The ARRL
Handbook for
the Radio Amateur

Published by the
AMERICAN RADIO RELAY LEAGUE
Newington, CT 06111 USA

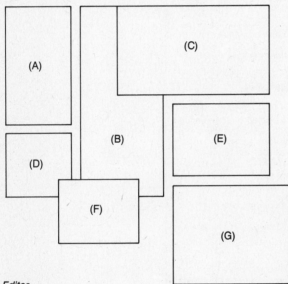

Cover photo credits

A—Antennas for 220 through 2304 MHz at the N6CA multi-multi operation on California's Mount Pinos during the 1988 ARRL June VHF QSO Party. (N6LL photo)

B—Ariane 4 takes to the air carrying AMSAT-OSCAR 13 on June 15, 1988. (Photo courtesy Arianespace)

C—This digital audio keyer, designed by W2IHY, is a sideband contester's dream come true. The details are in Chapter 29. (Meyers Studio photo)

D—AA6RX at the mic of the N6CA 6-meter operating position during the 1988 ARRL June VHF QSO Party. (N6LL photo)

E—The ground track of Fuji-OSCAR 12, as displayed by the Silicon Solutions GrafTrak® software. (WA2LQQ photo)

F—W6SFH with the 10-GHz dish antenna he used on July 19, 1987 to set a 414-mile DX record with N6GN. The previous record was 296 miles. (WA6KOD photo)

G—The WØMXY portable VHF station on a high ridge about 10 miles north of Limon, Colorado during the 1987 VHF QSO Party. (WØMXY photo)

Editor

Bruce S. Hale, KB1MW

Assistant Editor

Mark J. Wilson, AA2Z

Contributors

Michael Dinkelman, WA7UVJ
Gerald L. Hall, K1TD
Ed Hare, KA1CV
Albert Helfrick, K2BLA
Julius Jones, W2IHY
Andre Kesteloot, N4ICK
Zachary Lau, KH6CP
Domenic Mallozzi, N1DM
Thomas W. Miller, NK1P
David Newkirk, AK7M
Dick Stevens, W1QWJ

Production

Laird Campbell, W1CUT, Production Manager
Joel Kleinman, N1BKE, Deputy Production Manager
Michelle Chrisjohn, WB1ENT, Production Supervisor
Sue Fagan, cover
David Pingree
Leslie K. Bartoloth, KA1MJP
Steffie Nelson, KA1IFB
Meg Wise
Jean Wilson
Mark Kajpust

1989
Sixty-Sixth Edition

Foreword

The year 1989 marks the 75th anniversary of the founding of the ARRL. The League has been the national organization for radio amateurs since 1914, and ever since 1926 the ARRL *Handbook* has been *the* book to turn to for information about radio. This is the sixty-sixth edition of the *ARRL Handbook for the Radio Amateur* and the second edition to be offered exclusively in hardcover. We think you'll agree that this edition lives up to the reputation for excellence that the *Handbook* has developed over its long history. As always, this *Handbook* contains valuable information for both the beginner and the most advanced radio hobbyist.

As in previous editions, the 1989 *Handbook* has been updated to keep pace with progress in electronic technology. Rather than a complete revision of a few chapters, we have made a number of more subtle changes to much of the material. There are new sections on oscilloscopes, spectrum analyzers, digital frequency synthesis and phase-noise measurement. You will find several new construction projects, including a new 500-MHz frequency counter, a microprocessor-based memory keyer, a digital audio keyer and an inductance meter. Chapter 30 contains a new 1500-W amplifier design from Dick Stevens, W1QWJ, using the 3CX1200A7 triode.

We invite your comments on the *Handbook.* If you've developed a new technique or built a new project, let us know about it so we can share it with others. This is *your* book; help us make it better.

David Sumner, K1ZZ
Executive Vice President

Newington, Connecticut
October 1988

Schematic Symbols Used in Circuit Diagrams

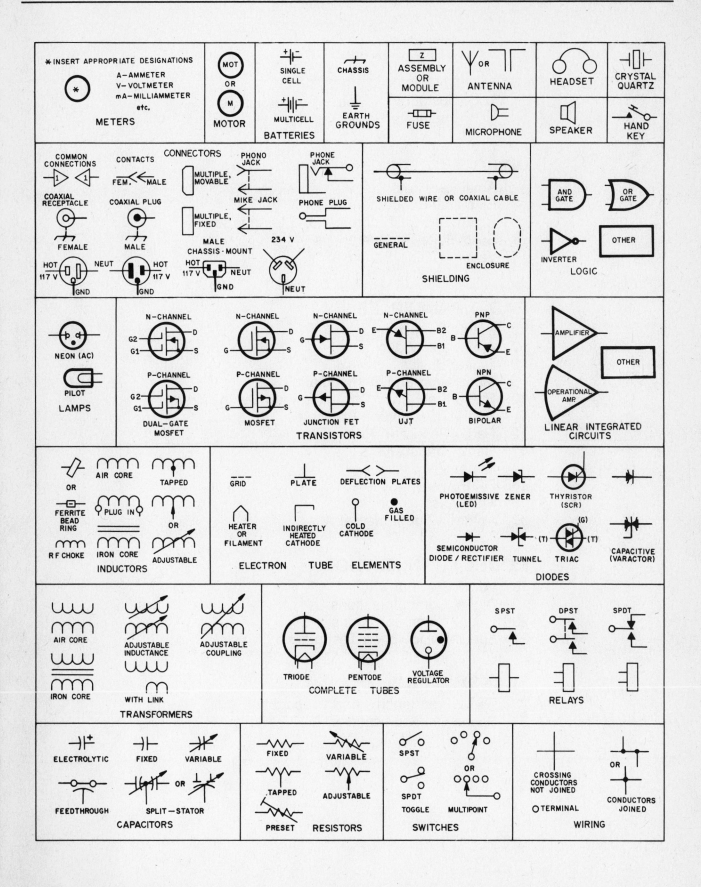

Contents

CONSTRUCTION and MAINTENANCE

ON THE AIR

U.S. Customary — Metric Conversion Factors

International System of Units (SI) — Metric Units

Prefix	Symbol	Multiplication Factor		
exa	E	10^{18}	=	1,000,000,000,000,000,000
peta	P	10^{15}	=	1,000,000,000,000,000
tera	T	10^{12}	=	1,000,000,000,000
giga	G	10^{9}	=	1,000,000,000
mega	M	10^{6}	=	1,000,000
kilo	k	10^{3}	=	1,000
hecto	h	10^{2}	=	100
deca	da	10^{1}	=	10
(unit)		10^{0}	=	1
deci	d	10^{-1}	=	0.1
centi	c	10^{-2}	=	0.01
milli	m	10^{-3}	=	0.001
micro	μ	10^{-6}	=	0.000001
nano	n	10^{-9}	=	0.000000001
pico	p	10^{-12}	=	0.000000000001
femto	f	10^{-15}	=	0.000000000000001
atto	a	10^{-18}	=	0.000000000000000001

Linear
1 meter (m) = 100 centimeters (cm) = 1000 millimeters (mm)

Area
$1\ m^2 = 1 \times 10^4\ cm^2 = 1 \times 10^6\ mm^2$

Volume
$1\ m^3 = 1 \times 10^6\ cm^3 = 1 \times 10^9\ mm^3$
$1\ liter\ (l) = 1000\ cm^3 = 1 \times 10^6\ mm^3$

Mass
1 kilogram (kg) = 1000 grams (g)
(Approximately the mass of 1 liter of water)
1 metric ton (or tonne) = 1000 kg

U.S. Customary Units

Linear Units

12 inches (in) = 1 foot (ft)
36 inches = 3 feet = 1 yard (yd)
1 rod = 5½ yards = 16½ feet
1 statute mile = 1760 yards = 5280 feet
1 nautical mile = 6076.11549 feet

Area
$1\ ft^2 = 144\ in^2$
$1\ yd^2 = 9\ ft^2 = 1296\ in^2$
$1\ rod^2 = 30¼\ yd^2$
$1\ acre = 4840\ yd^2 = 43{,}560\ ft^2$
$1\ acre = 160\ rod^2$
$1\ mile^2 = 640\ acres$

Volume
$1\ ft^3 = 1728\ in^3$
$1\ yd^3 = 27\ ft^3$

Liquid Volume Measure
$1\ fluid\ ounce\ (fl\ oz) = 8\ fluidrams = 1.804\ in^3$
1 pint (pt) = 16 fl oz
$1\ quart\ (qt) = 2\ pt = 32\ fl\ oz = 57¾\ in^3$
$1\ gallon\ (gal) = 4\ qt = 231\ in^3$
1 barrel = 31½ gal

Dry Volume Measure
$1\ quart\ (qt) = 2\ pints\ (pt) = 67.2\ in^3$
1 peck = 8 qt

$1\ bushel = 4\ pecks = 2150.42\ in^3$

Avoirdupois Weight
1 dram (dr) = 27.343 grains (gr) or (gr a)
1 ounce (oz) = 437.5 gr
1 pound (lb) = 16 oz = 7000 gr
1 short ton = 2000 lb, 1 long ton = 2240 lb

Troy Weight
1 grain troy (gr t) = 1 grain avoirdupois
1 pennyweight (dwt) or (pwt) = 24 gr t
1 ounce troy (oz t) = 480 grains
1 lb t = 12 oz t = 5760 grains

Apothecaries' Weight
1 grain apothecaries' (gr ap) = 1 gr t = 1 gr a
1 dram ap (dr ap) = 60 gr
1 oz ap = 1 oz t = 8 dr ap = 480 gr
1 lb ap = 1 lb t = 12 oz ap = 5760 gr

Multiply →

Metric Unit = Conversion Factor × U.S. Customary Unit

← Divide

Metric Unit ÷ Conversion Factor = U.S. Customary Unit

Metric Unit	=	Conversion Factor	×	U.S. Unit
(Length)				
mm		25.4		inch
cm		2.54		inch
cm		30.48		foot
m		0.3048		foot
m		0.9144		yard
km		1.609		mile
km		1.852		nautical mile
(Area)				
mm^2		645.16		$inch^2$
cm^2		6.4516		in^2
cm^2		929.03		ft^2
m^2		0.0929		ft^2
cm^2		8361.3		yd^2
m^2		0.83613		yd^2
m^2		4047		acre
km^2		2.59		mi^2
(Mass)		(Avoirdupois Weight)		
grams		0.0648		grains
g		28.349		oz
g		453.59		lb
kg		0.45359		lb
tonne		0.907		short ton
tonne		1.016		long ton

Metric Unit	=	Conversion Factor	×	U.S. Unit
(Volume)				
mm^3		16387.064		in^3
cm^3		16.387		in^3
m^3		0.028316		ft^3
m^3		0.764555		yd^3
ml		16.387		in^3
ml		29.57		fl oz
ml		473		pint
ml		946.333		quart
l		28.32		ft^3
l		0.9463		quart
l		3.785		gallon
l		1.101		dry quart
l		8.809		peck
l		35.238		bushel
(Mass)		(Troy Weight)		
g		31.103		oz t
g		373.248		lb t
(Mass)		(Apothecaries' Weight)		
g		3.387		dr ap
g		31.103		oz ap
g		373.248		lb ap

The Amateur's Code

ONE

The Amateur is Considerate. . . He never knowingly uses the air in such a way as to lessen the pleasure of others.

TWO

The Amateur is Loyal. . . He offers his loyalty, encouragement and support to his fellow radio amateurs, his local club and to the American Radio Relay League, through which Amateur Radio is represented.

THREE

The Amateur is Progressive. . . He keeps his station abreast of science. It is well-built and efficient. His operating practice is above reproach.

FOUR

The Amateur is Friendly. . . Slow and patient sending when requested, friendly advice and counsel to the beginner, kindly assistance, cooperation and consideration for the interests of others; these are marks of the amateur spirit.

FIVE

The Amateur is Balanced. . . Radio is his hobby. He never allows it to interfere with any of the duties he owes to his home, his job, his school, or his community.

SIX

The Amateur is Patriotic. . . His knowledge and his station are always ready for the service of his country and his community.

— PAUL M. SEGAL
ex-W3EEA, W9EEA

Chapter 1

Amateur Radio

Amateur Radio. You've heard of it. You probably know that Amateur Radio operators are also called hams. (Nobody knows quite why!) But who are these people and what do they do?

Every minute of every hour of every day, 365 days a year, radio amateurs all over the world communicate with each other. It's a way of discovering new friends while experimenting with different and exciting new ways to advance the art of their hobby. Ham radio is a global fraternity of people with common and yet widely varying interests, able to exchange ideas and learn more about each other with every on-the-air contact. Because of this Amateur Radio has the ability to enhance international relations as does no other hobby.

How is it possible to talk to an astronaut circling the earth in the Space Shuttle; a Tokyo businessman, a U.S. legislator, a Manhattan store owner, a camper in a Canadian national park, the head of state of a Mediterranean-area country, a student at a high school radio club in Wyoming, or a sailor on board a ship in the middle of the Pacific? And all without leaving your home! Only with Amateur Radio — that's how!

The way communication is accomplished is just as interesting as the people you get to meet. Signals can be sent around the world using reflective layers of the earth's ionosphere or relayed long distances via mountaintop radio repeaters. Orbiting satellites that hams built are used for two-way communication. Still other hams bounce their signals off the moon! Possibilities are almost unlimited. Not only do radio amateurs use international Morse code and voice for communication, but they also use radioteletype, facsimile and various forms of television. Many hams even have computers hooked up to their radios. As new techniques and modes of communication are developed, hams continue their long tradition of being among the first to use them.

Amateur Radio has flown two missions aboard the Space Shuttle. In the August 1985 mission, Tony England, W0ORE, carried along slow-scan television (SSTV) equipment and sent images back to Earth. Tony treated hams around the world to glimpses such as this of life aboard the Space Shuttle *Challenger* on the 2-meter downlink frequency.

What's in the future? Digital voice-encoding techniques? Three-dimensional TV? One can only guess. But if there is ever such a thing as a Star Trek transporter unit that makes it possible for people to physically travel via radio waves, hams will probably have one!

Once radio amateurs make sure that their gear does work, they look for things to do with the equipment and special skills they possess. Public service is a very large and integral part of the whole Amateur Radio Service. Hams continue this tradition by

becoming involved and sponsoring various activities in their communities.

Field Day, just one of the many public service-type activities, is an annual event occurring every June when amateurs take their equipment into the great outdoors (using electricity generated at the operating site) and test it for use in case of disaster. Not only do they test their equipment, but they make a contest out of the exercise and try to contact as many other hams operating emergency-type stations as possible (along with ordinary types). Often

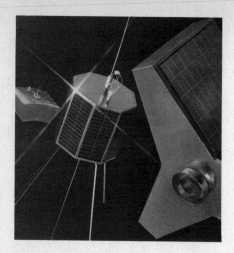

OSCAR (Orbiting Satellite Carrying Amateur Radio) is the name given to a series of satellites designed and built by hams.

they make Field Day a club social event while they are operating.

Traffic nets (networks) meet on the airwaves on a schedule for the purpose of handling routine messages for people all over the country and in other countries where such third-party traffic is permitted. By so doing, amateurs stay in practice for handling messages should any real emergency or disaster occur which would require operating skill to move messages efficiently. Nets also meet because the members often have common interests: similar jobs, interests in different languages, different hobbies (yes, some hams have hobbies other than ham radio!), and a whole barrelful of other reasons. It is often a way to improve one's knowledge and to share experiences with other amateurs for the good of all involved.

DX (distance) contests are popular, and awards are actively sought by many amateurs. This armchair travel is one of the more alluring activities of Amateur Radio. There are awards for Worked All States (WAS), Worked All Canadian Provinces (WAVE), Worked 100 Countries (DXCC), and many others.

Mobile operation (especially on the very high frequencies) holds a special attraction to many hams. It's always fun to keep in touch with ham friends over the local repeater (stations that receive your signal and retransmit it for better coverage of the area) or finding new friends on other frequencies while driving across the country. Mobile units are often the vital link in emergency communications, too, since they are usually first on the scene of an accident or disaster.

The OSCAR (Orbiting Satellite Carrying Amateur Radio) program is a relatively new challenge for radio amateurs. Built by hams from many countries around the world, these ingenious devices hitch rides as secondary payloads on space shots for commercial and governmental communications or weather satellites. OSCAR satellites receive

signals from the ground on one frequency and convert those signals to another frequency to be sent back down to earth. VHF (very-high frequency) and UHF (ultra-high frequency) signals normally do not bend around the earth much beyond the horizon, but when beamed to these satellites, a VHF/UHF signal's effective range is greatly increased to make global communication possible. These OSCAR satellites also send back telemetry signals either in Morse or radioteleprinter (RTTY) code, constantly giving information on the condition of equipment aboard the satellite.

Self-reliance has always been a trademark of the radio amateur. This is often best displayed by the many hams who design and build their own equipment. Many others prefer to build their own equipment from kits. The main point is that hams want to know their equipment, how it functions, what to do with it and how to fix it if a malfunction should occur. Repair shops aren't always open during hurricanes or floods and they aren't always out in the middle of the Amazon jungle, either. Hams often come up with variations on a circuit design in common use so that it may perform a special function, or a ham may bring out a totally original electronic design, all in the interest of advancing the radio art.

Looking Back

How did Amateur Radio become the almost unlimited hobby it is today? The beginnings are slightly obscure, but electrical experimenters around the turn of the century, inspired by the experiments of Marconi and others of the time, began duplicating those experiments and attempted to communicate among themselves. There were no regulatory agencies at that time, and much interference was caused by these amateur experimenters to other stations until governments the world over stepped in and established licensing, laws and regulations to control the problems involved in this new technology. Amateur experimenter stations were then restricted to the useless wavelengths of 200 meters and below. Amateurs suddenly found that they could communicate over longer distances than commercial stations on the longer wavelengths. Even so, signals often had to be relayed by intermediate amateur stations to get a message to the proper destination. Because of this, the American Radio Relay League was organized to establish routes of Amateur Radio communication and serve the public interest through Amateur Radio. But the dream of eventual transcontinental and even transoceanic Amateur Radio contact burned hot in the minds of radio amateur experimenters.

World War I broke out, and Amateur Radio, still in its infancy, was ordered out of existence until further notice. Many former Amateur Radio operators joined the armed services and served with distinc-

This emergency communications van, owned and operated by L'anse Creuse Radio Club (Michigan), demonstrates the commitment Amateur Radio operators have to being prepared for any emergency that may arise.

tion as radio operators, finding their skills to be much needed.

After the close of the "War to End All Wars," Amateur Radio was still banned by law, yet there were many hundreds of formerly licensed amateurs just itching to "get back on the air." The government had tasted supreme authority over the radio services and was half inclined to keep it. Hiram Percy Maxim, one of the founders of the American Radio Relay League, called the pre-war League's officers together and then contacted all the old members who could be found in an attempt to reestablish Amateur Radio. Maxim traveled to Washington, DC, and after considerable effort (and untold red tape) Amateur Radio was opened up again on October 1, 1919.

Experiments on shorter wavelengths were then begun with encouraging results. It was found that as the wavelength dropped (that is, frequency increased) greater distances were covered. The commercial stations were not about to miss out on this opportunity. They moved their stations to the new shorter wavelengths while the battle raged over who had the right to transmit in this new area. Usually, it turned out to be the station with the stronger signal, able to blot out everyone else.

National and international conferences were called in the twenties to straighten out the tangle of wavelength allocations. Through the efforts of ARRL officials, amateurs obtained frequencies on various bands similar to what we have today: 160 through 6 meters. When the amateur operators moved to 20 meters, the dream of coast-to-coast and transoceanic communication without relay stations was finally realized. (A more detailed history of the early days of Amateur Radio is con-

Ham operators are using computers more and more to do station "housekeeping," such as keeping a log of stations contacted.

tained in the ARRL publication *Two Hundred Meters and Down* by Clinton B. DeSoto.)

Public Service

Amateur Radio is a grand and glorious hobby, but this fact alone would hardly merit the wholehearted support given it by nearly all the world's governments at international conferences. There are other reasons. One of these is a thorough appreciation of the value of amateurs as sources of skilled radio personnel in time of war. Another asset is best described as public service.

The public service record of the amateur is a brilliant tribute to his work. These activities can be roughly divided into two classes: expeditions and emergencies. Amateur cooperation with expeditions began in 1923, when a League member, Don Mix, 1TS, accompanied MacMillan to the Arctic on the schooner Bowdoin with an amateur station. Amateurs in Canada and the U.S. provided the home contacts. The success of this venture was so outstanding that other explorers followed suit. During subsequent years Amateur Radio assisted perhaps 200 voyages and expeditions, the several explorations of the Antarctic being perhaps the best known. And this kind of work is not all in the distant past, either: In 1978 Japanese explorer Naomi Uemura, JG1QFW, became the first person to trek to the North Pole alone. Amateur Radio, through member stations of the National Capital DX Association at Washington, DC and the Polar Amateur Radio Club, VE8RCS, at Alert, NWT, Canada, provided important backup communications.

Sometimes Mother Nature goes on a rampage — with earthquakes such as those in Alaska in 1964, Peru in 1970, California in 1971, Guatemala in 1976 and Italy in 1980; floods like those in Big Thompson Canyon, Colorado, in 1976, Kentucky, Virginia, West Virginia, and Johnstown, Pennsylvania in 1977, Jackson, Mississippi in 1979; the big forest fires of California, particularly in 1977; and tornadoes, hurri-

canes, blizzards and typhoons, most anywhere, any year. In each of these disasters, amateurs were ready, with equipment not needing power from the electric company, to carry on communications for police, fire departments and relief organizations. The ability of radio amateurs to help the public in emergencies is one big reason Amateur Radio has survived and prospered.

Technical Developments

Amateurs started the hobby with spark-gap transmitters, which took up great hunks of frequency space. Then they moved on to tubes when these devices came along. Much later, transistors were used; now integrated circuits are a part of the everyday hardware in the Amateur Radio shack. This is because amateurs are constantly in the forefront of technical progress. Their incessant curiosity and eagerness to try anything new are two reasons. Another is that ever-growing Amateur Radio continually overcrowds its frequency assignments, spurring amateurs to the development and adoption of new techniques to permit the accommodation of more stations.

Amateurs have come up with ideas in their shacks while at home and then taken them to industry with surprising results. During World War II, thousands of skilled amateurs contributed their knowledge to the development of secret radio devices, both in government and private laboratories. Equally as important, the prewar technical progress by amateurs provided the keystone for the development of modern military communications equipment.

In the fifties, the U.S. Air Force was

This compact Amateur Radio station in a closet has been used by Ethan Winning, WD6GKF, to contact over 100 countries using a trap dipole antenna. Close the closet door, and the station disappears.

Greg Marshall, KA1DYT, generates electricity using his bicycle. This pedal power was used to contact five Amateur Radio stations.

faced with converting its long-range communications from Morse to voice; jet bombers had no room for skilled radio operators. At the time, amateurs had been using single sideband for about a decade, and were communicating by voice at great distances with both homemade and commercially built equipment. Generals LeMay and Griswold, both radio amateurs, hatched an experiment in which ham equipment was used to keep in touch with Strategic Air Command headquarters in Omaha, Nebraska, from an airplane traveling around the world. The system worked well; the equipment needed only slight modification to meet Air Force needs, and the expense and time of normal research and development procedures was saved.

Many youngsters build an early interest in Amateur Radio into a career. Later, as professionals, they may run into ideas which they try out in ham radio. A good example is the OSCAR series of satellites, initially put together by amateurs who worked in the aerospace industry, and launched as secondary payloads with other space shots. At this writing 11 Amateur Radio satellites have been launched by Western Hemisphere hams. OSCARs 9 and 10, portions of which were built by amateurs of several different countries, are currently in space relaying the signals of amateurs as well as scientific data. OSCAR can be heard on almost any 29-MHz receiver.

Development of third-generation Phase III satellites proceeds under the guidance of the Radio Amateur Satellite Corporation (AMSAT) with the assistance of Project OSCAR, Inc., the original nonprofit group, both affiliated with ARRL. The Phase III program was temporarily set back in May of 1980 when the first satellite in the series, Phase IIIA, ended up in the

Lunchtime at the Fontana (California) Junior High School Amateur Radio Club station involves the students making contact with other ham operators around the world. These kids have talked to places like Zaire and Finland using equipment they helped buy with money raised from candy sales and a balloon race.

Many Amateur Radio clubs hold theory and code classes to help prospective ham operators get licensed. ARRL Headquarters can provide you with the name and address of the club closest to you.

Atlantic Ocean because of a malfunctioning launch vehicle. It is a credit to its builders that up to the moment of its demise, the satellite functioned flawlessly. The thousands of hours of experience that went into the design and construction of Phase IIIA were used as the starting point for its successor, Phase IIIB. Then on June 16, 1983, Phase IIIB became OSCAR 10 when it was launched aboard a European Space Agency Ariane rocket from French Guiana. This OSCAR, placed in a highly elliptical orbit, provides a vast communications resource to amateurs throughout the world. The new Phase III satellites being built by AMSAT will continue the OSCAR program as older spacecraft are taken out of service.

The last week of November and the first week of December 1983 marked the first time a ham operator in space communicated with fellow hams in all parts of the world. Radio amateur Owen Garriott, W5LFL, made history aboard the space shuttle *Columbia*'s STS-9 mission by taking an amateur 2-meter-FM station along. Amateur stations throughout the world got the thrill of working an astronaut aboard an orbiting spacecraft. The second ham-in-space mission took place in July/August 1985. Mission Specialists Tony England, WØORE and John David Bartoe, W4NYZ beamed compatible-color SSTV signals to earth-bound stations, many of which were set up in schools.

THE ARRL

Thanks to the hard work of a few visionaries, a national Amateur Radio organization came into being in 1914. Since then, the continued hard work of dedicated men and women — ham radio operators — have made the American Radio Relay League what it is today. The American Radio Relay League (ARRL or *the League*)

is headquartered in the Hartford, Connecticut area. Its operations are, however, national and international in scope. Through its vast network of volunteers throughout the U.S. and Canada, plus a full-time, professional staff of about 125 employees in Newington, Connecticut, the ARRL promotes the advancement of the Amateur Radio Service.

The League operates strictly as a nonprofit, educational and scientific organization dedicated to the promotion and protection of the privileges that ham operators enjoy.

Of, by and for the radio amateur, ARRL numbers within its ranks the vast majority of active amateurs in North America. Approximately 120,000 licensed U.S. and Canadian amateurs are full members of the

Operating from New Hampshire, Tammy Keller, KA1HLG, is representative of the growing number of women who are becoming interested in electronics, communications and Amateur Radio.

League. An additional 11,000 foreign amateurs and unlicensed individuals are associate members. There is a place in the League for everyone who has an interest in Amateur Radio.

League Leadership

Full membership in ARRL gives one a voice in how the affairs of the organization are governed. Each year, half of the ARRL Board of Directors stands for election by full members in those areas or divisions represented by these directors. The Board of Directors, in turn, directs the operation of the League, including the hiring of an executive vice president to run ARRL Headquarters.

The ARRL Field Organization is headed by Section Managers. The Section Managers are elected by the membership. The Field Organization provides opportunities for members to promote Amateur Radio in such areas as *public service* (message handling, emergency preparedness, resolution of radio interference problems), *regulatory matters* (state government liaison, volunteer legal counsel, volunteer examiner, volunteer monitoring), *local clubs* (affiliated club coordinator, volunteer instructor), *public information* (public information officer, bulletin manager), *technical experimentation* (technical coordinator, technical advisor), and many others. The extent of involvement depends on one's interests and available time. But even if you don't have spare time to actively volunteer your services, you can nonetheless make a positive contribution by becoming a member of the ARRL.

ARRL Publications

A well-informed Amateur Radio community is crucial if ham radio is to continue as a trained and self-disciplined volunteer

The American Radio Relay League's station in Newington, Connecticut, W1AW, transmits bulletins of interest to radio amateurs and code practice that is heard throughout the world.

communications service worthy of the respect of federal, state and local government. At ARRL Headquarters, the League staff produces *QST*, a monthly journal that keeps the membership up to date on the technical and regulatory developments having an impact on Amateur Radio. Each month, the pages of *QST* report news of actions taken by the Federal Communications Commisssion (in Canada, the Department of Communications) that could have an impact on the future of Amateur Radio.

The monthly journal also keeps the membership informed of the actions the ARRL staff is taking, such as attendance at hearings and conferences. Monthly columns provide tips on contacting stations at great distances (called DXing), interfacing station equipment with computers, results of operating contests, and much more. Of course, you can also find the latest in building and constructing techniques, plus reviews of commerical equipment now available for the Amateur Radio market. *QST* is *the* publication devoted entirely to Amateur Radio. ARRL membership includes a subscription to *QST*.

QST requires the contributions of many people. In addition to the regular writers, members from all over submit items ranging from correspondence to feature technical project articles. Not all submissions are published, but if you have an original idea or approach to an Amateur Radio problem, why not share it with your fellow hams through the pages of *QST*?

The Headquarters staff also produces several technical publications (one of these is the very book you are now reading), plus books on operating, getting a license, the FCC Rules governing Amateur Radio and specialized communications. The ARRL *Satellite Experimenter's Handbook* is a good example of the latter. Experimentation with new or exotic modes of communications such as packet radio, spread spectrum and meteor burst is fostered by a special ARRL publication, *QEX, The ARRL Experimenter's Newsletter*. Volunteers in certain public service aspects of the Amateur Service can get another publication, *Field Forum,* to help prepare for their role in emergency situations.

Membership Services

Membership in ARRL means much more than receiving *QST* each month. Your membership dues also support many membership services. In addition to supporting the Amateur Radio Service through representation before government agencies and administrative support to the Field Organization, ARRL offers membership services on a personal level. One of these services is the ARRL DX QSL Bureau system. Radio amateurs, as a rule, exchange small postcards called QSL cards to confirm two-way radio contacts. For U.S. and Canadian amateurs, the most convenient way of sending cards to foreign stations is to use the ARRL Outgoing Overseas QSL Bureau. For $1 per pound (or portion thereof) of cards, ARRL members may send their cards to Headquarters for forwarding to QSL bureaus overseas.

Incoming cards destined for amateurs in the U.S. and Canada from amateurs overseas are handled by a far-ranging network of more than 400 hard-working volunteers. They are organized into call-area QSL bureaus. Each of these volunteers spends an average of 10 hours per month sorting cards, stuffing envelopes and mailing them to their destination stations in the U.S. and Canada. Out-of-pocket expenses for these incoming bureaus are borne by ARRL, but the key here is the time volunteered by sorters and handlers. This service is available to both members and non-members.

Recent cutbacks in the U.S. federal budget have led to amateurs taking over responsibility for administering Amateur Radio examinations. The ARRL Volunteer Examiner Program plays a crucial role as coordinator in the scheduling, printing and administration of amateur exams given by volunteers.

Other services performed by ARRL include low-cost insurance against equipment theft and certain other losses, assistance in resolving radio frequency interference and problems with local government restrictions on antennas, and compiling information that is useful to persons with disabilities.

Scholarships for bright, young radio amateurs and financial support for Amateur Radio satellite projects are just

Pete Demmer, KH6CTQ, has good reason to be proud of the portable windmill he designed and built. Assembled at the KH6IJ Field Day site in less than two hours, the windmill provided all the power needed to run the station.

two ways that the ARRL Foundation carries out its mission of advancing Amateur Radio. The ARRL Foundation gave financial assistance to the OSCAR 10 project and amateur efforts at WARC-79.

International Involvement

Do you plan on visiting a foreign country? ARRL Headquarters can provide you with information on how to apply for a reciprocal license to permit amateur operation from that country. ARRL serves as the international secretariat of a worldwide federation of national Amateur Radio societies, the International Amateur Radio Union (IARU). The IARU is comprised of societies like the League in 123 countries. The IARU has official observer status at the International Telecommunication Union (ITU), which is the branch of the United Nations in charge of world conferences and agreements concerning the radio spectrum. At the most recent World Administrative Radio Conference, held in Geneva, Switzerland in 1979, the IARU successfully defended Amateur Radio frequencies against other interests. At that conference, the member-societies of the IARU, after many years of careful planning and coordination, stunned the communications community by accomplishing what was earlier thought impossible — the acquisition of three new high-frequency bands for the use of radio amateurs.

Are You a Member?

No matter what particular aspect of Amateur Radio attracts you, ARRL membership is relevant and important. It has often been said that there would be no Amateur Radio as we know it today had it not been for the ARRL. The ARRL of today needs the broad-based support of radio amateurs and others interested in the continued existence of the Amateur Radio Service. An Amateur Radio license is not required for membership in the League, although full voting membership is granted only to licensed amateurs in the U.S. and Canada. Inquiries about membership are always welcome! For more information about the American Radio Relay League and answers to any questions you may have about Amateur Radio, write or call ARRL Headquarters, 225 Main St., Newington, CT 06111. Tel. 203-666-1541. Information on how to become a ham radio operator appears in Chapter 36.

Chapter 2

Electrical Fundamentals

Electricity and magnetism are familiar to everyone. The effects of static electricity on a dry, wintry day, the attraction of a compass needle to the Earth's magnetic poles, and the propagation and reception of radio waves are all examples of electromagnetic phenomena. Even the radiation of light and radiant heat from a stove are forms of electromagnetic radiation, governed by many of the same physical laws that apply to the simplest electronic circuit. Knowledge of these laws is fundamental to understanding almost every subject in Amateur Radio, and electronics in general.

Electromagnetic energy exists in the form of fields, a spatial and time-dependent pattern of energy. The analysis of electromagnetic fields, their interrelationships and their interaction with matter forms the basis of all electrical laws. Once a field problem is analyzed in a specific case, it is often possible to use the results over and over again for more general purposes. The field solution can be used to derive numerical formulas for such electrical properties as resistance, inductance and capacitance. These elements in turn form the building blocks for more complex configurations called networks or circuits.

ELECTROSTATIC FIELD AND POTENTIALS

Matter is made up of complex structures called atoms. Each atom is primarily composed of three fundamental particles called electrons, protons and neutrons. At the center of the atom is the nucleus formed by the combination of protons, particles with a positive charge, and neutrons, particles with no intrinsic charge themselves. Orbiting around the nucleus are the electrons, particles that have a negative charge. The unique combinations of these three atomic building blocks determine the chemical and electrical properties of the atom.

One of the fundamental laws of physics is the concept of electrostatic attraction and repulsion. Simply stated, particles with opposite charge (positive and negative) tend to attract, while particles with like charge (both positive or both negative) tend to repel. If charged particles are considered to exist at points in space, the force of attraction (or repulsion if the charges have like signs) is given by Coulomb's law

$$F = \frac{k_0 Q_1 Q_2}{r^2}$$

where Q_1 is the numerical value of one charge, Q_2 is the other charge value, r is the distance between particles, and F is the force charge Q_1 exerts on charge Q_2. The value of the proportionality constant, k_0, depends on the units used in specifying F, Q and r. In the metric system of units (SI — Systeme International d'Unites) the basic unit of charge is the coulomb, the unit of force is the Newton, and the unit of space is the meter. In the metric system, k_0 is then equal to

$$k_0 = \frac{1}{4\pi\epsilon_0}$$

where ϵ_0 has a value of

$$\frac{8.854 \times 10^{-12} \text{coul.}^2}{\text{Newton m}^2}$$

(usually referred to as "the permittivity of free space"). If a number of charged particles are present, the force on any one particle is a function of the positions and magnitudes of the other charges.

The fundamental forces inside an atom can be described by Coulomb's law. In the case of a complex atom, one having a number of protons in the nucleus, the repulsive forces between protons within the nucleus tend to force the protons apart. Balancing this is the attractive force between the electrons held in orbit about the nucleus and the protons in the nucleus. Although this description is highly simplistic, it does portray the actions within an atom to a surprisingly high degree of accuracy. The electron and the proton both have charges of equal magnitude — about 1.6×10^{-19} coulombs. The charge on the electron is negative, while the charge on the proton is positive. These values of charge may seem too minute to account for atomic

particle interaction, but consider the distances (r) involved. Almost all atoms have an outer diameter of approximately 10^{-10} meter. It is evident that substantial forces exist within the atom.

Analyzing the forces exhibited by multiple charges is overwhelmingly cumbersome, so the concept of electrostatic or electric field strength is a useful one to introduce. If a single charged particle is moved slowly around a fixed charged particle, everywhere the moving particle moves, a force is exerted on the moving particle. The force exerted on the moving particle displays the existence of a force field. If the moving particle is assigned a value Q_t for its charge, then by Coulomb's law, the force (F_t) on the moving particle is

$$F_t = \frac{Q_1 Q_2}{4\pi\epsilon_0 r^2}$$

Writing this as a force per unit charge gives

$$\frac{F_t}{Q_t} = \frac{Q_1}{4\pi\epsilon_0 r^2}$$

This expression describes the force on the moving particle totally as a function of the distance from Q_1 and the magnitude of Q_1; this is called the electric field intensity. For example, if a force of 1 newton existed on a test charge of 2 coulombs, the field intensity would be 0.5 newton/coulomb.

As described by Isaac Newton, whenever a force exists on an object, an expenditure of energy is needed to move the object against that force. In some instances, mechanical energy may be recovered (such as in a compressed spring) or the energy may be converted to another form of energy (such as heat produced by friction). In electrostatics, the energy is described in terms of potential. Potential is defined as the work required from some energy source in moving a unit positive charge between two points in an electric field. For example, if an energy expenditure of 5 newton-meters (5 joules) is needed to move a charge of 2 coulombs from a point of zero energy

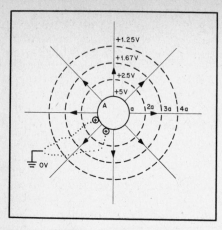

Fig. 1 — Field lines (solid) and potential lines (dashed) surrounding a charged sphere.

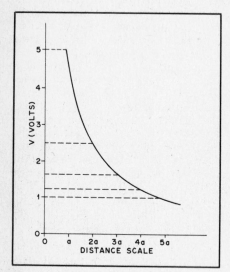

Fig. 2 — Variation of potential with distance for the charged sphere of Fig. 1.

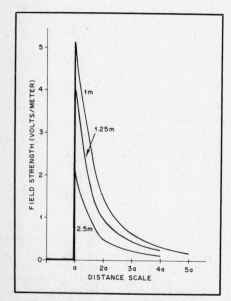

Fig. 3 — Variation of field strength with distance around a sphere charged to 5 volts for spheres of different radii.

to a given point, the potential at that point would be 2.5 joules/coulomb. The unit joules/coulomb is rather unwieldy, so another name is given to electrical potential: the volt. One joule per coulomb is equal to 1 volt. Notice that if the voltage is divided by length (meters), the dimensions of field intensity are obtained and a field strength of 1 newton/coulomb becomes 1 volt/meter.

The relationship between field intensity and potential is illustrated by the example shown in Fig. 1. A conductive sphere is gradually charged until its surface is at a potential of 5 volts. As charges arrive on the surface of a conductor, they tend to spread out into a uniform distribution; the outside of the sphere is said to constitute an equipotential surface. Since every point on the sphere has the same charge, and the charge spreads uniformly over the sphere's surface even as charge is added or removed, the same amount of energy is needed to bring a given amount of charge from a zero reference point to any one point on the sphere as would be required to bring the same amount of charge to any other point on the sphere. The amount of energy expended also is independent of the path traveled to get to the surface. For this particular sphere, it would require 5 joules of energy to bring a charge of 1 coulomb from a point of 0 voltage to any point on the sphere (as indicated by the dotted lines in Fig. 1).

The electromotive force applied to a charged particle at the surface of the sphere must be applied perpendicular to the surface for maximum potential to be realized. Any force applied from a direction other than normal to the surface will have two components: one is performing real work by moving the charge away from the sphere, and the other tends to pull the charge along the surface. Since the charges are free to move about the charged surface already, the energy source does no work in moving charges along the surface, only away from it; the potential difference between points on the sphere is zero. Once the charges are equally distributed along the sphere's surface, an equilibrium condition will be reached and any initial field component parallel to the surface becomes zero. The motion of charge under the influence of an electric field is a very important concept in electricity. The rate at which charge flows past a reference point is defined as the current. A rate of 1 coulomb per second is defined as 1 ampere.

In the case of the symmetrical sphere, the direction of the electromotive force and the electric field can be represented by the solid straight lines in Fig. 1. The arrows indicate the direction of the force for a positive charge. At points farther away from the sphere, less energy is required to move a charged particle from the zero reference. A series of concentric spherical shells, indicated by the dashed lines, can be used to define the equipotential surfaces around the

sphere. From the definition of potential difference, it can be shown through the use of integral calculus that the potential varies inversely to the distance from the center of the sphere. This relationship is indicated by the numbers in Fig. 1 and by the graph in Fig. 2.

The electric field defines the direction and magnitude of a force on a charged object. The field intensity is equal to the negative value of the slope of the curve in Fig. 2. The slope of a curve is the rate of change of some variable with distance; in this case the variable is the potential. This is why the electric field is sometimes called the potential gradient (gradient being equivalent to slope). In the case of a curve that varies as the inverse of the distance, the slope at any point is squared.

An examination of Fig. 1 suggests that the potential variation is dependent only on the shape of the conductor, and not its actual physical size. That is, once the value of the radius, a, of the sphere in Fig. 1 is specified, the potential at any other point a given distance from the sphere is also known. Thus, Fig. 1 can be used for any number of spheres with different radii. When it is changed by a certain percentage, all the other values would change by the same percentage, too. However, the amount of charge required to produce a given voltage, or voltage change, does depend on the size of the conductor, its shape, and its position in relation to other conductors and insulators. For a given conductor configuration, the voltage is related to the required charge by the formula

$$V = \frac{Q}{C}$$

where the entity C is defined as the capacitance. Capacitance will be discussed in more detail in a later section.

Since the electric field intensity is only related to the change in potential with distance, the rate at which it changes is unaffected by the absolute physical size of the conductor configuration. However, the exact numerical potential value at any point does depend on the dimensions of the configuration. This is illustrated in Fig. 3 for spheres with different radii. Note that for larger radii, the numerical value of the field strength at the surface of the sphere (distance equal to a) is less than it is for smaller radii. This effect is important in the design of transmission lines and capacitors. (A capacitor is a device for storing charge. In older terminology, it was sometimes called a condenser.)

Even when identical voltages are applied across the terminals of two different sized transmission lines or capacitors, the field strength between the conductors is higher for configurations of small physical size than for larger physical sizes. When the field strength becomes too high for a particular insulating material, the material (including air) can break down, allowing a conductive arc to pass through the insulation. This

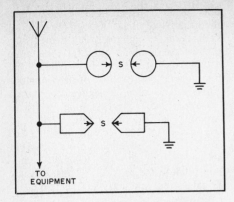

Fig. 4 — Spark gaps with sharp points break down at lower voltages than ones with blunt surfaces, even though the separation is the same.

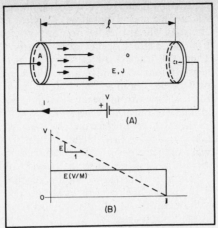

Fig. 5 — Potential and field strength along a current-carrying conductor.

effect is the basis for spark plugs in internal combustion engines. The gap between electrodes in the spark plug is adjusted to provide a calibrated breakdown of the insulator (the air-gas mixture) at a particular voltage. The arc is hot enough to ignite the fuel-air mixture.

Another example of high electric field breakdown is found in lightning arrestors. These are devices used to protect radio equipment connected to an antenna that might be subjected to high levels of atmospheric electricity. "Spark-gap" conductors are connected between the antenna transmission line and ground. The spark-gap conductors or electrodes are filed to sharp points. These needlepoints appear as conductors of very small radii, so the field strength between them is higher for the same applied potential than it would be for blunt electrodes (Fig. 4). For a given potential, this allows the separation between conductors to be relatively large so the radio frequency effects of the spark gap have minimal effect on normal circuit operation.

Increasing the size of a conductor reduces the electric field strength, thereby reducing the potential for insulation breakdown. A relatively common effect in high-strength electric fields is corona. Here the air insulation breaks down and actually glows. The effect is often seen on sharp tips of high power transmitting antennas. In order to reduce the localized field strength (and reduce the unwanted corona), a blunt electrode such as a sphere is often used on the sharp tip of a whip antenna.

Examination of Fig. 3 reveals that the field strength is zero for distances less than a (the sphere radius), which includes points actually inside the sphere. The implication here is that the effect of fields and charges cannot penetrate the conducting surface and disturb conditions inside the enclosure. The conducting sphere is said to form an electrostatic shield around the contents of the enclosure. The converse is not true, however. That is, charges inside the sphere do cause or induce a field on the outside surface. This is why it is very important

that enclosures designed to confine the effects of charges be connected to a point of zero potential. Such a point is often called a ground.

Fields and Currents

In the previous section, the motion of charged particles in the presence of an electric field was described in connection with charges placed on a conducting sphere, and the concept of current was introduced. In that example, it was assumed that charges could move around unimpeded on the surface of the sphere. In the real world using obtainable materials, this is not true. The movement of charge is dependent upon the transfer of energy between adjacent atoms in the conductor. Charges appear to collide with the constituent atoms as they move through the conductor while under the influence of the electric field of the conductor. The number of "collisions" depends on the chemical characteristics of the material used. Silver acts as a very good conductor with minimal opposition to the movement of charge; carbon and certain alloys of iron are rather poor conductors of charge flow. A measure of how easily charge can flow through a conductor is defined as the conductivity, and is denoted by σ.

The current density in a conductor, J, is the rate of charge flow or current through a given cross-sectional area. It is related to the electric field and conductivity by the formula

$$J = \sigma E$$

In general, the conductivity and electric field in a conductor do not remain constant over a large cross-sectional area. However, in order to simplify understanding of the following theoretical case, both are assumed to remain constant (Fig. 5).

A cylinder of a material with conductivity σ is inserted between two end caps of infinite conductivity. The end caps are connected to a voltage source, such as a bat-

tery or generator. (A battery consists of a number of cells that convert chemical energy to electrical energy, and a generator converts mechanical energy of motion to electrical energy.) The electric field is also considered to be constant along the length, ℓ, of the cylinder. As a consequence, the slope of the potential variation along the cylinder is also a constant. This is indicated by the dashed lines in Fig. 5B. Since the electric field is constant, the current density is also constant; the total current entering the end caps is the product of the current density and the cross-sectional area. The value of the electric field is the quotient of the total voltage and the length of the cylinder. Combining these results and introducing two new entities gives the following set of equations:

$$J = \sigma \frac{V}{\ell} \text{ since } J = \sigma E \text{ and } E = \frac{V}{\ell}$$

$$I = J(A) = \frac{\sigma A V}{\ell}$$

$$\rho = \frac{1}{\sigma} \text{ and } V = I\left(\frac{\rho \ell}{A}\right)$$

$$R = \frac{\rho \ell}{A} \text{ and } V = IR$$

where

ρ = the resistivity of the conducting material

R = the resistance

The final equation, called Ohm's Law, is a very basic one in circuit theory. Configurations similar to the one shown in Fig. 5 are very common in electrical circuits and are called resistors.

It will be shown in a later section that the power dissipated in a resistor is equal to the product of the resistance and the square of the current. Resistance is often an undesirable effect (such as in a wire carrying current from one location to another one), and in these situations must be reduced as much as possible. This can be accomplished by using a conductor with a low resistivity such as silver or copper (which is close to silver in resistivity, but is not as expensive) with a large cross-sectional area and as short a length as possible. The current-carrying capability decreases as the diameter of a conductor gets smaller.

Potential Drop and Electromotive Force

The application of the relationships between fields, potential and concepts similar to the physical configuration shown in Fig. 5 permitted the derivation of the formula that eliminated further consideration of the field problem. The idea of an electrical energy source was also introduced. A similar analysis involving mechanics and field theory would be required to determine the characteristics of an electrical generator, and an application of chemistry would be involved in designing a chemical cell. However, it will be assumed that this problem has been solved and that the energy source can be replaced with a symbol such as that used in Fig. 5.

Electrical Fundamentals   2-3

The term electromotive force (EMF) is applied to describe a source of electrical energy, and potential drop (or voltage drop) is used for a device that consumes electrical energy. A combination of sources and resistances (or other elements) that are connected in some way is called a network or circuit.

Charge Polarity and Electron Flow

The "+" and "−" symbols assigned to electromotive forces and potential drops are important in that they define the polarity of voltage and direction of current flow. These plus and minus symbols (representing positive and negative) were first used in the 18th century by Benjamin Franklin to describe two types of electric charge. Charged atoms (called ions) having more than the usual number of electrons have a negative charge, and those having less than the usual number are said to be positively charged. If a polarizing force is applied to some matter and then removed, the atoms will tend to revert to their natural states. This means that atoms deficient in electrons will attract the needed particles from those atoms having excess electrons. Thus, the transfer of electrons is from negative ions to positive ions. When the path for electrons is an electrical circuit, the current flows from negative to positive around the potential generator.

The direction of electron flow is important in applications of thermionic and semiconductor devices. The cathode of a vacuum tube is heated so that it will boil off electrons. Current will flow in the tube if and only if the anode is biased positive with respect to the cathode. This is known as the Edison effect.

Most modern circuits employ a chassis or ground plane or bus as a common conductor. This practice reduces the wiring or printed circuitry required and simplifies the schematic diagram. When the negative terminal of the power source is connected to this "ground" system, electrons flow from the negative terminal through the ground system and through the circuit elements to the positive terminal. While this is certainly a correct description of the action, it is more convenient to think of the common conductor as the return leg for all circuits. To accommodate this reasoning, electrical engineers have adopted a positive-to-negative convention. This convention is adhered to in most of the technical literature. The arrows in semiconductor schematic symbols point in the direction of this conventional current and away from actual electron flow.

In discussing network elements having one terminal connected to the "common," "ground" or "return" leg of the circuit, engineers use the terms source and sink to describe the current flow. A device is a current source if current flows away from the ungrounded terminal, and a current sink if current flows into the ungrounded terminal.

Alternating Currents

In considering current flow, it is natural to think of a single, constant force causing the electrons to move. When this is so, the electrons always move in the same direction through a path or circuit made up of conductors connected together in a continuous chain. Such a current is called a direct current, abbreviated dc. It is the type of current furnished by batteries and by certain types of generators.

It is also possible to have an EMF that periodically reverses. With this kind of EMF the current flows first in one direction through the circuit and then in the other. Such an EMF is called an alternating EMF, and the current is called an alternating current (abbreviated ac). The reversals (alternations) may occur at any rate from a few per second up to several billion per second. Two reversals make a cycle; in one cycle the force acts first in one direction, then in the other and then returns to the first direction to begin the next cycle. The number of cycles that occur in one second is called the frequency of the alternating current. The inverse of frequency, or the time duration of one cycle, is the period of the current.

The difference between direct current and alternating current is shown in Fig. 6. In these graphs, the horizontal axis indicates time, increasing toward the right. The vertical axis represents the amplitude or strength of the current, increasing in either the up or down direction away from the horizontal axis. If the graph is above the horizontal axis, the current is flowing in one direction through the circuit (indicated by the + sign). If it is below the horizontal axis the current is flowing in the

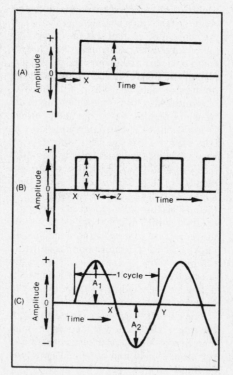

Fig. 6 — Three types of current flow. A — direct current; B — intermittent direct current; C — alternating current.

reverse direction through the circuit (indicated by the − sign). In Fig. 6A, assume that we close the circuit or make the path for the current complete at the time indicated by X. The current instantly takes the amplitude indicated by the height A. After that, the current continues at the same amplitude as time goes on. This is an ordinary direct current.

In Fig. 6B, the current starts flowing with the amplitude A at time X, continues at that amplitude until time Y and then instantly ceases. After an interval YZ the current again begins to flow and the same sort of start-and-stop performance is repeated. This is an intermittent direct current. We could get it by alternately closing and opening a switch in the circuit. It is a direct current because the direction of current flow does not change; the amplitude is always on the + side of the horizontal axis. The intermittent direct current illustrated has an ac component, however, that can be isolated by an electrical circuit called a filter. Filtering is discussed in greater detail in later sections.

In Fig. 6C the current starts at zero, increases in amplitude as time goes on until it reaches the amplitude A_1 while flowing in the + direction, then decreases until it drops to zero amplitude once more. At that time (X) the direction of the current flow reverses; this is indicated by showing the next part of the graph below the axis. As time goes on the amplitude increases, with the current now flowing in the − direction, until it reaches amplitude A_2. Then the amplitude decreases until finally it drops to zero (Y) and the direction reverses once more. This is an alternating current.

Waveforms

The type of alternating current shown in Fig. 6C is known as a sine wave. An electrodynamic machine called an alternator generates this waveshape because the current induced in the stator winding is pro-

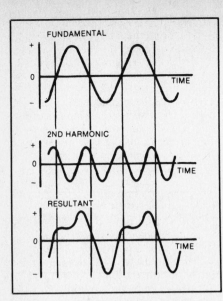

Fig. 7 — A complex waveform. A fundamental (top) and second harmonic (center) added together, point by point at each instant, result in the waveform shown at the bottom. When the two components have the same polarity at a selected instant, the resultant is the simple sum of the two. When they have opposite polarities, the resultant is the difference; if the negative-polarity component is larger, the resultant is negative at the instant.

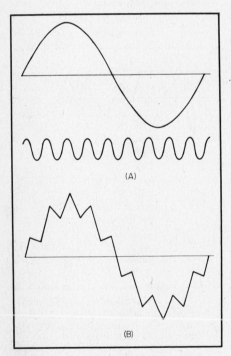

Fig. 8 — Two tones of dissimilar frequency and amplitude (A) are easily recognizable in the composite waveform (B).

portional to the sine of the angle that the winding makes with the magnetic flux lines produced by the rotating field. It is also possible to generate a sine wave electronically.

The variations in many ac waves are not so smooth, nor is one half cycle necessari-

ly just like the preceding one in shape. However, these complex waves can be shown to be the sum of two or more sine waves of frequencies that are exact integral (whole-number) multiples of some lower frequency. The lowest frequency is called the fundamental, and the higher frequencies are called harmonics.

Fig. 7 shows how a fundamental and a second harmonic (twice the fundamental) might add to form a complex wave. Simply by changing the relative amplitudes of the two waves, as well as the times at which they pass through zero amplitude, an infinite number of waveshapes can be constructed from just a fundamental and second harmonic. More complex waveforms can be constructed if more harmonics are used.

When two or more sinusoidal or complex signals that are not necessarily harmonically related are applied to a common load resistor, the resultant waveform is the sum of the instantaneous voltages. If the two signals have significantly different frequencies and amplitudes, they are easily distinguishable as components of a composite wave.

The illustration in Fig. 8 is an example of this phenomenon. Two signals having equal amplitudes and nearly equal frequencies combine to produce a composite wave that is not so simply analyzed. Shown in Fig. 9 are two signals having a frequency relationship of 1.5:1. When the positive peaks coincide, the resultant amplitude is twice that of either tone. Similarly, when the maximum negative excursion of one signal corresponds with the maximum positive excursion of the other, the resultant amplitude is the algebraic sum, or zero. The negative peaks never coincide; therefore this composite waveform is not symmetrical about the zero axis. Notice the periodic variation in the amplitude or envelope of the composite waveform. This variation has a frequency equal to the difference or beat between the two tones.

FREQUENCY AND WAVELENGTH

Frequencies ranging from about 20 to 20,000 cycles per second or hertz are called audio frequencies, because the vibrations of air particles that our ears recognize as sounds occur at a similar rate. Audio frequencies (abbreviated AF) are used to actuate loudspeakers and thus create sound waves.

Frequencies above about 20,000 hertz (Hz) are called radio frequencies (RF) because they are useful in radio transmission. Frequencies all the way up to and beyond 100,000,000,000 Hz have been used for radio purposes. At radio frequencies it becomes convenient to use a unit larger than the hertz. Three such units are the kilohertz, which is equal to 1000 Hz and is abbreviated kHz, the megahertz, equal to 1,000,000 hertz or 1000 kilohertz, and is abbreviated MHz, and the gigahertz, equal to 1,000,000,000 hertz or 1000 MHz

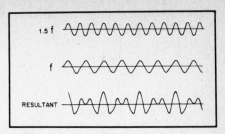

Fig. 9 — The graphic addition of equal-amplitude signals fairly close in frequency illustrates the phenomenon of beats. The beat-note frequency is 1.5f − f = 0.5f, and is visible in the resultant waveform.

Table 1

Conversion Factors for Fractional and Multiple Units

Change From	To	Divide By	Multiply By
Units	picounits		10^{12}
	nanounits		10^9
	microunits		10^6
	milliunits		10^3
	kilounits	10^3	
	megaunits	10^6	
	gigaunits	10^9	
Picounits	nanounits	10^3	
	microunits	10^6	
	milliunits	10^9	
	units	10^{12}	
Nanounits	picounits		10^3
	microunits	10^3	
	milliunits	10^6	
	units	10^9	
Microunits	picounits		10^6
	nanounits		10^3
	milliunits	10^3	
	units	10^6	
Milliunits	picounits		10^9
	nanounits		10^6
	microunits		10^3
	units	10^3	
Kilounits	units		10^3
	megaunits	10^3	
	gigaunits	10^6	
Megaunits	units		10^6
	kilounits		10^3
	gigaunits	10^3	
Gigaunits	units		10^9
	kilounits		10^6
	megaunits		10^3

and is abbreviated GHz. Table 1 shows how to convert between the various units in common use. The prefixes attached to the basic-unit name indicate the nature of the unit. These prefixes are

pico — one trillionth (abbreviated p)
nano — one billionth (abbreviated n)
micro — one millionth (abbreviated μ)
milli — one thousandth (abbreviated m)
kilo — one thousand (abbreviated k)
mega — one million (abbreviated M)
giga — one billion (abbreviated G)

Various radio frequencies are divided into classifications. These classifications constitute the frequency spectrum as far as it extends for radio purposes at the present time.

Frequency	Classification	Abbrev.
10 to 30 kHz	Very low freq.	VLF
30 to 300 kHz	Low freq.	LF
300 to 3000 kHz	Medium freq.	MF
3 to 30 MHz	High freq.	HF
30 to 300 MHz	Very high freq.	VHF
300 to 3000 MHz	Ultrahigh freq.	UHF
3 to 30 GHz	Superhigh freq.	SHF
30 to 300 GHz	Extremely high freq.	EHF

Wavelength

Radio waves travel at the same speed as light — about 300,000,000 meters per second or 186,000 miles per second in space. These waves can be set up by a radio-frequency current flowing in a circuit. This happens because the rapidly changing current sets up a magnetic field that changes in the same way, and the varying magnetic field in turn sets up a varying electric field. Whenever this happens, the two fields radiate at the speed of light.

Suppose an RF current has a frequency of 3,000,000 Hz. The field will go through complete reversals (one cycle) in 1/3,000,000 second. In that same period of time the field (the wave) moves 300,000,000/3,000,000 meters, or 100 meters. By the time the wave has moved that distance the next cycle has begun and a new wave has started out. In other words, the beginning of the first wave covers a distance of 100 meters before the beginning of the next, and so on. This distance is the wavelength.

The longer the time of one cycle (the lower the frequency) the greater the distance covered by each wave, and hence the longer the wavelength. The relationship between wavelength and frequency is shown by the formula

$$\lambda = \frac{300,000}{f\ (kHz)}$$

where
λ = wavelength in meters
f = frequency in kilohertz

or
$$\lambda = \frac{300}{f\ (MHz)}$$

where
λ = wavelength in meters
f = frequency in megahertz

Example: The wavelength corresponding to a frequency of 3650 kilohertz is

$$\lambda = \frac{300,000}{3650} = 82.2\ meters$$

PHASE

The term phase essentially means time, or the time interval between the instant when one thing occurs and the instant when a second related thing takes place. The later event is said to lag the earlier, while the one that occurs first is said to lead. In ac circuits the current amplitude changes continuously, so the concept of phase or time becomes important. Phase can be measured in ordinary time units, such as the second, but there is a more convenient method.

Since each ac or RF cycle occupies exactly the same amount of time as every other cycle of the same frequency, we can use the cycle itself as the time unit. Using the cycle as the time unit makes the specification or measurement of phase independent of the frequency of the current, as long as only one frequency is under consideration at a time. When two or more frequencies are to be considered, as in the case where harmonics are present, the phase measurements are made with respect to the lowest, or fundamental, frequency.

The time interval or phase difference under consideration usually will be less than one cycle. With a sine wave of current, the amplitude of the current at any instant is proportional to the sine of the fraction of the cycle completed at that moment from the instant the cycle began. Phase difference could be measured in decimal parts of a cycle or radian measure, but it is often more convenient to divide the cycle into 360 parts or degrees. A phase degree is therefore 1/360 of a cycle (Fig. 10).

Measuring Phase

The phase difference between two currents of the same frequency is the temporal or angular difference between corresponding parts of cycles of the two currents. This is shown in Fig. 11. The current labeled A leads the one marked B by 45°, since the A cycles begin 45° earlier in time. It is equally correct to say that B lags A by 45°.

Two important special cases are shown in Fig. 12. In the upper drawing B lags 90° behind A; that is, its cycle begins just one quarter cycle later than that of A. When one wave is passing through zero, the other is just at its maximum point.

In the lower drawing A and B are 180° out of phase. In this case it does not matter which one is considered to lead or lag. B is always positive while A is negative, and vice versa. The two waves are thus completely out of phase.

The waves shown in Figs. 10 and 11 could represent current, voltage or both. A and B might be two currents in separate circuits, or A might represent voltage and B current in the same circuit. If A and B represent two currents in the same circuit (or two voltages in the same circuit) the

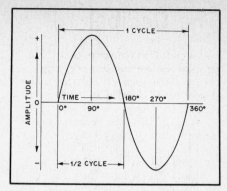

Fig. 10 — An ac cycle is divided into 360 degrees that are used as a measure of time or phase.

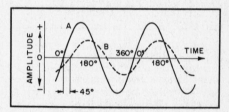

Fig. 11 — When two waves of the same frequency start their cycles at slightly different times, the time difference or phase difference is measured in degrees. In this drawing wave B starts 45° (one eighth cycle) later than wave A, and so lags 45° behind A.

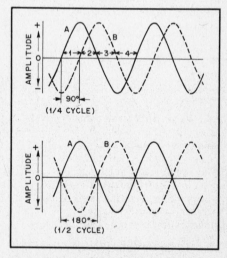

Fig. 12 — Two important special cases of phase difference. In the upper drawing, the phase difference between A and B is 90°; in the lower drawing the phase difference is 180°.

total or resultant current (or voltage) also is a sine wave, because adding any number of sine waves of the same frequency always gives a sine wave also of the same frequency.

The Decibel

It is useful to appraise signal strength in terms of relative loudness as registered by the ear. For example, if a person estimates that a signal is twice as loud when the transmitter power is increased from 10 watts to 100 watts, he or she will also estimate that a 1000-watt signal is twice as loud as a 100-watt signal. The human ear has a logarithmic response.

This fact is the basis for the use of the relative power unit called the decibel (dB). A decibel is one-tenth of a bel, the unit of sound named for Alexander Graham Bell. A change of 1 decibel in power level is just detectable as a change in loudness under ideal conditions. The number of decibels corresponding to a given power ratio is given by:

$$dB = 10 \log_{10}\left(\frac{P_2}{P_1}\right)$$

Note that the decibel is based on power ratios. Voltage or current ratios can be used, but only when the impedance is the same for both values of voltage, or current. The gain of an amplifier cannot be expressed correctly in decibels if it is based on the ratio of the output voltage to the input voltage unless both voltages are measured across the same value of impedance. When the impedance at both points of measurement is the same, the following formula may be used for voltage or current ratios:

$$dB = 20 \log\left(\frac{V_2}{V_1}\right) \text{ or } 20 \log\left(\frac{I_2}{I_1}\right)$$

where

V = voltage
I = current

If the voltage formula above is applied to an amplifier, where V_2 is the output voltage and V_1 is the input voltage, a positive decibel value indicates amplifier gain. On the other hand, applying the formula to a resistor network would result in a negative decibel value, signifying a loss.

When the decibel value is known, the numerical ratio can be calculated from:

$$\frac{P_2}{P_1} = \text{antilog}\left(\frac{dB}{10}\right) = 10^{\frac{dB}{10}}$$

or

$$\frac{V_2}{V_1} = \text{antilog}\left(\frac{dB}{20}\right) = 10^{\frac{dB}{20}}$$

Many mathematics textbooks contain tables of logarithms, or these numbers can be produced very quickly with a calculator. In any case, it is convenient to memorize the decibel values for a few of the common power and voltage ratios. For power changes, a numerical ratio of 2 is 3 dB, 4 is 6 dB, 10 is 10 dB, 100 is 20 dB, 1000 is 30 dB, and so on. When voltage changes are considered, doubling the voltage causes a 6-dB increase, a numerical ratio of 10 is worth 20 dB, 100 is 40 dB and so on. One can interpolate between known ratios to estimate a gain or loss within 1 decibel. Inverting a numerical ratio simply reverses the algebraic sign of the decibel value. For example, a voltage gain of 10 corresponds to 20 dB, while a gain of 1/10 (which is a loss of 10) corresponds to −20 dB.

In a system of cascaded gain and loss blocks where numerical ratios are specified for each block, the overall system gain or loss can be calculated this way: Convert the ratios to all gains or all losses (gains become fractional losses or losses become fractional gains). The overall numerical gain or loss will be the product of the individual figures, and the decibel value can be derived as before. If the individual gains and losses are given in decibels, the procedure is again to convert to all losses or gains. Only the algebraic signs need to be changed; that is, decibel losses become negative gains. Gains may also be treated as negative losses. The overall decibel gain or loss is the algebraic sum of the individual figures, and this can

be converted to a numerical ratio if desired. Fig. 13 illustrates both methods.

The decibel is a relative unit. When using decibels to specify an absolute voltage, current or power level, the decibel value must be qualified by a reference level. For example, in a discussion of sound intensity, a reference level of 1 dB corresponds to an acoustical field strength of 10^{-16} W / cm^2, the normal human hearing threshold at 600 Hz. A lion's roar at 20 feet might have a sound intensity of 90 dB, and the threshold of pain occurs at 130 dB. Thus, the human ear/brain has a dynamic range of 130 dB, or a ratio of 10 trillion to one.

In radio work, power is often rendered in dBW (decibels referenced to 1 watt) or dBm (decibels referenced to 1 milliwatt). With this system, 2 kilowatts equals +63 dBm or +33 dBW, and 5 microwatts equals −23 dBm or −53 dBW. Voltages are sometimes given as decibel values with respect to 1 volt or 1 microvolt; 2 millivolts equals +66 dBμV or −54 dBV. Antenna gain is specified with respect to some standard reference element such as isotropic radiator or a dipole. The measurement units are the dBi (gain over isotropic) and dBd (gain over a half-wave dipole). In spectrum analysis, noise, spurious signals and distortion products can be referenced to the carrier (if one exists), dBc. A certain frequency synthesizer might have a phase-noise specification of −40 dBc, 100 Hz removed from the carrier.

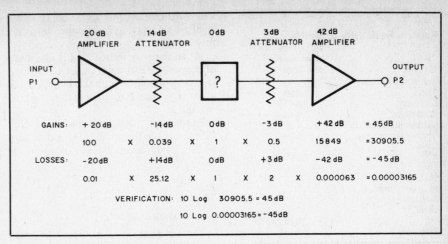

Fig. 13 — How to compute the composite gain of a system. Either numerical ratios or decibels may be used, but all must be consistent in gain or loss. The method is explained in the text.

Resistance and Conductance

Given two conductors of the same size and shape, but of different materials, the amount of current that will flow when a given EMF is applied will be found to vary with the resistance of the material. The

lower the resistance, the greater the current for a given value of EMF.

Resistance is measured in ohms (Ω). A circuit has a resistance of 1 ohm when an applied EMF of 1 volt causes a current of

1 ampere to flow. The resistivity of a material is the resistance, in ohms, of a cube of the material measuring 1 centimeter on each edge. One of the best conductors is copper, and in making resistance calcula-

Table 2

Relative Resistivity of Metals

Material	Resistivity Compared to Copper
Aluminum (pure)	1.60
Brass	3.7-4.90
Cadmium	4.40
Chromium	1.80
Copper (hard-drawn)	1.03
Copper (annealed)	1.00
Gold	1.40
Iron (pure)	5.68
Lead	12.80
Nickel	5.10
Phosphor bronze	2.8-5.40
Silver	0.94
Steel	7.6-12.70
Tin	6.70
Zinc	3.40

tions it is frequently convenient to compare the resistance of the material under consideration with that of a copper conductor of the same size and shape. Table 2 gives the ratio of the resistivity of various conductors to that of copper.

The longer the physical path, the higher the resistance of that conductor. For direct current and low-frequency alternating currents (up to a few thousand hertz) the resistance is inversely proportional to the cross-sectional area of the path the current must travel; that is, given two conductors of the same material and having the same length, but differing in cross-sectional area, the one with the larger area will have the lower resistance.

RESISTANCE OF WIRES

The problem of determining the resistance of a round wire of given diameter and length — or its converse, finding a suitable size and length of wire to provide a desired amount of resistance — can be solved easily with the help of the copper wire table given in Chapter 35. This table gives the resistance, in ohms per thousand feet, of each standard wire size.

Example: Suppose a resistance of 3.5 ohms is needed and some no. 28 wire is on hand. The wire table in Chapter 35 shows that no. 28 wire has a resistance of 66.17 ohms per thousand feet. Since the desired resistance is 3.5 ohms, the length of wire required is

$$\frac{3.5}{66.17} \times 1000 = 52.89 \text{ ft}$$

Suppose that the resistance of wire in a circuit must not exceed 0.05 ohm and that the length of wire required for making the connections totals 14 feet. Then

$$\frac{14}{1000} \times R = 0.05 \text{ ohm}$$

where R is the maximum allowable resistance in ohms per thousand feet. Rearranging the equation gives

$$R = \frac{0.05 \times 1000}{14} = 3.57 \text{ ohms/1000 ft}$$

The wire table shows that no. 15 is the smallest size having a resistance less than this value.

When the wire in question is not made of copper, the resistance values given in the wire table should be multiplied by the ratios given in Table 2 to obtain the resulting resistance.

Example: If the wire in the first example were made from nickel instead of copper, the length required for 3.5 ohms would be

$$\frac{3.5}{66.17 \times 5.1} \times 1000 = 10.37 \text{ ft}$$

Temperature Effects

The resistance of a conductor changes with its temperature. The resistance of practically every metallic conductor increases with increasing temperature. Carbon, however, acts in the opposite way; its resistance decreases when its temperature rises. It is seldom necessary to consider temperature in making resistance calculations for amateur work. The temperature effect is important when it is necessary to maintain a constant resistance under all conditions. Special materials that have little or no change in resistance over a wide temperature range are used in that case.

Resistors

A package of material exhibiting a certain amount of resistance made up into a single unit is called a resistor. Different resistors having the same resistance value may be considerably different in physical size and construction (Fig. 14). Current flowing through a resistance causes the conductor to become heated; the higher the resistance and the larger the current, the greater the amount of heat developed. Resistors intended for carrying large currents must be physically large so the heat can be radiated quickly to the surrounding air. If the resistor does not dissipate the heat quickly it may get hot enough to melt or burn.

Skin Effect

A conductor's effective resistance is not the same for alternating current as it is for direct current. With alternating current, other effects tend to force the current to flow mostly in the outer parts of the conductor. This decreases the effective cross-sectional area of the conductor, with the result that the resistance increases.

At low audio frequencies the increase in resistance is insignificant, but at radio frequencies above about 1 MHz, this "skin effect" is so pronounced that practically all the current flows within a few thousandths of an inch of a copper conductor surface. Above 10 MHz, only about the outer 0.0008 inch of the surface is used, and above 100 MHz the depth is less than 0.0003 inch. The RF resistance is consequently many times the dc resistance, and increases with frequency. In the RF range

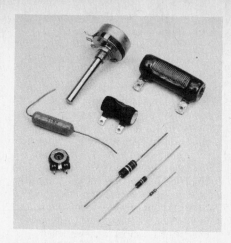

Fig. 14 — Examples of various resistors. In the right foreground are ¼-, ½- and 1-watt composition resistors. The three larger cylindrical components at the center are wire-wound power resistors. The remaining two parts are variable resistors, PC-board mount at the lower left and panel mount at the upper center.

a conductor of thin tubing will have just as low resistance as a solid conductor of the same diameter, because material not close to the surface carries practically no current.

Conductance

The reciprocal of resistance (1/R) is conductance. It is usually represented by the symbol G. A circuit having high conductance has low resistance, and vice versa. In radio work the term is used chiefly in connection with electron-tube and field-effect-transistor characteristics. The unit of conductance is the siemens, abbreviated S. A resistance of 1 ohm has a conductance of 1 siemens, a resistance of 1000 ohms has a conductance of 0.001 siemens, and so on. A unit frequently used in connection with electron devices is the microsiemens or one-millionth of a siemens. It is the conductance of a 1-million-ohm resistance.

OHM'S LAW

One of the simplest forms of electric circuit is a battery with a resistance connected to its terminals, as shown by the symbols in Fig. 15. A complete circuit must have an unbroken path so current can flow out of the battery, through the apparatus connected to it, and back into the battery. If a connection is removed at any point, the

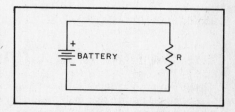

Fig. 15 — A simple circuit consisting of a battery and resistor.

circuit is broken, or open. A switch is a device for making and breaking connections and thereby closing or opening the circuit, either allowing current to flow or preventing it from flowing.

The values of current, voltage and resistance in a circuit are by no means independent of each other. The relationship between them is known as Ohm's Law. It can be stated as follows: The current flowing in a circuit is directly proportional to the applied EMF and inversely proportional to the resistance. Expressed as an equation, it is

$$I \text{ (amperes)} = \frac{E \text{ (volts)}}{R \text{ (ohms)}}$$

The equation gives the value of current when the voltage and resistance are known. It may be transposed so that each of the three quantities may be found when the other two are known:

$$E = IR$$

(that is, the voltage is equal to the current in amperes multiplied by the resistance in ohms) and

$$R = \frac{E}{I}$$

(or the resistance of the circuit is equal to the applied voltage divided by the current).

All three forms of the equation are used often in radio work. It must be remembered that the quantities are in volts, ohms and amperes; other units cannot be used in the equations without first being converted. For example, if the current is in milliamperes it must be changed to the equivalent fraction of an ampere before the value can be substituted in the equations.

The following examples illustrate the use of Ohm's Law:

The current flowing through a resistance of 20,000 ohms is 150 milliamperes. What is the voltage? Since the voltage is to be found, the equation to use is E = IR. The current must first be converted from milliamperes to amperes, and reference to Table 1 shows that to do so it is necessary to divide by 1000. Therefore,

$$E = \frac{150}{1000} \times 20{,}000 = 3000 \text{ volts}$$

When a voltage of 150 is applied to a circuit, the current is measured at 2.5 amperes. What is the resistance of the circuit? In this case R is the unknown, so

$$R = \frac{E}{I} = \frac{150}{2.5} = 60 \text{ ohms}$$

No conversion was necessary because the voltage and current were given in volts and amperes.

How much current will flow if 250 volts is applied to a 5000-ohm resistor? Since I is unknown

$$I = \frac{E}{R} = \frac{250}{5000} = 0.05 \text{ ampere}$$

Milliampere units would be more con-

venient for the current, and 0.05 ampere × 1000 = 50 milliamperes.

SERIES AND PARALLEL RESISTANCES

Very few actual electric circuits are as simple as Fig. 15. Commonly, resistances are found connected in a variety of ways. The two fundamental methods of connecting resistances are shown in Fig. 16. In the upper drawing, the current flows from the source of EMF (in the direction shown by the arrow, let us say) down through the first resistance, R1, then through the second, R2, and then back to the source. These resistors are connected in series. The current everywhere in the circuit has the same value.

In the lower drawing, the current flows to the common connection point at the top of the two resistors and then divides, one part of it flowing through R1 and the other through R2. At the lower connection point these two currents again combine; the total is the same as the current that flowed into the upper common connection. In this case the two resistors are connected in parallel.

Resistors in Parallel

In a circuit with resistances in parallel, the total resistance is less than that of the lowest value of resistance present. This is because the total current is always greater than the current in any individual resistor. The formula for finding the total resistance of resistances in parallel is

$$R = \cfrac{1}{\cfrac{1}{R1} + \cfrac{1}{R2} + \cfrac{1}{R3} + \cfrac{1}{R4} + \cdots}$$

where the dots again indicate that any number of resistors can be combined by the same method. For only two resistances in parallel (a very common case), the formula becomes

$$R = \frac{R1 \times R2}{R1 + R2}$$

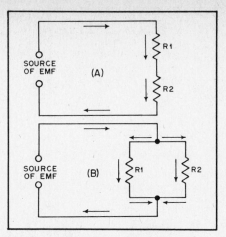

Fig. 16 — Resistors connected in series, A, and in parallel, B.

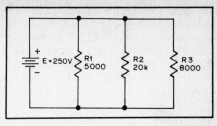

Fig. 17 — An example of resistors in parallel. The solution is worked out in the text.

Example: If a 500-ohm resistor is paralleled with one of 1200 ohms, the total resistance is

$$R = \frac{R1 \times R2}{R1 + R2} = \frac{500 \times 1200}{500 + 1200}$$

$$R = \frac{600{,}000}{1700} = 353 \text{ ohms}$$

Kirchhoff's First Law

There is another solution to the problem. Suppose the three resistors of the previous example are connected in parallel as shown in Fig. 17. The same EMF, 250 volts, is applied to all three of the resistors. The current in each can be found from Ohm's Law as shown below, I1 being the current through R1, I2 the current through R2 and I3 the current through R3.

For convenience, the resistance will be expressed in kilohms so the current will be in milliamperes.

$$I1 = \frac{E}{R1} = \frac{250}{5} = 50 \text{ mA}$$

$$I2 = \frac{E}{R2} = \frac{250}{20} = 12.5 \text{ mA}$$

$$I3 = \frac{E}{R3} = \frac{250}{8} = 31.25 \text{ mA}$$

The total current is

$$I = I1 + I2 + I3$$
$$I = 50 + 12.5 + 31.25$$
$$I = 93.75 \text{ mA}$$

This example illustrates Kirchhoff's current law: The current flowing into a node or branching point is equal to the sum of the individual currents leaving the node or branching point. The total resistance of the circuit is therefore

$$R = \frac{E}{I} = \frac{250}{93.75} = 2.667 \text{ kilohms}$$

$$R = 2667 \text{ ohms}$$

Resistors in Series

When a circuit has a number of resistances connected in series, the total resistance of the circuit is the sum of the individual resistances. If these are numbered R1, R2, R3 and so on, then

$$R \text{ (total)} = R1 + R2 + R3 + R4 + \cdots$$

where the dots indicate that as many

resistors as necessary may be added.

Example: Suppose that three resistors are connected to a source of EMF as shown in Fig. 18. The EMF is 250 volts. R1 is 5000 ohms, R2 is 20 kilohms (20,000 ohms) and R3 is 8000 ohms. The total resistance is then

R = R1 + R2 + R3
R = 5000 + 20,000 + 8000
R = 33,000 ohms.

The current flowing in the circuit is then

$$I = \frac{E}{R} = \frac{250}{33,000} = 0.00758 \text{ ampere}$$

I = 7.58 mA.

(We need not carry calculations beyond three significant figures; often, two will suffice because the accuracy of measurements is seldom better than a few percent.)

Kirchhoff's Second Law

Ohm's Law applies in any portion of a circuit as well as to the circuit as a whole. Although the current is the same in all three of the resistances in the example of Fig. 18, the total voltage divides among them. The voltage appearing across each resistor (the voltage drop) can be found from Ohm's Law.

Example: If the voltage across R1 is called E1, that across R2 is called E2, and that across R3 is called E3, then

E1 = IR1 = 0.00758 × 5000 = 37.9 volts
E2 = IR2 = 0.00758 × 20,000
   = 151.5 volts
E3 = IR3 = 0.00758 × 8000
   = 60.6 volts

Kirchhoff's second law accurately describes the situation in the circuit: The sum of the voltages in a closed current loop is equal to zero. The resistors are sinks of power, while the battery is a source of power, so the sign conversion described earlier causes the voltage potentials across the resistors to be opposite in sign from the battery voltage. Adding all the voltages yields zero. In the simple case of a single voltage source, simple algebraic manipulation implies that the sum of the individual voltage drops in the circuit must be equal to the applied voltage.

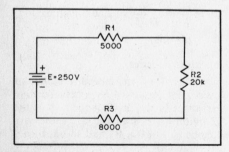

Fig. 18 — An example of resistors in series. The solution of the circuit is worked out in the text.

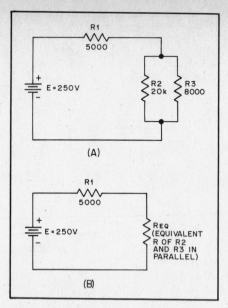

Fig. 19 — At A, an example of resistors in series-parallel. The equivalent circuit is shown at B. The solution is worked out in the text.

E = E1 + E2 + E3
E = 37.9 + 151.5 + 60.6
E = 250 volts

In problems such as this, when the current is small enough to be expressed in milliamperes, considerable time and trouble can be saved if the resistance is expressed in kilohms rather than in ohms. When the resistance in kilohms is substituted directly in Ohm's Law, the current will be milliamperes if the EMF is in volts.

Resistors in Series-Parallel

A circuit may have resistances both in parallel and in series, as shown in Fig. 19A. The method for analyzing such a circuit is as follows: Consider R2 and R3 to be the equivalent of a single resistor, R_{eq}, whose value is equal to R2 and R3 in parallel.

$$R_{eq} = \frac{R2 \times R3}{R2 + R3} = \frac{20,000 \times 8000}{20,000 + 8000}$$

$$R_{eq} = 5.71 \text{ k}\Omega$$

This resistance in series with R1 forms a simple series circuit, as shown in Fig. 19B. The total resistance in the circuit is then

R = R1 + R_{eq} = 5 kΩ + 5.71 kΩ
R = 10.71 kΩ

The current is

$$I = \frac{E}{R} = \frac{250 \text{ V}}{10.71 \text{ k}\Omega} = 23.3 \text{ mA}$$

The voltage drops across R1 and R_{eq} are

E1 = I × R1 = 23.3 × 5 = 117 volts
E2 = I × R_{eq} = 23.3 mA × 5.71 kΩ
E2 = 133 volts

with sufficient accuracy. These two voltage drops total 250 volts, as described by

Kirchhoff's first law. E2 appears across both R2 and R3 so,

$$I2 = \frac{E2}{R2} = \frac{133 \text{ V}}{20 \text{ k}\Omega} = 6.67 \text{ mA}$$

$$I3 = \frac{E2}{R3} = \frac{133 \text{ V}}{8 \text{ k}\Omega} = 16.67 \text{ mA}$$

where
  I2 = current through R2
  I3 = current through R3.

The sum of I2 and I3 is equal to 23.3 mA, conforming to Kirchhoff's second law.

Thevenin's Theorem

A useful tool for simplifying electrical networks is Thevenin's Theorem, which states that any two-terminal network of resistors and voltage or current sources can be replaced by a single voltage source and a series resistor. Such a transformation can simplify the calculation of current through a parallel branch. Thevenin's Theorem can be readily applied to the circuit of Fig. 19A, to find the current through R3.

In this example, R1 and R2 form a voltage divider circuit, with R3 as the load (Fig. 20A). The current drawn by the load (R3) is simply the voltage potential across R3, divided by its resistance. Unfortunately, the value of R2 affects the voltage potential across R3, just as the presence of R3 affects the potential appearing across R2. Some means of separating the two is needed; hence the Thevenin-equivalent circuit.

The voltage of the Thevenin-equivalent battery is the open circuit voltage as measured when no current is flowing from either terminal A or B. Without a load connected between A and B, the total current through the circuit is (from Ohm's Law)

$$I = \frac{E}{R1 + R2}$$

and the voltage between terminals A and B (E_{ab}) is

$$E_{ab} = I \times R2$$

By substituting the first equation into the second, a simplified expression for E_{ab} can be found to be

$$E_{ab} = \frac{R2}{R1 + R2} \times E$$

Using the actual values, this becomes

$$E_{ab} =$$

$$\frac{20,000}{5,000 + 20,000} \times 250 = 200 \text{ volts}$$

when nothing is connected to terminals A or B. With no current drawn, E then is equal to E_{ab}.

The Thevenin-equivalent resistance is the total net resistance between terminals A and B. The ideal voltage source, by definition, has zero internal resistance. As-

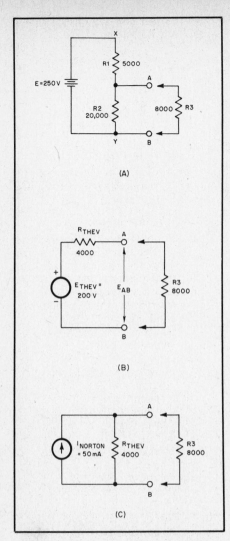

(A)

(B)

(C)

Fig. 20 — Thevenin-equivalent circuit for the circuit in Fig. 19.

Assuming the battery to be a close approximation of an ideal source, put a short between points X and Y in the circuit of Fig. 20A. R1 and R2 are then effectively placed in parallel, as viewed from terminals A and B. The Thevenin-equivalent resistance is then

$$R_{thev} = \frac{R1 \times R2}{R1 + R2} = 4000 \text{ ohms}$$

This gives the Thevenin-equivalent circuit as shown in Fig. 20B. The circuits of Figs. 19A and 19B as far as R3 is concerned are equivalent.

Once R3 is connected to terminals A and B, current will flow through R_{thev}, causing a voltage drop across R_{thev}, and reducing E_{ab}. However, the current through R3 is equal to

$$I = \frac{E_{thev}}{R_{total}} = \frac{E_{thev}}{R_{thev} + R3}$$

Substituting actual values gives

$$I = \frac{200}{4000 + 8000} = 16.67 \text{ mA}$$

This agrees with the value calculated earlier.

Norton's Theorem

Another tool for analyzing electrical networks is Norton's Theorem, which states that any two-terminal network of resistors and current or voltage sources can be replaced by a single current source and a parallel resistor. Norton's Theorem is to current sources what Thevenin's Theorem is to voltage sources. In fact, the Thevenin resistance as calculated previously is used as the equivalent resistance when using Norton's Theorem.

The circuit just analyzed by means of Thevenin's Theorem can be analyzed just as easily by Norton's Theorem. The equivalent Norton circuit is shown in Fig. 20C. The current I_{sc} of the equivalent current source is the short circuit current through terminals A and B. In the case of the voltage divider shown in Fig. 20A, the short circuit current is

$$I_{sc} = \frac{E}{R1} = \frac{250}{5000} = 50 \text{ mA}$$

The resulting Norton-equivalent circuit consists of a 50-mA current source placed in parallel with a 4000-ohm resistor. When R3 is connected to terminals A and B, by Kirchhoff's second law one third of the supply current flows through R3 and the remainder through R_{thev}. This gives a current through R3 of 16.67 mA, again agreeing with previous conclusions.

A Norton-equivalent circuit can be transformed into a Thevenin-equivalent circuit and vice versa. The equivalent resistor stays the same in both cases; it is placed in series with the voltage source in the case of a Thevenin-equivalent circuit, and in parallel with the current source in the case of a Norton-equivalent circuit. The voltage for a Thevenin-equivalent source is equal to the no load voltage appearing across the resistor in the Norton-equivalent circuit. The current for a Norton-equivalent source is equal to the short circuit current provided by the Thevenin source.

POWER AND ENERGY

Power, the rate of doing work, is equal to voltage multiplied by current. The unit of electrical power, called the watt, is equal to 1 volt multiplied by 1 ampere. The equation for power therefore is

$$P = EI$$

where

P = power in watts
E = EMF in volts
I = current in amperes.

Common fractional and multiple units for power are the milliwatt, one thousandth of a watt, and the kilowatt, or 1000 watts.

Example: The plate voltage on a transmitting vacuum tube is 2000 volts and the plate current is 350 milliamperes. (The

current must be changed to amperes before substitution in the formula, and so is 0.35 ampere.) Then

$$P = EI = 2000 \times 0.35 = 700 \text{ watts}$$

By substituting the Ohm's Law equivalent for E and I, the following formulas are obtained for power:

$$P = \frac{E^2}{R}$$

$$P = I^2R$$

These formulas are useful in power calculations when the resistance and either the current or voltage (but not both) are known.

Example: How much power will be used up in a 4000-ohm resistor if the potential applied to it is 200 volts? From the equation

$$P = \frac{E^2}{R} = \frac{200^2}{4000} = \frac{40,000}{4000}$$

$$P = 10 \text{ watts}$$

Now, suppose a current of 20 milliamperes flows through a 300-ohm resistor. Then

$$P = I^2R = 0.02^2 \times 300$$
$$P = 0.0004 \times 300$$
$$P = 0.12 \text{ watt}$$

Note that the current was changed from milliamperes to amperes before substitution in the formula.

Electrical power in a resistance is turned into heat. The greater the power, the more rapidly the heat is generated. Resistors for radio work are made in many sizes, the smallest being rated to dissipate (or carry safely) about 1/10 watt. The largest resistors commonly used in amateur equipment will dissipate about 100 watts.

When electrical energy is converted into mechanical energy, and vice versa, the following relationship holds: 1 horsepower = 746 watts. This formula assumes lossless transformation; practical efficiency is taken up shortly.

GENERALIZED DEFINITION OF RESISTANCE

Electrical power is not always turned into heat. The power used in running a motor, for example, is converted to mechanical motion. The power supplied to a radio transmitter is largely converted into radio waves. Power applied to a loudspeaker is changed into sound waves. But in every case of this kind, the power is completely used up — it cannot be recovered. Also, for proper operation of the device the power must be supplied at a definite ratio of voltage to current. Both of these features are characteristics of resistance, so it can be said that any device that dissipates power has a definite value of resistance.

This concept of resistance as something that absorbs power at a definite voltage-to-current ratio is very useful; it permits substituting a simple resistance for the load or power-consuming part of the device receiving power, often with considerable simplification of calculations. Of course, every electrical device has some resistance of its own in the more narrow sense, so a part of the power supplied to it is dissipated in that resistance and hence appears as heat even though the major part of the power may be converted to another form.

Efficiency

In devices such as motors and vacuum tubes, the objective is to obtain power in some form other than heat. Therefore power used in heating is considered to be a loss, because it is not useful power. The efficiency of a device is the useful power output (in its converted form) divided by the power input to the device. In a vacuum-tube transmitter, for example, the objective is to convert power from a dc source into ac power at some radio frequency. The ratio of the RF power output to the dc input is the efficiency of the tube. That is,

$$\text{Eff.} = \frac{P_o}{P_i}$$

where
Eff. = efficiency (as a decimal)
P_o = power output (watts)
P_i = power input (watts).

Example: If the dc input to the tube is 100 watts, and the RF power output is 60 watts, the efficiency is

$$\text{Eff.} = \frac{P_o}{P_i} = \frac{60}{100} = 0.6$$

Efficiency is usually expressed as a percentage; that is, it tells what percent of the input power will be available as useful output. The efficiency in the above example is 60%.

Suppose a mobile transmitter has an RF power output of 100 W at an efficiency of 52 percent at 13.8 V. The vehicular alternator system charges the battery at a 5-A rate at this voltage. Assuming an alternator efficiency of 68 percent, how much horsepower must the engine produce to operate the transmitter and charge the battery? Solution: To charge the battery, the alternator must produce 13.8 V × 5 A = 69 W. The transmitter dc input power is 100 W/0.52 = 192.3 W. Therefore, the total electrical power required from the alternator is 192.3 + 69 = 261.3 W. The engine load then is

$$\frac{261.3}{746 \times 0.68} = 0.515 \text{ horsepower}$$

Energy

In residences, the power company's bill is for electrical energy, not for power. What you pay for is the work that electricity does for you, not the rate at which that work is done. Electrical work is equal to power multiplied by time; the common unit is the watt-hour, which means that a power

of 1 watt has been used for one hour. That is,

W = PT

where
W = energy in watt-hours
P = power in watts
T = time in hours.

Other energy units are the kilowatt-hour and the watt-second (joule). These units are self-explanatory.

Energy units are seldom used in amateur practice, but it is obvious that a small amount of power used for a long time can eventually result in a power bill that is just as large as though a large amount of power had been used for a very short time.

AC Waveform Measurements

The time dependence of alternating current raises questions about defining and measuring values of voltage, current and power. Because these parameters change from one instant to the next, one might wonder, for example, which point on the cycle characterizes the voltage or current for the entire cycle. When you are viewing a single-tone signal (that is, a pure sine wave) on an oscilloscope, the easiest dimension to measure is the total vertical displacement, or peak-to-peak voltage.

This value, abbreviated P-P, is important in evaluating the signal-handling capability of a linear processing device such as an electronic amplifier or ferromagnetic transformer. If the steepest part of the waveform has a potential of zero, the signal has equal positive and negative excursions and no dc bias. The oscilloscope measurement of the maximum positive or negative excursion, or maximum instantaneous potential, is called the peak voltage, and in a symmetrical waveform it has half the value of the peak-to-peak amplitude. Insulators, air gaps and capacitor dielectrics must withstand the peak value of an ac voltage. In a well-designed ac-to-dc power supply the rectified dc output voltage will be nearly equal to the peak ac voltage.

When an ac voltage is applied to a resistor, the resistor will dissipate energy in the form of heat, just as if the voltage were dc. The dc voltage that would cause identical heating in the ac-excited resistor is the root-mean-square (RMS) value of the ac voltage. The RMS voltage of any waveform can be determined with the use of integral calculus, but for a pure sine wave the following relationships hold:

$$V_{peak} = V_{RMS} \times \sqrt{2} \approx V_{RMS} \times 1.414$$

and

$$V_{RMS} = \frac{V_{peak}}{\sqrt{2}} \approx V_{peak} \times 0.707$$

Unless otherwise specified or obvious from context, ac voltage is rendered as an RMS value. For example, the household

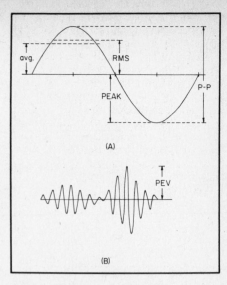

Fig. 21 — Ac voltage and current measurements. The sine-wave parameters are illustrated at A, while B shows the peak envelope voltage (PEV) for a composite waveform.

117-V ac outlet provides 117 V_{RMS}, 165.5 V_{peak} and 331 V_{P-P}.

An electrodynamic instrument such as a meter movement responds to the average value of an ac waveform. Again, integral calculus is required for computation of the average value of the general (complex) wave, but for a sinusoidal signal the peak, RMS and average voltages are related by the formulas:

$$V_{avg} = V_{peak} \times \frac{2}{\pi} \approx 0.63662 \ V_{peak}$$
$$\approx 0.9003 \ V_{RMS}$$

and

$$V_{peak} = \frac{V_{avg}}{0.636} = 1.5708 \ V_{avg}$$

Thus, our 117-V ac outlet provides an average voltage of 105.2.

Fig. 21A illustrates the four voltage parameters of a sine wave. The most accurate way to determine the RMS voltage of a complex wave is to measure the heat produced by applying the complex voltage to a known resistance and measure the dc voltage required to produce the same heat. However, some modern electronic voltmeters provide accurate RMS readings by performing mathematical operations on the waveform. The ratio of peak voltage to RMS voltage of an ac signal is called the crest factor. From the relationships presented earlier, the crest factor of a sine wave is the square root of 2.

The significant dimension of a multitone signal is the peak envelope voltage, shown in Fig. 21B. PEV is important in calculating the power in a modulated signal, such as that from an amateur SSB voice transmitter.

All that has been said about voltage measurements applies also to current (provided the load is resistive) because the waveshapes are identical. However, the terms RMS, average and peak have different meanings when they refer to ac power. The reason is that while voltage and current are sinusoidal functions of time, power is the product of voltage and current, and this product is a sine squared function. The mathematical operations that define RMS, average, and so on will naturally yield different results when applied to this new function. The relationships between ac voltage, current and power follow:

RMS voltage × RMS current = average power ≠ RMS power. The average power used to heat a resistor is equal to the dc power required to produce the same heat. RMS power is a mathematical curiosity only and has no physical significance.

Peak voltage × peak current = 2 × average power. Unfortunately, the definition given above for peak power conflicts with the meaning of the term when it is used in radio work. The peak power output of a radio transmitter is the power averaged over the RF cycle having the greatest amplitude. Modulated signals are

not purely sinusoidal because they are composites of two or more tones. However, the cycle-to-cycle variation is small enough that sine-wave measurement techniques produce accurate results. In the context of radio signals, then, peak power means maximum average power. Peak envelope power (PEP) is the parameter most often used to express the maximum signal-handling capability of a linear amplifier. To compute the PEP of a waveform such as that sketched in Fig. 21B, multiply the PEV by 0.707 to obtain the RMS value, square the result and divide by the load resistance.

Capacitance

Suppose two flat metal plates are placed close to each other (but not touching) and are connected to a battery through a switch, as illustrated in Fig. 22. At the instant the switch is closed, electrons are attracted from the upper plate to the positive terminal of the battery, and the same number are repelled into the lower plate from the negative battery terminal. Enough electrons move into one plate and out of the other to make the EMF between them the same as the EMF of the battery.

If the switch is opened after the plates have been charged in this way, the top plate is left with a deficiency of electrons and the bottom plate with an excess. Since there is no current path between the two, the plates remain charged despite the fact that the battery no longer is connected. If a wire is touched between the two plates (short-circuiting them), the excess electrons on the bottom plate flow through the wire to the upper plate, restoring electrical neutrality. The plates are discharged.

These two plates are representative of an electrical capacitor, a device possessing the property of storing electricity in the electric field between its plates. During the time the electrons are moving — that is, while the capacitor is being charged or discharged — a current flows in the circuit even though the circuit apparently is broken by the gap between the capacitor plates. However, the current flows only during the time of charge and discharge, and this time is usually very short. There can be no continuous flow of direct current through a capacitor. However, alternating current can "pass through" a capacitor. As fast as one plate is charged positively by the positive excursion of the alternating current, the other plate is being charged negatively. Positive current flowing into one plate causes a current to flow out of the other plate during one half of the cycle, while the reverse occurs during the other half of the cycle.

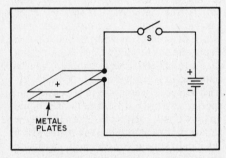

Fig. 22 — A simple capacitor.

The charge or quantity of electricity that can be held in the electric field between the capacitor plates is proportional to the applied voltage and to the capacitance of the capacitor:

$$Q = CV$$

where
- Q = charge in coulombs
- C = capacitance in farads
- V = potential in volts

The energy stored in a capacitor is also a function of potential and capacitance:

$$W = \frac{V^2 C}{2}$$

where
- W = energy in joules (watt-seconds)
- V = potential in volts
- C = capacitance in farads

The numerator of this expression can be derived easily from the definitions given earlier for charge, capacitance, current, power and energy. However, the denominator is not so obvious. It arises because the voltage across a capacitor is not constant, but is a function of time; the average voltage over the time interval determines the energy stored. The time dependence of the capacitor voltage is

Table 3

Dielectric Constants and Breakdown Voltages

Material	Dielectric Constant*	Puncture Voltage**
Air	1.0	21
Alsimag 196	5.7	240
Bakelite	4.4-5.4	300
Bakelite, mica filled	4.7	325-375
Cellulose acetate	3.3-3.9	250-600
Fiber	5-7.5	150-180
Formica	4.6-4.9	450
Glass, window	7.6-8	200-250
Glass, Pyrex	4.8	335
Mica, ruby	5.4	3800-5600
Mycalex	7.4	250
Paper, Royalgrey	3.0	200
Plexiglas	2.8	990
Polyethylene	2.3	1200
Polystyrene	2.6	500-700
Porcelain	5.1-5.9	40-100
Quartz, fused	3.8	1000
Steatite, low loss	5.8	150-315
Teflon	2.1	1000-2000

*At 1 MHz **In volts per mil (0.001 inch)

discussed in the section on time constant.

The larger the plate area and the smaller the spacing between the plates, the greater the capacitance. The capacitance also depends on the kind of insulating material between the plates; it is smallest with air insulation, and substitution of other insulating materials for air may increase the capacitance many times. The ratio of the capacitance with some material other than air between the plates to the capacitance of the same capacitor with air insulation is called the dielectric constant of that particular insulating material. The material itself is called a dielectric. The dielectric constants of a number of materials commonly used as dielectrics in capacitors are given in Table 3. If a sheet of polystyrene is substituted for air between the plates of

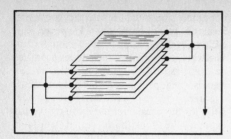

Fig. 23 — A multiple-plate capacitor. Alternate plates are connected together.

a capacitor, for example, the capacitance will be increased 2.6 times.

Units of Capacitance

The basic unit of capacitance is the farad, but this unit is much too large for practical work. Capacitance is usually measured in microfarads (abbreviated μF) or picofarads (pF). The microfarad is one millionth of a farad (10^{-6} F), and the picofarad is one millionth of a microfarad (10^{-12} F). Capacitors nearly always have more than two plates, the alternate plates being connected together to form two sets, as shown in Fig. 23. This makes it possible to attain a fairly large capacitance in a small space, since several plates of smaller individual area can be stacked to form the equivalent of a single large plate of the same total area. Also, all plates except the two on the ends are exposed to plates of the other group on both sides, and so are twice as effective in increasing the capacitance.

The formula for calculating capacitance is

$$C = \frac{0.224 \, KA}{d} \, (n - 1)$$

where

C = capacitance in pF

K = dielectric constant of material between plates
A = area of one side of one plate in square inches
d = separation of plate surfaces in inches
n = number of plates

If the plates in one group do not have the same area as the plates in the other, use the area of the smaller plates.

Capacitors in Radio

The types of capacitors used in radio work differ considerably in physical size, construction and capacitance. Representative types are shown in Fig. 24. In variable capacitors (almost always constructed with air for the dielectric) one set of plates is made movable with respect to the other set so the capacitance can be varied. Fixed capacitors (those having a single, nonadjustable value of capacitance) can also be made with metal plates and with air as the dielectric. However, fixed capacitors usually are constructed from plates of metal foil with a thin solid or liquid dielectric sandwiched between, so that a relatively large capacitance can be obtained in a small unit. The solid dielectrics commonly used are mica, paper and special ceramics. An example of a liquid dielectric is mineral oil. The electrolytic capacitor uses aluminum-foil plates with a semiliquid conducting chemical compound between them. The actual dielectric is a very thin film of insulating material that forms on one set of plates through electrochemical action when a dc voltage is applied to the capacitor. The capacitance obtained with a given plate area in an electrolytic capacitor is very large, compared with capacitors having other dielectrics, because the film is so thin — much less than any thickness practical with a solid dielectric.

The use of electrolytic and oil-filled capacitors is confined to power-supply filtering and audio-bypass applications because their dielectrics have high losses at the higher frequencies. Mica and ceramic capacitors are used throughout the frequency range from audio to several hundred megahertz.

Voltage Breakdown

When high voltage is applied to the plates of a capacitor, considerable force is exerted on the electrons and nuclei of the dielectric. The dielectric is an insulator; the electrons do not become detached from atoms the way they do in conductors. If the force is great enough the dielectric will break down. Usually failed dielectrics will puncture and offer a low resistance current path between the two plates.

The breakdown voltage a dielectric can withstand depends on the chemical composition and thickness of the dielectric, as shown in Table 3. Breakdown voltage is not directly proportional to the thickness; doubling the thickness does not quite double the breakdown voltage. Gas dielectrics break down, as evidenced by a spark or arc between the plates. Once the voltage is removed, the arc ceases and the capacitor is ready for use again. Solid dielectrics are permanently damaged by dielectric breakdown, and often will totally short out and melt or explode. Breakdown occurs at a lower voltage between pointed or sharp-edged surfaces than between rounded and polished surfaces; consequently, the breakdown voltage between metal plates of given spacing in air can be increased by buffing the edges of the plates.

A thick dielectric must be used to withstand high voltages, and since the capacitance is inversely proportional to dielectric thickness (plate spacing) for a given plate area, a high-voltage capacitor must have more plate area than a low-voltage one of the same capacitance. High-voltage, high-

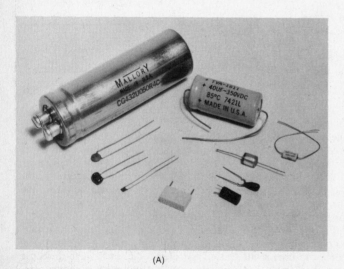

(A)

(B)

Fig. 24 — Fixed-value capacitors are shown at A. A large computer-grade unit is at the upper left. This 40-μF unit is an electrolytic capacitor. The smaller pieces are silver-mica, disc-ceramic, tantalum, polystrene and ceramic-chip capacitors. The small black unit (cylindrical) is a PC board-mount electrolytic. Variable capacitors are shown at B. A vacuum variable is at the upper left.

capacitance capacitors are physically large.

Capacitors in Series and Parallel

The terms parallel and series when applied to capacitance have a similar meaning as with resistances. When a number of capacitors are connected in parallel, as in Fig. 25A, the total capacitance of the group is equal to the sum of the individual capacitances:

$$C_{total} = C1 + C2 + C3 + C4 + \ldots$$

When two or more capacitors are connected in series, as in Fig. 25B, the total capacitance is less than that of the smallest capacitor in the group. The rule for finding the capacitance of a number of series-connected capacitors is the same as that for finding the resistance of a number of parallel-connected resistors. That is,

$$C_{total} = \cfrac{1}{\dfrac{1}{C1} + \dfrac{1}{C2} + \dfrac{1}{C3} + \dfrac{1}{C4} + \dfrac{1}{C5} + \ldots}$$

and, for only two capacitors in series,

$$C_{total} = \frac{C1 \times C2}{C1 + C2}$$

The same units must be used throughout; that is, all capacitances must be expressed in either μF or pF; both kinds of units cannot be used in the same equation.

Capacitors are usually connected in parallel to obtain a larger total capacitance than is available in one unit. The largest voltage that can be applied safely to a group of capacitors in parallel is the voltage that can be applied safely to the one having the lowest voltage rating.

When capacitors are connected in series, the applied voltage is divided up among them according to Kirchhoff's first law, and the situation is much the same as when resistors are in series and there is a voltage drop across each. The voltage that appears across each capacitor of a group connected in series is in inverse proportion to its capacitance, as compared with the capacitance of the whole group.

Example: Three capacitors having capacitances of 1, 2 and 4 μF, respectively, are connected in series as shown in Fig. 26. The total capacitance is

$$C = \cfrac{1}{\dfrac{1}{C1} + \dfrac{1}{C2} + \dfrac{1}{C3}} = \cfrac{1}{\dfrac{1}{1} + \dfrac{1}{2} + \dfrac{1}{4}}$$

$$C = \frac{1}{7/4} = \frac{4}{7} = 0.571 \ \mu F$$

The voltage across each capacitor is proportional to the total capacitance divided by the capacitance of the capacitor in question, so the voltage across C1 is

$$E1 = \frac{0.571}{1} \times 2000 = 1143 \text{ volts}$$

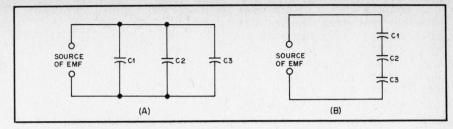

Fig. 25 — Capacitors in parallel, A, and in series, B.

Similarly, the voltages across C2 and C3 are

$$E2 = \frac{0.571}{2} \times 2000 = 571 \text{ volts}$$

$$E3 = \frac{0.571}{4} \times 2000 = 286$$

The sum of these three voltages equals 2000 volts, the applied voltage.

Capacitors are frequently connected in series to enable the group to withstand a larger voltage (at the expense of decreased total capacitance) than any individual capacitor is rated to stand. As shown by the previous example, the applied voltage does not divide equally among the capacitors (except when all the capacitances are precisely the same) and care must be taken to ensure that the voltage rating of no capacitor in the group is exceeded.

RC TIME CONSTANT

Connecting a source of EMF directly to the terminal of a capacitor charges the capacitor to the full EMF almost instantaneously. Any resistance added to the circuit as in Fig. 27A limits the amount of current flow, lengthening the time required for the EMF between the capacitor plates to build up to the same value as the EMF of the source. During this charging period, the current flowing from the source into the capacitor gradually decreases from its initial value; the increasing EMF stored in the capacitor's electric field offers increasing opposition to the steady EMF of the source.

The voltage potential between the terminals of the capacitor while being charged is an exponential function of time, and is given by

$$V_{(t)} = E \left(1 - e^{\frac{-t}{RC}} \right)$$

where
$V_{(t)}$ = capacitor EMF in volts at t
E = potential of charging source in volts
t = time in seconds after initiation of charging current
e = natural logarithmic base = 2.718
R = circuit resistance in ohms
C = capacitance in farads

Theoretically, the charging process is never really finished, but eventually the charging current drops to an unmeasurable value. When t = RC, the above equation becomes

$$V_{(RC)} = E(1 - e^{-1}) \approx 0.632E$$

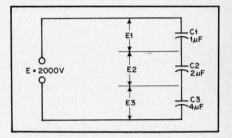

Fig. 26 — An example of capacitors connected in series. Finding the voltage drops, E1 through E3, is worked out in the text.

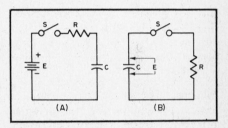

Fig. 27 — Illustrating the time constant of an RC circuit.

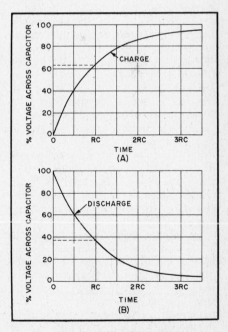

Fig. 28 — At A, how the voltage across a capacitor rises, with time, when charged through a resistor. The curve at B shows the way in which the voltage decreases across the capacitor terminals on discharging through the same value of resistance. From a practical standpoint, a capacitor may be considered as charged (or discharged) after 5RC times.

The product of R in ohms times C in farads is called the time constant of the circuit and is the time in seconds required to charge the capacitor to 63.2% of the supply voltage. After two time constants (t = 2RC) the capacitor charges another 63.2% of the difference between the capacitor voltage at one time constant and the supply voltage, for a total charge of 86.5%. After three time constants the capacitor reaches 95% of the supply voltage, and so on, as illustrated in the curve of Fig. 28A.

If a charged capacitor is discharged through a resistor, as indicated in Fig. 27B, the same time constant applies for the decay of the capacitor voltage. A direct short circuit applied between the capacitor's terminals would discharge instantly. R limits the current flow, so the capacitor voltage decreases only as rapidly as the capacitor can discharge itself through R. A capacitor discharging through a resistance exhibits the same time constant characteristics (calculated in the same way as above) as a charging capacitor. The voltage as a function of time while the capacitor is being discharged is given by

$$V_{(t)} = E\left(e^{\frac{-t}{RC}}\right)$$

where t = time in seconds after initiation of discharge.

Alternating Current in Capacitance

In Fig. 29 a sine-wave ac voltage having a maximum value of 100 is applied to a capacitor. In the period 0A, the applied voltage increases from 0 to 38; at the end of this period the capacitor is charged to that voltage. In interval AB the voltage increases to 71; that is, 33 volts additional. In this interval a smaller quantity of charge has been added than in 0A, because the voltage rise during interval AB is smaller. Consequently the average current during AB is smaller than during 0A. In the third interval, BC, the voltage rises from 71 to 92, an increase of 21 volts. This is less than the voltage increase during AB, so the quantity of electricity added is less; in other words, the average current during interval BC is still smaller. In the fourth interval, CB, the voltage increases only 8 volts; the charge added is smaller than in any preceding interval and therefore the current also is smaller.

By dividing the first quarter cycle into a very large number of intervals, it could be shown that the current charging the capacitor has the shape of a sine wave, just as the applied voltage does. The current is largest at the beginning of the cycle and becomes zero at the maximum value of the voltage, so there is a phase difference of 90° between the voltage and the current.

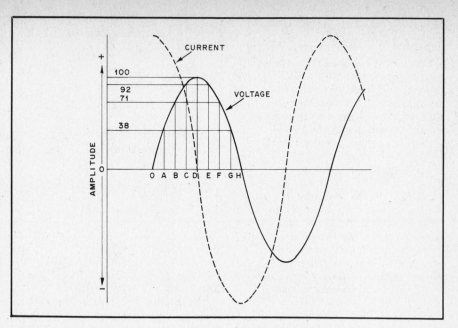

Fig. 29 — Voltage and current phase relationships when an alternating voltage is applied to a capacitor.

During the first quarter cycle the current is flowing in the normal direction through the circuit, since the capacitor is being charged. Hence the current is positive, as indicated by the dashed line in Fig. 29.

In the second quarter cycle — that is, in the time from D to H — the voltage applied to the capacitor decreases. During this time the capacitor loses its charge. Applying the same reasoning, it is evident that the current is small in interval DE and continues to increase during each succeeding interval. However, the current is flowing against the applied voltage because the capacitor is discharging into the circuit. The current flows in the negative direction during this quarter cycle.

The third and fourth quarter cycles repeat the events of the first and second, respectively, with this difference — the polarity of the applied voltage has reversed, and the current changes to correspond. In other words, an alternating current flows in the circuit because of the alternate charging and discharging of the capacitance. As shown in Fig. 29, the current starts its cycle 90° before the voltage, so the current in a capacitor leads the applied voltage by 90°.

Capacitive Reactance

The quantity of electric charge that can be placed on a capacitor is proportional to the applied EMF and the capacitance. This amount of charge moves back and forth in the circuit once each cycle, and so the rate of movement of charge (the current) is proportional to voltage, capacitance and frequency. When the effects of capacitance and frequency are considered together, they form a quantity that plays a part similar to that of resistance in Ohm's Law. This quantity is called reactance. The unit for reactance is the ohm, just as in the case of resistance. The formula for reactance is

$$X_C = \frac{1}{2\pi f C}$$

where

X_C = capacitive reactance in ohms
f = frequency in hertz
C = capacitance in farads
π = 3.1416

Although the unit of reactance is the ohm, there is no power dissipation in reactance. The energy stored in the capacitor in one portion of the cycle is simply returned to the circuit in the next.

The fundamental units for frequency and capacitance (hertz and farads) are too cumbersome for practical use in radio circuits. However, if the capacitance is specified in microfarads (μF) and the frequency is in megahertz (MHz), the reactance calculated from the previous formula retains the unit ohms.

Example: The reactance of a capacitor of 470 pF (0.00047 μF) at a frequency of 7.15 MHz is

$$X = \frac{1}{2\pi f c} = \frac{1}{2\pi \times 7.15 \times 0.00047}$$

$$X = 47.4 \text{ ohms}$$

Inductance

It is possible to show that the flow of current through a conductor is accompanied by magnetic effects; a compass needle brought near the conductor will deflect from its normal north-south position. The current sets up a magnetic field.

The transfer of energy to a magnetic field represents work performed by the source of EMF. Power is required for doing work, and since power is equal to current multiplied by voltage, there must be a voltage drop in the circuit while energy is being stored in the field. This voltage drop, exclusive of any voltage drop caused by resistance in the circuit, is the result of an opposing voltage induced in the circuit while the field is building up to its final value. Once the field becomes constant the induced EMF or back EMF disappears, since no further energy is being stored. The induced EMF opposes the EMF of the source and tends to prevent the current from rising rapidly when the circuit is closed. The amplitude of the induced EMF is proportional to the rate at which the current changes and to a constant associated with the circuit itself; the inductance of the circuit.

Inductance depends on the physical configuration of the conductor. If a conductor is formed into a coil, its inductance is increased. A coil of many turns will have more inductance than one of few turns, if both coils are otherwise physically similar. Furthermore, if a coil is placed around an iron core its inductance will be greater than it was without the magnetic core.

The polarity of an induced EMF is always such as to oppose any change in the current in the circuit. This means that when the current in the circuit is increasing, work is being done against the induced EMF by storing energy in the magnetic field. If the current in the circuit tends to decrease, the stored energy of the field returns to the circuit, and thus adds to the energy being supplied by the source of EMF. This tends to keep the current flowing even though the applied EMF may be decreasing or be removed entirely. The energy stored in the magnetic field of an inductor is given by

$$W = \frac{I^2 L}{2}$$

where

W = energy in joules
I = current in amperes
L = inductance in henrys

The unit of inductance is the henry. Values of inductance used in radio equipment vary over a wide range. In radio-frequency circuits, the inductance values used will be measured in millihenrys (a mH is one thousandth of a henry) at low frequencies, and in microhenrys (μH, one millionth of a henry) at medium frequencies and higher. Although coils for radio frequencies may be wound on special iron cores (ordinary iron is not suitable), many RF coils made and used by amateurs are of the air-core type, that is, wound on an insulating support consisting of non-magnetic material (Fig. 30).

Every conductor passing current has a magnetic field associated with it, and therefore inductance, even though the conductor is not formed into a coil. The inductance of a short length of straight wire is small, but it may not be negligible — if the current through it changes rapidly, the induced voltage may be appreciable. This is the case in even a few inches of wire when an alternating current having a frequency of the order of 100 MHz or higher is flowing. At much lower frequencies the inductance of the same wire could be ignored because the induced voltage would be negligible.

Calculating Inductance

The approximate inductance of a single-layer air-core coil may be calculated from the simplified formula

$$L \ (\mu H) = \frac{d^2 n^2}{18d + 40\ell}$$

where

L = inductance in microhenrys
d = coil diameter in inches
ℓ = coil length in inches
n = number of turns

The notation is explained in Fig. 31. This formula is a close approximation for coils having a length equal to or greater than 0.4d.

Example: Assume a coil has 48 turns wound at 32 turns per inch and a diameter of ¾ inch. Thus, d = 0.75 , ℓ = 48/32 = 1.5, and n = 48. Substituting,

$$L = \frac{0.75^2 \times 48^2}{18 \times 0.75 + 40 \times 1.5}$$

$$L = \frac{1296}{73.5} = 17.6 \ \mu H$$

To calculate the number of turns of a single-layer coil for a required value of inductance

$$n = \frac{\sqrt{L \ (18d + 40\ell)}}{d}$$

Example: Suppose an inductance of 10 μH is required. The form on which the coil is to be wound has a diameter of one inch and is long enough to accommodate a coil of 1¼ inches. Then d = 1, ℓ = 1.25 and L = 10. Substituting,

$$n = \frac{\sqrt{10 \ [(18 \times 1) + (40 \times 1.25)]}}{1}$$

$$n = \sqrt{680} = 26.1 \text{ turns}$$

A 26-turn coil would be close enough in practical work. Since the coil will be 1.25 inches long, the number of turns per inch will be 26.1 / 1.25 = 20.9. Consulting the wire table in Chapter 35, we find that no. 17 enameled wire (or anything smaller) can be used. The proper inductance is obtained by winding the required number of turns on the form and then adjusting the spacing between the turns to make a uniformly spaced coil 1.25 inches long.

Inductance Charts

Most inductance formulas lose accuracy when applied to small coils (such as are used in VHF work and in low-pass filters built for reducing harmonic interference to television) because the conductor thickness is no longer negligible in comparison with the size of the coil. Fig. 32 shows the measured inductance of VHF coils, and may be used as a basis for circuit design. Two curves are given; curve A is for coils wound to an inside diameter of ½ inch; curve B is for coils of ¾ inch inside diameter. In both curves the wire size is no. 12, winding pitch is eight turns to the inch (1/8 inch center-to-center turn spacing). The inductance values given include leads ½ inch long.

Machine-wound coils with the diameters and turns per inch given in Tables 4 and 5 are available in many radio stores, under the trade names of B&W Miniductor, Air-dux and Polycoil. Figs. 33 and 34 are used with Tables 4 and 5.

Fig. 30 — Assorted inductors. A rotary (continuously variable) coil is shown at the upper left. Slug-tuned inductors are visible in the lower foreground. An RF choke (three pi windings) is visible at the lower right.

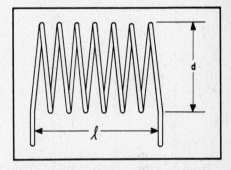

Fig. 31 — Coil dimensions used in the inductance formula. The wire diameter does not enter into the formula. The spacing has been exaggerated in this illustration for clarity. The formula is for closewound coils.

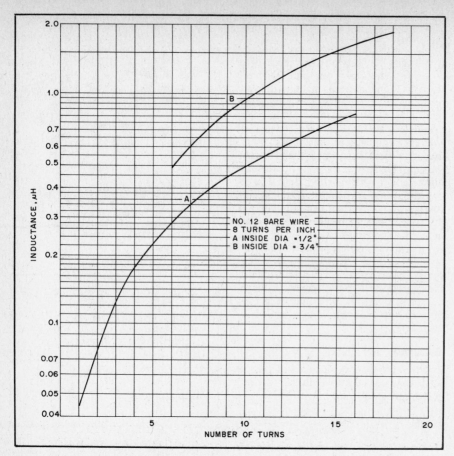

Fig. 32 — Measured inductance of coils wound with no. 12 bare wire, eight turns to the inch. The values include half-inch leads.

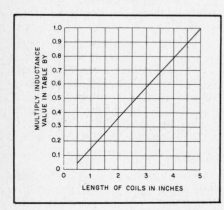

Fig. 33 — Factor to be applied to the inductance of coils listed in Table 4 for coil lengths up to 5 inches.

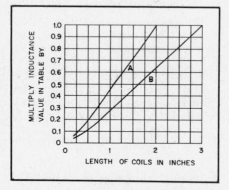

Fig. 34 — Factor to be applied to the inductance of coils listed in Table 5, as a function of coil length. Use curve A for coils marked A, and curve B for coils marked B.

Table 4
Machine-Wound Coil Specifications

Coil Dia Inches	Turns Per Inch	Inductance in μH
1¼	4	2.75
	6	6.3
	8	11.2
	10	17.5
	16	42.5
1½	4	3.9
	6	8.8
	8	15.6
	10	24.5
	16	63
1¾	4	5.2
	6	11.8
	8	21
	10	33
	16	85
2	4	6.6
	6	15
	8	26.5
	10	42
	16	108
2½	4	10.2
	6	23
	8	41
	10	64
3	4	14
	6	31.5
	8	56
	10	89

Table 5
Machine-Wound Coil Specifications

Coil Dia Inches	Turns Per Inch	Inductance in μH
½ (A)	4	0.18
	6	0.40
	8	0.72
	10	1.12
	16	2.9
	32	12
5/8 (A)	4	0.28
	6	0.62
	8	1.1
	10	1.7
	16	4.4
	32	18
¾ (B)	4	0.6
	6	1.35
	8	2.4
	10	3.8
	16	9.9
	32	40
1 (B)	4	1.0
	6	2.3
	8	4.2
	10	6.6
	16	16.9
	32	68

Forming a wire into a solenoid increases its inductance, and also introduces distributed capacitance. Since each turn is at a slightly different ac potential, each pair of turns effectively forms a parasitic capacitor. At some frequency the effective capacitance will have a reactance equal to that of the inductance, and the inductor will show self-resonance. (Reactance and resonance are treated later in this chapter.) Above the self-resonant frequency, a coil takes on the reactive properties of a capacitor instead of an inductor. The behavior of a coil with respect to frequency is illustrated in Fig. 35.

At low frequencies the inductance of a straight, round, nonmagnetic wire in free space is given by

$$L = 0.0002b\left[\left(\ln\frac{2b}{a}\right) - 0.75\right]$$

where

L = inductance in μH
a = wire radius in mm
b = wire length in mm
ln = natural logarithm = 2.303 × common logarithm (base 10)

If the dimensions are expressed in inches instead of mm, the equation may still be

used, except replace the 0.0002 value with 0.00508. Skin effect reduces the inductance at VHF and above. As the frequency approaches infinity, the 0.75 constant within the brackets approaches unity. As a practical matter, the skin effect doesn't reduce the inductance by more than a few percent.

As an example, find the inductance of

a wire that is 4 mm in diameter and 100 mm long. For the calculations, a = 2 mm (radius) and b = 100. Using the natural logarithm function, the problem is formulated as follows:

$$L = 0.0002 \, (100) \left[\ln\left(\frac{2 \times 100}{2}\right) - 0.75 \right]$$

$$L = 0.02 \, [\ln \, (100) - 0.75]$$
$$L = 0.02 \, (4.605 - 0.75)$$
$$L = (0.02) \, (3.855) = 0.077 \, \mu H$$

Fig. 36 is a graph of the inductance for wires of various radii as a function of length.

A VHF or UHF tank circuit can be fabricated from a wire parallel to a ground plane, with one end grounded. A formula for the inductance of such an arrangement is given in Fig. 37.

Suppose it is desired to find the inductance of a wire 100 mm long and 2 mm in radius, suspended 40 mm above a ground plane. (The inductance is measured between the free end and the ground plane, and the formula includes the inductance of the 40-mm grounding link.) A person skilled in the use of a sophisticated calculator could produce the answer with only a few keystrokes, but to demonstrate the use of the formula, begin by evaluating these quantities:

$$b + \sqrt{b^2 + a^2} = 100 + \sqrt{100^2 + 2^2}$$
$$= 100 + 100.02 = 200.02$$

$$b + \sqrt{b^2 + 4h^2}$$
$$= 100 + \sqrt{100^2 + 4 \times 40^2}$$
$$= 100 + 128.06$$
$$= 228.06$$

$$\frac{2h}{a} = \frac{2 \times 40}{2} = 40$$
$$\frac{b}{4} = \frac{100}{4} = 25$$

Substituting these figures into the formula yields:

$$L = 0.0004605 \times 100 \times$$

$$\log\left(\frac{2 \times 40}{2} \times \frac{200.02}{228.06}\right)$$

$$+ \, 0.0002 \, (128.06 - 100.02 + 25 - 80 + 2) = 0.066 \, \mu H$$

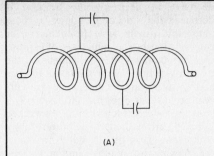

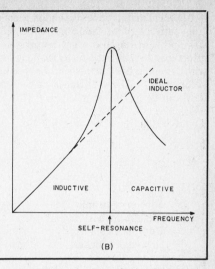

Fig. 35 — The proximity of the turns on a solenoid forms parasitic capacitors, as sketched in A. The net effect of the capacitors is called the distributed capacitance, and causes the coil to exhibit a self-resonance, illustrated in B.

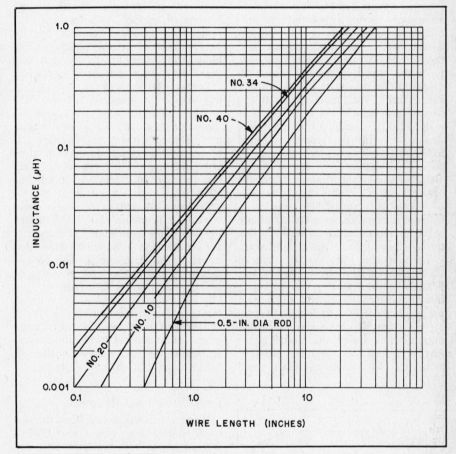

Fig. 36 — Inductance of various conductor sizes as straight wires.

$$L = 0.0004605b \left\{ \log_{10}\left[\frac{2h}{a}\left(\frac{b + \sqrt{b^2 + a^2}}{b + \sqrt{b^2 + 4h^2}} \right) \right] \right\} + 0.0002 \left(\sqrt{b^2 + 4h^2} - \sqrt{b^2 + a^2} + \frac{b}{4} - 2h + a \right)$$

where
L = inductance in μH
a = wire radius in mm
b = wire length parallel to ground plane in mm
h = wire height above ground plane in mm

Fig. 37 — Equation for determining the inductance of a wire parallel to a ground plane, with one end grounded. If the dimensions are in inches, the numerical coefficients become 0.0117 for the first term and 0.0508 for the second term.

Another conductor configuration that is frequently used for inductors is the flat strip. This arrangement has lower skin-effect loss at high frequencies than round wire because it has a higher surface-area to volume ratio. The inductance of such a strip can be found from

$$L = 0.00508b\left(\ln\frac{2b}{w + h} + 0.5 + 0.2235\frac{w + h}{b}\right)$$

where
L = inductance in microhenrys
b = length in inches
w = width in inches
h = thickness in inches

Permeability of Iron-Core Coils

Suppose the coil in Fig. 38 is wound on an iron core having a cross-sectional area of 2 square inches. When a certain current is sent through the coil, it is found that there are 80,000 lines of force in the core. Since the area is 2 square inches, the magnetic flux density is 40,000 lines per square inch. Now suppose that the iron core is removed and the same current is maintained in the coil. Also suppose the flux density without the iron core is found to be 50 lines per square inch. The ratio of these flux densities, iron core to air, is 40,000/50 or 800. This is called the permeability of the core. The inductance of the coil is increased 800 times by inserting the iron core, since the inductance will be proportional to the magnetic flux through the coils, other things being equal.

The permeability of a magnetic material varies with the flux density. At low flux densities (or with an air core), increasing the current through the coil will cause a proportionate increase in flux. But at very high flux densities, increasing the current may cause no appreciable change in the flux. When this is so, the iron is said to be saturated. Saturation causes a rapid

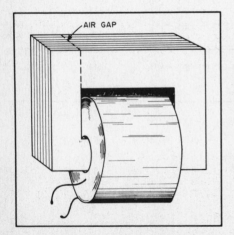

Fig. 38 — Typical construction of an iron-core inductor. The small air gap prevents magnetic saturation of the iron and thus maintains the inductance at high currents.

decrease in permeability, because it decreases the ratio of flux lines to those obtainable with the same current and an air core. Obviously, the inductance of an iron-core inductor is highly dependent on the current flowing in the coil. In an air-core coil, the inductance is independent of current because air does not saturate.

Iron-core coils such as the one sketched in Fig. 38 are used chiefly in power supply equipment. They usually have direct current flowing through the winding, and the variation in inductance with current is usually undesirable. It may be overcome by keeping the flux density below the saturation point of the iron. This is done by opening the core so there is a small air gap, indicated by the dashed lines in Fig. 38. The magnetic resistance introduced by such a gap is very large compared with that of the iron, even though the gap is only a small fraction of an inch. Therefore the gap, rather than the iron, controls the flux density. This reduces the inductance, but holds the value practically constant regardless of the magnitude of the current.

For radio-frequency work, the losses in iron cores can be reduced to a more useful level by grinding the iron into a powder and then mixing it with a "binder" of insulating material in such a way that the individual iron particles are insulated from each other. Using this approach, cores can be made that function satisfactorily even into the VHF range.

Because a large part of the magnetic path is through a nonmagnetic material (the "binder"), the permeability of the iron is low compared with the values obtained at power supply frequencies. The core is usually shaped in the form of a slug or cylinder for fit inside the insulating form on which the coil is wound. Despite the fact that the major portion of the magnetic path for the flux is in air, the slug is quite effective in increasing the coil inductance. By pushing the slug in and out of the coil, the inductance can be varied over a considerable range.

Eddy Currents and Hysteresis

When alternating current flows through a coil wound on an iron core an EMF is induced, as previously explained. Since iron is a conductor, a current flows in the core. Such currents are called eddy currents. Eddy currents represent lost power because they flow through the resistance of the iron and generate heat. Losses caused by eddy currents can be reduced by laminating the core (cutting the core into thin strips). These strips or laminations are then insulated from each other by painting them with some insulating material such as varnish or shellac.

Another type of energy loss is found in loaded inductors. Iron tends to resist any change in its magnetic state, so a rapidly changing current such as ac is continually forced to supply energy to the iron in order to overcome this "inertia." Losses caused

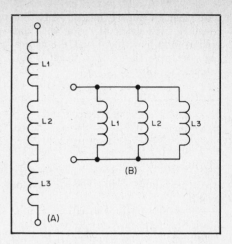

Fig. 39 — Inductances in series and parallel.

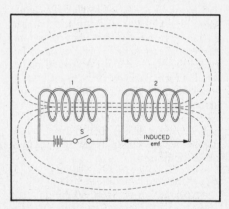

Fig. 40 — Mutual inductance. When S is closed, current flows through coil no. 1, setting up a magnetic field that induces an EMF in the turns of coil no. 2.

by this effect are referred to as hysteresis losses.

Eddy-current and hysteresis losses in iron increase rapidly as the frequency of the alternating current is increased. For this reason, ordinary iron cores can be used only at power-line and audio frequencies — up to approximately 15,000 hertz. Even then, a very good grade of iron or steel is necessary for the core to perform well at the higher audio frequencies. Iron cores of this type become completely useless at radio frequencies.

Inductances in Series and Parallel

When two or more inductors are connected in series (Fig. 39A), the total inductance is equal to the sum of the individual inductances, provided the coils are sufficiently separated so that coils are not in the magnetic field of one another. That is,

$$L_{total} = L1 + L2 + L3 + \ldots$$

If inductors are connected in parallel (Fig. 39B), and if the coils are separated sufficiently, the total inductance is given by

$$L_{total} = \frac{1}{\frac{1}{L1} + \frac{1}{L2} + \frac{1}{L3} + \ldots}$$

and for two inductances in parallel,

$$L = \frac{L1 \times L2}{L1 + L2}$$

Thus, the rules for combining inductances in series and parallel are the same for resistances, assuming the coils are far enough apart so that each is unaffected by another's magnetic field. When this is not so, the formulas given above cannot be used.

Mutual Inductance

When two coils are arranged with their axes on the same line, as shown in Fig. 40, current sent through coil 1 creates a magnetic field that cuts coil 2. Consequently, an EMF will be induced in coil 2 whenever the field strength is changing. This induced EMF is similar to the EMF of self-induction, but since it appears in the second coil because of current flowing in the first, it is a mutual effect and results from the mutual inductance between the two coils.

When all the flux set up by one coil cuts all the turns of the other coil, the mutual inductance has its maximum possible value. If only a small part of the flux set up by one coil cuts the turns of the other, the mutual inductance is relatively small. Two coils having mutual inductance are said to be coupled.

The ratio of actual mutual inductance to the maximum possible value that could theoretically be obtained with two given coils is called the coefficient of coupling between the coils. It is frequently expressed as a percentage. Coils that have nearly the maximum possible mutual inductance (coefficient = 1 or 100 percent) are said to be closely, or tightly, coupled. If the mutual inductance is relatively small the coils are said to be loosely coupled. The degree of coupling depends upon the physical spacing between the coils and how they are placed with respect to each other. Maximum coupling exists when they have a common axis and are as close together as possible (one wound over the other). The coupling is least when the coils are far apart or are placed so their axes are at right angles.

The maximum possible coefficient of coupling is closely approached when the two coils are wound on a closed iron core. The coefficient with air-core coils may run as high as 0.6 or 0.7 if one coil is wound over the other, but will be much less if the two coils are separated. Although unity coupling is suggested by Fig. 40, such coupling is possible only when the coils are wound on a closed magnetic core.

RL TIME CONSTANT

A comparable situation to an RC circuit exists when resistance and inductance are connected in series. In Fig. 41, first consider L to have no resistance and also assume that R is zero. Closing S sends a current through the circuit. The instantaneous transition from no current to a finite value, however small, represents a very rapid change in current, and a back

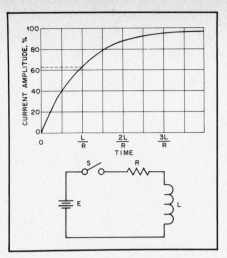

Fig. 41 — Time constant of an RL circuit.

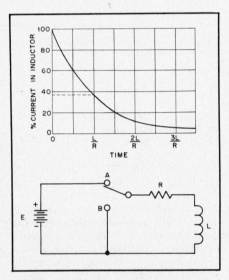

Fig. 42 — Instantaneously placing the switch in position B will discharge the inductor as shown in the graph. This is a theoretical model only; a mechanical switch cannot change state instantaneously.

EMF is developed by the self-inductance of L that is almost equal and opposite to the applied EMF. The resulting initial current is very small.

The back EMF depends on the change in current and would cease to offer opposition if the current did not continue to increase. With no resistance in the circuit (which would lead to an infinitely large current, by Ohm's Law) the current would increase forever, always growing just fast enough to keep the EMF of self-induction equal to the applied EMF.

When resistance in the circuit limits the current, Ohm's Law defines the value that the current can reach. The back EMF generated in L must only equal the difference between E and the drop across R, because the difference is the voltage actually applied to L. This difference becomes smaller as the current approaches the final Ohm's Law value. Theoretically, the back EMF never quite disappears and

so the current never quite reaches the Ohm's Law value, but practically the differences become unmeasurable after a time.

The current at any time after the switch in Fig. 41 has been closed can be found from

$$I_{(t)} = \frac{E}{R}\left(1 - e^{\frac{-tR}{L}}\right)$$

where

I_t = current in amperes at time t
E = power supply potential in volts
t = time in seconds after initiation of current
e = natural logarithmic base = 2.718
R = circuit resistance in ohms
L = inductance in henrys

The time in seconds required for the current to build up to 63.2% of the maximum value is called the time constant, and is equal to L/R, where L is in henrys and R is in ohms. After each time interval equal to this constant the circuit conducts an additional 63.2% of the remaining current. This behavior is graphed in Fig. 41.

An inductor cannot be discharged in the simple circuit of Fig. 41 because the magnetic field disappears as soon as current flow ceases. Opening S does not leave the inductor charged. The energy stored in the magnetic field returns instantly to the circuit when S is opened. The rapid disappearance of the field causes a very large voltage to be induced in the coil. Usually the induced voltage is many times larger than the voltage applied, because induced voltage is proportional to the rate that the field changes. The common result of opening the switch in such a circuit is that a spark or arc forms at the switch contacts at the instant of opening. When the inductance is large and the current in the circuit is high, large amounts of energy are released in a very short time. It is not at all unusual for the switch contacts to burn or melt under such circumstances. The spark or arc at the opened switch can be reduced or suppressed by connecting a suitable capacitor and resistor in series across the contacts. Such an RC combination is called a snubber network.

If the excitation is removed without breaking the circuit, as diagrammed in Fig. 42, the current will decay according to the formula

$$I_{(t)} = \frac{E}{R}\left[1 - \left(1 - e^{\frac{-tR}{L}}\right)\right]$$

or

$$I_{(t)} = \frac{E}{R}\left(e^{\frac{-tR}{L}}\right)$$

where t = time in seconds after removal of EMF.

After one time constant the current will lose 63.2% of its steady-state value (decay to 36.8% of the steady-state value). The graph in Fig. 42 shows the current decay waveform to be identical to the voltage discharge waveform of a capacitor. However, one should be careful about applying the terms charge and discharge to an inductive circuit. These terms refer to energy storage in an electric field. An inductor stores energy in a magnetic field.

Inductive Reactance

When an alternating voltage is applied to a pure inductance (one with no resistance — all practical inductors have some resistance) the current is 90° out of phase with the applied voltage. In this case the current lags 90° behind the voltage; the opposite of the capacitor current-voltage relationship (Fig. 43.)

The primary cause for current lag in an inductor is the back EMF generated in the inductance. The amplitude of the back EMF is proportional to the rate at which the current changes. This in turn is proportional to the frequency, so the amplitude of the current is inversely proportional to the applied frequency. Since the back EMF is proportional to inductance for a given rate of current change, the current is inversely proportional to inductance for a given applied voltage and frequency. (Another way of saying this is that just enough current flows to generate an induced EMF that equals and opposes the applied voltage.)

The combined effect of inductance and frequency is called inductive reactance, also expressed in ohms. The formula for it is

$$X_L = 2\pi fL$$

wnere

X_L = inductive reactance
f = frequency in hertz
L = inductance in henrys
π = 3.1416

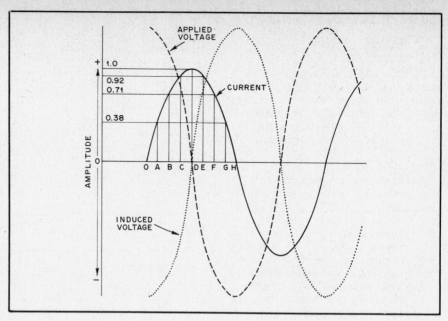

Fig. 43 — Phase relationships between voltage and current when an alternating voltage is applied to an inductance.

Example: The reactance of a coil having an inductance of 8 henrys, at a frequency of 120 hertz, is

$$X_L = 2\pi fL = 6.2832 \times 120 \times 8$$
$$X_L = 6032 \text{ ohms}$$

In radio-frequency circuits the inductance values usually are small and the frequencies are large. When the inductance is expressed in millihenrys and the frequency in kilohertz, the conversion factors for the two units cancel, and the formula for reactance may be used without first converting to fundamental units. Similarly, no conversion is necessary if the inductance is expressed in microhenrys and the frequency in megahertz.

Example: The reactance of a 15-microhenry coil at a frequency of 14 MHz is

$$X_L = 2\pi fL = 6.2832 \times 14 \times 15$$
$$X_L = 1319 \text{ ohms}$$

The resistance of the wire used to wind the coil has no effect on the reactance, but simply acts as a separate resistor connected in series with the coil.

Reactance and Impedance

Ohm's Law for Reactance

Ohm's Law equations for an ac circuit containing only reactance are

$$I = \frac{E}{X}, \qquad E = IX, \qquad X = \frac{E}{I}$$

where
E = EMF in volts
I = current in amperes
X = reactance in ohms

The reactance in the circuit may, of course, be either inductive or capacitive.

Example: If a current of 2 amperes is flowing through the capacitor of the earlier example (reactance = 47.4 ohms) at 7.15 MHz, the voltage drop across the capacitor is E = IX = 2 × 47.4 = 94.8 volts.

If 420 V at 120 hertz is applied to the 8-henry inductor of the earlier example, the current through the coil will be

$$I = \frac{E}{X} = \frac{420}{6032} = 0.0696A$$
$$= 69.6 \text{ mA}$$

Reactance Chart

The accompanying chart, Fig. 44, shows the reactance of capacitances from 1 pF to 100 μF, and the reactance of inductances from 0.1 μH to 10 henrys, for frequencies between 100 hertz and 100 megahertz. The approximate value of reactance can be read from the chart. More exact values can be calculated from the formulas.

Reactances in Series and Parallel

When reactances of the same kind are connected in series or parallel the resultant reactance is that of the resultant inductance or capacitance. This leads to the same rules that are used when determining the resistance when resistors are combined. That is, for series reactances of the same kind the resultant reactance is

$$X = X1 + X2 + X3 + X4 + \ldots$$

For reactances of the same kind in parallel the resultant is

$$X = \cfrac{1}{\dfrac{1}{X1} + \dfrac{1}{X2} + \dfrac{1}{X3} + \dfrac{1}{X4} + \ldots}$$

or for two in parallel,

$$X = \frac{X1 \times X2}{X1 + X2}$$

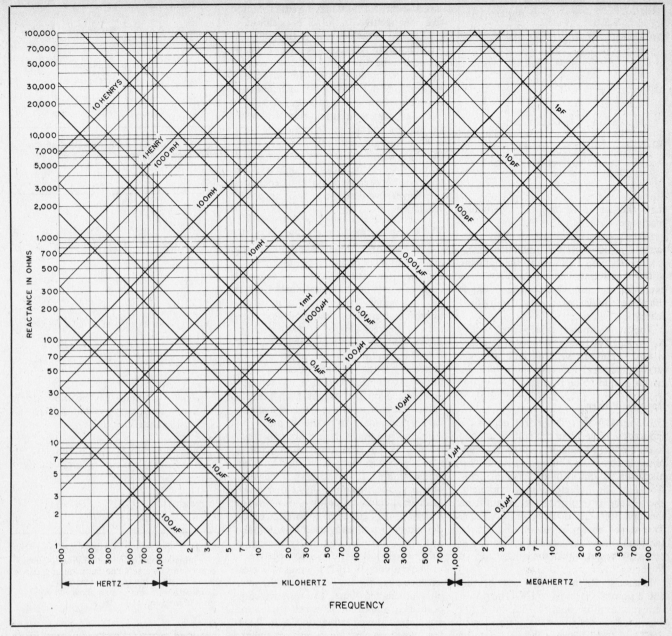

Fig. 44 — Inductive and capacitive reactance vs. frequency. Heavy lines represent multiples of 10, intermediate light lines multiples of five. For example, the light line between 10 μH and 100 μH represents 50 μH; the light line between 0.1 μF and 1 μF represents 0.5 μF, and so on. Intermediate values can be values within the chart range. For example, the reactance of 10 henrys at 60 Hz can be found by taking the reactance of 10 henrys at 600 Hz and dividing by 10 for the 10× times decrease in frequency.

The situation is different when reactances of opposite kinds are combined. Since the current in a capacitance leads the applied voltage by 90° and the current in an inductance lags the applied voltage by 90°, the voltage at the terminals of opposite types of reactance are 180° out of phase in a series circuit. (In a series circuit the current has to be the same through all elements.)

In a parallel circuit (in which the same voltage is applied to all elements), the currents in reactances of opposite types are 180° out of phase. The 180° phase relationship means the currents or voltages are of opposite polarity, so in the series circuit of Fig. 45A the voltage E_L across the induc-

tive reactance X_L is of opposite polarity to the voltage E_C across the capacitive reactance X_C. Thus if we call X_L "positive" and X_C "negative" (a common convention) the applied voltage E_{AC} is $E_L - E_C$. In the parallel circuit at B the total current, I, is equal to $I_L - I_C$, since the currents are 180° out of phase. In the series case, therefore, the resultant reactance of X_L and X_C is $X = X_L - X_C$ and in the parallel case (Fig. 45B),

$$X = \frac{-X_L X_C}{X_L - X_C}$$

Note that in the series circuit the total reactance is negative if X_C is larger than X_L; this indicates that the total reactance

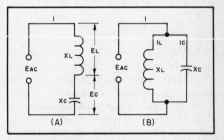

Fig. 45 — Series and parallel circuits containing opposite kinds of reactance.

is capacitive in such a case. The resultant reactance in a series circuit is always smaller

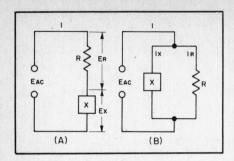

Fig. 46 — Series and parallel circuit containing resistance and reactance.

than the larger of the two individual reactances.

In the parallel circuit, the resultant reactance is positive (that is, inductive) if X_C is larger than X_L. This is just the opposite of the situation for reactances in series. In every case the resultant reactance is larger than the smaller of the two individual reactances.

There is one special case where $X_L = X_C$. The total reactance in this case is zero in the series circuit and infinitely large in the parallel circuit. In this case the circuit is said to be resonant.

Reactive Power

In Fig. 45A the voltage drop across the inductor is larger than the voltage applied to the circuit. This might seem to be an impossible condition, but it is not; the explanation is that while energy is being stored in the inductor's magnetic field, energy is being returned to the circuit from the capacitor's electric field, and vice versa. This stored energy is responsible for the fact that the voltages across reactances in series can be larger than the voltage applied to them.

In a resistance the current causes heating and a power loss equal to I^2R. The power in a reactance is equal to I^2X, but is not a loss; it is simply power that is transferred back and forth between the field and the circuit but not used up in heating anything. To distinguish this nondissipated power from the power actually consumed, the unit of reactive power is called the volt-ampere-reactive, or VAR, instead of watt. Reactive power is sometimes called wattless power.

Impedance

When a circuit contains both resistance and reactance the combined effect of the two is called impedance, symbolized by the letter Z. Impedance is thus a more general term than either resistance or reactance. The term is frequently used even for circuits that have only resistance or reactance, although usually with a qualification such as "resistive impedance" to indicate that the circuit has only resistance, for example.

The reactance and resistance comprising an impedance may be connected either in series or parallel, as shown in Fig. 46. In these circuits the reactance is shown as a

box to indicate that it may be either inductive or capacitive. In the series circuit at A the current is the same in both elements, with (generally) different voltages appearing across the resistance and reactance. In the parallel circuit at B the same voltage is applied to both elements, but different currents may flow in the two branches.

In a resistance the current is in phase with the applied voltage, while in a reactance it is 90° out of phase with the voltage. Thus, the phase relationship between current and voltage in the circuit as a whole may be anything between zero and 90°, depending on the relative amount of resistance and reactance.

The Complex Impedance Plane

The use of complex numbers in the analysis of ac circuits allows quicker solution without guess work. A complex number treats not only real numbers, but also "imaginary" numbers. An imaginary number is a number that cannot be plotted on the real axis of a graph such as shown in Fig. 47. On this graph the real axis is the horizontal axis, with positive values increasing to the right and negative values increasing to the left. Imaginary numbers are plotted on the Y axis with positive values rising from the origin and negative values descending from the origin.

In mathematical terms an imaginary number is an even root of a negative number. Since there is no real number when squared that will give a negative quantity, the imaginary system is used. Mathematicians define the value i as the square root of negative 1. In electronics, the symbol i is already used as a designation for current, so the symbol j is substituted; $j \times j$ is equal to negative 1.

The graph of Fig. 47, when used in electrical circuit analysis, is known as the complex impedance plane. The resistive components of the circuit have real values and are plotted along the real or X axis. Reactive values are plotted along the imaginary or Y axis. Inductive reactances, X_L, take on positive imaginary values because the voltage across an inductor leads the current through the inductor by 90 degrees. Capacitive reactances, X_C, take on negative imaginary values because the capacitive voltage lags the current by 90 degrees.

The familiar form, $R \pm jX$ is the rectangular coordinate form for a complex impedance as might be plotted on the complex impedance plane. For example, the value $27 - j8$ is plotted in Fig. 48A. The resistive value is 27 ohms and the capacitive reactance value is 8 ohms. So, 27 ohms positive and 8 imaginary ohms negative gives point A.

Fig. 48B shows another impedance value, where the resistance is 0.61 ohm and the inductive reactance is 1.15 ohms. The same impedance can be described in a secondary form called polar coordinate form. Here, instead of the point being

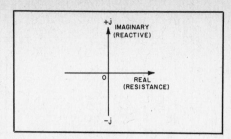

Fig. 47 — Real numbers are plotted on the X axis of the complex impedance plane; imaginary numbers are plotted on the Y axis.

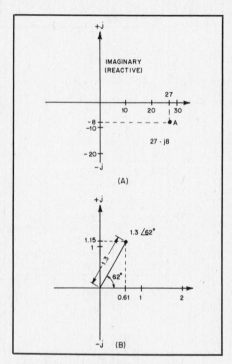

Fig. 48 — At A, the complex impedance $27 - j8$ is plotted graphically in rectangular coordinate form. At B, the complex impedance $0.61 + j 1.15$ is plotted in rectangular coordinate form and is also shown in polar coordinate form.

described by its X-Y values, its position is described by the magnitude of the distance from the origin and the angle in degrees rotated counterclockwise from the positive X-axis. This is shown in Fig. 48B. The impedance now is described in the form of some magnitude times a phase angle. For example, $1.3 \ \underline{/62°}$.

The values $0.61 + j1.15$ and $1.3 \ \underline{/62°}$ both describe the same impedance, only in different forms. The values are identical and each form can be easily converted to the other. There are advantages to each form. In rectangular form, complex numbers are easily added or subtracted. In polar form, complex numbers are easily multiplied and divided. The form used is largely dependent on the mathematical process to be performed on the complex number.

Addition and Subtraction of Complex Numbers

The rule for adding and subtracting com-

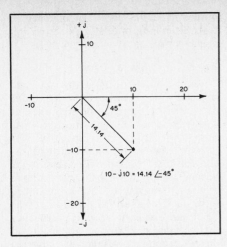

Fig. 49 — Converting from rectangular to polar form.

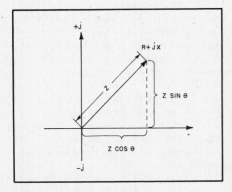

Fig. 50 — Converting from polar to rectangular form.

plex numbers in rectangular form is very simple: add or subtract the real and imaginary terms separately.

For example: $(15 + j32) + (10 - j8)$
$= 25 + j24$

Multiplication and Division of Complex Numbers

The rule for multiplying complex numbers in the polar form is: multiply the magnitudes and add the angles.

For example: $(8 \underline{/27°}) \times (2 \underline{/3°})$
$= 16 \underline{/30°}$

The rule for dividing complex numbers in polar form is: divide the magnitudes, but subtract the angle of denominator from the angle of the numerator.

For example: $(8 \underline{/27°}) / (2 \underline{/3°})$
$= 4 \underline{/24°}$

Converting from Rectangular to Polar Form

The easiest way to perform multiplication or division on a complex number is the use of polar form. Although it is possible to multiply complex numbers in their rectangular form, it is actually easier to first convert rectangular form to polar form and then perform the operations necessary. The magnitude of an impedance in polar form is equal to the distance the plotted impedance is from the origin. The

Pythagorean Theorem provides a simple means of calculating this distance. The magnitude is equal to

$$Z = \sqrt{R^2 + X^2}$$

This also is the formula for finding ac impedance.

The phase angle referenced to the positive real axis can also be found through simple trigonometry.

$$\theta = \arctan \frac{X}{R}$$

or

$$\theta = \arcsin \frac{X}{Z}$$

or

$$\theta = \arccos \frac{R}{Z}$$

For example: $10 - j10 = 14.14 \underline{/-45°}$ as shown in Fig. 49.

Converting from Polar to Rectangular Form

As shown in Fig. 50, the real term of an impedance plotted in polar form is equal to $Z \cos \theta$, and the imaginary term is equal to $Z \sin \theta$. Multiplying the magnitude of the impedance in polar form by the cosine of the phase angle gives the real term. Multiplying the magnitude of the impedance in polar form by the sine of the angle gives the reactive term.

For example: $12 \underline{/42°}$
$R = 12 \times \cos(42°) = 8.92$
$X = j \times 12 (\sin 42°) = j8.03$
$12 \underline{/42°} = 8.92 + j8.03$

Series Circuits

When resistance and reactance are in series, the impedance of the circuit is

$$Z = \sqrt{R^2 + X^2}$$

where

Z = impedance in ohms
R = resistance in ohms
X = reactance in ohms

The reactance may be either capacitive or inductive. If there are two or more reactances in the circuit they may be combined into a resultant by the rules previously given before substitution into the formula above; similarly for resistances.

Parallel Circuits

With resistance and reactance in parallel, as in Fig. 46B, the impedance is

$$Z = \frac{RX}{\sqrt{R^2 + X^2}}$$

where the symbols have the same meaning as for series circuits.

Just as in the case of series circuits, a number of reactances in parallel should be combined to find the resultant reactance before substitution in the formula above;

similarly for a number of resistances in parallel.

Equivalent Series and Parallel Circuits

The two circuits shown in Fig. 46 are equivalent if the same current flows when a given voltage of the same frequency is applied, and if the phase angle between voltage and current is the same in both cases. It is in fact possible to transform any given series circuit into an equivalent parallel circuit, and vice versa.

Transformations of this type often lead to simplification in the solution of complicated circuits. However, from the standpoint of practical work the usefulness of such transformations lies in the fact that the impedance of a circuit may be modified by the addition of either series or parallel elements, depending on which happens to be more convenient in the particular case. Typical applications are considered later in connection with tuned circuits and transmission lines.

A series RX circuit can be converted into its parallel equivalent by means of the formula

$$R_p = \frac{R_s^2 + X_s^2}{R_s}$$

$$X_p = \frac{R_s^2 + X_s^2}{X_s}$$

where the subscripts p and s represent the parallel- and series-equivalent values, respectively. If the parallel values are known, the equivalent series circuit can be found from

$$R_s = \frac{R_p}{1 + \left(\dfrac{R_p}{X_p}\right)^2}$$

and

$$X_s = \frac{R_s R_p}{X_p}$$

Ohm's Law for Impedance

Ohm's Law can be applied to circuits containing impedance just as readily as to circuits having resistance or reactance only. The formulas are

$$I = \frac{E}{Z}$$

$$E = IZ$$

$$Z = \frac{E}{I}$$

where

E = EMF in volts
I = current in amperes
Z = impedance in ohms

Fig. 51 shows a simple circuit consisting of a resistance of 75 ohms and a reactance of 100 ohms in series. From the formula previously given, the impedance is

$$Z = \sqrt{R^2 + Z_L^2} = \sqrt{75^2 + 100^2}$$
$$= 125$$

If the applied voltage is 250, then

$$I = \frac{E}{Z} = \frac{250}{125} = 2 \text{ amperes}$$

This current flows through both the resistance and reactance, so the voltage drops are

$$E_R = IR = 2 \times 75 = 150 \text{ volts}$$
$$EX_L = IX_L = 2 \times 100 = 200 \text{ volts}$$

The simple arithmetical sum of these two drops, 350 V, is greater than the applied voltage because the two voltages are 90° out of phase. Their actual resultant, when phase is taken into account, is

$$\sqrt{150^2 + 200^2} = 250 \text{ volts}$$

Power Factor

In the circuit of Fig. 51 an applied EMF of 250 V results in a current of 2 amperes, giving an apparent power of 250×2 = 500 watts. However, only the resistance actually consumes power. The power in the resistance is

$$P = I^2R = 2^2 \times 75 = 300 \text{ watts}$$

The ratio of the power consumed to the apparent power is called the power factor of the circuit. In this example the power factor would be 300/500 = 0.6. Power factor is frequently expressed as a percentage; in this case, it is 60 percent.

Real or dissipated power is measured in watts; apparent power, to distinguish it

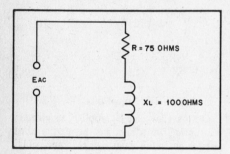

Fig. 51 — Circuit used as an example for impedance calculations.

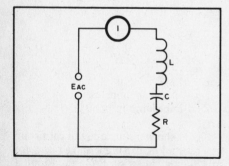

Fig. 52 — A series circuit containing L, C and R is resonant at the applied frequency when the reactance of C is equal to the reactance of L. The I in the circle is the schematic symbol for an ammeter.

from real power, is measured in volt-amperes (VA). It is simply the product of volts and amperes and has no direct relationship to the power actually used up unless the power factor of the circuit is known. The power factor of a purely resistive circuit is 100% or 1, while the power factor of a pure reactance is zero. In this illustration, the reactive power is

$$VAR = I^2X = 2^2 \times 100$$
$$VAR = 400 \text{ volt-amperes}$$

An equivalent definition of power factor is

$$PF = \frac{R}{Z} = \cos \theta$$

Since power factor is always rendered as a positive number, the value must be followed by the words leading or lagging to identify the phase of the voltage with respect to the current. Specifying the numerical power factor is not always sufficient. For example, many dc-to-ac power inverters can safely operate loads having a large net reactance of one sign but only a small reactance of the opposite sign.

Reactance and Complex Waves

It was pointed out earlier in this chapter that a complex wave (a nonsinusoidal wave) can be resolved into a fundamental frequency and a series of harmonic frequencies. When such a complex voltage wave is applied to a circuit containing reactance, the current through the circuit will not have the same wave shape as the applied voltage. This is because the reactance of an inductor and capacitor depend on the applied frequency. For the second-harmonic component of a complex wave, the reactance of the inductor is twice and the reactance of the capacitor is half their respective values at the fundamental frequency; for the third harmonic the inductive reactance is three times and the capacitive reactance is a third, and so on. Thus, the circuit impedance is different for each harmonic component.

Just what happens to the current wave shape depends on the values of resistance and reactance involved and how the circuit is arranged. In a simple circuit with resistance and inductive reactance in series, the amplitudes of the harmonic currents will be reduced because the inductive reactance increases in proportion to frequency. When capacitance and resistance are in series, the harmonic current is likely to be accentuated because the capacitive reactance becomes lower as the frequency is raised. When both inductive and capacitive reactance are present, the shape of the current wave can be altered in a variety of ways. The shape will depend on the circuit and the constants, or the relative values of L, C and R that are selected.

This property of nonuniform behavior with respect to fundamental and harmonics is an extremely useful one. It is the basis

of filtering, or the suppression of undesired frequencies in favor of a single desired frequency or group of such frequencies.

Resonance in Series Circuits

Fig. 52 shows a resistor, capacitor and inductor connected in series with a source of alternating current, the frequency of which can be varied over a wide range. At some low frequency the capacitive reactance will be much larger than the resistance of R, and the inductive reactance will be small compared with either the reactance of C or the resistance of R. (R is assumed to be the same at all frequencies.) On the other hand, at some very high frequency the reactance of C will be very small and the reactance of L will be very large. In either case the current will be small, because the net reactance is large.

At some intermediate frequency, the reactances of C and L will be equal. At this frequency the voltage drops across the coil and capacitor will be equal and 180° out of phase. Therefore, they cancel each other completely and the current flow is determined wholly by the resistance, R. At that frequency the current has its largest possible value, assuming the source voltage to be constant regardless of frequency. A series circuit in which the inductive and capacitive reactances are equal is said to be resonant.

The principle of resonance finds its most extensive application in radio-frequency circuits. The reactive effects associated with even small inductances and capacitances would place drastic limitations on RF circuit operation if it were not possible to cancel them out by supplying the right amount of reactance of the opposite kind — in other words, tuning the circuit to resonance.

Resonant Frequency

The frequency at which a series circuit is resonant is that at which $X_L = X_C$. Substituting the formulas for inductive and capacitance reactance gives

$$X_L = 2\pi fL = X_C = \frac{1}{2\pi fC}$$

$$2\pi fL = \frac{1}{2\pi fC}$$

$$f^2 = \frac{1}{4\pi^2 LC}$$

$$f = \frac{1}{2\pi \sqrt{LC}}$$

where
  f = frequency in hertz
  L = inductance in henrys
  C = capacitance in farads
  π = 3.14

These units are inconveniently large for radio-frequency circuits. A formula using more appropriate units is

$$f = \frac{10^6}{2\pi \sqrt{LC}}$$

where

f = frequency in kilohertz (kHz)
L = inductance in microhenrys (μH)
C = capacitance in picofarads (pF)
π = 3.14

Example: The resonant frequency of a series circuit containing a 5-μH inductor and a 35-pF capacitor is

$$f = \frac{10^6}{2\pi \sqrt{LC}} = \frac{10^6}{6.28 \times \sqrt{5 \times 35}}$$

$$f = \frac{10^6}{6.28 \times 13.23} = 12{,}037 \text{ kHz}$$

The resonant frequency is not affected by resistance in the circuit.

Resonance Curves

When a plot is drawn of the current in the circuit of Fig. 52 as the frequency is varied (the applied voltage being constant), it looks like one of the curves in Fig. 53. The shape of the resonance curve is determined by the ratio of reactance to resistance.

When the reactance of either the coil or capacitor is of the same order of magnitude as the resistance, the current decreases rather slowly as the frequency is moved in either direction away from resonance. Such a curve is said to be broad. Conversely, when the reactance is considerably larger than the resistance the current decreases rapidly as the frequency moves away from resonance and the circuit is said to be sharp. A sharp circuit will respond a great deal more readily to the resonant frequency than to frequencies quite close to resonance; a broad circuit will respond almost equally well to a group or band of frequencies centered around the resonant frequency.

Both types of resonance curves are useful. A sharp circuit gives good selectivity — the ability to respond strongly (in terms of current amplitude) at one desired frequency, and to discriminate against others. A broad circuit is used when the apparatus must give about the same response over a band of frequencies, rather than at a single frequency alone.

Q

Most diagrams of resonant circuits show only inductance and capacitance; no resistance is indicated. Nevertheless, resistance is always present. At frequencies up to perhaps 30 MHz this resistance is mostly in the wire of the coil. At higher frequencies energy loss in the capacitor (principally in the solid dielectric that must be used to form an insulating support for the capacitor plates) also becomes a factor. This energy loss is equivalent to resistance. When maximum sharpness or selectivity is needed, the objective of design is to reduce the inherent resistance to the lowest possible value.

The value of the reactance of either the inductor or capacitor at the resonant frequency of a series-resonant circuit, divided by the series resistance in the circuit, is called the Q (quality factor) of the circuit, or

$$Q = \frac{X}{R}$$

where

Q = quality factor
X = reactance of either coil or capacitor in ohms
R = series resistance in ohms

Example: The inductor and capacitor in a series circuit each has a reactance of 350 ohms at the resonant frequency. The resistance is 5 ohms. The Q is

$$Q = \frac{X}{R} = \frac{350}{5} = 70$$

The effect of Q on the sharpness of resonance of a circuit is shown by the curves of Fig. 54. In these curves the frequency change is shown in percentage above and below the resonant frequency. Q values of 10, 20, 50 and 100 are shown; these values cover much of the range commonly used in radio work. The unloaded Q of a circuit is determined by the inherent resistances associated with the components.

Voltage Rise at Resonance

When a voltage of the resonant frequency is inserted in series in a resonant circuit, the voltage that appears across either the inductor or capacitor is considerably higher than the applied voltage. The current in the circuit is limited only by the resistance and may have a relatively high value. However, the same current flows through the high reactances of the inductor and capacitor and causes large voltage drops.

The ratio of the reactive voltage to the applied voltage is equal to the ratio of reactance to resistance. This ratio is also the Q of the circuit. Therefore, the voltage across either the inductor or capacitor is equal to QE, where E is the voltage being applied to the series circuit. This fact accounts for the high voltages developed across the components of series-tuned antenna couplers.

Resonance in Parallel Circuits

When a variable-frequency source of constant voltage is applied to a parallel circuit of the type shown in Fig. 55, there is a resonance effect similar to that in a series circuit. In this case the line current (measured at the point indicated by the ammeter) is smallest at the frequency for which the inductive and capacitive reactances are equal. At that frequency the current through L is exactly canceled by the out-of-phase current through C, and only the current taken by R flows in the line. At frequencies below resonance the current

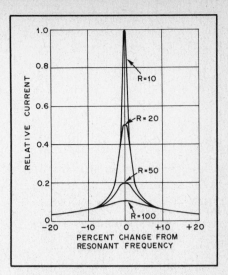

Fig. 53 — Current in a series-resonant circuit with various values of series resistance. The values are arbitrary and would not apply for all circuits, but represent a typical case. It is assumed that the reactances (at the resonant frequency) are 1000 ohms. Note that the current is substantially unaffected by the resistance in the circuit at frequencies more than plus or minus 10% away from the resonant frequency.

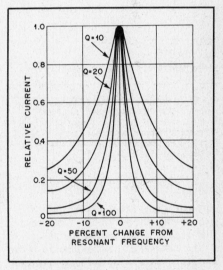

Fig. 54 — Current in series-resonant circuits having different Q values. In this graph the current at resonance is assumed to be the same in all cases. The lower the Q, the more slowly the current decreases as the applied frequency is moved away from resonance.

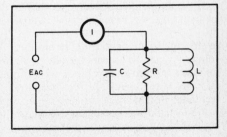

Fig. 55 — Circuit illustrating parallel resonance.

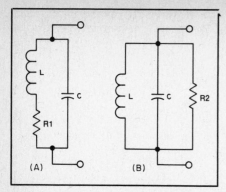

Fig. 56 — Series and parallel equivalents when the two circuits are resonant. The series resistance, R1 in A, is replaced in B by the equivalent parallel resistance R2, and vice versa.

$$R2 = \frac{X_L{}^2}{R1} = \frac{X_C{}^2}{R1}$$

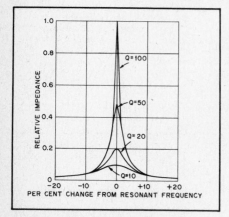

Fig. 57 — Relative impedance of parallel-resonant circuits with different Q values. These curves are similar to those in Fig. 54 for current in a series-resonant circuit. The effect of Q on impedance is most marked near the resonant frequency.

through L is larger than that through C, because the reactance of L is smaller and that of C is larger than at resonance.

There is only partial cancellation of the two reactive currents, and the line current therefore is larger than the current taken by R alone. At frequencies above resonance the situation is reversed and more current flows through C than through L, so the line current again increases. The current at resonance, being determined wholly by R, will be small if R is large, and large if R is small.

The R shown in Fig. 55 is not necessarily an actual resistor. In many cases it will be the series resistance of the coil transformed to an equivalent parallel resistance (see below). It may be antenna or other load resistance coupled into the tuned circuit. In all cases it represents the total effective resistance in the circuit.

Parallel and series resonant circuits are quite alike in some respects. For instance, when an external voltage is applied the circuits given at A and B in Fig. 56 will behave

identically, if (1) L and C are the same in both cases, and (2) R1 multiplied by R2, equals the square of the reactance (at resonance) of either L or C. When these conditions are met the two circuits will have the same Q. (These statements are approximate, but are quite accurate if the Q is 10 or more.) The circuit at A is a series circuit if it is viewed from within — that is, going around the loop formed by L, C and R1 — so its Q can be found from the ratio of X to R1.

Thus, a circuit like that of Fig. 56A has an equivalent parallel resistance (at resonance) of

$$R2 = \frac{X^2}{R1}$$

where X is the reactance of either the inductor or the capacitor. Although R2 is not an actual resistor, to the source of voltage the parallel-resonant circuit looks like a pure resistance of that value. It is pure resistance because the inductive and capacitive currents are 180° out of phase and are equal; thus there is no reactive current in the line. In a practical circuit with a high-Q capacitor, the parallel impedance at the resonant frequency is

$$Z_R = QX$$

where

   Z_R = resistive impedance at resonance
   Q = quality factor of inductor
   X = reactance (in ohms) of either the inductor or capacitor

Example: The parallel impedance of a circuit with a coil Q of 50 and having inductive and capacitive reactance of 300 ohms will be

$$Z_R = QX = 50 \times 300 = 15,000 \text{ ohms}$$

At frequencies off resonance the impedance is no longer purely resistive because the inductive and capacitive currents are not equal. The off-resonant impedance therefore is complex, and is lower than the resonant impedance for the reasons previously outlined.

The higher the circuit Q, the higher the parallel impedance. Curves showing the variation of impedance (with frequency) of a parallel circuit have the same shape as the curves showing the variation of current with frequency in a series circuit. Fig. 57 is a set of such curves. A set of curves showing the relative response as a function of the departure from the resonant frequency would be similar to Fig. 54. The −3 dB bandwidth (bandwidth at 0.707 relative response) is given by

$$\text{Bandwidth} -3 \text{ dB} = \frac{f_o}{Q}$$

where

   f_o = the resonant frequency
   Q = the circuit Q

It is also called the half-power bandwidth.

Parallel Resonance in Low-Q Circuits

The preceding discussion is accurate for Q values of 10 or more. When the Q is below 10, resonance in a parallel circuit having resistance in series with the coil, as in Fig. 56A, is not so easily defined. There is a set of values for L and C that will make the parallel impedance a pure resistance, but with these values the impedance does not have its maximum possible value. Another set of values for L and C will make the parallel impedance a maximum, but this maximum value is not a pure resistance.

Either condition could be called resonance, so with low-Q circuits it is necessary to distinguish between maximum impedance and resistive impedance when discussing parallel resonance. The difference between these L and C values and the equal reactances of a series-resonant circuit is appreciable when the Q is in the vicinity of 5, and becomes more marked with still lower Q values.

Q of Loaded Circuits

In many applications of resonant circuits the only power lost is that dissipated in the resistance of the circuit itself. At frequencies below 30 MHz most of this resistance is in the coil. Within limits, increasing the number of turns in the coil increases the reactance faster than it raises the resistance, so coils for circuits in which the Q must be high are made with relatively large inductance for the frequency.

However, when the circuit delivers energy to a load (as in the case of the resonant circuits used in transmitters) the energy consumed in the circuit itself is usually negligible compared with that consumed by the load. The equivalent of such a circuit is shown in Fig. 58A, where the parallel resistor represents the load to which power is delivered. If the power dissipated in the load is at least 10 times as great as the power lost in the inductor and capacitor, the parallel impedance of the resonant circuit itself will be so high compared with the resistance of the load that for all practical purposes the impedance of

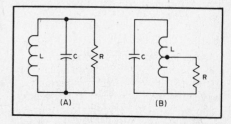

Fig. 58 — Equivalent diagram of a resonant circuit delivering power to a load. This resistor R represents the load resistance. At B the load is tapped across part of L, by which transformer action is equivalent to using a higher load resistance across the whole circuit.

the combined circuit is equal to the load resistance. Under these conditions the Q of a parallel resonant circuit loaded by a resistive impedance is

$$Q = \frac{R}{X}$$

where

R = parallel load resistance (ohms)
X = reactance (ohms)

Example: A resistive load of 3000 ohms is connected across a resonant circuit in which the inductive and capacitive reactances are each 250 ohms. The circuit Q is then

$$Q = \frac{R}{X} = \frac{3000}{250} = 12$$

The effective Q of a circuit loaded by a parallel resistance increases when the reactances are decreased. A circuit loaded with a relatively low resistance (a few thousand ohms) must have low-reactance elements (large capacitance and small inductance) to have reasonably high Q.

Impedance Transformation

An important application of the parallel-resonant circuit is as an impedance-matching device in the output circuit of an RF power amplifier. There is an optimum value of load resistance for each type of tube or transistor and set of operating conditions. However, the resistance of the load to which the active device is to deliver power usually is considerably lower than the value required for proper device operation.

To transform the actual load resistance to the desired value, the load may be tapped across part of the coil, as shown in Fig. 58B. This is equivalent to connecting a higher value of load resistance across the whole circuit, and is similar in principle to impedance transformation with an iron-core transformer. In high-frequency resonant circuits the impedance ratio does not vary exactly as the square of the turns ratio, because all the magnetic flux lines do not cut every turn of the coil. A desired reflected impedance usually must be obtained by experimental adjustment.

When the load resistance has a very low value (say below 100 ohms) it may be connected in series in the resonant circuit (as in Fig. 56A, for example), in which case it is transformed to an equivalent parallel impedance as previously described. If the Q is at least 10, the equivalent parallel impedance is

$$Z_R = \frac{X^2}{R}$$

where

Z_R = resistive parallel impedance at resonance
X = reactance (in ohms) of either the coil or the capacitor
R = load resistance inserted in series

If the Q is lower than 10 the reactance will have to be adjusted somewhat, for the reasons given in the discussion of low-Q circuits, to obtain a resistive impedance of the desired value.

While the circuit shown in Fig. 58B will usually provide an impedance step-up as with an iron-core transformer, the network has some serious disadvantages for some applications. For instance, the common connection provides no dc isolation, and the common ground is sometimes troublesome with regard to ground-loop currents. Consequently, a network with only mutual magnetic coupling is usually preferable. No impedance step-up will result unless the two coils are coupled tightly enough. The equivalent resistance seen at the input of the network will always be lower regardless of the turns ratio employed. However, such networks are still useful in impedance-transformation applications if the appropriate capacitive elements are used. A more detailed treatment of matching networks will be presented in a later section.

Unfortunately, networks involving reactive elements are usually narrowband in nature; it would be desirable if such elements could be eliminated in order to increase the bandwidth. With the advent of ferrites, this has become possible, and it is now relatively easy to construct actual impedance transformers that are both broadband and permit operation well up into the VHF portion of the spectrum. This is also accomplished in part by tightly coupling the two (or more) coils that make up the transformer, either by twisting the conductors together or winding them in a parallel fashion. The latter configuration is sometimes called a bifilar winding, as discussed in the section on ferromagnetic transformers.

Transformers

Two coils having mutual inductance constitute a transformer. The coil connected to the source of energy is called the primary coil, and the other is called the secondary coil.

The usefulness of the transformer lies in the fact that electrical energy can be transferred from one circuit to another without direct connection, and in the process can be readily changed from one voltage level to another. Thus, if a device to be operated requires, for example, 7-V ac and only a 440-V source is available, a transformer can be used to change the source voltage to that required. A transformer can be used only with ac, since no voltage will be induced in the secondary if the magnetic field is not changing. If dc is applied to the primary of a transformer, a voltage will be induced in the secondary only at the instant of closing or opening the primary circuit, since it is only at these

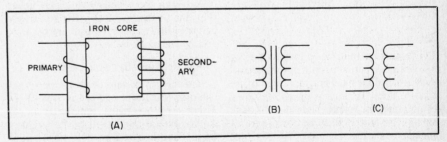

Fig. 59 — The transformer. A is a pictorial diagram. Power is transferred from the primary coil to the secondary by means of the magnetic field. B is a schematic diagram for an iron-core transformer, and C is for an air-core transformer.

times that the field is changing.

The Iron-Core Transformer

As shown in Fig. 59, the primary and secondary coils of a transformer may be wound on a core of magnetic material. This increases the inductance of the coils so that

a relatively small number of turns may be used to induce a given value of voltage with a small current. A closed core (one having a continuous magnetic path) such as that shown in Fig. 59 also tends to ensure that practically all of the field set up by the current in the primary coil will cut the turns of the secondary coil. However, the core introduces a power loss because of hysteresis and eddy currents, so this type of construction is normally practicable only at power and audio frequencies. The discussion in this section is confined to transformers operating at such frequencies.

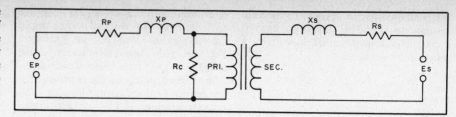

Fig. 60 — The equivalent circuit of a transformer includes the effects of leakage inductance and resistance of both primary and secondary windings. The resistance R_c is an equivalent resistance representing the core losses. Since these are comparatively small, their effect may be neglected in many approximate calculations.

Voltage and Turns Ratio

For a given varying magnetic field, the voltage induced in a coil in the field is proportional to the number of turns in the coil. When the two coils of a transformer are in the same field (which is the case when both are wound on the same closed core) it follows that the induced voltages will be proportional to the number of turns in each coil. In the primary the induced voltage practically equals and opposes, the applied voltage, as described earlier. Hence,

$$E_s = E_p \left(\frac{n_s}{n_p} \right)$$

where

E_s = secondary voltage
E_p = primary applied voltage
n_s = number of turns on secondary
n_p = number of turns on primary

Example: A transformer has a primary of 400 turns and a secondary of 2800 turns, and an EMF of 117 V is applied to the primary.

$$E_s = 117 \left(\frac{2800}{400} \right) = 117 \times 7$$

$$E_s = 819 \text{ volts}$$

Also, if an EMF of 819 V is applied to the 2800-turn winding (which then becomes the primary) the output voltage from the 400-turn winding will be 117 V.

Either winding of a transformer can be used as the primary, provided the winding has enough turns (enough inductance) to induce a voltage equal to the applied voltage without requiring an excessive current flow.

Secondary Current

The current that flows in the primary when no current is taken from the secondary is called the magnetizing current of the transformer. In any properly designed transformer the primary inductance is quite small. The power consumed by the transformer when the secondary is open (not delivering power) is only the amount necessary to overcome the losses in the iron core and in the resistance of the wire with which the primary is wound.

When power is taken from the secondary winding, the secondary current sets up a magnetic field that opposes the field set up by the primary current. For the induced voltage in the primary to equal the applied voltage, the original field must be maintained. The primary must draw enough additional current to set up a field exactly

equal and opposite to the field set up by the secondary current.

In practical calculations on transformers it may be assumed that the entire primary current is caused by the secondary load. This is justifiable because the magnetizing current should be very small in comparison with the primary load current at rated power output.

If the magnetic fields set up by the primary and secondary currents are to be equal, the primary current multiplied by the primary turns must equal the secondary current multiplied by the secondary turns. From this it follows that

$$I_p = I_s \left(\frac{n_s}{n_p} \right)$$

where

I_p = primary current
I_s = secondary current
n_p = number of turns on primary
n_s = number of turns on secondary

Example: Suppose the secondary of the transformer in the previous example is delivering a current of 0.2 ampere to a load. Then the primary current will be

$$I_p = I_s \left(\frac{n_s}{n_p} \right) = 0.2 \left(\frac{2800}{400} \right) = 0.2 \times 7$$

$$I_p = 1.4 \text{ amperes}$$

Although the secondary voltage is higher than the primary voltage, the secondary current is lower than the primary current, and by the same ratio.

Power Relationships and Efficiency

A transformer cannot create power; it can only transfer it and change the EMF. Hence, the power taken from the secondary cannot exceed that taken by the primary from the source of applied EMF. There is always some power loss in the resistance of the coils and in the iron core, so in all practical cases the power taken from the source will exceed that taken from the secondary. Thus,

$$P_o = nP_i$$

where

P_o = power output from secondary
P_i = power input to primary
n = efficiency factor

The efficiency, n, is always less than 1. It is usually expressed as a percentage; if n is 0.65, for instance, the efficiency is 65%.

Example: A transformer has an efficiency of 85% at its full-load output of 150 watts. The power input to the primary at full secondary load will be

$$P_i = \frac{P_o}{n} = \frac{150}{0.85} = 176.5 \text{ watts}$$

A transformer is usually designed to have the highest efficiency at the power output for which it is rated. The efficiency decreases with either lower or higher outputs. On the other hand, the losses in the transformer are relatively small at low output but increase as more power is taken. The amount of power that the transformer can handle is determined by its own losses, because these losses heat the wire and core. There is a limit to the temperature rise that can be tolerated, because too high a temperature can either melt the wire or cause the insulation to break down. A transformer can be operated at reduced output, even though the efficiency is low, because the actual loss will be low under such conditions. The full-load efficiency of small power transformers such as are used in radio receivers and transmitters usually lies between about 60 and 90 percent, depending on the size and design.

Leakage Reactance

In a practical transformer not all of the magnetic flux is common to both windings, although in well designed transformers the amount of flux that cuts one coil and not the other is only a small percentage of the total flux. This leakage flux causes an EMF of self-induction; consequently, there are small amounts of leakage inductance associated with both windings of the transformer. Leakage inductance acts in exactly the same way as an equivalent amount of ordinary inductance inserted in series with the circuit. It has, therefore, a certain reactance, depending on the amount of leakage inductance and the frequency. This reactance is called leakage reactance.

Current flowing through the leakage reactance causes a voltage drop. This voltage drop increases with increasing current, hence it increases as more power is taken from the secondary. Thus, the greater the secondary current, the smaller the secondary terminal voltage becomes. The resistances of the transformer windings also cause voltage drops when current is flowing; although these voltage drops are not in phase with those caused by leakage reactance, together they result in a lower secondary voltage under load than is indicated by the turns ratio of the transformer.

At power frequencies (60 Hz) the voltage at the secondary, with a reasonably well designed transformer, should not drop more than about 10% from open-circuit conditions to full load. The drop in voltage may be considerably more than this in a transformer operating at audio frequencies because the leakage reactance increases directly with the frequency. The various transformer losses are modeled in Fig. 60.

Impedance Ratio

In an ideal transformer — one without losses or leakage reactance — the following relationship is true:

$$Z_p = Z_s \left(\frac{n_p}{n_s} \right)^2$$

where

Z_p = impedance looking into primary terminals from source of power

Z_s = impedance of load connected to secondary

n_p/n_s = turns ratio, primary to secondary

That is, a load of any given impedance connected to the secondary of the transformer will be transformed to a different value looking into the primary from the source of power. The impedance transformation is proportional to the square of the primary-to-secondary turns ratio.

Example: A transformer has a primary-to-secondary turns ratio of 0.6 (primary has 6/10 as many turns as the secondary) and a load of 3000 ohms is connected to the secondary. The impedance at the primary then will be

$$Z_p = Z_s \left(\frac{n_p}{n_s} \right)^2 = 3000 \times (0.6)^2$$

$$Z_p = 3000 \times 0.36 = 1080 \text{ ohms}$$

By choosing the proper turns ratio, the impedance of a fixed load can be transformed to any desired value, within practical limits. If transformer losses can be neglected, the transformed (reflected) impedance has the same phase angle as the actual load impedance. Thus, if the load is a pure resistance, the load presented by the primary to the source of power will also be a pure resistance. The above relationship may be used in practical work even though it is based on an ideal transformer. Aside from the normal design requirements of reasonably low internal losses and low leakage reactance, the only requirement is

that the primary have enough inductance to operate with low magnetizing current at the voltage applied to the primary.

The primary terminal impedance of a transformer is determined wholly by the load connected to the secondary and by the turns ratio. If the characteristics of the transformer have an appreciable effect on the impedance presented to the power source, the transformer is either poorly designed or is not suited to the voltage and frequency at which it is being used. Most transformers will operate quite well at voltages from slightly above to well below the design figure.

Impedance Matching

Many devices require a specific value of load resistance (or impedance) for optimum operation. The impedance of the actual load that is to dissipate the power may differ widely from this value, so a transformer is used to change the actual load into an impedance of the desired value. This is called impedance matching. From the preceding,

$$\frac{n_p}{n_s} = \sqrt{\frac{Z_p}{Z_s}}$$

where

n_p/n_s = required turns ratio, primary to secondary

Z_p = primary impedance required

Z_s = impedance of load connected to secondary

Example: A transistor AF amplifier requires a load of 150 ohms for optimum performance, and is to be connected to a loudspeaker having an impedance of 4 ohms. The turns ratio, primary to secondary, required in the coupling transformer is

$$\frac{n_p}{n_s} = \sqrt{\frac{Z_p}{Z_s}} = \sqrt{\frac{150}{4}} = \sqrt{37.5} = 6.1$$

The primary therefore must have 6.1 times as many turns as the secondary.

Impedance matching implies adjusting the load impedance, by means of a transformer or otherwise, to a desired value. However, there is also another meaning. It is possible to show that any source of power will deliver its maximum possible output when the impedance of the load is equal to the internal impedance of the source. The impedance of the source is said to be matched under this condition. The efficiency is only 50% in such a case; just as much power is used up in the source as is delivered to the load. Because of the poor efficiency, this type of impedance matching is limited to cases where only a small amount of power is available and heating from power loss in the source is not important.

TRANSFORMER CONSTRUCTION

Transformers usually are designed so that the magnetic path around the core is

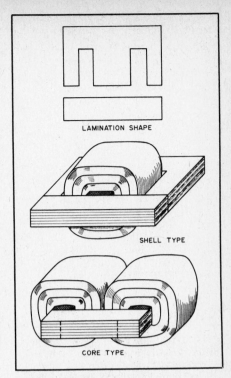

Fig. 61 — Two common types of transformer construction. Core pieces are interleaved to provide a continuous magnetic path.

as short as possible. A short magnetic path means that the transformer will operate with fewer turns, for a given applied voltage, than if the path were long. A short path also helps to reduce flux leakage and therefore minimizes leakage reactance.

Two core shapes are in common use, as shown in Fig. 61. In the shell type, both windings are placed on the inner leg, while in the core type the primary and secondary windings may be placed on separate legs, if desired. This is sometimes done when it is necessary to minimize capacitive effects between the primary and secondary, or when one of the windings must operate at very high voltage.

Core material for small transformers is usually silicon steel, called transformer iron. The core is built up of laminations, insulated from each other (by a thin coating of shellac, for example) to prevent eddy currents. The laminations are interleaved at the ends to make the magnetic path as continuous as possible and thus reduce leakage.

The number of turns required in the primary for a given applied EMF is determined by the size, shape and type of core material used, and the frequency. The number of turns required is inversely proportional to the cross-sectional area of the core. As a rough indication, windings of small power transformers frequently have about six to eight turns per volt on a core of 1-square-inch cross section and have a magnetic path 10 or 12 inches in length. A longer path or smaller cross section requires more turns per volt, and vice versa.

In most transformers the coils are wound

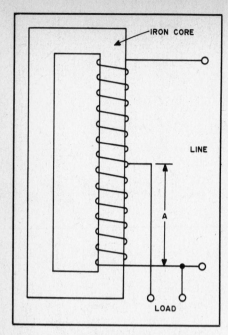

Fig. 62 — The autotransformer is based on the transformer principle, but uses only one winding. The line and load currents in the common winding (A) flow in opposite directions, so the resultant current is the difference between them. The voltage across A is proportional to the ratio of total turns to turns in A.

Fig. 63 — An assortment of toroid cores. A ferrite rod is placed at the top of the picture for comparison. The two light-colored, plastic-encased toroids at the upper left are tape-wound types (Hypersil steel), and are suitable for audio and dc-to-dc converter transformers. The wound toroid at the right center contains two toroid cores that have been stacked atop one another to increase the power capability.

in layers, with a thin sheet of treated-paper insulation between each layer. Thicker insulation is used between adjacent coils and between the first coil and the core.

Autotransformers

The transformer principle can be used with only one winding instead of two, as shown in Fig. 62; the principles just discussed apply equally well. A one-winding transformer is called an autotransformer. The current in the common section (A) of the winding is the difference between the line (primary) and the load (secondary) currents, since these currents are out of phase. Hence, if the line and load currents are

nearly equal, the common section of the winding may be wound with comparatively small wire. The line and load currents will be equal only when the primary (line) and secondary (load) voltages are not very different.

The autotransformer is used chiefly for boosting or reducing the power-line voltage by relatively small amounts. Continuously variable autotransformers are commercially available under a variety of trade names; Variac and Powerstat are typical examples.

Ferromagnetic Transformers and Inductors

The design concepts and general theory of transformers presented earlier in this chapter apply also to transformers wound on ferromagnetic-core materials (ferrite and powdered iron). As is the case with stacked cores made of laminations in the classic I and E shapes, the core material has a specific permeability factor that determines the inductance of the windings versus the number of wire turns used.

Both ferrite and powdered-iron materials are manufactured with a wide range of μ_i (initial permeability) characteristics. The value chosen by the designer depends on the intended operating frequency and the desired bandwidth of a given broadband transformer.

Core Types in Use at RF

For use in radio-frequency circuits especially, a suitable core type must be chosen to provide the Q required by the designer. The wrong core material destroys the Q of an inductor at RF.

Toroid cores are useful from a few hundred hertz well into the UHF spectrum. Tape-wound steel cores are employed in some types of power supplies — notably dc-to-dc converters. The toroid core is doughnut shaped, as shown in Fig. 63.

The principal advantage of this type of core is the self-shielding characteristic. Another feature is the compactness of a transformer or inductor. Therefore, toroids are excellent not only in dc-to-dc converters, but at audio and radio frequencies up to at least 1000 MHz, assuming the proper core material is selected for the range of operating frequencies. Toroid cores are available from microminiature sizes up to several inches in diameter. The latter can be used, as one example, to build a 20-kW balun for use in antenna systems.

Another form taken in ferromagnetic transformers and inductors is the pot-core or cup-core device. Unlike the toroid, which has the winding over the outer surface of the core material, the pot-core winding is inside the ferromagnetic material (Fig. 64). There are two cup-shaped halves to the assembly, both made of ferrite or powdered iron, which are connected tightly together by means of a screw that is passed through a center hole. The wire for the assembly is wound on an insulating bobbin that fits inside the

Fig. 64 — View of a disassembled pot core (left) and an assembled pot core (right).

Fig. 65 — A BC-band ferrite-rod loop antenna is shown at the top of the picture (J. W. Miller Co.). A blank ferrite rod is visible at the center, and a flat BC-band ferrite loop antenna in the lower foreground.

two halves of the pot-core unit.

The advantage to this type of construction is that the core permeability can be chosen to ensure a minimum number of wire turns for a given value of inductance. This reduces the wire resistance and increases the Q as opposed to an equivalent inductance wound on a core that has relatively low permeability. By virtue of the winding being contained inside the ferrite or powdered-iron pot core, shielding is excellent.

Still another kind of ferromagnetic-core inductor is the solenoidal type (Fig. 65). Transformers and inductors fabricated in this manner consist of a cylindrical, oval or rectangular rod of material over which the wire winding is placed. This variety of device does not have a self-shielding trait. Therefore, it must be treated in the same manner as any solenoidal-wound inductor (using external shield devices). An example of a ferrite-rod inductor is the built-in loop antennas found in portable radios and direction finders.

Core Size

The cross-sectional area of a ferromagnetic core should be chosen to prevent saturation from the load seen by the transformer. This means that the proper thickness and diameter are essential parameters to consider. For a specific core the maximum operational ac excitation can be determined by

$$B = \frac{E_{RMS} \times 10^8}{4.44\, f\, n_p\, A_e}$$

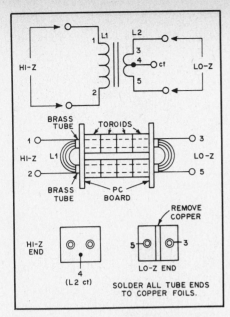

Fig. 66 — Schematic and pictorial representations of one type of conventional transformer. This style is used frequently at the input and output ports of RF power amplifiers that use transistors. The magnetic material consists of two rows of 950-μ toroid cores for use from 1.8 to 30 MHz. The primary and secondary windings are passed through the center holes of the toroid-stack rows as shown.

where

A_e = equivalent area of the magnetic path in sq. cm.
E_{RMS} = applied voltage
n_p = number of core turns
f = operating frequency in hertz
B = flux density in gauss

This equation is applicable to inductors that do not have dc flowing in the winding along with ac. When both ac and dc flow

$$B = \frac{E_{RMS} \times 10^8}{4.44 \ f \ n_p \ A_e} + \frac{N_p \ I_{dc} \ A_L}{10 A_e}$$

where

I_{dc} = the dc current through the winding
A_L = the manufacturer's index for the core being used, which can be obtained by consulting the manufacturer's data sheet.

Types of Transformers

The most common ferromagnetic transformers used in Amateur Radio work are the narrow-band, conventional, broadband and transmission-line varieties. Narrow-band transformers are used when selectivity is desired in a tuned circuit, such as an audio peaking or notching circuit, a resonator in an RF filter, or a tuned circuit associated with an RF amplifier. Broadband transformers are employed in circuits which must have a uniform response over a substantial spread of frequency, as in a 2- to 30-MHz broadband amplifier. In such an example the reactance of the windings should be at least four

times the impedance the winding is designed to look into. Therefore, a transformer that has a 300-ohm primary and a 50-ohm secondary load should have winding reactances (X_L) of at least 1200 ohms and 200 ohms, respectively. The windings, for all practical purposes, can be graded as RF chokes, and the same rules apply.

The permeability of the core material plays a vital role in designing a good broadband transformer. The performance of the transformer at the low-frequency end of the operating range depends on the permeability. That is, the μ_e (effective permeability) must be high enough in value to provide ample winding reactance at the low end of the operating range. As the operating frequency is increased, the effects of the core tend to disappear until there are scarcely any core effects at the upper limit of the operating range. For this reason it is common to find very low frequency core materials in transformers that are used in broadband circuits operating into the upper HF region, or even into the VHF spectrum. By way of simple explanation, at high frequencies the low-frequency core material becomes inefficient and tends to vanish electrically. This desirable trait makes possible the use of ferromagnetics in broadband applications.

Conventional transformers are those that are wound in the same manner as a power transformer. That is, each winding is made from a separate length of wire, with one winding being placed over the previous one with suitable insulation between (Figs. 66 and 67). A transmission-line transformer is, conversely, one that uses windings that are arranged to simulate a piece of transmission line of a specific impedance. This can be done by twisting the wires together a given number of times per inch, or by laying the wires on the core (adjacent to one another) at a distance apart which provides a two-wire line impedance of a particular value. In some applications these windings are called bifilar. A three-wire winding is known as a trifilar one, and so forth (Fig. 68).

It can be argued that a transmission-line transformer is more efficient than a conventional one, but in practice it is difficult to observe a significant difference in the performance characteristics. An interesting technical paper on the subject of toroidal broadband transformers was published by Sevick, W2FMI.[1] The classic reference work on the subject is by Ruthroff.[2]

Ferrite Beads

Another form of toroidal inductor is the ferrite bead. This component is available in various μ_i values and sizes, but most beads are less than 0.25 inch in diameter.

[1] J. Sevick, "Simple Broadband Matching Networks," QST, January 1976.
[2] Ruthroff, "Some Broadband Transformers," Proc. IRE, Vol. 47, August 1959, p. 137.

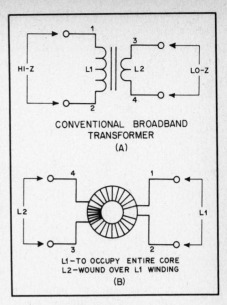

Fig. 67 — Another conventional transformer. Primary and secondary windings are wound over the outer surface of a toroid core.

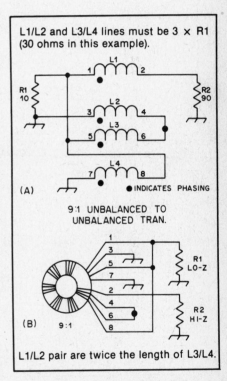

Fig. 68 — Schematic and pictorial presentations of a transmission-line transformer in which the windings need to be arranged for a specific impedance.

Ferrite beads are used principally as VHF and UHF parasitic suppressors at the input and output terminals of amplifiers. Another practical application is in decoupling networks that are used to prevent unwanted migration of RF energy from one section of a circuit to another. They are used also in suppressing RFI and TVI in hi-fi and television sets.

In some circuits it is necessary only to

Table 6

Powdered-Iron Toroidal Cores — A_L Values (μH per 100 turns)

Core Size	41-Mix Green $\mu = 75$	3-Mix Grey $\mu = .35$ 0.05-0.5 MHz	15-Mix Rd & Wh $\mu = 25$ 0.1-2 MHz	1-Mix Blue $\mu = 20$ 0.5-5 MHz	2-Mix Red $\mu = 10$ 1-30 MHz	6-Mix Yellow $\mu = 8$ 10-90 MHz	10-Mix Black $\mu = 6$ 60-150 MHz	12-Mix Gn & Wh $\mu = 3$ 100-200 MHz	0-Mix Tan $\mu = 1$ 150-300 MHz
T-200	755	360	NA	250	120	100	NA	NA	NA
T-184	1640	720	NA	500	240	195	NA	NA	NA
T-157	970	420	360	320	140	115	NA	NA	NA
T-130	785	330	250	200	110	96	NA	NA	NA
T-106	900	405	345	325	135	116	NA	NA	15.0
T-94	590	248	200	160	84	70	58	32	19.0
T-80	450	180	170	115	55	45	32	22	10.6
T-68	420	195	180	115	57	47	32	21	8.5
T-50	320	175	135	100	49	40	31	18	7.5
T-44	229	180	160	105	52	42	33	NA	6.4
T-37	308	120	90	80	40	30	25	15	6.5
T-30	375	140	93	85	43	36	25	16	4.9
T-25	225	100	85	70	34	27	19	13	6.0
T-20	175	90	65	52	27	22	16	10	4.5
T-16	130	61	NA	44	22	19	13	8	3.5
T-12	112	60	50	48	20	17	12	7.5	3.0

NA — Not available in that size.
$$\text{Turns} = 100 \sqrt{L_{\mu H} \div A_L \text{ value (above)}}$$
All frequency figures optimum.

Number of Turn vs. Wire Size and Core Size

Approximate maximum of turns — single-layer-wound enameled wire

Wire Size	T-200	T-130	T-106	T-94	T-80	T-68	T-50	T-37	T-25	T-12
10	33	20	12	12	10	6	4	1		
12	43	25	16	16	14	9	6	3		
14	54	32	21	21	18	13	8	5	1	
16	69	41	28	28	24	17	13	7	2	
18	88	53	37	37	32	23	18	10	4	
20	111	67	47	47	41	29	23	14	6	1
22	140	86	60	60	53	38	30	19	9	1
24	177	109	77	77	67	49	39	25	13	2
26	223	137	97	97	85	63	50	33	17	4
28	281	173	123	123	108	80	64	42	23	7
30	355	217	154	154	136	101	81	54	29	9
32	439	272	194	194	171	127	103	68	38	13
34	557	346	247	247	218	162	132	88	49	17
36	683	424	304	304	268	199	162	108	62	23
38	875	544	389	389	344	256	209	140	80	30
40	1103	687	492	492	434	324	264	178	102	51

Physical Dimensions

Core Size	Outer Dia (in)	Inner Dia (in)	Height (in)	Cross Sect. Area cm²	Mean Length cm	Core Size	Outer Dia (in)	Inner Dia (in)	Height (in)	Cross Sect. Area cm²	Mean Length cm
T-200	2.000	1.250	0.550	1.330	12.97	T-50	0.500	0.303	0.190	0.121	3.20
T-184	1.840	0.950	0.710	2.040	11.12	T-44	0.440	0.229	0.159	0.107	2.67
T-157	1.570	0.950	0.570	1.140	10.05	T-37	0.375	0.205	0.128	0.070	2.32
T-130	1.300	0.780	0.437	0.733	8.29	T-30	0.307	0.151	0.128	0.065	1.83
T-106	1.060	0.560	0.437	0.706	6.47	T-25	0.255	0.120	0.096	0.042	1.50
T-94	0.942	0.560	0.312	0.385	6.00	T-20	0.200	0.088	0.067	0.034	1.15
T-80	0.795	0.495	0.250	0.242	5.15	T-16	0.160	0.078	0.060	0.016	0.75
T-68	0.690	0.370	0.190	0.196	4.24	T-12	0.125	0.062	0.050	0.010	0.74

Courtesy of Amidon Assoc., N. Hollywood, CA 91607 and Micrometals, Inc.

place one or more beads over a short length of wire to obtain ample inductive reactance for creating an RF choke. A few turns of small-diameter enameled wire can be looped through the larger beads to increase the effective inductance. Ferrite beads are suitable as low-Q base impedances in solid-state VHF and UHF amplifiers. The low-Q characteristic prevents self-oscillation that might occur if a high-Q solenoidal RF choke were used in place of one made from beads. Miniature broadband transformers are sometimes fashioned from ferrite beads. For the most part, ferrite beads can be regarded as small toroid cores.

Number of Turns

The number of wire turns used on a toroid core can be calculated by knowing the A_L of the core and the desired inductance. The A_L is simply the inductance index for the core size and permeability being used. Table 6 provides information of interest concerning a popular assortment of powdered-iron toroid cores. The complete number for a given core is composed of the core-size designator in the upper left column, plus the corresponding mix number. For example, a half-inch diameter core with a no. 2 mix would be designated as a T-50-2 unit. The A_L would be 49 and the suggested operating frequency would be from 1 to 30 MHz. The μ_i for that core is 10.

The required number of wire turns for a specified inductance on a given type of core can be determined by

$$\text{Turns} = 100 \sqrt{\text{desired } L \ (\mu H) \div A_L}$$

where A_L is obtained from Table 6. The table also indicates how many turns of a

Table 7
Ferrite Toroids — A_L Chart (mH per 1000 turns) Enameled Wire

Core Size	63-Mix $\mu = 40$	61-Mix $\mu = 125$	43-Mix $\mu = 950$	72-Mix $\mu = 2000$	75-Mix $\mu = 5000$
FT-23	7.9	24.8	189.0	396.0	990.0
FT-37	17.7	55.3	420.0	884.0	2210.0
FT-50	22.0	68.0	523.0	1100.0	2750.0
FT-82	23.4	73.3	557.0	1172.0	2930.0
FT-114	25.4	79.3	603.0	1268.0	3170.0

Number turns = $1000 \sqrt{\text{desired L (mH)} \div A_L \text{ value (above)}}$

Ferrite Magnetic Properties

Property	Unit	63-Mix	61-Mix	43-Mix	72-Mix	75-Mix
Initial perm. (μ_i)		40	125	950	2000	5000
Maximum perm.		125	450	3000	3500	8000
Saturation flux density @ 13 oer	Gauss	1850	2350	2750	3500	3900
Residual flux density	Gauss	750	1200	1200	1500	1250
Curie temp.	°C	500	300	130	150	160
Vol. resistivity	ohm/cm	1×10^8	1×10^8	1×10^5	1×10^2	5×10^2
Opt. freq. range	MHz	15-25	0.2-10	0.01-1	0.001-1	0.001-1
Specific gravity		4.7	4.7	4.5	4.8	4.8
Loss factor	$\frac{1}{\mu_o}$	9.0×10^{-5} @ 25 MHz	2.2×10^{-5} @ 2.5 MHz	2.5×10^{-5} @ 0.2 MHz	9.0×10^{-6} @ 0.1 MHz	5.0×10^{-6} @ 0.1 MHz
Coercive force	Oer.	2.40	1.60	0.30	0.18	0.18
Temp. Coef. of initial perm.	%/°C 20-70°C	0.10	0.10	0.20	0.60	—

Ferrite Toroids — Physical Properties

Core Size	OD	ID	Height	A_e	I_e	V_e	A_s	A_w
FT-23	0.230	0.120	0.060	0.00330	0.529	0.00174	0.1264	0.01121
FT-37	0.375	0.187	0.125	0.01175	0.846	0.00994	0.3860	0.02750
FT-50	0.500	0.281	0.188	0.02060	1.190	0.02450	0.7300	0.06200
FT-82	0.825	0.520	0.250	0.03810	2.070	0.07890	1.7000	0.21200
FT-114	1.142	0.748	0.295	0.05810	2.920	0.16950	2.9200	0.43900

OD — Outer diameter (inches)
ID — Inner diameter (inches)
Hgt — Height (inches)
A_w — Total window area (in)^2

A_e — Effective magnetic cross-sectional area (in)^2
I_e — Effective magnetic path length (inches)
V_e — Effective magnetic volume (in)^3
A_s — Surface area exposed for cooling (in)^2

Courtesy of Amidon Assoc., N. Hollywood, CA 91607

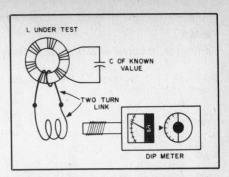

Fig. 69 — Method for checking the inductance of a toroid winding by means of a dip meter, known capacitance value and a calibrated receiver. The self-shielding properties of a toroidal inductor prevent dip meter readings when the instrument is coupled directly to the toroid. Sampling is done by means of a coupling link as illustrated.

particular wire gauge can be close wound to fill a specified core. For example, a T-68 core will contain 49 turns of no. 24 enameled wire, 101 turns of no. 30 enameled wire, and so on. Generally speaking, the larger the wire gauge the higher the unloaded Q of the toroidal inductor. The inductance values are based on the winding covering the entire circumference of the core. When there is space between the turns of wire, some control over the net inductance can be effected by compressing the turns or spreading them. The inductance will increase if compression is used and will decrease when the turns are spread farther apart.

Table 7 contains data for ferrite cores. The number of turns for a specified inductance in mH versus the A_L can be determined by

Turns = $1000 \sqrt{\text{desired L (mH)} \div A_L}$

where the A_L for a specific core can be taken from Table 7. Thus, if you required a 1-mH inductor and chose a no. FT-82-43 toroid core, the number of turns would be

Turns = $1000 \sqrt{1 \div 557}$
= $1000 \sqrt{0.001795}$
= $1000 \times 0.0424 = 42.4$ turns

For an FT-82 size core, no. 22 enameled wire would be suitable as indicated in Table 6 (using the T-80 core size as the nearest one to an FT-82). If the toroid core has rough edges (untumbled), it is suggested that insulating tape (3M glass epoxy tape or Mylar tape) be wrapped through the core before the wire is added. This will prevent the rough edges of the core from abrading the enameled wire.

Checking RF Toroidal Devices

The equations given previously will provide the number of wire turns needed for a particular inductance, plus or minus 10%. However, slight variations in core permeability may exist from one production run to another. Therefore, for circuits that require exact values of inductance, it is necessary to check the toroid winding by means of an RCL bridge or an RX meter. If these instruments are not available, close approximations can be had by using a dip

meter, standard capacitor (known value, stable type, such as a silver mica) and a calibrated receiver against which to check the dipper frequency.

Fig. 69 shows how to couple a dip meter to a completed toroid for testing. The coupling link in the illustration is necessary because the toroid has a self-shielding characteristic. That characteristic makes it difficult, and often impossible, to obtain a dip in the meter reading when coupling the instrument directly to the toroidal inductor or transformer. The inductance can be determined by X_L since $X_L = X_C$ at resonance. Therefore,

$$X_C = \frac{1}{2\pi fC} \text{ and } L_{(\mu H)} = \frac{X_L}{2\pi f}$$

where
X_C = the reactance of the known capacitor value
f = frequency in MHz
C = capacitance in μF

Using an example, where f is 3.5 MHz (as noted on a dip meter) and C is 100 pF, L is determined by

$$X_C = \frac{1}{6.28 \times 3.5 \times 0.0001} = 454.7 \text{ ohms}$$

L = X_C at resonance, where

$$L_{(\mu H)} = \frac{454.7}{6.28 \times 3.5} = 20.7 \ \mu H$$

It is assumed, for the purpose of accuracy, that the dip-meter signal is checked for precise frequency by means of a calibrated receiver.

Practical Considerations

Amateurs who work with toroidal inductors and transformers are sometimes confused by the winding instructions given in construction articles. For the most part, winding a toroid core with wire is less complicated than winding a cylindrical single-layer coil.

When many turns of wire are required,

Table 8
Ferrite Toroid Cores — Size (Inches) Cross-Reference

OD	ID	Thickness	Amidon	Fair-Rite	Indiana General	Ferroxcube	Magnetics, Inc
0.100	0.050	0.050	—	—	—	—	40200TC
0.100	0.070	0.030	—	701	F426-1	—	—
0.155	0.088	—	—	801	F2062-1	—	40502
0.190	0.090	0.050	—	—	—	213T050	—
0.230	0.120	0.060	FT-23	101	F303-1	1041T060	40601
0.230	0.120	0.120	—	901	—	—	—
0.300	0.125	0.188	—	—	F867-1	—	40705
0.375	0.187	0.125	FT-37	201	F625-9	266T125	41003
0.500	0.281	0.188	FT-50	301	—	768T188	—
0.500	0.312	0.250	—	1101	F627-8	—	41306
0.500	0.312	0.500	—	1901	—	—	—
0.825	0.520	0.250	FT-82	601	—	—	—
0.825	0.520	0.468	—	501	—	—	—
0.870	0.500	0.250	—	401	—	—	—
0.870	0.540	0.250	—	1801	F624-19	846T250	42206
1.000	0.500	0.250	—	1501	F2070-1	—	42507
1.000	0.610	0.250	—	1301	—	—	—
1.142	0.748	0.295	FT-114	1001	—	K300502	42908
1.225	0.750	0.312	—	1601	—	—	—
1.250	0.750	0.375	—	1701	F626-12	—	—
1.417	0.905	0.591	—	—	—	K300501	—
1.417	0.905	0.394	—	—	—	K300500	—
1.500	0.750	0.500	—	—	—	528T500	43813
2.000	1.250	0.750	—	—	—	400T750	—
2.900	1.530	0.500	—	—	—	144T500	—
3.375	1.925	0.500	—	—	F1707-15	—	—
3.500	2.000	0.500	—	—	F1707-1	—	—
5.835	2.50	0.625	—	—	F1824-1	—	—

Table 9
Ferrite Toroid Cores — Permeability Cross-Reference

μ_o	Amidon	Fair-Rite	Indiana General	Ferroxcube	Magnetics, Inc.
16	—	—	Q3	—	—
20	—	68	—	—	—
40	FT-63	63, 67	Q2	—	—
100	—	65	—	—	—
125	FT-61	61	Q1	4C4	—
175	—	62	—	—	—
250	FT-64	64	—	—	—
300	—	83	—	—	—
375	—	31	—	—	—
400	—	—	G	—	—
750	—	—	—	3D3	A
800	—	33	—	—	—
850	FT-43	43	H	—	—
950	—	—	TC-3	—	—
1200	—	34	—	—	—
1400	—	—	—	—	C
1500	—	—	TC-7	—	—
1800	FT-77	77	—	3B9	—
2000	FT-72	72	TC-9	—	S, V, D
2200	—	—	05	—	—
2300	—	—	—	3B7	G
2500	FT-73	73	TC-12	—	—
2700	—	—	—	3E (3C8)	—
3000	—	—	05P	3C5	F
4700	—	—	06	—	—
5000	FT-75	75	—	3E2A	J
10,000	—	—	—	—	W
12,500	—	—	—	3E3	—

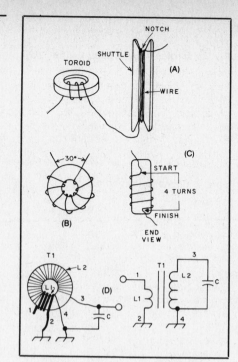

Fig. 70 — (A) Illustration of a homemade winding shuttle for toroids. The wire is stored on the shuttle and the shuttle is passed through the center hole of the toroid, again and again, until the required number of turns is in place. (B) It is best to leave a 30° gap beteen the ends of the toroid winding. This will reduce the distributed capacitance considerably. (C) Edgewise view of a toroid core, illustrating the method for counting the turns accurately. (D) The low-impedance winding of a toroidal transformer is usually wound over the large winding, as shown. For narrow-band applications the link should be wound over the cold end of the main winding (see text).

a homemade winding shuttle can be used to simplify the task. Fig. 70A illustrates this method. The shuttle can be fashioned from a piece of circuit-board material. A stiff piece of plastic works as well. The wire is wound on the shuttle after determining how many inches are required to provide the desired number of toroid turns. (A sample turn around the toroid core will reveal the wire length per turn.) Once the shuttle is loaded, it is passed through the toroid center again and again until the winding is completed. The edges of the shuttle should be kept smooth to prevent abrasion of the wire insulation.

How to Wind Toroids

The effective inductance of a toroid coil or a transformer winding depends in part on the distributed capacitance between the coil turns and between the ends of the winding. When a large number of turns are used (for example, 500 or 1000), the distributed capacitance can be as great as 100 pF. Ideally, there would be no distributed or parasitic capacitance, but this is not possible. Therefore, the unwanted capacitance must be kept as low as possible in order to take proper advantage of the A_L factors discussed earlier in this section. The greater the distributed capacitance, the more restrictive the transformer or inductor becomes when applied in a broadband circuit. In the case of a narrow-band application, the Q can be affected by the distributed capacitance.

The pictorial illustration at Fig. 70B shows the inductor turns distributed uniformly around the toroid core, but a gap of approximately 30° is maintained between the ends of the winding. This method is recommended to reduce the distributed capacitance of the winding. The closer the ends of the winding are to one another, the greater the unwanted capacitance. Also, in order to closely approximate the desired toroid inductance when using the A_L formula, the winding should be spread over the core as shown. When the turns of the winding are not close wound, they can be spread apart to decrease the effective inductance (this lowers the distributed C). Conversely, as the turns are pushed closer together, the effective inductance is increased by virtue of the greater distributed capacitance. This phenomenon can be used to advantage during final adjustment of narrow-band circuits in which toroids are used.

The proper method for counting the turns on a toroidal inductor is shown in Fig. 70C. The core is shown as it would appear when stood on its edge with the narrow dimension toward the viewer. In this example a four-turn winding has been placed on the core.

Some manufacturers of toroids recommend that the windings on toroidal transformers be spread around all of the core in the manner shown in Fig. 70B. That is, the primary and secondary windings should each be spread around most of the core. This is a proper method when winding conventional broadband transformers. However, it is not recommended when narrow-band transformers are being built. It is better to place the low-impedance winding (L1 of Fig. 70D) at the cold or grounded end of L2 on the core.

This is shown in pictorial and schematic form at Fig. 70D. The windings are placed on the core in the same rotational sense, and L1 is wound over L2 at the grounded end of L2. The purpose of this winding method is to discourage unwanted capacitive coupling between the windings, which aids in the reduction of spurious energy (harmonics, and so on) that might be present in the circuit where the transformer is used.

In circuits that have a substantial amount of voltage present in the transformer windings, it is good practice to use a layer of insulating material between the toroid core and the first winding. Alternatively, the wire itself can have high dielectric insulation, such as Teflon. This procedure prevents arcing between the winding and the core. Similarly, a layer of insulating tape (3-M glass tape, Mylar or Teflon) can be placed between the primary and secondary windings of the toroidal transformer (Fig. 70D). Normally, these precautions are not necessary at impedance levels under a few hundred ohms for RF power levels below 100 watts.

Once the inductor or transformer is wound and tested for proper performance, a coating or two of high-dielectric cement should be applied to the winding(s) of the toroid. This protects the wire insulation from abrasion, holds the turns in place and seals the assembly against moisture and dirt. Polystyrene Q Dope is excellent for the purpose.

The general guidelines given for toroidal components can be applied to pot cores and rods when they are used as foundations for inductors or transformers. The important thing to remember is that all of the powdered-iron and ferrite core materials are brittle. They break easily under stress.

Radio-Frequency Circuits

The designer of amateur equipment needs to be familiar with radio-frequency circuits and the various related equations. This section provides the basic data for most amateur circuit development.

COUPLED CIRCUITS AND FILTERS

Two circuits are said to be coupled when a voltage or current in one network produces a voltage or current in the other. The network where the energy originates is often called the primary circuit, and the network that receives the energy is called the secondary circuit. Such coupling is often desirable since, in the process, unwanted frequency components or noise may be rejected or isolated, and power may be transferred from a source to a load with greatest efficiency. On the other hand, two or more circuits may be coupled inadvertently and undesirable effects produced. While a great number of coupling-circuit configurations are possible, one very important class covers so many practical applications that it will be analyzed in detail. This class includes ladder networks and filters.

Ladder Networks

Any two circuits that are coupled can be drawn schematically as shown in Fig. 71A. A voltage source represented by E_{ac} with a source resistance R_p and a source reactance X_p is connected to the input of the coupling network, thus forming the primary circuit. At the output, a load reactance X_s and a load resistance R_s are con-

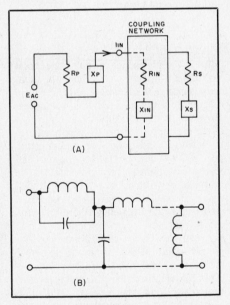

Fig. 71 — A representative coupling circuit (A) and ladder network (B).

nected as shown to form the secondary circuit. The circuit in the box could consist of an infinite variety of resistors, capacitors, inductors and even transmission lines. However, it will be assumed that the network can be reduced to a combination of series and shunt elements consisting only of inductors and capacitors, as indicated by the circuit shown in Fig. 71B. For obvious reasons, the circuit is often called a ladder network. In addition, if no resistive elements are present or if such elements can be neglected, the network is said to be dissipationless. It will consume no power.

If a network is dissipationless, all the power delivered to the input of the network will be dissipated in the load resistance, R_s. This effect leads to important simplifications in computations involved in coupled networks. The assumption of a dissipationless network is usually not valid with transmitting circuits since even a small network loss (0.5 dB) will result in considerable heating at the higher power levels used in amateur applications. On the other hand, coupled circuits used in some receiving stages may have considerable loss. This is because the network may have some advantage and its high loss can be compensated by additional amplification in another stage. However, such devices form a relatively small minority of coupled networks commonly encountered, and only the dissipationless case will be considered in this section.

Effective Attenuation and Insertion Loss

The most important consideration in any coupled network is the amount of power delivered to the load resistance, R_s from the source, E_{ac} with the network present. Rather than specify the source voltage each time, a comparison is made with the maximum available power from any source with a given primary resistance, R_p. The value of R_p might be considered as the impedance level associated with a complex combination of sources, transmission lines, coupled networks and even antennas. Typical values of R_p are 52, 75, 300 and 600 ohms. The maximum available power is given by

$$P_{max} = \frac{E_{ac}^2}{4 R_p}$$

If the network is also dissipationless, the power delivered to the load resistance, R_s is just the power dissipated in R_{in}. This power is related to the input current by

$$P_o = I_{in}^2 R_{in}$$

and the current in terms of the other variables is

$$I_{in} = \frac{E_{ac}}{\sqrt{(P_\pi + R_{in})^2 + (X_p + X_{in})^2}}$$

Combining these expressions gives a very useful formula for the ratio of power delivered to a load in terms of the maximum available power. This ratio, expressed in decibels, is given by

$$\text{Attn} = -10 \log \left(\frac{P_o}{P_{in}} \right)$$

$$= -10 \log \left[\frac{4 R_{in} R_p}{(R_p + R_{in})^2 + (X_p + X_{in})^2} \right]$$

and is sometimes called the effective attenuation.

In the special case where X_p and X_s are either zero or can be combined into a coupling network, and where R_p is equal to R_s the effective attenuation is also equal to the insertion loss of the network. The insertion loss is the ratio of the power delivered to the load (with the coupling network in the circuit) to the power delivered to the load with the network absent. Unlike the effective attenuation, which is always positive when defined by the previous formula, the insertion loss can take on negative values if R_p is not equal to R_s or if X_p and X_s are not zero. In effect, the insertion loss would represent a power gain under these conditions. The interpretation of this effect is that maximum available power does not occur with the coupling network out of the circuit because of the unequal source and load resistances and the nonzero reactances.

With the network in the circuit, the resistances are now matched, and the reactances are said to be "tuned out." The action of the coupling network in this instance is very similar to that of a transformer (which was discussed in a previous section), and networks consisting of pure inductors and capacitors are often used for this purpose. Such circuits are often called matching networks. On the other hand, it is often desired to deliver the greatest amount of power to a load at some frequencies while rejecting energy at other frequencies. A device that accomplishes this action is called a filter. In the case of unequal source and load resistance, it is often possible to combine the processes of filter-

ing and matching into one network.

Solving Ladder-Network Problems

From the previous section it is evident that if the values of R_{in} and X_{in} of Fig. 71A can be determined, the effective attenuation and possibly the insertion loss are also easily found. Being able to solve this problem has wide applications in RF circuits. For instance, design formulas for filters often include a simplifying assumption that the load resistance is constant with frequency. In the case of many circuits, this assumption is not true. However, if the values of R_s and X_s at any particular frequency are known, the attenuation of the filter can be determined even though it is terminated improperly.

Unfortunately, while the solution to any ladder problem is possible from a theoretical standpoint, practical difficulties are encountered as the network complexity increases. Many computations with a high degree of accuracy may be required, making the process a tedious one. Consequently, the use of a calculator or similar computing device is recommended. The approach used here is adapted readily to any calculating method including the use of inexpensive pocket calculators.

SUSCEPTANCE AND ADMITTANCE

The respective reactances of an inductor and a capacitor are given by

$$X_L = 2\pi fL \text{ and } X_c = \frac{-1}{2\pi fC}$$

In a simple series circuit, the total resistance is just the sum of the individual resistances in the network, and the total reactance is the sum of the reactances. However, it is important to note the sign of the reactance. Since capacitive reactance is negative and inductive reactance is positive, it is possible that the sum of the reactances might be zero even though the individual reactances are not zero. Recall that a series-circuit network is said to be resonant at the frequency where the reactances cancel.

A complementary condition exists in a parallel combination of circuit elements, and it is convenient to introduce the concepts of admittance, conductance and susceptance. In the case of a simple resistance, the conductance is merely the reciprocal. That is, the conductance of a 50-ohm resistance is 1/50 or 2×10^{-2}. The unit of conductance is the siemens, the reciprocal of the ohm. For simple inductances and capacitances, the formulas for the respective reciprocal entities are

$$B_L = \frac{-1}{2\pi fL} \text{ and } B_C = 2\pi fC$$

These entities are defined as susceptances. In a parallel combination of conductances and susceptances, the total conductance is the sum of the individual conductances, and the total susceptance is the

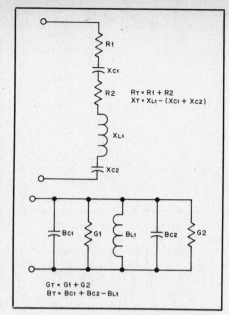

Fig. 72 — Resistances and reactances add in series circuits while conductances and susceptances add in parallel circuits.

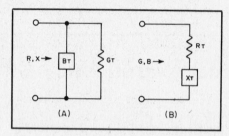

Fig. 73 — Conversion formulas can be used to transform a shunt conductance and susceptance to a series-equivalent circuit, A. The converse is illustrated at B.

sum of the individual susceptances, taking the respective susceptance signs into account. A comparison between the way resistance and reactance add and the manner in which conductance and susceptance add is shown in Fig. 72. An entity called admittance may be considered as the reciprocal of impedance. Admittance can be defined in terms of the total conductance and total susceptance by the formula

$$Y = \sqrt{G_T^2 + B_T^2}$$

where
  Y = the admittance
  G_t = total conductance
  B_t = total susceptance

If the impedance of a circuit is known, the admittance is merely the reciprocal. Likewise, if the admittance of a circuit is known, the impedance is the reciprocal of the admittance. However, conductance, reactance, resistance and susceptance are not so simply related. If the total resistance and total reactance of a series circuit are known, the conductance and susceptance of the circuit are related to them by the formulas

$$G = \frac{R_T}{R_T^2 + X_T^2} \text{ and}$$

$$B = \frac{-X_T}{R_T^2 + X_T^2}$$

On the other hand, if the total conductance and total susceptance of a parallel combination are known, the equivalent resistance and reactance can be found from the formulas

$$R = \frac{G_T}{G_T^2 + B_T^2} \text{ and}$$

$$X = \frac{-B_T}{G_T^2 + B_T^2}$$

The relationships are illustrated in Fig. 73. While the derivation of the mathematical expressions will not be given, the importance of the sign change cannot be stressed too highly. Solving network problems with a calculator is merely a matter of bookkeeping, and failure to take the sign change associated with the transformed reactance and susceptance into account is the most common source of error.

A Sample Problem

The following example illustrates the manner in which the theory just discussed can be applied to a practical problem. A filter with the schematic diagram shown in Fig. 74A has an insertion loss of 3 dB at 6 MHz when connected between a 52-ohm load and a source with a 52-ohm primary resistance (both X_p and X_s are zero). Find R_{in} and X_{in}. Since this is a case where the effective attenuation is equal to the insertion loss, the previous formula for effective attenuation applies.

Starting at the output, the values for the conductance and susceptance of the parallel RC circuit must be determined first. The conductance is just the reciprocal of 52 ohms and the previous formula for capacitive susceptance gives the values shown in parentheses in Fig. 74A. The next step is to apply the formulas for resistance and reactance in terms of the conductance and susceptance, and the results give a 26-ohm resistance in series with a −26-ohm capacitive reactance, indicated in Fig. 74B. The reactance of the inductor can now be added to give a total reactance of 78.01 ohms.

The conductance and susceptance formulas can now be applied; the results of both of these operations is shown in Fig. 74C. Finally, adding the susceptance of the 510.1-pF capacitor (Fig. 74A) and applying the formulas once more gives the value of R_{in} and X_{in} (Fig. 74F). If the latter values are substituted into the effective attenuation formula, the insertion loss and effective attenuation are 3.01 dB, which is very close to the value specified. The reader might verify that the insertion loss is 0.167, 0.37 and 5.5 dB at 3.5, 4.0 and 7.0 MHz,

respectively. If a plot of insertion loss versus frequency was constructed it would give the frequency response of the filter.

Frequency Scaling and Normalized Impedance

Often, it is desirable to change a coupling network at one frequency and impedance level to another one. For example, suppose it was desired to move the 3-dB point of the filter in the preceding illustration from 6 to 7 MHz. An examination of the reactance and susceptance formulas reveals that multiplying the frequency by some constant k and dividing both the inductance and capacitance by the same value of k leaves the equations unchanged. Thus, if the capacitances and inductance in Fig. 74A are multiplied by 6/7, all the reactances and susceptances in the new circuit will now have the same value at 7 MHz that the old one had at 6 MHz.

It is common practice, especially with many filter tables, to present all the circuit components for a number of designs at some convenient frequency. Translating the design to a desired frequency is accomplished simply by multiplying all the components by some constant factor. The most common frequency used is the value of f such that $2\pi F$ is equal to 1.0. This is sometimes called a radian frequency of 1.0 and corresponds to 0.1592 Hz. To change a 1-radian filter to a new frequency f_o (in Hz), all that is necessary is to multiply the inductances and capacitances by $0.1592/f_o$.

In a similar manner, if one resistance (or conductance) is multiplied by some factor n, all the other resistances (or conductances) and reactances (or susceptances) must be multiplied by the same factor to preserve the network characteristics. For instance, if the secondary resistance, R_s is multiplied by n, all circuit inductances must be multiplied by n and the circuit capacitances divided by n (since capacitive reactance varies as the inverse of C). If, in addition to converting the filter of Fig. 74A to 7 MHz from 6 MHz, it was also desired to change the impedance level from 52 to 600 ohms, the inductance would have to be multiplied by $(6/7) \times (600/52)$ and the capacitances by $(6/7) \times (52/600)$.

USING FILTER TABLES

In a previous example, it was indicated that the frequency response of a filter could be derived by solving for the insertion loss of the ladder network for a number of frequencies. The question might be asked if the converse is possible. That is, given a desired frequency response, could a network be found that would have this response? The answer is a qualified yes, and the technical nomenclature for this sort of process is network synthesis. Frequency responses can be cataloged and, if a suitable response can be found, the corresponding network elements can be determined from an associated table. Filters

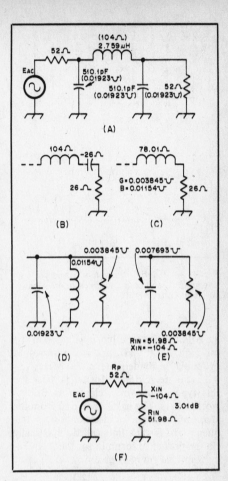

Fig. 74 — Problem illustrating network reduction to find insertion loss.

derived by network synthesis and similar methods (such as optimized computer designs) are often referred to as modern filters even though the theory has been in existence for years. The term is useful in distinguishing such designs from those of an older approximate method called image-parameter theory.

Butterworth Filters

Filters can be grouped into four general categories, as illustrated in Fig. 75A. Low-pass filters have zero insertion loss up to some critical frequency (f_c) or cutoff frequency, and then provide high rejection above this frequency. (The cutoff condition is indicated by the shaded lines in Fig. 75.) Band-pass filters have zero insertion loss between two cutoff frequencies, with high rejection outside of the prescribed bandwidth. Band-stop filters reject a band of frequencies while passing all others. And high-pass filters reject all frequencies below some cutoff frequency.

The attenuation shapes shown in Fig. 75A are ideal and can only be approached or approximated in practice. For instance, if the filter in the preceding problem was used for low-pass purposes in an 80-meter transmitter to reject harmonics on 40 meters, its performance would leave a lot to be desired. While insertion loss at

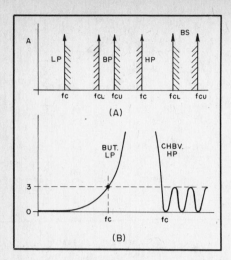

Fig. 75 — Ideal filter response curves are shown at A, and characteristics of practical filters are shown at B.

Table 10
Prototype Butterworth Low-Pass Filters

Fig. 76A	C1	L2	C3	L4	C5	L6	C7	L8	C9	L10
Fig. 76B	L1	C2	L3	C4	L5	C6	L7	C8	L9	C10
k										
1	2.0000									
2	1.4142	1.4142								
3	1.0000	2.0000	1.0000							
4	0.7654	1.8478	1.8478	0.7654						
5	0.6180	1.6180	2.0000	1.6180	0.6180					
6	0.5176	1.4142	1.9319	1.9319	1.4142	0.5176				
7	0.4450	1.2470	1.8019	2.0000	1.8019	1.2470	0.4450			
8	0.3902	1.1111	1.6629	1.9616	1.9616	1.6629	1.1111	0.3902		
9	0.3473	1.0000	1.5321	1.8794	2.0000	1.8794	1.5321	1.0000	0.3473	
10	0.3129	0.9080	1.4142	1.7820	1.9754	1.9754	1.7820	1.4142	0.9080	0.3129

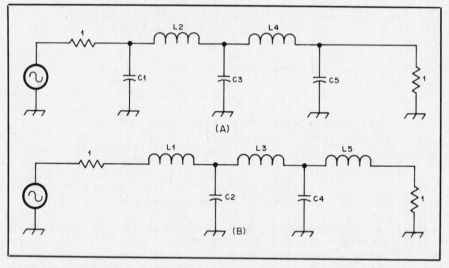

Fig. 76 — Schematic diagram of a Butterworth low-pass filter. (See Table 10 for element values).

3.5 MHz was acceptable, it would likely be too high at 4.0 MHz, and rejection would probably be inadequate at 7.0 MHz.

Fortunately, design formulas exist for this type of network, classed as Butterworth filters. The name is derived from the shape of the curve for insertion-loss vs. frequency and is sometimes called a maximally flat response. A formula for the frequency response curve is given by

$$A = 10 \log_{10} \left[1 + \left(\frac{f}{f_c} \right)^{2k} \right]$$

where

f = the frequency for an insertion loss of "A" dB

f_c = the frequency for an insertion loss of 3.01 dB

k = the number of circuit elements

The shape of a Butterworth low-pass filter is shown in the left-hand portion of Fig. 75B. (Another type that is similar but allows some ripple in the passband, is also shown in Fig. 75B. Here, a high-pass characteristic illustrates a Chebyshev response.)

As can be seen from the formula, increasing the number of elements will result in a filter that approaches the ideal low-pass shape. For instance, a 20-element filter designed for a 3.01-dB cutoff frequency of 4.3 MHz would have an insertion loss of 0.23 dB at 4 MHz, and a loss of 84.7 dB at 7 MHz. However, practical difficulties would make such a filter very hard to construct. Some compromises are always required between a theoretically perfect frequency response and ease of construction.

Element Values

Once the number of elements, k, is determined, the next step is to find the network configuration corresponding to k. Filter tables sometimes have sets of curves that enable the user to select the desired frequency response curve rather than use a formula. Once the curve with the fewest number of elements for the specified passband and stop-band insertion loss is found, the filter is then fabricated around the corresponding value of k. Table 10 gives normalized element values for values of k from 1 to 10. This table is for 1-ohm source and load resistances (reactance equal zero) and a 3.01-dB cutoff frequency of 1 radian/second (0.1592 Hz). There are two possible circuit configurations, and these are shown in Fig. 76. Here, a five-element filter is given as an example with either a shunt element next to the load (Fig. 76A) or a series element next to the load (Fig. 76B). Either filter will have the same response.

After the values for the 1-ohm, 1-radian/second prototype filter are found, the corresponding values for the actual frequency/impedance level can be determined (see the section on frequency and impedance scaling). The prototype inductance and capacitance values are multiplied by the ratio (0.1592/f_c) where f_c is the actual 3.01-dB cutoff frequency. Next, this number is multiplied by the load resistance in the case of an inductor and divided by the load resistance if the element is a capacitance. For instance, the filter in the preceding example is for a three-element design (k equal to 3), and the reader might verify the values for the components for an f_c of 6 MHz and load resistance of 52 ohms.

The formulas for change of impedance and frequency from the 1-ohm, 1-radian/second prototype to some desired level can also be conveniently written as

$$L = \frac{R}{2\pi f_c} L_{prototype}$$

$$C = \frac{1}{2\pi f_c R} C_{prototype}$$

where

R = the load resistance in ohms

f_c = the desired 3.01-dB frequency in Hz

Then L and C give the actual circuit-element values in henrys and farads in terms of the prototype element values from Table 10.

High-Pass Butterworth Filters

The usefulness of the low-pass prototype does not end here. If the following set of equations is applied to the prototype values, circuit elements for a high-pass filter can be obtained. The filters are shown

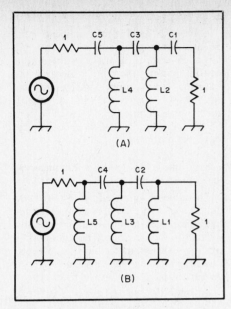

Fig. 77 — Network configuration of a Butterworth high-pass filter. The low-pass prototype can be transformed as described in the text.

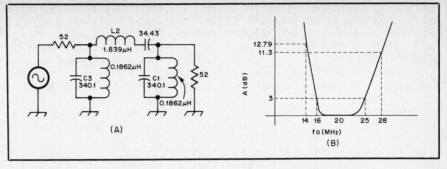

Fig. 78 — A Butterworth band-pass filter. Capacitance values are in picofarads.

in Fig. 77A and B, which correspond to Fig. 76A and B in Table 10. The equations for the actual high-pass circuit values in terms of the low-pass prototype are given by

$$C = \frac{1}{R2\pi f_c C_{prot.}}$$

$$L = \frac{R}{2\pi f_c L_{prot.}}$$

and the frequency response curve can be obtained from

$$A = 10 \log \left[1 + \left(\frac{f_c}{f} \right)^{2k} \right]$$

For instance, a high-pass filter with three elements, a 3.01-dB f_c of 6 MHz and a load resistance of 52 ohms, has a C1 and C3 of 510 pF and an L2 of 0.6897 μH. The insertion loss at 3.5 and 7 MHz would be 14.21 and 1.45 dB, respectively.

Butterworth Band-Pass Filters

Band-pass filters can also be designed through the use of Table 10. Unfortunately the process is not as straightforward as it is for low- and high-pass filters if a practical design is to be obtained. In essence, a low-pass filter is resonated to some center frequency with the 3.01-dB cutoff frequency being replaced by the filter bandwidth. The ratio of the bandwidth to center frequency must be relatively large; otherwise component values tend to become unmanageable.

While there are many variations of specifying such filters, a most useful approach is to determine an upper and lower frequency for a given attenuation. The center frequency and bandwidth are then given by

$$f_o = \sqrt{f_1 f_2}$$

$$BW = f_2 - f_1$$

If the bandwidth specified is not the 3.01-dB bandwidth (BW_c) this bandwidth can be determined from

$$BW_c = \frac{BW}{(10^{0.1A} - 1)^{\frac{1}{2k}}}$$

in the case of a Butterworth response or from tables of curves. A is the required attenuation at the cutoff frequencies. The upper and lower cutoff frequencies (f_{cu} and f_{cl} are then given by

$$f_{cl} = \frac{-BW_c + \sqrt{(BW_c)^2 + 4f_o^2}}{2}$$

and

$$f_{cu} = f_{cl} + BW_c$$

A somewhat more convenient method is to pick a 3.01-dB bandwidth (the wider the better) around some center frequency and compute the attenuation at other frequencies of interest by using the transformation:

$$\frac{f}{f_c} = \left| \left(\frac{f}{f_o} - \frac{f_o}{f} \right) \frac{f_o}{BW_c} \right|$$

which can be substituted into the insertion-loss formula or table of curves.

As an example, suppose you want to build a band-pass filter for the 15-meter Novice band in order to eliminate the possibility of radiation on the 14- and 28-MHz bands. For a starting choice, 16 and 25 MHz will be selected as the 3.01-dB points, giving a 3-dB bandwidth of 9 MHz. For these two points, f_o will be 20 MHz. It is common practice to equate the number of branch elements or filter resonators to certain mathematical entities called poles, and the number of poles is just the value of k for purposes of discussion here. For a three-pole filter (k of 3) the insertion loss will be 12.79 and 11.3 dB at 14 and 28 MHz, respectively.

C1, C3 and L2 are then calculated for a 9-MHz low-pass filter and the elements for this filter are resonated to 20 MHz as shown in Fig. 78A. The response shape is plotted in Fig. 78B, and it appears to be unsymmetrical about f_o. In spite of this fact, such filters are called symmetrical band-pass filters and f_o is the center frequency. If the response is plotted with a

logarithmic frequency scale, the symmetry will become apparent. Consequently, using a logarithmic plot is helpful in designing filters of this type.

Examination of the component values reveals that while the filter is practical, it is a bit untidy from a construction standpoint. Rather than using a single 340.1-pF capacitor, paralleling a number of smaller value units would be advisable. Encountering difficulty of this sort is typical of most filter designs; consequently, some tradeoffs among performance, complexity and ease of construction are usually required.

Additional Modern Filters

Butterworth filters are used where a flat passband response, well-behaved phase shift characteristics and exact impedance matching are important. They are subject to two limitations, however. For a given number of elements, the rate of attenuation (roll-off) in the transition band is not as great as can be obtained with some other designs of the same complexity. In addition, the ratio of capacitances is fixed at a number that may not result in standard component values. The values can be synthesized by connecting several capacitors in parallel, but this practice may set up parasitic resonances that greatly distort the band-pass and band-stop characteristics.

In applications requiring sharper cutoff or where some response ripple and passband impedance mismatch can be tolerated, a Chebyshev design should be considered. A variety of capacitance ratios is possible, depending on the allowable ripple. Many of these ratios result in standard-value capacitors, which greatly simplifies and economizes the circuit.

Another type of modern filter, the elliptic filter, is used in situations requiring extremely sharp cutoff with one or more infinitely deep notches in the stopband. Like the Chebyshev filter, the design parameters can be manipulated to produce circuits having at least some standard-value capacitors.

PASSIVE LC FILTER DESIGN

The following text by Ed Wetherhold, W3NQN, is adapted from a paper entitled "Simplified Passive LC Filter Design for the EMC Engineer." That work first appeared in the record of the 1985 IEEE International Symposium on Electromagnetic Compatibility, held August 20-22, 1985, in Wakefield, MA, pp. 575-584.© 1985 IEEE.

The filter design procedure presented here uses eight computer-calculated tables of performance parameters and component values for 5- and 7-branch Chebyshev and 5-branch elliptic 50-ohm filters. The tables permit the quick and easy selection of an equally-terminated passive LC filter for applications where the attenuation response is of primary interest. All of the capacitors in the Chebyshev designs and the three non-resonating capacitors in the elliptic designs have standard, off-the-shelf values to simplify construction. Although the tables cover only the 1- to 10-MHz frequency range, a simple scaling procedure gives standard-value capacitor (SVC) designs for any impedance level and virtually any cutoff frequency.

Introduction

The radio amateur frequently needs an equally-terminated inductor-capacitor (LC) filter for nonstringent applications such as harmonic attenuation or wideband preamplifier preselection where the amplitude response of sinusoidal signals is of primary concern. Suitable designs are usually obtained by scaling normalized low-pass designs for the desired frequency and impedance. These designs are published in several well-known references by Geffe, White and others.[1,2,3,4,5] The normalized tables in the first two references are based on specific levels of maximum passband ripple attenuation, while tables in the last three references are based on regular increments of reflection coefficient. Although the scaling and transformation procedures used with the normalized low-pass tables are relatively straightforward, the transformation and scaling calculations provide undesired opportunities for error. The calculated capacitor values will invariably be nonstandard, thus complicating the construction.

For most nonstringent filtering applications, it is not necessary that the actual cutoff frequency exactly match the desired cutoff frequency, and a deviation of 5% or so between the actual and desired cutoff frequencies is acceptable. In this case, it is feasible and more convenient to use design tables based on standard capacitor values instead of passband ripple attenuation or reflection coefficient.

Standard Values Used in Filter Design Calculations

Capacitors are commercially available in special series of preferred values having designations of E12 (10% tolerance) and E24 (5% tolerance).[6] The reciprocal of the E-number is the power to which 10 is raised to give the step multiplier for that particular series. By calculating normalized Chebyshev and elliptic component values based on many ratios of standard capacitor

values, and by using a 50-ohm impedance level, the parameters of a large number of designs can be calculated and tabulated to span the 1-10 MHz decade. Because of the large number of standard-value capacitor (SVC) designs in this decade, the increment in cutoff frequency from one design to the next is sufficiently small so that virtually any cutoff frequency requirement can be satisfied. Using such a table, the selection of an appropriate design consists of merely scanning the cutoff frequency column to find a design having a cutoff frequency that most closely matches the desired cutoff frequency.

Chebyshev and Elliptic SVC Design Tables

Low-pass and high-pass 5- and 7-element Chebyshev and 5-branch elliptic designs were selected for tabulation because they are easy to construct and will satisfy the majority of nonstringent filtering requirements where the amplitude response is of primary interest. The precalculated 50-ohm designs are presented in eight tables of five low-pass types (Tables 11-15) and three high-pass types (Tables 16-18) with cutoff frequencies covering the 1-10 MHz decade. The applicable filter configuration and attenuation response curve accompany each table. In addition to the component values, attenuation versus frequency data and SWR are also included in the table. The passband attenuation ripples are so low in amplitude that they are swamped by the filter losses and are not measurable. For this reason, they are not shown in the response curves.

Low-pass Tables

Tables 11 and 12 are for the low-pass 5- and 7-element Chebyshev capacitor input/output configuration. This configuration is generally preferred to the alternate inductor input/output configuration to minimize the number of inductors, and where a decreasing input impedance with increasing frequency in the stopband presents no problems. Tables 13 and 14 are also for low-pass applications, but with an inductor input/output configuration. This configuration is useful when the filter input impedance in the stopband must rise with increasing frequency. For example, some RF transistor amplifiers may become unstable when terminated in a low-pass filter having a stopband response with a decreasing input impedance. In this case, the inductor-input configuration may eliminate the instability.[7] Because only one capacitor value is required in the designs of Table 13, it was feasible to have the inductor value of L1 and L5 also be a standard value. Table 15 is for the low-pass 5-branch elliptic filter with the capacitor input/output configuration in which the nonresonating capacitors (C1, C3 and C5) are standard values. The alternate inductor input/output elliptic configuration is seldom used, and therefore it is not included.

High-pass Tables

Tables 16, 17 and 18 are for the high-pass 5- and 7-element Chebyshev and the 5-branch elliptic capacitor input/output configurations, respectively. Because the inductor input/output configuration is seldom used, it was not included. As with the low-pass elliptic filter, only capacitors C1, C3 and C5 in the high-pass elliptic filter have standard values.

Scaling the Tables to Other Frequencies and Impedances

The tables shown are for the 1-10 MHz decade and for a 50-ohm equally terminated impedance. The designs are easily scaled to other frequency decades and to other equally terminated impedance levels, however, making the tables a universal design aid for these specific filter types.

Frequency Scaling

To scale the frequency and the component values to the 10-100 or 100-1000 MHz decades, multiply all tabulated frequencies by 10 or 100, respectively, and divide all C and L values by the same number. The A_s and SWR data remain unchanged. To scale the filter tables to the 1-10 kHz, 10-100 kHz or the 0.1-1 MHz decades, divide the tabulated frequencies by 1000, 100 or 10, respectively, and multiply the component values by the same number. By changing the "MHz" frequency headings to "kHz" and the "pF" and "μH" headings to "nF" and "mH," the tables are easily changed from the 1-10 MHz decade to the 1-10 kHz decade, and the table values may be read directly. Because the impedance level is still at 50 ohms, the component values may be awkward, but this can be corrected by increasing the impedance level by ten times using the impedance scaling procedure described below.

Impedance Scaling

All the tabulated designs are easily scaled to impedance levels other than 50 ohms while keeping the convenience of standard-value capacitors and the "scan mode" of design selection. If the desired new impedance level differs from 50 ohms by a factor of 0.1, 10 or 100, the 50-ohm designs are scaled by inspection by shifting the decimal points of the component values. The other data remain unchanged. For example, if the impedance level is increased by ten or one hundred times (to 500 or 5000 ohms), the decimal point of the capacitor is shifted to the left one or two places and the decimal point of the inductor is shifted to the right one or two places. With increasing impedance the capacitor values become smaller and the inductor values become larger. The opposite is true if the impedance decreases.

When the desired impedance level differs from the standard 50-ohm value by a factor such as 1.2, 1.5 or 1.86, the following scaling procedure is used:

1. Calculate the impedance scaling ratio: $R = Z_x/50$, where Z_x is the desired new impedance level in ohms.

2. Calculate the cutoff frequency (F_{50co}) of a "trial" 50-ohm filter, where $F_{50co} = R \times F_{xco}$. R is the impedance scaling ratio and F_{xco} is the desired cutoff frequency of the filter at the new impedance level.

3. From the appropriate SVC table select a design having its cutoff frequency closest to the calculated F_{50co} value. The tabulated capacitor values of this design are taken directly, but the frequency and inductor values must be scaled to the new impedance level.

4. Calculate the exact F_{xco} values, where

$$F_{xco} = \frac{F'_{50co}}{R}$$

and F'_{50co} is the tabulated cutoff frequency of the selected design. Calculate the other frequencies of the design in the same way.

5. Calculate the inductor values for the new filter by multiplying the tabulated inductor values of the selected design by the square of the scaling ratio, R.

For example, assume a 600-ohm elliptic low-pass filter is desired with a cutoff frequency of 1.0 kHz. The elliptic low-pass table is frequency scaled to the 1-10 kHz decade by changing the table headings to kHz, nF and mH. A suitable design is then selected for scaling to 60 ohms. The 60-ohm design is then scaled to 600 ohms by shifting the decimal point to complete the scaling procedure. The calculations for this example follow, using the five steps outlined above:

1. $R = Z_x / 50 = 60/50 = 1.2$
2. $F_{50co} = 1.2 \times 1.0$ kHz $= 1.2$ kHz
3. From the elliptic low-pass table (Table 15), designs 5 and 10 have cutoff frequencies closest to the F_{50co} of 1.2 kHz, and either design is suitable. Design 5 is selected because of its better selectivity. The tabulated capacitor values of 2200 nF, 3900 nF, 1800 nF, 271 nf and 779 nF are copied directly.

4. All frequencies of the final design are calculated by dividing the tabulated frequencies (in kHz) of design 5 by the impedance scaling ratio of 1.2:

$F_{co} = 1.27/1.2 = 1.06$
$F_{3\,dB} = 1.45/1.2 = 1.21$
$F_{As} = 2.17/1.2 = 1.88$

Note that a cutoff frequency of 1.0 kHz was desired, but a 1.06 kHz cutoff frequency will be accepted in exchange for the convenience of using an SVC design.

5. The L2 and L4 inductor values of design 5 are scaled to 60 ohms by multiplying them by the square of the impedance ratio where $R = 1.2$ and $R^2 = 1.44$:

L2 $= 1.44 \times 7.85$ mH $= 11.3$ mH
L4 $= 1.44 \times 6.39$ mH $= 9.20$ mH

The 60-ohm design is now impedance scaled to 600 ohms by shifting the decimal

points of the capacitor values to the left and the decimal points of the inductor values to the right. The final scaled component values for the 600-ohm filter are:

C1 = 0.22 µF	C4 = 77.9 nF
C3 = 0.39 µF	L2 = 113 mH
C5 = 0.18 µF	L4 = 92.0 mH
C2 = 27.1 nF	

How to Use the Filter Tables

50-ohm impedance level: Before selecting a filter design, the important parameters of the filter must be known, such as type (low-pass or high-pass), cutoff frequency, impedance level, preferred input element (for low-pass only), and an approximation of the required stopband attenuation. It is obvious which tables to use for low-pass or high-pass applications, but it is not so obvious which one design of the many possible choices is optimum for the intended application. Generally, the Chebyshev will be preferred over the elliptic because the Chebyshev does not require tuning of the inductors; if the relatively gradual attenuation rise of the Chebyshev is not satisfactory, however, then the elliptic should be considered. For audio filtering, the elliptic designs with high values of SWR are preferred because these designs have a much more abrupt attenuation rise than the Chebyshev. For RF applications, SWR values less than 1.2 are recommended to minimize undesired reflections. Low SWR is also important when cascading high-pass and low-pass filters to achieve a band-pass response more than two octaves wide. Each filter will operate as expected if it is correctly terminated, but this will occur only if both designs have the relatively constant terminal impedance that is associated with low SWR.

Knowing the filter type and the response needed, select the table of designs most appropriate for the application on a trial basis. From the chosen table, scale the 1-10 MHz data to the desired frequency decade, and search the cutoff frequency column for a value nearest the desired cutoff frequency. After finding a possible design, check the stopband attenuation levels to see if they are satisfactory. Then check the SWR to see if it is appropriate for the application. Finally, check the component values to see if they are convenient. For example, in the audio-frequency range, the capacitor values probably will be in the microfarad range, and capacitors in this size are available only in the E12 series of standard values. Then connect them in accordance with the diagram shown in the table from which the design was selected.

Impedance levels other than 50 ohms: First calculate a "trial" filter design using the impedance scaling procedure previously explained. Then search the appropriate table for the best match to the trial filter and scale the selected design to the desired impedance level. In this way, the convenient scan mode of filter selection is used regardless of the desired impedance level.

References

[1] P. Geffe, *Simplified Modern Filter Design* (New York: John F. Rider, a division of Hayden Publishing Co., 1963).

[2] *A Handbook on Electrical Filters* (Rockville, Maryland: White Electromagnetics, 1963).

[3] R. Saal, *The Design of Filters Using the Catalog of Normalized Lowpass Filters* (Western Germany: Telfunken, 1966).

[4] A. Zverev, *Handbook of Filter Synthesis* (New York: John Wiley & Sons, 1967).

[5] A. B. Williams, *Electronic Filter Design Handbook* (New York: McGraw-Hill, 1981).

[6] *Reference Data for Radio Engineers,* Sixth edition, Table 2, p. 5-3. (Indianapolis, Indiana: Howard W. Sams & Co., 1981).

[7] R. Frost, "Large-scale S Parameters Help Analyze Stability," *Electronic Design,* May 24, 1980.

Selected Filter Bibliography

The following texts and articles are recommended for more in-depth study of filter theory and application.

Allen, W. H. "Modern Filter Design for the Radio Amateur." *Radio Communication,* August 1971.

Blinchikoff, H. and A. Zverev. *Filtering in the Time and Frequency Domains.* New York: John Wiley and Sons, 1976.

Christian, E. *LC-Filters: Design, Testing, and Manufacturing.* New York: John Wiley and Sons, 1983.

Geffe, P. *Simplified Modern Filter Design.* Rochelle Park, NJ: Hayden Publishing Co., 1963.

Graf, R. *Electronic Databook,* 3rd Ed., Passive LC Filter Design, pp. 117-143, Blue Ridge Summit, PA: TAB Books 1983.

Hayward, W. "The Peaked Lowpass: A Look at the Ultraspherical Filter," *Ham Radio,* June 1984.

Niewiadomski, S. "Elliptic Lowpass Audio Filter Design Using Miniature Preferred Value Components," *Radio Communication,* October 1984.

Webb, J. "High-Pass Filters for Receiving Applications." *QST,* October 1983.

Wetherold, E. "Modern Filter Design for the Radio Amateur." *QST,* September 1969.

Wetherold, E. "Low-Pass Filters for Amateur Radio Transmitters." *QST,* December 1979.

Wetherold, E. "Low-Pass Chebyshev Filters Use Standard-Value Capacitors." Engineer's notebook, *Electronics,* June 19, 1980.

Wetherold, E. "Design 7-element Low-Pass Filters Using Standard-Value Capacitors." *EDN,* Vol. 26, No. 1, January 7, 1981.

Wetherold, E. "Elliptic Lowpass Filters for Transistor Amplifiers." *Ham Radio,* January 1981.

Wetherold, E. "High-Pass Chebyshev Filters Use Standard-Value Capacitors." Engineer's notebook, *Electronics,*

Table 11

5-Element Chebyshev Low-pass Filter Designs
50-ohm Impedance, C-In/Out
For Standard E24 Capacitor Values

Filter No.	F_{co}	3 dB	20 dB	40 dB	Max. SWR	C1,5 (pF)	L2,4 (µH)	C3 (pF)
1	1.01	1.15	1.53	2.25	1.355	3600	10.8	6200
2	1.02	1.21	1.65	2.45	1.212	3000	10.7	5600
3	1.15	1.29	1.71	2.51	1.391	3300	9.49	5600
4	1.10	1.32	1.81	2.69	1.196	2700	9.88	5100
5	1.25	1.41	1.88	2.75	1.386	3000	8.67	5100
6	1.04	1.37	1.94	2.94	1.085	2200	9.82	4700
7	1.15	1.41	1.95	2.92	1.155	2400	9.37	4700
8	1.32	1.50	2.01	2.96	1.332	2700	8.29	4700
9	1.13	1.50	2.12	3.22	1.081	2000	9.00	4300
10	1.26	1.54	2.13	3.19	1.157	2200	8.56	4300
11	1.39	1.61	2.18	3.21	1.276	2400	7.88	4300
12	1.05	1.62	2.38	3.66	1.028	1600	8.35	3900
13	1.23	1.65	2.34	3.55	1.076	1800	8.19	3900
14	1.39	1.70	2.35	3.51	1.159	2000	7.75	3900
15	1.55	1.79	2.41	3.55	1.295	2200	7.05	3900
16	1.17	1.76	2.57	3.94	1.033	1500	7.70	3600
17	1.27	1.77	2.55	3.88	1.057	1600	7.64	3600
18	1.46	1.82	2.54	3.81	1.135	1800	7.28	3600
19	1.65	1.92	2.59	3.83	1.268	2000	6.64	3600
20	1.88	2.08	2.73	3.97	1.497	2200	5.70	3600
21	1.43	1.94	2.77	4.21	1.068	1500	6.96	3300
22	1.54	1.97	2.77	4.17	1.109	1600	6.79	3300
23	1.76	2.07	2.81	4.17	1.238	1800	6.21	3300
24	2.02	2.25	2.96	4.31	1.470	2000	5.31	3300
25	1.31	2.10	3.11	4.79	1.022	1200	6.43	3000
26	1.48	2.12	3.06	4.68	1.046	1300	6.39	3000
27	1.75	2.19	3.05	4.57	1.135	1500	6.07	3000
28	1.89	2.25	3.08	4.57	1.206	1600	5.77	3000
29	2.19	2.45	3.23	4.71	1.440	1800	4.92	3000
30	1.51	2.34	3.44	5.29	1.026	1100	5.78	2700
31	1.70	2.36	3.40	5.17	1.057	1200	5.73	2700
32	1.87	2.40	3.38	5.10	1.104	1300	5.57	2700
33	2.20	2.56	3.46	5.11	1.268	1500	4.98	2700
34	2.39	2.69	3.56	5.21	1.406	1600	4.53	2700
35	1.75	2.63	3.85	5.91	1.033	1000	5.14	2400
36	1.99	2.67	3.81	5.78	1.072	1100	5.05	2400
37	2.19	2.74	3.81	5.71	1.135	1200	4.85	2400
38	2.40	2.84	3.86	5.73	1.227	1300	4.55	2400
39	1.89	2.87	4.21	6.47	1.030	910	4.71	2200
40	2.14	2.91	4.16	6.31	1.068	1000	4.64	2200
41	2.39	2.99	4.16	6.23	1.135	1100	4.45	2200
42	2.64	3.11	4.22	6.25	1.238	1200	4.14	2200
43	2.93	3.29	4.36	6.39	1.398	1300	3.71	2200
44	2.05	3.16	4.64	7.13	1.028	820	4.28	2000
45	2.36	3.20	4.57	6.94	1.068	910	4.22	2000
46	2.63	3.28	4.57	6.86	1.135	1000	4.05	2000
47	2.93	3.43	4.65	6.89	1.251	1100	3.73	2000
48	3.29	3.67	4.85	7.07	1.440	1200	3.28	2000
49	2.34	3.51	5.14	7.88	1.033	750	3.85	1800
50	2.63	3.56	5.08	7.71	1.069	820	3.79	1800
51	2.96	3.66	5.09	7.62	1.145	910	3.61	1800
52	3.30	3.84	5.19	7.67	1.268	1000	3.32	1800
53	3.76	4.15	5.45	7.93	1.497	1100	2.85	1800
54	2.70	3.96	5.76	8.82	1.039	680	3.42	1600
55	3.06	4.03	5.71	8.63	1.086	750	3.34	1600
56	3.38	4.14	5.73	8.57	1.159	820	3.18	1600
57	3.82	4.39	5.89	8.67	1.311	910	2.86	1600
58	2.77	4.21	6.18	9.48	1.030	620	3.21	1500
59	3.14	4.26	6.10	9.26	1.067	680	3.17	1500
60	3.51	4.38	6.10	9.14	1.135	750	3.03	1500
61	3.88	4.56	6.20	9.17	1.241	820	2.82	1500
62	4.46	4.95	6.51	9.48	1.473	910	2.41	1500
63	3.39	4.88	7.08	10.8	1.044	560	2.77	1300
64	3.84	4.98	7.02	10.6	1.097	620	2.70	1300
65	4.26	5.14	7.08	10.5	1.181	680	2.55	1300
66	4.79	5.46	7.29	10.7	1.341	750	2.28	1300
67	3.61	5.28	7.68	11.8	1.039	510	2.56	1200
68	4.06	5.36	7.61	11.5	1.083	560	2.51	1200
69	4.55	5.54	7.65	11.4	1.167	620	2.37	1200
70	5.07	5.84	7.84	11.5	1.304	680	2.16	1200
71	3.96	5.76	8.38	12.8	1.041	470	2.35	1100
72	4.39	5.84	8.31	12.6	1.079	510	2.31	1100
73	4.88	6.01	8.33	12.5	1.152	560	2.20	1100
74	5.50	6.34	8.54	12.6	1.293	620	1.99	1100
75	4.40	6.34	9.20	14.1	1.043	430	2.13	1000
76	4.91	6.45	9.13	13.8	1.087	470	2.09	1000
77	5.38	6.62	9.17	13.7	1.154	510	2.00	1000
78	6.00	6.95	9.37	13.8	1.282	560	1.83	1000
79	4.81	6.97	10.1	15.5	1.042	390	1.94	910
80	5.43	7.09	10.0	15.2	1.091	430	1.89	910
81	6.00	7.31	10.1	15.1	1.167	470	1.80	910
82	6.60	7.64	10.3	15.2	1.283	510	1.66	910
83	4.86	7.69	11.4	17.5	1.023	330	1.76	820
84	5.51	7.76	11.2	17.1	1.052	360	1.74	820
85	6.07	7.89	11.1	16.8	1.095	390	1.70	820
86	6.77	8.17	11.2	16.7	1.184	430	1.60	820
87	7.54	8.61	11.5	17.0	1.327	470	1.45	820
88	5.26	8.40	12.4	19.2	1.022	300	1.61	750
89	6.04	8.49	12.2	18.7	1.052	330	1.59	750
90	6.70	8.64	12.2	18.4	1.101	360	1.55	750
91	7.33	8.89	12.3	18.3	1.175	390	1.48	750
92	8.24	9.42	12.6	18.5	1.327	430	1.33	750
93	6.69	9.36	13.5	20.6	1.054	300	1.44	680
94	7.48	9.56	13.4	20.2	1.110	330	1.40	680
95	8.25	9.89	13.6	20.2	1.196	360	1.32	680
96	9.10	10.4	13.9	20.4	1.328	390	1.20	680
97	7.21	10.2	14.8	22.6	1.048	270	1.32	620
98	8.18	10.5	14.7	22.2	1.107	300	1.28	620
99	9.11	10.9	14.9	22.1	1.203	330	1.19	620
100	10.1	11.5	15.3	22.5	1.355	360	1.08	620
101	7.82	11.3	16.4	25.1	1.042	240	1.19	560
102	9.02	11.6	16.3	24.6	1.105	270	1.16	560
103	8.66	12.4	18.0	27.6	1.044	220	1.09	510
104	9.64	12.6	17.9	27.1	1.088	240	1.06	510
105	9.22	13.5	19.6	30.0	1.039	200	1.00	470
106	9.85	14.7	21.5	33.0	1.034	180	0.919	430

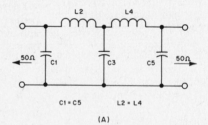

(A)

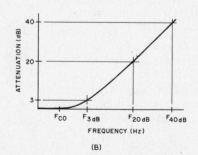

(B)

The schematic for a 5-element capacitor input/output Chebyshev low-pass filter is shown at A. At B is the typical attenuation response curve.

Table 12

7-Element Chebyshev Low-pass Filter Designs
50-ohm Impedance, C-In/Out
For Standard E24 Capacitor Values

Filter No.	F_{co}	3 dB	20 dB	40 dB	Max. SWR	C1,7 (pF)	L2,6 (µH)	C3,5 (pF)	L4 (µH)
1	1.02	1.10	1.31	1.65	1.254	3300	11.2	6200	12.6
2	1.04	1.16	1.40	1.79	1.142	2700	10.9	5600	12.6
3	1.13	1.23	1.45	1.84	1.264	3000	10.1	5600	11.3
4	1.05	1.23	1.51	1.96	1.071	2200	10.3	5100	12.3
5	1.12	1.26	1.53	1.96	1.123	2400	10.0	5100	11.7
6	1.23	1.34	1.59	2.01	1.247	2700	9.29	5100	10.4
7	1.03	1.30	1.63	2.15	1.030	1800	9.52	4700	11.9
8	1.12	1.33	1.64	2.13	1.064	2000	9.50	4700	11.4
9	1.21	1.37	1.66	2.13	1.119	2200	9.27	4700	10.8
10	1.29	1.42	1.70	2.16	1.200	2400	8.82	4700	10.0
11	1.10	1.41	1.79	2.36	1.023	1600	8.68	4300	11.0
12	1.21	1.45	1.79	2.33	1.058	1800	8.71	4300	10.5
13	1.31	1.49	1.81	2.33	1.114	2000	8.50	4300	9.91
14	1.42	1.56	1.86	2.36	1.202	2200	8.06	4300	9.14
15	1.54	1.65	1.93	2.43	1.336	2400	7.39	4300	8.18
16	1.25	1.57	1.97	2.59	1.031	1500	7.90	3900	9.85
17	1.32	1.59	1.97	2.57	1.050	1600	7.91	3900	9.62
18	1.44	1.64	1.99	2.56	1.109	1800	7.73	3900	9.04
19	1.57	1.72	2.05	2.60	1.205	2000	7.30	3900	8.27
20	1.44	1.73	2.14	2.78	1.056	1500	7.29	3600	8.82
21	1.52	1.76	2.15	2.78	1.086	1600	7.22	3600	8.54
22	1.66	1.84	2.20	2.81	1.176	1800	6.86	3600	7.83
23	1.83	1.96	2.30	2.90	1.327	2000	6.22	3600	6.90
24	1.51	1.86	2.32	3.05	1.037	1300	6.70	3300	8.27
25	1.68	1.93	2.35	3.03	1.099	1500	6.58	3300	7.72
26	1.77	1.98	2.38	3.05	1.147	1600	6.40	3300	7.37
27	1.96	2.11	2.49	3.14	1.294	1800	5.83	3300	6.50
28	1.56	2.02	2.56	3.38	1.021	1100	6.04	3000	7.68
29	1.68	2.05	2.56	3.35	1.042	1200	6.09	3000	7.47
30	1.79	2.09	2.57	3.33	1.073	1300	6.05	3000	7.21
31	1.99	2.20	2.64	3.37	1.176	1500	5.72	3000	6.52
32	2.11	2.28	2.70	3.42	1.257	1600	5.42	3000	6.08
33	1.75	2.25	2.84	3.75	1.023	1000	5.45	2700	6.89
34	1.89	2.29	2.84	3.71	1.048	1100	5.48	2700	6.68
35	2.02	2.34	2.86	3.70	1.086	1200	5.41	2700	6.40
36	2.15	2.41	2.90	3.72	1.141	1300	5.26	2700	6.06
37	2.44	2.61	3.07	3.86	1.327	1500	4.66	2700	5.18
38	2.01	2.54	3.20	4.21	1.027	910	4.86	2400	6.09
39	2.17	2.59	3.20	4.17	1.056	1000	4.86	2400	5.88
40	2.33	2.66	3.24	4.17	1.104	1100	4.77	2400	5.59
41	2.49	2.76	3.30	4.21	1.176	1200	4.57	2400	5.22
42	2.67	2.88	3.41	4.30	1.282	1300	4.27	2400	4.77
43	2.15	2.76	3.49	4.60	1.024	820	4.44	2200	5.61
44	2.35	2.82	3.49	4.55	1.053	910	4.46	2200	5.41
45	2.52	2.89	3.52	4.54	1.099	1000	4.38	2200	5.15
46	2.72	3.01	3.60	4.59	1.176	1100	4.19	2200	4.78
47	2.94	3.16	3.73	4.70	1.294	1200	3.88	2200	4.33
48	2.38	3.04	3.84	5.06	1.025	750	4.04	2000	5.09
49	2.57	3.09	3.84	5.01	1.050	820	4.06	2000	4.93
50	2.78	3.18	3.88	5.00	1.100	910	3.98	2000	4.68
51	2.99	3.31	3.96	5.05	1.176	1000	3.81	2000	4.35
52	3.26	3.50	4.12	5.19	1.308	1100	3.50	2000	3.89
53	2.67	3.38	4.26	5.61	1.027	680	3.64	1800	4.57
54	2.89	3.45	4.27	5.56	1.056	750	3.65	1800	4.41
55	3.09	3.54	4.31	5.55	1.100	820	3.59	1800	4.21
56	3.35	3.69	4.42	5.62	1.188	910	3.40	1800	3.87
57	3.65	3.92	4.60	5.80	1.327	1000	3.11	1800	3.45
58	3.07	3.82	4.80	6.30	1.033	620	3.24	1600	4.03
59	3.30	3.90	4.81	6.25	1.064	680	3.24	1600	3.88
60	3.55	4.02	4.87	6.26	1.120	750	3.15	1600	3.67
61	3.81	4.18	4.99	6.34	1.204	820	3.00	1600	3.39
62	3.16	4.05	5.12	6.75	1.024	560	3.03	1500	3.82
63	3.45	4.13	5.12	6.68	1.053	620	3.04	1500	3.69
64	3.69	4.24	5.17	6.66	1.097	680	2.99	1500	3.51
65	3.99	4.41	5.28	6.73	1.176	750	2.86	1500	3.26
66	4.31	4.64	5.48	6.91	1.297	820	2.64	1500	2.94
67	3.81	4.72	5.90	7.74	1.036	510	2.64	1300	3.26
68	4.10	4.82	5.93	7.69	1.070	560	2.62	1300	3.14
69	4.43	4.98	6.02	7.72	1.133	620	2.54	1300	2.94
70	4.78	5.21	6.19	7.85	1.230	680	2.39	1300	2.70
71	4.13	5.11	6.39	8.38	1.035	470	2.43	1200	3.01
72	4.40	5.20	6.41	8.33	1.064	510	2.43	1200	2.91
73	4.72	5.35	6.49	8.34	1.116	560	2.37	1200	2.76
74	5.12	5.60	6.67	8.48	1.214	620	2.23	1200	2.52
75	4.49	5.57	6.97	9.15	1.035	430	2.23	1100	2.76
76	4.82	5.68	7.00	9.09	1.066	470	2.22	1100	2.66
77	5.12	5.83	7.07	9.10	1.112	510	2.18	1100	2.54
78	5.52	6.07	7.24	9.21	1.196	560	2.07	1100	2.35
79	4.93	6.12	7.67	10.1	1.034	390	2.03	1000	2.51
80	5.33	6.26	7.70	10.0	1.069	430	2.02	1000	2.41
81	5.69	6.44	7.80	10.0	1.122	470	1.97	1000	2.29
82	6.08	6.68	7.97	10.1	1.198	510	1.88	1000	2.13
83	6.63	7.09	8.32	10.5	1.343	560	1.71	1000	1.89
84	5.48	6.75	8.43	11.0	1.038	360	1.85	910	2.28
85	5.84	6.87	8.46	11.0	1.068	390	1.84	910	2.20
86	6.28	7.09	8.58	11.0	1.126	430	1.79	910	2.07
87	6.75	7.39	8.80	11.2	1.213	470	1.69	910	1.91
88	5.68	7.39	9.37	12.4	1.020	300	1.65	820	2.10
89	6.17	7.52	9.36	12.2	1.043	330	1.66	820	2.04
90	6.60	7.68	9.41	12.2	1.079	360	1.65	820	1.96
91	7.01	7.89	9.53	12.2	1.131	390	1.61	820	1.86
92	7.59	8.27	9.82	12.5	1.233	430	1.51	820	1.70
93	6.72	8.21	10.2	13.4	1.042	300	1.52	750	1.87
94	7.23	8.40	10.3	13.3	1.080	330	1.51	750	1.79
95	7.72	8.66	10.4	13.4	1.138	360	1.46	750	1.69
96	8.24	9.00	10.7	13.6	1.222	390	1.39	750	1.57
97	7.36	9.04	11.3	14.8	1.039	270	1.38	680	1.70
98	7.98	9.27	11.4	14.7	1.082	300	1.37	680	1.62
99	8.58	9.59	11.6	14.8	1.148	330	1.32	680	1.52
100	9.23	10.1	11.9	15.1	1.247	360	1.24	680	1.39
101	7.91	9.86	12.4	16.2	1.032	240	1.26	620	1.56
102	8.67	10.1	12.4	16.1	1.075	270	1.25	620	1.49
103	9.39	10.5	12.7	16.2	1.145	300	1.20	620	1.39
104	8.86	11.0	13.7	18.0	1.036	220	1.14	560	1.40
105	9.49	11.2	13.8	17.8	1.068	240	1.13	560	1.35
106	9.72	12.0	15.0	19.7	1.036	200	1.03	510	1.28

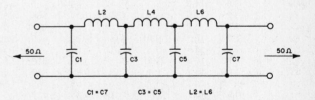

The schematic for a 7-element Chebyshev low-pass filter. See Table 11 for the attenuation response curve.

Table 13
5-Element Chebyshev Low-pass Filter Designs
50-ohm Impedance, L-In/Out
For Standard-Value L and C

Filter No.	F_{co}	3 dB	20 dB	40 dB	Max. SWR	L1,5 (μH)	C2,4 (pF)	L3 (μH)
1	0.744	1.15	1.69	2.60	1.027	5.60	4700	13.7
2	0.901	1.26	1.81	2.76	1.055	5.60	4300	12.7
3	1.06	1.38	1.94	2.93	1.096	5.60	3900	11.8
4	1.19	1.47	2.05	3.07	1.138	5.60	3600	11.2
5	1.32	1.58	2.17	3.23	1.192	5.60	3300	10.6
6	0.911	1.39	2.03	3.12	1.030	4.70	3900	11.4
7	1.08	1.50	2.16	3.29	1.056	4.70	3600	10.6
8	1.25	1.63	2.30	3.48	1.092	4.70	3300	9.92
9	1.42	1.77	2.46	3.68	1.142	4.70	3000	9.32
10	1.61	1.92	2.63	3.90	1.209	4.70	2700	8.79
11	1.05	1.64	2.41	3.72	1.025	3.90	3300	9.63
12	1.29	1.80	2.60	3.96	1.054	3.90	3000	8.83
13	1.54	1.99	2.80	4.22	1.099	3.90	2700	8.15
14	1.80	2.19	3.03	4.53	1.164	3.90	2400	7.57
15	1.99	2.35	3.20	4.75	1.222	3.90	2200	7.23
16	1.34	2.00	2.93	4.49	1.034	3.30	2700	7.89
17	1.68	2.25	3.20	4.84	1.077	3.30	2400	7.15
18	1.92	2.43	3.40	5.11	1.118	3.30	2200	6.73
19	2.16	2.63	3.62	5.40	1.174	3.30	2000	6.35
20	1.65	2.46	3.59	5.51	1.035	2.70	2200	6.43
21	1.99	2.70	3.86	5.85	1.069	2.70	2000	5.93
22	2.34	2.97	4.15	6.24	1.118	2.70	1800	5.50
23	2.71	3.27	4.49	6.68	1.188	2.70	1600	5.13
24	2.92	3.43	4.67	6.92	1.233	2.70	1500	4.97
25	2.01	3.01	4.39	6.74	1.034	2.20	1800	5.26
26	2.52	3.37	4.80	7.27	1.077	2.20	1600	4.76
27	2.78	3.57	5.02	7.56	1.107	2.20	1500	4.55
28	3.34	4.02	5.52	8.21	1.190	2.20	1300	4.18
29	2.36	3.61	5.29	8.14	1.029	1.80	1500	4.38
30	3.12	4.14	5.89	8.92	1.080	1.80	1300	3.88
31	3.51	4.45	6.23	9.36	1.118	1.80	1200	3.67
32	3.93	4.78	6.60	9.85	1.169	1.80	1100	3.48
33	4.37	5.15	7.01	10.4	1.233	1.80	1000	3.31
34	3.10	4.51	6.56	10.0	1.041	1.50	1200	3.51
35	3.65	4.90	6.99	10.6	1.073	1.50	1100	3.27
36	4.21	5.34	7.47	11.2	1.118	1.50	1000	3.06
37	4.75	5.77	7.95	11.9	1.173	1.50	910	2.89
38	3.53	5.41	7.94	12.2	1.029	1.20	1000	2.92
39	4.30	5.94	8.53	13.0	1.060	1.20	910	2.69
40	5.09	6.53	9.18	13.8	1.106	1.20	820	2.49
41	5.73	7.04	9.75	14.6	1.155	1.20	750	2.35
42	6.42	7.61	10.4	15.4	1.219	1.20	680	2.23
43	4.40	6.60	9.65	14.8	1.033	1.00	820	2.40
44	5.27	7.20	10.3	15.7	1.064	1.00	750	2.22
45	6.15	7.87	11.1	16.7	1.108	1.00	680	2.07
46	6.95	8.51	11.8	17.6	1.160	1.00	620	1.95
47	7.80	9.22	12.6	18.6	1.227	1.00	560	1.85
48	5.23	7.96	11.7	17.9	1.030	0.82	680	1.99
49	6.33	8.72	12.5	19.0	1.061	0.82	620	1.83
50	7.45	9.56	13.4	20.3	1.106	0.82	560	1.70
51	8.44	10.3	14.3	21.4	1.158	0.82	510	1.60
52	9.28	11.0	15.1	22.4	1.211	0.82	470	1.53
53	6.41	9.66	14.1	21.7	1.032	0.68	560	1.64
54	7.75	10.6	15.2	23.1	1.064	0.68	510	1.51
55	8.83	11.4	16.1	24.3	1.100	0.68	470	1.42
56	9.97	12.3	17.1	25.6	1.148	0.68	430	1.34

Table 14
7-Element Chebyshev Low-pass Filter Designs
50-ohm Impedance, L-In/Out
For Standard-Value Capacitors

Filter No.	F_{co}	3 dB	20 dB	40 dB	Max. SWR	L1,7 (μH)	C2,6 (pF)	L3,5 (μH)	C4 (pF)
1	1.01	1.18	1.44	1.87	1.081	5.89	4300	13.4	5100
2	1.09	1.29	1.60	2.08	1.059	5.06	3900	12.0	4700
3	1.03	1.09	1.26	1.58	1.480	10.1	4300	17.1	4700
4	1.20	1.40	1.73	2.24	1.071	4.81	3600	11.2	4300
5	1.16	1.23	1.44	1.81	1.383	8.34	3900	14.6	4300
6	1.33	1.54	1.88	2.43	1.087	4.58	3300	10.3	3900
7	1.42	1.68	2.07	2.70	1.064	3.95	3000	9.27	3600
8	1.34	1.41	1.63	2.04	1.506	7.98	3300	13.4	3600
9	1.53	1.85	2.31	3.02	1.045	3.36	2700	8.32	3300
10	1.50	1.59	1.86	2.33	1.406	6.57	3000	11.4	3300
11	1.63	2.06	2.59	3.41	1.029	2.83	2400	7.41	3000
12	1.69	1.81	2.13	2.68	1.317	5.36	2700	9.70	3000
13	1.86	2.27	2.83	3.70	1.042	2.71	2200	6.78	2700
14	1.91	2.07	2.46	3.12	1.238	4.31	2400	8.19	2700
15	2.14	2.52	3.11	4.04	1.064	2.63	2000	6.18	2400
16	2.01	2.11	2.45	3.06	1.506	5.32	2200	8.91	2400
17	2.29	2.78	3.46	4.52	1.045	2.24	1800	5.54	2200
18	2.25	2.39	2.79	3.49	1.406	4.38	2000	7.61	2200
19	2.45	3.09	3.88	5.11	1.029	1.89	1600	4.94	2000
20	2.53	2.71	3.19	4.02	1.317	3.57	1800	6.47	2000
21	2.85	3.37	4.15	5.39	1.064	1.97	1500	4.64	1800
22	2.86	3.11	3.69	4.68	1.238	2.88	1600	5.46	1800
23	3.13	3.84	4.79	6.27	1.039	1.59	1300	4.00	1600
24	3.27	4.12	5.18	6.81	1.029	1.41	1200	3.70	1500
25	3.47	3.90	4.70	6.02	1.140	2.01	1300	4.17	1500
26	3.99	4.61	5.64	7.28	1.087	1.53	1100	3.43	1300
27	4.27	5.05	6.22	8.09	1.064	1.32	1000	3.09	1200
28	4.01	4.22	4.90	6.11	1.506	2.66	1100	4.45	1200
29	4.63	5.53	6.85	8.91	1.056	1.17	910	2.81	1100
30	4.49	4.77	5.57	6.98	1.406	2.19	1000	3.81	1100
31	5.05	6.11	7.60	9.92	1.047	1.03	820	2.53	1000
32	4.93	5.23	6.10	7.64	1.416	2.02	910	3.49	1000
33	5.58	6.70	8.31	10.8	1.052	0.954	750	2.31	910
34	5.54	5.94	6.99	8.80	1.326	1.65	820	2.97	910
35	6.23	7.41	9.16	11.9	1.059	0.881	680	2.10	820
36	5.92	6.24	7.26	9.06	1.476	1.76	750	2.98	820
37	6.79	8.12	10.0	13.1	1.055	0.796	620	1.91	750
38	6.64	7.07	8.27	10.4	1.379	1.45	680	2.54	750
39	7.46	8.97	11.1	14.5	1.051	0.711	560	1.73	680
40	7.21	7.63	8.89	11.1	1.438	1.40	620	2.41	680
41	8.18	9.85	12.2	15.9	1.050	0.645	510	1.57	620
42	8.10	8.66	10.2	12.8	1.345	1.15	560	2.05	620
43	9.21	10.8	13.2	17.1	1.074	0.633	470	1.46	560
44	8.78	9.31	10.9	13.6	1.425	1.14	510	1.96	560
45	10.1	11.8	14.4	18.7	1.081	0.589	430	1.34	510

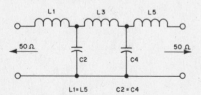

The schematic for a 5-element inductor-input/output Chebyshev low-pass filter. See Table 11 for the attenuation response curve.

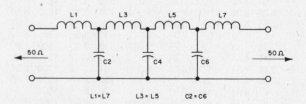

The schematic for a 7-element inductor input/output Chebyshev low-pass filter. See Table 11 for the attenuation response curve.

Table 15
5-Branch Elliptic Low-Pass Filter Designs
50-ohm Impedance
Standard E12 Capacitor Values for C1, C3 and C5

Filter No.	F_{co}	F_{3dB}	F_{As}	A_s	Max.	C1	C3	C5	C2	C4	L2	L4	F2	F4
	---------(MHz)---------			(dB)	SWR	----------------------(pF)------------------------					-----(µH)------		-----(MHz)-----	
1	0.795	0.989	1.57	47.4	1.092	2700	5600	2200	324	937	12.1	10.1	2.54	1.64
2	1.06	1.20	1.77	46.2	1.234	2700	4700	2200	341	982	9.36	7.56	2.82	1.85
3	1.47	1.57	2.15	45.4	1.586	2700	3900	2200	364	1045	6.32	4.88	3.32	2.23
4	0.929	1.18	1.91	48.0	1.077	2200	4700	1800	257	743	10.2	8.59	3.11	1.99
5	1.27	1.45	2.17	46.7	1.215	2200	3900	1800	271	779	7.85	6.39	3.45	2.26
6	1.69	1.82	2.54	45.9	1.489	2200	3300	1800	287	821	5.64	4.42	3.96	2.64
7	1.12	1.44	2.41	49.8	1.071	1800	3900	1500	192	549	8.45	7.25	3.95	2.52
8	1.49	1.73	2.70	48.8	1.183	1800	3300	1500	200	570	6.75	5.62	4.33	2.81
9	2.11	2.27	3.27	47.8	1.506	1800	2700	1500	213	604	4.55	3.64	5.12	3.40
10	1.28	1.66	2.63	46.3	1.064	1500	3300	1200	192	561	7.20	6.00	4.28	2.74
11	1.79	2.06	2.99	44.8	1.195	1500	2700	1200	204	592	5.52	4.42	4.75	3.11
12	2.52	2.70	3.63	43.8	1.525	1500	2200	1200	220	636	3.71	2.82	5.58	3.76
13	1.56	2.08	3.55	50.1	1.055	1200	2700	1000	127	363	5.88	5.07	5.83	3.71
14	2.23	2.59	4.04	48.8	1.183	1200	2200	1000	133	380	4.50	3.75	6.50	4.22
15	3.17	3.41	4.90	47.8	1.506	1200	1800	1000	142	402	3.03	2.42	7.68	5.10
16	1.94	2.52	4.15	48.4	1.064	1000	2200	820	115	331	4.79	4.06	6.78	4.34
17	2.73	3.14	4.73	47.0	1.199	1000	1800	820	121	348	3.66	2.99	7.56	4.93
18	3.73	4.02	5.63	46.2	1.491	1000	1500	820	129	368	2.56	2.01	8.76	5.85
19	2.39	3.11	5.20	49.4	1.065	820	1800	680	89.3	256	3.91	3.35	8.51	5.44
20	3.26	3.79	5.85	48.2	1.185	820	1500	680	93.6	267	3.07	2.54	9.39	6.10
21	4.83	5.17	7.30	47.2	1.569	820	1200	680	100	286	1.95	1.54	11.4	7.58
22	2.85	3.71	6.15	48.8	1.063	680	1500	560	76.6	220	3.26	2.78	10.1	6.43
23	4.16	4.74	7.14	47.3	1.221	680	1200	560	81.3	233	2.40	1.97	11.4	7.44
24	5.72	6.13	8.58	46.5	1.547	680	1000	560	86.3	246	1.65	1.30	13.3	8.91
25	3.67	4.69	7.95	50.5	1.076	560	1200	470	57.6	164	2.59	2.23	13.0	8.31
26	5.02	5.77	9.01	49.4	1.212	560	1000	470	60.3	171	2.01	1.68	14.5	9.40
27	7.18	7.68	11.1	48.6	1.582	560	820	470	64.1	181	1.32	1.06	17.3	11.5
28	4.40	5.60	9.24	49.3	1.079	470	1000	390	51.4	147	2.16	1.84	15.1	9.66
29	6.17	7.01	10.6	48.0	1.236	470	820	390	54.2	155	1.63	1.34	17.0	11.1
30	8.63	9.20	12.9	47.3	1.604	470	680	390	57.6	164	1.09	0.857	20.1	13.4
31	5.47	6.91	11.8	51.3	1.086	390	820	330	38.5	109	1.76	1.52	19.3	12.3
32	7.55	8.59	13.5	50.2	1.242	390	680	330	40.4	114	1.34	1.12	21.7	14.1
33	10.9	11.5	16.8	49.5	1.659	390	560	330	42.8	120	0.862	0.695	26.2	17.4
34	6.59	8.17	13.0	47.7	1.096	330	680	270	39.0	112	1.46	1.22	21.1	13.6
35	9.10	10.2	15.0	46.5	1.267	330	560	270	41.2	118	1.09	0.881	23.7	15.6
36	12.4	13.2	18.1	45.8	1.635	330	470	270	43.9	125	0.741	0.573	27.9	18.8

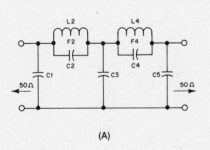

(A)

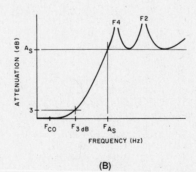

(B)

The schematic for a 5-branch elliptic low-pass filter is shown at A. At B is the typical attenuation response curve.

Table 16

5-Element Chebyshev High-pass Filter Designs
50-ohm Impedance, C-In/Out
For Standard E24 Capacitor Values

Filter No.	F_{co}	3 dB	20 dB	40 dB	Max. SWR	C1,5 (pF)	L2,4 (µH)	C3 (pF)
1	1.04	0.726	0.501	0.328	1.044	5100	6.45	2200
2	1.04	0.788	0.554	0.366	1.081	4300	5.97	2000
3	1.17	0.800	0.550	0.359	1.039	4700	5.85	2000
4	1.07	0.857	0.615	0.410	1.135	3600	5.56	1800
5	1.17	0.877	0.616	0.406	1.076	3900	5.36	1800
6	1.33	0.890	0.609	0.397	1.034	4300	5.26	1800
7	1.12	0.938	0.686	0.461	1.206	3000	5.20	1600
8	1.25	0.974	0.693	0.461	1.109	3300	4.86	1600
9	1.38	0.994	0.691	0.454	1.057	3600	4.71	1600
10	1.54	1.00	0.683	0.444	1.028	3900	4.67	1600
11	1.14	0.978	0.723	0.490	1.268	2700	5.09	1500
12	1.28	1.03	0.738	0.492	1.135	3000	4.64	1500
13	1.43	1.06	0.738	0.486	1.068	3300	4.44	1500
14	1.61	1.07	0.730	0.476	1.033	3600	4.38	1500
15	1.21	1.08	0.812	0.555	1.398	2200	4.82	1300
16	1.35	1.14	0.841	0.567	1.227	2400	4.29	1300
17	1.55	1.20	0.853	0.566	1.104	2700	3.94	1300
18	1.75	1.23	0.848	0.555	1.046	3000	3.81	1300
19	1.28	1.15	0.871	0.597	1.440	2000	4.57	1200
20	1.45	1.24	0.909	0.614	1.238	2200	3.99	1200
21	1.60	1.29	0.923	0.616	1.135	2400	3.71	1200
22	1.84	1.32	0.921	0.605	1.057	2700	3.54	1200
23	2.14	1.34	0.906	0.588	1.022	3000	3.50	1200
24	1.57	1.34	0.989	0.669	1.251	2000	3.69	1100
25	1.75	1.40	1.01	0.672	1.135	2200	3.40	1100
26	1.93	1.44	1.01	0.664	1.072	2400	3.27	1100
27	2.27	1.46	0.992	0.645	1.026	2700	3.21	1100
28	1.71	1.47	1.08	0.734	1.268	1800	3.39	1000
29	1.93	1.54	1.11	0.739	1.135	2000	3.09	1000
30	2.15	1.58	1.11	0.730	1.068	2200	2.96	1000
31	2.41	1.60	1.10	0.714	1.033	2400	2.92	1000
32	1.66	1.50	1.14	0.783	1.473	1500	3.54	910
33	1.82	1.59	1.18	0.803	1.311	1600	3.18	910
34	2.09	1.69	1.22	0.812	1.145	1800	2.83	910
35	2.36	1.74	1.22	0.802	1.068	2000	2.70	910
36	2.68	1.76	1.20	0.783	1.030	2200	2.66	910
37	2.12	1.81	1.33	0.898	1.241	1500	2.73	820
38	2.28	1.86	1.35	0.902	1.159	1600	2.58	820
39	2.61	1.93	1.35	0.890	1.069	1800	2.43	820
40	3.01	1.96	1.33	0.866	1.028	2000	2.39	820
41	2.17	1.90	1.42	0.970	1.341	1300	2.67	750
42	2.57	2.06	1.48	0.985	1.135	1500	2.32	750
43	2.76	2.10	1.48	0.978	1.086	1600	2.25	750
44	3.21	2.14	1.46	0.952	1.033	1800	2.19	750
45	2.45	2.13	1.58	1.08	1.304	1200	2.36	680
46	2.69	2.23	1.62	1.09	1.181	1300	2.17	680
47	3.17	2.33	1.63	1.07	1.067	1500	2.01	680
48	3.44	2.35	1.62	1.06	1.039	1600	1.99	680
49	2.70	2.34	1.74	1.18	1.293	1100	2.14	620
50	2.99	2.46	1.78	1.19	1.167	1200	1.96	620
51	3.28	2.53	1.79	1.19	1.097	1300	1.87	620
52	3.93	2.59	1.76	1.15	1.030	1500	1.81	620
53	3.02	2.60	1.93	1.31	1.282	1000	1.92	560
54	3.37	2.74	1.97	1.32	1.152	1100	1.75	560
55	3.72	2.81	1.98	1.31	1.083	1200	1.67	560
56	4.10	2.85	1.97	1.29	1.044	1300	1.64	560
57	3.31	2.86	2.12	1.44	1.283	910	1.75	510
58	3.69	3.00	2.17	1.45	1.154	1000	1.60	510
59	4.11	3.09	2.17	1.44	1.079	1100	1.52	510
60	4.59	3.14	2.15	1.41	1.039	1200	1.49	510
61	3.49	3.05	2.28	1.55	1.327	820	1.66	470
62	3.95	3.24	2.35	1.57	1.167	910	1.49	470
63	4.39	3.34	2.36	1.56	1.087	1000	1.41	470
64	4.94	3.40	2.34	1.53	1.041	1100	1.38	470
65	3.81	3.34	2.49	1.70	1.327	750	1.52	430
66	4.24	3.52	2.56	1.72	1.184	820	1.38	430
67	4.77	3.65	2.58	1.71	1.091	910	1.29	430
68	5.36	3.72	2.56	1.68	1.043	1000	1.26	430
69	4.20	3.68	2.75	1.87	1.328	680	1.38	390
70	4.72	3.89	2.83	1.90	1.175	750	1.24	390
71	5.22	4.02	2.84	1.88	1.095	820	1.17	390
72	5.93	4.10	2.82	1.85	1.042	910	1.14	390
73	4.48	3.95	2.96	2.02	1.355	620	1.30	360
74	5.01	4.18	3.05	2.05	1.196	680	1.16	360
75	5.60	4.34	3.08	2.04	1.101	750	1.09	360
76	6.23	4.42	3.07	2.01	1.052	820	1.06	360
77	4.79	4.25	3.20	2.19	1.391	560	1.22	330
78	5.44	4.55	3.33	2.24	1.203	620	1.07	330
79	6.03	4.72	3.36	2.23	1.110	680	1.00	330
80	6.77	4.82	3.35	2.20	1.052	750	0.970	330
81	7.70	4.87	3.30	2.14	1.023	820	0.962	330
82	5.28	4.68	3.53	2.41	1.386	510	1.10	300
83	5.94	4.99	3.65	2.46	1.212	560	0.978	300
84	6.66	5.20	3.70	2.46	1.107	620	0.910	300
85	7.43	5.31	3.68	2.42	1.054	680	0.882	300
86	8.56	5.36	3.62	2.35	1.022	750	0.875	300
87	6.05	5.31	3.97	2.70	1.332	470	0.956	270
88	6.69	5.58	4.07	2.74	1.196	510	0.870	270
89	7.43	5.78	4.11	2.73	1.105	560	0.817	270
90	8.39	5.91	4.08	2.68	1.048	620	0.792	270
91	7.07	6.09	4.51	3.06	1.276	430	0.818	240
92	7.84	6.38	4.61	3.08	1.155	470	0.752	240
93	8.59	6.55	4.62	3.06	1.088	510	0.719	240
94	9.64	6.66	4.58	3.00	1.042	560	0.702	240
95	7.61	6.60	4.90	3.33	1.295	390	0.760	220
96	8.53	6.95	5.02	3.36	1.157	430	0.690	220
97	9.43	7.15	5.04	3.33	1.085	470	0.658	220
98	10.4	7.26	5.01	3.28	1.044	510	0.644	220
99	7.58	6.83	5.19	3.56	1.470	330	0.776	200
100	8.53	7.33	5.42	3.67	1.268	360	0.678	200
101	9.36	7.64	5.52	3.70	1.159	390	0.628	200
102	10.4	7.88	5.54	3.66	1.081	430	0.596	200
103	8.55	7.67	5.81	3.98	1.440	300	0.685	180
104	9.69	8.24	6.06	4.09	1.238	330	0.597	180
105	10.7	8.57	6.15	4.10	1.135	360	0.556	180
106	9.80	8.73	6.58	4.50	1.406	270	0.595	160

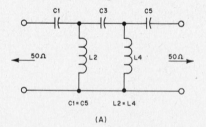

(A)

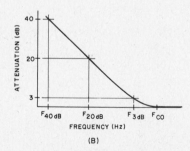

(B)

The schematic for a 5-element capacitor input/output Chebyshev high-pass filter, is shown at A. At B is the typical attenuation response curve.

Table 17
7-Element Chebyshev High-pass Filter Designs
50-ohm Impedance, C-In/Out
For Standard E24 Capacitor Values

Filter No.	F_{co}	3 dB	20 dB	40 dB	Max. SWR	C1,7 (pF)	L2,6 (µH)	C3,5 (pF)	L4 (µH)
1	1.02	0.826	0.660	0.504	1.036	5100	6.16	2000	4.98
2	1.00	0.880	0.724	0.563	1.109	3900	5.67	1800	4.86
3	1.08	0.905	0.732	0.563	1.058	4300	5.55	1800	4.60
4	1.16	0.922	0.734	0.558	1.030	4700	5.55	1800	4.45
5	1.00	0.924	0.780	0.617	1.257	3000	5.53	1600	4.93
6	1.09	0.971	0.806	0.630	1.147	3300	5.15	1600	4.48
7	1.16	1.00	0.819	0.634	1.086	3600	4.99	1600	4.22
8	1.23	1.02	0.824	0.632	1.050	3900	4.93	1600	4.05
9	1.34	1.04	0.825	0.625	1.023	4300	4.95	1600	3.92
10	1.03	0.958	0.815	0.648	1.327	2700	5.43	1500	4.89
11	1.13	1.02	0.853	0.669	1.176	3000	4.92	1500	4.31
12	1.22	1.06	0.871	0.676	1.099	3300	4.70	1500	4.01
13	1.30	1.09	0.879	0.675	1.056	3600	4.63	1500	3.83
14	1.39	1.11	0.880	0.670	1.031	3900	4.63	1500	3.71
15	1.22	1.13	0.954	0.755	1.282	2400	4.57	1300	4.09
16	1.34	1.20	0.994	0.776	1.141	2700	4.17	1300	3.62
17	1.45	1.24	1.01	0.780	1.073	3000	4.03	1300	3.38
18	1.57	1.27	1.02	0.775	1.037	3300	4.00	1300	3.24
19	1.31	1.21	1.03	0.816	1.294	2200	4.25	1200	3.81
20	1.41	1.28	1.07	0.836	1.176	2400	3.94	1200	3.45
21	1.55	1.34	1.09	0.845	1.086	2700	3.74	1200	3.16
22	1.68	1.37	1.10	0.841	1.042	3000	3.70	1200	3.01
23	1.41	1.32	1.12	0.887	1.308	2000	3.93	1100	3.53
24	1.54	1.39	1.16	0.912	1.176	2200	3.61	1100	3.16
25	1.65	1.44	1.19	0.921	1.104	2400	3.46	1100	2.95
26	1.80	1.49	1.20	0.919	1.048	2700	3.39	1100	2.78
27	1.97	1.52	1.20	0.907	1.021	3000	3.41	1100	2.68
28	1.54	1.44	1.22	0.971	1.327	1800	3.62	1000	3.26
29	1.70	1.53	1.28	1.00	1.176	2000	3.28	1000	2.87
30	1.82	1.59	1.31	1.01	1.099	2200	3.14	1000	2.67
31	1.95	1.63	1.32	1.01	1.056	2400	3.08	1000	2.55
32	2.15	1.67	1.32	1.00	1.023	2700	3.10	1000	2.45
33	1.85	1.67	1.40	1.10	1.188	1800	3.01	910	2.64
34	2.00	1.75	1.44	1.11	1.100	2000	2.85	910	2.43
35	2.15	1.80	1.45	1.11	1.053	2200	2.81	910	2.31
36	2.31	1.83	1.45	1.10	1.027	2400	2.81	910	2.24
37	1.91	1.77	1.50	1.19	1.297	1500	2.91	820	2.61
38	2.03	1.85	1.55	1.22	1.204	1600	2.74	820	2.42
39	2.22	1.94	1.59	1.24	1.100	1800	2.57	820	2.19
40	2.41	2.00	1.61	1.23	1.050	2000	2.53	820	2.08
41	2.61	2.03	1.61	1.22	1.024	2200	2.54	820	2.01
42	2.26	2.04	1.71	1.34	1.176	1500	2.46	750	2.16
43	2.38	2.10	1.73	1.35	1.120	1600	2.38	750	2.04
44	2.60	2.17	1.76	1.35	1.056	1800	2.31	750	1.91
45	2.83	2.22	1.76	1.34	1.025	2000	2.32	750	1.84
46	2.40	2.20	1.85	1.46	1.230	1300	2.31	680	2.05
47	2.69	2.34	1.92	1.49	1.097	1500	2.13	680	1.81
48	2.82	2.39	1.94	1.49	1.064	1600	2.10	680	1.75
49	3.11	2.45	1.94	1.47	1.027	1800	2.10	680	1.67
50	2.66	2.43	2.04	1.61	1.214	1200	2.08	620	1.84
51	2.84	2.52	2.09	1.63	1.133	1300	1.98	620	1.71
52	3.16	2.64	2.13	1.63	1.053	1500	1.91	620	1.58
53	3.33	2.67	2.13	1.62	1.033	1600	1.91	620	1.54
54	2.73	2.55	2.17	1.73	1.343	1000	2.05	560	1.85
55	2.98	2.71	2.27	1.79	1.196	1100	1.86	560	1.64
56	3.19	2.82	2.32	1.81	1.116	1200	1.77	560	1.52
57	3.39	2.89	2.35	1.81	1.070	1300	1.73	560	1.45
58	3.81	2.98	2.36	1.79	1.024	1500	1.73	560	1.37
59	3.27	2.97	2.49	1.96	1.198	1000	1.70	510	1.49
60	3.53	3.10	2.55	1.99	1.112	1100	1.61	510	1.38
61	3.76	3.18	2.58	1.99	1.064	1200	1.58	510	1.31
62	4.01	3.24	2.59	1.98	1.036	1300	1.57	510	1.27
63	3.51	3.21	2.69	2.12	1.213	910	1.58	470	1.40
64	3.79	3.35	2.76	2.15	1.122	1000	1.49	470	1.28
65	4.07	3.45	2.80	2.16	1.066	1100	1.45	470	1.21
66	4.35	3.52	2.81	2.14	1.035	1200	1.45	470	1.17
67	3.79	3.47	2.93	2.31	1.233	820	1.46	430	1.30
68	4.12	3.65	3.02	2.35	1.126	910	1.37	430	1.18
69	4.42	3.76	3.06	2.36	1.069	1000	1.33	430	1.11
70	4.77	3.85	3.07	2.34	1.035	1100	1.33	430	1.07
71	4.20	3.85	3.24	2.55	1.222	750	1.32	390	1.17
72	4.52	4.02	3.32	2.59	1.131	820	1.24	390	1.07
73	4.89	4.15	3.37	2.60	1.068	910	1.21	390	1.01
74	5.27	4.24	3.39	2.58	1.034	1000	1.20	390	0.969
75	4.48	4.13	3.48	2.75	1.247	680	1.24	360	1.10
76	4.86	4.33	3.59	2.80	1.138	750	1.15	360	1.00
77	5.20	4.47	3.65	2.82	1.079	820	1.12	360	0.942
78	5.64	4.58	3.67	2.80	1.038	910	1.11	360	0.899
79	4.87	4.49	3.79	2.99	1.254	620	1.14	330	1.01
80	5.26	4.71	3.91	3.05	1.148	680	1.06	330	0.924
81	5.67	4.87	3.98	3.07	1.080	750	1.03	330	0.864
82	6.07	4.98	4.00	3.06	1.043	820	1.02	330	0.829
83	5.32	4.91	4.15	3.28	1.264	560	1.04	300	0.930
84	5.80	5.18	4.30	3.36	1.145	620	0.965	300	0.838
85	6.22	5.36	4.37	3.38	1.082	680	0.933	300	0.787
86	6.71	5.49	4.40	3.36	1.042	750	0.923	300	0.752
87	7.25	5.58	4.40	3.33	1.020	820	0.931	300	0.731
88	5.98	5.50	4.64	3.66	1.247	510	0.926	270	0.824
89	6.46	5.77	4.78	3.74	1.142	560	0.867	270	0.752
90	6.98	5.97	4.87	3.76	1.075	620	0.837	270	0.703
91	7.50	6.11	4.89	3.74	1.039	680	0.831	270	0.675
92	6.39	5.97	5.08	4.04	1.336	430	0.873	240	0.787
93	6.94	6.32	5.29	4.16	1.200	470	0.798	240	0.704
94	7.41	6.55	5.41	4.21	1.123	510	0.762	240	0.656
95	7.95	6.75	5.48	4.22	1.068	560	0.742	240	0.620
96	8.61	6.90	5.50	4.19	1.032	620	0.740	240	0.595
97	7.56	6.88	5.77	4.54	1.202	430	0.733	220	0.646
98	8.11	7.16	5.91	4.60	1.119	470	0.697	220	0.599
99	8.63	7.35	5.98	4.61	1.071	510	0.681	220	0.570
100	9.28	7.51	6.00	4.58	1.036	560	0.677	220	0.548
101	7.70	7.19	6.11	4.86	1.327	360	0.723	200	0.652
102	8.30	7.56	6.34	4.99	1.205	390	0.667	200	0.589
103	8.97	7.90	6.51	5.06	1.114	430	0.632	200	0.542
104	9.59	8.11	6.58	5.07	1.064	470	0.618	200	0.515
105	8.72	8.09	6.86	5.44	1.294	330	0.637	180	0.571
106	9.42	8.51	7.11	5.57	1.176	360	0.590	180	0.517

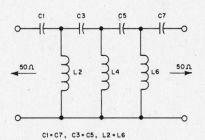

C1 = C7, C3 = C5, L2 = L6

The schematic for a 7-element capacitor input/output Chebyshev high-pass filter. See Table 16 for the attenuation response curve.

Table 18
5-Branch Elliptic High-Pass Filter Designs
50-ohm Impedance
Standard E12 Capacitor Values for C1, C3 and C5

Filter No.	F_{co}	F_{3dB} (MHz)	F_{As}	A_s (dB)	Max. SWR	C1	C3 (nF)	C5	C2	C4	L2 (µH)	L4	F2 (MHz)	F4
1	1.01	0.936	0.670	45.9	1.489	2.7	1.8	3.3	20.7	7.24	6.58	8.40	0.431	0.646
2	1.14	0.976	0.608	50.4	1.186	3.3	1.8	3.9	32.3	11.4	5.53	6.54	0.377	0.582
3	1.30	1.01	0.604	49.4	1.071	3.9	1.8	4.7	35.8	12.5	5.19	6.07	0.369	0.578
4	1.19	1.11	0.810	45.4	1.543	2.2	1.5	2.7	16.4	5.71	5.65	7.28	0.523	0.780
5	1.38	1.20	0.797	46.8	1.199	2.7	1.5	3.3	22.0	7.66	4.61	5.65	0.499	0.765
6	1.56	1.19	0.685	51.6	1.064	3.3	1.5	3.9	33.7	11.9	4.32	4.97	0.417	0.655
7	1.51	1.40	1.01	45.9	1.489	1.8	1.2	2.2	13.8	4.82	4.39	5.60	0.646	0.968
8	1.75	1.51	1.00	46.6	1.180	2.2	1.2	2.7	17.7	6.14	3.65	4.47	0.627	0.961
9	2.02	1.52	0.920	48.3	1.055	2.7	1.2	3.3	23.4	8.09	3.44	4.04	0.562	0.880
10	1.78	1.65	1.15	47.8	1.506	1.5	1.0	1.8	12.7	4.47	3.71	4.64	0.733	1.10
11	2.07	1.80	1.20	46.8	1.199	1.8	1.0	2.2	14.7	5.11	3.07	3.77	0.749	1.15
12	2.38	1.83	1.13	47.8	1.064	2.2	1.0	2.7	18.6	6.43	2.87	3.40	0.689	1.08
13	2.22	2.08	1.55	43.7	1.531	1.2	0.82	1.5	8.19	2.83	3.05	4.02	1.01	1.49
14	2.52	2.17	1.39	48.7	1.186	1.5	0.82	1.8	13.5	4.73	2.51	3.01	0.865	1.33
15	2.89	2.23	1.36	48.2	1.065	1.8	0.82	2.2	15.5	5.37	2.36	2.78	0.833	1.30
16	2.57	2.40	1.68	47.8	1.560	1.0	0.68	1.2	8.40	2.96	2.60	3.27	1.08	1.62
17	3.05	2.68	1.85	44.7	1.215	1.2	0.68	1.5	8.77	3.02	2.10	2.64	1.17	1.78
18	3.48	2.66	1.57	49.9	1.063	1.5	0.68	1.8	14.1	4.94	1.96	2.28	0.957	1.50
19	3.17	2.96	2.13	46.1	1.554	0.82	0.56	1.0	6.31	2.21	2.13	2.72	1.37	2.05
20	3.62	3.16	2.05	48.6	1.210	1.0	0.56	1.2	8.93	3.14	1.74	2.10	1.28	1.96
21	4.19	3.30	2.11	46.1	1.076	1.2	0.56	1.5	9.30	3.19	1.61	1.94	1.30	2.02
22	4.30	3.79	2.55	46.9	1.233	0.82	0.47	1.0	6.69	2.33	1.48	1.82	1.60	2.45
23	4.89	3.84	2.31	49.7	1.079	1.0	0.47	1.2	9.34	3.27	1.36	1.59	1.41	2.21
24	5.87	3.89	2.31	47.4	1.021	1.2	0.47	1.5	9.71	3.32	1.35	1.58	1.39	2.20
25	4.44	4.17	3.01	46.5	1.618	0.56	0.39	0.68	4.37	1.53	1.54	1.97	1.94	2.90
26	5.14	4.52	2.99	48.0	1.236	0.68	0.39	0.82	5.88	2.06	1.23	1.50	1.87	2.87
27	5.88	4.67	2.90	48.0	1.085	0.82	0.39	1.0	7.05	2.45	1.13	1.34	1.78	2.78
28	5.99	5.34	3.60	47.1	1.269	0.56	0.33	0.68	4.63	1.62	1.06	1.31	2.27	3.46
29	6.81	5.48	3.37	49.0	1.096	0.68	0.33	0.82	6.15	2.15	0.961	1.13	2.07	3.22
30	8.07	5.50	3.17	49.3	1.026	0.82	0.33	1.0	7.33	2.54	0.945	1.09	1.91	3.02
31	6.38	5.99	4.26	47.3	1.609	0.39	0.27	0.47	3.18	1.12	1.06	1.34	2.74	4.10
32	7.34	6.47	4.18	49.2	1.241	0.47	0.27	0.56	4.33	1.53	0.856	1.03	2.61	4.01
33	8.39	6.73	4.17	48.4	1.092	0.56	0.27	0.68	4.90	1.71	0.784	0.930	2.57	4.00
34	7.92	7.36	4.98	49.6	1.522	0.33	0.22	0.39	3.05	1.08	0.828	1.02	3.17	4.79
35	9.21	8.05	5.27	48.1	1.217	0.39	0.22	0.47	3.40	1.19	0.686	0.832	3.30	5.06
36	10.4	8.18	4.84	50.5	1.077	0.47	0.22	0.56	4.56	1.60	0.636	0.740	2.95	4.62

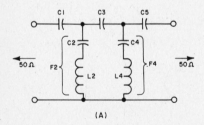

(A)

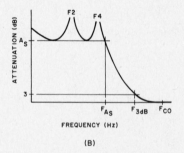

(B)

The schematic for a 5-branch elliptic high-pass filter is shown at A. At B is the typical attenuation response curve.

January 27, 1981.

Wetherhold, E. "Low-Pass Filters — Table of Precalculated Chebyshev Low-Pass Filters with Inductive Input and Output." *RF Design,* July/August, Sept./Oct. 1981.

Wetherhold, E. "Practical 75- and 300-Ohm High-Pass Filters." *QST,* February 1982.

Wetherhold, E. "Table Picks Standard Capacitors for Low-Pass Elliptic Filters." Designer's casebook, *Electronics,* November 30, 1982.

Wetherhold, E. "Table Picks Standard Capacitors for High-Pass Elliptic Filters." Designer's casebook, *Electronics,* February 24, 1983.

Wetherhold, E. "Simplified Elliptic Lowpass Filter Construction using Surplus 88-mH Inductors." *Radio Communication.* April 1983.

Wetherhold, E. "Low-pass Filters for Attenuating RF Amplifier Harmonics, Parts 1 and 2," *Short Wave Magazine,* Dec. 1983 and Jan. 1984.

Wetherhold, E. "Elliptical Lowpass Audio Filter Design (Using Surplus Inductors)," *Ham Radio,* Feb. 1984.

Wetherhold, E. "Practical LC Filter Design, Parts 1-6," *Practical Wireless,* July-Oct., Dec. 1984 and Jan. 1985.

Wetherhold, E. "A C.W. Filter for the Radio Amateur Newcomer," *Radio Communication,* Jan. 1985.

Wetherhold, E. "How to Build a C.W. Filter for the Novice Operator, Parts 1 and 2" *CQ* Magazine, Feb. and March 1985.

Wilkinson, J. "An Introduction to Elliptic Filters for the Radio Amateur." *Radio Communication,* February 1983.

Williams, A. B. *Electronic Filter Design Handbook.* New York: McGraw-Hill Book Co., 1981.

Zverev, A. *Handbook of Filter Synthesis.* New York: John Wiley and Sons, 1967.

COUPLED RESONATORS

Coupled resonators are frequently encountered in RF circuits. Applications include simple filters, oscillator tuned circuits and even antennas. The circuit shown in Fig. 79A illustrates the basic principles involved. A series RLC circuit and the external terminals ab are coupled through a common capacitance, C_m. Applying the formulas for conductance and susceptance in terms of series reactance and resistance gives

$$G_{ab} = \frac{R_r}{R_r^2 + X^2}$$

$$B_{ab} = B_{cm} - \frac{X}{R_r^2 + X^2}$$

The significance of these equations can be understood with the aid of Fig. 79B. At some point, the series inductive reactance will cancel the series capacitive reactance (at a point slightly below f_o where the conductance curve reaches a peak). Depending on the value of the coupling susceptance,

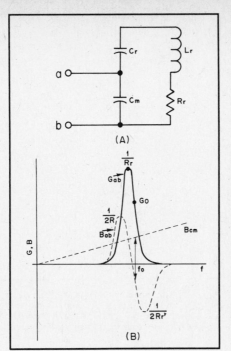

Fig. 79 — A capacitively coupled resonator is shown at A. See text for explanation of figure shown at B.

B_m, it is possible that a point can be found where the total input susceptance is zero. The input conductance at this frequency, f_o is then G_o.

Since G_o is less than the conductance at the peak of the curve, $1/G_o$ or R_o is going to be greater than R_r. This effect can be applied when it is desired to match a low-value load resistance (such as found in a mobile whip antenna) to a more practical value. Suppose R_r and C_r in Fig. 79A are 10 ohms and 21 pF, respectively, and represent the equivalent circuit of a mobile antenna. Find the value of L_r and C_m that will match this antenna to a 52-ohm feed line at a frequency of 3900 kHz. Substituting the above values into the formulas for input conductance gives

$$\frac{1}{52} = \frac{10}{10^2 + X^2}$$

Solving for X (the total series reactance) gives a value of the square root of 420 or 20.49 ohms. The reactance of a 21-pF capacitor at 3900 kHz is 1943.3 ohms so the inductive reactance must be 20.49 + 1943.3 = 1963.8 ohms. While either a positive or negative reactance will satisfy the equation for G_{ab} a positive value is required to tune out B_{cm}. If the coupling element was a shunt inductor, the total reactance would have to be capacitive or negative in value. Thus, the required inductance value for L_r will be 80.1 μH. In order to obtain a perfect match, the input susceptance must be zero. The value of B_{cm} can be found from

$$0 = B_{cm} - \frac{20.49}{10^2 + 20.49^2}$$

giving a susceptance value of 0.03941 siemens, which corresponds to a capacitance of 1608 pF.

Coefficient of Coupling

If the solution to the mobile whip antenna problem is examined, it can be seen that for given values of frequency, R_r, L_r, and C_r only one value of C_m results in an input load that appears as a pure resistance. While such a condition might be defined as resonance, the resistance value obtained is not necessarily the one required for maximum transfer of power.

A definition that is helpful in determining how to vary the circuit elements in order to obtain the desired input resistance is called the coefficient of coupling. The coefficient of coupling is defined as the ratio of the common or mutual reactance to the square root of the product of two specially defined reactances. If the mutual reactance is capacitive, one of the special reactances is the sum of the series capacitive reactances of the primary mesh (with the resonator disconnected) and the other one is the sum of the series capacitive reactances of the resonator (with the primary disconnected). Applying this definition to the circuit of Fig. 79A, the coefficient of coupling, k, is given by

$$k = \sqrt{\frac{C_r}{C_r + C_m}}$$

How meaningful the coefficient of coupling will be depends on the particular circuit configuration under consideration and on which elements are being varied. For example, suppose the value of L_r in the mobile-whip antenna problem was fixed at 100 μH, and suppose C_m and C_r were allowed to vary. (Recall that C_r is 21 pF and represents the antenna capacitance. However, the total resonator capacitance could be changed by adding a series capacitor between C_m and the antenna. Thus, C_r could be varied from 21 pF to some lower value but not a higher one.)

A calculated plot of k versus input resistance, R_{in} is shown in Fig. 80. Notice the rapid change in k for resistance values between 10 ohms and slightly higher resistances.

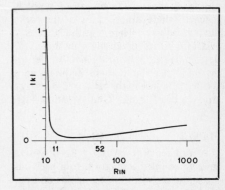

Fig. 80 — Variation of k with input resistance for the circuit of Fig. 79.

Similar networks can be designed to work with any ratio of input resistance and load resistance, but it is evident that small ratios are going to pose difficulties. For larger ratios, component tolerances are less critical. For instance, C_m might consist of switchable fixed capacitors with C_r being variable. With the given load resistance, C_m essentially sets the value of the reactance and thus the input resistance, while C_r and L_r provide the required reactance for the conductance formula. However, if L_r is varied, k varies also. Generally speaking, higher values of L_r (and consequently circuit Q) require lower values of k.

At this point, the question arises as to the significance and even the merit of such definitions as coefficient of coupling and Q. If the circuit element values are known, and if the configuration can be resolved into a ladder network, important properties such as input impedance and attenuation can be computed directly for any frequency. On the other hand, circuit information might be obscured or even lost by attempting to attach too much importance to an arbitrary definition. For example, the plot in Fig. 80 merely indicates that C_m and C_r are changing with respect to each other. But it doesn't illustrate how they are changing. Such information is important in practical applications, and even a simple table of C_m and C_r vs. R_{in} for a particular R_r would be much more valuable than a plot of k.

Similar precautions must be taken with the interpretation of circuit Q. Selectivity and Q are simply related for single resonators and circuit components, but the situation rapidly changes with complex configurations. For instance, adding loss or resistance to circuit elements would seem to contradict the idea that low-loss or high-Q circuits provide the best selectivity. However, this is actually done in some filter designs to improve frequency response. In fact, the filter with the added loss has identical characteristics to one with pure elements. The method is called predistortion and is very useful in designing filters where practical considerations require the use of circuit elements with parasitic or undesired resistance.

As the frequency of operation is increased, discrete components must become physically smaller. Eventually a point is reached where other forms of networks have to be used. Here, entities such as k and Q are sometimes the only means of describing such networks. Another definition of Q that is quite useful in this instance is that it is equal to the ratio of 2π (energy stored per RF cycle)/(energy lost per RF cycle).

Mutually Coupled Inductors

A number of very useful RF networks involve coupled inductors. In a previous section, there was some discussion on iron-core transformers which represent a special case of coupled inductance. The formulas

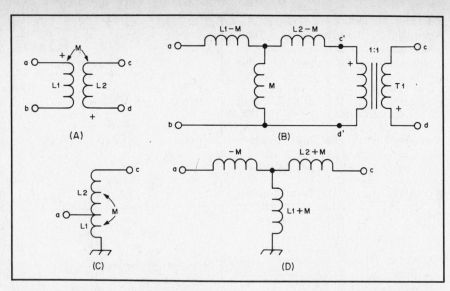

Fig. 81 — Two types of magnetically coupled circuits. At A, only mutual magnetic coupling exists while the circuit at C contains a common inductance also. Equivalents of both circuits at B and D permit the application of the ladder-network analysis discussed in this section. (If the sign of voltage is unimportant, T1 can be eliminated.)

presented apply to instances where the coefficient of coupling is very close to 1.0. While it is possible to approach this condition at frequencies in the RF range, many practical circuits work at values of k that are considerably less than 1.0. A general solution is rather complex, but many practical applications can often be simplified and solved through use of the ladder-network method. In particular, the sign of the mutual inductance must be taken into account if there are a number of coupled circuits or if the phase of the voltage between two coupled circuits is important.

The phase consideration is illustrated in Fig. 81. An exact circuit for the two mutually coupled coils at A is shown at B. T1 is an ideal transformer that provides the isolation between terminals ab and cd. If the polarity of the voltages between these terminals can be neglected, the transformer can be eliminated and just the circuit before the terminals cd substituted. A second circuit is shown in Fig. 81C and D. Here, it is assumed that the winding sense doesn't change between L1 and L2. If so, then the circuit of Fig. 81D can be substituted for the tapped coil shown in Fig. 81C.

Coefficients of coupling for the circuits in Fig. 81 are given by

$$k = \frac{M}{\sqrt{L1\ L2}}$$

and

$$k = \frac{L1 + M}{\sqrt{L1\ (L1 + L2 + 2M)}}$$

If L1 and L2 do not have the same value, an interesting phenomenon takes place as the coupling is increased. A point is reached where the mutual inductance exceeds the inductance of the smaller coil. The interpretation of this effect is illustrated in Fig. 82. While all the flux lines (as indicated

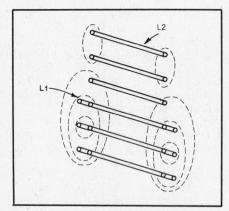

Fig. 82 — Diagram illustrating how M can be larger than one of the self inductances. This represents the transition from lightly coupled circuits to conventional transformers, since an impedance step-up is possible without the addition of capacitive elements.

by the dashed lines) associated with L1 also encircle turns of L2, there are additional ones that encircle extra turns of L2, also. Thus, there are more flux lines for M than there are for L1. Consequently, M becomes larger than L1. Normally, this condition is difficult to obtain with air-wound coils, but the addition of ferrite material greatly increases the coupling. As k increases so that M is larger than L1 (Fig. 82), the network begins to behave more like a transformer. For a k of 1, the equivalent circuit of Fig. 81B yields the transformer equations of a previous section. On the other hand, for small values of k, the network becomes merely three coils arranged in a T fashion. One advantage of the circuit of Fig. 81A is that there is no direct connection between the two coils. This property is important from an isolation standpoint and can be used to suppress unwanted currents that are often responsible for RFI difficulties.

Piezoelectric Crystals

A somewhat different form of resonator consists of a quartz crystal between two conducting plates. If a voltage is applied to the plates, the resultant electric field causes a mechanical stress in the crystal. Depending on the size and cut of the crystal, the crystal begins to vibrate at a certain frequency. The effect of this mechanical vibration is to simulate a series RLC circuit as in Fig. 79A. A capacitance associated with the crystal plates which appears across the terminals (C_m in Fig. 79A). Consequently, this circuit can also be analyzed with the aid of Fig. 79B. At some frequency (f1 in Fig. 83), the series reactance is zero and G_{ab} in the preceding formula will simply be $1/R_r$. Typical values for R_r range from 10 kΩ and higher. However, the equivalent inductance of the mechanical circuit is normally extremely high (over 10,000 henrys in the case of some low-frequency units), which results in a very high circuit Q (30,000).

Above f1 the reactance is inductive, and

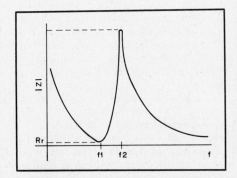

Fig. 83 — Frequency response of a quartz crystal resonator. The minimum value is only approximate since holder capacitance is neglected.

at f2 the susceptance of the series resonator is just equal to the susceptance of the crystal holder, B_{cm}. Here the total susceptance is zero. Since B_{cm} is usually very small, the equivalent series susceptance is also small. This means the value for X in the susceptance formula will be very large, and consequently G_{ab} will be small. This corresponds to a high input resistance. A plot of the magnitude of the impedance is shown in Fig. 83. The dip at f1 is called the series-resonant mode and the peak at f2 is referred to as the parallel-resonant or anti-resonant mode.

When specifying crystals for oscillator applications, the type of mode must be given along with the external capacitance across the holder or type oscillator circuit to be used. Otherwise, considerable difference in actual oscillator frequency will be observed. The effect can be used to advantage, however. The frequency of a crystal oscillator can be pulled with an external reactive element, or even frequency modulated with a device that converts voltage or current fluctuations into changes in reactance.

Matching Networks

In addition to filters, ladder networks are frequently used to match one impedance value to another one. While there are many circuits, a few of them offer such advantages as simplicity of design formulas or minimum number of elements. Some of the more popular ones are shown in Fig. 84. Shown at Fig. 84A and 84B, are two variations of an L network. These networks are relatively simple to design. The situation is somewhat more complicated for the circuits shown at Fig. 84C and 84D. For a given value of input and output resistance, there are many networks that satisfy the conditions for a perfect match. The difficulty

can be resolved by introducing the dummy variable labeled N.

From a practical standpoint, N should be selected in order to optimize circuit component values. Either values of N that are too low or too high result in networks that are hard to construct.

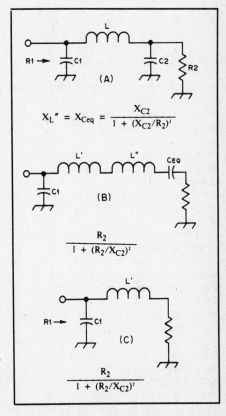

Fig. 85 — Illustration of the manner in which the network of Fig. 84C can be reduced to the one of Fig. 84A, assuming C2 is assigned some arbitrary value. (The formulas shown are for numerical reactance values.)

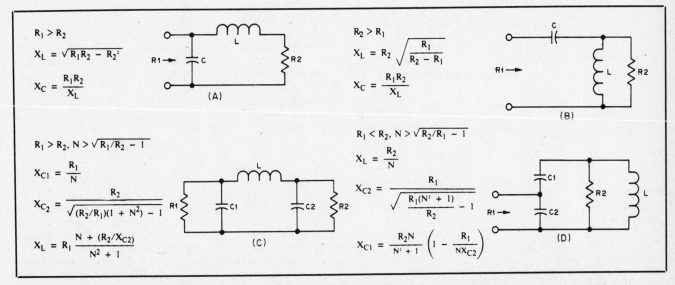

Fig. 84 — Four matching networks that can be used to couple a source and load with different resistance values. (Although networks are drawn with R1 appearing as the source resistance, all can be applied with R2 at the source end. Also, all formulas with capacitive reactance are for the numerical or absolute value.)

The reason for this complication is as follows. Only two reactive elements are required to match any two resistances. Consequently, adding a third element introduces a redundancy. This means one element can be assigned a value arbitrarily and the other two components can then be found. For instance, suppose C2 in Fig. 84C is set to some particular value. The parallel combination of C2 and R2 may then be transformed to a series equivalent (see Fig. 85). Then, L may be found by breaking it down into two components, L' and L". One component (L") would tune out the remaining capacitive reactance of the output series equivalent circuit. The network is then reduced to the one shown in Fig. 84A and the other component of L (L') along with the value of C1, could be determined from formulas (Fig. 84A). Adding the two inductive components gives the actual inductive reactance required for a match in the circuit of Fig. 84C.

As mentioned before, it is evident that an infinite number of networks of the form shown in Fig. 84C exist since C2 can be assigned any value. Either a set of tables or a family of curves for C1 and L in terms of C2 could then be determined from the method discussed earlier and illustrated in Fig. 87. However, similar data along with other information can be obtained by approaching the problem somewhat differently. Instead of setting one of the element values arbitrarily and finding the other two, a third variable is contrived, and in the case of Figs. 84C and 84D is labeled N. All three reactances are then expressed in terms of the variable N.

The manner in which the reactances change with variation in N for two representative circuits of the type shown in Fig. 84C is shown in Fig. 86. The solid curves are for an R1 of 3000 ohms and R2 equal to 52 ohms. The dashed curves are for the same R2 (52 ohms) but with R1

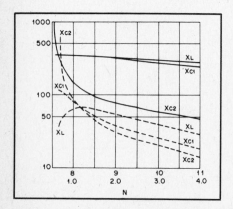

Fig. 86 — Network reactance variation as a function of dummy variable N. Solid curves and values of N from 8 to 11 are for an input resistance of 3000 ohms and an output resistance of 52 ohms. The dashed curves are for a similar network with an input and output resistance of 75 and 52 ohms, respectively. Values of N from 1 to 4 are for the dashed curves.

equal to 75 ohms. For values of N very close to the minimum specified by the inequality (Fig. 84C), X_{C2} becomes infinite, which means C2 approaches zero. As might be expected, the values of X_L and X_{C1} at this point are approximately those of an L network (Fig. 84A) and could be determined by means of the formulas in Fig. 84A for the corresponding values of R1 and R2.

The plots shown in Fig. 86 give a general idea of the optimum range of component values. The region close to the left-hand portion should be avoided since there is little advantage to be gained over an L network, while an extra component is required. For very high values of N, the

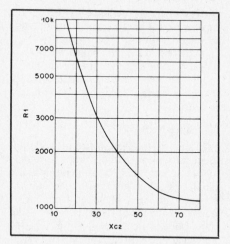

Fig. 87 — Input resistance vs. output reactance for an output resistance of 52 ohms. The curve is for a fixed inductor of 219 ohms (Fig. 84C). X_c varies from 196 to 206 ohms.

capacitance values become large without producing any particular advantage either. A good design choice is an N a few percent above the minimum specified by the inequality.

Often, one of the elements is fixed, with either or both of the remaining two elements variable. In many amateur transmitters, it is the inductor that remains fixed (at least for a given band), while C1 and C2 (Fig. 84C) are made variable. While this system limits the bandwidth and matching capability somewhat, it is still a very useful approach. For instance, the plot shown in Fig. 87 indicates the range of input resistance values that can be matched for an R2 of 52 ohms. The graph is for an inductive reactance of 219 ohms. X_{C1} varies from 196 to 206 ohms over the entire range of R1 (or approximately 20 percent). However, X_{C2} varies from 15 to almost 100 ohms, as indicated in the graph.

Since C2 more or less sets the transformed resistance, it is often referred to as the loading control on transmitters using the network of Fig. 84C, with C1 usually labeled TUNE. While the meaning of the latter term should be clear, the idea of

loading in a matching application perhaps needs some explanation. For small values of X_{C2} (very large C2), the transformed resistance is very high. Consequently, a source that was designed for a much lower resistance would deliver relatively little power. However, as the resistance is lowered, increased amounts of current will flow, resulting in more power output. Then the source is said to be loaded more heavily.

Similar considerations such as those discussed for the network of Fig. 84C also exist for the circuit of Fig. 84D. Only the limiting L network for the latter is the one shown in Fig. 84B. The circuit of Fig. 84C is usually called a pi network and, as pointed out, it is used extensively in the output stage of transmitters. The circuit of Fig. 84D has never been given any special name, but is quite popular in both antenna and transistor matching applications.

The plots shown in Fig. 86 are for fixed input and output resistances with the reactances variable. Similar figures can be plotted for other combinations of fixed and variable elements. An interesting case is for X_L and R1 fixed with R2, X_{C1} and X_{C2} variable. A lower limit for N also exists for this plot, except that instead of an L network, the limiting circuit is a network of three equal reactances. A feature of this circuit is that the output resistance is the ratio of the square of the reactance and the input resistance.

An analogous situation exists with a quarter-wavelength transmission-line transformer. The output resistance is the ratio of the square of the characteristic impedance of the line and the input resistance. Consequently, the special case in which all the reactances are equal in the circuit of Fig. 84C is the lumped-constant analog of the quarter-wavelength transformer. It has identical phase shift (90°) along with the same impedance-transforming properties.

Frequency Response

In many instances, a matching network performs a dual role in transforming a resistance value while providing frequency rejection. Usually the most important considerations are matching versatility, component values and the number of elements. But a matching network might also be able to provide sufficient selectivity for some application, thus eliminating the need for a separate circuit such as a filter.

Recall that Q and selectivity are closely related for simple RLC series and parallel circuits. Bandwidth and the parameter N of Fig. 84 are approximately related in this manner. For values of N much greater than the minimum specified by the inequality, N and Q can be considered to mean the same thing for all practical purposes. However, the frequency response of networks that are more complex than simple RLC types is usually more complicated. Consequently, some care is required in the interpretation of N or Q in regard to frequency rejection. For instance, a simple circuit has a frequency response that results in increasing attenuation for increasing excursions from resonance.

That is not true for the pi network, as can be seen from Fig. 88. For slight frequency changes below resonance, the attenuation increases as in the case of a simple RLC network. At lower frequencies, the attenuation decreases and approaches 2.55 dB. This plot is for a resistance ratio of 5:1, and the low-frequency loss is caused only by the mismatch in source and load resistance. Thus, while increasing N improves the selectivity near resonance, it has little effect on response for frequencies much farther away.

A somewhat different situation exists for the circuit of Fig. 84D. At frequencies far from resonance, either a series capacitance provides decoupling at the lower frequen-

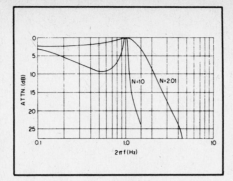
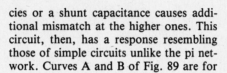

Fig. 88 — Frequency response of the network of Fig. 84C for two values of N.

cies or a shunt capacitance causes additional mismatch at the higher ones. This circuit, then, has a response resembling those of simple circuits unlike the pi network. Curves A and B of Fig. 89 are for

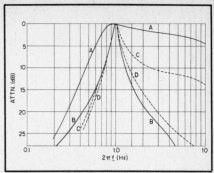

Fig. 89 — Frequency response of the circuit of Fig. 84D (see text).

a resistance ratio of 5:1, with N equal to 2.01 for curve A. Curve B is for an N of 10. Curves C and D are for a resistance ratio of 50:1, with N equal to 7.04 and 10, respectively.

Chapter 3

Radio Design Technique and Language

Many amateurs desire to construct their own radio equipment, and some knowledge of design procedures becomes important. Even when some commercially manufactured equipment is used, these techniques may still be required in setting up peripheral equipment. Also, an applicant for an Amateur Radio license might be tested on material in this area.

"Pure" vs. "Impure" Components

In the chapter on electrical laws and circuits, it is assumed that the components in an electrical circuit consist solely of elements that can be reduced to a resistance, capacitance or inductance. However, such elements do not exist in nature. An inductor always has some resistance associated with its windings and a carbon-composition resistor becomes a complicated circuit as the frequency of operation is increased. Even conductor resistance must be taken into account if long runs of cable are required.

In many instances, the effects of these "parasitic" components can be neglected and the actual device can be approximated by a "pure" element such as a resistor, capacitor, inductor or a short circuit in the case of an interconnecting conductor. In other cases, the unwanted component must be taken into account. However, it may be possible to break the element down into a simple circuit consisting of single elements alone. Then the actual circuit may be analyzed by means of the basic laws discussed in the previous chapter. It may also be possible to make a selection so the effects of the residual element are negligible.

There are times when unwanted effects may not only be difficult to remove but will affect circuit operation adversely as well. In fact, such considerations often set a limit on how stringent a design criterion can be tolerated. For instance, it is a common practice to connect small-value capacitors in various parts of a complex circuit, such as a transmitter or receiver, for bypassing purposes. A bypass capacitor permits energy below some specified frequency to

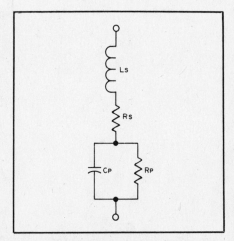

Fig. 1 — Equivalent circuit of a capacitor.

pass a given point while providing rejection to energy at higher frequencies. In essence, the capacitor is used in a single form of filter. In more complicated filter designs, capacitors may be required for complex functions (such as matching) in addition to providing a low reactance to ground.

An equivalent circuit of a capacitor is shown in Fig. 1. Normally, the series resistance, R_s, can be neglected. On the other hand, the upper frequency limit of the capacitor is limited by the series inductance, L_s. In fact, above the point where L_s and C_p form a resonant circuit, the capacitor actually appears as an inductor at the external terminals. This is why it is common practice to use two capacitors in parallel for bypassing, as shown in Fig. 2. At first inspection, this might appear as superfluous duplication. But the self-resonant frequency of a capacitor is lower for high-capacitance units than it is for smaller value ones. Thus, C1 in Fig. 2 provides a low reactance for low frequencies such as those in the audio range while C2 acts as a bypass for frequencies above the self-resonant frequency of C1.

RF Leakage

Although the capacitor combination shown in Fig. 2 provides a low-impedance path to ground, it may not be very effective in preventing RF energy that travels along the conductor at point 1 from reaching point 2. At dc and low-frequency applications, a circuit must always form a closed path for a current to flow. Consequently, two conductors are required if power is to be delivered from a source to a load. In many instances, one of the conductors may be common to several other circuits and constitutes a local ground.

As the frequency of operation is increased, however, a second type of coupling mechanism is possible. Power may be transmitted along a single conductor. (Although the same effect is possible at low frequencies, unless circuit dimensions are extremely large, such transmission effects can be neglected.) The conductor acts as

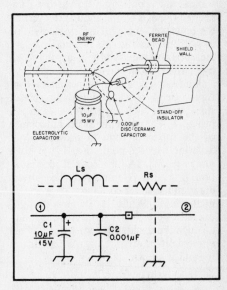

Fig. 2 — A bypassing arrangement that affords some measure of isolation (with the equivalent circuit shown.) Dashed lines indicate a mode of wave travel that permits RF energy to leak past the bypass circuit and that should be taken into account when more stringent suppression requirements are necessary. (L_s and R_s in the equivalent circuit represent the ferrite bead.)

a waveguide in much the same manner that a large conducting surface, such as the earth, will permit propagation of a radio wave close to its boundary with the air. A mode similar to ground-wave propagation that can travel along the boundary of a single conductor is illustrated by the dashed lines in Fig. 2. As with the wave traveling close to the earth, a poor conducting boundary will cause attenuation. This is why a ferrite bead is often inserted over the exit point of a conductor from an area where RF energy is to be contained or excluded. In addition to loss (particularly in the VHF range), the high permeability of the ferrite introduces a series-inductive reactance as well. Finally, the shield wall provides further isolation.

While the techniques shown in Fig. 2 get around some of the deficiencies of capacitors that are used for bypass purposes, the resulting suppression is inadequate for a number of applications. Examples would be protection of a VFO to surrounding RF energy, a low-frequency receiver with a digital display, and suppression of radiated harmonic energy from a transmitter. In each of these cases, a very high degree of isolation is required. For instance, a VFO is sensitive to voltages that appear on dc power supply lines and a transmitter output with a note that sounds "fuzzy" or rough may result.

Digital displays usually generate copious RF energy in the low-frequency spectrum. Consequently, a receiver designed for this range presents a situation where a strong source of emission is in close proximity to very sensitive receiving circuits. A similar case exists with transmitters operating on a frequency that is a submultiple of a fringe-area TV station. In the latter two instances, the problem is not so severe if the desired signal is strong enough to override the unwanted energy. Unfortunately, this is not the case normally and stringent measures are required to isolate the sensitive circuits from the strong source.

A different type of bypass-capacitor configuration is often used with associated shielding for such applications, as shown in Fig. 3. To reduce the series inductance of the capacitor, and to provide better isolation between points 1 and 2, either a disc type (Fig. 3A) or a coaxial configuration (Fig. 3B) is employed. The circuit diagram for either configuration is shown in the inset. While such "feedthrough" capacitors are always connected to ground through the shield, this connection is often omitted on drawings. Only a connection to ground is shown, as in Fig. 3C.

Dielectric Loss

Even though capacitors are usually high-Q devices, the effect of internal loss can be more severe than in the case of a coil. This is because good insulators of electricity are also usually good insulators of heat. Therefore, heat generated in a capacitor must be conducted to the outside via the

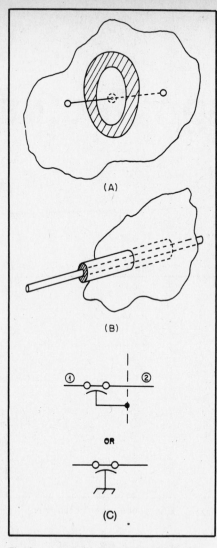

Fig. 3 — A superior type of bypassing arrangement to that shown in Fig. 2. Concentric conductors provide a low-inductance path to ground and better rejection of unwanted single-wire wave modes.

conducting plates to the capacitor leads. In addition, most capacitors are covered with an insulating coating that further impedes heat conduction. The problem is less severe with capacitors using air as a dielectric for two reasons. The first advantage of air over other dielectrics is that the loss in the presence of an alternating electric field is extremely small. Second, any heat generated by currents on the surface of the conducting plates is either drawn away by air currents or through the mass of the metal.

The dielectric loss in a capacitor can be represented by R_p, as shown in Fig. 1. However, if a dc ohmmeter were placed across the terminals of the capacitor, the reading would be infinite. This is because dielectric loss is an ac effect. Whenever an alternating electric field is applied to an insulator, there is a local motion of the electrons in the individual atoms that make up the material. Even though the electrons are not displaced as they would be in a con-

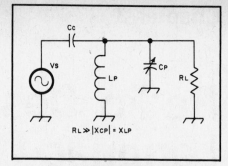

Fig. 4 — Consideration of capacitor voltage and current ratings should be kept in mind in moderate power applications.

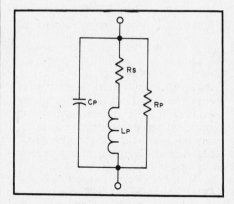

Fig. 5 — Equivalent circuit of an inductor.

ductor, this local motion requires the expenditure of energy and results in a power loss.

Consequently, some care is required in the application of capacitors in moderate to high-power circuits. The applied voltage should be such that RF current ratings are not exceeded for the particular frequency of operation. This is illustrated in Fig. 4. A parallel-resonant circuit consisting of L_p, C_p and R1 is connected to a voltage source, V_s, through a coupling capacitor, C_c. It is also assumed the Rl is much greater than either the inductive or capacitive reactance taken alone. This condition would be typical of that found in most RF power amplifier circuits using vacuum tubes.

Since the inductive and capacitive reactance of L_p and C_p cancel at resonance, the load presented to the source would be just Rl. This would mean the current through C_c would be much less than the current through either C_p or L_p. The effect of such current rise is similar to the voltage rise at resonance discussed in the previous chapter. Even though the current at the input of the parallel-resonant circuit is small, the currents that flow in the elements that make up the circuit can be quite large.

The requirements for C_c, then, would be rather easy to satisfy with regard to current rating and power dissipation. On the other hand, C_p would ordinarily be restricted to air-variable types, although

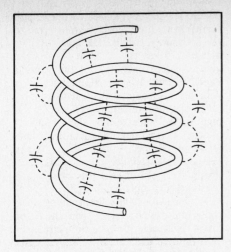

Fig. 6 — Distributed capacitance (indicated by dashed lines) affects the operation of a coil at high frequencies.

some experiments have been successful using Teflon as a dielectric. Generally speaking, the coupling capacitor should have a low reactance (at the lowest frequency of operation) in comparison to the load presented by the tuned circuit. The effect of the coupling-capacitor reactance could then be compensated by slightly retuning the parallel-resonant circuit.

Inductors

Similar considerations to those discussed in the previous sections exist with inductors also, as shown in Fig. 5. Since an inductor usually consists of a coil of wire, there will be a resistance associated with the wire material and this component is represented by R_s. In addition, there is always a capacitance associated with conductors in proximity as illustrated in Fig. 6. While such capacitance is distributed through the coil, it is a convenient approximation to consider that an equivalent capacitance, C_p, exists between the terminals (Fig. 5). Inductors are often wound on materials that have high permeability in order to increase the inductance. Thus, it is possible to build an inductor with fewer turns and that is smaller in size than an equivalent coil with an air core.

Unfortunately, high-permeability materials presently available have considerable loss in the presence of an RF field. It will be recalled that a similar condition existed with the dielectric in a capacitor. Consequently, in addition to the wire resistance, a loss resistance is associated with the core and represented by R_p in Fig. 5. Since this loss is more or less independent of the current through the coil but dependent on the applied voltage, it is represented by a parallel resistance.

RF Transformer

Although the term "transformer" might be applied to any network that transforms a voltage or an impedance from one level to another one, the term is usually reserved

for circuits incorporating mutual magnetic coupling. Examples would be IF transformers, baluns, broadband transformers and certain antenna matching networks (see Chapter 2). Of course, many devices used at audio and power frequencies are also transformers in the sense used here and have been covered in a previous chapter.

Networks that use mutual magnetic coupling exclusively have attractive advantages over other types in many common applications. A principal advantage is that there is no direct connection between the input and output terminals. Consequently, dc and ac components are separated easily, thus eliminating the need for coupling capacitors. Even more importantly, it is also possible to isolate RF currents because of the lack of a common conductor. Often, an HF receiver in an area where strong local broadcast stations are present will suffer from "broadcast harmonics" and possibly even rectified audio signals getting into sensitive AF circuits. In such cases, complicated filters sometimes prove ineffective, while a simple tuned RF transformer will clear up the problem completely. This is because the unwanted BC components are prevented from flowing on the receiver chassis, along with being rejected by the tuned transformer filter characteristic.

A second advantage of circuits using mutual magnetic coupling exclusively is that analysis is relatively simple compared to other forms of coupling, although exact synthesis is complicated. That is, finding a network with some desired frequency response would be difficult in the general case.

However, circuits using mutual magnetic coupling usually have very good out-of-band rejection characteristics when compared to networks incorporating other forms. (A term sometimes applied to transformer or mutual magnetic coupling is indirect coupling. Circuits with a single resistive or reactive element for the common impedance are called direct-coupled networks. Two or more elements in the common impedance are said to comprise complex coupling.) For instance, relatively simple band-pass filters are possible with mutual magnetic coupling and are highly recommended for VHF transmitter-multiplier chains. For receiving, such filters are often the main source of selectivity. Standard AM and FM broadcast receivers would be examples where intermediate-frequency (IF) transformers derive their band-pass characteristics from mutually coupled inductors.

A third advantage of mutually coupled networks is that practical circuits with great flexibility are possible particularly in regard to matching capabilities. For this reason, variable-coupling matching networks or those using link coupling have been popular for many years. In addition to matching flexibility, these circuits are good band-pass filters and can also provide isolation be-

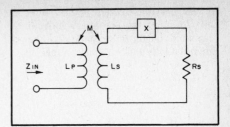

Fig. 7 — Basic magnetically coupled circuit.

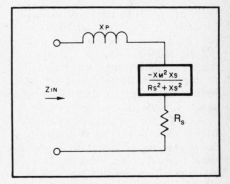

Fig. 8 — Equivalent single-mesh network of the two-mesh circuit of Fig. 7.

tween antenna circuits and those of the transmitter.

Design Formulas

A basic two-mesh circuit with mutual magnetic coupling is shown in Fig. 7. The reactance, X, is arbitrary and could be either inductive or capacitive. However, it is convenient to combine it with the secondary reactance (X_{LS}) since this makes the equations more compact. Hence, the total secondary reactance is defined by

$$X_s = 2\pi f L_s + X$$

The primary reactance and mutual reactance are also defined respectively as

$$X_p = 2\pi f L_p$$
$$X_m = 2\pi f M$$

A set of equations for the input resistance and reactance is given by

$$R_{in} = \frac{R_s X_m^2}{R_s^2 + X_s^2}$$

$$X_{in} = X_p - \frac{X_m^2 X_s}{R_s^2 + X_s^2}$$

This permits reducing the two-mesh circuit of Fig. 7 to the single-mesh circuit of Fig. 8.

Double-Tuned Circuits

A special case occurs if the value of X_s is zero. This could be accomplished easily by tuning out the inductive reactance of the secondary with an appropriate capacitor or by varying the frequency until a fixed capacitor and the secondary inductance resonated. Under these conditions, the input resistance and reactance would be

$$R_{in} = \frac{X_m^2}{R_s}$$

$$X_{in} = X_p$$

Then, in order to make the input impedance purely resistive, a second series capacitor could be used to cancel the reactance of X_p. The completed network is shown in Fig. 9 with C1 and C2 being the primary and secondary series capacitors.

If X_m could be varied, it is evident that the secondary resistance could be transformed to almost any value of input resistance. Usually, the desired resistance would be made equal to the generator resistance, R_g, for maximum power transfer. It might also be selected to satisfy some design goal, not necessarily related to maximum power transfer. This brings up a minor point but one that can cause considerable confusion. Normally, in transmitter circuits, the unloaded Q of the reactive components would be very high and the series parasitic resistances (discussed in a previous section) could be neglected. However, if it is not desired to do so, how should these resistances be taken into account? If maximum power transfer is the goal, the series resistance of the primary coil would be added to the generator resistance, R_g, and the transformed secondary resistance would be made equal to this sum.

On the other hand, a more common case requires the total input resistance to be equal to some desired value. For instance, an amplifier might provide optimum efficiency or harmonic suppression when terminated in a particular load resistance. Transmission lines also require a given load resistance in order to be matched. In such cases, the series resistance of the primary coil would be subtracted from the actual resistance desired and the transformed resistance made equal to this difference. As an example, suppose an amplifier required a load resistance of 3000 ohms, and the primary-coil resistance was 100 ohms. Then the transformed resistance must be equal

to 2900 ohms, (In either case, the secondary coil resistance is merely added to the secondary load resistance and the sum substituted for R_s.)

Coefficient of Coupling

Although the equations for the input impedance can be solved in terms of the mutual reactance, the transforming mechanism involved becomes clearer if the coefficient of coupling is used instead. The coefficient of coupling, k, in terms of the corresponding reactances of inductances is

$$k = \frac{X_m}{\sqrt{X_p X_s}} = \frac{L_m}{\sqrt{L_p L_s}}$$

Then, the input resistance becomes

$$R_{in} = \frac{k^2 X_p X_s}{R_s}$$

The primary and secondary Q values are defined as

$$Q_p = \frac{X_p}{R_g}$$

$$Q_s = \frac{X_s}{R_s}$$

where a loaded Q is assumed. This would mean R_s includes any secondary-coil loss. For maximum power transfer, R_g would be the total primary resistance, which consists of the generator and coil resistance.

The coefficient of coupling under these conditions reduces to a simple formula

$$k_c = \frac{1}{\sqrt{Q_p Q_s}}$$

However, if it is desired to make the input resistance some particular value (as in the case of the previous example), the coefficient of coupling is then

$$k_c = \sqrt{\frac{R_{in} - R_p}{X_p Q_s}}$$

If the primary "loss" resistance is zero, both formulas are identical.

At values of k less than k_c, the input resistance is lower than either the prescribed value or for conditions of maximum power transfer. Higher values of k result in a higher input resistance. For this reason, k_c is called the critical coefficient of coupling. If k is less than k_c the circuit is said to be undercoupled, and for k greater than k_c an overcoupled condition results. A plot of attenuation vs. frequency for the three cases is shown in Fig. 10. Critical coupling gives the flattest response, although greater bandwidth can be obtained by increasing k to approximately 1.5 k_c. At higher values, a pronounced dip occurs at the center or resonant frequency.

In the undercoupled case, a peak occurs at the resonant frequency of the primary and secondary circuit but the transformed resistance is too low and results in a mis-

match. As the coupling is decreased still further, very little power is transferred to the secondary circuit and most of it is dissipated in the primary-loss and generator-source resistances. On the other hand, an interesting phenomenon occurs with the overcoupled case. It will be recalled that the transformed resistance is too high at resonance because the coefficient of coupling is greater than the critical value. However, a special case occurs if the primary and secondary circuits are identical, which also means the transformed resistance R_{in} must equal R_s.

The behavior of the circuit under these conditions can be analyzed with the aid of Fig. 8. Assuming the Q of both circuits is high enough, the reactance, X_s, increases very rapidly on either side of resonance. If this variation is much greater than the variation of X_m with frequency, a frequency exists on each side of resonance where the ratio of X_m^2 to $R_s^2 + X_s^2$ is 1.0. Consequently, R_{in} is equal to R_s and the transformed reactance is $-X_s$. Since the primary and secondary resonators are identical, the reactances cancel because of the minus sign. The frequency plot for a k of 0.2 (k_c is 0.1) is shown in Fig. 10. If the primary and secondary circuits are not identical, a double-humped response still occurs, but at the points where the transformed resistance is equal to the desired

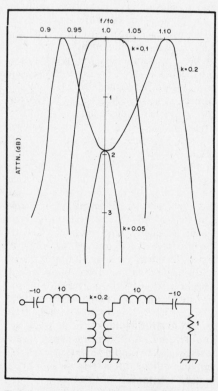

Fig. 10 — Response curves for various degrees of coupling coefficient, k. The critical coefficient of coupling for the network shown in the circuit diagram is 0.1. Lower values give a single response peak (but less than maximum power transfer) while "tighter" coupling results in a double-peak response.

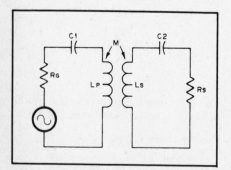

Fig. 9 — Double-tuned series circuits with magnetic coupling.

value, the reactances are not the same numerically. Consequently, there is attenuation at peaks, unlike the curve of Fig. 10.

Other Circuit Forms

While the coupled network shown in Fig. 9 is the easiest to analyze, it is not commonly encountered in actual circuits. As the resistance levels are increased, the corresponding reactances become very large also. In transmitter circuits, extremely high voltages are then developed across the coils and capacitors. For high-impedance circuits, the circuit shown in Fig. 11 is often used.

Although the frequency response for the circuit of Fig. 11 is somewhat different than the circuit of Fig. 9 (in fact, the out-of-band rejection is greater), a matching network can be designed based upon the previous analysis for the series circuit. This is accomplished by changing the parallel primary and secondary circuits to series equivalents. (It should be emphasized that this transformation is good at one frequency only.) The equivalent circuit of the one shown in Fig. 11 is illustrated in Fig. 12, where the new resistance and reactance of the secondary are given by

$$R_{eq}(S) = \frac{R_s}{1 + \gamma^2}$$

$$X_{eq}(C_s) = \frac{-R_s\gamma}{1 + \gamma^2}$$

$$\gamma = R_s / X_c$$

A similar set of transformations exists for the primary circuit also. In most instances, where one high-impedance load is matched to another, R_s in Fig. 11 is much greater than the reactance of C_s and C_p. This simplifies the transformations, and approximate relations are given by

$$R_{eq}(S) \approx \frac{X_c^2}{R_s}$$

$$X_{eq}(C_s) \approx X_c$$

As an example, suppose it was desired to match a 3000-ohm load to a 5000-ohm source using a coupled inductor with a 250-ohm (reactance) primary and secondary coil. Assume the coupling can be varied. Determine the circuit configuration and the critical coefficient of coupling.

Since the load and source resistance have a much higher numerical value than the reactance of the inductors, a parallel-tuned configuration must be used. In order to tune out the inductive reactance, the equivalent series capacitive reactance must be −250 ohms. Since both R_s and R_p are known, the exact formulas could be solved for γ and R_{eq}. However, because the respective resistances are much greater than the reactance, the simplified approximate formulas can be used. This means the primary and secondary equivalent

capacitive reactances are −250 ohms. The equivalent secondary resistance is $250^2/3000$ or 20.83 ohms, resulting in a secondary Q of 250/20.83 or 12. (A formula could be derived directly for the Q from the approximate equations.) The equivalent primary resistance and Q are 12.5 ohms and 20, respectively. Substituting the values for Q into the formula for the critical coefficient of coupling gives $1/\sqrt{(20)(12)}$ or 0.065.

Double-tuned coupled circuits of the type shown in Fig. 11 are used widely in radio circuits. Perhaps the most common example is the IF transformer found in AM and FM BC sets. Many communications receivers have similar transformers, although the trend has been toward somewhat different circuits. Instead of obtaining selectivity by means of IF transformers (which may require a number of stages), a single filter with quartz-crystal resonators is used instead. The subject of receivers is treated in a later chapter.

Single-Tuned Circuits

In the case of double-tuned circuits, separate capacitors are used to tune out the inductive components of the primary and secondary windings. However, examination of the equivalent circuit of the coupled coil shown in Fig. 8 suggests an alternative. Instead of a separate capacitor, why not detune a resonant circuit slightly and reflect a reactance of the proper sign into the primary in order to tune out the primary inductance. Since the transformation function (shown in the box in Fig. 8) reverses the sign of the secondary reactance, it is evident that X_s must be *inductive* in order to tune out the primary inductance.

This might seem to be a strange result but it can be explained with the following reasoning. From a mathematical point of view, the choice of the algebraic sign of the transformed reactance is perfectly arbitrary. That is, a set of solutions to the equations governing the coupled circuit is possible, assuming either a positive or negative sign for the transformed reactance. However, if the positive sign is chosen, the transformed resistance would be negative. But from a physical point of view, this is a violation of the law of conservation of energy since it would imply the secondary resistance acts as a source of energy rather than an energy sink. Consequently, the solution with the negative resistance does not result in a physically realizable network.

This phenomenon has implications for circuits one might not normally expect to be related to coupled networks. For instance, consider coil 1 (Fig. 13) in proximity to the one-turn shorted coil 2. A time-varying current in coil 1 will induce a current in coil 2. In turn, the induced current will set up a magnetic field of its own. The question is, will the induced field aid or oppose the primary field. Since the energy in

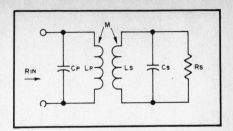

Fig. 11 — Coupled network with parallel-tuned circuits of IF transformer.

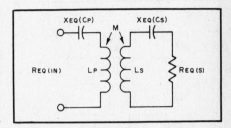

Fig. 12 — Equivalent series circuit of the parallel network shown in Fig. 11. This transformation is valid only at a single frequency and must be revalued if the frequency is changed.

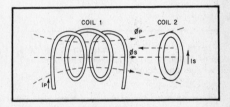

Fig. 13 — A coil coupled magnetically to a shorted turn provides insight to coils near solid shield walls.

a magnetic field is proportional to the square of the flux, the induced field must oppose the primary field. Otherwise the principle of the conservation of energy would be violated as it was with the "negative" resistance. Consequently, the induced current must always be in a direction such that the induced field opposes changes in the generating field. This result is often referred to as Lenz's Law.

If, instead of a one-turn loop, a solid shield wall was substituted, a similar phenomenon would occur. Since the total flux (for a given current) would be less with the shield present than it would in the absence of the shield, the equivalent coil inductance is decreased. That is why it is important to use a shield around a coil that is big enough to reduce the effect of such coupling. Also, a shield made from a metal with high conductivity such as copper or aluminum is advisable; otherwise a loss resistance will be coupled into the coil as well.

Link Coupling

An example of a very important class of

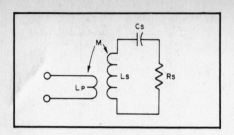

Fig. 14 — Link coupling can be used to analyze a number of important circuits.

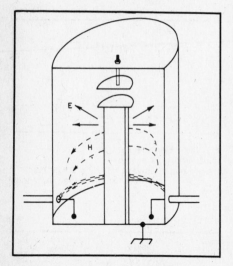

Fig. 15 — A VHF/UHF circuit which can be approximated by a link-coupled network using conventional components.

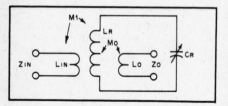

Fig. 16 — Equivalent low-frequency analog of the circuit shown in Fig. 15.

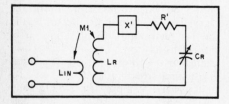

Fig. 17 — The network of Fig. 16 can be reduced with the transformation shown in Fig. 8.

single-tuned circuits is shown in Fig. 14. The primary inductor consists of a small coil either in close proximity to or wound over one end of a large coil. Two resonators can be coupled in this manner, although there may be considerable separation (and no mutual coupling between the larger coils). Hence the term, link coupling. While this particular method is seldom used nowadays, the term is still applied to the basic configuration shown in Fig. 14. Ap-

plications would be antenna-matching networks, output stages for amplifiers, and especially important, many circuits used at VHF that have no direct HF equivalent.

The cavity resonators used in repeater duplexers are one form of VHF circuit that uses link coupling. A cross-sectional view of a representative type is shown in Fig. 15. Instead of ordinary coils and capacitors, a section of coaxial transmission line comprises the resonant circuit. The frequency of the resonator may be varied by adjusting the tuning screw that changes the value of the capacitor. Energy is coupled into and out of the resonator by means of two small, one-turn loops. Current in the input loop causes a magnetic field (shown by dashed lines). If the frequency of the generating field is near one of the resonant modes of the configuration, an electric field will also be generated (shown by solid lines). Finally, energy may then be coupled out of the resonator by means of a second loop.

A low-frequency equivalent circuit of the resonator is shown in Fig. 16. However, the circuit can be used only to give an approximate idea of the actual frequency response of the cavity. At frequencies not close to the resonant frequency, the mathematical laws governing resonant circuits are different from those of discrete components used at HF. Over a limited frequency range, the resonator can be approximated by the series LC circuit shown in Fig. 16.

By applying the formulas for coupled networks shown in Fig. 8 to the two-link circuit of Fig. 16, the output link and load can be transformed to an equivalent series resistance and reactance. This is shown in Fig. 17. In most instances the reactance, X_s in the formula, is just the reactance of the output link. Since the two-link network has been reduced to a single coupled circuit, the formulas can be applied again to find the input resistance and reactance.

Analysis of Single-Tuned Circuits

Single-tuned circuits are very easy to construct and adjust experimentally. The tuned circuit consisting of L_s, C_s, and perhaps the load, R_s, can be constructed first and tuned to the natural resonant frequency

$$f_o = \frac{1}{2\pi \sqrt{L_s C_s}}$$

Then the primary inductor, which may be a link or a larger coil, is brought into proximity of the resonant circuit. The resonant frequency will usually shift upward. For instance, a coil and capacitor combination was tuned to resonance by means of a dip oscillator (see the chapter on measurements) at a frequency of 1.8 MHz. When a two-turn link was wound over the coil and coupled to the dip oscillator, the resonant frequency had increased to 1.9 MHz. A three-turn link caused a change to 2 MHz.

Often an actual load may be an unknown

quantity, such as an antenna, and insight into the effects of the various elements is helpful in predicting single-tuned circuit operation. Usually, as in the case of most matching networks, R_s (Fig. 8) and the input resistance are specified with the reactive components being the variables. Unfortunately, the variables in the case of mutually coupled networks are not independent of each other, which complicates matters.

Examination of the equivalent circuit shown in Fig. 8 would indicate the first condition is that the reactance reflected from the secondary into the primary be sufficient to tune out the primary reactance. Otherwise, even though the proper resistance transformation is obtainable, a reactive component would always be present. A plot of the reflected reactance as a function of X_s is shown in Fig. 18. From mathematical considerations it can be shown that the maximum and minimum of the curve have a value equal to $X_m^2/2R_s$. Consequently, this value must be greater than or equal to X_p so a value of X_s may exist such that the reflected reactance will cancel X_p. In the usual case where $X_m^2/2R_s$ is greater than X_p, it is interesting to note that two values of X_s exist where X_p and the reflected reactance cancel. This means there are two cases where the input impedance is purely resistive, and R_s could be matched to either one of two source resistances if so desired. The value of X_s at these points is designated as X_{s1} and X_{s2}.

On the other hand, a high value of R_s requires X_m to be large also. This could be accomplished by increasing the coefficient of coupling or by increasing the turns on the secondary coil. Increasing the turns on the primary also will cause X_m to be higher, but X_p will increase also. This is self-defeating, since X_{m2} is proportional to X_p.

An alternative approach is to use the parallel configuration of Fig. 19. The approximate equivalent series resistance of the parallel combination is then $X(C_s)^2/R_s$ and the reactance is approximately $X(C_s)$. (See diagram and text for Fig. 12.) This approach is often used in multiband antenna systems. On some frequencies the impedance at the input of the feed line is high, so the circuit of Fig. 18 is employed. This is referred to as parallel tuning. If the impedance is very low, the circuit of Fig. 14 is used and is called series tuning.

As an example, suppose a single-tuned circuit is to be used to match a 1-ohm load to a 50-ohm source, as shown in Fig. 20. It might be pointed out here that networks using mutual magnetic coupling can be scaled in the same manner that filter networks are scaled (as discussed in Chapter 2). For instance, the circuit of Fig. 20 could be scaled in order to match a 50-ohm load to a 2500-ohm source merely by multiplying all the reactances by a factor of 50.

The input resistance and reactance of the

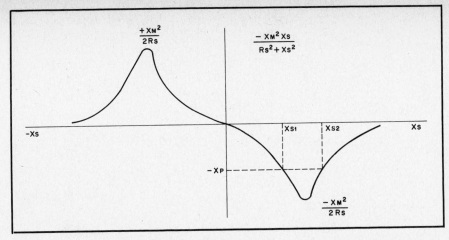

Fig. 18 — Reflected reactance into the primary of a single-tuned circuit places restraints on resistances that can be matched. This gives rise to a general rule that high-Q secondary circuits require a lower coefficient of coupling than low-Q ones.

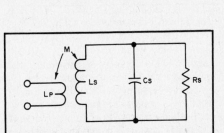

Fig. 19 — Single-tuned circuit with a parallel RC secondary.

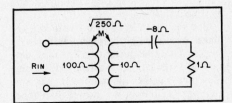

Fig. 20 — Example of a single-tuned circuit.

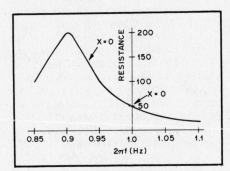

Fig. 21 — Input resistance of the circuit of Fig. 20 as a function of frequency.

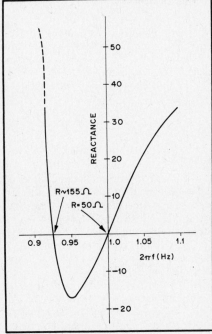

Fig. 22 — Input reactance of the network of Fig. 20. Note two resonant frequencies (where reactance is zero).

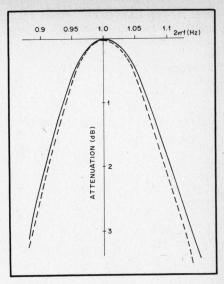

Fig. 23 — Response of the circuit shown in Fig. 20.

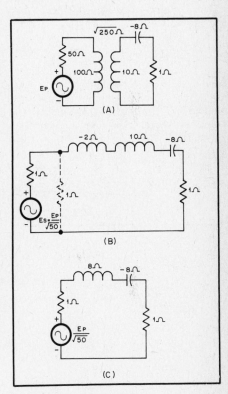

Fig. 24 — The transformation of Fig. 8 applied to the primary side of the circuit of Fig. 20.

circuit of Fig. 20 are plotted in Figs. 21 and 22, respectively. As pointed out earlier, there are two possible points where the reactance is zero, and this circuit could be used to match the 1-ohm load to either a 50-ohm or 155-ohm source. Assuming a 50-ohm source is being used, the attenuation plot as a function of frequency is given by the solid curve of Fig. 23.

With slight modification to include the effect of the source, the transformation of Fig. 8 can be applied to the primary side of the coupled circuit shown in Fig. 20. This is illustrated in Fig. 24. The complete circuit is shown at Fig. 24A and the network with the transformed primary resistance and reactance is shown in Fig. 24B.

In a lossless transformer, the maximum available power at the secondary must be the same as that of the original source on the primary side, neglecting the effects of reactance. That is, the power delivered to a 1-ohm resistance (shown as a dashed line in Fig. 24B) must be the same as that delivered to a 50-ohm load in Fig. 24A.

This assumes that the rest of the circuit has been disconnected in either case. In order to fulfill this requirement, the original source voltage must be multiplied by the square root of the ratio of the new and old source resistance.

The single-mesh transformed network is shown in Fig. 24C. It is interesting to compare the response of an RLC series circuit that actually possesses these element values at resonance with the circuit of Fig. 20. For comparison, the response of such a circuit is shown in Fig. 23 as a dashed curve; this curve differs only slightly from the coupled-circuit curve (solid line). The reason for the similarity is that even though the transformation of the primary resistance and reactance also changes with frequency, the effect is not that great in the present case.

Broadband RF Transformers

The sensitivity of the frequency characteristic of the transformation shown in Fig. 8 depends mostly on the ratio of X_s to R_s. However, if X_s is much greater than R_s, the transformed reactance can be approximated by

$$\frac{-X_m^2 X_s}{R_s^2 + X_s^2} \approx \frac{-X_m^2}{X_s}$$

and the resistance becomes

$$\frac{R_s X_m^2}{R_s^2 + X_s^2} \approx R_s \frac{X_m^2}{X_s^2}$$

Applying this approximation to the general coupled circuit shown in Fig. 25A results in the transformed network of Fig. 25B. The coefficient of coupling for the circuit of Fig. 25A is

$$k = \frac{X_m}{\sqrt{X_1 X_2}}$$

and the network shown in Fig. 25B in terms of the coefficient of coupling is illustrated in Fig. 25C. If k equals 1.0, the input reactance is zero and the input resistance is given by

$$R_{in} = \left(\frac{X_1}{X_2}\right)R_2 = \left(\frac{L_1}{L_2}\right)R_2 \approx \left(\frac{N_1}{N_2}\right)R_2$$

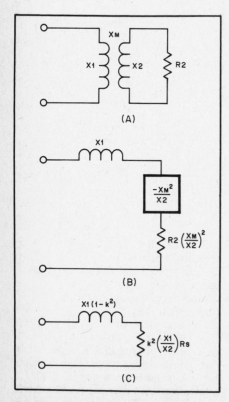

Fig. 25 — Equivalent-circuit approximation of two coupled coils.

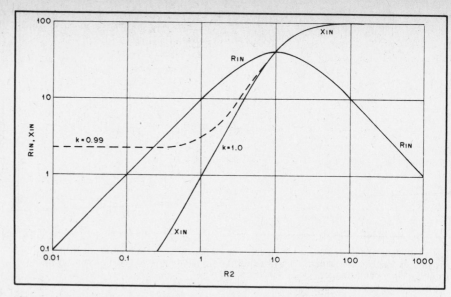

Fig. 26 — Input resistance and reactance as a function of output load resistance for X_1, and X_2 equal to 100 ohms and 10 ohms, respectively (Fig. 25).

where N_1 and N_2 are the number of turns on coil 1 and 2, respectively. From maximum-power transfer considerations such as those discussed for the circuit of Fig. 24, the voltage transfer ratio becomes

$$e_2 = \left(\frac{N_2}{N_1}\right)e_1$$

It will be recalled the above equations occurred in the discussion of the "ideal transformer" approximation in Chapter 2. It was assumed then that the leakage reactance and magnetizing current were negligible. The effects on circuit operation of these variables are shown in Fig. 26. The curves were computed for various load resistances (R_s) using the exact equations shown in Fig. 8.

X_1 and X_2 are assumed to be 100 and 10 ohms, respectively, with the solid curves for a k of 1.0 and the dashed reactance curve for k equal to 0.99 (the resistance curve for the latter values is the same as the one for k equal to 1.0). The ideal-transformer representation can be modified slightly to approximate the curve of Fig. 26 as shown in Fig. 27. The shunt reactance, X_{mag}, is called the magnetizing reactance and X_L is referred to as the leakage reactance.

Unfortunately, the two reactances are not independent of each other. That is, attempts to change one reactance so that its effect is suppressed causes difficulties in eliminating the effects of the other reactance. For instance, increasing X_l, X_m, and X_2 will increase X_{mag}, which is desirable. However, examination of Fig. 25C reveals that the coefficient of coupling, k, will have to be made closer to 1.0. Otherwise, the leakage reactance increases since it is proportional to X_1.

High-Permeability Cores

As a consequence of the interaction between the leakage reactance and the magnetizing reactance, transformers that approach ideal conditions are extremely difficult (if not impossible) to build using techniques common in air-wound or low-permeability construction. In order to build a network that will match one resistance level to another one over a wide range of frequencies, ideal transformer conditions have to be approached quite closely. Otherwise, considerable inductive reactance will exist along with the resistance component, as shown in Fig. 26.

One approach is to use a core with a higher permeability than air. Familiar examples would be power transformers and similar types common to the AF range. However, when an inductor configuration contains materials of more than one permeability, the analysis relating to Fig. 25C has to be modified somewhat. The manner in which the core affects the circuit is a bit complicated although even a qualitative idea of how such transformers work is very useful.

First, consider the coupled coils shown in Fig. 28. For a given current, I_1 a number of flux lines are generated that link both coil 1 and coil 2. Note that in coil 1, not all of the flux lines are enclosed by all the turns. The inductance of a coil is equal to the ratio of the sum of flux lines linking each turn and the generating current or

$$L_1 = \frac{\Lambda_{TOTAL}}{I_1}$$

where for the example shown in Fig. 28, Λ_{TOTAL} is given by

$$\Lambda_{TOTAL} = \Lambda_1 + \Lambda_2 + \Lambda_3 + \Lambda_4 + \Lambda_5$$

Counting up the number of flux linkages in coil 1 gives

$$\Lambda_{TOTAL} = 5 + 5 + 7 + 7 + 5 = 29$$

If all the flux lines linked all the turns,

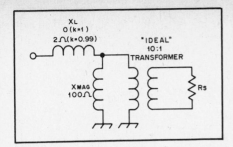

Fig. 27 — Approximate network for the curves of Fig. 26.

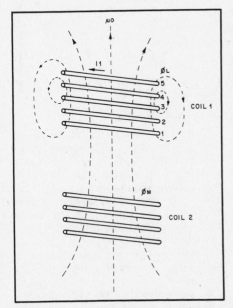

Fig. 28 — Coupled coils showing magnetic flux lines.

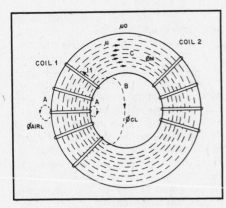

Fig. 29 — Toroidal transformer.

Λ_{TOTAL} would be 35 so L_1 is 29 / 35 or 83 percent of its maximum possible value. Likewise, if all the flux (7 lines) generated in coil one linked all the turns of coil 2, the maximum number of flux linkages would be the number of turns on coil 2 times 7 or 28. Since only three lines link coil 2, the mutual inductance is $3 \times 4/28$ or 43 percent of maximum.

Assuming both coils are perfect, if a current I_1 produced 7 flux lines in a five-turn

coil, then the same current in a four-turn coil would produce (4/5)(7) flux lines, since the flux is proportional to the magnetizing current times the number of turns. Consequently, the maximum flux linkages in coil 1 from a current of the same value as I_1 but in coil 2 instead would be (4/5)(7)(5) or 28. Therefore, it can be seen that the mutual inductance is independent of the choice of coil used for the primary or secondary. That is, a voltage produced in one coil by a current in the other one would be the same if the coils were merely interchanged. (This result has been used implicitly on a number of previous occasions without proof.) In addition, the maximum flux linkages in coil 2 produced by a current, I_1, would be (4/5)(7)(4). As an exercise, substitute the maximum inductance values into the formula for the coefficient of coupling and show that k is 1.0.

The next step is to consider the effect of winding coils on a form with a magnetic permeability much higher than that of air. An example is illustrated in Fig. 29, and the configuration shown is called a toroidal transformer. Since the flux is proportional to the product of the permeability and the magnetizing current, the flux in the core shown in Fig. 29 will be much greater than the coil configuration of Fig. 28. However, not all of the flux is confined to the core. As can be seen in Fig. 29, some of the flux lines never penetrate the core (see lines marked A in Fig. 29) while others enclose all the windings of coil 1 but not coil 2 (see line marked B). The significance of these effects is as follows. The total flux linkage produced by the current, I_1, is

$$\Lambda_{\text{TOTAL}} = \Lambda_{\text{air}} + \Lambda_{\text{core}}$$

and dividing both sides of the equation by I_1 gives

$$L_T = L_{\text{air}} + L_{\text{core}}$$

Consequently, the circuit of Fig. 25 can be represented as shown in Fig. 30A. For values of X_2 much greater than the load resistance, the approximate network of Fig. 30B can replace the one of Fig. 30A.

At first sight, it might seem as though little advantage has been gained by introducing the core since the formulas are much the same as those of Fig. 25C. However, the reactances associated with the core can be made very high by using a material with a high permeability. Also, even though there may be some leakage from the core as indicated by line B in Fig. 29, it is ordinarily low and the coefficient of coupling in the core can be considered 1.0 for all practical purposes. This is especially true at AF and power frequencies with transformers using iron cores where the permeability is extremely high. This means the magnetizing reactance can be made very high without increasing the leakage reactance accordingly as is the case with the circuit in Fig. 25C. Therefore, ideal transformer conditions are considered to exist in the core, and the final circuit can

be approximated by the one shown in Fig. 30C.

Bifilar and Twisted-Pair Windings

Although the core helps alleviate some of the problems with leakage and magnetizing reactance, the residual parasitic elements must still be made as low as possible. This is especially important in matching applications, as the following example illustrates. A transformer has a primary and secondary leakage reactance of 1 ohm and 0.1 ohm, respectively, with a coefficient of coupling of 1.0 in the core. X_1 and X_2 are 1000 ohms and 100 ohms.

A plot similar to the one of Fig. 26 is shown in Fig. 31, along with a curve for voltage standing wave ratio (VSWR). These results are based on the exact equations and it can be seen that the approximate relations shown in Fig. 30C are valid up to 1 ohm or so. Curve A (Fig. 31) includes only the effect of the secondary reactance and illustrates the manner in which the reactance is transformed. Curve B shows the total input reactance, which merely requires the addition of 1 ohm. The VSWR curve includes the effect of the input reactance. The useful range of the transformer is between 1 and 10 ohms, with rapid deterioration in VSWR outside of these values. (The VSWR curve is for a characteristic impedance equal to 10 times the secondary resistance. For instance, the transformer would be useful in matching a 5-ohm load to a 50-ohm line.)

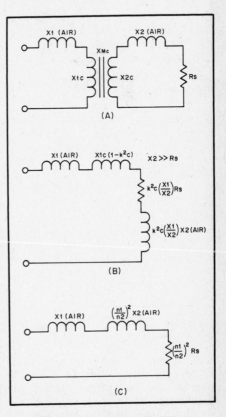

Fig. 30 — Effect of a high-permeability core on transformer equivalent circuit.

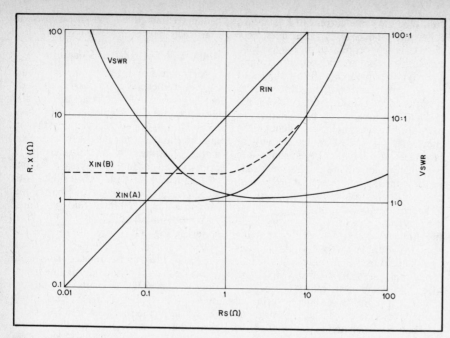

Fig 31 — Curve for transformer problem discussed in the text.

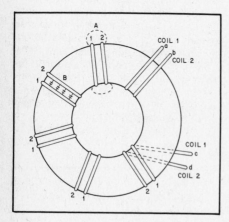

Fig. 32 — Bifilar-wound transformer on toroidal core.

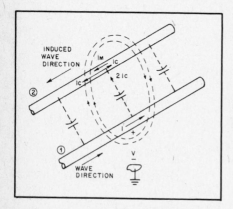

Fig. 33 — Effect of distributed capacitance on transformer action.

As mentioned previously, these difficulties are less pronounced at audio frequencies since the permeabilities normally encountered in iron-core transformers are so high. The actual inductance of the winding itself is therefore small in comparison to the component represented by the core. That is, a small number of turns of wire wound on a core may actually be the equivalent of a very large coil. However, materials suitable for RF applications have much lower permeabilities, and a narrower range of matching values is likely to be the result (such as in the example of Fig. 31). Therefore, other means are required in keeping the parasitic elements as low as possible. Either that, or less conventional transformer designs are used.

One approach is shown in Fig. 32. Instead of separating the windings on the core as shown in Fig. 29, they are wound in parallel fashion. This is called a bifilar winding, although a more common approach to achieve the same purpose is to twist the wires together. Either way, there are a number of advantages to be gained (and some disadvantages). With reference to Fig. 28, the fact that not all the flux lines linked all of the turns of a particular coil meant the self inductance was lower than if all the turns were linked. Since the separation between turns of a particular coil is quite large in the configuration of Fig 32, the flux linkage between turns is quite low. This means the corresponding leakage inductance is reduced accordingly. However, the coupling between both coils is increased because of the bifilar winding (flux line A) in Fig. 32 which also tends to reduce the leakage inductance of either coil.

On the other hand the capacitance between windings is increased considerably as indicated by B in Fig. 32. As a result, the coupling between windings is both electrical and magnetic in nature. Generally speaking, analysis of the problem is complicated. However, a phenomenon usually associated with such coupling is that it

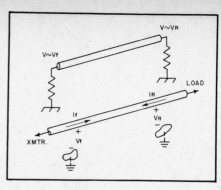

Fig. 34 — Basic configuration for a directional-coupler type of VSWR detector.

tends to be directional. That is, energy transferred from one winding to the other propagates in a preferred direction rather than splitting equally.

Directional Coupling

Two conductors are oriented side by side over a conducting plane as shown in Fig. 33. The current I in conductor 1 will induce a current I_m in conductor 2 because of magnetic coupling. The actual value of the current will depend on the external circuitry attached to the conductors but it will be assumed that the two of them extend to infinity in both directions.

Since capacitive coupling exists also, a second set of current components denoted by I_c will flow. The result is that a wave traveling toward the right in conductor 1 will produce a wave traveling toward the left in conductor 2. Such coupling is called contradirectional coupling, since the induced wave travels in the opposite direction to the generating wave.

This is the principle behind many practical devices and ones that are quite common in amateur applications. In adjusting a load such as an antenna, it is desirable to ensure that energy is not reflected back to the transmitter. Otherwise, the impedance presented to the transmitter output may not be within range of permissible values. A directional coupler is useful in determining how much power is reflected, as indicated in Fig. 34. Energy originating from the transmitter and flowing to the right causes a voltage to be produced across the resistor at the left. On the other hand, a wave traveling from the right to the left produces a voltage across the right-hand resistor. If both of these voltages are sampled, some idea of the amount of power reflected can be determined. (The subject of reflected power is taken up in more detail in the chapter on transmission lines.)

It some situations, the coupling described can be very undesirable. For instance, the lines shown in Fig. 34 might be conductors on a circuit board in a piece of equipment. As a result, the coupling between lines can cause feedback and, because of its directional nature, can be very difficult to sup-

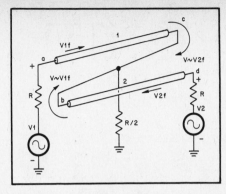

Fig. 35 — Directional-coupler hybrid combiner.

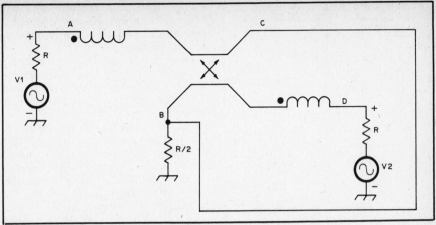

Fig. 37 — Equivalent circuit of transmission-line transformer in the presence of the core. Dots indicate winding sense of coils. A positive current into a dotted end of one coil will produce a voltage in the other coil because of mutual coupling. The polarity of this voltage will be such that dotted end of the "secondary" coil will be positive. The crossed-arrows symbol in the middle of the parallel lines is the standard symbol for a directional coupler.

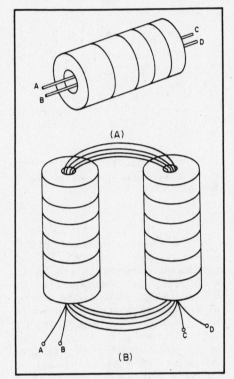

Fig. 36 — Transmission-line transformer with ferrite cores.

press with conventional methods. Therefore, it is good design practice to use double-sided board (board with conductive foil on both sides) so a ground plane of metal is in close proximity to the conductors. This tends to confine the fields to the region in the immediate vicinity of the wires.

Transmission-Line Transformers

In effect, sections of transmission line in close proximity act as transformers with the unique feature that the coupling is directional. For instance, if only magnetic coupling were present in the configuration of Fig. 34, power would be divided equally between the resistors at either end of the secondary section of transmission line. As another example of directional effects, the network shown in Fig. 35 can be used to couple two sources to a common load without cross-coupling of power from one

source to the other one. (This assumes the sources have the same frequency and phase. Otherwise, a resistance of value 2R must be connected from point a to d.) Such a configuration is called a hybrid combiner and is often used to combine the outputs of two solid-state amplifiers in order to increase the power-handling capability. This permits the use of less expensive low-power devices rather than very expensive high-power ones. Even though more devices are required, it is still simpler since the difficulties in producing a high-power transistor increase in a greater proportion as the power level is raised.

The manner in which the circuit shown in Fig. 35 operates is as follows. A wave from the generator on the left end of line 1 travels toward the right and induces a wave in line 2 that travels toward the left and on into the load. No wave is induced in line 2 that travels toward the right except for a small fraction of power.

A similar situation exists with the second generator connected at the right end of line 2. A wave is induced in line 1 that travels toward the right. Since the load is also connected to the right end of line 1, power in the induced wave will be dissipated here with little energy reaching the generator at the left end of line 1. In order to simulate a single load (since there are two generators involved), the value of the load resistance must be half of the generator resistance. Assuming that two separate resistors of value R were connected to the ends of the line, it would be possible to connect them together without affecting circuit operation. This is because the voltage across both resistors is of the same phase and amplitude. Consequently, no additional current would flow if the two resistors were paralleled or combined into a single resistor of R/2.

Extending the Low-Frequency Range

As might be expected, the coupling

mechanism illustrated in Figs. 32 through 34 is highly dependent on dimensions such as conductor spacing and line length. For instance, maximum coupling of power from the primary wave to the induced wave occurs when the secondary line is a quarter-wavelength long[1] or some odd multiple of a quarter-wavelength. This would normally make such couplers impractical for frequencies in the HF range. However, by running the leads through a ferrite core as shown in Fig. 36, lower frequency operation is possible. Although the transformer of Fig. 36A is seldom used, it illustrates the manner in which the conductors are employed electrically in the more complicated configurations of Fig. 32 and 36B. Also, the relationship between the parallel-line coupler in Fig. 35 and the loaded version of Fig. 36A is easier to visualize.

As covered in an earlier problem discussion (Fig. 29), a set of coupled coils wound on a high-permeability core can be broken down into combinations of two series inductances. One inductance represents the path in air while the other includes the effects of the flux in the core. As before, it is assumed that the coefficient of coupling in the core is 1.0.

If the hybrid combiner of Fig. 35 is wound on a core (such as those of Fig. 32 or Fig. 36), the low-frequency range of the entire system is increased considerably. The equivalent circuit showing the effect of the core on the air-wound coupler is illustrated in Fig. 37. (The symbol in the middle of the parallel lines is the standard for a directional coupler.) At the higher frequencies, most core materials decrease in permeability. The operation approaches that of the original air-wound coupler, and

[1]Oliver, "Directional Electromagnetic Couplers," *Proceedings of the I.R.E.*, Vol. 42, pp. 1686-1692, November 1954.

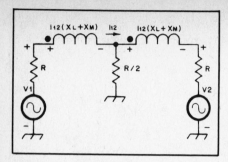

Fig. 38 — Low-frequency equivalent circuit of hybrid combiner showing isolation of sources.

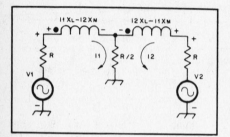

Fig. 39 — Desired coupling mode of hybrid combiner.

the inductance produced by the core can be neglected. At the low end of the frequency range, the line lengths are usually too short to provide much coupling or isolation. Therefore, the circuit can be represented by the set of coupled coils shown in Figs. 38 and 39.

For a current I_{12} flowing from a source 1 over to the mesh that includes source 2, the mutual-reactance components add to the self-inductance of each coil. Consequently, a large reactance appears in series between the two sources, which effectively isolates them. On the other hand, currents from both sources that flow through the load resistor, R/2, produce fluxes that cancel, and the voltages produced by the self- and mutual-reactance terms subtract. If both sources have the same amplitude and phase, currents I_1 and I_2 must be identical because of the symmetry involved. However, if the coefficient of coupling is 1.0, the self- and mutual-reactance must be equal. Therefore the voltage across either coil is zero, since the terms subtract, and a low-impedance path exists between both sources and the load.

Other Transformer Types

The hybrid combiner is only one application of a combination transmission-line or directional-coupler transformer and conventional coupled-coil arrangement. With other variations, the low-frequency isolation is accomplished in the same manner. Mutual-reactance terms add to the self-reactance to provide isolation for some purpose with cancellation of reactive components in the path for the desired coupling. Very good bandwidth is possible

with a range from BC frequencies to UHF in the more esoteric designs. Models that cover all the amateur HF bands can be constructed easily.[2]

Unfortunately, there is also a tendency to expect too much from such devices. Misapplication or poor design often results in inferior performance. For instance, as indicated in an earlier example (Fig. 31), actual impedance levels are important, along with the desired transforming ratio. Using a transformer for an impedance level that was not intended results in undesirable reactive components and an improper transforming ratio. However, when applied properly, the transformers discussed in the previous sections can provide bandwidth characteristics that are obtainable in no other way.

Another transformer type is shown in Fig. 40A. The windings of the coils are such that the voltages across the inductors caused by the desired current are zero. This is because the induced voltages produced by the current in the mutual-reactance terms just cancel the voltage drop caused by the current flowing in the self-reactances of either coil, (assuming the coefficient of coupling is 1.0) However, an impedance connected to ground at point c would be in series with the self-reactance (X_L) of the coil connected between points a and c. But there would be no induced voltage to counter the voltage drop across this coil. Therefore, if X_L is large, very little current would flow in the impedance Z and it would be isolated effectively from the source.

In fact, terminal c could be grounded as shown in Fig. 40B. The voltage drop across the coil from a to c would then be equal to V_1. However, the induced voltage in the coil connected between points b and d would also be V_1, assuming unity coupling (k equal to 1.0). Although the voltage drop produced by the inductors around the mesh through which I_1 flows is still zero, point d is now at potential $-V1$ and a phase reversal has taken place. For this reason, the configuration shown in Fig. 40B is called a phase-reversal transformer.

Baluns

The circuit shown in Fig. 40A is useful in isolating a load from a grounded source. This is required in many applications, and the device that accomplishes this goal is called a balun (balanced to unbalanced) transformer. Baluns may also be used to transform impedances and to provide isolation. A 1:1 balun such as the one shown in Fig. 40A means the impedance at the input terminals ab will be the same as the load connected across terminals cd. Other transforming ratios, such as 4:1, are possible with the appropriate circuit connections.

[2]Ruthoff, "Some Broad-Band Transformers," *Proceedings of the I.R.E.*, Vol. 97, pp. 1337-1342, August 1959.

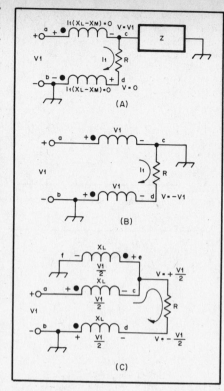

Fig. 40 — Applications of transmission-line transformers. See text.

One disadvantage of the network of Fig. 40A is that although the load is isolated from the source, the voltages at the output are not balanced. This is important in some applications such as diode-ring mixers, where a push-pull input is required, and so the circuit of Fig. 40C is used. A third coil connected between points e and f is wound on the same core as the original transformer (Fig. 40A). This coil is connected so a voltage across it produces a flux that adds to that produced by the coil between a and c. Assuming both coils are identical, the voltage drop across either one must be the same, or half the applied voltage. However, since the coil between b and d is also coupled to this combination (and is an identical coil), the induced voltage must also be $-V_1/2$. Consequently, the end of the load connected to points c and e is at a potential of $+V_1/2$ with respect to ground, while point d is $-V_1/2$ with respect to ground when the input voltage has the polarity shown. Therefore, this circuit not only isolates the load from the source but provides a balanced voltage.

Either the circuit of Fig. 40A or Fig. 40C can be used if only isolation is desired. However, the network shown in Fig. 40C is more difficult to design and construct since the reactance of the coils between points a and f must be very high throughout the frequency range of the transformer. With both transformers, the coefficient of coupling must also be very close to 1.0 in order to prevent undesirable reactance in series with the load. This problem can be offset somewhat by reducing X_L slightly (by fewer turns) but this is counter to the requirement of a large X_L in the circuit of Fig. 40C. Isolation is reduced in both cases, although no detrimental effect on input impedance results in the transformer of Fig. 40A by reducing X_L.

Twisted Pairs — Impedance and Attenuation

Twisted pairs of wire are often used in the construction of broadband RF transformers. The question often arises as to what size conductors and what number of twists per inch should be used. To help answer these questions the information contained in Tables 1 and 2 was developed. Table 1 illustrates the approximate impedance for various size conductors with different numbers of twists per inch. These values are based on measurements in the ARRL laboratory, and should be accurate to within an ohm or two. Enameled copper wire was used for each pair. The information shown in Table 2 is the attenuation per foot for the same twisted pairs of wire. Information is not included for twists per inch greater than 7½ for the no. 20 wire

Table 1
Impedance of Various Two-Conductor Lines

Wire Size	Twists per Inch				
	2½	5	7½	10	12½
no. 20	43	39	35		
no. 22	46	41	39	37	32
no. 24	60	45	44	43	41
no. 26	65	57	54	48	47
no. 28	74	53	51	49	47
no. 30			49	46	47

Measured in ohms at 14.0 MHz

This chart illustrates the impedance of various two-conductor lines as a function of the wire size and number of twists per inch.

Table 2
Attenuation per Foot for Lines in Table 1

Wire Size	Twists per Inch				
	2½	5	7½	10	12½
no. 20	0.11	0.11	0.12		
no. 22	0.11	0.12	0.12	0.12	0.12
no. 24	0.11	0.12	0.12	0.13	0.13
no. 26	0.11	0.13	0.13	0.13	0.13
no. 28	0.11	0.13	0.13	0.16	0.16
no. 30			0.25	0.27	0.27

Measured in decibels at 14.0 MHz

Attenuation in dB per foot for the same lines as shown in Table 1.

since this results in an unusable tight pair. Likewise, the information for twists per inch less than 7½ for no. 30 wire is omitted since these pairs are extremely loose.

As a general rule the wire size can be selected based on the core size to be used and the required number of turns. The number of twists per inch can be selected according to the impedance level of transmission line that is needed. For applications where moderate levels of power are to be handled (such as in the low- and medium-level stages of a solid-state transmitter), smaller wire sizes should be avoided. For receiver applications, very small wire can be used. It is not uncommon to find transformers wound with pairs of no. 32 wire and smaller.

Nonlinear and Active Networks

Almost all the theory in previous sections has dealt with so-called passive components. Passive networks and components can be represented solely by combinations of resistors, capacitors and inductors. As a consequence, the power output at one set of terminals in a passive network cannot exceed the total power input from sources connected to other terminals in the circuit. This assumes all the sources are at one frequency. Similar considerations hold true for any network, but, it is possible for energy to be converted from one frequency (including dc) to other ones. While the total power input must still be equal to the total power output, it is convenient to consider certain elements as controllable sources of power. Such devices are called amplifiers and are part of a more general class of circuits called active networks. An active network generally possesses characteristics that are different from those of simple RLC circuits, although the goal in many instances is to attempt to represent them in terms of passive elements and generators.

Nonlinearity

Two other important attributes of passive RLC elements are that they are linear and bilateral. A two-terminal element such as a resistor is said to be bilateral since it doesn't matter which way it is connected in a circuit. Semiconductor and vacuum-tube devices such as triodes, diodes, transistors and integrated circuits (ICs) are all examples where the concept of a bilateral element no longer holds true. The manner in which the device is connected in a circuit and the polarity of the voltages involved are very important.

An implication of the failure to satisfy the bilateral requirements is that such devices are nonlinear in the strictest sense. Linearity means that the amplitude of a voltage or current is related to other voltages and currents in a circuit by a single proportionality constant. For instance, if all the voltages and currents in a circuit were doubled, a single remaining voltage or current would be doubled also. That is, it couldn't change by a factor of one half, or three, no matter how complex the network might be. Likewise, if all the polarities of the currents and voltages in a circuit are reversed, the polarity of a remaining voltage or current must be reversed also. Finally, if all the generators or sources in a linear network are sine waves at a single frequency, any voltage or current produced by these sources must also be a sine wave at the same frequency.

Consequently, if a device is sensitive to the polarity of the voltage applied to its terminals, it doesn't meet the requirements of either a bilateral element or a linear one. Because of the extreme simplicity of the

mathematics of linear circuits as compared to the general nonlinear case, however, there is tremendous motivation in being able to represent a nonlinear circuit by a linear approximation. Many devices exhibit linear properties over part of their operating range or may satisfy some but not all of the requirement of linear circuits. Such devices in these categories are sometimes termed piece-wise linear (or; simply, as linear). For instance, a linear mixer doesn't satisfy the rule that a voltage or current must be at the same frequency as the generating source(s). Since the desired output voltage (or current) varies in direct proportion to the input voltage (or current), however, the term linear is applied to distinguish the mixer from types without this quasi-linear property.

Harmonic Frequency Generation

In a circuit with nothing but linear components, the only frequencies present are those generated by the sources themselves. This is not true with nonlinear elements. One of the properties of nonlinear networks mentioned earlier is that energy at one frequency (including dc) may become converted to other frequencies. In effect, this is how devices such as transistors and vacuum tubes are able to amplify radio signals. Energy from the dc power supply is converted to energy at the desired signal

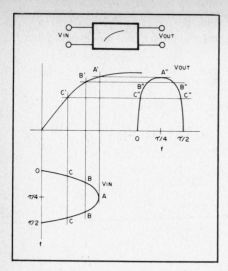

Fig. 41 — Nonlinear transfer charcteristic (see text discussion).

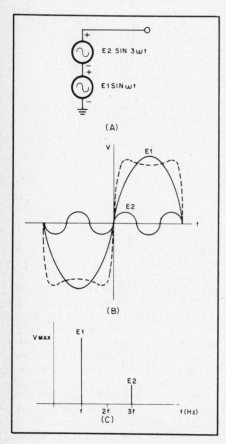

Fig. 42 — Harmonic analysis and spectrum.

frequency. Therefore, a greater amount of signal power is available at the output of the network of an active device than at the input.

On the other hand, such frequency generation may be undesirable. For instance, the output of a transmitter may have energy at frequencies that could cause interference to nearby receiving equipment. Filters and similar devices must be used to suppress this energy as much as possible.

The manner in which this energy is produced is shown in Fig. 41. A sine wave at the input of a nonlinear network (V_{in}) is "transformed" into the output voltage waveform (V_{out}) illustrated. If the actual device characteristic is known, the waveform could be constructed graphically. It could also be tabulated, if the output voltage as a function of input voltage was available in either tabular or equation form. (Only one-half of the period of a sine wave is shown in Fig. 41 for clarity.)

Although the new waveform retains many of the characteristics of the original sine wave, some transformations have taken place. It has zero value when t is either 0 or T/2, and attains a maximum at T/4. However, the fact that the curve is flattened somewhat means energy at the original sine-wave frequency has been converted to other frequencies. It will be recalled that the sum of a number of sine waves at one frequency result in another sine wave at the same frequency. Therefore, the waveform of Fig. 41 must have more than one frequency component present since it is no longer a sine wave.

One possible model for the new waveform is shown in Fig. 42A. Instead of one sine wave at a single frequency, there are two generators at three times the "fundamental" frequency ω, where ω is $2\pi f(Hz)$. If the two sine waves are plotted point by point, the dashed curve of Fig. 42B results. While this curve doesn't resemble the one of Fig. 41 very closely, the general symmetry is the same. It would take an infinite number of generators to represent the desired curve exactly, but it is evident that all the frequencies must be odd multiples of the fundamental. Even multiples would produce a lopsided curve that might be useful for representing other types of waveforms.

In either case, the multiples have a specific name and are called harmonics. There is no first harmonic. By definition, the second, third and fourth multiples are designated as the second, third and fourth harmonics. Thus the dashed curve of Fig. 42 is the sum of the fundamental and third harmonic.

Analyzing waveforms such as those of Fig. 41 is a very important procedure. A plot of harmonic amplitude such as that shown in Fig. 42C is called the spectrum of the waveform and can be displayed on an instrument called a spectrum analyzer. If the mathematical equation or other data for the curve is known, the harmonics can also be determined by means of a mathematical process called Fourier analysis.

Linear Approximations of Nonlinear Devices

Nonlinear circuits may have to be analyzed graphically as in the previous example. There are many other instances where only a graphical method may be practical, such as in power-amplifier prob-

lems. However, a wide variety of applications permit a different approach. A model is derived from the nonlinear characteristics using linear elements to approximate the more difficult nonlinear problem. This model is then used in more complicated networks instead of the nonlinear characteristics, which simplifies analysis considerably.

The following example illustrates how this is accomplished. Although a vacuum-tube application is considered, a similar process is employed in solving semiconductor problems as well. However, there are some additional factors involved in semiconductor design that do not apply to vacuum tubes. In fact, much of the analysis required with vacuum tubes is unnecessary with modern solid-state components since many of the problems have already been "solved" before the device leaves the counter at the radio store. That is, amplifiers such as those in integrated circuits have the peripheral elements built in, and there is no need to determine the gain or other parameters such as the values of bias resistors.

The Triode Amplifier

A simple network using a triode vacuum tube is shown in Fig. 43A, and a typical set of characteristic curves is illustrated in Fig. 44A. The first chore in finding a suitable linear approximation for the triode is to determine an optimum operating point. Generally speaking, a point in the center of the set of curves is desirable and is indicated by point Q in Fig. 44A. (Other areas are aften picked for power-amplifier operation, but the goal here is to find a point where the maximum voltage swing is possible without entering regions where the nonlinearities affect the linear approximation.)

In the particular operating point chosen, the cathode to grid voltage is -3, the cathode to plate voltage is 280, and the plate current is 10 mA. It is assumed that the input signal source in Fig. 43A is a short-circuit at dc, and a 3-V battery connected as shown results in a dc voltage of -3 being applied to the grid at all times. Such a battery is called a bias battery or bias supply.

The next step is to determine how the plate voltage varies with grid voltage (e_g) for a constant plate current. Assuming that the characteristic curves were completely linear, this would permit evaluation of an equivalent ac voltage generator as shown in Fig. 43B. For a constant plate current of 10 mA, the plate voltage changes from 325 (point b) to 230 (point a) when the grid voltage is changed from -4 to -2 (Fig. 44A).

These numbers can be used to compute the amplification factor (μ) of the triode which is

$$\mu = \frac{325 - 230}{(-4) - (-2)} = -47.5$$

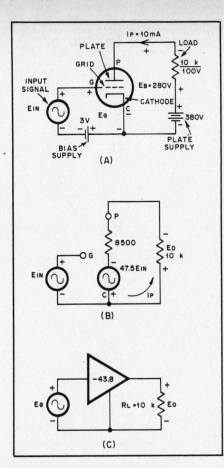

Fig. 43 — Basic triode amplifier and equivalent circuit.

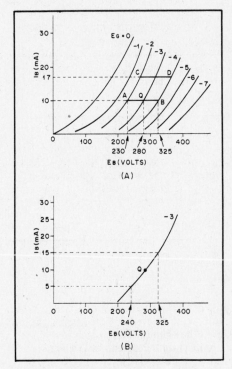

Fig. 44 — Triode characteristics and derivations of small-signal parameters.

Quite often a set of characteristics will not be published for a triode, and only the amplification factor will be given along with a typical operating point. However, note that the amplification factor is negative. This means that for an increase in the signal voltage (e_{in}), the controlled generator decreases in voltage. Consequently, there is a 180° phase shift between the input voltage and the controlled source. (Note the polarity of the generator shown in Fig. 43B.)

In order to complete the equivalent generator circuit, the source impedance must be computed. This is accomplished by determining how the plate voltage varies with plate current at constant grid voltage as shown in Fig. 44B. The plate resistance is then

$$r_p = \frac{325 - 240}{(15 - 5) \times 10^{-3}} = 8500 \text{ ohms}$$

which must be considered to be in series with the controlled source of Fig. 43B.

It should be pointed out that the reasoning why this procedure is valid has not been presented. That is, why was the amplification factor defined as the ratio of a change in plate voltage to change in grid voltage at constant current? Unfortunately, the calculations involved, although not difficult, are somewhat sophisticated. Some knowledge of the subject of partial differential equations is required for the theoretical derivation of these parameters. However, an intuitive idea can be obtained from the following.

If the characteristics were completely linear, instead of being nonlinear as shown, the equivalent generator would not be affected by changes in plate current, but only by changes in grid voltage. For instance, if the plate current was increased from 10 to 17 mA (Fig. 44A), the amplification factor would be the equivalent of the change in voltage represented by the line cd divided by −2. However, since the length of cd is almost the same as that of ab (the difference in plate voltage for a −2-V change at 10 mA), it can be concluded μ doesn't change very much — at least not in the center region of the characteristics.

Similar considerations hold for the plate resistance, r_p. It wouldn't matter if the curve for −4 or −2 V was picked (Fig. 44B), since the change in plate voltage vs. plate current would be approximately the same. Entities such as μ and r_p are often called incremental or small-signal parameters. This means they are valid for small ac voltages or currents around some operating point, but less so for large variations in signal or for regions removed from the specified operating point. Also, such parameters are not closely related to dc voltage characteristics. For instance, a static plate resistance could be defined as the ratio of plate voltage to plate current. For the −3-V operating point chosen, the static plate resistance would be 280 divided by 10×10^{-2} or 28 kΩ. This is considerably different from the small-signal

plate resistance determined previously, which was 8500 ohms.

Amplifier Gain

The ratio of the variation in voltage across the load resistance to change in input voltage is defined as the gain of the amplifier. For the equivalent circuit shown in Fig. 43B, this ratio would be

$$A = \frac{e_o}{e_{in}}$$

To solve for the gain, the first step is to determine the incremental plate current. This is just the source voltage divided by the total resistance of the circuit mesh or

$$i_p = \frac{47.5 e_{in}}{10 + 8.5} \text{mA}$$

The output voltage is then

$$e_o = i_p 10.$$

Combining these two equations gives

$$A = \frac{e_o}{e_{in}} = \frac{(47.5)(10)}{10 + 8.5} = 25.67$$

It is inconvenient to have the input and output voltages defined with opposite polarities as shown in Fig. 43B. Therefore, the gain becomes negative as illustrated in the triangle in Fig. 43C. A triangle is the standard way of representing an amplifier stage in block diagram form. The amplifier gain depends on the load resistance, Rl. A general formula for the gain of the circuit of Fig. 43B is

$$A = \frac{-\mu R_L}{r_p + R_L}$$

Feedback

Being able to eliminate the equivalent circuit and use only one parameter such as the gain permits analysis of more complicated networks. A very important application occurs when part of the output energy of an amplifier is returned to the input circuit and gets amplified again. Since energy is being fed back into the input, the general phenomenon is called feedback. The manner in which feedback problems are analyzed is illustrated in Fig. 45. The output voltage is sampled by a network in the box marked beta and is multiplied by this term. This transformed voltage then appears in series with the input voltage, e_{in} which is applied to the input terminals of the amplifier (triangle with A_o). A_o is defined as the open-loop gain. It is the ratio of the voltage that appears between terminals 3 and 4 when a voltage is applied to terminals 1 and 2. The circuit of Fig. 45 is an example of voltage feedback; a similar analysis holds for networks incorporating current feedback.

The closed-loop gain, A_c, can then be found by inspection of Fig. 45. From the

diagram, the output voltage must be

$$e_o = A_o(e_{in} + \beta e_o)$$

Rearranging terms gives

$$e_o(1 - \beta A_o) = A_o e_{in}$$

The closed-loop gain is defined by

$$A_c = \frac{A_o}{1 - \beta A_o} = \frac{e_o}{e_{in}}$$

Cathode Bias

As an application of the feedback concept, consider the amplifier circuit shown in Fig. 46. It will be recalled that a bias battery was required in the previous example. A method of eliminating this extra source is to insert a small-value resistor in series with the cathode lead to ground (Fig. 46A). In terms of the amplifier block diagram, the circuit of Fig. 46B results. The next task is to evaluate the open-loop gain and the value of β.

With the exception of the cathode resistor, the circuit of Fig. 46 is the same as that of Fig. 43. Consequently, the ac plate current must be

$$i_p = \frac{-\mu e_{12}}{r_p + R_L + R_c}$$

The open-loop gain can then be determined from

$$\frac{e_o}{e_{12}} = A_o = \frac{-\mu R_L}{r_p + R_L + R_c}$$

Next, β is determined from the expression for output voltage

$$e_o = i_p R_L$$

and the feedback voltage, which is

$$e_f = i_p R_c$$

Then β is

$$\beta = \frac{e_f}{e_o} = \frac{i_p R_c}{i_p R_L} = \frac{R_c}{R_L}$$

Note that β is positive since if the path 1 to 2 is considered, the feedback voltage is added to the input signal. Substituting the values of β and A_o into the feedback equation gives

$$A_c = \frac{A_o}{1 - \frac{R_c}{R_L} A_o}$$

which after some manipulation becomes

$$A_c = \frac{-\mu R_L}{r_p + R_L + (1 + \mu)R_c}$$

Comparison of this equation with the one for the previous circuit with no cathode resistor reveals that the gain has decreased

because of the term $(1 + \mu)R_c$ in the denominator. Such an effect is called negative or degenerative feedback.

On the other hand, if the feedback was such that the gain increased, regenerative or positive feedback would result. Positive feedback can be either beneficial or detrimental in nature, and the study of feedback is an important one in electronics. For instance, frequency generation is possible in a circuit called an oscillator. But on the other hand, unwanted oscillation or instability in an amplifier is very undesirable.

Oscillators

A special case of feedback occurs if the term $1 - \beta A_o$ becomes zero. This would mean the closed-loop gain would become infinite. An implication of this effect is that a very small input signal would be amplified and fed back and amplified again until the output voltage became infinite. Either that, or amplifier output would exist with no signal input. Random noise would trigger the input into producing output.

Of course, an infinite output voltage is a physical impossibility and circuit limitations such as the nonlinearities of the active device would alter the feedback equation. For instance, at high output voltage swings, the amplifier would either saturate (be unable to supply more current) or limit (be cut off because the grid was too negative), and A_o would decrease.

It should be stressed that it is the product of βA_o that must be 1.0 for oscillations to occur. In the general case, both β and A_o may be complex numbers unlike those of the cathode-bias problem just discussed. That is, there is a phase shift associated with A_o and β with the phase shift of the product being equal to the sum of the individual phase shifts associated with each entity.

Therefore, if the total phase shift is 180° and if the amplitude of the product is 1.0, oscillations will occur. At low frequencies, these conditions normally are the result of the effect of reactive components. A typical example is shown in Fig. 47; the configuration is called a tuned-plate tuned-grid oscillator. If the input circuit consisting of L1 and C1 is tuned to a frequency f_o, with the output circuit (L2, C2) tuned to the same frequency, a high impedance to ground will exist at the input and output of the amplifier. Consequently, a small capacitance value represented by C_f is capable of supplying sufficient voltage feedback from the plate to the grid.

At other frequencies, or if either circuit is detuned, oscillations may not occur. For instance, off-resonant conditions in the output tank will reduce the output voltage and in effect, reduce the open-loop gain to the point where oscillations will cease. On the other hand, if the input circuit is detuned far from f_o, it will present a low impedance in series with the relatively high reactance of C_f. The voltage divider thus formed will result in a small value for β and

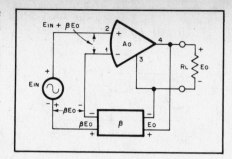

Fig. 45 — Network illustrating voltage feedback.

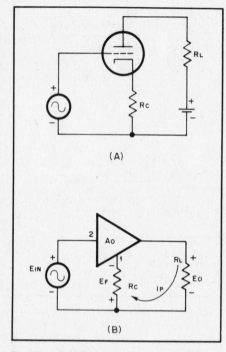

Fig. 46 — Feedback example of an amplifier with cathode bias.

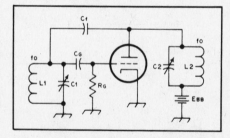

Fig. 47 — Tuned-plate tuned-grid oscillator.

the conditions for oscillations will not be fulfilled. However, for conditions near f_o, both the amplitude and phase of the βA_o product will be correct for oscillations to occur.

Under some conditions, the voltage across the tank circuit may be sufficient to cause the grid to be driven positive with respect to the cathode, and grid current will flow through C_g. During the rest of the RF cycle, C_g will discharge through R_g, causing a negative bias voltage to be ap-

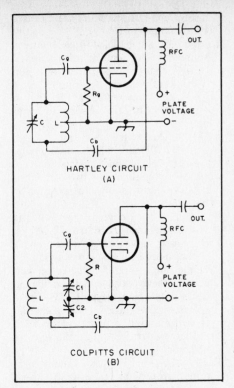

Fig. 48 — Hartley and Colpitts oscillators.

plied to the grid. This bias voltage sets the operating point of the oscillator and prevents excessive current flow.

Two other common type of oscillators are shown in Fig. 48. In Fig. 48A, feedback voltage is applied across a tapped inductor while in Fig. 48B, the voltage is applied across a capacitor instead. Quite often, a tuned plate circuit is not employed and an RF choke coil provides a high impedance load instead.

Classes of Amplifier Service

The amplifiers discussed so far have been described mostly in terms of ideal linear gain blocks. Class-A amplifiers, with or without negative feedback, reproduce the input signal with the greatest fidelity or least distortion. They accomplish this at the expense of poor power efficiency, meaning the signal power delivered to the load is usually only a small fraction of the dc power taken from the power supply. The remainder must be dissipated as heat. The maximum theoretical efficiency for a class-A amplifier is 50 percent, with 25 percent being a more typical figure. One factor preventing the theoretical maximum efficiency from being realized is that a practical amplifying device cannot be a perfect conductor — it will have a finite minimum voltage drop that prevents the full power supply potential from being impressed across the load. The operating point for a class-A amplifier is chosen to permit power supply current to flow over the entire 360° of the input signal cycle, and the average supply current does not vary with the signal level.

An advantage of a class-A amplifier, aside from nearly perfect input waveform reproduction (linearity), is that it requires minimal energy from the signal source. So while the power efficiency is poor, the power gain can be extremely high. The amplifier described in Figs. 43 and 44 operates in class-A.

An amplifier operating in class-B is biased so that no power supply current flows without an input signal applied. Current is taken from the supply over exactly 180° of the input signal cycle. A single-ended class-B amplifier that works only into a resistor for a load is, in effect, a half-wave rectifier (see Chapter 6), because one half of the input cycle is duplicated in the output and the other half is absent. This type of circuit finds some application as a rectifier or detector in radio work, but the distortion would be intolerable if it were used as an audio amplifier.

Class-B amplifiers can be used in low-distortion applications, however. Two class-B devices can be combined in a push-pull circuit. The two input terminals are driven 180° out of phase, so the amplifying devices conduct on opposite half cycles of the input signal. The output terminals are similarly coupled out of phase (usually by means of a transformer) to reproduce the input waveform faithfully.

A single-ended class-B stage is suitable for linear RF amplification if the load is a tuned circuit having a Q of five or better. The flywheel (ringing) effect of the tank circuit supplies the missing half of the output cycle. The efficiency of this type of amplifier is typically 60 percent.

An amplifier that is biased so as to conduct over the entire cycle for small input signals but at a reduced angle for large signals is a class-AB amplifier. The operating point is adjusted for the desired compromise among efficiency, power gain and linearity.

The output terminal of a single-device class-A, AB or B amplifier operated without negative feedback acts as a current source. This means that the instantaneous power supply current is essentially independent of the supply potential and is determined by the input signal amplitude. When an amplifying device is driven into saturation, that is, increasing the input amplitude causes no further increase in output, the output terminal no longer approximates a current source. Instead, the output terminal becomes "ohmic," meaning that the supply current varies approximately linearly with the supply voltage.

The signal output voltage also varies linearly with the supply voltage. This effect is highly useful in an important mode of radiotelephony. The saturated amplifier just described is called class-C. Its operating point is selected to allow power supply current to flow only in narrow pulses corresponding to the peaks of the input signal. Class-C amplifiers are extremely

nonlinear, and the harmonics produced in the wave form distortion process make the class-C amplifier useful as a frequency multiplier. High-Q tank circuits are required to suppress unwanted frequency components generated by a class-C stage.

Apart from its utility as an amplitude modulated amplifier or a frequency multiplier, the principal asset of a class-C amplifier is high power efficiency. The practical efficiency in straight amplifier service can run as high as 85 percent, but considerable drive power is required from the signal source, so the power gain is not as high as can be obtained in class-A or B. In multiplier service the maximum theoretical efficiency is the reciprocal of the harmonic number.

Still higher efficiency can be obtained from class-D and class-E amplifiers. In these circuits the active devices act as saturated switches that are controlled by the input signal. The class-E RF circuit is a recent development (1970s) by WA1HQC that yields high efficiency in an interesting way: The single active device switches the supply current through the load at critical points of the waveform. The output waveform is such that the electronic switch does not conduct current and withstand voltage simultaneously, resulting in very little energy waste. As in the class-C case, this output waveform is rich in harmonics, which must be filtered carefully.

Glossary of Radio Terms

active — As used in active filter or active device: A device or circuit that requires an operating voltage. (See passive.)

analog — A term used in computer work, meaning a system which operates with numbers represented by directly measurable quantities (analog readout mechanical dial system. See digital).

attenuator — A passive network that reduces the power level of a signal without introducing appreciable distortion.

balun — Balanced to unbalanced-line transformer.

bank wound — Pertaining to a coil (inductor) that has two or more layers of wire, each being wound over the top of the preceding one. (See solenoid.)

bandpass — A circuit or component characteristic that permits the passage of a single band of frequencies while attenuating those frequencies above and below that frequency band.

band-reject — A circuit or device that rejects a specified frequency band while passing those frequencies that lie above and below the rejected band (opposite of band-pass). Sometimes called band-rejection, as applied to a filter.

bandwidth — The frequency width of circuit or component, such as a band-pass filter or tuned circuit. Usually measured at the half-power points of the response curve (−3 dB points).

base loading — Applies to vertical antennas for mobile and fixed-location use; an inductance placed near the ground end of a vertical radiator to change the electrical length. With variations the inductor aids in impedance matching.

bifilar — Two conducting elements used in parallel; for example, two parallel wires wound on a coil form.

bilateral — Having two symmetrical sides or terminals; a filter (as one example) that has a 50-ohm characteristic at each port, with either port suitable as the input or output one.

bias — To influence current to flow in a specified direction by means of dc voltage; forward bias on a transistor stage, or grid bias on a tube type of amplifier.

blanker — A circuit or device that momentarily removes a pulse or signal so it is not passed to the next part of a circuit; a noise blanker. Not to be confused with clipper, which clips part of a pulse or waveform.

bridge — An electrical instrument used for measuring or comparing inductance, impedance, capacitance or resistance by comparing the ratio of two opposing voltages to a known ratio; to place one component in parallel with another; to join to conductors or components by electrical means.

broadband — A device or circuit that is broadband has the capability of being operated over a broad range of frequencies. A broadband antenna is one example.

cascade — One device or circuit that directly follows another; two or more similar devices or circuits in which the output of one is fed to the input of the succeeding one (tandem).

cascode — Two-stage amplifier having a grid-driven (or common-emitter or source) input circuit and a grounded-grid (or common base or gate) output circuit.

chip — Slang term for an integrated circuit, meaning a chip of semiconductor material upon which an IC is formed.

clamp — A circuit that maintains a predetermined characteristic of a wave at each occurrence so that the voltage or current is clamped or held at a specified value.

clipper — A device or circuit that limits the instantaneous value of a wave form or pulse to a predetermined value (see blanker).

closed loop — A signal path that includes a forward route, a feedback path, and a summing point which provides a closed circuit. In broad terms, an amplifying circuit providing voltage or power gain while being terminated correctly at the input and output ports, inclusive of feedback.

cold end — The circuit end of a component that is connected to ground or is bypassed for ac or RF voltage (the grounded end of a coil or capacitor).

common-mode signal — The instantaneous algebraic average of two signals applied to a balanced circuit, both signals being referred to a common reference.

composite — Made of a collection of distinct components; a complete (composite) circuit rather than a discrete part of an overall circuit.

conversion loss/gain — Relating to a mixer circuit from which less output energy is taken than is supplied at the input-signal port (loss); when a mixer delivers greater signal output than is supplied to the input-signal port (gain).

converter — A circuit used to convert one frequency to another. In a receiver the converter stage converts the incoming signals to the intermediate frequency.

core — An element made of magnetic material, serving as part of a path for magnetic flux.

damping — A progressive reduction in amplitude of a wave with respect to time (usually referenced to microseconds or milliseconds); a device or network added to a circuit to damp unwanted oscillations.

decay time — The period of time during which the stored energy or information decays to a specified value less than its initial value, such as the discharge time of a timing network.

decibel (dB) — A logarithmic expression of power ratio; also used as an expression of voltage ratios, provided the voltages appear across identical values of resistance or impedance. A decibel is 1/10 of a bel, and represents the smallest change in the power level of an audio tone that the human ear can detect.

decoder — A device or circuit that translates a combination of signals into a single signal representing that combination. One such circuit is that used for decoding the output signal of a DTMF pad.

differential amplifier — An amplifier with an output signal that is proportional to the algebraic difference between two input signals (sometimes called a difference amplifier).

digital — Refers to signals having or circuits that work with a finite number of discrete levels or states, as opposed to an analog signal which may have an infinite number of levels.

diplexing — The simultaneous transmission or reception of two signals while using a common antenna, made possible by using a device called a diplexer. Used in TV broadcasting to transmit visual and aural carriers by means of a single antenna.

discrete — A single device or circuit (a transistor as opposed to an IC). (See composite.)

dish — An antenna reflector for use at VHF and higher that has a concave shape. For example, a part of a sphere or paraboloid.

Doppler — The phenomenon evidenced by the change in the observed frequency of a wave in a transmission system caused by a time rate of change in the effective length of the path of travel between source and the point of observation.

drift — A change in component or circuit operation over time.

drive — RF energy applied at the input of an RF amplifier (RF driving power or voltage).

DTMF — Dual-tone multifrequency, such as a DTMF pad, which is sometimes called a Touch-Tone® pad.

duplex — Simultaneous two-way independent transmission and reception in both directions.

duplexer — A device that permits simultaneous transmission and reception of related signal energy while using a common antenna (see diplexing).

dynamic range — Difference in decibels between the overload level and minimum discernible signal level (MDS) in a system, such as a receiver. Parameters include desensitization point and distortion products as referenced to the receiver noise floor.

EME — Earth-moon-earth. Communications carried on by bouncing signals off the lunar surface. Commonly referred to as moonbounce.

empirical — Not based on mathematical procedures; experimental endeavor during design or modification of a circuit. Founded on case-history experience or intuition.

enabling — The preparation of a circuit for a subsequent function (enabling pulse or signal).

encoder — A circuit or device which changes discrete inputs into coded combinations of outputs. A DTMF pad is an encoder.

excitation — Signal energy used to drive a transmitter stage (see drive). Voltage applied to a component to actuate it, such as the field coil or a relay.

Faraday rotation — Rotation of the plane of polarization of an electromagnetic wave when traveling through a magnetic field. In space communications this effect occurs when signals traverse the ionosphere.

feedback — A portion of the output voltage being fed back to the input of an amplifier. Description includes ac and dc voltage which can be used separately or together, depending on the particular circuit.

feedthrough — Energy passing through a circuit or component, but not usually desired. A type of capacitor which can be mounted on a chassis or panel wall to permit feeding through a dc voltage while bypassing ac or RF to ground. Sometimes called a coaxial capacitor.

ferromagnetic — Material that has a relative permeability greater than unity and requires a magnetizing force. (Ferrite and powdered-iron rods and toroids).

finite — Having a definable quantity; a

finite value of resistance or other electrical measure.

flip-flop — An active circuit or device that can assume either of two stable stages at a given time, as dictated by the nature of the input signal.

floating — A circuit or conductor that is above ac or dc ground for a particular reason. Example: A floating ground bus which is not common to the circuit chassis.

gate — A circuit or device, depending on the nature of the input signal, which can permit the passage or blockage of a signal or dc voltage.

GDO — Abbreviation for a grid-dip or gate-dip oscillator (test instrument).

ground loop — A circuit-element condition (PC board conductor, metal chassis or metal cabinet wall) that permits the unwanted flow of ac from one circuit point to another.

half-power point — The two points on a response curve that are 3 dB lower in level than the peak power. Sometimes called the 3dB bandwidth.

Hall effect — The change of the electric conduction caused by the component of the magnetic field vector normal to the current density vector, which instead of being parallel to the electric field, forms an angle with it.

high end — Refers generally to the hot (RF or dc) end of a component or circuit; the end opposite the grounded or bypassed end (see cold end).

high level — The part of a circuit which is relatively high in power output and consumption as compared to the small signal end of a circuit. Example: A transmitter PA stage is the high-level amplifier, as might be the driver also.

high pass — Related mainly to filters or networks which are designed to pass energy above a specified frequency, but attenuate or block the passage of energy below that frequency.

high-Z — The high-impedance part of a circuit; a high-impedance microphone; a high-impedance transformer winding.

hot end — see high end.

hybrid — A combination of two generally unlike things; a circuit that contains transistors and tubes, for example.

ideal — A theoretically perfect circuit or component; a lossless transformer or device that functions without any faults.

insertion loss — That portion of a signal current or voltage that is lost as it passes through a circuit or device. The loss of power through a filter or other passive network.

interpolate — To estimate a value between two known values.

leakage — The flow of signal energy beyond a point at which it should not be present. Example: Signal leakage across a filter because of poor layout (stray coupling) or inadequate shielding.

linear amplification — The process by which a signal is amplified without

altering the characteristic of the input waveform. Class-A, AB and B amplifiers are generally used for linear amplification.

load — A circuit or component that receives power; the power delivered to such a circuit or component. Example: A properly matched antenna is a load for a transmitter.

loaded — A circuit is said to be loaded when the desired power is being delivered to a load.

logic — Decision-making circuitry of the type found in computers.

long wire — A horizontal wire antenna that is one wavelength or greater in size. A long piece of wire (less than one wavelength) does not qualify as a long wire.

low end — see cold end.

low level — Low-power stage or stages of a circuit as referenced to the higher-power stages (see high level).

low pass — A circuit property that permits the passage of frequencies below a specified frequency but attenuates or blocks those frequencies above that frequency (see high pass).

low-Z — Low impedance (see high-Z).

mean — A value between two specified values; an intermediate value.

master oscillator — The primary oscillator for controlling a transmitter or receiver frequency. Can be a VFO (variable-frequency oscillator), VXO (variable crystal oscillator), PTO (permeability-tuned oscillator), PLL (phase-locked loop), LMO (linear master oscillator) or frequency synthesizer.

microwaves — AC signals having frequencies greater than about 1000 MHz.

modulation index — The ratio of the frequency deviation of the modulated wave to the frequency of the modulating signal.

narrowband — A device or circuit that can be operated only over a narrow range of frequencies. Low-percentage bandwidth.

network — A group of components connected together to form a circuit that will conduct power, and in most examples effect an impedance match. Examples: An LC matching network between stages of a transistorized transmitter.

noise figure — Of a two-port transducer the decibel ratio of the total noise power to the input noise power, when the input termination is at the standard temperature of 290 K.

nominal — A theoretical or designated quantity that may not represent the actual value. Sometimes referred to as the ball-park value.

op amp — Operational amplifier. A high-gain, feedback-controlled amplifier. Performance is controlled by external circuit elements.

open loop — A signal path that does not contain feedback (see closed loop).

parameter — The characteristic behavior of a device or circuit, such as the current

gain of a 2N5109 transistor.

parametric amplifier — Synonym for reactance amplifier. An inverting parametric device for amplifying a signal without frequency translation from input to output. Used for low-noise UHF and microwave amplification.

parasitic — Unwanted condition or quantity, such as parasitic oscillations or parasitic capacitance; additional to the desired characteristic.

passive — Operating without an operating voltage. Example: An LC filter that contains no amplifier, or a diode mixer.

PEP — Peak envelope power; maximum amplitude that can be obtained with any combination of signals.

permeability — A term used to express relationships between magnetic induction and magnetic force.

pill — Slang expression for a transistor or an IC.

PL — Private Line, such as a repeater accessed by means of a specified tone.

PLL — Phase-locked loop.

port — The input or output terminal of a circuit or device.

prototype — A first full-scale working version of a circuit design.

Q_L — Loaded Q of a circuit.

Q_u — Unloaded Q of a circuit.

quagi — An antenna consisting of both full-wavelength loops (quad) and Yagi elements.

resonator — A general term for a high-Q resonant circuit, such as an element of a filter.

return — That portion of a circuit which permits the completion of current flow, usually to ground; a ground return.

ringing — The generation of an audible or visual signal by means of oscillation or pulsating current; the annoying sound developed in some audio filters when the Q is extremely high.

ripple — Pulsating current. Also, the gain depressions that exist in the flat portion of a band-pass response curve (above the −3 dB points on the curve). Example: Passband ripple in the nose of an IF filter response curve.

rise time — The time required for a pulse or waveform to reach a specified value from some smaller specified value. The specified values are typically 10 and 90 percent of the peak amplitude.

RMS — Root mean square. The square root of the mean of the square of the voltage or current during a complete cycle.

rotor — A moving rotary component within a rotation-control device. Examples include the moving plates of a variable capacitor and the armature of an alternator. Not to be confused with an antenna rotator, which is the total assembly.

saturation — A condition that exists when a further change in input produces no additional output (a saturated amplifier).

selectivity — A measure of circuit capability to separate the desired signal from those at other frequencies.

shunt — A device placed in parallel with or across part of another device. Examples: meter shunts, shunt-fed vertical antennas and a capacitor placed (shunted) across another capacitor.

solenoidal — A single-layer coil of wire wound to form a long cylinder.

spectral purity — The relative freedom of an emission from harmonics, spurious signals and noise.

standing-wave ratio — The ratio of the maximum to minimum voltage or current on a transmission line at least a quarter-wavelength long.

strip — General term for two or more stages of a circuit that in combination perform a particular function. Examples: a local-oscillator strip, an audio strip or an IF strip.

subharmonic — A frequency that is an integral submultiple of a frequency to which it is referred. A misleading term that implies that subharmonic energy can be created along with harmonic energy (not true). More aptly, a 3.5-MHz VFO driving a 40-meter transmitter, with 3.5-MHz leakage at the output, qualifying as a subharmonic.

tank — A circuit consisting of inductance and capacitance, capable of storing electrical energy over a band of frequencies continuously distributed about a single frequency at which the circuit is said to be resonant, or tuned.

toroidal — Doughnut-shaped physical format, such as a toroid core.

transducer — A device which is used to transport energy from one system (electrical, mechanical or acoustical) to another. Example: A loudspeaker or phonograph pickup.

transceiver — A combination transmitter and receiver that uses some parts of the circuit for both functions.

Transmatch — An LC network used to effect an impedance match between a transmitter and a feed line to an antenna.

transmission line — One or more conductors used to convey ac energy from one point to another, as from a radio station to its antenna.

transverter — A converter that permits transmitting and receiving at a specified frequency apart from the capability of the transceiver to which it is connected as a basic signal source. Example: a 2-meter transverter used in combination with an HF-band transceiver.

trap — A device consisting of L and C components that permits the blockage of a specified frequency while allowing the passage of other frequencies. Example: a wave trap or an antenna trap.

trifilar — Same as bifilar, but with three parallel conductors.

trigger — To initiate action in a circuit by introducing an energy stimulus from an external source, such as a scope trigger.

U — Symbol for unrepairable assembly, such as an integrated circuit (U1, U2, etc.).

unloaded — The opposite condition of loaded.

varactor — A two-terminal semiconductor device (diode) which exhibits a voltage-dependent capacitance. Used primarily as a tuning device or frequency multiplier at VHF and UHF.

VCO — Voltage-controlled oscillator. Uses tuning diodes that have variable dc applied to change their junction capacitances.

VSWR — Voltage standing-wave ratio. (See standing-wave ratio).

VU — Volume unit.

VXO — Variable crystal oscillator.

wave — A periodically varying electromagnetic field radiated from a conductor.

waveguide — A hollow conducting tube used to convey microwave energy.

wavelength — The distance between the two points of corresponding phase of two consecutive cycles of an electromagnetic signal.

X — The symbol for reactance.

Zener diode — A diode used to regulate voltage or function as a clamp or clipper. Named after the inventor.

Z — Symbol for a device or circuit that contains two or more components. Example: A parasitic suppressor that contains a resistor and an inductor in parallel (Z1, Z2, etc.). Z is also the symbol for impedance.

Chapter 4

Solid-State Basics

Solid-state devices are also called semiconductors. The prefix semi means half. They have electrical properties between those of conductors and those of insulators.

The conductivity of a material is proportional to the number of free electrons in the material. Pure germanium and pure silicon crystals have relatively few free electrons. If, however, carefully controlled amounts of impurities (materials having a different atomic structure, such as arsenic or antimony) are added, the number of free electrons, and consequently the conductivity, is increased. When certain other impurities are introduced (such as aluminum, gallium or indium), an electron deficiency, or hole, is produced. As in the case of free electrons, the presence of holes encourages the flow of electrons in the semiconductor material, and the conductivity is increased. Semiconductor material that conducts by virtue of the free electrons is called N-type material; material that conducts by virtue of an electron deficiency is called P-type.

Electron and Hole Conduction

If a piece of P-type material is joined to a piece of N-type material as at A in Fig. 1 and a voltage is applied to the pair as at B, current will flow across the boundary or junction between the two (and also in the external circuit) when the battery has the polarity indicated. Electrons, indicated by the minus symbol, are attracted across the junction from the N material through the P material to the positive terminal of the battery. Holes, indicated by the plus symbol, are attracted in the opposite direction across the junction by the negative potential of the battery. Thus, current flows through the circuit by means of electrons moving one way and holes the other.

If the battery polarity is reversed, as at C, the excess electrons in the N material are attracted away from the junction and the holes in the P material are attracted by the negative potential of the battery away from the junction. This leaves the junction region without any current carriers; consequently, there is no conduction.

In other words, a junction of P- and N-type materials constitutes a rectifier. It differs from the vacuum-tube diode rectifier in that there is a measurable, although comparatively very small, reverse current. The reverse current results from the presence of some carriers of the type opposite to those that principally characterize the material.

With the two plates separated by practically zero spacing, the junction forms a capacitor of relatively high capacitance. This places a limit on the upper frequency at which semiconductor devices of this construction will operate, as compared with vacuum tubes. Also, the number of excess electrons and holes in the material depends on temperature, and since the conductivity in turn depends on the number of excess holes and electrons, the device is more temperature sensitive than is a vacuum tube.

Capacitance may be reduced by making the contact area very small. This is done by means of a point contact, a tiny P-type region being formed under the contact point during manufacture when N-type material is used for the main body of the device.

SEMICONDUCTOR DIODES

The vacuum-tube diode has been replaced in modern equipment designs. Semiconductor diodes are more efficient because they do not consume filament power. They are much smaller than tube diodes. In low-level applications they operate cooler than tubes do. Solid-state diodes are superior to tube types in that solid-state diodes operate into the microwave region, while most vacuum-tube diodes are not practical at frequencies above 50 MHz.

Semiconductor diodes fall into two main categories, structurally. Although they can be made from silicon or germanium crystals, they are usually classified as PN junction diodes or point-contact diodes. These formats are illustrated in Fig. 2. Junction diodes are used from dc to the microwave region, but point-contact diodes are intended primarily for RF applications. The internal capacitance of a point-contact diode is considerably less than that of a junction diode designed for the same circuit application. As the operating frequency is increased the unwanted internal and external capacitance of a diode becomes

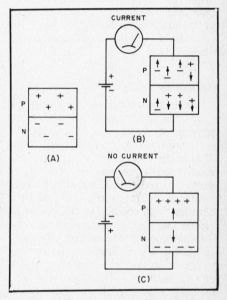

Fig. 1 — A PN junction (A) and its behavior when conducting (B) and nonconducting (C).

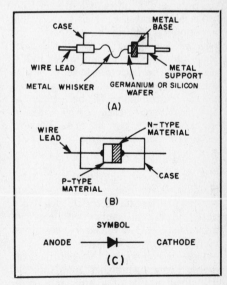

Fig. 2 — A point-contact type of diode is shown at A. A junction diode is depicted at B and the diode symbol is at C. The cathode end is usually indicated by a band on the body of the device.

more troublesome to the designer. Where a given junction type of diode may exhibit a capacitance of several picofarads, the point-contact device will have an internal capacitance of 1 pF or less.

Selenium Diodes

Power rectifiers made from selenium were in common use in ac power supplies before 1965. Selenium diodes are characterized by high forward voltage drop (which increases with diode age) and high reverse leakage current. The voltage drop causes the device to dissipate power, and a typical rectifier stack has large cooling fins. An additional shortcoming of selenium rectifiers is that they sometimes emit toxic fumes when they burn out. When replacing selenium diodes with silicon units, be certain that the filter capacitors (and the entire equipment) can withstand the higher output voltage. Some early solar cells were made of selenium, but silicon devices have taken over this area, too.

Germanium Diodes

The germanium diode is characterized by a relatively large current flow when small amounts of voltage are applied in the forward direction (Fig. 1). Small currents will flow in the reverse (back) direction for much larger applied voltages. A representative curve is shown in Fig. 3. The dynamic resistance in either the forward or back direction is determined by the change in current that occurs, at any given point on the curve, when the applied voltage is changed by a small amount. The forward resistance will vary somewhat in the region of very small applied voltages. However, the curve is mostly straight, indicating a relatively constant dynamic resistance. For small applied voltages the resistance is on the order of 200 ohms or less.

The back resistance exhibits considerable variation and depends on the specific applied voltage during the test. It may vary from a few thousand ohms to well over a megohm. The back resistance of a germanium diode is considerably lower than that of a silicon diode. The back resistance for silicon is greater than a megohm in most instances, but the germanium diode is normally less than a megohm.

Common Silicon Diodes

Apart from the fact that silicon P and N materials are used in the formation of a silicon junction diode, the characteristics of these devices are similar to those of germanium diodes. The voltage/current curves of Fig. 3 are representative.

The junction barrier voltage for silicon diodes is somewhat higher (approximately 0.7 volt) than that of a germanium diode (about 0.3 volt). The majority of the diodes in use today fall into the silicon class. They are rugged and reliable from RF small-signal applications to dc power use.

Silicon diodes are available in ratings of 1000 volts (PIV) or greater. Many of these diodes can accommodate dc in excess of 100 amperes. The primary rule in preventing damage to any diode is to operate the device within the maximum ratings specified by the manufacturer. The device temperature is one of the important parameters. Heat sinks are used with diodes that must handle large amounts of power, thereby holding the diode junction temperature at a safe level.

The behavior of junction diodes under varying temperatures is of interest to designers of circuits that must perform over some temperature range. The relationship between forward bias current, forward bias voltage and temperature is defined by the classic diode equation:

$$I_f = I_s \left(e^{\frac{qV}{kt}} - 1 \right)$$

where q is the fundamental electronic charge (1.6×10^{-19} coulombs), V is the bias potential, k is Boltzmann's constant (1.38×10^{-23} joules/kelvin), kelvin = °Celsius + 273), t is the junction temperature in kelvins, I_s is the reverse-bias saturation current, I_f is the forward-bias current and e is the natural logarithmic base (2.718). The ratio q/k is approximately 11,600, so the diode equation can be written:

$$I_f = I_s \left(e^{\frac{11,600V}{t}} - 1 \right) \qquad \text{(Eq. 1)}$$

It is useful to have an expression for the voltage developed across the junction when the forward current is held constant. To obtain such an expression we must solve the diode equation for V. Expanding the right side of Eq. 1 yields:

$$I_f = I_s e^{\frac{11,600V}{t}} - I_s \qquad \text{(Eq. 2)}$$

Adding I_s to both sides gives:

$$I_f + I_s = I_s e^{\frac{11,600V}{t}} \qquad \text{(Eq. 3)}$$

Dividing through by I_s produces:

$$\frac{I_f}{I_s} + 1 = e^{\frac{11,600V}{t}} \qquad \text{(Eq. 4)}$$

which implies

$$\frac{11,600V}{t} = \ln \left(\frac{I_f}{I_s} + 1 \right) \qquad \text{(Eq. 5)}$$

Multiplying each term by $\frac{t}{11,600}$ leaves:

$$V = \frac{t}{11,600} \ln \left(\frac{I_f}{I_s} + 1 \right) \qquad \text{(Eq. 6)}$$

The undetermined quantity in Eq. 6 is I_s, the reverse saturation current. In ordinary silicon signal diodes this current approximately doubles with each 4.5 kelvin-temperature increase. A mathematical expression for this behavior as a function of temperature is:

$$I_{s(t)} = 2I_{s(t-4.5)} \qquad \text{(Eq. 7)}$$

At room temperature (300 kelvins), the reverse saturation current is on the order of 10^{-13} amperes. Eq. 7 describes a phenomenon similar to radioactive decay, where the 4.5-kelvin current-doubling interval is analogous to the half-life of a radioactive substance. This equation with the given initial condition sets up an initial-value problem, the solution of which is:

$$I_{s(t)} = 10^{-13} e^{\frac{(t-300)\ln 2}{4.5}} \qquad \text{(Eq. 8)}$$

Substituting this expression for I_s into Eq. 6 produces the diode voltage drop as a function of temperature for a constant current:

$$V_{(t)} = \frac{t}{11,600} \times$$
$$\ln \left[\frac{I_f}{10^{-13} e^{\frac{(t-300)\ln 2}{4.5}}} + 1 \right] \quad \text{(Eq. 9)}$$

The temperature coefficient of the junction potential can be obtained from the partial derivative of V with respect to t, but it's a simple matter (with the aid of a pocket calculator) to extract the information directly from Eq. 9. If the forward current is fixed at 1 milliampere, the diode drop at room temperature is 0.5955 volts. This potential decreases at an initial rate of 2 millivolts per kelvin. The temperature coefficient gradually increases to 3 millivolts per kelvin at 340 kelvins. While the temperature curve isn't linear, it is gradual enough to be considered linear over small intervals. When the bias current is increased to 100 milliamperes, the room temperature junction potential increases to 0.7146 volts as might be expected, but the temperature coefficient stays well-behaved. The initial potential decrease is 1.6 millivolts per

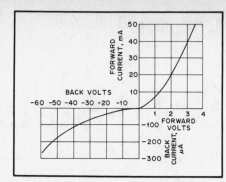

Fig. 3 — Typical point-contact diode (germanium) characteristic curve. Because the back current is much smaller than the forward current, a different scale is used for back voltage and current.

kelvin, and this value increases to 2.5 millivolts per kelvin at 340 kelvins.

The significance of the very minor dependence of temperature coefficient on bias current is that it isn't necessary to use an elaborate current regulator to bias diodes used in temperature-compensation applications. The equations defining the behavior of junction diodes are approximations. Some of the voltages were expressed to five significant figures so the reader can verify his calculations, but this much precision exceeds the accuracy of the approximations.

Diodes as Switches

Solid-state switching is accomplished easily by using diodes or transistors in place of mechanical switches or relays. The technique is not a complicated one at dc and audio frequencies when large amounts of power are being turned on and off, or transferred from one circuit point to another.

Examples of shunt and series diode switching are given in Fig. 4. The illustration at A shows a 1N914 RF-switching diode as a shunt on-off element between C1 and ground. When + 12 volts is applied to D1 through R1, the diode saturates and effectively adds C1 to the oscillator tank circuit. R1 should be no less than 2200 ohms in value to prevent excessive current flow through the diode junction (I_{max} = 12 V / 2200 Ω).

Series diode switching is shown in Fig. 4B. In this example the diode, D1, is inserted in the audio signal path. When S1 is in the ON position the diode current path is to ground through R2, and the diode saturates to become a closed switch. When S1 is in the OFF state R1 is grounded and + 12 volts is applied to the diode cathode. In this mode D1 is back biased (cut off) to prevent audio voltage from reaching the transistor amplifier. This technique is useful when several stages in a circuit are controlled by a single mechanical switch or relay. RF circuits can also be controlled by means of series diode switching.

A significant advantage to the use of diode switching is that long signal leads are eliminated. The diode switch can be placed directly at the circuit point of interest. The dc voltage that operates it can be at some convenient remote point. The diode recovery time (switching speed) must be chosen for the frequency of operation. In other words, the higher the operating frequency the faster the switching speed required. For dc and audio applications one can use ordinary silicon power-supply rectifier diodes.

Diodes as Gates

Diodes can be placed in series with dc leads to function as gates. Specifically, they can be used to allow current to flow in one direction only. An example of this technique is given in Fig. 5A.

A protective circuit for the solid-state transmitter is effected by the addition of

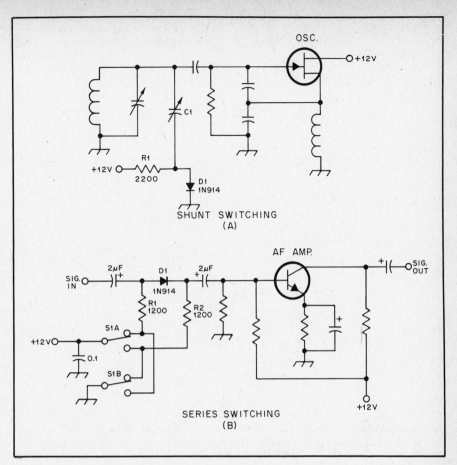

Fig. 4 — A silicon-switching diode, D1, is used at A to place C1 in the circuit. At B is shown a series switch with D1 in the signal path.

D1 in the 13.6-volt dc line to the equipment. The diode allows the flow of positive current, but there will be a drop of approximately 0.7 volt across the diode, requiring a supply voltage of 14.3. Should the operator mistakenly connect the supply leads in reverse, current will not flow through D1 to the transmitter. In this application the diode acts as a gate. D1 must be capable of passing the current taken by the transmitter without overheating.

A power type of diode can be used in shunt with the supply line to the transmitter for protective purposes. This method is illustrated in Fig. 5B. If the supply polarity is crossed accidentally, D1 will draw high current and cause F1 to open. This is sometimes referred to as a crowbar protection circuit. The primary advantage of circuit B over circuit A is that there is no voltage drop between the supply and the transmitter.

Diodes as Voltage References

Zener diodes are discussed later in this chapter. They are used as voltage references or regulators. Conventional junction diodes can be used for the same purposes by taking advantage of their barrier-voltage characteristics. The greater the voltage needed, the higher the number of diodes

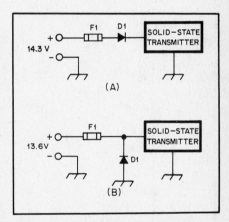

Fig. 5 — D1 at A protects the equipment if the supply leads are cross-polarized in error. At B the fuse will blow if the power supply is connected for the wrong polarity.

used in series. Some examples of this technique are given in Fig. 6. At A the diode (D1) establishes a fixed value of forward bias (0.7 V) for the transistor, thereby functioning as a regulator. R1 is chosen to permit a safe amount of current to flow through the diode junction while it is conducting at the barrier voltage.

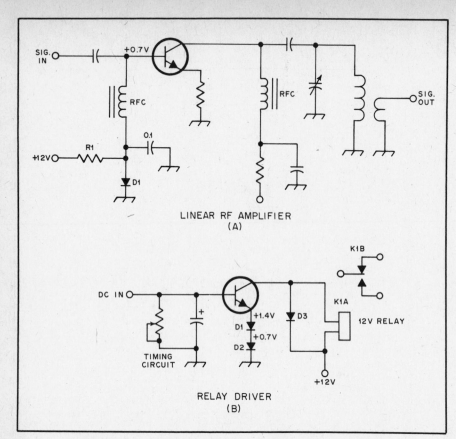

Fig. 6 — D1 establishes a 0.7-volt bias reference at A. Approximately 1.4 volts of emitter bias is established by connecting D1 and D2 in series at illustration B.

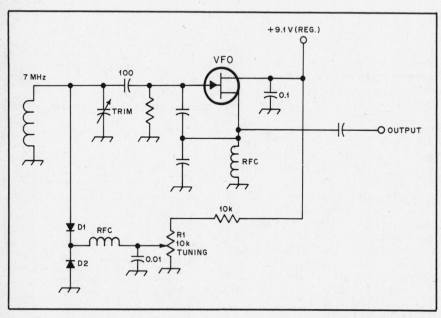

Fig. 7 — High-speed switching diodes of the 1N914 variety can be connected back to back and used as tuning diodes. As the reverse voltage is varied by means of R1, the internal capacitance of the diodes will change.

The circuit of Fig. 6B shows two diodes inserted in the emitter return of a relay-driver transistor. D1 and D2 set up a cutoff voltage of approximately 1.4. This reduces the static current of the transistor when forward bias is not provided at the transistor base. If too much static current flows, the relay may not drop out when the forward bias decays across the timing network. The more sensitive the relay the greater the

chance for such a problem. D1 and D2 prevent relay dropout problems of this variety. D3 is used as a transient suppressor. A spike will occur when the relay coil field collapses. If the amplitude of the spike is great enough, the transient, while following the dc bus in a piece of equipment, can destroy transistors and diodes elsewhere in the circuit. In this application the diode (D3) can be regarded as a clamp, since it clamps the spike at approximately 0.7 volt.

Using Diodes as Capacitors

Later in this chapter there is a discussion about VVC (voltage-variable capacitor) diodes. They are known also as tuning diodes and Varicap diodes. It is possible, however, to use ordinary silicon diodes as voltage-variable capacitors. This is accomplished by taking advantage of the inherent changes in diode junction capacitance as the reverse bias applied to them is changed. The primary limitation in using high-speed switching diodes of the 1N914 variety is a relatively low maximum capacitance. At a sacrifice to low minimum capacitance, diodes can be used in parallel to step up the maximum available capacitance.

An example of two 1N914 silicon diodes in a diode tuning circuit is given in Fig. 7. As R1 is adjusted to change the back bias on D1 and D2, there will be a variation in the junction capacitance. That change will alter the VFO operating frequency. The junction capacitance increases as the back bias is lowered. In the circuit shown here the capacitance will vary from roughly 5 pF to 15 pF as R1 is adjusted. The diodes used in circuits of this kind should have a high Q and excellent high-frequency characteristics. Generally, tuning diodes are less stable than mechanical variable capacitors are. This is because the diode junction capacitance will change as the ambient temperature varies. This circuit is not well suited to mobile applications because of its lack of stability.

Diode Clippers and Clamps

The previous mention of diode clamping action (D3 in Fig. 6) suggests that advantage can be taken of the characteristic barrier voltage of diodes to clip or limit the amplitude of a sine-wave. Although there are numerous applications in this general category, diode clippers are more familiar to the amateur in noise limiter, audio limiter and audio compressor circuits. Fig. 8 illustrates some typical circuits that use small-signal diodes as clamps and clippers. D1 in Fig. 8A functions as a bias clamp at the gate of the FET. It limits the positive sine-wave swing at approximately 0.7 V. Not only does the diode tend to regulate the bias voltage, it limits the transconductance of the FET during the positive half of the cycle. This action restricts changes in transistor junction capacitance. As a result, frequency stability of the oscillator is enhanced and the generation of harmonic

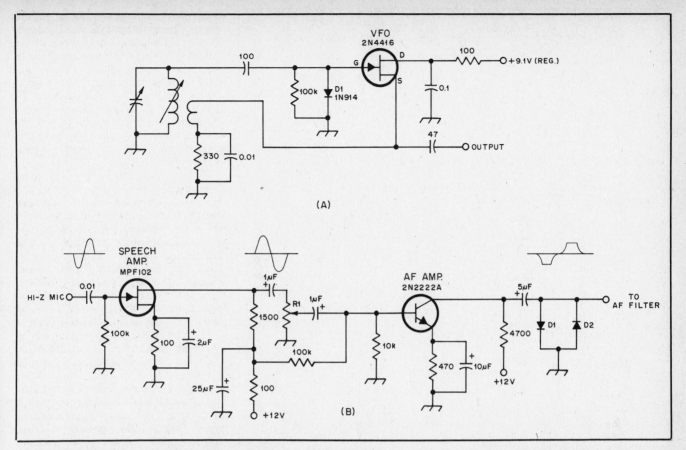

Fig. 8 — D1 serves as a bias stabilization device at A (see text). At B, D1 and D2 are employed as clippers to flatten the positive and negative AF peaks. Clipping will occur at roughly 0.7 volt if silicon diodes are used. Audio filtering is required after the clipper to remove the harmonic currents caused by the diode action.

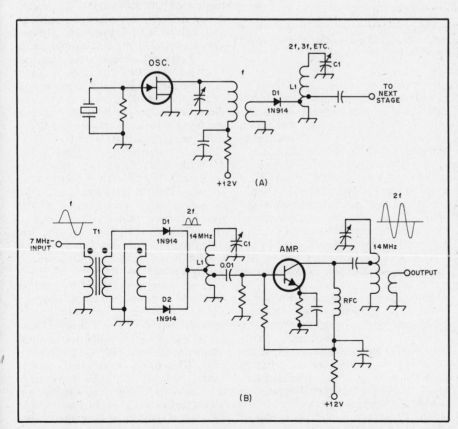

Fig. 9 — A simple diode frequency multiplier is shown at A. A balanced diode frequency doubler is at B. T1 is a trifilar-wound broadband toroid transformer.

currents is greatly minimized.

The circuit of Fig. 8B shows how a pair of diodes can be connected in back-to-back fashion for the purpose of clipping the negative and positive sine-wave peaks in an audio amplifier. If germanium diodes are used at D1 and D2 (1N34As or similar) the audio will limit at roughly 0.3 V. With silicon diodes (1N914 or rectifier types) the voltage will not exceed 0.7 V. R1 serves as the clipping-level control. An audio gain control is normally used after the clipper filter, along with some additional gain stages. The output of the clipper must be filtered to restore the sine wave if distortion is to be avoided. Diode clippers generate considerable harmonic currents, thereby requiring an RC or LC type of audio filter.

Diode Frequency Multipliers

Designers of RF circuits use small-signal diodes as frequency multipliers when they want to minimize the number of active devices (tubes or transistors) in a circuit. The primary disadvantage of diode multipliers is a loss of signal compared to that available from an active multiplier. Fig. 9 contains examples of diode frequency multipliers. The circuit at A is useful for obtaining odd or even multiples of the driving voltage. The efficiency of this circuit is not high, requiring that an amplifier

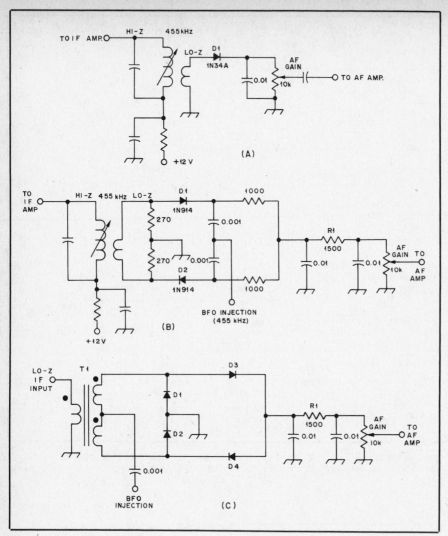

Fig. 10 — D1 at A is used as a simple AM detector. Two versions of diode product detectors are illustrated at B and C. BFO injection for B and C should be approximately 19 volts P-P for best detector performance.

two-diode product detector. R1 and the two bypass capacitors serve as an RF filter to keep signal and BFO energy out of the following AF amplifier stage. A four-diode product detector is illustrated at C. T1 is a trifilar-wound broadband transformer. The characteristic input impedance of T1 is 50 ohms. An RF filter follows this detector also. BFO injection voltage for the detectors at B and C should be between 8 and 10 volts P-P for best detector performance.

Circuits for typical diode mixers are given in Fig. 11. Product detectors are also mixers except for the frequencies involved. The output energy is at audio frequencies rather than at some RF intermediate frequency. The examples at A and B can be compared to those at C and D for the purpose of illustrating the similarity between balanced modulators and mixers. It is evident that product detectors, balanced modulators and mixers are of the same family. The diodes in all examples can be hot-carrier types or matched silicon switching diodes of the 1N914 class.

C1 and C2 in Fig. 11C and D are used for balancing purposes. They can be employed in the same manner with the circuits at A and B. The transformers in each illustration are trifilar-wound toroidal types. They provide a broadband circuit characteristic.

Hot-Carrier Diodes

One of the more recent developments in the semiconductor field is the hot-carrier diode, or HCD. It is a metal-to-semiconductor, majority-carrier conducting device with a single rectifying junction. The carriers are typically high-mobility electrons in an N type of semiconductor material. The HCD is particularly useful in mixers and detectors at VHF and higher. Notable among the good features of this type of diode are its higher operating frequency and lower conduction voltage compared to a PN junction diode such as the 1N914.

When compared to a point-contact diode, the HCD is mechanically and electrically superior. It has lower noise, greater conversion efficiency, larger square-law capability, higher breakdown voltage and lower reverse current. The internal capacitance of the HCD is markedly lower than that of a PN junction diode and it is less subject to temperature variations.

Fig. 12 shows how the diode is structured internally. A typical set of curves for an HCD and a PN junction diode is given in Fig. 13. The curves show the forward and reverse characteristics of both diode types.

Fig. 14 illustrates the noise figure and conversion loss of an HCD with no bias applied. When forward bias is applied to the diode, the noise figure will change from that shown in Fig. 14. Curves for various bias amounts are given in Fig. 15. The numbers at the ends of the curves signify the amount of current (in milliamperes) flowing into the test circuit at point A.

be used after the diode multiplier in most applications. Resonator L1/C1 must be tuned to the desired output frequency.

A diode frequency doubler is shown at B in Fig. 9. It functions like a full-wave power-supply rectifier, where 60-Hz energy is transformed to 120-Hz by virtue of the diode action. This circuit will cause a loss of approximately 8 dB. Therefore, it is shown with a succeeding amplifier stage. If reasonable circuit balance is maintained, the 7-MHz energy will be down some 40 dB at the output of D1 and D2 — prior to the addition of L1 and C1. Additional suppression of the driving energy is realized by the addition of resonator L1/C1. T1 is a trifilar-wound toroidal transformer. At this frequency (7 MHz) a 0.5-inch-diameter ferrite core (permeability of 125) will suffice if the trifilar winding contains approximately 10 turns. Additional information on this subject is given in the ARRL publication, *Solid State Design for the Radio Amateur*.

Diode Detectors and Mixers

Diodes are effective as detectors and mixers when circuit simplicity and strong-signal-handling capability are desired. Impedance matching is an important design objective when diodes are used as detectors and mixers. The circuits are lossy, just as is the case with diode frequency multipliers. A diode detector or mixer will exhibit a conversion loss of 7 dB or more in a typical example. Therefore, the gain before and after the detector or mixer must be chosen to provide an acceptable noise figure for the overall circuit in which the diode stage is used. This is a particularly critical factor when diode mixers are used at the front end of a receiver.

A significant advantage in the use of diode mixers and detectors is that they are broadband in nature, and they provide a wide dynamic range. Hot-carrier diodes are preferred by some designers for these circuits, but the 1N914 class of switching diodes provides good performance if they are matched for a similar resistance before being placed in the circuit.

Fig. 10 illustrates some examples of diode detectors. A basic AM detector is shown at A. The circuit at B is that of a

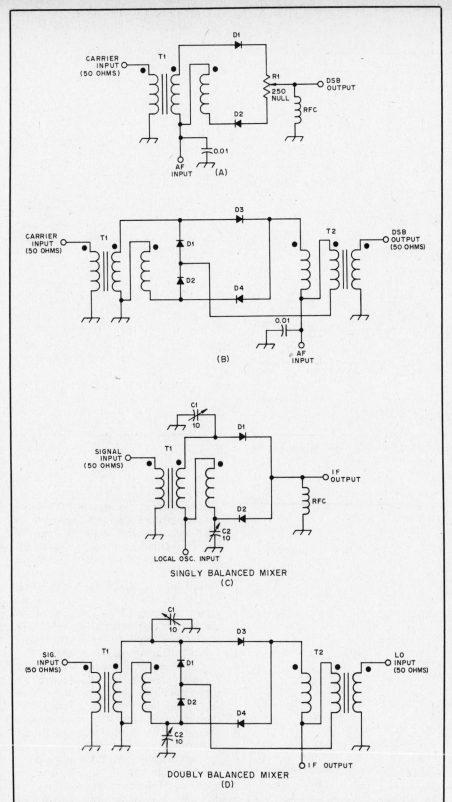

Fig. 11 — The circuits at A and B are for use in balanced modulators. The similarity between these and balanced mixers is shown at C and D.

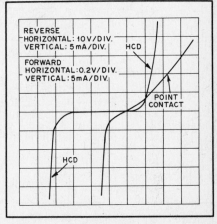

Fig. 12 — Cross-sectional representation of a hot-carrier diode (HCD).

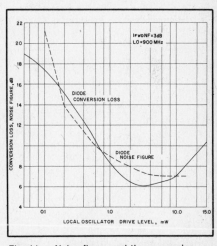

Fig. 13 — Forward and reverse characteristics of a hot-carrier diode as compared to a PN junction diode.

Fig. 14 — Noise figure and the conversion loss of a typical HCD that has no bias applied.

A set of curves showing conversion loss versus LO drive for an HCD mixer are given in Fig. 16. The test circuit used for the curve of Fig. 15 applies. The curve numbers indicate milliamperes measured at point A.

Varactor Diodes

Mention was made earlier in this chapter of diodes being used as voltage-variable capacitors, wherein the diode junction capacitance can be changed by varying the reverse bias applied to the diode. Manufac-

turers have designed certain diodes for this application. They are called Varicaps (variable capacitor diodes) or varactor diodes (variable reactance diodes). These diodes depend on the change in capacitance that occurs across their depletion layers. They are not used as rectifiers.

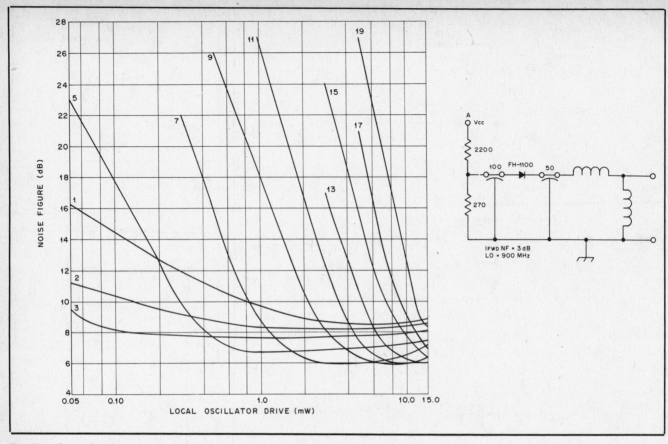

Fig. 15 — Curves for hot-carrier diode noise figure versus local-oscillator drive power. The bias currents are in milliamperes as measured at point A in the representative test circuit.

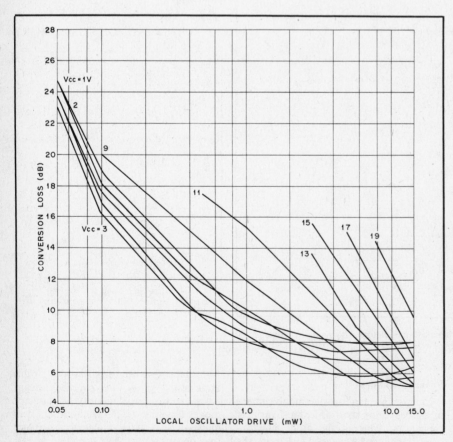

Fig. 16 — Local-oscillator drive power versus conversion loss for a specified bias amount. Bias currents are in milliamperes as measured at point A of the circuit in Fig. 15.

Varactors are designed to provide various capacitance ranges from a few picofarads to more than 100 pF. Each one has a specific minimum and maximum capacitance, and the higher the maximum amount the greater the minimum value. Therefore, the amateur finds it necessary to tailor circuits for the midrange of the capacitance curve. Ideally, he will choose the most linear portion of the curve. Fig. 17A shows typical capacitance-voltage curves for three varactor diodes.

A representative circuit of a varactor diode is presented in Fig. 18. In this equivalent circuit the diode junction consists of C_j (junction capacitance) and R_j (junction resistance). The bulk resistance is shown as R_s. For the most part R_j can be neglected. The performance of the diode junction at a particular frequency is determined mainly by C_j and R_s. As the operating frequency is increased, the diode performance degrades, owing to the transit time established by C_j and R_s.

An important characteristic of the varactor diode is the Q, or figure of merit. The Q of a varactor diode is determined by the ratio of its capacitive reactance (X_j) and its bulk resistance, R_s, just as is true of other circuit elements such as coils and capacitors, where $Q = X/R_s$ at a specified frequency. Fig. 17B characterizes the Q of three Motorola varactor diodes (versus reverse bias) at 50 MHz.

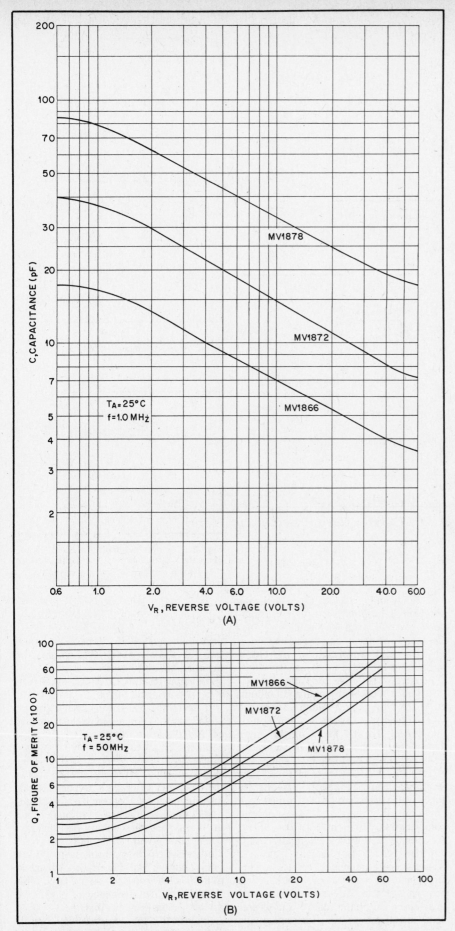

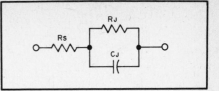

Fig. 18 — Representative circuit of a varactor diode showing case resistance, junction resistance and junction capacitance.

Fig. 17 — Reverse voltage respective to diode capacitance of three Motorola varactor diodes (A). Reverse voltage versus diode Q for the varactors at A are shown at B.

Present-day varactor diodes operate into the microwave part of the spectrum. They are quite efficient as frequency multipliers at power levels as great as 25 watts. The efficiency of a correctly designed varactor multiplier exceeds 50 percent in most instances. Fig. 19 illustrates the basic circuit of a frequency multiplier that contains a varactor diode. D1 is a single-junction device that serves as a frequency tripler in this example. FL1 is required to ensure reasonable purity of the output energy. It is a high-Q strip-line resonator. Without FL1 in the circuit there would be considerable output energy at 144, 288 and 864 MHz. Similar circuits are used as doublers, quadruplers and higher.

A Motorola MV104 tuning diode is used in the circuit of Fig. 20. It contains two varactor diodes in a back-to-back arrangement. The advantage in using two diodes is reduced signal distortion, as compared to a one-diode version of the same circuit. Reverse bias is applied equally to the two diodes in the three-terminal device. R1 functions as an RF isolator for the tuned circuit. The reverse bias is varied by means of R2 to shift the operating frequency. Regulated voltage is as important to the varactor as it is to the FET oscillator if reasonable frequency stability is to be ensured. Varactor diodes are often used to tune two or more circuits at the same time (receiver RF amplifier, mixer and oscillator), using a single potentiometer to control the capacitance of the diodes.

It is worth mentioning that some Zener diodes and selected silicon power-supply rectifier diodes will work effectively as varactors at frequencies as high as 144 MHz. If a Zener diode is used in this manner it must be operated below its reverse breakdown voltage point. The stud-mount variety of power supply diodes (with glass headers) are reported to be the best candidates as varactors, but not all diodes of this type will work effectively; experimentation is necessary.

Gunn Diodes

Gunn diodes are named after the developer, J. B. Gunn, who was studying carrier behavior at IBM Corp. in 1963. Basically, the Gunn effect is a microwave oscillation that occurs when heavy current is passed through bulk semiconductor material. A Gunn device is a diode only in

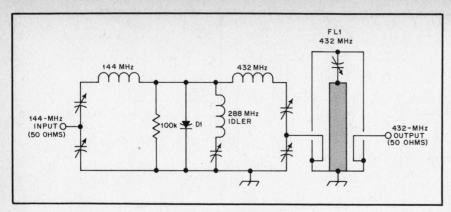

Fig. 19 — Typical circuit for a varactor-diode frequency tripler.

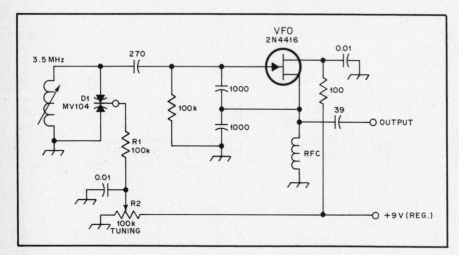

Fig. 20 — Example of a varactor-tuned VFO. D1 contains two varactors, back to back (see text).

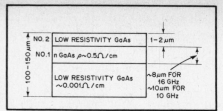

Fig. 21 — Cross-sectional illustration of Gunn diode.

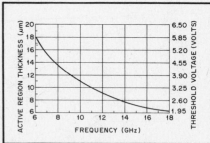

Fig. 22 — Active region thickness versus frequency of a Gunn diode.

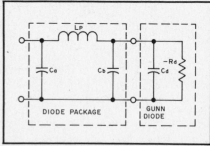

Fig. 23 — Equivalent circuit of a Gunn diode. The parasitic reactances of the diode package are included.

the sense that it has two electrodes; it has no rectifying properties.

Amateurs have been using Gunn oscillators at 10 and 24 GHz, but these devices are employed commercially from 4 to 100 GHz and beyond. In high-power (>100 mW) or relatively narrow-bandwidth applications, a varactor diode is used to adjust the oscillation frequency. Gunn oscillators in octave-bandwidth systems such as laboratory sweep generators use YIG tuning. In this method, the volume of a resonant cavity is varied by the magnetic expansion of an yttrium iron garnet sphere. The magnetic force for tuning is generated by a coil external to the cavity, and the oscillation frequency is very nearly a linear function of the coil current.

YIG tuning produces a much cleaner harmonic and close-in oscillation spectrum than that obtainable with varactor tuning, because of the higher Q. The output power of YIG-tuned oscillators must be limited to about 100 mW. This is because the YIG sphere becomes a nonlinear circuit element at higher powers and can cause spurious emissions. An in-depth treatment of the

technology and physics of Gunn devices is provided in the *Gunn Diode Circuit Handbook* by Microwave Associates, Inc. Chapter 32 of this book contains a section on practical applications of Gunn technology for radio amateurs.

Fig. 21 shows a cross-sectional representation of a slice of the material (gallium arsenide) from which Gunn diodes are made. Layer no. 1 is the active region of the device. The thickness of this layer depends on the chosen frequency of operation. For the 10-GHz band it is approximately 10 μm (10^{-5} meters) thick. The threshold voltage is roughly 3.3 volts. At 16 GHz the layer would be formed to a thickness of 8 μm, and the threshold voltage would be about 2.6 volts.

Layer no. 2 is grown epitaxially and is doped to provide low resistivity. This layer is grown on the active region of the semiconductor, but it is not essential to the primary operation of the diode. It is used to ensure good ohmic contact and to prevent metalization from damaging the N layer of the diode.

The composite wafer of Fig. 21 is

metalized on both sides to permit bonding into the diode package. This process of metalization also ensures a low electrical and thermal resistance. The completed chip is bonded to a gold-plated copper pedestal, with layer no. 2 next to the heat sink. A metal ribbon is connected to the back side of the diode to provide for electrical contact.

The curve in Fig. 22 shows the relationship of the diode active-region thickness to the frequency of operation. The curve illustrates an approximation because the actual thickness of the active region depends on the applied bias voltage and the particular circuit used. The input power to the diode must be 20 to 50 times the desired output power. Thus the efficiency from dc to RF is on the order of two to five percent.

The resonant frequency of the diode assembly must be higher than the operating frequency to allow for parasitic C and L components which exist. Fig. 23 shows the equivalent circuit of a packaged Gunn

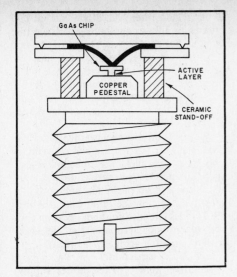

Fig. 24 — Illustration of a packaged Gunn diode as presented in literature from Microwave Associates.

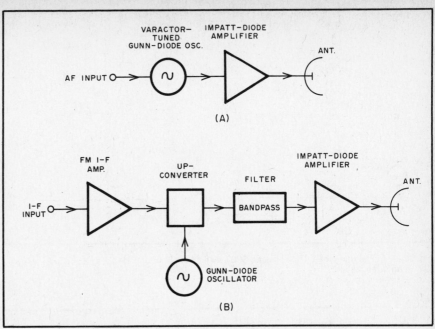

Fig. 25 — Block diagram of a simple Gunn-diode transmitter for FM (A) and an up-converter Gunn-diode transmitter (B).

diode. Assuming a diode natural resonant frequency of 17 GHz, the following approximate values result: L_p = 0.25 nH, C_a = 0.15 pF and C_b = 0.15 pF. Additional components exist within the diode chip. They are represented by C_d (capacitance) and $-R_d$ (negative resistance). These quantities, plus the stray resonances in the diode holder and bias leads in the microwave cavity, have a direct bearing on the electrical behavior of the Gunn oscillator. A cross-sectional representation of a packaged Gunn diode is shown at Fig. 24.

Presently, Gunn diodes are useful for generating powers between 0.1 and 1 watt. As the technology advances these power limits will increase. IMPATT (impact-avalanche transit time) diodes are useful as microwave amplifiers after a Gunn diode signal source. IMPATT diodes are also capable of providing power output in the 0.1- to 1-watt class.

Fig. 25 shows block diagrams of two Gunn-diode systems. In each example an IMPATT diode is used as an amplifier. Fig. 25A shows a direct FM transmitter which employs a varactor-deviated Gunn-diode oscillator as a signal source. FM is provided by applying audio to the bias lead of the varactor diode. The latter is coupled to the Gunn-diode cavity. Fig. 25B illustrates a microwave relay system in which a Gunn diode is used as an LO source. Essentially, the equipment is set up as a heterodyne up-converter transmitter. The upper sideband from the mixer is amplified at microwave frequency by means of an IMPATT diode.

PIN Diodes

A PIN diode is formed by diffusing heavily doped P+ and N+ regions into an almost intrinsically pure silicon layer, as illustrated in Fig. 26. In practice it is impossi-

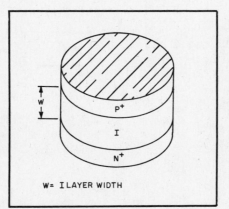

Fig. 26 — The PIN diode is constructed by diffusing P+ and N+ regions into an almost intrinsically pure silicon layer. Thus the name PIN diode.

ble to obtain intrinsically pure material, and the I layer can be considered to be a lightly doped N region. Characteristics of the PIN diode are primarily determined by the thickness, area and semiconductor nature of the chip, especially that of the I region. Manufacturers design for controlled-thickness I regions having long carrier lifetime and high resistivity. Carrier lifetime is basically a measure of the delay before an average electron and hole recombine. In a pure silicon crystal the theoretical delay is on the order of several milliseconds, although impurity doping can reduce the effective carrier lifetime to microseconds or nanoseconds.

When forward bias is applied to a PIN diode, holes and electrons are injected from the P+ and N+ regions into the I region. These charges do not immediately recom-

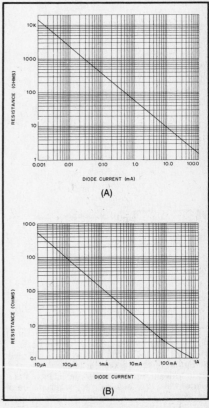

Fig. 27 — At A is a graph comparing diode resistance to forward-bias current for a PIN diode intended for low-level receiver applications. At B is a similar graph for a diode capable of handling over 100 watts of RF.

bine. Rather, a finite quantity of charge always remains stored and results in a lowering of the I-region resistivity. The amount of stored charge depends on the

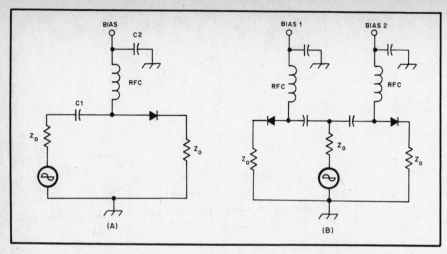

Fig. 28 — At A a PIN diode is used as an SPST switch. At B, two diodes form an SPDT switching arrangement.

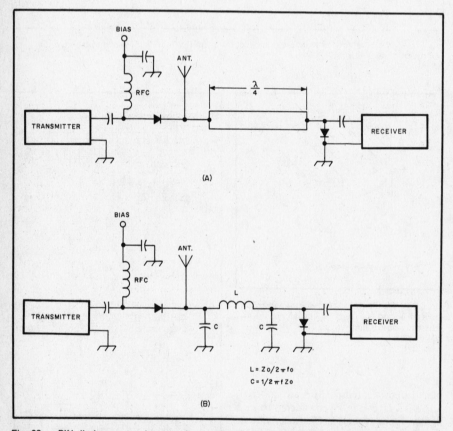

Fig. 29 — PIN diodes are used to transfer a common antenna to either a transmitter or receiver. A voltage applied to the bias terminal will switch the system to the transmit mode, connecting the output of the transmitter to the antenna. At the same time the diode across the receiver input is forward biased to a low-resistance state to protect the input stage of the receiver. The quarter-wave line isolates the low resistance of the receiver diode from the common antenna connection. At B the quarter-wave line is replaced with a lumped-element equivalent.

recombination time (carrier lifetime) and the level of the forward-bias current. The resistance of the I region under forward-bias conditions is inversely proportional to the charge and depends on the I-region width and mobility of the holes and electrons of the particular semiconductor material. Representative graphs of

resistance vs. forward bias level are shown in Fig. 27A and B for low-level receiving and high-power transmitting PIN diodes.

When a PIN diode is at zero or reverse bias, there is essentially no charge, and the intrinsic region can be considered as a low-loss dielectric. As with an ordinary PN junction there is a reverse breakdown or

Zener region where the diode current increases rapidly as the reverse voltage increases. For the intrinsic region to remain in a low-loss state, the maximum instantaneous reverse or negative voltage must not exceed the breakdown voltage. Also, the positive voltage excursion must not cause thermal losses to exceed the diode dissipation rating.

At high radio frequencies, when a PIN diode is at zero or reverse bias, the diode appears as a parallel plate capacitor, essentially independent of reverse voltage. It is the value of this capacitance that limits the effective isolation the diode can provide. PIN diodes intended for high isolation and not power-handling capability are designed with as small a geometry as possible to minimize the capacitance.

Manufacturers of PIN diodes supply data sheets with all necessary design data and performance specifications. Key parameters are diode resistance (when forward biased), diode capacitance, carrier lifetime, harmonic distortion, reverse voltage breakdown and reverse leakage.

PIN diodes are used in many applications, such as RF switches, attenuators and various types of phase-shifting devices. Our discussion will be confined to switch and attenuator applications since these are the most likely to be encountered by the amateur. The simplest type of switch that can be created with a PIN diode is the series SPST type. The circuit is shown in Fig. 28A. C1 functions as a dc blocking capacitor and C2 is a bypass capacitor. In order to have the signal from the generator flow to the load, a forward bias must be applied to the bias terminal. The amount of insertion loss caused by the diode is determined primarily by the diode bias current.

Fig. 28B illustrates an SPDT switch arrangement that uses essentially two SPST switches with a common connection. For a generator current to flow into the load resistor at the left, a bias voltage is applied to bias terminal 1. For signal to flow into the load at the right a bias must be applied to terminal 2. In practice it is usually difficult to achieve more than 40 dB isolation with a single diode switch at UHF and microwaves. Better performance, in excess of 100 dB, is achievable using compound switches. Compound switches are made up of two or more diodes in a series/shunt arrangement. Since not all diodes are biased for the same state, some increase in bias-circuit complexity results.

One general class of switches used in connection with transceive applications requires that a common antenna be connected to either the receiver or transmitter during the appropriate receive or transmit states. When PIN diodes are used as switching elements in these applications, higher reliability, better mechanical ruggedness and faster switching speeds are achieved relative to the electromechanical relay.

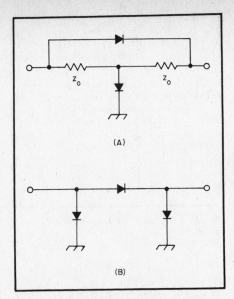

Fig. 30 — Two types of PIN diode attenuator circuits. The circuit at A is called a bridged T and the circuit at B is a pi type. Both exhibit very broadband characteristics.

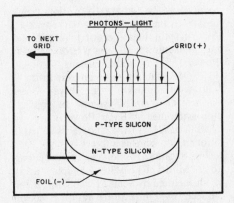

Fig. 31 — A solar-energy diode cell. Electrons flow when light strikes the upper surface. The bottom of the cell is coated with foil to collect current for the load, or for the succeeding cell in series-connected arrays of cells.

A basic approach is shown in Fig. 29A and B where a PIN diode is used in series with the transmit line and another diode in shunt with the receive line. A single bias supply is used to turn on the series diode during transmit while also turning on the shunt diode to protect the receiver. The quarter-wave line between the two diodes is necessary to isolate the low resistance of the receiver diode from the antenna connection. During receive periods both diodes are effectively open circuited, allowing signal energy to be applied to the receiver. At B is the same basic circuit, although the quarter-wave line has been replaced with a lumped element section.

Two of the more common types of attenuators using PIN diodes are shown in Fig. 30. The type at A is referred to as a bridged T, while the circuit at B is the com-

mon pi type. Both are useful as very broadband devices. It is interesting to note that the useful upper frequency of these attenuators often depends on the bias circuit isolation rather than the PIN diode characteristics.

Light-Emitting Diodes (LEDs)

The primary component in optoelectronics is the LED. This diode contains a PN junction of crystal material that produces luminescence around the junction when forward bias current is applied. LED junctions are made from gallium arsenide (GaAs), gallium phosphide (GaP), or a combination of both materials (GaAsP). The latter depends on the color and light intensity desired. Today, the available LED colors are red, green and yellow.

Some LEDs are housed in plastic affixed to the base header of a transistor package. Other LEDs are contained entirely in plastic packages that have a dome-shaped head at the light-emitting end. Two wires protrude from the opposite end (positive and negative leads) for applying forward bias to the device.

There are countless advantages to the use of LEDs. Notable among them are the low current drain, long life (sometimes 50 years, as predicted) and small size. They are useful as visual indicators in place of incandescent panel lamps. One of their most common applications is in digital display units, where arrays of tiny LEDs are arranged to provide illuminated segments in numeric-display assemblies.

The forward bias current for a typical LED ranges between 10 and 20 mA for maximum brilliance. An applied voltage of 1½ to 2 is also typical. A 1000-ohm resistor in series with a 12-volt source will permit the LED to operate with a forward current of approximately 10 mA (IR drop = 10 V). A maximum current of 10 mA is suggested in the interest of longevity for the device.

LEDs are also useful as reference diodes, however unique the applications may seem. They will regulate dc at approximately 1.5 V.

The following are definitions and terms used in optics to characterize the properties of an LED.

Incident flux density is defined as the amount of radiation per unit area (expressed as lumens/cm² in photometry; watts/cm² in radiometry). This is a measure of the amount of flux received by a detector measuring the LED output.

Emitted flux density is also defined as radiation per unit area and is used to describe light reflected from a surface. This measure of reflectance determines the total radiant luminous emittance.

Source intensity defines the flux density that will appear at a distant surface and is expressed as lumens/steradian (photometry) or watts/steradian (radiometry).

Luminance is a measure of photometric brightness and is obtained by dividing the luminous intensity at a given point by the

projected area of the source at the same point. Luminance is a very important rating in the evaluation of visible LEDs.

While luminance is equated with photometric brightness, it is inaccurate to equate luminance as a figure of merit for brightness. The only case in which this rating is acceptable is when comparing physically identical LEDs. Different LEDs are subject to more stringent examination. Manufacturers do not use a set of consistent ratings for LEDs (such as optical flux, brightness and intensity). This is because of the dramatic differences in optical measurements between point- and area-source diodes. Point-source diodes are packaged in a clear epoxy or set within a transparent glass lens. Area-source diodes must employ a diffusing lens to spread the flux over a wider viewing area and hence have much less point intensity (luminance) than the point-source diodes.

Solar-Electric Diodes

Sunlight can be converted directly into electricity by a process known as photovoltaic conversion. A solar cell is used for this purpose. It relies on the photoelectric properties of a semiconductor. Practically, the solar cell is a large-area PN junction diode. The greater the area of the cell, the higher the output current will be. A dc voltage output of approximately 0.5 is obtained from a single cell. Numerous cells can be connected in series to provide 6, 9, 12, 24 or whatever low voltage is required. In a like manner, cells can be connected in parallel to provide higher output current, overall.

The solar diode cell is built so that light can penetrate into the region of the PN junction, Fig. 31. Most modern solar cells use silicon material. Impurities (doping) are introduced into the silicon material to establish excess positive or negative charges which carry electric currents. Phosphorous is used to produce N-type silicon. Boron is used as the dopant to produce P-type material.

Light is absorbed into the silicon to generate excess holes and electrons (one hole/electron pair for each photon absorbed). When this occurs near the PN junction, the electric fields in that region will separate the holes from the electrons. This causes the holes to increase in the P-type material. At the same time the electrons will build up in the N-type material. By making direct connection to the P and N regions by means of wires, these excess charges generated by light (and separated by the junction) will flow into an external load to provide power.

Approximately 0.16 A can be obtained from each square inch of solar-cell material exposed to bright sunlight. A 3½-inch-diameter cell can provide 1.5 A of output current. The efficiency of a solar cell (maximum power delivered to a load versus total solar energy incident on the cell) is typically 11 to 12 percent.

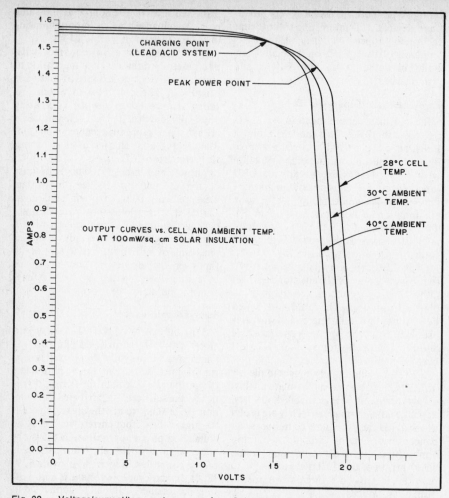

Fig. 32 — Voltage/current/temperature curve for a Solar Power Corp. array that contains 36 solar-electric cells in series. The curves are for a model E12-01369-1.5 solar panel.

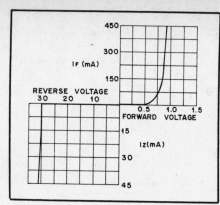

Fig. 34 — Typical characteristics of a Zener diode (30 V).

Arrays of solar cells are manufactured for all manner of practical applications. A storage battery is used as a buffer between the solar panel and the load. A PN junction diode should be used between the solar-array output and the storage battery to prevent the battery from discharging back into the panel during dark periods. An article on the subject of solar cells and their amateur applications (DeMaw "Solar Power for the Radio Amateur," August 1977 *QST)* should be of interest to those who wish to utilize solar power. *Solar Electric Generator Systems,* an application pamphlet by Solar Power Corp. of N. Billerica, MA 01862, contains valuable information on this subject.

Fig. 32 shows the voltage/current curves to a Model E12-01369-1.5 solar array manufactured by Solar Power Corp. It can be seen that temperature has an effect on the array performance.

Tunnel Diodes

One type of semiconductor diode having no rectifying properties is called a tunnel diode. The bidirectional conduction of the device is a result of heavily doped P and N regions with a very narrow junction. The Fermi level lies within the conduction band for the N side and within the valence band for the P side. A typical current-vs.-voltage curve for a tunnel diode is sketched in Fig. 33. When the forward bias potential exceeds about 30 mV, increasing the voltage causes the current to decrease, resulting in a negative resistance characteristic. This effect makes the tunnel diode capable of amplification and oscillation.

At one time tunnel diodes were expected to dominate in microwave applications, but other devices soon surpassed tunnel diodes in performance. The two-terminal oscillator concept had great fad appeal, and some amateurs built low-power transmitters based on tunnel diodes. In the 1960s the Heath Company marketed a dip meter that used a tunnel diode oscillator. Tunnel diodes are not widely used in new designs;

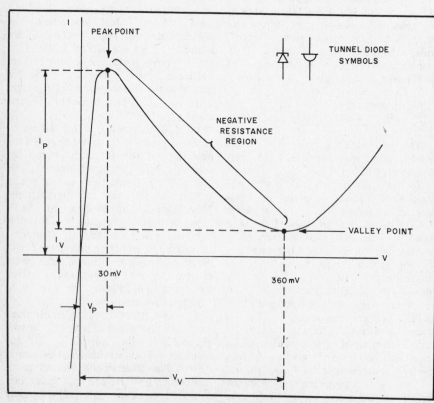

Fig. 33 — Schematic symbol and current-vs.-voltage characteristic for a tunnel diode.

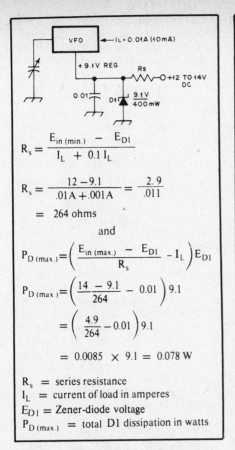

$$R_s = \frac{E_{in\,(min.)} - E_{D1}}{I_L + 0.1\,I_L}$$

$$R_s = \frac{12 - 9.1}{.01\,A + .001\,A} = \frac{2.9}{.011}$$

$$= 264 \text{ ohms}$$

and

$$P_{D\,(max.)} = \left(\frac{E_{in\,(max.)} - E_{D1}}{R_s} - I_L\right) E_{D1}$$

$$P_{D\,(max.)} = \left(\frac{14 - 9.1}{264} - 0.01\right) 9.1$$

$$= \left(\frac{4.9}{264} - 0.01\right) 9.1$$

$$= 0.0085 \times 9.1 = 0.078 \text{ W}$$

R_s = series resistance
I_L = current of load in amperes
E_{D1} = Zener-diode voltage
$P_{D\,(max.)}$ = total D1 dissipation in watts

Fig. 35 — Example of how a shunt type of Zener diode regulator is used. The equations show how to calculate the value of the series resistor and the diode power dissipation. In this example a 400-mW Zener diode will suffice (D1).

this material is included only for completeness.

Zener Diodes

Zener diodes have, for the most part, replaced the gaseous regulator tube. They have been proved more reliable than tube types of voltage regulators, are less expensive and far smaller in size.

These diodes fall into two primary classifications: voltage regulators and voltage-reference diodes. When they are used in power supplies as regulators, they provide a nearly constant dc output voltage even though there may be large changes in load resistance or input voltage. As a reference element the Zener diode utilizes the voltage drop across its junction when a specified current passes through it in the reverse-breakdown direction (sometimes called the Zener direction). This "Zener voltage" is the value established as a reference. Therefore, if a 6.8-volt Zener diode was set up in this manner, the resultant reference voltage would be 6.8.

At the present time it is possible to purchase Zener diodes rated for various voltages between 2.4 and 200. The power ratings range from ¼ to 50 watts. Fig. 34 shows the characteristics of a Zener diode designed for 30-volt operation.

Fig. 35 shows how to calculate the series

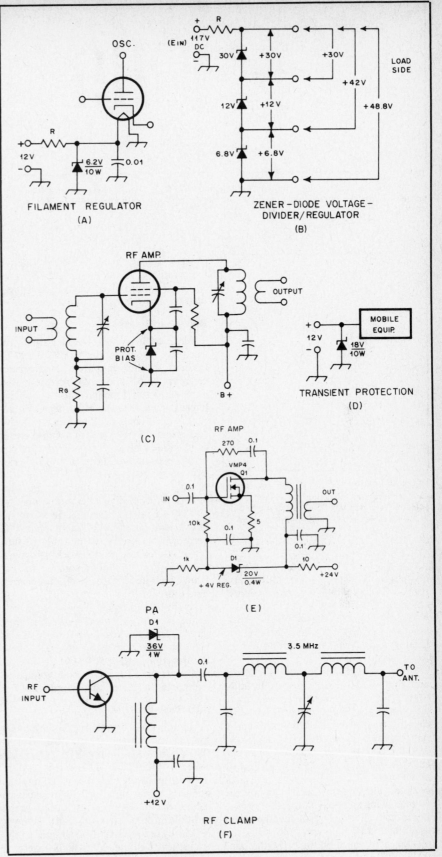

Fig. 36 — Practical examples of Zener diode applications. The circuit at A is useful for stabilizing the filament voltage of oscillators. Zener diodes can be used in series to obtain various levels of regulated voltage (B). Fixed-value bias for transmitter stages can be obtained by inserting a Zener diode in the cathode return (C). At D an 18-volt Zener diode prevents voltage spikes from harming a mobile transceiver. A Zener-diode series regulator (20-V drop) is shown at E and an RF clamp is shown at F. D1 in the latter circuit will clamp at 36 volts to protect the PA transistor from dc voltage spikes and extreme sine-wave excursions at RF. This circuit is useful in protecting output stages during no-load or short-circuit conditions.

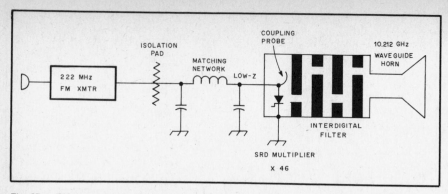

Fig. 37 — Step-recovery-diode frequency multiplier for 10 GHz. The matching network elements are represented as lumped components but would take the form of a microstripline in an actual design.

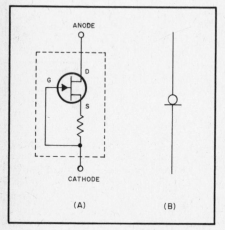

Fig. 38 — At A, an N-channel JFET connected as a constant-current source. At B, the schematic symbol for the circuit in A when it is packaged as a two-terminal device.

resistance needed in a simple shunt regulator which employs a Zener diode. An equation is included for determining the wattage rating of the series resistor. Additional data on this subject is given in Chapter 7 of *Solid State Design for the Radio Amateur.*

Some practical applications for Zener diodes are illustrated in Fig. 36. In addition to the shunt applications given in the diagram, Zener diodes can be used as series elements when it is desired to provide a gate that conducts at a given voltage. These diodes can be used in ac as well as dc circuits. When they are used in an ac type of application they will conduct at the peak voltage value or below, depending on the voltage swing and the voltage rating of the Zener diode. For this reason they are useful as audio and RF clippers. In RF work the reactance of the diode may be the controlling factor above approximately 10 MHz with respect to the performance of the RF circuit and the diode.

Most Zener diodes which are rated higher than 1 watt in dissipation are contained in stud-mount packages. They should be affixed to a suitable heat sink to

prevent damage from excessive junction temperatures. The mounting techniques are the same as for power rectifiers and high-wattage transistors.

Reference Diodes

While ordinary Zener diodes are useful as voltage regulators, they don't exhibit the thermal stability required in precision reference applications. A reverse-biased semiconductor junction has a positive temperature coefficient of barrier potential. A forward-biased junction has a negative coefficient. The way to temperature-compensate a Zener diode is to connect one or more common silicon diodes in series with it. When this is done as part of the manufacturing process, the resulting component is termed a reference diode. A 1N3499 6.2-volt reference diode will maintain a temperature coefficient of 0.0005 percent per degree over the range of 0 to 75° C. Reference diodes work best when operated at a few milliamperes of current from a high-impedance or constant-current bias source.

As the name implies, these diodes aren't suited for circuits where power is taken directly from the device. Reference diodes can't be tested with an ohmmeter because two junctions are back-to-back — the instrument can't supply enough voltage to overcome the Zener barrier potential.

Step-Recovery Diodes

One device characterized by extremely low capacitance and short storage time is the step-recovery diode (SRD), sometimes called a "snap" diode. These diodes are used as frequency multipliers well into the microwave spectrum. Switching the device in and out of forward conduction is the multiplication mechanism, and the power efficiency is inversely proportional to the frequency multiple. Very high orders of multiplication are possible with step-recovery diodes. One use for this feature is in a comb generator, an instrument used to calibrate the frequency axis of a spectrum analyzer.

A single harmonic of the excitation frequency can be selected by an interdigital

filter or cavity resonator. A 1-watt, 220-MHz FM transmitter could drive a snap diode multiplier (×46) and filter combination to an output of about 10 mW in the 10-GHz band — a typical and effective power level at that frequency. A representative system of this variety is suggested in Fig. 37. The exciter should be well isolated from the SRD and its matching network to prevent parasitic oscillations.

Current-Regulator Diodes

A JFET with its gate shorted to its source or connected below a source resistor will draw a certain current whose value is almost entirely independent of the applied potential. The current (I_{DSS} in FET terminology) is also quite stable with temperature. Semiconductor manufacturers take advantage of those properties and package the JFET circuit of Fig. 38A in a two-terminal package and call it a constant-current diode. A special symbol, given in Fig. 38B, is assigned to this type of diode. The 1N5305 diode approaches an ideal current generator, in that it draws 2 mA over the range of 1.8 to 100 volts. Constant-current diodes find application in ohmmeters, ramp generators and precision voltage references.

BIPOLAR TRANSISTORS

The word "transistor" was chosen to describe the function of a three-terminal PN junction device that is able to amplify signal energy (current). The inherent characteristic is one of "transferring current across a resistor." The transistor was invented by Shockley, Bardeen and Brattain at Bell Labs in 1947 and has become the standard amplifying device in electronic equipment. In RF and audio applications it is practical to obtain output power in excess of 1000 watts by using several amplifier blocks and hybrid power combiners. The primary limitation at the higher power levels is essentially a practical or economic one: Low voltage, high-current power supplies are required, and the cost can exceed that of a high-voltage moderate-current supply of the variety that would be employed with a vacuum-tube amplifier of comparable power. The primary advantages of solid-state power amplifiers are compactness and reliability.

In small-signal applications the transistor outweighs the vacuum-tube in performance. There remains in some quarters a belief that transistors are hard to tame, are noisier than tubes and are subject to damage at the flick of a switch. None of this is true. A transistorized circuit designed and operated correctly is almost always capable of outperforming an equivalent vacuum-tube circuit in all respects. An understanding of how transistors function will help to prevent poor circuit performance. The fundamentals outlined in this chapter are provided for the amateur designer so the common pitfalls can be avoided.

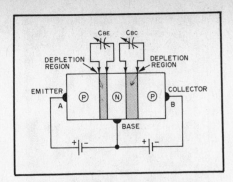

Fig. 39 — Illustration of a junction PNP transistor. Capacitances C_{be} and C_{bc} vary with changes in operating and signal voltage (see text).

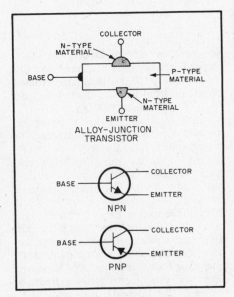

Fig. 40 — Pictorial and schematic representations of junction transistors. The base, collector and emitter can be compared to the grid, plate and cathode of a triode tube, respectively.

Fig. 39 shows a "sandwich" made from two layers of P-type semiconductor material with a thin layer of N-type between. There are in effect two PN junction diodes back-to-back. If a positive bias is applied to the P-type material at the left, current will flow through the left-hand junction, the holes moving to the right and the electrons from the N-type material moving to the left. Some of the holes moving into the N-type material will combine with the electrons there and be neutralized, but some of them also will travel to the region of the right-hand junction.

If the PN combination at the right is biased negatively, as shown, there would normally be no current flow in this circuit. However, there are now additional holes available at the junction to travel to point B and electrons can travel toward point A, so a current can flow even though this sec-

tion of the sandwich is biased to prevent conduction. Most of the current is between A and B and does not flow out through the common connection to the N-type material in the sandwich.

A semiconductor combination of this type is called a transistor, and the three sections are known as the emitter, base and collector. The amplitude of the collector current depends principally on the amplitude of the emitter current; that is, the collector current is controlled by the emitter current.

Between each PN junction exists an area known as the depletion, or transition, region. It is similar in characteristics to a dielectric layer, and its width varies in accordance with the operating voltage. The semiconductor materials either side of the depletion region constitute the plates of a capacitor. The capacitance from base to emitter is shown as C_{be} (Fig. 39), and the base-collector capacitance is represented as C_{bc}. Changes in signal and operating voltages cause a nonlinear change in these junction capacitances, which must be taken into account when designing some circuits. A base-emitter resistance, rb', also exists. The junction capacitance, in combination with rb', determines the useful upper frequency limit (f_T or f_a) of a transistor by establishing an RC time constant.

Power Amplification

Because the collector is biased in the back direction the collector-to-base resistance is high. On the other hand, the emitter and collector currents are substantially equal, so the power in the collector circuit is larger than the power in the emitter circuit ($P = I^2R$, so the powers are proportional to the respective resistances, if the currents are the same). In practical transistors emitter resistance is on the order of a few ohms while the collector resistance is hundreds or thousands of times higher, so power gains of 20 to 40 dB or even more are possible.

Transistor Types

The transistor may be one of the types shown in Fig. 40. The assembly of P- and N-type materials may be reversed, so that PNP and NPN transistors are both possible.

The first two letters of the NPN and PNP designations indicate the respective polarities of the voltages applied to the emitter and collector in normal operation. In a PNP transistor, for example, the emitter is made positive with respect to both the collector and the base, and the collector is made negative with respect to both the emitter and the base.

Manufacturers are constantly working to improve the performance of their transistors — greater reliability, higher power and higher frequency ratings, and improved uniformity of characteristics for any given type number. One such development provided the overlay transistor, whose

emitter structure is made up of several emitters that are joined together at a common case terminal. This process lowers the base-emitter resistance, rb', and improves the transistor input time constant. The time constant is determined by rb' and the junction capacitance of the device.

The overlay transistor is extremely useful in VHF and UHF applications. It is capable of high-power operation well above 1000 MHz. These transistors are useful as frequency doublers and triplers, and are able to provide an actual power gain in the process.

Another multi-emitter transistor that has been developed for use from HF through UHF should be of interest to the radio amateur. It is called a balanced-emitter transistor (BET), or "ballasted" transistor. The transistor chip contains several triode semiconductors whose bases and collectors are connected in parallel. The various emitters, however, have built-in emitter resistors (typically about 1 ohm) that provide a current-limiting safety factor during overload periods, or under conditions of significant mismatch. Since the emitters are brought out to a single case terminal the resistances are effectively in parallel, thus reducing the combined emitter resistances to a fraction of an ohm. (If a significant amount of resistance were allowed to exist it would cause degeneration in the stage and would lower the gain of the circuit.)

Most modern transistors are of the junction variety. Various names have been given to the several types, some of which are junction alloy, mesa and planar. Though their characteristics may differ slightly, they are basically of the same family and simply represent different physical properties and manufacturing techniques.

Transistor Characteristics

An important characteristic of a transistor is its beta (β), or current-amplification factor, which is sometimes expressed as H_{FE} (static forward-current transfer ratio) or h_{fe} (small-signal forward-current transfer ratio). Both symbols relate to the grounded-emitter configuration. Beta is the ratio of the collector current to the base current

$$\beta = \frac{I_c}{I_b}$$

Thus, if a base current of 1 mA causes the collector current to rise to 100 mA the beta is 100. Typical betas for junction transistors range from as low as 10 to as high as several hundred.

A transistor's alpha (α) is the ratio of the collector to the emitter current. Symbols h_{FB} (static forward-current transfer ratio) and h_{fb} (small-signal forward-current transfer ratio), common-base hookup, are frequently used in connection with gain. The smaller the base current, the closer the collector current comes to being equal to that of the emitter, and the closer alpha

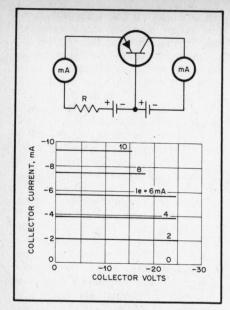

Fig. 41 — Typical collector-current versus collector-voltage characteristics of a junction transistor for various emitter-current values. Because the emitter resistance is low, a current-limiting resistor (R) is placed in series with the source current. The emitter current can be set at a desired value by adjustment of this resistance.

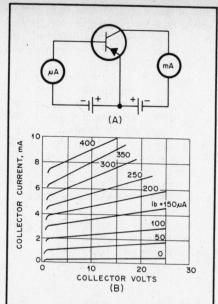

Fig. 42 — Collector current versus collector voltage for various values of base current in a junction transistor. The illustration at A shows how the measurements are made. At B is a family of curves.

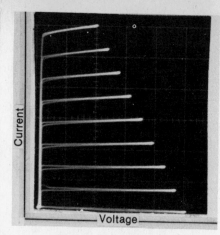

Fig. 43 — Curve-tracer display of a small-signal transistor characteristics.

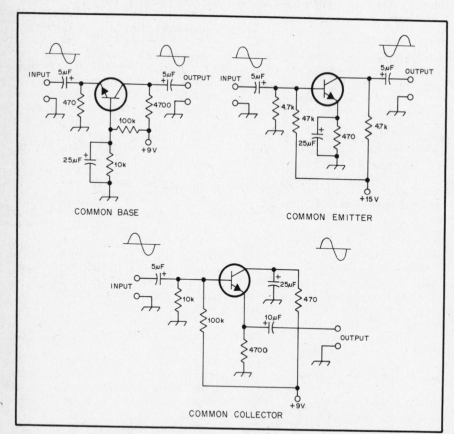

COMMON BASE

COMMON EMITTER

COMMON COLLECTOR

Fig. 44 — Basic transistor amplifiers. Observe the input and output phase relationships for the various configurations.

comes to being 1. Alpha for a junction transistor is usually between 0.92 and 0.98.

Transistors have frequency characteristics that are important to circuit designers.

Symbol f_T is the gain bandwidth product (common-emitter) of the transistor. This is the frequency at which the gain becomes unity, or 1. The expression "alpha cutoff" is frequently used to express the useful upper-frequency limit of a transistor, and this relates to the common-base hookup. Alpha cutoff is the point at which the gain is 0.707 its value at 1000 Hz.

Another factor that limits the upper-frequency capability of a transistor is its transit time. This is the period of time required for the current to flow from emitter to collector, through the semiconductor base material. The thicker the base material, the greater the transit time. Hence, the thicker the base material the more likelihood there will be of phase shift of the signal passing through it. At frequencies near and above f_T or alpha cutoff, partial or complete phase shift can occur. This will give rise to positive feedback because the internal capacitance, C_{be}, feeds part of the in-phase collector signal back to the base. The positive feedback can cause instability and oscillation, and in most cases will interlock the input and output tuned circuits of an RF amplifier so it is almost impossible to tune them properly. This form of feedback can be corrected by using what is termed "unilateralization." Conventional positive feedback can be nullified by using neutralization, as is done with vacuum-tube amplifiers.

Characteristic Curves

The operating principles of transistors can be shown by a series of characteristic curves. One such set of curves is shown in Fig. 41. It shows the collector current vs. collector voltage for a number of fixed values of emitter current. Practically, the collector current depends almost entirely on the emitter current and is independent of the collector voltage. The separation between curves is quite uniform, indicating that almost distortionless output can be obtained over the useful operating range of the transistor.

Another type of curve is shown in Fig. 42, together with the circuit used for obtaining it. This also shows collector current vs. collector voltage, but for a number of different values of base current. In this case the emitter element is used as the common point in the circuit. The collector current is not independent of collector voltage with this type of connection, indicating that the output resistance of the device is fairly low. The base current also is quite low, which means the resistance of the base-emitter circuit is moderately high with this method of connection. This may be contrasted with the high values of emitter current shown in Fig. 41. An actual oscillograph of a characteristic family of curves for a small-signal transistor is shown in Fig. 43. It was obtained by means of a curve tracer.

Transistor Amplifiers

Amplifier circuits used with transistors fall into one of three types, known as the common-base, common-emitter and common-collector circuits. These are shown in Fig. 44 in elementary form. The three circuits correspond approximately to the grounded-grid, grounded-cathode and cathode-follower circuits, respectively, used with vacuum tubes.

The important transistor parameters in these circuits are the short-circuit current transfer ratio, the cutoff frequency, and the input and output impedances. The short-circuit current transfer ratio is the ratio of a small change in output current to the change in input current that causes it, the output circuit being short-circuited. The cutoff frequency was discussed earlier in this chapter. The input impedance is that which a signal source working into the transistor would see, and the output impedance is the internal output impedance of the transistor (corresponding to the plate resistance of a vacuum tube, for example).

Common-Base Circuit

The input circuit of a common-base amplifier must be designed for low impedance, since the emitter-to-base resistance is of the order of $26/I_e$ ohms, where I_e is the emitter current in milliamperes. The optimum output load impedance, R_L, may range from a few thousand ohms to 100,000, depending on the requirements.

In this circuit the phase of the output (collector) current is the same as that of the input (emitter) current. The parts of these currents that flow through the base resistance are likewise in phase, so the circuit tends to be regenerative and will oscillate if the current amplification factor is greater than one.

Common-Emitter Circuit

The common-emitter circuit shown in Fig. 44 corresponds to the ordinary grounded-cathode vacuum-tube amplifier. As indicated by the curves of Fig. 42, the base current is small and the input impedance is therefore fairly high — several thousand ohms in the average case. The collector resistance is some tens of thousands of ohms, depending on the

signal source impedance. The common-emitter circuit has a lower cutoff frequency than does the common-base circuit, but it gives the highest power gain of the three configurations.

In this circuit the phase of the output (collector) current is opposite to that of the input (base) current so such feedback as occurs through the small emitter resistance is negative and the amplifier is stable.

Because the common-emitter amplifier circuit illustrated in Fig. 44 is one of the most often seen applications for a bipolar transistor, a brief analysis and discussion of the design network biasing the base supplies an open-circuit potential of 1.36 V. Connecting the base has little loading effect if the transistor beta is high (a fair assumption). A beta of 100, for example, causes the 470-ohm emitter resistor to present a dc base resistance of 47 kΩ, which is a negligible shunt. The base-emitter junction drops 0.6 V, so the potential across the emitter resistor is $1.36 - 0.6 = 0.76$ V. This voltage causes 1.6 mA to flow in the emitter and the 470-ohm resistor. Since the transistor alpha is nearly unity, the collector current is also 1.6 mA, which drops 7.52 volts across the collector load resistor. The quiescent (no signal) collector voltage is therefore $15 - 7.52 = 7.48$ V. This value allows the maximum undistorted output voltage swing between cutoff and saturation.

Assuming the emitter bypass capacitor has negligible reactance at the operating frequency, the emitter is at ac ground. Because of the high alpha mentioned earlier, any emitter current variation caused by an input signal will also appear in the collector circuit. Since the current variation is the same, the voltage gain is the ratio of the collector load resistance to the (internal) emitter resistance. For small signals, this emitter resistance can be approximated by $R_e = 26/I_e$, where R_e is the emitter-to-base junction resistance, and I_e is the emitter current in milliamperes. In our example, $I_e = 1.6$, so $R_e = 16.25$ Ω. The voltage gain, then, is 289, which is 49 dB. The ac base impedance is given by R_e. Using the previous values for beta and emitter resistance results in a base impedance of 1625 Ω. The circuit input impedance is found by shunting the base impedance with the bias resistors. The result in this case is about 1189 Ω.

If the emitter bypass capacitor is omitted, the external resistor dominates the gain equation, which becomes: $A_v = R_L/R_E$, where A_v is the voltage gain, R_L is the collector load resistance as before, and R_E is the unbypassed emitter resistance. Without the bypass capacitor, the common-emitter circuit in Fig. 44 exhibits a numerical voltage gain of 10 (20 dB). The base impedance becomes βR_E, or 47 kΩ. This value is swamped by the bias network for a circuit input impedance of 3.91 kΩ. The emitter resistor has introduced 29 dB of degenerative feedback to the circuit,

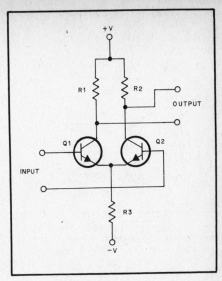

Fig. 45 — Differential amplifier. This arrangement can be analyzed as a composite of the common-collector and common-base circuits.

stabilizing the gain and impedance values over a wide frequency range. A dc beta of 100 was assumed in this example, and for convenience this value was also assigned at the operating frequency. In reality, however, beta decreases with increasing frequency, as noted in the section on transistor characteristics. Degenerative feedback overcomes this effect to a large extent.

Common-Collector Circuit (Emitter-Follower)

Like the vacuum-tube cathode follower, the common-collector transistor amplifier has high input impedance and low output impedance. The latter is approximately equal to the impedance of the signal input source multiplied by $(1 - \alpha)$. The input resistance depends on the load resistance, being approximately equal to the load resistance divided by $(1 - \alpha)$. The fact that input resistance is directly related to the load resistance is a disadvantage of this type of amplifier if the load is one whose resistance or impedance varies with frequency.

The current transfer ratio with this circuit is

$$\frac{1}{1 - \alpha}$$

and the cutoff frequency is the same as in the grounded-emitter circuit. The output and input currents are in phase.

Differential Amplifier Circuit

An important variation of the fundamental amplifier types is the differential amplifier, drawn in Fig. 45. The output voltage is proportional to the difference (with respect to ground) between the

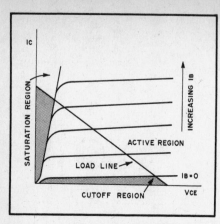

Fig. 46 — Typical characteristic for the collector of an NPN transistor that shows the three primary regions involved during switching.

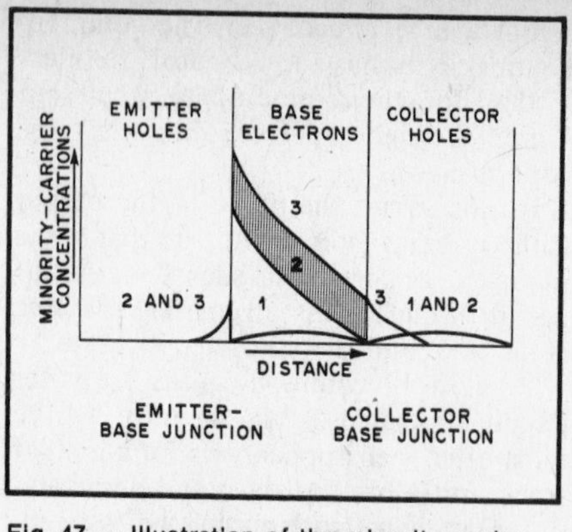

Fig. 47 — Illustration of the minority-carrier concentrations in an NPN transistor. No. 1 shows the cutoff region. No. 2 is the active region at the threshold of the saturation region. No. 3 is in the saturation region.

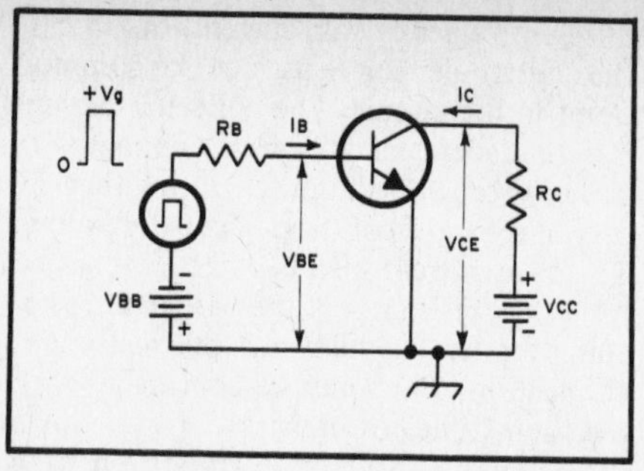

Fig. 48 — Circuit for a transistor switching circuit (saturated).

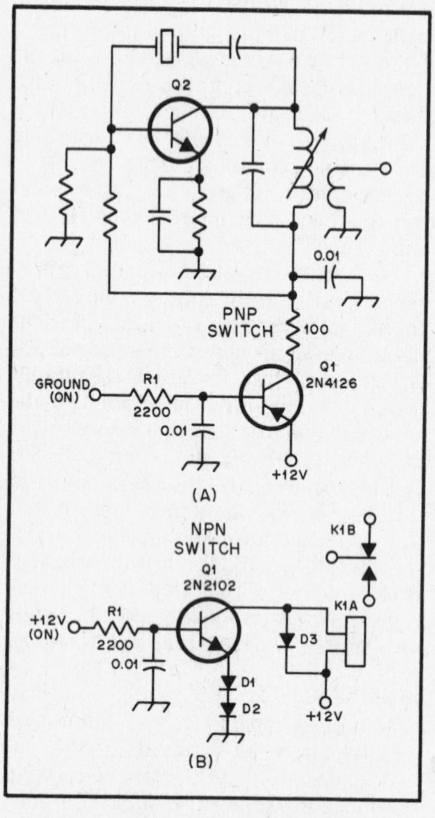

Fig. 49 — Examples of practical switching circuits. A PNP switch is used to key an oscillator at A. When R1 is grounded the switching transistor is forward biased to saturation, permitting current to flow from the dc supply line to Q2. The circuit at B shows Q1 as a relay-driver NPN switch. When +12 volts is applied to the base of Q1 it is forward biased to saturation, permitting current to flow through the field coil of K1A. D1 and D2 are included to reduce the static collector current of Q1, which in some instances could cause K1A to remain closed after forward bias was removed from Q1. D3 serves as a spike suppressor when the field of K1A collapses.

voltages applied to the input terminals. With the proper choice of operating conditions, several differential amplifier stages of the type shown can be cascaded directly. Fig. 45 shows the circuit in its classic balanced form, but many circuits use differential amplifiers in a single-ended configuration. When only a single input and output terminal is required, R1 could be a short circuit and the Q2 base could be grounded. Under these circumstances the differential amplifier can be understood as an emitter-follower driving a common-base stage. The output is taken between the Q2 collector and ground. R3 establishes the current in Q1 and Q2, which should be equal under static conditions.

Differential amplifiers work best when R3 is replaced by some type of constant-current source. One type of current regulator has been discussed in the diode section, and current sources made from bipolar transistors are covered later.

With a current source biasing Q1 and Q2 the input signal cannot modulate the total collector current; only the ratio of the currents varies. One beneficial result of the constant-current bias is that a higher impedance is presented to the driving signal.

Bipolar Transistor Dissipation

Apart from the characteristics mentioned earlier, it is necessary to consider the matters of collector dissipation, collector voltage and current, and emitter current. Variations in these specifications are denoted by specific parameter symbols which appear later in the chapter. The maximum dissipation ratings of transistors, as provided on the manufacturer's data sheets, tend to confuse some amateurs. An acceptable rule of thumb is to select a transistor that has a maximum dissipation rating of approximately twice the dc input power of the circuit stage. That is, if a 5-watt dc input is contemplated, choose a transistor with a 10-watt or greater rating.

When power levels in excess of a few hundred milliwatts are necessary there is a need for heat sinking. A sink is a metal device that helps keep the transistor cool by virtue of heat transfer from the transistor case to the sink. At power levels below 5 watts it is common practice to employ clip-on heat sinks of the crown variety. For powers greater than 5 watts it is necessary to use large-area heat sinks fashioned from extruded aluminum. These sinks have cooling fins on one or more of their surfaces to hasten the cooling process. Some high-power, solid-state amplifiers employ cooling fans from which the air stream is directed on the metallic heat sink. Regardless of the power level or type of heat sink used, silicone heat-transfer compound should always be used between the mating surfaces of the transistor and the heat sink. Another rule of thumb is offered: If the heat-sink-equipped transistor is too warm to touch with comfort, the heat sink is not large enough in area.

Excessive junction heat will destroy a transistor. Prior to destruction the device may go into thermal runaway. During this condition the transistor becomes hotter and its internal resistance lowers. This causes an increase in emitter/collector and emitter/base current. This increased current elevates the dissipation and further lowers the internal resistance. These effects are cumulative: Eventually the transistor will be destroyed. A heat sink of proper size will prevent this type of problem. Excessive junction temperature will eventually cause the transistor to become open. Checks with an ohmmeter will indicate this condition after a failure.

Excess collector voltage will also cause immediate device failure. The indication of this type of failure, as noted by means of an ohmmeter, is a shorted junction.

Bipolar Transistor Applications

Silicon transistors are the most common types in use today, although a few germanium varieties are built for specific applications. Collector voltages as great as 1500 can be accommodated by some of the high-power silicon transistors available now. Most small-signal transistors will safely handle collector voltages of 25 or greater. Generally speaking, transistors in the small-signal class carry dissipation ratings of 500 mW or less. Power transistors are normally classed as 500-mW and higher devices.

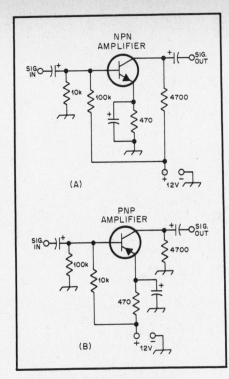

Fig 50 — Examples of PNP and NPN amplifiers operating from a power supply with a negative ground.

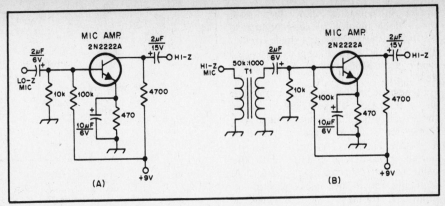

Fig. 51 — RC and transformer-coupled audio amplifiers suitable for high- and low-impedance microphones.

The practical applications for all of these semiconductors range from dc to the microwave spectrum.

Bipolar Transistor Switches

Our present-day technology includes the use of solid-state switches as practical alternatives to mechanical switches. When a bipolar transistor is used in a switching application it is either in an "on" or "off" state. In the "on" state a forward bias is applied to the transistor, sufficient in level to saturate the device. The common-emitter format is used for nearly all transistor switches. Switching action is characterized by large-signal nonlinear operation of the device. Fig. 46 shows typical output characteristics for an NPN switching transistor in the common-emitter mode.

There are three regions of operation — cutoff, active and saturation. In the cutoff region the emitter-base and collector-base junctions are reverse biased. At this period the collector current is quite small and is comparable to the leakage current, I_{ceo}, I_{cev} or I_{cbo}.

Fig. 47 illustrates the minority-carrier concentration relative to an NPN transistor. During cutoff the concentration is zero at both junctions because they are reverse biased (curve 1).

The emitter-base junction is forward biased in the active region. At this time the collector-base junction is reversed biased. Fig. 46 shows a load line along which switching from the cutoff to the active region is done. The transit time (speed)

through the active region depends on the transistor frequency-response characteristics. Thus, the higher the frequency rating of the device, the faster the switching time. Curve 2 in Fig. 47 depicts the minority-carrier concentration of the active region.

In the saturation region the emitter-base and collector-base junctions are forward biased. During this period the forward voltage drop across the emitter-base junction $V_{BE(sat)}$ is larger than it is across the collector-base junction. This results in a collector-emitter voltage termed $V_{CE(sat)}$. Series resistances present in the emitter and collector legs of the circuit contribute to the determination of $V_{CE(sat)}$. Since the collector in this state is forward biased, additional carriers are injected into the base. Some also reach the collector. Curve 3 in Fig. 47 shows this minority-carrier concentration. Fig. 48 contains the circuit for a basic saturated-transistor switch.

It is extremely important to make certain that none of the transistor voltage ratings is exceeded during the "off" period. The minimum emitter-base breakdown voltage, $V_{(BR)EBO}$, must not exceed $V_{BE(off)}$. Also, the minimum collector-base breakdown voltage, $V_{(BR)CBO}$, should be no greater than $V_{cc} + V_{BE(off)}$. Finally, the minimum collector-to-emitter breakdown voltage, $V_{(BR)CERL}$, must be greater than V_{cc}. As is true in any transistor application, the junction temperature must be maintained at a safe value by whatever means necessary.

A transistor switch can be turned on by means of a pulse (Fig. 48) or by application of a dc forward bias. Typical circuits for the latter are given in Fig. 49. The circuit at A illustrates how a PNP transistor can be used as a low-power switch to turn oscillator Q2 on and off. In the "on" state R1 is grounded. This places the bipolar switch, Q1, in a saturated mode, thereby permitting current to flow to Q2. A transistor switch of the type shown at A of Fig. 49 can be used to control more than one circuit stage simultaneously. The primary criterion is that the switching transistor be

capable of passing the combined currents of the various stages under control. The method shown at A is often used in keying a transmitter.

An NPN transistor switch is shown in Fig. 49B. If desired, it can be "slaved" to the circuit of Fig. 49A by attaching R1 of circuit B to the collector of Q1 in circuit A. Because an NPN device is used at B, a positive forward bias must be applied to the base via R1 to make the transistor saturate. When in that state, current flows through the relay (K1A) field coil to actuate the contacts at K1B. D3 is connected across the relay coil to damp inductive spikes that occur when the relay-coil field collapses.

D1 and D2 may not be necessary. This will depend on the sensitivity of the relay and the leakage current of Q1 in the off state. If there is considerable leakage, K1 may not release when forward bias is removed from Q1. D1 and D2 will elevate the emitter to approximately 1.4 volts, thereby providing sufficient reverse bias to cut off Q1 in the off state. It can be seen from the illustrations in Fig. 49 that either NPN or PNP transistors can be used as electronic switches.

Transistor Audio Amplifiers

Bipolar transistors are suitable for numerous audio-amplifier applications from low-level to high power. It is common practice to use all NPN or all PNP devices, regardless of the polarity of the power supply. In other circuits a mixture of the two types may be found, especially when direct-coupled or complimentary-symmetry stages are included. Fig. 50 shows how PNP or NPN stages can be used with power supplies that have positive or negative grounds. The essential difference in the circuits concerns returning various elements to the negative or positive sides of the power supply. The illustrations show that all one needs to do to use either type of device with the same power supply is to interchange the resistor connections. The same principle applies when using NPN or PNP transistors with a power supply that has a positive ground. Knowledge of how this is done enables the designer to mix NPN and PNP devices in a single circuit. This basic technique is applicable to any type of transistor circuit — RF, audio or dc.

Some basic low-level audio amplifiers are shown in Fig. 51. These stages operate in the Class A mode. The input impedance of these circuits is low — typically between 500 and 1500 ohms. For the most part the output impedance is established by the value to the collector load resistor. A matching transformer can be used at the input of these stages (Fig. 51B) when it is necessary to use, for example, a high-impedance microphone with one of them. T1 serves as a step-down transformer.

Some direct-coupled audio amplifiers are shown in Fig. 52. The circuit at A combines PNP and NPN devices to provide a compatible interface between them. Three NPN stages are in cascade at B to provide high gain. This circuit is excellent for use in direct-conversion receivers, owing to the need for very high gain after the detector. At C is a Darlington pair, named after the person who developed the configuration. The principal advantages of this circuit are high gain, high input impedance and low output impedance.

Transistor RF Amplifiers

In most respects small-signal RF amplifiers are similar in performance to those used in audio applications. However, to effect maximum stable amplification some important design measures are necessary. Furthermore, the matter of proper impedance matching becomes more important than it is in simple audio amplifiers. Other considerations are noise figure, purity of the amplified signal and dynamic range.

Although bipolar transistors can be used as RF amplifiers for receiver front ends, they are not found there in most of the high-performance receivers: Field-effect transistors are more often the designer's choice because of their high input impedance and good dynamic-range traits. A correctly designed bipolar RF input stage can exhibit good dynamic range, however. It is necessary to operate a fairly husky low-noise transistor in Class A, using a relatively high standing collector current — 50 to 100 mA, typically.

Some RF and IF amplifier circuits that use bipolar transistors are shown in the examples of Fig. 53. When used with the appropriate L and C networks they are suitable for either application. At A in Fig. 53 the transistor base is tapped near the cold end of the input tuned circuit to provide an impedance match. The collector is tapped down on the output tuned circuit to provide a proper match. If it is desired, the base and collector taps can be moved even farther down on the tuned circuits. This will result in a deliberate mismatch. The technique is sometimes used to aid

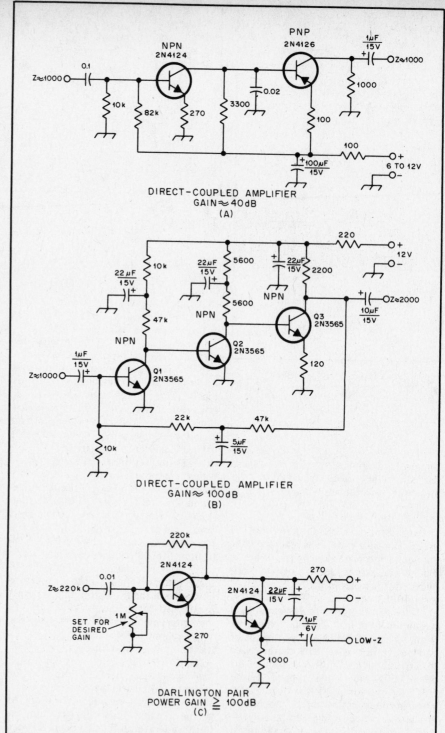

Fig. 52 — Practical examples of direct-coupled audio amplifiers.

stability and/or lower the stage gain. The circuit at B is operated in the common-base mode. Taps are shown on the input and output coils for impedance-matching purposes.

Broadband amplifiers with heavy negative feedback are useful as small-signal RF and IF amplifiers. An example is shown at C in Fig. 53. Not only is negative feedback applied (collector to base), but

degenerative feedback is obtained by virtue of the unbypassed 10-ohm emitter resistor. The use of feedback ensures an unconditionally stable stage. As the operating frequency is decreased the negative feedback increases because the feedback-network reactance becomes lower. This is important if reasonably constant gain is desired over a wide range of frequencies, say from 1.8 to 30 MHz.

This form of gain compensation is necessary because as the operating frequency of a given transistor is decreased, the gain increases. Typically, the gain will increase on the order of 6 dB per octave. Therefore, the probability of instability (self-oscillations) becomes a major consideration at low frequencies in an uncompensated RF amplifier.

The circuit of Fig. 53C operates stably and has a characteristic input and output impedance of approximately 50 ohms. The broadband 4:1 transformer in the collector circuit is required to step down the collector impedance to 50 ohms. Design information on this type of circuit is provided in the ARRL book, *Solid State Design for the Radio Amateur*. A band-pass filter is needed at the amplifier input. Another can be used at the output of the 4:1 transformer if desired. The transistor used in any of the amplifiers of Fig. 53 should have an f_T that is 5 to 10 times greater than the highest operating frequency of the stage. The 2N5179 has an f_T in excess of 1000 MHz, making it a good device up to 148 MHz for this application.

Transistor RF Power Amplifiers

RF power amplifiers that use bipolar transistors fall into two general categories — Class C and linear. The latter is used for AM and SSB signal amplification and the class of operation is A or AB. These amplifiers are designed for narrow or wideband applications, depending on the purpose for which the stage or stages will be used. Class C bipolar-transistor amplifiers are used for FM and CW work.

Most wideband amplifiers contain ferrite-loaded broadband transformers at the input and output ports. The output transformer is followed by a multipole low-pass filter for each band of operation. This is necessary to attenuate harmonic currents so they will not be radiated by the antenna system. Although this type of filtering is not always needed with a narrow-band amplifier (the networks provide reasonable selectivity), filters should be used in the interest of spectral purity. Two-section filters of the half-wave or low-pass T variety are entirely suitable for harmonic reduction at the 50-ohm output ports of amplifiers.

One of the principal difficulties encountered by amateurs who design and build their own high-power, solid-state amplifiers is instability at some point in the power range. That is, an amplifier driven to its maximum rated output may be stable when terminated properly, but when the drive level is reduced it is apt to break into self-oscillation at the operating frequency, at VHF or perhaps at very low frequencies. Part of the problem is caused by an increase in beta as the collector current is decreased. This elevates the amplifier gain to encourage instability. Also, solid-state amplifiers are designed for a specific network impedance at a specified power-output level. When the drive is reduced the

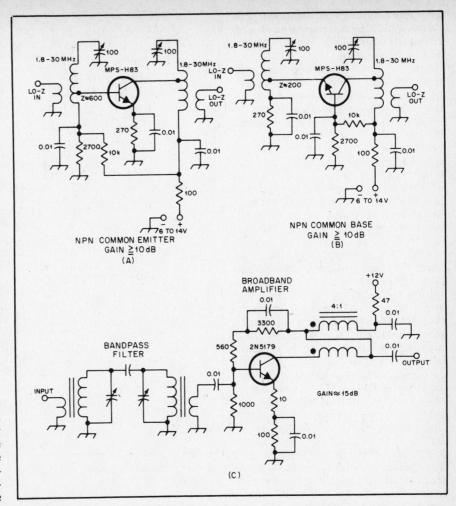

Fig. 53 — Illustrations of common-emitter and common-base RF amplifiers for narrow-band use. The circuit at C is that of a fed-back broadband amplifier that has a band-pass filter at the input.

collector and base impedances increase. This causes a mismatch. An increase in the loaded Q of the networks may also result — a situation that encourages instability. Therefore, it is best to design for a specified power output and adhere closely to that level during operation.

Solid-state power amplifiers should be operated just below their saturation points for best efficiency and stability. That is the point which occurs when no additional RF output can be obtained with increased driving power. Some designers recommend that, for example, a 28-volt transistor be used for 12-volt operation. Saturation will take place at a level where the transistors are relatively safe from damage if a significant output mismatch is present. Stability is usually better under these same conditions, although the gain of the transistors will be considerably lower than would be the case if equivalent types of 12-volt devices were used.

Fig. 54 shows two single-ended amplifiers of typical design. At A a broadband input transformer steps down the 50-ohm source to the lower base impedance of Q1. (Most power stages have a base im-

pedance of 5 ohms or less.) Although a number of suitable tuned networks can be used to effect the desired impedance match, the use of T1 eliminates components and the sometimes complex calculations required for the design of a proper network. When the actual base impedance of Q1 is unknown (it varies with respect to drive level and operating frequency), empirical adjustment of the T1 turns ratio will permit close matching. An SWR indicator can be used between T1 and the signal source to indicate a matched condition. This test should be made with the maximum intended drive applied.

To continue the discussion relating to Fig. 54A, a 10-ohm resistor (R1) is bridged across the secondary of T1 to aid stability. This measure is not always necessary. It will depend on the gain of the transistor, the layout and the loaded Q of T1. Other values of resistance can be used. A good rule of thumb is to employ only that value of resistance which cures instability. It must be remembered that R1 is in parallel with the transistor input impedance. This will have an effect on the turns ratio of T1. When excessive driving power is available, a deliberate mismatch can be introduced at the input to Q1 by reducing the number of secondary turns. If that is done, R1 can often be eliminated.

The shortcoming resulting from this technique is that the driving source will not be looking into a 50-ohm termination. T1 is normally a ferrite-loaded transformer — toroidal or solenoidal. The core material for operation from 1.8 to 30 MHz is typically the 950-μ_i (initial permeability) type. The primary winding of T1 (and other broadband transformers) should be approximately four times the terminal impedance with respect to reactance. Therefore, for a 50-ohm load characteristic the primary-winding reactance of T1 should be roughly 200 ohms.

Two RF chokes are shown in Fig. 54A. These are necessary to assure ample dc-lead decoupling along with the related bypass capacitors. The upper RF choke serves also as a collector load impedance. The reactance should be four or five times the collector impedance. Three values of bypass capacitors are used to ensure effective decoupling at VHF, HF and LF. If the decoupling is inadequate, RF from the amplifier can flow along the 12-volt bus to other parts of the transmitter, thereby causing instability of one or more of the stages. A simple low-pass T network is used for matching the collector to the 50-ohm load. It also suppresses harmonic energy. The loaded Q of this general type of matching network should be kept below 4 in the interest of amplifier stability. Information on this and other types of tuned matching networks is given in the *ARRL Electronics Data Book*. Data are also given in that volume concerning broadband transformer design.

Fig. 54B shows the same general amplifier. The difference is in the biasing. The circuit at A is set up for Class C. It is driven into the cutoff region during

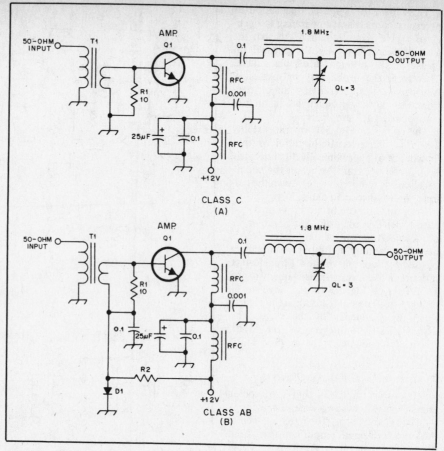

Fig. 54 — Circuits for RF power amplifiers. At A is a Class C type. The circuit at B is biased for linear amplification.

operation. At B there is a small amount of forward bias applied to Q1 (approximately 0.7 volt) by means of the barrier voltage set by D1, a silicon power diode. D1 also functions as a simple bias regulator. R2 should be selected to provide fairly substantial diode current. The forward bias establishes linearity for the amplifier so

SSB and AM driving energy can be amplified by Q1 with minimum distortion.

Although some transistors are designed especially for linear amplification, any power transistor can be used for the purpose. Once the proper biasing point is found for linear amplification with a Class C type of transistor, an investigation of

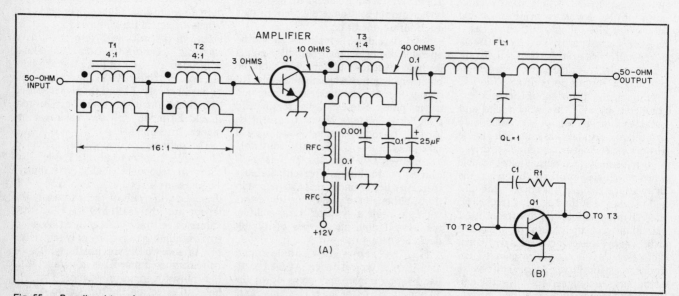

Fig. 55 — Broadband transformers are employed at A for impedance matching. FL1 suppresses harmonic currents at the amplifier output. In the examples at B are feedback components C1 and R1 (see text).

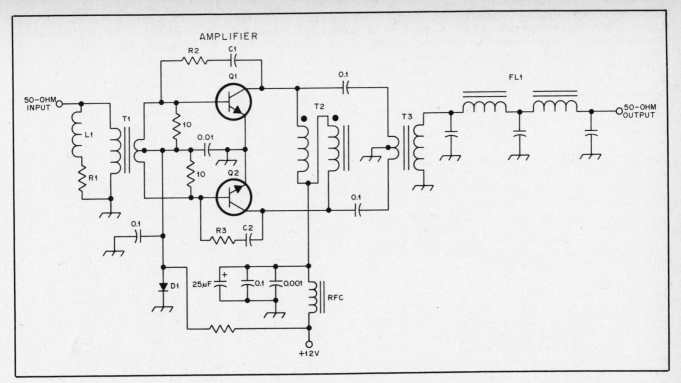

Fig. 56 — Example of a fed-back, push-pull, RF power amplifier set up for broadband service from 1.8 to 30 MHz. The circuit is biased for linear amplification.

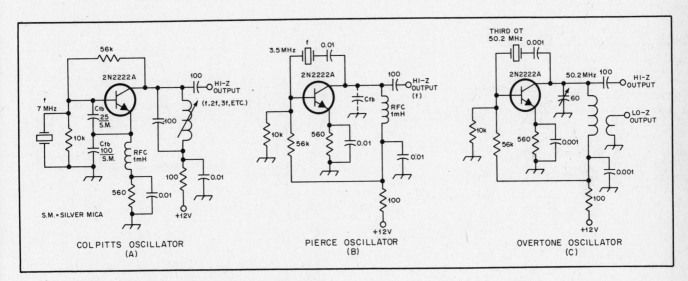

Fig. 57 — Examples of transistor oscillators that use crystal control.

linearity versus output power should be undertaken. Some Class C transistors are incapable of delivering as much power (undistorted) in Class AB as they can under Class C conditions. Most power transistors intended for linear amplification have built-in, degenerative-feedback resistors at the emitter sites. This technique aids linearity. Depending on the package style of a Class C type of transistor, an emitter-feedback resistor can be added externally. Such resistors are usually on the order of 1 ohm.

A broadband Class C amplifier is shown in Fig. 55A. T1 and T2 are 4:1 broadband transformers connected in series to provide an impedance step down of 16:1. For most

applications this arrangement will provide an acceptable match between 50 ohms and the base impedance of Q1. In the example we have assumed a base impedance of approximately 3 ohms.

T3 serves as a collector load and a step-up transformer. It is useful to use a step-up transformer when the collector impedance is low (25 ohms or less). This enables the designer to work with filter-component values (FL1) that are more practical than would be the case if an attempt were made to match 10 ohms to 50 ohms with the filter network. FL1 in this example is a double pi-section low-pass type (half-wave filter). It is designed to

match 40 ohms to 50 ohms and has a loaded Q of 1.

Feedback can be applied to stabilize the amplifier. This is seen in Fig. 55B. C1 and R1 are chosen to reduce the amplifier gain by whatever amount is necessary to provide stability and the broadband characteristics desired. C1 serves as a dc blocking capacitor.

A push-pull broadband linear amplifier is illustrated at Fig. 56. When additional frequency compensation is desired (beyond that available from a negative-feedback network) L1 and R1 can be added across the amplifier input. They are selected to roll off the driving power toward the low end of the amplifier operating range. As the frequency is reduced, L1 represents a lower reactance, thereby permitting some of the drive power to be dissipated in R1.

T1 is a conventional broadband transformer (not a transmission-line type) with a turns ratio set for matching 50 ohms to the base load presented by Q1 and Q2. T2, another broadband transformer, is used to provide balanced dc feed to the collectors. T3 is another broadband transformer which is wound for lowering the collector-to-collector impedance to 50 ohms. FL1 is designed for a bilateral impedance of 50 ohms in this example.

Bipolar-Transistor Oscillators

Transistors function well as crystal-controlled or LC oscillators. RC oscillators are also practical when a bipolar transistor is used as the active element. The same circuits used for tube-type oscillators apply when using transistors. The essential difference is that transistor oscillators have lower input and output impedances, operate at low voltages and deliver low output power — usually in the milliwatt range. The greater the oscillator power, the greater the heating of the transistor junction and other circuit elements. Therefore, in the interest of oscillator stability it is wise to keep the dc input power as low as practical. The power level can always be increased by means of subsequent amplifier stages at minor cost.

Some representative examples of crystal-controlled oscillators are provided in Fig. 57. At A is an oscillator that can be used to obtain output at f (the crystal frequency), or at multiples of f. The circuit at B illustrates a Pierce type of oscillator for fundamental output at 3.5 MHz. C_{fb} may be necessary with some crystals to provide ample feedback to cause oscillation. The value of C_{fb} will depend on the operating frequency and the gain of the transistor. Typically for 1.8 to 20-MHz crystals (fundamental mode) the capacitance value ranges from 25 to 100 pF. The higher values are typical at the lower end of the frequency range. In Fig. 57C is an overtone oscillator. The collector tuned circuit must be able to resonate slightly above the crystal overtone frequency in order to ensure oscillation. Low-impedance output can be had by means of the link shown. Alternatively, a capacitive divider can be placed across the inductor to provide a low-Z tap-off point. The trimmer should be retained in parallel with the inductor to permit resonating the circuit.

Some typical RF and audio oscillators are shown in Fig. 58. The circuit at A obtains feedback by means of the emitter tap on the tuned circuit. Approximately 25 percent of the oscillator RF power is used as feedback. The tap point on this type of oscillator is between 10 and 25 percent of the total coil turns. The designer should use the smallest amount of feedback that will provide reliable oscillator performance with the load connected.

Fig. 58B illustrates a series-tuned Colpitts oscillator, although this general circuit is often referred to as a "series-tuned Clapp" oscillator. It is very stable when polystyrene capacitors are used in the feedback and tuned circuits. Silver-mica capacitors can be used as substitutes at a slight sacrifice in drift stability (long term).

A twin-T audio oscillator is shown at C in Fig. 58. It is a very stable type of circuit that delivers a clean sine-wave output. Mylar or polystyrene capacitors should be used for best stability.

A simple feedback circuit is effected by means of T1 in Fig. 58D. T1 is a small transistor output transformer with a center-tapped primary and an 8-ohm secondary. This circuit is excellent for use as a code-practice or side-tone oscillator. All of the RF oscillators described in these examples should be followed by one or more buffer

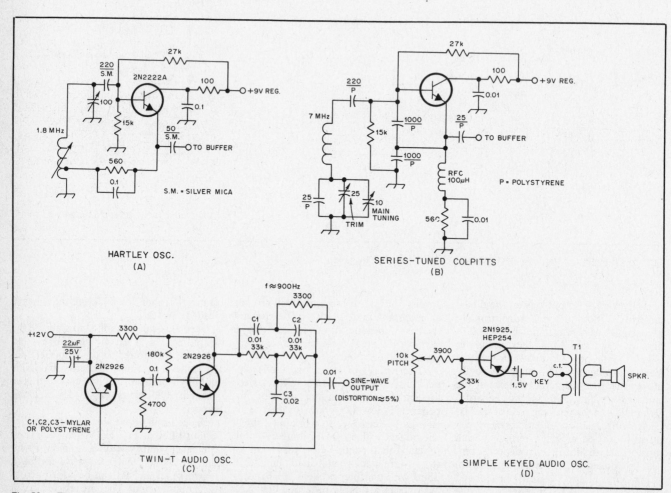

Fig. 58 — The circuits at A and B are VFOs for use in transmitters or receivers. Audio oscillators are shown at C and D.

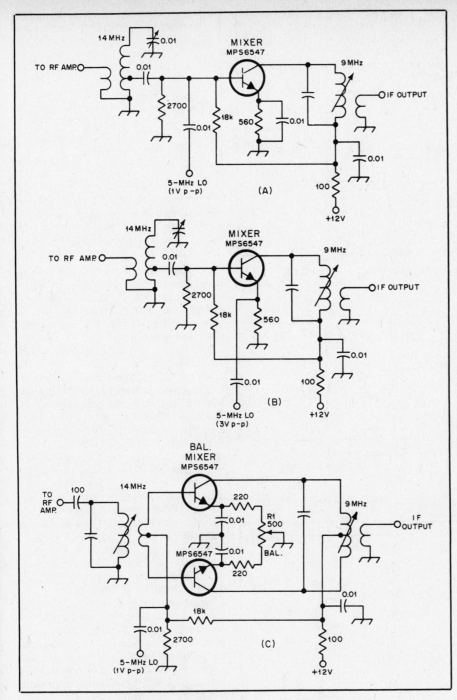

Fig. 59 — Some typical bipolar-transistor mixers. Their characteristics are discussed in the text.

mixers of Fig. 59 can be used without RF amplifier stages ahead of them for frequencies up to and including 7 MHz. The noise in that range (ambient from the antenna) will exceed that of the mixer.

The primary limitation in the performance of the mixer of Fig. 59A is that the local-oscillator voltage is injected at the base. This does not afford good LO/input-signal isolation. The unfavorable result can be oscillator "pulling" with input load changes, and/or radiation of the LO energy via the antenna if the front-end selectivity is marginal or poor. The advantage of the circuit is that it requires less injection voltage than the one at B, where emitter injection is used.

At Fig. 59B is the same basic mixer, but with LO voltage applied to the emitter. This technique requires slightly higher levels of LO energy, but affords greater isolation from the mixer input port.

A singly balanced bipolar-transistor mixer is illustrated in Fig. 59C. R1 is adjusted to effect balance. This circuit could be modified for emitter injection by changing R1 to 1000 ohms, replacing the 220-ohm resistors with 1-mH RF chokes, and injecting the LO output at the junction of the two 0.01-μF capacitors. The center tap of the input transformer (base winding) would then be bypassed by means of a 0.01-μF capacitor.

Other Uses for Bipolar Transistors

It is possible to take advantage of the junction characteristics of small-signal transistors for applications that usually employ diodes. One useful technique is using transistors as voltage-variable capacitors (varactors). This method is depicted in Fig. 60. The collector-base junction of Q1 and Q2 serve as diodes for tuning the VFO. In this example the emitters are left floating. A single transistor could be used, but by connecting the pair in a back-to-back arrangement they never conduct during any part of the RF cycle. This minimizes loading of the oscillator. The junction capacitance is varied by adjusting the tuning control, R1. In this circuit the tuning range is approximately 70 kHz.

A bipolar-transistor junction can be used as a Zener diode in the manner shown in Fig. 61. Advantage is taken of the reverse-breakdown characteristic of Q1 to establish a fixed reference level. Most transistors provide Zener-diode action between 6 and 9 volts. The exact value can be determined experimentally.

When the base and collector of a bipolar transistor are connected and a forward bias applied to the base-emitter junction, a superdiode results. If the collector were left open, the base-emitter junction would behave like an ordinary diode. With the collector tied to the base, the diode current rises much more rapidly with applied voltage because of the amplification provided by the transistor action. Two cross-

stages to prevent frequency changes resulting from load variations occurring after the oscillator chain.

Transistor Mixers

Much of the modern equipment used by amateurs contains mixers that use FETs or diode rings, both of which offer good dynamic range. However, there is no reason why a bipolar mixer can't be used to obtain satisfactory results if care is taken with the operating parameters and the gain distribution in the receiver and transmitter where they are used. The bipolar transistors used in receiver mixers should be selected

according to noise figure (low) and dynamic range (high). The signal applied to it should be kept as low as possible, consistent with low-noise operation. Most semiconductor manufacturers specify certain transistors for mixer service. Although this does not mean other types of bipolar transistors can't be used for mixing, it is wise to select a device designed for that class of service.

Fig. 59 contains examples of three basic types of transistor mixers. At A is the most common one. It is found in simple circuits such as transistor AM broadcast-band receivers. As an aid to dynamic range, the

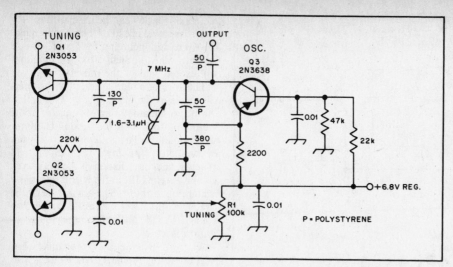

Fig. 60 — Bipolar transistors serve as varactor tuning diodes in this circuit (Q1 and Q2).

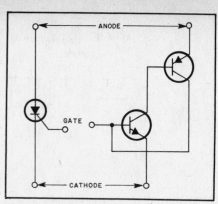

Fig. 64 — An SCR and its discrete functional near-equivalent.

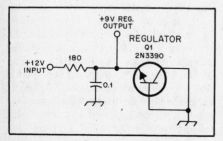

Fig. 61 — A bipolar transistor will function as a Zener diode when connected as shown here.

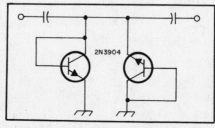

Fig. 62 — A peak clipper circuit using bipolar transistors connected in the superdiode configuration.

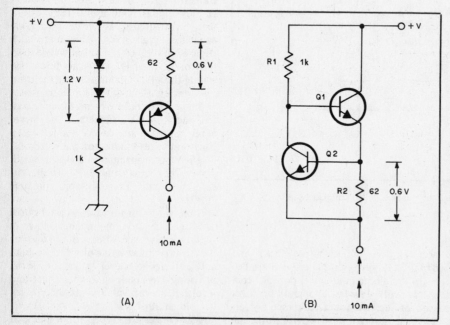

Fig. 63 — Constant-current generators made with bipolar transistors. In A, the reference voltage established by the diodes is converted to a current by the emitter resistor. A two-transistor feedback arrangement is employed at B. The functions of both circuits are explained in the text.

connected super-diodes form the basis for a highly effective peak clipper or hard limiter. Fig. 62 illustrates the application. NPN transistors are shown, but PNP units will yield identical performance.

Constant-Current Generators

The curves in Figs. 41 to 43 show that the collector current of a bipolar transistor is essentially independent of the collector-to-emitter potential when the device is biased in its active region. Fig. 63A illustrates a constant-current source (or sink, if actual electron flow is considered) using a PNP transistor. A fairly constant 1.2-volt potential drop is maintained across the diode string. The base-emitter junction introduces a diode drop, so the EMF applied to the 62-Ω resistor is a constant 0.6 volt. A constant voltage across a resistor forces a constant current. This current flows in the emitter, and the high alpha causes the collector current to be nearly the same.

The circuit of Fig. 63B works in a similar manner. R1 biases Q1 into conduction. When the EMF developed by R2 reaches 0.6 volt, Q2 begins to conduct, shunting base drive away from Q1 and limiting its collector current.

A device that passes an arbitrary current independent of the applied voltage presents an infinite dynamic impedance to the driving signal. This feature makes the constant-current generator valuable in several applications. One use for the circuits of Fig. 63 is in the bias control circuit of a differential amplifier. Either configuration can be used to establish the proper amplifier current while providing the tightest possible coupling between the emitters of the differential pair.

Another way to employ a constant-current circuit is to use it as an active load for the collector of a transistor amplifier stage. The infinite dynamic impedance of the current source causes the amplifier to exhibit very high voltage gain. When the amplifier is an NPN transistor, the current source must be a PNP device, and vice-versa.

Thyristors

Two complementary bipolar transistors connected as in Fig. 64 form the solid-state analog of the latching relay; a trigger pulse applied to the base of Q2 will initiate current flow in both devices. This current is limited only by the external circuit resistance and continues independent of the trigger signal until the main source is interrupted. Four-layer semiconductors

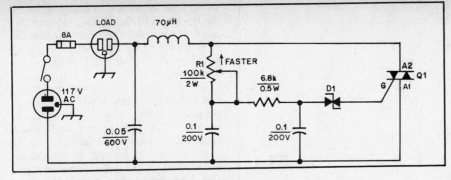

Fig. 65 — Schematic diagram of motor-speedcontrol.
Q1 — Triac (silicon bidirectional thyristor), 8A, 200 V (Motorola MAC2-4 or HEP340 or equiv.).

D1 — Diac (silicon bilateral trigger), 2 A, 300 mW.

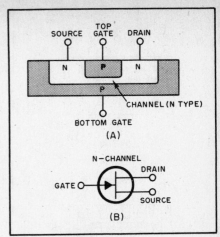

Fig. 67 — Profile and symbol for an N-channel junction field-effect transistor. In a P-channel device, all polarities are reversed and the gate arrow points away from the substrate.

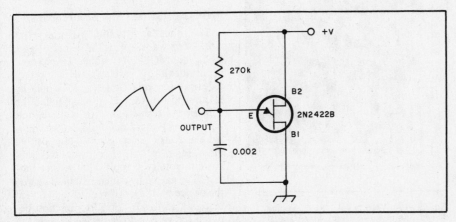

Fig. 66 — A relaxation oscillator based on a unijunction transistor. The frequency of oscillation is approximately 1500 Hz.

(PNPN or NPNP) having this property are known as thyristors or silicon controlled rectifiers (SCRs). SCRs find use in power supply overvoltage protection circuits (crowbars), electronic ignition systems, alarms, solid-state commutating systems for dc motors and a host of other applications.

Two complementary SCRs fabricated in parallel, with a common gate terminal, form a triac. These are used to switch alternating currents. The most common application of the triac is in incandescent light dimmers. Triacs have sensitive gates, and prolonging the trigger signal or injecting excessive gate current can cause excessive heating. In circuits operating on 117-volt ac, a diac is used to trigger a triac. A diac is a bidirectional current-limiting diode. Structurally, it can be compared to a triac without a gate. A motor speed control illustrating the use of triacs and diacs are drawn in Fig. 65.

Unijunction Transistors

An unusual three-terminal semiconductor device is the unijunction transistor (UJT), sometimes called a double-base diode. The elements of a UJT are base 1, base 2 and emitter. The single rectifying junction is between the emitter and the silicon substrate. The base terminals are ohmic contacts, meaning the current is a

linear function of the applied voltage. Current flowing between the bases sets up a voltage gradient along the substrate. In operation, the direction of flow causes the emitter junction to be reverse biased.

The relaxation oscillator circuit (the most common UJT application) of Fig. 66 illustrates the function of the UJT. When the circuit is energized, the capacitor charges through the resistor until the emitter voltage overcomes the reverse bias. As soon as current flows in the emitter, the resistance of the base 1 region decreases dramatically, discharging the capacitor. The decreased base 1 resistance alters the voltage distribution along the substrate, establishing a new bias point for the emitter junction. As more and more emitter current flows, the majority carrier injection builds a space charge in the base 1 region, which causes the emitter current to cease. Current is again available to charge the capacitor and the cycle repeats.

If the resistor were replaced by a constant-current source, the output waveform would be a linear ramp instead of a sawtooth. The UJT schematic symbol resembles that of an N-channel JFET — the angled emitter distinguishes the unijunction transistor.

Field-Effect Transistors

Field-effect transistors are assigned that

name because the current flow in them is controlled by varying electric field that is brought about through the application of a voltage that controls the electrode known as the gate. By contrast, in a bipolar transistor, output current flow is controlled by the current applied to the base electrode.

There are two types of field-effect transistors (FETs) in use today — the junction FET (JFET) and the MOSFET. The JFET has no insulation between its elements, as is the case with a bipolar transistor. The MOSFET has a thin layer of oxide between the gate or gates and the drain-source junction. The term MOSFET is derived from metal-oxide silicon field-effect transistor. The basic characteristic of the two types are similar — high input impedance and good dynamic range. These characteristics apply to small-signal FETs. Power FETs, which will be treated later, have different characteristics.

Although some MOSFETs have but one gate, others have two gates. Single-gate FETs can be equated to a triode vacuum tube. The gate represents the grid, the anode is similar to the drain and the cathode is like the source. The input impedance of FETs is a megohm or greater. Their noise figure is quite low, making them ideal as preamplifiers for audio and RF well into the UHF region. Nearly all of the MOSFETs manufactured today have built-in gate-protective Zener diodes. Without this provision the gate insulation can be perforated easily by small static charges on the user's hands or by the application of excessive voltages. The protective diodes are connected between the gate (or gates) and the source of FET.

The Junction FET

As was stated earlier, field-effect transistors are divided into two main groups: Junction FETs and MOSFETs. The basic JFET is shown in Fig. 67.

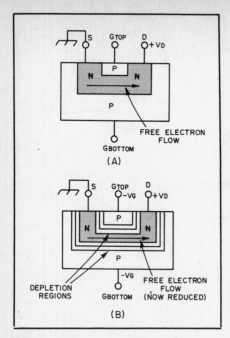

Fig. 68 — Operation of a JFET under applied bias. A depletion region (light shading) is formed, compressing the channel and increasing the resistance to current flow.

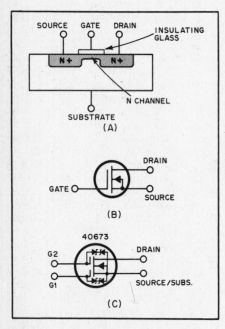

Fig. 69 — Profile and symbol for a MOSFET.

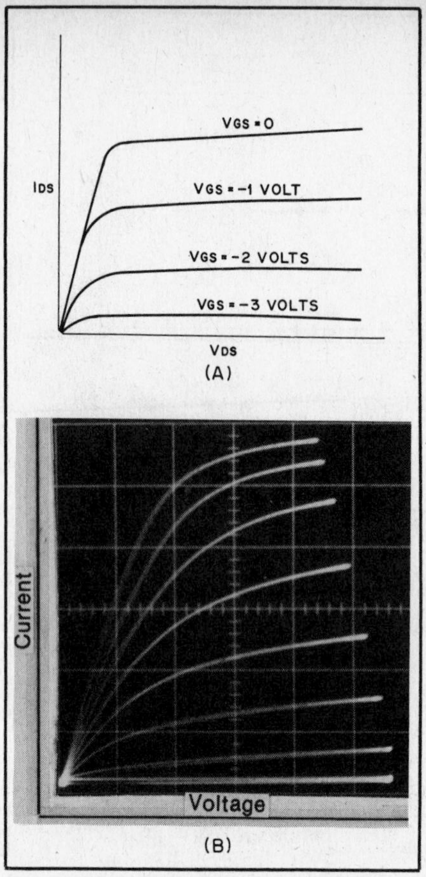

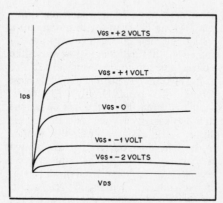

Fig. 70 — At A are typical JFET characteristic curves. The picture at B shows an actual oscillograph of the family of curves produced by a curve tracer.

Fig. 71 — Typical characteristic curves for a MOSFET.

The reason for the terminal names will become clear later. A dc operating condition is set up by starting a current flow between source and drain. This current flow is made up of free electrons since the semiconductor is N-type in the channel, so a positive voltage is applied at the drain. This positive voltage attracts the negatively charged free electrons and the current flows (Fig. 68). The next step is to apply a gate voltage of the polarity shown in Fig. 68. Note that this reverse-biases the gates

with respect to the source, channel and drain. This reverse-bias gate voltage causes a depletion layer to be formed, which takes up part of the channel. Since the electrons now have less volume in which to move, the resistance is greater and the current between source and drain is reduced. If a large gate voltage is applied, the depletion regions meet, causing pinch-off; consequently, the source-drain current is reduced nearly to zero. Since the large source-drain current changes with a relatively small gate

voltage, the device acts as an amplifier. In the operation of that JFET, the gate terminal is never forward biased, because if it were the source-drain current would all be diverted through the forward-biased gate junction diode.

The resistance between the gate terminal and the rest of the device is very high, since the gate terminal is always reverse biased, so the JFET has a very high input resistance. The source terminal is the source of current carriers, and they are drained out of the circuit at the drain. The gate opens and closes the amount of channel current that flows in the pinch-off region. The operation of an FET thus closely resembles the operation of the vacuum tube with its high grid-input impedance.

MOSFETs (Metal-Oxide Semiconductors)

The other large family that makes up field-effect transistors is the insulated-gate FET, or MOSFET, which is pictured schematically in Fig. 69. In order to set up a dc operating condition, a positive polarity is applied to the drain terminal. The substrate is connected to the source, and both are at ground potential, so the channel electrons are attracted to the positive drain. In order to regulate this source-drain current, voltage is applied to the gate contact. The gate is insulated from the rest of the device by a layer of very thin dielectric material, so this is not a PN junction between the gate and the device — thus the name insulated gate. When a negative gate polarity is applied, positive-charged holes from the P-type substrate are attracted toward the gate and the conducting channel is made more narrow; thus, the source-drain current is reduced.

When a positive gate voltage is connected, the holes in the substrate are repelled, the conducting channel is made larger and the source-drain current is increased. The MOSFET is more flexible since either a positive or negative voltage can be applied to the gate. The resistance between the gate and the rest of the device is extremely high because they are separated by a layer of thin dielectric. Thus, the MOSFET has an extremely high input impedance. In fact, since the leakage through the insulating material is generally much smaller than through the reverse-biased PN gate junction in the JFET, the MOSFET has a much higher input impedance. Typical values of R_{in} for the MOSFET are over a million megohms.

Both single-gate and dual-gate MOSFETs are available. The latter has a signal gate, gate 1, and a control gate, gate 2. The gates are effectively in series, making it an easy matter to control the dynamic range of the device by varying the bias on gate 2. Dual-gate MOSFETs are widely used as AGC-controlled RF and IF amplifiers, as mixers and product detectors and as variable attenuators. The isolation between the gates is relatively high in mixer service. This reduces oscillator "pulling"

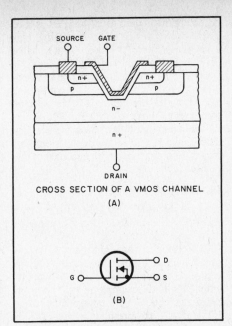

Fig. 72 — Profile and symbol for a power FET (VMOS enhancement-mode type).

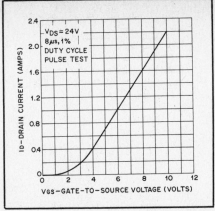

Fig. 73 — Curve showing relationship between gate-source voltage and drain current of a power FET.

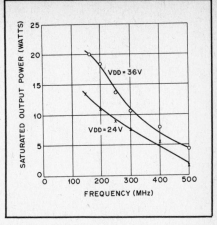

Fig. 74 — Curves for 24- and 36-volt operation of a power FET.

and reduces oscillator radiation. The forward transadmittance (transconductance, or G_m) of dual-gate MOSFETs is as high as 40,000 microsiemens.

FET Characteristics

The characteristic curves for the FETs described above are shown in Figs. 70 and 71. The drain-source current is plotted against drain-source voltage for given gate voltages.

The dynamic characteristics of an FET are most heavily influenced by dynamic mutual conductance or transconductance. This parameter is defined as the ratio of drain current change to the small gate-to-source voltage change that caused it. Mathematically, the relationship is expressed $G_m = \Delta I_D / \Delta E_{GS}$, where Δ represents a small change or increment. Typical general-purpose JFETs for small-signal RF and audio work have G_m values in the neighborhood of 5000 microsiemens, while some units designed for CATV service feature transconductance over 13,000 μS. As mentioned earlier, 40,000 μS (40 mS) is the transconductance figure for the "hottest" dual-gate MOSFETs. Some JFETs intended for analog switching also have 40 mS or more of transconductance to achieve low ON resistance. The newer power FETs boast transconductance figures on the order of 1 siemen.

Transconductance is of great importance in calculating the gain and impedance values of FET circuits. In common-source and common-gate amplifiers with no degeneration, the numerical voltage gain is given by $A_v = G_m R_L$, where A_v is the gain, G_m is the transconductance in siemens and R_L is the drain load resistance

in ohms. Also, the source impedance of a common-gate or common-drain (source follower) amplifier is approximately $1/G_m$.

Classifications

Field-effect transistors are classified into two main groupings for application in circuits — enhancement mode and depletion mode. The enhancement-mode devices are those specifically constructed so they have no channel. They become useful only when a gate voltage is applied that causes a channel to be formed. IGFETs (insulated-gate FET) can be used as enhancement-mode devices since both polarities can be applied to the gate without the gate becoming forward biased and conducting.

A depletion-mode unit corresponds to Figs. 66 and 67, shown earlier, where a channel exists with no gate voltage applied. For the JFET we can apply a gate voltage and deplete the channel, causing the current to decrease. With the MOSFET we can apply a gate voltage of either polarity so the device can be depleted (current decreased) or enhanced (current increased).

To sum up, a depletion-mode FET is one that has a channel constructed; thus, it has a current flow for zero gate voltage. Enhancement-mode FETs are those that have no channel, so no current flows with zero gate voltage.

Power FETs

FETs capable of handling substantial amounts of power are available for use from dc through the VHF spectrum. They are known under more than one name — vertical FETs, MOSPOWER FETs and VMOS FETs. The power FET was introduced in 1976 by Siliconix, Inc. The device enabled designers to switch a current of 1 ampere in less than 4 nanoseconds. The transfer characteristic of the power FET is a linear one. It can be employed as a linear power amplifier or a low-noise, small-signal amplifier with high dynamic range. With

this kind of FET there is no thermal runaway, as is the case with power types of bipolar transistors. Furthermore, there is no secondary breakdown or minority-carrier storage time. This makes them excellent for use in switching amplifiers (Class D service). Of particular interest to amateurs is the immunity of power FETs to damage from a high SWR (open or short condition). These devices can be operated in Class A, AB, B or C. Zero bias results in Class C operation.

Fig. 72 illustrates the manner in which a power FET is formed. These devices operate in the enhancement mode. The current travels vertically. The source is on top of the chip but the drain is on the bottom of the chip. In this vertical structure there are four layers of material (N+, P, N− and N+). This device offers high current density, high source/drain breakdown capability and low gate/drain feedback capacitance. These features make the transistor ideal for HF and VHF use.

Fig. 73 depicts the drain current as being linearly proportional to the gate-to-source voltage. The more conventional JFET exhibits a square-law response (drain current being proportional to the square of the gate-to-source voltage).

As an example, the M/A COM PHI VMP-4 power FET can provide a power just short of 20 watts (saturated) at 160 MHz. Fig. 74 shows curves for this device respective to saturated output power versus frequency. In this case both the input and output impedances of the transistor are matched conjugately. An advantage to this device over the power bipolar transistor is that these impedances are barely affected by the drive levels applied. In wideband amplifier service the power FET can be operated with complete stability. In-depth data on these devices is given in M/A COM application note AN80-2.

GaAsFETs

For low-noise amplification at UHF and

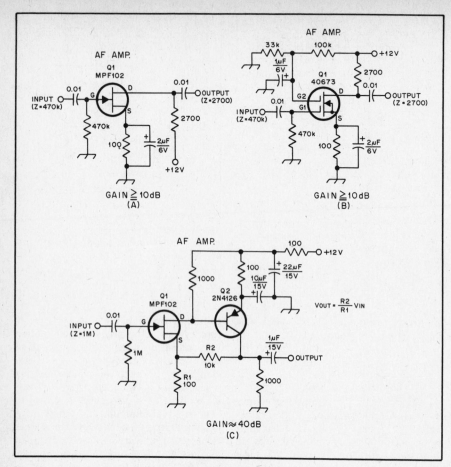

Fig. 75 — Some typical audio amplifiers that employ FETs.

microwaves, the state of the art is defined by field-effect transistors fabricated from gallium arsenide. Also used in LEDs and microwave diodes, gallium arsenide is semiconductor compound, as opposed to silicon and germanium, which are elements. This compound exhibits greater carrier mobility (the electrons can move more freely) than silicon or germanium; hence, the transit time is reduced and high-frequency performance improved in GaAsFETs. GaAsFETs are classified as depletion-mode junction devices. The gate is made of gold or aluminum, (which is susceptible to damage from static charges).

GaAsFETs are available for both small-signal and power applications. The power devices have noise figures almost as low as those specified for the small-signal types, and naturally exhibit greater dynamic range and ruggedness. Several semiconductor manufacturers offer gallium-arsenide field-effect transistors in various noise figure, frequency and power ratings. In the U.S., Hewlett-Packard and Microwave Semiconductor Corp. feature units usable up through frequency of Ku-band. Representative type numbers are HFET-2201 and MSC H001, respectively. In Great Britain,

the Plessey GAT5 and GAT6 devices feature low-noise performance up to 14 GHz. The Nippon Electric Company of Japan is also competing strongly for leadership in the GaAsFET market. A practical GaAsFET preamplifier is featured in Chapter 32. For more background information on GaAsFETs, see Wade, "Introduction to GaAs Field-Effect Transistors," *Ham Radio,* January 1978, and Wade and Katz, "Low-Noise GaAsFET UHF preamplifiers," *QST,* June 1978.

Practical FET Circuits

Small-signal FETs can be used in the same general types of circuits given earlier for bipolar transistors. The primary obstacle in some types of amplifier circuits is instability. Certain precautions are necessary to prevent unwanted self-oscillations, but they do not differ markedly from those techniques applied when working with triode tubes.

In Fig. 75 are examples of FET audio amplifiers. The circuit at A shows a simple RC coupled stage with a gain of 10 dB or greater. The input and output impedances are set by the gate and drain resistors. The circuit at B in Fig. 75 is similar to that at

A, except that a dual-gate MOSFET is used as the active device (Q1). A positive bias is supplied to gate 2 by means of a resistive divider. In the circuit of Fig. 75C, a PNP transistor is combined with a JFET to provide a direct-coupled pair. This configuration provides high gain. The amount of gain is set by the ratio of R1 and R2. Again, the input and output impedances are largely determined by the values of the input and output resistors, 1 megohm and 1000 ohms, respectively.

RF and IF Amplifiers

Small-signal RF and IF amplifiers that use FETs are capable of good dynamic range and will exhibit a low noise figure. It is for these reasons that many designers prefer them to bipolar transistors. Fig. 76 contains examples of FET RF or IF amplifiers. In the example at A the gate and drain elements of Q1 are tapped down on L2 and L3 to provide stability. This represents an intentional mismatch, which causes a slight sacrifice in stage gain. The 10-ohm drain resistor (R1) is used only if VHF parasitic oscillations occur.

Fig. 76B depicts a common-gate FET amplifier. The source is tapped well down on the input tuned circuit to effect an impedance match. This circuit is characterized by its excellent stability, provided the gate lead is returned to ground by the shortest path possible. This type of circuit will have slightly less gain capability than the common-source example at A.

In Fig. 76 at C is an illustration of a dual-gate MOSFET amplifier. Provided the input and output tuned circuits are well isolated from one another, there is less chance for self-oscillation than with a JFET. A positive bias is applied to gate 2, but AGC voltage can be used in place of a fixed-value voltage if desired. This circuit can provide up to 25 dB of gain.

Fig. 76D shows the configuration of a cascode RF amplifier in which a dual JFET (Siliconix U257) is specified. The advantage in using the dual FET is that both transistors have nearly identical characteristics, because they are fabricated on a common substrate. Two separate JFETs can be used in this circuit if the one nearest to V_{DD} has an I_{DSS} higher than its mate. This ensures proper dc bias for cascode operation. Unmatched FETs require special forward-biasing techniques and ac-coupling measures that aren't seen in this circuit.

Cascode amplifiers are noted for their high gain, good stability and low noise figure. With the circuit shown the noise figure at 28 MHz is approximately 1.5 dB. Short leads are necessary, and shielding between the tuned circuits is recommended in the interest of stability. Careful layout will permit the use of toroidal inductors at L2 and L3. These components should be spaced apart and mounted at right angles to one another in order to reduce unwanted infringing magnetic fields. AGC can be applied to this amplifier by routing the control voltage to the gate of Q1B.

FET Mixers

There are three types of FET mixers in common use today — single-ended, singly balanced and doubly balanced. In all cases there is an advantage to using active devices in place of passive ones (diodes). This ensures a conversion gain that helps minimize the number of gain stages required in a given circuit.

A single-ended JFET mixer requires 0 dBm of LO injection power. It can provide several decades of bandwidth and has a good IMD characteristic. The latter is far superior to most bipolar single-ended mixers. The major shortcoming is very poor isolation between the three mixer ports (RF, LO and IF). A typical single-ended mixer using a JFET is shown in Fig. 77A. Optimum trade-off between conversion gain and IMD occurs near the point where the self-bias is 0.8 V. LO injection voltage will be on the order of 1 (P-P) to provide good mixer performance. Conversion gain with this mixer will be approximately 10 dB.

Fig. 77B illustrates a singly balanced JFET mixer. A broadband transformer (T1) provides a low-impedance source for the LO and supplies injection voltage in push-pull to the gates of Q1 and Q2. The latter should be matched FETs or a dual FET such as the U430 by Siliconix. This mixer provides between 10 and 20 dB of isolation between the mixer ports. The signal is applied in parallel across the sources of Q1 and Q2 by means of broadband transformer T2. Output at the IF is taken from a balanced tuned circuit.

A doubly balanced FET mixer is shown at C in Fig. 77. Broadband transformers are used throughout, with FL1 and FL2 providing low-pass selectivity at the mixer output. The filters also provide an impedance stepdown between the drains of Q1 and T4 (1700 ohms to 100 ohms). LO injection is supplied to the gates and signal input is to the sources. Port-to-port isolation with this mixer is on the order of 30 dB or greater. Bandwidth is one octave. Indepth information on this type of circuit is given in Siliconix application note AN-73-4.

Fig. 78 contains the circuit of a typical dual-gate MOSFET single-ended mixer. Its performance characteristics are similar to those of the mixer at Fig. 77A. The primary exception is that the port-to-port isolation is somewhat better by virtue of the gate 2 isolation from the remainder of the electrodes. This mixer and all other active FET mixers require a fairly low drain-load impedance in the interest of good IMD. If the drain tuned circuit is made high in terms of impedance (in an effort to improve conversion gain) the drain-source peak signal swing will be high. This will lead to a change in junction capacitance (varactor effect) and the generation of harmonic currents. The result is distortion.

Of primary significance is the condition called "drain-load distortion." This malady occurs when excessive signal levels overload the drain circuit. The result is degraded IMD and cross-modulation effects. R1 in Fig. 78 is used to decrease the drain-load impedance by means of swamping. A value of 10,000 ohms is suitable for a 40673 MOSFET mixer. Some JFETs require a lower drain load for optimum performance. Values as low as 5000 ohms are not unusual. This form of

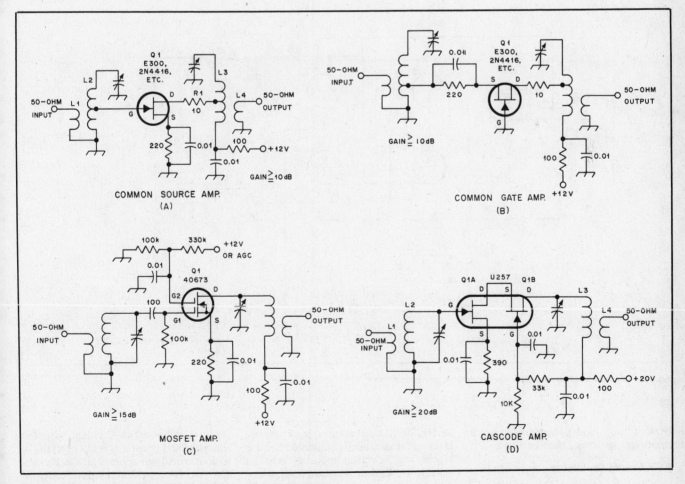

Fig. 76 — Examples of FET amplifiers suitable for RF or IF applications.

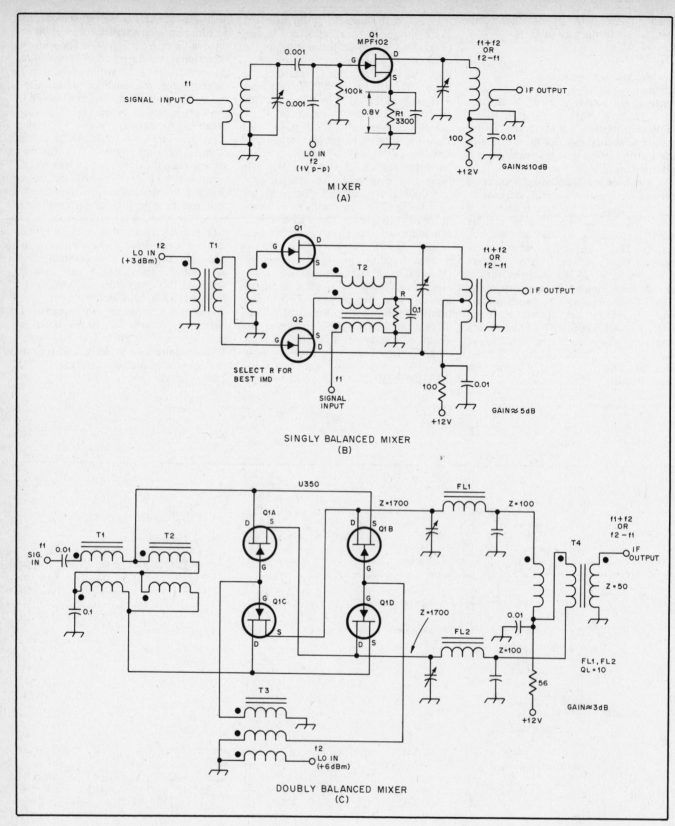

Fig. 77 — Various JFET mixers. See text for data.

overloading is more pronounced at low dc drain-voltage levels, such as 6 to 8.

FET Crystal Oscillators

A group of crystal-controlled FET oscillators is presented in Figs. 79 and 80.

At Fig. 79A is an overtone type. The tuned circuit in the drain is resonated slightly higher than the crystal frequency to ensure reliable oscillation. The circuit at B is a variation of the one at A, but performs the same function. A Pierce triode oscillator is shown in Fig. 80 at A. It is suitable for use with fundamental types of crystals. A Colpitts oscillator appears at B in Fig. 80. C_{fb} in these circuits are feedback capacitors. C_{fb} in the circuit at C is chosen experimentally. Typically, it will be from

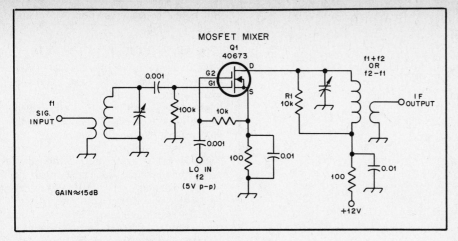

Fig. 78 — A dual-gate MOSFET single-ended mixer.

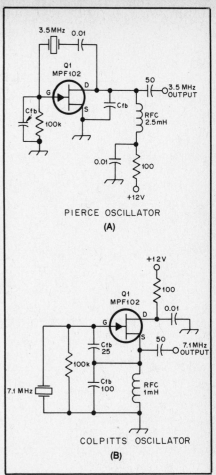

PIERCE OSCILLATOR
(A)

COLPITTS OSCILLATOR
(B)

Fig. 80 — Fundamental-mode FET crystal oscillators.

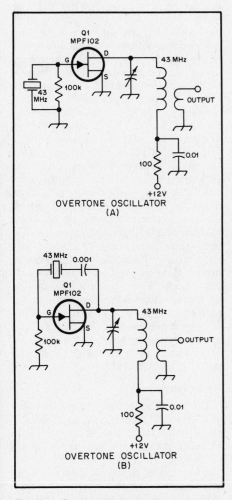

OVERTONE OSCILLATOR
(A)

OVERTONE OSCILLATOR
(B)

Fig. 79 — Overtone crystal oscillators using FETs.

100 to 500 pF, depending on the transistor characteristics and the crystal activity.

FET VFOs

The principle of operation for FET VFOs is similar to that discussed in the section on bipolar transistor oscillators. The notable difference is the impedance level at the device input. The circuits of Fig. 81 all have high-impedance gate terminals. Furthermore, fewer parts are needed than is true of bipolar transistor equivalent circuits: There is no resistive divider for applying forward bias.

The circuits of Fig. 81A and B are identical except for the biasing of gate 2 at B. Both circuits illustrate oscillators. The source tap on L1 should be selected to provide approximately 25 percent of the oscillator power as feedback. D1 in each example is used to stabilize the gate bias. It acts as a diode clamp on positive-going excursions of the signal. This aids oscillator stability and reduces the harmonic output of the stage. The latter is reduced as a result of the positive swing of the sine-wave being limited by D1, which in turn limits the device transconductance on peaks. This action reduces changes in junction capacitance, thereby greatly restricting the varactor action that generates harmonic currents. D1 is most effective when source-bias resistors are included in the circuit (R1).

Shown in Fig. 81 at C is a series-tuned Colpitts VFO that uses a JFET. This is an exceptionally stable VFO if careful design and component choice is applied. All of the fixed-value capacitors in the RF parts of the circuit should be temperature-stable. Polystyrene capacitors are recommended, but dipped silver-mica capacitors will serve adequately as a second choice. L1 should be a rigid air-wound inductor. A slug-tuned inductor can be used if the coil Q is high. In such cases the slug should occupy the least amount of coil space possible: Temperature changes have a marked effect on ferrite or powdered-iron slugs, which can change the coil inductance markedly. Capacitors C_{fb} of Fig. 81C are on the order of 1000 pF each for 3.5-MHz operation. They are proportionally lower in capacitance as the operating frequency is increased, such as 680 pF at 7 MHz, and so on.

Power FET Examples

Fig. 82 contains examples of three amplifiers that use power FETs. The circuit at A is an audio amplifier that can deliver 4 watts of output. At 3 watts of output the distortion is approximately two percent. Feedback is employed to aid the reduction of distortion.

A Class C amplifier is shown at B in Fig. 82. The VN67AJ is capable of a saturated output near 15 watts at 30 MHz. In this circuit the power output is considerably less. A medium output power of 7 to 10 watts is suggested. The gain is approximately 8 dB over the frequency range specified (with the appropriate drain tank). If proper layout techniques are used, this amplifier is unconditionally stable.

A broadband HF linear amplifier is shown at C. A narrow-band linear VHF power amplifier is shown at D in Fig. 82. Power output is 5 watts PEP. IMD is −30 dB. It is interesting to realize that this same amplifier is suitable as a high-dynamic-range preamplifier for a VHF receiver. In this application the noise figure is on the order of 2.5 dB and the gain is 11 dB.

Other FET Uses

Fig. 83 contains illustrations of additional practical uses for JFETs. The circuit

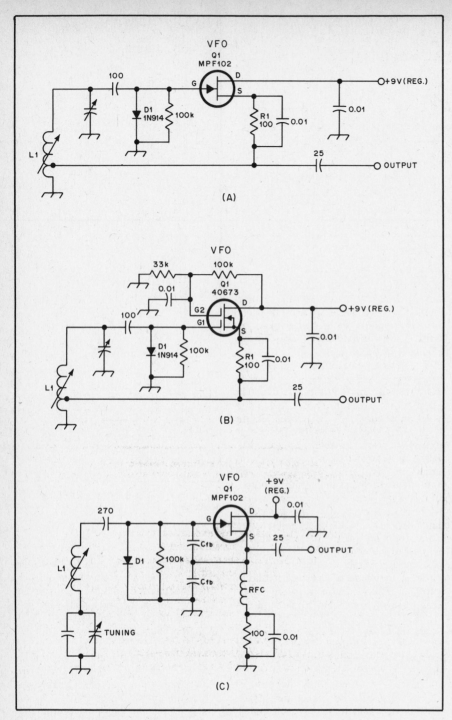

Fig. 81 — Three examples of VFOs in which FETs are used.

eliminate frequency component f at the doubler output.

LINEAR INTEGRATED CIRCUITS

There are two general types of ICs (integrated circuits). The first variety, which we are addressing at the moment, are called linear ICs. The other group are known generally as logic ICs. These devices will be discussed later in the chapter.

ICs are characterized by the term "microcircuit." In essence they are composed of numerous — sometimes hundreds — of bipolar and/or field-effect transistors on a single silicon chip (substrate). Along with the individual transistors formed on the substrate are diodes, capacitances and resistances. Some ICs contain only diodes. Others may contain only resistors.

The principal advantages of ICs are their compactness over an equivalent number of discrete transistors, and the fact that all of the devices on the substrate are evenly matched in characteristics. That is the result of the manufacturing process, whereby all of the IC transistors are formed from a single slice of semiconductor material under the same environmental conditions. This provides an inherent balance in their performance traits — a condition nearly impossible to realize with closely matched discrete transistors. Therefore, when changes in IC temperature take place, the parameters of the transistors on the chip change in unison — a distinct advantage when the IC is used in, say, a balanced modulator, mixer or push-push doubler.

Most of the theory given earlier for bipolar transistors applies to ICs, so it will not be repeated here. Rather, the text will provide data on practical applications of ICs. Linear ICs are so-called because in most applications where they are used the performance mode is a linear one. This does not mean, however, that they can't be used in a nonlinear mode, such as Class C. The biasing will determine the operating mode, Class A through Class C.

IC Structures

The basic IC is formed on a uniform chip of N-type or P-type silicon. Impurities are introduced into the chip, their depth into it being determined by the diffusion temperature and time. The geometry of the plane surface of the chip is determined by masking off certain areas, applying photochemical techniques and providing a coating of insulating oxide. Certain areas of the oxide coating are then opened up to allow the formation of interconnecting leads between sections of the IC. When capacitors are formed on the chip, the oxide serves as the dielectric material.

Fig. 84 shows a representative three-component IC in both pictorial and schematic form. Most integrated circuits are housed in TO-5 type cases, or in flat-pack epoxy blocks. ICs may have as many as 12 or more leads that connect to the various elements on the chip.

at A shows a Schmitt trigger. It is emitter-coupled and provides a comparator function. Q1 places very light loading on the measured input voltage. Q2 has high beta to enable the circuit to have a fast transition action and a distinct hysteresis loop. Additional applications of this type are found in *Linear Applications* by National Semiconductor Corp.

A simple FET dc voltmeter with high input impedance is shown in Fig. 83B. Multiplier resistances are given for a full-scale range of 2 or 20 volts. Meter accuracy is quite good, with a linear reading provided by M1.

A push-pull frequency doubler is shown at C in Fig. 83. The input frequency (f) is applied to the gates of Q1 and Q2 in push-pull. Output from the doubler is taken with the connected drains in parallel. R1 is adjusted for best waveform purity at 2f. The efficiency of this Class C doubler is on par with that of a straight-through Class C amplifier. Careful adjustment will nearly

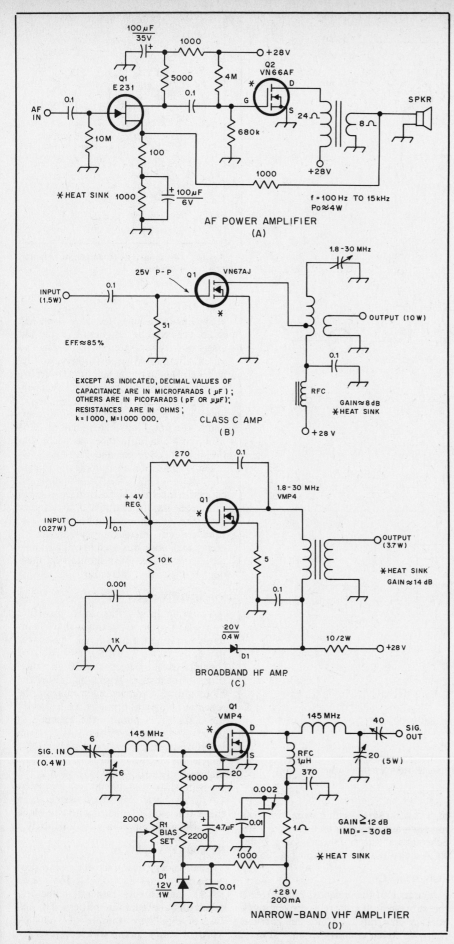

Fig. 82 — Examples of power FETs in three amplifier circuits.

Some present-day ICs are called LSI (large-scale integration) chips. Such devices may contain the equivalent of several conventional ICs, and can have dozens of dual-in-line package (DIP) connector pins. LSI ICs are used in electronic organs, digital clocks, electronic calculators and so on. Essentially, they are just super-size ICs.

Some Practical Considerations

In circuits where slight extra lead lengths can be tolerated, it is prudent to install the ICs in sockets rather than solder them into the PC board directly. In amateur work there is an occasional need to replace an IC during circuit development for a typical one-shot design. This is particularly pertinent when bargain-house ICs are purchased: Many have defects, and the task of removing an IC that is not in a socket is a task no builder finds delightful.

When using ICs for RF work it is best to install them in a low-profile type of IC socket (minimum lead length type). The thicker sockets are suitable for dc and audio applications, where lead length is not likely to be a critical factor. Excessive lead length can cause instability. This is brought on by having numerous high-gain devices packaged physically close to one another on the common substrate: High gain and stray lead coupling set the stage for self-oscillation!

CMOS ICs

The term CMOS means the IC is a complementary metal-oxide silicon type of integrated circuit. Essentially, the internal workings of the device are similar to the MOSFETs that were treated earlier in this chapter: MOSFETS are formed on the CMOS IC substrates.

CMOS devices consume very low power — especially advantageous in battery-operated equipment. The transit time (propagation delay) through the FET gates of a CMOS IC is very short — ideal in logic circuits. It ranges from 25 to 50 ns in most devices. This does not imply that CMOS ICs aren't useful in linear applications. Some are designed primarily for the linear amplification of audio and RF energy (CA3600E, for one).

Another salient feature of CMOS chips is low noise. Because FETs are used in these ICs the input impedance is high, making them more suitable than bipolar ICs for interfacing with comparable impedance levels outside the IC package. Fig. 85 shows the diagram of a CA3600E CMOS IC along with the block-symbol circuit for its use as a high-gain audio amplifier.

Array ICs

One branch of the linear-IC family is known as the IC array group. A short course on these and other linear ICs was given by DeMaw in *QST* for January through March 1977. Basically, the IC array is a substrate which contains a number of individual diodes or NPN

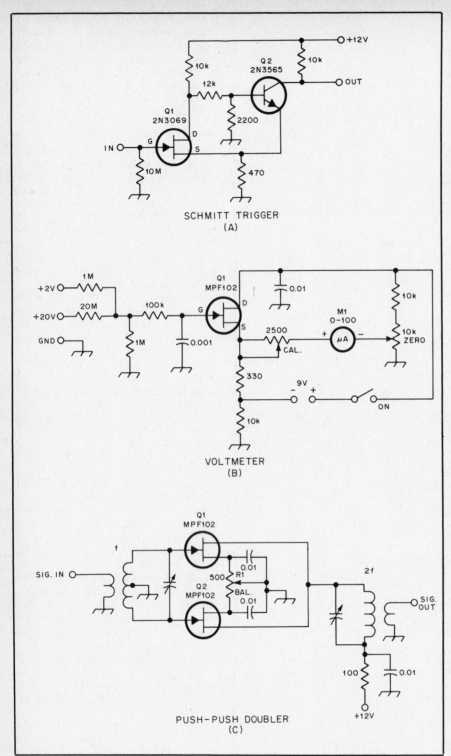

SCHMITT TRIGGER
(A)

VOLTMETER
(B)

PUSH-PUSH DOUBLER
(C)

Fig. 83 — JFETs are useful in additional kinds of circuits. Here are three examples.

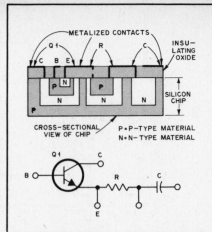

Fig. 84 — Pictorial and schematic representation of a simple IC.

active-device circuitry for an FM or AM radio receiver. The designer need only add essential outboard components (front-end tuned circuits, IF transformers, tuning meter and audio power amplifier) to realize a composite piece of equipment. Other subsystem ICs may contain only the IF amplifiers, product detectors, AGC loops and audio amplifiers. This style of IC is sold by such manufacturers as RCA, National Semiconductor and Plessey.

Fig. 87 illustrates an example of this kind of device — the RCA CA3089E, designed expressly for use in wide- or narrow-band FM receivers. It features a quadrature detector, and contains amplifiers, limiters, squelch circuit, metering circuit and an AF amplifier. Those interested in compact portable amateur receivers should find these devices especially interesting.

Practical Examples

The main disadvantage in the use of IC symbols in circuit diagrams is that the internal workings are not shown. This makes the designer work with a collection of "magic boxes." Fortunately, IC manufacturers publish data books that show the block symbols, pin arrangements and the schematic diagrams of the active devices on the chips. This permits the amateur to understand what the circuit configuration is before the design work is started. It is impractical to include the schematic diagrams of the ICs used in this book, but we will show the circuit of the RCA CA3028A, because it is used frequently in the next section. Fig. 88 contains the block and schematic representation of this IC.

RF and IF Amplifiers

Nearly every manufacturer of ICs produces chips that are suitable for use as RF/IF amplifiers, mixers, detectors, oscillators and audio amplifiers. The circuits of Fig. 89 are examples of CA3028A RF or IF amplifiers to which AGC is applied. Maximum gain occurs when the

bipolar transistors. They differ from conventional ICs by virtue of having each of the transistors independent from one another. Each transistor base, emitter and collector is brought out of the IC package by means of its own single pin. This enables the designer to treat each transistor as a discrete device, with the advantage that each transistor has nearly identical electrical characteristics (f_T, beta, dissipation rating, etc.). Some array ICs have f_T ratings as

high as 1200 MHz, with maximum collector dissipation ratings as high as 1 watt. Schematic illustrations of some popular RCA array ICs are shown in Fig. 86.

Subsystem ICs

A branch of the linear-IC family tree is the subsystem IC. It is a conventional-package integrated circuit, but contains nearly as much circuitry as an LSI chip. Some of these devices represent the entire

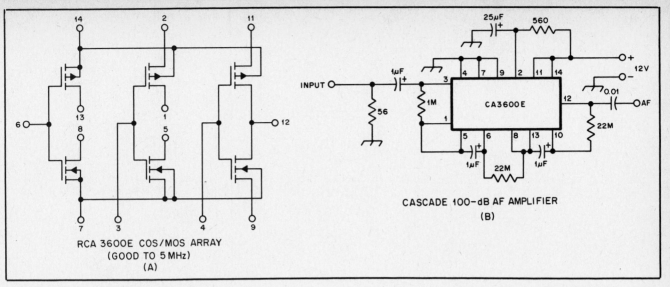

Fig. 85 — The diagram at A shows the internal workings of a CMOS IC. A 100-dB audio amplifier that employs the CA3600E is shown at B.

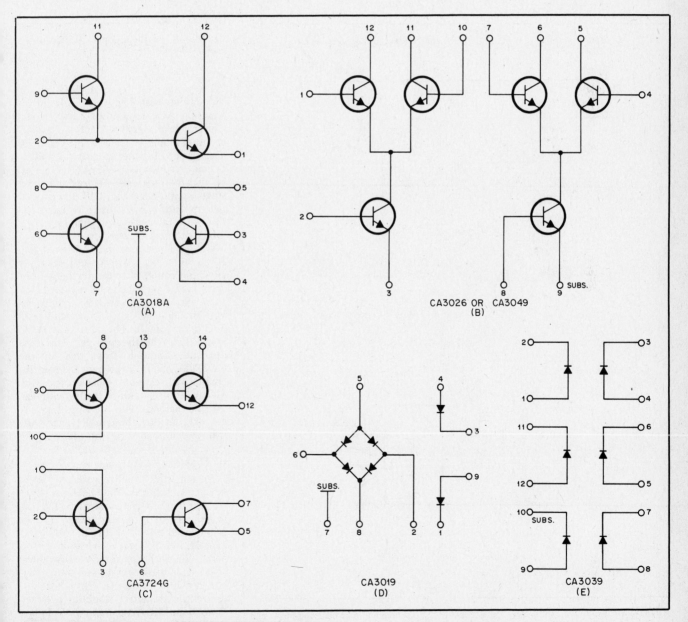

Fig. 86 — Various transistor and diode-array ICs. The configurations suggest a variety of amateur applications.

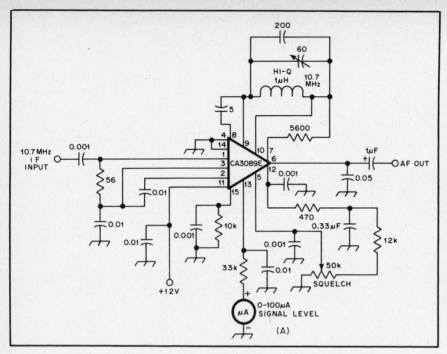

Fig. 87 — Example of a subsystem IC used as the heart of a narrow-band FM receiver.

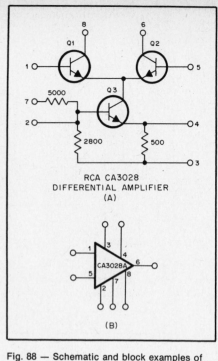

Fig. 88 — Schematic and block examples of an RCA CA3028A IC.

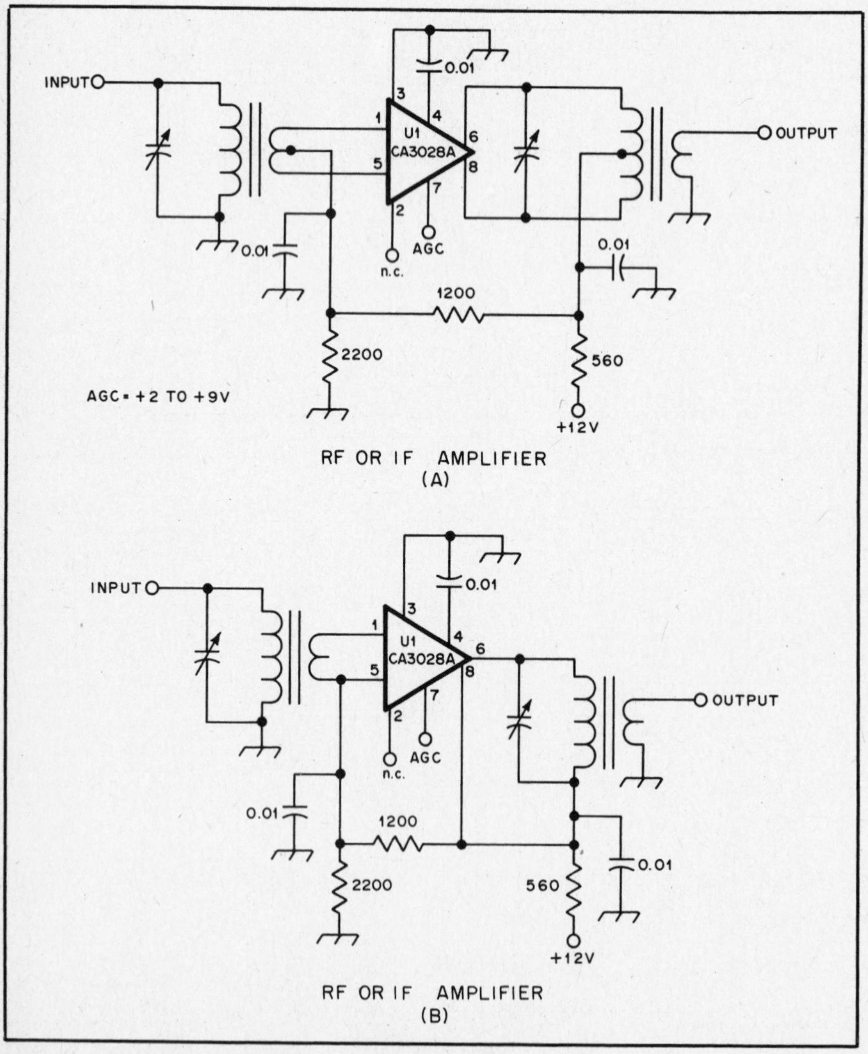

Fig. 89 — RF and IF amplifiers using the CA3028A IC. The example at A is balanced for ac and dc, whereas the circuit at B is balanced only for dc conditions.

AGC voltage (IC forward bias) is at its highest potential. The IC is nearly cut off when the AGC level drops below 2 volts. The circuit of Fig. 89A functions as a differential amplifier, as does the one at B. The basic difference is that dc and ac balance are featured at A, whereas only dc balance is effected at B. The gain of either stage is approximately 40 dB. Pin 2 of U1 is left floating, but is used for LO injection when the CA3028A is employed as a mixer or product detector. A Motorola MC1550G is similar to the IC shown in Fig. 89. A MC1590G is a more suitable IC for IF amplification when greater amounts of stage gain or AGC control are desired.

An example of an MC1590G amplifier is given in Fig. 90A. It is shown with AGC applied to pin 2. The lower the AGC voltage the higher the stage gain. This is the opposite condition from that of the CA3028A of Fig. 89, where the gain increases with elevated AGC voltage. The MC1350P of Fig. 90B is the low-cost version of the MC1590G. It is shown with manual control of the gain (R1), but AGC voltage can be applied instead.

IC Mixers

Examples of IC active mixers are given in Fig. 91. At A is a singly balanced mixer formed by the differential transistor pair in a CA3028A. A doubly balanced mixer is illustrated at B in Fig. 91. The MC1496G contains two differential transistor pairs to permit the doubly balanced configuration. This circuit does not exactly follow the suggested one by Motorola. It has been optimized for use as a transmitting mixer by W7ZOI and KL7IAK *(Solid State Design for the Radio Amateur,* 1st edition, page

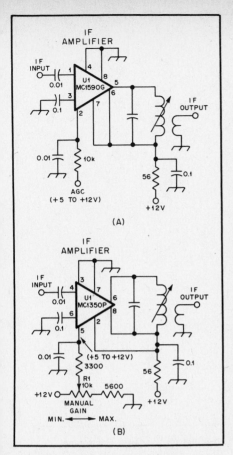

Fig. 90 — Circuit examples for Motorola IC IF amplifiers.

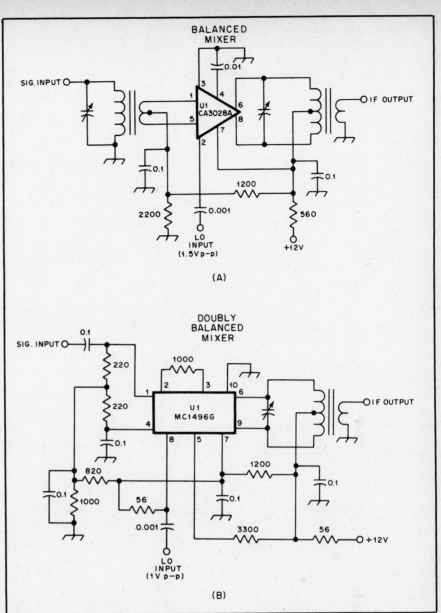

Fig. 91 — Two types of ICs are shown as mixers. The one at A is a singly balanced mixer.

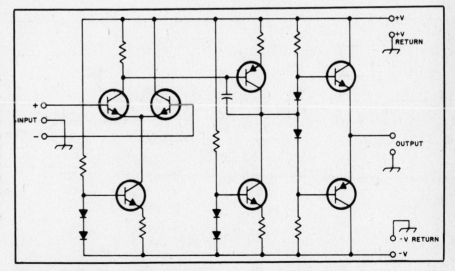

Fig. 92 — An operational amplifier assembled from discrete components. IC op amps contain more transistors (for current limiting and other peripheral functions), but the circuit topology is similar to that shown here.

204). Numerous other ICs can be used as mixers. Examples of many practical circuits are given in the ARRL book just referenced.

The circuit arrangements for product detectors and balanced modulators are similar to those shown in Fig. 91. They will not be described in this text, because the primary difference between them and a regular mixer lies in the frequencies of the signals mixed (AF versus RF) and the frequency of the resulting output energy.

IC Audio Amplifiers

Practically every IC manufacturer offers a line of audio ICs. Some are for use as low-noise preamplifiers and others are capable of delivering up to a few watts of output to a loudspeaker. Most of the audio-power ICs are designed for looking directly into an 8- or 16-ohm load without the need for a matching transformer. Because these circuits are relatively mundane they shall not be offered here as examples. Practical applications for audio ICs can be found in the construction projects elsewhere in this *Handbook*. Manufacturer's data sheets also provide definitive information on the use of these devices.

Operational Amplifiers

An operational amplifier (op amp) is a high-gain, direct-coupled differential amplifier whose characteristics are chiefly

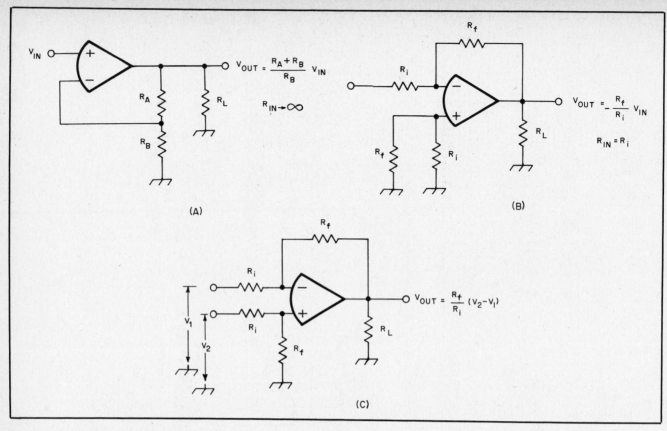

Fig. 93 — The standard negative-feedback op-amp circuits with their transfer equations. At A is a noninverting amplifier, at B an inverting amplifier, and at C is a differential amplifier.

determined by components external to the amplifier unit. Op amps can be assembled from discrete transistors, but better thermal stability results from fabricating the circuit on a single silicon chip. Integrated circuit op amps are manufactured with bipolar, JFET and MOSFET devices, either exclusively or in combination.

A design based on discrete components is shown in Fig. 92. Circuits of this variety were in common use before the advent of inexpensive IC-fabrication technology. The input stage consists of a differential pair biased by a constant current source. The terminal marked " – " is the inverting input and the one marked " + " is the noninverting input. The next stage, the PNP transistor, provides most of the voltage gain. High gain is realized through the use of a constant current source for the PNP collector load. The frequency response is determined by the collector-to-base capacitance of the PNP stage. This capacitance is fixed internally in some IC op amps and connected externally to others. A pair of emitter followers in a complementary symmetry arrangement forms the output buffer. A more comprehensive discussion of operational amplifiers is given by Woodward in "A

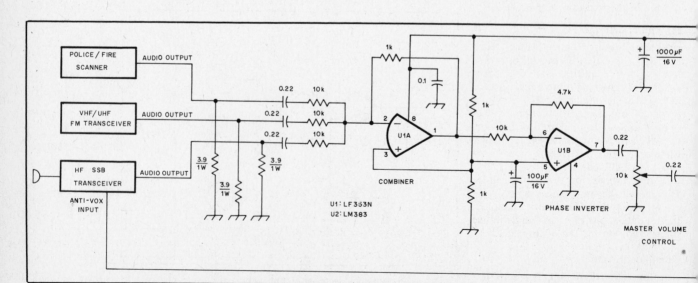

Fig. 94 — An amateur application for a summing amplifier — an audio combiner.

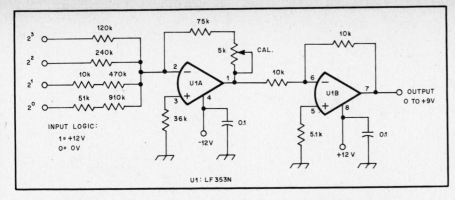

Fig. 95 — BCD D/A converter suitable for connection to a B-series CMOS driving source.

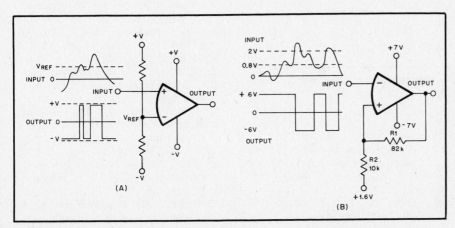

Fig. 96 — (A) Differential voltage comparator. Either inverting or noninverting circuits may be used. (B) Schmitt trigger. The constants shown here are suitable for connecting +5-V TTL to ±7-V CMOS logic.

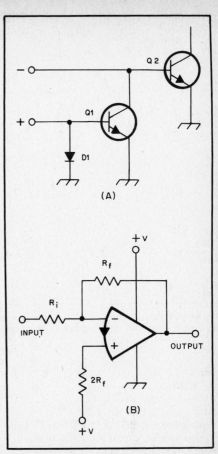

Fig. 97 — (A) Input circuit of a Norton operational amplifier. (B) Norton op amp connected as an inverting amplifier. Note the special symbol used to denote a Norton IC.

Beginner's Look at Op Amps,'' April and June 1980 QST.

The most common application for op amps is in negative feedback circuits operating from dc to perhaps a few hundred kilohertz. Provided the device has sufficient open-loop gain, the amplifier transfer function is determined almost solely by the external feedback network. The

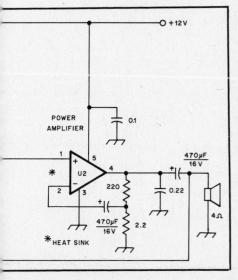

differential inputs allow for both inverting and noninverting circuits. Fig. 93 shows these configurations and gives their transfer equations. R_L does not appear in the equations, implying that the output impedance is zero. This condition results from the application of heavy negative feedback. Most IC op amps have built-in current limiting. This feature protects the IC from damage caused by short circuits, but also limits the values of load resistance for which the output impedance is zero. Most op amps work best with load resistances of at least 2 kΩ.

Since the op amp magnifies the difference between the voltages applied to its inputs, applying negative feedback has the effect of equalizing the input voltages. In the inverting amplifier configuration the feedback action combined with Kirchhoff's current law establishes a zero impedance, or virtual ground at the junction of R_f and R_i. The circuit input impedance is just R_i. Negative feedback applied to the noninverting configuration causes the input impedance to approach infinity.

The virtual ground at the inverting input terminal of an inverting operational amplifier circuit allows several currents to be summed without interaction. This principle can be used to advantage by the amateur wishing to simplify his or her station control system. An example of a summing amplifier is given in Fig. 94. The cir-

cuit shown allows the operator to monitor the outputs of several receivers with one loudspeaker. The 3.9-Ω resistors simulate the loudspeaker in each receiver. An inverter follows the summing amplifier to restore the antivox signal to the proper phase. Fig. 95 shows another application for a summing amplifier, a D/A converter. An FET-input operational amplifier can operate with the high-value resistors required by CMOS digital ICs while maintaining low offset and drift errors.

ICs intended for op-amp service can also be used in open-loop or positive feedback applications. Connecting one input to a fixed reference voltage as in Fig. 96A forms a comparator. The open-loop gain of the IC is so high that it acts more like a switch than an amplifier. When the voltage applied to the free input terminal is less than the reference voltage, the IC output stays near one of the power rails. If the input voltage exceeds the reference, the output swings to the opposite rail. A comparator with positive feedback, or hysteresis, is called a Schmitt trigger. A Schmitt trigger is illustrated in Fig. 96B. The potential on the noninverting input terminal depends on the output state as well as the reference voltage.

The Norton Amplifier

An unusual type of op amp is the

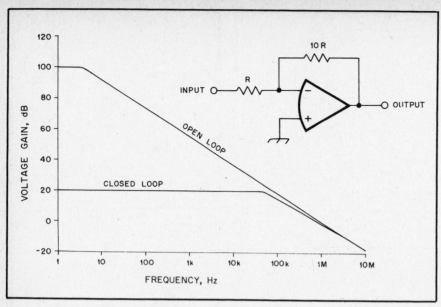

Fig. 98 — Open-loop gain and closed-loop gain as a function of frequency. The vertical distance between the curves is the feedback or gain margin.

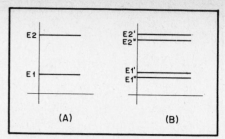

Fig. 99 — Energy-level diagram of a single atom is shown at A. At B, the levels split when two atoms are in close proximity.

Norton, named for the network theorem on which its operation is based. Fig. 97 shows a simplified diagram of the input stage of a Norton amplifier. The noninverting input makes use of D1 and Q1 in a current mirror configuration. When input current is applied to Q1, it steals base drive from Q2, the inverting input. This amplifier must have input current to operate, hence it is not a high-impedance device. In the inverting-amplifier configuration the numerical voltage gain is R_f/R_i, but the noninverting input terminal must be returned to the positive supply through a resistance of $2R_f$ to equalize the input currents. Any attempt to use this type of IC as a voltage follower is doomed to failure — the input stage will be destroyed by excessive current. The chief use of Norton amplifiers is for single-supply applications where the dc level of the signal is very near ground.

Important Op Amp Specifications

Construction projects in the amateur literature call for the 709 and 741 more than any other type of operational amplifier because until recently they were the only ones commonly available to the electronics hobbyist. A much wider selection of op amps is available today, and the amateur designer can choose the components best suited to the application. Also, the performance of some existing circuits can be upgraded by replacing 709s and 741s with improved devices. To this end, a brief survey of op amp specifications is in order.

Offset voltage is the potential between the amplifier input terminals in the closed-loop condition. Ideally, this voltage would be zero. Offset results from imbalance between the differential input transistors. Values range from millivolts in ordinary consumer-grade devices to only nanovolts in premium Milspec units. The temperature

coefficient of offset voltage is drift. Drift is usually considered in relation to time. Heat generated by the op-amp itself or by associated circuitry will cause the offset voltage to change over time. A few microvolts per degree Celsius (at the input) is a typical drift specification.

There are two types of noise associated with operational amplifiers. Burst or popcorn noise is a low-frequency pulsing, usually below 10 Hz. The amplitude of this noise is approximately an inverse function of temperature. The other noise is sometimes called flicker, and is a wideband signal whose amplitude varies inversely with frequency. For some analytical purposes, drift is considered as a very low frequency noise component. Op amps that have been optimized for offset, drift and noise are called instrumentation amplifiers. The latest instrumentation amplifier is the National Semiconductor LM10, designed by Robert Widlar, the acknowledged "father of the IC op amp." The architecture of the LM10 is different from any other device, but the practical applications are the same.

The small-signal bandwidth of an op amp is the frequency range over which the open-loop voltage gain is at least unity. This specification depends mostly on the frequency compensation scheme (for example, the capacitor in Fig. 92). Fig. 98 shows how the maximum closed-loop gain varies with the frequency. The power bandwidth of an operational amplifier is a function of slew rate, and is always less than the small-signal value. Slew rate is a measurement of output voltage swing per unit time. Values from 0.8 to 13 volts per microsecond are typical of modern devices.

The hobbyist should maintain a supply of inexpensive 741 and 301 op amps for breadboarding, but should also be prepared to use improved devices in the final design. In an active filter for example, a 741 will

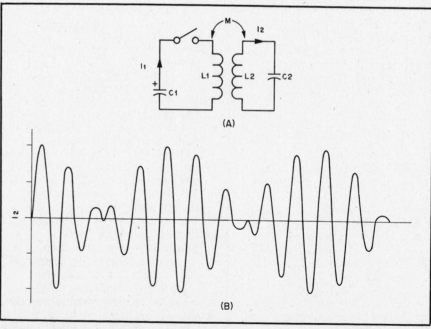

Fig. 100 — Electrical-circuit analog of coupled atoms.

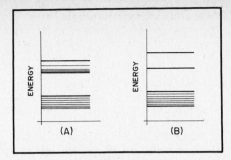

Fig. 101 — The energy level of a conductor is illustrated at A. A similar level for an insulator is depicted at B.

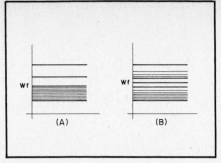

Fig. 102 — Semiconductor energy-level representation.

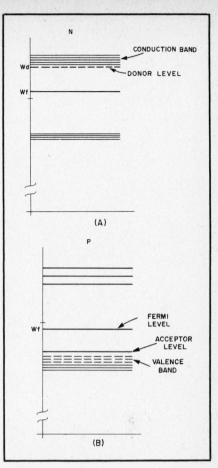

Fig. 103 — The effects on the energy level if impurity atoms are introduced.

demonstrate whether or not the circuit is working, but a low-noise, wide-bandwidth device will give higher performance, especially in receiving service. An abbreviated table of operational amplifier specifications is given in Chapter 35. Most of the devices listed are available from hobby electronics stores or the mail-order firms listed in Chapter 35.

APPENDIX

The Physics of Solid-State Devices

The electrical characteristics of solid-state devices such as diodes and transistors depend on phenomena that take place at the atomic level. While semiconductors can be employed without a complete knowledge of these effects, some understanding is helpful in various applications. Electrons, which are the principal charge carriers in both vacuum tubes and semiconductors, behave much differently in either of the two circumstances. In free space, an electron can be considered as a small charged solid particle. On the other hand, the presence of matter affects this picture greatly. For instance, an electron attached to an atom has many properties similar to those of RF energy in tuned circuits. It has a frequency and wavelength that depend on atomic parameters just as the frequency associated with electrical energy in a tuned circuit depends on the values of inductance and capacitance.

A relation between the energy of an electron in an atomic orbit and its associated frequency is given by

$$f(Hz) = \frac{E \text{ (joules)}}{6.625 \times 10^{-34}}$$

where the expression in the denominator is called Planck's constant. This equation is quite important when an electron is either raised or falls between two different energy states. For instance, when an electron drops from one level to a lower one, energy is emitted in the form of electromagnetic radiation. This is the effect that gives the characteristic glow to neon tubes, mercury-vapor rectifiers and light-emitting diodes. The frequency of the emitted radiation is given by the formula where E is the difference in energy. However, if an electron

receives enough energy such that it is torn from an atom, a process called ionization is said to occur (although the term is also loosely applied to transitions between any two levels). If the energy is divided by the charge of the electron (-1.6×10^{-19} coulombs), the equivalent in voltage is obtained.

A common way of illustrating these energy transitions is by means of the energy-level diagram shown in Fig. 99A. It should be noted that unlike ordinary graphical data, there is no significance to the horizontal axis. In the case of a single atom, the permitted energy can only exist at discrete levels (this would be characteristic of a gas at low pressure where the atoms are far apart). However, if a single atom is brought within close proximity of another one of similar type, the single energy levels split into pairs of two that are very close together (Fig. 99B). The analogy between tuned circuits and electron energy levels can be carried even further in this case.

Consider the two identical circuits that are coupled magnetically as shown in Fig. 100A. Normally, energy initially stored in C1 would oscillate back and forth between L1 and C1 at a single frequency after the switch was closed. However, the presence of the second circuit consisting of L2 and C2 (assume L1 equals L2 and C1 equals C2) results in the waveform shown in Fig. 100B. Energy also oscillates back and forth between the two circuits and the current then consists of components at two slightly different frequencies. The effect is similar to the splitting of electron energy levels when two atoms are close enough to interact.

Conductors, Insulators and Semiconductors

Solids are examples of large numbers of atoms in close proximity. As might be expected, the splitting of energy levels continues until a band structure is reached. Depending on the type of atom, and the physical arrangement of the component atoms in the solid, three basic conditions can exist. In Fig. 101A, the two discrete energy levels have split into two bands. All the states in the lower band are occupied by electrons while the ones in the higher energy band are only partially filled.

In order to impart motion to an electron, the expenditure of energy is required. This means an electron must then be raised from one energy state to a higher one. Since there are many permitted states in the upper level of Fig. 101A that are both unoccupied and close together, electrons in this level are relatively free to move about. Consequently, the material is a conductor.

In Fig. 101B, all the states in the lower level are occupied, there is a big gap between this level and the next higher one and the upper level is empty. This means if motion is to be imparted to an electron, it must be raised from the lower level to the upper one. Since this requires considerable energy, the material is an insulator. (The energy-level representation gives an insight into the phenomenon of breakdown. If the force on an electron in an insulator becomes high enough because of an applied field, it can acquire enough energy to be raised to the upper level. When this happens, the material goes into a conducting state.)

A third condition is shown in Fig. 102. In the material associated with this diagram, the upper level is unoccupied but is very close to the occupied one. Hence, under conditions where the random electron motion is low (low temperature), the material acts as an insulator (Fig. 102A). However, as the random or thermal mo-

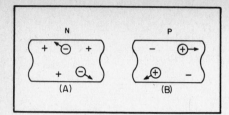

Fig. 104 — N- and P-type semiconductors.

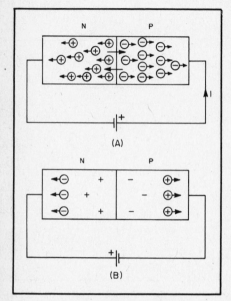

Fig. 105 — Elementary illustration of current flow in a semiconductor diode.

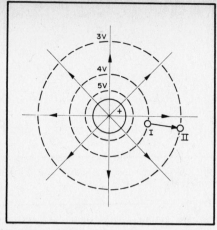

Fig. 106 — Potential diagram of an electron in atomic orbit.

tion increases, some electrons acquire enough energy to move up to states in the upper level. Consequently, both levels are partially occupied, as shown in Fig. 102B.

The line marked W_f represents a statistical entity related to the average energy of electrons in the material and is called the Fermi Level. At absolute zero (no thermal motion), W_f is just at the top of the lower energy level. As electrons attain enough energy to move to the upper level, W_f is approximately halfway between the two levels.

The PN Junction

The material for the diagram shown in Fig. 102 is called an intrinsic semiconductor. Examples are the elements germanium and silicon. As such, the materials do not have any rectifying properties by themselves. However, if certain elements are mixed into the intrinsic semiconductor in trace amounts, a mechanism for rectification exists. This is shown in Fig. 103A. If an element with an occupied energy level such as arsenic is introduced into germanium, a transformation in conductivity takes place. Electrons in the new occupied level are very close to the upper partially filled band of the intrinsic germanium. Consequently, there are many extra charge

carriers available when thermal energy is sufficient to raise some of the electrons in the new level to the partially filled one. Germanium with an excess of mobile electrons is called an N-type semiconductor.

By introducing an element with an empty or unoccupied energy level near the lower partially filled level (such as boron), a somewhat different transformation in conductivity occurs. This is shown in Fig. 103B. Electrons from the lower level can move into the new unoccupied level if the thermal energy is sufficient. This means there is an excess of unoccupied states in the germanium lower energy level. Germanium treated this way is called P-type semiconductor.

A physical picture of both effects is shown in Fig. 104. The trace elements or impurities are spread throughout the intrinsic crystal. Since the distance of separation is much greater for atoms of the trace elements than it is for ones of the intrinsic crystal, there is little interaction between the former. Because of this lack of coupling, the distribution of energy states is a single level rather than a band. In Fig. 104A, atoms of the trace element are represented by the + signs since they have lost an electron to the higher energy level. Consequently, such elements are called donors. In Fig. 104B, the impurity atoms that have trapped an electron in the new state are indicated by the − signs. Atoms of this type are called acceptor impurities.

While it is easy to picture the extra free electrons by the circled minus charges in Fig. 104A, a conceptual difficulty exists with the freed positive charges shown in Fig. 104B. In either case, it is the motion of electrons that is actually taking place and the factor that is responsible for any current. However, it is convenient to consider that a positive charge carrier, called a hole, exists. It would seem as though a dislocation in the crystal-lattice structure was moving about and contributing to the total current.

If a section of N-type material is joined

to another section made from P-type, a one-way current flow results. This is shown in Fig. 105. A positive potential applied to the P-type electrode attracts any electrons that diffuse in from the N-type end. Likewise, holes migrating from the P-type end into the N-type electrode are attracted to the negative terminal. Note that the diagram indicates not all the carriers reach the terminals. This is because some carriers combine with ones of the opposite sign while enroute. In the case of a diode, this effect doesn't present much of a problem since the total current remains the same. Other carriers take the place of those originally injected from the opposite regions. However, such recombination degrades the performance of transistors considerably and will be discussed shortly.

If a voltage of the opposite polarity to that of Fig. 105A is applied to the terminals, the condition in Fig. 105B results. The mobile charge carriers migrate to each end as shown leaving only the fixed charges in the center near the junction. Consequently, little current flows and the PN junction is back biased. It can be seen that the PN junction constitutes a diode since current can flow readily only in one direction. While this simple picture suffices for introductory purposes, proper treatment of many important effects in semiconductors requires a more advanced analysis than the elementary model affords. Returning to Figs. 100, 101 and 102, it would be convenient if the diagrams were in terms of voltage rather than energy. As pointed out earlier, the relation between energy and voltage associated with an electron is given by

$$W = eV = (-1.6 \times 10^{-19}) V$$

Because the electron has been assigned a minus charge, a somewhat upside-down world results. However, if it is kept in mind that it requires the expenditure of energy to move an electron from a point of higher potential to one at a lower value, this confusion can be avoided. As an illustration, suppose an electron is moved from an atomic orbit indicated by I in Fig. 106 to orbit II. This would mean the electron would have had to have been moved against the force of attraction caused by the positive nucleus resulting in an increase in potential energy. (In other words, orbit II is at a higher energy level than orbit I.) However, note that the electrostatic potential around the nucleus decreases with distance and that orbit II is at a lower potential than orbit I.

Consequently, the energy-level diagram in terms of voltage becomes inverted as shown in Fig. 107. It is now possible to approach the problem of the PN junction diode in terms of the energy-level diagrams presented previously. If a section of N-type and P-type material is considered separately, the respective energy (or voltage) levels would be the same. However, if the two

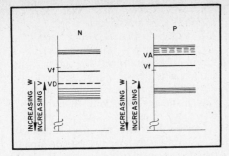

Fig. 107 — Energy-level diagram in terms of potential.

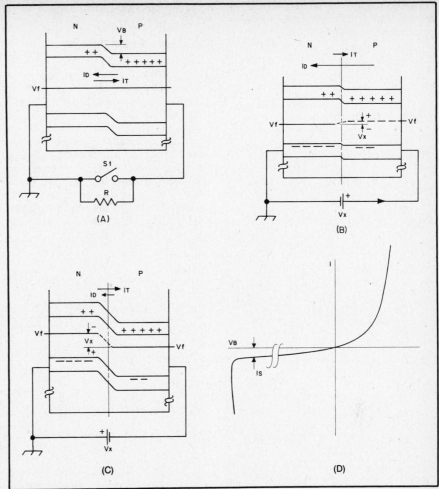

Fig. 108 — Energy-level diagrams for unbiased (A), forward-biased (B) and reversed-biased diode (C). Illustration D shows the resultant characteristics of the diode.

sections were joined together and connected by an external conductor as shown in Fig. 108, a current would flow initially. This is because the voltage corresponding to the statistical entity referred to previously (Fermi Level) is not the same for P- and N-type materials at the same temperature.

At the Fermi Level, the probability that a particular energy state is occupied is one half. For N-type material, the Fermi Level is shifted upward toward the conduction band (Fig. 108A). In a P-type material, it is shifted downward toward the valence band. Although the theory behind the Fermi Level and definitions concerning the conduction and valence bands won't be dealt with here, it is sufficient to know that the band structure shifts so that the Fermi Levels are the same in both parts of the joined sections (Fig. 108).

The reasoning behind this effect is as follows. Consider conditions for hole flow only for the moment. Since there is an excess of holes in the P region (Fig. 108), there is a tendency for them to move over into the adjacent N region because of diffusion. The process of diffusion is demonstrated easily. If a small amount of dye is dropped into some water, it is concentrated in a small area at first. However, after a period of time has passed, it spreads out completely through the entire volume.

Once the holes diffuse into the N region, they recombine with the electrons present and produce a current in the external terminals denoted by I_D (Fig. 108). But a paradox results because of this current. If S1 is opened so that I_D flows through R, where does the energy that is transferred (irreversibly) to this resistance come from? In effect, it represents a perpetual-motion dilemma or else the semiconductor will cool down since the diffusion process is the result of a form of thermal motion. Both conclusions are against the laws of physics, so a third alternative is necessary. It is then assumed that the Fermi Levels align so that the potential across the terminals becomes zero, and no current will flow in the external circuit.

However, if the Fermi Levels are the same, the conduction and valence bands in either section will no longer align. As a consequence, a difference in potential between the two levels exists and is indicated by V_B

in Fig. 108. The formation of this junction or barrier voltage is of prime importance in the operation of PN-junction devices. Note that holes in the P region must overcome the barrier voltage that impedes the flow of the diffusion current. It will also be recalled that both holes and electrons were generated in the intrinsic semiconductor because of thermal effects (Fig. 102B). The addition of either donor or acceptor atoms modifies this effect somewhat. If donor atoms are present (N-type material), fewer holes are generated. On the other hand, if acceptor atoms represent the impurities, fewer electrons are generated in comparison to conditions in an intrinsic semiconductor. In the case of P-type material, holes predominate and are termed the majority carriers. Since there are fewer electrons in P-type material they are termed the minority carriers.

Referring to Fig. 108A, there are some holes in the N region (indicated by the + signs) because of the thermal effects mentioned earlier. Those near the junction will experience a force caused by the electric field associated with the barrier voltage. This field will produce a flow of holes into

the P region and the current is denoted by I_T. Such a current is called a drift current as compared to the diffusion current I_D. Under equilibrium conditions, the two currents are equal and just cancel each other. This is consistent with the assumption that no current flows in the external circuit because of the fact that the Fermi Levels are the same and no voltage is produced.

So far, only conditions for the holes in the upper (or conduction) band have been considered, but identical effects take place with the motion of electrons in the lower energy band (valence band). Since the flow of charge carriers is in opposition, but because holes and electrons have opposite signs, the currents add.

The Forward-Biased Diode

If an external EMF is applied to the diode terminals as shown in Fig. 108B, the equilibrium conditions no longer exist and the Fermi Level voltage in the right-hand region is shifted upward. This means the barrier voltage is decreased and considerable numbers of carriers may now diffuse across the junction. Consequently, I_D becomes very large while I_T decreases in value because of the decrease in barrier voltage. The total current under forward-bias conditions then becomes

$$I = I_s \left(e^{\frac{qv_x}{kt}} - 1 \right)$$

where

q = 1.6×10^{-19} coulombs (the fundamental charge of an electron)
k = 1.38×10^{-23} joules/kelvin (Boltzmann's constant)
t = junction temperature in kelvins
e = 2.718 (natural logarithmic base)
V_x = applied EMF and I_s = reverse-bias saturation current.

This equation is discussed in greater detail in the section dealing with common silicon diodes.

The Reverse-Biased Diode

If the source, V_x, is reversed as shown in Fig. 108C, the barrier voltage is increased. Consequently, charge carriers must overcome a large potential hill and the diffusion current becomes very small. However, the drift current caused by the thermally generated carriers returns to the value it had under equilibrium conditions. For large values of V_x, the current approaches I_s, defined as the reverse saturation current, I_s is the sum of I_T and its counterpart in the lower or "valence" band. Finally, the characteristic curves of the forward- and reversed-bias diode can be constructed and are shown in Fig. 108D.

It is obvious that I_s should be as small as possible in a practical diode since it would only degrade rectifier action. Also, since it is the result of the generation of thermal carriers, it is quite temperature sensitive, which is important when the diode is part of a transistor. If the reverse voltage is increased further, an effect called avalanche breakdown occurs as indicated by the sudden increase in current at V_b. In such an instance, the diode might be damaged by excessive current. However, the effect is also useful for regulator purposes and devices used for this purpose are called Zener diodes.

Chapter 5

Vacuum-Tube Principles

This chapter covers the principles of vacuum-tube operation. Additional material on vacuum tubes may be found in Chapter 15. The outstanding difference between the vacuum tube and most other electrical devices is that the electric current does not flow through a conductor, but through empty space — a vacuum. This is only possible when free electrons — that is, electrons that are not attached to atoms — are somehow introduced into the vacuum. Free electrons in an evacuated space will be attracted to a positively charged object within the same space, or will be repelled by a negatively charged object. The movement of the electrons under the attraction or repulsion of such charged objects constitutes the current in the vacuum.

Thermionic Emission

The most practical way to introduce a sufficiently large number of electrons into the evacuated space is by thermionic emission. If a piece of metal is heated to incandescence in a vacuum, electrons near the surface are given enough energy of motion to fly off into the surrounding space. The higher the temperature, the greater the number of electrons emitted. In a vacuum tube, the electron-emitting material is called the cathode.

If the cathode is the only electrode in the vacuum, most of the emitted electrons stay in its immediate vicinity, forming a cloud about the cathode. The reason for this is that the electrons in the space, being negative electricity, form a negative charge (space charge) in the region of the cathode. The space charge repels those electrons nearest the cathode, tending to make them fall back on it.

Now suppose a second conductor is introduced into the vacuum, but not connected to anything else inside the tube. If this second conductor is given a positive charge by connecting a voltage source between it and the cathode, as indicated in Fig. 1, electrons emitted by the cathode are attracted to the positively charged conductor. An electric current then flows through the circuit formed by the cathode, the charged conductor and the voltage source. In Fig. 1 this voltage source is a battery (B battery); a second battery (A battery) is also indicated for heating the cathode to the

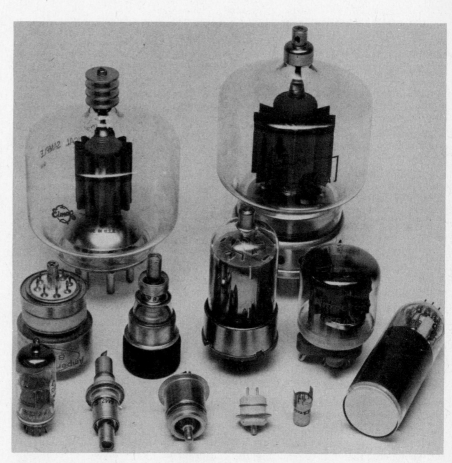

Transmitting tubes are shown in the back and center rows. Receiving tubes are in the front row (l to r): miniature, pencil, planar triode (two), Nuvistor and 1-inch-diameter cathode-ray tube.

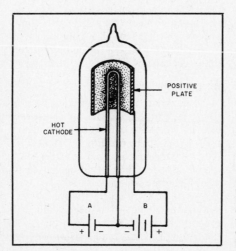

proper operating temperature.

The positively charged conductor is usually a metal plate or cylinder (surrounding the cathode) and is called the anode or plate. Like the other working parts of a tube, the plate is a tube element or electrode. The tube shown in Fig. 1 is

Fig. 1 — Conduction by thermionic emission in a vacuum tube. The A battery is used to heat the cathode to a temperature that will cause it to emit electrons. The B battery makes the plate positive with respect to the cathode, thereby causing the emitted electrons to be attracted to the plate. Electrons captured by the plate flow back through the B battery to the cathode.

Labels in Fig. 1: POSITIVE PLATE, HOT CATHODE, A, B

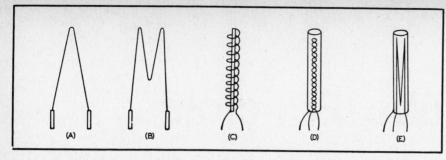

Fig. 2 — Types of cathode construction. Directly heated cathodes or filaments are shown at A, B and C. The inverted V filament is used in small receiving tubes, the M in both receiving and transmitting tubes. The spiral filament is a transmitting tube type. The indirectly heated cathodes at D and E show two types of heater construction, one a twisted loop and the other bunched heater wires. Both types tend to cancel the magnetic fields set up by the current though the heater.

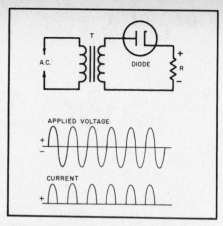

Fig. 4 — Rectification in a diode. Current flows only when the plate is positive with respect to the cathode, so that only half cycles of current flow through the load resistor, R.

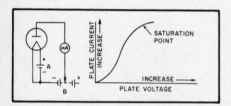

Fig. 3 — The diode, or two-element tube, and a typical curve showing how the plate current depends on the voltage applied to the plate.

a two-element or two-electrode tube (a diode), one element being the cathode and the other the anode or plate.

Since electrons have a negative charge, they will be attracted to the plate only when the plate is given a positive charge. If the plate is given a negative charge, the electrons will be repelled back to the cathode and no current will flow. The vacuum tube therefore can conduct only in one direction.

Cathodes

Before electron emission can occur, the cathode must be heated to a high temperature. However, it is not essential that the heating current flow through the actual material that does the emitting; the heating filament or heater can be electrically separate from the emitting cathode. Such a cathode is said to be indirectly heated, while an emitting filament is called a directly heated cathode. Fig. 2 shows both types in the forms they commonly take.

Much greater electron emission can be obtained, at relatively low temperatures, by using special cathode materials rather than pure metals. One of these is thoriated tungsten (tungsten in which thorium is dissolved). Still greater efficiency is obtained in the oxide-coated cathode, a cathode in which rare-earth oxides form a coating over a metal base.

Although the oxide-coated cathode has the highest efficiency, it can be used successfully only in tubes that operate at low plate voltages. Its use is therefore confined to receiving-type tubes and to the smaller varieties of transmitting tubes. The thoriated filament, on the other hand, will

operate well in high-voltage tubes.

Plate Current

If there is only a small positive voltage on the plate, the number of electrons reaching it will be small because the space charge (which is negative) prevents those electrons nearest the cathode from being attracted to the plate. As the plate voltage is increased, the effect of the space charge is increasingly overcome and the number of electrons attracted to the plate becomes larger. That is, the plate current increases with increasing plate voltage.

Fig. 3 shows a typical plot of plate current vs. plate voltage for a two-element tube or diode. A curve of this type can be obtained with the circuit shown, if the plate voltage is increased in small steps and a current reading is taken by means of the current-indicating instrument (a milliammeter) at each voltage. The plate current is zero with no plate voltage and the curve rises until a saturation point is reached. This is where the positive charge on the plate has substantially overcome the space charge and almost all the electrons are going to the plate. At higher voltages the plate current stays at practically the same value.

The plate voltage multiplied by the plate current is the power input to the tube. In a circuit like that of Fig. 3 this power is all used in heating the plate. If the power input is large, the plate temperature may rise to a very high value (the plate may become red or even white hot). The heat developed in the plate is radiated to the bulb of the tube, and in turn to the surrounding air.

Rectification

Since current can flow through a tube in only one direction, a diode can be used to change alternating current into direct current. It does this by permitting current flow only when the anode is positive with respect to the cathode. There is no current flow when the plate is negative.

Fig. 4 shows a representative circuit. Alternating voltage from the secondary of the transformer, T, is applied to the diode tube in series with a load resistor, R. The

voltage varies as is usual with ac, but current flows through the tube and R only when the plate is positive with respect to the cathode — that is, during the half cycle when the upper end of the transformer winding is positive. During the negative half cycle there is simply a gap in the current flow. This rectified alternating current therefore is an intermittent direct current.

The load resistor, R, represents the actual circuit in which the rectified alternating current does work. All tubes work with a load of one type or another; they must cause power to be developed in a load in order to serve a useful purpose. Also, to be efficient, most of the power must do useful work in the load rather than in heating the plate of the tube. Thus, the voltage drop across the load should be much higher than the drop across the diode.

With the diode connected as shown in Fig. 4, the polarity of the current through the load, R, is as indicated. If the diode were reversed, the polarity of the voltage developed across the load would be reversed.

VACUUM-TUBE AMPLIFIERS
Triodes

If a third element — called the control grid or, simply, grid — is inserted between the cathode and plate as in Fig. 5, it can be used to control the effect of the space charge. If the grid is given a positive voltage with respect to the cathode, the positive charge will tend to neutralize the negative space charge. The result is that, at any selected plate voltage, more electrons will flow to the plate than if the grid were not present. On the other hand, if the grid is made negative with respect to the cathode the negative charge on the grid will add to the space charge. This will reduce the number of electrons that can reach the plate

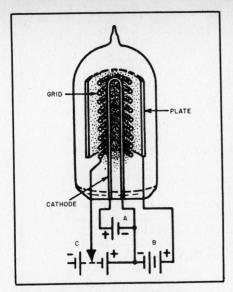

Fig. 5 — Construction of a basic triode vacuum tube, showing the directly heated cathode (filament), grid (with an end view of the grid wires) and plate. The relative density of the space charge is indicated roughly by the dot density.

Fig. 6 — Curves of grid voltage vs. plate current at various fixed values of plate voltage (e_b) for a typical small triode. Characteristic curves of this type can be taken by varying the battery voltages in the circuit shown.

is one with an amplification of perhaps 30 or more. Medium-μ tubes have amplification factors in the approximate range 8 to 30, and low-μ tubes in the range below 7 or 8. The μ of a triode is useful in computing stage gains.

The best all-around indication of the effectiveness of a tube as an amplifier is its grid-plate transconductance — also called mutual conductance or g_m. It is the change in plate current divided by the change in grid voltage that caused the change; it can be found by dividing the amplification factor by the plate resistance. Since current divided by voltage is conductance, transconductance is measured in the unit of conductance, the siemens.

Practical values of transconductance are very small, so the microsiemens (one millionth of a siemens) is the commonly used unit. Different types of tubes have transconductance values ranging from a few hundred to several thousand. The higher the transconductance the greater the possible amplification.

Amplification

The way in which a tube amplifies is best shown by a type of graph called the dynamic characteristic. Such a graph, together with the circuit used for obtaining it, is shown in Fig. 7. The curves are taken with the plate supply voltage fixed at the desired operating value. The difference between this circuit and the one shown in Fig. 6 is that in Fig. 7 a load resistance is connected in series with the plate of the tube. Fig. 7 thus shows how the plate current will vary, with different grid voltages, when the plate current is made to flow through a load and thus do useful work.

The several curves in Fig. 7 are for various values of load resistance. When the resistance is small (as in the case of a 5-kΩ load) the plate current changes rapidly with a given change in grid voltage. If the load resistance is high (as in the 100,000-kΩ curve), the change in plate current for the same grid-voltage change is relatively small; also, the curve tends to be straighter.

at any selected plate voltage.

The grid is inserted in the tube to control the space charge and not to attract electrons to itself, so it is made in the form of a wire mesh or spiral. Electrons then can pass through the open spaces in the grid to reach the plate.

Characteristic Curves

For any particular tube, the effect of the grid voltage on the plate current can be shown by a set of characteristic curves. A typical set of curves is shown in Fig. 6, together with the circuit that is used for determining them. For each value of plate voltage, there is a value of negative grid voltage that will reduce the plate current to zero; that is, there is a value of negative grid voltage that will cut off the plate current.

The curves could be extended by making the grid voltage positive as well as negative. When the grid is negative, it repels electrons and therefore none of them reaches it; in other words, no current flows in the grid circuit. When the grid is positive, however, it attracts electrons and a current (grid current) flows, just as current flows to the positive plate. Whenever there is grid current there is an accompanying power loss in the grid circuit, but as long as the grid is negative no power is used.

The grid can act as a valve to control the flow of plate current. Actually, the grid has a much greater effect on plate current flow than does the plate voltage. A small change in grid voltage is just as effective in bringing about a given change in plate current as is a large change in plate voltage.

The fact that a small voltage change on the grid is equivalent to a large voltage change on the plate indicates the possibility of amplification with the triode tube. The

many uses of the electron tube are nearly all based on this amplifying characteristic. The amplified output is not obtained from the tube itself, but from the voltage source connected between its plate and cathode. The tube simply controls the power from this source, changing it to the desired form.

To utilize the controlled power, a load must be connected in the plate or output circuit, just as in the diode case. The load may be either a resistance or an impedance. The term impedance is frequently used even when the load is purely resistive.

Tube Characteristics

The physical construction of a triode determines the relative effectiveness of the grid and plate in controlling the plate current. The control of the grid increases by moving it closer to the cathode or by making the grid mesh finer.

The plate resistance of a vacuum tube is the ac resistance of the path from cathode to plate. For a given grid voltage, it is the quotient of a small change in plate voltage divided by a resultant change in plate current. Thus, if a 1-volt change in plate voltage caused a plate-current change of 0.01 mA (0.00001 or 1×10^{-5} ampere), the plate resistance would be 100,000 ohms.

The amplification factor (usually designated by the Greek letter μ) of a vacuum tube is defined as the ratio of the change in plate voltage to the change in grid voltage to effect equal changes in plate current. If, for example, an increase of 10 plate volts raised the plate current 1.0 mA, and an increase in (negative) grid voltage of 0.1 volt were required to return the plate current to its original value, the amplification factor of this triode tube would be 100. The amplification factors of triode tubes range from 2 to 100 or so. A high-μ tube

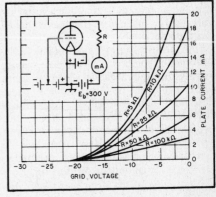

Fig. 7 — Dynamic characteristics of a small triode with various load resistances from 5 kΩ to 100 kΩ.

Fig. 8 shows the same type of curve, but the circuit is arranged so that a source of alternating voltage (signal) is inserted between the grid and the grid battery (C battery in this example). The voltage of the grid battery is fixed at −5 volts, and from the curve it may be seen that the plate current at this grid voltage is 2 milliamperes. This current flows when the load resistance is 50,000 × 0.002 = 100 volts, leaving 200 volts between the plate and cathode.

When a sine-wave signal having a peak value of 2 volts is applied in series with the bias voltage in the grid circuit, the instantaneous voltage at the grid will swing to −3 volts when the signal reaches its positive peak, and to −7 volts when the signal reaches its negative peak. The maximum plate current will occur at the instant the grid voltage is −3 volts. As shown by the graph, it will have a value of 2.65 milliamperes. The minimum plate current occurs at the instant the grid voltage is −7 volts, and has a value of 1.35 mA. At intermediate values of grid voltage, intermediate plate current values will occur.

The instantaneous voltage between the plate and cathode of the tube is also shown on the graph. When the plate current is maximum, the instantaneous voltage drop in R_p is 50,000 × 0.00265 = 132.5 volts; when the plate current is minimum the instantaneous voltage drop in R_p is 50,000 × 0.00135 = 67.5 volts. The actual voltage between plate and cathode is the difference between the plate supply potential, 300 volts, and the voltage drop in the load resistance. The plate-to-cathode voltage is therefore 167.5 volts at maximum plate current and 232.5 volts at minimum plate current.

This varying plate voltage is an ac voltage superimposed on the steady plate-cathode potential of 200 volts (as previously determined for no-signal conditions). The peak value of this ac output voltage is the difference between either the maximum or minimum plate-cathode voltage and the no-signal value of 200 volts. In the illustration this difference is 232.5 − 200 or 200 − 167.5; that is, 32.5 volts in either case. Since the grid signal voltage has a peak value of 2 volts, the voltage-amplification ratio of the amplifier is 32.5/2 or 16.25. That is, approximately 16 times as much voltage is obtained from the plate circuit as is applied to the grid circuit.

As shown by the drawings in Fig. 8, the alternating component of the plate voltage swings in the negative direction (with reference to the no-signal value of plate-cathode voltage) when the grid voltage swings in the positive direction, and vice versa. This means that the alternating component of plate voltage (that is, the amplified signal) is 180 degrees out of phase with the signal voltage on the grid.

Bias

The fixed negative grid voltage (called

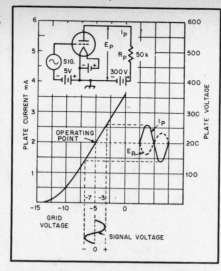

Fig. 8 — Amplifier operation. When the plate current varies in response to the signal applied to the grid, a varying voltage drop appears across the load, R_p, as shown by the dashed curve, E_p. I_p is the plate current.

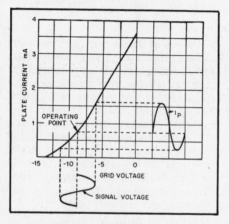

Fig. 9 — Distortion resulting from the choice of an operating point on the curved part of the tube characteristic. The lower half cycle of plate current does not have the same shape as the upper half cycle.

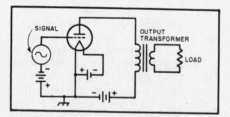

Fig. 10 — An elementary power amplifier circuit in which the power-consuming load is coupled to the plate circuit through an impedance-matching transformer.

grid bias) in Fig. 8 serves a very useful purpose. One purpose of the type of amplification shown in this drawing is to obtain, from the plate circuit, an alternating voltage that has the same wave shape as the signal voltage applied to the grid. To do so, an operating point on the straight part of

the curve must be selected. The curve must be straight in both directions from the operating point at least far enough to accommodate the maximum value of the signal applied to the grid. If the grid signal swings the plate current back and forth over a part of the curve that is not straight as shown in Fig. 9, the shape of the ac wave in the plate circuit will not be the same as the shape of the grid-signal wave. In such a case the output wave shape will be distorted.

A second reason for using negative grid bias is that any signal whose peak positive voltage does not exceed the fixed voltage on the grid cannot cause grid current to flow. With no current flow there is no power consumption, so the tube will amplify without taking any power from the signal source. (However, if the positive peak of the signal does exceed the negative bias, current will flow in the grid circuit during the time the grid is positive.)

Distortion of the output wave shape that results from working over a part of the curve that is not straight (that is, a nonlinear part of the curve) has the effect of transforming a sine-wave grid signal into a more complex waveform. As explained in an earlier chapter, a complex wave can be resolved into a fundamental frequency and a series of harmonics. In other words, distortion from nonlinearity causes the generation of harmonic frequencies — frequencies that are not present in the signal applied to the grid. Harmonic distortion is undesirable in most amplifiers, although there are occasions when harmonics are deliberately generated and used.

Class-A Amplifiers

An amplifier in which voltage gain is the primary consideration is called a voltage amplifier. Maximum voltage gain is obtained when the load resistance or impedance is made as high as possible in comparison with the plate resistances of the tube. In such a case, the major portion of the voltage generated will appear across the load.

Voltage amplifiers belong to a group called class-A amplifiers. A class-A amplifier is one operated so that the wave shape of the output voltage is the same as that of the signal voltage applied to the grid. If a class-A amplifier is biased so that the grid is always negative, even with the largest signal to be handled by the grid, it is called a class-A_1 amplifier. Voltage amplifiers are always class-A_1 amplifiers.

Power Amplifiers

Fig. 10 shows an elementary power-amplifier circuit. It is simply a transformer-coupled amplifier with the load connected to the secondary. Although the load is shown as a resistor, it actually would be some device, such as an antenna, that employs the power usefully. Every power tube requires a specific value of load resistance from plate to cathode, usually some thousands of ohms, for optimum operation. The resistance of the actual load is rarely the right value for matching this optimum load resistance, so the turns ratio of the transformer is chosen to reflect the power value of resistance into the primary. The transformer may be either step up or step down, depending on whether the actual load resistance is higher or lower than the tube wants.

The power amplification ratio of an amplifier is the ratio of the power output obtained from the plate circuit to the power required from the ac signal in the grid circuit. There is no power lost in the grid circuit of a class-A_1 amplifier, so such an amplifier has an infinitely large power amplification ratio. However, it is quite possible to operate a class-A amplifier in such a way that current flows in the grid circuit during at least part of the cycle. In such a case, power is used up in the grid circuit and the power amplification ratio is not infinite. A tube operated in this fashion is known as a class-A_2 amplifier. It is necessary to use a power amplifier to drive a class-A_2 amplifier, because a voltage amplifier cannot deliver power without serious distortion of the wave shape.

Another term used in connection with power amplifiers is power sensitivity. In the case of a class-A_1 amplifier, the term means the ratio of power output to the grid signal voltage. If grid current flows, the term usually means the ratio of plate power output to grid power input.

The ac power that is delivered to a load by an amplifier tube has to be paid for in power taken from the source of plate voltage and current. In fact, there is always more power going into the plate circuit of the tube than is coming out as useful output. The difference between the input and output power is used up in heating the plate of the tube, as explained previously. The ratio of useful power output to dc plate input is called the plate efficiency. The higher the plate efficiency, the greater the amount of power that can be taken from a tube having a given plate dissipation rating.

Parallel and Push-Pull Amplifiers

When it is necessary to obtain more power output than one tube is capable of giving, two or more similar tubes may be connected in parallel. In this case the similar elements in all tubes are connected together. This method is shown in Fig. 11 for a transformer-coupled amplifier. The power output is in proportion to the number of tubes used; the grid signal or exciting voltage required, however, is the same as for one tube. If the amplifier operates in such a way as to consume power in the grid circuit, the grid power required is in proportion to the number of tubes used.

An increase in power output also can be had by connecting two tubes in push-pull.

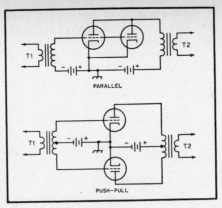

Fig. 11 — Parallel and push-pull AF amplifier circuits.

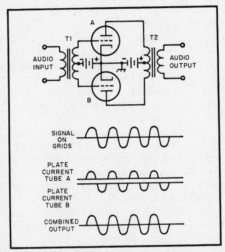

Fig. 12 — Class-B amplifier operation.

In this case the grids and plates of the two tubes are connected to opposite ends of a balanced circuit, as shown in Fig. 11. At any instant the ends of the secondary winding of the input transformer, T1, will be at opposite polarity with respect to the cathode connection, so the grid of one tube is swung positive at the same instant that the grid of the other is swung negative. Hence, in any push-pull connected amplifier the voltages and currents of one tube are out of phase with those of the other tube.

In push-pull operation the even-harmonic distortion (second, fourth, etc.) is balanced out in the plate circuit. This means that for the same power output the distortion will be less than with parallel operation.

The excitation voltage measured between the two grids must be twice that required for one tube. If the grids consume power, the driving power for the push-pull amplifier is twice that taken by either tube alone.

Cascade Amplifiers

It is readily possible to take the output of one amplifier and apply it as a signal to the grid of a second amplifier, then take the second amplifier's output and apply it to a third, and so on. Each amplifier is called a stage, and stages used successively are said to be in cascade.

Class-B Amplifiers

Fig. 12 shows two tubes connected in a push-pull circuit. If the grid bias is set at the point where the plate current is just cut off (when no signal is applied) then a signal can cause plate current to flow in either tube only when the signal voltage applied to that particular tube is positive with respect to the cathode. Since in the balanced grid circuit the signal voltages on the grids of the two tubes always have opposite polarities, plate current flows only in one tube at a time.

The graphs show the operation of such an amplifier. The plate current of tube B is drawn inverted to show that it flows in the opposite direction, through the primary of the output transformer, to the plate current of tube A. Thus each half of the output-transformer primary works alternately to induce a half cycle of voltage in the secondary. In the secondary of T2, the original waveform is restored. This type of operation is called class-B amplification.

The class-B amplifier has considerably higher plate efficiency than the class-A amplifier. Furthermore, the dc plate current of a class-B amplifier is proportional to the signal voltage on the grids, so the power input is small with small signals. The dc plate power input to a class-A amplifier is the same whether the signal is large, small or absent altogether; therefore the maximum dc plate input that can be applied to a class-A amplifier is equal to the rated plate dissipation of the tube or tubes. Two tubes in a class-B amplifier can deliver approximately 12 times as much audio power as the same two tubes in a class-A amplifier.

A class-B amplifier usually is operated in such a way as to provide the maximum possible power output. This requires large values of plate current. To obtain these currents, the signal voltage must completely overcome the grid bias during at least part of the cycle, so grid current flows and the grid circuit consumes power. While the power requirements are fairly low (as compared with the power output), the fact that the grids are positive during only part of the cycle means the load on the preceding amplifier or driver stage varies in magnitude during the cycle. The effective load resistance is high when the grids are not drawing current, and is relatively low when they do take current. This must be allowed for when designing the driver.

Certain types of tubes have been designed specifically for class-B service and can be operated without fixed or other form of grid bias (zero-bias tubes). The amplification factor is so high that the plate current is small without signal. Because there is no fixed bias, the grids start drawing current immediately whenever a signal is applied, so the grid-current flow is continuous throughout the cycle. This makes the load on the driver much more constant than is the case with tubes of lower μ biased to plate-current cutoff.

Class-B amplifiers used at radio frequencies are known as linear amplifiers, because they are adjusted to operate in such a way that the power output is proportional to the square of the RF excitation voltage. This permits amplification of a modulated RF signal without distortion. Push-pull is not required in this type of operation; a single tube can be used equally well.

Class-AB Amplifiers

Class-AB audio amplifiers are push-pull amplifiers with higher bias than would be normal for pure class-A operation, but less than the cutoff bias required for class-B. At low signal levels the tubes operate as class-A amplifiers, and the plate current is the same with or without signal. At higher signal levels, the plate current of one tube is cut off during part of the negative cycle of the signal applied to its grid, and the plate current of the other tube rises with the signal. The total plate current for the amplifier also rises above the no-signal level when a large signal is applied.

In a properly designed class-AB amplifier the distortion is as low as with a class-A stage, but the efficiency and power output are considerably higher than with pure class-A operation. A class-AB amplifier can be operated either with or without driving the grids into the positive region. A class-AB_1 amplifier is one in which the grids are never positive with respect to the cathode; therefore no driving power is required — only voltage. A class-AB_2 amplifier is one that has grid-current flow during part of the cycle if the applied signal is large; it takes a small amount of driving power. The class-AB_2 amplifier will deliver somewhat more power (using the same tubes), but the class-AB_1 amplifier avoids the problem of designing a driver that will deliver power without distortion into a load of highly variable resistance.

Conduction Angle

Inspection of Fig. 12 shows that either of the two vacuum tubes is working for only half the ac cycle and idling during the other half. It is convenient to describe the amount of time during which plate current flows in terms of electrical degrees. In Fig. 12 each tube has 180-degree excitation, a half cycle being equal to 180 degrees. The number of degrees during which plate current flows is called the conduction angle of the amplifier. From the description given above, it should be clear that a class-A amplifier has 360-degree excitation, because plate current flows during the whole cycle. In a class-AB amplifier the conduction angle is between 180 and 360 degrees (in each tube), depending on the particular operating conditions chosen. The

greater the amount of negative grid bias, the smaller the conduction angle becomes.

A conduction angle of less than 180 degrees leads to a considerable amount of distortion, because there is no way for the tube to reproduce even a half cycle of the signal on its grid. Using two tubes in push-pull, as in Fig. 12, would merely put together two distorted half cycles. A conduction angle of less than 180 degrees therefore cannot be used if distortionless output is wanted.

Class-C Amplifiers

In power amplifiers operating at radio frequencies, distortion of the RF waveform is relatively unimportant. Tuned circuits are usually used to filter out the RF harmonics resulting from distortion.

A radio-frequency power amplifier therefore can be used with a conduction angle of less than 180 degrees. This is called class-C operation. The advantage is that the plate efficiency is increased because the loss in the plate is proportional, among other things, to the amount of time during which the plate current flows. This time is reduced by decreasing the operating angle.

Depending on the type of tube, the optimum load resistance for a class-C amplifier ranges from about 1.5 kΩ to 5 kΩ. Tuned-circuit arrangements are normally used to transform the resistance of the actual load to the value required by the tube. The grid is driven well into the positive region, so grid current flows and power is consumed in the grid circuit. The smaller the conduction angle, the greater the driving voltage and the larger the grid driving power required to develop full output in the load resistance. The best compromise between driving power, plate efficiency and power output usually results when the minimum plate voltage is just equal to the peak positive grid voltage. These conditions must exist at the peak of the driving cycle, when the plate current reaches its highest value. Under such conditions the conduction angle is usually between 120 and 150 degrees, and the plate efficiency lies in the range of 60 to 80 percent. While higher plate efficiencies are possible, attaining them requires excessive driving power and grid bias, together with higher plate voltage than is normal for the particular tube type.

Feedback

It is possible to take a part of the amplified energy in the plate circuit of an amplifier and insert it into the grid circuit. This is called feedback.

If the voltage that is inserted in the grid circuit is 180 degrees out of phase with the signal voltage acting on the grid, the feedback is called negative or degenerative. On the other hand, if the voltage is fed back in phase with the grid signal, the feedback is called positive or regenerative.

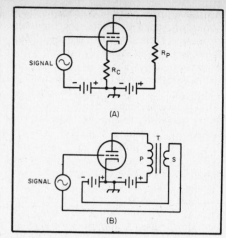

Fig. 13 — Simple circuits for producing feedback.

Negative Feedback

With negative feedback the voltage that is fed back opposes the signal voltage. This decreases the amplitude of the voltage acting between the grid and cathode and thus has the effect of reducing the voltage amplification. That is, a larger exciting voltage is required for obtaining the same output voltage from the plate circuit.

The greater the amount of negative feedback (when properly applied), the more independent the amplification becomes of tube characteristics and circuit conditions. This tends to make the frequency-response characteristic of the amplifier flat — that is, the amplification tends to be the same at all frequencies within the range for which the amplifier is designed. Also, any distortion generated in the plate circuit of the tube tends to "buck itself out." Amplifiers with negative feedback therefore have lower harmonic distortion. These advantages are worthwhile if the amplifier otherwise has enough voltage gain for its intended use.

In the circuit shown at A in Fig. 13 resistor R_c is in series with the regular plate resistor, R_p and this is a part of the load for the tube. Therefore, part of the output voltage will appear across R_c. R_c also is connected in series with the grid circuit, however, so the output voltage that appears across R_c is in series with the signal voltage. The output voltage across R_c opposes the signal voltage, so the actual ac voltage between the grid and cathode is equal to the difference between the two voltages.

The circuit shown at B in Fig. 13 can be used to give either negative or positive feedback. The secondary of a transformer is connected back into the grid circuit to insert a desired amount of feedback voltage. Reversing the terminals of either transformer winding (but not both simultaneously) will reverse the phase.

Positive Feedback

Positive feedback increases the amplification because the feedback voltage adds to the original signal voltage, and the resulting larger voltage on the grid causes a larger output voltage. The amplification tends to be greatest at one frequency (which depends on the particular circuit arrangement) and harmonic distortion is increased. If enough energy is fed back, a self-sustaining oscillation will be set up — one in which energy at essentially one frequency is generated by the tube itself. In such cases all the signal voltage on the grid can be supplied from the plate circuit. No external signal is needed because any small irregularity in the plate current (and there are always some irregularities) will be amplified and thus give the oscillation an opportunity to build up. Positive feedback finds a major application in oscillators, and in addition is used for selective amplification at both audio and radio frequencies, the feedback being kept below the value that causes self-oscillation.

Interelectrode Capacitances

Each pair of elements in a tube forms a small capacitor. There are three such capacitances in a triode — that between the grid and cathode, that between the grid and plate, and that between the plate and cathode. The capacitances are very small, only a few picofarads at most, but they frequently have a pronounced effect on the operation of an amplifier circuit.

Input Capacitance

It was explained previously that the ac grid voltage and ac plate voltage of an amplifier having a resistive load are 180 degrees out of phase, using the cathode of the tube as a reference point. However, these two voltages are in phase going around the circuit from plate to grid as shown in Fig. 14. This means that their sum is acting between the grid and plate — that is, across the grid-plate capacitance of the tube.

As a result, a capacitive current flows around the circuit, its amplitude being directly proportional to the sum of the ac grid and plate voltages and to the grid-plate capacitance. The source of the grid signal must furnish this amount of current, in addition to the capacitive current that flows in the grid-cathode capacitance. Hence the signal source sees an effective capacitance that is larger than the grid-cathode capacitance. This is known as the Miller effect.

The greater the voltage amplification, the greater the effective input capacitance. The input capacitance of a resistance-coupled amplifier is given by the formula

$$C_{input} = C_{gk} + C_{gp} (A + 1)$$

where

C_{gk} is the grid-to-cathode capacitance
C_{gp} is the grid-to-plate capacitance
A is the voltage amplification.

The input capacitance may be as much as several hundred picofarads when the

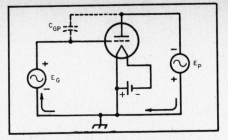

Fig. 14 — The ac voltage appearing between the grid and plate of the amplifier is the sum of the signal voltage and the output voltage, as shown by this simplified circuit. Instantaneous polarities are indicated.

voltage amplification is large, even though the interelectrode capacitances are quite small.

Output Capacitance

The principal component of the output capacitance of an amplifier is the actual plate-to-cathode capacitance of the tube. The output capacitance usually need not be considered in audio amplifiers, but becomes of importance at radio frequencies.

Tube Capacitance at RF

At radio frequencies the reactances of even very small interelectrode capacitances drop to very low values. A resistance-coupled tube amplifier gives very little amplification at RF, for example, because the reactances of the interelectrode capacitors are so low that they practically short circuit the input and output circuits, and thus the tube is unable to amplify. This is overcome at radio frequencies by using tuned circuits for the grid and plate, making the tube capacitances part of the tuning capacitances. In this way the circuits can have the high resistive impedances necessary for satisfactory amplification.

The grid-plate capacitance is important at radio frequencies because its reactance, relatively low at RF, offers a path over which energy can be fed back from the plate to the grid. In practically every case the feedback is in the right phase and of sufficient amplitude to cause self-oscillation, so the circuit becomes useless as an amplifier.

Special neutralizing circuits can be used to prevent feedback but they are, in general, not too satisfactory when used in radio receivers. They are used in transmitters, however.

Tetrodes

The grid-plate capacitance can be greatly reduced by inserting a second grid between the control grid and the plate, as indicated in Fig. 15. The second grid, called the screen grid, acts as an electrostatic shield to reduce capacitive coupling between the control grid and plate. It is made in the form of a grid or coarse screen so electrons can pass through it. A tube having a

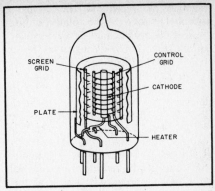

Fig. 15 — Representative arrangement of elements in a screen-grid tetrode, with part of plate and screen cut away. This shows single-ended construction with a button base, typical of miniature receiving tubes. To reduce capacitance between control grid and plate, the leads from these elements are brought out at opposite sides. Actual tubes probably would have additional shielding between these leads.

cathode, control grid, screen grid and plate (four elements) is called a tetrode.

Because of the shielding action of the screen grid, the positively charged plate cannot easily attract electrons from the cathode as it does in a triode. In order to enhance electron flow to the plate, it is necessary to apply a positive voltage to the screen (with respect to the cathode). The screen then attracts electrons much as does the plate in a triode tube. In traveling toward the screen the electrons acquire such velocity that most of them shoot between the screen wires and then are attracted to the plate. A small proportion does strike the screen, however, with the result that some current also flows in the screen-grid circuit.

To be a good shield, the screen grid must be connected to the cathode through a circuit that has low impedance at the frequency being amplified. A bypass capacitor from screen grid to cathode, having a reactance of not more than a few hundred ohms, is generally used.

Pentodes

When an electron traveling at appreciable velocity through a tube strikes the plate, it dislodges other electrons, which splash from the plate into the interelement space. This is called secondary emission. In a triode the negative grid repels the secondary electrons back into the plate and they cause no disturbance. In the screen-grid tube, however, the positively charged screen attracts the secondary electrons, causing a reverse current to flow between screen and plate.

To overcome the effects of secondary emission, a third grid, called the suppressor grid, may be inserted between the screen and plate. This grid, negatively changed with respect to the screen and plate, acts as a shield between the screen grid and plate so the secondary electrons cannot be attracted by the screen grid. They are hence attracted back to the plate without appreciably obstructing the regular plate-current flow. A five-element tube of this type is called a pentode.

Although the screen grid in either the tetrode or pentode greatly reduces the influence of the plate voltage on plate-current flow, the control grid still can control the plate current in essentially the same way that it does in a triode. Consequently, the grid-plate transconductance (or mutual conductance) of a tetrode or pentode will be of the same order of value as in a triode of corresponding structure. On the other hand, since a change in plate voltage has very little effect on the plate-current flow, both the amplification factor and plate resistance of a pentode or tetrode are very high. Because of the high plate resistance, the actual voltage amplification possible with a pentode is very much less than the large amplification factor might indicate. A voltage gain in the vicinity of 50 to 200 is typical of a pentode stage.

In practical screen-grid tubes, the grid-plate capacitance is only a small fraction of a picofarad. This capacitance is too small to cause an appreciable increase in input capacitance by means of the Miller effect, so the input capacitance of a screen-grid tube is equal to the capacitance between the plate and screen.

The chief function of the screen grid is to serve as an accelerator of electrons, so that large values of plate current can be drawn at relatively low plate voltages. Tetrodes and pentrodes have high power sensitivity compared with triodes of the same power output, although harmonic distortion is somewhat greater.

Beam Tubes

A beam tetrode is a four-element screen-grid tube constructed in such a way that the electrons are formed into concentrated beams on their way to the plate. Additional design features overcome the effects of secondary emission so that a suppressor grid is not needed. The beam construction makes it possible to draw large plate currents at relatively low plate voltages, and increases the power sensitivity of the tube.

For power amplification at radio frequencies, beam tetrodes have largely supplanted the nonbeam types. This is because large power outputs can be secured with very small amounts of grid driving power.

Input and Output Impedances

The input impedance of a vacuum-tube amplifier is the impedance seen by the signal source when connected to the input terminals of the amplifier. In the types of amplifiers previously discussed, the input impedance is the impedance measured between the grid and cathode of the tube with operating voltages applied. At audio frequencies the input impedance of a class-A1 amplifier is for all practical purposes the input impedance of the stage. If the tube

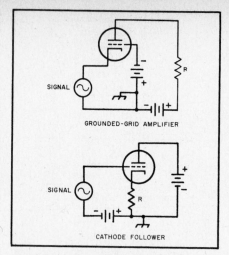

Fig. 16 — In each circuit shown, R represents the load. In the upper circuit, the grid is the junction point between the input and output circuits. In the lower drawing, the plate is the junction. In either case the output is developed in the load resistor, and may be coupled to a following amplifier by the usual methods.

is driven into the grid-current region, there is in addition a resistance component in the input impedance. The resistance has an average value equal to E^2/P, where E is the RMS driving voltage and P is the power in watts consumed in the grid. The resistance usually will vary during the ac cycle, because grid current may flow only during part of the cycle. Also, the grid-voltage/grid-current characteristic is seldom linear.

The output impedance of amplifiers of this type consists of the plate resistance of the tube shunted by the output capacitance.

At radio frequencies, when tuned circuits are employed, the input and output impedances are usually pure resistances; any reactive components are tuned out in the process of adjusting the circuits to resonance at the operating frequency.

Other Types of Amplifiers

In the amplifier circuits discussed so far, the signal has been applied between the grid and cathode, and the amplified output has been taken from the plate-to-cathode circuit. That is, the cathode has been the meeting point for the input and output circuits. However, it is possible to use any one of the three principal elements as the common point. This leads to two additional kinds of amplifiers, commonly called the grounded-grid amplifier and the cathode follower.

These two circuits are shown in simplified form in Fig. 16. In both circuits, R represents the load into which the amplifier works. The actual load may be resistance-capacitance coupled, transformer coupled, or may be a tuned circuit if the amplifier operates at radio frequencies. Also, in both circuits the batteries that supply grid bias and plate power are assumed to have such

negligible impedance that they do not enter into the operation of the circuits.

Grounded-Grid Amplifier

In the grounded-grid amplifier the input signal is applied between the cathode and grid, and the output is taken between the plate and grid. The grid is thus the common element. The ac component of the plate current has to flow through the signal source to reach the cathode. The source of signal is in series with the load through the plate-to-cathode resistance of the tube, so some of the power in the load is supplied by the signal source. In transmitting applications this fed-through power is of the order of 10 percent of the total power output, using tubes suitable for grounded-grid service.

The input impedance of the grounded-grid amplifier consists of a capacitance in parallel with an equivalent resistance that would absorb the power furnished by the driving source to the grid and to the load. This resistance is of the order of a few hundred ohms. The output impedance, neglecting the interelectrode capacitances, is equal to the plate resistance of the tube. This is the same as in the case of the grounded-cathode amplifier.

The grounded-grid amplifier is widely used at radio frequencies. With a triode tube designed for this type of operation, an RF amplifier can be built that is free from the type of feedback that causes oscillation. This requires that the grid act as a shield between the cathode and plate, reducing the plate-cathode capacitance to a very low value.

Cathode Follower

The cathode follower uses the plate of the tube as the common element. The input signal is applied between the grid and plate (assuming negligible impedance in the batteries) and the output is taken between cathode and plate. This circuit is degenerative; in fact, all of the output voltage is fed back into the input circuit out of phase with the grid signal. The input signal therefore has to be larger than the output voltage; that is, the cathode follower gives a loss in voltage, although it gives the same power gain as other circuits under equivalent operating conditions.

An important feature of the cathode follower is its low output impedance, which is given by the formula (neglecting interelectrode capacitances)

$$Z_{out} = \frac{r_p}{1 + \mu}$$

where

r_p is the tube plate resistance
μ is the amplification factor.

Low output impedance is a valuable characteristic in an amplifier designed to cover a wide band of frequencies. In addition, the input capacitance is only a fraction of the grid-to-cathode capacitance of the tube, a feature of further benefit in a wide-band amplifier. The cathode follower is useful as a step-down impedance transformer, since the input impedance is high and the output impedance is low.

Cathode Circuits and Grid Bias

Most of the equipment used by amateurs is powered by the ac line. This includes the filaments or heaters of vacuum tubes. Supplies for the plate (and sometimes the grid) are usually ac that is rectified and filtered to give pure dc — that is, direct current that is constant and without a superimposed ac component. The relatively large currents required by filaments and heaters however, usually make a rectifier-type dc supply impractical. Therefore, alternating current is almost always used to heat the tube cathode.

Filament Hum

Alternating current is just as good as direct current from the heating standpoint, but some of the ac voltage is likely to be coupled to the grid and cause ac hum to be superimposed on the output. Hum troubles are worst with directly heated cathodes (filaments), because with such

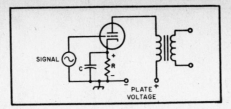

Fig. 18 — Cathode biasing. R is the cathode resistor and C is the cathode bypass capacitor.

cathodes there has to be a direct connection between the source of heating power and the rest of the circuit.

Hum can be minimized by either of the connections shown in Fig. 17. In both cases the grid- and plate-return circuits are connected to the electrical midpoint (center tap) of the filament supply. Thus, as far as the grid and plate are concerned, the voltage and current on one side of the filament are balanced by an equal and opposite voltage and current on the other side. The balance is never quite perfect, however, so filament-type tubes are never completely hum free. For this reason directly heated filaments are employed for the most part in power tubes, where the hum introduced is extremely small in comparison with the power-output level.

With indirectly heated cathodes the chief problem is the magnetic field set up by the heater. Occasionally, also, there is leakage between the heater and cathode, allowing a small ac voltage to reach to the grid. If hum appears, grounding one side of the heater supply usually will help to reduce it, although sometimes better results are obtained if the heater supply is center-tapped and the center-tap grounded, as shown in Fig. 17.

Cathode Bias

In the simplified amplifier circuits discussed in this chapter, grid bias has been supplied by a battery. In equipment that operates from the power line, however, cathode bias is almost universally used for tubes that are operated in class-A (constant dc input).

The cathode-bias method uses a resistor (cathode resistor) connected in series with the cathode, R in Fig. 18. The direction of plate-current flow is such that the end of the resistor nearest the cathode is positive. The voltage drop across R therefore places a negative voltage on the grid with respect to the cathode. This negative bias is obtained from the steady dc plate current.

If the alternating component of plate current flows through R when the tube is

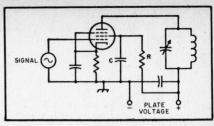

Fig. 19 — Screen-voltage supply for a pentode tube through a dropping resistor, R. The screen bypass capacitor, C, must have low enough reactance to bring the screen to ground potential for the frequency or frequencies being amplified.

amplifying, the voltage drop caused by the ac will be degenerative (note the similarity between this circuit and that of Fig. 13A). To prevent this, the resistor is bypassed by the capacitor, C, that has very low reactance compared with the resistance of R. Depending on the type of tube and the particular kind of operation, R may be between about 100 and 3000 ohms. For good bypassing at radio frequencies, capacitances of about 100 pF to 0.1 μF are used; the small values are sufficient at very high frequencies and the largest at low and medium frequencies. In the range 3 to 30 megahertz a capacitance of 0.01 μF is satisfactory.

The value of cathode resistor for an amplifier having negligible dc resistance in its plate circuit (transformer or impedance coupled) can be calculated from the known operating conditions of the tube.

Screen Supply

In practical circuits using tetrodes and pentodes, the voltage for the screen frequently is taken from the plate supply through a resistor. A typical circuit for an RF amplifier is shown in Fig. 19. R is the screen dropping resistor, and C is the screen bypass capacitor. In flowing through R, the screen current causes a voltage drop in R that reduces the plate-supply voltage to the proper value for the screen. When the plate-supply voltage and the screen current are known, the value of R can be calculated from Ohm's Law.

The reactance of the screen bypass capacitor, C, should be low compared with the screen-to-cathode impedance. For radio-frequency applications between 3 and 30 MHz, a capacitance in the vicinity of 0.01 μF is amply large.

In some vacuum-tube circuits the screen voltage is obtained from a voltage divider connected across the plate supply. The design of voltage dividers is discussed in Chapter 6.

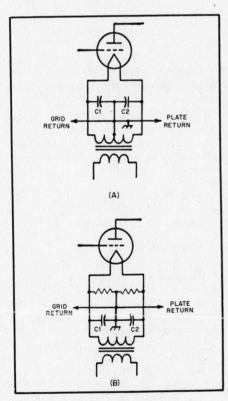

Fig. 17 — Filament center-tapping methods for use with directly heated tubes.

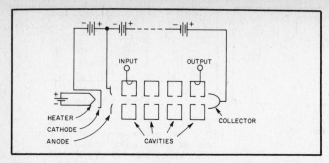

Fig. 20 — Simplified diagram of a klystron amplifier. Not shown is the focusing magnet used to keep the electron beam from striking the cavities or body of the tube.

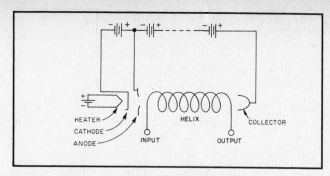

Fig. 21 — Simplified diagram of a traveling-wave-tube (TWT) amplifier. Like the klystron, this tube requires a focusing magnet.

Tubes for UHF and Microwaves

Earlier discussions in this chapter have assumed that the time it takes for the electrons to travel from the cathode to the plate does not affect the performance of vacuum-tube operation. As the frequency of operation is raised, this time becomes increasingly important. Called the transit time, this time depends on the voltage from the cathode to the plate and the spacing between them. The higher the voltage and the smaller the spacing, the shorter the transit time. This is why tubes designed for VHF and UHF work have very small interelectrode spacings. The power handling capabilities also get smaller as the spacing decreases, however, so there is a frequency limit above which ordinary triode and pentode tubes cannot be operated efficiently.

Many different tubes have been developed which actually use transit-time effects to an advantage. Velocity modulation of the electron stream in a klystron is one example. A small voltage applied across the gap in a reentrant cavity resonator either retards or accelerates an electron stream by means of the resultant electric field. Initially, all the electrons are traveling at the same velocity, and the current in the beam is uniform. After the velocity fluctuations are impressed on the beam, the current is still uniform for a while, but then the electrons that were accelerated begin to catch up with the slower ones that passed through when the field was zero. These faster electrons are also catching up with ones that passed through the gap earlier but were retarded. The result is that the current in the beam is no longer uniform, but consists of a series of pulses.

If the beam now passes through another cavity gap, a current will be induced in the cavity walls and an electric field also will be set up across the gap. If the phase of the electric field is right, the electron pulses or bunches pass through the gap and are retarded, thus giving up energy to the electric field. When the electric field reverses, it would normally accelerate the same number of electrons and give back the

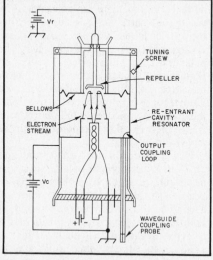

Fig. 22 — Cross-sectional view of a typical reflex klystron. The frequency of the cavity resonator is changed by varying the spacing between the grids using a tuning mechanism and a flexible bellows.

energy, but fewer electrons now pass through the gap and the energy is from the beam to the cavity. If the voltage produced across the output cavity is greater than that across the input cavity, amplification results (assuming the two cavity impedances are the same).

A four-cavity klystron is shown in Fig. 20. In this example, a pair of intermediate cavities has been added. These additional cavities reinforce the bunching of electrons within the electron beam, which in turn increases the gain.

Not shown in Fig. 20 is the magnet that must be used to keep the electron beam focused down the center of the tube. Beam electrons must be returned through the collector and not by way of the cavities or body of the tube.

Traveling Wave Tube

As you can see in Fig. 21, a helix traveling wave tube (TWT) looks much like a klystron. The major difference is that the klystron cavities have been replaced with a helix.

A signal traveling along the helix moves more slowly than the electrons in the beam. The signal on the helix tends to retard beam electrons. As beam electrons slow down they give up energy to the signal on the helix.

Another type of TWT uses a series of coupled cavities instead of a helix. Table 1 compares TWTs and klystrons.

Microwave Oscillator Tubes

Another type of klystron is the reflex klystron oscillator, shown in Fig. 22. Here the input and output cavity are the same. The electron stream makes one pass through, becomes velocity modulated, and is turned around by the negative charge on an element called the repeller. During the second pass through, the stream is now bunched and delivers some of its energy to the cavity. The dissipated beam is then picked up by the cavity walls and the circuit is completed.

Other microwave oscillator wave tubes are the back wave oscillator (BWO) and the magnetron. The BWO works on the same principle as the TWT. The magnetron uses a magnetic field in conjunction with an electric field in its operation. Home microwave ovens use magnetrons to produce RF energy for cooking.

Table 1
Comparison of Klystrons and TWTs

Type	Gain (dB)	Efficiency (%)	Bandwidth (%)
Klystron	40-60	30-70	1-5
Helix TWT	30-50	20-40	30-120
Cavity TWT	30-50	20-40	5-40

Chapter 6

Power Supplies

In most residential systems, three wires are brought in from the outside to the house distribution panel, although there are only two wires in older systems. In the three-wire system, the third wire is neutral and is grounded. The voltage between the two wires normally is 234. Half of this voltage appears between each of these wires and neutral, as indicated in Fig. 1A.

In systems of this type, the 117-volt household load is divided as evenly as possible between the two sides of the circuit. Half of the load is connected between one wire and neutral, and the other half of the load is connected between the other wire and neutral. Heavy appliances, such as electric stoves and heaters, are designed for 234-volt operation and are connected across the two ungrounded wires.

While both ungrounded wires should be fused, a fuse or switch should never be used in the neutral wire. The reason for this is that opening the neutral wire does not disconnect the equipment. It simply leaves the equipment on one side of the 234-volt circuit in series with whatever load may be across the other side of the circuit. Furthermore, with the neutral open, the voltage will then be divided between the two sides in inverse proportion to the load resistance. The voltage will drop below normal on one side and soar on the other side, unless the loads happen to be equal. More information on 117- and 234-V ac wiring for amateur stations may be found in Chapter 37.

The electrical power required to operate Amateur Radio equipment is usually taken from the alternating current (ac) lines when the equipment is operated where power is available. For mobile operation, the source of power is almost always the car storage battery.

Direct current (dc) voltages used in transmitters, receivers and other related equipment are derived from the commercial ac lines by using a transformer-rectifier-filter system. The transformer changes the ac voltage to a suitable value, and the rectifier converts the ac to pulsating dc. A filter is used to smooth out these pulsations to an acceptably low level. Essentially pure dc is required to prevent 60- or 120-Hz hum in most amateur equipment. If a constant voltage is required under conditions of changing load or ac

line voltage, a regulator is used following the filter.

Fusing

All transformer primary circuits should be fused properly. To determine the approximate current rating of the fuse or circuit breaker to be used, multiply each current being drawn from the supply in amperes by the voltage at which the current is being drawn. Include the current taken by bleeder resistors and voltage dividers. In the case of series resistors, use the source voltage, not the voltage at the equipment end of the resistor. Include filament power if the transformer is supplying filaments.

After multiplying the various voltages and currents, add the individual products. Then divide by the line voltage and add 10 or 20 percent. Use a fuse or circuit breaker with the nearest larger current rating.

When selecting fuses and circuit breakers for use in motor circuits (for example, blower lines) a slow-blow type should be used. When the motor is started, it can draw up to 10 times its normal operating current until the shaft reaches its normal operating speed. This can take up to 30 seconds on a large blower. Such an overload will usually open a fast- or normal-blow fuse.

For low-power semiconductor circuits, use fast-blow fuses. As the name implies, such fuses open very quickly once the cur-

rent rating is exceeded by more than 10%.

TRANSFORMERS

Volt-Ampere Rating

The number of volt-amperes (VA) delivered by a transformer depends on the type of filter used (capacitor or choke input), and on the type of rectifier used (full-wave center tap or full-wave bridge). With a capacitive-input filter, the heating effect in the secondary is higher because of the high ratio of peak-to-average current. The volt-amperes handled by the transformer may be several times the watts delivered to the load. The primary volt-amperes will be somewhat higher because of transformer losses.

Broadcast and Television Replacement Transformers

Small power transformers sold for replacement in broadcast and television receivers are usually designed for service in terms of use for several hours continuously with capacitor-input filters. In the usual type of amateur transmitter service, where most of the power is drawn intermittently for periods of several minutes with equivalent intervals in between, the published ratings can be exceeded without excessive transformer heating.

With a capacitor-input filter, it should be safe to draw 20 to 30 percent more current than the rated value. With a choke-input filter, an increase in current of about

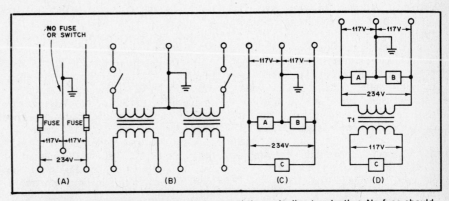

Fig. 1 — Three-wire power-line circuits. At A, normal three-wire-line termination. No fuse should be used in the grounded (neutral) line. At B, a switch in the neutral line would not remove voltage from either side of the line; C, connections for both 117- and 234-volt transformers; D, operating a 117-volt plate transformer from the 234-volt line to avoid light blinking. T1 is a 2:1 step-down transformer.

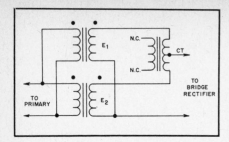

Fig. 2 — One method of equalizing two transformers connected in parallel.

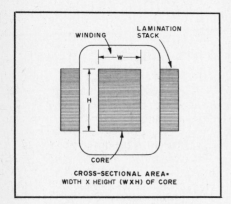

Fig. 3 — Cross-sectional drawings of a typical power transformer. Multiplying the height (or thickness of the laminations) by the width of the central core area in inches gives the value to be applied to Fig. 4.

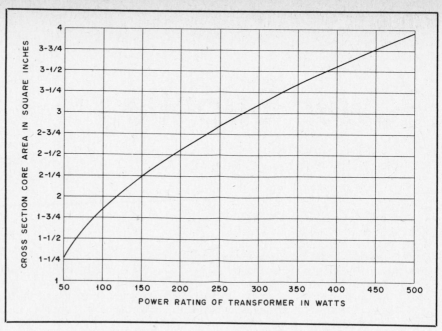

Fig. 4 — Power-handling capability of a transformer versus cross-sectional area of core.

50 percent is permissible. If a bridge rectifier is used, the output voltage will be approximately doubled. In this case, it should be possible in amateur transmitter service to draw the rated current, thus obtaining about twice the rated output power from the transformer. This does not apply, of course, to amateur transmitter plate transformers, which usually are rated for intermittent service.

Transformers in Parallel

The secondary windings of power transformers can be connected in parallel to obtain a greater current rating than can be had with a single transformer. A common application might be a high-current, 12-V supply for a transceiver. One potential problem is equalizing the current between the two transformers.

A timeless scheme for paralleling nominally-equal transformers is shown in Fig. 2. An inductor, in this case the center-tapped secondary winding of a low-voltage filament transformer, is used to equalize the current. This winding will only see the difference between voltages E1 and E2.

In the ideal case (E1 = E2), the currents are equal but opposite in the center-tapped winding. There is no net flux, and therefore no inductive voltage drop. Ideally, there is no loss.

When E1 is not equal to E2, the open-circuit voltage at the center tap is E2 + 1/2(E1 − E2). If the difference between E1 and E2 is very small, the currents will be approximately equal, and there will be essentially only the normal transformer losses.

Rewinding Power Transformers

Although the home winding of power transformers is a task that few amateurs undertake, the rewinding of a transformer secondary to give some desired voltage is not difficult. It involves a matter of only a small number of turns, and the wire is large enough to be handled easily. Often a receiver power transformer with a burned-out high-voltage winding or the power transformer from a discarded TV set can be converted into an entirely satisfactory transformer without great effort and with little expense.

The average TV power transformer for a 17-inch or larger vacuum-tube set is capable of delivering from 350 to 450 watts, continuous duty. If an amateur transmitter is being powered, the service is not continuous, so the ratings can be increased by 40 or 50 percent without danger of overloading the transformer.

The primary volt-ampere rating of the transformer to be rewound, if known, can be used to determine its power-handling capability. The secondary volt-ampere rating will be 10 to 20 percent less than the primary rating. The power rating may also be determined approximately from the cross-sectional area of the core that is inside the windings. Fig. 3 shows the method of determining the area, and Fig. 4 may be used to convert this information into a power rating.

Before disconnecting the winding leads from their terminals, each should be marked for identification. In removing the core laminations, care should be taken to note how the core is assembled, so the reassembling will be done in the same manner. Most transformers have secondaries wound over the primary, while in some the order is reversed. In case the secondaries are on the inside, the turns can be pulled out from the center after slitting and removing the fiber core.

The turns removed from one of the original filament windings of known voltage should be carefully counted as the winding is removed. This will give the number of turns per volt; the same figure should be used in determining the number of turns for the new secondary. For instance, if the old filament winding was rated at 5 volts and had 15 turns, this is 15/5 = 3 turns per volt. If the new secondary is to deliver 18 volts, the required number of turns on the new winding will be 18 × 3 = 54 turns.

In winding a transformer, the size of wire is an important factor in the heat developed in operation. A cross-sectional area of 1000 circular mils per ampere (cmil/A) is conservative. A value commonly used in amateur-service transformers is 700 cmil/A. The larger the cmil/A figure, the cooler the transformer will run. The current rating in amperes of various wire sizes is shown in the copper-wire table in Chapter 35. If the transformer being rewound is a filament transformer, it may be necessary to choose the wire size carefully to fit the small available space. On the other hand, if the transformer is a power unit with the high-voltage winding removed, there should be plenty of room for a size of wire that will conservatively handle the required current.

After the first layer of turns is put on during rewinding, secure the ends with cellulose tape. Each layer should be insulated from the next. Ordinary household waxed paper can be used; a single layer is adequate. Sheets cut to size beforehand may be secured over each layer with tape. Be sure to bring all leads out the same side of the core so the covers will go in place when the unit is completed. When the last layer of the winding is put on, use two sheets of waxed paper, and then cover those with vinyl electrical tape, keeping the tape as taut as possible. This will add mechanical strength to the assembly.

The laminations and housing are assembled in just the opposite sequence to that followed in disassembly. Use a light coating of shellac between each lamination. During reassembly, the lamination stack may be compressed by clamping in a vise. If the last few lamination strips cannot be replaced, it is better to omit them than to force the unit together.

RECTIFIERS

Half-Wave Rectifier

Fig. 5 shows a simple half-wave rectifier circuit. As explained in Chapter 4, a rectifier (in this case a semiconductor diode) will conduct current in one direction but not the other. During one half of the ac cycle, the rectifier will conduct and current will flow through the rectifier to the load (indicated by the solid line in Fig. 5B). During the other half cycle, the rectifier is reverse biased and no current will flow (indicated by the broken line in Fig. 5B) to the load. As shown, the output is in the form of pulsed dc, and current always flows in the same direction. A filter can be used to smooth out these variations and provide a higher average dc voltage from the circuit. This idea will be covered in the section on filters.

The average output voltage — the voltage read by a dc voltmeter — with this circuit (no filter connected) is 0.45 times the RMS value of the ac voltage delivered by the transformer secondary. Because the frequency of the pulses is low (one pulsation per cycle), considerable filtering is required to provide adequately smooth dc output. For this reason the circuit is usually limited to applications where the current required is small, as in a transmitter bias supply.

The peak inverse voltage (PIV), the voltage that the rectifier must withstand when it isn't conducting, varies with the load. With a resistive load, it is the peak ac voltage ($1.4 E_{RMS}$). With a capacitor filter and a load drawing little or no current, it can rise to $2.8 E_{RMS}$. The reason for this is shown in Figs. 5C and 5D. With a resistive load as shown at C, the amount of voltage applied to the diode is that voltage on the lower side of the zero-axis line, or $1.4 E_{RMS}$. A capacitor connected to the circuit (shown at D) will store the peak positive voltage when the diode conducts on the positive pulse. If the circuit is not supplying any current, the voltage across the capacitor will remain at that

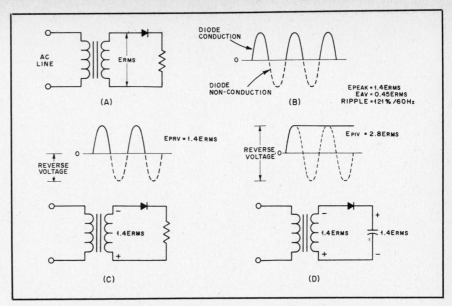

Fig. 5 — Half-wave rectifier circuit. A illustrates the basic circuit, and B displays the diode conduction and nonconduction periods. The peak-inverse voltage impressed across the diode is shown at C and D with a simple resistor load at C and a capacitor load at D. E_{PIV} is 1.4 E_{RMS} for the resistor load and 2.8 E_{RMS} for the capacitor load.

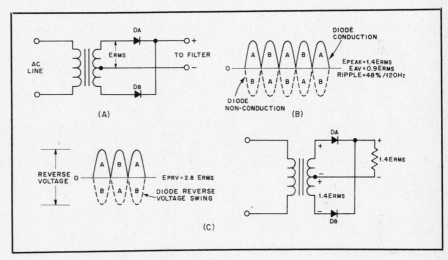

Fig. 6 — Full-wave center-tap rectifier circuit. A illustrates the basic circuit. Diode conduction is shown at B with diodes A and B alternately conducting. The peak-inverse voltage for each diode is 2.8 E_{RMS} as depicted at C.

same level. The peak inverse voltage impressed across the diode is now the sum of the voltage stored in the capacitor plus the peak negative swing of voltage from the transformer secondary. In this case the PIV is 2.8 E_{RMS}.

Full-Wave Center-Tap Rectifier

A commonly-used rectifier circuit is shown in Fig. 6. Essentially an arrangement in which the outputs of two half-wave rectifiers are combined, it makes use of both halves of the ac cycle. A transformer with a center-tapped secondary is required with the circuit.

The average output voltage is 0.9 times the RMS voltage of half the transformer secondary; this is the maximum that can be obtained with a suitable choke-input filter. The peak output voltage is 1.4 times

the RMS voltage of half the transformer secondary; this the maximum voltage that can be obtained from a capacitor-input filter.

As can be seen in Fig. 6C, the PIV impressed on each diode is independent of the type of load at the output. This is because the peak inverse voltage condition occurs when diode A conducts and diode B does not conduct. The positive and negative voltage peaks occur at precisely the same time, a condition different from that in the half-wave circuit. As the cathodes of diodes A and B reach a positive peak (1.4 E_{RMS}), the anode of diode B is at a negative peak, also 1.4 E_{RMS}, but in the opposite direction. The total peak inverse voltage is therefore 2.8 E_{RMS}.

Fig. 6B shows that the frequency of the output pulses is twice that of the half-wave

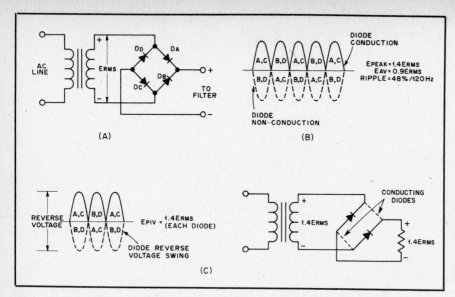

Fig. 7 — Full-wave bridge rectifier circuit. The basic circuit is illustrated at A. Diode conduction and nonconduction times are shown at B. Diodes A and C conduct on one half of the input cycle while diodes B and D conduct on the other. C displays the peak inverse voltage for one-half cycle. Since this circuit reverse-biases two diodes essentially in parallel, 1.4 E_{RMS} is applied across each diode.

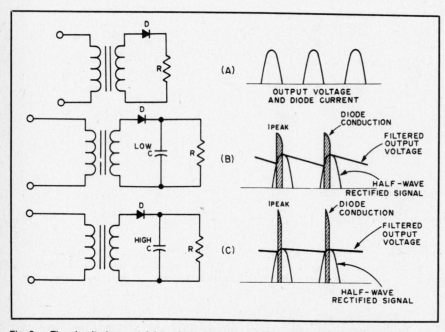

Fig. 8 — The circuit shown at A is a simple half-wave rectifier with a resistive load. The waveform shown to the right is that of output voltage and diode current. B illustrates how the diode current is modified by the addition of a capacitor filter. The diode conducts only when the rectified voltage is greater than stored capacitor voltage. Since this time period is usually only a short portion of a cycle, the peak current will be quite high. C shows an even higher peak current. This is caused by the larger capacitor, which effectively shortens the conduction period of the diode.

rectifier. Comparatively less filtering is required. Since the rectifiers work alternately, each handles half of the load current. The current rating of each rectifier need be only half the total current drawn from the supply.

Full-Wave Bridge Rectifier

Another commonly used rectifier circuit is illustrated in Fig. 7. In this arrangement, two rectifiers operate in series on each half of the cycle, one rectifier being in the lead to the load, the other being the return lead. As shown in Fig. 7A and B, when the top lead of the transformer secondary is positive with respect to the bottom lead, diodes A and C will conduct while diodes B and D are reverse biased. On the next half cycle, when the top lead of the transformer is negative with respect to the bottom, diodes B and D will conduct while diodes A and C are reverse biased.

The output wave shape is the same as that from the simple full-wave center-tap rectifier circuit. The maximum output voltage into a resistive load or choke-input filter is 0.9 times the RMS voltage delivered by the transformer secondary; with a capacitor filter and a light load, the output voltage is 1.4 times the secondary RMS voltage.

Fig. 7C shows the inverse voltage to be 1.4 E_{RMS} for each diode. When an alternate pair of diodes (such as D_A and D_C) is conducting, the other diodes are essentially connected in parallel in a reverse-biased direction. The reverse stress is then 1.4 E_{RMS}. Each pair of diodes conducts together on alternate half cycles with the full load current flowing through each diode during its conducting half cycle. Since each diode is not conducting during the other half cycle the average current is one half the total load current drawn from the supply.

Comparing the full-wave center-tap rectifier circuit and the full-wave bridge-rectifier circuit, it can be seen that both circuits have almost the same rectifier requirement since the center tap has half the number of rectifiers as the bridge, but these rectifiers have twice the inverse voltage rating requirement of the bridge diodes. The diode current ratings are identical for the two circuits. The bridge makes better use of the transformer's secondary than the center tap, since the transformer's full winding supplies power during both half cycles, while each half of the center tap's secondary provides power only during its positive half-cycle. The bridge rectifier often takes second place to the full-wave center tap rectifiers in high-current low-voltage applications, however, since the two forward-conducting series-diode drops in the bridge introduce an additional volt or more of voltage loss, and thus more heat to be dissipated, than does the single diode drop of the center tap.

Semiconductor Rectifier Ratings

Silicon rectifiers are being used almost exclusively in power supplies designed for amateur equipment. They are available in a wide range of voltage and current ratings. In peak inverse voltage (PIV) ratings of 600 or less, silicon rectifiers carry current ratings as high as 400 amperes. At 1000 PIV, the current ratings may be several amperes. The extreme compactness of silicon rectifiers makes feasible the stacking of several units in series for higher voltages. Stacks are available commercially that will handle up to 10,000 PIV at a dc load current of 1 A or more.

Protection of Silicon Power Diodes

The important specifications of a silicon diode are:

1) PIV, the peak inverse voltage.

2) I_o, the average dc current rating.

3) I_{REP}, the peak repetitive forward current.

4) I_{SURGE}, a nonrepetitive, peak half-sine wave of 8.3 ms duration (one-half cycle

of 60 Hz line frequency).

The first two specifications appear in most catalogs. The last two often do not, but they are very important.

Because the rectifier never allows current to flow more than half the time, when it does conduct it has to pass at least twice the average direct current. With a capacitor-input filter, the rectifier conducts much less than half the time, so that when it does conduct, it may pass as much as 10 to 20 times the average dc current, under certain conditions. This is shown in Fig. 8. At A is a simple half-wave rectifier with a resistive load. The waveform to the right of the drawing shows the output voltage along with the diode current. At B and C there are two periods of operation to consider. After the capacitor is charged to the peak-rectified voltage, a period of diode nonconduction elapses while the output voltage discharges through the load. As the voltage begins to rise on the next positive pulse, a point is reached where the rectified voltage equals the stored voltage in the capacitor. As the voltage rises beyond that point, the diode begins to supply current. The diode will continue to conduct until the waveform reaches the crest, as shown. Since the diode must pass a current equal to that of the load over a short period of a cycle, the current will be high. The larger the capacitor for a given load, the shorter the diode conduction time and the higher the peak repetitive current (I_{REP}).

When the supply is first turned on, the discharged input capacitor looks like a dead short, and the rectifier passes a very heavy current. This is I_{SURGE}. The maximum I_{SURGE} rating is usually for a duration of one-half cycle (at 60 Hz), or about 8.3 milliseconds.

If a manufacturer's data sheet is not available, an educated guess about a diode's capability can be made by using these rules of thumb for silicon diodes of the type commonly used in amateur power supplies:

Rule 1) The maximum I_{REP} rating can be assumed to be approximately four times the maximum I_o rating.

Rule 2) The maximum I_{SURGE} rating can be assumed to be approximately 12 times the maximum I_o rating. (This figure should provide a reasonable safety factor. Silicon rectifiers with 750-mA dc ratings, for example, seldom have 1-cycle surge ratings of less than 15 amperes; some are rated up to 35 amperes or more.) From this it can be seen that the rectifier should be selected on the basis of I_{SURGE} and not on I_o ratings.

Thermal Protection

The junction of a diode is quite small; hence it must operate at a high current density. The heat-handling capability is, therefore, quite small. Normally, this is not a prime consideration in high-voltage, low-current supplies. Use of high-current rectifiers at or near their maximum ratings

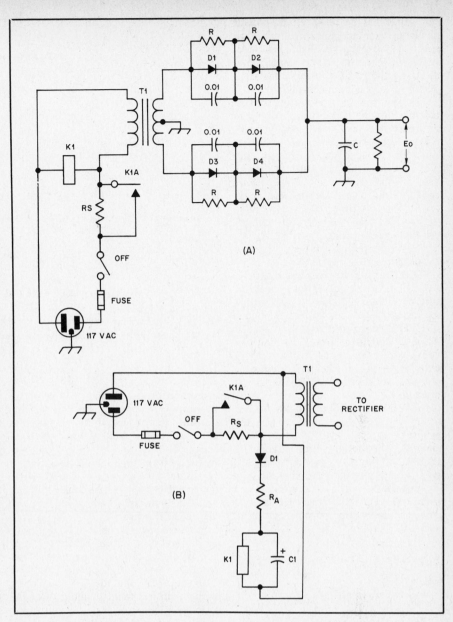

(A)

(B)

Fig. 9 — The primary circuit of T1 at A shows how a 117-volt ac relay and a series-dropping resistor, R_S, can provide surge protection while C charges. At B is an alternative primary circuit using an RC time delay. When silicon rectifiers are connected in series for high-voltage operation, the inverse voltage does not divide equally. The inverse voltage drops can be equalized with resistors, as shown in the secondary circuit. To protect against voltage spikes that may damage an individual rectifier, each rectifier should be bypassed by a 0.01-μF capacitor. Connected as shown, two 400-PIV silicon rectifiers can be used as an 800-PIV rectifier, although it is preferable to include a safety factor and call it a 750-PIV rectifier. The rectifiers, D1 through D4, should be of the same type number and ratings.

(usually 2-ampere or larger stud-mount rectifiers) requires some form of heat sinking. Frequently, mounting the rectifier on the main chassis — directly, or with thin mica insulating washers — will suffice. If insulated from the chassis, a thin layer of silicone grease should be used between the diode and the insulator, and between the insulator and the chassis, to assure good heat conduction. Large, high-current rectifiers often require special heat sinks to maintain a safe operating temperature. Forced-air cooling is sometimes used as a further aid. Safe case temperatures are usually given in the manufacturer's data sheets and should be observed if the maxi-

mum capabilities of the diode are to be realized. See the thermal design section presented later in this chapter for more information.

Surge Protection

Each time the power supply is activated, assuming the input filter capacitor has been discharged, the rectifiers must look into what represents a dead short. Some form of surge protection is usually necessary to protect the diodes until the input capacitor becomes nearly charged, unless the diodes used have a very high surge-current rating (several hundred amperes). Although the dc resistance of the transformer secondary can sometimes be relied on to provide ample surge-current limiting, this is seldom true on high-voltage power supplies. Series resistors can be installed between the secondary and the rectifier strings, but are a deterrent to good voltage regulation.

The best method of protecting the rectifiers is to apply low primary voltage to the transformer for a short period (typically 0.5 to 1 second) until the capacitors charge. After the capacitors have charged, full primary voltage may be applied. A variable autotransformer (Variac or Powerstat are common trade names) may be used to increase the voltage gradually. Some transformers are wound with multiple primaries, and a relay-switching scheme may be employed to step the primary up to full potential.

Fig. 9 illustrates several ways to limit inrush current automatically at turn-on. In each case, R_S is a 25-watt resistor with a resistance of between 15 and 50 ohms, depending on power-supply characteristics. At power supply turn-on, R_S introduces a voltage drop into the primary feed to T1 until the coil of K1 is energized, closing the contacts and shorting out R_S for normal operation.

In the circuit shown in Fig. 9A, K1 has an ac coil rated for the same voltage as the primary of T1. As C becomes nearly charged, the voltage drop across R_S lessens, allowing K1 to pull in. At Fig. 9B, an RC time constant determines the time between application of primary voltage and closure of K1. In this circuit, K1 has a dc coil. The delay time may be calculated using the equation

$$\tau = R_{TH}C$$

where

  t   = time in seconds

$$R_{TH} = \frac{R_A\, R_C}{R_A + R_C}$$

R_A = resistance of R_A in ohms
R_C = coil resistance of K1 in ohms
C = capacitance of C1 in farads

A third method of energizing K1 is with a fixed time-delay relay having appropriate ratings.

Transient Problems

A common cause of trouble is transient voltages on the ac power line. These are short spikes, mostly, that can temporarily increase the voltage seen by the rectifier to values much higher than the normal transformer voltage. They come from distant lightning strokes, electric motors turning on and off, and so on. Transients cause unexpected, and often unexplained, failure of silicon rectifiers.

It's always wise to suppress line transients, and it can be done easily. Fig. 10A shows one way. C1 looks like 280 kΩ at 60 Hz, but to a sharp transient (which has only high-frequency components), it is an

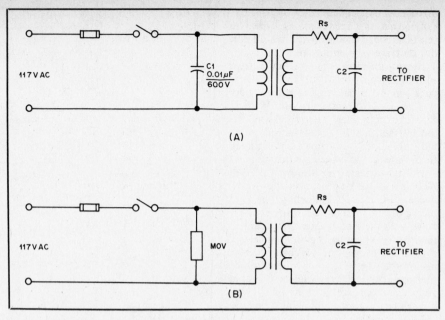

Fig. 10 — Methods of suppressing line transients. See text.

effective bypass. C2 provides additional protection on the secondary side of the transformer. It should be 0.01 μF for transformer voltages of 100 or less, and 0.001 μF for high-voltage transformers.

Fig. 10B shows another transient-suppression method using metal-oxide varistor (MOV) suppressors. The MOVs do not conduct unless the peak voltage becomes abnormally high. Then they clip the transient peaks.

Transient voltages can go as high as twice the normal line voltage before the suppressor diodes clip the peaks. Capacitors cannot give perfect suppression either. Thus, it is a good idea to use power supply rectifiers rated at about twice the expected PIV.

Diodes in Series

When the PIV rating of a single diode is not sufficient for the application, similar diodes may be used in series. (Two 500-PIV diodes in series will withstand 1000 PIV, and so on.) When this is done, a resistor and a capacitor should be placed across each diode in the string to equalize the PIV drops and to guard against transient voltage spikes, as shown in Fig. 11A. Even though the diodes are of the same type and have the same PIV rating, they may have widely different back resistances when they are cut off. The inverse voltage divides according to Ohm's Law, and the diode with the higher back resistance will have the higher voltage developed across it. The diode may break down.

If we put a swamping resistor across each diode, R as shown in Fig. 11A, the resultant resistance across each diode will be almost the same, and the back voltage will divide almost equally. A good rule of thumb for resistor size is this: Multiply the PIV rating of the diode by 500 ohms. For

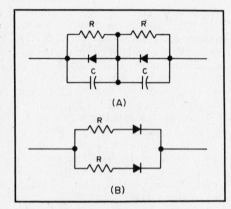

Fig. 11 — (A) Diodes connected in series should be shunted with equalizing resistors and spike-suppressing capacitors. (B) Diodes connected in parallel should have series current equalizing resistors.

example, a 500-PIV diode should be shunted by 500 × 500, or 250,000 ohms. Metal film resistors are generally rated for working voltage as follows: ¼ W — 250 V (RMS); ½ W — 350 V; 1 W — 500 V; 2 W — 750 V.

The shift from forward conduction to high back resistance does not take place instantly in a silicon diode. Some diodes take longer than others to develop high back resistance. To protect the fast-switching diodes in a series string until all the diodes are properly cut off, a 0.01-μF capacitor should be placed across each diode. Fig. 11A shows the complete series-diode circuit. The capacitors should be noninductive (ceramic disc, for example) and should be well matched. Use 10-percent-tolerance capacitors if possible.

Diodes in Parallel

Diodes can be placed in parallel to

increase current-handling capability. Equalizing resistors should be added as shown in Fig. 11B. Without the resistors, one diode may take most of the current. The resistors should be selected to have about a 1-volt drop at the expected peak current.

FILTERING

The pulsating dc waves from the rectifiers are not sufficiently constant in amplitude to prevent hum corresponding to the pulsations. Filters are required between the rectifier and the load to smooth out the pulsations into an essentially constant dc voltage. The design of the filter depends to a large extent on the dc voltage output, the voltage regulation of the power supply and the maximum load current rating of the rectifier. Power-supply filters are low-pass devices using series inductors and shunt capacitors.

Load Resistance

In discussing the performance of power-supply filters, it is sometimes convenient to express the load connected to the output terminals of the supply in terms of resistance. The load resistance is equal to the output voltage divided by the total current drawn, including the current drawn by the bleeder resistor.

Voltage Regulation

The output voltage of a power supply always decreases as more current is drawn, not only because of increased voltage drops in the transformer and filter chokes, but also because the output voltage at light loads tends to soar to the peak value of the transformer voltage as a result of charging the first capacitor. Proper filter design can eliminate the soaring effect. The change in output voltage with load is called voltage regulation, and is expressed as a percentage.

$$\text{Percent regulation} = \frac{100\,(E_1 - E_2)}{E_2}$$

where
E_1 = the no-load voltage
E_2 = the full-load voltage

A steady load, such as that represented by a receiver, speech amplifier or unkeyed stages of a transmitter, does not require good (low) regulation as long as the proper voltage is obtained under load conditions. However, the filter capacitors must have a voltage rating safe for the highest value to which the voltage will soar when the external load is removed.

A power supply will show more (higher) regulation with long-term changes in load resistance than with short temporary changes. The regulation with long-term changes is often called the static regulation, to distinguish it from the dynamic regulation (short temporary load changes). A load that varies at a syllabic or keyed rate, as represented by some audio and RF

amplifiers, usually requires good dynamic regulation (15 percent or less) if distortion products are to be held to a low level. The dynamic regulation of a power supply can be improved by increasing the value of the output capacitor.

When essentially constant voltage regardless of current variation is required (for stabilizing an oscillator, for example), special voltage-regulating circuits described later in this chapter are used.

Bleeder

A bleeder resistor is a resistance connected across the output terminals of the power supply. Its functions are to discharge the filter capacitors as a safety measure when the power is turned off and to improve voltage regulation by providing a minimum load resistance. When voltage regulation is not of importance, the resistance may be as high as 100 ohms per volt. The resistance value to be used for voltage-regulating purposes is discussed in later sections. From the consideration of safety, the power rating of the resistor should be as conservative as possible, since a burned-out bleeder resistor is dangerous!

Ripple Frequency and Voltage

Pulsations at the output of the rectifier can be considered to be the result of an alternating current superimposed on a steady direct current. From this viewpoint, the filter may be considered to consist of shunt capacitors that short circuit the ac component while not interfering with the flow of the dc component. Series chokes will readily pass dc but will impede the flow of the ac component.

The alternating component is called ripple. The effectiveness of the filter can be expressed in terms of percent ripple, which is the ratio of the RMS value of the ripple to the dc value in terms of percentage.

$$\text{Percent ripple (RMS)} = \frac{100\,E_1}{E_2}$$

where
E_1 = the RMS value of ripple voltage
E_2 = the steady dc voltage

Any multiplier or amplifier supply in a CW transmitter should have less than five percent ripple. A linear amplifier can tolerate about three percent ripple on the plate voltage. Bias supplies for linear amplifiers should have less than one-percent ripple. VFOs, speech amplifiers and receivers may require a ripple reduction to 0.01 percent.

Ripple frequency is the frequency of the pulsations in the rectifier output wave — the number of pulsations per second. The frequency of the ripple with half-wave rectifiers is the same as the frequency of the line supply — 60 Hz with 60-Hz supply. Since the output pulses are doubled with a full-wave rectifier, the ripple frequency is doubled — to 120 Hz with a 60-Hz supply.

The amount of filtering (values of inductance and capacitance) required to give adequate smoothing depends on the ripple frequency. More filtering is required as the ripple frequency is lowered.

TYPES OF FILTER

Power supply filters fall into two classifications: capacitor input and choke input. Capacitor-input filters are characterized by relatively high output voltage with respect to the transformer voltage. Capacitor-input filters are especially useful with silicon rectifiers.

The output voltage of a properly de-

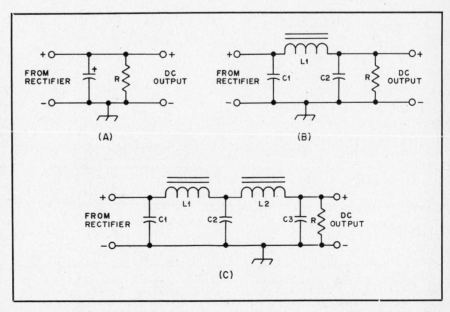

Fig. 12 — Capacitor-input filter circuits. At A is a simple capacitor filter. B and C are single- and double-section filters, respectively.

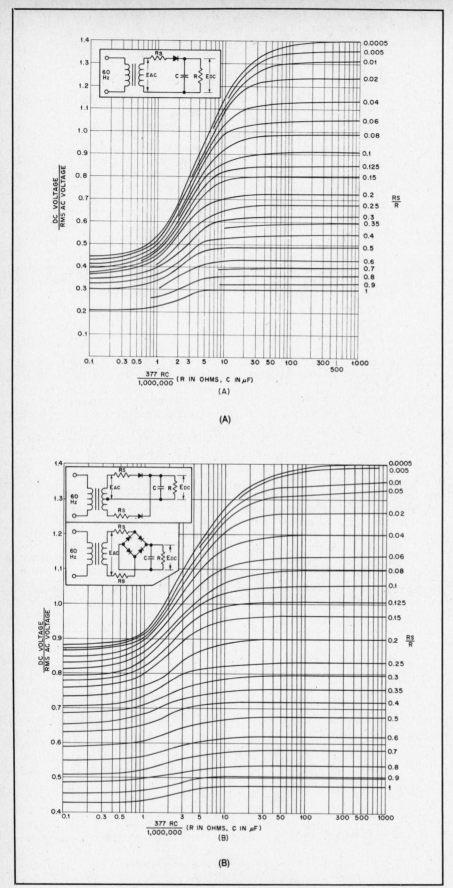

Fig. 13 — Dc output voltages from a half- and full-wave rectifier circuit as a function of the filter capacitance and load resistance (half-wave shown at A and full-wave shown at B). R_s includes transformer winding resistance and rectifier forward resistance. For the ratio R_s/R, both resistances are in ohms; for the RC product, R is in ohms and C is in microfarads. These curves are adapted from those published by Otto H. Schade in "Analysis of Rectifier Operation," *Proceedings of the I.R.E.*, July 1943.

signed choke-input power supply is less than would be obtained with a capacitor-input filter from the same transformer. Generally speaking, a choke-input filter will permit a higher load current to be drawn from a thermionic rectifier without exceeding the peak rating of the rectifier. This is a result of the greatly reduced peak current that the rectifier is subjected to.

Capacitor-Input Filters

Capacitor-input filter systems are shown in Fig. 12. Disregarding voltage drops in the chokes, all have the same characteristics except with respect to ripple. Better ripple reduction will be obtained when LC sections are added as shown in Figs. 12B and C.

Output Voltage

To determine the approximate dc voltage output when a capacitor-input filter is used, the graphs shown in Fig. 13 will be helpful. An example of how to use the graph is given below.

Example:
Full-wave rectifier (use graph at B)
Transformer RMS voltage = 350
Load resistance = 2000 ohms
Series resistance = 200 ohms
Input capacitance = 20 μF

$$\frac{R_s}{R} = \frac{200}{2000} = 0.1$$

$$\frac{377 \, RC}{1,000,000} = \frac{377 \times 2000 \times 20}{1,000,000} = 15.08$$

From curve 0.1 and 377 RC/1,000,000 = 15, the dc voltage is (350 × 1.04) = 364.

In many cases it is desirable to know the amount of capacitance required for a power supply, given certain performance criteria. This is especially true when designing a power supply for an application such as a solid-state transceiver. The following example shows the builder how to arrive at circuit values for a power supply using a single capacitor filter.

Fig. 14 is the circuit diagram of the power supply to be used.

Requirements:
Output voltage = 12.6
Output current = 1 ampere
Maximum ripple = 2 percent
Load regulation = 5 percent

The RMS secondary voltage of T1 must be the desired output voltage plus the voltage drops across D2 and D4 divided by 1.41.

$$E_{SEC} = \frac{12.6 + 1.4}{1.41} = 9.93$$

In practice, the nearest standard transformer (10 V) would work fine. Alternatively, the builder could wind his own transformer, or remove secondary turns from a 12-volt transformer to obtain the desired RMS secondary voltage.

A two-percent ripple referenced to 12.6 volts is 0.25 V RMS. The peak-to-peak value is therefore 0.25 × 2.8 = 0.7 V. This value is necessary to calculate the required capacitance for C1.

Also needed for determining the value of C1 is the time interval (t) between the full-wave rectifier pulses, which is calculated as follows:

$$t = \frac{1}{f_{(Hz)}} = \frac{1}{120} = 8.3 \times 10^{-3}$$

where

t = the time between pulses
f = the ripple frequency in Hz

Since the circuit makes use of a full-wave rectifier, a pulse occurs twice during each cycle. With half-wave rectification, a pulse would occur only once a cycle. Thus 120 Hz is used as the frequency for this calculation.

C1 is calculated from

$$C_{(\mu F)} = \left[\frac{I_L t}{E_{rip}(P\text{-}P)} \right] 10^6$$

$$= \left[\frac{1 A \times 8.3 \times 10^{-3}}{0.7} \right] 10^6$$

$$= 11,857 \ \mu F$$

where I_L is the current taken by the load. The nearest standard capacitor value is 12,000 μF. This value is acceptable, although the tolerance of electrolytic capacitors is rather loose, so the builder may elect to use the next larger standard value.

Diodes D1-D4, inclusive, should have a PIV rating of at least two times the transformer secondary peak voltage. Assuming a transformer secondary RMS value of 10 volts, the PIV should be at least 28 volts. Four diodes rated at 50 volts will provide a margin of safety. Although the minimum current rating needed is one-half the load current, the forward current rating of the diodes should be at least twice the load current for safety. For a 1-A load, the diodes should be rated for at least 2 A.

The load resistance, R_L, is determined by E_o/I_L, which in this example is 12.6/1 = 12.6 ohms. This factor must be known in order to find the necessary series resistance for five-percent regulation. Calculate as follows:

$$R_{s(max)} = \text{load regulation} \times R_L$$

$$= 0.05 \times 12.6 = 0.63 \text{ ohm}$$

Therefore, the transformer secondary dc resistance should be no greater than 0.63 ohm. The secondary current rating should be equal to or greater than the I_L = 1 ampere. C1 should have a minimum working voltage of 1.4 times the output voltage. In this power supply the capacitor should be rated for at least 18 volts.

Choke-Input Filters

Better voltage regulation results when a choke-input filter, as shown in Fig. 15, is

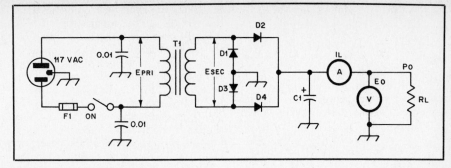

Fig. 14 — This figure illustrates how to design a simple unregulated power supply. See text for a thorough discussion.

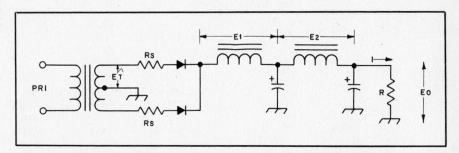

Fig. 15 — Diagram showing various voltage drops that must be taken into consideration in determining the required transformer voltage to deliver the desired output voltage.

used. Choke input permits better utilization of thermionic rectifiers, because a higher load current can be drawn without exceeding the peak current rating. Choke-input filters are generally not used with silicon rectifiers.

Minimum Choke Inductance

A choke-input filter will tend to act as a capacitor-input filter unless the input choke has at least a certain minimum value of inductance called the critical value. For full-wave 60-Hz supplies, this critical value is given by

$$L_{crit} \text{ (henrys)} = \frac{E \text{ (volts)}}{I \text{ (mA)}}$$

where

E = the supply output voltage
I = the current being drawn through the filter

If the choke has at least the critical value, the output voltage will be limited to the average value of the rectified wave at the input to the choke when the current drawn from the supply is small. This is in contrast to the capacitor-input filter in which the output voltage tends to soar toward the peak value of the rectified wave at light loads.

Minimum-Load — Bleeder Resistance

From the above formula for critical inductance, it is obvious that if no current is drawn from the supply, the critical inductance will be infinite. So that a practical value of inductance may be used, some current must be drawn from the supply at all

times the supply is in use. From the formula we find that this minimum value of current is

$$I(mA) = \frac{E(volts)}{L_{crit}}$$

In the majority of cases, it will be most convenient to adjust the bleeder resistance so that the bleeder will draw the required minimum current. From the formula, it may be seen that the value of critical inductance becomes smaller as the load current increases.

Swinging Chokes

Less costly chokes are available that will maintain at least the critical value of inductance over the range of current likely to be drawn from practical supplies. These chokes are called swinging chokes. As an example, a swinging choke may have an inductance rating of 5/25 H and a current rating of 200 mA. If the supply delivers 1000 volts, the minimum load current should be 1000/25 = 40 mA. When the full load current of 200 mA is drawn from the supply, the inductance will drop to 5 H. The critical inductance for 200 mA at 1000 volts is 1000/200 = 5 H. Therefore the 5/25-H choke maintains the critical inductance at the full current rating of 200 mA. At all load currents between 40 mA and 200 mA, the choke will adjust its inductance to the approximate critical value.

Output Voltage

Provided the input-choke inductance is at least the critical value, the output voltage may be calculated quite closely by

$$E_o = 0.9E_t - (I_B + I_L) \times (R1 + R2) - E_r$$

where

E_o = output voltage
E_t = RMS voltage applied to the rectifier (RMS voltage between center-tap and one end of the secondary in the case of the center-tap rectifier)
I_B = bleeder current (A)
I_L = load current (A)
R1 = first filter choke resistance
R2 = second filter choke resistance
E_r = voltage drop across the rectifier

The various voltage drops are shown in Fig. 15. At no load, I_L is zero; hence the no-load voltage may be calculated on the basis of bleeder current only. The voltage regulation may be determined from the no-load and full-load voltages using the formulas previously given.

Output Capacitance

Whether the supply has a choke- or capacitor-input filter, the reactance of the output capacitor should be low for the lowest audio frequency if it is intended for use with a class-A AF amplifier. For vacuum-tube loads, 16 μF or more is usually adequate. When the supply is used with a class-B amplifier (for modulation or for SSB amplification) or a CW transmitter, increasing the output capacitance will result in improved dynamic regulation of the supply. However, a region of diminishing returns can be reached, and 20 to 30 μF will usually suffice for any supply subjected to large changes at a syllabic (or keying) rate.

Resonance

Resonance effects in the series circuit across the output of the rectifier, formed by the first choke and first filter capacitor, must be avoided, because the ripple voltage would build up to large values. This not only is the opposite action to that for which the filter is intended, but also may cause excessive rectifier peak currents and abnormally high peak-inverse voltages. For full-wave rectification, the ripple frequency will be 120 Hz for a 60-Hz supply; resonance will occur when the product of choke inductance in henrys times capacitor capacitance in microfarads is equal to 1.77. At least twice this product of inductance and capacitance should be used to ensure against resonance effects. With a swinging choke, the minimum rated inductance of the choke should be used. If too high an LC filter product is used, the resonance may occur at the radio-telegraph keying or voice syllabic rate, and large voltage excursions (filter bounce) may be experienced at that rate.

Ratings of Filter Components

In a power supply using a choke-input filter and properly designed choke and bleeder resistor, the no-load voltage across

the filter capacitors will be about nine-tenths of the ac RMS voltage. Nevertheless, it is advisable to use capacitors rated for the peak transformer voltage. This large safety factor is suggested because the voltage across the capacitors can reach this peak value if the bleeder should burn out and there is no load on the supply.

In a capacitor-input filter, the capacitors should have a working-voltage rating at least as high, and preferably somewhat higher, than the peak voltage from the transformer. Thus, in the case of a center-tap rectifier having a transformer delivering 550 volts each side of the center tap, the minimum safe capacitor voltage rating will be 550 × 1.41, or 775 volts. An 800-volt capacitor should be used, or preferably a 1000-volt unit.

Filter Capacitors in Series

Filter capacitors are made in several different types. Electrolytic capacitors, which are available for peak voltages up to about 800, combine high capacitance with small size. This is possible because the dielectric is an extremely thin film of oxide on aluminum foil. Capacitors of this type may be connected in series for higher voltages, although the filtering capacitance will be reduced to the resultant of the two capacitances in series.

If this arrangement is used, it is important that each capacitor be shunted with a resistor of about 100 ohms per volt of supply voltage applied to the individual capacitors, with an adequate power rating.

These resistors may serve as all or part of the bleeder resistance. Capacitors with higher voltage ratings are usually made with a dielectric of thin paper impregnated with oil. The working voltage of a capacitor is the voltage that it will withstand continuously.

Filter Chokes

Filter chokes or inductances are wound on iron cores, with a small gap in the core to prevent magnetic saturation of the iron at high currents. When the iron becomes saturated, its permeability decreases, and consequently the inductance also decreases. Despite the air gap, the inductance of a choke usually varies to some extent with the direct current flowing in the winding; hence it is necessary to specify the inductance at the current which the choke is intended to carry. Its inductance with little or no direct current flowing in the winding will usually be considerably higher than the value when full load current is flowing.

Negative-Lead Filtering

For many years it has been almost universal practice to place filter chokes in the positive leads of plate power supplies. This means that the insulation between the choke winding and its core (which should be grounded to chassis as a safety measure) must be adequate to withstand the output voltage of the supply. This voltage requirement is removed if the choke is placed in the negative lead, as shown in Fig. 16. With this connection, the capacitance of the

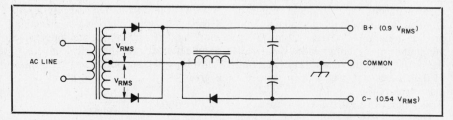

Fig. 16 — In most applications, the filter choke may be placed in the negative instead of the positive side of the circuit. This reduces the danger of a voltage breakdown between the choke winding and core. The ripple voltage developed across the choke can be rectified to provide a free negative bias supply.

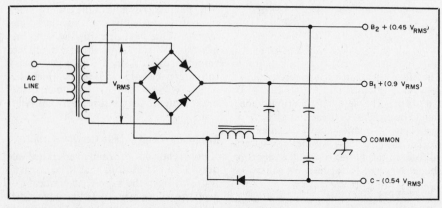

Fig. 17 — The "economy" power-supply circuit is a combination of the full-wave and bridge-rectifier circuits.

transformer secondary to ground appears in parallel with the filter chokes, tending to bypass the chokes. However, this effect will be negligible in practical application except in cases where the output ripple must be reduced to a very low figure. Choke terminals, negative capacitor terminals and the transformer center-tap terminal should be well protected against accidental contact, since these will assume full supply voltage to chassis should a choke burn out or the chassis connection fail.

A free negative supply of about 60% of the plate potential can be obtained by rectifying the ripple voltage developed across the filter choke. The voltage across the choke has an effective value of the peak secondary voltage (measured with respect to the center tap) minus the main dc output voltage. Rectification multiplies the choke voltage by 1.4.

The Economy Power Supply

In many transmitters of the 100-watt class, an excellent method for obtaining plate and screen voltages without wasting power in voltage-dropping resistors is use of the economy power supply circuit. Shown in Fig. 17, it is a combination of the full-wave and bridge-rectifier circuits. The voltage at B2 is the normal voltage obtained with the full-wave circuit, and the voltage at B1 is that obtained with the bridge circuit. The total dc power obtained from the transformer is, of course, the sum of the power used at each voltage. In CW and SSB applications, additional power can usually be drawn without excessive heating, especially if the transformer has a rectifier filament winding that isn't being used. When negative-lead choke filtering is used, the free negative bias supply scheme described above can be incorporated, making all of a transmitting tube's operating potentials available from a single transformer. The entire peak secondary voltage is used to calculate the resulting bias voltage.

VOLTAGE MULTIPLICATION

Half-Wave Voltage Doubler

Fig. 18 shows the circuit of a half-wave voltage doubler. B, C and D of Fig. 18 illustrate the circuit operation. For clarity, assume the transformer voltage polarity at the moment the circuit is activated is that shown at B. During the first negative half cycle, D_A conducts (D_B is in a nonconductive state), charging C1 to the peak rectified voltage (1.4 E_{RMS}). C1 is charged with the polarity shown at B. During the positive half cycle of the secondary voltage, D_A is cut off and D_B conducts, charging capacitor C2. The amount of voltage delivered to C2 is the sum of peak secondary voltage of the transformer plus the voltage stored in C1 (1.4 E_{RMS}). On the next negative half cycle, D_B is nonconducting and C2 will discharge into the load. If no load is connected across C2, the

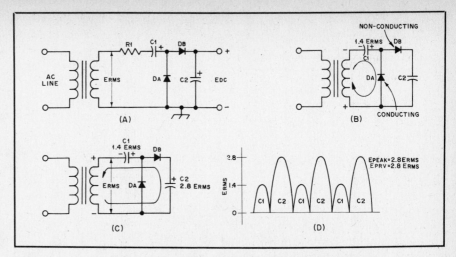

Fig. 18 — Illustrated at A is a half-wave voltage-doubler circuit. B displays how the first half cycle of input voltage charges C1. During the next half cycle (shown at C), capacitor C2 is charged with the transformer secondary voltage plus that voltage stored in C1 from the previous half cycle. D illustrates the levels to which each capacitor is charged throughout the cycle.

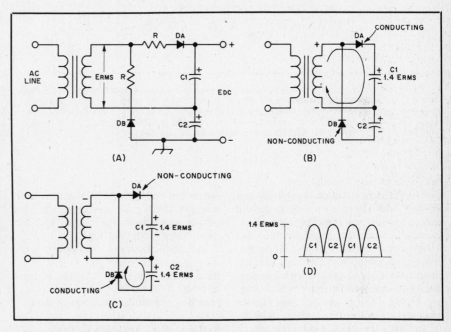

Fig. 19 — A full-wave voltage doubler is displayed at A. One half cycle is shown at B and the next half cycle at C. Each capacitor receives a charge during every cycle of input voltage. D illustrates how each capacitor is charged alternately.

capacitors will remain charged — C1 to 1.4 E_{RMS} and C2 to 2.8 E_{RMS}. When a load is connected to the output of the doubler, the voltage across C2 drops during the negative half cycle and is recharged up to 2.8 E_{RMS} during the positive half cycle.

The output waveform across C2 resembles that of a half-wave rectifier circuit in that C2 is pulsed once every cycle. Fig. 18D illustrates the levels to which the two capacitors are charged throughout the cycle. In actual operation, the capacitors will not discharge all the way to zero as shown.

Full-Wave Voltage Doubler

Shown in Fig. 19 is the circuit of a full-wave voltage doubler. The circuit operation

can best be understood by following B, C and D of Fig. 19. During the positive half cycle of transformer secondary voltage, as shown at B, D_A conducts charging capacitor C1 to 1.4 E_{RMS}. D_B is not conducting at this time.

During the negative half cycle, as shown at C, D_B conducts, charging capacitor C2 to 1.4 E_{RMS}, while D_A is nonconducting. The output voltage is the sum of the two capacitor voltages, which will be 2.8 E_{RMS} under no-load conditions. Fig. 19D illustrates that each capacitor alternately receives a charge once per cycle. The effective filter capacitance is that of C1 and C2 in series, which is less than the capacitance of either C1 or C2 alone.

Resistors R in Fig. 19A are used to limit

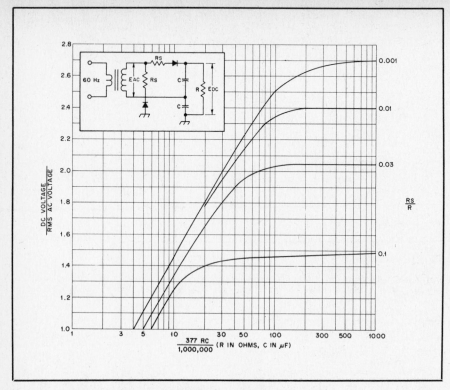

Fig. 20 — Dc output voltages from a full-wave voltage-doubling circuit as a function of the filter capacitances and load resistance. For the ratio R_s/R and for the RC product, resistances are in ohms and capacitance is in microfarads. Equal resistance values for R_s and equal capacitance values for C are assumed. These curves are adapted from those published by Otto H. Schade in "Analysis of Rectifier Operation," *Proceedings of the I.R.E.,* July 1943.

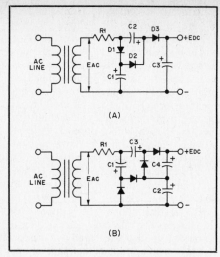

Fig. 21 — Voltage-multiplying circuits with one side of the transformer secondary used as common. A, voltage tripler; B, voltage quadrupler. Capacitances are typically 20 to 50 μF depending on output current demand. Dc ratings of capacitors are related to E_{peak} (1.4 E_{ac}).

C1 — Greater than E_{peak}
C2 — Greater than $2E_{peak}$
C3 — Greater than $3E_{peak}$
C4 — Greater than $2E_{peak}$

the surge current through the rectifiers. Their values are based on the transformer voltage and the rectifier surge-current rating, since at the instant the power supply is turned on the filter capacitors look like a short-circuited load. Provided the limiting resistors can withstand the surge current, their current-handling capacity is based on the maximum load current from the supply. Output voltages approaching twice the peak voltage of the transformer can be obtained with the voltage doubling circuit shown in Fig. 19. Fig. 20 shows how the voltage depends upon the ratio of the series resistance to the load resistance, and the load resistance times the filter capacitance. The peak inverse voltage across each diode is 2.8 E_{RMS}.

Voltage Tripling and Quadrupling

A voltage-tripling circuit is shown in Fig. 21A. On one half of the ac cycle, C1 and C3 are charged to the source voltage through D1, D2 and D3. On the opposite half of the cycle, D2 conducts and C2 is charged to twice the source voltage, because it sees the transformer plus the charge in C1 as its source (D1 is cut off during this half cycle.) At the same time, D3 conducts, and with the transformer and the charge in C2 as the source, C3 is charged to three times the transformer voltage.

The voltage-quadrupling circuit of Fig. 21B works in substantially similar

fashion. In either of the circuits of Fig. 21, the output voltage will approach an exact multiple of the peak ac voltage when the output current drain is low and the capacitance values are high.

In the circuits shown, the negative leg of the supply is common to one side of the transformer. The positive leg can be made common to one side of the transformer by reversing the diodes and capacitors.

VOLTAGE STABILIZATION

Zener Diode Regulation

A Zener diode (named after Dr. Carl Zener) can be used to maintain the voltage applied to a circuit at a practically constant value, regardless of the voltage regulation of the power supply or variations in load current. The typical circuit is shown in Fig. 22A. Note that the cathode side of the diode is connected to the positive side of the supply. The electrical characteristics of a Zener diode under conditions of forward and reverse voltage are given in Chapter 4.

Zener diodes are available in a wide variety of voltages and power ratings. The voltages range from less than two to a few hundred, while the power ratings (power the diode can dissipate) run from less than 0.25 watt to 50 watts. The ability of the Zener diode to stabilize a voltage depends on the conducting impedance of the diode. This can be as low as one ohm or less in a low-voltage, high-power diode or as high

as a thousand ohms in a high-voltage, low-power diode.

Diode Power Dissipation

Zener diodes of a particular voltage rating have varied maximum current capabilities, depending on the power ratings of each of the diodes. The power dissipated in a diode is the product of the voltage across it and the current through it. Conversely, the maximum current a particular diode may safely conduct is equal to its power rating divided by its voltage rating. Thus, a 10-V, 50-W Zener diode, if operated at its maximum dissipation rating, would conduct 5 amperes. A 10-V, 1-W diode, on the other hand, could safely conduct no more than 0.1 A, or 100 mA. The conducting impedance of a diode is its voltage rating divided by the current flowing through it. In the above examples, the conducting impedance would be 2 ohms for the 50-W diode, and 100 ohms for the 1-W diode. Disregarding small voltage changes which may occur, the conducting impedance of a given diode is a function of the current flowing through it, varying in inverse proportion.

The power-handling capability of most Zener diodes is rated at 25 °C, or approximately room temperature. If the diode is operated in a higher ambient temperature, its power capability must be derated. A typical 1-watt diode can safely dissipate only ½ watt at 100 °C.

Part C of Fig. 22 illustrates a method for multiplying the effective power-handling capability of a small Zener diode. When the diode conducts, it supplies base current to a power transistor. This causes collector

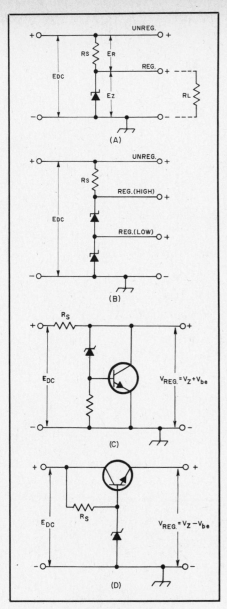

Fig. 22 — Zener-diode voltage regulation. The voltage from a negative supply may be regulated by reversing the power-supply connections and the diode and transistor polarities. The operation of these circuits is discussed in the text.

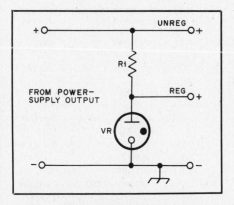

Fig. 23 — Voltage stabilization circuit using a VR tube. A negative-supply output may be regulated by reversing the polarity of the power-supply connections and the VR-tube connections from those shown here.

current to be drawn through the series-limiting resistance (see next section). This current drops the diode voltage ($V_Z + V_{be}$) to the conduction threshold. When load current is drawn through R_S, the output voltage tends to drop below the conduction threshold. But when the diode ceases to conduct, the transistor collector ceases to draw current. This negative feedback, or servo action, stabilizes the current in the limiting resistor and regulates the output voltage at $V_Z + V_{be}$. With this approach, the conducting impedance of the Zener diode is divided by the transistor beta.

When high load current is required, a Zener diode shunt regulator is grossly inefficient because of the voltage dropped and power dissipated by the current-limiting resistor. This deficiency can be overcome by using the emitter follower arrangement of Fig. 22D. A Zener diode establishes a reference voltage for the transistor base, and the load current flows in the collector-emitter circuit. The limiting resistor is selected to provide the maximum load current divided by the minimum transistor beta, plus the desired Zener current. As with the shunt current boosting circuit described above, the output impedance of this voltage regulator will be the Zener impedance divided by the transistor beta. The output voltage is $V_Z - V_{be}$.

Limiting Resistance

The value of R_S in Fig. 22 is determined by the load requirements. If R_S is too large, the diode will be unable to regulate at large values of I_L (the current through R_L). If R_S is too small, the diode dissipation rating may be exceeded at low values of I_L. The optimum value for R_S can be calculated by

$$R_S = \frac{E_{dc(min)} - E_Z}{1.1\ I_{L(max)}}$$

When R_S is known, the maximum dissipation of the diode, P_D, may be determined by

$$P_D = \left[\frac{E_{dc(max)} - E_Z}{R_S} - I_{L(min)}\right]E_Z$$

In the first equation, conditions are set up for the Zener diode to draw 1/10 the maximum load current. This assures diode regulation under maximum load.

Example: A 12-volt source is to supply a circuit requiring 9 volts. The load current varies between 200 and 350 mA.

$E_Z = 9.1$ V (nearest available value)

$$R_S = \frac{12 - 9.1}{1.1 \times 0.35} = \frac{2.9}{0.385} = 7.5\ \text{ohms}$$

$$P_D = \left[\frac{12 - 9.1}{7.5} - 0.2\right]9.1$$

$$= 0.185 \times 9.1 = 1.7\ \text{W}$$

The nearest available dissipation rating above 1.7 W is 5; therefore, a 9.1-V,

5-W Zener diode should be used. Such a rating, it may be noted, will cause the diode to be in the safe dissipation range even though the load is completely disconnected [I_L (min) = 0].

Obtaining Other Voltages

Fig. 22B shows how two Zener diodes may be used in series to obtain regulated voltages not normally obtainable from a single Zener diode, and also to give two values of regulated voltage. The diodes need not have equal breakdown voltages, because the arrangement is self equalizing. However, the current-handling capability of each diode should be taken into account. The limiting resistor may be calculated as above, taking the sum of the diode voltages as E_Z, and the sum of the load currents as I_L.

Gaseous Regulator Tubes

Although at first glance it may seem inappropriate to discuss gaseous voltage-regulator tubes (commonly called VR tubes) in a modern Handbook, they do offer advantages over their solid-state counterparts in some applications. Tube-type amplifier screen supplies often call for regulated voltages between 200 and 400. To obtain these voltages with Zener diodes, series combinations of four to six devices are often required. The cost can add up quickly. In addition to being far less expensive than solid-state devices, gaseous regulator tubes are much more forgiving of line transients, application of overvoltage and reverse screen current.

Gaseous regulator tubes (0A2/VR150, 0B2/VR105, etc.) can be used to good advantage. The voltage drop across such tubes is constant over a moderately wide current range. Tubes are available for regulated voltages near 150, 105, 90 and 75 volts.

The fundamental circuit for a gaseous regulator is shown in Fig. 23. The tube is connected in series with a limiting resistor, R1, across a source of voltage that must be higher than the starting voltage. The starting voltage is about 30 to 40 percent higher than the operating voltage. The load is connected in parallel with the tube. For stable operation, a minimum tube current of 5 to 10 mA is required. The maximum permissible current with most types is 40 mA; consequently, the load current cannot exceed 30 to 35 mA if the voltage is to be stabilized over a range from zero to maximum load. A single VR tube may also be used to regulate the voltage to a load current of almost any value as long as the variation in the current does not exceed 30 to 35 mA. If, for example, the average load current is 100 mA, a VR tube may be used to hold the voltage constant provided the current does not fall below 85 mA or rise above 115 mA.

The value of the limiting resistor must lie between that which just permits minimum tube current to flow and that which just passes the maximum permissible tube current when there is no load current. The latter value is generally used. It is given by the equation

$$R = \frac{E_s - E_r}{I}$$

where

R = limiting resistance in ohms
E_s = voltage of the source across which the tube and resistor are connected
E_r = rated voltage drop across the regulator tube
I = maximum tube current in amperes (usually 40 mA, or 0.04 A)

Two tubes may be used in series to give a higher regulated voltage than is obtainable with one, and also to give two values of regulated voltage. Regulation of the order of one percent can be obtained with these regulator tubes when they are operated within their proper current range. The capacitance in shunt with a VR tube should be limited to 0.1 μF or less. Larger values may cause the tube drop to oscillate between the operating and starting voltages.

Do not use any of the unconnected tube-socket pins as circuit tie points even though such pins are unconnected. They penetrate the tube envelope and terminate abruptly. Depending on the voltage applied to the pin, intermittent breakdown may be experienced.

ELECTRONIC VOLTAGE REGULATION

When extremely low ripple is required, or when the supply voltage must be constant with large fluctuations of load current and line voltage, a closed-loop amplifier is used to regulate the supply. There are two main categories of electronic regulators: linear regulators, in which the condition of a control element is varied in direct proportion to the line voltage or load current; and switching regulators, in which the control device is switched on and off, with the duty cycle proportional to the line or load conditions. Each system has relative advantages and disadvantages, and there are applications for both types in Amateur Radio equipment.

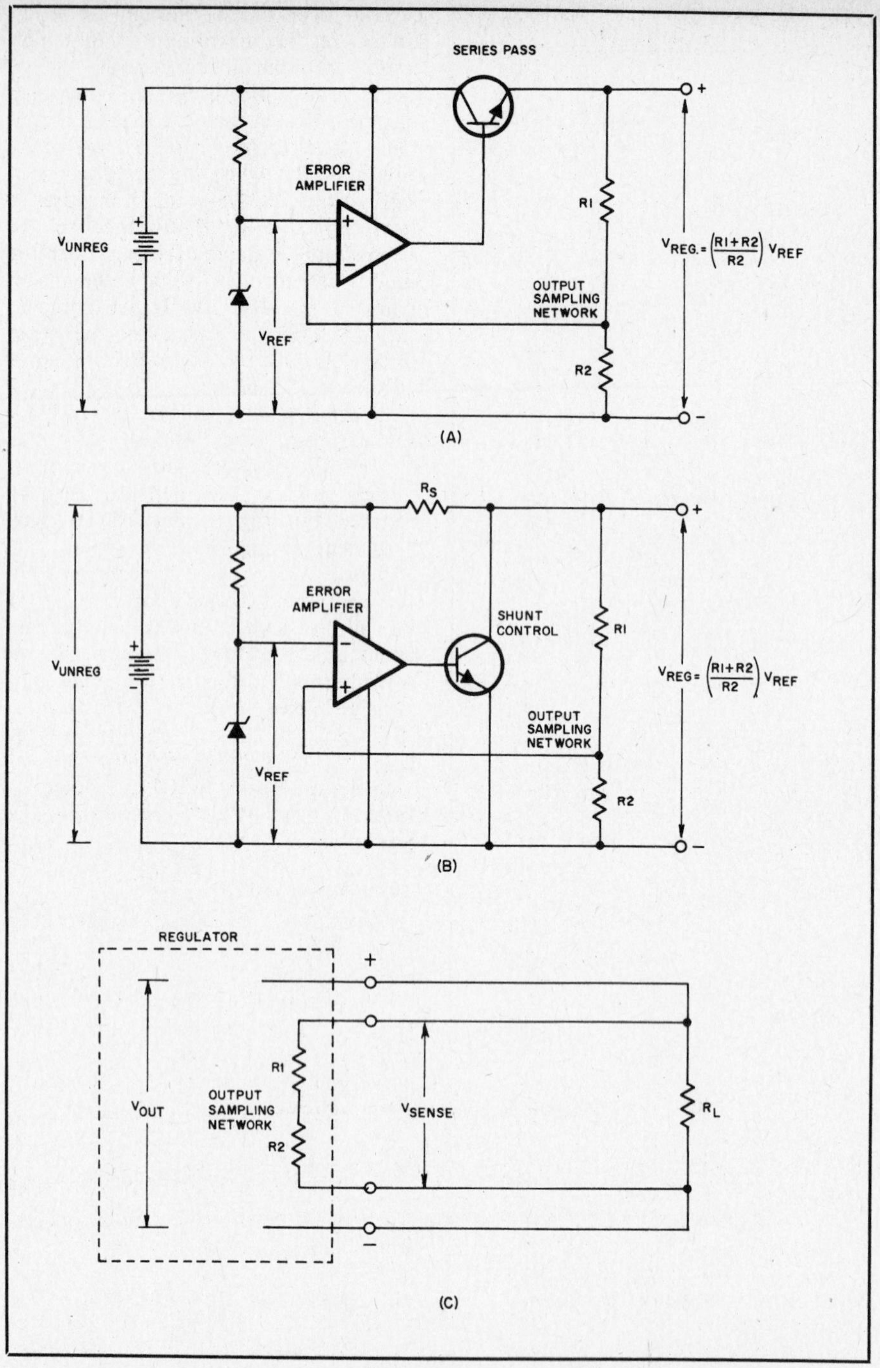

Linear Regulators

The two basic forms of linear regulators are sketched in Fig. 24. A series regulator, shown at A, is used in applications requiring efficient use of the primary power source. A stable reference voltage is established by a Zener diode. In critical applications a temperature-compensated reference diode (discussed in Chapter 4) would be used. The output voltage is sampled by the error amplifier, which compares the output (usually attenuated by a voltage divider) to the reference. If the scaled-down output voltage is higher than

Fig. 24 — Linear electronic voltage regulator circuits. In these diagrams batteries represent the unregulated input voltage source. A transformer, rectifier and filter would serve this function in most applications. (A) Series regulator. (B) Shunt regulator. (C) Remote sensing overcomes poor load regulation caused by IR drop in the connecting wires by bringing them inside the feedback loop.

the reference voltage, the error amplifier reduces the drive current to the pass transistor. Conversely, if the load pulls the output voltage below the value called for, the amplifier drives the pass transistor into increased conduction. The series pass transistor forms a voltage divider with the load resistance. In a series regulator, the power dissipation of the pass transistor is directly proportional to the load current and input-output voltage differential. The series-pass element can be located in either leg of the supply. Either NPN or PNP

devices can be used, depending on the ground polarity of the unregulated input.

Fig. 24B shows a simple shunt regulator circuit. Such a system would be used when the load on the unregulated voltage source must be kept constant. The operation is similar to that of the series regulator, except that the control element is in parallel with the output load. As the load current varies, the control transistor changes resistance in opposition to the load such that the resistance of the parallel combination remains constant, drawing a constant current through R_S. Note that the input polarity of the error amplifier is reversed with respect to the series regulator configuration. This is a result of the inversion caused by the common-emitter control transistor. The power dissipation in the control transistor is greatest when the output is unloaded.

The "stiffness" or tightness of regulation of a linear regulator depends on the gain of the error amplifier and the ratio of the output scaling resistors. The resistor scaling is usually the most significant because high-gain IC op amps can be used in the error amplifier. If the reference potential is 1 volt and the scaling resistors are selected to produce an output of 10 volts, the loop gain is reduced (hence output impedance increased) by a factor of 10. Increasing the reference potential to 5 volts and rescaling the sampling divider for the same 10-volt output will improve the regulation by a factor of five compared to the previous example. Thus, if the 10-volt supply using a 1-volt reference produced an output change of 5 mV for a certain load variation, the supply having the 5-volt reference will produce only a 1-mV change. The same rule applies to output ripple. However, this characteristic can be controlled independently of the load regulation by shunting R1 in Fig. 24 (A and B) with a capacitor. This allows the regulator to function at maximum loop gain for the ac component of the output.

The above discussion leads to the conclusion that the reference voltage in a linear regulator should be as high as possible. However, solid-state physics and fabrication technology dictate that the best temperature stability is obtained with reference diodes having operating potentials in the 6-volt region.

In any regulator, the output is cleanest and regulation stiffest at the point where the sampling network or error amplifier is connected. If heavy load current is drawn through long leads, the voltage drop can degrade the regulation at the load. To combat this effect, the feedback connection to the error amplifier can be made directly to the load. This technique, called remote sensing, moves the point of best regulation to the load by bringing the connecting loads inside the feedback loop. This is shown in Fig. 24C.

Linear voltage regulators are called for in applications requiring tight regulation with very low output noise and ripple. The trade-offs for these traits are bulk, weight and heat dissipation. Several examples of linear regulator design are given in Chapter 27.

Safe Operating Area

In choosing a suitable transistor for use as the series pass device in a linear regulator, care must be taken to ensure that the transistor chosen will survive any possible combination of voltage and current stress that may occur during normal or

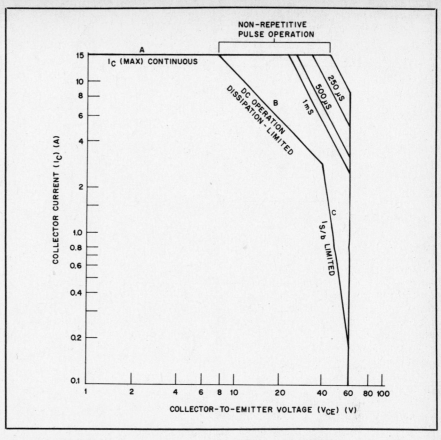

Fig. 25 — Graph of the safe operating area of a 2N3055 transistor. See text for details. This graph is based on material presented in RCA *Power Devices Data Book* SSD-220C.

short-circuit operation. Although the transistor's listed wattage rating is an indicator of its power-handling capability, it must be understood that this rating usually applies only to low forward voltage drops and high collector currents. As the collector-to-emitter voltage is increased, the power-handling capability of the transistor can decrease. This decrease, known as the forward-biased second-breakdown limit, depends greatly upon the processes used in the manufacture of the transistor.

To assist the circuit designer in choosing a transistor that will meet power ratings, transistor manufacturers provide curves relating collector current and voltage limits. These curves are referred to as safe-operating-area (SOAR) curves.

A typical SOAR curve (for a 2N3055 transistor) is shown in Fig. 25. Line A is maximum emitter current that can be handled by the device without damage to the emitter lead. Line B shows the constant power region of operation, which is the transistor's maximum power rating. Finally, line C is the second breakdown limit. Also shown are plots that represent short-term or pulsed-power limitations for various pulse widths.

The power-supply designer should consult the data sheets for his choice of transistor and check to make sure that the safe operating area limits will not be exceeded during normal and short-circuit operation

of the power supply. The addition of a current limiter or foldback limiter will help to protect the pass transistor against destruction during overload conditions.

Current Limiting for Discrete-Component Regulators

Damage to a pass transistor can occur when the load current exceeds the safe amount. Fig. 26 illustrates a simple current-limiter circuit that will protect Q1. All of the load current is routed through R1. A voltage difference will exist across R1; the value will depend on the exact load current at a given time. When the load current exceeds a predetermined safe value, the voltage drop across R1 will forward bias Q2 and cause it to conduct. Because Q2 is a silicon transistor, the voltage drop across R1 must exceed 0.6 V to turn Q2 on. This being the case, R1 is chosen for a value that provides a drop of 0.6 V when the maximum safe load current is drawn. In this instance, 0.6 volts will be seen when IL reaches 0.5 A. R2 protects the base-emitter junction of Q2 from current spikes, or from destruction in the event of failure of Q1 under short-circuit conditions.

When Q2 turns on, some of the current through R_S flows through Q2, thereby depriving Q1 of some of its base current. This action, depending upon the amount of Q1 base current at a precise moment, cuts off Q1 conduction to some degree,

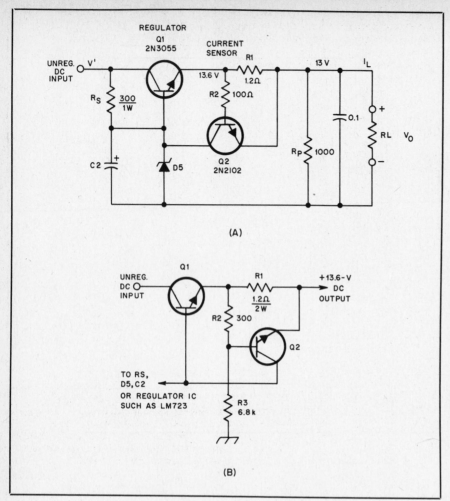

(A)

(B)

Fig. 26 — Overload protection for a regulated supply can be implemented by addition of a current-overload protective circuit, as shown at A. At B, the circuit has been modified to employ current foldback limiting.

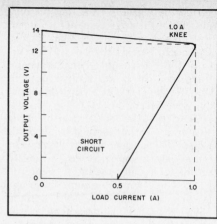

Fig. 27 — The 1-A regulator shown in Fig. 26B will fold back to 0.5 A under short-circuit conditions. See text.

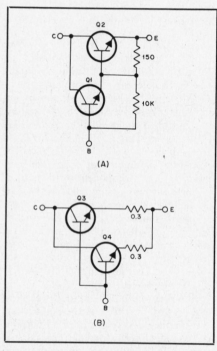

(A)

(B)

Fig. 28 — At A, a Darlington-connected pair for use as the pass element in a series-regulating circuit. At B, the method of connecting two or more transistors in parallel for high current output. Resistances are in ohms. The circuit at A may be used for load currents from 100 mA to 5 A, at B for currents from 6 to 10 A.

Q1 — Motorola MJE 340 or equiv.
Q2-Q4 — Power transistor such as 2N3055 or 2N3772.

thus limiting the flow of current through it.

Foldback Limiting

Under short-circuit conditions, the previously described current limiter must withstand the full source voltage and short-circuit current simultaneously, which can impose a very high power dissipation or second breakdown stress on the series pass transistor. For example, a 12-volt regulator with current limiting set for 10 amperes and having a source of 16 volts will have a dissipation of 40 watts [(16 V − 12 V) × 10 A] at the point of current limiting (knee). But its dissipation will rise to 160 watts under short-circuit conditions (16 volts × 10 amps).

A modification of the limiter circuit can cause the regulator's output current to decrease with decreasing load resistance after the knee. With the output shorted, the output current is only a fraction of the knee current value, which protects the series pass transistor from excessive dissipation and possible failure. Using the previous example of the 12-volt, 10-amp regulator, if the short-circuit current is designed to be 3 amperes (knee is still 10 amps), the tran-

sistor dissipation with short circuit will be only 16 V × 3 A = 48 watts.

Fig. 26B shows how the current limiter example given in the previous section would be modified to incorporate foldback limiting. The divider string formed by R2 and R3 provides a negative bias to the base of Q2, which prevents Q2 from turning on until this bias is overcome by the drop in R1 caused by load current. Since this hold-off bias decreases as the output voltage drops, Q2 becomes more sensitive to current through R1 with decreasing output voltage. See Fig. 27.

The circuit is designed by first calculating the value of R1 for short-circuit current. For example, if 0.5 ampere is chosen, the value for R1 is simply 0.6 volts/0.5 amp = 1.2 ohms (with the output shorted, the amount of hold-off bias supplied by R2 and R3 is very small and can be neglected). The knee current is then chosen. For this example, the selected value will be 1.0 ampere. The divider string is then proportioned to provide a base voltage at the knee that is just sufficient to turn on Q2 (a value of 13.6 volts for 13.0 volts output). With 1.0 ampere flowing through

R1, the voltage across the divider will be 14.2 V. The voltage dropped by R2 must then be 14.2 V − 13.6 V, or 0.6 volt. Choosing a divider current of 2 milliamperes, the value of R2 is then 0.6 V/0.002 A = 300 ohms. R3 is calculated to be 13.6 V/0.002 A = 6800 ohms.

High-Current-Output Regulators

A simple Zener diode reference or IC op-amp error amplifier may not be able to source enough current to a pass transistor that must conduct heavy load current. The Darlington configuration of Fig. 28A multiplies the pass transistor beta, thereby extending the control range of the error amplifier. If the Darlington arrangement is implemented with discrete transistors, resistors across the base-emitter junctions may be necessary to prevent collector-to-base leakage currents in Q1 from being amplified and turning on the transistor pair. These resistors are contained in the envelope of a monolithic Darlington device.

When a single pass transistor is not available to handle the current required from a regulator, the current-handling capability may be increased by connecting two or more pass transistors in parallel. The circuit at B of Fig. 28 shows the method of connection. The resistances in the emitter leads of each transistor are necessary to equalize the currents.

IC Voltage Regulators

The modern trend in regulators is toward the use of three-terminal devices commonly referred to as three-terminal regulators. Inside each regulator is a reference, a high-gain error amplifier, temperature-compensated voltage-sensing resistors and transistors and a pass element. Many currently available units have thermal shutdown, overvoltage protection and current foldback, making them virtually destruction-proof. Several supplies using these ICs are featured in Chapter 27.

Three-terminal regulators (a connection for unregulated dc input, regulated dc output and ground) are available in a wide range of voltage and current ratings. Fairchild, National and Motorola are perhaps the three largest suppliers of these regulators at present. It is easy to see why regulators of this sort are so popular when one considers the low price and the number of individual components they can replace. The regulators are available in several different package styles, depending on current ratings. Low-current (100 mA) devices frequently use the plastic TO-92 and DIP-style cases. TO-220 packages are popular in the 1.5-A range, and TO-3 cases house the larger 3-A and 5-A devices.

Three-terminal regulators are available as positive or negative types. In most cases, a positive regulator is used to regulate a positive voltage and a negative regulator a negative voltage. However, depending on the systems ground requirements, each regulator type may be used to regulate the "opposite" voltage.

Fig. 29A and B illustrate how the regulators are used in the conventional mode. Several regulators can be used with a common-input supply to deliver several voltages with a common ground. Negative regulators may be used in the same manner. If no other common supplies operate off the input supply to the regulator, the circuits of Figs. 29C and D may be used to regulate positive voltages with a negative

regulator and vice versa. In these configurations the input supply is floated; neither side of the input is tied to the system ground.

Manufacturers have adopted a system of family numbers to classify three-terminal regulators in terms of supply polarity, output current and regulated voltage. For example, National Semiconductor uses the number LM7805C to describe a positive 5-V, 1.5-A regulator; the comparable unit from Texas Instruments is a UA7805KC. LM7812C describes a 12-V regulator of similar characteristics. LM7905C denotes a negative 5-V, 1.5-A device. There are many such families with widely varied ratings available from manufacturers. Fixed-voltage regulators are available with output ratings in most common values between 5 and 28 V. Other families include devices that can be adjusted from 1.25 to 50 V.

Regulator Specifications

When choosing a three-terminal regulator for a given application, the most important specifications to consider are device output voltage, output current, input-to-output differential voltage, line regulation, load regulation and power dissipation. Output voltage and current requirements will be determined by the load the supply will ultimately be used with.

Input-to-output differential voltage is one of the most important three-terminal regulator specifications to consider when designing a supply. The differential value (the difference between the voltage applied to the input terminal and the voltage on the output terminal) must be within a specified range. The minimum differential value, usually about 2.5 V, is called the dropout voltage. If the differential value is less than the dropout voltage, no regulation will take place. At the other end of the scale, maximum input-output differential voltage is generally about 40 V. If this differential value is exceeded, maximum device dissipation may be reached.

Increases in either output current or differential voltage produce proportional increases in device power consumption. By employing a safety feature called current foldback, some manufacturers have ensured that maximum dissipation will never be exceeded in normal operation. Fig. 30 shows the relationship between output current, input-output differential and current limiting for a three-terminal regulator nominally rated for 1.5-A output current. Maximum output current is available with differential voltages ranging from about 2.5 V (dropout voltage) to 12 V. Above 12 V, the output current decreases, limiting the device dissipation to a safe value. If the output terminals are accidentally short circuited, the input-output differential will rise, causing current foldback, and thus preventing the power-supply components from being over-stressed. This protective feature makes three-terminal regulators

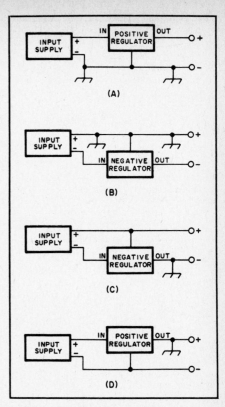

Fig. 29 — A and B illustrate the conventional manner in which three-terminal regulators are used. C and D show how one polarity regulator can be used to regulate the opposite polarity voltage.

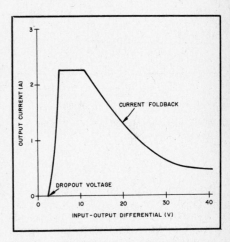

Fig. 30 — Effects of input-output differential voltage on three-terminal regulator output current.

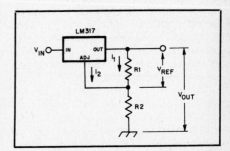

Fig. 31 — By varying the ratio of R2 to R1 in this simple LM317 schematic diagram, a wide range of output voltages is possible. See text for details.

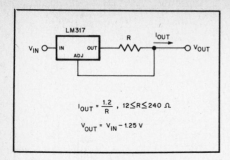

$$I_{OUT} = \frac{1.2}{R} \quad , \quad 12 \le R \le 240 \ \Omega$$

$$V_{OUT} = V_{IN} - 1.25 \ V$$

Fig. 32 — The basic LM317 voltage regulator is converted into a constant-current source by adding only one resistor.

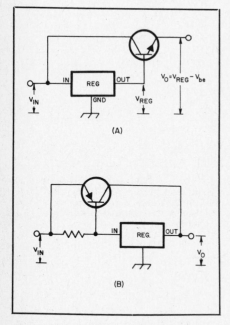

Fig. 33 — Two methods for boosting the current output capacity of an IC voltage regulator. (A) NPN emitter follower. (B) PNP "wraparound." Operation of these circuits is explained in the text.

particularly attractive in simple power supplies.

When designing a power supply around a particular three-terminal regulator, input-output voltage characteristics of the regulator should play a major role in selecting the transformer secondary and filter capacitor component values. The unregulated voltage applied to the input of the three-terminal device should be higher than the dropout voltage, yet low enough that the regulator does not go into current limiting caused by an excessive differential voltage. If, for example, the regulated output voltage of the device shown in Fig. 30 was 12, then unregulated input voltages of between 14.5 and 24 would be acceptable if maximum output current is desired.

In use, all but the lowest current regulators generally require an adequate external heat sink because they may be called on to dissipate a fair amount of power.

Also, because the regulator chip contains a high-gain error amplifier, bypassing of the input and output leads is essential for stable operation.

Most manufacturers recommend bypassing the input and output directly at the leads where they protrude through the heat sink. Tantalum capacitors are usually recommended because of their excellent bypass capabilities into the VHF range.

In addition to fixed-output-voltage ICs, high-current, adjustable voltage regulators are available. These ICs require little more than an external potentiometer for an adjustable output range from 5 to 24 V at up to 5 A. The unit price on these items is only a few dollars, making them ideal for testbench power supplies. A very popular low-current, adjustable-output-voltage three-terminal regulator, the LM317, is shown in Fig. 31. It develops a steady 1.25-V reference, V_{ref}, between the output and adjustment terminals. By installing R1 between these terminals, a constant current, I1, is developed, governed by the equation $I1 = V_{ref}/R1$. Both I1 and a 100-μA error current, I2, flow through R2, resulting in output voltage V_o. Vo can be calculated using the equation

$$V_o = V_{ref} \left(1 + \frac{R2}{R1} \right) + I_2 R_2$$

Any voltage between 1.2 and 37 volts may be obtained with a 40-volt input by changing the ratio of R2 to R1.

Fig. 32 shows one of many flexible applications for the LM317. By adding only one resistor, the voltage regulator can be changed into a constant current source, capable of charging NiCd batteries, for example. Design equations are given in the figure. The same precautions should be taken with adjustable regulators as with the fixed-voltage units. Proper heat sinking and lead bypassing are essential for proper circuit operation.

When the maximum current output from an IC voltage regulator is insufficient to operate the load, discrete power transistors may be connected to increase the current capability. Fig. 33 shows two methods for boosting the current output of a positive regulator, although the same techniques can be applied to negative regulators.

In A, an NPN transistor is connected as an emitter follower, multiplying the output current capacity by the transistor beta. The shortcoming of this approach is that the base-emitter junction is not inside the feedback loop. The result is that the output voltage is reduced by the base-emitter drop, and the load regulation is degraded by variations in this drop.

The circuit at B has a PNP transistor "wrapped around" the regulator. The regulator draws current through the base-emitter junction, causing the transistor to conduct. The IC output voltage is un-

changed by the transistor because the collector is connected directly to the IC output (sense point). Any increase in output voltage is detected by the IC regulator, which shuts off its internal pass transistor, and this stops the boost-transistor base current.

Most IC regulators are internally configured as linear series regulators. However, the manufacturer's applications literature shows how to use these devices in shunt regulators and even in switching regulators.

Thermal Design

Just as important as power-supply electrical design is power-supply thermal design. Linear power supplies are inefficient. Rectifiers, regulators and pass transistors all dissipate heat, and this heat must be safely carried away to avoid device damage. Although the black-magic rule of thumb "bigger is better" applies to heat sinks, this section will show you how to calculate heat-sink requirements and choose the right cooler for any power supply. This material was prepared by ARRL TA Dick Jansson, WD4FAB.

Creating the thermal design for a properly cooled power supply is a very well-defined process. To some amateurs experienced only in the electronic arts, heat transfer is a subject shrouded with mystery. The following information will remove these veils and show what is needed to make a good electronic design behave properly in the thermal world. An outline of the steps to be taken will show the logic to be applied:

1) Determine the expected power dissipation.
2) Identify the requirements for the dissipating elements.
3) Estimate heat-sinking requirements.
4) Rework the electronic device to meet the thermal requirements.
5) Select the heat exchanger.

To illustrate this discussion, a 28-V, 10-A power supply will be used as an example. Construction details of this supply may be found in Chapter 27. The steps described here were actually followed during the design of that supply.

The first step is to estimate the filtered, unregulated supply voltage under full load. A schematic diagram of this supply is shown in Fig. 34. The transformer has a secondary output of 32-V ac. Since this is the input to a full-wave bridge rectifier, the filtered dc output from C1 will follow those output characteristics shown in Fig. 7, presented earlier in this chapter. As a first guess, estimate 40 V as the filtered dc output at a 10-A load. The filter capacitor is 22,000 μF, so the output is reasonably well filtered. We do not know the value of the series resistance/load resistance, R_s/R, but if the power supply is going to be worth its salt, that value needs to be low! Bracketing the series resistance, R_s, would

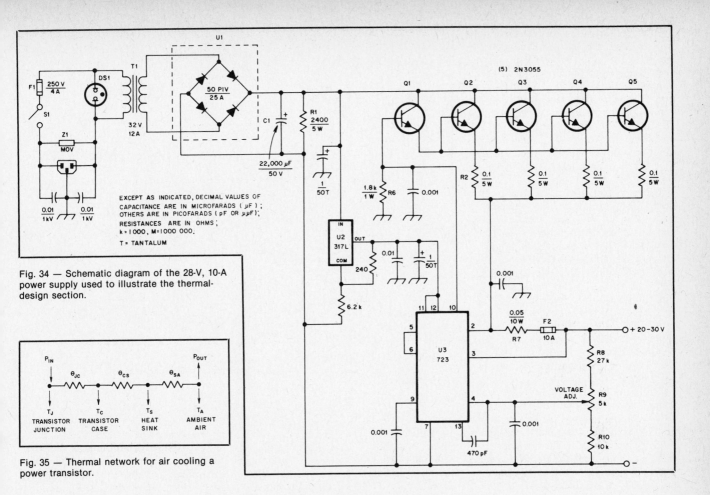

Fig. 34 — Schematic diagram of the 28-V, 10-A power supply used to illustrate the thermal-design section.

Fig. 35 — Thermal network for air cooling a power transistor.

place the dc output of the filter between 34 V and 44 V. The estimate of 40 V output still looks good, so continue to use that value as the filtered dc value that will feed the regulator stage at a 10-A load.

The next step is to estimate the total pass-transistor power dissipation under full load. Allowing for some small voltage drops in the power-transistor emitter circuitry, the output of the series pass transistors (Q2-Q5) is about 29 V for a delivered 28 V under a 10-A load. With an unregulated voltage of 40-V, the total pass-transistor dissipation, in the form of heat, is (40 V − 29 V) × 10 A, or 110 W. The heat sink for this power supply must be able to handle that amount of dissipation and still keep the transistor junctions below the specified safe operating temperature limits.

The next consideration is the ratings of the pass transistors to be used. This supply calls for four 2N3055 pass transistors, even though a data sheet for that transistor shows that each 2N3055 is rated for 15-A service and 115-W dissipation. Why so many 2N3055s? Here we need to get past the big, bold type at the top of the specification data sheet, and on to such subtle characteristics as the junction-to-case thermal resistance, θ_{jc}, and the maximum allowable junction temperature, T_j. Unfortunately, some of these characteristics are not contained in most elementary listings of transistor properties. Don't be

deluded by the specified maximum continous device dissipation, P_d, as that value is the computed maximum dissipation at a case temperature of 25 °C. It is not practical to try to keep the case of a power transistor at 25 °C when room temperature can easily exceed that.

The Greek letter θ is used to express values of thermal resistance, while R is used to express values of electrical resistance. Thermal resistance can be equated to electrical resistance, and thermal networks that look like electrical networks can be constructed. In fact, the basic equations are the same. Heat, measured in watts, flows in the network as does electrical current, measured in amperes. Temperature differences, in degrees Celsius, are equivalent to electrical potential in volts. Resistance to heat flow is expressed in °C/W, and is equivalent to resistance measured in ohms.

The concept of thermal resistance can be understood by reviewing basic heat-flow phenomena. A thermal network that describes the flow of heat from a transistor junction to the ambient air is shown in Fig. 35. The ambient air is considered here as the ultimate heat sink. θ_{jc} describes the thermal resistance from the transistor junction to its case. θ_{cs} is the resistance of the mounting interface between the transistor case and the heat sink. θ_{sa} is the resistance of the heat sink to the ambient air.

Checking the specification sheet for

the 2N3055 shows that it is rated for $\theta_{jc} = 1.52$ °C/W, and a maximum case (and junction) temperature of 220 °C. Industry experience has shown that silicon transistor reliability suffers substantially when junctions are operated at those elevated temperatures. Most commercial and military specifications will not permit design junction temperatures to exceed 125 °C, and in some cases 110 °C. A transistor operated above 125 °C does not "go away" instantly, but will degrade slowly because of excessive thermal migration in the silicon material.

At some point, you need to decide how the power supply is going to be used. Thermal requirements (as well as other design requirements) vary significantly, depending on expected duty cycle. Continuous commercial service (CCS) is exactly what it sounds like: "Brick-on-the-key" operation, such as FM, Baudot RTTY or ATV. Amateur equipment designed primarily for CW and SSB operation is often designed for Intermittent Commercial and Amateur Service (ICAS), nominally considered to be a duty cycle of 33% of the CCS. To provide a power supply capable of performing any chore within its power output capability, the designer of the example supply has chosen to design for CCS operation, even though many amateur operators may never stress the supply to that extent. This philosophy is conservative and safe.

Other considerations in how the supply will be used include the maximum air temperature that the unit will see in service and the maximum junction temperature to be allowed for the design. A lot of deliberation of these subjects was done, with the conclusion that an ambient air temperature (T_a) of 35 °C (95 °F), $T_j = 125$ °C for ICAS operation and $T_j = 150$ °C for CCS operation were safe assumptions. These assumptions allow design flexibilities, as will be shown later.

The electrical design for this supply originally called for two 2N3055 transistors in the regulator pass stage. This design was based on electrical characteristics only, and did not take thermal considerations into account. The equivalent thermal network for a two-transistor supply is shown in Fig. 36. Each of the 2N3055s would have to dissipate 55 W.

The total thermal resistance between the junction and the heat sink is equal to the junction-to-case thermal resistance (θ_{jc}), plus the thermal resistance associated with mounting of the transistor to the heat sink (θ_{cs}).

Proper mounting of most TO-3 power transistors requires that they have an electrical insulator between the transistor case and the heat sink. This electrical insulator must, however, provide for very good thermal contact between the two. To achieve a quality mounting, you should use thin polyimid or mica formed washers and a suitable thermal compound to exclude air from the interstitial space. "Thermal greases" are commonly available for this function. Silicone grease may be used, but

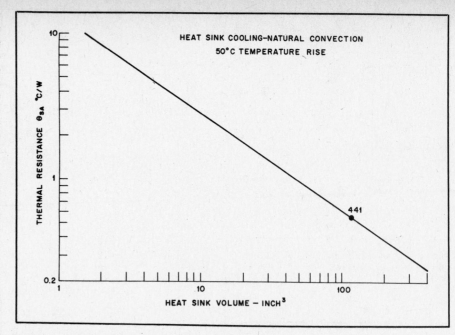

Fig. 38 — Graph of thermal resistance vs. heat-sink volume for natural-convection cooling of a heat sink with a 50 °C temperature rise. This graph is based on engineering data from EG&G Wakefield.

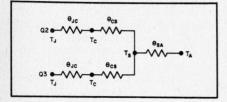

Fig. 36 — Thermal network for a power supply with two pass transistors.

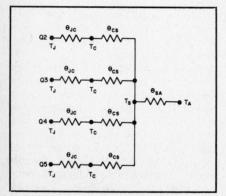

Fig. 37 — Thermal network for a power supply with four pass transistors.

filled silicone oils made specifically for this purpose are better.

A good starting value for mounting thermal resistance is $\theta_{cs} = 0.50$ °C/W. Lower values of θ_{cs} are possible, but the techniques needed to achieve them are expensive and not generally available to the average amateur. Furthermore, serious work on lowering θ_{cs} is not warranted in light of the somewhat higher value of θ_{jc}, which cannot be lowered without going to a somewhat more exotic pass transistor.

The total value of the junction-to-sink resistances in this example is $\theta_{js} = 2.02$ °C/W ($\theta_{jc} = 1.52 + \theta_{cs} = 0.5$). At a power dissipation (P_d) of 55 W, the transistor junction can be expected to operate 111 °C (55 W × 2.02 °C/W) above the heat-sink temperature (T_s). If the maximum value for T_j is 150 °C, then the heat-sink temperature could rise no higher than $T_s = 39$ °C. With a ambient air temperature (T_a) estimated at 35 °C (worst case), it would require an impossibly large heat sink to keep T_s within 4 °C of T_a to keep the two 2N3055s cool. The conclusion to be drawn from this analysis is that use of two 2N3055s is not feasible from a thermal standpoint, although from an electrical standpoint two transistors are more than adequate.

At this point, the power-supply design was modified to use four 2N3055 pass transistors. This change divides the dissipation load by four, making $P_d = 27.5$ W for each transistor. The equivalent thermal network for four devices is shown in Fig. 37. Going through the steps described earlier, the junction-to-sink temperature difference is now $T_j - T_s = 55.6$ °C. Initially, it was hoped to keep the target maximum junc-

tion temperature below 125 °C. At 27.5 W per transistor, that means a sink temperature of $T_s = 69$ °C. This temperature is possible to achieve with two large heat sinks (two devices per sink), or forced-air cooling of a single large heat sink.

Forced air cooling of a heat sink requires the use of a fan or blower to move the air. That means a "moving" part and thus a potential reduction in supply reliability. Also, blowers are noisy. If at all possible, the power-supply design should make use of natural-convection cooling and not rely on a blower. By reasoning that most operation would not be CCS, and that during occasional CCS operation junction temperatures of $T_j = 150$ °C are tolerable, a design is possible that offers heat-sink temperatures of $T_s = 94$ °C and a sink-to-air temperature difference $T_s - T_a = 59$ °C. It is possible to implement these parameters with a single heat sink of the "right" design. Please note that a heat-sink temperature of 94 °C is substantially hotter than the allowable "finger-touch thermometer" temperature of 52 °C. Do not attempt to use the finger thermometer with this design.

Now comes the big question: What is the "right" heat sink to use? We have established its requirements. It must be capable of dissipating 110 W, and it must allow a temperature rise no greater than 59 °C, or $\theta_{sa} = 0.54$ °C/W (59 °C divided by 110 W).

From Fig. 38, the relationship of heat-sink volume and thermal resistance for a natural-convection operation can be seen. This relationship presumes the use of suitably spaced fins (0.35 inch or greater)

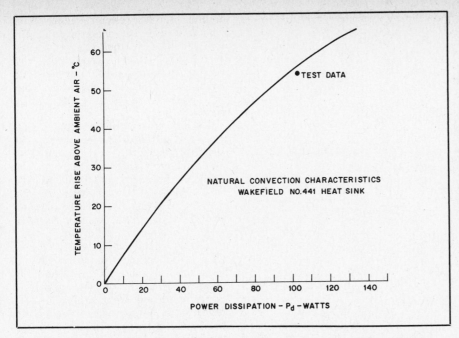

(A)

Fig. 39 — Graph of the natural convection characteristics under load of the EG&G Wakefield model no. 441 heat sink. This graph is based on actual testing performed by ARRL TA Dick Jansson, WD4FAB.

(B)

Fig. 40 — The EG&G Wakefield 441 heat sink is a "three-dimensional" design that makes excellent use of space. The photo at A shows the heat sink with four 2N3055 transistors in place. At B, the wiring to the transistors is visible.

and provides a "rough order-of-magnitude" value for sizing a heat sink. The needs of the supply used for this example could be met with a heat-sink volume of approximately 120 in³, which is within the range of available heat sinks, although it certainly is not small. Unlike a heat sink for use on an RF amplifier, this application can use a heat sink where only "wire access" to the transistor connections is available. A heat sink with a "three-dimensional shape" can be used, allowing the selection of a more efficient design. Most RF designs require the transistor mounting surface to be completely exposed so that the PC board can be mounted close to the transistors.

A quick consultation with several manufacturer's catalogs reveals that Wakefield Engineering Co. model nos. 441 and 435 heat sinks meet the needs of this application.[1] A Thermalloy model no. 6441 is suitable as well. Data published in the catalogs of these manufacturers show that in natural-convection service, the expected temperature rise for 100 W dissipation would be just under 60°C, an almost perfect fit for this application. See Fig. 39. Moreover, the no. 441 heat sink can easily mount four TO-3-style 2N3055 transistors, as shown in Fig. 40.

[1]There are numerous manufacturers of excellent heat sinks. References to any one manufacturer are not intended to exclude the use of products of any others, nor to indicate any particular predisposition to one manufacturer. The catalogs and products referred to here are from: EG&G Wakefield Engineering, 60 Audubon Rd., Wakefield, MA 01880; and Thermalloy, Inc., 2021 West Valley View La., Dallas, TX 75234.

Thermal power-loading tests of a no. 441 heat sink showed a very close agreement with the manufacturer's data — a 58°C temperature rise for 110-W power dissipation. It is not necessary to test heat sinks used for every application. The tests were done just to illustrate that the data of reliable suppliers of heat sinks for transistor cooling can be believed. Once you determine the heat-sink requirements for your application, you can make use of the manufacturer's data to select the right cooler. Remember: heat sinks should be mounted with the fins and transistor mounting area vertical to promote convection cooling.

Another point here is that very satisfactory results can be obtained through the good application of "handbook and catalog engineering." There is absolutely no need for you to conduct a lengthy, detailed computer thermal analysis of individual heat sinks. Such analysis is not needed if you use existing heat-sink designs.

Switching Regulators

Just as the transistor has replaced the inefficient vacuum tube, the linear voltage regulator is rapidly being replaced by the switching regulator. This transition is being assisted by the availability of components made especially for switching supplies, such as high-speed switching transistors and diodes, capacitors, transformer cores, and so on. The introduction of switching-regulator integrated circuits has also lowered the complexity of switchers to a degree that rivals linear regulators, and new circuit design techniques have reduced regulator noise and ripple to acceptable levels for communications equipment.

The switching regulator (also known as a switching-mode regulator, or SMR) offers several advantages over the linear type:

1) Much higher efficiency resulting from minimal regulator dissipation. Very desirable for battery-powered applications.

2) Lighter in weight as a result of less heat sink material.

3) Smaller in size.

These advantages are offset, however, by slower response to load transients, increased circuit complexity and higher cost.

The simplest type of SMR somewhat resembles a linear regulator in form and is known as the voltage-bucking (buck) con-

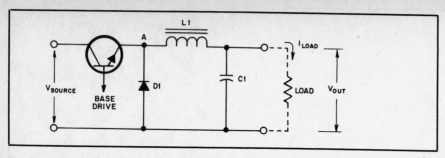

Fig. 41 — Schematic diagram of a simple switching regulator.

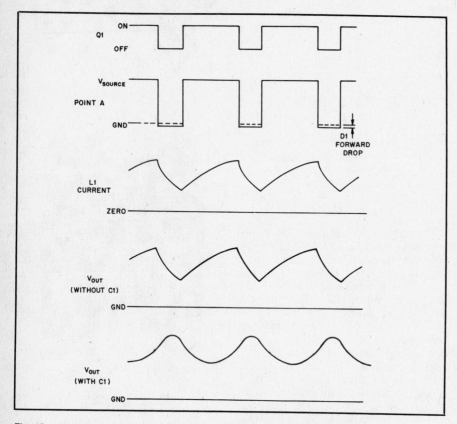

Fig. 42 — Waveforms seen at various points in the switching regulator shown in Fig. 41.

figuration. The circuit reduces (or "bucks") the input voltage to produce a lower value of output voltage. By re-arrangement of circuit components, the reverse can be made to occur; the output voltage can actually be higher than the input voltage. This is known as a boost configuration.

The basic elements of the buck regulator are shown in Fig. 41. They are switch transistor Q1, inductor L1, capacitor C1 and diode D1. A positive source voltage is applied to the collector of Q1, and the load is connected as shown. Some control circuitry for applying base drive to Q1 is also needed; for the present, assume that we can turn Q1 fully on and off at will.

Initially, ignore the presence of C1. As Q1 is turned on, all of the source voltage instantaneously appears across L1 since the initial current in L1 is zero. With time, the current in L1 begins to increase and V_{OUT} begins to rise; V_{OUT} is allowed to continue to rise until it reaches the desired output voltage. When V_{OUT} exceeds the desired voltage, switch Q1 is turned off and point A in Fig. 41 immediately goes negative in an effort to maintain the current flowing in L1, thereby causing D1 to conduct. As the energy which has become stored in L1 dissipates into the load through D1, the current in L1 decreases, as does the output voltage. When V_{OUT} drops to slightly less than the desired value, Q1 is turned on again, bringing point A to the source voltage level, and the cycle repeats.

Obviously, in this circuit, Q1 must switch on and off at an extremely fast rate to minimize the swing in output voltage around the desired value, since this voltage variation is directly related to the on and off times of Q1. This situation can be improved by adding C1, which supports the output voltage while the current decreases in L1, and reduces overshoot by absorbing the excess current in L1 while Q1 is on. See

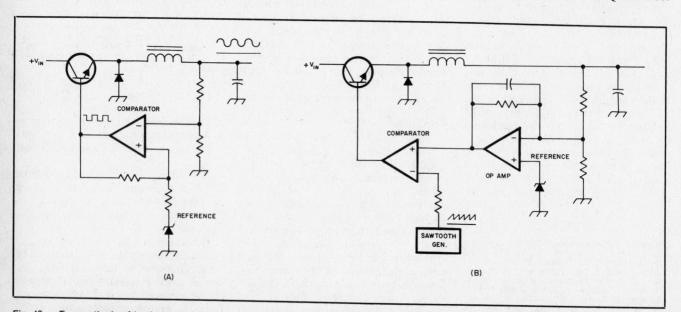

Fig. 43 — Two methods of implementing a pulse-width modulator. The circuit at A uses the ripple modulation technique, while the circuit at B employs fixed-frequency (duty-cycle) modulation.

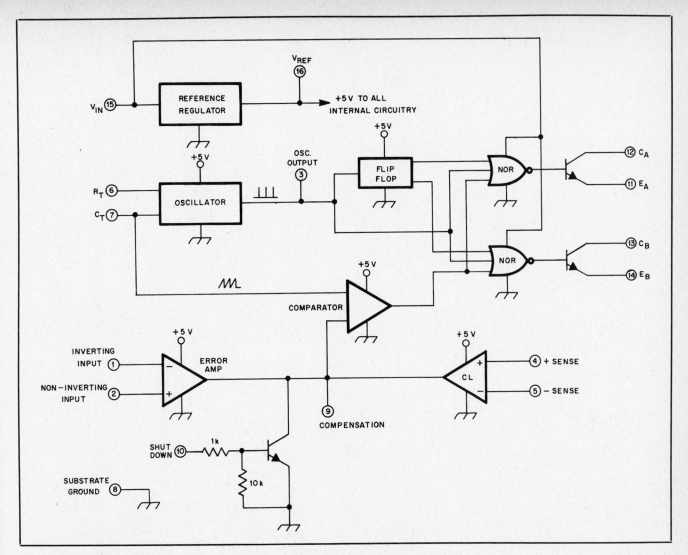

Fig. 44 — Block diagram of the Silicon General SG3524 regulating pulse-width modulator IC. This drawing is based on information presented in the *Silicon General Product Catalog.*

the waveforms in Fig. 42. The transistor can then be switched at a slower speed, and the inductance of L1 reduced because of the smoothing effect of C1.

In this example, power is only transmitted from the source when Q1 is on, but this power is distributed to the load continuously during the entire cycle. The power delivered to the load is $V_{OUT} \times I_{LOAD}$, which must equal the power obtained from the source while Q1 conducts. Since Q1 conducts for only a portion of the cycle at an average current during conduction of I_{LOAD}, the output voltage is then equal to the duty cycle $\times V_{SOURCE}$ or

$$V_{OUT} = \frac{t_{ON}}{t_{CYCLE}} \times V_{SOURCE}$$

In actual practice, the losses in D1, forward drop in Q1, resistance in L1, and various other factors combine to affect the above formula, but except for very-low-output-voltage, high-current regulators, these losses can be neglected since the control circuits will automatically compensate for them.

The control circuit that must be added to the basic circuit in Fig. 41 has to be capable of turning the switch transistor on and off quickly and providing the proper ratio of on-time to off-time to maintain regulation. This portion of a switching regulator is referred to as a pulse-width modulator (PWM).

The simplest PWM is a comparator that has the dc output of the regulator applied through a divider to one input and a reference voltage applied to the alternate input. See Fig. 43A. A small amount of the comparator's output voltage is fed back to its non-inverting input to provide hysteresis, thereby ensuring that the comparator's output is either high or low. With the ripple-modulated PWM, there must be sufficient output ripple to overcome the hysteresis, which is why the term "ripple modulation" is used.

By forcing the regulator to switch at a fixed frequency that is much higher than the ripple modulator would produce, the output LC filter can become much more efficient. This can be done by injecting a

fixed-frequency sawtooth waveform to the comparator, as shown in Fig. 43B. The output voltage is now compared to the reference at the input to an op amp. Op-amp response is slowed by a capacitor between its output and inverting terminals so that it will not respond to switching ripple. As the output of the op amp rises and falls with output voltage variations, the comparator "chops off" more or less of the sawtooth that varies the duty cycle of the drive to the switching transistor's base.

Integrated circuits are now available that provide the op amp, comparator, reference and drive circuits all in one package. A block diagram of one of these ICs, a Silicon General SG3524, is shown in Fig. 44.

Dc-to-Dc Conversion

Dc-to-dc conversion is probably the oldest form of switching technology in the power-supply field, dating back to the vibrator supplies used in automobile radios and low-power surplus gear. The noisy and unreliable mechanical vibrators were eventually replaced by transistors or SCRs.

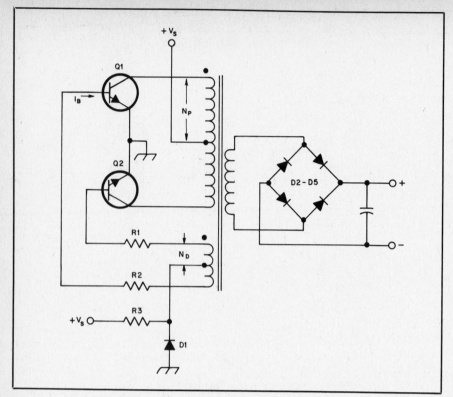

Fig. 45 — Schematic diagram of a basic dc-to-dc converter.

Power is delivered to the load during each half-cycle. Neglecting diode drops and transistor saturation voltages, the output voltage is

$$V_{OUT} = \frac{N_S}{N_P} \times V_S$$

This type of self-excited, single-transformer converter design is known as a Royer Oscillator. The frequency of oscillation is dependent upon the time required to saturate the core with the applied voltage and is given by the formula

$$F_{(Hz)} = \frac{V_S \times 10^8}{4 \times N_P \times B_M \times A_C}$$

where

V_S = voltage applied at the primary center tap

N_P = no. of turns from primary center tap to one end

B_M = saturation flux density of the core in gauss

A_C = cross-sectional area of the core in cm^2

The best cores for the Royer oscillator are tape-wound toroids using orthonol or square permalloy-type alloys. Some ferrite cores are also specifically made for this application.

Since the tape-wound toroids used for self-excited dc-dc converters become very lossy at frequencies above the audible range, the self-excited configuration is rarely used above about 15 to 20 kHz. For higher frequency use, an electronically driven type is usually employed. This consists of an electronic square-wave oscillator operating at the desired frequency to drive the bases of the switching transistors. The output transformer is made with a core material that does not have sharp saturation characteristics and is designed to be operated well within the maximum flux density rating to avoid saturation. High-flux-density ferrite is usually chosen for the core.

Push-Pull Converter/Regulator

By reducing the width of the drive pulses applied to the transistor to less than 50% duty cycle, the push-pull converter just described can be made to regulate. Integrated circuits that contain most of the components necessary for the construction of a push-pull regulator are now available from several manufacturers (Silicon General, Unitrode, Texas Instruments). The popular Silicon General type SG3524 is one example.

These integrated circuits work on the basis of a constant switching frequency with variable duty cycle, and can also be used for making a much improved version of the single transistor buck regulator described earlier.

One of the big advantages of the regulating converter is the transformer isolation that it provides between the

The dc-to-dc converter works on the principle of chopping the dc power source to produce an ac component and then using a transformer to obtain a different voltage level. The new voltage can then be rectified to produce a dc output voltage higher or lower than the source. In the case of the old vibrator supplies, the 6-volt or 12-volt car battery voltage was chopped, transformed up to several hundred volts ac, and rectified to provide plate voltage for the equipment. Another example, still used today, eliminated the output rectifiers and used a transformer wound to produce 120-V ac to power ac equipment. This type of circuit is often referred to as a dc-to-ac inverter, although nothing is really "inverted."

The operation of a dc-dc converter is as follows. Refer to Fig. 45. Q1 and Q2 are initially biased on by R3. One transistor (assume Q1), draws slightly more current than its mate (Q2). Because of the autotransformer effect of the transformer primary, the collector of Q1 goes toward ground, while the collector of Q2 goes more positive. This primary voltage swing is reflected into the base drive winding, which applies increased base drive to the heavily-conducting transistor (Q1) while removing bias from Q2. When Q1 is saturated, the voltage across the drive winding is equal to the ratio of drive winding turns to primary turns, multiplied by the source voltage.

$$V_{DRIVE\ WINDING \atop (END\ TO\ CENTER\ TAP)} = \frac{N_D}{N_P} \times V_S$$

Since the base of Q1 does not permit the

forward-biased base voltage to be more than about 1 volt, the center tap of the drive winding is pushed negative until D1 conducts, again at less than 1 volt (negative). The remainder of the winding voltage across N_D is dropped across R1, and it is the current resulting from that drop across R1 that determines the drive current

$$I_B = \frac{V_{ND} - V_{BE}\ (Q1) - V_{FD}\ (D1)}{R1}$$

Usually V_{BE} and V_{FD} are about 0.8 volts, therefore

$$I_B \approx \frac{V_{ND} - 1.6}{R1}$$

Also, V_{ND} is usually designed to be 5 volts or less in order to prevent reverse conduction of the base-emitter junction of Q2.

Q1 remains in conduction, producing output power, until the core of the transformer saturates. At that point, it is incapable of transferring power, and the base drive to Q1 reduces. The collector current rises abruptly, pulling Q1 out of saturation, since the saturated transformer's primary looks like a short circuit to Q1.

As the base drive winding swings through zero volts, the base of Q2 is turned on and collector current in Q2 begins to flow. This helps Q1 to turn off and swings the excitation in the transformer primary in the opposite direction. Q2 saturates, with Q1 off, and remains in saturation until the transformer core once again saturates, initiating another cycle.

primary and secondary circuits. Since the transformer is operating at high frequency, its size and weight can be much less than a 60-Hz transformer handling the same power. To take advantage of this, the 117-V ac line is bridge rectified and capacitive-input filtered to produce about 170-V dc. This voltage is then applied to a regulating converter to produce the desired regulated and isolated dc output voltage. This application is usually referred to as off-line conversion.

Battery Power

The availability of solid-state equipment makes practical the use of battery power under portable or emergency conditions. Hand-held transceivers and instruments are obvious applications, but even fairly powerful transceivers (100 W or so output) may be practical users of battery power. Solid-state kilowatt mobile amplifiers exist, but these are intended for operation from an auxiliary battery that is constantly charged. The lower-power equipment can be powered from two types of batteries. The primary battery is intended for one-time use; the storage (or secondary) battery may be recharged many times.

A battery is a group of chemical cells, usually series-connected to give some desired multiple of the cell voltage. Each assortment of chemicals used in the cell gives a particular nominal voltage; this must be taken into account to make up a particular battery voltage.

Primary Batteries

The most common primary cell is the carbon-zinc flashlight type, in which chemical oxidation converts the zinc into salts and electricity. When there is no current flow, the oxidation stops until the next time current is required. Some chemical action does continue, so stored batteries eventually will degrade or dry out to the point where the battery will no longer supply the desired current. The time taken for degradation without battery use is called shelf life.

The carbon-zinc battery has a nominal voltage of 1.5 volts, as does its heavy-duty or industrial brother. These latter types are capable (for a given size) of producing more milliampere hours and less voltage drop than a carbon-zinc battery of the same size. They also have longer shelf life. Alkaline primary batteries have even better characteristics and will retain more capacity at low temperatures. Nominal voltage is 1.5 volts.

Lithium primary batteries have a nominal voltage of about 3 volts per cell and by far the best capacity, discharge, shelf-life and temperature characteristics. Their disadvantages are high cost and the fact that they cannot be readily replaced by other types in an emergency.

Silver oxide (1.5 V) and mercury (1.4 V) batteries are very good where nearly constant voltage is desired at low currents for long periods. Their main use (in subminiature versions) is in hearing aids, though they may be found in other mass-produced devices such as household smoke alarms.

Rechargeable or Storage Batteries

Many of the chemical reactions in primary batteries are theoretically reversible if current is passed through the battery in the reverse direction. For instance, zinc may be plated back onto the negative electrode of a zinc-carbon battery. Recharging of primary batteries should not be done for two reasons: It may be dangerous because of heat generated within sealed cells, and even in cases where there may be some success, both the charge and life are limited. In the zinc-carbon example, the zinc may not replate in the locations that had been oxidized. Pinholes in the case result, with consequent fluid leakage that will damage the equipment powered by the batteries.

One type of alkaline battery is rechargeable, and is so marked. If the recommended charging rate is not marked on such a battery, the manufacturer's advice should be asked. In a number of cases, the manufacturer markets chargers and recommends that only those should be used.

The most common small-storage battery is the nickel-cadmium (NiCd) type, with a nominal voltage of 1.2 V per cell. Carefully used, these are capable of 500 or more charge and discharge cycles, compared to 50 or so for alkaline types. The NiCd battery must not be fully discharged for best life. Where there is more than one cell in the battery, the most-discharged cell may suffer polarity reversal, resulting in a short circuit, or seal rupture. All storage batteries have discharge limits, and nickel-cadmium types should not be discharged to less than 1.0 V per cell.

The most widely used storage battery is the lead-acid type. In automotive service, the battery is usually expected to discharge partially at a very high rate, and then to be recharged promptly while the alternator is also carrying the electrical load. If the conventional auto battery is allowed to discharge fully from its nominal 2 V per cell to 1.75 V per cell, only about 50 cycles of charge and discharge may be expected, with reduced storage capacity.

The most attractive battery for extended high-power electronic applications is the so-called "deep-cycle" battery (intended for such use as powering electrical fishing motors and the accessories in recreational vehicles). The size 24 and 27 batteries furnish a nominal 12 volts and are about the size of small and medium automotive batteries. These batteries may furnish between 1000 and 1200 watt-hours per charge at room temperature. When properly cared for, they may be expected to last more than 200 cycles. They often have lifting handles and screw terminals, as well as the conventional truncated-cone automotive terminals. They may also be fitted with accessories such as plastic carrying cases, with or without built-in chargers.

Lead-acid batteries are also available with gelled electrolyte. Commonly called gel cells, these types may be mounted in any position if sealed, but some vented types are position sensitive.

Lead-acid batteries with liquid electrolyte usually fall into one of three classes — conventional with filling holes and vents, permitting the addition of distilled water lost from evaporation or during high-rate charge or discharge; maintenance-free, from which gas may escape but water cannot be added; and sealed. Generally, the deep-cycle batteries have filling holes and vents.

Battery Capacity

The common rating of battery capacity is ampere hours (Ah), the product of current drain and time. The symbol "C" is commonly used; C/10, for example, would be the current available for 10 hours con-

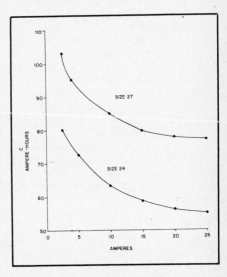

Fig. 46 — Output capacity as a function of discharge rate for two sizes of lead-acid batteries.

tinuously. The value of C changes with the discharge rate and might be 110 at 2 amperes but only 80 at 20 amperes. Fig. 46 gives capacity-to-discharge rate for two standard-size lead-acid batteries. Capacity may vary from 35 mA hours for some of the small hearing-aid batteries to more than 100 ampere hours for a size 28 deep-cycle storage battery.

The primary cells, being sealed, usually benefit from intermittent (rather than continuous) use. The resting period allows completion of chemical reactions needed to dispose of by-products of the discharge.

All batteries will fall in output voltage as discharge proceeds. "Discharged" condition for a 12-V lead-acid battery, for instance, should not be less than 10.5 volts. It is also good to keep a running record of hydrometer readings, but the conventional readings of 1.265 charged and 1.100 discharged apply only to a long, low-rate discharge. Heavy loads may discharge the battery with little reduction in the hydrometer reading.

Batteries that become cold have less of their charge available, and some attempt to keep a battery warm before use is worthwhile. A battery may lose 70% or more of its capacity at cold extremes, but it will recover with warmth. All batteries have some tendency to freeze, but those with full charges are less susceptible. A full-charged lead-acid battery is safe to $-30\,°F$ ($-26\,°C$) or colder. Storage batteries may be warmed somewhat by charging. Blowtorches or other flame should never be used to heat any type of battery.

A practical limit of discharge occurs when the load will no longer operate satisfactorily on the lower output voltage near the "discharged" point. Much gear intended for "mobile" use may be designed for an average of 13.6 V and a peak of perhaps 15 V, but will not operate well below 12 V. For full use of battery charge, the gear should operate well (if not at full power) on as little as 10.5 V with a nominal 12 to 13.6-V rating.

Somewhat the same condition may be seen in the replacement of carbon-zinc cells by nickel-cadmium storage cells. Eight of the former will give 12 V, while 10 of the same size nickel-cadmium units are required for the same voltage. If a 10-cell battery holder is used, the equipment should be designed for 15 V in case the carbon-zinc units are plugged in.

Deep-cycle and nickel-cadmium storage batteries will tolerate a light continual charge (trickle). The sealed nickel-cadmium types tolerate a near-full charging rate continuously.

Discharge Planning

Transceivers usually drain a battery at two or three rates: one for receiving, one for transmit standby and one for key-down or average voice transmit. Considering just the first and last of these (assuming the transmit standby equal to receive), average

two-way CW communication would require the low rate ¾ of the time and the high rate ¼ of the time. The ratio may vary somewhat with voice. The user may calculate the percentage of battery charge used in an hour by the combination (sum) of rates. If, for example, 20% of the battery capacity is used, the battery will provide five hours of communications per charge. In most actual traffic and DX-chasing situations the time spent listening should be much greater than that spent transmitting.

Caring for Storage Batteries

In addition to the precautions given above, the following are recommended. (Your manufacturer's advice will probably be more applicable.)

Gas escaping from storage batteries may be explosive. Keep flame away.

Dry-charged storage batteries should be given electrolyte and allowed to soak for at least half an hour. They then should be charged at perhaps a 15 A rate for 15 minutes or so. The capacity of the battery will build up slightly for the first few cycles of charge and discharge, and then have fairly constant capacity for many cycles. Slow capacity decrease may then be noticed.

No battery should be subjected to unnecessary heat, vibration or physical shock. The battery should be kept clean. Frequent inspection for leaks is a good idea. Leaking or spraying electrolyte should be cleaned from the battery and surroundings. The electrolyte is chemically active and electrically conductive, and may ruin electrical equipment. Acid may be neutralized with sodium bicarbonate (baking soda), and alkalies may be neutralized with a weak acid such as vinegar. Both neutralizers will dissolve in water, and should be quickly washed off. Do not let any of the neutralizer enter the battery.

Keep a record of the battery usage, and include the last output voltage and (for lead-acid storage batteries) the hydrometer reading. This allows prediction of useful charge remaining, and the recharging or procuring of extra batteries, thus minimizing failure of battery power during an excursion or emergency.

Charging Rechargeable Batteries

The rated full charge of a battery, C, is expressed in ampere-hours. No battery is perfect, so more charge than this must be offered to the battery for a full-charge. If, for instance, the charge rate is 0.1 C (the 10-hour rate), 12 or more hours may be needed for the charge.

Basically nickel-cadmium batteries differ from the lead-acid types in the methods of charging. It is important to note these differences, since improper charging can drastically shorten the life of a battery.

Nickel-cadmium cells have a flat voltage-vs.-charge characteristic until full charge is reached; at this point the charge voltage rises abruptly. With further charging, the

electrolyte begins to break down and oxygen gas is generated at the positive (nickel) electrode and hydrogen at the negative (cadmium) electrode.

Since the cell should be made capable of accepting an overcharge, battery manufacturers typically prevent the generation of hydrogen by increasing the capacity of the cadmium electrode. This allows the oxygen formed at the positive electrode to reach the metallic cadmium of the negative electrode and reoxidize it. During overcharge, therefore, the cell is in equilibrium. The positive electrode is fully charged and the negative electrode less than fully charged, so oxygen evolution and recombination "wastes" the charging power being supplied.

In order to ensure that all cells in a NiCd battery reach a fully charged condition, NiCd batteries should be charged by a constant current at about a 0.1 C current level. This level is about 50 mA for the AA-size cells used in most hand-held radios. This is the optimum rate for most NiCds since 0.1 C is high enough to provide a full charge, yet it is low enough to prevent overcharge damage and provide good charge efficiency.

Although fast-charge-rate (3 to 5 hours typically) chargers are available for hand-held transceivers, they should be used with care. The current delivered by these units is capable of causing the generation of large quantities of oxygen in a fully-charged cell. If the generation rate is greater than the oxygen recombination rate, pressure will build in the cell, forcing the vent to open and the oxygen to escape. This can eventually cause drying of the electrolyte, and then cell failure. The cell temperature can also rise, which can shorten cell life. To prevent overcharge from occurring, fast-rate chargers should have automatic charge-limiting circuitry which will switch or taper the changing current to a safe rate as the battery reaches a fully-charged state.

By contrast, the gelled-electrolyte lead-acid battery cannot tolerate overcharging, even at very low charge rates. For this reason, constant current or trickle charging should not be used for charging gel cells unless the battery voltage is monitored and charging is terminated when full charge is reached. Voltage-limited charging is best for these batteries. The proper charger would charge the battery at a safe current (C/10 or C/20) until a per-cell voltage of about 2.3 volts is reached (13.8 V for a 12-V battery). At this point, the charge current will taper to only enough to maintain the full-charge voltage. This makes the gel cell the optimum choice for "floating" across a regulated 13.8 volt power system to provide battery backup in the event of power failure.

Deep-cycle lead-acid cells are best charged at a slow rate, while automotive and some nickel-cadmium types may safely be given quick charges. This depends on the amount of heat generated within each

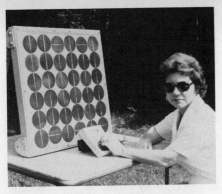

Fig. 47 — Solar-electric arrays are excellent for short- or long-term field and emergency use to power amateur stations. A 14-volt, 1.5-A solar panel and two automobile batteries in parallel can provide many after-dark hours of operation with typical 100-watt HF-band transceivers of the solid-state variety.

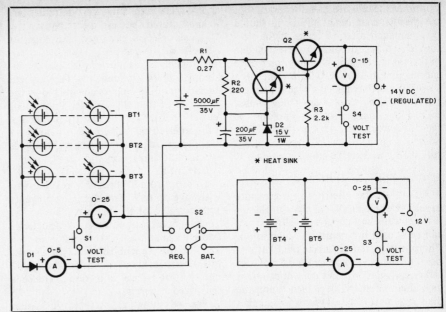

cell, and cell venting to prevent pressure build-up. Some batteries have built-in temperature sensing, used to stop or reduce charging before the heat rise becomes a danger. Quick and fast charges do not usually allow gas recombination, so some of the battery water will escape in the form of gas. If the water level falls below a certain point, acid hydrometer readings are no longer reliable. If the water level falls to plate level, permanent battery damage may result.

Overcharging in moderation causes little loss of battery life. However, continuous overcharge may generate a voltage depres-

Fig. 48 — Schematic diagram of a solar power supply. Note that the battery-charging circuit does not employ a regulator or switch to shut off charging current once the storage battery reaches full charge state. Because the output of the solar panels is, at most, 1½ A and the storage batteries are full-size automobile batteries, the danger of damage from overcharging is not great. Anyone contemplating higher current solar batteries or smaller storage batteries should give serious consideration to a regulator and/or an automatic cutoff switch for the charging circuit. (See Fig. 49.)

BT1, BT2, BT3 — 20-V, ½-A solar panels by Spectrolab.
BT4, BT5 — 12-V lead-acid automobile batteries.
D1 — 1N5401 or any diode with at least 2-A capacity and with at least 50 PIV.
Q1 — NPN silicon 90-W transistor, power switching, TIP31, Radio Shack 276-2020 or equiv.

Q2 — NPN silicon 115 W transistor, power switching, 2N3055 or equiv.
R1 — 0.27 Ω, 1 W.
R2 — 220 Ω, 1 W.
R3 — 2.2 Ω, 1 W.
S1, S2, S3 — SPST, momentary-contact switch.
S2 — DPDT knife switch.

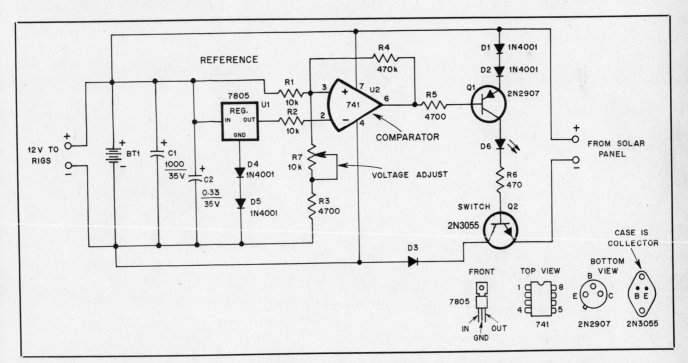

Fig. 49 — Schematic diagram of the electronic switch. Resistances are in ohms; k = 1000; capacitance values are in microfarads (µF).

BT1 — Automotive storage battery, lead-acid type.
C1 — 1000-µF, 35-V electrolytic.
C2 — 0.33-µF, 35-V.
D3 — Silicon diode, PIV of 50 or more, current

rating sufficient to pass full output of the solar panel.
D6 — Light-emitting diode, any type.
Q2 — Low-frequency power transistor; 2N3055, or equiv. Use heat sink of 9 in²

or more.
R7 — 10 kΩ, ½ W, carbon control, linear taper, PC mount.
U1 — 3-terminal, 5-volt regulator.
U2 — Op amp, any of the 741 family usable.

sion when the cells are later discharged. For best results, charging of NiCd cells should be terminated after 15 hours at the slow rate. Better yet, circuitry may be included in the charger to stop charging, or reduce the current to about 0.02 C when the 1.43-V-per-cell terminal voltage is reached. For lead-acid batteries, a timer may be used to run the charger to make up for the recorded discharge, plus perhaps 20%. Some chargers will switch over automatically to an acceptable standby charge.

SOLAR-ELECTRIC POWER

Although solar-electric arrays are quite expensive when purchased new, surplus individual cells and groups of cells (arrays) can be bought inexpensively on occasion. Photons from the sun strike the PN junctions of the cells to generate 0.5 volt per cell (see Chapter 4). The current rating of an individual cell is dependent upon the diameter of the cell. Typical production units deliver 100 mA, 600 mA, 1 A or 1.5 A. Cells with higher current ratings are manufactured, but are quite costly.

A solar-electric panel generally contains 36 cells wired in series. This provides approximately 18 volts dc under no-load conditions at peak sunlight. The current capability of the panel is determined by the diameter of the cells. Greater amounts of current output can be had by paralleling like panels. That is, two 1.5-A panels can be operated in parallel to deliver 3 amperes of current, and so on.

The usual operating system has the array output routed through a regulator to a storage battery. The regulator prevents overcharging of the battery. The station equipment takes its power from the battery. Most automotive 12-volt batteries are suitable for use with solar-electric panels. NiCd batteries are also satisfactory. Fig. 47 shows a solar array in a frame. The cells are wired in series.

Fig. 48 shows a solar-electric system suitable for low- or high-power operation. If the current drain is less than the capacity of the solar bank (1.5 A in this case), the load can be powered from the solar cells through the regulator circuit. For heavier loads, the current is taken from the storage batteries, which are charged by the solar array. The circuit of Fig. 48 was designed by John Akiyama, W6PQZ, and was described by John Halliday, W5PIZ in August 1980 QST. In the same issue, Doug Blakeslee, N1RM, described an electronic switch to automatically disconnect storage batteries from a solar system when full charge (13.5 V) is reached. The circuit is shown in Fig. 49. U1, D4 and D5 establish a 6.2-V reference for comparator U2. A voltage divider composed of R1, R3 and R7 scales the battery voltage down to the reference value, while R4 provides hysteresis to prevent oscillation. When the battery potential exceeds the comparator threshold, U2 goes high, turning off Q1 and Q2. The LED, D6, indicates that the battery is being charged.

Chapter 7

Audio and Video

Of the five senses we humans possess — sight, smell, taste, hearing and feeling — three are used for communication. We communicate visually, through actions and through body language such as facial expressions. We communicate aurally, through speech. On occasion we may communicate through feeling by touching; a tap on the shoulder, for example. Aural and visual communications can be transmitted electronically, and are therefore pertinent to Amateur Radio. This chapter deals with the fundamentals involved.

SPEECH

Human speech consists predominantly of two types of sounds — voiced and unvoiced. Voiced sounds originate when air passes from the speaker's lungs through the larynx (voice box), a passage in the human throat with the opening obstructed by vocal cords. As air travels past these cords they vibrate, causing puffs of air to escape into the aural cavity, which consists of the throat, nasal cavity and mouth. Studies indicate that the acoustic waveform produced by the vocal cords has many harmonics of the fundamental vibration. Because of the irregular shape of the aural cavity, the spectral-amplitude distribution of the harmonics tends to show peaks at distinct points. As speech is produced, changes occur in the aural cavity shape, thus changing the spectral location of these peaks.

Fig. 1 shows a spectrogram or voice print of the utterance, "digital communications." The darkness of the bands indicates amplitude or voice strength. The fine structure of amplitude peaks very close together in the horizontal dimension is a measurement of vocal-cord vibration frequency or fundamental.

Notice the rather strong amplitude con-centrations below 4000 Hz. These are the spectral peaks referred to above. They are called *formants*. The first three formants are shown in Fig. 1 at the beginning of the utterance. Fig. 1 also indicates a range of frequencies between the first and second formants, from approximately 600 to 1500 Hz, where little energy is present. This is identified in the figure as a spectral gap.

Unvoiced sounds in speech occur when there is no vocal track excitation. They are caused by the speaker using his or her tongue, lips and teeth to cause clicks, hisses and popping sounds. These sounds, or evidence of their occurrence by formant extensions into or from an unvoiced sound, are very important in the intelligibility of speech. The spectral amplitude distribution of unvoiced sounds is generally above 1500 Hz and is "noise-like" in that very little periodic structure is present.

One other important aspect of speech is

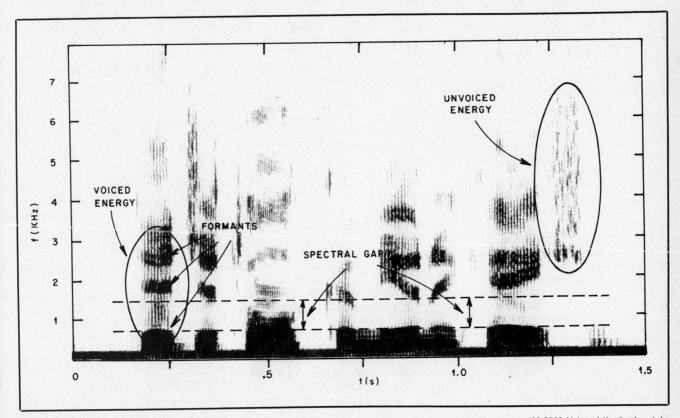

Fig. 1 — A speech spectrogram of the utterance, "digital communications." The vertical axis represents frequency (80-8000 Hz) and the horizontal axis represents time (0-1.5 seconds).

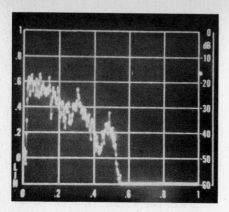

Fig. 2 — Long-time audio spectrum of speech. The horizontal scale is 1000 Hz per division or 0 to 5000 Hz. The vertical scale is 10 dB per division.

the pause between acoustic sounds. Juncture pauses carry meaning and cannot be eliminated without impairing intelligibility. Some long pauses can, however, be shortened and thereby reduce message length. Briefly, then, speech is the continuous production of voiced and unvoiced sounds with appropriate pauses to add clarity and distinctness.

Frequency and Amplitude Considerations

Measurements of the voices of different persons indicate that the first three formants, identified in Fig. 1, lie predominantly below 2500 Hz. Speech consisting of these three formants is of good quality both from an intelligibility and listenability standpoint. Sufficient information as to the existence of some unvoiced sounds appears to lie in this range also. For example, to produce s or hissing sounds the frequency range must extend to about 4000 Hz. This is not usually required for intelligibility, however, since contextual clues provide sufficient evidence for the listener to "hear" an s.

Fig. 2 shows a long-time amplitude spectrum of speech. Note the presence of the first three formants, appearing as peaks in the pattern. Note also that speech components above 2500 Hz (mid-scale on the horizontal axis) are 20 to 25 dB lower than those below 2500 Hz. From a frequency point of view, evidence from theory and practice indicates a bandwidth of 300 to 2500 Hz is adequate for good-quality speech. This is a bandwidth of 2.2 kHz.

From an amplitude standpoint, tests conducted with single coherent tone interference indicate that a dynamic range of 40 dB is quite adequate for good-quality speech. This is reinforced by the information displayed in Fig. 2. Many communications channels allow only 10 to 30 dB of signal-to-noise ratio, so equipment designed to preserve more than 40 dB of dynamic range is not warranted.

SOUND WAVES

Sound waves travel through the air at the

speed of approximately 1190 feet per second. They propagate as pressure waves, or alternating condensations and rarefactions of the air. The amplitudes of the pressure changes that make up sound waves from human speech are quite small. Sound power is measured in microjoules, a metric unit. One microjoule equals 10 ergs. When a sentence of a few words is spoken by a typical voice, the potential energy in the resultant sound wave is approximately 30 or 40 microjoules. Compare this to the 10^8 microjoules per second required to power a 100-watt lamp. It has been calculated that it would take 500 people talking continuously for a year to produce enough energy to heat a cup of tea. Because of the extremely low level of human speech, sensitive microphones and speech amplifiers are required in equipment intended for radio communications.

Sound intensity levels are ordinarily expressed in decibels. The reference level in power units for these decibel figures, established by the American Standards Association, is 10^{-16} watt per square centimeter. This equates approximately to 0.0002 dyne per square centimeter in pressure units, a figure that is more commonly used. (There is variation in the precise relationship between power and pressure units because of temperature and atmospheric pressure changes. However, agreement between the above values will be within 1 dB under most conditions.) The sound intensity level in decibels is calculated from the usual equation, 10 times the common logarithm of the power ratio, while using 10^{-16} watt per square centimeter as the reference. This is the same relationship as 20 times the logarithm of the pressure ratio.

Instruments are available to read sound pressure levels. These are calibrated in decibels, and are called sound level meters. Typical sound levels are 35 dB in a hospital, 40 dB in a classroom, 45 dB in offices, and 50 dB in banks, stores and restaurants. Intensity levels in factories may be as high as 80 dB.

Human speech intensities vary from one talker to another, and also vary for the same talker depending on the environment and distance between the talker and the listener. The average intensity level for talkers, both men and women, is 66 dB. Softer talkers speaking in conversational tones were measured to be as low as 54 dB, while loud talkers were as high as 75 dB. When one whispers the level is about 26 dB, while a loud shout produces an intensity level of about 86 dB. Thus, the human voice has a dynamic range of approximately 60 dB. These figures are for intensity levels averaged over a period of time. In actual speech, the instantaneous power varies from zero during pauses and intervals between words to peak values as high as 100 times the average value. Men tend to speak with greater intensity levels than women. There are other differences, too.

The pitch of a woman's voice is about one octave higher than that of a man, and a woman's voice is richer in frequencies above 3000 Hz.

HEARING

Pressure variations in the sound wave produce mechanical vibrations in the eardrum. These vibrations are transmitted to the inner ear, where nerve endings are excited. The ear consists of three parts: the outer ear, the middle ear and the inner ear. The outer ear is the external portion and the auditory canal. The middle ear consists of the ear drum and three small bones that are linked mechanically to the cochlea in the inner ear. In the cochlea are the nerves that give us the sense of hearing. The inner ear also contains parts that have no function in the mechanism of hearing, but do serve as an organ of balance or equilibrium.

Every person has a threshold of hearing, a level below which he or she cannot detect any sound, and that threshold level is quite dependent on frequency. If the sound is a single-frequency tone, as is commonly used for acuity measurements, an intermittent tone can be heard at levels 10 to 20 dB below that for a continuous tone. Frequencies of 20 Hz and 20 kHz represent the probable limits of human hearing, although not many persons can hear sounds at these extremes. A person with acute hearing has a threshold of approximately 20 dB in the mid-frequency range, 2 to 3 kHz.

As a person becomes older, hearing acuity decreases, particularly for the high frequencies. The loss below 1 kHz is usually less than 10 dB, but the losses with age run from 10 dB at 40 years to 40 dB at 60 years. (These decibel losses are referenced to normal hearing.) Hearing losses also occur with continued exposure to very loud sounds. The loss is usually permanent with continued exposure to intensities in excess of 85 dB. Men hear low frequencies better than women at comparable ages, and women hear high frequencies better, with the range from 1.5 to 2 kHz being heard about equally well.

MICROPHONES

A microphone is the transducer that converts sound waves into electrical energy. All voice transmissions begin with a microphone. Mics, as they are commonly called, have a number of basic characteristics including distortion, frequency response, impedance, output level, mechanical design, and directional or noise-canceling ability.

Distortion in a microphone refers to a nonlinear response to sound pressure waves. Distortion is not usually a problem in mics of modern design, except the carbon type.

Frequency Response

The frequency response of a mic is the bandwidth over which it responds to sound waves. In amateur transmissions, only the portion of the speech band from about 200 to 3500 Hz needs to be transmitted for good intelligibility. Mics usually have bandwidths greater than this, so the speech amplifiers in transmitters are often designed to limit the audio bandwidth to the minimum necessary.

A microphone should have a response that is as uniform as possible over the desired speech range, but this is the most difficult microphone characteristic to achieve. The inexpensive types often have peaks and valleys in their response as much as 20 dB above and below the nominal output level. This peaky response will increase the peak-to-average ratio of the speech waveform, which reduces the average level of audio power transmitted.

Microphones are generally omnidirectional, responding to sound from all directions, or unidirectional, picking up sound from one direction. If a microphone is to be used close to the operator's mouth, an omnidirectional microphone is ideal. If, however, speech is generated a foot or more from the microphone, a unidirectional microphone will reduce reverberation by a factor of 1.7:1. Some types of unidirectional microphones have a proximity effect in that low frequencies are accentuated when the microphone is too close to the mouth.

Impedance

A microphone is a voltage generator that has an optimum load impedance. Mics are available in two broad categories of impedance, high and low. The high-Z mics are used within a short distance of the transmitter, as the connecting cable cannot be made very long without compensation or matching transformers. Shielding of the microphone cable is necessary to prevent hum and noise pickup. The actual impedance of a "high-Z" mic will vary with manufacturer and type, but values between 100 kΩ and 500 kΩ are typical.

Low-Z microphones are sold in three popular impedances, 50, 150 and 600 ohms. The 50-ohm types are used by the commercial communications services, while the 150- and 600-ohm mics are usually found in broadcast, telephone and public address applications. The amateur's choice of mic impedance is usually dictated by the requirements of the transmitter, unless an impedance-matching transformer is used.

Output Level

Related to impedance, output level varies widely with different types of microphones. There are several systems of specifying microphone output level in use, so direct comparison of various manufacturers' specifications can lead to erroneous conclusions. In addition, the test level, the reference level and the load impedances used in the specification of high-Z and low-Z mics are different, so the two cannot be compared directly.

The popular rating for high-Z microphones is the output voltage, expressed in

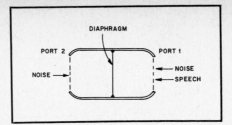

Fig. 3 — Simplified center chamber of a noise-canceling microphone.

decibels below 1 volt, measured across an open circuit with 1 dyne per square centimeter test signal at 1000 Hz applied to the microphone. Low-Z mics use a different system with the rating in output power, expressed in decibels below 1 milliwatt, measured across a 600-ohm load and using a test tone of 10 dynes per square centimeter at 1000 Hz. Several other test procedures are in use, so direct comparison of manufacturers' specifications will be valid only if the same measurement standards have been used.

Noise Cancellation

Noise cancellation is the ability to cancel or reject noise while not attenuating the desired voice signal. A noise-canceling mic can be used to advantage in a noisy environment. The principle of noise cancellation (perhaps more properly called noise attenuation, as it is impossible to get rid of all noise) is quite simple.

See Fig. 3. The noise originating at some distant point enters the mic through two ports and is applied to both sides of the diaphragm. If the sound pressure is equal on both sides, the diaphragm will not move and the mic has no output. Speech is allowed to enter through only one port; the diaphragm moves in response to this sound pressure and output is produced. The shape of the diaphragm, the material it is made from, and the mounting are all important design considerations. A problem area is the tendency of the diaphragm to "lock up" under heavy noise pressure, distorting the desired speech signal. Considerable progress has been made in this area, and a good design will take very heavy noise without producing appreciable speech distortion.

The measurement of a mic's ability to attenuate noise is not standard, and most microphones or elements are sold simply with the label, "noise canceling," with no information on how well the mic actually cancels background noise. A good mic should attenuate noise by at least 20 dB across the speech band.

In noise-canceling mics designed for hand-held operation, the mic element itself must be physically small and light. This means the two sampling ports cannot be spaced very far apart. If the voice is allowed to enter both ports, it too will be at-

tenuated. For this reason, noise-canceling mics are inherently close-talk devices. They should be held within ½ inch of the lips for best results.

TYPES OF MICROPHONES

Microphones are available in many types. Those commonly available or popular for amateur use are discussed in the sections that follow.

Carbon Microphones

The carbon microphone consists of a metal diaphragm placed against a cup of loosely packed carbon granules. As the diaphragm is actuated by the sound pressure, it alternately compresses and decompresses the granules. When current is flowing through the element, a variable direct current corresponds to the movement of the diaphragm. The carbon mic is used almost universally by the telephone company, and until quite recently was quite popular with the military. Carbon elements are low impedance (usually between 30 and 80 ohms), and have very high output — 0.2 V or more, depending on the dc voltage across the element. They differ from most other popular types in that the transmitter (or some other source) must provide an operating voltage for the element. Practically, the voltage requirement limits the use of the element to transmitters designed for carbon mics.

The carbon mic has two principal disadvantages, a high level of distortion compared to other types and a problem when the element is exposed to moisture for long periods. Moisture in the element will pack the carbon granules together, reducing the output and increasing distortion. In some instances the element stops working altogether. The mic can sometimes be restored by heating it for several hours in the oven or with a sun lamp, and then tapping it lightly to loosen the granules.

Piezoelectric Microphones

Piezoelectric microphones make use of the phenomenon by which certain materials produce a voltage by mechanical stress or distortion of the material. A diaphragm is coupled to a small bar of material such as Rochelle salt or ceramic made of barium titanate or lead zirconium titanate. The diaphragm motion is thus translated into electrical energy. These microphones are of high impedance. The Rochelle salt or crystal mic is of a fragile nature, and is susceptible to high temperatures, high humidity and extreme dryness. For this reason is is not suitable for mobile or portable work. A crystal mic has high output and a wide frequency response.

The ceramic type is impervious to temperature and humidity. This type has become very popular with manufacturers who supply mics with their amateur and CB transceivers. The reason is that the ceramics sell for low prices when mass produced. The impedance is high, and the output of

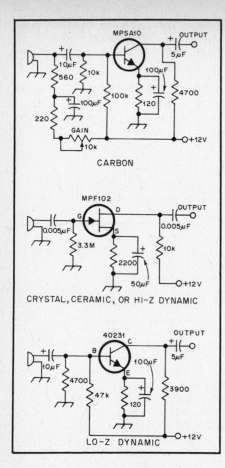

Fig. 4 — Speech circuits for use with standard-type microphones. Typical parts values are shown.

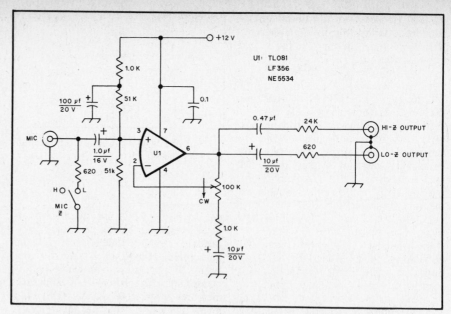

Fig. 5 — A speech amplifier suitable for microphone or interstage use. The input and output impedances can be tailored to match a wide range of loads. Maximum gain of this circuit is 40 dB.

a modern ceramic microphone is generally enough to drive most popular rigs. However, a ceramic mic is a low-output device, and operation can be marginal with a transmitter that is lacking gain in the speech amplifier. An interesting characteristic of the ceramic microphone is that if it is terminated in a lower-than-optimum resistance (optimum being 100 kΩ to 250 kΩ), the low-frequency output is attenuated — a simple way to eliminate the unwanted frequencies below 200 Hz. This frequency characteristic is also true of microphones made with Rochelle salts, another piezoelectric material.

Many of the inexpensive ceramic mics show the effects of mass production — excessive peaks in their frequency response. Manufacturers rate the nominal frequency response for these mics, but never mention the limits of deviation from the nominal output level. It is generally true that the cost of materials for all types of microphones is quite low unless put into a very fancy housing. The yield, testing, rework and retesting that a manufacturer goes through to ensure that a microphone meets a specified frequency response within specified limits is what adds up to the price of a "good" microphone.

Dynamic Microphones

The dynamic microphone resembles a dynamic loudspeaker. A lightweight coil, usually made of aluminum wire, is attached to a diaphragm. This coil is suspended in a magnetic field. When sound pressure impinges on the diaphragm, it moves the coil through the magnetic field, generating an alternating voltage. Dynamic mics are one of the most popular types for radio service. Dynamic elements are low impedance; matching transformers are built in if a high-Z output is desired.

Dynamics range all the way in price until they rival the cost of a communications receiver. The very expensive types are generally intended for recording and broadcast work. The dynamic element can be made noise canceling or directional. The mounting of the diaphragm is such that sound can easily be applied to both the front and the rear of the diaphragm to produce a desired pattern or effect. Because of its excellent speech quality and reliability, the military services have been changing over to this type of mic for use on field radio sets.

Electret Microphones

The electret microphone has recently appeared as a feasible alternative to the carbon, piezoelectric or dynamic microphone. An electret is an insulator that has a quasi-permanent static electric charge trapped in or on it. The electret operates in a condenser fashion, using a set of biased plates. Motion caused by air pressure variations creates a changing capacitance and an accompanying change in voltage. The electret acts as the plates would, and being charged, it requires no bias voltage. A low voltage provided by a battery used for an FET impedance converter is the only power required to produce an audio signal.

Electrets traditionally have been suscep-

tible to damage from high temperatures and high humidity. New materials and different charging techniques have lowered the chances of damage, however. Only in extreme conditions (such as 120 °F at 49% humidity) are problems present. The output level of a typical electret is higher than that of a standard dynamic microphone.

SPEECH AMPLIFIERS

The purpose of a speech amplifier is to raise the level of audio output from a microphone to that required by the modulator of a transmitter. In SSB and FM transmitters the modulation process takes place at low levels, so only a few volts of audio is necessary. One or two simple voltage-amplifier stages will suffice. AM transmitters often use high-level plate modulation requiring considerable audio power, compared to SSB and FM. The microphone-input and audio voltage-amplifier circuits are similar in all three types of phone transmitters, however.

When designing speech equipment it is necessary to know (1) the amount of audio power the modulation system must furnish, and (2) the output voltage developed by the microphone when it is spoken into from a normal distance with ordinary loudness. It then becomes possible to choose the number and type of amplifier stages needed to generate the required audio power without overloading or undue distortion anywhere in the system.

The circuit immediately following the audio input establishes the signal-to-noise ratio of the transmitter. General-purpose ICs such as 741 op amps are widely used in speech amplifiers, but they are fairly noisy, so it is best to precede them with a lower noise discrete device (FET or bipolar transistor). The circuits

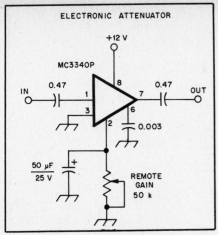

Fig. 6 — A dc voltage controls the gain of this IC, eliminating the need for shielded leads to the gain control.

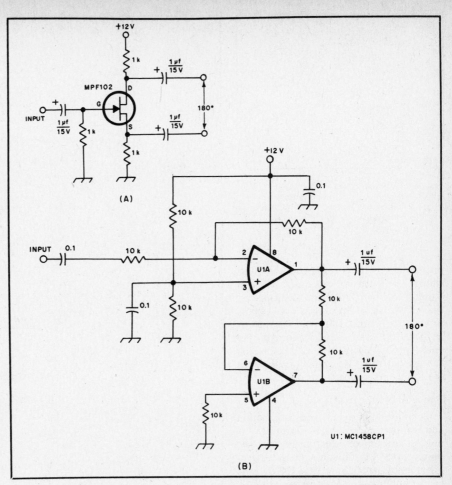

Fig. 7 — Phase splitter circuits using (A) a JFET and (B) a dual operational amplifier.

in Fig. 4 fulfill this requirement.

Voltage Amplifiers

The important characteristics of a voltage amplifier are its voltage gain, maximum undistorted output voltage and its frequency response. The voltage gain is the voltage-amplification ratio of the stage. The output voltage is the maximum AF voltage that can be obtained from the stage without distortion. The amplifier frequency response should be adequate for voice reproduction; this requirement is easily satisfied.

The voltage gain and maximum undistorted output voltage depend on the operating conditions of the amplifier. The output voltage is in terms of *peak* voltage, rather than RMS; this makes the rating independent of the waveform. Exceeding the peak value causes the amplifier to distort, so it is more useful to consider only peak values in working with amplifiers.

A circuit suitable for use as a microphone preamplifier or the major gain block of a speech system is shown in Fig. 5. The response rolls off below 200 Hz to reduce hum pickup. Ordinary 741 op amps can be used in stages following the preamp, provided the voltage gain is held to about 20 (26 dB).

Gain Control

A means for varying the overall gain of the amplifier is necessary for keeping the final output at the proper level for modulating the transmitter. The common method of controlling the gain is to adjust the value of ac voltage applied to the input of one of the amplifiers by means of a voltage divider or potentiometer.

The gain-control potentiometer should be near the input end of the amplifier, at a point where the signal voltage level is so low there is no danger that the stages ahead of the gain control will be overloaded by the full microphone output. In a high-gain amplifier it is best to operate the first stage

at maximum gain, since this gives the best signal-to-hum ratio. The control is usually placed in the input circuit of the second stage.

Remote gain control can be accomplished with an electronic attenuator IC, such as the Motorola MC3340P. A dc voltage varies the gain of the IC from +6 to −85 dB, eliminating the need for shielded leads to a remotely located volume control. A typical circuit is shown in Fig. 6.

Phase Inversion

Some balanced modulators and phase shifters require push-pull audio input. The obvious way to obtain push-pull output from a single-ended stage is to use a transformer with a center-tapped secondary. Phase-inverter or phase-splitter circuits can accomplish the same task electronically. A differential amplifier can be used to convert a single-ended input to a push-pull output. Two additional phase-splitter circuits are shown in Fig. 7.

Speech-Amplifier Construction

Once a suitable circuit has been selected for a speech amplifier, the construction problem resolves itself into avoiding two difficulties — excessive hum and unwanted feedback. For reasonably humless operation, the hum voltage should not exceed about one percent of the maximum audio output voltage — that is, the hum and noise should be at least 40 dB below the output level.

Unwanted feedback, if negative, will reduce the gain below the calculated value. If positive, feedback is likely to cause self-oscillation or howls. Feedback can be minimized by isolating each stage with decoupling resistors and capacitors, by avoiding layouts that bring the first and last stages near each other, and by shielding of "hot" points in the circuit, such as high-impedance leads in low-level stages.

If circuit-board construction is used, high-impedance leads should be kept as short as possible. All ground returns should be made to a common point. A good ground between the circuit board and the metal chassis is necessary. Complete shielding from RF energy is always required for low-level solid-state audio circuits. The microphone input should be decoupled for RF with a filter, as shown in Fig. 8. At A, an RF choke with a high impedance over the frequency range of the transmitter is used. For high-impedance inputs, a resistor may be used in place of the choke, shown in Fig. 8B.

When using paper capacitors as by-

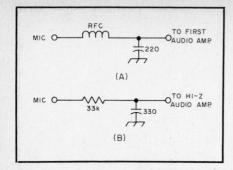

Fig. 8 — RF filters using LC components, A, and RC components, B. These filters are used to prevent feedback caused by RF pickup on the microphone lead.

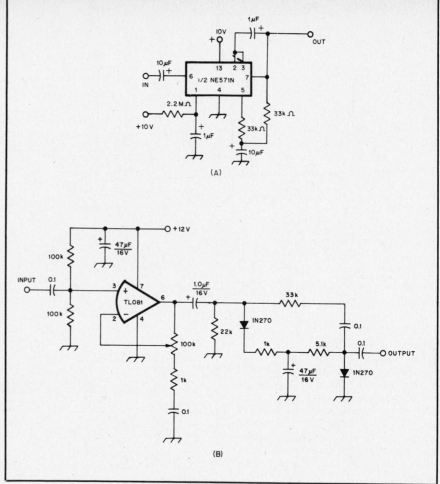

Fig. 9 — Typical solid-state compressor circuits. The circuit at A works on the AGC principle, while that at B is a forward-acting compressor.

passes, be sure the terminal marked "outside foil," often indicated with a black band, is connected to ground. This utilizes the outside foil of the capacitor as a shield around the "hot" foil. When paper or Mylar capacitors are used for coupling between stages, always connect the outside foil terminal to the side of the circuit having the lower impedance to ground.

Driver and Output Stages

Commonly used balanced modulators and transmitting mixers have power outputs too low for consistently effective communications. Most modern grounded-grid linear amplifiers require 30 to 100 watts of exciter output power to drive them to their rated power output. An exciter output amplifier serves to boost the output power to a useful level while providing additional selectivity to reject spurious mixing products.

Two stages are usually required to obtain the necessary power. The stage preceding the output amplifier is called the driver. Some tubes that work well as drivers are the 6CL6, 12BY7, 6EH7 and 6GK6. Since all of these tubes are capable of high gain, instability is sometimes encountered in their use. Parasitic suppression should be included as a matter of course. Some form of neutralization is recommended. Driver stages should be operated in Class A or AB1 to minimize distortion. The higher quiescent dissipation can be handled easily at these power levels. VMOS power FETs are also well suited to SSB driver circuits.

SPEECH PROCESSING

Four basic systems, or a combination thereof, can be used to reduce the peak-to-average ratio, and thus raise the average power level of an SSB signal. They are compression or clipping of the AF wave before it reaches the balanced modulator, and compression or clipping of the RF waveform after the SSB signal has been generated. One form of RF compression, commonly called automatic level control or ALC, is used almost universally in amateur SSB transmitters. Audio processing is also

used to increase the level of audio power contained in the sidebands of an AM transmitter and to maintain constant deviation in an FM transmitter. Both compression and clipping are used in AM systems, while most FM transmitters employ only clipping.

Volume Compression

Although it is obviously desirable to keep the voice level as high as possible, it is difficult to maintain constant voice intensity when speaking into the microphone. To overcome this variable output level, it is possible to use automatic gain control that follows the *average* (not instantaneous) variations in speech amplitude. This can be done by rectifying and filtering some of the audio output and applying the resultant dc to a control electrode in an early stage in the amplifier.

The circuit of Fig. 9A works on this AGC principle. One section of a Signetics NE571N IC is used. The other section can be connected as an expander to restore the dynamic range of received signals that have been compressed in transmission. Operational transconductance amplifier ICs such

as the CA3080 are also well suited for speech-compression service.

When an audio AGC circuit derives control voltage from the output signal, the system is a closed loop. If short attack time is necessary, the rectifier-filter bandwidth must be opened up to allow syllabic modulation of the control voltage. This allows some of the voice-frequency signal to enter the control terminal, causing distortion and instability. Because the syllabic frequency and speech-tone frequencies have relatively small separation, the simpler feedback AGC systems compromise fidelity for fast response.

Problems with loop dynamics in audio AGC systems can be sidestepped by eliminating the loop and using a forward-acting system. The control voltage is derived from the *input* of the amplifier, rather than from the output. Eliminating the feedback loop allows unconditional stability, but the trade-off between response time and fidelity remains. Care must be taken to avoid excessive gain between the signal input and control voltage output. Otherwise, the transfer characteristic can reverse; that is, an increase in input level can cause a

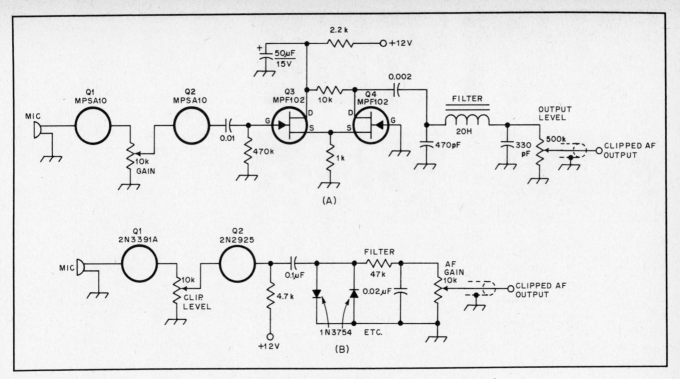

Fig. 10 — This diagram illustrates the use of JFETs or silicon diodes to clip positive and negative voice peaks.

decrease in output. A simple forward-acting compressor is shown in Fig. 9B.

To test or adjust a speech compressor, one would apply a single steady tone to the input and measure the input and output signal levels over the dynamic range of the instrument. The single-tone test may indicate gross distortion, but one cannot judge the speech performance by this result. Provided the release time is longer than the time between syllables, the compressor will generate distortion only during brief signal peaks.

Speech Clipping and Filtering

In speech waveforms the average power content is considerably less than in a sine wave of the same peak amplitude. If the high amplitude peaks are clipped off, the remaining waveform will have a considerably higher ratio of average power to peak amplitude. Although clipping distorts the waveform and the result therefore does not sound exactly like the original, it is possible to obtain a worthwhile increase in audio power without sacrificing intelligibility. Once the system is properly adjusted *it will be impossible to overdrive* the modulator stage of the transmitter because the maximum output amplitude is fixed.

By itself, clipping generates high-order harmonics and therefore will cause splatter. To prevent this, the audio frequencies above those needed for intelligible speech must be filtered out *after* clipping and *before* modulation. The filter required for this purpose should have relatively little attenuation below about 2500 Hz, but high attenuation for all frequencies above 3000 Hz.

There is a loss in naturalness with "deep" clipping, even though the voice is highly intelligible. With moderate clipping levels (6 to 12 dB) there is almost no change in "quality" but the voice power is increased considerably.

Before drastic clipping can be used, the speech signal must be amplified several times more than is necessary for normal modulation. Also the hum and noise must be much lower than the tolerable level in ordinary amplification, because the noise in the output of the amplifier increases in proportion to the gain.

In the circuit of Fig. 10B a simple diode clipper is shown following a two-transistor preamplifier section. The 1N3754s conduct at approximately 0.7 volt of audio and provide positive- and negative-peak clipping of the speech waveform. A 47-kΩ resistor and a 0.02-µF capacitor follow the clipper to form a simple RC filter for attenuating the high-frequency components generated by the clipping action, as discussed earlier. Any silicon diodes may be used in place of the 1N3754s. Germanium diodes (type 1N34A) can also be used, but will clip at a slightly lower peak audio level.

SSB Speech Processing

Compression and clipping are related, as both have fast attack times. And when the compressor release is made quite short, the effect on the waveform approaches that of clipping. Speech processing is most effective when accomplished at radio frequencies, although a combination of AF clipping and compression can produce worthwhile results. The advantage of an outboard audio speech processor is that no in-

ternal modifications are necessary to the SSB transmitter with which it will be used.

To understand the effect of SSB speech processing, consider that basic RF waveforms without processing have high peaks but low average power. With proper processing, the amount of average power is raised considerably. Fig. 11 shows a comparison of received signal-to-noise ratio with different forms of signal processing. An advantage of several decibels for RF clipping (for 20 dB of processing) is shown over any other type of processing.

Investigations by W6JES, reported in January 1969 *QST,* showed that, for a transmitted signal using 15 dB of audio clipping from a remote receiver, the intelligibility threshold was improved nearly 4 dB over a signal with no clipping. Increasing the AF clipping level to 25 dB gave an additional 1.5 dB improvement in intelligibility. Audio compression was found to be valuable for maintaining relatively con-

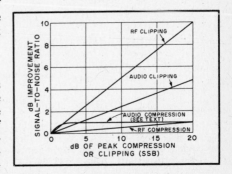

Fig. 11 — The improvement in received signal-to-noise ratio obtained by the simple forms of signal processing.

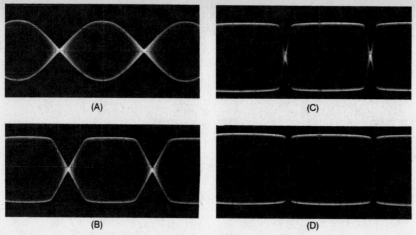

(A)

(B)

(C)

(D)

Fig. 12 — Two-tone envelope patterns with various degrees of RF clipping. All envelope patterns are formed using tones of 600 and 1000 Hz. At A, clipping threshold; B, 5 dB of clipping; C, 10 dB of clipping; D, 15 dB of clipping.

stant average-volume speech, but such a compressor added little to the intelligibility threshold at the receiver — only about 1 to 2 dB.

Evaluation of RF clipping from the receive side with constant-level speech, and filtering to restore the original bandwidth, resulted in an improved intelligibility threshold of 4.5 dB with 10 dB of clipping. Raising the clipping level to 18 dB gave an additional 4 dB improvement at the receiver, or 8.5 dB total increase. The improvement of the intelligibility of a weak SSB signal at a distant receiver can thus be substantially improved by RF clipping. The effect of such clipping on a two-tone test pattern is shown in Fig. 12.

Automatic level control, although a form of RF speech processing, has found its primary application in maintaining the

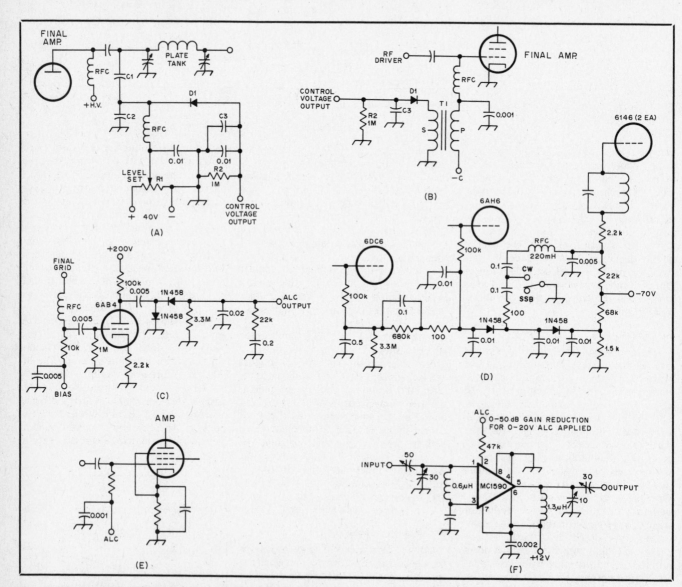

Fig. 13 — At A, control voltage obtained by sampling the RF output voltage of the final amplifier. The diode back bias, 40 volts or so maximum, may be taken from any convenient positive voltage source in the transmitter. R may be a linear control having a maximum resistance on the order of 50 kΩ. D1 may be a 1N34A or similar germanium diode.

At B, control voltage obtained from the grid circuit of a Class AB1 tetrode amplifier. T1 is an interstage audio transformer having a turns ratio, secondary to primary, of 2 or 3 to 1. An inexpensive transformer may be used, since the primary and secondary currents are negligible. D1 may be a 1N34A or similar type. Time constant R2-C3 is discussed in the text.

At C, control voltage is obtained from the grid of a Class AB1 tetrode amplifier and amplified by a triode audio stage.

At D, ALC system used in the Collins 32S-3 transmitter. At E, applying control votage to the tube or, at F, to the linear IC-controlled amplifier.

peak RF output of an SSB transmitter at a relatively constant level, desirably below the point at which the final amplifier is overdriven during peaks in the audio input. These typical ALC systems, shown in Fig. 13, by the nature of their design time constants offer a limited increase in the ratio of transmitted average power to peak envelope power. A value in the region of 2 to 5 dB is typical. An ALC circuit with shorter time constants will function as an RF syllabic compressor, producing up to 6 dB improvement in the intelligibility threshold at a distant receiver. The Collins Radio Company used an ALC system with dual time constants (Fig. 13D) in their S/Line transmitters, and this has proved to be quite effective.

Heat is an extremely important consideration in the use of any speech processor that increases the ratio of average to peak power. Many transmitters, in particular those using television sweep tubes, simply are not built to withstand the effects of increased average input, either in the final-amplifier tube(s) or in the power supply. If heating in the final tube is the limiting factor, adding a cooling fan may be a satisfactory answer.

Compandoring

Compressing the audio signal prior to modulation for more efficient use of the transmitter power has been widely used by amateurs. However, by compressing only the amplitude peaks, background noise increases relative to the peaks. A significant advantage can be obtained by expanding the compressed audio at the receiver. The term "compandor" is a fusion of the terms "compressor" and "expandor." The positive effects of compandoring are illustrated in Fig. 14. A short segment of the speech time waveform is shown in Fig. 14A. When compressed and transmitted, this waveform becomes noisy because of compression, and also from the added circuit and radio channel noise. This is shown in Fig. 14B. Notice the increased noise between the two passages of speech. After the waveform is processed by an amplitude expander (the receiving portion of an amplitude compandor), the waveform appears as shown in Fig. 14C. The noise during the quiet passage has been reduced relative to the high-level passages. Although the S/N ratio during the loud passage is not as good as the input waveform, the ratio for the overall passage is much better than it would be without expansion. Noise during loud passages is not nearly as objectionable as noise between passages of speech.

Using a two-to-one amplitude compandor* (Signetics NE571N), tests performed for the FCC indicated a measurable 12- to 15-dB improvement with a full

amplitude compandor (compression on transmit and expansion on receive). When only compression was used, significant S/N ratio reduction resulted, even though a higher average transmitter power was being used.

An amplitude expander circuit is shown in Fig. 15. It is important to realize that the amplitude compandor will improve the S/N ratio, but only when sufficient signal is present for use as a reference. Thus, it has a thresholding effect. As long as the received signal is a few decibels above the noise, the expander can expand on the basis of the reference to provide a communications channel with a cleaner signal over a wider dynamic range. However, when the signal level drops into the noise, the expander will not operate properly. The S/N improvement is limited, but the improvement over the useful range is worthwhile. If we consider that a certain S/N ratio and corresponding voice quality were required, use of an amplitude compandor would allow the achievement of this goal with less transmitter power. In this respect the amplitude compandor can achieve the same quality S/N ratio with 12-15 dB less transmitted power. This is a significant power savings and can have a definite impact on the quality of communications per watt of transmitter power.

Amplitude-Compandored Single Sideband

Work has been taking place to develop an amplitude-compandored single-sideband system (ACSSB) in the land-mobile communications industry. The purpose is to reduce the spectrum requirements per channel from that presently required for FM. Ordinary SSB operation was not found to be satisfactory for operators who were accustomed to noise-free FM and who were not trained communicators. The "clarifier" or receiver incremental tuning control was a nuisance, as was signal fading that never occurred on FM. So to get the spectrum savings of SSB over FM, the engineers have come up with a system that gives the land mobile users everything they are accustomed to with FM, including signaling.

A conventional SSB transmission has a suppressed carrier. In order to demodulate the signal at the receiver, the missing carrier must be replaced. At the higher frequencies this becomes difficult. In ACSSB, a pilot tone is added to the audio, sufficiently separated in frequency and filtered so the two do not interfere. Although there is some variation among manufacturers, the pilot tone is usually 3 or 3.1 kHz. The pilot power level is about 10 dB below the voice peak power. Thus, with no voice modulation the transmitter has an output power of approximately 1/10th that when there is voice modulation.

At the receiver, the pilot tone is compared with an internal reference oscillator in a phase-locked loop circuit. The difference voltage thus produced is used to

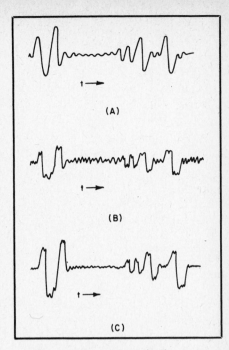

Fig. 14 — Speech waveforms for the amplitude compandor, showing amplitude (vertical axis) vs. time, t. At A is the input waveform, at B the compressed waveform with additional circuit noise as might be received, and at C the expanded waveform. Note the improvement in signal-to-noise ratio upon expansion after reception.

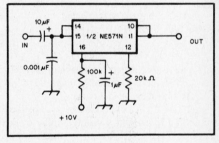

Fig. 15 — Amplitude expander circuit.

shift the receiver local oscillator until the frequency error is eliminated. The time constant of the phase-locked loop is short enough to be ignored by the user and to be compatible with tone decoders used for signal purposes.

The compandor thresholding effect provides an operator with a sensation similar to the capture effect in FM. Thus, when the signal strengths are high, FM and ACSSB give comparable results with noticeably less flutter on ACSSB than on FM.

ACSSB is a means of packing about four times more voice transmissions in a given band than FM. The pilot tones and compandoring compensate for some inherent SSB shortcomings, but with additional complexity. This appears to be an area for Amateur Radio experimentation.

MORSE CODE

Previous sections have been concerned

*A two-to-one amplitude compandor yields a compression of 1 dB out for every 2 dB in and an expansion of 2 dB out for every 1 dB in.

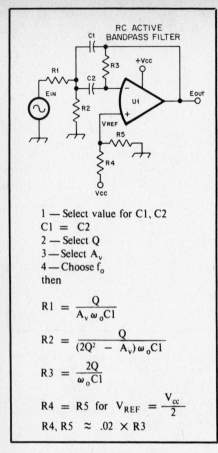

RC ACTIVE BANDPASS FILTER

1 — Select value for C1, C2
C1 = C2
2 — Select Q
3 — Select A_v
4 — Choose f_o
then

$$R1 = \frac{Q}{A_v \omega_o C1}$$

$$R2 = \frac{Q}{(2Q^2 - A_v) \omega_o C1}$$

$$R3 = \frac{2Q}{\omega_o C1}$$

$$R4 = R5 \text{ for } V_{REF} = \frac{V_{cc}}{2}$$

$$R4, R5 \approx .02 \times R3$$

Fig. 16 — Basic circuit for an RC active band-pass filter. One pole is shown along with the fundamental equations for finding the resistance values needed.

$$f_o = 900 \text{ Hz} \qquad A_v = 1 \qquad \omega_o = 2\pi f_{o(Hz)}$$

$$Q = 5 \qquad C1, C2 = 0.00068 \, \mu F$$

$$R1 = \frac{5}{(1) (6.28 \times 900) (0.00068 \times 10^{-6})}$$

$$= 1,300,948 \text{ ohms}$$

$$R2 = \frac{5}{[2(25) - 1] [6.28 \times 900 \, (0.00068 \times 10^{-6})]} = 26,550 \text{ ohms}$$

$$R3 = \frac{10}{(6.28 \times 900) (0.00068 \times 10^{-6})} = 2,601,896 \text{ ohms}$$

$$R4, R5 \approx 2,601,896 \times 0.02 = 52,018 \text{ ohms}$$

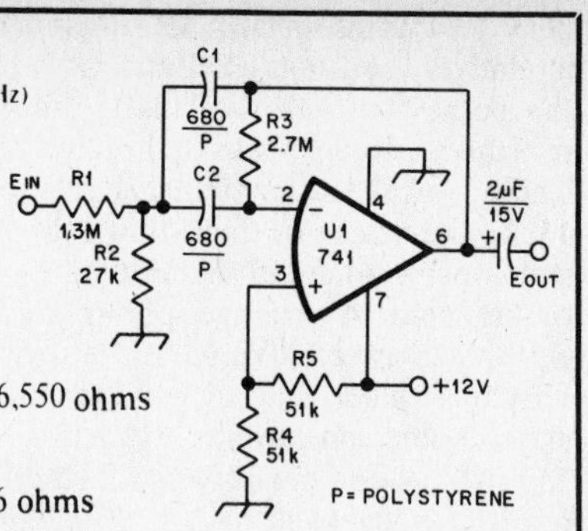

Fig. 17 — A design example based on the circuit of Fig. 16.

primarily with voice communications, but Morse code is equally important in Amateur Radio. With computers or other logic circuits, Morse code can be copied by machine to present a visual display of the text being received, but by far the majority of operators copy code "by ear." To do this, of course, it is necessary to *hear* the received code.

Most operators prefer to receive code signals in a frequency range from 500 to 800 Hz. The ear is better able to distinguish two tones of fixed frequency separation at low pitch rather than at higher pitches. In other words, if an undesired signal is present at a frequency 200 Hz removed from a desired signal, the ear can better distinguish the desired signal if the tones are 500 Hz and 700 Hz, instead of 1100 and 1300 Hz. Thus, the ear is its own form of selective filter. For this reason, some operators prefer to adjust the receiver tuning for lower pitches yet, in the order of 200 to 300 Hz for the desired signal.

Morse code signals occupy significantly less bandwidth than do voice signals. Therefore, a much narrower filter may be used for code reception than is possible for receiving voice transmissions. Decreasing the received bandwidth improves the signal-to-noise ratio under a given set of conditions, as well as improving the selectivity.

If a receiver lacks IF selectivity, an outboard filter may be used ahead of the speaker or headphones.

Op Amps as Audio Filters

One of the more common uses to which operational amplifiers (op amps) are put can be seen in the RC active audio-filter field. Op amps have the distinct advantage of providing gain and variable parameters when used as audio filters. Passive filters that contain L and C elements are generally committed to some fixed frequency, and they exhibit an insertion loss. Finally, op amps contribute to the attainment of miniaturization that is seldom possible with bulky inductors in a passive type of audio filter.

Although RC active filters can be built with bipolar transistors, the modern approach is to use operational amplifiers. The use of an op-amp IC, such as a type 741, results in a compact filter pole that will provide stable operation. Only five connections are made to the IC, and the gain of the filter section, plus the frequency characteristic, is determined by the choice of components external to the IC.

Although there are numerous applications for RC active filters, their principal use in amateur work is that of establishing selectivity at audio frequencies. One or two poles may be used as a band-pass or low-pass section for improving the passband characteristics during SSB or AM reception. Up to four filter poles are frequently used to acquire selectivity for CW or RTTY reception. The greater the number of poles, up to a practical limit, the sharper the skirt response of the filter. Not only does a well-designed RC filter help to reduce QRM, but it improves the signal-to-noise ratio in some receiving systems.

Considerable design data is found in the National Semiconductor Corp. application note. A thorough treatment of Norton amplifiers is given, centering on the LM3900 current-differencing type of op amp. Design information is included for high-, low- and band-pass types of RC active audio filters. The simplified design data

presented here are based on the technique used in AN72-15.

Fig. 16 shows a single band-pass filter pole and gives the equations for obtaining the desired values for the resistors if the gain, Q, f_o and C1-C2 values are chosen. C1 and C2 are equal in value and should be high-Q, temperature-stable components. Polystyrene capacitors are excellent for use in this part of the circuit. Disc-ceramic capacitors *are not recommended*. R4 and R5 are equal in value and are used to establish the op-amp reference voltage. This is $V_{cc}/2$.

C1 and C2 should be standard values of capacitance. The filter design is less complicated when C1 and C2 serve as the starting point for the equations. Otherwise some awkward values for C1 and C2 might result. The resistance values can be "fudged" to the nearest standard value after the equations have been worked. The important consideration is that matched values must be used when more than one filter pole is employed. For most amateur work it will be satisfactory to use five percent, ½-watt, composition resistors. If the resistor and capacitor values are not held reasonably tight in tolerance for a multipole filter, the f_o for each pole may be different, however slight. The result is a wide nose for the response, or even some objectionable passband ripple.

Fig. 17 illustrates the design of a single-pole band-pass filter. An arbitrary f_o of 900 Hz has been specified, but for CW reception the operator may prefer something much lower — 200 to 700 Hz. An A_v (gain) of 1 (unity) and a Q of 5 are stated. Both the gain and the Q can be increased for a single-section filter if desired, but for a multisection RC active filter it is best to restrict the gain to 1 or 2 and use a maximum Q of 5. This will help prevent unwanted filter "ringing" and audio instability.

C1 and C2 are 680-pF polystyrene capacitors. Other standard values can be used from, say, 500 to 2000 pF. The limiting factor will be the resultant resistor values. For certain design parameters and

C1-C2 values, unwieldy resistance values may result from the equations. If this happens, select a new value for C1 and C2.

The resistance values assigned to R1 through R5, inclusive, are the nearest standard values to those obtained from the equations. The principal effect from this is a slight alteration of f_o and A_V.

In a practical application the RC active filter should be inserted in the low-level audio stages. This will prevent overloading the filter during the reception of strong signals. The receiver AF gain control should be used between the audio preamplifier and the input of the RC active filter for best results. If audio-derived AGC is used in the receiver, the RC active filter will give best performance when it is contained within the AGC loop. Information on other types of active filters is given by Bloom in July 1980 *QST*.

LOUDSPEAKERS AND HEADPHONES

The counterpart for the microphone at the transmitter is the loudspeaker or headphones at the receiver. These devices are transducers that convert electrical energy into vibrations of a diaphragm, thereby setting up sound waves.

Dynamic loudspeakers, or simply speakers, use a diaphragm of heavy paper or other thin material. This diaphragm is formed in the shape of a shallow cone, with its rim attached to the speaker frame. At the apex of the cone is a small core on which a coil of fine wire is wound. The core and the coil are free to move, and are suspended in the field of a permanent magnet. When a direct current is passed through the coil, the apex of the cone or diaphragm is deflected in one direction. When the current is removed, the diaphragm returns to its resting position. A direct current of the opposite polarity will deflect the diaphragm in the opposite direction. An alternating current causes the diaphragm to move back and forth at the frequency of the current and thereby set up sound waves, assuming the frequency is not too high for the diaphragm to follow.

Speakers are available from an inch or less to many inches in diameter. The larger speakers are more efficient at reproducing lower frequencies. For the best audio quality, a speaker should be mounted in a cabinet or some form of baffle. Even a good quality speaker, if operated in open air, will have a "tinny" sound — reproducing only intermediate frequencies. Low frequencies will be conspicuously absent. A baffle or cabinet serves to reinforce or enhance the low frequencies. Small speakers mounted at the bottom of a small chassis leave something to be desired also when it comes to audio fidelity.

Commonly available loudspeakers are designed with an impedance in the range of 4 to 16 Ω. Another impedance range available is 45 Ω; speakers in this range are often used in intercommunication stations. In this application the speaker doubles as a microphone for talking — sound pressure waves move the diaphragm and the attached coil in the magnetic field, thereby inducing a voltage into the coil.

Headphones operate on the same general principle as dynamic speakers. In fact, some high-quality headsets for hi-fi or stereo listening use small dynamic speakers. A thin, flat, metal diaphragm is used in most headsets. In operation the diaphragm is set in motion by a varying magnetic field caused by current flowing through the headset coils, which are wound on a magnetic core. Headphone impedances of 4 to 8 Ω are the most popular because a reasonable amount of power can be coupled to the phones with relatively low audio voltages. Headphones are also available in an impedance range of 500 to 600 Ω, and in the 2000-Ω range. Those in the 2000-Ω range are referred to as high-impedance headphones. They may be operated with relatively low currents, but do require a higher voltage amplitude than phones of lower impedance.

Video

For the sighted, much human communication is visual. In personal communication we watch faces, hands and bodies for visual cues and clues. Your grandparents used to listen to the radio — today the family watches television. Several hours a day is average. Visual communication is a part of Amateur Radio.

QST began carrying articles on TV in 1925. Those early experiments used crude mechanical methods. Nevertheless, radio amateurs were transmitting pictures. As early as 1928 a *QST* author pointed out the possibility of electronic television. The days of the motor-driven scanning disc were numbered. Introduction of the moderately priced iconoscope in 1940 made electronic television a practical reality for the amateur. By 1960, amateurs were transmitting color-TV signals.

Vision

Light is physical energy in the form of electromagnetic radiation. The visible light range falls between microwaves and X rays, as shown in Fig. 18. The human eye is sensitive to only a relatively narrow band of electromagnetic radiation. That band falls between the wavelengths of 400 and 720 nanometers, approximately.

Solar radiation contains all the visible wavelengths. Other light sources have different wavelength distributions. Incandescent sources tend to have more orange or yellow light; fluorescent light tends to be more bluish.

Physical objects selectively absorb light energy. Reflected light, therefore, has a different energy distribution than that of the original light. It is that difference that makes vision possible. Differences in reflected light account for the perception of brightness, texture and color.

Light enters the eye, and is focused by the lens onto the retina. The amount of light entering the eye is controlled by the iris. See Fig. 19. Two types of light sensors make up the retina — rods and cones.

Rods are sensitive only to black, white and shades of gray. Cones allow us to perceive color. Rods are more sensitive than cones — that is they work at lower light levels. That is why we can see color in the sky at twilight or dawn, but we can only perceive black, white and gray for the objects on the ground about us. The rela-

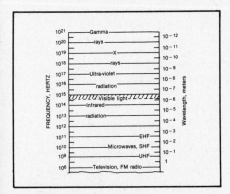

Fig. 18 — The electromagnetic spectrum from VHF to gamma rays.

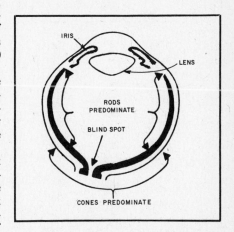

Fig. 19 — Cross-sectional view of the human eye. The retina contains rods and cones (see text). There are neither rods nor cones in the blind spot — that is where the optic nerve enters the eye.

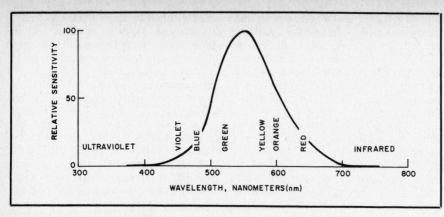

Fig. 20 — Relative sensitivity of the human eye to light at various wavelengths.

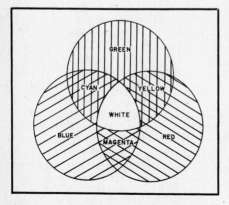

Fig. 21 — The three primary colors are red, green and blue. The primaries combine, two at a time, to form the secondary colors, yellow, cyan and magenta. White is a combination of red, green and blue.

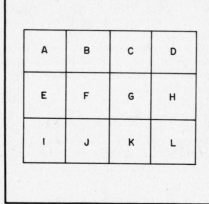

Fig. 22 — A picture can be described by dividing it into parts and then describing the parts.

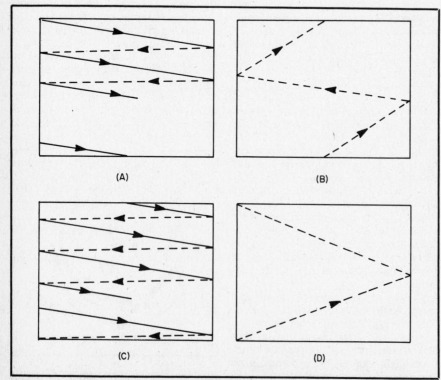

Fig. 23 — Interlaced scanning as used in TV. In field one, 262½ lines are scanned (A). At the end of field one, the electron scanning beam is returned to the top of the picture area (B). Scanning lines in field two (C) fall between the lines of field one. At the end of field two, the scanning beam is again returned to the top where scanning continues with field one (D).

tive sensitivity of the human eye to light varies with wavelength. See Fig. 20.

Color

Impulses from the rods and cones in the retina travel along the optic nerve to the central nervous system. There the impulses are "seen" as light and color. Any number of different energy distribution combinations may be perceived as the same color. The eyes do not sort out wavelengths like a prism. Cones are of three "kinds"; each kind being more sensitive to light in a particular portion of the visible spectrum. The areas of senstivity are known as red, blue and green light. Color is not a property of the electromagnetic energy that we call light. Rather, color is a psychological experience and a perceptual response to that energy.

It is possible to reproduce any color by combining in proper proportions three primary colors. The primaries are red, green and blue, as shown in Fig. 21. Combinations of equal amounts of two of the primaries form the secondary colors: yellow, cyan and magenta. White is a combination of all three primaries in the proper proportions.

Combining light is an additive process. A green light and a red light falling on a white wall make a yellow glow. This should not be confused with paints or pigments — they combine in a subtractive process.

This trichromatic system of color perception makes the transmission of color TV pictures possible. Light from the scene being transmitted is separated into red, green and blue components. These are transmitted to the receiver where they are simultaneously displayed on a screen. Through the eye and the central nervous system, the color picture is "reassembled" by the viewer.

Flicker

A flickering light can appear to be continuous. When light reaches the eye for a brief instant, the eye responds, sending the information to the brain. However, when the light stops, the eye continues to send to the brain for an additional fraction of a second. If you have ever experimented with a strobe light or viewed a motion picture, you should know that light does not have to be continuous for the eye to perceive it as such. The fact that the eye "retains" an image for a fraction of a second after it has gone makes television possible. This retention of an image is called persistence. For high-intensity light, a rate of 50 to 60 flashes per second results in no visible flicker.

PICTURE TRANSMISSION

If you wish to transmit a picture to another location, it is neither practical nor necessary to send all the information in that picture all at one time. Television and facsimile transmission systems send pictures "a piece at a time." In Fig. 22 an area has been divided into a mosaic of 12 squares. If the total area contained a picture, the scene could be described by detailing the contents of the individual squares in sequence (such as A through L). That is the concept behind television. Television (and facsimile) systems "divide" a picture in many more than 12 portions. In a workable system, each portion would be so small that it would contain only a single value of brightness (and color). Each tiny portion becomes a picture element or pixel, as it is commonly called.

The number of horizontal lines and the number of "pieces" along a line varies. There are three standards used worldwide for television broadcasting. Amateurs may use commercial transmission standards for TV. They are not limited to commercial standards, however. Amateurs use slow-scan TV and many experiment with other standards and systems. See Chapter 20 for information on Amateur TV and facsimile transmission.

This chapter will cover the theory of television. Discussion details are based on U.S. commercial standards — fast-scan TV (FSTV).

The Scanning Process

Scanning is the way a picture is divided sequentially into pieces for transmission or presentation (viewing). A total of 525 scan lines comprise a frame (complete picture) in the U.S. television system. Thirty frames are generated each second. Each frame consists of two fields, each field containing 262-½ lines. Sixty fields are generated each second. Scan lines from one field fall between (interlace) lines from the other field. The scanned area is called the television raster.

Fig. 23 illustrates the principle involved in scanning an electron beam (in the pickup or receiving device) across and down to produce the television raster. Scanning of field one begins in the upper left of the picture area. The electron beam is swept across the picture to the right side. At the end of the line, the beam is "turned off" or blanked and returned to the left side where the process repeats. In the meantime, the beam has also been moved slightly downward. At the end of 262½ horizontal scans (lines), the beam is blanked and rapidly returned to the top of the picture area. At that point, scanning of field two begins. Notice that this time the beam starts scanning from the top in the middle of the picture. For that reason, the scanning lines of field two will fall half way between the lines of field one. At the end of 262½ lines the beam is rapidly returned to the top of the picture again. Scanning continues — this time field one of the next frame.

Because the picture area is scanned top to bottom 60 times each second, there is no flicker. Because the entire picture area is scanned only 30 times each second, bandwidth is reduced.

It should be obvious that the horizontal

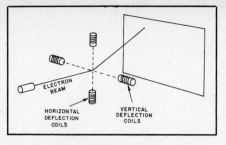

Fig. 24 — Electron beam deflection in TV cameras and receivers is usually accomplished by using two sets of coils. This is called electromagnetic deflection.

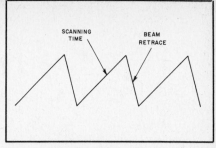

Fig. 25 — Current waveform used in electromagnetic deflection coils for deflection of the electron scanning beam.

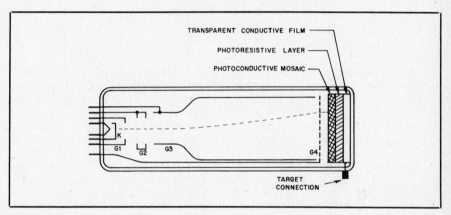

Fig. 26 — Cross-sectional view of a vidicon.

and vertical oscillators that control the electron beam movement must be "locked together" for the two fields to interlace properly. If the frequencies of these oscillators are not locked, proper interlace will be lost and vertical resolution or detail will be degraded.

Deflection

In most TV applications, the electron beam is scanned (deflected) by means of two pairs of coils. Because the deflection of the electron beam is accomplished magnetically, coils for horizontal deflection are located above and below the beam. Vertical deflection coils are located on either side of the beam. See Fig. 24.

The electron beam is deflected as a result of a "sawtooth" current passing through the deflection coils (Fig. 25). The frequency of the horizontal sawtooth current is 15,750 Hz. A similar waveform with a frequency of 60 Hz causes vertical deflection. The electron beam is turned off during beam retrace time by a process called blanking. Blanking is only associated with the electron beam and does not affect the deflection coil current or the resulting magnetic fields.

THE TELEVISION CAMERA

Most modern TV cameras use vidicon tubes or vidicon derivatives. It is in the vidicon that optical energy is converted into a video (electrical) signal. The construction of a typical vidicon can be seen in Fig. 26.

Details, such as tube size, number of grids and the type of focus and deflection used, vary among the different types.

The cathode is the source of the electrons that form the electron beam. Grid 1 is a tubular element mounted concentric to the beam that controls the quantity of beam electrons, or the beam intensity. Grid 2, another tubular element, accelerates the beam electrons toward the target. Grid 3 operates in conjunction with an external focus coil to concentrate the beam into a focused electron stream at the target. Grid 4 is a screen mesh that is used to decelerate beam electrons and ensure that the beam strikes the target perpendicularly. Grid 4 also collects any excess beam electrons that are not attracted to the target.

The photosensitive element is made up of layers that are bonded to the faceplate glass of the tube. The layer directly adjacent to the faceplate glass is a transparent coating that is a conducting material. This coating is the signal electrode of the vidicon. Deposited on top of this layer are two more layers, the three forming the vidicon target. Adjacent to the first layer is a photoresistive layer that is backed by a conductive mosaic. A positive voltage is applied to the target to attract the beam electrons.

The target acts as a "leaky capacitor" at each point where the electron beam strikes the target. In other words the target can be viewed as a large number of small capacitors side by side distributed over its

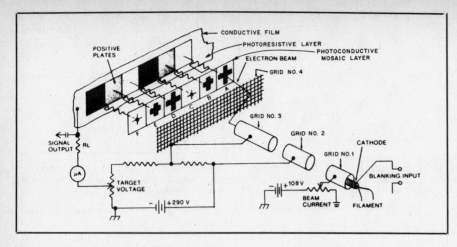

Fig. 27 — Vidicon operation depicted. See text.

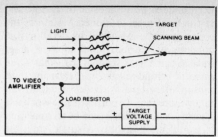

Fig. 28 — Vidicon equivalent circuit.

surface. On one side the capacitors are all joined together with a common connection for the output signal; on the other side the capacitors are left floating where they are exposed to the electron beam. The dielectric of these capacitors is the photoresistive layer of the target. Resistance of the middle layer is reduced in proportion to the intensity of light falling on it. This creates, in effect, a "leaky" capacitance with the amount of leakage proportional to the light falling on the target.

Target operation is depicted in Fig. 27. The light image consists of a series of black, gray and white areas. Where the image is black, the resistance of the photoresistive layer is very high. Where the image is white, the resistance is low. In the white areas, the positive potential on the target attracts many electrons from the photoconductive mosaic — that leaves an area with a high positive potential. That is shown in Fig. 27 by a large plus sign on the photoconductive mosaic. In the gray area fewer electrons are attracted across the "leaky capacitor." That leaves a less positively charged area than for white. In the black portion, the charge is essentially zero.

Any given element of the target is scanned by the electron beam once for each frame (two fields) or every 33 ms. The charge on the floating side of the target is neutralized each time the beam scans it. As electrons are added to the back of the target, other electrons are drawn off through the target connection. This causes varying amounts of current to flow through the target connection as the scanning beam sweeps the picture area. The amount of current flow is proportional to the amount of light in a given area. In the black areas no change in resistance occurs and the actual resistance is the static resistance of the photoresistive material (Fig. 28). Some current will flow, however, and this is called the dark current of the vidicon.

Electrons flowing from the target through the load resistor create the signal voltage or video signal that corresponds to the optical image on the face of the vidicon. The signal polarity across the load resistor is positive going for low light areas and negative going for the high light areas as shown in Fig. 29. A properly adjusted vidicon will produce approximately 0.3 μA in the high light areas and as little as 0.02 μA (dark current) in the low light areas. The difference between the dark current and the maximum current is the useful signal current.

Blanking pulses are applied to the vidicon cathode to turn off the electron beam during retrace time of the scanning beam. Usually the pulse shape at the output of the vidicon is not as clean as shown in Fig. 29. The signal requires processing.

Good signal-to-noise ratio is a function of several factors. Two factors you can control are plenty of light and only enough beam current to discharge the white highlights.

It is wise to keep scene contrast to a minimum. In other words, keep the lighting flat for best results. About 100 foot candles should be enough illumination. Too little light results in noisy video. Too much light

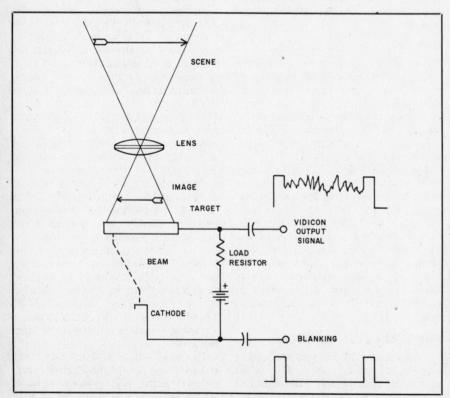

Fig. 29 — The vidicon output is a combination of video and blanking. See text.

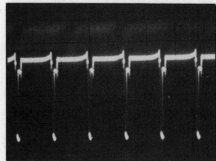

Fig. 30 — Waveform showing horizontal sync pulses.

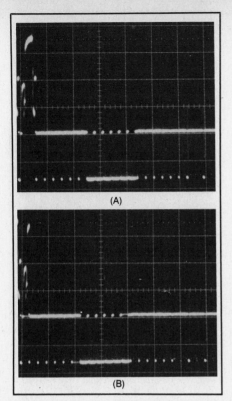

Fig. 31 — Vertical sync. Field 1 is shown at A, field 2 at B.

Table 1
Standard Video Levels

	IEEE Units	%PEV	%PEP
Zero carrier	120	0	0
White	100	12.5	1.6
Black	7.5	70	50
Blanking	0	75	56
Sync tip	−40	100	100

runs up the electrical bill and can be uncomfortable — even with air conditioning.

Synchronizing Pulses

To ensure a stable picture at the receiver, the scanning process at the transmitting and receiving ends must be synchronized. Fig. 30 shows the pulses used to control the horizontal scan. They are called horizontal sync pulses, and occur once per line — a frequency of 15,750 Hz. Horizontal sync pulses occur during the horizontal blanking interval.

Vertical scanning is controlled by vertical sync pulses, which can be seen in Fig. 31. The vertical sync pulse is preceded and followed by six equalizing pulses. The vertical sync pulse has a duration of three line times and is slotted or serrated at the end of each half-line time. Blanking turns off the electron beam during vertical retrace and sync.

The equalizing pulses are twice the frequency and approximately half the duration of horizontal sync pulses (2 equalizing pulses per line). The equalizing pulses

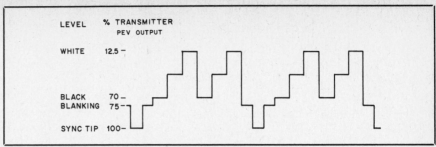

Fig. 32 — Composite video waveform. For transmission, sync is positive and white is negative.

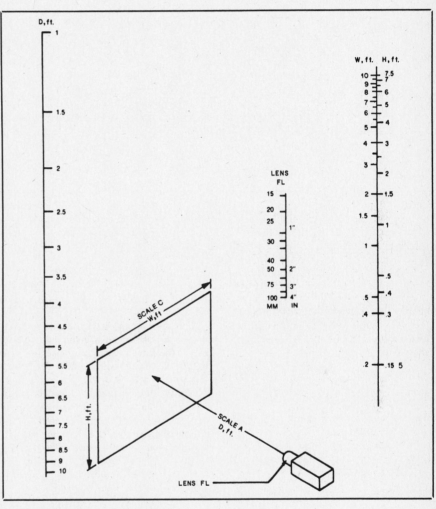

Fig. 33 — Field of view nomograph.

ensure that there is always a pulse to lock the horizontal sweep to at the beginning of each line. Line 1 of field 1 begins at the first equalizing pulse and vertical sync starts at line 4. In field 2, line 1 begins at the second equalizing pulse and vertical sync starts at line 3½. This creates the ½-line offset that generates the interlaced scanning pattern shown in Fig. 23.

Composite Video

Video, as it comes from the camera shown in Fig. 29 is not suitable for transmission. Horizontal and vertical sync must be added first. The composite video signal contains video, blanking and sync. See Table 1. Fig. 32 shows these three elements combined for two horizontal lines of the visual image of Fig. 27.

Compact, hand-held cameras today require only power and light to generate a composite video signal. Some older cameras and most TV studio cameras require horizontal and vertical sync signals from an external source. Those signals can be provided by a sync generator.

Camera Optics

The camera lens is used primarily to focus the scene image on the face of the pickup tube. The two basic characteristics of lenses are focal length (FL) and diameter

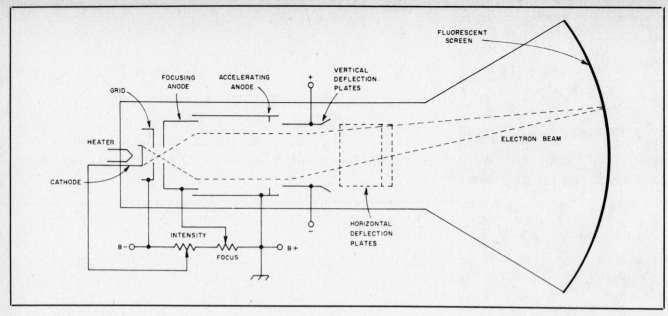

Fig. 34 — Cathode ray tube using electrostatic deflection and focusing.

of the lens opening (related to f-stop).

Lens FL is the distance from the optical center of the lens to the point behind the lens where the image comes into focus. For example, if an object at infinity comes into focus 1 inch behind a lens the FL is 1 in.

Focal length governs the magnification of the image and therefore the field of view of the lens. Long focal length lenses bring objects up closer and the field of view is less. The field of view of a lens used with a camera is related to the scanned area by the ratio:

$$FL/w = D/W \qquad \text{(Eq. 1)}$$

where

FL = focal length of the lens
w = width of the scanned area in the same units as FL
D = distance to the subject
W = width of the field in the same units as D

The image size of the scanned area of a typical vidicon is 0.375 in high × 0.50 in wide. If a 1-in lens is used, the width of the field of view would be:

$$W = D \times 0.5 \text{ in}/1 \text{ in}$$

If the subject were 5 ft away from the camera the field of view would be 2.5 ft wide (and 1.875 ft high).

This same equation can be used to determine the position of a given sized object to fill the screen. Example: Use Eq. 1 to determine how far from the camera with the 1-inch lens a 2-ft wide card should be placed to fill the screen. The answer is 4 ft.

By rearranging Eq. 1, you can determine the focal length of the lens required to view a scene of a given width at a given distance. For example, if you wish to view a scene 10 ft wide at a distance of 10 feet the lens must have a FL of 0.5 in.

$$FL = w \times D/W = 0.5 \times 10/10 = 0.5$$

Fig. 33 is a nomograph that can be used to determine these relationships easily.

F Stops

The lens iris of a TV camera operates like the iris of the human eye. If you suddenly look into a bright light, the iris of your eye closes down to admit less light. Since camera tubes operate over a limited light level range, the camera lens iris must be used to control the amount of light striking the pickup tube.

The f stop of a lens is a function of the FL and lens diameter with the iris wide open:

$$f = FL/d \qquad \text{(Eq. 2)}$$

where

f = f-stop number
FL = focal length of the lens
d = diameter of the iris opening in the same units as FL

Use Eq. 2 to prove that a lens with an 1-in FL and an iris opening of 0.66 in has an f-stop number of 1.5. Lenses are referred to as "fast lenses" or "slow lenses." These are expressions that refer to lens performance based on f-stop number. As the f-stop number of a lens increases, the opening size decreases and less light passes through the lens. The amount of light passing through the lens is inversely proportional to the square of the f-stop numbers. For example, the amount of light passing through an f/1.5 lens compared to an f/3 lens is:

$$(f - 3/f - 1.5)^2 = 4$$

The f/1.5 lens passes four times as much

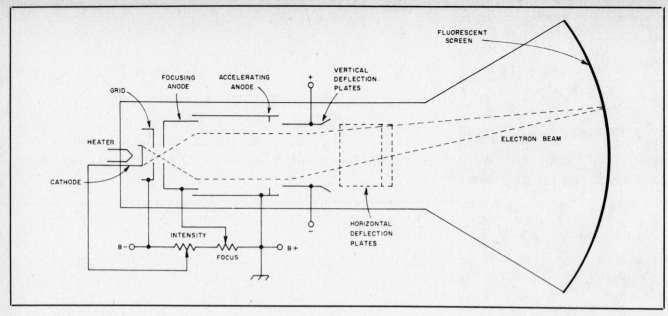

(A)   (B)

Fig. 35 — Electromagnetic focus is shown at A and deflection at B. The drawings show the physical location of the windings.

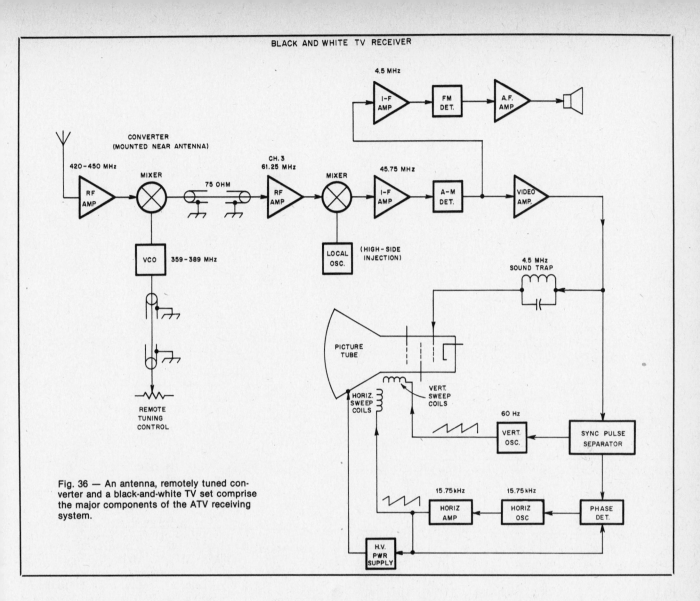

Fig. 36 — An antenna, remotely tuned converter and a black-and-white TV set comprise the major components of the ATV receiving system.

light as the f/3 lens. To put it another way, when you set an f/1.5 lens iris for f/3, light passing through the lens is reduced to one quarter of the original amount.

Depth of Field

Depth of field describes how much of the scene, from front to rear, is in focus. The larger the f-stop number used with a lens the greater is the depth of field. Also, the shorter the focal length of a lens the more depth of field it has.

There is no substitute for a good lens — beware of bargains! In choosing a lens, many other qualities should be considered. A discussion of those qualities is beyond the scope of this *Handbook*.

The TV Picture Tube

A picture or cathode-ray tube (CRT) similar to those used in oscilloscopes is shown in Fig. 34. The heater and cathode supply a stream of electrons that is controlled by the grid. Focus of the stream to a single point at the screen is realized by adjusting the voltage on the focusing anode. Voltage differences between the deflection plates control the deflection of

the electron stream. In Fig. 34 the top vertical deflection plate is positive with respect to the bottom plate. Therefore, the negatively charged electron stream is deflected upward. When the electrons strike the coating on the back of the screen there is a resulting fluorescence. Persistence and color of the fluorescence will depend on the material that coats the back of the screen.

Unlike an oscilloscope CRT, the modern TV picture tube uses electromagnetic deflection and focus. This is illustrated in Fig. 35. Notice that the top and bottom deflection coils are used to sweep the beam horizontally. That is because they are using the magnetic rather than the electric field for deflection.

Fig. 36 is a block diagram of an ATV receiving station. For best results the converter should be mounted near the antenna. Any TV set may be used after the converter. See Chapter 20 for more information on assembling and operating an ATV station.

TV Sound

In a regular TV channel, the sound is transmitted on an FM carrier 4.5 MHz

higher than the picture carrier. A separate transmitter can be used to generate the sound carrier, but that requires a hard-to-build combiner or a second antenna. It is easy to frequency-modulate the visual carrier, but that requires special circuitry for receiving. An attractive alternative to these methods is to generate an FM sound subcarrier at 4.5 MHz and combine that with the picture information at the video modulator input. Perhaps the easiest way for amateurs to transmit TV sound is to use an FM simplex channel on another band.

In many areas a 2-meter FM simplex channel is used for ATV coordination. This helps minimize interference. Another signal 50 dB weaker can cause noticeable interference to an ATV picture. When the ATV station is sending in-channel sound, the receiving station can transmit on the coordinating frequency; the result is full duplex sound — a most pleasant benefit. Most CQ calls for ATV are heard on the coordinating frequency.

COLOR TV

A color TV camera typically splits light, by means of a prism, into the three primary

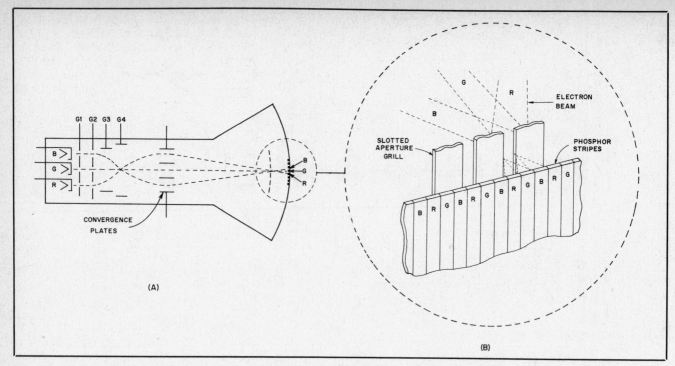

Fig. 37 — Modern single-gun color picture tube. Earlier color CRTs used three guns and a shadow mask. The advantage of the single gun type of tube is that it is easier to converge. Convergence occurs when the RGB electron beams illuminate the same aperture in the slotted grill at the same time. That ensures that the constituent parts of the color image are aligned with one another.

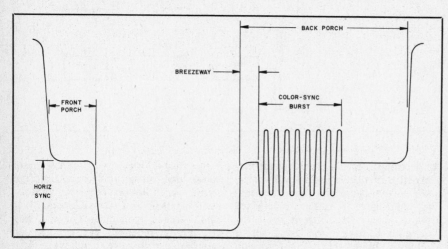

Fig. 38 — Details of horizontal sync and blanking for color TV.

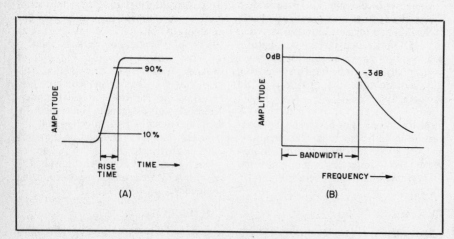

Fig. 39 — Rise time is depicted at A, bandwidth at B. See text.

colors — red, green and blue (RGB). A separate tube is used to detect each primary. The RGB outputs are combined to give a luminance (brightness) signal, which is transmitted in the normal way. In addition, difference signals are used to phase-modulate a 3.58-MHz subcarrier (actually 3,579,545 Hz) with the color, or chroma, information.

A color TV receiver is similar to the set shown in Fig. 36. The major difference is in the video detection and display. A simple envelope detector will not do for color reception; the phase-encoded chroma information must be extracted from the 3.58-MHz subcarrier. RGB plus luminance signals are detected and are then used to control three separate electron beams (one per primary) in the picture tube. A typical color picture tube can be seen in Fig. 37.

Eight to 10 cycles of color subcarrier are sent after horizontal sync. See Fig. 38. These are the color-sync burst, or color burst, as it is commonly called. Color burst is used for a subcarrier phase reference in the receiver color detector circuitry synchronous demodulator. Also given in Fig. 38 are the names for the various portions of the horizontal sync and blanking waveform.

Video Frequencies and Bandwidth

The lowest video frequency component is determined by the time rate of change of average luminance of the scene. For a picture that is uniform and unchanging, the lowest frequency component is zero — dc. It would be very difficult to pass a dc signal

Table 2

Relationship of Rise Time and Bandwidth

Rise time (µs)	Bandwidth (MHz)
0.35	1
0.18	2
0.117	3
0.087	4
0.07	5
0.058	6
0.05	7
0.044	8
0.039	9
0.035	10

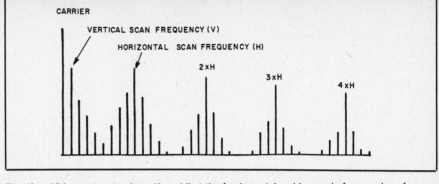

Fig. 40 — Video energy tends to "bunch" at the fundamental and harmonic frequencies of horizontal and video sync.

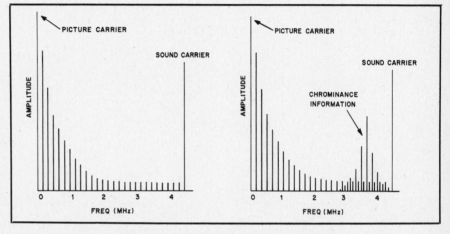

Fig. 41 — With the proper selection of sync and chrominance-subcarrier frequencies, chrominance and luminance information "bunches" fall between each other.

from the camera, through the transmitter and to the TV picture tube. The first blocking capacitor would spell and end to 0-Hz frequency response! Fortunately, that kind of response is not necessary.

A reference (dc) level can be set for each video scan line through a technique known as clamping. Clamping sets or holds a level at a reference point. Sync and blanking are present for each line of the video frame. The sync tip or the blanking level can be used to "clamp" or set a dc level at the beginning of each line time.

Maximum video frequency must be considered for three different points. First, the maximum frequency generated by the camera. Second, the maximum video frequency passed through the transmitter. Third, the maximum frequency present at the receiver picture tube.

The maximum camera frequency is determined by the size of the electron scanning beam and the bandwidth of the video amplifiers. Bandwidth limitation is based on the ability to reproduce an abrupt level change. Consider a black-to-white transition. It will take some finite amount of time for the level to shift from black to white. The time it takes for levels to shift between the 10 and 90 percent points in the level shift is called the rise time. See Fig. 39A. The bandwidth required to make that shift is depicted in Fig. 39B. The relationship of bandwidth and rise time is:

$$BW = 0.35/T \qquad \text{(Eq. 3)}$$

where
  BW = bandwidth in hertz
  T = rise time in seconds

The relationship between rise time and bandwidth for some specific values is given in Table 2.

Video cameras typically have a bandwidth of 8 MHz or so. Simple black-and-white cameras may have less, and sophisticated color cameras have more.

The maximum video passed through a transmitter is usually limited by the transmitter circuitry. Broadcast TV transmitters are carefully designed and adjusted to pass video frequencies according to strict standards. Amateurs do not have the same restraints. A little over 4-MHz bandwidth is required to pass carrier and one sideband of a color-TV signal.

A number of factors determine the maximum video frequency to reach the TV receiver picture tube. IF bandwidth is usually the limiting factor. On-screen resolution of the CRT is also limited by the design and construction of the electron-

beam focus and deflection coils.

TV Energy Distribution

The square-wave sync pulses are transmitted at maximum power level. For that reason, power distribution in the video passband is concentrated at the fundamental and harmonics of the horizontal scanning frequency. Surrounding these are sidebands related to the vertical scanning frequency and its harmonics. That distribution is depicted in Fig. 40.

For color TV, chrominance information must be added to the luminance information. Fig. 41 shows how this is done to minimize interference between the two signal components. The chrominance subcarrier has a nonintegral relationship to the horizontal scanning frequency. The energy "bunches" from color and luminance fall between each other.

Chapter 8

Digital Basics

Digital electronics is an important aspect of Amateur Radio. Everything from simple digital circuits to sophisticated microcomputers has found application in modern Amateur Radio systems. Applications include digital communications, code conversion, signal processing, station control, frequency synthesis, amateur satellite telemetry, message handling, word processing and other information-handling operations.

The fundamental principle underlying digital electronics is that a device can have only a finite number of states. In *binary* digital systems there are two discrete states, represented in base-2 arithmetic by the numerals 0 and 1. Although binary digital systems are the most common, trinary (3 states), quaternary (4 states) and other digital systems exist. Digital systems, with their discrete states, are contrasted with analog systems, in which voltages or other quantities vary in a continuous manner. Fig. 1 compares a sine wave with a digital approximation of the same waveform. There are situations in which a designer is faced with a choice between digital technology and analog technology. Some problems are best solved by digital circuits, some by analog circuits, and others by circuits using both digital and analog components. To understand Amateur Radio equipment, you will need a good working knowledge of both analog and digital circuits. This chapter examines digital electronics from the simplest digital building blocks to the complex microcomputer.

Binary Quantities

As stated above, most digital systems are binary. When discussing these systems, each binary digit, or *bit,* is represented by a 1 or a 0. The binary states described as 1 and 0 may represent on and off, a punched hole and the absence of a hole in paper tape or card, or a mark and space in a communications transmission. In most electronic digital logic systems, bits are represented by voltage levels. For example, in one common circuit family, 0 volts is a binary 0, and +5 volts is a binary 1. Because it is not always possible to achieve exact voltages in practical circuits, digital circuits consider the signal to be a 1 or 0

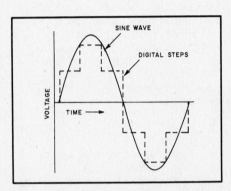

Fig. 1 — A sine wave and a digital approximation of that waveform. Note that the sine wave has continuously varying voltage and that the digital waveform is composed of discrete steps.

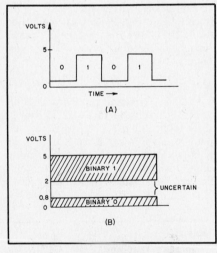

Fig. 2 — A typical digital logic signal is shown in A. B illustrates possible ranges of recognition for binary 1 and 0.

if the voltage comes within certain bounds, as shown in Fig. 2.

In electronic circuits, changes of state do not occur instantly. The time it takes to go from a 0 to a 1 state is called *rise time. Fall time* is the time it takes to change from a 1 to a 0. Fig. 3 shows how rise and fall times effect the shape of a binary signal, or *pulse.* Rise and fall times describe a signal at a specific point in a circuit. The

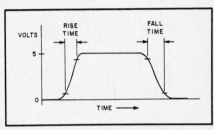

Fig. 3 — Shape of a typical digital pulse.

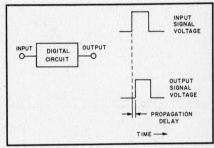

Fig. 4 — Propagation delay in a digital circuit.

distributed inductances and capacitances in a wire or PC-board trace cause rise and fall times to lengthen as a pulse gets farther from its source. Rise and fall times are typically in the microsecond or nanosecond range and vary with the logic family used.

Digital signals take a measurable amount of time to pass through a circuit. This phenomenon is called *propagation delay,* and results in an output signal being slightly delayed from an input signal, as shown in Fig. 4. Propagation delay is the result of transistor switching delays, reactive element charging times and the time needed for signals to travel through wires. In complex circuits, different propagation delays through different paths can cause problems when pulses must arrive somewhere at exactly the same time.

COMBINATIONAL LOGIC

In binary digital circuits, a combination of binary inputs results in a specific binary output or combination of binary outputs. These circuits are used to implement everything from simple switches to

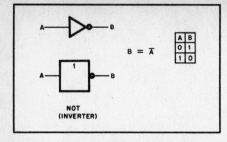

Fig. 5 — Inverter. The distinctive triangular symbol is the standard used in ARRL and other U.S. publications. The square symbol is used in some other countries.

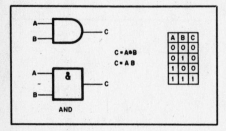

Fig. 6 — Two-input AND gate.

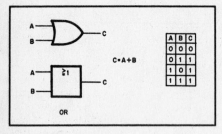

Fig. 7 — Two-input OR gate.

Table 1

Useful Properties and Theorems in Boolean Algebra

Commutative Property	$AB = BA$
	$A + B = B + A$
Associative Property	$A(BC) = (AB)C$
	$A + (B + C) = (A + B) + C)$
Distributive Property	$A(B + C) = AB + AC$
	$A + (BC) = A + B(A + C)$
Theorems	$A \cdot 0 = 0$
	$A \cdot 1 = A$
	$A \cdot A = A$
	$A \cdot \overline{A} = 0$
	$\overline{\overline{A}} = A$
	$A + 0 = A$
	$A + 1 = 1$
	$A + A = A$
	$A + \overline{A} = 1$

---

powerful computers. Some binary circuits are called *combinational logic networks.* The output of a combinational logic network is dependent only on its inputs, and its output changes immediately when its input changes. While there are more complex types of binary circuits, these circuits are usually composed of combina-

tional logic elements. To understand complex digital circuits, one must understand combinational logic.

BOOLEAN ALGEBRA

Just as there are mathematical theories and equations used in designing analog circuits, there is a special branch of mathematics, *Boolean algebra,* that is used to describe and design binary digital circuits. In standard algebra there are several basic operations: addition, subtraction, multiplication and division. In Boolean algebra, the basic operations are *logical operations:* AND, OR and NOT. Boolean algebra is used to describe complex logical functions (often digital circuits) as combinations of these logical operators.

A logical function or operator can be defined by describing its output for all possible inputs. Such a list of inputs and outputs is called a *truth table.* In a truth table, a T represents a true or a 1 signal and an F represents a false or a 0 signal.

Boolean Operators

Figs. 5 through 7 show the truth tables, circuit symbols and Boolean algebra symbols for the three basic Boolean operations, NOT, AND and OR. All combinational logic functions, no matter how complex, can be described in terms of these three operators.

NOT

The NOT operation is called inversion, negation or complement. The Boolean notation for NOT is a bar over a variable or expression. As can be seen from the truth table in Fig. 5, NOT 1 is 0 and NOT 0 is 1. The circuit that implements the NOT function is called an *inverter* or *inverting buffer.*

AND

The AND operation results in a 1 only when both of its operands are 1. That is, if the inputs are called A and B, the output is 1 only if A *and* B are 1. In Boolean notation, AND is represented by a middle dot (•) between the variables, or nothing between the variables. Both of these forms

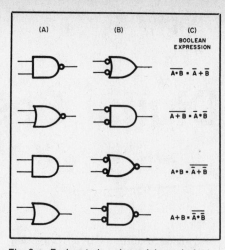

Fig. 8 — Each gate in column A is equivalent to the gate opposite it in column B. The Boolean expressions in column C formally state the equivalences.

are shown in Fig. 6. The circuit that implements the AND function is called an AND *gate.* Gates are combinational logic elements with two or more inputs and one output state.

OR

The OR operation results in 1 if one or the other of its operands is 1 (Fig. 7). This also includes the case where both operands are 1; because of this, OR is sometimes called the INCLUSIVE OR. The Boolean notation for OR is + between the two operands. The circuit that implements the OR function is called an OR gate.

Boolean Identities

Algebraic identities, such as $A \times B = B \times A$, help us solve complicated problems in mathematics. There are several identities in Boolean algebra that are useful for solving complex logic problems. These identities are shown in Table 1.

DeMORGAN'S THEOREM

One of the Boolean identities deserves

Table 2

DeMorgan's Theorem

(A) $\overline{A \cdot B} = \overline{A} + \overline{B}$

(B) $\overline{A + B} = \overline{A} \cdot \overline{B}$

(1)	(2)	(3)	(4)	(5)	(6)	(7)	(8)	(9)	(10)
A	B	$\overline{A}$	$\overline{B}$	$A \cdot B$	$\overline{A \cdot B}$	$A + B$	$\overline{A + B}$	$\overline{A} \cdot \overline{B}$	$\overline{A} + \overline{B}$
0	0	1	1	0	1	0	1	1	1
0	1	1	0	0	1	1	0	0	1
1	0	0	1	0	1	1	0	0	1
1	1	0	0	1	0	1	0	0	0

(A) and (B) are statements of DeMorgan's Theorem. The truth table at (C) is proof of these statements; (A) is proven by the equivalence of columns 6 and 10; (B) is proven by the fact that columns 8 and 9 are the same.

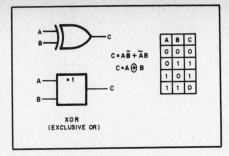

Fig. 9 — Two-input XOR gate.

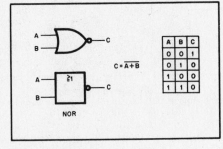

Fig. 10 — Two input NOR gate.

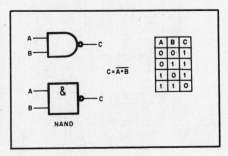

Fig. 11 — Two-input NAND gate.

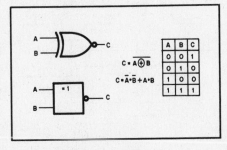

Fig. 12 — Two-input XNOR gate.

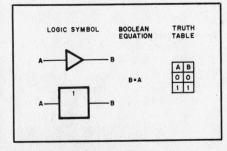

Fig. 13 — Noninverting buffer.

special attention. This identity is called DeMorgan's Theorem. DeMorgan's Theorem consists of two identities: $A \cdot B = \overline{A} + \overline{B}$ and $\overline{A + B} = \overline{A} \cdot \overline{B}$. The truth table in Table 2 proves these identities. The important consequence of DeMorgan's Theorem is that any logical function can be implemented using either inverters and AND gates or inverters and OR gates. The theorem gives the circuit designer freedom, because it allows AND gates and inverters to be substituted for OR gates, and vice versa. The gate substitutions shown in Fig. 8 illustrate some uses of DeMorgan's Theorem. Such substitutions can lead to simplification of design or to more efficient use of available gates.

Other Common Gates

While we have shown that any logical function can be implemented with only two logic elements (AND and NOT or OR and NOT), there are several other logic elements that are easily implemented and commonly available. Using these available gates (illustrated in Figs. 9 through 13) can simplify logic circuits.

EXCLUSIVE OR

In an EXCLUSIVE OR (XOR) gate, the output is 1 only if a single input is 1. If both inputs are 1 or both inputs are 0, the output is 0. The symbol $\oplus$ in a Boolean expression represents the XOR operation.

NOR

NOR means NOT OR. A NOR gate is a circuit that produces a 0 at its output if any or all of its inputs are 1.

NAND

NAND is a contraction of NOT AND. A NAND gate has a 0 output only when all of its inputs are 1.

EXCLUSIVE NOR

The EXCLUSIVE NOR (XNOR) gate produces 0 output only if a single input is 1. Note that if both of the inputs are 1, there is a 1 on the output.

Noninverting Buffer

The *noninverting buffer* does not implement a Boolean function. Its output is simply equal to its input. While not useful for logical operations, a noninverting buffer has several uses: (a) providing sufficient current to drive a number of gates, (b) in-

terfacing between two logic families, (c) obtaining a desired pulse rise time and (d) providing a slight delay to make pulses arrive at the proper time.

Polarity

In Boolean algebra, gates operate on the logical values 1 (true) and 0 (false). When implementing gates as electronic circuits, we need to assign voltage levels to 1 and 0. The intuitively acceptable convention is to use a higher voltage for 1 and a lower voltage for 0. This convention is called *positive logic,* and the standard logic symbols illustrated above are symbols for positive logic elements. In *negative logic,* the higher voltage represents 0 and the lower voltage represents 1. A negative logic input or output is represented by a small circle on the negative logic terminal. One can also think of these circles as representing inverted inputs or outputs. While logic polarity can be confusing, being able to think in both positive and negative logic sometimes makes circuit design easier.

SEQUENTIAL LOGIC

The previous sections discussed combinational logic, which has its outputs *immediately* dependent on its *inputs only*. There is another class of logic called *sequential logic*. In sequential logic, a network's outputs are dependent not only on its current inputs, but also on its current outputs. This is a type of feedback and requires memory. The logic element that supplies memory is generally called a *flip-flop*.

A flip-flop (also known as a *bistable multivibrator* or an *Eccles-Jordan circuit*) is a binary sequential logic element with two stable states. One is the *set* state (1 state); the other is the *reset* state (0 state). The schematic symbol for a flip-flop is a rectangle with the letters FF. These letters may be omitted if there is no ambiguity.

Flip-flop inputs and outputs are normally identified by a single letter. A letter followed by a subscripted letter (such as D_C), means that the input is dependent on the input of the subscripted letter (input D_C is dependent on input C). Flip-flop outputs are labelled Q and $\overline{Q}$ (Table 3).

State Transition Tables

Conventional truth tables are difficult to apply to flip-flops, because flip-flop outputs depend on current outputs as well as on current inputs. Also, a certain input

---

Table 3
Flip-Flop Out Designations

Output	Action	Restrictions
Q (Set)	Normal output	Only two output states are
$\overline{Q}$ (Reset)	Inverted output	possible: Q = 1, $\overline{Q}$ = 0; and Q = 0, $\overline{Q}$ = 1.

Notes
1) $\overline{Q}$ is the complement of Q.
2) The normal output is normally marked Q or unmarked.
3) The inverted output is normally marked $\overline{Q}$. If so, if there is a 1 state at Q, there will be a 0 state at $\overline{Q}$.

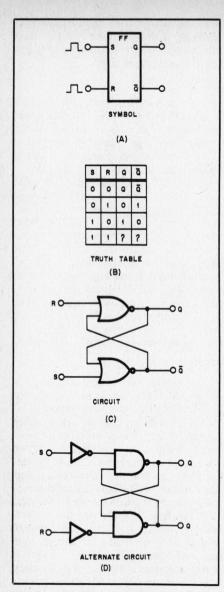

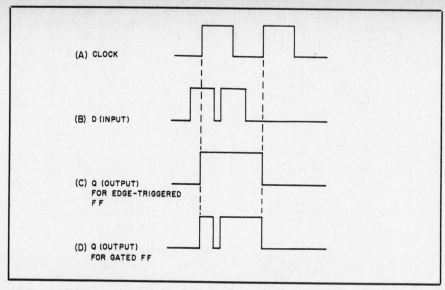

Fig. 15 — This figure illustrates the different outputs that may result when a D flip-flop is edge triggered (C) or gated (D). Notice that the short negative pulse on the input (B) is not reproduced by the edge-triggered flip-flop.

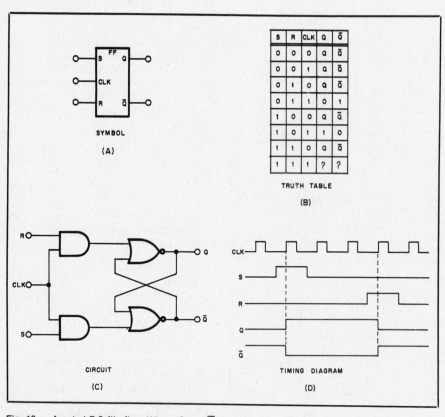

Fig. 16 — A gated R-S flip-flop. Where Q and $\overline{Q}$ are shown in the table, the previous states are retained. A question mark (?) indicates an unpredictable output state.

Fig. 14 — An unclocked R-S flip-flop. Where Q and $\overline{Q}$ are shown in the truth table, the previous states are retained. A question mark (?) indicates that the output state is unpredictable. C shows an R-S flip-flop made from two NOR gates. Circuit D is another implementation using two NAND gates and two inverters.

condition may not have a unique output state. The simpler sequential circuits can be defined by state tables showing input and output states as they are now, and output states as they will be after the next input pulse. These tables are called *state transition tables*. State transition tables are used for designing sequential logic networks.

Unclocked R-S Flip-Flop

One of the simplest circuits for storing a bit of information is the R-S (or S-R) flip-flop. Its inputs are R (reset) and S (set). Fig. 14 shows how an R-S flip-flop can be made using combinational logic elements.

When S = 0 and R = 0 the output will stay the same as it was at the last input pulse. This is indicated by the letter Q in the truth table.

If S = 1 and R = 0, Q will change to 1. If S = 0 and R = 1, Q will change to 0.

If both the R input and the S input are 1, the flip-flop output assumes an unpredictable state. This state varies from circuit to circuit, and may be Q = $\overline{Q}$ = 1 or Q = $\overline{Q}$ = 0. While Q = $\overline{Q}$ is a logical impossibility, real flip-flops may present this output state. Avoid the R = S = 1 input, and make no assumptions about the resulting output.

Synchronous Flip-Flops

The R-S flip-flop is an *asynchronous* device; its outputs change immediately to reflect changes on its inputs. *Synchronous* flip-flops change state (to reflect input conditions) only at specific times. These times are dictated by a *clock, toggle* or *gate* input.

Synchronizing inputs can take two

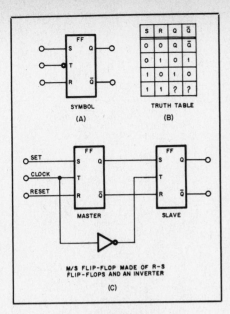

Fig. 17 — A master/slave flip-flop.

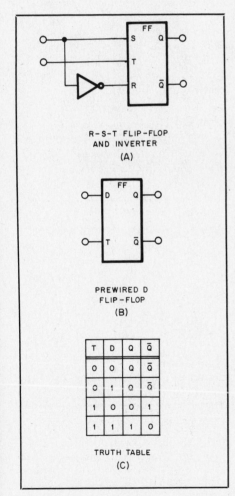

Fig. 18 — The D flip-flop. When T = 0, Q and Q̄ states don't change. When T = 1, the output states change to reflect the input (D).

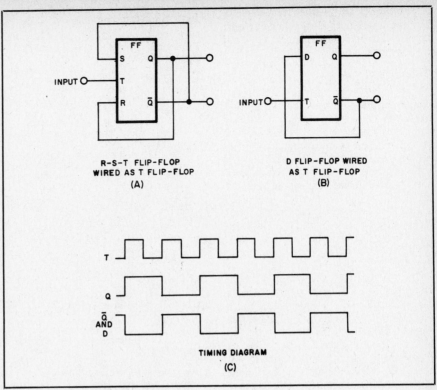

Fig. 19 — The T flip-flop. The output frequency is half of the input frequency.

a specified level (1 or 0). Dynamic or *edge triggered* inputs allow the flip-flop to change state only when the clock *changes* level. Dynamic inputs that react to a transition from 0 to 1 are called *positive-edge triggered* inputs, and are indicated by a small isosceles triangle inside the flip-flop at the clock input. Inputs that are sensitive to the transition from 1 to 0 are *negative-edge triggered,* and are indicated by the triangle and a negation symbol (a small circle outside the flip-flop rectangle). It is important to remember that input levels affect gated flip-flops whenever the clock is in a specific state, while inputs affect edge triggered flip-flops only when the clock is in transition. Fig. 15 shows how the same input to a gated and an edge-triggered flip-flop results in two different outputs.

Gated R-S Flip-Flop

The gated R-S flip-flop, or *gated latch,* has R, S and clock inputs. It is also called the R-S-T flip-flop, the "T" indicating a *toggle* input. Although not often used, the R-S-T flip-flop is important because it illustrates a step between the R-S flip-flop and the J-K flip-flop. The inputs R and S produce the same results as those on an unclocked R-S flip-flop, but no change in output will occur unless the toggle input is high. A gated R-S flip-flop is illustrated in Fig. 16 along with a timing diagram to show the effect of the clock input.

Master/Slave Flip-Flop

The importance of synchronous flip-flops is illustrated in the construction of the

master/slave flip-flop. The master/slave (M/S) flip-flop solves a problem known as *race.* Race occurs when flip-flops are connected in series and triggered by the same clock pulse or when a flip-flop uses its own output as an input. When an input changes just as an output is about to change, an erroneous output may result.

This problem is solved by building a flip-flop that samples and stores its inputs before changing its outputs. Such a flip-flop can be built by placing two gated R-S flip-flops in series. The first, or *master,* flip-flop is active when the clock is positive; it samples and stores the inputs when the clock is 1. The second, or *slave,* flip-flop takes its input from the output of the master, and acts when the clock is negative. Hence, when the clock is 1, the input is sampled, and when the clock is 0, the output is generated. Changes in the output are now isolated from the input, and there is no race condition.

An M/S flip-flop can be made by wiring two R-S flip-flops in series and connecting an inverter between the T inputs. Thus when the T input is 1, the master (first flip-flop) is enabled. When the T input is 0, the slave (second flip-flop) is enabled and the master disabled. An R-S M/S flip-flop is shown in Fig. 17.

D Flip-Flop

In a D flip-flop, the *data* input (D) is transferred to the outputs (Q and Q̄) when the flip-flop is enabled by the T input. Fig. 18 shows a typical D flip-flop and its state transition table. The logic level at D

forms, *static* or *dynamic.* Static, *gated,* or *level triggered* inputs allow the flip-flop to change state whenever the clock input is at

is transferred to Q when the clock is positive. The Q output will retain this logic level, regardless of any changes at D, until the next positive clock pulse.

Toggle Flip-Flop

Fig. 19 shows a *toggle,* or T flip-flop. The timing diagram shows that the flip-flop output changes state with each positive clock pulse. Because of this, the T flip-flop is also called a complementing flip-flop. Examination of the pulses in Fig. 19C shows that the output frequency of the T flip-flop is one half of the input frequency. Thus a T flip-flop is a 2:1 (also called a *modulo-2* or *radix-2*) frequency divider. Two T flip-flops connected in series form a 4:1 divider, and so on.

J-K Flip-Flop

The J-K flip-flop is a common flip-flop. It has no invalid input combinations (a disadvantage of the R-S flip-flop), and it can complement (divide by 2). The J-K flip-flop is illustrated in Fig. 20.

On receipt of a clock pulse, inputs of J = 1, K = 0 sets Q = 1. J = 0, K = 1 resets Q = 0. Simultaneous J = 1, K = 1 causes Q to change state (complement). If J = 0 and K = 0, Q will not change state when the flip-flop is clocked.

Asynchronous Inputs

Most commercially available synchronous flip-flops have some asynchronous inputs. These inputs allow you to change the state of the flip-flop regardless of the state of the clock input. Asynchronous inputs usually found on flip-flops are the *direct reset,* RD, and *direct set,* SD. As their names suggest, these inputs allow the flip-flop to be set or reset asynchronously. Such inputs are used to establish known conditions in a circuit or to interrupt an ongoing process.

COUNTERS

A *counter, divider* or *divide-by-n* counter is a circuit that stores pulses and produces an output pulse when a specified number, n, of pulses are stored. In a counter consisting of flip-flops connected in series, a change of state of the first stage affects the second stage and so on.

A *ripple, ripple-carry* or *asynchronous* counter passes the count from stage to stage; each stage is clocked by the preceding stage. In a *synchronous* counter, each stage is controlled by a common clock.

Weighting is the count value assigned to each counter output. In a four-stage binary ripple counter, the output of the first stage has a weight of 1, the second 2, the third 4 and the fourth 8. Base 10 (decade) counters consist of a divide-by-2 and divide-by-5 counters (encoded in binary-coded decimal or BCD code, as shown in Table 4). A base 12 counter has a divide-by-2 and a divide-by-6 counter.

Most counters have the ability to *clear*

Digital Number Systems

Decimal

The familiar *decimal* or *base-10* system is comprised of 10 symbols (digits or numerals): 0, 1, 2, 3, 4, 5, 6, 7, 8 and 9. The value of each depends on its position in the number. For example: In the decimal number 123, 1 = 100, 2 = 20 and 3 = 3 units.

The first position carries the greatest weight and is known as the *most significant digit* (MSD), while right-most position is referred to as the *least significant digit* (LSD). The positional values relative to the decimal point can be expressed in terms of *powers of 10,* also known as scientific notation.

Decimal position values of the number decimal 234.56 in powers of 10 are:

100	10	1		0.1	0.01	Decimal value
10^2	10^1	10^0		10^{-1}	10^{-2}	Power of 10 value
2	3	4	.	5	6	Decimal number
↑					↑	
MSD		Decimal point			LSD	

Binary

In the *binary* or *base-2* system only two symbols are used: 0 and 1. As in the decimal system, the position of a binary digit determines its weight which is expressed in powers of 2. The left-most position is the *most significant bit* (MSB), and the right-most position is the *least significant bit* (LSB).

Position values of the binary number 1101011 are:

64	32	16	8	4	2	1	Decimal value
2^6	2^5	2^4	2^3	2^2	2^1	2^0	Power of 2 value
1	1	0	1	0	1	1	Binary number
↑						↑	
MSB					LSB		

Octal

In the *octal* or *base-8* system, 8 symbols are used: 0, 1, 2, 3, 4, 5, 6 and 7. Each positional value can be expressed in powers of 8. Position values for the octal number 7654 are:

512	64	8	1	Decimal value
8^3	8^2	8^1	8^0	Power of 8 value
7	6	5	4	Octal number
↑			↑	
MSD			LSD	

Hexadecimal

In the *hexadecimal* (hex) or *base-16* system 16 symbols are used: 0, 1, 2, 3, 4, 5, 6, 7, 8, 9, A, B, C, D, E and F. Each position value can be expressed in powers of 16. Position values for the hexadecimal number FF07 are:

4096	256	16	1	Decimal value
16^3	16^2	16^1	16^0	Power of 16 values
F	F	0	7	Hexadecimal number
↑			↑	
MSD			LSD	

Binary-Coded Decimal

In the binary-coded decimal (BCD) system the decimal numbers 0 through 9 are represented by 4-bit binary patterns:

BCD	Decimal	BCD	Decimal
0000	0	1000	8
0001	1	1001	9
0010	2	1010	
0011	3	1011	Not
0100	4	1100	Used
0101	5	1101	
0110	6	1110	
0111	7	1111	

The decimal number 789 would be expressed in BCD as:

7	8	9	Decimal number
0111	1000	1001	BCD code

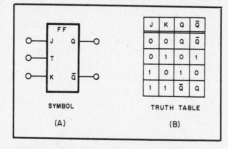

Fig. 20 — A J-K flip-flop. The truth table shows that if J = 0 and K = 0, the output does not change, and if J = 1 and K = 1, the output state is inverted.

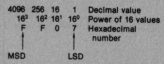

Table 4
Binary-Coded Decimal (BCD) Code

Decimal Number	Binary Weights			
	8	4	2	1
0	0	0	0	0
1	0	0	0	1
2	0	0	1	0
3	0	0	1	1
4	0	1	0	0
5	0	1	0	1
6	0	1	1	0
7	0	1	1	1
8	1	0	0	0
9	1	0	0	1

Note: BCD does not use hexadecimal (base 16) numbers A (1010), B (1011), C (1100), D (1101), E (1110) or F (1111).

the count to 0. Some counters can be *preset* to a desired count.

Counters may either count up (increment) or down (decrement). *Up* counters count in increasing weighted values (for example: 0000, 0001, 0010,..., 1111). *Down* counters work in the reverse order. There are also *up/down* counters that can be commanded to count either up or down.

REGISTERS

Groups of flip-flops may be used to temporarily store data. Such a data-storage device is called a *register.* There are two classes of registers: *shift registers* and *word* or *byte registers.* Registers are used in counters, data-storage devices and sequence generators.

Shift Registers

A shift register is a group of flip-flops connected so that when a clock pulse occurs, each flip-flop assumes the state of the preceding flip-flop. The first flip-flop of the group usually receives its input from

Table 5
Number Conversion

Decimal	Binary	Octal	Hexa-decimal	BCD
0	0	0	0	0000
1	1	1	1	0001
2	10	2	2	0010
3	11	3	3	0011
4	100	4	4	0100
5	101	5	5	0101
6	110	6	6	0110
7	111	7	7	0111
8	1000	10	8	1000
9	1001	11	9	1001
10	1010	12	A	0001 0000
11	1011	13	B	0001 0001
12	1100	14	C	0001 0010
13	1101	15	D	0001 0011
14	1110	16	E	0001 0100
15	1111	17	F	0001 0101
16	10000	20	10	0001 0110
17	10001	21	11	0001 0111
18	10010	22	12	0001 1000
19	10011	23	13	0001 1001
20	10100	24	14	0010 0000
21	10101	25	15	0010 0001
22	10110	26	16	0010 0010
23	10111	27	17	0010 0011
24	11000	30	18	0010 0100
25	11001	31	19	0010 0101
26	11010	32	1A	0010 0110
27	11011	33	1B	0010 0111
28	11100	34	1C	0010 1000
29	11101	35	1D	0010 1001
30	11110	36	1E	0011 0000
31	11111	37	1F	0011 0001
32	100000	40	20	0011 0010
33	100001	41	21	0011 0011
34	100010	42	22	0011 0100
35	100011	43	23	0011 0101
36	100100	44	24	0011 0110
37	100101	45	25	0011 0111
38	100110	46	26	0011 1000
39	100111	47	27	0011 1001
40	101000	50	28	0100 0000
41	101001	51	29	0100 0001
42	101010	52	2A	0100 0010
43	101011	53	2B	0100 0011
44	101100	54	2C	0100 0100
45	101101	55	2D	0100 0101
46	101110	56	2E	0100 0110
47	101111	57	2F	0100 0111
48	110000	60	30	0100 1000
49	110001	61	31	0100 1001
50	110010	62	32	0101 0000
51	110011	63	33	0101 0001
52	110100	64	34	0101 0010
53	110101	65	35	0101 0011
54	110110	66	36	0101 0100
55	110111	67	37	0101 0101
56	111000	70	38	0101 0110
57	111001	71	39	0101 0111
58	111010	72	3A	0101 1000
59	111011	73	3B	0101 1001
60	111100	74	3C	0110 0000
61	111101	75	3D	0110 0001
62	111110	76	3E	0110 0010
63	111111	77	3F	0110 0011
64	1000000	100	40	0110 0100
65	1000001	101	41	0110 0101
66	1000010	102	42	0110 0110
67	1000011	103	43	0110 0111
68	1000100	104	44	0110 1000
69	1000101	105	45	0110 1001
70	1000110	106	46	0111 0000
71	1000111	107	47	0111 0001
72	1001000	110	48	0111 0010
73	1001001	111	49	0111 0011
74	1001010	112	4A	0111 0100
75	1001011	113	4B	0111 0101
76	1001100	114	4C	0111 0110
77	1001101	115	4D	0111 0111
78	1001110	116	4E	0111 1000
79	1001111	117	4F	0111 1001
80	1010000	120	50	1000 0000
81	1010001	121	51	1000 0001
82	1010010	122	52	1000 0010
83	1010011	123	53	1000 0011
84	1010100	124	54	1000 0100
85	1010101	125	55	1000 0101
86	1010110	126	56	1000 0110
87	1010111	127	57	1000 0111
88	1011000	130	58	1000 1000
89	1011001	131	59	1000 1001
90	1011010	132	5A	1001 0000
91	1011011	133	5B	1001 0001
92	1011100	134	5C	1001 0010
93	1011101	135	5D	1001 0011

Decimal	Binary	Octal	Hexa-decimal	BCD
94	1011110	136	5E	1001 0100
95	1011111	137	5F	1001 0101
96	1100000	140	60	1001 0110
97	1100001	141	61	1001 0111
98	1100010	142	62	1001 1000
99	1100011	143	63	1001 1001
100	1100100	144	64	0001 0000 0000
101	1100101	145	65	0001 0000 0001
102	1100110	146	66	0001 0000 0010
103	1100111	147	67	0001 0000 0011
104	1101000	150	68	0001 0000 0100
105	1101001	151	69	0001 0000 0101
106	1101010	152	6A	0001 0000 0110
107	1101011	153	6B	0001 0000 0111
108	1101100	154	6C	0001 0000 1000
109	1101101	155	6D	0001 0000 1001
110	1101110	156	6E	0001 0001 0000
111	1101111	157	6F	0001 0001 0001
112	1110000	160	70	0001 0001 0010
113	1110001	161	71	0001 0001 0011
114	1110010	162	72	0001 0001 0100
115	1110011	163	73	0001 0001 0101
116	1110100	164	74	0001 0001 0110
117	1110101	165	75	0001 0001 0111
118	1110110	166	76	0001 0001 1000
119	1110111	167	77	0001 0001 1001
120	1111000	170	78	0001 0010 0000
121	1111001	171	79	0001 0010 0001
122	1111010	172	7A	0001 0010 0010
123	1111011	173	7B	0001 0010 0011
124	1111100	174	7C	0001 0010 0100
125	1111101	175	7D	0001 0010 0101
126	1111110	176	7E	0001 0010 0110
127	1111111	177	7F	0001 0010 0111
128	10000000	200	80	0001 0010 1000
129	10000001	201	81	0001 0010 1001
130	10000010	202	82	0001 0011 0000
131	10000011	203	83	0001 0011 0001
132	10000100	204	84	0001 0011 0010
133	10000101	205	85	0001 0011 0011
134	10000110	206	86	0001 0011 0100
135	10000111	207	87	0001 0011 0101
136	10001000	210	88	0001 0011 0110
137	10001001	211	89	0001 0011 0111
138	10001010	212	8A	0001 0011 1000
139	10001011	213	8B	0001 0011 1001
140	10001100	214	8C	0001 0100 0000
141	10001101	215	8D	0001 0100 0001
142	10001110	216	8E	0001 0100 0010
143	10001111	217	8F	0001 0100 0011
144	10010000	220	90	0001 0100 0100
145	10010001	221	91	0001 0100 0101
146	10010010	222	92	0001 0100 0110
147	10010011	223	93	0001 0100 0111
148	10010100	224	94	0001 0100 1000
149	10010101	225	95	0001 0100 1001
150	10010110	226	96	0001 0100 0000
151	10010111	227	97	0001 0101 0001
152	10011000	230	98	0001 0101 0010
153	10011001	231	99	0001 0101 0011
154	10011010	232	9A	0001 0101 0100
155	10011011	233	9B	0001 0101 0101
156	10011100	234	9C	0001 0101 0110
157	10011101	235	9D	0001 0101 0111
158	10011110	236	9E	0001 0101 1000
159	10011111	237	9F	0001 0101 1001
160	10100000	240	A0	0001 0110 0000
161	10100001	241	A1	0001 0110 0001
162	10100010	242	A2	0001 0110 0010
163	10100011	243	A3	0001 0110 0011
164	10100100	244	A4	0001 0110 0100
165	10100101	245	A5	0001 0110 0101
166	10100110	246	A6	0001 0110 0110
167	10100111	247	A7	0001 0110 0111
168	10101000	250	A8	0001 0110 1000
169	10101001	251	A9	0001 0110 1001
170	10101010	252	AA	0001 0111 0000
171	10101011	253	AB	0001 0111 0001
172	10101100	254	AC	0001 0111 0010
173	10101101	255	AD	0001 0111 0011
174	10101110	256	AE	0001 0111 0100
175	10101111	257	AF	0001 0111 0101
176	10110000	260	B0	0001 0111 0110
177	10110001	261	B1	0001 0111 0111
178	10110010	262	B2	0001 0111 1000
179	10110011	263	B3	0001 0111 1001
180	10110100	264	B4	0001 1000 0000
181	10110101	265	B5	0001 1000 0001
182	10110110	266	B6	0001 1000 0010
183	10110111	267	B7	0001 1000 0011
184	10111000	270	B8	0001 1000 0100
185	10111001	271	B9	0001 1000 0101
186	10111010	272	BA	0001 1000 0110
187	10111011	273	BB	0001 1000 0111

Decimal	Binary	Octal	Hexa-decimal	BCD
188	10111100	274	BC	0001 1000 1000
189	10111101	275	BD	0001 1000 1001
190	10111110	276	BE	0001 1001 0000
191	10111111	277	BF	0001 1001 0001
192	11000000	300	C0	0001 1001 0010
193	11000001	301	C1	0001 1001 0011
194	11000010	302	C2	0001 1001 0100
195	11000011	303	C3	0001 1001 0101
196	11000100	304	C4	0001 1001 0110
197	11000101	305	C5	0001 1001 0111
198	11000110	306	C6	0001 1001 1000
199	11000111	307	C7	0001 1001 1001
200	11001000	310	C8	0010 0000 0000
201	11001001	311	C9	0010 0000 0001
202	11001010	312	CA	0010 0000 0010
203	11001011	313	CB	0010 0000 0011
204	11001100	314	CC	0010 0000 0100
205	11001101	315	CD	0010 0000 0101
206	11001110	316	CE	0010 0000 0110
207	11001111	317	CF	0010 0000 0111
208	11010000	320	D0	0010 0000 1000
209	11010001	321	D1	0010 0000 1001
210	11010010	322	D2	0010 0001 0000
211	11010011	323	D3	0010 0001 0001
212	11010100	324	D4	0010 0001 0010
213	11010101	325	D5	0010 0001 0011
214	11010110	326	D6	0010 0001 0100
215	11010111	327	D7	0010 0001 0101
216	11011000	330	D8	0010 0001 0110
217	11011001	331	D9	0010 0001 0111
218	11011010	332	DA	0010 0001 1000
219	11011011	333	DB	0010 0001 1001
220	11011100	334	DC	0010 0010 0000
221	11011101	335	DD	0010 0010 0001
222	11011110	336	DE	0010 0010 0010
223	11011111	337	DF	0010 0010 0011
224	11100000	340	E0	0010 0010 0100
225	11100001	341	E1	0010 0010 0101
226	11100010	342	E2	0010 0010 0110
227	11100011	343	E3	0010 0010 0111
228	11100100	344	E4	0010 0010 1000
229	11100101	345	E5	0010 0010 1001
230	11100110	346	E6	0010 0011 0000
231	11100111	347	E7	0010 0011 0001
232	11101000	350	E8	0010 0011 0010
233	11101001	351	E9	0010 0011 0011
234	11101010	352	EA	0010 0011 0100
235	11101011	353	EB	0010 0011 0101
236	11101100	354	EC	0010 0011 0110
237	11101101	355	ED	0010 0011 0111
238	11101110	356	EE	0010 0011 1000
239	11101111	357	EF	0010 0011 1001
240	11110000	360	F0	0010 0100 0000
241	11110001	361	F1	0010 0100 0001
242	11110010	362	F2	0010 0100 0010
243	11110011	363	F3	0010 0100 0011
244	11110100	364	F4	0010 0100 0100
245	11110101	365	F5	0010 0100 0101
246	11110110	366	F6	0010 0100 0110
247	11110111	367	F7	0010 0100 0111
248	11111000	370	F8	0010 0100 1000
249	11111001	371	F9	0010 0100 1001
250	11111010	372	FA	0010 0101 0000
251	11111011	373	FB	0010 0101 0001
252	11111100	374	FC	0010 0101 0010
253	11111101	375	FD	0010 0101 0011
254	11111110	376	FE	0010 0101 0100
255	11111111	377	FF	0010 0101 0101

Decimal	Power of 2	Binary	Octal	Hexa-decimal
256	2^8	100000000	400	100
512	2^9	1000000000	1000	200
1024	2^{10}	10000000000	2000	400
2048	2^{11}	100000000000	4000	800
4096	2^{12}	1000000000000	10000	1000
8192	2^{13}	10000000000000	20000	2000
16384	2^{14}	100000000000000	40000	4000
32768	2^{15}	1000000000000000	100000	8000
65536	2^{16}	10000000000000000	200000	10000

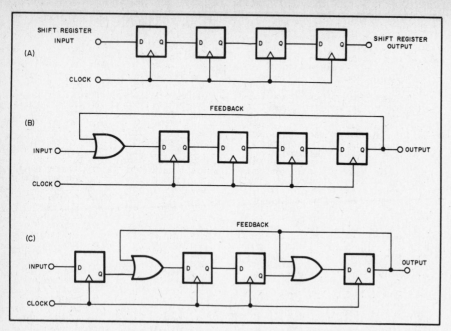

Fig. 21 — Shift registers. Four D flip-flops connected as a simple shift register are shown at A. B shows the same flip-flops used as a ring counter. Notice that the feedback is ORed with the input. In C, the feedback is ORed with the inputs to the third and fourth flip-flops, creating a pseudorandom number generator.

an external logic signal. Shift registers may provide parallel outputs (an output from each flip-flop), and they may provide a means to directly set the state of each flip-flop (parallel loading) synchronously or asynchronously. Fig. 21A illustrates four D flip-flops connected as a shift register.

Shift registers can be used to delay or synchronize data. In this application, the register is constructed so that a desired bit is at the output of the register during a specified clock pulse. Any output of a shift register can be fed back to its input to produce an almost infinite array of possibilities. The *ring counter* (Fig. 21B) is a shift register with its output fed back to its input. When a clock pulse is applied to

a ring counter, the bit pattern stored in the register is rotated either left or right. Shift registers with several feedback paths can be used as *pseudorandom-number generators* (Fig. 21C). The output of such a generator is a sequence of bits that meets one or more criteria of true randomness. Pseudorandom-number generators have applications in noise generators and spread-spectrum key generators.

Byte and Word Registers

A byte register or a word register is simply a group of flip-flops that is thought of as a single register. Such registers are used in computers for temporary data storage. The number of flip-flops in a word

register depends on the word length of the computer. Word length and computer architecture will be more thoroughly discussed in later sections of this chapter.

ONE-SHOT MULTIVIBRATOR

A *one-shot* or *monostable* multivibrator is a circuit that has one stable state and an unstable state. When the one-shot is *triggered* by a clock pulse, it assumes its unstable state for a period of time determined by circuit components. Thus, the one-shot multivibrator puts out a pulse of some predetermined duration (T). If a one-shot multivibrator is triggered while it is in the unstable state, two things may happen. A *non-retriggerable* multivibrator is not affected by the second trigger pulse. A *retriggerable* multivibrator will start "counting" its pulse duration (T) from the most recent trigger pulse. Both types of one-shots are common.

A 555 timer IC connected as a one-shot multivibrator is shown in Fig. 22. The one-shot is activated by negative-going pulse between the trigger input and ground. The trigger pulse causes the output (Q) to go positive and capacitor C to charge through resistor R. When the voltage across C reaches two-thirds of V_{cc}, the capacitor is quickly discharged to ground and the output returns to 0. The output remains at logic 1 for a time determined by

$$T = 1.1 \ RC \qquad \text{(Eq. 1)}$$

where
    R = resistance in ohms
    C = capacitance in farads

ASTABLE MULTIVIBRATOR

An *astable* or *free-running* multivibrator is a circuit that alternates between two unstable states. It can be synchronized by an input signal of a frequency that is slightly higher than the astable multivibrator free-running frequency. An astable

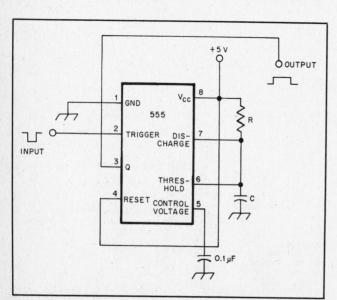

Fig. 22 — A 555 timer connected as a one-shot multivibrator. See text for a formula to calculate values of R and C.

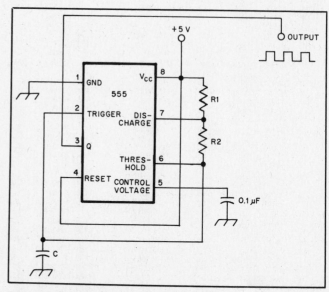

Fig. 23 — A 555 timer connected as an astable multivibrator. See text for formula to calculate values of R1, R2 and C.

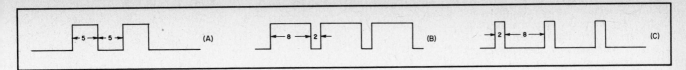

Fig. 24 — Pulse sequences showing 50% (A), 80% (B) and 20% (C) duty cycles.

multivibrator creates a sequence of pulses at a desired frequency and width.

An astable multivibrator circuit using the 555 timer is shown in Fig. 23. The capacitor C charges to two-thirds V_{cc} through R1 and R2 and discharges to one-third V_{cc} through R2. The ratio of R1:R2 sets the *duty cycle* of the pulses. The duty cycle is the ratio of the time that the output is high to the total period of the output cycle (Fig. 24). The output frequency is determined by

$$f = \frac{1.46}{(R_1 + 2R_2)\,C} \qquad \text{(Eq. 2)}$$

where
  R = resistance in ohms
  C = capacitance in farads

SUMMARY

Digital logic is playing an increasingly important role in Amateur Radio. Most of this logic is binary and can be described and designed using Boolean algebra. Using the NOT, AND and OR gates of combinational logic, designers can build sequential logic circuits that have memory and feedback. The simplest memory circuit is called a flip-flop, or latch, and these latches can be combined to form such useful circuits as counters, shift registers and byte registers.

Digital Integrated Circuits

Integrated circuits are the cornerstone of digital logic devices. An integrated circuit is a miniature electronic module of components and conductors manufactured as a single unit. Today's complex digital equipment would be impossible with vacuum tubes or even discrete transistors as active elements.

As one generation of ICs outshined the previous one, it became necessary to classify them according to the number of gates on a single chip. These classifications are defined below.

Small-scale integration (SSI) — 10 or fewer gates on a chip.
Medium-scale integration (MSI) — 10-100 gates.
Large-scale integration (LSI) — 100-1000 gates.
Very-large-scale integration (VLSI) — 1000 or more gates.

COMPARING LOGIC FAMILIES

When selecting logic devices for a circuit, a designer is faced with the choice between many families and subfamilies of logic ICs. Which subfamily is right for the application at hand is often a function of several parameters. The characteristics that are usually important are logic speed, power consumption and fan-out.

Impedance

One parameter that is discussed in the following sections is gate impedance. The ideal gate would have low output impedance, enabling it to deliver high current without dropping considerable voltage in its output stages. The ideal gate would have infinite input impedance and would, therefore, draw no current. Unfortunately, such a gate does not exist, and the designer must compromise on input and output impedances.

Speed

Logic devices operate at widely-varying clock speeds. Standard TTL devices can only operate up to a few MHz, while some ECL ICs can operate at several GHz. Gate propagation delay determines the maximum clock speed at which an IC can be operated; the clock period must be long enough for all signals within the IC to propagate to their destinations. ICs that use capacitively coupled inputs have minimum, as well as maximum, clock rates. While the designer might want to use the fastest available ICs, higher speed usually results in higher power consumption.

Power Consumption

In some applications, power consumption by logic gates is a critical design consideration. This power consumption can be divided into two parts: *static power* and *dynamic power*. Gates are either holding a state (high or low), or they are changing states, and each of these conditions has different power requirements. The power that is consumed when a gate changes state is called dynamic power. The power consumed when a gate is holding a state is called static power. Calculating the total power that will be required by an IC with several gates performing diverse functions is a complex task. Nominal power requirements, however, can be used to compare logic subfamilies.

Fan-out

A gate's output can only supply a limited amount of current. Therefore, a single output can only drive a limited number of in-

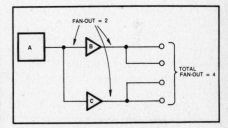

Fig. 25 — Gate A (fan-out = 2) is connected to the two buffers (B and C), each with a fan-out of 2, to achieve total fan-out of 4.

puts. The measure of driving ability is called *fan-out*. Fan-out is expressed as the number of inputs (of the same subfamily) that can be driven by a single output. If a logic family that is otherwise desirable does not have sufficient fan-out, consider using noninverting buffers to increase fan-out (Fig. 25).

Other Considerations

The parameters mentioned above are the basic considerations that influence the selection of a logic subfamily for a specific application. These considerations have complex interactions that come into play in demanding low-current, high-speed or high-complexity circuits. If you are designing a circuit, consult the application notes for the logic devices that you are going to use. Application notes contain speed, power-consumption and fan-out specifications, as well as other useful information.

DIGITAL-LOGIC IC FAMILIES

A number of families of digital logic ICs fall under the two broad categories of *bipolar* and *metal-oxide semiconductor*

(MOS). Numerous manufacturing techniques have been developed using both bipolar and MOS technologies. Each family has its own inherent advantages and disadvantages, and each of the surviving families has found its special niche in the market.

BIPOLAR LOGIC FAMILIES

Bipolar semiconductor ICs employ NPN and PNP junction transistors. While early bipolar logic was faster and had higher power consumption than MOS logic, these distinctions have been blurred as manufacturing technology has developed. There are several families of bipolar logic devices, and within some of these families there are subfamilies.

Direct-coupled-transistor logic (DCTL) was the first commercial IC family. It is now obsolete. It had poor noise immunity and high current consumption.

Resistor-transistor logic (RTL) was the earliest popular IC logic family. It suffered from low speed, poor noise immunity and low fan-out. *Diode-transistor logic* (DTL) reduced power requirements but was slower than other bipolar logic families. Both RTL and DTL are no longer used in new designs and are available for replacement purposes only. *High-Threshold Logic (HTL)* is a variation of DTL that uses higher supply voltages (around 15 V) and Zener diodes at gate inputs to achieve greater noise immunity. *Resistor-capacitor-transistor logic* (RCTL) is a variation of RTL. RCTL gates have bypass capacitors across their input resistors to increase speed.

Transistor-Transistor Logic (TTL)

TTL (also known as T²L) has seen widespread acceptance because it is faster and more immune to noise than RTL and DTL. The TTL family is divided into standard TTL and variations. The transistors in standard TTL circuits saturate, thus reducing operating speed. The variations cure this by clamping the transistors with *Schottky* diodes to prevent saturation or by using gold dope to reduce transistor recovery time. Both manufacturing processes are used, but Schottky-clamped TTL is faster.

Most TTL ICs are identified by numbers beginning with 54 or 74. The 54 prefix denotes a military temperature range of −55 to 125 °C, while 74 indicates a commercial temperature range of 0 to 70 °C. Standard TTL ICs have a 2, 3 or 4-digit device identification number immediately following the prefix (for example, 5400 and 74174). Standard TTL devices are usable up to 35 MHz. One or two letters in the middle of an ID number indicate the special TTL subfamily:

AS — Advanced Schottky
ALS — Advanced low-power Schottky
H — High speed (up to 50 MHz)
L — Low power (up to 3 MHz)
LS — Low-power Schottky (up to 45 MHz)

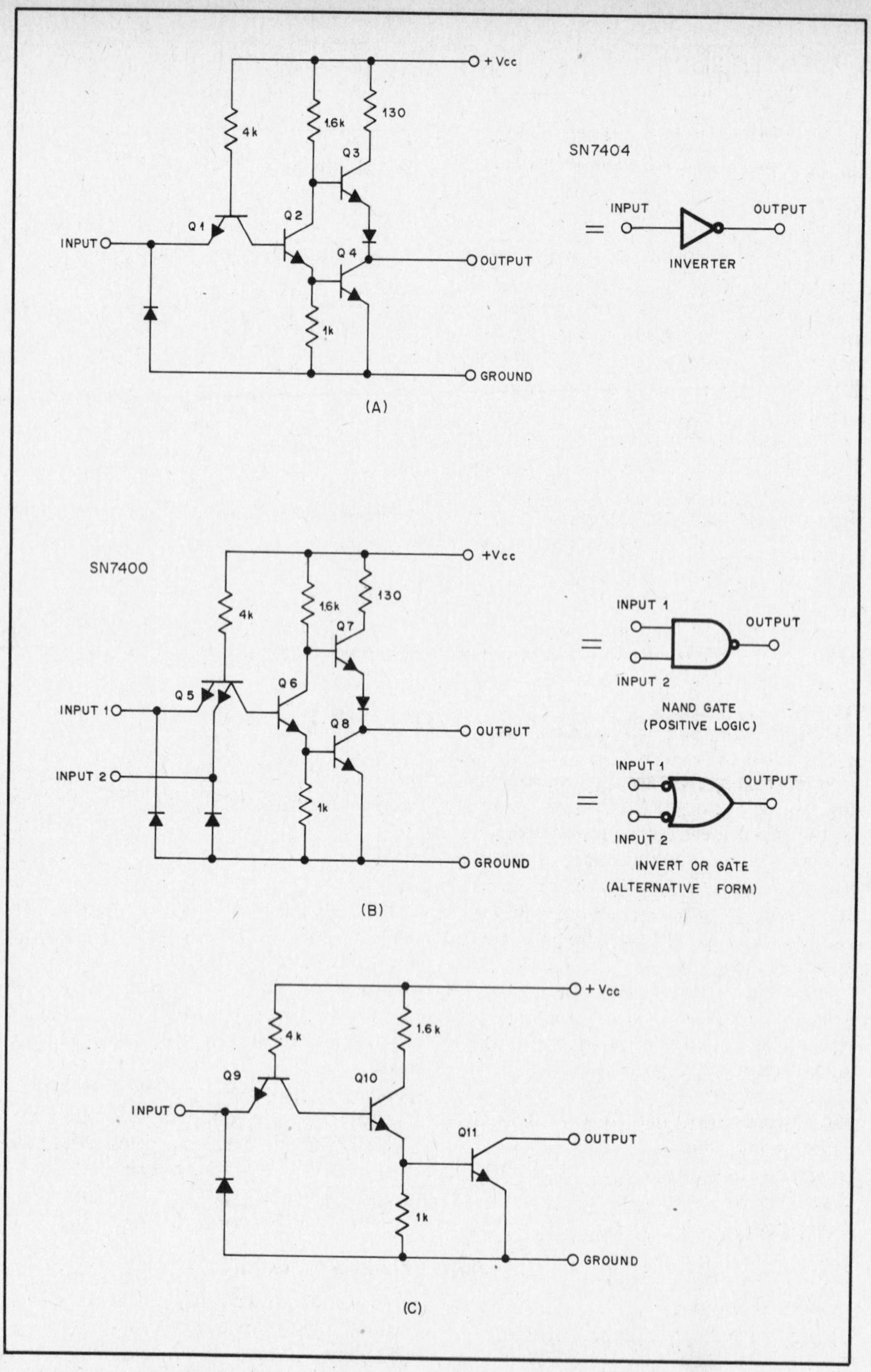

Fig. 26 — TTL circuits and their equivalent logic symbols. The circuits in A and B have totem-pole outputs, while C has an open-collector output. Indicated resistor values are typical. Identification of transistors is for text reference only; these are *not* discrete components.

S — Schottky (some up to 125 MHz)

TTL Circuits

One section of a 7404 hex inverter is represented schematically in Fig. 26A. A low applied to the input will cause Q1 to conduct. This will cause Q2 to be near cutoff, in turn biasing Q3 into saturation and Q4 near cutoff. As a result, the output will be high, about 1 V below V_{cc}. If the signal at the input is high, the conduction state of each transistor reverses, and the output drops nearly to ground potential (low). The input diode protects the circuitry by clamping any negative potential to approximately −0.7 V, limiting the current in Q1 to a safe value. The output diode at the emitter of Q3 is required to ensure that Q3 is cut off when the output is low.

The circuit in Fig. 26B, one section of a 7400 gate, is very similar to that of A. The difference is that Q5 is a multiple-emitter transistor with one input to each emitter. A low at either input will turn on Q5, causing the output to go high. Both inputs must be high to produce a low out-

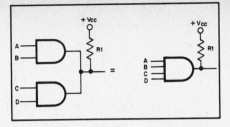

Fig. 27 — The outputs of two open-collector-output AND gates are shorted together (wire ANDed) to produce an output the same as would be obtained from a four-input AND gate.

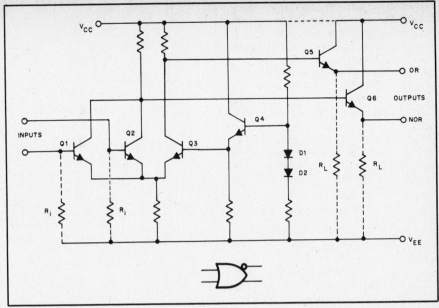

Fig. 28 — Circuit topology of the ECL family. The functions of the various components are explained in the text. The availability of the complimentary output is indicated by the modified logic symbol.

put. This is a NAND gate.

TTL Outputs

In the TTL circuit of Fig. 26A, transistors Q3 and Q4 are arranged in a totem-pole configuration. This output circuit has a low source impedance, allowing the gate to supply or sink substantial output current. The 130-ohm resistor between the collector of Q3 and $+V_{cc}$ limits the current through Q3, protecting Q3 if the gate output is short-circuited.

Some TTL gates employ *open-collector* outputs, as shown in Fig. 26C. Open-collector outputs have an advantage over totem-pole outputs: they can be *wire-ANDed*. In a wire-AND circuit, several open-collector outputs can be connected to a single *pull-up* resistor. When all of the outputs are high, the point where they are connected will be high. If any of the outputs are low, the connection will be low. This circuit implements the logical AND function, as shown in Fig. 27. Open-collector outputs are also useful for interfacing TTL gates to gates from other logic families. TTL outputs have a minimum high level of 2.4 V and a maximum low level of 0.4 V. A pull-up resistor, typically 2.2 kΩ, can be used to raise the high voltage level to 5 V, which may be needed to drive a non-TTL circuit. Although +5 is the recommended power-supply voltage for TTL, the pull-up resistor on an open collector can be connected to a different voltage, providing a high logic level that is not 5 V.

There is another type of output that is available in TTL and other logic families. This output is called the *three-state output*. Three-state output devices are used to connect two or more outputs to a single input in such a way that *only one* of the outputs affects the input. To accomplish this, each output can take on a third, high-impedance, state (as well as the standard high and low states). When the output is in the third state, it is essentially disconnected from the circuit. Outputs in the high-impedance state do not affect the input to which they are connected; only the output that is in a normal high or low state will affect the input. Care must be taken to ensure that all of the outputs but one are in the high-impedance state. Three-state devices are used in computers to allow several

devices to share a common set of wires (called a *bus*).

Typically, TTL devices have a fan-out of 10 within a subfamily. Buffers may be used to increase fan-out.

TTL Inputs

If TTL inputs are left open, they assume a high logic state. However, greater noise immunity will be realized if pull-up resistors are used. A pull-up resistor is used to connect an input to the supply voltage. When operated with a +5-V supply, any input level between 2.0 and 5.5 V is defined to be high. A level less than 0.8 V is an input low.

Decoupling

When a TTL gate changes state, the amount of current that it draws changes rapidly. These changes in current, called switching transients, appear on the power supply line and can cause false triggering of other devices. For this reason, the power bus should be adequately decoupled. For proper decoupling, use 0.1 μF from V+ to ground at each standard TTL IC and 0.01 μF at each L, LS or S device.

Emitter-Coupled Logic (ECL)

ECL, also called current-mode logic (CML), has exceptionally high speed, high input impedance and low output impedance. Some ECL devices can operate at frequencies higher than 1.2 GHz. This family is different from the other bipolar-logic families; the transistors operate in a nonsaturated mode that is analogous to that of some linear devices. ECL devices are used in UHF frequency counters, UHF synthesizers and high-speed computers.

A significant feature of ECL gates is that complementary output functions are

available from each circuit. For example, the circuit of Fig. 28 provides both an OR output and a NOR output. Q1 or Q2, together with Q3 forms a differential amplifier. When the Q2 collector goes high, the Q3 collector goes low. These levels then appear at the emitters of the output buffers, Q5 and Q6. Q4, D1 and D2 and associated circuitry form a bias generator. The reference voltage established at the base of Q3 determines the input switching threshold.

The typical logic swing for ECL devices is only 800 mV. ECL devices are characterized for use with a −5.2-V power supply, but operation from other supplies is possible. If the V_{cc} terminal is connected to +2.0 V and the V_{ee} terminal connected to −3.2 V, the device can drive a 50-ohm load. The power output obtained this way is about 1 mW. Heat sinking is sometimes necessary because ECL ICs dissipate a great deal of power.

Several ECL subfamilies are now being produced. High speed and low power dissipation are largely conflicting goals, so different subfamilies are offered to allow the designer to choose the trade-offs. ECL subfamilies are compatible, but only over a limited temperature range. The differences between subfamilies are mostly in resistance values and the presence or absence of input and output pull-down resistors.

Other Bipolar Logic Families

There are several bipolar logic families that are used *within* ICs to increase functional density and to give some analog function to bipolar ICs. These logic families are generally *not* present at the inputs and outputs of ICs, and are not encountered by the user.

Integrated Injection Logic (I[2]L) has increased density and decreased power consumption at speeds near those of TTL. It can handle both digital and analog operations on a single chip. Current-Hogging Logic (CHL) is a saturated-logic variation of I[2]L and has higher noise immunity.

Complementary-Constant-Current Logic (C[3]L) operates at faster than I[2]L speeds without sacrificing high chip density.

Triple-Diffused Emitter-Follower Logic (3D-EFL) is an unsaturated bipolar logic family that takes its name from the triple-diffusion fabrication process used to simplify production.

METAL-OXIDE SEMICONDUCTOR (MOS) LOGIC FAMILIES

While bipolar devices use PNP and NPN junction transistors, MOS devices are composed of N-channel or P-channel field-effect transistors (FETs). MOS is characterized by simple device structure, small size (high density) and ease of LSI fabrication. MOS families are used extensively in microprocessors, digital watches and calculators.

P-Channel MOS (PMOS)

PMOS, in which P-type dopant is used for the source and drain, is the oldest and simplest MOS process. Electrical current in a PMOS device is conducted by the flow of positive charges (holes). PMOS power consumption is lower than that of bipolar logic, but its operating speed is slower. The only extensive use of PMOS is in calculators and watches, where low speed is acceptable and low power consumption and cost are desirable.

N-Channel MOS (NMOS)

In NMOS, N-type material is used for FET sources and drains, and electrons carry current. Since electrons flow faster than holes, NMOS has at least twice the speed of PMOS. NMOS also has greater gain than PMOS and supports greater packaging density through the use of smaller transistors.

Complementary MOS (CMOS)

CMOS uses both P-channel and N-channel devices on the same substrate to achieve high noise immunity and low power consumption (less than 1 mW per gate and negligible power during standby). This accounts for the widespread use of CMOS in battery-operated equipment.

A notable feature of CMOS devices is that the logic levels swing to within a few millivolts of the supply voltages. The input switching threshold is approximately one half the supply voltage ($V_{dd} - V_{ss}$). This characteristic contributes to high immunity to noise on the input signal or power supply lines. CMOS input-current drive requirements are miniscule, so their fan-out is great, at least in low-speed systems. (For high-speed systems, the input capacitance increases the dynamic

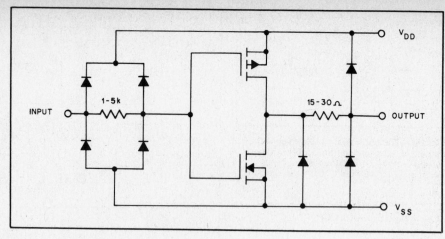

Fig. 29 — Internal structure of a CMOS gate.

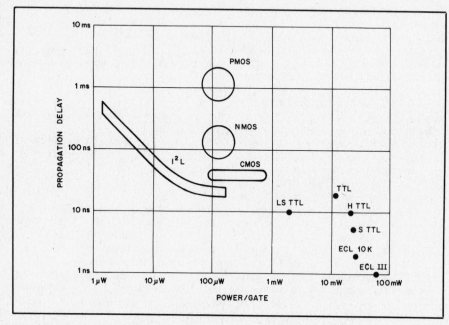

Fig. 30 — Comparison of speed and power capabilities of various logic families.

power dissipation and limits the fan-out.) There are now a number of CMOS logic IC subfamilies in production.

The 4000A series is the original commercial line. This series operates with supplies of 3 to 12 V.

A subfamily having some improved characteristics is the 4000B (for buffered) series. The B series can be powered from supplies of up to 18 V. This feature makes the devices especially attractive for automotive applications. The output impedance of buffered ICs is independent of the input state. An unbuffered series, designated the 4000UB, meets all the B-series specifications except that the logic outputs are not buffered. Also, input logic levels must be within 20 percent of the supply voltages. Several trade-offs must be considered when choosing between buffered and unbuffered ICs. Buffered devices have greater noise immunity and drive capability but lower speed than unbuffered

types. Some special-purpose 4000-series ICs have three-state output circuits. The third state is neither high or low, but is a high-impedance condition, which allows several outputs to be paralleled for wire ORing or multiplexing. (See preceding discussion of TTL outputs.)

The 54C/74C series is designed to be a CMOS plug-in replacement for low-power TTL devices in some applications. The 54HC/74HC is a subfamily of high-speed CMOS devices using a silicon-gate CMOS process to attain speeds similar to Schottky TTL. 54HC/74HC devices were designed specifically to replace 54LS/74LS devices. These have input, output and supply parameters tailored to the 54LS/74LS series, resulting in the relatively high speed of bipolar logic and low power consumption and high noise immunity of CMOS.

Special Considerations

A simplified diagram for a CMOS logic

inverter is given in Fig. 29. Some of the diodes in the input- and output-protection circuits are an inherent part of the manufacturing process. Even with the protection circuits, CMOS ICs are susceptible to damage from static charges. To protect against damage from static, the pins should not be inserted in Styrofoam™ as is commonly done with TTL ICs. Instead a spongy conductive material is available for this purpose. Before removing a CMOS IC from its protective material, make certain that your body is grounded. A conductive bracelet connected to the ground terminal of a 3-wire ac outlet through a 10-megohm resistor is adequate for this purpose.

All CMOS inputs should be tied to an input signal. + supply voltage or ground. Unterminated CMOS inputs, even on unused gates, may cause gate outputs to oscillate. Oscillating gates draw high current and overheat.

The low power consumption of CMOS ICs made them attractive for satellite applications, but standard CMOS devices have proven to be sensitive to low levels of radiation — cosmic rays, gamma rays and X rays. Now, radiation-hardened CMOS ICs are able to tolerate 10^6 rads, making them suitable for space applications.

Silicon-on-Sapphire (SOS)

SOS is a fabrication technique that takes its name from depositing a thin silicon film on an insulating sapphire substrate. This minimizes the parasitic capacitance normally found between the source, drain and a silicon substrate. SOS technology has been used to produce CMOS ICs with greatly reduced gate propagation times.

SUMMARY

There are many types of logic ICs, each with its own advantages and disadvantages. If you want low power consumption, you should probably use CMOS. If you want ultra-high-speed logic, you will have to use ECL. The graphs in Fig. 30 should give you some idea of what logic family and subfamilies are acceptable in your circuit. Whatever the application, consult up-to-date literature when designing logic circuits. IC data books and applications notes are usually available from IC manufacturers and distributors.

Digital Interfacing

The term *interface* is used to describe a boundary between, or a physical or logical interconnection of, two devices, circuits or systems. *Physical interfaces* may be electrical and/or mechanical. Electrical interfaces must match such parameters as signal level and circuit impedance. *Mechanical interfaces* involve the mating of connectors and specific pin assignments within those connectors. *Logical interfaces* provide communication between different modules in software.

Different interfacing tasks call for different levels of formality and standardization. Designers are free to define their own interfaces between logic families on a circuit board or circuit boards within a piece of equipment. When interfacing between two computers or between a terminal and a modem, however, existing interface standards should be followed.

INTERFACING LOGIC FAMILIES

Each semiconductor logic family has its own advantages in particular applications. For example, the highest frequency stages in a UHF counter or a frequency synthesizer would use ECL. After the frequency has been divided down to less than 25 MHz, the speed of ECL is unnecessary, and its expense and power dissipation are unjustified. TTL or CMOS are better choices at lower frequencies.

Each logic family has its own input voltage and current requirements, so they can't be mixed with satisfactory results. The term *interface* is used to describe the point at which one logic family ends and another begins.

There are a number of ICs intended especially for mating different logic families. The CD4049UB and CD4050B hex buffers are designed to drive TTL gates from CMOS input signals. TTL-to-ECL and ECL-to-TTL conversion can be implemented with the N1017 and N1068 ICs manufactured by Signetics and others. Unfortunately, these components aren't always conveniently available to the small-quantity purchaser, so logic interface sometimes must be accomplished by other circuitry.

A knowledge of the internal circuits of integrated circuits and their input/output (I/O) characteristics will allow the designer to devise reliable digital interfaces. Typical internal structures have been illustrated for each common logic family. The I/O characteristics of the common logic families are listed in Table 6. This information was compiled from various industrial publications and is intended only as a guide. Certain ICs may have characteristics that vary from the values given.

The following section discusses some specific logic conversions. Often more than one conversion scheme is possible, depending on whether the designer wishes to optimize power consumption or speed. Usually one must be traded off for the other. Where an electrical connection between two logic systems isn't possible, an optical isolator can sometimes be used.

TTL Driving CMOS

A CMOS gate is easily driven by a TTL device when both are powered by a + 5-V source. However, the totem-pole output structure of most TTL ICs prevents a high output level of sufficient potential to properly activate a CMOS input. A pull-up resistor connected from the interface point to the power bus will remedy this problem. The maximum usable value for this resistor is 15 kΩ, but the circuit capacitance, coupled with this large resistance, will reduce the maximum possible speed of the CMOS gate. Lower resistance values will generate a more favorable RC product at the expense of increased power dissipation. A standard TTL gate can drive a pull-up resistor of 330 Ω, but a low-power version is limited to a 1.2 kΩ minimum. The resistor pull-up technique is illustrated in Fig. 31A.

When the CMOS device is operating on a power supply other than + 5 V, the TTL interface is more complex. One fairly simple technique uses a TTL open-collector output connected to the CMOS input and a pull-up resistor, the other end of which is connected to the CMOS supply. Although this method works in many applications, the common-base level shifter of Fig. 31B will translate a TTL output signal to a + 15-V CMOS signal while preserving the full noise immunity of both gates. An operational amplifier configured as a comparator, as in Fig. 31C, makes an excellent converter from TTL to CMOS using dual power supplies. An FET op amp is shown because its output voltage can usually swing closer to the rails (+ and − supply voltages) than a bipolar unit. Where the pulse rate is below 10 kHz or so, a 741 op amp may be used.

CMOS Driving TTL

The 4049UB and 4050B devices already mentioned can drive two standard TTL loads when a common + 5-V supply is used. Most A-series CMOS ICs can't sink enough current to drive TTL gates to a reliable low input state. Gates from the B series can drive one low-power TTL load directly. The 74C00 family is capable of

Table 6

Electrical Characteristics of Common Logic Families

	Standard TTL (active pull-up) $V_{cc} = +5.0$ V	Schottky TTL (74S) $V_{cc} = +5.0$ V	High-speed TTL (74H) $V_{cc} = +5.0$ V	Low-power TTL (74L) $V_{cc} = +5.0$ V	Schottky low-power TTL (74LS) $V_{cc} = +5.0$ V	ECL III (1600 series) $V_{cc} = 0$ V, $V_{EE} = -5.2$ V	ECL 10 k
minimum HIGH input voltage	2.0 V	2.0 V	2.0 V	2.0 V	2.0 V	−1.095 V	−1.105 V
maximum HIGH input current	40 μA	50 μA	50 μA	20 μA	20 μA	—	—
maximum LOW input voltage	0.8 V	0.8 V	0.8 V	0.8 V	0.8 V	−1.485 V	−1.475 V
maximum LOW input current	1.6 mA	2.0 mA	2.0 mA	400 μA	400 μA	—	—
minimum HIGH output voltage	2.4 V	2.5 V	2.4 V	2.4 V	2.5 V	−0.9 V (30 mA)	−0.825 V (30 mA)
maximum HIGH output current	800 μA	1.0 mA	1.0 mA	400 μA	400 μA	40 mA	50 mA
maximum LOW output voltage	0.4 V	0.5 V	0.4 V	0.3 V	0.5 V	−1.75 V	−1.725 V
maximum LOW output current	16 mA	20 mA	20 mA	4.0 mA	8.0 mA	Open emitter — pull-down resistor required	

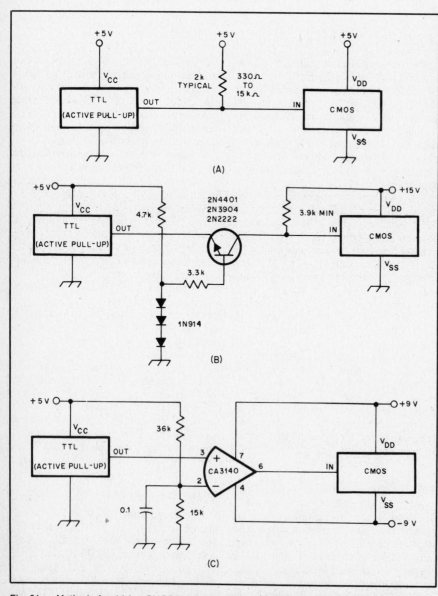

Fig. 31 — Methods for driving CMOS loads from TTL sources. The circuit complexity depends on the power supply voltages. The operation of these circuits is discussed in the text.

direct connection to low-power TTL with a fan-out of two. The drive capability of CMOS gates can be increased by connecting identical gates in parallel, but this practice is not recommended unless all the gates are contained in a single IC package.

Fig. 32A shows a simple method for driving a TTL load from a CMOS source operating with a higher-voltage power supply. The diode blocks the high voltage from the CMOS gate when it is in the high output state. A germanium diode is used because its lower forward-voltage drop provides higher noise immunity for the TTL device in the low state. The 68-kΩ resistor pulls the input high when the diode is back biased.

Standard TTL inputs draw 1.6 mA in the low state. A pull-down resistor for this purpose can be no larger than 220 Ω. To pull a 220-Ω resistor up to an acceptable high level requires 10 mA, which is beyond the capabilities of most CMOS devices. When a pull-down resistor is used, a dual-gate MOSFET having high transconductance makes a good buffer between CMOS and TTL circuits. This scheme is diagrammed in Fig. 32B. The CMOS power supply voltage isn't critical when this system is used, because the output impedance of the CMOS device is high compared to the pull-down resistance, and the protective diodes in the FET can sink more current than the CMOS IC can source. In fact, this circuit can also be used with split supplies, provided that the positive CMOS output excursion is at least 5 V.

TTL Driving ECL

When a common power supply is used, the resistor network of Fig. 33A will allow a standard TTL gate to drive an ECL input at the maximum TTL speed. Although shown with V_{cc} connected to +5 V and V_{ee} grounded, the same circuit will work

CMOS (4000A) V_{cc}		CMOS (4000B) V_{cc}			CMOS (74C00) V_{cc}		
+5.0 V	+10.0 V	+5.0 V	+10.0 V	+15.0 V	+5.0 V	+10.0 V	+15.0 V
3.5 V	7.0 V	4.0 V	8.0 V	12.5 V	3.5 V	8.0 V	—
—	—	—	—	1.0 µA	—	—	1.0 µA
1.5 V	3.0 V	1.0 V	2.0 V	2.5 V	1.5 V	2.0 V	—
—	—	—	—	1.0 µA	—	—	1.0 µA
4.95 V	9.95 V	4.95 V	9.95 V	14.95 V	2.4 V (360 µA)	9.0 V (10 µA)	—
(--------------------------------------- no load --------------------------------------)							
300 µA (2.5 V)	250 µA (9.5 V)	1.6 mA (2.5 V)	1.3 mA (9.5 V)	3.4 mA (13.5 V)	1.75 mA (0 V)	8.0 mA (0 V)	—
0.05 V	0.05 V	0.05 V	0.05 V	0.05 V	0.4 V (360 µA)	1.0 V (360 µA)	—
(--------------------------------------- no load --------------------------------------)							
300 µA (0.4 V)	600 µA (0.5 V)	500 µA (0.4 V)	1.3 mA (0.5 V)	3.4 mA (1.5 V)	1.75 mA (5 V)	8.0 mA (10 V)	—

with V_{cc} connected to ground and V_{ee} (and the "ground" terminal of the TTL device) connected to −5 V. This arrangement provides full noise immunity for the ECL circuit. Where speed is not a consideration, a TTL output can be connected to an ECL input with a pull-up resistor.

Independent TTL and ECL circuits can be coupled by the interfaces in Fig. 33B and C. In B, the TTL gate is divorced from the voltage-divider network when the output is high. In this state the junction of the 1.2-kΩ and 12-kΩ resistors assumes a potential of nearly +5 V. When the TTL output goes low, the anode end of the diode string is pulled down to about +2.5 V. This 2.5-V logic swing is attenuated and shifted to the proper nonsaturating ECL levels by the resistor network.

An emitter-follower stage is used in Fig. 33C. The −1.8-V potential at the ECL input established by the resistor network prevents the transistor from turning on when the TTL output is low. A germanium diode provides a stiff voltage reference in the low state and prevents excessive conduction in the upper transistor of the IC output structure. The voltage translation process is similar to that in Fig. 33B. Returning the collector to +5 V rather than ground keeps the transistor well out of saturation.

ECL Driving TTL

The complementary output of ECL gates can be used to advantage in converting to TTL levels. Modern ECL ICs have emitter-follower outputs that are ideal for switching the base-emitter junctions of bipolar transistors. For coupling logic circuits having a common 5- or 5.2-V power supply, the PNP-transistor and pull-down-resistor combination of Fig. 34A may be used. A positive supply is shown, but the circuit will also work with negative supplies. The circuit in Fig. 34B will condition −5.2-V ECL

signals to drive +5-V TTL gates. Transposing the OUT and OUT connections of the ECL device will effect a logic inversion with the translation. This technique can also be applied to flip-flops, which have Q and Q outputs.

CMOS Driving ECL

Speed is rarely a consideration when mating a relatively slow logic family to one that is very fast — the system cannot be faster than the slowest logic element used. The speed of ECL ICs comes from keeping the transistors out of saturation, and it is for this reason that the defined input-logic swing is only about 400 mV. However, the input levels can be anywhere within the range of the power supply without damaging the device. Negligible input current is required for either logic state, so when a common 5- or 5.2-V power supply is used, CMOS can drive ECL directly. A variety of circuits can be used between CMOS and ECL systems having different power supplies. The scheme illustrated in Fig. 35A is useful when a split power supply is used for the CMOS logic. The advantage of using a MOSFET converter is that the fan-out (to other CMOS devices) is not compromised. Fig. 35B shows a +15-V CMOS system driving a −5.2-V ECL gate through a PNP transistor. Altering some of the resistance values will make this circuit work with split-supply CMOS as well. This conversion method results in a logic inversion, but that problem usually can be remedied at the ECL output.

ECL Driving CMOS

Some voltage amplification is required if an ECL gate is to drive CMOS. When the ECL supply is negative and the CMOS supply is positive, the circuit of Fig. 34B illustrated for ECL-to-TTL conversion may be used. All of the resistors can be made much larger with CMOS for reduced power consumption.

The differential comparator arrangement

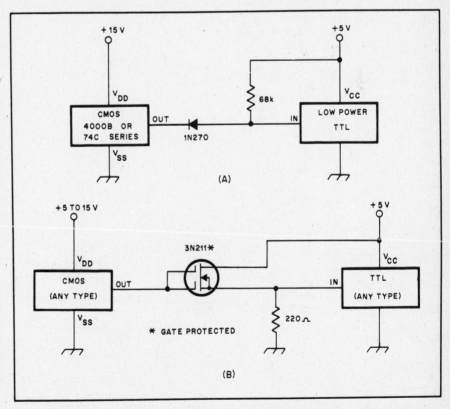

Fig. 32 — CMOS-to-TTL interface circuits. When both devices operate from a +5-volt supply, the diode in A can be eliminated. The circuit in B will work with a wide range of supply voltages and logic subfamilies.

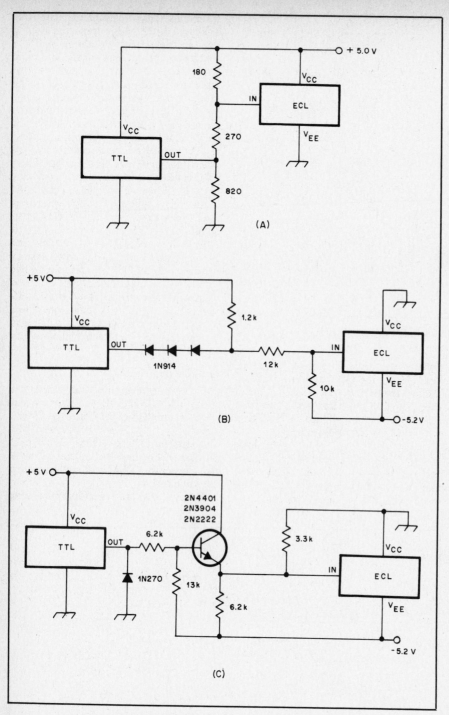

Fig. 33 — These circuits allow TTL gates to drive ECL systems using various power supply arrangements. Each is discussed in the text.

in Fig. 36A is another good translation method. If the CMOS circuit has split power supplies, the − V terminal of the op amp should be returned to V_{ss}. With split CMOS supplies the op amp can be connected directly to the CMOS input; the resistor and diode are unnecessary. If complementary ECL outputs aren't available, one of the comparator inputs should be biased to some potential between the two ECL logic levels.

Fig. 36B shows a way to obtain the required CMOS logic swing when both families are powered from the same sup-

ply. This NPN-PNP saturated amplifier will also work when a common negative supply is used.

Summary

Most logic families and subfamilies can be interconnected. As the schematics in the preceding sections show, interface circuits are not complex. As in the initial logic family selection, high speed and low power consumption must be juggled when designing interface circuits. You will construct successful interfaces if you remember to stay within the current and voltage limits

for gate inputs and do not draw excessive current from gate outputs.

OPTOCOUPLERS

An *optocoupler,* or *optoisolator,* is an LED and a phototransistor in a single IC package. Because they use light instead of an electrical connection, optoisolators provide a clean interface between circuits using widely differing voltages. Most LEDs in optocouplers are infrared emitters, although some operate in the visible light spectrum.

Fig. 37A shows the schematic diagram of a typical optocoupler. In this case, the base of the phototransistor is brought outside the package. Access to the base lead is useful for controlling the transistor when the LED is not energized.

The figure of merit for an optocoupler is the current transfer ratio (CTR), the ratio of I_E to I_F, where I_E is the phototransistor emitter current and I_F is the forward current through the LED. A Darlington phototransistor is used in some devices to establish a more favorable transfer characteristic. The schematic of an optocoupler with a Darlington phototransistor is shown in Fig. 37B.

In optocouplers with an internal light path, the light is transmitted from the LED to the phototransistor detector by means of a plastic light pipe or air gap. There are other configurations such as ones with an external reflective light path and others with an external light gap. An *optical shaft-encoder* is an array of open optocouplers chopped by a rotating wheel.

It is possible to make an optocoupler from a separate emitter and photodetector for applications such as a punched-tape or punched-card reader. Also, separate emitters and photodetectors are used for transmission of digital pulses over considerable distances. Amateur Radio applications of this type include use of light pipes to connect different pieces of equipment in the presence of high RF fields. Amateurs should consider optical coupling to minimize radiofrequency interference between digital and analog equipment such as computers and radios.

An optoisolator packaged with a triac forms a *solid-state relay* that can replace electromechanical units in many applications. The advantages of this scheme include freedom from contact bounce, arcing, mechanical wear and noise. Some solid-state relays are capable of switching 10 A at 117 V from CMOS signals.

Optical isolators can solve problems that might otherwise require complex circuits. Using optical isolators, low-voltage logic circuits can control high-voltage "real-world" devices, such as radio transceivers. Thus, optical isolators are critical to many digital control circuits. When designing circuits with optical isolators, check the current transfer ratio, isolation voltage and

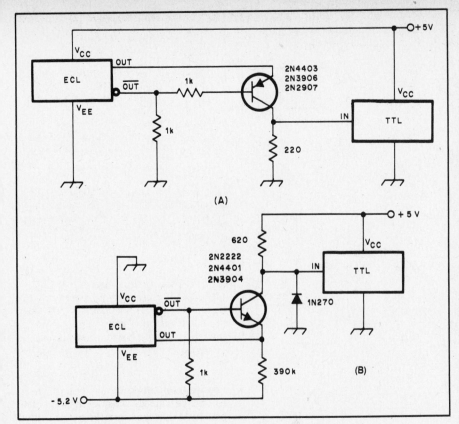

Fig. 34 — Bipolar transistors are used in these ECL-to-TTL converters.

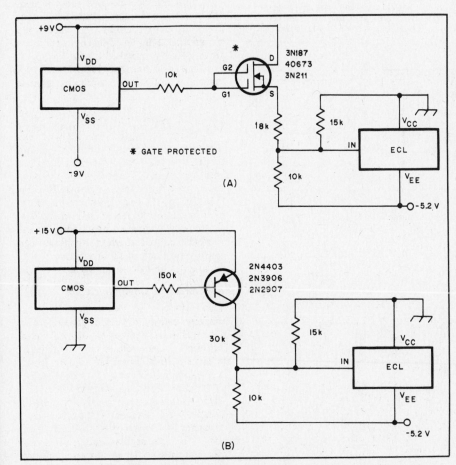

Fig. 35 — Split-supply CMOS logic can drive ECL through a MOSFET, as in A. When the CMOS system is powered from a single high-voltage supply, the bipolar transistor circuit of B can be used. Note that this results in a logic inversion.

switching speed specifications.

INTERFACING EXTERNAL DEVICES

There are numerous considerations in design of an interface for connection with an external device. If the external device already exists and cannot be changed, the interface design task is largely a matter of meeting the established requirements of the external device. Key questions are: What type of connector is used, and what are the pin assignments? What voltage or current levels are required? What is the data rate? What is the signaling format?

If the external device can be modified or has yet to be designed, it is prudent to investigate existing standards and conventions for interfacing such devices. To do otherwise would be not only "reinventing the wheel" but limiting the possibilities of interfacing with other similar devices. An interface standard encountered often in Amateur Radio is the Electronic Industries Association (EIA) RS-232-C and its International Telegraph and Telephone Consultative Committee (CCITT) counterparts V.24 and V.28. Details of these and other established standards are presented in Chapter 19. To understand these standards, it is necessary to know something about how data is communicated among devices.

Parallel and Serial Signaling

Within a piece of digital equipment, it is normal to handle a number of associated bits in *parallel* — all the bits in a group (byte or word) are handled at the same time. Typically, but not always, there are 8 associated bits, and 8 conductors to move the information from one circuit to another. The requirement for 8 separate conductors is not a problem within a piece of equipment. When transmitting digital information over long distances, however, providing 8 wires or 8 simultaneous communications channels may be too expensive for the application.

The solution to this problem is *serial* signaling, in which each of the bits is sent in turn, in agreed sequence, over a single channel or wire. Continuing with 8 bits as an example, it is convenient to name the bits b_0, b_1, b_2, b_3, b_4, b_5, b_6 and b_7. (The subscripted figures represent each bit's binary weight.) One of the things that must be specified for serial communications is whether b_0 (the least-significant bit or LSB) or b_7 (the most-significant bit or MSB) is sent first. Serial signaling makes it possible to send multi-bit binary-coded data over a single digital channel.

A graphic illustration of parallel and serial signaling is given in Fig. 38. In most amateur digital communications (either via radio or telephone lines), serial transmission is used to minimize cost and complexity.

Operational Mode

The number of channels needed for either serial or parallel signaling depends

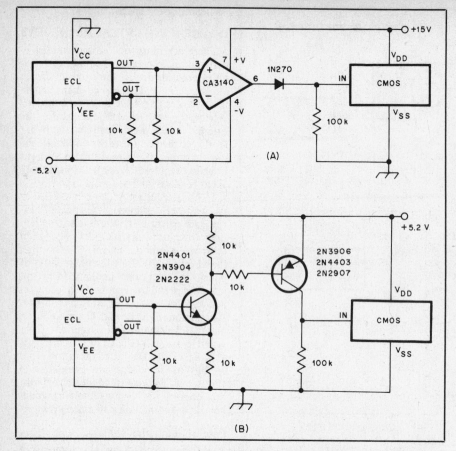

Fig. 36 — ECL-to-CMOS interface circuits. A method useful when different power supplies are used is illustrated in A. The diode prevents the −5.2-volt LOW level from damaging the CMOS device. When a common supply is available, the two-transistor amplifier/translator shown at B may be used.

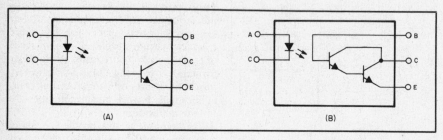

Fig. 37 — Optocouplers consisting of an LED and phototransistor detector. The device shown at A uses a typical single-transistor detector, while B has a Darlington phototransistor for improved transfer ratio.

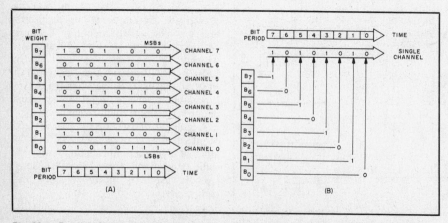

Fig. 38 — Parallel (A) and serial (B) signaling. Parallel signaling in this example uses 8 channels and is capable of transfering 8 bits per bit period. Serial transfer uses only one channel and can send only one bit per bit period.

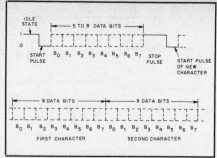

Fig. 39 — Serial data transmission format. In asynchronous signaling at A, a start pulse of one bit period is followed by the data bits and a stop pulse of at least one bit period. In synchronous signaling at B, the data bits are sent continuously without start or stop pulses.

on the operational mode.

Simplex operation involves only one-way transmission. So only those channels needed to get the information from the sender to the receiver are necessary.

Half-duplex operation is similar to simplex except that the direction of transmission can be reversed. This permits two-way communication without the need for a separate channel each way. The penalty is that the channel must be shared and cannot provide full-time communications in both directions.

Full-duplex operation is simultaneous two-way signaling. For parallel transfer of 8 bits in both directions, 16 channels are required. Full-duplex serial communication requires two channels. Full-duplex operation has the advantage of being able to provide communications in both directions at once.

Synchronization

In order to correctly receive data, the receiving interface must know when data bits will occur — it must be *synchronized* with the sender. There are two types of synchronization. In *asynchronous* communication, the receiver synchronizes on each incoming character. In *synchronous* communication, data is sent in long blocks, without start and stop bits or gaps between characters. Each of these schemes has its advantages and disadvantages.

Asynchronous Communications

In asynchronous communications, each character transmitted begins with a *start bit* and ends with a *stop bit* (Fig. 39). The start bit (usually a zero) tells the receiver to begin receiving a character. The stop bit (usually a one) signals the end of a character. Between characters, the transmitting circuit sends the stop-bit state (steady one or zero). Because the receiver is always told when a character begins and ends, characters need not be sent at regular intervals. This is especially useful when there is a person typing the input characters; the typist is usually slower than the data communications equipment and may not type at an even pace. One disadvantage of asynchronous communications is the inclusion of the start and stop bits, which are not "useful" data; if you are transmitting 8 data bits, a start bit and a stop bit, 2 of the 10 bits transmitted, or 20% of the bits, are *overhead,* used only for synchronization. On the other hand, asynchronous communications are rather simple. Because the receiver is newly synchronized at the beginning of each character, it does not need complex circuits to keep it synchronized. Because the characters need not be sent in a steady stream, no stringent demands are made on the person or process generating the characters. Amateur RTTY is usually asynchronous.

Synchronous Communications

In synchronous communications, data is sent in *blocks,* usually longer than a single character (Fig. 39). At the beginning of each block, the sender transmits a special sequence of bits that the receiver uses for initial synchronization. After becoming synchronized at the beginning of a block, the receiver must stay synchronized until the end of the block. Because the sender and receiver may be using slightly different clock frequencies, it is usually not adequate for the receiver to merely get synchronized initially and then depend on its internal clock. There are several ways for the receiver to stay synchronized. The transmitter may send the clock signal on a separate channel, but this is wasteful. The modulation technique used on the communications channel may convey clock information (Fig. 40C), or the clock might be implicit in the transmitted data (Fig. 40D).

One disadvantage of synchronous transmission is that the data must be sent as a continuous stream; characters must be placed in a buffer until there are enough to make a block. Also, while errors in asynchronous signaling usually only affect one character (the receiver can get resynchronized at the beginning of the next character), error recovery on synchronous channels may be a longer process involving several lost characters or an entire lost block. The major advantage of synchronous signaling is that it does not impose 20% overhead (the start and stop bits) on each character. This is an important consideration during large data transfers. As AMTOR and packet radio become popular, synchronous communication will play a major role in Amateur Radio.

Data Rate

There are limitations on how fast data can be transferred. First, the sending equipment has an upper speed at which it can produce the data in a continuous stream. Second, the receiving equipment has an upper limit on how fast it can accept and process data. Third, the signaling channel itself has a speed limit, often based on how fast data can be sent without errors. Final-

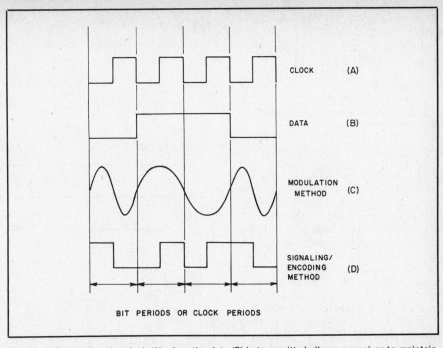

Fig. 40 — Recovering the clock (A) when the data (B) is transmitted allows a receiver to maintain synchronization during synchronous communication. The modulation method shown at C results in a transition of the received carrier at the beginning and end of each clock period. The encoding method shown at D results in a data transition in the middle of each clock period. Either of these methods provides enough information for clock recovery.

ly, standards and the need for compatibility with other equipment may have a strong influence on the data rate used.

Data rate can be expressed in a number of ways, but the basic method is to refer to the number of events or symbols per second. The usual symbol is the bit, and data rate is often given in bits per second (abbreviated bit/s). The *baud* (abbreviated Bd) is a unit of symbols per second; in simple signaling systems, bauds equal bits per second.

Within a given piece of equipment, it is desirable to use the highest possible data rate. When external devices are interfaced, it is normal practice to select the highest standard signaling rate at which both the sending and receiving equipment can operate. Standard speeds used in 8-bit serial data transmission include 110, 150, 300, 600, 1200, 2400, 4800 and 9600 bits per second.

Error Detection in Data Transfers

Whenever data transfers are subject to errors, it is important to include some method of detecting and correcting errors. In data communications, there are numerous techniques for error detection and correction. The technique employed in any specific circumstance depends on the types of errors likely to be encountered in the communications media. Some of these techniques are discussed in Chapter 19.

Parity Check

For some data transfers, *parity check* provides adequate error detection. To use the parity check, a *parity bit* is transmitted along with the data bits. In systems

using *odd parity,* the parity bit is selected such that the transmitted character (data bits and parity bit) has an odd number of ones. In *even parity* systems, the parity bit is chosen to give the character an even number of ones. For example, if the data 1101001 (which has an even number of ones already) is to be transmitted in an even parity system, the parity bit should be a 0 (maintaining even parity). If the same data is to be transmitted in an odd parity system, the parity bit should be a 1, resulting in a total of 5 ones. When a character is received, the receiver checks parity by counting the ones in the character. If the parity is correct (odd in an odd system, even in an even system), the data is assumed to be correct. If the parity is wrong, an error has been detected.

Parity checking only detects a small fraction of possible errors. This can be intuitively understood by noting that a randomly chosen word has a 50% chance of having even parity and a 50% chance of having odd parity. Fortunately, on relatively error-free channels, single-bit errors are the most common, and parity checking will always detect a single bit in error. Parity checking is a simple error-detection strategy, and because it is easily implemented, it is frequently used.

Parallel Input/Output Interfacing

There are a number of parallel input/output (PIO) ICs available. Typically they have eight data lines and one or more *handshaking* lines. Fig. 41 shows a simplified diagram of a PIO chip and its lines to an external device. The term

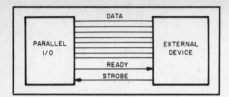

Fig. 41 — Parallel interface with READY and STROBE handshaking lines.

Table 7

Centronics Parallel-Printer Interface

Pin	Name	Source	Meaning
1	STROBE	Computer	Data lines are valid
2-9	D_0-D_7	Computer	Data to be printed
10	ACKNOWLEDGE	Printer	Data has been accepted
11	BUSY	Printer	Printer can't receive data
12	PAPER OUT	Printer	No paper in printer
31	INPUT-PRIME	Computer	Resets printer
32	ERROR	Printer	Printer error
20-27	RETURNS	—	Grounds for D_0-D_7

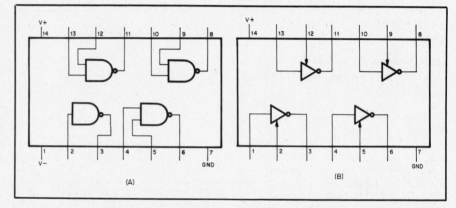

Fig. 42 — RS-232-C transmission-line drivers and receivers. A is a 1488 quad line driver. B is a 1489 quad line receiver. Note that pins 2, 5, 9 and 12 of the 1489 are response-control lines that are typically connected via 330-pF capacitors to ground.

"handshaking" includes a number of functions needed to coordinate the data transfer. One of these is the READY line, which, when it makes the transition from 0 to 1, indicates that data is now available on all 8 data lines. If only the ready line is used, the receiver may not be able to keep up with the data. In that case, a STROBE line is added to enable the receiver to tell the transmitter that it is ready to receive the next character. There are many other handshaking strategies that may be used during parallel signaling. The Centronics standard (Table 7) is commonly used for sending parallel data to printers.

Serial Input/Output Interfacing

Serial I/O (SIO) interfacing is more complex than parallel I/O. Fortunately there are a number of ICs available to handle the job. The advantages of serial data transfers and the general formats for asynchronous and synchronous serial transmission have been discussed earlier in this chapter. What is left to understand SIO is a discussion of the conversion process.

Within computers and other digital circuits, data is usually operated on, stored and transmitted in parallel. So, at the terminals of a serial interface, data must be converted from parallel to serial and back again. Serial-to-parallel and parallel-to-serial conversion is normally done with a shift register. The *universal asynchronous receiver/transmitter* (UART) is a shift-register device designed to handle these data conversions. The UART is capable of operating in the full-duplex communications mode. Furthermore, it can simultaneously send and receive data. UARTs can be programmed to select the number of data bits and whether the parity is odd, even, or not used. The UART transmitter accepts data in parallel form and shifts it out in asynchronous serial format. The transmitted serial format includes a start pulse, data pulses and stop pulses of various lengths. The receiver operates by shifting the serial pulses into a register. Then, when the specified number of bits have been stored in the shift register and a stop bit is detected, the data is made available in parallel form.

The *universal synchronous receiver/transmitter* (USRT) is similar to the UART but is for synchronous operation. A more popular device is the *universal synchronous/asynchronous receiver transmitter* (USART), which combines both a UART and a USRT on one chip.

Signaling on Line

Inside equipment and for short runs of wire between equipment, the normal practice is to use neutral keying; that is, simply to key a voltage such as +5 V on and off. In neutral keying, the off condition is considered to be 0 V. Over longer runs of wire, the capacitance between conductors and the inductance of the conductors becomes of concern. As the capacitance increases, more time is needed to make the transition from 0 to 1 or vice versa. This decreases the maximum speed at which data can be transferred on the wire, and may also cause the 1s and 0s to be different lengths. This is known as *bias distortion*. Also, on longer lines, there is a possibility of picking up noise, which can make it difficult for the receiver to decide exactly when the transition takes place.

Because of these problems, *polar* or *bipolar* keying is used on longer lines. Polar keying uses one polarity (for example +) for a logical 1 and the other (− in this example) for a 0. This means that the decision threshold at the receiver is 0 V for polar keying. Thus, any positive voltage can be taken as a 1, and any negative voltage becomes a 0.

Since neutral keying is usually used inside equipment, and polar keying is desirable for lines leaving the equipment, signals must be converted between polar and bipolar. The old way of doing this was to use an electromechanical polar relay.

Today, op-amp circuits handle this conversion. Even more convenient than the op amp, however, are ICs called *line drivers* and *line receivers*. There are a number of different types available, but the most popular ones are the 1488 quad line driver and the 1489 quad line receiver. "Quad" signifies that there are four drivers or receivers in the same IC package. Diagrams of the 1488 and 1489 with their pin assignments are shown in Fig. 42. The 1488 is capable of converting four data streams at standard TTL logic levels to output levels that meet EIA RS-232-C or CCITT V.24 standards. The 1488 can also be used to interface DTL/TTL to MOS, DTL/TTL to HTL, and DTL/TTL to RTL. The 1489 has four receivers that can convert RS-232-C or V.24 levels to TTL/DTL levels.

Summary

There are many types of digital interface, but a given interface can be categorized by answering several questions: Is it a parallel or a serial interface? Is it internal to a device, or is it between two devices? Does the interface signal asynchronously or synchronously? What recognized standards might apply to the interface? Once these broad questions have been answered, you can begin to study the interface in detail.

DIGITAL AND ANALOG INTERFACING

Until now, we have confined our discussion to digital electronics — the generation

A/D and D/A Experimentation in Amateur Radio

Analog-to-digital and digital-to-analog conversion have found important applications in the commercial and consumer worlds. These widespread applications have made A/D and D/A technology available to amateur experimenters. Two examples of this consumer-oriented digital signal processing are digital voice transmission and digital high-fidelity audio.

The telephone companies pioneered the digital transmission of voice on a large scale in the early 1960s. Most digital voice transmissions in North America use the AT&T "mu-law" standard — an 8-kHz sampling rate and 8 bits per sample. The resulting signal has an audio bandwidth of 4 kHz and requires a transmission speed of 64,000 bits per second. Mu-law encoding differs from standard A/D conversion in that the steps defined by the 256 quantizing levels are not evenly spaced. This technique causes the quantizing noise to vary with signal level, giving a nearly constant signal-to-noise ratio over a wide range of speech amplitudes. Because of the widespread use of digital speech transmission within the telephone network, mu-law coder-decoder (CODEC) ICs containing A/D and D/A converters and filters have become available at reasonable prices from several manufacturers. These ICs make excellent building blocks for experiments in digital voice communication.

Another consumer product that employs digital signal processing is the high-fidelity Compact Disk Digital Audio System, commonly referred to as the CD. A CD player uses a laser to scan a 12-cm disk that holds up to 74 minutes of very high fidelity digitized audio. The two audio channels are each sampled at a rate of 44.1 kHz, 16 bits per sample. Based on the rule of thumb 6 dB/bit, the quantizing signal-to-noise ratio is approximately 96 dB! As CD technology becomes more popular and more manufacturers produce inexpensive ICs that perform the necessary functions, experimenters will be able to adapt some of these parts to amateur applications. In particular, the CD system uses a sophisticated form of forward error correction (FEC), and ICs developed for this purpose may be very useful in improving the error performance of high speed amateur digital transmissions.

As amateurs gain experience with digital audio, the prospect of integrating this technology with packet radio will become attractive. Voice packet switching represents a way in which amateurs with diverse interests might join forces to build an "integrated voice/data network" capable of satisfying a wide range of communication needs. The basic technology needed is here today; all we need are enough creative and motivated amateurs to make this a reality.
— *Phil Karn, KA9Q*

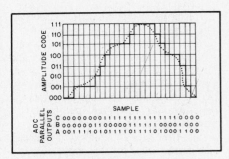

Fig. 43 — Analog-to-digital conversion of a complex wave. In this illustration, the voltage of the analog waveform is converted to the closest digital voltage level at the sample times. At each sample time, the digital code corresponding to the digital voltage level appears on output on the parallel outputs of the converter.

and manipulation of bits. It will come as no surprise, however, that it is an analog world. Virtually all measurable values in the physical universe are continuously variable and are thus analog. How can the many advantages of digital processing be applied to the analog world? We must employ *analog-to-digital conversion* (ADC) and *digital-to-analog conversion* (DAC).

Why Convert?

There are many reasons for using digital techniques on analog signals. Analog signals have the undesirable characteristic that they often pick up noise, and it is difficult or impossible to the remove the noise from the signal. Digital signals, on the other hand, are not likely to pick up noise (errors), and once noise has been introduced, it can often be detected and cor-

rected. When an analog signal has been *digitized,* it is subject to the whole spectrum of logical and computerized manipulation. It can be sent from place to place in digital form. It can be stored without corruption. It can be digitally filtered. In short, anything that can be done to or with a collection of bits can now be done to or with the original analog information. Digital signal processing is becoming common as more and more digital computers are connected to the analog "real world."

Analog to Digital Conversion

The object of analog to digital conversion is to change an analog signal into a collection of bits that accurately represents that signal. This is done by measuring the amplitude of the analog signal and then expressing that amplitude as a binary number (Fig. 43). As we have seen previously, binary numbers can be easily expressed and manipulated by electronic digital circuits. The process of measuring an analog signal and converting it to a binary number is called *sampling.* Correct sampling involves

two things: *sampling rate* and *sampling resolution.*

Sampling Rate

Each digital sample represents the amplitude of the analog signal at a specific time. As the amplitude of the signal changes, the binary number representing the amplitude must change. The rate at which the binary number is updated is called the sampling rate. Fig. 44 shows that if the sampling rate is too low, the digital amplitude values will represent a low-frequency *alias,* as well as the original analog signal. This condition is called *under sampling.* In order for the digital samples to correctly represent the analog signal, the sampling frequency must be at least two times the highest frequency component of the analog signal. This rule and its mathematical proof are the *Nyquist Sampling Theorem,* and the minimum sampling rate is called the *Nyquist rate.*

To apply the Nyquist Theorem, you must know the highest frequency component in the signal that you want to sample. For example, if you want to sample a signal that has components as high as 3 kHz, the theorem specifies a minimum sampling rate of 6 kHz. In practice, it is desirable to sample above the Nyquist rate. This is because the output of an A/D converter contains not only the original signal, but "double sideband" images of the original signal, centered at multiples of the sampling frequency.

Fig. 45 shows the spectrum of an analog signal and the spectrum at the output of an A/D converter. For clarity, only the first image spectrum (centered at the sampling frequency) is shown. The image must be eliminated if the digital sample is to correctly represent the analog signal. If the sampling frequency is below the Nyquist rate, the image overlaps the desired signal (Fig. 45A). Because of the overlap, the image cannot be eliminated without eliminating some of the desired signal. This is called *aliasing,* and is the result of under sampling. When the sampling rate is equal to the Nyquist rate, the image is just adjacent to the desired signal (Fig. 45B). In this case, an ideal filter (with vertical skirts) could eliminate the image.

Only when the sampling rate is above the Nyquist rate is there some separation be-

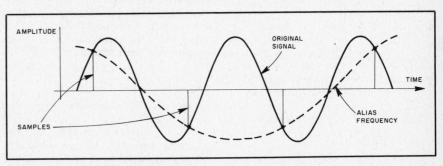

Fig. 44 — A sine wave sampled below the Nyquist rate. Notice that the samples specify an alias signal as well as the original sine wave.

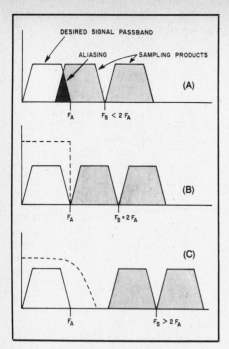

Fig. 45 — The unfiltered spectral output of an A/D converter. The unshaded spectrum is that of the analog signal being sampled. F_A is the highest frequency in the analog signal. The shaded spectrum is the first set of sampling products, centered at the sampling frequency (F_S). A shows undersampling, B shows sampling at the Nyquist rate, and C shows reasonable oversampling. The dotted lines in B and C represent the filter passbands that would separate the desired signal from the sampling products.

Table 8
Signal-to-Noise Ratios for Ideal Analog-to-Digital Conversion

Number of Bits	Amplitude Levels	Signal-to-Noise Ratio (dB)
3	8	18
4	16	24
5	32	30
6	64	36
7	128	42
8	256	48
9	512	54
10	1024	60
11	2048	66
12	4096	72
13	8192	79
14	16384	84
15	32768	90
16	65536	96

tween the desired signal and the image spectrum (Fig. 45C). With this gap between the two spectra, real filters can filter out the image and leave the desired signal unattenuated. The higher the sampling rate with respect to the highest frequency in the analog signal, the farther the undesired image is from the desired signal.

Sampling Resolution

Analog signals are infinitely variable; they can take on an infinite number of amplitudes. An A/D converter with n out-

put bits can only represent 2^n amplitudes (Table 8). Thus, the A/D converter cannot perfectly represent the amplitude of an analog signal. The number of bits in an A/D converter is the *resolution* of the converter. When an analog signal looks like a straight line, the output of an A/D converter has a "stair step" characteristic — it is *quantized* (Fig. 46). Fig. 46 shows the difference between an analog signal and a three-bit digital representation of that signal. At each sampling instant, there is a difference between the analog signal and the closest available digital value. This difference is called *quantization error*. Quantization error introduces noise, called *quantization noise,* to the sampled signal. The higher the resolution of the A/D converter, the lower the quantization error and the smaller the quantization noise. The relationship between resolution (in bits) and quantization noise (for an ideal A/D converter) can be expressed as

$$S/N = -20 \log \left(\frac{1}{2^n} \right) \qquad \text{(Eq. 3)}$$

where

n = the resolution of the A/D converter in bits

S/N = the signal-to-noise ratio in dB

This relationship can also be approximated as:

$$S/N = 6 \times n.$$

Practical Considerations

From the preceding discussions of sampling rate and resolution, we see that large, frequent samples are ideal for A/D conversion. In practical circuits, however, these characteristics conflict with each other and with external constraints. Large samples require longer sampling times, imposing a maximum sampling rate. Samples must be transmitted to the device that is to process them, and communications channels can only handle a limited number of bits per second. Finally, the device that is processing the samples has a maximum processing speed. With all of these varying constraints and the fixed constraints of the Nyquist Theorem and S/N equations, what is an acceptable A/D sample? Two industry standards are worth mentioning: the standard for digital telephony is 8-bit samples and a 8-kHz sampling frequency; the Compact Disk Digital Audio System takes 16-bit samples at 44.1 kHz. These systems, and their impact on amateur experimenters, are explored in the sidebar.

Digital to Analog Conversion

Once an analog signal has been turned into a digital bit-stream and manipulated, stored or communicated, it is sometimes converted back into an analog signal. Frequency and voice synthesizers must also convert digital information into analog signals. This process is called digital-to-

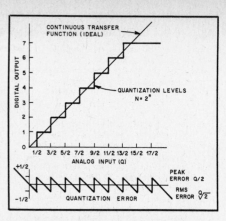

Fig. 46 — Comparison of a linear, analog transfer function with a quantized, digital transfer function. The difference between the two transfer functions results in quantization error, as shown at the bottom of the figure.

analog (D/A) conversion. In D/A conversion, each bit in a digital sample is converted to a voltage or current that is weighted to correspond to the bit's binary value. The resulting analog signals are summed to obtain an analog representation of the binary number. Fig. 47 shows a D/A converter using switched current sources. The output of a D/A converter is called a *digitally synthesized waveform*. Digitally synthesized waveforms are quantized, and the abrupt changes that characterize quantized signals appear on D/A converter outputs. These abrupt changes manifest themselves as unwanted high-frequency components in the output of the D/A converter. They must be removed by filtering. The correct filter for this application is a low-pass filter with cutoff frequency at or below one-half the sampling frequency.

Summary

D/A and A/D conversion allow digital circuits to act upon and generate analog signals. Correct A/D conversion requires sampling at or above the Nyquist rate and enough bits per sample to yield a good signal-to-noise ratio. D/A conversion produces an analog signal from a sequence of digitally encoded amplitudes.

HYBRID ANALOG AND DIGITAL ICs

There is a growing number of integrated circuits that have both analog and digital inputs and outputs. Some of these chips switch analog signals by means of digital control signals. Others are signal processors.

Analog Switches and Multiplexers

Bilateral switches are CMOS devices that act as digitally-controlled, SPST, solid-state switches. The signals on the input/output ports of these switches may be either analog or digital. The switches are controlled by digital signals. The block diagram of a quad bilateral switch is shown in Fig. 48.

Analog multiplexers are devices that use

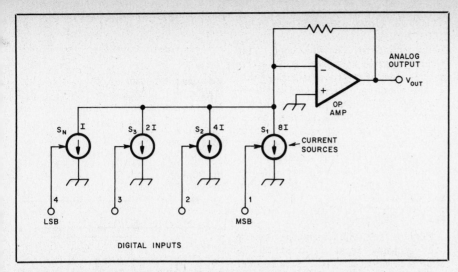

Fig. 47 — A D/A converter using four appropriately weighted current sources and a summing current-to-voltage converter. Notice that the MSB, representing $2^3 = 8$, generates 8 times the current generated by the LSB, which represents $2^0 = 1$.

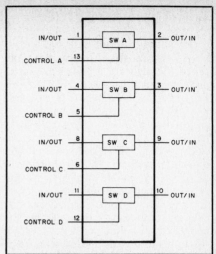

Fig. 48 — Block diagram of a CMOS 4066 quad bilateral switch. Each switch has its own control input.

bilateral switches to implement single-pole multiple-throw switches. Analog multiplexers are bilateral; pins are not strictly inputs or outputs, but can be either inputs or outputs. Unlike bilateral switches, analog multiplexers do not use a single binary control for each switch. Instead, analog multiplexers use binary-coded control inputs. For example, the analog multiplexer represented in Fig. 49 selects one of 8 IN/OUT lines using three digital-control inputs.

Analog bilateral switches and multiplexers can be used to switch signals, control the gain of an amplifier, select among resistive and reactive components, and perform numerous other tasks. Typically, these devices can handle signals up to 15 V peak-to-peak, provided that the power supply can handle the signal excursions. Complex analog multiplexing schemes can be realized by using digital logic to control analog switches.

Sample-and-Hold Amplifiers

A sample-and-hold (S/H) amplifier uses a capacitor to store a sample of input voltage. The charge on this capacitor tracks or samples the input voltage until the amplifier is put into the HOLD mode. At this time, the capacitor stops tracking and holds the input voltage (Fig. 50). S/H amplifiers have numerous applications in A/D and D/A converters. They are also used in data-acquisition and synchronous-demodulation applications.

Switched-Capacitor Filters

The switched-capacitor filter (SCF) is an IC that can be used as a low-pass, high-pass, band-pass or notch filter with as many as five poles on a single chip. SCFs are easy to use: filter type, Q and bandwidth are determined by external resistors, and the center frequency of the filter is determined by an external clock signal. Furthermore, frequency and bandwidth

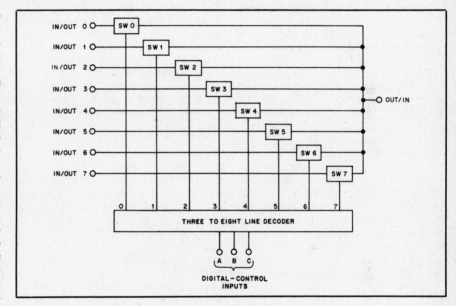

Fig. 49 — Block diagram of an 8-channel analog multiplexer/demultiplexer.

can be digitally controlled. Currently-available SCFs have dynamic ranges of around 80 dB and Qs up to 50, and they are usable at frequencies up to 250 kHz. Radio amateurs are using SCFs as CW and RTTY audio-bandpass filters and in DTMF and modem waveform generators.

SUMMARY

Interfaces take many forms. There are interfaces between IC families, interfaces between circuit boards, interfaces between digital devices and interfaces between digital and analog domains. Each type of interface places different demands on circuit designers. It is important to know what kind of interface you are designing, what its power requirements and restrictions are, what speed it will operate at, how noise is likely to be introduced into it and what standards might apply to the inter-

face. Only when you understand these characteristics can you design and build a successful interface.

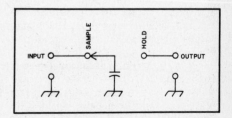

Fig. 50 — Simplified diagram of a sample-and-hold (S/H) amplifier. When the switch is in the SAMPLE position, the voltage across the capacitor is equal to the input voltage. When the switch is in the HOLD position, the capacitor retains its charge, presenting a sample of the input voltage to the S/H output.

Digital Basics 8-23

Memory Devices

Computers and other digital circuits usually rely on stored information. This stored information may be data to be acted upon or instructions to direct circuit actions. Digital information is stored in *memory devices.*

A memory device consists of a large number of *memory locations* or *words.* Each word, in turn, is composed of some number of *memory cells,* each capable of remembering one bit of binary information. The number of bits in each word, equal to the number of memory cells per memory location, is variable. Eight-, 16- and 32-bit words are common.

In order to store data in, or retrieve data from, a memory device, it is necessary to gain access to specific memory cells. The words *write* and *read* are used to signify storage and retrieval, respectively.

Memories can be divided into broad categories based on what memory cells can be accessed at a given instant. *Sequential-access memories* (SAMs) must be accessed by stepping past each memory location until the desired location is reached. Magnetic tapes implement SAM; to reach information in the middle of the tape, the tape head must pass over all of the information on the beginning of the tape. Two special types of SAM are the *queue* and the *push-down stack.* In a queue, also called a *first-in, first-out* (FIFO) memory, locations must be read in the order that they were written. The queue is a "first-come, first-served" device, like a line at a ticket window. The push-down stack is also called *last-in, first-out* (LIFO) memory. In LIFO memory, the location written most recently is the next location read. LIFO can be visualized as a stack; items are added to and removed from the "top" of a push-down stack.

The second category of memory is *random-access memory* (RAM). Random-access memory allows any memory cell to be accessed at any instant, with no time wasted stepping past "preceding" locations. Random-access memory is like a bookcase; any book can be pulled out at any time.

It is usually faster to access a desired word in random-access memory than it is to access a word in SAM. Also, all words in random-access memory have the same access time, while each word in a SAM has a different access time. Generally, the semiconductor memory devices internal to computers are random-access memories. Magnetic devices, such as tapes and disks, have at least some sequential access characteristics. We will leave tapes and disks for a later section and concentrate here on random-access, solid-state memories.

Addressing

An *address* is the identifier, or name, given to a particular location in memory. Within a memory IC, each word is given a unique address. The first word in the memory is given the address 0. To access a given word, the binary address of that word is placed on the chip's *address lines.* For example, to access the 11th word in a memory with eight address lines, the address lines would be set to 00001010. For ease of notation, programmers and circuit designers do not write addresses in binary form, but use octal (base-8) or hexadecimal (base-16) notation. In *hex,* 00001010 is written "0A."

Memory Chip Size

Because, internally, memories are addressed using binary arithmetic, the number of words in an IC and the number of bits in a word are usually powers of 2. An IC with 8 address lines has 2^8 or 256 uniquely addressable words. The size of a memory IC is usually expressed as "M × N," where M is the number of words in the IC, and N is the number of bits in each word. Memory chips as small as 16 × 4 and as large as 256,000 × 1 are common. To understand how memory ICs are organized into large computer memory systems, one must first understand how small memory chips work. We will use a 256 × 1 IC as an example.

Memory Architecture

Memory chips, no matter how large or small, have several things in common. Each chip must have *address lines, data lines* and *control lines* (Fig. 51). A memory IC must have enough address lines to uniquely address each of its words. A 256 × 1 memory must, therefore, have 8 address lines. There must be as many data lines on a chip as there are bits per word in the chip, so the 256 × 1 memory has 1 data line. Data lines are both inputs to the chip (when writing) and outputs from the chip (when reading). Therefore, there must be a control line that distinguishes read operations from write operations. This line is labeled *R/W* in Fig. 51. When R/W is 1, the chip is in the write mode.

The final control line on this simple memory is the chip select (CS). When CS is 1, the chip is selected, and it acts on the R/W, address and data information presented to it. When CS is low, the data line enters a high-impedance state, so that the deselected chip does not effect operation of devices or circuits attached to it. On actual ICs, CS and R/W may be inverted or edge-triggered inputs, and there may be other lines on a memory chip. The chip in Fig. 51 does, however, provide a realistic basis for understanding memory chips.

Writing to Memory

If we want to write a 1 to the 11th word of the 256 × 1 memory in Fig. 51, we must execute the following steps:

1) Place the correct address (in this case

Table 9

Comparison of Memory Chips

Type of Memory	Volatile	Frequent Writes	Erasable	Erasable in System
RAM	yes	yes	yes	yes
NVRAM	no	yes	yes	yes
ROM	no	no	no	no
PROM	no	no	no	no
EPROM	no	no	yes	no
EEPROM	no	no	yes	yes
Bubble	no	yes	yes	yes

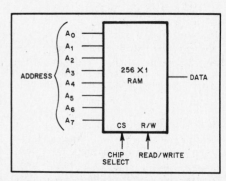

Fig. 51 — Simplified 256 × 1 memory IC. The DATA line is an input during write operations and an output when data is being read from the memory. All other signals are inputs to the IC.

0A hex or 00001010 binary) on the address lines.

2) Place a 1 on the data line.

3) Set the R/W control line to 1 (write).

4) Set CS to 1.

This will result in the data on the data line (1) being written to the address on the address lines (00001010).

Reading from Memory

To read the contents of the 11th word:

1) Place the address 00001010 on the address lines.

2) Set the R/W control line to read (0).

3) Set CS to 1.

When all of this has been done, the data from address 0A (hex) will be placed on the data line.

Timing

The above discussion ignores subtle timing requirements. These requirements are strict, and usually insist that data being written to the memory be present for a minimum *setup time* before and *hold time* after the CS and R/W signals. Timing specifications also specify how long address lines must be stable and when data being read from the memory is valid. Consult manufacturers' data sheets and applications notes for timing information.

Larger Words

We have shown how a one-bit-wide

memory works. Usually, though, we want a wider memory. Low-cost home computers use 8-bit words, and 16- and 32-bit computers are becoming popular. One way to get wider memory is to use several 1-bit-wide memory chips, as shown in Fig. 52. The address and control lines go to each chip, and each chip's data is used as a single bit in the large word. It is easy to see that when reading from address 0A (hex), the data lines D_0 through D_3 contain the data from address 0A of chips 0 through 3. The same is true when writing to an address. An address placed on the shared address lines (called an *address bus)* now specifies an entire word of data. If all three memory chips were put in a single package, they would make a 256 × 4 IC. This IC would look like the IC in Fig. 51, except that it would have 4 data lines. It would perform like the circuit in Fig. 52. In fact, ICs with 4- and 8-bit words are quite common, and even larger words are available.

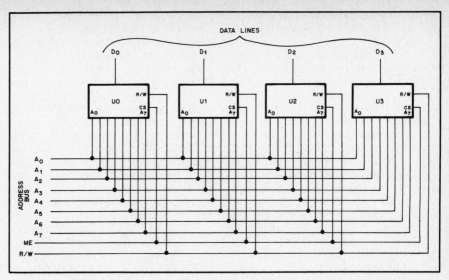

Fig. 52 — A 256 × 4 memory built with four of the ICs in Fig. 51. Address and control signals are decoded by the four ICs, resulting in four bits of data (one bit from each IC) being read or written by the circuit.

More Address Space

Most tasks require more than 256 words of memory. It is not uncommon for a personal computer to address 256,000 8-bit words. To understand how such large memories are built and addressed, we will look at a 1024 × 8 (1-kbyte) memory built from four 256 × 8 memory chips (Fig. 53).

Ten address lines are needed to address 1024 locations (2^{10} = 1024). Eight of the 10 address lines ($A_0 - A_7$) are used as a normal address bus for chips 0 through 3. The remaining 2 address lines (A_8 and A_9) are used to choose among the 4 memory chips. A_8 and A_9 are run through a *2-to-4-line decoder.* For each of the four combinations of A_8 and A_9, the decoder places a 1 on a different one of its outputs. When employed in this manner, the 2-to-4-line decoder is called an *address decoder.* To assert the CS input for one of the memory chips, ME must be 1 and the correct output of the 2-to-4-line decoder must also be 1. When an address is placed on A_0 through A_9, a single memory chip is selected by ME, A_8 and A_9, and the other 8 address lines address a single word from that chip. The three chips that are not selected enter a high-impedance state and do not affect the data lines. This example shows how large memories are built from small memory ICs.

Summary

All memories built with more than one memory IC employ the techniques described above. Using the proper memory ICs and address decoding, any size memory with any word length can be built.

MEMORY TYPES

The concepts described above are applied to several types of random-access, semiconductor memory. Semiconductor memories are categorized by the ease and speed with which they can be written and their ability to "remember" when power is removed

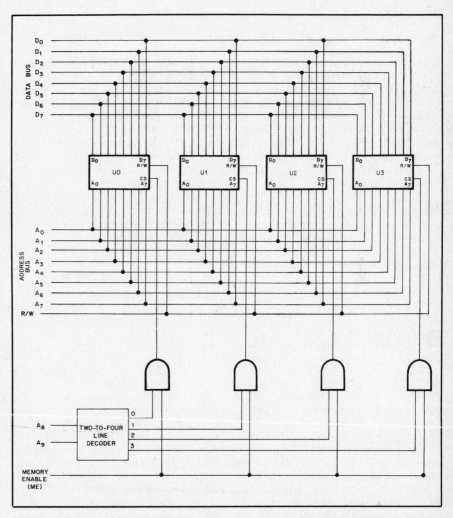

Fig. 53 — A 1024 × 8 memory using four 256 × 8 memory ICs. The high address bits, A_8 and A_9, select one of the memory ICs. The remaining seven address bits address a byte from the selected IC. Address and data lines shared in parallel by several chips are called a *bus.*

from the circuit.

RAM

The term "RAM", standing for "random-access memory," is usually used for semiconductor memories that can be written to or read from with equal ease and speed. It is usually safe to assume that

RAMs are *volatile,* meaning that information stored in RAM will be lost if power is removed.

Two types of RAM are common: *static RAM* and *dynamic RAM.* Dynamic RAMs (DRAMs) store a bit as the presence or absence of charge on MOSFET gate-substrate capacitance. Because the capacitance is not perfect, the charge leaks or degrades slowly and must be refreshed periodically (typically every few milliseconds). This memory refresh is usually performed by a *dynamic RAM controller* IC. A static RAM (SRAM) is an array of flip-flops. When a bit is stored in a flip-flop, the bit will remain until power is removed or another bit replaces it. SRAMs do not need to be refreshed. While SRAMs may seem more desirable, they are neither as simple to manufacture nor as dense as DRAMs. DRAMs are often 1.5 to 4 times more dense than SRAMs. Increased density and ease of manufacture make DRAMs significantly cheaper than SRAMS.

Both bipolar and MOS semiconductors are used to make RAMs. While all bipolar RAMs are SRAMs, MOS techniques can produce both static and dynamic RAMs. Cost, power consumption and access speed are the factors that help designers select static or dynamic, MOS or bipolar RAMs. Generally, MOS RAMs have lower power consumption than bipolar RAMs. Access speed varies widely within and between RAM catagories. To select the best RAM for a given application, consult manufacturers' data sheets.

Read-Only Memory

Read-only memory (ROM) is nonvolatile; its contents are not lost when power is removed from the memory. Although the name "read-only memory" implies that ROMs can never be written, all ROMs can be written or *programmed* at least once. *Mask ROMs* are programmed by having ones and zeros etched into their semiconductors at manufacturing time.

There are other types of ROMs, called *programmable ROMs* (PROMs), that can be written to after manufacture. *Fusible-link PROMs* are written by burning away specific diodes or transistors in a semiconductor memory array. This type of PROM can be written only once. Two types of PROMs can be erased and reprogrammed. Erasing is done by ultraviolet light for UV PROMs (EPROMs), and electronically for electronically erasable PROMs (EEPROMs). The erasure cycle for EEPROMs is long, and UV PROMs must be removed from operation for reprogramming. These requirements mean that data cannot be written as quickly and easily to ROMs as it can to RAMs.

What (besides their being nonvolatile) distinguishes PROMs from RAMs? Read and write times for RAMs are nearly equal, but PROMs have slow write times (in the millisecond range), and RAM-like read times (measured in nanoseconds). Two other factors make it hard to write to PROMs. First, PROMs must be erased before they can be reprogrammed. Second, PROMs often require a programming voltage higher than their operating voltage. ROMs are practical only for storing data or programs that do not change frequently and must survive when power is removed from the memory. The programs that start up a computer when it is first turned on or the call sign in a repeater IDer are prime candidates for ROM.

Nonvolatile RAM

For some tasks, the ideal memory would be as nonvolatile as ROM but as easy to write to as RAM. Low-power RAMs can be used in such applications if they are supplied with NiCd or lithium cells for backup power. A more elegant and durable solution to this memory problem is called a *nonvolatile RAM* (NVRAM, or Xicor NOVRAM™). An NVRAM chip includes both RAM and PROM. When power is first applied or the chip is reset, the contents of the ROM are copied into the RAM. The system using the NVRAM reads and writes to the RAM as though the ROM did not exist. When requested, perhaps before power is turned off, logic in the NVRAM copies the entire contents of the RAM into the ROM for nonvolatile storage. The RAM in a NVRAM is called *shadow RAM,* and it is standard volatile RAM. The backup ROM in NVRAM is EEPROM, and each bit in the ROM can be reversed several thousand times. Since the RAM is only copied into the ROM on command, NVRAM lasts long enough for many applications. NVRAM fills the gap between easily written memory and nonvolatile memory.

SUMMARY

With RAM, ROM and NVRAM available in many sizes and packages, you should be able to find a memory IC that meets your design requirements. By using correct address decoding, relatively small memory ICs can be used to build large memory systems. Like most IC technology, memories are changing quickly — getting cheaper, faster, more efficient and smaller. When you are designing memory systems, be sure to consult the *latest* literature.

Digital Computers

Until the 1970s, digital computers were strictly the tools of industry and education; there were no home computers or hand-held calculators as we know them today. Today, only 10 years later, everything from watches to automobiles is computerized. This revolutionary increase in the availability of computers was made possible by several simultaneous changes in technology that decreased the size, increased the complexity and decreased the price of semiconductors. These advances made possible the construction of the *microprocessor* (μP) IC. This single chip performs all of the control, logical and mathematical functions necessary for computation. A microprocessor, memory and I/O circuits are all that is needed to make a *microcomputer* (μC).

Modern Amateur Radio transceivers, scanners, digital-code converters, packet-radio controllers and television systems contain *special purpose computers.* These are dedicated microcomputers that perform specialized control tasks and are not available for other uses. Many amateurs also have *general-purpose computers* that they use for Amateur Radio applications. The obvious result of the increase in use of microcomputers by amateurs has been a substantial increase in computerized RTTY. There are many other applications of computers in Amateur Radio; CW sending and receiving, contest duping and logging, computer-aided design (CAD), computer-aided instruction (CAI), satellite tracking, propagation prediction and record keeping are just a few. Amateurs have found numerous opportunities to use microcomputers, and as pocket computers and single-chip microcomputers become less expensive, they will be found in most amateur stations.

COMPUTER ORGANIZATION

A computer consists of a number of interconnected subsystems, each of which performs some specific task. These subsystems may be spread among many chips, or they may be concentrated in a single-chip microcomputer. Fig. 54 illustrates the organization of a typical computer.

Input Device

Input devices are used to feed the computer both *data* to work on and *programs* to tell it what to do. Some digital input devices are: switches, keyboards, magnetic disks and tapes, paper cards and tapes, demodulators and analog-to-digital converters.

Output Device

Output devices are used to present the results of computer operations to the user or to some other system. Output devices also allow the computer to control external devices. Printers, video displays, LEDs, LCDs, modems, speech synthesizers, plotters, relays, and magnetic disks and tapes are common output devices.

Memory

The memory subsystem usually has several thousand words of storage, both temporary (RAM) and nonvolatile (ROM). It holds computer programs as well as input, intermediate and output data.

Arithmetic Logic Unit

The arithmetic logic unit (ALU) performs logical operations (AND, OR, etc.) and arithmetic operations (addition, subtraction, multiplication and division). The ALU may depend on the control unit to supply its data and send its results to memory, or it may be able to access memory itself.

Control Unit

The control unit directs the operation of the computer. It takes instructions from memory and executes them. In the course of executing an instruction, the control unit may call on the ALU, perform I/O and access memory to read or write data. The control unit maintains the *program counter*, which contains the address of the next instruction that the control unit will fetch for execution. The program counter usually moves from lower addresses to higher addresses, but some instructions cause the control unit to change the contents of the program counter. This is called looping or branching. The control unit and the ALU are usually both in the microprocessor IC.

Registers

Microprocessor chips have some internal memory locations that are used by the control unit and the ALU. Because they are inside the microprocessor IC, these *registers* can be accessed much more quickly than other memory locations. Some registers are purely internal and cannot be accessed directly by programs. Others can be accessed directly. The registers that can be programmed are divided into *general purpose* and *special purpose* registers. General purpose registers can be used as though they were RAM locations, although they are referred to by name rather than by address. The special purpose registers usually include the program counter, the *accumulator*, the *stack pointer* and a *flag register*. The program counter, as discussed before, contains the address of the next instruction that the computer will execute. The accumulator contains the result of the

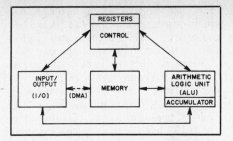

Fig. 54 — Major subsystems in a microcomputer. Not all of the communications paths shown on this diagram are present in every computer. DMA stands for direct memory access.

latest ALU operation. For program control, the microprocessor uses a section of RAM as a push-down stack. The stack pointer contains the address of the "top" of this stack. The flag register holds status and flag bits that indicate, among other things, whether the result of the latest ALU operation was zero, positive or negative.

Timing

The computer understands a few relatively simple instructions: add A to B, move the contents of X to Y, send C to the output device, etc. For the execution of these simple instructions to result in meaningful operations, thousands of instructions must be executed each second. The rate at which instructions are executed and the synchronization of the systems in a microcomputer are controlled by a single clock. Usually, a quartz crystal generates the clock signal. The output of this clock goes to the microprocessor and to any other ICs that need to be synchronized with the microprocessor.

Summary

The subsystems discussed above may exist in several ICs or a single-chip microcomputer. At the least, a microprocessor *central processing unit* (CPU) contains the control subsystem and the ALU.

Selecting a Microprocessor

Microprocessors can be classified by the number of bits their accumulators will handle at one time. Four, 8, 12, 16 and 32-bit microprocessors are available. Amateurs usually use 8- or 16-bit microprocessors. Some microcomputers now have both an 8-bit and a 16-bit processor on the same PC board, permitting operation of either processor or simultaneous operation of the two.

One important consideration when selecting a microprocessor is the availability of desirable software for that processor. Familiarity of the designer with programming the microprocessor, or an earlier model with a similar instruction set, may also be an important consideration. Some microprocessors are *upwardly compatible* with older microprocessors of the same family. Upward compatibility means that

programs written for the older microprocessor will run on the newer microprocessor. Upward compatibility eases the transition from one processor to another, and allows a designer to gain the advantages of a new microprocessor without having to replace a complete software library.

Speed is another important consideration in selecting a microprocessor. Clock rates typically run from 1 to 10 MHz, with the 1- to 6-MHz range being suitable for most Amateur Radio applications. The clock frequency defines the duration of the *machine cycle* — the basic unit of time for a microprocessor. In one machine cycle, the microprocessor can perform one basic operation. For example, it might take one machine cycle for the control unit to fetch a number from memory. Each instruction that the microprocessor can execute is made up of several basic operations and will take several machine cycles to execute. Manufacturers' data sheets may refer only to average instruction execution times. For this reason, designers wishing to select the fastest microprocessor for a given job may develop a *benchmark* program to test how fast different microprocessors execute a desired task.

When selecting a microprocessor, it is important to know whether it will be supported by the manufacturer and/or other sources. Microprocessor manufacturers usually offer an entire family of support chips, such as a parallel I/O, serial I/O (UART), DMA controller and timer/counter. It is best to consider the entire chip set before making the decision to select a particular processor.

Documentation is also vital to the designer. Manufacturers may offer an array of data sheets, application notes, programmer's guides and references and other design aids. It will be helpful to know whether the manufacturer will answer technical questions over the phone. Manufacturers may avoid documenting idiosyncracies that could be design hurdles in particular applications. There are quite a few books on how to work with specific microprocessors. Maintaining contact with other designers, either directly or via users' groups, is another way of finding design information and learning about undocumented features.

Power requirements are important in some applications and may be the primary reason for selecting a given microprocessor. Early microprocessors used as many as four voltages, complicating power-supply design. Most designers prefer chip sets that use one voltage, usually +5 V. CMOS microprocessors are normally chosen for satellites and other battery-operated equipment in which power is at a premium. If supply voltage varies widely, the voltage range over which a microcomputer will operate is also important.

Future cost and availability of a microprocessor are difficult to estimate.

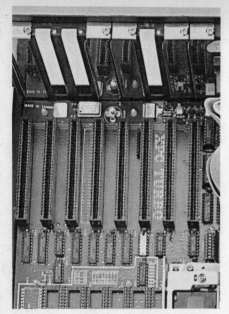
Fig. 55 — The backplane of a popular personal computer has several sockets for accessory PC boards such as video adapters, serial and parallel interface cards, or disk-drive controllers.

Nevertheless, there are numerous indicators such as reputation of the manufacturer, the amount of support available and *second sourcing* (arrangements made with other manufacturers to produce the same chip). When a chip is first introduced, its cost may be high, perhaps $50 or more. If it is used in large-quantity products, the unit price will drop, probably to less than $10. Many amateurs are simply concerned with the current price of the chips they use. Manufacturers pay less attention to the current cost of a chip and more to their estimate of what it will cost when they have to order production quantities. If you are designing equipment for other amateurs to build (for example, if you plan to write an article for *QST*), try to project cost and availability ahead to the probable publication date. The chips that are cheapest at the moment may be phased out by the time people want to duplicate your circuit.

MICROCOMPUTER BUSES

Most microprocessors have a single set of data, address and control pins; most of the rest of the microcomputer must be connected, in parallel, to these pins. The address, control and data circuits that connect the subsystems of a microcomputer are called a *bus*. The bus usually also carries power throughout the computer. One popular mechanism for routing the bus to various parts of the microcomputer is to print the bus as a set of parallel PC traces on a PC board called the *mother board*. Multi-conductor connectors, placed at intervals along these traces, are used to connect individual memory, I/O or microprocessor PC boards to the bus (Fig. 55). Common home computers like the Apple II

series and the IBM PC have the microprocessor and some memory on the mother board and several bus connectors, called slots, available for system expansion.

One popular bus was introduced with the the Altair 8800 computer in 1975. Because it was a 100-wire bus, it became known as the *S-100* bus. The S-100 bus accommodates PC boards measuring 5 by 10 inches. The S-100 bus is widely supported, and S-100 PC boards for almost every microcomputer function are available. Because of the widespread use of S-100 bus, the Institute of Electrical and Electronics Engineers (IEEE) ultimately developed an S-100 bus standard known as IEEE-696-1983. The S-100 bus computers are popular among radio amateurs who need versatile, expandable computers.

Another bus worthy of note is the IEEE-488-1978 standard. The 488 bus was originated by Hewlett-Packard under the name HPIB and has also been called GPIB. Its main use is for computer control of electronic test equipment, but it can also be used as a microcomputer peripheral bus. Most professional test equipment can be controlled by the 488 bus. Amateur applications of the 488 bus are expected to grow, particularly as bus controllers and test equipment with the 488 bus become available as surplus.

For an overall view of microcomputer buses, see *The S-100 & Other Micro Buses,* by Poe and Goodwin, 2nd Ed., 1981, Howard W. Sams & Co., Inc.

I/O Transfers

The microcomputer would be useless without I/O. Input and output allow the computer to react to and affect the outside world. Microcomputers transfer data to and from peripheral devices using several techniques.

Program-Controlled I/O

Under *program-controlled,* or *polled, I/O,* all input and output events are initiated by the program that is running on the microcomputer. The program must wait until the I/O device is ready to accept or deliver data and then execute the instruction that actually sends or receives the data. An example of program-controlled I/O might be a program waiting for a keyboard entry:

Has a key been pressed?

If not, then check again.

If yes, then find out what key was pressed and continue.

One of the disadvantages of program-controlled I/O is that the program must spend its time checking the status of the I/O device. If the program must perform the I/O before it can continue anyway, time spent checking I/O status is not wasted. If, on the other hand, the program could be performing other tasks while waiting for I/O, time spent checking status is wasted. Program-controlled I/O is easily written and debugged. In home computers using the BASIC language, program-

controlled I/O is often the only kind of I/O used.

Interrupt-Driven I/O

Program-controlled I/O might be understood by imagining a cook returning to the oven every few minutes to see if a meal is ready to be served. In *interrupt-driven I/O,* the cook would go do something else, and the meal would interrupt him when it was ready to come out of the oven. An *interrupt* is a temporary break in the normal execution of a program. Under interrupt-driven I/O, external devices that are ready to accept output or generate input interrupt the program that is executing. When the microprocessor is interrupted, it suspends whatever program it was processing, branches to an *interrupt-service* program, takes care of the device that generated the interrupt and then resumes execution of the interrupted program. The major disadvantage of interrupt-driven I/O is that program flow can get very confusing; what if an interrupt-service program gets interrupted? If a program does not work, is there a problem with the program itself, or is there a problem in an interrupt-service program? These problems are not impossible to solve, and interrupt-driven I/O is common. Programs that rely on interrupt-driven I/O do not have to constantly check the status of I/O devices. Also, each interrupting device can be given a priority, so that I/O events get serviced in order of importance. Interrupt-driven I/O, despite its complexity, is often necessary or desirable in control and communications programs.

Memory-Mapped I/O

In *memory-mapped I/O,* addresses that are treated like RAM by the microprocessor are actually I/O devices. Thus, a command that would usually be used to read or write to a memory location might actually result in an I/O operation. Memory mapping is an addressing technique and might be used with interrupt-driven or program-controlled I/O.

Direct-Memory Access (DMA)

DMA is used with I/O devices that need rapid transfer of data, such as magnetic disk drives or high-speed data-communications equipment. When a device that uses DMA is ready to transfer data (either to or from memory), a *DMA controller* makes the transfer directly between memory and the I/O device without interrupting or using the main control unit. Thus, DMA can perform I/O while another program is running, without interrupting that program. Although some microprocessors have built-in DMA controllers, DMA usually requires a special controller IC.

PERIPHERAL DEVICES

Peripherals are devices separate from the central-processing unit that provide additional facilities such as outside communication. Tape recorders, printers, video-display terminals and modems are common peripherals.

MOVING MEDIA

The term *moving memory* refers to the mechanical motion of the storage medium past a read or write transducer (often called a *head*). The earliest forms of moving memory used hole patterns punched in paper cards or tapes. Punched cards (also known as IBM™ cards), once widely used in the computer industry, have had virtually no application in Amateur Radio. On the other hand, punched tapes have seen Amateur Radio use. The tape format that gained the most popularity among amateur radioteletype operators was the 5-level punched tape, using the Baudot code. The tapes were prepared on a perforator (or reperforator) and read on a transmitter-distributor (TD).

Magnetic Media

Magnetic tape is widely used for storing digital information. Inexpensive audio cassettes, designed primarily for analog recording, are used by amateurs for digital storage. Most low-end microcomputers use cassettes as an inexpensive magnetic storage medium. Unfortunately, for proprietary reasons, there is little standardization of the tones and data formats used by different microcomputer manufacturers. The Kansas City standard, developed in the mid-70s was an attempt to get all manufacturers to agree on a data rate of 1200 bits per second and the tones 1200 and 2400 Hz. Although it met with little commercial success, the Kansas City standard is now used for data transmission from the UOSAT amateur

satellites and by the Radio Netherlands BASICODE project.

Streamer is the term given to a relatively new type of digital magnetic 4-track, ¼-inch tape-cartridge recorders. These recorders were designed to back up hard disks.

The *floppy disk* (also called *diskette*) is the peripheral most widely used in microcomputers to store large amounts of data. The diskette itself is a paper-thin mylar disk upon which a thin layer of magnetic material has been deposited. This disk, which is like a sheet of magnetic tape, can be 3 to 4, 5¼ or 8 inches in diameter. It is enclosed in an envelope that protects the magnetic coating while allowing a read/write head access to the disc (Fig. 56). Floppy disks are read and written by a *disk drive*. The drive spins the disk (within its envelope) much like a record player spins a record. The read/write head may access any spot on the disk by moving to the correct point in the *head-access window* and waiting for the data to spin "into view" (Fig. 57). Because the disk makes several revolutions per second, data on diskette can be accessed much more quickly than data on magnetic tape.

Disks and disk drives may be either *single-sided* (SS) or *double-sided* (DS). Double-sided disks are prepared for reading and writing on both sides, and double-sided drives have read/write heads for both sides of the disk. Depending on the *disk controller* used to access the disk drive, data may be written in either a *single-density* (SD) or *double-density* (DD) format. Using double-density results in nearly twice the storage capacity of single-density but requires higher-quality diskettes. The

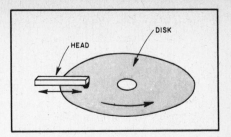

Fig. 57 — The read/write head in a disk drive can be positioned over any track on a disk. Disk rotation will then bring the desired sector beneath the head.

initials specifying the number of sides and the density of a disk system are often combined, as in "SSSD," meaning "single sided, single density."

Fig. 56 shows that data is stored on a disk in a number of concentric bands, called *tracks*. Some disks have a track density of 48 tracks per inch (tpi), while others use 96 tpi. Each track is divided into *sectors* — the smallest addressable units of information on a disk. A photoelectric sensor in the disk drive may be used in conjunction with the *index hole* in the disk envelope to sense the radial position of the disk. *Hard-sectored* disks have holes in the disk for each sector. More common *soft-sectored* disks have only one hole. Depending on the number of tracks, sectors, bytes per sector and sides used, a floppy disk can store anywhere from 80 kilobytes to several megabytes.

There is standardization in 8-inch floppy disk formats. The most common format, developed by IBM, has 77 concentric tracks with track 0 on the outside and track 76 inside near the hub. Each track is broken up into 26 equal sectors, each holding 128 bytes. There is a "write-protect" notch on the side of the protective envelope. On 8-inch disks, the disk cannot be written to unless the write-protect notch is covered.

There has not been much standardization of 5¼-inch floppy disk formats. The write-protect convention for 5¼-inch disks is just the opposite of that for 8-inch disks — expose the notch to write to the disk.

Standardization efforts are now underway for smaller 3- to 4-inch *microfloppy* disks. The American National Standards Institute (ANSI) X3B8 Microfloppy Industry Committee (MIC) has proposed adoption of a 3½-inch standard with 80 tracks. The unformatted capacity is 500 kbytes for single-sided 3½-inch disks and 1000 for double-sided disks.

It is important to remember that the read/write head is in contact with the surface of the floppy disk, and the disk is rotating at several revolutions per second. Dust and dirt on the disk and the imperfections in the disk surface gradually damage both the disk and the head. This means that disks eventually wear out. When a disk wears out, the data stored on that disk will probably be lost. It is prudent, therefore, to make back-up copies of your disks.

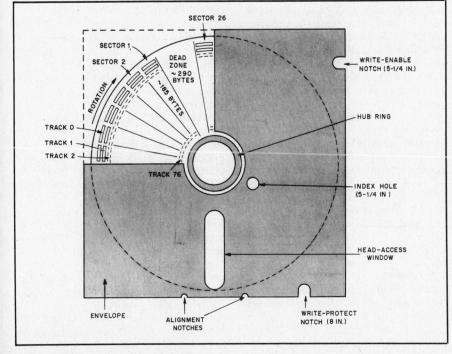

Fig. 56 — Construction of a floppy disk.

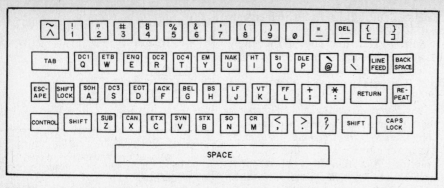

Fig. 58 — Typical computer-terminal keyboard layout. Two- and three-letter abbreviations shown at the top of letter and underline keys designate control functions. The CONTROL key must be pressed at the same time as the other keys to enter a control code. The complete ASCII-coded character set is listed in Chapter 19.

Back-up disks, stored in a clean, dry, cool place, will help you avoid disaster when your disks wear out.

In *hard disk* systems, the disk is rigid, and the read/write head does not contact the disk. The absence of friction between the head and the disk allows finer head positioning and higher disk speeds. For these reasons, hard disks hold more data and are accessed more quickly than floppy disks. Hard disks currently used for microcomputers have 5¼- or 8-inch disks capable of storing from 2 to 84 megabytes of data. Although the common Winchester hard disk drives do not, some hard disk drives have removable, interchangable disks. Hard disk systems are high-precision mechanical devices, and while they do not fail frequently, even one failure can result in the loss of megabytes. To avoid this, you should frequently make back-up copies of your hard disks. Using floppy disks for hard-disk back up can take a lot of diskettes, and streamer tapes (mentioned earlier) make more economical hard-disk back ups.

Summary

Some type of moving memory is usually included in a general-purpose computer system. Inexpensive home computers use cassette tapes, but floppy disks are more convenient. Advancing technology has been pushing the storage capacity of floppy disks up, while bringing prices down. High-capacity hard disks are also coming within the reach of the amateur budget. Remember, whatever moving medium you use, always make back-up copies of your important data.

Keyboards

A *keyboard* is an input device consisting of numerous key switches that, when pressed, cause the generation of a particular bit pattern. Most computer keyboards have keys arranged in the QWERTY order, similar to a typewriter keyboard layout. However, the similarity ends with the location of the 26 letters of the Latin alphabet and the 10 Arabic numerals. The placement of punctuation, other printable symbols and *control keys* varies. A typical computer-terminal keyboard layout is shown in Fig. 58.

Computer keyboards have keys not traditionally found on typewriters. While typewriters have one RETURN key that issues both a carriage return (CR) and a line feed (LF) for a new line, computer keyboards usually have separate CR and LF keys. The CAPS LOCK key on a computer keyboard puts all letters in upper case but preserves the action of the SHIFT key over keys with numerals, punctuation and other symbols. Some keyboards include a special numeric keypad seperate from the main keyboard. This keypad makes it easy for operators familiar with calculators or adding machines to enter lists of numbers into the computer. Other special keys often found on computer keyboards include a CONTROL key, an ESCAPE key and perhaps some *function keys*.

Most computer keyboards will generate the entire 128-character ASCII set. (Chapter 19 examines the ASCII characters and their meanings.) Ninety-six of the ASCII characters are numerals, alphabetics and printable punctuation marks. These characters are usually generated by single keys or shifted keys. The other 32 characters are *control characters*, and are generated by pressing alphabetic and numeric keys while holding down the CONTROL key. Because control characters are nonprinting, various authors have adopted many different styles to tell the reader to enter a control character. For example, variations on how to hit the BEL character include "CONTROL-G," "CTRL-G," "<control G>," "[G]," or "∧G." Some control characters can be generated by single keystrokes. In particular, the CR character is generated by a key marked RETURN or ENTER. Some keyboards can generate 8-bit *extended ASCII* characters. These characters are generated by a process similar to that used for control keys, except that another special key (sometimes marked ALT) is held down instead of the CONTROL key. Using CONTROL and the ALT keys, a keyboard with a reasonable number of keys can generate the complete, 256-character extended ASCII character set.

Software, in conjunction with keys that generate control characters, can create special keyboard effects. When a function key is typed it can be recognized by a special program and expanded to have the effect of several keystrokes. *Escape sequences*, initiated by the ESC character, open up further possibilities. Typing the ESC key alerts software that the next character typed has a special meaning. Unlike the CONTROL key, the ESC key is not held down while the second key is typed. There has been an attempt to standardize the meanings of escape sequences, but there is still a wide variation between manufacturers.

There are many different types of key switches in use. Many provide the travel, acoustic feedback and "feel" of a typewriter keyboard. Some *membrane* keyboards are deficient in this respect but have other special features such as compactness and protection against moisture and dust. With few exceptions, individual keys simply make and break a single electrical contact. Unfortunately, the make is usually not just one clean pulse but a series of disorderly pulses resulting from mechanical contact bounce. Key bounce was a serious problem with some early keyboards. To *debounce* a keyboard, circuits that verify intended key closures and ignore key bounces and noise are added.

Computer keyboards vary from one model of computer to the next. Key placement, the meaning of escape sequences and control characters, and even the availability of specific characters will change if you change computers. The method of entering control characters and escape sequences, however, remains the same.

Screen Displays

A video display terminal (VDT) consists of a keyboard, as discussed above, and a *monitor*. Most monitors are *cathode-ray tubes*, either video monitors with no RF circuits or standard TV sets. Portable computers often use flat-screen, *liquid-crystal displays* (LCDs).

Whether CRT or LCD, the monitor is usually treated as a dot-matrix display with some number of picture elements *(pixels)* or dots horizontally and some number of pixels vertically. Characters are formed on the display by turning on specific pixels in a *character matrix*. Most CRTs employ *raster scanning*, a method of covering the screen by writing one row of pixels at a time. The electron beam in the CRT writes the top row of pixels first and then continues line- by-line to the bottom of the screen. Then, a *vertical retrace* brings the beam back to the top of the screen to begin again. To display characters on a raster-scanned screen, the electron beam must be turned on and off at precisely the times needed to illuminate the pixels that form the characters.

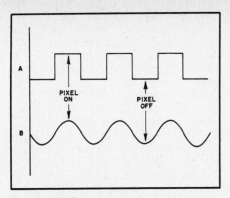

Fig. 59 — Control signals for turning on every-other pixel on a CRT. A shows the ideal square-wave control signal. The signal at B is a sine-wave approximation of the signal at A.

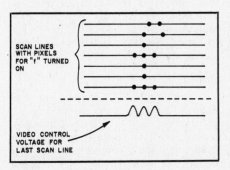

Fig. 60 — Portion of a raster-scanned video display.

Like any other signal, the signal that turns the electron beam on and off occupies bandwidth within the monitor circuits. To calculate the bandwidth needed for a specific display, we must consider the worst-case characteristics of the electron-beam control signal. This worst (highest bandwidth) case is when all of the pixels on the screen are set such that adjacent pixels have different states. The ideal control signal for such a display is shown in Fig. 59A. The frequency (F) of this pulse train can be calculated by

$$F = \frac{(N/2)}{T} Hz \qquad (Eq.\ 4)$$

where
N = the number of pixels on the screen
T = the time taken to complete a whole-screen scan

If the control signal were approximated by a sine wave (Fig. 59B), the bandwidth needed to pass the signal would be equal to the frequency of the sine wave. Unfortunately, the sine wave does not turn the beam on and off abruptly and would produce blurred pixels. To get good definition, components up to the 3rd harmonic of the fundamental square-wave frequency (F) must be accommodated. Thus, a minimum bandwidth of $3 \times (N/2)/T$ Hz is necessary to scan N pixels in T seconds. Example 1 shows how to calculate the bandwidth needed to display 25 lines, with 80 characters/line and a 10 × 7-pixel matrix

used for each character.

Example 1:
(10 × 7 matrix) = 70 pixels/character
Characters/line = 80
Lines/screen = 25
Pixels/screen (N) = 25 × 80 × 70
    = 140,000
Refresh time (T) = 1/60 seconds
Worst case frequency (F) = 140,000/2 × 60 = 4,200,000 Hz
Bandwidth required = 3 × 4.2 MHz = 12.6 MHz

Raster scanning is used in television receivers, which may be used for displaying computer text. Television receivers have limited IF bandwidth — on the order of 3.5 MHz for color, 3 MHz or less for black and white. The example above shows that such bandwidth results in a poor display if 25 lines of 80 characters are desired. Computer monitors used to display 80 × 25 screens have a video bandwidth up to 25 MHz. What can be displayed on a TV receiver? Example 2 calculates the bandwidth needed for 16 lines with 40 9 × 7 characters on each line.

Example 2:
Pixels/character = 9 × 7 = 63
    Characters/line = 40
Lines/screen = 16
Pixels/screen (N) = 16 × 40 × 63
    = 40320
Scan time (T) = 1/60 seconds
Worst case frequency (F) = 40320/2 × 60 = 1,209,600 Hz
Necessary bandwidth = 3 × 1.2 MHz = 3.6 MHz

Thus, this example shows about as much text as can be displayed on a normal TV screen. This screen format (40 × 16) is used by some popular microcomputers to permit use of a TV receiver as an inexpensive display. The individual characters are formed as shown in Fig. 60.

An alternative to raster scanning of a CRT is the *vector mode*. In this mode, the user can draw a straight line by specifying only two points. Vector-mode displays are used primarily for graphics.

Character Displays

Rather than having a full screen display, individual characters can be presented on single-character displays. Some of the formats available are shown in Fig. 61. A popular display is the seven-segment version, as shown in Fig. 62, which is an excellent choice for displaying numbers.

Character displays are built using several different manufacturing technologies. LEDs are common because they are easy to read when there is low ambient light. LCDs are now coming into use and are especially useful in low-current-drain battery operation. Unlike the LEDs, LCDs are visible in bright ambient light. However, since LCDs do not emit light, but reflect it, they are not visible in low-light environments. LCDs are also sluggish at low

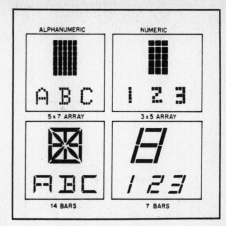

Fig. 61 — Various character display formats.

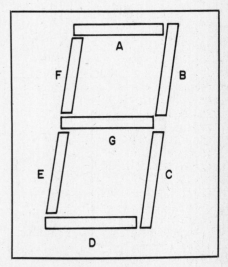

Fig. 62 — Segment identification and layout for a seven-segment display.

temperatures. LED and LCD characters are available in several sizes: 0.3, 0.6 inches and larger.

A numeral to be displayed on seven-segment display is usually encoded in BCD form, and a combinational logic circuit activates the proper segments of the display. This logic circuit is called a *decoder*. Various seven-segment decoders are manufactured to drive common-cathode and common-anode displays. One decoder chip available to hobbyists is the 7447A. This is an open-collector TTL device designed to pull down common-anode displays through external current-limiting resistors. A 7447A will also drive common-cathode displays if external transistors are used. Fig. 63 shows the connections for both types of display.

The dc illumination method shown in Fig. 63 is the easiest to implement, but higher light output with lower energy consumption can be obtained by pulsing the display voltage. A pulse rate of 100 Hz is imperceptible because of the persistence of human vision.

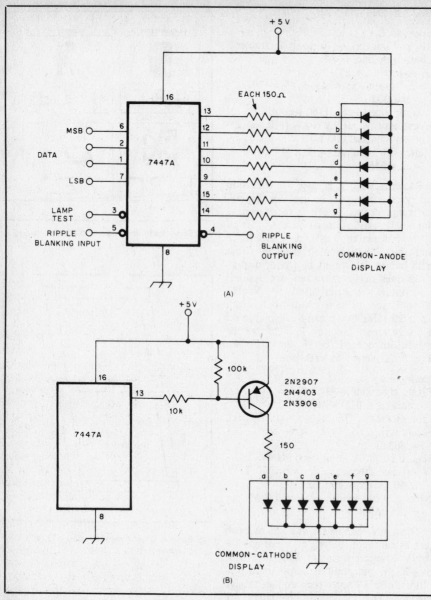

Fig. 63 — A shows a 7447 decoder/driver connected to a common-anode LED display. At B is a method for using the same chip with a common-cathode device.

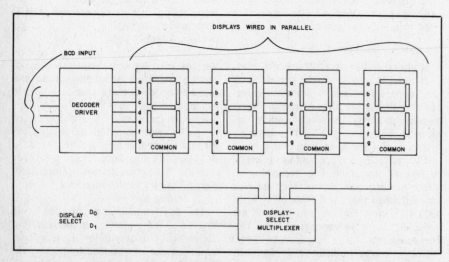

Fig. 64 — Multiplexed character displays. BCD data for a particular display is placed at the input of the decoder/driver, and the desired display is selected by placing its "address" on the inputs to the display-select multiplexer.

As more digits are added to a display, using a decoder/driver for each digit becomes unfeasible from an economic and PC-board real-estate point of view. A technique called *multiplexing* (Fig. 64) allows a single decoder/driver IC to be shared by several digits. The digits are wired in parallel; that is, all of the "a" segments are connected together and so on. The multiplexer logic sends input data for a digit into the decoder and enables the common element of the correct digit. This process is repeated for each digit. Only when both a segment and its associated common lead are selected will that segment be energized. With this system, only one digit is energized at any instant, a factor that greatly reduces power-supply requirements. To maintain the brightness of each digit, the current to each display segment must be increased. When implementing a multiplexed display, you should be sure not to exceed the peak- and average-current specifications for the display.

Printers

It is often necessary to have a printed copy or "hard copy" of communications, programs or data that are stored in a computer. Printers are used to produce such hard copy. Some printers produce *letter quality* text, with characters that look like they were typed on a typewriter. Other printers produce *data-processing* quality type, used mostly for program listings. Printers are categorized by their printing speed and the quality of type they produce.

In the mid 1970s, a number of amateurs used ITA2 (Baudot) teleprinters as a computer output device. Since these teleprinters had a 5-bit code, they did not print all the characters of ASCII. Certain characters were either ignored or converted to substitutes. These machines operated at speeds in the 45- to 75-baud range (or 6 to 10 char/s) — quite slow by computer standards. ASCII teleprinters, such as the Teletype Corporation Model 33, were only slightly better. These teleprinters operated at 110 bauds or 10 char/s.

Most printers currently used with microcomputers have the full ASCII coded character set. They are called *character printers* because they print only one character at a time, not a whole line at once as *line printers* do. Printers used with personal computers are described below.

Fully formed character printers are usually *impact printers*; they print by striking a character-shaped hammer against a ribbon. These printers may use interchangeable type elements (called daisy wheels or thimbles). Many character fonts are available. The characters are fully formed and match those of a typewriter. While characters are printed slowly (less than 80 characters/second), they can be sent to the printer at a much higher rate. The characters are stored in a FIFO buffer, and hardware or software *flow control* (*handshaking*) between the printer input and the computer output port keeps the computer from overfilling the buffer.

Dot-matrix printers print each character as a matrix of dots, much the same way characters are displayed on a raster-scanned monitor. A 5-by-7 matrix (measured 5 dots wide by 7 dots high) is just adequate for the standard ASCII character set. However, such a matrix makes the normal Latin letters look ragged. At least a 7-by-9 matrix is needed to make Latin letters look better and is considered minimum for some foreign alphabets. While dot-matrix printers do not form perfect letters, they are fast (greater than 100 chars/sec), they can print graphics, and they can be used as plotters. Some can print multicolor graphics. Also, dot-matrix printers are generally less expensive than impact printers.

While most printers used on home computers are either dot-matrix or impact printers, several other types of printers are beginning to gain popularity. The *ink-jet* printer has been too expensive for most amateurs, but lower-cost versions are now available. The ink-jet printer has some of the advantages of both letter-quality and dot-matrix printers. *Electrostatic* printers use the process commonly associated with photocopy machines, where a dye is deposited onto charged paper. *Laser* printers are just becoming available for home computers. They are very fast (only several seconds per page), and can produce both graphics and letter quality text.

SUMMARY

Peripherals allow computers to interact with the outside world. General-purpose computers need both a monitor, to present results to the operator, and a keyboard, to accept commands and data from the operator. Mass storage devices, usually cassette tapes or floppy disks, are used to store programs and data. Printers are used to make hard copies of computer outputs. Peripherals often cost more than the microcomputer to which they are attached; when selecting a microcomputer, pay attention to the availability and price of necessary peripherals.

COMPUTER LANGUAGES

There are two broad classifications of computer languages: low level and high level. Generally, a low-level language is one that is designed to communicate with a computer in its native tongue, while a high-level language is intended to be more like a widely used human language.

LOW-LEVEL LANGUAGES

Programmers who wish to converse with a microprocessor directly must do so in *machine language*. The symbols used in machine language are 1s and 0s. Every microprocessor has a built-in instruction set, and each instruction in that set can be represented by a binary word. Few pro-

grammers actually write machine language. When they do, they express machine instructions as octal or hexadecimal numbers rather than as binary numbers. The instruction to move the contents of register B into register A, understood by the microprocessor as "01111000," would be written as "78" in hexadecimal (hex).

Most programmers cannot remember what all the numerical machine-language instructions stand for, so *assembly language* is used to substitute a mnemonic abbreviation for each instruction. In the above example, "78" would become "MOV A,B" in assembly language. Once you get a program written in assembly language, it must be *assembled* to convert it to machine language. This is done with an assembler program that converts commands from mnemonics to binary words; "MOV A,B" would be converted to the byte "01111000." A single assembly-language instruction generates a single machine-language instruction; programming in assembly language creates programs that are as fast and efficient as if they were written in machine language.

Assembly language programs and programs in high-level languages are usually stored as text. Such text is called *source code*, while binary files of machine-language instructions are called *object code*. A single source code instruction, such as "MOV A,B," is called a *statement*.

Both machine and assembly languages are usually different for each microprocessor. However, once you learn assembly-language programming on a particular chip, you can easily work with others in the same family. Even when switching families, you will get the hang of the new mnemonics in no time. Sometimes old habits are hard to break, and you will yearn for your favorite special feature that is not in the new instruction set. In general, you end up doing the same operations but have to learn new mnemonics. A programmer's reference guide or card can help you remember all the mnemonics until they become second nature through repetition.

Many experienced programmers prefer to work in assembly language because it gives them step-by-step control of the processing of data. Programming in assembly language usually results in smaller object code and faster execution than can be achieved with higher-level languages. This is especially useful in high-speed code-conversion and other signal-processing applications.

HIGH-LEVEL LANGUAGES

High-level languages are oriented toward the user or toward a specific task rather than toward the microprocessor. In order to appear familiar, high-level languages are typically designed to look like a subset of English. Plain-English programming is not yet possible, because English has ambiguities and numerous shades of meaning and

subtleties. We must be content with high-level computer languages in which words have specific meanings (strict *semantics*) and must be used according to rigid rules (*syntax*).

Many high-level languages are extensions of the assembly-language concept — that of substituting something easy to remember for something that is easy to forget. Instead of simply having a few mnemonic letters represent a few numbers, though, high-level languages set up more complex relationships. A single English word (instruction or statement) in a high-level language can cause the generation and execution of several machine-language instructions. Thus, high-level languages provide a way of expressing complex computing processes in notation that is easy to understand and remember.

Like assembly-language source statements, high-level-language statements must be converted to machine language. Conversion is performed by an *interpreter* or a *compiler*. An interpreter translates a statement to machine code and then executes the code immediately. It does not create an object program. A compiler takes an entire high-level source program and translates it into a machine-language object program. The object program may be executed immediately or saved for later execution. If a particular statement is executed several times, an interpreter will translate that statement over and over, while a compiler translates it only once. Thus, compiled programs usually run faster than interpreted programs. However, during program development it is time consuming to recompile a program after every small change is made. In this case, the immediate execution offered by an interpreter can be convenient. Interpreters and compilers each have their advantages and disadvantages. Most home computers are provided with a built-in interpreter for the computer language BASIC, and BASIC compilers can be purchased for some of these computers.

There are hundreds of high-level computer languages. Some are used for specific tasks, some for a wide range of tasks. The languages described below are common or of some specific interest to radio amateurs.

BASIC

The Beginner's All-purpose Symbolic Instruction Code (BASIC) was developed at Dartmouth College as a language similar to, but easier to learn than, FORTRAN II. BASIC is by far the most common high-level language used by radio amateurs and personal-computer users. Numerous dialects of BASIC exploit the special features of different computers. Many of these dialects implement a common set of BASIC instructions and then add *extensions* that provide new features. Fortunately, many BASIC dialects for different computers were written by the same software

Glossary of Digital Terms

Access — To read or write to memory or an I/O device.

Accumulator (ACC) — The register in a computer that holds the result of ALU operations.

Address — An expression, either numeric or symbolic, that specifies a memory location.

Algorithm — A set of directions describing a way to perform a certain task.

Architecture — Organizational structure of a system.

Arithmetic Logic Unit (ALU) — The part of a computer that performs mathematical operations.

Assemble — To convert a symbolic assembly-language program into machine language.

Assembly Language — A symbolic language representing machine language in mnemonic form.

Asynchronous — Communication in which data characters are individually synchronized and may be sent at any time.

Baud — A bit per second.

Benchmark — A program used to compare the performance of two or more computer systems.

Binary — The base-two number system, using the numerals 0 and 1.

Bipolar — A broad category of semiconductors based on PNP or NPN junction transistors. See MOS.

Bit — The basic element of binary information, usually represented by 0 and 1; a contraction of "binary digit."

Bug — An error that causes hardware or software to malfunction.

Byte — A group of bits, usually 8.

Central Processing Unit (CPU) — The part of a computer that interprets and executes instructions.

Chip — An integrated circuit.

Clear — To make a value or state equal to 0.

CMOS — Complementary metal-oxide semiconductor, an integrated-circuit logic subfamily employing both n-channel and p-channel FETs; noted for its low power consumption.

Compiler — Software that converts a high-level-language program into a machine- or assembly-language-program.

Cross Assembler — An assembler that translates code from one machine-language to another.

Debug — To remove errors from hardware or software.

Decimal — The base-ten number system, using the numerals 0 through 9.

Display — Presentation of information in a form to be sensed by humans, usually visual.

Documentation — Reference material detailing specifications and technical information on systems, hardware or software.

ECL — Emitter-coupled logic, an integrated circuit logic family capable of operating at frequencies above 1.2 GHz.

Edge-triggered — A circuit that responds to data or clock transitions, rather than responding to data or clock levels.

Execute — To interpret instructions and perform the indicated operations.

Fan-out — The number of parallel loads of a given family that can be driven by one output of that family.

Flip-Flop — A circuit with two stable states, usually 0 and 1, capable of storing a single bit of information.

Floppy disk — A flexible magnetic recording medium for data storage. Also called a diskette.

Flowchart — A graphic representation of the structure of a program or process.

Handshaking — Exchange of prearranged signals between devices to determine their readiness to establish or maintain communications.

Hard disk — A rigid magnetic recording medium for data storage.

Hardware — Physical equipment.

Hexadecimal — The base-16 number system, using the numerals 0 through 9, and A through F.

Instruction — A statement that specifies a computer operation.

Instruction set — The repertoire of machine-language instructions for a microprocessor.

Interface — The common boundary between two devices or software modules.

Integrated circuit (IC) — A miniature electronic module of components and conductors manufactured as a single unit.

Interpreter — A program that translates and then immediately executes statements in a high-level language.

I/O — Input/output.

Keyboard — A set of key-switches used to enter data or commands.

Label — One or more characters used to identify a memory location or a statement or an item of data in a computer program.

Latch — A flip-flop capable of storing one bit of information. Also used as a verb meaning to hold state.

LSI — Large-scale integration, ICs with more than 1000 devices per chip.

Machine — A computer.

---

house, which has had the effect of minimizing unnecessary variations.

Although there is a common core of BASIC that is *transportable* from one computer to another, do not expect to type in a BASIC program listing written for one computer and run it on another without modification. Books like *The Basic Handbook* by David A. Lien discuss translating BASIC from one dialect to another (see the Selected Bibliography at the end of this section).

Here are a few common *commands* and *statements* in BASIC:

END: Stops execution of the program.

LET: Assigns values to variables. LET is often optional; "LET X = 1" and "X = 1" usually mean the same thing.

LIST: Displays the program.

PRINT: Means to display whatever follows, such as a variable (X in the line 40 below) or anything enclosed in quotes.

REM: Remarks or comments. A computer will ignore lines beginning with REM.

Example:
20 PRINT "Join the computer revolution."
30 LET X = 1
40 PRINT X
50 END

Pascal

Pascal is a structured programming language; structured programs are written as systems of interacting subprograms. Large programming tasks can be broken into many small tasks, and each of these small tasks can be translated into a Pascal subprogram. (Although BASIC can be used to write such structured programs, it was not designed for easy, efficient structured programming.) Pascal is taught in many engineering courses, since it provides a good basis for understanding other programming languages. Pascal software is available to amateurs for most microcomputers at modest cost.

C

C is a structured language that, like assembly language, generates compact, fast object programs. Unlike Pascal and many other high-level languages, C allows programs to interact freely with computer hardware. Because of these characteristics, C is becoming popular for Amateur Radio applications.

IPS

The Interpreter for Process Structure (IPS) is a derivative of the high-level language, FORTH. IPS is a "threaded-code" language written by Karl Meinzer, DJ4ZC, to control the integrated housekeeping unit (IHU) 1802-based computer in the AMSAT-OSCAR Phase III satellites.

OPERATING SYSTEMS

An *operating system* (OS) is a group of computer programs that supervises the sequencing and processing of other computer programs. The programs supervised by an OS are usually called *applications programs*. Operating systems are also responsible for I/O management — the creation and assignment of files, the maintainance of file directories, and so on. The OS allows applications programs to use I/O devices without knowing the precise characteristics of those devices. In computers that have disk drives, the OS is usually called the disk operating system (DOS). Concurrent or *multi-tasking* operating systems allow several applications programs to share a single CPU. A good OS will provide simple but powerful I/O services to applications programs and an understandable, complete set of commands for human users.

CP/M® is an operating system popular with microcomputer users and manufacturers. CP/M stands for control program/microcomputer, and is a registered trademark of Digital Research Inc. (DRI). CP/M-80 runs on the 8080 family of microprocessors, and CP/M-86 runs on the 8086 family. As supplied by DRI, CP/M must be customized to fit specific microcomputer hardware. CP/M is popular because it provides a well-defined

Machine Language — The numeric form of specifying instructions used by a computer.

Microcomputer (uC) — A computer having a microprocessor CPU.

Microprocessor (uP) — One or a small set of chips capable of performing the functions of a CPU.

Mnemonic — A symbol chosen to assist the human memory. The symbolic name used to represent instructions, registers and memory locations.

Monitor — A display device, usually a cathode-ray tube. A permanent program resident in a computer to perform basic operations.

MOS — Metal-oxide semiconductor. Semiconductors based on field-effect transistors (FETs), rather than bipolar-junction transistors. See bipolar.

MSI — Medium-scale integration, ICs with 10 to 100 devices per chip.

Nyquist rate — In analog-to-digital conversion, the lowest sampling rate that will result in correct reproduction of the analog signal. This rate is twice that of the highest frequency component in the original analog signal.

Object program — A machine-language program.

Octal — The base-eight number system, using numerals 0 through 7.

Open collector — A TTL output that is not internally pulled up through a resistor or transistor.

Operating system — A group of programs that supervise I/O and the execution of other programs in a computer.

Parallel — Simultaneous handling of several bits, with a separate channel for each bit. See serial.

Port — A set of I/O connections.

Printer — An output device that produces a paper copy.

Program — A collection of instructions that perform a certain task.

Program counter (PC) — A register that specifies the address of the next instruction to be executed.

Pull-down resistor — A resistor connected from the input or output of a logic element to ground.

Pull-up resistor — A resistor used to connect the input or output of a logic element to a supply voltage.

Rails — The positive and negative supply voltages.

Random-Access Memory (RAM) — Volatile storage. Storage in which any address can be accessed as easily and quickly as any other.

Read — To copy the contents of a memory location or peripheral storage medium.

Read-Only Memory (ROM) — Storage that can be easily read but not easily written to; usually nonvolatile.

RTL — Resistor-transistor logic, an early bipolar integrated-circuit logic family.

Schottky — A TTL manufacturing process that provides high-speed by eliminating transistor saturation.

Sequential-access memory — Memory in which the first n addresses must be accessed in order to access address $n + 1$.

Serial — A process in which a group of bits are handled one at a time over a single channel.

Software — Computer programs.

SSI — Small-scale integration, ICs with 10 or fewer devices on a chip.

Start bit — The bit transmitted to indicate the beginning of an asynchronous character. See stop bit.

State — A mode or condition of a circuit or program.

State diagram — A chart showing all states in a device or process and the events leading to and resulting from each state.

Stop bit — The bit transmitted at the end of an asynchronous character and between such characters. See start bit.

Subroutine — A section of a program used to divide the work of a large program into small segments.

Synchronous — A method of communication in which data is sent as a continuous, exactly-timed stream, with special synchronization information sent only at the beginning of long data blocks.

Totem pole — A TTL output circuit in which a transistor is used to pull up the collector of another transistor. See open collector.

TTL — Transistor-transistor logic, a widely used bipolar IC logic family.

UART — Universal asynchronous receiver/transmitter, an IC that converts parallel data into asynchronous serial data, or vice versa.

USART — Universal synchronous/asynchronous receiver/transmitter, an integrated circuit that converts parallel data into either synchronous or asynchronous serial data, or vice versa.

VLSI — Very-large-scale integration, ICs with 1000 or more devices on a chip.

Volatile memory — Storage that does not retain its contents when its power is removed.

Write — To place data in memory or peripheral storage.

---

set of I/O subprograms. Applications programs can use these subprograms without knowing what hardware-specific I/O instructions are actually executed.

Customizing CP/M involves writing a *basic input/output system* (BIOS) that permits the generic I/O subprograms to access existing hardware devices. Once you have a customized BIOS for your computer, you will be able to run all CP/M programs — it is not necessary to customize every application that you wish to run.

Many CP/M programs are in the public domain, meaning that they can be freely copied and used without violating copyrights. A catalog of public-domain software for CP/M is available from NYACC, the New York Amateur Computer Club, Inc., P.O. Box 106, New York, NY 10008.

MS DOS

MS DOS, written by Microsoft, is quickly becoming as popular as CP/M. A version of this DOS written specifically for the IBM PC is called PC DOS. Because of the prestige of IBM, PC DOS is possibly the fastest-growing microcomputer DOS on the market. Not long after the introduction of the IBM PC, many applications programs written for CP/M and other operating systems became available for PC DOS. A catalog of public-domain software for the IBM PC is also available from NYACC.

Other Operating Systems

There are nearly as many microcomputer operating systems as there are microcomputer manufacturers. Some manufacturers supply a proprietary OS with specific computer models. These operating systems are intended to make best use of hardware features or limitations. When you choose a particular personal computer, you will usually not have many operating systems to choose from. You may, however, decide to select a certain computer *because* it uses an OS that you are familiar with.

NOISE PRECAUTIONS

Noise is an important consideration in designing digital circuits. Noise on a signal or power line may cause logic circuits to perform erratically. Wires and printed-circuit traces may act as antennas that radiate RF noise. These same "antennas" can receive energy, especially when immersed in high RF fields near a radio transmitter.

Pulse rise and fall times determine the frequency of some of the electromagnetic interference radiated or conducted from a digital circuit. You can find the approximate upper frequency of the interference by dividing the constant 0.34 by the time the pulse takes to rise from 10% to 90% of its full value. For example, a rise time of 15×10^{-9} seconds between 10% and 90% levels would yield RF outputs up to 22.6 MHz. High-speed logic devices may be appealing for certain digital applications but can cause needless RFI because of fast rise and fall times. This can be avoided by using logic that is no faster than necessary.

Digital circuits also produce energy at the repetition rates of the pulses. For example, in digital equipment using a 3.579545-MHz clock, there will be RF energy at that frequency and its harmonics. Whether that energy emanates from the digital equipment and causes interference to radio reception depends largely on the design and construction of the digital equipment.

Designers of *computing devices* for sale in the U.S. are subject to compliance with Part 15 of the FCC rules. A computing device, in the FCC definition, is any electronic system that generates pulses in excess of 10 kHz and uses digital techniques. Class A devices are for commercial or industrial use. Amateurs should be most concerned with Class B devices, which are for use in a residential environment. Class B devices must not radiate more than 100 μV/m (30-88 MHz), 150 μV/m (88-216 MHz) or 200 μV/m (216-1000 MHz) at a distance of 3 meters, nor conduct to the power line more than 250 μV (0.45-30 MHz). These restrictions are sufficient to minimize RFI to neighbors but inadequate to prevent interference to nearby Amateur Radio receivers. Clearly, digital equipment for the shack must be designed to meet more stringent requirements than those of Part 15 of the FCC rules!

Shielding of digital equipment is often neglected. Nonmetallic enclosures may save money initially but can allow unnecessary interference to nearby radio and television receivers. Whenever possible, use a metal enclosure that has solid electrical contact among all of its parts. Paint should be removed where parts meet. To ensure that the entire surface makes contact, you may also need to add some sheet-metal screws or sandwich some copper wire braid between the mating surfaces. Copper or aluminum screen may be used to cover holes without blocking ventilation. If a plastic or wooden enclosure cannot be avoided, it may be possible to add internal shielding using heavy-duty aluminum foil. Otherwise, consider an electrically conductive coating such as nickel-filled acrylic. Coating materials are available in spray cans or may be applied commercially.

Line filters should be used at the ac input of power supplies for digital circuits to minimize interference conducted on the power lines. Three-contact ac-power receptacles of the standard CEE-22 type with built-in line filters are recommended.

All signal and dc-power lines leaving enclosures should be bypassed or filtered. Connectors can be purchased with built-in bypass capacitors on every contact, but they are expensive. The use of individual bypass capacitors (with the shortest possible leads) on all lines capable of causing interference is an economical approach. Large ferrite toroids can be used to decouple cables, and individual wires can be decoupled with ferrite beads.

The wires in cables that carry digital signals between enclosures may act as antennas. One solution to this problem is to avoid unnecessary external cabling by mounting different digital components in one shielded enclosure. Another way of avoiding the problem is to use optical fibers to transmit digital signals between enclosures. Commercial fiber-optic cabling systems are expensive, but they are expected to come down in price. There is also an opportunity for amateur innovation in inexpensive fiber-optic cabling.

Where external wire cabling is necessary, shielding is recommended. Each conductor may be separately shielded to minimize coupling between signal lines. In long shielded cables with a radiation problem, consider adding toroids along the cable to break up the resonances.

A low-impedance RF ground is as important for digital equipment as it is for radio equipment. Ensure that all digital equipment enclosures, frames, mother boards and cable shields are properly grounded. Use any space left on a PC board for an adequate ground plane.

Take design precautions to avoid power-supply noise from current spikes. Power-supply wiring should be as thick as practical, and wide PC-board power traces should be used to keep the line impedance low. This is particularly true for +5-V lines, which may be the only supply voltage or the one carrying the largest current in a circuit using more than one supply voltage. Ground traces should also be as large possible, since they often carry as much current as power-supply traces. Different voltage traces should be arranged for minimum mutual coupling by keeping them as far apart as possible. Where they cross, try to arrange them at right angles. An on-board voltage regulator is recommended to help isolate noise to and from the rest of the system. Good engineering practice suggests being generous with decoupling capacitors. Typical decoupling practice for +5-V lines includes: a 20-μF electrolytic capacitor at the input to the board, a 6.8-μF electrolytic capacitor in parallel with a 0.1- to 1.5-μF tantalum capacitor near every 1 to 8 devices and a 0.01- to 0.5-μF ceramic capacitor for every 1 or 2 adjacent devices.

Digital Basics Selected Bibliography

Deem, Muchow, and Zeppa, *Digital Computer Circuits and Concepts,* 2nd ed., Reston Publishing Co., Inc., Reston, VA, 1977.

Fink, *Electronics Engineers' Handbook,* 2nd ed., McGraw-Hill, Inc., 1982.

ITT, *Reference Data for Radio Engineers,* 6th ed., Howard W. Sams & Co., Inc., Indianapolis, IN, 1975.

Lien, *The BASIC Handbook,* Compusoft™ Publishing, San Diego, CA, 1978.

Poe and Goodwin, *The S-100 & Other Micro Buses,* Howard W. Sams & Co., Inc., 1981.

Sippl, *Data Communications Dictionary,* Van Nostrand Reinhold Co., 1976.

Soucek, *Microprocessors & Microcomputers,* John Wiley & Sons, Inc., 1976.

Tocci, *Digital Systems,* Prentice-Hall, Inc., Englewood Cliffs, NJ, 1980.

Williams, *Designer's Handbook of Integrated Circuits,* McGraw-Hill, Inc., 1984.

Additional reading material on digital basics and microcomputers is available from a variety of sources. *EDN, Electronics Week, Computer Design, Electronic Engineering Times* and other trade journals regularly print late news of semiconductor development. Semiconductor manufacturers, or their field representatives, are the best sources for data sheets and application notes on specific devices. *Byte* and a number of other magazines cover the general field of personal computers; there are other magazines oriented toward a specific line or type of computer. Amateur computer clubs are excellent sources of hardware and software information. *QST* and *QEX* regularly publish articles on digital subjects.

Chapter 9

Modulation and Demodulation

Modulation is the process of varying some characteristic of a *carrier* wave in accordance with a signal to be conveyed. A carrier is a wave having at least one characteristic that can be varied from a known reference by modulation. That known reference is typically a steady sine wave but could be a recurring series of pulses in which no signal is present between pulses. A *modulator* is a circuit or device in which the carrier and modulating signal come together to produce a modulated carrier, or one that processes the modulating signal and presents it to the circuit or device to be modulated. That is, modulation may or may not occur in the modulator, depending how the term is applied in practice. Modulation is normally associated with transmitters. The modulator may be built into the transmitter or exist as a separate unit.

Demodulation is the process of deriving the original modulation signal from a modulated carrier wave. But that's not the whole story. The received signal usually contains noise and distortion picked up along the way. For the original modulating signal to be reproduced faithfully, the effects of noise and distortion need to be dealt with effectively during reception and demodulation. Demodulation is part of the receiving process and is performed in a *demodulator*. While a demodulator is a normal part of a radio receiver, external demodulators are used for some types of communications.

The *baseband* is the range of frequencies occupied by the signal before it modulates the carrier wave. The signal in the baseband is usually at frequencies that are substantially lower than that of the carrier. On the low end, the baseband may approach or include dc (0 Hz). The high end depends on the rate at which information occurs as well as the presence of subcarriers or other special signals within the baseband. There is a baseband for all types of signals, whether analog or digital. Also, it should be realized that the term baseband is relative to the specific modulation being considered. There can be more than one baseband in a complete modulation process. For example, to a transmitter, a keyed tone going into the microphone input is its (analog) baseband; to the tone keyer, the keying pulses are its (digital) baseband.

This chapter presents the fundamentals of the modulation and demodulation processes. Subsequent chapters provide details concerning the generation of modulated signals and demodulation circuitry for types of modulation used in the Amateur Radio Service.

DESIGNATION OF EMISSIONS

One result of the World Administrative Radio Conference held in 1979 (WARC '79) was the adoption of a new system of designating emissions according to their necessary bandwidth and their classification. These new designations are used throughout this chapter with secondary reference to the pre-WARC '79 designations.

The new designations are not without problems. The new designations take into account the method of generating a signal rather than simply what the signal looks like when transmitted. Thus, the new system would have you use one designation for direct modulation of the main carrier and a different one if you modulate a subcarrier of a single-sideband suppressed-carrier transmitter. Also, if you key a radioteletype modem with a direct-printing teletypewriter then switch to using a computer to do the same thing, the modulation designation changes under the new system.

On November 27, 1984, the FCC released the Third Report and Order for General Docket No. 80-739 to implement the new emission designators in all radio services, effective January 3, 1985. The appendix amended the Amateur Radio Service rules by deleting the old designations and replacing them with new designations. The amended rules omitted numerous types of emission that were permitted, or implicit, in the rules prior to the change. As this is being written, the FCC is considering a further amendment to either add the missing designators or generalize the wording to allow various types of emission.

Bandwidth

Whenever a carrier is modulated, sidebands are produced. Sidebands are the frequency bands on both sides of a carrier resulting from the baseband signal varying some characteristic of the carrier. The modulation process creates two sidebands: the *upper* sideband (USB) and the *lower* sideband (LSB). The width of each sideband is generally equal to the highest frequency component in the baseband signal. In some modulation systems, the width of the sidebands may greatly exceed the highest baseband frequency component. The USB and LSB are mirror images of each other and carry identical information. Some modulation systems transmit only one sideband and partially or completely suppress the other in order to conserve bandwidth.

Occupied bandwidth is that between lower and upper limits of the signal where the mean power is 0.5% (-23 dB) of the total mean power. In some cases a different relative power level may be specified; for example, -26 dB (0.25%) is used in section 97.69 of the FCC rules governing digital communications. Occupied bandwidth is not always easy for the amateur to determine. It can be measured on a spectrum analyzer — a piece of test equipment that is not generally available to most amateurs. Occupied bandwidth can also be calculated, but these calculations require an understanding of the mathematics of information theory and are not covered in this chapter.

Necessary bandwidth is that part of the occupied bandwidth sufficient to ensure transmission of the information at the rate and with the quality required. Formulas used by the International Radio Consultative Committee (CCIR) and the FCC for calculation of necessary bandwidths are included in this chapter's discussion of specific modulation systems.

Assigned frequency band is the necessary bandwidth plus twice the absolute frequency tolerance. *Frequency tolerance* (expressed in parts per 10^6, percentage, or in hertz) is the maximum permissible departure by a frequency from its correct frequency.

Prior to WARC '79, the CCIR and FCC expressed necessary bandwidth in terms of kilohertz. As examples, 0.1 was used to mean 100 Hz, and 6000 for 6 MHz. Spectrum managers had no taste for ragged computer-generated columns with different numbers of places, so a more elegant technique was clearly needed. The WARC '79 delegates agreed on a new system of expressing necessary bandwidth by three numerals and one letter. The letter occupies the position of the decimal point and represents the unit of bandwidth. The letters H (Hz), K (kHz), M (MHz) and G (GHz) are used, but neither 0 nor K, M nor G may appear in the first character. Numerical values of more than three significant places are rounded off to three places. Some examples taken from the CCIR Radio Regulations are listed in the accompanying table.

MAJOR MODULATION SYSTEMS

The broadest category of modulation

Examples of CCIR Necessary Bandwidths

0.002 Hz = H002	6 kHz = 6K00	1.25 MHz = 1M25
0.1 Hz = H100	12.5 kHz = 12K5	2 MHz = 2M00
25.3 Hz = 25H3	180.4 kHz = 180K	10 MHz = 10M0
400 Hz = 400H	180.5 kHz = 181K	202 MHz = 202M
2.4 kHz = 2K40	180.7 kHz = 181K	5.65 GHz = 5G65

systems is how the main carrier is modulated. The major types are amplitude modulation, angle modulation and pulse modulation.

Amplitude Modulation

Amplitude modulation (AM) covers a class of modulation systems in which the amplitude is the characteristic that is varied. It is possible to amplitude modulate a main carrier or subcarrier. However, in the context of the new emission symbols, it is the modulation of the *main* carrier that determines the first symbol (type of

modulation of the main carrier).

Amplitude modulation is often simplistically described as varying the amplitude of the carrier from a zero power level to a peak power level. This chapter includes some photographs of oscilloscope patterns that seem to prove this notion. In fact the oscilloscope pattern is a composite of the energy produced over a range of frequencies. The carrier itself stays at the same amplitude when modulated by an analog (such as voice) baseband signal. The modulation itself produces sidebands. They are bands of frequencies on both sides of

---

Emission Classifications

Emissions are classified by the following basic characteristics:

(1) First symbol — type of modulation of the main carrier

(1.1) Emission of an unmodulated carrier ... N

(1.2) Emission in which the main carrier is amplitude modulated (including cases where subcarriers are angle modulated)

  (1.2.1) Double sideband ... A

  (1.2.2) Single sideband, full carrier ... H

  (1.2.3) Single sideband, reduced or variable-level carrier ... R

  (1.2.4) Single sideband, suppressed carrier ... J

  (1.2.5) Independent sidebands ... B

  (1.2.6) Vestigial sideband ... C

(1.3) Emission in which the main carrier is angle modulated

  (1.3.1) Frequency modulation ... F

  (1.3.2) Phase modulation ... G

(1.4) Emission in which the main carrier is amplitude and angle modulated either simultaneously or in a pre-established sequence. ... D

(1.5) Emission of pulses[1]

  (1.5.1) Sequence of unmodulated pulses ... P

  (1.5.2) A sequence of pulses

    (1.5.2.1) modulated in amplitude ... K

    (1.5.2.2) modulated in width/duration ... L

    (1.5.2.3) modulated in position/phase ... M

    (1.5.2.4) in which the carrier is angle modulated during the period of the pulse ... Q

    (1.5.2.5) which is the combination of the foregoing or is produced by other means ... V

(1.6) Cases not covered above, in which an emission consists of the main carrier modulated, either simultaneously or in a pre-established sequence, in a combination of two or more of the following modes: amplitude, angle, pulse ... W

(1.7) Cases not otherwise covered ... X

(2) Second symbol — nature of signal(s) modulating the main carrier

(2.1) No modulating signal ... 0

(2.2) A single channel containing quantized or digital information without the use of a modulating subcarrier[2] ... 1

(2.3) A single channel containing quantized or digital information with the use of a modulating subcarrier[2] ... 2

(2.4) A single channel containing analog information ... 3

(2.5) Two or more channels containing quantized or digital information ... 7

(2.6) Two or more channels containing analog information ... 8

(2.7) Composite system with one or more channels containing quantized or digital information, together with one or more channels containing analog information ... 9

(2.8) Cases not otherwise covered ... X

(3) Third symbol — type of information to be transmitted[3]

(3.1) No information transmitted ... N

(3.2) Telegraphy — for aural reception ... A

(3.3) Telegraphy — for automatic reception ... B

(3.4) Facsimile ... C

(3.5) Data transmission, telemetry, telecommand ... D

(3.6) Telephony (including sound broadcasting) ... E

(3.7) Television (video) ... F

(3.8) Combination of the above ... W

(3.9) Cases not otherwise covered ... X

The following two optional characteristics may be added for a more complete description of an emission:

(4) Fourth symbol — detail of signal(s)

(4.1) Two-condition code with elements of differing numbers and/or durations ... A

(4.2) Two-condition code with elements of the same number and duration without error correction ... B

(4.3) Two-condition code with elements of the same number and duration with error correction ... C

(4.4) Four-condition code in which each condition represents a signal element (of one or more bits) ... D

(4.5) Multi-condition code in which each condition represents a signal element (of one or more bits) ... E

(4.6) Multi-condition code in which each condition or combination of conditions represents a character ... F

(4.7) Sound of broadcasting quality (monophonic) ... G

(4.8) Sound of broadcasting quality (stereophonic or quadraphonic) ... H

(4.9) Sound of commercial quality (excluding categories K and L below) ... J

(4.10) Sound of commercial quality with the use of frequency inversion or band-splitting ... K

(4.11) Sound of commercial quality with separate frequency-modulated signals to control the level of demodulated signal ... L

(4.12) Monochrome ... M

(4.13) Color ... N

(4.14) Combination of the above ... W

(4.15) Cases not otherwise covered ... X

(5) Fifth symbol — nature of multiplexing

(5.1) None ... N

(5.2) Code-division multiplex ... C

(5.3) Frequency-division multiplex ... F

(5.4) Time-division multiplex ... T

(5.5) Combination of frequency-division multiplex and time-division multiplex ... W

(5.6) Other types of multiplexing ... X

Notes

[1]Emissions where the main carrier is directly modulated by a signal that has been coded into quantized form (for example, pulse code modulation) should be designated under (1.2) or (1.3).

[2]This excludes time-division multiplex.

[3]In this context the word "information" does not include information of a constant, unvarying nature such as is provided by standard-frequency emissions, continuous-wave and pulse radars, and so forth.

[4]This includes bandwidth-expansion techniques.

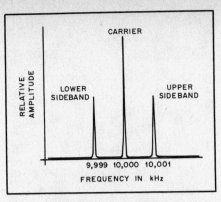

Fig. 1 — A 10-MHz carrier modulated by a 1-kHz sine wave.

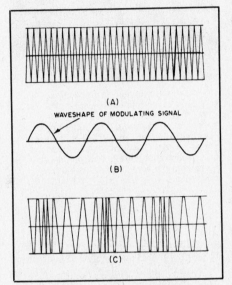

Fig. 2 — Graphical representation of frequency modulation. In the unmodulated carrier at A, each RF cycle occupies the same amount of time. When the modulating signal, B, is applied, the radio frequency is increased and decreased according to the amplitude and polarity of the modulating signal.

the carrier frequency. AM is basically a process of heterodyning or nonlinear mixing. As in any mixer, when a carrier and baseband modulation are combined, there are three products in the frequency range of interest: (1) the carrier, (2) the lower sideband (LSB), and (3) the upper sideband (USB). Thus, if a carrier of 10 MHz were modulated by a 1-kHz sine wave, the outputs would be as shown in Fig. 1.

The bandwidth of the modulated signal in this example would be 2 kHz, the difference between the lowest and highest frequencies. In AM, the difference between the carrier and farthest component of the sideband is determined by the highest frequency component contained in the baseband modulating signal.

Angle Modulation

Two particular forms of angle modulation are *frequency modulation* (FM) and *phase modulation* (PM).

The term frequency modulation is often used to cover various forms of angle modulation. Amateurs use FM when referring to angle-modulated VHF/UHF radios which may use either FM or PM. The pre-WARC '79 emission symbol, F, was used for both FM and PM. However, the post-WARC '79 symbols are F for FM and G for PM.

Frequency and phase modulation are not independent, since the frequency cannot be varied without also varying the phase, and vice versa.

The communications effectiveness of FM and PM depends almost entirely on the receiving methods. If the receiver will respond to frequency and phase changes but is insensitive to amplitude changes, it will discriminate against most forms of noise, particularly impulse noise such as that from ignition systems.

Frequency Modulation

Fig. 2 is a representation of frequency modulation. When a modulating signal is applied, the carrier frequency is increased during one half cycle of the modulating signal and decreased during the half cycle of the opposite polarity. This is indicated in the drawing by the fact that the RF cycles occupy less time (higher frequency) when the modulating signal is positive, and more time (lower frequency) when the modulating signal is negative.

The change in the carrier frequency (*frequency deviation*) is proportional to the instantaneous amplitude of the modulating signal. Thus, the deviation is small when the instantaneous amplitude of the modulating signal is small, and is greatest when the modulating signal reaches its peak, either positive or negative. The drawing shows the amplitude of the RF signal does not change during modulation. This is an oversimplification and is true only in the overall sense, as the amplitude of both the carrier and sidebands varies with frequency modulation. FM is capable of conveying dc levels, as it can maintain a specific frequency.

Phase Modulation

In phase modulation, the characteristic varied is the carrier phase from a reference value. In PM systems, the demodulator responds only to instantaneous changes in frequency. PM cannot convey changes in dc levels unless special phase-reference techniques are used. The amount of frequency change, or deviation, is directly proportional to how rapidly the phase is changing and the total amount of the phase change. The rapidity of the phase change is directly proportional to the frequency of the modulating signal and to the instantaneous amplitude of the modulating signal. This means that in a PM system, deviation increases with both the instantaneous amplitude and frequency of the modulating signal. Thus, PM modulators have a built-in *pre-emphasis*, where devia-

tion increases with modulating frequency. By way of contrast, deviation in FM systems is proportional only to the amplitude of the modulating signal. Apart from this characteristic it is difficult to distinguish between the two.

Angle-Modulation Sidebands

In AM there is a single set of sidebands above and below the carrier frequency. FM and PM produce many sets of sidebands that occur at integral multiples of the modulating frequency on both sides of the carrier. An angle-modulation signal therefore inherently occupies a wider channel than an AM signal.

In angle modulation, the number of sidebands depends on the relationship between the modulating frequency, frequency deviation and *modulation index*. For angle modulation with a sinusoidal modulating function

$$\chi = \frac{D}{m} = \phi \qquad \text{(Eq. 1)}$$

where
- χ = modulation index
- D = peak deviation (half the difference between the maximum and minimum values of the instantaneous frequency
- m = modulation frequency in hertz
- ϕ = phase deviation in radians (a radian = $180/\pi$ or approximately 57.3 degrees)

For example: The peak frequency deviation of an FM transmitter is 3000 Hz either side of the carrier frequency. The modulation index when modulated by a sine wave of 1000 Hz is

$$\text{modulation index} = \frac{3000}{1000} = 3$$

When modulated with a 3000-Hz sine wave at the same deviation the index would be 1; at 100 Hz the index would be 30, and so on.

Given a constant input level to the modulator, in PM the modulation index is constant regardless of the modulating frequency; in FM it varies with the modulating frequency, as shown in the above example. In an FM system the ratio of the *maximum* carrier-frequency deviation to the *highest* modulating frequency is called the *deviation ratio*. Thus

$$\text{deviation ratio} = \frac{D}{M} \qquad \text{(Eq. 2)}$$

where
- D = peak deviation
- M = maximum modulation frequency in hertz

For example: The deviation ratio for narrow-band FM is 5000 Hz (peak deviation) divided by 3000 Hz (maximum modulation frequency) and is 1.67.

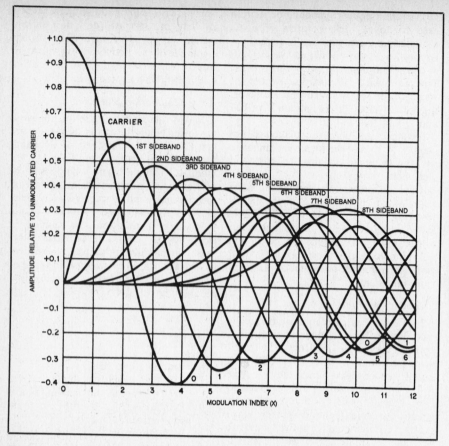

Fig. 3 — Amplitude variation of the carrier and sideband pairs with modulation index. This is a graphical representation of mathematical functions developed by F. W. Bessel. Note that the carrier completely disappears at modulation indexes of 2.405, 5.52 and 8.654.

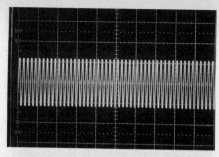
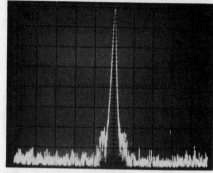

Fig. 4 — The top photo is an oscilloscope pattern of an unmodulated RF carrier. The vertical (Y) axis shows amplitude, and the horizontal (X) axis time. The bottom photo is a spectrum-analyzer display, where amplitude is shown on the Y axis and frequency on the X axis. Notice that there is only one line that represents the amplitude at the carrier frequency.

Fig. 3 shows how the amplitudes of the carrier and the various sidebands vary with the modulation index. This is for single-tone modulation; the first-order sidebands (a pair, one above and one below the carrier frequency) are displaced from the carrier by an amount equal to the modulating frequency, the second is twice the modulating frequency away from the carrier, and so on. For example, if the modulating frequency is 2000 Hz and the carrier frequency is 29,500 kHz, the first sideband pair is at 29,498 and 29,502 kHz, the second pair is at 29,496 and 29,504 kHz, the third at 29,494 and 29,506 kHz, and so on. The amplitudes of these sidebands depend on the modulation index, not on the frequency deviation.

Note that as shown in Fig. 3, the carrier strength varies with the modulation index. (In amplitude modulation the carrier strength is constant; only the sideband amplitude varies.) At a modulation index of 2.405, the carrier disappears entirely. It then becomes "negative" at a higher index, meaning that its phase is reversed compared to the phase without modulation. In angle modulation, the energy that goes into the sidebands is taken from the carrier, the *total* power remaining the same regardless of the modulation index.

Since there is no change in total amplitude with modulation, an FM or PM signal can be amplified without distortion by an ordinary class-C amplifier. The modulation can take place in a very low-level stage, and the signal can then be amplified by either frequency multipliers or straight-through amplifiers.

If the modulated signal is passed through one or more frequency multipliers, the modulation index is multiplied by the same factor as the carrier frequency. For example, if modulation is applied at 3.5 MHz and the final output is on 28 MHz, the total frequency multiplication is eight times, so if the frequency deviation is 500 Hz at 3.5 MHz it will be 4000 Hz at 28 MHz.

Pulse Modulation

FCC rules restrict pulse modulation to the microwave region. Because of this it is not in common usage among amateurs. Pulse-modulated signals are usually sent as a series of short pulses separated by relatively long stretches of time with no signal being transmitted. (A typical pulse transmission might use 1-μs pulses with a 1000-Hz pulse rate. Thus, the peak power of a pulse transmission is usually much greater than its average power.

The pre-WARC '79 emission symbol for all pulse-type emissions was P. Since WARC '79, P is used only for a sequence of unmodulated pulses. Other types of pulse modulation are:
• Emission symbol K — pulse-amplitude modulation (PAM)
• Emission symbol L — pulse-width modulation (PWM) or pulse-duration modulation (PDM)
• Emission symbol M — pulse position/phase modulation (PPM)
• Emission symbol Q — angle modulation during the pulse
• Emission symbol V — combination of the above or by other means

AMATEUR RADIO MODULATION SYSTEMS

Specific modulation systems of general interest to radio amateurs are discussed below. Pre- and post-WARC '79 emission designations and graphic illustrations are included.

UNMODULATED CARRIER

Just as number systems include the digit 0, RF carriers have their unmodulated state. The emission symbol for an unmodulated carrier is N0N (formerly A0 or F0). N0N emission is permitted in all amateur bands above 51 MHz. In addition, it is permitted for beacon operation in bands above 28.2 MHz specified by FCC rules. N0N may also be used on any amateur band below 51 MHz for short

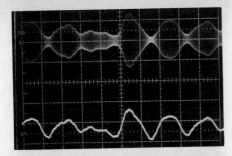

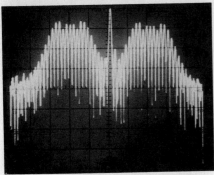

Fig. 5 — Voice-modulated AM DSB signal. On top is a dual-trace oscilloscope pattern of amplitude (Y axis) vs. time (X axis). The lower trace is a baseband voice signal. The upper trace is the carrier 100% modulated by the voice signal. The bottom photo is the spectrum-anaylzer display of amplitude (Y axis) vs. frequency (X axis).

periods of time when required for authorized remote control purposes or for experimental purposes (such as for transmitter tuning).

The bandwidth of an N0N emission is 0 Hz except for the transient sidebands that occur when the carrier is turned on or off. Fig. 4 shows an oscilloscope amplitude vs. time display and a spectrum analyzer amplitude vs. frequency display of an unmodulated carrier.

ANALOG TELEPHONY

Modulation systems for transmission of a single analog channel of telephony (including sound broadcasting) are identified by the post-WARC '79 symbols — 3E. The dash is replaced by a letter representing the type of modulation of the main carrier. Types of modulation for speech transmission that are commonly used in the Amateur Radio Service follow.

Double-Sideband Amplitude-Modulated Telephony

Double-sideband (DSB) amplitude-modulated (AM) telephony is generated by varying the amplitude of the carrier by a voice-frequency analog speech signal. Its post-WARC '79 designation is A3E (formerly A3). The full designation for commercial or amateur quality is 6K00A3EJN, meaning: 6 kHz bandwidth, DSB AM, telephony, sound of commercial quality, no multiplexing. In this case, the highest modulating frequency is 3 kHz,

resulting in 3-kHz sidebands above and below the carrier.

While DSB AM was the primary phone modulation system used in the early days of Amateur Radio, very few amateurs use it today. DSB AM has been virtually replaced by single-sideband amplitude modulation (SSB AM) in the HF bands. Angle (frequency or phase) modulation and SSB telephony are used on the VHF and UHF bands. Nevertheless, the most popular way of producing an SSB signal is to first generate a DSB AM signal and filter out the unwanted sideband.

Fig. 5 shows an oscilloscope amplitude vs. time display and a spectrum-analyzer amplitude vs. frequency display of a voice-modulated AM DSB signal. The signal in Fig. 5 is said to be 100% modulated on peaks, which is the maximum that the FCC permits. Modulation above 100% is unproductive because it adds distortion and generates unwanted sidebands, called *splatter*. Modulation with peaks much below 100% is undesirable because signal-to-noise ratio suffers.

At 100% modulation, voltage of each of the sidebands is half that of the carrier. Therefore, the power in each sideband is $(\frac{1}{2})^2$ times or ¼ that of the carrier. For a transmitter with a carrier output power of 100 watts, the power in each sideband is 25 watts, and the total sideband power is 50 watts. Fig. 6 illustrates the power in the carrier and sidebands of a 100%-modulated AM DSB signal. However, the same 100-watt AM DSB transmitter produces instantaneous peaks of 400 watts as illustrated in the vector diagram of Fig. 7. The bandwidth of a double-sideband AM telephony signal is calculated

$$B_n = 2M \qquad \text{(Eq. 3)}$$

where

B_n = necessary bandwidth in hertz
M = maximum modulation frequency in hertz

Receiving DSB AM Telephony

A double-sideband voice-modulated signal with carrier can be demodulated to audio by means of a half-wave or full-wave rectifier; simply one or more diodes. When the modulated signal passes through a nonlinear device, the carrier mixes with the two sidebands and produces an audio output.

On HF, multipath propagation causes AM DSB signals to fade in amplitude. Fading can be overcome to some extent by using an automatic gain control (AGC) loop circuit to keep the audio output at a constant level. Another multipath effect is selective fading, which has the effect of periodically distorting the demodulated audio. This type of distortion may be virtually eliminated by independent-sideband detection of the upper and lower sidebands and by representing the USB and LSB outputs to separate earphones. Simple elec-

trical combination after detection will reintroduce the distortion.

Single-Sideband Suppressed-Carrier AM Telephony

As both sidebands of a DSB AM signal convey the same information, a single sideband (SSB) may be transmitted and received. Since the carrier contains no intelligence, it too may be suppressed prior to transmission and reinserted in the receiver. SSB transmission and reception adds complexity and cost to radio equipment. However, SSB has a number of advantages:

• *spectrum conservation* — two SSB signals can occupy the space of one DSB signal.

• *more talk power for a transmitter final stage* — the same transmitter illustrated in Figs. 6 and 7 that ran 100 W carrier, 400 W peak and 50 W of talk power (USB + LSB) in DSB AM service can run 400 W of peak SSB talk power. Thus for a given transmitter power amplifier, SSB is said to have a 400/50 = 8/1 = 9-dB advantage over DSB AM, considering transmitter peak power alone.

• elimination of carrier interference — this eliminates the possibility of a carrier heterodyning with an adjacent signal.

• reduction in multipath distortion — SSB does not suffer the audio distortion

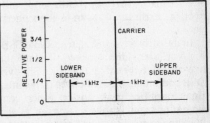

Fig. 6 — Example of a carrier 100% modulated by a single 1-kHz tone.

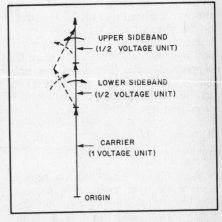

Fig. 7 — Vector diagram of 100% modulation of an AM carrier at the instant when peak conditions exist. The broken vectors show the relationships at an instant when the modulating signal is somewhat below its peak.

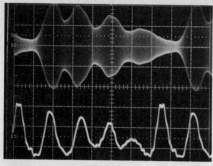

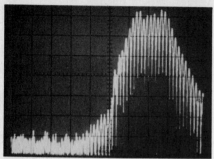

Fig. 8 — Single-sideband suppressed-carrier voice modulation. At top is a dual-trace oscilloscope pattern of amplitude (Y axis) vs. time (X axis). The bottom trace is a baseband voice signal. The upper trace is the SSBSC RF telephony signal. The lower photo is the spectrum-analyzer display of amplitude (Y axis) vs. frequency (X axis).

300 to 3000 Hz is 2K70J3EJN. On the end, J means commercial quality, and N signifies no multiplexing.

Receiving SSBSC Telephony

SSBSC signals can be demodulated by an ordinary AM detector and by using a beat-frequency oscillator to reinsert the carrier within a few hertz of where it should be relative to the sidebands. Better performance can be realized by using a product detector, which is explained in Chapter 18.

Amplitude-Compandored Single-Sideband Telephony

Amplitude *compression* and limiting are common speech-processing techniques used in most SSB transmission systems to limit peak amplitudes and keep the average speech level high. In some SSB receiving systems, an *expansion* circuit is included to restore the compressed speech to its full dynamic range. A circuit that handles both transmitting compression and receiving expansion is known as a compandor. Unless there is some coordination between the compressor and expander, the expander will not know when and how much to expand. In amplitude-compandored single sideband (ACSSB) the control information is sent as amplitude modulation of a 3.1-kHz pilot tone. ACSSB offers great potential for use in the VHF and UHF amateur bands, and is explained in Chapter 18.

There is no special post-WARC '79 emission symbol for ACSSB. It would fall into the SSBSC category of J3E (formerly A3J), however. The necessary bandwidth would be calculated by

$$B_n = M - L \qquad \text{(Eq. 5)}$$

where

B_n = necessary bandwidth in hertz.
M = maximum modulation frequency in hertz, which in this case is that of the pilot tune
L = lowest modulation frequency in hertz.

Single-Sideband Full-Carrier Telephony

This type of modulation is the same as SSBSC except that a full carrier has been added. It is also known as *compatible AM* (CAM) because it can be received on an ordinary AM receiver with no beat-frequency oscillator. CAM has a number of additional advantages:

• receiver tuning is less critical than for SSBSC.
• bandwidth is half that of DSB AM.
• it is easy to generate using an SSBSC transmitter by the simple insertion of the carrier.

Disadvantages are:

• transmission of the full carrier reduces the peak power available for the intelligence-bearing sideband.
• carriers produce unnecessary beat notes in the phone bands.

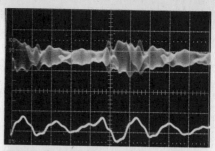

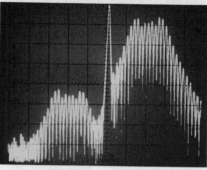

Fig. 9 — Displays of a single-sideband, full-carrier telephony, or compatible-AM (CAM), signal. At top is a dual-trace oscilloscope pattern of amplitude (Y axis) vs. time (X axis). The bottom trace is a baseband voice signal. The upper trace is the CAM RF signal. The lower photo is the spectrum-analyzer display of amplitude (Y axis) vs. frequency (X axis).

• reception with an ordinary DSB AM receiver will be less effective than a receiver designed for SSB reception.

The post-WARC '79 emission symbol for single-sideband full-carrier telephony is H3E (formerly A3H). The full CCIR emission designation 3K00H3EJN would mean a bandwidth of 3 kHz, full-carrier, single-sideband telephony, commercial quality and no multiplexing. Necessary bandwidth is calculated by

$$B_n = M \qquad \text{(Eq. 6)}$$

where

B_n = necessary bandwidth in hertz
M = maximum modulation frequency in hertz

Fig. 9 shows oscilloscope and spectrum-analyzer displays of a sample CAM signal.

Angle-Modulated Telephony

Below 29.0 MHz, FCC rules limit the bandwidth of frequency- or phase-modulated telephony to that of an AM transmission having the same audio characteristics. The full emission designation for such a signal would be 6K00F3EJN for frequency modulation and 6K00G3EJN for phase modulation. These designations break into: 6-kHz bandwidth, F for frequency modulation or G for phase modulation, single-channel analog, telephony, sound of commercial quality and no multiplexing.

(mushiness) heard when receiving a DSB AM signal where the USB and LSB propagate through separate ionospheric paths.

In the early days of Amateur Radio SSB, a favorite pastime was to calculate the overall effective advantage of SSB over DSB communications. Besides transmitter final amplifier peak power, other considerations are amplifier average power handling capability, the effects of impulse noise on SSB and DSB reception, fading and distortion resulting from multipath propagation, and reception in the presence of different types of interference. However, at risk of resurrecting an old controversy, the overall advantage of SSB over DSB is on the order of 12 dB.

Fig. 8 shows amplitude vs. time and amplitude vs. frequency displays of an SSB suppressed-carrier (SC) signal. The emission symbol is J3E (formerly A3J) for SSBSC. The necessary bandwidth is calculated

$$B_n = M - L \qquad \text{(Eq. 4)}$$

where

B_n = necessary bandwidth in hertz
M = maximum modulation frequency in hertz
L = lowest modulation frequency in hertz

The full emission designation for an SSBSC telephony signal modulated from

Because of the limitation not to exceed the bandwidth of an AM transmission of the same characteristics, the bandwidth is restricted to 2M (twice the maximum modulation frequency) or 2 × 3000 = 6000 Hz. You cannot use 5-kHz or even 3-kHz deviation without greatly exceeding the 6-kHz bandwidth limit. In fact, the usual FM bandwidth formulas will not work in this case, and it is necessary to examine the power in the second pair of sidebands and limit it to about 1% of the total. If this were done assuming sinusoidal modulation at 3000 Hz, the highest modulation index would be 0.4, and deviation would be limited to 1.2 kHz. In practice, however, the energy distribution in a complex (voice) wave is such that the modulation index for any one frequency component is reduced as compared to the index with a sine wave having the same peak amplitude as the voice wave. Thus, for voice, a modulation index of 0.6 and a 1.8-kHz deviation will ensure that the second pair of sidebands will be 27 dB below the unmodulated carrier level.

To obtain the benefits of this mode, a good FM receiver is required. As shown in Fig. 3, at an index of 0.6 the amplitude of the first pair of sidebands is about 25 percent of the unmodulated-carrier amplitude; this compares with a sideband amplitude of an AM transmitter modulated 100 percent.

Above 29.0 MHz, the standard FM deviation is 5 kHz. The formula for necessary bandwidth is

$$B_n = 2M + 2DK \qquad \text{(Eq. 7)}$$

where
- B_n = necessary bandwidth in hertz
- M = maximum modulation frequency in hertz
- D = peak deviation in hertz
- K = 1

For a maximum modulation frequency of 3000 Hz and a peak deviation of 5000 Hz, the necessary bandwidth is 16,000 Hz. The emission designation would be 16K0F3EJN for 16-kHz bandwidth, frequency modulation, analog, telephony, commercial quality, and no multiplexing. In the case of phase modulation, the designation would be 16K0G3EJN.

In the early 1970s, 15-kHz deviation was used extensively by amateurs. Using Eq. 7, the bandwidth would be 36 kHz, and the emission designations would be 36K0F3EJN and 36K0G3EJN for frequency modulation and phase modulation, respectively. Currently, 15-kHz deviation is used mostly on frequencies above 450 MHz. One exception is that some fast-scan ATV groups apply 15-kHz FM audio modulation to the video carrier on repeater input frequencies.

TV Sound

The sound channel for U.S. TV broadcasting has a maximum modulation frequency of 15 kHz and uses a peak deviation of 25 kHz. Using Eq. 7, necessary bandwidth is calculated to be 2M + 2DK = (2 × 15) + (2 × 25 × 1) = 80 kHz. The emission designation would be 80K0F3EGN, meaning: 80-kHz necessary bandwidth, frequency modulation, analog, telephony, sound of broadcasting quality (monophonic) and no multiplexing. This is of interest to amateurs who transmit ATV for reception by ordinary TV receivers. Amateurs may not be interested in transmitting the full 50- to 15,000-Hz audio range and can reproduce high-quality speech with a 250- to 6000-Hz audio pass-band. In this case, the necessary bandwidth would be reduced to 62 kHz. If the maximum modulation frequency were reduced to 3000 Hz, the necessary bandwidth would be 56 kHz.

FM Broadcasting

The characteristics of FM sound broadcasting are of interest to amateurs for special applications. Chapter 32 illustrates the use of an FM broadcast receiver as a demodulator in a Gunnplexer™ microwave communication system.

FM sound broadcasting uses a peak deviation of 75 kHz and a maximum modulation frequency of 15 kHz. Again using Eq. 7, the necessary bandwidth is 2M + 2DK = 2 × 15 + 2 × 75 × 1 = 180 kHz. The emission designation is 180KF3EGN, which breaks down to: 180-kHz necessary bandwidth, frequency modulation, analog, telephony, sound of broadcasting quality (monophonic), and no multiplexing. For stereophonic or quadraphonic sound, the G would be replaced by H. N would become F to indicate frequency-division multiplexing.

AMATEUR IMAGE COMMUNICATIONS MODULATION

Image communications systems used in the Amateur Radio Service are outlined in Chapter 20. Below is a discussion of the most popular modulation systems for these modes.

Fast-Scan Television

Two types of amplitude modulation are presently in use by Amateur Radio fast-scan television (FSTV) operators. One is double-sideband, and the other is vestigial sideband. Although most FSTV is in color, some monochrome (black and white) FSTV is still in use.

The simplest way to transmit FSTV is monochrome DSB AM. In this case, the necessary bandwidth can be calculated from

$$B_n = 2M \qquad \text{(Eq. 8)}$$

where
- B_n = necessary bandwidth in hertz
- M = maximum modulation frequency in hertz

Typically, monochrome FSTV has a video bandwidth (M in Eq. 8) of 2.5 MHz. The complete CCIR emission designation would be 5M00A3FMN for: 5-MHz necessary bandwidth, double-sideband AM, single-channel analog information, television, monochrome, and no multiplexing. This emission designation does not account for any audio channel — the possibilities include sending the audio independently or multiplexing it with the video transmission.

Using the NTSC color system, the video bandwidth needs to be about 4 MHz. Again using Eq. 8, the DSB AM necessary bandwidth would be 8 MHz. The emission designation becomes 8M00A3FNM, which stands for: an 8-MHz necessary bandwidth, double sideband AM, single-channel analog information, television, color and no multiplexing. Again, no audio channel is included.

Double-sideband AM transmission of either monochrome or color TV occupies a considerable amount of spectrum. So, it is desirable to use vestigial-sideband (VSB) transmission wherever possible for FSTV to reduce bandwidth. This is particularly important for FSTV repeater outputs because repeaters are generally well sited for wide-area coverage and may run several hundred watts of power. For vestigial-sideband FSTV, the necessary bandwidth of the visual transmission can be calculated from

$$B_n = M + V \qquad \text{(Eq. 9)}$$

where
- B_n = necessary bandwidth in hertz
- M = maximum modulation frequency in hertz
- V = bandwidth of the vestigial sideband (typically 1.25 MHz)

For example, the full emission designation for a VSB color TV transmission might be 5M25C3FNN, meaning: 5.25-MHz necessary bandwidth, vestigial-sideband AM, single-channel analog information, television, color, and no multiplexing. For monochrome VSB TV, the emission designation might be 4M00C3FMN, meaning: 4-MHz necessary bandwidth, vestigial-sideband AM, single-channel analog, television, monochrome and no multiplexing. Neither of these designations includes an audio channel which would most likely be 4.5 MHz above the video carrier and use 80K0F3EGN emission as discussed earlier in this chapter.

Analog Slow-Scan Television or Facsimile

Slow-scan television (SSTV) or facsimile (FAX) is typically transmitted by frequency modulating an audio oscillator and feeding that into a single-sideband suppressed-carrier transmitter. The bandwidth has to take into account both the frequency range over which the oscillator frequency is swung (such as from 1200 to

2300 Hz) and the number of picture elements (pixels) transmitted per second. The necessary bandwidth can be calculated from

$$B_n = 2M + 2DK \qquad \text{(Eq. 10)}$$

where
B_n = necessary bandwidth in hertz
$M = N/2$
N = number of pixels per second
$K = 1.1$ (typically)
D = peak deviation (half the difference between the maximum and minimum values of the instantaneous frequency)

For example, with a tone that can be swung from 1200 to 2300 Hz (a difference of 1100 Hz), D = half the difference, or 550 Hz. If the horizontal sweep rate is 15 Hz and there are 100 pixels per line, N = (15 × 100) = 1500 pixels/s, and M = 1500/2 = 750. Pouring these numbers in Eq. 10, B_n = (2 × 750) + (2 × 550 × 1.1) = 1500 + 1210 = 2710 Hz. If this were an SSTV transmission, the emission designator could be 2K71J3FMN, for: 2.71-kHz necessary bandwidth, single-sideband suppressed-carrier, single-channel analog information, monochrome and no multiplexing. For facsimile, it would be 2K71J3CMN — the C designating facsimile. Note that it is the SSBSC modulation of the main carrier that determines the basic type of emission (J), not the fact that the subcarrier is frequency modulated. On the other hand, if you directly frequency-modulate the main carrier, the J in the above designators would change to F.

Quantized Slow-Scan Television or Facsimile

Many amateurs are now using computers to generate *quantized* (rather than continuously varying signals). The bandwidth of quantized SSTV or FAX may be calculated using Eq. 15 provided that the transitions between quantized steps are smooth. For abrupt transitions between steps where image contouring is to be preserved, larger values of K should be used. If the modulation system is as described under Analog Slow-Scan Television or Facsimile, above, the emission designators for quantized SSTV or FAX would be changed by substituting the digit 2 (meaning quantized modulation of a tone) for the digit 3 (analog) in the third symbol.

TELEGRAPHY

These modulation systems include telegraphy for aural reception (Morse code) and telegraphy for automatic reception (radioteletype).

Continuous-Wave Morse Telegraphy

Telegraphy by *on-off keying* (OOK) of a carrier is the oldest radio modulation system. It is also known as CW (for continuous wave). While CW is used by

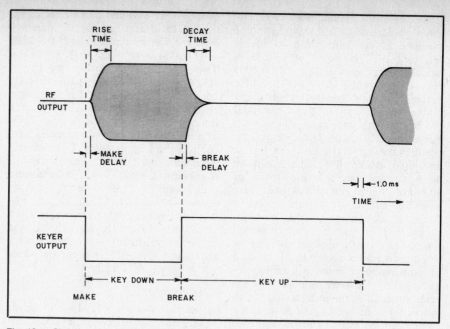

Fig. 10 — Shaping and timing in CW keying systems.

amateurs and other communicators to mean OOK telegraphy by Morse code, parts of the electronics industry use CW to signify an unmodulated carrier. OOK telegraphy is also known as *amplitude-shift keying* (ASK).

The emission symbol for Morse OOK of the main carrier is A1A (formerly A1). The first letter is A because keying a carrier on and off produces double (upper and lower) sidebands corresponding to the frequency components of the keying pulse. The amplitude of the RF signal rises from zero to full strength some time after the key is closed, and returns to zero some time after the key is opened. The lengths of these transition intervals and their wave shapes determine the bandwidth occupied by the signal.

In Fig. 10, the envelope rises from zero to full amplitude in 5 ms. The waveform during the rise time resembles 90 degrees of sine wave having a period of 20 ms, or a frequency of 50 Hz. Amplitude-modulation theory tells us that the occupied bandwidth will be 100 Hz; that is, 50 Hz for the lower sideband (LSB) and 50 Hz for the upper sideband (USB). If the waveform deviates from sinusoidal, harmonics are present, and the bandwidth will be some integral multiple of 100 Hz. The fall time is just as important as the rise time; whichever is shorter or has the greatest harmonic content will determine the bandwidth of the emission.

League publications have long promoted 5-ms rise and fall times for CW keying envelopes. This figure was not chosen arbitrarily. Appendix 6 of the CCIR Radio Regulations and section 2.202 of the FCC rules define the bandwidth for an A1A emission as

$$B_n = B K \qquad \text{(Eq. 11)}$$

where
B_n = necessary bandwidth in hertz
B = telegraph speed in bauds
K = an overall numerical factor that varies according to the emission and that depends on the allowable signal distortion.

The CCIR and FCC recommend $K = 3$ for nonfading circuits and $K = 5$ for fading circuits when using telegraphy for aural reception. If $K = 3$ is chosen, that means the frequency of the keying envelope rise is the third harmonic of the keying frequency. $K = 3$ is the minimum needed for aural reception because a $K = 1$ (although useful for machine recognition) sounds like the keying is too soft. $K = 5$ is more appropriate for fading circuits because the faster rise time helps the ear sense the transitions. Further, $K = 5$ says that the third *and* fifth harmonic of the keying pulses are present in both sidebands. Thus, if any of these harmonics are lost through *selective fading* caused by *multipath* propagation, the rest of the signal stands a good chance of getting through and being interpreted correctly by the receiving operator. Selective fading isn't usually a problem on HF at speeds below 50 bauds because the upper and lower sidebands are fairly close in frequency and are usually *correlated* (they fade together). Every now and then you can hear the exception, which sounds like the signal is in a barrel.

In Morse telegraphy a dot occupies one unit of time, as does a space. For a string of dots having a duration of 20 ms, there will be 25 such dots in a second, and the keying speed will be 50 bauds. We are more accustomed to words per minute (WPM)

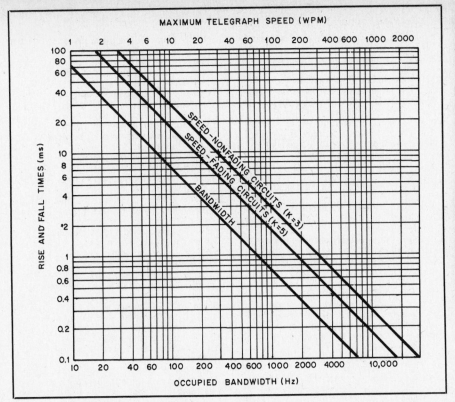

Fig. 11 — Curves for determining 23-dB bandwidths and maximum telegraph speeds for various keying rise and fall times.

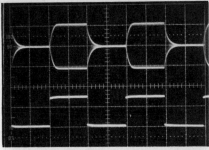
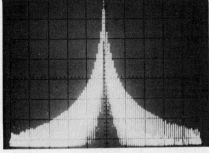

Fig. 12 — Displays of a CW OOK signal. At top is a dual-trace oscilloscope pattern of amplitude (Y axis) vs. time (X axis). The bottom trace is the CW keying input before shaping at 25 dots/s. The upper trace is the keyed RF carrier after keying has been shaped for approximately 5-ms rise and fall times. The bottom photo is the spectrum-analyzer display of amplitude (Y axis) vs. frequency (X axis) for the same signal.

than bauds when dealing with Morse code, so let's apply the following formulas.

WPM = 2.4 × dot/s
WPM = 1.2 B

where

WPM = telegraph speed in words per minute
2.4 = a constant calculated by comparing dots per second with plain-language Morse code sending the word PARIS
1.2 = a constant calculated by comparing the signaling rate in bauds with plain-language Morse code sending the word PARIS
B = telegraph speed in bauds

Thus a keying speed of 25 dot/s or 50 bauds is equal to 60 WPM. Returning to Eq. 11, on a nonfading circuit the necessary bandwidth will be 50 × 3 = 150 Hz. This value differs from the 100 Hz derived earlier because the formula is based on a larger energy spectrum and recognizes that the keying waveform isn't likely to be purely sinusoidal. Also, we defined the rise and fall times to include the complete transition interval for ease of illustration. Measuring the rise and fall times according to the 10% to 90% convention shortens the times somewhat and tends to further reconcile the two bandwidth values. Given a 150-Hz bandwidth, how fast can we communicate over a fading path? Setting B_n = 150 and K = 5, and solving Eq. 11 for B produces a telegraph speed of

30 bauds, which is 36 WPM (PARIS). You may conclude that 5-ms rise and fall times are suitable for up to 36 WPM on fading circuits and 60 WPM on nonfading circuits. These numbers can be scaled up or down according to your needs as shown in Fig. 11.

Fig. 12 shows an oscilloscope amplitude vs. time display and a spectrum-analyzer amplitude vs. frequency display of an OOK signal.

To complete the picture, the full CCIR emission designation for the Morse code keying just discussed is 150HA1AAN. The figure 150 is the necessary bandwidth in hertz (H); A = double-sideband amplitude modulation; 1 = a single channel containing quantized or digital information without the use of a modulating subcarrier; the second A = telegraphy — for aural reception; the third A = two-condition code with elements of differing numbers and/or durations; and N = no multiplexing.

Receiving Morse Telegraphy

Proper reception of a Morse-code transmission requires that the receiver bandwidth be at least that of the necessary bandwidth plus any frequency error. Thus, if you have 150-Hz receiver bandwidth, it would be necessary for you to carefully tune your receiver to receive a 150-Hz-bandwidth transmission. Or, if you want to have a fixed-frequency receiver, probably one that's crystal controlled, you will need to increase the 150-Hz bandwidth

by the combined error of both the transmitter and receiver frequencies.

For aural reception, a Morse-code OOK radio signal is not completely demodulated to its original dc pulse, since only thumping would be heard. Instead, it consists of simply heterodyning the RF signal down to an AF signal somewhere in the vicinity of 700 Hz. If the IF bandwidth is much greater than 150 Hz, the overall bandwidth can be reduced by adding an audio filter in the receiver AF circuits. These concepts and practical circuits are developed in Chapter 28.

Single-Sideband Suppressed Carrier Tone-Modulated Morse

Rather than keying a main carrier on and off, many modern transmitters generate a CW signal by keying an audio tone of about 700 or 800 Hz and feeding it to an SSBSC transmitter. The transmitter output looks just the same as though the main carrier had been keyed except for the suppressed carrier 700 or 800 Hz away. In fact, the carrier is so well suppressed that it cannot be heard any appreciable distance away from the transmitter. "If it looks like a duck, walks like a duck and quacks like a duck..." Well, not in this case. The transmitted signal is not classified as A1A but as J2A (formerly A2J). Eq. 11 can still be used to calculate bandwidth. Using the 5-ms rise-time example discussed under continuous-wave Morse telegraphy, the bandwidth would remain 150 Hz. Thus, the

Glossary Of Modulation, Multiplexing And Demodulation Terms

ACSB® — Amplitude-compandored single sideband; registered trademark of Sideband Technology Inc. (See ACSSB.)

ACSSB — Amplitude-compandored single sideband.

ADM — Adaptive delta modulation.

AFSK — Audio-frequency-shift keying.

AM — Amplitude modulation.

Analog — Information in continuously variable quantities. (Opposite digital. Also see quantized.)

ANBFM — Adaptive narrow-band frequency modulation.

Apparent carrier — The frequency, f_0, at which the signal maximum appears.

APK — Amplitude phase keying. (Same as QAM.)

APCM — Adaptive pulse-code modulation.

ASK — Amplitude-shift keying.

AWGN — Additive white Gaussian noise.

Baseband — The band of frequencies of a signal (a) prior to modulation of the main carrier or (b) after demodulation of the main carrier.

Baud (Bd) — A unit of signaling speed equal to the number of discrete conditions or events per second. (If the duration of a pulse is 20 ms, the signaling rate is 50 bauds or the reciprocal of 0.02.)

Bit/s — A measure of the number of bits of data transmitted per second. (The prefixes k, M, or G may be added to denote higher speeds; for example 1 kbit/s is 1000 bits per second.)

BPSK — Binary phase-shift keying.

Carrier — A continuous radio frequency capable of being modulated by a baseband signal or subcarrier.

CDM — Companded delta modulation.

CDMA — Code-division multiple access.

CFSK — Coherent frequency-shift keying.

C/N — Carrier-to-noise ratio. (Expressed in dB.)

Compandor — A contraction of compressor-expander.

CPFSK — Continous-phase frequency-shift keying.

CPM — Continous-phase modulation.

CVSD — Continuously variable slope delta modulation. (Used to convert analog speech into compressed digital form.)

CW — Continuous wave. (In Amateur Radio, CW usually means a continuous carrier wave keyed on and off by Morse code.)

DAV — Data above voice.

DBPSK — Differentially binary phase-shift keying.

DCPSK — Differentially coherent phase-shift keying.

Demodulation — The process of deriving the original modulation signal from a modulated carrier wave.

DEBPSK — Differentially encoded binary phase-shift keying, with carrier recovery.

DECPSK — Differential encoded coherent phase-shift keying.

DFSK — Double frequency-shift keying.

Dibit — A group of two bits. (The possible states are 00, 01, 10 and 11. In phase modulation each dibit translates to one of four specific phases or transitions.)

Digital — Information in discrete states. (Opposite analog.)

DIV — Data in voice.

DLL — Delay-locked loop.

DPCM — Differential pulse-code modulation.

DPMM — Digitally processed multimode modulation.

DPSK — Differential phase-shift keying.

DS — Direct sequence. (A technique used in spread spectrum.)

DSB — Double sideband.

DSBSC — Double-sideband suppressed carrier.

DUV — Data under voice.

Envelope delay — The propagation time of the envelope of a wave.

Equalizer — A circuit with fixed or variable parameters for correction of amplitude or phase distortion.

Eye diagram — (Also called eye pattern). An oscilloscope display of a digital signal stream, triggered at the transmission bit rate, to show the received waveform. (The displayed waveform resembles an eye.)

FAX — Facsimile.

FDM — Frequency-division multiplexing. (Typically used to transmit a number of sub-carriers [each having its own frequency] over one carrier.)

FDMA — Frequency-division multiple access.

FFSK — Fast frequency-shift keying. (Same as MSK.)

FH — Frequency hopping. (A modulation technique used in spread spectrum.)

FM — Frequency modulation.

FSK — Frequency-shift keying.

FSTV — Fast-scan television.

GMSK — Gaussian-filtered minimum-shift keying.

Group delay — The rate of change with frequency of the phase shift during propagation from one point to another.

ICW — Interrupted continuous wave.

ISB — Independent sideband.

Isochronous distortion — The peak-to-peak phase jitter, which may be expressed in percentage of the unit interval.

Intersymbol interface (ISI) — Pulse stretching of one pulse into the next pulse.

Jitter — The peak-to-peak time difference between the ideal time and the actual time for a pulse transition.

LPC — Linear predictive coding. (A technique for converting analog speech to digital form using a small number of bits.)

LSB — Lower sideband.

---

complete emission designation would be 150HJ2AAN.

Amplitude Tone-Modulated Morse Telegraphy

This type of modulation, often called *modulated* CW (MCW), was lumped under an ambiguous pre-WARC '79 emission symbol of A2. MCW or A2 meant that a Morse-keyed tone modulated a double-sideband AM transmitter. The ambiguity was whether only the sidebands were keyed or the sidebands and the carrier. Carrier-keyed MCW is permitted in the Amateur Radio Service only on frequencies between 50.1 and 54 MHz and above 144.1 MHz. MCW produced by keying the tone but not the carrier can be operated on the frequencies just mentioned and as specified in §97.65(b) of the FCC rules: "Whenever code practice, in accordance with §97.91 (d), is conducted in bands authorized for A3 emission tone modulation of the radio telephone transmitter may be utilized when interspersed with appropriate voice instructions."

The post-WARC '79 emission symbol for MCW is A2A. Again, no distinction is made in the emission symbol to tell whether the carrier is keyed. The formula for necessary bandwidth is

$$B_n = BK + 2M \qquad \text{(Eq. 12)}$$

where

B_n = necessary bandwidth in hertz
B = modulation rate in bauds
M = maximum modulation frequency in hertz
K = 5 for fading circuits, 3 for non-fading circuits

For a 25-WPM (B = 20) transmission with M = 1000 and K = 5, the bandwidth will be 2.1 kHz. The full emission designation in this case will be 2K10A2AAN. Broken down, the designation stands for: 2.1 kHz bandwidth, double-sideband AM, a single channel containing quantized or digital information, telegraphy for aural reception, two-condition code with elements of differing numbers and/or durations, and no multiplexing.

Frequency-Shift-Keyed Telegraphy

On-off keying is effective for aurally received Morse-code transmissions because the ear is able to recognize the difference between the presence of a signal and noise (the absence of signal). Demodulators for teletypewriters and computers do much better with two positive conditions, each representing a specific logic state (1 or 0). Frequency-shift keying (FSK) provides the two positive states just mentioned. The transmitter shifts between two predetermined frequencies, called *mark* (1) and *space* (0). In Amateur Radio Baudot and AMTOR systems, the standard shift is now 170 Hz. Deviation (usually abbreviated D) is measured from an imaginary center frequency and the mark or space frequencies, and is half the shift. Thus a 170-Hz shift equals a deviation of 85 Hz.

Sometimes you will see the abbreviation CP used in front of FSK. CP, meaning *continuous phase,* says that the mark and space frequencies as well as the dc pulses that switch between them are clocked in a way that ensures a continuous sinusoidal function. This avoids generating unwanted sidebands that would have resulted from waveform discontinuities.

To date, virtually all amateur FSK has been NCFSK, the NC standing for *non-coherent.* You can tell this quickly by looking at the signaling rate (in bauds) and the shift: If one is not an integral multiple of the other, the FSK will be noncoherent. For example, a 170-Hz-shift FSK system keyed at 45.45 bauds is noncoherent because 170 divided by 45.45 equals 3.740374 — far from a round number. Unless special design precautions are taken, an NCFSK signal could contain many unwanted sideband components (key clicks),

MAP — Maximum *a posteriori* probability. (A mathematical estimative technique used in data demodulation. See ML.)

Mark — The presence of a signal. (See space.)

M-ary — M is a variable representing the number of states (in modulation).

MCW — Modulated continuous wave. (In Amateur Radio, MCW usually means a tone-modulated carrier keyed on and off by Morse code.)

MASK — Multiple amplitude-shift keying.

MFSK — Multiple frequency-shift keying.

ML — Maximum likelihood. (A mathematical estimative technique used in data demodulation. See MAP.)

MPSK — Multiple phase-shift keying.

Modem — Modulator-demodulator.

Modulation — The process of varying some characteristic of a carrier wave to convey intelligence.

Modulation index — In frequency modulation, the number of sideband pairs for a given modulating frequency.

MSK — Minimum-shift keying. (Same as FFSK.)

MSTV — Medium-scan television.

MUX — Multiplex; multiplexer.

NBFM — Narrow-band frequency modulation.

NBPSK — Narrow-band phase-shift keying.

NCFSK — Noncoherent frequency-shift keying.

Necessary bandwidth — The bandwidth just sufficient to ensure the transmission of information at the rate and with the quality required.

NLA-QAM — Nonlinear amplified QAM.

NLF-OQPSK — Nonlinearly filtered QPSK; also known as Feher's QPSK.

Occupied bandwidth — The bandwidth the upper and lower limits of which have mean powers equal to a specified pecentage (0.5% unless otherwise specified) of the total mean power.

OOK — On-off keying.

OQPSK — Offset (or staggered) quadrature phase-shift keying.

PAM — Pulse-amplitude modulation.

PCM — Pulse-code modulation.

PDM — Pulse-duration modulation. (Same as PWM.)

PFM — Pulse-frequency modulation.

PLL — Phase-locked loop.

PM — Phase modulation.

PSK — Phase-shift keying.

Pulse spreading — Elongation of a received pulse as a result of multipath propagation.

PWM — Pulse-width modulation. (Same as PDM.)

QAM — Quadrature amplitude modulation. (Same as APK.)

QASK — Quadrature amplitude-shift keying.

Q-M/PSK — Quadrature suppressed-carrier AM added to quadrature carrier and hard limited.

QPPM — Quantized pulse-position modulation.

QPRS — Quadrature-partial-response modulation system.

QPSK — Quadrature phase-shift keying (quadraphase).

Quadbit — A group of 4 bits. In phase modulation each quadbit translates to one of 16 specific phases or transitions.)

Quantize — To divide the range of variables into a finite number of specific values or steps.

Raised cosine (RC) — A method of shaping a keying waveform to reduce bandwidth by following the cosine function with the negative peak of the wave raised (or biased) to 0-V dc.

Reduced carrier (RC) — A method of transmission wherein the carrier is reduced by a predetermined number of decibels.

SCPC — Single channel per carrier.

Sideband — The group of frequencies above or below the main carrier frequency which contain the product of modulation.

S/N — Signal-to-noise ratio. (Expressed in dB.)

Space — The absence of a signal. (See mark.)

Splatter — Generation of unnecessary sidebands by exceeding 100% modulation in an AM system or by flat-topping.

SQAM — Staggered (or offset) quadrature amplitude modulation.

SQPSK — Staggered (or offset) quaternary phase-shift keying.

SSB — Single sideband.

SSBSC — Single-sideband suppressed carrier.

SSMA — Spread-spectrum multiple access.

SSTV — Slow-scan television.

Suppressed carrier (SC) — A method of transmission wherein the carrier is greatly attenuated.

TADI — Time assignment digital interpolation. (A technique for sending bursts of data during pauses in speech.)

TDM — Time-division multiplexing.

TDMA — Time-division multiple access.

TFM — Tamed frequency modulation.

TH — Time hopping. (A technique used in spread spectrum.)

Unit interval — The duration of the shortest signal element. (Expressed in seconds, the unit interval is the reciprocal of the signaling speed in bauds.)

USB — Upper sideband.

VF — Voice frequencies.

VFCT — Voice-frequency carrier telegraphy. (A technique for combining multiple telegraph channels in the frequency spectrum normally occupied by one voice channel.)

VSB — Vestigial sideband.

VSDM — Variable-slope delta modulation.

WBFM — Wideband frequency modulation.

WGN — White Gaussian noise.

WPM — Words per minute. (A unit of speed in telegraph systems.)

8PSK — Eight-ary or octal phase-shift keying.

---

which would interfere with nearby channels.

The pre-WARC '79 emission symbol for FSK was F1. The new designation still begins with F1 but includes a third letter (and the fourth and fifth letters if you care to add them). For conventional 170-Hz FSK 45.45-baud Baudot RTTY, the short-form emission symbol is F1B, the B standing for telegraphy — for automatic reception. The complete emission designation would be 249HF1BBN. The optional fourth letter means a two-condition code with elements of the same number and duration without error correction; the fifth letter N is for no multiplexing. For Baudot RTTY, the necessary bandwidth is calculated

$$B_n = B + 2DK \qquad \text{(Eq. 13)}$$

where

B_n = necessary bandwidth
B = modulation rate in bauds
D = peak deviation (half the frequency shift) in hertz
K = 1.2 (typically)

The complete emission designation for AMTOR FSK RTTY is 304HF1BCN. The bandwidth 304H is calculated using Eq. 13, above; in this case, B is 100 bauds. The optional fourth and fifth letters are: C for two-condition code with elements of the same number and duration with error correction and N for no multiplexing.

For both Baudot and AMTOR RTTY, the third letter B, meaning telegraphy for automatic reception, implies *direct-printing* telegraphy. In CCIR lingo, direct-printing describes the traditional teletypewriter-to-teletypewriter circuit where what you type on one end comes out on a typewriter-like printer on the other end. However, when you depart from direct-printing telegraphy and enter the world of computers to generate FSK, you are entering the "twilight zone" of CCIR terminology. Thus, if you are transmitting ASCII from teletypewriter to teletypewriter, you are transmitting F1B emission. If, on the other hand, you are transmitting *data* (in this case loosely meaning that a computer generates it and it is probably received by a computer-like device at the other end), then the emission symbol is F1D. We suggest using the third symbol B for all narrow-band direct-printing telegraphy whether sent by a teletypewriter or a computer.

Audio-Frequency-Shift-Keyed Telegraphy

Audio-frequency-shift keying (AFSK) is more popular than direct FSK of the main carrier for both direct-printing telegraphy and data transmission in the Amateur Radio Service. AFSK into an SSBSC transmitter is common for both the HF bands and via satellite. For Baudot and AMTOR, the emission designations of AFSK-generated signals would be 249HJ2BBN and 304HJ2BCN, respectively, using the same parameters as in the FSK examples given earlier.

Above 50 MHz, radioteletype transmissions are often sent by frequency-shift keying an audio tone and feeding the AFSK tone into the microphone input of an FM transmitter. Necessary bandwidth calculations get a bit tricky because of the double modulation. First, let us determine the baseband bandwidth of the frequency-shifted tones. That is done by using Eq. 13, which will yield bandwidths of 249 Hz for 45-baud Baudot, as calculated above. If the AFSK tones used are 2125 and 2295 Hz, you can determine the highest modulation frequency by finding the center frequency and adding half the bandwidth to that. As 2125 and 2295 Hz differ by 170 Hz, the center frequency is 2210 Hz. Half the 249-Hz bandwidth is 125 Hz (124.5 Hz rounded up), so 2210 + 125 = 2335. Thus the highest modulation frequency is 2335 Hz. To complete the bandwidth calculation, we then use the general necessary bandwidth formula for the main FM carrier

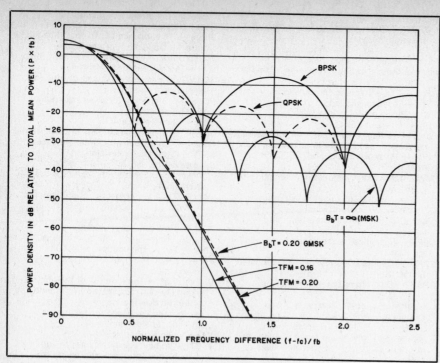

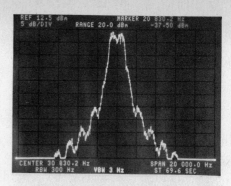

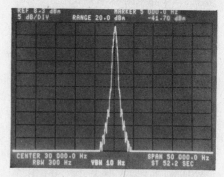

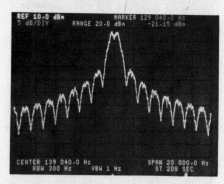

Fig. 13 — Power spectral density of BPSK, QPSK, MSK, GMSK and TFM relative to the mean power of the total emission. (After CCIR Rep. 767-1 and Steber, March 1984)

B_bT = normalized bandwidth of premodulation Gaussian band-pass filter (GMSK)
f_c = center frequency
f_b = bandwidth (Hz)
P = spectral density (W/Hz)

Fig. 14 — Wide spectral displays of 1200-baud data transmissions. At top is a display of amplitude (Y axis) vs. frequency (X axis) of a 1200-baud signal produced by a Bell 202 modulator, which has a frequency shift of 1000 Hz. At center is a similar display of the output of a minimum-shift-keying (MSK) modulator, which has a shift of 600 Hz. At bottom is a display of a binary-phase-shift-keying (BPSK) modulator. Note that the top and bottom displays have spans of 20 kHz, while the center display span is 50 kHz.

$$B_n = 2M + 2DK \qquad \text{(Eq. 10)}$$

where

B_n = necessary bandwidth in hertz
M = maximum modulation frequency in hertz
D = peak deviation in hertz
K = 1

Thus, if a frequency deviation of 5 kHz is used, $B_n = (2 \times 2335) + (2 \times 5000 \times 1) = 4670 + 10,000 = 14,670$ Hz.

DATA, TELEMETRY AND TELECOMMAND

Data, telemetry and telecommand are identified by D as the third symbol of the emission designator. Included in this category are computer-to-computer data transfers, packet radio, telemetry from satellites, and remote control of models and other objects. Excluded from this category is narrow-band direct-printing telegraphy (conventional radioteletype).

Efficient Data Communications

A measure of data communication efficiency is the number of hertz or frequency shift (in an FSK system) per baud (Hz/Bd) — this is nothing more than FM modulation index. Narrow-band direct-printing RTTY transmissions use low signaling rates and small frequency shifts in absolute terms (Hz). Typically an RTTY signal uses a frequency shift of 170 hertz and signals at a rate of 45 bauds. This amounts to 170 Hz/45 Bd, which reduces to a modulation index of 3.78 (Hz/Bd). If we were to use the same modulation index for a 1200-baud data transmission, the frequency shift would be $3.78 \times 1200 = 4536$ Hz. Further, applying Eq. 13, the necessary bandwidth would be

$$B_n = B + 2DK = 1200 + (4536 \times 1.2)$$
$$= 1200 + 5443 = 6643 \text{ Hz.}$$

Clearly, it would not be good spectrum efficiency to use more than 6 kHz of bandwidth when existing technology permits transmission of a 1200-baud data signal in a much smaller bandwidth. Through this analysis, we have established that conventional RTTY, while not occupying much spectrum in absolute terms, is spectrally inefficient with respect to its signaling rate. Also, it can be concluded that data transmission at higher speeds must use spectrally efficient modulation techniques in order to find room in crowded Amateur Radio bands.

The FCC recognized this matter of spectrum efficiency in the rules, which state: "the ... frequency shift (difference between the frequency for the "mark" signal and that for the "space" signal), as appropriate, shall not exceed 1000 Hz. When these emissions are used on frequencies above 50 MHz, the frequency shift, in hertz, shall not exceed the sending speed, in baud(s), of the transmission, or 1000 Hz, whichever is greater." The frequency shift (in Hz) not exceeding the sending speed (in Bd) means a limit of 1 Hz/Bd (a modulation index of 1) for signaling rates of 1000 Bd or more.

For 1200-Bd packet radio, it is common to use surplus Bell 202 compatible modems. As Bell 202 tones (1200 and 2200 Hz) shift 1000 Hz, the result is 1000/1200 or a modulation index of 0.83 (Hz/Bd). On HF packet radio, there is activity using Bell 103 tones (originate tones of 1070 and 1270 Hz or answer tones of 2025 and 2225 Hz) at a signaling speed of 300 bauds. This amounts to a modulation index of (200/300) = 0.67 (Hz/Bd). A frequency shift of 600 Hz at a signaling speed of 1200 bauds would give a modulation index of

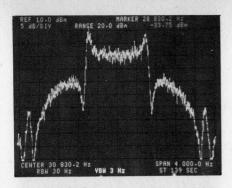

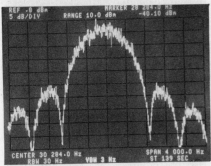

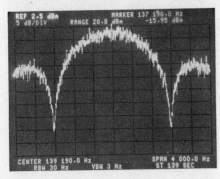

Fig. 15 — Close-in spectral displays of same types of modulation as in Fig. 14. Bell 202 FSK (1000-Hz shift) at top; MSK (600-Hz shift) at center; and BPSK (180° phase shift) at bottom.

$(600/1200) = 0.5$ and would be called *minimum-shift keying* (MSK) or, less frequently, *fast frequency-shift keying* (FFSK).

Eq. 13 is considered a good rule of thumb for modulation indexes greater than 1. For modulation indexes below 1, Eq. 13 is inaccurate and yields excessively large necessary bandwidth figures. In such cases, we recommend either a literature search for

the specific type of modulation or actual measurement with a spectrum analyzer to determine occupied bandwidth. For your convenience, Fig. 13 summarizes the spectral densities of a number of digital modulation schemes.

Multilevel Modulation

It is now commonplace to use more than one modulation system to generate a radio signal. As an example, you can use a digital stream to frequency modulate an audio tone that frequency-modulates the transmitter carrier. This technique is used for transmission of RTTY at VHF. In some texts, this complex modulation technique is abbreviated FM-FM (or FM/FM). The first term is the modulation applied to the subcarrier, while the second is the modulation of the main carrier.

Here are some examples of double modulation systems used in the Amateur Radio Service:

AFSK-SSB — A digital stream (Baudot, AMTOR, packet radio) modulating an audio-frequency-shift-keyed tone that modulates a single-sideband transmitter. (Emission designation: J2B or J2D.)

FM-SSB — A signal frequency modulating a subcarrier that modulates a single-sideband transmitter. (This is commonly used for SSTV and FAX on high-frequency transmitters; emission designations would be J3F and J3C, respectively.)

FM-VSB — A signal frequency modulating a subcarrier that modulates a vestigial-sideband transmitter. (Example: The aural subcarrier injected into a UHF fast-scan TV transmitter.)

FURTHER READING ON MODULATION, MULTIPLEXING AND DEMODULATION

Sandro Belline, et al., "Noncoherent Detection of Tamed Frequency Modulation," *IEEE Trans. on Communications,* March 1984.

CCIR, *Radio Regulations,* International Radio Consultative Committee, International Telecommunication Union, Geneva, 1982.

CCIR, *Recommendations and Reports of the CCIR, 1982,* "Efficient Digital Satellite Communications," Volume VIII, Rep. 767-1, International Radio Consultative Committee, International Telecommunication Union, Geneva, 1982.

Kah-Seng Chung, "Generalized Tamed Frequency Modulation and its Application for Mobile Radio Communications," *IEEE Trans.* on Vehicular Technology, August 1984.

Dennis Connors, KD2S, "Bibliography on Minimum-Shift Keying," *QEX,* March 1983.

John P. Costas, K2EN, "Synchronous Communications," *IRE Proc.,* December 1956.

FCC, "Part 2 — Frequency Allocations and Radio Treaty Matters; General Rules and Regulations, and Part 97 — Amateur Radio Service," *Rules and Regulations.*

Kamilo Feher, *Digital Communications Microwave Applications,* Prentice-Hall, Inc., Englewood Cliffs, NJ, 1981.

Kamilo Feher, *Digital Communications Satellite/Earth Station Engineering,* Prentice-Hall, Inc., Englewood Cliffs, NJ, 1983.

Roger L. Freeman, W1HIA, *Telecommunication Transmission Handbook,* John Wiley & Sons, New York, 1975.

Roger L. Freeman, *Reference Manual for Telecommunications Engineering,* John Wiley & Sons, New York, 1985.

Robert M. Gagliardi, *Satellite Communications,* Lifetime Learning Publications, Belmont, California, 1984.

NTIA, *Manual of Regulations and Procedures for Federal Radio Frequency Management,* Washington, DC, revised 1984.

Frank F. E. Owen, *PCM and Digital Transmission Systems,* McGraw-Hill Book Company, New York, 1982.

E. W. Pappenfus, et al., *Single Sideband Principles and Circuits,* McGraw-Hill Publications Company, New York, 1964.

Mischa Schwarz, *Information Transmission, Modulation, and Noise,* McGraw-Hill Publishing Company, New York, 1980.

James J. Spilker, Jr., *Digital Communications by Satellite,* Prentice-Hall, Inc., Englewood Cliffs, NJ, 1977.

William D. Stanley, *Electronic Communications Systems,* Reston Publishing Company, Inc., Reston, Virginia, 1982.

J. Mark Steber, "Understanding PSK Demodulation Techniques," *Microwaves & RF,* March and April 1984.

Chapter 10

Radio Frequency Oscillators and Synthesizers

RF oscillators and synthesizers are fundamental building blocks used in radio equipment of all kinds. The first stage in almost all simple transmitters is an oscillator. This stage determines on what particular frequency the transmitter will operate. In a similar manner, the various oscillators used in a superheterodyne communications receiver determine the frequency the receiver is tuned to. The simplest transmitter is a crystal-controlled oscillator. A full-feature transceiver may use a microprocessor-controlled frequency synthesizer to determine transmit and receive frequencies.

An oscillator is actually a special type of amplifier. The transistor or other active device used in the oscillator amplifies the signal applied to its input just as any amplifier circuit does. The basic difference is that a portion of the output from the device is fed back to the input of the amplifier so that it aids the applied signal. This is called positive feedback or regeneration. An example of this is the howling that sometimes occurs in a public-address system. Negative feedback or degeneration tends to oppose the input signal.

Crystal Oscillators

A crystal-controlled oscillator uses a piece of quartz that has been ground to a particular thickness, length and width. For the most part, the thickness determines the frequency at which the crystal oscillates, irrespective of the stray capacitance in the immediate circuit of the crystal. Stray capacitance has some effect on the operating frequency, but overall the effect is minor. The maximum power available from such an oscillator is restricted by the heat (caused by circulating RF current) the crystal can withstand before fracturing. The circulating current is determined by the amount of feedback required to ensure excitation. Excessive heating of the crystal causes frequency drift. The extent of the drift is related to the manner in which the quartz crystal is cut. It is for these reasons that the amount of feedback used should be held to only that level which provides quick oscillator starting and reliable opera-

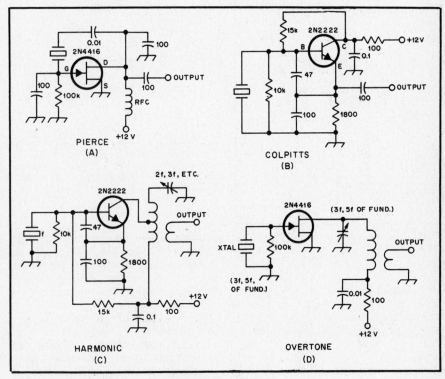

Fig 1—Four types of popular solid-state crystal oscillators.

tion under load. The power necessary to excite a successive stage properly can be built up inexpensively by low-level amplifiers.

Although a wide variety of crystal-controlled oscillator circuits can provide acceptable performance in amateur equipment, only a few of the popular ones will be highlighted here. In the circuits offered as illustrations, the feedback must be sufficient to assure quick starting of the oscillator. Some circuits function well without the addition of external feedback components (internal capacitance within the transistor or tube being adequate). Other circuits need external feedback capacitors. Poor-quality (sluggish) crystals generally require larger amounts of feedback to provide operation comparable to that of lively crystals. Some surplus crystals are sluggish, as can be the case with those

which have been reground or etched for a different operating frequency. Therefore, some experimentation with feedback voltage may be necessary when optimizing a given circuit. As a rule of thumb it is necessary to use one-fourth of the oscillator output power as feedback power to ensure oscillation.

The active element in an oscillator can be a tube, transistor or IC. Some common examples of solid-state oscillators are shown in Fig 1. A Pierce oscillator that employs a JFET is illustrated at A. A bipolar transistor is used at B to form a Colpitts oscillator. The example in Fig 1C shows how to extract the harmonic of a crystal by tuning the collector circuit to the desired harmonic. Unless a band-pass filter is used after the tuned circuit, various harmonics of the crystal frequency will appear

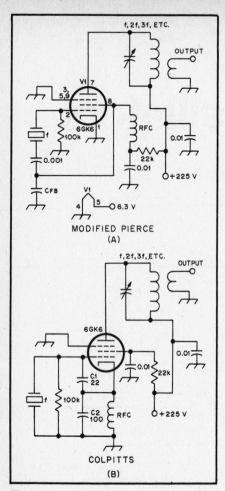

Fig 2—Two common tube types of crystal oscillators.

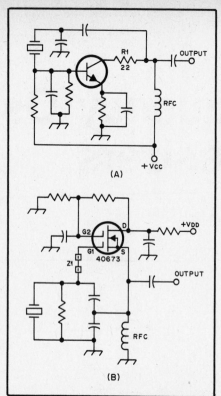

Fig 3—Two methods for suppressing VHF and UHF parasitic oscillations. R1 at A damps the parasitics and Z1 at B (ferrite beads) serves that purpose.

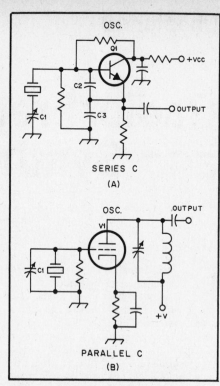

Fig 4—The crystal-oscillator operating frequency can be shifted slightly by means of trimmer capacitors as shown at A and B. A series hookup (A) is used with transistors to help compensate for the relatively high input capacitance of the transistor.

in the output. Therefore, if good spectral purity is desired it is necessary to use a double-tuned collector tank to obtain a band-pass characteristic, or to employ the tank circuit shown and follow it with a harmonic filter.

An overtone oscillator is depicted in Fig 1 at D. The crystal oscillates near an odd multiple of the fundamental cut—usually the third or fifth harmonic. In this example the drain tank is tuned approximately to the desired overtone. Oscillation will begin when the tank is tuned slightly above the overtone frequency. A high-Q tuned circuit is necessary.

Vacuum-tube crystal oscillators are presented in Fig 2. A modified Pierce oscillator is shown at A. In this case the screen grid of V1 functions as the plate of a triode tube. Feedback is between the screen and control grids. C_{fb} will range from 10 to 100 pF for oscillators operating from 1.8 to 20 MHz. At lower frequencies the feedback capacitor may require a higher value.

A Colpitts style of tube oscillator is illustrated in Fig 2B. The feedback occurs between the grid and cathode by means of a capacitive divider (C1 and C2). The plate tank can be tuned to the crystal frequency or its harmonics. In the interest of good

oscillator stability, the supply voltage to the circuits of Figs 1 and 2 should be regulated. This is especially significant in the case of harmonic or overtone oscillators where small amounts of drift are multiplied by the chosen harmonic factor.

The usual cause of erratic oscillation, or no oscillation at all, is excessive loading on the oscillator output by the succeeding stage of circuit, insufficient feedback or a sluggish crystal. Concerning the latter, a crystal not ground to a uniform thickness and feathered carefully around the edges may be difficult to make oscillate. Attempts by amateurs inexperienced in grinding their own crystals may lead to this condition.

Some crystal oscillators develop unwanted VHF self-oscillation (parasitics) even though the circuits may be functioning normally otherwise. The result will be a VHF waveform when the RF voltage is viewed by means of an oscilloscope. Parasitics can cause TVI and specific problems elsewhere in the circuit in which the oscillator is used. Two simple methods for preventing VHF parasitics are shown in Fig 3. The technique at A calls for the insertion of a low-value resistance (R1) in the collector lead as close to the transistor body as possible. Typical resistance values are 10 to 27 ohms. The damping action of the resistor inhibits VHF oscillation. An alternative to the use of resistance for swamping VHF oscillation is illustrated at B in Fig 3. One or two high-mu miniature

ferrite beads (μ_i = 950) are placed near the transistor body in the lead to gate 1. The beads can be used in the drain lead when a tuned circuit or RF choke is used in that part of the circuit. Ferrite beads can be used in the base or collector lead of the circuit of Fig 3A, rather than employing R1. Similarly, R1 can be used at gate 1 of the oscillator.

Adjusting Frequency

It is necessary in some crystal oscillator applications to ensure spot accuracy of the operating frequency. Various reactances are present in most oscillator circuits, causing the operating frequency to differ somewhat from that marked on the crystal case. Addition of a trimmer capacitor will permit "rubbering" the crystal to a specified frequency within its range. This procedure is sometimes referred to as "netting" a crystal.

Fig 4 shows two circuits in which a trimmer capacitor might be used to compensate for differences in the operating frequency of the oscillator. At A the series capacitor (C1) is connected between the low side of the crystal and ground. The series hookup is used to help offset the high input capacitance of the oscillator. The input capacitance consists of the series value of feedback capacitors C2 and C3 plus the input capacitance (C_{in}) of Q1. Conversely, the input capacitance of the circuit at B in Fig 4 is quite low because a triode tube is employed. In this kind of circuit the trimmer capacitor is connected in parallel. The choice between series and parallel trimming will depend on the active device used and the amount of input capacitance present. This rule applies to tube oscillators as well as those which use transistors.

Fig 5 shows four practical oscillator circuits of this type. C1 is included for adjusting the crystal to the frequency for which it has been ground. In circuits where considerable shunt capacitance is present (Fig 5A and C) the trimmer is usually connected in series with the crystal. When there is minimal parallel capacitance (approximately 6 pF in the circuits at B and D, Fig 5) the netting trimmer can be placed in parallel with the crystal. Whether a series or parallel trimmer is used will depend also on the type of crystal used (load capacitance and other factors).

Feedback capacitance (C_{fb}) for the circuit at B in Fig 5 must be found experimentally. Generally, a value of 100 pF will suffice for operation from 3.5 to 20 MHz. As the operating frequency is lowered it may require additional capacitance. The drain RF choke should be self-resonant below the operating frequency.

A third-overtone crystal is illustrated at Fig 5D. Satisfactory operation can be had by inserting the crystal as shown by the dashed lines. This method is especially useful when low-activity crystals are used in the overtone circuit. However, C1 will have little effect if the crystal is connected from gate to drain, as shown. C2-L1 is adjusted slightly above the desired overtone frequency to ensure fast starting of the oscillator. The circuits shown in Fig 5 can be used with dual-gate MOSFETs also, assuming that gate 2 is biased with a positive 3 to 4 volts.

Crystal Switching

Although several crystals for a single oscillator can be selected by mechanical means, a switch must be contained in the RF path. This can impose severe restrictions on the layout of a piece of equipment. Furthermore, mechanical switches normally require that they be operated from the front panel of the transmitter or receiver. That restriction complicates the remote operation of such a unit. Also, the switch leads can introduce unwanted reactances in the crystal circuit. A better technique is illustrated in Fig 6, where D1 through D3—high-speed silicon switching diodes—are used to select one of three or more crystals from some remote point. As operating voltage is applied to one of the diodes by means of S1, it is forward biased into "hard" conduction, thereby completing the circuit between the crystal trimmer and ground. Some schemes actually call for reverse-biasing the unused diodes when they are not activated. This ensures almost complete cutoff, which may not be easy to achieve in the circuit shown because

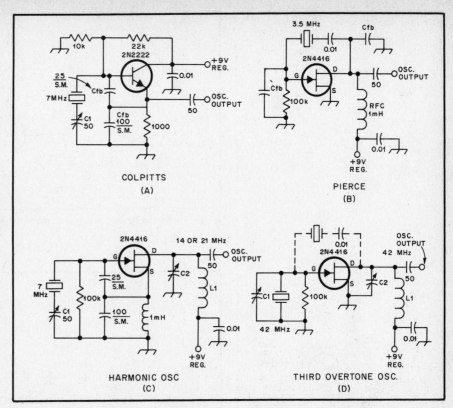

Fig 5—Practical examples of crystal-controlled oscillators that can be frequency trimmed.

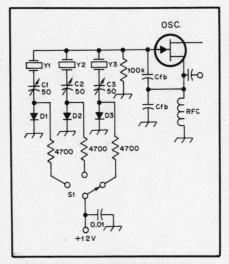

Fig 6—Method for changing crystals by means of diode switching.

of the existing RF voltage on the diode anodes.

VXO Circuits

Variable-frequency crystal oscillators (VXOs) are useful in place of conventional crystal oscillators when it is necessary to "rubber" the crystal frequency a few kilohertz. AT-cut crystals in HC-6/U type holders seem to provide the greatest frequency change when used in a VXO. To obtain maximum frequency shift it is vital to reduce stray circuit capacitance to the

smallest possible amount. This calls for low-capacitance switches, low minimum-capacitance variable capacitors, and the avoidance of crystal sockets. The crystals should be spaced well away from nearby metal surfaces and circuit components to further reduce capacitance effects. The higher the crystal fundamental frequency, the greater the available frequency swing. For example, a 3.5-MHz crystal might be moved a total of 3 kHz, whereas a 7-MHz crystal could be shifted 10 kHz. Although shifts as great as 50 kHz at 7 MHz may be possible, they should be avoided. Under those conditions, the high-stability traits of a crystal oscillator are lost.

Fig 7 contains a simple VXO circuit at A. By adjusting X_L the operator can shift the crystal frequency. The range will start at the frequency for which the crystal is cut and move lower. In both circuits, D1 is included to stabilize the FET bias and reduce the transistor junction capacitance during the peak of the positive RF-voltage swing. It acts as a clamp, thereby limiting the transistor g_m at peak-voltage periods. This lowers the junction capacitance and provides greater VXO swing. D1 also reduces harmonic output from the VXO by restricting the nonlinear change in transistor junction capacitance—a contributing factor to the generation of harmonic currents. Clamp diodes are used for the same purpose in conventional FET VFOs. The circuit of Fig 7A will provide a swing of approximately 5 kHz at 7 MHz.

An improved type of VXO is presented

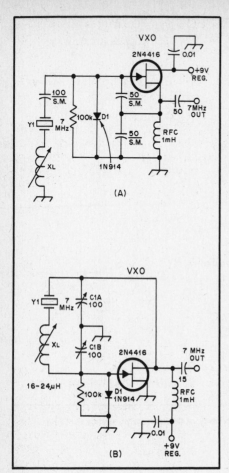

Fig 7—Circuits for two types of VXOs.

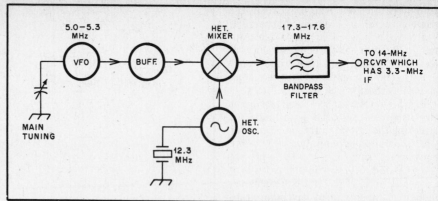

Fig 8—Technique for heterodyne frequency generation in a receiver.

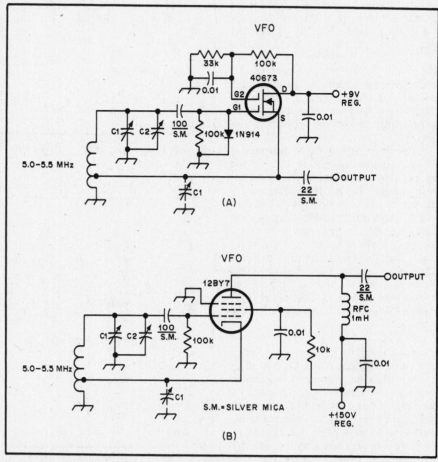

Fig 9—Examples of Hartley VFOs. A MOSFET version is shown at A; a vacuum-tube type at B.

at B in Fig 7. Depending on the exact characteristics of the crystal used at Y1, swings as great as 15 kHz are possible at 7 MHz. X_L is set initially for a reactance value that will provide the maximum possible frequency shift when C1 is tuned through its range. The frequency shift should be only that which corresponds to true VXO control, even though greater range can be had after the circuit ceases to be a highly stable one. X_L is not adjusted again. A buffer stage should be used after either of the VXO circuits to prevent frequency pulling during load changes.

VXOs of this general type are useful in portable transmitters and receivers when full band coverage is sacrificed in exchange for stability and simplicity. Output from VXOs can be multiplied several times to provide LO energy for VHF and UHF receivers and transmitters. When that is done it is possible to realize 100 kHz or more of frequency change at 144 MHz.

VFO Circuits

Variable-frequency oscillators are similar to VXOs. The essential difference is that greater frequency coverage is possible, and no crystals are used. The practical upper frequency limits for good CW or SSB stability is between 7 and 10 MHz. For operation at higher frequencies, it is better to employ a heterodyne-type VFO. This calls for a VFO operating at, say, 5 MHz. The VFO output is heterodyned in a mixer with energy from a crystal-controlled oscillator. This provides a resultant or difference frequency at the desired LO-chain output frequency. A block diagram is given at Fig 8 to illustrate the concept as it might appear in a receiver. The same scheme could be used in a transmitter. The heterodyne oscillator has a crystal for each amateur band accommodated by the receiver. The crystals and appropriate band-pass filters are switched by a panel-mounted control. The band-pass filter (Fig 8) is desirable to prevent 5- and 12.3-MHz energy from reaching the receiver mixer. A doubly balanced mixer is recommended if minimum unwanted energy is desired at the mixer output.

Some typical VFOs are shown in Figs. 9 and 10. A vacuum-tube Hartley is compared to a similar one that utilizes a dual-gate MOSFET in Fig 9. The capacitor shown in dashed lines (C1) can be used in that part of the circuit rather than at the

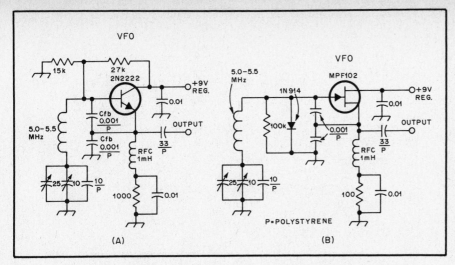

Fig 10—Colpitts VFOs. A bipolar transistor type is seen at A, while an FET version is given at B.

low end of the tank coil if greater bandspread is desired. C1 is the main-tuning capacitor and C2 is the padder for calibrating the oscillator. The coil tap is approximately 25 percent of the total number of turns for proper feedback.

A comparison is drawn in Fig 10 between a bipolar transistor and JFET version of a series-tuned Colpitts oscillator (sometimes called a "series-tuned Clapp"). These VFO circuits can be scaled to other operating frequencies by using the values shown to determine the reactances of the capacitors. This information will enable the designer to select approximate values in picofarads for other frequencies.

Fundamental Stability Considerations

There are some specific measures that must be taken when designing a VFO. The form on which the VFO coil is wound is of special significance with regard to stability. Ideally, the use of metallic core material should be avoided. Therefore, powdered iron, brass, copper and ferrite slugs, or toroid cores for that matter, are not recommended when high stability is required. The reason is that the properties of such core materials are affected by changes in temperature and can cause a dramatic shift in the value of inductance that might not occur if an air-core coil was used. Furthermore, some styles of slug-tuned inductors are subject to mechanical instability in the presence of vibration. This can cause severe frequency jumping and a need for frequent recalibration of the VFO readout.

Regardless of the format selected for the VFO coil, the finished product should be coated with two or three applications of polystyrene cement (Q dope) or similar low-loss dopant. This will keep the coil turns secured in a permanent position—an aid to mechanical stability.

The VFO coil should be mounted well away from nearby conducting objects (cabinet walls, shield cans, and so on) to prevent frequency shifts which are likely to occur if the chassis or cabinet are stressed during routine handling or mobile operation. Movement of the chassis, cabinet walls and other nearby conductive objects can (if the coil is close by) change the coil inductance. Furthermore, the proximity effects of the conductive objects present an undefined value of capacitance between the coil and these objects. Changes in spacing will alter that capacitance, causing frequency shifts of an abrupt nature.

It follows that all forms of mechanical stability are of paramount importance if the VFO is to be of "solid" design. Thus, the trimmer or padder capacitors used in the circuit should be capable of remaining at their preset values despite temperature changes and vibration. For this reason it is not wise to use ceramic or mica trimmers. Air-dielectric variable capacitors of the PC-board-mount subminiature type are recommended.

The main tuning element (capacitor or permeability tuner) needs to have substantial rigidity; it should be mounted in a secure manner. Variable capacitors used as main-tuning elements should be of the double-bearing variety. They should rotate easily (minimum torque) to minimize mechanical stress of the VFO assembly when they are adjusted. Variable capacitors with plated brass plates are preferred over those with aluminum plates. Compared with brass, aluminum is more subject to physical changes in the presence of temperature variations. The VFO tuning-capacitor rotor must be grounded at both ends as a preventive measure against instability. Some designers have found that a 1/8- to 1/4-inch-thick piece of aluminum or steel plate serves as an excellent base for the

VFO assembly. It greatly reduces instability that can be caused by stress on the main chassis of the equipment. The VFO module can be installed on shock mounts to enhance stability during mobile operation.

Electrical Stability

Apart from the mechanical considerations just discussed, the relative quality of the components used in a VFO circuit is of great importance. Fig 11 contains three illustrations of basic solid-state tunable oscillators suggested for amateur applications. The numbered components have a direct bearing on the short- and long-term stability of the VFO. That is, the type of component used at each specified circuit point must be selected with stability foremost in mind. The fixed-value capacitors, except for the drain bypass, should be temperature-stable types. NP0 ceramic capacitors are recommended for frequencies up to approximately 10 MHz. A second choice is the silver-mica capacitor (dipped or plain versions). Silver micas tend to have some unusual drift characteristics when subjected to changes in ambient temperature. Some increase in value while others decrease. Still others are relatively stable.

It is often necessary to experiment with several units of a given capacitance value in an effort to select a group of capacitors that are suitably temperature stable. The same is not true of polystyrene capacitors. When they are used with the commonly available slug-tuned coils, the temperature characteristics of the polystyrene capacitors and those of the inductor tend to cancel each other. This results in excellent frequency stability. If toriodal cores are used, they should be made of SF-type powdered iron material (Amidon mix 6). This material has a low temperature coefficient and when used with NP0-type ceramic capacitors produces very low drift oscillators. Ordinary disc-ceramic capacitors are unsuitable for use in stable VFOs. Those with specified temperature characteristics (N750 and similar) are useful, however, in compensating for drift.

The circuit of Fig 11A is capable of very stable operation if polystyrene capacitors are used at C3 through C8. A test model for 1.8 to 2.0 MHz exhibited only 1 Hz of drift from a cold start to a period some two hours later. Ambient temperature was $25°$ C. Q1 can be any high-g_m JFET for use at VHF or UHF. Capacitors C1 through C4 are used in parallel as means of distributing the RF current among them.

A single fixed-value capacitor in that part of the circuit would tend to change value because of the RF heating within it. Therefore, a distinct advantage exists when several capacitors can be used in parallel at such points in a VFO circuit. The same concept is generally true of C5, C6 and C7. In the interest of stability, C5 should be the smallest value that will permit reliable oscillation. Feedback capacitors C6 and C7 are typically the same value and have an X_c of roughly 60 Ω. Therefore, a suitable value for a 1.9-MHz VFO would be 1500 pF.

C8 of Fig 11A should be the smallest capacitance value practical with respect to ample oscillator drive to the succeeding stage. The smaller the value of C8, the less the chance for oscillator pulling during load changes. D1 is a gate-clamping diode for controlling the bias of the FET. Basically, it limits the positive swing of the sine wave. This action restricts the change in Q1 junction capacitance to minimize harmonic generation and changes in the amount of C associated with L1.

The reactance of RFC can be on the order of 10 kΩ. The drain bypass, C9, should have a maximum X_c of 10 ohms to ensure effective bypassing at the operating frequency. Ideally, an X_c of 1 ohm would be used (0.1 μF at 1.5 MHz). D2 is used to provide 9.1 volts, regulated, at the drain of Q1. Lower operating voltages aid stability through reduced RF-current heating, but at the expense of reduced oscillator output.

A Hartley oscillator is shown in Fig 11B. This circuit offers good stability also, and is one of the better circuits to use when the tank is parallel tuned. The tap on L1 is usually between 10 and 25 percent of the total coil turns, tapped above the grounded end. This ensures adequate feedback for reliable oscillation. The higher the FET g_m, the lower the feedback needed. Only that amount of feedback necessary to provide oscillation should be used: Excessive feedback will cause instability and prohibitive RF heating of the components. Most of the rules for the circuit of Fig 11A apply to the oscillator in Fig 11B.

Parallel tuning of the kind used in Fig 11B and C are suitable for use below 6 MHz, although the circuit at B can be used successfully into the VHF region. However, the Colpitts oscillators of A and C in Fig 11 have large amounts of shunt capacitance caused by C6 and C7 of A, and C5 and C6 of C. The smaller the coupling capacitor between L1 and the gate, the less pronounced is this effect. The net result is a relatively small value of inductance at L1, especially with respect to Fig 11C, which lowers the tank impedance and may prevent oscillation (high C-to-L ratio). The series-tuned circuit of Fig 11A solves the shunt-C problem nicely by requiring considerably greater inductance at L1 than would be acceptable in the circuit of Fig 11C. The circuit at A resembles the popular "Clapp" circuit of the early 1950s.

A suitable transistor for Q1 of Fig 11C is an RCA 40673. The Texas Instruments 3N211 is also ideal, as it has an extremely high g_m—approximately 30,000 microsiemens (μS). Dual-gate MOSFETs are suitable for the circuits of Fig 11A and B if biased as shown at C. Also, they can be used as single-gate FETs by simply connecting gates 1 and 2 together. No

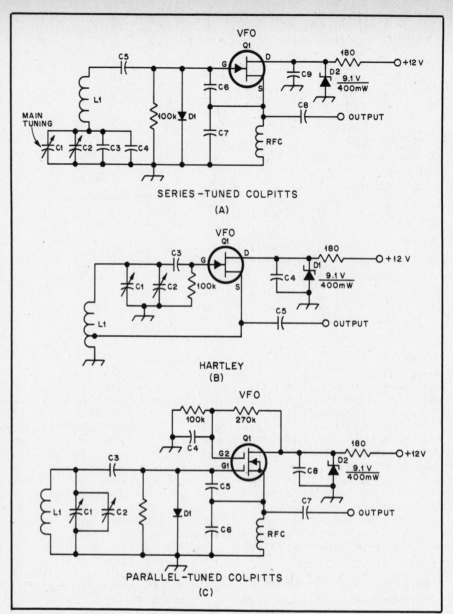

Fig 11—Three common types of VFOs for use in receivers and transmitters.

external bias is required if this is done. Gate 2 of Q1 (Fig 11C) should by bypassed with a low-reactance capacitor (C4), as is the rule for the drain bypassing of all three examples given in Fig 11.

Bipolar transistors are satisfactory for use in the three VFOs just discussed. The principal disadvantage of bipolars in these circuits is the low base impedance and higher device input capacitance. Most FETs exhibit an input C of approximately 5 pF, but many bipolar transistors have a substantially higher capacitance, which tends to complicate a VFO design for the higher operating frequencies. The UHF small-signal transistors, such as the 2N5179, are best suited to the circuits under discussion.

Load Isolation for VFOs

Load changes after the oscillator have a pronounced effect on the operating frequency. Therefore, it is imperative to provide some form of buffering (isolating stage

or stages) between the oscillator and the circuit to which it will be interfaced. The net effect of load changes, however minor, is a change in reactance which causes phase shifts. The latter affects the operating frequency to a considerable degree. Therefore, the more isolating stages which follow the oscillator (up to a practical number, of course), the less likelihood of load shifts being reflected back to the oscillator.

Buffer stages can perform double duty by affording a measure of RF amplification, as needed. But, care must be taken to avoid introducing narrow-band networks in the buffer/amplifier chain if considerable frequency range is planned, for example, 5.0 to 5.5 MHz. If suitable broadband characteristics are not inherent in the design, the oscillator-chain output will not be constant across the desired tuning range. This could seriously affect the conversion gain and dynamic range of a receiver mixer, or lower the output of a transmitter in some parts of a given band.

Fig 12A illustrates a typical RC coupled VFO buffer with broadband response. C1 is selected for minimum coupling to the oscillator, consistent with adequate drive to Q1. Q1 and Q2 should have high f_T and medium beta to ensure a slight RF-voltage gain. Devices such as the 2N2222A and 2N5179 are suggested.

Q2 of Fig 12A operates as an emitter-follower. The RF-voltage output will be approximately 0.9 of that supplied to the base. In a typical VFO chain, using an oscillator such as the one in Fig 11A, this buffer strip will deliver approximately 1 volt P-P across the 470-Ω emitter resistor of Q2.

A better circuit is offered in Fig 12B. Q1 is a JFET that has a high input impedance (1 MΩ or greater) by virtue of the FET-device characteristic. This minimizes loading of the oscillator. RFC1 is chosen to resonate broadly with the stray circuit capacitance (roughly 10 pF) at the mid-range frequency of the LO chain. Although this does not introduce significant selectivity, it does provide a rising characteristic in the RF-voltage level at the source of Q1.

Q2 functions as a fed-back amplifier with shunt feedback and source degeneration. The feedback stabilizes the amplifier and makes it broadband. The drain tank is designed as a pi network with a loaded Q of 3. The transformation ratio is on the order of 20:1 (1000-ohm drain to 50-ohm load). R1 is placed across L1 to further broaden the network response. The 50-ohm output level is recommended in the interest of immunity to load changes: The higher the output impedance of a buffer chain the greater the chance for oscillator pulling with load changes. P-P output across C3 should be on the order of 3 volts when using the oscillator of Fig 11A.

Fig 13 illustrates a composite VFO that

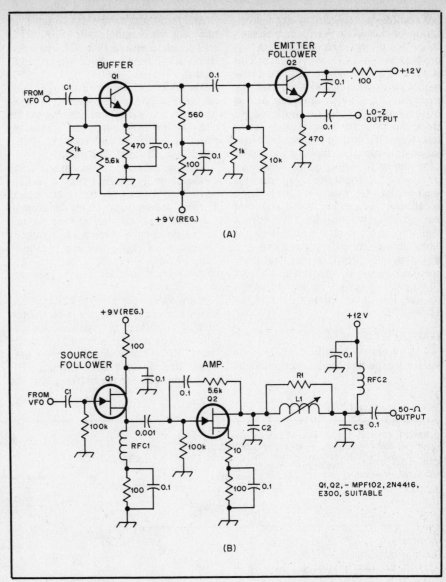

(A)

(B)

Fig 12—VFO buffer and buffer/amplifier sections, which provide isolation between the oscillator and the VFO-chain load. The circuit at B is recommended for most applications.

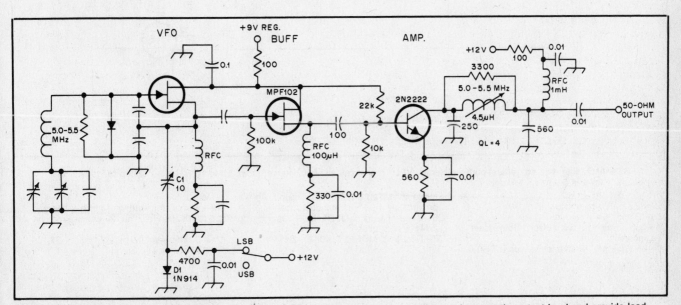

Fig 13—Suggested circuit for a stable series-tuned Colpitts VFO. Buffering follows the oscillator to increase the output level and provide load isolation.

has a buffer stage followed by an amplifier. D1 can be included to provide the necessary frequency offset when switching the receiver from upper to lower sideband. This is necessary to eliminate the need to readjust the calibration dial when changing sidebands. C1 is adjusted for the desired offset amount. The pi-network output from the amplifier stage is designed to transform 500 ohms to 50 ohms. The low-impedance output is desirable in the interest of minimum frequency pulling from load changes. A 3300 ohm swamping resistor is used across the pi-network inductor to broadband the tuned circuit and to prevent any tendency toward instability when a high-impedance load is attached to the circuit. Long-term drift measurements with this type of circuit at the frequency specified indicated a maximum shift of 60 Hz over a three-hour period. Output was measured at approximately 1 volt RMS across 50 ohms.

Other VFO Criteria

Apart from the stability considerations just treated, purity of emissions from VFOs is vital to most designers. It is prudent to minimize the harmonic output of a VFO chain and to ensure that VHF parasitic energy is not being generated within the LO system.

The pi-network output circuit of Fig 12B helps reduce harmonics because it is a low-pass network. Additional filtering can be added at the VFO-chain output by inserting a half-wave filter with a loaded Q of 1 (X_L and X_C = 50 for a 50-Ω line).

VHF parasitics are not uncommon in the oscillator or its buffer stages, especially when high f_T transistors are employed. The best preventive measures are keeping the signal leads as short as possible and adding parasitic suppressors as required. The parasitic energy can be seen as a superimposed sine wave riding on the VFO output waveform when a high-frequency scope is used.

A low-value resistor (10 to 22 ohms) can be placed directly at the gate or base of the oscillator transistor to stop parasitic oscillations. Alternatively, one or two ferrite beads (950 μ_i) can be slipped over the gate or base lead to resolve the problem. If VHF oscillations occur in the buffer stages, the same preventive measures can be taken.

VFO noise should be minimized as much as possible. A high-Q oscillator tank will normally limit the noise bandwidth adequately. Resistances placed in the signal path will often cause circuit noise. Therefore, it is best to avoid the temptation to control the RF excitation to a given LO stage by inserting a series resistor. The better method is to use small-value coupling capacitors.

A Hartley VFO circuit is shown in Fig 14. Operating frequency range with the components shown is 5.0 to 5.5 MHz, but the circuit will function well from 2.5 to 10 MHz. The design guidelines offered in the previous section will be useful for altering the circuit for other frequency ranges. More information about this VFO circuit can be found in the High-Performance Communications Receiver project in Chapter 30, and another practical VFO circuit is shown in the Band-Imaging Receiver project in Chapter 30. Other VFO circuits and design information appear in the ARRL publication *Solid-State Design for the Radio Amateur*.

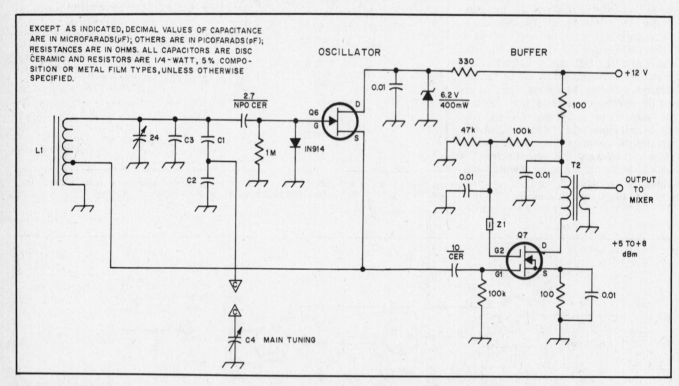

Fig 14—Schematic diagram of a general-purpose Hartley VFO. The circuit will function well from 2.5 to 10 MHz; the tuning range with the components shown is 5.0 to 5.5 MHz.

C1—100-pF NP0 ceramic.
C2—82-pF NP0 ceramic.
C3—126-pF NP0 ceramic.
C4—365-pF air variable capacitor (broadcast replacement type).
L1—35 turns no. 28 enamel wire on a T-50-6

toroid core. Tap 8 turns from ground. Approximately 4.9-μH total inductance.
Q6—General-purpose JFET, MPF-102, 2N4416 or similar.
Q7—Dual-gate MOSFET, 40673, 3N140 or

similar.
T2—Ferrite transformer, 18-turn primary, 5-turn secondary, no. 28 enamel wire on an FT-37-43 core.
Z1—Ferrite bead on gate 2 lead of Q7. FB43-101 or similar.

Radio Frequency Synthesizers

Frequency synthesis has become a feature in almost every type of amateur equipment. The first amateur application was in 2-meter FM transceivers. The frequency accuracy and stability required by repeaters rules out VFOs; they tend to fall short in stability and frequency display accuracy. When 2-meter FM activity became heavy, the cost of crystals to fully use a 2-meter transceiver became prohibitive, and synthesizers began appearing.

The frequency synthesizer can be viewed as a simple system or, as it is often called in the electrical engineering world, a "black box." This black box is designed to provide an output signal at some convenient level on some desired frequency. The desired output signal frequency is an input to the black box. That input may come from electrical logic signals or mechanical switches.

The range of frequencies available from a synthesizer could span from a few percent to several decades depending on the application. The smallest frequency change that can be accommodated by a frequency synthesizer is called the resolution. This can vary from a few hertz to several megahertz again depending on the application. Usually the resolution is a power of 10, that is 1, 10 or 100 Hz, etc.

Another basic characteristic of a synthesizer is the lock-up time. This is the amount of time from the instant that a new frequency is called for until the output has assumed that frequency. Lock-up time can vary from less than one cycle of the current output frequency to several seconds or even minutes for highly accurate synthesizers and frequency standards. This is a critical characteristic when it is necessary to change frequencies very quickly—such as when the same synthesizer is used to control both the receiver and the transmitter frequency. When changing from receive to transmit, the synthesizer must provide the proper output frequency quickly.

A synthesizer should produce a single output frequency without distortion or other spectral impurity. Harmonics and other far-removed signals are not usually a problem in most systems; they can be easily removed with a low-pass filter. Spectral components close to the desired signal are not easily removed by simple filtering.

One type of close-in spectral impurity is known as phase noise. This is residual random variation of the phase difference between the synthesizer output and a perfect sine wave of the same frequency. Phase noise tends to reduce the ability of a receiver to separate nearby signals in spite of a good IF filter because it tends to "spread" the local oscillator signal. This phenomenon is known as reciprocal mix-

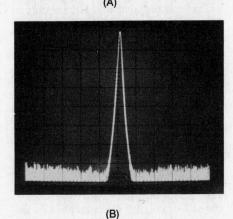

(A)

(B)

Fig 15—Spectrum analyzer photographs of a VCO. The VCO is disconnected from the phase-locked loop at A. The reduction of noise sidebands provided by the loop is shown at B.

ing. A receiver that has good selectivity, sensitivity and dynamic range can be destroyed by a local oscillator with excessive phase noise. Likewise, a transmitted signal will appear "broad" if the frequency control system has excessive phase noise.

Phase noise is generated within the oscillators of the synthesizer. A crystal oscillator has little phase-noise. Synthesizers that use only crystal oscillators usually have excellent phase-noise characteristics. Some synthesizers use voltage-controlled oscillators (VCOs). VCOs have considerably higher amounts of phase noise. VCO phase noise can be reduced in a properly designed synthesizer. Fig 15 shows a spectrum analyzer display of a typical VCO before and after reduction of phase noise. Just how much phase noise can be tolerated depends on the type of operation intended. Narrow-band modes require better phase-noise characteristics than wider modes. Phase noise is measured as the power contained in a specific bandwidth at a frequency removed from the carrier. For example, synthesizer phase noise could be specified as −90 dB from

the carrier in a 10-Hz bandwidth 1 kHz from the carrier. A move detailed discussion of phase noise appears later in this chapter.

Frequency accuracy can be an important requirement in some synthesizer applications. Accuracy and resolution should not be confused. Just because a synthesizer displays a lot of digits and has very narrow frequency resolution does not mean that the actual frequency is the one displayed to the last digit. Frequency inaccuracy is the difference between actual and displayed frequency.

An attribute of frequency synthesizers is stability. Generally, frequency synthesizers are in some fashion or another locked to the frequency of a crystal oscillator. In that case, the stability of the synthesizer is the stability of the crystal oscillator. There are some designs where more than one crystal oscillator or other non-crystal-controlled oscillators are used. In these cases, the stability of the synthesizer is a function of the oscillators involved.

Another important characteristic is the amount of signal radiated from the synthesizer. Most systems have numerous internal signals of various frequencies. It would be nearly impossible for all of these frequencies to never cause any harmful interference. Therefore, careful shielding is important in the design of any frequency synthesizer. Typically all programming, power and other control lines are well bypassed and shielded. The output of the synthesizer is usually band-pass filtered to eliminate the escape of any internal signals.

DIRECT NUMERICAL SYNTHESIS

A common technique for producing sine waves from digital sources is to construct the sine wave from a series of samples, as shown in Fig 16. In this example, a sine wave is constructed from 16 samples. The wave is divided into 16 equally-spaced samples, resulting in an approximation of the actual sine wave. The raggedness of the sine wave reconstructed from the samples can be smoothed by low-pass filtering; the result is shown at C in Fig 16.

The output frequency of the digitally generated sine wave is the rate at which the samples are generated divided by the number of samples for a complete sine wave. For example, for the sine wave in Fig 16, 16 samples are generated per cycle at a 16 kHz rate. This means we are producing a 1-kHz sine wave.

An interesting twist on this method of sine wave generation is to supply only every second, third, fourth (or Nth sample where N is any integer). This is done by making the next sample the current sample plus N. For example, for N equal to 4, if the first

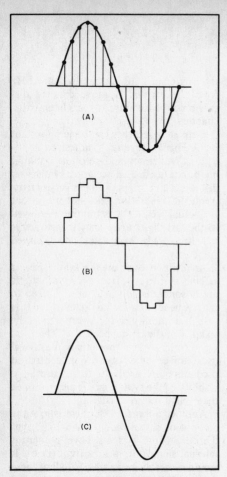

Fig 16—A sine wave can be created digitally using discrete samples. If the sine wave shown at A is sampled at each point on the curve, the samples can be used to recreate the wave; this is shown at B. The roughness in the recreated wave can be smoothed out by low-pass filtering, resulting in the smooth curve shown at C.

sample is number zero, the next would be 4, then 8, then 12 and so on. When N is greater than one, fewer samples are used to produce a complete sine wave than the total number of samples available; the number of samples used per cycle is equal to the total number of available samples divided by N. The output frequency is still the rate at which samples are generated

divided by the number of samples per complete sine wave. To use the previous example, for N equal to 4, only four of the available 16 samples are used per cycle, but these samples are produced at one-fourth the 16 kHz sampling rate, again resulting in a 1-kHz sine wave.

The actual number of samples used to generate the sine wave is the total number of available samples divided by N. The actual output frequency is the rate at which the samples are generated divided by the number of samples used for each output cycle:

$$F_{out} = f_{clock} / P \times N$$

where P is the total number of samples available.

If f_{clock} / P is selected to be a convenient number, such as 1 Hz, the output frequency is directly proportional to N in steps of 1 Hz. The lower limit to the number of samples is two; the familiar Nyquist sampling limit states that the sample rate must be at least twice the maximum frequency to be sampled.

Fig 17 shows a block diagram of a direct numerical synthesizer. The values of the samples are stored in a read-only memory (ROM) which drives a D/A converter to create an analog output equal to the value of the sample. A latch holds the address of the ROM until the next sample is desired. The value of the following sample is determined by adding a fixed number to the current ROM address. This is done with an adder using the programmed input frequency under control of the clock.

The synthesized sine-wave output is not perfect; there are several factors that contribute to the distortion of the sine wave. First, the number of bits employed in the A/D converter will limit just how accurate the samples used for the sine function can be. This means that the samples can be close but not always precise. The fact that the sine wave is sampled at finite intervals will cause distortion in spite of the fact that most of the distortion can be removed with low-pass filtering.

Another type of distortion arises because the number of samples used to generate the

sine function does not have to be an integer. In the previous example 16 samples were used with an N of 4. This resulted in exactly 4 samples per output cycle. If N is 5, there will be 3.2 samples per output cycle. This means that the points on the sine wave where the sampling occurs are at different points on the following cycle. Using the example of N equal to 5, if the first sample is at 0 degrees, the second occurs at 75 degrees, then 150, 225 and 300 degrees. The first sample of the next output cycle will be at 15 degrees. This sort of thing will continue until the fifth cycle when the sampling is back to 0, 75, 150 and so on again. This produces some low-level sidebands that have complex relationships to the programmed frequency.

This method of frequency synthesis is used mostly for the generation of audio frequency signals and RF signals in the lower HF region. The method offers excellent resolution without high cost. Using high-quality components, the level of the complex sidebands can be kept low enough for many applications.

When this technique is used at HF and higher, very fast logic, ROMs and D/A converters are required. The sidebands become more difficult to control and the simplicity of the system is overshadowed by the relative cost of the components. Improved components are making the technique more attractive, however.

Indirect Synthesis: Phase-Locked Loop

The two synthesizers described earlier are examples of "direct" synthesis. Their output frequency is derived directly from one or more oscillators. Another method of frequency synthesis, perhaps the most widely used method, is called the "indirect" method. This method uses the phase-locked loop (PLL). The description "indirect" refers to the fact that the output frequency is generated by stabilizing a voltage-controlled oscillator (VCO).

A simple phase-locked loop synthesizer block diagram is shown in Fig 18. Five elements make up the simple loop; a VCO, a programmable divider, a phase detector, a loop filter and a reference frequency source.

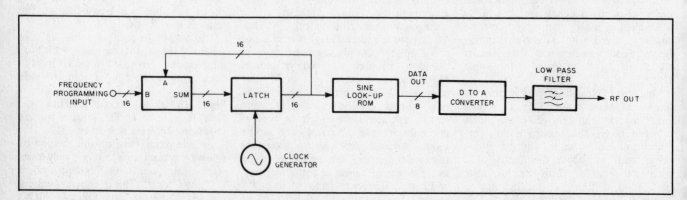

Fig 17—A block diagram of a direct numerical synthesizer. See text for details.

The purpose of the PLL synthesizer is to take a signal from a VCO, divide the frequency by an integer, and compare that result to a precise and stable reference frequency in a phase detector. The phase detector produces an electrical output that indicates positive or negative phase difference. The phase detector output is fed back to the VCO through a filter. The loop works to adjust the output of the phase detector to zero automatically. That means the output of the programmable divider is precisely on the same frequency as the reference. This can be described mathematically as:

$$F = NFr$$

where F is the output frequency, N is the value of the programmable divider, which is always an integer, and Fr is the reference frequency.

Depending on the reference frequency, phase comparison can take place at great speed. Frequency corrections can be made at such a rate that small and rapid deviations in frequency such as phase noise can be corrected. A control voltage correction can only be made once every cycle of the reference frequency. Therefore, the correction at the output of the phase detector is not a continuous function but rather a sampled function. The sample frequency is equal to the reference frequency. This signal is filtered to remove some of the energy at the reference frequency. Not all of the reference frequency energy can be removed; what remains is applied to the VCO control voltage. Any variation of the control voltage frequency modulates the VCO. This modulation causes sidebands to be generated. These sidebands appear not only at frequencies removed from the carrier by the reference frequency, but also at harmonics of the reference frequency. The amount of reference energy that is applied to the VCO is a function of the loop filter and the phase detector.

The loop filter has an effect not only on the reference sidebands of the PLL synthesizer, but also on the time and nature of the lock up. Fig 19 shows the three basic paths for PLL frequency change. In the first path the loop changes from the original to the new frequency directly without overshoot and in minimum time. This is called critically damped and is the usual path chosen in synthesizer design, especially if the frequency must change between transmit and receive. Curve B in Fig 19 shows the frequency rapidly changing to the desired frequency, overshooting the mark and continuing to oscillate around the desired frequency before becoming stable. This characteristic is called underdamped.

It may appear that this transient characteristic has no value in the design of synthesizers. A certain amount of underdamping can make the loop filter design easier for realizing low sideband levels in the system; this approach may be used

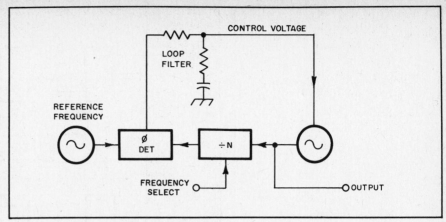

Fig 18—A simple single-loop PLL synthesizer.

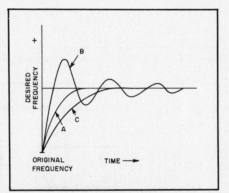

Fig 19—Three transient frequency-change characteristics: critically damped at A, underdamped at B, overdamped at C.

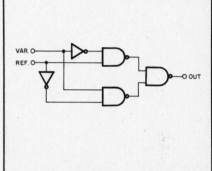

Fig 20—Exclusive-OR type phase detector.

when the synthesizer is not required to change frequency rapidly. Curve C in Fig 19 shows the overdamped case. In this case the frequency changes from the old frequency to the new frequency without overshoot, but the time required is considerably longer than in the critically damped case.

Several varieties of phase detectors are suitable for use in PLL systems. One very simple type of phase detector is the exclusive-OR circuit shown in Fig 20. The signals presented to the exclusive-OR phase detector must have exactly 50 percent duty cycle—a condition not difficult to generate when using IC dividers.

Fig 21 shows the input and output waveforms found in the exclusive-OR phase detector. When the phase angle between the two inputs is 90 degrees (Fig 21B), the average value of the output is 50 percent. When the phase angle is less, the average value of the output changes depending on which input is leading. Furthermore, the output is a pulse train with a frequency equal to twice the input frequency. It is important when using the simple exclusive-OR phase detector that the loop filter have sufficient rejection of the phase detector output frequency.

The exclusive-OR phase detector has one serious drawback that limits its use in certain types of frequency synthesizers: If

there is any harmonic relationship between the two inputs to the phase detector, the output will allow the loop to lock up on an incorrect output frequency. This can be used to advantage for PLL systems where harmonics are desired. In a PLL where only one frequency is desired, however, this possibility must be eliminated. There would be no problem if the frequency range of the VCO were limited to a ratio of less than 2 to 1. This limitation would preclude the VCO from ever locking on a harmonic because the VCO cannot generate the fundamental and the harmonic of any frequency.

A phase detector specifically designed for phase-locked loops is shown in Fig 22. This phase detector is not susceptible to harmonically related inputs; for that reason it is more accurately called a phase/frequency detector. The phase/frequency detector also has a three-state output. If the phase or frequency difference is such that a positive output is required, the upper (P-channel) FET conducts supplying a current path from V_{DD} to the output. If the input phase relationship is such that a negative output is required, the lower (N-channel) FET conducts and provides a current path from the output to V_{SS}. If the phase angle is very close neither FET conducts and the output floats. Because the output has the capability of two "on"

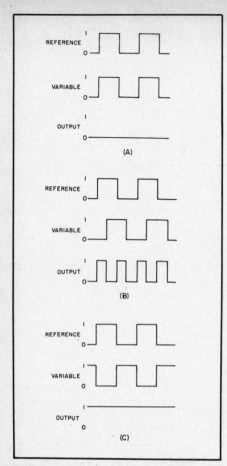

Fig 21—Waveforms associated with an exclusive-OR phase detector. Variable and reference are in phase at A. Variable leads reference by 90° at B. At C, variable and reference are 180° out of phase.

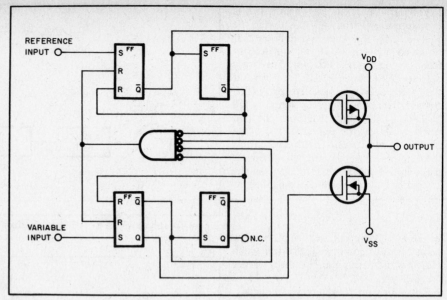

Fig 22—A phase/frequency detector specifically designed for phase-locked loop synthesizers.

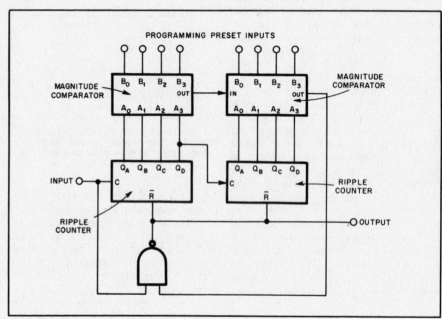

Fig 23—A programmable divider based on a ripple counter.

states as well as an "off" state, the amount of reference ripple in the output is much lower than with the exclusive-OR phase detector. The only significant disadvantage of this phase detector is the definition of where the phase angle between the two inputs is small enough for both output transistors to be off. In operation the loop will "bounce" between the two on states and cause a low frequency noise that can be objectionable. The usual solution to this problem is to inject a small amount of current at the phase detector to force the phase detector out of this "dead zone."

Programmable Dividers

Programmable dividers are generally constructed from IC counters. The counter can be binary or decade depending on the source of programming information. Binary counters are suited to microprocessor programming where there is a binary output. For simple systems, where information is accepted from mechanical switches, binary-coded-decimal programming (BCD) is required. Therefore, both binary and BCD programmable dividers are used in synthesizer systems.

Fig 23 shows a programmable divider using ripple BCD counters as the dividers.

A magnitude comparator accepts information from the programming switches and compares this input to the state of the counter. When the counter reaches the selected state, the "equals" output of the magnitude comparator provides a reset pulse to the counter. This type of programmable divider is frequency limited to only a few megahertz.

A higher frequency alternative is to use a synchronous counter with internal zero detect as shown in Fig 24. This programmable divider is preset to a number and then counts down to a value of one and then is preset to the same number. Normally, IC counters are provided with internal zero detection for the purpose of cascading counters. For use as a programmable divider, however, the detection of

the count of one is required. This can be done simply by detecting the zero state for all of the cascaded counters except the least significant digit that detects the count of one.

The advantage of this type of programmable divider is a significant increase in operating speed. First, the propagation delay within the counter is much less than that of a ripple counter. Second, only one state of the counter must be decoded, a value of one, and the magnitude decoder required for that task is simpler and therefore, faster than a full-function magnitude comparator. A typical TTL programmable divider using synchronous down-counters can operate reliably at 15 MHz.

There are special programmable dividers using CMOS technology with special look-

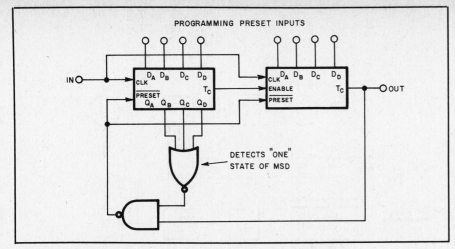

Fig 24—A programmable divider based on a synchronous down counter.

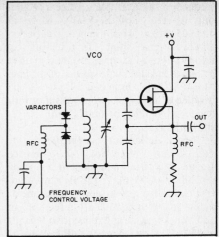

Fig 25—A Colpitts oscillator with varactor tuning suitable for use as a VCO.

ahead decoding that can further increase the speed to a point where CMOS logic speed can approach that of TTL. These specialized circuits are typically used in synthesizer LSI chips.

Voltage-Controlled Oscillators

The VCO used for a PLL can be practically any one of the standard oscillator circuits with a varactor diode in the tuned circuit to allow voltage tuning. Fig 25 shows a Colpitts oscillator that has a varactor diode. The VCO must not only provide the output from the synthesizer, but also must provide drive for the programmable divider. Usually several buffer amplifier stages are used to isolate the VCO from the output of the synthesizer as well as the programmable divider. Changes in the VCO load can cause the generation of unwanted phase noise.

The noise level of a synthesizer will depend to a great extent on the design of the VCO. One method of reducing VCO phase noise is to use two varactor diodes back to back, as shown in Fig 25. This prevents the RF voltage at the VCO tuned circuit from being rectified in the varactor diode and changing the tuning voltage.

SYNTHESIZERS FOR HIGH-FREQUENCY COMMUNICATIONS SYSTEMS

The synthesizer required for high frequency communications systems presents several design problems. First, the operating frequency span is large, from 3 to 30 MHz for general coverage. Also, the frequency resolution must be fine; and 100 Hz is considered the absolute maximum for SSB or CW operation and then only if a clarifier or receiver-incremental tuning (RIT) control is provided for SSB operation. For RTTY operation or SSB without the use of a clarifier control even better resolution is required, typically on the order of 25 Hz or less. Because 25-Hz steps are not easily displayed by the frequency dis-

play, some designs use 10-Hz resolution, which is sufficiently small to accommodate any type of operation. Narrow resolution and wide frequency coverage are two characteristics that are not complementary. That is, it would be easy to design a phase-locked loop with a narrow frequency range and narrow resolution. It is difficult to design a synthesizer that has a broad frequency range and narrow frequency resolution, while maintaining good noise performance and quick lock-up time.

Typical amateur HF operation involves continuous tuning for signals, as opposed to VHF FM where operation takes place on well-defined frequencies. Therefore, the time it takes the synthesizer to lock on the proper frequency is an important parameter and must be taken into account in the design of a high-frequency synthesizer. All of these requirements taken together dictate a system more complicated than a single-loop synthesizer.

The modern trend in receiver (and transceiver) design is up conversion, a scheme in which the first IF is higher than the highest received frequency. For coverage to 30 MHz, the minimum IF must be above 30 MHz. That places the local oscillator at 30.1 to 60 MHz for a 0.1- to 30-MHz input. In practice the first IF is 45 MHz, or higher, which dictates first local-oscillator coverage from 45 to 75 MHz—a span of less than an octave.

A ratio of two to one between the highest and the lowest frequency generated, or an octave, is about the maximum range that can be covered reliably by a single VCO. Greater frequency ratios are obtained by using several VCOs or band-switching the VCO with transistors or PIN diodes. It is especially important to minimize the VCO frequency range in the design of low-noise synthesizers. As previously mentioned, the exclusive-OR phase detector requires an input range of less than two to one. Essentially, a simple synthesizer cannot operate with a maximum to minimum frequency ratio of more than two to one.

For a frequency resolution of 50 Hz (the maximum acceptable for tuning SSB) a conventional, single-loop synthesizer requires a 50-Hz reference. The loop bandwidth must be considerably less than 50 Hz, in order to reduce the level of reference sidebands. Lock up time in this case would be too slow for manual tuning of amateur bands. Any frequency excursion of more than a few kilohertz would find the loop dragging behind the frequency display. To counteract this, multiple loop synthesizers are used. The output frequency of these synthesizers is a function of two phase-locked loops.

Two-Loop Synthesizers

Fig 26 shows an example of a two-loop synthesizer for use with an up-converting receiver or transceiver with a 45-MHz first IF. In this example, a coarse-tune PLL provides an output at 50 to 80 MHz. This loop has a reference frequency of 100 kHz and a fast lock-up time. Because of the high reference frequency, this loop can provide a very clean output signal. The coarse-tune VCO output is mixed with the main VCO output. The output of the mixer is tuned to 5 MHz. For each frequency of the coarse-tune synthesizer, there are two possible frequencies of the main VCO that will provide a 5-MHz output; the sum and the difference.

To prevent the VCO from operating at the wrong frequency, a coarse-tune voltage is applied to the main VCO. This voltage sets the VCO to approximately the correct frequency. The mixer output is phase compared to the fine-tune PLL synthesizer output. The output of this phase detector is fed through a loop amplifier to the main VCO. Thus, the frequency difference between the coarse tune loop and the main VCO is precisely equal to the output frequency of the fine-tune loop.

To avoid the lock-up time problems, the fine-tune loop does not operate in the 5-MHz range, but at 50 times that frequency, or 250 MHz. The required output frequency is 5.0000 MHz to 4.90005 MHz. This frequency variation is the reverse from the programming information; as the programming data is increased in value the frequency of the fine-tune VCO is decreased. This can be handled either by a microprocessor or with logic within the fine-tune synthesizer as shown in Fig 27.

Because the loop operates at 50 times the desired output frequency, the reference frequency is 50 times the desired frequency resolution—2500 Hz for 50-Hz resolution. It is not difficult to achieve a lock time of a few ms with a 2500-Hz reference. In addition, the noise and reference sidebands at 250 MHz are reduced 33 dB because of the frequency division.

Practically any division factor can be used for the fine-tune loop. For example, a standard ECL divide-by-40 chip could be used. The VCO would operate in the 200-MHz region; the reference would be 2000 Hz. Lock-up time would be slightly longer, but the programmable divider ratios would remain exactly the same as in the previous example.

General-coverage receivers and transceivers increase the need for frequency synthesis in HF equipment. A typical amateur-band-only transceiver of a few years ago used a crystal oscillator in conjunction with a conventional VFO. The VFO operated in the megahertz region and the crystal oscillator operated in the low-VHF region. This scheme provides reasonable noise and stability performance. When more than a few frequency bands are covered, however, the cost of the required crystals becomes prohibitive. For a general-coverage receiver with the range divided into 500-kHz bands, approximately 59 crystals are required; that is totally unacceptable.

Partial Synthesis

One solution to the crystal problem is the partially synthesized transceiver (see Fig 28). Partially synthesized means that the operating frequency is set by both a synthesizer and a free-running VFO. This synthesizer consists of a single PLL. The 500-kHz reference frequency ensures that phase noise will be sufficiently low for SSB, RTTY and CW operation; the VFO ensures that the resolution is nearly infinite.

A VCO (actually three separate switchable oscillators) is the first local oscillator. It operates 48.05 MHz above the input frequency. The output of this VCO is mixed with a 45.05- to 45.55-MHz signal to produce a 3.0- to 33.0-MHz output. This output feeds the programmable divider. The 45.05- to 45.55-MHz signal is generated by mixing a 40-MHz oscillator with a 5.05- to 5.55-MHz VFO.

Output from the programmable divider feeds a phase detector, which also receives a 500-kHz reference frequency. The programmable divider works in a range

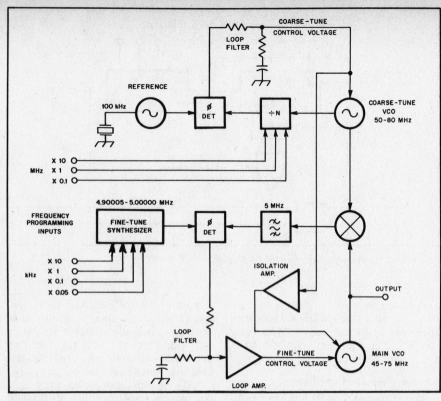

Fig 26—A two-loop synthesizer used in a high-frequency receiver or transceiver with a 45-MHz IF.

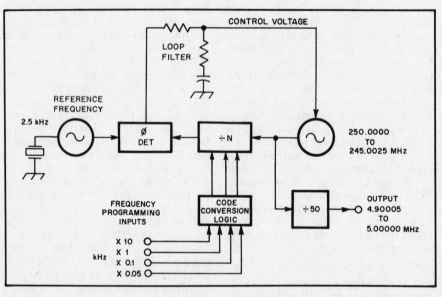

Fig 27—Logic circuit for use with the fine-tune synthesizer within the system shown in Fig 26.

from 6 to 65; that corresponds to frequency coverage from 0 to 29.5 MHz. The programmable divider sets the operating frequency of the transceiver in 500-kHz steps while the VFO adds from 0 to 500 kHz to the basic synthesizer frequency.

There are some significant disadvantages to this type of system. First, the frequency cannot be fully programmed externally. Because of the mechanically tuned VFO, only the 500-kHz bands can be programmed. Second, the frequency stability is not better than that of the VFO. Because the VFO is operating in the low MHz region, however, the stability will be acceptable for all but the most demanding operation. Finally, the frequency readout is a combination of digital inputs to the synthesizer and the dial of a VFO. This can be improved by using a frequency counter to display the frequency electronically.

VHF SYNTHESIZER APPLICATIONS

The requirements of a frequency synthesizer for VHF operation are generally less stringent than those for HF. Normally the synthesizer covers only one amateur band. That means a span that is a small

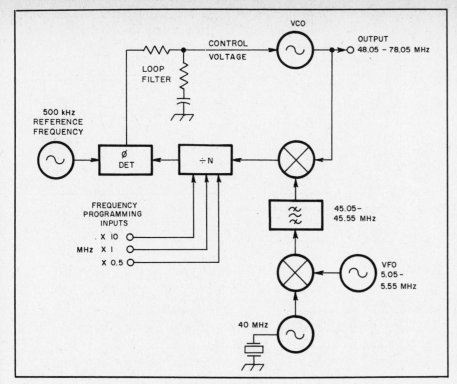

Fig 28—Frequency control for a partially synthesized transceiver for high-frequency operation.

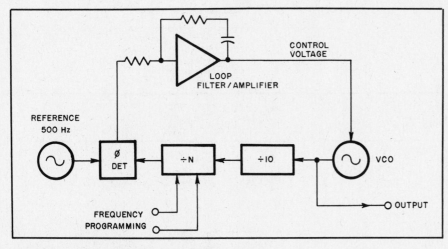

Fig 29—VHF synthesizer using a fixed-modulus prescaler.

percentage of the operating frequency. Resolution or channel separation is typically 5 or 10 kHz for FM operation and perhaps 100 Hz for SSB operation. A simple synthesizer would appear to suit the VHF transceiver. The problem is that conventional logic cannot operate at frequencies of 150 MHz. Even emitter-coupled logic (ECL) with propagation delays as short as a few nanoseconds cannot be used to construct programmable dividers to operate at 150 MHz.

One method of constructing a VHF PLL synthesizer is shown in Fig 29. In this example, the VCO output is divided by 10 in an ECL prescaler. ECL logic is fast enough to operate as a fixed divider at 150 MHz, but fails as a conventional programmable divider.

The major disadvantage of this syn-

thesizer design method is that the loop reference frequency must be reduced by the same factor as the fixed-modulus divider. As an example, if a synthesizer for a 2-meter transceiver were designed with 5-kHz channel resolution using a simple phase-locked loop, a 5-kHz reference frequency could be used. If a fixed-modulus prescaler with a division of 10 were placed between the VCO and the programmable divider, the reference frequency would have to be reduced to 500 Hz.

It is important that the highest reference frequency possible be used in a PLL synthesizer for best noise performance and lowest reference sidebands. In this example, the output of the prescaler can be used to drive relatively low-speed logic in the programmable divider. However, even 15 MHz requires fast logic in a program-

mable divider. Fixed-modulus prescaler divisions of more than 10 are typically required with some low-power logic families. This means a less than 500-Hz reference frequency in the 2-meter synthesizer example. That would greatly compromise the synthesizer performance.

Dual-Modulus Programmable Dividers

One technique used in VHF synthesizers with conventional logic without reducing the reference to an unacceptable frequency is the dual-modulus programmable divider. The heart of the divider is the dual-modulus prescaler shown in Fig 30. This ECL chip can divide by two factors that differ by 1. The dual-modulus prescaler shown in Fig 30 can divide by either 10 or 11, depending on the state of the modulus control input. The 10/11 ratio has important applications in BCD-programmed synthesizers. They are available in a variety of frequency capabilities up to more than 500 MHz. Some standard ratios are: 5/6, 8/9, 15/16, 20/21, 32/33, 40/41 and 80/81. Other ratios can be created with one of these chips and some external logic.

The dual-modulus programmable divider consists of three parts: the main counter, the auxiliary counter and the dual-modulus prescaler (see Fig 31). Both the main and auxiliary counters are presettable; they are often down counters. At the beginning of the divider cycle both the main and auxiliary counters are preset to the respective programmed value. Both counters are clocked together from the output of the dual-modulus prescaler. If the auxiliary counter is preset to a value other than zero, the dual-modulus prescaler starts counting by the larger of its two divisors $(N + 1)$ at the beginning of the divider cycle. When the auxiliary counter reaches zero, the mode of the dual-modulus prescaler is changed to the smaller of the two divisors (N). The main counter continues to count toward zero, while the auxiliary counter stops counting. When the main counter reaches zero both the main and auxiliary counters are preset, and the divider cycle starts anew.

The total number of VCO cycles required for a complete divider cycle can be calculated in the following manner. Assume the main counter is preset to the value, M, and the auxiliary counter is set to the value of A. Because the dual-modulus prescaler divides by $N + 1$ until the auxiliary counter reaches zero, $(N + 1)A$ VCO cycles are required to reach this point. The mode is then switched and the dual-modulus prescaler divides by N until the main counter reaches zero. The main counter requires $(M - A)$ counts or $N(M - A)$ VCO cycles to reach zero. The total cycles of the VCO required for the complete divider cycle is:

$$(N + 1) A + N (M - A) = NM + A$$

From the equation, it can be seen that M has more effect on the division value of the dual-modulus divider than A. In the case of a 10/11 prescaler, the effect of M is 10 times that of A. That ratio is an important advantage of the 10/11 prescaler. In BCD-programmed synthesizers such as those programmed directly from switches, some of the switches can program the main counter while others are used to program the auxiliary counter.

There are some restrictions on the possible values of M and A. Because the auxiliary counter must reach zero before or concurrent with the main counter, the value of M must be equal to or greater than A. There is also a limit on the minimum division factor. That limit is equal to N(N − 1). For example, the minimum division factor for a dual-modulus divider using a 10/11 prescaler is 90. As a proof, try to determine what numbers must be set into the main and auxiliary counters to create a division of 89. It is not possible to find two numbers that will work and satisfy the requirement that M be equal to or greater than A. This minimum does not usually cause any problems in VHF synthesizers in which the minimum division factor is usually quite high.

Integrated Circuits

One significant advancement in synthesizer design is the LSI chip made specifically for dual-modulus synthesizers. Chips such as the Motorola MC145156 and MC145152, when used with suitable prescalers, allow two-chip synthesizer designs to operate at more than 500 MHz. A block diagram of a Motorola chip is shown in Fig 32. This chip is unique; the frequency-programming information is entered into the chip serially to reduce the package size. This could be a disadvantage in some systems, but it can be handled easily by a microprocessor. The MC145152 is a parallel entry version of the MC145156; the '145152 is housed in a 24-pin package, while a 16-pin package serves for the '145156.

Another important advance in dual-modulus VHF synthesizer design came when low power ECL prescalers became available. Perhaps the most significant disadvantage of the divider was the cost and the power requirements of the dual-modulus prescaler. An early 150-MHz ECL dual-modulus prescaler required as much as 100 mA at 5 volts; the power requirement for a modern prescaler is as low as 7 mA. This makes the dual-modulus prescaler a possible contender for portable, battery-operated equipment.

Frequency Modulation

FM is accomplished in a PLL by modulating the oscillator control voltage. The loop normally will tend to remove any frequency variation. To prevent removing all of the applied modulation, the response of the loop is restricted to frequencies below the voice-frequency band or approximately 300 Hz. Unfortunately, this prevents the loop from removing mechanically induced FM of the VCO known as

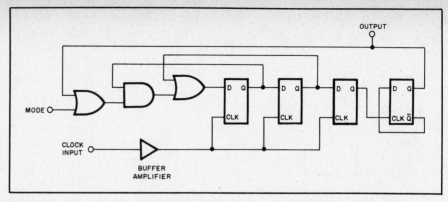

Fig 30—Block diagram of a dual-modulus prescaler suitable for VHF synthesizers.

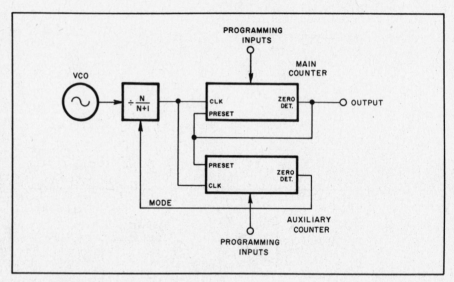

Fig 31—Block diagram of a dual-modulus divider.

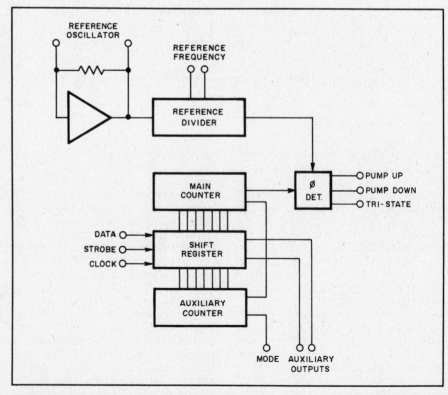

Fig 32—Block diagram of a Motorola LSI synthesizer chip.

microphonics. When the loop frequency response is so limited, the VCO must have few microphonics. This requires rigid shields and components secured with adhesives or potting compounds.

When modulation is applied to the VCO control voltage the amount of frequency deviation depends on the VCO constant. In VHF synthesizers where the bandwidth is a small percentage of the operating frequency, the variation in VCO constant versus operating frequency is small. When the VCO frequency range is large compared to the operating frequency, such as the 4-MHz span of the 50-MHz band, VCO constant can be a significant problem.

Mixing in VHF Synthesizers

VHF synthesizers can also use some form of mixing. Mixing can achieve two goals simultaneously. It can reduce the VCO frequency to a value that can be handled by conventional logic. Mixing can also perform some of the synthesizer arithmetic such as providing the IF offset for receive.

Fig 33 shows a block diagram of a VHF synthesizer using a mixing technique. In this example, the VCO operates in two bands, 133.3 to 137.3 MHz which is the receiver local-oscillator range, and the transmitter range from 144 MHz to 147.995 MHz. The same VCO serves as the receiver local oscillator and the transmitter oscillator. The VCO is mixed with a 125.3-MHz oscillator for receive and a 136-MHz oscillator for simplex transmit. In Fig 33, the programmable divider division factor goes from 1600 at 144 MHz to 2399 at 147.995 MHz. No special code conversion is required for this synthesizer, and Fig 34 shows the logic diagram of the programmable divider. On simplex, the VCO frequency is changed by 10.7 MHz between transmit and receive, by switching crystals in the offset oscillator. For repeater operation, the crystal oscillator is changed to provide a transmit frequency 600 kHz up or down from the receive frequency. More crystals are required to operate with other splits. The disadvantage of this synthesizer is the number of required crystals. A 2-meter synthesizer with simplex, 600-kHz positive and negative repeater offsets and 1-MHz repeater split requires seven crystals including the reference-oscillator crystal. The output frequency is directly dependent on each crystal, unlike the dual-modulus synthesizer, in which frequency accuracy depends on only one crystal.

The VCO must shift approximately 10 MHz between transmit and receive. This requires the synthesizer to lock up quickly. The transmitter must be inhibited while the synthesizer is out of lock. To facilitate the frequency shift, a fixed capacitor can be switched in and out of the VCO tank circuit.

A variation of this synthesizer uses only two crystal oscillators, including the 5-kHz reference. VCO shifts, transmit/receive switching and repeater offsets are con-

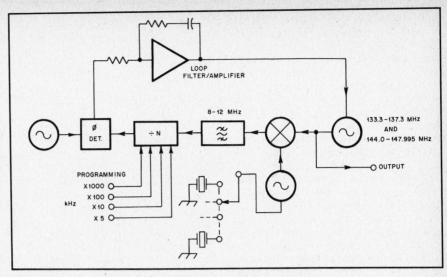

Fig 33—Block diagram of a VHF synthesizer using mixing techniques.

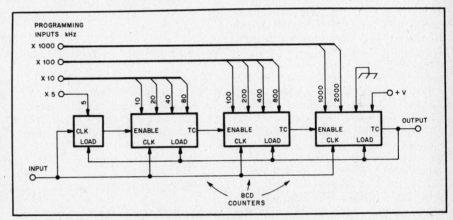

Fig 34—Logic diagram of the programmable divider for the synthesizer shown in Fig 33.

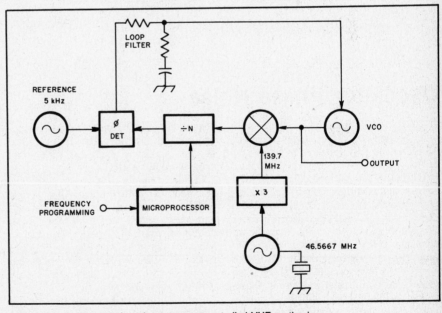

Fig 35—Block diagram of a microprocessor-controlled VHF synthesizer.

trolled by a microprocessor. A block diagram of this synthesizer is shown in Fig 35. The VCO must still shift for transmit and receive; it shares this problem with the previous example. However, only two crystals are required—a significant advantage over the previous case.

One example of a synthesizer design that

does not require a VCO shift between transmit and receive is shown in Fig 36. In this design, the VCO serves as the receiver local oscillator at all times. On transmit, the VCO is mixed with a crystal-controlled, frequency-modulated oscillator operating at the receiver IF. Repeater offsets are handled by the programmable divider. For repeater operation, the VCO has to shift between transmit and receive, but only 600 kHz or 1 MHz, depending on the offset. That is an improvement over the previous examples, where the VCO was required to shift an amount equal to the receiver IF plus any repeater offset.

Like any heterodyning transmitter there is a possibility of transmitting spurious signals. To minimize this chance, the receiver IF should be as high as practical. In the example shown, the receiver IF is 16.9 MHz. The local oscillator range is 127.1 to 131.095 MHz. LO output is mixed with a 16.9-MHz, frequency-modulated crystal oscillator to provide the transmitted frequency. Care must be taken to prevent local oscillator radiation. The 16.9-MHz signal is filtered easily. To effectively eliminate the local oscillator from the transmit signal requires a sharp tuned circuit. This circuit can be adjusted automatically with a varactor diode, tuned from the VCO control voltage. This technique usually reduces the spurious signal level to an acceptable limit.

Frequency synthesis at UHF uses similar techniques to those used at VHF, with some minor changes. The frequencies involved in this range limit the use of the dual-modulus synthesizer and make mixing schemes difficult. One acceptable method of UHF synthesis is to use a fixed-modulus

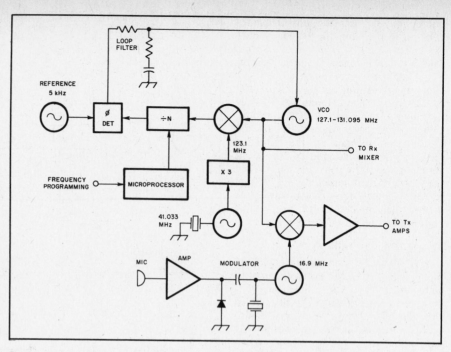

Fig 36—Block diagram of a synthesizer requiring only minimal VCO frequency shifting.

prescaler. Because most modes of operation at UHF are broadband, the presence of the reference frequency in the PLL is not a significant problem. Simple frequency dividers such as divide by 2, 4 or 8 are available for operating frequencies as high as 2 GHz and are expected to become available for even higher frequencies.

Synthesizer Bibliography

Helfrick, A. D. *Amateur Radio Equipment Fundamentals,* Englewood Cliffs: Prentice-Hall, 1982, pp. 111-119.

Helfrick, *Practical Repair and Maintenance of Communications Equipment,* Englewood Cliffs: Prentice-Hall, 1983, chapter 5.
Helfrick, "A High Performance Synthesized 2-Meter Transmitter," *QST,* September 1980.
Helfrick, "The Universal Synthesizer," *QST,* September 1981.
Helfrick, "A Modern Up-Converting General-Coverage Receiver," *QST,* December 1981.
Helfrick, "Versatile Communications Receiver," *Ham Radio,* July 1982.
Helfrick, "A Modern Synthesizer for Portable VHF Transceivers," *QST,* April 1982.
Rohde, Ulrich L, *Digital PLL Frequency Synthesizers,* Englewood Cliffs: Prentice-Hall, 1983.

Oscillator Phase Noise

The following material on phase noise was written by John Grebenkemper, KI6WX, and first appeared as a two-part article in March and April 1988 *QST.*

Oscillators have formed the basis of most radio communications systems since spark-gap transmitters were retired. Originally, these oscillators used piezoelectric crystals or high-Q LC tuned circuits as resonant elements. More recently, *phase-locked loops* (PLLs) have been used in radios with synthesized local oscillators. There is no theoretical reason that prevents a phase-locked oscillator from having better phase-noise characteristics than a free-running oscillator using an LC tuned circuit.

An ideal oscillator would generate a sine wave at a given frequency with no deviation in amplitude or phase over time. If we looked at this signal with an ideal spectrum analyzer, we would see all of the energy

concentrated at the fundamental frequency and none at any other frequency. If ideal oscillators existed, problems with phase noise wouldn't exist.

Oscillator signals vary in both amplitude and phase as a function of time. These variations are referred to as *amplitude noise* and *phase noise.* Amplitude noise is an undesired variation in the *amplitude* of an oscillator signal. If we use an oscilloscope to look at oscillator output, we can see that the amplitude peaks of the sine wave vary with time (Fig 37).

Phase noise is an undesired variation in the phase of the signal. In this case, an oscilloscope shows that the time between zero crossings of the signal varies over time when compared to the zero crossings of an ideal sine wave. An exaggerated example of phase noise is shown in Fig 38.

Several factors combine to make ampli-

tude noise much less important than phase noise. The first is that most oscillators tend to produce less amplitude noise than phase noise. This occurs because the active devices in most oscillators are operated in a gain-saturated condition. Gain saturation occurs when increasing the input to a device or stage results in no further increase in output. If a device or stage is sufficiently gain saturated, small changes in the amplitude of the input signal (amplitude noise, in other words) produce no change in the output level.

Amplitude noise is secondary in importance to phase noise for another reason. Most modern ham rigs use mixers in their signal-generation schemes. Because mixers are usually operated with their local-oscillator input ports gain saturated, any amplitude noise on the input signal generally does not appear on the mixer output.

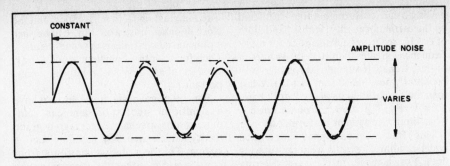

Fig 37—Oscillator amplitude noise effects as viewed on an oscilloscope. The broken line shows the output of an ideal oscillator; the solid line shows the output waveform of an oscillator with a large amount of amplitude noise.

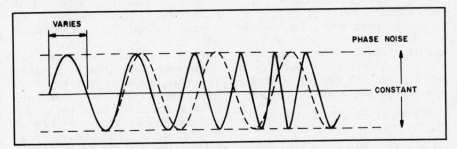

Fig 38—Oscillator phase-noise effects as viewed on an oscilloscope. The broken line shows the output of an ideal oscillator; the solid line shows the output waveform of an oscillator with a large amount of phase noise.

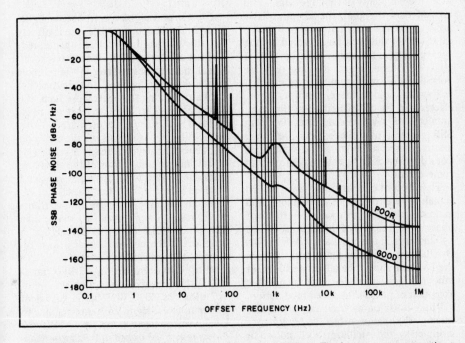

Fig 39—Phase-noise characteristics of two synthesized oscillators. The lower curve represents the output of a well-designed synthesized oscillator. The upper curve represents that of a poorly designed synthesized oscillator. Note the broad peak at 1 kHz, and the spikes visible near 60 and 120 Hz, and 11 and 22 kHz. (See text.)

SPECIFYING PHASE NOISE LEVELS

There are many ways to display oscillator phase noise. The most common presentation is called single-sideband (SSB) phase noise. This is simply a display of the phase-noise characteristics on one side of the carrier. The SSB phase-noise presentation has probably achieved its widespread use because SSB phase noise is easy to measure and interpret for the average person who is familiar with RF measurement techniques. The display of SSB phase noise is exactly what you would see if the oscillator was observed using an ideal spectrum analyzer with a 1-Hz sampling bandwidth.

The vertical axis is the SSB phase noise measured in units of "dBc/Hz." The "dBc" means that we are measuring the power in the phase noise relative to the power in the carrier. The "Hz" implies that we are making this measurement in a 1-Hz bandwidth. (Phase noise is usually not actually *measured* in a 1-Hz bandwidth, for reasons of instrumentation limitations.)

The horizontal axis represents the difference between the measurement and oscillator frequencies, and is usually referred to as the *offset frequency*. Offset frequency is measured in Hz, and it is usually plotted on a logarithmic scale. An offset of 1 kHz means that we are measuring the phase noise 1 kHz from the carrier; an offset of 10 kHz is 10 kHz from the carrier, and so on. A 10-kHz offset from a 10-MHz carrier means that we are making measurements at 9.99 MHz or 10.01 MHz.

Two typical phase-noise measurements are shown in Fig 39. At an offset frequency of 0 Hz, the SSB phase noise is always 0 dBc/Hz, because we are measuring the carrier power, which is the reference for these measurements. At any other offset, the SSB phase noise is always a negative dB value.

Phase noise in a communications system can be divided into two different regions. The *in-band* or *close-in* phase noise is the component that is *within the modulation bandwidth* of the signal. The *out-of-band* or *far-out* phase noise is the component that is *beyond* the modulation bandwidth of the signal. The modulation bandwidth of an amateur SSB signal is 3 kHz. Therefore, we can define the close-in phase noise of an SSB signal as the noise at offsets of less than 3 kHz, and the far-out phase noise as the phase noise at offsets greater than that. The difference between close-in and far-out phase noise is similarly defined for modes with different modulation bandwidths.

The example used in Fig 39 shows the phase noise for two different synthesized oscillators. The phase-noise plot labeled "good" is the phase noise from an excellent synthesized local oscillator. These phase-noise characteristics are considered the best presently obtainable for oscillators used in HF receivers. The phase noise from this oscillator rapidly decreases with increasing offset frequency, reaching −160 dBc/Hz at an offset frequency of 100 kHz. The slight flattening of the phase-noise plot near 1 kHz results from the natural frequency of the PLL (the natural frequency produces optimum loop response). (It is normal for a synthesized oscillator to show flattening or a few dB of peaking of the phase noise at the natural frequency of the PLL.) This oscillator would present only minor phase-noise problems in amateur applications.

The phase-noise plot labeled "poor" is typical of what one would expect from an inferior synthesized oscillator. The characteristics of such an oscillator include strong spurious signals, an underdamped PLL (lack of sufficient negative feedback), and a high phase-noise floor. Large spurious signals are shown at harmonics of 60 Hz and 11 kHz. These signals result from frequency modulation of the oscillator in the PLL, and are generally caused by inadequate shielding or poor loop design. The broad peak at 1 kHz results from an underdamped PLL. Any peaking greater than this is indicative of a poorly designed phase-locked oscillator.

The phase noise of the poorer oscillator in Fig 39 is only -130 dBc/Hz at a 100-kHz offset. This is generally caused by deficient design of the phase-locked oscillator. Any of the above forms of degradation in the phase noise of an oscillator compromise the performance of a transmitter (and/or receiver).

EFFECTS OF OSCILLATOR PHASE NOISE

In order to understand the effects of phase noise, we have to keep some basic principles in mind. The term *communications system* used here refers to the receiver, the transmitter to which the receiver is tuned, and other nearby transmitters. The term *system* is used to convey that phase-noise effects can result from *both the receiver and transmitters*, not just the affected receiver.

Principle 1: Whenever a carrier is passed through a mixer, the phase noise of the oscillator driving that mixer is *added* to the carrier. This means that the oscillator SSB phase-noise power at a given offset frequency is added to the phase noise already present on the carrier at the same offset frequency. For instance, suppose we pass an ideal carrier (one with no phase noise) through two mixers, each driven by an oscillator with an SSB phase noise of -130 dBc/Hz at a 10-kHz offset. After passing the carrier through the first mixer, the SSB phase noise on the carrier is -130 dBc/Hz. After passing the carrier through the second mixer, the phase noise on the carrier is -127 dBc/Hz. This is the *sum* of the phase-noise *power* of the carrier and local oscillators at the 10-kHz offset frequency. (Summing two equal powers is a doubling of either power, or a 3-dB increase. -127 dBc is twice as much power as -130 dBc.) This example illustrates that *the oscillator with the worst phase-noise characteristics limits the performance of the system*. This principle holds for both transmitters and receivers.

Principle 2: *Phase noise on a transmitted signal causes effects identical to phase noise generated in a receiver*. (Of course, transmitter-generated phase noise can affect many users of the radio spectrum, while receiver-generated phase noise only affects the person using that receiver!)

Principle 3: Passing an oscillator signal through a filter *reduces* the phase noise in accordance with the filter bandwidth and attenuation characteristics. For instance,

passing a carrier through a filter (centered on the carrier frequency) with a bandwidth of 20 kHz reduces the phase noise for offset frequencies greater than 10 kHz. As a result of this, transmitters and receivers have an *apparent phase noise* that is usually less than the sum of their local oscillator phase-noise levels. The *apparent* phase noise is what limits the *system* performance.

This effect must be applied with a good understanding of exactly what the filtering does. For instance, in receivers, filtering *before* a mixer determines the apparent phase noise; in transmitters, filtering *after* a mixer determines apparent phase noise. The improvement in the apparent phase-noise floor achievable with filtering depends on the power level of the signal being filtered. A normal filter at room temperature generates thermal-noise power of about -170 dBm/Hz. A 10-dBm signal passed through this filter can achieve a noise floor no lower than -160 dBc/Hz.

CLOSE-IN VERSUS FAR-OUT PHASE NOISE

The close-in and far-out phase-noise components act on a communications system in different ways. Close-in phase noise limits the performance of the system even when there is plenty of signal present. With an AM or FM signal, the close-in phase noise limits the maximum signal-to-noise ratio (SNR) that can be achieved by the system. With a frequency-shift-keyed or a phase-shift-keyed signal, the close-in phase noise limits the maximum bit error rate that the system can achieve. Both of these effects can be quantified once the communications system is defined. With an SSB voice signal, the effects are much harder to predict, but excessive phase noise does degrade SSB signal intelligibility to some extent.

The effect of close-in phase noise on FM signals is relatively easy to calculate, and makes a good case study of the effects of phase noise, because of the wide use of FM by amateurs in the VHF and UHF bands. Oscillator phase-noise limits derived from the FM case should be adequate for most other modulation modes, with the possible exception of phase-shift keyed modulation.

Phase noise on an oscillator signal has exactly the same effect as frequency modulating the oscillator with noise. In fact, given the SSB phase noise of an oscillator, it is possible to calculate the equivalent frequency deviation. This deviation is called the *incidental frequency modulation* (IFM) of the oscillator.

The phase noise in an FM communications system causes a constant noise level to appear at the receiver output, regardless of the strength of the received signal. This noise imposes a maximum SNR that the communications system can achieve.

It takes an oscillator with a very high level of close-in phase noise to degrade amateur communications systems. Most amateur communications are accomplished

with marginal signal to noise ratios, and the close-in phase noise would have to be *very* strong to cause problems in such situations. Therefore, for most amateur applications, close-in phase noise is relatively unimportant.

Far-out phase noise, however, can have a significant impact on amateur communications. Its main effect is to degrade the dynamic range of a communications system. This degradation can result from phase noise in either the receiver or the transmitter. If the transmitter has excessive phase noise, transmitted signals are partly composed of broadband FM noise. The phase-noise emissions of some current amateur equipment are capable of significantly increasing the noise level on an entire amateur band! The effect of this is generally experienced only by amateurs living within a few miles of the offending transmitter, because the power level of the phase noise is quite low. A kilowatt transmitter with a high level of phase noise produces only a few microwatts in a typical amateur receiving bandwidth. As low as this may seem, however, it can be quite a problem for nearby receivers when you consider that a typical HF receiver is capable of detecting signals of less than 10^{-15} watts!

Far-out phase noise in a receiver oscillator is less destructive, because it affects only the receiver in which it is generated. The net effect, however, is the same: Any signal that reaches a mixer in the receiver is modulated by the phase noise in the local oscillator driving that mixer. As such, the signal appears to have at least as much phase noise as the local oscillator. Thus, sufficiently strong signals *off* the receiving frequency can degrade receiver sensitivity by raising the noise floor *at* the receiving frequency. Receiver dynamic range is reduced as the noise floor rises.

As an example, suppose that a transmitter has a power output of 1 kW and an SSB phase noise of -100 dBc/Hz at a given offset frequency. At that offset frequency, the transmitter radiates noise at a level of -40 dBm/Hz. In a 2-kHz bandwidth, this is equivalent to a noise level of -7 dBm, or 200 μW. This is sufficient power to cause significant interference to local receivers. *The effect on a receiver is the same as if the transmitter had low phase noise and the receiver had an SSB phase noise of* -100 dBc/Hz. Fortunately, the oscillators in most amateur equipment don't have this much SSB phase noise at offset frequencies greater than a few kilohertz.

The effect of phase noise on an amateur communications system can be calculated with reasonable accuracy if a few simple assumptions are made. From the calculated effects, we can calculate the oscillator performance necessary to meet our requirements. The assumptions are:

1) The transmitter is located 1 mile from the receiver.

2) The EIRP (effective isotropic radiated power) of the transmitter is 100 W (+50 dBm).

3) The loss from the transmit antenna to the receive antenna is the same as in free space. (This isn't strictly true for signals propagated above ground, but it gives an approximation of the expected signal attenuation. The actual attenuation is almost impossible to calculate because of the effects of the local terrain.)

4) The receive antenna gain is 0 dBi in all directions.

5) Our goal is a phase-noise level that does not significantly degrade the performance of the communications system. That is, the received noise resulting from oscillator phase noise should be equal to or less than the background noise normally picked up by an antenna on a given amateur band. (Background noise is composed of radio emissions from our galaxy, as well as man-made noise and atmospheric noise, and varies with frequency.)

Table 1 shows the results of these calculations. The table shows values for both 14 MHz (representative of the HF bands) and 144 MHz (representative of the VHF bands). As can be seen from these calculations, the phase noise must be less than −140 dBc/Hz if there is to be no noticeable interference. A higher EIRP or less transmitter-to-receiver antenna separation would require even better phase-noise performance to guarantee no significant interference. The graph of Fig 40 shows phase-noise interference generated at 14 MHz as a function of transmitter-to-receiver antenna separation, and the apparent SSB phase noise of the system. The interference level is plotted in S units, assuming S9 is equal to 50 μV and each S unit is 6 dB. As you can see from the graph, an SSB phase noise of −120 dBc/Hz at a distance of 500 feet produces nearly an S9 signal. This graph may be applied to local situations to estimate the interference potential that could result from oscillator phase noise.

PHASE NOISE LEVELS IN AMATEUR EQUIPMENT

Equipment cost limits the extent to which we can minimize phase noise in communications systems; with this in mind, we can set rough phase-noise performance standards for equipment in various price ranges.

Curves showing recommendations for such standards appear in Fig 41. These curves represent tools for comparison of phase-noise performance in amateur equipment.

Four curves are shown: "excellent," "good," "fair" and "poor." When the phase-noise characteristics of a radio are known, they can be compared to the curves in Fig 41. The worst curve that a radio's phase-noise characteristics pass through should be considered the performance of that radio. The "excellent" curve approx-

Table 1
Typical Maximum Tolerable Phase-Noise Levels

	14 MHz	144 MHz
EIRP	+50 dBm	+50 dBm
Loss from transmitter to receiver (distance = 1 mile)	60 dB	80 dB
Signal power at receiving antenna (0-dBi gain receiving antenna)	−10 dBm	−30 dBm
Noise power at receiving antenna (background noise)	−150 dBm/Hz	−172 dBm/Hz
Equivalent SSB phase noise	−140 dBc/Hz	−142 dBc/Hz

This table shows maximum phase-noise levels tolerable in a communications system before dynamic range is significantly degraded. In both cases, EIRP is 100 W; separation between transmitting and receiving antennas is 1 mile; the receiving antenna is omnidirectional and has 0 dBi gain; and the receiver is sensitive enough to detect the noise.

At 14 MHz, path loss and received noise power allow the receiver to detect any phase noise stronger than −140 dBc/Hz. At 144 MHz, background noise is lower and path loss is higher, resulting in a tolerable equivalent phase-noise level of −142 dBc/Hz at this frequency. See text for additional information.

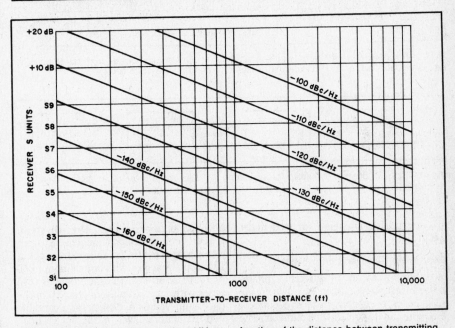

Fig 40—Phase-noise interference at 14 MHz as a function of the distance between transmitting and receiving antennas. The curves are based on a transmitter EIRP of 100 W and a receiving bandwidth of 2 kHz. The path loss from the transmit antenna to the receive antenna is assumed to follow the free-space attenuation curve. S9 is equivalent to 50 μV.

imates the best performance that can be expected from modern HF and VHF synthesized oscillators. For amateur equipment, this curve is approached by free-running LC and crystal oscillators, and by the very best transceivers.

The "good" curve represents what is achieved by most well-designed amateur equipment. It doesn't cost a great deal more to design a radio that meets this curve (as opposed to the "poor" or "fair" curves). A radio with phase-noise performance that only meets the "fair" curve hasn't been designed with much regard for minimizing phase noise. The "poor" curve represents a design that is so poor that a transmitter with this performance shouldn't be used on the amateur bands because of the great amount of interference it can generate.

The close-in phase-noise characteristics shown in the curves of Fig 41 are relatively

easy for most oscillators to achieve. The close-in regions of the curves were derived mainly from the effects of IFM on the limiting FM SNR. Table 2 lists the close-in phase-noise IFM integrated from 10 Hz to 10 kHz for each of these curves. (Even a radio generating the IFM level corresponding to phase noise matching the "poor" curve provides adequate voice communications capability.)

MEASURING PHASE NOISE

Equipment capable of measuring low levels of far-out phase noise is quite expensive. It is not unusual for commercial enterprises to invest $100,000 in good phase-noise measurement equipment. Inexpensive techniques are, however, available for measuring the apparent phase noise of communications receivers and transmitters. The equipment necessary to measure *receiver* phase noise can be built

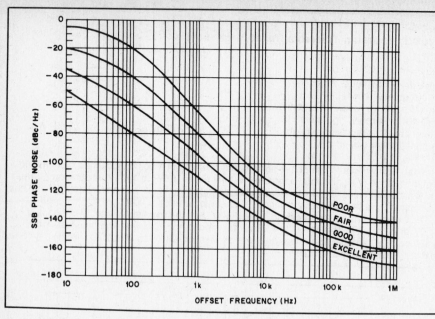

Fig 41—Phase-noise performance levels for amateur equipment. The curves from top to bottom represent poor, fair, good and excellent phase-noise performance. (See text.)

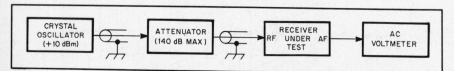

Fig 42—Block diagram of test setup for measuring the SSB phase noise of an SSB or CW receiver.

Table 2
Incidental Frequency Modulation

Curve	IFM
Excellent	0.4 Hz
Good	2.5 Hz
Fair	20 Hz
Poor	160 Hz

Incidental frequency modulation (IFM) resulting from the phase-noise levels shown in Fig 41. The IFM was calculated for the frequency range from 10 Hz to 10 kHz.

by enterprising amateurs for well under $100.

Measuring Receiver Phase Noise

A block diagram of a receiver phase-noise measurement setup is shown in Fig 42. The test setup consists of an oscillator, a step attenuator and an ac voltmeter. The output of the oscillator is fed through the attenuator to the receiver's RF input. The ac voltmeter is connected to the audio output of the receiver. The receiver is operated in SSB or CW mode with the AGC disabled.

The oscillator and step attenuator must be carefully constructed (well shielded, minimum necessary lead lengths and so on) for proper operation of this test setup. A simple oscillator circuit suitable for use in this application appears on page 126 of the ARRL publication *Solid State Design for the Radio Amateur*. The oscillator provides the reference signal for measuring phase noise. The phase noise of the oscillator must be less than that of the receiver if we are to measure receiver phase noise accurately. The oscillator should have a power output of at least +10 dBm (10 mW) into 50 Ω. Less power than this may limit the accuracy of the measurement setup at low phase-noise levels. The oscillator should be constructed in a metal box and powered by an internal 9-V battery to minimize unwanted signal leakage.

The step attenuator must have a maximum attenuation of at least 140 dB. Prefer-

ably, it should be adjustable in 1-dB steps. It is difficult to design a single attenuator circuit that provides this much attenuation, so a step attenuator with a maximum of 60 or 70 dB of attenuation should be used in series with fixed attenuators to obtain the maximum value. A design for a suitable step attenuator is given in Chapter 25. Each of the fixed attenuators can be constructed in a shielded enclosure using standard Pi or T resistor configurations to achieve the desired attenuation. Use BNC connectors on the oscillator and step attenuator; they have better shielding properties than UHF connectors. Commercial step attenuators and in-line coaxial attenuators are sometimes available on the surplus market. If you choose to build attenuators, use 5% (or better) precision resistors. The accuracy of the phase-noise measurements depend to a great extent on the accuracy of the attenuators.

The ac voltmeter is used to measure the audio output level from the receiver. The voltmeter should measure true RMS voltage, but commonly available peak-reading ac voltmeters are adequate for this application.

Phase-noise measurements are done on the band for which the oscillator is built. Measurements should be made in each amateur band, since phase-noise levels of many receivers vary from one amateur band to another. Briefly, the measurement procedure is as follows: A reference audio output level is established by tuning the

receiver to the oscillator frequency and measuring the audio output voltage. The receiver is then tuned above (or below) the oscillator frequency in increments of the desired offset frequency. The attenuation is then decreased until the noise voltage at the audio output is the same as the reference level. The difference in the two attenuator settings is then corrected for the receiver bandwidth to give the SSB phase noise at that offset frequency. The process is then repeated for several different offset frequencies. The step-by-step measurement procedure follows.

Step-By-Step Measurement Procedure

1) Connect the test equipment as shown in Fig 42. Set the step attenuator to its maximum attenuation. Switch the oscillator out of the line. Set the receiver for SSB or CW reception with the AGC disabled. Preamplifiers or input attenuators in the receiver should be disabled, if possible. Use the narrowest IF bandwidth available in the receiver. Set the RF and AF gain controls to maximum, unless this causes the audio amplifier to overload from receiver noise. In case of audio-amplifier overload, reduce the audio gain until there is a minimum of 10 dB of headroom in the audio amplifier. This can be done by decreasing the AF gain (while monitoring audio output voltage) by 10 dB after all signs of gain compression disappear. In other words, decrease the AF gain to 10 dB below the point at which the audio output voltage begins to vary linearly with adjustment of the AF gain control.

2) Determine the 6-dB bandwidth of the receiver. An easy way to do this is to enable the oscillator, tune the oscillator signal through the receiver passband and note the frequencies at which the audio voltage is 6 dB below the peak response. The difference between the upper and lower frequencies is the receiver's bandwidth. Alternatively, the published 6-dB bandwidth for the receiver can be used. The bandwidth measurement doesn't have to be very accurate—a 25% error in the bandwidth factor affects the phase-noise measurement by only 1 dB. Record the receiver bandwidth (Δf).

3) Switch the oscillator into the line. Center the oscillator signal in the receiver passband. Switch the oscillator out of the line and measure and record the audio output voltage. Switch the oscillator into the line again, and adjust the step attenuator until the audio voltage increases by 41% (3 dB) from the no-signal value. The signal

is now at the receiver's MDS (minimum discernible signal) level. This setting is a compromise: A higher setting allows more precise measurement of the audio output voltage; a lower setting decreases the possibility of overloading the receiver front end. Record the frequency to which the receiver is tuned (fo), the setting of the step attenuator (A_0), and the audio output voltage (V_0).

4) Measure the phase noise by first tuning the receiver to the desired offset frequency, given by fo + f, where f is the offset frequency. Then, adjust the attenuator until the audio output voltage is as close as possible to (V_0). Record the total attenuation (A_1) and the audio voltage (V_1). The SSB phase noise at this offset frequency is found by

$$L(f) = A_1 - A_0 - 10 \log (\Delta f)$$

where

$L(f) = $ SSB phase noise in dBc/Hz

5) Determine if the receiver is overloaded at this offset frequency as follows: Decrease the attenuation by 3 dB. Note the audio voltage at the new attenuator setting (V_2). The audio output voltage of the receiver should increase by approximately 22% (1.7 dB) from V_1. If it increases by less than 18% (1.4 dB), it is likely that some stage in the receiver is overloaded, and the SSB phase-noise measurement is inaccurate. If the overload is in the audio stages, it can be eliminated by decreasing the audio gain and repeating the measurements. If the overloading is occurring in the RF or IF stages, the receiver's blocking dynamic range has been exceeded. The overload may be eliminated by reducing the receiver's RF gain and repeating the measurements. The existence of this problem indicates that, at this offset frequency, the receiver's performance is limited by blocking dynamic range, not phase noise.

6) Repeat steps 4 and 5 for several offset frequencies. Ideally, measurements should be made at offset frequencies separated by no more than half of the receiver's bandwidth, in order to make sure that no discrete phase-noise components are overlooked. This is tedious at the greater offset frequencies. Generally, a coarser sampling is adequate to plot the phase-noise curve, as long as you check the intermediate offset frequencies to make sure that no discrete signals are present. The measurements can start at offset frequencies as close as a few times the receiver's bandwidth, if the receiver has good IF filters. You can determine if the offset is too small by disconnecting the audio voltmeter from the receiver and listening to the audio. If you can hear the carrier signal at all, the offset is too small. All that should be present is noise if the measurement is to be accurate. Increase the offset until no trace of carrier remains in the audio to be sure of the measurement.

The technique outlined in the steps above is quite accurate. SSB phase-noise levels of less than −160 dBc/Hz can be measured

Table 3
SSB Phase Noise of ICOM IC-745 Receiver Section

Oscillator output power = −3 dBm (0.5 mW)
Receiver bandwidth (Δf) = 1.8 kHz
Audio noise voltage = 0.070 V
Audio reference voltage (V_0) = 0.105 V
Reference attenuation (A_0) = 121 dB

Offset Frequency (kHz)	Attenuation (A_1) (dB)	Audio V_1 (volts)	Audio V_2 (volts)	Ratio V_2/V_1	SSB Phase Noise (dBc/Hz)
4	35	0.102	0.122	1.20	−119
5	32	0.104	0.120	1.15	−122*
6	30	0.104	0.118	1.13	−124*
8	27	0.100	0.116	1.16	−127*
10	25	0.106	0.122	1.15	−129*
15	21	0.100	0.116	1.16	−133*
20	17	0.102	0.120	1.18	−137
25	14	0.102	0.122	1.20	−140
30	13	0.102	0.122	1.20	−141
40	10	0.104	0.124	1.19	−144
50	8	0.102	0.122	1.20	−146
60	6	0.104	0.124	1.19	−148
80	4	0.102	0.126	1.24	−150
100	3	0.102	0.126	1.24	−151
150	3	0.102	0.124	1.22	−151
200	0	0.104			−154
250	0	0.100			−154
300	0	0.098			< −154
400	0	0.096			< −154
500	0	0.096			< −154
600	0	0.097			< −154
800	0	0.096			< −154
1000	0	0.096			< −154

Asterisks indicate measurements possibly affected by receiver overload (see text).

on a receiver that has good sensitivity and a high gain-compression point (good dynamic range), as determined in step 5. It is difficult to construct an oscillator that has a noise floor much below this level, and, as mentioned earlier, the oscillator phase-noise level limits the accuracy of the measurement setup.

This measurement technique has advantages and disadvantages. One of its major advantages is that it measures the *apparent* phase noise of the receiver. Apparent phase noise includes the effects of filters; it is the phase noise that actually limits a receiver's dynamic range.

One disadvantage of this measurement technique is that it doesn't allow *direct* measurement of the phase noise of an oscillator. Therefore, it can't be used to directly measure the phase noise of a transmitter. This limitation can be overcome by using a receiver with known phase-noise characteristics to measure the phase noise of a transmitter. The transmitter is substituted for the oscillator in the test setup, and its output is attenuated to the power level of the oscillator. As long as the transmitter phase noise is greater than that of the receiver, measurements of the transmitter's phase noise will be accurate.

Another disadvantage is that, at some offset frequencies, the receiver's dynamic range may make phase-noise measurement impossible. Precision in this measurement technique requires that the oscillator out-

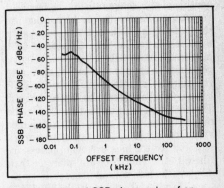

Fig 43—Measured SSB phase noise of an ICOM IC-745 transceiver (serial number 01528).

put be variable from a very low level to a very high level, in small, precise steps.

Table 3 and Fig 43 show phase-noise characteristics measured in the receiver section of an ICOM IC-745 transceiver using this method. Table 3 includes the attenuation and voltage values to show their relationships to phase-noise levels in the IC-745 over this range of offset frequencies. The data obtained from the ratio of V_2 to V_1 indicates that the receiver was probably overloaded at offset frequencies from 5 kHz to 15 kHz. This measurement was limited to an SSB phase-noise floor of −154 dBc/Hz because of the low power-output level of the oscillator. Fig 43

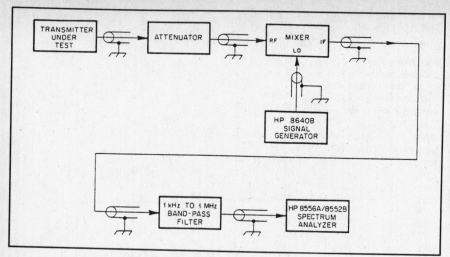

Fig 44—ARRL transmitter phase-noise measurement setup.

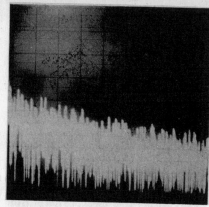

Fig 45—ICOM IC-745 (serial number 03101) phase-noise characteristics. Measurement frequency: 3.5 MHz, power output: 100 W.

is based on the far-out phase noise data obtained from Table 3, as well as the close-in phase noise measured using laboratory test equipment and the same IC-745. The data obtained using the method described earlier and that obtained from the laboratory test equipment closely track each other. The phase-noise performance of this transceiver generally exceeds that indicated by the "good" curve in Fig 41.

TRANSMITTER PHASE-NOISE MEASUREMENT IN THE ARRL LAB

The system used to measure transmitter phase noise in the ARRL Lab essentially consists of a direct-conversion receiver with very good phase-noise characteristics. As shown in Fig 44, an attenuator is used after the transmitter, a Mini-Circuits ZAY-1 mixer, a Hewlett-Packard 8640B signal generator, a bandpass filter and an audio-frequency spectrum analyzer (HP 8556A/8552B) to make the measurements. The transmitter signal is mixed with the output of the signal generator, and signals produced in the mixing process that are not required for the measurement process are filtered out. The spectrum analyzer then displays the transmitted phase-noise spectrum. The 100 mW output of the HP 8640B is barely enough to drive the mixer—the setup would work better with 200 or even 400 mW of drive. To test the phase noise of an HP 8640B, we use a second '8640B as a reference source. It is quite important to be sure that the phase noise of the reference source is lower than that of the signal under test, because what is actually being measured is the combined phase-noise output of the signal generator and the transmitter. It would be quite embarrassing to publish phase-noise plots of the reference generator instead of the transmitter under test! The HP 8640B has much cleaner spectral output than most transmitters.

Fig 45 shows a phase noise photograph for ICOM 745 serial number 03101. This photograph was taken directly from the spectrum-analyzer display, using the test

setup shown in Fig 44. The photo does not necessarily reflect the phase-noise characteristics of all ICOM 745 transceivers.

The carrier is not shown in the photograph because it is nulled by heterodyning and bandpass filtering. The log reference level (the top horizontal line on the scale in the photos) represents − 60 dBc/Hz. It is common in industry to use a 0-dBc log reference, but such a reference level would not allow measurement of phase-noise levels below − 80 dBc/Hz. The actual measurement bandwidth used on the spectrum analyzer is 100 Hz, but the reference is scaled for a 1-Hz bandwidth. This allows phase-noise levels to be read directly from the display in dBc/Hz. Because each vertical division represents 10 dB, the photos show the noise level between − 60 dBc and − 140 dBc. The horizontal scale is 2 kHz per division.

What do the Phase-Noise Pictures Mean?

Although they are useful for comparing different radios, phase-noise photographs can also be used to calculate the amount of interference you may receive from a nearby transmitter with known phase-noise characteristics. An approximation is given by

Interfering signal level = NL + 10 × log BW

where

Interfering signal level is in dBc

NL is the noise level on the receiving frequency

BW is receiver IF bandwidth in Hz

For instance, if the noise level is − 90 dBc and you are using a 2.5-kHz SSB filter, the interfering signal will be − 56 dBc. In other words, if the transmitted signal is 20 dB over S9, and each S unit is 6 dB, the interfering signal will be as strong as an S3 signal.

The measurements made in the ARRL Lab apply only to transmitted signals. It is reasonable to assume that the phase-noise characteristics of most transceivers are similar on transmit and receive, because the

same oscillators are generally used in the local-oscillator (LO) chain.

In some cases, the receiver may have better phase-noise characteristics than the transmitter. Why the possible difference? The most obvious reason is that circuits often perform less than optimally in strong RF fields, as anyone who has experienced RFI problems can tell you. A less-obvious reason results from the way that many high dynamic-range receivers work. To get good dynamic range, a sharp crystal filter is often placed immediately after the first mixer in the receive line. This filter removes all but a small slice of spectrum for further signal processing. If the desired filtered signal is a product of mixing an incoming signal with a noisy oscillator, signals far away from the desired one can end up in this slice. Once this slice of spectrum is obtained, however, unwanted signals cannot be reintroduced, no matter how noisy the oscillators used in further signal processing. As a result, some oscillators in receivers don't affect phase noise.

The difference between this situation and that in transmitters is that crystal filters are seldom used for reduction of phase noise in transmitting because of the high cost involved. Equipment designers have enough trouble getting smooth, click-free break-in operation in transceivers without having to worry about switching crystal filters in and out of circuits at 40-WPM keying speeds!

CONCLUSION

Close-in phase noise generally has little effect on the performance of amateur communications systems. However, far-out phase noise *can* significantly reduce the dynamic range of a receiver. Far-out phase-noise performance has effects just as critical as blocking dynamic range and two-tone dynamic range performance of receivers. Ideally, these measurements should be made for a range of offset frequencies. Far-out phase noise in receivers can be measured with relatively inexpensive test equipment, as long as care is taken to perform the measurements properly.

Chapter 11

Radio Transmitting Principles

Even though some modern transmitters and transceivers contain only solid-state devices, it is still practical to use a hybrid circuit that contains a mixture of tubes and active semiconductor stages. Typically, the unit has transistors, diodes and ICs up to the driver stage of the transmitter. At that point one will find a tube driver used to supply RF power to a vacuum-tube amplifier. The amplifier might consist of a pair of 6146Bs.

The principal advantage of tube amplifiers is that they are somewhat less subject to damage from excessive drive levels and mismatched loads. However, a properly designed solid-state driver and PA section should be immune to output mismatch damage, provided an SWR-protection circuit has been included in the transmitter. A solid-state amplifier is slightly more difficult to design and have work correctly than is a tube amplifier of equivalent power. This is because purity of emissions is harder to achieve when transistor power stages are employed. Transistors generate considerably more harmonic energy than tubes do, and are prone to self-oscillation at LF, VLF and audio frequencies unless some careful design work is done. This is not generally true of tube amplifiers.

If one is to ignore these problems and concentrate mainly on cost and convenience, transistors may have the edge over tubes. A 13.6-volt design can be operated directly from an automotive or solar-electric supply, whereas a tube amplifier requires a high-voltage power supply for mobile, portable and fixed-station use. When an ac power source is required, the cost of a high-voltage, medium-current supply for tubes versus a low-voltage, high-current power source for transistors is similar, provided new components are used in both. At power-output levels in excess of approximately 150 watts, the transistor-amplifier power supply becomes rather expensive because of complex regulator-

circuit requirements. This is why most amateurs use vacuum tubes in high-power HF and VHF amplifiers. The number of power transistors required (plus combiners) to generate a l-kW signal may run considerably higher in cost than a tube or tubes for an amplifier of equivalent power. The price of large heat sinks versus a cooling fan may place the solid-state amplifier in a prohibitive price class also.

The decision to buy or build a transmitter is founded on some basic considerations: cost compared to features; professional appearance contrasted to that of homemade equipment; the knowledge and satisfaction gained from building equipment, as weighed against buying commercially made gear and simply becoming an operator. The judgment must fit the amateur's objectives and affluence. Home-built transmitters are usually easier to service than commercial ones because the builder knows the circuit layout and how each stage functions. Futhermore, the cost of maintenance is markedly lower for homemade equipment than for most factory-built gear. But the greatest significance to home-built circuits is the knowledge gained from constructing a project and the pride that goes with using it on the air!

Not all parts of a transmitter are covered in depth in this chapter. A discussion of frequency-generating circuits is found in Chapter 10. For information on RF power amplifiers, see Chapter 15.

Above 50 MHz

The frequencies above 50 MHz were once a world apart from the rest of Amateur Radio — in equipment required, in modes of operation and in results obtained. Today these worlds blend increasingly.

The almost universal use of SSB for voice work in the HF range has had a major impact on equipment design for the VHF and even UHF bands. Many amateurs have

a considerable investment in HF sideband gear. This equipment provides accurate frequency calibration and good mechanical and electrical stability. It is effective in CW as well as SSB communication. These qualities are attractive to VHF operators, so it is natural for them to look for ways to use HF gear on frequencies above 50 MHz.

Use of VHF accessory devices, both ready made and homebuilt, is increasing. This started years ago with the VHF receiving converter. Similar conversion equipment for transmitting has been widely used since SSB has been on the HF bands. Today the HF trend is to one-package stations, transceivers. The obvious move for many VHF operators is a companion box to perform both transmitting and receiving conversion functions. Known as transverters, these are offered by several manufacturers. They are also relatively simple to build, and are thus attractive projects for the home-builder of VHF gear.

CW, SSB and FM are the predominant modes on the VHF bands. Each method of communication has certain transmitter requirements. In general equipment used for FM is of the oscillator-multiplier type. The multiplier stages are needed to realize the desired deviation.

Because SSB cannot be passed satisfactorily through a frequency-multiplication stage, generation of VHF SSB signals requires the use of one or more mixer stages. VHF CW may be generated either by mixing or by frequency multiplying. Recently, manufacturers of amateur FM transceivers have been using a combination of both approaches. In the multimode VHF transceiver, which offers the operator a choice of CW, SSB, FM and often AM, we find both approaches to signal generation.

Transmitter Basics

The most basic type of transmitter is one that contains a single stage, is crystal con-

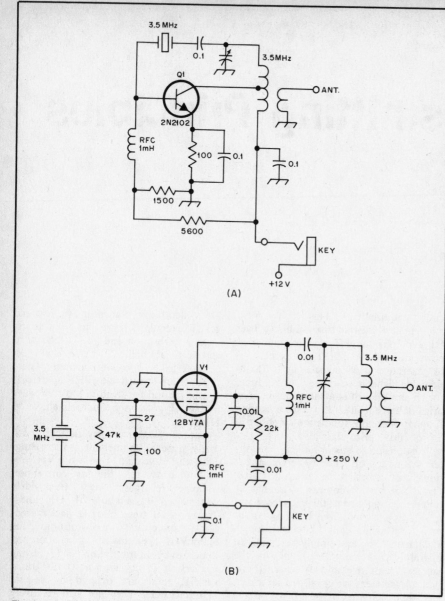

Fig. 1 — A transistor oscillator is shown at A. The example at B illustrates a type of crystal oscillator.

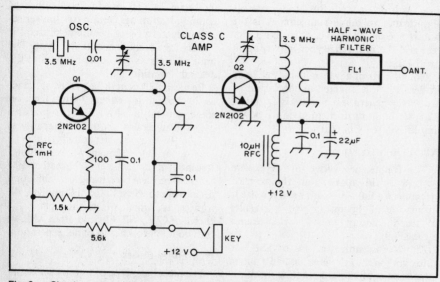

Fig. 2 — Circuit example of a simple, solid-state CW transmitter.

trolled, and is designed for CW operation. A circuit example is given in Fig. 1. This kind of transmitter is not especially suitable for use on the air because it is inefficient and is prone to generating a chirpy CW signal unless loaded lightly. But the same circuit is entirely acceptable when followed by an isolating stage (buffer-amplifier) as shown in Fig. 2. The second stage not only builds up the power level, but also gives the oscillator a relatively constant load to look into. The constant load helps to prevent oscillator pulling and attendant chirping of the CW note.

Fig. 3 illustrates the basic types of transmitters for CW and RTTY work. The drawing at A represents the general circuit given in Fig. 2. Illustration B is an expansion of that circuit and inclues a frequency multiplier. A heterodyne type of generator, which is currently popular for multiband transmitters and tranceivers, is shown as the exciter section of a transmitter in drawing C. A frequency synthesizer is shown as the RF generator at D.

For operation on AM, any of the lineups given in Fig. 3 is suitable, provided a modulator is added. It is used to modulate the operating voltage to the PA stage, or in some designs the operating voltage to the PA and the stage immediately before it.

The block diagram of Fig. 4 outlines the functional stages of an SSB transmitter. Zl can be a simple VFO, a heterodyne generator (Fig. 3C) or a frequency synthesizer. The essential difference between this type of transmitter and one that would be used for CW/RTTY is that the RF amplifiers must operate in Class A, AB or B (linear), rather than in Class C, which is suitable for CW work. However, linear amplifiers are entirely satisfactory for any transmission mode at a sacrifice in efficiency. Once the SSB signal is generated it cannot be passed through a frequency multiplier. All post-filter stages must operate straight through. Class C amplifiers are generally used in FM transmitters as well as in CW and RTTY transmitters.

Fig. 5 illustrates two approaches to VHF FM transmitters. Shown at A is a transmitter that uses phase modulation, also known as indirect FM. Multiplier stages bring the signal to the operating frequency and increase the deviation to the desired range. A frequency-synthesized direct FM transmitter is shown at B. Fewer multiplier stages are required for this method because greater deviation can be realized at the modulator. Audio is amplitude limited and the frequency response is shaped before modulation in both transmitters.

Crystal Switching

Although several crystals for a single oscillator can be selected by mechanical means, a switch must be contained in the RF path. This can impose severe restrictions on the equipment layout. Furthermore, mechanical switches normally require that they be operated from the front panel of the transmitter or receiver. That type of format complicates the remote operation of such a unit. Also, the switch leads can introduce unwanted reactances in the crystal circuit. A better technique is illustrated in Fig. 6, where D1 and D2 — high-speed silicon switching diodes — are used to select one of two or more crystals from some remote point. As operating voltage is applied to one of the diodes by means of S1, it is forward biased into "hard" conduction, thereby completing the circuit between the crystal trimmer and ground. Some schemes actually call for reverse-biasing the unused diode or diodes when they are not activated. This ensures almost complete cutoff, which may not be easy to obtain in the circuit shown because of the existing RF voltage on the anodes of D1 and D2.

VARIABLE-FREQUENCY OSCILLATORS

The theory and general application of variable-frequency oscillators is treated in Chapter 10. The circuit principles are the same regardless of the VFO application.

Some additional considerations pertain to the use of VFOs in transmitters as compared to a VFO in a receiver. Generally, heating of the interior of a transmitter cabinet is greater than in a receiver. This is because considerably more power is being dissipated in the transmitter. Therefore, greater care must be given to oscillator long-term stability. Temperature-compensating capacitors are often needed in the frequency-determining portion of the oscillator to level off the long-term stability factor. Some oscillators are designed for use with a temperature-control oven for the purpose of maintaining a relatively constant ambient temperature in the oscillator compartment — even while the equipment is otherwise turned off. Another design

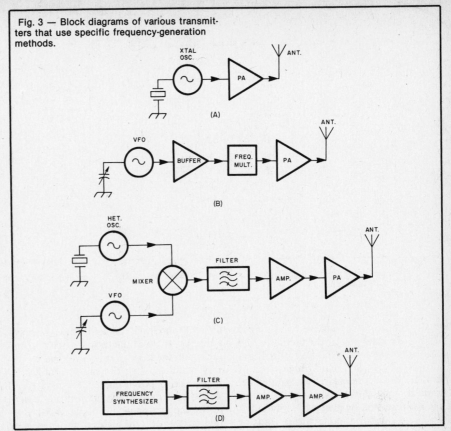

Fig. 3 — Block diagrams of various transmitters that use specific frequency-generation methods.

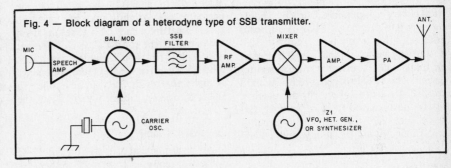

Fig. 4 — Block diagram of a heterodyne type of SSB transmitter.

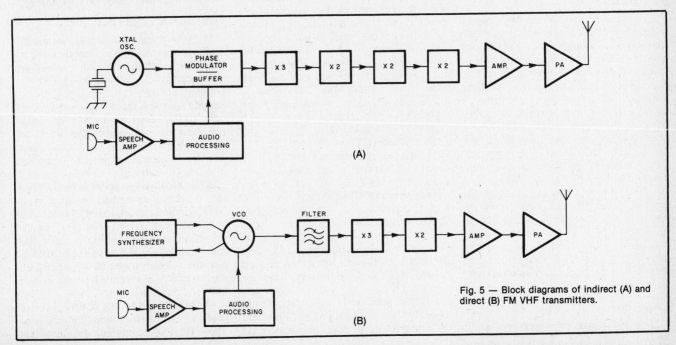

Fig. 5 — Block diagrams of indirect (A) and direct (B) FM VHF transmitters.

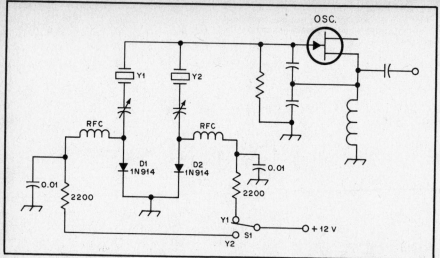

Fig. 6 — A method for selecting one of two (or several) crystals by means of diode switching. D1 and D2 are the switches.

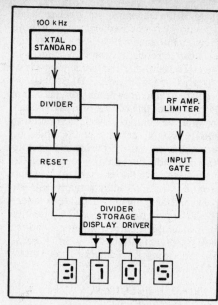

Fig. 7 — Frequency counter block diagram.

matter related to a transmitter-contained VFO is RF shielding of the oscillator and the attendant low-level buffer-isolation stages that follow it. Fairly high levels of stray RF can be present in a transmitter and some of that energy may migrate to the oscillator section by means of stray radiation or conduction along wiring leads or circuit-board elements. Thus, it is important to provide as much physical and electrical isolation as possible. The VFO should be housed in a rigid metal box. All dc leads entering the enclosure require RF-decoupling networks that are effective at all frequencies involved in the transmitter design. The VFO box needs to be fastened securely to the metal chassis on which it rests to ensure good electrical contact. Excessive stray RF entering the VFO circuitry can cause severe instability and erratic oscillator operation.

VFO Dials

One of the tasks facing an amateur builder is the difficulty of finding a suitable dial and drive assembly for a VFO. A dial should provide a sufficiently slow rate of tuning — 10 to 25 kHz per knob revolution is considered optimum — without backlash. Planetary drives are popular because of their low cost; however, they often develop objectional backlash after a short period of use. Several types of two-speed drives are available. They are well suited to home-made amateur equipment. The Eddystone 898 precision dial has long been a favorite with amateurs, although the need to elevate the VFO far above the chassis introduces some mechanical stability problems. If a permeability-tuned oscillator (PTO) is used, one of the many types of turn counters made for vacuum variable capacitors or rotary inductors may be employed.

Linear Readout

If a linear-dial frequency readout is

desired, the variable capacitor must be only a small portion of the total capacitance in the oscillator tank. Capacitors tend to be very nonlinear near the ends of rotation. A gear drive providing a 1.5:1 reduction should be employed so that only the center of the capacitor range is used. Then, as a final adjustment, the plates of the capacitor must be filed until linear readout is achieved. In a PTO, the pitch of the oscillator coil winding may be varied so that a linear frequency change results from the travel of the tuning slug. A different approach is to use a variable-capacitance diode (Varicap) as the VFO tuning element. A meter that reads the voltage applied to the Varicap is calibrated to indicate the VFO frequency.

Electronic Dials

An electronic dial consists of a simplified frequency counter that reads either the VFO or operating frequency of a transmitter or receiver. The advantage of an electronic dial is the excellent accuracy (to 1 hertz, if desired) and the fact that VFO tuning does not have to be linear. The readout section of the dial may use neon-glow tubes called Nixies® or a seven-segment display using incandescent lamps, filament wires in a vacuum tube, LCDs (liquid crystal display) or LEDs (light-emitting diodes). The use of MSI and LSI circuits, some containing as many as 200 transistors on a single chip, reduces the size required for an electronic dial to a few square inches of circuit-board space.

A typical counter circuit is given in Fig. 7. The accuracy of the counter is determined by a crystal standard, often referred to as a clock. The output from a 100-kHz calibration oscillator, the type often used in receivers and transceivers, may be employed if an accuracy of 100 Hz is sufficient. For readout down to 1 Hz, a 1- to 10-MHz AT-cut crystal should be chosen, because this type of high-accuracy crystal

exhibits the best temperature stability. The clock output energy is divided in decade-counter ICs to provide the pulse that opens the input gate of the counter for a preset time.

The number of RF cycles that pass through the gate while it is open are counted and stored. Storage is used so the readout does not blink. At the end of each counting cycle, the stored information activates the display LEDs, which present the numbers counted until another count cycle is complete. A complete electronic dial arranged to be combined with an existing transmitter or receiver was described in October 1970 *QST*. Also, Macleish et al. reported an adapter that allows a commercially made frequency counter to be mated with ham gear so the counter performs as an electronic dial (May 1971 *QST*).

Frequency Multiplication

If it is necessary to use frequency multipliers at some point after the VFO or other frequency generator in a transmitter, the circuits of Fig. 8 can be applied. Of course, vacuum-tube multipliers are entirely suitable if the design is not one that uses semiconductors. The fundamental principles for frequency multiplication are applicable to tubes and transistors alike. The requirement is to operate the devices in Class C. Although a transistor circuit may be operated with forward bias applied to a frequency multiplier, the stage must be driven hard enough to override the bias and operate Class C. Forward bias is sometimes used in a multiplier stage (solid state) to lower the excitation requirements. Negative voltage (reverse bias) is often used on the grid of a vacuum-tube multiplier, but forward bias is not.

The circuit of Fig. 8A is probably the least suitable for frequency multiplication. Typically, the efficiency of a doubler of this type is 50 percent, a tripler is 33 percent and a quadrupler is 25 percent. Additionally, harmonics other than the one to which the output tank is tuned will appear in the output unless effective band-pass filtering is applied. The collector tap on L1 of Fig. 8A is placed at a point that offers a reasonable compromise between power output and spectral purity. The lower the tap with respect to V_{cc}, the lighter the collector loading on L1 and the greater the filtering action of the tuned circuit. The trade-off is, however, a reduction in output power as the mismatch of the collector to the load increases.

A push-push doubler is shown at Fig. 8B. Because of the conduction angle of this type of circuit the efficiency is similar to that of a straight amplifier operating in Class C. Also, the driving frequency (f) will be well attenuated at the doubler output if electrical balance and component symmetry are ensured. A 12AU7A tube will work nicely in this type of circuit well into the VHF region. T1 in this example is a trifilar-wound, broad-band toroidal transformer. It drives the gates of Q1 and Q2 in push-pull (opposite phase). The drains are in parallel and are tuned to 2f. R1 is used to establish electrical balance between Q1 and Q2. R1 is set while the doubler is being fully driven. Diode doublers can be used in a similar circuit; see Chapter 4.

A push-pull tripler is illustrated in Fig. 8C. The matter of electrical balance and symmetry is important to proper operation. The circuit discriminates against even harmonics, aiding spectral purity. The efficiency is somewhat better than a tripler using the circuit of Fig. 8A. If vacuum tubes are used in the circuits of Fig. 8, the input ports should employ high-impedance tuned circuits for best performance.

The Oscillator-Multiplier VHF Transmitter

This type of transmitter, which may be used for FM or CW, generally starts with a crystal oscillator operating in the HF range, followed by one or more frequency-multiplier stages and at least one amplifier stage. While relatively simple to construct, such transmitters can cause much grief unless the builder takes precautions to prevent undesired multiples of the oscillator and the multiplier stages from being radiated. For frequencies below 450 MHz the transmitting mixer is not difficult to construct and is recommended for most applications. Spurious-signal radiation is much easier to prevent with the mixer, although it does not lend itself to compact FM equipment design. For operation on the amateur UHF bands, the oscillator-multiplier approach offers definite advantages. Fig. 9 shows how the harmonics of a 144-MHz signal may be multiplied to permit operation on amateur UHF and microwave bands. Stability at 144 MHz is easy to realize with current technology, making stable microwave signals simple to generate. Varactor diodes are used as frequency-multiplying devices. They are installed in resonant cavities. Operation will probably be crystal controlled, as even the best transceiver/transverter combination used to generate the 144-MHz signal may create problems when the output is multiplied in frequency 40 times! A frequency synthesizer with a stable reference oscillator may be used to generate the 144-MHz signal, but its output should be well filtered to eliminate noise.

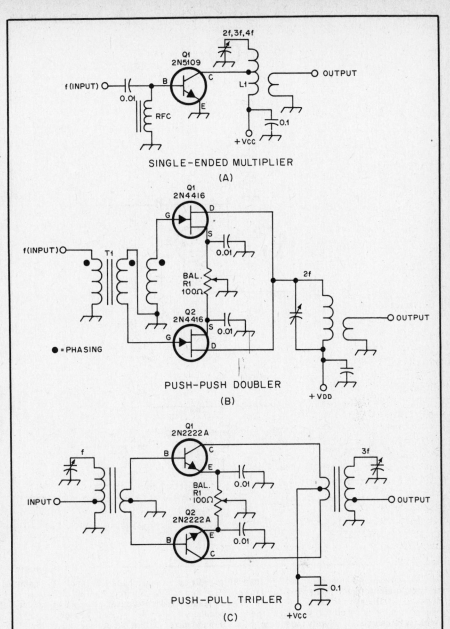

Fig. 8 — Single-ended multiplier (A), push-push doubler (B) and push-pull tripler (C).

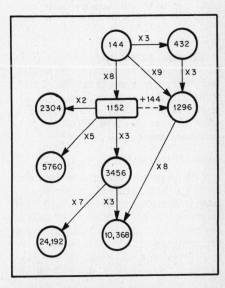

Fig. 9 — Multiplier scheme for generating RF on weak-signal communication frequencies at UHF and above. All frequencies are in megahertz. An 1152-MHz LO can be used to drive microwave multiplier circuits or serve as an LO for a 1296-MHz transverter with a 144-MHz IF.

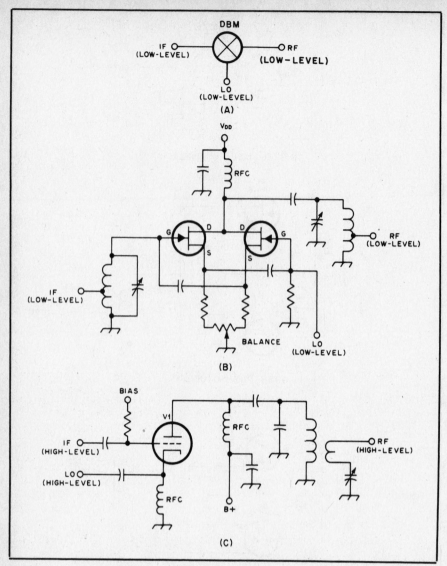

Fig. 10 — A trio of commonly used VHF transmitting mixers. At A, perhaps the simplest, a commercially available diode doubly balanced mixer. RF output is CW, requiring the use of several stages of amplification to reach a useful level. At B, a singly balanced mixer using FETs. Adjustment of this circuit is critical to prevent the local-oscillator signal from leaking through. A mixer of this type can supply slightly more output than a diode mixer (3 dBm, as opposed to 1 dBm for the mixer shown at A). At C, a high-level mixer using a vacuum-tube triode is shown. In this case, V1 might be a type 2C39 or 7289 tube. With the correct circuit constants this mixer could provide an SSB output of 15 watts on 1296 MHz. Power input would be about 100 watts! In addition, 10 watts of LO and 5 watts of IF drive would be needed. Despite these requirements, such a circuit provides a relatively low-cost means of generating high-level microwave SSB. Spurious outputs at the LO and image frequencies will be quite strong. To attenuate them, a strip-line or cavity filter should be used at the mixer output.

Although spurious outputs of the various multiplier stages may not cause harmful interference, that is no excuse for not removing them. In most cases, the Q of successive cavities will suffice. A band-pass filter may be used to filter the final multiplier stage.

Transmitting Mixers

With the possible exception of the power levels involved, there is no reason to consider transmitting mixers differently than their receiving counterparts. One thing to keep in mind is that many deficiencies in the transmitter mixer will show up on the air. Receiver mixer troubles are *your* problem. Transmitter mixer troubles

become everyone's problem!

A trio of popular types of transmitting mixers is shown is Fig. 10. The doubly balanced diode mixer at A may be built using discrete components or the phase relationship between ports may be established using etched-circuit strip lines. Miniature DBMs are available at low cost from several manufacturers. They offer an almost foolproof method of generating VHF SSB. Another popular mixer uses a pair of FETs in a singly balanced configuration. If care is taken in construction and adjustment, local-oscillator rejection will be adequate with this circuit. To be safe, a series-tuned trap, designed to attenuate the LO leakage even further,

should follow this stage. A typical FET balanced mixer is shown in Fig. 10B.

Finally, we see a typical vacuum-tube mixer (Fig. 10C). Because it can handle more power, the tube mixer has endured at VHF. Its higher output, when compared to most solid-state mixers, reduces the number of subsequent amplifier stages needed to reach a specific power level. This is the only advantage of using them as mixers, at least on the lower VHF bands.

High-Level Transmitting Mixers

When you are designing a transmitting converter for VHF, the trade-offs between the advantages of mixing at a low power level, such as in a diode-ring mixer, and using several stages of linear amplification must be weighed against the cost of amplifying devices. On 432 MHz and above, it may be desirable to mix the IF and local-oscillator signals at a fairly high level. This method makes it unnecessary to use costly linear devices to reach the same power level. High-level mixing results in a slightly distorted signal, so it should be used only when essential.

Fig. 11 gives the schematic diagram of a typical 432-MHz high-level mixer. V1 is the final amplifier tube of a retired commercial 450-MHz FM transmitter. The oscillator and multiplier stages now produce local-oscillator injection voltage, which is applied to the grid as before. The major change is in the cathode circuit. Instead of being directly at ground, a parallel LC circuit is inserted and tuned to the IF. In this case 10-meters was chosen. With the exception of 144 MHz, any amateur band could serve as the IF. A 144-MHz IF is unsuitable because the third harmonic would appear at the output, where it would combine with the desired signal. For that reason, some additional output filtering is needed with this circuit. The original crystal in the transmitter is replaced with one yielding an output at the desired local-oscillator frequency, then the intermediate stages are retuned.

One disadvantage of the high-level mixer is the relatively large amount of local-oscillator injection required. In many cases it is simpler to mix at a lower level and use linear amplifiers than to construct the local-oscillator chain. On the higher bands, it may be feasible to generate local-oscillator energy at a lower frequency and use a passive varactor mixer to reach the injection frequency. In this case the previous statements about diode-multiplier spurious outputs apply. If the local-oscillator injection is impure, the mixer output will be also. For further information on mixers see Chapters 4 and 12 of this publication and *Solid State Design for the Radio Amateur,* an ARRL publication.

Output Filtering

Output purity from oscillators, multipliers and amplifiers is of paramount importance to the performance of numerous circuits. In the interest of compliance with current FCC regulations, wherein all spurious emissions from a transmitter must be 40 dB or greater below the peak power of the desired signal at HF and 60 dB or greater between 30 and 225 MHz, filtering is important. The type of filter used — band-pass, notch, low-pass or high-pass — will depend on the application. Band-pass filters afford protection against spurious responses above and below the amateur band for which they have been designed. Low-pass filters attenuate energy above the desired output frequency, while high-pass filters reduce energy below the band of interest.

It is common practice to include a harmonic filter at the output of a VFO chain to ensure purity of the driving voltage to a mixer or amplifier stage. The filter bandwidth must be adequate for the tuning range of the VFO in order to prevent attenuation of the output energy within the desired band. For this reason, some designers prefer to use a low-pass type of filter instead of a band-pass one.

The information contained in Figs. 12 through 14 and in Tables 1 through 5 will allow the builder to select an appropriate Chebyshev filter design to fulfill a particular need. Information is included for both high-pass and low-pass filters with 1, 0.1, 0.01 and 0.001 dB band-pass ripple. These figures correspond to SWRs of 2.66, 1.36, 1.10 and 1.03, respectively. Additionally, information is provided for both T and pi types of filter configurations.

The filters are nomalized to a frequency of 1 MHz and an input and output impedance of 50 ohms. To translate the designs to other frequencies, simply divide the component values by the new frequency in megahertz. (The 1-MHz value represents a "cutoff" frequency. That is, the attenuation increases rapidly above this frequency for the low-pass filter or below this frequency for the high-pass filter. This effect should not be confused with the variations in attenuation in the passband.) For instance, if it is desired to reduce harmonics from a VFO at frequencies above 5 MHz (the new cutoff frequency), the inductance values would be divided by 5 and the capacitance values multiplied by 5.

Other impedance levels can also be used by multiplying the inductor values by the ratio $Z_o/50$ and the capacitor values by $50/Z_o$, where Z_o is the new impedance. This factor should be applied in addition to the ones for frequency translation.

In order to select a suitable filter design the builder must determine the amount of attenuation required at the harmonic frequencies (for the low-pass case) or subharmonic frequencies (for the high-pass application). Additionally, the builder must determine the maximum permissible amount of band-pass ripple and therefore the SWR of the filter. With this information the builder can refer to Table 1 to select an appropriate filter design. The at-

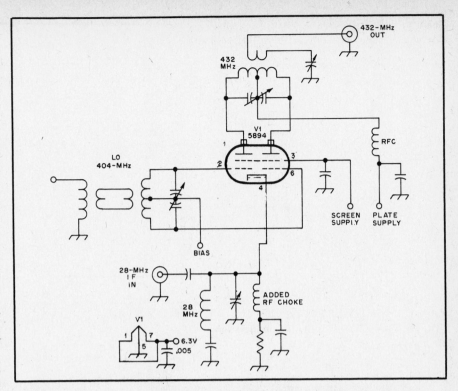

Fig. 11 — Partial schematic diagram of a 70-cm (432-MHz) mixer, built from a converted FM transmitter. The original oscillator-multiplier-driver stages of the unit now provide LO injection. A stripline filter should be used at the output of the mixer to prevent radiation of spurious products.

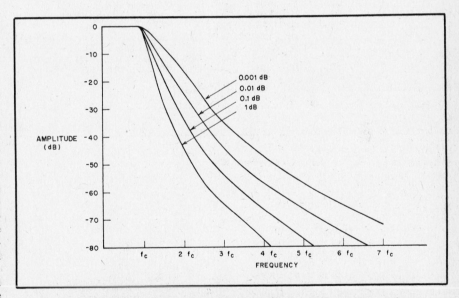

Fig. 12 — A representative drawing of the amplitude vs. frequency response that can be expected from a 5-element low-pass filter designed from the information contained in Tables 2 or 3. The exact amount of attenuation (theoretical) can be obtained from Table 1. This drawing shows how passband ripple and roll-off slope are interrelated.

tenuation values given here are theoretical and assume perfect components, no coupling between filter sections and no signal leakage around the filter. A "real life" filter should follow these values fairly close down to the 60- or 70-dB attenuation level. At this point the theoretical response will likely be degraded somewhat by the factors just mentioned. Once the filter design has been selected the builder

can refer to Tables 2 through 5 to obtain the normalized component values.

In many cases the calculated capacitor values will be sufficiently close to a standard value so that the standard-value item may be used. Alternatively, a combination of fixed-value silver-mica capacitors and mica compression trimmers can be used in parallel to obtain the chart values. Toroidal inductors, because of their self-shielding

Table 1
Chebyshev High-Pass and Low-Pass Filters — Attenuation (dB)

No. elements	Ripple, dB	SWR	$2f_c$	$3f_c$	$4f_c$	$5f_c$	$6f_c$	$7f_c$
3	1	2.66	22.46	34.05	41.88	47.85	52.68	56.74
3	0.1	1.36	12.24	23.60	31.42	37.39	42.22	46.29
3	0.01	1.10	4.08	13.73	21.41	27.35	32.18	36.24
3	0.001	1.03	0.63	5.13	11.68	17.42	22.20	26.25
5	1	2.66	45.31	64.67	77.73	87.67	95.72	102.50
5	0.1	1.36	34.85	54.21	67.27	77.21	85.26	92.04
5	0.01	1.10	24.82	44.16	57.22	67.17	75.22	82.00
5	0.001	1.03	14.94	34.16	47.22	57.16	65.22	71.99
7	1	2.66	68.18	95.29	113.57	127.49	138.77	148.26
7	0.1	1.36	57.72	84.83	103.11	117.03	128.31	137.80
7	0.01	1.10	47.68	74.78	93.07	106.99	118.27	127.75
7	0.001	1.03	37.68	64.78	83.06	96.98	108.26	117.75
9	1	2.66	91.06	125.91	149.42	167.32	181.82	194.01
9	0.1	1.36	80.60	115.45	138.96	156.86	171.36	183.55
9	0.01	1.10	70.56	105.41	128.91	146.81	161.31	173.51
9	0.001	1.03	60.55	95.40	118.91	136.91	151.31	163.50

Note: For high-pass filter configuration $2f_c$ becomes $f_c/2$, etc.

Fig. 13 — A photograph of a 7-element low-pass filter designed with the information contained in Table 3. The filter is housed in a small aluminum Minibox.

Table 2
Chebyshev Low-Pass Filter — T Configuration

No. elements	Ripple, dB	L1	L2	L3	L4	L5	C1	C2	C3	C4
3	1	16.10	16.10				3164.3			
3	0.1	8.209	8.209				3652.3			
3	0.01	5.007	5.007				3088.5			
3	0.001	3.253	3.253				2312.6			
5	1	16.99	23.88	16.99			3473.1	3473.1		
5	0.1	9.126	15.72	9.126			4364.7	4364.7		
5	0.01	6.019	12.55	6.019			4153.7	4153.7		
5	0.001	4.318	10.43	4.318			3571.1	3571.1		
7	1	17.24	24.62	24.62	17.24		3538.0	3735.4	3538.0	
7	0.1	9.400	16.68	16.68	9.400		4528.9	5008.3	4528.9	
7	0.01	6.342	13.91	13.91	6.342		4432.2	5198.4	4432.2	
7	0.001	4.690	12.19	12.19	4.690		3951.5	4924.1	3951.5	
9	1	17.35	24.84	25.26	24.84	17.35	3562.5	3786.9	3786.9	3562.5
9	0.1	9.515	16.99	17.55	16.99	9.515	4591.9	5146.2	5146.2	4591.9
9	0.01	6.481	14.36	15.17	14.36	6.481	4542.5	5451.2	5451.2	4542.5
9	0.001	4.854	12.81	13.88	12.81	4.854	4108.2	5299.0	5299.0	4108.2

Component values normalized to 1 MHz and 50 ohms, L in µH; C in pF.

Table 3
Chebyshev Low-Pass Filter — Pi Configuration

No. elements	Ripple, dB	C1	C2	C3	C4	C5	L1	L2	L3	L4
3	1	6441.3	6441.3				7.911			
3	0.1	3283.6	3283.6				9.131			
3	0.01	2002.7	2002.7				7.721			
3	0.001	1301.2	1301.2				5.781			
5	1	6795.5	9552.2	6795.5			8.683	8.683		
5	0.1	3650.4	6286.6	3650.4			10.91	10.91		
5	0.01	2407.5	5020.7	2407.5			10.38	10.38		
5	0.001	1727.3	4170.5	1727.3			8.928	8.928		
7	1	6896.4	9847.4	9847.4	6896.4		8.85	9.34	8.85	
7	0.1	3759.8	6673.9	6673.9	3759.8		11.32	12.52	11.32	
7	0.01	2536.8	5564.5	5564.5	2536.8		11.08	13.00	11.08	
7	0.001	1875.7	4875.9	4875.9	1875.7		9.879	12.31	9.879	
9	1	6938.3	9935.8	10,105.	9935.8	6938.3	8.906	9.467	9.467	8.906
9	0.1	3805.9	6794.5	7019.9	6794.5	3805.9	11.48	12.87	12.87	11.48
9	0.01	2592.5	5743.5	6066.3	5743.5	2592.5	11.36	13.63	13.63	11.36
9	0.001	1941.7	5124.6	5553.2	5124.6	1941.7	10.27	13.25	13.25	10.27

Component values normalized to 1 MHz and 50 ohms, L in µH; C in pF.

It is a 7-element, low-pass type of pi configuration. The unit is housed in a small aluminum Minibox and makes use of BNC connectors for the input and output connections.

Driver Stages

The choice between tubes and transistors in low-level amplifier and driver stages will depend on the nature of the composite transmitter. Some designs contain a mixture (hybrid) of tubes and semiconductors, while other circuits have no vacuum tubes at all. If tubes are used in a hybrid circuit, they are generally restricted to the driver and PA sections of the transmitter.

There is no reason why tubes should be used in preference to power transistors for output powers up to say, 150 watts, despite the prevailing myth that tubes are more rugged, operate more stably and produce less spurious output. It is true that transistors are less tolerant than tubes to SWR levels in excess of 2:1, but a correctly designed transistor amplifier can be operated safely if SWR-protection circuitry is included. Furthermore, spectral purity can be just as good from a solid-state amplifier as it is from a vacuum-tube amplifier. A harmonic filter normally follows a solid-state power stage, whereas this measure may not be required when tubes are used. Amplifier IMD (third- and fifth-order products) in solid-state power stages which operate linearly is fully as acceptable as that observed in most tube types of linear amplifiers. Typically, if a design is correct, the IMD will be on the order of -33 dB from the reference power value.

The major area of concern when designing a solid-state driver or PA section is to prevent low-frequency self-oscillations. Such parasitics tend to modulate the carrier and appear as spurious responses within the amplifier passband. The low-frequency parasitics occur as a result of the extremely high gain exhibited by HF and VHF transistors at the low-frequency end of the spectrum. The theoretical gain increase for a given transistor is 6 dB per octave as the operating frequency is lowered. The same is not true of vacuum tubes. Therefore, it is necessary to employ

properties, are ideal for use in these filters. Miniductor stock can also be used. However, it is much bulkier and will not offer the same degree of shielding between filter sections. Disc ceramic or paper capacitors are not suitable for use in RF filters. Standard mica or silver-mica types are recommended.

Fig. 13 shows a filter that was designed with the information contained in Table 3.

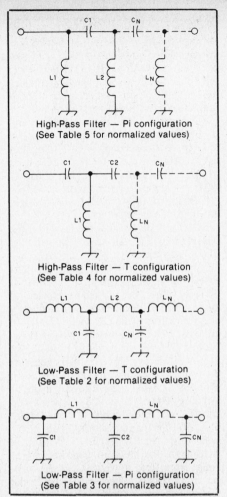

High-Pass Filter — Pi configuration
(See Table 5 for normalized values)

High-Pass Filter — T configuration
(See Table 4 for normalized values)

Low-Pass Filter — T configuration
(See Table 2 for normalized values)

Low-Pass Filter — Pi configuration
(See Table 3 for normalized values)

Fig. 14 — Shown here are the four filter types discussed in the text and Tables 1 through 5.

Table 4
Chebyshev High-Pass Filter — T Configuration

No. elements	Ripple, dB	C1	C2	C3	C4	C5	L1	L2	L3	L4
3	1	1573.0	1573.0				8.005			
3	0.1	3085.7	3085.7				6.935			
3	0.01	5059.1	5059.1				8.201			
3	0.001	7786.9	7786.9				10.95			
5	1	1491.0	1060.7	1491.0			7.293	7.293		
5	0.1	2775.6	1611.7	2775.6			5.803	5.803		
5	0.01	4208.6	2018.1	4208.6			6.098	6.098		
5	0.001	5865.7	2429.5	5865.7			7.093	7.093		
7	1	1469.2	1028.9	1028.9	1469.2		7.160	6.781	7.160	
7	0.1	2694.9	1518.2	1518.2	2694.9		5.593	5.058	5.593	
7	0.01	3994.1	1820.9	1820.9	3994.1		5.715	4.873	5.715	
7	0.001	5401.7	2078.0	2078.0	5401.7		6.410	5.144	6.410	
9	1	1460.3	1019.8	1002.7	1019.8	1460.3	7.110	6.689	6.689	7.110
9	0.1	2662.2	1491.2	1443.3	1491.2	2662.2	5.516	4.922	4.922	5.516
9	0.01	3908.2	1764.1	1670.2	1764.1	3908.2	5.576	4.647	4.647	5.576
9	0.001	5218.3	1977.1	1824.6	1977.1	5218.3	6.657	4.780	4.780	6.657

Component values normalized to 1 MHz and 50 ohms, L in μH; C in pF.

Table 5
Chebyshev High-Pass Filter — Pi Configuration

No. elements	Ripple, dB	L1	L2	L3	L4	L5	C1	C2	C3	C4
3	1	3.932	3.932				3201.7			
3	0.1	7.714	7.714				2774.2			
3	0.01	12.65	12.65				3280.5			
3	0.001	19.47	19.47				4381.4			
5	1	3.727	2.652	3.727			2917.3	2917.3		
5	0.1	6.939	4.029	6.939			2321.4	2321.4		
5	0.01	10.52	5.045	10.52			2439.3	2439.3		
5	0.001	1.466	6.074	1.466			2837.3	2837.3		
7	1	7.159	5.014	5.014	7.159		1469.2	1391.6	1469.2	
7	0.1	6.737	3.795	3.795	6.737		2237.2	2023.1	2237.2	
7	0.01	9.985	4.552	4.552	9.985		2286.0	1949.1	2286.0	
7	0.001	13.50	5.195	5.195	13.50		2564.1	2057.7	2564.1	
9	1	3.651	2.549	2.507	2.549	3.651	2844.1	2675.6	2675.6	2844.1
9	0.1	6.656	3.728	3.608	3.728	6.656	2206.5	1968.9	1968.9	2206.5
9	0.01	9.772	4.410	4.176	4.410	9.772	2230.5	1858.7	1858.7	2230.5
9	0.001	13.05	4.943	4.561	4.943	13.05	2466.3	1911.8	1911.8	2466.3

Component values normalized to 1 MHz and 50 ohms, L in μH; C in pF.

quality decoupling and bypassing in the circuit. It is similarly important to use low-Q, low-inductance RF chokes and matching networks to discourage low-frequency tuned-base, tuned-collector oscillations. The suppression concepts just discussed are illustrated in Fig. 15 at B and C. In the circuit at B there are two 950-mu ferrite beads added over the pigtail of RFC1 to swamp the Q of the choke. Three bypass capacitors (0.001, 0.01 and 0.1 μF) are used with RFC2 of Fig. 15B to provide effective RF decoupling from VHF to MF. A 22-μF capacitor is used near RFC2 to bypass the +V_{cc} line at low frequency and audio. This method is recommended for each high-gain solid-state stage in a transmitter.

Driver Circuits

The circuits of Fig. 15A and 15B are typical of those that would be employed to excite a tube-type PA stage. The 6GK6 tube driver at A can be biased for Class C or Class AB operation, making it suitable for CW or SSB service. Of course, the AB mode would be suitable for CW and SSB, and would require considerably less excitation power than would the same stage

operating in Class C. Other tubes that perform well in this circuit are the 6CL6, 12BY7A and 5763. The output tank is designed for high impedance so it will interface properly with the high-impedance grid of the PA. It may be necessary to include a neutralization circuit with this type of amplifier, especially if care is not exercised in the layout. The high transconductance of the 6GK6 series encourages self-oscillation near the operating frequency. Z1 is a parasitic choke that should be included as a matter of course to prevent VHF parasitics.

A transistor amplifier suitable for driving a Class C tube PA is presented in Fig. 15B. Q1 operates Class C, so it is not satisfactory for amplifying SSB energy. However, forward bias (approximately 0.7 volt) can be added between the emitter and ground to move the operating curve into the Class AB (linear) region, thereby making the stage suitable for SSB signal amplification. A 1.5-ohm resistor can be added between the emitter and ground to help prevent thermal runaway and to introduce degeneration

(feedback) for enhancing stability. No bypass capacitor would be used from emitter to ground if this were done. T1 is a narrow-band toroidal RF transformer that has a turns ratio suitable for transforming the collector impedance to the grid impedance (determined by the value of the grid resistor of the PA) of the final amplifier. The secondary winding of T1 is tuned to resonance at the operating frequency. Approximately 1 watt of output can be taken from Q1 in the HF region when a 12-volt V_{cc} is used. This is ample power for driving a pair of 6146B tubes in Class AB1.

A broadband solid-state driver is shown in Fig. 15C. The trade-off for broadband operation (1.8 to 30 MHz in this example) is a reduction in maximum available gain. Therefore, the output from Q1 of Fig. 15C will be less than 1 watt. The stage operates Class A, making it linear. The emitter is unbypassed to provide emitter degeneration. Shunt feedback is used between the base and collector to enhance stability and contribute to the broadband characteristic the

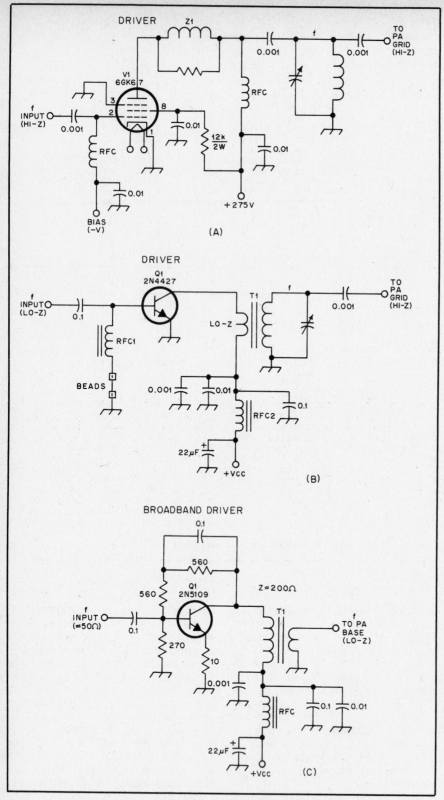

Fig. 15 — Circuit examples of transistor and tube driver stages for use in transmitters.

or peaked by means of a front-panel control. The transistor selected for broadband service should have very high f_T and beta ratings. Transistors designed for UHF service are excellent as HF amplifiers when broadbanding is contemplated. Neutralization is not necessary when using bipolar-transistor amplifiers.

A practical three-stage broadband amplifier strip is shown schematically in Fig. 16. With an input level of 10 mW, it is possible to obtain 1.4 watts of output from 3.5 to 29 MHz. A keying transistor (Q4) is included for turning the amplifier off by means of a VOX, or for keying it during CW operation.

RMS and dc voltages are noted on the diagram of Fig. 16 to aid in troubleshooting. Overall gain for the strip at 7 MHz is 31 dB, with slight gain variations elsewhere in the passband. T1 consists of 30 turns of no. 28 enameled wire (primary) on an FT50-43 toroid core. The secondary has four turns of no. 28 wire. T2 uses 16 turns of no. 28 enameled wire (primary) looped through an Amidon BLN-43-302 ferrite balun core. The secondary contains four turns of no. 28 wire. RFC1, RFC2 and RFC3 are 250-μH units. They are made by winding 20 turns of no. 28 enameled wire on FT37-43 toroid cores. D1 and D2 are 1-A, 50-PIV rectifier diodes. This driver was designed to excite a Motorola MRF449A PA stage to a power-output range from 15 to 30 watts. C1 at the emitter of Q1 can be selected to provide the overall gain needed in this strip. The value given at C1 proved suitable for the ARRL version of this amplifier. The final value will depend on the gain of the individual transistors acquired for this circuit.

Coupling Between Transmitter Stages

Correct impedance matching between a stage and its load provides maximum transfer of power. The load can be an antenna or a succeeding stage in a transmitter. Thus, the output impedance of a stage must be matched to the input of the following stage. Various forms of coupling networks are popular for use in tube or transistor circuits. The choice will depend on a number of considerations — available driving power versus tolerable mismatch, the selectivity required and impedance matching can be approximated by

$$Z = \frac{V_{cc}^2}{2P_o}$$

where Z is in ohms and P_o is the power output from the stage. However, determining the input impedance of the base of the following stage is difficult to do without expensive laboratory equipment. Generally, when the PA delivers more than 2 watts of output power, the base impedance of that stage will be less than 10 ohms — frequently just 1 or 2 ohms. For this reason, some kinds of LC matching networks do not lend themselves to the application. Furthermore, with the precise input impedance

circuit. T1 is a conventional broadband transformer wound on a toroid core. The turns ratio is adjusted to match the approximate 200-ohm collector impedance to the base impedance of the transistor PA stage — typically less than 5 ohms. Heat sinks are required for the transistors of Fig. 15B and C. The primary of T1 should have

a reactance of roughly four times the collector impedance at the lowest proposed operating frequency. Therefore, for 1.8 MHz the primary winding would be 70 μH (X_L = 800 ohms). This can be achieved easily by using an FT50-43 Amidon core. The primary advantage to a broadband driver is that it need not be band-switched

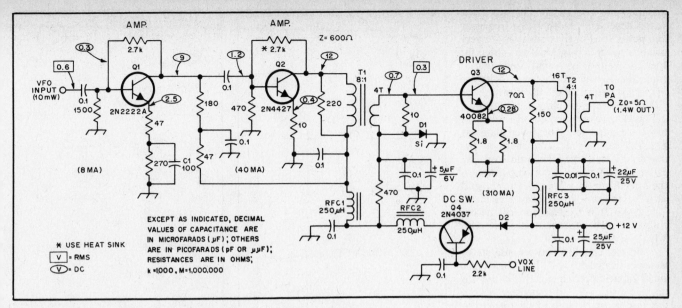

Fig. 16 — Practical circuit for a three-stage broadband amplifier/driver. See text.

of a transistor power amplifier not known, it becomes desirable to use what is sometimes referred to as a "sloppy" matching network. This is an LC network in which both the inductance and capacitance elements are variable to allow wide range of adjustment for matching. On the other hand, some designers purposely introduce a mismatch between stages to control the power distribution and aid stability. When this technique is used, it is necessary to have more driving power than would be needed under a matched condition. An intentional mismatch results in a trade-off between gain and the desired end effect of introducing a mismatch.

In the interest of stability, it is common practice to use low-Q networks between stages in a solid-state transmitter. The penalty for using a low-Q resonant network is poor selectivity; there is little attenuation of harmonic or other spurious energy. Conversely, tube stages operate at relatively high plate and grid impedance levels and can be neutralized easily (not true of transistors). This permits the use of high-Q networks between stages, which in turn provide good selectivity. Most solid-state amplifiers use matching networks with loaded Qs of 5 or less. Tube stages more commonly contain networks with loaded Qs of 10 to 15. The higher the Q, up to a practical limit, the greater the attenuation of frequencies other than the desired ones. In all cases, the input and output capacitances of tubes and transmitters must be included in the network constants, or to use the engineering vernacular, "absorbed" into the network. The best source of information on the input and output capacitances of power transistors is the

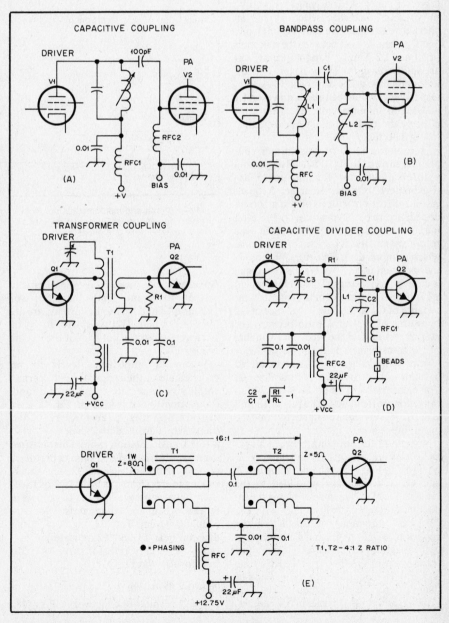

Fig. 17 — Typical coupling methods for use between amplifier stages. See text.

manufacturer's data sheet. The impedance values depend on the operating frequency and power level of the transistor — a very complex set of curves. Most data sheets list specific values of input and output capacitance, which do not vary with the operating frequency or power level.

The interstage coupling method shown in Fig. 17A is a common one when vacuum tubes are employed. The driver plate has a tuned circuit that is resonant at the operating frequency. A low-value coupling capacitor (100 pF in this example) routes the drive from the plate of V1 to the grid of V2 across a high-impedance element, RFC2. RFC1 is used as part of the decoupling network for the supply voltage to V1.

Band-pass coupling between tube stages is demonstrated at Fig. 17B. C1 has a very small capacitance value and is chosen to provide a single-hump response when the two resonators (L1 and L2) are peaked to the operating frequency. The principal advantage to this circuit over that of Fig. 17A is greater purity of the driving energy to V2 by virtue of increased selectivity. As an alternative to capacitive coupling (C1), link coupling can be used between the cold ends of L1 and L2. Similar band-pass networks are applicable to transistor stages. The collector and base of the driver and PA stages, respectively, would be tapped down on L1 and L2 to minimize loading. This helps preserve the loaded Q of the tuned circuits, aiding selectivity.

A common form of transformer coupling is shown in Fig. 17C. T1 is usually a toroidal inductor for use up to approximately 30 MHz. At higher frequencies it is often difficult to obtain a correct impedance ratio. Depending on the total number of transformer turns used, the secondary might call for less than one turn, which is impractical. However, for most of the spectrum up to 30 MHz this technique is entirely satisfactory. The primary tap on T1 is chosen to transform the collector impedance of Q1 to the base impedance of Q2 by means of the turns ratio between the tapped section and the secondary winding of the transformer. If there is a tendency toward self-oscillation, R1 may be added in shunt with the secondary to stabilize Q2. The value used will be in the 5- to 27-ohm range for most circuits. The rule of thumb is to use just enough resistance to tame the instability.

A method for coupling between stages by means of a capacitive divider is illustrated in Fig. 17D. The net value of C1 and C2 in series must be added to the capacitance of C3 when determining the inductance required for resonance with L1. The basic equation for calculating the capacitance ratio of C1 and C2 is included in the diagram. RFC1 serves as a dc return for the base of Q2. The Q of the RF choke is degraded intentionally by the addition of two 950-mu ferrite beads. This aids stability, as discussed earlier in this chapter.

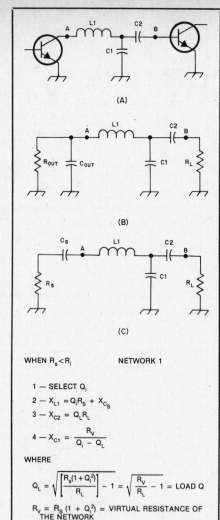

Fig. 18 — Circuit and mathematical solution for matching network no. 1. Actual circuit (A), parallel equivalent (B) and series equivalent (C).

An advantage to using this type of circuit is that VHF and UHF parasitics are discouraged and harmonic currents are attenuated when C2 is fairly high in capacitance. This is not true of the circuit in Fig. 17C.

When the impedance levels to be matched are of the proper value to permit employing specific-ratio broadband transformers, the circuit of Fig. 17E is useful. In this example, two 4:1 transformers are used in cascade to provide a 16:1 transformation ratio. This satisfies the match between the 80-ohm collector of Q1 and the 5-ohm base of Q2. The shortcoming of this technique is the lack of selectivity between stages, but the advantage is in the broadband characteristic of the coupling system. The phasing dots on the diagram near T1 and T2 indicate the correct electrical relationship of the transformer windings.

Network Equations

The three networks shown in Figs. 18,

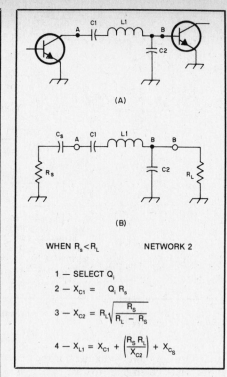

Fig. 19 — Circuit and equations for network no. 2. Actual circuit (A) and series equivalent (B).

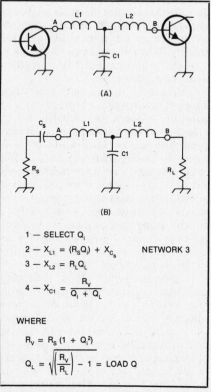

Fig. 20 — Network-solution equations and circuit for network no. 3. Actual circuit (A) and series equivalent (B).

19 and 20 will provide practical solutions to many of the impedance-matching problems encountered by amateurs. In each of the figures it is assumed that the output

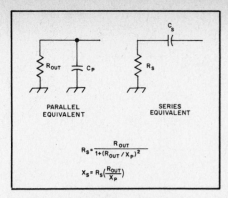

Fig. 21 — Parallel and series equivalent circuits and the formulas used for conversion.

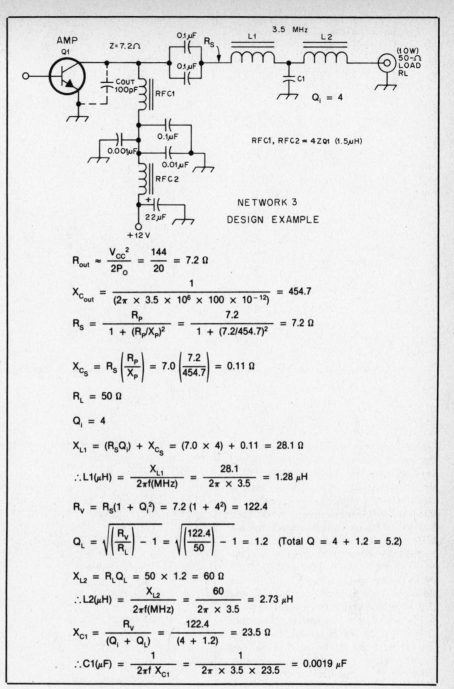

Fig. 22 — A practical example of network no. 3 and the solution to the network design.

impedance being matched is lower than the input impedance of the following stage. If this is not the case, the network (the circuit elements between the points marked A and B) can be turned around to provide the correct transformation.

Normally, the output impedance of a transistor is given as a resistance in parallel with a capacitance, C_{out}. To use the design equations for these three networks, the output impedance must first be converted from the parallel form (R_{out} and C_{out}) to the equivalent series form (R_s and C_s). These equivalent circuits and the equations for conversion are given in Fig. 21. Often the output capacitance is small enough that it may be neglected; the resulting error is compensated for by using variable components in the network.

The low-pass T network (Fig. 20) has the advantage of matching a wide range of impedances with practical component values. Some designers feel that of the various networks used in solid-state work, the T network is best in terms of collector efficiency. The harmonic suppression afforded by the T network varies with the transformation ratio and the total Q of the network. For stages feeding an antenna, additional harmonic suppression will normally be needed. This is also true for networks 1 and 2. These three networks are covered in detail in *Motorola Application Note AN-267.* Another excellent paper on the subject was written by Becciolini, *Motorola Application Note AN-271.*

The equations for networks 1, 2 and 3 were taken from AN-267. That paper contains computer solutions to these networks and others, with tabular information for various values of Q and source impedances. A fixed load value of 50 ohms is the base for the tabular data.

A design example for network 3 is given in Fig. 22. The solutions for the other two networks follow the same general trend, so examples for networks 1 and 2 will not be given. In Fig. 22 the component "C_{out}" is taken from the manufacturer's data sheet. If it is not available, it can be ignored at the expense of a slight mathematical error in the network determination. By making

C1 variable the network can be made to approximate the correct transformation ratio. At the lower frequencies C1 will be fairly large in value. This may require a fixed-value silver-mica capacitor in parallel with a mica compression trimmer to obtain the exact value of capacitance needed. The equations will seldom yield standard values of capacitance.

L1 and L2 of Fig. 22 can be wound on powdered-iron toroid cores of suitable cross-sectional area for the power involved. This is explained in an earlier chapter of this book. L1 and L2 should be separated by mounting them apart and at right angles. Alternatively, a shield can be used

between the inductors. This will prevent unwanted capacitive and inductive coupling effects between the input and output terminals of the network. Despite the self-shielding nature of toroidal inductors, some coupling is possible when they are in close proximity.

Broadband Transformers

Toroidal broadband transformers are very useful in a great many amateur applications. Some of the more popular transformer configurations are presented here for those who wish to use them in matching networks associated with solid-state devices and tubes. It is important to

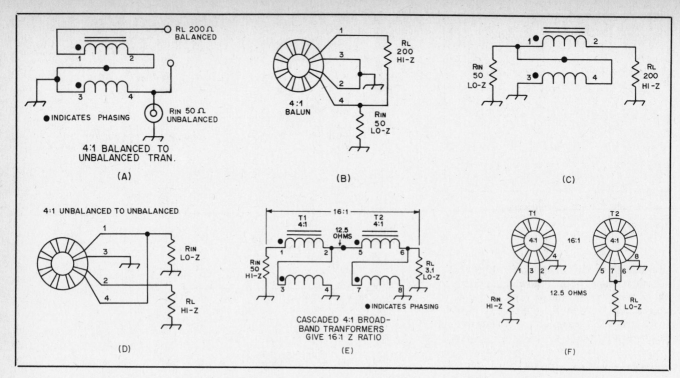

Fig. 23 — Circuit illustrations of 4:1 broadband transformers.

realize that broadband transformers are best suited to low-impedance applications, say, up to a few hundred ohms. They should be thought of as devices that can transform one impedance to another, in terms of their transformation ratio. They should not be regarded as devices that are built for some specified pair of impedances, such as 200 ohms to 50 ohms in the case of a 4:1 transformer. The term "balun" pertains only to a broadband transformer that converts a balanced condition to one that is unbalanced, or vice versa.

The broadband transformers illustrated in Figs. 23, 24 and 25 are suitable for use in solid-state circuits, as matching devices between circuit modules and in antenna-matching networks. For low power levels the choice of core material is often ferrite. Powdered-iron is more often the designer's preference when working with high power levels. The primary reason to avoid ferrite at high power is the potential for damage to the core material during saturation and overheating. This can alter the permeability factor of the core material permanently. Powdered-iron is more tolerant in this regard.

Fig. 23 shows two types of 4:1 transformers, plus a method for connecting two of them in series to effect a 16:1 transformation. The circuit at E is often used between a 50-ohm source and the base of an RF power transistor.

Two styles of 9:1 transformer are shown in Fig. 24 at A and C. They are also found at the input to transistor amplifiers and between the collector and the load. The variable-ratio transformer of Fig. 24C is ex-

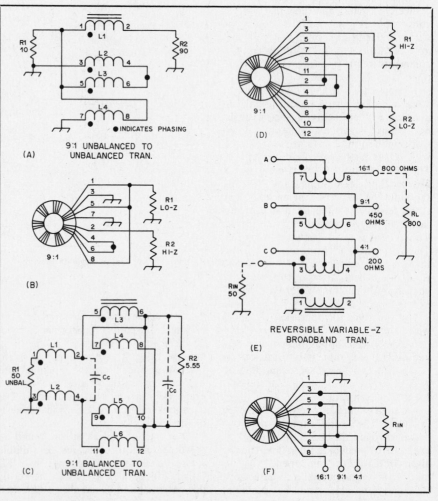

Fig. 24 — Circuit examples of 9:1 broadband transformers (A and C) and a variable-impedance transformer (E).

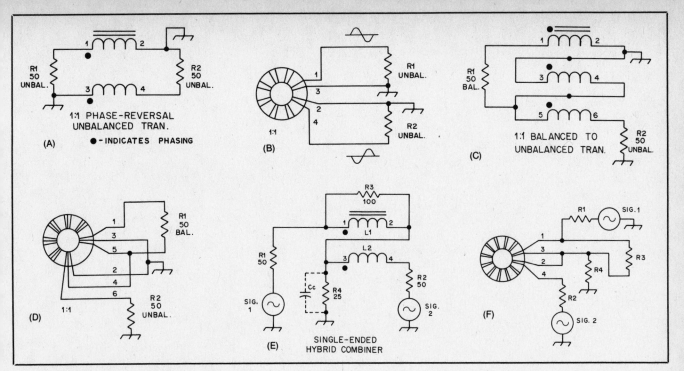

Fig. 25 — Assorted broadband transformers.

cellent for obtaining a host of impedance transformations. This transformer was developed by W2FMI for use in matching ground-mounted vertical antennas.

Phase-reversal, 1:1 balun and hybrid-combiner transformers are shown in Fig. 25. The circuit at E of Fig. 25 is useful when it is necessary to feed two signals to a single load. When the input signals are on different frequencies, the power is split evenly between R3 and R4. When the input voltages are on the same frequency (as with two transistor amplifiers feeding a single load), with the amplitudes and phase identical, all of the power is delivered to R4.

Chapter 12

Radio Receiving Principles

How good should receiver performance be? A suitable answer might be, "As good as is possible, consistent with the state of the amateur art and the money available to the purchaser." That opens up a wide area for debate, but the statement is not meant to imply that a receiver has to be costly or complex to provide good performance: Some very basic, inexpensive homemade receivers offer outstanding performance.

For many years the evolution of commerical amateur receivers seemed to stagnate except for the window dressing and frills added to the front panels. Emphasis was placed on "sensitivity" (whatever was really meant by that term) in the advertising. Some amateurs concluded (as a result of the strong push for sensitive receivers) that the mark of a good unit was seen when atmospheric noise on the HF bands could push the S-meter needle up to an S2 or S3. Very little thought, if any, was given to the important parameters of a receiver — high dynamic range, fine readout resolution and frequency stability. Instead, receivers were placed on the market with 5- or 10-kHz dial increments and excessive amounts of front-end gain. The excessive gain caused the mixer (or mixers) to collapse in the presence of moderate and strong signals.

Double-conversion superheterodyne receivers were for a long time the choice of manufacturers and amateurs. The second IF was often 100 or 50 kHz, thereby enabling the designer to get fairly reasonable orders of selectivity by means of high-Q IF transformers. That concept predated the availability of crystal-lattice and mechanical filters. The low-frequency second IF dictated the use of a double-conversion circuit in order to minimize image responses.

Single-conversion receivers offer much cleaner performance in terms of spurious responses and dynamic range. Although their performance is acceptable in many instances, there is considerable room for improvement. At least there is only one mixer to cause intermodulation-distortion (IMD) and overloading problems in a single-conversion superheterodyne receiver. A strong doubly balanced mixer (DBM) and careful gain distribution in such a receiver can yield superb performance. Of course, the local oscillator should be stable and low in noise components to further enhance performance. Thus far, not many commerically built amateur receivers meet these criteria. In terms of dynamic range, some manufactured receivers exhibit an MDS (minimum discernible signal) of − 145 dBm (referenced to the noise floor); blocking of the desired signal does not occur (1 dB of compression) until the adjacent test signal is some 116-dB above the noise floor, and the two-tone IMD dynamic range is on the order of 85 dB. Greater detail concerning this measurement technique is given in Chapter 25. A receiver with the approximate figures just given is considered to be an acceptable one for use where fairly strong signals prevail. However, improvements over those numbers have been achieved by amateurs who designed and built their own receivers.

VHF and UHF

Adequate receiving capability is essential in VHF and UHF communications. This is true whether the station is a transceiver or a combination of separate transmitting and receiving units, regardless of the modulation system used. Transceivers and FM receivers are treated separately in this Handbook, but their performance involves basic principles that apply to all receivers for frequencies above 30 MHz. Important attributes are good signal-to-noise ratio (low noise figure), adequate gain, stability and freedom from overloading and other spurious responses.

A VHF station often has an HF-band communications receiver, with a crystal-controlled converter for the desired VHF band ahead of it. The HF receiver serves as a tunable IF system, complete with detector, noise limiter, BFO and audio amplifier.

Choice of a suitable communications receiver for use with converters should not be made lightly. A good receiver for this purpose has all the characteristics of a good HF receiver. The special requirements of FM phone are discussed in Chapter 18. Good mechanical design and frequency stability are important. Image rejection should be high in the converter output range. This may rule out 28 MHz with single-conversion receivers having 455-kHz IF systems.

This suggests strongly that amateurs consider designing and building their own receivers. Certainly, this is within the capability of many experimenters. The satisfaction derived from such an effort can't be measured. The following sections of this chapter are written for those who wish to acquire a better understanding of how a practical receiver operates. Design data and related philosophy are included for those who are inspired toward developing a homemade receiver.

Sensitivity

One of the least understood terms among amateurs is sensitivity. In a casual definition the word refers to the ability of a receiver to respond to incoming signals. It is proper to conclude from this that the better the sensitivity, the more responsive the receiver will be to weak signals. The popular misconception is that the greater the receiver front-end gain, the higher the sensitivity. An amateur who subscribes to this concept can ruin the performance of a good receiver by installing a high-gain preamplifier ahead of it. Although this will cause the S meter to read much higher on all signals, it can actually degrade the receiver sensitivity if the preamplifier is of inferior design (noisy).

A true measure of receiver sensitivity is obtained when the input signal is referenced to the noise generated within the receiver. Since the significant noise generated inside a receiver of good design originates in the RF and mixer stages (sometimes in the post-mixer amplifier), a low-noise front end is vital to high sensitivity. The necessary receiver gain can be developed after the mixer — usually in the IF amplifier section. The internal noise is generated by the thermal agitation of electrons inside the tubes, transistors or ICs. It is evident from this discussion that a receiver of high sensitivity could be one with relatively low front-end gain. This thought should be kept in

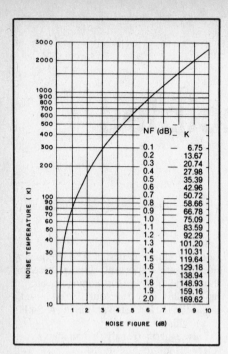

Fig. 1 — Relationship between noise figure and noise temperature.

mind as we enter the discussion of dynamic range and noise figure.

Noise Temperature, Noise Factor and Noise Figure

The lower the receiver noise figure (NF), the more sensitive it is. Receiver noise figures are established primarily in the RF amplifier and/or mixer stages. Low-noise active devices (tubes or transistors) should be used in the receiver front end to help obtain a low noise figure. The unwanted noise, in effect, masks the weaker signals and makes them difficult to copy. Noise generated in the receiver front end is amplified in the succeeding stages along with the signal energy. Therefore, in the interest of sensitivity, internal noise should be kept as low as possible.

Don't confuse external noise (man-made and atmospheric, which comes in on the antenna) with receiver noise during discussions of noise figure. Although the ratio of the external noise to the incoming signal level has a lot to do with reception, external noise does not relate to this general discussion. It is because external noise levels are quite high on 160, 80, 40 and 20 meters that emphasis is seldom placed on a low receiver noise figure for those bands. However, as the operating frequency is increased from 15 meters up through the microwave spectrum, the matter of receiver noise becomes a primary consideration. At these higher frequencies the receiver noise almost always exceeds that from external sources, especially at 2 meters and above.

Receiver noise is produced by the movement of electrons in any substance (such as resistors, transistors and FETs) that has

a temperature above absolute zero ($-273°$ C or 0 K). Electrons move in a random fashion colliding with relatively immobile ions that make up the bulk of the material. The final effect is that in most substances there is no net current in any particular direction on a long-term average, but rather a series of random pulses. These pulses produce what is called thermal agitation noise, thermal noise or Johnson noise.

As the currents caused by electron movement increase with temperature, so does the noise power. Also, as the pulses are random, they spread out over a broad frequency spectrum. As it turns out, if we examine the power contained in a given passband, the value of that power is independent of the center frequency of the passband. This is expressed as

$$p = kTB$$

where p is the thermal noise power, k is Boltzmann's constant (1.38×10^{-23} joule per K), T is absolute temperature in kelvins and B is the bandwidth in hertz. Notice that the power is directly proportional to temperature, and at 0 K the noise power is zero.

Active devices normally exhibit noise temperatures different from their ambient temperatures. The thermal noise produced by a semiconductor device will limit its ability to respond to input signals below the level of the internally generated noise. Noise temperature, noise factor and noise figure are all measures of this device noise. The results are expressed in terms of temperature, ratios and decibels, respectively.

Consider a 50-ohm termination connected to the input of a device with the termination cooled to absolute zero. There would be no noise produced by this source, and the noise output from the device would be that of the internally generated noise. If the termination were now heated to a temperature that would raise the output noise of the device by 3 dB (thermal agitation noise equal to the internally generated noise of the device) and the temperature of the termination measured, the effective input noise temperature (T_E) of the device would be this value. The noise temperature specification is independent of bandwidth and is directly proportional to noise power. For example, if we were to halve the noise temperature we would double the signal-to-noise ratio.

In order to convert a noise temperature measurement to noise figure, an intermediate calculation is required — noise factor (f). Noise factor is by definition the ratio of the total output noise power to the input noise power when the termination is at the standard temperature of 290 K (17° C). The noise power caused only by the input noise of the termination is simply the noise power of the source multiplied by the gain of the device. Mathematically

$$N_{power\ input} = GkBT_o$$

where G is the gain of the device and T_o is 290 K. The total noise caused by the input noise of the termination and the internally generated noise is simply the sum of the two noise sources multiplied by the gain of the device, or

$$N_{power\ total} = GkB(T_o + T_E)$$

where T_E is the effective input noise temperature and T_o is 290 K. The noise factor (f) is calculated as

$$f = \frac{N_{power\ total}}{N_{power\ input}}$$

where T_E is the effective input noise temperature and T_o is 290 K. Noise figure can then be calculated as follows:

$$NF = 10 \log_{10} f = 10 \log_{10} \left(1 + \frac{T_E}{T_o}\right)$$

where noise figure is expressed in decibels. Should the noise figure of the device be known and it is desired to find the noise temperature, the equation can be rearranged as follows:

$$T_E = 290\ [antilog(NF/10) - 1]K$$

where noise figure is expressed in decibels. A graph illustrating the relationship between noise figure and noise temperature is given in Fig. 1.

Noise factor can also be represented in terms of signal-to-noise ratios as

$$f = \frac{S/N\ at\ input}{S/N\ at\ output}$$

and noise figure can be found from

$$NF = 10 \log_{10} f$$
$$= 10 \log_{10} \frac{S/N\ at\ input}{S/N\ at\ output}$$

A receiving system consists of an interconnection of individual stages, some noisier than others. Each stage's noise contribution to the reduction of signal-to-noise ratio can be expressed as a noise figure. How much the noise figure of a particular stage affects system noise figure depends on the gain of the stages between that stage and the antenna. That is, if a stage's gain is sufficiently large, its noise figure will tend to override or "mask" the noise contribution of the stage following it. Mathematically, the noise factor of a receiving system can be expressed as

$$F = f_1 + \frac{f_2 - 1}{G_1} + \frac{f_3 - 1}{G_1\ G_2} +$$
$$\cdots + \frac{f_n - 1}{G_n \ldots G_2\ G_1}$$

where

f_n = noise factor of the nth stage

G_n = gain of the nth stage.

Brief analysis of this equation shows that the first stage of a receiving system is the most important with regard to noise figure. If the gain of this and succeeding stages is greater than unity, the denominator of each successive term becomes greater. The numerical value of terms beyond the second or third approaches zero and can be ignored.

It might seem that the more gain an RF amplifier has, the better the signal-to-noise ratio and therefore the better the reception. This is not necessarily true. The primary function of an RF amplifier is to establish the noise figure of the system. One good RF stage is usually adequate unless the mixer is a passive type with loss instead of gain. Two RF stages are the usual maximum requirement.

Once the system noise figure is established, any further gain necessary to bring a signal to audible levels may be obtained from intermediate-frequency stages or in the audio channel. Use the minimum gain necessary to set the overall receiver noise figure; this will help avoid overloading and spurious signals in subsequent stages.

Further examination of the equation points out the desirability of mounting the first stage of the receiver system at the antenna for VHF and UHF. The transmission line from the antenna to the receiver can be considered as a stage in the receiving system. The first stage of a receiving system makes the major contribution of noise figure to the system, so it is highly desirable that the first stage be a low-noise amplifier with gain. A transmission line is a "lossy" amplifier, and if placed at the first stage of a receiving system, automatically limits the system noise figure to that of the transmission line, at best. If the first RF amplifier is placed before the lossy transmission line stage, at the antenna, the amplifier gain will tend to mask the noise added by the transmission line.

Noise-Figure Measurements

Amateurs can use a thermal noise source for determining receiver noise figure. The resistance of the noise-generator output must match that of the receiver input, 50 ohms to 50 ohms, for example. Fig. 2 shows a setup for making these measurements. The first reading is taken with the noise generator turned off. The receiver audio gain is adjusted for a convenient noise reading in decibels, as observed on the audio power meter. The noise generator is turned on next, and its output is increased until a convenient power ratio, expressed by N2/N1, is observed. The ratio N2/N1 is referred to as the "Y-factor", and this noise figure measurement technique is commonly called the Y-factor method. From the Y-factor and output power of the noise generator, the noise figure can be calculated

$$NR = ENR - 10 \log_{10} (Y - 1)$$

where
NF = noise figure in decibels
ENR = excess noise ratio of the noise generator in decibels
Y = output noise ratio, N2/N1.
The excess noise ratio of the generator is

$$ENR = 10 \log_{10} \left(\frac{P2}{P1} - 1 \right)$$

where
P2 = noise power of the generator
P1 = noise power from a resistor at 290 kelvins.

If a thermionic diode such as a 5722 tube is used as the noise source, and if the circuit is operated in the "temperature-limited mode" (portion of the tube curve where saturation occurs, dependent on cathode temperature and plate voltage), the (excess) decibels can be calculated by

$$(\text{excess}) \ dB = 10 \log_{10} (20 \ Rd \ Id)$$

where
Rd = the noise source output resistance
Id = the diode current in amperes.

Most manufacturers of amateur communications receivers rate the noise characteristics with respect to signal input. A common expression is S + Noise/Noise, or the signal-to-noise ratio. Usually the sensitivity is given as the number of microvolts required for an S + Noise/Noise ratio of 10 dB. Sensitivity can also be expressed in terms of the minimum discernible signal (MDS) or noise floor of the receiver. (See Chapter 25.)

Typically, the noise figure for a good receiver operating below 30 MHz runs about 5 to 10 dB. Lower noise figures can be obtained, but they are of no real value because of the external noise arriving from the antenna. It is important to remember also that optimum noise figure in an RF amplifier does not always coincide with maximum stage gain, especially at VHF and higher. This is why actual noise measurements must be used to peak for best noise performance.

Selectivity

Many amateurs regard the expression "selectivity" as equating to the ability of a receiver to separate signals. This is a fundamental truth, particularly with respect to IF selectivity which has been established by means of high-Q filters (LC, crystal, monolithic or mechanical). But in a broader sense, selectivity can be employed to reject unwanted signal energy in any part of a receiver — the front end, IF section, audio circuit or local-oscillator chain. Selectivity is a relative term, since the degree of bandwidth can vary from a few hertz to more than a megahertz, depending on the design objectives. Therefore, it is not uncommon to hear terms like "broadband

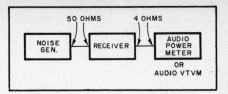

Fig. 2 — Block diagram of a noise measurement setup.

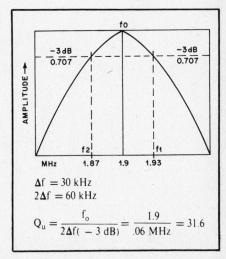

Fig. 3 — A curve and equation for determining the unloaded Q of a tuned circuit.

filter" or "narrow-band filter."

The degree of selectivity is determined by the bandwidth of a filter network. The bandwidth is normally specified for the minus 3-dB points on the filter response curve; the frequencies where the filter output power is half the peak output elsewhere in the passband. The difference in frequency between a minus 3-dB point and the filter center frequency is known as Δ f. The bandwidth of the filter then becomes 2 Δ f. Fig. 3 illustrates this principle and shows how the unloaded Q of a tuned circuit or resonator relates to the bandwidth characteristic.

If a tuned circuit is used as a filter, the higher its loaded Q, the greater the selectivity. To make the skirts of the response curve steeper, several high-Q resonators can be used in cascade. This aids the selectivity by providing greater rejection of signals close in frequency to the desired one. The desirable effect of cascaded filter sections can be seen in Fig. 4. The circuit is that of a tunable Cohn type of three-pole filter for use in the front end of a 160-meter receiver. The response curve is included to illustrate the selectivity obtained.

An ideal receiver with selectivity applied to various significant parts of the circuit might be structured something like this:

a) Selective front end for rejecting out-of-band signals to prevent overloading and spurious responses.

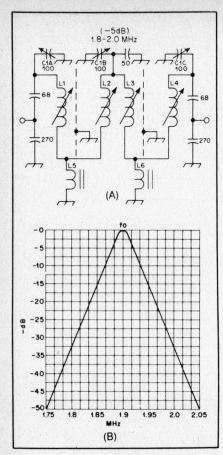

Fig. 4 — A tunable Cohn-type filter is shown at A. L5 and L6 are the bottom-coupling inductors (1.45 µH). L1 and L4 are 70 pH and L2, L3 are 140 µH. A response curve for the tunable filter is given at B.

b) Selective IF circuit (two IF filters: one for 2.4-kHz SSB bandwidth and one for 300-Hz CW and RTTY bandwidth).

c) RC active or passive audio filter for audio selectivity to reduce wideband noise and provide audio selectivity in the range from 300 to 2700 Hz (SSB), 2100 to 2300 Hz (or 1250 to 1450 Hz), for RTTY and a very narrow bandwidth, such as 750 to 850 Hz, for CW.

d) Selective circuits or filters in the local oscillator chain to reject all mixer injection energy other than the desired frequency.

This illustrates clearly that selectivity means more than the ability of a receiver to separate one amateur signal from another that is nearby in frequency. Specifically, it means that selectivity can be used to select one frequency or band of frequencies while rejecting others.

Selectivity at VHF

Ever-increasing occupancy of the radio spectrum brings with it a parade of receiver overload and spurious responses. Overloading problems can be minimized by the use of high-dynamic-range receiving techniques, but spurious responses such as

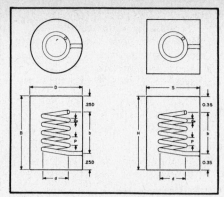

Fig. 5 — Round and square helical resonators, showing principal dimensions. Diameter, D (for side, S) is determined by the desired unloaded Q. Other dimensions are expressed in terms of D (or S) as described in the text.

the image frequency must be filtered out before mixing occurs. Conventional tuned circuits cannot provide the selectivity necessary to eliminate the plethora of signals found in most urban and many suburban neighborhoods. Other filtering techniques must be used.

Although some amateurs use quarter-wavelength coaxial cavities on 50, 144 and 220 MHz, the helical resonators shown in Fig. 5 are usually a better choice as they are smaller and easier to build. In the frequency range from 30 to 100 MHz, where it's difficult to build high-Q inductors, and because coaxial cavities are very large, the helical resonator is an excellent choice. At 50 MHz, for example, a capacitively tuned, quarter-wavelength coaxial cavity with an unloaded Q of 3000 would be about 4 inches in diameter and nearly 5 feet long. On the other hand, a helical resonator with the same unloaded Q is about 8.5 inches in diameter and 11.3 inches long. Even at 432 MHz, where coaxial cavities are common, the use of a helical resonator will result in substantial size reductions. The following design information on helical resonators originally appeared in a June 1976 QST article by W1HR.

The helical resonator has often been described simply as a coil surrounded by a shield, but it is actually a shielded, resonant section of helically wound transmission line with relatively high characteristic impedance and low axial propagation velocity. The electrical length is about 94 percent of an axial quarter wavelength, or 84.6 electrical degrees. One lead of the helical winding is connected directly to the shield, and the other end is open circuited as shown in Fig. 5. Although the shield may be any shape, only round and square shields will be considered in this section.

Design

The unloaded Q of a helical resonator is determined primarily by the size of the shield. For a round resonator with a copper coil on a low-loss form, mounted in a copper shield, the unloaded Q is given by

$$Q_u = 50 D \sqrt{f_o}$$

where D = inside diameter of the shield in inches and f_o = frequency in MHz.

If the shield can is square, assume D to be 1.2 times the width of one side. This formula, which includes the effects of losses and imperfections in practical materials, yields values of unloaded Q which are easily attained in practice. Silver plating of the shield and coil will increase the unloaded Q by about three percent over that predicted by the equation. At VHF and UHF, however, it is more practical to increase slightly the shield size (i.e., increase the selected Q_u by about three percent before making the calculation). The fringing capacitance at the open-circuited end of the helix is about 0.15D pF (i.e., approximately 0.3 pF for a shield 2 inches in diameter).

Once the required shield size has been determined, the total number of turns, N, winding pitch, P, and characteristic impedance, Z_o, for round and square helical resonators with air dielectric between the helix and shield, are given by

$$N = \frac{1908}{f_o D} \quad P = \frac{f_o D^2}{2312} \quad Z_o = \frac{99,000}{f_o D}$$

$$N = \frac{1590}{f_o S} \quad P = \frac{f_o S^2}{1606} \quad Z_o = \frac{82,500}{f_o S}$$

In these equations, dimensions D and S are in inches, and f_o is in megahertz. The design nomograph for round helical resonators in Fig. 6, which can be used with slide-rule accuracy, is based on these formulas.

Although there are many variables to consider when designing helical resonators, certain ratios of shield size and length, and coil diameter and length, will provide optimum results. For helix diameter, d = 0.55D, or d = 0.66S. To determine helix length, b = 0.825D, or b = 0.99S. For shield length, B = 1.325D and H = 1.60S.

Calculation of these dimensions is simplified by the design chart of Fig. 7. Note that these ratios result in a helix with a length 1.5 times its diameter, the condition for maximum Q. The shield is about 60 percent longer than the helix — although it can be made longer — to completely contain the electric field at the top of the helix and the magnetic field at the bottom.

It should be mentioned that the winding pitch, P, is used primarily to determine the required conductor size. During actual construction the length of the coil is adjusted to that given by the equations for helix length. Conductor size ranges from 0.4P to 0.6P for both round and square resonators and is plotted graphically in Fig. 8.

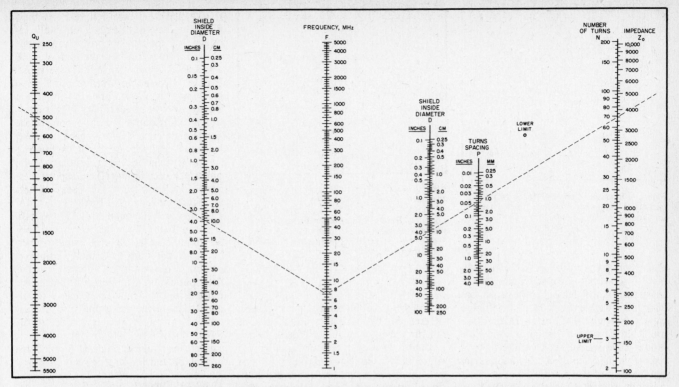

Fig. 6 — Design nomograph for round helical resonators. After selecting unloaded Q_u, required shield diameter is indicated by index line from Q_u scale to frequency scale (dashed index line here indicates a shield of about 3.8 inches for an unloaded Q of 500 at 7 MHz). Number of turns, N, winding pitch, P, and characteristic impedance, Z_o, are determined by index line from the frequency scale through previously determined shield diameter on right-hand side of the chart (index line indicates P = 0.047 inch, N = 70 turns, and Z_n = 3600 ohms).

Obviously, an area exists (in terms of frequency and unloaded Q) where the designer must make a choice between a conventional cavity (or lumped LC circuit) and a helical resonator. At the higher frequencies, where cavities might be considered, the choice is affected by shape factor: a coaxial resonator is long and relatively small in diameter, while the length of a helical resonator is not much greater than its diameter. A second consideration is that point where the winding pitch, P, is less than the radius of the helix (otherwise the structure tends to be nonhelical). This condition occurs when the helix has fewer than three turns ("upper limit" on the design nomograph of Fig. 6).

Construction

To obtain as high an unloaded Q as possible, the shield should should not have any seams parallel to the axis of the helix. This is usually not a problem with round resonators because large-diameter copper tubing is used for the shield, but square resonators require at least one seam and usually more. However, the effect on unloaded Q is minimal if the seam is silver soldered carefully from one end to the other.

Best results are obtained when little or no dielectric is used inside the shield. This is usually no problem at VHF and UHF because the conductors are large enough that a supporting coil form is not required. The lower end of the helix should be soldered to the inside of the shield at a

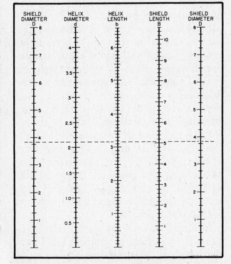

Fig. 7 — Helical resonator design chart. After the shield diameter has been determined, helix diameter, d, helix length, b, and shield length, B, can be determined with this graph. Index line indicates that a shield diameter of 3.8 inches requires helix mean diameter of 2.1 inches, helix length of 3.1 inches, and shield length of 5 inches.

point directly opposite from the bottom of the coil.

Although the external field is minimized by the use of top and bottom covers, the top and bottom of the shield may be left open with negligible effect on frequency or unloaded Q. If covers are provided, however, they should make good electrical contact with the shield. In those resonators where the helix is connected to the bottom

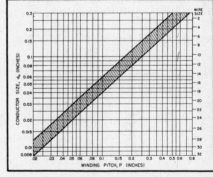

Fig. 8 — Helix conductor size vs. winding pitch, P. A winding pitch of 0.047 inch for example, dictates a conductor diameter between 0.019 and 0.028 inch (number 22 or 24 AWG).

cover, that must be soldered solidly to the shield to minimize losses.

Tuning

A helical resonator designed from the nomograph of Fig. 6, if carefully built, will resonate very close to the design frequency. Resonance can be adjusted over a small range by slightly compressing or expanding the helix. If the helix is made slightly longer than that called for in Fig. 7, the resonator can be tuned by pruning the open end of the coil. However, neither of these methods is recommended for wide frequency excursions because any major deviation in helix length will degrade the unloaded Q of the resonator.

Most helical resonators are tuned by

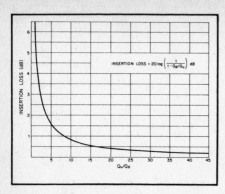

Fig. 9 — Insertion loss of all tuned resonant circuits is determined by the ratio of loaded to unloaded Q as shown here.

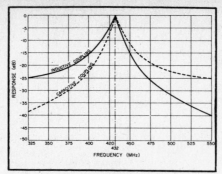

Fig. 10 — Response curve for a single-resonator 432-MHz filter showing the effects of capacitive and inductive input/output coupling. Response curve can be made symmetrical on each side of resonance by combining the two methods (inductive input and capacitive output, or vice versa).

means of a brass tuning screw or high quality air-variable capacitor across the open end of the helix. Piston capacitors also work well, but the Q of the tuning capacitor should ideally be several times the unloaded Q of the resonator. Varactor diodes have sometimes been used where remote tuning is required, but varactors can generate unwanted harmonics and other spurious signals if they are excited by strong, nearby signals.

When a helical resonator is to be tuned by a variable capacitor, the shield size is based on the chosen unloaded Q at the operating frequency. Then the number of turns, N, and the winding pitch, P, are based on resonance at $1.5f_o$. Tune the resonator to the desired operating frequency, f_o.

Insertion Loss

The insertion loss (dissipation loss), IL, in decibels, of all resonant circuits is given by

$$IL = 20 \log_{10} \left(\frac{1}{1 - Q_d/Q_u} \right) dB$$

where
 Q_d = loaded Q
 Q_u = unloaded Q.

This is plotted in Fig. 9. For the most practical cases ($Q_d > 5$), this can be closely approximated by IL ≈ 9.0 (Q_d/Q_u) dB. The selection of a loaded Q for a tuned circuit is dictated primarily by the required selectivity of the circuit. However, to keep dissipation loss to 0.5 dB or less (as is the case for low-noise VHF receivers), the unloaded Q must be at least 18 times the loaded Q. Although this may be difficult to achieve in practice, it points up the necessity of considering both selectivity and insertion loss before choosing the unloaded Q of any resonant tuned circuit.

Coupling

Signals may be coupled into and out of helical resonators with inductive loops at the bottom of the helix, direct taps on the coil or any combination of these. Although the correct tap point can be calculated easily, coupling by loops and probes must be determined experimentally. When only one resonator is used, the input and output coupling is often provided by probes. For maximum isolation the probes are positioned on opposite sides of the resonator.

When coupling loops are used, the plane of the loop should be perpendicular to the axis of the helix and separated a small distance from the bottom of the coil. For resonators with only a few turns, the plane of the loop can be tilted slightly so it is parallel with the slope of the adjacent conductor. Helical resonators with inductive coupling (loops) will exhibit more attenuation to signals above the resonant frequency (as compared to attenuation below resonance), whereas resonators with capacitive coupling (probes) exhibit more attenuation below the passband, as shown for a typical 432-MHz resonator in Fig. 10. This characteristic may be a consideration when choosing a coupling method. The passband can be made more symmetrical by using a combination of coupling methods (inductive input and capacitive output, for example).

If more than one helical resonator is required to obtain a desired band-pass characteristic, adjacent resonators may be coupled through apertures in the shield wall between the two resonators. Unfortunately, the size and location of the aperture must be found empirically, so this method of coupling is not very practical unless you're building a large number of identical units.

Since the loaded Q of a resonator is determined by the external loading, this must be considered when selecting a tap (or position of a loop or probe). The ratio of this external loading, R_b, to the characteristic impedance, Z_o, for a quarter-wavelength resonator is calculated from

$$K = \frac{R_b}{Z_o} = 0.785 \left(\frac{1}{Q_d} - \frac{1}{Q_u} \right)$$

Even when filters are designed and built properly, they may be rendered totally ineffective if not installed properly. Leakage around a filter can be quite high at VHF and UHF, where wavelengths are short. Proper attention to shielding and good grounding is mandatory for minimum leakage. Poor coaxial cable shield connection into and out of the filter is one of the greatest offenders with regard to filter leakage. Proper dc lead bypassing throughout the receiving system is good practice, especially at VHF and above. Ferrite beads placed over the dc leads may help to reduce leakage even further.

Proper termination of a filter is a necessity if minimum loss is desired. Most VHF RF amplifiers optimized for noise figure do not have a 50-ohm terminating input impedance. As a result, any filter attached to the input of an RF amplifier optimized for noise figure will not be properly terminated, and the filter's loss may rise substantially. As this loss is directly added to the RF amplifier's noise figure, prudent consideration should be made of filter choice and placement in the receiver.

Dynamic Range

Here is another term which seems to confuse some amateurs and even some receiver manufacturers. The confusion concerns true dynamic range (as treated briefly at the start of this chapter) and the AGC control range in a receiver. That is, if a receiver AGC circuit has the capability of controlling the overall receiver gain by some 100 dB from a no-signal to a large-signal condition, a misinformed individual might claim that the dynamic range of the receiver is 100 dB. A receiver with a true dynamic range of 100 dB would be a very fine piece of equipment, indeed!

Dynamic range relates specifically to the amplitude levels of multiple signals that can be accommodated during reception. This is expressed as a numeric ratio, generally in decibels. The present state of the receiver art provides optimum dynamic ranges of up to 100 dB. This is the maximum dynamic range attainable when the distortion products are at the sensitivity limit of the receiver. Simply stated, dynamic range is the decibel difference (or ratio) between the largest tolerable receiver input signal (without causing audible distortion products) and the minimum discernible signal (sensitivity).

Poor dynamic range can cause a host of receiving problems when strong signals appear within the front-end passband. Notable among the maladies is cross modulation of the desired signal. Another effect is desensitization of the receiver from a strong unwanted signal. Spurious signals may appear in the receiver tuning range when a strong signal is elsewhere in the band. This is caused by IMD products from the mixer. Clearly, strong signals cause undesired interference and distortion of the desired signal when a receiver's dynamic range is poor. Design features of importance to high-dynamic-range receivers will appear in this chapter.

TRF Receivers

Tuned-radio-frequency receivers have little value in Amateur Radio today, but in the early days they were suitable for the reception of spark and AM signals. They consisted mainly of a couple of stages of selective RF amplification, an AM detector and an audio amplifier. Variations were developed as regenerative and super-regenerative receivers.

The straight regenerative detector was simply a self-oscillating detector that provided increased sensitivity (similar in function to a product detector) and a beat note for CW reception. Amplitude-modulated signals could be copied, if they were loud, when the regeneration control was set for a nonoscillating condition. For weak-signal AM reception the regeneration control was advanced to increase the detector sensitivity, and the signal was tuned in at zero beat, thereby eliminating the heterodyne from the carrier. Present-day uses for the TRF receiver are restricted mainly to reception of AM broadcast signals, for hi-fi reception and for field-strength indicators of CW or AM signals.

Superregenerative receivers were quite popular among VHF and UHF amateurs in the '30s, '40s and early '50s. The principle of operation was an oscillating detector that had its oscillation interrupted (quenched) by a low-frequency voltage slightly above the audible range (20 to 50 kHz being typical). Some superregenerative detectors employed a so-called self-quenching trait, brought about by means of an RC network of the appropriate time constant. The more esoteric "supergenny" or "rushbox" detectors used an outboard quench oscillator. This circuit was more sensitive than the straight regenerative detector, but was best suited for reception of AM and wide-band FM signals. Because of the quenching action and frequency, the detector response was extremely broad, making it unsuitable for narrow-band signals versus audio recovery. High-Q input tuned circuits helped make them more selective, but a typical superregenerative receiver that used a tuned cavity at the detector input could accommodate only ten 1000-μV, 30-percent-modulated AM signals in a range from 144 to 148 MHz without signal overlap. These tests were performed in the ARRL laboratory with the 10 signals separated from one another by equal amounts.

A major problem associated with the use of regenerative and superregenerative receivers was oscillator (detector) radiation. The isolation between the detector and the antenna was extremely poor, even when an RF amplifier was employed ahead of the detector. In many instances the radiated energy could be heard for several miles, causing intense interference to other

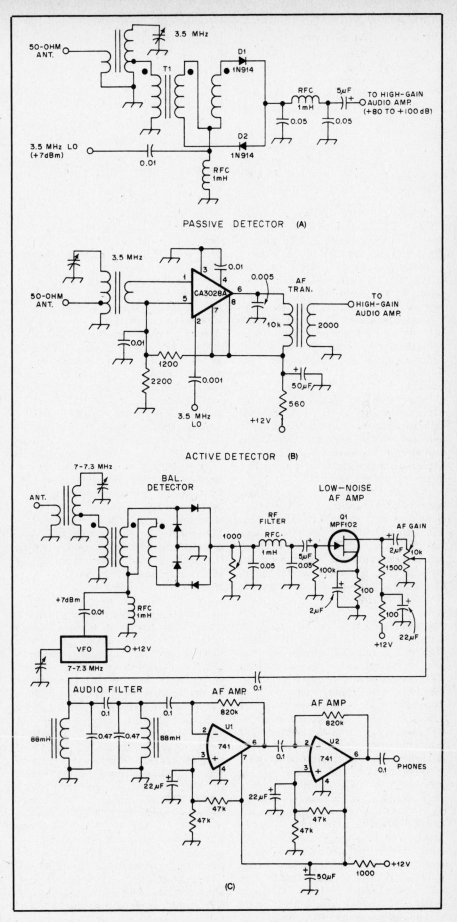

Fig. 11 — Typical detectors that can be used in the front ends of direct-conversion receivers. A passive diode is shown in A. The active detector (B) provides considerable conversion gain. An example of a practical direct-conversion receiver is shown in C.

amateurs in the community.

Direct-Conversion Receivers

A more satisfactory simple receiver for CW or SSB is called the direct-conversion or synchrodyne type. Although there is a distinct possiblity of signal radiation, it is considerably lower in level than with regenerative receivers. This results from better isolation between the antenna and the source of oscillation. A modern direct-conversion receiver uses a separate oscillator and a balanced or doubly balanced detector. Both features help to reduce unwanted radiation.

The detection stage of this receiver is actually a product detector that operates at the desired signal frequency. The product-detector circuits described in Chapter 18 are suitable in this kind of receiver. A tuned RF amplifier is useful ahead of the detector at 14 MHz and higher, but it is seldom necessary from 160 through 40 meters. This is because the atmospheric and man-made noise from the resonant antenna usually exceeds that of the detector below 14 MHz. When no RF stage is used, it is desirable to include a tuned network ahead of the detector.

Fig. 11 shows typical front ends for direct-conversion receivers. One circuit (A) employs a passive detector. The other (B) contains an active detector. Circuit B is desirable in the interest of increased gain. More information about product detectors can be found in Chapter 18.

The circuit of Fig. 11A shows a singly balanced passive detector. Front-end selectivity is provided by a tuned circuit. T1 is a broadband, trifilar-wound toroid transformer. It is tapped on the input tuned circuit at the approximate 50-ohm point. An RF filter is used after D1 and D2 to prevent LO energy from being passed on to the audio amplifier.

Fig. 11B illustrates an active singly balanced IC detector. The input impedance across pins 1 and 5 is roughly 1000 ohms. However, the secondary winding of the input tuned circuit can be made lower than 1000 ohms to reduce the signal amount to the detector. This will ensure improved dynamic range through a deliberate mismatch. Such a practice is useful when an RF amplifier precedes the detector. For maximum sensitivity when no RF amplifier is included, it is more practical to use a 1000-ohm transformation from the 50-ohm antenna (larger link at the detector input). An audio transformer is used at the detector output. The primary winding should have low dc resistance to provide dc balance between the collectors of the differential-amplifier pair in the IC. Alternatively, a center-tapped primary can be used. If this is done, pin 8 should be connected to one end of the winding and the B+ fed to the center tap. The impedance between pins 6 and 8 is approximately 8000 ohms.

In order to obtain ample headphone volume during reception of weak signals, it is necessary to use an audio amplifier that has between 80 and 100 dB of gain. The first AF amplifier should be low-noise type, such as a JFET. The audio-gain control should follow the first audio amplifier. Selectivity for SSB and CW reception can be had by including a passive or RC active audio filter after the gain control. Fig. 11C contains a circuit that shows a typical direct-conversion receiver in its entirety. As was stated earlier, the detector is operating as a product detector rather than a mixer, and the VFO is serving as a BFO. The difference frequency between the incoming 7-MHz signal and the 7-MHz BFO injection voltage is at audio frequency (zero IF). This is amplified by Q1, filtered through a passive LC audio network, then amplified by two 40-dB op-amp stages. It is possible to copy AM signals with this type of receiver by tuning the signal in at zero beat.

Direct-conversion receivers of the type illustrated in Fig. 11 provide double-signal reception; that is, a CW beat note will appear either side of the zero beat. This is useful during sideband reception, wherein the upper sideband is received on one side of zero beat and the lower sideband will appear on the opposite side of zero beat. QRM will be greater, of course, with this kind of receiver because there is no rejection of the unwanted sideband. Some designers have contrived elaborate circuits that, by means of phasing networks, provide single-signal reception. Unfortunately, the circuit becomes nearly as complex as that of superheterodyne. The benefits obtained are probably not worth the effort.

Direct-conversion receivers are not especially suitable above 14 MHz because it is difficult to secure adequate BFO stability at so high a frequency. A practical solution to the problem is the employment of a heterodyne BFO chain in which a 5-MHz VFO is heterodyned with crystal-controlled oscillators. Direct-conversion receivers are ideal for use in simple transceivers because the BFO can be used also as the frequency source for the transmitter, provided the appropriate frequency offset is included between transmit and receive to permit copy of SSB and CW signals without readjusting the BFO.

Characteristic Faults

A major difficulty connected with direct-conversion receivers is microphonics. The effect is noted when the operating receiver is bumped or moved. An annoying ringing sound is heard in the receiver output until the mechanical vibration ceases. The simple act of peaking the front end or adjusting the volume control can set off a microphonic response. This trait is caused by the extremely high gain needed in the audio amplifier. Slight electrical noises in the receiver front end, caused by small vibrations, are amplified many times by the audio channel. They are quite loud by the time they reach the speaker or phones. The best precautionary measure to reduce microphonics is to make all of the detector and BFO circuit leads and components as rigid as possible. Addition of an RF amplifier stage ahead of the detector will also help by virtue of increasing the front-end gain. This reduces the amount of audio gain needed to copy a signal, thereby diminishing the loudness of the microphonics.

The other common problem inherent in direct-conversion receivers is hum (Fig. 12). The fault is most pronounced when an ac power supply is used. The hum becomes progressively worse as the operating frequency is increased. For the most part, this is caused by ac ground loops in the system. The ac modulates the BFO voltage, and the hum-modulated energy is introduced directly into the detector, as well as being radiated and picked up by the antenna. The most practical steps toward a cure are to affix an effective earth ground to the receiver chassis and power supply, use a battery power supply and feed the antenna with coaxial cable. End-fed wire antennas increase the possiblity of hum if they are voltage fed (high impedance at the receiver end). Decoupling the ac power supply leads (dc leads to the receiver) is also an effective preventive measure for hum. The cure described by Wes Hayward, W7ZOI, is to add a toroidal decoupling choke, bifilar wound, in the plus and minus dc leads from the power supply. This will prevent high-impedance RF paths between the power supply and receiver. The effect is to prevent BFO energy from entering the power supply, being modulated by the rectifier diodes and reradiated by the ac line. This form of buzz is called "common-mode hum."

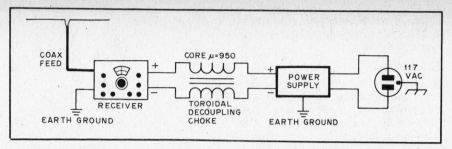

Fig. 12 — A method for eliminating common-mode hum in a direct-conversion receiver, as described by W7ZOI.

Superheterodyne Receivers

Nearly all of the present-day communications receivers are structured as superheterodyne types. Fig. 13 shows a simple block diagram of a single-conversion superheterodyne circuit. This basic design has been popular since the 1930s, and only a few general circuit enhancements have been introduced in recent years. Sophisticated versions of this type of receiver use alternatives to the circuits indicated in the block diagram. The local oscillator, for example, might utilize a phase-locked loop or synthesizer LO chain rather than a straight VFO. Digital readout is used in many models in place of the more traditional analog readout method. RF noise blankers (often very complex) are chosen by some designers in preference to simple shunt audio noise limiters. An assortment of techniques is being used to improve the overall selectivity of these receivers — elaborate IF filtering, RC active or LC passive audio filters. However, the basic circuit concept remains unchanged. The advancement of greatest significance in recent years is the changeover from vacuum tubes to semiconductors. This has increased the life span of the equipment, improved overall efficiency, aided stability (through reduced heating) and contributed to greater ruggedness and miniaturization.

Some manufacturers still produce double- or multiconversion superheterodyne receivers, but the circuits are similar to that of Fig. 13. Multiconversion receivers have a second mixer and LO chain for the purpose of making the second IF lower than the first. This helps to increase the overall selectivity in some designs, but it often degrades the receiver dynamic range through the addition of a second mixer. Multiconversion receivers are more prone to spurious responses than are single-conversion designs, owing to the additional oscillator and mixing frequencies involved. The "cleanest" performance is obtained from properly designed single-conversion receivers.

Circuit Function

In the example of Fig. 13 it is assumed that the receiver is adjusted to receive the 20-meter band. Front-end selectivity is provided by the resonant networks before and after the RF amplifier stage. This part of the receiver is often called the preselector, meaning that it affords a specific degree of front-end selectivity at the operating frequency. The RF amplifier increases the level of the signal from the antenna before it reaches the mixer. The amount of amplification is set by the designer, consistent with the overall circuit requirements (gain distribution). Generally, the gain will be from a few decibels to as much as 25 dB.

When the incoming signal reaches the mixer it is heterodyned with the local-oscillator frequency to establish an IF (intermediate frequency). The IF can be the sum or the difference of the two frequencies. In the example given, the IF is the difference frequency, or 9 MHz.

An IF filter (crystal lattice or ceramic monolithic) is used after the mixer. At low intermediate frequencies (455 kHz and similar), mechanical filters are often used. The IF filter sets the overall receiver selectivity. For SSB reception it is usually 2.4 kHz wide at the 3-dB points of the filter response curve. For CW reception it is be-

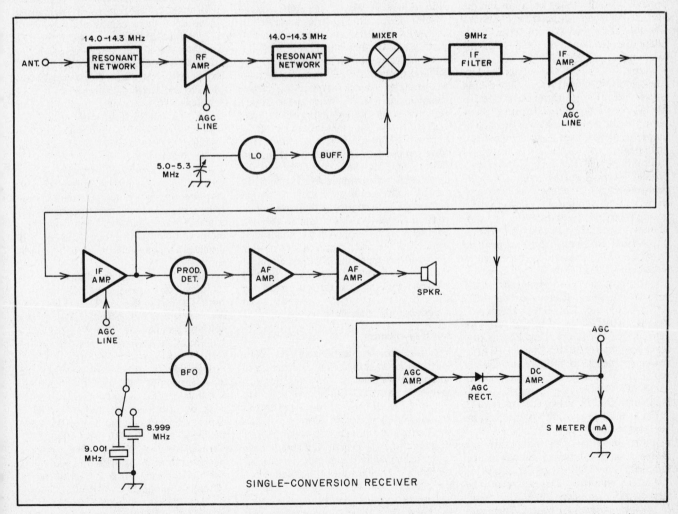

Fig. 13 — Block diagram of a single-conversion superheterodyne receiver for 20 meters. The arrows indicate the direction of signal and voltage components.

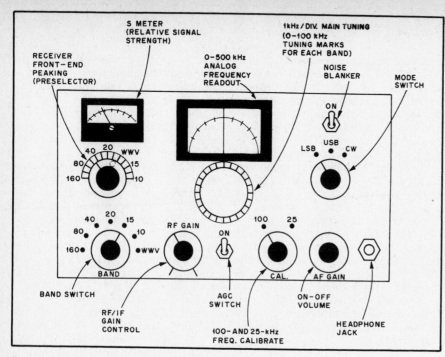

Fig. 14 — Layout of a typical amateur superheterodyne receiver.

tween 200 and 500 Hz in bandwidth, depending on the design objective and the operator preference. A good bandwidth for CW and RTTY reception is 300 Hz. Wider filters are available for AM and FM reception.

Output from the IF filter is increased by one or more amplifier stages. The overall gain of most IF strips varies from 50 to 100 dB. The amount of signal gain is determined by the design objective, the type of amplifier devices used and the number of gain stages.

The amplified IF energy is routed to a product detector, where it is mixed with the beat-frequency oscillator output. This produces an audio frequency voltage, which is amplified and fed to a speaker or headphones. The BFO is adjusted for reception of the upper or lower sideband, depending on which is appropriate at the time. In either case the BFO frequency is offset slightly from the center frequency of the IF filter. For SSB reception it is usually offset approximately 1.5 kHz, in which case it falls on the slope of the IF response curve. For CW reception the BFO is offset approximately 800 Hz from the IF filter center frequency to produce an 800-Hz peak audio tone in the speaker. Other values of CW offset are common, but 800 Hz is preferred by many CW operators.

The overall gain of the receiver can be adjusted manually (by means of a panel-mounted control) or automatically (by means of an AGC circuit). Energy can be sampled from the IF amplifier output or the audio amplifier. Depending on the method used, the resultant AGC is called *IF derived* or *audio derived*. There are many arguments pro and con about which

method is best. They shall not be considered here. In Fig. 13 the AGC voltage is sampled from the IF strip, amplified by the AGC amplifier and then rectified to provide a dc control voltage. A dc amplifier is used to drive the AGC terminals of the RF and IF amplifiers. It can also be used to operate an S meter for observing relative signal-strength levels. When the incoming signal is weak the gain-controlled stages operate fully. As the incoming signal becomes stronger the AGC circuit starts lowering the gain of the RF and IF stages, thereby leveling the audio output at the speaker. A well-designed AGC system will provide a uniform level of audio output (at a given AF-gain control setting) over an incoming signal-level variation of 100 dB. The net effect is to prevent overloading of some of the receiver stages and to protect the operator from the startling effect of tuning from a weak signal to an extremely loud one. Fig. 14 shows the front panel and controls for a typical amateur-band superheterodyne receiver.

Local Oscillators

A good communications receiver contains oscillators that operate in a stable and spectrally pure manner. Poor oscillator performance can spoil the best of receivers even though all other parts of the circuit are functioning in elegant fashion. Not only should the oscillator be stable with regard to short- and long-term drift, it should have minimum noise in the output (at least 80 dB below the peak value of the fundamental energy) and be reasonably free of spurious responses. Concerning the latter, it is not difficult to design an oscillator that has all harmonics attenuated by 60 or 70

dB. Another important characteristic of an oscillator is quick starting when operating voltage is applied.

Oscillator instability can result from a host of poor design practices. To improve the stability characteristics it is useful to observe the following:

1) Use well-filtered regulated operating voltages.

2) Use temperature-stable, fixed-value capacitors in the frequency-determining part of the circuit. Polystyrene and silver mica capacitors are recommended.

3) Ensure that all mechanical and electrical components are secured rigidly in their part of the circuit. This will lessen the chance for mechanical instability.

4) Build the oscillator on firm, flex-free material.

5) When practical, enclose the oscillator in its own shield compartment and use RF filtering in the dc supply leads. The more constant the ambient temperature surrounding the oscillator, the greater will be the frequency stability.

Precautions should be taken to ensure that the oscillator in a receiver looks into a constant load impedance. Even minute load changes will cause phase shifts which can affect the oscillator frequency. The effect is more pronounced with VFOs than it is with crystal-controlled oscillators. Because of these conditions it is good design practice to couple very lightly to the oscillator stage. The power level can be increased by adding one or more buffer/amplifiers before the oscillator signal is applied to the mixer or detector.

Changes in operating voltage will result in frequency shifts. It is for this reason that regulated voltage is recommended for oscillators. Zener diodes are adequate for the purpose.

Magnetic cores, such as those in slug-tuned coils, change their properties with variations in ambient temperature, thereby causing inductance changes that can severely affect the oscillator frequency. Furthermore, mechanical instability can result if the slugs are not affixed securely in the coil forms.

Oscillator noise can be held to an acceptable level by employing high-Q tuned circuits. The higher the tank Q, the narrower the bandwidth and, hence, the lower the noise output voltage. Excessive LO noise will have a serious effect on mixer performance.

High amounts of harmonic current in the LO-chain output can cause unwanted mixer injection. If the receiver front-end selectivity is not of high magnitude, spurious signals from outside the band of interest will be heard along with the desired ones. Harmonic energy can degrade the performance of some kinds of mixers, making it worthwhile to use suitable filtering at the LO-chain output.

Receiver Front Ends

The designer has a number of options

available when planning the input section of a receiver. The band-pass characteristics of the input tuned circuits are of considerable significance if strong out-of-band signals are to be rejected — an ideal design criterion. Many of the commerical receivers available to the amateur use tuned circuits that can be adjusted from the front panel of the equipment. The greater the network Q, the sharper the frequency response and, hence, the better the adjacent-frequency rejection. For a given network design the bandwidth doubles for each octave higher. That is, an 80-meter front-end network may have a 3-dB bandwidth of 100 kHz for a given Q and load factor. At 40 meters the same type of network would be 200-kHz wide at the 3-dB points of the response curve. This is the reason that most receivers have a tunable front-end section (preselector). If fixed-tuned filters were used, at least two such filters would be necessary to cover from 3.5 to 4 MHz or 1.8 to 2.0 MHz. This would complicate the design and increase the equipment cost.

Fig. 15 shows the two concepts just discussed. The circuit at A covers all of the 80-meter band, and if selective enough offers some in-band rejection. A pair of Butterworth band-pass filters might be used at FL1 and FL2 of Fig. 15B to cover all of the 80-meter band. A lot of additional components would be required, and the in-band rejection of unwanted signals would be less than in the case of circuit A. The principal advantage of the circuit at B is that front-panel peaking adjustments would not be necessary once the trimmers in the filters were set for the desired response. A similar tuned circuit for either example in Fig. 15 would be used between the RF amplifier and the mixer.

Regardless of the type of LC input network used, a built-in step attenuator is worth considering. It can be used for measuring changes in signal level, or to reduce overloading effects when strong signals appear in the receiver passband. Fig. 16 shows how this can be done. The example at A is suitable for simple receivers when calibration in decibels is not a requisite, and when maintaining an impedance match between the tuned circuit and the antenna is not vital. The circuit at B is preferred because the pads are of 50-ohm impedance. In the circuit shown three steps are available: 6, 12 and 18 dB, depending on how the switches are thrown. The resistance values specified are the closest standard ones to the actual values needed to provide precisely 6 or 12 dB of attenuation. For amateur work the accuracy is adequate. Front-end attenuators are useful when VHF converters are used ahead of the station receiver. If the converters have a significant amount of overall gain they can degrade the dynamic range of the main receiver when strong signals are present. The attenuators can be set to simulate a condition of unity gain through the converter, thereby aiding receiver dynamic

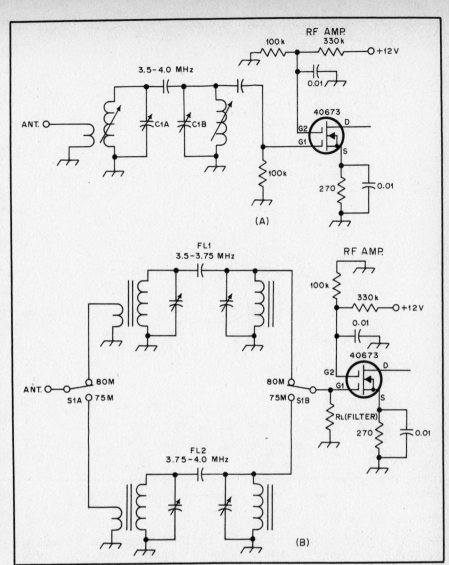

Fig. 15 — Method for selecting band-pass filters for 75 and 80 meters at the input to an RF amplifier.

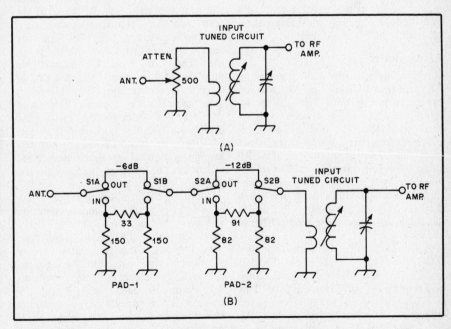

Fig. 16 — Front-end attenuators. A simple type is given at A and a step-attenuator version is shown at B.

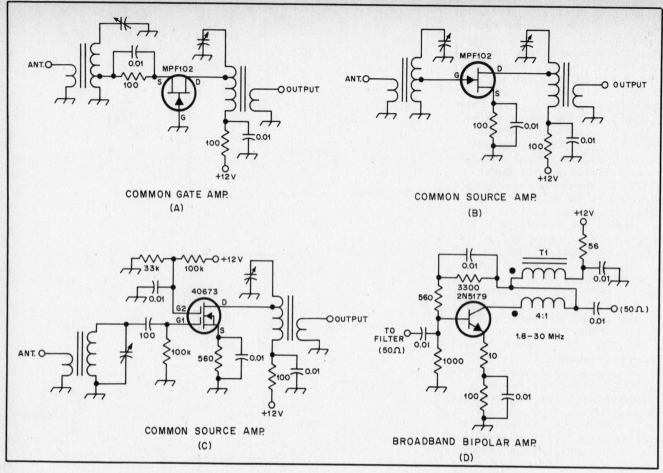

Fig. 17 — Narrowband RF amplifiers are shown from A to C. A fed-back broadband RF amplifier is shown at D.

RF Amplifiers

range. It is worth remembering, however, that an attenuator used at the input of a receiver when no converter is attached will degrade receiver sensitivity and noise figure. A receiver used frequently for antenna and received-signal decibel measurements might have several 3-dB pads included, thereby providing greater resolution during measurements.

RF Amplifiers

It was implied earlier in the chapter that RF amplifiers are useful primarily to improve the receiver noise figure. When atmospheric and man-made noise levels exceed that of the mixer it should be possible to realize better dynamic range by not having an RF amplifier. The gain of the RF stage, when one is used, should be set for whatever level is needed to override the mixer noise. Sometimes that is only a few decibels. A good low-noise active device should be employed as the RF amplifier in such instances.

A well-designed receiver should not have AGC applied to the RF amplifier. The best noise figure and RF-stage dynamic range will result when AGC is not applied. This is because the AGC voltage changes the operating characteristics of the RF amplifier from Class A to a less-linear mode.

Fig. 17 shows some typical RF amplifiers for use in amateur receivers. Tube-type circuits have not been included because they do not offer any particular advantage over solid-state amplifiers.

The circuit at A in Fig. 17 is likely to be the least subject to self-oscillaton of the four examples given. The common-gate hookup helps to ensure stability if the gate lead is kept as short as is physically possible. The gain from a common-gate amplifier of this type is lower than that of a common-source amplifier. However, gains up to 15 dB are typical. The drain of the FET need not be tapped down on the drain coil, but if it is there will be less loading on the tuned circuit, thereby permitting somewhat greater tuned-circuit selectivity: The lower the drain tap, the less the stage gain.

All of the FET amplifiers in Fig. 17 are capable of providing low-noise operation and good dynamic range. The common-source circuits at illustrations B and C can provide up to 25 dB of gain. However, they are more prone to instability than is the circuit at A. Therefore, the gates are shown tapped down on the gate tank: Placing the input at a low impedance point on the tuned circuit will discourage self-oscillation. The same is true of the drain tap. JFETs will hold up under considerable

RF input voltage before being damaged. Laboratory test of the MPF102 showed that 80 volts P-P (gate to source) was required to destroy the device. However, in the interest of good operating practices the P-P voltage should be kept below 10. Tapping the gate down on the input tuned circuit will result in lower levels of P-P input voltage, in addition to aiding stability.

A broadband bipolar-transistor RF amplifier is shown in Fig. 17 at D. This type of amplifier will yield approximately 16 dB of gain up to 148 MHz, and it will be unconditionally stable because of the degenerative feedback in the emitter and the negative feedback in the base circuit. A broadband 4:1 transformer is used in the collector to step the impedance down to approximately 50 ohms at the amplifier output. A 50-ohm characteristic exists at the input to the 2N5179 also. A band-pass filter should be used at the input and output of the amplifier to provide selectivity. The 4:1 transformer helps to ensure a collector load of 200 ohms, which is preferred in an amplifier of this type. This style of amplifier is used in CATV applications where the transformation from collector to load is 300 to 75 ohms.

Stability

Excessive gain or undesired feedback

may cause amplifier instability. Oscillation may occur in unstable amplifiers under certain conditions. Damage to the active device from overdissipation is only the most obvious effect of oscillation. Deterioration of noise figure, spurious signals generated by the oscillation, and reradiation of the oscillation through the antenna, causing RFI to other services, can also occur from amplifier instability.

Neutralization or other forms of feedback may be required in RF amplifiers to reach stability. Amplifier neutralization is achieved by feeding energy from the amplifier output circuit back to the input in such an amount and phase as to cancel out the effects of device internal capacitance and other unwanted input-output coupling. Care in termination of both the input and output can produce stable results from an otherwise unstable amplifier. Attention to proper grounding and proper isolation of the input from the output by means of shielding can also yield stable operating conditions.

Overloading and Spurious Signals

Normally, the RF amplifier is not a significant contributor to overloading problems in VHF receiving systems. The RF amplifiers in the first or second stage of a receiving system operate in a linear service, and if properly designed require a substantial signal input to cause deviation from linearity. Overloading usually occurs in the naturally nonlinear mixer stages. Images and other responses to out-of-band signals can be reduced or eliminated by proper filtering at the amplifier input.

In general, unwanted spurious signals and overloading increase as the signal levels rise at the input to the offending stage. Consequently, minimum gain prior to the stage minimizes overloading. Since noise figure may suffer at reduced gain, a compromise between optimum noise figure and minimum overloading must often be made. Especially in areas of high amateur activity, sacrificing noise figure somewhat may result in increased weak-signal reception effectiveness if the lower noise-figure system is easily overloaded.

Typical Circuits for VHF and Above

Common circuits for the RF amplifiers used at VHF and UHF are illustrated in Figs. 18 through 21. The termination impedances of both the input and output of these examples are low (50 ohms), suiting them well to preamplifer service. Preamplifers are useful for improving the noise figure of existing equipment.

The choice of active device has a profound effect on the weak-signal performance of an RF amplifier. Although tubes can be used on the VHF and UHF bands, their use is seldom seen, as solid-state devices provide far better performance at lower cost. Bipolar transistors can provide excellent noise figures up through 4 GHz if chosen and used properly. The

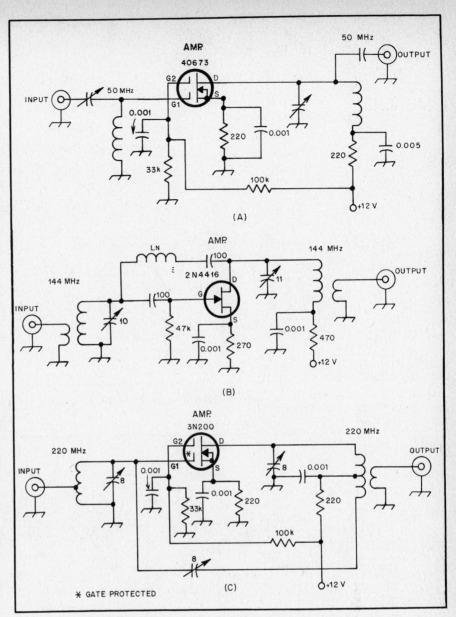

Fig. 18 — Typical grounded-source RF amplifiers. The dual-gate MOSFET, A, is useful below 500 MHz. The junction FET, B, and neutralized MOSFET, C, work well on all VHF bands. Except where given, component values depend on frequency.

JFET is usable through 450 MHz, although the most commonly available ones drop off in performance quickly beyond that frequency. Dual-gate MOSFETs also are usable through 450 MHz. The GaAsFET provides superior noise figures past 1300 MHz.

Most RF amplifiers for use below 225 MHz use FETs rather than bipolars. Unless bipolar transistors are run at relatively high standing currents they are prone to overloading from strong signals. Additionally, their lower terminating impedances can present somewhat awkward design considerations to the builder. The FET minimizes these problems while presenting an acceptable noise figure.

Above 225 MHz, inexpensive FETs cannot provide the low noise figure attainable from bipolars. The wavelength at these higher frequencies also allows the con-

venient use of tuned lines rather than conventional coils, easing the possible design difficulties of the lower terminating impedances of bipolars.

The input network of an RF amplifier should be as low in loss as possible to assure a low noise figure, since any loss before the first stage is effectively added to the noise figure. High-selectivity circuits often have significant losses and should be avoided at the front end. L networks usually provide the least loss while achieving proper impedance matching. High-quality components should also be employed in the input circuit to further reduce losses.

It should be pointed out that the terminating impedance of transistors for optimum noise figure is usually not the same as that for optimum power transfer (gain). This complicates the design and tuning procedures somewhat, but careful

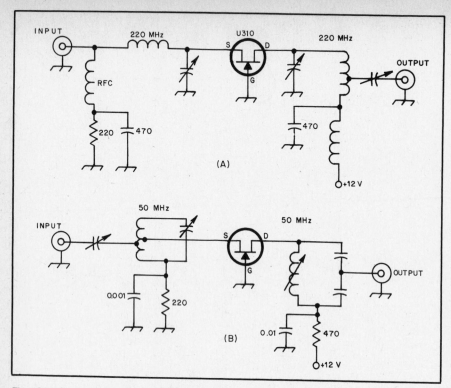

Fig. 19 — Grounded-gate-FET preamplifier tends to have lower gain and broader frequency response than other amplifiers described.

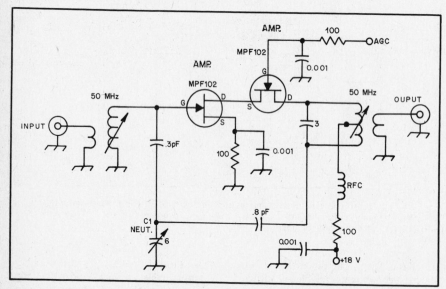

Fig. 20 — Cascode amplifier circuit combines grounded-source and grounded-gate stages, for high gain and low noise figure. Though JFETs are shown, the cascode principle is useful with MOSFETs as well.

measurements and adjustment can compensate for these shortcomings. The dual-gate MOSFET has different internal geometry, so optimum noise match is virtually identical to optimum gain match. This means that adjusting a dual-gate MOSFET amplifier for maximum gain usually provides best noise figure.

Some examples of common-source amplifiers are shown in Fig. 18. Many properly designed dual-gate MOSFET amplifiers do not require neutralization to achieve stability and best noise figure. An

example of this approach is shown in Fig. 18A. Neutralization may be required; Fig. 18C shows capacitive neutralization applied to a dual-gate-MOSFET amplifier. Common-source JFET amplifiers usually require neutralization to attain satisfactory operation. Inductive neutralization as shown in Fig. 18B is commonly used.

Using the gate as the common stage element introduces negative feedback and eliminates the need for neutralization in a common-gate amplifier, as shown in Fig. 19. The feedback reduces the stage

gain and lowers the input impedance, increasing the bandwidth of the stage. An additional benefit of common-gate amplifiers is reduced susceptibility to overload as compared to common-source amplifiers.

The cascode circuit of Fig. 20 combines the common-source and the common-gate amplifiers, securing some of the advantages of each. Increased gain over a single stage is its greatest asset.

Fig. 21 shows typical bipolar amplifiers for the UHF range. Fig. 21A illustrates a common-emitter amplifier, analogous to the common-source FET amplifier. The common-base amplifier of Fig. 21B can similarly be compared to a common-gate FET amplifier.

Front-End Protection

The first amplifier of a receiver is susceptible to damage or complete burnout through application of excessive voltage to its input element by way of the antenna. This can be the result of lightning discharges (not necessarily in the immediate vicinity), RF leakage from the station transmitter through a faulty send-receive relay or switch, or RF power from a nearby transmitter and antenna system. Bipolar transistors often used in low-noise UHF amplifiers are particularly sensitive to this trouble. The degradation may be gradual, going unnoticed until the receiving sensitivity has become very poor.

No equipment is likely to survive a direct hit from lightning, but casual damage can be prevented by connecting diodes back-to-back across the input circuit. Either germanium or silicon VHF diodes can be used. Both have the thresholds of conduction well above any normal signal level, about 0.2 volt for germanium and 0.6 volt for silicon. The diodes used should have fast switching times. Diodes such as the 1N914 and hot-carrier types are suitable. A check on weak-signal reception should be made before and after connection of the diodes.

MIXERS

Conversion of the received energy to a lower frequency, so it can be amplified more efficiently than would be possible at the signal frequency, is a basic principle of the superheterodyne receiver. The stage in which this is done may be called a converter or frequency converter, but we will use the more common term, mixer, to avoid confusion with converter, as applied to a complete VHF receiving accessory. Mixers perform similar functions in both transmitting and receiving circuits.

The mixer is one of the most important parts of a high-performance receiver. It is at this point where the greatest consideration for dynamic range exists. For best receiver performance the mixer should receive only enough preamplification to overcome the mixer noise. When excessive amounts of signal energy are permitted to reach the mixer, there will be desensitization, cross-modulation and IMD products in the mixer. When these effects are severe enough the receiver can be rendered useless. Therefore, it is advantageous to utilize what is often called a "strong mixer." That is, one which can handle high signal levels without being affected adversely.

Generally speaking, diode-ring passive mixers fare the best in this regard. However, they are fairly noisy and require considerably more LO injection than is the case with active mixers. For the less sophisticated types of receivers, it is adequate to use single-ended active mixers, provided the gain distribution between the antenna and mixer is proper for the mixer device used. Field-effect transistors are preferred by most designers; bipolar-transistor mixers are seldom used.

The primary advantage of an active mixer is that it has conversion gain rather than loss. This means that the stages following the mixer need not have as much gain as when diode mixers are used. A typical doubly balanced diode mixer will have a conversion loss of some 8 dB, whereas an FET active mixer may exhibit a conversion gain as great as 15 dB. The cost of gain stages is relatively small. This consideration easily justifies the use of strong passive mixers in the interest of high dynamic range.

An LC, crystal-lattice, or mechanical type of band-pass filter is almost always used after the mixer or the post-mixer amplifier. This helps to establish the overall selectivity of the receiver. It also rejects unwanted mixer products that fall outside the passband of the filter.

In the interest of optimum mixer performance, the LO energy supplied to it should be reasonably clean with respect to unwanted frequencies. Many designers, for this reason, use a band-pass filter between the LO output and the mixer input. Excessive LO noise will seriously degrade receiver performance. LO noise should be 80 dB or more below the peak output level. Excessive noise will appear as noise sidebands in the receiver output.

A receiver for 50 MHz or higher usually has at least two mixer stages: one in the VHF or UHF converter, and usually two or more in the communications receiver that follows it. We are primarily concerned here with the first mixer.

The ideal mixer would convert any input signal to another chosen frequency with no distortion, and would have a noise figure of 0 dB. Unfortunately, a mixer such as that exists only in a dream world. The mixer that has a 0-dB noise figure (or equivalent loss) has yet to be conceived. This means that the proper use of RF amplification and perhaps post-mixer amplification is necessary for maximum receiver performance with regard to sensitivity. Poor sensitivity is the least difficult of the mixer failings to mend.

Because the mixer operates in a nonlinear mode, reduction of distortion becomes a major design problem. As the mixer input

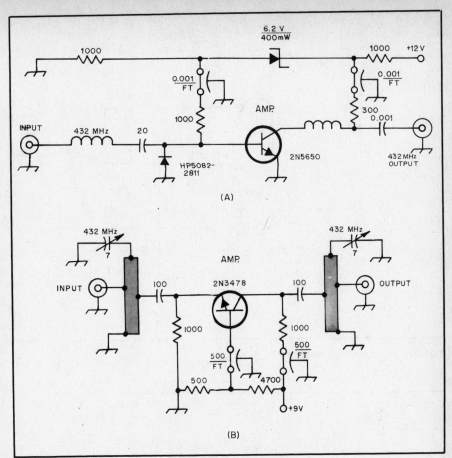

Fig. 21 — Examples of VHF amplifiers using bipolar transistors.

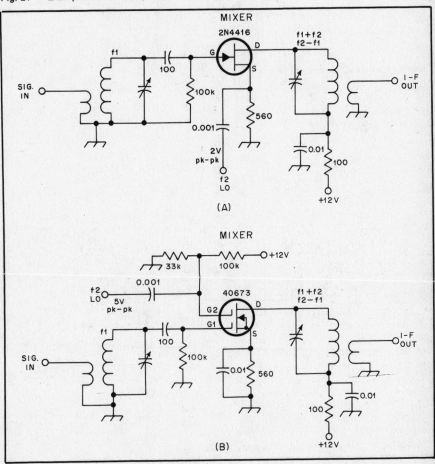

Fig. 22 — Two styles of active mixers using FETs.

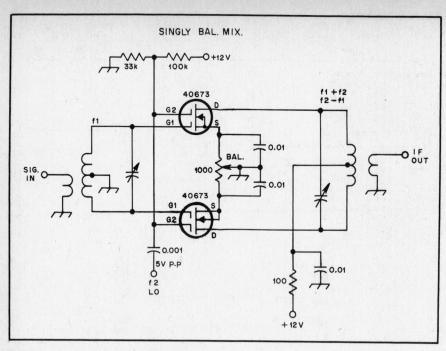

Fig. 23 — An active singly balanced FET mixer.

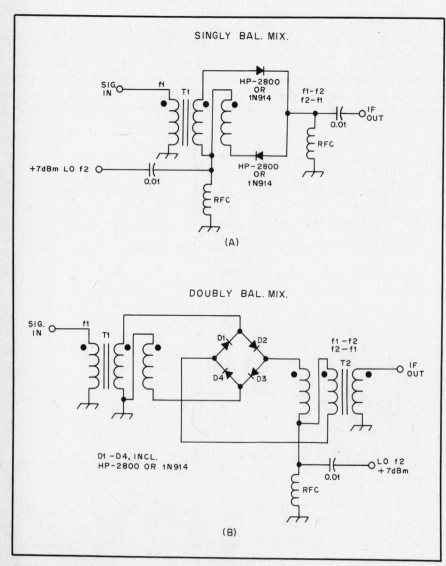

Fig. 24 — Singly and doubly balanced diode mixers.

level is increased, a point is reached where the output no longer increases linearly with input. A phenomenon known as compression occurs. When the compression point is reached, the sensitivity of the mixer is reduced for every signal in the passband. This is manifested as desensing. Different types of mixers characteristically reach their compression points at different input levels, so proper mixer choice can minimize this type of distortion. Any amplifier before the mixer will increase the input levels to the mixer, lowering the point where the input level to the receiving system will cause compression. Therefore, a builder should not use more gain than is necessary to establish system noise figure prior to the mixer.

If more than one signal is present in the passband going into the mixer, they may mix to produce spurious responses known as intermodulation distortion (IMD) products. As the input levels increase further, higher-order IMD products may appear, seemingly filling the passband. Proper mixer operating conditions will alleviate IMD problems and also reduce gain-compression problems.

A third type of distortion is cross modulation. This is most readily observed on AM signals. When the carrier is on, cross modulation is evidenced by modulation characteristics of another signal being superimposed on the received carrier. Techniques to improve IMD characteristics also improve cross-modulation performance.

A problem inherent in all mixing systems is image generation. Whenever two signals are mixed, components are produced at the sum and difference of the two signal frequencies and at multiples of these frequencies. For receiving applications amateurs typically want to detect only one of the mixing products, usually the first-order mixing product. Filtering must be applied to separate the desired signal from the rest. Post-mixer filtering is not adequate, as input images can be mixed to the same intermediate frequency as the desired signal. Input filtering discriminates against these images and prevents unwanted out-of-band signals from overloading the mixer.

Typical HF Mixer Circuits

Fig. 22 shows two single-ended active mixers that offer good performance. The example at A employs a JFET with LO injection supplied to the source across a 560-ohm resistor. This injection mode requires somewhat more LO power than would be used if injection was done at the gate. However, there is less occasion for LO pulling when source injection is used, and there is better isolation between the LO and antenna than would be the case with gate injection.

The circuit at B in Fig. 22 is similar in general performance to that at A. The major difference is that a dual-gate MOSFET is used to permit injection of the LO energy at gate 2. Since there is considerable signal isolation between gates 1 and 2, LO pulling is minimized, and antenna-to-LO isolation is good.

A singly balanced active mixer is illustrated in Fig. 23. Two 40673 dual-gate MOSFETs are connected in push-pull, but with the LO frequency injected in parallel at gate 2 of each device. A potentiometer is used in the sources of the transistors to permit circuit balance. This mixer offers performance superior to that of the mixers shown in Fig. 22.

One of the least complicated and inexpensive mixers is the two-diode version (singly balanced) seen in Fig. 24A. A trifilar-wound broadband toroidal transformer is used at the mixer input. The shortcoming of this mixer over the one at B is that signal isolation between all three mixer ports is not possible. A better version is that at B in Fig. 24. In this case all three mixer ports are well isolated from one another. This greatly reduces the probability of spurious responses in the receiver. Conversion loss with these mixers is approximately 8 dB. The impedance of the mixer ports is approximately 50 ohms.

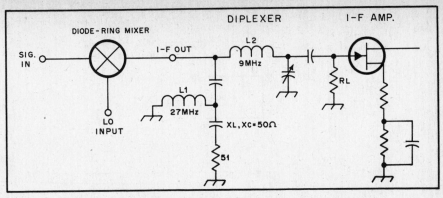

Fig. 25 — Method for diplexing the mixer output to improve the IMD characteristics.

Improved IMD characteristics can be had from a diode-ring mixer, by placing a diplexer after the mixer as shown in Fig. 25. The diplexer consists of a high-pass network (L1) and a low-pass one (L2). L2 is tuned to the IF and serves as a matching network between 50 ohms and R_L, the FET gate resistor. L1 and the associated series capacitors are tuned to three times the IF and terminated in 50 ohms. This gives the mixer a proper resistive termination without degrading the 9-MHz IF. The high-pass network has a loaded Q of 1.

IC Mixers

Although numerous ICs are available for use as mixers, only three are shown here. Fig. 26 shows a CA3028A singly balanced active mixer. The diagram at B shows the inner workings of the IC. The LO is in-

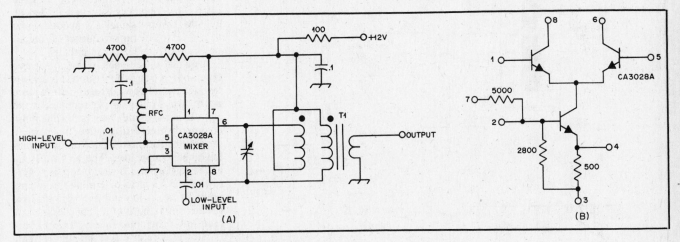

Fig. 26 — CA3028A singly balanced mixer. The circuit for the IC is given at B.

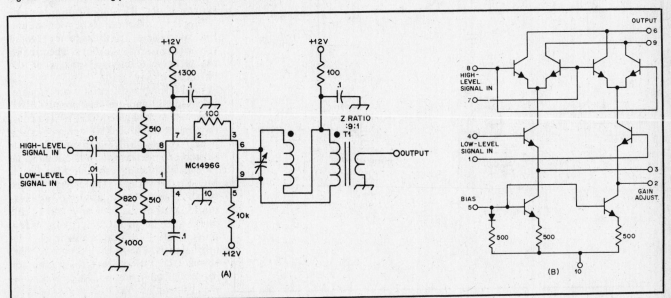

Fig. 27 — MC1496G doubly balanced mixer and circuit of the IC.

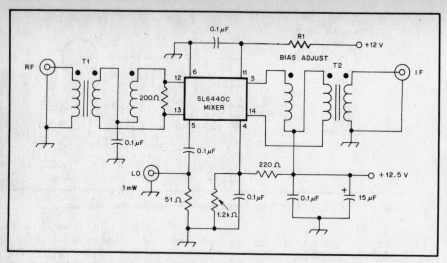

Fig. 28 — Plessey SL6400C doubly balanced mixer. R1 is selected for a bias current of about 12 mA. T1 and T2 are broadband transformers would on ferrite toroidal cores.

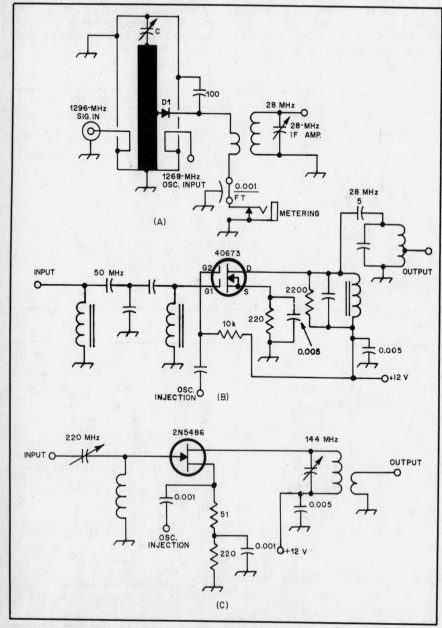

Fig. 29 — Examples of single-ended mixers. The diode mixer, A, is usable through the microwave region. FET mixers, B and C, offer conversion gain and low noise figure.

jected at pin 2 of the IC. Conversion gain is on the order of 15 dB.

Fig. 27A illustrates a doubly balanced IC active mixer that employs an MC1496G. A schematic diagram of the IC is shown at B. The performance of this mixer is excellent, but is it not as strong a mixer as that of Fig. 24. However, it has good conversion gain and a fairly low noise figure.

The Plessey SL6400C programmable high-level doubly balanced mixer is shown in Fig. 28 in a broadband circuit. This configuration produces a conversion gain of 8 dB and a third-order output intercept of approximately +22 dBm. The single-sideband noise figure was found to be 11 dB. If the mixer input and output ports are terminated in 50 ohms, rather than 200 ohms as shown in Fig. 28, the typical conversion gain will be −1 dB.

Single-ended Mixers for VHF and UHF

The simplest type of mixer is the diode mixer. The signal and the heterodyning frequency are fed into the mixer, and the mixer output includes both the sum and the difference frequencies of the two. In the case of the circuit shown in Fig. 29A the difference frequency is retained, so the 1296-MHz input signal is converted down to 28 MHz. The sum frequency is filtered out by the 28-MHz tuned circuits.

A quality diode (such as the hot-carrier type) has a fairly low noise figure up through the microwave region. Since most active mixers fall off in performance above 500 MHz, the diode mixer is the one most commonly found in amateur microwave service. Unfortunately, all diode mixers have conversion loss. The loss must be added to the noise figure of the stage following the mixer to determine the system noise figure. A low-noise stage following the mixer is necessary for good weak-signal reception. The noise figure of most communications receivers is far higher than what is needed for a low-noise-figure system; if not, RF amplification is used.

Bipolar transistors are not good square-law type devices, and thus are not favored for single-ended applications. Their major use is in switching-type mixers of the balanced variety.

Field-effect transistors have good square-law response and are very popular VHF mixers. The dual-gate MOSFET is a common mixer found in VHF amateur equipment. The MOSFET can provide considerable conversion gain, while at the same time maintaining a reasonable noise figure. MOSFET overload characteristics are suitable for the vast majority of applications. Local-oscillator energy can be applied at one of the MOSFET gates, effectively isolating the local oscillator from the other signals. The gate impedance is high, so relatively little injection is needed for maximum conversion gain. A typical example is shown in Fig. 29B.

JFETs are close to the MOSFET in mixer performance but are more difficult to apply in practical hardware. As with the MOSFET, input impedance to a JFET mixer is high, and substantial conversion gain is available. JFET bias for mixer service is critical and must be adjusted for best results. The output impedance of a JFET is lower than a dual-gate MOSFET; typically around 10 kΩ. Although other possibilities exist, local-oscillator injection should be made at the JFET source for best results. The source is a low-impedance point, so considerably more local-oscillator power is required than if a dual-gate MOSFET were used as the mixer. Noise figures as low as 4 dB are possible with circuits like that shown in Fig. 29C.

The injection level of the local oscillator affects mixer performance. Raising the LO level increases conversion gain in an FET mixer. The local-oscillator signal should be as large as possible without pushing the FET into its pinchoff region. The gate junction of the FET should never conduct in mixer applications. Increased IMD products result from either of these conditions and should be avoided. The local-oscillator energy should be as pure as possible. Distorted injection energy not only increases IMD production but also increases stage noise figure.

Proper termination of the output of an FET mixer optimizes overload performance. If the impedance seen at the drain of an FET mixer is too high at any of the mixer product frequencies, large voltage excursions can occur on the FET drain. If the voltage excursion on the drain is large enough, output distortion will be evident. Often these high-voltage excursions occur at frequencies outside the desired passband, causing distortion from signals not even detectable by the receiver. A resistor within the output matching network may be used to limit the broadband impedance to a suitably low level.

Balanced Mixers

Use of more than one device in either a singly or doubly balanced mixer offers many advantages over a single-ended mixer. The balance prevents energy injected into a mixer port from appearing at another port. The implications of this are significant when minimum mixer distortion is sought. The port-to-port isolation inhibits any signals other than the mixing products from reaching subsequent stages where they might be mixed and cause undesirable products. The usually large local-oscillator signal is kept away from the RF amplifier stages where it might cause gain compression because of its magnitude. Any amplitude-modulated noise found on the local-oscillator signal is suppressed at the mixer output, where it might be later detected. In a singly balanced mixer only one port, usually the local-oscillator input, is isolated from the other two. A doubly balanced mixer isolates all three ports from each other.

The most common balanced mixer uses

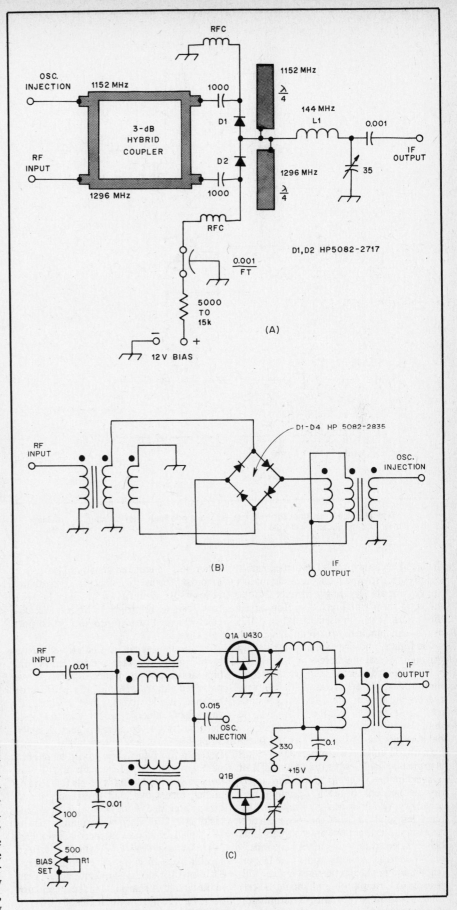

Fig. 30 — Balanced mixers for VHF and UHF. The singly balanced mixer, A, provides isolation of the local oscillator from the output. The doubly balanced diode mixer, B, has all ports isolated from each other, and is broadband throughout VHF. A special dual JFET is used at C to give high dynamic range with low noise figure.

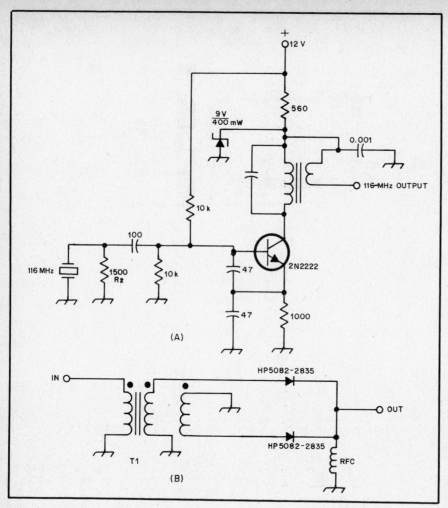

Fig. 31 — Typical crystal oscillator for VHF use, A. The diode frequency doubler, B, provides good rejection of the fundamental signal.

diodes. The disadvantages presented earlier with respect to single-ended diode mixers apply here also. A singly balanced diode mixer is shown in Fig. 30A. Hot-carrier diodes are normally used for D1 and D2, as they can handle high currents, have a low noise figure, and are available for use up through several gigahertz.

The doubly balanced mixer is more common today. Commercial modules, very reasonably priced, are often used instead of homemade circuits. Large-scale manufacturing can usually offer electrical balance not so easily attained with the homemade units. Isolation of 35 to 40 dB is typical at VHF, with only 6 to 7 dB of conversion loss. High local-oscillator injection is needed to reach optimum performance with these mixers. Proper broadband termination of all the mixer ports is necessary to prevent unwanted signals from being reflected into the mixer "rat race," only to emerge at another port. The IF port (shown in Fig. 30B) is the most critical with respect to termination and should be terminated at 50 + j0 ohms. Transmission line transformers provide the necessary phase shift, as half the bridge is fed 180 degrees out of phase with respect to the

other half. These can be wound on ferrite toroidal forms to effect a broadband response. Careful winding of the transformers improves balance in the circuit, which in turn improves port-to-port isolation.

Active devices can be used very effectively in balanced mixers. Both FETs and bipolars can be used successfully. Active balanced mixers offer all the benefits of balanced diode mixers, plus the added advantage of conversion gain rather than loss. Because of this conversion gain, less RF amplification is needed to establish a lower system noise figure than would be needed with a diode mixer. Low gain prior to the mixer keeps mixer-input levels low, maximizing mixer overload resistance. High-dissipation active devices can be used, yielding better mixer performance than is available from diode balanced mixers. Fig. 30C shows a dual FET which has been specially designed for mixer applications. R1 allows for electrical balance adjustment in the circuit. A sharp null in local oscillator output at the mixer output can be observed when R1 is set to the optimum point, showing that electrical balance has been achieved.

Injection Stages

Oscillator and multiplier stages supplying heterodyning energy to the mixer should be as stable and free of unwanted frequencies as possible. Proper application of crystal control gives the stability needed. Two major factors influence oscillator stability: temperature and operating voltage. As the temperature of a component changes, its internal geometry is altered somewhat as the constituent materials expand or contract. This typically results in internal capacitance changes that affect the resonant frequency of the tuned circuits controlling the oscillator frequency. Use of quality components that have good temperature characteristics helps in this regard. Minimum power should be extracted from the oscillator as excessive heat dissipation within either the crystal or the transistor will cause internal capacitance changes in those devices, moving the resonant frequency. Voltage to the transistor should be regulated for best stability. Simple Zener-diode regulation is sufficient, or a three-terminal regulator IC can be used.

Any unwanted injection frequencies will mix with signals present in the mixer, creating spurious outputs at the mixer output. A clean local oscillator will prevent these unwanted outputs. The oscillator chain output can be heavily filtered to cut down the harmonic content of the oscillator. Good planning and design will minimize the unwanted energy, making the filtering job less demanding. A high-frequency crystal in the oscillator minimizes the number of times that the fundamental oscillator frequency has to be multiplied to reach the converter injection frequency. Proper use of doublers rather than triplers can eliminate any odd oscillator frequency multiples, so a low-pass filter at the output has to filter only the fourth harmonic and beyond. A band-pass filter would be needed at the output of a tripler to eliminate the second harmonic and the higher ones. Finally, good shielding and power-line filtering should be used throughout to prevent any line noise from reaching the mixer or causing RFI problems elsewhere.

Fig. 31A shows a typical circuit useful for providing the 116-MHz injection energy necessary to convert a 144-MHz signal down to 28 MHz. R_z dampens the crystal action somewhat, assuring that the proper overtone is the actual oscillation frequency. The collector tank network is parallel tuned and can be wound on a toroidal core to reduce radiation. The output is link coupled from the tank, minimizing harmonic coupling. This oscillator would be followed by a buffer to bring the signal up to that level needed and to purify the oscillator signal further.

A similar oscillator could be used in a 220-MHz converter. Since crystals are not available at 192 MHz, the frequency required for conversion to 28-MHz converter output, the most logical approach is to use a 96-MHz oscillator and double its output. Fig. 31B shows a diode frequency doubler suitable for the application. The phase-shifting transformer can be made from a trifilar winding on a ferrite core. Hot-carrier diodes allow the use of a doubler like this up through at least 500 MHz. There is a loss of about 8 dB through the doubler, so amplification is needed to raise the injection signal to the appropriate level. Fundamental energy is down by as much as 40 dB from the second harmonic with a balanced diode doubler such as this. All of the odd harmonics are well down in amplitude also, without using tuned circuits. A low-pass filter can be used to eliminate the undesired harmonics from the output.

DOUBLY BALANCED MIXERS

Advances in technology have provided the amateur builder with many new choices of hardware to use in the building of receivers, converters or preamplifiers. The broadband doubly balanced mixer (DBM) package is a fine example of this progress. As amateurs gain an understanding of the capabilities of this device, they are incorporating this type of mixer in many pieces of equipment, especially receiving mixers. The combined mixer/amplifier described here was presented originally in March 1975 *QST* by K1AGB.

Mixer Comparisons

Is a DBM really better than other types? What does it offer, and what are its disadvantages? To answer these questions, another look at more conventional active (voltages applied) mixing techniques and some of their problems is in order. Briefly, common single-device active mixers with gain at VHF and UHF are beset with problems of noise, desensitization and insufficient local-oscillator (LO) isolation from the RF and IF "ports." As mixers, most devices have noise figures in excess of those published for them as RF amplifiers and will not provide sufficient

sensitivity for weak-signal work. To minimize noise, mixer-device current is generally maintained at a low level. This can reduce dynamic range and increase overload potential. Gain contributions of RF amplifiers (used to establish a low system noise figure) further complicate the overload problem. LO-noise leakage to the RF and IF ports adversely affects system performance. Mixer dynamic range can be limited by conversion of this noise to the IF, placing a lower limit on mixer system sensitivity. Generally 20 dB of mixer mid-band interport isolation is required, and most passive DBMs can offer greater than 40 dB.

A commercially manufactured doubly balanced diode mixer offers performance predictability, circuit simplicity and flexibility. Closely matched Schottky-barrier hot-carrier diodes, commonly used in most inexpensive mixers of this type, provide outstanding strong-signal mixer performance (up to about 0 dBm at the RF input port) and add little (0.5 dB or so) to the mixer noise figure. Essentially, diode conversion loss from RF to IF, listed in Table 1, represents most of the mixer contribution to system noise figure. Midband isolation between the LO port and the RF and IF ports of a DBM is typically > 35 dB — far greater than that achievable with conventional single-device active-mixing schemes. This isolation is particularly advantageous in dealing with low-level local-oscillator harmonic and noise content. Of course, selection of LO devices with low audio noise figures, and proper RF filtering in the LO output, will reduce problems from this source.

Often-listed disadvantages of a diode DBM are (a) conversion loss, (b) LO power requirements, and (c) IF-interface problems. The first two points are closely interrelated. Conversion loss necessitates some low-noise RF amplification to establish a useful weak-signal system noise figure. Additional LO power is fairly easy to generate, filter and measure. If we accept the fact that more LO power is

necessary for the DBM than is used in conventional single-device active mixing circuits, we leave only two real obstacles to be overcome in the DBM, those of conversion loss and IF output interfacing.

To minimize conversion loss in a DBM, the diodes are driven by the LO beyond their square-law region, producing an output spectrum which in general includes the terms[2]

1) Fundamental frequencies fLO and fRF
2) All of their harmonics
3) The desired IF output, fLO $\pm$ fRF
4) All higher order products of nfLO $\pm$ mfRF, where n and m are integers.

The DBM, by virtue of its symmetry and internal transformer balance, suppresses a large number of the harmonic modulation products. In the system described here, f LO is on the low side of fRF; therefore, numerically, the desired IF output is fRF − fLO. Nonetheless, the term fLO + fRF appears at the IF output port equal in amplitude to the desired IF signal, and this unused energy must be effectively terminated to obtain no more than the specified mixer-conversion loss. This is not the image frequency, fLO − fIF, which will be discussed later.

In any mixer design, all RF port signal components must be bypassed effectively for best conversion efficiency (minimum loss). Energy not "converted" by mixing action will reduce conversion gain in active systems, and increase conversion loss in passive systems such as the diode DBM. RF bypassing also prevents spurious resonances and other undesired phenomena from affecting mixer performance. In this sytem, RF bypassing at the IF output port will be provided by the input capacitance of the IF interface. The DBM is not a panacea for mixing ills, and its effectiveness can be reduced drastically if all ports are not properly terminated.

[2]See appendix at the end of this section on mixers.

Table 1
Doubly Balanced Mixers

Manufacturer	Relcom	Anzac	MCL	MCL	MCL	MCL
Model	M6F	MD-108	SRA-1	SRA-1H	RAY-1	MA-1
Frequency Range (MHz)						
LO	2-500	5-500	5-500	5-500	5-500	1-2500
RF	2-500	5-500	51-500	5-500	5-500	1-2500
IF	DC-500	DC-500	DC-500	DC-500	DC-500	1-1000
Conversion loss	9 dB max.	7.5 dB max.	6.5 dB typ.	6.5 dB typ.	7.5 dB typ.	8.0 dB typ.
Mid-range Isolation, LO-RF	34-40 dB min.	40 dB min.	45 dB typ.	45 dB typ.	40 dB typ.	40 dB typ.
Mid-range LO-IF	25-35 dB min.	35 dB-min.	40 dB typ.	40 dB typ.	40 dB typ.	40 dB typ.
Total input power	50 mW	400 mW	500 mW	500 mW	1W	50 mW
LO power requirement	+7 dBm (5 mW)	+7 dBm (5 mW)	+7 dBm (5 mW)	+17 dBm (50 mW)	+23 dBm (200 mW)	+10 dBm (10 mW)
Signal 2-dB compression level	Not spec.	Not spec.	+1 dBm	+10 dBm	+15 dBm	+7 dBm
Impedance, all ports	50 ohms	50 ohms	50 ohms	50 ohms	50 ohms	50 ohms

Relcom, Division of Watkins-Johnson, 2525 N. First St., San Jose, CA 95131, tel. 408-262-1411
Anzac Electronics, 39 Green Street, Waltham, MA 02154, tel. 617-891-6220
MCL — Mini-Circuits, P. O. Box 166, Brooklyn, NY 11235 tel. 718-934-4500

All specifications apply only at stated LO power level.

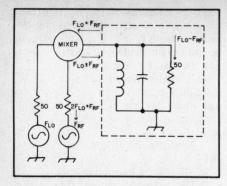

Fig. 32 — The IF port of a doubly balanced mixer, matched at fLO − fRF and reactive at fLO + fRF. In this configuration conversion loss, RF compression and desensitization levels can vary ±3dB while harmonic modulation and third-order IMD products can vary ±20 dB.

DBM-Port Termination

Most DBM performance inconsistencies occur because system source and load impedances presented to the mixer are not matched at all frequencies encountered in normal operation. The terminations (attenuator pads) used in conjunction with test equipment by manufacturers to measure published performance characteristics are closely matched over a wide range of frequencies. Reactive mixer terminations can cause system problems, and multiple reactive terminations can usually compound these problems to the point where performance is very difficult to predict. Let's see how we can deal with reactive terminations.

The IF Port

The IF port is very sensitive to mismatch conditions. Reflections from the mixer/IF-amplifier interface (the pi network in Fig. 33) can cause the conversion loss to vary as much as 6 dB. Also greatly affected are third-order intermodulation-product ratio and the suppression of spurious signals, both of which may vary ±10 dB or more. It is ironic that the IF port is the most sensitive to a reactive termination, as this is a receiving system point where sharp-skirted filters are often desired.

Briefly, here is what happens with a reactive IF port termination. Fig. 32 shows a DBM with ''high side'' LO injection and an IF termination matched at fLO − fRF but reactive to fLO + fRF. The latter term re-enters the mixer, again combines with the LO, and produces terms that exit at the RF port, namely 2 fLO + fRF, a dc term, and fLO + fRF − fLO (the original RF-port input frequency). This condition affects conversion loss, as mentioned earlier, in addition to RF-port VSWR, depending on the phase of the reflected signal. The term 2 fLO + fRF also affects the harmonic spectrum resulting in spurious responses.

One solution to the IF interface problem is the use of a broadband 50-ohm resistive termination, like a pad, to minimize reflec-

tions. In deference to increased post-conversion system noise figure, it seems impractical to place such a termination at the mixer IF output port. While a complementary filter or diplexer (high-pass/low-pass filters appropriately terminated) can be used to terminate both fRF + fLO and fRF − fLO[†], a simpler method can be used if fRF + fLO is less than 1 GHz and $(f$RF + fLO$)/(f$RF − fLO$) \geq 10$. Place a short-circuit termination to fRF + fLO, like a simple lumped capacitance, directly at the mixer IF terminal. This approach is easiest for the amateur to implement and duplicate, so a form of it was tried — with success. In our circuit, C1 serves a dual purpose. Its reactance at fRF + fLO is small enough to provide a low-impedance ''short-circuit'' condition to this term for proper mixer operation. Additionally, it is part of the input reactance of the mixer IF amplifier interface. Fortunately, the network impedance-transformation ratio is large enough, and in the proper direction, to permit a fairly large amount of capacitance (low reactance) at the mixer IF-output port. The capacitor, in its dual role, must be of good quality at VHF/UHF (specifically fRF + fLO), with short leads, to be effective. The mixer condition $(f$RF + fLO$)/(f$RF − fLO$) \geq 10$ is met at 432 and 220 MHz with a 404/192 MHz LO (28-MHz IF) and on 144 MHz with a 130-MHz LO (14-MHz IF). At 50 MHz, with a 36-MHz LO, we are slightly shy of the requirement, but no problems were encountered in an operating unit. The pi-type interface circuit assures a decreasing impedance as IF operation departs from mid-band, lessening IMD problems.

The LO Port

The primary effect of a reactive LO source is an increase in harmonic modulation and third-order IMD products. If the drive level is adequate, no effect is noted on conversion loss, RF compression and desensitization levels. A reactive LO source can be mitigated by simply padding the LO port with a 3- or 6-dB pad and increasing the LO drive a like amount. If excess LO power is not available, matching the LO source to the mixer will improve performance. This method is acceptable for single-frequency LO applications, when appropriate test equipment is available to evaluate matching results. For simplicity, a 3-dB pad was incorporated at the LO-input port as an interface in both versions of the mixer. Thus the LO port is presented with a reasonably broadband termination, and is relatively insensitive to applied frequency, as long as it is below about 500 MHz. This implies that frequencies other than amateur assignments may be covered — and such is indeed the case when ap-

[†]Presentation and calculation format of these terms is based on ''low-side'' LO injection; see the appendix for explanation.

propriate LO frequencies and RF amplifiers are used. Remotely located LOs, when adjusted for a 50-ohm load, can be connected to the mixer without severe SWR and reflective-loss problems in the transmission line.

Broadband mixers exhibit different characteristics at different frequencies, owing to circuit resonances and changes in diode impedances resulting from LO power-level changes. Input impedances of the various ports are load dependent, even though they are isolated from each other physically, and by at least 35 dB electrically. At higher frequencies, this effect is more noticeable, since isolation tends to drop as frequency increases. For this reason, it is important to maintain the LO power at its appropriate level, once other ports are matched.

The RF Port

A reactive RF source is not too detrimental to system performance. This is good, since the output impedance of most amateur preamplifiers is seldom 50 ohms resistive. A 3-dB pad is used at the RF port in the 50- and 144- to 14-MHz mixers, and a 2-dB pad is used in the 220/432- to 28-MHz units, although they add directly to mixer noise figure. RF inputs between about 80 and 200 MHz are practical in the 14-MHz-IF-output model, while the 28-MHz-output unit is most useful from 175 to 500 MHz. Mixer contribution to system noise figure will be almost completely overcome by a low-noise RF amplifier with sufficient gain and adequate image rejection.

Image Response

Any broadband mixing scheme will have a potential image-response problem. In most amateur VHF/UHF receiver systems (as in these units), single-conversion techniques are employed. The LO is placed below the desired RF channel for noninverting down-conversion to IF. Conversion is related to both IF and LO frequencies. Because of the broadband nature of the DBM, input signals at the RF image frequency (numerically $fLO - fRF$ in our case) will legitimately appear inverted at the IF-output port, unless proper filtering is used to reduce them at the mixer RF-input port. For example, a 144-MHz converter with a 28-MHz IF output (116-MHz LO) will have RF image-response potential in the 84- to 88-MHz range. TV channel 6 wideband-FM audio will indeed appear at the IF-output port near 28 MHz unless appropriate RF- input filtering is used to eliminate it. While octave-bandwidth VHF/UHF "imageless mixer" techniques can improve system noise performance by about 3 dB (image noise reduction), and image signal rejection by 20 dB — and much greater with the use of a simple gating scheme — such a system is a bit esoteric for our application. Double- or multiple-conversion techniques can be used to advantage, but they further complicate an otherwise simple system. Image noise and signal rejection will depend on the effectiveness of the filtering provided in the RF-amplifier chain.

Mixer Selection

The mixer used in this system is a Relcom M6F, with specifications of the M6F and suitable substitutes given in Table 1. While mixers are available with connectors attached; they are more expensive. The MF6 and substitutes are designed for printed-circuit applications. Their short leads assure a proper interface between the mixer and the IF amplifier. The combining of mixer and IF amplifier in one converter package was done for that reason. Along these lines, the modular approach permits good signal isolation and enables the mixer-amplifier/IF system to be used at a variety of RF and LO-input frequencies, as mentioned earlier.

Most commonly available, inexpensive DBMs are not constructed to take advantage of LO powers much above +10 dBm (10 mW). To do so requires additional circuitry, which could degrade other mixer characteristics, specifically conversion loss and interport isolation. The advantage of higher LO power is primarily one of improved strong-signal-handling performance. At least one manufacturer advertises a moderately priced "high-level" receiving DBM that can use up to +23 dBm (200 mW) LO power, and still retain the excellent conversion loss and isolation characteristics shown in Table 1. The usefulness of mixers with LO power requirements above the commonly available +7 dBm (5 mW) level in amateur receiving applications may be a bit moot, as succeeding stages in most amateur receivers will likely overload before the DBM does. Overdesign is not necessary.

In general, mixer selection is based on the lowest practical LO level requirement that will meet the application, as it is more economical and results in the least LO leakage within the system. As a first-order approximation, LO power should be 10 dB greater than the highest anticipated input-signal level at the RF port. Mixers with LO requirements of +7 dBm are adequate for amateur receiving applications.

Application Design Guidelines

While the material just presented only scratches the surface in terms of DBM theory and utilization in amateur VHF/UHF receiving systems, some practical solutions to the mixer-port mismatch problem have been offered. To achieve best performance from most commerically manufactured broadband DBMs in amateur receiver service, the following guidelines are suggested:

1) Choose IF and LO frequencies that will provide maximum freedom from interference problems. Don't guess; go through the numbers!

2) Provide a proper IF-output termination (most critical).

3) Increase the LO-input power to RF-input power ratio to a value that will provide the required suppression of any in-band interfering products. The specified LO power (+7 dBm) will generally accomplish this.

4) Provide as good an LO match as possible.

5) Include adequate premixer RF-image filtering at the RF port.

When the mixer ports are terminated properly, performance usually in excess of published specifications will be achieved — and this is more than adequate for most amateur VHF/UHF receiver mixing applications.

THE COMBINED DBM/IF AMPLIFIER

A low-noise IF amplifier (2 dB or less) following the DBM helps ensure an acceptable system noise figure when the mixer is preceded by a low-noise RF amplifier.

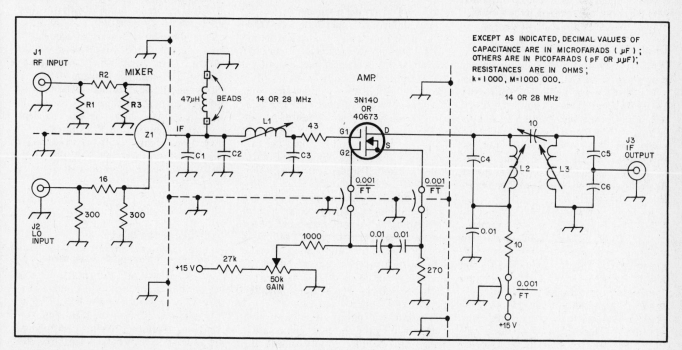

Fig. 33 — A schematic diagram for the doubly balanced mixer and IF post amplifier. The IF can be either 14 or 28 MHz. Parts values are given in Table 2.

A pi-network matching system used between the mixer IF-output port and gate 1 of the 3N140 transforms the nominal 50-ohm mixer-output impedance to a 1500-ohm gate-input impedance (at 28 MHz) specifically for best noise performance. The network forms a narrow-band mixer/IF-output circuit, which serves two other importance functions: It helps achieve the necessary isolation between RF and IF signal components, and serves as a 3-pole filter, resulting in a monotonic decrease in match impedances as the operating IF departs from midband. This action aids in suppression of harmonic-distortion products.

The combined DBM/IF amplifier is shown schematically in Fig. 33 and pictorially in the photographs. In the 14-MHz model, the 3N140 drain is tapped down on its associated inductance to provide a lower impedance for better strong-signal-handling ability. The 3N140 produces about 19 dB gain across a 700-kHz passband, flat within 1 dB between 13.8 and 14.5 MHz. A 2-MHz passband is used for the 28-MHz model, and the device drain is connected directly to the high-impedance end of its associated inductance. Both amplifiers were tuned independently of their mixers, and checked for noise figure as well as gain. With each IF amplifier pretuned and connected to its mixer, signals were applied to the LO- and RF-input ports. The pi-network inductance in the IF interface was adjusted carefully to see if performance had been altered. No change was noted. IF gain is controlled by the externally accessible potentiometer. Passband tuning adjustments in the drain circuit are best made with a sweep generator, but single-signal tuning techniques will be adequate. While there should be no difficulty with the non-gate-protected 3N140, a 40673 may be substituted directly if desired.

DBM/IF Amplifier IMD Evaluation

Classical laboratory IMD measurements were made on the DBM/IF amplifier using two − 10 dBm signals, closely spaced in the 144-MHz range. The LO power used was + 7 dBm. Conversion loss was 5 dB and the calculated third-order output intercept point was + 15 dBm. In operation, as simulated by these test conditions, equivalent output signal levels at J3 would be strong enough to severely overload most amateur receivers.

A high-performance, small-signal, VHF/UHF receiving amplifier optimized for IMD reduction and useful noise figure is only as good as any succeeding receiving-system stage, in terms of overload. The DBM/IF-amplifier combination presented significantly reduces common first-mixer overload problems, leaving the station receiver as the potentially weak link in the system. When properly understood and employed, the broadband DBM followed by a selective low-noise IF amplifier can be a useful tool for the amateur VHF/UHF

receiver experimenter.

APPENDIX

Mixer Terminology

fRF — RF input frequency
fLO — local-oscillator input frequency
fIF — IF output frequency

By convention, mixing signals and their products are referred to the LO frequency for calculations. In the mixer system, fRF is always above fLO, so we will refer our signals to fRF, with the exception of Fig. 32 which used the fLO reference.

Overload

A generic term covering most undesired operating phenomena associated with device nonlinearity.

Harmonic Modulation Products

Output responses caused by harmonics of fLO and fRF and their mixing products.

RF Compression Level

The absolute single-signal RF input-power level that causes conversion loss to increase by 1 dB.

RF Desensitization Level

The RF input power of an interfering signal that causes the small-signal conversion loss to increase by 1 dB, i.e., reducing a weak received signal by 1 dB.

Intermodulation Products

Distortion products caused by multiple RF signals and their harmonics mixing with each other and the LO, producing new output frequencies.

Mixer Intermodulation Intercept Point

Because mixers are nonlinear devices, all signals applied will generate others. When two signals (or tones), F1 and F2, are applied simultaneously to the RF-input port, additional signals are generated and appear in the output as fLO $\pm$ (nF1 + mF2). These signals are most troublesome when $n + m$ is a low odd number, as the resulting product will lie close to the desired output. For $n − 1$ (or 2) and $m − 2$ (or 1), the result is three (3), and is called the two-tone/third-order intermodulation products. When F1 and F2 are separated by 1 MHz, the third-order products will lie 1 MHz above and below the desired output. Intermodulation is generally specified under anticipated operating conditions since performance varies over the broad mixer-frequency ranges. Intermodulation products may be specified at levels required (i.e., 50 dB below the desired outputs for two 0-dBm input signals) or by the intercept point.

The intercept point is a ficitious point determined by the fact that an increase of level of two input tones by 10 dB will cause the desired output to increase by 10 dB, but the third-order output will increase by 30

dB. If the mixer exhibited no compression, there would be a point at which the level of the desired output would be equal to that of the third-order product. This is called the third-order-intercept point.

Noise Figure

Noise figure (NF) is a relative measurement based on excess noise power available from a termination (input resistor) at a particular temperature (290 K). When measuring the NF of a doubly balanced mixer with an automatic system, such as the HP-342A, a correction may be necessary to make the meter reading consistent with the accepted definition of receiver noise figure.

In a broadband DBM, the actual noise bandwith consists of two IF passbands, one on each side of the local-oscillator frequency (fLO + fIF and fLO − fIF). This double-sideband (DSB) IF response includes the RF channel and its image. In general, only the RF channel is desired for further amplification. The image contributes nothing but receiver and background noise.

When making an automatic noise-figure measurement using a wideband noise source, the excess noise is applied through both sidebands in a broadband DBM. Thus the instrument meter indicates NF as based on both sidebands. This means that the noise in the RF and image sidebands is combined in the mixer IF-output port to give double contribution (3 dB greater than under SSB conditions). For equal RF-sideband responses, which is a reasonable assumption, and in the absence of preselectors, filters or other image rejection elements, the automatic NF meter readings are 3 dB lower than the actual NF for DBM measurements.

The noise figure for receivers (and most DBM) is generally specified with only one sideband for the useful signal. As mentioned in the text, most DBM diodes add no more than 0.5 dB (in the form of NF) to conversion loss, which is generally measured under single-signal RF-input (SSB) conditions. Assuming DBM conversion efficiency (or loss) to be within specifications, there is an excellent probability that the SSB NF is also satisfactory. Noise-figure calculations in the text were made using a graphical solution of the well known noise-figure formula:

$$f_T = f_1 + \frac{f_2 - 1}{g_1}$$

converted to decibels.

Improved Wideband IF Responses

The following information was developed in achieving broad-band performance in the mixer-to-amplifier circuitry. In cases where only a small portion of a band is of interest the original circuit values are adequate. For those who need to receive over a considerable portion of a band, say 1 to 2 MHz, a change of some component will provide improved performance over a broad range while maintaining an acceptable noise figure.

The term "nominal 50-ohm impedance" applied to diode DBM ports is a misnomer, as their reflective impedance is rarely 50 ohms + j_0, and a VSWR of 1 is almost never achieved. Mixer performance specified by the manufacturer is measured in a 50-ohm broadband system, and it is up to the designer to provide an equivalent termination to ensure that the unit will meet specifications. Appropriate matching techniques at the RF and LO ports will reduce conversion loss and LO-power requirements. Complex filter synthesis can improve the IF output match. However, if one does not have the necessary equipment to evaluate these efforts, they may be wasted. Simple, effective, easily reproduced reduced circuitry was desired as long as the trade-offs were acceptable. Measurements indicate this to be the case.

The most critical circuit in the combined unit is the interface between mixer and IF amplifier. It must be low-pass in nature to satisfy VHF signal component bypassing requirements at the mixer IF port. For best mixer IMD characteristics and low conversion loss, it must present to the IF port a nominal 50-ohm impedance at the desired frequency, and this impedance value must not be allowed to increase as IF operation departs from midband. The impedance at the IF-amplifier end of the interface network must be in the optimum region for minimum cross-modulation and low noise. A dual-gate device offers two important advantages over most bipolars. Very little, if any, power gain is sacrificed in achieving best noise figure, and both parameters (gain and NF) are relatively independent of source resistance in the optimum region. As a result, the designer has a great deal of flexibility in choosing a source impedance. In general, a 3:1 change in source resistance results in only a 1-dB change in NF. With minimum cross-modulation as a prime system consideration, this 3:1 change (reduction) in source resistance implies a 3:1 improvement in cross-modulation and total harmonic distortion.

Tests in the 3N201 dual-gate MOSFET have shown device noise performance to be excellent for source impedances in the 1000- to 2000-ohm region. For optimum noise and good cross-modulation performance, the nominal 50-ohm mixer-IF-output impedance is stepped up to about 1500 ohms for IF-amplifier gate 1, using the familiar low-pass pi network. This is a mismatched condition for gate 1, as the device input impedance for best gain in the HF region is on the order of 10 kΩ. Network loaded-Q values in the article are a bit higher than necessary, and a design for lower Q_L is preferred. Suggested modified component values are listed in Fig. 34. High-frequency attenuation is reduced somewhat, but satisfactory noise and bandwidth perfor-

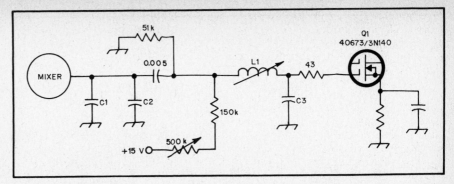

Fig. 34 — Suggested changes in the mixer-to-3N140 pi-network interface circuit, producing lower Q_L and better performance.

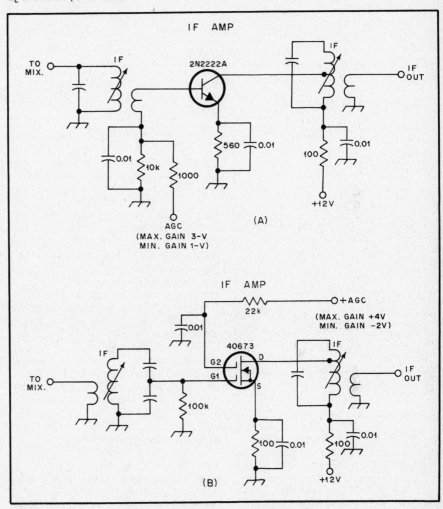

Fig. 35 — Methods for applying AGC to a bipolar IF amplifier (A) and a dual-gate-MOSFET IF stage (B).

mance are more easily obtained. Coil-form size is the same, so no layout changes are required for the modification. Components in the interface must be of high Q and few in number to limit their noise contribution through losses. The 28-MHz values provide satisfactory interface network performance over a 2-MHz bandwidth. A higher Q_L in the 28-MHz interface can be useful if one narrows the output network and covers only a few hundred kilohertz bandwidth, as is commonly done in 432-MHz weak-signal work.

Device biasing and gain control methods

were chosen for simplicity and adequate performance. Some sort of gain adjustment is desirable for drain-circuit overload protection. It is also a handy way to "set" the receiver S meter. A good method for gain adjustment is reduction of the gate-2 bias voltage from its initial optimum-gain bias point (greater than +4 V dc), producing a remote-cutoff characteristic (a gradual reduction in drain current with decreasing gate bias). The initial gain-reduction rate is higher with a slight forward bias on gate 1, than for $Vg_1s = 0$. Input- and output-circuit detuning resulting from gain reduc-

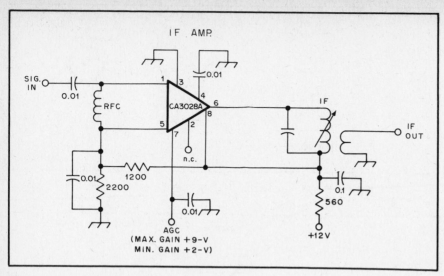

Fig. 36 — An IC type of IF amplifier with AGC applied.

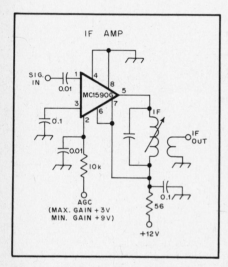

Fig. 37 — AGC is applied to an MC1590G IC.

tion (Miller effect) are inconsequential as the gate-1 and drain susceptances change very little over a wide range of V_{g_2}s and I_D at both choices of IF. Best intermodulation figure for the 3N201 was obtained with a small forward bias on gate 1. The bias-circuit modification shown in Fig. 34 may be tried, if desired.

IF Amplifiers

The amount of amplification used in a receiver will depend on how much signal level is available at the input to the IF strip. Sufficient gain is needed to ensure ample audio output consistent with driving head-phones or a speaker. Another consideration is the amount of AGC-initiated IF gain range. The more IF stages used (a maximum of two is typical), the greater the gain change caused by AGC action. The range is on the order of 80 dB when two CA3028A ICs are used in the IF strip. A pair of MC1590G ICs will provide up to 120 dB of gain variation with AGC applied.

Nearly all modern receiver circuits utilize ICs as IF amplifiers. Numerous types of ICs are available to provide linear RF and IF amplification at low cost. The CA3028A and MC1590G ICs are the most popular ones for amateur work because they are easy to obtain and are inexpensive. With careful layout techniques either device will operate in a stable manner. Bypassing should be done as near to the IC pins as possible. Input and output circuit elements must be separated to prevent mutual coupling which can cause unstable operation. If IC sockets are used they should be of the low-profile variety with short socket conductors.

Fig. 35 contains examples of bipolar transistor and FET IF amplifiers. Typical component values are given. A CA3028A IC, connected for differential-amplifier operation, is shown in Fig. 36 as an IF amplifier. Up to 40 dB of gain is possible with this circuit. The IC is useful up to 120 MHz and has a low noise figure.

A Motorola MC1590G IC will provide up to 50 dB of gain when used as an IF amplifier. An example of the circuit is given in Fig. 37. AGC operates in the reverse of that which is applied to a CA3028A. With the CA3028A, the gain will be maximum with maximum AGC voltage. An MC1590G delivers maximum gain at the low AGC voltage level.

With both amplifiers (Figs. 36 and 37) the input impedance is on the order of 1000 ohms. The output impedance is close to 4000 ohms. These values are for single-ended operation, as shown. The values are doubled when either device is operated in push-pull with respect to input and output tuned circuits.

Choice of Frequency

The selection of an intermediate frequency is a compromise between conflicting factors. The lower the IF the higher the selectivity and gain, but a low IF brings the image nearer the desired signal and hence decreases the image ratio. A low IF also increases pulling of the oscillator frequency. A high IF improves the image ratio and reduces pulling, but the gain is lowered and selectivity is harder to obtain by simple means.

An IF of the order of 455 kHz gives good selectivity and is satisfactory from the standpoint of image ratio and oscillator pulling at frequencies up to 7 MHz. The image ratio is poor at 14 MHz when the mixer is connected to the antenna, but adequate when there is a tuned RF amplifier between the antenna and mixer. At 28 MHz and at very high frequencies, the image ratio is very poor unless several RF stages are used. Above 14 MHz, pulling is likely to be bad without very loose coupling between mixer and oscillator. Tuned-circuit shielding also helps.

With an IF of about 1600 kHz, satisfactory image ratios can be secured on 14, 21 and 28 MHz with one RF stage of good design. For frequencies of 28 MHz and higher, a common solution is to use double conversion, choosing one high IF for image reduction (9 MHz is frequently used) and a lower one for gain and selectivity.

In choosing an IF it is wise to avoid frequencies on which there are strong signals, such as broadcast bands, since such signals may get into the IF amplifier by conduction or radiation pickup. Shifting the IF or better shielding are solutions to this interference problem.

Fidelity: Sideband Cutting

If the selectivity is too great to permit uniform amplification over the band of frequencies occupied by the modulated signal, some of the sidebands are "cut." While sideband cutting reduces fidelity, it is frequently done to favor communications effectiveness.

IF Selectivity

Assuming that all stages are properly designed, IF selectivity is the most significant selectivity in a receiver. This selectivity separates signals and reduces QRM. At intermediate frequencies above 500 kHz it is common practice to use crystal filters. These can be designed with just one crystal (Fig. 38A), or with two or more crystals. Fig. 38B illustrates a two-crystal, half-lattice filter. A cascaded half-lattice filter is shown at C of Fig. 38. Table 4 lists various crystal filters manufactured by Spectrum International.

The single-crystal example shown at A of Fig. 38 is best suited for simple receivers intended mainly for CW use. C1 is adjusted to provide the bandpass characteristic shown adjacent to the circuit. When the BFO frequency is placed on the part of the low-frequency slope (left) which gives the desired beat note respective to f_o (approximately 800 Hz), single-signal reception will result. To the right of f_o in Fig. 38A the response drops sharply to reduce output on the unwanted side of zero beat, thereby making single-signal reception possible. If no IF filter were used, or if the BFO frequency fell at f_o, nearly equal response would exist on either side of zero beat (double-signal response) as is the case with direct-conversion receivers. QRM on the unwanted-response side of the IF passband would interfere with reception. The single-crystal filter shown is capable of at least 30 dB of rejection on the high-frequency side of zero beat. The filter termination, R_T, has a marked effect on the response curve. It is necessary to experiment with the resistance value until the desired response is obtained. Values can range from 1500 to 10,000 ohms.

A half-lattice filter is shown at B in Fig. 38. The response curve is symmetrical, and there is a slight dip at center frequency. The dip is minimized by proper selection of R_T. Y1 and Y2 are separated in frequency by the amount needed to obtain CW or SSB selectivity. The bandwidth at the 3-dB points will be approximately 1.5 times the crystal-frequency spacing. For upper- or lower-sideband reception, Y1 and Y2 would be 1.5 kHz apart, yielding a 3-dB bandwidth of approximately 2.25 kHz. For CW work, a crystal spacing of 0.4 kHz would result in a bandwidth of roughly 600

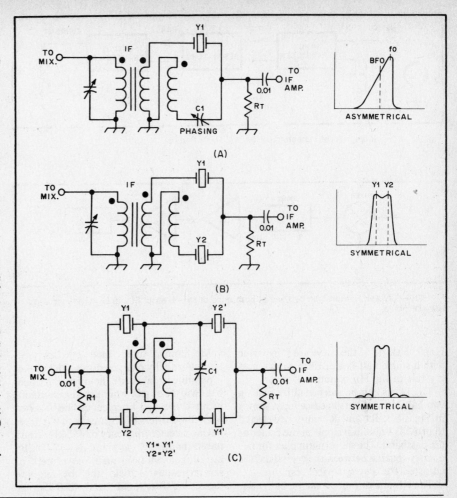

Fig. 38 — A comparison among crystal IF filters. The selectivity is increased as crystals are added.

Table 3

Collins Radio Mechanical Filters

Part & Type Numbers	Min. 3 dB BW @ 25° (kHz)	Min. 4 dB CBW OTR (kHz)	Max. 60 dB BW @ 25°C (kHz)	Max. 60 dB BW OTR (dB)	Max. RV @ 25°C (dB)	Max. RV OTR (dB)	Max. IL @ 25°C (dB)	Max. IL COTR (dB)	Min. 60 dB SBR (kHz)	S & L −5% ohms	Res. Cap. +5% (pF)
526-9689-010 F455FD-04	0.375	0.375	3.5	4.0	3.0	4.0	10.0	12.0	455-F60L F6OH-465	2000 350	350
526-9690-010 F455FD-12	1.2	1.2	8.7	9.5	3.0	4.0	10.0	12.0	445-F6OL F6OH-465	2000 350	350
F526-9691-010 F455FD-19	1.9	1.9	5.4	5.9	3.0	4.0	10.0	12.0	445-F6OL F6OH-465	2000	330 350
526-9692-010 F455FD-25	2.5	2.5	6.5	7.0	3.0	4.0	10.0	12.0	445-F60L F6OH-465	2000 510	510
526-9693-010 F455FD-29	2.9	2.9	7.0	8.0	3.0	4.0	10.0	12.0	445-F6OL F6OH-465	2000 510	510
526-9694-010 F455FD-38	3.8	3.8	9.0	10.0	3.0	4.0	10.0	12.0	445-F6OL F6OH-465	2000 1000	1000
526-9695-010 F455FD-58	5.8	5.8	14.0	15.0	3.0	4.0	10.0	12.0	445-F6OL F6OH-465	2000 1100	1100

OTR = Operating Temperature Range, RV = Ripple Voltage, IL = Insertion loss, SBR = Stop Band Range, S & L = Source and Load
Courtesy of Collins Radio Co.

Table 4

Crystal Filters

Application	SSB Tran.	SSB Rec.	CW or Digital Data	AM	AM	CW	FM
Filter Type	XF-9A	XF-9B	XF-9NB	XF-9C	XF-9D	XF-9M	XF-9E
No. of crystals	5	8	8	8	8	4	8
6-dB of bandwidth	2.5 kHz	2.4 kHz	0.5 kHz	3.75 kHz	5.0 kHz	0.5 kHz	12 kHz
Passband ripple	<1 dB	<2 dB	<0.5 dB	<2 dB	<2 dB	<1 dB	<2 dB
Insertion loss	<3 dB	<3.5 dB	<6.5 dB	<3.5 dB	<3.5 dB	<5 dB	<3 dB
Term. impedance	500 Ω	500 Ω	500 Ω	500 Ω	500 Ω	500 Ω	1200 Ω
Ripple capacitors	30 pF	30 pF	30 pF	30 pF	30 pF	30 pF	30 pF
Shape factor	6:50 dB 1.7	6:60 dB 1.8	6:60 dB 2.2	6:60 dB 1.8	6:60 dB 1.8	6:60 dB 4.4	6:60 dB 1.8
Stop-band atten.	>45 dB	>100 dB	>90 dB	>100 dB	>100 dB	>90 dB	>90 dB

Courtesy of Spectrum International

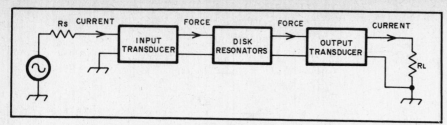

Fig. 39 — Block diagram of a mechanical filter (Collins Radio).

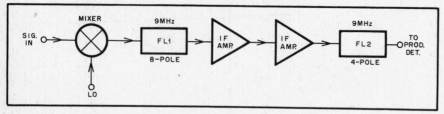

Fig. 40 — Crystal IF filters can be used at both ends of the IF strip. FL2 greatly reduces wideband IF noise.

Hz. The skirts of the curve are fairly wide with a single half-lattice filter, which uses crystals in the HF region.

The skirts can be steepened by placing two half-lattice filters in cascade, as shown in Fig. 38C. R1 and R_T must be selected to provide minimum ripple at the center of the passband. The same rule applies for frequency spacing between the crystals. C1 is adjusted for a symmetrical response.

The circuits of Fig. 38 can be built easily and inexpensively by amateurs. The transformers shown are tuned to center frequency. They are wound bifilar or trifilar on ferrite or powdered-iron cores of appropriate frequency characteristics.

A technique that has recently gained favor with amateurs is that of building crystal filters in the form of a ladder network, in which all crystals are cut to the same frequency. TV color-burst and CB crystals are commonly used to fabricate low-cost filters having high performance. The design procedure involves analysis of a quartz crystal in terms of its lumped-constant equivalent circuit. These lumped-constant values are then used to design a modern Chebyshev or Butterworth band-pass filter. Wes Hayward, W7ZOI, described the procedure in detail in "A Unified Approach to the Design of Crystal Ladder Filters" in May 1982 *QST* (pp. 21-27).

An illustration of how a mechanical filter operates is provided in Fig. 39. Perhaps the most significant feature of a mechanical filter is the high Q of the resonant metallic disks it contains. A Q figure of 10,000 is the nominal value obtained with this kind of resonator. If L and C constants were employed to acquire a bandwidth equivalent to that possible with a mechanical filter, the IF would have to be below 50 kHz. Table 3 lists various mechanical filters that are manufactured by Collins Radio Co.

Mechanical filters have excellent frequency-stability characteristics. This makes it possible to fabricate them for fractional bandwidths of a few hundred hertz. Bandwidths down to 0.1 percent can be obtained with these filters. This means that a filter having a center frequency of 455 kHz could have a bandwidth as small as 455 Hz. By inserting a wire through the centers of several resonator disks, thereby coupling them, the fractional bandwidth can be made as great as 10 percent of the center frequency. The upper limit is governed primarily by occurrence of unwanted spurious filter responses adjacent to the desired passband.

Mechanical filters can be built for center frequencies from 60 to 600 kHz. The main limiting factor is disk size. At the low end of the range, the disks become prohibitively large. At the high limit of the range, the disks become too small to be practical.

The principle of operation is shown in Fig. 39. As the incoming IF signal passes through the input transducer, the signal is converted to mechanical energy. This energy is passed through the disk resonators to filter out the undesired frequencies, then through the output transducer where the mechanical energy is converted back to the original electrical form.

The transducers serve a second function: They reflect the source and load impedances into the mechanical portion of the circuit, thereby providing a termination for the filter.

Mechanical filters require external resonating capacitors, which are used across the transducers. If the filters are not resonated, there will be an increase in insertion loss and degradation of the passband characteristics. Concerning the latter, there will be various unwanted dips in the nose response (ripple), which can lead to undesirable effects. The exact amount of

shunt capacitance will depend on the filter model used. The manufacturer's data sheet specifies the proper capacitor values.

Collins mechanical filters are available with center frequencies from 64 to 500 kHz and in a variety of bandwidths. Insertion loss ranges from 2 dB to 12 dB, depending on the style of filter used. Of greatest interest to amateurs are the 455-kHz mechanical filters with the prefix F455. They are available in bandwidths of 0.375, 1.2, 1.9, 2.5, 2.9, 3.8 and 5.8 kHz. Maximum insertion loss is 10 dB, and the characteristic impedance is 2000 ohms. Different values of resonating capacitance are required for the various models, ranging from 350 to 1100 pF. Although some mechanical filters are terminated internally, this series requires external source and load termination of 2000 ohms. The F455 filters are the least expensive of the Collins line.

Most modern receivers have selectable IF filters to provide suitable bandwidths for SSB and CW. Most commercial receivers use a 250- to 600-Hz bandwidth filter for CW and a 1.8- to 2.7-kHz bandwidth for SSB. CW or SSB filters are used to receive RTTY. The input and output ends of a filter should be well isolated from one another if the filter characteristics are to be realized. Leakage across a filter will negate the otherwise good performance of the unit. The problem becomes worse as the filter frequency is increased. Mechanical switches are not recommended above 455 kHz for filter selection because of leakage across the switch wafers and sections. Diode switching is preferred by most designers. The switching diodes for the filter that is out of the circuit are usually back-biased to ensure minimum leakthrough.

In the interest of reducing wideband noise from the IF amplifier strip, it is worthwhile to use a second filter that has exactly the same center frequency as the first. The second filter is placed at the end of the IF strip, ahead of the product detector. This is shown in Fig. 40. The technique was described by W7ZOI in March and April 1974 *QST*. The second filter, FL2, has somewhat wider skirts than the first, FL1. An RC active audio filter after the product detector has a similar effect, but the results are not quite as spectacular as when two IF filters are used. The overall signal-to-noise ratio is enhanced greatly by this method.

Automatic Gain Control

Automatic regulation of the gain of the receiver in inverse proportion to the signal strength is an operating convenience in reception, because it tends to keep the output level of the receiver constant regardless of input-signal strength. The average rectified dc voltage, developed by the received signal across a resistance in a detector circuit, is used to vary the bias on the RF and IF amplifier stages. Since this voltage is proportional to the average amplitude of the signal, the gain is reduced as the signal strength increases. The control will be more complete and the output more constant as the number of stages to which the AGC bias is applied is increased. Control of at least two stages is advisable.

Various schemes from simple to extravagant have been conceived to develop AGC voltage in receivers. Some perform poorly because the attack time of the circuits is wrong for CW work, resulting in "clicky" or "pumping" AGC. The first significant advance toward curing the problem was presented by Goodman, W1DX, "Better AVC for SSB and Code Reception," January 1957 *QST*. He coined the term "hang" AVC, and the technique has been adopted by many amateurs who have built their own receiving equipment. The objective is to make the AGC take hold as quickly as possible to avoid the ailments mentioned earlier.

For best receiver performance the IF filters should be contained within the AGC loop, which strongly suggests the use of RF-derived AGC. Most commercial receivers follow this rule. However, good results can be obtained with audio-derived AGC, despite the tendency toward a clicky response. If RC active audio filters are used to obtain receiver selectivity, they should be contained within the audio-AGC loop if possible.

Fig. 41 illustrates the general concept of an AGC circuit. RF energy is sampled from the output of the last IF by means of light coupling. This minimizes loading on the tuned circuit of the IF amplifier. The IF energy is amplified by the AGC amplifier, then convert to dc by means of an AGC rectifier. R1 and C1 are selected to provide a suitable decay time constant (about 1 second for SSB and CW). Q1 and Q2 function as dc amplifiers to develop the voltage needed for AGC control of the IF (and sometimes, RF) amplifier stages. The developed AGC voltage can be used to drive an S meter. A level control can be placed at the input of the AGC amplifier to establish the signal input level (receiver front end), which turns on the AGC system. Most designers prefer to have this happen when a received signal level is between 0.25 and 1 μV. The exact parameters are based somewhat on subjectivity.

An AF-derived AGC loop is shown in Fig. 42. It is suitable for use with CA3028A IF-amplifier ICs. Provision is made for manual IF-gain control. D1 functions as a gating diode to prevent the manual-control circuitry from affecting the normal AGC action. This circuit was first used in a receiver described by DeMaw in June and July 1976 *QST*.

An RF-derived AGC system is seen in Fig. 43. It operates on a principle similar to that shown in Fig. 42, except that an op amp is used in place of the discrete bipolar dc amplifiers. Current changes are sampled across the 10-kΩ FET source resistor by

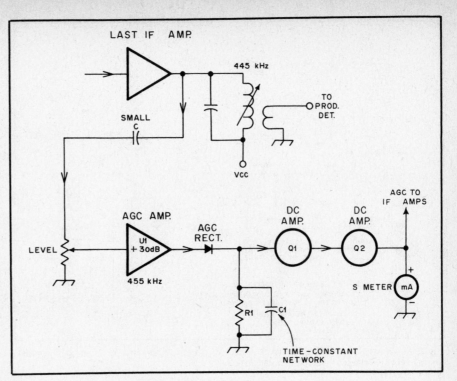

Fig. 41 — A system for developing receiver AGC voltage.

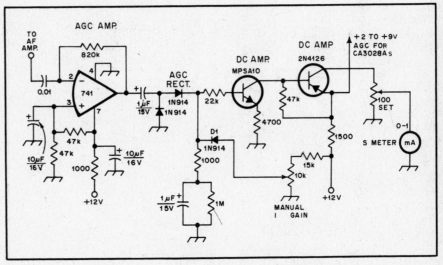

Fig. 42 — A practical circuit for developing AGC voltage for a CA3028A IF amplifier.

means of the op amp difference amplifier. With the values of resistance given, the output dc swing of the op amp is the desired +2 to +9 volts for controlling CA3028A IF amplifiers. This system was also used in the DeMaw receiver.

It certainly is not essential to have AGC in a receiver. If the operator is willing to adjust the gain manually, good performance is certain to result. AGC is mainly an operator convenience: It prevents loud signals from blasting out of the speaker or headphones when the operator tunes the band at a given AF-gain setting.

Beat-Frequency Oscillators

The circuits given for crystal-controlled oscillators in Chapter 10 are suitable for use in BFO circuits. A beat oscillator generates energy that is supplied to a product detector for reception of CW and SSB signals. The BFO frequency is offset by the appropriate amount with respect to the center frequency of the IF filter. For example, a BFO used during CW reception is usually some 800 Hz above or below the IF center frequency. During SSB reception the offset is slightly more — approximately 1.5 kHz above or below the IF center frequency, depending upon the need for upper or lower sideband operation. Typically, the BFO is placed roughly 20 dB down on the slope of the IF passband curve for SSB reception or transmission.

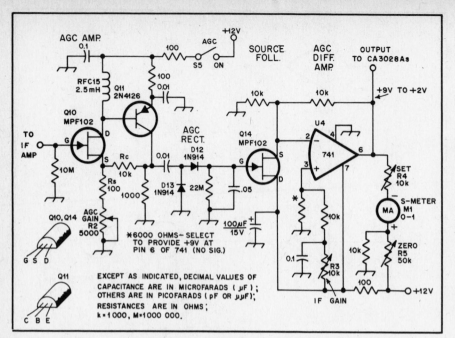

Fig. 43 — An AGC system for CA3028A IF amplifiers. An op amp is used as difference amplifier to provide AGC voltage while operating an S meter.

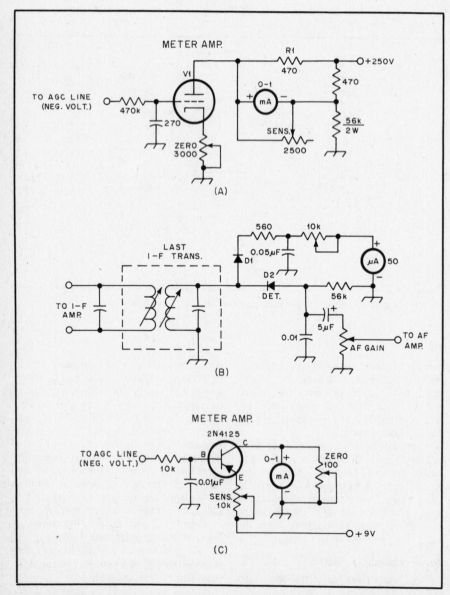

A BFO need not be crystal controlled. It can use a VFO circuit, or it can be tuned by means of a Varactor diode. Elimination of the crystals represents a cost savings to the builder, but frequency stability may not be as good as when crystal control is employed.

When the BFO is operated at frequencies above 3 MHz it is helpful to use a buffer stage after the oscillator to minimize the effects of pulling. Furthermore, if a passive product detector is used in the receiver, a substantial amount of BFO output power will be required — approximately +7 dBm. The buffer/amplifier helps to boost the oscillator output to satisfy this requirement.

S Meters

Signal-strength meters are useful when there is a need to make comparative readings. Such might be the case when another operator asks for a comparison between two antennas being tested. Because S meters are relative-reading instruments, signal reporting based on the amount of needle deflection is generally without meaning. No two receivers render the same reading a given signal, unless by coincidence. This is because the gain distribution within an amateur receiver varies from band to band. Since most S meters are activated from the AGC line in a receiver, what might be S9 on one ham band could easily become S6 or 10 dB over S9 on another band. A receiver that rendered accurate readings on each band it covered would be extremely esoteric and complex.

An attempt was made by at least one receiver manufacturer in the early 1940s to establish some significant numbers for S meters. S9 was to be equivalent to 50 μV, and each S unit would have been equal to 6 dB. The scale readings above S9 were given in dB. The system never took hold in the manufacturing world, probably for the reasons given earlier.

In addition to the example shown in Fig. 43, some typical S-meter circuits are offered in Fig. 44. The example at C can be used with RF- or audio-derived AGC.

NOISE REDUCTION

The only known approach to reducing tube, transistor and circuit noise is through

Fig. 44 — Various methods for using an S meter. At A, V1 is a meter amplifier. As the AGC voltage increases, the plate current decreases to lower the voltage drop across R1. An up-scale meter reading results as the current through the meter increases. At B, the negative-going AGC voltage is inverted and amplified by an operational amplifier circuit. R1 establishes the dB-per-S-unit sensitivity; R2 is the meter-zeroing adjustment. R3 and D1 compress the readings over S9. This circuit is well suited to an IF amplifier having a gain control characteristic that is linear in dB/volt, such as one using dual-gate MOSFETs. At C, the negative AGC voltage forward biases the transistor to cause an increase in collector current, thereby deflecting the meter upward with signal increases.

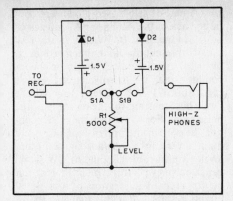

Fig. 45 — A simple audio limiter/clipper. R1 sets the bias on the diodes for the desired limiting·level.

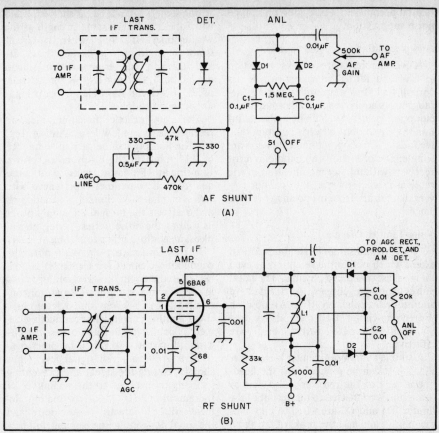

Fig. 46 — Examples of RF and audio ANL circuits. Positive and negative clipping takes place in both circuits. The circuit at A is self-adjusting.

the choice of low-noise, front-end, active components and through more overall selectivity.

In addition to the active-device and circuit noise, much of the noise interference experienced in reception of high-frequency signals is caused by domestic or industrial electrical equipment and by automobile ignition systems. The interference is of two types in its effects. The first is "hiss," consisting of overlapping pulses similar in nature to the receiver noise. It is largely reduced by high selectivity in the receiver, especially for code reception. The second is the "pistol-shot" or "machine-gun" type, consisting of separated impulses of high amplitude. The hiss type of interference usually is caused by commutator sparking in dc and series-wound ac motors, while the shot type results from separated spark discharges (ac power leaks, switch and key clicks, ignition sparks and the like).

Impulse Noise

Impulse noise, because of the short duration of the pulses compared with the time between them, must have high amplitude to contain much average energy. Hence, noise of this type strong enough to cause

much interference generally has an instantaneous amplitude much higher than that of the signal being received. The general principle of devices intended to reduce such noise is to allow the desired signal to pass through the receiver unaffected, but to make the receiver inoperative for amplitudes greater than that of the signal. The greater the amplitude of the pulse compared with its time of duration, the more successful the noise reduction.

Another approach is to "silence" (render

inoperative) the receiver during the short duration time of any individual pulse. The listener will not hear the " hole" because of its short duration, and very effective noise reduction is obtained. Such devices are called "blankers" rather than "limiters."

In passing through selective receiver circuits, the time duration of the impulses is increased, because of the bandwidth of the circuits. Thus, the more selectivity ahead of the noise-reducing device, the more dif-

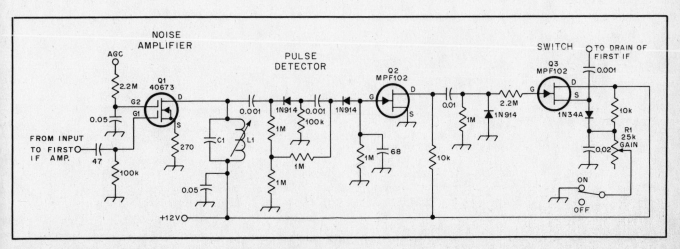

Fig. 47 — Diagram of a noise blanker. C1 and L1 are tuned to the receiver IF.

ficult it becomes to secure good pulse-type noise suppression.

Audio Limiting

A considerable degree of noise reduction in CW and RTTY reception can be accomplished by amplitude-limiting arrangements applied to the audio-output circuit of a receiver. Such limiters also maintain the signal output nearly constant during fading. These output-limiter systems are simple, and are readily adaptable to most receivers without any modification of the receiver itself. However, they cannot prevent noise peaks from overloading previous stages.

Noise-Limiter Circuits

Pulse-type noise can be eliminated to an extent which makes the reception of even the weakest of signals possible. The noise pulses can be clipped or limited in amplitude, at either an RF or AF point in the receiver circuit. Both methods are used by receiver manufacturers; both are effective.

A simple audio noise limiter is shown at Fig. 45. It can be plugged into the headphone jack of the receiver and a pair of headphones connected to the output of the limiter. D1 and D2 are wired to clip both the positive and negative peaks of the audio signal, thus removing the high spikes of pulse noise. The diodes are back-biased by 1.5-volt batteries permitting R1 to serve as a clipping-level control. This circuit also limits the amount of audio reaching the headphones. When tuning across the band, strong signals will not be ear shattering and will appear to be the same strength as the weaker ones. S1 is open when the circuit is not in use to prevent battery drain. D1 and D2 can be germanium or silicon diodes, but 1N34As or 1N914s are generally used. This circuit is usable with only high-impedance headphones.

The usual practice in communication receivers is to use low-level limiting, Fig. 46. The limiting can be carried out at RF or AF points in the receiver, as shown. Limiting at RF does not cause poor audio quality as is sometimes experienced with series or shunt AF limiters. Limiting at audio affects the normal AF signal peaks as well as the noise pulses, giving an unpleasant audio quality to strong signals.

In a series-limiting circuit, a normally conducting element (or elements) is connected in the circuit in series and operated in such a manner that it becomes nonconductive above a given signal level. In a shunt limiting circuit, a nonconducting element is connected in shunt across the circuit and operated so that it becomes conductive above a given signal level, thus short-circuiting the signal and preventing its being transmitted to the remainder of the amplifier. The usual conducting element will be a forward-biased diode, and the usual nonconducting element will be a back-biased diode. In many applications, the value of bias is set manually by the operator; usually the clipping level will be set at about 1 to 10 volts.

The AF shunt limiter at A and the RF shunt limiter at B operate in the same manner. A pair of self-biased diodes are connected across the AF line at A, and across an RF inductor at B. When a steady CW signal is present the diodes barely conduct, but when a noise pulse rides in on the incoming signal, it is heavily clipped because capacitors C1 and C2 tend to hold the diode bias constant for the duration of the noise pulse. For this reason the diodes conduct heavily in the presense of noise and maintain a fairly constant signal output level. Considerable clipping of CW signal peaks occurs with this type of limiter, but no apparent deterioration of the signal quality results. L1 at B is tuned to the IF of the receiver. An IF transformer with a conventional secondary winding could be used in place of L1, the clipper circuit being connected to the secondary winding; the plate of the 6BA6 would connect to the primary winding in the usual fashion.

IF Noise Silencer

The IF noise silencer circuit shown in Fig. 47 is designed to be used ahead of the high-selectivity section of the receiver. Noise pulses are amplified and rectified, and the resulting negative-going dc pulses are used to cut off an amplifier stage during the pulse. A manual "threshold" control is set by the operator to a level that only permits rectification of the noise pulses that rise above the peak amplitude of the desired signal. The clamp transistor, Q3, short circuits the positive-going pulse "overshoot." Running the 40673 controlled IF amplifier at zero gate 2 voltage allows the direct application of AGC voltage.

Chapter 13

Radio Transceivers

Transceivers have taken the place of separate transmitter/receiver combinations in nearly every Amateur Radio station. Some of the reasons for this are cost, compactness, portability and ease of operation.

The three previous chapters (10, 11 and 12) covered the fundamentals of frequency generation, transmitting and receiving principles. The transceiver brings these concepts together in an integrated package.

Circuit Considerations

Every transceiver, whether simple or complex, uses some of the same circuitry for transmitting and receiving. Shared stages frequently include oscillators, mixers, amplifiers, filters and tuned circuits. Usually, sharing is achieved by switching the input and output of a particular stage from the receiver to the transmitter. Shared circuitry may result in more economical and compact equipment.

The designer must determine whether any transmitting and receiving stages are sufficiently alike to permit their replacement by a single circuit. To do this, three conditions must be satisfied.

First, the stages must operate in the same frequency range. Depending on the circuit, that range may be wide or narrow.

Second, input and output power levels must be comparable. Often, the low level of IF signal processing that takes place in SSB transmitters allows one or more of the same IF amplifiers to be used by the receiver. However, there are other stages, such as the receiver RF amplifier and transmitter final power amplifier, that are never replaced by a single circuit. Because linearity, efficiency and heat dissipation are major concerns in SSB power amplifier design, the transistors chosen for these stages have internal geometries that optimize their power-handling capability. Receiver RF amplifiers, however, process much smaller signal levels, so power-handling capability is less important. Instead, great care is taken to ensure that the stage exhibits low distortion and good noise figure.

Finally, impedances must be matched. This is often done by designing each stage

The Ten-Tec Corsair II offers 160-10 meter coverage. Power output is 100 W on all bands. For the CW enthusiast, this radio offers a wide range of optional crystal filters, as well as full break-in operation. A memory keyer is built in.

for an input and output impedance of 50 Ω. Even when sharing is possible, it may not always be desirable. The cost and complexity of the required switching circuitry and the additional effort that may be needed to align the transceiving stage must be taken into account. The designer can weigh these factors against the reduction in size, weight and component count that the transceiving approach makes possible.

Historical Developments

In the early days of radio, great constructional differences existed between transmitters and receivers. The spark-gap transmitter had little in common with the crystal or electrolytic receiver-detector.

The invention of the vacuum tube and development of continuous-wave (CW) transmission led to transmitters and receivers that used amplification. However, differences in power levels and frequency schemes prohibited circuit sharing. Fig. 1 shows the block diagram of a typical AM transmitter. Each block represents a different stage employing one or more tubes. A stable, low-frequency signal is generated by the crystal oscillator, and is isolated from later stages by a buffer amplifier. This signal is then raised to the desired frequency using a nonlinear multiplying stage. Finally, the multiplier output is filtered and then amplified by an RF amplifier. Audio from

the microphone is amplified and then delivered to the RF power amplifier, where voice information is impressed on the carrier. Audio circuits use linear (Class A, AB, or B) amplifiers to avoid distortion. RF amplifiers and frequency multipliers use nonlinear (Class C) stages.

In contrast, a typical AM receiver appears in Fig. 2. There are several important differences. Instead of nonlinear multiplying, the receiver uses heterodyning methods for frequency conversion. Also, frequencies and power levels are generally different from those used in the AM transmitter. The only stages performing the same function are the audio stages. Depending on transmitter power, a microphone amplifier or even the modulator could be used for audio amplification in the receiver.

The development of circuitry for single-sideband suppressed-carrier communication led to many similarities between transmitters and receivers. The AM receiver of Fig. 3A requires only a BFO to demodulate SSB signals. The block diagram of an SSB transmitter, shown in Fig. 3B, more closely resembles that of a heterodyne receiver than an AM transmitter. In fact, the major difference between block diagrams of an SSB heterodyne transmitter and receiver is the direction of signal flow. Both use similar oscillator frequencies, mixing schemes and power levels to pro-

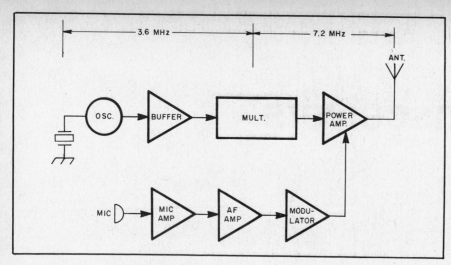

Fig. 1 — Block diagram of a conventional AM transmitter.

Both transmitter and receiver are simultaneously tuned by a 5.1-MHz VFO. On transmit, the VFO signal and the 3-MHz single-sideband signal are mixed together to produce an 8-MHz IF signal. A band-pass filter attenuates all RF energy falling outside this 500-kHz-wide IF. On receive, a signal entering the second receiver mixer from this same band-pass filter is mixed with the VFO signal. This converts the received signal from the 8-MHz IF down to 3 MHz, where it can be "cleaned up" by the narrow 2.1-kHz crystal filter.

A crystal-controlled heterodyne oscillator is used for mixing in a manner similar to the VFO. But instead of producing a variable frequency, this oscillator establishes a fixed frequency, one per band, to convert that band to the 8-MHz IF. This method allows every block to the left of the heterodyne oscillator to use the same frequencies regardless of the particular band of operation. The receiver IF circuitry passes signals between 8.4 and 8.9 MHz. Depending on the setting of the VFO, only one particular signal in this 500-kHz-wide IF is converted to 3 MHz, then passed by the narrow crystal filter and demodulated into audio. A similar process is used by the transmitter to convert audio into an 8-MHz single-sideband signal.

All circuitry to the right of the heterodyne oscillator operates at RF. As a result, these blocks must employ band-switching to resonate the proper LC combinations for each band of operation. The driver grid and driver plate circuits provide RF selectivity to both the transmitter and the receiver.

cess SSB signals, leading to many opportunities for circuit sharing.

Block Diagram Basics

Analyzing a transceiver reveals how the concept of block sharing has been realized. By minimizing circuitry, size, cost and energy consumption can be reduced. Fig. 4 shows how compatible signal levels and frequency schemes have been successfully used to eliminate duplicate circuitry in the Heath HW-101. Proceeding from left to right, the first shared circuit is the crystal-controlled carrier oscillator/BFO. This circuit supplies RF to the balanced modulator, for production of a double sideband, suppressed carrier signal, and to the product detector, for converting the receiver second IF signal to audio. The oscillator supplies both signals simultaneously, eliminating switching, impedance-matching and frequency-pulling problems.

The product detector and balanced modulator are mixers. In this transceiver, the transmitter requires a balanced mixer to null out the unwanted carrier signal, while the receiver uses a simple, active product detector. There is no reason why a singly or doubly balanced product detector might not be employed, instead. Such a circuit could serve a dual role as both product detector and balanced modulator.

The crystal filter, an expensive item, is also shared. On transmit, it is used to pass only one sideband of the double-sideband signal and to provide additional carrier suppression. On receive, the steep-skirted 2.1-kHz passband response of the filter provides sharp selectivity. Output from the crystal filter is applied to an adjustable-gain IF amplifier. Gain in this stage is controlled by either an automatic level control (ALC), or an automatic gain control (AGC) voltage, depending on mode. On transmit, an ALC bias voltage developed from the signal in the final amplifier stage ensures maximum transmitter output without overloading. On receive, an AGC voltage similarly reduces the IF gain to maintain a relatively consistent volume level.

Design Considerations

Circuit sharing in a transceiver helps to streamline a design, but the switching circuitry is sometimes rather extensive. To show just how complicated the switching task may become, let us refer once again

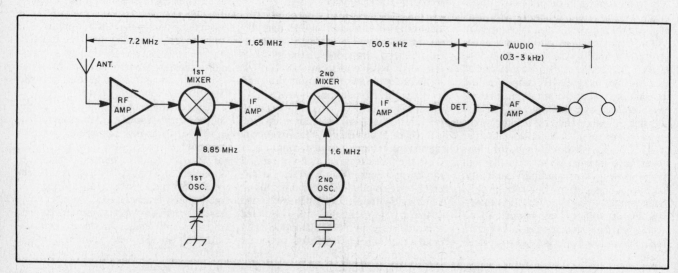

Fig. 2 — A superheterodyne AM receiver.

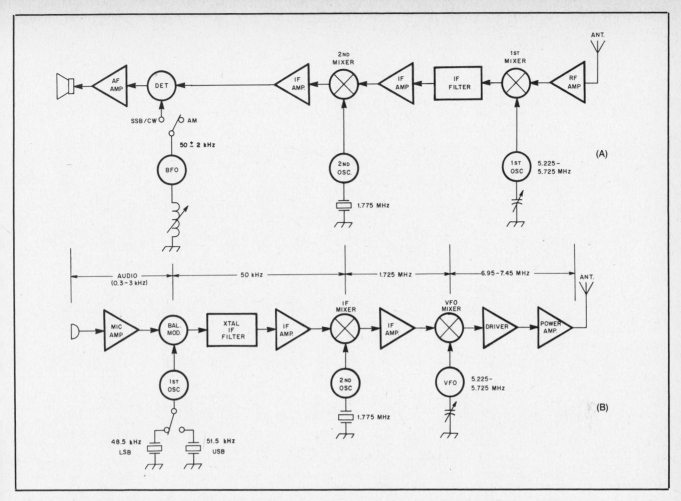

Fig. 3 — Block diagrams show similarities between a superheterodyne AM/SSB receiver (A) and a heterodyne SSB transmitter (B).

to Fig. 4. In this tube/transistor hybrid, voltage is switched from the screens of V2, V7, V8 and V9 in the transmit mode to V4, V10 and V11 in the receive mode. Bias voltage used for cutting off the receiver audio amplifier and second mixer in transmit is grounded to allow reception, while simultaneously, cutoff bias is applied to the transmitter mixers and driver stage. Switching circuitry also protects the sensitive small-signal stages of the receiver. An especially large negative bias applied to the grids of the receiver RF amplifier is necessary to ensure that this stage remains cut off. Otherwise, peaks in the driver RF envelope might turn this tube on, resulting in spikes on the transmitted waveform.

Occasionally, it may be necessary to actually change circuitry within a common stage to achieve proper transceive operation. In the transceiver of Fig. 4, additional capacitance must be switched into the driver preselector circuit when receiving, to compensate for tube stray capacitance. Otherwise, the driver preselector control setting that produces a peak on receive will not produce a coinciding peak on transmit.

Most transceivers use a combination of relays and switching diodes to perform the required switching. PIN diodes, which can be biased to appear as either open or closed

switches to RF energy, are especially useful. By placing them at the required circuit location, long and potentially troublesome RF interconnections may be avoided.

Most transceiver circuits require some form of sequencing to ensure that the receiver is properly muted before the transmitter is energized. This involves shutting off and turning on various transmitting and receiving stages in a predetermined sequence. Yaesu uses such time-delay sequencing circuitry in their FT-ONE transceiver. Fig. 5 shows a section of this circuitry. During CW operation, key closures are detected by the KEY line and applied to two inverter gates, U14e and U14f. The RC time constant on U14f delays the keying signal before it can pass through buffer U13d to actuate Q20, the transmit bias switch. The TX SIG line, whose logical state changes at the same time as the KEY line, has many functions. It supplies the third input to the transmit bias switch (U14d) and provides a proper voltage to the RC network on the input of Q12. This RC time constant delays operation of the receiver zero-voltage reference switch (RX 0). The TX SIG line supplies inputs to several other sequentially delayed gates. These control the transmit zero-voltage reference switch (Q11), both transmit and

receive 13.5-volt supply switches (Q19 and Q22) and the receiver bias switches (Q15, Q16). The antenna relay, powered by relay drivers Q25 and Q26, is also controlled by the TX SIG line.

TRANSCEIVER FEATURES

The basic transceiver is fine for casual sideband and CW work, but it has limitations in more rigorous operation, such as DX chasing and contesting. Fortunately, most transceivers today offer built-in features to help increase their flexibility.

RIT and XIT

Receiver incremental tuning (RIT) has become an almost universal transceiver feature that allows adjustment of the receive frequency without changing the transmit frequency. One important use of this feature is correcting for an operator who tunes to a beat note that differs from the transceiver's built-in frequency offset on CW or who prefers to listen to slightly different-sounding voice characteristics on SSB. By using the RIT control instead of the VFO dial, the pitch of the received audio can be changed while leaving the transmit frequency undisturbed. Otherwise, if during a QSO one operator adjusts the VFO to fine tune the receiver, he will also

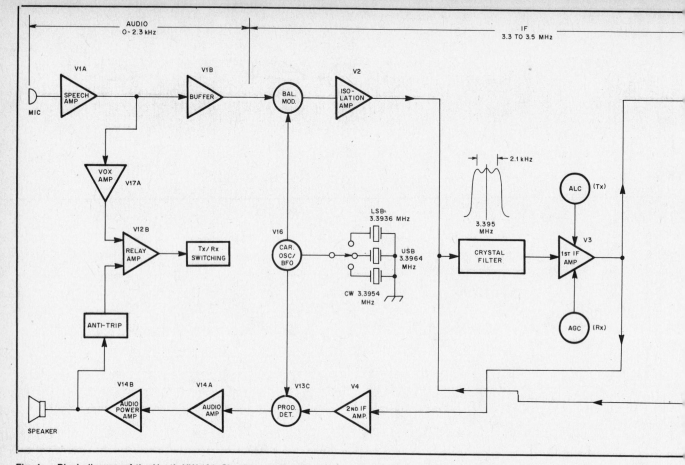

Fig. 4 — Block diagram of the Heath HW-101. Circuitry used exclusively for transmitting is shown at the top. Receiver circuitry is at the bottom. The middle section contains blocks used for both transmitting and receiving.

Ten-Tec's Argosy is a SSB/CW QRP transceiver. This unit allows for operator selection of input power levels of 10 or 100 watts, QSK (break-in) CW operation and frequency coverage of 80-10 meters.

adjust the transmit frequency. The other operator will likely compensate for this frequency shift by similarly adjusting the VFO. Before long, such frequency leap-frogging may move both stations far enough to cause interference with other ongoing communications.

Some RIT controls permit the receive frequency to be shifted as much as 10 kHz. These controls can be used as a second VFO, allowing split-frequency operation.

XIT (Transmitter Incremental Tuning) is closely related to RIT, permitting the transmitter frequency to be shifted a few

kilohertz away from the receive frequency. Many transceivers use the same knob for controlling both the RIT and XIT functions to permit listening on the transmitter offset frequency.

Fig. 6A is a schematic diagram of an RIT similar to that used in the Kenwood TS-520S and TS-820S transceivers. D1 is a varactor diode. Because it forms parts of the VFO tuning circuitry, any change of dc bias on D1 will cause a shift in VFO frequency. Potentiometer R1 is used as the RIT frequency-shift control. It forms part of a voltage divider with R3 and R4 when the RIT is switched on, allowing a variable bias to be applied to D1 (Fig. 6B). When the RIT is off, or when the transceiver is in transmit, however, a fixed bias is applied to D1 (Fig. 6C). When R1 is at the center of its range, the transmit and receive frequencies should coincide. R2, an internal trimmer potentiometer, ensures this by providing limited adjustment of the fixed varactor dc bias, eliminating any frequency difference.

XIT circuits operate in the same manner, except that the potentiometer that applies dc bias to the varactor diode is switched into operation during transmit instead of receive. Generally, RIT and XIT circuits

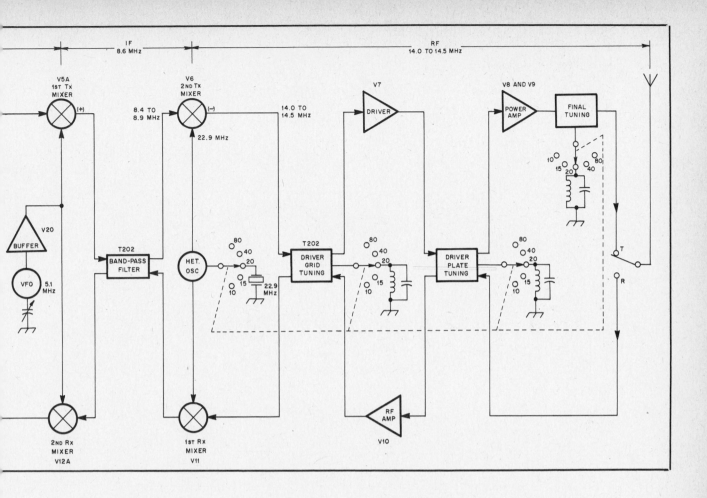

use relays or solid-state devices for switching.

IF Shift

This useful receiver feature, sometimes called passband tuning, effectively changes the center frequency of the receiver passband without changing the frequency to which the receiver is tuned. By leaving the pitch of the received signal or signals undisturbed, IF shift permits optimum placement of the available receiver selectivity around an SSB or CW signal.

Although IF shift can be viewed as changing the center frequency of the receiver IF filter, this filter operates on only one fixed frequency. Instead, the IF signal is shifted above and below this passband. Fig. 7 is a block diagram of a simple IF shift system. A tunable oscillator provides an 8.455-MHz signal to the mixers. The first mixer uses this signal to convert the incoming 455-kHz IF to the 8.000-MHz filter frequency. The second mixer then takes this 8.000-MHz signal and converts it back down to 455 kHz. By varying the oscillator frequency, the position of the desired signal within the filter passband can be changed.

When properly tuned, a received signal

The ICOM IC-735 offers SSB, CW, AM and FM modes, a 100-W transmitter, versatile memory system and general-coverage receiver in a compact package.

will pass through the center of the IF filter. Interference to one side of this desired signal may be removed by several methods. Fig. 8A shows such an interfering signal and its relationship to the passband filter. If the transceiver is tuned upward 250 Hz (Fig. 8B), the interfering signal will no longer fall within the passband. This is an effective and practical interference cure on CW, requiring only that the operator listen

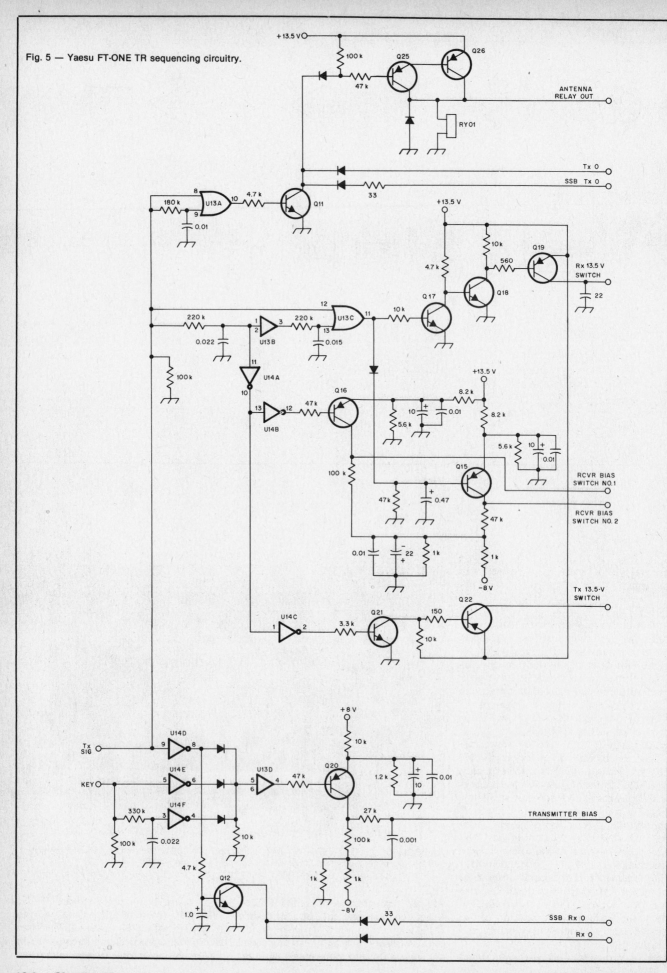

Fig. 5 — Yaesu FT-ONE TR sequencing circuitry.

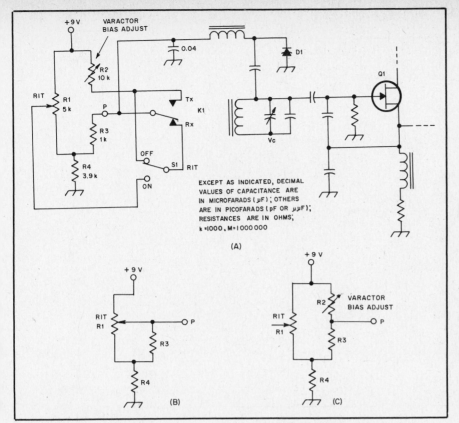

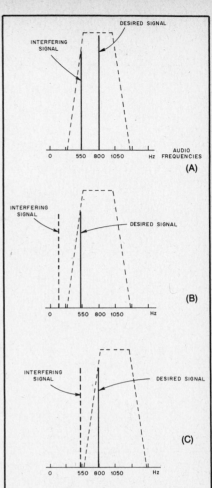

Fig. 6 — RIT circuit similar to that used in the Kenwood TS-520S and TS-820S series transceivers (A). At B is the equivalent circuit formed when the RIT is switched on. At C is the circuit when RIT is off or during transmit.

Fig. 8 — At A, receiver tuning places desired signal in center of the 500-Hz-wide CW passband filter. Another signal 250 Hz away will also fall within the passband, causing interference. At B, interfering signal has been removed by tuning receiver up 250 Hz. Desired signal will now have a 250-Hz lower pitch, however. At C, IF shift permits the passband to be moved, eliminating the interfering signal without changing the pitch of the desired signal.

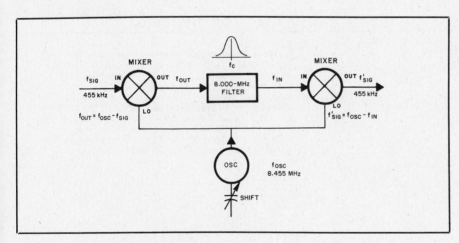

Fig. 7 — Block diagram of a simple IF-shift circuit.

to a CW note 250 Hz lower than normal. On SSB, however, this solution is impractical, because it leaves the voice signal unintelligible. By using the IF shift control, the interfering signal may still be eliminated (Fig. 8C), and signal intelligibility retained.

In practice, an IF shift circuit may not be as simple as the one shown in Fig. 7. The Kenwood TS-820 HF transceiver incorporates an IF shift circuit that is also part of the phase-locked-loop (PLL) frequency synthesizer. In the block diagram (see Fig. 9) operation on 14.250 MHz is shown.

In the synthesizer, the heterodyne oscillator supplies a fixed 19.5-MHz signal to the heterodyne mixer. This is mixed with a 23.08-MHz signal from the voltage controlled oscillator (VCO) to produce a difference frequency of 3.58 MHz. With the IF shift off, this 3.58-MHz difference frequency mixes with an 8.8315-MHz carrier oscillator signal in the carrier mixer to produce a 5.2515-MHz difference frequency.

The phase detector compares this signal with the 5.2515-MHz signal generated by the VFO. If any phase difference between these two signals is detected, an appropriate dc voltage will be produced by the phase detector and applied to the voltage-controlled oscillator. The VCO will respond by changing frequency slightly, in a direction that reduces this phase difference. The net effect is that the VCO remains locked to the 5.2515-MHz VFO signal.

The PLL has two outputs. One supplies the 23.08-MHz VCO signal to the RF mixer. This mixes with the desired RF input signal (14.250 MHz) to produce an 8.83-MHz IF signal. The other output is at 8.8315 MHz; it mixes with the filtered 8.83-MHz IF signal in a product detector, producing audio at 1.5 kHz.

As the VFO is moved up in frequency, an error voltage will be produced in the phase detector, causing the VCO frequency to shift down by the same amount. This relationship is a result of the mixing scheme, which uses difference frequencies instead of sum frequencies. The result is a lower frequency being tuned in by the transceiver.

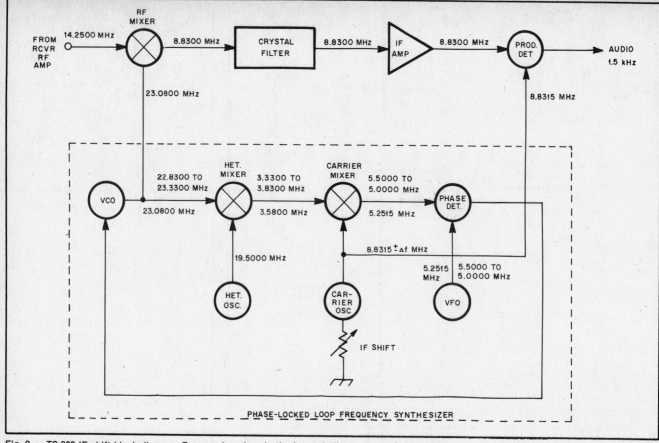

Fig. 9 — TS-820 IF shift block diagram. Frequencies given in the boxes indicate ranges for the parts of the synthesizer. For further discussions see text.

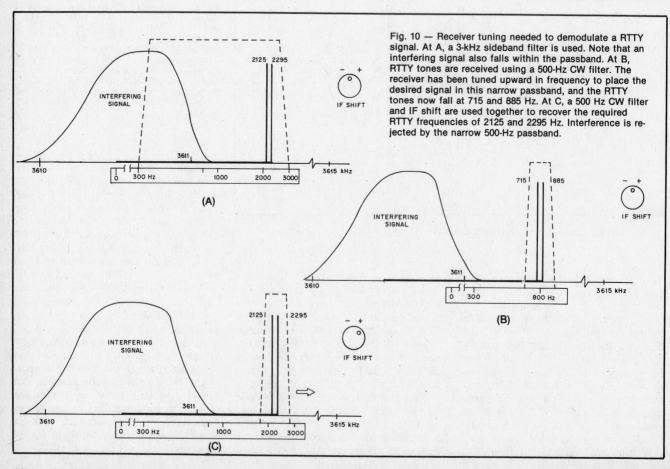

Fig. 10 — Receiver tuning needed to demodulate a RTTY signal. At A, a 3-kHz sideband filter is used. Note that an interfering signal also falls within the passband. At B, RTTY tones are received using a 500-Hz CW filter. The receiver has been tuned upward in frequency to place the desired signal in this narrow passband, and the RTTY tones now fall at 715 and 885 Hz. At C, a 500 Hz CW filter and IF shift are used together to recover the required RTTY frequencies of 2125 and 2295 Hz. Interference is rejected by the narrow 500-Hz passband.

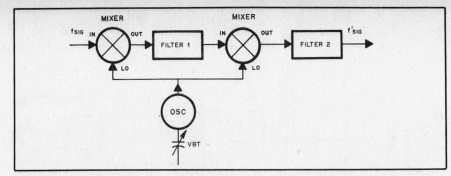

Fig. 11 — Variable-bandwidth tuning can be obtained by placing a second filter at the input or output of the IF shift circuit shown in Fig. 7.

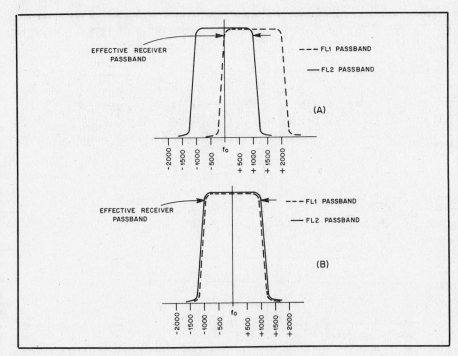

Fig. 12 — Variable-bandwidth tuning requires the use of two passbands. By moving the center frequency of one passband, the overall bandwidth is reduced.

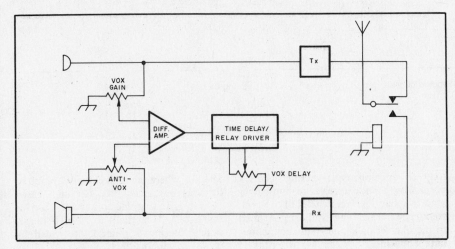

Fig. 13 — A basic VOX system.

Consider the transceiver once again tuned to 14.25 MHz. As the IF shift control is adjusted to raise the carrier oscillator 500 Hz to 8.832 MHz, a higher difference frequency will be produced by the carrier mixer. The phase detector will in turn detect this frequency difference and force the VCO up 500 Hz, to 23.0805 MHz, to compensate. At this point, the higher VCO frequency injected into the RF mixer will cause a portion of the 20 meter amateur band 500-Hz higher to pass through the crystal filter, just as a 500-Hz adjustment of the VFO would. (Keep in mind, however, that the VFO frequency has not been changed.)

After filtering, the resultant IF signal is mixed in the product detector with the IF-shifted carrier oscillator signal. This 500 Hz higher carrier-oscillator signal effectively compensates for the 500 Hz higher VCO signal, producing demodulated, IF-shifted audio. If the carrier oscillator had not been shifted up 500 Hz, the net effect would be exactly as if the VFO had been used to shift the receiver frequency up 500 Hz to 14.2505 MHz.

IF shift is also useful for improving reception of radioteletype (RTTY) and other data signals. Ordinarily, a reasonably wide SSB filter is used to recover the standard RTTY modem or terminal unit (TU) audio tones of 2295 and 2125 Hz. Unfortunately, BFO placement relative to the center frequencies of many such SSB filters allows recovery of audio frequencies from 300 to 3000 Hz (Fig. 10A). Only the higher frequencies are important here, so the lower 300 to 2000 Hz response represents unneeded bandwidth that only makes the receiver more susceptible to interference from adjacent signals.

To get around this, a 500-Hz CW filter might be used, but such filters are generally used with BFO placements allowing an 800-Hz beat note with signal at filter center. The two RTTY tones separated by only 170 Hz would still be heard, but their recovered audio frequencies would be centered about 800 Hz (885 and 715 Hz, respectively) — totally unacceptable to the RTTY demodulators (Fig. 10B). The solution, IF shift, allows the recovered-audio pitch at filter center to be varied at will — from 800 to 2210 Hz, for example. This allows reception of audio tones at 2295 and 2125 Hz, and provides a narrow passband response (Fig. 10C).

Frequently, an SSB or a special FSK mode is used by transceivers for RTTY transmission and reception. When poor band conditions make reception difficult with the SSB filter, a narrow CW filter and IF shift can improve copy. However, in most transceiver designs, the CW filter can be used only in the CW mode. This situation requires the operator to switch back and forth between modes when transmitting and receiving RTTY.

Variable-Bandwidth Tuning

To avoid the expense of equipping a transceiver with several IF filters, variable-bandwidth tuning (VBT) has been developed. This circuit is similar to IF shift. However, instead of shifting the IF passband, VBT permits adjustment of the IF width to optimize reception under various operating conditions. VBT also provides reasonable receiver CW selectivity in transceivers equipped only with SSB filters.

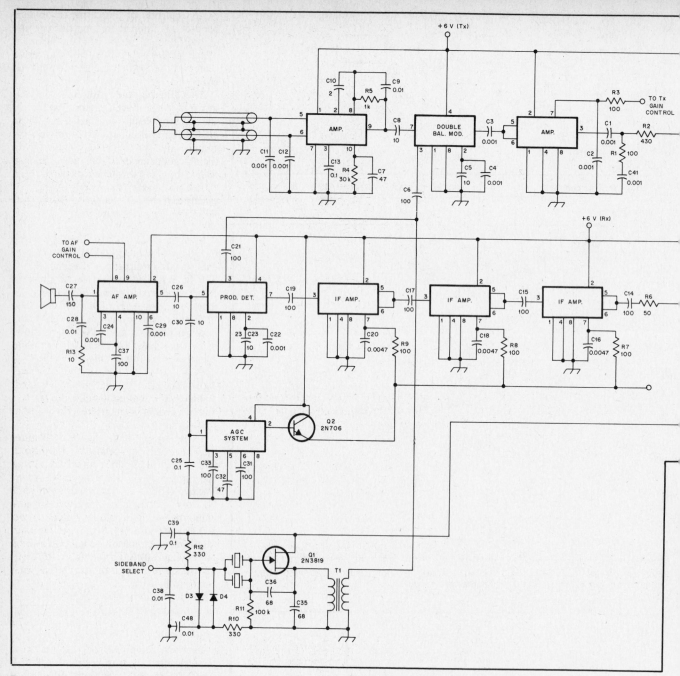

Fig. 14 — Schematic diagram of a transceiver IF module constructed from integrated-circuit building blocks.

One way of implementing VBT is by taking the simple IF-shift system of Fig. 7 and placing a second IF filter at its output. The resulting VBT system is shown in Fig. 11. By shifting the passband of the first filter with respect to the second, the effective system bandwidth may be reduced (Fig. 12A). When two equal-bandwidth filters are exactly aligned, the overall system bandwidth will be at a maximum (Fig. 12B).

Notch Filtering

Another way of reducing interference is to notch, or reject, the offending signal. A notch filter reduces receiver response over a narrow band of frequencies. The notch frequency is adjustable, so it can be tuned to the interfering signal frequency, eliminating unwanted CW or AM carrier tones, while preserving the desired signal's audio response.

Today, two types of notch filters are found in amateur transceivers. One type, the IF notch, operates at the receiver IF. The other type is the audio notch filter. IF notch filters are most often employed at relatively low frequencies, such as 455 or

50 kHz, where the high circuit Q required for a narrow bandwidth notch can be easily obtained. In a single-conversion receiver that uses an IF in the 8- to 10-MHz range, notch filtering is generally provided by means of audio filters.

Both filter types can be effective, although some audio notch filters do not provide the attenuation or notch depth afforded by typical IF notch filters. Typical values of attenuation range from 30 to 60 dB. If a filter has only a 30-dB notch, it will attenuate a 20-dB-over-S9 signal to a level equivalent to about S7. A 60-dB notch

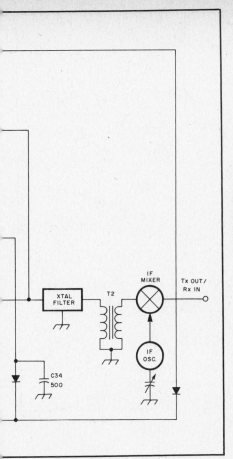

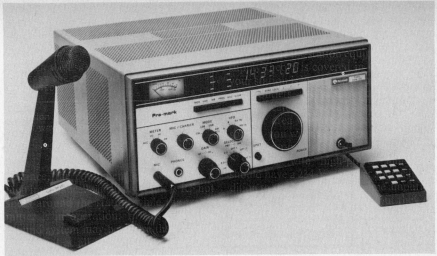

The Rockwell/Collins KWM-380. Extended 160-10 meter frequency coverage is provided. The optional 16-button keypad offers alternative methods of frequency control. A combination ac/dc power supply is self-contained.

would reduce the same signal to approximately S2.

Audio Peak Filters

Another type of filter found in some receivers is the audio peak filter (APF). These filters are narrow-bandwidth bandpass filters. Usually, the bandwidth is a few hundred hertz (or less), and the center frequency is adjustable.

An APF can offer a significant improvement in overall selectivity, even in a receiver equipped with a narrow bandwidth IF filter. There may be some problems, however, if an APF is used instead of an IF crystal filter for CW reception. Because the APF is located in the receiver audio section, a strong signal outside the APF passband can still be amplified to a very high level in the receiver IF stages. If fact, the signal level can overload some receiver stages, producing distortion and reducing receiver sensitivity. An IF filter prevents this problem by removing the undesired signal before it reaches the IF amplifiers. Even a receiver that is not overloaded may suffer from an undesired signal activating the AGC circuit and reducing receiver sensitivity to the point where a weak signal will not be heard. Sometimes this problem

can be overcome by disabling the AGC circuitry.

VOX

A transceiver equipped with VOX (voice operated switch) may be changed from receive to transmit by merely speaking into the microphone. Fig. 13 is a block diagram of a basic VOX system. It consists of a differential amplifier, variable time-delay circuit and transmit-receive (TR) switch. Audio energy from the microphone, controlled by the VOX-gain potentiometer, is applied to one of the two input terminals of a differential amplifier. Audio input level is adjusted so the operator's normal voice intensity will energize the TR switch. The time delay circuit keeps the TR switch

energized during short pauses in speech. Without the time delay, the switch would alternate rapidly between transmit and receive. Proper adjustment of the VOX delay prevents this annoying condition.

A problem may develop when audio from the speaker reaches the microphone. If it is as loud as the operator's voice, it will turn on the transmitter. A solution might be to place the speaker and microphone at opposite ends of the shack, or to wear headphones. Fortunately, a more convenient solution, anti-VOX, makes this unnecessary. With anti-VOX, a portion of the receiver audio is applied to the other input of the differential amplifier. Signal level is determined by the anti-VOX (or anti-trip) potentiometer. When the anti-VOX po-

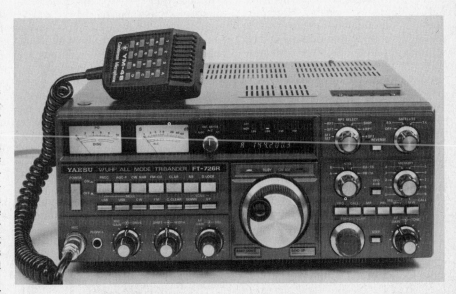

Yaesu's innovative FT-726 VHF/UHF all-mode transceiver also covers the 10- and 15-meter bands. Plug-in modules are installed for the bands of interest. Up to three such modules may be installed at any time; the 2-meter module is supplied as standard equipment. The '726 can be thought of as a 10-MHz IF unit with separate transverter modules for the bands covered. The power supply is built in.

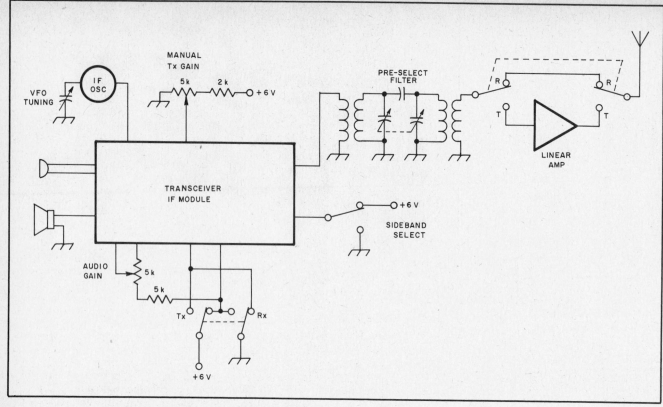

Fig. 15 — A complete transceiver, using the IF module of Fig. 14.

tentiometer is set properly, any speaker audio picked up by the microphone will be cancelled by this anti-VOX signal, preventing the receiver from keying the transmitter.

FUTURE TRENDS

The latest transceiver designs are strongly influenced by recent developments in RF and computer technology. Complete receiver subsystems are now available in single IC packages. Improved fabrication techniques promise to increase device reliability while decreasing cost. These changes are sure to have an effect on the Amateur Radio equipment of tomorrow.

Building Block Basics

It is already possible to assemble a transceiver from IC building blocks, adding only a few discrete components (individual resistors, capacitors, transistors, etc.). The schematic diagram in Fig. 14 bears strong resemblance to a block diagram, since the circuitry needed for each stage function (mixer, amplifier), is entirely contained within IC modules.

This single-conversion design is described in the RSGB *Radio Communications Handbook*. It can be used as the foundation for either a single- or multi-band HF transceiver. Audio at the transmitter microphone input will produce a 70-mV RMS, 9-MHz SSB output. Similarly, the receiver will accept a 9-MHz SSB input signal and convert it down to audio, providing 65 mW of power to a speaker or external power amplifier. Fig. 15 is a schematic diagram showing how the basic transceiver may be expanded into a fully operational HF communications system.

This circuit design illustrates an important shift away from conventional transceiver design philosophy. There is practically no sharing of circuitry. This approach may well be more economical than trying to use a single chip for both transmitting and receiving, for several reasons. First of all, it is much easier to merely switch supply voltage from the receiver circuitry to the transmitter circuitry than to develop extensive RF switching arrangements for each shared chip. Second, the cost of switching circuitry might easily add up to more than the cost of the chips they were designed to

Trio-Kenwood's TS-940S is a top-notch radio that offers SSB, CW, AM, FM and FSK operation on the 160-10 meter amateur bands, as well as a general-coverage receiver. The power supply is built in. Options include an automatic antenna tuner and several crystal-filter combinations.

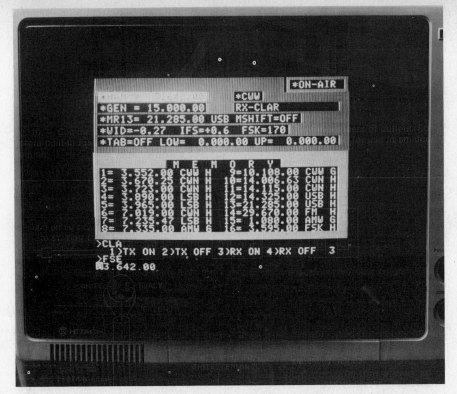

Fig. 16 — Computer graphic display from the Yaesu FT-980 transceiver controller program.

The '980 can be connected to a personal computer through an appropriate interface. This arrangement makes it possible to tune the radio and store and manipulate frequencies from the computer keyboard. Control of such important operating features as RIT, IF shift, mode control, IF width, FSK shift, and even TR switching may be performed by the computer.

Fig. 16 is the graphic display from an FT-980 controller program. This particular program, as supplied by Yaesu, is designed to be run on an Apple II computer. Complete transceiver diagnostics are given, and any listed parameter may be changed at will, making the transceiver front panel unnecessary. In fact, practically every front panel control is disabled when the computer program runs.

A more complex program might not only control the transceiver, but also format and send RTTY text. Or, the computer might serve as a processor for voice signals, permitting fingertip control of compression, frequency response, and so on. The possibilities are limited only by the skill and imagination of the software designer.

Two relatively new forms of digital communications, AMTOR and packet radio, make computer control necessary both for generating and interpreting data and for transmit/receive mode switching. Close synchronization is required among the various hardware and software elements within each system, and among all active systems comprising a network. A transceiver used in an AMTOR system, for example, must exhibit quick recovery time (on the order of 10 ms) when switching from receive to transmit. Care must be taken to ensure that circuit time constants are short enough to permit this type of high-speed operation.

Tomorrow's Transceiver

As digital methods improve and digital logic increases in speed, it may only be a matter of time before amateur signals are

Digital Aspects

Digital electronics plays an extensive role in Amateur Radio today. For the amateur transceiver, digital methods are used both externally, for generating and interpreting various modes of digital communications, and internally, for performing such circuit functions as stabilizing the operating frequency, storing and recalling control set-tings from memory, and preventing out-of-band operation.

The Yaesu FT-980, for example, has a built-in microprocessor that allows up to 16 frequencies to be stored and instantly recalled from memory. Mode and VFO information for each channel is simultaneously programmed and stored. Other features of the digital frequency controller include push-button frequency tuning and tab setting. This tab function permits an operator to partition an amateur band to avoid accidental transmissions on unauthorized frequencies.

eliminate! Additionally, alignment is sure to be quicker and easier when transmitting and receiving circuitry can be isolated from one another.

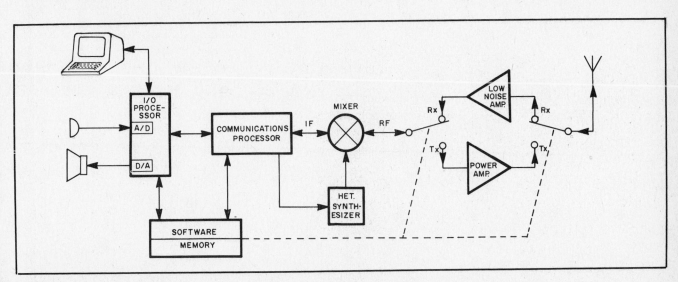

Fig. 17 — A transceiver of the future?

exclusively digitally processed. The transceiver of tomorrow may employ a receiver consisting of an analog-to-digital (A/D) converter at the antenna or IF input and a digital-to-analog (D/A) converter/ amplifier or a line driver at the audio/ baseband output to accommodate voice or data signals, respectively (Fig. 17). In between, all information would be processed as digital data. When the transceiver is transmitting, the process would simply be reversed, with audio from a microphone applied to an analog-to-digital converter, digitally processed up to the operating frequency, and transmitted using conventional RF techniques. With the ever-increasing speeds of microcomputers, the transceiver of today may tomorrow be reduced to little more than packaged software, with the necessary RF A/D and D/A conversion tasks handled by small plug-in interface modules.

Chapter 14

Repeaters

Repeaters, particularly FM repeaters, have had a great impact on Amateur Radio. Find a copy of any Amateur Radio magazine from the early 1970s and compare the amount of advertising for FM equipment with the amount of advertising in a current issue. Dramatic isn't it? Advertising from manufacturers and dealers is a good barometer of buyers' interest.

Why the widespread popularity of FM repeater operation? Repeaters are ideal for increasing the range of mobile and hand-held portable radios. Telephone interconnects (autopatches) provide instant access to local police, fire and medical assistance during emergencies. The repeater provides a local "party line" for club members. These are some of the reasons why repeaters have become so popular.

A repeater is a well situated automatic retransmission station used to extend the range of other stations. Repeaters often serve hand-held, portable and mobile stations in a local area. The repeater station consists of a receiver, control circuitry, interconnect circuitry, a transmitter, power supply(ies), at least one antenna and various other associated items (Fig 1).

Repeaters are as much a product of FCC rules, regulations and definitions as they are of radio technology. Unlike other less-regulated aspects of Amateur Radio, the repeater operator must calculate the effective radiated power (ERP; definition follows) instead of simply measuring the output power. Furthermore, FCC rules require the repeater operator to calculate the height above average terrain (HAAT; definition follows).

The FCC has valid reasons for treating repeaters more strictly than other amateur stations. It is usually difficult, if not impossible, for two repeaters to share the same service area and frequency(ies). In most stations, the operator (person controlling the equipment) is also the most frequent user of the station; not so for repeaters. In a typical repeater group, one or two individuals may be the operators, but tens (hundreds) of other stations will use the facilities of the repeater.

This chapter is meant to provide the reader with an overview of the dual nature

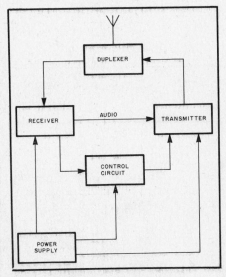

Fig 1—Block diagram of simple dual-frequency repeater.

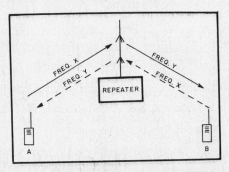

Fig 2—Pictorial diagram of frequency usage in a dual-frequency repeater system. A transmits on frequency X, which is the repeater receive frequency. The repeater retransmits A's signal on frequency Y, which is B's receive frequency. When it is B's turn to transmit, the paths are reversed, with B transmitting on X and A receiving on Y.

of repeaters—the technology and the rules. Most of the circuits serve to illustrate the concept as simply as possible and, therefore, may not be the best design for those looking for a construction project. For those in the process of building a repeater station, we suggest you consult recent issues of *QST, Ham Radio* or *73* for the latest in practical circuits, and ARRL's *FCC Rule Book* for the latest in repeater regulations.

In the amateur service at VHF and UHF, there are basically two approaches to producing a station that functions as a repeater. The first repeater format receives the signal on one frequency, demodulates it, processes it, applies it to the input of the repeater transmitter and retransmits it (usually on a second frequency in the same band), as depicted in Fig 2. Most repeater stations in the US are designed to operate with FM in the 2-meter band. But FM voice repeaters can be found on the 28, 50, 220, 450-MHz and higher bands as well. Some repeaters are designed for other modes besides voice, such as RTTY, slow-scan TV, fast-scan TV and packet radio.

Packet repeaters usually use the second fundamental format for repeater operation (Fig 3). Here, only one frequency is used: The repeater station receives and transmits on the same frequency. *It does not, however, transmit and receive at the same time.* With digital data there is no problem storing the received information and retransmitting it. Thus, one frequency can be used, and many of the problems associated with dual-frequency systems are not present. Obviously, voice conversations do not lend themselves to this store-and-forward technique. Single-frequency repeaters are fundamentally different from dual-frequency repeaters and have radically different problems associated with them. The balance of this chapter is devoted primarily to dual-frequency systems.

In some instances, a voice repeater may also be used for another mode, such as RTTY. If this is accomplished with audible tones, its operation is not different than that of normal voice repeaters. Some groups have successfully integrated auxiliary modes with the primary voice mode using techniques that make the secondary communications transparent to the voice users. Such techniques are beyond the scope of this chapter.

Isolation

The primary consideration in dual-frequency repeaters is isolation. Everything else is secondary. Isolation means keeping the output of the repeater transmitter and other unwanted transmitters from affecting the input of the repeater receiver. No

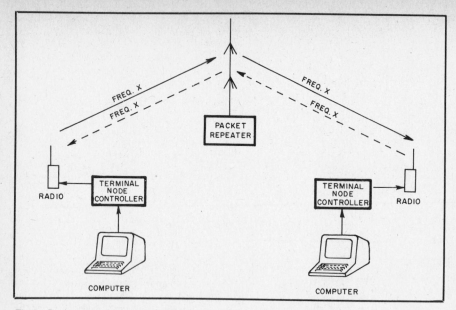

Fig 3—Packet repeaters use only one frequency for transmitting and receiving, but they do not transmit and receive at the same time.

matter what options are added to a repeater, the performance will be unsatisfactory if the transmitter "desenses" the receiver.

Desensitization is the reduction in receiver sensitivity caused by noise or RF overload of the receiver front end. Although desensitization from the repeater transmitter may be the most common form of this malady, it can be caused by any strong signal or noise reaching the receiver front end.

The least expensive—and usually the least effective—method of providing the needed isolation is to use separate antennas for the transmitter and receiver. To be effective, the antennas must have low angles of radiation with virtually no lobes of power high in the vertical plane (Fig 4). Usually in such installations, the receiver antenna is mounted directly above the transmitter antennna. The greater the vertical separation, the better the isolation, but the greater the separation, the greater the difference in the patterns of the two antennas. If the receive and transmit antenna patterns do not match closely, the system will seem to perform erratically for stations in certain locations (Fig 5). For instance, a weak hand-held might be able to access the repeater, but a directional gain antenna and preamplifier might be required to receive it. Thus, one is forced to compromise between isolation and consistent performance.

One way of minimizing the effect of the transmitter on the receiver in such installations is to keep the transmitter power to a minimum. Such a system might be marginally useful when the transmitter operates at a power of 1- or 2-watts output, and totally useless when the transmitter output power increases to 25 or 30 watts. Receiver desense can often be identified in a repeater

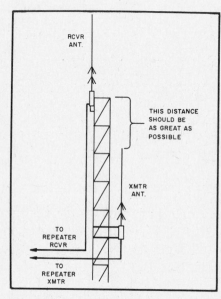

Fig 4—Pictorial diagram of a typical two-antenna repeater installation. The vertical separation should be at least four wavelengths.

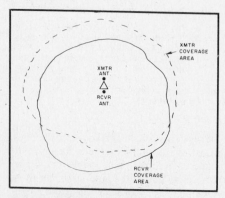

Fig 5—Use of two antennas can cause disparity in the coverage areas of the receiver and transmitter. The solid line represents the boundaries of the receiver coverage area, while the broken line represents the coverage area of the transmitter.

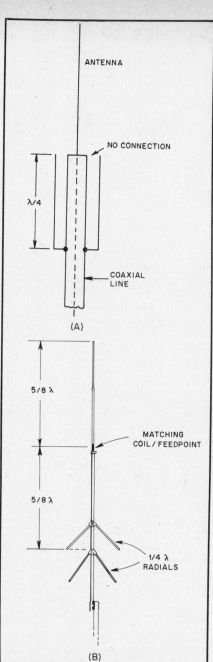

Fig 6—At A, the coaxial sleeve, or "bazooka," balun for isolating the feed line from the antenna. At B, a typical 2-meter FM antenna with two sets of radials spaced ¼-wavelength apart.

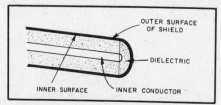

Fig 7—Cutaway view of a section of coaxial Hardline cable. At VHF and UHF, virtually all currents flow at the surface of the conductor; therefore, the signal path is along the outer surface of the inner conductor and the inner surface of the shield. Current flowing on the outer surface of the shield can create fields that combine with the antenna fields causing a change in the radiation pattern of the antenna.

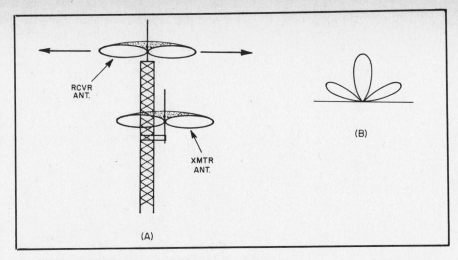

Fig 8—At A, pictorial diagram of main lobes of radiation for receiving and transmitting antennas. At B, representation of a distorted radiation pattern, which would cause problems if it were to replace the transmitter pattern in A.

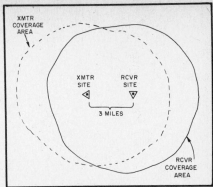

Fig 9—Disparity in coverage area caused by using two sites.

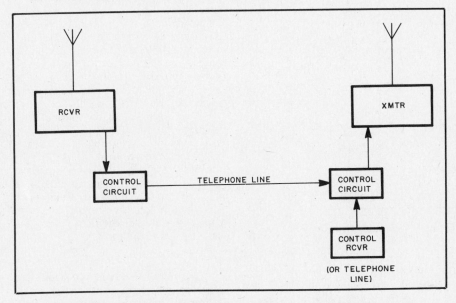

Fig 10—Block diagram of two-site system using telephone line to send audio and control functions from the receiver to the transmitter.

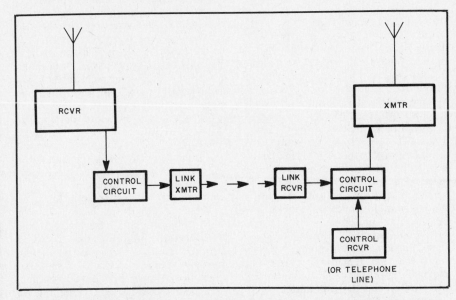

Fig 11—Block diagram of two-site system using RF link between receiver and transmitter sites.

system using this principle, whether the system is using vertically separated antennas or some other isolation procedure. If a weak station is able to break the receiver squelch, but disappears when the repeater transmitter is turned on, desense is indicated. To confirm this condition, reduce the output power of the repeater or simply disable the transmitter; the weak signal should now be able to capture and hold the repeater receiver.

Antennas that are well decoupled from the transmission line and the use of Hardline coaxial cable are two means of improving the performance of the two-antenna system. Decoupling the feed line from the antenna means preventing the antenna field from inducing current flow on the outside of the coaxial cable shield. Current flowing on the outer surface of the shield will create fields that combine with and distort those of the antenna, which may produce lobes in undesired directions.

Various devices can be used to accomplish decoupling; they create a very high impedance point somewhere along the surface of the outer conductor that effectively blocks the current, and they provide another very low impedance path that effectively short circuits the current and prevents it from flowing farther along the transmission line. Fig 6 shows antennas with a "bazooka" at A and two sets of quarter-wave radials at B. The "bazooka" is ¼-wavelength-long conductive sleeve attached to the shield at the bottom and insulated at the top. Because it is ¼-wavelength long, it has a very low impedance at the point that it is attached, but a very high impedance at the top where it is insulated from the shield. Since the "bazooka" and the corresponding portion of the outer surface of the shield form a section of transmission line, the shield also has a very high impedance at this point and blocks the RF current. The ¼-wavelength radials in B provide a low-impedance path for the current. Two sets of radials spaced ¼-wavelength apart are substantially more effective than one set.

Since at VHF and higher most of the RF current tends to flow on the surface of a

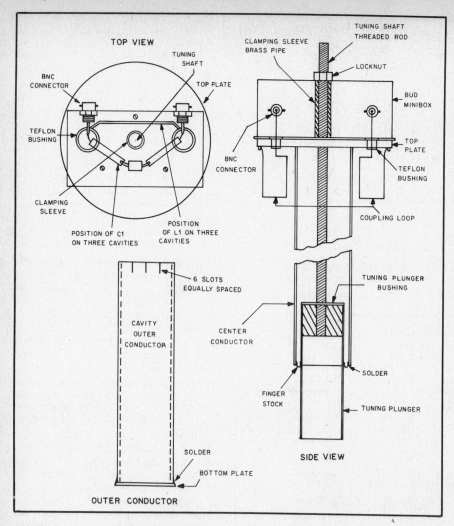

TOP VIEW

Fig 12—Details of a 146-MHz cavity to be used as part of a duplexer. To minimize loss, the conducting surfaces should be silver plated.

conductor, a coaxial cable can be considered to have three conductors: the outer surface of the inner conductor, the inner surface of the shield and the outer surface of the shield. In a perfect system the signal travels on the outer surface of the inner conductor and the inner surface of the shield (Fig 7). If the outer shield has holes in it (for example, gaps in braid), fields on the outside of the coaxial cable can influence the field existing on the inside of the conductor. Because Hardline coaxial cable has a shield made of a conductive tubing (usually copper or aluminum), the holes are entirely eliminated. The only possibility of an outside field invading the internal field is at the junction points (connectors). If braided-shield type coaxial cable is to be used, the double-shielded variety is preferable.

A little reflection on Fig 8 will give the reader some insight into the significance of Hardline and decoupling. Since the receiver antenna is mounted above the transmitter antenna, it is obvious that the receiver transmission line must run through the major lobes of the transmitter antenna field. The limitations of the vertically

separated, two-antenna configuration often outweigh the benefits of cost-savings.

Two Sites

A second means for providing isolation between the repeater receiver and transmitter is to place the two at different sites and provide some means for linking the output of the repeater receiver to the input of the repeater transmitter. Where we talk of vertical separation in terms of a few feet or a few wavelengths, we talk of horizontal separation in terms of miles.

The pitfalls of a two-site system are many. To the system users, the most significant may be the disparity in coverage between the receiver and transmitter (Fig 9). Where this is a minor consideration for vertically separated antenna systems, it becomes a major concern for the two-site system. Two sites require two separate installations and the expenses that go with them. Amateur Radio groups are often required to pay rent for the space on a mountaintop, tower or tall building. Both the receiver and the transmitter require some means of providing electrical power. All this spells added expense for the group.

For the repeater operator, controlling two sites and interconnections are the big problems. Somehow the audio output of the receiver must be linked to the input of the transmitter. This can be accomplished in different ways. Probably the easiest, and in the long run the most expensive, is to rent a dedicated line from the local telephone company (Fig 10). A second method is to use an RF link consisting of a transmitter located at the receiver site and a receiver located at the transmitter site (Fig 11).

The equipment of the link is frequently chosen to operate on a band other than the one the repeater operates on. Although the initial expense may be greater, equipment can be selected with a life expectancy of several years. If the equipment is amortized over its expected life, the cost will often be found to be lower than that of renting a telephone line for a comparable period of time.

In addition to getting the audio from the receiver to the transmitter, the transmitter must be told to turn on when a signal is present at the output of the receiver. Typically, some form of encoding will be used to accomplish this task. For instance, a tone or current (dc) may be used on the telephone line to signal the transmitter to turn on.

A further complicating factor comes when service is required. It may not be obvious to the operator whether a problem exists at the receiver, over the link, or at the transmitter. Unless a technician is available for both sites, several trips back and forth may be required to bring the system back to optimum.

Duplexers

A duplexer is a device that allows the repeater transmitter and receiver to be connected to the same antenna. If properly constructed and installed, a duplexer will allow the signal from the antenna to reach the receiver while blocking energy from the transmitter. We live in a world governed by the axiom, "There's no such thing as a free lunch." The price of including a duplexer in a repeater system is insertion loss. Properly constructed, a duplexer will provide adequate isolation with an insertion loss of less than 1 dB.

What is a duplexer? How does it work? A duplexer is a device that provides a high Q passband for both the receiver and the transmitter. On the one hand, the passband circuit connected to the receiver passes energy at the receive frequency with a minimum of attenuation, while providing a high degree of attenuation (a notch) for energy at the transmitter frequency. On the other hand, the passband circuit connected to the transmitter passes energy at the transmit frequency, while rejecting or notching energy at the receive frequency.

Some FM transmitters lacking in good design produce a broad spectrum of energy as well as energy at the desired output

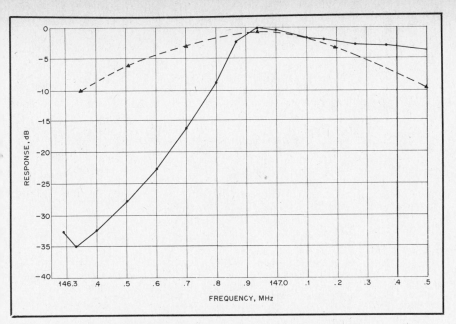

Fig 13—Typical frequency response of a single cavity of the type used for duplexers. The dashed line represents one cavity alone, while the solid line is for a cavity with a shunt capacitor connected between input and output. An inductance connected in the same manner will cause the rejection notch to be above the frequency that the cavity is tuned to.

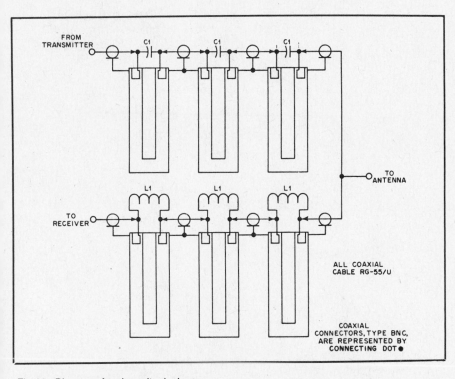

Fig 14—Diagram of a six-cavity duplexer.

frequency. This broad spectrum of energy is often referred to as white noise. Practically any transmitter will produce some white noise, but in good designs that noise will be at least 80 dB lower than the carrier. Although this may seem like an insignificant amount of power, it is more than enough to cause desense when the white noise appears on the frequency the receiver is tuned to. In poorly designed transmitters, the white noise is not likely to be down as much as 80 dB. A duplexer has the added advantage of preventing white noise energy

on the receive frequency from passing from the transmitter to the receiver.

In most VHF/UHF applications, resonant coaxial cavities are used to form the high-Q circuits needed for the duplexer (Fig 12). The cavity functions as a series-resonant circuit. When a capacitor or inductor is connected across a series-resonant circuit, an antiresonant notch is produced, and the resonant frequency is shifted. If a capacitor is added, the notch appears below the resonant frequency. Adding an inductance will make the notch

appear above the resonant frequency. The value of either component will determine the spacing between the notch and the resonant frequency.

Fig 13 shows the band-pass characteristics of the cavity with and without shunt elements. This example shows a cavity tuned to 146.94 MHz. Note that without the shunt capacitor, the cavity would provide only about 10 dB of attenuation 600 kHz below the primary frequency, which is the standard offset for the 2-meter band in the US. With the shunt element, the attenuation is approximately 35 dB—an increase of nearly 25 dB. The insertion loss in both cases is less than ½ dB, which is an acceptable level.

Typically, a duplexer in the Amateur Radio Service will consist of three cavities in the receive leg and three cavities in the transmitter leg (Fig 14). Using cavities similar to those described, we see that the receive leg will provide 100 dB of rejection of the transmitter signal, while having no more than 1.5 dB of insertion loss. Both of the these figures are acceptable and attainable. Commercially available six-cavity duplexers may have less than 1 dB of insertion loss per leg, while providing over 100 dB of isolation. Measurements in the ARRL lab have found insertion losses as high as 5 dB per leg in homemade duplexers manufactured from circuit-board material. Although far from ideal, these duplexers are probably still preferable to using separate antennas.

A second type of duplexer uses the hybrid ring circuit of Fig 15. The isolation provided by a hybrid ring depends on the arrival of two signals of equal magnitude but 180 degrees out of phase, at the output terminal. The signal is split into two parts at the input. One signal path is made one-half wavelength longer than the other to achieve the required 180-degree phase shift.

For duplex operation, the hybrid ring must be made to pass the desired frequency while attenuating the unwanted signal. A high-Q cavity is placed in the ring to act as a switch to take one of the signal paths out of the circuit. The cavity is resonant at the pass frequency of the hybrid ring, and thus will act as a short circuit at this frequency. Quarter-wavelength lines isolate the cavity from the signal so the desired energy is not attenuated.

At the frequency to be canceled, the cavity is not resonant and, therefore, it acts as a high impedance in parallel with one branch of the hybrid ring. Because the cavity does contribute some phase shift, a stub tuner is included in the opposite leg as a means to adjust the ring for maximum attenuation of the unwanted signal.

Common Practices

Even with a good-quality duplexer installed, there is no guarantee that the repeater receiver will be isolated from the transmitter. It is common practice to locate the transmitter and receiver in the same

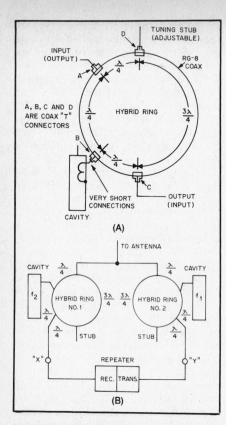

Fig 15—At A, a hybrid ring with a tuned cavity attached. At B, two hybrid-ring and cavity assemblies may be combined to serve as a duplexer. For additional isolation between the receiver and transmitter, another ring and cavity assembly may be inserted at points X and Y.

Rear view of Maggiore Hi Pro Mk II 220-MHz repeater. The transmitter and receiver strips are housed inside the cast aluminum boxes mounted to the chassis. Feedthrough capacitors are used for all leads going to the transmitter and receiver strips. The control circuitry is located beneath the chassis. Audio signals are at line level, minimizing the possibility of RF interference. These measures represent an acceptable level of isolation between the receiver and transmitter.

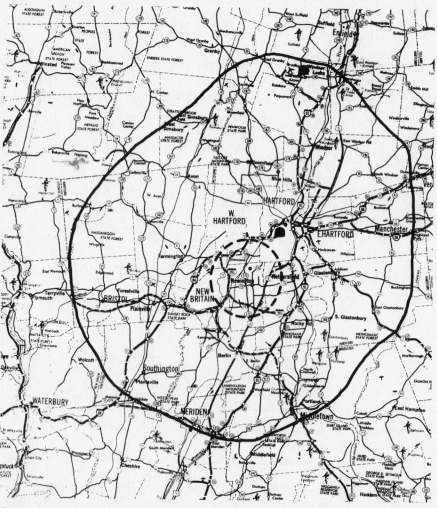

Fig 16—Map used to define coverage area and to estimate ''center'' area where search for repeater site is to be conducted.

cabinet. This is not mandatory, but given the cramped space at most repeater sites, it is quite common and often necessary. Care must be taken to properly shield the receiver from the transmittter. Most commercially available repeaters designed for the amateur service house the transmitter and receivers in RF-tight metal boxes. Equipment for the commercial services, which is usually more costly, often has even more elaborate shielding. Power, audio and control leads are fed into the transmitter and receiver by way of feed-through capacitors and other decoupling devices. It simply does not make good sense to spend hundreds of dollars for a duplexer, and then turn around and defeat its purpose with sloppy construction practices.

Another potential problem point is the coaxial cables that connect the duplexer to the transmitter and receiver. If RF leaks out of or into these cables, the duplexer will likely be compromised. Simply keeping the cables as far away from each other as possible will help. These jumper cables should be made of high-quality double-shielded coaxial cable as a minimum. Hardline would be preferable from this point of view, but it sometimes is not practical because a fair amount of flexibility is required.

If double-shielded cable is not conveniently available, it can be fabricated from ordinary single-shielded cable and braid removed from another section of cable. The coaxial cable is inserted into and pushed through the center of the braid—it will probably be necessary to do this before attaching the connectors. The braid is then secured to the connectors with ordinary hose clamps. Although not pretty, this procedure is functional. Some hams have even used soft-drawn copper tubing as an

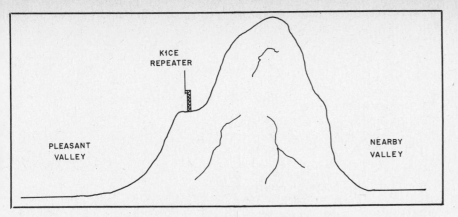

Fig 17—Location of K1CE repeater halfway up the mountain gives desired coverage of Pleasant Valley, while reducing co-channel interference from signals originating in Nearby Valley.

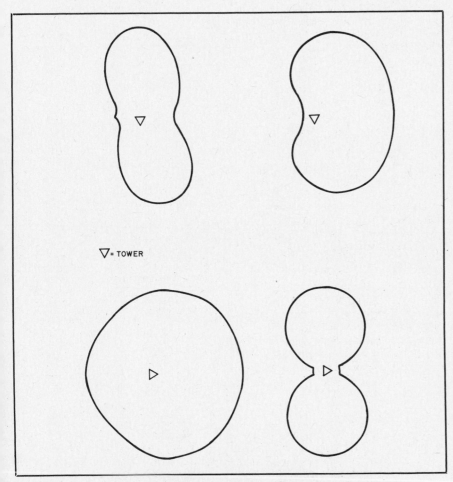

▽ = TOWER

Fig 18—Special radiation patterns obtained by varying installation of "omnidirectional" Stationmaster antenna, as suggested in Phelps Dodge sales literature.

outer shield when flexibility was not of prime importance.

Repeater Sites

Usually hams are interested in finding a repeater site that will provide good coverage of the community. This roughly translates into the the highest building, hill or mountain in the area. The site must be accessible, and some means for providing electrical power must exist. For control purposes and possibly autopatch, it is desirable to have access to telephone lines also.

In the early days of Amateur Radio repeaters, the commonly accepted definition of the best site was simply the highest point on the highest hill or building. Those were the days when repeaters were few and far between. In most urban areas, particularly along the East and West coasts of the US, the 2-meter band is approaching complete occupancy—and in some areas this is even true of the 220- and 450-MHz bands. Thus, it is not uncommon to have mobile stations accessing more than one repeater with the slightest band opening.

It is time to reexamine our definition of "best site." A 2-meter repeater located in Lewistown, Montana would probably be expected to have a much wider service area than one located in Hartford, Connecticut. Although it may be advantageous for the Lewistown repeater to cover an area with a radius of 100 miles, it would be irresponsible for a repeater in Hartford to do the same. Such a service area for the Lewistown repeater would not even include any of the major cities of Montana, whereas a similar sized service area centered in Hartford would take in both Boston and New York City.

To find the ideal site, the first thing that should be considered is the service area—specifically, what area should the repeater cover. For crowded urban areas, this probably should be no greater than a 60-mile diameter. Using topographical maps and taking the geographical idiosyncrasies into consideration, locate the center of the service area (Fig 16) and begin the search for a suitable site in that vicinity. A moderately high hill may provide better coverage of the service area than the highest hill—simply because the system will receive less co-channel interference.

Sometimes geographical peculiarities dictate the selection of a site at one extreme of the service area, or at least somewhat removed from the center of the service area. If the service area is to be a valley, it may be wise to locate the repeater on one of the mountains bordering the valley (Fig 17). Rather than choosing the peak of the mountain, the repeater operator should locate the station somewhat below the crest to minimize coverage to the other side of the mountain.

Directional antennas can also be used to shape the pattern of coverage also. For more information on this topic, consult *The ARRL Antenna Book* and literature available from various VHF/UHF antenna manufacturers. For instance, the Phelps Dodge Stationmaster series of antennas are fiberglass-enclosed collinear arrays exhibiting (usually) an omnidirectional horizontal pattern. Fig 18 displays a variety of horizontal radiation patterns that can be achieved by varying the installation of a Stationmaster. Exposed multiple-dipole arrays (Fig 19) and beam antennas offer even more possibilities.

Businesses are often willing to pay rent for the right to install their radio equipment at these sites. Some landlords are willing to rent space at a reduced rate or even donate it to an Amateur Radio group. When you discuss financial matters with the landlord, it may be good strategy to emphasize emergency communications and other forms of public service—if you are prepared to make good on any promises. Good negotiating and salesmanship skills on the part of the site committee may mean

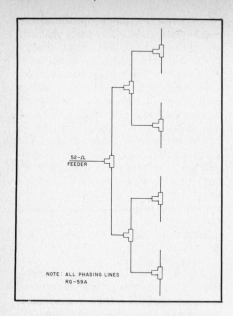

Fig 19—Typical dipole array for repeater use. Various patterns can be obtained by varying the alignment of each element.

NOTE: ALL PHASING LINES RG-59A

52-Ω FEEDER

the difference between a good site (high altitude, low rent) and a poor one. Increased cost may be the least of the problems caused by the proximity to other radio stations, though.

Living With Your New Neighbors

With the proliferation of radio transmitters in service, even a mediocre site is likely to have several transmitters. The repeater receiver, in this case, not only contends with the RF from the repeater transmitter, it must also face the onslaught of all the other RF at the site. Two problems are likely to develop. The RF fields in and around the site may be so strong that they simply overload the front end of the receiver. A more insidious problem exists when fundamental and harmonic signals from the various transmitting devices mix (heterodyne) somewhere outside the repeater producing a resultant signal on the frequency the receiver is tuned to. This problem is called intermodulation distortion, or simply intermod.

With simple front-end overload, cavities often can be used to eliminate the problem. Instead of placing the cavity in series with the transmission line, it is placed in parallel. Since it is a series resonant device, it now acts as a very sharp notch filter. Thus, a cavity across the transmission line can sharply attenuate an interfering signal (Fig 20).

The front end of an FM receiver can be further fortified with the addition of helical resonators, if they are not already installed. Helical resonators can be built exhibiting Qs of 1000 or more at VHF and UHF. Because the Q is so high, it is possible to design front-end tuned circuits using helical resonators that provide a high degree of selectivity with low insertion loss. Various coupling designs may be used. The walls

of the cavity should be seamless, if possible. Receivers for the commercial services often come with six or more resonators constructed in a cast-aluminum housing. Such design and construction practices provide for "bulletproof" front ends, but they are expensive to produce. Some of the equipment manufactured for the Amateur Service uses less expensive—and less effective—designs.

If the intermod signal is being generated somewhere other than the front end of the repeater receiver, cavities and bulletproof front ends will be of no use in combating it (Fig 21). The first step is to determine the frequency of various transmitters operating at or near the site. Remember, broadcast radio and TV signals, which may be very powerful at your site, can beat with other signals to produce the intermod. Once the likely culprits have been identified it will be necessary to work with the technicians who service the equipment causing the problem. (See May 1983 *QST*, pp 17-18, for a primer on intermod.)

At this stage of the game, human relations skills may be more essential than engineering skills in resolving the problem. Some form of filtering applied to one or more of the offending pieces of equipment will be necessary to clean up the problem. Although technically and legally it may be the responsibility of the owners of this equipment to pay for the filters and their installation, it may be difficult or impossible to force them to take action. In such cases, the expedient solution is to provide the filtering for them.

Antennas, Gain, ERP and HAAT

The most commonly used type antenna for repeater service is the vertically polarized omnidirectional gain antenna. *(The ARRL Antenna Book* contains a discussion of polarization and patterns.)* In short, an antenna does not create or amplify power or energy, but it does concentrate it in some directions at the expense of others. The ideal repeater antenna will have most of its energy in the vertical plane concentrated near the horizon. If you conceive of the antenna as being represented by a pencil standing on its end on a table top, and of the radiation from the antenna as being a doughnut surrounding the pencil, increasing the gain toward the horizon is accomplished by flattening out the doughnut (Fig 22). The doughnut has not been "created" or "amplified;" it merely has been rearranged or concentrated.

Sales literature for antennas from reputable manufacturers usually express gain figures as compared to either an isotropic radiator or to a half-wavelength dipole. The isotropic radiator is imaginary, and thus the gain figures can only be calculated instead of measured. An isotropic radiator is a single point in space and radiates energy in a pattern that can be visualized as a sphere. The dipole, of

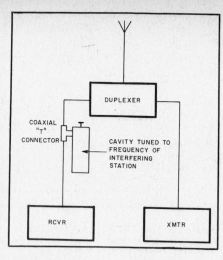

Fig 20—Single cavity used to notch out interfering signal.

DUPLEXER

COAXIAL "T" CONNECTOR

CAVITY TUNED TO FREQUENCY OF INTERFERING STATION

RCVR

XMTR

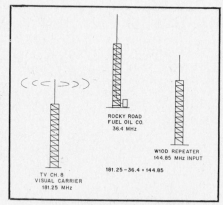

Fig 21—Pictorial diagram representing creation of intermod signal on repeater input frequency, when energy from the channel 8 visual carrier combines with output signal of the Rocky Road Fuel Oil Co. transmitter in the output stage of the transmitter (181.25 MHz – 36.4 MHz = 144.85 MHz). Some form of filtering must be applied to the Rocky Road transmitter to eliminate the intermod.

TV CH. 8 VISUAL CARRIER 181.25 MHz

ROCKY ROAD FUEL OIL CO. 36.4 MHz

W1OD REPEATER 144.85 MHz INPUT

181.25 – 36.4 = 144.85

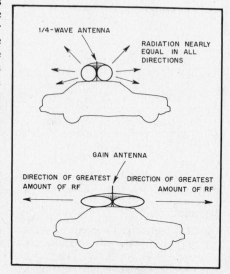

Fig 22—At A, radiation nearly equal in all directions from the ¼-wavelength vertical antenna. At B, the gain antenna has concentrated the signal toward the horizon at the expense of high angle radiation.

1/4-WAVE ANTENNA

RADIATION NEARLY EQUAL IN ALL DIRECTIONS

GAIN ANTENNA

DIRECTION OF GREATEST AMOUNT OF RF

DIRECTION OF GREATEST AMOUNT OF RF

course, is a real antenna producing the familiar doughnut pattern.

In practice, measurements on the antenna being tested usually are compared with measurements made on a half-wavelength dipole. The latter should be at the same height and have the same polarization as the antenna under test, and the reference field—that from the half-wavelength dipole comparison antenna—should be measured in the most favored direction of the dipole. The data can be obtained either by measuring the field strengths produced at the same distance from both antennas when the same power is supplied to each, or by measuring the power required in each antenna to produce the same field strength at the same distance.

A half-wavelength dipole has a theoretical gain of 2.14 dB over an isotropic radiator. Thus the gain of an actual antenna over the dipole can be referenced to the isotropic by adding 2.14 dB to the measured gain. If the gain is expressed over an isotropic antenna, it can be referenced to a half-wavelength dipole by subtracting 2.14 dB. With respect to FCC rules, antenna gain is based on a reference half-wavelength dipole. If an antenna is rated at 9-dB gain over as isotropic radiator, the gain of the antenna by FCC standards is 9 − 2.14, or 6.86 dB.

Repeater output power is usually specified in terms of effective radiated power (ERP), which is a term not widely used in most other Amateur Radio activities. Commercial services used ERP for many years before it came to be used in the Amateur Service.

It is not difficult to determine ERP for *a given installation*. First you need to know the exact power output from the transmitter—usually measured with a calibrated wattmeter. Next, deduct from this figure any losses caused by the duplexer. If this figure is not known, it is acceptable to measure the power coming out of the duplexer. Subtract the power lost in the feed line. This figure can be obtained from a table of coaxial cable characteristics, such as those found in Chapter 16. This gives the power reaching the feed point of the antenna. Combine this figure with the gain of the antenna; the result is the effective radiated power. In short, ERP is a measure of the power aimed in a direction that should be useful to the system.

For instance, assume the power output of a repeater transmitter is 200 watts. The duplexer has a loss of 3 dB. This means that 100 watts of power is reaching the feed line input. The feed line also has a loss of 3 dB, resulting in 50 watts reaching the antenna input. The antenna has a gain of 13 dB, which means the ERP of the repeater station is 1000 watts. (Note: These figures were chosen to simplify computation.)

Another important repeater specification is the *height above average terrain* (HAAT). HAAT is a geomathematical construct that means what it sounds like.

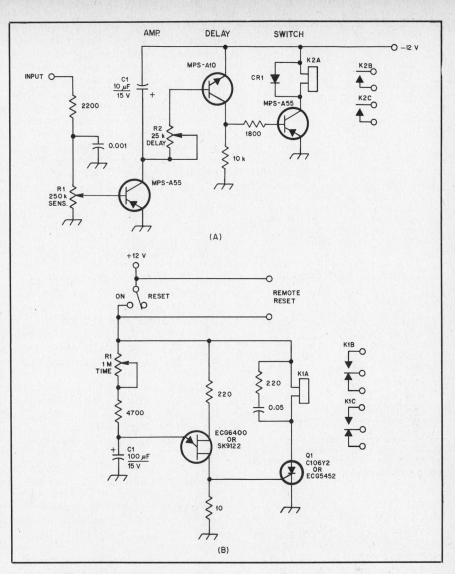

Fig 23—Elementary COR circuits.

To determine HAAT, follow this procedure: On a US Geological Survey map having a scale of 1:250,000, lay out eight evenly spaced radials (0, 45, 90, 135, 180, 225, 270 and 315 degrees) extending from the transmitting site to a distance of 10 miles. By reference to the map contour lines, establish the ground elevation above mean sea level (AMSL) at 2, 4, 6, 8 and 10 miles from the antenna structure along each radial. If no elevation figure or contour line exists for any particular point, use the nearest contour line elevation available. Now, compute the height of average terrain by adding the individual elevations at each of the 40 points and dividing that sum by 40. Thus the HAAT is the height AMSL of the transmitting antenna's center of radiation minus the height of average terrain.

Basic Repeater Controls

For a repeater to operate automatically, some circuit must determine when to turn the transmitter on and off — it's against FCC rules (and common sense) to simply let the transmitter stay on the air con-

tinuously without a signal to be retransmitted. The circuit that handles this function is usually called a carrier-operated relay or COR (Fig 23). Other names such as squelch relay, squelch-operated relay, squelch gate or signal-operated relay may be used. The name is not important, but the function is. In fact, in modern solid-state equipment, the COR function is often accomplished without benefit of a relay in the circuit, but the name persists.

An early COR used a portion of the limiter grid voltage in the receiver to drive a dc amplifier. The dc amplifier, sometimes called a relay driver, activated the relay, thereby controlling the transmitter. Such a simple system has some flaws, a major one being that anything causing limiter grid current to flow will activate the COR. Atmospheric noise, white noise, off-channel signals, a change in temperature or a change in voltage may cause such a simple system to turn the transmitter on.

A more reliable COR is one that is derived from the high-frequency noise component in the discriminator passband. This is the hissing sound you hear when an

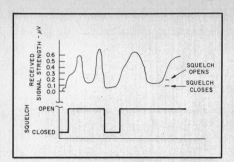

Fig 24—The function of hysteresis in a COR circuit. The upper line represents variations in the signal strength, while the lower line represents the squelch action. Notice that it takes a stronger signal to open the squelch than it does to keep it open.

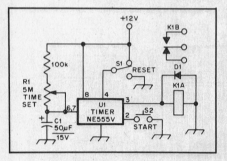

Fig 25—Elementary timer circuit using 555 IC. S1 and S2 should be parallel with the corresponding elements of the COR to provide start and reset functions.

FM receiver squelch is open but no carrier is present. A carrier "quiets" the receiver, causing the noise to disappear. Depending on the receiver design, this noise band may extend well beyond the range of human hearing. Any such ultrasonic component will also be quieted by a carrier. Sophisticated squelch circuits detect and act on the quieting of this ultrasonic component. Such squelch circuits are more reliable than those based on grid (or collector) current.

A second feature that improves the COR is hysteresis in the squelch circuit. Hysteresis is the property wherein it takes a stronger signal to open the squelch than it takes to keep it open (Fig 24). This is advantageous when a mobile station is on the fringe of the service area. Without this feature, the squelch will open and close rapidly as the signal level fluctuates. The result is audio that clips in and out, which is usually rather frustrating to operators attempting to listen.

Another improvement to the basic COR is to add a circuit that produces a "hangtime." Hangtime is simply the amount of time that the COR holds the transmitter on after the signal has disappeared from the receiver input. Such a feature prevents wear and tear on the equipment, particularly relays, if any are involved. It is also desirable during periods when the received signal is fading or "picket fencing."

A VOX circuit, similar to those commonly used in HF SSB equipment, can be connected to the COR control. To activate the transmitter, a voice or audio tone, as well as a carrier, must be received. This prevents the transmitter from being held on for a long time by an unmodulated carrier.

FCC rules require that some means be incorporated for disabling the transmitter if a carrier holds the repeater receiver squelch open for more than three minutes. The circuit that performs this function is often referred to as the time-out timer. In the early days, time-out timers were based on the electromechanical devices or simple RC circuits. Such procedures are archaic by today's standards.

One inexpensive modern approach is to use a 555 timer IC (Fig 25). The 555 contains two voltage comparators, a flip-flop and an output stage. The chief drawback is that large-value capacitors are required for durations in excess of a few seconds. The leakage of such capacitors often degrades the precision and accuracy of such circuits. Also, as the components age, some adjustments to the circuit will likely be necessary.

A second inexpensive solution is to use digital clock and counter circuitry. The common problems with the 555-type circuits are overcome at the cost of increased component count and design complexity. A simple digital timer circuit is shown in Fig 26. An MM5369 counter-timer IC is used to provide a stable 60-Hz reference signal. The 60-Hz signal is divided by 3600 by U1-U4, providing a 1-minute clock (1/60 Hz) at the input of U5. The last divide-by-ten counter (U5) can provide outputs at one-minute intervals from one minute to ten minutes. The selected output is wrapped around to the ENABLE input of the counter; this holds the count when the selected output is reached. As long as the RESET input of U5 is open, the counter is disabled. This input must be LOW to allow the timer to run. This circuit can be used either as a three-minute time-out timer or as part of a Morse or voice identifier circuit (See Fig 27).

More important than the particular circuit is the functioning of the timer. Rules require the timer to time out no longer than three minutes after the signal appears and remains on frequency. Some means must be provided for the timer to reset automatically during the course of a QSO. For the convenience of the users, most operators design the time-out timer to reset on the disappearance of a received carrier at the repeater receiver. One problem with this approach is that intermittent noise at the repeater input can cause the time-out timer to reset before the transmitter carrier drops. The result is that the transmitter can stay on indefinitely. This problem can be solved by resetting the time-out timer only after the transmitter carrier drops, but this method is often annoying to users.

Courtesy tones are frequently added to

the timer circuit. These short "beeps" notify the user that the timer has been reset and that it is okay to begin transmitting. Sometimes a tone with a slightly different pitch is used to inform listeners that the transmitting station has ceased transmitting. Although such devices have the beneficial effect of encouraging courtesy in a Pavlovian sort of way, an operator who neglects careful analysis of tone and timing when he installs them on his repeater runs the risk of having his repeater sound like a "Pong" game.

Identification

Another function required by FCC rules is that the repeater be identified every ten minutes during use. Some of the first "IDer" circuits used electromechanical devices to key Morse code into the repeater transmitter. These archaic mechanical monsters required frequent adjustment. Few repeater operators lamented their passing.

A simple solid-state identifier circuit is shown in Fig 27. An RS flip-flop (U1A and U1B) latches the input COR signal. The flip-flop gates a divide-by-ten counter (U2) which counts one-minute (1/60 Hz) input signals from a divider circuit like the one shown in Fig 26. When the selected time is reached by the counter, the output goes high, resetting the RS flip-flop and a binary counter (U4). When the binary counter is reset, its clock (U3A) is allowed to run. The binary counter clocks the address lines of a memory chip (U5), and the data programmed into the EPROM appears at pin 9. This data is used to gate the tone oscillator (U3B) to produce the Morse code ID. When U4 reaches binary 10000000, its Q8 output is high, and this stops the U3A clock until the counter is again reset by U2.

Only the least-significant bit of each data word in the EPROM is used in this application. The ID is programmed into the EPROM by burning a one for tone on and a zero for tone off. Three ones in a row makes a dash, and three zeros makes a character space. Six or seven zeros in a row can be used as a word space. Programming for "de AA2Z" is shown in Fig 28. The remaining seven bits in the data word are ignored.

The beauty of this approach is its flexibility. Other IDs or messages could be programmed into the EPROM, using the seven unused data bits. A simple logic circuit could then be used to select which bit of the data word is used to gate the tone oscillator. Other circuit modifications will probably be obvious to more advanced builders.

Ideally, an identifier circuit should incorporate means for the operator to vary the speed, pitch and amplitude of the Morse code being sent. FCC rules require that the code be sent at a speed no greater than 20 WPM and that it have sufficient deviation to be understandable when a voice is simultaneously being transmitted.

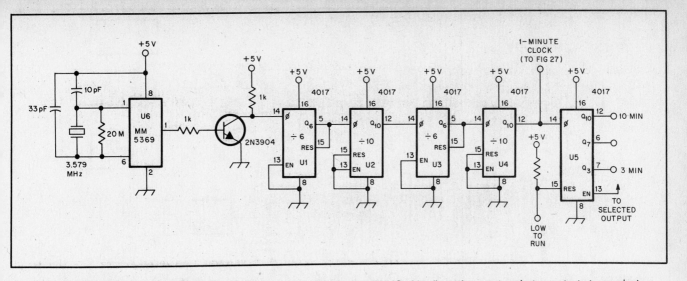

Fig 26—A simple digital timer circuit. Ten output lines are provided by the last 4017 IC; this allows the user to select an output at one-minute intervals from one to ten minutes. The three-minute output is useful for time-out timers, while the seven-, nine- or ten-minute outputs could be used to time an identifier circuit.

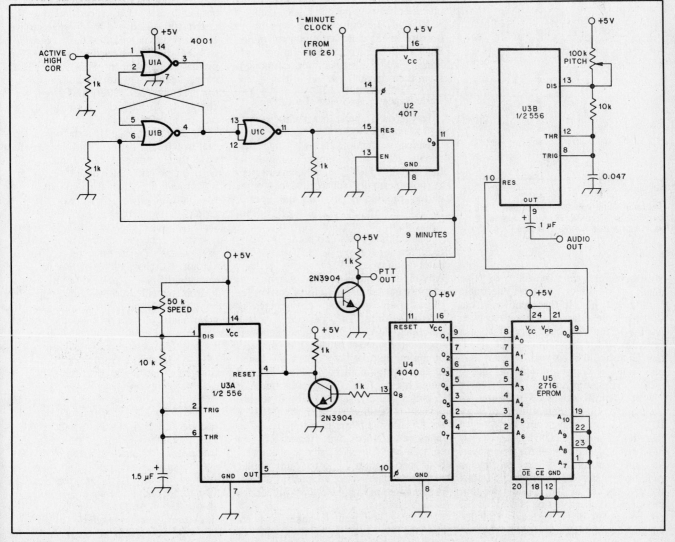

Fig 27—A simple Morse code identifier circuit. See text for details.

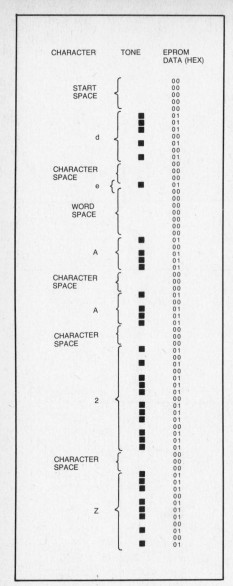

CHARACTER		TONE	EPROM DATA (HEX)
START SPACE	{		00 00 00 00 00
d		■ ■ ■	01 01 01 00 01 00 01
CHARACTER SPACE	{		00 00 01
e		■	00 01 00
WORD SPACE			00 00 00 00 00
A		■ ■ ■	01 00 01 01 00
CHARACTER SPACE	{		00 00 01
A		■ ■	00 01 01 00
CHARACTER SPACE	{		00 01
2	{	■ ■ ■ ■ ■	00 00 01 01 00 01 01 00 01 01 00 01 01
CHARACTER SPACE	{		00 00 01
Z	{	■ ■ ■ ■ ■	01 01 00 01 01 01 00 01 00 01

Fig 28—The message "de AA2Z" programmed into an EPROM for use in the identifier circuit of Fig 27. A one means "tone on" while a zero means "no tone."

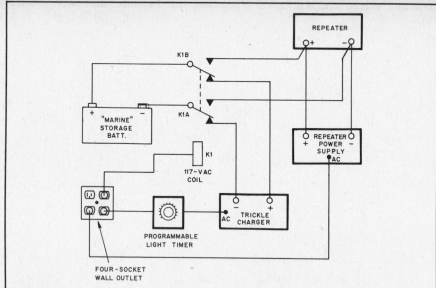

Fig 29—Simple circuit to isolate battery and trickle charger until commercial power fails. The line cord connected to the 117-V relay coil should be plugged into the same outlet as the repeater power supply. Values are not critical as long as the relay contacts will handle the current drawn by the repeater. The timer is any convenient programmable light timer.

Most operators prefer the ability to "fine tune" such things. Another desirable feature will inhibit the identifier from starting while a carrier is being received. Assuming the time-out timer of such a system is set for three minutes, the identifier timer should be set for no greater than seven minutes to ensure strict adherence to FCC identification rules.

Voice identification is also permitted. In the past, the most common form of voice ID used a tape player and prerecorded cassettes or cartridges. Obviously, circuitry must be employed to start the tape player, key the transmitter while the tape is running, and shut off the player and release the transmitter at the end of the tape message. A modern version of the voice ID relies on digital speech synthesis techniques, which is beyond the scope of this chapter.

In its most basic form, the voice repeater requires a COR, a time-out timer and an identifier circuit to operate legally. And this is all that is required for a repeater to operate in the automatic control mode. Such a repeater would be operating "bare bones"; many users want more functions from their "machines."

Basic Repeater Accessory Features

Few repeater operators are satisfied with a "bare bones" system. If the repeater is intended to function during emergencies, an emergency power supply is one of the first options added. Typically, an emergency power supply consists of a storage battery capable of powering the repeater for at least several hours during times that commercial power is interrupted.

One common mistake made by many repeater operators is to install an automobile-type lead-acid storage battery. Automobile batteries are designed to supply tremendous amounts of current for short durations—perhaps 200 amperes for 2 to 10 seconds. They perform this task quite well. A repeater emergency supply is apt to be required to provide a few amperes on an intermittent basis over a few hours or days. Automobile batteries are not well suited for this task. The deep discharge or "marine" battery—also a lead-acid storage device—is far better suited for this type service. Marine batteries can be found in many retail stores that cater to boating enthusiasts. The price, size and ampere-hour specifications are often quite similar to their automotive counterparts, but their performance in emergency back-up service is far superior.

Another common mistake is over-charging the battery to the point of destroying it. Particularly during hot

summer months, constant charging, even with a "trickle" charger, can boil away the electrolyte. A lead-acid battery may survive one or two such episodes, but its capacity will be diminished and it will require premature replacement. Some commercial repeater controllers provide diode switching for the emergency battery and a "trickle" charge. If the "trickle" is too great, the battery may be destroyed.

Fig 29 shows a simple, inexpensive device that isolates the battery from the repeater until the commercial power mains fail. Once the power is interrupted, the battery is automatically switched in and remains on line until the power is restored. Since the battery is isolated from the repeater, a separate trickle charger will be required. The trickle charger can be connected to a programmable light timer to control the amount of charge the battery receives each day. Some experimentation with a hydrometer will be needed to determine the minimum charging time each day to maintain a fully charged battery without harming it. Of course, after an extended period of use, the emergency battery will have to be recharged with a regular battery charger, because a trickle charger is not capable of restoring the charge quickly.

Another feature that many repeaters add is a Dual-Tone Multi-Frequency (DTMF) decoder. (DTMF decoders and encoders are often referred to by Western Electric's trademark, "Touch Tone.") A DTMF encoder produces two tones simultaneously—seven distinct tones are required for 12-number pads and eight are required for 16-number pads. DTMF devices are, for this reason, sometimes referred to as 2-of-7 or 2-of-8 equipment. Table 1 shows the standard tones produced for each of the 16

Table 1
Touch-Tone Audio Frequencies

Low Tone (Hz)	High Tone			
	1209 Hz	1336 Hz	1477 Hz	1633 Hz
697	1	2	3	F_o
770	4	5	6	F
852	7	8	9	I
941	*	0	#	P

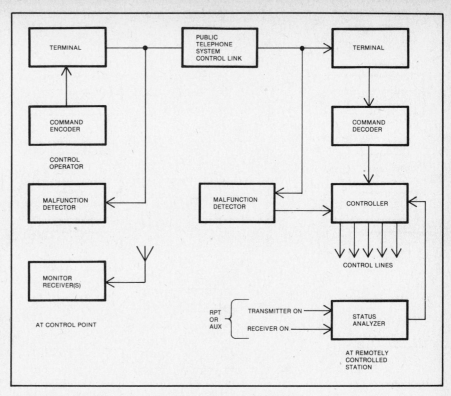

Fig 30—Block diagram of a public telephone system control link.

possible numbers and symbols.

When originally developed by Bell Labs, DTMF encoders and decoders were constructed using precision inductors and capacitors. For audio frequencies, such components are rather bulky, but they are capable of close tolerances and good stability over long periods of time.

In the last few years, advances in solid-state technology have produced single-IC encoders that derive the audio tones digitally from crystal controlled oscillators. These encoders are much smaller than their inductor-capacitor counterparts, which makes them ideal for use with miniaturized equipment, such as hand-held transceivers. Several companies manufacture complete pads that require only mounting and connecting three wires. As an alternative, the amateur can obtain the IC and custom design his own encoder. One minor drawback is that these encoders are slightly more susceptible to RF interference than the inductor-capacitor types.

Surplus telephone company DTMF decoders have never been widely available, probably because the telephone system requires far fewer of them than it does of the encoders. But they are available for those who wish to invest the time and effort involved in locating them. As with the encoders, these inductor-capacitor decoders are durable and provide reliable service.

When the 567 tone-decoder IC was introduced it received widespread attention in the Amateur Radio community and quickly became the nucleus of a number of DTMF decoders. As with many other popular ICs from that period, mass production and mass marketing has brought the consumer price down to under $1 per IC from many surplus dealers. The 567 was readily available, easy to work with and required no special components or techniques.

Most repeater operators soon learned that there is another side to decoders built with 567s. Resistors used to set the decoder operating frequency often change value as they age or as temperature or humidity change. The result is that the operator must frequently "fine tune" the decoder. These decoders have a tendency to "false"—indicate the presence of a signal that really is not there.

DTMF tone-decoder ICs have improved dramatically in the last few years. Where eight 567 ICs are required to decode all 16 digits, new devices require only one IC to accomplish this task. The Silicon System SSI204 14-pin single-chip DTMF decoder is typical of these new devices. A complete one-of-sixteen DTMF decoder can be built with the addition of only a few external components. One circuit using this device is shown in Chapter 34. Any repeater operator considering a DTMF decoder would be well advised to opt for one of the modern single-IC decoders over the 567 types.

Once the DTMF decoder is functioning, the repeater operator may install a number of options for his convenience in controlling the repeater and for that of the users. Remote sensors to provide information about the state and condition of various circuits can be included. Tape recorders may be connected to provide various forms of bulletins and general-interest information. The repeater may be linked to transmitters and receivers on other bands (crossband links). One thing that may not be done legally is to connect a receiver tuned to another radio service frequency—such as a NOAA weather station—to the repeater transmitter. The automatic retransmission of signals from other radio services by an Amateur Radio station is illegal. The one exception to this rule is that the FCC has granted a permanent waiver for amateurs to retransmit NASA space shuttle signals during shuttle missions.

Controlling the Repeater

As with any amateur station, a station in repeater operation must be under the control of a licensed amateur called the control operator. He and the station licensee are both responsible for the transmissions of the repeater. There are three kinds of control according to FCC thinking—local, remote and automatic.

Local control means that the control operator is on duty at a control point, which is the same location as the transmitter. The control point is defined in the rules as the operating position of an Amateur Radio station where the control operator function is performed.

Remote control means that the control operator is on duty at a control point located at a place other than the transmitter site. Remote control takes place over a control link, such as telephone line or radio link. A phone line (Fig 30) can be used in the normal manner, with the operator dialing an unlisted number that rings at the repeater site. Some device at the repeater site must answer the ring. He then sends a signal which either activates or deactivates the repeater functions. Another kind of control involves a dedicated line, which is constantly connected to the repeater. Again some signal must pass from the control operator enabling or disabling a function.

Remote control over a radio link (Fig 31) is possible by transmitting command signals to a special receiver interfaced with the repeater to accomplish the control function. (DTMF encoders and decoders are most often used for these signaling functions.) The repeater cannot be legally controlled on the repeater receive (input) frequency. A signal to control any Amateur Radio station—including repeater stations

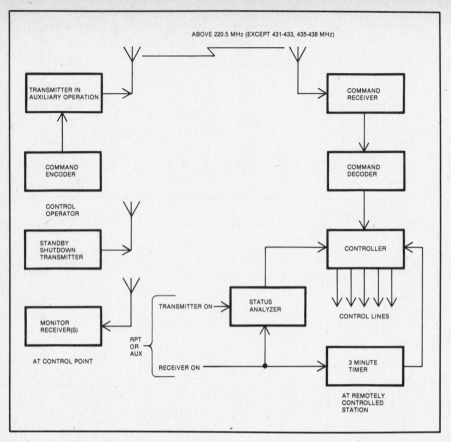

ABOVE 220.5 MHz (EXCEPT 431-433, 435-438 MHz)

TRANSMITTER IN AUXILIARY OPERATION

COMMAND RECEIVER

COMMAND ENCODER

COMMAND DECODER

CONTROL OPERATOR

STANDBY SHUTDOWN TRANSMITTER

CONTROLLER

MONITOR RECEIVER(S)

STATUS ANALYZER

TRANSMITTER ON

CONTROL LINES

AT CONTROL POINT

RPT OR AUX

RECEIVER ON

3 MINUTE TIMER

AT REMOTELY CONTROLLED STATION

Fig 31—Block diagram of an RF control link.

—must be transmitted on a frequency where auxiliary operation (radiocommunication for the purpose of remotely controlling another Amateur Radio station) is permitted: 220.5 MHz and above, except 431 to 433 and 435 to 438 MHz.

Automatic control means that the repeater is controlling itself, without a control operator being at a control point, local or remote. Under such operating conditions third-party traffic is prohibited, and control features must be built in to ensure that the repeater transmitter will be taken off the air if the repeater malfunctions. If such safeguards are built in, locally or remotely controlled repeaters may be placed in automatic control for short periods of time when it is not convenient to have a control operator on duty—such as the wee hours of the morning. Once the control operator is back on duty, the repeater is no longer considered under automatic control, and functions that permit third-party traffic are again legal.

Autopatch

Autopatch is a function that allows the repeater to be connected to the commercial telephone lines, thus permitting phone calls to be placed by repeater users. This is the most common form of third-party traffic found in repeater usage. Most repeaters use DTMF signaling to accomplish this function. Usually the user transmits an access code, which enables the patch. He then stops transmitting and waits to hear the dial tone. Once he hears the tone, he transmits again and dials the number using his DTMF encoder. Audio from the repeater receiver is passed on to the phone line, and the telephone switching office detects the tones originated in the user's encoder as it would tones coming from a standard telephone.

Noise from the radio link may interfere with the switching office's ability to decode the tones. Distortion in the radio link and the interface may force the tones outside the tolerances in the switching equipment. In such simple autopatch designs, we often find cases where A's encoder always "works," and B's doesn't. The solution is to provide circuitry at the repeater that decodes and stores the number being encoded by the user (Fig 32). The repeater then uses an onboard encoder to dial the number into the telephone system. In areas having only pulse dialing available, it will be necessary to use a tone-to-pulse converter.

Care should be taken not to design an autopatch that will encourage abuse. Calls should be limited to short durations. The patch should be disabled when a control operator is not on duty. Calls that might be construed as business in nature should be avoided. Devices can be added that automatically lock out long-distance dialing. Autodialer features can be added, that automatically dial prestored emergency numbers when the user dials a short command (for example, *0 to connect to the State Police Department).

Microprocessors As Controllers

Properly interfaced, microprocessors make marvelous repeater controllers, because the programming can be tailored to suit the needs and desires of the operator. Functions that might have required a vast array of logic ICs can be easily handled by the microprocessor (and a few support ICs). Success or failure depends largely on the software.

If the operator wants to change a repeater function, he has but to change the programming. Where discrete logic controllers encourage a set-it-and-forget-it attitude with respect to repeater control parameters, the microprocessor encourages the "fine tuning" instinct.

The programming must be stored in some kind of memory device. If the programming is stored in ROM or PROM, then changing the software is a relatively involved procedure. A new IC must be "burned" each time changes are to be made. If the programming is to be stored in RAM, it is much easier to make changes. In fact, it is quite possible to reprogram the controller remotely in such a system. The drawback here is that a nearby lightning strike or a power-line glitch may reprogram the controller, also.

Advanced Computer Controls, which manufactures the ACC RC-850 computer controller, has gotten around these problems by using three different forms of memory in their controller. The basic operating system of the microprocessor is stored in EPROMs. This programming is permanent; ACC does from time to time update and add features by exchanging new EPROMs for old. When power is removed from the controller, this basic operating programming is there to "reboot" the microprocessor when power is restored.

A second form of memory found in the RC-850 is RAM "scratchpad," which permits the operators to make minor changes to optimize the repeater functions for the particular application. Any or all of dozens of various parameters can be changed locally or remotely by means of DTMF commands. Any programming stored in this scratchpad will be lost if power is removed from the system. This might be annoying if it happened often.

ACC gets around this potential pitfall by adding a third type of memory, Electrically Erasable Programmable Read Only Memories (EEPROMs). Like PROMs and EPROMs, EEPROMs retain memory when removed from a power source. Identifier messages, "permanent" changes in operating parameters, and such are programmed into the EEPROMs by means of the onboard EEPROM programmer. Thus, if the repeater system loses power, the operator

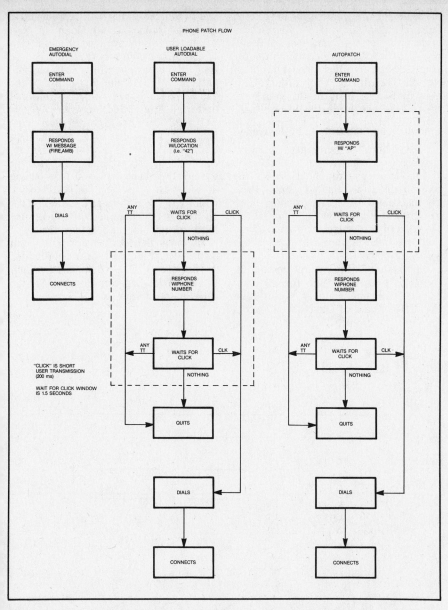

PHONE PATCH FLOW

Fig 32—Flow sequence for ACC's RC-850 autopatch. It is actually three separate patches in one. The emergency autodialer rapidly dials stored emergency numbers when accessed. Both the user loadable autodialer and the standard autopatch provide the user with a great deal of control. All tones are regenerated by the RC-850.

Table 2
EIA Standard Subaudible Tone Frequencies

Code†	Freq. (Hz)	Code†	Freq. (Hz)
XZ	67.0	4Z	136.5
WZ	69.3	4A	141.3
XA	71.9	4B	146.2
WA	74.4	5Z	151.4
XB	77.0	5A	156.7
WB	79.7	5B	162.2
YZ	82.5	6Z	167.9
YA	85.4	6A	173.8
YB	88.5	6B	179.9
ZZ	91.5	7Z	186.2
ZA	94.8	7A	192.8
ZB	97.4	8Z	206.5
1Z	100.0	9Z	229.1
1A	103.5	M1	203.5
1B	107.2	M2	210.7
2Z	110.9	M3	218.1
2A	114.8	M4	225.7
2B	118.8	M5	233.6
3Z	123.0	M6	241.8
3A	127.3	M7	250.3
3B	131.8	0Z	254.1

†Motorola designator

gained popularity as a means of minimizing *co-channel interference*, also. Nonetheless, many amateurs reject it without giving consideration to its merits.

A CTCSS tone is any one of the standard tones (Table 2) established by the EIA that range from 67.0 to 254.1 Hz. This group is generally referred to as the subaudible tone group—subaudible because the tones are set at a very low deviation (usually around 500 Hz in a 5-kHz system) and because the corresponding receiver generally has a high-pass filter installed in the audio chain to prevent the tone from reaching the speaker.

At the transmitter end, the tone is activated when the transmitter begins transmitting and stays on for the duration of the transmission. The receiver is equipped with a decoder that is set to recognize the presence of that particular tone. The presence or absence of the tone operates the squelch (COR) function instead of the standard squelch. In a typical commercial installation, the operators at the receiver and transmitter have switches that allow them to choose either CTCSS or standard squelch operation.

Some amateur groups have shown more creativity in the use of CTCSS in their repeater systems. For instance, one group has both squelch systems wired in parallel on their repeater receiver, but the standard squelch is set to require a signal strength 20 dB greater than that of the CTCSS to operate. Thus, no one is excluded from the repeater if he has a strong signal. Weak signals must have the right tone, though. Frequent users with low-powered equipment or those who operate from the fringes have installed CTCSS encoders. Transient mobile stations can access the repeater. Basically, the only signals excluded are those coming from great distances away during marginal band openings.

can rest assured that the repeater will function "normally" when it returns to the air. Such an elaborate system provides the repeater operator with tremendous versatility without the risk of frequent major reprogramming efforts. The drawbacks to such a system are primarily ones of cost and increased complexity.

Home computers offer repeater operators another avenue for developing flexible computerized controllers. The same problems with memory storage exist for these devices as for the custom-designed microprocessor controllers. One way around this is to add a disc drive to the system. Of course, the functioning of the controller will largely depend on the creativity and programming skill of the designer/operator.

CTCSS

Continuous Tone Coded Squelch System (CTCSS) is probably one of the most controversial features to be found in modern FM and repeater equipment. Often referred to by Motorola's trademark, "PL," CTCSS, like many other features and procedures found in Amateur Radio FM and repeater communications, was adapted to the Amateur Radio Service after being developed for the commercial services. Its earliest use in the Amateur Radio Service was to keep uninvited stations from using private closed repeaters. In the commercial services, it is primarily used to minimize co-channel interference. As the Amateur Radio bands have become more crowded, CTCSS has

In the early days, CTCSS encoders and decoders used special circuits that required an expensive "reed"—an electromechanical vibrator much like a tuning fork—for each individual tone. Besides being expensive, these devices are bulky by today's standards. Several manufacturers make miniature encoders and decoders that use digital techniques and IC technology to reduce the size to a point that they will fit in a small hand-held transceiver. The user programs the desired tone by setting switches or soldering jumpers. The basic encoder units cost approximately $30. These encoders can hardly be considered a financial burden for most hams.

Several other systems can be employed to minimize co-channel interference. In digital CTCSS, the sine-wave tone is replaced by a 21-bit digital character string. Tone-burst requires a short (approximately 200 millisecond) audible tone to be sent at the beginning of each transmission or each series of transmissions. A DTMF code can be required to set a timer that will automatically turn the repeater transmitter off at the end of a specified time. Hams often call this "Touch Tone Up." Like CTCSS, these systems serve to limit the on-channel signals that will turn on the transmitter.

Coordination, Standards and Cooperation

Early efforts to coordinate repeaters were confined to limited geographical areas around major metropolitan areas. The coordinating bodies that grew out of these efforts evolved differently from one place to another. No national group was involved, and as a result, little communication or standardization existed from one group to the next.

But the popularity of repeaters mushroomed, and the need for national standards became obvious. At this point the ARRL Board of Directors stepped in and began consulting with representative groups and individuals around the country. The band plans found in the current edition of *The ARRL Repeater Directory* are the result of this ongoing effort. Some regional variation will be found for all except the 220-MHz band plan, which is followed uniformly around the country.

The FCC has begun to recognize these band plans and the local coordinating bodies in administrative decisions. Although not specifically endorsed in Part 97, these band plans and coordinating decisions carry the weight of law. Anyone desiring to place a repeater station on the air must get prior authorization from the local coordinator and follow the band plans, or run the risk of incurring the wrath of the FCC if interference to coordinated activity occurs. Consult the latest edition of *The ARRL Repeater Directory* for the coordinator in your area.

Linear Translators and Transponders

A linear translator is similar to a repeater in many ways. Each is a combination of a receiver and a transmitter that is used to extend the range of mobile, portable and fixed stations. A typical FM voice repeater receives on a single frequency or channel and retransmits what it receives on another channel. By contrast, a typical translator receiver passband is wide enough to include many channels. Received signals are amplified, shifted to a new frequency range and retransmitted by the translator. Fig 33 shows block diagrams for a simple voice repeater and a simple translator.

By comparing block diagrams, you can see that the major hardware difference between a repeater and a translator is signal detection. In a repeater, the signal is reduced to baseband (audio) before it is retransmitted. In a translator, signals in the passband are moved to an IF for amplification and retransmission.

A Multi-Mode Machine

Operationally, the contrast is much greater. An FM voice repeater is a one signal, one mode input and one signal, one mode output device. A translator can receive several signals at once and convert (or translate) them to a new range. Further, a linear translator can be thought of as a multi-mode repeater. It is said of computers, "garbage in—garbage out." In a similar manner, whatever goes into a linear translator is what comes out. A linear translator can be used for AM, SSB, ACSB, FM, SSTV and CW—either single or mixed modes and in any combination.

Translator Applications

For a variety of reasons, linear trans-

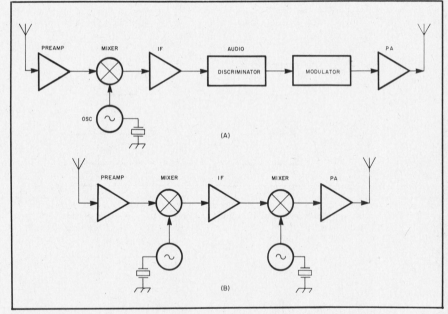

Fig 33—Block diagram of an FM voice repeater at A; of a linear translator at B.

lators have not been used very often in commercial applications. For that reason, there is not much surplus equipment and practical knowledge to be found. This is one area where the amateur community has taken the lead, especially in satellite communications.

Some work has been done with linear translators for terrestrial Amateur Radio communication, particularly on the West Coast of the USA. For most of us, however, the more familiar application is in the amateur communication satellites (OSCARs and RS).

Transponders

By convention, any linear translator that is installed in a satellite is called a transponder. Because of the universal availability of satellite communications to amateurs, the remainder of this chapter will refer to linear translators as transponders. For more information on communicating through satellites, see Chapter 23.

A transponder aboard a Phase II satellite may have 130-dB gain and an output power of 1 watt. In that regard, it is much like a communications receiver. The main difference is that the receiver has an audio output and the transponder an RF output.

All stations using a transponder must share the limited power available. For that reason, high-duty-cycle modes such as RTTY and SSTV are limited to specific calling frequencies.

Repeaters are often referred to by their operating frequencies. For example, a .34/.94 machine receives on 146.34 MHz and transmits on 146.94 MHz. A different convention applies to transponders: The input band is given, followed by the corresponding output band. For example, a 2-m/10-m transponder would have an input passband centered near 146 MHz and an output passband centered near 29 MHz. Mostly, transponders are identified by mode—not mode of transmission such as SSB or CW. Mode has an entirely different meaning in this case. The various frequency band combinations used for uplink and downlink are referred to as modes. See Table 3. For example, OSCAR 10 has Mode B and Mode L transponders.

A block diagram of a simple transponder is shown in Fig 34. For several reasons, transponders used aboard satellites are more complex than the one shown. Just as a receiver design provides band-pass filtering, image rejection, AGC and overall gain, a transponder design must meet similar considerations. For that reason, transponder designers use multiple-frequency conversions. A block diagram of the basic Mode-A transponder used on OSCAR 8 is shown in Fig 35.

Inverting and Noninverting Transponders

In any conversion stage, the LO may be above or below the incoming frequency. The LO frequencies in a single- or multi-conversion transponder may be chosen so that entering signals are inverted. That type design changes upper-sideband signals into lower-sideband (and vice versa), transposes relative mark-space placement in RTTY, and so on. When this scheme is used, the result is called an inverting transponder.

Table 3
Bands Used for Satellite Communications

Mode	Uplink	Downlink
A	2 m	10 m
B	70 cm	2 m
J	2 m	70 cm
K	15 m	10 m
L	23 cm	70 cm
S	70 cm	13 cm
T	15 cm	2 cm

The major benefit of an inverting transponder is that Doppler shifts on the uplink and downlink are in opposite directions and will, to a limited extent, cancel. With the 146/29-MHz Mode A link combination, Doppler is not serious; transponders using this frequency combination are typically noninverting. Transponders using higher frequency combinations are usually inverting.

Power and Bandwidth

The power output, bandwidth and operating frequencies of a transponder must be compatible. When a transponder is fully loaded with equal-strength signals, each signal should provide an adequate signal-to-noise ratio at the ground.

Low-altitude (800-1500 km) satellites perform well with from 1 to 4 watts of PEP

at frequencies between 29 and 435 MHz, using a 100-kHz-wide transponder. More sophisticated, high-altitude (35,000 km) provide acceptable performance with 35 watts PEP using a 500-kHz-wide transponder downlink at 146 or 435 MHz. A gain budget for Phase III, Mode B can be found in Chapter 23.

Dynamic Range

Dynamic-range requirements for a transponder may appear to be less severe than for an HF receiver. An HF receiver will handle signals differing in level by as much as 100 dB. A low-altitude satellite will encounter signals in its passband differing by perhaps 40 dB. That, however, is only part of the story. The HF receiver filters out all but the desired signal before introducing most of the gain. All transponder users must be accommodated simultaneously—quite a different situation. In one sense, all stations share a single channel. Transponder gain can, therefore, be limited by the strongest signal in the passband.

When a strong station is driving the satellite transponder to full power output, stations more than 22 dB lower than that level will not be heard on the downlink. That is true even if the weaker stations are perfectly readable when the strong signal is not present! User stations must cooperate by using only the minimum power necessary for communication through the satellites.

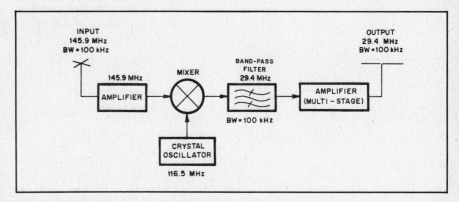

Fig 34—Block diagram of a simple 2-m/10-m linear transponder.

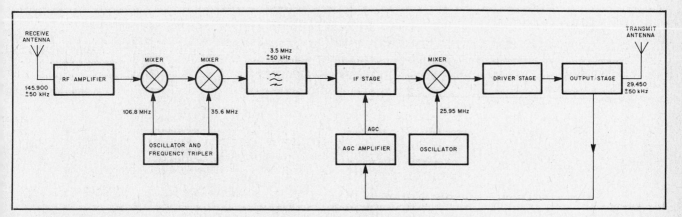

Fig 35—Block diagram of AMSAT-OSCAR 8 Mode A transponder.

Chapter 15

RF Power Amplifiers

RF power amplifiers present significant design challenges to the radio amateur. A power amplifier in an amateur station may be called upon to develop as much as 1500 watts of RF output power, the legal maximum in the United States. The voltages and currents needed to perform this feat are much higher than those found in other amateur equipment; the voltage and current levels are potentially lethal, in fact. Every component in an RF power amplifier must be carefully selected to endure these high electrical power levels without failing. Large quantities of heat are produced in the amplifier and have to be dissipated safely. Spurious signal generation must be minimized, not only for legal reasons, but also for sound neighbor relations. Every one of these challenges has to be overcome to produce a loud, clear signal from a safe and reliable amplifier.

TYPES OF POWER AMPLIFIERS

Power amplifiers are categorized by their power level, intended frequencies of operation, device type, class of operation and circuit configuration. Within each of these categories there usually is a wide range of circuit options. Choosing from almost all the possibilities to arrive at a suitable combination is the entire concept of design.

Solid State versus Vacuum Tubes

Today, most items of commercially manufactured amateur equipment use solid-state devices exclusively. A few manufacturers still offer HF transceivers with tubes in the final transmitter output stage, but their popularity is dwindling. Using transistors exclusively offers the manufacturers several advantages:

1) Compact design — Even with their heat sinks, transistors are smaller than tubes, allowing smaller packages.
2) Power-supply simplification — Tube amplifiers need a multitude of voltages to function. The filament requires a low voltage, high current supply; the plate circuit, several hundred volts. To meet these needs requires additional power supply components, beyond what the rest of the transceiver needs.
3) "No-tune-up" operation — By their nature transistors lend themselves to low impedance, broadband operation. Harmonic energy suppression can be performed by fixed-tuned, low-pass filters; these can be switched when changing bands. Tube amplifiers must be retuned on each band, and even for significant frequency movement within a band.
4) Long life — Nothing in a transistor wears out. Unless the device is defective, or damage is inflicted on it because its ratings are exceeded, transistors can last virtually forever. Vacuum tubes wear out as their filaments deteriorate from continued operation.
5) Manufacturing ease — Transistors are ideally suited for printed-circuit-board fabrication. The low voltages and low impedances encountered with transistor circuitry work very well on printed circuits (some circuits use the circuit board traces themselves as circuit elements); the high impedances found with vacuum tubes do not. The transistor's physical size and shape also works well with printed circuits; the device can be soldered right to the board, an impossible task with a vacuum tube. These advantages in fabrication mean reduced manufacturing costs.

Based on all this reasoning, it seems there would be no place for vacuum tubes in a transistorized world. Transistors do have significant limitations, however, especially in a practical sense. Individual transistors available today cannot develop more than approximately 150-W output. Pairs of transistors or even pairs of pairs are usually employed in practical designs, even at the 100-W level. Present-day transistors just cannot withstand the high currents needed for high power levels, so the current must be divided among several devices. Beyond the 300-W output level, somewhat exotic (at least for radio amateurs) techniques of power combination from multiple amplifiers must be used. Although this has been successfully done, it is an expensive proposition.[1] Even if one ignores the challenge of the RF portions of a high power transistor amplifier, there is the dc power supply to consider. A solid-state amplifier capable of delivering 1 kilowatt of RF output might require regulated 50 volts at more than 40 amperes. Developing that much current is a formidable and expensive task. These limitations considered, solid-state amplifiers are practical up to a couple of hundred watts output. Beyond that point, the vacuum tube still reigns.

CLASSES OF OPERATION

The class of operation of an amplifier stage is defined by its output current conditions. This, in turn, determines the amplifier's efficiency, linearity and operating impedances.

• Class A: The RF drive level and dc bias are set so there is device output current all the time. The conduction angle is 360° (see Fig. 1A). (The conduction angle of an amplifier is defined as the angular portion of a drive cycle over which there is output current. Conduction angle is measured in degrees; a full sine-wave cycle is 360°.) The active device in a Class A amplifier acts as a variable resistor. Output voltage is generated by the varying current flowing through the load resistance. Maximum linearity is achieved in a Class A amplifier, but the efficiency of the stage is low. In theory, maximum efficiency is 50%, but in practice it usually runs about 25 to 30%.

• Class AB: Drive level and dc bias are adjusted so output current flows appreciably more than half the drive cycle, but less than the whole drive cycle. The

[1]Granberg, "One KW — Solid-State Style," *QST*, April 1976, p. 11; May 1976, p. 28.

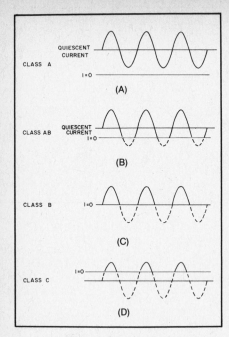

Fig. 1 — Amplifying device output current for various classes of operation. All assume a sinusoidal drive signal.

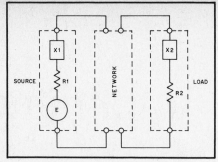

Fig. 2 — The fundamental circuit for a generator and a load includes the internal power source resistance (R_1), the load resistance (R_2), and associated reactances (X_1 and X_2). A network is used to "match" the impedances.

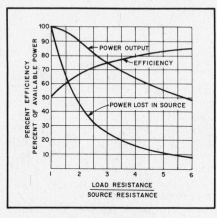

Fig. 3 — Efficiency is dependent on the ratio of the load resistance to the source resistance.

operating angle is much greater than 180° but less than 360° (see Fig. 1B). Efficiency is much better than Class A, averaging over 50%. Class AB linearity is not as good as that found in Class A, but is very acceptable for even the most rigorous SSB applications.

• Class B: Bias and RF drive are at a level where the device is just cut off (see Fig. 1C) and output current flows during one half of the drive cycle (conduction angle = 180°). Improved efficiency is attained, up to 65%, and acceptable linearity is still maintained.

• Class C: The dc bias is adjusted so that the device is cut off. Output current flows only during peaks in the drive cycle (see Fig. 1D). The output current consists of pulses at the drive frequency; the conduction angle is considerably less than 180°. Since the output device is cut off for more than half the operating cycle, efficiency is relatively high, up to 80%. Linearity is extremely poor, though.

There are other amplifier classes not usually found in amateur service. Most amplifiers that use Class D through H operate at audio frequencies. Their prime virtue is high efficiency. Manufacturers of audio equipment use these amplifier classes to reduce power supply requirements and the associated high costs in high-power stereo amplifiers. Sophisticated techniques would be required to implement these operating classes in radio-frequency amplifiers, particularly at high power levels. The additional complexity and cost is difficult to justify for amateur RF service.

The class of operation is independent of device type and circuit configuration. The active amplifying device and the circuit itself must be uniquely applied for each operating class, but amplifier linearity and efficiency are limited by the class of operation. Clever amplifier design cannot improve on these fundamental limits. Poor design and implementation, though, can certainly prevent an amplifier from approaching its potential in efficiency and linearity.

IMPEDANCE MATCHING

All power sources, such as the tube or transistor in a power amplifier, have an internal impedance. The impedance usually consists of a resistance in combination with some reactance. Current flowing through this internal resistance in the process of generating power is converted to heat. The electrical power consumed by the internal resistance is unavailable to the load, and only goes to heating up the power source. It can be proven algebraically that the maximum power available from a power source is transferred to the load resistance when the load resistance is equal to the source resistance. If the source has reactance associated with it, then the load impedance must be equal to the complex conjugate of the source impedance.

Usually, neither the source impedance nor the load impedance can be changed. In order to achieve maximum power transfer, a matching network is used between the source and the load to transform the load impedance to the source impedance (and vice versa). This is illustrated in Fig. 2.

The total power generated by the source is

$$P_{IN} = P_{OUT} + P_D$$

where
P_{IN} = the source power input
P_{OUT} = the power delivered to the load
P_D = the power dissipated in the source resistance.

The efficiency (the ratio of output power to total power generated) is

$$\text{Eff.} = \frac{P_{OUT}}{P_{IN}} \text{ or } \frac{P_{OUT}}{P_{OUT} + P_D}$$

Efficiency is usually expressed as a percentage.

When the source impedance is matched to the load impedance, each impedance consumes the same amount of power (simple Ohm's Law: $P = I^2R$, the current being the same through both series impedances). It then follows that half the power generated is transferred to the load, and half is dissipated by the source. Efficiency is 50% when the source resistance is equal to the load resistance.

One of the main objectives in power amplifier design is to transfer as much fundamental frequency power to the load as possible, without exceeding the amplifier's active device ratings. In other words, maximum efficiency is needed. As illustrated in Fig. 3, efficiency improves as the load resistance to source resistance ratio increases. The usual practice in RF power amplifier design is to select an optimum load resistance that will provide the highest power output, while staying within the amplifying device's power rating.

The optimum load resistance is determined by the amplifying device's current transfer characteristics, and in the case of vacuum tubes, by the amplifier's class of operation. For a transistor amplifier, the optimum load resistance is

$$R_L = \frac{V_{cc}^2}{2P_O}$$

Where R_L (the load resistance) is equal to the dc collector voltage (V_{cc}) squared, divided by two times P_O (the amplifier power output in watts). Vacuum tubes have more complex current transfer characteristics and each class of operation produces different RMS values of RF current through the load impedance. The optimum load resistance for vacuum tube amplifiers can be approximated by the ratio of the dc plate voltage to the dc plate current, divided by a constant, appropriate to each class of operation (and this in turn determines the maximum efficiency each class of operation can provide). The optimum tube load resistance is

$$R_L = \frac{V_P}{KI_P}$$

where
    R_L = the appropriate load resistance
    V_P = the dc plate potential in volts
    I_P = the dc plate current in amperes
    K = a constant that approximates the RMS current to dc current ratio appropriate for each class. For the different classes of operation
        Class A, $K \approx 1.3$
        Class AB, $K \approx 1.5$
        Class B, $K \approx 1.57$
        Class C, $K \approx 2$

The load resistance for an RF power amplifier usually is a transmission line connected to an antenna, or the input of another amplifier. It isn't practical, or usually even possible, to modify either of these load resistances to the optimum value needed for high-efficiency operation. A matching network is normally used to transform the true load resistance to the optimum load resistance for the amplifying device. The matching network is also used to cancel any reactance in the two impedances (source and load). Two types of matching networks are found in RF power amplifiers: tank circuits and transformers.

TANK CIRCUITS

Parallel resonant circuits and their equivalents have the ability to store energy. Capacitors store electrical energy in the electric field between their plates; inductors store energy in the magnetic field induced by the coil winding. These circuits are referred to as tank circuits, since they act as storage "tanks" for RF energy.

The energy stored in the individual tank circuit components varies with time. At some point the capacitor in a tank circuit is fully charged, and the current through the capacitor is zero. All the energy in the tank is stored in the capacitor's electric field. Since the inductor is connected directly across the capacitor, the capacitor then starts to discharge through the inductor (since the inductor has no energy stored in it, there is no magnetic field and the voltage potential across the inductor is zero).

The current flowing into the inductor creates a magnetic field; energy now is being stored in the inductor's magnetic field. Assuming there is no resistance in the tank circuit, the increase in energy stored in the inductor's magnetic field is balanced by the decrease in the energy stored in the capacitor's electric field. The total energy stored in the tank circuit stays constant; some is stored in the inductor, some in the capacitor. As the energy level rises in the inductor, the voltage potential across the

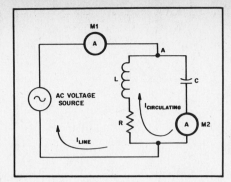

Fig. 4 — There are two currents in a tank circuit: the line current (I_{LINE}) and the circulating current ($I_{CIRCULATING}$). The circulating current is dependent on tank Q.

inductor will climb from its original zero voltage condition. According to Lenz's law, the voltage produced is in opposition to the increasing current flowing into the inductor. Eventually, the induced voltage is large enough to totally oppose the current created by the transfer of energy from the capacitor to the inductor. The capacitor's electric field is depleted at this point, and all the tank circuit's energy is stored in the magnetic field of the inductor.

The voltage potential across the inductor, which is in parallel with the capacitor, then starts to recharge the capacitor (the polarity is now reversed from the original condition). Eventually, the capacitor charges fully, and all the energy is then stored in the capacitor's electric field instead of in the inductor's magnetic field. Again, the capacitor discharges into the inductor, where a voltage in opposition to the current develops across the inductor. Once the potential is high enough, the capacitor again recharges. The flow of energy into the inductor from the capacitor and back again is an alternating current. It can be mathematically shown that this alternating current is sinusoidal, and has a frequency of

$$F = \frac{1}{2\pi\sqrt{LC}}$$

which of course is the resonant frequency of the tank circuit. In the absence of a load or any losses to dissipate tank energy, the tank circuit current would oscillate forever.

A typical tank circuit is shown in Fig. 4. The values for L and C are chosen so that the reactance at L (X_L) is equal to the reactance of C (X_C) at the frequency of the signal generated by the ac voltage source. Since X_L is equal to X_C, the line current, I_{LINE}, as measured by meter M1 is close to zero; only the leakage through C is measurable. However, the current, $I_{circulating}$, in the loop defined by L, R and C is definitely not zero. Examine what would happen if the circuit were suddenly broken at points A and B. The circuit is now comprised of L, C, R and M2, all in

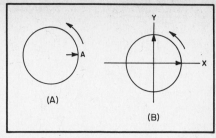

Fig. 5 — The rotation of a point on a flywheel defines a sine wave.

series. X_L is equal to X_C, so they are series resonant. The current in the circuit (if some voltage were applied between points A and B), the circulating current, is only limited by the resistance R. If R is equal to zero, the circulating current would be infinite!

The Flywheel Effect

A tank circuit can be likened to a flywheel — a mechanical device for storing energy. The energy in a flywheel is stored in the angular momentum of the wheel. Most flywheels are very heavy, and concentrate their mass as close to their outer rim as possible to maximize angular momentum. A perfect, unloaded flywheel is unaffected by friction or air resistance. It would spin forever once given an initial shot of energy to start it turning. No energy is expended heating up bearings or moving air; all the energy is stored in the flywheel's rotation.

But, as soon as a load of some sort is attached, the wheel starts to slow or even stop. Some of the energy stored in the spinning flywheel is now being transferred to the load. In order to keep the flywheel turning at a constant speed, the energy drained by the load has to be replenished. Energy has to be added to the flywheel from some external source. If sufficient energy is added to the flywheel, it maintains its constant rotation.

In the real world, of course, flywheels suffer from the same fate as tank circuits; there are system losses that dissipate some of the stored energy without performing any useful work. Air resistance and friction in the bearings slow the flywheel. In a tank circuit, resistive losses drain energy.

The comparison between a flywheel and a tank circuit is more accurate than first appearances might suggest. Fig. 5A shows the side view of a flywheel. This particular flywheel has no load attached, and happens to be in an environment where there is no friction or air resistance (perhaps in outer space). It is already spinning. Since there are no losses and no load draining energy from it, the wheel rotates at a constant speed and probably has been doing so for a long time.

At some particular time, point A is exactly at the three o'clock position. The time it takes for the wheel to make one complete revolution so that point A is again at ex-

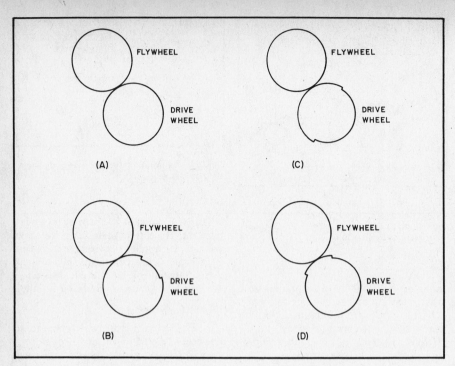

Fig. 6 — Equivalent amounts of energy can be coupled into a flywheel over varying operating angles.

actly three o'clock is called the period of rotation. The frequency of rotation is the reciprocal of this; instead of the number of time units it takes for the flywheel to rotate once, the motion is expressed as the number of rotation units the wheel makes in one time unit. In order to measure this properly, some reference points are required. A set of axes, passing through the shaft of the flywheel, is convenient. One is labeled X, one is Y. When point A is exactly at three o'clock, it lies exactly on the X axis. At that instant, point A has a Y coordinate of zero, because it is at the same point along the Y axis as the origin of the flywheel shaft. One quarter revolution later, point A now is exactly on the Y axis. The flywheel happens to have a radius of 1 unit, so the Y coordinate of A at one quarter revolution (90° of rotation) is 1.

One quarter turn later, point A is again on the X axis, but opposite from where it started. It has rotated one half turn or 180°. The Y coordinate is again zero. Another quarter turn of rotation later, point A has again rotated 90°, for a total of 270°. It's again on the Y axis, but directly opposite its position one half turn ago. The Y coordinate is now −1, to indicate that point A's position is below the flywheel shaft rather than above it. Another one-quarter rotation, for a total of 360° since first measured, and point A is back to its position along the X axis. The Y coordinate is again zero, since point A is again vertically level with the flywheel shaft. All further rotations of the wheel follow precisely the same pattern.

By careful measurement, the Y coordinate of point A could be plotted for the entire rotation of the flywheel, rather than just every quarter turn. The plot would be of a sine wave, just like the circulating current in a tank circuit. This is no mere coincidence; the circulating current in the tank circuit is exactly analogous to the rotation of the flywheel. The circulating current in the tank circuit is the result of the rotation of an electromagnetic field vector as energy is transferred back and forth between the tank circuit's inductor and capacitor.

Fig. 6 illustrates a variety of techniques that can be used to replace the energy transferred to a load from a flywheel. A drive wheel, powered by an electric motor of some type, is coupled to the flywheel. The coupling can be through gear teeth, friction drive or any number of other ways. In this case, the flywheel and drive wheel are coupled together by a 100% efficient friction drive. The drive wheel is shown in Fig. 6A. With no load attached to the flywheel, the energy in the (perfect) flywheel remains constant, demanding no additional energy from the drive wheel and the drive motor. The drive motor uses no electrical current.

As soon as a load is attached, energy flows from the flywheel into the load. In order to maintain the flywheel's energy and thus its rotation, this energy must be replaced. The only source of energy available to the flywheel is the drive wheel, so energy flows from the drive wheel into the flywheel. The drive wheel derives its energy from its drive motor, which in turn takes externally generated electricity and converts it to mechanical energy. A loss in available energy anywhere along this chain causes a similar loss for the remainder of the

linkage. In scientific terms, energy must be conserved.

Conversely, if additional energy becomes available anywhere along the chain, perhaps caused by an increase in electrical power at the drive motor, the additional energy is transferred down the line to the load. This implies that controlling the energy supplied to the drive motor affects the energy levels throughout the linkage, and the energy ultimately supplied to the load. The energy available to the load can then be varied, or modulated, by varying the energy supplied to the drive motor.

Another drive system that requires less energy from the drive motor is shown in Fig. 6B. Part of the drive wheel has been cut away so that the drive wheel engages the flywheel for only a part of the drive wheel's rotation. When the drive wheel contacts the flywheel, it supplies energy to the flywheel. The drive motor draws electrical current since it has to provide the energy for the drive wheel. During the portion of the rotation when the cutout on the drive wheel faces the flywheel (that portion of the rotation when the drive wheel would still contact the flywheel if there were no cutout), the drive motor can coast. The motor draws no current. No energy is being transferred from the drive wheel to the flywheel, and the drive motor coasts until the cutout is passed. When the wheels again engage, the drive wheel transfers energy to the flywheel and the drive motor again draws current.

Since the drive motor draws current only during a portion of the flywheel rotation (because of the cutout), the average current consumption per flywheel rotation is reduced. System efficiency improves, but with one drawback — the drive wheel has no control over the flywheel while the cutout faces the flywheel. However, since the uncontrolled portion of the flywheel rotation is short compared to the overall flywheel rotational period, any corrections can be made during the much longer time frame when the two wheels are in contact.

Fig. 6C represents a drive wheel-flywheel system in which the drive wheel contacts and drives the flywheel for only half of the flywheel rotation. During the other half, the drive wheel coasts and drive motor current goes to zero. The low drive-motor duty cycle makes this scheme the most efficient so far. The bad part is that control of the flywheel is lost for half the rotation period.

A limiting case of this efficiency concept is shown in Fig. 6D. Here the drive wheel only "kicks" the flywheel for a short portion of the flywheel rotation period. The duty cycle is low, but there is almost no control of the flywheel except during the short contact period.

The flywheel drive systems just described are analogous to the different classes under which an amplifier can be run. Fig. 6A is just like Class A operation. During the entire flywheel rotation period, the drive wheel provides energy to the flywheel, just as the output device in a Class A amplifier transfers current into its tank circuit. The efficiency is not very good, but the control over the flywheel, or tank circuit circulating current, is as tight as it can be. For an amplifier, this translates to faithful reproduction of the drive signal with low distortion. Classes AB, B, and C are represented by Figs. 6B, 6C and 6D, respectively.

Progressively with each class, the drive wheel actually drives the flywheel for shorter segments of the flywheel rotation period. In Fig. 6C the contact is for one half the rotation; the conduction angle is 180°. So it is also in an amplifier. In a Class B amplifier, device output current flows for half the drive cycle or 180°. What is important in both the flywheel system and the RF power amplifier is that the drive system supplies enough energy to keep the flywheel turning or the tank current circulating. It is also important to note that as the conducting angle decreases, control of the flywheel or tank circuit is given up. The amount of energy transferred from the flywheel or tank circuit into the load is decreasingly controlled by the drive system. Reproduction of changes in the drive energy level at the load (also referred to as modulation), deteriorates as the conduction angle decreases. In an RF power amplifier this is called distortion.

Tank Circuit Q

In order to quantify the ability of a tank circuit or even a flywheel to store energy, a quality factor, Q, is defined. Q is the ratio of energy stored in a system to energy lost.

$$Q = 2\pi \frac{W_S}{W_L}$$

where

W_S = is the energy stored by the flywheel or tank circuit
W_L = the energy lost to heat and the load.

By algebraic substitution and appropriate integration, the Q for a tank circuit can be expressed as

$$Q = \frac{X}{R}$$

where

X = the reactance of either the inductor or the capacitor
R = the series resistance.

Since both circulating current and Q are proportional to $1/R$, circulating current is therefore proportional to Q. The tank circulating current is equal to the line current multiplied by Q. If the line current is 100 mA, and the tank Q is 10, then the circulating current through the tank is 1 ampere. (This also implies, according to Ohm's Law, that the voltage potentials across the components in a tank circuit are also proportional to Q.)

When there is no load connected to the tank, the only resistances contributing to

R are the losses in the tank circuit. The unloaded Q_U in that case is

$$Q = \frac{X}{R_{Loss}}$$

where

X = the reactance of either the inductor or capacitor
R_{Loss} = the loss resistance in the circuit.

To the tank circuit, a load acts exactly the same as circuit losses. Both consume energy. It just happens that energy consumed by the circuit losses becomes heat. The only difference is the use the energy is put to. When energy is coupled out of the tank circuit into a load resistor, the unloaded tank circuit Q (Q_L) is

$$Q_L = \frac{X}{R_{Loss} + R_{Load}}$$

where R_{Load} is the load resistance. Energy dissipated in R_{Loss} is wasted as heat; ideally, all the tank circuit energy should be dissipated in R_{Load}. This implies that R_{Loss} should be as small as possible; the unloaded Q will rise accordingly.

Tank Circuit Efficiency

The efficiency of a tank circuit is the ratio of power dissipated in the load resistance, R_{Load} (the useful power) to the total power dissipated in the tank circuit (by R_{Load} and R_{Loss}). Within the tank circuit, R_{Load} and R_{Loss} are in series, and the circulating current flows through both. The power dissipated by each is therefore proportional to their resistance. The loaded tank efficiency can therefore be defined as

$$\text{Tank efficiency} = \frac{R_{Load}}{R_{Load} + R_{Loss}} \times 100$$

where efficiency is measured as a percentage. By algebraic substitution, the loaded tank efficiency can also be expressed as

$$\text{Tank Efficiency} = \left(1 - \frac{Q_L}{Q_U}\right) \times 100$$

where

Q_L = the tank circuit loaded Q
Q_U = the unloaded Q of the tank circuit.

It follows then that tank efficiency can be maximized by keeping Q_L low, which keeps the circulating current low and the I^2R losses down. The unloaded Q should be maximized for best efficiency; this means keeping the circuit losses low.

The selectivity provided by a tank circuit helps suppress harmonic currents generated by the amplifier. The amount of harmonic suppression is dependent upon circuit Q, so a conflict exists for the amplifier designer. A low Q_L is desirable for best tank efficiency, but low Q_L yields poor harmonic suppression. High Q_L keeps amplifier harmonic levels low, but tank ef-

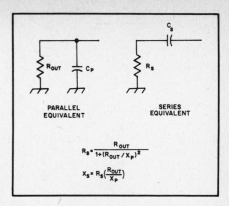

Fig. 7 — Parallel and series equivalent circuits and the formulas used for conversion.

ficiency suffers. At HF, Q_L can usually be chosen as a compromise value where tank efficiency remains high and harmonic suppression is also reasonable. At the higher frequencies the tank Q_L is not as controllable. However, unloaded Q can always be optimized, regardless of frequency, by keeping circuit losses low.

Matching Circuits

Tank circuit type matching networks need not take the form of a capacitor directly connected in parallel with an inductor. A number of equivalent circuits can be used to match the impedances normally encountered in a power amplifier. Most are operationally more flexible than a parallel resonant tank. Each has it virtues and failings, but the final choice usually is based on the actual component values needed to implement a particular network. Different networks may require unreasonably high or low inductance or capacitance values. In that case, a change to another type of network or to a different value of Q_L may be required. Several different networks may have to be investigated before the final design is reached.

Normally, the impedances of active amplifying devices are given as a resistance in parallel with the reactance. The equivalence between parallel impedance and series impedance is the basis for all matching network design. The equivalent circuits and the equations for conversion are given in Fig. 7. In order to use most design equations for computing matching networks, the parallel impedance must first be converted to its equivalent series form.

The Q_L of a parallel impedance can be converted from the series form as well. Substitution of the usual formula for calculating Q into the equations from Fig. 7 gives

$$Q_L = \frac{R_P}{X_P}$$

where

R_P = the parallel equivalent resistance
X_P = the parallel equivalent reactance.

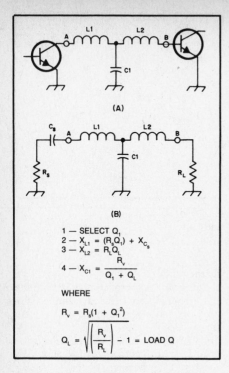

Fig. 8 — Network-solution equations and circuit for a T network. Actual circuit (A) and series equivalent (B).

1 — SELECT Q_1

2 — $X_{L1} = (R_sQ_1) + X_{C_s}$

3 — $X_{L2} = R_LQ_L$

4 — $X_{C1} = \dfrac{R_v}{Q_1 + Q_L}$

WHERE

$R_v = R_s(1 + Q_1^2)$

$Q_L = \sqrt{\left(\dfrac{R_v}{R_L}\right) - 1} = $ LOAD Q

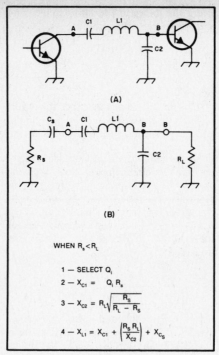

Fig. 10 — Circuit and equations for a series C, L network. Actual circuit (A) and series equivalent (B).

WHEN $R_s < R_L$

1 — SELECT Q_1

2 — $X_{C1} = Q_1 R_s$

3 — $X_{C2} = R_L\sqrt{\dfrac{R_s}{R_L - R_s}}$

4 — $X_{L1} = X_{C1} + \left(\dfrac{R_s R_L}{X_{C2}}\right) + X_{C_s}$

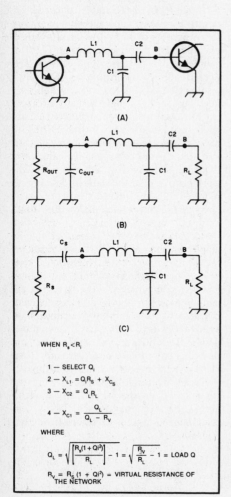

Fig. 9 — Circuit and mathematical solution for a shunt C, L network. Actual circuit (A), parallel equivalent (B) and series equivalent (C).

WHEN $R_s < R_i$

1 — SELECT Q_i

2 — $X_{L1} = Q_iR_s + X_{C_s}$

3 — $X_{C2} = Q_LR_L$

4 — $X_{C1} = \dfrac{Q_L}{Q_L - R_v}$

WHERE

$Q_L = \sqrt{\left[\dfrac{R_s(1 + Qi^2)}{R_L}\right] - 1} = \sqrt{\dfrac{R_v}{R_L} - 1} = $ LOAD Q

$R_v = R_s (1 + Qi^2) = $ VIRTUAL RESISTANCE OF THE NETWORK

A low-pass T network is shown in Fig. 8. It is capable of matching a wide range of impedances and practical component values and is often used in solid state work. The harmonic suppression afforded by the T network varies with the transformation ratio and the Q_L of the network. The networks shown in Figs. 9 and 10 also provide good harmonic suppression. These three networks are covered in detail in Motorola Application Note AN-267. Computer generated tables of solutions for all three networks are presented in that Application Note. The *Motorola RF Data Manual* contains a copy of AN-267 as well as a number of other very useful technical resources. The harmonic suppression of each network can be calculated as described earlier in this Handbook, once the component values are chosen.

The pi network may also be used in amplifier matching. Harmonic suppression in a pi network is a function of the impedance transformation ratio and the Q_L of the circuit. Second harmonic attenuation is approximately 35 dB for a load impedance of 2000 Ω in a pi network with a Q_L of 10. The third harmonic is typically 10 dB lower and the fourth harmonic approximately 7 below that. Again, the techniques described earlier in the book can be used to calculate exact theoretical values for different impedances and Q_L values. A typical pi network as used in the output circuit of a tube amplifier is shown in Fig. 11. The component values for a pi network with a Q_L of 12 are stated in Table 1. Other values of Q_L can be used by applying the formulas

$$\frac{Q_N}{12} = \frac{C_N}{C_{12}} \quad \text{and} \quad \frac{Q_N}{12} = \frac{L_{12}}{L_N}$$

where

Q_N = the new Q_L value for the network
C_{12} and L_{12} = the different component values taken from Table 1
C_N and L_N = the component values for a pi network with a Q_L of Q_N.

The pi-L network is a combination of a pi network followed by an L network. The pi network transforms the load resistance to an intermediate impedance level called the image impedance. Typically, the image impedance is chosen to be between 300 and 700 Ω. The L section then transforms from the image impedance down to 50 Ω. The output capacitor of the pi network is combined with the input capacitor for the L network, as shown in Fig. 12. The pi-L configuration attenuates harmonics better than a pi network. Second harmonic levels for a pi-L network with a Q_L of 10 is approximately 52 dB down from the fundamental. The third harmonic is attenuated 65 dB and the fourth harmonic approximately 75 dB. Component values for a pi-L network with a Q_L of 12 are given in Table 2. Other component values can be calculated for different values of Q_L by using the formulas described previously for the pi network.

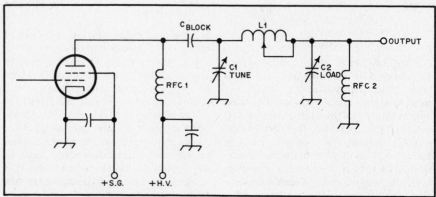

Fig. 11 — A pi matching network used at the output of a tetrode power amplifier. RFC2 is used for protective purposes in the event C_{block} fails.

The values for L and C in Tables 1 and 2 are based on purely resistive impedances. Any reactance modifies these values. Stray circuit reactances as well as tube or capacitor reactances all should be included as part of the matching network. Very often these reactances will render most of the matching circuits impractical. Either unacceptable loaded Q values or unrealistic component values are required. A compromise must be made once all the matching network alternatives have been investigated and found unacceptable.

Above 30 MHz, transistor and tube reactances completely dominate circuit impedances. At the lower impedances found in transistor circuits, the standard networks can be applied as long as suitable components are used. Capacitors rarely exhibit their marked values above 50 MHz because of internal reactances in addition to lead inductance, and must be compensated for. Tuned circuits frequently take the form of strip lines or other transmission lines in order to obtain "pure" inductances. The choice of components is often more significant than the type of network used.

The high impedances encountered in VHF tube amplifier plate circuits are not easily matched with the typical networks. The tube output capacitance is usually of such a high value that most matching networks are unsuitable. The usual practice is to resonate the tube output capacitance with a low-loss inductance connected in parallel. The result is a very high Q tank circuit. Component losses must be kept to an absolute minimum in order to achieve reasonable tank efficiency. Output impedance transformation is usually performed by a link inductively coupled to the tank circuit or by a parallel transformation of the output resistance through a series capacitor.

Transformers

Broadband transformers are often used in matching to the input impedance or optimum load impedance in a power amplifier. Multioctave power amplifier performance can be achieved by appropriate application of these transformers. The input and output transformers are two of the most critical components in a broadband amplifier. Amplifier efficiency, gain flatness, input SWR and even linearity are all affected by the transformer design and use. There are two transformer types, as described earlier in this Handbook; the conventional transformer and the transmission-line transformer.

The conventional transformer is wound much the same way as a power transformer. The primary and secondary windings are wound around a high-permeability core, usually made from a ferrite or powdered-iron material. Coupling between the secondary and primary is made as tight as possible to minimize leakage inductance. At low frequencies, the coupling between windings is predominantly

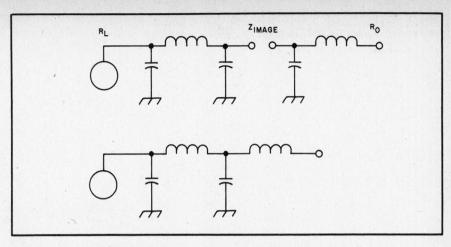

Fig. 12 — The pi-L network uses a pi network to transform the input impedance (R_L) to the image impedance (Z_{IMAGE}). An L network transforms Z_{IMAGE} to R_o.

Table 1
Pi-Network Values for Various Plate Impedances with a Loaded Q of 12

	MHz	1500(12)	2000(12)	2500(12)	3000(12)	3500(12)	4000(12)	5000(13)	6000(14)	8000(16)
C1	1.8	708	531	424	354	303	264	229	206	177
	3.5	364	273	218	182	156	136	118	106	91
	7	182	136	109	91	78	68	59	53	46
	14	91	68	55	46	39	34	30	27	23
	21	61	46	36	30	26	23	20	18	15
	28	46	34	27	23	20	17	15	13	11
C2	1.8	3413	2829	2415	2092	1828	1600	1489	1431	1392
	3.5	1755	1455	1242	1076	940	823	766	736	716
	7	877	728	621	538	470	411	383	368	358
	14	439	364	310	269	235	206	192	184	179
	21	293	243	207	179	157	137	128	123	119
	28	279	182	155	135	117	103	96	92	90
L1	1.8	12.81	16.6	20.46	24.21	27.90	31.50	36.09	39.96	46.30
	3.5	6.59	8.57	10.52	12.45	14.35	16.23	18.56	20.55	23.81
	7	3.29	4.29	5.26	6.22	7.18	8.12	9.28	10.26	11.90
	14	1.64	2.14	2.63	3.11	3.59	4.06	4.64	5.14	5.95
	21	1.10	1.43	1.75	2.07	2.39	2.71	3.09	3.43	3.97
	28	0.82	1.07	1.32	1.56	1.79	2.03	2.32	2.57	2.98

These component values are for use with the circuit of Fig. 11 and were provided by W6FFC and K1JX.

magnetic. As the frequency rises, the permeability in the core decreases and the leakage inductance increases; transformer losses increase as well.

Typical examples of conventional transformers are shown in Fig. 13. In Fig. 13A, the primary windings consist of brass or copper tubes, inserted into ferrite sleeves. The tubes are shorted together at one end by a piece of copper-clad printed circuit board material. The secondary winding is threaded through the tubes. Since the low-impedance winding is only a single turn, the transformation ratio is limited to the squares of integers — 1, 4, 9, 16, etc. The lowest effective transformer frequency is determined by the inductance of the one-turn winding. It should have a reactance at least four times greater than the impedance it is connected to, for extended low-frequency response.

The coupling coefficient between the two windings is a function of the primary tube diameter and its length, and the diameters and insulation thickness of the wire used in the high-impedance winding. High impedance ratios, greater than 36:1, should use large-diameter secondary windings. Miniature coaxial cable (using only the braid as the conductor) works very well. Another use for coaxial cable braid is illustrated in Fig. 13B. Instead of using tubing for the primary winding, the secondary winding is threaded through copper braid. Performance of the two units is almost identical.

The cores used must be large enough so the core material will not saturate at the power level applied to the transformer. Core saturation can cause permanent changes to the core permeability, as well as overheating. Transformer nonlinearity also develops at core saturation. Harmonics and other distortion products are produced, clearly an undesirable situation. Multiple cores can be used to increase the power capabilities of the transformer. The core size required can be calculated through the use of formulas presented in an earlier chapter of this Handbook in the discussion of ferromagnetic transformers and inductors.

Table 2
Pi-L-network Values for Various Plate Impedance and Frequencies

These values are based on a loaded Q of 12.

Zin (Ohms)	Freq. (MHz)	C1 (pf)	L1 (µH)	C2 (pF)	L2 (µH)	Zin (Ohms)	Freq. (MHz)	C1 (pF)	L1 (µH)	C2 (pF)	L2 (µH)
1500	1.80	784.	14.047	2621.	8.917	3500	14.35	38.	4.118	206.	1.259
1500	2.00	636.	14.047	1982.	8.917	3500	21.00	27.	2.755	136.	0.843
1500	3.50	403.	7.117	1348.	4.518	3500	21.45	25.	2.755	138.	0.843
1500	4.00	318.	7.117	991.	4.518	3500	28.00	21.	1.989	106.	0.609
1500	7.30	174.	3.900	543.	2.476	3500	29.70	18.	1.989	99.	0.609
1500	7.00	188.	3.900	596.	2.476	4000	1.80	297.	32.805	1841.	8.917
1500	14.00	93.	1.984	292.	1.259	4000	2.00	238.	32.805	1412.	8.917
1500	14.35	89.	1.984	276.	1.259	4000	3.50	153.	16.621	947.	4.518
1500	21.00	62.	1.327	191.	0.843	4000	4.00	119.	16.621	706.	4.518
1500	21.45	59.	1.327	185.	0.843	4000	7.00	71.	9.107	418.	2.476
1500	28.00	48.	0.959	152.	0.609	4000	7.30	65.	9.107	387.	2.476
1500	29.70	43.	0.959	134.	0.609	4000	14.00	35.	4.633	204.	1.259
2000	1.80	591.	17.933	2355.	8.917	4000	14.35	33.	4.633	197.	1.259
2000	2.00	478.	17.933	1788.	8.917	4000	21.00	23.	3.099	137.	0.843
2000	3.50	304.	9.086	1211.	4.518	4000	21.45	22.	3.099	132.	0.843
2000	4.00	239.	9.086	894.	4.518	4000	28.00	18.	2.238	107.	0.609
2000	7.00	142.	4.978	534.	2.476	4000	29.70	16.	2.238	95.	0.609
2000	7.30	131.	4.978	490.	2.476	5000	1.80	239.	40.011	1696.	8.917
2000	14.00	70.	2.533	264.	1.259	5000	2.00	190.	40.011	1316.	8.917
2000	14.35	67.	2.533	249.	1.259	5000	3.50	123.	20.272	872.	4.518
2000	21.00	47.	1.694	173.	0.843	5000	4.00	95.	20.272	658.	4.518
2000	21.45	45.	1.694	167.	0.843	5000	7.00	57.	11.108	387.	2.476
2000	28.00	36.	1.224	135.	0.609	5000	7.30	52.	11.108	360.	2.476
2000	29.70	32.	1.224	120.	0.609	5000	14.00	29.	5.651	186.	1.259
2500	1.80	474.	21.730	2168.	8.917	5000	14.35	27.	5.651	183.	1.259
2500	2.00	382.	21.730	1654.	8.917	5000	21.00	19.	3.780	125.	0.843
2500	3.50	244.	11.010	1115.	4.518	5000	21.45	18.	3.780	123.	0.843
2500	4.00	191.	11.010	827.	4.518	5000	28.00	15.	2.730	95.	0.609
2500	7.00	114.	6.033	493.	2.476	5000	29.70	13.	2.730	89.	0.609
2500	7.30	105.	6.033	453.	2.476	6000	1.80	200.	47.118	1612.	8.917
2500	14.00	56.	3.069	240.	1.259	6000	2.00	160.	47.118	1242.	8.917
2500	14.35	53.	3.069	230.	1.259	6000	3.50	103.	23.873	829.	4.518
2500	21.00	38.	2.053	158.	0.843	6000	4.00	80.	23.873	621.	4.518
2500	21.45	36.	2.053	154.	0.843	6000	7.00	48.	13.081	368.	2.476
2500	28.00	29.	1.483	127.	0.609	6000	7.30	44.	13.081	340.	2.476
2500	29.70	26.	1.483	111.	0.609	6000	14.00	24.	6.655	172.	1.259
3000	1.80	397.	25.466	2026.	8.917	6000	14.35	22.	6.655	173.	1.259
3000	2.00	318.	25.466	1554.	8.917	6000	21.00	16.	4.452	117.	0.843
3000	3.50	204.	12.903	1042.	4.518	6000	21.45	15.	4.452	116.	0.843
3000	4.00	159.	12.903	777.	4.518	6000	28.00	13.	3.215	87.	0.609
3000	7.00	94.	7.070	468.	2.476	6000	29.70	11.	3.215	84.	0.609
3000	7.30	87.	7.070	426.	2.476	8000	1.80	152.	61.119	1453.	8.917
3000	14.00	47.	3.597	222.	1.259	8000	2.00	120.	61.119	1138.	8.917
3000	14.35	44.	3.597	217.	1.259	8000	3.50	78.	30.967	747.	4.518
3000	21.00	32.	2.406	146.	0.843	8000	4.00	60.	30.967	569.	4.518
3000	21.45	30.	2.406	145.	0.843	8000	7.00	36.	16.968	337.	2.476
3000	28.00	24.	1.738	115.	0.609	8000	7.30	33.	16.968	312.	2.476
3000	29.70	21.	1.738	105.	0.609	8000	14.00	18.	8.632	165.	1.259
3500	1.80	338.	29.155	1939.	8.917	8000	14.35	17.	8.632	159.	1.259
3500	2.00	272.	29.155	1476.	8.917	8000	21.00	12.	5.775	104.	0.843
3500	3.50	174.	14.772	997.	4.518	8000	21.45	11.	5.775	106.	0.843
3500	4.00	136.	14.772	738.	4.518	8000	28.00	9.	4.171	86.	0.609
3500	7.00	81.	8.094	444.	2.476	8000	29.70	8.	4.171	77.	0.609
3500	7.30	75.	8.094	404.	2.476						
3500	14.00	40.	4.118	215.	1.259						

Operating Q — 12. Output load — 52 ohms. Computer data provided by Bill Imamura, JA6GW and Clarke Greene, K1JX.

(A)

(B)

Fig. 13 — The two methods of constructing the transformers as outlined in the text. At (A), the one-turn loop is made from brass tubing; at (B), a piece of coaxial cable braid is used for the loop.

Transmission line transformers are similar to conventional transformers, but can be used over wider frequency ranges. In a conventional transformer, the high-frequency losses are caused by leakage inductance, rising with frequency. In a transmission line transformer, the windings are arranged so there is tight capacitive coupling between the two. A high coupling coefficient is maintained up to a high frequency.

Fig. 14 shows two types of 4:1 transformers, and a method for connecting two of them in series to effect a 16:1 transformation. The circuit at E is often used between a 50-Ω source and the base of an RF power transistor.

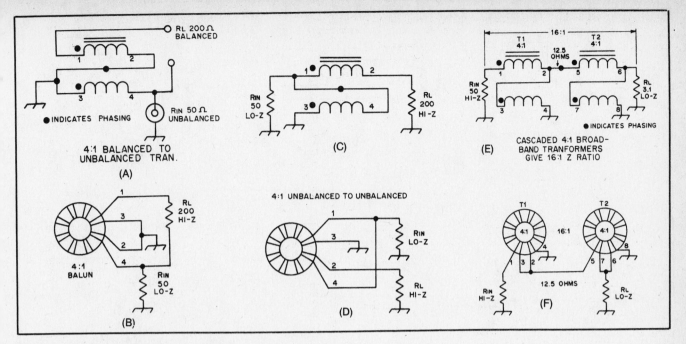

Fig. 14 — Circuit illustrations of 4:1 broadband transformers.

Two types of 9:1 transformers are illustrated in Fig. 15. Phase reversal, 1:1 balanced to unbalanced and hybrid combiner transformers are shown in Fig. 16. The circuit at E is used to combine the output of two simultaneously driven amplifiers into a common load. When the amplitude and phase of the signal sources are identical, all of the power is delivered to the load, R4. The amplifiers are effectively isolated from each other.

Output Filtering

Filtering of the amplifier output is sometimes necessary to meet spurious-signal requirements. Broadband amplifiers, by definition, provide little if any suppression of harmonic energy. Amplifiers using tank circuits in their output often require further attenuation of undesired harmonics. High level signals, particularly at multiple transmitter sites, can travel down the feedline and mix in a power amplifier, causing spurious outputs. For example, an HF transceiver signal radiated from a triband beam is picked up by a VHF FM antenna on the same mast. The signal saturates the low-power FM transceiver output stage, even with power off, and is reradiated by the VHF antenna. Proper use of filters can reduce spurious energy considerably.

The type of filter used will depend on the application and the level of spurious attenuation needed. Band-pass filters attenuate spurious signals above and below the band for which they are designed. Low-pass filters attenuate signals above the desired output frequency, while high-pass filters reduce energy below the design frequency.

The filters shown in Fig. 17 are low-pass and high-pass versions of Chebyshev filters.

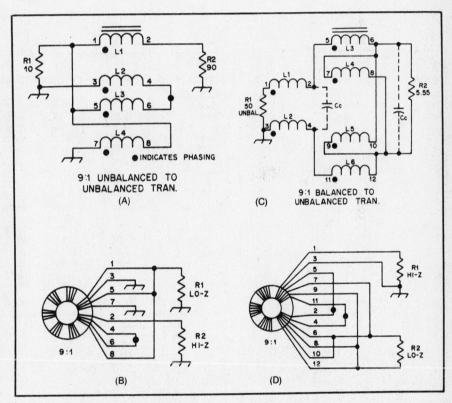

Fig. 15 — Circuit examples of 9:1 broadband transformers (A and C).

The filters may be realized in either a T or pi configuration. In this design, the number of filter circuit elements is determined by the desired stop-band frequency roll-off and the tolerable passband ripple. Steeper roll-off requires more circuit elements as does lower passband ripple. Passband ripple manifests itself by increased passband SWR.

In order to select a suitable filter design, the builder must determine the amount of attenuation required in the stopband. Additionally the builder must determine the maximum amount of passband ripple, and therefore the SWR of the filter. Table 3 and Fig. 18 show attenuation values achieved with the various filters and passband ripples. The attenuation levels are quoted at

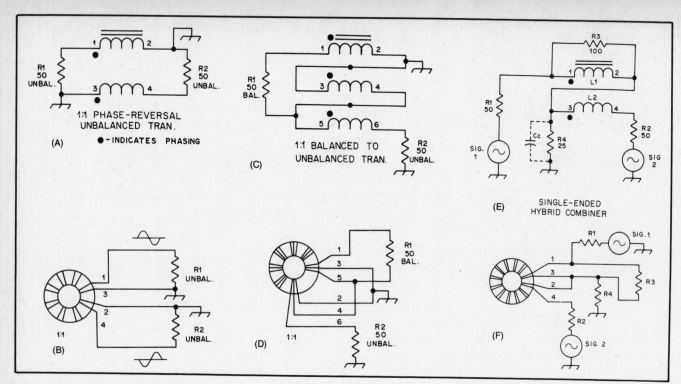

Fig. 16 — Assorted broadband transformers.

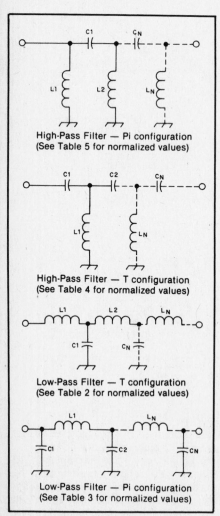

High-Pass Filter — Pi configuration
(See Table 5 for normalized values)

High-Pass Filter — T configuration
(See Table 4 for normalized values)

Low-Pass Filter — T configuration
(See Table 2 for normalized values)

Low-Pass Filter — Pi configuration
(See Table 3 for normalized values)

Fig. 17 — The four filter types discussed in
the text and Tables 3 through 8.

Table 3
Chebyshev High-Pass and Low-Pass Filters — Attenuation (dB)

No. elements, ripple, dB	SWR	2 f_c	3 f_c	4 f_c	5 f_c	6 f_c	7 f_c
3, 1	2.66	22.46	34.05	41.88	47.85	52.68	56.74
3, 0.1	1.36	12.24	23.60	31.42	37.39	42.22	46.29
3, 0.01	1.10	4.08	13.73	21.41	27.35	32.18	36.24
3, 0.001	1.03	0.63	5.13	11.68	17.42	22.20	26.25
5, 1	2.66	45.31	64.67	77.73	87.67	95.72	102.50
5, 0.1	1.36	34.85	54.21	67.27	77.21	85.26	92.04
5, 0.01	1.10	24.82	44.16	57.22	67.17	75.22	82.00
5, 0.001	1.03	14.94	34.16	47.22	57.16	65.22	71.99
7, 1	2.66	68.18	95.29	113.57	127.49	138.77	148.26
7, 0.1	1.36	57.72	84.83	103.11	117.03	128.31	137.80
7, 0.01	1.10	47.68	74.78	93.07	106.99	118.27	127.75
7, 0.001	1.03	37.68	64.78	83.06	96.98	108.26	117.75
9, 1	2.66	91.06	125.91	149.42	167.32	181.82	194.01
9, 0.1	1.36	80.60	115.45	138.96	156.86	171.36	183.55
9, 0.01	1.10	70.56	105.41	128.91	146.81	161.31	173.51
9, 0.001	1.03	60.55	95.40	118.91	136.91	151.31	163.50

Note: For high-pass filter configuration 2f_c becomes $f_c/2$, etc.

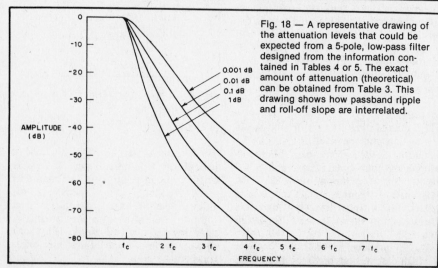

Fig. 18 — A representative drawing of
the attenuation levels that could be
expected from a 5-pole, low-pass filter
designed from the information contained in Tables 4 or 5. The exact
amount of attenuation (theoretical)
can be obtained from Table 3. This
drawing shows how passband ripple
and roll-off slope are interrelated.

integer multiples of the fundamental frequency (for the low-pass case) or integer submultiples of the fundamental frequency (for the high-pass application). The values given are theoretical and assume perfect components, no coupling between filter sections and no signal leakage around the filter. A working model should follow these values down to the 60- to 70-dB attenuation level. Beyond this point, the theoretical response will likely be degraded somewhat by the factors just mentioned.

Once the filter is chosen, the component values can be found. The filters listed in Tables 4 through 7 are normalized to a frequency of 1 MHz and an input and output impedance of 50 ohms. The 1 MHz value represents a cut-off frequency. The attenuation increases rapidly above or below this frequency into the stopband. In order to translate these designs to other frequencies, it is necessary to divide the component values by the new frequency in megahertz. For example, if a low-pass filter is desired to reduce harmonics from the 10-meter FM transceiver, the cut-off frequency should be 29.7 MHz. The inductance and capacitance values shown in the tables would then be divided by 29.7. Other impedance levels can also be used by multiplying the inductor values by the ratios $Z_0/50$ and the capacitor values by $50/Z_0$, where Z_0 is the new impedance. The correction factor should be applied in addition to any factors for frequency translation.

In many cases the calculated capacitor values will be such that a standard-value unit may be used. Alternatively, a combination of fixed-value silver mica capacitors and mica compression trimmers can be used in parallel to obtain the chart values. Toroidal inductors, because of their self-shielding properties, are ideal for use in these filters. Miniductor stock can also be used. However, it is much bulkier and will not offer the same degree of shielding between filter sections. Disc ceramic or paper capacitors are not suitable for use in RF filters. Standard mica or silver mica types are recommended.

Fig. 19 shows a filter that was designed with the information contained in Table 5. It is a 7-element, low-pass type of pi configuration. The unit is housed in a small aluminum Minibox and makes use of BNC connectors for the input and output connections.

TRANSMITTING DEVICE RATINGS

Plate Dissipation

The ultimate determination of the power-handling capability of a tube is its plate dissipation. It is the measure of how many watts of heat the tube can safely dissipate without exceeding critical temperatures. Excessive temperature can weaken or even melt internal tube connections or the seals that retain vacuum in the tube. The result — tube failure. The same tube may have different ratings, depending on the conditions under which it is operated, but all the ratings are based primarily upon the heat that the tube can safely dissipate.

Table 4
Chebyshev Low-Pass Filter — T Configuration

No. elements, ripple, dB	L1	L2	L3	L4	L5	C1	C2	C3	C4
3, 1	16.10	16.10				3164.3			
3, 0.1	8.209	8.209				3652.3			
3, 0.01	5.007	5.007				3088.5			
3, 0.001	3.253	3.253				2312.6			
5, 1	16.99	23.88	16.99			3473.1	3473.1		
5, 0.1	9.126	15.72	9.126			4364.7	4364.7		
5, 0.01	6.019	12.55	6.019			4153.7	4153.7		
5, 0.001	4.318	10.43	4.318			3571.1	3571.1		
7, 1	17.24	24.62	24.62	17.24		3538.0	3735.4	3538.0	
7, 0.1	9.400	16.68	16.68	9.400		4528.9	5008.3	4528.9	
7, 0.01	6.342	13.91	13.91	6.342		4432.2	5198.4	4432.2	
7, 0.001	4.690	12.19	12.19	4.690		3951.5	4924.1	3951.5	
9, 1	17.35	24.84	25.26	24.84	17.35	3562.5	3786.9	3786.9	3562.5
9, 0.1	9.515	16.99	17.55	16.99	9.515	4591.9	5146.2	5146.2	4591.9
9, 0.01	6.481	14.36	15.17	14.36	6.481	4542.5	5451.2	5451.2	4542.5
9, 0.001	4.854	12.81	13.88	12.81	4.854	4108.2	5299.0	5299.0	4108.2

Component values normalized to 1 MHz and 50 ohms. L in µH; and C in pF.

Table 5
Chebyshev Low-Pass Filter — Pi Configuration

No. elements, ripple, dB	C1	C2	C3	C4	C5	L1	L2	L3	L4
3, 1	6441.3	6441.3				7.911			
3, 0.1	3283.6	3283.6				9.131			
3, 0.01	2002.7	2002.7				7.721			
3, 0.001	1301.2	1301.2				5.781			
5, 1	6795.5	9552.2	6795.5			8.683	8.683		
5, 0.1	3650.4	6286.6	3650.4			10.91	10.91		
5, 0.01	2407.5	5020.7	2407.5			10.38	10.38		
5, 0.001	1727.3	4170.5	1727.3			8.928	8.928		
7, 1	6896.4	9847.4	9847.4	6896.4		8.85	9.34	8.85	
7, 0.1	3759.8	6673.9	6673.9	3759.8		11.32	12.52	11.32	
7, 0.01	2536.8	5564.5	5564.5	2536.8		11.08	13.00	11.08	
7, 0.001	1875.7	4875.9	4875.9	1875.7		9.879	12.31	9.879	
9, 1	6938.3	9935.8	10,105.	9935.8	6938.3	8.906	9.467	9.467	8.906
9, 0.1	3805.9	6794.5	7019.6	6794.5	3805.9	11.48	12.87	12.87	11.48
9, 0.01	2592.5	5743.5	6066.3	5743.5	2592.5	11.36	13.63	13.63	11.36
9, 0.001	1941.7	5124.6	5553.2	5124.6	1941.7	10.27	13.25	13.25	10.27

Component values normalized to 1 MHz and 50 ohms. L in µH; C in pF.

Table 6
Chebyshev High-Pass Filter — T Configuration

No. elements, ripple, dB	C1	C2	C3	C4	C5	L1	L2	L3	L4
3, 1	1573.0	1573.0				8.005			
3, 0.1	3085.7	3085.7				6.935			
3, 0.01	5059.1	5059.1				8.201			
3, 0.001	7786.9	7786.9				10.95			
5, 1	1491.0	1060.7	1491.0			7.293	7.293		
5, 0.1	2775.6	1611.7	2775.6			5.803	5.803		
5, 0.01	4208.6	2018.1	4208.6			6.098	6.098		
5, 0.001	5865.7	2429.5	5865.7			7.093	7.093		
7, 1	1469.2	1028.9	1028.9	1469.2		7.160	6.781	7.160	
7, 0.1	2694.9	1518.2	1518.2	2694.9		5.593	5.058	5.593	
7, 0.01	3994.1	1820.9	1820.9	3994.1		5.715	4.873	5.715	
7, 0.001	5401.7	2078.0	2078.0	5401.7		6.410	5.144	6.410	
9, 1	1460.3	1019.8	1002.7	1019.8	1460.3	7.110	6.689	6.689	7.110
9, 0.1	2662.2	1491.2	1443.3	1491.2	2662.2	5.516	4.922	4.922	5.516
9, 0.01	3908.2	1764.1	1670.2	1764.1	3908.2	5.576	4.647	4.647	5.576
9, 0.001	5218.3	1977.1	1824.6	1977.1	5218.3	6.657	4.780	4.780	6.657

Component values normalized to 1 MHz and 50 ohms. L in µH; and C in pF.

Efficiency for amplifiers runs between 25 and 75 percent, depending on the operating class. The efficiency indicates how much of the stage's power is useful RF; the rest

Table 7
Chebyshev High-Pass Filter — Pi Configuration

No. elements, ripple, dB	L1	L2	L3	L4	L5	C1	C2	C3	C4
3, 1	3.932	3.932				3201.7			
3, 0.1	7.714	7.714				2774.2			
3, 0.01	12.65	12.65				3280.5			
3, 0.001	19.47	19.47				4381.4			
5, 1	3.727	2.652	3.727			2917.3	2917.3		
5, 0.1	6.939	4.029	6.939			2321.4	2321.4		
5, 0.01	10.52	5.045	10.52			2439.3	2439.3		
5, 0.001	1.466	6.074	1.466			2837.3	2837.3		
7, 1	7.159	5.014	5.014	7.159		1469.2	1391.6	1469.2	
7, 0.1	6.737	3.795	3.795	6.737		2237.2	2023.1	2237.2	
7, 0.01	9.985	4.552	4.552	9.985		2286.0	1949.1	2286.0	
7, 0.001	13.50	5.195	5.195	13.50		2564.1	2057.7	2564.1	
9, 1	3.651	2.549	2.507	2.549	3.651	2844.1	2675.6	2675.6	2844.1
9, 0.1	6.656	3.728	3.608	3.728	6.656	2206.5	1968.9	1968.9	2206.5
9, 0.01	9.772	4.410	4.176	4.410	9.772	2230.5	1858.7	1858.7	2230.5
9, 0.001	13.05	4.943	4.561	4.943	13.05	2466.3	1911.8	1911.8	2466.3

Component values normalized to 1 MHz and 50 ohms. L in µH; and C in pF.

Fig. 19 — A photograph of a 7-pole low-pass filter designed with the information contained in Table 5. The filter is housed in a small aluminum Minibox.

is wasted power in the form of heat, dissipated through the plate. By knowing the plate-dissipation limit of the tube, and the efficiency expected from the chosen class of operation, the maximum power output and input levels can be determined. The maximum power output possible is

$$P_{OUT} = \frac{P_D N_P}{100 - N_P}$$

where

P_{OUT} = the power output in watts
P_D = the plate dissipation in watts
N_P = the efficiency (as a percentage).

The dc input power would simply be

$$P_{IN} = \frac{100\, P_{OUT}}{N_P}$$

Almost all tube-type power amplifiers in amateur service today operate as linear amplifiers (Class AB or B) with an efficiency of approximately 60%. That means a useful power output of approximately 1.5 times the plate dissipation can be reliably expected from a linear amplifier (assuming everything else works right).

The amplifier duty cycle determines the dissipation rating. Some types of operation are less efficient than others, meaning that the tube must dissipate more heat. Some forms of modulation, such as CW or SSB are intermittent in nature, causing less average heating than modulation formats in which there is continuous transmission, such as RTTY. Power tube manufacturers use two different rating systems to allow for the variations in service. CCS — Continuous Commercial Service — is the more conservative rating and is used for specifying tubes that are in constant use. The second rating system is based on intermittent, low-duty-cycle operation, and is known as ICAS — Intermittent Commercial and Amateur Service. ICAS ratings are normally used by amateurs who wish to obtain maximum power output with reasonable tube life in CW and SSB service. The conservative CCS ratings should be used for FM, RTTY and SSTV applications. (Plate power transformers for amateur service are also rated in CCS and ICAS terms.)

Maximum Ratings

Tube manufacturers publish sets of maximum values for the tubes they produce. No single maximum value should ever be used, unless all other ratings are simultaneously held substantially below their maximum values. As an example, a tube may have a maximum plate-voltage rating of 2000, a maximum plate-current rating of 300 mA, and a maximum plate power-input rating of 400 W. Therefore, if the maximum plate voltage of 2000 is used, then the plate current should be limited to 200 mA (not 300 mA) to stay within the maximum power-input rating of 400 W.

The manufacturers also publish sets of typical operating values that should result in good efficiency and long tube life. These values for many tubes can also be found in Chapter 35.

TRANSISTOR POWER DISSIPATION

RF power amplifier transistors are limited in power-handling capability by the amount of heat the device can safely dissipate. Power dissipation for a transistor is expressed symbolically as P_D. The maximum rating is based on maintaining a case temperature of 25 °C. If greater temperatures are anticipated, the device has to be derated in milliwatts per °C, as specified by the manufacturer for that particular device. The efficiency considerations described earlier in reference to plate dissipation apply here, also. A rule of thumb for selecting a P_D rating suitable for a given RF power output level is to choose a transistor that has a maximum dissipation of twice the desired output power.

Maximum Transistor Ratings

Transistor data sheets specify the maximum operating voltage for several conditions. Of particular interest is the V_{CEO} specification (collector to emitter voltage, with the base open). In RF amplifier service the collector to emitter voltage can rise to twice the dc supply potential. Thus, if a 12-V supply is used, the transistor should have a V_{CEO} of 24 or greater to preclude damage.

The maximum collector current is also specified by the manufacturer. This specification is actually limited by the current-carrying capabilities of the internal bonding wires. Of course, the collector current must stay below the level that generates heat higher than the allowable device power dissipation. Many transistors are also rated for the load mismatch they can safely withstand. A typical specification might be for a transistor to tolerate a 30:1 SWR at all phase angles.

Transistor manufacturers publish data sheets that describe all the appropriate device ratings. Typical operating results are also given in these data sheets. In addition, many manufacturers publish application notes illustrating the use of their devices in practical circuits. Construction details are usually given. Perhaps owing to the popularity of Amateur Radio among electrical engineers, most of the notes describe applications especially suited to the Amateur Service. Specifications for some of the more popular RF power transistors are found in Chapter 35.

PASSIVE COMPONENT RATINGS

Output Tank Capacitor Ratings

The tank capacitor in a high-power amplifier should be chosen to allow sufficient spacing between plates to preclude high-voltage breakdown. The peak RF voltage present across a tank circuit under load, but without modulation, may be taken conservatively as being equal to the dc plate or collector voltage. If the dc supply voltage also appears across the tank capacitor, this must be added to the peak RF voltage, making the total peak voltage twice the dc supply voltage. At the higher voltages, it is usually desirable to design the tank circuit so that the dc supply voltages do not appear across the tank capacitor, thereby allowing the use of a smaller capacitor with less plate spacing. Capacitor manufacturers usually rate their products in terms of the peak voltage between plates. Typical plate spacings are given in Table 8.

Output tank capacitors should be mounted as close to the tube as temperature considerations will permit, to make possible the shortest capacitive path from plate to cathode. Especially at the higher frequencies, where minimum circuit capacitance becomes important, the capacitor should be mounted with its stator plates well spaced from the chassis or other shielding. In circuits in which the rotor must be insulated from ground, the capacitor should be mounted on ceramic insulators of a size commensurate with the plate voltage involved and — most important of all, from the viewpoint of safety to the operator — a well-insulated coupling should be used between the capacitor shaft and the dial. The section of the shaft attached to the dial should be well grounded. This can be done conveniently through the use of panel shaft-bearing units.

High-Frequency Tank Coils

Tank coils should be mounted at least their diameter away from shielding to prevent a marked loss in Q. Except perhaps at 28 MHz, it is not important that the coil be mounted quite close to the tank capacitor. Leads up to 6 or 8 inches are permissible. It is more important to keep the tank capacitor, as well as other components, out of the immediate field of the coil. For this reason, it is preferable to mount the coil so that its axis is parallel to the capacitor shaft, either alongside the capacitor or above it.

Many factors must be taken into consideration in determining the size of wire (see Table 9) that should be used in winding a tank coil. The considerations of form factor and wire size which will produce a coil of minimum loss are often of less importance in practice than the coil size that will fit into available space or that will handle the required power without excessive heating. This is particularly true in the case of input circuits for screen-grid tubes where the relatively small driving power required can be easily obtained even if the losses in the driver are quite high. It may be considered preferable to take the power loss if the physical size can be kept down by making the coils small.

The wire sizes in Table 9 are larger than those typically used in modern commercial and home-built equipment. That is because their values were derived for AM service, where there is a continuous carrier and modulation peaks have a PEP four times the carrier value. Thus, the no. 10 wire required for 1000-W input on 75-meter AM would be adequate for at least 4-kW PEP in SSB service.

Table 8
Typical Tank-Capacitor Plate Spacings

Spacing Inches	Peak Voltage	Spacing Inches	Peak Voltage	Spacing Inches	Peak Voltage
0.015	1000	0.07	3000	0.175	7000
0.02	1200	0.08	3500	0.25	9000
0.03	1500	0.125	4500	0.35	11000
0.05	2000	0.15	6000	0.5	13000

Larger wire size than required for current handling capabilities is often used to maximize unloaded Q. Particularly at higher frequencies where skin depth considerations increase losses, the greater surface area of large diameter windings can be beneficial. Small-diameter copper tubing, up to 0.25-in outer diameter, can be used successfully for tank coils up through the lower VHF range. Silver plating the tubing further reduces losses. This is especially true as the tubing ages, and oxidizes. Silver oxide is a much better conductor than copper oxides, so silver-plated tank coils maintain their low-loss characteristics even after years of use.

At VHF and above, tank circuits take on forms less reminiscent of the familiar coil inductor. The inductances required to resonate tank circuits of reasonable Q at these higher frequencies is small enough that only strip lines or coaxial lines are practical. Since these are constructed from sheet metal or large-diameter tubing, current-handling capabilities normally are not a relevant factor.

RF Chokes

The characteristics of any RF choke will vary with frequency, from characteristics resembling those of a parallel-resonant circuit of high impedance, to those of a series-resonant circuit, where the impedance is lowest. In between these extremes, the choke will show varying amounts of inductive or capacitive reactance.

In series-feed circuits, these characteristics are of relatively small importance because the RF voltage across the choke is negligible. In a parallel-feed circuit, however, the choke is shunted across the tank circuit, and is subject to the full tank RF voltage. If the choke does not present a sufficiently high impedance, enough power will be absorbed by the choke to burn it out. To avoid this, the choke must have a sufficiently high reactance to be effective at the lowest frequency, and yet have no series resonances near the higher frequency bands. These potential resonances should be carefully investigated with a dip meter. A resonant choke failure in a high power amplifier can be very dramatic and damaging!

Blocking Capacitors

A series capacitor is usually used at the input of every amplifier output circuit. Its purpose is to block dc from appearing on

Table 9
Wire Sizes for Transmitting Coils for Tube Transmitters

Power Input (Watts)	Band (MHz)	Wire Size
1000	28-21	6
	14.7	8
	3.5-1.8	10
500	28-21	8
	14-7	12
	3.5-1.8	14
150	28-21	12
	14-7	14
	3.5-1.8	18
75	28-21	14
	14-7	18
	3.5-1.8	22
25 or less*	28-21	18
	14-7	24
	3.5-1.8	28

*Wire size limited principally by consideration of Q.

any of the matching circuit components and the antenna. As mentioned in the section on tank capacitors, output circuit component ratings can be relaxed somewhat when only the ac component of the output voltage is present.

The voltage rating for a blocking capacitor should be equal to at least twice the dc voltage applied. The peak voltage in a tank circuit can approach this value. A large safety margin is desirable, since blocking capacitor failure can bring catastrophic results.

The blocking capacitor should be low impedance at the amplifier operating frequencies. Since all the amplifier RF output current flows through the capacitor, a low impedance is necessary to avoid dissipation of RF power in the capacitor. Not only is that power wasted, but also the power-dissipation capabilities of most capacitors is minimal. A reasonable capacitance value for a blocking capacitor is one that limits its dissipation to less than 0.1% of the amplifier output power. This implies a capacitive reactance less than one thousandth the circuit impedance.

The current flowing through the blocking capacitor is dependent on circuit loaded Q. High levels of current flow in typical power amplifier output circuits. The capacitor must be capable of withstanding these currents. Below a couple of hundred watts at the high frequencies, suitable voltage rating disc ceramic capacitors work well in high-impedance tube amplifier output circuits. Above that power level and at VHF, ceramic doorknob capacitors are needed for their low losses and high current handling capabilities. The so-called "TV doorknob" breaks down at high RF current levels and should be avoided.

The very high values of Q_L found in many VHF and UHF tube-type amplifier tank circuits often require custom fabrication of the blocking capacitor. This can usually be accommodated through the use of a Teflon "sandwich" capacitor. Here, the blocking capacitor is formed from two parallel plates separated by a thin layer of Teflon. This capacitor often is part of the tank circuit itself, forming a very low-loss blocking capacitor. Teflon is rated for a minimum breakdown voltage of 2000 volts per mil of thickness, so voltage breakdown should not be a factor in any practically realized circuit.

The capacitance formed from such a Teflon sandwich can be calculated from the information presented earler in this Handbook (use a dielectric constant of 2.1 for Teflon). In order to prevent any potential irregularities caused by dielectric thickness variations (including air gaps), Dow-Corning DC-4 silicone grease should be evenly applied to both sides of the Teflon dielectric. This grease has similar properties to Teflon, and will fill in any surface irregularities that might cause problems.

The very low impedances found in transistorized amplifiers present special problems. In order to achieve the desired low blocking capacitor impedance, large-value capacitors are required. Special ceramic chips and mica capacitors are available that meet the requirements for high capacitance, large current carrying capability, and low associated inductance. These capacitors are more costly than standard disc-ceramic or silver-mica units, but their level of performance easily justifies their price. Most of these special-purpose capacitors can be identified by the fact that they are either leadless or come with wide straps instead of normal wire leads. Disc-ceramic and other wire-lead capacitors are generally not suitable for transistor power amplifier service.

SOURCES OF OPERATING VOLTAGES

Tube Filament or Heater Voltage

The heater voltage for the indirectly heated cathode-type tubes found in low-power classifications may vary 10 percent above or below rating without seriously reducing the life of the tube. The voltage of the higher-power, filament-type tubes should be held closely between the rated voltages as a minimum and five percent above rating as a maximum. Because of internal tube heating at UHF and higher, the manufacturers' rated filament voltage often is reduced at these higher frequencies. The derated filament voltages should be fol-

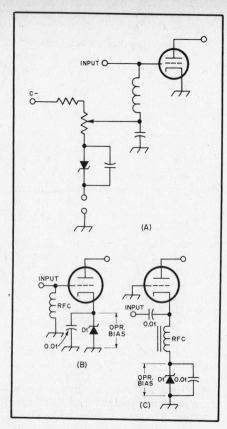

Fig. 20 — Various techniques for providing operating bias with tube amplifiers.

lowed carefully to maximize tube life. Series dropping resistors may be required in the filament circuit to attain the correct voltage. The voltage should be measured at the filament pins of the tube socket while the amplifier is running. The filament choke and the interconnecting wiring all have voltage drops associated with them. The high current drawn by a power tube heater circuit causes substantial voltage drops to occur across even a small resistance. Also, make sure that the plate power drawn from the power line does not cause a drop in filament voltage below the proper value when plate power is applied.

Thoriated-type filaments lose emission when the tube is overloaded appreciably. If the overload has not been too prolonged, emission sometimes may be restored by operating the filament at rated voltage with all other voltages removed for a period of 10 minutes, or at 20% above rated voltage for a few minutes.

Vacuum Tube Plate Voltage

Dc plate voltage for the operation of RF amplifiers is most often obtained from a transformer-rectifier-filter system (see Chapters 6 and 27) designed to deliver the required plate voltage at the required current. It is not unusual for a power tube to arc over internally. This is usually caused by the intrusion of a charged subatomic particle into the tube. The flashover by itself is not normally dangerous to the tube, as long as the event is short lived. Damage,

to both the high-voltage supply and the tube, usually occurs when the flashover totally discharges the power supply capacitor bank. The subsequent high-current discharge into the tube not only damages the tube, but also frequently destroys the rectifiers in the power supply.

A good preventative measure against this is the inclusion of a 50- to 100-Ω, high-wattage power resistor in series with the plate voltage circuit. The additional resistance in the line lengthens the time constant associated with the capacitor-load resistance circuit. Short-duration flashovers are dampened by the RC low-pass action of the plate circuit, once the resistor is added. Any future flashovers probably will go undetected. Even at high plate-current levels, the addition of the resistor does little to affect the dynamic regulation of the plate supply. The wattage rating of the resistor may be calculated from the expected maximum dc plate current drawn by the tube. A safety factor of at least two should be applied to ensure resistor reliability.

Grid Bias

The grid bias for a linear amplifier should be highly filtered and very well regulated. Any ripple or other voltage change in the bias circuit modulates the amplifier. This causes hum to appear on the signal as well as distortion. Since most linear amplifiers draw only small amounts of grid current, these bias supply requirements are not hard to meet.

Fixed bias is usually obtained from a variable-voltage regulated supply. Voltage adjustment allows setting bias level to give the desired resting plate current. Fig. 20 shows a simple Zener-diode-regulated bias supply. The dropping resistor is chosen to allow approximately 10 mA of Zener current. Bias is then regulated for all drive conditions up to a grid current level of about 5 mA. The potentiometer allows the bias level to be adjusted between V_{Zener} to approximately 15 volts higher. This range is usually adequate to allow for variations in the operating conditions between different tubes. Under stand-by conditions, when it is desirable to cut off the tube entirely, the Zener ground return is interrupted so the full bias supply voltage is applied to the grid.

In Fig. 20B and C, bias is obtained from the voltage drop across a Zener diode in the cathode (or filament center-tap) lead. Operating bias is obtained by the voltage drop across D1 as a result of plate (and screen) current flow. The diode voltage drop effectively raises the cathode potential relative to the grid. The grid is therefore lower in potential than the cathode by the Zener voltage of the diode. Assigning an arbitrary potential of zero to the cathode gives a grid bias equal to the Zener drop. The Zener-diode wattage rating should be twice the product of the maximum cathode current times the developed bias. Therefore, a tube requiring 15 volts of bias during a maximum cathode current flow of 100 mA would dissipate 1.5 W in the Zener diode. The diode rating, to allow a suitable safety factor, would be 3 W or greater. The circuit of Fig. 20C illustrates how D1 would be used with a cathode driven (grounded grid) amplifier as opposed to the grid-driven example at B.

In all cases the Zener diode should be bypassed by a 0.01 µF capacitor of suitable voltage. Current flow through any type of diode generates shot noise. If not bypassed, this noise would modulate the amplified signal, causing distortion in the amplifier output.

Screen Voltage For Tubes

Power tetrode screen current varies widely with both excitation and loading. The current may be either positive or negative. In a linear amplifier the screen voltage should be well regulated for all values of screen current. The power output from a tetrode is very sensitive to screen voltage, and any dynamic change in the screen potential can cause distorted output. Voltage regulator tubes or Zener diodes are usually used for screen regulation.

Fig. 21 shows a typical example of a regulated screen supply for a power tetrode amplifier. The voltage from a fixed dc supply is dropped to the Zener stack voltage by the current-limiting resistor. A screen bleeder resistor is connected in parallel with the Zener stack to allow for the negative screen current developed under certain tube operating conditions. The bleeder current is chosen to be somewhat greater than the expected maximum negative screen current. For external anode tubes in the 4CX250 family, a typical value would be 20 mA. For the 4CX1000 family, a screen bleeder current of 70 mA is required. The idling current for the Zeners is set to be approximately 10 mA higher than the expected maximum screen current, plus the screen bleeder-resistor current. The screen voltage is then regulated for all values of current between maximum negative screen current and maximum positive screen current.

A tetrode should never be operated without plate voltage and load; otherwise the screen would act like an anode and draw excessive current. Supplying the screen through a series dropping resistor from the plate supply affords a measure of protection, since the screen voltage only appears when there is plate voltage. Alternatively, a fuse can be placed between the regulator and the bleeder resistor. The fuse should not be installed between the bleeder resistor and the tube, because the tube should never be operated without a load on the screen. Without a load, the screen potential tends to rise to the anode voltage. Any screen bypass capacitors or other associated circuitry would likely be damaged by this high voltage.

In Fig. 21, a varistor is connected from

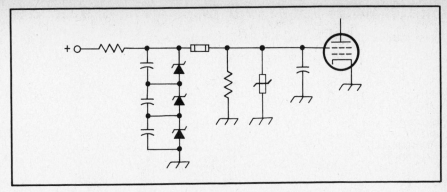

Fig. 21 — A Zener-regulated screen supply for use with a tetrode. Protection is provided by a fuse and a varistor.

screen to ground. If, because of some circuit failure, the screen voltage should rise substantially above its nominal level, the varistor will conduct and clamp the screen voltage to ground. This usually causes a fuse to blow in the power supply, and power is interrupted before any damage occurs. The varistor voltage should be chosen to be approximately 50 percent higher than normal screen voltage.

Transistor Biasing

Solid-state power amplifiers generally operate in Class C or AB. When some bias is desired during Class C operation (Fig. 22A), a resistance of the appropriate value can be placed in the emitter return as shown. Most transistors will operate in Class C without adding bias externally, but in some instances the amplifier efficiency can be improved by means of emitter bias. Reverse bias supplied to the base of the Class C transistor should be avoided because it will lead to internal breakdown of the device during peak drive periods. The damage is frequently a cumulative phenomenon, leading to gradual destruction of the transistor junction.

A simple method for Class AB biasing is shown in Fig. 22B. D1 is a silicon diode that acts as a bias clamp at approximately 0.7 V. The forward bias establishes linear-amplification conditions. That value of bias is not always optimum for a specified transistor in terms of IMD. Variable bias of the type illustrated in Fig. 22C permits the designer sufficient variance to locate the operating point for best linearity. The diode clamp or the reference for another type of regulator is usually thermally bonded to the power transistor. The bias level then tracks the thermal characteristics of the output transistor. Since a transistor's current transfer characteristics are a function of temperature, thermal tracking of the bias is necessary to maintain device linearity and to prevent thermal runaway and the subsequent destruction of the transistor.

AMPLIFIER COOLING

Tube Cooling

Vacuum tubes must be operated within

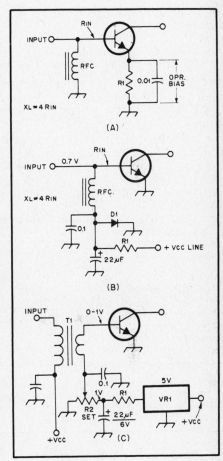

Fig. 22 — Biasing methods for use with transistor amplifiers.

the temperature range specified by the manufacturer if long tube life is to be achieved. Tubes with glass envelopes rated at up to 25 W plate dissipation may be run without forced-air cooling, if a moderate amount of cooling by convection can be arranged. If a perforated metal enclosure is used, and a ring of ¼-inch-diameter holes are placed around the tube socket, normal air flow can be relied on to remove excess heat at room temperatures.

For tubes with greater plate dissipation, or those operated with plate currents in excess of the manufacturer's ratings, forced-

Table 10

Specifications of Some Popular Tubes, Sockets and Chimneys

Tube	CFM	Back Pressure (inches)	Socket	Chimney
3-400Z/8163	13	0.13	SK-400, SK-410	SK-416
3-500Z	13	0.082	SK-400, SK-410	SK-406
3CX800A7	19	0.35	EIMAC P/N 154353	
3-1000Z/8164	25	0.38	SK-500, SK-510	SK-516
3CX1500/8877	35	0.41	SK-2200, SK-2210	SK-2216
4-250A/5D22	2	0.1	SK-400, SK-410	SK-406
4-400A/8438	14	0.25	SK-400, SK-410	SK-406
4-1000A/8166	20	0.6	SK-500,SK-510	SK-506
4CX250R/7850W	6.4	0.59	SK-600, SK-600A, SK-602A, SK-610, SK-610A, SK-611, SK-612, SK-620, SK-620A SK-621, SK-630	SK-606 SK-626
4CX300A/8167	7.2	0.58	SK-700, SK-710, SK-711A, SK-712A, SK-740, SK-760, SK-761, SK-770	SK-606
4CX350A/8321	7.8	1.2	Same as 4CX250R	
4CX1000A/8168	25	0.2	SK-800B, SK-810B, SK-890B	SK-806
4CX1500/8660				
8874	8.6	0.37		

These values are for sea-level elevation. For locations well above sea level (Denver, Colorado, for example), add an additional 20% to the figure listed.

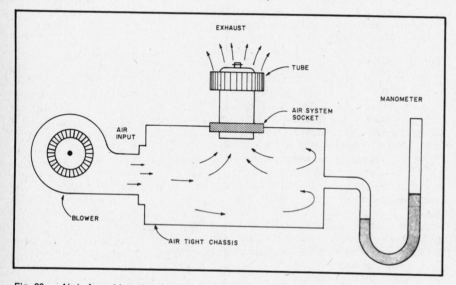

Fig. 23 — Air is forced into the chassis by the blower and exits through the tube socket. The manometer is used to measure system back pressure, which is an important factor in determining the proper size blower.

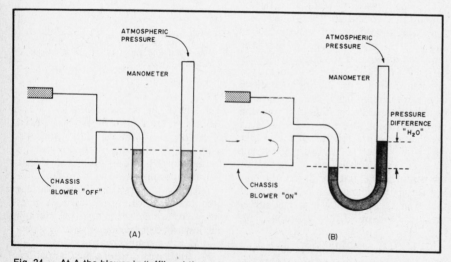

Fig. 24 — At A the blower is "off" and the water will seek its own level in the manometer. At B the blower is "on" and the amount of back pressure in terms of inches of water can be measured as indicated.

air cooling with a fan or blower is needed. Most manufacturers rate tube cooling requirements for continuous-duty operation. Their literature will indicate the required volume of air flow, in cubic feet per minute (CFM), at some particular back pressure.

Back pressure is the pressure that is built up inside an imperfect air passage when pressurized by an air source such as a blower. A perfect air passage offers no resistance to air flow, so the volume of air entering the passage flows out at the same rate. Any practical air passage is finite in size, and restricts the volume of air that can flow through. The difference between the potential volume of air that is available for flow through the air passage and the volume that actually flows manifests itself as back pressure. Anything that represents resistance to the air flow develops back pressure.

Tubes and their associated components are obstacles to the free passage of air and so develop back pressure to an air stream. The exact amount of pressure will depend on the blower, tube socket, tube and chimney characteristics. Blowers vary in their ability to work against back pressure so the matter of blower selection should not be taken lightly.

Values of CFM and back pressure for some of the more popular tubes, sockets and chimneys are given in Table 10. The back pressure is specified in inches of water and can be measured easily in an operational air passage as indicated in Figs. 23 and 24. The pressure differential between the air passage and the atmospheric pressure is measured with a device called a manometer. A manometer is nothing more than a piece of clear tubing, open at both ends and fashioned in the shape of a "U." The manometer is temporarily connected to the chassis and is removed after the measurements are completed. As shown in the diagrams, a small amount of water is placed in the tube. At Fig. 24A the blower is "off" and the water will seek its own level, because the air pressure (ordinary atmospheric pressure) is the same at both ends of the manometer tube. At B, the blower is "on" (socket, tube and chimney in place) and the pressure difference, in terms of inches of water, is measured. For most applications a standard scale can be used for the measurement and the results will be sufficiently accurate.

Table 11 illustrates the performance specifications for Dayton blowers, which are available through W. W. Grainger outlets throughout the U.S. Blowers having wheel diameters similar to those in Table 11 will likely have similar flow and back-pressure characteristics. If in doubt about specifications, consult the manufacturer. The setup of Fig. 24 is the authoritative determinant of cooling adequacy. As an example, assume that an amplifier is to be built using a 3-1000Z tube. A blower capable of supplying 25 CFM at a back pressure of 0.38 inch of water is required.

It appears (from Table 11) that the second blower listed would be suitable, although it may be marginal since it can supply only 25 CFM into a back pressure of 0.4 inch of water. The next larger size would provide a margin of safety.

When a pair of tubes is used, the CFM rating is doubled, but the back pressure remains the same as that for one tube. A pair of 3-1000Z tubes, for example, would require 50 CFM at a back pressure of 0.38 inch of water. In this case the fifth blower listed in the table would be suitable since it can supply 85 CFM at a back pressure of 0.4 inch of water. Always choose a blower that can supply at least the required amount of air. Smaller blowers will almost certainly lead to shortened tube life.

One method for directing the flow of air around a tube envelope or through tube cooling fins involves the use of a pressurized chassis. This system is shown in Fig. 23. A blower is attached to the chassis and forces air up through the tube socket and around the tube. A chimney (not shown in this drawing) is used to guide the air around the tube as it leaves the socket. A chimney will prevent the air from being dispersed as it hits the envelope or cooling fins, concentrating the flow for maximum cooling.

A less conventional approach is shown in Fig. 25. Here the anode compartment is pressurized by the blower. A special chimney is installed between the anode heat exchanger and a ventilation hole in the compartment cover. The blower pressurizes the anode compartment, and the only paths for air flow are through the anode and its chimney, and through the air system socket. Measurements by Dick Jansson, WD4FAB, indicate a nearly even split of the air flow between the anode and the air system socket. The pressure losses in a test made on an amplifier using a 4CX250 indicates a back pressure of only 0.33 inch of water with this scheme. This compares to a measured back pressure of 0.76 inch of water encountered by conventional base-to-anode cooling. Blower requirements are reduced considerably.

Table 10 also contains the part numbers for air-system sockets and chimneys to be used with the tubes that are listed. The builder should investigate which of the sockets listed for the 4CX250R, 4CX300A, 4CX1000A and 4CX1500A best fit the circuit needs. Some of the sockets have certain tube elements grounded internally through the socket. Others have elements bypassed to ground through capacitors that are integral parts of the sockets.

Depending on one's design philosophy and tube sources, some compromises in the cooling system may be appropriate. For example, if glass tubes are available inexpensively as broadcast pulls, a shorter lifespan may be acceptable. In such a case, an increase of convenience and a reduction in cost, noise and complexity can be had by using a pair of "muffin" fans. One fan may be used for the filament seals and one

Table 11
Blower Performance Specifications

Wheel Dia.	Wheel Width	RPM	Free Air	Back Pressure (inches)					Cutoff	Stock No.
				0.1	0.2	0.3	0.4	0.5		
2"	1"	3160	15	13	4	—	—	—	0.22	2C782
3"	1-15/32"	3340	54	48	43	36	25	17	0.67	4C012
3"	1-7/8"	3030	60	57	54	49	39	23	0.60	4C440
3"	1-7/8"	2880	76	70	63	56	45	8	0.55	4C004
3-13/16"	1-7/8"	2870	100	98	95	90	85	80	0.80	4C443
3-13/16"	2-1/2"	3160	148	141	135	129	121	114	1.04	4C005

for the anode seal, dispensing with a blower and air-system socket and chimney. The air flow path for this scheme is not as uniform as with the use of a chimney. The tube envelope cylinder mounted in a cross flow has flow stagnation points and low heat transfer in certain regions of the cylinder. These points become hotter than the rest of the envelope. The use of multiple fans to disturb the cross air flow can significantly reduce this problem. Many amateurs have used this cooling method successfully in low-duty-cycle CW and SSB operation, but it is not recommended for AM, SSTV or RTTY service.

The true test of the effectiveness of a forced air cooling system is the amount of heat conducted away from the tube by the air stream. The power dissipated can be calculated from the air flow temperatures. The dissipated power is

$$P = 169Q_A [(T_2/T_1) - 1]$$

where

P = the dissipated power in watts
Q_A = the air flow in CFM
T_1 = the inlet air temperature (normally room temperature)
T_2 = the amplifier exhaust temperature.

The exhaust temperature can be measured with a cooking thermometer at the air outlet. The thermometer should not be placed inside the anode compartment because of the high voltage present.

Transistor Cooling

Transistors used in power amplifiers dissipate significant amounts of power, and the heat so generated must be effectively removed to maintain acceptable device temperatures. Some bipolar power transistors have the collector connected directly to the case of the device, as the collector must dissipate most of the heat generated when the transistor is in operation. Others have the emitter connected to the case. However, even the larger case designs cannot conduct heat away fast enough to keep the operating temperature of the device within the safe area, the maximum temperature that a device can stand without damage. Safe area is usually specified in a device data sheet, often in graphical form. Germanium power transistors may be operated at up to 100°C while the silicon

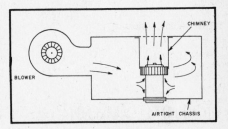

Fig. 25 — Anode compartment pressurization is more efficient than grid compartment pressurization. Air exits through the tube anode, through a chimney, and through the tube socket.

types may be run at up to 200°C. Leakage currents in germanium devices can be very high at elevated temperatures; thus, silicon transistors are preferred for power applications.

A thermal sink, properly chosen, will remove heat at a rate that keeps the transistor junction temperature in the safe area. For low-power applications a simple clip-on heat sink will suffice, while for 100-W input power a massive cast-aluminum finned radiator will be necessary. The appropriate size heat sink can be calculated based on the thermal resistance of the transistor case to ambient temperature. The first step is to calculate the total power dissipated by the transistor:

$$P_D = P_{DC} + P_{RFin} - P_{RFout}$$

where

P_D = the total power dissipated by the transistor in watts
P_{DC} = the dc power into the transistor in watts
P_{RFin} = the RF power into the transistor in watts
P_{RFout} = the RF output power from the transistor in watts.

The value of P_D is then used to obtain the Θ_{CA} value from

$$\Theta_{CA} = \frac{T_C - T_A}{P_D}$$

where

Θ_{CA} = the thermal resistance of the device case to ambient
T_C = the device case temperature
T_A = the ambient temperature (room temperature).

A suitable heat sink can then be chosen from the manufacturer's specifications for Θ_{CA}. The heat generated by the transistor must be radiated to the ambient by the heat sink so a low Θ_{CA} is needed. A well designed heat sink system minimizes thermal path lengths and has a large cross-sectional area. The thermal bonding of the transistor to the heat sink should have minimum thermal loss. In general, the heat sink mounting surface must be flat and the transistor firmly attached to the heat sink so intimate contact is made between the two. The use of thermal silicone based heat sink compounds offer considerable improvement in thermal transfer. The thermal resistance of this grease is considerably less than that of air, but is not nearly as good as that of aluminum. The quantity of grease should be kept to an absolute minimum. Enough should be used only to fill in any air gaps between the transistor and the heat sink surface.

The maximum temperature rise in the transistor junction may easily be calculated by using the equation

$$T_J = (\Theta_{JC} + \Theta_{CA}) P_D + T_A$$

where
    T_J = the transistor junction temperature

Θ_{JC} = the manufacturer's published thermal resistance of the transistor from junction to case
Θ_{CA} = the thermal resistance of the device case to ambient
P_D = the power dissipated by the transistor
T_A = the ambient temperature.

The value of T_J must be kept below the manufacturer's recommended junction temperature to prevent transistor failure. Measured values of the ambient temperature and the device case temperature can be used in the preceding formulas to calculate junction temperature.

Design Guidelines and Examples

Most of the design problems facing an amplifier builder are not theoretical, but have to do with real-world component limitations. There is no such thing as a "pure" resistance or "pure" inductance or capacitance. The tubes and transistors used in RF amplifiers have stray reactances and resistances associated with every element of their construction. Every passive component has stray reactance and resistance to go along with its "advertised" function. Some of these factors can be ignored at some frequencies, but successful design and construction accounts for all of these "facts of life" and their effects.

A simplified equivalent schematic of an amplifying device is shown in Fig. 26A. The input is represented by a resistance in parallel with a capacitance. The output consists of a current generator in parallel with a resistance and capacitance. This is an accurate description of both a transistor and a vacuum tube, regardless of circuit configuration (as demonstrated in Fig. 26B and C). Both the input and output impedances have a resistive component in parallel with a reactive component.

The amplifier input and output matching networks must transform the complex impedances found in the amplifying device to the transmission line impedance (usually 50 ohms). Other impedances associated with the amplifier circuit, such as the impedance of a dc-supply choke, must also be included in the matching-network design. The matching networks and the associated components are influenced by each other's presence, and proper consideration must be made to account for their effects.

Perhaps the best way to explain the design considerations necessary to implement a particular amplifier function is through example. The following examples are representative of the various problems associated with power amplifier design. They are not intended to be detailed con-

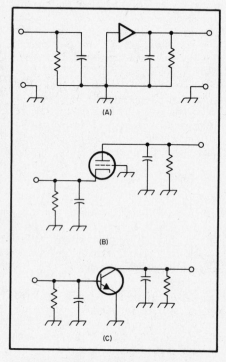

Fig. 26 — The electrical equivalents for power amplifiers. In A, the input is represented by a resistor in parallel with a capacitor and the output is a current source in parallel with another resistor and capacitor. These effects are applied to tubes and transistors in B and C.

struction plans; they only demonstrate the approach one would take in designing similar projects.

DESIGN EXAMPLE 1: A HIGH-POWER VACUUM TUBE HF AMPLIFIER

Most popular HF transceivers produce approximately 100-W output. According to the manufacturer's specifications, the EIMAC 8877 is a suitable candidate for use in a power amplifier capable of the maximum legal power output with this much ex-

citation. The 8877 can develop 1500-W output for approximately 60-W of drive when used in a grounded grid circuit. Grounded-grid operation is usually the easiest tube amplifier circuit to implement. The input impedance is relatively low, not too different from 50 Ω. The negative feedback inherent in the grounded-grid configuration reduces the likelihood of amplifier instability. Fewer supply voltages are needed in this configuration compared to others; only high-voltage dc for the plate and low-voltage ac for the filament are needed.

The first step in the amplifier design process is to verify that the tube is actually capable of producing the anticipated results, and still stay within manufacturer's ratings. The plate dissipation expected during normal operation of the amplifier is computed first. Since the amplifier will be used for SSB operation among others, a class of operation producing linear amplification must be used. Class AB operation is a very good compromise between linearity and good efficiency, with an expected efficiency of about 60%. Based on efficiency formulas presented earlier, an input power of 2500 W is needed to produce the desired 1500-W output. Operated under these conditions, the tube dissipates 1000 W — well within the manufacturer's specifications.

The specification for maximum grid current must be adhered to closely in high-mu triodes, for maximum tube life. The grid structure in these tubes is the most fragile element, since it is closely spaced to the cathode and carefully aligned to achieve high current gains. For a fixed level of drive, grid current tends to decrease with increasing plate potential. The maximum plate voltage allowable for the 8877 is specified to be 4000 V. A plate potential of 3500 V is chosen to minimize grid current, which is well within the published guidelines. In order to reach the 2500-W input power, 714 mA of plate current must be drawn at this plate voltage.

The next step in the design process is to calculate the optimum plate load resistance at this plate voltage and current for Class AB operation. From the earlier equations, R_L is calculated to be 3268 Ω. This result is approximate since the value for K in Class AB is only approximately 1.5. For ease in further calculations, the plate resistance can be rounded off to 3500 Ω.

The first decision that has to be made is to choose input and output matching networks. Almost any network can be used at the output to transform the 3500-Ω plate impedance down to 50 Ω, but experience shows that the pi and pi-L networks yield the most practical component values. Additionally, these two networks provide reasonable harmonic attenuation. The pi-L does give significantly greater harmonic attenuation than the pi, so it is usually a better choice. The penalty of using a pi-L network is the additional burden of the "L" inductor and band-switching this inductor in multiband amplifiers.

The input impedance of a grounded-grid 8877 is specified to be approximately 54 Ω. While this is close enough to 50 Ω to cause only minimal SWR at the input, the impedance varies with drive level. The fluctuating impedance can present problems for the exciter output matching network, increasing the intermodulation distortion level from the exciter. A tank circuit matching network at the amplifier input can stabilize the impedance through its "flywheel effect." The matching circuit must have a Q_L of at least two for this to occur. The pi network is a suitable choice for this application.

Fig. 27 illustrates these input and output networks applied in the amplifier circuit. The schematic shows the major components in the amplifier RF section, but with band switching omitted. C1 and C2 and L1 form the input pi network. C3 is a blocking capacitor to keep the cathode dc potential isolated from the exciter.

The filament in an 8877 is in close proximity to the cathode — it has to be to adequately heat the cathode. A capacitance of several picofarads is formed between the two. Particularly at high frequencies where these few picofarads represent a relatively low reactance, RF drive intended for the cathode can be capacitively coupled into the filament, where it becomes heat. To keep the drive where it belongs in the cathode, the filament must be kept at a high RF impedance above ground. The high impedance minimizes RF current flow in the filament circuit; RF dissipated in the filament becomes virtually zero. The dc resistance should be kept to a minimum to lessen voltage drops in the high-current filament circuit.

The best type of choke for use in this application is a pair of heavy gauge bifilar windings over a ferrite rod. The ferrite core raises the inductive reactance throughout the HF region so that a minimum of wire

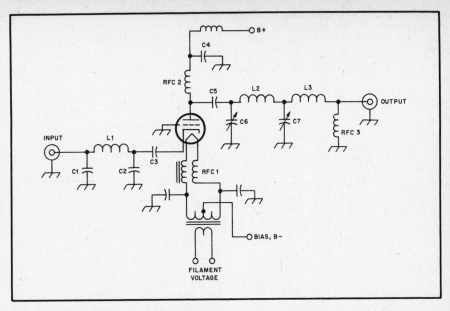

Fig. 27 — A simplified schematic of a grounded-grid amplifier using a pi network input and a pi-L network output.

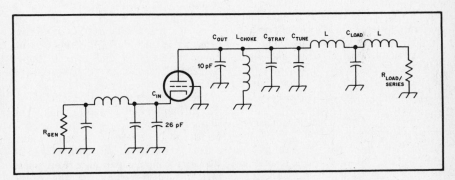

Fig. 28 — The effective reactances for the amplifier in Fig. 27.

is needed, keeping filament-circuit voltage drops low. The bifilar winding technique adds to the bandwidth of the choke and assures that both filament terminals are at the same RF potential.

Plate voltage is supplied to the tube through RFC2. C5 is the plate blocking capacitor. The output pi-L network is composed of tuning capacitor C6, loading capacitor C7, pi coil L2 and L coil L3. RFC3 is a high-inductance RF choke placed at the output for safety purposes. Its value, usually 1 to 2 millihenrys, is high enough so that it appears as an open circuit across the output connector for RF. However, should the plate blocking capacitor fail and allow high voltage onto the output matching network, RFC3 would short the dc to ground and blow the power-supply fuse. This prevents dangerous high voltage from appearing on the feed line or antenna.

The next calculations to be made involve the design of the input and output matching networks. The resistances to be matched have already been defined. The tube input resistance is specified by the manufacturer to be 54 Ω. The calculated plate resistance

is 3500 Ω. Both have to be transformed to 50 Ω, the impedance of the connecting cables. Unfortunately, neither of these impedances is purely resistive. As shown in Fig. 26, the tube input and output impedances are composed of both resistance and internal capacitances. These capacitances must be taken into account when designing the matching networks. Additionally, any reactance introduced into the networks by dc supply line chokes and interconnecting leads must be accounted for.

Fig. 28 shows the various reactances in the amplifier circuit. At the tube input, the impedance is 54 ohms in parallel with C_{IN}, 26 pF. In the output circuit C_{OUT} is the tube output capacitance, 10 pF. The reactance presented by the plate choke is shown as L_{CHOKE}. C_{STRAY} represents the combined stray capacitances of the interconnecting leads in the output circuit, the enclosure metal work in proximity to the tube, and the parasitic reactances in the tuning capacitor. The rest of the components C_{TUNE}, C_{LOAD} and the two Ls form the pi-L network.

Design of the input matching circuit is straightforward. The pi network circuit equations from Chapter 2 are used to calculate component values. A Q_L of between two and three should be used. Higher Q values reduce the network's bandwidth, perhaps requiring the inclusion of a front-panel control for use across an entire amateur band. The purpose of this input network is to stabilize the impedance presented to the exciter, not to offer selectivity. The value of the capacitor at the tube input end of the pi network should be reduced by 26 pF; that value of capacitance is automatically added in parallel to the pi-network capacitor by the tube input impedance.

The output matching network is usually the most difficult section of an amplifier to design. The tables presented earlier in the chapter greatly simplify the effort. Both the pi and pi-L design tables were calculated around a loaded Q value of 12. Loaded Qs lower than 10 do not provide adequate harmonic suppression; values much higher than 18 increase matching network losses caused by high circulating currents. A Q_L value of 12 is a good compromise between harmonic suppression and circuit losses. Theoretical matching network component values can be taken right from the chart for the 3500-Ω plate impedance. These values can then be adjusted to allow for the previously described circuit reactances.

First, the low frequency component values should be examined. At 3.5 MHz the tuning capacitor value from Table 2 is 174 pF. Fig. 28 shows three reactances directly in parallel with C_{TUNE}. The tube output capacitance, C_{OUT}, adds 10 pF directly in parallel to C_{TUNE}. Strays in the circuit, represented by C_{STRAY}, also are directly added in parallel to C_{TUNE}. An exact value for C_{STRAY} is difficult to define, so an educated guess must be made. In a well constructed, carefully thought out power amplifier, C_{STRAY} can be estimated to be approximately 10 pF. Between C_{OUT} and C_{STRAY}, 20 pF of pi-L network input capacitance is already accounted for. The inductance of the plate choke is also in parallel with C_{TUNE}. A popular plate choke used in many amplifiers has an inductance of 90 μH. At 3.5 MHz this is an inductive reactance of 1979 Ω. When combined in parallel with the 174 pF (261 Ω at 3.5 MHz) at the input of the pi-L network, an effective capacitance of 151 pF results. Since the pi-L network requires an effective capacitance of 174 pF at its input, the value of C_{TUNE} must be increased to compensate for the plate choke inductance. A total capacitance at the pi-L input of 197 pF, in parallel with the 90-μH plate choke, gives an effective capacitance of 174 pF. Of the 197 pF required, 10 pF comes from circuit strays, 10 pF comes from C_{OUT}, and 177 pF must come from the tuning capacitor, C_{TUNE}. A tuning capacitor with a maximum capacitance

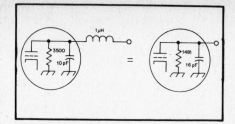

Fig. 29 — The effect of a series inductance in vacuum tube output circuit. In this case, the 3500-Ω plate resistance is transformed down to 1481 ohms.

around 200 pF should be used to allow for loads with higher than a 1:1 SWR.

The component values for the high end of the amplifier frequency range should also be examined. At 29.7 MHz, Table 2 calls for an input capacitance of 18 pF. Since C_{OUT} and C_{STRAY} already account for 20 pF of pi-L input capacitance, a problem exists. Nothing can be done to reduce C_{OUT}; C_{STRAY} presumably is at a minimum also. Additionally, the tuning capacitor probably has a minimum capacitance around 15 pF. The minimum circuit capacitance found at the pi-L input is already 35 pF, nearly twice the required value.

There are potentially three solutions to this dilemma. The first is probably the simplest to implement. A higher loaded Q for the pi-L network will require more capacitance at the input. Raising Q_L to 24 changes the input capacitance requirement to 36 pF, which is within the range of the circuit as described. Many amplifier builders, both commercial and amateur, take this approach. The major drawback is substantially increased tank circuit losses caused by the increased Q.

Another potential solution is to reduce the minimum capacitance provided by C_{TUNE}. Vacuum variable capacitors have minimum capacitances as low as three to five pF, and can be used at C_{TUNE}. This reduces the minimum effective circuit capacitance to 25 pF. A pi-L network with a loaded Q of 17 needs only 26 pF at its input. While not optimum, a Q_L of 17 is perfectly acceptable from a tank circuit loss perspective. Unfortunately, vacuum variables are very expensive and cumbersome mechanically.

A third possibility is the use of an additional reactance in the output circuit. An inductance connected in series between the tube and the tuning capacitor will act as an L network in conjunction with C_{OUT} and transform the plate load resistance to a lower value. This is shown in Fig. 29. Lower input impedance-matching networks require higher input capacitances while maintaining a Q_L of 12. For example, a matching network designed for a plate impedance of 1500 Ω requires 43 pF of input capacitance at 29.7 MHz. In order to

calculate the L network, the series equivalent impedance for the tube must be computed. The plate impedance is 3500 Ω in parallel with 10 pF (C_{OUT}). At 10 meters, this is 3500 Ω in parallel with −j535 ohms.

Using the equivalence formulas described earlier, this converts to an equivalent series impedance of 80 −j523 Ω. A 1-μH inductor, having an impedance of j187 ohms at 29.7 MHz, can be added as the series inductance for the L network. A series equivalent resistance of 80 −j523 plus j187 or 80 −j336 Ω results. This converts to an equivalent parallel impedance of 1491 Ω in parallel with −j355 Ω, which is equal to 1491 ohms (rounded to 1500 Ω) in parallel with 15 pF at 29.7 MHz. This new impedance is what the pi-L must now match. C_{OUT} is now effectively 15 pF, and when combined with the 10 pF of C_{STRAY} and the 15 pF minimum capacitance of C_{TUNE}, gives an effective input capacitance of 40 pF. C_{TUNE} must be adjusted to a value of 18 pF to give the required pi-L input capacitance of 43 pF for this new impedance level. The L inductor should be a high-Q coil wound from copper tubing to keep losses low. The L inductor has a decreasing yet significant effect on progressively lower frequencies. A similar calculation to the above should be made on each band to determine the transformed equivalent plate impedance, before referring to Table 2.

The 90-μH plate choke still remains in parallel with C_{TUNE}. At 29.7 MHz its impedance is in excess of 16,000 ohms and has very little effect on the RF characteristics of the circuit.

An analysis like this should be performed on every band the amplifier will operate on. Only then can the actual impedance levels be calculated and acted upon. Since the other components in the circuit, L2, L3 and C7, have no other reactances associated with them, their use is straightforward.

This amplifier is made operational on multiple bands by switching in additional inductances for L2 and L3 as the operating frequency is lowered. The usual practice is to use a shorting switch that will short out the unused portions of the total inductors for the band in use. The wiring to the switch and the switch itself add stray inductance and capacitance to the circuit. To minimize these effects at the higher frequencies, the unswitched 10-meter L2 should be placed closest to the high-impedance end of the network at C6. The effects of stray reactance diminish as the impedance level goes down. Reactances associated with the switch are effectively placed in parallel with C7 where the impedance level is between 300 and 700 ohms. The coil taps are best determined in an empirical fashion.

The impedance match in both the input and output networks can be checked without applying dc voltage, once the amplifier is built. In actual operation, the tube input and output resistances are the product of current flow through the tube. Without dc present, these resistances don't exist; the tube is an open circuit for dc. The input and output resistances can be easily simulated by ordinary quarter- or half-watt resistors (not wirewound, though; they are more inductive than resistive at RF). Resistors having the same value as the input and output resistances, connected in parallel with the tube input and output, present the same termination resistances to the matching networks as the tube does in operation. C_{IN} and C_{OUT} are physical properties of the tube, and are there whether dc is applied or not.

With the termination resistors in place, the impedance matches achieved by the matching networks can be measured by means of a noise bridge or return loss measurement set-up. For measuring the input match, the bridge is connected to the amplifier input. The matching network can then be adjusted for the desired match. To measure the output match, the bridge is connected to the amplifier output. Since the output matching network is bilateral (it works in both directions), the matching effectiveness of the network can be measured in either direction; obviously it's easier to measure from the 50-Ω end of the network. The taps on the inductors are then determined experimentally, adjusting the taps and the variable capacitors for best match. Once the best tap points are found, the connecting wires can be soldered in place and the termination resistors removed. The amplifier is now ready for full-voltage testing.

Design Example 2: A Medium Power 144-MHz Amplifier

For several decades the 4CX250 family of power tetrodes has been used successfully up through 500 MHz. The latest versions in the family are still state-of-the-art products. They are relatively inexpensive, produce high gain and lend themselves to relatively simple amplifier designs. In amateur service at VHF, the 4CX250 is a particularly attractive choice for an amplifier. Most VHF exciters used now by amateurs are solid-state and usually develop 10 W or less output. The drive requirement for the 4CX250 in grounded cathode, Class AB operation ranges between 2 and 8 watts for full power output, depending on frequency. At 144 MHz, manufacturer's specifications suggest an available output power of over 300 W. This is clearly a substantial level improvement over 10 W, so a 4CX250B will be used in this amplifier.

The first design step is the same as performed in the previous example — verify that the proposed tube will perform as desired while staying within the manufacturer's ratings. Again assuming a basic amplifier efficiency of 60% for Class AB operation, 300 W of output requires a plate input power of 500 W. Tube dissipation is

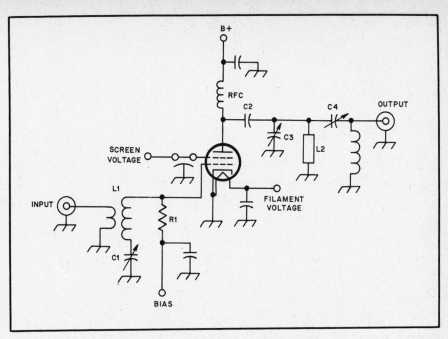

Fig. 30 — The simplified schematic for a VHF power amplifier using a power tetrode. The output circuit is a parallel-tuned tank circuit with series reactance output matching.

rated at 250 W, so plate dissipation is apparently not a problem as the tube will only be dissipating 200 W in this amplifier. The recommended maximum plate potential is 2000 V; the tube will be running near its maximum ratings. Plate current for 500-W input is then 250 mA, which is within the manufacturer's ratings. The plate load resistance can now be calculated. Using the same formula as before, the value is determined to be 5333 Ω.

The next step is to investigate the output circuit. The manufacturer's specification for C_{OUT} is 4.7 pF. The inevitable circuit strays along with the tuning capacitor add to the circuit capacitance. A carefully built amplifier might only have 7 pF of stray capacitance, and a specially-made tuning capacitor can be fabricated to have a mid-range value of 3 pF. The total circuit capacitance adds up to about 15 pF. At 144 MHz this represents a capacitive reactance of only 74 Ω. The Q_L of a tank circuit with this reactance in parallel with a plate load resistance of 5333 Ω is over 72. A pi output matching network would be totally impractical. The L required would be extremely small and circuit losses would be prohibitive. The best solution is the simplest: Connect an inductor in parallel with the circuit capacitance to form a parallel-resonant tank circuit.

To keep tank circuit losses low with such a high Q_L, a very high Q inductor must be used. The lowest loss inductors are formed from transmission line sections. These can take the form of either coaxial lines or strip lines. Both have their advantages and disadvantages, but the strip line is so much easier to fabricate that it is almost exclusively used in VHF tank circuits today.

The reactance of a terminated transmis-

sion line section is a function of both its characteristic impedance and its length (see Chapter 16). The reactance of a line terminated in a short circuit is

$$X_{in} = Z_o \tan \ell$$

where
  X = is the circuit reactance
  Z_o = the line impedance
  ℓ = the transmission line length in degrees.

For lines shorter than a quarter wavelength (90°) the circuit reactance is inductive.

In order to resonate with the tank circuit capacitive reactance the transmission line reactance must be the same value, but inductive. This requires an inductive reactance of 74 ohms. Examination of the formula for transmission-line circuit reactance suggests that a wide range of lengths can yield the same inductive reactance as long as the line Z_o is appropriately scaled. Based on circuit Q considerations, the best bandwidth in a tank circuit results when the ratio of Z_o to X_{in} is between one and two. This implies that transmission line lengths between 26.5° and 45° give the best bandwidth. Between these two limits, and with some adjustment of Z_o, practical transmission lines can be designed. A transmission line length of 35° is 8 inches long at 144 MHz, which is a workable dimension mechanically. Substitution of this value into the transmission line equation gives a Z_o of 105 Ω for the transmission line.

The width of the strip line and its placement relative to the ground planes determine the line impedance. Other stray capacitances such as mounting standoffs

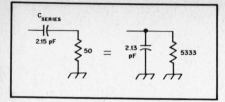

Fig. 31 — Series reactance matching as applied to the amplifier in Fig. 30.

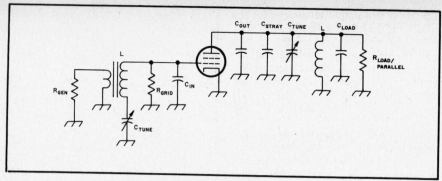

Fig. 32 — The reactances and resistances for the amplifier in Fig. 30.

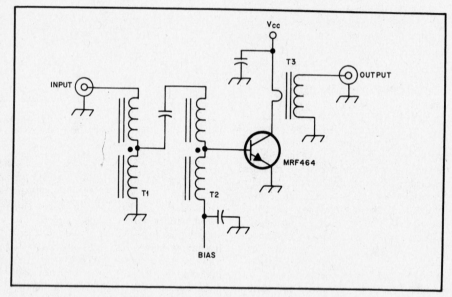

Fig. 33 — A simplified schematic of a broadband HF transistorized power amplifier. T1 and T2 are 4:1 broadband transformers to match the low input impedance of the transistor.

also affect the impedance. Accurate calculation of the line impedance for most physical configurations requires extensive application of Maxwell's equations and is beyond the scope of this book. The specialized case in which the strip line is parallel to and located halfway between two ground planes has been documented in *ITT Reference Data for Radio Engineers.* According to charts presented in that book, a 105-Ω strip line impedance is obtained by placing a line with a width of approximately 0.4 times the ground plane separation halfway between the ground planes. Assuming the use of a standard 3-inch-deep chassis for the plate compartment, this yields a strip line width of 1.2 in. A strip line 1.2 in. wide located 1.5 in above the chassis floor and grounded at one end has an inductive reactance of 74 ohms at 144 MHz.

The resulting amplifier schematic diagram is shown in Fig. 30. L2 is the strip-line inductance just described. C3 is the tuning capacitor, made from two parallel brass plates whose spacing is adjustable. One of the plates is connected directly to the strip line while the other is connected to ground through a wide, low-inductance strap. C2 is the plate blocking capacitor. This can be either a ceramic doorknob capacitor such as the Centralab 850 series or a homemade "Teflon sandwich." Both are equally effective at 144 MHz.

Impedance matching from the plate resistance down to 50 ohms can be either through an inductive link or through capacitive reactance matching. Mechanically, the capacitive approach is simpler to implement. Fig. 31 shows the development of reactance matching through a series capacitor (C4 in Fig. 30). By using the parallel equivalent of the capacitor in series with the 50-Ω load, the load resistance can be transformed to the 5333-Ω plate resistance. Substitution of the known values into the parallel-to-series equivalence formulas reveals that a 2.15 pF capacitor at C4 matches the 50-Ω load to the plate resistance. The resulting parallel equivalent for the load is 5333 Ω in parallel with 2.13 pF. The 2.13-pF capacitor is effectively in parallel with the tank circuit.

A new plate line length must now be calculated to allow for the additional capacitance. The equivalent circuit diagram containing all the various reactances is shown in Fig. 32. The total circuit capacitance is now just over 17 pF, which

is a reactance of 64 Ω. Keeping the strip-line width and thus its impedance constant at 105 Ω dictates a new resonant line length of 31°. This calculates to be 7.14 in for 144 MHz.

The alternative coupling scheme is through the use of a inductive link. The link can be either tuned or untuned. The length of the link can be estimated based on the amplifier output impedance, in this case 50 Ω. For an untuned link, the inductive reactance of the link itself should be approximately equal to the output impedance, 50 Ω. For a tuned link, the length depends on the link Q_L. The link Q_L should generally be greater than two, but usually less than five. For a Q_L of three this implies a capacitive reactance of 150 Ω, which at 144 MHz is just over 7 pF. The self inductance of the link should of course be such that its impedance at 144 MHz is 150 Ω (0.166 μH) at 144 MHz. Adjustment of the link placement determines the transformation ratio of the circuit line. Some fine adjustment of this parameter can be made through adjustment of the link series tuning capacitor. Placement of the

link relative to the plate inductor is an empirical process.

The input circuit is shown in Figs. 30 and 32. C_{IN} is specified to be 18.5 pF for the 4CX250. This is only 60 Ω at 2 meters, so the pi network again is unsuitable. Since a surplus of drive is available with a 10-W exciter, circuit losses at the amplifier input are not as important as at the output. An old-fashioned "split stator" tuned input can be used. L1 in Fig. 30 is series tuned by C_{IN} and C1. The two capacitors are effectively in series (through the ground return). A 20-pF variable at C1 set to 18.5 pF gives an effective circuit capacitance of 9.25 pF. This will resonate at 144 MHz with an inductance of 0.13 μH at L1. L1 can be wound on a toroid core for mechanical convenience. The 50-Ω input impedance is then matched by link coupling to the toroid. The grid impedance is primarily determined by the value for R1, the grid bias feed resistor.

Design Example 3: A Broadband HF Solid-State Amplifier

Linear power amplifier design using

transistors at HF is a fundamentally simple process. An appropriate transistor meeting the desired design specifications is selected on the basis of dissipation and power output. The transistor manufacturers greatly simplify the design by specifying each RF power transistor according to its frequency range and power output. The amplifier designer need only match to the device input and output and provide appropriate dc bias currents to the transistor.

The Motorola MRF464 is an RF power transistor capable of 80-W PEP output with low distortion. Its usable frequency range is up through 30 MHz. At a collector potential of 28 V, a collector efficiency of 40% is possible.

Fig. 33 shows the schematic diagram of a 2- to 30-MHz broadband linear amplifier using the MRF464. The input impedance of the transistor is specified by the manufacturer to be $1.4 - j0.30\ \Omega$ at 30 MHz and drops down to $9.0 - j5.40$ at 2 MHz. Transformers T1 and T2 match the 50-ohm amplifier input impedance to an approximate median value of the transistor input impedance. They are both 4:1 step-down ratio transmission-line transformers. A single 16:1 transformer could be used in place of T1 and T2, but 16:1 transformers are more difficult to fabricate for broadband service.

The manufacturer specifies the transistor output resistance to be approximately 6 ohms (in parallel with an equivalent output capacitance) across the frequency range. T3 is a ferrite-loaded conventional transformer with a step-up ratio of approximately 8:1. This matches the transistor output to 50 Ω.

The amplifier has a falling gain characteristic with rising frequency. To flatten out the gain across the frequency range, negative feedback could be applied. However, most power transistors have highly reactive input impedances and large phase errors would occur in the feedback loop. Instability could potentially occur.

A better solution is to use an input correction network. This network is used as a frequency-selective attenuator for amplifier drive. At 30 MHz, where transistor gain is least, the input power loss is designed to be minimal (less than 2 dB). The loss increases lower in frequency to compensate for the increased transistor gain. The MRF464 has approximately 12 dB more gain at 1.8 MHz than at 30 MHz; the compensation network is designed to have 12 dB loss at 1.8 MHz. A properly designed compensation network will result in an overall gain flatness of approximately 1 dB.

AMPLIFIER STABILIZATION

Stable Operating Conditions

Purity of emissions and the useful life of the active devices in a tube or transistor circuit depend heavily on stability during operation. Oscillations can occur at the operating frequency or far from it because of undesired positive feedback in the amplifier. Unchecked, these oscillations pollute the RF spectrum and can lead to tube or transistor overdissipation and subsequent failure. Each type of oscillation has its own cause and its own cure.

A linear amplifier operates with its input and output circuits tuned to the same frequency. Unless the coupling between these two circuits is kept to a minimum, some energy from the output will be coupled in phase back to the input and the amplifier will oscillate. Care should be used in arranging components and wiring of the two circuits so that there will be negligible opportunity for coupling external to the tube or transistor itself. Complete shielding between input and output circuits usually is required. All RF leads should be kept as short as possible, and particular attention should be paid to the RF return paths from input and output tank circuits to emitter or cathode.

In general, the best arrangement using a tube is one in which the input circuit and the output circuit are on opposite sides of the chassis. Individual compartments for the input and output circuitry add to the isolation. Transistor circuits are somewhat more forgiving, since all the impedances are relatively low. However, the high currents found on most amplifier circuit boards can easily couple into unintended circuits. Proper layout, the use of double-sided circuit board material (with one side used exclusively as a ground plane and low-inductance ground return), and heavy doses of bypassing on the dc supply lines can keep most solid-state amplifiers from oscillating.

VHF and UHF Parasitic Oscillations

Z1 of Fig. 34 is a parasitic choke. This network will help dampen oscillations at

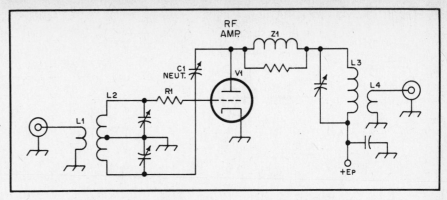

Fig. 34 — Single-ended, neutralized RF amplifier. Z1 is a parasitic suppressor, see text.

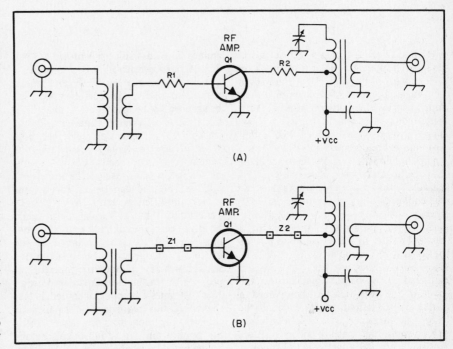

Fig. 35 — Suppression methods for VHF and UHF parasitics in solid-state amplifiers.

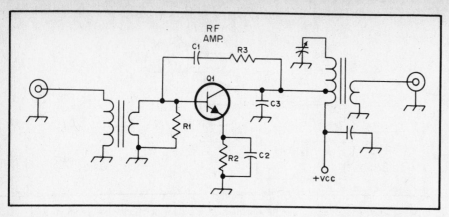

Fig. 36 — Illustration of shunt feedback in a transistor amplifier. C1 and R3 comprise the feedback network.

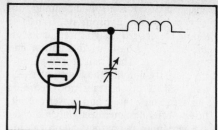

Fig. 37 — The equivalent feedback path due to the internal capacitance of the tube grid plate structure in a power amplifier.

VHF and UHF. It acts as a high series impedance at these frequencies to break up the unwanted VHF/UHF circuit path causing oscillation. Z1 consists of a non-inductive resistor, between 51 and 100 Ω, with a coil wound around the resistor body. This forms a broadband RF choke that presents a high impedance at VHF and higher, but looks like a low reactance in the HF region and lower. The resistance in parallel with the inductor lowers the Q of the choke, and dissipates undesired VHF energy.

A typical parasitic suppressor for a power level up to 150 W uses 6 to 8 turns of no. 20 wire wound around a 56-Ω, 1-W composition resistor. The coil ends are soldered to the resistor pigtails near the body of the resistor. Z1 is then placed in series with the output circuit, as close to the tube plate pin or cap as possible. For higher RF powers a high-wattage Globar resistor, or a 25-W noninductive (NIT) power resistor in parallel with the minimum inductance that will stop the oscillations should be used. Larger inductances will affect operation at the upper end of HF (particularly 10 meters). A few turns of no. 12 or 14 wire around the resistor body should suffice. Additional parasitic suppression can be attained, if needed, by connecting a low value resistor (10 to 51 Ω) in series with the tube input, near the tube socket. This is illustrated by R1 of Fig. 34.

Parasitic oscillations can be dampened in low-power solid-state amplifiers by using a small amount of resistance in series with the base or collector lead as shown in Fig. 35A. The value of R1 or R2 typically should be between 10 and 22 Ω. The use of both resistors is seldom necessary, but an empirical determination must be made. R1 or R2 should be located as close to the transistor as practical.

At power levels in excess of approximately 0.5 W, the technique of parasitic suppression shown in Fig. 35B is effective. The voltage drop across a resistor would be prohibitive at the higher power levels, so one or more ferrite beads placed over

connecting leads can be substituted (Z1 and Z2). A bead permeability of 125 presents a high impedance at VHF and above without affecting HF performance. The beads need not be used at both circuit locations. Generally, the terminal carrying the least current is the best place for these suppression devices. This suggests that the resistor or ferrite beads should be connected in the base lead of the transistor.

C3 of Fig. 36 can be added to some power amplifiers to dampen VHF/UHF parasitic oscillations. The capacitor should be low in reactance at VHF and UHF, but must present a high reactance at the operating frequency. The exact value selected will depend upon the collector impedance. A reasonable estimate is to use an X_c of 10 times the collector impedance at the operating frequency. Silver-mica or ceramic chip capacitors are suggested for this application. An additional advantage is seen in the bypassing action for VHF and UHF harmonic energy in the collector circuit. C3 should be placed as close to the collector terminal as possible, using short leads. The effects of C3 in a broadband amplifier are relatively insignificant at the operating frequency. However, when a narrow band collector network is used, the added capacitance of C3 must be absorbed into the network design in the same manner as the C_{OUT} of the transistor.

Low-Frequency Instability

Bipolar transistors exhibit a rising gain characteristic as the operating frequency is lowered. To preclude low-frequency instabilities because of the high gain, shunt and degenerative feedback are often used. In the regions where low-frequency self-oscillations are most likely to occur, the feedback increases by nature of the feedback network, reducing the amplifier gain. In the circuit of Fig. 36, C1 and R3 provide negative feedback, which increases progressively as the frequency is lowered. The network has a small effect at the desired operating frequency but has a pronounced effect at the lower frequencies.

The values for C1 and R3 are usually chosen experimentally. C1 will usually be between 220 pF and 0.0015 µF for HF-band amplifiers while R3 may be a value from 51 to 5600 Ω.

R2 of Fig. 36 develops emitter degeneration at low frequencies. The bypass capacitor, C2, is chosen for adequate RF bypassing at the intended operating frequency. The impedance of C2 rises progressively as the frequency is lowered, thereby increasing the degenerative feedback caused by R2. This lowers the amplifier gain. R2 in a power stage is seldom greater than 10 Ω, and may be as low as 1 Ω. It is important to consider that under some operating and layout conditions R2 can cause instability. This form of feedback should be used only in those circuits in which unconditional stability can be achieved.

R1 of Fig. 36 is useful in swamping the input of an amplifier. This reduces the chance for low-frequency self-oscillations, but has an effect on amplifier performance in the desired operating range. Values from 3 to 27 Ω are typical. When connected in shunt with the normally low base impedance of a power amplifier, the resistors lower the effective device input impedance slightly. R1 should be located as close to the transistor base terminal as possible, and the connecting leads must be kept short to minimize stray reactances. The use of two resistors in parallel reduces the amount of inductive reactance introduced compared to a single resistor.

Although the same concepts can be applied to tube-type amplifiers, the possibility of self-oscillations at frequencies lower than VHF is significantly lower than in solid-state amplifiers. Tube amplifiers will usually operate stably as long as the input-to-output isolation is greater than the stage gain. Proper shielding and dc power lead bypassing essentially eliminate feedback paths, except for those through the tube itself.

Amplifier Neutralization

Depending on the inner electrode capacitances of tubes, a neutralization circuit may be necessary. Output energy can be capacitively coupled back to the input as shown in Fig. 37. Neutralization involves coupling a small amount of the output energy (which by the tube phase relationships is out of phase with the input energy) and feeding it back to the amplifier input. There the neutralization circuit energy cancels the unwanted in-phase (positive) feedback. A typical circuit is given in Fig. 38. L2 provides a 180° phase reversal because it is center tapped. C1 is connected between the plate and the lower half of the grid tank. C1 is then adjusted so that the energy coupled from the tube output through the neutralization circuit is equal in amplitude and exactly 180° out of phase with the energy coupled from the output back through the tube. The two signals then cancel and the oscillation ceases.

The easiest way to adjust a neutralization circuit is to connect an RF source to the amplifier output tuned to the amplifier operating frequency. An RF detector of some kind is then connected to the amplifier input. The amplifier should remain turned off for this test. The amplifier tuning and loading controls, as well as any input network adjustments should then be peaked for maximum indication on the RF detector connected at the input. C1 is then adjusted for minimum response on the detector. This null indicates that the neutralization circuit is canceling energy coupled back to the amplifier input through the tube (or transistor) internal capacitances.

Screen-Grid Tube Neutralizing Circuits

The plate-to-grid capacitance in a screen-grid tube is reduced to a fraction of a picofarad by the interposed grounded screen. Nevertheless, the power sensitivity of these tubes is so great that only a very small amount of feedback is necessary to start oscillation. To assure a stable tetrode amplifier, it is usually necessary to load the grid circuit, or to use a neutralizing circuit.

A capacitive neutralizing system for screen-grid tubes is shown in Fig. 39. C1 is the neutralizing capacitor. The value of C1 should be chosen so that at some adjustment of C1,

$$\frac{C1}{C3} =$$

$$\frac{\text{Tube grid-plate capacitance (or } C_{gp})}{\text{Tube input capacitance (or } C_{IN})}$$

The grid-to-cathode capacitance must include all strays directly across the tube capacitance, including the capacitance of the tuning capacitor stator to ground. This may amount to 5 to 20 pF.

Grid Loading

The use of a neutralizing circuit may often be avoided by loading the grid circuit if the driving stage has some power capacity to spare. Loading by tapping the grid down on the grid tank coil or by a resistor from grid to cathode is effective in stabiliz-

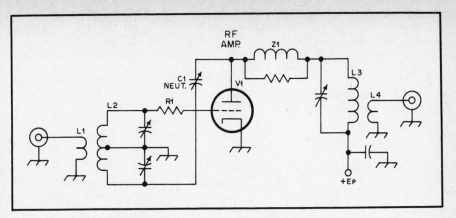

Fig. 38 — Example of neutralization of a single-ended RF amplifier.

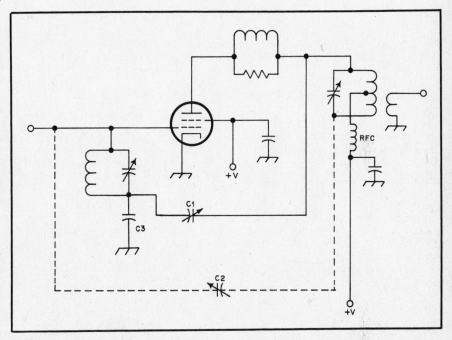

Fig. 39 — A neutralization circuit may use either C1 or C2 to cancel the effect of the tube internal capacitance.

ing an amplifier. Either measure reduces the gain and the sensitivity of the amplifier somewhat, lessening the possibility of oscillation. Substitution of a resistor for the grid circuit bias supply choke accomplishes the same goal.

BIBLIOGRAPHY

Belcher, "RF Matching Techniques, Design and Example," *QST,* October 1972, pp. 24-30

Feynman, *Lectures on Physics,* Vol. I, Addison-Wesley Publishing Co., 1977.

Goodman, "My Feed Line Tunes My Antenna," *QST,* April 1977, pp. 40-42.

Granberg, "Build This Solid-State Titan," *QST,* June 1977, pp. 27-31 (Part 1);

QST, July 1977, pp. 27-29 (Part 2).

Granberg, "One KW-Solid-State Style," *QST,* April 1976, pp. 11-14 (Part 1); *QST,* May 1976, pp. 28-30 (Part 2).

Hejhall, "Broadband Solid-State Power Amplifiers for SSB Service," *QST,* March 1972, pp. 36-43.

Johnson and Artigo, "Fundamentals of Solid-State Power-Amplifier Design," *QST,* September 1972, pp. 29-36 (Part 1); *QST,* November 1972, pp. 16-20 (Part 2); *QST,* April 1973, pp. 28-34 (Part 3).

Johnson, "Heat Losses in Power Transformers," *QST,* May 1973, pp. 31-34.

Knadle, "A Strip-line Kilowatt Amplifier for 432 MHz," *QST,* April 1972, pp. 49-55.

Meade, "A High-Performance 50-MHz Amplifier," *QST,* September 1975, pp. 34-38.

Meade, "A 2-KW PEP Amplifier for 144 MHz," *QST,* December 1973, pp. 34-38.

Olsen, "Designing Solid-State RF Power Circuits," *QST,* August 1977, pp. 28-32 (Part 1); *QST,* September 1977, pp. 15-18 (Part 2); *QST,* October 1977, pp. 22-24 (Part 3).

Orr, *Radio Handbook,* 22nd Ed., Howard W. Sams & Co., Inc., 1981.

Potter and Fich, *Theory of Networks and Lines,* Prentice-Hall, Inc., 1963.

Reference Data for Radio Engineers, ITT, Howard & Sams Co., Inc.

RF Data Manual, Motorola, Inc., 1982.

Simpson, *Introductory Electronics for Scientists and Engineers,* Allyn and Bacon, Inc., 1975.

Solid State Power Circuits, *RCA Designer's* Handbook, 1972.

White, "Thermal Design of Transistor Circuits," *QST,* April 1972, pp. 30-34.

Chapter 16

Transmission Lines

The place where RF power is generated is very frequently not the place where it will be used. A transmitter and its antenna are a good example. To radiate well, the antenna should be high above the ground and should be kept clear of trees, buildings and other objects that might absorb energy. The transmitter itself, however, is most conveniently installed indoors, where it is readily accessible.

A *transmission line* is the means by which RF energy is conveyed from one point to another. The types of transmission lines include simple two-conductor configurations, such as the familiar coaxial cable and TV parallel-wire line. Such lines are useful from power-line frequencies to well into the microwave region, and they form the most commonly used class. Waveguide is representative of a second type. Here the conductor configuration is complex, and ordinary concepts such as voltage and current tend to become obscure. As a consequence, various waveguide parameters are expressed in terms of the electric and magnetic fields associated with the line. The propagation of electromagnetic energy through waveguides is closely related to similar phenomena in space itself.

Transmission Lines and Circuits

A transmission line differs from an ordinary circuit in one very important aspect. *Propagation delays* exist from one end of the transmission line to the other. The speed at which electrical energy travels, while tremendously high when compared to mechanical motion, is not infinite. Propagation times are often of no significance in circuit design, as the layout dimensions are small in terms of signal wavelength. This is not true for transmission-line considerations, where finite propagation time becomes a factor of paramount importance. This is illustrated in Fig. 1. A transmission line connects a source at point S to a load at point L, separated by a distance ℓ.

Suppose that the distance ℓ is 300 meters (nearly 1000 feet). Also suppose for a moment that the source is a battery, rather than the RF generator shown at the left in Fig. 1. At the instant the battery is connected to the two wires of the transmission line, an electric current can be detected in

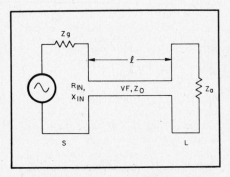

Fig. 1—Source and load connected by means of a transmission line. VF (velocity factor) and Z_0 (characteristic impedance) are properties of the line, as discussed in the text.

the wires near the battery terminals. This current does not flow instantaneously throughout the length of the wires, however.

The electric field that causes the current cannot travel faster than the speed of light, so a measurable interval of time elapses before the currents become evident even a relatively short distance away. The current could not be detected at the load, 300 meters away, until one microsecond (one millionth of a second) after the connection was made. By ordinary standards this is a very short length of time, but in terms of radio frequencies it represents the time for one complete cycle of a 1-MHz radio signal. It also represents the time for 10 complete cycles of a 10-MHz signal.

In a practical transmission line the energy travels somewhat slower than the speed of light (typically from 65 to 97 percent of light speed), depending on the type of transmission line. This line characteristic is called the *velocity factor* (VF) of the line.

The transmission-line current charges the capacitance between the two wires. The conductors of this "linear" capacitor also have inductance, however. The transmission line may be thought of as being composed of a whole series of small inductors and capacitors connected as shown in Fig. 2. Each inductor represents the inductance of a very short section of one wire, and each capacitor represents the capacitance between two such short sections.

CHARACTERISTIC IMPEDANCE

Consider an infinitely long chain of inductors and capacitors connected as shown in Fig. 2. All the small inductors have the same value, as do all the small capacitors. This line has an important property. To an electrical impulse applied at one end, the combination appears to have an impedance. This is called the *characteristic impedance* or *surge impedance*, often abbreviated Z_0. Its value is approximately equal to $\sqrt{L/C}$, where L and C are the inductance and capacitance per unit length.

In the equivalent transmission line of Fig. 2 there are no resistors, and thus no power is lost in the line. In this case, Z_0 becomes a pure resistance, R_0. (Lost power is represented by I^2R, or power dissipated in the line.) With no resistance, there is no power loss no matter what the line length may be. This may not seem consistent with calling the impedance a pure resistance, because the word *resistance* implies that some power will be dissipated if current flows. But with an infinitely long line, the effect (as far as the source of power is concerned) is exactly the same as if the line were replaced by a pure resistance. This is because the energy leaves

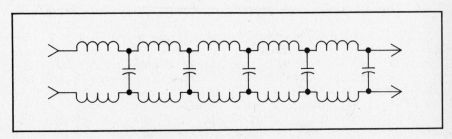

Fig. 2—Equivalent of a transmission line in lumped circuit constants.

the source and travels outward forever along the line.

The characteristic impedance of a line determines the amount of current that can flow when a given voltage is applied to an infinitely long line. The impedance limits the flow of current in the same way that a definite value of actual resistance limits current flow when a voltage is applied.

All practical transmission lines exhibit some power loss. These losses occur in the resistance that is inherent in the conductors that make up the line, and from leakage currents flowing in the dielectric material between the conductors. Because of these losses, the characteristic impedance (or Z_0) of practical lines contains a small reactive component. This component is usually capacitive, but it is small enough that its effects may be disregarded in Amateur Radio applications.

The inductance and capacitance per unit length of line depend on the size of the conductors and the spacing between them. The smaller the spacing between the two conductors and the greater their diameter, the higher the capacitance and the lower the inductance. A line with closely spaced large conductors will have low impedance, while one with widely spaced small conductors will have relatively high impedance.

Matched Lines

Actual transmission lines do not extend to infinity, but have a definite length. In use they are connected to, or *terminate* in, a load at the output end of the line, as illustrated in Fig. 1. If the load is a pure resistance of a value equal to the characteristic impedance of the line, the line is said to be *matched*. To current traveling along the line, such a load just looks like still more transmission line of the same characteristic impedance.

In other words, a short line terminated in a purely resistive load equal to the characteristic impedance of the line acts just as if the line was infinitely long. In a matched transmission line, energy travels outward along the line from the source until it reaches the load, where it is completely absorbed.

RF on Lines

Assume now that a very short burst of power is emitted from the source in Fig. 1. This is represented by the vertical line at the left of the series of lines in Fig. 3. As the pulse voltage appears across the load Z_a of Fig. 1, all the energy may be absorbed, or part of it may be reflected in much the same manner that energy in a water wave is reflected as the wave hits a steep breakwater or the end of a container. This reflected wave is represented by the second line in the series, and the arrow above the line indicates the direction of travel. As the second wave reaches the source, the process is repeated, with all

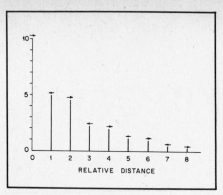

Fig. 3—Component magnitudes for forward and reverse traveling waves produced by a short pulse on a transmission line.

of the energy being either absorbed or partially reflected.

Even though the back-and-forth cycle is actually infinite, after a few reflections the intensity of the traveling wave becomes very small. If, instead of a short pulse, a continuous RF voltage is applied to the terminals of a transmission line, the voltage at any point along the line will consist of a sum of voltages—the composite of waves traveling toward the right and waves traveling toward the left. The sum of the waves traveling toward the right is called the *forward wave* or incident wave, while the one traveling toward the left is called the *reflected wave*. If Z_a contains resistance, there will be a net flow of energy from the source to the load.

Line Factors and Equations

In practical applications of transmission lines, one factor is very important: the manner in which the source and load are connected. Some "impedance lump" effects may arise from connectors and from the method of connecting the source and the load. These effects may cause the line to "see" a different impedance than would be indicated if the measurements were made directly at the load terminals. This is shown in Fig. 1 by the abrupt change in dimensions at each end of the line, ℓ. Even though the short line connecting the generator to the main transmission line (and the one connecting the load to the line) might have the same characteristic impedance, a mismatch will still occur if the physical sizes are different, unless the adapter connecting the lines is tapered to effect a smooth change between the different dimensions. This effect can usually be neglected at HF, but it becomes important as the frequency of operation is extended into the VHF region and above.

In the example shown in Fig. 3, the ratio of the voltage in the reflected wave to the voltage in the incident wave is defined as the voltage reflection coefficient. This coefficient is designated by the Greek letter rho (ρ). The relationship between the out-

put resistance (R_a), the output reactance (X_a), the line impedance (Z_0) and the magnitude of the reflection coefficient is

$$\rho = \sqrt{\frac{(R_a - R_0)^2 + X_a{}^2}{(R_0 + R_a)^2 + X_a{}^2}} \qquad \text{(Eq. 1)}$$

Note that if R_a is equal to R_0 and X_a is 0, the reflection coefficient is 0. This represents a matched condition where all the energy in the incident wave is transferred to the load. In effect, it is as if there is an infinite line of characteristic impedance Z_0 connected at A. On the other hand, if R_a is 0, regardless of the value of X_a, the reflection coefficient is 1.0. This means that *all* the power is reflected, in much the same manner as radiant energy is reflected from a mirror.

If there are no reflections from the load, the voltage distribution along the line is constant or "flat." A line operating under these conditions is often referred to as a *flat line*. If reflections exist, a voltage standing-wave pattern will result. The ratio of the maximum voltage on the line to the minimum value (provided the line is longer than a quarter wavelength) is defined as the *voltage standing wave ratio*, or VSWR. Reflections from the load also result in a standing-wave pattern of current flowing in the line. The ratio of maximum to minimum current, or ISWR, is identical to the VSWR in a given line. In amateur literature the abbreviation SWR is commonly used for standing-wave ratio, as the results are identical when taken from proper measurements of either current or voltage. The SWR is related to the reflection coefficient by

$$SWR = \frac{1 + \rho}{1 - \rho} \qquad \text{(Eq. 2)}$$

and

$$\rho = \frac{SWR - 1}{SWR + 1} \qquad \text{(Eq. 3)}$$

The definition of Eq. 3 is a more general one, valid for any line length. Often, the actual load impedance is unknown. An alternative way of expressing the reflection coefficient is

$$\rho = \sqrt{\frac{P_r}{P_f}} \qquad \text{(Eq. 4)}$$

where

P_r = the power in the reflected wave
P_f = the power in the forward wave

The parameters are relatively easy to measure with a directional RF wattmeter (power meter), either homemade or purchased. It should be obvious, however, that there can be no other power sources at the load if the definition above is to hold. For instance, assume the reflection coefficient (ρ) of the generator in the example shown in Fig. 3 is 0.9. This value could be obtained by substituting the generator resistance and reactance into Eq. 1, but not by measurement if the source is activated.

It is possible to determine the input resistance and reactance of a terminated line if the load resistance and reactance are known. The line length and characteristic impedance must also be known. The equations are

$$R_{in} = \frac{R_a (1 + \tan^2\ell)}{\left(1 - \frac{X_a}{Z_0} \tan\ell\right)^2 + \left(\frac{R_a}{Z_0} \tan\ell\right)^2}$$

(Eq. 5)

$$X_{in} = \frac{X_a (1 - \tan^2\ell) + \left(Z_0 - \frac{R_a{}^2 - X_a{}^2}{Z_0}\right)\tan\ell}{\left(1 - \frac{X_a}{Z_0} \tan\ell\right)^2 + \left(\frac{R_a}{Z_0} \tan\ell\right)^2}$$

(Eq. 6)

where

R_{in} = resistive component of the impedance at the input end of the line
X_{in} = reactive component of the impedance at the input end of the line
R_a = load resistance
X_a = load reactance
ℓ = electrical line length in degrees or radians

To determine the value of the tangent function, the line length, along with the frequency in MHz and the velocity factor (VF), can be substituted into the following expression:

$$\ell \text{ (degrees)} = 0.366f_{(MHz)} \times \ell_{(feet)}/VF$$

(Eq. 7)

Lost Power

Often it is mistakenly assumed that power reflected from a load represents power that is lost in some way. With proper matching at the input end of the line, this is true only if there is considerable loss in the line itself. In addition to the forward loss, some reflected power is dissipated on the way back to the source.

If the terminating resistance is zero, the input reactance consists of a pure reactance, which is given by

$$X_{in} = Z_0 \tan\ell$$

(Eq. 8)

If the line termination is infinite in value (an open circuit), the input reactance is given by

$$X_{in} = Z_0 \cot\ell$$

(Eq. 9)

A short length of line (less than a quarter wavelength) with a short circuit as a terminating load appears as an inductance, while an open-circuited line appears as a capacitance.

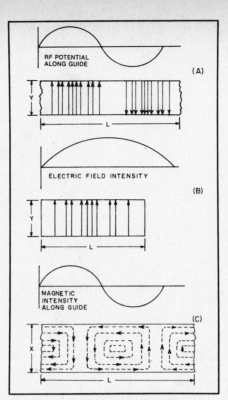

Fig. 4 — Field distribution in a rectangular waveguide. The $TE_{1,0}$ mode of propagation is depicted.

Waveguides

A waveguide is a conducting tube through which energy is transmitted in the form of electromagnetic waves. The tube is not considered as carrying a current in the same sense that the wires of a two-conductor line do, but rather a boundary which confines the waves to the enclosed space. Skin effect prevents any electromagnetic effects from being evident outside the guide. The energy is injected at one end, either through capacitive or inductive coupling or by radiation, and is received at the other end. The waveguide merely confines the energy of the fields, which are propagated through it to the receiving end by means of reflections against its inner walls.

Analysis of waveguide operation is based on the assumption that the guide material is a perfect conductor of electricity. Typical distributions of electric and magnetic fields in a rectangular guide are shown in Fig. 4. It will be observed that the intensity of the electric field is greatest (as indicated by closer spacing of the lines of force) at the center along the X dimension, Fig. 4B, diminishes to zero at the end walls. Zero field intensity is a necessary condition at the end walls, since the existence of any electric field parallel to the walls at the surface would cause an infinite current to flow in a perfect conductor. This represents an impossible situation.

Modes of Propagation

Fig. 4 represents a relatively simple

Table 1
Wavelength Formulas for Waveguide

	Rectangular	Circular
Cut-off wavelength	2X	3.41R
Longest wavelength transmitted with little attenuation	1.6X	3.2R
Shortest wavelength before next mode becomes possible	1.1X	2.8R

distribution of the electric and magnetic fields. An infinite number of ways exist in which the fields can arrange themselves in a guide, as long as there is no upper limit to the frequency to be transmitted. Each field configuration is called a mode. All modes may be separated into two general groups. One group, designated TM (transverse magnetic), has the magnetic field entirely transverse to the direction of propagation, but has a component of electric field in that direction. The other type, designated TE (transverse electric) has the electric field entirely transverse, but has a component of magnetic field in the direction of propagation. TM waves are sometimes called E waves, and TE waves are sometimes called H waves. The TM and TE designations are preferred, however.

The particular mode of tranmission is identified by the group letters followed by subscript numberals; for example $TE_{1,0}$, $TM_{1,1}$, etc. The number of possible modes increases with frequency for a given size of guide. There is only one possible mode (called the dominant mode) for the lowest frequency that can be transmitted. The dominant mode is the one normally used in practical work.

Waveguide Dimensions

In the rectangular guide the critical dimension in Fig. 4C is X; this dimension must be more than one-half wavelength at the lowest frequency to be transmitted. In practice, the Y dimension usually is made about equal to ½ X to avoid the possibility of operation at other than the dominant mode. Other cross-sectional shapes than the rectangle can be used, the most important being the circular pipe. Many of the same considerations apply as in the rectangular shape.

Wavelength formulas for rectangular and circular guides are given in Table 1, where X is the width of a rectangular guide and R is the radius of a circular guide. All figures are in terms of the dominant mode.

Coupling to Waveguides

Energy may be introduced into or extracted from a waveguide or resonator by means of either the electric or magnetic field. The energy transfer takes place frequently through a coaxial line. Two methods for coupling are shown in Fig. 5. The probe at A is simply a short extension

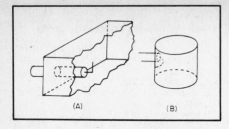

Fig. 5 — Coupling to waveguide and resonators.

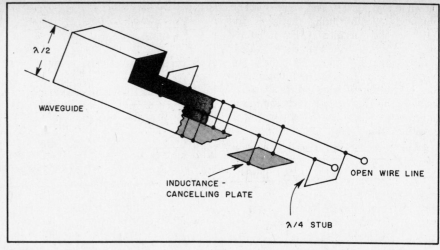

Fig. 6 — At its cutoff frequency a rectangular waveguide can be analyzed as a parallel two-conductor transmission line supported from top and bottom by an infinite number of quarter-wavelength stubs.

of the inner conductor of the coaxial line, oriented so it is parallel to the electric lines of force. The loop shown at B is arranged to enclose some of the magnetic lines of force. The point at which maximum coupling will be obtained depends on the particular mode of propagation in the guide or cavity; the coupling will be maximum when the coupling device is in the most intense field.

Coupling can be varied by rotating the probe or loop through 90 degrees. When the probe is perpendicular to the electric lines the coupling will be minimum; similarly, when the plane of the loop is parallel to the magnetic lines, the coupling will be minimum.

Evolution of a Waveguide

Suppose an open-wire line is used to carry VHF energy from a generator to a load. If the line has any appreciable length, it must be well insulated from the supports to avoid high losses. Since high-quality insulators are difficult to make for microwave frequencies, it is logical to support the transmission line with quarter-wavelength stubs, shorted at the far end. The open end of such a stub presents an infinite impedance to the transmission line, provided the shorted stub is nonreactive. However, the shorting link has finite length and, therefore, some inductance. This inductance can be nullified by making the RF current flow on the surface of a plate rather than through a thin wire. If the plate is large enough, it will prevent the magnetic lines of force from encircling the RF current.

An infinite number of these quarter-wave stubs may be connected in parallel without affecting the standing waves of voltage and current. The transmission line may be supported from the top as well as the bottom, and when infinitely many supports are added, they form the walls of a waveguide at its cutoff frequency. Fig. 6 illustrates how a rectangular waveguide evolves from a two-wire parallel transmission line. This simplified analysis also shows why the cutoff dimension is a half wavelength.

While the operation of waveguides is usually described in terms of fields, current flows on the inside walls, just as fields exist between the conductors of a two-wire transmission line. At the waveguide cutoff frequency, the current is concentrated in the

center of the walls, and disperses toward the floor and ceiling as the frequency increases.

MATCHING THE ANTENNA TO THE LINE

The load for a transmission line may be any device capable of dissipating RF energy. When lines are used in applications involving transmitters, the most common load is an antenna. When a transmission line is connected between an antenna and a receiver, the receiver input circuit is the load (not the antenna), because the power taken from a passing wave is delivered to the receiver.

Whatever the application, the conditions existing at the load, and *only* the load, determine the standing-wave ratio on the line. If the load is purely resistive and equal to the characteristic impedance of the line, there will be no standing waves. If the load is not purely resistive, or is not equal to the line Z_0, there will be standing waves. No adjustments can be made at the input end of the line to change the VSWR. Neither is the VSWR affected by changing the line length.

Only in a few special cases is the load inherently of the proper value to match a practicable transmission line. In all other cases it is necessary either to operate with a mismatch and accept the VSWR that results, or else to take steps to bring about a proper match between the line and load by means of transformers or similar devices. Impedance-matching transformers may take a variety of physical forms, depending on the circumstances.

Note that it is essential, if the VSWR is to be made as low as possible, that the load at the point of connection to the transmission line be purely resistive. In general, this requires that the load be tuned to resonance. If the load itself is not resonant at the operating frequency, the tuning can

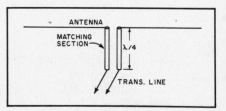

Fig. 7 — "Q" matching section, a quarter-wave impedance transformer.

sometimes be accomplished in the matching system.

The Antenna as a Load

Every antenna system, no matter what its physical form, will have a definite value of impedance at the point where the line is connected. The problem is to transform this antenna input impedance to the proper value to match the line. In this respect there is no "best" type of line for a particular antenna system, because it is possible to transform impedances in any desired ratio. Consequently, any type of line may be used with any type of antenna. There are frequently reasons other than impedance matching that dictate the use of one type of line in preference to another, such as ease of installation, inherent loss in the line, and so on, but these are not considered in this section.

Although the input impedance of an antenna system is seldom known accurately, it is often possible to make a close estimate of its value. Matching circuits can be built with ordinary coils and capacitors, but these are not used extensively because they must be supported at the antenna and must be weatherproofed. The systems described below use *linear transformers*.

The Quarter-Wave Transformer or "Q" Section

As mentioned previously, a quarter-wave

transmission line may be used as an impedance transformer. See Fig. 7. If the antenna impedance and the characteristic impedance of the line to be matched are known, the characteristic impedance of a matching section is

$$Z = \sqrt{Z_1 Z_0}$$

where

Z_1 = antenna impedance
Z_0 = characteristic impedance of the line to which it is to be matched.

Example: To match a 600-ohm line to an antenna presenting a 72-ohm load, the quarter-wave matching section would require a characteristic impedance of

$$\sqrt{72 \times 600} = \sqrt{43,200} = 208 \text{ ohms}$$

The spacing between conductors and the conductor size determine the characteristic impedance of the transmission line. As an example, for the 208-ohm transmission line required above, the line could be made from ½-inch-diameter tubing spaced 1.5 inches between conductors.

The length of the quarter-wave matching section may be calculated from

$$\text{Length (feet)} = \frac{246 \text{ VF}}{f}$$

$$\text{Length (meters)} = \frac{75 \text{ VF}}{f}$$

where

VF = velocity factor
f = frequency in MHz

Example: A quarter-wave transformer of RG-11 is to be used at 28.7 MHz. From Table 2, VF = 0.66.

$$\text{Length} = \frac{246 \times 0.66}{28.7} = 5.66 \text{ feet}$$

$$= 5 \text{ feet } 8 \text{ inches}$$

The antenna must be resonant at the operating frequency. Setting the antenna length by formula is accurate enough with single-wire antennas, but in other systems, particularly close-spaced arrays, the antenna should be adjusted to resonance before the matching section is connected.

When the antenna input impedance is not accurately known, it is advisable to construct the matching section so the spacing between conductors can be changed. The spacing may then be adjusted to give the lowest SWR on the transmission line.

Series-Section Transformers

The *series-section transformer* has advantages over the quarter-wave transformer. The material in this section is based on a *QST* article by Frank A. Regier, OD5CG/WA1ZQC (Ref. 1 of the bibliography at the end of this chapter). The transformer illustrated in Fig. 8 closely resembles the ¼-wave transformer. (Actually, the ¼-wave transformer is a special case of the series-section transformer.) The important differences are: first, that the matching section need not be located exactly at the load; second, that it may be less than a quarter wavelength long; and third, that there is great freedom in the choice of the characteristic impedance of the matching section.

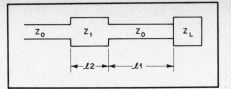

Fig. 8 — Series section transformer Z_1 for matching transmission-line Z_0 to load, Z_L.

In fact, the matching section can have any characteristic impedance that is not too close to that of the main line. Because of this freedom, it is almost always possible to find a length of commercially available line that will be suitable as a matching section. As an example, consider a 75-Ω line, a 300-Ω matching section, and a purely resistive load. It can be shown that such a section may be used to match any resistance between 5 Ω and 1200 Ω to the main line.

The design of a series-section transformer consists of determining (1) the length $\ell2$ of the series or matching section and (2) the distance $\ell1$ from the load to the point where the section should be inserted into the main line. Three quantities must be known: the characteristic impedances of the main line and of the matching section, both assumed purely resistive, and the complex load impedance, Z_L.

The two lengths, $\ell1$ and $\ell2$, are to be determined from the characteristic impedances of the main line and the matching section, Z_0 and Z_1, respectively, and from the load impedance ($Z_L = R_L + jX_L$). The first step is to determine the normalized impedances.

$$n = \frac{Z_1}{Z_0}$$

$$r = \frac{R_L}{Z_0}$$

$$x = \frac{X_L}{Z_0}$$

Next, $\ell2$ and $\ell1$ are determined from

$$\ell2 = \arctan B$$
$$\ell1 = \arctan A$$

where

$$B = \pm \sqrt{\frac{(r-1)^2 + x^2}{r\left(n - \frac{1}{n}\right)^2 - (r-1)^2 - x^2}}$$

$$A = \frac{\left(n - \frac{r}{n}\right)B + x}{r + xnB - 1}$$

Lengths $\ell2$ and $\ell1$ thus determined are electrical lengths in degrees. Actual lengths are obtained by dividing by 360° and multiplying by the wavelength measured along the line (main line or matching

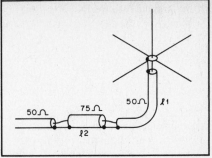

Fig. 9 — Example of series-section matching. A 38-Ω antenna is matched to 50-Ω coax by means of a length of 75-Ω cable.

section, as the case may be), taking the velocity factor of the line into account.

The sign of B may be chosen either positive or negative, but the positive sign is preferred because it results in a shorter matching section. The sign of A may not be chosen, but may turn out to be either positive or negative. If a negative sign occurs and an electronic calculator is used to determine $\ell1$, a negative electric length will result. If this happens, add 180°. The resultant electrical length will be correct both physically and mathematically.

In calculating B, if the quantity under the radical is negative, an imaginary value for B results. This means that Z_1, the impedance of the matching section, is too close to Z_0 and should be changed.

Limits on the characteristic impedance of Z_1 may be calculated in terms of the standing-wave ratio produced by the load on the main line without matching. For matching to occur, Z should either be greater than $Z_0\sqrt{VSWR}$ or less than $Z_0\sqrt{VSWR}$.

An Example

Suppose we want to feed a 29-MHz ground-plane vertical antenna with RG-58-type foam-dielectric coax (Fig. 9). We'll assume the antenna impedance to be 38 ohms pure resistance, and use a length of RG-59 foam-dielectric coax as the series section.

Z_0 is 50 ohms, Z_1 is 75 ohms, and both cables have a velocity factor of 0.79. (From above, Z_1 must have an impedance greater than 57.4 Ω or less than 43.6 Ω.) The design steps are as follows.

From the earlier equations,
n = 1.5, r = 0.76 and x = 0.
Further, B = 0.3500 (positive sign chosen), $\ell2$ = 19.29° and A = −1.4486. Calculating $\ell1$ yields −55.38°. Adding 180° to obtain a positive result gives $\ell1$ = 124.62°.

To find the physical lengths $\ell1'$ and $\ell2'$, we first find the free-space wavelength.

$$\lambda_o = \frac{984}{f_{MHz}} \text{ feet}$$

and the transmission-line wavelength

$$\lambda = \lambda_o \times \text{velocity factor}$$

In the present case we find $\lambda = 26.81$ ft. Finally, we have

$$\ell1' = \frac{\ell1 \times \lambda}{360} = 9.28 \text{ ft, and}$$

$$\ell2' = \frac{\ell2 \times \lambda}{360} = 1.44 \text{ ft.}$$

This completes the calculations. Construction consists of cutting the main coax at a point 9.28 ft from the antenna and adding a 1.44-ft length of the 75-Ω cable.

The Quarter-Wave Transformer

The antenna in the preceding example could have been matched by a ¼-λ transformer at the load. Such a transformer would have a characteristic impedance of 43.6 Ω. It is interesting to see what happens in the design of a series-section transformer if this value is chosen as the characteristic impedance of the series section.

Following the same steps as before, we find n = 0.872, r = 0.76 and x = 0.

From these values we find B = ∞ and $\ell2 = 90°$. Further, A = 0 and $\ell1 = 0°$. These results represent a quarter-wave section at the load, and indicate that, as stated earlier, the quarter-wave transformer is indeed a special case of the series-section transformer.

Folded Dipoles

If a half-wave antenna element is split into two or more parallel conductors, it can be made to match various line impedances. The transmission line is attached at the center of only one conductor. Various forms of such "folded dipoles" are shown in Fig. 10. Currents in all conductors are in phase in a folded dipole and, since the conductor spacing is small, the folded dipole is equivalent in radiating properties to an ordinary single-conductor dipole. However, the current flowing into the input terminals of the antenna from the line is the current in one conductor only, and the entire power from the line is delivered at this value of current. This is equivalent to saying that the input impedance of the antenna has been raised by splitting it up into two or more conductors.

The ratio by which the input impedance of the antenna is stepped up depends not only on the number of conductors in the folded dipole but also on their relative diameters, since the distribution of current between conductors is a function of their diameters. (When one conductor is larger than the other, as in Fig. 10C, the larger one carries the greater current.) The ratio also depends on the spacing between the conductors, as shown by the graphs of Figs. 11 and 12. An important special case is the two-conductor dipole with conductors of equal diameter; as a simple antenna, not a part of a directive array, it has an input impedance close enough to 300 ohms to afford a good match to 300-ohm twin-lead.

Fig. 11 indicates the required ratio of conductor diameters in a 2-conductor

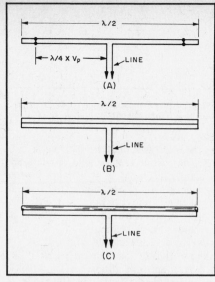

Fig. 10 — The folded dipole, a method for using the antenna element itself to provide an impedance transformation. Depending on the velocity factor of the line used for the flat top, the conductors should be shorted part-way in from the end. This is because the sections must be resonant both as quarter-wave transmission-line stubs and quarter-wavelength single-wire radiators. A flat top made from 300-ohm twin-lead would have shorting straps at about 85 percent of the distance from center to end. However, no straps are necessary when open-wire line is used in the flat top. This is because the velocity factor of the line (about 97 percent) is very close to the percentage of a physical half wavelength needed for electrical resonance in the "thick" antenna thus created.

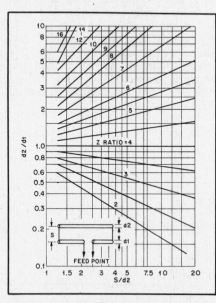

Fig. 11 — Impedance transformation ratio, two-conductor folded dipole. The dimensions d1, d2 and s are shown on the inset drawing. Curves show the ratio of the impedance (resistive) seen by the transmission line to the radiation resistance of the resonant antenna system.

dipole to give a desired impedance ratio. Similar information for a three-conductor dipole is given in Fig. 12. This graph applies if all the three conductors are in the same plane. The two conductors not con-

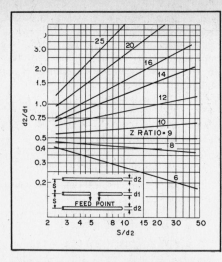

Fig. 12 — Impedance transformation ratio, three-conductor folded dipole. The dimensions d1, d2 and s are shown on the inset drawing. Curves show the ratio of the impedance (resistive) seen by the transmission line to the radiation resistance of the resonant antenna system.

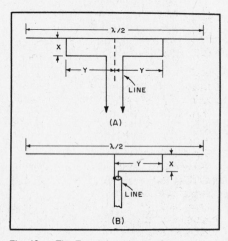

Fig. 13 — The T match and gamma match.

nected to the transmission line must be equally spaced from the fed conductor, and must have equal diameters. The fed conductor may have a different diameter, however. The unequal-conductor method has been found particularly useful in matching to low-impedance antennas such as directive arrays using close-spaced parasitic elements.

The antenna element should be approximately self-resonant at the median operating frequency. The length is seldom critical, because a folded dipole tends to have the characteristics of a "thick" antenna and thus has a relatively broad frequency-response curve.

T and Gamma Matching Sections

The method of matching shown in Fig. 13A is based on the fact that the impedance between any two points equidistant from the center along a resonant antenna is resistive, and has a value that depends on the spacing between the two points. It is therefore possible to choose a pair of points between which the impedance will have the right value to match a transmission line. In practice, the line cannot be connected directly at these points because the distance between them is much greater than the conductor spacing of a practical transmission line. The T arrangement in Fig. 13 overcomes this difficulty by using a second conductor paralleling the antenna to form a matching section to which the line may be connected.

The T is particularly suited to use with a parallel-conductor line. The operation of this system is somewhat complex. Each T conductor (Y in the drawing) forms a short section of transmission line with the antenna conductor opposite it. Each of these transmission-line sections can be considered to be terminated in the impedance that exits at the point of connection to the antenna. Thus, the part of the antenna between the two points carries a transmission-line current in addition to the normal antenna current. The two transmission-line matching sections are in series, as seen by the main transmission line.

If the antenna by itself is resonant at the operating frequency, its impedance will be purely resistive. In this case the matching-section lines are terminated in a resistive load. As transmission-line sections, these matching sections are terminated in a short, and are shorter than a quarter wavelength. Thus their input impedance, i.e., the impedance seen by the main transmission line looking into the matching-section terminals, will be reactive as well as resistive. This prevents a perfect match to the main transmission line, since its load must be purely resistive for perfect matching. The reactive component of the input impedance must be tuned out before a proper match can be obtained.

One way to do this is to detune the antenna just enough, by changing its length, to cause reactance of the opposite kind to be reflected to the input terminals of the matching section, thus canceling the reactance introduced. Another method, which is considerably easier to adjust, is to insert a variable capacitor in series with each matching section where it connects to the transmission line, as shown in Chapter 19. The capacitors must be protected from the weather.

The method of adjustment commonly used is to cut the antenna for approximate resonance and then make the spacing X some value that is convenient constructionally. The distance Y is then adjusted, while maintaining symmetry with respect to the center, until the VSWR on the transmission line is as low as possible. If the VSWR is not below 2:1 after this adjustment, the antenna length should be changed slightly and the matching section taps adjusted again. This procedure may be continued until the VSWR is close to 1:1.

When the series-capacitor method of reactance compensation is used, the anten-

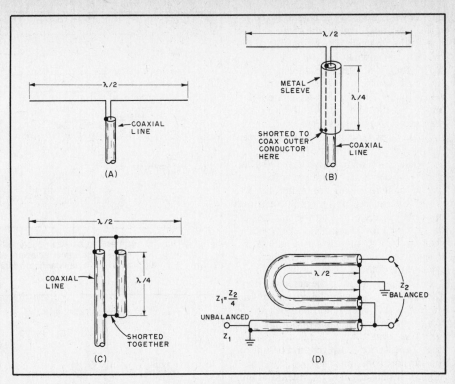

Fig. 14 — Radiator with coaxial feed (A) and methods of preventing unbalanced currents from flowing on the outside of the transmission line (B and C). The half-wave phasing section shown at D is used for coupling between an unbalanced circuit when a 4:1 impedance ratio is desired or can be accepted.

na should be the proper length for resonance at the operating frequency. Trial positions of the matching-section taps are then taken, each time adjusting the capacitor for minimum VSWR, until the standing waves on the transmission line are brought down to the lowest possible value. The unbalanced (gamma) arrangement in Fig. 13B is similar in principle to the T, but is adapted for use with single coax line. The method of adjustment is the same.

LOADS AND BALANCING DEVICES

The most important practical load for a transmission line is an antenna, which in most cases will be "balanced" — that is, constructed symmetrically with respect to the feed point. A balanced antenna should be fed with a balanced feeder system in order to preserve symmetry with respect to ground. This will avoid difficulties with unbalanced currents on the line and consequent undesirable radiation from the transmission line itself. Line radiation can be prevented by a number of devices that detune or decouple the line for "antenna" currents, and thus greatly reduce their amplitude.

Often the antenna is fed through coaxial line, which is inherently unbalanced. In such a case, some method should be used for connecting the line to the antenna without upsetting the symmetry of the antenna itself. This requires a circuit that will isolate the balanced load from the unbalanced line while providing efficient power transfer. Devices for doing this are

called baluns (a contraction for "balanced to unbalanced").

An antenna with open ends, of which the half-wave dipole is an example, is inherently a balanced radiator. When the antenna is opened at the center and fed with a parallel-conductor line, this balance is maintained throughout the system, as long as external causes of unbalance are avoided.

If the antenna is fed at the center through a coaxial line, as indicated in Fig. 14A, this balance is upset because one side of the radiator is connected to the shield while the other is connected to the inner conductor. On the side connected to the shield, current can flow down over the outside of the coaxial line. The fields thus set up cannot be canceled by the fields from the inner conductor because the inside fields cannot escape through the shielding of the outer conductor. Hence these "antenna" currents flowing on the outside of the line will be responsible for radiation.

Linear Baluns

Fig. 14B shows a balun arrangement known as a *bazooka*, which uses a sleeve over the transmission line. The sleeve, with the outside of the outer conductor, forms a shorted quarter-wave line section. As described earlier in this chapter, the impedance looking into the open end of such a section is very high, so the end of the outer conductor of the coaxial line is effectively isolated from the part of the line below the sleeve. The length is an electrical quarter wave, and may be physically shorter if the insulation between the sleeve and the line is other than air. The bazooka (and the balun form shown in Fig. 14C) have no effect on antenna impedance at the frequency where the ¼-wave sleeve (and the line section in C are resonant. However, the sleeve and line section both add inductive shunt reactance at frequencies lower, and capacitive shunt reactance at frequencies higher than the ¼-wave-resonant frequency.

Another method that gives an equivalent effect is shown at C. Since the voltages at the antenna terminals are equal and opposite (with reference to ground), equal and opposite currents flow on the surfaces of the line and second conductor. Beyond the shorting point, in the direction of the transmitter, these currents combine to cancel out. The balancing section "looks like" an open circuit to the antenna, since it is a quarter-wave parallel-conductor line shorted at the far end, and thus has no effect on the normal antenna operation. This is not essential to the line-balancing function of the device, however, and baluns of this type are sometimes made shorter than a quarter wavelength to provide the shunt inductive reactance required in certain matching systems.

Fig. 14D shows a third balun, in which equal and opposite voltages, balanced to ground, are taken from the inner conductors of the main transmission line and half-wave phasing section. Since the voltages at the balanced end are in series while the voltages at the unbalanced end are in parallel, there is a 4:1 step-down in impedance from the balanced to the unbalanced side. This arrangement is useful for coupling between a 300-ohm balanced line and a 75-ohm coaxial line, for example.

Coil Baluns

The coil balun is based on the principles of a linear-transmission-line balun as shown in the upper drawing of Fig. 15. Two transmission lines of equal length having a characteristic impedance Z_0 are connected in series at one end and in parallel at the other. At the series-connected end the lines are balanced with respect to ground and will match an impedance equal to $2Z_0$. At the parallel-connected end the lines will be matched by an impedance equal to $Z_0/2$. One side may be connected to ground at the parallel-connected end, provided the two lines have a length that is an odd multiple of a quarter wavelength.

A definite line length is required only for decoupling purposes and, as long as there is adequate decoupling, the system will act as a 4:1 impedance transformer regardless of line length. If each line is wound into a coil, as in the lower drawing, the inductances so formed will act as choke coils and will tend to isolate the series-connected end from any ground connection that may be placed on the parallel-connected end. Balun coils made in this way will operate over a

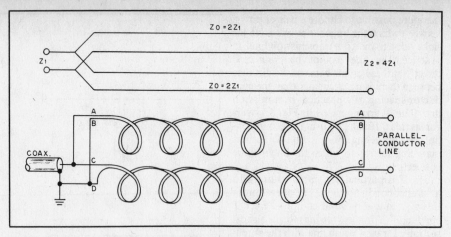

Fig. 15 — Baluns for matching between push-pull and single-ended circuits. The impedance ratio is 4:1 from the push-pull side to the unbalanced side. Coiling the lines (lower drawing) increases the frequency range over which satisfactory operation is obtained.

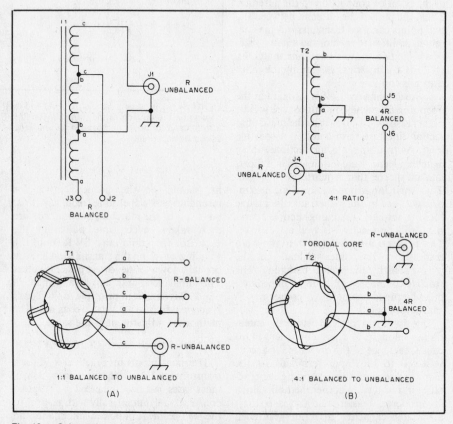

Fig. 16 — Schematic and pictorial representations of balun transformers. The windings are labeled a, b and c to show the relationship between the pictorial and schematic illustrations. Construction information on baluns of this type is given in Chapter 33.

wide frequency range, since the choke inductance is not critical. The lower frequency limit is where the coils are no longer effective in isolating one end from the other; the length of line in each coil should be about equal to a quarter wavelength at the lowest frequency to be used.

The principal application of such coils is in going from a 300-ohm balanced line to a 75-ohm coaxial line. This requires that the Z_0 of the lines forming the coils be 150 ohms.

A balun of this type, when matched, is simply a fixed-ratio transformer. It cannot compensate for inaccurate matching

elsewhere in the system. With a "300-ohm" line on the balanced end, for example, a 75-ohm coaxial cable will not be matched unless the 300-ohm line actually is terminated in a 300-ohm load.

Toroidal Baluns

Air-wound balun transformers are somewhat bulky when designed for operation in the 1.8- to 30-MHz range. A more compact broadband transformer can be made by using toroidal ferrite core material as the foundation for bifilar-wound coil balun transformers.

In Fig. 16 at A, a 1:1 ratio balanced-to-

Fig. 17 — An RF choke formed by coiling the feed line at the point of connection to the antenna. The inductance of the choke isolates the antenna from the remainder of the feed line.

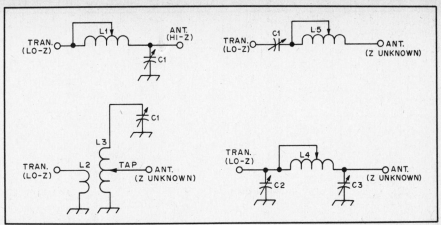

Fig. 18 — Networks for matching a low-impedance transmitter output to a random-length end-fed wire antennas.

unbalanced-line transformer is shown. This transformer is useful in converting a 50-ohm balanced line condition to one that is 50 ohms, unbalanced. Similarly, the transformer will work between balanced and unbalanced 75-ohm impedances. A 4:1 ratio transformer is illustrated in Fig. 16 at B. This balun is useful for converting a 200-ohm balanced condition to one that is 50 ohms, unbalanced. In a like manner, the transformer can be used between a balanced 300-ohm point and a 75-ohm unbalanced line.

Low-loss high-frequency ferrite core material is used for the toroidal forms. Power losses in the core are unavoidable, however, and the core size and wire size must be suitably chosen for the highest level of power that will be used. Similarly, the core must be chosen for the type of material that is suitable for the frequency range of operation.

Coil baluns may also be used as impedance transformers, as previously mentioned for the 4:1 balun of Fig. 16B. Be aware, however, that core saturation may occur at high power levels, thus nullifying a portion of the impedance-transformation capabilities. *Mismatches are masked at high powers because of this effect.* For this reason, power limitations and VSWR limitations are imposed on coil baluns. Further, a coil balun introduces an impedance transfer error, which causes skewing of the VSWR and impedance plots. Stated another way, the input impedance of a 1:1 coil balun that is terminated in a 50-ohm resistance will not be precisely 50 ohms of resistance. Depending on the balun construction, there will usually be an inductive reactance in addition to the resistance. But if the limitations of coil baluns are understood and respected, such baluns can be quite useful in an antenna system. Construction information on this type of balun is presented in the chapter on antenna projects.

Choke Baluns

The unbalanced coupling that results from connecting coaxial line to a balanced antenna may be nullified by choking off the current from flowing on the outside of the feed line. Strictly speaking, the baluns of Fig. 14B and 14C are a form of choke

balun. These baluns, by their nature, are frequency sensitive, however.

A more direct approach to the objective of choking off outside currents on a coaxial feeder is shown in Fig. 17, where the line itself is formed into a coil at the antenna feed point. Choke baluns of this type are broadband in nature. Ten turns of coaxial line coiled at a diameter of 6 or 8 inches form an inductor with enough series reactance to minimize unwanted currents at frequencies from 14 to 30 MHz. (The turns may be held in place with electrical tape.) The effectiveness of a choke of this type decreases at the higher frequencies because of the distributed capacitance among the turns. At lower frequencies, too much coiled-up feed line would be required to yield enough isolation of the remainder of the line from the antenna.

In addition to broad bandwidth, choke baluns avoid the core saturation that occurs at high power levels with toroidal baluns. Also, there are no limitations on the maximum power or VSWR to which the balun may be subjected. Further, there is no skewing of the impedance with a choke balun.

The frequency range of the choke balun can be extended to well below 2 MHz by using a core of high-permeability ferrite instead of air. With higher core permeability, the choke inductance increases dramatically, thereby increasing the reactance at lower frequencies. Of great importance, no core saturation occurs at high power levels in the choke balun because there is no high level of primary current that exists in coil baluns.

Another type of choke balun that is very effective was originated by M. Walter Maxwell, W2DU. (See Ref. 2 of the bibliography at the end of this chapter.) One or more ferrite beads may be placed over the outer shield of the coax where it is connected to the antenna. The beads present a high impedance to RF currents that would otherwise tend to flow on the outer conductor. If beads are stacked, the total impedance at a given frequency is in approximate proportion to the number of

beads. For a given number of beads, the impedance generally increases with frequency, but this depends on the type of bead material.

Bead materials of various sizes and RF characteristics are available. By using a stack or a "sleeve" of appropriate beads over ordinary coaxial line, a choke balun may be made to cover a broad frequency range, such as from 2 to 250 MHz.

Nonradiating Loads

Not all loads terminating transmission lines are radiators. Typical examples of nonradiating loads are the grid circuit of a power amplifier, the input circuit of a receiver and another transmission line. This last case includes the "antenna tuner." (Some prefer the term "antenna coupler" because such a device is used to couple a transmission line to a transmitter. ARRL publications use the word *Transmatch,* a coined word covering devices that match the transmitter output to whatever load it sees.)

Coupling to a Receiver

A good match between an antenna and its transmission line does not guarantee a low standing-wave ratio on the line when the antenna system is used for receiving. The VSWR is determined wholly by what the line "sees" at the receiver antenna-input terminals. For minimum VSWR the receiver input circuit must be matched to the line. The rated input impedance of a receiver is a nominal value that varies over a considerable range with frequency. Most HF receivers are sensitive enough that exact matching is not necessary. The most desirable condition is when the receiver is matched to the Z_0 of the line, which in turn is matched to the antenna. This transfers maximum power from the antenna to the receiver with the least transmission line loss.

Coupling to Random-Length Antennas

Several Transmatch schemes are shown in Fig. 18, permitting random-length wires to be matched to normal low-Z transmit-

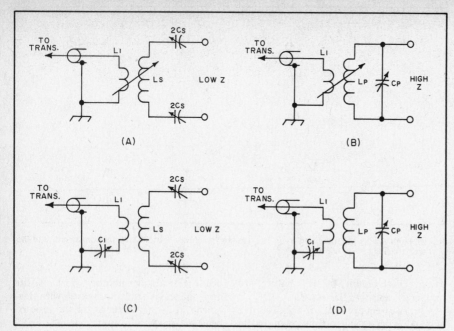

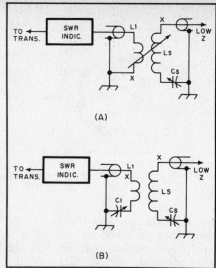

Fig. 19 — Simple circuits for coupling a transmitter to a balanced line that presents a load different than the transmitter output impedance. A and B respectively are series- and parallel-tuned circuits using variable inductive coupling between coils. C and D are similar but use fixed inductive coupling and a variable series capacitor, C1. A series-tuned circuit works well with a low-impedance load; the parallel circuit is better with high-impedance loads (several hundred ohms or more).

Fig. 20 — Coupling from a transmitter designed for 50- to 75-ohm output to a coaxial line with a 3 or 4:1 VSWR is readily accomplished with these circuits. Essential difference between the circuits is adjustable inductive coupling at A, and fixed inductive coupling with variable series capacitor at B. In either case the circuit can be adjusted to give a 1:1 VSWR in the line to the transmitter. The coil ends marked X should be adjacent, for minimum capacitive coupling.

ter outputs. The circuit to be used will depend on the length of the antenna wire and its impedance at the desired operating frequency. Ordinarily one of the four methods shown will provide a suitable impedance match to an end-fed random wire, but the configuration will have to be determined experimentally. A VSWR bridge should be used as a match indicator.

Coupling the Transmitter to the Line

The type of coupling system needed to transfer maximum power from the final RF amplifier to the transmission line depends almost entirely on the input impedance of the line. The input impedance is determined by the standing-wave ratio, the line length and the Z_0 of the line. The simplest case is that in which the line is terminated in its characteristic impedance. In this case the VSWR is 1:1 and the input impedance is equal to the Z_0 of the line, regardless of line length. For this condition the line is said to be "flat."

Coupling systems that will deliver power into a flat line can be readily designed. For practical purposes the line can be considered flat if the VSWR is no greater than about 1.5:1. That is, a coupling system designed to work into a pure resistance equal to the line Z_0 will have enough leeway to take care of the small variations in input impedance that will occur when the line length is changed, if the VSWR is higher than 1:1 but no greater than 1.5:1.

Current practice in transmitter design is to provide an output circuit that will work into such a line, usually a coaxial line of 50 to 75 ohms characteristic impedance. If the input impedance of the transmission line differs appreciably from the transmitter output design impedance, then a matching network must be inserted between the transmitter and the line input terminals.

Impedance-Matching Circuits for Transmission Lines

As shown earlier in this chapter, the input impedance of a line that is operating with a high standing-wave ratio can vary over quite wide limits. The simplest type of circuit that will match such a range of impedances to 50 or 75 ohms is a simple series- or parallel-tuned circuit, approximately resonant at the operating frequency. If the load presented by the line at the operating frequency is low (below a few hundred ohms), a series-tuned circuit should be used. When the load is higher than this, the parallel-tuned circuit is easier to use.

Typical simple circuits for coupling between the transmitter with 50- to 75-ohm coaxial-line output and a balanced transmission line are shown in Fig. 19. The inductor L1 should have a reactance of about 60 ohms when adjustable inductive coupling is used (Figs. 19A and 19B). When a variable series capacitor is used, L1 should have a reactance of about 120 ohms. The variable capacitor, C1, should have a reactance at maximum capacitance of about 100 ohms.

On the secondary side, L_s and C_s should be capable of being tuned to resonance at about 80 percent of the operating frequency. In the series-tuned circuits, for a given low-impedance load, looser coupling can be used between L1 and L_s as the L_s-to-C_s ratio is increased. In the parallel-tuned circuits, for a given high-impedance load, looser coupling can be used between L1 and L_p as the C_p-to-L_p ratio is increased. The constants are not critical; the rules of thumb are mentioned to assist in correcting a marginal condition where sufficient transmitter loading cannot be obtained.

Coupling to a coaxial line that has a high VSWR, and that consequently may present the transmitter with a load it cannot couple to, is done with an unbalanced version of the series-tuned circuit, as shown in Fig. 20. The rule given above for coupling ease and L_s-to-C_s ratio applies to these circuits as well.

The most satisfactory way to set up any of the circuits of Fig. 19 or 20 initially is to connect a coaxial VSWR indicator in the line to the transmitter, as shown in Fig. 20. The RF-wattmeter type of bridge, which can handle the full transmitter power and may be left in the line for continuous monitoring, is excellent for this purpose. A simple resistance bridge is perfectly adequate, however, requiring only that the transmitter output be reduced to a very low value so the bridge will not be overloaded. To adjust the circuit, make a trial setting of the coupling (coil spacing in Figs. 19A and B and 20A, C1 setting in others) and adjust C_s or C_p for minimum VSWR as indicated by the bridge. If the VSWR is not 1:1, readjust the coupling and retune C_s or C_p, continuing this procedure until the VSWR is at or near 1:1. The settings may then be logged for future reference.

In the series-tuned circuits of Figs. 19A

and C, the two capacitors should be adjusted for similar settings. The "2C$_s$" indicates that a balanced series-tuned coupler requires twice the capacitance in each of two capacitors as does an unbalanced series-tuned circuit, all other things being equal.

It is possible to use circuits of this type without initially setting them up with a VSWR bridge. In such a case it is a matter of cut and try until adequate power transfer between the amplifier and main transmission line is obtained. Initial tuning can be done by listening on a receiver tuned to the intended operating frequency, and adjusting for maximum band noise or received signal strengths. Tuning without an SWR bridge, however, frequently results in high VSWR in the link with consequent power loss, "hot spots" in the coaxial cable and tuning that is critical with frequency. The VSWR indicator method is simple and gives the optimum operating conditions quickly and with certainty.

THE CONJUGATE MATCH AND THE Z$_0$ MATCH

One purpose of matching impedances is to enable a power source or generator, having an optimum load impedance, to deliver its maximum available power to a load having a different impedance, usually through a transmission line having still another impedance. The same principles apply, and the same wave actions and reflections occur, whether the matching device is a stub on a line, a quarter-wave line transformer, or a network of lumped reactances such as a pi-network tank circuit or a Transmatch. (In ARRL literature, a Transmatch is considered to be a matching device used between the transmitter and the feed line.) To aid in understanding the principles of matching, we shall consider transmission lines and reactance components of the Transmatch as lossless elements, and then treat the effects of loss later.

The conjugate match is defined by the National Bureau of Standards as the condition for maximum power absorption by a load, in which the impedance seen looking toward the load at a point in the circuit is the complex conjugate of that seen looking toward the source. The Z$_0$ match is the condition in which the impedance seen looking into a transmission line is equal to the characteristic impedance of the line.

If a generator, a load, and an ideal lossless line connecting them all have the same Z$_0$, both a Z$_0$ match and a conjugate match exist. (A Z$_0$ match is also a special condition of the conjugate match.) Under these conditions the generator delivers its maximum available power into the line, which transfers it to the load where it is completely absorbed.

If the Z$_0$ load is now replaced by another, such as an antenna having an impedance $Z_L = R + jX \neq Z_0$, we have two mismatches at the load, a Z$_0$ mismatch and a conjugate mismatch. The Z$_0$ mismatch creates a reflection having a magnitude

$$\rho = \frac{Z_L - Z_0}{Z_L + Z_0}$$

causing a reflection loss ρ^2 that is referred back along the line to the generator. This in turn causes the generator to see the same magnitude of Z$_0$ mismatch at the line input. This referred Z$_0$ mismatch causes the generator to deliver less than its maximum available power by the amount equal to the reflection loss. There is no power lost in the line — only a reduction of power delivered. And all of the reduced power delivered is absorbed by the load.

If a matching device is now inserted between the line and the load, the device provides a conjugate match at the load by supplying a reflection gain which cancels the reflection loss, so the line sees a Z$_0$ match at the input of the matching device. Since the line is terminated in a Z$_0$ match, the generator now sees a Z$_0$ match at the line input, and will again deliver its maximum available power into the line, to be completely absorbed by the load.

However, if a Transmatch is inserted between the generator and the line instead of a matching device at the load, the Transmatch provides the conjugate match at the line input, and the generator sees a Z$_0$ match at the Transmatch input. At the load, the line now sees a conjugate match, but also sees a Z$_0$ mismatch. The Z$_0$ mismatch again causes a reflection loss that is referred back to the line input as before. With the Transmatch at the line input, the reflection gain, which again cancels the reflection loss, shields the generator from seeing the referred Z$_0$ mismatch again appearing at the line input.

The reflection gain of the Transmatch also creates the conjugate match at the load, and thus enables the load to absorb the maximum available power delivered by the generator. Since the generator sees a Z$_0$ match at the Transmatch input, it continues delivering its maximum available power into the line through the Transmatch, and the power is still completely absorbed by the load because of the conjugate match.

The amount of forward power flowing on a lossless line is determined by the expression

$$\frac{1}{1 - \rho^2}$$

where ρ = voltage coefficient of reflection.

On a real line with attenuation, forward power at the conjugate matching point is

$$\frac{1}{1 - \rho^2 e^{-4\alpha}}$$

where

α = line attenuation in nepers
  = dB/8.686

e = 2.71828..., the base of natural logarithms

The power arriving at the mismatched load with attenuation in the line is

$$\frac{e^{-2\alpha}}{1 - \rho^2 e^{-4\alpha}}$$

and the power absorbed in the load is

$$\frac{(1 - \rho^2)\, e^{-2\alpha}}{1 - \rho^2\, e^{-4\alpha}}$$

The power reflected is

$$\frac{\rho^2 e^{-2\alpha}}{1 - \rho^2 e^{-4\alpha}}$$

The Conjugate Matching Theorem

To understand how a conjugate match is created at the load with the Transmatch at the line input, examine the Conjugate Matching Theorem and the wave actions which produce the conjugate match. It is the conjugate match that enables the load to absorb the maximum available power from the generator despite the Z$_0$ load mismatch.

"If a group of four-terminal networks containing only reactances (or lossless lines) are arranged in tandem to connect a generator to its load, then if at any junction there is a conjugate match of impedances, there will be a conjugate match of impedances at every other junction in the system."

To paraphrase from the NBS definition, "conjugate match" means that if in one direction from a junction the impedance has the dimensions R + jX, then in the opposite direction the impedance will have the dimensions R − jX. And according to the theorem, when a conjugate match is accomplished at any of the junctions in the system, any reactance appearing at any junction is canceled by an equal and opposite reactance, which also includes any reactance appearing in the load. This reactance cancellation results in a net system reactance of zero, establishing resonance in the entire system. In this resonant condition the generator delivers its maximum available power to the load, which is why an antenna operated away from its natural resonant frequency is tuned to resonance by a Transmatch connected at the input to the transmission line.

Conjugate matching is obtained through a controlled wave interference between two sets of reflected waves. One set of waves is reflected by the load mismatch, and the other by a complementary mismatch introduced by the reactances in the Transmatch. The mismatch introduced by the reactances of the Transmatch creates reflected waves of voltage and current at the matching point (the Transmatch input). These waves are equal in magnitude but opposite in phase relative to those arriving from the load mismatch. The two sets of reflected voltage and current waves created by the two mismatches combine at the matching point to produce resultant waves of voltage and current that are respectively at 0° and 180° in phase relative to the source wave.

The phase relationship of 180° between the resultant voltage and current waves creates a virtual open circuit to waves traveling toward the source, which totally re-reflects both sets of reflected waves back into the line. This re-reflection prevents the reflected waves from traveling beyond the matching point, which is the reason an SWR indicator connected at the input of the Transmatch shows zero reflected power at the conclusion of the tune-up procedure. On re-reflection the voltage and current components of the reflected waves emerge in phase with the corresponding components of the source wave, so by superposition, all of the power contained in the reflected waves is added to the power in the source wave. Consequently, the forward power in the line is greater than the source power whenever the line is terminated in a Z_0 mismatch. This is why the forward-power wattmeter connected at the output of the Transmatch indicates a value higher than the source power, by the amount equal to the reflected power.

This enlargement of the forward power is called "reflection gain." Reflection gain cancels the reflection loss, creating the conjugate match that enables the Z_0-mismatched load to absorb the maximum available power delivered by the generator. When the source power enlarged by the reflected power reaches the Z_0-mismatched load, Z_L, the power previously reflected and added to the source power is again subtracted by the Z_0 mismatch reflection, leaving the source power to be completely absorbed by the load. This is why all power entering the line is absorbed in the load, regardless of the mismatch or SWR.

In lines having attenuation, all power entering the line is absorbed in the load, except for that lost because of attenuation. When the matching is performed at the line input, the attenuation increases as the Z_0 load mismatch increases because, in addition to the attenuation of the forward power, the reflected power is also attenuated in the same proportion. However, when the matched-line attenuation is low, as it is in typical amateur installations, the additional loss because of Z_0 mismatch is small. The additional losses are so low that even with moderate to high SWR values, the difference in power radiated compared to that with a 1:1 SWR is too small to be discerned by the receiving station. Information for calculating the additional power lost versus SWR is available from the graph in Fig. 23, Chapter 16 of this *Handbook,* and from Fig. 9 in December 1974 *QST,* p. 12.

THE CONJUGATE MATCH AND THE PI-NETWORK TANK

In transceivers having tubes in the RF output stage, the pi network serves as a conjugate matching network as well as a harmonic-controlling tank circuit. When

Table 2

Characteristics of Commonly Used Transmission Lines

Type of line	Z_0 Ohms	Vel. %	pF per foot	OD	Diel. Material	Max. Operating Volts (RMS)
RG-6	75	75	18.6	0.266	Foam PE	400
RG-8X	52.0	75	26.0	0.242	Foam PE	300
RG-8	52.0	66	29.5	0.405	PE	4000
RG-8 foam	50.0	80	25.4	0.405	Foam PE	600
RG-8A	52.0	66	29.5	0.405	PE	5000
RG-9	51.0	66	30.0	0.420	PE	4000
RG-9A	51.0	66	30.0	0.420	PE	4000
RG-9B	50.0	66	30.8	0.420	PE	5000
RG-11	75.0	66	20.6	0.405	PE	4000
RG-11 foam	75.0	80	16.9	0.405	Foam PE	600
RG-11A	75.0	66	20.6	0.405	PE	5000
RG-12	75.0	66	20.6	0.475	PE	4000
RG-12A	75.0	66	20.6	0.475	PE	5000
RG-17	52.0	66	29.5	0.870	PE	11,000
RG-17A	52.0	66	29.5	0.870	PE	11,000
RG-55	53.5	66	28.5	0.216	PE	1900
RG-55A	50.0	66	30.8	0.216	PE	1900
RG-55B	53.5	66	28.5	0.216	PE	1900
RG-58	53.5	66	28.5	0.195	PE	1900
RG-58 foam	53.5	79	28.5	0.195	Foam PE	200
RG-58A	53.5	66	28.5	0.195	PE	1900
RG-58B	53.5	66	28.5	0.195	PE	1900
RG-58C	50.0	66	30.8	0.195	PE	1900
RG-59	73.0	66	21.0	0.242	PE	2300
RG-59 foam	75.0	79	16.9	0.242	Foam PE	300
RG-59A	73.0	66	21.0	0.242	PE	2300
RG-62	93.0	86	13.5	0.242	Air space PE	750
RG-62A	93.0	86	13.5	0.242	Air space PE	750
RG-62B	93.0	86	13.5	0.242	Air space PE	750
RG-133A	95.0	66	16.2	0.405	PE	4000
RG-141	50.0	70	29.4	0.190	PTFE	1900
RG-141A	50.0	70	29.4	0.190	PTFE	1900
RG-142	50.0	70	29.4	0.206	PTFE	1900
RG-142A	50.0	70	29.4	0.206	PTFE	1900
RG-142B	50.0	70	29.4	0.195	PTFE	1900
RG-174	50.0	66	30.8	0.1	PE	1500
RG-213	50.0	66	30.8	0.405	PE	5000
RG-214*	50.0	66	30.8	0.425	PE	5000
RG-215	50.0	66	30.8	0.475	PE	5000
RG-216	75.0	66	20.6	0.425	PE	5000
RG-223*	50.0	66	30.8	0.212	PE	1900
9913 (Belden)*	50.0	84	24.0	0.405	Air Space PE	—
9914 (Belden)*	50.0	78	26.0	0.405	Foam PE	—

Aluminum Jacket, Foam Dielectric

1/2 inch	50.0	81	25.0	0.5		2500
3/4 inch	50.0	81	25.0	0.75		4000
7/8 inch	50.0	81	25.0	0.875		4500
1/2 inch	75.0	81	16.7	0.5		2500
3/4 inch	75.0	81	16.7	0.75		3500
7/8 inch	75.0	81	16.7	0.875		4000

Open wire	—	97	—	—		—
75-ohm transmitting twin lead	75.0	67	19.0	—		—
300-ohm twin lead	300.0	82	5.8	—		—
300-ohm tubular	300.0	80	4.6	—		—

Open Wire, TV Type

1/2 inch	300.0	95	—	—		—
1 inch	450.0	95	—	—		—

Dielectric Designation	Name	Temperature Limits
PE	Polyethylene	−65° to + 80° C
Foam PE	Foamed polyethylene	−65° to + 80° C
PTFE	Polytetrafluoroethylene (Teflon)	−250° to + 250° C

*Double shield

properly adjusted, the pi network matches the optimum load resistance of the tubes to the impedance seen at the input of the coaxial line connected to the output of the network.

In the ever-present quest for the perfect 1:1 SWR, many amateurs are unaware that the pi network is usually capable of matching the optimum tube load resistance to a wider range of output load impedances than defined by a 2:1 SWR; some will match to as high as 3:1, and sometimes even 4:1. During tune-up, if a resonance dip in plate current can be obtained at the proper loaded value of current within the range of the tuning and loading capacitors in the pi network, the network has correctly matched the tubes to whatever impedance

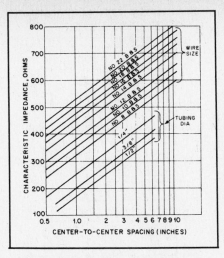

Fig. 21 — Chart showing the characteristic impedance of spaced-conductor parallel transmission lines with air dielectric. Tubing sizes are given for outside diameters.

is seen at the output. This is true no matter how high the SWR. When correctly tuned to the plate-current dip, the tubes see the desired resistive (nonreactive) load, the network has supplied a conjugate match, and any reflected power as shown by an SWR indicator has been re-reflected in the same manner as described above for the Transmatch. In other words, the pi network can serve as the Transmatch for whatever SWR it can load properly.

TRANSMISSION LINE CHARACTERISTICS

Each type of line has a characteristic velocity factor, related to its insulating-material properties. The velocity factor must be taken into account when cutting a transmission line to a specific part of a wavelength — such as with a quarter-wavelength transformer. For example, if RG-8A were used to make a quarter-wavelength line at 3.5 MHz, the line dimension should be $0.66 \times 246/f$ (MHz), because RG-8A has a velocity factor of 0.66. Thus, the line would be 46.4 feet long instead of the free-space length of 70.3 feet. Table 2 shows various velocity factors for the transmission lines commonly used by amateurs. Open-wire line has a velocity factor of essentially unity because it lacks a substantial amount of solid insulating material. Conversely, molded 300-ohm TV line has a velocity factor of 0.80 to 0.82.

Amateurs can construct their own parallel transmission lines by following the chart contained in Fig. 21. With wire conductors it is easy to fabricate open-wire feed lines. Spacers made of high-quality insulating material are used at appropriate distances apart to maintain the spacing between the wires.

Characteristic Impedance

Too great a distance between insulating spacers may yield variations in line impedance because of nonconstant conductor spacing, and may permit the conductors to short out. These considerations are more important in outdoor installations in windy weather. The characteristic impedance of an air-insulated parallel-conductor line, neglecting the effect of the insulating spacers, is given by

$$Z_0 = 276 \log \frac{2S}{d}$$

where

Z_0 = characteristic impedance
S = center-to-center distance between conductors
d = diameter of conductor (in same units as S)

The characteristic impedance of an air-insulated coaxial line is given by the formula

$$Z_0 = 138 \log \frac{b}{a}$$

where

Z_0 = characteristic impedance
b = inside diameter of outer conductors
a = outside diameter of inner conductor (in same units as b)

It does not matter what units are used for a and b as long as they are the same units. The characteristic impedance of any transmission line can be expressed as a function

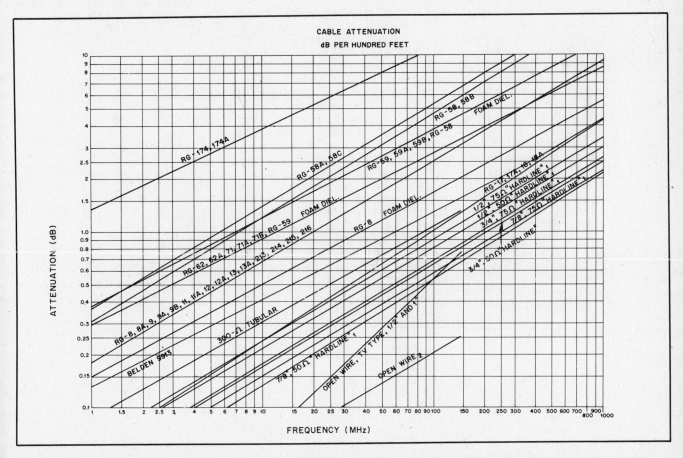

Fig. 22 — This graph displays the attenuation in decibels per 100-foot lengths of many popular transmission lines. The vertical axis represents attenuation and the horizontal axis frequency. Note that these loss figures are only accurate for properly matched transmission lines.

of the distributed capacitance and inductance:

$$Z_0 = \sqrt{L/C}$$

where

- Z_0 = characteristic impedance
- L = inductance (henrys) per unit length
- C = capacitance (farads) per unit length

Line Losses

The power loss in a transmission line is not directly proportional to the line length, but varies logarithmically with the length. That is, if 10% of the input power is lost in a section of line having a certain length, then 10% of the remaining power will be lost in the next section of the same length. For this reason it is customary to express line losses in terms of decibels per unit length, because the decibel is a logarithmic unit. Calculations are simple because the total loss in a line is merely the loss per unit length multiplied by the total number of units. Line loss is usually expressed in decibels per hundred feet. It is necessary to specify the frequency for which the loss applies, because the loss varies with frequency.

Fig. 22 shows the loss (or attenuation) per hundred feet versus frequency for several common types of lines, both coaxial and balanced types.

Effect of VSWR

The power lost in a line is lowest when the line is terminated in a resistance equal to its characteristic impedance, and increases with an increase in the standing-wave ratio. This is because the average values of both current and voltage become larger as the VSWR becomes greater. The increase in effective current raises the I^2R (ohmic) losses in the conductors, and the increase in effective voltage increases the E^2/R losses in the dielectric.

The increased loss caused by a VSWR greater than 1 may or may not be serious. If the VSWR at the load is not greater than 2, the *additional* loss caused by the standing waves, as compared with the loss when the line is perfectly matched, does not amount to more than about ½ dB even on very long lines. Since ½ dB is an undetectable change in signal strength a VSWR of 2 or less is every bit as good as a perfect match, as far as losses are concerned.

The effect of VSWR on line loss is shown in Fig. 23. The horizontal axis is the attenuation, in decibels, of the line when

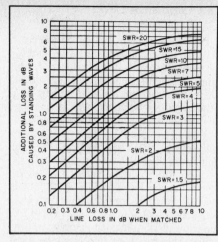

Fig. 23 — Increase in line loss because of standing waves (VSWR measured at the load). To determine the total loss in decibels in a line having a VSWR greater than 1, first determine the loss for the particular type of line, length and frequency, on the assumption that the line is perfectly matched (Table 2). Locate this point on the horizontal axis and move up to the curve corresponding to the actual VSWR. The corresponding value on the vertical axis gives the additional loss in decibels caused by the standing waves.

perfectly matched. The vertical axis gives the *additional* attenuation. For example, if the loss in a certain line is 1 dB when perfectly matched, a VSWR of 3 on that same line will cause an additional loss of approximately 0.5 dB. The total loss on the poorly matched line is therefore 1 + 0.5 = 1.5 dB. If the VSWR were 10 instead of 3, the additional loss would be 2.5 dB, and the total loss 1 + 2.5 = 3.5 dB.

It is important to note that the curves in Fig. 23 that represent VSWR are the values that exist *at the load*. In most cases of amateur operation, this will be at the antenna end of a length of transmission line. The VSWR as measured at the input or transmitter end of the line will be less, depending on the line attenuation. It is not always convenient to measure VSWR directly at the antenna. By using the graph shown in Fig. 24, however, you can obtain the VSWR at the load by measuring it at the input to the transmission line and using the known (or estimated) loss of the transmission line. For example, if the VSWR at the transmitter end of a line is measured as 3 to 1 and the line is known to have a total attenuation (under matched con-

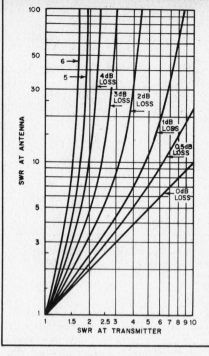

Fig. 24 — VSWR at input end of transmission line vs. VSWR at load end for various values of matched-line loss.

ditions) of 1 dB, the VSWR at the load end of the line will be 4.4 to 1. From Fig. 23, the additional loss is nearly 1 dB because of the presence of the VSWR. The total line loss in this case is 2 dB.

Bibliography

[1]Regier, F. A., "Series-Section Transmission Line Impedance Matching," *QST*, July 1978, pp. 14-16.
[2]Maxwell, M. W., "Some Aspects of the Balun Problem," *QST*, March 1983, pp. 38-40.
[3]Maxwell, M. W., "Another Look at Reflections," Part 4, *QST*, October 1973, pp. 22-29.
[4]Everitt, W. L., *Communication Engineering*, 2nd Edition (New York: McGraw-Hill), 1937.
[5]Smith, P. H., *Electronic Applications of the Smith Chart in Waveguide, Circuit and Component Analysis* (New York: McGraw Hill).
[6]Friis, H. T., and Schelkunoff, S. A., *Antennas: Theory and Practice* (New York: John Wiley and Sons), 1952, p. 258.
[7]Maxwell, M. W., "Another Look at Reflections," Part 6, *QST*, December 1974, p. 12, Fig. 9.
[8]Maxwell, M. W., "Another Look at Reflections," Part 7, *QST*, August 1976, pp. 15-20.
[9]Maxwell, M. W., "Wave Reflections in Attenuators, Filters and Matching Networks," Technical Correspondence, *QST*, November 1981, pp. 47-48.
[10]*The ARRL Antenna Book*, 14th edition, ed. G. L. Hall (Newington: American Radio Relay League), 1982, p. 3-12, Fig. 23.
[11]Maxwell, M. W., "How Does a Transmatch Work?," Technical Correspondence, *QST*, March 1985, pp. 45-46.

Chapter 17

Antenna Fundamentals

An antenna system is comprised of all the components used between the transmitter or receiver and the actual radiator. Therefore, such items as the antenna proper, transmission line, matching transformers, baluns and Transmatch qualify as parts of an antenna system.

In a well designed system, only the antenna does the radiating. It is noteworthy that any type of feed line can be utilized with a given antenna, provided the following conditions are met. A suitable matching device must be used to ensure a low standing-wave ratio (SWR) between the feed line and the antenna, and another between the feed line and the transmitter (or receiver).

Some antennas possess an input impedance at the feed point close to that of certain transmission lines. For example, a half-wavelength, center-fed dipole, placed at a correct height above ground, will have a feed-point impedance of approximately 75 ohms. In such a case it is practical to use a 75-ohm coaxial or balanced line to feed the antenna. But few amateur half-wavelength dipoles actually exhibit a 75-ohm impedance. This is because at the lower end of the high-frequency spectrum the typical height above ground is rarely more than ¼ wavelength. The 75-ohm characteristic is most likely to be realized in a practical installation when the horizontal dipole is approximately one-half, three-quarters or one wavelength above ground. At lower heights the approximate feed-point resistance may be obtained from Fig. 1, but values may vary at different locations because of differences in ground conductivity.

Fig. 1 shows the difference between the effects of perfect ground and real earth at low antenna heights. The effect on the resistance of a horizontal half-wave antenna is negligible only as long as the height of the antenna is greater than 0.2 wavelength. Below this height, while decreasing rapidly to zero over perfectly conducting ground, the resistance decreases less rapidly with height over actual ground. At lower heights the resistance stops decreasing at around 0.15 wavelength, and thereafter increases as height decreases further. The reason for the increasing resistance is that more and more of the induction field of the antenna is absorbed by

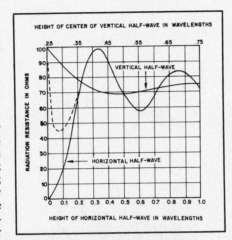

Fig. 1 — Curves showing the radiation resistance of vertical and horizontal half-wavelength dipoles at various heights above ground. The broken-line portion of the curve for a horizontal dipole shows the resistance over real earth, the solid line for perfectly conducting ground.

the lossy ground as the height drops below ¼ wavelength.

For a half-wave vertical antenna, the center of which is ¼ wavelength or more above the surface, differences between the effects of perfect ground and real earth on the impedance is negligible. The resistance of a vertical dipole at various heights above real and perfectly conducting ground is shown in Fig. 1.

The Antenna Choice

The amount of available space should be high on the list of factors to consider when selecting an antenna. Those who live in urban areas often must accept a compromise type of antenna for the HF bands because the city lot won't accommodate full-size wire dipoles, end-fed systems or high supporting structures. Other constrictions are imposed by the amount of money available for an antenna system (including supporting hardware), the number of amateur bands to be worked and local zoning ordinances.

Finally, the operation objective comes into play; to dedicate one's self to DXing, or to settle for a general type of operation that will yield short- and long-haul QSOs during periods of good propagation.

Because of these influences, it is impossible to suggest one type of antenna system over another. A rule of thumb might be to erect the biggest and best collection of antennas that space and finances will allow. If a modest system is the order of the day, then use whatever is practical and accept less than optimum performance. Practically any radiator will enable the operator to make good contacts under some conditions of propagation, assuming the radiator is able to accept power and radiate it at some useful angle respective to earth.

In general, the height of the antenna above ground is the most critical factor at the higher end of the HF spectrum — 14 through 30 MHz. This is because the antenna should be clear of conductive objects such as power lines, phone wires, gutters and the like, plus high enough to have a low radiation angle. This is not nearly as important at 2 to 10 MHz, but lower frequency antennas still should be well away from conductive objects and as high above ground as possible in the interest of good performance. The exception is a ground-mounted vertical antenna. Ground-plane verticals, however, should be installed as high above ground as possible so their performance will not be degraded by conductive objects.

Antenna Polarization

Most HF-band antennas are either vertically or horizontally polarized, although circular polarization is possible, just as it is at VHF and UHF. The polarization is determined by the position of the radiating element or wire with respect to earth. Thus, a radiator that is parallel to the earth radiates horizontally, while an antenna at a right angle to the earth (vertical) radiates a vertical wave. If a wire antenna is slanted above earth, it radiates waves that are between vertical and horizontal in nature.

During line-of-sight communications, maximum signals will exist when the antennas at both ends of the circuit have the same polarity; cross polarization results in many decibels of signal reduction. During propagation via the ionosphere (sky wave), however, it is not essential to have the same polarization as the station at the opposite end of the circuit. This is because the radiated wave is bent and it tumbles considerably during its travel through the atmospheric layer from which it is refracted. At the far end of the communications path the wave may be horizontal, vertical or somewhere in between at a given instant. On multihop transmissions, in which the signal is refracted more than once from the atmosphere, and similarly reflected from the earth's surface during its travel, considerable polarization shift will occur. Therefore, the main consideration for a good DX antenna is a low angle of radiation rather than the polarization. It should be said, however, that most DX antennas for HF work are horizontally polarized. The major exception is the ground-plane vertical and phased vertical arrays.

Impedance

The impedance at a given point in the antenna is determined by the ratio of the voltage to the current at that point. For example, if there were 100 RF volts and 1.4 amperes of current at a specified point in an antenna and if they were in phase, the impedance would be approximately 71 ohms.

The impedance is significant with regard to matching the feeder to the feed point: Maximum power transfer takes place under a perfectly matched condition. As the mismatch increases, so does the reflected power. If the feed line is not too lossy or long, good performance can be had at HF when the standing-wave ratio (SWR) is 3:1 or less. When the feeder loss is very low — as with open-wire transmission line — much higher SWR is not detrimental to performance if the transmitter is able to work into the mismatched condition satisfactorily. In this regard, a Transmatch (matching network between the transmitter and the feed line) is often employed to compensate for the mismatch condition, enabling the operator to load the transmitter to its full rated power.

Antenna impedance can be either resistive or complex (containing resistance and reactance). This will depend on whether or not the antenna is resonant at the operating frequency. Many operators mistakenly believe that a mismatch, however small, is a serious matter, and that their signals won't be heard well even if the SWR is as low as 1.3:1. This unfortunate fallacy has cost much wasted time and money among some amateur groups as individuals attempted to obtain a perfect match. A perfect match, however ideal the concept may be, is not necessary. The significance of a perfect match becomes more pronounced only at VHF and higher, where feed-line losses are a major problem.

Antenna Bandwidth

The bandwidth of an antenna refers generally to the range of frequencies over which the antenna can be used to obtain good performance. The bandwidth is usually referenced to some SWR value, such as, "The 2:1 SWR bandwidth is 3.5 to 3.8 MHz." Some more specific bandwidth terms are used also, such as the gain bandwidth and the front-to-back ratio

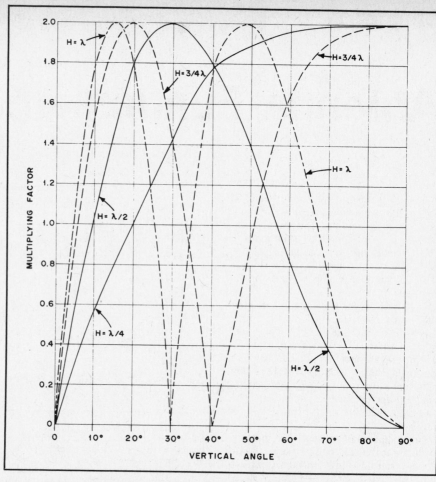

Fig. 2 — Effect of ground on the radiation of horizontal antennas at vertical angles for four antenna heights. These data are based on perfectly conducting ground.

bandwidth. The gain bandwidth is significant because the higher the antenna gain, the narrower the gain bandwidth will be for a given gain-bandwidth product.

For the most part, the lower the operating frequency of a given antenna design, the narrower the bandwidth. This follows the rule that the bandwidth of a resonant circuit doubles as the frequency of operation is increased by one octave (or doubled), assuming the Q is the same for each case. Therefore, it is often difficult to obtain sufficient bandwidth to cover all of the 160- to 80-meter bands with a dipole antenna cut for each of those bands. The situation can be aided by applying broadbanding techniques, such as fanning the far ends of a dipole to simulate a conical type of dipole.

Radiation Angle

The vertical angle of maximum radiation is of primary importance, especially at the higher frequencies. It is advantageous, therefore, to erect the antenna at a height that will take advantage of ground reflection so as to reinforce the space radiation at the most desirable angle. Since low angles usually are most effective, this generally means that the antenna should be high. The height should be at least one-half

wavelength at 14 MHz, and preferably three-quarters or one wavelength. At 28 MHz the height should be at least one wavelength, and preferably higher.

The physical height required for a given height in wavelengths decreases as the frequency is increased, so that acceptable heights are not impractical; a half wavelength at 14 MHz is only 35 feet, while the same height represents a full wavelength at 28 MHz. At 10 MHz and lower, the higher radiation angles are effective, so a useful antenna height is not difficult to attain. But greater height is important at 10 MHz and below when it is desired to work DX consistently. Heights between 35 and 70 feet are suitable for the upper bands, the higher figures being preferable. It is well to remember that most simple horizontally polarized antennas do not exhibit the directivity they are capable of unless they are one-half wavelength or more above ground. Therefore, with dipole-type antennas it is not important to choose a favored broadside direction unless the antenna is at least one-half wavelength above ground.

Imperfect Ground

Fig. 2 is based on ground having perfect conductivity, whereas the earth is not a perfect conductor. The principal effect of actual ground is to make the curves inaccurate at the lowest angles; appreciable high-frequency radiation at angles smaller than a few degrees is practically impossible to obtain over horizontal ground. Above 15 degrees, however, the curves are accurate enough for all practical purposes, and may be taken as indicative of the result to be expected at angles between 5 and 15 degrees.

The effective ground plane — that is, the plane from which ground reflections can be considered to take place — seldom is the actual surface of the ground. Instead, it is several inches to a few feet below it, depending on the characteristics of the soil.

Current and Voltage Distribution

When power is fed to an antenna, the current and voltage vary along its length. The current is maximum (loop) at the center and nearly zero (node) at the ends. The opposite is true of the RF voltage. The current does not actually reach zero at the current nodes, because of the end effect. Similarly, the voltage is not zero at its node because of the resistance of the antenna, which consists of both the RF resistance of the wire (ohmic resistance) and the radiation resistance. The radiation resistance is the equivalent resistance that would dissipate the power the antenna radiates, with a current flowing in it equal to the antenna current at a current loop (maximum). The ohmic resistance of a half-wavelength antenna is ordinarily small enough, compared with the radiation resistance, to be neglected for all practical purposes.

Conductor Size

The impedance of the antenna also depends on the diameter of the conductor in relation to the wavelength, as indicated in Fig. 3. If the diameter of the conductor is increased, the capacitance per unit length increases and the inductance per unit length decreases. Since the radiation resistance is affected relatively little, the decreased L/C ratio causes the Q of the antenna to decrease so that the resonance curve becomes less sharp. Hence, the antenna is capable of working over a wide frequency range. This effect is greater as the diameter is increased, and is a property of some importance at the very high frequencies where the wavelength is small.

THE HALF-WAVELENGTH ANTENNA

A fundamental form of antenna is a single wire whose length is approximately equal to half the transmitting wavelength. It is the unit from which many more complex forms of antennas are constructed and is known as a dipole antenna.

The length of a half-wave in free space is

$$\text{Length (ft)} = \frac{492}{f \text{ (MHz)}} \quad \text{(Eq. 1)}$$

The actual length of a resonant half-

wavelength antenna will not be exactly equal to the half wavelength in space, but depends on the thickness of the conductor in relation to the wavelength. The relationship is shown in Fig. 3, where K is a factor that must be multiplied by the half wavelength in free space to obtain the resonant antenna length. An additional shortening effect occurs with wire antennas supported by insulators at the ends because of the capacitance added to the system by the insulators (end effect). The following formula is sufficiently accurate for wire antennas for frequencies up to 30 MHz.

$$\text{Length of half-wave antenna (ft)} =$$

$$\frac{492 \times 0.95}{f \text{ (MHz)}} = \frac{468}{f \text{ (MHz)}} \quad \text{(Eq. 2)}$$

Example: A half-wave antenna for 7150 kHz (7.15 MHz) is 468/7.15 = 65.45 ft, or 65 ft 5 in.

Above 30 MHz the following formulas should be used, particularly for antennas constructed from rod or tubing. K is taken from Fig. 3.

$$\text{Length of half-wave antenna (ft)} =$$

$$\frac{492 \times K}{f \text{(MHz)}} \quad \text{(Eq. 3)}$$

$$\text{Length (in)} = \frac{5904 \times K}{f \text{ (MHz)}} \quad \text{(Eq. 4)}$$

Example: Find the length of a half-wavelength antenna at 28.7 MHz, if the antenna is made of ½-inch-diameter tubing. At 28.7 MHz, a half wavelength in space is

$$\frac{492}{28.7} = 17.14 \text{ ft}$$

from Eq. 1. The ratio of half wavelength to conductor diameter (changing wavelength to inches) is

$$\frac{(17.14 \times 12)}{0.5 \text{ in.}} = 411$$

From Fig. 3, K = 0.97 for this ratio. The length of the antenna, from Eq. 3 is

$$\frac{492 \times 0.97}{28.7} = 16.63 \text{ ft}$$

or 16 feet 7½ inches. The answer is obtained directly in inches by substitution in Eq. 4

$$\frac{5904 \times 0.97}{28.7} = 199.5 \text{ inches}$$

The length of a half wavelength antenna is affected also by the proximity of the dipole ends to nearby conductive and semiconductive objects. In practice, it is often necessary after cutting the antenna to the computed length to do some experimental "pruning" of the wire, lengthening or shortening it in increments to obtain a low

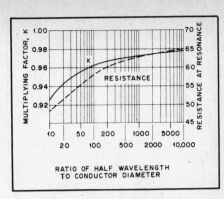

Fig. 3 — Effect of antenna diameter on length for half-wavelength resonance, shown as a multiplying factor, K, to be applied to the free-space, half-wavelength equation (Eq. 1 of text). The effect of conductor diameter on the center feed-point impedance is shown here also.

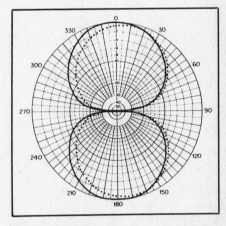

Fig. 4 — Response of a dipole antenna in free space, looking at the plane of the conductor, solid line. If the currents in the halves of the dipole are not in phase, some distortion of the pattern will occur, broken line. (Pattern calculation courtesy of "Annie.")

SWR. This can be done by applying RF power through an SWR indicator and observing the reflected-power reading. When the lowest SWR is obtained for the desired part of an amateur band, the antenna is resonant at that frequency. The value of the SWR indicates the quality of the match between the antenna and the feed line. With feed-line impedances of 50 or 75 ohms, the SWR at resonance should fall between 1.1:1 and 1.7:1 for antennas of average height that are clear of nearby conductive objects. If the lowest SWR obtainable is too high for use with solid-state rigs, a Transmatch or line-input matching network may be used, as described in Chapter 16.

Radiation Characteristics

The classic radiation pattern of a dipole antenna is most intense perpendicular to the wire. A figure-8 pattern (Fig. 4) can be

'Annie, a commercially available program for plotting antenna patterns on the Apple II series of computers, is available from Sonnet Software, Dept. QH, 4397 Luna Course, Liverpool, NY 13088.

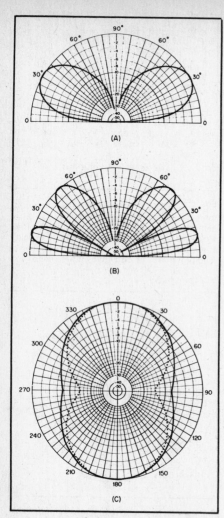

(A)

(B)

(C)

Fig. 5 — At A, elevation-plane response of a dipole antenna placed ½ wavelength above a perfectly conducting ground, and B, the pattern for the same antenna when it is raised to one wavelength height. C shows the azimuth patterns of the dipole for the two heights at the most favored elevation angle, the solid-line plot for the ½-λ height at an elevation angle of 30 degrees, and the broken-line plot for the 1-λ height at an elevation angle of 15 degrees. (Pattern calculation courtesy of "Annie.")

assumed off the broad side of the antenna (bidirectional pattern) if the dipole is ½ wavelength or greater above earth and is not degraded by nearby conductive objects. This assumption is based also on a symmetrical feed system. In practice, a coaxial feed line may distort this pattern slightly, as shown in Fig. 4. Minimum horizontal radiation occurs off the ends of the dipole. This discussion applies to a half-wavelength antenna that is parallel to the earth. If the dipole is erected vertically, however, uniform radiation in all compass directions will result in a doughnut pattern if it could be viewed from above the antenna.

Many beginners assume that a dipole antenna will exhibit a broadside pattern at any height above ground. In fact, as the antenna is brought closer to ground, the radiation pattern deteriorates until the antenna is, for the most part, an omnidirec-

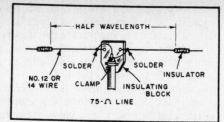

Fig. 6 — Method for affixing the feed line to the center of a dipole antenna. A plastic block is used as a center insulator. The coaxial cable is held in place by means of a metal clamp.

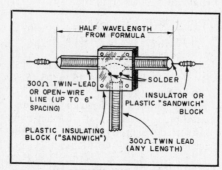

Fig. 7 — Construction details for a folded-dipole antenna. TV ribbon line is used as the dipole and feed line. Two pieces of plastic form an insulator/sandwich at the center to hold the conductor junction secure. When ribbon line is used for the flat top, the two conductors should be shorted at about 82% of the distance from the feed point to the ends.

tional radiator of high-angle waves. Many are tempted to use any convenient height, such as 20 or 30 feet above ground, for an 80-meter dipole, only to learn that the system is effective in all directions over a relatively short distance (out to 500 or 1000 miles under good conditions).

It can be seen from this that height above ground is important for a host of reasons. Fig. 5 illustrates clearly the advantage gained from antenna height. The radiation angle of Fig. 5A is 30 degrees, whereas at a height of one wavelength (Fig. 5B) the lobes split and the lower ones provide a good DX-communications angle of 15 degrees. The directivity of the antenna at the two heights is shown in Fig. 5C. The solid line shows the azimuth pattern at a radiation angle of 30 degrees for a ½-λ dipole height, and corresponds to the plot in Fig. 5A. The broken line shows the pattern for a radiation angle of 15 degrees and a 1-λ height, corresponding to the plot of Fig. 5B. Fig. 5C illustrates that there is significant radiation off the ends of a low horizontal dipole, even at the most favored elevation angle. For the ½-λ height (solid line), the radiation off the ends is only 7.6 dB lower than that in the broadside direction.

The higher angle lobes of the pattern of Fig. 5B (50 degrees) are useful for short-haul communications and compare favorably in practice with the lobe angle

shown in Fig. 5A. At heights appreciably lower than ½ wavelength, the lobe angle becomes higher. Eventually the two lobes converge to create the discrete "ball of radiation" which has a very high-angle nature (poor for long-distance communications).

Feed Methods

Most amateur single-wire dipole antennas (half wavelength) have a feed-point impedance between 50 and 75 ohms, depending on the installation. Therefore, standard coaxial cable is suitable for most installations. The smaller types of cable (RG-58 and RG-59) are satisfactory for power levels up to a few hundred watts if the SWR of the system is low. For high-power stations (RG-8 or RG-11) larger cables should be used.

These cables can be connected at the center of the antenna, as shown in Fig. 6. A plastic insulating block is used as central reinforcement for the cable and the dipole wires. The coax shield braid is connected to one leg of the dipole and the center conductor is soldered to the remaining leg. The exposed end of the cable should be sealed against dirt and moisture to prevent degradation of the transmission line.

Symmetrical feed can be obtained by inserting a 1:1 balun transformer at the dipole feed point. If one is not used, it is unlikely that the slight pattern skew resulting from nonsymmetrical feed will be noticed. The effects of unbalanced feed are most significant in beam antennas at VHF and higher. The narrower the beam pattern, the more annoying the condition will be.

The characteristic impedance of a dipole antenna can be increased by using a two-wire or folded dipole of the type shown in Fig. 7. This antenna offers a good match to 300-ohm TV ribbon. Alternatively, two pieces of wire can be used to form the equivalent of the TV-line dipole. If this is done it will be necessary to locate insulating spacers every few feet along the length of the dipole to keep the wires spaced apart uniformly and to prevent short circuiting. Open-wire TV ladder line is excellent for use in a 300-ohm folded dipole antenna, both for the radiator and the feed line. Feeder losses with this type of construction will be very low as opposed to molded TV twin-lead.

A dipole antenna can be used as an all-band radiator by using tuned open wire feed line. This principle is shown in Fig. 8A. In this example the dipole is cut to a half wavelength for the lowest desired amateur band. It is operated on its harmonics when used for the other chosen amateur bands. A typical antenna of this type might be utilized from 80 through 10 meters. This style of radiator is known by some amateurs as the center-fed Zepp. An end-fed version (end-fed Zepp) is shown in Fig. 8 at B. This version is not quite as

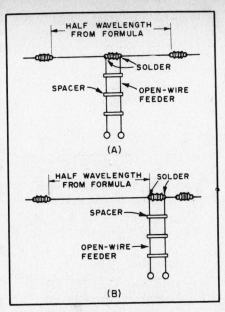

Fig. 8 — Center fed Zepp antenna (A) and an end-fed Zepp at B.

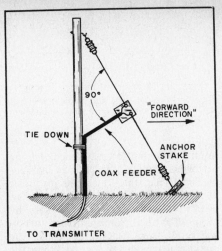

Fig. 9 — Example of a sloping half-wavelength dipole. On the lower HF bands over poor to average earth, maximum radiation is off the sides and in the "forward direction" indicated here, if a nonconductive support is used. A metal support will alter this pattern by acting as a parasitic element, depending on its electrical height.

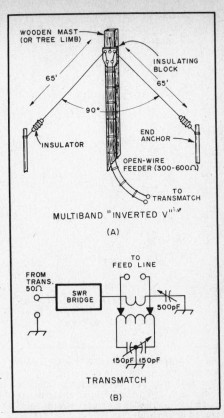

Fig. 10 — At A, details for a drooping dipole which can be used for multiband HF operation. A Transmatch is shown at B, suitable for matching the antenna to the transmitter over a wide frequency range.

desirable as the center-fed version because the feed system is not symmetrical. This can cause feeder radiation and a distortion of the antenna radiation pattern.

Both types of Zepp antenna require a matching network (Transmatch) at the transmitter end of the line to convert the feeder impedance to 50 ohms, and to change the balanced condition to an unbalanced one. Although the feed line may be anything from 200 ohms up to 600 ohms (not critical), losses will be insignificant when open-wire line is used. This is true despite the variations in antenna feed-point impedance from band to band. The feed-point impedance for the dipole will be high at even harmonics and will be low at the lowest operating frequency and at odd harmonics thereof. For example, if the dipole is cut for 40 meters, the feed-point impedance will be low on 40 and 15 meters, but it will be high at 20 and 10 meters.

With a dipole antenna the feed line should be routed away from the antenna at a right angle for as great a distance as possible. This will help prevent current imbalance in the line caused by RF pickup from the dipole. A right-angle departure of ¼ wavelength or greater is suggested.

Under some circumstances it may be necessary to experiment with the length of the open-wire feeders when using an all-band Zepp. This is because at some operating frequencies the line may present an "awkward" impedance to the Transmatch, making it impossible to obtain a suitable load condition for the transmitter. This will depend on the capability of the Transmatch being used.

Dipole Variations

The physical application of dipoles can be varied to obtain radiation properties that differ from those of the more conventional horizontal dipole. Furthermore, the nature and amount of property at the site will often dictate departures from the conventional means of erecting a dipole.

A sloping dipole can be useful for DX work on the lower HF bands because of its low angle of radiation. A sloping dipole is shown in Fig. 9. Excellent results can be had when the ground end of the antenna is only a few feet above the earth. When the antenna is supported with a nonconducting mast over poor to average earth, the radiation pattern tends to be cardioid shaped, but without a sharp null at the back. Maximum response is off the sides and in the "forward direction" as indicated in Fig. 9. The front-to-back ratio of the sloping dipole depends on the frequency and the conductivity of the earth, as well as the amount of slope and the vertical elevation angle of concern. Surprisingly, the sloping dipole becomes an omnidirectional radiator over a perfectly conducting earth.

If a metal mast or tower is used, it will alter the shape of the radiation pattern by acting as a parasitic element. Best results as a reflector seem to be obtained when the support is grounded and is approximately a quarter wave in physical height, supporting rotary beam or other antennas at the top. Some amateurs install four slopers for a given amateur band, spaced equidistantly around the tower. A feed-line switching system is used to obtain directivity in the chosen direction.

Drooping Dipole (Inverted V)

Another popular type of dipole antenna is the so-called inverted V, or drooping dipole. Because it has no characteristics of a true V antenna, the name inverted V is misleading. A drooping dipole is shown in Fig. 10. Newcomers to Amateur Radio are frequently led to regard this antenna as a panacea, but there is nothing magical or superior about a drooping dipole. In fact, its performance is inferior to that of a horizontal dipole at the same height as the apex. The main attribute is that it requires only one supporting structure. As with a horizontal dipole, it can be used for single or multiband operation.

Contrary to popular belief, the radiation off the ends of the drooping dipole is essentially no greater than that of a horizontal dipole at the same height as the apex. This myth probably arises because the drooping dipole is not an efficient radiator in the broadside direction, so it merely seems by comparison to its broadside performance that it is a good radiator off the ends. As is true of a horizontal dipole, the higher the drooping dipole is mounted above the ground, the better it will perform respective to low-angle, long-distance communications.

Best results are obtained when the enclosed angle of the drooping dipole is 90° or more. At angles less than 90°, considerable cancellation takes place, and reduced antenna performance results. At angles greater than 120° the antenna begins to function as a horizontal dipole.

When the ends of the drooping dipole are relatively close to the earth, pruning of the dipole legs may be necessary to compensate for the capacitive effect to ground.

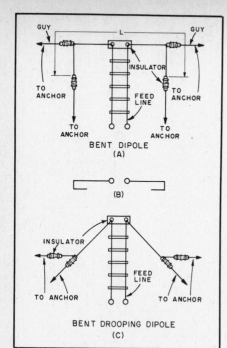

Fig. 11 — When limited space is available for a dipole antenna the ends can be bent downward as at A, or back on the radiator as shown at B. The drooping dipole at C can be erected with the ends bent parallel with the ground when the available supporting structure is not high enough to permit an enclosed angle of approximately 90°.

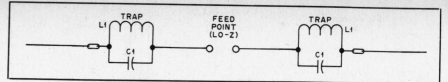

Fig. 12 — Example of a trap dipole antenna. L1 and C1 can be tuned to the desired frequency by means of a dip meter before they are installed in the antenna.

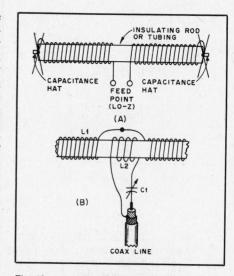

Fig. 13 — A helically wound dipole is illustrated at A. As shown, the radiation resistance will be very low and will require a broadband matching transformer. The coupling method shown at B is more satisfactory for providing a matched condition.

Thus, if the dipole is cut to length by means of the standard half-wavelength equation, it may be too long as a drooping dipole. Incremental trimming of each end of the dipole can be done while using an SWR indicator. This will show when the lowest SWR is obtained. A 50-ohm feed line offers a good match for drooping dipoles when single-band operation is desired. For multiband use, the antenna can be fed with open-wire line and matched to the transmitter with a Transmatch.

The most ideal supporting structure for a drooping dipole is a wooden or other nonconductive one. This type of support will have the least effect on the antenna pattern and performance.

Drooping dipoles are effective when used for a single band, the one for which it has been cut. Good results can also be had for two-band operation. An 80-meter dipole can be fed with open-wire line and used also on 40 meters as two half waves in phase. On 40 meters its performance at low radiation angles off the ends will be slightly superior to that of a 40-meter *half-wave* horizontal dipole at the same height as the apex.

Bent Dipoles and Trap Dipoles

When there is insufficient real estate to permit the erection of a full-size horizontal or drooping dipole, certain compromises are possible in the interest of getting an antenna installed. The voltage ends (far ends) of a dipole can be bent downward toward earth to effect resonance, and the performance will not be reduced markedly. Fig. 11 illustrates the technique under discussion. At A the dipole ends are bent downward and secured to anchors by means of guy line. Some pattern distortion will result from bending the ends. The dipole ends can also be bent back over the wire halves of the antenna, as shown at B in Fig. 11. This causes some signal cancellation (more severe than with the system of Fig. 11A), so it is not a preferred technique.

Fig. 11C demonstrates a bending technique for drooping dipoles when the available supporting mast or tree is too short to permit normal installation. The ends of the dipole are guyed off by means of insulators and wires, as shown. Alternatively, but not preferred, is the fold-back method at B in Fig. 11.

All of the shortening systems highlighted in Fig. 11 will have an effect on the overall length of the dipole. Therefore, some cutting and testing will be necessary to ensure a low SWR in the favored part of the amateur band for which the antenna is built. If open-wire feeders and a Transmatch are used, the dipole length will not be critical, provided it is close to the length required for a fully extended half-wavelength dipole. Pruning will be required if single-band operation with coaxial feed line is planned.

Trap dipoles offer one solution to multiband operation with a shortened radiator. The concept is shown in Fig. 12. In this example the dipole is structured for two-band use. Assuming that the antenna is made for operation on 80 and 40 meters, the overall radiator (including the traps) must be resonant at the center of the chosen section of the 80-meter band. The traps add loading to the dipole, so the length from the feed point to the far end of each leg will be somewhat shorter than normal. During 40-meter operation the traps present a high impedance to the signal and divorce the wires beyond the traps. Therefore, the wire length from the feed point to each trap is approximately what it would be if the dipole were cut for just 40 meters, with no traps in the line.

This principle can be extended for additional bands, using a new set of traps for each band. Since there is considerable interaction between the various segments of a multiband trap dipole, considerable experimentation with the wire lengths between the traps and beyond will be necessary.

The trap capacitors (C1) should be high-voltage and high-current units. Transmitting mica capacitors offer good performance. Transmitting ceramic capacitors are usable, but change value with extreme changes of temperature. Therefore, they are more suitable for use in regions where the climate is fairly constant throughout the year. The coils (L1) should be reasonably heavy wire gauge to minimize I^2R losses. The X_L and X_C values in traps are not critical. Generally the reactance can be on the order of 100 to 300 ohms. The traps should be checked for resonance before they are installed in the antenna system. This can be done with a dip meter and a calibrated receiver. Weatherproofing should be added to the traps as a measure against detuning and damage from ice, snow and dirt.

Helically Wound Dipoles

The overall length of a half-wavelength dipole can be reduced considerably by using helically wound elements. Fig. 13A shows the general form taken with this type of antenna. A length of insulating rod or tubing (fiberglass or phenolic) is used to contain the wire turns of the dipole. The material should be of high dielectric quality. Varnished bamboo has been used successfully by some in lieu of the more expensive materials. A hardwood pole from a lumber yard can be used after being coated one or more times with exterior spar varnish.

To minimize losses, the wire used should be of the largest diameter practical. The turns can be close-wound or spaced apart, with little difference in performance. The ends of the helical dipole should contain capacitance hats (disks or wire spokes preferred) of the largest size practical. The hats will lower the Q of the antenna and broaden its response. If no disks are used, extremely high RF voltage can appear at the ends of the antenna. At medium power levels and higher the insulating material can burn when no hats are used. The voltage effect is similar to that of a Tesla coil.

The feed-point impedance of helical dipoles or verticals is quite low. Therefore, it may be necessary to use some form of matching network to interface the antenna with 50-ohm coaxial cable. A broadband, variable-impedance transformer is convenient for determining the turns ratio of the final transformer used. The feed method shown at B of Fig. 13 can be used to obtain a matched condition. L2 is wound over L1, or between the two halves of L1, as illustrated. C1 is adjusted for an SWR of 1 at the center of the desired operating range. The bandwidth of this type of antenna is quite narrow. A 40-meter version with an 18-foot overall length exhibited a 2:1 SWR bandwidth of 50 kHz. The capacitance hats on that model were merely 18-inch lengths (whiskers) of no. 8 Copperweld wire. Greater bandwidth would result with larger capacitance hats.

To obtain half-wavelength performance it is necessary to wind approximately one wavelength of wire on the tubing. Final pruning can be done while observing an SWR indicator placed in the transmission line. Proximity to nearby conductive objects and the earth will have a significant effect on the resonance of the antenna. Ideally, final adjustments should be made with the antenna situated where it will be during use. Marine spar varnish should be painted on the elements after all tuning is finished. This will protect the antenna from the weather and will lock the turns in place so that detuning will not occur later.

A reasonably linear current and voltage distribution will result when using a helically wound dipole or vertical. The same is not true of center-, mid- or end-loaded (lumped inductance) dipoles. The efficiency of the helically wound antenna will be somewhat less than a full-size dipole. The performance will suffer as the helices are made shorter. Despite the gain-length trade-off, this antenna is capable of good performance when there is no room for a full-size dipole.

LONG-WIRE ANTENNAS

An antenna will be resonant so long as an integral number of standing waves of current and voltage can exist along its length; in other words, so long as its length

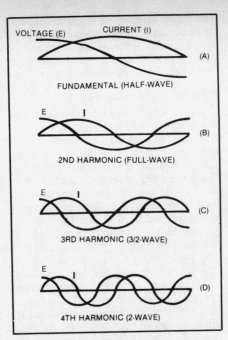

Fig. 14 — Standing-wave current and voltage distribution along an antenna when it is operated at various harmonics of its fundamental resonant frequency.

is some integral multiple of a half wavelength. When the antenna is more than one wavelength long it usually is called a long-wire antenna, or a harmonic antenna.

Current and Voltage Distribution

Fig. 14 shows the current and voltage distribution along a wire operating at its fundamental frequency (where its length is equal to a half wavelength) and at its second, third and fourth harmonics. For example, if the fundamental frequency of the antenna is 7 MHz, the current and voltage distribution will be as shown at A. The same antenna excited at 14 MHz would have current and voltage distribution as shown at B. At 21 MHz, the third harmonic of 7 MHz, the current and voltage distribution would be as in C; and at 28 MHz, the fourth harmonic, as in D. The number of the harmonic is the number of half waves contained in the antenna at the particular operating frequency.

The polarity of current or voltage in each standing wave is opposite to that in the adjacent standing waves. This is shown in the figure by drawing the current and voltage curves successively above and below the antenna (taken as a zero reference line), to indicate that the polarity reverses when the current or voltage goes through zero. Currents flowing in the same direction are in phase; in opposite directions, out of phase.

It is evident that one antenna may be used for harmonically related frequencies, such as the various amateur bands. The long-wire or harmonic antenna is the basis of multiband operation with one antenna.

Physical Length

The length of a long-wire antenna is not

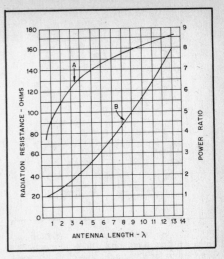

Fig. 15 — Curve A shows the variation in radiation resistance with antenna length when the antenna is fed at a current maximum. Curve B shows power in the lobes of maximum radiation for long-wire antennas as a ratio to the maximum radiation for a half-wave antenna.

an exact multiple of that of a half-wave antenna because the end effects operate only on the end sections of the antenna; in other parts of the wire these effects are absent, and the wire length is approximately that of an equivalent portion of the wave in space. The formula for the length of a long-wire antenna, therefore is

$$\text{Length (feet)} = \frac{492 (N - 0.05)}{f \text{ (MHz)}}$$

where N is the number of half waves on the antenna.

Example: An antenna 4 half waves long at 14.2 MHz would be

$$\frac{492 (4 - 0.05)}{14.2} = \frac{492 (3.95)}{14.2}$$

= 136.9 feet, or 136 feet 10 inches.

It is apparent that an antenna cut as a halfwave for a given frequency will be slightly off resonance at exactly twice that frequency (the second harmonic), because of the decreased influence of the end effects when the antenna is more than one-half wavelength long. The effect is not important, except for a possible unbalance in the feeder system and consequent radiation from the feed line. If the antenna is fed in the exact center, no unbalance will occur at any frequency, but end-fed systems will show an unbalance on all but one frequency in each harmonic range.

Impedance and Power Gain

The radiation resistance as measured at a current loop becomes higher as the antenna length is increased. Also, a long-wire antenna radiates more power in its most favorable direction than does a half-wave antenna in its most favorable direction.

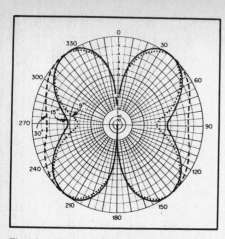

Fig. 16 — Horizontal patterns of radiation from a full-wave antenna. The solid line shows the pattern for a vertical angle of 15°; broken lines show deviation from the 15° pattern at 9° and 30°. All three patterns are drawn to the same relative scale; actual amplitudes will depend on the height of the antenna.

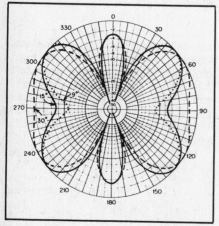

Fig. 17 — Horizontal patterns of radiation from an antenna three half-waves long. The solid line shows the pattern for a vertical angle of 15°; broken lines show deviation from the 15° pattern at 9° and 30°. Minor lobes coincide for all three angles.

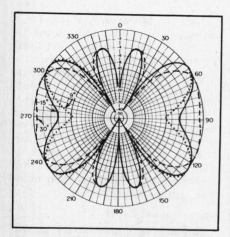

Fig. 18 — Horizontal patterns of radiation from an antenna 2 wavelengths long. The solid line shows the pattern for a vertical angle of 15°; broken lines show deviation from the 15° pattern at 9° and 30°. The minor lobes nearly coincide for all three angles.

This power gain is obtained at the expense of radiation in other directions. Fig. 15 shows how the radiation resistance and the power in the lobe of maximum radiation vary with the antenna length.

Directional Characteristics

As the wire is made longer in terms of the number of half wavelengths, the directional effects change. Instead of the doughnut pattern of the half-wave antenna, the directional characteristic splits up into lobes which make various angles with the wire. In general, as the length of the wire is increased the direction in which maximum radiation occurs tends to approach the line of the antenna itself.

Directional characteristics for antennas $1\,\lambda$, $1\tfrac{1}{2}\,\lambda$, and $2\,\lambda$ are given in Figs. 16, 17 and 18 for three vertical angles of radiation. Note that, as the wire length increases, the radiation along the line of the antenna becomes more pronounced. Still longer antennas can be considered to have practically end-on directional characteristics, even at the lower radiation angles.

When a long-wire antenna is fed at one end or at the current loop closest to that end, the radiation is most pronounced from the long section. This unidirectional pattern can be accentuated by terminating the far end in a load resistance to ground. The load resistor will dissipate energy that would ordinarily be radiated toward the feed point. Depending on the pattern symmetry of the unterminated antenna, this resistor must handle up to half the power delivered to the feed point. The exact resistance must be determined empirically, but the voltage-to-current ratio at a current node is a good starting value. A quarter-wavelength wire beyond the resistor can serve as a pseudo ground for the system. Low-angle radiation from a long wire can be enhanced by sloping the wire down toward the favored direction.

Methods of Feeding

In a long-wire antenna, the currents in adjacent half-wave sections must be out of phase, as shown in Fig. 14. The feeder system must not upset this phase relationship. This is satisfied by feeding the antenna at either end or at any current loop. A two-wire feeder cannot be inserted at a current node, however, because this invariably brings the currents in two adjacent half-wave sections in phase. A long-wire antenna is usually made a half wavelength at the lowest frequency and fed at the end.

Long-Wire Directive Arrays

Two long wires can be combined in the form of a horizontal V, in the form of a horizontal rhombus, or in parallel, to provide a long-wire directive array. In the V and rhombic antennas the main lobes reinforce along a line bisecting the acute angle between the wires; in the parallel antenna the reinforcement is along the line of the lobe. This reinforcement provides both

gain and directivity along the line, since the lobes in other directions tend to cancel. When the proper configuration for a given length and height above ground is used, the power gain depends on the length (in wavelengths) of the wires.

Rhombic and V antennas are normally bidirectional along the bisector line mentioned above. They can be made unidirectional by terminating the ends of the wires away from the feed point in the proper value of resistance. When properly terminated, V and rhombic antennas of sufficient length work well over a three-to-one or four-to-one frequency range and hence are useful for multiband operation.

Antenna gains of the order of 10 to 15 dB can be obtained with properly constructed long-wire arrays. The pattern is rather sharp with gains of this order, however, and rhombic and V beams are not used by amateurs as commonly as they once were, having been displaced by the rotatable multielement Yagi beam. Futher information on these antennas can be found the *The ARRL Antenna Book*.

BEAMS WITH DRIVEN ELEMENTS

By combining individual half-wave antennas into an array with suitable spacing between the antennas (called elements) and feeding power to them simultaneously, it is possible to make the radiation from the elements add up along a single direction and form a beam. In other directions the radiation tends to cancel, so a power gain is obtained in one direction at the expense of radiation in other directions.

There are several methods of arranging the elements. If they are strung end to end, so that all lie on the same straight line, the elements are said to be collinear. If they are parallel and all lying in the same plane, the elements are said to be broadside when the phase of the current is the same in all, and end-fire when the currents are not in phase.

Collinear Arrays

Simple forms of collinear arrays, with current distribution, are shown in Fig. 19. The two-element array at A is popularly known as a Franklin array, two half-waves in phase or a double Zepp antenna. It will be recognized as simply a center-fed dipole operated at its second harmonic.

By extending the antenna, as at B, the additional gain of an extended double Zepp antenna can be obtained. Carrying the length beyond that shown will result in an X-shaped pattern that no longer has the maximum radiation at right angles to the wire.

Collinear arrays may be mounted either horizontally or vertically. Horizontal mounting gives increased azimuthal directivity, while the vertical directivity remains the same as for a single element at the same height. Vertical mounting gives the same horizontal pattern as a single element, but improves the low-angle radiation.

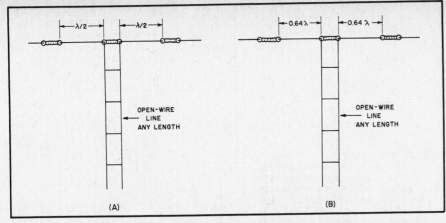

Fig. 19 — Collinear antennas in phase. The system at A is known as two half waves in phase and has a gain of 1.8 dB over a half-wave antenna. By lengthening the antenna slightly, as in B, the gain can be increased to 3 dB. Maximum radiation is at right angles to the antenna. The antenna at A is sometimes called a double Zepp antenna, and that of B is known as an extended double Zepp.

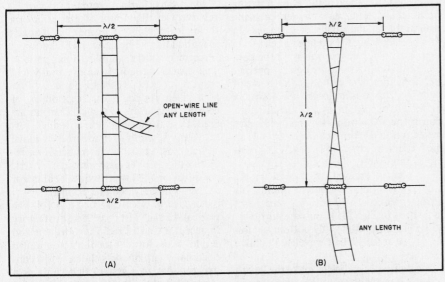

Fig. 20 — Simple broadside array using horizontal elements. By making the spacing S equal to 3/8 wavelength, the antenna at A can be used at the corresponding frequency and up to twice that frequency. Thus when designed for 14 MHz it can also be used on 18, 21, 25 and 28 MHz. The antenna at B can be used on only the design band. This array is bidirectional, with maximum radiation broadside or perpendicular to the antenna plane (perpendicularly through this page). Gain varies with the spacing S, running from 2½ to almost 5 dB. (See Fig. 22.)

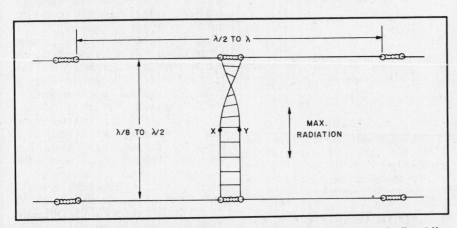

Fig. 21 — Top view of a horizontal end-fire array. The system is fed with an open-wire line at X and Y; the line can be of any length. Feed points X and Y are equidistant from the two insulators, and the feed line should drop down vertically from the antenna. The gain of the system will vary with the spacing, as shown in Fig. 22, and is a maximum at 1/8 wavelength. By using a length of 33 feet and a spacing of 8 feet, the antenna will work on 20, 17, 15, 12 and 10 meters.

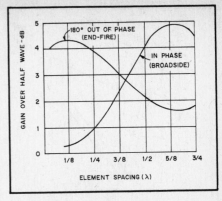

Fig. 22 — Gain vs. spacing for two parallel half-wave elements combined as either broadside or end-fire arrays.

Broadside Arrays

Parallel antenna elements with currents in phase may be combined as shown in Fig. 20 to form a broadside array, so named because the direction of maximum radiation is broadside to the plane containing the antennas. The gain and directivity depend on the spacing of the elements.

Broadside arrays may be suspended either with the elements all vertical or with them horizontal and one above the other (stacked). In the case of vertical elements, the horizontal pattern becomes quite sharp, while the vertical pattern is the same as that of one element alone. If the array is suspended horizontally, the horizontal pattern is equivalent to that of one element while the vertical pattern is sharpened, giving low-angle radiation.

Broadside arrays may be fed either by tuned open-wire lines or through quarter-wave matching sections and flat lines. In Fig. 20B, note the crossing over of the phasing section, which is necessary to bring the elements into proper phase relationship.

End-Fire Arrays

Fig. 21 shows a pair of parallel half-wave elements with currents out of phase. This is known as an end-fire array because it radiates best along the plane of the antennas, as shown. The end-fire principle was first demonstrated by John Kraus, W8JK, and 2-element arrays of this type are often called "8JK" antennas.

The end-fire array may be used either vertically or horizontally (elements at the same height), and is well adapted to amateur work because it gives maximum gain with relatively close element spacing. Fig. 22 shows how the gain varies with spacing. End-fire elements may be combined with additional collinear and broadside elements to give a further increase in gain and directivity.

Either tuned or untuned lines may be used with this type of array. Untuned lines preferably are matched to the antenna through a quarter-wave matching section or phasing stub.

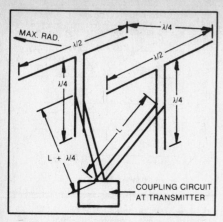

Fig. 23 — Unidirectional two-element end-fire array and method of obtaining 90° phasing.

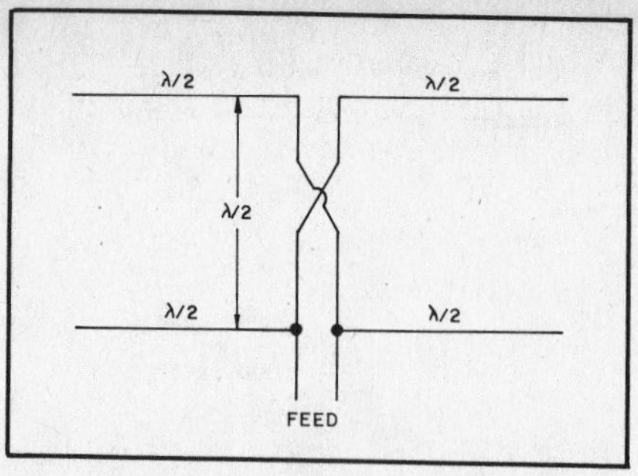

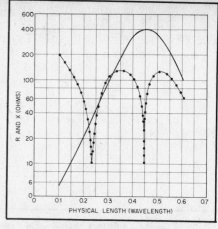

Fig. 24 — A four-element combination broadside collinear array, popularly known as the lazy-H antenna. A closed quarter-wave stub may be used at the feed point to match into an untuned transmission line, or tuned feeders may be attached at the point indicated. The gain over a half-wave antenna is 5 to 6 dB.

Fig. 25 — Radiation resistance (solid curve) and reactance (dotted curve) of vertical antennas as a function of physical height.

Unidirectional End-Fire Arrays

Two parallel elements spaced ¼ wavelength apart and fed equal currents 90° out of phase will have a directional pattern in the plane of the array at right angles to the elements. The maximum radiation is in the direction from the element in which the current lags. In the opposite direction the fields from the two elements cancel.

One way in which the 90° phase difference can be obtained is shown in Fig. 23. Each element must be matched to its transmission line, the two lines being of the same type except that one is an electrical quarter wavelength longer than the other. The length L can be any convenient value. Open quarter-wave matching sections could be used instead. The two transmission lines are connected in parallel at the transmitter coupling circuit.

When the currents in the elements are neither in phase nor 180° out of phase the radiation resistances of the elements are not equal. This complicates the problem of feeding equal currents to the elements. If the currents are not equal, one or more minor lobes will appear in the pattern and decrease the front-to-back ratio. The adjustment process is likely to be tedious and requires field-strength measurements in order to get the best performance.

More than two elements can be used in unidirectional end-fire arrays. The requirement for unidirectivity is that there must be a progressive phase shift in the element currents. The shift must equal the spacing between the elements in electrical degrees, and the current amplitudes also must be properly related. This requires binominal current distribution — i.e., the ratios of the currents in the elements must be proportional to the coefficients of the binominal series. In the case of three elements, this requires that the current in the center element be twice that in the two outside elements, for 90° (quarter-wave spacing and element current phasing. This antenna has an overall length of ½ wavelength.

Combined Arrays

Broadside, collinear and end-fire arrays may be combined to give both horizontal and vertical directivity, as well as additional gain. The lower angle of radiation resulting from stacking elements in the vertical plane is desirable at the higher frequencies. In general, doubling the number of elements in an array by stacking will raise the gain from 2 to 4 dB, depending on the spacing.

Although arrays can be fed at one end as in Fig. 20B, it is not especially desirable in the case of large arrays. Better distribution of energy between elements, and hence better overall performance, will result when the feeders are attached as nearly as possible to the center of the array.

A 4-element array, known as the lazy-H antenna, is a popular configuration. This arrangement is shown with the feed point indicated, in Fig. 24. (Compare with Fig. 20B.) For best results, the bottom section should be at least a half wavelength above ground.

It will usually suffice to make the length of each element equal to that given by the dipole formula. The phasing line between the parallel elements should be of open-wire construction and its length can be calculated from:

Length of half-wave line (feet)

$$= \frac{480}{f \text{ (MHz)}}$$

Example: A half-wavelength phasing line for 28.8 MHz would be

$$\frac{480}{28.8} = 16.67 \text{ feet} = 16 \text{ feet } 8 \text{ inches}$$

The spacing between elements can be made equal to the length of the phasing line. No special adjustments of line or element length or spacing are needed, provided the formulas are followed closely.

THE VERTICAL ANTENNA

One of the more popular amateur antennas is the vertical type. With this style of antenna it is possible to obtain low-angle radiation for ground-wave and DX work.

Additionally, vertical antennas occupy a relatively small amount of space, making them ideal for city-lot property and apartment buildings. The principal limitation in performance is the omnidirectional pattern. This means that QRM can't be nulled out from the directions that are not of interest at a given period. The exception is, of course, when arrays of vertical elements are used. Despite the limitation of a single vertical element with a ground screen or radial system, cost versus performance is an incentive that inspires many antenna builders.

For use on the lower frequency amateur bands — notably 160 and 80 meters — it is not always practical to erect a full-size vertical. In such instances it is satisfactory to accept a shorter radiating element and employ some form of loading to obtain an electrical length of one's choice. Most constructors design a system that contains a ¼-wavelength driven element. However, good results and lower radiation angles are sometimes realized when using a 3/8- or ½-wavelength vertical. At the lower amateur frequencies the larger verticals become prohibitive, especially in urban areas where zoning ordinances may exist, and where limited acreage may rule out the installation of guy-wire systems.

Fig. 25 provides curves for the physical height of verticals in wavelength versus radiation resistance and reactance. The plots are based on perfectly conducting ground, a condition seldom realized in practical installations. It can be seen that the shorter the radiator, the lower the radiation resistance — with 6 ohms being typical for a 0.1-wavelength antenna. The lower the radiation resistance, the more the antenna efficiency depends on ground conductivity. Also, the bandwidth decreases markedly as the length is reduced toward the left of the scale in Fig. 25. Difficulty is also experienced in developing a suitable matching network when the radiation resistance is very low.

Illustrations of various vertical-antenna radiation patterns are given in Fig. 26. The example at A is for a quarter-wavelength radiator over a theoretically ideal ground. The dashed lines show the current distribution, including the image portion below ground. The image can be equated to one half of a dipole antenna, with the vertical radiator representing the remaining dipole half.

The illustration at B characterizes the pattern of a half-wavelength vertical. It can be seen that the radiation angle is somewhat lower than that of the quarter-wavelength version at A. The lower angles enhance the DX capability of the antenna. Two half wavelengths in phase are shown in Fig. 26 at C and D. From a practical point of view, few amateurs could erect such an antenna unless it was built for use on the higher HF bands, such as 20, 15 or 10 meters. The very low radiation angle is excellent for DXing, however.

Full-Size Vertical Antennas

When it is practical to erect a full-size vertical antenna, the forms shown in Fig. 27 are worthy of consideration. The example at A is the well-known vertical ground plane. The ground system consists of four or more above-ground radial wires against which the driven element is worked. The length of the radials and the driven element is derived from the standard equation

$$L \text{ (feet)} = \frac{234}{f \text{ (MHz)}}$$

It has been established generally that with four equidistant radial wires drooped at approximately 45° (Fig. 27A) the feed-point impedance is roughly 50 ohms. When the radials are at right angles to the radiator (Fig. 27B) the impedance approaches 36 ohms. The major advantage in this type of vertical antenna over a ground-mounted type is that the system can be elevated well above nearby conductive objects (power lines, trees, buildings, etc.). When drooping radials are utilized they can be used as guy wires for the mast that supports the antenna. The coaxial cable shield braid is connected to the radials, and the center conductor is common to the driven element.

The Marconi antenna shown in Fig. 27C is the classic form taken by a ground-mounted vertical. It can be grounded at the base and shunt fed, or it can be isolated from ground, as shown, and series fed. This antenna depends on an effective ground system for efficient performance. The subject of ground screens is treated later in this chapter. If a perfect ground were located below the antenna, the feed impedance would be near 36 ohms. In a practical case, owing to imperfect ground, the impedance is more apt to be in the vicinity of 50 to 75 ohms.

A gamma feed system for a grounded ¼-wavelength vertical is presented in Fig. 27D. Some rules of thumb for arriving at workable gamma-arm and capacitor di-

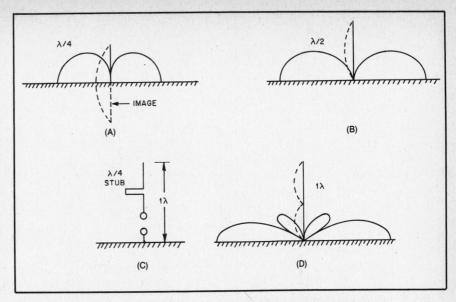

Fig. 26 — Elevation plane responses for a quarter-wavelength vertical antenna (A), a ½-wavelength type (B) and two half waves in phase (C and D). It can be seen that the examples at B and D provide more radiated power at low radiation angles than the version at A.

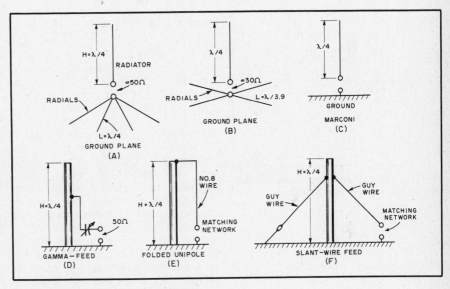

Fig. 27 — Various types of vertical antennas.

mensions are to make the rod length 0.04 to 0.05 wavelength, its diameter ⅓ to ½ that of the driven element and the center-to-center spacing between the gamma arm and the driven element roughly 0.007 wavelength. The capacitance of C1 at a 50-ohm matched condition will be some 7 pF per meter. The absolute value at C1 will depend on whether the vertical is resonant and on the precise value of the radiation resistance. Generally, best results can be had when the radiator is approximately three percent shorter than the resonant length.

Amateur antenna towers lend themselves well to use as shunt-fed verticals, even though an HF band beam antenna may be mounted on the tower. The overall system should be close to resonance at the desired operating frequency if gamma feed is to be used. The HF-band beam will contribute somewhat to top loading of the tower. The natural resonance of such a system can be checked by dropping a no. 12 or 14 wire down from the top of the tower (making it common to the tower top) to form a folded unipole (Fig. 27E). A four- or five-turn link can be inserted between the lower end of the drop wire and the ground system, then a dip meter inserted in the link to observe the resonant frequency. If the tower is equipped with guy wires, they should be broken up with strain insulators to prevent unwanted loading of the vertical. In such cases where the tower and beam antennas are not able to provide ¼-wavelength resonance, portions of the top guy wires can be used as top-loading capaci-

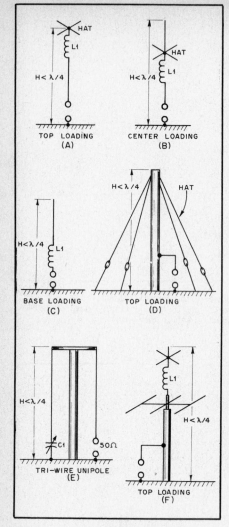

Fig. 28 — Vertical antennas that are less than one quarter wavelength in height.

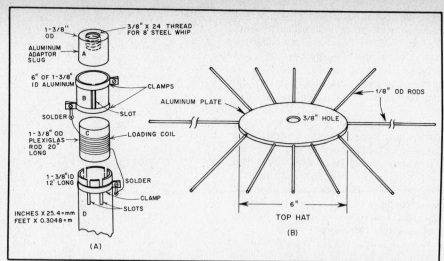

Fig. 29 — At A are the details for the tubing section of the loading assembly. Illustration B shows the top hat and its spokes. The longer the spokes, the better.

tance. It will be necessary to experiment with the guy-wire lengths (using the dipmeter technique) while determining the proper dimensions.

A folded-unipole type of vertical is depicted at E of Fig. 27. This system has the advantage of increased feed-point impedance. Furthermore, a Transmatch can be connected between the bottom of the drop wire and the ground system to permit operation on more than one band. For example, if the tower is resonant on 80 meters, it can be used as shown on 160 and 40 meters with reasonable results, even though it is not electrically long enough on 160. The drop wire need not be a specific distance from the tower, but spacings between 12 and 30 inches are suggested.

The method of feed shown at Fig. 27F is commonly referred to as "slant-wire feed." The guy wires and the tower combine to provide quarter-wave resonance. A matching network is placed between the lower end of one guy wire and ground and adjusted for an SWR of 1. It does not matter at which level on the tower the guy wires are connected, assuming that the Transmatch is capable of effecting a match to 50 ohms.

Physically Short Verticals

A group of short vertical radiators is presented in Fig. 28. Illustrations A and B are for top and center loading. A capacitance hat is shown in each example. The hat should be as large as practical to increase the radiation resistance of the antenna and improve the bandwidth. The wire in the loading coil is chosen for the largest gauge consistent with ease of winding and coil-form size. The larger wire diameters will reduce the I^2R losses in the system. The coil-form material should be of the medium or high dielectric type. Phenolic or fiberglass tubing is entirely adequate.

A base-loaded vertical is shown at C of Fig. 28. Since this is the least effective method of loading in terms of antenna performance, it should be used only as a last choice. The primary limitation is that the current portion of the vertical exists in the coil rather than the driven element. With center loading the portion of the antenna below the coil carries current, and with the top loading version the entire vertical element carries current. Since the current part of the antenna is responsible for most of the radiating, base loading is the least effective of the three methods. The radiation resistance of the coil-loaded antennas shown is usually less than 16 ohms.

A method for using guy wires to top load a short vertical is illustrated in Fig. 28 at D. This system works well with gamma feed. The loading wires are trimmed to provide an electrical quarter wavelength for the overall system. This method of loading will result in a higher radiation resistance and greater bandwidth than the systems shown at A through C of Fig. 28. If any HF band or VHF array is atop the tower, it will simply contribute to the top loading.

A tri-wire unipole is shown at E of Fig.

28. Two no. 8 drop wires are connected to the top of the tower and brought to ground level. The wires can be spaced any convenient distance from the tower — normally 12 to 30 inches from one side. C1 is adjusted for an SWR of 1. This type of vertical has a fairly narrow bandwidth, but because C1 can be motor driven and controlled from the operating position, QSYing is accomplished easily. This technique will not be suitable for matching to 50-ohm line unless the tower is less than an electrical quarter wavelength high.

A different method for top loading is shown at F of Fig. 28. W9UCW described this system in "The Minooka Special," December 1974 *QST*. An extension is used at the top of the tower to effect an electrical quarter-wavelength vertical. L1 is a loading coil with sufficient inductance to provide antenna resonance. This type of antenna lends itself nicely to operation on 160 meters.

A method for effecting the top-loading shown in Fig. 28F is illustrated in the drawing of Fig. 29. Pipe section D is mated with the mast above the HF-band beam antenna. A loading coil is wound on solid Plexiglas rod or phenolic rod (item C), then clamped inside the collet (B). An aluminum slug (part A) is clamped inside item B. The top part of A is bored and threaded for a 3/8 inch × 24 thread stud. This will permit a standard 8-foot stainless-steel mobile whip to be threaded into item A above the loading coil. The capacitance hat (Fig. 29B) can be made from a 1/4-inch-thick brass or aluminum plate. It may be round or square. Lengths of 1/8-inch brazing rod can be threaded for a 6-32 format to permit the rods to be screwed into the edge of the aluminum plate. The plate contains a row of holes along its perimeter, each having been tapped for a 6-32 thread. The capacitance hat is affixed to item A by means of the 8-foot whip antenna. The whip will increase the effective height of the vertical antenna.

Cables and Control Wires on Towers

Most vertical antennas of the type shown in Fig. 28 consist of towers and HF or VHF beam antennas. The rotator control wires and the coaxial feeders to the top of the tower will not affect antenna performance adversely. In fact, they become a part of the composite antenna. To prevent unwanted rf currents from following the wires into the shack, simply dress them close to the tower legs and bring them to ground level. This decouples the wires at RF. The wires should then be routed along the earth surface (or buried underground) to the operating position. It is not necessary to use bypass capacitors or RF chokes in the rotator control leads if this is done, even when maximum legal power is employed.

Variations in Verticals

A number of configurations qualify for use as vertical antennas even though the radiators are fashioned from lengths of wire. Fig. 30A shows a flat-top T vertical. Dimension H should be as tall as possible for best results. The horizontal section, L, is adjusted to a length which provides resonance. Maximum radiation is polarized vertically despite the horizontal top-loading wire. A variation of the T antenna is depicted at B of Fig. 30. This antenna is commonly referred to as an inverted L. Vertical member H should be as long as possible. L is added to provide an electrical quarter wavelength overall. Some amateurs believe that a 3/8-wavelength version of this antenna is more effective, since the current portion of the wire is elevated higher above ground than is the case with a quarter-wavelength wire.

Half-Sloper Antenna

A basic-half sloper antenna is shown at C of Fig. 30. The wire portion represents one half of a dipole. The feed point is between the tower leg and the upper end of the slope wire, with the coax shield braid being connected to the tower. The feed-point impedance is generally between 30 and 60 ohms, depending on the length of the wire, the tower height and the enclosed angle between the slope wire and the tower. The tower constitutes a portion of this antenna, and will have a voltage maximum somewhere between the feed point and ground. The half sloper has been found useful for close-in work as well as for DXing. A VSWR as low as 1.5:1 can be obtained (using 50-ohm coax feed line) with the system of Fig. 30C by experimenting with the slope-wire length and the enclosed angle. This assumes there are no guy wires that are common to the tower. If guys are used they should be insulated from the tower and broken into nonresonant lengths for the band of operation.

Many amateurs are likely to have towers that support HF-band Yagis or quad

beams. The conductors situated above the sloper feed point have a marked effect on the tuning of the sloper system. The beam antenna becomes a portion of the overall tower/sloper system. Although the presence of the beam seems to have no appreciable effect on the radiation pattern of the sloper, it does greatly affect the VSWR. If a low VSWR can't be obtained for a given amateur antenna installation, the system illustrated in Fig. 30D can be applied.

The sloper is erected in the usual manner, but a second wire (L2) is added. It is connected to the tower at the antenna feed point, but insulated at the low end. L2 plus the tower constitutes the missing half of the dipole. This half is effectively a fanned element (similar to a conical element) and tends to increase the bandwidth of the sloper system over that which can be obtained with the version at C. The "compensating wire" (L2) is spaced approximately 90° from L1, and is roughly 30° away from the side of the tower. The VSWR can be brought to a low value (usually 1:1) by varying the L2/tower and L1/L2 enclosed angles. The degree of these angles will depend on the tower height, the beam atop the tower and the ground condition below the system.

Tests indicate the sloper to be effective on its harmonics for DX and local work. If harmonic operation is planned, the 50-ohm coaxial feeder should be replaced with open-wire line, which is coupled to the transmitter by means of a Transmatch. With open-wire feed, the L2 compensating wire of Fig. 30D will not be necessary.

Best performance will be had when the base of the tower is well grounded. Buried radials are highly recommended as part of the ground system. Operation on 160 meters can be had by feeding the low end of L2. The bandwidth of the system in Fig. 30 will be approximately 50 kHz on 160 meters, 100 kHz on 80 meters, 200 kHz on 40 meters, and so on, between 2:1 VSWR points.

Ground Systems

The importance of an effective ground system for vertical antennas cannot be emphasized too strongly. However, is it not always possible to install a radial network that approaches the ideal. A poor ground is better than no ground at all, and therefore the amateur should experiment with whatever is physically possible rather than ruling out vertical antennas. It is often possible to obtain excellent DX results with practically no ground system at all.

Although the matter of less-than-optimum ground systems could be debated almost endlessly, some practical rules of thumb are in order for those wishing to erect vertical antennas. Generally a large number of shorter radials offers a better ground system than a few longer ones. For example, 8 radials of 1/8 λ are preferred over 4 radials of ¼ λ. If the physical height

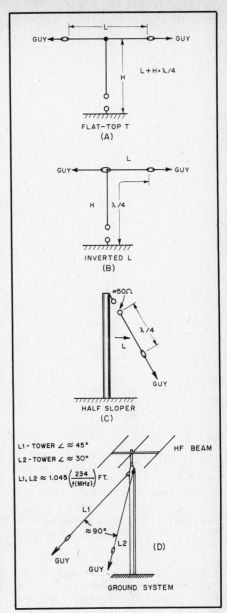

Fig. 30 — Some variations in vertical antennas that offer excellent performance.

of the vertical is an eighth wavelength, the radial wires should be of the same length and dispersed uniformly from the base of the tower.

The conductor size of the radials is not especially significant. Wire gauges from no. 4 to no. 20 have been used successfully by amateurs. Copper wire is preferred, but where soil is low in acid or alkali, aluminum wire can be used. The wires can be bare or insulated, and they can be laid on the earth's surface or buried a few inches below ground. The insulated wires will have greater longevity by virtue of reduced corrosion and dissolution from soil chemicals.

If time and expense are not prime considerations, the amateur should bury as much ground wire as possible. Some operators have literally miles of wire buried radially beneath their vertical antennas.

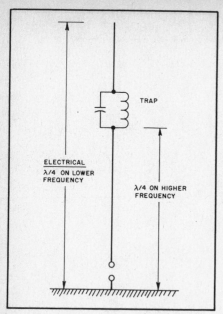

Fig. 31 — A two-band trap vertical antenna. The trap should be resonated as a parallel resonant circuit at the center of the operating range for the higher frequency band; typical component reactances range from 100 to 300 ohms. At the lower frequency the trap will act as a loading inductor, adding electrical length to the total antenna.

When property dimensions do not allow a classic installation of equally spaced radial wires, they can be placed in the ground wherever space will permit. They may run away from the antenna in only one or two compass directions. Results will still exceed those of when no ground system is used.

A single ground rod, or group of them bonded together, is seldom as effective as a collection of random-length radial wires. In some instances a group of short radial wires can be used in combination with ground rods driven into the soil near the base of the antenna. Bear in mind, though, that RF currents at MF and HF seldom penetrate the earth more than several inches. The power company ground can be tied in, and if a metal fence skirts the property it can also be used as part of the ground system. A good rule is to use anything that will serve as a ground when developing a radial ground system.

All radial wires must be connected together at the base of the vertical antenna. The electrical bond needs to be of low resistance. Best results will be obtained when the wires are soldered together at the junction point. When a grounded vertical is used, the ground wires should be affixed securely to the base of the driven element. A lawn edging tool is excellent for cutting slits in the soil when laying radial wires.

Trap Verticals

Although a full-size, single-band antenna is more effective than a lumped-constant one, there is justification for using trap types of multiband antennas. The concept is especially useful to operators who have limited antenna space on their property. Multiband compromise antennas are also appealing to persons who engage in portable operation and are unwilling to transport large amounts of antenna hardware to the field.

The 2-band trap vertical antenna of Fig. 31 operates in much the same manner as a trap dipole or trap-style Yagi. The notable difference is that the vertical is one half of a dipole. The radial system (in-ground or above ground) functions as a ground plane for the antenna, and represents the missing half of the dipole. Therefore, the more effective the ground system, the better the antenna performance.

Trap verticals are adjusted as quarter-wavelength radiators. The portion of the antenna below the trap is adjusted as a quarter-wavelength radiator at the higher proposed operating frequency. That is, a 20/15-meter trap vertical would be a resonant quarter wavelength at 15 meters from the feed point to the bottom of the trap. The trap and that portion of the antenna above the trap (plus the 15-meter section below the trap) constitute the complete antenna during 20-meter operation. But because the trap is in the circuit, the overall physical length of the vertical antenna will be slightly less than that of a single-band, full-size 20-meter vertical.

Traps

The trap functions as the name implies: It traps the 15-meter energy and confines it to the part of the antenna below the trap. During 20-meter operation it allows the RF energy to reach all of the antenna. Therefore, the trap in this example should be tuned as a parallel resonant circuit to 21 MHz. At this frequency it divorces the top section of the vertical from the lower section because it presents a high-impedance (barrier) at 21 MHz. Generally, the trap inductor and capacitor have a reactance of 100 to 300 ohms. Within that range it is not critical.

The trap is built and adjusted separately from the antenna. It should be resonated at the center of the portion of the band to be operated. Thus, if one's favorite part of the 15-meter band is between 21,000 and 21,100 kHz, the trap would be tuned to 21,050 kHz.

Resonance is checked by using a dip meter and detecting the dipper signal in a calibrated receiver. Once the trap is adjusted it can be installed in the antenna, and no further adjustment will be required. It is easy, however, to be misled after the system is assembled: Attempts to check the trap with a dip meter will suggest that the trap has moved much lower in frequency (approximately 5 MHz lower in a 20/15-meter vertical). This is because the trap has become absorbed into the overall antenna, and the resultant resonance is that of the total antenna. Ignore this phenomenon.

Multiband operation for three or four bands is quite practical by using the appropriate number of traps and tubing sections. The construction and adjustment procedure is the same, regardless of the number of bands covered. The highest frequency trap is always closest to the feed end of the antenna, and the next to lowest frequency trap is always the farthest from the feed point. As the operating frequency is progressively lowered, more traps and more tubing sections become a functional part of the antenna.

The trap should be weatherproofed to prevent moisture from detuning it. Several coatings of high dielectric compound, such as Polystyrene Q Dope, are effective. Alternatively, a protective sleeve of heat-shrink tubing can be applied to the coil after completion. The coil form for the trap should be of high dielectric quality and be rugged enough to sustain stress during periods of wind.

The trap capacitor must be capable of withstanding the RF voltage developed across it. The amount of voltage present will depend on the operating power of the transmitter. Fixed-value ceramic transmitting capacitors are suitable for most power levels if they are rated at 5000 to 10,000 volts. A length of RG-58/U or RG-59/U coax cable can be used successfully up to 200 watts. (Check to see how many picofarads per foot your cable is rated at before cutting it for the trap.) RG-8/U or RG-11/U cable is recommended for the trap capacitor at powers in excess of 200 watts. The advantage of using coax cable is that it can be trimmed easily to adjust the trap capacitance.

Large-diameter copper magnet wire is suggested for the trap coil. The heavier the wire gauge the lower the trap losses and the higher the Q. The larger wire sizes will reduce coil heating.

YAGI AND QUAD DIRECTIVE ANTENNAS

Most of the antennas described earlier in this chapter have unity gain or just slightly more. For the purpose of obtaining gain and directivity it is convenient to use the Yagi-Uda or cubical quad types of HF-band beam antennas. The former is commonly called a Yagi and the latter is referred to as a quad in the amateur vernacular.

Most operators prefer to erect these antennas for horizontal polarization, but they can be used as vertically polarized arrays as well merely by rotating the elements by 90°. In effect, the beam antenna is turned on its side for vertical polarity. The number of elements employed will depend on the gain desired and the capability of the supporting structure to contain the array safely. Many amateurs obtain satisfactory results with only two elements in a beam antenna, while others have several

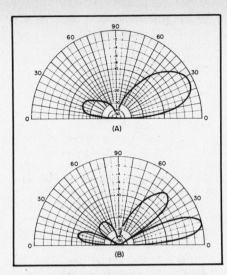

Fig. 32 — Elevation-plane response of a three-element Yagi placed ½ wavelength above a perfect ground (A) and the same antenna spaced one wavelength above ground (B). Pattern calculations courtesy of "Annie." [2]

elements operating for a single amateur band.

Regardless of the number of elements used, the height-above-ground considerations discussed earlier for dipole antennas remain valid with respect to the angle of radiation. This is demonstrated in Fig. 32 at A and B where a comparison of radiation characteristics is given for a three-element Yagi at one-half and one wavelength above a perfectly conducting ground. It can be seen that the higher antennas (Fig. 32B) has a lobe that is more favorable for DX work (roughly 15°) than the larger lobe of Fig. 32A (approximately 30°). The pattern at B shows that some useful high-angle radiation exists also, and the higher lobe is suitable for short-skip contacts when propagation conditions dictate the need.

A free-space azimuth pattern for the same antenna is provided in Fig. 33. The back-lobe pattern reveals that most of the power is concentrated in the forward lobe. The power difference dictates the front-to-back ratio in decibels. It is infrequent that two 3-element Yagis with different element spacings will yield the same lobe patterns. The pattern of Fig. 33 is shown only for illustrative purposes.

Parasitic Excitation

In most of these arrangements the additional elements receive power by induction or radiation from the driven element and reradiate it in the proper phase relationship to give the desired effect. These elements are called parasitic elements, as contrasted to the driven elements, which receive power directly from the transmitter through the transmission line.

The parasitic element is called a director when it reinforces radiation on a line

[2]See footnote 1 earlier in this chapter.

pointing to it from the driven element, and a reflector when the reverse is the case. Whether the parasitic element is a director or reflector depends on the parasitic-element tuning, which usually is adjusted by changing its length.

Gain vs. Spacing

The gain of an antenna with parasitic elements varies with the spacing and tuning of the elements. Thus, for any given spacing, there is a tuning condition that will give maximum gain at this spacing. The maximum front-to-back ratio seldom, if ever, occurs at the same condition that gives maximum forward gain. The impedance of the driven element also varies with the tuning and spacing, and thus the antenna system must be tuned to its final condition before the match between the line and the antenna can be completed. The tuning and matching may interact to some extent, however, and it is usually necessary to run through the adjustments several times to ensure that the best possible tuning has been obtained.

Two-Element Beams

A two-element beam is useful where space or other considerations prevent the use of the larger structure required for a three-element beam. The general practice is to tune the parasitic element as a reflector and space it about 0.15 wavelength from the driven element, although some successful antennas have been built with 0.1-wavelength spacing and director tuning. Gain vs. element spacing for a two-element antenna is given in Fig. 34 for the special case where the parasitic element is resonant. It is indicative of the performance to be expected under maximum-gain tuning conditions.

Three-Element Beams

A theoretical investigation of the three-element case (director, driven element and reflector) has indicated a maximum gain of slightly more than 7 dB. A number of experimental investigations has shown that the optimum spacing between the driven element and reflector is in the region of 0.15 to 0.25 wavelength, with 0.2 wavelength representing probably the best overall choice. With 0.2-wavelength reflec-

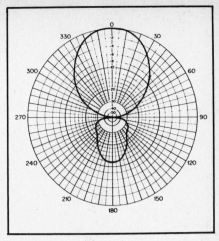

Fig. 33 — Azimuth-plane pattern of a three-element Yagi in free space. Pattern calculation courtesy of "Annie."

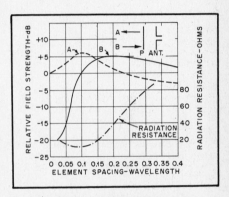

Fig. 34 — Gain vs. element spacing for an antenna and one parasitic element. The reference point, 0 dB, is the field strength from a half-wave antenna alone. The greatest gain is in the direction A at spacings of less than 0.14 wavelength, and in direction B at greater spacings. The front-to-back ratio is the difference in decibels between curves A and B. Variation in radiation resistance of the driven element is also shown. These curves are for a self-resonant parasitic element. At most spacings the gain as a reflector can be increased by slight lengthening of the parasitic element; the gain as a director can be increased by shortening. This also improves the front-to-back ratio.

tor spacing, Fig. 35 shows that the gain variation with director spacing is not especially critical. Also, the overall length of the array (boom length in the case of a

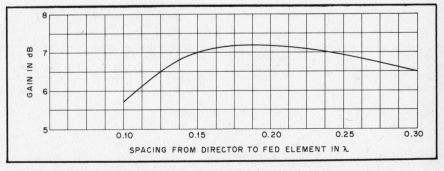

Fig. 35 — General relationship of gain of three-element Yagi versus director spacing, the reflector being fixed at 0.2 wavelength.

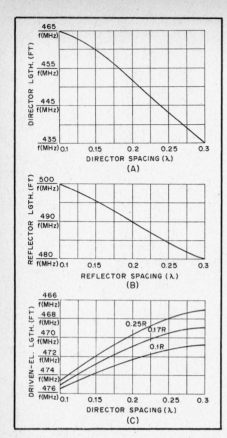

Fig. 36 — Element lengths for a three-element beam. These lengths will hold closely for the tubing elements supported at or near the center.

rotatable antenna) can be anywhere between 0.35 and 0.45 wavelength with no appreciable difference in gain.

Wide spacing of both elements is desirable not only because it results in high gain but also because adjustment of tuning or element length is less critical and the input resistance of the driven element is higher than with close spacing. A higher input resistance improves the efficiency of the antenna and makes a greater bandwidth possible. However, a total antenna length, director to reflector, of more than 0.3

wavelength at frequencies of the order of 14 MHz introduces considerable difficulty from a construction standpoint. Lengths of 0.25 to 0.3 wavelength are therefore used frequently for this band, even though they are less than optimum.

In general, the antenna gain drops off less rapidly when the reflector length is increased beyond the optimum value than it does for a corresponding decrease below the optimum value. The opposite is true of a director. It is therefore advisable to err, if necessary, on the long side for a reflector and on the short side for a director. This also tends to make the antenna performance less dependent on the exact frequency at which it is operated: An increase above the design frequency has the same effect as increasing the length of both parasitic elements, while a decrease in frequency has the same effect as shortening both elements. By making the director slightly short and the reflector slightly long, there will be a greater spread between the upper and lower frequencies at which the gain starts to show a rapid decrease.

When the overall length has been decided on, the element lengths can be found by referring to Fig. 36. The lengths determined by these charts will vary slightly in actual practice with the element diameter and the method of supporting the elements. The tuning of a beam should always be checked after installation. However, the lengths obtained by the use of the charts will be close to correct in practically all cases, and they can be used without checking if the beam is difficult to access.

Table 1 can be used to determine the lengths needed for 4-element Yagis. Both CW and phone lengths are included for the three bands, 20, 15 and 10 meters. The 0.2-wavelength spacing will provide greater bandwidth than the 0.15 spacing. Antenna gain is essentially the same with either spacing. The element lengths given will be the same whether the beam has two, three or four elements. It is recommended that plumbers delight type construction be used, where all the elements are mounted direct-

ly on and grounded to the boom. This puts the entire array at dc ground potential, affording better lightning protection. A gamma section can be used for matching the feed line to the array.

Tuning Adjustments

The preferable method for checking the beam is by means of a field-strength meter, or the S meter of a communications receiver, used in conjunction with a dipole antenna located at least 10 wavelengths away and as high or higher than the beam that is being checked. A few watts of power fed into the antenna will give a useful signal at the observation point, and the power input to the transmitter should be held constant for all the readings.

Preliminary matching adjustments can be done on the ground. The beam should be set up so the reflector element rests on the earth, with the remaining elements in a vertical configuration. In other words, the beam should be aimed straight up. The matching system is then adjusted for a 1:1 SWR between the feed line and driven element. When the antenna is raised to its operating height, only slight touch-up of the matching network should be required.

A great deal has been printed about the need for tuning elements of a Yagi-type beam. However, experience has shown that for Yagi arrays made from metal tubing, the lengths given in Fig. 36 and Table 1 are close enough to the desired length that no further tuning should be required.

Simple Systems — The Rotary Beam

Two- and three-element systems are popular for rotary-beam antennas, where the entire antenna system is rotated, to permit its gain and directivity to be utilized for any compass direction. The antennas may be mounted either horizontally (with the plane containing the elements parallel to the earth) or vertically.

A four-element beam will give still more gain than a three-element one, provided the support is sufficient for about 0.2 wavelength spacing between elements. The

Table 1

Element Lengths for 20, 15 and 10 Meters, Phone and CW

Freq. (kHz)	Driven Element		Reflector		First Director		Second Director	
	A	B	A	B	A	B	A	B
14,050	33' 5-3/8"	33' 8"	35' 2-1/2"	35' 5-1/4'	31' 9-3/8"	31' 11-5/8"	31' 1-1/4"	31' 3-5/8"
14,250	32' 11-3/4"	33' 2-1/4"	34' 8-1/2"	34' 11-1/4"	31' 4"	31' 6-3/8"	30' 8"	30' 10-1/2"
21,050	22' 4"	22' 5-5/8"	23' 6"	23' 7-3/4"	21' 2-1/2"	21' 4"	20' 9-1/8"	20' 10-7/8"
21,300	22' 3/4"	22' 2-3/8"	23' 2-5/8"	23' 4-1/2"	20' 11-1/2"	21' 1"	20' 6-1/4"	20' 7-3/4"
28,050	16' 9"	16' 10-1/4"	17' 7-5/8"	17' 8-7/8"	15' 11"	16'	15' 7"	15' 9-1/2"
28,600	16' 5-1/4"	16' 6-3/8"	17' 3-1/2"	17' 4-3/4"	15' 7-1/4"	15' 8-1/2"	15' 3-3/8"	15' 4-1/2"

A
| 0.2 | 0.2 | 0.2 |

B
| 0.15 | 0.15 | 0.15 |

These lengths are for 0.2- or 0.15-wavelength element spacing.

To convert ft to meters multiply ft × 0.3048.
Convert in to mm by multiplying in × 25.4.

tuning for maximum gain involves many variables, and complete gain and tuning data are elusive.

The elements in close-spaced arrays (less than ¼-wavelength element spacing) preferably should be made of tubing of ½ to 1 inch diameter. A conductor of large diameter not only has less ohmic resistance but also has lower Q; both these factors are important in close-spaced arrays because the impedance of the driven element usually is quite low compared to that of a simple dipole antenna. With three- and four-element close-spaced arrays the radiation resistance of the driven element may be so low that ohmic losses in the conductor can consume an appreciable fraction of the power.

Feeding the Rotary Beam

Any of the usual methods of feed (described later under Impedance Matching) can be applied to the driven element of a rotary beam. The popular choices for feeding a beam are the gamma match with series capacitor, the T match with series capacitors, and a half-wavelength phasing section, as shown in Fig. 37. These methods are preferred over any others because they permit adjustment of the matching and the use of coaxial-line feed. The variable capacitors can be housed in small plastic cups for weatherproofing; receiving types with close spacing can be used at powers up to a few hundred watts. Maximum capacitance required is usually 140 pF at 14 MHz and proportionally less at the higher frequencies.

If physically possible, it is better to adjust the matching device after the antenna has been installed at its ultimate height, since a match made with the antenna near the ground may not hold for the same antenna in the air.

Sharpness of Resonance

Peak performance of a multielement parasitic array depends on proper phasing or tuning of the elements, which can be exact for one frequency only. Close-spaced arrays usually are quite sharp-tuning because of the low radiation resistance. The frequency range over which optimum results can be obtained is only of the order of one or two percent of the resonant frequency, or up to about 500 kHz at 28 MHz. However, the antenna can be made to work satisfactorily over a wider frequency range by adjusting the director or directors to give maximum gain at the highest frequency to be covered, and by adjusting the reflector to give optimum gain at the lowest frequency. This sacrifices some gain at all frequencies but maintains more uniform gain over a wider frequency range.

The use of large-diameter conductors will broaden the response curve of an array, because a larger diameter lowers the Q. This causes the reactances of the elements to change rather slowly with frequency, with the result that the tuning stays near

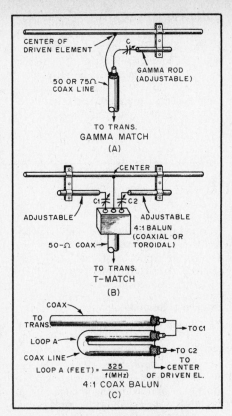

Fig. 37 — Illustrations of gamma and T matching systems. At A, the gamma rod is adjusted along with C until the lowest SWR is obtained. A T match is shown at B. It is the same as two gamma-match rods. The rods and C1 and C2 are adjusted alternately for a 1:1 SWR. A coaxial 4:1 balun transformer is shown at C. A toroidal balun can be used in place of the coax model shown. The toroidal version has a broader frequency range than the coaxial one. The T match is adjusted for 200 ohms and the balun steps this balanced value down to 50 ohms, unbalanced. Or the T match can be set for 300 ohms, and the balun used to step this down to 75 ohms unbalanced. Dimensions for the gamma and T match rods are not given by formula. Their lengths and spacing will depend on the tubing size used, and the spacing of the parasitic elements of the beam. Capacitors C, C1 and C2 can be 140 pF for 14-MHz beams. Somewhat less capacitance will be needed at 21 and 28 MHz.

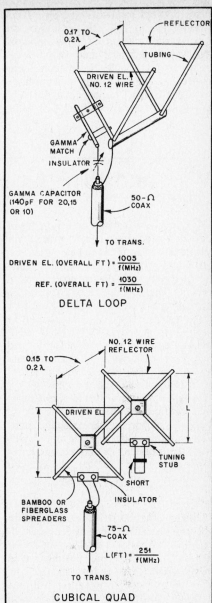

Fig. 38 — Information on building a quad or a delta-loop antenna. The antennas are electrically similar, but the delta-loop uses plumber's delight construction.

optimum over a considerably wider frequency range than with wire conductors.

Delta Loops and Quad Beams

One of the more effective DX arrays is called the cubical quad, or simply, quad antenna. It consists of two or more square loops of wire, each supported by a bamboo or fiberglass cross-arm assembly. The loops are a quarter wavelength per side (full wavelength overall). One loop is driven and the other serves as a parasitic element — usually a reflector. A variation of the quad is called the delta loop. The electrical properties of both antennas are the same, generally speaking, though some operators report better DX results with the delta loop. Both antennas are shown in Fig. 38. They differ mainly in their physical properties,

one being of plumber's delight construction, while the other uses insulating support members. One or more directors can be added to either antenna if additional gain and directivity is desired, though most operators use the two-element arrangement.

It is possible to interlace quads or "deltas" for two or more bands, but if this is done the formulas given in Fig. 38 may have to be changed slightly to compensate for the proximity effect of the second antenna. For quads the length of the full-wave loop can be computed from

$$\text{Full-wave loop (ft)} = \frac{1005}{f \text{ (MHz)}}$$

If multiple arrays are used, each antenna should be tuned separately for maximum forward gain, as noted on a field-strength meter. The reflector stub on the quad should be adjusted for this condition. The delta loop gamma match should be adjusted for a 1:1 SWR. No reflector tuning is needed. The delta loop antenna has a broader frequency response than the quad, and holds at an SWR of 1.5:1 or better across the band for which it is constructed.

The resonance of the quad antenna can be found by checking the frequency at which the lowest SWR occurs. The element length (driven element) can be adjusted for resonance in the most-used portion of the band by lengthening or shortening it.

A two-element quad or delta loop antenna compares favorably with a three-element Yagi array in terms of gain (see *QST,* May 1963 and January 1969, for additional information). The quad and delta-loop antennas perform very well at 50 and 144 MHz. A discussion of radiation patterns and gain, quads vs. Yagis, was presented by Lindsay in May 1968 *QST.*

VHF and UHF Antennas

Improving an antenna system is one of the most productive moves open to the VHF enthusiast. It can increase transmitting range, improve reception, reduce interference problems and bring other practical benefits. The work itself is by no means the least attractive part of the job. Even with high-gain antennas, experimentation is greatly simplified at VHF and UHF because an array is a workable size, and much can be learned about the nature and adjustment of antennas. No large investment in test equipment is necessary.

Whether we buy or build our antennas, we soon find that there is no one "best" design for all purposes. Selecting the antenna best suited to our needs involves much more than scanning gain figures and prices in a manufacturer's catalog. The first step should be to establish priorities.

Gain

Shaping the pattern of an antenna to concentrate radiated energy, or received-signal pickup, in some directions at the expense of others is the only possible way to develop gain. This is best explained by starting with the hypothetical isotropic antenna, which would radiate equally in all directions. A point source of light illuminating the inside of a globe uniformly, from its center, is a visual analogy. No practical antenna can do this, so all antennas have gain over isotropic (dBi). A half-wave dipole in free space has a gain of 2.1 dBi. If we can plot the radiation pattern of an antenna in all planes, we can compute its gain, so quoting it with respect to isotropic is a logical base for agreement and understanding. It is rarely possible to erect a half-wave antenna that has anything approaching a free-space pattern; this fact is responsible for much of the confusion about true antenna gain.

Radiation patterns can be controlled in various ways. One is to use two or more driven elements, fed in phase. Such collinear arrays provide gain without markedly sharpening the frequency response, compared to that of a single element. More gain per element, but with a sacrifice in frequency coverage, is obtained by placing parasitic elements, longer and shorter than the driven one, in the plane of the first element, but not driven from the feed line. The reflector and directors of a Yagi array are highly frequency sensitive and such an antenna is at its best over frequency changes of less than one percent of the operating frequency.

Frequency Response

Ability to work over an entire VHF band may be important in some type of work. The response of an antenna element can be broadened somewhat by increasing the conductor diameter, and by tapering it to something approximating a cigar shape, but this is done mainly with simple antennas. More practically, wide frequency coverage may be a reason to select a collinear array, rather than a Yagi. On the other hand, the growing tendency to channelize operations in small segments of our bands tends to place broad frequency coverage low on the priority list of most VHF stations.

Radiation Pattern

Antenna radiation can be made omnidirectional, bidirectional, practically unidirectional, or anything between these conditions. A VHF net operator may find an omnidirectional system almost a necessity, but it may be a poor choice otherwise. Noise pickup and other interference problems tend to be greater with such antennas, and those having some gain are especially bad in these respects. Maximum gain and low radiation angle are usually prime interests of the weak-signal DX aspirant. A clean pattern, with lowest possible pickup and radiation off the sides and back, may be important in high-activity areas, or where the noise level is high.

Height Gain

In general, the higher an antenna is installed, the better in VHF antenna installations. If raising the antenna clears its view over nearby obstructions, it may make dramatic improvements in coverage. Within reason, greater height is almost always worth its cost, but height gain must be balanced against increased transmission-line loss. Line losses are considerable at VHF, and they increase with frequency. The best available line may be none too good, if the run is long in terms of wavelength. Give line-loss information, shown in table form in Chapter 16, close scrutiny in any antenna planning.

Physical Size

A given antenna design for 432 MHz will have the same gain as one for 144 MHz, but being only one-third the size it will intercept only one-third as much energy in receiving. Thus, to be equal in communication effectiveness, the 432-MHz array should be at least equal in size to the 144-MHz one, which will require roughly three times as many elements. With all the extra difficulties involved in going higher in frequency, it is well to be on the big side in building an antenna for the higher bands.

DESIGN FACTORS

Having sorted out objectives in a general way, we face decisions on specifics, such as polarization, type of transmission line, matching methods and mechanical design.

Polarization

Whether to position the antenna elements vertically or horizontally has been a moot point since early VHF pioneering. Tests show little evidence on which to set up a uniform polarization policy. On long paths there is no consistent advantage, either way. Shorter paths tend to yield higher signal levels with horizontal in some kinds of terrain. Man-made noise, especially ignition interference, tends to be lower with horizontal. Verticals are markedly simpler to use in omnidirectional systems and in mobile work.

Early VHF communication was largely vertical, but horizontal gained favor when directional arrays became widely used. The major trend to FM and repeaters, particularly in the 144-MHz band, has tipped the balance in favor of verticals in mobile work and for repeaters. Horizontal predominates in other communication on 50 MHz and higher frequencies. It is well to check in advance in any new area in which you expect to operate, however, as some localities still use vertical almost exclusively. A circuit loss of 20 dB or more can be expected with cross-polarization.

Transmission Lines

There are two main categories of

transmission lines, balanced and unbalanced. Balanced lines include open-wire lines separated by insulating spreaders, and twin-lead, in which the wires are embedded in solid or foamed insulation. Line losses result from ohmic resistance, radiation from the line and deficiencies in the insulation. Large conductors, closely spaced in terms of wavelength, and using a minimum of insulation, make the best balanced lines. Impedances are mainly 300 to 500 ohms. Balanced lines are best in straight runs. If bends are unavoidable, the angles should be as obtuse as possible. Care should be taken to prevent one wire from coming closer to metal objects than the other. Wire spacing should be less than 1/20 wavelength.

Properly built open-wire line can operate with very low loss in VHF and even UHF installations. A total line loss under 2 dB per hundred feet at 432 MHz is readily obtained. A line made of no. 12 wire, spaced ¾ inch or less with Teflon spreaders, and running essentially straight from antenna to station, can be better than anything but the most expensive coax, at a fraction of the cost. This assumes the use of baluns to match into and out of the line, with a short length of quality coax for the moving section from the top of the tower to the antenna. A similar 144-MHz setup could have a line loss under 1 dB.

Small coax such as RG-58/U or -59/U should never be used in VHF work if the run is more than a few feet. Half-inch lines (RG-8/U or -11/U) work fairly well at 50 MHz, and are acceptable for 144-MHz runs of 50 feet or less. If these lines have foam rather than solid insulation they are about 30 percent better. Aluminum-jacket lines with large inner conductors and foam insulation are well worth their cost. They are readily waterproofed, and can last almost indefinitely. Beware of any "bargains" in coax. Lost transmitter power can be made up to some extent by increasing power, but once lost, a weak signal can never be recovered in the receiver.

Effects of weather should not be ignored. A well-constructed open-wire line works well in nearly any weather, and it stands up well. Twin-lead is almost useless in heavy rain, wet snow or icing. The best grades of coax are impervious to weather. They can be run underground, fastened to metal towers without insulation, or bent into any convenient position, with no adverse effects on performance.

Impedance Matching

Theory and practice in impedance matching are given in detail in earlier chapters, and theory, at least, is the same for frequencies above 50 MHz. Practice may be similar, but physical size can be a major modifying factor in choice of methods.

Universal Stub

As its name implies, the double-

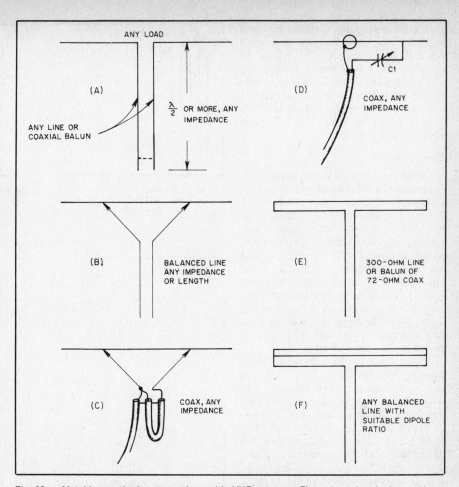

Fig. 39 — Matching methods commonly used in VHF antennas. The universal stub, A, combines tuning and matching. The adjustable short on the stub and the points of connection of the transmission line are adjusted for minimum reflected power in the line. In the delta match, B and C, the line is fanned out to tap on the dipole at the point of best impedance match. Impedances need not be known in A, B and C. The gamma match, D, is for direct connection of coax. C1 tunes out inductance in the arm. Folded dipole of uniform conductor size, E, steps up antenna impedance by a factor of four. Using a larger conductor in the unbroken portion of the folded dipole, E, gives higher orders of impedance transformation.

adjustment stub of Fig. 39A is useful for many matching purposes. The stub length is varied to resonate the system, and the transmission line and stub impedances are equal. In practice this involves moving both the sliding short and the point of line connection for zero reflected power, as indicated on an SWR indicator connected in the line.

The universal stub allows for tuning out any small reactance present in the driven part of the system. It permits matching the antenna to the line without knowledge of the actual impedances involved. The position of the short yielding the best match gives some indication of how much reactance is present. With little or no reactive component to be tuned out, the stub will be approximately a half wavelength from load to short.

The stub should be stiff bare wire or rod, spaced no more than 1/20 wavelength. Preferably it should be mounted rigidly, on insulators. Once the position of the short is determined, the center of the short can be grounded, if desired, and the portion of

the stub no longer needed can be removed.

It is not necessary that the stub be connected directly to the driven element. It can be made part of an open-wire line, as a device to match into or out of the line with coax. It can be connected to the lower end of a delta match, or placed at the feed point of a phased array. Examples of these uses are given later.

Delta Match

Probably the first impedance match was made when the ends of an open line were fanned out and tapped onto a half-wave antenna at the point of most efficient power transfer, as in Fig. 39B. Both the side length and the points of connection either side of the center of the element must be adjusted for minimum reflected power in the line, but as with the universal stub, the impedances need not be known. The delta makes no provision for tuning out reactance, so the universal stub is often used as a termination for it, to this end.

Once thought to be inferior for VHF applications because of its tendency to radiate if adjusted improperly, the delta has come back to favor now that we have good methods for measuring the effects of matching. It is very handy for phasing multiple-bay arrays with open lines, and its dimensions in this use are not particularly critical. It should be checked out carefully in applications like that of Fig. 39C which have no tuning device.

Gamma Match

The gamma match is shown in Fig. 39D. There being no RF voltage at the center of a half-wave dipole, the outer conductor of the coax is connected to the element at this point, which may also be the junction with a metallic or wooden boom. The inner conductor, carrying the RF current, is tapped out on the element at the matching point. Inductance of the arm is canceled by means of C1, resulting in electrical balance. Both the point of contact with the element and the setting of the capacitor are adjusted for zero reflected power, with a bridge connected in the coaxial line.

The capacitor can be made variable temporarily, then replaced with a suitable fixed unit when the required capacitance value is found, or C1 can be mounted in a waterproof box. Maximum capacitance should be about 100 pF for 50 MHz and 35 to 50 pF for 144. The capacitor and arm can be combined with the arm connecting to the driven element by means of a sliding clamp, and the inner end of the arm sliding inside a sleeve connected to the inner conductor of the coax. One can be constructed from concentric pieces of tubing, insulated by plastic sleeving. RF voltage across the capacitor is low, once the match is adjusted properly, so with a good dielectric, insulation presents no great problem, if the initial adjustment is made with low power. A clean, permanent, high-conductivity bond between arm and element is important, as the RF current is high at this point.

Folded Dipole

The impedance of a half-wave antenna broken at its center is 72 ohms. If a single conductor of uniform size is folded to make a half-wave dipole, as shown in Fig. 39E, the impedance is stepped up four times. Such a folded dipole can thus be fed directly with 300-ohm line with no appreciable mismatch. Coaxial line of 70 to 75 ohms impedance may also be used if a 4:1 balun is added. (See balun information in Chapter 16.) Higher impedance step-up can be obtained if the unbroken portion is made larger in cross-section than the fed portion, as in Fig. 39F. For design information see Chapter 16.

Baluns and Transmatches

Conversion from balanced loads to unbalanced lines, or vice versa, can be performed with electrical circuits, or their equivalents made of coaxial line. A balun made from flexible coax is shown in Fig. 40A. The looped portion is an electrical

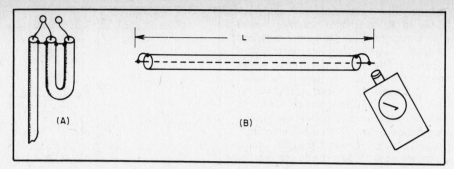

Fig. 40 — Conversion from unbalanced coax to a balanced load can be done with a half-wave coaxial balun, A. Electrical length of the looped section should be checked with a dip meter, with ends shorted, B. The half-wave balun gives a 4:1 impedance step up.

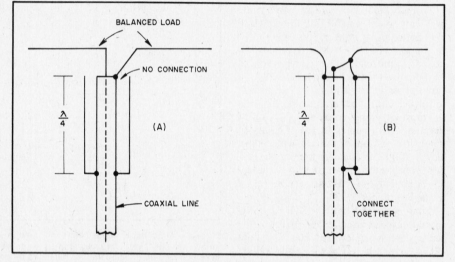

Fig. 41 — The balun conversion function, with no impedance change, is accomplished with quarter-wave lines, open at the top and connected to the coax outer conductor at the bottom. The coaxial sleeve, A, is preferred.

half-wavelength. The physical length depends on the propagation factor of the line used, so it is well to check its resonant frequency, as shown at B. The two ends are shorted, and the loop at one end is coupled to a dip-meter coil. This type of balun gives an impedance step-up of 4:1 in impedance, 50 to 200 ohms, or 75 to 300 ohms typically.

Coaxial baluns giving a 1:1 impedance transfer are shown in Fig. 41. The coaxial sleeve, open at the top and connected to the outer conductor of the line at the lower end (A) is the preferred type. A conductor of approximately the same size as the line is used with the outer conductor to form a quarter-wave stub, in B. Another piece of coax, using only the outer conductor, will serve this purpose. Both baluns are intended to present an infinite impedance to any RF current that might otherwise tend to flow on the outer conductor of the coax.

The functions of the balun and the impedance transformer can be handled by various tuned circuits. Such a device, commonly called an antenna coupler or Transmatch, can provide a wide range of impedance transformations. Additional

selectivity inherent to the Transmatch can reduce RFI problems.

Stacking Yagis

Where suitable provision can be made for supporting them, two Yagis mounted one above the other and fed in phase may be preferable to one long Yagi having the same theoretical or measured gain. The pair will require a much smaller turning space for the same gain, and their lower radiation angle can provide interesting results. On long ionospheric paths a stacked pair occasionally may show an apparent gain much greater than the 2 to 3 dB that can be measured locally as the gain from stacking.

Optimum spacing for Yagis of five elements or more is one wavelength, but this may be too much for many builders of 50-MHz antennas to handle. Worthwhile results can be obtained with as little as one-half wavelength (10 feet), but 5/8 wavelength (12 feet) is markedly better. The difference between 12 and 20 feet may not be worth the added structural problems involved in the wider spacing, at 50 MHz at least. The closer spacings give lowered

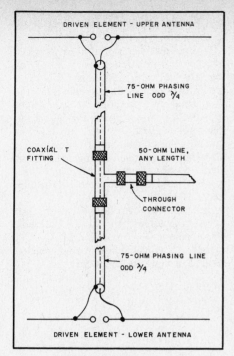

Fig. 42 — A method for feeding a stacked Yagi array.

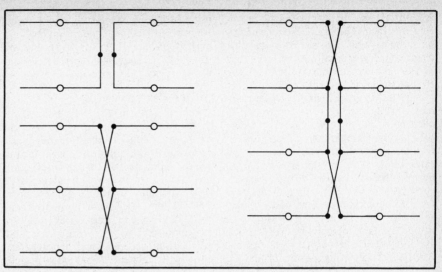

Fig. 43 — Element arrangements for 8, 12 and 16-element collinear arrays. Parasitic reflectors, omitted here for clarity, are five percent longer and 0.2 wavelength in back of the driven elements. Feed points are indicated by black dots. Open circles are recommended support points. The elements can run through wood or metal booms, without insulation, if supported at their centers in this way. Insulators at the element ends (points of high RF voltage) tend to detune and unbalance the system.

measured gain, but the antenna patterns are cleaner (less power in the high-angle lobes) than will be obtained with one-wavelength spacing. The extra gain with wider spacings is usually the objective on 144 MHz and higher bands, where the structural problems are not severe.

One method for feeding two 50-ohm antennas, as might be used in a stacked Yagi array, is shown in Fig. 42. The transmission lines from each antenna to the common feed point must be equal in length and an odd multiple of a quarter wavelength. This line acts as an impedance transformer and raises the feed impedance of each antenna to 100 ohms. When the two antennas are connected in parallel at the coaxial T fitting, the resulting impedance is close to 50 ohms.

COLLINEAR ANTENNAS

Information given thus far is mainly on parasitic arrays, but the collinear antenna has much to recommend it. Inherently broad in frequency response, it is a logical choice where coverage of an entire band is wanted. This tolerance also makes a collinear array easy to build and adjust for any VHF application. The use of many driven elements is popular in very large phased arrays, such as may be required for moonbounce (EME) communication.

Large Collinear Arrays

Bidirectional curtain of arrays of four, six and eight half-waves in phase are shown in Fig. 43. Usually reflector elements are added, normally at about 0.2 wavelength in back of each driven element, for more gain and a unidirectional pattern. Such

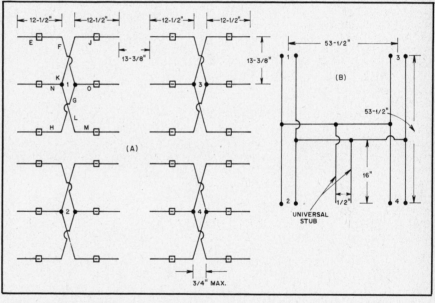

Fig. 44 — Large collinear arrays should be fed as sets of no more than eight driven elements each, interconnected by phasing lines. This 48-element array for 432 MHz (A) is treated as if it were four 12-element collinears. Reflector elements are omitted for clarity. Phasing harness is shown at B.

parasitic elements are omitted from the sketch in the interest of clarity.

When parasitic elements are added, the feed impedance is low enough for direct connection to open line or twin-lead, connected at the points indicated by black dots. With coaxial line and a balun, it is suggested that the universal stub match, Fig. 39A, be used at the feedpoint. All elements should be mounted at their electrical centers, as indicated by open circles in Fig. 43. The framework can be metal or insulating material, with equally good results. The metal supporting structure is entirely in back of the plane of the reflector

elements. Sheet-metal clamps can be cut from scraps of aluminum to make this kind of assembly, which is very light in weight and rugged as well. Collinear elements should always be mounted at their centers, where RF voltage is zero — never at their ends, where the voltage is high and insulation losses and detuning can be very harmful.

Collinear arrays of 32, 48, 64 and even 128 elements can be made to give outstanding performance. Any collinear should be fed at the center of the system, for balanced current distribution. This is very important in large arrays, which are treated as sets of six or eight driven elements each. The sets are fed through a balanced harness, each section of which is a resonant length, usually of open-wire line. A 48-element collinear array for 432 MHz, Fig. 44, illustrates this principle.

A reflecting plane, which may be sheet metal, wire mesh, or even closely spaced elements of tubing or wire, can be used in place of parasitic reflectors. To be effective, the plane reflector must extend on all sides to at least a quarter wavelength beyond the area occupied by the driven elements. The plane reflector provides high front-to-back ratio, a clean pattern, and somewhat more gain than parasitic elements, but large physical size rules it out for amateur use below 420 MHz. An interesting space-saving possibility lies in using a single plane reflector with elements for two different bands mounted on opposite sides. Reflector spacing from the driven element is not critical. About 0.2 wavelength is common.

Circular Polarization

Polarization is described as "horizontal" or "vertical," but these terms have no meaning once the reference of the earth's surface is lost. Many propagation factors can cause polarization change — reflection or refraction and passage through magnetic fields (Faraday rotation), for example. Polarization of VHF waves is often random, so an antenna capable of accepting any polarization is useful. Circular polarization, generated with helical antennas or with crossed elements fed 90 degrees out of phase, will respond to any linear polarization.

The circularly polarized wave, in effect, threads its way through space, and it can be left- or right-hand polarized. These polarization senses are mutually exclusive, but either will respond to any plane polarization. A wave generated with right-hand polarization, when reflected from the moon, comes back with left-hand, a fact to be borne in mind in setting up EME circuits. Stations communicating on direct paths should have the same polarization sense.

Both senses can be generated with crossed dipoles, with the aid of a switchable phasing harness. With helical arrays, both senses are provided with two antennas, wound in opposite directions.

Chapter 18

Voice Communications

T he history of radiotelephony is indeed interesting. Way back in 1878, Bell and Tainter devised a system to voice-modulate a light beam! Called the photophone, it relied on a silvered plane mirror mounted so that it could be vibrated by sound waves. Detection was by a light-sensitive selenium cell in the receiver. By using a focusing lens and a parabolic mirror, this method spanned 250 yards.

Radio voice communications owe credit to Fessenden. In 1902, he developed a system to modulate continuous waves with the human voice. Before this time, most voice transmissions were attempts at modulating spark transmitters, with generally poor results.

Frequency modulation (FM) was first technically addressed by John R. Carson in *Proceedings of the IRE*, in February 1922. By mathematical analysis, he "proved" FM inferior to AM on two counts: bandwidth requirements and distortion.

The Carson analysis held until May 1936, when another paper on FM appeared in the same journal. In this work, Major Edwin H. Armstrong set the stage for viable FM communications. Some of his ideas are still in use today!

Eleven years later, single-sideband suppressed-carrier (SSB) came on the amateur scene. On September 21, 1947, Oswald G. Villard, W6QYT, operating W6YX, worked Winfield G. Wagener, W6VQD on the 75-meter band. The SSB era had begun.

Recently, a new single-sideband technique, Amplitude Compandored Single Sideband (ACSSB) has been introduced. This mode, basically designed for the Land Mobile Service, combines the advantages of FM and SSB. ACSSB may very well find a future in the crowded amateur repeater subbands, and for satellite work.

Single Sideband

On the high-frequency amateur bands, single sideband is the most widely used radiotelephony mode. Since SSB is a sophisticated (or simplified, depending on one's point of view) form of amplitude modulation, it is worthwhile to take a brief look at some AM fundamentals. Modulation is a mixing process. When RF and AF signals are combined in a standard AM transmitter (such as one used for commercial broadcasting) four output signals are generated: the original carrier or RF signal, the original AF signal, and two sidebands, whose frequencies are the sum and difference of the original RF and AF signals, and whose amplitudes are proportional to that of the original AF signal. The sum component is called the upper sideband. It is erect, in that increasing the frequency of the modulating audio signal causes a corresponding increase in the frequency of the RF output signal. The difference component is called the lower sideband, and is inverted, meaning an increase in the modulating frequency results in a decrease in the output frequency. The amplitude and frequency of the carrier are unchanged by the modulation process, and the original AF signal is rejected by the RF output network. The RF envelope (sum of sidebands and carrier), as viewed on an oscilloscope, has the shape of the modulating waveform.

Fig. 1B shows the envelope of an RF

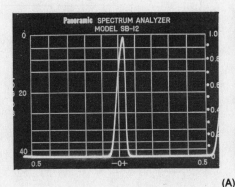
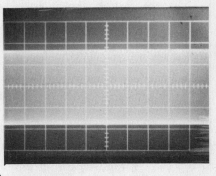

(A)

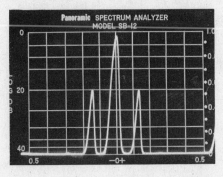
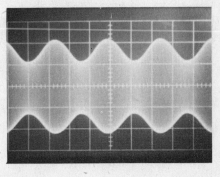

(B)

Fig. 1 — Electronic displays of AM signals in the frequency and time domains. (A) Unmodulated carrier or single-tone SSB signal. (B) Full-carrier AM signal with single-tone sinusoidal modulation.

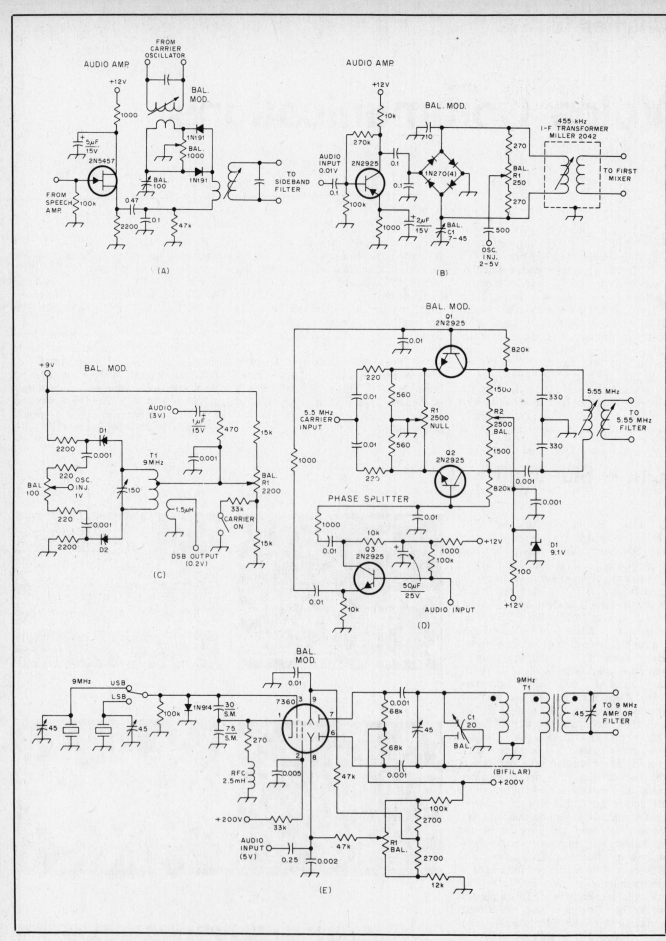

Fig. 2 — Typical balanced modulator circuits. The representative parts values should serve as a basis for designing one's own equipment.

signal that is modulated 20 percent by an AF sine wave. The envelope varies in amplitude because it is the vector sum of the carrier and the sidebands. A spectrum analyzer or selective receiver will show the carrier to be constant. The spectral photograph also shows that the bandwidth of an AM signal is twice the highest frequency component of the modulating wave.

An amplitude-modulated signal cannot be frequency multiplied without special processing because the phase/frequency relationship of the components of the modulating waveform would be severely distorted. For this reason, once an AM signal has been generated, it can be moved in frequency only by heterodyning.

All of the intelligence is contained in the sidebands, but two-thirds of the RF power is in the carrier. The carrier serves only to demodulate the signal in the receiver. If this carrier is suppressed in the transmitter and reinserted in the proper phase in the receiver, several significant communications advantages accrue. If the reinserted carrier is strong compared to the incoming double-sideband signal, exalted carrier reception is achieved in which distortion caused by frequency-selective fading is reduced greatly. A refinement of this technique, called synchronous detection, uses a phase-locked loop to enhance the rejection of interference. Also, the lack of a transmitted carrier eliminates the heterodyne interference common to adjacent AM signals. Perhaps the most important advantage of eliminating the carrier is that the overall efficiency of the transmitter is increased. The power consumed by the carrier can be put to better use in the sidebands. The power in the carrier is continuous, and an AM transmitter requires a heavy-duty power supply. An SSB transmitter having the same power output as an AM transmitter can use a much lighter power supply because the duty cycle of voice operation is low.

Balanced Modulators

The carrier can be suppressed or nearly eliminated by using a balanced modulator or an extremely sharp filter. In SSB transmitters it is common practice to use both devices. The basic principle of any balanced modulator is to introduce the carrier in such a way that only the sidebands will appear in the output. The type of balanced-modulator circuit chosen by the builder will depend on the constructional considerations, cost, and the active devices to be employed.

In any balanced-modulator circuit there will be no output with no audio signal. When audio is applied, the balance is upset, and one branch will conduct more than the other. Since any modulation process is the same as "mixing," sum and difference frequencies (sidebands) will be generated. The modulator is not balanced for the

Table 1

Forward-to-Reverse Resistance Ratio for Different Classes of Solid-State Diodes

Diode Type	Ratio M = 1,000,000	
Point-contact germanium (1N98)	500	
Small-junction germanium (1N270)	0.1	M
Low-conductance silicon (1N457)	48	M
High-conductance silicon (1N645)	480	M
Hot-carrier (HPA-2800)	2000	M

sidebands, and they will appear in the output.

A rectifier-type balanced modulator is shown in Fig. 2, at A and B. If they have equal forward resistances, the diode rectifiers are connected so that no RF can pass from the carrier source to the output circuit by either of the two possible paths. The net effect is that no RF energy appears in the output. When audio is applied, it unbalances the circuit by biasing the diode (or diodes) in one path, depending on the instantaneous polarity of the audio, and hence some RF will appear in the output. The RF energy in the output will appear as a double-sideband suppressed-carrier signal.

For minimum distortion in any diode modulator, the RF voltage should be at least six to eight times the peak audio voltage. The usual operation involves a fraction of a volt of audio and several volts of RF. Desirable diode characteristics for balanced modulator and mixer service include: low noise, low forward resistance, high reverse resistance, good temperature stability and fast switching time (for high-frequency operation).

Table 1 lists the different classes of diodes, giving the ratio of forward-to-reverse resistance of each. This ratio is an important criterion in the selection of

diodes. Also, the individual diodes used should have closely matched forward and reverse resistances; an ohmmeter can be used to select matched pairs or quads.

One of the simplest diode balanced modulators in use is that of Fig. 2A. Its use is usually limited to low-cost portable equipment in which a high degree of carrier suppression is not vital. A ring balanced modulator, shown in Fig. 2B, offers good carrier suppression at low cost. Diodes D1 through D4 should be well matched and can be 1N270s or similar. C1 and R1 are adjusted for best RF balance as evidenced by maximum carrier null. It may be necessary to adjust each control several times alternately to secure optimum carrier suppression.

Varactor diodes are part of the unusual circuit shown in Fig. 2C. This arrangement allows single-ended input of nearly equal levels of audio and carrier. Excellent carrier suppression, 50 dB or more, and a simple method of unbalancing the modulator for CW operation are features of this design. D1 and D2 should be rated at 20 pF for a bias of −4 V. R1 can be adjusted to cancel any mismatch in the diode characteristics, so it isn't necessary that the varactors be well matched. T1 is wound on a small-diameter toroid core. The tap on the primary winding of this transformer is at the center of the winding.

A bipolar-transistor balanced modulator is shown in Fig. 2D. This circuit is similar to one used by Galaxy Electronics and uses closely matched transistors at Q1 and Q2. A phase splitter (inverter), Q3, is used to feed audio to the balanced modulator in push-pull. The carrier is supplied to the circuit in parallel and the output is taken in push-pull. D1 is a Zener diode and is used to stabilize the dc voltage. Controls R1 and R2 are adjusted for best carrier suppression.

The circuit at E offers superior carrier suppression and uses a 7360 beam-deflec-

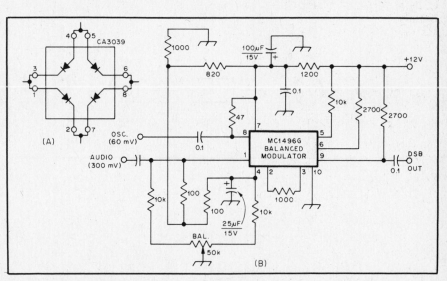

Fig. 3 — Additional balanced-modulator circuits in which integrated circuits are used.

Table 2

Guidelines for Amateur SSB Signal Quality

Parameter	Suggested Standard
Carrier suppression	At least 40-dB below PEP
Opposite-sideband suppression	At least 40-dB below PEP
Hum and noise	At least 40-dB below PEP
Third-order intermodulation distortion	At least 30-dB below PEP
Higher-order intermodulation distortion	At least 35-dB below PEP
Long-term frequency stability	At most 100-Hz drift per hour
Short-term frequency stability	At most 10-Hz P-P deviation in a 2-kHz bandwidth

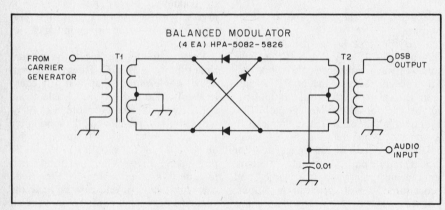

Fig. 4 — Balanced-modulator design using hot-carrier diodes.

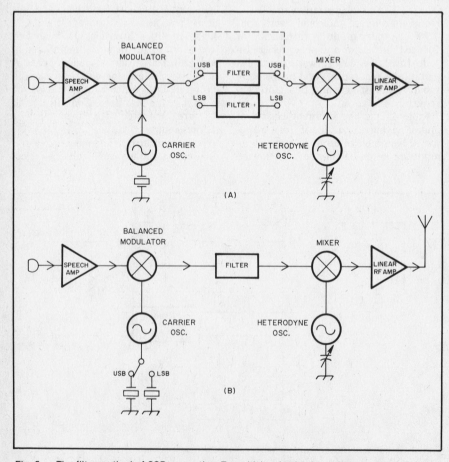

Fig. 5 — The filter method of SSB generation. Two sideband selection schemes are commonly used.

tion tube as a balanced modulator. This tube is capable of providing as much as 60 dB of carrier suppression. When it is used with mechanical or crystal-lattice filters, the total carrier suppression can be as great as 80 dB. Most well-designed balanced modulators can provide between 30 and 50 dB of carrier suppression. The primary of transformer T1 should be bifilar wound for best results.

Vacuum-tube balanced modulators can be operated at high power levels and the double-sideband output can be used directly into the antenna. Past issues of *QST* have given construction details on such transmitters.

IC Balanced Modulators

Integrated circuits (ICs) are presently available for use in balanced-modulator and mixer circuits. A diode array such as the RCA CA3039 is ideally suited for use in circuits such as that of Fig. 3A. Since all diodes are formed on a common silicon chip, their characteristics are extremely well matched. This fact makes the IC ideal in a circuit where good balance is required. The hot-carrier diode also has closely matched characteristics and excellent temperature stability. Using broadband toroidal-wound transformers, it is possible to construct a circuit similar to that of Fig. 4 that will have 40 dB of carrier suppression without the need for balance controls. T1 and T2 consist of trifilar windings, 12 turns of no. 32 enam. wire wound on a ½-inch toroid core. Another device with good inherent balance is the special IC made for modulator/mixer service, such as the Motorola MC1496G or Signetics S5596. A sample circuit using the MC1496 is shown in Fig. 3B. R1 is adjusted for best carrier balance. The amount of energy delivered from the carrier generator affects the level of carrier suppression; 100 mV of injection is about optimum, producing up to 55 dB of carrier suppression.

SINGLE-SIDEBAND EMISSION

A further improvement in communications effectiveness can be obtained by transmitting only one of the sidebands. When the proper receiver bandwidth is used, a single-sideband signal will show an effective gain of up to 9 dB over an AM signal of the same peak power. Because the redundant information is eliminated, the required bandwidth of an SSB signal is half that of a comparable AM or DSB emission. Unlike DSB, the phase of the local carrier generated in the receiver is unimportant.

Generating the SSB Signal: The Filter Method

If the DSB signal from the balanced modulator is applied to a band-pass filter, one of the sidebands can be greatly attenuated. Because a filter cannot have infinitely steep skirts, to obtain adequate suppression of the unwanted sideband the response of the filter must begin to roll off within about 300 Hz of the phantom carrier. This effect limits the ability to transmit bass frequencies, but these frequencies have little communications value. The filter rolloff can be used to obtain an additional 20 dB of carrier suppression. The bandwidth of an SSB filter is selected for the specific application. For voice communications, typical values are 1.8 to 3.0 kHz.

Fig. 5 illustrates two variations of the filter method of SSB generation. The heterodyne oscillator is represented as a simple VFO, but may be a premixing system or synthesizer. The scheme at B is perhaps less expensive than that of A, but the heterodyne-oscillator frequency must be shifted when changing sidebands if the dial calibration is to be maintained. The ultimate sense (erect or inverted) of the final output signal is influenced as much by the relationship of the heterodyne-oscillator frequency to the fixed SSB frequency as by the filter or carrier frequency selection. The heterodyne-oscillator frequency must be chosen to allow the best image rejection. This consideration requires that the heterodyne-oscillator frequency be above the fixed SSB frequency on some bands and below it on others. To reduce circuit complexity, early amateur filter types of SSB transmitters did not include a sideband selection switch. The result was that the output was LSB on 160, 75 and 40 meters, and USB on the higher bands. This convention persists despite the flexibility of most modern amateur SSB equipment.

Filter Types

For carrier frequencies in the 50- to 100-kHz region, a satisfactory filter can be made up of lumped-constant LC sections. High quality components and careful adjustment are required for good results with this type of filter. An alternate possibility is a "synthesized" filter comprised of high-performance operational amplifiers used as gyrators or "active inductors." A further drawback of SSB generation in this frequency range is that multiple conversion is necessary to reach the desired output frequency with adequate suppression of spurious mixing products.

Mechanical filters are an excellent choice for SSB generation in the 400- to 500-kHz region. These filters are described in some detail in the receiving chapter. For wide-dynamic-range receiving applications, the more modern types using piezoelectric transducers are preferred for lowest intermodulation distortion. In transmitters, where the signal levels can be closely predicted, the types using magnetostrictive transducers are entirely suitable.

Quartz crystal filters are commonly used in systems in which the SSB signal is generated in the high-frequency range. Some successful amateur designs have also employed crystals at 455 kHz. Generally, four or more crystal elements are required to obtain adequate selectivity for SSB

transmission. Crystal-filter design is a sophisticated subject, and the more esoteric aspects are beyond the scope of this *Handbook*. The discussion of piezoelectric crystal theory in Chapter 4 is sufficient background material for the general understanding of the concepts outlined in this section.

A fundamental crystal-filter section is the half-lattice, shown in Fig. 6. The passband of this type of filter is slightly wider than the frequency spacing between the crystals. The antiresonant (parallel resonant) frequency or pole of the low-frequency crystal must be equal to the series-resonant frequency or zero of the high-frequency crystal. Such a filter is useful for casual receiving purposes, but the ultimate stopband attenuation is poor, and numerous spurious responses will exist just outside the passband. Cascading two of these sections back to back, as in Fig. 7, will greatly suppress these parasitic resonances and steepen the skirts without materially affecting the passband. An important factor in the design of this type of filter is the coefficient

of coupling between the two halves of the transformer. The coupling must approach unity for proper operation. A twisted bifilar winding on a high-permeability ferrite core most nearly approximates this ideal. Some crystal filters have tuned input and output transformers. The flatness of the passband is heavily dependent on the terminating resistances. Lattice filters exhibit fairly symmetrical response curves and can be used for LSB or USB selection by means of placing the carrier frequency on the upper or lower skirt.

An asymmetrical filter is shown in Fig. 8. Using this approach, good unwanted sideband suppression can be obtained with only two crystals. The crystals are ground for the same frequency. The potential bandwidth here is only half that obtained with a half-lattice design. The maximum bandwidth of almost any crystal filter can be increased by using plated crystals intended for overtone operation.

The home construction of crystal filters can be very time consuming, if not expensive. The reason for this is that one must

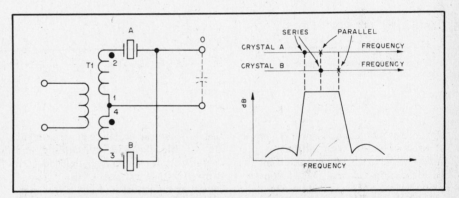

Fig. 6 — The half-lattice crystal filter. Crystals A and B should be chosen so that the parallel-resonant frequency of one is the same as the series-resonant frequency of the other. Very tight coupling between the two halves of the secondary of T1 is required for optimum results. The theoretical attenuation-vs.-frequency curve of a half-lattice filter shows a flat passband between the lower series-resonant frequency and higher parallel-resonant frequency of the pair of crystals.

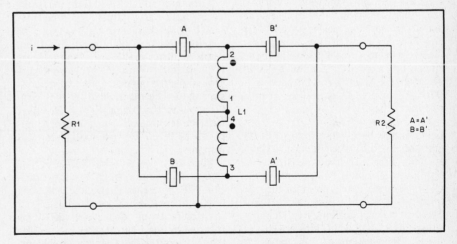

Fig. 7 — Half-lattice filters cascaded in a back-to-back arrangement. The theoretical curve of such a filter has increased skirt selectivity and fewer spurious responses, as compared with a simple half lattice, but the same passband as the simple circuit.

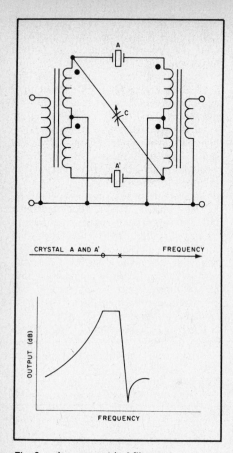

Fig. 8 — An asymmetrical filter and theoretical attenuation curve.

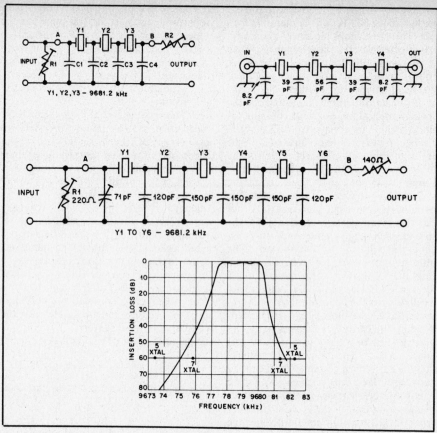

Fig. 9 — Some ladder filters based on CB crystals, with the response that can be expected from the 6-pole unit.

experiment with a large number of crystals to produce a filter with satisfactory performance. Crystal grinding and etching can be a fascinating and highly educational activity, but most home builders would prefer to spend their time on other aspects of equipment design. High-quality filters are available from several manufacturers. Most amateurs who build their own SSB equipment adopt a "systems engineering" approach and design their circuits around filters of known performance. Some filter suppliers are listed in the Component Data chapter. It is still worthwhile to have an appreciation for the basic design ideas, however, for many of the less expensive filters can be improved markedly by the addition of a couple of crystals external to the package. The technique is to steepen the skirts by grouping sharp notches on either side of the passband.

An important exception is the ladder filter. Although this type of filter is treated in textbooks, it has received attention in the amateur literature only recently. The significant feature of ladder filters is that all of the crystals are ground for the same frequency. Low-cost CB crystals are ideally suited to this application. Representative designs by F6BQP and G3JIR are given in Fig. 9. Filter sections of this type can be cascaded for improved shape factor with very little effect on the 3-dB bandwidth.

Ladder filters having six or more elements are suitable for SSB transmitting and receiving service. In general, the bandwidth is inversely proportional to the values of the shunt capacitors and directly proportional to the terminal impedances. Table 3 lists the frequencies of the CB channels. Overtone crystals for CB service have fundamental resonance at approximately one third of the listed frequency.

Filter Applications

The important considerations in circuits using band-pass filters are impedance matching and input/output isolation. The requirements for isolation are less severe in transmitting applications than they are for receiving. With proper layout and grounding, the opposite sideband suppression should be determined by the shape factor rather than signal leakage. The filter must be terminated with the proper impedances to ensure a smooth bandpass response.

Fig. 10A shows a typical SSB generator using a KVG crystal filter. The grounded-gate JFET presents a broadband 50-ohm termination to the balanced modulator and transforms the impedance to the 500 ohms required by the filter. The DC return for the source of the JFET is through the output transformer of the modulator. The tank circuit is broadly resonant at 9 MHz

and rejects any spurious signals generated in the modulator that might be propagated through the filter. Crystal filters should be isolated from any dc voltages present in the circuit.

A circuit using a Collins mechanical filter is illustrated in Fig. 10B. The IF transformer prevents spurious responses and

Table 3
CB Frequencies

Channel	Frequency (MHz)	Channel	Frequency (MHz)
1	26.965	21	27.215
2	26.975	22	27.225
3	26.985	23	27.255
4	27.005	24	27.235
5	27.015	25	27.245
6	27.025	26	27.265
7	27.035	27	27.275
8	27.055	28	27.285
9	27.065	29	27.295
10	27.075	30	27.305
11	27.085	31	27.315
12	27.105	32	27.325
13	27.115	33	27.335
14	27.125	34	27.345
15	27.135	35	27.355
16	27.155	36	27.365
17	27.165	37	27.375
18	27.175	38	27.385
19	27.185	39	27.395
20	27.205	40	27.405

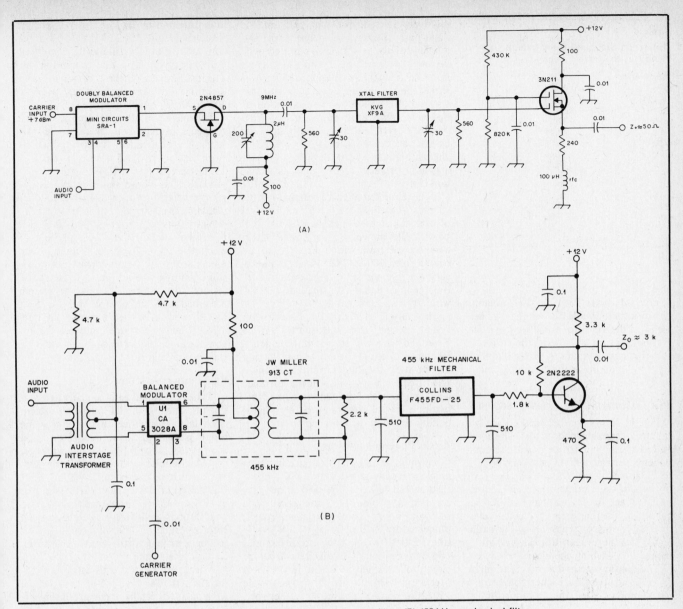

Fig. 10 — Connecting a packaged filter into an SSB generator. (A) 9-MHz crystal filter, (B) 455-kHz mechanical filter.

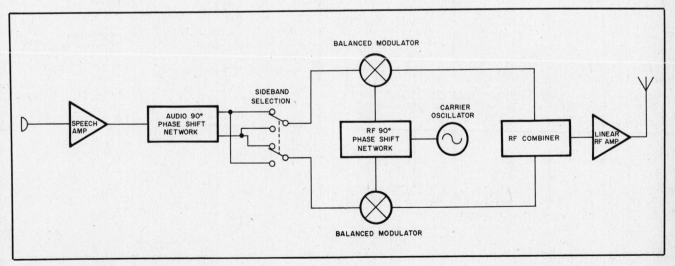

Fig. 11 — The phasing system of SSB generation.

Table 4

Unwanted Sideband Suppression as a Function of Phase Error

Phase Error (deg.)	Suppression (dB)
0.125	59.25
0.25	53.24
0.5	47.16
1.0	41.11
2.0	35.01
3.0	31.42
4.0	28.85
5.0	26.85
10.0	20.50
15.0	16.69
20.0	13.93
30.0	9.98
45.0	6.0

removes dc bias. The output terminating network does double duty as the bias network for the transistor amplifier stage. The filter output transformer is the dc return for the base circuit. This technique is legitimate so long as the current is limited to 2 mA.

SSB Generation: The Phasing Method

Fig. 11 shows another method for obtaining a single-sideband signal. The audio and carrier signals are each split into components separated 90° in phase and applied to balanced modulators. When the DSB outputs of the modulators are combined, one sideband is reinforced and the other is cancelled. The figure shows sideband selection by means of transposing the audio leads, but the same result can be had by means of switching the carrier leads.

The phasing method was used in many pre-1960 amateur SSB exciters, but became less popular after the introduction of relatively inexpensive high-performance band-pass filters. The phase shift and amplitude balance of the two channels must be very accurate if the unwanted sideband is to be adequately attenuated. Table 4 shows the required phase accuracy of one channel (AF or RF) for various levels of opposite sideband suppression. The numbers given assume perfect amplitude balance and phase accuracy in the other channel. It can be seen from the table that a phase accuracy of ±1° must be maintained if the signal quality is to satisfy the criteria tabulated at the beginning of this chapter. It is difficult to achieve this level of overall accuracy over the entire speech band. Note, however, that speech has a complex spectrum with a large gap in the octave from 700 to 1400 Hz. The phase-accuracy tolerance can be loosened to ±2° if the peak deviations can be made to occur within the spectral gap.

The major advantage of the phasing system is that the SSB signal can be generated at the operating frequency without the need for heterodyning. Phasing can be used to good advantage even in fixed-frequency systems. A loose-tolerance (±4°) phasing exciter followed by a simple two-pole crystal filter can generate a high-quality signal at very low cost.

Audio Phasing Networks

It would be difficult to design a two-port network having a quadrature (90°) phase relationship between input and output with constant-amplitude response over a decade of bandwidth. A practical approach, pioneered by Robert Dome, W2WAM, is to use two networks having a differential phase shift of 90°. This differential can be closely maintained in a simple circuit if precision components are used. The 350/2Q4 audio phase shift network

manufactured by Barker and Williamson is such a circuit. The 2Q4 is a 1950 vintage component but it is still useful. A modern design using this device is given in Fig. 12. The insertion loss of the 2Q4 is 30 dB, and the phase-shift accuracy is ±1.5° over the 300-3000 Hz speech band.

The tolerances of the components can be relaxed considerably if several phase-shift sections are cascaded. A sixth-order network designed by HA5WH is shown in Fig. 13. Using common ±10% tolerance components, this phase shifter provides approximately 60 dB of opposite-sideband attenuation over the range of 300 to 3000 Hz.

Numerous circuits have been developed to synthesize the required 90° phase shift electronically. Active-filter techniques are used in most of these systems, but precision components are needed for good results. An interesting phasing system described in *Electronics* for April 13, 1978, makes use of a tapped analog delay line. These "bucket brigade" devices are becoming available at reasonable prices on the surplus market.

RF Phasing Networks

If the SSB signal is to be generated at a fixed frequency, the RF phasing problem is trivial; any method that produces the proper phase shift can be used. If the signal is produced at the operating frequency, problems similar to those in the audio networks must be overcome.

A differential RF phase shifter is shown in Fig. 14. The amplitudes of the quadrature signals won't be equal over an entire phone band, but this is of little consequence as long as the signals are strong enough to saturate the modulators.

Where percentage bandwidths are small, such as in the 144.1- to 145-MHz range, the RF phase shift can be obtained convenient-

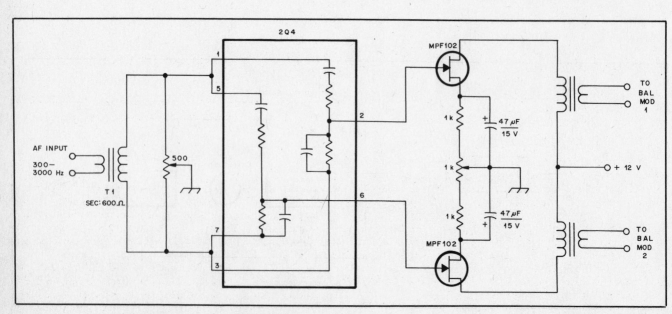

Fig. 12 — A circuit using the B&W 2Q4 audio phase shift network.

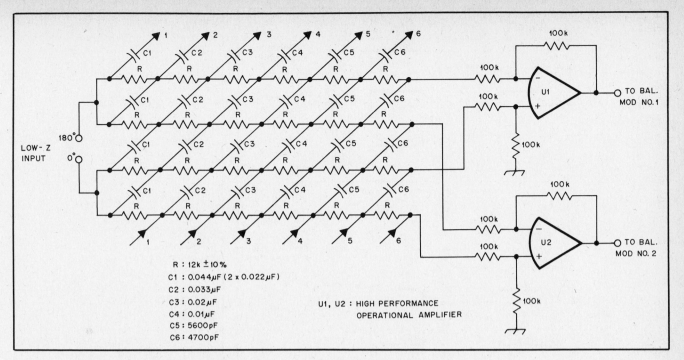

Fig. 13 — A high-performance audio phase shifter made from ordinary loose-tolerance components.

R : 12k ±10%
C1 : 0.044 μF (2 x 0.022 μF)
C2 : 0.033 μF
C3 : 0.02 μF
C4 : 0.01 μF
C5 : 5600 pF
C6 : 4700 pF

U1, U2 : HIGH PERFORMANCE OPERATIONAL AMPLIFIER

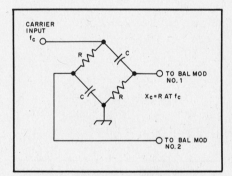

Fig. 14 — A simple RF phase shifter. One of the capacitors can be variable for precise alignment.

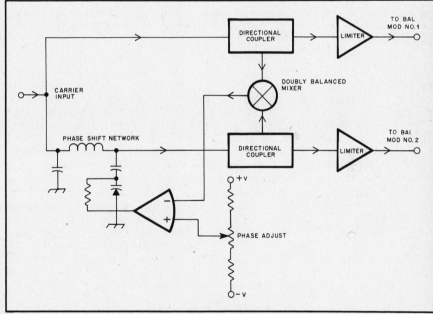

Fig. 15 — Block diagram of a phase-locked-loop phase-shifting system capable of maintaining quadrature over a wide bandwidth. The doubly balanced mixer is used as a phase detector.

ly by means of transmission-line methods. If one balanced-modulator feed line is made an electrical quarter wavelength longer than the other, the two signals will be 90° out of phase. It is important that the cables be properly terminated.

One method for obtaining a 90° phase shift over a wide bandwidth is to generate the quadrature signals at a fixed frequency and heterodyne them individually to any desired operating frequency. Quadrature hybrids having multioctave bandwidths are manufactured commercially, but they cost hundreds of dollars. Another practical approach is to use two VFOs in a master-slave phase-locked-loop system. Many phase detectors lock the two signals in phase quadrature. A doubly balanced mixer has this property. One usually thinks of a phase-locked loop as having a VCO locked to a reference signal, but a phase dif-ferential can be controlled independently of the oscillator. The circuit in Fig. 15 illustrates this principle. Two digital phase shifters are sketched in Fig. 16. If ECL ICs are used, this system can work over the entire HF spectrum.

DETECTORS

Diode Detectors

The simplest detector for AM is the diode. A germanium or silicon crystal is an imperfect form of diode (a small current can usually pass in the reverse direction), but the principle of detection in a semiconductor diode is similar to that in a vacuum-tube diode.

Circuits for both half-wave and full-wave diodes are given in Fig. 17. The simplified half-wave circuit at Fig. 17A includes the RF tuned circuit, L2C1, a coupling coil, L1, from which the RF

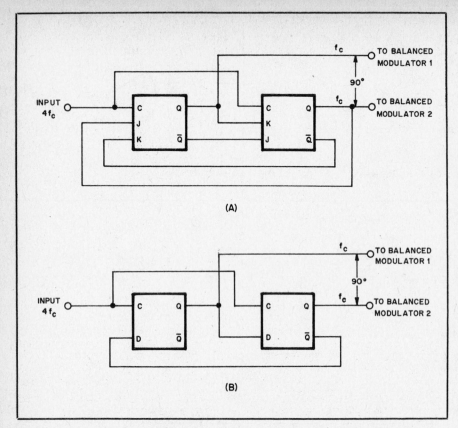

Fig. 16 — Digital RF phase shift networks. The circuit at A uses JK flip flops, and the circuit at B uses D flip flops. In each case the desired carrier frequency must be quadrupled before it is processed by the phase-shift network.

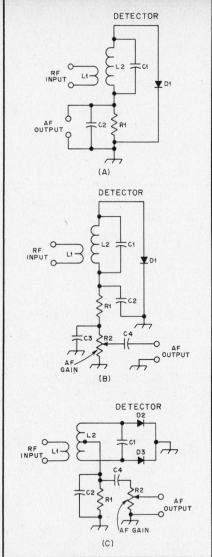

Fig. 17 — Simplified and practical diode detector circuits. A, the elementary half-wave diode detector; B, a practical circuit with RF filtering and audio output coupling; C, full-wave diode detector, with output coupling indicated. The circuit, L2C1, is tuned to the signal frequency; typical values for C2 and R1 in A and C are 250 pF and 250 kΩ, respectively; in B, C2 and C3 are 100 pF each; R1, 50 Ω, and R2 250 Ω. C4 is 0.1 pF.

energy is fed to L2C1, and the diode, D1, with its load resistance, R1, and bypass capacitor, C2.

The progress of the signal through the detector or rectifier is shown in Fig. 18. A typical modulated signal as it exists in the tuned circuit is shown at A. When this signal is applied to the rectifier, current will flow only during the part of the RF cycle when the anode is positive with respect to cathode, so that the output of the rectifier consists of half-cycles of RF. These current pulses flow in the load circuit comprised of R1 and C2. C2 charges to the peak value of the rectified voltage on each pulse and retains enough charge between pulses so that the voltage across R1 is smoothed out, as shown in C. C2 thus acts as a filter for the radio-frequency component of the output of the rectifier, leaving a dc component that varies in the same way as the modulation on the original signal. When this varying dc voltage is applied to a following amplifier through a coupling capacitor (C4 in Fig. 17), only the variations in voltage are transferred, so the final output signal is ac, as shown in D.

In the circuit at 17B, R1 and C2 have been divided for the purpose of providing a more effective filter for RF. It is important to prevent the appearance of any RF voltage in the output of the detector, because it may cause overloading of a succeeding amplifier stage. The audio-frequency variations can be transferred to another circuit through a coupling capacitor, C4. R2 is usually a "potentiometer" so the audio volume can be adjusted to a desired level.

Coupling from the potentiometer (volume control) through a capacitor also avoids any flow of dc through the moving contact of the control. The flow of dc through a high-resistance volume control often tends to make the control noisy (scratchy) after a short while.

The full-wave diode circuit at Fig. 17C differs in operation from the half-wave circuit only in that both halves of the RF cycle are utilized. The full-wave circuit has the advantage that RF filtering is easier than in the half-wave circuit. As a result, less attenuation of the higher audio frequencies will be obtained for any given degree of RF filtering.

The reactance of C2 must be small compared to the resistance of R1 at the radio frequency being rectified, but at audio frequencies must be relatively large compared to R1. If the capacitance of C2 is too large, response at the higher audio frequencies will be lowered.

Compared with most other detectors, the gain of the diode is low, normally running around 0.8 in audio work. Since the diode consumes power, the Q of the tuned circuit is reduced, bringing about a reduction in selectivity. The loading effect of the

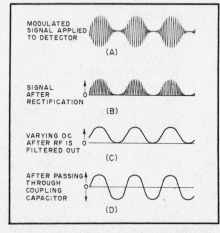

Fig. 18 — Illustrations of the detection process.

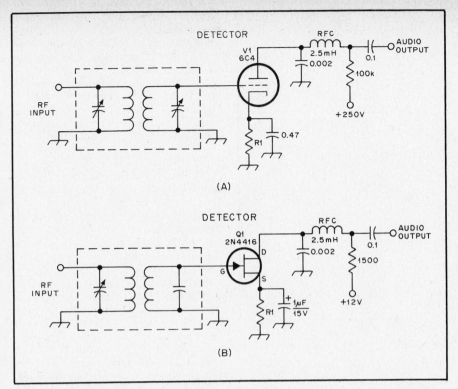

Fig. 19 — Plate-detection circuits. In each example the input circuit is tuned to the signal frequency. Typical R1 values for the tube circuit at A are 1000 to 5600 ohms. For the FET circuit at B, R1 is on the order of 100 to 3900 ohms.

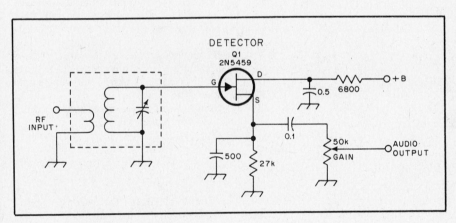

Fig. 20 — An infinite-impedance detector.

diode is close to one half the load resistance. The detector linearity is good, and the signal-handling capability is high.

Plate Detectors

The plate detector is arranged so that rectification of the RF signal takes place in the plate circuit of the tube or the drain of the FET. Sufficient negative bias is applied to the grid to bring the plate current nearly to the cutoff point, so that application of a signal to the grid circuit causes an increase in average plate current. The average plate current follows the changes in the signal in a fashion similar to the rectified current in a diode detector.

In general, transformer coupling from the plate circuit of a plate detector is not satisfactory, because the plate impedance of any tube is very high when the bias is near the plate-current cutoff point. The same is true of a JFET or MOSFET. Impedance coupling may be used in place of the resistance coupling shown in Fig. 19. Usually 100 henrys or more of inductance is required.

The plate detector is more sensitive than the diode detector because there is some amplifying action in the tube or transistor. It will handle large signals, but is not as tolerant as the diode. Linearity, with the self-biased circuits shown, is good. Up to

the overload point the detector takes no power from the tuned circuit, and so does not affect its Q and selectivity.

Infinite-Impedance Detector

The circuit of Fig. 20 combines the high signal-handling capabilities of the diode detector with the low distortion and, like the plate detector, does not load the tuned circuit it connects to. The circuit resembles that of the plate detector, except that the load resistance, 27 kΩ, is connected between source and ground and thus is common on both gate and drain circuits, giving negative feedback for the audio frequencies. The source resistor is bypassed for RF but not for audio, while the drain circuit is bypassed to ground for both audio and radio frequencies. An RF filter can be connected between the source and the output coupling capacitor to eliminate any RF that might otherwise appear in the output.

The drain current is very low with no signal but increases with signal as in the case of the plate detector. The voltage drop across the source resistor consequently increases with signal. Because of this and the large initial drop across this resistor, the signal usually cannot drive the gate positive with respect to the source.

Product Detectors

A product detector is similar in function to a balanced or product modulator. It is also similar to a mixer. In fact, the latter is sometimes called a "first detector" in a receiver circuit. Product detectors are used principally for SSB, CW and RTTY FSK signal detection. Essentially, it is a detector whose output is equal to the product of the beat-frequency oscillator (BFO) and the RF signals applied to it. Output from the product detector is at audio frequency. Some RF filtering is necessary at the detector output to prevent unwanted IF or BFO voltage from reaching the audio amplifier which follows the detector. LC or RC RF decoupling networks are satisfactory, and they need not be elaborate. Fig. 22 illustrates this type of filtering.

Diode Product Detectors

The product detectors shown in Fig. 21 are called "passive." The term means that the devices used do not require an operating voltage. Active devices (transistors, ICs and tubes) do require an operating voltage. Passive mixers and detectors exhibit a conversion loss, whereas active detectors can provide conversion gain. Passive detectors usually require a substantially greater level of BFO injection voltage than is the case with active detectors. Therefore, the primary drawbacks to the use of diodes in these circuits are the loss in gain and the high injection level required. A typical conversion loss for a two-diode detector (Fig. 21A) is 5 dB. The four-diode detectors have a loss of approximately 8 dB. The BFO injection level for each of the diode detectors shown in Fig. 21 is +13 dBm, or 20 mW.

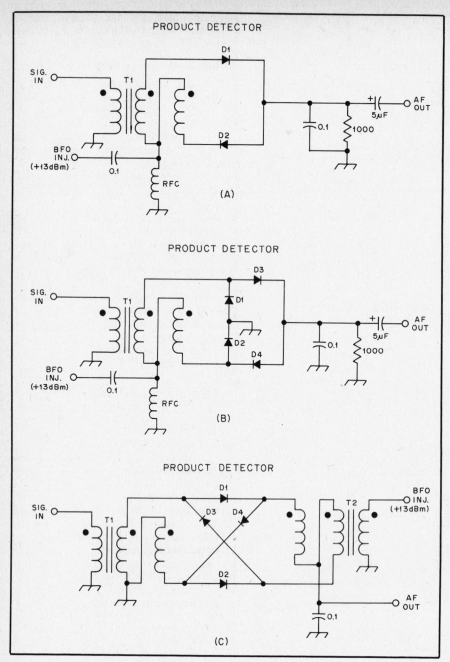

Fig. 21 — Examples of diode product detectors. Singly balanced types are shown at A and B. A doubly balanced version is illustrated at C.

Since the terminal impedance of the detector is roughly 50 ohms, an RMS BFO voltage of 1, or P-P voltage of 2.8, is required.

The advantages to the use of diodes in a product detector are circuit simplicity, low cost, broadband characteristics, low noise figure and good port-to-port signal isolation. This type of detector is excellent at the input of a direct-conversion receiver (to be treated later in the chapter).

The transformers shown in the circuits of Fig. 21 are broadband, toroidally wound types. The black dots near the windings of T1 and T2 indicate the phasing required. The core material is ferrite, and the windings are trifilar. Core permeability can be 950 for most applications, although some designers use cores with less initial permeability. An Amidon FT-50-43 is entirely suitable as a transformer core for the circuits shown. Fifteen trifilar turns are ample for each transformer.

High-speed silicon switching diodes are satisfactory for use in the circuits of Fig. 21. They should be as closely matched as possible for forward and back resistance. Closely matched diodes can be had by using a diode-array IC, such as the RCA CA3019 or 3039. Hot-carrier diodes are excellent for the circuits shown. Matched 1N914s are the choice of many amateur designers.

A singly balanced detector is shown at A in Fig. 21. An improved singly balanced detector is shown at B. Two diodes have been added to improve the circuit balance while presenting a more symmetrical load to the BFO. The result is better isolation between the BFO and IF input ports.

Two broadband transformers are used to provide the doubly balanced detector of Fig. 21C. The advantage with this configuration is that all three ports are effectively isolated from one another.

Simple Active Product Detectors

Fig. 22 contains two examples of single-ended active detectors that employ FETs. They are quite acceptable for use in simple receivers that do not require high performance characteristics. The circuit at A uses a JFET which has BFO injection voltage supplied across the source resistor. Because the source is not bypassed, instability can occur if the circuit is used as a mixer which has an IF that is close to the signal frequency. This problem is not apt to become manifest when the output is at audio frequency. Slightly more injection power is needed for circuit A than is necessary for the detector at B. An RMS voltage of roughly 0.8 is typical (6.5 mW).

The detector of Fig. 22B operates in a similar fashion to that of A, but the BFO is injected on control gate no. 2. Approximately 1 volt RMS is needed (0.1 mW). FETs with proper injection levels and moderate signal-input amounts have excellent IMD characteristics. Generally, they are preferred to single-ended, bipolar-transistor detectors. The circuits at A and B contain RF chokes and bypass capacitors in the drain leads to minimize the transfer of BFO energy to the succeeding audio stage. The bypass capacitors are also useful for rolling off the unwanted high-frequency audio components.

Active Balanced Product Detectors

Examples of active IC product detectors are given in Fig. 23. A singly balanced version is shown at A. It uses an RCA differential-pair IC. Except for its conversion gain, it performs similarly to the singly balanced diode detector of Fig. 21B. Doubly balanced active detectors are seen at B and C of Fig. 23. These ICs contain two sets of differential amplifiers each. The "diffamps" are cross-connected in the examples shown to obtain double balanced circuits. The virtues of these detectors are similar to the equivalent four-diode types, but they exhibit several decibels of conversion gain. The MC1496G is made by Motorola, and the CA3102E is an RCA device.

Other SSB Modes

An SSB transmitter is simply a frequency translator. Any frequency- or amplitude-varying signal (within the bandwidth capabilities of the transmitter) applied to the input will be translated intact (although frequency inversion takes place in LSB) to the chosen radio frequency. If amplitude-

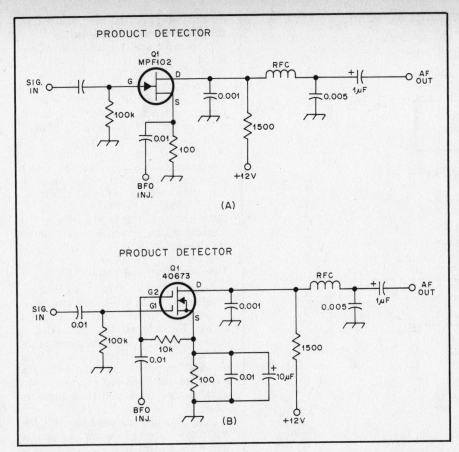

Fig. 22 — Active product detectors. A JFET example is provided at A and a dual-gate MOSFET type is at B.

limited tones corresponding to the video information of a slow-scan television picture are fed into the microphone input, F3F emission will result. Two alternating tones from an AFSK RTTY keyer will cause the transmitter to produce an F1B signal. A keyed audio tone will be translated into an A1A signal. This technique is a perfectly legitimate way to operate CW with an SSB transceiver, and is simpler than the more traditional method of upsetting the balanced modulator for carrier leakage. One can vary the transmitting frequency independently of the receiving frequency by means of changing the audio tone. The strength of the tone determines the transmitter power output. Good engineering practice requires that the tone be frequency-stable and that the total harmonic distortion be less than 1%. Also, the carrier and opposite sideband must be suppressed at least 40 dB. Of course the rise and decay times of the audio envelope must be controlled to avoid key clicks.

Independent-Sideband Emission

If two SSB exciters, one USB and the other LSB, share a common carrier oscillator, two channels of information can be transmitted from one antenna. Methods for ISB generation in filter and phasing transmitters are shown in Fig. 24. May 1977 *QST* carried an article on converting the popular Drake TR-4C to ISB.

Many commercially manufactured SSB transceivers have provisions for controlling the transmit or receive frequency with an external VFO or receiver. With slight modification it should be possible to slave two transceivers to a single VFO for ISB operation. The oscillators in the transceivers must be aligned precisely.

The most obvious amateur application for independent sideband is the transmission of slow-scan television with simultaneous audio commentary. On the VHF bands, other combinations are possible, such as voice and code or SSTV and RTTY.

Power Ratings of SSB Transmitters

Fig. 25A is more or less typical of a few voice-frequency cycles of the modulation envelope of a single-sideband signal. Two amplitude values associated with it are of particular interest. One is the maximum peak amplitude, the greatest amplitude reached by the envelope at any time. The other is the average amplitude, which is the average of all the amplitude values contained in the envelope over some significant period of time, such as the time of one syllable of speech.

The power contained in the signal at the maximum peak amplitude is the basic transmitter rating. It is called the peak-envelope power, abbreviated PEP. The peak-envelope power of a given transmitter is intimately related to the distortion considered tolerable. The lower the signal-to-distortion ratio the lower the attainable peak-envelope power as a general rule. For splatter reduction, an S/D ratio of 25 dB is considered a border-line minimum, and better figures are desirable.

The signal power, S, in the classical definition of S/D ratio is the power in one tone of a two-tone test signal. This is 6 dB below the peak-envelope power in the same signal. Manufacturers of amateur SSB equipment usually base their published S/D ratios on PEP, thereby getting an S/D ratio that looks 6 dB better than one based on the classical definition. *QST* product reviews also use the PEP reference. In comparing distortion-product ratings of different transmitters or amplifiers, first make sure the ratios have the same base.

Peak vs. Average Power

Envelope peaks occur only sporadically during voice transmission, and have no direct relationship to meter readings. The meters respond to the amplitude (current or voltage) of the signal averaged over several cycles of the modulation envelope. (This is true in practically all cases, even though the transmitter RF output meter may be calibrated in watts. Unfortunately, such a calibration means little in voice transmission since the meter can be calibrated in watts only by using a sine wave — which a voice-modulated signal definitely is not.)

The ratio of peak-to-average amplitude varies widely with voices of different characteristics. In the case shown in Fig. 25 the average amplitude, found graphically, is such that the peak-to-average ratio of amplitudes is almost 3:1. The ratio of peak power to average power is something else again. There is no simple relationship between the meter reading and actual average power, for the reason mentioned earlier. See Chapter 2 for a complete discussion of computing PEP.

Testing a Sideband Transmitter

There are three commonly used methods for testing an SSB transmitter. These include the wattmeter, oscilloscope and spectrum-analyzer techniques. In each case, a two-tone test signal is fed into the mic input to simulate a speech signal. From the measurements, information concerning such quantities as PEP and intermodulation-distortion-product (IMD) levels can be obtained. Depending on the technique used, other aspects of transmitter operation (such as hum problems and carrier balance) can also be checked.

As might be expected, each technique has both advantages and disadvantages and the suitability of a particular method will depend on the desired application. The wattmeter method is perhaps the simplest one but it also provides the least amount of in-

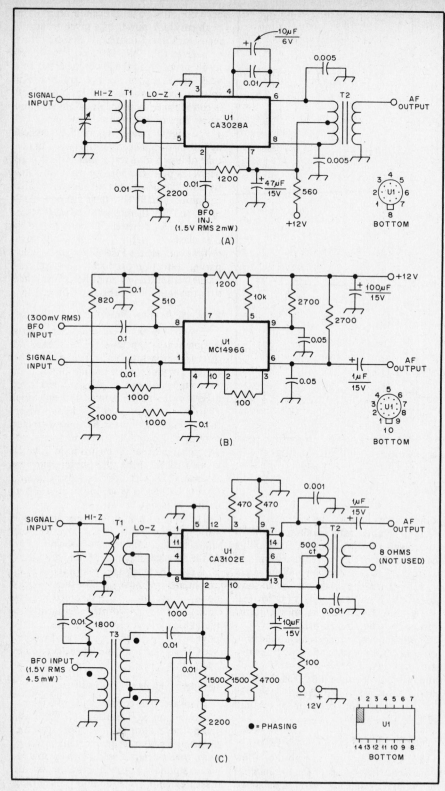

Fig. 23 — Examples of IC product detectors. At A is a singly balanced version, while those at B and C are doubly balanced.

method (using a two-tone test input signal), the power output is multiplied by two.

A spectrum analyzer is capable of giving the most information (of the three methods considered here), but it is also the most costly method and the one with the greatest chance of misinterpretation. Basically, a spectrum analyzer is a receiver with a readout that provides a plot of signal amplitude vs. frequency. The readout could be in the form of a paper chart but usually it is presented as a trace on a CRT. For a spectrum analyzer to provide accurate information about a signal, that signal must be well within the linear dynamic range of the analyzer. For an explanation of the function and application of this instrument see Chapter 25.

Two-Tone Tests and Scope Patterns

A very practical method for amateur applications is to use a two-tone test signal (usually audio) and sample the transmitter output. The sample is then applied directly to the vertical-deflection plates in an oscilloscope. An alternative method is to use an RF probe and detector to sample the waveform and apply the resulting audio signal to the vertical-deflection amplifier input.

If the amplifier has good linearity, the resulting envelope will approach a perfect sine wave pattern (see Fig. 26A). As a comparison, a spectrum-analyzer display for the same transmitter and under the same conditions is shown in Fig. 26B. In this case, spurious products approximately 30 dB below the amplitude of each of the tones can be seen.

As the distortion increases, so does the level of the spurious products, and the resulting waveform departs from a true sine-wave function. This can be seen in Fig. 26C. One of the disadvantages of the scope and two-tone test method is that a relatively high level of IMD-product voltage is required before the waveform seems distorted to the eye. For instance, the waveform in Fig. 26C doesn't seem too much different from the one in Fig. 26A but the IMD level is only 17 dB below the level of the desired signal (see analyzer display in Fig. 26D). A 17- to 20-dB level corresponds to approximately 10% distortion in the voltage waveform. Consequently, a "good" waveform means the IMD products are at least 30 dB below the desired tones. Any noticeable departure from the waveform in Fig. 26A should be suspect, and the transmitter operation should be checked.

The relation between the level at which distortion begins for the two-tone test signal and an actual voice signal is a rather simple one. The maximum deflection on the scope is noted (for an acceptable two-tone test waveform) and the transmitter is then operated such that voice peaks are kept below this level. If the voice peaks go above this level, a type of distortion called "flattopping" will occur. Results are shown for a two-tone test signal in

formation. RF wattmeters suitable for single-tone or CW operation may not be accurate with a two-tone test signal. A suitable wattmeter for the latter case must have a reading that is proportional to the actual power consumed by the load. The reading must be independent of signal waveform. A thermocouple ammeter connected in series with the load would be a typical example of such a system. The output power would be equal to I^2R, where I is the current in the ammeter and R is the load resistance (usually 50 ohms). In order to find the PEP output with the latter

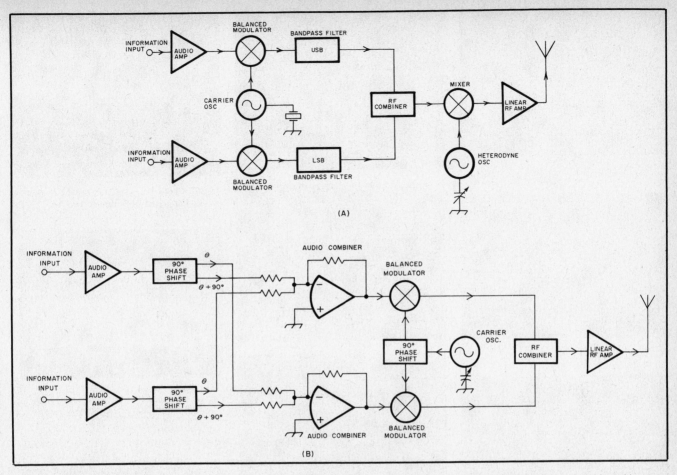

Fig. 24 — Independent-sideband generators. (A) Filter system. (B) Phasing system. The block marked "RF combiner" can be a hybrid combiner or a summing amplifier.

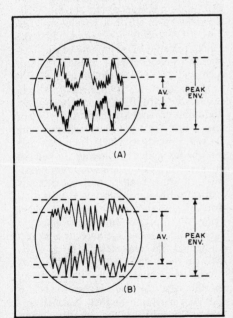

Fig. 25 — (A) Typical SSB voice-modulated signal might have an envelope of the general nature shown, where the RF amplitude (current or voltage) is plotted as a function of time, which increases to the right horizontally. (B) Envelope pattern after speech processing to increase the average level of power output.

Fig. 26E. IMD-product levels rise very rapidly when flattopping occurs. For instance, third-order product levels will increase 30 dB for every 10-dB increase in desired output as the flattopping region is approached, and fifth-order terms will increase by 50 dB.

Interpreting Distortion Measurements

Unfortunately, considerable confusion has grown concerning the interpretation and importance of distortion in SSB gear. Distortion is a very serious problem when high spurious-product levels exist at frequencies removed from the passband of the desired channel. In this case, such distortion may cause needless interference to other channels ("splatter"), and should be avoided. This can be seen quite clearly in Fig. 26F, where the flattopping region is approached and the fifth and higher order terms increase dramatically. The situation is less serious if such products fall within the bandwidth of operation.

On the other hand, attempting to suppress in-band products more than necessary is not only difficult to achieve but may not result in any noticeable increase in signal quality. In addition, measures required to suppress in-band IMD often cause problems at the expense of other qualities such as efficiency. This can lead to serious difficulties such as shortened tube life or transistor heat-dissipation problems.

The two primary causes of distortion can be seen in Fig. 27. While the waveform is for a single-tone input signal, similar effects occur for the two-tone case. As the drive signal is increased, a point is reached at which the output current (or voltage) cannot follow the input, and the amplifier saturates. As mentioned earlier, this condition is often referred to as flattopping. It can be prevented by ensuring that excessive drive doesn't occur; the usual means of accomplishing this is by ALC action. The ALC provides a signal that is used to lower the gain of earlier stages in the transmitter.

The second type of distortion is called "crossover" distortion and occurs at low signal levels. (See Fig. 27.) Increasing the idling plate or collector current is one way of reducing the effect of crossover distortion in regards to producing undesirable components near the operating frequency. Instead, the components occur at frequencies considerably removed from the

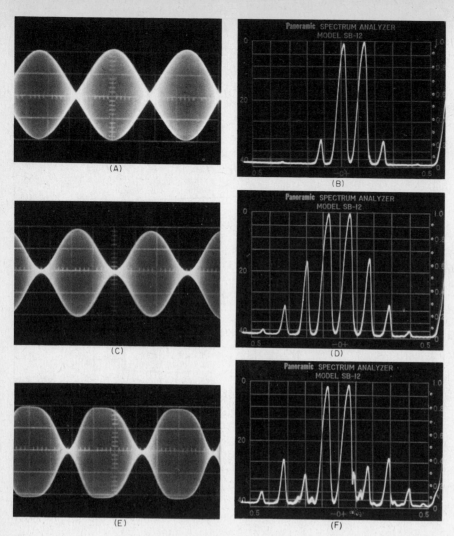

Fig. 26 — Scope patterns for a two-tone test signal and corresponding spectrum-analyzer displays. The pattern in A is for a properly adjusted transmitter and consequently the IMD products are relatively low as can be seen on the analyzer display. At C, the PA bias was set to zero idling current and considerable distortion can be observed. Note how the pattern has changed on the scope and the increase in IMD level. At E, the drive level was increased until the flattopping region was approached. This is the most serious distortion of all since the width of the IMD spectrum increases considerably, causing splatter (F).

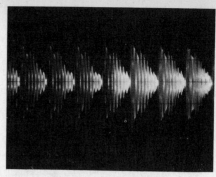

Fig. 28 — Speech pattern of the word "X" in a properly adjusted SSB transmitter.

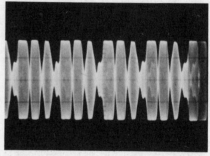

Fig. 29 — Severe clipping (same transmitter as Fig. 22 but with high drive and ALC disabled).

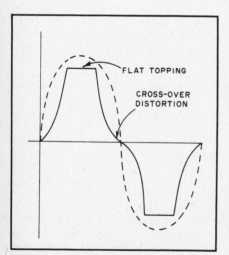

Fig. 27 — Waveform of an amplifier with a single-tone input showing flattopping and crossover distortion.

operating frequency and can be eliminated by filtering.

As implied in the earlier discussion, the effect of distortion is to generate frequency components that add or subtract in order to make up the complex waveform. A more familiar example would be the harmonic generation caused by the nonlinearities often encountered in amplifiers. However, a common misconception that should be avoided is that IMD is caused by fundamental-signal components beating with harmonics. Generally speaking, no such simple relation exists. For instance, single-ended stages have relatively poor second-harmonic suppression but with proper biasing to increase the idling current, such stages can have very good IMD-suppression qualities.

However, a definite mathematical relation does exist between the desired components in an SSB signal and the "distortion signals." Whenever nonlinearities exist, products between the individual components which make up the desired signal will occur. The mathematical result of such multiplication is to generate other signals of the form $(2f1 - f2)$, $(3f1)$, $(5f2 - f1)$ and so on. Hence the term inter-modulation-distortion products. The "order" of such products is equal to the sum of the multipliers in front of each frequency component. For instance, a term such as $(3f1 - 2f2)$ would be called a fifth-order term, since $3 + 2$ is equal to 5. In general, the third, fifth, seventh and similar "odd-order" terms are the most important ones since some of these fall near the desired transmitter output frequency and can't be eliminated by filtering. As pointed out previously, such terms do not normally result from fundamental components beating with harmonics. An exception would be when the fundamental signal along with its harmonics is applied to another nonlinear stage such as a mixer. Components at frequencies identical to the IMD products will result.

When two equal tones are applied to an amplifier and the result is displayed on a spectrum analyzer, the IMD products appear as "pips" off to the side of the main signal components (Fig. 26). The amplitudes associated with each tone and the IMD products are merely the decibel difference between the particular product and one tone. However, each desired tone is 3 dB down from the average power output and 6 dB down from the PEP output.

Since PEP represents the most important quantity as far as IMD is concerned, relating IMD-product levels to PEP is one logical way of specifying the "quality" of a transmitter or amplifier in regard to low distortion. For instance, IMD levels are referenced to PEP in "Product Review" specifications of commercially made gear in *QST*. PEP output can be found by multiplying the PEP input by the efficiency of the amplifier. The input PEP for a two-tone test signal is given by

$$PEP = \frac{E_p I_p}{n} \left[1.57 - 0.57 \frac{I_o}{I_p} \right]$$

where
E_p = plate voltage

I_p = average plate current
I_o = idling (zero signal) plate current
n = duty cycle of test tones ($= 1.0$ for steady two-tone test, 0.5 for 50% duty cycle, etc)

The inclusion of a factor for test-tone duty cycle allows calculation of amplifier PEP input with the application of *pulsed* test tones. (Pulsed tones result in lower average power input, and this allows longer test periods if the amplifier under test cannot tolerate continuous operation at maximum PEP input. See Goodman, "Pulsed Signals Through S.S.B. Transmitters," *QST*, Sept 1965, pp 18-19.) If continuous test tones are used, $n = 1.0$.

Generally speaking, most actual voice

patterns will look alike (in the presence of distortion) except in the case where severe flattopping occurs. This condition is not too common since most rigs have an ALC system, which prevents overdriving the amplifiers. However, the voice pattern in a properly adjusted transmitter usually has a "Christmas tree" shape when observed on a scope; an example is shown in Fig. 28.

More Information

Further discussion on measurements, including FCC power measurement procedures, can be found in Chapter 25. For a discussion of voice processing see Chapter 7. Equipment and accessory construction projects are presented in Chapters 27 through 34.

Amplitude Compandored Single Sideband

There has not been much said about Amplitude Compandored Single Sideband (ACSSB) in Amateur Radio literature. However, amateur experimenters should be aware of ACSSB developments now underway in the land-mobile communications industry.

SSB is the standard mode for HF voice communications. Amateurs have used SSB at VHF and UHF with good success but usually between fixed locations and experienced operators. When SSB was tried in the Land Mobile Service, several problems were noticed. One was that the users (who are not trained operators) couldn't master the control known as CLARIFIER to land-mobile people and receiver incremental tuning or RIT to amateurs. In addition, the users were annoyed by the fading and noise experienced with SSB over the FM they were familiar with. So, to get the spectrum savings of SSB over FM, the land-mobile engineers had to come up with a form of SSB that gave the land-mobile users everything that they were accustomed to with FM, including signaling.

Pilot Tone

A conventional SSB transmission has a suppressed carrier. In order to demodulate the signal at the receiver, the missing carrier is replaced. At higher frequencies it is unrealistic to expect the original carrier at the transmitter and that reinserted at the receiver to be exactly the same. As mentioned above, adding a knob to compensate for the error is unacceptable for land-mobile service. So manufacturers have added a pilot tone to the audio, sufficiently separated in frequency and filtered so that the two do not interfere with each other. The pilot tone is usually 3.1 kHz. As shown

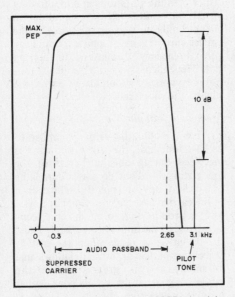

Fig. 30 — Bandwidth of an ACSSB signal is 3.1 kHz.

in Fig. 30, the pilot power level is about 10 dB below the voice peak power. Thus, the transmitter has an output power of approximately 1/10th of maximum when there is no voice modulation.

At the receiver, the pilot tone is compared with an internal reference oscillator in a phase-locked-loop circuit. The difference voltage thus produced is used to shift the receiver local oscillator until the frequency error is eliminated. The time constant of the phase-locked loop is short enough to be ignored by the user and to be compatible with tone decoders used for signaling purposes.

The pilot tone serves other functions than locking the receiver on frequency. It is used to control squelch and AGC circuits

as well. The pilot tone can also be modulated for signaling or telemetry transmission — a useful feature in logic-controlled systems.

Amplitude Compandoring

Compression of voice signals is common in Amateur Radio SSB. What is not common in amateur SSB is to employ an expandor at the receiving end. The term "compandor" is a fusion of the terms "compressor" and "expandor." The compandor circuitry provides an operator with a sensation similar to the "capture effect" in FM. Thus, when the signal strengths are high, FM and ACSSB give comparable results with noticeably less flutter on ACSSB than on FM. A detailed discussion of compandor operation is included in Chapter 7.

In Operation

The FCC has made tests of ACSSB in direct comparison with FM under actual operating conditions. Preliminary results indicate that with equal power (PEP and carrier), range and capture effect are about the same. ACSSB is quieter until lock is lost; in the fringe area, noise begins to appear on voice peaks (because of compandoring).

ACSSB is a means of packing about four times more voice transmission in a given band than FM. The pilot tones and compandoring compensate for some inherent SSB shortcomings but with additional complexity. This appears to be an interesting area for Amateur Radio experimentation.

Further Reading on ACSSB

J. S. Belrose, VE2CV, "On SSB Radio Communications," *QEX*, July 1982, pp. 3-4.

F. B. Childs, "ACSB — Filling the Gaps Between FM Channels," *Mobile Radio Technology,* January/February 1983, pp. 60-66.

B. Snyder, W9GT, "What ACSB Means to You," *Business Radio Action,* June 1982, pp. 13-17.

Haller and Van Deursen, "Amplitude Compandored Sideband Compared to Conventional Frequency Modulation for VHF Mobile Radio: Laboratory and Field Testing Results," *OST Technical Memorandum FCC/OST TM83-7,* Federal Communications Commission, Washington, DC, October 1983.

Frequency Modulation

Methods of radiotelephone communication by frequency modulation were developed in the 1930s by Major Edwin Armstrong in an attempt to reduce the problems of static and noise associated with receiving AM broadcast transmissions. The primary advantage of FM, the ability to produce a high signal-to-noise ratio when receiving a signal of only moderate strength, has made FM the mode chosen for mobile communications services and quality broadcasting. The disadvantages (the wide bandwidth required and the sometimes poor results obtained when an FM signal is propagated via the ionosphere because of phase distortion), have limited the use of frequency modulation to the 10-meter band and the VHF/UHF section of the spectrum.

FM has some impressive advantages for VHF operation, especially when compared to AM. With FM, the modulation process takes place in a low-level stage and remains the same, regardless of transmitter power. The signal may be frequency multiplied after modulation, and the PA stage can be operated Class C for best efficiency, as the "final" need not be linear.

In recent years there has been increasing use of FM by amateurs operating around 29.6 MHz. The VHF spectrum now in popular use includes 52 to 54 MHz, 146 to 148 MHz, 222 to 225 MHz, and 440 to 450 MHz.

Frequency and Phase Modulation

It is possible to convey intelligence by modulating any property of a carrier, including its frequency and phase. When the frequency of the carrier is varied in accordance with the variations in a modulating signal, the result is frequency modulation (FM). Similarly, varying the phase of the carrier current is called phase modulation (PM). Frequency and phase modulation are not independent, since the frequency cannot be varied without also varying the phase, and vice versa.

Frequency Modulation

When a modulating signal is applied to an FM modulator, the carrier frequency is increased during one half cycle of the modulating signal and decreased during the half cycle of opposite polarity. The change in the carrier frequency (frequency deviation) is proportional to the instantaneous amplitude of the modulating signal. Thus, the deviation is small when the instantaneous amplitude of the modulating signal is small, and is greatest when the modulating signal reaches its peak, either positive or negative.

Phase Modulation

If the phase of the current in a circuit shifts, there is an instantaneous frequency change during the time that the phase is shifting. The amount of frequency change, or deviation, is directly proportional to how rapidly the phase is shifting and the total amount of the phase shift. The rapidity of the phase shift is directly proportional to the frequency of the modulating signal. Further, in a properly operating PM system the amount of phase shift is proportional to the instantaneous amplitude of the modulating signal. This means that the amount of frequency change, or deviation, caused by a phase modulator, is directly proportional to both the instantaneous voltage and the frequency of the modulating signal. This is the outstanding difference between FM and PM, since in FM the frequency deviation is proportional only to the amplitude of the modulating signal. See Chapter 9 for further discussion of this and related topics.

FREQUENCY MODULATION METHODS

Direct FM

A simple, satisfactory device for producing FM in an amateur transmitter is the reactance modulator. This is a vacuum tube or transistor connected to the RF tank circuit of an oscillator in such a way as to act as a variable inductance or capacitance.

Fig. 32A is a representative circuit. Gate 1 of the modulator MOSFET is connected across the oscillator tank circuit, C1/L1, through resistor R1 and blocking capacitor C2. C3 represents the input capacitance of the modulator transistor. The resistance of R1 is large compared to the reactance of C3, so the RF current through R1/C3 will be practically in phase with the RF voltage appearing at the terminals of the tank circuit. However, the voltage across C3 will lag the current by 90 degrees. The RF current in the drain circuit of the modulator will be in phase with the grid voltage, and consequently is 90 degrees behind the current through C3, or 90 degrees behind the RF tank voltage. This lagging current is drawn through the oscillator tank, giving the same effect as though an inductance were connected across the tank. The frequency increases in proportion to the amplitude of the lagging plate current of the modulator. The audio voltage, introduced through a radio-frequency choke, varies the transconductance of the transistor and thereby varies the RF drain current.

The modulated oscillator usually is operated on a relatively low frequency, so a high order of carrier stability can be secured. Frequency multipliers are used to raise the frequency to the final frequency desired.

A reactance modulator can be connected to a crystal oscillator as well as to the self-controlled type as shown in Fig. 32B. However, the resulting signal can be more phase-modulated than it is frequency-

Fig. 31 — Portability and reliability make 2-meter FM hand-held operation a natural for supplying communications during emergencies, particularly when normal lines of communication are disrupted and access to the affected area is limited. (*KB4GJH photo*)

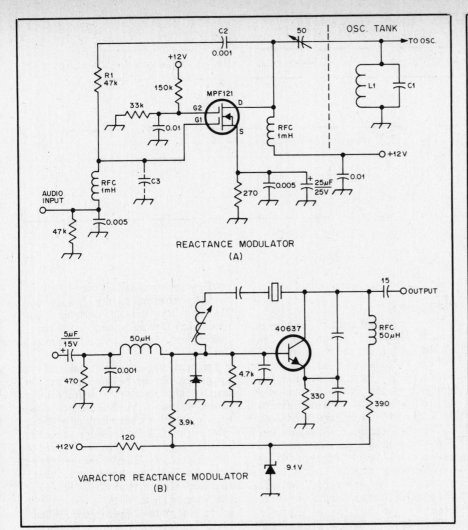

Fig. 32 — Reactance modulators using (A) a high-transconductance MOSFET and (B) a varactor diode.

Fig. 33 — (A) The phase-shifter type of phase modulator. (B) preemphasis and (C) deemphasis circuits.

modulated, because the frequency deviation that can be secured by varying the frequency of a crystal oscillator is quite small.

The sensitivity of the modulator (frequency change per unit change in modulating voltage) depends on the transconductance of the modulator transistor. It increases when R1 is made smaller in comparison with C3. It also increases with an increase in LC ratio in the oscillator tank circuit. However, for highest carrier stability it is desirable to use the largest tank capacitance that will permit the desired deviation to be secured while keeping within the limits of linear operation.

A change in any of the voltages on the modulator transistor will cause a change in RF drain current, and consequently a frequency change. Therefore it is advisable to use a regulated power supply for both modulator and oscillator.

Indirect FM

The same type of reactance-tube circuit that is used to vary the tuning of the oscillator tank in FM can be used to vary the tuning of an amplifier tank and thus vary the phase of the tank current for PM.

Hence the modulator circuit of Fig. 32A or 33A can be used for PM if the reactance transistor or tube works on an amplifier tank instead of directly on a self-controlled oscillator. If audio shaping is used in the speech amplifier, as described above, an FM-compatible signal will be generated by the phase modulator.

The phase shift that occurs when a circuit is detuned from resonance depends on the amount of detuning and the Q of the circuit. The higher the Q, the smaller the amount of detuning needed to secure a given number of degrees of phase shift. If the Q is at least 10, the relationship between phase shift and detuning (in kilohertz either side of the resonant frequency) will be substantially linear over a phase-shift range of about 25 degrees.

From the standpoint of modulator sensitivity, the tuned circuit Q on which the modulator operates should be as high as possible. On the other hand, the effective Q of the circuit will not be very high if the amplifier is delivering power to a load since the load resistance reduces the Q. There must therefore be a compromise between modulator sensitivity and RF power out-

put from the modulated amplifier. An optimum Q figure appears to be about 20; this allows reasonable loading of the modulated amplifier, and the necessary tuning variation can be secured from a reactance modulator without difficulty. It is advisable to modulate at a low power level.

Reactance modulation of an amplifier stage usually results in simultaneous amplitude modulation because the modulated stage is detuned from resonance as the phase is shifted. This must be eliminated by feeding the modulated signal through an amplitude limiter or one or more saturating stages — that is, amplifiers that are operated Class C and driven hard enough so that variations in the amplitude of the input excitation produce no appreciable variations in the output amplitude.

For the same type of reactance modulator, the speech-amplifier gain required is the same for PM as for FM. However, since the actual frequency deviation increases with the modulating audio frequency in PM, it is necessary to cut off the frequencies above about 3000 Hz before modulation takes place. If this is not done,

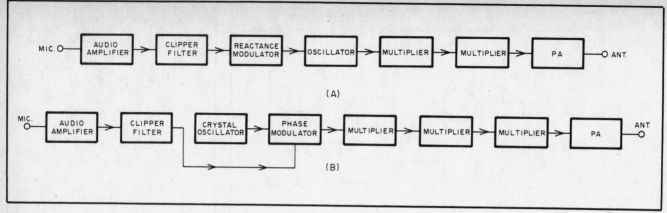

Fig. 34 — Block diagrams of typical FM exciters.

unnecessary sidebands will be generated at frequencies considerably removed from the carrier.

Speech Processing for FM

The speech amplifier that feeds the modulator is of ordinary design, except that no power is taken from it and the AF voltage required by the modulator input usually is small — only a volt or two for transistors. Because of these modest requirements, only a few speech stages are needed; a two-stage amplifier consisting of two bipolar transistors, both resistance-coupled, will more than suffice for crystal ceramic or Hi-Z dynamic microphones.

Several forms of speech processing produce worthwhile improvements in FM system performance. It is desirable to limit the peak amplitude of the audio signal applied to an FM or PM modulator, so that the deviation of the FM transmitter will not exceed a preset value. This peak limiting is usually accomplished with a simple audio clipper placed between the speech amplifier and modulator.

The clipping process produces high-order harmonics which, if allowed to pass through to the modulator stage, would create unwanted sidebands. Therefore, an audio low-pass filter with a cut-off frequency between 2.5 and 3 kHz is needed at the output of the clipper. Excess clipping can cause severe distortion of the voice signal. An audio processor consisting of a compressor and a clipper has been found to produce audio with a better sound (less distortion) than a clipper alone.

To reduce the amount of noise in some FM communication systems, an audio shaping network called preemphasis is added at the transmitter to proportionally attenuate the lower audio frequencies, giving an even spread to the energy in the audio band. This results in an FM signal of nearly constant energy distribution. Preemphasis applied to an FM transmitter will give the emission the deviation characteristics of PM. The reverse process, called deemphasis, is accomplished at the receiver to restore the audio to its original

relative proportions. See Fig. 33.

FM Exciters

FM exciters and transmitters take two general forms. One, shown at Fig. 34A, consists of a reactance modulator which shifts the frequency of an oscillator to generate an FM signal directly. Successive multiplier stages provide output on the desired frequency, which is amplified by a PA stage. This system has a disadvantage in that, if the oscillator is free running, it is difficult to achieve sufficient stability for VHF use. If a crystal-controlled oscillator is employed, because the amount that the crystal frequency is changed is kept small, it is difficult to achieve equal amounts of frequency swing.

The indirect method of generating FM shown in Fig. 34B is currently popular. Shaped audio is applied to a phase modulator to generate FM. Since the amount of deviation produced is very small, a large number of multiplier stages is needed to achieve wide-band deviation at the operating frequency. In general, the system shown at A will require a less complex circuit than that at B, but the indirect method (B) often produces superior results.

Testing an FM Transmitter

Accurate checking of the operation of an FM or PM transmitter requires different methods than the corresponding checks on an AM or SSB set. This is because the common forms of measuring devices either indicate amplitude variations only (a milliammeter, for example), or because their indications are most easily interpreted in terms of amplitude.

The quantities to be checked in an FM transmitter are the linearity and frequency deviation and the output frequency, if the unit uses crystal control. The methods of checking differ in detail.

Frequency Checking

Channelized FM operation requires that a transmitter be held within a few hundred hertz of the desired channel even in the wide-band system. Having the transmitter

on the proper frequency is particularly important when operating through a repeater. The rigors of mobile and portable operation make a frequency check of a channelized transceiver a good idea at three-month intervals.

Frequency meters generally fall into two categories, the heterodyne type and the digital counter. Today the digital counter is used almost universally; units counting to over 500 MHz are available at relatively low cost in kit form. Even less expensive low-frequency counters can be employed by using a prescaler, a device that divides an input frequency by a preset ratio, usually 10 or 100. Many prescalers may be used at 148 MHz or higher, using a counter with a 2-MHz (or more) upper frequency limit. If the counting system does not have a sufficient upper frequency limit to measure the output of an FM transmitter directly, one of the frequency-multiplier stages can be sampled to provide a signal in the range of the measurement device. Alternatively, a crystal-controlled converter feeding an HF receiver that has accurate frequency readout can be employed, if a secondary standard is available to calibrate the receiving system.

Deviation and Deviation Linearity

A simple deviation meter can be assembled following the diagram of Fig. 35A. This circuit was designed by K6VKZ. The output of a wide-band receiver discriminator (before any deemphasis) is fed to two amplifier transistors. The output of the amplifier section is transformer coupled to a pair of rectifier diodes to develop a dc voltage for the meter, M1. There will be an indication on the meter with no signal input because of detected noise, so the accuracy of the instrument will be poor on weak signals.

To calibrate the unit, signals of known deviation will be required. If the meter is to be set to read 0-15 kHz, then a 7.5-kHz deviation test signal should be employed. R1 is then adjusted until M1 reads half scale, 50 μA. To check the peak deviation of an incoming signal, close both S1 and

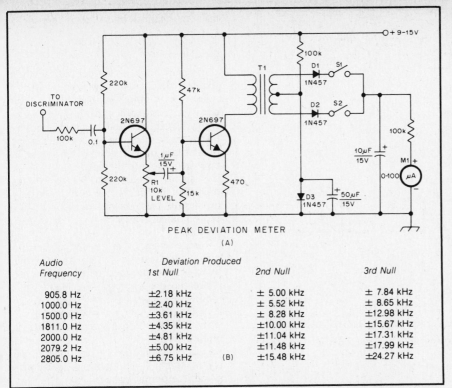

PEAK DEVIATION METER

(A)

Audio Frequency	Deviation Produced		
	1st Null	2nd Null	3rd Null
905.8 Hz	±2.18 kHz	± 5.00 kHz	± 7.84 kHz
1000.0 Hz	±2.40 kHz	± 5.52 kHz	± 8.65 kHz
1500.0 Hz	±3.61 kHz	± 8.28 kHz	±12.98 kHz
1811.0 Hz	±4.35 kHz	±10.00 kHz	±15.67 kHz
2000.0 Hz	±4.81 kHz	±11.04 kHz	±17.31 kHz
2079.2 Hz	±5.00 kHz	±11.48 kHz	±17.99 kHz
2805.0 Hz	±6.75 kHz (B)	±15.48 kHz	±24.27 kHz

Fig. 35 — (A) Schematic diagram of the deviation meter. Resistors are ½-watt composition, and capacitors are ceramic, except those with polarity marked, which are electrolytic. D1-D3, incl., are high-speed silicon switching diodes. R1 is a linear-taper composition control, and S1, S2 are SPST toggle switches. T1 is a miniature audio transformer with 10-kΩ primary and 20-kΩ center-tapped secondary (Triad A31X). (B) Chart of audio frequencies that will produce a carrier null when the deviation of an FM transmitter is set for the values given.

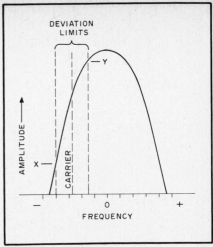

Fig. 36 — FM detector characteristics. Slope detection, using the sloping side of the receiver's selectivity curve to convert FM to AM for subsequent detection.

S2. Then, read the meter. Opening first one switch and then the other will indicate the amount of positive and negative deviation of the signal, a check of deviation linearity.

Measurement of Deviation Using Bessel Functions

Using a mathematical relationship known as the Bessel Function it is possible to predict the points at which, with certain audio-input frequencies and predetermined deviation settings, the carrier output of an FM transmitter will disappear completely. Thus, by monitoring the carrier frequency with a receiver, it will be possible to identify aurally the deviation at which the carrier is nulled. A heterodyne signal at either the input or receiver IF is required so the carrier will produce a beat note that can be identified easily. Other tones will be produced in the modulation process, so some concentration is required when making the test. With an audio tone selected from the chart (Fig. 35B), advance the deviation control slowly until the first null is heard. If a higher-order null is desired, continue advancing the control further until the second, and then the third, null is heard. Using a carrier null beyond the third is generally not practical.

For example, if a 905.8-Hz tone is used, the transmitter will be set for 5-kHz deviation when the second null is reached. The second null achieved with a 2805-Hz audio

input will set the transmitter deviation at 15.48 kHz. The Bessel-function approach can be used to calibrate a deviation meter, such as the unit shown in Fig. 35A.

Reception of FM Signals

Receivers for FM signals differ from others principally in two features — there is no need for linearity preceding detection (it is, in fact, advantageous if amplitude variations in signal and background noise can be minimized). Also, the detector must be capable of converting frequency variations of the incoming signal into amplitude variations.

Frequency-modulated signals can be received after a fashion on any ordinary receiver. The receiver is tuned to put the carrier frequency partway down on one side of the selectivity curve. When the frequency of the signal varies with modulation it swings as indicated in Fig. 36, resulting in an AM output varying between X and Y. This is then rectified as an AM signal.

With receivers having steep-sided selectivity curves, the method is not very satisfactory because the distortion is quite severe unless the frequency deviation is small, since the frequency deviation and output amplitude is linear over only a small part of the selectivity curve.

The FM Receiver

Block diagrams of an AM/SSB and an

FM receiver are shown in Fig. 37. Fundamentally, to achieve a sensitivity of less than 1 μV, an FM receiver requires a gain of several million — too much total gain to be accomplished with stability on a single frequency. Thus, the use of the superheterodyne circuit has become standard practice. Three major differences will be apparent from a comparison of the two block diagrams. The FM receiver employs a wider-bandwidth filter and a different detector, and has a limiter stage added between the IF amplifier and the detector. Otherwise the functions, and often the circuits, of the RF, oscillator, mixer and audio stages will be the same in either receiver.

In operation, the noticeable difference between the two receivers is the effect of noise and interference on an incoming signal. From the time of the first spark transmitters, QRN has been a major problem for amateurs. The limiter and discriminator stages in an FM set can eliminate a good deal of impulse noise, except noise that manages to acquire a frequency-modulation characteristic. Accurate alignment of the receiver IF system and phase tuning of the detector are required to achieve good noise suppression. FM receivers perform in an unusual manner when QRM is present, exhibiting a characteristic known as the capture effect. The loudest signal received, even if it is only two or three times stronger than other stations on the same frequency, will be the only transmission demodulated. By comparison, an S9 AM or CW signal can suffer noticeable interference from an S2 carrier.

Bandwidth

Most FM sets that use tubes achieve IF selectivity by using a number of over-coupled transformers. The wide bandwidth and phase-response characteristic needed in the IF system dictate careful design and

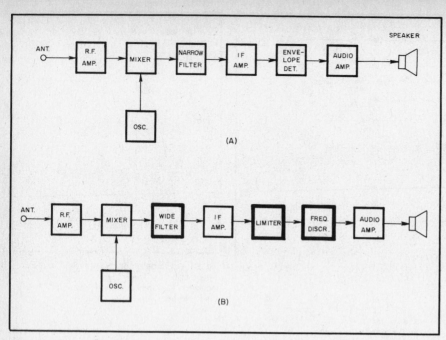

Fig. 37 — Block diagrams of (A) an AM, (B) an FM receiver. Dark borders outline the sections that are different in the FM set.

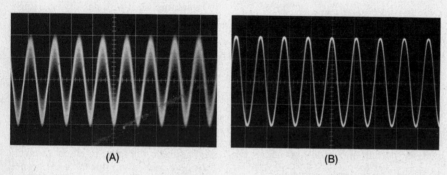

(A)   (B)

Fig. 38 — (A) Input wave form to a limiter stage shows AM and noise. (B) The same signal, after passing through two limiter stages, is devoid of AM components.

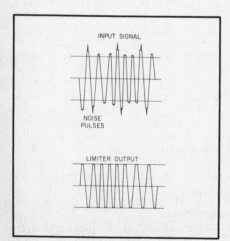

Fig. 39 — Representation of limiter action. Amplitude variations on the signal are removed by the diode action of the grid- and plate-current saturation.

adjacent-channel interference, especially in congested areas.

Limiters

When FM was first introduced, the main selling point for the new mode was that noise-free reception was possible. The circuit in the FM receiver that has the task of chopping off noise and amplitude modulation from an incoming signal is the limiter. Most types of FM detectors respond to both frequency and amplitude variations of the signal. Thus, the limiter stages preceding the detector are included so only the desired frequency modulation will be demodulated. This action can be seen in Fig. 38.

Limiter stages can be designed using tubes, transistors or ICs. For an amplifier to act as a limiter, the applied voltages are chosen so the stage will overload easily, even with a small amount of signal input. As shown in Fig. 39, the input signal limits when it is of sufficient amplitude so that diode action of the input voltage and output current saturation clip both sides of the input signal, producing a constant-amplitude output voltage.

Obviously, a signal of considerable strength is required at the input of the limiter to assure full clipping, typically 1 volt for transistors and several hundred microvolts for ICs. Limiting action should start with an RF input of 0.2 μV or less, so a large amount of gain is required between the antenna terminal and the limiter stages. For example, many commercial solid-state receivers use nine transistor stages for sufficient gain before the first limiter. The new ICs offer some simplification of the IF system, as they pack a lot of gain into a single package.

When sufficient signal arrives at the receiver to start limiting action, the set quiets — that is, the background noise disappears. The sensitivity of an FM receiver is rated in terms of the amount of input signal required to produce a given amount of quieting, usually 20 dB. Use of solid-state devices allows receivers to achieve 20 dB quieting with 0.15 to 0.5 μV of input signal.

A single transistor amplifier stage will not provide good limiting over a wide range of input signals. Two stages, with different input time constants, are a minimum requirement. The first stage is set to handle impulse noise satisfactorily while the second is designed to limit the range of signals passed on by the first. At frequencies below 1 MHz it is useful to employ untuned RC-coupled limiters that provide sufficient gain without oscillation.

Fig. 40A shows a two-stage limiter using transistors in two stages biased for limiter service. The base bias on either transistor may be varied to provide limiting at a desired level. The input-signal voltage required to start limiting action is called the limiting knee. This refers to the point at which collector current ceases to rise with

alignment of all interstage transformers.

For the average ham, the use of a high-selectivity filter in a homemade receiver offers some simplification of the alignment task. Following the techniques used in SSB receivers, a crystal or ceramic filter should be placed in the circuit as close as possible to the antenna connector — at the output of the first mixer, in most cases. Table 5 lists a number of suitable filters available to amateurs. Experimenters who wish to build their own can use surplus HF crystals or ceramic resonators.

One item of concern to every amateur FM user is the choice of IF bandwidth for the receiver. Deviation of 5 kHz is now standard on the amateur bands above 29 MHz. A wide-band receiver can receive narrow-band signals, suffering only some loss of audio level in the detection process. Naturally, it also will be subject to

Table 5
FM-bandwidth Filters Available to Amateurs

Manufacturer	Model	Center Frequency	Nominal Bandwidth	Ultimate Rejection	Impedance Ω In	Out	Insertion Loss	Crystal Discriminator
KVG (1)	XF-9E	9.0 MHz	12 kHz	90 dB	1200	1200	3 dB	XD9-02
KVG (1)	XF-107A	10.7 MHz	12 kHz	90 dB	820	820	3.5 dB	XD107-01
KVG (1)	XF-107B	10.7 MHz	15 kHz	90 dB	910	910	3.5 dB	XD107-01
KVG (1)	XF-107C	10.7 MHz	30 kHz	90 dB	2000	2000	4.5 dB	XD107-01
CDI (2)-		21.5 MHz	15 kHz	90 dB	550	550	3 dB	—
CDI (2)-		21.5 MHz	30 kHz	90 dB	1100	1100	2 dB	—
Clevite (3)	TCF4-12D3CA	455 kHz	12 kHz	60 dB	40 k	2200	6 dB	—
Clevite (3)	TCF4-18G45A	455 kHz	18 kHz	50 dB	40 k	2200	6 dB	—
Clevite (3)	TCF6-30D55A	455 kHz	30 kHz	60 dB	20 k	1000	5 dB	—

Manufacturers' addresses are as follows: (1) Spectrum International, P.O. Box 1084, Concord, MA 01742, tel. 617-263-2145; (2) Crystal Devices, Inc. (CDI), 6050 N. 52nd Ave., Glendale, AZ 85301, tel. 602-934-5234; (3) Semiconductor Specialists, Inc., P.O. Box 66125, O'Hare International Airport, Elmhurst, IL 60126, tel. 312-279-1000.

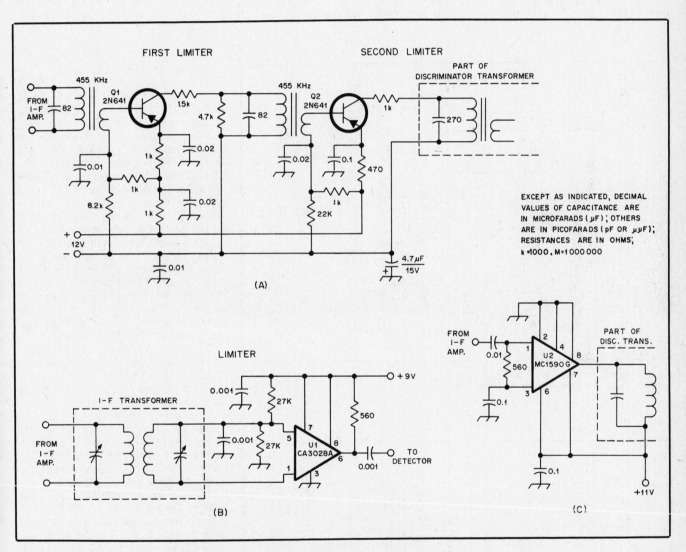

Fig. 40 — Typical limiter circuits using (A) transistors, (B) a differential IC, (C) a high-gain linear IC.

increased input signal. Modern ICs have limiting knees of 100 mV for the circuit shown in Fig. 40C, using the RCA CA3028A or Motorola MC1550G, or 200 mV for the MC1590G of Fig. 40D. Because the high-gain ICs such as the CA3076 and MC1590G contain as many as six or eight active stages that will saturate with sufficient input, one of these devices pro-

vides superior limiter performance compared to a pair of transistors.

Detectors

The first type of FM detector to gain popularity was the frequency discriminator. The characteristic of such a detector is shown in Fig. 41. When the FM signal has no modulation, and the carrier is at point

0, the detector has no output. When audio input to the FM transmitter swings the signal higher in frequency, the rectified output increases in the positive direction. When the frequency swings lower, the output amplitude increases in the negative direction. Over a range where the discriminator is linear (shown as the straight portion of the line), the conversion

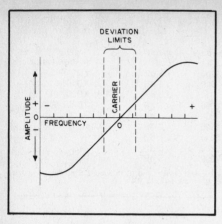

Fig. 41 — The characteristic of an FM discriminator.

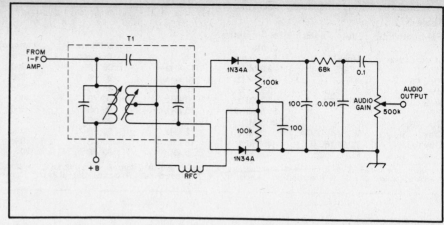

Fig. 42 — Typical frequency-discriminator circuit used for FM detection. T1 is a Miller 12-C45.

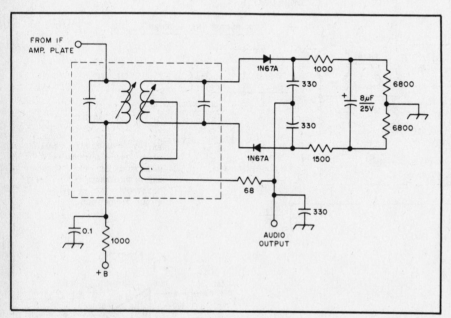

Fig. 43 — A ratio detector of the type often used in entertainment radio and TV sets. T1 is a ratio-detector transformer such as the Miller 1606.

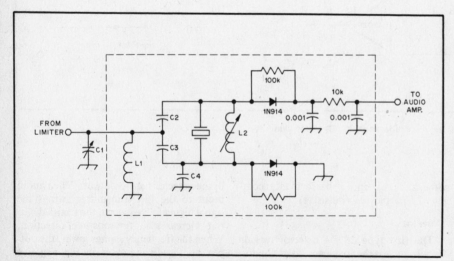

Fig. 44 — Crystal discriminator, C1 and L1 are resonant at the intermediate frequency. C2 is equal in value to C3. C4 corrects any circuit imbalance so equal amounts of signal are fed to the detector diodes.

of FM to AM will also be linear.

A practical discriminator circuit is shown in Fig. 42. The voltage induced in the T1 secondary is 90 degrees out of phase with the current in the primary. The primary signal is introduced through a center tap on the secondary, coupled through a capacitor. The secondary voltages combine on each side of the center tap so that the voltage on one side leads the primary signal while the other side lags by the same amount. When rectified, these two voltages are equal and of opposite polarity, resulting in zero-voltage output. A shift in input frequency causes a shift in the phase of the voltage components that results in an increase of output amplitude on one side of the secondary, and a corresponding decrease on the other side. The differences in the two changing voltages, after rectification, constitute the audio output.

In the search for a simplified FM detector, RCA developed a circuit that became standard in entertainment radios; it eliminated the need for a preceding limiter stage. Known as the ratio detector, this circuit is based on the idea of dividing a dc voltage into a ratio that is equal to the ratio of the amplitudes from either side of a discriminator-transformer secondary. With a detector that responds only to ratios, the input signal may vary in strength over a wide range without causing a change in the level of output voltage — FM can be detected, but not AM. In the actual ratio detector, Fig. 43, the required dc voltage is developed across two load resistors, shunted by an electrolytic capacitor. Other differences include the two diodes, which are wired in series aiding rather than series opposing, as in the standard discriminator circuit. The recovered audio is taken from a tertiary winding that is tightly coupled to the primary of the transformer. Diode-load resistor values are selected to be lower (5000 ohms or less) than for the discriminator.

The sensitivity of the ratio detector is one-half that of the discriminator. In general, however, the transformer design values for Q, primary-secondary coupling

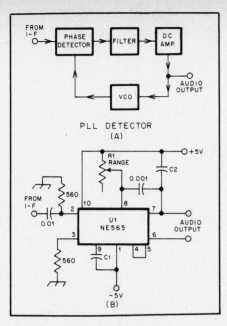

Fig. 45 — (A) Block diagram of a PLL demodulator. (B) Complete PLL circuit.

and load will vary greatly, so the actual performance differences between these two types of FM detectors are usually not significant. Either circuit can provide excellent results. In operation, the ratio detector will not provide sufficient limiting for communications service, so this detector also is usually preceded by at least a single limiting stage.

Other Detector Designs

The difficulties often encountered in building and aligning LC discriminators have inspired research that has resulted in a number of adjustment-free FM detector designs. The crystal discriminator utilizes a quartz resonator, shunted by an inductor, in place of the tuned-circuit secondary used in a discriminator transformer. A typical circuit is shown in Fig. 44. Some commercially made crystal discriminators have the input-circuit inductor, L1, built in (C1 must be added), while in other types both L1 and C2 must be supplied by the builder. Fig. 44 shows typical component values; unmarked parts are chosen to give the desired bandwidth. Sources for crystal discriminators are listed in Table 5.

The PLL

Since the phase-locked loop (PLL) was reduced to a single IC package, it has made significant impact on receiver design. Signetics first introduced a PLL in a single flat-pack IC. Motorola and Fairchild are making the PLL in separate building-block ICs. These allow a builder to assemble an FM receiving system with a minimum of bother.

A basic phase-locked loop consists of a phase detector, a filter, a dc amplifier and a voltage-controlled oscillator (VCO). The VCO runs at a frequency close to that of an incoming signal. The phase detector produces an error voltage if any frequency difference exists between the VCO and the IF signal. This error voltage is applied to the VCO. Any changes in the frequency of the incoming signal are sensed at the detector and the error voltage readjusts the VCO frequency so it remains locked to the intermediate frequency. The bandwidth of the system is determined by a filter on the error-voltage line. See Fig. 45.

Because the error voltage is a copy of the audio variations originally used to shift the frequency of the transmitter, the PLL functions directly as an FM detector. The sensitivity achieved with the Signetics NE565 PLL is good — about 1 mV for a typical circuit. No transformers or tuned circuits are required. The PLL bandwidth is usually 2 to 10 percent of the IF for FM detection. Components R1/C1 set the VCO close to the desired frequency. C2 is the loop-filter capacitor, which determines the capture range — that range of frequencies over which the loop will acquire lock with an unlocked input signal. The NE565 has an upper frequency limit of 500 kHz.

Chapter 19

Digital Communications

There is a growing trend away from analog transmission methods and toward digital communications. This is not limited to systems, such as computers, that use digital codes in their native language. Digital types of transmission are in increasing use for signals that are analog in origin and must be delivered to the ultimate user in analog form.

This chapter covers not only the types of digital communication covered in §97.69 of the FCC rules (Baudot, AMTOR, ASCII and other radioteletype [RTTY] or computer-to-computer communications) but the granddaddy of them all — Morse code. It may be convenient to label these different digital codes and tuck them away in separate categories. To do so, however, limits one's thinking to what already has been done.

The future of digital communication will see rapid expansion. Distinctions between the present categories will become increasingly blurred, and there will be a scramble for new terminology to keep up with technological change. Looking to the future, the FCC has given the Amateur Radio Service wide latitude for the use of digital communication. The preamble of §97.69 will give you the flavor: "...an amateur radio communication may include digital codes which represent alphanumeric characters, analogue measurements or other information. These digital codes may be used for such communications as (but not limited to) radio teleprinter, voice, facsimile, television, communications to control amateur radio stations, models and other objects, transference of computer programs or direct computer-to-computer communications, and communications in various types of data networks (including so-called "packet switching" systems); provided that such digital codes are not intended to obscure the meaning of, but are only to facilitate, the communications..."

The term "radioteletype" has been used to mean communications between typewriterlike *machines* using a telegraph code known as Baudot (or Murray). There was no serious movement to use other codes,

Fig. 1 — A family portrait of telegraph keyers used over this century. At front left is a hand key or straight key. Next is a semiautomatic key, usually called a *bug*, which produced a series of dots automatically but not dashes (which had to be formed individually by the operator). At front right is an electronic paddle keyer that produced both dots and dashes automatically when initiated by the operator. At rear left, a keyboard-operated electronic keyer. At right is a personal computer that has been programmed to generate and receive Morse code.

and the FCC rules permitted only Baudot code for RTTY communications. Before 1975, few amateurs thought about any terminal other than a teletypewriter.

The year 1975 marked the availability of the first affordable personal computers. A number of radio amateurs acquired computers (which used an eight-bit code) and immediately thought of interfacing them to Baudot teleprinters (which used five information bits). One application was to use a teleprinter to produce a paper printout (hard copy) for their computer. Another was to use the computer to send and receive Baudot RTTY and CW on the amateur bands.

Regulatory Changes

On September 15, 1978, the Canadian Department of Communications (DOC) announced rules creating the Amateur Digital Radio Operator's Certificate and establishing regulations for packet radio

(covered later in this chapter).

On March 17, 1980 the FCC legalized Amateur Radio transmission of the American National Standard Code for Information Interchange (ASCII). U.S. amateurs were then permitted to use either Baudot at 45, 50, 56.25 or 75 bauds or ASCII at 300 bauds below 28 MHz, 1200 bauds between 28 and 225 MHz and 19,600 bauds above 420 MHz.

In response to an ARRL petition (RM-3788), on October 28, 1982, the FCC amended its rules to permit any digital code to be used above 50 MHz. It raised the permissible speeds for ASCII transmission to 19.6 kilobauds between 50 and 220 MHz and 56 kilobauds above 220 MHz. Rather than set maximum speeds for other digital codes, the FCC specified maximum bandwidths of 20 kHz between 50 and 220, and 100 kHz between 220 and 1215 MHz.

In response to another ARRL petition (RM-4122) on February 8, 1983, the FCC amended its rules to permit amateurs to use Amateur Teleprinting Over Radio (AMTOR) in the high-frequency bands. Officially, AMTOR was described in terms of Recommendation No. 476-2 (1978) of the International Radio Consultative Committee (CCIR). In the same order, they made the maximum speeds for AMTOR and Baudot the same as specified for ASCII.

Another FCC order, which was effective on June 15, 1983, made several other changes that affected digital communications. The maximum audio- or radio-frequency shift was redefined as not to exceed 1000 Hz (between the mark and space frequencies) below 50 MHz, also that the frequency shift in hertz shall not exceed the sending speed in bauds or 1000 Hz, whichever is greater. The same order eliminated the need for CW or voice identification of digital transmissions and authorized stations to identify in Baudot, AMTOR or ASCII, whichever code is used for all or part of the communication below 50 MHz or when any digital code is used above 50 MHz. The order also added that for AMTOR either CCIR Recommendation 476-2 (1978) or 476-3 (1982) may be used.

These regulatory changes have done much to clear the way for significant changes in Amateur Radio digital communications in the years ahead. No doubt, there will be more changes designed to facilitate new forms of digital communications.

Morse Code

Radiotelegraphy by Morse code is the original modulation method used in Amateur Radio. The code is named after its inventor, Samuel F. B. Morse, 1791-1872, an American artist and promoter of the telegraph. Unlike codes designed for teletypewriters, the Morse code consists of elements of unequal length. The short pulse is called a dot (often sounded *dit*), and the long pulse is a dash (sounded *dah*). There are unique combinations of dots and dashes for the letters of the alphabet, numerals, punctuation marks and procedure signals (prosigns). A comprehensive list of the code combinations and their meanings appears in Table 1. The list shows code combinations for languages that use the Latin alphabet and includes accented letters, which are needed in some languages.

The listing of all of the code combinations does not imply that amateurs should learn and use all of them. In fact, to satisfy FCC testing requirements and for general use on the ham bands, you need to learn only the 26 letters, 10 numerals, the period, the comma, the question mark, fraction bar ($\overline{DN}$), $\overline{AR}$, $\overline{SK}$ and $\overline{BT}$. If you do not already

---

Table 1

Morse Code Character Set[1]

A	didah	• —	I	didit	••	S	dididit	•••
B	dahdididit	— •••	J	didahdahdah	• — — —	T	dah	—
C	dahdidahdit	— • — •	K	dahdidah	— • —	U	dididah	•• —
D	dahdidit	— ••	L	didahdidit	• — ••	V	didididah	••• —
E	dit	•	M	dahdah	— —	W	didahdah	• — —
F	dididahdit	•• — •	N	dahdit	— •	X	dahdididah	— •• —
G	dahdahdit	— — •	O	dahdahdah	— — —	Y	dahdidahdah	— • — —
H	didididit	••••	P	didahdahdit	• — — •	Z	dahdahdidit	— — ••
			Q	dahdahdidah	— — • —			
			R	didahdit	• — •			

1	didahdahdahdah	• — — — —	4	didididah	•••• —	8	dahdahdahdidit	— — — ••
2	dididahdahdah	•• — — —	5	didididit	•••••	9	dahdahdahdahdit	— — — — •
3	didididahdah	••• — —	6	dahdidididit	— ••••	0	dahdahdahdahdah	— — — — —
			7	dahdahdididit	— — •••			

Period [.]:	didahdidahdidah	• — • — • —	$\overline{AAA}$	Fraction bar [/]:	dahdidididahdit	— •• — •	$\overline{DN}$
Comma [,]:	dahdahdididahdah	— — •• — —	$\overline{MIM}$	Quotation marks ["]:	didahdididahdit	• — •• — •	$\overline{AF}$
Question mark				Dollar sign [$]:	dididididahdididah	••• — •• —	$\overline{SX}$
or request for				Apostrophe [']:	didahdahdahdahdit	• — — — — •	$\overline{WG}$
repetition [?]:	dididahdahdidit	•• — — ••	$\overline{IMI}$	Paragraph [¶]:	didahdidahdidit	• — • — ••	$\overline{AL}$
Error:	dididididididit	••••••••	$\overline{HH}$	Underline [__]:	dididahdahdidah	•• — — • —	$\overline{IQ}$
Hyphen or dash [—]:	dahdididididah	— •••• —	$\overline{DU}$	Starting signal:	dahdidahdidah	— • — • —	$\overline{KA}$
Double dash [=]:	dahdidididah	— ••• —	$\overline{BT}$	Wait:	didahdididit	• — •••	$\overline{AS}$
Colon [:]:	dahdahdahdididit	— — — •••	$\overline{OS}$	End of message or cross [+]:	didahdidahdit	• — • — •	$\overline{AR}$
Semicolon [;]:	dahdidahdidahdit	— • — • — •	$\overline{KR}$	Invitation to transmit [K]:	dahdidah	— • —	K
Left parenthesis [(]:	dahdidahdahdit	— • — — •	$\overline{KN}$	End of work:	dididahdidah	••• — • —	$\overline{SK}$
Right parenthesis [)]:	dahdidahdahdidah	— • — — • —	$\overline{KK}$	Understood:	dididahdit	••• — •	$\overline{SN}$

Notes

1. Not all Morse characters shown are used in FCC code tests. License applicants are responsible for knowing, and may be tested on, the 26 letters, the numerals 0 to 9, the period, the comma, the question mark, $\overline{AR}$, $\overline{SK}$, $\overline{BT}$ and fraction bar [$\overline{DN}$].

2. The following letters are used in certain European languages which use the Latin alphabet:

Ä, Ą	didahdidah	• — • —	Ö, Ő, Ó	dahdahdahdit	— — — •	
Á, Å, À, Â	didahdahdidah	• — — • —	Ñ	dahdahdidahdah	— — • — —	
Ç, Ć	dahdidahdidit	— • — ••	Ü	dididahdah	•• — —	
É, È, Ę	dididahdidit	•• — ••	Ż	dahdididahdit	— •• — •	
È	didahdidahdah	• — • — —	Ź	dahdahdididah	— — •• —	
Ê	dahdididahdit	— •• — •	CH, Ş	dahdahdahdah	— — — —	

3. Special Esperanto characters:

Ĉ	dahdidahdidit	— • — ••
Ĝ	dididahdit	••• — •
Ĥ	didahdahdahdit	• — — — •
Ĵ	dahdahdahdidit	— • — — •
Ŭ	didahdah	•• — —

4. Signals used in other radio services:

Interrogatory	dididandidah	•• — • —	$\overline{INT}$
Emergency silence	dididididahdah	•••• — —	$\overline{HM}$
Executive follows	dididahdididah	••• — •• —	$\overline{IX}$
Break-in signal	dahdahdahdahdah	— — — — —	$\overline{TTTTT}$
Emergency signal	dididadahdidididit	••• — — •••	$\overline{SOS}$
Relay of distress	dahdididahdididahdidit	— •• — •• — ••	$\overline{DDD}$

know these Morse characters, it is advisable to learn the code in an orderly manner as outlined in Chapter 36, preferably with the help of an instructor. After you have learned the basic characters and have some on-the-air experience, you may wish to expand your vocabulary of Morse code characters as you hear them or have need to send them.

Table 1 will also prove useful to those wishing to develop computer programs to send and receive Morse code. Some such programs have been flawed by omissions of valid characters and bogus combinations resulting from an incomplete listing of existing assignments.

Table 2 lists Morse code combinations for languages that are not based on the Latin alphabet. While most Amateur Radio CW communications use the International Morse Code (Table 1), you can occasionally hear hams native in the same language chat with one another in the codes listed in Table 2.

Timing of Morse Code

The basic unit of time in Morse code is

Table 2
Morse Codes for Other Languages

Code	Japanese	Korean	Arabic	Hebrew	Russian	Greek
•	ヘ he	ㅏ a		ו vav	Е,Э E	Ε epsilon
—	ム mu	ㅓ ŏ	ت ta	תּ tav	Т T	Τ tau
• •	〃 nigori	ㅑ ya	ﺱ ya	׳ yod	И I	Ι iota
• —	イ i	ㅗ o	ا alif	א aleph	А A	Α alpha
— •	タ ta	ㅛ yo	ن noon	נן nun	Н N	Ν nu
— —	ヨ yo	ㅁ m	م meem	םמ mem	М M	Μ mu
• • •	ラ ra	ㅕ yŏ	س seen	שׁש shin	С S	Σ sigma
• • —	ウ u		ط ta	ט tet	У U	ΟΥ omicron ypsilon
• — •	ナ na	ㅠ yu	ر ra	ר reish	Р R	Ρ rho
• — —	ヤ ya	ㅂ p(b)	و waw	צץ tzadi	В V	Ω omega
— • •	ホ ho		د dal	ד dalet	Д D	Δ delta
— • —	ワ wa	o -ng	ك kaf	ךכ chaf	К K	Κ kappa
— — •	リ ri	ㅅ s	غ ghain	ג gimmel	Г G	Γ gamma
— — —	レ re	ㅍ p'	خ kha	ה heh	О O	Ο omicron
• • • •	ヌ nu	ㅜ u	ح ha	ח chet	Х H	Η eta
• • • —	ク ku	ㄹ r-(-l)	ض dad		Ж J	ΗΥ eta ypsilon
• • — •	チ ti	ㄴ n	ف fa	ףפ feh	Ф ,F	Φ phi
• • — —	ノ no				Ю yu	ΑΥ alpha ypsilon
• — • •	カ ka	ㄱ k(g)	ل lam	ל lamed	Л L	Λ lambda
• — • —	ロ ro		ع ain		Я ya	ΑΙ alpha iota
• — — •	ツ tu	ㅈ ch(j)		פ peh	П P	Π pi
• — — —	ヲ wo	ㅎ h	ج jeem	ע ayen	Й Y	ΥΙ ypsilon iota
— • • •	ハ ha	ㄷ t(d)	ب ba	בּב bet	Б B	Β beta
— • • —	マ ma	ㅋ k'	ص sad		Ь,ь mute	Ξ xi
— • — •	ニ ni	ㅊ ch'	ث tha	ם samech	Ц TS	Θ theta
— • — —	ケ ke	ㅔ e	ظ za		Ы I	Υ ypsilon
— — • •	フ hu	ㅌ t'	ذ dhal	ז zain	З Z	Ζ zeta
— — • —	ネ ne	ㅐ ae	ق qaf	ק kof	Щ SHCH	Ψ psi
— — — •	ソ so		ز zay		Ч CH	ΕΥ epsilon ypsilon
— — — —	コ ko		غ sheen		Ш SH	Χ khi
• • — • •	ト to		ء he			
• • — • —	ミ mi					
• • — — •	° han-nigori					
• — • • •	オ o					
• — • • —	ヰ (w)i					
• — • — •	ン n					
• — • — —	テ te					
• — — • •	ヱ (w)e					
• — — • —	- hyphen					
• — — — •	セ se					
— • • • —	メ me					
— • • — •	モ mo					
— • • — —	ユ yu					
— • — • •	キ ki					
— • — • —	サ sa					
— • — — •	ル ru					
— • — — —	エ e					
— — • • —	ヒ hi					
— — • — •	シ si					
— — — • —	ア a					
— — — • — —	ス su					
• — • • • —						

ﻳ lam-alif

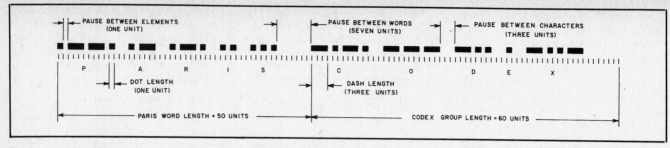

Fig. 2 — Timing of Morse code elements and spaces. The word PARIS is used as the typical word in English plain-language text. CODEX is typical of random-letter groups.

the length of the dot. The duration of the dash is three times that of the dot. The term *element* is used to include both dots and dashes. The space between two elements forming the same character is the same as for the dot. The space between characters is equal to three dots. The space between words or groups is equal to seven dots. These relationships are illustrated in Fig. 2.

Note that the length of characters varies. The letter E is the shortest because it is used most frequently in English plain-language text. T and I are next in line, and the codes get longer for letters that are less frequently used.

Analysis of English plain-language text indicates that the average word (including the space between words) is 50 units long. The word PARIS is of this length and is used to accurately set transmission speed. For example, if you wish to transmit at 10 WPM, adjust the keying speed until the word PARIS is sent 10 times in one minute. If you were to set the speed for PARIS but then send random-letter groups for code practice, you would find that the number of groups sent per minute is smaller. The reason is that the average length of random-letter groups is 60 units — the length of the group CODEX. The code speed for plain-language English can be found from

$$\text{speed (WPM)} = \frac{\text{dots/min}}{25}$$

$$= 2.4 \times \text{dots/s}$$

For example, a properly adjusted keyer gives a string of dots that are counted at 10 dots per second: speed = 2.4 × 10 = 24 WPM. For these keyers the code speed may be determined directly from the clock frequency

$$\text{speed (WPM)} = 1.2 \times \text{clock freq. (Hz)}$$

For a quick and simple means of determining the code speed, send a continuous string of dashes and count the number of dashes that occur in a five-second period. This number, to a close approximation, is the code speed in words per minute. A method for checking the speed of a Morse keyboard is to send a continuous string of zeros, with proper interletter spacing. Most keyboards will automatically insert the correct space if the key is released and reactivated before the end of the character. If zeros are sent for one minute, the speed is approximately

$$\text{speed (WPM)} = (\text{zeros/min}) \times 0.44.$$

Cutting Numbers

You will notice that the numerals 0 through 9 in Morse code are longer than letters. That's fine as long as numbers are only an incidental part of the transmission. Radio services that send lots of numeric text, such as weather information, usually cut the length of the codes for numbers. Table 3 lists both long numbers and cut number substitutions. Cut numbers should be used sparingly whenever in doubt that the receiving operator will understand them. Also, only the long numbers are legal for identification purposes. Cut numbers

are not in widespread use by amateurs except for sending RST reports. Even there, usually only the numeral 9 is sent in cut form. Another exception is that the cut form of zero is frequently used in ragchewing. The long form of zero should be sent in call signs and in message headings.

MORSE KEYING AND SWITCHING

Two basic problems involved with Morse code radiotelegraphy are keying the transmitter on and off and switching between transmit and receive states. If the transmitter is switched on and off too slowly, the resultant RF signal will be difficult to read because the keying sounds too soft. If the transitions are too fast, extraneous sidebands, called *key clicks*, are generated. It is the responsibility of the transmitter operator to ensure that unnecessary key clicks are not transmitted.

Ideally, a transceiver should quickly change from receive to transmit states when the key is pressed. Similarly, it should revert to the receiving state as soon as the key is opened. This is called *break-in keying* and allows the sending operator to hear between keying elements in the event that the receiving station is trying to interrupt. This is an important feature for the serious CW operator and is attainable by careful transceiver design.

Power-Supply Regulation for CW Keying

It is impractical to control the shape of the RF keying envelope unless the transmitter power supply is properly filtered and regulated. Improper regulation could cause the effect shown in Fig. 3D. This type of wave is even richer in harmonics than a square wave. The severity of the problem is a function of power supply load regulation. Note that in a class-C amplifier the change in power output related to plate or collector voltage regulation is proportional to the *square* of the voltage change. The power supply for a solid-state transmitter may be regulated easily and inexpensively by electronic means, but such a scheme applied to a high-power tube type of transmitter is costly and dissipates considerable power. Fortunately, the plate voltage waveform can be corrected with a passive circuit (see Dome, May 1977 *QST*). Note also that the power source must be nearly pure dc to ensure that the transmitter output

Table 3
Morse Cut Numbers

Numeral	Long Number		Cut Number		Equivalent Character
1	didahdahdahdah	·————	didah	·—	A
2	dididahdahdah	··———	dididah	··—	U
3	didididahdah	···——	didididah	···—	V
4	dididididah	····—	dididididah	····—	4
5	didididdit	·····	didididdit	····· or ·	5 or E
6	dahdidididit	—····	dahdidididit	—····	6
7	dahdahdidididit	——···	dahdididit	—···	B
8	dahdahdahdidit	———··	dahdidit	—··	D
9	dahdahdahdahdit	————·	dahdit	—·	N
0	dahdahdahdahdah	—————	dah	—	T

Note: These cut numbers are not legal for use in call signs. They should be used only where there is agreement between operators and when no confusion will result.

signal is not broadened by hum modulation.

RF Envelope Shaping

On-off keying is a form of amplitude modulation and, as such, generates sidebands whose spacing from the carrier is a function of envelope rise and fall times, which are the highest-frequency components of the keying waveform. An untreated keying waveform approaches *square-wave* modulation, consisting of the keying frequency plus *all of its odd harmonics*. These harmonics create key clicks extending many kilohertz either side of the carrier. Adjusting the rise and fall times of the keying waveform to about 5 ms controls key clicks yet allows a wide range of practical keying speeds, as explained in Chapter 9.

Once the power supply voltage has been brought under control, it is a simple matter to shape the keying envelope with an RC network. The figures in this section illustrate the application of time-constant circuits to various keying methods.

When a circuit carrying current is opened or closed mechanically, a spark is generated. This spark causes the circuit to radiate energy throughout the electromagnetic spectrum. When a transmitter is keyed manually or through a relay, the spark at the contacts can cause local BCI, *but this spark has no effect on the RF output signal*. A simple filter (0.01-μF capacitor in series with 10 ohms) across the key or relay contacts will usually reduce the local clicks to a tolerable level. Solid-state switching methods significantly reduce the current and voltage that must be switched mechanically, thereby reducing local clicks and enhancing operator safety. Modern transistorized transmitters incorporate this type of keying. With proper device selection, solid-state keying may be implemented in older tube designs as well.

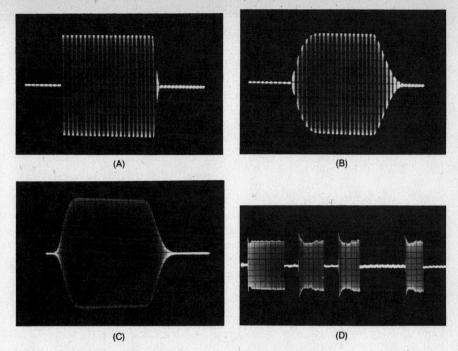

(A) (B) (C) (D)

Fig. 3 — These photos show CW signals as observed on an oscilloscope. At A is a dot generated at 46 bauds with no intentional shaping, while at B the shaping circuits have been adjusted for approximately 5-ms rise and decay times. Vertical lines are from a 1-kHz signal applied to the Z or intensity axis for timing. Shown at C is a shaped signal with the intensity modulation of the pattern removed. For each of these photos, sampled RF from the transmitter was fed directly to the deflection plates of the oscilloscope.

A received signal having essentially no shaping is shown at D. The spike at the leading edge is typical of poor power-supply regulation, as is also the immediately following dip and rise in amplitude. The clicks were quite pronounced. This pattern is typical of many observed signals, although not by any means a worst case. The signal was taken from the receiver's IF amplifier (before detection) using a hand-operated sweep circuit to reduce the sweep time to the order of one second. (Photos from October and November 1966 QST.)

Amplifier tubes may be keyed in the cathode (filament transformer center-tap for directly heated types), grid-bias supply or screen. Transistors should be keyed in one leg of the collector supply. The low impedance of RF power transistor circuits usually requires the emitter to be grounded as directly as possible; therefore, no solid-state analog of cathode keying exists. Similarly, blocked-grid keying has no transistor equivalent, because a reverse bias sufficient to cut the stage off in the presence of heavy excitation would cause breakdown of the base-emitter junction.

Mechanical contacts frequently bounce several times before stabilizing in the closed

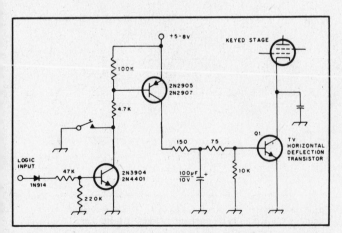

Fig. 4 — Cathode keying. Envelope shaping is accomplished by means of the RC network. Q1 must be able to withstand the plate voltage of the keyed stage. Some suitable types are: DTS-423, 2N6457 (400V), SDT 13305 (500V), DTS-801 (800V), MJ12010 (950V), 2SC1308K, ECG 238 (1500V). These are high-energy devices and are capable of switching any value of plate current the tube is likely to draw. For plate voltages below about 350, the 2N3439 is adequate (and much less expensive).

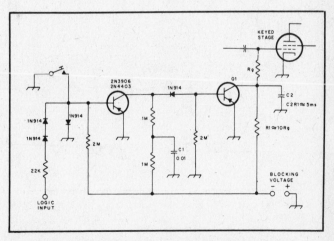

Fig. 5 — Blocked-grid keying. The rise time of the keying pulse is determined by C1 and its associated network. The decay time is governed by the R1C2 product. R_g is the existing grid leak. Typical values for R1 and C2 are 220 kΩ and 0.022 μF. Some transistors suitable for Q1 are: 2N5415 (200 V), MM4003 (250 V), MJE350 (300 V), 2N5416, RCS882 (350 V), 2N6213 (375 V), 2N6214 (425 V).

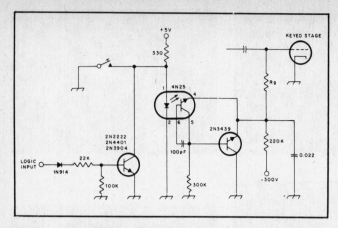

Fig. 6 — If a suitable high-voltage PNP transistor cannot be obtained, an NPN unit can be used with an optical isolator. The rise time of the keying envelope is controlled by the "integrating" capacitor connected to the base of the phototransistor.

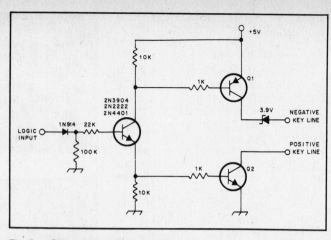

Fig. 7 — Circuit to interface digital logic with positive or negative key lines. Q1 and Q2 must be able to withstand the expected negative and positive keying voltages and currents.

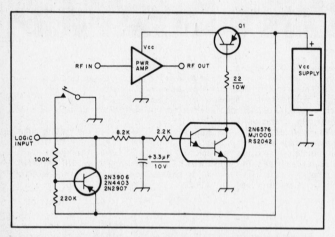

Fig. 8 — Keying circuit for solid-state class-C amplifiers in the 100-watt class, such as those sold for VHF FM service. Q1 must be able to pass the amplifier collector current without dropping too much voltage. Types 2N6246, SK3173 and RS2043 are good for currents up to 15 amperes.

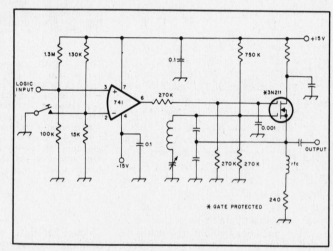

Fig. 9 — Keying a dual-gate MOSFET oscillator. The 741 op amp is used as a comparator. With the input resistors shown, the circuit can be triggered by any +5-volt logic device.

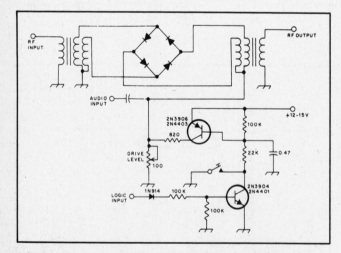

Fig. 10 — Keying a doubly balanced modulator in a CW-SSB transmitter.

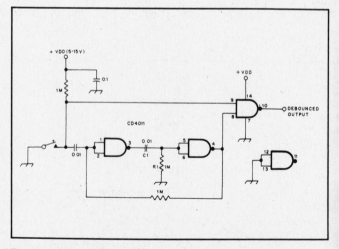

Fig. 11 — Debouncing circuit for hand keys and relay contacts. The minimum dot length is determined by the R1C1 product.

state. The beginnings of keying pulses formed by bouncing contacts are poorly defined. This defect can degrade the readability of a code signal under adverse conditions. Relays and semiautomatic keys are especially prone to this malady. The circuit of Fig. 11 will help clean up the pulses generated by mechanical contacts.

A satisfactory CW RF envelope can be amplified by means of a *linear* amplifier without affecting the keying characteristics. If, however, the signal is amplified by one or more nonlinear stages (e.g., a class-C multiplier or amplifier), the signal envelope will be modified, possibly introducing significant key clicks. It is possible to compensate for this effect by using longer-than-normal rise and decay times in the exciter and letting the amplifier modify the signal to an acceptable one. Any clicks generated by a linear amplifier are likely to be the result of low-frequency parasitic oscillations.

Oscillator Keying

A change in frequency at the beginning of a keying pulse is called a *chirp*. If the oscillator isn't keyed, chirp is the result of changing dc operating potentials or changing RF load conditions on the oscillator. The voltage to the oscillator can be regulated easily, as most transmitters use fairly low-power oscillators. If the oscillator frequency is pulled by the loading effect of subsequent keyed stages, better load isolation is indicated.

If break-in operation is desired (see below), it may be necessary to key the transmitter's oscillator. Oscillators may be keyed by the same methods used for amplifiers, but greater care is required to obtain good results. In general, the goals of clickless and chirpless oscillator keying are mutually exclusive. This is because a key-click filter will cause the operating voltage to be applied *slowly*, thereby creating a chirp. Crystal oscillators may be keyed satisfactorily if active crystals are used. A keyed oscillator may exhibit a continuous frequency change during a keying pulse. This defect is called a *yoop*. It is caused by RF heating in the tank circuit or crystal. Yoop is usually an indication of a faulty component or excessive oscillator power.

Break-in operation with a VFO-controlled transmitter usually dictates some form of *differential* keying. In this system, the oscillator is turned on as quickly as possible and the amplifier is keyed after the oscillator has stabilized. When the key is released, the oscillator operates until the amplifier output has decayed to zero. Shaping of the keying envelope is accomplished in the amplifier keying circuit. In the past, break-in operation was implemented with VR tubes and relays. The complexity of such systems frightened many hams away from this convenient mode. With modern circuitry, break-in can be simple, quiet and safe. One method of break-in keying is shown in Fig. 12.

A few notes concerning oscillator keying are in order. Do not attempt to key an oscillator unless it is stable while running free, at the same time the other stages are keyed. If the oscillator frequency is multiplied, chirps and yoops will be multiplied also. The transmitter signal should be checked at the highest operating frequency. It may be found that the signal quality is satisfactory on 160 meters but leaves something to be desired on 10. Modern transmitters and transceivers provide multi-

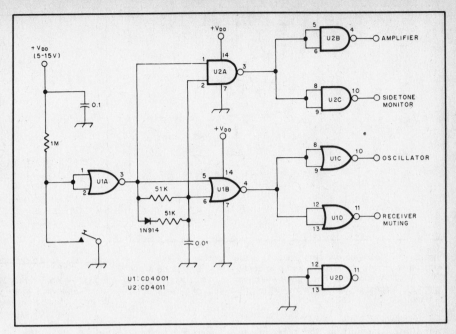

Fig. 12 — Differential (sequential) keying system for fast break-in with oscillator-multiplier transmitters.

band operation by the heterodyne method and should be stable on all bands.

Backwave

If the transmitter oscillators run continuously, they may be audible as a *backwave* between keying pulses. A strong backwave may indicate the need for neutralizing one or more transmitter stages. In general, if the backwave conforms to the −40 dB spurious signal rule, it won't be objectionable.

Practical Keying Circuits

The figures in this section illustrate methods by which various solid-state and thermionic devices may be keyed. In these circuits, the armature of the hand key is at ground potential and the voltage across the key is imperceptible. The current through the key is generally less than 1 milliampere. A neon bulb with a proper series resistor across the key will alert the operator to junction breakdown of the high-voltage transistors. As long as safety is given due consideration, the key-at-ground convention need not be followed, but this standardization is useful for equipment interconnections. Digital control is shown in all of the examples. This feature simplifies the simultaneous keying of transmitter stages, TR switches, sidetone oscillators and muting systems. The ICs used to perform the control functions are plentiful and inexpensive. These systems use a logic high to indicate a keydown condition.

Break-In

Break-in (QSK) is a system of radiotelegraph transmission in which the station receiver is sensitive to other signals between the transmitted keying pulses. This capability is very important to traffic

handlers, but can be used to great advantage in ragchewing as well. Break-in gives CW communication the dimension of more natural conversation.

Most commercially manufactured transceivers feature a "semi break-in" mode in which the first key closure actuates the VOX relay. The VOX controls are usually adjusted to hold the relay closed between letters. With proper VOX adjustment, it is possible for the other operator to break your transmission between words, but this system is a poor substitute for true break-in.

Separate Antennas

The simplest way to implement break-in is to use a separate antenna for receiving. If the transmitter output power is low (below 50 watts or so) and the isolation between transmitting and receiving antennas is good, this method can be satisfactory. Best isolation is obtained by mounting the antennas as far apart as possible and at right angles. Smooth break-in involves protecting the receiver from damage by the transmitter power and assuring that the receiver will "recover" fast enough to be sensitive between keying pulses. If the receiver recovers fast enough but the transmitter clicks are bothersome (they may be caused by receiver overload and so exist only in the receiver), their effect on the operator can be minimized through the use of an output limiter. The separate antenna method is most useful on the 160-, 80- and 40-meter bands, where the directional effects of the antennas aren't pronounced.

Switching a Common Antenna

When powers above about 50 watts are used, where two antennas are not available, or when it is desired to use the same an-

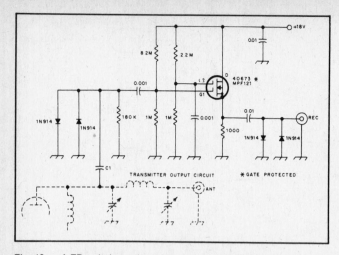

Fig. 13 — A TR switch can be connected to the input side of the transmitter pi network. For powers up to about 100 W, C1 can be a 5-10 pF, 1000-V mica unit. For high-power operation a smaller "gimmick" capacitor made from a short length of coaxial cable should be used.

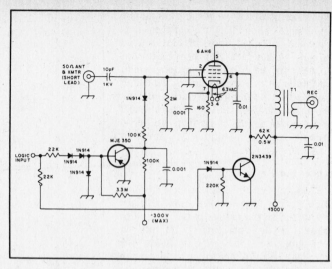

Fig. 14 — An external TR switch. The primary of T1 is 50 turns of no. 30 enameled wire on an FT37-43 toroid core. The secondary is 15 turns of no. 30 wound over the primary.

tenna for transmitting and receiving (a "must" when directional antennas are used), special treatment is required for quiet break-in operation on the transmitter frequency.

Vacuum relays or reed switches may be used to switch the antenna between the transmitter and receiver in step with the keying. This method is satisfactory for power levels up to the legal limit, but the relays are expensive and the system timing is critical.

Perhaps the most modern and elegant approach is the use of PIN diodes to switch the antenna. These devices are available in power ratings up to about 100 watts, but are quite expensive at present. There are no keying-speed constraints when PIN diodes are used, and if the proper devices are selected, the spectral purity of the output signal won't be affected. The important electrical parameter in this regard is *carrier lifetime*.

An easy and economical way to implement break-in with a single antenna is to use an electronic *TR switch*. With such a device the antenna is connected to the transmitter at all times. In the most common type of electronic TR switch, a tube is used to couple the antenna to the receiver. When the transmitter is keyed, the RF output causes the tube to draw grid current through a high-resistance grid leak. The high negative bias thus developed cuts off the plate current, limiting the signal delivered to the receiver.

Unfortunately, when the grid circuit is driven into rectification, harmonics are generated. A low-pass filter between the TR switch and the antenna can help to eliminate TVI caused by the harmonics, but the lower-order harmonics may cause interference to other communications. Another common shortcoming of TR switches is that the transmitter output circuit may "suck out" the received signal. In a transmitter having a high-impedance

tuned-output tank circuit, both of these problems can be circumvented by connecting the TR switch to the *input* side of the tank circuit. With this configuration, the grid rectification harmonics are suppressed and the received signal peaked by means of the tuned circuit. Fig. 13 shows a MOSFET TR switch that works on this principle.

A TR switch for use external to any transmitter is shown in Fig. 14. The tube is grid-block keyed, and the fixed-value bias prevents any grid current flow at power levels up to 800 watts. This power figure assumes a 50-ohm system with a unity SWR. The circuit can withstand peak RF voltages up to 300. The power capability must be derated if the impedance at the point of connection is higher than 50 ohms. Although the signal path has a diode, it is effectively "linearized" by the high-value series resistor, and should not significantly degrade the spectral purity of the transmitter output.

External TR switches should be well shielded and the power leads carefully filtered. In general, the coaxial cable to the transmitter should be as short as possible, but some experimenting may be necessary to eliminate "suck out" of the received signal. It is commonly stated that electronic TR switches are usable only with transmitters having class-C output stages because the "diode noise" generated by the resting current of a linear amplifier will mask weak signals. Actually, the class-of-service designation is not related to key-up conditions, so there is no reason a linear amplifier can't be biased off during key-up periods.

Reduction of Receiver Gain During Transmission

For absolutely smooth break-in operation with no clicks or thumps, means must be provided for momentarily reducing the gain through the receiver. A muting function completely disables the receiver audio during key-down periods. Assuming the transmitter signal at the receiver is held

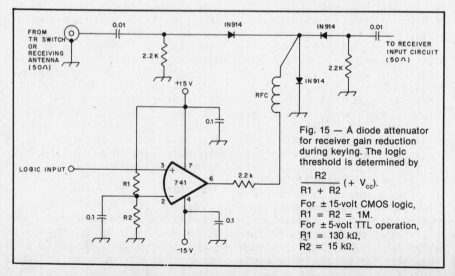

Fig. 15 — A diode attenuator for receiver gain reduction during keying. The logic threshold is determined by

$$\frac{R2}{R1 + R2} (+ V_{cc}).$$

For ± 15-volt CMOS logic, R1 = R2 = 1M.
For ± 5-volt TTL operation, R1 = 130 kΩ, R2 = 15 kΩ.

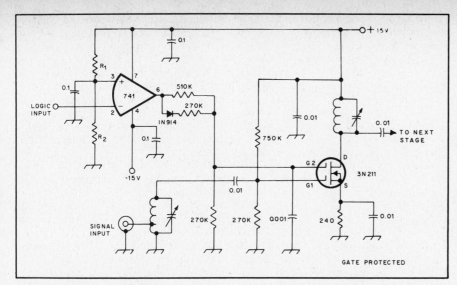

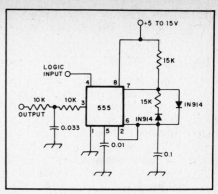

Fig. 17 — A 555 universal timer used as a sidetone generator. Pin 4 is taken to ground to interrupt the tone. The frequency of oscillation is about 500 Hz with the constants shown.

Fig. 16 — Gain-reduction circuit for receivers using a fixed-bias dual-gate MOSFET in the first stage. As much as 40 dB of attenuation is possible with this method. The logic threshold is calculated as in Fig. 15.

below the damage level, muting the audio output can be an effective means of achieving smooth break-in *provided no AGC is used*. AGC systems suitable for CW operation are characterized by long "hang" times. Unless the transmitter signal in the receiver is at a level similar to that of the other station, the AGC system will seriously desense the receiver, rendering the break-in system useless. A diode attenuator suitable for use with TR switches or separate antennas is shown in Fig. 15. If the receiver uses a dual-gate MOSFET with no AGC in the first stage, the method of Fig. 16 may be used.

Monitoring

If the receiver output is muted, an audio sidetone oscillator must be used to monitor one's sending. A 555 timer connected as an astable multivibrator is commonly used for this purpose. This device delivers rectangular output pulses, and the resulting signal often sounds quite raucous. A variation of the standard 555 circuit appears in Fig. 17. The diodes maintain the symmetry of the waveform independently of the pitch and the RF filter removes many of the objectional harmonics. A keying monitor can be powered by the RF output of the transmitter. Such a circuit is shown in Fig. 18. Keying monitors often have built-in loudspeakers, but it is less expensive and more convenient to inject the monitor signal into the audio output stage of the

receiver. With this system one always hears the sidetone from the same source (speaker or headphones) as the other station's signal.

If the audio output isn't muted, the receiver can be used to monitor one's keying, provided both stations are on the same frequency. Some DX operators transmit and listen on separate frequencies. When using your receiver as a monitor, you should be careful about drawing any conclusions concerning the quality of your signal. The signal reaching the receiver must be free of any line voltage effects induced by the transmitter. To be certain of your signal quality you should listen to your station from a distance. Trading stations with a nearby amateur is a good way to make signal checks.

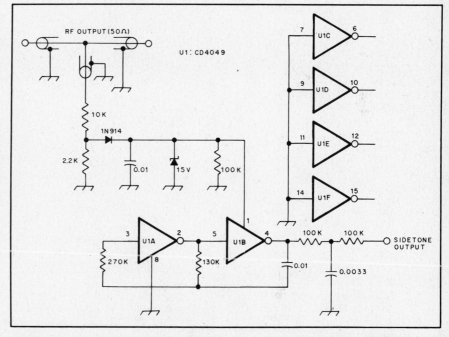

Fig. 18 — An RF-powered keying monitor suitable for power outputs from about 20 to 100 W. This circuit should be installed inside the transmitter or a shielded enclosure to minimize RFI.

Narrow-Band Direct-Printing Telegraphy

The term "narrow-band direct-printing telegraphy" (NBDP) is used by the ITU to mean communications between two teletypewriters. Isn't that what we've been calling "radioteletype" (RTTY) all along? Yes, but there may be a subtle distinction, or at least a clarification. "Direct-printing" makes it rather clear that typewriterlike printers are involved and not data communications between computers. The two direct-printing telegraph systems Baudot and AMTOR used in the Amateur Radio

Service are presented here. ASCII and packet radio are covered later in this chapter.

BAUDOT RADIOTELETYPE

The commercial- or military-surplus Baudot-encoded teletypewriter (TTY) was the mainstay of amateur RTTY operators from 1946 through around 1977. There are still numerous TTY machines in use, but many RTTY operators are now using computer-based terminals. In some cases, the terminal was designed from the outset as a multimode RTTY and CW terminal. In others, amateurs adapted personal computers to send and receive the Baudot code.

The Baudot Telegraph Code

One of the first data communications codes to receive widespread use had five information pulses (or levels) to present the alphabet, numerals, symbols and machine functions. In the United States, the current five-level code is commonly called the Baudot code. In Great Britain the almost-identical code is called the Murray code.

There are many variations in five-level coded character sets, principally to accommodate foreign-language alphabets.

Current FCC rules in §97.69(a) identify the Baudot code as the International Telegraph Alphabet Number 2 (ITA2). This code is recognized by the International Telegraph and Telephone Consultative Committee (CCITT) and allows for national variations. The ITA2 coded character set is shown in Table 4.

As ITA2 is a five-level code, and there are only two conditions for each level (A or Z, space or mark, binary 0 or 1). Therefore, a total of $2 \times 2 \times 2 \times 2 \times 2 = 2^{(5)} = 32$ different code combinations are possible. Because it is necessary to provide 26 Latin letters, 10 numerals and punctuations, the 32 code combinations are not sufficient. This problem is solved by using the codes twice; once in the *letters* (LTRS) case and again in the *figures* (FIGS) case. Two special characters, LTRS and FIGS are used to indicate whether subsequent characters will be in letters or figures case. The receiving terminal stores the last received LTRS or FIGS characters so it remains in the last-received case until changed. Control functions such as LTRS, FIGS, carriage return (CR), line feed (LF), space (SP) and blank are assigned to both the letters and figures cases so that they can be used in either case. The remaining 26 code combinations have different character meaning, depending on whether preceded by a LTRS or FIGS character.

FCC rules provide that ITA2 transmissions must be single channel and sent using start-stop pulses as illustrated in Fig. 19. The bits in Fig. 19 are arranged in a left-to-right order, as would be observed on an oscilloscope.

FCC rules also permit ITA2 figures case positions not used for numerals and the slant sign or fraction bar to be used for the remote control of receiving printers. They may also be used for other purposes indicated in §97.69 of the rules.

Baudot Teletypewriters

Mechanical teletypewriter equipment consists of these basic components:

A *printer* prints on a continuous supply of paper (in a roll or fan folded) about the width of letter-size paper. Although it has been known as a *page* printer, it prints a character at a time. A printer, often called a *Receive-Only* or *RO* terminal, will allow you to decode and print what RTTY operators are sending.

A *keyboard* is used by an operator to originate messages. Most mechanical keyboards have encoders to send out the character code in start-stop (asynchronous) sequence. A teleprinter with keyboard and printer may be known as *KSR*, which stands for *Keyboard Send-Receive* terminal.

A *perforator* is a unit that punches paper tape and may be controlled by a keyboard by direct mechanical connection. The term *reperforator (reperf)* signifies that the perforator punches tape from electrically received signals and usually prints the corresponding characters on the tape. A *Receive-Only Typing Reperforator (ROTR)* receives serial data and perforates and prints on the tape.

A *Transmitter-Distributor (TD)* set is a tape reader that senses the perforations in the tape and produces encoded start-stop pulses. A teletypewriter with a tape capability may be called *Automatic Send-Receive* or *ASR*.

Computer-Based Baudot Terminals

Electronic Baudot terminal equipment, as shown in Fig. 20, is similar to counterpart mechanical teletypewriter units.

A *microcomputer* is the foundation of the system and is used to encode and decode ASCII characters, convert them to and from the comparable Baudot codes, and provide for the start-stop code sequence for transmission and reception. The heart of the microcomputer is the

Table 4

ITA2 (Baudot) and CCIR Rec. 476 (AMTOR) Codes

Combination No.	ITA2[1] Code Bit No. 43210	Hex	CCIR 476[2] Code Bit No. 6543210	Hex	Letters Case	Figures Case ITA2	Figures Case U.S. TTYs[3]
1	00011	03	1000111	47	A	—	—
2	11001	19	1110010	72	B	?	?
3	01110	0E	0011101	1D	C	:	:
4	01001	09	1010011	53	D	[5]	$
5	00001	01	1010110	56	E	3	3
6	01101	0D	0011011	1B	F	[4]	!
7	11010	1A	0110101	35	G	[4]	&
8	10100	14	1101001	69	H	[4]	# or motor stop
9	00110	06	1001101	4D	I	8	8
10	01011	0B	0010111	17	J	BELL	'
11	01111	0F	0011110	1E	K	(	(
12	10010	12	1100101	65	L	)	)
13	11100	1C	0111001	39	M	.	.
14	01100	0C	1011001	59	N	,	,
15	11000	18	1110001	71	O	9	9
16	10110	16	0101101	2D	P	0	0
17	10111	17	0101110	2E	Q	1	1
18	01010	0A	1010101	55	R	4	4
19	00101	05	1001011	4B	S	'	BELL
20	10000	10	1110100	74	T	5	5
21	00111	07	1001110	4E	U	7	7
22	11110	1E	0111100	3C	V	=	;
23	10011	13	0100111	27	W	2	2
24	11101	1D	0111010	3A	X	/	/
25	10101	15	0101011	2B	Y	6	6
26	10001	11	1100011	63	Z	+	"
27	01000	08	1111000	78	← CR (Carriage return)		
28	00010	02	1101100	6C	≡ LF (Line feed)		
29	11111	1F	1011010	5A	↓ LTRS (Letter shift)		
30	11011	1B	0110110	36	↑ FIGS (Figure shift)		
31	00100	04	1011100	5C	SP (Space)		
32	00000	00	1101010	6A	BLK (Blank)		

Notes

[1] 1 represents the mark condition (shown as Z in ITU recommendations) which is the higher emitted radio frequency for FSK, the lower audio frequency for AFSK. 0 represents the space condition (shown as A in ITU documents). Bits are numbered 0 (least significant bit) through 4 (most-significant bit). The order of bit transmission is LSB first, MSB last. Symbols A and Z are defined in CCIR Rec. R.140.

[2] 1 represents the mark condition (shown as B in CCIR recommendations), which is the higher emitted radio frequency for FSK, the lower audio frequency for AFSK. 0 represents the space condition (shown as Y in CCIR recommendations). Bits are numbered 0 (LSB) through 6 (MSB). The order of bit transmission is LSB first, MSB last.

[3] Many U.S. teletypewriters have these figures case characters.

[4] At present unassigned. Reception of these signals, however, should not initiate a request for repetition.

[5] The pictorial representations of ⊠ or ✚ indicate WRU (Who are you?), which is used for an answer-back function in telex networks.

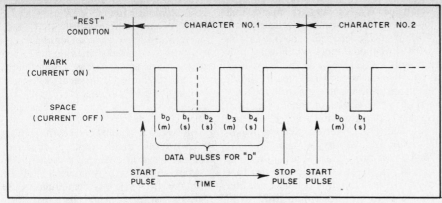

Fig. 19 — Time sequence of typical Baudot character, the letter D.

Fig. 20 — Many RTTY operators now use all-electronic systems.

Table 5
CCIR Rec. 476 Service Information Signals

Mode A (ARQ)	Bit No. 6543210	Hex	Mode B (FEC)
Control signal 1 (CS1)	1100101	65	
Control signal 2 (CS2)	1101010	6A	
Control signal 3 (CS3)	1001101	4D	
Idle signal β	0110011	33	
Idle signal α	0001111	0F	Phasing signal 1
Signal repetition	1100110	66	Phasing signal 2

Table 6
Baudot Signaling Rates and Speeds

Signaling Rate (bauds)	Data Pulse (ms)	Stop Pulse (ms)	Speed (WPM)	Common Name
45.45	22.0	22.0	65.00	Western Union
	22.0	31.0	61.33	"60 speed"
	22.0	33.0	60.61	45 bauds
50.00	20.0	30.0	66.67	European; 50 bauds
56.92	17.57	25.00	76.68	"75 speed"
	17.57	26.36	75.89	57 bauds
74.20	13.47	19.18	100.00	"100 speed"
	13.47	20.21	98.98	74 bauds
100.0	10.00	15.00	133.33	100 bauds

microprocessor. There is also *random-access memory (RAM)* for volatile (temporary) storage of data and *read-only memory (ROM)* for nonvolatile (permanent) storage. *Erasable programmable read-only memory (EPROM)* is nonvolatile but can be erased and reprogrammed.

The other units described below can be considered *peripherals* from the standpoint of the microcomputer.

A *keyboard* usually produces an ASCII-encoded parallel output.

A *video-display terminal* or *VDT* is the normal display device. Many VDTs use cathode-ray tubes (CRTs), but flat-screen character displays are coming into use.

A *character printer* is used when a paper (hard) copy is desired.

A *disk drive* may be added to the system to store text or software on flexible (floppy) diskettes. Alternatively, an audio cassette tape recorder may be used.

Baudot Speeds and Signaling Rates

The operating speed of mechanical teletypewriter machines is determined by motor speed and gearing ratios or electronic timing circuits. Common speeds are 60, 67, 75 or 100 WPM. See Table 6 for the relationship of speeds and signaling rates and pulse times.

The speed is given in the *approximate* number of five-letter-plus-space combinations transmitted in a *continuous sequence* in start-stop format over a one-minute

interval. Convenient choices of gearing ratios and motor-shaft speeds have resulted in noninteger WPM rates. Common usage, however, has rounded the exact speeds to easily remembered numbers. Thus, "60 speed" Baudot actually is sent at 61.33 WPM, and "75 speed" is really 76.67 WPM.

A problem occurs with the use of words per minute or characters per second as TTY speed specifications because of the varying length of stop pulses in use. For example, "60 speed" Baudot TTY has 22-ms-long start and data pulses and a 31-ms stop pulse (1.42 times the data pulse width); the Western Union "65 speed" also has 22-ms start and data pulses, but the stop pulse is also 22 ms long (a 1:1 ratio); some electronic terminals use 22-ms start and data pulses and 33-ms stop pulses (a 1.5:1 ratio).

In light of the above character-length variation from the different stop-pulse-to-data-pulse ratios in use, it is now common to refer to the number of shortest pulses per second. The *baud* is a unit of signaling speed equal to one pulse (event or symbol) per second. The signaling rate in bauds can be found by dividing the shortest pulse length into one; for example, $1/0.022 = 45.45$, commonly abbreviated to 45.5 or even 45 bauds.

Teletypewriter Loop Circuits

Teletypewriter printing mechanisms use selector magnets to sense the presence

(mark) or absence (space) of the loop current. The characters typed on the sending keyboard are encoded with proper mark and space pulses. Since the keyboards and selector magnets of both machines are series connected, text typed on one keyboard is reproduced on both printers. Connection of the keyboard directly to its associated printer is called a *local loop*, giving local copy of transmitted text.

Selector magnets used in older machines such as the Teletype Corp. model 15 were designed for a mark loop current of 60 mA dc. Newer machines use a 20-mA loop, and some electronic interface circuits accept a wide range of loop currents (10 to 120 mA for the HAL DS3100, for example).

Since the dc resistance of the selector magnets is rather low (100 to 300 ohms, typically), it would at first seem that a low-voltage loop supply could be used. However, the inductance of the magnet is usually quite high (on the order of 4 henrys for a model 15), causing a delay in the current rise time. This, in turn, delays the selector magnet response to a mark pulse, distorting the signal. This distortion can be severe enough to cause misprinting of received text, particularly if other forms of distortion are present (such as caused by variations in the radio signal). In general, the higher the loop voltage and loop resistance used, the lower the distortion. In practice, loop power-supply voltages between 100 and 300-V dc are common.

Newer loop supplies use a 150- to 200-volt loop supply and a 2000- to 3000-ohm loop resistor to set the 60-mA loop current. Amateur Baudot RTTY demodulators normally contain loop power supplies and current-limiting resistors.

Baudot RTTY Modems

The term *modem* is a contraction of *modulator-demodulator*. Some amateurs use the term *terminal unit* or *TU* to mean modem or sometimes just a demodulator.

It is possible to transmit teleprinter signals by on/off keying (OOK) as is used for regular Morse code CW transmission. In the early days of amateur RTTY, OOK was the only legal way of operating RTTY on the HF bands. OOK demodulators had problems distinguishing between signal and no signal because the no-signal condition contains both natural and man-made noise. This method of keying worked only when there was a sufficiently high signal-to-noise ratio.

In 1953, the FCC changed its rules to permit frequency-shift keying (FSK) in the HF amateur bands. FSK has the advantage of providing either a mark or space signal during transmission. Thus the receiving demodulator has the easier decision between two on-the-air signals rather than comparing the mark signal against noise. With proper demodulation, FSK transmission provides much better performance than OOK.

Frequency-Shift Keying

Below 50 MHz, FSK (F1B emission) is used for Baudot RTTY. The nominal transmitter frequency is the mark condition. The space condition causes the transmitter to shift downward in frequency, normally 170 Hz. The frequency shift on the HF bands is limited to a maximum of 1000 Hz. Direct frequency-shift keying of HF transmitters is possible using a diode and a variable capacitor or a voltage-variable-capacitance diode in an oscillator circuit. Some commercial transceivers have frequency-shift keyers of this type built in.

Audio-Frequency-Shift Keying

Above 50 MHz where A2B or F2B emissions are permitted, audio-frequency-shift keying (AFSK) is generally used. In this case the RF carrier stays on the air throughout the transmission and a modulating audio tone is shifted in frequency. U.S. amateurs traditionally have used an audio frequency of 2125 Hz for mark and a somewhat higher frequency for space. At one time, when 850-Hz shift was used by amateurs on HF, this same shift was used for VHF AFSK, making the space tone 2975 Hz. After the shift on HF was reduced to 170 Hz, the common VHF AFSK space frequency was changed to 2295 Hz.

AFSK of SSB Transmitters

Probably most modern RTTY stations simply feed AFSK tones to the microphone input of an SSB transmitter or transceiver. When properly designed and adjusted, this method of modulation, while technically J2B emission, cannot be distinguished from F1B emission on the air. The user should make certain that audio distortion, carrier and unwanted sidebands are not present to the degree of causing interference. The user should also make certain that *the equipment is capable of withstanding the 100-percent duty cycle* needed for the duration of an RTTY transmission. For safe operation, it is often necessary to reduce the transmitter power output to 25 to 50 percent of the power level that is safe for CW operation.

In the U.S. it has been customary to use the same modems for both VHF AFSK and HF via an SSB transmitter. This proved difficult when the 850-Hz shift tones of 2125 Hz mark and 2975 Hz space were used because of the narrow filtering used in most SSB equipment. So some amateurs, particularly those in IARU Region 1 (Europe and Africa), went to the "low tones" of 1275 Hz mark and 2125 Hz space to operate more in the center of the SSB filter passbands. When the shift was reduced to 170 Hz, 1275 Hz was retained mark, and 1445 Hz was used for space. Both high and low tones can be used interchangeably on the HF bands because only the amount of shift is important. The slight frequency difference is unnoticed on the air because each station tunes its transmitters and receivers for best results. On VHF AFSK, however, the high and low tone pairs are not compatible.

Because the convention on HF is that the *higher* frequency is the mark condition, and as the AFSK modem uses the *lower* frequency for mark, it is normal to use the *lower sideband* mode for RTTY on SSB radio equipment. If you want to tune to an exact RTTY frequency, remember that most SSB radio equipment will display the frequency of its (suppressed) carrier, not the frequency of the mark signal. For example, if you want to operate on 14,083 kHz and you are using a 2125-Hz AFSK mark frequency, your SSB radio (suppressed-carrier) frequency should be 14,083 + 2.125 = 14,085.125 kHz.

AFSK Modulators

The audio tone produced by the AFSK modulator must be nearly sinusoidal. A nonsinusoidal waveform contains harmonics of the fundamental frequency. If, for example, a nonsinusoidal (low) tone of 1275 Hz is used, the audio outputs will be 1275, 2550 and 3825 Hz. Any properly designed SSB transmitter can be expected to filter out 3825 Hz but may not be able to reduce 2550 Hz to an acceptable level, depending on the filter characteristics. Particularly when using the low tones, the harmonic distortion of the tones should be kept to a few percent.

Most modern AFSK generators are of the *continuous-phase (CPFSK)* type. Older types of noncoherent-FSK (NCFSK) generators had no provisions for phase continuity and produced sharp switching transients. The noise from phase discontinuity caused interference several kilohertz around the RTTY signal.

AFSK Demodulators

An AFSK demodulator takes the shifting tones from the audio output of a receiver and produces TTY keying pulses.

Many AFSK demodulators are of the FM type. In this type of demodulator, the signal is first sent through a band-pass filter to remove out-of-band interference and noise. It is then limited to remove amplitude variations. The signal is FM-demodulated (in a discriminator or a phase-locked loop or PLL). The output of the detector is run through a low-pass filter to remove noise at frequencies above the keying rate. The result is fed to a circuit that makes the ultimate decision between mark and space.

Other demodulator designs use AM (limiterless) detectors in place of a discriminator or PLL. These types of demodulators, when properly designed, permit continuous copy even when the mark or space frequency fades out completely. However, at 170-Hz shift, the mark and space frequencies tend to fade more at the same time than they do independently. In other words, when one fades chances are that the other will fade about the same time. For this reason, FM and AM types of demodulators are comparable at 170-Hz shift. However, at wider shifts (say 425 Hz and above), the independently fading mark and space can be used to achieve an *in-band frequency-diversity* effect if the demodulator is capable of processing it. To conserve spectrum, it is generally desirable to stay with 170-Hz shift for 45-baud Baudot and forego the possible in-band-frequency-diversity gain. However, the inband-frequency-diversity gain should be kept in mind for higher signaling rates that would justify greater shift.

Diversity Reception

Another type of diversity can be achieved by using two antennas, two receivers and a dual demodulator. This setup is not as far fetched as it may sound; some amateurs are using it with excellent results. One of the antennas would be the normal station antenna for that band. The second antenna could be either another antenna of the same polarization located at least 3/8-wavelength away, or an antenna of the opposite polarization located at the first antenna or anywhere nearby. A problem is to get both receivers on the same frequency without having to carefully tune both. While rare, some RTTY diversity enthusiasts have located slaved receivers on the surplus market. ICOM has produced the IC-7072 Transceiver Unit, which slaves an IC-720(A) transceiver to an IC-R70 receiver. Other methods could include a computer controlling two receivers so that both would track.

Two demodulators are needed for this type of diversity. Also, some type of diversity combiner or selector is needed. Many commercial or military RTTY demodulators are equipped for diversity reception.

The payoff for using diversity is a worthwhile improvement in copy. Depending on fading conditions, adding diversity may be equivalent to raising transmitter power sevenfold (8 dB).

Autostart

Mechanical teletypewriters cannot print until their motors are running at the proper speed. Rather than letting the motor run continuously, many Baudot RTTY stations employ an *autostart* circuit that senses the presence of a mark signal and turns on the motor.

Further Reading on Baudot RTTY

DATACOM, British Amateur Radio Teleprinter Group (BARTG), c/o Pat Beedie, GW6MOJ, "Ffynnonlas," Salem, Llandeilo, Dyfed, Wales SA19 6EW, published quarterly.

Hobbs, Yeomanson and Gee, *Teleprinter Handbook*, Radio Society of Great Britain, Alma House, Cranborne Rd., Potters Bar, Herts EN6 3JW, England, 1983.

Ingram, *RTTY Today*, Universal Electronics, Inc. 4555 Groves Rd., Suite 3, Columbus, OH 43232, 1984.

Kretzman, *The New RTTY Handbook*, Cowan Publishing Corp., 14 Vanderventer Ave., Port Washington, NY, 1962.

Nagle, "Diversity Reception: an Answer to High Frequency Signal Fading," *Ham Radio*, November 1979, pp. 48-55.

RTTY Journal, 9085 La Casita Ave., Fountain Valley, CA 92708

Schwartz, "An RTTY Primer," *CQ*, issues of August 1977, November 1977, February 1978, May 1978 and August 1978.

Tucker, *RTTY from A to Z*, Cowan Publishing Corp., 14 Vanderventer Ave., Port Washington, NY, 1970.

AMTOR

RTTY circuits are plagued with problems of fading and noise unless something is done to mitigate these effects. Frequency, polarization and space diversity are methods of providing two or more simultaneous versions of the transmission to compare at the receiving station. Another method of getting more than one opportunity to see a given transmission is time diversity. The same signal sent at different times will experience different fading and noise conditions. Time diversity is the basis of *AMTOR* or *Am*ateur *T*eleprinting *O*ver *R*adio.

AMTOR always uses two forms of time

Fig. 21 — W1AW AMTOR operating position.

diversity in either Mode A (ARQ or Automatic Repeat Request) or Mode B (FEC or Forward Error Correction). In Mode A, a repeat is sent only when requested by the receiving station. In Mode B, each character is sent twice. In both Mode A or Mode B, the second type of time diversity is supplied by the redundancy of the code itself.

This new type of RTTY was introduced to Amateur Radio by Peter Martinez, G3PLX, through several articles that appeared in RSGB's *Radio Communication*, *QST* and elsewhere. AMTOR was derived from a commercial system, SITOR, which was designed for use in the Maritime Mobile Service. It is used for ship-to-shore, ship-to-ship and between a ship and a subscriber of the (international) telex network. If you listen to shore station CW traffic lists, you will probably hear references to both TOR (SITOR) and TLX (telex).

Bill Meyn, K4PA, and a small group of U.S. amateurs operated AMTOR under a Special Temporary Authority (STA) that was granted by the FCC for tests that began in January 1982. His test report was published in the July 1983 *QST* Technical Correspondence column. Also in the same *QST* issue was "An Introduction to AMTOR," by Paul Newland, AD7I, one of the AMTOR experimenters under the STA. Additional references are listed at the end of the AMTOR part of this chapter.

As a result of a January 27, 1983 FCC order, AMTOR was made part of the U.S. Amateur Radio rules. The rule provided that "the code, baud rate and emission timing shall conform to the specifications of CCIR Recommendation 476-2 (1978) Mode A or B." In a later action, the FCC

also recognized Recommendation 476-3, a later document which corrected some inconsistencies in the earlier one. The latest (1986) CCIR documents are Rec. 476-4 and Rec. 625. Anyone interested in the design aspects of AMTOR should refer to these recommendations. You may obtain a complete reprint of Rec. 476-3 as part of the proceedings of the Third ARRL Amateur Radio Computer Networking Conference, available from ARRL Hq.

An Overview of AMTOR

In the Maritime Mobile Service, this type of system was devised as a means of improving communications between 5-unit, asynchronous teleprinters using the ITA2 (Baudot) code. The system converts the 5-unit code to a 7-unit code for transmission. There is a one-for-one correspondence between the 5- and 7-unit codes for all 32 combinations of the 5-unit code, as can be seen in Table 4. Ordinarily, a 7-unit code could have up to 2^7 or 128 possible combinations. In this case, there is a constant ratio of four marks (Bs or 1s) to three spaces (Ys or 0s). If a received character does not have this constant ratio, the receiving station knows it is erroneous. The constant ratio limits the number of usable combinations to 35. So 32 of the combinations equate to the 32 ITA2 combinations, and there are 3 left over for use as *service information signals* as shown in Table 5. These three unique combinations are Idle Signal Beta, Idle Signal Alpha and *RQ* or *Repeat Request*. In addition to these three unique service information signals, three others (CS1, CS2 and CS3) are not unique but are borrowed from the 32 combinations that equate to ITA2. They are not confused with the message characters because they are sent only in the reverse direction.

Mode A (ARQ)

This is a synchronous system, transmitting blocks of three characters from the *Information Sending Station (ISS)* to the *Information Receiving Station (IRS)*.

The station that initiates the QSO is known as the *Master Station (MS)*. The MS first sends the selective call of the called station in blocks of three characters, listening between blocks. Four-letter calls are used in AMTOR and are normally derived from the first character and the last three letters of the station call sign. As an example, for a call sign of W1AW the AMTOR call would be WWAW. The *Slave Station (SS)* recognizes its selective call and answers that it is ready. The MS now becomes the Information Sending Station and will send traffic as soon as the Information Receiving Station says it is ready.

After contact is established, the ISS sends its message in groups of three characters and pauses between groups for a reply from the IRS. Each character is sent at a rate of 100 bauds, amounting to 70 ms for one character or 210 ms for a three-character block. The block repetition cycle is every 450 ms, so there are 240 ms during each cycle that the ISS is not sending. This 240-ms period is taken up by propagation time from the ISS to the IRS, 70 ms for the IRS to send its service information signal, and the return trip back to the ISS. Turnaround or switching time associated with the modem and the radio transmitter and receiver (known as *equipment delay*, or t_E in CCIR Rec. 476-3) must occur within this 240-ms period. A receive-transmit turnaround time of less than 20 ms is desirable.

When an ISS is done sending, it can enable the other station to become the ISS by sending the three-character sequence FIGS Z B. A station may end the contact by sending an "end of communication signal" consisting of three Idle Signal Alphas.

On the air, AMTOR Mode A signals have a characteristic chirp-chirp sound to them. Because of the 210/240-ms on/off timing, Mode A can be used with some transmitters at full power levels.

Mode B (FEC)

When transmitting to no particular station (for example calling CQ, net or

bulletins) there is no (one) station to act as IRS, thus no one to send in the reverse direction. Even if there were one, his or her ability to receive properly may not be representative of others desiring to copy the signal.

Mode B uses a simple forward-error-control (FEC) technique of sending each character twice. If the repetition were sent immediately, a single noise burst or quick fade could mutilate both. Burst errors can be virtually eliminated by delaying the repetition for a period thought to exceed the duration of most noise bursts. In AMTOR, the first transmission (DX) of a character is followed by four other characters, after which the retransmission (RX) of the first character occurs. At 70 ms per character, this leaves 280 ms between the end of the first transmission and the beginning of the second. From the start of the first to the start of the second, the duration is 350 ms.

In Mode B, the receiving station tests for the constant 4-mark/3-space ratio and prints only unmutilated DX or RX characters, or prints an error symbol or space, if both are mutilated.

The sending station's transmitter must be capable of 100%-duty-cycle operation for Mode B. Thus it may be necessary to reduce transmitter power level to 25% to 50% of full rating.

AMTOR Operation

Hundreds of U.S. and overseas stations are now active on AMTOR. It no doubt will grow in popularity as soon as amateurs realize the virtually error-free operation that it offers.

The frequency shift is 170 Hz for both Maritime Mobile and amateur use of CCIR Rec. 476-3. Existing 170-Hz-shift (Baudot RTTY) modems may be used for AMTOR. However, some demodulators designed for 45- or 50-baud Baudot RTTY service have narrow filter bandwidths and may require modification before they are suitable for 100-baud AMTOR operation. Most modem manufacturers will have modification instructions or kits available. If you have built a modem according to someone else's design, an s.a.s.e. to the designer probably will bring you the modification information.

Some of the commercially built AMTOR converters have built-in modems (for example, the I.C.S. Electronics AMT-1 uses the low-tone pair of 1275 Hz and 1445 Hz).

AMT-1 AMTOR converters sold by AEA for marketing in the U.S. provide the high-tone pair of 2125 Hz and 2295 Hz. One commercial unit (the Hal Communications Corp. ARQ1000) has an optional 170-Hz-shift modem with a 1700-Hz center frequency as specified in CCIR Recommendation 476-3. Other manufacturers of AMTOR equipment are Digital Electronic Systems, Inc, Kantronics and Microlog. Most AMTOR activity is around 14,075 and 3637.5 kHz, which are used as calling frequencies. Operators usually move off the calling frequency to free it for others. However, there are some automatically controlled stations that just stay on these frequencies. At this writing, unattended automatic control of an AMTOR station in the U.S. requires an FCC Special Temporary Authority.

You can listen for W1AW AMTOR Mode B transmissions following the Baudot and ASCII bulletins. Check *QST* for the W1AW schedule.

Further Reading on AMTOR

CCIR, Recommendation 476-3, "Direct-Printing Telegraph Equipment in the Maritime Mobile Service." *Reprint available from ARRL Hq. as part of the proceedings of the Third ARRL Amateur Radio Computer Networking Conference.*

DATACOM, British Amateur Radio Teleprinter Group (BARTG, c/o Pat Beedie, GW6MOJ, "Ffynnonlas," Salem, Llandeilo, Dyfed, Wales SA19 7NP, published quarterly.

Martinez, "Amtor, An Improved RTTY System Using a Microprocessor," *Radio Communication,* RSGB, August 1979.

Martinez, "Amtor, The Easy Way," *Radio Communication,* RSGB, June/July 1980.

Martinez, "Amtor — a Progress Report," *Radio Communication,* RSGB, September 1981, p. 813.

Meyn, "Operating with AMTOR," Technical Correspondence, *QST,* July 1983, pp. 40-41.

Newland, "An Introduction to AMTOR," *QST,* July 1983.

Newland, "Z-AMTOR: An Advanced AMTOR Code Converter," *QST,* February 1984. *Summarized in Chapter 29 of this Handbook.*

Newland, "A User's Guide to AMTOR Operation," *QST,* October 1985.

American National Standard Code for Information Interchange (ASCII)

The American National Standard Code for Information Interchange (ASCII) is a coded character set used for information-processing systems, communications systems and related equipment. Current FCC regulations provide that amateur use of ASCII shall conform to ASCII as defined in American National Standards Institute (ANSI) Standard X3.4-1968. ANSI has issued a revision X3.4-1977, which differs only in terminology and elimination of two choices for certain graphics. Its international counterparts are International Organization for Standardization Standard ISO 646-1973 and International Alphabet No. 5 (IA5) as specified in CCITT Recommendation V.3.

ASCII uses seven bits to represent letters, figures, symbols and control characters. Unlike ITA2 (Baudot), ASCII has both upper- and lower-case letters. The ASCII coded character set is shown in Table 7. The bits in the table are arranged according to standard binary representation (b_6 through b_0). In the international counterpart codes, £ usually replaces #, and $ may be replaced by the international currency sign ¤.

Some early video display terminals and teleprinters, such as the Teletype Corp. Model 33, implemented only the upper-case letters. They usually display the upper-case letters when the lower-case letters are received. On newer terminal equipment, a CAPS-LOCK feature on the keyboard may be used to shift all letters to upper case.

Control Characters

ASCII has 32 *control characters* plus the special characters for space and delete. They are not consistently used as specified in the ANSI X3.4 standard. However, it will be helpful to know the standard uses. There are five groups in the control set: Logical Communication Control, Physical Communication, Device Control, Information Separators and Code Extension. Below is an explanation of each control character. The first line of each entry follows the following format:

Table 7
The ASCII Coded Character Set

Bit Number					6 5 4	0 0 0	0 0 1	0 1 0	0 1 1	1 0 0	1 0 1	1 1 0	1 1 1
3	2	1	0	Hex 1st / 2nd		0	1	2	3	4	5	6	7
0	0	0	0	0		NUL	DLE	SP	0	@	P	`	p
0	0	0	1	1		SOH	DC1	!	1	A	Q	a	q
0	0	1	0	2		STX	DC2	"	2	B	R	b	r
0	0	1	1	3		ETX	DC3	#	3	C	S	c	s
0	1	0	0	4		EOT	DC4	$	4	D	T	d	t
0	1	0	1	5		ENQ	NAK	%	5	E	U	e	u
0	1	1	0	6		ACK	SYN	&	6	F	V	f	v
0	1	1	1	7		BEL	ETB	'	7	G	W	g	w
1	0	0	0	8		BS	CAN	(	8	H	X	h	x
1	0	0	1	9		HT	EM	)	9	I	Y	i	y
1	0	1	0	A		LF	SUB	*	:	J	Z	j	z
1	0	1	1	B		VT	ESC	+	;	K	[	k	{
1	1	0	0	C		FF	FS	,	<	L	\	l		
1	1	0	1	D		CR	GS	−	=	M	]	m	}
1	1	1	0	E		SO	RS	.	>	N	^	n	~
1	1	1	1	F		SI	US	/	?	O	_	o	DEL

ACK	= acknowledge	FF	= form feed
BEL	= bell	FS	= file separator
BS	= backspace	GS	= group separator
CAN	= cancel	HT	= horizontal tab
CR	= carriage return	LF	= line feed
DC1	= device control 1	NAK	= negative acknowledge
DC2	= device control 2	NUL	= null
DC3	= device control 3	RS	= record separator
DC4	= device control 4	SI	= shift in
DEL	= (delete)	SO	= shift out
DLE	= data link escape	SOH	= start of heading
ENQ	= enquiry	SP	= space
EM	= end of medium	STX	= start of text
EOT	= end of transmission	SUB	= substitute
ESC	= escape	SYN	= synchronous idle
ETB	= end of block	US	= unit separator
ETX	= end of text	VT	= vertical tab

Notes
1. "1" = mark, "0" = space.
2. Bit 6 is the most-significant bit (MSB), bit 0 the least-significant bit (LSB).

ASCII Char.	Keyboard Character	Dec	Hex	Binary 6543210	Name	Graphic Symbol

Logical Communication Control Group

SOH Control A 1 01 0000001 Start of Heading ⌐
Used at the beginning of a message heading.

STX Control B 2 02 0000010 Start of Text ⊥
Used at the beginning of a message heading.

ETX Control C 3 03 0000011 End of Text ⌐|
Terminates text.

EOT Control D 4 04 0000100 End of Transmission ⤡
Indicates conclusion of a transmission, which may have contained one or more texts and any associated headings. It may be used to turn off the receiving device.

ENQ Control E 5 05 0000101 Enquiry ⊠
Requests a response from a remote station. The response may be the remote station's station of identification (Who are you?).

ACK Control F 6 06 0000110 Acknowledge ✓
Transmitted by a receiver as an affirmative response to a sender.

DLE Control P 16 10 0010000 Data Link Escape ⊟
Changes the meaning of a limited number of continuously following characters. It is used exclusively to provide supplementary controls in data-communication networks. DLE is normally terminated by a shift in (SI) character.

NAK Control U 21 15 0010101 Negative Acknowledgment ⤨
Transmitted by a receiver as a negative response to a sender.

SYN Control V 22 16 0010110 Synchronous Idle ⎍
Used by a synchronous transmission system in the absence of any other character to provide a signal from which synchronization may be achieved or retained.

ETB Control W 23 17 0010111 End of Transmission Block ⊣
Used to indicate the end of a block of data for communication purposes.

Physical Communication Group

NUL Control @ 0 00 0000000 Null ☐
Media or time fill. It may be inserted into, or removed from, a stream of data without affecting the information content of that stream.

CAN Control X 24 18 0011000 Cancel ⌧
Indicates that the data with which it is sent is in error or is to be disregarded. It usually means cancel the line (data back to the last CR).

EM Control Y 25 19 0011001 End of Medium ⬧
Identifies the physical end of the medium, or the end of the used or wanted portion of information recorded.

SUB Control Z 26 1A 0011010 Substitute ⸮
A substitute for another character in the same position to keep data fields correct.

Device Control Group

BEL Control G 7 07 000111 Bell ⌓
Used to call for operator attention. It may trigger an alarm or other attention devices.

BS Control H 8 08 0001000 Backspace ◄
A format effector that controls the movement of the printing position to one space backward on the same printing line. It is used for things such as: underlining, restriking for boldness and forming composite characters by over-striking. It is a very unpredictable character, that is implemented differently on printers and video-display terminals.

HT Control I 9 09 0001101 Horizontal Tab →
A format effector that controls the movement of the printing position to the next in a series of predetermined positions along the same printing line.

LF Control J 10 0A 0001010 Line Feed ≡
A format effector that controls the movement of the printing position to the next printing line.

VT Control K 11 0B 0001011 Vertical Tab ↓
A format effector that controls the movement of the printing position to the next in a series of predetermined printing lines.

FF Control L 12 0C 0001100 Form Feed ⤓
A format effector that controls the movement of the printing position to the first predetermined printing line on the next form or page.

CR Control M 13 0D 0001101 Carriage Return ◄—
A format effector that controls the movement of the printing position to the first printing position on the same printing line. In some cases, the CR will incorporate an automatic line feed (LF) as either a default or an option. This feature, called New Line (NL), is used to make computer keyboards more like those of typewriters.

DC1	Control Q	17	11	0010001	Device Control 1	
DC2	Control R	18	12	0010010	Device Control 2	
DC3	Control S	19	12	0010011	Device Control 3	
DC4	Control T	20	14	0010100	Device Control 4	

These are the control of ancillary devices associated with data processing or telecommunications systems, especially for switching devices on or off. While DC1 and DC3 are often used to start and stop the flow of data, these control signals are "wild cards" that are used for nearly any switching purposes.

Information Separators

FS	Control \	28	1C	0011100	File Separator	
GS	Control]	29	1D	0011101	Group Separator	
RS	Control ∧	30	1E	0011110	Record Separator	
US	Control —	31	1F	0011111	Unit Separator	

FS, GS, RS and US is a hierarchical group of information separators used for formatting and string processing. FS is the most inclusive, and US the least inclusive. A *file* is a grouping of all related records. A *group* is a collection of *records* within a file. A record is a collection of related fields of data. A *unit* is a number of fields 1 to 80 characters long.

Code Extension Group

SO   Control N   14   0E   0001110   Shift Out
Indicates that the code combinations that follow are to be interpreted as outside of the standard character set until a shift in (SI) character is reached.

SI   Control O   15   0F   0001111   Shift In
Indicates that the code combinations that follow are to be interpreted according to the standard code table.

ESC   ESC 7   1B   0011011   Escape
Intended to provide code extension (supplementary characters). The ESC character itself is a prefix affecting the interpretation of a limited number of continuously following characters. It may be terminated by a shift in (SI) character.

Parity

While not strictly a part of ASCII (ANSI X3.4), an eighth bit (P) may be added for *parity* checking. FCC rules permit optional use of the parity bit. The applicable U.S. standard (ANSI X3.16-1976) and international standard (CCITT Rec. V.4) recommend an *even* parity sense for asynchronous data communication and *odd* parity sense for synchronous data communication. However, there is very little standardization within the computer industry, and you will find five variations: (1) no parity, (2) always mark, (3) always space, (4) odd parity or (5) even parity. See Chapter 8 for details of parity checking.

Code Extensions

By sacrificing parity, the eighth bit can be used to extend the ASCII 128-character code to one having 256 possibilities. Work is now underway to produce an international standard on extended coded character sets for text communication that will provide supplementary sets of graphic characters. Use of these extended character sets by U.S. amateurs is permitted above 50 MHz but not below 50 MHz where only specified codes (Baudot, AMTOR and ASCII) are authorized by FCC rules.

ASCII Serial Transmission

Serial transmission of ASCII is covered in ANSI Standards X3.15 and X3.16, and CCITT Recommendations V.4 and X.4. They specify that the bit sequence for serial transmission of an ASCII character shall be least-significant bit first to most-significant bit or b_0 through b_6 (plus the parity bit P if used).

Serial transmission may be either *synchronous* or *asynchronous*. In synchronous transmissions, only the information bits (and optional parity bit) are sent, as shown in Fig. 22A.

Asynchronous (start-stop) serial transmission adds a *start pulse* at the beginning and a *stop pulse* at the end of each character. The start pulse is the same length as for the information pulses. The stop pulse may be one or two bit periods long. Although there is some variation, the convention is that the stop pulse is one bit period long except for 110-baud transmissions where mechanical teletypewriters may be involved.

ASCII Data Rates

Personal computers usually provide a choice of ASCII serial input/output data rates ranging from 110 to 19,200 bits per second, as shown in Table 8.

Data communications signaling rates depend largely on the medium and the state of the art when the equipment was selected. Numerous national and international standards that recommend different data rates, are listed in Table 9. The most-used rates tend to progress in 2:1 steps from 300 to 9600 bits per second and in 8000 bits-per-second increments from 16,000 bits per second. For Amateur Radio serial ASCII transmissions the data rates of 75, 110, 150, 300, 600, 1200, 2400, 4800, 9600, 16,000,

Table 8
ASCII Asynchronous Signaling Rates

Bits per Second	Data Pulse (ms)	Stop Pulse (ms)	CPS	WPM
110	9.091	18.182	10.0	100
150	6.667	6.667	15.0	150
300	3.333	3.333	30.0	300
600	1.667	1.667	60.0	600
1200	0.8333	0.8333	120	1200
2400	0.4167	0.4167	240	2400
4800	0.2083	0.2083	480	4800
9600	0.1041	0.1041	960	9600
19200	0.0520	0.0520	1920	19200

CPS = characters per second

$$= \frac{1}{START + 7 (DATA) = PARITY + STOP}$$

$$WPM = \text{words per minute} = \frac{CPS}{6} \times 60$$

= number of 5-letter-plus-space groups per minute

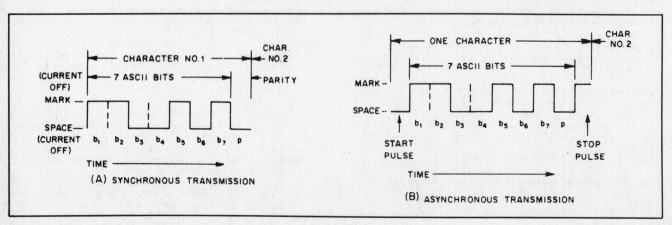

Fig. 22 — Time sequence of typical ASCII   character, the letter S sent serially.

19,200 and 56,000 bits per second are suggested.

Bauds vs. Bits Per Second

The term *baud* is a unit of signaling speed equal to one discrete condition or event per second. In single-channel transmission, such as the FCC prescribes for Baudot transmissions, the signaling rate in bauds equals the data rate in bits per second. However, the FCC does not limit ASCII to single-channel transmission. There are digital modulation systems that have more than two (mark and space) states. In *dibit* modulation two ASCII bits are sampled at a time. The four possible states for a dibit are 00, 01, 10 and 11. In four-phase modulation, each state is assigned an individual phase of 0, 90, 180 and 270 degrees, respectively. Using dibit phase modulation, the signaling speed in bauds is half the information-transfer rate in bits per second. As the FCC specifies the digital sending speed in bauds, amateurs may transmit ASCII at higher information rates by using digital modulation systems that encode more bits per signaling element. This technology is open for exploration by Amateur Radio experimenters.

Amateur ASCII RTTY Operations

On April 17, 1980 the FCC first permitted ASCII in the Amateur Radio Service. U.S. amateurs have been slow to abandon Baudot in favor of asynchronous serial ASCII.

Some of the reluctance to use it arises from the reasoning that asynchronous ASCII has two (or three with a parity bit added) more bits than asynchronous Baudot and is usually sent at higher speeds. Thus, it was felt that the greater data rates and increased bandwidth needed for ASCII would make its reliability lower than that of Baudot. This is true as far as it goes but does not exhaust the theoretical possibilities, which will be discussed below.

On the practical side, some amateurs tried ASCII on the air and experienced poor results. In some cases, this can be traced to the use of modems that were optimized for 45-baud operation. At 110 or 300 bauds, the 45-baud mark and space filters are too narrow. On the HF bands, speeds above 50 or 75 bauds are subject to intersymbol interference (ISI) (slurring one pulse into the next) from multipath propagation. Multiple paths can be avoided by operating at the maximum usable frequency (MUF) where there is only one ray path. The amount of multipath delay varies according to operating frequency with respect to the MUF and path distance. Paths in the 600- to 5000-mile range are generally less bothered by multipath than shorter or longer ones, with paths of 250 miles or less being difficult from a multipath standpoint. As a result, successful operation at the higher ASCII speeds depends on using the highest frequency band possible as well as having

Table 9
Data Transmission Signaling-Rate Standards

Standard	Signaling Rates (bit/s)	Tolerance
CCITT		
V.5	600, 1200, 2400, 4800	± 0.01%
V.6	Preferred: 600, 1200, 2400, 3600, 4800, 7200, 9600	± 0.01%
	Supplementary: 1800, 3000, 4200, 5400, 6000, 6600, 7800, 8400, 9000, 10200, 10800	
V.21	200	≤ 200 bit/s
	300 (where possible)	≤ 300 bit/s
V.23	600	≤ 600 bit/s
	1200	≤ 1200 bit/s
	75 (backward channel)	≤ 75 bit/s
V.35	Preferred: 48000	± 1 bit/s
	When necessary: 40800	± 1 bit/s
V.36	Recommended for international use: 48,000	
	Certain applications: 56000, 64000, 72000	
X.3	Packet assembly/disassembly speeds: 50, 75, 100, 110, 134.5, 150, 200, 300, 600, 1200, 1200/75, 1800, 2400, 4800, 9600, 19200, 48000, 56000, 64000	
ANSI		
X3.1	Serial: 75, 150, 300, 600, 1200, 2400, 4800, 7200, 9600	
	Parallel: 75, 150, 300, 600, 900, 1200	
X3.36	Above 9600 bit/s, signaling rates shall be in integral multiples of 8000 bit/s.	
	Selected standard rates: 16000, 56000, 1344000 and 1544000	
	Recognized for international use: 48000	
EIA		
RS-269-B	(Same as ANSI X3.1)	
FED STD		
-1001	(Same as ANSI X3.36) For foreign communications: 64000	
-1041	2400, 4800, 9600	

suitable modems at both ends of the circuit.

Returning to the theoretical comparison of Baudot and ASCII, recall that the FCC requires asynchronous (start-stop) transmission of Baudot. This means that the five information pulses must be sent with a start pulse and a stop pulse, usually of 1.42 times the length of the information pulse. Thus, an asynchronous Baudot transmitted character requires 7.42 units. In contrast, 7 bits of ASCII plus a parity bit, a start and a two-unit stop pulse has 11 units. However, it is possible to send only the 7 ASCII information bits synchronously (without start and stop pulses), making the number of units that must be transmitted (7 vs. 7.42) slightly smaller for ASCII than for Baudot. Or, it is possible to synchronously transmit 8 bits (7 ASCII bits plus a parity bit) and take advantage of the error-detection capability of parity. Also, there is nothing to prevent ASCII from being sent at a lower speed such as 50 or 75 bauds, to make it as immune to multipath as 45- or 50-baud Baudot RTTY is. So it is easy to see that ASCII can be as reliable as Baudot RTTY if care is used in system design.

While 45- or 50-baud RTTY circuits can

provide reliable communications, this range of signaling speeds does not make full use of the HF medium. Speeds ranging from 75 to 1200 bauds can be achieved on HF with error-detection and error-correction techniques similar to those used in AMTOR. Reliable transmission at these higher speeds can be accomplished by means of packet radio, which is described later in this chapter.

Bibliography on ASCII

ANSI X3.4-1977, "Code for information interchange," American National Standards Institute, 1430 Broadway, New York, NY 10018.

ANSI X3.15-1976, "Bit sequencing of the American national standard code for information interchange in serial-by-bit data transmission."

ANSI X3.16-1976, "Character structure and character parity sense for serial-by-bit data communication information interchange."

ANSI X3.25-1976, "Character structure and character parity sense for parallel-by-bit communication in American national standard code for information interchange."

Bemer, "Inside ASCII," *Interface Age*, May, June and July 1978.

CCITT V.3, "International Alphabet No. 5," International Telegraph and Telephone Consultative Committee, CCITT volumes with recommendations prefixed with the letters V and X are available from United Nations Bookstore, Room 32B, UN General Assembly Building, New York, NY 10017 or from International Telecommunication Union, General Secretariat, Sales Service, Place des Nations, CH 1211, Geneva 20, Switzerland.

CCITT V.4, "General structure of signals of International Alphabet No. 5 code for data transmission over the public telephone network."

ISO 646-1973 (E), "7-bit coded character set for information processing interchange," International Organization for Standardization, available from ANSI, 1430 Broadway, New York, NY 10018.

Mackenzie, *Coded Character Sets, History and Development*, Addison-Wesley Publishing Co., 1980.

Code Conversion

Conversion between International Morse, Baudot and ASCII is tricky business. In fact, it is virtually impossible to come up with a perfect conversion between any two of these codes, let alone all three. Anyone who has given this serious consideration will confirm this.

The root of the problem is that each code has a somewhat different character set. All three have all the numerals (0-9), a word space, certain punctuation marks and the Latin alphabet upper case in common. If you have original text with upper- and lower-case letters in ASCII, the case information will be lost when converting to Baudot or Morse, both of which assume that all letters are in upper case. Once lost, the case information cannot be regained except, perhaps, by a computer program designed to guess the case.

Another problem with code conversion is that the conversion tables cannot be bilateral. Using the case example, above, both upper- and lower-case ASCII will translate to upper-case Baudot and Morse, but Baudot and Morse letters cannot translate to both ASCII cases. There are some subtleties to this game. Thus, if you want to design a versatile code-conversion scheme, it is useful to build unilateral conversion tables, one for translating code a to b, another for converting code b to a.

The differences between the international and national implementations of Baudot add some further complications. On most U.S. teletypewriters, the figures-case characters for D, F, G and H are used for $, !, & and # (or motor stop), respectively. In ITA2, these characters are unassigned, meaning they are set aside for national use. So, you can't use these characters for international communications unless you have an agreement with the other station as to what each means. Perhaps even worse is that the figures-case characters for J, S, V and Z, while specified to be audible signal (bell), ', = and +, respectively, for ITA2, show up as ', bell, ; and " on many U.S. TTYs. You can use these characters internationally as long as you bear in mind that hams in other countries will probably interpret them within their ITA2 meanings.

There are some more complications when considering conversion to and from ASCII or its international versions, IA5 and ISO 646. The ASCII characters #, $, @, [, \,], ^ , `, `, {, |, } and ~ may be assigned to different graphic characters in other countries. Some languages need these wild cards for accented letters.

Morse code presents a few dilemmas too. One is that few hams know the complete Morse character set shown in Table 1. To limit the character set to just those known by most amateurs might be advisable for QSOs that will be aurally read by the average operator. However, it could be unnecessarily restrictive when sending to the expert operator or to a computer that knows the whole code. A code-conversion program that produces Morse might include a menu-selectable feature to vary the conversion table according to the receiving audience.

Code Conversion Standards

International standards, ISO/DIS 6936 and CCITT Rec. S.18, exist for conversion from ITA2 to ASCII and the reverse. They include direct conversion for all common characters and change both upper- and lower-case ASCII into upper-case ITA2 in a straightforward manner. When translating from ITA2 to ASCII, ITA2 characters for which there are no corresponding ASCII characters are translated to SUB. Going the other way around, SOH, STX, ETX, EOT, ACK, DLE, NAK, SYN, ETB and DEL are removed; the other nontranslatable characters are replaced by the question mark (?). Bear in mind that ISO/DIS 6936 is based on ITA2, not the version of Baudot that is found on many U.S. teletypewriters.

Limiting code conversion simply to direct equivalents may not be the best approach for Amateur Radio. In this respect, it is important to note that some characters (in Morse, Baudot and ASCII) change their meanings according to the context, or protocol, in which they are used. For example, the Morse code combination •-•- is an accented letter A for certain European languages but is also used as a separator in Amateur Radio message headings. The meanings of ASCII control characters change as different implementors of computer systems or data-communications networks reassign them to meet specific needs. What is ETX (end of text) to one system is Control C (crash the program) to another.

To go beyond the direct equivalents requires that the context(s) be carefully defined and standardized. In other words, if you know how a certain character (or sequence of characters) is used in one code using one protocol, there must be a corresponding character or set of characters in the other code and protocol.

Now that you're completely confused, let's look at how to say that a message is coming. ASCII has its SOH (start of heading) — just one control character can do the job. In Baudot it's a problem, but not insurmountable. CCITT Rec. S.4 (see Chapter 38) comes to the rescue by using ZCZC as the start of the heading. And in Morse — ah Morse! — we have two possibilities: HR (meaning here is, or hear the following message) and $\overline{KA}$, a starting signal. There are similar correspondences between ASCII EOT, Baudot (CCITT Rec. S.4) NNNN, and Morse $\overline{AR}$.

These examples illustrate the practicality of developing contextual code-conversion tables for Amateur Radio. This is not to suggest that everyone rush out there and implement one on the assumption that any right-thinking person would go along with your conversion table. Beware of the mines, quicksand, traps and low doorways! How many programmers made up their own ASCII character for the Morse code combination -•--• ($\overline{KN}$) when left parenthesis [(] was part of International Morse? Many other radio services, even other IARU Member Societies, have been using -•--• for left parenthesis all along.

The ARRL Digital Communications Committee will be working on standardized code-conversion tables for use by amateurs. This will dovetail with other work on standardizing message formats.

CODE-CONVERSION TABLES

Tables 10 through 14 are based on direct character-for-character translations between Morse, Baudot and ASCII. They are conservative and do not conflict with international standards. Conversions between ITA2 and ASCII comply with ISO/DIS 6936. The tables have numerous blanks where there are no exact equivalents in the other codes. The order of characters for

Table 10
Conversion from Morse to ASCII and Baudot

Int'l Morse Code	Char.	ASCII Code 6543210	Char.	Baudot Code ↓43210	Char.²
·	E	1000101	E	↓00001	E
—	T	1010100	T	↑10000	T
··	I	1001001	I	↓00110	I
·—	A	1000001	A	↓00011	A
—·	N	1001110	N	↓01100	N
——	M	1001101	M	↓11100	M
···	S	1010011	S	↓00101	S
··—	U	1010101	U	↓00111	U
·—·	R	1010010	R	↓01010	R
·——	W	1010111	W	↓10011	W
—··	D	1000100	D	↓01001	D
—·—	K	1001011	K	↓01111	K
——·	G	1000111	G	↓11010	G
———	O	1001111	O	↓11000	O
····	H	1001000	H	↓10100	H
···—	V	1010110	V	↓11110	V
··—·	F	1000110	F	↓01101	F
··——	Ü				
·—··	L	1001100	L	↓10010	L
·—·—	Ä				
·——·	P	1010000	P	↓10110	P
·———	J	1001010	J	↓01011	J
—···	B	1000010	B	↓11001	B
—··—	X	1011000	X	↓11101	X
—·—·	C	1000011	C	↓01110	C
—·——	Y	1011001	Y	↓10101	Y
——··	Z	1011010	Z	↓10001	Z
——·—	Q	1010001	Q	↓10111	Q
———·	Ö				
————	CH				
·····	5	0110101	5	↑10000	5
····—	4	0110100	4	↑01010	4
···——	S̄N				
···—·	3 / ÉÈẸ	0110011	3	↑00001	3
··—··	ĪNT				
··—·—					
··——·	2	0110010	2	↑10011	2
·—·—·	ĀS				
·—··—	ĀU				
·——·—	ĀR +	0101011	+	↑10001	+ (ITA)
·—·—·—	ÁÀÂ				
·————	1	0110001	1	↑10111	1
—····	6	0110110	6	↑10101	6
—···—	B̄T =	0111101	=	↑11110	= (ITA)
—··—·	D̄N /	0101111	/	↑11101	/
—·—··	Ç				
—·—··	K̄A				
—·——·	K̄N (	0101000	(	↑01111	(
——···	7	0110111	7	↑00111	7
——··—	Ź / Ñ				
———··	8	0111000	8	↑00110	8
————·	9	0111001	9	↑11000	9
—————	0	0110000	0	↑10110	0
····——	H̄M				
···—·—	S̄K				
··—··—	ĪX				
··——··	ĪMI ?	0111111	?	↑11001	?
··——·—	ĪQ _	10111*1	_		
·—··—·	ĀF "	0100010	"	↑10001	" (U.S.A.)
·—·—··	ĀL ¶				
·—·—·—	ĀAA .	0101110	.	↑11100	.
·————·	W̄G '	0100111	'	↑00101 / ↑01011	' (ITA) / ' (U.S.)
—····—	D̄U -	0101101	-	↑00011	.
—·—·—·	K̄R ;	0111011	;	↑11110	; (U.S.)
—·——·—	K̄K)	0101001	)	↑10010	)
——··——	M̄IM ,	0101100	,	↑01100	,
———···	ŌS :	0111010	:	↑01110	:
···—··—	S̄X $	0100100	$	↑01001	$ (U.S.)
········	Error				

Notes

1 In Baudot code, it is necessary to check to see if the terminal is in the correct case (↓ = letters, ↑ = figures). If not, a LTRS or FIGS must be inserted before the code shown.

2 Figures-case characters are the same for both ITA2 and U.S. teletypewriters except where indicated. If only (U.S.) is shown, this is an unassigned character in ITA2.

Table 11
Conversion from Baudot to ASCII and Morse

Baudot Code ¹ 43210	Char.²	ASCII Code 6543210	Char.	Int'l Morse Code	Char.
↓↑00000	Blank	0000000	NUL		
↓ 00001	E	1000101	E	·	E
↑ 00001	3	0110011	3	···——	3
↓↑00010	LF	0001010	LF		
↓ 00011	A	1000001	A	·—	A
↑ 00011	—	0101101	—	—····—	—
↓↑00100	SP	0100000	SP		space
↓ 00101	S	1010011	S	···	S
↑ 00101	' (ITA)	0100111	'	·————·	'
↑ 00101	Bell (U.S.)	0000111	BEL		
↓ 00110	I	1001001	I	··	I
↑ 00110	8	0111000	8	———··	8
↓ 00111	U	1010101	U	··—	U
↑ 00111	7	0110111	7	——···	7
↓↑01000	CR	0001101	CR		
↓ 01001	D	1000100	D	—··	D
↑ 01001	WRU (ITA)	0000101	ENQ		
↑ 01001	$ (U.S.)	0100100	$	···—··—	$
↓ 01010	R	1010010	R	·—·	R
↑ 01010	4	0110100	4	····—	4
↓ 01011	J	1001010	J	·———	J
↑ 01011	Bell (ITA)	0000111	BEL		
↑ 01011	' (U.S.)	0100111	'	·————·	'
↓ 01100	N	1001110	N	—·	N
↑ 01100	,	0101100	,	——··——	,
↓ 01101	F	1000110	F	··—·	F
↑ 01101	! (U.S.)	0100001	!		
↓ 01110	C	1000011	C	—·—·	C
↑ 01110	:	0111010	:	———···	:
↓ 01111	K	1001011	K	—·—	K
↑ 01111	(	0101000	(	—·——·	K̄N (
↓ 10000	T	1010100	T	—	T
↑ 10000	5	0110101	5	·····	5
↓ 10001	Z	1011010	Z	——··	Z
↑ 10001	+ (ITA)	0101011	+	·—·—·	ĀR +
↑ 10001	" (U.S.)	0100010	"	·—··—·	ĀF "
↓ 10010	L	1001100	L	·—··	L
↑ 10010	)	0101001	)	—·——·—	K̄K)
↓ 10011	W	1010111	W	·——	W
↑ 10011	2	0110010	2	··———	2
↓ 10100	H	1001000	H	····	H
↑ 10100	# (U.S.)	0100011	#		
↓ 10101	Y	1011001	Y	—·——	Y
↑ 10101	6	0110110	6	—····	6
↓ 10110	P	1010000	P	·——·	P
↑ 10110	0	0110000	0	—————	0
↓ 10111	Q	1010001	Q	——·—	Q
↑ 10111	1	0110001	1	·————	1
↓ 11000	O	1001111	O	———	O
↑ 11000	9	0111001	9	————·	9
↓ 11001	B	1000010	B	—···	B
↑ 11001	?	0111111	?	··——··	ĪMI ?
↓ 11010	G	1000111	G	——·	G
↑ 11010	& (U.S.)	0100110	&		
↓↑11011	FIGS				
↓ 11100	M	1001101	M	——	M
↑ 11100	.	0101110	.	·—·—·—	ĀAA .
↓ 11101	X	1011000	X	—··—	X
↑ 11101	/	0101111	/	—··—·	D̄N /
↓ 11110	V	1010110	V	···—	V
↑ 11110	= (ITA)	0111101	=	—···—	B̄T =
↑ 11110	; (U.S.)	0111011	;	—·—·—·	K̄R ;
↓↑11111	LTRS				

Notes

1 In Baudot code, it is necessary to check to see what the current case is before conversion (↓ = letters, ↑ = figures, ↓↑ = either).

2 Figures-case characters are the same for both ITA2 and U.S. teletypewriters except where indicated. If only (U.S.) is shown, this is an unassigned character in ITA2.

---

each table is determined by the digital value of the code from which conversion is to be made.

Conversion from Morse to ASCII and Baudot

Table 10 starts with the Morse E and progresses through the code to the longest characters. This progression has been used in code-conversion programs written in BASIC. All Morse Latin letters translate directly to Baudot letters and upper-case ASCII letters. All direct equivalents of punctuation marks are included.

Table 12

Conversion from ASCII to Morse and Baudot

ASCII Code 6543210	Char.	Int'l Morse Code	Char.	Baudot Code ¹ 43210	Char.²
0000000	NUL			↓↑ 00000	Blank
0000001	SOH				
0000010	STX				
0000011	ETX				
0000100	EOT				
0000101	ENQ			↑ 01001	WRU (ITA)
0000110	ACK				
0000111	BEL			↑ 01011	Bell (ITA)
				↑ 00101	Bell (U.S.)
0001000	BS				
0001001	HT				
0001010	LF			↓↑ 00010	LF
0001011	VT				
0001100	FF				
0001101	CR			↓↑ 01000	CR
0001110	SO				
0001111	SI				
0010000	DLE				
0010001	DC1				
0010010	DC2				
0010011	DC3				
0010100	DC4				
0010101	NAK				
0010110	SYN				
0010111	ETB				
0011000	CAN				
0011001	EM				
0011010	SUB				
0011011	ESC				
0011100	FS				
0011101	GS				
0011110	RS				
0011111	US				
0100000	SP	Space		00100	SP
0100001	!				
0100010	"	•–••–•	$\overline{AF}$ "	↑ 10001	" (U.S.)
0100011	#			↑ 10100	# (U.S.)
0100100	$	•••–••–	$\overline{SX}$ $	↑ 01001	$ (U.S.)
0100101	%				
0100110	&				
0100111	'	•––––•	$\overline{WG}$ '	↑ 00101	' (ITA)
				↑ 01011	' (U.S.)
0101000	(	–•––•–	$\overline{KN}$ (	↑ 01111	(
0101001	)	–•––•–	$\overline{KK}$)	↑ 10010	)
0101010	*				
0101011	+	•–•–•	$\overline{AR}$ +	↑ 10001	+ (ITA)
0101100	,	––••––	$\overline{MIM}$,	↑ 01100	,
0101101	-	–••••–	$\overline{DU}$ –	↑ 00011	–
0101110	.	•–•–•–	$\overline{AAA}$.	↑ 11100	.
0101111	/	–••–•	$\overline{DN}$ /	↑ 11101	/
0110000	0	–––––	0	↑ 10110	0
0110001	1	•––––	1	↑ 10111	1
0110010	2	••–––	2	↑ 10011	2
0110011	3	•••––	3	↑ 00001	3
0110100	4	••••–	4	↑ 01010	4
0110101	5	•••••	5	↑ 10000	5
0110110	6	–••••	6	↑ 10101	6
0110111	7	––•••	7	↑ 00111	7
0111000	8	–––••	8	↑ 00110	8
0111001	9	––––•	9	↑ 11000	9
0111010	:	–––•••	$\overline{OS}$:	↑ 01110	:
0111011	;	–•–•–•	$\overline{KR}$;	↑ 11110	; (U.S.)
0111100	<				
0111101	=	–•••–	$\overline{BT}$ =	↑ 11110	= (ITA)
0111110	>				
0111111	?	••––••	$\overline{IMI}$?	↑ 11001	?
1000000	@				
1000001	A	•–	A	↓ 00011	A
1000010	B	–•••	B	↓ 11001	B
1000011	C	–•–•	C	↓ 01110	C
1000100	D	–••	D	↓ 01001	D
1000101	E	•	E	↓ 00001	E
1000110	F	••–•	F	↓ 01101	F
1000111	G	––•	G	↓ 11010	G
1001000	H	••••	H	↓ 10100	H
1001001	I	••	I	↓ 00110	I
1001010	J	•–––	J	↓ 01011	J
1001011	K	–•–	K	↓ 01111	K
1001100	L	•–••	L	↓ 10010	L
1001101	M	––	M	↓ 11100	M
1001110	N	–•	N	↓ 01100	N
1001111	O	–––	O	↓ 11000	O
1010000	P	•––•	P	↓ 10110	P
1010001	Q	––•–	Q	↓ 10111	Q
1010010	R	•–•	R	↓ 01010	R
1010011	S	•••	S	↓ 00101	S
1010100	T	–	T	↓ 10000	T
1010101	U	••–	U	↓ 00111	U
1010110	V	•••–	V	↓ 11110	V
1010111	W	•––	W	↓ 10011	W
1011000	X	–••–	X	↓ 11101	X
1011001	Y	–•––	Y	↓ 10101	Y
1011010	Z	––••	Z	↓ 10001	Z
1011011	[				
1011100	\				
1011101	]				
1011110	^				
1011111	_	••––•–	$\overline{IQ}$ _		
1100000	`				
1100001	a	•–	A	↓ 00011	A
1100010	b	–•••	B	↓ 11001	B
1100011	c	–•–•	C	↓ 01110	C
1100100	d	–••	D	↓ 01001	D
1100101	e	•	E	↓ 00001	E
1100110	f	••–•	F	↓ 01101	F
1100111	g	––•	G	↓ 11010	G
1101000	h	••••	H	↓ 10100	H
1101001	i	••	I	↓ 00110	I
1101010	j	•–––	J	↓ 01011	J
1101011	k	–•–	K	↓ 01111	K
1101100	l	•–••	L	↓ 10010	L
1101101	m	––	M	↓ 11100	M
1101110	n	–•	N	↓ 01100	N
1101111	o	–––	O	↓ 11000	O
1110000	p	•––•	P	↓ 10110	P
1110001	q	––•–	Q	↓ 10111	Q
1110010	r	•–•	R	↓ 01010	R
1110011	s	•••	S	↓ 00101	S
1110100	t	–	T	↓ 10000	T
1110101	u	••–	U	↓ 00111	U
1110110	v	•••–	V	↓ 11110	V
1110111	w	•––	W	↓ 10011	W
1111000	x	–••–	X	↓ 11101	X
1111001	y	–•––	Y	↓ 10101	Y
1111010	z	––••	Z	↓ 10001	Z
1111011	{				
1111100						
1111101	}				
1111110	~				
1111111	DEL				

Notes

¹ In Baudot code, it is necessary to check to see what the current case is before conversion (↓ = letters, ↑ = figures, ↓↑ = either case).

² Figures-case characters are the same for both ITA 2 and U.S. teletypewriters except where indicated.

International Morse has specific code combinations for accented letters. However, they cannot be translated out of context to either Baudot or ASCII because individual national implementations of these codes place them differently.

The programmer is left to determine how to display nonconvertible Morse characters. For example, an incoming $\overline{SK}$ cannot be displayed or printed as such because an overlining capability is not usually provided in computer peripherals. Perhaps it could be shown as SK if there is an underlining capability or <END OF WORK> if not.

Alternatively, it could be ignored at the expense of losing that meaning.

When converting from Morse (or ASCII) to Baudot, the program has to check to see what case the Baudot is in. If, for example, you just received WB and the next character is 4, you have to check to see that

Table 13

Code Conversion, ITA1 through 4 (Notes 1 and 2)

Combination No.	ITA1 Bit No. 43210	Figure Case ITA1	Letter Case All Codes	Figure Case ITA2-4	ITA2 Bit No. 43210	ITA3 Bit No. 6543210	ITA4 Bit No. 543210
1	+ + + + −	1	A	—	00011	0101100	000110
2	+ − − + +	8	B	?	11001	1001100	110010
3	+ − + − +	9	C	:	01110	0011001	011100
4	+ − − − −	0	D	Note 4	01001	0011100	010010
5	+ + + − +	2	E	3	00001	0001110	000010
6	+ − − − +	Note 3	F	Note 4	01101	1100100	011010
7	+ − + − +	7	G	Note 4	11010	1000011	110100
8	+ − + − −	+	H	Note 4	10100	0100100	101000
9	+ + − − +	Note 3	I	8	00110	0000111	001100
10	+ − + + −	6	J	BELL	01011	1100010	010110
11	− − + + −	(	K	(	01111	1101000	011110
12	− − + − −	=	L	)	10010	0100011	100100
13	− − + − +	)	M	.	11100	1000101	111000
14	− − − − +	Note 3	N	,	01100	0010101	011000
15	+ + − − −	5	O	9	11000	0110001	110000
16	− − − − −	%	P	0	10110	0101001	101100
17	− − − + −	/	Q	1	10111	1011000	101110
18	− − + + +	−	R	4	01010	0010011	010100
19	− + + − −	.	S	'	00101	0101010	001010
20	− + + − −	Note 3	T	5	10000	1010001	100000
21	+ + − + −	4	U	7	00111	0100110	001110
22	− + − − −	'	V	=	11110	1001001	111100
23	− + − − +	?	W	2	10011	1010010	100110
24	− + + − +	,	X	/	11101	0110100	111010
25	+ + − + +	3	Y	6	10101	1010100	101010
26	− + + − −	:	Z	+	10001	1000110	100010
27	+ + + − −	Carriage return	Carriage return	Carriage return	01000	1100001	010000
28	− + + + −	Line feed	Line feed	Line feed	00010	0001101	000100
29	− + + + +	Letter blank (space)	Letter shift	Letter shift	11111	0111000	111110
30	+ − + + +	Figure blank (space)	Figure shift	Figure shift	11011	0110010	110110
31	− − + + +	Error	Space	Space	00100	0001011	001000
32	+ + + + +	Instrument at rest	Blank	Blank	00000	1110000	000001
— —				Phasing signal	—	—	110011
— —				Signal repetition	—	0010110	—
— —				Signal alpha	Note 5	1001010	000000
— —				Signal beta	Note 6	0011010	111111

Note 1: For complete specifications of these codes see the following International Telecommunication Union documents: ITA1 and 2 — Telegraph Regulations (Geneva Revision, 1958), ITA3 — CCITT Rec. S.13, ITA4 — CCITT Rec. R.44.

Note 2: In ITA1, + indicates positive current, − negative current. In ITA2 through ITA4, 1 represents mark condition (shown as Z in ITU recommendations, which is the higher emitted radio frequency for FSK, the lower for AFSK). 0 represents the space condition (shown as A in ITU recommendations). For meanings of A and Z see CCITT Rec. 140. The normal order of bit transmission is lowest significant bit (LSB) first.

Note 3: At the disposal of each administration for its internal service.

Note 4: At present unassigned. Reception of these signals, however, should not initiate a request for repetition. See CCITT Rec. S.4.

Note 5: Permanent 0 polarity.

Note 6: Permanent 1 polarity.

---

the device is in letters case before giving it the B. In this case it is so no LTRS function need be issued. Then, having checked or remembered that the device is in letters case for the B, a FIGS function must be sent before sending the Baudot device the 4.

Conversion from Baudot to ASCII and Morse

Table 11 is arranged by binary progression starting with 00000 (blank). Note that blank translates to an ASCII NUL, but the letters-shift (LTRS) character does not convert to an ASCII DEL. When a mistake is made during preparation of paper tape, the operator may backspace the tape and use LTRS or DEL to punch all hole positions to erase erroneous characters. If this is done when preparing Baudot tapes, the LTRS function will be read as a change from figure case (if it was there) to letter case. If letter case was not wanted, a figures-shift character must be inserted after the LTRS functions. DEL causes no case change, so LTRS and DEL are not equivalent.

The bell and apostrophe (') translate nicely from ITA2 to ASCII. But note that these codes in ITA2 are just exactly reversed in the code used in most U.S. Baudot TTYs. Because of this difference, it is usually desirable to send both characters in Baudot. This causes both ITA2 and U.S. machines to ring the bell and print an apostrophe. In that other stations may send one code, the other or both together, this makes an interesting puzzle for the code conversion programmer.

The WRU (Who are you?) character in ITA2 converts to ENQ (enquiry) in ASCII. They are both used to request the identity of the other terminal. However, ENQ may also be used to request the status of the other terminal.

Figures H (10100) is messy. It is unassigned in ITA2 and available for national use. On U.S. TTYs, it is probably a pound sign (#), and thus will translate # to an ASCII #. On older machines, it may be used as a motor stop.

When using figures case, you should be aware of a feature called *unshift on space* that may be implemented on some Baudot terminals. If you send a group of numbers and a space, the machine will print letters-case characters after the space if it has the unshift-on-space feature. To avoid this problem, it is a good idea always to send a FIGS just before each group of figures that are preceded by a space.

Some terminals have a "nonoverline" feature. An automatic LF is sent every time a CR is received. Others have a "wrap-around" feature that automatically moves to the first printing position of the new line when the old line is full. If your terminal has these features, don't assume everyone has them.

Conversion from ASCII to Morse and Baudot

This conversion is probably the most-

Table 14

Conversion Between ASCII, EBCDIC and Selectric®

Binary	Dec.	Hex.	ASCII	EBCDIC	Selectric Unshift	Selectric Shift
00000000	000	00	NUL	NUL	Space	Space
00000001	001	02	SOH	SOH	!	°
00000010	002	03	STX	STX	t	T
00000011	003	03	ETX	ETX	j	J
00000100	004	04	EOT	PF	4	$
00000101	005	05	ENQ	HT	o	O
00000110	006	06	ACK	LC	l	L
00000111	007	07	BEL	DEL	/	?
00001000	008	08	BS		5	%
00001001	009	09	HT	RLF	'	"
00001010	010	0A	LF	SMM	e	E
00001011	011	0B	VT	VT	p	P
00001100	012	0C	FF	FF		
00001101	013	0D	CR	CR		
00001110	014	0E	SO	SO		
00001111	015	0F	SI	SI		
00010000	016	10	DLE	DLE	2	@
00010001	017	11	DC1	DC1	.	.
00010010	018	12	DC2	DC2	n	N
00010011	019	13	DC3	DC3	=	+
00010100	020	14	DC4	RES	z	Z
00010101	021	15	NAK	NL		
00010110	022	16	SYN	BS		
00010111	023	17	ETB	IL		
00011000	024	18	CAN	CAN	6	¢
00011001	025	19	EM	EM	i	I
00011010	026	1A	SUB	CC	k	K
00011011	027	1B	ESC		q	Q
00011100	028	1C	FS	IFS	UC	UC
00011101	029	1D	GS	IGS	BS	BS
00011110	030	1E	RS	IRS		
00011111	031	1F	US	IUS	LC	LC
00100000	032	20	SP	DS	1	±
00100001	033	21	!	SOS	m	M
00100010	034	22	"	FS	x	X
00100011	035	23	#		g	G
00100100	036	24	$	BYP	0	)
00100101	037	25	%	LF	s	S
00100110	038	26	&	EOB/ETB	h	H
00100111	039	27	'	ESC/PRE	y	Y
00101000	040	28	(		7	&
00101001	041	29	)		r	R
00101010	042	2A	*	SM	d	D
00101011	043	2B	+		;	:
00101100	044	2C	,			
00101101	045	2D	–	ENQ	NL	NL
00101110	046	2E	.	ACK	LF	LF
00101111	047	2F	/	BEL	HT	HT
00110000	048	30	0		3	#
00110001	049	31	1		v	V
00110010	050	32	2	SYN	u	U
00110011	051	33	3		f	F
00110100	052	34	4	PN	9	(
00110101	053	35	5	RS	w	W
00110110	054	36	6	UC	b	B
00110111	055	37	7	EOT	–	–
00111000	056	38	8		8	•
00111001	057	39	9		a	A
00111010	058	3A	:		c	C
00111011	059	3B	;		,	,
00111100	060	3C	<	DC4	EOT	EOT
00111101	061	3D	=	NAK	IL	IL
00111110	062	3E	>			
00111111	063	3F	?	SUB		
01000000	064	40	@	SP		
01000001	065	41	A			
01000010	066	42	B			
01000011	067	43	C			
01000100	068	44	D			
01000101	069	45	E			
01000110	070	46	F			
01000111	071	47	G			
01001000	072	48	H			
01001001	073	49	I			
01001010	074	4A	J	¢		
01001011	075	4B	K	.		
01001100	076	4C	L	<		
01001101	077	4D	M	(		
01001110	078	4E	N	+		
01001111	079	4F	O				
01010000	080	50	P	&		
01010001	081	51	Q			
01010010	082	52	R			
01010011	083	53	S			
01010100	084	54	T			
01010101	085	55	U			
01010110	086	56	V			
01010111	087	57	W			
01011000	088	58	X			
01011001	089	59	Y			
01011010	090	5A	Z	!		
01011011	091	5B	[	$		
01011100	092	5C	\	*		

Binary	Dec.	Hex.	ASCII	EBCDIC
01011101	093	5D	]	)
01011110	094	5E	^	;
01011111	095	5F	_	–|
01100000	096	60	`	–
01100001	097	61	a	/
01100010	098	62	b	
01100011	099	63	c	
01100100	100	64	d	
01100101	101	65	e	
01100110	102	66	f	
01100111	103	67	g	
01101000	104	68	h	
01101001	105	69	i	
01101010	106	6A	j		
01101011	107	6B	k	,
01101100	108	6C	l	%
01101101	109	6D	m	_
01101110	110	6E	n	>
01101111	111	6F	o	?
01110000	112	70	p	
01110001	113	71	q	
01110010	114	72	r	
01110011	115	73	s	
01110100	116	74	t	
01110101	117	75	u	
01110110	118	76	v	
01110111	119	77	w	
01111000	120	78	x	
01111001	121	79	y	\
01111010	122	7A	z	:
01111011	123	7B	{	#
01111100	124	7C			@
01111101	125	7D	}	'
01111110	126	7E	~	=
01111111	127	7F	DEL	"
10000000	128	80		
10000001	129	81		a
10000010	130	82		b
10000011	131	83		c
10000100	132	84		d
10000101	133	85		e
10000110	134	86		f
10000111	135	87		g
10001000	136	88		h
10001001	137	89		i
10001010	138	8A		
10001011	139	8B		
10001100	140	8C		
10001101	141	8D		
10001110	142	8E		
10001111	143	8F		
10010000	144	90		
10010001	145	91		j
10010010	146	92		k
10010011	147	93		l
10010100	148	94		m
10010101	149	95		n
10010110	150	96		o
10010111	151	97		p
10011000	152	98		q
10011001	153	99		r
10011010	154	9A		
10011011	155	9B		
10011100	156	9C		
10011101	157	9D		
10011110	158	9E		
10011111	159	9F		
10100000	160	A0		
10100001	161	A1		~
10100010	162	A2		s
10100011	163	A3		t
10100100	164	A4		u
10100101	165	A5		v
10100110	166	A6		w
10100111	167	A7		x
10101000	168	A8		y
10101001	169	A9		z
10101010	170	AA		
10101011	171	AB		
10101100	172	AC		
10101101	173	AD		
10101110	174	AE		
10101111	175	AF		
10110000	176	B0		
10110001	177	B1		
10110010	178	B2		
10110011	179	B3		
10110100	180	B4		
10110101	181	B5		
10110110	182	B6		
10110111	183	B7		
10111000	184	B8		
10111001	185	B9		

Binary	Dec.	Hex.	ASCII	EBCDIC
10111010	186	BA		
10111011	187	BB		
10111100	188	BC		
10111101	189	BD		
10111110	190	BE		
10111111	191	BF		
11000000	192	C0		{
11000001	193	C1		A
11000010	194	C2		B
11000011	195	C3		C
11000100	196	C4		D
11000101	197	C5		E
11000110	198	C6		F
11000111	199	C7		G
11001000	200	C8		H
11001001	201	C9		I
11001010	202	CA		
11001011	203	CB		
11001100	204	CC		
11001101	205	CD		
11001110	206	CE		
11001111	207	CF		
11010000	208	D0		}
11010001	209	D1		J
11010010	210	D2		K
11010011	211	D3		L
11010100	212	D4		M
11010101	213	D5		N
11010110	214	D6		O
11010111	215	D7		P
11011000	216	D8		Q
11011001	217	D9		R
11011010	218	DA		
11011011	219	DB		
11011100	220	DC		
11011101	221	DD		
11011110	222	DE		
11011111	223	DF		
11100000	224	E0		
11100001	225	E1		
11100010	226	E2		S
11100011	227	E3		T
11100100	228	E4		U
11100101	229	E5		V
11100110	230	E6		W
11100111	231	E7		X
11101000	232	E8		Y
11101001	233	E9		Z
11101010	234	EA		
11101011	235	EB		
11101100	236	EC		
11101101	237	ED		
11101110	238	EE		
11101111	239	EF		
11110000	240	F0		0
11110001	241	F1		1
11110010	242	F2		2
11110011	243	F3		3
11110100	244	F4		4
11110101	245	F5		5
11110110	246	F6		6
11110111	247	F7		7
11111000	248	F8		8
11111001	249	F9		9
11111010	250	FA		
11111011	251	FB		
11111100	252	FC		
11111101	253	FD		
11111110	254	FE		
11111111	255	FF		

needed one because many amateurs are using, or would like to use, an ASCII keyboard to generate Morse and Baudot codes. Table 12 shows that most code combinations needed in Morse and Baudot can be translated from the ASCII character set. However, there are some Morse characters (accented letters, $\overline{SN}$, $\overline{INT}$, $\overline{AS}$, $\overline{AU}$, $\overline{HM}$, $\overline{SK}$, $\overline{IX}$, or $\overline{HH}$) that don't appear on an ASCII keyboard. Some of the control characters or graphic characters could be used until an Amateur Radio standard code conversion table is developed.

Code Conversion ITA1 through 4

Table 13 gives character-for-character code equivalents for these CCITT codes. ITA1 was used on cables. ITA2 is in standard use in the Amateur Service and in telex services worldwide. ITA3 is found in Table A-1 of Annex A to CCITT Recommendation S.13, "Use on Radio Circuits of 7-Unit Synchronous Systems Giving Error Correction by Automatic Repetition." ITA4 is specified in Annex A to CCIR Rec. R.44 entitled "6-Unit Synchronous Time-Division 2-3-Channel Multiplex Telegraph System for Use over FMVFT Channels Spaced at 120 Hz for Connection to Standardized Teleprinter Networks." The abbreviation FMVFT means frequency modulated voice-frequency telegraph. ITA1, 3 and 4 are not currently used in the Amateur Radio Service but would be legal in the U.S. above 50 MHz. They are included herein primarily for information.

Conversion Between ASCII, EBCDIC and Selectric

Although they are not common in the Amateur Radio Service, it is possible to transmit EBCDIC and IBM Selectric® codes in the U.S. above 50 MHz. It is more likely that surplus equipment using these codes would be used as peripherals in a computer setup. Table 14 is included for the individual who wants to write ASCII conversion tables for EBCDIC or Selectric devices.

Packet Radio — Integrated Digital Communications

The term *data communications* is virtually synonymous with transfer of information between computers. In fact, the term includes any types of signals that can be expressed in digital form. It is no secret that every major communications system is changing to digital transmission and phasing out many older analog systems. Analog communications systems can be made to work well over one or two relays without problems with background noise or distortion. However, if you add dozens of relays then the noise and distortion are not simply that of the weakest link but the accumulation of the ill effects.

With digital transmission, these effects are eliminated on a link-by-link basis by regenerating the digital information on each relay. Also, error-control techniques are brought into play to eliminate any missed or wrong bits. Thus the end-to-end noise and distortion are simply that of the digitizing process, not the transmission of the digital information.

The larger commercial and governmental communications carriers have embarked on *integrated* digital networks, that is, ones that will handle voice, images and data. Amateur Radio has taken bold steps in that direction under the banner of *packet radio*. The ultimate goal is to have a global amateur packet-radio network capable of automatically sending keyboard data, speech and images to other amateurs anywhere in the world. Much of the technology is now available to us, and agreement on standards is well underway. The ARRL has provided a framework for packet-radio experimenters to exchange ideas through its Amateur Radio Computer Networking Conferences held since 1981. The ARRL Ad Hoc Committee on Amateur Radio Digital Communication was established by the Board of Directors to develop standards. The committee has representation from the major packet-radio clubs and has been successful in developing cooperative standards.

The foundation for a global amateur packet-radio network has been laid. Current Amateur Radio packet technology will support direct connections (read that error-free two-way communications) between two stations with as many as eight digital repeaters in between. Packet radio is currently active on a local-area basis in a number of North American locations and some other countries. While most of the activity has been on VHF, 1984 became the year of growing packet-radio operation on HF and via amateur satellite. Also, a number of groups on the East and West Coasts of the United States have successfully chained a number of packet repeaters, which by 1986 should support relaying of packets from San Diego to Vancouver and from Miami to Montreal.

In the 1984 *Handbook*, we estimated that over 300 U.S. amateurs were active on packet radio. That number has grown to 3900 as the 1986 *Handbook* is being edited. Additional rapid growth will continue as more amateurs realize the error-free communications available now from packet radio. Further expansion is inevitable as more amateurs understand the potential services that a global amateur integrated data network will provide:

• Two-way terminal-to-terminal communications between your station and any other within the network.

• Computer file transfer between stations.

• If the desired station is not logged onto the network at the time you wish, you can leave a message on the computer-based message system (CBMS) used by that station.

• Host computers will be scattered through the network to provide a variety of services such as data bases with information of interest to amateurs.

• Access to the network will be open to those with even the most modest Amateur Radio station. If you can presently use a hand-held transceiver to work into a 2-meter repeater, you could use the same hand-held to get into a 2-meter packet repeater if there is one in your local area.

• Routing of packets throughout the network will be automatic and will take advantage of VHF/UHF terrestrial relays, meteor-scatter, HF and satellite packet-radio links.

• Third-party traffic, where permitted, will be automatically handled through the network.

• The latest bulletins will be available to you whenever you are signed on to the network.

Is all of this *blue sky*? Not really. Many of the capabilities mentioned above are already available in a growing number of North American metropolitan areas. However, internetworking between these locations is not yet a reality except for linking some packet repeaters up and down the East and West Coasts, and in several inland areas. Internetworking is contingent on some further developments, namely:

• Agreement on networking standards (now under development by the ARRL Digital Committee),

• Design and installation of high-speed (9.6- to 56-kbit/s) packet-radio stations for the terrestrial *backbone* that will interconnect metropolitan areas.

• Participation by more clubs with the technical and funding resources to establish packet-radio facilities in their areas and link with neighboring areas.

• Individual amateurs interested in getting their own station on packet radio and willing to help others do the same.

When 2-meter FM began, who would have guessed that today there would be over 9000 VHF/UHF FM repeaters in North America? A look through the ARRL *Repeater Directory* will show that there are few places within the United States and southern Canada that are not served by at least one FM repeater. Packet radio has the same potential but is in an early stage of its growth. A number of packet radio repeaters are shown in the *Repeater Directory*; others are not yet listed and more are planned by various groups. The eventual number of packet repeaters is left to speculation. However, it is a safe bet that the number will be in the thousands.

Amateur Packet Radio History

The Canadian Department of Communications authorized packet radio in 1978. Amateurs in Montreal were the first to get on the air. In Vancouver, BC, Doug Lockhart, VE7APU, began experimenting with *bit-oriented* packet radio based on the International Organization for Standardization (ISO) high-level link control (HDLC) protocol. Bit-oriented protocols permit different meanings to be assigned to each bit and are more efficient than *character-oriented* protocols, which need to send an entire character for each morsel of information.

Doug also designed a special-purpose computer called a terminal-node controller (TNC). He founded the Vancouver Amateur Digital Communications Group (VADCG) that arranged for the manufacture of over 500 TNC boards and sold them to amateurs in North America and to a few overseas amateurs. These TNCs and VADCG's newsletter, *The Packet*, helped many experimenters get started in packet radio. VADCG provided North American "packeteers" a way to check their TNC boards by transmitting packet beacon messages on 14.0765 MHz. They also cooperated with Ontario amateurs and exchanged packets via Canada's ANIK B satellite.

The Canadian Radio Relay League awarded a CRRL Certificate of Merit to Doug Lockhart in August 1984 in recognition of his packet-radio pioneering work.

Genesis in the United States

In 1980, two groups in the U.S. began experimenting with the VADCG TNC boards. One was near San Francisco where Hank Magnuski, KA6M, activated the first U.S. amateur packet-radio repeater on 2 meters on December 10, 1980. KA6M/R did not use the VADCG TNC but was equipped with a homemade TNC based on an STD-bus Z80® microprocessor and a Western Digital 1933 HDLC chip. Following the VADCG pattern, KA6M/R used a Bell 202 modem, chiefly because of their surplus availability. Hank founded the Pacific Packet Radio Society (PPRS), pioneered sending packets through the OSCAR 10 satellite, and has been an important contributor to the development of amateur packet radio in the United States.

In the Washington, DC area, the Amateur Radio research and Development Corporation (AMRAD) began spreading the word about packet radio through its *AMRAD Newsletter*. Bill Moran, W4MIB, was the first East Coast packeteer to get on the air on May 4, 1981. AMRAD used an unmodified 2-meter FM voice repeater instead of building a simplex (single-frequency) packet-only repeater as KA6M had done. A simplex repeater was later activated to serve the Washington, DC area (currently WB4JFI/R on 145.01 MHz).

Early witnesses of packet-radio operation were amazed to see the one- or two-second bursts turn into almost instant displays of error-free text across the CRT screen. It took no convincing for them to visualize a packet-radio network spanning North America if not the globe. It became clear that the local-area groups needed to be interconnected by means of terrestrial VHF/UHF *backbone* network, HF and satellite. Commercial experience strongly suggested that meteor-scatter communications would be excellent for packet radio over 600- to 1200-mile paths.

First Computer Networking Conference

Greater forces were needed to solve some immediate technical problems and to establish packet-radio activities in other areas. It was time to get the existing packet experimenters together to exchange ideas and to invite others to get involved. Thus, in October 1981, AMRAD and AMSAT organized the first ARRL Amateur Radio Computer Networking Conference, which convened at the National Bureau of Standards headquarters in Gaithersburg, Maryland. The conference was attended by 80 amateurs, many of whom presented their views on how an amateur packet-radio network should be designed. Participation by Canadian and Swedish amateurs made the conference an international event. The second day of the conference was held at the Goddard Space Flight Center as a prelude to the AMSAT annual meeting.

After the first conference, Den Connors, KD2S, moved to Tucson, Arizona and founded the Tucson Amateur Packet Radio Corporation (TAPR). This group designed three successive TNCs. The alpha test TNC was built in small quantities. TAPR then designed their beta test TNC and had 180 of them fully assembled for field testing in a number of geographical areas. The feedback that they received was used in the design of the TAPR TNC kit, over 2000 of which have been sold throughout the world. In recognition of TAPR's contribution to packet radio, TAPR President Lyle Johnson, WA7GXD, was given the Dayton ARA Technical Excellence Award in 1984.

Packet-radio protocols were the subject of discussion and diversity of views. The original software supplied by VADCG underwent some evolutionary changes. However, that software provided only one-byte addresses, meaning that the number of stations that could be addressed was quite small. Some thought that the packet-radio addressing system should be based on amateur call signs, while others proposed dynamic-addressing schemes wherein the specific address would be assigned at the time of signing onto the network. AMRAD and the Radio Amateur Telecommunications Society (RATS) of New Jersey held meetings to see whether the CCITT X.25 public packet-switched network protocol could become the basis for an Amateur Radio packet protocol. There were problems, but most were solved by Eric Scace, K3NA, who served on the CCITT committee and was one of the authors of X.25. What emerged from these meetings was called AX.25 (Amateur X.25). AX.25 was based on the use of Amateur Radio call signs for the address field of the HDLC frame. AX.25 was not instantly accepted.

AX.25 Agreement

By mid 1982, there was some concern that the various packet groups would not be able to agree on a common protocol by the time OSCAR 10 was launched. In October 1982, Tom Clark, W3IWI, as president of AMSAT, called a meeting of active U.S. packet-radio groups. His strategy was to lock them all in a room until they agreed. It worked. The outcome was a new version of AX.25 that would allow a repeater call sign as an optional third address in the HDLC frame.

At the October 1982 meeting mentioned, AMSAT's packet-satellite (PACSAT) project was started. The concept for a low-orbiting packet "mailbox" that would serve the whole earth was accepted by the AMSAT membership. KD2S was designated the project leader and later passed that role to Harold Price, NK6K. PACSAT is now scheduled to be launched in 1987.

Within weeks after the October 1982 meeting, KA6M and NK6K had early versions of AX.25 software written for the VADCG and TAPR TNCs. The VADCG and TAPR versions of AX.25 wouldn't talk to each other immediately, but some software patches took care of that in early 1983.

Second Computer Networking Conference

The second ARRL Amateur Radio Computer Networking Conference was held in March 1983, in San Francisco. This conference served to consolidate gains made thus far and to introduce others to packet radio.

In July 1983, a meeting was held at the home of Harold Price, NK6K, to talk about linking of packet repeaters from San Diego to San Francisco. WESTNET was born and with it a proposal to permit up to eight repeater call signs in the AX.25 frame. This proposal was accepted by the ARRL Digital Communications Committee in November 1983 and made part of AX.25 until a true networking protocol could be agreed to and implemented.

Word of WESTNET's creation prompted formation of EASTNET to link Washington, DC to the packet repeater in Lowell, Massachusetts operated by the New England Packet Radio Association (NEPRA). As a part of this network, ARRL HQ activated the W1AW packet repeater on February 18, 1984. The

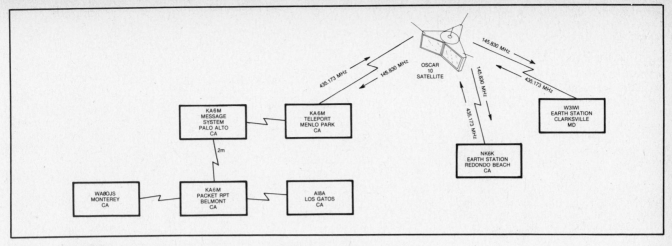

Fig. 23 — Link diagram of the first teleport experiments via OSCAR 10 on March 11, 1984.

southeast part of the country was subsequently served by SOUTHNET, operated by the Florida Amateur Digital Communications Association (FADCA). EASTNET and SOUTHNET are to be joined through the Carolinas.

Third Computer Networking Conference

A third ARRL Amateur Radio Computer Networking Conference was held in Trenton, New Jersey in April 1984. Incidentally, printed proceedings of each of the three conferences are available from ARRL Hq.; see the bibliography at the end of this section. The Dayton Hamvention followed in two weeks and held a packet-radio forum with a record turnout of over 300 in attendance. It was clear from the meetings in Trenton and Dayton that packet radio was no longer the exclusive province of bearded experimenters but was a proper topic for discussion in "polite company." At this time, it was estimated that about 1700 TNCs (or software equivalents) had been sold and that about 1300 of them were considered operational.

On November 14, 1984, the ARRL filled a Petition for Rule Making (RM-4879) with the FCC requesting permission for automatic control of an amateur station on frequencies above 30 MHz when using digital communications. This addition to the rules would clear away obstacles to computer based message system (CBMS) operation and various forms of networking under automatic control. In response, on April 5, 1985, the FCC adopted a Notice of Proposed Rule Making (PR Docket No. 85-105) in which they proposed that any Amateur Radio station be permitted to operate under automatic control, except when transmitting on frequencies below 29.5 MHz.

An FCC Order adopted March 25, 1985, eliminated the special requirement that a description of special digital codes used above 50 MHz be included in the station log and provided to the Commission upon request.

Teleport Experiments

On March 11, 1984, several stations demonstrated the interconnection of packet-radio stations through terrestrial repeaters and the teleport via the OSCAR 10 satellite. Those involved were Tom Clark, W3IWI, Hank Magnuski, KA6M, Harold Price, NK6K, Jim Tittsler, AI8A, and Ron McMurday, WA0OJS. The first contact was between W3IWI and WA0OJS. Then NK6K chatted with AI8A for about 10-15 minutes. The links were as shown in Fig. 23. During each contact text was transmitted in both directions. A total of five TNCs had to be properly connected for all this to work. Not suspecting that anything special was going on, WA0OJS and AI8A connected with KA6M to pick up their mail. Instead they found that they were connected to stations in remote geographic areas via satellite.

The FCC granted a joint ARRL/AMSAT request for a special temporary authority (STA) to permit automatic teleport operation of 24 stations for a six-month period beginning in October 1984. On October 28, 1984, a unique packet-radio test was conducted. Amateur packet-radio stations VE1PAC/VE6 in Alberta, K7PYK in Arizona, W4DAQ in Alabama, and N5AHD in Texas successfully used the W3IWI bulletin board via OSCAR 10. K7PYK maintained contact for about an hour and acquired about 50 kbytes of documentation from W3IWI.

HF Packet Operation

Around March 1984, Bob Bruninga, WB4APR, in Annapolis, Maryland, began regular 30-meter packet-radio operation with W9TD, W0RPK, N0AN, W9HLY and others. On weekends, WB4APR acted as a bridge between 30 and 2 meters to give midwest packeteers immediate access to VHF packet repeaters in the Washington, DC area.

Aeronautical Packet Repeater

In July and August 1984, Gordon Bass, K2UIR, operated what may be the first airborne amateur packet-radio repeater on weekend flights between Rochester, New York and Rockport, Maine. Operating under the call sign W2DUC, the repeater used a GLB Electronics PK-1 TNC at an altitude of about 10,000 feet.

Packet Bulletin Boards

In mid 1984, Hank Oredson, W0RLI, activated a "packet bulletin board system" (PBBS) using software written in Z80® assembly language, run on a Xerox 820 computer board with a TAPR TNC. The PBBS was designed to originate, receive and automatically forward messages in the National Traffic System (NTS) format. The W0RLI PBBS software is used by numerous stations.

Packet Meteor Scatter Experiments

On August 1, 1984, well before the peak of the Perseids meteor showers, Ralph Wallio, W0RPK, in central Iowa held 6-meter packet radio schedules with Bob Carpenter, W3OTC, in Maryland. W3OTC received about 2% of the packets sent by W0RPK. Four nights later, these two stations had what is believed to be the first two-way amateur packet-radio meteor-scatter contact.

Formal Approval of AX.25

At its meeting on September 15, 1984, the ARRL Ad Hoc Committee on Amateur Radio Digital Communication finalized the AX.25 protocol specification. To distinguish it from previous versions, the protocol was called Version 2.0.

AX.25 Version 2.0 was approved by the ARRL Board of Directors at their October 26, 1984 meeting. An article describing the

Board action and the protocol appeared in December 1984 *QST*. The protocol specification was printed in a booklet entitled *Amateur Packet-Radio Link Layer Protocol, AX.25*, which is available from ARRL HQ. The booklet gives the complete specification in sufficient detail for those wishing to write software implementations.

UoSAT-OSCAR 11 and PACSAT Progress

Packet-radio messages were successfully exchanged between a station assembled by WH6AMX, WA3ZIA and VE3FLL in Honolulu, NK6K in Los Angeles and the UoSAT command station in Surrey, England, using the data communications experiment (DCE) on board UoSAT-OSCAR 11, on January 16, 1985. The Hawaii end of this experiment was conducted at the Pacific Telecommunications Conference in Honolulu. The DCE was intended to prove the concept of worldwide store-and-forward packet communications as a precursor to PACSAT and JAS-1 satellites.

A critical design review for the PACSAT project was held in Arlington, Virginia in early March. The digital store-and-forward subassembly is to have 2 to 4 Mbytes of memory and will be designed by the Tucson Amateur Packet Radio group and the Ottawa Group. The University of Surrey is to fabricate the actual PACSAT spacecraft.

Fourth Computer Networking Conference

The Fourth ARRL Amateur Radio Computer Networking Conference was held in San Francisco on March 30, 1985. A significant paper was presented by Steve Goode, K9NG, on his 9600-bit/s FSK modem. A number of groups were eagerly awaiting the availability of a practical 9600-bit/s modem for terrestrial intercity linking as well as 6-meter meteor-scatter operation.

The ARRL Ad Hoc Committee for Amateur Radio Digital Communication met that same weekend. It approved work done by Doug Lockhart, VE7APU, to develop amateur standards based on CCITT X.3, X.28 and X.29 protocols. The committee made plans to develop message protocols after study of the CCITT X.400 series. At that meeting, Hank Magnuski, KA6M, announced that the Pacific Packet Radio Society "Golden Packet" award for stations participating in the first transcontinental terrestrial VHF/UHF/microwave linking was to be established.

Fifth Computer Networking Conference

On March 9, 1986, Orlando, Florida was the site of the Fifth ARRL Amateur Radio Computer Networking Conference. The number of packet-radio stations had grown to around 14,000 worldwide, and 2-meter frequency congestion had become a major problem. At the conference, working prototypes of two networking technologies were demonstrated: virtual circuit and datagram. Key players in these developments were J. Gordon Beattie, N2DSY,

Thomas Moulton, W2VY, Howard Goldstein, N2WX, Terry Fox, WB4JFI, and Philip Karn, KA9Q. Printed proceedings of the 5th conference are available from ARRL HQ.

OPEN SYSTEMS INTERCONNECTION REFERENCE MODEL

Introduced in 1978, the Open Systems Interconnection (OSI) Reference Model (RM) is a long-term project of the International Organization for Standardization (ISO). The OSI-RM was developed to promote compatible communications among a wide variety of systems. In it, the communications structure is specified in seven distinct levels. The terms *level* and *layer* are used almost synonymously in the OSI-RM context. More often than not, the term level is used when referring to the function, and the term layer is used when referring to the name. This usage can be seen in the following list of the seven levels that make up the OSI-RM:
• Level 7 — Application Layer (highest level)
• Level 6 — Presentation Layer
• Level 5 — Session Layer
• Level 4 — Transport Layer
• Level 3 — Network Layer
• Level 2 — Link Layer
• Level 1 — Physical Layer (lowest level)
Primary documents for the OSI Reference Model are ISO 7498 and CCITT Recommendation X.200. Protocols for each level are in various stages of development. ISO documents progress through three stages before adoption: draft proposal (DP), draft international standard (DIS) and international standard (ISO). At each level there are normally two parts: a service specification (defines *what* the layer is supposed to do) and a protocol specification (defines *how* the layer is supposed to do it).

The OSI-RM has won general international acceptance in the telecommunications industry and in amateur packet-radio circles. One advantage is that the levels are

modular. This allows different people to work on individual levels with some assurance that the various levels can be made to work together in a system. Modularity also permits more than one protocol at a level. Each level must communicate with the level(s) above and below it, and must follow agreed rules for these interlevel interfaces. In order for distant systems to communicate, *peer* levels (for example, the link layer on one system and the link layer on the other) must exchange data in an agreed protocol. The interlevel and peer-to-peer relationships between two systems in direct communication are shown in the "apartment-house" diagram in Fig. 24.

Level 1 — The Physical Layer

The Physical Layer is the lowest level in the OSI Reference Model. Its function is to send and receive *bits* (binary ones and zeros — marks and spaces). This level is concerned with the following:
• Physical connections — allows duplex or half-duplex transmission
• Physical service data units (PSDUs) — one bit in serial transmission, "n" bits in parallel transmission
• Circuit identification
• Bit sequencing
• Notification of fault conditions
• Deriving quality of service parameters
In addition, these functions are associated with the physical layer:
• Modulation and demodulation (modems)
• Signaling speed
• Transmission of data and handshaking signals
• Characterization of communications media (radio circuit)
The physical layer is the only level that maintains an actual electrical connection with its peers. The higher levels communicate with their peers through *logical* or *virtual* connections.

Level 2 — The Link Layer

The link layer (also known as the data-link control layer) arranges the bits into *frames*. The most common protocol is ISO

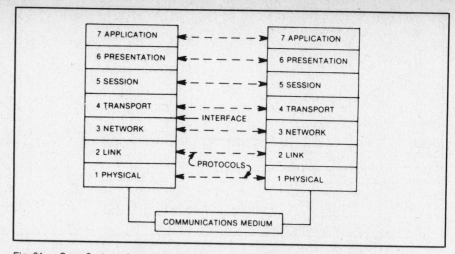

Fig. 24 — Open Systems Interconnection reference model protocol levels. The solid lines between layers, as shown between 3 and 4, are interfaces. The dashed lines indicate logical communications between peer layers.

High-Level Data Link Control Procedures (ISO 3309). The link layer performs the following services:

- Establishes and releases one or more link connections
- Exchanges data-link service data units (DLSDUs) — frames
- Identifies endpoints
- Keeps DLSDUs/frames in proper sequence
- Notifies network layer when errors are detected
- Controls data flow
- Selects optional quality-of-service parameters

Level 3 — The Network Layer

The network layer organizes data into *packets*, which are frames with network information added. In CCITT X.25 protocol, it is called the *packet layer*. Services provided by the network layer are:

- Network addressing and identifiers
- Network connections and release
- Transmission of network service data units (NSDUs) — packets
- Quality of service parameters
- Notifies transport layer of errors
- May provide sequenced delivery
- Flow control
- Expedited service through network

There are two basic types of network-layer protocols: connection-oriented and connectionless. Connection-oriented protocols set up a specific *virtual circuit* (VC) between the two end points. Connectionless protocols use the *datagram* (DG) with complete addressing information in each packet so they can follow any available route through a network.

Some view the network layer as two sublayers: Level 3 (or Level 3A) — the network (or *intranet*) sublayer and Level 3.5 (or Level 3B) — the internetworking (*internet*) sublayer. As an example, CCITT X.25 is an intranet protocol, while X.75 is an internet protocol.

Level 4 — The Transport Layer

The transport layer organizes data into a *transport protocol data unit* (TPDU), which is a packet with transport-layer data added. This layer ensures that all data sent is received completely and in proper sequence at the destination. The following functions are performed at all times during a transport connection:

- Transmission of TPDUs — messages.
- Multiplexing and demultiplexing — used to share a network connection between two or more transport connections
- Error detection — detect loss, garbling, duplication, misdelivery or missequencing of TPDUs
- Error recovery — recovery from detected errors.

Other functions are:

- Connection establishment
- Data transfer
- Release — disconnection of transport connection

The CCITT transport protocol in Recommendation X.224 specifies transport service in five classes, each of which is designed according to the error behavior of network connections. These are:

Class 0 — Simple class
Class 1 — Basic error-recovery class
Class 2 — Multiplexing class
Class 3 — Error-recovery and multiplexing class
Class 4 — Error-detection and recovery class

The amount of work to be done at the transport layer is dependent on the type of protocol (virtual circuit or datagram) uses at the network layer. As it is possible for individual datagrams to arrive via different routes through a network, the transport layer in a connectionless network needs to buffer messages to return them to their original order. Connection-oriented networks permit use of a leaner transport layer.

Level 5 — The Session Layer

The session layer organizes data into *session protocol data units* (SPDUs). This layer is concerned with:

- Dialog management
- Data-flow synchronization, and resynchronization
- Mapping addresses with names (so users may retain a name if they move)
- Graceful or abrupt disconnection
- Buffering data until delivery time

Session services are grouped into the following phases:

- Connection establishment
- Data transfer
- Connection release

Level 6 — The Presentation Layer

The presentation layer is responsible for terminal management. It performs the following services:

- Transfer of syntax for character sets, text strings, data display formats, graphics, file organization and data types
- Encoding, decoding and compaction of data
- Interpret character sets (example: ASCII)
- Code conversion

Level 7 — The Application Layer

The application layer serves as a window between the OSI communications environment and the application process. It is the only level that does not interface with a higher one. The application layer provides the following functions:

- Log in — identification of communications partners
- Password checks and authority to communicate
- Determine adequacy of resources
- Determine acceptable quality of service
- Synchronization of applications programs
- Selecting the dialog procedures
- Agreement on error-recovery responsibility

- Procedures for control of data integrity
- Identifying data syntax constraints

Application layer protocols are classified in five groups:

Group 1 — System-management protocols
Group 2 — Application-management protocols
Group 3 — System protocols
Group 4 — Industry-specific protocols
Group 5 — Enterprise-specific protocols

PHYSICAL-LAYER CONSIDERATIONS

Layer 1 of the OSI model defines the electrical and mechanical standards such as connector configuration and which pins are used for different signals. The physical layer is responsible for getting individual *bits* transferred from one piece of equipment through the communications medium to the other equipment.

DATA INTERFACE STANDARDS

It is common for digital communications equipment to have one or more serial binary data ports to connect to the next piece of equipment. Serial ports send and receive data one bit at a time. In Electronic Industries Association (EIA) and CCITT terminology, they are defined as *interchange circuits* at the interface between the DTE and DCE for the transfer of binary data, control and timing signals, and analog signals where appropriate. DTE is the abbreviation for *data terminal equipment*, meaning a video display terminal (VDT) or a computer that serves as a data terminal. DCE stands for *data circuit-terminating equipment* — typically a modulator-demodulator (modem) that terminates the analog communications circuit.

With some equipment configurations it is difficult to tell which is the DTE and which is the DCE. It is necessary to know one from the other in order to know how to interpret the standard. A fairly safe rule is that the DTE is the farthest from the communications channel. So in a setup with a terminal, a computer, a packet-assembler/disassembler (PAD) and a modem:

- at the terminal-computer interface, the terminal is the DTE, and the computer is the DCE.
- at the computer-PAD interface, the computer is the DTE, and the PAD is the DCE.
- at the PAD-modem interface, the PAD is the DTE, and the modem is the DCE.

Internationally, data-interchange interfaces are defined in CCITT Recommendations V.24 and V.28 for data rates up to 20,000 bit/s. V.35, which is primarily a 48 kbit/s modem specification, is finding use as an interchange-circuit standard for data rates above 20,000 bit/s.

In the United States, the most common standard is EIA RS-232-C for speeds up to 20,000 bit/s. EIA also has issued other standards, known as RS-422, RS-423 and RS-449, although these interfaces are not yet common in personal computers or Amateur Radio equipment. When looking at surplus military equipment, bear in mind that interfaces may be designed against military interface standards such as MIL-STD-188C (similar to RS-232-C), MIL-STD-188-114 (similar to RS-422), MIL-STD-188-144 (similar to RS-423), and MIL-STD-1397 (42 and 250 kbit/s).

Also, but not as often, some parallel ports are found in amateur digital equipment. In parallel interfaces, the eight bits in a byte move into and out of the equipment simultaneously on eight separate wires. Two parallel interface standards are of interest to amateurs. One is the Institute of Electronics and Electrical Engineers (IEEE) Standard 488, also known as the IEEE 488 bus, IEC 625, Hewlett-Packard Interface Bus (HPIB) and General Purpose Interface Bus (GPIB). The other is the Centronics parallel interface, a de facto standard interface for parallel printer ports (see Chapter 8 for pin assignments).

Fig. 25 — Electronic Industries Association RS-232-C data interface standard and several DB25 connectors.

Balanced vs. Unbalanced Interfaces

Both balanced and unbalanced circuits are used in digital interfaces, much the same as for RF circuits. In fact, data signals sent at speeds of megabytes per second *are* RF and need to be treated as such. The rise and fall transitions of much slower data rates may be in the RF range. Like any transmission line, if the line is not terminated at both ends in its characteristic impedance, the conductors will radiate. The line will also be susceptible to electromagnetic interference from nearby conductors.

An unbalanced circuit has one signal conductor, and perhaps one or more timing- and control-signal conductors; the return path for each is through a common signal ground. Yet, unbalanced circuits are commonly used for circuits operating at speeds below 20,000 bit/s and for wire lines not exceeding 50 feet or 15 meters, as in EIA RS-232-C. In this standard, longer cables may be used if the load capacitance does not exceed 2500 pF. In office buildings twisted-pair RS-232-C cables are sucessfully run as long as 500 feet, but the standard guarantees only 50 feet.

Balanced interface circuits use two conductors balanced to ground. When one wire is positive, the other is negative, and vice versa. Balanced circuits are a little more trouble from a design standpoint but offer higher data rates over longer distances with less tendency for interference. Balanced data circuits should be considered for data rates over 20,000 bit/s.

Bipolar vs. Neutral Interfaces

Most data interface circuits between equipment use bipolar signaling; that is, the circuit switches between a positive voltage (say 5-V dc) for a binary 0 and a negative voltage (say −5-V dc) for a binary 1. (For a discussion of polar and neutral keying, see Chapter 8.) A problem with polar

Table 15

Data Interface Connections

Pin	Ckt	RS-232-D Description	V.24 No.	Name	Common Abbr.*
1	—	Protective Ground	101	Frame Ground	FG
2	BA	Transmitted Data	103	Transmitted Data	TxD
3	BB	Received Data	104	Received Data	RxD
4	CA	Request to Send	105	Request to Send	RTS
5	CB	Clear to Send	106	Clear to Send	CTS
6	CC	Data Set Ready	107	Data Set Ready	DSR
7	AB	Signal Ground	102	Signal Ground	SG
8	CF	Received Line Signal Detector	109	Data Carrier Detect	CD
9	—	(Reserved for Data Set Testing)			
10	—	(Reserved for Data Set Testing)			
11	—	Unassigned			
12	SCF/CI	Sec. Rec'd Line Sig. Detector	122	Backward Channel Received Line Signal Detector	SCD
13	SCB	Sec. Clear to Send	121	Backward Channel Ready	SCTS
14	SBA	Sec. Transmitted Data	118	Transmitted Backward Channel Data	STxD
15	DB	Transmission Signal Element Timing (DCE Source)	114	Transmitter Signal Element Timing (DCE Source)	TxC
16	SBB	Sec. Received Data	119	Received Backward Channel Data	SRxD
17	DD	Receiver Signal Element	115	Receiver Signal Element	RxC
18	LL	Local Loopback			
19	SCA	Sec. Request to Send	120	Transmitted Backward Line Signal	SRTS
20	CD	Data Terminal Ready	108/2	Data Terminal Ready	DTR
21	RL/CG	Remote Loopback/Signal Quality Detector	110	Data Signal Quality Detector	SQ
22	CE	Ring Indicator	125	Calling Indicator	RI
23	CH/CI	Data Signal Rate Selector	111	Data Rate Selector	
			112	Data Rate Selector	
24	DA	Transmit Signal Element Timing (DTE Source)	113	Transmitter Signal Element Timing, (DTE Source)	ETxC
25	TM	Test Mode			

*Most abbreviations in this column are generally recognized by association with their full names. Exceptions are: ETxC = External Transmitter Clock, RxC = Receiver Clock and TxC = Transmitter Clock.

signaling is that the power supply must be able to supply both positive and negative voltages, thus adding complexity and cost. Another is that integrated circuits called line drivers and receivers are needed to convert the normal neutral signals used in most digital circuits to and from polar line signals.

Some designers avoid the extra power-supply and bipolar line-driver/receiver-chip requirements by simply transfering neutral signals between equipment, typically at so-called TTL levels of approximately 0 and + 5-V dc. Neutral TTL-level signaling has the disadvantages of poor noise immunity and signal distortion. This method of interfacing is viable only for short lines and where the designer is confident that the equipment at the other end of the line will be using TTL levels, not a standard interface circuit. It is considered to be an expedient, and as such, may cause incompatibility when equipment is used for different purposes.

RS-232-C, V.24 and V.28

EIA Standard RS-232-C has to be the ubiquitous interface standard of the age. Its DB25 connector, shown in Fig. 25, is manifest everywhere you go these days: on office equipment, on the customer side of point-of-sale terminals at check-out counters, on personal computers. It is not unknown for avid packeteers to have RS-232-C receptacles in various rooms of the house, even the bathroom.

RS-232-C's international counterparts are CCITT Recommendations V.24 and V.28. V.24 describes the functions of each circuit. V.28 specifies the electrical parameters: voltages, impedances, load capacitance and signal characteristics. Specifications for the 25-pin connector will not be found in these documents but in ISO 2210 which also assigns pin numbers to V.24 circuits. The commercial types are the TRW-Cinch D-Subminiature DB25P ad DB25S or equivalent by other manufacturers. The male pin connector (DB25P) is normally installed on the DTE and the female connector (DB25S) on the DCE. Pin assignments for RS-232-C and V.24 are shown in Table 15. Designers should find EIA Industrial Electronics Bulletin No. 9 useful, as it contains application notes for RS-232-C.

Signal Levels

The voltage ranges for RS-232-C and V.24 are shown in Fig. 26. Most signals more positive than +5 V and more negative than −5 V will operate the circuit successfully. It is common practice to supply +12-V and −12-V power to MC1488 quad line drivers, which are used in conjuction with MC1489 quad line receivers.

If you do much RS-232-C and V.24 interfacing, you will soon memorize the functions most-used pins: 1 through 8 and 20. These nine pins are virtually always used. The remaining pins may or may not be used depending on the circumstances. You will find RS-232-C cables with anywhere from the basic 9 to all 25 lines. Similarly, modems and other digital equipment advertised as "RS-232-C compatible" can be expected to use signal voltage levels specified in RS-232-C and V.28, and at least do not violate the standard. However, they may not implement the entire standard and may leave out certain control lines which may or may not be needed for your applications. Also, particularly where space is at a premium, a connector smaller than the DB25 may be used.

DB25 connectors come in many varieties — designed for mounting on panels or printed-circuit boards and for terminating cable assemblies. If possible, use shielded hoods and cables to reduce the possibility of electromagnetic interference. Keep an

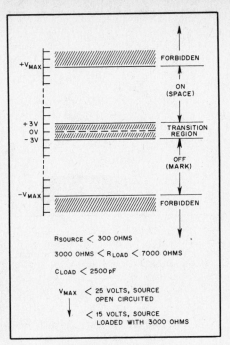

Fig. 26 — RS-232-C and V.28 voltage limits.

eye peeled for the types of DB25 hardware you need at flea markets.

Gender Changes

Although the standards are clear that the male DB25P goes on the DTE and the female DB25S on the DCE, many designers get them mixed up. It is not uncommon to see equipment with only the female DB25S connectors and cable assemblies with male DB25Ps. It is a good idea to have a gender changer handy when connecting equipment, just in case someone got it backwards. A gender changer can be easily made by physically mounting two DB25Ss or DB25Ps back to back and connecting correspondingly numbered pins together with ribbon cable or individual wires.

Null Modems

Another useful gadget to have handy is a so-called *null modem*. That is a pair of DB25 connectors of the appropriate gender(s) mounted back to back and wired pin for pin except that pins 2 (transmit data) and 3 (receive data) are crossed and pins 4 (request to send) and 5 (clear to send) are crossed. This may take care of the problem of connecting two pieces of equipment both wired as DCEs or DTEs. For example, a computer serial port intended for a printer is usally wired as a DCE, and the printer as a DTE. If one or the other is not wired that way, you may find that the equipment has an internal jumper to swap lines 2 and 3. If not, a null modem will do the trick.

RS-422-A, V.11 and X.27

The RS-422-A specifies differential, balanced circuits for rates up to 100 kilobauds at distances up to 4000 feet or 10 megabauds up to 40 feet. D-Subminiature DB37 (37-pin) connectors wired according to ISO 4902 are used. RS-422-A is compatible with CCITT Recommendations V.11 and X.27. However, it is not interoperable with RS-232-C, MIL-STD-188C or CCITT V.28 and V.35. It is not compatible with RS-423-A without some reconfiguration.

RS-423-A, V.10 and X.26

RS-423-A is an unbalanced, bipolar circuit capable of distances up to 4000 feet at 3 kilobauds or 40 feet at 300 kilobauds. It is compatible with CCITT Recommendations V.10 and X.26. RS-423-A can interoperate with RS-232-C circuits; EIA Industrial Electronics Bulletin No. 12 deals with implementation. RS-423-A and V.10/X.26 use D-Subminiature DB37 (37-pin) connectors wired according to ISO 4902.

RS-449

RS-449 is capable of signaling rates up to 2 Mbit/s at a cable length of 667 feet. The standard has 10 more interchange circuits than RS-232-C and specifies two D-Subminiature DB37 and DB9 (37- and 9-pin) connectors. EIA Industrial Electronics Bulletin No. 12 has application

Table 16
EIA RS-449 37-Pin Connector Assignments

Pin	Direction	Mnemonic	Circuit name
1	—	SHIELD	
2	from DCE	SI	Signaling rate indicator
3	—	SPARE	
4	to DCE	SD	Send data
5	from DCE	ST	Send timing
6	from DCE	RD	Receive data
7	to DCE	RS	Request to send
8	from DCE	RT	Receive timing
9	from DCE	CS	Clear to send
10	to DCE	LL	Local loopback
11	from DCE	DM	Data mode
12	to DCE	TR	Terminal ready
13	from DCE	RR	Receiver ready
14	to DCE	RL	Remote loopback
15	from DCE	IC	Incoming call
16	to DCE	SF/SR	Select frequency Signaling rate selector
17	to DCE	TT	Terminal timing
18	from DCE	TM	Test mode
19	—	SG	Signal ground
20	from DCE	RC	Receive common
21	—	SPARE	
22	to DCE	SD	Send data
23	from DCE	ST	Send timing
24	from DCE	RD	Receive data
25	to DCE	RS	Request to send
26	from DCE	RT	Receive timing
27	from DCE	CS	Clear to send
28	to DCE	IS	Terminal in service
29	from DCE	DM	Data mode
30	to DCE	TR	Terminal ready
31	from DCE	RR	Receiver ready
32	to DCE	SS	Select standby
33	from DCE	SQ	Signal quality
34	to DCE	NS	New signal
35	to DCE	TT	Terminal timing
36	from DCE	SB	Standby indicator
37	to DCE	SC	Send common

Table 17
EIA RS-449 9-Pin Connector Assignments

Pin	Direction	Mnemonic	Circuit name
1	—	SHIELD	
2	from DCE	SRR	Secondary receiver ready
3	to DCE	SSD	Secondary send data
4	from DCE	SRD	Secondary receive data
5	—	SG	Signal ground
6	—	RC	Receive common
7	to DCE	SRS	Secondary request to send
8	from DCE	SCS	Secondary clear to send
9	to DCE	SC	Send common

Fig. 27 — ISO 2593 rack-and-panel connector. This connector is usually identified with V.35 and is found on modems operating at speeds in excess of 20,000 bit/s.

Table 19
IEEE 488 Bus Pin Allocations

Pin	Mnemonic	Signal
1	D101	Data line 1
2	D102	Data line 2
3	D103	Data line 3
4	D104	Data line 4
5	EOI(24)	End or identify
6	DAV	Data valid
7	NRFD	Not ready for data
8	NDAC	Not data accepted
9	IFC	Interface clear
10	SRQ	Service request
11	ATN	Attention
12	SHIELD	Shield
13	D105	Data line 5
14	D106	Data line 6
15	D107	Data line 7
16	D108	Data line 8
17	REN(24)	Remote enable
18	GND(6)	Ground for pin 6
19	GND(7)	Ground for pin 7
20	GND(8)	Ground for pin 8
21	GND(9)	Ground for pin 9
22	GND(10)	Ground for pin 10
23	GND(11)	Ground for pin 11
24	GND(logic)	Ground for pins 5 and 27

notes on interconnection between RS-449 and RS-232-C.

V.35

CCITT Recommendation V.35 was written as a standard for modems operating at 48 kbit/s. However, its interface specifications and the 34-pin Winchester rack-and-panel connector (specified in ISO 2593, not V.35) have become standard in the United States for interchange circuits operating at 48, 56 and 64 kbit/s. The ISO 2593 connector is shown in Fig. 27.

IEEE 488 Parallel Interface

This is a parallel bus for limited distances on the order of 67 feet. It is of interest to amateurs as a control bus for test equipment and some communications equipment. The Commodore PET personal computer used the IEEE 488 bus for control of its peripherals (disk drive, printer, etc.). Most new test instruments being purchased for industrial use have the IEEE 488 bus so that testing can be computer controlled. Since instruments with IEEE 488 are rela-

tively new, they are not generally available at prices that amateurs can afford. However, that should change as laboratories get newer equipment and older equipment with the IEEE 488 bus becomes available on the surplus market. This bus uses a unique 24-pin connector as shown in Fig. 28.

OPERATIONAL MODES

There are a number of terms that describe the operation of circuits or connections, grouped under the category variously called *operational*, *communication* or *transmission* modes. There are some minor differences between data-communications and telephone-company usage within the United States, and between U.S. and international usage. Since packet radio is used internationally, it is important to remember that different interpretations of the same terms may cause problems in understanding terminology between packeteers in different

countries. Also, some confusion can occur between the operational mode of the communications and how many radio frequencies are being used. For example, it is possible to have a half-duplex data exchange using one frequency (often called simplex) or two frequencies (called duplex).

Although this subject is treated here under physical-layer considerations, there are parallel uses of the terms defined below at higher ISO protocol levels.

Fig. 29 will help in understanding the terms discussed below.

Simplex (SX)

According to the IEEE, simplex operation is "A method of operation in which communication between two stations takes place in one direction at a time." The ITU

Table 18
ISO 2593 Pin Allocations for V.35 Interfaces

Pin	Circuit	Direction	Function	Pin	Circuit	Direction	Function
A	101	Common	Protective ground or earth	AA	114	to DTE	Transmitter signal element timing B-wire
B	102	Common	Signal ground or common return				
C	105	from DTE	Request to send	P	103	from DTE	Transmitted data A-wire
D	106	to DTE	Ready for sending	S	103	from DTE	Transmitted data B-wire
E	107	to DTE	Data set ready	U	113	from DTE	Transmitter signal element timing A-wire
F	109	from DTE	Data channel received line signal detector	Z	—	—	F_3
H	108/1	from DTE	Connect data set to line	W	113	from DTE	Transmitter signal element timing B-wire
	108/2	from DTE	Data terminal ready				
	125	to DTE	Calling indicator	BB	—	—	F_3
K	—	—	F_1	CC	—	—	F_4
L	—	—	F_2	DD	—	—	F_5
M	—	—	F_1	EE	—	—	F_4
N	—	—	F_2	FF	—	—	F_5
R	104	to DTE	Received data A-wire	HH	—	—	N_1
T	104	to DTE	Received data B-wire	JJ	—	—	N_2
V	115	to DTE	Receiver signal element timing A-wire	KK	—	—	N_1
X	115	to DTE	Receiver signal element timing B-wire	LL	—	—	N_2
Y	114	to DTE	Transmitter signal element timing A-wire	MM	—	—	F
				NN	—	—	F

N = Pins permanently reserved for national use.
Pins HH, JJ and KK are used in the U.K. for transmitter-clock control.

F = Pins reserved by ISO, not for national use.
Subscripts indicate pins to form pairs.

Table 20
American and International Interface Standards

American	Circuits	Electrical	Connector
EIA RS-232-C	CCITT V.24	CCITT V.28	ISO 2110
EIA RS-449/432-A	CCITT V.24	CCITT V.10	ISO 4902
EIA RS-449/422-A	CCITT V.24	CCITT V.11	ISO 4902
(no equivalent)	CCITT V.24	CCITT V.35	ISO 2593

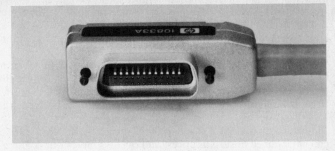

Fig. 28 — 24-pin connector used for IEEE 488 cables.

Radio Regulations define it as an "Operating method in which transmission is made possible alternately in each direction of a *telecommunication* channel..." A footnote says that simplex operation may use either one or two frequencies. CCIR Rec. R.140 says "simplex (circuit) permitting the transmission of signals in either direction, but not simultaneously."

Radio amateurs often use the term simplex to indicate that only one frequency is used for a VHF FM circuit. This is the current amateur digital transmission mode, where two or more stations involved in a contact operate on exactly the same frequency and take turns transmitting and receiving. Simplex operation keeps equipment requirements to a minimum by permitting sharing of some equipment for both transmitting and receiving. A transceiver (in contrast with separate receiver and transmitter) and a common antenna may be used.

In some other uses of the term, simplex transmission is in only one direction with no reverse direction possible. CCIR R.140 calls this a *unidirectional connection*. One-way operation is seldom used in data communications since there is no return channel or way of reversing the channel to detect and correct errors. Nevertheless, there are some Amateur Radio one-way transmission applications, usually point to multipoint, such as for transmission of bulletins and beacon operation.

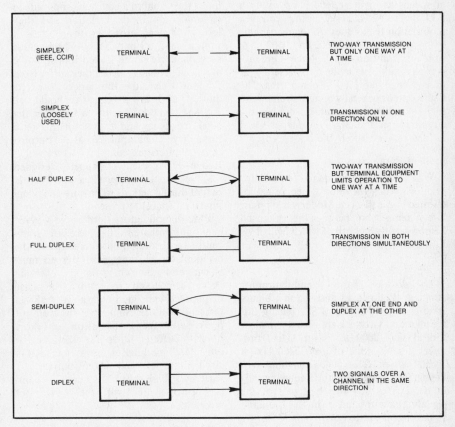

Fig. 29 — Operational modes. U.S. and international usage of these terms varies. See discussion in text.

Half Duplex (HDX)

A half-duplex circuit can transmit in either direction, but only one direction at a time; so sayeth the IEEE and other references. That sounds just like simplex, doesn't it? The clue to understanding the difference between simplex and half-duplex operation is in the CCITT R.140 definition: "half duplex circuit (or connection) — A circuit (or connection) capable of duplex operation, but which, on account of the nature of its termination, can be operated alternately only." In plain English, the transmission channel is capable of simultaneous operation in both directions but the terminal equipment can only send or receive, not do both at the same time.

Full Duplex (FDX)

In a full-duplex circuit, information can flow both ways simultaneously. The IEEE and CCIR definitions agree on this, although CCIR calls it simply *duplex* operation. The CCIR Radio Regulations include a footnote to the effect that "duplex operation require(s) two frequencies in radiocommunication." CCIR Rec. R.140 has another term: "duplex, two way simplex (connection) — A circuit permitting the exchange of signals in both directions."

Full-duplex operation is not yet in general use in packet radio. But it will be required for high-speed packet trunks where simplex or half-duplex circuits cannot provide sufficient peak capacity. Full duplex is inherently more complex than simplex because the transmitter and receiver at the same site must operate simultaneously. This requires separation or duplexing of receiving and transmitting antennas as is normal practice in FM voice repeaters.

Semi-Duplex

This is a term found in the ITU Radio Regulations to describe "A method which is *simplex operation* at one end and *duplex operation* at the other."

Diplex

CCIR Rec. R.140 defines this as "Permitting simultaneously and in the same direction, the transmission or reception of two signals over a circuit or channel."

MODEMS

Modem is a contraction of *modulator-demodulator*. In telephone terminology, it is called a *data set*. A modem is used to convert digital baseband signals to and from analog form for transmission over analog media. Analog communications systems designed to transmit speech have an analog baseband bandwidth of about 3 kHz. There is a relationship between the analog bandwidth and the data rates that can be accommodated. It is relatively easy to design a modem that can handle up to 1200 bit/s in a 3-kHz analog channel. Modems operating at 300 and 1200 bit/s can be obtained at surplus prices, and can be inexpensively fabricated by the amateur. The state of the art permits transmission at speeds as high as 9600 bit/s within a 3-kHz analog channel using complex modulation techniques. Modems capable of operating at 2400, 4800 and 9600 bit/s over unconditioned telephone lines are commercially available but presently at greater cost than 1200-bit/s modems. The higher speeds generally require circuits with higher signal-to-noise ratios; so when the circuit is not the best, the higher-speed modems have slower *fallback* speeds.

Modem Parameters

When they work, modems are taken for granted. When they don't, people get interested in the subject. Modems live on the line of demarcation between the digital and analog worlds, in both of which Murphy's Law claims sovereignty.

Modulation

The most-used form of modulation for Amateur Radio data transmission is binary frequency-shift keying (FSK). Using this technique, a binary 1 is translated to a mark tone at (say) 1200 Hz, while a binary 0 produces a space tone of (say) 2200 Hz. A special case of FSK is called minimum-shift keying (MSK) in which the frequency shift in hertz is exactly one half that of the signaling rate in bauds. Thus a 1200-baud signal that is shifted 600 Hz is MSK.

Phase-shift keying (PSK) is another popular form of angle modulation used for data transmission. Its use in Amateur Radio, thus far, has been limited to satellite telecommand and telemetry, but interest is growing in using various forms of PSK for high-speed packet radio. While binary phase-shift keying (BPSK) is robust, it is not particularly spectrum efficient. *M-ary* PSK schemes such as 4-ary PSK (4PSK) or 8-ary PSK (8PSK) permit transmission of more bits per second at a lower baud rate and are used in many high-speed modems designed for telephone use.

Amplitude-shift keying (ASK) is not presently used in amateur packet radio but is a possibility for the future, particularly in combination with PSK, to achieve high speeds within relatively narrow bandwidths.

Some additional information on digital modulation techniques is presented in Chapter 9.

Demodulation

The function of a demodulator in a data modem is to convert the analog signal back to its digital baseband. Designers will testify that the modulator is the easy part of modem design. Demodulators are the downfall of most modem projects. They take even longer than the usual project-completion rule of multiplying the time estimate by 2 and using the next larger time increment.

FSK demodulators are among the simpler ones. There are two main design approaches: the FM type and the AM type. The FM type of demodulator first hard-limits the signal to a constant amplitude to eliminate amplitude variations. It then looks for zero crossings to produce a square-wave output. Many FM demods nowadays use phase-locked loops. While adequate in most respects, they suffer from dropouts when either the mark or space frequency fade below a certain threshold level. An AM demodulator has a filter for the mark frequency and a filter for the space frequency. The amplitudes of the outputs of these filters are compared to see which frequency is present. A properly designed AM type demod can produce good digital output during periods when only one signal (mark or space) is present.

PSK demodulators need some way of determining phase shifts. There is no ready source of absolute phase reference at the demodulator, so some phase reference must be derived by the demodulator. In normal PSK, it is necessary to compare the phase of the received signal with a fixed phase-reference source. Some systems transmit phase-reference information in short bursts. Differential phase-shift keying (DPSK) demodulation does not need a fixed phase reference signal but decides whether a phase shift has occurred by comparing one symbol with the last symbol received.

Synchrony

Most of the lower-speed modems are asynchronous and operate at a variable rate. You will note this when seeing a speed reference such as "0-300" bit/s. Of course, a 0-bit/s modem is nothing but a power supply, but this designation tells you that the modem is capable of maintaining a 0 or 1 logic state indefinitely. When each character is started and when transitions may occur therein are functions of the data keying the modulator. When receiving data, the demodulator will recognize changes of state that occur at any time, perhaps within certain limits to reject noise.

Higher-speed modems are usually synchronous. The modulator clocks out data at a fixed rate. The demodulator duplicates or regenerates the clock from the received data. Bit-synchronous demodulators have the advantage of knowing when a bit is to occur and can use various decision strategies to determine which of two binary states (or which of 4 dibit states, which of 8 tribit states, etc.) was received. Synchronous demodulators are usually assigned a 3-dB S/N

advantage over asynchronous demodulators.

Frequency Modes: Originate and Answer

Modems designed for full-duplex telephone-line communications divide the audio-frequency spectrum into high and low frequency bands. Transmission in one direction is possible by modulating a tone (or tones) in the higher-frequency band, while transmission in the opposite direction is possible using lower-frequency tone(s). To eliminate confusion on the telephone, the calling (originating) station (say) transmits on the low tone(s), and the called (answering) station transmits on the high tone(s). The two stations stay with the tone(s) they chose when the circuit was established; the tones are not changed as communications go back and forth.

In the early days of time-shared computers, many terminals were equipped with *originate-only* modems because they only called the computer center, never answered calls. This was a way of cutting costs at the terminal stations. Similarly, computer centers that only answered calls minimized cost by installing *answer-only* modems. These one-way modems saved money, but when they became available on the surplus market were a source of frustration for personal-computer users who wanted to both originate and answer calls. Amateur Radio operators found that originate-only or answer-only modems could be modified to change the tones to the same frequency band (high or low) for packet radio simplex operation.

Carrier Detection

Most demodulators have a carrier-detect (CD) output. It is more formally called "Received Line Signal Detector." In telephone modems, a 0 condition on the CD output means that no signal is being received or that the signal being received is unsuitable for demodulation. In a half-duplex circuit, whenever there is a 1 output on the CD line, the modem is inhibited from transmitting. That way your modem knows that there is a live modem of its own species on the other end of the circuit.

In Amateur Radio, our need for this signal is different from that of the telephone system. Generally, we want the modem to know when there is anyone else using the channel. Increasingly, VHF packet-radio channels are coordinated for that use and not voice, so anyone else on the channel is likely another packet station. However, there are situations where VHF voice and packet communications are mixed, as in some 2-meter repeaters. Unfortunately, modem CD outputs will not always go to 1 when there are voice signals on the line. In this case, squelch output should be used to inhibit packet transmission. That way, the packet transmissions avoid stepping on voice users of the repeater.

Similarly on HF, packet operations usually share frequencies with RTTY activities. If both sets of users equip their modems/transceivers with a carrier-detect feature, then the number of collisions will be reduced.

Transmission Impairments

Most modems now in use are intended for use with radios designed for voice communications. The human ear is very tolerant of distortion. Thus, you may not even suspect that your trusty hand-held transceiver has audio that would gag a data modem. Of course, we are talking about a whole system, not just your radio. The radio at the other end of the circuit and any artifacts introduced by the propagational medium need to be taken into account. When using voice radios for packet radio, one way to find out if it will work is to simply try it. Most amateurs do not have the test equipment to accomplish all the testing that is needed to check out a system piece by piece. We would hope, however, that such testing would be performed before offering radio equipment for general data-communications use.

Frequency Response

Many radios have nonuniform amplitude-vs.-frequency responses. Fig. 30 illustrates an audio-frequency response that might be seen through a radio system. This may be the result of roll-off high frequencies in an audio amplifier or using an IF filter that is too narrow. Low-frequency losses could be caused by use of too small a value capacitor used for interstage coupling in an amplifier. An IF filter with excessive ripple in the passband could wreak havoc with the frequency response. Similar problems could be the fault of poor IF alignment. Virtually all voice radios can be used for packet operation — many without modification of any kind. Others may require some minor surgery to make them suitable for data communications.

Phase-Shift and Envelope-Delay Distortion

Propagation through an audio communications channel should vary uniformly with frequency. If the propagation time is not uniform, the components of a complex waveform (for example, the fundamental frequency and several harmonics) will arrive out of their proper phase relationships. This will result in a distorted waveform. Fig. 31 illustrates linear and nonlinear phase-shift characteristic, as well as the corresponding envelope delays.

Amplitude Nonlinearity

This is not usually a serious problem with using voice radios for data communications systems with binary amplitude states. However, any nonlinear distortion in the voice-frequency channel adds harmonics to the fundamental modulating signal. Amateur Radio transmitter microphone-input circuits typically have substantial amounts of compression with time constants optimized for speech. Unwanted

products may be generated, especially if a data signal is at a level high enough to be well into the compression or clipping region.

Nonlinear distortion becomes serious in certain high-speed modems which use quantized amplitude modulation, possibly in combination with phase-shift keying. In such modems nonlinear response in the channel will make the amplitude differences between adjacent levels uneven and could result in falsing, particularly in the presence of noise.

Equalization

Various forms of distortion, such as nonuniform frequency response and envelope-delay distortion can be corrected by equalization. An equalizer is a circuit designed to compensate for undesired amplitude-frequency and/or phase-frequency characteristics. Equalizers can be designed to overcome the poor response in a radio receiver or an entire system if the distortion at the transmitter and that introduced by the medium are known. An equalizer works by setting the gains of an audio-frequency filter. *Automatic equalization* and *adaptive equalization* are techniques that can automatically compensate for circuit distortion. Automatic equalizers adjust to circuit conditions upon reception of a special bit sequence, called *training*, sent just before transmission of data. Adaptive equalizers make continuous adjustments during reception.

Frequency-Translation Error

When an FM voice radio is used with an AF modem, the received tones are at the exact frequencies produced by the modulator at the transmitting station. However, if the modulator and demodulator are not tuned to precisely the same frequencies, some distortion in the received data should be expected, particularly under poor S/N conditions.

Data transmission with AF modems through SSB voice radios is a different matter. Even small mistuning of the receiver relative to the transmitter frequency can cause distortion. Doppler shift, if not corrected, can also corrupt received data. Under these circumstances accurate frequency control and a demodulator tuning indicator are recommended.

Noise

Periodic or random noise will affect the performance of a digital demodulator. Some types of noise, particularly transients, may simply be an annoyance to voice communications. It takes only one bit error to corrupt a packet. Impulse noise on the order of 1-ms duration and decaying over 4 ms will likely have little effect on a 45-baud RTTY signal with a 22-ms data pulse. But at 1200 bauds, that same pulse could wipe out 1 to 4 bits.

Just as in any other radio application, it is necessary to minimize the noise level

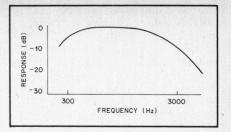

Fig. 30 — Amplitude-vs.-frequency response of a radio system designed for voice communications. Note the attenuation of both highs and lows.

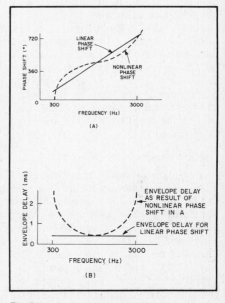

Fig. 31 — At A is the phase-shift-vs.-frequency response of a radio system. B shows the corresponding envelope delay distortion.

presented to the data demodulator. Proper siting, use of directive antennas with minimal noise pickup and low-noise receiving preamps are desirable for packet-radio stations. These considerations are of great importance for meteor-scatter and satellite reception. Careful selection of demodulator input filter shape and bandwidth can contribute to reducing the impact of the noise received by the modem. While most demodulators perform flawlessly in the absence of noise, some fail sooner than others as the signal-to-noise ratio is decreased. What separates the best radio modems from the pack is their ability to perform in the presence of interference.

Jitter

Jitter comes in many flavors and may be introduced almost anywhere throughout a system. *Phase jitter* consists of abrupt variations in phase or unwanted angle modulation. It can be described as the received data signal zero crossings occurring at the wrong time. Phase jitter can be introduced in the modulation process. It may be introduced by power-supply hum modulating the master oscillator or by the data modulator clocking on the wrong pulse in a train. Jitter on the order of 15 degrees or more can be expected to cause misinterpretation of data by a demodulator, depending on how sampling is performed. If the demodulator is designed to look for edges of pulses any jitter is likely to introduce errors. If the demodulator samples in the middle of the pulse period, or integrates energy over the entire pulse period, jitter may not be a serious problem. In addition to phase jitter, which may occur throughout a transmission, *phase hits* are a transient unwanted angle modulation that may last 4 ms or longer, according to the telephone definition.

Amplitude jitter is not usually a problem for angle-modulated data communications but could be for systems using quantized amplitude modulation. In telephone terminology, *gain hits* are transient phenomena lasting 4 ms or more where changes in amplitude exceed 12 dB. Gain hits in the negative direction are called *dropouts*.

Modem Standards

Two organizations that have written the lion's share of modem standards are the AT&T and CCITT. A summary of these standards is given in Table 16. Many modems for different applications were developed by AT&T and are known by their Bell designations (for example, Bell 103). CCITT modems closely parallel Bell types but are not necessarily interoperable with them. For example, both the Bell 103 and the CCITT V.21 can operate at 300 bit/s, and are alike in many respects, but the two use different sets of tone frequencies and cannot intercommunicate. There is the same similarity and incompatibility between Bell 202 and CCITT V.23. While international technocrats have been having their disagreements, IC designers have been designing single-chip modems, such as the Am7910 that will do Bell 103, V.21, Bell 202 and V.23. "The man wants a Bell 202, pull MC1 high!" See the construction project, An LSI Modem for Amateur Radio, in Chapter 29.

Bell and CCITT Modems

Bell 103/113 Family

The Bell 103/113, usually abbreviated Bell 103, is of particular interest to amateurs for both telephone-line and packet-radio communications. Many personal computers use Bell 103-compatible modems for 300 bit/s phone-line data transmission, although the Bell 212A (1200-bit/s) modems are gradually replacing them.

These modems are part of the Bell 100-series data sets and are also called the Bell 103/113 family. Model number 103 designates an originate/answer modem, whereas the letter suffix denotes the configuration. Model number 113 can be either

Table 21
Bell 103/113 Modem Frequency Scheme

Mode	Originate	Answer
Transmit	F1M 1270 Hz mark	F2M 2225 Hz mark
	F1S 1070 Hz space	F2S 2025 Hz space
Receive	F2M 2225 Hz mark	F1M 1270 Hz mark
	F2S 2025 Hz space	F1S 1070 Hz space

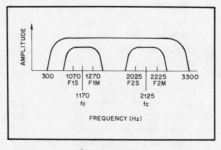

Fig. 32 — Bell 103/133 modem frequencies. The call-originating station transmits on the lower-frequency band and receives on the higher-frequency band. The called (answering) station does the opposite.

originate or answer, as differentiated by the letter suffix. Specific models are:

103A: Older originate/answer, single-channel modem for installation in Model 33 or 35 Teletype® equipment.

103E (Computer Site): Originate/answer, up to 40 in a cabinet.

103F: For leased lines only.

103G: Originate/answer, single-channel modem that includes an integrated telephone set.

103H: Originate/answer, single-channel modem for direct installation in Model 37 Teletype sets.

103J, 103JR: Originate/answer, single-channel modem. The 103J was the model used prior to the FCC Registration Program. The "R" suffix indicates registration under Part 68 of the FCC rules.

113A: Originate-only, single-channel "bare-bones" modem for terminal installations. Early 113A-L1 and -L1A versions did not require ac power supplies as they were powered from the telephone line. Later models with L1/2 and L1A/2 suffixes require ac power supplies.

113AR: Registered version of the 113A.

113C: Full-featured originate-only, single-channel modem used prior to registration.

113CR: Registered version of 113C.

113B: Answer only, in 20-channel increments. Older modem used in multiline computer installations.

113D: Unregistered version of 113DR.

113DR: Answer-only registered modem in current use by the Bell System. Available in individual and multiline configurations.

The Bell 103/113 family operates at

speeds up to 300 bit/s. Although used since its introduction for 110-bit/s ASCII teleprinters, it is used nowadays almost exclusively for computers at 300 bit/s.

The spectrum is divided into two bands. Each band has a carrier which uses frequency-shift keying (FSK) between two frequencies. The two lower-band frequencies are called F1M (frequency 1 mark) and F1S (space). The frequencies in the higher band are called F2M and F2S. The originating station transmits on F1 and receives on F2. The answering station transmits on F2 and receives on F1. See Table 21 and Fig. 32.

Bell 103/113 modems normally use the EIA RS-232-C interface. The Clear to Send circuit (from modem to DTE) is on when the modem has established a connection. The Transmitted Data circuit can be used only when Clear to Send is on. Received Data (modem to DTE) is the circuit that delivers the data received from the other modem. When on, the Carrier Detect (modem to DTE) shows that the data carrier is being received from the modem at the other end of the circuit. When Carrier Detect is off, this circuit is held in a marking condition. An on condition on the Data Set Ready circuit (modem to DTE) indicates that the modem is connected to the telephone line and in the data mode. An on condition on the Ring Indicator (modem to DTE) circuit tells the DTE that a ringing signal has been received. Data Terminal Ready (DTE to modem), when on, permits the modem to answer incoming calls automatically. An off condition tells the modem to disconnect from the line when the call is completed.

CCITT V.21

This is the modem used for low-speed (110, 200 and 300-bit/s) asynchronous data transmission in Europe and elsewhere outside of North America. The frequency scheme for the V.21 is shown in Table 22 and Fig. 33. V.21 is similar to, but uncompatible with, Bell 103/113. It is unfortunate that you cannot use a Bell 103 modem on the telephone to call a computer in Europe using a V.21 modem, or vice versa. However, you *can* use a Bell 103 to communicate with a V.21 modem when using HF packet radio. Although the specific frequencies are different, the mark and space frequencies for both modems are 200 Hz apart.

Bell 202

The Vancouver Amateur Digital Communications Group (VADCG) selected the Bell 202 modem for their 2-meter packet-radio network because some were available on the surplus market. It is widely used in North America for VHF packet radio using AFSK on an FM carrier and has been used experimentally on 10 meters using FSK (AFSK to an SSB transceiver).

The Bell 202 is is a half-duplex modem designed for asynchronous, binary, serial

Table 22
CCITT V.21 Modem Frequency Scheme

Mode	Channel 1 (Calling station)	Channel 2 (Called station)
Transmit	F_A 1180 Hz 0 bit F_Z 980 Hz 1 bit	F_A 1850 Hz 0 bit F_Z 1650 Hz 1 bit
Receive	F_A 1850 Hz 0 bit F_Z 1650 Hz 1 bit	F_A 1180 Hz 0 bit F_Z 980 Hz 1 bit

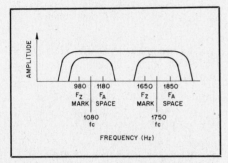

Fig. 33 — CCITT V.21 modem frequencies. The call-originating station transmits on the lower-frequency band and receives on the higher-frequency band similar to Bell 103/113 modems.

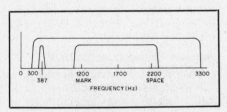

Fig. 34 — Bell 202 modem frequencies. The 387-Hz reverse channel on some Bell 202 modems is not used for amateur packet radio.

data transmission at speeds up to 1200 bit/s. The maximum bit rate is also a function of the transmission circuit; Bell 202 modems can handle up to 1800 bit/s on conditioned lines. It uses FSK modulation with the mark and space frequencies of 1200 Hz and 2200 Hz, respectively. See Fig. 34. Some models have a 5-bit/s on/off-keyed (OOK) reverse channel on 387 Hz used primarily for a break signal to signal when to turn the circuit around and for certain forms of error control. To defeat echo suppressors found on North American long-distance telephone lines, the modem sends a 2025-Hz echo-suppressor disabling tone for 300 ±50 ms. Because echo suppressors turn on after 100 ms of no signal, the modem is designed to keep its forward- or reverse-channel transmitter on at all times. Bell 202 modems use EIA RS-232-A, -B or -C interfaces, according to when they were designed.

Specific Bell model numbers are: 202A and 202B: Superseded by 202C and 202D.

Table 23
Summary of Bell and CCITT Modem Characteristics

Standard	Speed in bit/s Forward	Fallback	Reverse	Sync/async	Modulation	Transmitting frequencies
Bell 103/113	0-300			Async	FSK	Originate: 1070 Hz space 1270 Hz mark Answer: 2025 Hz space 2225 Hz mark
CCITT V.21	0-300			Async	FSK	Originate: 980 Hz mark 1080 Hz space Answer: 1650 Hz mark 1850 Hz space
Bell 202	1200			Async	FSK	Forward: 1200 Hz mark 2200 Hz space
			5		OOK	Reverse: 387 Hz
CCITT V.23	1200			Async	FSK	Forward: 1300 Hz mark 2100 Hz space
		600			FSK	Fallback: 1300 Hz mark 1700 Hz space
			75		FSK	Reverse: 390 Hz mark 450 Hz space
Bell 212A	1200			Sync	4DPSK	Originate: 1200 Hz Answer: 2400 Hz
		300			FSK	Same as Bell 103
CCITT V.22	1200			Both	4DPSK	Modes I and II: Originate: 1200 Hz Answer: 2400 Hz
		600			2DPSK	Modes II and IV: Originate: 1200 Answer: 2400
		300			FSK	Same as V.21
Bell 201CR	4800			Sync	4DPSK	Carrier: 1800 Hz
CCITT V.22bis	2400			Both	QAM	Originate: 1200 Hz Answer: 2400 Hz
		1200		Both	4DPSK	Same as V.22
CCITT V.26	2400			Sync	4DPSK	Carrier: 1800 Hz
			75	Async	FSK	Same as V.23
CCITT V.26bis	2400			Sync	4DPSK	Carrier: 1800 Hz
		1200		Sync	2DPSK	
			75	Async	FSK	Same as V.23
CCITT V.26ter	2400			Sync	4DPSK	Carrier: 1800 Hz
		1200		Sync	2DPSK	
			75	Async	FSK	Same as V.23
Bell 208BR	4800			Sync	8DPSK	Carrier: 1800 Hz
CCITT V.27	4800			Sync	8DPSK	Carrier: 1800 Hz
			75	Async	FSK	Same as V.23
CCITT V.27bis	4800			Sync	8DPSK	Carrier: 1800 Hz
		2400		Sync	4DPSK	
			75	Async	FSK	Same as V.23
CCITT V.27ter	4800			Sync	8DPSK	Carrier: 1800 Hz
		2400		Sync	4DPSK	
			75	Async	FSK	Same as V.23
Bell 203	3600 4800 5400 6400 7200 9600 10800			Sync		
Bell 209A	9600			Sync	QAM	
CCITT V.29	9600			Sync	QAM	Carrier: 1700 Hz
		7200		Sync	QAM	
		4800		Sync	QAM	
CCITT V.32	9600			Sync	QAM	Carrier: 1800
		4800		Sync	QAM	
		2400		Sync	QAM	(Further study)
CCITT V.35	48000			Sync	SSBSC	Carrier: 100 kHz
		40800				
CCITT V.36	48000 56000 64000 72000			Sync	SSBSC	Carrier: 100 kHz
CCITT V.37	72000-160000					

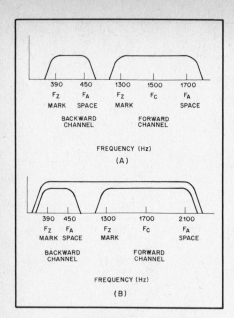

Fig. 35 — CCITT V.23 modem frequencies. Those for Mode 1 (600 bauds) are at A. Mode 2 (1200-baud) frequencies are at B.

202C: Older modem with handset, single channel, self contained. Switched lines — 0-1200 bit/s. Leased lines — 0-1800 bit/s.

202C5, 9 or 11: Without reverse channel.

202C6, 10 or 12: With reverse channel.

202D: Same as 202C without handset; requires 804A auxiliary set.

202E: Older transmit-only data set with telephone set. Reverse channel is optional.

202R: 0-1800 bit/s modem similar to 202D, no auto answer.

202S: 0-1200 bit/s auto answer before FCC registration.

202SR: Registered version of 202S.

202T: 0-1800 bit/s, for leased lines, 1 to 8 in housing.

You may see some newer Bell 202-compatible modems with 75-baud or 150-baud reverse channels. This feature has been added to some modems for leased-line and videotex service. The Am7911 is a single-chip modem with a 150-baud reverse channel in both Bell 202 and V.23 modes.

CCITT V.23

The CCITT V.23 is similar to the Bell 202. Although V.23 1200-baud mark and space frequencies are 100 Hz different than those of the 202, the two type of modems are interoperable under certain conditions. In addition, V.23 has a fallback speed of 600 bauds and uses a different space frequency for that speed. The fallback speed is used when the channel is noisy. Another difference is that the V.23 backward channel uses FSK and is capable of data rates up to 75 bauds. V.23 frequencies are shown in Fig. 35. V.23 also has an echo-suppressor disabling tone of 2100 Hz. V.23 modems use the V.24 interface. Pin 23, data signaling rate selector (circuit no. 111) is used to

tell the modem to change speed.

Bell 212A

In late 1976, the Bell System introduced a full-duplex 1200-bit/s modem — the Bell 212A. Bell 212AR denotes the FCC-registered version now in use. To achieve this within the audio passband of a telephone channel, the bandwidth of each 1200-bit/s signal was reduced by using *dibit*-encoded, differential phase-shift keying (DPSK). In dibit encoding of a PSK signal, two bits are used to determine the phase of the transmitted signal:

Dibit	Phase
00	90°
01	0°
10	180°
11	270°

The originating station transmits on 1200 Hz, whereas the answering station transmits on 2400 Hz, as shown in Fig. 36. In addition, the 212A has a low-speed FSK modem for communication with Bell 103 modems. The operating speed is selected by the modem originating the call. An answering 212A can tell which type of modem it has connected with and automatically adjusts to the right speed and protocol.

The Bell 212A employs a bit-sequence encoder, called a *scrambler*, to prevent transmitting a pathological bit sequence such as a long string of 0s. The scrambler operates according to a fixed algorithm designed to produce a pseudorandom bit pattern, thus having control over the shape of the spectrum produced by keying. Without scrambling, the demodulator would lose synchronization with the dibit clock.

Bell 212A modems are becoming popular with amateurs for telephone-line use as most computer services and some bulletin boards now use 212A modems. Little or no experimentation with the 212A for packet-radio is known as this is written.

CCITT V.22

The CCITT V.22 is similar to the Bell 212A and can interoperate synchronously at 1200 bit/s. V.22 uses the same frequencies as the 212A, namely 1200 Hz (channel 1) for the calling station (originate) and 2400 Hz (channel 2) for the called station (answer). Whenever channel 2 is in use, V.22 modems transmit an 1800-Hz guard tone at all times to suppress in-band telephone signaling.

V.22 has the following alternative modes:

Mode	Operation
I	1200 bit/s sync
II	1200 bit/s async
III	600 bit/s sync
IV	600 bit/s async
V	1200 and 0-300 bit/s async

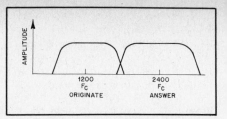

Fig. 36 — Bell 212A modem frequencies for 1200-bit/s full-duplex operation. See Fig. 32 for frequencies when in the Bell 103 mode.

Like the 212A, V.22 uses dibit DPSK. Below are phase changes for the various modes:

Mode	00	01	10	11	(dibits)
I, II	90°	0°	270°	180°	
V	270°	180°	90°	0°	

Mode	0	1	(bits)
III, V	90°	270°	

V.22 modems are supplied in various configurations. Alternative A is synchronous, B is A plus asynchronous, and C is B with Mode V added. Some V.22 modems do not have the 600-bit/s fallback speed.

The V.22 is becoming more widely used in Europe. There is no known packet-radio use known at this time.

Amateur Packet-Radio Modems

The Bell 202 modem predominates 1200-baud VHF packet-radio operation. It has proven to be a good initial choice for VHF FM packet radio because of its surplus availability. It has some problems, however. It is North American standard, not international, although it is to some extent interoperable with CCITT V.23 modems. Bell 202 modems have been used to copy UoSAT-OSCAR 11 signals using the "Kansas City standard", which uses 1200- and 2400-Hz tones. There is the growing belief that the Bell 202 is too slow and should be replaced, initially for intercity trunks, with a modem capable of 9600 bit/s. Also, while the Bell 202 is usable for communications through the OSCAR 10 satellite with its limited signal-to-noise ratio, a more robust form of modulation is needed to reduce the bit error rate. On HF, Bell 202 frequency shift of 1000 Hz, while legal in the U.S., is excessive for operation at 300 bauds — the speed limit below 28 MHz. Even at 1200 bauds, 1000-Hz shift is not necessary, and a narrower shift of 800 Hz (CCITT V.23) or 600 Hz (minimum-shift keying — MSK) would be in the interest of spectral efficiency.

Bell 103 modems are used for 300-baud HF packet-radio communications because of their ready availability and reasonably narrow 200-Hz frequency shift (compared to 1000 Hz for Bell 202 used for VHF packet radio). Before the Bell 103 caught on for this application, some packet transmissions were made using 170-Hz-shift RTTY modems. They were not always successful, however, because the RTTY modems have filters optimized for lower speeds (45 to 100 bauds). Widening RTTY demodulator filters has proven difficult in some cases. Some have found it possible to carry on packet-radio contacts between one station with a Bell 103 200-Hz shift modem and another using an RTTY 170-Hz shift modem if carefully tuned.

Receiver tuning when using a telephone-type modem can be difficult as there is usually no tuning indicator of any kind except for a Carrier Detect LED. External tuning indicators have been added to some packet-radio modems to facilitate HF operation. An oscilloscope is generally superior to LED tuning indicators and can be configured to display circles, flags or "crossed bananas" as in RTTY demodulators or eye patterns described in this chapter.

1200-Baud Modems

The VADCG terminal-node controller (TNC) was designed without a modem, as the plan was to use surplus Bell 202 modems. In late 1982, VADCG produced a Bell 202-like modem in both kit and PC-board-only form. It used an Exar XR-2206 function generator as the modulator and an XR-2211 phase-locked loop as the demodulator. The VADCG radio modem is no longer available.

When TAPR designed their TNC 1, they decided to add the modem to the TNC board to simplify getting on the air. The same Exar chips were used, but with a National MF10 switched-capacitor filter (SCF) for equalization. TAPR found that equalization was required to compensate for the audio responses of different VHF radios. A modem-disconnect arrangement was built in to permit use of an external modem. This same modem design was carried over to the TAPR TNC 2. The TAPR modem can be reconfigured for 300-baud, 200-Hz FSK.

A number of packeteers have built 1200-baud modems around the Advanced Micro Devices Am7910 chip. A modem built around the Am7910 is described in Chapter 29. The following modes are programmable:

- Bell 103 originate, 300 bit/s, full duplex
- Bell 103 answer, 300 bit/s, full duplex
- Bell 202 1200-bit/s, half duplex
- Bell 202 1200-bit/s, half duplex, with equalizer
- CCITT V.21 originate, 300 bit/s, full duplex
- CCITT V.21 answer, 300 bit/s, full duplex
- CCITT V.23 Mode 2, 1200 bit/s, half duplex
- CCITT V.23 Mode 2, 1200 bit/s, half duplex, with equalizer
- CCITT V.23 Mode 1, 600 bit/s, half duplex
- Loopback in above modes.

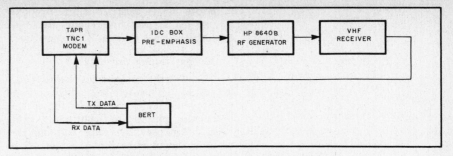

Fig. 37 — K9NG's bit-error-rate-test setup for the TAPR TNC 1 modem.

A number of manufacturers have introduced so-called "single-chip" modems, and more are anticipated in the years ahead. The "single-chip" description is more wishful thinking than reality in that additional support chips are usually needed. Nevertheless, they represent a significant breakthrough in modem-design simplification. Switched-capacitor-filter chips are also part of most new modem designs.

Packet Adaptive Modem

A multispeed HF packet-radio modem designed by Robert Watson and W4RI is currently being tested. It uses a standard 600-Hz shift at all speeds (75, 150, 300, 600 and 1200 bauds). At 1200 bauds, a 600-Hz shift FSK modem becomes MSK. The 600-Hz shift is wide enough to take advantage of in-band frequency diversity. Design details were included in the proceedings of the Second ARRL Amateur Radio Computer Networking Conference.

9600-Baud Modems

As this is being written, there is a widespread need for a modem that will operate at 9600 bauds for terrestrial trunking, meteor-scatter and satellite communications. This speed seems to be a natural next step as it is a worthwhile speed increase over 1200 bauds and because existing TNCs will function at 9600 bit/s. Several amateurs have taken different design approaches.

Phil Karn, KA9Q, designed a 9600-baud modem using BPSK primarily for use in PACSAT. The design proved difficult to reproduce, and it has been included with other candidate designs being investigated at the University of Surrey, U.K. BPSK is considered to be a robust modulation method, but occupies more bandwidth than some newer schemes. Filtering to restrict bandwidth introduces amplitude modulation. If a filtered BPSK signal is amplified by a nonlinear power amplifier, the AM is removed, and the full bandwidth is restored.

Gary Field, WA1GRC, has been developing a 9600-baud FSK 21.4-MHz RF modem with 9600-Hz shift. He is also working on a transverter board to convert 21.4 MHz to a frequency in the 220-MHz band. Twenty units are planned for alpha testing.

At the Fourth ARRL Amateur Radio Computer Networking Conference, Steve Goode, K9NG, introduced a new 9600-baud FSK RF modem that uses a 3-kHz shift. Circuit details are printed in the conference proceedings. The Tucson Amateur Packet Radio Corporation is producing PC boards for the K9NG modem.

Higher-Speed Modems

Spectrally efficient modem designs will be needed for speeds in excess of 9600 bit/s. Current thinking is that the next jump in speed will be to 56 kbit/s. Commercial 56-kbit/s modems are available but at prices beyond most amateurs. If an amateur 56-kbit/s modem were completed now, it would not be useful until a packet assembler/disassembler capable of that same speed is available. There is the future possibility of even higher speeds, such as 1.544 Mbit/s, for amateur packet radio. This entire area of high-speed modems and PADs is a challenging one for Amateur Radio development.

Modem Testing

Bit Error Rate

Bit-error-rate (BER) testing (BERT) is done by sending a known pseudorandom bit sequence, comparing the received bit sequence with the known one and analyzing the results. BER is the basic, but not the only, measure of overall performance of a digital communications system. The bit error rate is the probability of not properly receiving a bit. BER is expressed in terms of the number of errors $\times 10^{-n}$ or in a percentage.

A test setup described by Steve Goode, K9NG, in the August 1983 issue of *QEX* is shown in Fig. 37. It was used to test the BER performance of the TAPR TNC 1 modem, the results of which are presented in Fig. 38. BERT results of the K9NG 9600-bit/s modem are shown in Fig. 39.

Block Error Rate

Block error rate (BLER, also seen as BKER) is probably more meaningful in packet radio than BER. "Block" refers to a transmission block, which is the length of one transmission — in this case the length of a packet. Thus BLER testing (BLERT) tells you how many transmission blocks can be sent without any errors. A

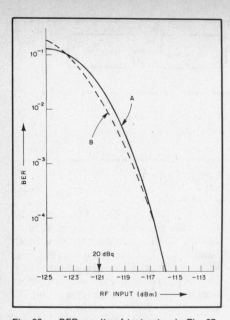

Fig. 38 — BER results of test setup in Fig. 37. BER for 1200 bit/s and 300 bit/s are shown at A and B, respectively. In both tests, an RF input signal of − 121 dBm represents 20 dBq (decibels of quieting).

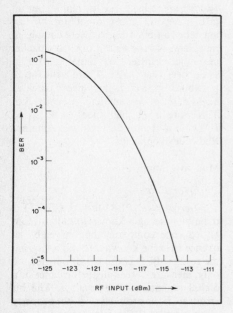

Fig. 39 — BER results of testing the K9NG 9600-bit/s modem.

typical VHF FM radio channel with a full-quieting signal could be expected to allow a BER of 1×10^{-5} or higher. That means that, on average, an error would occur every 100,000 bits. If a 2000-bit packet is sent repeatedly, on average, one out of every $(100,000/2000 =) 50$ packets would be rejected because of a bit error introduced by the channel. Such a circuit would have a BLER of $(1/50 =) 2 \times 10^{-2}$.

It might seem at first that bit errors would be fairly evenly spread throughout various blocks. While largely random, bit errors tend to occur in bursts. Three bit

errors may be clustered and destroy only one block, not three. Thus, BER and BLER are not necessarily proportional.

Modem Testing Techniques

Some telephone modems have a *self-test* feature. The self-test may activate automatically with power on, or the operator pushes a test button, to activate an integral word generator. A comparison of the output of the word generator and the same data sent through the modem is made, and the operator is signaled the results.

Loopback testing connects the output of the modulator to the input of the demodulator, either inside the modem or through a full-duplex communications channel. In the case of full-duplex modems which use separate originate and answer tones, the modem must be programmed to send and receive the same tones. For example, the modem is set up to send on its originate tones and receive on its answer tones, and vice versa.

In packet radio, *remote testing* of not just the modem but the entire packet-radio system can be accomplished by "connecting to yourself" through a digipeater. A simple way to do this is to make the connection then send a test sequence on the keyboard and observe the results on the screen. If any errors occur in transmission, it is extremely unlikely that the sequence will be repeated by the digipeater.

Another packet-radio testing method for equipment and the propagational path has provided some interesting results. In it, two stations agree that one station is to send a series of brief packets. In order to tell which packets get through and which do not, the data sent in the packet is simply the time in hours, minutes and seconds. The receiving station saves the times received on disk and can use the times for statistical analysis of the system over the path. A simple analytical technique is to print all times beginning with the same hour and minute on one line. The hard copy becomes a histogram of the experimental results.

RS-232-C Break-Out Box

Aside from the a bit-error-rate tester, the most powerful everyday diagnostic tool for modem testing is the break-out box. Staring at the modem and cables for a long time or incantations won't help much. BOB — An RS-232-C Break-Out Box in Chapter 29 is one you can build for a reasonable price. Once you use a BOB, you will never want to be without one. Even if you don't have anything to troubleshoot, watching its blinking LEDs can be fine entertainment when the bands are dead.

In addition to modem testing, a break-out box is useful for checking any RS-232-C interfacing problem. Sooner or later you will run into an interfacing situation that you can't seem to figure out by flipping lines 2 and 3 and performing

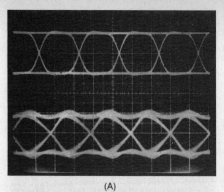

(A)

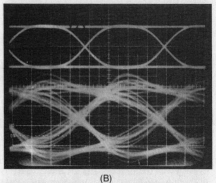

(B)

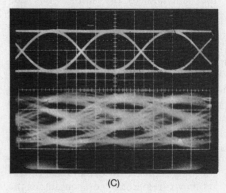

(C)

Fig. 40 — K9NG phototgraphs of eye patterns using his 9600-bit/s modem. Upper trace on each of the above represents the digital baseband signal before modulation. The lower trace on each photo is the received data after demodulation. Data rates were 9.6, 12 and 16 kbit/s at A, B and C, respectively.

gender-change operations on connectors. Sometimes the control lines, particularly pins 8 and 20, need to be reversed. Or, signals may not be appearing on various RS-232-C lines at the right times. A break-out box can also help you monitor what the modem is doing or not doing at various times. It helps to listen to transmitter and receiver audio at the same time.

Eye Patterns

An essential diagnostic technique for modem design and testing is the *eye pattern* or *eye diagram* display on an oscilloscope. It gets its name from the shape of the display. The eye-pattern diplay is a comparison of transmitted data and received data signals. Some eye patterns photographed by K9NG appear in Fig. 40.

To produce an eye pattern, connect the

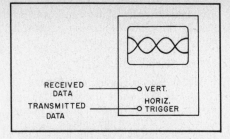

Fig. 41 — Oscilloscope connecions for eye patterns. A single-trace 'scope is shown. If available, a dual-trace 'scope can be used to display eye patterns prior to modulation and subsequent to demodulation.

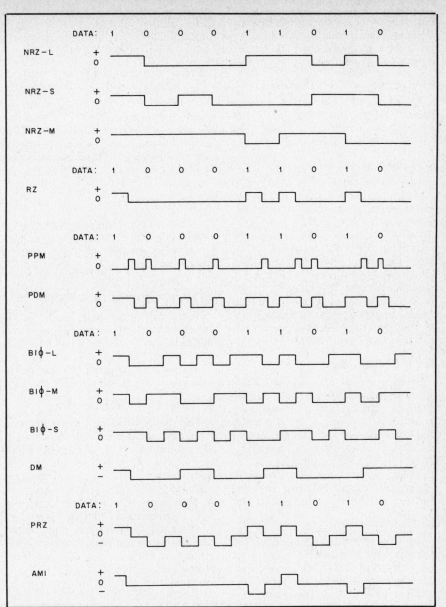

Fig. 42 — Pulses produced by various binary baseband codes.

transmitted data to the horizontal trigger input of the 'scope. The 'scope is then set up to begin a trace when triggered, and the horizontal sweep period is set to an integral multiple of the bit period. If you have only a single-trace 'scope then connect the signal you wish to display to the 'scope vertical input as shown in Fig. 41.

In the case of the K9NG eye patterns a dual-trace 'scope was used. The upper trace in each photograph displays the digital baseband signal prior to modulation. Note that the top trace in the photo at A has smooth transitions between the upper and lower rails but that there is some flat-topping on the rails. As the bit rate is increased, the data looks less like a square wave and becomes sinusoidal in the top trace at C. The lower traces were produced by feeding the received data out of the demodulator into the second vertical channel input of the 'scope. In the lower trace at A, the eye pattern looks good with only minor amplitude and phase jitter. Amplitude jitter is seen by thickening or separation of the trace(s) in the vertical direction, while phase jitter appears as thickening or separation of the line(s) horizontally. Note how both amplitude and phase jitter increase as the data rate through the modem is increased.

BINARY BASEBAND CODING

With so many experts working on things nothing remains simple. You didn't really think that 1s and 0s got translated directly to marks and spaces and that would be that, did you? Well, that's the way things were done in the early days of digital signaling. Experience with wire lines, missile telemetry and digital magnetic recording spawned the development of a number of different binary baseband encoding techniques, also called line coding, transmission coding, channel coding, baseband data format and pulse-code modulation (PCM).

A number of popular baseband codes are described below and illustrated in Fig. 42.

NRZ-L (Nonreturn-to-zero — level)

NRZ-L is often called simply NRZ. It is also known as NRZ (change). A 1 is represented by one level, a 0 by the other level. Baudot and AMTOR RTTY normally use NRZ-L coding. This type of coding is not suitable for transmission or recording systems incapable of handling dc levels. Without transitions, there is a tendency for the baseline to shift and cause signal-detection errors.

NRZ-S (Nonreturn-to-zero — space)

The familiar term in packet-radio circles is NRZI (nonreturn-to-zero — inverted). A 0 produces a change in level. A 1 causes no change. It does not matter whether a keying sequence is started on one tone or the other (usually called mark and space); only the transitions between the two tones are important. In NRZ-S, long sequences of 1 bits will cause no transitions, giving this code a strong dc component. However, in packet radio, *bit stuffing* or zero insertion after 5 continuous 1 bits limits the dc component to some extent. This is the predominant binary baseband code used for packet radio at this time. It is supported by numerous protocol-controller chips including the Intel 8273, Zilog 8530 and Western Digital 1933.

NRZ-M (Nonreturn-to-zero — mark)

This is also sometimes called NRZI, chiefly in digital recording, but it is just the opposite of NRZ-S. A 1 produces a change in level. A 0 causes no change. NRZ-M has a strong dc component whenever there are long sequences of 0 bits.

RZ (Return-to-zero)

A 1 produces a half-bit-period pulse. A 0 produces no pulse. RZ has a strong dc component since no pulses are produced for 0 bits. The half-bit-period pulses require twice the bandwidth of codes that transmit only full-bit-period pulses.

PPM (Pulse-position modulation)

A 1 produces a pulse in the middle of the symbol period. A 0 produces a pulse at the beginning of the symbol period. PPM has eliminated the dc component but requires greater bandwidth than codes that transmit only full-bit-period pulses.

PDM (Pulse-duration modulation)

A 1 causes a long pulse, a 0 a shorter pulse. PDM has no dc component but requires greater bandwidth than codes that transmit only full-bit-period pulses.

Biφ-L (Biphase — Level)

Biφ-L is often called "Manchester." More precisely, it is "Manchester II," as it is the second code developed at Manchester, England. It is also referred to as "split phase." A 1 produces a half-bit-period 1 pulse followed by a half-bit-period 0 pulse during the symbol period. A 0 produces a half-bit-period pulse sequence of 01. Biφ-L has eliminated the dc component but requires twice the bandwidth of codes that transmit only full-bit-period pulses. Biφ-L has been used for AMSAT data links and by a few packet-radio experimenters.

Biφ-M (Biphase-Mark)

Biφ-M is also called Manchester I, PE-M (phase encoding-mark), diphase and DF (double-frequency) and FM (frequency modulation) recording. There is a transition at the start of every bit period. A 1 causes a second transition halfway through the bit period. A 0 does not cause a second transition. Biφ-M has no dc component but requires twice the bandwidth of codes that transmit only full-bit-period pulses.

Biφ-S (Biphase-Space)

Biφ-S is also called phase encoding. There is a transition at the beginning of every bit period. A 0 causes a second transition halfway through the bit period. A 1 does not cause a second transition. Biφ-S has no dc component but requires twice the bandwidth of codes that transmit only full-bit-period pulses.

DM (Delay Modulation)

DM is often called *Miller Code*, also MFM (modified frequency modulation). A 1 causes a transition halfway through the bit period. A 0 causes no transition unless another 0 immediately follows, in which case a transition occurs at the end of the first bit period. DM is dc free and does not send pulses of less than one bit period.

PRZ (Polar return-to-zero)

PRZ is a pseudoternary or multilevel-binary (MLB) code. A 1 produces a half-bit-period pulse of one polarity. A 0 produces a half-bit-period pulse of the opposite polarity. PRZ has no dc component.

AMI (Alternate-mark-inversion)

AMI is also known as BP (bipolar). It is in the pseudoternary or multilevel-binary (MLB) class of codes. A 1 causes a positive pulse for the first half of the bit period and a 0 for the second half. A 0 produces a

0-level output. AMI has no dc component but lacks clocking information during long sequences of zeros.

HDBn (High-density bipolar codes)

To overcome timing problems associated with AMI, HDB*n* limits the maximum number of 0s to *n*. If *n* + 1 consecutive 0s occur in the data, the *n* + 1th 0 is replaced by a pulse of the same polarity as the pulse produced by the most-recent 1 bit. HDB3 (where *n* = 3) is the most popular form.

Other Codes

There are numerous other binary baseband codes, each having a peculiar combination of spectral and timing-recovery characteristics.

CIRCUIT TURNAROUND TIME

The circuit turnaround time is the interval needed between the last bit transmitted by one station and first bit the station receiver can properly process. It is comprised of two parts: (1) transceiver switching delays and (2) propagation delay.

Transceiver Switching Delay

Transceiver switching delay is the time it takes a transceiver to change from receive to transmit or vice versa and stabilize at full performance in the new condition. For design purposes, it is necessary to analyze the transceiver switching delay in terms of its four components: (1) transmitter ON time, (2) transmitter OFF time, (3) receiver ON time and (4) receiver OFF time. Worst-case measurements to date have been in the vicinity of 400 ms. Most HF and VHF radios can be modified to reduce the switching time to the 10- to 20-ms range. The requirements for HF/VHF packet radio are essentially the same as for AMTOR. All transceivers tested in the ARRL lab for the *QST* Product Review column will include switching times. For higher-speed packet radio (above 1200 bauds), a better design goal would be on the order of 1 ms.

HF Skywave Propagation Delay

The worst-case propagation delay for a skywave path to the other side of the earth (12,000 miles) is about 70 ms. The approximate delay in ms can be computed by dividing the distance in statute miles by 186 (in km by 300). Some additional delay needs to be added to account for the distance traveled by the wave from earth to the ionosphere and back to earth again. That distance is more significant than the distance along the ground for near vertical-incidence skywave paths and can amount to as much as 260 mi (1.4-ms delay).

VHF/UHF Terrestrial Propagation Delay

The VHF/UHF delay in ms can be calculated by dividing the distance in miles by 186. For example, a 32-mile path has a

propagation delay of 0.17 ms, which is usually insignificant compared to the equipment switching delay.

Meteor-Scatter Propagation Delay

Meteor-scatter paths are limited to around 1200 miles as a result of the curvature of the earth. Meteor trails range from about 50 to 80 miles above the earth. Propagation delays run from around 0.3 to 6.7 ms, depending on path length.

Satellite Propagation Delay

OSCAR 10 is in a highly elliptical orbit. The maximum round-trip delay at apogee is around 240 ms, or 33 ms at perigee.

CHANNEL ACCESS

ALOHA, named after early packet-radio experiments at the University of Hawaii, is a channel-access method wherein a station transmits without first checking to see whether the channel is free. When a *collision* occurs, the TNC waits for a random time interval and retries until it gets the frame through and an acknowledgment from the other station. Because of collisions, the maximal channel loading is about 18%, which is acceptable as long as traffic on a frequency is light.

There are more sophisticated access arbitration schemes that involve listening before transmitting. Current packet-radio TNCs look for the carrier-detect signal from the modem before transmitting. This isn't absolute insurance against collisions because two stations could decide to transmit at exactly the same moment. Also, propagation delay makes it impossible to tell if a distant station is transmitting until some milliseconds after transmission began.

THE LINK LAYER

The link layer defines the protocol for error-free transmission and reception between two points that are at two ends of a communications medium. In packet radio, that means two stations that work each other directly (not through a network). We have seen that the job of the physical layer is to get *bits* from one place through modems, radio transceivers and the propagation medium as reliably as possible. The function of the link layer is to move frames from one place to another through the physical layer and communications medium. In case the physical layer gets some bits wrong, the link layer will detect these errors and keep trying until a correct frame gets to the other station.

AX.25 LEVEL 2 PROTOCOL

On October 26, 1984, the ARRL Board of Directors approved the AX.25 Amateur Packet-Radio Link-Layer Protocol (Version 2.0). The AX.25 level 2 protocol follows, in principle, CCITT Recommendation X.25. Exceptions are that the address field has been extended to accommodate Amateur Radio call signs and that an Unnumbered Information (UI) frame feature has been added. It also follows the principles of CCITT Rec. Q.921 (LAPD) in the use of multiple links, distinguished by the address field, on a single shared channel. With the exception of an extended address, it could be considered a subset of the ANSI Advanced Data Comunications Control Protocol (ADDCP), balanced mode. It also follows the frame structure of High-Level Data Link Control Procedures (HDLC) per ISO 3309.

The AX.25 link-layer protocol formally specifies the format of a packet-radio frame and the actions a packet-radio station must taken when it transmits or receives such a frame.

At the link layer, data is sent in transmission blocks called *frames*. In addition to carrying data, each frame contains addressing, error-checking and control information. The addressing information tells what amateur station sent the frame, what station the frame is to be received by and, optionally, what stations should relay the frame. This addressing technique allows many packet-radio stations to share the same frequency. A station can monitor and display all the activity on a channel, regardless of to whom the frames are addressed. Or, a station may accept only the frames intended for it and ignore the rest. The error-checking information in the frame allows the receiving station to determine whether the frame contains any errors. If the frame is error-free, the receiving station accepts it and sends an acknowledgement to the transmitting station, if the two stations have established a *connection*. If the frame contains errors, the receiving station ignores it and waits for it to be retransmitted by the sending station.

AX.25 Link-Layer Format

Link-layer packet-radio transmissions are sent in *frames*. Each frame is divided into *fields*. Fig. 43 shows the frame format. Transmission of the frame is normally preceded by 16 bit reversals for synchronization. The frame consists of the beginning flag, address field, control field, a network protocol identifier (PID), information field, frame-check sequence (FCS) and an ending flag.

Flag Field

Each frame starts and ends with a *flag*. The flag has a peculiar bit pattern: 01111110. The pattern appears only at the beginning and end of frames. If five 1 bits show up somewhere else in the frame, a 0 bit is inserted by the sending station and removed by the receiving station. This is called *zero insertion* or *bit stuffing*, and is illustrated in Fig. 44.

Address Field

The *address field* consists of 2 to 10 specially encoded Amateur Radio call signs. The first address is that of the sta-

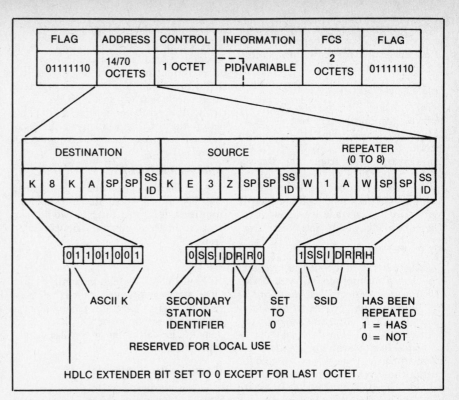

Fig. 43 — AX.25 frame format. The repeater address field is optional and may contain as many as eight repeater call signs.

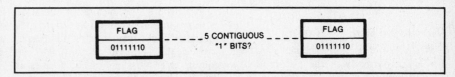

Fig. 44 — The HDLC controller chip uses zero-insertion or bit-stuffing to eliminate the possibility of flags appearing within a frame. The transmitter examines the content between flags and inserts a 0 bit after all sequences of contiguous 1 bits to ensure that a flag sequence is not simulated. The receiver discards any 0 bit that directly follows 5 contiguous 1 bits that occur between the flag sequences.

tion for which the frame is intended. The second is the source or station originating the frame. If the two stations are in direct contact, that's all there is to the address field. If not, the call signs of 1 to 8 *digipeaters* may appear.

Each call sign takes up six character spaces. Call signs are made up of upper-case alpha and numeric ASCII characters only; shorter call signs are left padded with ASCII spaces. A seventh character, called a *Secondary Station Identifier* (SSID), is added to permit up to 16 different packet-radio stations to operate under one radio call sign. For example, W1AW 5 is a digipeater, while W1AW 4 is a bulletin board station. The SSID also keeps track of which digipeaters have repeated, and which have yet to repeat, the frame. In Fig. 43, the two bits marked "R" are reserved and may be used in an agreed-upon manner in individual networks. When not implemented, they are set to 1.

The SSID *octet* (8 bits) in the repeater address, when used, differs from the others in that its last bit, "H" is used to indicate whether a frame has been repeated. This

is necessary to prevent the reception of identical frames, one direct and one from the repeater.

Control Field

The *control field* contains a bit pattern that tells what kind of frame it is (*information* or *supervisory*) and a frame number (0 to 7) for acknowledgments. It is used to signal a connection request, ready/not-ready conditions, frame numbering and the specific mode of operation.

Protocol Identifier Field

Strictly speaking, the *Protocol Identifier Field* (PID) is part of the information field and is shown as such in Fig. 43. It appears in information (I) and unnumbered information (UI) frames only. It identifies what kind of network-layer protocol, if any, is in use.

Information Field

The *information field* (I field) contains the data to be transmitted. This field can have any integral number of octets, up to 256, of information. As the network and higher protocol levels are implemented, some of the octets at the beginning of the I field will be used for addressing and control for those levels.

Frame Check Sequence Field

The *frame-check sequence* (FCS) is a 16-bit number calculated by both the sender and receiver of a frame. It follows an algorithm published in ISO 3309 (HDLC). Upon receipt of a frame, the receiver calculates the FCS on the basis of the received data then compares the answer with the FCS calculated by the sender. If the two match, that frame is acknowledged. The FCS is followed by the ending flag.

AX.25 Procedure

Packet-radio protocols are generally executed in a microprocessor-based device called a *packet assembler/disassembler* (PAD), also called a *terminal-node controller* (TNC). Protocols can also be implemented in software written for a microcomputer. In any event, the procedures are automatic and operate without the need for operator intervention. The frame format and procedures for handling the frame are link-layer matters. What you see on your computer screen may or may not be a close approximation of what goes on at the link layer, since the screen display is a presentation-layer matter.

Disconnected State

When power is applied, packet-radio TNCs are normally in the disconnected state or monitor mode. This permits displaying all activities on the channel. The TNC also looks for any connection requests from other stations and will respond by establishing the connection or ignoring the connection request, depending on circumstances.

Connection Establishment

When one station wishes to connect with another, it sends a command frame to the other station and starts a time-out timer. If the other station is on the air and able to connect, it sends an acknowledgment frame. If the called station doesn't respond before the timer runs out, the calling station will reinitiate the request a number of times.

Information Transfer

After the link connection is established, the TNCs will enter the information-transfer state. In this state, the two stations may exchange information and supervisory frames.

Disconnection

While in the information-transfer state, either station may send a request to disconnect. The disconnection occurs after a response from the other station or if no response is heard after several disconnect attempts.

Connectionless Operation

This is a procedure that permits round tables and bulletin transmissions via packet radio. The normal method of connection between two amateur stations is not practical when multiple stations are involved. The protocols allows this type of operation by using unnumbered information (UI) frames. Without frames being numbered, however, the TNCs do not automatically request retransmissions of bad frames as they do when connected.

Frame Acknowledgments

The control field of each frame sent includes the number (0 to 7) of the last frame correctly received from the other station. If the sending station had sent frame

number 5 but received acknowledgment for frame number 4, it knows to repeat frame number 5.

AX.25 Details

The protocol specification, "AX.25 Amateur Packet-Radio Link-Layer Protocol," is available from ARRL Hq. A tutorial based on an earlier version of the protocol appears in the proceedings of the Second ARRL Amateur Radio Computer Networking Conference, also available from ARRL Hq. See the end of this chapter for a bibliography.

PACKET ASSEMBLER/DISASSEMBLERS

The assembly and disassembly of frames (or packets) is performed automatically by a computer. Special-purpose computers, called terminal-node controllers (TNCs) or packet assembler/disassemblers (PADs) have been designed for Amateur Radio use.

The TNC was introduced to Amateur Radio by the Vancouver Amateur Digital Communications Group (VADCG). The Tucson Amateur Packet Radio Corporation (TAPR) produced several TNCs that made a great contribution to the growth of Amateur packet radio. The TAPR TNC 1 and TNC 2 have been widely "cloned" by U.S. and foreign manufacturers. In addition, a number of manufacturers have introduced their own innovative TNC designs. Packet-radio manufacturers are listed near the end of this chapter. Refer to *QST* for articles and advertisements on packet-radio products. *Gateway,* the ARRL Packet-radio Newsletter carries the latest news on new packet equipment and software.

DIGIPEATERS

A *digipeater* is a packet-radio station

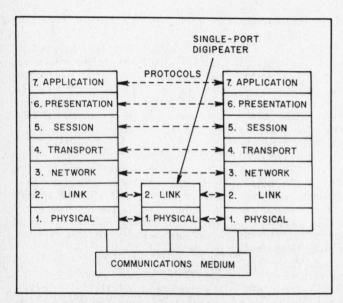

Fig. 45 — Open Systems Interconnection model of two packet-radio stations communicating through a single-port digipeater. The dashed lines represent logical communications between peer layers.

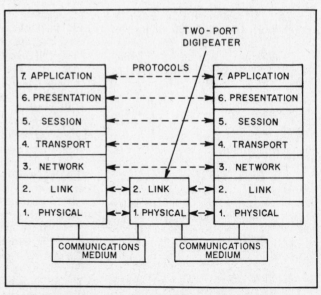

Fig. 46 — Open Systems Interconnection model of two packet-radio stations communicating through a two-port digipeater. The dashed lines represent logical communications between peer layers.

capable of recognizing and selectively repeating frames. An equivalent term used in industry is *bridge*. A digipeater can be either a single-port or multiport station. Virtually any TNC can be used as a single-port digipeater, provided that it has supporting software. TAPR TNCs and their commercial equivalents have software that includes a digipeating mode. Unmodified VADCG TNCs require a change of EPROMs for digipeater operation.

A multiport digipeater is necessary when one port operates at one set of Level 1 parameters (say 1200 bauds, Bell 202, on 2 meters) and a second port uses another set (say 9600 bauds, FSK, on 220 MHz).

The Xerox 820, without disk drives, monitor or keyboard, can be used as a multiport digipeater using software developed by KE3Z. The software can be used to link packet-radio stations operating on different frequencies, or can be used for a single-port digipeater. It uses the on-board SIO and two state machines of the type mentioned earlier in this chapter. To receive a copy of the software, send an 8-inch disk to ARRL Hq. with sufficient return postage.

An "apartment-house" diagram of how two packet-radio stations connect through a single-port or multiport digipeater is shown in Fig. 45.

TELEPORTS

A *teleport* capable of real-time relaying of packets between packet-radio stations on the ground and a satellite is a special form of a two-port digipeater. What goes in the terrestrial port immediately goes out the port to the satellite and vice versa.

Real-time operation is not the ultimate in teleports, since when the users want to operate and when the satellite is in view may not coincide. Future teleports will include message buffering and will operate on a store-and-forward basis. Store-and-forward teleports are expected to greatly increase the utility of satellites for the average packet radio station.

Automatically operated teleports presently require a special temporary authority (STA) from the FCC. However, the rules may be changed to permit automatic operation by the time you read this.

THE NETWORK LAYER

The purpose of the network layer is to get packets to their destination through a network. No networking standard exists as this edition of the *Handbook* is being finalized. Nor is there agreement among packet-radio experimenters on what form the network layer should take — virtual connections or datagrams, both of which are discussed below.

Fig. 47A illustrates the classical Open Systems Interconnection model of how two computers communicate through a network. Note the peer-to-peer (for example transport-to-transport) protocol relationships between the two end points and how

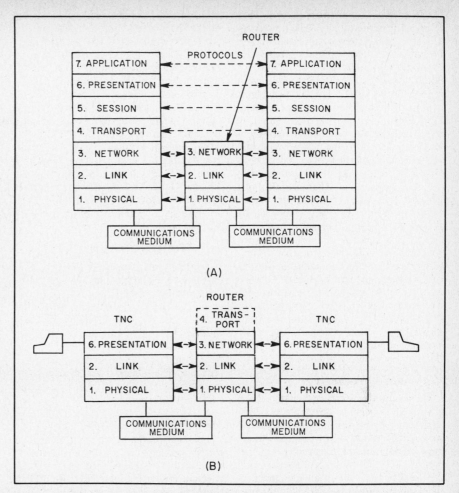

Fig. 47 — Possible packet-radio networking models. A shows the classical OSI model of two stations communicating through one or more network routers. This model will likely apply only to hosts that provide a service to the network. B represents a more modest idea of how users with present TNCs may communicate through a network. Dotted lines around the transport-layer box indicate that a transport-layer protocol may be required at the router if a datagram networking protocol is adopted. Dashed lines show logical communications between peer levels of protocol.

the physical, link and network-layer protocols involve the relay station, called a *router*. The idea is that both end points have computers with software for each of the protocol layers. The full 7-story apartment-house protocol structure may indeed become the pattern for *hosts* — those stations that will provide substantial computer services to the network.

At the moment, however, most packet-radio stations are equipped with TNCs that have only a modest capability. They implement the physical layer (modem) and the link layer (frame assembly and disassembly), and have a rather basic presentation layer. The presentation-layer code takes care of the commands and responses between the terminal and the TNC. The network, transport and application layers are skipped — often called *null layers*. Some of the future TNCs (communications processors, or whatever they will be called) may eventually implement all protocol levels. But for the moment, none do.

There is general agreement that however elaborate the network may become, net-

work designers should take care not to disenfranchise the user with a present-day TNC. Furthermore, such users should be able to enjoy the full capabilities of the worldwide packet-radio network. Just how this will be accomplished is a matter of speculation. Yet, it stands to reason that if all the fancy protocols are not in the user's TNC or personal computer, then they are likely to be in some super station nearby.

Fig. 47B may be closer to reality for amateur packet radio — certainly for the near future. The dashed lines forming the network- and transport-layer boxes in the router indicate some uncertainty of whether only the network layer or both the network and transport layers will be implemented. That depends largely on the outcome of the controversy between virtual connection and datagram approaches described below. If virtual connection is the approach chosen, only the network layer need be implemented. If datagram, then both need be included.

The data-communications industry uses

the term *local-area network* (LAN). That term, while perhaps okay when describing a 2-meter packet network, is inappropriate for packet-radio nets that might span half the globe, as they can on HF or via satellite. So, we have coined the term *intranet* to mean a (level 3) network with one net controller (router) and a bunch of user stations. The term *internet*, on the other hand, is widely used to denote communication between (intra)nets. If the network goes the virtual-connection route, there will be both intranets and internets. If it goes datagram, there may or may not be intranets. Regardless of the outcome, it is hoped that we will end up with one worldwide standard. To do otherwise would mean that high-level *gateways* have to be established between networks using different protocols. Unfortunately, when this is done, the unique advantages and enhancements in one network cannot be translated to networks not having the protocols to support them. In other words, communication between dissimilar networks becomes a matter of the least common denominator — something to avoid like grid current.

NETWORK PROTOCOL

The protocol(s) for the network layer are now in the process of development. Two approaches are being studied by the ARRL Ad Hoc Committee on Amateur Radio Digital Communication: virtual circuit and datagram. This is a subject of considerable controversy both within amateur packet-radio circles and in the data communications industry. The Committee agreed to a period of parallel experimentation with the two approaches prior to settling on one or the other.

Operating prototypes of both virtual-circuit and datagram network protocols were unveiled at the Fifth ARRL Amateur Radio Computer Networking Conference in Orlando on March 9, 1986. Field trials in several locations across the U.S. were planned throughout 1986.

Virtual-Circuit Protocol Proposal

A proposal based on CCITT Recommendation X.25 packet-layer protocol (PLP) was presented by Terry Fox, WB4JFI, at the Third ARRL Amateur Computer Networking Conference and elaborated upon in the Fourth Conference. Software written by Howie Goldstein, N2WX, was demonstrated at the Fifth Conference.

The virtual-circuit (VC) approach has the advantage of less overhead in the packet header and making the network (not the end user) responsible for acknowledgement of packets. In a virtual circuit, only the first packet establishing a virtual connection has the complete information needed by the network to route packets in the network header. Subsequent packets have an abbreviated network header.

Datagram Protocol Proposal

Phil Karn, KA9Q, has proposed use of a datagram type of network protocol based on the IP (Interned Protocol) developed by the Defense Advanced Research Projects Agency (DARPA). In a datagram protocol, each packet contains complete network addressing/routing information. There is more overhead per packet, but the advantage is that packets will reach the destination via any route still open even in an unreliable network.

Arguments in favor of a DARPA IP-based datagram protocol were given in a paper by KA9Q in the proceedings of the Fourth ARRL Amateur Radio Computer Networking Conference. Software written by KA9Q was demonstrated at the Fifth Conference.

THE TRANSPORT LAYER

The transport layer makes sure that the data passed to and from higher layers is all there and put in the right order. In datagram protocols, it is possible that one packet sent before another one could beat the other to the destination. The transport layer has to be able to buffer incoming packets and reassemble them in the correct sequence. The transport header adds some overhead to the packet to carry out the necessary transport layer functions.

Transport Protocol Proposals

The transport and network layers are closely associated. If a CCITT X.25 PLP-based virtual-circuit protocol is adopted, it is likely that a subset of the ISO TP-4 transport protocol now in the form of a draft international standard (DIS) will be used. On the other hand, if a DARPA IP-based network protocol adopted, it is likely that the DARPA TCP-based transport control protocol will be used.

THE SESSION LAYER

The session layer is responsible for connection establishment and termination. In fact, it can set up multiple connections to the same or several different places. This protocol is not yet under active study.

THE PRESENTATION LAYER

The presentation layer comes close to a three-ring circus. It talks to the session layer to send and receive messages from distant computers. It deals with data files, code conversion and text strings, and is the traffic cop for computer peripherals. This layer also serves all the programs that might be running at the application layer.

Standardizing TNC Commands and Responses

In 1979, the VADCG TNC original TIP (terminal interface program) had only two commands: CONTROL-X followed by a call sign meant "connect" and CONTROL-Y meant "disconnect." Commands were changed and added in other versions of the TIP. The software used in the TAPR TNC included a more-elaborate set of commands and responses. Commercial variants of the TAPR TNC used the TAPR commands and responses with only slight deviation.

In 1984, the ARRL Ad Hoc Committee on Amateur Radio Digital Communication circulated information on the latest versions of the following CCITT Recommendations:

X.3   Packet assembly/disassembly facility (PAD) in a public data network.

X.28   DTE/DCE interface for a start/stop mode data terminal equipment accessing the packet assembly/disassembly facility (PAD) in a public data network situated in the same country.

In simplified terms, X.3 specifies what a PAD (TNC) is supposed to do, and X.28 standardizes the commands and responses between the DTE (terminal) and the PAD (TNC). They are presentation-level protocols.

Doug Lockhart, VE7APU, studied the latest X.3 and X.28 documents for the Committee, and presented a paper on his findings at the Fourth ARRL Amateur Radio Computer Networking Conference. He proposed that an amateur version of these protocols should be adopted to standardize TNC commands and responses. A draft protocol proposal will be considered by the Committee.

Videotex

Another presentation-level protocol can-

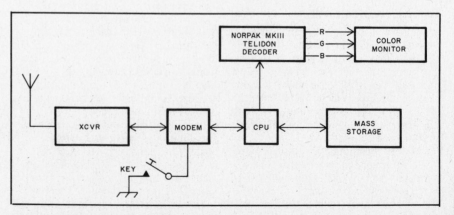

Fig. 48 — Block diagram of the station set up used for the Telidon tests.

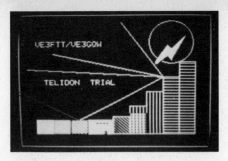

Fig. 49 — Photograph of one trial picture, as received by VE3FTT. The picture was received error free

didate is videotex. An article in the September 1983 issue of *QST* documented some early experiments with videotex using Telidon picture-transmission techniques. A block diagram of the test setup used by VE3FFT is shown in Fig. 48. A photograph of a test picture received by VE3FFT is reproduced in Fig. 49.

Two major videotex standards exist:
ANSI X3.110-1983/CSA T500-1983, Videotex/Teletext Presentation Level Protocol Syntax
CCITT S.100, International Information Exchange for Interactive Videotex.
Both of these standards have a rich repertoire of graphic options, such as: alphanumeric characters, alphamosaic option, alphageometric option, alphaphotographic option and dynamically redefinable character sets (DRCS) option. Amateur experimentation with videotex will be extremely interesting in the years ahead.

THE APPLICATION LAYER

This is the layer that runs one or more application programs — the real work of a computer in the user's view. Some computers will be set up to run one program that will offer a service to the packet-radio network. Others will be user oriented and will use the network to communicate needed information to and from other computers in the network.

Message Protocols

Computer-based message systems (CBMSs) are high-interest applications for Amateur Radio. Probably, every metropolitan area within the network will have a CBMS to serve local amateurs. A number of packet-radio CBMSs are already in operation. Message systems help to sustain interest in packet radio by providing a reason to check in every day. They are also helpful in keeping users well informed.

It is unfortunate that the protocols used in various CBMSs are different and incompatible. Message formats used for CBMSs differ from those of the ARRL National Traffic System (NTS), the Military Affiliate Radio System (MARS) and electronic mail services. This may be of little immediate concern to many amateurs at the moment but causes a great deal of work for amateur stations that refile messages between net-

works. Lip service has been given to standardizing message formats. It seems that everyone has been willing to standardize if the rest of the world will make the first move. That may have happened. After years of discussion, the CCITT has completed work on message-handling standards and has published their recommendations in documents in the X.400 series. They will be studied closely for application to Amateur Radio.

Host Protocols

A number of packet stations have set up host computers to provide a general or specific service to packet-radio users. The general-purpose type have included RCPM systems that allow *remote* use of a computer running CP/M® .

Except for brief demonstrations, permanent data-base hosts have not yet emerged. This is expected to be an important application for packet radio. Imagine that you're at your word processor one evening and need to have the title and author's name of an article on solar batteries. You don't know which Amateur Radio magazine it was in but think it was published in the fall of 1982 (or was it early 1983?). If an individual or club has gone to the trouble of building such a data base, you could log on the network and get the information you need within a few minutes. Or, you want to get ready for a DX contest and would like an up-to-date propagation prediction for East Asia. Again, how about signing on the network and asking for the latest word on MUFs and LUFs for the path.

Unlike message protocols, host protocols for other purposes may not need to be standardized for quite some time to come. For the present, this would be an area for wide-ranging experimentation to learn different ways of better serving the Amateur Radio community.

Weather Applications

Packet radio is expected to make a great contribution to transmission of weather data via Amateur Radio. A paper entitled "Packet Radio and the National Hurricane Center" by Joel Kandel, KI4T, is available in the proceedings of the Fourth ARRL Amateur Radio Computer Networking Conference.

A weather-reporting network around the Minneapolis-St. Paul area is being developed. A contact for this activity is Dr. Noel Petit, WBØVGI, Suite 220-B, 511 11th Ave. S, Minneapolis, MN 55415.

ADVANCES IN PACKET-RADIO NETWORKING

As this edition of *The ARRL Handbook* is being prepared, several network-layer schemes are under consideration and are being tested in the field. (The network layer handles routing for the user.) Each of the network-layer protocols under considera-

tion is described briefly in the following section.

The ARRL Ad Hoc Committee on Amateur Radio Digital Communication has decided that networking protocols must compete in field trials until such time as one demonstrates its superiority over the others. Once the facts are in, the Digital Committee will make a recommendation to the ARRL Board of Directors on a networking standard.

NET/ROM

The W6AMT network of digipeaters in California is testing a new firmware program for the TNC 2. Called "NET/ROM," the new firmware supports networking capabilities (commonly referred to in packet-radio circles as "layer three" and "layer four").

Developed by Ron Raikes, WA8DED, and Mike Busch, W6IXU, of Software 2000, Inc, NET/ROM runs on a standard TAPR TNC 2 terminal node controller, or on any of the commercially available TNC 2 "clones." NET/ROM is distributed in the form of a 27C256 EPROM which simply plugs into the ROM socket of the TNC 2 in place of the standard TAPR firmware ROM. NET/ROM is intended for use primarily at wide-coverage digipeater sites. It is not appropriate for end-user or mailbox stations.

A NET/ROM node provides the normal functions of an ordinary AX.25 digipeater, plus a set of sophisticated higher-level networking capabilities. A NET/ROM node user may display a list of other known network nodes; establish a transport-level circuit to a distant node; and connect to another end-user or mailbox in the vicinity of the distant node. Compared with conventional AX.25 multihop digipeating, NET/ROM's true store-and-forward packet switching technology can provide an order-of-magnitude improvement in throughput, especially over long paths. Routing from the local node to the distant node is handled automatically, and even includes alternate routing to circumvent network outages.

NET.EXE

Another approach to the implementation of higher-level protocols is that taken by Phil Karn, KA9Q, in his NET.EXE program for the IBM PC. Essentially, Phil argues that there is no need to tie ourselves down to the AX.25 link layer protocol when a layer-four (transport) protocol is operating to ensure end-to-end data integrity. Phil's approach, as embodied in his NET.EXE program, allows several possible link-layer protocols (including AX.25).

NET.EXE executes the Defense Advanced Research Projects Agency (DARPA) suite of protocols. Included in these are IP, the Internet Protocol; TCP, the Transmission Control Protocol; ARP, the Address Resolution Protocol; FTP, the

File Transfer Protocol and SMTP, the Simple Mail Transfer Protocol. All of these are above the link layer in the hierarchy of protocols. At the link layer, NET.EXE supports simple serial data transfer via SLIP, the Serial Line Interface Protocol, and non-protocol serial I/O.

One of the most interesting aspects of the NET.EXE system is that it can do more than one job concurrently. TCP, which is used to provide end-to-end data integrity, can support multiple connections. NET.EXE also supports multiple applications processes. Using the multi-connect protocols, you could, for example, initiate a file transfer to another computer using FTP, and while that transfer is taking place you can chat with the operator of the remote computer using TELNET (the terminal-to-terminal "chat" protocol). Or perhaps chat with the operator of a *different* remote computer. Or receive forwarded mail. Or receive a file. Or... (You get the idea.)

TEXNET

This networking protocol uses datagrams with node-to-node acknowledgments. User selection of node resources is performed on the basis of SSIDs. For example, connecting to the node with the node's call sign and an SSID of 1 will attach the user to the National Weather Service interface. Connecting to SSID 4 will let the user access the network, at which point the user can command the network node to list the other known network nodes, establish a circuit through the network, display node-activity statistics or access the message system.

The TEXNET network nodes under construction are two-port devices, with a 2-meter port for user access and a 70-cm port for node-to-node linking. The nodes will be operated from a battery and charger, providing emergency communication capabilities. Linking between nodes will be performed at 9600 bit/s.

The TPRS development team includes George Baker, W5YR, Tom McDermott, N5EG, and Tom Aschenbrenner, WB5PUC. TPRS expects to make printed-circuit boards and software available.

Virtual Circuit Networking

A virtual circuit (VC) networking proposal based on CCITT X.25 was introduced by Terry Fox, WB4JFI, at the Third ARRL Amateur Radio Computer Networking Conference in 1984. Other early VC supporters were Gordon Beattie, N2DSY, Tom Moulton, W2VY, and Howard Goldstein, N2WX. Advocates of this approach point out that X.25 is an international standard in widespread use in commercial and government-owned packet-switched networks. Furthermore, unlike a datagram network, the network (not the endpoint) takes responsibility for delivering all packets where they are addressed and in the right order. One advantage of VC is that once a virtual circuit is established through a network, subsequent packets have less overhead in the packet headers, whereas datagrams contain the full address every transmission.

It was Howie who wrote the code for an X.25-based VC network protocol resident in a TAPR TNC 2 that was demonstrated at the Fifth Networking Conference at Orlando in 1986. A number of groups are experimenting with this networking protocol, but a final, field-tested package has yet to be released. Howie has written dual-port networking code for the PAC-COMM dual-port TNC, which will be released soon.

MORE INFORMATION ON PACKET RADIO

There are numerous sources of additional information on packet radio. The International Telegraph and Telephone Consultative Committee (CCITT), books and trade magazines are the best sources for packet switching in industry. The Amateur Radio literature on packet radio is rapidly growing. The bibliography that follows includes many useful articles and papers, which, in turn, will lead you to other material. Packet-radio clubs and their newsletters are also sources of new information.

PACKET-RADIO SOURCES

There are numerous sources of packet-radio products and information. The sources listed below, while not exhaustive, are representative of what is available and will, in turn, lead you to other material.

BIBLIOGRAPHY

ARRL, *ARRL Amateur Radio Computer Networking Conference 1-4,* pioneer papers on packet radio 1981-1985. Proceeding available from ARRL HQ.

ARRL, *ARRL Amateur Radio 5th Computer Networking Conference,* Orlando, Florida—March 9, 1986. Proceedings available from ARRL HQ.

ARRL, *ARRL Amateur Radio 6th Computer Networking Conference,* Redondo Beach, CA—August 29, 1987. Proceedings available from ARRL HQ.

Borden, D., and P. Rinaldo, "The Making of an Amateur Packet- Radio Network," *QST,* October 1981.

Fox, T., *AX.25 Amateur Packet-Radio Link-Layer Protocol* (Version 2.0), ARRL, October 1984.

Grubbs, J., *Get *** Connected to Packet Radio,* QSKY Publishing, 1986. Available from ARRL HQ.

Horzepa, S., "On Line" column, *QST.*

Horzepa, S., "Packet Radio" chapter of *The ARRL Operating Manual,* Third Ed.

Horzepa, S., *Your Gateway to Packet Radio,* ARRL 1987.

Jahnke, B., *The ARRL Repeater Directory,* 1988-1989 Edition, ARRL, 1988.

Johnson, L., "Join the Packet-Radio Revolution," *73,* November 1983 through January 1984.

Magnuski, H. "National Standards for Amateur Packet Radio Networks," Conference Proceedings of the Eighth (1983) West Coast Computer Faire.

Mayo, J., *Packet Radio Handbook,* Tab Books 1987.

McLanahan, D., "A Packet Radio Primer," *Ham Radio,* December 1985.

Meijer, A., and P. Peeters, *Computer Network Architectures,* Computer Science Press, 1982.

Morrison, M., D. Morrison, and L. Johnson, "Amateur Packet Radio," *Ham Radio,* July and August 1983.

Price, H., "What's All This Racket About Packet," *QST,* July 1985.

Price, H., "Packet Radio—A Closer Look." *QST,* August 1985.

Rinaldo, P., "ARRL Board Approves AX.25 Packet-Radio Link-Layer Protocol," *QST,* December 1984.

Stallings, W., *Tutorial: COMPUTER COMMUNICATIONS: Architectures, Protocols, and Standards,* IEEE Computer Society, 1985.

Sumner, D., "It Seems to Us...: Packet Fever," *QST,* April 1986.

Sumner, D., "It Seems to Us...: Good News for Packeteers," *QST,* May 1986.

Tanenbaum, A., *Computer Networks,* Prentice-Hall, Inc, Englewood Cliffs, NJ 1981.

PACKET-RADIO PERIODICALS

Amateur Radio Research and Development Corp (AMRAD), monthly *AMRAD Newsletter* ($15). AMRAD, PO Drawer 6148, McLean, VA 22106-6148.

ARRL, biweekly, *Gateway—The ARRL Packet-Radio Newsletter* ($6 members, $9 nonmembers). ARRL, 225 Main St, Newington, CT 06111.

ARRL, monthly *QEX—The ARRL Experimenter's Exchange* ($8 members, $16 nonmembers). ARRL, 225 Main St, Newington, CT 06111.

British Amateur Radio Teleprinter Group (BARTG), quarterly *Datacom,* Pat & John Beedie, GW6MOJ/GW6MOK, "Ffynnonlas," Salem, Llandeilo, Dyfed, SA19 7NP, Wales.

New England Packet Radio Association (NEPRA), monthly *NEPRA PacketEar* ($15). NEPRA, PO Box 208, East Kingston, NH 03827.

Northwest Amateur Packet Radio Association (NAPRA), *Zero Retries,* c/o John Gates, N7BTI, 13304 131st St. KPN, Gig Harbor, WA 98335.

Rocky Mountain Packet Radio Association (RMPRA), *RMPRA > Packet* (quarterly), c/o Bob Gobrick, WA6ERB, 14311 W. Virginia Dr., Lakewood, CO 80228.

Sydney Amateur Digital Communications Group, *The Australian Packeteer* (SADCG), PO Box 231, French's Forrest, NSW 2086, Australia.

Tucson Amateur Packet Radio (TAPR), *Packet Status Register* (included with TAPR membership, $15 US, $18 Canada/ Mexico, $25 elsewhere), PO Box 22888, Tucson, AZ 85734.

Utah Packet Radio Association, *UPRA Connect,* UPRA, 4382 Cherryview Drive, West Valley City, UT 84120.

Vancouver Amateur Digital Communications Group (VADCG), *The Packet,* ($10, indefinite), c/o Doug Lockhart, VE7APU, 9531 Odlin Rd, Richmond, BC V6X 1E1, Canada.

Wisconsin Amateur Packet Radio Association, *The Wisconsin Packeteer,* (WAPRA), PO Box 1215, Fond Du Lac, WI 54935.

PACKET-RADIO PRODUCT SOURCES

Advanced Electronic Applications, Inc, PO Box C-2160, Lynnwood, WA 98036.

Applied Digital Technologies, 2056 E Sutter Pl, Oxnard, CA 63033.

Brincomm Technology, 2980 Wayward Dr, Marietta, GA 30066.

GLB Electronics, 151 Commerce Parkway, Buffalo, NY 14224.

Hamilton Amateur Packet Network (HAPN), Box 4466, Station D, Hamilton, ON L8V 4S7, Canada.

Heathkit, Benton Harbor, MI 49022.

Kalt & Associates, 2440 E Tudor Rd Suite 138, Anchorage, AK 99507.

Kantronics, 1202 E 23rd St, Lawrence, KS 66046.

MFJ Enterprises, Box 494, Mississippi State, MS 39762.

Pac-Comm Packet Radio Systems, Inc, 3652 W. Cypress St., Tampa, FL 33607.

Packeterm, PO Box 835, Amherst, NH 03031.

Richcraft Engineering Ltd, Drawer 1065, #1 Wahmeda Industrial Park, Chautauqua, NY 14722.

Tono Corporation, 98 Motosoja-Machi, Maebashi-Shi, 371 Japan.

Tucson Amateur Packet Radio (TAPR), PO Box 22888, Tucson, AZ 85734.

Ward Co Ltd, 471-8 Imagome, Higashiosaka City 578, Japan.

Glossary of Digital Communications Terminology

ACK — Acknowledgment, the control signal sent to indicate the correct receipt of a transmission block.

Address — A character or group of characters that identifies a source or destination.

AFSK — Audio frequency-shift keying.

ALOHA — A channel-access technique wherein each packet-radio station transmits without first checking to see if the channel is free; named after early packet-radio experiments at the University of Hawaii.

AMICON — AMSAT International Computer Network — Packet-radio operation on SSC L1 of AMSAT-OSCAR 10 to provide networking of ground stations acting as gateways to terrestrial packet-radio networks.

AMRAD — Amateur Radio Research and Development Corporation, a nonprofit organization involved in packet-radio development (P.O. Drawer 6148, McLean, VA 22106).

AMTOR — Amateur Teleprinting Over Radio, an amateur radioteletype transmission technique employing error correction as specified in CCIR 476-2 or 476-3. CCIR Rec. 476-3 is reprinted in the proceedings of the Third ARRL Amateur Radio Computer Networking Conference, available from ARRL Hq.

ANSI — American National Standards Institute (1430 Broadway, New York, NY 10018.)

Answer — The station intended to receive a call. In modem usage, the *called* station or modem tones associated therewith.

ARQ — Automatic Repeat Request, an error-control technique in which a sending station, after transmitting a data block, awaits a reply (ACK or NAK) to determine whether to repeat the last block or proceed to the next.

ASCII — American National Standard Code for Information Interchange, a code consisting of seven information bits.

ASR — An Automatic Send-Receive teletypewriter.

Autostart — A device that automatically senses presence of a signal and turns on a printer motor.

AX.25^{SM pend.} — Amateur packet-radio link-layer protocol approved by ARRL Board of Directors in October 1984. Copies of protocol specification are available from ARRL Hq for $8 U.S., $9 Canada and elsewhere.

Backwave — An unwanted signal emitted between the pulses of an on/off-keyed signal.

Balanced — A relationship in which two stations communicate with one another as equals; that is, neither is a primary (master) or secondary (slave).

Baud — A unit of signaling speed equal to the number of discrete conditions or events per second. (If the duration of a pulse is 20 ms, the signaling rate is 50 bauds or the reciprocal of 0.02.) (abbreviated Bd)

Baudot code — A coded character set in which five bits represent one character. Used in the U.S. to refer to ITA2.

Bell 103 — A 300-baud full-duplex modem using 200-Hz-shift FSK of tones centered at 1170 and 2125 Hz.

Bell 202 — A 1200-baud modem standard with 1200-Hz mark, 2200-Hz space, used for VHF FM packet radio.

BERT — Bit-error-rate test.

BER — Bit error rate.

Bit — Binary digit, a single symbol, in binary terms either a one or zero.

Bit stuffing — Insertion and deletion of 0s in a frame to preclude accidental occurrences of flags other than at the beginning and end of frames.

BLER — Block error rate.

BLERT — Block-error-rate test.

Break-in — The ability to hear between elements or words of a keyed signal.

Byte — A group of bits, usually eight.

Carrier Detect — Formally, Received Line Signal Detector, a physical-level interface signal that indicates that the receiver section of the modem is receiving tones from the distant modem.

CCIR — International Radio Consultative Committee, an International telecommunication Union agency.

CCIR Rec. 476-3 — The CCIR Recommendation used as the basis of AMTOR and incorporated by reference into the FCC rules.

CCITT — International Telegraph and Telephone Consultative Committee, an ITU agency. CCIR and CCITT recommendations are available from the U.N. Bookstore, United Nations Building, New York, NY 10017.

Chirp — Incidental frequency modulation of a carrier as a result of oscillator instability during keying.

Clear to Send — A physical-level interface circuit generated by the DCE that when ON indicates the signals presented on the Transmitted Data circuit can be transmitted.

Collision — A condition that occurs when two or more transmissions occur at the same time and cause interference to the intended receivers.

Connection — A logical communication channel established between peer levels of two packet-radio stations.

Contention — A condition on a communications channel that occurs when two or more stations try to transmit at the same time.

Control field — An 8-bit pattern in an HDLC frame containing commands or responses, and sequence numbers.

CRC — Cyclic Redundancy Check, a mathematical operation in which the results are sent with a transmission block to enable the receiving station to check the integrity of the data.

CSMA — Carrier Sense Multiple Access, a channel access arbitration scheme in which packet-radio stations listen on a channel for the presence of a carrier before transmitting a frame.

Cut numbers — In Morse code, shortening of codes sent for numerals.

DARPA — Defense Advanced Research Projects Agency; formerly ARPA, sponsors of ARPANET.

Datagram — A mode of packet networking in which each packet contains complete addressing and control information. (compare Virtual Circuit)

Data Set — Modem.

Destination — In packet radio, the station that is the intended receiver of the frame sent over a radio link either directly or via a repeater.

DCE — Data Circuit-Terminating Equipment, the equipment (for example, a modem) that provides communication between the DTE and the line radio equipment.

Digipeater — A link-level gateway station capable of repeating frames. The term "bridge" is used in industry.

Domain — In packet radio, the combination of a frequency and a geographical service area.

DTE — Data Terminal Equipment, for example a VDU or teleprinter.

DXE — In AX.25, Data Switching Equipment, a peer (neither master nor slave) station in balanced mode at the link layer.

EASTNET — A series of digipeaters along the U.S. East Coast.

EIA — Electronic Industries Association. (2001 Eye St., N.W., Washington, DC 20006.)

Envelope-delay distortion — In a complex waveform, unequal propagation delay for different frequency components.

Equalization — Correction for amplitude-frequency and/or phase-frequency distortion.

Eye pattern — An oscilloscope display in the shape of one or more eyes for observing the shape of a serial digital stream and any impairments.

FADCA — Florida Amateur Digital Communications Association, c/o Gwyn Reedy, W1BEL, 812 Childers Loop, Brandon, FL 33511.

FCS — Frame Check Sequence. (See CRC.)

FEC — Forward Error Correction, an error-control technique in which the transmitted data is sufficiently redundant to permit the receiving station to correct some errors.

Field — In packet radio, at the link layer, a subdivision of a frame, consisting of one or more octets.

Flag — In packet switching, a link-level octet (01111110) used to initiate and terminate a frame.

Frame — In packet radio, a transmission block consisting of opening flag, address, control, information, frame-check-sequence and ending flag fields.

FSK — Frequency-Shift Keying.

Gateway — In packet radio, an interchange point.

HDLC — High-level Data Link Control Procedures as specified in ISO 3309.

Host — As used in packet radio, a computer with applications programs accessible by remote stations.

IA5 — International Alphabet Number 5, a 7-bit coded character set, CCITT version of ASCII.

Information field — Any sequence of bits containing the intelligence to be conveyed.

ISI — Intersymbol interference; slurring of one symbol into the next as a result of multipath propagation.

ISO — International Organization for Standardization.

ITA2 — International Telegraph Alphabet Number 2, a CCITT 5-bit coded character set commonly called the Baudot or Murray code.

JAS-1 — JAMSAT/JARL amateur satellite having packet-radio capability.

Jitter — Unwanted variations in amplitude or phase in a digital signal.

Key clicks — Unwanted transients beyond the necessary bandwidth of a keyed radio signal.

KSR — Keyboard Send-Receive teletypewriter.

LAP — Link Access Procedure, CCITT X.25 unbalanced-mode communications.

LAPB — Link Access Procedure, Balanced, CCITT X.25 balanced-mode communications.

Layer — In communications protocols, one of the strata or levels in a reference model.

Level 1 — Physical layer of the OSI reference model.

Level 2 — Link layer of the OSI reference model.

Level 3 — Network layer of the OSI reference model.

Level 4 — Transport layer of the OSI reference model.

Level 5 — Session layer of the OSI reference model.

Level 6 — Presentation layer of the OSI reference model.

Level 7 — Application layer of the OSI reference model.

Loopback — A test performed by connecting the output of a modulator to the input of a demodulator.

LSB — Least-significant bit.

Mode A — In AMTOR, an Automatic Repeat Request transmission method.

Mode B — In AMTOR, a Forward Error Correction transmission method.

Modem — Modulator-Demodulator, a device that connects between a data terminal and communication line (or radio). Also called Data Set.

MSB — Most-significant bit.

MSK — Frequency-shift keying where the shift in Hertz is equal to half the signaling rate in bits per second.

NAK — Negative Acknowledge. (opposite of ACK)

NAPLPS — ANSI X3.110-1983 Videotex/Teletext Presentation Level Protocol Syntax.

NBDP — Narrow-band direct-printing telegraphy.

NEPRA — New England Packet Radio Association, P.O. Box 15, Bedford, MA 01730.

Node — A point within a network, usually where two or more links come together, performing switching, routine and concentrating functions.

NRZI — Nonreturn to zero. A binary baseband code in which output transitions result from data 0s but not from 1s. Formal designation is NRZ-S (nonreturn-to-zero — space).

Null modem — A device to interconnect two devices both wired as DCEs or DTEs; in EIA RS-232-C interfacing, back-to-back DB25 connectors with pin-for-pin connections except that Received Data (pin 3) on one connector is wired to Transmitted Data (pin 3) on the other.

Octet — A group of eight bits.

OOK — On-off keying.

Originate — The station initiating a call. In modem usage, the *calling* station or modem tones associated therewith.

OSI-RM — Open Systems Interconnection Reference Model specified in ISO 7498 and CCITT Rec. X.200.

Packet Radio — A digital communications technique involving radio transmission of short bursts (frames) of data containing addressing, control and error-checking information in each transmission.

PACSAT — Proposed AMSAT packet-radio satellite with store-and-forward capability.

PAD — Packet Assembler/Disassembler, a device that assembles and disassembles packets (frames). It is connected between a data terminal (or computer) and a modem in a packet-radio station. (see also TNC)

Parity Check — Addition of noninformation bits to data, making the number of ones in a group of bits always either even or odd.

PID — Protocol identifier. Used in AX.25 to specify the network-layer protocol used.

PPRS — Pacific Packet Radio Society, P.O. Box 51562, Palo Alto, CA 94303.

Primary — The master station in a master-slave relationship; the master maintains control and is able to perform actions that the slave cannot. (Compare secondary.)

Protocol — A formal set of rules and procedures for the exchange of information within a network.

PSK — Phase-Shift Keying.

RAM — Random Access Memory.

Received Data — Physical-level signals generated by the DCE are sent to the DTE on this circuit. (abbreviated RxD)

Request to Send — Physical-level signal used to control the direction of data transmission of the local DCE. (abbreviated RTS)

RO — Receive-Only teletypewriter.

ROM — Read-Only Memory.

ROTR — Receive-Only Typing Reperforator Set.

Router — A network packet switch. In packet radio, a network-level relay station capable of routing packets.

RS-232-C — An EIA standard physical-level interface between DTE (terminal) and DCE (modem), using 25-pin connectors.

RTTY — Radioteletype.

Secondary — The slave in a master-slave relationship. Compare primary.

SOFTNET — An experimental packet-radio network at the University of Linköping, Sweden.

Source — In packet radio, the station transmitting the frame over a direct radio link or via a repeater.

SSID — Secondary Station Identifier. In AX.25 link-layer protocol, a multipurpose octet to identify several packet-radio stations operating under the same call sign.

SOUTHNET — A series of digipeaters along the U.S. Southeast Coast.

TAPR — Tucson Amateur Packet Radio Corporation, a nonprofit organization involved in packet-radio development (P.O. Box 22888, Tucson, AZ 85734).

TD — Tape Transmitter-Distributor Set, a paper tape reader.

Teleport — A radio station that acts as a relay between terrestrial radio stations and a communications satellite.

Teleprinter — Trademark of ITT/Creed. A printing teletypewriter.

Teletype — Trademark of the Teletype Corporation, usually used to refer to teletypewriters.

Teletypewriter — Generic term for printing telegraph equipment. (For specific types see ASR, KSR, RO, ROTR and TD.) (abbreviated TTY)

TNC — Terminal Node Controller, a device that assembles and disassembles packets (frames). (used interchangeably with PAD)

TR switch — Transmit-receive switch to allow automatic selection between receive and transmitter for one antenna.

Transmitted Data — Physical-level data signals transferred on a circuit from the DTE to the DCE. (abbreviated TxD)

TTY — Teletypewriter.

TU — Terminal Unit, a radioteletype modem or demodulator.

Turnaround time — The time required to reverse the direction of a half-duplex circuit, required by propagation, modem reversal and transmit-receive switching time of a transceiver.

UI — Unnumbered Information frame.

VADCG — Vancouver Amateur Digital Communications Group. (c/o Doug Lockhart, VE7APU, 9531 Odlin Rd., Richmond, BC V6X 1E1, Canada.)

VDT — Video-display terminal.

VDU — Video Display Unit, a device used to display data, usually provided with a keyboard for data entry.

Videotex — A presentation-layer protocol for two-way transmission of graphics.

Virtual Circuit — A mode of packet networking in which a logical connection that emulates a point-to-point circuit is established. (compare datagram)

V.24 — A CCITT standard defining physical-level interface circuits between a DTE (terminal) and DCE (modem), equivalent to EIA RS-232-C.

V.28 — A CCITT standard defining electrical characteristics for V.24 interface.

WESTNET — A series of digipeaters along the U.S. West Coast.

Window — In packet radio at the link layer, the range of frame numbers within the control field used to set the maximum number of frames that the sender may transmit before it receives an acknowledgment from the receiver.

X.25 — CCITT packet-switching protocol.

Chapter 20

Image Communications

Judging by the upsurge of amateur interest in systems that can send and receive pictures, many hams must be taking to heart the old Chinese proverb, "one picture is worth a thousand words." This, along with the availability of inexpensive video components, may explain why so many hams are motivated to use and experiment actively with systems capable of exchanging pictures.

There are three basic image communications systems in popular use by the amateur community. Each approach has unique features, capabilities and applications:

• *Fast-scan television (FSTV)*: Moving pictures are displayed on a standard TV set; includes sound. Its performance is similar to commercial broadcast TV pictures. It is used predominately in the UHF bands (70 and 23 cm) to provide local area coverage.

• *Slow-scan television (SSTV)*: Low-resolution still pictures are displayed on a standard TV set. Used in the HF bands, it provides worldwide coverage.

• *Facsimile (FAX)*: High-resolution still picture produced on paper or photographic film. It is used for weather-satellite reception and in the HF bands to provide worldwide coverage.

The recent amateur interest in image communications essentially reflects the current video orientation of our society. This is evidenced by the huge sales of home consumer items such as color TV sets, video cassette recorders, TV cameras, video games and computer graphics displays. Many of these consumer devices can be used in your shack as a foundation for building a complete amateur image communications system.

Fast-Scan Amateur Television

Fast-scan TV (FSTV), also referred to simply as amateur television (ATV), uses a transmission format fully compatible with video equipment designed for the home consumer market. The video is amplitude modulated; the audio is frequency modulated. Each video frame consists of 525 horizontal lines with 30 frames transmitted each second. (See Chapter 7 for more details on the scanning process.) Translated into simple terms, this means that an FSTV picture displays full motion, has a simultaneous sound channel, can be in color and has excellent detail just like commercial TV pictures.

Amateur television offers a major advantage over broadcast TV, though, and that is rooted in our ability to communicate interactively or two ways. Amateur television people have been communicating in round table nets for many years — long before industry discovered the benefits of interactive video or, as it's called, "teleconferencing."

Because the signals occupy several megahertz of bandwidth, the FCC does not permit FSTV on bands below 420 MHz. Most ATV is performed in the 420- to 440-MHz and 1240- to 1294-MHz segments of the 70- and 23-cm bands. Some ATV activity can be found in the 902-928 MHz band. These frequencies generally limit FSTV range to line of sight with extended

Fig. 1 — Fast-scan ATV station assembled using circuits described in this chapter plus a home-video camera and standard TV set.

coverage possible when ducting conditions exist or repeaters are used.

A Technician class or higher license is required to transmit FSTV below 1240 MHz. Novices may operate ATV in the 1270-1295 MHz portion of the 23-cm band. All you need is a home-video type camera, microphone, ATV transmitter and antenna. Depending on your application, a personal computer terminal or video tape recorder can be used instead of a camera. You don't even have to provide external synchronization. Sync pulses are part of the composite video signal from the camera, video cassette recorder or computer. No

modification to these devices is necessary.

To receive, all you need is an antenna, RF converter (also called a downconverter, it shifts the received signal to a standard VHF TV channel) and an unmodified home TV set. No scan converters are required. The basic elements of a fast-scan ATV station are shown in Fig. 1.

In keeping with the ham tradition for both public service and inventiveness in communications, ATVers have been using their live action video systems in many exciting applications. For example, the Southern California ATVers annually help coordinate and monitor the Tournament of Roses Parade and local marathons. During the Christmas season, amateurs have taken their systems into hospitals to permit children to see and talk with Santa Claus over live TV. Warren Weldon, W5DFU, uses his ATV cameras on a tall tower to search the Oklahoma skies for "killer tornadoes" while transmitting the pictures directly to the Weather Service to provide a valuable early-warning capability. Amateurs have transmitted ATV from Civil Air Patrol airplanes and even hot-air balloons. ATV has also been used in radio-controlled model airplanes and robots to provide the necessary feedback to aid the operator. Most recently, the FCC has permitted ATVers to retransmit the spectacular Space Shuttle video provided on

Fig. 2 — Aft end of STS-9 Shuttle cargo bay as seen via fast-scan ATV.

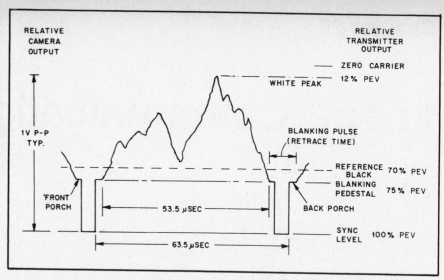

Fig. 3 — TV Waveform. Camera and corresponding transmitter RF output power levels during one horizontal line scan for black-and-white TV. (A color camera would generate a "burst" of 8 cycles at 3.58 MHz on the back porch of the blanking pedestal.) Note that "black" corresponds to a higher transmitter output power than "white."

NASA's public information channel on Satcom 1R (Fig. 2).

On a day-to-day basis, there are many down-to-earth uses for ATV. You can show off your shack, your family, home movies, video tapes and travel slides. You do run the risk of putting fellow ATV watchers to sleep! Some amateurs exchange video tape copies of their contacts with hams in other countries.

Your home computer system and ATV station make an ideal marriage. By simply taking the video output of your home computer into the video input of your ATV transmitter, the computer may be used to show still and animated graphics, teach computer programming and debugging, repeat W1AW RTTY bulletins and display bulletin-board messages. Serial data can also be placed on the TV sound subcarrier for exchanging computer programs.

Since ATVers share the bands with other types of emissions and amateur interests, frequencies have evolved through formal frequency-coordination organizations that minimize mutual interference to all band users in any given area. Before you start transmitting, check with local ATVers and the *ARRL Repeater Directory*. The *Repeater Directory* also lists your area frequency-coordination committee.

TV Waveforms

Thanks to the availability of off-the-shelf cameras, and receiving converters that preclude the need to perform surgery on your home TV set, it is not necessary to understand the inner workings of a camera or TV set. It is only necessary for you to know what the waveform at the output of a camera, computer or video tape recorder will look like to help you adjust your ATV station properly.

Fig. 3 shows what a single horizontal line of video looks like at the camera and transmitter output. Two basic elements can be seen in the waveform: the synchronization pulses and the video.

• Synchronization pulses — These are used in the TV receiver to synchronize the CRT electron beam scan with that of the camera (see Chapter 7). Although only "horizontal" pulses are shown in the figure, there are also "vertical" pulses,

which occur 60 times a second to ensure that the top of the received picture lines up properly with the top of image as seen by the camera. It is only important for you to know that the proper amplitude of these synchronization pulses must be maintained in the transmission process. The transmitter and linear amplifier must be adjusted properly to ensure that the sync is not overly compressed or clipped.

• Video — Between the synchronization pulses lies the video information that controls the intensity of the beam as it scans across the receiver picture tube. The signal level determines whether an individual picture element (pixel) shall be white, black or a given shade of gray. This concept is shown in Figs. 4 and 5.

Color TV is more complicated than black-and-white TV. The transmitting and receiving equipment is the same for black and white transmission, except that a color camera and TV receiver are required. The corresponding frequency spectrum of the color TV signal is shown in Fig. 6.

A brief discussion of color TV can be found in Chapter 7. An excellent source of more complete information on the basics of both monochrome and color television techniques may be found in the book *Television Handbook for the Amateur* by Biagio Presti. See the bibliography at the end of this section.

FSTV System Elements

As previously emphasized, standard, unmodified, off-the-shelf video components are used to generate and display the ATV pictures. A block diagram of a typical FSTV station is shown in Fig. 7. The most important part of the system is the antenna and coax. Since vegetation greatly attenuates UHF signals, mounting the antenna above tree tops is highly desirable. DX

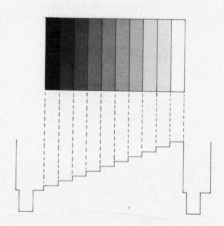

Fig. 4 — The video signal resulting from viewing a gray-scale pattern ranging from black to white level.

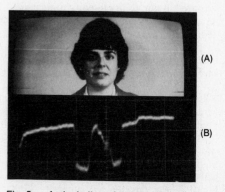

(A)

(B)

Fig. 5 — A single line of the TV picture at A has its voltage waveform shown at B.

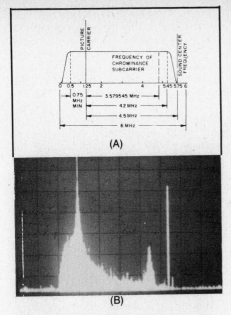

(A)

(B)

Fig. 6 — Frequency spectrum of a color TV signal shown in diagram at (A) and on a spectrum analyzer at (B). Each vertical division represents 10 dB; horizontal divisions are 1 MHz. Spectrum power density will vary with picture content, but typically 90% of sideband power is within the first megahertz.

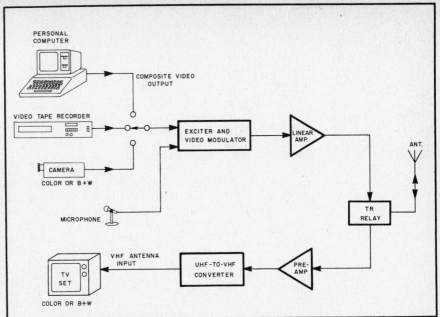

Fig. 7 — Fast-scan ATV station.

will vary with terrain, trees and temperature inversion characteristics (ducting). Because the bandwidth of an FSTV system is considerably wider than a narrow-band voice circuit operating on the same frequency at the same power level, it is necessary to compensate for the accompanying decrease in received signal-to-noise by using relatively high-gain antennas, low-loss transmission lines and low-noise preamplifiers. One should not compare ATV performance with UHF broadcast TV reception. Commercial and public stations transmit a million, or more, watts of effective radiated power. Amateurs use considerably lower power levels, of course.

Antennas

The antenna must have both high gain and broad bandwidth. For the home-brew enthusiast, the 15-element quagi is popular (see Chapter 33). The MBM48/70 J Beam and several KLM models are good, high-gain, commercially constructed beams that are specially designed for broadband work. For portable applications, a small Yagi antenna, or omnidirectional antennas such as the AEA Isopole are used.

Polarization used by ATVers varies throughout the country. In general, most DX operations use horizontal, while most repeaters use vertical polarization. Check with local ATVers to determine the polarization used in your area before setting up an antenna.

Transmission Line

At 420 MHz, cable loss can be quite high, especially for long cable runs. Types such as RG-58 are very lossy at UHF (15-dB loss/100 feet at 420 MHz) and are not used for ATV work except for short equipment interconnects in the shack. Up to 50 feet, RG-8, -213 or -214 (5 dB/100 ft) are acceptable, but for longer runs, ½-inch Hardline such as Phelps-Dodge PD-FX 12-50H (2.1 dB/100 ft) or low-loss coax such as Belden 8214, Belden 9913, Belden 9914, or Saxton 8285 (3.8 dB/ 100 ft) is recommended. The exception to this is when the receive preamp can be mast-mounted at the antenna. The type of connector used (UHF or N) is not as important as proper installation and weatherproofing.

Receiver

The purpose of the receive converter is to shift the UHF signal to a frequency that is tunable on a standard TV set. The output is fed into the VHF antenna terminals on the set. No modification to the TV is required.

A preamplifier may be included as part of the converter or as a separate mast-mounted unit. The use of GaAsFET front ends for preamps, by virtue of their wide dynamic input range and related immunity from self-generated intermodulation components, has reduced the requirements for filtering ahead of the preamp. A popular commercial preamp for ATV use is Advanced Receiver Research model P432VDG.

Video Transmitter

A video transmitter is very similar in design to a UHF FM voice transmitter except that the constant dc voltage on the final RF amplifier stage is replaced with the amplitude-varying output of a video modulator to produce amplitude modulation. The principle is identical to plate (or collector) modulation prevalent in the pre-SSB era.

The transmitter is modulated with the video and synchronization signals provided by a camera. The modulator gain must be adjusted to provide the proper modulation level. Too little gain will produce a picture lacking contrast; too much gain may cause the picture to "white out" or clip off sync. Proper set-up of modulation level is just as important as transmitter power level. A strong received signal exhibiting poor contrast may look worse than a weak signal with excellent contrast. The modulator must have the bandwidth to pass the 3.58-MHz color-burst signal provided by the camera and for a 4.5-MHz sound subcarrier. Most solid-state UHF transmitters have output matching circuits that pass the wide-bandwidth ATV signals.

Power Amplification

To increase the power output from the video transmitter, a linear amplifier is used. Tube amplifiers generally provide better linearity than solid-state units, while solid-state amplifiers tend to have broader bandwidths.

Proper sync-to-video ratio and picture linearity are the main concerns when using a high-power transistor amplifier. A sync stretcher in the video modulator is used to compensate for gain compression (nonlinearity at high power levels) typically found in solid-state amplifiers. By exaggerating the sync level at the amplifier input, the proper sync-to-video relationship will result at the amplifier output as shown in Fig. 8.

Transistor amplifiers, by virtue of their low loaded Qs, have sufficient bandwidth for ATV. This is generally not the case with

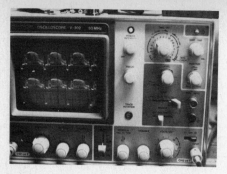

Fig. 8 — Oscilloscope used to observe the video waveform. The lower trace is the video signal as it comes out of the sync stretcher. The upper trace is the signal from the Mirage D1010-N amplifier.

tube amplifiers. At UHF, many tetrode high-power amps using 4CX250-type tubes have too high a loaded Q at the grid and plate to pass the color and sound subcarriers without significant attenuation. This can be overcome by control-grid modulation of the final stage and increased loading of the plate circuit. Grounded-grid triodes like the 7289 are less bandwidth restrictive in the linear mode because of the lower impedances associated with cathode drive. However, direct cathode modulation is usually preferred and much simpler. Sync stretching is normally not necessary with tube amplifiers, as their linearity is excellent.

For more discussion on the special considerations for linear amplifiers used in a television role, see "The Care and Feeding of Linear Amplifiers for ATV," *QST*, August 1982, p. 24.

Audio Transmission Techniques

Sound can be transmitted on FSTV using two main approaches. One method is to FM audio modulate the AM video carrier frequency. This permits a rather simple transmitter design, but the penalty is paid at the receiving end since a standard TV will not be able to pick up the audio. A separate FM receiver, tuned to the video frequency, must be used to recover the audio.

A more popular approach is to provide FM audio at a frequency 4.5 MHz above the video carrier using a separate transmitter or using a subcarrier generator that feeds the input of the video modulator. The advantage of this approach is that no separate receiver is required since a conventional TV set includes the circuitry to demodulate the 4.5-MHz subcarrier.

In addition to either technique, a separate 2-meter circuit has proven highly beneficial for talk back and coordination. By using 2 meters to establish a full-duplex audio link, the receiving station can provide valuable picture-quality reports to the TV transmitting station. Also, a casual 2-meter listener is intrigued when hearing the comments about a picture that cannot be seen on 2 meters, and is surprised to even

find out that there is ATV activity in the area. Many 2-meter enthusiasts have been drawn into ATV in this manner.

ATV Repeaters

There are two types of ATV repeaters on the air. Most are in-band machines with input and output in the 70-cm band. The other type of repeater is cross band. The input is in the 70-cm band and the output is in the 23-cm band. Exact frequencies are listed in the *ARRL Repeater Directory*. Cross band has an advantage of full-duplex monitoring of the received repeater video since receive system desense, a normal problem with in-band repeaters, is virtually eliminated. This allows an operator to make picture adjustments without relying on descriptions from other station operators.

Vertical polarization is predominant for repeater antennas because it is much easier to build or purchase commercial high-gain omnidirectional vertical antennas than horizontal types. Also it is easier for communication with mobile units which almost exclusively use vertical whip antennas.

To conserve transmitted signal bandwidth and prevent receiver desense, ATV repeaters use vestigial sideband, which is also used in commercial broadcasting. A filter at the output of the transmitter is used to suppress the lower sideband beyond 1.25 MHz while allowing up to 4.75 MHz on the upper sideband. This creates an ATV signal with 6-MHz bandwidth. Filtering can be accomplished using an interdigital filter of the type described by Fisher, "Interdigital Bandpass Filters for Amateur VHF/UHF Applications," *QST*, March 1968. (A commercially built interdigital filter is available from Spectrum International, model PSF-ATV.) Filtering can also be performed using a group of cavities tuned to provide a flat low-loss response over several megahertz but having very sharp skirts to perform the necessary out-of-band attenuation. (Such a multicavity filter is commercially available from TX-RX Systems Inc., model 19-66-79154-2.)

Amateur TV repeater design is described in more detail in Brown, "The WR4AAG ATV Repeater," *QST*, October 1975, p. 98; also Pasternak/Morris, *The Practical Handbook of Amateur Radio FM and Repeaters*, Tab Books, 1980; and *Everything You Always Wanted To Know About Amateur Television*, QCD Publications.

Constructing a Basic ATV System

An ATV station can be assembled using four basic circuit boards. (These boards may be purchased built and tested from PC Electronics, 2522 S. Paxson La., Arcadia, CA 91006.) This approach will enable the transmission and reception of full-color television with sound.

Shown in Figs. 9 through 12, all boards run from a common 13.8-V dc, 3 A regulated power source. The complete system can be mounted in a 12 × 8 × 3 inch

chassis. At the builder's option, the receive converter and TR relay can be mounted at the antenna in an outside weatherproof electrical box and controlled from the transmitter location.

Shielded connections between modules are made with the small RG-174 50-ohm coax. At UHF, the center conductor of the coax should not be more than ¼ inch past the shield for all board connections to minimize SWR and radiation. Power and signal connections are made with no. 24 vinyl-covered wire except to the 10-watt power module, which is no. 18.

RECEIVING PREAMP AND CONVERTER

The converter, shown in Fig. 10, tunes the 420- to 450-MHz band and shifts the received signal to TV Channel 3 (video carrier at 61.25 MHz) or Channel 2 or 4 if Channel 3 is used in your area. Trimmer capacitor, C3, which sets the varactor-diode-controlled oscillator range, sets the received frequency. The lead length from the board to the 10-kΩ frequency-tuning pot is not critical. This allows remote tuning if desired. The board, however, must be in a shielded enclosure to minimize stray pickup of unwanted adjacent TV channels.

A Mini-Circuits Lab SBL-1 doubly balanced mixer module is followed by an 80-MHz low-pass filter to reduce intermodulation from other strong signals. Ahead of the mixer is a low-noise MRF901 or NE64535 preamp stage. The MRF901 has a typical noise figure of 1.7 dB with 15-dB gain, while the NE64535 exhibits a 0.9-dB NF and 18-dB gain.

Coaxial cable with an impedance of 75 ohms is recommended to connect the unit to the TV-set VHF antenna input. Twin lead can act as an antenna bringing in strong adjacent-channel interference.

To align the converter, preset the three trimmer capacitors and the 10-kΩ VCO pot to their center positions. Set the 5-kΩ preamp transistor bias pot for 0.5-volt drop across the 100-Ω collector resistor if the MRF901 is used, or 0.8 volt with the NE64535. Have a nearby ATV station transmit a picture to you or use a signal generator. Use an insulated tuning tool to carefully adjust C3 for best reception of the signal. Watch your TV set as you adjust C3. You may find reception possible in two positions — one on the high side of the local oscillator injection frequency, the other on the low side. You want the low side. To do this, select the position with the most capacity on C3 (plates meshed). Next, fine tune with the 10-kΩ pot for best picture and finally peak C1 and C2.

TRANSMITTING CIRCUITS

Three circuit boards are used to transmit video and sound at approximately 12-watts PEP output. The exciter/modulator generates 80 milliwatts of AM, which drives a sealed broad-band power module.

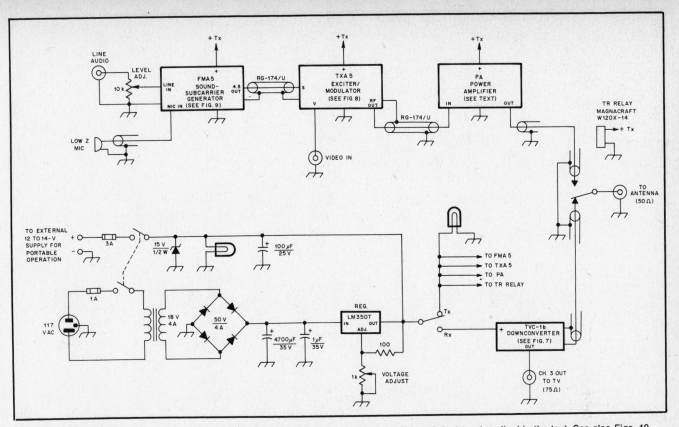

Fig. 9 — ATV system-interconnection and power-supply schematic diagram using circuit board designs described in the text. See also Figs. 10, 11 and 12. The boards are also available fully built from P. C. Electronics. Output from the converter connects to a standard TV set. Video input is from a standard 1-volt peak-to-peak, 75-ohm source such as camera, video tape recorder or computer.

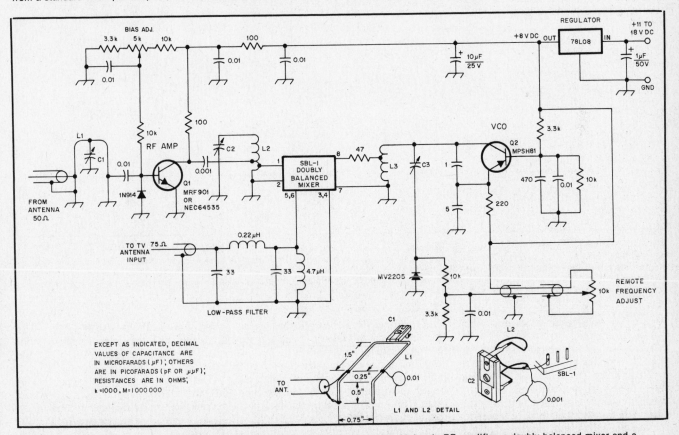

Fig. 10 — The circuit of the ATV receiving converter (TVC-1b) consists of a low-noise, high-gain RF amplifier, a doubly balanced mixer and a varicap-tuned VCO. Remote tuning of the converter permits it to be mounted at the antenna if desired. Variable capacitors are Arco 400, 1-10 pF, or equiv. L1 consists of a horseshoe-shaped loop of no. 18 wire, 4.5-inches long and 0.75 inch wide. Loop ends are soldered to a ground plane; a right-angle bend 0.5 inch from the grounded ends causes the rest of the loop to be parallel to the ground plane. Taps are made 0.75 inch from the ends. L2 consists of 1½ turns of no. 18 wire, 0.375-inch diameter, tapped at 0.5 inch from the lower end. L3 is a hairpin loop, 0.5 inch across the bottom and 0.625 inch high. It is made with no. 18 bus wire.

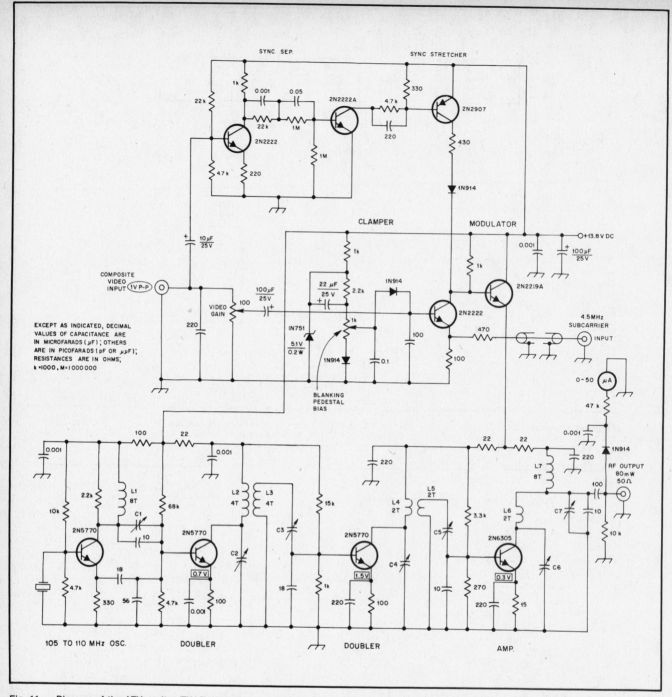

Fig. 11 — Diagram of the ATV exciter (TXA5). Variable capacitors are Arco 400, 1- to 10-pF, or equiv. All coils are wound from no. 24 bus wire, using a 10-24 screw for a winding form. Coupled coil pairs are placed 1/16 inch apart. Each stage is tuned with a voltmeter monitoring and peaking the next stage's emitter voltage. If peaked correctly, the voltages should be at or above those noted at each emitter. The 1-kΩ blanking pedestal bias pot is backed down to give 75% of the maximum unmodulated RF power output for proper video to sync ratio setup.

Another module generates the 4.5-MHz sound subcarrier.

The exciter/modulator, shown in Fig. 11, contains a fifth overtone crystal oscillator in the 100-MHz range. Following the oscillator are two doublers, the last of which is collector modulated along with the final. The bypass capacitors are chosen to remove 400-MHz energy but not attenuate the video below 8 MHz. This gives high-resolution and color even for 64-character-per-line computer video. The pedestal bias pot controls the blanking clamp level and

sets the proper video-to-sync ratio. A sync stretcher in the modulator circuit is used to compensate for nonlinearities in the final amplifier module. The stretcher circuit pulls the applied voltage on the modulated stages up to almost full V_{cc} during sync times. This is necessary since the input-vs.-output characteristic curves of UHF transistor amplifiers generally show gain compression in their upper power range. Typically, the last 3-dB increase in power output takes 6 dB of input power increase. Without sync stretching to compensate,

power output would have to be set to less than half the rated capability to maintain sync level and picture stability at the receiver.

The power amplifier consists of an MHW 710-2 or Amprex BGY41B power module with a heat sink and printed-circuit board for the line filtering parts (Fig. 9). Filtering of V_{cc} at the RF amplifier module under video modulation is critical. The supply impedance must be very low at video frequencies up to 5 MHz where the resistance and inductance of the supply

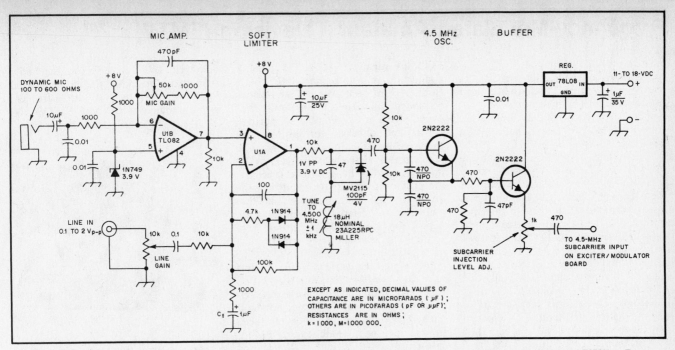

Fig. 12 — This ATV sound-subcarrier generator (FM-A5) permits both voice and video to be transmitted. U1 is either a National LF353N or Texas Instruments TL082CP. The MV2115 is a Motorola varicap of 100 pF at 4 volts.

leads are significant. The power modules are designed for 15 watts using FM, and are sufficiently linear for ATV.

The subcarrier sound generator, shown in Fig. 12, has a high-gain dynamic microphone amplifier consisting of a dual-section, FET-input op amp and soft limiter. A second input is used for mixing high-level audio from a VCR or color camera microphone. A 4.5-MHz oscillator is varactor modulated by the output of the audio amplifier. The deviation is fixed at 25 kHz by soft limiting back-to-back diodes. The 1-kΩ pot on the emitter follower adjusts the sound subcarrier injection level. The sound subcarrier is usually set about 15 dB below the video carrier level. This corresponds to around midpoint on the pot or 0.5 volt P-P on a scope. Too much injection will give a beat pattern in a color picture, and not enough will give a low audio level. The oscillator frequency should be set by adjusting the 18-μH inductor slug to within 3 kHz of 4.5 MHz using a counter connected to the 1-kΩ pot output.

Transmitter Tuning

Tune-up requires a voltmeter, UHF RF power meter and a low-SWR antenna or dummy load. Adjust each trimmer on the exciter/modulator board for peak emitter voltage on the following stage. There should be no video or subcarrier connected during the RF tune up. Set the pedestal clamp bias pot for maximum voltage on the emitter of the 2N2219A modulator transistor. When some power is indicated by

the wattmeter, carefully peak C6 and C7 followed by C4 and C5 for maximum power. With 13.8-V dc, the PEP output should be at least 10 watts. To establish the proper video-to-sync ratio (Fig. 8), reduce the measured power output by 75% by adjusting the pedestal bias pot. For example, if the maximum power output is 12 watts, the blanking pedestal power setting should be adjusted for a 9-watt output. The sync stretcher will pull the power back up to 12 watts during sync time when video is applied. An average-reading wattmeter has little significance when measuring video-modulated RF since the reading will fluctuate greatly depending upon video gain setting, camera setup and image content. As previously described, the wattmeter reading is valid for adjusting the blanking pedestal and determining the PEP level with no video applied.

Connect any source of composite video (typically 1 volt peak-to-peak) to the input and adjust the 100-Ω video-gain pot for best picture at another ATVer's location, preferably at least 1 mile away. Picture monitoring on a TV set in your own shack can give a false indication due to overload and reflections. Use of a two-meter talk-back channel and a distant viewer will facilitate a rapid and accurate adjustment. It is best to use a scene or display with lots of white for proper modulation depth adjustment. Increase the gain until the white areas begin to smear; then back off until the smearing stops. Overmodulation will cut off the carrier during white areas of the

picture causing undesirable buzzing in the received audio.

Proper video-to-sync ratio and picture linearity are major concerns when adding a high-power transistor linear amplifier, such as the Mirage D1010N. Setup is basically similar to that previously described using the 10-watt PEP module. However, the blanking pedestal pot should be adjusted for an output not exceeding 75 watts even though 100 watts is possible. With 75 watts measured at the amplifier output, the drive power to the final amplifier will be about 4 watts. The sync stretcher will still put the sync tip at over 10 watts drive to compensate for the high-power amplifier gain compression. The end result is to provide a 100-watt sync tip and the proper video-to-sync ratio at the amplifier output as shown in Fig 8.

FSTV Bibliography

Presti, *Television Handbook for the Amateur*, Aptron Labs, 1978; Available from QCD Publications, PO Box H, Lowden, IA 52255-0408.

Everything You Always Wanted To Know About Amateur Television, 3rd Edition, 1982, QCD Publications.

SPEC-COM, QCD Publications.

Brown and Wood, *Amateur Television Handbook Volume 1*, 1981, British Amateur Television Club

Brown, *Amateur Television Handbook Volume 2*, 1982, British Amateur Television Club

Frequency-Modulated Amateur Television (FMTV)

In the previous section, we described the use of amplitude modulation for fast-scan television. Another method just beginning to make inroads into the amateur community is frequency-modulated (fast-scan) television (FMTV). There are performance advantages possible with FMTV.

In the U.S., FMTV is used above 420 MHz, with the majority of activity found in the 1240-1260 MHz segment of the 24-cm band. Despite the additional path losses encountered at these higher frequencies, excellent performance can be achieved because of (1) the overall processing gain advantages afforded by FM; (2) the ability to use relatively compact high-gain antennas; and (3) the availability of interference-free channels.

Frequency-modulated television offers several exciting opportunities for amateurs to experiment with new modulation enhancement techniques, and also motivates fresh exploration into the relatively untouched microwave regions. In this section, the primary focus will be on 24-cm operation; however, the FMTV principles can be readily transferred to the higher-frequency bands.

History

The concept of transmitting fast-scan television pictures via FM is not exactly new. In fact, the possibilities of using FM for television were suggested as early as 1929.[1] The first quantitative analysis was reported in 1940 and was immediately followed by hardware development focused on using FM to relay commercial broadcast television between cities.[2,3,4] During World War II, the military sponsored research into FMTV and in the late 40s

FCC investigated FMTV for possible application in commercial UHF broadcasting.[5] During the 1950s and 60s research continued into various techniques such as pre-emphasis and de-emphasis, improved limiter and discriminator designs, and extensive investigations into bandwidth requirements. In the late 60s and early 70s interest in FMTV heightened as it became the primary mode for transmitting television through communications satellites and also in terrestrial communication applications for sending pictures from commercial news-gathering remote units back to their TV stations.[6]

In the amateur world, European hams have been at the forefront of FMTV experimentation and equipment development. Virtually all commercially-built amateur FMTV equipment is produced in Europe, with patches of FMTV just beginning to sprout in the United States.

24-cm FMTV Transmitters

The two basic techniques for generating a 24-cm FMTV signal are shown in Fig. 13. One method is to use a varactor to triple the amplified signal from a voltage-controlled 416-433 MHz oscillator that has been video modulated. The other approach is to simply amplify the output from a video-modulated 24-cm VCO. In either case, a standard composite video signal provided from a TV camera, video tape machine or computer is processed through a pre-emphasis filter which accentuates the high video frequencies to improve the overall system signal-to-noise ratio. The video signal is then combined with an audio subcarrier signal, typically at 5.8 or 6.0 MHz. With AM FSTV, the audio subcarrier is at 4.5 MHz. Experimentation has shown that a higher subcarrier frequency is required for good FMTV per-

(A)

(B)

Fig. 14 — FMTV transmitter components using a varactor tripler technique. At A, a Wood & Douglas 416-MHz voltage-controlled oscillator (UFM01) is used to drive a 10-W UHF amplifier (70LIN10). At B, a varactor tripler is used to produce approximately 3 W at 24 cm. This technique requires an external audio subcarrier generator and video pre-emphasis circuit. The boards are available fully assembled but require mounting and heat sinking.

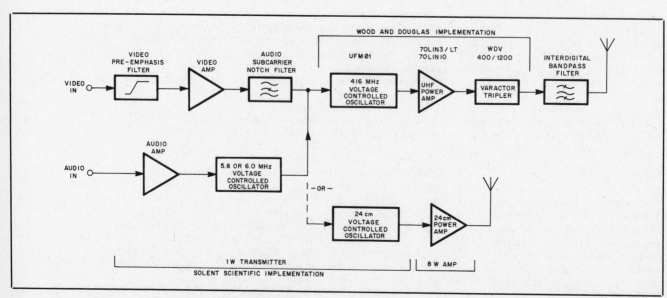

Fig. 13 — Two methods of generating a 24-cm FMTV signal.

Fig. 15 — This 1-W FMTV transmitter kit from Solent Scientific employs a 24-cm VCO and amplifier. The design includes an on-board audio subcarrier generator and video pre-emphasis filter. The output may be amplified to 8 W using an optional power amplifier kit.

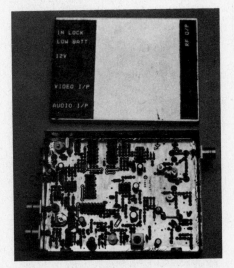

Fig. 16 — This experimental 24-cm transmitter from Wood and Douglas uses a phase-locked loop and reference crystal oscillator for precise frequency stability. The unit which produces a 20-mW output, has a built-in audio subcarrier generator but does not use video pre-emphasis.

formance. The audio subcarrier design shown in Fig. 13 can be easily set on 5.8 or 6.0 MHz for this purpose. The amplitude-varying signal, consisting of the FM audio riding the video envelope, controls the frequency of a voltage-controlled oscillator to generate the video FM signal. The signal is amplified (and tripled, if it was originally generated in the lower portion of the 400-MHz band) and then transmitted. Figs. 14, 15 and 16 show FMTV hardware using both transmission techniques.

24-cm FMTV Receivers

While a 24-cm FM transmitter is really not much more complicated than an AM transmitter, this is not the case with the receiver. The additional complexity in the receiver design is readily apparent in the block diagram in Fig. 17. For AM FSTV, it was only necessary to employ a converter which could produce a signal in the frequency range that could be tuned by a standard TV set. The TV possessed all of the necessary circuitry to demodulate the video and audio. With FM, much of this circuitry must be provided external to the TV set.

A converter, like the one shown in Fig. 18, is used to translate the signal to an intermediate frequency. Depending on the receiver design, the IF may be anywhere from 45 to 70 MHz. Typically, a phase-locked loop or doubly-balanced mixer is used as a discriminator to demodulate the signal to baseband video where the audio subcarrier is removed and further demodulated to baseband audio. This function is usually contained on an IF/video demodulator board, as shown in Fig. 19. The video portion is de-emphasized (high noise component attenuated) and made available for display on a video

Fig. 18 — This Wood and Douglas 24-cm to VHF converter (1250DC50) is fully built and packaged.

Fig. 19 — This Wood and Douglas video IF board (VIDIF) is used to demodulate the signal at the IF to baseband video. An external subcarrier demodulator and de-emphasis unit is required, and the board requires mounting in a suitable enclosure, as shown.

monitor. A Channel 3 or 4 RF modulator may be used to upconvert the baseband audio and video signals into a format viewable on a standard TV set.

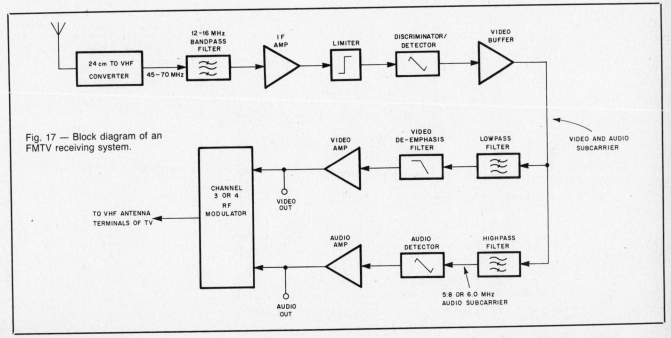

Fig. 17 — Block diagram of an FMTV receiving system.

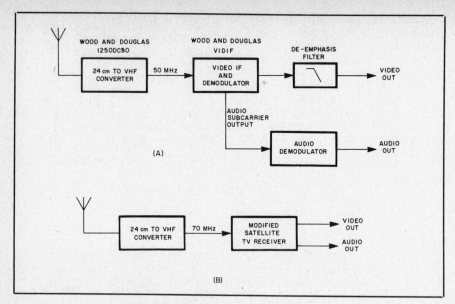

Fig. 20 — Two approaches to 24-cm FMTV receiver design. The design at A uses components designed for amateur use. At B, a modified 70-MHz satellite TV IF strip is used. See text.

Fig. 21 — This complete 24-cm FMTV receiver contains a preamplifier, converter, video IF strip, audio subcarrier demodulator, video de-emphasis filter and 12-V power supply.

Two possible receiver configurations based on commercially-available equipment are shown in Fig. 20. A complete 24-cm FMTV receiver corresponding to the Fig. 20A design is shown in Fig. 21. The schematic for the sound- subcarrier demodulator is shown in Fig. 22 while the circuit for the de-emphasis network will be described later in the text.

It is of interest to note that the operation of the FMTV receiver is almost identical to that used for home satellite TV reception except for the lower microwave carrier frequency and lower FM deviation levels typically used by amateurs operating in the 24-cm band. The 70-MHz satellite TV IF strips that are just beginning to become available on the surplus market can be

adapted for amateur use. If the strip is to be used on 24 cm to receive a nominal 4-MHz peak deviation signal, there are two basic modifications that are required. First, the surface acoustic wave (SAW) bandpass filter in the IF, typically exhibiting a 28 MHz bandpass, must be changed to one 12 to 16 MHz wide to better match the lower bandwidth occupied by amateur signals. This will improve sensitivity and reduce interference. Secondly, the overall gain of the discriminator and/or video amplifier circuits must be increased to compensate for the substantially lower amateur deviation levels to maintain the 1 V peak to peak video output needed for driving a standard TV monitor. If the strip is to be used to receive a signal exhibiting an 8-12 MHz deviation, which may be more common with 10-GHz operation, no modification will be required.

There are some satellite units called "block receivers" that directly cover the 950-1450 MHz range. At first glance, it would seem that this receiver would be ideal for amateurs since it could directly tune across the 24-cm amateur band; however, experimentation has shown that this receiver class has very poor sensitivity. (In normal satellite applications, the receiver depends on a high-gain, low-noise front end to convert the 3.7-4.2 GHz satellite band to 950-1450 MHz.) Unless a very high-gain and low noise figure pre-amplifier, along with a 12 to 16 MHz wide 24-cm bandpass filter, can be used at its front end, this type of receiver is not recommended.

24-cm Antennas and Transmission Line

Generally, the main requirement for an antenna system is that it must possess sufficient bandwidth to support FMTV. Most commercial and homebrew designs are adequate as long as they are cut for

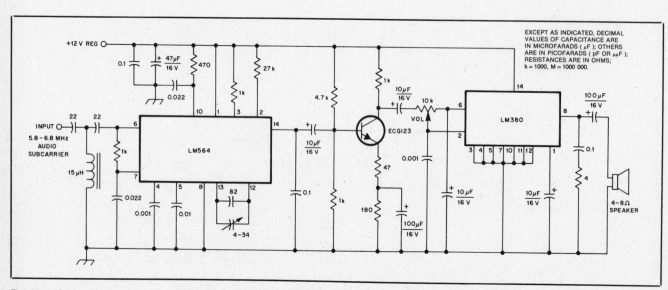

Fig. 22 — Schematic diagram of an audio subcarrier demodulator circuit that converts 5.8-6.8 MHz audio subcarrier to baseband audio.

resonance at the transmit carrier frequency. A very popular antenna for ATV use world-wide is the French-built F9FT or TONNA antenna shown in Fig. 23. It is available built for three different frequencies: 1296, 1269 and 1248 MHz. Another possibility is the loop Yagi described in Chapter 32.

Transmission line losses at 24 cm can be quite formidable. If it is not possible to mount the receiver preamplifier, transmitter power amp or varactor tripler directly at the antenna mast, very low-loss transmission line must be used. For long runs, 7/8-inch Hardline (2.1 dB attenuation per 100 feet for foam dielectric and 1.5 dB/100 feet for air dielectric) should be used. For shorter lengths, ½-inch Hardline (3.5 dB attenuation per 100 feet for foam dielectric and 2.9 dB for air) or Belden 9913 (4.6 dB/100 feet) can be used. For very short hookups between equipment in the shack, ¼-inch foam dielectric, flexible Hardline (7 dB/100 feet) is ideal, but RG-8/-214 (12 dB/100 ft) or RG-58 (22 dB/100 feet) can be used as a more practical, cost-effective alternative.

Test Equipment

Good low-cost test equipment for 24 cm is generally scarce. One exception is the availability of 24-cm test generators that produce a video-modulated signal in the 10-80 mW range. Fig. 24 shows some commercially available units. These devices are not only invaluable in checking out FMTV receivers, but are also small enough to be built into a camera or installed in model airplanes to provide a compact mobile transmit capability over short ranges. For measuring frequencies in the 24-cm band, the Ramsey PS-10B prescaler has proven to be a highly cost-effective adjunct to a low-frequency counter.

Comparison of 24-cm FM-ATV with 70-cm FSTV

Given some of the apparent complexities of FM and some of the drawbacks of transmission on frequencies above 1 GHz, why use FMTV rather than stick with the venerable AM FSTV using readily available 70-cm gear? To fully examine the pros and cons of these systems, it is useful to first examine the relative merits of FM versus AM video and then compare 24-versus 70-cm propagation. Finally, we will look at the combined picture.

FM versus AM TV

A comparison between these two modulation modes has been the subject of university research and numerous studies in industry and the military. The specific advantages of FM television include:
• Better linearity and greater average transmitted power. Amplifiers can be operated in class-C mode and operated at full power using FM. Thus it is possible to obtain two to three times greater average output from a solid state amplifier oper-

Fig. 23 — The F9FT TONNA antenna is popular worldwide for FMTV work. Versions are available for 1248, 1269 and 1296 MHz.

(A)

(B)

Fig. 24 — 24-cm video test generators are helpful for checking out receivers. The 10-mW transmitter shown at A is fully built and packaged by Solent Scientific. The 80-mW transmitter board from PC Electronics, shown at B, requires an additional video inverter board to obtain the proper transmitted signal polarity.

ating in FM rather than AM. Another advantage is that sync compression, which is a common malady with improperly adjusted AM systems, is not a major problem when communicating with FMTV. These systems are therefore very easy to initially set up and operate.
• Signal-to-noise ratio improvement at demodulator output. With amplitude-modulated systems, the resulting video output signal-to-noise ratio will generally be the same as the receiver input carrier-to-noise ratio. In the case of FM, the video output signal-to-noise ratio will be significantly greater than the input carrier-to-noise ratio. The actual level depends on the input carrier-to-noise ratio being above a

"threshold level," typically 7-12 dB, and also on the value of the deviation index (the ratio of the peak deviation to the highest modulating frequency). Several studies have specifically examined the gain advantages that FM can provide over AM television.[7-10] Additional studies have examined tradeoffs of various parameters such as bandwidth and deviation to further optimize system performance.[11-16]
• Increased signal-to-noise performance through pre-emphasis and de-emphasis. As is shown in Fig. 6B, the high-frequency components of a video signal are considerably lower in amplitude than the low-frequency components. These higher frequencies, therefore, contribute very little to the overall transmitted bandwidth. Studies and analyses have demonstrated that it is useful to emphasize these higher-frequency components prior to transmission and then de-emphasize them at the receiver to restore the signal to its original composition.[17-20] While performing this de-emphasis, the higher-frequency components of the noise are attenuated, which effectively increases signal-to-noise ratio. (A good discussion of the overall concept of pre-emphasis and de-emphasis can be found in Reference 21.)
• Improved co-channel interference immunity. Studies have clearly demonstrated that FM television is more resistant than AM to a wide variety of interference sources.[22-24]
• Immunity from fading. Because the FM receiver uses hard limiting, any amplitude variations in the received signal caused by flutter normally associated with ATV mobile operations are reduced. This greater immunity from fast fading provides more solidly locked pictures.

Of course, with every set of advantages, you can always expect some disadvantages. The specific disadvantages of FMTV are:
• Equipment complexity. As previously described, the receiving equipment is more complex and expensive than that required to receive AM.
• Greater signal bandwidth. The transmitted signal bandwidth is greater than for vestigial AM FSTV. The microwave frequencies that can accommodate this bandwidth exhibit greater propagation losses.
• Poor weak-signal performance. FMTV performance for very weak signals will be worse than AM because of the FM receiver threshold effect.
• Degradation from multi-path interference. FMTV can be more susceptible to degradation from severe multi-path effects. While an amplitude-modulated TV shows ghosting in the face of multipath, FMTV degradation manifests itself as picture distortion such as a portion of the image appearing like a photographic negative.

70 cm versus 24 cm
• Although path losses are greater on 24 cm than 70 cm, these losses can be somewhat compensated by the fact that higher-gain 24-cm antennas can be constructed for the

same boom length as a 70-cm design.

• Vegetation and buildings tend to attenuate 24 cm more than 70 cm, making 24-cm antenna placement more critical.

• As 70 cm becomes more populated with narrowband voice repeaters that can put fundamentals and intermodulation products in the video receiver passband (degrading picture quality), system performance may become limited by interference more than by receiver noise figure. On the other hand, the amateur microwave frequencies are relatively free of mutual amateur interference sources. The major problem besetting the 23/24-cm amateur allocation is FAA Air Route Surveillance Radars which operate at 4 megawatts ERP and occupy an 8 MHz bandwidth in a variety of locations and center frequencies throughout the country.

• Test equipment for 24 cm is generally expensive. As amateur interest in the microwave regions begins to grow, however, new low-cost equipment should become available.

On balance, then, what does this all mean? In short, the higher path losses of 24 cm can be compensated for through higher-gain antennas, by the inherent system processing gains afforded through the use of FM and by the availability of more interference-free channels. In practical use, given paths that are at or near line-of-sight, and comparing transmitters operating at equal peak power levels, 24-cm FM-ATV will usually equal or outperform 70-cm FSTV. In cases involving large amounts of vegetation, rough terrain or man-made obstacles, or in situations normally involving very weak signal work, the 70-cm FSTV will be better.

FMTV Operating Parameters

Amateur standards for FMTV using

Table 1

Typical FMTV Amateur Operating Parameters

1. *Signal Polarity:* A black-to-white transition in the video causes an instantaneous increase in transmitted carrier frequency.
2. *Transmit Pre-emphasis:* In accordance with CCIR Recommendation 405-1 for 525 line systems
3. *Receiver De-emphasis:* In accordance with CCIR Recommendation 405-1 for 525 line systems
4. *Deviation:* 4 MHz Peak (8 MHz Peak-to-Peak)
5. *Sound subcarrier frequency:* 5.8 MHz or 6 MHz
6. *Sound subcarrier level:* 10 dB below video

A listing of countries using the 525-line format is provided in Table 2.

Table 2

Countries Using 525-Line Television Systems

United States	Mexico
Canada	Netherlands Antilles
Japan	Panama
Korea	Peru

525-line systems are beginning to evolve. Table 1 shows typical operating parameters.

The purpose of developing standards is to foster interoperability between stations, to promote operation with equipment exhibiting uniformly good performance, and to aid equipment manufacturers and builders in developing equipment that can best suit the needs of the amateur community. The local nature of microwave operation permits each local amateur group to establish its own operating parameters and standards to best suit local needs. The operating parameters shown in Table 1 represent a typical system. The values were derived based on both amateur experimentation in the U.S. and research into previous industrial and military investigations.

Tradeoffs in performance, bandwidth and cost must be considered when selecting operating parameters. A major goal is to operate with the minimum bandwidth practical for good performance. Minimum bandwidth is desirable not only from the standpoint of spectrum conservation, but also from concern to avoid sideband overlap with the 8-MHz-wide FCC Air Route Surveillance Radars, the 1240-MHz band edge, 1296-MHz DX/EME stations and 1269-MHz satellite uplinks. Careful selection of operating parameters also permits the effective use (with appropriate modifications) of low-cost off-the-shelf FMTV equipment available in Europe and satellite-TV 70-MHz IF receiver strips available in the U.S.

The specific rationale for the parameters in Table 1 is as follows:

• Signal Polarity: A "Black-to-White/ Low-to-High Frequency" polarity was selected because this is the predominant mode for commercial FMTV satellite systems and amateur FMTV systems used in Europe. No modification to the video polarity of these units is needed.

• Transmit pre-emphasis and de-emphasis was selected in accordance with the International Telecommunication Union (ITU) International Radio Consultative Committee (CCIR) Recommendation 405-1. Pre-emphasis and de-emphasis is very inexpensive to implement, yet is a highly effective means of substantially improving the performance of FMTV systems. The circuit shown in Fig. 25, which uses readily-available parts, is a practical and low-cost implementation of the CCIR 405-1 pre-emphasis network suitable for 525-line Amateur Radio FMTV systems. Table 2 shows the countries using a 525-line standard. The performance characteristics of the circuit are shown in Fig. 26. The actual network losses are provided for design guidance; however, it is the relative losses, i.e., curve shape, that primarily affect system performance. The

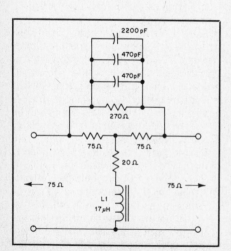

Fig. 25 — Schematic diagram of a practical 525-line video pre-emphasis network for transmitter input, based on CCIR Recommendation 405-1. The capacitors should be plastic-film or mica for good stability. L1 is 56t no. 30 enam. wire on a T-50-2 core, or a Coilcraft SLOT-10-1-09.

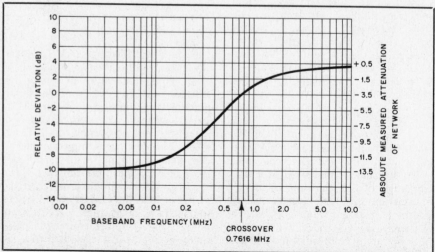

Fig. 26 — Characteristic response curve for a video pre-emphasis network.

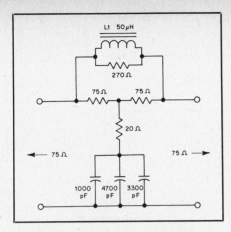

Fig. 27 — Schematic diagram of a 525-line video de-emphasis network for receiver output. The capacitors should be plastic-film or mica for good stability. L1 is 52t no. 30 on a T-44-3 core, or a Coilcraft SLOT-10-1-12.

matching de-emphasis network for use at the receiver is shown in Fig. 27.

• A 4-MHz peak deviation is fairly standard world-wide for terrestrial (non-satellite) FMTV communications.[25] This deviation provides good picture quality at a lower bandwidth and deviation than is used by satellite TV systems. If a smaller deviation was used, the receiver processing gain advantages would rapidly diminish. A greater deviation, although improving system performance, would occupy excessive bandwidth; particularly for the 1240-1300 MHz (23/24-cm) amateur band. For some of the higher microwave regions

where sufficient bandwidth exists, particularly at 3.3 GHz and above, experimentation with greater deviations would be not only practical but highly desirable. Using higher deviation levels would permit the use of unmodified 70-MHz IF strips in the receiver design.

• A sound subcarrier frequency of 5.8 MHz (used by satellites) or 6.0 MHz (used by European FMTV hams) is high enough to avoid interference with the chroma signal yet low enough to minimize total transmit bandwidth.

• The −10 dB sound subcarrier level will allow good audio reception for P3.5 and better pictures. This level is referenced to the video level at the output of the pre-emphasis filter with a 1-V peak-to-peak 762-kHz sine wave applied to the transmitter video input. Increasing the subcarrier level too much can degrade picture quality while setting it too low may prevent the sound from being received under less than P5 conditions.

Practical FMTV Measurement Techniques

To assist in setting up the FMTV transmit parameters shown in Table 1 using test equipment affordable by amateurs, the following procedures are provided.

Transmit Video Deviation

To properly set the deviation of a transmitter using pre-emphasis, attention must be paid to using the proper input test frequency as well as voltage level. The following procedure may be used to establish a 4-MHz peak deviation with a 1 V

Table 3

Derivation of the Test Voltage Level for the 4 MHz Peak Deviation Adjustment:

For FM signals:

$$\text{modulation index} = \frac{\text{peak deviation}}{\text{modulating frequency}}$$

A Bessel function chart shows the relation between carrier and sideband amplitudes of the modulated signal as a function of the modulation index. At a specified value of the modulation index, the carrier and various sidebands go to zero.[26, 27]

By theory:

modulation index = 2.40 for first carrier null

At the pre-emphasis crossover frequency of 762 kHz, the peak deviation of the first carrier null will be:

$$2.40 \times 0.762 \text{ MHz} = 1.8288 \text{ MHz}$$

Since the desired transmitter performance is 4 MHz peak deviation with a 1 V peak to peak input and assuming that the transmitter VCO performance is linear such that input voltage is directly proportional to transmit deviation, the results can be scaled:

$$\frac{1 \text{ V}_{P-P}}{4 \text{ MHz}} = \frac{X \text{ V}_{P-P}}{1.8288 \text{ MHz}}$$

cross multiplying:

$$1.8288 = 4X$$

and dividing:

$$\frac{1.8288}{4} = X$$

$$0.457 \text{ V} = X$$

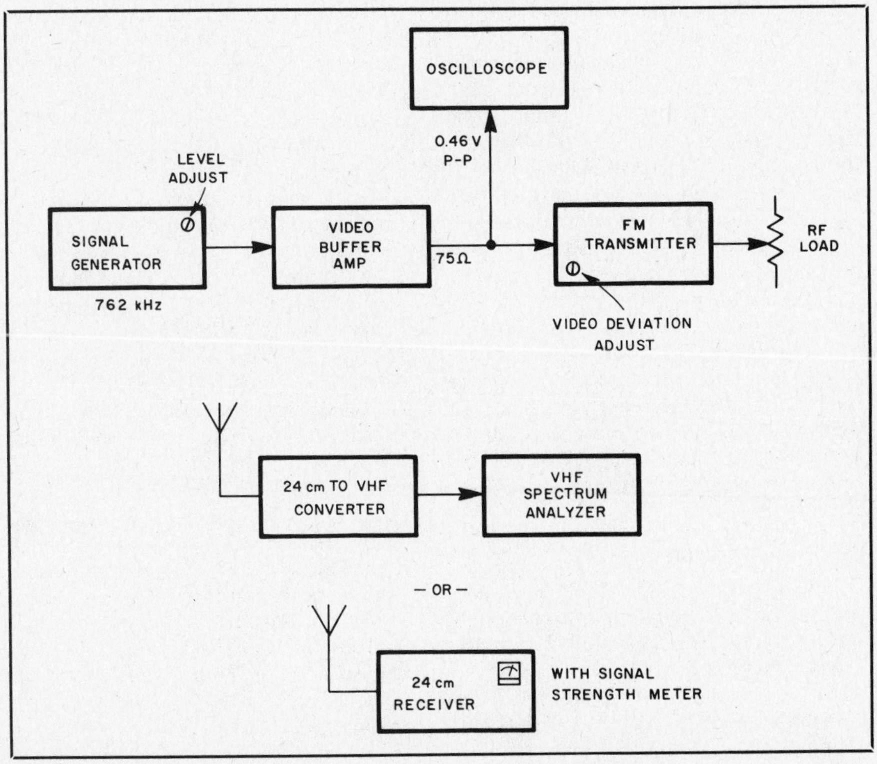

peak to peak input for transmitters employing CCIR Rec 405-1 525-line pre-emphasis. It is based on the Bessel function null measurement technique and is founded on the theory that the transmitted signal carrier for an FM signal will be nulled when the ratio of the peak deviation to the modulating frequency is equal to 2.40.[26,27]

Step 1 — Set the signal generator to the pre-emphasis crossover frequency of 762 kHz and the level to 0.46 V peak to peak (See Table 3) using the set-up shown in Fig. 28.

Step 2 — Set the spectrum analyzer (or a receiver with a signal strength meter) to monitor the transmitted signal. Use a resolution bandwidth of 200 kHz or less. Starting with the transmit deviation control at the lowest possible level, slowly increase the deviation. The carrier amplitude will increase and rise above the upper and lower sidebands as shown in Fig. 29A.

Step 3 — Eventually the carrier will begin to drop with increasing deviation. Set the deviation control at the point the carrier is lowest (null). See Fig. 29B. The deviation is now properly set for 4-MHz peak. You may note that if you continue increasing the deviation, there will be other settings

Fig. 28 — Block diagram of the test configuration for setting video deviation.

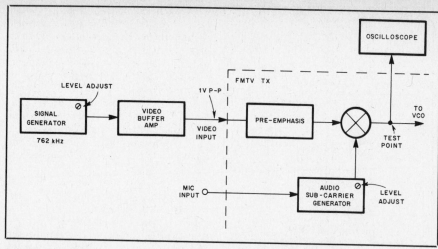

Fig. 30 — A block diagram of the test configuration for measuring audio-subcarrier level relative to the video level.

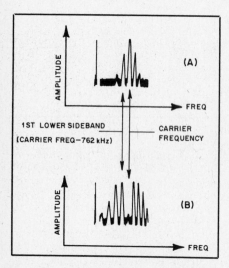

Fig. 29 — Spectrum pictures of transmitter output observed on the Science Workshop "Poor Man's Analyzer" while adjusting transmitter deviation. As the deviation is first increased, the carrier and sideband levels will increase in amplitude, as shown at A; just before the desired deviation is reached, however, the carrier level will begin to drop. At the proper deviation level, the carrier will be nulled, as shown at B.

that will also achieve a null. The setting that is desired is the one with the lowest deviation; i.e., the first null.

Step 4 — To set up additional transmitters with the properly adjusted unit, send a test signal with the calibrated transmitter and then note the resulting peak-to-peak-amplitude output on a receiver. For another transmitter, deviation can be rapidly set by adjusting its deviation control to produce the same receiver output levels.

When transmitting a normal camera-produced picture, if the transmitter deviation exceeds the dynamic range of the receiver demodulator, you will note the presence of undesirable "sparklies" on the screen. These are shimmering black spots which appear in the white areas of the scene. If the receiving station cannot eliminate the sparklies through careful tuning, it may be necessary for the transmitting station to slightly reduce deviation.

Transmit Audio Level

The test set-up for measuring the audio subcarrier level with reference to the video level is shown in Fig. 30.

Step 1 — With the audio subcarrier turned off, apply a 1-V peak-to-peak 762-kHz signal at the video input to the transmitter's pre-emphasis filter. Measure the resulting peak-to-peak voltage level at the test point where the video and audio are summed, typically at the output of the pre-emphasis filter. This establishes the video reference level.

Step 2 — Turn off the signal generator and activate the audio subcarrier generator. Measure the peak-to-peak voltage level.

Step 3 — Divide the audio subcarrier level by the video reference level. This will yield the relative audio subcarrier level. An audio level 10 dB down from video will correspond to a voltage ratio of 0.32.

FMTV Bibliography

[1] J.H.O. Harries, "Can Frequency and Phase-Change Modulation Reduce Interference?," *Television,* Vol. 2, no. 18, Aug. 1929, pp. 310-314.

[2] C.W. Carnahan, "F-M Applied to a Television System," *Electronics,* Feb. 1940, pp. 29-32.

[3] F.H. Kroger et al, "A 500-Megacycle Radio-Relay Distribution System for Television," *RCA Review,* Vol. 5, Jul. 1940, pp. 31-50.

[4] C.W. Hansell, "Development of Radio Relay Systems," *RCA Review,* 1947, pp. 367-384.

[5] W.C. Boese, "Survey Report on the use of Frequency Modulation for Television Transmission," *Technical Research and Technical Information Reports,* Part 2, Federal Communications Commission, May 25, 1950.

[6] L.S. Golding and J.E.D. Ball, "Satellite Television Covers the World," *IEEE Spectrum,* Vol. 10, No. 8, Aug. 1973, pp. 24-31.

[7] H. Drysdale, "A Comparison of the Performance of Television Transmission Systems Using Amplitude Modulation and Frequency Modulation," Royal Aircraft Establishment, *Technical Memorandum No. WE1201(E),* Feb. 1966.

[8] K.G. Johannsen and E.A. Enriquez, "Investigation of Surface Educational Television Distribution," *IEEE Transactions of Broadcasting,* Vol. BC-22, No. 2, Jun. 1976, pp. 45-52.

[9] D.O. Cummings, "Some Considerations for Amplifier Spacing for AM and FM Supertrunks," *IEEE Transactions on Cable Television,* Vol. CATV-1, No. 1, Oct. 1976, pp. 63-73.

[10] P.R.J. Court, "A Frequency Modulation System for Cable Transmission of Video or Other Wideband Signals," *IEEE Transactions on Cable Television,* Vol. CATV-3, No. 1, Jan. 1978, pp. 24-35.

[11] G.W. Beakley, "Television to Small Earth Stations," *IEEE Transactions on Broadcasting,* Vol. BC-22, No. 3, Sep. 1976, pp. 96-100.

[12] L. Clayton, "FM Television Signal-to-Noise," *IEEE Transactions on Cable Television,* Vol. CATV-1, No. 1, Oct. 1976, pp. 25-30.

[13] K.G. Johannsen et al, "Television Sound Subcarrier Transmission in Space Communication," *IEEE Transactions*

on *Broadcasting,* Vol. BC-20, No. 3, Sep. 1974, pp. 42-48.

[14]D.W. Halayko and R.W. Huck, "Small Television Receive-Only 12 GHz Ground Terminals," *IEEE Transactions on Cable Television,* Vol. CATV-3, No. 3, Jul. 1978, pp. 112-119.

[15]"Broadcasting-Satellite Service (Sound and Television)," CCIR Documents of the XVth Plenary Assembly, Geneva, 1982, Volumes X and XI, Part 2, Recommendation 215-1.

[16]"Transmission de la Television en Modulation de Frequence," France Cables sous Marine et de Radio, Paris, 1984.

[17]C. Loo, "S/N Improvement of an FM Signal At or Near Threshold Due to De-Emphasis and Noise Weighting," *IEEE Transactions on Broadcasting,* Vol. BC-23, No. 3, Sep. 1977, pp. 65-70.

[18]"Pre-Emphasis Characteristics for Frequency Modulation Radio-Relay Systems for Television," CCIR Documents of the XVth Plenary Assembly, Geneva, 1982, Vol. 4, Recommendation 405-1.

[19]C. Loo, "Optimum Pre-Emphasis Network for Satellite Transmission of an FM TV Signal Near or at Threshold," *IEEE Transactions on Broadcasting,* Vol. BC-28, No. 2, Jun. 1982, pp. 37-43.

[20]W. F. Thorn, "FM Transmission of Television from Rockets and Balloons," Northeastern University, Electronics Research Laboratory, Boston, Scientific Report No. 3, 11 Jul. 1979.

[21]Mischa Schwartz, "Information Transmission, Modulation, and Noise," Polytechnic Institute of Brooklyn, 1959, pp. 305-311.

[22]P.R. Groumpos and B.D. Dimitriadis, "On the Subjective Evaluation of Protection-Ratios' Measurements for Co-Channel FM-TV Interference Under Variable (S/N) Ratios," *IEEE Transactions on Broadcasting,* Vol. BC-29, No. 3, Sep. 1983, pp. 101-105.

[23]"Broadcasting-Satellite Service (Sound and Television), Measured Interference Protection Ratios for Planning Television Broadcasting Systems," CCIR Documents of the XVth Plenary Assembly, Geneva, 1982, Volumes X and XI, Part 2, Recommendation 634-2.

[24]S. P. Barnes and E. F. Miller, "Carrier-Interference Ratios for Frequency Sharing Between Frequency-Modulated and Amplitude-Modulated-Vestigial Sideband Television Systems," Lewis Research Center, Cleveland, *NASA Technical Paper 1264,* Aug. 1978.

[25]"Frequency Deviation and the Sense of Modulation for Analogue Radio-Relay Systems for Television," CCIR Documents of the XVth Plenary Assembly, Geneva, 1982, Vol. 4, Recommendation 276-2.

[26]Murray G. Crosby, "A Method of Measuring Frequency Deviation," *RCA Review,* 1940, pp. 473-477.

[27]"Spectrum Analysis Amplitude and Frequency Modulation," Hewlett Packard Spectrum Analyzer Application Note No. 150-1, Nov. 1971.

Slow-Scan Television

Slow-scan television (SSTV) started in 1958 as the effort of a small band of Amateur Radio operators led by Copthorne Macdonald, VE1BFL. The in-intention of the project was to send television pictures on the HF amateur bands. By virtue of HF radio design and propagation characteristics, the SSTV system design is tailored to allow operation in a 3-kHz bandwidth — considerably less than the 6-MHz width of a commercial TV channel.

To meet the objective, bandwidth was decreased by reducing the information rate. Instead of 30 pictures a second, SSTV takes 8 seconds to send one picture. Instead of 525 lines, an SSTV picture is composed of 120 lines. The video is sent as a frequency-modulated subcarrier, which varies between 1500 Hz for black elements and 2300 Hz for white elements. Vertical and horizontal synchronizing signals are sent as bursts of 1200-Hz tones. Table 5 summarizes the SSTV standards for black-and-white TV operation.

In the early years of SSTV, the picture was displayed on surplus P7 radar tubes that could hold a slowly decaying picture for about 8 seconds when viewed in a darkened room. Special shuttered vidicons or flying spot scanners were used as the video sources. By 1974, several amateurs had built digital scan converters that permitted viewing a nondecaying picture in full daylight. The digital scan converter was designed so that the SSTV image could be stored in solid-state memory, and displayed on an unmodified black-and-white TV set. Fig. 31 shows an example of SSTV picture

Table 5
SSTV Standards

Frame Time	8 seconds
Lines per frame	120
Duration of line	67 ms
Duration of V sync	30 ms
Duration of H sync	5 ms
H or V sync frequency	1200 Hz
Black frequency	1500 Hz
White frequency	2300 Hz

quality using this technique. The converter also can be used to change the video output from a fast-scan camera into an SSTV-compatible signal. Thus, the scan converter is an interface between the slow-scan TV and fast-scan TV formats. A few years later, several commercial companies began producing these rather complex converters for amateur SSTV use.

Station Operation

As shown in Fig. 32, a typical SSTV station uses a standard transceiver, scan converter, TV camera and TV set. The TV receiver can be used for monitoring both the received signal and image to be transmitted.

Reception of SSTV is quite simple. The SSB voice of the transmitting amateur is tuned in so it sounds "natural." When the SSTV transmission begins, the signal should be in tune. The picture can be improved by using a transceiver that has passband tuning. Under ideal conditions the audio highs can be boosted to give some-

Fig. 31 — SSTV picture as seen on a standard TV set using digital scan converter.

what higher resolution. When poorer conditions exist, the passband can be tuned to minimize interference.

A General class or higher license is required to operate SSTV. It can be used on every band. The most popular frequency band is 20 meters, especially on 14.230 MHz, where an SSTV net meets each Saturday afternoon. Operators on the lower HF bands often experience poor quality because of multipath distortion, while the VHF and UHF bands limit the operational range.

Scan-Converter Design

The scan converter plays an integral part in both receiving and transmitting SSTV. The heart of the SSTV scan converter is the memory system. Most present systems use dynamic RAM ICs that contain enough bits

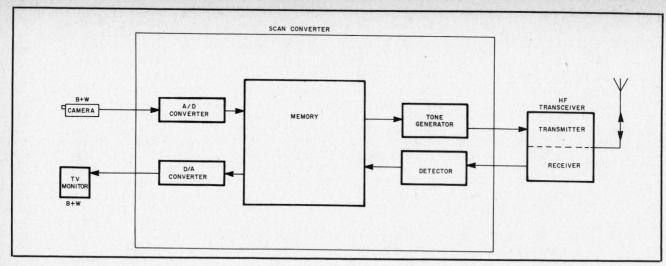

Fig. 32 — Basic SSTV station.

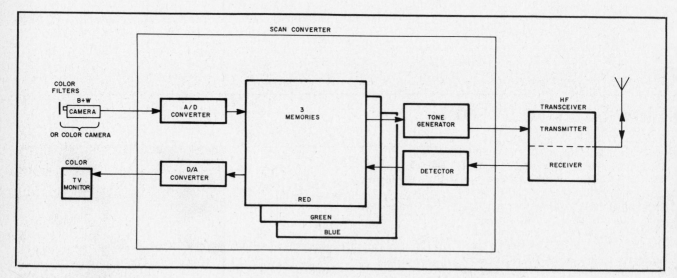

Fig. 33 — Color SSTV station.

to adequately resolve the picture elements and their gray scales. Sixty-five thousand bits (65 kbits) is the minimum considered acceptable for a single black-and-white picture.

Receiving

The converter has input detection circuitry to translate the SSTV audio-frequency variations from the receiver audio output into digital pulses. These contain horizontal and vertical synchronization as well as binary values representing the gray scales being sent. The pulses are stored in memory. The information is then sent to a digital-to-analog converter, which converts the stored data into an analog video signal fully compatible with a standard TV monitor to produce a nondecaying image.

Transmitting

Output from a black-and-white TV camera is converted into horizontal and vertical pulses along with digitized gray-scale values. The image is "snapped" in

1/60 of a second from the TV camera. The information is stored in memory. The digital output from memory goes to circuitry that produces the audio-frequency-modulated signal conforming to the SSTV standards. This SSTV output is then fed to the microphone input of the transceiver. Since the SSTV FM audio signal to the transmitter is continuous, the transceiver/linear amplifier RF power should be reduced to safe dissipation levels under this high-duty cycle operation.

Color SSTV

SSTVers have been experimenting with color since before the advent of the scan converters. In the initial experiments, red, green and blue filters were placed in front of the black and white TV camera and the face of the P7 monitor. The output of the filtered P7 monitor was captured on a color Polaroid® camera. The transmitting station sent a set of three 8-second SSTV frames that were received and recorded on film as a color picture, presuming the critical

photographic camera alignment was not disturbed.

The scan converter offers a vast improvement over that interesting, but error-prone, technique for color SSTV. If two more memories are added to the black and white scan converter (195 kbits total), as shown in Fig. 33, each memory can be dedicated to a color. A black and white camera with a Red no. 25 filter, a Green no. 58 filter and a Blue no. 47B filter successively placed in front of the lens then sends one or more SSTV frames of each color. (A color TV camera is often used — that avoids the filter changing, but generally at a higher cost!) The receiving SSTV station switches the output of the SSTV detector to load the appropriate "color" memory. This is referred to as multiframe color SSTV. The outputs of the three color memories are translated from their digital gray-scale values into analog voltage levels capable of directly driving the gun drivers of a color TV monitor. Other designs translate the digital gray-scale values into composite-

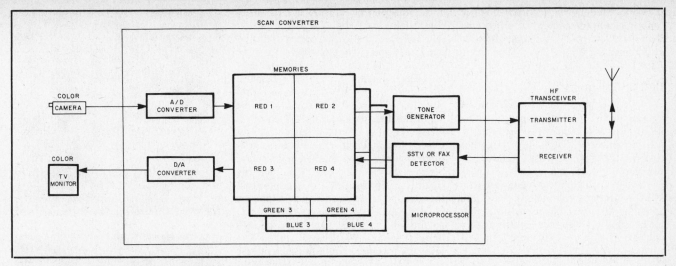

Fig. 34 — Color SSTV station capable of storing multiple images.

Fig. 35 — Microprocessor-based color SSTV station of WØLMD.

(A)

(B)

Fig. 36 — Noise caused by adjacent-channel interference is shown at A. Picture improvement when noise is removed by digital noise clipping is shown at B.

video levels and sends them to a conventional color TV set. The composite TV method simplifies the cabling and avoids TV set modification, at the cost of reduced bandwidth and resolution.

Recent Developments

Since the advent of the microprocessor, and because of the interest in more applications of the SSTV scan converters, systems are continuing to evolve in complexity and capability.

Multiple Image Storage

Since memories are continually getting larger and cheaper, present scan converters are being designed to hold four or more color pictures simultaneously (Fig. 34). This means a good received picture stored in a single memory does not have to be destroyed to make way for a new picture. If memories with 256 kbit capacity are used, the whole scan-converter memory can be built from three ICs. (A similar design 10 years ago would have required 768 of the 1-kbit memories used at that time.)

Increasing Resolution

The large array of memory opens up several design possibilities. The memory can be used for double vertical and double horizontal resolution if a single high-resolution picture for reception of schematics or weather satellite images is required.

Improved Color SSTV

In color SSTV applications, a microprocessor can be used to assign automatically the incoming SSTV color frame to the correct memory. The microprocessor also opens the way for even better transmission schemes such as sending all of the color information required for each line before the next line is sent. This results in a single-frame color SSTV system. Several techniques have been proposed for transmitting the color information. One system is described in "8 and 12 Second Single-Frame Color SSTV," *A5 Magazine*, November 1982, p. 32; another is described in "Robot Model 450-C Composite Single Frame Color Specification," *A5 Magazine*, February 1984, p. 28. A major objective is to maintain compatibility with existing black and white SSTV systems.

FAX Compatibility

FAX and weather satellite images can be seen on an SSTV scan converter if a format translator is attached to the input of the scan converter. Facsimile and weather satellite pictures typically use over 1000 picture elements per line and over 1000 lines per picture. Since even high-resolution SSTV converters only have 256 × 256 resolution, the full quality of facsimile cannot be achieved. Nevertheless, the quality will be adequate for most experimenters.

Image Processing

The microprocessor may also be utilized for digital image processing to remove transmission noise from received pictures, and to boost the contrast of weak pictures. Since SSTV is much slower in data rate than conventional fast-scan ATV, the microprocessor has a chance to operate on the received images without resorting to very expensive high-speed digital signal-processing hardware. A highly sophisticated microprocessor-controlled color SSTV station is shown in Fig. 35.

A picture is often distorted significantly when it arrives at the receive station after HF transmission. A picture may have "buckshot noise" from adjacent-signal interference, as shown in Fig. 36A. A digital signal-processing algorithm known as a threshold noise clipper may be utilized to reduce the visibility of the noise interference. This algorithm compares the gray-scale value of each picture element (pixel) with the eight picture elements surrounding it. If the gray-scale value is significantly different from the average of the eight neighbors, then the average gray-scale value of the eight neighbors is taken instead of the original value of the center pixel. The restored picture is shown in Fig. 36B.

Another example of digital image processing is with SSTV pictures received with weak contrast as shown in Fig. 37A. This means that the picture did not go either completely black or completely white. Often the problem is also compounded by marginal-signal conditions that inject considerable noise into the already weak picture. The restored picture in Fig. 37B shows what can be expected after a digital contrast-enhancing algorithm is used. The algorithm first derives a histogram of all the shades of gray that it finds in the received picture. Then it calculates the gray shade that represents the 10% point of the blackest picture elements, and the gray shade that represents the 10% point of all the white picture elements. Next a look-up table of expanded gray-scale values is made so that the weak shades are expanded linearly to full black and white. The microprocessor now modifies each picture element according to the expanded gray-scale values. While the effect is impressive on black-and-white pictures, it is very spectacular on color SSTV pictures. The microprocessor can also be used to electronically title an SSTV picture, as shown in the figures.

Future SSTV

The design of the scan converters is continuing to evolve. Now that image-processing techniques are becoming available in commercial SSTV scan converters, the next step will be scan-converter systems that are multitasking. This means that the scan converter can be simultaneously processing one picture, receiving a new picture and retrieving a picture from hard-disk memory for later transmission. A split-screen image could monitor all activities or concentrate on any one of them.

Another area of development will be simultaneous voice and video. This will involve sending the SSTV image during intervals of silence in the operator's voice transmission.

Standard digital silence-detecting algorithms can produce the effect of a perfect voice-control system, and instantaneously select voice or video modes along with a mode-change header.

Several amateurs in Australia and New Zealand have been experimenting with low-resolution, narrow-bandwidth motion

Weak Contrast
(A)

Contrast Enhanc
(B)

Fig. 37 — Weak-contrast picture (A) and the same picture after digital contrast enhancement (B).

systems. These systems are designed to operate in voice grade bandwidth, making them applicable for long-distance HF transmission. Techniques such as these may permit the visual transmission of sign language images by Amateur Radio operators with speech or hearing difficulties.

SSTV today represents a wide open field for the new breed of experimenter. Today's SSTV systems involve microprocessors and software development. For those wanting to experiment with uncharted fields, the digital imaging possibilities within SSTV offer a tremendous challenge.

Facsimile

The earliest operational system for image transmission, facsimile or FAX networks date back to the late 20s and early 30s. Like any other imaging system, FAX compromises in some areas to gain advantages in others. Conventional fast-scan TV is capable of handling moving images of moderate resolution, but the rapid image repetition rate requires several megahertz of signal bandwidth, which confines it to the UHF range. SSTV sacrifices resolution and motion to transmit relatively low resolution images in an 8-second frame within the bandwidth constraints of conventional SSB transceivers. The more advanced SSTV formats have resolution capabilities approaching fast scan but, given bandwidth restrictions, this is achieved by lengthening the transmission time for still frames to 17 or 34 seconds.

By comparison, FAX represents a high-resolution service with pictures ranging from 800 to several thousand lines per frame. This high resolution is achieved within the constraints of audio bandwidths by using relatively long transmission times ranging from 3.3 to 15 minutes.

FAX is used widely in a number of services. Weather chart distribution is perhaps the most common, using both HF and landline links to weather service and forecast centers throughout the world. News photographs, sometimes called "wirephotos," are also distributed throughout the world in a similar manner. A wide variety of weather satellites transmit cloud-cover pictures back to earth using FAX. Various systems are becoming increasingly common in business offices and law-enforcement agencies to facilitate the transfer of documents, signatures, fingerprints and photographs.

Amateur FAX work has been limited, primarily because of historical regulatory factors. During the post-WW II years, the 11-meter band was the only amateur HF assignment authorized for FAX operation. The subsequent loss of 11 meters left the amateur fraternity with no regular FAX operating privileges on HF. It was only with the recent authorization of FAX on all amateur bands that interest has been renewed.

Image Transmission

A typical FAX transmitter consists of a drum driven by a synchronous motor operating from a crystal or tuning-fork frequency standard. The material to be transmitted is wrapped around the drum, which rotates at a constant speed. A small spot of light is focused on the printed material (photo, text, map, etc.), and the light reflected from the subject is picked up by a photocell, photomultiplier or phototransistor. The rotation of the drum provides the equivalent of the "horizontal scanning" of the material. A carriage, supporting the light source and pickup assembly, is moved along the drum at a constant rate providing the "vertical scanning." The carriage is usually operated by a threaded rod driven by another synchronous motor. As the carriage travels from one end of the drum to the other, the entire picture area is scanned. The voltage variations from the light pickup are amplified and used to modulate an audio subcarrier, using either amplitude or frequency modulation. For positive AM, the modulation is proportional to brightness, which typically ranges from 4% for black through 90 to 100% for white. Negative modulation reverses this relationship with minimum modulation level corresponding to white, and maximum to black.

Amplitude-modulated-subcarrier systems are typically used with VHF and microwave FM links (weather satellites using positive modulation) or weather chart and photo transmission via landline (typically negative modulation). FM-subcarrier operation uses the voltage variations from the light pickup to shift the subcarrier frequency. Black material results in a 1500 Hz frequency, rising in linear fashion to 2300 Hz for white. The FM format is used with all FAX systems that operate on HF, reducing problems caused by the fading that is inevitable on any long-distance short-wave transmission circuit.

Image Display

The options for image display are far more varied than those for transmission. The device that displays the picture is known as the "recorder" and is found in a wide variety of configurations. Drum-type recorders hold a single sheet of the recording medium on a drum that is quite similar to that used in the transmitter. By contrast, continuous-feed FAX recorders are designed to display pictures using material fed from a roll. Such recorders do not require that material be reloaded for each picture but are typically more complex than drum-type recorders.

Facsimile recorders also differ widely in the nature of the recording medium. Photographic systems use film or photographic paper that is exposed by a modulated-light source and lens. Such systems require a darkroom and photographic-processing chemicals and are not well suited for high-volume use. The most advanced photographic systems use a dry silver paper and modulated laser light source. These are usually continuous read-out machines in which the paper is heat processed prior to emerging from the recorder. Such systems provide superb image quality and are convenient to use. However, they are very expensive to purchase new and are not yet available on the surplus market.

Weather chart recorders are typically of the continuous-feed type and employ a damp, chemically treated paper that is "exposed" using an electrolytic process that creates a dark line when current is passed through the paper. These systems are ideal for maps and charts and can provide a reasonable gray scale for continuous-tone images.

Another medium is electrosensitive paper in which printing is accomplished by applying a high voltage to a printing wire or stylus. The paper is dry and has a black base layer with a white coating. The applied voltage will cause varying amounts of the surface material to be burned away, thus creating a trace that can be varied from white (no coating removed) to black (all the coating burned away). With the proper choice of paper and printing electronics a fine gray scale can be obtained with images approaching photographic quality.

Yet another recording medium involves the use of a combination of carbon and white paper with the printing electronics driving a mechanical system that exerts varying pressure on the carbon to produce a trace on the white paper beneath. The image quality is marginal, not to mention all of the wasted carbon paper that is discarded. Long-term availability of the specialized paper may also be a problem.

Standards

The essential standards for a FAX unit are defined by three parameters: drum speed (helix speed in the case of continuous read-out recorders), carriage or paper feed rate, and modulation characteristics. To be suitable for use in a given class of service,

a piece of equipment must match the characteristics of the service.

Drum Speed

Drum speed is expressed in terms of revolutions per minute (r/min) which is equivalent to the scanning line frequency (line/min). Several "standard" speeds may be found in surplus commercial equipment with some units having multiple speed options:
- 60 line/min — standard speed for old wirephoto and chart recorders. Few modern services use this line rate and it is unsuitable for amateur use because of the extended periods required for image transmission with standard frame rates.
- 100 line/min — This speed was used for a considerable period with various wirephoto services and some machines still on the surplus market. Not used by current services so conversion to 120 or 240 line/min would be required.
- 120 line/min — Standard speed for HF and landline weather charts, some wirephoto services and weather satellite use. This line rate is marginal for amateur use since transmission times will range from 7 to 15 minutes depending on the other system specifications.
- 180 line/min — Used by some office FAX systems such as the Xerox Model 400. It is acceptable for point-to-point links on VHF but for general HF use, weather charts and satellite reception, conversion to 120 or 240 line/min will be required.
- 240 line/min — Standard for weather satellite use and used by more modern wirephoto and chart transmissions. This is a good rate for amateur HF use since detailed pictures (800 lines) can be transmitted in as little as 3.3 minutes.

Other speeds can be found in specialized gear, and all will require conversion to either 120 or 240 line/min to be useful for amateur work.

Scanning Density

Scanning density is usually expressed by the number of lines per inch (line/in). There are two different ways in which the scanning density can be used.

The first case is where the frame-time parameters of a signal format are known and you wish to obtain the appropriate scanning density for proper image display. An example might be the WEFAX weather satellite format in which the image has a 1:1 aspect ratio (square), a line rate of 240 line/min, and a frame time of 200 seconds, producing an 800-line image (4 lines/s × 200 seconds). Obviously the recorder should have a 240-r/min drum speed, but in choosing a recorder the proper scanning density must be determined. To compute the required scanning density, you need to know the width of the image format on the recorder. If you have some sample printouts, you can simply measure the width. In the case of a drum-type system, the width can be approximated by the relationship:

$$L = 3.14 D$$

where

L = length of the scanning line (width)
D = drum diameter

Since domestic recorders are calibrated in terms of lines/inch, your measurements should be in inches or converted to inches from the metric equivalents. In the case of a continuous readout system, the image width is usually slightly less than the width of the paper.

Another parameter is the aspect ratio of the image format (ratio of image width to image height). For the WEFAX example, the aspect ratio (A) is 1:1 or simply 1. The required scanning density can be computed from:

$$F = (tR)/(L/A)$$

where

F = scanning density in line/in
t = frame time in minutes
R = scanning line frequency in line/min
L = length of scanning line in inches
A = aspect ratio

In the case of our WEFAX example, the scanning density for an 11-inch recorder is:

$$F = (3.33 \times 240)/(11/1) = 800/11 = 72.7$$

The scanning line frequency is not a particularly critical factor. If the value is within 10% of nominal, the distortion in the aspect ratio will not be noticeable. In the case of 240-line/min recorders, a scanning density of 75 line/in is commonly available and will produce an essentially perfect picture. Many recorders to find their way onto the surplus market recently use 96 line/in. This rate will compress the image vertically — i.e., it will not be as tall as it is wide — but the picture is acceptable.

A second use for the scanning density is the computation of the transmission time using a specific format or the computation of the line/in requirement when using a recorder with a different image width. As an example, we might consider computing the transmission time for a 19-inch weather chart recorder operating with a scanning frequency of 120 r/min and a 96-line/in scanning density, which is standard for weather chart service. Since width (19 in) and scanning density (96 line/in) are given and if we specify the aspect ratio of the image (A), we can compute the transmission/reception time (t) in minutes by rearranging the previous equation as:

$$t = ((L/A) F)/R$$

where

t = transmission or reception time in minutes
L = length of scanning line (width) in inches
A = aspect ratio
F = scanning density in line/min
R = scanning line frequency in r/min

If the image is square (A = 1), then the time can be computed from:

$$t = ((19/1) \times 96)/120 = (19 \times 96)/240$$
$$= 1824/240 = 7.6 \text{ minutes}$$

If A is greater than 1 (width > height), the transmission time will be reduced. Likewise if A is less than 1 (width < height), the transmission time will be increased.

If we know the drum speed and scanning density specifications for a specific format for one recorder, we can readily compute the required scanning density (line/in) setting for other recorders. In the previous case, we noted that 96 line/in is the standard setting for 120-r/min weather charts using a 19-inch recorder. The scanning density for other recorder widths can be calculated as follows:

$$F' = (FL)/L'$$

where

F' = unknown scanning density in line/in
F = reference scanning density in line/in
L = reference line width
L' = line width of new format

If we had access to an 11-inch recorder and wished to copy 120-r/min weather charts (F = 96 and L = 19), the required scanning density can be computed as follows:

$$F' = (96 \times 19)/11 = 1824/11 = 165.8$$

If you were to check the specification of the 11-inch recorder, you would find a 166-line/in scanning density switch setting. That should work fine!

Modulation Format

The various modulation formats have already been briefly discussed. Chart recorders designed for landline use will usually have negative amplitude modulation. Weather satellite systems require positive amplitude modulation, while HF systems use FM subcarrier modulation. Some units will have several switch-selectable modulation formats. Ideally, a recorder should have complete selection for AM, FM, and positive and negative formats.

Surplus Equipment

Surplus commercial equipment is an excellent way to acquire FAX hardware since the systems are well constructed (often to MIL specs), and the cost is low. The ideal unit will have multiple drum speeds, scanning densities and demodulation formats. In practice, some conversion will probably be needed. Before you consider conversion, factors such as documentation, circuit type and media should be considered. The modification process will require information about motor drive frequencies, control signals, circuit configurations, etc. A good technical manual containing schematics is a must! Unless you are going to use the system only for the mechanics (with a complete replacement of the electronics), you should not consider a unit for which a manual is unobtainable.

The nature of the electronic circuits is another important factor. Tube-type units are typical surplus items because they have been phased out of service in favor of solid-state equipment. Although the price may look appealing, consider carefully the present cost and availability of replacement tubes. You should not necessarily reject a tube system, but you can expect to do some circuit replacement as tubes become either too expensive or unavailable.

Finally, check out the recording medium. Electrolytic and electrostatic papers of various sorts will be around for quite some time, but some specialized media such as carbon-paper systems may be in short supply in the near future.

Given that a specific surplus unit is deemed suitable in most ways, the three primary conversion concerns are likely to be the drum speed, scanning density and modulation formats. Each of these will now be discussed.

Drum Speed Changes

Virtually all FAX systems on the surplus market use synchronous motors for control of the drum speed. These motors are ac units that require a specific drive frequency to operate at their rated output speed. Typically, a highly stable reference signal will be derived from a crystal standard or an electrical tuning fork. This signal is amplified in a class-A or -B amplifier to a voltage and power level sufficient to operate the motor. The actual motor speed is usually quite high (several thousand r/min) with internal or external gearing used to achieve the desired drum speed. Major speed changes (such as doubling or halving the drum speed) are usually best accomplished by changing gears. Often the manufacturer has or had such gear sets available. If they can be located, conversion is simple. Occasionally, standard instrument gear sets can be made to fit in the space available.

Less drastic changes in drum speed can often be accomplished by changing the reference or drive frequency to the motor. The relationship of the original drive frequency and the new drive frequency is as follows:

$$f' = (fR')/R$$

where

f' = new drive or frequency
f = original drive frequency
R' = new drum speed
R = original drum speed

For example, the vintage D-611-P wirephoto recorder operates at 100 line/min with a drive frequency of 1000 Hz. The new drive frequency that will cause the motor to operate at 120 line/min is:

$$f' = (1000 \times 120)/100 = 120,000/100$$
$$= 1200 \text{ Hz}$$

The motor amplifier can be driven by a 1200-Hz signal. That signal can be derived from a 1.2-MHz oscillator with the output divided by 1000. The motor will operate the drum at the proper 120-line/min rate.

Another example is the Xerox 400 TELECOPIER. This unit has a drum speed of 180 line/min derived from a 368.64-Hz crystal standard. The drive frequency to operate this recorder at 240 r/min is:

$$f' = (368.64 \times 240)/180 = 884,736/180$$
$$= 491.52 \text{ kHz}$$

Simply substituting a 491.52-kHz crystal for the original will complete the conversion.

Office FAX machines such as the Xerox 400 use square-wave-driven motors. Most systems, though, like the D-611-P, use true synchronous motors that employ one or more capacitors. The capacitors serve to resonate the motor inductance at the drive frequency. Any significant change in the drive frequency will require a change in the motor capacitor(s). The new capacitor value can be determined if the original capacitor value, original drive frequency and new drive frequency are known:

$$C' = C (f/f')^2$$

where

C' = new capacitor value
C = original capacitor value
f = original drive frequency
f' = new drive frequency

As an example, the Muirhead K-550 wirephoto recorder normally operates at 100 r/min with 50-Hz motor drive. From our earlier examples, you could compute the value for the new drive frequency for 120-r/min operation. The result is 60 Hz. This unit uses a pair of 320-μF motor capacitors. The new capacitance value is:

$$f' = 320 \times ((50 \times 50)/(60 \times 60))$$
$$= 320 \times (2500/3600) = 320 \times 0.6944$$
$$= 222.22 \ \mu F$$

A pair of 220-μF capacitors should do the job.

Proper drum speed is one of the most critical factors in an effective FAX recorder or transmitter. Precision frequency sources must be used! Although the K-550 machine requires a 60-Hz drive, the 60-Hz power-line frequency is not stable enough for proper sync. A suitable source would be a 6-MHz crystal oscillator, followed by five decade counters (100,000 total division).

Scanning-density Changes

If the surplus recorder does not have the precise scanning density (lines per inch) desired, it is possible to alter the scanning density in several ways. In the case of a drum-type system, where the print head is typically driven by a threaded lead screw, the scanning density can be altered by changing the thread pitch, gear drive (if used), or rod drive motor speed (motor substitution or drive-frequency change). In continuous-feed systems, the paper feed rate is motor-controlled to drive rollers through gears or pulleys. The drive roller size can be altered, gears or pulleys

Table 6

Standards For Various Classes of FAX Service

Service	Drum Speed (r/min or line/min)	Size (in)	Scan Density (line/in)
WEFAX			
Satellite	240	11	75
APT			
Satellite	240	11	166
Weather			
Charts	120	19	96
Wirephotos	90	11	96
Wirephotos	180	11	166

changed, or drive motor speed shifted. Given the considerable latitude for acceptable scanning density ranges with most image material, you can usually get by with no changes or, at least, minimal changes in the carriage or paper drive.

Modulation Format

Most surplus units will have amplitude-modulation capability, usually with both positive and negative formats. In some systems, shifting polarity is as simple as reversing diodes and changing the bias on a video stage. Addition of an FM capability to an AM-only unit can be achieved by using a standard SSTV demodulator with a suitable interface buffer to meet your circuit requirements.

Equipment Sources

Conversion requirements for surplus equipment can range from nothing to complete replacement of electronics and even mechanical changes. You should carefully study the requirements of the mode that you are interested in and evaluate the cost and conversion requirements of surplus equipment. The equipment can be found at the larger hamfests, through dealers or directly from the manufacturers. Three reliable sources of equipment are:

Alden Electronic and Impulse Recording Equipment, Inc.
Washington St.
Westboro, MA 01581

Muirhead Inc.
1101 Bristol Rd.
Mountainside, NJ 07092

Atlantic Surplus
3730 Nautilus Ave.
Brooklyn, NY 11224

Home Construction

Effective FAX systems can be built from scratch with performance equal to commercial equipment. Such projects are an interesting combination of electronics and mechanics. Reference 1 in the FAX bibli-

ography at the end of this chapter describes a complete direct-readout recorder for use with satellite imagery. This system is completely solid state and uses dry electrostatic paper. The documentation includes circuit changes for photographic printing if desired. Reference 2 describes a continuous-read-out electrolytic recorder for weather charts. Either system could be adapted for other services with the addition of different demodulators and drive motors. Anyone seriously considering building the FAX mechanical subsystem should obtain the Hurst catalog (Hurst Manufacturing Company, 1551 E. Broadway, Princeton, IN 47670). Hurst is one of the largest manufacturers of synchronous motors and carries a large line of stock and built-to-order units.

Services

Services of potential interest to amateurs include weather satellites, weather chart HF transmissions, HF press photos and two-way amateur operation. Table 6 summarizes the standardized recorder requirements for each service.

Satellites

Amateurs have been very active in this area since the inception of the satellite program in the '60s and have made many valuable contributions. Reception and display of both geostationary and polar orbiting satellite signals is entirely practical, but does involve antennas, receivers, tracking and a knowledge of the various satellite systems in addition to FAX techniques. Reference 1 in the FAX bibliography provides a complete guide to setting up a station including construction projects. Reference 3 also provides a system overview with valuable information on the conversion of surplus FAX gear. References 4 and 5 provide some additional sources of information as well as a very broad overview of amateur satellite activities.

Weather Charts

Weather charts are transmitted on literally dozens of HF channels throughout the day. Reception requires a stable general-coverage SSB receiver or a ham receiver with suitable frequency converter.

Press Photos

A very large number of press photos are sent daily on HF. The precise frequencies are not a matter of public record, although some are included in various specialized frequency guides. These pictures can be copied quite easily but are not transmitted in a broadcast service. The press agencies are covered by the nondisclosure provisions of the Communications Act of 1934. You can copy them for your enjoyment; it would be illegal to publish samples, however. Once you learn to recognize FAX signals, you can locate the stations with a little patient searching. For starters, try combing the region between 15 and

Fig. 38 — A test chart such as this can be used to check out a TV camera or an entire ATV transmitter system. It provides patterns for horizontal and vertical resolution and gray-scale testing. An 11 × 14-inch high-quality version printed on heavy paper is available from ARRL Hq. for $3 postpaid. You can personalize the ARRL test chart by adding your call letters, location and a photograph in the white space provided.

20 MHz during daylight hours.

Two-way Amateur FAX

One of the greatest difficulties for two-way operation is locating FAX transmitters. While FAX recorders are available from surplus, FAX transmitters are much rarer. Unless a unit is found, you may have to build a transmitter that involves a simple drum and traverse system with the addition of a light source, photo transistor, an op-amp buffer and an SSTV modulator. See bibliography reference 1. Office FAX recorders, although sometimes requiring conversion, are attractive because they usually incorporate a transmitter and receiver in the same package.

For VHF operation, any kind of modulation system can be used as long as the units are compatible. Operation at HF requires the use of an FM subcarrier. At present, there are no uniform standards for HF FAX. It is anticipated that standards can be developed within the amateur community in the not-too-distant future.

A critical element in such considerations is the transmission time. Any format requiring more than 10 minutes for a single frame transmission is illegal because of the requirement for identification. An acceptable format should be reasonably short, recognizing crowded band conditions, propagation, and avoiding excessive wear and tear on RF transmitter finals caused by extended duration, high-duty-cycle service. Recorders exhibiting 120 line/min are marginal in this respect since the transmission of quality pictures requires 6 or more minutes. As a starting point, a 240-r/min standard (75 line/in for 11-inch recorders or 42 line/in for 19-inch recorders) would seem workable. This is the essence of the WEFAX satellite format but using FM modulation. Such an approach would transmit an 800-line high-quality image in just 3.3 minutes.

With the availability of computers, we may wish to evolve our own standards as amateurs have done with SSTV. For the moment, a 240-line/min standard would appear to be a good starting point for discussion and experimentation. A national amateur FAX net is conducted every Sunday on 14.245 MHz at 2000 UTC.

Computers

Digital scan converters have revolutionized SSTV and are about to do the same with FAX. The use of home computers offers an exciting challenge for use as "intelligent" scan converters. An example of such a system is one developed by WB8DQT and K6AEP for use with the Radio Shack TRS-80C Color Computer.[6] The system has four elements: a FAX interface circuit, a high-resolution display board, the computer with 64-kbyte RAM and a software package. The input signals are processed by the interface board and

Fig. 39 — A typical FAX picture.

the computer, samples the image, loads it into memory and passes the picture information to a display board for presentation on a standard TV monitor. The picture is displayed in a 256 × 256 format with 16 gray-scale levels.

WEATHER SATELLITES

Since the early days of weather satellites, radio amateurs have been active in the reception of satellite imagery. Assembling a successful weather satellite receiving station can provide an exciting challenge to the interested amateur. While the technical challenges to be overcome are significant, the step-by-step improvement of the amateur's station is rewarded by continual improvements in the imagery received. There are challenges to be met in the areas of antennas and receivers, storage and processing of the signal received, image production and finally image interpretation. Some unsuspecting radio amateurs have become passably good meteorologists quite by accident! In addition there is the excitement of monitoring an operational satellite system on a day-to-day basis. In some cases radio amateurs have alerted the satellite operators to problems on board the spacecraft before the operators became aware of the problems. Amateurs are continually upgrading their imaging systems and, in some cases, the imagery received challenges that produced by commercial systems.

Modern weather satellites generate two types of images. First is the high-speed digital imagery that is designed to efficiently transfer all the data generated by the imaging instruments. These wide-band digital signals are usually transmitted at microwave frequencies and require complex receiving stations. A few amateurs with special technical skills have succeeded

in receiving and displaying this imagery, but these signals require station complexity currently beyond the reach of most amateurs.

The second type of imagery available on modern meteorological satellites is the "Automatic Picture Transmission" or APT imagery. Early in the meteorological satellite programs it became apparent that for users in remote areas to have access to the satellite imagery on a timely basis, the satellite would have to transmit directly to the remote user. The APT format was developed to meet this need. It was first carried on the US Tiros VIII spacecraft (launched Dec. 21, 1963) and has continued to be supported for the same reasons originally proposed in the early 1960s.

Construction of an APT station is much less complicated than assembly of a digital imaging station. Because of the narrower bandwidth and relatively slow rate of data transfer, standard medium bandwidth VHF receivers can be used for reception of the satellite signal and home-type audio recording equipment can be used to store the satellite signal. The VHF receivers, antennas, preamplifiers and transmission lines described elsewhere in this *Handbook* can be used on the satellite bands with appropriate modifications to change the frequency range covered.

APT image resolution is about 2-3 km for polar orbiting spacecraft and 8-12 km for the geostationary spacecraft. Clear images, completely adequate for meteorological purposes, can be obtained. APT offers higher resolution pictures than those received on commercial television broadcasts, for example. Images are much higher in resolution than those received on amateur SSTV. Under optimal lighting conditions, satellite images will show medium-sized cities, moderately large rivers and forest areas, and even, at times, interstate highways! Fortunately, the same APT format has been used by the weather satellites of the United States, the Soviet Union and the European Space Agency (ESA). Because of this international cooperation, the same equipment (with minor adjustments) can be used for reception of imagery from all of these spacecraft.

APT Signals

The APT mode, like facsimile, allows the transmission of a relatively high-resolution image in a moderate bandwidth by using a longer time for image transmission. With APT, the detected FM signal consists of an audio-frequency subcarrier that has been amplitude modulated. The amplitude peaks on the subcarrier correspond to white areas on the image, and the low amplitude portions correspond to black areas on the image. The image is transmitted at either 120 or 240 lines per minute

(see Table 7).

The United States spacecraft and the ESA Meteosat use 2400 Hz as the subcarrier frequency. With these spacecraft the signal is always modulated at least 5%. If a noise-free signal is available, the user can employ a phase-locked loop circuit to track the subcarrier and use that information for detection or line synchronization. Soviet spacecraft do not use the 2400 Hz subcarrier frequency, although the subcarrier they use is close to that frequency (apparently about 2500 Hz). On some Soviet spacecraft, the subcarrier modulation falls below 5% during the scan line. For this reason it is not possible to "lock on" to the subcarrier of these spacecraft.

The term "WEFAX" describes a special type of APT format. This format consists of separate image frames consisting of 800 lines of imagery, with defined borders on all four sides. Data is transmitted at 240 lines per minute. In addition, clearly identified start and stop tones make it easier to automate image reception. Since the imagery is prepared at the spacecraft command station, some of the "images" may consist of text bulletins and weather charts.

There are some important differences between the APT format (including WEFAX) and the facsimile mode used for HF transmissions. It is probably more helpful to think of APT as a completely different mode. It should not be assumed that surplus facsimile equipment will automatically function in the APT mode, although many successful conversions have been made.

Spacecraft

Currently active spacecraft can be divided into two categories. First are the polar-orbiting spacecraft. This group includes the Tiros-N series operated by the United States and the Meteor 2 and Meteor 3 series spacecraft operated by the Soviet Union. Other countries, including the People's Republic of China, have plans for launching and operating weather satellites of this type. Currently active polar-orbiting meteorological spacecraft all transmit APT imagery in the form of scanning radiometer imagery. To produce this imagery the spacecraft views the earth below through a rotating mirror which scans a narrow strip of the earth below and then transmits that one line of the image obtained. See Fig. 40. As the mirror rotates, the spacecraft moves through space; each rotation of the mirror covers a new portion of the earth. In this way the spacecraft produces a long, continuous image with clearly defined side borders, but no definite top or bottom.

The second group of meteorological satellites are the geostationary satellites. This group includes the GOES series op-

Table 7

Currently Active Meteorological Spacecraft with APT Service

Spacecraft	Operating Country	Orbit Type	Inclination (Degrees)	Orbital Period	Frequencies (MHz)	Image Format	Line Rate (lines/min)
Tiros-N Series	USA-NOAA	Polar	98	102 min.		S.R.*	120
NOAA 6 (#11416)					137.500		(visible & IR)
NOAA 9 (#15427)					137.620		
Meteor 2 Series	USSR	Polar	82	104 min.		S.R.*	120
Meteor 2-14 (#16735)					137.300		(visible only)
					137.400**		
					137.850**		
Meteor 3 Series	USSR	Polar	82	110 min.			120
Meteor 3-1 (#16191)					137.850		(visible only)
GOES Series	USA-NOAA	Geostat. 75°W	0	24 hrs.	1691.00	WEFAX	240
GOES-E ***							
GOES-5 (#12472)							
GOES-C ***		114°W					
GOES-2 (#10061)							
GOES-W ***		135°W					
GOES-3 (#10953)							
METEOSAT	ESA	Geostat. 0°W	0	24 hrs.	1694.50	WEFAX	240
METEOSAT 2 (#12544)							

*Scanning radiometer format.
**Alternate frequencies for Meteor 2 series spacecraft.
*** GOES-E, GOES-W and GOES-C refer to spacecraft locations. GOES-2, GOES-3 and GOES-5 refer to particular spacecraft.

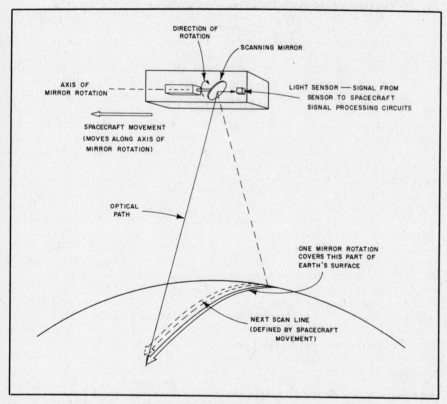

Fig. 40 — Meteorological satellites view the Earth through rotating mirrors to produce APT images.

erated by the United States and the Meteosat spacecraft operated by the European Space Agency. Both of these spacecraft carry APT data in the WEFAX format, which differs from the scanning-radiometer data obtained from the polar orbiters. In these spacecraft the high resolution digital data is transmitted to a ground command station, computer processed and changed to the WEFAX format. In some cases geographical boundaries and titles are overlayed on the images. The image is then transmitted up to the spacecraft from where it is retransmitted to WEFAX receiving stations, including amateur stations. In this case, although the spacecraft carries the imaging instrument it is serving as a transponder in the WEFAX mode; imagery from one spacecraft may be transmitted on a different spacecraft altogether. Spacecraft with non-functioning imaging instruments have, at times, been used to retransmit APT data from other spacecraft.

A third, somewhat unpredictable, source of APT data is APT instruments carried on experimental Soviet spacecraft. Spacecraft in the Meteor-Priroda series have carried APT instruments in the past and transmitted in the 137 MHz band. These are usually 240 line-per-minute scanning-radiometer images and have been popular among amateurs because of the high resolution offered. In addition, some of the Cosmos radar imaging spacecraft (Cosmos 1500 and 1602) have transmitted imagery in the APT format. These spacecraft seldom transmit over North America although occasionally imagery may be available. European amateurs can frequently receive APT signals from these spacecraft as they transmit data to receiving stations within the Soviet Union.

RF Signals

The two types of weather satellites transmit imagery in different frequency ranges. The polar-orbiting meteorological spacecraft use the 137 MHz VHF band for APT data. The geostationary spacecraft have used the S-band microwave frequencies of 1691 MHz (GOES) and 1694.5 MHz (Meteosat). In both cases the signal is sent as a medium bandwidth FM signal (about 30 kHz). Conventional FM receivers can be used as long as proper system band-

Fig. 41 — This APT image from a Soviet Meteor-Priroda (experimental Meteor) shows the southeastern United States.

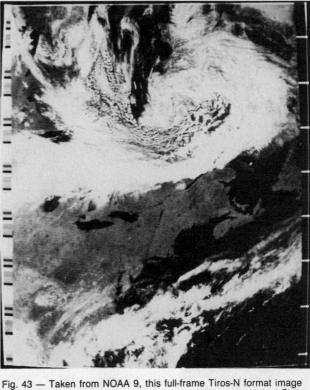

Fig. 43 — Taken from NOAA 9, this full-frame Tiros-N format image shows dramatic weather patterns over the northeastern United States and the Canadian Maritime provinces.

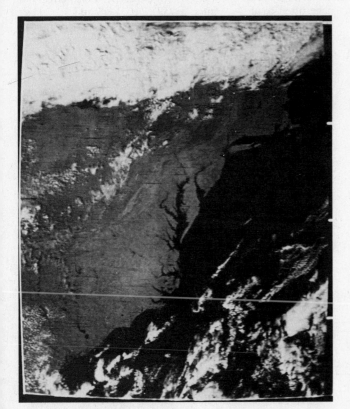

Fig. 42 — Much of the east coast of the United States is visible in this expanded Tiros-N frame.

width is maintained and good dynamic range is preserved.

Receiving Systems

Two frequency ranges are of interest for APT users. First is the 136-138 MHz satellite band. Weather satellites are assigned to the frequencies between 137 and 138 MHz. Commonly used frequencies are shown in Table 7. The frequencies used by the NOAA series spacecraft have not changed for many years. The three frequencies noted here in use by Meteor series spacecraft have been the primary channels for their operational Meteor 2 series spacecraft during the past 6 years. Other frequencies have been used intermittently by Soviet experimental spacecraft, but these are variable and change without notice. If crystal control of the receiver is anticipated, these five frequencies are the most active and should be included. Typical antennas for this frequency range include the turnstile antenna as described for OSCAR use, the multiturn helical antenna, and crossed Yagi antennas with right-hand circular polarization. An antenna-mounted preamplifier is recommended, but not absolutely required. If a high performance, low noise preamplifier is available, good imagery may be obtained with an omnidirectional antenna such as a short helix or a discone antenna.

The other frequencies of interest are 1691 MHz and 1694.5 MHz in the microwave S band. These frequencies are used by the geostationary spacecraft operated by the U.S. and ESA. Interdigital or stripline downconverters can be built to convert these frequencies to the desired intermediate frequency. Several commercial units converting 1691 MHz down to 137.5 MHz are now available. A parabolic dish antenna (usually about 6 feet in diameter) is a typical antenna, although smaller dish antennas can be used if high performance, low noise preamplifiers are available.

Imaging Systems

Several types of instruments for displaying satellite imagery in are common use at this time. First are the "drum" type machines such as that described many years ago in *QST* by Wendell Anderson. With this type of unit a sheet of light-sensitive photographic paper is wrapped around a long cylinder or (drum) which is then spun at exactly the same rate as the satellite image line rate. If the satellite transmits data at 240 lines per minute, for example, the drum must be rotated at precisely 240 revolutions per minute. The sensitized paper is exposed one line at a time by an intensity-modulated light source which is slowly drawn along the length of the drum. In this way the entire surface of the paper is exposed. After exposure the paper is removed from the drum and is developed using standard photo-finishing techniques. The image can then be examined. A variation of this method is the use of an electrostatic paper on an office-type facsimile machine. The Xerox model 400 and the older Western Union Desk-Fax machines have been used successfully for APT after modifications.

A second widely used method of preparing satellite imagery is with the use of a dedicated CRT. With this method the CRT is controlled with a very slow sweep rate to match the line rate as transmitted by the spacecraft. The image is drawn on the face of the CRT slowly, one line at a time. Because of the length of time required to draw an entire image on the face of the CRT, a camera is needed to photograph the screen of the CRT in order to see the detail in the image.

A third approach is the modification of commercial facsimile equipment for print-out of satellite images. Some models of commercial equipment work well in this capacity. The amateur is cautioned to become familiar with the equipment and select a particular machine carefully. This equipment is still fairly expensive, even as surplus. Some units are designed to print only black and white (as for weather maps and line drawings); these may not work well in an application (such as APT) where a good gray scale is important.

More recently, with the availability of microcomputers and digital memory, the prospect of using scan converters for APT imagery has become a practical possibility. The use of this technique has lagged behind its use in SSTV because of the concern for the loss of detail in the scan conversion process. Since the APT image has a great many more pixels than an SSTV image, more of the image will be sacrificed for a given framestore size. For example, while 256 pixels by 256 pixels may be perfectly adequate for an SSTV image it would be considered a practical minimum for APT use. Even at this level, at least half of the image detail will be lost.

There are three general approaches to the development of a scan converter (digital framestore device) for APT use. The first is to sample the APT image, store the data in the memory of a personal computer and display the image on the screen of the personal computer. This has been done with Commodore 64, Apple II, and IBM personal computers among others. The results of this approach are usually disappointing. Since the amount of memory available for data storage may be limited in a PC, a relatively low resolution image often results. The second approach involves the use of a microcomputer to control a separate block of RAM with additional hardware to send the display to a dedicated monitor. See references 6 and 7. The third approach is the development of a dedicated hardware-controlled scan converter without the use of a microcomputer. While this approach is less flexible it does offer the option of producing imagery without the expense of a microcomputer.[8]

FAX Bibliography and Text References

[1]Taggart, R. E., *The New Weather Satellite Handbook*, Wayne Green Inc., Peterborough, NH, 132 pages.

[2]Lee, C. B., "Build a FAX From Scratch," *73*, July 1976, pp. 44-50.

[3]Summers, R. J.; Gotwald, T. *Teachers' Guide for Building and Operating Weather Satellite Ground Stations*, NASA EP-184, Office of Public Affairs, NASA Goddard Space Flight Center, Greenbelt, MD, 83 pages.

[4]Dean, R., "Facsimile For the Radio Amateur," *73*, September 1971, pp. 66-77.

[5]Davidoff, Martin R., *The Satellite Experimenter's Handbook*, ARRL, 1984, Chapter 11.

[6]Abrams, C. and Taggart, R., "A Color Computer SSTV/FAX System," *73*, Nov. and Dec. 1984.

[7]Abrams C., "In Search of the Perfect Picture," *QST*, Dec. 1985 and Jan. 1986.

[8]Vidmar, M. "A Digital Storage and Scan Converter for Weather Satellite Images," *VHF Communications*, Winter 1982 and Spring 1983.

Chapter 21

Special Modulation Techniques

A mateur Radio is not limited to sending telegraphy, voice or images by conventional modulation systems. There are numerous uses for telecommand (remote control) and telemetry (distant reading of a sensor), both of which may use a variety of signaling techniques. In some applications, telecommand or telemetry signals are sent over a radio link dedicated to that purpose. In others, signaling is sent along with telegraph, voice or image communications in a secondary or incidental way, usually to facilitate communications. This chapter will cover some of the signaling techniques of interest to amateurs. Also, this chapter contains information about uses of experimental transmission techniques by radio amateurs.

RADIO COMMAND AND CONTROL

Amateur Radio may be used for remotely controlling an object or another radio station. Although command and control transmissions may not be used indiscriminately on the amateur bands, FCC rules include a number of special provisions for different applications. Before designing or attempting to operate new telecommand systems, it is a good idea to check *The FCC Rule Book*, available from the ARRL.

According to §97.3(l) of the FCC rules, *radio control operation* is one-way radio communication for remotely controlling objects or apparatus other than Amateur Radio stations. Reading further in §97.3(m), *control* means techniques used for accomplishing the immediate operation of an Amateur Radio station. Control includes one or more of the following: *local control, remote control* and *automatic control*. Remote control is defined by §97.3(m) (2) as manual control, with the other control operator monitoring the operation on duty at a control point located elsewhere than at the station transmitter, such that the associated operating adjustment(s) are accessible through a *control link*. §97.3(n) defines a control link as apparatus for effecting remote control be-

Fig. 1 — A typical radio-controlled model airplane with its accompanying control transmitter. The operator works the joysticks, and his or her actions are translated into commands that operate servos to control the airplane.

tween a control point and a remotely controlled station.

Whether controlling objects, apparatus or other Amateur Radio stations, bear in mind that N0N (formerly A0 or F0) emission (an unmodulated carrier) is not permitted below 51 MHz. This would apply to control transmissions that send control signals only occasionally but with the carrier left on continuously.

RADIO CONTROL OF A REMOTE MODEL AIRCRAFT OR VEHICLE

Radio control (R/C) of remote model aircraft or vehicles by hobbyists was started in the 1930s. ARRL's Ross Hull with R. B. Bourne, Clinton DeSoto, Byron Goodman, Harner Selvidge and H. M. Plummer did some early work, as did Walt and William Good. The Good brothers used a radio system to steer the free-flight type of model airplanes back to the launch point. The system was big, and not totally reliable, but it showed definite potential. A large tube-type transmitter on the ground sent a steering pulse on command to a tube-type receiver in the model airplane.

All of the early experimenting was carried on by Amateur Radio operators because amateur frequencies were the only frequencies available. Later, in the early 1950s, Citizen Band spots at 27.255 and 465 MHz were opened by the FCC for shared use by R/C hobbyists. This caused rapid growth of the R/C hobby due to the simpler license.

When the Citizens Band was formed from the old 11-meter amateur band, certain frequencies were set aside for or shared by the R/C enthusiasts. However, since many of the early receivers were wideband superregenerative types, many fliers found their airplanes crashing because of adjacent CB interference. Many became amateurs as a result, and operate to this day at the top of the 6-meter band, a territory unused by most amateurs because of its proximity to TV Channel 2.

Recent pressure by 6-meter FM repeater operators to use this upper area has resulted in the suggested R/C frequencies being moved to the bottom end of 6 meters. Some amateur R/C operators use 10 meters by obtaining new crystals and retuning the

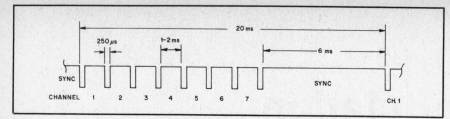

Fig. 2 — Sample signal from an R/C transmitter.

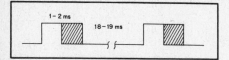

Fig. 3 — The signal seen by each servo.

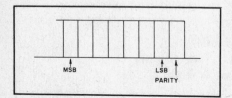

Fig. 4 — Example of a PCM-encoded signal.

older 27-MHz CB R/C equipment. The bulk of R/C operation occurs on the new license-free radio-control bands at 72-73 and 75-76 MHz.

The license-free frequencies do get overcrowded at many flying fields, so most amateurs who are R/C hobbyists purchase 6-meter R/C transmitters and receivers.

Uses

The majority of R/C systems are used for controlling the flight of model airplanes, as shown in Fig. 1. Other applications include R/C cars and R/C motorboats and sailboats. The airplanes vary in size from about a pound in weight with a 1-foot wingspan to gigantic multi-engine models weighing over 50 pounds with 15-foot wingspans. Many modelers fly large-wingspan gliders. R/C cars may have electric motors or be gas-engine driven.

The operator controls the model from a transmitter that has control sticks. Miniature receivers are built into the model to receive signals from the transmitter. The action of the control sticks is sent through the transmitter and receiver to servos located in the model. These translate the stick movement into some mechanical movement which produces the desired control. The operator can control movements such as ailerons, elevator, rudder or throttle in an airplane. The operator may control the steering, throttle, rudder or a sail winch in a model car or boat. Often, more sophisticated scale models have other systems controlled remotely, such as retracting wheels, "bomb dropping," flaps, battleship guns that turn and fire, anchors that raise and lower, and many other interesting actions.

Because R/C systems are widely available at low cost, amateurs can use them for many unique applications around the station that may require some kind of mechanical adjustment. The amateur TV station can have its TV cameras panned, zoomed or focused using R/C. Potentiometers or tuning knobs can be turned by R/C. Miniature antennas could be turned by some of the special high-thrust servos intended for very large model airplanes. Several amateurs active on SSTV use R/C servos to swing in color filters over their black-and-white TV cameras.

Design

Most current R/C designs consist of a low-power transmitter that sends out a repeating series of width-modulated pulses to the R/C receiver. Each width-modulated pulse represents a single channel, for example the rudder channel. An R/C system is rated by the number of channels it supports, based on how many of these width-modulated pulses are involved. See Fig. 2. For example, systems are called 4-channel systems, 7-channel systems, and so on. An R/C system also may be sending out more than "4 channels," but only have four servos or four control sticks included with the system, so it is still known as a "4 channel." The series of pulses repeat about 50 times a second, so that any control stick movement can quickly result in an appropriate servo movement.

Each channel's servo monitors a positive pulse that varies in duration between 1 and 2 milliseconds, centered at 1.5 ms. See Fig. 3. The servo's electronics are designed so a 1.5-ms pulse will cause the motor to move the control arm to a centered position. If the pulse width changes to 1 ms, the servo control arm will turn 45 degrees in one direction. If the pulse width had changed to 2 ms, the servo control arm would have turned 45 degrees in the other direction. The servo quickly reacts to any change in the transmitter control stick, and maintains position if no new stick movement has occurred. Depending on manufacturer and type, the precise pulse duration and amount may vary slightly.

Some servos are very small and light, with little torque. Others are larger with considerable torque. Most of the newer servos have three leads — a ground or common lead, a +4.8-volt lead, and the pulse lead. Prices vary from $10 to $50 per servo.

Some recent systems utilize pulse-coded modulation (PCM) instead of the traditional pulse-width modulation scheme. See Fig. 4. PCM sends the absolute position of each channel, encoded as bits (similar to ASCII), plus a check bit. The receiver is then able to reject any faulty or corrupted signal, and use the last good position setting. If no good positions arrive over a predetermined short period of time, the system can reset the servos to an assumed safe postion, such as low throttle and neutral control surfaces, until the system once again obtains acceptable PCM codes.

Most of the existing R/C transmitters use discrete components to encode the pulses. Fig. 1 shows a prototype transmitter that uses a CMOS microprocessor for the encoding task. The block diagram of such a transmitter is shown in Fig. 5. This transmitter has the ability to have totally new control features and options added by simply replacing the program in EPROM. This transmitter can retain the flight trim settings of four different airplanes, or it could provide an unlimited number of channels for unique applications such as scale boats or robotics.

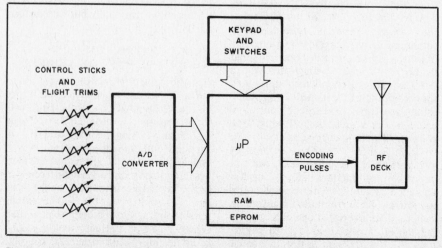

Fig. 5 — Simplified block diagram of a microprocessor-encoded R/C transmitter.

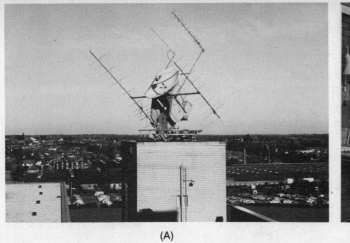 

(A)   (B)

Fig. 6 — UoSAT-OSCAR command station, G3YJO, at the University of Surrey, Guildford, Surrey, England. At A is the 2-m, 70-cm and 23-cm az-el antenna array. Inside the station, at B, are the digital command and display consoles alongside the transmitting and receiving equipment.

Because of the potentially lethal nature of R/C airplanes flying often in excess of 100 mi/h, 6-meter repeater operators on 50 MHz or 53 MHz should coordinate their frequencies with the local R/C clubs. The most commonly used Amateur Radio R/C frequencies are: 53.1, 53.2, 53.3, 53.4, 53.5, 53.6, 53.7, 53.8 and 53.9 MHz. Most current Amateur Radio R/C activities occur on 53.1 to 53.5. These frequencies are expected to be vacated by 1990.

New frequencies are:

Frequency	Channel No.
50.80	00
50.82	01
50.84	02
50.86	03
50.88	04
50.90	05
50.92	06
50.94	07
50.96	08
50.98	09

Recommended practice is to use the even-numbered channels until 1990 and all channels thereafter, when the selectivity of R/C receivers is expected to improve.

RADIO REMOTE CONTROL OF AN AMATEUR RADIO STATION

Radio communication for remotely controlling another Amateur Radio station is considered *auxiliary operation* by the FCC. All Amateur Radio bands above 220.5 MHz (excluding 431 to 433 and 435 to 438 MHz) are available for auxiliary operation. FCC rules specify that the remotely controlled station must have provisions to limit transmission to a period of no more than 3 minutes in the event of malfunction in the control link.

Remote Control of a Repeater

When remotely controlling a station in *repeater operation*, the control link must use frequencies above 220.5 MHz and may not use the input (receiving) frequencies of the repeater, according to FCC rules. For

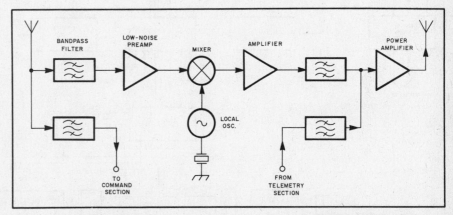

Fig. 7 — Functional block diagram of a typical satellite transponder showing telecommand and telemetry channels.

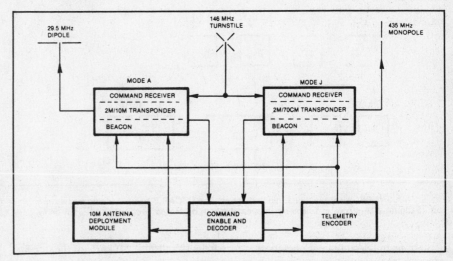

Fig. 8 — AMSAT-OSCAR 8 satellite functional block diagram. OSCAR 8 telemetry was sent using 20-WPM Morse code. A complete set of data required about 20 seconds to send.

a discussion of repeater-control methods see Chapter 14.

Telecommand of a Satellite

The FCC defines *telecommand operation* as earth-to-space Amateur Radio communication to initiate, modify or terminate functions of a station in space operation (that is, an Amateur Radio communications satellite).

The following frequencies are available for telecommand operation: 7.0-7.1, 14.0-14.25, 21.0-21.45, 24.89-24.99, 28.0-29.7, 144-146, 435-438 and 24,000-24,050 MHz.

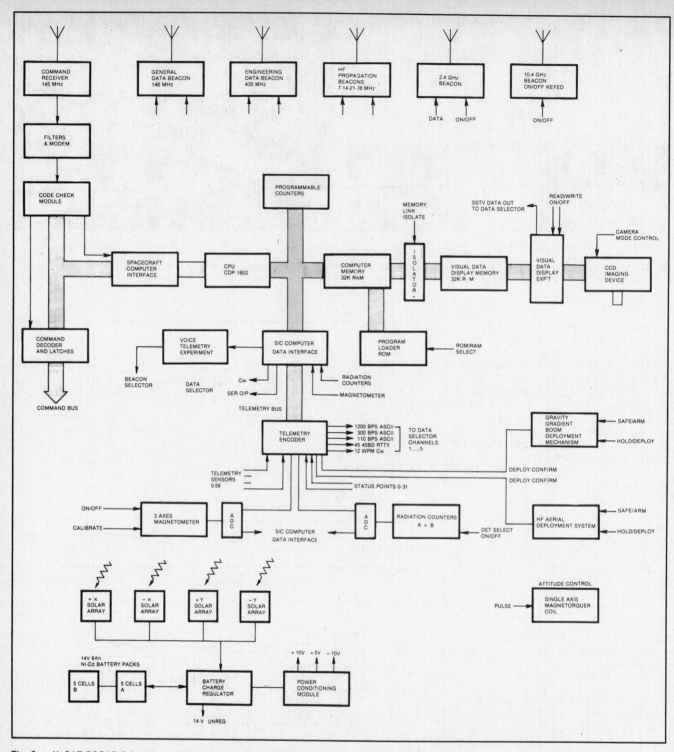

Fig. 9 — UoSAT-OSCAR 9 functional block diagram. The OSCAR 9 telemetry system has provisions for monitoring 60 analog sensor channels and 45 digital status points. Telemetry data can be sent on Morse code, RTTY (Baudot code) or ASCII.

Stations in telecommand operation are exempt from station identification requirements and may transmit special codes intended to obscure the meaning of command messages to the satellite, according to §97.421 of the FCC rules.

Telecommand links are used to switch OSCAR satellites from one operating mode to another and to turn off a malfunctioning transmitter that might cause interference. In addition, the ability to telecommand subsystems prolongs the useful life of the satellite by turning off parts of the satellite to conserve energy. Command frequencies, codes and transmission formats are confidential and provided only to AMSAT-approved command stations.

There are volunteer command stations in more than 8 countries. Fig. 6 shows the UoSAT-OSCAR command station at the University of Surrey, England.

TELEMETRY

Telemetry is the process of measuring a quantity, transmitting the result to a distant station, and there indicating or recording the measured quantity. It is strongly associated with rocketry and satellites but has many other applications. In many instances where radio is used to telecommand or control a distant radio station, a telemetry link is needed to monitor conditions at the remote station. When a command is sent, telemetry provides the feedback necessary to see whether the command was carried out and to monitor conditions afterwards to ensure the desired results.

AMATEUR-SATELLITE TELEMETRY

In §97.403(d) of the FCC rules, telemetry is defined as space-to-earth transmissions, by a station in space operation, of results of measurements made in the station, including those relating to the function of the station. The FCC considers telemetry transmissions by amateur satellites to be permissible one-way communications and says that such transmissions may consist of specially coded messages intended to facilitate communications.

Telemetered information typically includes satellite temperature, solar-cell current, battery temperature and voltage, and other parameters that describe the electrical and mechanical conditions of the satellite. Telemetry data may also include the results of scientific observations in the case of amateur satellites with a research mission.

Telemetry is transmitted by means of a beacon, usually just outside the satellite's communications passband. Beacon transmitters typically have output power levels of 50 mW to 1 W depending on the altitude of the spacecraft. Telemetry encoding methods vary, but OSCAR satellites have used Morse code, Baudot radioteletype, "advanced encoding techniques" (ASCII for computer processing) and digitized speech.

SIGNALING

Signaling systems for command, control and telemetry must be designed with high reliability in mind. Many early signaling systems used analog transmission techniques to represent continuously variable quantities by the amplitude, frequency or duration of a tone. Most new systems use some type of digital transmission technique and often include a degree of redundancy for detection and correction of errors.

Multiplexing

Most signaling systems are capable of sending more than one piece of information in the same transmission. For example, the telemetry from a satellite might routinely include a complete system check with readings from numerous sensors. Combining two or more streams of information on one radio carrier is called *multiplexing*.

Frequency-division multiplexing (FDM) is a method of sending each of the information streams over separate subcarriers, each at a different frequency. The subcarriers could be modulated in various ways, according to the nature of the information. Fig. 10 illustrates the baseband of an FDM signal.

Time-division multiplexing (TDM) uses only one channel but assigns the different data streams to different time slots. For example, telemetry from sensor 1 could be followed in turn by that of sensors 2, 3, and

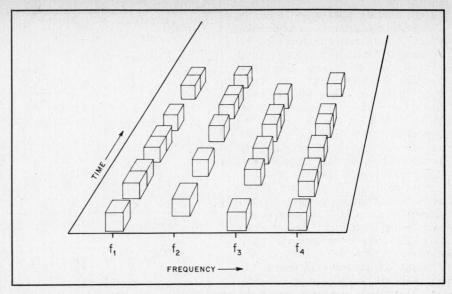

Fig. 10 — Frequency-division multiplexing (FDM). Each block represents a logical 1 being transmitted. The spaces in between are 0s. Four streams of data are being sent, each on a separate frequency.

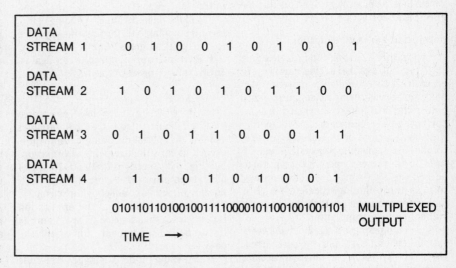

Fig. 11 — Time-division multiplexing (TDM). Four data streams are interleaved to produce one multiplexed data stream at four times the individual data rate.

so forth, as shown in Fig. 11.

Coding

Essentially every signaling system, whether for control or telemetry, requires some type of encoding at the sending station and decoding at the receiving station. A voltage or current reading cannot be transmitted in its original form but could be used to frequency-modulate an audio tone suitable for transmission. The signal can be demodulated by a frequency-to-voltage converter. Alternatively, the original signal could have been fed to an analog-to-digital converter to produce a digital code that, in turn, could frequency-shift key an audio tone. This signal could be received by an FSK demodulator and decoded by a digital-to-analog converter.

In general, FCC rules permit codes where they facilitate communication rather than obscure meaning. A careful reading of *The*

FCC Rule Book is advised before using specific coding and modulation systems.

Error Control

When designing a signaling system, particularly one for command and control, serious attention should be given to the possibility of errors and the consequences of those errors causing undesirable results. There is no such thing as an error-free communications channel, although some approach the level of perfection. Amateur Radio channels are subject to noise, interference and fading, so some errors may be encountered from time to time. It is best to ensure that the communications channel has a basic reliability well into the 90% range before considering it for control purposes.

It is worthwhile to consider at least some type of *error detection* scheme to prevent the decoder from accepting erroneous commands or falsing on no signal. To give the decoder something to work on for detecting errors, it is necessary to add some redundancy to the encoding process prior to transmission. For example, a single-bit parity check is often added to the ASCII code. There are several ways of using parity, but using the "odd parity" system, the parity bit is set to make sure that the number of 1 bits in the character is always an odd number. If, on reception, the number of 1s is even, the decoder knows that the command contains an error.

Error-detection systems are simply methods of confirming the (probable) correctness of a signal. More sophisticated techniques provide error correction, often called *forward error correction (FEC)*. It takes more than 1 bit of redundancy for a 7-bit code to implement FEC. Several additional bits will be needed to provide sufficient redundancy to determine what the correct signal should have been. The exact ratio of parity bits to original coding bits varies according to the specific type of FEC system and the percentage of errors that the system can correct on its own.

Encryption and Authentication

Encryption is a technique of encoding each symbol against a changing key in order to obscure meaning or prevent someone else from knowing what command to send. The FCC has recognized its importance as a safeguard against tampering with satellites and permits its use for telecommand of amateur spacecraft. The FCC rules do not presently permit encryption to be used for other Amateur Radio purposes, such as controlling a repeater. Yet, it is undesirable to use a control system that is vulnerable to manipulation.

Where tampering is a threat, an *authentication* technique should be considered. Using this approach, you can send the same command each time but have a changing element appended to the end for authentication purposes. For example, if the command for turning on the autopatch is OA, then the first transmission of that command might be OA7, the next one OAR, the third OAM, and so forth, the third character being selected from a sequence known only to the encoder and decoder.

SPECIFIC SIGNALING TECHNIQUES

Before settling on a specific signaling technique, it is useful to consider some of the trade-offs. Here is a suggested checklist:

• What is the original form of the data to be transmitted?
• Does it need to be converted to a transmissible format?
• Will the modulation fit within a standard voice channel, on-off or frequency-shift-key a transmitter, or does it need a special modulator?
• What type of demodulator is required?
• Does the data need to be converted after demodulation?

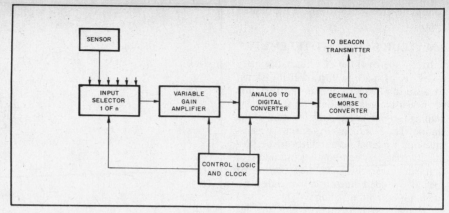

Fig. 12 — Block diagram of a typical AMSAT-OSCAR Morse code telemetry encoder.

• Is it legal within the FCC rules for the frequency purposes?
• Is frequency coordination needed?
• Will the encoding/modulating equipment be accessible to the stations transmitting the signals?
• Will the demodulating/decoding equipment be available to the stations expected to receive the signals?
• Have precautions been taken against errors, tampering or simple loss of control?

Morse Code

Morse code has been used for transmitting telemetry from OSCAR satellites because no special decoding equipment is needed to copy the telemetry information. See Fig. 12. Speeds of 10 and 20 WPM have been used to ensure that a large number of people could copy the telemetry. The chief disadvantage of this application is that speeds that are easily copied aurally do not permit the transmission of massive data.

If sent using on-off keying, Morse-code transmission may not be reliable enough for command and control purposes. Use of frequency-shift keying would overcome that problem. Computers can be used for both the generation and decoding of Morse characters.

Baudot Code

ITA2 (commonly called Baudot) is normally a direct-printing telegraph system. However, §97.69 of the FCC rules provides that ITA2 figures-case characters except the numerals and slant bar may be used for remote control of receiving printers and other objects. The purpose of doing so must be to facilitate communication, not to obscure meaning. The codes available for control purposes are the figures-case positions corresponding to the following letters: A, S, D, F, G, H, J, K, L, Z, C, V, B, N and M.

For telemetry purposes, the complete ITA2 character set may be used. Numerous amateurs throughout the world are equipped to receive and print these transmissions.

The complete ITA2 character set is given in Chapter 19.

AMTOR

What was said about ITA2 applies to AMTOR, as AMTOR uses the ITA2 character set. However, AMTOR has two optional error-control modes (ARQ and FEC), which may be valuable for control or telemetry transmissions.

ASCII

ASCII has a larger character set than ITA2. Further, it is usually sent at speeds much higher than ITA2 or AMTOR. Also, it is the native code for most microcomputers, making code conversion unnecessary. These advantages make ASCII suitable for many telemetry applications.

The same FCC rule provision for controlling printers or other objects mentioned above for Baudot applies even more so in this case. The reason is that ASCII has a much larger number of control characters already built into the code. See Chapter 19 for the complete ASCII character set.

Digitized Speech

UoSAT-OSCAR 9 first used digitized speech for telemetry. It has the advantage that anyone who knows the (English) language can instantly decode the meaning. It is particularly useful for educational demonstrations involving students in the lower grades. Its low data rate for easy listening makes it unsuitable for most telemetry applications.

Amateur Television

It may seem strange to consider amateur television (both slow scan and fast scan) as a signaling system, but don't overlook it as a possibility. Image communications systems are excellent for some types of telemetry. TV could be used to look at meter faces, confirm that an antenna system is still there and pointing in the right direction, or could look at the weather situation at a remote location. The camera can be remotely controlled so it can view

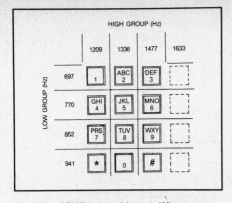

Fig. 13 — DTMF key pad layout. Whenever a key is pressed, the pad produces two AF tones, one from the low group and one from the high group. Twelve-button key pads are standard. Optional 16-button key pads are available on some hand-held radios.

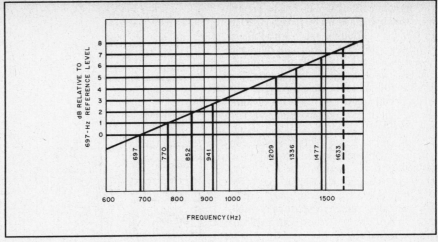

Fig. 14 — The effect of pre-emphasis on DTMF tones.

different scenes, zoom in on items of interest, and so forth. Fast-scan ATV might also be the "eyes" for remote control of models.

Dual-Tone Multifrequency Signaling

In the late 1940s, Bell Laboratories developed the Touch-Tone® signaling system to replace telephone pulse dialing. The dual-tone multifrequency (DTMF) signaling system, as it is known generically, became widely used in Amateur Radio with the introduction of autopatches to 2-meter FM repeaters. Repeater operators also found many control applications for DTMF. These aspects of DTMF signaling are covered in Chapter 14.

The fact that many amateurs own VHF/UHF transceivers with built-in DTMF encoders makes DTMF an attractive candidate for a variety of control purposes. However, this is simply a one-way advantage (that is, from a hand-held transceiver to a base unit) because the hand-held radios do not presently include DTMF decoders. Such decoders would be used for selective-calling purposes in order to mute the hand-held receiver until a prearranged tone sequence is received.

The standard 12-button DTMF key pad is a 2-out-of-7 tone encoder, meaning that the encoder selects one tone from the low group of 697, 770, 852 or 941 Hz and a second from the high group of 1209, 1336 or 1477 Hz. An optional 16-button DTMF key pad is also available; it is a 2-out-of-8 tone encoder. The 16-button variety uses the additional high-group frequency of 1633 Hz. The key pad layout with tones related to character assignments is shown in Fig. 13.

A number of amateurs have puzzled over how to encode the 26 Latin letters on a DTMF pad. There is an encoding scheme that was originally devised to enable the deaf to use a DTMF pad on a telephone to report a fire to the local fire department. In this system, two buttons are pressed for each letter. First, push the button with the letter printed on it — that would be the "2"

button for A, B and C. The result would be ambiguous, so press *, 0 or # to indicate whether it is A, B or C, respectively. The only problem left is what to do with Q or Z, which do not appear anywhere on the key pad. The solution is to send ** for Q and ## for Z. They're easy to remember because they fall directly below the keys (PRS) and (WYX) where they were omitted. As an example, 225 Main St. would be encoded

$$\frac{2}{2}\ \frac{2}{2}\ \frac{5}{5}\ \frac{M}{6^*}\ \frac{A}{2^*}\ \frac{I}{4\#}\ \frac{N}{60}\ \frac{S}{7\#}\ \frac{T}{8^*}$$

DTMF signals may be encoded with either a manual key pad or an automatic dialer. With manual keying, the duration of the tone depends on how long the key is pressed, and the length of the wait between tones is a function of how long the operator takes to press the next key. A minimum dwell time of 30 to 40 ms is necessary to ensure that most DTMF decoders will properly recognize the tones. Typical decoders need at least 10 ms delay between tones and may reset themselves if the next tones are not received within 2.5 seconds. Thus, a decoder designed for 35 ms dwell time and 15 ms of spacing could decode up to 20 digits per second. Automatic dialers are programmed to transmit one or more telephone numbers in rapid sequence. Repertory dialers are automatic dialers that store a large quantity of telephone numbers, any one of which can be recalled by a few key strokes.

Twist or tone differential refers to the relative amplitude difference between the low-group and high-group tones, expressed in decibels. DTMF tone encoders are usually designed to produce a twist of 3 dB, where the high group is at the higher amplitude. Some DTMF key pads have an internal adjustment that can vary the low or high group as much as ±5 dB. DTMF decoders can tolerate a certain amount of twist, sometimes as much as 12 dB, without malfunctioning, in order to allow latitude for misadjustment of the encoder and fre-

quency distortion in the communications channel.

When DTMF tones are fed into the audio input stage of an FM transmitter, the levels of all tones should be about equal at that point. The DTMF tones should emerge from the audio-processing stage of the transmitter without clipping. Any clipping at this point will generate intermodulation products that could cause interference and may hinder proper decoding at the receiving station. In a normal FM transmitter, the modulation is pre-emphasized by 6 dB per octave. This means that a frequency of 1394 Hz would be 6 dB higher after pre-emphasis than a 697-Hz tone, assuming that both tones were of the same amplitude before pre-emphasis. Fig. 14 shows the relative amplitudes of the DTMF tones after pre-emphasis. It can be seen that the combination of the highest tone from the low group (941 Hz) and the highest from the high group (1477 Hz on a 12-button key pad and 1633 on a 16-button key pad) will produce the greatest frequency deviation. So when you set the peak deviation for DTMF signaling on an FM transmitter, it is necessary to do so at the highest tones from the low and high groups. On a 12-digit key pad use the pound sign (#) to adjust the deviation to about 3 kHz (for a transmitter designed for 5-kHz deviation on voice peaks).

At the receiving end, 6-dB-per-octave de-emphasis does the job of restoring the DTMF tones to an equal level.

SPREAD-SPECTRUM COMMUNICATIONS

The common rule of thumb for judging the efficiency of a modulation scheme is to examine how tightly it concentrates the energy of the signal for a given rate of information. While the compactness of the signal appeals to the conventional wisdom, spread-spectrum modulation techniques take the exact opposite approach — that of spreading the signal out over a very wide bandwidth.

Communications signals can be greatly increased in bandwidth by factors of 10 to 10,000 by combining them with binary sequences using several techniques that will be described later. The result of this spreading has two beneficial effects. The first effect is dilution of the signal energy so that while occupying a very large bandwidth, the amount of power density present at any point within the spread signal is very slight. The amount of signal dilution depends on several factors such as transmitting power, distance from the transmitter and the width of the spread signal. The dilution may result in the signal being below the noise floor of a conventional receiver, and thus invisible to it, while it can be received with a spread-spectrum receiver!

The second beneficial effect of the signal spreading process is that the receiver can reject strong undesired signals—even those much stronger than the desired spread-spectrum signal power density. This is because the desired receiver has a copy of the spreading sequence and uses it to "despread" the signal. Nonspread signals are then suppressed in the processing. The effectiveness of spread spectrum's interference-rejection property has made it a popular military antijamming technique.

Conventional signals such as narrow-band FM, SSB and CW are rejected, as are other spread-spectrum signals not bearing the desired pseudonoise (PN) coding sequence. The result is a type of private channel, one in which only the spread-spectrum signal using the same pseudonoise sequence will be accepted by the spread-spectrum receiver. A two-party conversation can take place, or if the code sequence is known to a number of people, net-type operations are possible.

The use of different binary sequences allows several spread-spectrum systems to operate independently of each other within the same band. This is a form of sharing called code-division multiple access. If the system parameters are chosen judiciously and if the right conditions exist, conventional users in the same band space will experience very little interference from spread-spectrum users. This allows more signals to be packed into a band; however, each additional signal, conventional or spread spectrum, will add some interference to all users.

Benefits for Spectrum Managers

Effective spectrum management attempts to use a band as fully as possible while keeping interference to a minimum. A limit exists as to how many signals can be put in a band. When the allocation is used up, additional stations that use conventional modes may cause interference that degrades or blocks communications by other users. Additional spread-spectrum signals, however, may not cause severe interference; they just raise the background noise level. The limit to the number of spread-spectrum signals that can occupy a

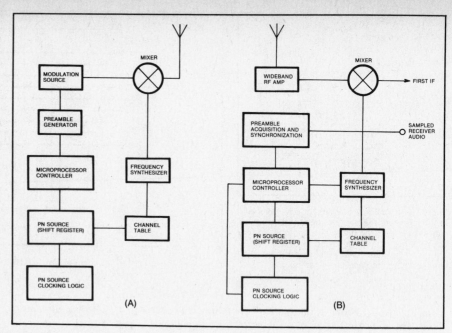

Fig. 15 — The block diagram of a frequency-hopping (FH) transmitter is shown at A. The channel table contains frequencies that the FH transmitter will visit as it hops through the band. These channels are selected to avoid interference to fixed band users (such as repeaters). The preamble employs the falling edge of an audio tone to trigger hopping. Conventional modulation, such as SSB, is employed. The block diagram at B is for an FH receiver. Preamble acquisition and synchronization trigger the beginning of the hopping mode and keep the receiver channel changes in step with the transmitter.

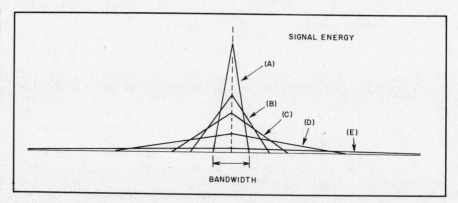

Fig. 16 — A graphic representation of the distribution of power as the signal bandwidth increases. The unspread signal (A) contains most of its energy around a center frequency. As the bandwidth increases (B), the power about the center frequency falls. At C and D, more energy is being distributed in the spread signal's wider bandwidth. At E, the energy is diluted as the spreading achieves a very wide bandwidth. Bandwidth is roughly twice the bit speed of the PN code generator.

band is sometimes called a "soft" limit because the effects of over allocation are not as severe as the rapid degradation caused by over allocation of conventional users.

Overlay is a spectrum-management concept that takes advantage of the spread-spectrum signal rarification and interference-rejection effects to share a band with conventional modulation users. In bands that are channelized and are fully allocated by assigned users such as repeaters, there are few ways to accommodate new users. Yet, viewing such a band on a spectrum analyzer reveals that much of the spectrum is only lightly used because of the intermittent use of many repeaters and the presence of numerous guard bands between fixed channels.

In the overlay concept, the spread-

spectrum signal is continuously spread over a shared band. It can exist in both the unused guard bands and on intermittently used repeater channels. It is, in effect a form of frequency diversity that takes advantage of whatever unused spectrum space is available within its spreading width.

Spread spectrum possesses a number of additional advantages. Spread spectrum provides a degree of protection against fading (a large variation in received signal strength caused by reflections, as in TV "ghosts"). Frequency-selective fading typically affects a relatively narrow band of frequencies. The spread carrier offers a form of frequency diversity that can offset the faded frequencies by those within the spread-spectrum signals that do not experience the same fading. The effects of reflective or multipath interference, which result from several parts of the same signal arriving at the spread-spectrum receiver at slightly different times, can be largely reduced. Signals that arrive late at the spread-spectrum receiver will not match the spreading code currently being used to decode the signal. Hence they are rejected as interference.

Finally, spread spectrum can be used to construct very precise ranging and radar systems. The spread carrier, modulated with the PN sequence, permits the receiver to measure very precisely the time the signal was sent; thus, spread spectrum can be used to time the distance to a transmitter for ranging, or to an object as in the case of a radar reflection. Both applications have been commonly used in the aerospace field for many years.

Types of Spread Spectrum

There are numerous ways to cause a carrier to spread; however, all spread-spectrum systems can be viewed as two modulation processes. First, the information to be transmitted is applied. A conventional form of modulation, either analog or digital, is commonly used for this step. Second, the carrier is modulated by the spreading code, causing it to spread out over a large bandwidth. Four spreading techniques are commonly used in military and space communications, but amateurs are currently authorized to use only two of the four techniques.

Frequency Hopping

Frequency hopping (FH) is a form of spreading in which the center frequency of a conventional carrier is altered many times a second in accordance with a pseudo-random list of channels. See Fig. 15. The amount of time the signal is present on any single channel is called the dwell time. To avoid interference both to a conventional user and from conventional users, the dwell time must be very short, commonly less than 10 milliseconds.

Direct Sequence

Direct sequence (DS) is a second form of spreading in which a very fast binary bit stream is used to shift the phase of an RF carrier. This binary sequence is designed to appear to be random (that is, a mix of approximately equal numbers of zeroes and ones), but is generated by a digital circuit. This binary sequence can be duplicated and synchronized at the transmitter and receiver. Such sequences are called pseudo-noise or PN. See Fig. 16. Each PN code bit is called a *chip*. The phase shifting is commonly done in a balanced mixer that typically shifts the RF carrier between 0 and 180 degrees; this is called binary phase-shift keying (BPSK). Other types of phase-shift keying are also used. For example, quadrature phase-shift keying (QPSK) shifts between four different phases.

DS spread spectrum is typically used to transmit digital information. See Figs. 17 and 18. A common practice in DS systems is to mix the digital information stream with the PN code. The result of this mixing causes the PN code to be either inverted for a number of PN chips for an information bit of one or left unchanged for an information bit of zero. This modulation process is called bit-inversion modulation for obvious reasons. The resulting PN code is mixed with the RF carrier to produce the DS signal.

Chirp

The third spreading method employs a carrier that is swept over a range of frequencies. This method is called *chirp* spread spectrum and finds its primary application in ranging and radar systems.

Time Hopping

The last spreading method is called *time hopping*. In a time-hopped signal, the carrier is on-off keyed by the PN sequence, resulting in a very low duty cycle. The speed of keying determines the amount of signal spreading.

A hybrid system is formed by combining two or more forms of spread spectrum into a single system. Typically, a hybrid system combines the best points of two or more spread-spectrum systems. The performance

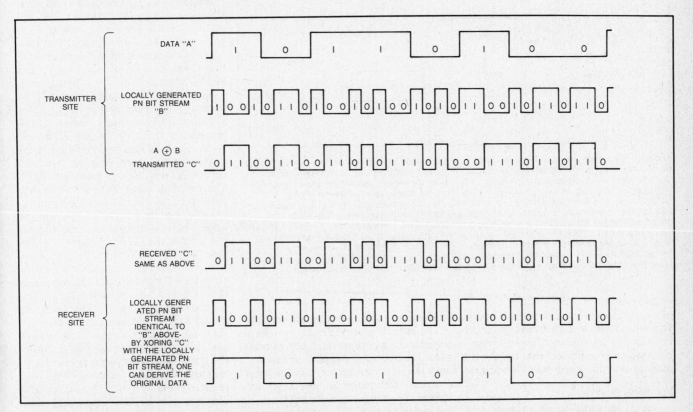

Fig. 17 — In bit-inversion modulation, a digital information stream is combined with a PN bit stream, which is clocked at four times the information rate. The combination is the exclusive-or sum of the two. Notice that an information bit of one inverts the PN bits in the combination, while an information bit of zero causes the PN bits to be transmitted without inversion. The combination bit stream has the speed characteristics of the original PN sequence, so it has a wider bandwidth than the information stream.

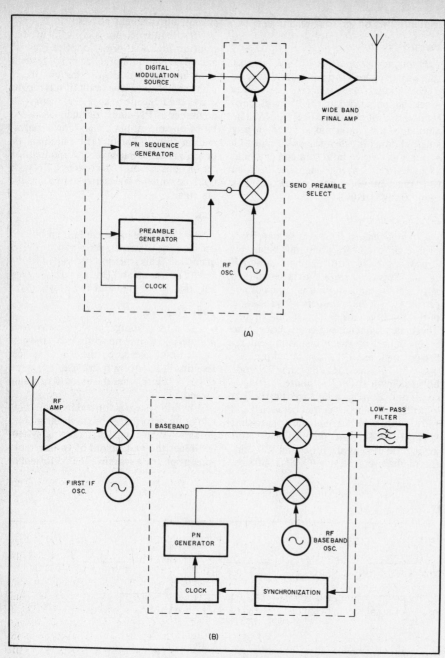

Fig. 18 — The block diagram of a direct sequence transmitter is shown at A. The digital modulation source is mixed with a combination of the PN sequence mixed with the carrier oscillator. The PN sequence is clocked at a much faster rate than the digital modulation. A very fast composite signal emerges as a result of the mixing. The preamble is selected at the start of transmission. Part B shows a direct sequence receiver. The wideband signal is translated down to a baseband (common) frequency. The processes form a correlator, mixing a baseband oscillator with the PN source and then mixing the result against the incoming baseband RF. The synchronization process keeps the PN sequence in step by varying the clock for optimal lock. After mixing, the information is contained as a digital output signal and all interference is spread to noise. The low-pass filter removes some of this noise. Notice that the transmitter and receiver employ very similar designs, one to perform spreading, the other to despread.

of a hybrid system is usually better than can be obtained with a single spread-spectrum technique for the same cost. The most common hybrids combine both frequency-hopping and direct-sequence techniques.

Spreading Sequences

One of the most important aspects of spread spectrum is the PN sequence used to spread the RF carrier. The spreading sequence determines how well the various

properties of spread spectrum will perform.

The spreading sequence takes a form geared to the type of spread spectrum being used. For frequency hopping, the spreading code is a stream of numbers that represent the channels to which the frequency hopper will travel. In a DS system, the spreading code is a very fast bit stream generated by a digital circuit. In both cases, the PN code is generated to resemble random activity and passes many of the

tests devised to identify random sequences. Codes that have random-like properties are called pseudorandom or pseudonoise sequences.

Correlation

Correlation is a fundamental process in a spread-spectrum system and forms a common method of receiving signals. Correlation measures how alike two signals are; that is, how similar in appearance they are to each other. The degree of likeness is often expressed as a number between zero and one. A perfect match is typically indicated by a one, while no match may be indicated by a zero. Partial matches yield values between one and zero, depending upon likeness.

In a spread-spectrum receiver, correlation is often used to identify a signal that has been coded with a desired PN sequence. Correlation is usually done with a circuit known as a correlator. A correlator is typically composed of a mixer followed by a low-pass filter that performs averaging. The mixer is where the two signals to be compared are multiplied together. A match yields a high value of output; but if the two mixed signals differ, the output will be lower depending on how different the signals are. The averaging circuit reports the average output of the mixer. This value is therefore the average likeness of the two signals.

In a DS system, the correlator is used to identify and detect signals with the desired spreading code. Signals spread with other PN codes, or signals not spread at all, will differ statistically from the desired signal and give a lower output from the correlator. The desired signal will have a strong match with locally generated PN code and yield a larger output from the correlator.

Notice that the averaging circuit of the correlator gives the average mixer value over time. If noise or interference is present, some of the received signal will be corrupted. After mixing, the interfering signals are spread and resemble noise, while the desired signal is despread and narrowband. The averaging circuit of the correlator then performs a low-pass filter function, thereby reducing the noise while passing the desired narrowband information. This is the heart of the DS interference-rejection process.

Correlation action in an FH system is implemented differently, but the concept is the same. In a frequency-hopping system, the transmitter carrier frequency is being moved about many times a second according to the spreading sequence. The receiver uses the same spreading sequence to follow the transmitter, moving from channel to channel in exact step with the signal. If the receiver is out of step with the signal, it cannot recover the information being transmitted.

FH signals that are under the control of a different PN sequence will be received only randomly, by the chance that both the desired and undesired PN codes have a channel in common at that moment. Narrowband signals will be visited occasionally by the hopping signal and should not be a cause of interference.

Despreading and Detection

In DS systems, collapsing the spread-spectrum signal by removing the effects of the spreading sequence is called despreading. If the DS signal is viewed as a signal with two types of modulation impressed on it (one for spreading and the other containing information) then despreading is a demodulation step aimed at the spreading sequence. What remains after the removal of the spreading sequence is the digital information stream.

The despreading process is in fact performed by a correlator containing a mixer and an averaging circuit. In the correlator, like signals produce a high value while unlike signals produce lower values. Thus signals that are the same produce high outputs because the signals reinforce each other. Signals that are unlike cannot reinforce each other and therefore form lower-valued products.

Bit-inversion modulation is detected by this correlator action since an information bit of "1" caused the PN code to be inverted, while a "0" leaves the PN code unchanged. In the correlator, a "0" information bit generates an uninverted PN code stream that closely correlates with the local PN code. Comparatively, a "1" results in a complete decorrelation since the PN code is inverted for this information bit. Through this action the correlator recovers the transmitted information.

Undesired signals in the DS receiver passband are not correlated with the local PN code. Within the correlator, the undesired signals randomly fall in and out of match on a bit-by-bit basis. Here the mixer output resembles noise that is filtered by the correlator low-pass filter. In comparison, the mixer will produce a stream of ones for an uninverted PN sequence or a stream of zeros for the inverted PN sequence.

Thus, the mixing process despreads the desired spread-spectrum signal and causes the undesired signals to spread to noise. It is mainly through the mixing process that interference is rejected within a spread-spectrum receiver.

Spreading Sequence Generators

Generation of the PN sequence is commonly done with linear-feedback shift registers (LFSR). See Fig. 19. The shift register consists of a number of one-bit memory registers that shift their contents to the right with each clock pulse. New values are introduced at the left-most stage while output is commonly taken from the right-most stage.

The shift register can be used to generate a wide variety of sequences. Generation of a linear sequence is typically done by tapping the values of certain cells and com-

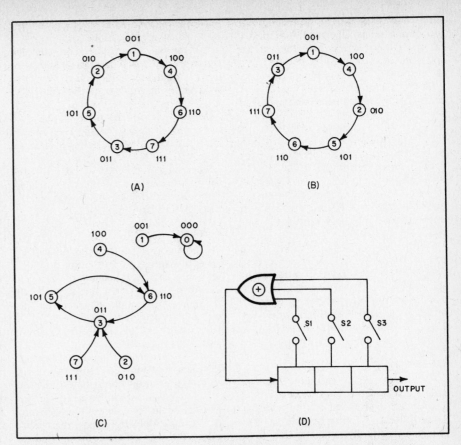

Fig. 19 — Various PN sequences. At A, a three-stage shift register with (1,3) stages tapped. There are seven unique code words, each as wide as the shift register. At B, a three-stage shift register with (2,3) stages tapped produces a somewhat different arrangement of code words. At C, two nonmaximal-length cycles result from the tap choices of (1,2) for the three-stage register. A and B are maximal-length cycles. At D is a three-stage linear feedback shift register. The stages to be combined with exclusive-OR to produce feedback are selected by S1-S3.

bining the values using an exclusive-OR (one-bit addition without carry) gate. The combined output is then fed back to the input of the left-most cell.

The LFSR sequence is a repeating sequence whose maximun length is related to the number of stages (N) in the shift register by the equation: length = $2^N - 1$. This equation determines the maximum length; not all tap choices produce the longest sequence, however. There are two cases to consider: the maximal-length sequence or m-sequence and the nonmaximal-length or composite sequences.

The length of an m-sequence is given by the equation $2^N - 1$ and is the longest possible for a given register length. Since the output repeats after producing $2^N - 1$ bits, the entire sequence can be viewed as a cycle (see Fig. 19).

Initializing the register with any sequence of consecutive bits from the cycle will generate the entire cycle from that point. A cycle is commonly started by an initial value of all zeros followed by a one such as 0000001 for a seven-stage register. Other starting values still generate the same sequence but are said to be phase shifted from the zeros followed by a one initial starting point. A run of bits selected from the sequence that is the same length as the generating shift register is called a code

word of the sequence. All code words from an m-sequence are unique with no repeats, and no code word contains all zeros.

Comparatively, the length of the nonmaximal sequence is always shorter than $2^N - 1$, and its outputs form a number of distinct short cycles. Code words are not unique for this type of sequence. The nonmaximal sequences are called a composite sequence because a composite sequence can be expressed as a combination of shorter maximal-length sequences. The nonmaximal-sequence length is in fact the product of the length of the smaller m-sequences of which it is composed.

Determining whether a given tap sequence will produce a maximal or composite sequence is quite complicated. Most efforts have been carried out by computer search methods, and a number of excellent tables are available for the designer.

A maximal-length sequence seven-stage shift register can easily be built with two ICs, as shown in Fig. 20. A more complete description of this shift register appeared in October 1986 QEX.

Orthogonality

An important consideration in choosing a sequence for a spread-spectrum system is the amount of statistical similarity a sequence has with conventional signals and with sequences employed by other spread-spectrum systems. The greater the degree of similarity, the less able the spread-spectrum receiver will be able to reject interference.

The ideal sequence will show a very low correlation when compared with undesired sequences. Moreover, the PN sequence should also have a low correlation with shifted versions of itself. A measure of a sequence's self-correlation is measured by its autocorrelation function (ACF). The ACF measures how much interference will be received from other spread-spectrum units using the same PN sequence but with different starting points within the PN code cycle.

Two code families that have found common usage are the m-sequences and Gold codes. Gold codes are a family of spreading codes generated by exclusive-OR combining the output of two "preferred" m-sequence generators. Many other PN codes can be used with spread spectrum — Dixon lists a number of such codes.

Synchronization

Synchronization is the most difficult issue for spread-spectrum systems. For a spread-spectrum receiver to demodulate the desired signal, it must be able to synchronize the locally generated PN code reference with the one used by the transmitter. This operation commonly takes place at very high speeds. It is usually viewed as two processes: rough synchronization which searches and acquires the receiver's PN code within one bit or channel of the transmitter's, and fine synchronization which maintains bit or channel dwell timing.

In rough synchronization, the receiver attempts to line up the local PN code as close as possible to the transmitters. There are two methods. The first is called epoch synchronization, in which the transmitter periodically sends a special synchronization sequence. Commonly, this synchronization sequence is a unique short bit sequence chosen because it is easily detected by a simple correlator known as a digital matched filter. This filter consists of a shift register that clocks the received sequence by one bit at a time and compares it against the unique sequence. The output of the matched filter is highest when the unique word is found in the input bit stream.

A second method of synchronization is known as phase synchronization. Here, the receiver attempts to determine which of the $2^N - 1$ phases the m-sequence could be in. To determine the phase, and thereby the synchronization, at least N bits would have to be received without error. Noisy signals may require more than N bits depending on the signal-to-noise ratio.

PN codes with short cycle times will exhibit several repetitions in a short amount of time. A digital matched filter can be constructed to signal the presence of a unique code word from this short spread-

ing code sequence. This is essentially a form of epoch synchronization.

Longer sequences require more-complex synchronization methods. One method uses a sequence based on the time of day. The receiver sets up a digital matched filter with a value several seconds ahead of the current time and waits for this unique code word to appear. The procedure may be useful for net operations in which stations can enter or leave a net at will.

There are a number of specially developed sequences available for rapid synchronization; however, they fall beyond the scope of this brief discussion.

Preamble

A preamble signal is commonly sent by the spread-spectrum transmitter immediately before the transmitter enters the spread mode. The preamble signal alerts the receiver to set up its synchronizing procedure to acquire the spread signal. The format of the preamble is different from system to system.

In slow frequency hopping, which has hop rates of less than 100 times a second, a tone appearing on a prearranged frequency or home channel can be used. The falling edge of the tone signals the beginning of hopping. At faster speeds, the precise instant the tone falls may be difficult to measure accurately; other preamble methods are typically employed. One such method calls for the frequency hopping receiver to examine a specific set of channels continually. A frequency-hopping transmitter would hop on these channels a prearranged number of times to allow the receiver to synchronize.

In direct-sequence transmissions the preamble performs three functions. First, the RF carrier must be acquired by the receiver. This may be done by transmitting a run of all zero bits, which reveals the carrier's center frequency to the receiver. Second, local clock synchronization is established by transmitting a sequence of alternating ones and zeros. The receiver will detect these transitions and derive bit timing from them. Last, the spreading code itself must be synchronized. This can be

done by an epoch synchronization procedure.

Once rough synchronization is established, the spread-spectrum receiver must track and maintain bit timing or channel dwell timing. There are several techniques to accomplish fine synchronization; however, the underlying mechanism is a feedback loop that attempts to minimize the timing error between the transmitter and receiver PN code at the fraction of a bit time.

The feedback loop itself commonly consists of a differencing circuit that calculates the actual bit difference and a feedback path that adjusts the receiver spreading code clock. In one technique called dithering, the spreading-code clock speed is continually rocked back and forth around the synchronization point, causing the receiver to periodically move in and out of synchronization a small amount. The point at which best synchronization is achieved is used to synchronize the receiver clock. Dithering can track a spreading code whose timing may be changing, perhaps because the transmitter is mobile, or in the case of HF spread spectrum, with changes in the height of the reflecting layers of the ionosphere.

Spectrum of Spread Spectrum

The spectrum of each type of spread-spectrum signal depends on several factors, such as the speed at which the spreading code is clocked, the type of spreading code used, whether frequency hopping or direct sequence is being used, the modulation bandwidth and the method of modulation.

Fig. 21 is the spectrum of a BPSK DS sequence. The signal is symmetric around the center frequency and contains several peaks that are called lobes. The main lobe is maximum at the center frequency but falls rapidly. The point at which the main lobe falls to its low point is called the first zero; subsequent lobes are called spectral sidelobes. The main lobe of a DS signal contains the majority of power, about 90 percent, while the remaining 10 percent is distributed over the side lobes. In many systems, the side lobes are clipped since

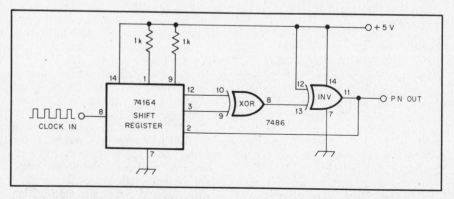

Fig. 20 — A maximal-length sequence seven-stage shift register. A complete description of this circuit appeared in October 1986 *QEX*.

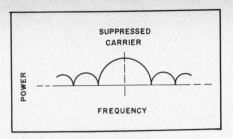

Fig. 21 — Power vs. frequency for a direct-sequence-modulated spread-spectrum signal. The envelope assumes the shape of a sin x^2 divide by x curve. With proper modulating techniques, the carrier is suppressed.

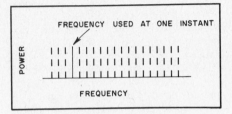

Fig. 22 — Power vs. frequency for frequency-hopping spread-spectrum signals. Emissions jump around in pseudorandom fashion to discrete frequencies.

they tend to extend out over a large span of spectrum but carry little of the DS signal's power.

The chip rate determines the overall spread of the DS signal. The overall size of the DS carrier expands as the chip rate increases, while lower chip rates collapse the signal into a nonspread conventional PSK signal. In addition, the chip rate determines the size of the DS main lobe and the location of the zeros. The main lobe is 2/chip-rate wide, centered about the DS carrier's center frequency. Zeros occur at multiples of 1/chip-rate, symmetrically on either side of the center frequency.

The spectrum of an FH signal is shown in Fig. 22. The spectrum structure is not as complex as in the DS signal, but depends on the spreading bandwidth and type of modulation used. The spectrum of FH signals consists of a carrier that moves pseudorandomly among many channels. As the speed of hopping increases, channels are visited more frequently with less time spent on a channel. At very fast hopping speeds, significant sidebands can be observed on the FH signal. These sidebands are generated from the pulse-like behavior that the FH signal exhibits at high speeds.

The amount of power the frequency-hopping transmitter delivers per channel is related to how often the channel is visited and the dwell time. The amount of power per channel is greater with fewer channels; conversely, a large number of channels decreases the frequency of visits, hence a lower power per channel. In sum, the larger the number of channels, the lower the signal power per channel.

Last, the method of modulation has a

marked effect on the spectrum of an FH signal. For example NBFM produces a constant-amplitude signal and generates a relatively flat spectrum. SSB, however, emits energy only when there is information to transmit, and hence less energy is emitted overall.

Near vs. Far-Field Strength

Spread-spectrum signals exhibit some unusual but logical signal effects. Close to the transmitter, a spread-spectrum signal may be observed readily on a conventional receiver; however, at a distance (50-100 miles) the signal may be noticed only with careful measurement. This results in a property called *low probability of intercept* (LPI). Some authors dispute the notion of LPI, saying that spread-spectrum signal energy is in fact intercepted by the receiving station's antennas and receivers; however, there is a low probability that the listener will recognize the presence of spread spectrum. The term *low probability of recognition* (LPR) has been suggested in place of LPI.

This radical difference in signal strength between close-in and distant observers is quite logical. The power of a narrow-band signal is typically concentrated about a center frequency; hence, a conventional narrow-band receiver will be in a position to collect much of the original power subject to path loss. In the DS spread-spectrum case, the same power is distributed over a band of frequencies, giving less power per hertz. A narrow-band receiver can collect only as much of this distributed power as the width of its IF passband; consequently, it registers a much weaker signal. A spread-spectrum receiver's passband is quite wide, allowing it to receive the entire bandwidth containing the spread-spectrum signal. This power is concentrated by the despreading process, making the output signal-to-noise ratio at least equal to the narrow-band signal's.

In the case of FH spread spectrum, a full-power narrow-band carrier is hopped among many channels. The concentrated power of the carrier can be observed at a distance equal to that of a nonhopping narrow-band signal; however, a conventional receiver will only catch a glimpse of the FH signal as it briefly visits the currently received channel.

FH Frequency Generation

Generation of FH frequencies is commonly done with a synthesizer that operates in one of several ways. The most popular amateur synthesizer technique uses a phase-locked loop (PLL) in which a voltage-controlled oscillator (VCO) generates the frequency of interest. The VCO is typically locked to a multiple of a reference crystal that also determines the channel spacing. The loop is established by sampling the VCO output frequency and dividing it down so its divided frequency matches the reference-oscillator frequency.

The error voltage generated by the phase comparetor holds the VCO on the desired multiple of the reference oscillator.

This approach is limited because of the time required for the loop to settle. Various modifications are possible such as replacing the loop's low-pass filter with an active low-pass filter; however, the settling-time problem is difficult to overcome. One successful method has been to use two or more synthesizers, alternately allowing one to settle while using the other to control the current channel. The increase in hopping speed can be found by calculating one minus the number of synthesizers multiplied by the ratio of dwell to settling time.

Another enhancement to the basic PLL synthesizer comes by inserting a D/A converter into the PLL. The D/A takes a digital word from a computer which represents a voltage feed to the VCO. The voltage has been selected so the VCO will produce a desired frequency. The PLL and reference oscillator are used to lock the VCO output to a precise channel; however, the large settling time, attributed mainly to the loop tracking large jumps in frequency, is absent.

A different approach to FH generation can be found in the direct-synthesis approach. Here, a large number of samples are taken of a sine wave that is stored in a ROM. A signal is generated by feeding these samples into a wide-band D/A converter, producing an RF sine wave. The frequency of the RF sine wave is determined by the rate at which the samples are fed to the D/A. A synthesizer of this type is capable of hopping many thousands of times per second and commonly has a resolution of 1-hertz steps.

Self Policing and Enforcement

Spread spectrum belongs to the general class of wideband signals that are new to many Amateur Radio operators. Traditionally, amateur experience has been with narrow-band signals that can be easily received with conventional receiving equipment. Wideband signals require larger IF bandwidths and some new ways of thinking.

Spread spectrum employs a carrier with a very wide bandwidth. In addition, the carrier is coded with a PN sequence. These two factors make conventional amateur receivers unsuitable for spread-spectrum reception. There are, however, many techniques available that recognize, locate and recover the transmitted information for amateur spread-spectrum users.

Amateur operators are interested in experiments that enhance communications; hence, system parameters are chosen to facilitate communications. Amateur systems are designed so they can be received without too much difficulty. Conversely, military users are interested in signal hiding — to prevent reception by unauthorized users — so military system parameters are set with this objective in mind.

When a spread-spectrum signal is above the noise level, it can be received with a wideband receiver, and direction-finding techniques can be used to locate the signal source. Interference to other amateur operations or TVI complaints are resolved easily because a signal strong enough to cause interference is far enough above the noise that it can be located with conventional DF techniques. In particular, a directional antenna can be connected to the receiver that is being interfered with. Rotation of the antenna can be used to help locate the source. Moreover, the periodic identification required of all amateur signals can aid the interference-determination process.

Currently, the FCC has authorized two types of spread-spectrum communications for amateurs. The user must identify with narrow-band transmissions on one frequency in the band being used. Alternatively, the station ID may be transmitted while in spread-spectrum operation if the transmission is changed so that CW, SSB or narrow-band FM receivers can be used to identify the station. One proposed identification procedure is for the stations involved to send call signs by on-off keying of the spread-spectrum carrier. This will provide a method to readily identify an amateur spread-spectrum user.

Sometimes it is necessary to monitor the contents of an amateur spread-spectrum transmission to determine if the amateur station is being used properly. This is possible when the signal is above the noise; a wideband receiver and appropriate detector to demodulate the signal are necessary. Other signals in the passband can cause interference, so a sharp beam antenna is required to reduce the effects of the undesired signal.

In DS spread-spectrum transmissions, both the spreading PN code and the transmitted information are digital in nature and are often mixed before being used to modulate the RF carrier. The signal must be despread before the digital information can be recovered; however, the PN code must be known in advance. In the spread-spectrum rules, the FCC defines three m-sequence PN codes for amateur use. These codes are generated by three shift-register configurations: a seven-stage register tapped at stages (7,1); a 13-stage register tapped at stages (13,4,3,1); and a 19-stage register tapped at stages (19,5,2,1). These three sequences are the only ones that amateurs are allowed to use. It is not difficult to test all three sequences against a received amateur DS signal to determine which is in use. Once the sequence is identified and synchronized, the transmitted digital information can be recovered.

Amateur SS Experimentation

Spread-spectrum development for jamproof military communications began in the late 1940s. John P. Costas, W2CRR, was the first person to recognize non-

military applications for SS. Costas presented a paper entitled "Poisson, Shannon, and the Radio Amateur," in the *Proceedings of the IRE* for December 1959. (Poisson and Shannon developed mathematical models for communications systems — all analytical studies in communication theory and information theory are based on their results.)

In this paper, Costas explained that congested band operation presents an interesting problem in analysis that can be solved by statistical methods. He showed that in spite of the bandwidth economy of SSB, there are definite advantages to using very broadband techniques. Costas concluded that broadband techniques would result in more efficient use of the spectrum and an increase in the number of available channels. At the time, Costas' statements were revolutionary, for they challenged the conventional theory that congestion in the radio spectrum can be relieved only by the use of smaller transmission bandwidths.

In 1980, Dr. Michael Marcus of the FCC Office of Science and Technology (OST) suggested that radio amateurs experiment with spread-spectrum modulation techniques. The rationale was that (a) the civil radio services could take advantage of the spread-spectrum pioneering of the military, (b) design of spread-spectrum systems by the private sector was slow because of the high cost of development vs. return on investment, (c) more experimentation was needed in areas such as designing for low-cost and on-the-air testing in congested frequency bands, and (d) radio amateurs could perform useful experiments without the need for either governmental or industrial research and development money.

The Amateur Radio Research and Development Corporation (AMRAD) requested, and the FCC granted, a Special Temporary Authority (STA) to permit spread-spectrum tests in the amateur bands by a small number of amateurs, for one year beginning in March 1981. Under the STA, the first Amateur Radio SS tests were conducted by W4RI in McLean, Virginia and K2SZE in Rochester, New York. Later, WA3ZXW in Annapolis, Maryland ran additional on-the-air tests with K2SZE. The equipment used was capable of hopping over a frequency range up to 100 kHz at rates of 1, 2, 5 and 10 hops per second. RF power output levels of 100 and 500 watts were used into dipole antennas.

These particular radios functioned best at 5 hops per second. This was subjectively judged on the basis of least-bothersome interference from the various signals at the different hopping frequencies. It was observed that frequency hopping was more successful in the presence of heavy CW interference than it was in the presence of heavy SSB interference. In comparison, conventional SSB usually provided better communications than frequency-hopped SSB whenever a single clear channel could be found for the conventional SSB.

However, the conventional SSB could be disrupted by strong interference on that channel. While hampered by cyclic interference when busy frequencies were revisited, the frequency-hopped link could be maintained despite band congestion.

Although the tests were announced beforehand in Amateur Radio publications and on the air from W1AW, no correspondence was received indicating that the frequency-hopping tests either interfered with, or were heard by, other Amateur Radio stations. The only exception was that several amateurs in the Northern Virginia area could recognize the presence of the frequency-hopped transmissions on conventional SSB receivers after learning what the signal sounded like. All were within five miles of W4RI and were able to hear both ends.

AMRAD member Chuck Phillips, N4EZV, built a VHF frequency-hopping radio in 1980. The design consisted of a modified VHF-Engineering scanner board capable of switching between four different crystal oscillators. By tripling the clocking speed, the transmitter was made to hop at 12 hops per second. The oscillators were arranged to operate continuously, and were switched in and out by using high-isolation solid-state switches. Sometime in 1981, the Motorola synthesizer chip showed up in a GLB add-on synthesizer board. N4EZV modified a couple of those boards and managed to reach 15 hops per second. Later that year, N4EZV tried a new synthesizer, working at lower frequencies, by modifying some Heathkit HW2031s. The synthesizer signal was then heterodyned to the desired output frequency. Both the transmitter and receiver were equipped with identical frequency "look-up tables" stored in EEPROMs. When the push-to-talk microphone switch was depressed, a short audio tone burst was transmitted and then decoded at the receiver. At the end of the burst, both transmitter and receiver started hopping to the frequencies pre-loaded in the look-up tables. The hopping rate was derived by dividing down from the reference crystal used to pilot the whole setup. When the microphone push-to-talk button was released at the end of a transmission, another tone was sent, which stopped the hoppers. If the receiver detected only noise, (caused by interference), it would automatically revert to its "home frequency" where it would wait for another transmission. This system seems to operate well up to about 100 hops per second over a 4-megahertz frequency range.

Experiments run by AMRAD in 1985 used a Commodore computer to control the frequency of an ICOM IC-2AT. Very little interference to conventional 2-meter users was created by the spread spectrum transmissions. The signals were inaudible to amateurs involved in regular QSOs, but the transmissions did cause some interference by keying some repeaters that did not have a carrier-sense activation delay.

This interference affected only a few repeaters out of many in the Northern Virginia area, and the problem was cured by removing the appropriate repeater input frequencies from the table of frequencies used in the frequency-hopping scheme.

In 1987, AMRAD member Elton Sanders, WB5MMB, used the direct-synthesis oscillator described by Fred Williams in the February 1985 issue of QST to build a frequency hopper. The frequency of the direct-synthesis oscillator was determined by a program written by David Borden, K8MMO, to run on an IBM® PC-XT clone, and the oscillator output was limited to frequencies within the 6.15 to 6.25 MHz band. This output was then multiplied by 72, and fed to a 70-centimeter output amplifier. (Hopping rates ranging between 250 and 1000 hops per second were obtained by this method.)

Initial discussions with the FCC concerning the second STA revealed that the FCC Field Operations Bureau was interested in conducting monitoring and direction-finding exercises against amateur spread-spectrum transmissions. In February 1985, FCC personnel conducted a "fox hunt" to find an amateur spread-spectrum transmitter operating at an undisclosed site. Using the standard FCC spectrum analysis van and enforcement car, FCC personnel located the transmitter within 25 minutes, proving that they can DF amateur spread-spectrum transmissions if necessary.

On May 9, 1985, the FCC amended the rules to permit SS operation on the amateur bands above 420 MHz at power output levels no greater than 100-W PEP for domestic communications. These rules became effective in June 1986. The amateur community is working toward establishing a set of standards for amateur SS operation. The current rules authorize three PN sequences, and both FH and DS are allowed. Hybrid spread-spectrum techniques are not permitted. Bandwidths may be as large as the amateur band of operation. Also, SS users are specifically prohibited from causing interference to other users of the band.

Under these new rules, amateurs engaging in SS operation must maintain a log with the following information.

1) A technical description of the transmitted signal.

2) Signal parameters including the frequency or frequencies of operation, the chip rate, the code, the code rate, the spreading function, the transmission protocol(s) including the method of achieving synchronization, and the modulation type.

3) A general description of the type of information being conveyed; for example, voice, text, memory dump, facsimile, television and so on.

4) The method and, if applicable, the frequencies used for station identification.

5) The beginning date and ending date of use of each type of transmitted signal.

Throughout 1986 and 1987, AMRAD continued to devise and run experiments designed to bring this spread-spectrum technology within the reach of the average radio amateur, by utilizing only parts and technologies readily available to everyone. In June 1986, AMRAD member Andre Kesteloot, N4ICK, demonstrated a way to derive clock signals (suitable for frequency-hopping) by using broadcast TV signals. The complete experiment is described in October 1986 QEX, pp 4-7. With the equipment described, the trailing edge of an audio tone burst was used to derive original synchronization, and the two sets hopped at 15 hops per second for more than one hour without losing sync.

In September 1986, N4ICK demonstrated an experimental board allowing for the transmission of data with direct-sequence spread spectrum. Synchronization was obtained by using a sliding correlator, fully described in the December 1986 issue of QEX. In 1987, N4ICK developed a method for deriving precise clock signals from AM radio broadcast transmitters (see October 1987 QEX) which was used in an interesting experiment conducted in January 1988. This experiment (see May 1988 QEX) can be summarized as follows (see Fig. 23): N4ICK used two hand-held transceivers operating on 445 MHz. At both sites, PN generators were driven by clock pulses derived from the same AM radio transmitter. These two PN sequences were thus driven by synchronous clocks (it was still necessary to correlate them). At the receiver end, additional pulses were fed into the clock pulse stream, making that clock operate slightly faster than the transmitting clock. The receiver output was connected to a DTMF decoder. In the absence of a valid tone, the receiver clock would continue to operate faster than the transmitter clock. If a DTMF tone was sent at the transmitter for a certain duration of time, there would necessarily be a moment when the transmitting and the receiving PN sequences would be in phase. At that moment, the output of the receiver's doubly balanced mixer would be correlated, and a signal would reach the DTMF decoder. A valid output from the DTMF decoder would be used to slow the receiver clock-pulse generator down to its normal speed. Although the overall synchronization process was relatively slow

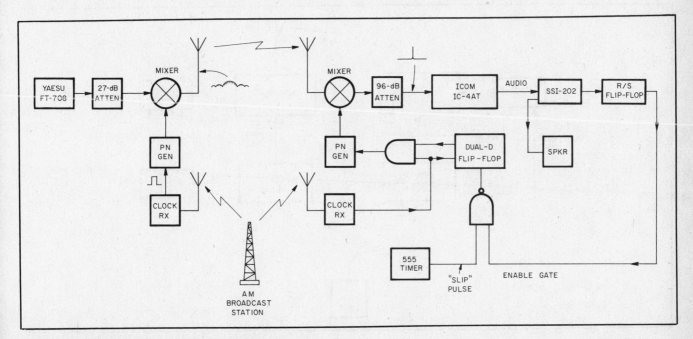

Fig. 23 — Block diagram of one of N4ICK's spread-spectrum experiments.

(it could take up to 45 seconds to reach sync) once synchronization was reached both stations would remain in lock for hours.

The above AMRAD experiments all relied on a clock derived from an external reference. Eventually circuitry will be devised so that it will be possible to derive a clock directly from the received signal. Present investigations at AMRAD are directed toward this goal, and although promising results have been obtained at 144 MHz at power levels of the order of the milliwatt, it remains a formidable challenge to translate these experiments into actual QSOs at 440 MHz. (Interested readers should write directly to AMRAD, PO Box 6148, McLean, VA 22106.)

Selected SS Bibliography

Reading material on spread spectrum may be difficult to obtain for the average amateur. Below are references that can be mail ordered. Spread-spectrum papers have also been published in *IEEE Transactions on Communications, on Aerospace and Electronic Systems* and *on Vehicular Technology.*

Dixon, *Spread Spectrum Systems*, second edition, 1984, Wiley Interscience, 605 Third Ave., New York, NY 10016.

Dixon, *Spread Spectrum Techniques*, IEEE Service Center, 445 Hoes La., Piscataway, NJ 08854.

Golomb, *Shift Register Sequences*, 1982, Aegean Park Press, Laguna Hills, CA.

Hershey, *Proposed Direct Sequence Spread Spectrum Voice Techniques for Amateur Radio Service*, 1982, U.S. Department of Commerce, NTIA Report 82-111.

Holmes, *Coherent Spread Spectrum Systems*, 1982, Wiley Interscience, New York.

The *AMRAD Newsletter* carries a monthly column on spread spectrum and reviews ongoing AMRAD experiments. In addition, the following articles on SS have appeared in Amateur Radio publications:

Feinstein, "Spread Spectrum — A report from AMRAD," *73*, November 1981.

Feinstein, "Amateur Spread Spectrum Experiments," *CQ*, July 1982.

Kesteloot, "Practical Spread Spectrum: A Simple Clock Synchronization Scheme," *QEX*, October 1986.

Kesteloot, "Experimenting with Direct Sequence Spread Spectrum," *QEX*, December 1986.

Kesteloot, "Extracting Stable Clock Signals from AM Broadcast Carriers for Amateur Spread Spectrum Applications," *QEX*, October 1987.

Kesteloot, "Practical Spread Spectrum: Achieving Synchronization with the Slip-Pulse Generator," *QEX*, May 1988.

Rhode, "Digital HF Radio: A Sampling of Techniques," *Ham Radio*, April 1985.

Rinaldo, "Spread Spectrum and the Radio Amateur," *QST*, November 1980.

Sabin, "Spread Spectrum Applications in Amateur Radio," *QST*, July 1983.

Williams, "A Digital Frequency Synthesizer," *QST*, April 1984.

Williams, "A Microprocessor Controller for the Digital Frequency Synthesizer," *QST*, February 1985.

COHERENT CW

While spectrum management has received much attention in the recent Amateur Radio literature, the problems and possibilities of "more QSOs per kilohertz" were first recognized more than half a century ago. The late Frederick Emmons Terman, 6FT, presented his vision of narrow-band communications in "Some Possibilities of Intelligence Transmission When Using a Limited Band of Frequencies," published in *Proceedings of the Institute of Radio Engineers*, January 1930.

As early as 1927, the Bell Telephone Company had reported successful experiments with 200-WPM Baudot TTY communications in a 50-Hz bandwidth over undersea cables. The bandwidth reduction resulted from synchronization of the transmitter and receiver.

Technology made giant leaps in the next 45 years. In September 1975 *QST*, Raymond Petit, W6GHM, described the experiments of some radio amateurs with a mode he called "coherent CW." Petit did not acknowledge Terman's paper, so we must conclude that he rediscovered the wheel.

In CCW, the receiver is designed to respond to characteristics that differentiate the transmitted signal from received noise. The transmitted signal has three distinguishing characteristics: precise frequency, precise pulse length and predetermined pulse sequencing (pulse phase).

Fig. 24 shows a block diagram of a CCW system. Notice that a precise frequency standard is used at both the receiver and the transmitter. These frequency standards, synchronized to WWVB (and by WWVB to each other) are the source of the "coherence" in CCW. Digital techniques are used to derive timing signals from the frequency standard; these timing signals set the precise frequency of the transmitter or receiver, determine the basic pulse length and establish the exact instant at which each pulse begins and ends. The transmitter uses these timing signals to generate a signal with pulse length and phase exactly synchronized to the master standard. The

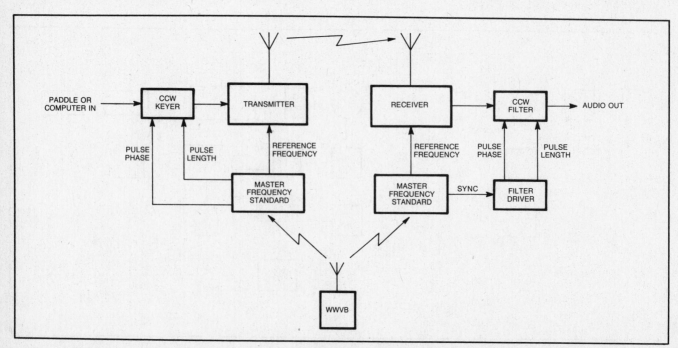

Fig. 24 — Block diagram of a coherent CW system. Precise timing signals from WWVB are used to synchronize the transmitter and receiver.

Fig. 25 — This is the first complete amateur station to be built for coherent CW operation. Assembled by Andy McCaskey, WA7ZVC, it consists of a modified Ten-Tec PM-2 transceiver and homemade modules that provide for the control and processing of signals as required for coherent CW operation.

receiver samples the received signal in pulses synchronized by its master oscillator. In this way, the receiver and transmitter are both synchronized to the same master frequency standard (WWVB) and their operation is "coherent."

Using CCW techniques allows a very narrow filter to be used at the receiver. The bandwidth required for transmitting a radiotelegraph signal is directly proportional to the keying rate. For a speed of 12 WPM the unit pulse length is 0.1 second. Since a dot and a space each require 0.1

second, a string of dots at 12 WPM is a square wave having a fundamental frequency of 5 Hz. To preserve the square-wave characteristic of the emission, an SSB transmission bandwidth of at least 15 Hz is required. Terman reported that with synchronization techniques, the receiver bandwidth could be reduced to 1.5 to 2.0 times the keying rate. In practical CCW systems, a combination of digital and analog techniques is used to produce a filter with a 3-dB bandwidth of 9 Hz, which is within the range predicted by Terman.

Noise bursts and strong adjacent-channel signals result in an occasional extra dot or an elongated dash, but are otherwise unnoticed. At the 12-WPM keying speed used by CCW experimenters, a signal-to-noise ratio improvement of about 20 dB can be realized over the bandwidths typically used for CW. Faster speeds are possible, but the bandwidth must be increased at the expense of signal-to-noise ratio.

To establish CCW contact, one station sends a preamble of dots to allow the receiving operator to synchronize his filter. Experience thus far indicates that once the filter has been synchronized, it usually won't need adjustment for several hours. Fig. 25 depicts a typical CCW station. The early experimenters built their stations around simple QRP equipment to dramatize the communications advantages

offered by the mode and to emphasize the accessibility of the necessary technology. The simple gear requires some add-on circuitry to allow oscillator stabilization.

The more modern synthesized transceivers can be outfitted for CCW more easily — replacing the internal reference oscillator with an external standard is all that's required. To send CCW, the paddle-actuated clock in the keyer must be replaced by a continuous pulse train from the frequency standard. Coordinating one's paddle movements with the "metronome" requires a different keying technique. A buffered keyboard (controlled by the standard) is the ideal CCW sending instrument.

When more stations have CCW capability, the mode may prove highly useful for emergency communications. Another possible use for CCW is in EME work.

Coherent CW Bibliography

Petit, "Coherent CW— Amateur Radio's Newest State of the Art?," *QST*, September 1975.

Weiss, "Coherent CW—The CW of the Future," Part 1, *CQ*, June 1977, Part 2, *CQ*, July 1977.

Woodson, "Coherent CW—The Concept, *QST*, May 1981.

Woodson, "Coherent CW—The Practical Aspects," *QST*, June 1981.

Chapter 22

Radio Frequencies and Propagation

Although great advances in understanding the many modes of radio wave propagation have been made in recent years, the number of variables affecting long distance communication are very complex and not entirely predictable. Amateur attempts to schedule operating times and frequencies for optimum results may not always succeed, but familiarity with the nature of radio propagation can improve your chances for success and add greatly to your enjoyment of the pursuit of distant communications. Indeed, personal computer technology now allows the average radio amateur to more accurately assess how day-to-day propagation changes will influence his operation and provides guidelines for frequency and operating time selections.

The sun, ultimate source of life and energy on Earth, influences all radio communication beyond ground-wave or line-of-sight ranges. Conditions vary with such obvious sun-related cycles as time of day and season of the year. Since these conditions differ for appreciable changes in latitude and longitude, and everything is constantly changing as the Earth rotates, almost every communications circuit has unique features with respect to the band of frequencies that are useful and the quality of signals in portions of that band.

Historically, propagation forecasting has been somewhat of a mystery, consisting of a combination of experience, "old wives' tales" and reference manuals. Today the radio amateur has new tools to perform rational frequency management, so frequency selection need not be a hit or miss proposition. In this chapter we will look at sky wave propagation and what influences it. We will also look at some ways that radio amateurs can reduce the mystery and even use the idiosyncrasies of propagation to their advantage.

As we study the various modes and conditions that affect the propagation of radio waves, we need to be sure that we understand the terms being used. Reflection and refraction are two words that often seem to be used interchangeably, even though they describe quite different phenomena.

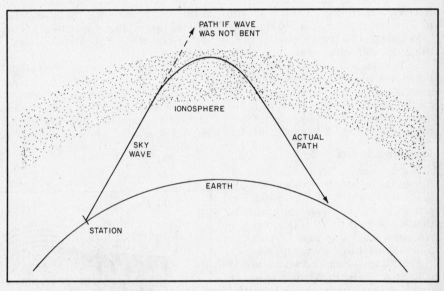

Fig. 1 — Because radio waves are bent in the ionosphere, they return to Earth far from their origin. Without refraction in the ionosphere, radio waves would pass into space.

Reflection occurs at any boundary between materials with different dielectric constants. Familiar examples with light are reflections from water surfaces, window glass and mirrors. Both water and glass are transparent for light, but their dielectric constants are very different from that of air. Some of the light energy that hits either material will bounce off, or be reflected. Radio waves, having much longer wavelengths than light, are practically unaffected by a thin layer of glass, but their behavior upon encountering water may vary, depending on the purity of the water. Distilled water is a good insulator; salt water is a relatively good conductor.

Depending on their wavelength (or frequency), radio waves may be reflected by buildings, trees, vehicles, the ground, water, ionized layers in the outer atmosphere, or at boundaries between air masses having different temperatures and moisture content. Some of the radio energy will be absorbed in the medium that the wave hits, and some of it may pass into (or through) the material. *Medium* is the term used to refer to a material that a wave is traveling through.

Refraction is the bending of a wave as it passes from one medium into another. (If the wave is moving in a direction that is at a right angle to the boundary between the two materials, the wave will not be bent.) This bending occurs because the wave travels at a different speed in the new material. If you place a straight object into water at an angle, the object will appear to be bent at the water's surface. (You can try this with a pencil and a glass of water.) This is an example of refraction of light waves.

Radio waves bend when they pass from one material into another in the same way that light waves do. The degree of bending depends on the difference in speeds at which the waves move through the two materials. The amount of bending also increases at higher frequencies. As radio waves travel through different areas of the atmosphere, there may be gradual changes in the speed of the waves as the temperature, air density and levels of ionization change. This will result in a gradual bending of the radio waves. Most amateur communication on the HF bands depends on this bending of radio waves, so it is a very important concept. Fig. 1 illustrates the refraction of a radio wave as it is returned to Earth.

THE IONOSPHERE

On frequencies below 30 MHz, long distance communication is the result of refraction (bending) of the wave in the ionosphere. The ionosphere is a region 30 to 260 miles above the Earth's surface where free ions and electrons exist in sufficient quantity to affect the direction of wave travel. Depending on the frequency used and the time of day, the ionosphere can support communications from very short ranges of less than 60 miles (called Near Vertical Incidence Signals — NVIS) to distances of greater than 6000 miles.

Ionization of the upper atmosphere is attributed to ultraviolet radiation from the sun. The result is not a single ionized region, but several layers of varying densities at various heights surrounding the Earth. Each layer has a central region of maximum electron density, which tapers off both above and below that altitude.

Ionospheric Layers

The ionospheric layers that most influence HF communications are the D, E, E_s, F_1 and F_2 layers. Of these, the D layer acts as a large RF sponge that absorbs signals passing through it. Depending on frequency and time of day, the remaining four are useful to the communicator.

The D layer, which is between 45 and 55 miles above the Earth, is most pronounced during daylight hours with its ionization being directly proportional to how high the sun is in the sky. Because absorption is inversely proportional to frequency, wave energy in the two lowest amateur bands (1.8 and 3.5 MHz) is almost completely absorbed by this layer during daylight hours. The rise and fall of the D layer dictates the lowest usable frequency (LUF) over a given path. While the effects of the D region are reasonably well known, the exact chemistry and physics of how this layer behaves are not well known or defined.

The lowest region of the ionosphere useful for returning radio signals to the Earth is the E layer. Its average height of maximum ionization is from 65 to 75 miles. The atmosphere here is still dense enough so that ions and electrons set free by solar radiation do not have to travel far before they meet and recombine to form neutral particles. For this reason the layer can maintain its ability to bend radio waves only in the presence of sunlight. Ionization is thus greatest around local noon, and it practically disappears after sundown.

A nomadic cousin of the E region is called sporadic-E. The sporadic-E region consists of relatively dense patches of ionization that literally drift around about 70 miles above the Earth. The effects of sporadic-E become confused with those of other types of ionization on the lower amateur frequencies, but they stand out above 21 MHz and into the VHF region. This is especially true during a solar minimum, when the bands above 21 MHz are rarely open. Sporadic-E propagation is covered in more detail in the VHF/UHF section of this chapter.

The region of ionization mainly responsible for long distance communication is the F-layer. Its altitudes of ionization range from about 90 to 250 miles. It ionizes very rapidly at sunrise, reaching peak electron density early in the afternoon at the middle of the propagation path. The ionization decays very slowly after sunset, reaching the minimum value just before sunrise. This period, called the pre-sunrise depression, is the time when the lowest MUF over the path is observed. During the day, the F region is split into two layers, the F_1 and the F_2.

The F_1 layer is not an important propagation medium. Its refracting heights are between 90 and 120 miles, and it forms and decays in direct correlation to the passage of the sun. After sunset, the F_1 layer decays and it is replaced by a broadened F_2 layer.

The F_2 region is the primary medium supporting amateur HF communications. The thickness of this layer ranges from a relatively thin layer (50 miles) during the day to a broad layer (95 miles) centered at about 200 miles above the Earth at night. The daytime layer is very dense and has very complex combination/recombination characteristics. Because of Earth/ionospheric geometry, the maximum range of a single hop off the F_2 region is about 2500 miles. Fig. 2 summarizes day to night variations of the layers and shows approximate heights for them.

Traditionally, the ionospheric layers are characterized as nice, well behaved, stratified layers. Recent technologies allow very high resolution measurements of the medium to be made, and indicate that the ionospheric layers are in continual horizontal and vertical motion. You should recognize that these variations are a routine situation that must be dealt with. Periods of very high solar activity, with their disruptions, may present the opportunity for unusual propagation periods. Use the winter solar minimum periods to exploit 160 meters and use the uncertainties of HF propagation to your advantage.

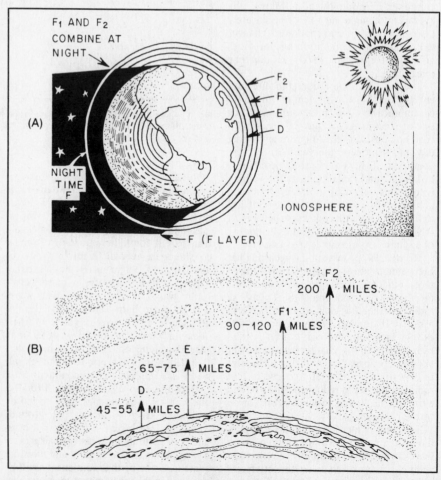

Fig. 2 — The ionosphere consists of several layers of ionized particles at different heights above the Earth. At night, the D and E layers disappear and the F_1 and F_2 layers combine to form a single F layer.

Virtual Height

An ionospheric layer is a region of considerable depth, but for practical purposes it is convenient to think of each layer as having a definite height. The height from which a simple reflection from the layer would give the same effects (observed from the ground) as the effects of the gradual bending that actually takes place is called the virtual height of that layer. Fig. 3 illustrates the virtual height of a layer in the ionosphere.

The virtual height of an ionospheric layer for various frequencies is determined with a variable-frequency sounding device (called an ionosonde) that directs energy vertically and measures the time required for the round-trip path. As frequency is increased, a point is reached where no energy will be returned from the ionosphere. The frequency where radio waves will penetrate a layer of the ionosphere instead of returning to Earth is called the critical frequency for that layer.

TYPES OF PROPAGATION

Depending on the propagation method, radio waves can be classified as ionospheric waves (sky waves), tropospheric waves or ground waves (surface waves). The sky wave is that main portion of the total radiation leaving the antenna at angles above the horizon. Without the refracting (bending) qualities of the ionosphere, these signals would be lost in space. The tropospheric wave is that portion of the radiation kept close to the Earth's surface as a result of bending in the lower atmosphere. This normally occurs only at the higher HF or lower VHF frequencies. The ground wave is that portion of the radiation directly affected by the surface of the Earth. It has two components — an Earth-guided surface wave and the space wave. Ground wave propagation is vertically polarized and its absorption increases with frequency. At equal power levels, a surface-wave signal over water will travel tens to hundreds of miles farther than the same signal over land.

Nearly all amateur communication on frequencies below 30 MHz is by means of sky waves. After leaving the transmitting antenna, this type of wave travels from the Earth's surface at an angle that would send it out into space if its path were not bent enough to bring it back to Earth. As the radio wave travels outward from the Earth, it encounters ionized particles in the ionosphere. The ionosphere refracts (or bends) radio waves, and at some frequencies the radio waves are refracted enough so they return to Earth at a point far from the originating station.

Because most amateur operation uses the sky wave mode of propagation, let's try to visualize just what is happening over a typical HF circuit. The *propagating bandwidth* is that band of frequencies which are propagating or usable between two points. This band of frequencies is

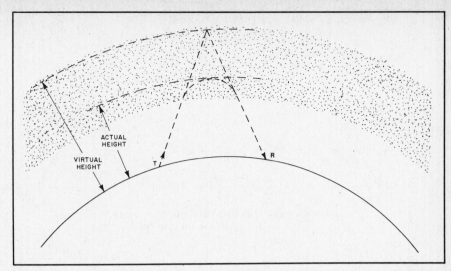

Fig. 3 — The virtual height of a layer in the ionosphere is the height at which a simple reflection would return the wave to the same point as the gradual bending that actually takes place.

constantly changing throughout the day, as shown in Fig. 4. This day/night plot (called a diurnal plot) shows how the band of propagating frequencies changes as a function of day and night. The upper boundary is called the maximum usable frequency (MUF).

The MUF is a function of how highly ionized the F region is. During a solar maximum, the MUF can rise above 30 MHz during daylight hours. Fig. 4 was developed from data taken during a solar minimum between Cycles 20 and 21.

The low-frequency boundary on Fig. 4 is called the lowest usable frequency (LUF). The LUF is a function of absorption, signal-to-noise ratio, power and transmission mode. Note that the change in LUF directly correlates to the movement of the sun over the path, and peaks at noon at the midpoint of the path. The vertical axis in Fig. 4 shows relative signal amplitude. The irregularity is due to the various multihop modes that can exist between the MUF and LUF.

It is important to remember that the definitions of LUF and MUF are for a specific communications path. For example, the frequencies useful for communications from your location to western Europe may be quite different from the frequencies useful for communications to Australia.

Multihop Propagation

When an ionospherically propagated wave returns to Earth, it can be reflected upward from the ground, travel again to the ionosphere, and be refracted to Earth. This process can be repeated several times under ideal conditions, leading to very long distance communications. Ordinarily, ionospheric absorption and ground-reflection losses exact tolls in signal level and quality, so multiple-hop propagation usually yields lower signal levels and more distorted modulation than single hop

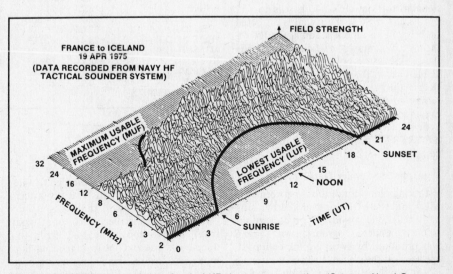

Fig. 4 — Day-Night characteristics of typical HF sky wave propagation. (Courtesy Naval Ocean Systems Center.)

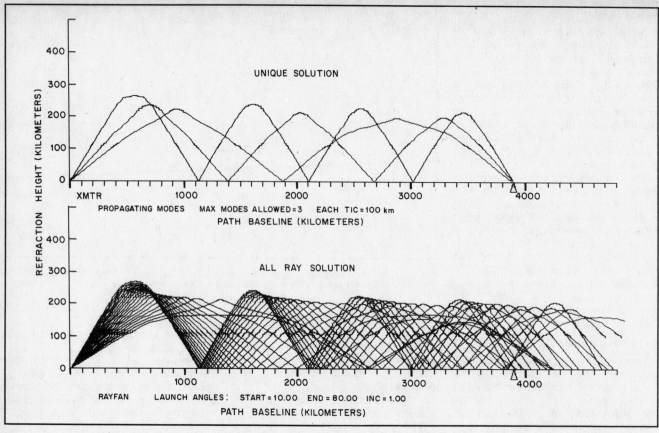

Fig. 5 — Ray trace of 14.0-MHz Signals over the Hawaii to California path at 1800 UTC on February 21, 1973. (Courtesy Naval Ocean Systems Center.)

signals. There are exceptions, however, and under ideal conditions even communications using long path signals are possible. These signals are discussed in the next section.

The same signal can be propagated over a given path via several different numbers of hops. The example in Fig. 5 shows that 2-hop, 3-hop, and 4-hop F_2 modes existed at the same time for a 14-MHz path. The receiver sees these three signals all arriving at very slightly different times, causing signal distortion. On CW, there is no real problem. On SSB, the signal is distorted slightly due to time spreading. On RTTY the distortion may be sufficient to degrade the signal to the point that it is useless. Table 1 shows the difference in these three signals, their angles of arrival, path loss and signal to noise ratio. This data was prepared using a sophisticated computer analysis known as Advanced PROPHET, a FORTRAN program run on mainframe computers.

Ionospheric Sounder Data

Propagation conditions between the MUF and the LUF are explored by using an ionospheric sounder. The oblique incidence sounder ionogram shown in Fig. 6 was produced by sweeping the entire HF spectrum at the transmitter and measuring all the propagating energy at the receiver. This shows that six modes of propagation

exist for this particular path, each covering a different part of the HF spectrum with some overlapping. For example, if 14.3 MHz was used on this path between California and Hawaii, the signal would arrive at three discreet times, one signal from the two-hop mode, one from the three-hop mode and one from the four-hop mode. At the receiver, the three modes arriving at slightly different times would cause distortion of the signal. It is im-

portant to note that the characteristics seen in Fig. 6 are unique to this HF sky wave circuit. Each and every different point to point circuit will have its own mode structure as a function of time of day. This means that every HF propagation problem has a totally unique solution.

Depending on how intensely the F region is ionized, an HF signal will be refracted in the ionosphere and reflected from the ground several times before it gets to the

Table 1

Multimode Parameters — Advanced PROPHET Ray Trace Synopsis

DATE: 2/17/73   TIME: 18:00 UT   ATMOSPHERIC NOISE: YES   BWIDTH: 3,000 KHz

FREQ: 14.0   SSN: 126.4   PK: 1.0   MAN-MADE NOISE: QM   SNR REQD: 12.0 db

XMTR: HONO   LAT: 21.4   LON: 158.2°   ANT: 101 @ *OMNI*   PWR: 100.00 W

RCVR: SFRAN   LAT: 38.1°   LON: 122.3°   ANT: 161 @ 261.9°   RANGE: 3893.6 KM

IONOSPHERE: FOE = 2.6 MHz   FOF1 = 3.6 MHz   FOF2 = 8.2
    HMF2 = 326 KM   YMF2 = 108.9 KM

NHOP	2	3	4	0	0	0
MODE	3300000	3330000	3333000	0000000	0000000	0000000
ANGLE	13.75	19.55	29.10	0.00	0.00	0.00
DELAY (MSEC)	13.930	14.302	15.498	0.000	0.000	0.000
LOSS (DB)	163.08	161.98	152.04	0.00	0.00	0.00
1HZ SNR (DB)	28.44	29.53	39.48	0.00	0.00	0.00
ADJ SNR (DB)	−6.33	−5.24	4.71	0.00	0.00	0.00
HT1 (KM)	312.74	298.03	356.96	0.00	0.00	0.00
HT2 (KM)	345.78	270.95	304.13	0.00	0.00	0.00
HT3 (KM)	0.00	254.85	282.81	0.00	0.00	0.00
(VIRTUAL HEIGHTS)						

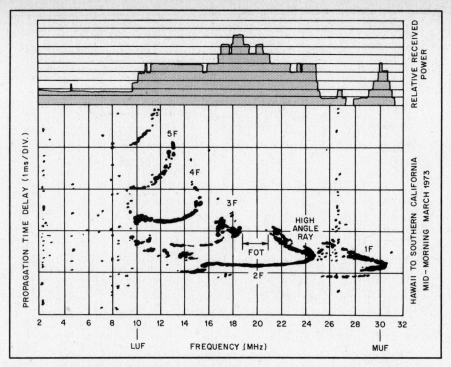

Fig. 6 — HF oblique sounder ionogram. This photo shows a typical chirpsounder measurement on a 2500-mile path during midmorning in March.

desired receiver. Fig. 5 illustrates how several propagation modes can be expected, depending on the vertical radiation pattern of the transmitting antenna. The height of the refracting medium, the distance to the desired receiver and the signal radiation angle from the transmitter create the geometry of a path. If a receiver is too close to the transmitter, it will be in the skip zone where the ionosphere is not ionized enough to return very steep angles of signal radiation. There may be places along the transmission path where the signal cannot be heard because of the *geometry* of the situation, even though the transmission frequency is between the MUF and LUF.

The higher the radiation angle, the deeper into the ionosphere the signal penetrates before it is bent back to Earth. In order to establish a successful communications link, a user must choose a frequency between the MUF and the LUF, the antenna radiation pattern must meet the path geometry requirements and the signal must exceed the required signal-to-noise ratio.

SOLAR PHENOMENA

Man's interest in the sun is older than recorded history. Records of sunspot observations translatable into modern terms go back nearly 300 years. The first complete sunspot cycle to be recorded began in 1755, and is referred to as Cycle 1. Sunspot cycles have been numbered consecutively since then, so we are at the beginning of Cycle 22 now.

Current observations are statistically *smoothed*, or averaged, to maintain a

continuous record in the form of the International Sunspot Number (ISN). (The smoothed number means that it is an average of the values for six months before and six months after the month in question — International Sunspot Numbers are always six months behind the present date. For example, the latest ISN on January 1, 1987 is for July 1, 1986.) This number is the basis for traditional HF propagation predictions. In this section we will describe solar radiation, how it influences propagation, and how radio amateurs can use the knowledge of its day to day variations to select operating times and frequencies.

Solar Cycles

The periodic rise and fall of sunspot

numbers had been studied for many years before their correlation with radio-propagation variations was well known. Sunspot cycles average 10.7 years in length (we usually refer to an 11-year cycle), but have been as short as 7.3 years and as long as 17.1 years. The highs and lows also vary greatly. Cycle 19 peaked in 1958 with a sunspot number of over 200. Cycle 20, of more average intensity, reached 120 in 1969, while Cycle 21 rose to 164.5 in December 1979. By contrast, one of the lowest cycles, Cycle 14, peaked at only 60 in 1907. Several cycle lows, called solar minimums, have not reached zero levels on the International scale for any appreciable period, while little or no activity occurs for several months during the minimum of some cycles. (The minimum period of 1985 is a good example of a time with almost no activity.)

What does this mean to the radio amateur? At solar maxima (high sunspot number), the radiation from all the active regions around the sunspots makes the ionosphere capable of returning higher-frequency radio signals to the Earth instead of allowing them to pass through. This means that frequencies up to 40 MHz or higher are usable for long-distance communication. During periods of minimum solar activity, the amount of radiation is reduced, and frequencies above 20 MHz become unreliable for long-distance communication on a day-to-day basis. For this reason, the radio amateur needs to be aware of the current solar cycle status.

Navy scientists have recently revised their predictions for the next five solar cycles, which extends through the year 2040. Fig. 7 shows a graph of sunspot numbers for Solar Cycles 17 through 21 and predictions for Cycles 22 through 26. Fig. 7 predicts that the next solar maximum will occur around 1991-1992 with a peak level similar to Cycle 20 (1969).

Note that the cycles do not have a sinewave shape. The rise time is shorter than the decay time, but neither is clearly defined. There are no absolutes at either solar maximum or solar minimum; periods

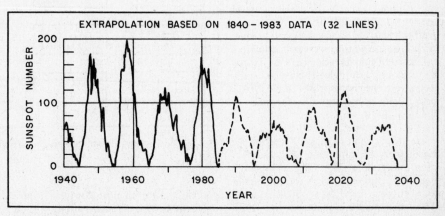

Fig. 7 — Solar sunspot number prediction for the next five sunspot cycles, from the Naval Ocean Systems Center.

of extremely intense activity have been observed from a single sunspot region during solar minimum. This can cause short term openings on the higher frequencies. Large numbers of very low radiation sunspots have also been observed during a solar maximum. This allows unexpected propagation conditions to exist at either solar minimum or maximum.

A useful indication of overall solar activity is the solar-flux index, a number associated with the measured radio noise coming from the sun at 2800 MHz. One such measurement is made at 1700 UTC daily in Ottawa, Canada. The latest solar-flux index is transmitted at 18 minutes past every hour by WWV. Because it is a current measurement for solar activity, it tends to be more useful in predicting propagation conditions than the more traditional sunspot number. This solar flux, more commonly called *10.7-cm solar flux* or *2800-MHz solar flux* can be directly related to sunspot number using the chart shown in Fig. 8.

Solar Radiation

There are two types of radiation from the sun that influence propagation; electromagnetic and particle emissions. The electromagnetic emissions include X-rays, ultraviolet (UV) and extreme ultraviolet (EUV). Each has an impact on a different part of the ionosphere. Particle emissions come in two forms: high-energy particles (high-energy protons and alpha particles) and low-energy particles (low-energy protons and electrons). The different types of radiation travel from the Earth to the sun at different speeds and have varying effects on sky-wave propagation. The time it takes for each type of radiation to travel from the sun to the Earth is shown in Fig. 9. The effects that result from each radiation type are also summarized in this figure.

EUV ionizes the ionospheric F region and is always present at some level. During solar maxima, the increased number of active regions on the sun provide for increased EUV, which in turn increases the F-region ionization and allows the use of higher frequencies for long-distance communication. UV and X-ray emissions ionize the D region, which absorbs HF energy. During solar flares, UV and X-ray emissions increase, causing increased signal loss on those HF circuits facing the sun. It takes about 8 minutes for electromagnetic emissions from the sun to travel to the Earth, because this energy travels at the speed of light.

High-energy particle radiation travels more slowly than light, and reaches the Earth from 15 minutes to several hours after a large solar flare. The high-energy particle emissions cause much higher absorption in the Earth's polar regions. They also create a radiation hazard to satellite systems and personnel orbiting the Earth in spacecraft. The lower-energy

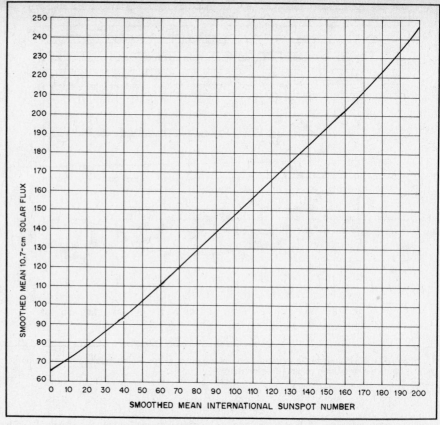

Fig. 8 — Relationship between the smoothed mean sunspot number and the 2800-MHz (10.7-cm) solar flux. In the low months of a solar cycle, flux values run between about 66 and 86. Intermediate years may see values of 85 to 150. The peak years of Cycle 21 brought readings between 140 and 380, in 1979 and 1980.

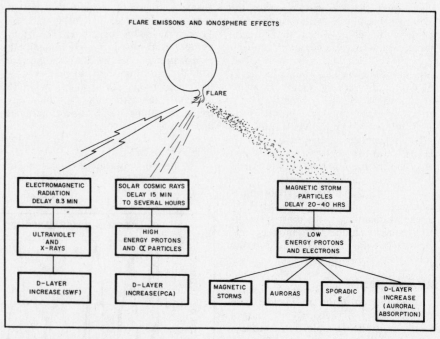

Fig. 9 — A summary of solar flare emissions and ionospheric effects.

particles travel even more slowly, and cause magnetic disturbances, auroras, sporadic E and increased polar-region absorption. These emissions arrive 20 to 40 hours after the solar flare has occurred.

It is difficult to compare solar activity to ionospheric variations because the chief ionizing agent for the F-region of the ionosphere is EUV, and this makes the traditional solar sunspot number (SSN) a poor short-term measure of solar effects on the ionosphere. EUV must be measured from satellite platforms, so relatively little is known about its variability. There is evidence, however, that EUV varies in much the same way as the 10.7-cm solar flux, which can be measured at ground level. For that reason, the solar-flux index is replacing the SSN as a short-term measure of solar activity.

Variations in the level of solar radiation can be gradual, as with the passage of sunspot groups and other long-lived activity centers across the solar surface, or the variations may be sudden and abrupt as with solar flares. Because the rotational period of the sun is approximately 28 days, a region of high activity that is on the disk (that region of the sun facing the Earth) will be in a position of affecting propagation for about 14 days and then hidden for about 14 days. A very active region can produce short-term radio blackouts for the full 14 days of its transit across the disk, but will produce particle events such as magnetic storms and radio aurora for only about seven days. Generally, until an active region crosses the central meridian of the sun, a strong flare cannot produce particle disturbances on the Earth because of the Earth/sun trajectory.

Absorption

When moving particles collide with each other, the one with the higher amount of energy will generally lose some of its energy to the other particle. As a radio wave travels through the ionosphere, it bumps into ionized particles and gives up some of its energy to those particles as it sets them in motion. The ionosphere absorbs energy from the radio wave. Such absorption is greater at lower frequencies, because it is proportional to the inverse square of frequency. It is this absorption that defines the LUF of a circuit. (If too much energy is absorbed, the signal will be too weak to be heard at the receiving station.)

Absorption is caused by X-rays and UV radiation, and it increases markedly during solar flares and auroral disturbances. A signal loses strength every time it passes through the D region. For this reason, a three-hop signal will have more path loss, (or less signal at the destination), than a one-hop signal. In order to minimize absorption, amateurs should use frequencies as near the MUF as possible.

EFFECTS OF THE EARTH'S MAGNETIC FIELD

The ionosphere has been discussed thus far in terms of simple bending, or refraction. An understanding of long-distance propagation must also take the Earth's magnetic field into account. Because of the Earth's magnetic field, the ionosphere is a double refracting medium that breaks up plane-polarized waves into what are known as the ordinary and extraordinary waves. This helps to explain the dispersal of plane polarization encountered in most ionospheric communication.

The Earth's magnetic field is disturbed by particle emissions from the sun. These emissions cause the magnetic field to vary, which causes loss of signals in the polar regions, increased aurora, and depleted F_2 ionization (called an ionospheric storm) at mid latitudes. Ionospheric storms tend to migrate southward from the polar region 36 to 72 hours after the actual solar event. This depletion of the ionosphere tends to affect most ham bands, resulting in generally poor operating conditions for the duration of the storm. An indicator of this happening is when the K index, also announced at 18 minutes after the hour on WWV, rises above 3. Remember that K_p is on a scale of 0 to 9, with 9 being the most disturbed condition.

RADIATION ANGLE AND SKIP DISTANCE

The lower the angle above the horizon at which a wave leaves the antenna, the less refraction in the ionosphere is required to bring it back to Earth, or to maintain useful signal level. This is the basis for the emphasis on low radiation angles in the pursuit of DX on the HF and VHF bands.

Some of the effects of radiation angle are illustrated in Fig. 10. The high-angle waves from the transmitter, T, are bent only slightly in the ionosphere, and so pass through it. The wave at a somewhat lower angle is just capable of being returned by the ionosphere.

This radiation angle from the antenna is called the critical angle. Radiation at angles more than the critical angle above the horizon will not be returned to Earth, while radiation at angles less than the critical angle will be returned farther from the transmitter. In daylight, if the frequency is low enough, the wave might be returned by the E layer, reaching R1. If the wave is returned from the F_2 layer, it will reach R2. The higher angle wave is returned closer to the transmitting point, T, than is the lowest-angle wave, which returns to R3.

The area between R1 and the outer reaches of the ground wave, near the transmitter, is called the skip zone. If R2 is at the shortest distance where energy returned from the F_2 layer is usable, the distance between R2 and T is the skip distance. The actual distances to both R1 and R2 depend on the ionization density, the radiation angle at T, and on the frequency in use. Table 2 has been included to provide a general indication of typical F-layer skip distances for the MF and HF amateur bands. Because these values can vary as much as 50% or more, they should be used only for comparison.

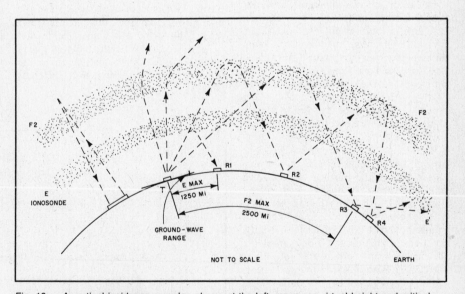

Fig. 10 — A vertical incidence sounder, shown at the left, measures virtual height and critical frequency of the F_2 layer. The transmitter, T, is shown radiating signals at four different angles. At the two highest radiation angles, the waves pass through the ionosphere after slight refraction. The lower-angle wave is returned to Earth by the E layer, if the frequency is low enough, at a maximum distance of 1250 miles. The signals are returned from the F_2 layer at a maximum distance of about 2500 miles, depending on the radiation angle. The signal may make a double hop, going from R2 to R4, which is beyond the maximum single-hop range. The lowest-angle wave reaches the maximum practical single-hop distance at R3.

LONG-PATH PROPAGATION

Propagation between any two points on the Earth's surface is usually by the shortest direct route, which is a great-circle path between the two points. A great circle is an imaginary line drawn around the Earth, formed by a plane passing through the center of the Earth. The diameter of a great circle is equal to the diameter of the Earth. You can find a great-circle path between two points by stretching a string tightly between those two points on a globe. If a rubber band is used to mark the entire great circle, you can see that there are really two great-circle paths. See Fig. 11.

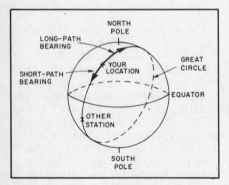

Fig. 11 — A sketch of the Earth, showing a great-circle path between two stations. Short-path and long-path bearings are shown from the northern-hemisphere station.

One of those paths will usually be longer than the other, and it may be useful for communications when conditions are favorable. Of course you must have a beam antenna that you can point in the desired direction to make effective use of long-path propagation. The station at the other end of the path must also point his or her antenna in the long-path direction to your station to make the best use of this propagation mode.

The long- and short-path directions always differ by 180°. Since the circumference of the Earth is about 24,000 miles, short-path propagation is always over a path length of less than 12,000 miles. The long-path distance is 24,000 miles minus the short-path distance for a specific communications circuit. For example, if the distance from Gordon, Pennsylvania to the Canary Islands is 3510 miles at a bearing of 85°, then the long-path circuit would be a distance of 20,490 miles at a bearing of 265°. (Chapter 16 of *The ARRL Antenna Book* contains information about distance and bearing calculations.)

Normally, radio signals propagate most effectively along the great-circle path that provides the least absorption. Since the D layer of the ionosphere is most responsible for signal absorption, the amount of sunlight that the signal must travel through

helps determine how strong the signals will be. So at times, your signal will be stronger at the receiving station when it travels over the nighttime side of the Earth.

For paths less than about 6000 miles, the short-path signal will almost always be stronger, because of the increased losses caused by multiple-hop ground-reflection losses and ionospheric absorption over the long path. When the short path is more than 6000 miles, however, long-path propagation will usually be observed either along the "gray line" between darkness and light, or over the nighttime side of the Earth.

While signals will generally travel best along the great-circle path, there is often some deviation from the exact predicted beam heading. Ionospheric conditions can cause radio signals to reflect or refract in unexpected ways. So it is always a good idea to rock your rotator control back and forth around the expected beam heading, and listen for the peak signal strength. If the strongest signals come from a direction that is significantly different from the great-circle headings, then we say the signals are following a crooked path. Anytime signals are not following a great-circle path, you have crooked-path propagation.

GRAY-LINE PROPAGATION

The *gray line* (sometimes called the *twilight zone*) is a band around the Earth that separates the daylight from darkness. Astronomers call this the terminator. It is a somewhat fuzzy region because the Earth's atmosphere tends to diffuse the light into the darkness. Fig. 12 illustrates the gray line around the Earth. Notice that on one side of the Earth, the gray line is coming into daylight (sunrise), and on the other side it is coming into darkness (sunset).

Propagation along the gray line is very efficient. One major reason for this is that the D layer, which absorbs HF signals, disappears rapidly on the sunset side of the gray line, and it has not yet built up on the sunrise side.

The gray line runs generally north and south, but varies as much as 23° either side of the north-south line. This variation is caused by the tilt of the Earth's axis relative to its orbital plane around the sun. The gray line will be exactly north and south at the equinoxes (March 21 and September 21). On the first day of summer in the northern hemisphere (June 21) it is tilted a maximum of 23° one way and on the first day of winter (December 21) it is tilted a maximum of 23° the other way. Fig. 13 illustrates the changing gray-line tilt. The tilt angle will be between these extremes during the rest of the year.

One way to describe the gray-line tilt is as an angle measured upward from the equator, looking east. It is important to note that if you measure an angle greater than 90° at sunrise, then you will measure an angle less than 90° at sunset on the same day. For example, at sunrise on April 16 the gray line makes an angle of approximately 99° with the equator, and at sunset it makes an angle of 81°. This means you can work into a different area of the world using gray-line propagation at sunset than you could at sunrise on the same day.

FADING

Changes in signal level, called fading, arise from a variety of phenomena, some natural and some man-made. Two or more parts of the wave may follow different paths, causing phase differences between wave components at the receiving end. Total field strength may be greater or less than the strength of one component. This

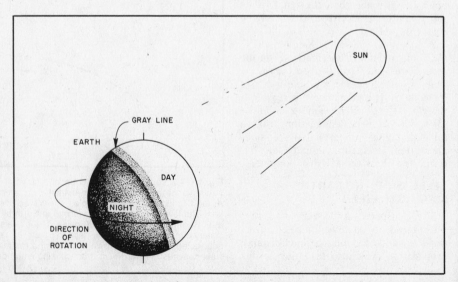

Fig. 12 — The gray line is a transition region between daylight and darkness. One side of the Earth is coming into sunrise, and the other side is just past sunset.

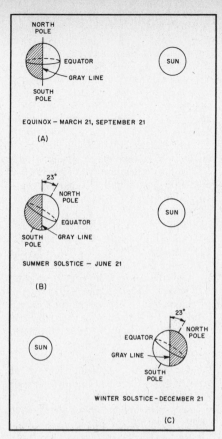

Fig. 13 — The angle with which the gray line crosses the equator depends on the time of year, and on whether it is at sunset or sunrise. A shows that the gray line is perpendicular to the equator twice a year, at the vernal equinox and the autumnal equinox. B indicates that the north pole is tilted toward the sun at an angle of 23° at the summer solstice, and C shows that it is tilted away from the sun at an angle of 23° at the winter solstice. These changes in the sun's position relative to the poles cause the gray line to shift position.

type of fading is called multipath interference. Fluctuating signal levels also result from the changing nature of the wave path, as in the case of moving air-mass boundaries in tropospheric propagation on the higher frequencies. Man-made sources of fading include reflections from aircraft and ionospheric discontinuities produced by exhaust from large rocket engines. Routinely, most fading comes from natural sources, like multipath interference.

Under some circumstances the wave path may vary with very small changes in frequency, so that modulation sidebands arrive at the receiver out of phase, causing mild or severe distortion. Called selective fading, this problem increases with signal bandwidth. Double-sideband AM signals suffer much more from selective fading than do single- sideband, suppressed-carrier signals.

THE SCATTER MODES

Much long-distance propagation can be described in terms of discrete reflection, though the analogy is never precise since true reflection would be possible only with perfect mirrors in a vacuum. All electromagnetic wave propagation is subject to scattering influences that alter idealized patterns to a great degree. The Earth's atmospheric and ionospheric layers, and objects in the wave path all act to scatter the signal. Strong returns are thought of as reflections and weaker ones as scattering, but both influence wave propagation. Scatter modes are useful in many kinds of communications.

Forward Scatter

We describe the skip zone as if there were no signal heard between the end of useful ground-wave range and the points R1 or R2 of Fig. 10. But actually, the transmitted signal can be detected over much of the skip zone, if sufficiently sensitive devices and methods are used. A small portion of the transmitted energy is scattered back to Earth in several ways, depending on the frequency.

Tropospheric scatter extends the local communications range to an increasing degree with frequencies above 20 MHz, becoming most useful in the VHF range. *Ionospheric scatter,* mostly used from the height of the E region, is most marked at frequencies up to about 60 or 70 MHz. *VHF tropospheric scatter* is usable within the limits of amateur power levels and antenna techniques, out to nearly 500 miles. *Ionospheric forward scatter* is discernible in the skip zone at distances up to 1200 miles or so.

A major component of ionospheric scatter is that contributed by short-lived columns of ionized particles formed around meteors entering the Earth's atmosphere. This can be anything from very short burst of little communications value to sustained periods of usable signal level, lasting up to a minute or more. Most common in the early morning hours, meteor scatter can be an interesting adjunct to amateur communication at 21 MHz and higher, especially in periods of low solar activity. It is at its best during major meteor showers.

Backscatter

A complex form of scatter is readily observed when working near the maximum usable frequency for the F layer over a particular path at a given time. The transmitted wave is refracted back to Earth at some distant point, which may be an ocean area or a land mass. A small portion of the transmitted signal may be reflected back into the ionosphere toward the transmitter when it reaches the Earth. The reflected wave helps fill in the skip zone, as shown in Fig. 14.

Backscatter signals are generally very weak, but there is no evidence of fading. Voice signals have a hollow, barrel-like sound. With optimum equipment, such signals are usable at distances from just beyond the reliable local range out to several hundred miles. Under ideal conditions, backscatter communications are possible over 3000 miles or more, though the term *sidescatter* is more descriptive of what happens on such long paths.

The scatter modes contribute to the usefulness of the higher parts of the DX spectrum, especially during periods of low solar activity when the normal ionospheric modes are often less available.

MF AND HF PROPAGATION

The amateur MF and HF bands (160 meters being the sole MF entry) are distinct in their propagation characteristics, but share one common bond: All of them utilize the F_2 layer for the primary mode of long-distance propagation. Here are capsule descriptions of propagation on those bands:

• The 1.8-MHz (160-meter) band suffers from extreme daytime D-layer absorption. Therefore, the top band is primarily useful at night. Only daytime signals entering the ionosphere at very high angles will be returned to Earth, which limits communications to 75 miles or so. At night, the D layer dissipates, and lower-angle signals are propagated. Under these conditions, communications over several thousands miles are possible.

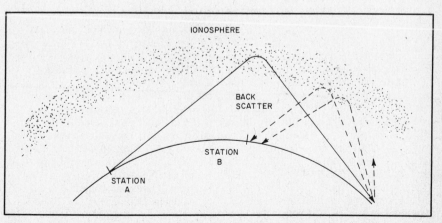

Fig. 14 — When radio waves pass into the ionosphere, some energy may be scattered back into the skip zone.

Another factor affecting propagation on this band is noise, both atmospheric and man-made. Since warm weather and noise producing thunderstorms often go hand-in-hand, winter is generally the best time to work DX on 160 meters.

• The 3.5-MHz (80-meter) band is similar to the 1.8-MHz band in many respects. Daytime absorption is significant, but not quite as extreme as on 1.8 MHz. Remember, absorption is proportional to the inverse square of frequency so as frequency is increased, D-layer effects are reduced. Typically, daytime communications range is approximately 250 miles. During nighttime, signals are often propagated halfway around the world. As on 1.8 MHz, atmospheric noise tends to be high, making winter the most attractive season for the 80 meter DXer.

• The 7-MHz (40-meter) band is the lowest frequency band that shows appreciable skip distance (see Table 2). During the day, a typical amateur station can cover a radius of approximately 500 miles. At night, reliable world-wide communications are possible on 40 meters. Atmospheric noise is somewhat more diminished, compared to 80 meters, but still presents a problem during the summer months. However, 40-meter DX signals are often of sufficient amplitude to override even high-level summer static. It is for that reason that 40 meters is the lowest-frequency amateur band considered reliable for year-round DX communication.

Table 2

Approximate Minimum Skip Distances for the Amateur MF and HF Bands

Band	Noon*	Midnight*
160 m	0	0
80 m	0	0
40 m	0	300 mi
30 m	200 mi	600 mi
20 m	500 mi	1000 mi
17 m	750 mi	(Daytime only)
15 m	800 mi	(Daytime only)
12 m	1000 mi	(Daytime only)
10 m	1200 mi	(Daytime only)

*Local time at the midpoint of the path.

• The 10-MHz (30-meter) band is unique in that it shares characteristics of both daytime and nighttime bands. Communications of up to 1000 miles are typical during the daytime, with this distance extending halfway around the world at night. In this respect, the band is generally useful on a 24-hour basis. During a solar minimum, the MUF on some DX paths may drop below 10 MHz during the night. Under these conditions, 30 meters adopts the characteristics of the higher frequency, daytime bands described below. Because the 30-meter band is a transitional band that is usable during both the day and night

for long-distance communications, it also shows the least variation in conditions during the swing between solar maximum and solar minimum.

• The 14-MHz (20-meter) band is traditionally regarded as the amateurs' primary long-haul DX band. Regardless of our position in the 11-year cycle, 20 meters can be depended on for world-wide propagation during the daylight hours. During solar maximum periods, 20 meters will often stay open to distant locations throughout the night. Skip distance is appreciable on the band (see Table 2), and is always present to some degree. Atmospheric noise is not too much of a consideration on the band, even in the summer. Because of its popularity, 20 meters tends to be very congested during the daylight hours.

• The 18-MHz (17-meter) band is similar to the 15-meter band in many respects, but the effects of fluctuating solar activity are not quite as pronounced. During the years of high solar activity, 17 meters is reliable for daytime and early-evening long-range communications, often lasting well after sunset. During moderate years, this band resembles 15 meters, opening during sunlight hours and closing shortly after sunset. During solar minimum periods, 17 meters will open to middle and equatorial latitudes, but only for short periods during midday on north-south paths.

• The 21-MHz (15-meter) band has characteristics similar to the 20-meter band, but shows greater fluctuation with changing solar activity. During peak years, 15 meters is reliable for daytime long-distance communications, and will often stay open well into the night. Twenty-four hour, continuous DX operation is possible on occasion, although less often than on 20 meters, and only during the peak of a solar maximum. In periods of moderate solar activity, 15 meters is basically a daytime-only band, closing shortly after sunset. During solar minimum periods, 15 meters may not open at all except for infrequent north-south transequatorial circuits. Sporadic-E is observed occasionally in early summer and mid-winter, although this is not common and the effects are not as pronounced as on the higher frequencies.

• The 24-MHz (12-meter) band offers propagation that combines the best of the 10- and 15-meter bands. While 12 meters is primarily a daytime band during low and moderate sunspot years, periods of solar activity find the band open well after sunset, often over long paths. During years of moderate solar activity, 12 meters opens to the low and middle latitudes during the daytime hours, but seldom remains open after sunset. Unlike 10 meters, periods of low solar activity seldom cause this band to go completely dead except at higher latitudes. Occasional daytime openings, especially in the lower latitudes, have been noted over north-south paths. The main sporadic-E season on 24 MHz lasts from late spring through summer and short

openings may be observed in mid-winter.

• The 28-MHz (10-meter) band is well known for extreme variations in characteristics. During a solar maximum, long-distance propagation is so efficient that very low powers can produce very loud signals half-way around the globe. DX is abundant with very modest equipment. Many a new DXer has been born on 10 meters. During a solar maximum, 10 meters is a daytime band remaining open a few hours past sunset. During periods of moderate solar activity, 10 usually opens only to low and transequatorial latitudes around midday. During a solar minimum, 10 meters is completely dead all day. There are anomalous openings (scatter, meteor burst, sporadic-E) during solar minimum that go unnoticed because most of the amateur population has migrated to lower frequencies. This lack of occupancy tends to enhance the dead appearance of the band. Amateurs who specialize in 10-meter operation make contacts throughout the solar cycle.

Sporadic-E propagation is fairly common on 10 meters, occurring between late April and early August. This phenomenon occurs regardless of F-layer conditions, and provides single-hop communications out to 1300 miles and multiple-hop communications to 2600 miles.

PROPAGATION PREDICTION

To the casual amateur, part of the thrill in ham radio is the challenge of finding an open frequency to operate on. The more serious amateur, interested in DX or traffic handling, wants to know, in advance, the band that will have the best stability and hearability in the geographical area of interest. Today's amateur has several options in planning his or her operations. These include the propagation charts in *The ARRL Operating Manual,* home computer predictions and the use of HF reference beacons. When these three techniques are used together, they can take a lot of mystery out of propagation conditions.

In December 1982, *QST* published the first MUF prediction program, called MINIMUF, for the ham microcomputer. Response throughout the amateur community was positive and the program quickly proliferated through many of the common home computers. In the early 1980's, a new generation of synthesized, digital HF receivers and transceivers provided full coverage for the entire HF spectrum causing renewed interest in shortwave broadcast listening. With these resources, today's enlightened amateur can schedule his or her operating activities using the long term predictions in *The ARRL Operating Manual,* fine tune them with short term predictions from a home computer and calibrate these signal predictions made in advance with on-the-air observations of known shortwave broadcasters. *The ARRL Operating Manual* contains an extensive set of propagation prediction charts, with complete information on how to use them to make a realistic estimate of propagation conditions

MUF Prediction on the Home Computer Using MINIMUF

Traditionally, the prediction of the maximum usable frequency (MUF) is done by a large, complex computer model, nominally consisting of 150,000 to 200,000 bytes of computer code. In 1977, a simple model was developed to emulate the dynamics of the MUF and its sensitivity to solar activity and seasonal changes. Simple is an understatement; the original model consisted of 80 BASIC program steps! Since that time many military, industrial, commercial and amateur HF users have successfully implemented and used MINIMUF.

The initial verification was done by comparing the prediction with oblique-incidence sounder data, which is the only way to observe the actual MUF boundary. The original sounder data base encompassed 196 path months (4074 test points) of observed maximum usable frequencies measured over 23 different HF sounder paths. MINIMUF was found to have an rms error of ± 3.8 MHz. Current users find it useful from 2 MHz to 50 MHz for MUF predictions out to 6000 miles. However, accuracy degrades for ranges of less than 250 miles.

As one can imagine, anything as simple as MINIMUF invites tinkering. Numerous experimenters have made attempts to improve the model with such features as adding an E and F_1 layer (MINIMUF is a single-layer model), changing constants to reflect local conditions and giving it more diurnal variation. All of these revisions, when compared against oblique-sounder data, degraded the accuracy and made the program more complicated. These exercises only serve to prove the old adage, "If it works, don't fix it." The version of MINIMUF described in this *Handbook* was first published in 1978 (see bibliography) and is still the principal version in use.

Application Tips

As solar activity varies, user interest in updating the MINIMUF prediction to reflect current conditions also increases. The updating method found to be most effective was to vary the sunspot number input parameter as a function of the 10.7-cm solar flux. Because of the lag in F-layer response to a rapid increase in solar activity, it is best to use either a 5-day, 15-day or 90-day running average of the 10.7-cm flux. The type of application will determine which is best. The 5-day mean is a short term, more dynamic input while the 90-day mean is more applicable for long term planning. As was indicated in a previous section, the current 10.7-cm solar flux information is transmitted at 18 minutes after every hour on WWV. The

conversion curve to translate the 10.7-cm flux to sunspot number is shown in Fig. 8.

Two other points are borne out by field testing. First, MINIMUF is an F-region approximation. Any intervention by E-region modes of propagation, either as multiple E or EF modes is not predictable by MINIMUF. Such operational situations are proving to represent only a small percentage of the total, however. Second, MINIMUF has the greatest accuracy within the one- and three-hop ranges between 300 miles and 7500 miles. Predictions for transmission paths longer than this should be used with some caution.

Many different conversions of the original MINIMUF article in December 1982 *QST* were received by the Program Exchange Department at ARRL Headquarters. In addition, input was received from numerous amateurs on its modification (see bibliography). A printed listing of an enhanced version of MINIMUF, which includes an LUF-plotting routine, written in BASIC, is available for $2.00 and an SASE. Write to ARRL Headquarters, Technical Department, 225 Main St., Newington, CT 06111. A table of benchmark data is also included to assure correct program loading.

Further Notes on Use

After you have a version of MINIMUF running on your computer, there are countless ways to test it yourself. One method is to observe the passage of the MUF through a frequency where a reference transmitter operates. WWV, WWVH and many shortwave broadcast signals will do nicely. Normally, ± 30 minutes from the predicted time of MUF passage (either the signal appears or goes away) is a good window.

Do not be mislead by comparing your prediction to published propagation charts. All HF prediction programs do well on some paths and poorly on others (and not necessarily the same paths). Comparing one HF prediction program against another is not valid unless all of the assumptions within the models are approached in the same manner. Such comparisons should be used with great caution because they are comparisons of apples and strawberries.

For those that are interested in calculating the Frequency of Optimum Transmission (FOT) and the LUF, two simplified approaches may suffice. The FOT may be approximated by 0.85 × MUF. This is the frequency which is supposed to have the highest circuit reliability and the least multipath interference. The LUF may be approximated by 0.25 × MUF with an upper limit of 12 MHz. After sunset, the LUF drops to below 2 MHz and remains there until sunrise.

HF Reference Beacons

Another method to assist day to day operations is establishing the habit of monitoring an HF signal that regularly

appears on the same frequency, at the same time, and maintains fairly constant transmission characteristics. The three most available sources of signals to monitor are (1) HF shortwave broadcasters; (2) time standard stations; and (3) amateur beacons. This method of environmental assessment is made even more attractive by a new generation of fully synthesized, programmable, general-coverage receivers, both in receive-only and transceive configurations. In general, these receivers are state-of-the-art, sensitive and very frequency agile. Stations to be monitored may be programmed in any order, with almost any revisit time and dwell time. This allows you to browse around the HF spectrum. You will find a whole new world of interesting signals; commercial, military, business, affairs of state, and broadcast; signals most amateurs don't normally listen to.

The most prolific HF signals that are useful for routinely monitoring conditions are those of the high power broadcasters. Most countries in the world maintain a foreign radio service to transmit news, music, and shows of interest from that country to large portions of the world. These signals, mostly double-sideband, amplitude modulation (AM), have 100 kW or more power and use large directive antennas. It takes effective radiated powers such as this to maintain broadcast quality at HF. The main advantage of using these signals is that they transmit on several frequencies simultaneously, allowing a listener to estimate the propagating bandwidth (LUF to MUF) over a given reference path.

While there are literally hundreds of these stations, some of the stronger signals to listen for in the USA are Voice of America (USA), BBC (UK), Radio Moscow (USSR), Deutsche-Welle (FDR), Radio Canada International (Canada), Radio South Africa (South Africa), Radio Nacional Buenos Aires (Argentina) and Radio Australia (Australia). The most useful frequencies to listen on are 5.9 to 6.2 MHz, 7.1 to 7.3 MHz, 9.5 to 9.775 MHz, 11.7 to 11.975 MHz, 15.1 to 15.4 MHz, 17.7 to 17.9 MHz, 21.4 to 21.75 MHz, and 25.6 to 26.1 MHz.

To use these broadcasts, one must establish routine listening habits and monitor the same transmissions at the same time each day. Develop a sense of trend. Are conditions improving or degrading? Monitoring several different signals can yield a picture of global conditions. Are BBC's signals making it over the Polar regions? Can you hear both the 6.0- MHz and 9.5-MHz signals from Radio Australia, and are they of equal strength?

In general, there are HF broadcasters in almost any part of the world an amateur may be interested in. If they are properly monitored they can provide an excellent assessment of global propagation. You should be aware, however, that the programs from some of these stations are transmitted from locations outside of the country. So if you hear a program from Radio Moscow, the Voice of America or the BBC, for example, you may not be listening to radio signals coming from Moscow, the USA or Great Britain. Shortwave listener's guides can help you learn where the broadcast is coming from, and program schedules from the stations themselves will often list the transmitter locations.

Time Standard Stations

Probably the most-used HF reference signals are the US Time Standard signals from WWV and WWVH. The Department of Commerce, National Oceanic and Atmospheric Administration, Environmental Research Laboratories in Boulder, Colorado maintain time standard stations at Ft. Collins, Colorado (WWV) and Kuaui, Hawaii (WWVH). These stations transmit standard "tic" pulses each second, with minute announcements and special bulletins periodically (such as 10.7 cm flux at 18 minutes after the hour on WWV). WWV transmits on 1.5, 5, 10, 15 and 20 MHz with very precise time and frequency stability. For amateurs in the USA, these signals are very useful to calibrate MINIMUF predictions, and to assess day to day communications reliability over one-hop ranges.

Amateur Beacon Stations

Historically, DX clubs and experimenters have established special interest beacons to alert DX operators of new long-range openings. There is a series of beacons on 10 meters between 28.1 and 28.5 MHz. These beacons have been established to alert 10-meter fans to long-haul openings or a sporadic-E opening. DX organizations are now establishing beacons at lower frequencies to optimize their DX-search operations.

Because these beacons emerge and leave the ham bands on a sporadic basis, no attempt will be made to list them here. A current list of beacons appears periodically in the "How's DX?" column of QST.

SELECTED HF-PROPAGATION REFERENCES

Anderson, John E., "MINIMUF for the Ham and the IBM Personal Computer," QEX, Nov. 1983, pp. 7-14.

Bramwell, Denton, "Technical Correspondence — More MINIMUF Mods," QST, May 1983, p. 43.

Horzepa, Stan, "On Line — Translating BASIC," QST, Aug. 1983, p. 64.

Paul, Dr. Adolf K., "NOSC Scientist Predicts Sunspot Activity," Technical Briefs, Naval Ocean System Center, 17 Feb. 1986.

Rose, Robert B., J. N. Martin and P. H. Levine, "MINIMUF-3: A Simplified HF MUF-Prediction Algorithm," Naval Ocean Systems Center Technical Report TR-186, Feb. 1, 1978.

Rose, Robert B., "MINIMUF: A Simplified MUF-Prediction Program for Microcomputers," QST, Dec. 1982, pp. 36-38.

Rose, Robert B., "Technical Correspondence — MINIMUF Revisited," QST, Mar. 1984, p. 46.

Rose, Robert B., "New Initiatives in HF-Signal Coverage Assessment," Ionospheric Effects Symposium, Naval Research Laboratory, Washington, DC, June 1984.

Shallon, Sheldon C., "Technical Correspondence — MINIMUF on Polar Paths," QST, Oct. 1983, p. 48.

THE WORLD ABOVE 50 MHz

50-54 MHz (6-meter band)

The lowest amateur VHF band shares many of the characteristics of both lower and higher frequencies. Nearly every form of propagation is found in the 6-meter band, which contributes greatly to its popularity. In the absence of any favorable propagation conditions, well-equipped 50-MHz stations work regularly over a radius of at least 300 miles, depending on terrain, power, receiver capabilities, and antenna. Weak-signal tropo scatter allows the very best stations to make 500-mile contacts nearly any time. Weather effects may extend the normal range by a few hundred miles, especially during the summer months, but true ducting is rare.

Sporadic-E is probably the most common and certainly the most popular form of propagation on the 6-meter band. Single-hop E-skip openings may last many hours for contacts in the 400- to 1300-mile range, primarily during the spring and early summer. Multiple-hop sporadic-E provides transcontinental contacts several times a year, and contacts between the U.S. and South America, Europe, and Japan via multiple-hop E-skip have been made on rare occasions. During the peak of the 11-year sunspot cycle, world-wide 50-MHz DX is possible due to F_2 reflections during daylight hours. F_2 backscatter provides an additional propagation mode for contacts up to 2000 miles when the maximum usable frequency (MUF) is just below 50 MHz.

Other types of ionospheric propagation make 6 meters an exciting band. Aurora effects in the temperate latitudes allow contacts up to 1300 miles and sometimes farther. In the early morning hours, meteor scatter is a popular 50-MHz endeavor that provides brief contacts in the 500- to 1300-mile range. Transequatorial spread F (TE) creates 50-MHz paths as long as 5000 miles across the equator during peak sunspot years. Propagation by field-aligned irregularities (FAI) may provide additional hours of contacts at distances of up to 1300 miles.

144-148 MHz (2-meter band)

Ionospheric effects are significantly reduced at 144 MHz, but they are far from absent. F-layer propagation is unknown except for TE, which is responsible for the current 144-MHz terrestrial DX record of nearly 5000 miles. Sporadic-E occurs as high as 144 MHz about a tenth as often as at 50 MHz, but the maximum single-hop distance is the same, about 1300 miles. Sporadic-E contacts greater than 2000 miles have been made across the U.S. and Canada, as well as in Europe, but they are rare.

Auroral propagation is quite similar to that found on 50 MHz, except that signals tend to be weaker and more distorted. Meteor-scatter contacts up to 1300 miles are limited primarily to the periods of the great annual meteor showers. Contacts have been made via FAI on 144 MHz, but its potential has not been fully explored.

Tropospheric effects improve with increasing frequency, and 144 MHz is the lowest VHF band at which weather plays an important propagation role. Weather-induced enhancements may extend the normal 300- to 400-mile range of well equipped stations to 500 miles and more, especially in the summer and early fall. Tropospheric ducting makes it possible for contacts of more than 1200 miles over the continent and at least 2500 miles over water.

220-225 MHz (135-cm band)

The 135-cm band shares many characteristics with the 2-meter band. The normal working range of better 220-MHz stations is nearly as good as comparably equipped 144-MHz stations. The 135-cm band is slightly more sensitive to tropospheric effects, but ionospheric modes are more difficult to use.

Aurora and meteor-scatter signals are somewhat weaker than on 144 MHz. Sporadic-E contacts on 220 MHz are extremely rare, but may become more common as use of this band increases. TE is well within the possibilities of 220 MHz, and FAI may be possible as well. Increased activity on the 135-cm band will surely reveal the extent of 220-MHz propagation modes.

420-450 MHz (70-cm band)

The lowest amateur UHF band marks the highest frequency on which ionospheric propagation is commonly observed. Aurora signals are much weaker, and the range is usually less, than on 144 or 220 MHz. Meteor scatter is much more difficult than on the lower bands because signals are significantly weaker and of much shorter duration. Although sporadic-E and FAI are unknown as high as 432 MHz and probably impossible, TE may be useful. TE reception has been observed on this band, but no two-way 432-MHz contacts have been reported.

Well equipped 432-MHz stations can expect to work over a radius of at least 300 miles in the absence of any propagation enhancement. Tropospheric bending is more pronounced at 432 MHz and provides the most frequent and useful means of extended range contacts. Tropospheric ducting supports contacts of 1000 miles and farther over land and more than 2500 miles over water. The current 432-MHz terrestrial DX record of 2550 miles was accomplished via ducting over water. Reflections from airplanes, mountains and other stationary objects may also be useful adjuncts to 70-cm propagation.

902-928 MHz (33-cm band) and Higher

Ionospheric modes of propagation are nearly unknown in the UHF bands above 902 MHz. Aurora scatter may be within the range of amateur power levels at 902 and 1296 MHz, but signal levels will be well below those at 432 MHz. Doppler shift and distortion will be considerable, and the signal bandwidth may be quite wide. No other ionospheric propagation modes are likely.

Almost all extended distance work in the UHF and microwave bands is accomplished with the aid of tropospheric enhancements. The frequencies above 902 MHz are very sensitive to changes in the weather. Tropospheric ducting is more common than in the VHF bands. Contacts up to 2500 miles have been made on 1296 MHz and more than 1000 miles on 10 GHz by ducting. Limited amateur experiences have shown that well equipped stations can work reliably up to 200 miles and farther on 1296 MHz, but in general, normal working ranges are reduced with increasing frequency. Above 10 GHz, attenuation caused by atmospheric water vapor and oxygen become the most significant limiting factors in long-distance communications.

VHF/UHF/MICROWAVE PROPAGATION MODES

There is a wide variety of means by which radio frequencies above 50 MHz may be propagated beyond the horizon. They can be divided into three broad categories: propagation modes involving the ionosphere, such as F_2-layer reflections and meteor scatter; those limited to the troposphere (the weather-producing lower 10 miles of the atmosphere), such as ducting; and reflections from large solid objects, such as mountains and airplanes. Each of these is discussed in turn. Table 3 summarizes the characteristics of these propagation effects.

F2-Layer Reflections

Most long-distance radio communications in the HF range (3-30 MHz) result from one or more reflections in the F_2-layer of the ionosphere, 150 to 250 miles in altitude. The MUF of F_2 reflections reaches only as high as 70 MHz during the peak years of the sunspot cycle, making 6 meters the only useful VHF band for this type of communications. The skip distance at 50 MHz is usually between 1800 and 2500 miles, but multiple-hop paths make world-wide DX possible. The MUF may exceed 50 MHz during a three-year period around the peak of the 11-year sunspot maximum. The last solar peak was in 1980 and the next maximum is predicted for 1991. World-wide 50-MHz F_2 DX may be possible as early as the winter of 1989.

The MUF follows seasonal and daily cycles as well as the sunspot cycle. Even during the peak sunspot years, the MUF may reach 50 MHz only during the fall and winter. 50-MHz F_2 contacts are generally completed only over paths entirely in daylight, since the MUF declines sharply after sunset. Stations running as few as ten watts and simple antennas can work F_2 DX easily.

It is possible to forecast F_2-layer MUF using the 2800-MHz solar flux values broadcast by WWV at 18 minutes after each hour. The MUF approaches 50 MHz when the solar flux rises to 175 (equivalent to a sunspot number of 125). The MUF can also be followed more directly from the high frequencies through 50 MHz, using a suitable general coverage receiver.

Just before the MUF reaches 50 MHz F_2 backscatter is often evident on the

Table 3
VHF, UHF and Microwave Propagation

Propagation Mode	Useful Frequency	Expected Distances	Most Favorable Times
Ionospheric Modes			
F_2 Backscatter	50 MHz	100-2000 miles	Daylight during peak sunspot years
F_2 Reflection	50 MHz	1800-12,500 miles	Daylight during peak sunspot years
Transequatorial Spread-F	50-432 MHz	2000-5000 miles	1700-2200 local time during peak sunspot years
Sporadic-E	50-220 MHz	400-2500 miles and farther	Anytime, but common late mornings and early evenings May-July and December-January
Field-aligned irregularities	50-144 MHz, possibly 220 MHz	400-1300 miles	During and just after intense sporadic-E sessions
Aurora	50-432 MHz, possibly to 1296 MHz	100-1300 miles	Anytime, but more likely late afternoons during March-April and September-October
Meteor scatter	50-432 MHz	500-1300 miles	Early mornings, especially during one of the great annual meteor showers
Tropospheric Modes			
Line of sight	50 MHz-300 GHz	1-200 miles	Anytime, but cold weather better for bands above 10 GHz
Tropo scatter	50 MHz-300 GHz	50-500 miles	Anytime
Rain scatter	1296 MHz-10 GHz	5-100 miles	During periods of heavy rain
Ducting	50 MHz-300 GHz	100-2500 miles and farther	During favorable weather conditions
Reflection and diffraction by solid objects			
Stationary object reflections	144 MHz-300 GHz	5-150 miles	Anytime. Objects must be above the radio horizon for both stations
Airplane reflection	144 MHz-300 GHz	5-400 miles	During passage of an airplane between two stations
Earth-moon-earth reflection	50 MHz-2304 MHz, and higher	100-12,500 miles	When the moon is at perigee and within sight of the two stations
Knife-edge diffraction	432 MHz-300 GHz	5-200 miles	Anytime. Hills or mountains must lie between the two stations

6-meter band. Signals transmitted in the direction of an F_2 region with an MUF just below 50 MHz will be scattered back toward the transmitter over a wide area. Backscatter may provide another useful way of making contacts up to 2000 miles and farther. Backscattered signals are generally weak and have a characteristic hollow or ringing sound to them. High power and directive antennas are usually required to make backscatter contacts.

Transequatorial Spread-F

Discovered in 1947, *transequatorial spread-F (TE)* propagation makes it possible for contacts between 3000 and 5000 miles across the equator on frequencies as high as 432 MHz. Stations attempting TE contacts must be nearly equidistant from the geomagnetic equator. Many contacts have been made on 50 and 144 MHz between Europe and South Africa, Japan and Australia, and the Caribbean region and South America. Fewer contacts have been made on the 220-MHz band. TE signals have been heard on 432 MHz, but so far, no two-way contacts have resulted.

Unfortunately, for most continental U.S. stations, the geomagnetic equator dips south of the geographic equator in the western hemisphere, making only the most southerly portions of Florida and Texas within TE range. TE contacts from the southeast part of the country may be possible with Argentina, Chile, and even South Africa.

Limited experience with TE indicates that it peaks between 1700 and 2200 local time during years with high sunspot numbers. Signals have a rough aurora-like note. High power and large antennas are not required to work TE, as stations with 100 watts and single long Yagis have been successful.

The physics of TE propagation is not well understood, but a popular hypothesis suggests that during peak sunspot years, the F_2 layer near the equator bulges and intensifies slightly. The MUF late in the day may increase to 1.5 times its normal level 15° either side of the magnetic equator. VHF and UHF signals are refracted twice over the magnetic equatorial region at angles that normally would be insufficient to reflect the signals back toward Earth. The geometry is such, however, that two shallow reflections in the F_2 layer can create north-south terrestrial paths up to 5000 miles.

Sporadic-E

"Short skip," long familiar on the 10-meter band during the summer months, also affects the VHF bands as high as 220 MHz. *Sporadic-E,* as it is more properly called, propagates 50- and 144-MHz signals between 400 and 1300 miles. Multiple-hop sporadic-E has supported contacts up to 5000 miles on 50 MHz and more than 2000

miles on 144 MHz. The first confirmed 220-MHz sporadic-E contact was made in June 1987, but such contacts are likely to be very rare in the future.

Although sporadic-E is most common during the months of May, June and July, with a less-intense season at the end of December and early January, it can appear anytime. It is apparently unaffected by the sunspot cycle, and its association with the weather is highly uncertain. It is most likely to occur from 0900 to 1200 and again early in the evening between 1700 and 2000 local time.

Efforts to predict sporadic-E openings have not been highly successful, probably because its causes are not well understood. Studies have demonstrated that small and unusually dense patches of ionization in the E-layer, between 50 and 60 miles in altitude, are responsible for sporadic-E reflections. These sporadic-E clouds, as they are called, may form suddenly, move quickly from their birthplace, and dissipate within a few hours.

Sporadic-E clouds also exhibit an MUF that can be followed from 28 MHz through the 50-MHz band and higher. When the skip distance on 28 MHz is as short as 300 or 400 miles, it is a good sign that the MUF has reached 50 MHz. Contacts at the maximum sporadic-E distance, about 1300 miles, should then be possible at 50 MHz. E-skip contacts as short as 400 miles on 50 MHz, in turn, may be a sign that 144-MHz contacts in the 1200-mile range can be completed. Backscatter from sporadic-E clouds may also be evident on frequencies just below the MUF of the cloud.

The alert VHF operator will monitor

6-meter openings for signs that the MUF may be approaching 144 MHz. Hearing a 50-MHz station work progressively shorter paths suggests that the MUF is rising rapidly and gives clues to possible 144-MHz paths. See Fig. 15. The MUF can also be followed through TV channels 2 through 6 (56-88 MHz), the FM broadcast band (88-108 MHz), and into the aircraft band above that. Sporadic-E openings occur about a tenth as often on 144 MHz as on 50 MHz and for much shorter periods. It is common for the 6-meter band to be open for hours via sporadic-E, but 2-meter openings typically last only a matter of minutes. A few spectacular 2-meter sporadic-E sessions have lasted for longer than an hour. Sporadic-E may also affect the frequencies above the 2-meter band; this can be followed on TV channels 6 through 13 (174-216 MHz). The MUF for sporadic-E has been observed to exceed 200 MHz on a few occasions, but 220-MHz contacts are still very rare events.

Field-Aligned Irregularities

Amateurs have experimented with a little-known scattering mode known as *field-aligned irregularities (F AI)* on 50 and 144 MHz since 1978. Propagation via FAI accompanies sporadic-E openings and may persist for several hours after E-skip as disappeared. There are indications that 2000 to 2400 local time may be the most productive for FAI. Stations attempting FAI contacts direct their antennas toward a common scattering region that corresponds to a known sporadic-E reflection point. They need not be along a single great-circle path. Stations in southern

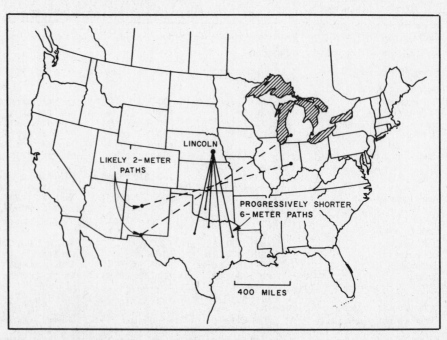

Fig. 15 — Sporadic-E contacts of 400 miles on 50 MHz indicate that 1200-mile paths may be open on 144 MHz. Hearing a 50-MHz station in Lincoln, Nebraska, work progressively shorter paths into Texas and Oklahoma (solid lines) suggests that 144-MHz contacts (dashed lines) may be possible. The midpoint of all paths is the same area of high-MUF, sporadic-E activity, in this case just north of the Kansas-Oklahoma border.

Florida, for example, have completed 144-MHz FAI contacts with Dallas, Texas, when all participating stations were beamed toward a common scattering region over northern Alabama.

FAI-propagated signals are weak and fluttery, reminiscent of aurora signals. Doppler shifts of as much as 3 kHz have been observed in some tests. Stations running as little as 100 watts and a single Yagi should be able to complete FAI contacts during the most favorable times, but higher power and larger antennas may yield better results. Contacts have been made on 50 and 144 MHz, and 220-MHz FAI seems probable as well. Expected maximum distances should be similar to other forms of E-layer propagation — up to 1300 miles.

Aurora

Radio signals with frequencies as high as 3000 MHz have been scattered by the aurora borealis or northern lights (aurora australis in the Southern Hemisphere), but amateur *aurora* contacts have been reported only through 432 MHz. By pointing directional antennas toward the center of aurora activity (generally north in the Northern Hemisphere) contacts up to 1300 miles distant are possible. Stations north of 42° latitude experience many aurora openings each year, while those in the Gulf Coast states may hear aurora signals no more than once a year, if that often.

The appearance of aurora is closely linked to solar activity. During massive solar storms, high energy streams of electrons and protons spew out from the sun. These high-energy particles are trapped by the Earth's magnetic field in the polar regions. They ionize the gases of the E-layer and higher, giving rise to spectacular visual displays as well as affecting radio communications. Aurora generally absorbs radio signals below 20 MHz, while those above 20 MHz are scattered and reflected.

Auroras occur most often around the spring and fall equinoxes (March-April and September-October), but aurora may appear in any month. Radio aurora activity generally peaks between 1600 and 2000 local time and may reappear later in the evening. Aurora may be anticipated by following the A- and K-index reports on WWV at 18 minutes after each hour. A K-index of 4 or greater and an A-index of at least 30 are indications that a geomagnetic solar storm is in progress and an aurora likely. The probability, intensity, and southerly extent of aurora increase as the two index numbers rise. Weak and watery signals on the AM broadcast band through the 40-meter band late in the afternoon may be an indication that an aurora is already in progress. The 20-meter band may close down altogether. Satellite operators have also noticed that 144-MHz downlink signals are often weak and distorted when satellites pass near the polar regions during magnetic storms.

You do not have to able to see an aurora to make aurora contacts. Useful aurora may be as distant as 650 miles and well below the visual horizon. Auroras scatter radio signals back toward the transmitter over a very wide area, making possible contacts with stations up to 1300 miles away. Antennas should be pointed generally north and then used to probe east and west to peak signals. Offsets from north are usually greatest when the aurora is closest. There may be some advantage to antennas that can be elevated, especially when auroras are nearby. High power and large antennas are not necessary to work aurora. Stations with as little as 10 watts output have made aurora contacts on frequencies as high as 432 MHz.

Aurora-scattered signals are easy to identify. On 50-MHz SSB, signals sound very distorted and somewhat wider than normal; on 144 MHz and above, the distortion may be so severe that only CW is useful. Aurora CW signals have a distinctive note variously described as a buzz, hiss, or mushy sound. Doppler shifts of 1 kHz may be evident on 144 MHz and several kHz on 432 MHz. Signal strength dramatically decreases with higher frequencies, while Doppler shift and distortion increase. Especially strong aurora may evolve into a mode known as auroral-E on 50 MHz. Distortion disappears and path lengths may increase well beyond the usual 1300-mile limit for aurora.

Meteor Scatter

Contacts between 500 and 1300 miles distance can be made on 50 through 432 MHz via reflections from the ionized trails left by meteors as they travel through the ionosphere. The kinetic energy of meteors no larger than peas, and more commonly the size of grains of sand, are sufficient to ionize a column of air up to 12 miles long in the E-layer. The particle itself evaporates and never reaches the ground, but the ionized column may persist for a few seconds to a minute or more before it dissipates. This is enough time to make very brief contacts by reflections from the ionized trails. Hundreds of thousands of meteors enter the Earth's atmosphere every day, but relatively few have the required size, speed, and orientation to the Earth to make them useful for *meteor-scatter* propagation.

Radio signals in the 30- to 100-MHz range are reflected best by meteor trails, making the 50-MHz band prime for meteor-scatter work. The early morning hours around dawn are the most productive, because the Earth's rotation sweeps up incoming meteors and contributes to their apparent velocity. Meteor contacts ranging from a second or two to more than a minute can be made nearly any morning on 50 MHz. Meteor-scatter contacts on 144 MHz and higher are more difficult because reflected signal strength and duration drop sharply with increasing frequency. A meteor trail that provides 30 seconds of communication on 50 MHz will last only 3 seconds on 144 MHz and less than 1 second on 432 MHz.

Meteor scatter may be somewhat better during July and August because the average number of meteors entering the Earth's atmosphere peaks during those months. The very best times are during one of the great annual meteor showers, when the number of useful meteors may increase 50-fold over the normal rate of one or two per hour. See Table 4. A meteor shower occurs when the Earth passes through a relatively dense stream of particles, thought to be the remnants of a comet, that are also in orbit around the sun. The greatest showers are relatively consistent from year to year, both in their recurrence and meteor density.

Table 4
Prominent Annual Meteor Showers

Name	Peak dates	Rate, meteors/hour
Quadrantids	January 3	50
Arietids	June 7-8	60
Perseids	August 12-13	50
Geminids	December 12-13	60

Because meteors provide only fleeting moments of communication even during one of the great meteor showers, especially on 144 MHz and above, special operating techniques are often used to increase the chances of completing a contact. Prearranged schedules between two stations establish times, frequencies, and precise operating standards. Usually, each station transmits on alternate 15-second periods until enough information is pieced together a bit at a time to confirm contact. Nonscheduled random meteor contacts are quite common on 50 MHz and 144 MHz during meteor showers, but short transmissions and alert operating habits are required.

It is helpful to run several hundred watts to a large antenna, but meteor-scatter can be used by modest stations under optimal conditions. During the best showers, a few watts and a small Yagi are sufficient on 50 MHz. On 144 MHz, at least 100 watts output and a long Yagi are needed for consistent results. Proportionately higher power is required for 220 and 432 MHz even under the best conditions.

Line of Sight

At one time it was thought that communications in the VHF range and higher would be restricted to "line of sight" paths. Although this has not proven to be the case even in the microwave region, the concept

of line of sight is still useful in understanding propagation. In the absence of
any kind of enhancement, radio waves
traveling through the weather-producing
lower few thousand feet of the atmosphere,
called the troposphere, do not travel in
straight lines. Radio waves are normally
refracted, or bent, slightly earthward. The
reason is not hard to understand. The
normal drop in temperature, pressure, and
water vapor content of air with increasing
altitude affects the refractive properties of
the atmosphere. Just as light is refracted
as it crosses the boundary between two
dissimilar media, such as water and air,
radio waves are refracted as they pass
through an atmosphere whose refractive
properties change with altitude. Under
average conditions, radio waves are
refracted toward Earth enough to make the
horizon appear 1.34 times farther away
than the visual horizon.

A simple formula can be used to estimate
the distance to the radio horizon under
average conditions:

$$d = \sqrt{2h}$$

where
    d = distance to the radio horizon
        in miles
    h = height above average terrain in
        feet

The distance to the radio horizon for an
antenna 100 feet above average terrain is
thus 14.1 miles; a station on top of a
5000-foot mountain would have a radio
horizon of 100 miles. The line-of-sight
distance for any pair of stations is their
combined distance to a common horizon.
Stations on top of two 5000-foot mountains
no more than 200 miles distant would be
within each other's radio horizon.

Tropospheric Scatter

Contacts beyond the horizon are made
every day on the VHF bands and higher
without the aid of obvious propagation
enhancement. "Ground wave," a term that
is more appropriate to propagation on the
frequencies below 3 MHz, plays no
important role above 50 MHz. Beyond-the-
horizon communication in the VHF range
and higher is possible because a tiny
portion of the transmitted signal is
scattered by small changes in the index of
refraction of air, and by dust, clouds, and
other naturally occurring particles in the
troposphere. Most amateurs are unaware
that they use *tropo scatter,* even though it
plays an essential role in most ordinary
VHF and UHF communication. The
normal ranges for well equipped stations
provided in the propagation summaries of
each band, for example, are really the
maximum distances commonly worked via
tropo scatter.

The theoretical maximum distance that
can be worked via tropo scatter is limited
ultimately by the line-of-sight distance two
stations have of the same scattering region

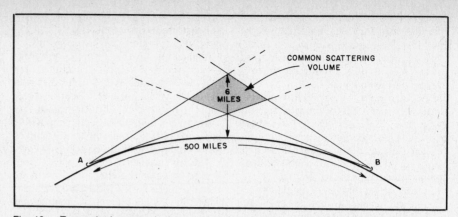

Fig. 16 — Tropospheric scatter is the most common form of propagation in the bands above
50 MHz. A small portion of transmitted signals is scattered by the atmosphere, but only that
portion scattered within sight of two stations — the common scattering volume — provides the
vital tropo-scatter link. The maximum theoretical range of 500 miles for tropo scatter is limited
by the highest part of the atmosphere useful for scattering, about 6 miles in altitude. At 6 miles,
the radio horizon is 250 miles distant.

of the troposphere. See Fig. 16. The highest
altitude for which scattering is efficient at
amateur power levels is about 6 miles. An
application of the distance-to-horizon
formula shows that 500 miles is the
maximum distance for stations relying only
on tropo scatter at any frequency.

In practice, tropo scatter paths are
shorter than 500 miles because most
stations are less than optimal, and scattering path losses increase with frequency.
Almost all extended-distance tropo scatter
work involves weak signals and considerable fading. Scatter may be worked any time
or season. Weather affects scattering
efficiency, but its effects are nearly
impossible to predict over any given path.

Rain scatter is a special case of tropo
scatter that is practical in the 1296 MHz to
10 GHz range. Stations simply point their
antennas toward a common area of rain.
A certain portion of radio energy is
scattered by the raindrops, making over-
the-horizon or obstructed-path contacts
possible, even with low power. The
theoretical range for rain scatter is as great
as 400 miles, but the experiences of
amateurs in the microwave bands suggests
that expected ranges are less than 100 miles.
Snow and hail make less efficient scattering
media unless the ice particles are partially
melted. Smoke and dust particles are too
small for extraordinary scattering, even in
the microwave bands.

Tropospheric Enhancement and Ducting

Weather dramatically affects tropospheric refraction, and specific weather
conditions can be linked to enhanced
propagation. Enhancement refers to
tropospheric refraction that is greater than
normal but insufficient to bend radio waves
back to the surface of the Earth. Increased
signal strength and extended path distances
to the limits of tropo scatter, or about
500 miles, result from enhanced tropospheric conditions. Ducting takes place
when refraction is so great that radio

signals are bent all the way back to the
surface of the Earth. When ducting conditions in the troposphere exist over a wide
geographic area, signals may remain very
strong over distances of 1000 miles or
more. Higher frequencies are more sensitive to tropospheric refraction, and effects
may be observed in the microwave bands
before they affect the lower frequencies.

Enhancement and ducting occur when
the normal lapse in air temperature with
altitude is reversed, that is when temperature increases with altitude. This condition
is known as a temperature inversion.
Temperature inversions are often accompanied by unusual drying of air with
altitude, which contributes to increased
refractivity. A large number of weather
conditions are known to create temperature
inversions. Some of the more common and
most useful inversion-producing weather
conditions are discussed in turn.

Radiation inversions are probably the
most common and widespread of the
various weather conditions useful to
propagation. Radiation inversions form
only over land after sunset as a result of
progressive cooling of the air near the
Earth's surface. As the Earth cools by
radiating heat into space, the air just above
the ground is cooled in turn. At higher
altitudes, the air remains relatively warmer,
thus creating the inversion. The cooling
process may continue through the evening
and predawn hours, creating inversions
that extend as high as 1500 feet. Radiation
inversions are most common during clear,
calm, summer evenings. They are more
distinct in dry climates, in valleys, and over
open ground. Their formation is inhibited
by wind, wet ground, and cloud cover.
Although such inversions are common and
widespread, they are rarely strong enough
to cause true ducting. The enhanced
conditions so often observed on summer
evenings are usually a result of radiation
inversions.

Large, sluggish, *high-pressure systems*

(or anticyclones) create the most dramatic and widespread tropospheric ducts. Inversions in high-pressure systems are created by the sinking and compressing of air within the high. Useful inversions appear between 1000 and 5000 feet. Ducts intensify during the evening and early morning hours. The longest and strongest radio paths usually lie to the south and southwest of the high-pressure centers. See Fig. 17.

Over continental areas, sluggish high-pressure systems that are likely to contain strong temperature inversions are most common in late summer over the eastern half of the U.S. They move very slowly out of Canada, generally southeastward, and may linger for days, providing many hours of extended propagation. The southeastern part of the country, including the Gulf and the lower Midwest, experiences the most high-pressure openings; the upper Midwest and East Coast somewhat less frequently; the mountain states and the West Coast rarely.

Semipermanent high-pressure systems — nearly constant climatic features in certain parts of the world — have created the longest and most exciting ducting paths. The Eastern Pacific High, which migrates northward off the coast of California during the summer, has been responsible for the longest ducting paths reported to date, more than 2500 miles on 144 MHz through 1296 MHz between California and Hawaii. The Bermuda High is a nearly permanent feature in the Caribbean area. During the summer, it too moves north and often covers the southeastern U.S. It has produced contacts in excess of 1800 miles from Florida and the Carolinas to the West Indies, but its potential has not been exploited. Other semipermanent highs lie in the Indian Ocean, the western Pacific, and off the coast of west Africa.

The *wave cyclone* is a more dynamic weather system seen primarily during the spring months over the middle part of the American continent. The wave begins as a disturbance in a long continental front, which creates two distinct parts. On the western leg, a cold front forms and moves rapidly eastward; to the east, a warm front moves slowly northward. When the wave is in its open position, as shown in Fig. 18, north-south radio paths 1000 miles and longer may be possible in the warm sector — that area to the east of the cold front and south of the warm front. East-west paths nearly as long may also open up in the southerly parts of the warm sector. Evening and early morning are usually the best times. Wave cyclones are rarely productive for more than 24 to 36 hours, because the eastward moving cold front eventually closes off the warm sector. Temperature inversions associated with wave cyclones are created by a fast stream of warm, dry air above 3000 feet that covers relatively cooler and moister Gulf air flowing northward near the Earth's surface in the warm sector.

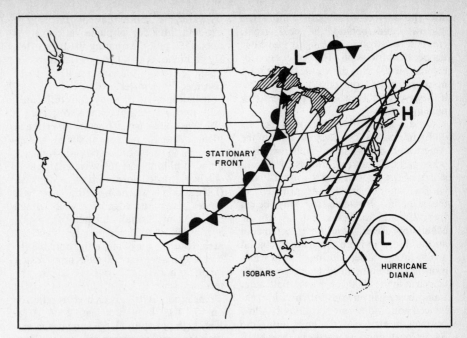

Fig. 17 — Sprawling, slow-moving, high-pressure systems, a common feature in the eastern half of the country during the late summer, provide the majority of extended tropospheric ducts. Shown here are the surface weather features on September 8, 1984, when many contacts of 1000 miles or more were made on all bands from 144 MHz through 1296 MHz. Solid lines represent typical paths reported during the early evening hours. Note that long paths appeared to the southwest of the high-pressure center. For more details about this tropospheric event, see January 1985 *QST*, pp. 74-77; and March 1986 *QEX*, pp. 5-12.

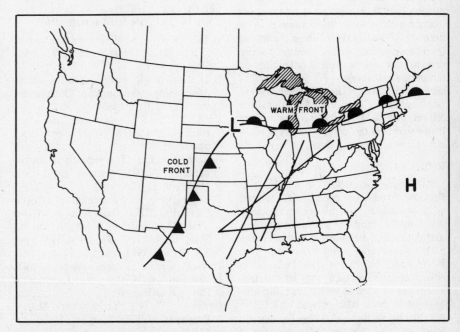

Fig. 18 — The wave cyclone also creates tropospheric ducts that support contacts in excess of 1000 miles in the bands above 144 MHz. Shown here is a wave that formed on June 2, 1980. Solid lines show typical contacts made on 144 MHz and 432 MHz. Similar contacts were made on the following day, before the wave collapsed. Another wave formed along the same continental front on June 6-7 and that provided an additional period of VHF and UHF DX.

Warm fronts and *cold fronts* sometimes bring enhanced conditions, but rarely true ducting. A warm front marks the surface boundary between a mass of warm air flowing over an area of relatively cooler and more stationary air. Inversion conditions may be stable enough for several hundred miles ahead of the warm front to bring enhancement and sometimes ducting. A cold front marks the surface boundary between a mass of cool air that is trying to wedge itself under more stationary warm air. The warmer air is pushed aloft in a narrow band behind the cold front, creating a strong but highly unstable temperature inversion. The best chance for enhancement occurs during the two or three hours after the front passes. Long paths, if they are possible, will open parallel to the passing front.

Certain kinds of wind may also create useful inversions. The *Chinook wind* of the American high plains (and similar winds in other parts of the world) floods the eastern slopes of the Rockies and the Great Plains with warm and very dry air, primarily in the springtime. Inversion effects may be felt from Canada to Texas and east to the Mississippi River. *Land breezes,* the cool air that blows in from the ocean in coastal areas after sunset during the warmer months, is accompanied by a return flow of warm air at 1000 feet or so in altitude. Land breeze inversions often bring enhanced conditions and occasionally allow contacts in excess of 500 miles along the East Coast and other coastal areas. In southern Europe, a hot, dry wind known as the *sirocco* that blows northward from the Sahara Desert across the Mediterranean has probably been responsible for record-breaking microwave contacts in excess of 1000 miles.

Over warm water areas, such as the Caribbean and other tropical waters, *evaporation inversions* may create ducts that are useful only in the microwave region between 3.3 and 24 GHz. These are among the few inversions that depend on a sharp drop in water-vapor content rather than on temperature inversion. Air just above the surface of warm bodies of water is saturated because of evaporation. Within 10 to 30 feet in altitude, the water-vapor content drops significantly, creating a very shallow but stable and strong duct.

Reflections

At frequencies above 144 MHz, reflections from a variety of large objects, such as buildings, airplanes, mountains, water towers, and the like can provide a useful means of extending over-the-horizon paths to several hundred miles. Two stations need only beam toward a common reflector, whether stationary or moving. The reflector must be within the radio line-of-sight of both stations and be large in terms of wavelength. Contrary to common sense notions, the best position for a reflector is not midway between two stations. Signal strength increases as the reflector approaches one end of the path; the reflector may even be positioned on the far side of one station with good result.

The maximum range is limited by the line-of-sight distance of both stations to the reflector and to a lesser extent by reflector size. A water tower on top of a hill 500 feet above average terrain could extend reliable microwave coverage of a station in a near-by valley to more than 30 miles. The extreme limit for airplane reflections is about 550 miles, assuming the largest jets fly no higher than 40,000 feet. Actual airplane reflection contacts are likely to be considerably shorter.

In addition to making fair reflectors, hills and mountains can also serve as effective knife-edge diffractors at VHF and higher. A small portion of signals traveling over the crest of a mountain range or a line of hills at least 100 wavelengths long will be diffracted downward on the far side. (Diffraction is a bending of the wave as it passes across the edge of a well defined object.) Only a small portion of the signal energy will be diffracted into the "shadow" area, but it may make it possible to complete paths up to 100 miles or more that might otherwise be obstructed by the terrain.

The moon has also been used as reflector on 50 MHz through at least 2304 MHz. Very high power and large antennas, along with the very best receivers, are required to overcome the extreme losses involved in Earth-Moon-Earth (EME) paths. Contacts from one side of the Earth to the other, 12,500 miles, are possible. The techniques of EME communication are specialized enough to be discussed in detail elsewhere in the *Handbook.*

SELECTED VHF/UHF PROPAGATION BIBLIOGRAPHY

Bain, Walter F., "VHF Propagation by Meteor-Trail Ionization," *QST*, May 1974, pp. 41-47, 176.

Bean, Bradford R. and E. J. Dutton, *Radio Meteorology* (Washington: Government Printing Office, 1968).

Collier, James S., "Upper-Air Conditions for Two-Meter DX," *QST*, Sept. 1955, pp. 16-18.

Cracknell, R. G., "Transequatorial Propagation of V.H.F. Signals,"*QST*, Dec. 1959, pp. 11-17.

Cracknell, Ray, Fred Anderson, and Costas Fimerelis, "The Euro-Asia to Africa VHF Transequatorial Circuit During Solar Cycle 21,"*QST*, Nov. 1981, pp. 31-36; Dec. 1981, pp. 23-27.

Davies, Kenneth, *Ionospheric Radio Propagation* (Washington: Government Printing Office, 1965).

Gannaway, J. N., "Tropospheric Scatter Propagation," *QST*, Nov. 1983, pp. 43-48.

Greene, Clarke, "Meteor-Scatter Communications," *QST*, Jan. 1986, pp. 14-17.

Hall, Gerald L., ed., *ARRL Antenna Book* (Newington: ARRL, 1984).

Jessop, G. R., ed., *VHF/UHF Manual*, 4th edition (Hertfordshire, England: Radio Society of Great Britain, 1983).

Kneisel, Thomas F., "Ionospheric Scatter by Field-Aligned Irregularities at 144 MHz," *QST*, Jan. 1982, pp. 30-32.

Miller, Richard, "Radio Aurora," *QST*, Jan. 1985, pp. 14-18.

Owen, Michael R., "VHF Meteor Scatter — An Astronomical Perspective," *QST*, June 1986, pp. 14-20.

Owen, Michael R., "The Great Spordic-E Opening of June 14, 1987," *QST*, May 1988, pp. 21-29.

Pocock, Emil, "The Weather that Brings VHF DX," *QST*, May 1983, pp. 11-16.

Pocock, Emil, "Sporadic-E Propagation at VHF: A Review of Progress and Prospects." *QST,* April 1988, pp. 33-39.

Reisert, Joseph H. and Gene Pfeffer, "A Newly Discovered Mode of VHF Propagation,"*QST*, Oct. 1978, pp. 11-14.

Reisert, Joseph, "Improving Meteor Scatter Communications," *Ham Radio* June 1984, pp. 82-92.

Tilton, Edward P., *The Radio Amateur's VHF Manual*, 3rd ed. (Newington: ARRL, 1972). (Out of print.)

Wilson, Melvin S., "Midlatitude Intense Sporadic-E Propagation," *QST*, Dec. 1970, pp. 52-55 and Mar. 1971, pp. 54-57.

NEWSLETTERS WITH VHF/UHF PROPAGATION NEWS

Midwest VHF Report, Roger A. Cox, WBØDGF, 3451 Dudley St., Lincoln, NE 68503 (monthly, $10/year).

Northeast VHF News, Lewis D. Collins, W1GXT, 10 Marshall Terr., Wayland, MA 01778 (bimonthly, $3/year).

Pack Rats' Cheese Bits, Mt. Airy VHF Radio Club, Harry B. Stein, W3CL, 2087 Parkdale Ave., Glenside, PA 19038 (monthly, $3/year).

Sidewinders on Two Bulletin, Howard Hallman, WD5DJT, 3230 Springfield, Lancaster, TX 75134.

Six Shooter, Six Meter International Radio Klub, Ray Clark, K5ZMS, 7158 Stone Fence, San Antonio, TX 78227 (every 3 months, $3/year).

220 Notes, Walt Altus, WD9GCR, 215 Villa Rd., Streamwood, IL 60103 (bimonthly, $5/year).

VHF/UHF and Above Information Exchange, Rusty Landes, KAØHPK, Box 270, West Terre Haute, IN 47885.

VHF/UHF Newsletter, Radio Society of Great Britain, Alma House, Cranborne Rd., Potters Bar, Hertfordshire, EN6 3JW England.

West Coast VHFer, Bob Cerasuolo, WA6IJZ, Box 2041, Oxnard, CA 93034 (monthly, $10/year).

See *QST*, Feb 1987, p 62, for addresses of other VHF/UHF newsletters.

COLUMNS IN AMATEUR JOURNALS

CQ: VHF: Principles, Practices, and Products

Ham Radio: VHF/UHF World

QST: World Above 50 MHz and The New Frontier

Chapter 23

Space Communications

Terrestrial communication range is limited by the spherical shape of the earth. Various propagation mechanisms can be used to transmit a signal *around* the earth. Beyond line-of-sight distances, however, reliable communication may require the use of higher effective radiated power and may not be possible at all.

When a space object is visible from a point on earth, there is nothing in the path to block or scatter the signal. When that object is visible at two points on earth, communication between these two points is possible using the object as a reflector (passive relay) or a transponder (active relay). The signal trip to space and back avoids the greater attenuation involved in sending a signal around the earth.

Amateur Radio space communication has two major facets: artificial satellites and earth-moon-earth (EME). Both are rich with technical and operational achievements by radio amateurs. Together, they make the VHF and higher frequencies usable for amateur transcontinental communications and push today's technology to the limit.

SATELLITE COMMUNICATIONS

The excitement of global VHF/UHF DX can be within the reach of nearly every amateur, and it can be experienced with small antennas, low power and comparatively simple equipment. Amateur satellite communications is a mature program that has been developing for more than a quarter of a century. It began in 1961 and has been enhanced with every launch. The rich history of the amateur satellite program is well covered in *The Satellite Experimenter's Handbook,* published by the ARRL. Presented here is detailed information on setting up a ground station for this fascinating facet of Amateur Radio. Along with the satellite tracking and operating techniques presented in *The ARRL Operating Manual,* amateurs now have the most up-to-date information available on satellite communications.

Spacecraft

Amateurs currently enjoy communications through the thirteenth OSCAR (Orbiting Satellite Carrying Amateur Radio). OSCAR 13 is the second phase III

Table 1
OSCAR Operating Modes

Mode	Uplink Band	Downlink Band
A	2 m (145 MHz)	10 m (29 MHz)
B	70 cm (435 MHz)	2 m (145 MHz)
J	2 m (145 MHz)	70 cm (435 MHz)
L	24 cm (1269 MHz)	70 cm (436 MHz)
S	70 cm (436 MHz)	13 cm (2401 MHz)
JL	2 m/23 cm	70 cm
K	15 m	10 m
T	15 m	2m

Table 2
Spacecraft Frequencies

OSCAR 9

HF Beacons	7.050, 14.002, 21.002, and 29.510 MHz. CW keyed telemetry
VHF Beacon	145.825 MHz NBFM ± 5 kHz. ASCII, Baudot, voice, AFSK and Morse telemetry and bulletins.
UHF Beacon	435.025 MHz NBFM ± 5 kHz. ASCII, Baudot, voice, AFSK and Morse telemetry and bulletins.
S-Band Beacon	2401.0 MHz NBFM ± 10 kHz. ASCII, Baudot, voice, AFSK and Morse telemetry and bulletins.
X-Band Beacon	10.470 GHz steady carrier, LHCP

OSCAR 10

	Uplink (MHz)	Downlink (MHz)	Beacon (MHz)
Mode B	435.025-435.175	145.975-145.825	General 145.810 Engrg. 145.990
Mode L	1269.050-1269.850	436.950-436.150	General 436.040 Engrg. 436.020

OSCAR 11

VHF Beacon	145.825 MHz NBFM FSK Telemetry
UHF Beacon	435.025 MHz NBFM FSK, PSK Telemetry
S-Band Beacon	2401.5 MHz AFSK, PSK

OSCAR 13

	Uplink (MHz)	Downlink (MHz)	Beacon (MHz)
Mode B	435.425-435.575	145.975-145.825	General 145.8125 Engrg. 145.975
Mode L	1269.575-1269.325	435.725-435.975	General 435.650 Engrg. 436.020
Mode J	145.820-145.860 144.440-144.480	435.930-435.970 (Note 1) 435.930-435.970	
Digital Mode L	1269.675	435.675 (Note 2)	
Mode S	435.625	2401.337 (Note 3)	2401.267

Notes:
1) Both Mode J options are expected to be implemented with software control of the uplink frequency selection.
2) RUDAK is a digital-only transponder that will use PSK encoding.
3) Mode S will be a soft-limited FM transponder that can also accommodate up to four SSB signals.
4) Although minor revisions in frequencies may be anticipated, the frequencies shown are those currently planned.

FO-12

	Uplink (MHz)	Downlink (MHz)
Linear Service Mode J	145.900-146.000	435.900-435.800
Packet Digital Service Mode J	145.850, 145.870, 145.890, and 145.910	435.910 (1 channel)
Beacon		435.795

satellite. The first successful phase III satellite, OSCAR 10, demonstrated a noteworthy capability for global communications since it was first made available in August 1983. While it had a few problems, they were in the shadow of the success that introduced many thousands of amateurs to satellite communications. This success provided a host of new transceivers, preamplifiers and antennas for the amateur world.

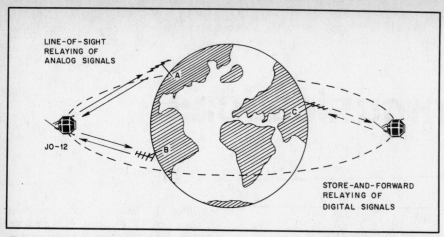

Fig. 1 — FO-12 provides for conventional amateur communications through its transponders, as well as a digital store-and-forward capability.

As this *Handbook* is being prepared, the Phase III-C spacecraft (OSCAR 13) has been successfully launched, with improved capabilities taken from lessons learned from AO-10. In addition to Mode B, OSCAR 13 has an improved Mode L transponder that should be much more sensitive than that aboard AO-10. See Tables 1 and 2. This Mode L transponder is coupled with a Mode J unit. Both transponders offer QRN-free 70-cm downlinks. The world of packet digital communications is be served by the RUDAK Mode L transponder using PSK, Manchester II modulation. OSCAR 13 will also be an experimental test bed to prove the feasibility of 2.4 GHz Mode S transponders for use in future OSCARs.

Amateurs in more countries are participating in the process of designing and

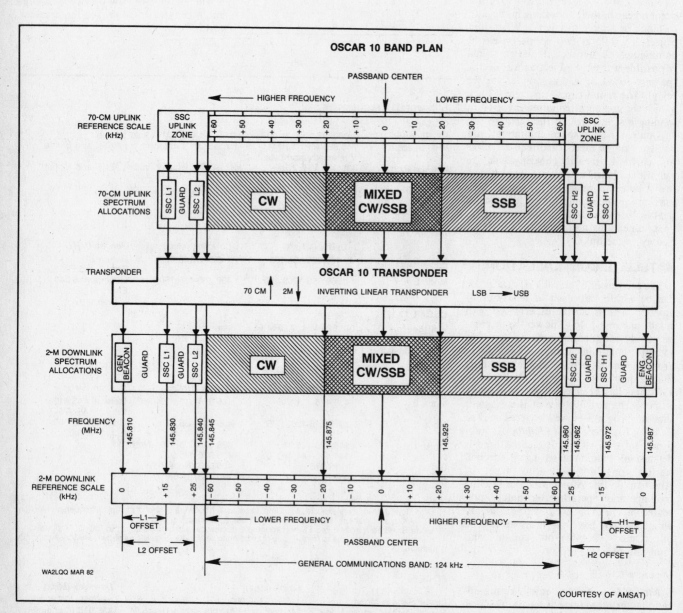

Fig. 2 — The OSCAR 10 band plan allows for CW only, mixed CW/SSB and SSB only operation. Courteous operators observe this voluntary band plan at all times.

constructing spacecraft. In addition to the launch of OSCAR 13 in 1988, there is the first Japanese OSCAR, called FO-12. This is a joint effort between the Japan Amateur Radio League (JARL) and the Japanese AMSAT (JAMSAT) organization. It was launched by the Japanese National Space Development Agency (NASDA) as the payload on an experimental launch vehicle H-1 — truly a Japanese product in total. FO-12 is a Phase II spacecraft in a 1500 km circular orbit. It carries a Mode J transponder that serves two functions: linear SSB service and an advanced packet store-and-forward global message service with a 1.5 megabyte RAM storage bank. See Fig. 1.

Russian spacecraft designs have concentrated more on HF communications links to accommodate the kinds of equipment readily available in that country. RS 10/11, two separate "satellites" on the same spacecraft, include mode A and mode K transponders. The spacecraft is in a stable circular orbit at fairly low altitude.

Operating Frequencies

Early OSCARs used Mode A transponders that were affected by F2-layer ionization phenomena during periods of high solar activity. As shown in Table 1, Mode A uses a 2-meter uplink (ground-to-spacecraft frequency) and a 10-meter downlink (spacecraft-to-ground). When the ionosphere is active, 10-meter satellite downlinks can be seriously affected.

Newer classes of amateur satellites have been designed to use higher-frequency amateur bands for a number of reasons. In addition to avoiding ionospheric effects, spectrum capacity and attainable antenna directivity are among some of the other strong reasons for moving to the UHF region for future satellites. OSCAR 10 (also called AO-10) features a combination of Mode B and Mode L transponders. OSCAR 13 carries a combination of Modes B, J, L, and S (see Table 2).

Future satellites will make use of higher frequencies. They will be placed in highly elliptical orbits, like OSCAR 13, or into the geostationary orbits. From these high altitudes, more than 45% of the Earth's surface can be viewed at one time, so the potential number of users is tremendous. A span of several hundred kilohertz is needed to afford sufficient spectrum for these users, and this much free space can only be found in the UHF region.

Another reason for using UHF with OSCARs is that high-gain, directional antennas are needed to make communications possible over such long paths. Since satellites are powered by batteries and solar cells, there are limitations on the amount of power the satellite transmitter can consume. The most efficient way to get the needed ERP is through the use of high-gain, narrow beamwidth antennas. Such

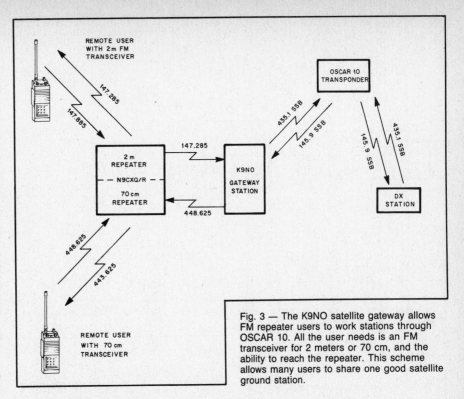

Fig. 3 — The K9NO satellite gateway allows FM repeater users to work stations through OSCAR 10. All the user needs is an FM transceiver for 2 meters or 70 cm, and the ability to reach the repeater. This scheme allows many users to share one good satellite ground station.

Amateur Satellites: A History

OSCAR I, the first of the Phase I satellites, was launched on December 12, 1961. The 0.10 W transmitter onboard discharged its batteries after only three weeks.

OSCAR II was launched on June 2, 1962. Identical to OSCAR I, Amateur Radio's second venture into space lasted 18 days.

OSCAR III, launched on March 9, 1965, was the first satellite for amateur communications. During its two-week life, more than 100 amateurs in 16 countries communicated through the linear transponder.

OSCAR IV was launched on December 21, 1965. This satellite had a Mode J linear transponder. A launch vehicle defect placed OSCAR IV into poor orbit, preventing widespread amateur use.

OSCAR 5, built by students at Melbourne University in Australia, transmitted telemetry on both 2 m and 10 m for more than a month.

OSCAR 6, the first of the Phase II satellites, was launched on October 15, 1972. It carried a Mode A linear transponder and lasted nearly five years.

OSCAR 7, built by hams from many countries was launched on November 15, 1974. It carried Mode A and Mode B linear transponders and served the amateur community for six years.

OSCAR 8, again a cooperative international effort, was launched on March 5, 1978. Mode A and Mode J transponders were carried on board. The spacecraft lasted six years.

Radio Sputniks 1 and 2, launched from the Soviet Union on October 26, 1978, each carried a sensitive Mode A transponder. Their useful lifetimes were only a few months.

OSCAR Phase III-A, the first of a new satellite series, was launched on May 23, 1980, but it failed to achieve orbit because of a failure in the launch vehicle.

OSCAR 9, built at the University of Surrey in England, was launched in October 1981. This is a scientific/educational low-orbit satellite containing many experiments and beacons, but no amateur transponders. UoSAT-OSCAR 9 is fully operational as this is written.

Radio Sputniks 3-8 were simultaneously launched aboard a single vehicle in December 1981. Several carried Mode A transponders and two carried a device nicknamed "Robot" that could automatically handle a CW QSO.

Iskra 2 was launched manually from the Salyut 7 space station in May 1982 and sported a Mode K HF transponder. It was destroyed upon re-entering the atmosphere a few weeks after launch.

Iskra 3, launched in November 1982 from Salyut 7, was even shorter lived than its predecessor.

OSCAR 10, the second Phase III satellite, was launched on June 16, 1983, aboard an ESA Ariane rocket, and was placed in a elliptical orbit. OSCAR 10 carries Mode B and Mode L transponders. OSCAR 10 is operational only part time as this is written. Listen to W1AW bulletins for operating schedules.

OSCAR 11, another scientific/educational low-orbit satellite like OSCAR 9, was built at the University of Surrey in England and launched on March 1, 1984. This spacecraft has also demonstrated the feasibility of store-and-forward packet digital communications and is fully operational as this is written.

OSCAR 12, built in Japan and launched by the Japanese in August 1986, is a low-orbit (Phase II) satellite carrying Mode J and Mode JD (digital store-and-forward) transponders.

Radio Sputnik 10/11, launched in June 1987, is two separate "satellites" on the same low-orbit spacecraft. The satellites carry Mode A, K and T transponders and CW QSO "Robots."

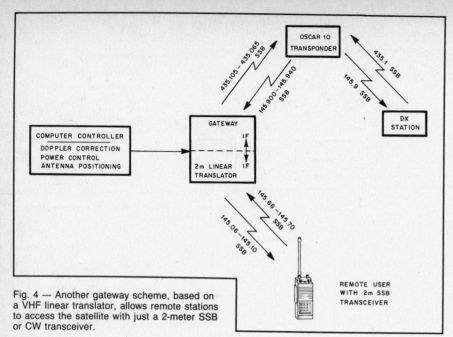

Fig. 4 — Another gateway scheme, based on a VHF linear translator, allows remote stations to access the satellite with just a 2-meter SSB or CW transceiver.

antennas would be too large and heavy at HF and VHF for the strict size and mass constraints imposed by the launch vehicle.

Communications Methods

Amateurs who use satellites for communications must be alert to the needs of using that resource efficiently. Only narrow bandwidth modulation methods, such as CW and SSB, should be used for general QSOs. NBFM is *not* an efficient use of the limited satellite transponder spectrum. Also, the 100% duty cycle of FM voice is a very inefficient use of transponder RF power. Conventional RTTY is allowed on AO-10 on the calling frequency of 145.880. Packet radio is the preferred data-transmission mode, though, because the information-rate/duty-cycle ratio is much higher.

Most satellites have recommended use patterns for their transponders. Attention is placed on devoting portions of the bandwidth of the transponder to various uses. Typical band segments include CW only, mixed CW and SSB, SSB only, channels for specialized or experimental modes, and special channels for telemetry and bulletins. Fig. 2 shows the OSCAR 10 band plan. Satellite users should consult *The ARRL Operating Manual* for operating information for amateur satellites.

Another interesting use of AO-10's global communications capability is the service provided by ''gateway'' stations. A gateway station provides an interface between differing services. In the case of AO-10, some amateur communities of VHF FM operators have been served by gateways that translate their FM repeater signals into satellite SSB signals, and vice-versa. Global and transcontinental contacts have been made through AO-10 by operators using VHF hand-held transceivers.

Fig. 3 shows the block diagram of a manually operated gateway that has been active from Illinois. Another interesting form of gateway is the land-based linear translator that provides a portion of the AO-10 passband for operators with limited resources for satellite operation. One particular translator gateway provides a 40-kHz-wide band segment for SSB service as shown in Fig. 4. Both of these gateways operate in half-duplex for the remote user. Other schemes permit a remote user to operate full duplex, but at the expense of more equipment on his part.

A variant of SSB holds promise for satellite users. Amplitude compandored single sideband, ACSSB, is spectrally and duty-cycle efficient. It can enhance the signal-to-noise ratio of satellite signals, all of which are relatively weak. ARRL, AMSAT and Project OSCAR are jointly sponsoring a test program called Project Companion to introduce ACSSB to Amateur Radio by way of OSCAR 10. The initial trials are very encouraging, but the program needs devoted participants to keep it on track.

COMMUNICATIONS REQUIREMENTS

Most long-distance communications depend on natural propagation phenomena. Some days, a kilowatt and high Yagi are needed to establish communications; at other times, 5 W and a long wire will do the trick. Stations designed to take advantage of ionospheric propagation are largely a matter of personal preference. Bigger stations usually mean bigger signals.

Operation through a satellite is somewhat different. Fixed conditions such as transponder receiver sensitivity and power output, and variable conditions such as spacecraft attitude and path losses, are the dominating factors in establishing minimum ground station requirements. Once the minimum requirements are met, you can have reliable communications any time the satellite is in view regardless of changing propagation phenomena. The communications requirements for OSCAR 10 have been pretty well defined since the spacecraft became operational in 1983. Unlike other areas of amateur interest, an inadequate station won't merely limit your ability to communicate — it may well prohibit communications entirely.

Satisfactory satellite communications might be possible with an existing station, but planning the ground station around certain parameters will enhance its effectiveness. Ground stations for Phase III satellites such as OSCAR 10 need to consider such subtleties as low-loss feed line, a single-transistor low-noise GaAsFET preamplifier, and high-gain VHF/UHF antennas that can be pointed accurately.

The parameters that should be considered for an OSCAR ground station are:

1) 146-MHz antenna gain of at least 13 dBi, circularly polarized. Right-hand circular polarization is a must for effective operation. More antenna gain is always desirable for better performance. Antenna gain of much more than 18 dBi, however, is not cost effective — satellite transponder noise floors will limit further benefits.

2) 435-MHz antenna gain of at least 13 dBi, circularly polarized. For this band, it is highly desirable to be able to switch between right-hand circular polarization (RHCP) and left-hand circular polarization (LHCP). Gains of 14 to 18 dBi are preferred for OSCAR 10, Mode B. For OSCAR 10, Mode L, antenna gain around 20 dBi is necessary. Future Mode L operations are expected to be satisfactory with the 14- to 18-dBi class antenna.

3) 1269-MHz antenna gain of 20+ dBi is required for OSCAR 10, Mode L. Current AO-10 users are having success with linear antenna polarization, even though the spacecraft antenna uses RHCP. Future operations will probably demand a switchable circular polarization. Future Mode L operations are also expected to be less demanding in the antenna department. Gain in the range of 16 to 20 dBi should yield satisfactory results.

4) Circularly polarized antennas must be mounted on nonmetallic booms so that the antenna performance is not degraded by a metal boom in the antenna field. Similarly, coaxial feed lines must be routed axially rearward past the antenna reflector to avoid pattern distortion.

5) Feed-line loss of less than 2 dB is needed; less than 1 dB is better. This can be achieved without expensive Hardline through the use of antenna mounted preamplifiers and other equipment.

6) 435-MHz EIRP (effective isotropic radiated power) of no more than 500 W. Most operations can be conducted with 100-W EIRP, and at times it's possible to use only 10-W EIRP.

7) 1269-MHz EIRP of 3000 to 25,000 W for OSCAR 10, Mode L. Future satellite requirements expected to be in the 500 to 3000 W category.

8) 146-MHz receive system noise figure no greater than 2 dB. Noise figures less than 1 dB are not warranted because of the limitations imposed by terrestrial noise sources.

9) 435-MHz receive system noise figure of less than 1 dB. Terrestrial noise levels are not a limiting factor on this band.

10) Frequency readout resolution of 1 kHz on transmit and receive. Accurate frequency readout allows you to find your signal on the downlink without causing needless QRM.

11) An accurate means of determining antenna direction at all times during an orbit. This means knowing where the antennas are pointed, and where to point them. Accurate graphical plotting or computer printout of pointing angles is required.

Like many other aspects of amateur radio, tradeoffs can be made to minimize expenditure of time and money. Up to a point, antenna gain is easier to obtain than transmitter power. It is easier to put up a bigger Yagi to increase your EIRP than to build or buy an amplifier. For example, a commercially available 40-element 435 MHz crossed Yagi makes enough EIRP for AO-10, Mode B communications with only 25 W of 435-MHz RF at the antenna. Often, only 5 to 15 W are all that's necessary.

Better antennas also improve receive performance. All things considered, it is much more cost effective — and desirable from a performance standpoint — to optimize the receiving set-up before investing in higher transmit EIRP. Don't be an alligator!

AMSAT and Project OSCAR have jointly produced a very useful manual that defines the details of the satellite link characteristics. This manual is recommended reading for all interested amateurs.

Circular Polarization

In the HF bands, polarization differences between antennas are not really noticeable because of the nonlinearities of ionospheric reflections. On the VHF and UHF bands, however, there is little ionospheric reflection. Cross-polarized stations (one using a vertical antenna, the other a horizontal antenna) often find considerable difficulty, with upwards of 20-dB loss. Such linearly polarized antennas are "horizontal" or "vertical" in terms of the antenna's position relative to the surface of the Earth, a reference that loses its meaning in space.

The use of circularly polarized (CP) antennas for space communications is well established. If spacecraft antennas used linear polarization, ground stations would not be able to maintain polarization alignment with the spacecraft because of changing orientation. Ground stations

Table 3
Polarization and Gain of the Phase III-B Antennas

Frequency (MHz)	High Gain Antennas Polarization	Gain (dBi)	Omni Antennas Polarization	Gain (dBi)
146	RHCP	9.0	Linear	0.0
436	RHCP	9.5	Linear	2.1
1269	RHCP	12.0	RHCP	0.0

(adapted from *The AMSAT-Phase III Satellite Operations Manual* prepared by AMSAT and Project OSCAR)

Fig. 5 — This diagram shows the far-field radiation pattern (undeformed) for the 2-meter high-gain antenna on OSCAR 10. (Adapted from *The AMSAT-Phase III Satellite Operations Manual* prepared by AMSAT and Project OSCAR.)

using CP antennas are not sensitive to the polarization motions of the spacecraft antenna, and therefore will maintain a better communications link.

All AO-10 gain antennas (for 2 m, 70 cm, and 24 cm) are configured for RHCP operation along their maximum gain direction. See Table 3 and Fig. 5. Since this direction is also the main antenna lobe along the spacecraft +Z axis, the best communications with AO-10 will also be along that direction. Since the AO-10 radiations are RHCP, ground stations should also be RHCP for optimum communications.

There are times, however, when LHCP provides a better satellite link. AO-10 is designed so that the main antenna lobe is oriented toward the center of the Earth when the satellite is at apogee. It is only so oriented for up to a few hours either side of apogee, however. As the satellite moves away from apogee, the orientation changes and the RHCP main lobe is not centered on Earth. At other times, ground command stations have to orient the spacecraft to odd angles to maintain proper sunlight on the solar cells that are used for prime power. Again, at such times the satellite antenna's main lobe is not pointed directly at Earth. During the periods when the main lobe is directed away from Earth, communications are still possible through the use of OSCAR 10's antenna side lobes. CP antenna patterns, however, exhibit side lobes that can be polarized in the opposite sense to the main lobe. At times when only LHCP side lobes are available, ground stations must be able to switch to LHCP to maintain optimum communications.

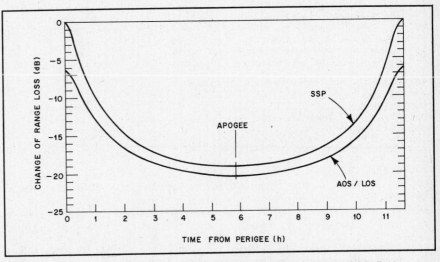

Fig. 6 — Range loss of signal strength during the 11.7 hour elliptical orbit of OSCAR 10. (Adapted from *The AMSAT-Phase III Satellite Operations Manual* prepared by AMSAT and Project OSCAR.)

Table 4

Summary of One-Way Transmission Losses for Communications Paths Between Earth and PHASE III Spacecraft at Apogee in an 11.7 Hour Elliptical Orbit, as Viewed Near AOS-LOS Range

Loss Mechanism	Attenuation (dB)		
	2m	70cm	24cm
Path Loss	168.07	177.57	186.86
Tropo/Ionospheric Refraction	0.002	0.0003	0.0002
Tropo Absorption	0.1	0.7	1.65
Ionospheric Absorption (by category)			
D-Layer	0.12	0.013	0.002
F-Layer	0.12	0.013	0.002
Aurora	0.13	0.014	0.002
Polar Cap Absorption	0.47	0.053	0.006
(A Rare D-Layer Event)			
Field-Aligned Irregularities	†	†	†

†Little data available, characterized by rapid amplitude and phase fluctuations.

(adapted from *The AMSAT-Phase III Satellite Operations Manual* prepared by AMSAT and Project OSCAR)

Table 5

Link Budget Computations for Mode B and Mode L Transponders on OSCAR 10 at Apogee and AOS-LOS Range

Ground Station Uplink

Symbol	Parameter	Mode B (70 cm Uplink)	Mode L (24 cm Uplink)
N_f	Sat. Receiver Noise Floor	−136.7 dBm	−137.4 dBm
SNR	Avg. Signal-to-Noise Ratio	15.0 dB	15.0 dB
P_s	Signal at Satellite	−121.7 dBm	−122.4 dBm
G_r	Satellite Antenna Gain	9.0 dBi	12.0 dBi
P_a	Signal at Sat. Antenna	−130.7 dBm	−134.4 dBm
P_L	Path Loss	177.57 dB	186.86 dB
$L_{i,t}$	Iono/Tropo Loss	0.73 dB	1.65 dB
$L_{p,p}$	Pointing/Polarization Loss	1.5 dB	1.5 dB
$P_t + G_t$	Reqd. Gnd. Stn. avg. ERP	19.10 dBW	25.61 dBW
	or:	81.3 W	363.9 W
	Gnd. Stn PEP ERP	25.10 dBW	31.61 dBW
	or:	323.6 W	1448.8 W
G_t	Gnd. Station Antenna Gain	16.0 dBi	19.0 dBi
P_t	Gnd. Stn. PEP Power Output	9.1 dBW	13.6 dBW
	or:	8.1 W	18.2 W

Satellite Downlink

Symbol	Parameter	Mode B (2 m Downlink)	Mode L (70 cm Downlink)
P_t	Satellite Max. PEP Output	16.99 dBW	16.99 dBW
	Average Output	10.99 dBW	10.99 dBW
L_m	Multi-user Load Share	−15.0 dB	−20.0 dB
G_s	Satellite AGC Compression	0.0 dB	0.0 dB
G_t	Satellite Antenna Gain	9.0 dBi	9.5 dBi
P_u	Average User ERP Output	34.99 dBm	30.49 dBm
P_L	Path Loss	168.07 dB	177.57 dB
$L_{i,t}$	Iono/Tropo Loss	0.33 dB	0.73 dB
$L_{p,p}$	Pointing/Polarization Loss	1.5 dB	1.5 dB
G_r	Receiving Antenna Gain	12.0 dBi	16.0 dBi
P_s	Received Signal Level	−122.91 dBm	−133.31 dBm
P_n	Noise in Rcv. Bandwidth	−137.76 dBm	−146.67 dBm
SNR	Avg. Signal-to-Noise Ratio	14.8 dB	13.4 dB

Link computations are based on the following parameters:

11.7 hour elliptical orbit
Eccentricity = 0.6
Slant range = 41,395 km at AOS/LOS
2.4 kHz SSB Signal
6 dB Peak (PEP) to Average signal ratio
Ground Station Receiving Noise Characteristics:
   at 2m: T = 505 K, NF = 4.4 dB
   at 70cm: T = 65 K, NF = 0.9 dB

(adapted from *The AMSAT-Phase III Satellite Operations Manual* prepared by AMSAT and Project OSCAR)

Independently switchable RHCP/LHCP antenna circularity is necessary, especially for 70 cm. It is expected to be required for 24 cm operation on Phase III-C as well.

Operators who use AO-10, Mode L have determined that the uplink can be affected by LHCP antenna lobes when the satellite is located off-axis to the ground station. Switchable RHCP/LHCP on 2 m may be convenient, but it is rarely needed.

Spin Modulation

An AO-10 design characteristic is that the 2-m high-gain array pattern has three side lobes located along the spin axis that are only about 3 dB weaker than the primary axial lobe. The effective radiation pattern intended for this antenna is shown in Fig. 5. The axial radiation lobes of the 70-cm and 24-cm high-gain antennas are similarly symmetrical about the spin axis. Lobes in the off-axis antenna patterns produce amplitude modulation of signal strength as the spacecraft spins about its axis. Off-axis operations of AO-10 are an established fact of life, creating spin modulation frequencies of about 0.5 Hz from a 10 r/min spin rate.

Stations using CP antennas are much less prone to be affected by spin modulation that those using linearly polarized antennas. Good signals are maintainable even far off-axis from the main lobe, if the ground station is using CP.

Path Loss

In free space, a station located at a distance from an RF source will receive an average power flow from that source that is inversely proportional to the square of the distance. Doubling the range will reduce the signal by 6 dB. Since Phase III satellites have substantially elliptical orbits, the changes in path length (slant range) and loss is correspondingly sizable. Fig. 6 illustrates the changes in path length of AO-10 signals for all conditions from apogee to perigee and for direct overhead passes (SSP location) to the far-limb viewing at AOS or LOS.

Other physical processes introduce additional losses in signals traveling to and from AO-10 through the troposphere and ionosphere. While many of these effects are frequency dependent, the additional path losses for AO-10 signals are small compared to the slant range path losses. Table 4 summarizes the path losses for each band used on AO-10.

Link Budget

Link budget computations can be deceiving to the unwary observer. The problems arise from the application of wide-bandwidth spacecraft transponders to handle a wide variety of complex, time-averaged, narrow-bandwidth QSOs. The analysis presented in Table 5 employs a somewhat lower signal-to-noise ratio than the source information. Actual on-the-air experience has shown these lower values to be workable. The computation is also performed on the basis of averaged RF power levels. The conversion to PEP is done at the end, assuming a 6 dB relationship between PEP and average.

In the link computations, no allowances have been made for ground station transmission line attenuation. Information for attenuation of a variety of popular feed lines is shown in Chapter 16. If you measure RF power near the antenna and have a tower-mounted, low-noise preamp, the link computations can be used as presented. The computation is based on two possible worst-case conditions: with the maximum slant range values at AOS/LOS, and with the spacecraft at apogee. An important assumption that the spacecraft antenna pattern is on-axis to the communications path, a highly unusual spacecraft orientation condition. Of course, an infinite number of cases could be presented in such a tabular assessment.

Another highly variable parameter that must be taken into account is pointing and polarization loss. With the wide variety of OSCAR 10 offset pointing situations that have been seen in the first few years of AO-10 operation, the pointing and polarization losses can easily achieve values of 10 dB or more. Also witnessed in AO-10 operations is the very sharp degradation of the communications link when the satellite is at very low elevation angles (essentially on the horizon). Nevertheless, the presentation of Table 5 is in the ballpark for Mode B operations for those power level and received signal conditions seen in practice.

Mode L operations on OSCAR 10 have not achieved the performance values seen in Table 5 owing to a faulty voltage regulator in the Mode L transponder. Consistent Mode L performance, with the spacecraft at apogee, can be attained with uplink power in the range of 10 to 25 kW ERP. Also required is a good receiving setup, with 70-cm downlink antenna gain about 22 dBi and a low-noise GaAsFET preamp at the antenna. When the satellite is at lower altitudes, successful OSCAR 10, Mode L QSOs can be made with uplink power as low as 3 kW ERP.

Table 5 also provides a good guide for expected link operations for Phase III-C. Mode JL transponder link performance can be determined easily, provided there are no unusual equipment failures when the spacecraft is launched.

EQUIPPING A STATION

Perhaps the biggest difference between terrestrial and satellite communications is that the latter is full-duplex operation. This means that you transmit and receive simultaneously. You can hear your own downlink signal while transmitting, as well as that of the station being worked. Full duplex provides the opportunity for a fully interactive conversation, as if the other station is in the very same room.

Successful satellite operation demands that you can locate and hear your own signal from the spacecraft. Choose equipment with this goal in mind. Equipping a station for full-duplex operation is not too difficult because the transmitter is on a different band than the receiver. Satellite ground-station configurations vary ac-

cording to the satellite communications "mode" being used. Figs. 7, 8 and 9 show several different configurations for Mode B, Mode L and Mode J. Each of these options will be discussed.

Receivers

Receiving requirements for OSCAR 10 are demanding, but pleasurable results can be achieved with the right kind of equipment. Do not expect to find 60-dB-over-S9 signals on the downlink. OSCAR operation is a weak-signal situation where contacts can be made with signals that are 4 dB stronger than the noise. Conversational quality can be assured with signals that are 6-9 dB or greater out of the noise.

The old adage "You can't work 'em if you can't hear 'em" especially applies to satellite work. The first step that should be taken toward gearing up for OSCAR 10 is to assemble the best receiving setup possible. There is no point in getting transmitting capability until the satellite signals can be comfortably heard.

There are a number of options for starting a satellite station from scratch. You may wish to try building almost any part of the station from a simple controller, all the way up to a complete station. Alternatively, the busy operator may wish to purchase everything. All of the necessary components are readily available from commercial suppliers.

Amateurs active on 2 meters with a multimode transceiver already have the basic building block for receiving Mode B and transmitting on Modes L and J. If you currently have no VHF equipment, consider that a multimode transceiver will also allow you to explore the exciting world of terrestrial 2-meter SSB operation. The basic requirements are that the rig includes SSB and CW modes and that it covers the entire 2-meter band. A multimode transceiver

also makes an excellent replacement for an FM-only 2-meter rig.

The major equipment manufacturers listed in Table 6 all make suitable transceivers. The current crop of base-station rigs includes the Kenwood TS-711A, ICOM IC-271A and Yaesu FT-726R. There are also several compact multimode radios intended for mobile use that will be quite usable. These include the Yaesu FT-480R, Kenwood TR-9130 and ICOM IC-290H. In addition, there are often good buys on the used market, if you're interested in an older radio. Gear such as the Kenwood TS-700 series, Yaesu FT-225RD and ICOM IC-251 are still popular. Many of these transceivers have been reviewed in *QST*.

An excellent solution to receiving satellite signals can be found in the form of receiving converters used with your quality HF transceiver or receiver. The receiving converter consists of a mixer and a local oscillator and may contain a preamplifier. The local oscillator frequency is usually chosen so that signals will be converted for reception by any receiver that covers the 10-meter band. In addition, a number of manufacturers offer transverters that include a receiving converter and transmitting converter in the same package.

Receiving converters are available commercially from several suppliers listed in Table 6. For those who enjoy building equipment, there are several suitable construction projects in Chapter 31.

There are several advantages to using a receiving converter. Modern HF transceivers and receivers most likely have excellent frequency stability, a frequency readout in 1 kHz or smaller steps, good SSB and CW crystal filters, an effective noise blanker and high dynamic range. Chances are good that a multimode VHF transceiver will offer some, but not all, of these features. Cost is another factor. If

The ICOM IC-271A is typical of the 2-meter multimode transceivers on the market today. It can be used for reception on the OSCAR 10 downlink. The IC-471A, virtually identical in appearance, offers a matching 70-cm, 25-W signal for the uplink.

Table 6

Suppliers of Equipment of Interest to Satellite Operators

Multimode VHF and UHF Transceivers and Specialty Equipment
ICOM America, Inc., 2380-116th Ave. NE, Bellevue, WA 98004.
Kenwood USA Corp., 2201 East Dominguez St., Long Beach, CA 90810.
Ten-Tec Inc., Sevierville, TN 37862.
Yaesu Electronics Corp, 17210 Edwards Road, Cerritos, CA 90701

Converters, Transverters and Preamplifiers
Advanced Receiver Research, Box 1242, Burlington, CT 06013.
Angle Linear, P.O. Box 35, Lomita, CA 90717 (preamps only).
Hamtronics, Inc., 65-E Moul Rd., Hilton, NY 14468.
Henry Radio, 2050 S. Bundy Dr., Los Angeles, CA 90025.
The PX Shack, 52 Stonewyck Dr., Belle Mead, NJ 08502.
Radio Kit, PO Box 973, Pelham, NH 03076.
Spectrum International, P.O. Box 1084, Concord, MA 01742.
Transverters Unlimited, Box 6286, Station A, Toronto, ON M5W 1P3, Canada.

Power Amplifiers
Alinco Electronics, P.O. Box 20009, Reno, NV 89515.
Communications Concepts, Inc., 2648 North Aragon Ave., Dayton, OH 45420.
Down East Microwave, Box 2310-RR 1, Troy, ME 04987.
Encomm, 1506 Capitol Ave., Plano, TX 75074.
Falcon Communications, P.O. Box 8979, Newport Beach, CA 92658.
Mirage Communications, P.O. Box 1000, Morgan Hill, CA 95037.
TE Systems, P.O. Box 25845, Los Angeles, CA 90025.

Antennas
Cushcraft Corp., 48 Perimeter Rd., Manchester, NH 03108.
Down East Microwave, Box 2310-RR 1, Troy, ME 04987.
KLM Electronics, Inc., P.O. Box 816, Morgan Hill, CA 95037.
Telex Communications, Inc., 9600 Aldrich Ave. S, Minneapolis, MN 55420

Note: This is a partial list. The ARRL does not endorse specific products.

you already own an HF rig, but are not interested in terrestrial VHF/UHF SSB operation (you don't need 2-meter transmit capability for Mode B), the cost of building or buying a superior receiving converter will be significantly less than that of even an older multimode transceiver.

Experience has often shown that daytime noise will often raise the practical 2-meter receiver noise floor by 10 to 20 dB, thus making OSCAR 10, Mode B, daytime communications difficult, at best. Weak downlink signals are often no match for the noise. In general, noise is not a problem on 70-cm (for Mode J and Mode L reception), but in some areas interference from airport radar can be troublesome. In addition, local FM repeaters may be heard in the satellite passband of the ground-based receiver because the VHF transceiver may offer poor rejection of strong nearby signals. Use of a high-dynamic-range receiving converter with a good HF transceiver has been shown to solve both of these problems. The lesson here is that many VHF transceiver noise blankers are inadequate for VHF/UHF satellite operation, and that some VHF transceivers do not work well in areas with many strong, nearby signals. Better results may be had with a receiving converter than with a VHF

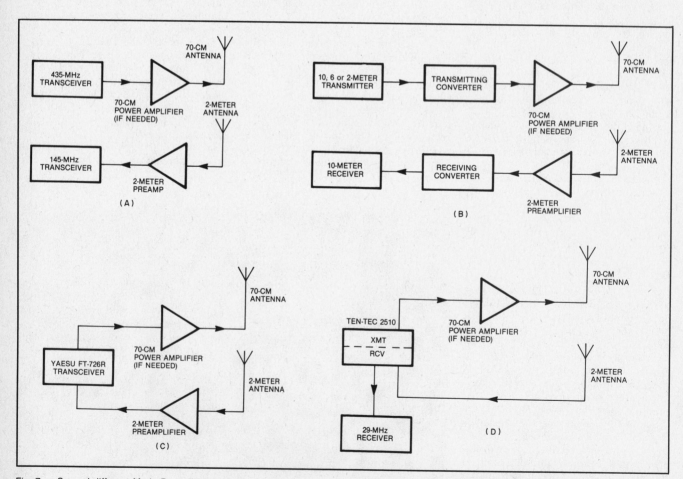

Fig. 7 — Several different Mode B satellite-station configurations are shown here. At A, separate VHF/UHF multimode transceivers are used for transmitting and receiving. The configuration shown at B uses transmitting and receiving converters or transverters with HF equipment. At C, the Yaesu FT-726R can perform both transmitting and receiving functions, full duplex, in one package. The Ten-Tec 2510 shown at D contains a 435-MHz transmitter and a 2-meter to 10-meter receiving converter.

For amateurs with HF receivers or transceivers who have no desire for transceive operation on 2 meters, a receiving converter such as this Advanced Receiver Research R144VDA may be the answer. Shown with matching low-noise GaAsFET preamplifier, this unit converts 2-meter OSCAR 10 downlink signals for reception on any receiver that can tune to 29 MHz.

multimode transceiver.

Preamplifiers

No discussion of satellite receiving systems would be complete without mentioning preamplifiers. Good, low-noise preamplifiers are essential for receiving weak downlink signals. Multimode rigs and most transverters will hear much better with the addition of a GaAsFET preamplifier ahead of the receiver front end, albeit at the expense of a considerable reduction in the third-order intercept point of the receiver. While a preamplifier can be added right at the receiver in the station, it may not do much good there. Considerably better results can be obtained if the preamp is mounted at the antenna. Indeed, antenna mounting of a preamp is essential for UHF operation. Losses in the feed line will seriously degrade the noise figure of even the best preamplifier mounted at the receiver.

Table 6 lists several sources of commercially-built preamplifiers. These are available in several configurations. Some models are designed to be mounted in a receive-only line, for use with a receiving converter or transverter. Others, designed with multimode transceivers in mind, have built-in relays and circuitry that automatically switch the preamplifier out of the antenna line during transmit. Still others are housed, with relays, in weatherproof enclosures that mount right at the antenna. For the equipment builder, several suitable designs appear in Chapter 31 and in *The Satellite Experimenter's Handbook*.

Transmitters

The OSCAR 10, Mode B uplink requires 5 to 20 W of 435.1-MHz RF at the antenna. This assumes a good antenna, which will be discussed later. Feed-line losses in a typical 435-MHz installation can easily run 3 dB, so you'll need anywhere between 10 and 50 W output from your transmitter. OSCAR 10, Mode L, presents rather high transmitting power requirements (approximately 10 to 25 kW ERP), caused by the failure of a single transistor. The Phase III-C spacecraft is expected to have a more normal 2 kW ERP requirement.

Since the possible number of combinations of transmitter power and antenna gain needed to give a satisfactory signal through OSCAR 10 is infinite, satellite users generally talk about their uplink capability in terms of effective radiated power (ERP). ERP takes into account antenna gain, feed-line loss and RF output power. For example, a 10-W signal into a 3-dB-gain antenna will have an ERP of 20 W (3 dB greater than, or twice as strong as, 10 W). This assumes no loss in the feed line; all 10 W from the transmitter reaches the antenna. If the signal is 10 W into a 10-dB-gain antenna, the ERP is 100 W. The same 100-W ERP can be achieved with a 50-W transmitter and a 3-dB-gain antenna.

Stations with an uplink ERP as low as 10 W can be copied through OSCAR 10, Mode B, but ERP levels of 100 to 400 W are the norm. No matter what your ERP, your signal on the downlink should never be stronger than the AO-10 general beacon at 145.81 MHz. You must have a way of adjusting your uplink signal so that it is no stronger than the beacon. This point is discussed in detail in *The ARRL Operating Manual*.

If the Mode B satellite ground station has a 10-W transmitter, a short run of low-loss feed line and good antenna gain, an additional amplifier would probably not be needed. If losses and gains do not add up to enough station ERP, a 30- to 40-W amplifier may be needed. Some operators have 100-W amplifiers, but with the antennas available today, use of that much power is guaranteed to create an uplink signal that far exceeds the beacon level. This is considered by good operators to be an antisocial action. Considerate operators with the 100-W amplifiers quickly reduce drive power to lower the ERP to acceptable levels.

Most satellite operators use UHF multimode transceivers to generate Mode B uplink signals. The manufacturers listed in Table 6 all make 70-cm multimode transceivers that are similar to the 2-meter units described earlier. Although most of these transceivers provide 10-W output, some can deliver 25 W or more.

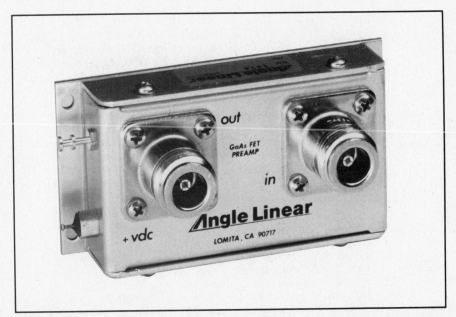

A low-noise GaAsFET preamplifier such as this Angle Linear model is essential if you want to maximize your downlink reception. The preamplifier should be mounted at the antenna for best results.

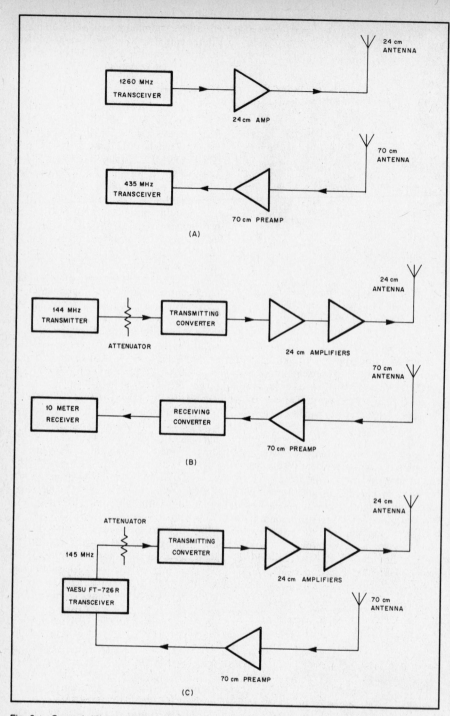

Fig. 8 — Several different Mode L satellite-station configurations are shown here. At A, separate VHF/UHF multimode transceivers are used for transmitting and receiving. The configuration shown at B uses transmitting and receiving converters or transverters with HF equipment. At C, the Yaesu FT-726R can be used for full duplex receiving and transmitting (with the addition of a 2-meter to 24-cm transmitting converter).

Unless you want to try 70-cm terrestrial communication (and that can be fun, too), there is no need for a complete UHF transceiver. Transmitting converters for use with HF transceivers are available from suppliers listed in Table 6. In addition, some of those manufacturers make transverters that are suitable for satellite work. If more 435-MHz power is needed, there are a number of solid-state amplifiers on the market. See Table 6. Choose carefully; you don't need a rock crusher.

For Mode L transmitting, there are several transmitting converters, transverters, amplifiers and even a multimode transceiver available from the suppliers listed in Table 6. Chapter 32 details several construction projects that will get you on Mode L. AO-10, Mode L operation will require the power levels offered by a tube-type amplifier, while it is expected that Phase III-C will only need power levels that

can be generated by transistors.

Specialty Equipment

Separate transceivers or transmitting and receiving converters are no longer the only way to go. Modern equipment offerings by Yaesu and Ten-Tec, tailored for the satellite user, do it all in one package.

The Yaesu FT-726R starts out as a 2-meter multimode transceiver. It is, however, expandable to work on other bands with the addition of optional modules. The Mode-B satellite operator would most likely be interested in an FT-726R with the stock 144-MHz and optional 430-MHz modules. These same RF modules will also serve well for Mode J directly, and Mode L using an outboard 23-cm transmitting converter. To tie it all together, Yaesu offers an optional satellite module that allows the amateur to transmit on one band for the uplink while receiving on another band for the downlink. This is full duplex operation; the effect is the same as having two separate radios in one box.

Ten-Tec's 2510 is tailored specifically for Mode-B satellite operation. This unusual piece of equipment includes a hot receiving converter that converts 145-MHz signals to 10 meters for reception on any HF receiver or transceiver. A low-noise GaAsFET preamplifier is built in, so no external preamp is required. For the uplink, the 2510 has a complete 10-W, 435-MHz SSB and CW transmitter. The 2510 has only one frequency tuning control for the receiver and the transmitter. The receiver automatically tracks the transmitter, an exceptionally useful feature for busy satellite operators.

Antennas

The best antennas for OSCAR 10 have circular polarization (CP). Helical antennas like the ones shown in Fig. 10 were the way to go for Phase II satellites. An eight-turn helical for 70 cm and a huge six-turn helical for 2 meters provided excellent results for OSCAR 8. For OSCAR 10, however, more gain is needed than a reasonably sized helical antenna can supply. In addition, successful OSCAR 10 operation requires the ability to conveniently switch from RHCP to LHCP.

The present trend in satellite arrays uses a different method to achieve circular polarization. These antennas, shown in Fig. 11, are essentially two complete Yagis mounted perpendicular to each other on the same boom. One set of elements is mounted ¼ wavelength ahead of the other. The antennas are fed in phase and are switchable from RHCP to LHCP. These particular antennas, manufactured by KLM, have proved to be excellent performers. Cushcraft and Telex/HyGain also manufacture crossed-Yagi satellite antennas for OSCAR service. Emphasis must be placed on achieving CP operation on single boom antennas, rather than multiple separate Yagis mounted orthogonal to

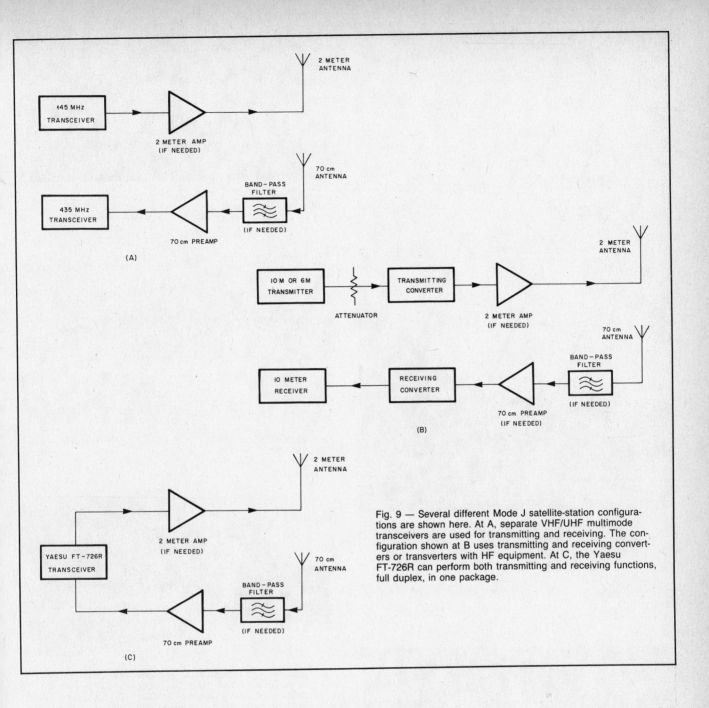

Fig. 9 — Several different Mode J satellite-station configurations are shown here. At A, separate VHF/UHF multimode transceivers are used for transmitting and receiving. The configuration shown at B uses transmitting and receiving converters or transverters with HF equipment. At C, the Yaesu FT-726R can perform both transmitting and receiving functions, full duplex, in one package.

each other. The latter can lead to somewhat uncertain circularity.

Satellite antennas should be mounted as close to the station as possible. Height above ground makes no difference for satellite work, except that the antennas must be mounted high enough that trees and other obstructions do not block the view of the satellite at low elevations. A low mount allows use of shorter feed lines (lower losses) and often reduces QRN pickup by the antennas. Many operators are able to set up their antennas on a 10 to 15-foot mast right next to the shack and have only 20 feet of feed line. Plan to use good-quality, low-loss coaxial cable from the start, such as Belden 9913. Even better is a run of Hardline.

Mode L transmitting antennas have taken numerous forms, mostly based on

the technology needed for EME communications. Loop Yagis, such as those presented in Chapter 32, are popular, but you must stack several in an array to achieve the 24 dBi gain necessary for AO-10, Mode L. Also popular are the large parabolic reflector antennas seen in EME service. For Mode L, and especially for the upcoming Phase III-C satellite, some of these higher gain antennas are overkill. While higher gain means lower transmitter power, the narrow beamwidths require the operator to reposition the antenna more often.

Although a practical CP Yagi for 24 cm has not yet been demonstrated, such an approach may be feasible. Active Mode L operators have found that a small parabolic dish (4 to 6 feet in diameter) with a circularly polarized feed makes a fine Mode

L antenna. A number of reasonably priced TVRO dishes are available, and they require only the addition of a suitable feed for Mode L service. A low-cost, home built parabolic Mode L antenna is presented later in this chapter.

STATION ACCESSORIES

Satellite communications, like any other specialized facet of Amateur Radio, requires some specialized station accessories. Having the best equipment does not necessarily guarantee success. There are a number of "hints-and-kinks" type ideas that can make OSCAR operation far more satisfying. These items convert a basic receiver, transmitter and a pair of antennas into a very workable operating position. Some of the items discussed here will

Fig. 10 — Helical antennas such as these gave excellent performance for low-orbiting Phase II satellites.

The Kenwood TS-811A is a 70-cm multimode transceiver that can be used to generate a 10-W, 435-MHz signal for the uplink. A similar unit, the TS-711A, may be used for the 2-meter downlink.

Another means of generating a 435-MHz uplink signal is with an HF transceiver and a 10-meter to 70-cm transverter such as this SSB Electronics TV28-432.

Usually, a little more than 10-W uplink power is required for a good downlink signal. Solid-state "brick" amplifiers such as this Mirage D1010 provide the extra power needed. Care must be taken, however, to use only the minimum power necessary to maintain reliable communications.

Fig. 11 — A very popular commercially manufactured antenna array for OSCAR 10, Mode B is a pair of KLM crossed Yagis. The large box on the mast contains a 2-meter preamplifier and a 70-cm power amplifier, as well as power-supply circuitry.

W1INF, the ARRL Laboratory station, has OSCAR 10 Mode-B capability. The uplink is a 2-meter multimode transceiver driving a 2-meter to 70-cm transverter, while the downlink is a receiving converter and 10-meter receiver. A GaAsFET preamplifier is mounted at the antenna.

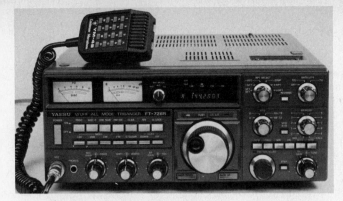

Yaesu's FT-726R is a favorite among satellite users because it can work on both 2 meters and 70 cm. With the optional satellite module that allows full-duplex operation, the effect is practically the same as having two separate transceivers in one box.

The Ten-Tec 2510 Mode-B satellite station is designed specifically for Mode-B operation. The box contains a 435-MHz SSB and CW transmitter, as well as a 145- to 28-MHz receiving converter with a low-noise front end.

provide capabilities beyond that of satellite operation. They also apply to VHF and UHF terrestrial work. Design your station to suit your own needs.

Antenna Rotators

Unlike stations located on the surface of the Earth, OSCAR 10 will be found somewhere in the sky above. Operators commonly aim antennas toward another station by changing the pointing angle, or azimuth (sometimes called az). Aiming antennas toward OSCAR 10 requires the control of antenna elevation (el). Satellite antennas must be able to rotate from side to side and up and down simultaneously. See Fig. 12. While the use of electrically controlled antenna rotators will be discussed here, it might be noted that OSCAR 10's motions are slow enough that hand-operated, "armstrong" antenna control is feasible. At times, the antennas do not need to be repositioned for periods of up to four hours.

Azimuth Rotators

Azimuth rotators are commonly used for positioning terrestrial HF and VHF antennas. Antennas for OSCAR 8 and other low-orbit satellites were on the

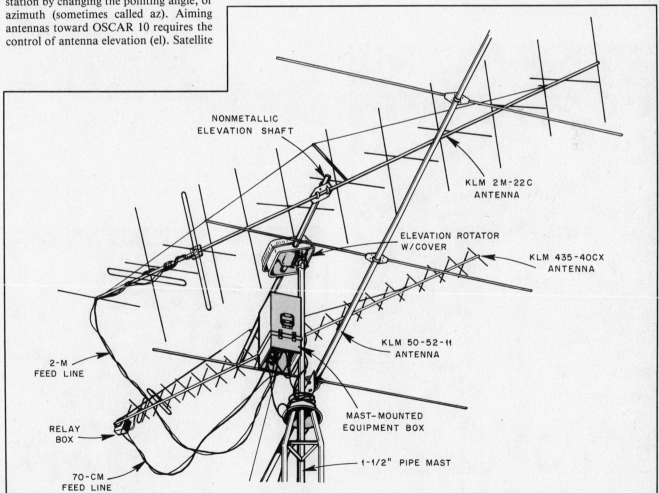

NONMETALLIC ELEVATION SHAFT

KLM 2M-22C ANTENNA

ELEVATION ROTATOR W/COVER

KLM 435-40CX ANTENNA

KLM 50-52-11 ANTENNA

2-M FEED LINE

RELAY BOX

MAST-MOUNTED EQUIPMENT BOX

70-CM FEED LINE

1-1/2" PIPE MAST

Fig. 12 — Some of the finer points of this OSCAR array are discussed in the text.

smaller and lighter side, so light-duty TV-antenna rotators such as those sold by Alliance, Channel Master, Radio Shack and others could be used for the azimuth rotator. Today's high-gain satellite arrays are a bit large for these light-duty rotators. Look for something more robust, such as a rotator recommended for turning a small HF beam or VHF array. Various models manufactured by Alliance, Daiwa, Kenpro, Telex and others are routinely advertised in *QST*.

Elevation Rotators

Elevation rotator selection is somewhat more limited, but there are some interesting things that can be done. Commercially manufactured models are available. The Kenpro KR500, designed specifically for elevating small- to medium-size VHF or UHF arrays, is quite popular among satellite operators. Other offerings include the Dynetic Systems DR10 and Kenpro KR5400, both of which are combined az-el rotators.

A lower cost, commercially manufactured alternative is the Alliance U110 TV-antenna rotator. Rotators of this type have been used by satellite operators for quite a few years. Despite its relatively light construction, the U110 will handle antenna loads weighing up to 80 pounds or more. The key to success is to achieve a static

balance of the antenna mass so that the rotator does not have to elevate a "dead" load. A highly attractive feature of the elevation rotators noted previously is that the cross boom to be rotated passes completely through the rotator. This allows the mounting of one antenna on each side of center and the adjustment of their respective positions for a side-to-side balanced load.

Figs. 13 and 14 show one particular method of mounting the U110. The rotator is clamped to the mast (the one that the azimuth rotator turns) with a plate that permits it to mount 90° from its normal orientation. With the rotator mounted in this position, it is not protected from rain or snow as well as it is in the normal position. A cover (an appropriately sized plastic dishpan or bucket is ideal) is added to afford protection from the elements. A problem with polyethylene plastics, commonly used in kitchenware, is that solar radiation quickly deteriorates their polymeric structure and causes the plastic to break apart. As shown, the plastic cover has been protected from the sun with an aluminum baking pan.

Antenna Cross Boom Construction

One requirement not commonly discussed is that of using a nonmetallic elevation axis boom for antennas that have their boom-to-mast mounting hardware in the center of the boom. A metal cross boom will seriously distort the beam pattern of a circularly polarized antenna, so it is important to make the cross boom from nonmetallic material.

From a structur-

al standpoint, the best nonmetallic material for this job is glass-epoxy composite tubing, because its stiffness is excellent. Lengths of this material may be found at an industrial supply house that specializes in plastics. Also, KLM sells lengths of 1½ inch OD fiberglass masting for this purpose, and Telex/HyGain includes fiberglass masting with their OSCAR antennas. If you have a rotator that will accept a 1½ inch elevation boom, then your best bet is to use a single piece of this tubing.

A less expensive alternative is to make the cross boom from a combination of metallic and nonmetallic tubing. For strength and stiffness, use a short length of steel or aluminum tubing through the middle of the rotator. Let the metal tubing extend for about 6 inches on each side of the rotator. Then install nonmetallic masting, such as common PVC pipe, over the steel stubs.

The elevation boom pictured in Figs. 12 to 14 was constructed with this method. The center piece that fits through the U110 is a 2-foot section of 1.33-inch-OD steel tubing that originally was part of the top support rail of a chain-link fence. Attached to the steel stub on each side of the rotator is a 4-foot length of 1¼ inch, schedule-40 PVC pipe. PVC pipe is specified by the nominal ID, here 1¼ inches. There are several varieties of 1¼ inch PVC pipe — schedule 40 indicates a thick-wall, heavy-duty version. This pipe slipped nicely over the center stub. The fit is perfect — no machining was needed.

Unlike glass-epoxy tubing, PVC pipe is not very stiff; it needs help. The secret to making PVC pipe capable of supporting satellite Yagis is to insert a wooden dowel into the PVC pipe, along its entire length. The finished dimension of 1-3/8 inch wooden clothes-rod dowel (the kind you might hang inside a closet) is just perfect for a slide fit into the pipe. This material is available from most lumber yards. Add a few ¼ inch bolts to each side to secure the

Fig. 13 — The elevation rotator, an Alliance U110, is protected from the elements by plastic and aluminum covers. The large white box holds tower-mounted equipment.

Fig. 14 — Close-up of the U110 showing how it mounts to the mast. Note the PVC pipe that slides over the steel stub protruding from the rotator.

Fig. 15 — Details of the position potentiometer mounting and weighted arm.

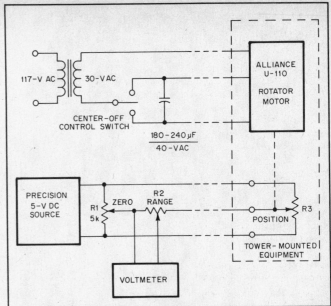

Fig. 16 — Block diagram of the elevation rotator direction control and position readout.

pieces, and you've got a sturdy, inexpensive, nonmetallic elevation boom.

Position Indicators

You may notice additional hardware around the elevation rotator shown in the photos. With the use of high-directivity antennas, the accuracy of the U110 control box is questionable. Adding a single-turn, 1-kΩ precision potentiometer provides the ability to closely control and position the elevation boom. See Fig. 15. The potentiometer (a large surplus instrumentation model) is attached to the elevation boom with a U bolt and angle bracket. A metal arm with lead weights at the end is attached to the potentiometer shaft. The weighted arm and gravity hold the shaft still while the potentiometer body turns with the elevation boom.

A simple circuit, shown in Fig. 16, is all that is needed to control the U110 and to use signals from the precision potentiometer (R3) for an accurate position indicator. R1 is used to zero the scale, while R2 calibrates the range of the indicator. The value of R2 will depend on the value of the voltmeter you use. For example, a 0- to 100-mV meter might be used, adjusted so that 10 mV equals 10° elevation, 30 mV equals 30° elevation, and so on.

Tower-Mounted Preamplifiers

To get the most out of your satellite station, you'll need to mount a low-noise preamplifier on the tower or mast, near the antenna, so that feed-line losses do not degrade low-noise performance. Feed-line losses ahead of the preamplifier add directly to receiver noise figure. A preamp with a 0.5-dB noise figure will not do you much good if there is 3 dB of feed-line loss between it and the antenna.

There is a variety of good preamplifiers available. Chapters 31 and 32 detail several suitable construction projects, and Table 6 lists many commercial suppliers.

Mast mounting of sensitive electronic equipment has been a fact of life for the serious VHF/UHFer for years, although it may seem to be a strange or difficult technology for many HF operators. A mast-mounted preamp is not difficult to construct, and there is a growing number of commercially available models as well. See the list of equipment suppliers in Table 6.

Take another look at Fig. 13. The large white box located below the elevation rotator can't be missed. Fig. 17 shows the interior of this box: It contains a lot of items besides a simple preamplifier. The box holds two racks of equipment. On the right are two dc voltage regulators with their pass-transistor heat sinks. These regulators provide on-site-regulated 28-V dc and 12-V dc from an unregulated 50-V dc supply located in the shack. Below the regulators is a 24-cm transmitting converter for the Mode L uplink. Opposite the regulators on the other rack panel is a 70-cm solid-state power amplifier for the

Mode B uplink, a 70-cm preamplifier for the Mode L downlink, and relays.

Fig. 18 is another view that details the area below the 24-cm transmitting converter. A close look shows a 2-meter preamplifier for the Mode B downlink and relays to switch it in and out of the line to the antenna. This setup is overkill to beat excessive feed-line losses. Normal installations require only the receiving preamplifier (2 meters for Mode B and 70 cm for Mode L) to be mounted on the mast.

Station Control Circuitry

Fig. 19 is a schematic diagram of the control circuitry for the tower-mounted rack. Parts of this diagram will be helpful, even if only the preamp is just mounted at the antenna. Note that this circuit is designed around the surplus coaxial relays that were available at the time. Your version will probably be different and will depend on the relays available to you. Switching requirements for coaxial relays were the subject of a comprehensive discussion by Joe Reisert, W1JR. Fig. 19 is one version of his concepts.

This circuitry performs several functions. For starters, it places the preamp in the line only during receiving periods and takes it out of the line during transmitting periods as well as at those times when the station is not in use. This is needed if the satellite array is used for terrestrial transceive operation as well. The switching arrangement shown also protects the preamp from stray electromagnetic pulses (EMP), such as lightning strokes, when the station is not in use. EMP protection is desirable even if the antenna and preamp are used only for receiving OSCAR 10 signals.

Fig. 19 is a bit more complicated than the average mast-mounted preamp setup

Fig. 17 — Interior of the tower-mounted equipment rack with the cover removed. The 70-cm equipment is on the left, while power-supply regulators, a 24-cm transmit converter and a 2-meter preamp are mounted on the right.

Fig. 18 — Close-up of the 2-meter preamplifier and relays.

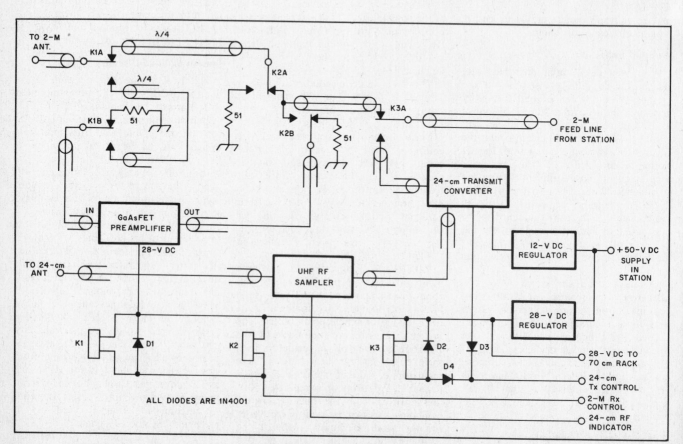

Fig. 19 — Control circuitry for the mast-mounted 2-meter preamplifier. K1-K3 are surplus coaxial relays.

because it also allows 2-meter RF to drive a 24-cm Mode L transmitting converter. An extra relay (K3) is used to switch between 2-meter and 24-cm operation. K1, a DPDT transfer relay, switches the input of the preamp between the antenna and a 50-ohm termination. K2, another DPDT relay, switches the preamp output between a 50-ohm termination and the feed line to the shack. The coaxial cable used for connections between the relays is cut to 0.1 to 0.2 electrical wavelength to achieve maximum isolation between the transmitted RF and the preamp input. The relays are connected so that they must be energized to place the preamp in line.

Tower-Mounted Equipment Shelters

A great many amateurs seem apprehensive about placing their valuable radio equipment outdoors. Such fears are unfounded if adequate care is taken to protect the equipment from the elements. The equipment shown in the photos has been outdoors for years without any adverse effects.

The present mast-mounted enclosure shown earlier is a welded aluminum box purchased from a surplus dealer. It was used because it was available and the price was right. You don't really need a big box like this if you just want to protect a preamp and relays.

Fig. 20 shows the basic scheme for weatherproofing tower-mounted equipment. The fundamental concept is to provide a cover to shelter equipment from rain. A 2-inch-deep aluminum cake pan is about the minimum acceptable cover. A trip to the housewares section of the local department store will reveal a variety of plastic and aluminum trays and pans that can make suitable rain covers. As mentioned before, polyethylene plastic must be protected from sunlight. Clear polystyrene refrigerator containers work better than those made of polyethlene, and aluminum is best of all. Choose a cover that is large enough for your equipment; remember to leave room for connecting cables.

The bottom of the rain cover is open to the elements. This is done on purpose and will not cause any problems. Do *not* try to hermetically seal the enclosure. By leaving the bottom open, adequate ventilation will prevent accumulation of water condensation. Just make sure that water cannot run into the enclosure by way of cables coming from above. Form the cables as shown in Fig. 20 to provide drip loops. Adding a piece of window screen over the opening should be considered to avoid infestations of nesting insects, such as wasps.

Transmitting Accessories

Fig. 21 shows the 70-cm rack from the mast-mounted equipment box. The 70-cm power amplifier (built on the heat sink at the left of the photo) is mounted in the box near the antenna to avoid feed-line losses.

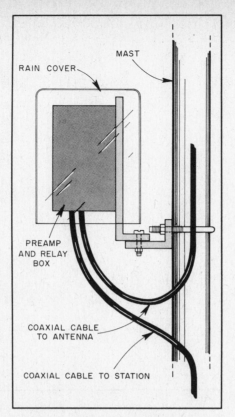

Fig. 20 — Protection for tower-mounted equipment need not be elaborate. Be sure to dress the cables as shown so that water drips off the cable jacket before it reaches the enclosure.

Tower mounting of a transmitter is probably unnecessary for 70 cm, but becomes more important for higher frequencies. Feed-line losses at 70 cm are generally twice those of 2 meters, while those at 24 cm are about twice those of 70 cm, or four times those of 2 meters. For an 80 foot feed line, the losses at 24 cm can easily reach 6 dB for even the best flexible coaxial cable. A 100-W amplifier in the shack will only yield 25 W at the antenna. A good alternative is to place the transmitting converter and a 20-W solid-state amplifier in the tower mounted box near the antennas. The results are nearly the same, and you avoid the time and money needed to generate high power that will just be lost in the feed line anyway.

A coaxial RF sampler is connected to the output of the 70-cm amplifier shown also in Fig. 21. It's nice to be able to monitor the power at the antenna to be sure that everything is working properly. To the right of the power amplifier is a 70-cm preamp and coaxial relays used for OSCAR Mode L reception and terrestrial operation.

One very important aspect of using GaAsFET preamps with transmitting equipment is getting everything to switch in the proper sequence. If transmitters, amplifiers and antenna relays are keyed simultaneously, it's likely that RF will be applied to the feed line before the relays are fully closed. Such *hot switching* can easily arc the contacts on expensive coaxial relays. In addition, if the TR relay is not fully closed, RF may be applied to the preamplifier. Such bursts of RF energy are guaranteed to destroy the GaAsFET in the preamplifier. Many pieces of transmitting equipment (especially multimode transceivers) emit a short burst of RF power when switched on or off, so there is the risk of transmitting into your preamp even if you are careful to pause before keying.

Ideally, keying of a transmitter should follow a sequence that will ensure the safety of the equipment. When you switch into transmit from receive, the coaxial relays change state to remove the preamplifier from the line. Next, the power amplifier is keyed on. The last thing that happens is that the transmitter RF is enabled. When switching back to receive, the sequence is just the opposite. First, the transmitter RF is switched off, the power amplifier is disabled, and then the TR relays change state to place the preamp back in service.

Fig. 21 — The 70-cm equipment panel holds a power amplifier, preamp and TR relays.

Solid-state sequencers to control station TR switching are shown in Chapters 31 and 32. In addition, a unit designed specifically for use with the Yaesu FT-726R is presented later in this chapter.

Receiving Accessories

The only additional useful equipment applies to those stations using a receiving converter and an HF receiver for downlink reception. An in-line switchable attenuator can be installed between the converter output and the antenna jack of the 10-meter receiver. Such an attenuator can be used to lower the AGC level and improve the perceived signal-to-noise ratio. In addition, by adjusting the attenuator so that the S meter on the HF rig rests at zero at no signal, more accurate signal reports can be given. The attenuator circuit is shown in Chapter 25. A modified form has been found useful so that there are only three steps: 5, 10 and 20 dB. These three settings allow attenuation in 5-dB steps from 0 to 35 dB.

Antenna Accessories

Long-boom Yagis, such as the KLM antennas shown in Fig. 11, can suffer from boom sag that might cause pattern distortion and pointing errors. In addition, a boom support is desirable in areas where high winds or ice are a problem. To avoid possible interference with the antenna pattern, the boom brace should be made from nonconductive material such as Phillystran™ HPTG2100 guy cable.

Details are shown in Fig. 22. The vertical boom-brace support member is also nonconductive, made from a fiberglass fishing rod blank. A short piece of threaded stainless-steel rod inserted in the top of the tube is used to adjust tension on the boom brace. A 2-inch piece of 5/16-inch copper tubing brazed across the threaded rod in a "T" fashion holds the Phillystran cable in place. Jam nuts secure the threaded rod once the boom is straight.

Experience with the exposed relays on the circularity switchers used on some commercial antennas has shown that they are prone to failure caused by an elusive mechanism known as "diurnal pumping." The relay is covered with a plastic case, and the seam between the case and PC board is sealed with a silicone sealant. It is not hermetically sealed, however. As a result, the day/night temperature swings pump air and moisture in and out of the relay case. Under the right conditions of temperature and moisture content, moisture from the air will condense inside the relay case. Water builds up inside the case, promoting extensive corrosion and unwanted electrical conduction, seriously degrading relay performance in a short time.

A good rain shelter keeps water from falling directly on the equipment, yet allows the enclosure to "breathe," eliminating water buildup. Equipment in such an enclosure will stay dry and healthy, save

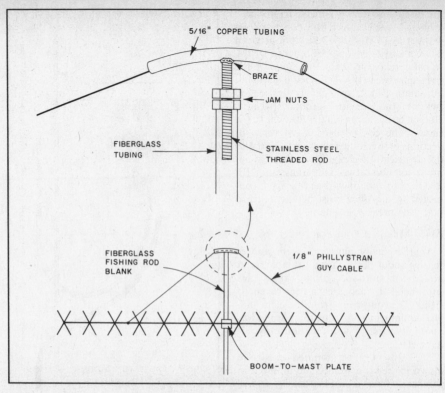

Fig. 22 — A boom brace may be desirable for long Yagis. This arrangement is made from nonconductive material to prevent undesirable effects on the antenna pattern.

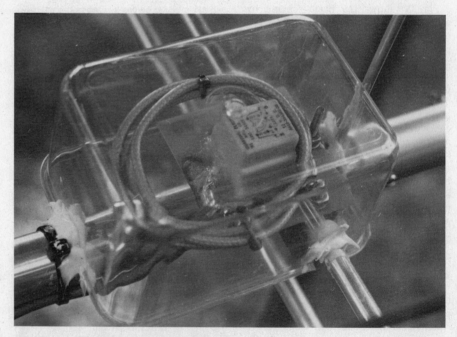

Fig. 23 — KLM 2M-22C antenna CP switcher relay with relocated balun and protective cover.

perhaps for minor corrosion on unprotected steel.

If you have antennas with sealed plastic relays, such as the KLM CX series, you can avoid problems by making the modifications shown in Fig. 23. Relocate the 4:1 balun as shown and place a clear polystyrene plastic refrigerator container over the relay. Notch the container edges for the driven element and the boom so that the container will sit down over the relay, sheltering it from the elements. Bond the container in place with a few dabs of RTV adhesive sealant.

Position the antenna in an "X" orientation, so that neither set of elements is parallel to the ground. The switcher board should now be canted at an angle, and one

Fig. 24 — This low-profile control unit houses a TR sequencer and provides instant control of many station functions.

side of the relay case should be lower than the other. Carefully drill, by hand, a pair of 3/32-inch holes through the low side to vent the relay case. See Fig. 23. The added cover keeps rain water off the relay, and the holes prevent any condensation buildup inside the relay case.

Station Control

Depending on the complexity of your satellite station, you might want to combine most of the switching and control circuitry into a single box so that it provides ready access to all controls. Fig. 24 shows a Minibox cut to a low profile (small enough to fit underneath a transceiver) that contains all of the switches needed to control the station. It houses a TR se-

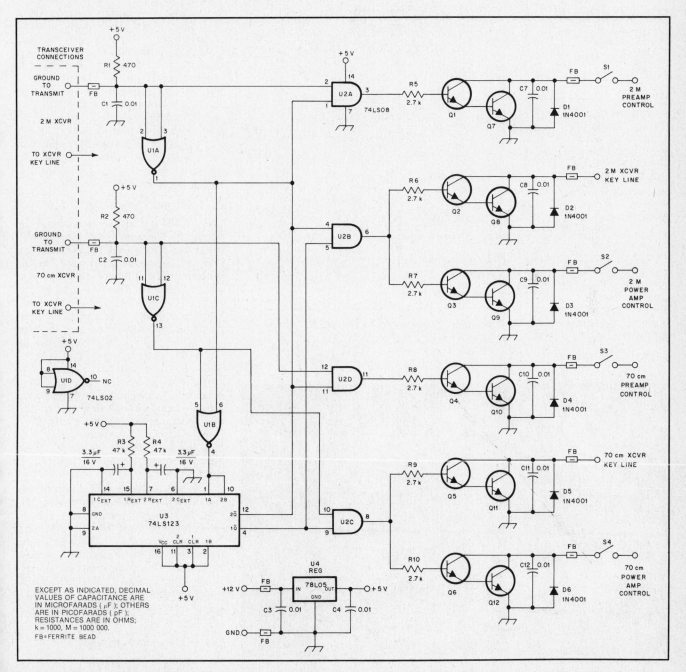

Fig. 25 — Schematic diagram of the digital TR sequencer. All resistors are ¼ W unless noted. Capacitors are disc ceramic unless noted. Capacitors marked with polarity are electrolytic. The output switching transistors (Q1-Q12) can be just about anything that will handle the current requirements.

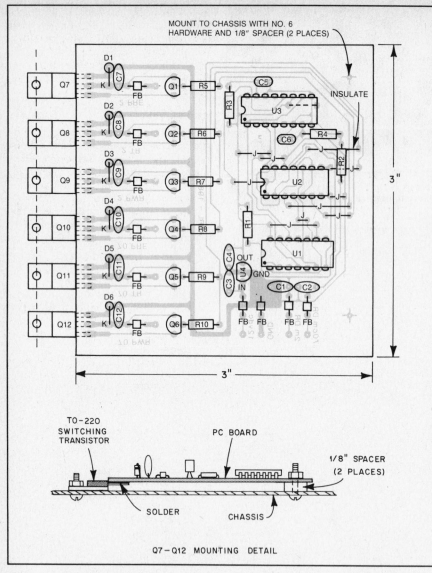

Fig. 26 — Parts placement diagram for the digital TR sequencer. A full-size etching pattern is given at the back of this book.

Fig. 27 — The digital TR sequencer is built on a PC board. The output switching transistors can be mounted to the chassis for improved thermal dissipation.

quencer circuit board as well. This box controls the following functions: change antenna polarization from RHCP to LHCP on 2 meters and 70 cm; switch the 2-meter and 70-cm preamplifiers in and out of the circuit; and switch the power amplifier in or out of the line. Also provided is the option for control of the PTT. The microphone connects to the box, which allows the PTT line to be intercepted before going to the transceiver. The mic PTT switch activates the sequencer, which controls all the other switching, including transmitter PTT.

A DIGITAL TR SEQUENCER FOR SATELLITE WORK

The need for properly controlling the sequence of events in switching a station from transmit to receive is covered in detail in Chapters 31 and 32. The main advantage of sequencing, rather than just switching everything simultaneously, is to protect equipment from damage. For example, sequencing will prevent you from transmitting into a preamplifier or hot switching relays. The TR sequencer described here can be used for other aspects of VHF and UHF operation, but it was built with full duplex (on two bands) satellite operation in mind. Dick Jansson, WD4FAB, designed and built this sequencer for use with a Yaesu FT-726R transceiver.

Circuit Details

Fig. 25 is a schematic diagram of the sequencer. The component count is quite low. Most of the discrete components have been added to protect the circuit from RFI problems. U1 and U2 provide control signal conditioning and digital switching logic, while U3 provides both the key-down and key-release timing. A pair of Darlington-connected transistors were used for each relay driver (Q1-Q12) so that practically any garden-variety transistors can be used. The tabs of the TO220 transistors are mounted to a heat sink for conservative operation. Although most of the switching needs in a modern amateur station require transistors that can handle a few milliamperes of current, some surplus coaxial relays require 500 mA or more. These "generic" control transistors will handle all jobs.

This sequencer provides a 50-ms delay for the transmitter and power amplifier outputs, giving the preamp relays time to settle. Another 50-ms delay is provided for returning the preamps to the line after the transmitter and power amplifier have been turned off. These time delays are conservative, but not at all noticeable in operation. If 50 ms is too long, the 3.3 μF timing capacitors could be reduced to 2.2 μF to get into the 30 ms timing range.

Construction and Installation

Construction is straightforward. All components are mounted on a PC board as shown in Fig. 26. Etched circuit boards

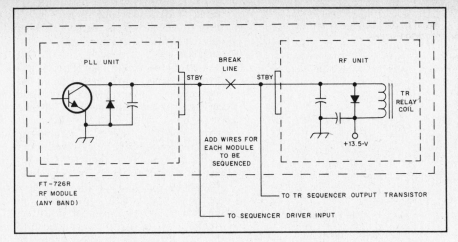

Table 7

Design Considerations for Parabolic Dish Antennas

Dish Size			
Beamwidth	4†	5†	6†
(Deg)	13.2	10.5	8.7
Gain (dBi)	21.8	23.7	25.3
Output Power (W)	EIRP	EIRP	EIRP
100 W	16.2 kW	25.4 kW	36 kW
50 W	8 kW	12.75 kW	18.3 kW
30 W	4.8 kW	7.65 kW	11 kW
15 W	2.4 kW	3.75 kW	5.4 kW
10 W	1.6 kW	2.5 kW	3.6 kW
5 W	800 W	1.25 kW	1.8 kW
2.5 W	400 W	625 W	900 W

†Based upon 50% reflection efficiency and 10-dB taper feedhorn for 0.4 f/D dish

Fig. 28 — The Yaesu FT-726R can be modified so that the TR sequencer can be used to switch it into transmit.

and partial parts kits are available from A&A Engineering (see Chapter 35 parts suppliers list). The output transistors are mounted so that they hang over the edge of the board and can be mounted to the chassis to dissipate heat. See the detail in Fig. 26. Fig. 27 is a photo of the finished unit.

Although this unit was designed with the FT-726R in mind, it can be used with any equipment. If you wish to use it with the FT-726R, Fig. 28 shows a good way to connect everything. This modification allows you to intercept the TR switching for each FT-726R RF module, but it requires no drilling or irreversible modifications. Each RF module in the Yaesu rig has a TR relay and driver transistor that are on separate PC boards within the module. A small three wire cable and connector join them. Simply break the lead and run wires out through the louvers on the bottom of the transceiver.

A MODE-L PARABOLIC ANTENNA AND FEEDHORN FOR OSCAR-10

OSCAR 10, Mode L requires a high-gain antenna for the uplink. This parabolic dish antenna, first described by Eugene F. Ruperto, W3KH, in May 1986 *QST*, provides adequate gain for successful Mode L QSOs. It can be built at home for a small investment in time and money, and is made from readily available materials.

Design Strategy

Some Mode L stations are using commercially built TV dishes. Considering the work needed to put them together, cover them with screen, build a feed antenna and supports, add a mount, and live with a deep f/D (usually 0.375), the buyer basically gets only a preformed wire grid for his money. The custom-design approach advocated here offers the builder some measure of control over the finished product. It's possible to design and construct parabolic reflectors of any reasonable size for any amateur endeavor.

This dish was designed with a three-step approach. First, the dish should be effective, which means that a study of size versus power must be considered. Calculations indicate that a gain between 21 dBi and 27 dBi would be sufficient, provided that the transmitter power is enough to reach the target value of 3 kW EIRP (see Table 7). Second, the dish should be light so that it can be mounted at the mast with other satellite antennas. This prevents an aiming discrepancy later on (such as the downlink antenna pointing at the bird and the uplink antenna pointing toward some meaningless position in space). This becomes more apparent as dish size increases, which brings up the third requirement — beamwidth. A wide-beamwidth dish is nice to have for a moving satellite, but it means a sacrifice in gain. On the other hand, too large a dish will narrow the beamwidth sufficiently to make tracking a chore. Although the Phase III satellites don't require much beam pointing near apogee, the Mode-L operating period is sometimes changed to a lower mean anomaly for operational reasons, and the ground track can cover a considerable angle in terms of antenna movement. It's a nice feeling to not have to aim the antenna more than once every 10 or 15 minutes during the Mode-L period. Calculations indicate that a five-foot (1.52-m) dish, with 12 watts of power at the feed horn, should do the job.

Materials

The dish ribs are made from ¼-inch diameter steel pencil rod (a mild steel). With a weight of 0.167 pounds per foot, it allows the antenna to be light enough to rotate with other antennas. This pencil stock is usually sold in 20-foot lengths, called "joints," and at this length they are extremely flexible. To get them home, plan to cut them in half with a hacksaw or carry them on an extension ladder on top of the pickup. Keeping the original length will minimize waste when making your cuts. The cost of a 20-foot joint is less than

$2.00. Roughly figured, five joints will be needed for a 5-foot dish. Excluding the price of a piece of pipe and three hose clamps, the steel needed for the 5-foot dish will set you back about $10, which is a lot less than the cost of your average catalog TV dish.

The dish covering is common aluminum screen-door material, which is a lot easier to work with than hardware cloth. The door screen material mesh size is much smaller than the one-tenth wavelength required to be a perfect reflector at this frequency, but it also presents greater wind loading, with the probability of ice and snow accumulation. The fact that the dish can be pointed to a stored position when not in use, however, minimizes these effects. A 16-foot-long by 3-foot-wide piece of aluminum screen, which can be obtained at most hardware stores for less than $10, is more than ample for the 5-foot dish.

A very neat, taut covering can be accomplished by using Liquid Nail™ or similar material to glue the screen to the ribs. Use a caulking tube (price less than $2) to apply the glue. Make sure the glue is waterproof since it will be exposed to the weather. A question arises concerning possible galvanic action between the steel rod and the aluminum screen. The glue matrix acts as a buffer between the metals and, for the most part, as an insulator between the two. As a consequence, only very small areas will occasionally make contact. Though these could disintegrate over a period of time, the glued surface area will be much larger and still retain the original strength. The entire screen surface may be removed with a pair of scissors or a sharp knife. After grinding the rib surfaces free of glue, a new screen surface can be applied.

Dish Construction

A ¼-inch-thick piece of scrap steel plate, 2 × 4 feet, is used as a jig to build the ribs. The parabolic X and Y coordinates (see Table 8) are transferred to the plate, spaced roughly every 3 inches, by tracing the figure on the plate. Short pieces of ¼ inch steel strap are tack welded along the curve to

Table 8

Detailed Design for Three Parabolic Dish Antenna Sizes

Dish Size (Feet)	Y, X Coord[†] (In)	Focal Length (In)	Depth of Dish (In)	Gain (dBi)	Beamwidth (Degrees)
	0, 0				
	3 0.117				
	6 0.468				
	9 1.054				
4	12 1.875	19.2	7.5	21.8	13.8
	15 2.929				
	18 4.218				
	21 5.742				
	24 7.5				
	0, 0				
	3 0.039				
	6 0.375				
	9 0.843				
	12 1.5				
5	15 2.343	24	9.375	23.76	10.54
	18 3.375				
	21 4.594				
	24 6.0				
	27 7.60				
	30 9.375				
	0, 0				
	3 0.078				
	6 0.312				
	9 0.703				
	12 1.25				
	15 1.953				
6	18 2.812	28.8	11.25	25.35	8.78
	21 3.828				
	24 5.0				
	27 6.328				
	30 7.812				
	33 9.453				
	36 11.250				

[†]Cartesian coordinates derived from $y^2 = 4FX$.
y = radius of dish, x = axis of focus, 0, 0 = dish center.

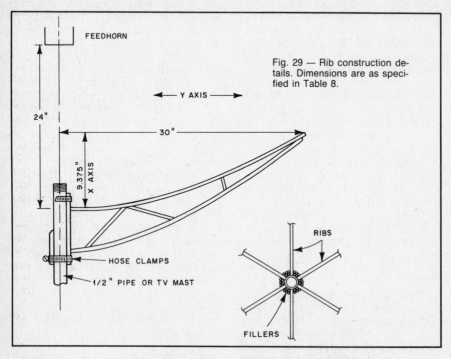

Fig. 29 — Rib construction details. Dimensions are as specified in Table 8.

FEEDHORN

Y AXIS

24"

30"

9.375" X AXIS

HOSE CLAMPS

1/2" PIPE OR TV MAST

RIBS

FILLERS

hold the pencil rod along the front of the curve; then similar pieces are tack welded along the back to hold the pencil rod at the desired shape (see Figs. 29 and 30). The process is repeated for the back rib member as well as the piece of pencil rod that forms the hub portion of the rib. The hub pieces extend about 2 inches beyond the rib dimensions so that when completed, a hose clamp can be fitted on both front and back of the rib to clamp around a slightly longer piece of 1/2 inch pipe, threaded at both ends, that forms the central hub.

Once the jig is completed, and the first rib welded and checked, the remaining five ribs can be constructed in a short time. The rib accuracy is checked using a plywood curve gauge cut out on a bandsaw. Because of the six-rib design, a sloppy fit results when the ribs are clamped to the hub piece. Add additional 9-inch pieces of pencil rod as fillers to make a tighter fit when attaching the ribs to the ½ inch pipe hub. Place the ribs and pipe face down to space the ribs evenly. The dish shown used a joint of pencil rod, bent to a diameter of five feet around a tractor tire, that was welded to the outer tip of each rib before tightening the hose clamps. Alternatively, single pieces of rod may be cut to fit between the ribs at the edge of the dish to form the rim. The completed dish, with center pipe hub, weighs approximately 12 pounds. More ribs could be added, at the expense of additional weight.

Cut sectors of aluminum screen that are slightly larger than the rib spacing so that the sectors will overlap. Run a bead of Liquid Nail, or other glue, on the front side of two ribs and the included part of the external ring and press the screen sector in place. The glue dries fairly rapidly at room temperature. Run a strip of thin wood over the screen surface to tighten it and allow the glue to penetrate and hold the screen surface. By the time this operation is complete, the next sector can be applied in the same manner.

Feed-horn Construction

Many tin-can feed horn designs have been described in past articles. Designs that use motor oil and coffee cans seem to work well enough, but require some modification, usually in the length, to perform optimally at the design frequency. In addition, they also have a tendency to disintegrate after a short stay in the elements, even when painted. The feed horn used here was custom built from 26-gauge galvanized steel capped and soldered at one end. It has only one seam running longitudinally along the feed axis. A piece of copper tubing, threaded to accommodate a no. 8-32 bolt for tuning, is soldered to an N connector (see Fig. 31). The N connector is then soldered with a propane torch into a 5/8-inch hole, located as shown in Fig 31. Apply two coats of paint to the outside of the horn for protection.

The feed used here is linear-polarized for several reasons. Most of the stations heard on Mode L are using linear polarization, and the downlink signals sound okay, primarily because of favorable scheduling of OSCAR 10. Feed horns can be easily replaced, so you can try a switchable left- or right-hand circularly polarized feed at a later date. Such a feed horn is described later. The design described here has an SWR of 1.15:1 at 1269.5 MHz, measured

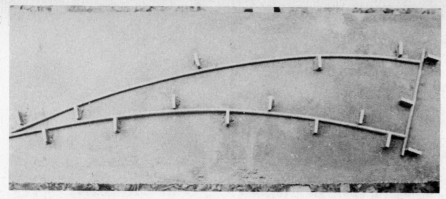

Fig. 30 — Rib assembly jig.

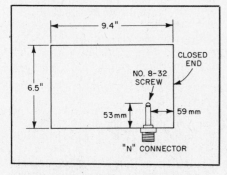

Fig. 31 — Feedhorn construction dimensions. See text for details.

at the feed horn.

The feed horn is mounted on a small plywood saddle attached to a length of ¼-inch pipe by two small U clamps. The ¼-inch pipe is offset at the center of the dish with a ¾-inch, 90° elbow, a short nipple, another 90° elbow, another short nipple, and a reducer from ¾ to ½ inch (see Fig. 32).

Conclusion

The rib design is very strong. During initial testing, the dish took an unscheduled trip across the back yard during a severe windstorm. It suffered only one small deformation on the outer ring and a bent feed horn mount, both of which were easily repaired. The antenna is now tower mounted, as shown in Fig. 33. With 40 W at the feed horn, performance is satisfactory and the pointing angles for tracking are reasonable.

MODE L FEED HORN WITH CIRCULAR POLARIZATION

The feed horn described here provides circular polarization, which might well prove to be the most effective feed method for Mode L use. It was originally described by Dr. John L. DuBois, W1HDX, in the March/April 1983 issue of the AMSAT journal, *Orbit*.

A popular feed among 1296 MHz EME operators is a circular horn with two quarter-wavelength stub radiators inside, set at right angles and fed 90 degrees out of phase. This produces the required circular polarization and is inherently unbalanced for simple coaxial cable feed. The feed horn described here is based on that design, but scaled for 1269 MHz.

The feed horn is basically a piece of circular waveguide operated above the TE11 mode cutoff and below the TM01 mode cutoff. It is excited by stub radiators approximately ¼ wavelength long and ½ wavelength from the closed end of the guide. A # 10 food can, generally used for institutional quantities of fruits or vegetables, is 6.1 inches in diameter and 6.85 inches long. This gives a TE11 cutoff of 1135 MHz and TM01 cutoff of 1484 MHz, neatly positioning 1269 MHz in between. The length should be approximately one wavelength long. This length turned out to be extremely important in tuning the feed for 1269 MHz and determining bandwidth. The diameter and length of the feed stubs are also important tuning factors, determining primarily the input impedance.

Much time was spent optimizing these dimensions with an HP swept return-loss measurement system. The result for a single feed stub (linear polarization) is 20 dB return loss at 1269 MHz over a 20 MHz bandwidth in a 50 ohm system. This corresponds to a 1.2:1 SWR.

Dimensions of the feed horn and placement of the feed stubs are shown in Fig. 34. Two full length #10 cans and part of a third are soldered together. It is necessary to file or steel brush the rims very clean before soldering to get good solder flow. The feed stubs are made from a two inch length of 1/8 inch hobby brass tubing soldered to the center pin of a chassis-mount Type N connector. Cut a hole in the assembled feed can to clear the rear insulator diameter of the N connector and scrape paint away from the edges of the hole for about one inch diameter. Then solder the connector shell (already assembled to the brass stub) directly to the outside of the can. This will take a very hot iron and some patience, but you should end up with a solder bead uniformly around the edge forming a weather seal. Be sure to measure carefully to place the second connector and probe exactly 90 degrees around the can and at the same distance from the open end as the first one. After this, the can should be coated with plastic spray to prevent rusting.

The last required detail is a method of splitting the 50 ohm transmitter power into two lines with a phase difference of 90 degrees to feed the two radiator stubs. This is ideally handled by a 3-dB quadrature hybrid. The fourth terminal of the hybrid is terminated in 50 ohms and absorbs any reflections from the antenna. It could also be fed to a power meter to continuously monitor SWR.

Fig. 32 — Completed dish assembly. Note the feedhorn mounting scheme.

Fig. 33 — The completed Mode L dish is mounted on an elevation cross boom with other satellite antennas.

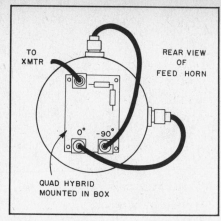

Fig. 36 — Follow this wiring method for right-hand circular polarization. Reversing the connections at the hybrid would yield left-hand circular polarization.

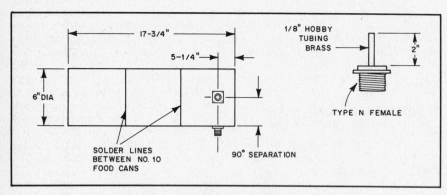

Fig. 34 — Feed horn and radiator probe details. All dimensions are in inches.

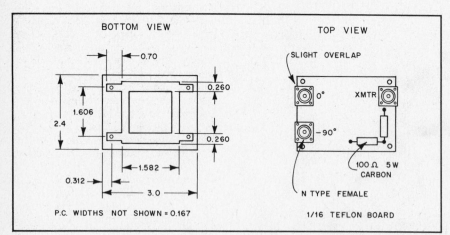

Fig. 35 — 1269 MHz quadrature hybrid. All dimensions are in inches.

In this case, the hybrid and termination were designed to handle up to 100 W forward power. The hybrid was designed on 1/16 inch, double-sided, Teflon circuit board. It is a single square design shown in Fig. 35. The shells of the N connectors connectors are soldered directly to the top foil of the board. The post passes through the board and is soldered to the hybrid pattern on the other side. The termination is made from two 5-W, 100-ohm, carbon-composition (not wirewound!) resistors in parallel. These resistors are mounted on top of the board and the grounded ends are soldered to the top ground foil with extremely short lead length. The other leads pass through the board and are soldered to the fourth hybrid arm. Keep lead length to the absolute minimum.

Cover the resistors with a liberal amount of silicone sealant and mount the board as the cover of a 2½- × 3- × 1¾-inch Pomona box. Four screws secure the board at the corners and all edges are covered with silicone sealant for a weather tight seal.

Secure the box to the rear of the feed horn. Two coaxial cables of equal length (about 16 inches long) connect the hybrid to the feed stubs in the can. Use RG-213, Belden 9913 or similar low-loss cable. Follow the connection pattern in Fig. 36 for right-hand circular polarization. Reversing the feed cables would yield left-hand circular polarization. The exact details of mounting this feed to your dish will depend on dish design.

Selected References

Satellite Equipment Selection

D. DeMaw, "Trio-Kenwood TS-700S 2-Meter Transceiver," *QST*, Feb. 1978, pp. 31-32.

C. Hutchinson, "TEN-TEC 2510 Mode B Satellite Station," *QST*, Oct. 1985, pp. 41-43.

D. Ingram, "The Ten-Tec 2510 OSCAR Satellite Station/Converter," *CQ Magazine*, Feb. 1985, pp. 44-46.

J. Kleinman, "ICOM IC-290H All-Mode 2-Meter Transceiver," *QST*, May 1983, pp. 36-37.

J. Lindholm, "ICOM IC-471A 70-cm Transceiver," *QST*, Aug. 1985, pp. 38-39.

M. Wilson, "ICOM IC-271A 2-Meter Multimode Transceiver," *QST*, May 1985, pp. 40-41.

M. Wilson, "Yaesu Electronics Corp. FT-726R VHF/UHF Transceiver," *QST*, May 1984, pp. 40-42.

M. Wilson, "Yaesu FT-480R 2-Meter Multimode Transceiver," *QST*, Oct. 1981, pp. 46-47.

H. Winard and R. Soderman, "A Survey of OSCAR Station Equipment," *Orbit*, No. 16, Nov.-Dec. 1983, pp. 13-16 and No. 18, Mar.-Apr. 1984, pp. 12-16.

Antennas

M. Davidoff, "Off-Axis Circular Polarization of Two Orthognal Linearly Polarized Antennas," *Orbit*, No. 15, Sep.- Oct. 1983, pp. 14-15.

J. L. DuBois, "A Simple Dish for Mode-L," *Orbit*, No. 13, Mar.-Apr. 1983, pp. 4-6.

B. Glassmeyer, "Circular Polarization and OSCAR Communications," *QST*, May 1980, pp. 11-15.

G. L. Hall, ed., *The ARRL Antenna Book* (Newington: ARRL, 1982). Available from your local radio store or from ARRL for $8.00 ($8.50 outside US). Add $2.50 ($3.50 UPS) per order for shipping and handling.

R. Jansson, "Helical Antenna Construction for 146 MHz," *Orbit*, May-June 1981, pp. 12-15.

R. Jansson, "KLM 2M-22C and KLM 435-40CX Yagi Antennas," *QST*, Oct. 1985, pp. 43-44.

R. Jansson, "70-Cm Satellite Antenna Techniques," *Orbit*, No. 1, Mar. 1980, pp. 24-26.

C. Richards, "The Chopstick Helical," *Orbit*, No. 5, Jan.- Feb. 1981, pp. 8-9.

V. Riportella, "Amateur Satellite Communications," *QST*, May 1986, pp. 70-71.

G. Schrick, "Antenna Polarization," *The ARRL Antenna Compendium*, Vol. 1 (Newington: ARRL 1985), pp. 152-156.

A. Zoller, "Tilt Rather Than Twist," *Orbit*, No. 15, Sep.- Oct. 1983, pp. 7-8.

General Texts

M. Davidoff, *The Satellite Experimenter's Handbook* (Newington: ARRL, 1984). Available from your local radio store or from ARRL for $10 ($11 outside US). Add $2.50 ($3.50 UPS) per order for shipping and handling.

R. Halprin, ed., *The ARRL Operating Manual* (Newington: ARRL 1985).

The AMSAT-Phase III Satellite Operations Manual, prepared by Radio Amateur Satellite Corp. and Project OSCAR, Inc., 1985. Available from AMSAT, 850 Sligo Ave., Silver Spring, MD 20910, for $15.00.

Glossary of Satellite Terminology

AMSAT — A registered trademark of the Radio Amateur Satellite Corporation, a non-profit scientific/educational organization located in Washington, DC. It builds and operates Amateur Radio satellites and has sponsored the OSCAR program since the launch of OSCAR 5. (AMSAT, P.O. Box 27, Washington, DC 20044.)

Anomalistic period — The elapsed time between two successive perigees of a satellite.

AO-# — The designator used for AMSAT OSCAR spacecraft in flight, by sequence number (AO-5 through AO-10).

AOS — Acquisition of signal — The time at which radio signals are first heard from a satellite, usually just after it rises above the horizon.

Apogee — The point in a satellite's orbit where it is farthest from Earth.

Area coordinators — An AMSAT corp of volunteers who organize and coordinate amateur satellite user activity in their particular state, municipality, region, or country. This is the AMSAT grass roots organization set up to assist all current and prospective OSCAR users.

Argument of perigee — The polar angle that locates the perigee point of a satellite in the orbital plane; drawn between the ascending node, geocenter, and perigee; and measured from the ascending node in the direction of satellite motion.

Ascending node — The point on the ground track of the satellite orbit where the sub-satellite point (SSP) crosses the equator from the Southern Hemisphere into the Northern Hemisphere.

Az-el mount — An antenna mount that allows antenna positioning in both the azimuth and elevation planes.

Azimuth — Direction (side-to-side in the horizontal plane) from a given point on Earth, usually expressed in degrees. North = 0° or 360°; East = 90°; South = 180°; West = 270°.

Circular polarization (CP) — A special case of radio energy emission where the electric and magnetic field vectors rotate about the central axis of radiation. As viewed along the radiation path, the rotation directions are considered to be right-hand (RHCP) if the rotation is clockwise, and left-hand (LHCP) if the rotation is counterclockwise.

Decending node — The point on the ground track of the satellite orbit where the sub-satellite point (SSP) crosses the equator from the Northern Hemisphere into the Southern Hemisphere.

Desense — A problem characteristic of many radio receivers in which a strong RF signal overloads the receiver, reducing sensitivity.

Doppler effect — An apparent shift in frequency caused by satellite movement toward or away from your location.

Downlink — The frequency on which radio signals originate from a satellite for reception by stations on Earth.

Eccentricity — The orbital parameter used to describe the geometric shape of an elliptical orbit; eccentricity values vary from e = 0 to e = 1, where e = 0 describes a circle and e = 1 describes a straight line.

EIRP — Effective isotropic radiated power — same as ERP except the antenna reference is an isotropic radiator.

Elliptical orbit — Those orbits in which the satellite path describes an ellipse with the Earth at one focus.

Elevation — Angle above the local horizontal plane, usually specified in degrees. (0° = plane of the Earth's surface at your location; 90° = straight up, perpendicular to the plane of the Earth).

Epoch — The reference time at which a particular set of parameters describing satellite motion (Keplerian elements) are defined.

EQX — The reference equator crossing of the ascending node of a satellite orbit, usually specified in UTC time and degrees of longitude of the crossing.

ERP — Effective radiated power — System power output after transmission-line losses and antenna gain (referenced to a dipole) are considered.

ESA — European Space Agency — A consortium of European governmental groups pooling resources for space exploration and development.

FO-12—Japanese Amateur Radio satellite, Fuji-OSCAR 12.

Geocenter — The center of the Earth.

Geostationary orbit — A satellite orbit at such an altitude (approximately 22,300 miles) over the equator that the satellite appears to be fixed above a given point.

Ground station — A radio station, on or near the surface of the earth, designed to transmit or receive to/from a spacecraft.

Groundtrack — The imaginary line traced on the surface of the Earth by the subsatellite point (SSP).

Inclination — The angle between the orbital plane of a satellite and the equatorial plane of the Earth.

Increment — The change in longitude of ascending node between two successive passes of a specified satellite, measured in degrees West per orbit.

Iskra — Soviet low-orbit satellites launched manually by cosmonauts aboard Salyut missions. Iskra means "spark" in Russian.

JAMSAT — Japan AMSAT organization.

Keplerian Elements — The classical set of six orbital element numbers used to define and compute satellite orbital motions. The set is comprised of inclination, Right Ascension of Ascending Node (RAAN), eccentricity, argument of perigee, mean anomaly and mean motion, all specified at a particular epoch or reference year, day and time. Additionally, a decay rate or drag factor is usually included to refine the computation.

LHCP — Left hand circular polarization.

LOS — Loss of signal — The time when a satellite passes out of range and signals from it can no longer be heard. This usually occurs just after the satellite goes below the horizon.

Mean anomaly (MA) — An angle that increases uniformly with time, starting at perigee, used to indicate where a satellite is located along its orbit. MA is usually specified at the reference epoch time where the Keplerian elements are defined. For AO-10 the orbital time is divided into 256 parts, rather than degrees of a circle, and MA (sometimes called phase) is specified from 0 to 255. Perigee is therefore at MA = 0 with apogee at MA = 127.

Mean motion — The Keplerian element to indicate the complete number of orbits a satellite makes in a day.

NASA — National Aeronautics and Space Administration, the U.S. space agency.

Nodal period — The amount of time between two successive ascending nodes of a satellite orbit.

Orbital elements — See Keplerian Elements.

Orbital plane — An imaginary plane, extending throughout space, that contains the satellite orbit.

OSCAR — Orbiting Satellite Carrying Amateur Radio.

OSCARLOCATOR — A graphical satellite tracking device used to locate a satellite in its orbit and aid in pointing antennas. Available from ARRL.

PACSAT — A proposed AMSAT packet radio satellite with store-and-forward capability.

Pass — An orbit of a satellite.

Passband — The range of frequencies handled by a satellite translator or transponder.

Perigee — The point in a satellite's orbit where it is closest to Earth.

Period — The time required for a satellite to make one complete revolution about the Earth. See anomalistic period and nodal period.

Phase I — The term given to the earliest, short-lived OSCAR satellites that were not equipped with solar cells. When their batteries were depleted, they ceased operating.

Phase II — Low altitude, long-lived OSCAR satellites. Equipped with solar panels that powered the spacecraft systems and recharged their batteries, these satellites have been shown to be capable of lasting up to five years (OSCARs 6, 7 and 8, for example).

Phase III — Extended-range, high-orbit OSCAR satellites with very long-lived solar power systems (OSCAR 10 and Phase IIIC).

Phase IV — Proposed OSCAR satellites in geostationary orbits.

Precession — An effect that is characteristic of AO-10 and Phase III orbits. The satellite apogee SSP will gradually change over time.

Project OSCAR — The California-based group, among the first to recognize the potential of space for Amateur Radio; responsible for OSCARs I through IV.

QRP days — Special orbits set aside for very low power uplink operating through the satellites.

RAAN — Right Ascension of Ascending Node — The Keplerian element specifying the angular distance along the celestial equator, between the vernal equinox and the ascending node of a spacecraft. This can be simplified to mean roughly the longitude of the ascending node.

Radio Sputnik — Soviet Amateur Radio satellites (see RS #).

Reference orbit — The orbit of Phase II satellites beginning with the first ascending node during that UTC day.

RHCP — Right-hand circular polarization.

RS # — The designator used for most Soviet Amateur Radio satellites (RS-1 through RS-8, for example).

Satellipse — A graphical tracking device, similar to OSCARLOCATOR, designed to be used with satellites in elliptical orbits.

Satellite pass — Segment of orbit during which the satellite ''passes'' nearby and in range of a particular ground station.

Sidereal day — The amount of time required for the Earth to rotate exactly 360 degrees about its axis with respect to the ''fixed'' stars. The sidereal day contains 1436.07 minutes (see Solar day).

Solar day — The solar day, by definition, contains exactly 24 hours (1440 minutes). During the solar day, the Earth rotates slightly more than 360 degrees about its axis with respect to ''fixed'' stars (see Sidereal day).

Spin modulation — Periodic amplitude fade-and-peak resulting from the rotation of a satellite's antennas about its spin axis, rotating the antenna peaks-and-nulls.

SSC — Special service channels — Frequencies in the downlink passband of AO-10 that are set aside for authorized, scheduled use in such areas as education, data exchange, scientific experimentation, bulletins and official traffic.

SSP — Subsatellite point — Point on the surface of the Earth directly between the satellite and the geocenter.

Telemetry — Radio signals, originating at a satellite, that convey information on the performance or status of onboard subsystems. Also refers to the information itself.

Transponder — A device onboard a satellite that receives radio signals in one segment of the spectrum, amplifies them, translates (shifts) their frequency to another segment of the spectrum and retransmits them. Also called linear translator.

UoSAT-OSCAR (UO #) — Amateur Radio satellite built under the coordination of Radio Amateurs and educators at the University of Surrey, England.

Uplink — The frequency at which signals are transmitted from ground stations to a satellite.

Window — Overlap region between acquisition circles of two ground stations referenced to a specific satellite. Communication between two stations is possible when the subsatellite point is within the window.

Earth-Moon-Earth

Popularly known as moonbounce, EME is the most popular method of space communication after OSCAR. The concept is straightforward: Stations that can simultaneously see the moon communicate by reflecting VHF and UHF signals off the lunar surface. Unlike OSCAR, though, the two stations have a relatively stable target and may be separated by virtually 180 degrees of arc on the earth's surface, which translates to more than 11,000 miles.

There is a trade-off, though; since the moon's mean distance from earth is 239,000 miles, path losses are huge when compared to "local" VHF work. Thus, each station on an EME circuit demands the most out of the transmitter, antenna, receiver and operator skills. Even with all those factors in an optimum state, the signal in the headphones may be barely perceptible above the noise. Nevertheless, for any type of amateur communication over a distance of 500 miles or more at 432 MHz, for example, moonbounce comes out the winner over terrestrial methods when various factors are figured on a balance sheet.

EME thus presents amateurs with the ultimate challenge in strengthening radio systems. Before amateur involvement, the only other known moon relay circuit was operated by the U.S. Navy between Washington, DC and Hawaii. Their 400 megawatts of effective radiated power carried four multiplexed RTTY channels. The first two-way amateur link took place between the EIMAC Radio Club, W6HB, and the Rhododendron Swamp VHF Society, W1BU, on 1296 MHz in July 1960.

Only a few amateurs heard anything more than their own echoes during the next few years. Hams at government and private institutions began conducting tests with other hams by using very large arrays such as the 150-foot steerable dish at WA6LET (Stanford University) or the 1000-foot parabolic surface at KP4BPZ (Arecibo). Amateur-to-amateur contacts did not become established until the early '70s, a notable effort being between VE7BBG and WA6HXW. Activity spread to all continents — except South America. In July 1976, the Mt. Airy VHF Club of Philadelphia (Pack Rats) staged an expedition to Barranquilla, Colombia, which allowed K2UYH to become the first amateur to work all continents on 432 MHz.

Amateurs have even used the 2300-MHz band for EME work, starting with a QSO between W3GKP and W4HHK in 1970. There are presently active stations in half a dozen countries on that band.

Most of the components for an EME station on 144, 220, 432 or 1296 MHz are now commercially available. Whether a prospective EMEer chooses that route or builds all the gear, some design considerations must be taken because it is weak-signal work.

1) Transmissions must be made on CW or SSB with as close to the maximum legal output as possible.

2) The antenna should have at least 20 dB of gain over a dipole.

3) As with an OSCAR antenna system, rotators are needed for both azimuth and elevation. Since the half-power beamwidth of a high-gain antenna is quite sharp, the rotators must have appropriate accuracy.

4) Transmission-line losses should be held to a minimum.

5) The receiving systems should have a very low noise figure.

Don't let these requirements scare you! Most EMEers started out as listeners. In the ARRL International EME Competition, operators with nothing more than a simple Yagi, preamplifier and multimode transceiver hear the stronger stations.

Calculating EME Capabilities

EME path loss can be determined from Fig. 37. The received signal-to-noise ratio can be calculated from the formula:

$$S/N = P_o - L_t + G_t - P_l + G_r - P_n \quad \text{(Eq. 1)}$$

where

P_o = transmitter output power (dBW)
L_t = transmitter feed-line loss (dB)
G_t = transmitting antenna gain (dBi)
P_l = total path loss (dB)
G_r = receiving antenna gain (dBi)
P_n = receiver noise power (dBW)

Receiver noise power, P_n, is determined by the following:

$$P_n = 10 \log_{10} K \times B \times T_s \quad \text{(Eq. 2)}$$

where

$K = 1.38 \times 10^{-23}$ (Boltzmann's constant)
B = bandwidth (Hz)

T_s = receiving system noise temperature (K)

Receiving system noise temperature, T_s, can be found from:

$$T_s = T_a + (L_r - 1) T_l + L_r T_r \quad \text{(Eq. 3)}$$

where

T_a = antenna temperature (K)
L_r = receiving feed-line loss (ratio)
T_l = physical temperature of feed line (normally 290 K)
T_r = receiver noise temperature (K)

The meaning of these formulas becomes clear if we substitute appropriate values and calculate S/N for a typical 432-MHz EME communications link:

P_o = +30 dBW (1000 W)
L_t = 0.5 dB
G_t = 23.0 dBi ($8 \times 4.2 - \lambda$ NBS Yagi)
P_l = 262 dB
G_r = 23.5 dBi (15-ft parabolic)
B = 100 Hz
T_a = 100 K
L_r = 1 (0 dB, preamp at antenna)
T_l = 290 K
T_r = 50.72 K (NF = 0.7 dB)
S/N = +0.8 dB

It is obvious that there is no place for compromise in assembling the EME station. Even relatively sophisticated equipment gives marginal results.

EME Scheduling

The best days to schedule are usually when the moon is at perigee (closest to the earth) since the path loss is typically 2 dB less than when the moon is at apogee (farthest from the earth). The moon's perigee and apogee dates may be determined from publications such as *The*

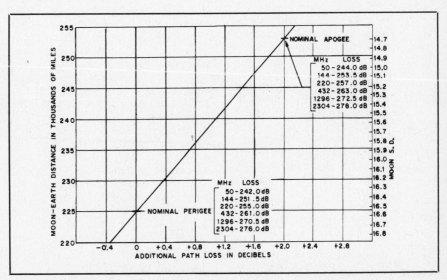

Fig. 37 — Variations in EME path loss can be determined from this graph. S.D. refers to the semi-diameter of the moon, which is indicated for each day of the year in *The Nautical Almanac*.

Nautical Almanac[1] by inspecting the section of the tables headed "S.D." (semi-diameter of the moon in minutes of arc). An S.D. of 16.53 equates to an approximate earth-to-moon distance of 225,000 miles, typical perigee, and an S.D. of 14.7 to an approximate distance of 252,500 miles, typical apogee. If the semi-diameters are located on Fig. 37, the EME path losses in decibels may be determined for the most popular amateur frequencies.

The moon's orbit is slightly elliptical. Hence, the day-to-day path-loss changes at apogee and perigee are minor. The greatest changes take place at the time when the moon is traversing between apogee and perigee. However, several other factors must be considered for optimum scheduling aside from the path losses.

If perigee occurs near the time of a new moon, one to two days will be unusable since the sun behind the moon will cause increased sun-noise pickup. Therefore, schedules should be avoided when the moon is within 10° of the sun (and farther if your antenna has a wide beam or strong side lobes). The moon's orbit follows a cycle of 18 to 19 years, so the relationships between perigee and new moon will not be the same from one year to the next.

Low moon declinations and low aiming elevations generally produce poor results and should be avoided if possible. Conversely, high moon declinations and high elevation angles should yield best results. Good results are usually obtained when both stations are using similar elevation angles, since then both stations are looking through comparable electron densities. Generally, low elevation angles increase antenna-noise pickup and increase tropospheric absorption, especially above 420 MHz, where the galactic noise is very low. This situation cannot be avoided when one station is unable to elevate the antenna above the horizon or when there is a great terrestrial distance between stations. Ground gain (gain obtained when the antenna is aimed at the horizon) has been used very effectively at 144 and 432 MHz. While potentially useful, ground gain is a complex phenomenon and difficult to predict.

Usually, signals are stronger in the fall and winter months and weaker in the summer. Also, signals are generally better at night than during the day. This may be attributable to decreased ionization or less Faraday rotation.

Whenever the moon crosses the galactic plane (twice a month for three to five days each occurrence), the sky temperature will be higher. Areas of the sky to avoid are the constellations Orion and Gemini at northern declinations and Scorpius and Sagittarius at southern declinations. Positions of the moon with respect to these constellations can be checked with *Sky and Telescope* magazine or *The Nautical Almanac*. The galactic plane is biased toward southern declinations, which will cause southerly declinations to be less desirable (with respect to noise) than are northern declinations.

Finally, the time of the day and the day of the week must be considered since most of us have to work for a living and cannot always be available for schedules. Naturally, weekends and evenings are preferred, especially when perigee occurs on a weekend.

General Considerations

It helps to know your own EME window as accurately as possible. The term "window" means the period of time that a station can "see" the moon. This can be determined with the help of information contained in a later section of this chapter. Most EME operators determine their local window and translate it into GHA (Greenwich hour angle) and declination. This information is a constant, so once it is determined it is usable by other stations just as one would use UTC. Likewise, it helps to know the window of the station to be scheduled. Most EME stations are limited in some way by local obstructions, antenna-mounting constraints, geographical considerations, and the like. Therefore, the accuracy of each station's EME window is very important for locating common windows and setting schedule times.

A boresight of some type is practically mandatory in order to align your antenna accurately with the moon. Most antenna systems exhibit some pattern skewing which must be accounted for. A simple calibration method is to peak your antenna on received sun noise and then align the boresight tube on the sun. The boresight of the antenna is now calibrated and can be used to aim the antenna at the moon. Readers are cautioned against using a telescope or other device employing lenses as a boresight device! Even the best of optical filters will not eliminate the hazard from solar radiation when viewed directly. A simple piece of small diameter tubing, two or three feet long can serve this purpose. A symmetrical spot of light cast upon a piece of paper near the back end of the tube will indicate alignment.

A remote readout is a highly recommended convenience. Accuracies of ±2° are usually necessary and can be attained with syncros. A remote readout is particularly important for scheduling when the moon is within 45° of the sun or when the sky is overcast. There are very few areas where the moon is not occasionally obscured by cloud cover. Aiming an antenna blindly seldom pays off.

Locating the Moon

The moon orbits the earth once in approximately 28 days, a lunar month.

Because the plane of the moon's orbit is tilted from the earth's equatorial plane by approximately 23.5°, the moon swings in a sine-wave pattern both north and south of the equator. The angle of departure of the moon's position at a given time from the equatorial plane is termed declination. Declination angles of the moon, which are continually changing (a few degrees a day), indicate the latitude on the earth's surface where the moon will be at zenith. For this presentation, positive declination angles are used when the moon is north of the equator, and negative angles when south.

The longitude on the earth's surface where the moon will be at zenith is related to the moon's Greenwich Hour Angle, abbreviated G.H.A. or GHA. "Hour angle" is defined as the angle in degrees to the west of the meridian. If the GHA of the moon were zero degrees, it would be directly over the Greenwich meridian. If the moon's GHA were 15 degrees, the moon would be directly over the meridian which is designated as 15° W longitude on a globe. As one can readily understand, the GHA of the moon is continually changing, too, because of both the orbital velocity of the moon and the earth's rotation inside the moon's orbit. The moon's GHA changes at the rate of approximately 347° per day.

GHA and declination are terms that may be applied to any celestial body. *The Nautical Almanac* and other publications list the GHA and decl. of the sun and moon (as well as for other celestial bodies that may be used for navigation) for every hour of the year. This information may be used to point an antenna when the moon is not visible. Almanac tables for the sun may be useful for calibrating remote-readout systems.

Using the Almanac

Books such as *The Nautical Almanac* and other almanacs show the GHA and declination of the sun or moon at hourly intervals for every day of the period covered by the book. Instructions are included in such books for interpolating the positions of the sun or moon for any time on a given date. The orbital velocity of the moon is not constant, and therefore precise interpolations are not linear.

Fortunately, linear interpolations from one hour to the next, or even from one day to the next, will result in data that is entirely adequate for Amateur Radio purposes. If linear interpolations are made from 0000 UTC on one day to 0000 UTC on the next, worse-case conditions exist when apogee or perigee occurs near midday on the next date in question. Under such conditions, the total angular error in the position of the moon may be as much as a sixth of a degree. Because it takes a full year for the earth to orbit the sun, the similar error for determining the position of the sun will be no more than a few hundredths of a degree.

If a polar mount (a system having one

[1]*The Nautical Almanac for the Year*****, where **** indicates the calendar year for the data. This annual publication is printed by the U.S. Printing Office, Washington, DC. It is available from the Superintendent of Documents and from many dealers of marine products.

axis parallel to the earth's axis) is used, information from the Almanac may be used directly to point the antenna array. The local hour angle (LHA) is simply the GHA plus or minus the observer's longitude (plus if east long., minus if west). The LHA is the angle west of the observer's meridian at which the celestial body is located. LHA and declination information may be translated to an EME window by taking local obstructions and any other constraints into account.

Azimuth and Elevation

An antenna system that is positioned in azimuth (compass direction) and elevation (angle above the horizon) is called an az-el system. For such a system, some additional work will be necessary to convert the almanac data into useful information. The GHA and decl. information may be converted into azimuth and elevation angles with the mathematical equations that follow. An electronic calculator or computer that treats trigonometric functions may be used. CAUTION: Most almanacs list data in degrees, minutes, and either decimal minutes or seconds. Computer or calculator programs generally require this information in degrees and decimal fractions, so a conversion may be necessary before the almanac data is entered.

Determining az-el data from equations follows a procedure similar to calculating great-circle bearings and distances for two points on the earth's surface. There is one additional factor, however. Visualize two observers on opposite sides of the earth who are pointing their antennas at the moon. Imaginary lines representing the boresights of the two antennas will converge at the moon at an angle of approximately 2°. Now assume both observers aim their antennas at some distant star. The boresight lines now may be considered to be parallel, each observer having raised his antenna in elevation by approximately 1°. The reason for the necessary change in elevation is that the earth's diameter in comparison to its distance from the moon is significant. The same is not true for distant stars, or for the sun.

Equations for az-el calculations are:

$$\sin E = \sin L \sin D + \cos L \cos D \cos LHA \qquad \text{(Eq. 4)}$$

$$\tan F = \frac{\sin E - K}{\cos E} \qquad \text{(Eq. 5)}$$

$$\cos C = \frac{\sin D - \sin E \sin L}{\cos E \cos L} \qquad \text{(Eq. 6)}$$

where

E = elevation angle for the sun
L = your latitude (negative if south)
D = declination of the celestial body
LHA = local hour angle = GHA plus or minus your longitude (plus if east long., minus if west long.)
F = elevation angle for the moon
K = 0.01657, a constant (see text that follows)

C = true azimuth from north if sin LHA is negative; if sin LHA is positive, then the azimuth = 360 minus C

Assume our location is 50° N. lat., 100° W. long. Further assume that the GHA of the moon is 140° and its declination is 10°. To determine the az-el information we first find the LHA, which is 140 minus 100 or 40°. Then we solve Eq. 4:

$$\sin E = \sin 50 \sin 10 + \cos 50 \cos 10 \cos 40$$

$$\sin E = 0.61795 \text{ and } E = 38.2°$$

Solving Eq. 5 for F, we proceed. (The value for sin E has already been determined in Eq. 4.)

$$\tan F = \frac{0.61795 - 0.06175}{\cos 38.2}$$

$$= 0.76489$$

From this, F, the moon's elevation angle, is 37.4°.

We continue by solving Eq. 6 for C. (The value of sin E has already been determined.)

$$\cos C = \frac{\sin 10 - 0.61795 \sin 50}{\cos 38.2 \cos 50}$$

$$= -0.59308$$

C therefore equals 126.4°. To determine if C is the actual azimuth, we find the polarity for sin LHA, which is sin 40° and has a positive value. The actual azimuth then is 360 − C = 233.6°.

If az-el data is being determined for the sun, use of Eq. 5 must be omitted; Eq. 5 takes into account the nearness of the moon. The solar elevation angle may be determined from Eq. 4 alone. In the above example, this angle is 38.2°.

The mathematical procedure is the same for any location on the earth's surface. Remember to use negative values for southerly latitudes. If solving Eq. 4 or 5 yields a negative value for E or F, this indicates the celestial body is below the horizon.

These equations may also be used to determine az-el data for man-made satellites, but a different value for the constant, K, must be used. K is defined as the ratio of the earth's radius to the distance from the earth's center to the satellite.

The value for K as given above, 0.01657 is based on an average earth-moon distance of 239,000 miles. The actual earth-moon distance varies from approximately 225,000 to 253,000 mi. When this change in distance is taken into account, it yields a change in elevation angle of approximately 0.1° when the moon is near the horizon. For greater precision in determining the correct elevation angle for the moon, the moon's distance from the earth may be taken as:

$$D = -15,074.5 \times \text{S.D.} + 474,332$$

where

D = moon's distance in miles

S.D. = moon's semi-diameter, from the almanac.

Calculator and Computer Programs

As has been mentioned, a calculator or computer may be used in solving the equations for azimuth and elevation. Tables in this section list suitable programs. The program of Table 9 is for Hewlett-Packard HP-25 and similar calculators using reverse Polish notation (RPN). With this program, the GHA and declination of a celestial body are entered for a particular time of day, and the calculator computes the azimuth and elevation for that time. Calculations must then be repeated for a different time of day, by using different GHA and declination values, as appropriate. For EME work, it is convenient to calculate az-el data at 30-minute intervals or so, and to keep the results of all calculations handy during the EME window. Necessary antenna-position corrections can then be made periodically.

Table 10 is a BASIC language program for the IBM PC. This program provides azimuth and elevation information for half-hour intervals during a UTC day when the celestial body is above the horizon. The program makes a linear interpolation of GHA and declination values (discussed earlier) during the period of the UTC day.

With the program of Table 10 in operation, the computer first asks for various data to be entered via the keyboard, including GHA and declination for 2400 UTC for the date of the calculations. This data, of course, will be the same as for 0000 UTC for the following day from the almanac. After all data is entered, the computer performs calculations for each half hour. Only those results for which the body is above the horizon are displayed, so there may be a period of 15 seconds or more during which nothing appears to be happening. This is normal. The program may be modified to print the results of calculations on paper if a line printer is available, or to adapt it for use on other computers with variations in BASIC language.

Either of these programs may be used for determining the azimuth and elevation for more distant bodies, by using program steps 26 and 27 for the sun in Table 9, or by entering SUN at the keyboard for statement 130 in Table 10. Further, Table 9 may be used for determining the positions of man-made satellites for a given time, if the appropriate value for the constant K is stored in register 3. For this application, the coordinates of the sub-satellite point are used in place of GHA and declination.

Libration Fading of EME Signals

One of the most troublesome aspects of receiving a moonbounce signal, besides the enormous path loss and Faraday rotation fading, is libration fading. This section will deal with libration (pronounced lie-brayshun) fading, its cause and effects, and possible measures to minimize it.

Table 9

Calculator Program for Determining Azimuth and Elevation of Celestial Bodies

	Line	Key Entry	
	01	RCL 1	
	02	—	
	03	STO 4	
	04	f cos	
	05	x ≥ y	
	06	STO 5	
	07	f cos	
	08	X	
	09	RCL 0	
	10	f cos	
	11	STO 6	
	12	X	
	13	RCL 0	
	14	f sin	
	15	STO 7	
	16	RCL 5	
	17	f sin	
	18	STO 5	
	19	X	
	20	+	
Use these	21	STO X 7	Use these pro-
program steps	22	Enter ↑	gram steps for
for the moon	23	g sin⁻¹	the sun and
	24	f cos	distant bodies
	25	STO X 6	

26	g 1/x	26	f Last x
27	x ≥ y	27	GTO 32

	28	RCL 3
	29	—
	30	X
	31	g tan⁻¹
	32	RCL 5
	33	RCL 7
	34	—
	35	RCL 6
	36	÷
	37	g cos⁻¹
	38	RCL 4
	39	f sin
	40	g x ≥ 0
	41	GTO 44
	42	R↓
	43	GTO 00
		(or R/S)
	44	R↓
	45	CHS
	46	RCL 2
	47	+
	48	GTO 00 (or R/S)

Instructions

1) Load the program, selecting lines 26 and 27 for either the moon or for more distant bodies. Switch to RUN.
2) Initialize: f PRGM; f FIX 1; g DEG
3) Store constants: 360 STO 2; 0.01657 STO 3
4) Store data
Your latitude (degrees and decimal; negative if east) STO 0
Your longitude (degrees and decimal; negative if east) STO 1
5) Input data:
Decl. of celestial body (degrees and decimal) Enter ↑
GHA of celestial body (degrees and decimal) R/S
The result displayed after a few seconds is the azimuth or bearing in degrees clockwise from north.
Depressing x ≥ y displays the elevation angle above the horizon. (The body is below the horizon if a negative angle is displayed.)
6) For another az-el calculation from the same location, go to step 5.
7) For az-el calculations from a different location on the earth's surface, go to step 4 using new latitude and longitude.

Table 10

Computer Program for Determining Azimuth and Elevation of Celestial Bodies

```
10 REM * * * MOONTRAK.BAS * * *
20 CLS:PRINT"PROGRAM TO CALCULATE AZ-EL DATA FOR THE SUN OR MOON"
30 PRINT:PRINT"Program by J. Hall, K1TD, ARRL Hq., Rev 1.1, June 1983":PRINT
40 PRINT"This program may be reproduced without prior permission"
50 PRINT"provided The ARRL Handbook is credited.":PRINT
60 B$="###.#":I=57.2958:K=.01657
70 PRINT"Enter negative values for southerly latitudes."
80 PRINT"Enter negative values for easterly longitudes."
90 PRINT:INPUT"Your latitude (degrees and decimal)";A
100 INPUT"Your longitude (degrees and decimal)";L1
110 D=SIN(A/I):F=COS(A/I)
120 INPUT"UTC date (no comma, please)";A$
130 INPUT"Data for which, sun or moon";C$:PRINT
140 PRINT A$" GHA of ";C$;" at 0000 UTC":INPUT"   (degrees and decimal)";L2
150 PRINT A$" declination at 0000 UTC":INPUT"   (degrees and decimal)";B1
160 PRINT A$" GHA at 2400 UTC":INPUT"   (degrees and decimal)";L3
170 PRINT A$" declination at 2400 UTC":INPUT"   (degrees and decimal)";B2
180 GI=(L3+360-L2)/24
190 BI=(B2-B1)/24
200 J%=0:GOSUB 330
210 FOR A%=0 TO 48:G=SIN(B1/I):L=L2-L1
220 E=D*G+F*COS(B1/I)*COS(L/I):C=(G-D*E)/F:J=E
230 IF E>=1 THEN E=1.5708:GOTO 250
240 IF E<=-1 THEN E=-1.5708 ELSE E=ATN(E/SQR(ABS(1-E*E)))
250 C=C/COS(E):IF C>=1 THEN C=0:GOTO 270
260 IF C<=-1 THEN C=180 ELSE C=I*(-ATN(C/SQR(ABS(1-C*C))))+1.5708
270 IF LEFT$(C$,1)="M" OR LEFT$(C$,1)="m" THEN GOSUB 430
280 E=I*E:IF J%=8 AND E>=0 THEN GOSUB 410
290 IF E>=0 THEN GOSUB 360
300 B1=B1+BI/2:L2=L2+GI/2:IF L2>360 THEN L2=L2-360
310 NEXT:PRINT:PRINT"Data for ";A$;" is completed for the ";C$
320 INPUT"To continue, press enter";J%:CLS:PRINT"Next ";:GOTO 120
330 CLS:PRINT"Data for the ";C$" from "A;"deg. lat., ";L1;"deg. long."
340 PRINT"for ";A$:PRINT
350 PRINT"Time, UTC","Azimuth","Elevation":PRINT:RETURN
360 IF INT(A%/2)=A%/2 THEN D$=STR$(50*A%) ELSE D$=STR$(50*A%-20)
370 D$=RIGHT$(D$,LEN(D$)-1)
380 IF LEN(D$)<4 THEN D$="0"+D$:GOTO 380
390 PRINT D$,:IF SIN(L/I)<=0 THEN PRINT USING B$;C; ELSE PRINT USING B$;360-C;
400 PRINT TAB(32);:PRINT USING B$;E:J%=J%+1:RETURN
410 PRINT:INPUT"For more data press enter";J%
420 J%=0:GOSUB 330:RETURN
430 J=(J-K)/COS(E):E=ATN(J):RETURN
440 REM NOTE:  % = integer
```

Libration fading of an EME signal is characterized in general as fluttery, rapid, irregular fading not unlike that observed in tropospheric scatter propagation. Fading can be very deep, 20 dB or more, and the maximum fading will depend on the operating frequency. At 1296 MHz the maximum fading rate is about 10 Hz, and scales directly with frequency.

On a weak CW EME signal, libration fading gives the impression of a randomly keyed signal. In fact on very slow CW telegraphy the effect is as though the keying is being done at a much faster speed.

On very weak signals only the peaks of libration fading are heard in the form of occasional short bursts or "pings."

Fig. 38 shows samples of a typical EME echo signal at 1296 MHz. These recordings, made at W2NFA, show the wild fading characteristics with sufficient S/N ratio to record the deep fades. Circular polarization was used to eliminate Faraday fading; thus these recordings are of libration fading only. The recording bandwidth was limited to about 40 Hz to minimize the higher sideband-frequency components of libration fading that exist but are much smaller

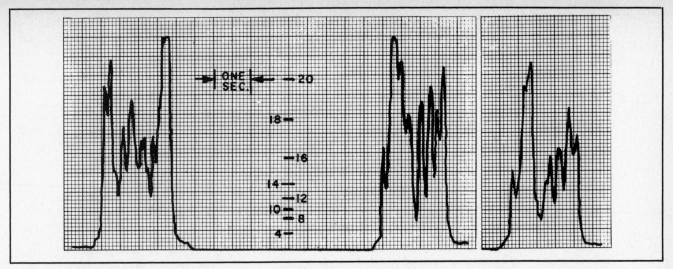

Fig. 38 — Chart recording of moon echoes received at W2FNA on July 26, 1973, at 1630 UTC. Antenna gain 44 dBi, transmitting power 400 watts and system temperature 400 K.

in amplitude. For those who would like a better statistical description, libration fading is Raleigh distributed. In the recordings shown by Fig. 38, the average signal-return level computed from path loss and mean reflection coefficient of the moon is at about the +15 dB S/N level.

It is clear that enhancement of echoes far in excess of this average level is observed. This point should be kept clearly in mind when attempting to obtain echoes or receive EME signals with marginal equipment. The probability of hearing an occasional peak is quite good since random enhancement as much as 10 dB is possible. Under these conditions, however, the amount of useful information that can be copied will be near zero. Enthusiastic newcomers to EME communications will be stymied by this effect since they know they can hear the signal strong enough on peaks to copy but can't make any sense out of what they try to copy.

What causes libration fading? Very simply, multipath scattering of the radio waves from the very large (2000-mile diameter) and rough moon surface combined with the relative motion between earth and moon called librations.

To understand these effects, assume first that the earth and moon are stationary (no libration) and that a plane wave front arrives at the moon from your earth-bound station as shown in Fig. 39A.

The reflected wave shown in Fig. 39B consists of many scattered contributions from the rough moon surface. It is perhaps easier to visualize the process as if the scattering were from many small individual flat mirrors on the moon that reflect small portions (amplitudes) of the incident wave energy in different directions (paths) and with different path lengths (phase). Those paths directed toward the moon arrive at your antenna as a collection of small wave fronts (field vectors) of various amplitudes

and phases. The vector summation of all these coherent (same frequency) returned waves (and there is a near-infinite array of them) takes place at the feed point of your antenna (the collecting point in your antenna system). The level of the final summation as measured by a receiver can, of course, have any value from zero to some maximum. Remember that we assumed the earth and moon were stationary, which means that the final summation of these multipath signal returns from the moon will be one fixed value. The condition of zero relative motion between earth and moon is a rare event that will be discussed later in this section.

Consider now that the earth and moon are moving relative to each other (as they are in nature), so the incident radio wave "sees" a slightly different surface of the moon from moment to moment. Since the lunar surface is very irregular, the reflected wave will be equally irregular, changing in amplitude and phase from moment to moment. The resultant continuous summation of the varying multipath signals at your antenna feed point produces the effect called libration fading of the moon-reflected signal.

The term libration is used to describe small perturbations in the movement of celestial bodies. Each libration consists mainly of its diurnal rotation; moon libration consists mainly of its 28-day rotation, which appears as a very slight rocking motion with respect to an observer on earth. This rocking motion can be visualized as follows: Place a marker on the surface of the moon at the center of the moon disc, which is the point closest to the observer, as shown in Fig. 40. Over time, we will observe that this marker wanders around within a small area. This means the surface of the moon as seen from the earth is not quite fixed but changes slightly as different areas of the periphery are exposed because

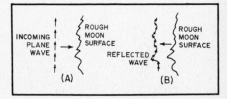

Fig. 39 — How the rough surface of the moon reflects a plane wave as one having many field vectors.

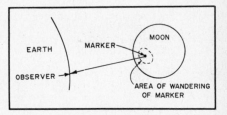

Fig. 40 — The moon appears to "wander" in its orbit about the earth. Thus a fixed marker on the moon's surface will appear to move about in a circular area.

of this rocking motion. Moon libration is very slow (on the order of 10^{-7} radians per second) and can be determined with some difficulty from published moon ephemeris tables.

Although the libration motions are very small and slow, the larger surface area of the moon has nearly an infinite number of scattering points (small area). This means that even slight geometric movements can alter the total summation of the returned multipath echo by a significant amount. Since the librations of the earth and moon are calculable, it is only logical to ask if there ever occurs a time when the total libration is zero or near zero. The answer is yes, and it has been observed and verified

Table 11

Signal Reports Used on 144-MHz EME

T — Signal just detectable
M — Portions of call copied
O — Complete call set has been received
R — Both "O" report and call sets have been received
$\overline{SK}$ — End of contact

Table 12

Signal Reports Used on 432-MHz EME

T — Portions of call copied
M — Complete calls copied
O — Good signal—solid copy (possibly enough for SSB work)
R — Calls and reports copied
$\overline{SK}$ — End of contact

Table 13

144-MHz Procedure — 2-Minute Sequence

Period	1½ minutes	30 seconds
1	Calls (W6XXX DE W1XXX)	
2	W1XXX DE W6XXX	T T T T
3	W6XXX DE W1XXX	O O O O
4	RO RO RO RO	DE W1XXX K
5	R R R R R	DE W6XXX K
6	QRZ? EME	DE W1XXX K

Table 14

432-MHz Procedure — 2½-Minute Sequence

Period	2 minutes	30 seconds
1	VE7BBG DE K2UYH	
2	K2UYH DE VE7BBG	
3	VE7BBG DE K2UYH	T T T
4	K2UYH DE VE7BBG	M M M
5	RM RM RM RM	DE K2UYH K
6	R R R R R	DE VE7BBG $\overline{SK}$

experimentally on radar echoes that minimum fading rate (not depth of fade) is coincident with minimum total libration. Calculation of minimum total libration is at best tedious and can only be done successfully by means of a computer. It is a problem in extrapolation of rates of change in coordinate motion and in small differences of large numbers.

EME Operating Techniques

Many EME signals are near the threshold of readability, a condition caused by a combination of path loss, Faraday rotation and libration fading. This weakness and unpredictability of the signal has led to the development of techniques for exchange of EME information that differs from those used for normal VHF work — the usual RST reporting would be jumbled and meaningless for many EME contacts. Dashes are often chopped into pieces, a string of dots would be incomplete, and complicated words would make no sense at all.

Unfortunately, there is no universal agreement as to procedures for all the bands, although there is similarity. Two-meter operators generally use the "T M O R " system, while those on 432 MHz use a similar system but applied at somewhat different levels of readability. The meanings and typical use of each part of the sequence are given in Tables 11 through 14.

At the moment, there is no uniform system in use for EME operation on all bands. The 432-MHz procedure is used on 432 MHz and above. It has also become the preferred system on 220 MHz, although the 144-MHz procedure is sometimes used on this band.

In the schedule sequence for both 144 and 432 MHz, the initial period starts on the hour, but because of the difference in sequence lengths for the two bands, schedules starting on the half hour will not be the same. On 2 meters, there are 15 sequence periods to the half hour, which make the period 0030 to 0032 an "even" sequence. This could make a difference, depending on which operator was assigned an "odd" or "even" sequence. Note that odd or even refers to the sequence number, not the minutes designated within that sequence.

On 432 MHz, there are 12 sequence periods to the half hour. The eastern-most station from the international date line usually calls first, and since two 2½-minute periods fill a 5-minute space, it works out conveniently that the eastern (or first) station will call starting with every five minute mark, and start listening 2½-minutes later. Thus a schedule starting at 0030 would be an "odd" period, although operators on 432 MHz seldom label them as such. It is convenient for the operators simply to start with the eastern-most station calling on the hour or half hour, unless arranged otherwise.

Of course there is much room for change in these arrangements, but they do serve as vital guidelines for schedules. As signals become stronger, the rules can be relaxed to a degree, and after many contacts, stations can often ignore them completely, if the signals are strong enough.

Calls are often extremely difficult to hear in their entirety. A vital dot or dash can be missing, which can render a complex call unreadable. To copy both calls completely requires much patience and a good ear. Both calls must be copied, because even though most work is by timed schedules, there can be last-minute substitutions because of equipment trouble at one station, unexpected travel, changes in plans or the like, which make it impossible for the scheduled station to appear. Thus, rather than have one station spend the entire period listening, only to find that no one is there, a system of standby stations is becoming more popular. This is good, because nothing will demoralize a newcomer faster than several one-sided schedules.

An exchange of signal reports is a useful and required bit of information: useful because it helps in evaluating your station performance and the conditions at the time, required because it is a "non-prearranged" exchange, thereby requiring that you copy what was sent as part of the contact. Obviously, other things could be included in an "exchange of unknown information," and when conditions permit stronger signals, many operators do include names, elaborate on the signal reports, arrange next schedule times, and so on. Unfortunately, such exchanges are rare.

Confirmation is essential for completion of the exchange. There is no way that you can be sure that the other operator copies what you sent until you hear him say so. The final R or ROGER means that he has copied your information, and your two-way contact is complete.

Sending speed is usually in the 10- to 13-WPM range, although it can be adjusted according to conditions and operator skill. Characters sent too slowly tend to become chopped up and confusing. High-speed CW is hard to copy at marginal signal levels for most amateurs, and the fading that is typical of an EME path can make it nearly impossible to decipher the content.

Other Modes

Only a few stations have the capability of sending (and receiving) signals of a strength sufficient to allow experimentation with other than CW. SSB contacts are now common. In general, only the stations with large antennas try the more difficult modes for EME work. Such installations are often "borrowed" from some research program for the amateur endeavors.

Frequencies

EME contacts are generally made randomly or by prearranged schedule. Many stations, especially those with marginal capability, prefer to set up a specific time and frequency in advance so that they will have a better chance of finding each other. The larger stations, especially on 144 and 432 MHz where there is a good amount of activity, often call CQ during evenings and weekends when the moon is at perigee and listen for random replies. Most of the work on 220, 1296 and 2304 MHz, where activity is light, is done by schedule.

Most amateur EME work on 144 and 220 MHz takes place near the low edge of the band. Activity is found 50 kHz or higher in the band during peak hours. Generally, random activity and CQ calling take place in the lower 10 kHz or so, and schedules are run higher in the band.

Formal schedules (that is, schedules arranged well in advance and published in the *432 and Above EME Newsletter* published by Al Katz, K2UYH) are run on 432.000, 432.025 and 432.030 MHz. Other schedules are normally run on 432.035 and up to 432.070. For this band, the EME random calling frequency is 432.010 with random activity spread out between 432.005 and 432.020. Random SSB CQ calling is at 432.015. Terrestrial activity is centered on 432.100 and is, by agreement, limited to 432.075 and above in North America.

Moving up in frequency, formal schedules are run on 1296.000 and 1296.025 MHz. The EME random calling frequency is 1296.010 with random activity spread out between 1296.005 and 1296.020. Terrestrial activity is centered at 1296.100. There is some EME activity on 2304 MHz. Specific frequencies are dictated by equipment availability and are arranged by the stations involved.

For EME SSB contacts on 144 and 432 MHz, contact is usually established on CW, and then the stations move up 100 kHz from the CW frequency. (This method was adopted because of the U.S. requirement for CW only below 144.1 MHz.)

Of course, it is obvious that as the number of stations on EME increases, the frequency spread must become greater. Since the moon is in convenient locations only a few days out of the month, and only a certain number of stations can be scheduled for EME during a given evening, the answer will be in the use of simultaneous schedules, spaced a few kilohertz apart. The time may not be too far away — QRM has already been experienced on each of our three most active EME frequencies.

EME Net Information

An EME net meets on weekends on 14.345 MHz for the purpose of arranging schedules and exchanging pertinent information. The net meets at 1600 UTC concerning EME operation on 432 MHz and above. At 1700 UTC the net is for 2-meter EME operation. Both nets carry information on 220-MHz EME operation.

Antenna Requirements

The tremendous path loss incurred over an EME circuit places stringent requirements on station performance. Low-noise receiving equipment, maximum legal power and large antenna arrays are required for successful EME operation. Although it may be possible to copy some of the better-equipped stations while using a single high-gain Yagi antenna, it is doubtful whether such an antenna could provide reliable two-way communication. Antenna gain of at least 20 dB is required for reasonable success. Generally speaking, more antenna gain will yield the most noticeable improvement in station performance, as the increased gain will aid both the received and transmitted signals.

Several types of antennas have become popular among EME enthusiasts. Perhaps the most popular antenna for 144-MHz work is an array of either four or eight long-boom (14- to 15-dB gain) Yagis. The four-Yagi array would provide approximately 20-dB gain, and the eight-antenna system would show an approximate 3-dB

Fig. 41 — The 2-meter EME antenna system used at DL8DAT — 16 × 24-element Yagis.

increase over the four-antenna array. At 432 MHz, eight or 16 long-boom Yagis are used. Yagi antennas are available commercially or can be constructed from readily available materials. Information on maximum-gain Yagi antennas is presented in Chapter 33. The dimensions presented are based on figures developed by the National Bureau of Standards for Yagi design. At least one manufacturer has used the NBS design information for their latest series of high-performance antennas.

A moderately sized Yagi array has the advantage that it is relatively easy to construct and can be positioned in azimuth and elevation with commercially available equipment. Matching and phasing lines present no particular problems. The main disadvantage of a Yagi array is that the polarization plane of the antenna cannot be conveniently changed. Polarization shift at 144 MHz is fairly slow, and the added complexity of the cross-polarized antenna

system may not be worth the effort. At 432 MHz, where the shift is at a somewhat faster rate, an adjustable polarization system offers a definite advantage over a fixed one.

Quagi antennas (made from both quad and Yagi elements) are also popular for EME work. Slightly more gain per unit boom length is possible as compared to the conventional Yagi. Additional information on the quagi is presented in Chapter 33.

The collinear is another popular type of antenna for EME work. A 40-element collinear array has approximately the same frontal area as an array of four Yagis. The collinear array would produce approximately 1- to 2-dB less gain. Of course the depth dimension of the collinear array is considerably less than for the long-boom Yagis. An 80-element collinear would be marginal for EME communications, providing approximately 19-dB gain. Many operators choosing this type of antenna use 160-element or larger systems. As with Yagi and quagi antennas, the collinear cannot be easily adjusted for polarity changes. From a constructional standpoint there may be little difference in complexity and material costs between the collinear and Yagi arrays.

The parabolic dish is another antenna used extensively for EME work. Unlike the other antennas described, the major problems associated with dish antennas are mechanical. Dishes 12 feet in diameter are required for successful EME operation on 432 MHz. With structures of this size, wind and ice loading place a severe strain on the mounting and positioning systems. Extremely rugged mounts are required for large dish antennas, especially when used in windy locations. Several aspects of the parabolic dish antennas make the extra mechanical problems worth the effort. For example, the dish antenna is inherently broadband and may be used on several different bands by simply changing the feed.

The graph at Fig. 42 relates antenna

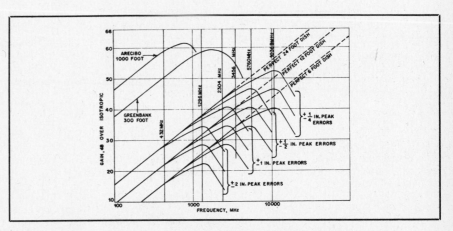

Fig. 42 — Parabolic antenna gain versus size, frequency and surface errors. All curves assumed 60-percent aperture efficiency and 10-dB power taper. Reference: J. Ruze, Britsh IEE.

Fig. 43 — A newcomer to EME stands in awe of the K2UYH 28-foot dish.

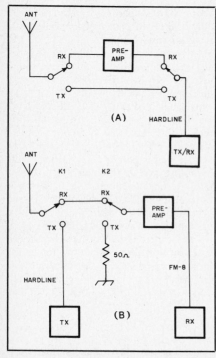

Fig. 44 — Two systems for switching a preamplifier in and out of the receive line. At A, a single length of cable is used for both the transmit and receive line. At B is a slightly more sophisticated system that uses two separate transmission lines. See text for details.

gain, frequency and size. As can be seen, an antenna that is suitable for 432-MHz work is also usable for each of the higher amateur bands. Additional gain is available as the frequency of operation is increased. Another advantage of this antenna is in the feed system. The polarization of the feed, and therefore the polarization of the antenna, can be adjusted with little difficulty. It should be relatively easy to devise a system whereby the feed could be rotated remotely from the shack. Changes in polarization of the signal could be compensated for at the operating position! As polarization changes can account for as much as 30 dB of signal attenuation, the rotatable feed could make the difference between working a station and not. A photograph of the parabolic dish antenna used at K2UYH is shown in Fig. 43. More information on parabolic dish antennas is available in *The ARRL Antenna Book*.

Antennas suitable for EME work are by no means limited to the types described thus far. Rhombics, quad arrays, helixes and others have not gained the popularity of the Yagi, quagi, collinear and parabolic dish, however.

Receiver Requirements

A low-noise receiving setup is essential for successful EME work. Since many of the signals to be copied on EME are barely, but not always, out of the noise, a low-noise-figure receiver is a must. The mark to shoot for at 144 MHz is something under 0.5 dB, as the cosmic noise will then be the limiting factor in the system. Noise figures of this level are relatively easy to achieve, even with available inexpensive devices.

As low a noise figure as can be attained will be usable at 432 MHz. Noise figures on the order of 0.5 dB are possible with GaAsFETs. Other builders choose the rugged bipolar transistor, which offers a noise figure just under 1 dB.

The loss in the transmission line that connects the antenna to the preamplifier adds directly to the system noise figure. Not only does the line have loss, it also is a source of noise, which further deteriorates the system noise figure. For those reasons, most serious EME operators mount a preamplifier at the top of the tower or directly at the antenna. If an exceptionally good grade of transmission line is available, it is possible to obtain almost as good results with the preamplifier located in the shack.

Two relay/preamplifier switching systems are sketched in Fig. 44. The system at A makes use of two relays and a single transmission line for both transmit and receive. The preamplifier is simply switched "in" for receive and "out" for transmit.

The system outlined at Fig. 44B also uses two relays, but the circuit is somewhat more sophisticated. Two transmission lines are used, one for the receive line and one for the transmit line. In addition, a 50-ohm termination is provided. Since relays with high isolation in the VHF/UHF range are difficult and expensive to obtain, two relays with a lower-isolation factor may be used.

When the relays are switched for the transmit mode, K1 connects the antenna to the transmit line, K2 switches the preamplifier into the 50-ohm termination. Hence, two relays provide the isolation between the transmitter connection and the preamplifier. If independent control of K2 is provided for, the preamplifier can be switched between the 50-ohm termination and the antenna during receive. This feature is especially useful when making sun-noise measurements to check system performance. For this measurement the antenna is directed toward the sun and the preamplifier is alternately switched between the 50-ohm load and the antenna. The decibel difference can be recorded and used as a reference when checking system improvements. A less convenient, but better, check is to compare sun noise with cold sky noise. The complete circuit for this relay system is presented later in this chapter.

As the preamplifier is mounted ahead of the transmission line to the receiver, a cable of mediocre performance can be used. The loss of the cable, as long as it is within reason, will not add appreciably to the system noise figure. Information contained in the Receiving chapter of this book explains how to calculate system noise figures. Foam-type RG-8 cable is acceptable for runs up to 100 feet at 144 MHz. See Chapter 31 for examples of practical preamplifiers and remote-mounted relay/preamp projects.

It is important to get as much transmitter power as possible to the antenna. For this reason rigid or semirigid low-loss cable is specified for the transmit line.

Transmitter Requirements

In many EME installations the antenna gain is not much above the minimum required for communications. It is highly likely that the maximum legal limit of power will be required for successful EME work on up through 432 MHz. Since many contacts may require long, slow sending, the transmitter and amplifier should have adequate cooling. An amplifier with some power to spare rather than an amplifier running "flat out" is desirable. This is especially important should SSB communication be attempted. An amplifier run all out on SSB will likely produce large amounts of odd-order IMD products that fall within the band. While the splatter produced will not affect your communications, it will certainly affect that of others close in frequency!

Chapter 24

Construction Techniques

While a better job can be done with a variety of tools, by taking little care it is possible to turn out a fine piece of equipment with only a few common hand tools. A list of tools that are indispensable in the construction of electronic equipment is found on this page. With these tools it should be possible to prepare panels and metal chassis for assembly and wiring. It is an excellent idea for the amateur who builds gear to add tools from time to time as finances permit.

RECOMMENDED TOOLS AND MATERIALS

Long-nose pliers, 6- and 4-inch
Diagonal cutters, 6- and 4-inch
Combination pliers, 6-inch
Screwdriver 6- to 7-inch, 1/4-inch blade
Screwdriver, 4- to 5-inch, 1/8-inch blade
Phillips screwdriver, 6- to 7-inch
Phillips screwdriver, 3- to 4-inch
Long-shank screwdriver with holding clip on blade
Scratch awl or scriber for marking metal
Combination square, 12-inch, for layout work
Hand drill, 1/4-inch chuck or larger
Soldering pencil, 30-watt, 1/8-inch tip
Soldering iron, 200-watt, 5/8-inch tip
Hacksaw and 12-inch blades
Hand nibbling tool, for chassis-hole cutting
Hammer, ball-peen, 1-lb head
Heavy-duty jackknife
File set, flat, round, half-round, and triangular. Large and miniature types recommended
High-speed drill bits, no. 60 through 3/8-inch diameter
Set of Spintite socket wrenches for hex nuts
Adjustable wrenches, 6- and 10-inch
Machine-screw taps, 4-40 through 10-32 thread
Socket punches, 1/2 in, 5/8 in, 3/4 in, 1-1/8 in, 1-1/4 in, and 1-1/2 in
Tapered reamer, T-handle, 1/2-inch maximum width
Bench vise, 4-inch jaws or larger
Medium-weight machine oil
Tin shears, 10-inch
Motor-driven emery wheel for grinding

Table 1
Numbered Drill Sizes

No.	Diameter (Mils)	Will Clear Screw	Drilled for Tapping from Steel or Brass
1	228.0	12-24	—
2	221.0	—	—
3	213.0	—	14-24
4	209.0	12-20	—
5	205.0	—	—
6	204.0	—	—
7	201.0	—	—
8	199.0	—	—
9	196.0	—	—
10	193.5	—	—
11	191.0	10-24 10-32	—
12	189.0	—	—
13	185.0	—	—
14	182.0	—	—
15	180.0	—	—
16	177.0	—	12-24
17	173.0	—	—
18	169.5	—	—
19	166.0	8-32	12-20
20	161.0	—	—
21	159.0	—	10-32
22	157.0	—	—
23	154.0	—	—
24	152.0	—	—
25	149.5	—	10-24
26	147.0	—	—
27	144.0	—	—
28	140.0	6-32	—
29	136.0	—	8-32
30	128.5	—	—
31	120.0	—	—
32	116.0	—	—
33	113.0	4-40	—
34	111.0	—	—
35	110.0	—	—
36	106.5	—	6-32
37	104.0	—	—
38	101.5	—	—
39	099.5	3-48	—
40	098.0	—	—
41	096.0	—	—
42	093.5	—	—
43	089.0	—	4-40
44	086.0	2-56	—
45	082.0	—	—
46	081.0	—	—
47	078.5	—	3-48
48	076.0	—	—
49	073.0	—	—
50	070.0	—	2-56
51	067.0	—	—
52	063.5	—	—
53	059.5	—	—
54	055.0	—	—

Solder, *rosin core only*
Contact cleaner, liquid or spray can
Duco cement or equivalent
Electrical tape, vinyl plastic

Radio-supply houses, mail-order retail stores and most hardware stores carry the various tools required for building or servicing Amateur Radio equipment. While power tools (electric drill or drill press, grinding wheel, etc.) are very useful and will save a lot of time, they are not essential.

Twist Drills

Twist drills are made of either high-speed steel or carbon steel. The latter type is more common and will usually be supplied unless a specific request is made for high-speed drills. The carbon drill will suffice for most equipment construction work and costs less than the high-speed type.

While twist drills are available in a number of sizes, those listed in bold type in Table 1 will be most commonly used in construction of amateur equipment. It is usually desirable to purchase several of each of the commonly used sizes rather than a standard set, most of which will be used infrequently, if at all.

Although Table 1 lists drills down to no. 54, the series extends to no. 80. No. 68 and no. 70 are useful for drilling printed-circuit boards for component leads.

Care of Tools

The proper care of tools is not only a matter of pride to a good worker. Energy is saved and annoyance avoided by possessing a full kit of well-kept, sharp-edged tools.

Drills should be sharpened at frequent intervals so grinding is kept at a minimum each time. This makes it easier to maintain the rather critical surface angles required for best cutting with least wear. Occasional oilstoning of the cutting edges of a drill or reamer will extend the time between grindings.

The soldering iron can be kept in good condition by keeping the tip well tinned with solder and not allowing it to run at

Fig. 1 — A compact assembly of commonly available items, this soldering station sanitizes the electronics assembly process. Miniature toggle switches are used because of the minimal force required to manipulate them. The force required to operate standard-size switches could destabilize the unit.

Fig. 2 — View of the chassis underside with the bottom plate removed. No. 24 hookup wire is adequate for all connections. Use sleeving wherever the possibility of a short circuit exists. The diode may be installed in either direction.

full voltage for long periods when it is not being used. After each period of use, the tip should be removed and cleaned of any scale that may have accumulated. An oxidized tip may be cleaned by dipping it in sal ammoniac (ammonium chloride) while hot and then wiping it clean with a rag. If a copper tip becomes pitted, it should be filed until smooth and bright and then tinned immediately with solder. Most modern soldering iron tips are iron-clad and cannot be filed.

Useful Materials

Small stocks of various miscellaneous materials will be required in constructing radio apparatus. Most of these are available from hardware or radio-supply stores. A representative list follows:

Sheet aluminum, solid and perforated, 16 or 18 gauge, for brackets and shielding.

Aluminum angle stock, 1/2 × 1/2-inch

1/4-inch-diameter round brass or aluminum rod for shaft extensions.

Machine screws: Round-head and flat head, with nuts to fit. Most useful sizes: 4-40, 6-32 and 8-32, in lengths from 1/4-inch to 1-1/2 inches. (Nickel-plated steel will be found satisfactory except in strong RF fields, where brass should be used.)

Bakelite, Lucite, polystyrene and copper-clad PC-board scraps.

Soldering lugs, panel bearings, rubber grommets, terminal-lug wiring strips,

varnished-cambric insulating tubing, heat-shrinkable tubing.

Shielded and unshielded wire.

Tinned bare wire, nos. 22, 14 and 12.

Machine screws, nuts, washers, soldering lugs, etc., are most reasonably purchased in quantities of a gross. Many radio-supply stores sell small quantities and assortments that come in handy.

A DELUXE SOLDERING STATION

The simple device shown on this page can enhance the versatility and longevity of a soldering iron as well as make electronic

assembly more convenient. Fig. 1 depicts the obvious convenience features — a protective heat sink and case, and a tip-cleaning sponge rigidly attached to a sturdy base for efficient one-handed operation. Inside the chassis are some electrical refinements that justify the sophisticated name soldering station.

Soldering iron tips and heating elements last longer if operated at a lower-than-maximum temperature when idling. In the unit described here, temperature reduction is accomplished by halving the duty cycle of the applied ac voltage. D1 in Fig. 3 con-

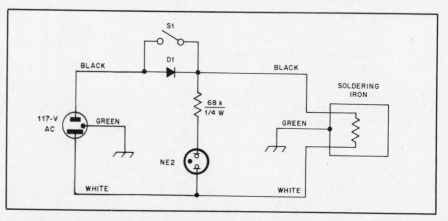

Fig. 3 — Schematic diagram of the soldering station. D1 is a silicon power rectifier, 1-A, 400-PIV. S1 is a miniature SPST toggle switch rated 3 at 125 V. This current is satisfactory for use with irons having power ratings up to 100 W.

ducts only when the hot ac line is positive with respect to neutral. If the diode were reversed, the soldering iron would be heated only on the negative half cycles, but the result would be the same. This is one of the rare applications of rectifier diodes where the polarity is not important. With current flowing only in one direction, only one electrode of the neon bulb will glow. Closing S1 short-circuits the diode and applies full power to the soldering iron, igniting both bulb electrodes brightly.

CMOS ICs are prone to damage by static charges, so they should be soldered with an iron having a grounded tip. This requirement is fulfilled by most irons having 3-wire power cords.

The base for the unit is a 2 × 6 × 4-inch (HWD) aluminum chassis (Bud AC-431 or equivalent). An Ungar model 8000 soldering iron fits neatly on the chassis top. The holder has two mounting holes in each foot. A sponge tray nests between the feet and the case. In this model, a sardine tin is used for the sponge tray. A suitable watertight enclosure can also be fabricated from strips of copper-clad circuit-board material. The tray and iron holder are secured to the chassis by 6-32 × 1/2-inch pan-head machine screws and nuts, with flat washers under the screw heads (sponge tray) and lock washers under the nuts (chassis underside). One of these nuts fastens a 6-lug tie point strip to the chassis bottom. Use the soldering iron holder base as a template for drilling the chassis and sponge tray. The floor of the sponge tray must be sealed around the screw heads to prevent moisture from leaking into the electrical components below the chassis. RTV compound was used for this purpose in the unit pictured.

Purchase a separate 3-wire cord for the power input. Merely splicing the soldering station into existing soldering iron cord will shorten the operating radius of the iron and make it awkward to use. Heyco bushings were used to anchor both cords in the unit decribed. If these aren't available, grommets and cable clamps will work as well. Knotting the cords inside the chassis is a simple expedient that normally provides adequate strain relief.

The underchassis assembly is shown in Fig. 2. The neon bulb is forced through a 3/16-inch-ID grommet. The leads are sleeved to prevent short circuits. If you mount the bulb in a fixture or socket, use a clear lens to ensure that the electrodes are distinctly visible. Fit a cover to the bottom of the chassis to prevent accidental contact with the live ac wiring. Stick-on rubber feet will ensure a skid-free unit that won't mar your work surface.

SOLDERING-IRON TEMPERATURE CONTROL

Greater flexibility can be realized by installing a temperature rather than an idle control for your soldering iron. An incandescent-light dimmer can be used to

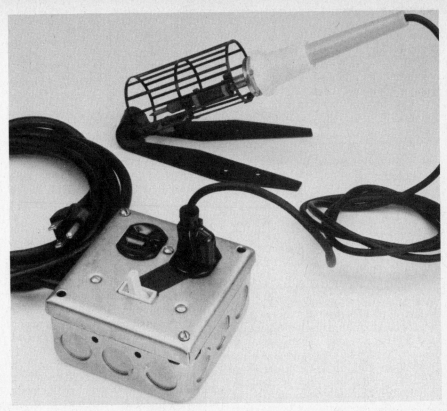

Fig. 4 — An incandescent-light dimmer controls soldering iron tip temperature. Only one of the duplex outlets is connected through the dimmer.

control the working temperature of the tip. Shown in Fig. 4 is a temperature control that can be built into an electrical box. A dimmer and a duplex outlet are mounted in the box; the wiring diagram is shown in Fig. 5. Be sure to break the link that connects the two hot terminals of the duplex outlet. The hot terminal is narrower than the neutral one. Neutral terminals may remain interconnected.

The dimmer shown in Fig. 4 was purchased at Radio Shack. It looks like a switch but it is really a variable control. Finer adjustment could be realized with a knob-controlled unit, but this one works fine.

When soldering static-sensitive devices, two or three jumpers can be used to ground you, the work and the iron. If the iron does not have a ground wire in the power cord, clip a jumper from the metal part of the iron near the handle to the metal box that houses the temperature control. Another jumper connects from the box to the work. Finally, a jumper goes from the box to your metal watch band. 3M makes an elastic wrist band for static grounding. This wrist band is equipped with a snap-on ground lead. Also, a 1-MΩ resistor is built into the snap of the strap to protect the user should a live circuit be contacted while using the strap. A similar resistor should be built into any homemade ground strap to protect the operator. It's not good practice to work on live circuits, however. This resistor will pro-

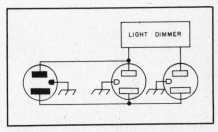

Fig. 5 — Schematic diagram of the soldering-iron temperature control.

tect you, but you should make sure that all power is removed from a circuit before working on it. Remember: even 24 V can be lethal! This technique works well for soldering CMOS ICs and GaAsFETs.

Chassis Working

With a few essential tools and proper procedure, building radio gear on a metal chassis is a relatively simple matter. Aluminum is preferred to steel, not only because it is a superior shielding material, but because it is much easier to work and provides good chassis contacts.

The placement of components on the chassis is shown clearly in the photographs in this *Handbook*. Aside from certain essential dimensions, which usually are given in the text, exact duplication is not necessary.

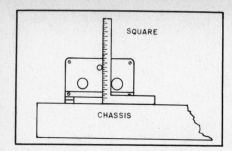

Fig. 6 — Method of measuring the heights of capacitor shafts. If the square is adjustable, the end of the scale should be set flush with the face of the head.

Much trouble and energy can be saved by spending sufficient time in planning the job. When all details are worked out beforehand, the actual construction is greatly simplified.

Cover the top of the chassis with a piece of wrapping paper, or, preferably, cross-section paper. Fold the edges down over the sides of the chassis and fasten with adhesive tape. Then assemble the parts to be mounted on top of the chassis and move them about until a satisfactory arrangement has been found, keeping in mind any parts to be mounted underneath, so interference in mounting can be avoided. Place capacitors and other parts with shafts extending through the panel first, and arrange them so the controls will form the desired pattern of the panel. Be sure to line up the shafts squarely with the chassis front. Locate any partition shields and panel brackets next, and then sockets and any other parts. Mark the mounting-hole centers of each accurately on the paper. Watch out for capacitors whose shafts are off center and do not line up with the mounting holes. Do not forget to mark the centers of socket holes and holes for wiring leads. The small holes for socket-mounted screws are best located and center-punched, using the socket itself as a template, after

the main center hole has been cut.

Use the square to indicate the centers of shafts. These should be extended to the chassis front and marked on the panel at the chassis line, the panel being fastened on temporarily. The hole centers may then be punched in the chassis with the center punch. After drilling, the parts that require mounting underneath may be located and the mounting holes drilled, making sure by trial that no interferences exist with parts mounted on top. Mounting holes along the front edge of the chassis should be transferred to the panel by once again fastening the panel to the chassis and marking it from the rear.

Next, mount on the chassis the capacitors and any other parts with shafts extending to the panel, and measure accurately the height of the center of each shaft above the chassis, as illustrated in Fig. 6. The horizontal displacement of shafts having already been marked on the chassis line on the panel, the vertical displacement can be measured from this line. The shaft centers may now be marked on the back of the panel, and the holes drilled. Holes for any other panel equipment coming above the chassis line may then be marked and drilled, and the remainder of the apparatus mounted. Holes for terminals and other parts of the rear edge of the chassis should also be marked at this time.

Drilling and Cutting Holes

When drilling holes in metal with a hand drill it is important that the centers first be located with a center punch, so the drill point will not "walk" away from the center when starting the hole. When the drill starts to break through, special care must be used. Often it is an advantage to shift a two-speed drill to low gear at this point. Holes more than ¼-inch in diameter should be started with a smaller drill and reamed out with the larger drill.

The chuck on the usual type of hand drill is limited to ¼-inch drills. The ¼-inch hole may be filed out to larger diameters with a round file. Another method possible with limited tools is to drill a series of small holes with the hand drill along the inside of the circumference of the large hole, placing the holes as close together as possible. The center may then be knocked out with a cold chisel and the edges smoothed with a file.

Taper reamers which fit into the carpenter's brace will make the job easier. A large rat-tail file clamped in the brace makes a good reamer for holes up to the diameter of the file.

For socket holes and other large holes in an aluminum chassis, socket-hole punches should be used. They require first drilling a guide hole to pass the bolt that is turned to squeeze the punch through the chassis. The threads of the bolt should be oiled occasionally.

Large holes in steel panels or chassis are best cut with an adjustable circle cutter ("flycutter"). Occasional application of machine oil in the cutting groove will help. The cutter first should be tried out on a block of wood or scrap material to make sure it is set for the right diameter.

The burrs or rough edges that usually result after drilling or cutting holes may be removed with a file, or sometimes more conveniently with a sharp knife or chisel. It is a good idea to keep an old wood chisel sharpened and available for this purpose.

Rectangular Holes

Square or rectangular holes may be cut out by making a row of small holes as previously described, but is more easily done by drilling a ½-inch hole inside each corner, as illustrated in Fig. 7, and using these holes for starting and turning the hacksaw. The socket-hole punch and the square punches that are now available also may be of considerable assistance in cutting out large openings.

Semiconductor Heat Sinks

Homemade heat sinks can be fashioned from brass, copper or aluminum stock with ordinary workshop tools. The dimensions of the heat sink will depend on the type of device used and the amount of heat that must be conducted away from the body of the semiconductor.

Fig. 8 shows the order of progression for forming a large heat sink from aluminum or brass channels of near-equal height and depth. The width is lessened in parts B and C so each channel will fit into the preceding one as shown in the completed model at D. The three pieces are bolted together with 8-32 screws and nuts. Dimensions given are for illustrative purposes only.

Heat sinks for smaller transistors can be

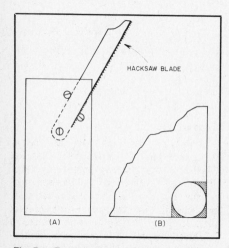

Fig. 7 — To cut rectangular holes in a chassis corner, holes may be filed out as shown in the shaded portion of B, making it possible to start the hacksaw blade along the cutting line. A shows how a single-ended handle may be constructed for a hacksaw blade.

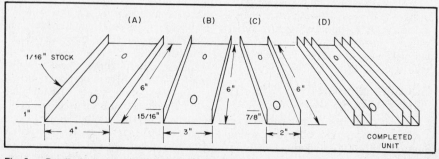

Fig. 8 — Details for forming channel-type heat sinks.

fabricated as shown in Fig. 9. Select a drill bit that is one size smaller than the diameter of the transistor case and form the heat sink from 1/16-inch-thick brass, copper or aluminum stock as shown in steps A, B and C. Form the stock around the drill bit by compressing it in a vise (A). The completed heat sink is press-fitted over the body of the semiconductor as illustrated at D. The larger the heat sink area, the greater the amount of heat conducted away from the transistor body. In some applications, such as power transistor stages, the heat sinks shown in Fig. 9 may be 2 or 3 inches in height.

Another technique for making heat sinks for TO-5 type transistors and larger models is shown in Fig. 10. This style of heat sink will dissipate considerably more heat than will the type shown in Fig. 9. The main body of the sink is fashioned from a piece of 1/8-inch-thick aluminum angle bracket — available from most hardware stores. A hole is bored in the angle stock to allow the transistor case to fit *snugly* into it. The transistor is held in place by a small metal plate whose center hole is slightly smaller in diameter than the case of the transistor. Details are given in Fig. 10.

A thin coating of silicone grease, available from most electronic supply houses, can be applied between the case of the transistor and the part of the heat sink with which it comes in contact. The silicone grease will aid the transfer of heat from the transistor to the sink. Electrically conductive and nonconductive grease is available. Do not use conductive grease for applications where the device is insulated from the heat sink. This practice can be applied to all models shown here. In the example given in Fig. 8, the grease should be applied between the three channels before they are bolted together, as well as between the transistor and the channel it contacts.

Construction Notes

If a control shaft must be extended or insulated, a flexible shaft coupling with adequate insulation should be used. Satisfactory support for the shaft extension, as well as electrical contact for safety, can be provided by means of a metal panel bushing made for the purpose. These can be obtained singly for use with existing shafts, or they can be bought with a captive extension shaft included. In either case the panel bushing gives a solid feel to the control. The use of fiber washers between ceramic insulation and metal brackets, screws or nuts will prevent the ceramic parts from breaking.

Cutting and Bending Sheet Metal

If a metal sheet is too large to be cut conveniently with a hacksaw, it may be marked with scratches as deep as possible along the line of the cut on both sides of the sheet, and then clamped in a vise and worked back and forth until the sheet breaks at the line. Do not carry the bending too far until

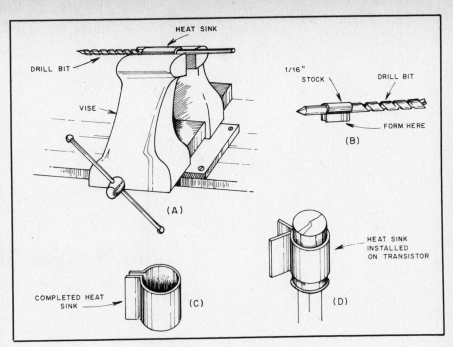

Fig. 9 — Steps used in constructing heat sinks for small transistors.

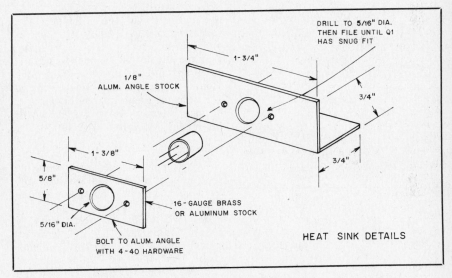

Fig. 10 — Layout and assembly details of another homemade heat sink. The completed assembly can be insulated from the main chassis of the transmitter by using insulating washers.

the break begins to weaken; otherwise the edge of the sheet may become bent. A pair of iron bars or pieces of heavy angle stock, as long or longer than the width of the sheet, to hold it in the vise, will make the job easier. C clamps may be used to keep the bars from spreading at the ends. Bends may be made similarly. The rough edges may be smoothed with a file or by placing a large piece of emery cloth or sandpaper on a flat surface and running the edge of the metal back and forth over the sheet.

Much of the tedium of sheet metal work can be relieved by using copper-clad printed-circuit board material wherever possible. Copper-clad stock is manufac-

tured with phenolic, FR-4 fiberglass and Teflon base materials in thicknesses up to 1/8 inch. While it is manufactured in large sheets for industrial use, some hobby electronics stores and surplus outlets market usable scraps at reasonable prices. PC-board stock is easily cut with a small hacksaw. Because the nonmetallic base material isn't malleable, it can't be bent in the usual way. However, corners are easily formed by holding two pieces at right angles and soldering the seam. Excellent RF-tight enclosures can be fabricated in this manner. Many projects in this Handbook were constructed using this technique. If mechanical rigidity is required of a large

Table 2
Standard Metal Gauges

Gauge No.	American or BS†	U.S. Standard††	Birmingham or Stubs†††
1	0.2893	0.28125	0.300
2	0.2576	0.265625	0.284
3	0.2294	0.25	0.259
4	0.2043	0.234375	0.238
5	0.1819	0.21875	0.220
6	0.1620	0.203125	0.203
7	0.1443	0.1875	0.180
8	0.1285	0.171875	0.165
9	0.1144	0.15625	0.148
10	0.1019	0.140625	0.134
11	0.09074	0.125	0.120
12	0.08081	0.109375	0.109
13	0.07196	0.09375	0.095
14	0.06408	0.078125	0.083
15	0.05707	0.0703125	0.072
16	0.05082	0.0625	0.065
17	0.04526	0.05625	0.058
18	0.04030	0.05	0.049
19	0.03589	0.04375	0.042
20	0.03196	0.0375	0.035
21	0.02846	0.034375	0.032
22	0.02535	0.03125	0.028
23	0.02257	0.028125	0.025
24	0.02010	0.025	0.022
25	0.01790	0.021875	0.020
26	0.01594	0.01875	0.018
27	0.01420	0.0171875	0.016
28	0.01264	0.015625	0.014
29	0.01126	0.0140625	0.013
30	0.01003	0.0125	0.012
31	0.008928	0.0109375	0.010
32	0.007950	0.01015625	0.009
33	0.007080	0.009375	0.008
34	0.006350	0.00859375	0.007
35	0.005615	0.0078125	0.005
36	0.005000	0.00703125	0.004
37	0.004453	0.006640626	—
38	0.003965	0.00625	—
39	0.003531	—	—
40	0.003145	—	—

†Used for aluminum, copper, brass and nonferrous alloy sheets, wire and rods.
††Used for iron, steel, nickel and ferrous alloy sheets, wire and rods.
†††Used for seamless tubes; also by some manufacturers for copper and brass.

copper-clad surface, stiffening ribs may be soldered at right angles to the sheet.

Finishing Aluminum

Aluminum chassis, panels and parts may be given a sheen finish by treating them in a caustic bath. An enameled or plastic container, such as a dishpan or infant's bathtub, should be used for the solution. Dissolve ordinary household lye in cold water in a proportion of one-quarter to one-half can of lye per gallon of water. The stronger solution will do the job more rapidly. Stir the solution with a stick of wood until the lye crystals are completely dissolved. Be very careful to avoid skin contact with the solution. It is also harmful to clothing. Sufficient solution should be prepared to cover the piece completely. When the aluminum is immersed, a very pronounced bubbling takes place and ventilation should be provided to disperse the escaping gas. A half hour to two hours in the bath should be sufficient, depending on the strength of the solution and the desired surface.

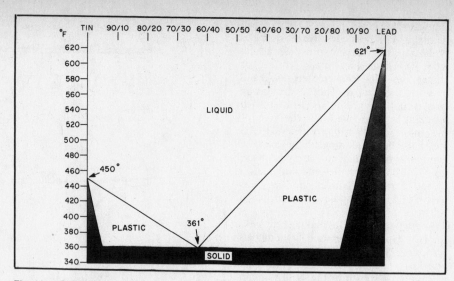

Fig. 11 — State diagram for tin-lead solder alloys. Courtesy of Litton Industries, Inc., Kester Solder Co.

Raw aluminum can be prepared for painting by abrading the surface with medium-grade sandpaper, making certain the strokes are applied in the same direction (not circular or random). This process will create tiny grooves on the otherwise smooth surface. As a result, paint or lacquer will adhere well. Before painting, wash the abraded aluminum with alcohol or soap and hot water, then dry thoroughly. Avoid touching the prepared surface before painting it. One or two coats of zinc chromate primer applied before the finish paint will ensure good adherence.

Soldering

The secret of good soldering is to use the right amount of heat. Solders have different melting points, depending on the ratio of tin to lead. Tin melts at 450 °F and lead at 621 °F (see Fig. 11). Solder with 63% tin and 37% lead melts at 361 °F, the lowest melting point for a tin and lead mixture. Called 63-37, this ratio also provides the most rapid solid to liquid transition and best stress resistance.

As shown in Fig. 11, solder that is not 63-37 goes through a *plastic* state. If the solder is deformed while it is in the plastic state, that deformation will remain when the solder freezes into the solid state. Any stress or motion applied to "plastic solder" will result in poor solder joint.

A 60-40 solder has the best *wetting* qualities. Wetting is the ability to spread rapidly and alloy uniformly. Soldering is not like glueing. The solder does more than bind pieces of metal together and provide an electrically conductive path between them. In the soldering process, the materials being joined and the solder combine to form an alloy. This effect is illustrated in Fig. 12. Because of its low melting point, small plastic area and excellent wetting characteristics, 60-40 solder is the most commonly used alloy in electronics.

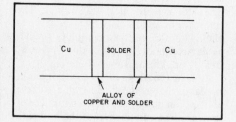

Fig. 12 — An alloy layer exists in a good solder joint.

Various solder alloys are used for special purposes. Other tin-lead alloys are 50-50 and 40-60. A low-melting-point solder, 50-32-18 (tin-lead-cadmium) has a melting temperature of under 300 °F. This alloy is used for heat sensitive components.

Some connections carrying very high current can't be made with ordinary tin-lead solder because the heat generated by the joint resistance would melt the solder. Automotive starter brushes and UHF transmitter tank circuits are two cases in which this situation can occur. High-melting-point silver solder prevents this condition. A typical alloy contains 5-93.5-1.5 (tin-lead-silver) and the melting point is about 600 °F.

To prevent the silver plating on some components from leaching into the solder, some silver is added to the solder. A typical alloy contains 62-36-2 (tin-lead-silver) and melts at 354 °F. Another silver solder is 96.3-3.7 (tin-silver); it contains no lead and melts at 430 °F.

Always use rosin-core solder; never acid-core. The rosin is a flux. Flux removes oxide by suspending it in solution and floating it to the top. Flux is not a cleaning agent! The work must be cleaned before soldering. Flux is not a part of a soldered connection — it merely aids the soldering process. After soldering, any remaining flux should be removed. Alcohol is a good solvent for the job. A toothbrush is an excellent tool for applying the alcohol and scrubbing the excess flux away.

The two key factors in quality soldering are time and temperature. Generally, rapid heating is desired. If heat is applied too long, the flux may become used up and surface oxidation can become a problem. Further, thin coating may dissolve from leads and components, or components may be harmed. The soldering-iron tip should be hot enough to readily melt the solder without burning, charring or discoloring components, PC boards or wires. Usually, a tip temperature about 100°F above the solder melting point is about right for mounting components on PC boards.

When soldering, apply the solder to the joint — not to the iron. A small amount of solder on the tip will improve heat transfer. The iron and the solder should be applied simultaneously to the joint. Keep the iron clean by brushing the tip with a cloth towel or a moist sponge. Allow the joints to cool in air at room temperature. Make sure there is no movement or stress on joints during cooling and solidification of the solder.

When you solder transistors, crystal diodes or small resistors, the lead should be gripped with a pair of pliers up close to the unit so that the heat will be conducted away. Overheating of a transistor or diode while soldering can cause permanent damage. Also, mechanical stress will have a similar effect. Therefore, a small unit should be mounted so there is no appreciable mechanical strain on the leads.

Trouble is sometimes experienced soldering to the pins of coil forms or male cable plugs. It helps if the pins are first cleaned on the inside with a suitable twist drill and then tinned by flowing rosin-core solder into them. Immediately clear the surplus solder from each hot pin by a whipping motion or by blowing through the pin from the inside of the form or plug. Before inserting the wire in the pin, file the nickel plate from the pin tip. After soldering, remove excess solder with a file.

When soldering to the pins of polystyrene coil forms, hold the pin to be soldered with a pair of heavy pliers to form a heat sink and ensure that the pin does not heat enough in the coil form to loosen and become misaligned.

Just as important as soldering is desoldering to remove components. Commercially made wicking material (braid) will soak up excess solder from a joint. Another useful tool is an air-suction solder remover. Or, you can heat the joint and "flick" the wet solder off. Keep a large cardboard box handy to flick the solder into, or you'll get solder splashes everywhere.

Wiring

The wire used in connecting amateur equipment should be selected by considering both the maximum current it will be

called on to handle and the voltage its insulation must stand without breakdown. Also, from the consideration of TVI, the power wiring of all transmitters should be done with wire that has a braided shielding cover. Receiver and audio circuits may also require the use of shielded wire at some points for stability or the elimination of hum.

No. 20 stranded wired is commonly used for most receiver wiring (except for the high-frequency circuits) where the current does not exceed 2 or 3 amperes. For higher-current heater circuits, no. 18 is available. Wire with cellulose acetate insulation is good for voltages up to about 500. For higher voltages, Teflon-insulated or other special HV wire should be used. Teflon insulation has the additional advantage of not melting when a soldering iron is applied. This makes it particularly helpful in tight places or large harnesses. Although Teflon-insulated wire is more expensive, it is often available from industrial surplus houses. Inexpensive wire strippers make the removal of insulation from hookup wire an easy job.

When power leads have several branches in the chassis, it is convenient to use fiber-insulated multiple tie points as anchorages of junction points. Strips of this type are also useful as insulated supports for resistors, RF chokes and capacitors. Exposed points of high-voltage wiring should be held to a minimum; those that cannot be avoided should be made as inaccessible as possible to accidental contact or short-circuit.

Where shielded wire is called for and capacitance to ground is not a factor, Belden type 8885 shielded grid wire may be used. If capacitance must be minimized, it may be necessary to use a piece of car radio low-capacitance lead-in wire or coaxial cable.

For wiring high-frequency circuits, rigid wire is often used. Bare soft-drawn tinned wire, size 22 to 12 (depending on mechanical requirements) is suitable. Kinks can be avoided by stretching a piece 10 or 15 feet long and then cutting it into short lengths that can be handled conveniently. RF wiring should be run directly from point to point with a minimum of sharp bends and the wire kept well spaced from the chassis or other grounded metal surfaces. Where the wiring must pass through the chassis or a partition, a clearance hole should be cut and lined with a rubber grommet. If insulation is necessary, varnished cambric tubing (spaghetti) can be slipped over the wire.

In transmitters in which the peak voltage does not exceed 2500, the shielded grid wire mentioned above should be satisfactory for power circuits. For higher voltages, Belden type 8656, Birnbach type 1820, or shielded ignition cable can be used. In the case of filament circuits carrying heavy current, it may be necessary to use no. 10 or 12 bare or enameled wire, slipped through

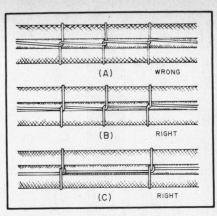

Fig. 13 — Methods of lacing cables. The method shown at C is more secure, but takes more time than the method of B. The latter is usually adequate for most amateur requirements.

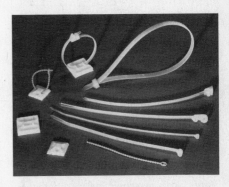

Fig. 14 — Plastic products are used to form and secure wires into a cable.

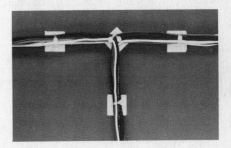

Fig. 15 — Adhesive-backed blocks and plastic wraps used to make wiring cables. Some wraps are reusable.

spaghetti, and then covered with copper braid pulled tightly over the spaghetti. If the shielding is simply slid back over the insulation and solder flowed into the end of the braid, the braid usually will stay in place without the necessity for cutting it back or binding it in place. The braid should be cleaned first so solder will take with a minimum of heat. RF wiring in transmitters usually follows the methods described above for receivers with due respect to the voltages involved.

Where power or control leads run together for more than a few inches, they will present a better appearance when bound together in a single cable. The correct technique is illustrated in Fig. 13; both

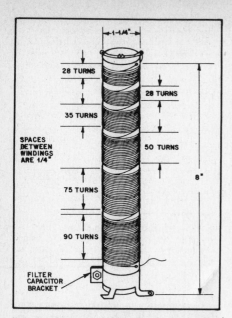

Fig. 16 — Pictorial diagram of a heavy-duty plate choke for the HF bands. No. 26 enameled wire is used for the winding. Low-loss material should be used for the form.

plastic and waxed-linen lacing cords are available. A variety of plastic devices are available to bundle wires into cables and to clamp or secure them in place. Figs. 14 and 15 show some of the products and techniques that you might use. Check with your local electronic parts supplier for items that are in stock.

To give a commercial look to the wiring of any unit, run any cabled leads along the edge of the chassis. If this isn't possible, the cabled leads should then run parallel to an edge of the chassis. Further, the generous use of the tie points mounted parallel to an edge of the chassis, for the support of one or both ends of a resistor or fixed capacitor, will add to the appearance of the finished unit. In a similar manner, arrange the small components so that they are parallel to the panel or sides of the chassis.

Winding Coils

Close-wound coils are readily wound on the specified form by anchoring one end of the length of wire (in a vise or to a doorknob) and the other end to the coil form. Straighten any kinks in the wire and then pull to keep the wire under slight tension. Wind the coil to the required number of turns while walking toward the anchor,

always maintaining a slight tension on the wire.

To space-wind the coil, wind the coil simultaneously with a suitable spacing medium (heavy thread, string or wire) in the manner described above. When the winding is complete, secure the end of the coil to the coil-form terminal and then carefully unwind the spacing material. If the coil is wound under suitable tension, the spacing material can be easily removed without disturbing the winding. Finish the space-wound coil by judicious applications of Duco cement to hold the turns in place.

The cold end of a coil is the end at or close to chassis or ground potential. Coupling links should be wound on the cold end of a coil to minimize capacitive coupling.

RF chokes must often present a high impedance over a broad frequency range. This requirement calls for the avoidance of series resonances within the range. Such resonances can be avoided in single-layer solenoids by separating the winding into progressively shorter sections. A practical choke suitable for HF amateur service at plate impedances up to 51-kΩ and currents up to 600 mA is shown in Fig. 16.

Another way to build a broadband choke is to wind a small number of turns on a high-permeability ferrite rod (such as the type used for antennas in some portable radios). The magnetic core supplies a large inductance with a small winding. Keeping the number of turns small reduces the distributed capacitance and raises the self-resonant frequency of the choke. Ferrite chokes are best suited to low-impedance applications. A bifilar (this term is explained in the following paragraphs) filament choke for grounded-grid kilowatt HF amplifiers appears in Fig. 17.

Toroidal inductors and transformers are specified for many projects in this *Handbook*. The advantages of this type of winding include compactness and a self-shielding property. Figs. 18 and 19 illustrate the proper way to wind and count the turns on a toroidal core.

A *bifilar* winding is one that has two identical lengths of wire, which when placed on the core result in the same number of turns for each wire. The two wires can be put on the core side by side at the same time, just as if a single winding were being applied. An easier and more popular method is to twist the two wires (8 to 15 times per inch will suffice), then

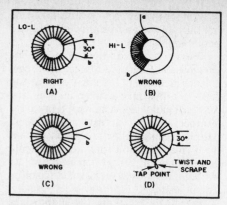

Fig. 18 — The suggested winding method for a single-layer toroid as shown at A. A 30° gap is recommended (see text). Wrong methods are shown at B and C. At D is a method for placing a tap on the coil.

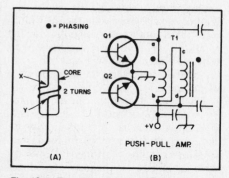

Fig. 19 — The view at A shows how the turns on a toroid should be counted. The large black dots in the diagram at B are used to indicate the polarity (phasing) of the windings.

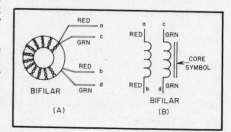

Fig. 20 — Schematic and pictorial presentation of a bifilar-wound toroidal transformer.

wind the pair on the core. The wires can be twisted handily by placing one end of the length of two wires in a bench vise. The remaining wire-pair ends are tightened into the chuck of a small hand drill, and the twisting is done.

A *trifilar* winding has three wires, and a *quadrifilar* winding has four. The procedure for preparation and winding is otherwise the same as for a bifilar winding. Fig. 20 shows a bifilar type of toroid in schematic and pictorial form. The wires have been twisted together prior to placing them on the core. It is helpful, though by no means essential, to use wires of different color when multifilar-winding a core. The more wires used, the more perplexing it is

Fig. 17 — Bifilar filament choke wound on a ferrite rod. Heat-shrink tubing will help anchor the winding. If enameled wire is used, the form should be insulated before winding.

to identify the end of the windings correctly once the core has been wound. There are various colors of enamel insulation available, but it is not easy for amateurs to find this wire locally or in small-quantity lots. This problem can be solved by taking lengths of wire (enameled magnet wire), cleaning them to remove dirt and grease, then spray-painting them. Ordinary aerosol-can spray enamel works fine. Spray lacquer is not as satisfactory because it is brittle when dry and tends to flake off the wire.

The winding sense of a multifilar toroidal transformer is important in most circuits. Fig. 19B illustrates this principle. The black dots (called phasing dots) at the top of the T1 windings indicate polarity. That is, points a and c are both start or finish ends of their respective windings. In this example, points a and d are of opposite phase (180° phase difference) to provide push-pull voltage feed to Q1 and Q2.

CIRCUIT-BOARD FABRICATION

Modern-day builders prefer the neatness and miniaturization made possible by the use of etched printed-circuit boards. There are additional benefits to be realized from the use of circuit boards: Low lead inductances, excellent physical stability of the components and interconnecting leads, and good repeatability of the basic layout of a given project. The repeatability factor makes the use of circuit boards ideal for group projects.

Planning and Layout

The constructor should first plan the physical layout of the circuit by sketching a pictorial diagram on paper, drawing it to scale. Once this has been done, the interconnecting leads can be inked in to represent the copper strips that will remain on the etched board. The Vector Company sells layout paper for this purpose. It is marked with the same patterns used on their perforated boards.

After the basic etched-circuit design has been completed, the designer should go over the proposed layout several times to insure against errors. When that has been done, the pattern can be painted on the copper surface of the board to be etched. Etch-resistant solutions are available from commercial suppliers and can be selected from their catalogs. Some builders prefer to use India ink for this purpose. Perhaps the most readily available material for use in etch-resist applications is ordinary exterior enamel paint. The portions of the board to be retained are covered with a layer of paint applied with an artist's brush, duplicating the pattern that was drawn on the layout paper. The job is made a bit easier by tracing over the original layout with a ballpoint pen and carbon paper while the pattern is taped to the copper side of the unetched circuit board. The carbon paper is placed between the pattern and the

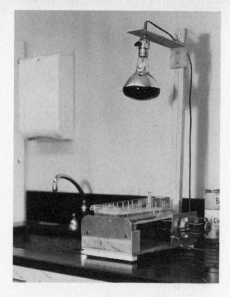

Fig. 21 — A homemade stand for processing etched-circuit boards. The heat lamp maintains the etchant-bath temperature between 90 and 115°F (32 and 46°C) and is mounted on an adjustable arm. The tray for the bath is raised and lowered at one end by the action of a motor-driven eccentric disc, providing the necessary agitation of the chemical solution. A darkroom thermometer monitors the temperature of the bath.

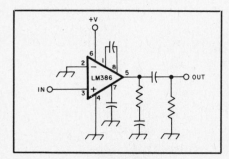

Fig. 22 — Schematic diagram of an audio amplifier.

circuit board. After the paint has been applied, it should be allowed to dry for at least 24 hours prior to the etching process. The Vector Company produces a rub-on transfer material that can also be used as etch-resist when laying out circuit-board patterns. Thin strips of ordinary masking tape, cut to size and firmly applied, serve nicely as etch-resist material too.

To make a single PC board, it is convenient to cover the copper surface with masking tape, transfer the circuit pattern by means of carbon paper, then cut out and remove the sections of masking tape where the copper is to be etched away. An X-acto® hobby knife is excellent for the purpose. Masking tape, securely applied, serves as a fine etch-resist material.

Many magazine articles feature printed-circuit layouts. The more-complex patterns (those containing ICs and high component densities) are difficult to duplicate accurately by hand. A photographic process is the most efficient way to transfer a layout from a magazine page to a circuit board. A Thermofax transparency-producing machine (most schools have these) will copy the circuit on a clear plastic sheet for use as a negative. Pressing this negative against a photosensitive copper-clad board with a piece of glass and exposing the assembly to sunlight for about 90 seconds will deactivate the etchant resist on the exposed part of the board. The portion of the copper that is shielded from the light by the negative will resist etching. Photosensitive PC-board material is manufactured by Kepro Company.

The Etching Process

Almost any strong acid bath will serve as an etchant, but the two chemical preparations recommended here are the safest to use. A bath can be prepared by mixing one part ammonium persulphate crystals with two parts clear water. A normal quantity of working solution for most Amateur Radio applications is composed of one cup of crystals and two cups of water. To this mixture add ¼ teaspoon of mercuric chloride crystals; this serves as an activator for the bath. Ready-made etchant kits that use these chemicals are available from Vector. Complete kits that contain circuit boards, etchant powders, etch-resist transfers, layout paper, and plastic etchant bags are also available from Vector at moderate prices.

Another chemical bath that works satisfactorily for copper etching is made up from one part ferric chloride crystals and two parts water. No activator is required with this bath. Ready-made solutions (one-pint and one-gallon sizes) are available through some mail-order houses at low cost. They are manufactured by Kepro Company and carry stock numbers E-1PT and E-1G, respectively.

Etchant solutions become exhausted after a certain amount of copper has been processed. Therefore, it is wise to keep a quantity of the bath on hand if frequent use is anticipated. With either chemical bath, the working solution should be maintained at a temperature between 90 and 115°F. A heat lamp can be directed toward the bath during the etching period, its distance set to maintain the required temperature. A darkroom thermometer is handy for monitoring the temperature of the bath.

While the circuit is immersed in the solution, it should be agitated continuously to permit uniform reaction to the chemicals. This action will also speed up the etching process somewhat. Normally, the circuit board should be placed in the bath with the copper side facing down, toward the bottom of the tray. The tray should be nonmetallic, preferably a Pyrex dish or a photographic darkroom tray.

Fig. 21 shows a homemade etching stand made up from a heat lamp, some lumber, and an 8-r/min motor. An eccentric disc has been mounted on the motor shaft and butts against the bottom of the etchant tray. As the motor turns, the eccentric disc raises and lowers one end of the tray, thus providing continuous agitation of the solution. The heat lamp is mounted on an adjustable, slotted wooden arm. Its height above the solution tray is adjusted to provide the desired bath temperature. Because the etching process takes between 15 minutes and one hour — dependent on the strength and the temperature of the bath — such an accessory is convenient.

After the etching process is completed, the board is removed from the tray and washed thoroughly with fresh, clear water. The etch-resist material can then be rubbed off by applying a few brisk strokes with medium-grade steel wool. WARNING: *Always use rubber gloves when working with etchant powders and solutions. Should the acid bath come in contact with the body, immediately wash the affected area with clear water. Protect the eyes when using acid baths.*

No-Etch PC Boards

Some would-be builders express revulsion at the prospect of PC-board fabrication. The distaste for chemical processes should not deter a person, however, for several alternatives exist.

A construction technique that is practically indistinguishable from true printed circuitry is excavating a copper-clad board with hand-held grinding tool, such as the Moto-tool manufactured by the Dremel Company. The simpler circuits can be cut out of the board with an X-acto knife. Unwanted copper can be removed by heating it with a soldering iron and lifting it off with a knife while it is still hot.

Alternative Construction Methods

Many methods can be used to construct a circuit. Say, for example, you wish to build the audio amplifier shown in Fig. 22. What construction method would you use? Many possibilities exist — some of them will be covered in this section.

You can build the audio amplifier on an etched PC board. That technique was described earlier. The results might look like the board shown in Fig. 23. For a complicated project, a lot of work can go into getting one perfect PC board. However, after that first one is perfected, dozens more can be made with relative ease.

Solderless Prototyping Board

An etched PC board is not always the best way to construct a circuit. That is especially true when you are developing a circuit. It can be frustrating to have to make another PC board for every circuit variation you want to try. A better method for experimental circuit construction is to use a solderless prototyping board. Several models are available from many sources. One of these is illustrated in Fig. 24.

(A)

(B)

Fig. 23 — Audio amplifier built on a PC board. Top view at A; bottom view at B.

Fig. 24 — The audio amplifier of Fig. 22 has been built on this solderless prototyping board.

Metal strips inside the board grip inserted component leads and hookup wire. The strips are arranged in short columns along the center of the board and in long rows at the top and bottom. Circuit components and hookup wire are used to interconnect the strips and make the circuit. This is a fast and easy way to test an idea and work on modifications. Component and interconnection changes can be readily made.

There are some minor disadvantages to these solderless boards. Because of the capacitance between the finger-stock strips, the boards are not good for building RF circuits. Also, to prevent deforming the

finger stock when using components with heavy leads, you must add short lead extensions of small diameter wire.

Perforated Construction Board

Perhaps the least complicated approach to circuit-board fabrication is the use of unclad perforated board. Perforated board is available with many different hole patterns. Your choice will depend on your needs.

A great deal of time can be spent laying out and etching a single PC board. Circuit construction on perforated board is much easier. Start by placing the components on the board and moving them about until a satisfactory layout is obtained. Component leads are then pushed through the holes, connected with bus or hookup wire and soldered. The audio amplifier of Fig. 22 is shown constructed with this technique in Fig. 25.

Perforated boards and accessories are manufactured by several companies. Accessories include mounting hardware and a variety of connection terminals for solder and solderless construction. These products are widely available.

Utility PC Boards

Utility PC boards offer a hybrid alternative to the custom-designed etched PC board. They offer the flexibility of perforated board construction and the mechanical and electrical advantages of etched circuit connection pads. Utility PC boards can be used to build anything from simple passive filter circuits to computers.

(A)

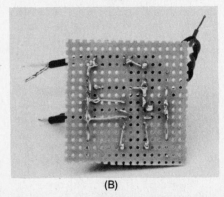

(B)

Fig. 25 — Audio amplifier built on perforated board. Top view at A; bottom view at B.

Fig. 26 — Etching pattern for the universal breadboard. Black represents copper; the pattern is shown actual size. You can etch your own from this pattern or purchase a board from Circuit Board Specialists, P.O. Box 969, Pueblo, CO 81002.

Fig. 27 — Utility PC boards like these are available from Oak Hills Research, 4061 N. Douglas Rd., Luther, MI 49656.

(A)

(B)

Fig. 28 — Audio amplifier built on a multipurpose PC breadboard. Top view at A; bottom view at B.

Circuits can be built on boards on which the copper cladding has been divided into connection pads. Power supply voltages can be distributed on bus strips. The etching pattern for a board of this type can be found in Fig. 26. Boards like those shown in Fig. 27 are commercially available.

The example audio amplifier is shown in Fig. 28. This time the amplifier was built on a Radio Shack board that is designed for building circuits with ICs. Component leads are inserted into the board and soldered to the etched pads. Wire jumpers connect the pads together to complete the circuit.

Utility boards with one or more etched plugs for use in computer-bus, interface and general purpose applications are widely available. Connectors, mounting hardware and other accessories are also available. Check with your parts supplier for details.

Plug-in boards facilitate maintenance and repair.

Wire Wrap

Wire-wrap construction can be used to construct a circuit quickly and without solder. Low- and medium-speed digital circuits are often assembled on a wire-wrap board. The technique is not limited to digital circuits, however. Fig. 29 shows the example audio amplifier built using wire wrap. Circuit changes can be easily made, and yet the method is suitable for permanent assemblies.

Wire wrap basically consists of twisting or wrapping a wire around a post to make each connection. The wrapping tool resembles a thick pencil. Powered wire-wrap guns are used when many connections must be made. The wire is almost always no. 30 and has a thin insulation on it. Two wire-wrap methods are used: the standard

and the modified wrap (Fig. 30). The modified wrap is more secure and should be used for no. 30 wire. The wrap-post terminals are usually square; they should be long enough for at least two connections. For proper wire-wrap technique, see Figs. 30 and 31.

When wire wrapping, small components can be mounted on an IC header plug. The plug is then inserted into a wire-wrap IC socket, as shown in Fig. 29. The large capacitor in that figure has its leads soldered to Y-shaped wrap posts. Wire wrap works only on posts with sharp corners!

Ugly or Dead-Bug Construction

The term "ugly construction" was coined by Wes Hayward, W7ZOI. "Dead-bug construction" gets its name from the appearance of an IC with its leads sticking up in the air. This technique uses copper-clad circuit-board material as a foundation and ground plane on which to build a circuit using point-to-point wiring.

This method is flexible. The builder may use the parts on hand, something that is often difficult to do with the projects that use etched boards. The circuit may be changed easily; that aids experimentation. Speed is the greatest virtue of ugly construction. A moderately complex project (like a QRP transmitter) can be built in half the time, or less, than that required for the

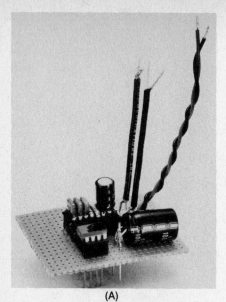

(A)

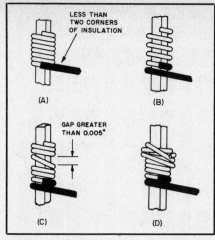

Fig. 31 — Improper wire-wrap connections. Insufficient insulation for modified wrap is shown at A; a spiral wrap at B; an open wrap at C and an over wrap at D.

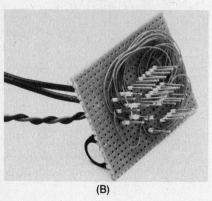

(B)

Fig. 29 — Audio amplifier built using wire-wrap techniques.

Fig. 32 — Audio amplifier built using ugly or dead bug construction.

Fig. 34 — Cordwood modules make compact subassemblies. This broad-band amplifier consists of two transistors, two capacitors and six resistors.

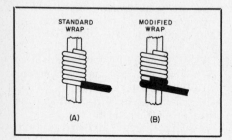

Fig. 30 — Wire-wrap connections. Standard wrap is shown at A; modified wrap at B. The modified wrap should be used with no. 30 wire.

stand-off insulators. One resistor lead is soldered to the copper ground plane, the other lead is used as a circuit connection point. You can use ¼- or ½-watt resistors in values from 220 kΩ to several megohms. The high-ohmage resistors act more like *insulators* than true resistors. Therefore, the higher the resistance the better for this application. As a rule of thumb, the resistor being used as a stand-off insulator should have a value that is at least 10 times the circuit impedance or value or resistance used at that circuit point. For example, if a resistor is used as a tie point at the 50-Ω part of a circuit, the resistor should be 500 Ω or greater.

Fig. 33A illustrates pictorially how one might apply the stand-off technique to wire the circuit shown at C of Fig. 33. Illustration B demonstrates how the resistor pigtails are bent before the component is

same project using boards that have to be etched and drilled.

Components are mounted just above the copper-clad board. (See Fig. 32.) Copper on the board becomes the circuit ground. Ground connections are made directly, thereby minimizing component lead lengths. Short lead lengths and a low-impedance ground conductor help prevent circuit instability. For that reason, ugly construction is particularly good for RF circuits — and for sensitive circuits, better than etched circuit boards.

High-ohmage resistors can be used as

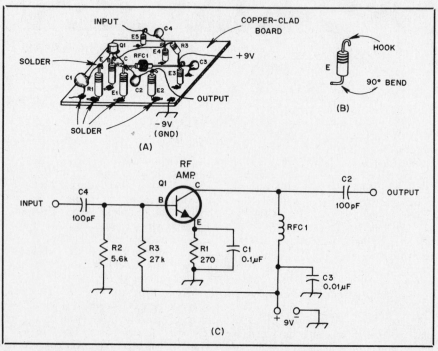

Fig. 33 — Pictorial view of a circuit board that uses ugly construction is shown at A. A closeup view of one of the standoff resistors is shown at B. Note how the pigtails are bent. The schematic diagram at C shows the assembled circuit displayed at A.

soldered to the PC-board material. Components E1 through E5 are resistors that are used as stand-off insulators. They do not appear in the schematic diagram. The base circuitry at Q1 of Fig. 33A is stretched out excessively. This was done to reduce clutter in the drawing. In a practical circuit all of the signal leads should be kept as short as possible. E4 would, therefore, be placed much closer to Q1 than the drawing indicates.

R1 and R2 of Fig. 33 actually serve two purposes: They are not only the normal circuit resistances, but function as stand-off posts as well. This practice should be followed wherever a capacitor or resistor can be employed in the dual role.

Space Savers.

Compact assemblies can be made using cordwood modules. A cordwood module resembles a sandwich. On the outside are the end pieces. In the center are the circuit

Fig. 35 — Surface-mounted components take up little space on this thick-film module. *(Courtesy Tactical RF Inc., Norfolk, Massachusetts.)*

components. The end pieces have holes for the component leads to pass through. Etched traces or point-to-point wiring can be used for circuit connections. Component leads can be used to make connections to the module. This construction method is particularly useful for making small

subassemblies. A broad-band amplifier with two transistors, two capacitors and six resistors built with this technique is shown in Fig. 34.

Another space-saving technique is the use of surface-mounted or chip components. Most UHF and microwave enthusiasts know about chip resistors and capacitors. Other devices, such as transistors and diode arrays, are also available in this space-saving format.

Surface-mounted components take up very little space on a PC board. Even less space is required when circuit traces and components are placed on a piece of ceramic substrate material to make a thick-film module. It is possible for the advanced amateur builder to construct thick-film modules at home. The techniques used today are beyond the scope of this book. Fig. 35 shows what can be done using thick-film technology for making miniature circuits.

Chapter 25

Test Equipment and Measurements

Measurement and testing go hand in hand, but it is useful to make a distinction between measuring and testing. Measurement is commonly considered to give a meaningful quantitative result, while for testing a simple indication of satisfactory or unsatisfactory may suffice. In any event, the accurate calibration associated with real measuring equipment is seldom necessary for simple test apparatus.

Certain items of measuring equipment that are useful to amateurs are readily available in kit form, at prices that represent a genuine saving over the cost of identical parts. Included are volt-ohm-milliammeter combinations, vacuum-tube and transistor voltmeters, oscilloscopes and the like. The coordination of electrical and mechanical design, components and appearance make it far preferable to pur-

chase such equipment than to attempt to build one's own.

However, some test gear can easily be built at home. This chapter considers the principles of the more useful types of measuring equipment and concludes with the descriptions of several pieces that not only can be built satisfactorily at home but also will facilitate operation of the amateur station.

The Direct-Current Instrument

In measuring instruments and test equipment suitable for amateur purposes, the ultimate readout is generally based on a measurement of direct current. A meter for measuring dc uses electromagnetic means to deflect a pointer over a calibrated scale in proportion to the current flowing through the instrument.

In the D'Arsonval type, a coil of wire to which the pointer is attached is pivoted between the poles of a permanent magnet. When current flows through the coil, it sets up a magnetic field that interacts with the field of the magnet to cause the coil to turn. The design of the instrument normally makes the pointer deflect in direct proportion to the current.

A less expensive type of instrument is the moving-vane type, in which a pivoted soft-iron vane is pulled into a coil of wire by the magnetic field set up when current flows through the coil. The farther the vane extends into the coil, the greater the magnetic pull on it for a given change in current. This type of instrument thus does not have linear deflection. The intervals of equal current change are crowded together at the low-current end and spread out at the high-current end of the scale.

Current Ranges

The sensitivity of an instrument is usually expressed in terms of the current required for full-scale deflection of the pointer.

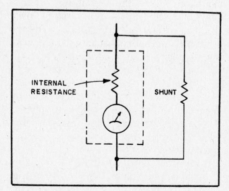

Fig 1—Use of a shunt to extend the calibration range of a current-reading instrument.

Although a very wide variety of ranges is available, the meters of interest in amateur work have basic movements that will give maximum deflection with currents measured in microamperes or milliamperes. They are called microammeters and milliammeters, respectively.

Thanks to the relationships between current, voltage and resistance expressed by Ohm's Law, it becomes possible to use a single low-range instrument—e.g., 1 milliampere or less full-scale pointer deflection—for a variety of direct-current measurements. Through its ability to measure current, the instrument can also be used indirectly to measure voltage. Likewise, a

measurement of both current and voltage will obviously yield a value of resistance. These measurement functions are often combined in a single instrument—the volt-ohm-milliammeter or VOM, a multirange meter that is one of the most useful pieces of measuring and test equipment an amateur can possess.

Accuracy

The accuracy of a dc meter of the D'Arsonval type is specified by the manufacturer. A common specification is 2 percent of full scale, meaning that a 0-100 microammeter, for example, will be correct to within 2 microamperes at any part of the scale. There are very few cases in amateur work where accuracy greater than this is needed. However, when the instrument is part of a more complex measuring circuit, the design and components can all cause error, reducing the overall accuracy of the complete device.

Extending the Current Range

Because of the way current divides between two resistances in parallel, it is possible to increase the range (more specifically, to decrease the sensitivity) of a dc microammeter or milliammeter. The meter itself has an inherent resistance—its internal resistance—which determines the full-rated current passing through it when its rated voltage is applied. (This rated voltage is on

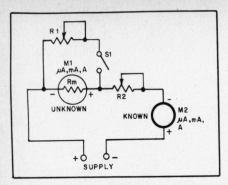

Fig 2—A safe method for determining the internal resistance of a meter.

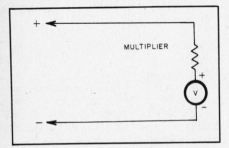

Fig 3—A voltmeter is a current-indicating instrument in series with a high resistance, the "multiplier."

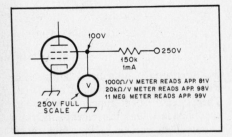

Fig 4—Effect of voltmeter resistance on accuracy of readings. It is assumed that the dc resistance of the screen circuit is constant at 100 kilohms. The actual current and voltage without the voltmeter connected are 1 mA and 100 volts. The voltmeter readings will differ because the different types of meters draw different amounts of current through the 150-kΩ resistor.

the order of a few millivolts.) By connecting an external resistance in parallel with the internal resistance, as in Fig 1, the current will divide between the two, with the meter responding only to that part of the current that flows through the internal resistance of its movement. Thus, it reads only part of the total current; the effect is to make more total current necessary for a full-scale meter reading. The added resistance is called a shunt.

It is necessary to know the meter's internal resistance before the required value for a shunt can be calculated. It may vary from a fraction of an ohm to a few thousand ohms, with the higher resistance values associated with higher sensitivity. When this resistance is known, it can be used in the formula below to determine the re-

quired shunt for a given multiplication:

$$R = \frac{R_m}{n - 1}$$

where
    R = the shunt
    R_m = internal resistance
    n = the factor by which the original meter scale is to be multiplied.

Often the internal resistance of a particular meter will be unknown. This is usually the case when the meter is purchased at a flea market or is obtained from a commercial piece of equipment. Unfortunately, the internal resistance of the meter can not be measured directly with a VOM or VTVM without risk of damage to the meter movement.

Most modern digital multimeters *can* be used to measure the internal resistance of a meter directly.

Fig 2 shows another method that can be used safely to determine the internal resistance of a meter. A calibrated meter capable of measuring the same current as the unknown meter is required. The system works as follows: S1 is placed in the open position and R2 is set for maximum resistance. A supply of constant voltage is connected to the terminals + and − (a battery will work fine) and R2 is adjusted so that the unknown meter reads exactly full scale. Note the current shown on M2. Close S1 and alternately adjust R1 and R2 so that the unknown meter (M1) reads exactly half scale and the known meter (M2) reads the same value as in the step above. At this point the current in the circuit flows through M1 and half through R1. To determine the internal resistance of the meter, simply open S1 and read the resistance of R1 with a VTVM, VOM or digital volt-ohmmeter.

The values of R1 and R2 will depend on the meter sensitivity and the supply voltage. The maximum resistance value for R1 should be approximately twice the expected internal resistance of the meter. For highly sensitive meters (10 μA and less) 1 kilohm should be adequate. For less sensitive meters 100 ohms should suffice. Use no more supply voltage than necessary.

The value for minimum resistance at R2 can be calculated using Ohm's Law. For example, if the meter is a 0-1 mA type and the supply is a 1.5-volt battery, the minimum resistance required at R2 will be

$$R2 = \frac{1.5}{0.001}$$

R2(min) = 1500 ohms

In practice a 2- or 2.5-kΩ potentiometer would be used.

Making Shunts

Homemade shunts can be constructed from any of various special kinds of resistance wire, or from ordinary copper wire if no resistance wire is available. The copper wire table in this handbook gives

the resistance per 1000 feet for various sizes of copper wire. After computing the resistance required, determine the smallest wire size that will carry the full-scale current (250 circular mils per ampere is a satisfactory figure for this purpose). Measure off enough wire to provide the required resistance. A high-resistance 1- or 2-watt carbon-composition resistor makes an excellent form on which to wind the wire. It is recommended that you use the largest size copper wire that is practical to keep the effects of the temperature coefficient of copper to a minimum.

The Voltmeter

If a large resistance is connected in series with a current-reading meter, as in Fig 3, the current multiplied by the resistance will be the voltage drop across the resistance. This is known as a multiplier. An instrument used in this way is calibrated in terms of the voltage drop across the multiplier resistor and is called a voltmeter.

Sensitivity

Voltmeter sensitivity is usually expressed in ohms per volt, meaning that the meter full-scale reading multiplied by the sensitivity will give the total resistance of the voltmeter. For example, the resistance of a 1-kΩ-per-volt voltmeter is 1000 times the full-scale calibration voltage, and by Ohm's Law the current required for full-scale deflection is 1 milliampere. A sensitivity of 20 kΩ per volt, a commonly used value, means that the instrument is a 50-micro-ampere meter.

The higher the resistance of the voltmeter, the more accurate the measurements, especially in high-resistance circuits. Current flowing through the voltmeter will cause a change in the voltage between the points where the meter is connected, compared with the voltage with the meter absent. This is illustrated in Fig 4.

Multipliers

The required multiplier resistance is found by dividing the desired full-scale voltage by the current, in amperes, required for full-scale deflection of the meter alone. Strictly, the internal resistance of the meter should be subtracted from the calculated value but this is seldom necessary (except perhaps for very low ranges), since the meter resistance will be negligible compared with the multiplier resistance. An exception is when the instrument is already a voltmeter and is provided with an internal multiplier, in which case the multiplier resistance required to extend the range is

$$R = R_m (n - 1)$$

where
    R_m = total resistance of the instrument
    n = factor by which the scale is to be multiplied

For example, if a 1-kΩ-per-volt voltmeter

having a calibrated range of 0-10 volts is to be extended to 1000 volts, R_m is 1000 $\times$ 10 = 10 kΩ, n is 1000/10 = 100, and R = 10,000 (100 − 1) = 990 kΩ.

When extending the range of a voltmeter or converting a low-range meter into a voltmeter, the rated accuracy of the instrument is retained only when the muliplier resistance is precise. Precision wire-wound resistors are used in the mulipliers of high-quality instruments. These are relatively expensive, but the home constructor can do quite well with 1-percent-tolerance composition resistors. They should be derated when used for this purpose—that is, the actual power dissipated in the resistor should be not more than ¼ to ½ the rated dissipation—and care should be used to avoid overheating the body of the resistor when soldering to the leads. These precautions will help prevent permanent change in the resistance of the unit.

Ordinary composition resistors are generally furnished in 10- or 5-percent-tolerance ratings. If possible errors of this order can be accepted, resistors of this type may be used as multipliers. They should be operated below the rated power dissipation figure, in the interest of long-term stability.

DC MEASUREMENT CIRCUITS

The Voltmeter

A current-measuring instrument should have very low resistance compared with the resistance of the circuit being measured; otherwise, inserting the instrument will cause the current to differ from its value with the instrument out of the circuit. The resistance of many circuits in radio equipment is high and the circuit operation is affected little, if at all, by adding as much as a few hundred ohms in series. In such cases the voltmeter method of measuring current, shown in Fig 5, is frequently convenient. A voltmeter (or low-range milliammeter provided with a multiplier and operating as a voltmeter) having a full-scale voltage range of a few volts is used to measure the voltage drop across a suitable value of resistance acting as a shunt.

The value of shunt resistance must be calculated from the known or estimated maximum current expected in the circuit (allowing a safe margin) and the voltage required for full-scale deflection of the meter with its multiplier.

Power

Power in direct-current circuits is determined by measuring the current and voltage. When these are known, the power is equal to the voltage in volts multiplied by the current in amperes. If the current is measured with a milliammeter, the reading of the instrument must be divided by 1000 to convert it to amperes.

The setup for measuring power is shown in Fig 6, where R is any dc load, not necessarily an actual resistor.

Resistance

Obviously, if both voltage and current are measured in a circuit such as that in Fig 6, the value of resistance R (in case it is unknown) can be calculated from Ohm's Law. For accurate results, the internal resistance of the ammeter or milliammeter, mA, should be very low compared with the resistance, R, being measured. This is because the voltage read by the voltmeter, V, is the voltage across mA and R in series. The instruments and the dc voltage should be chosen so that readings are in the upper half of the scale, if possible, since the percentage error is less in this region.

The Ohmmeter

Although Fig 6 suffices for occasional resistance measurements, it is inconvenient when frequent measurements over a wide range of resistance are to be made. The device generally used for this purpose is the ohmmeter. This consists fundamentally of a voltmeter (or milliammeter, depending on the circuit used) and a small battery, the meter being calibrated so the value of an unknown resistance can be read directly from the scale. Typical ohmmeter circuits are shown in Fig 7. In the simplest type, Fig 7A, the meter and battery are connected in series with the unknown resistance. If a given deflection is obtained with terminals A-B shorted, inserting the resistance to be measured will cause the meter reading to decrease. When the resistance of the voltmeter is known, the following formula can be applied.

$$R = \frac{eR_m}{E} - R_m$$

where
R = resistance to be found
e = voltage applied (A-B shorted)
E = voltmeter reading with R connected, and
R_m = resistance of the voltmeter.

The circuit of Fig 7A is not suited to measuring low values of resistance (below a hundred ohms or so) with a high-resistance voltmeter. For such measurements the circuit of Fig 7B can be used. The unknown resistance is

$$R = \frac{I_2 R_m}{I_1 - I_2}$$

where
R = the unknown resistance
R_m = the internal resistance of the milliammeter
I_1 = current with R disconnected from terminals A-B
I_2 = current with R connected

The formula is based on the assumption that the current in the complete circuit will be essentially constant whether or not the

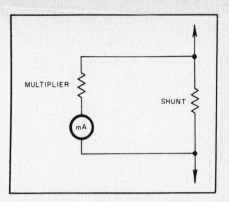

Fig 5—Voltmeter method of measuring current. This method permits using relatively large values of resistance in the shunt, standard values of fixed resistors frequently being usable. If the multiplier resistance is 20 (or more) times the shunt resistance, the error in assuming that all the current flows through the shunt will not be of consequence in most practical applications.

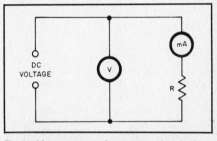

Fig 6—Measurement of power requires both current and voltage measurements; once these values are known, the power is equal to the product P = EI. The same circuit can be used for measurement of an unknown resistance.

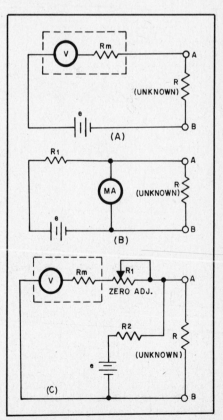

Fig 7—Ohmmeter circuits. Values are discussed in the text.

$\longrightarrow$

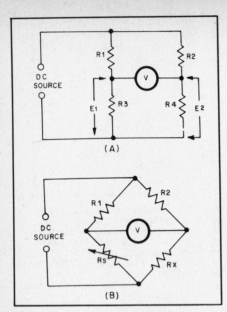

Fig 8—The Wheatstone bridge circuit. It is frequently drawn as at B for emphasizing its special function.

Fig 9—The IC voltmeter mounted in a small plastic case. This basic instrument measures only dc voltages, but with an RF probe as described later in this chapter, it can be used for RF measurements as well. It may also be used for resistance measurements, by using techniques described in the previous section. DS1, mounted between the binding posts, is a power-on indicator.

unknown terminals are short-circuited. This requires that R1 be very large compared with R_m, e.g., 3000 ohms for a 1-mA meter having an internal resistance of perhaps 50 ohms. A 3-volt battery would be necessary in this case in order to obtain a full-scale deflection with the *unknown* terminals open. R1 can be an adjustable resistor, to permit setting the open-terminal current to exact full scale.

A third circuit for measuring resistance is shown in Fig 7C. In this case a high-resistance voltmeter is used to measure the voltage drop across a reference resistor, R2, when the unknown resistor is connected so that current flows through it, R2 and the battery in series. By suitable choice of R2 (low values for low-resistance, high values for high-resistance unknowns), this circuit will give equally good results on all resistance values in the range from one ohm to several megohms. That is provided that the voltmeter resistance, R_m, is always very high (50 times or more) compared with the resistance of R2. A 20-kΩ-per-volt instrument (50-μA movement) is generally used. Assuming that the current through

the voltmeter is negligible compared with the current through R2, the formula for the unknown is

$$R = \frac{eR2}{E} - R2$$

where

R and R2 are shown in Fig 7C
e = the voltmeter reading with A-B open circuited
E = voltmeter reading with R connected.

The zero adjuster, R1, is used to set the voltmeter reading exactly to full scale when the meter is calibrated in ohms. A 10-kΩ variable resistor is suitable with a 20-kΩ-per-volt meter. The battery voltage is usually 3 volts for ranges up to 100 kΩ or so and 6 volts for higher ranges.

Bridge Circuits

An important class of measurement circuits is the bridge. A desired result is obtained by balancing the voltages at two different points in the circuit against each other so there is zero potential difference between them. A voltmeter bridged between the two points will read zero (null) when this balance exists, but will indicate some definite value of voltage when the bridge is not balanced.

Bridge circuits are useful both on direct current and on ac of all frequencies. The majority of amateur applications are at radio frequencies, as shown later in this chapter. However, the principles of bridge operation are most easily introduced in

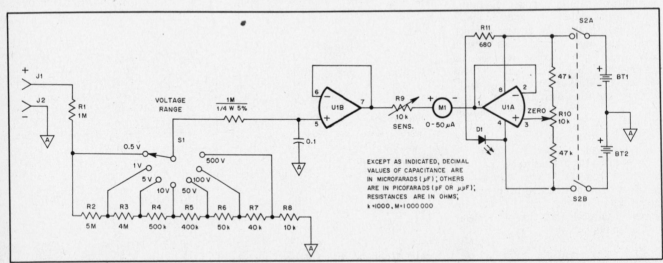

Fig 10—A high impedance voltmeter need not be complex. This circuit uses a single IC. BT1 and BT2 can be 2 AAA (or AA) cells in a holder or 9-V transistor radio batteries (see text).

D1—LED, 5-V, 20 mA; Radio Shack 276-041 or equiv.
J1, J2—Banana jacks, 500-V insulation (see text), RS 274-662 or equiv.
M1—50-μA dc meter.
R1—1.0-MΩ, 1/2-W, 5% resistor.
R2—4.7-MΩ and 300-kΩ, 1/4-W, 5% resistors in series.
R3—3.9-MΩ and 100-kΩ, 1/4-W, 5% resistors in series.
R4—470-kΩ and 30-kΩ, 1/4-W, 5% resistors in series.

R5—390-kΩ and 10-kΩ, 1/4-W, 5% resistors in series.
R6—47-kΩ and 3-kΩ, 1/4-W, 5% resistors in series.
R7—39-kΩ and 1-kΩ, 1/4-W, 5% resistors in series.
R8—10-kΩ, 1/4-W, 5% resistor.
R9—10-kΩ, 1/4-W PC-mount potentiometer, RS 271-218 or equiv.
R10—10-kΩ, panel-mount potentiometer, RS 271-1722.

R11—Current-limiting resistor; 680 Ω for 3-V batteries at BT1 and BT2, or 1.6 kΩ for 9-V batteries.
S1—1-pole, 7-position rotary switch (see text), RS 275-1385 or equiv.
S2—2-pole, 2-position toggle switch, RS 275-614 or equiv.
U1—LF353N dual JFET op amp, RS 276-1715 or equiv.

terms of dc, where the bridge takes its simplest form.

The Wheatstone Bridge

The simple resistance bridge, known as the Wheatstone bridge, is shown in Fig 8. All other bridge circuits—some of which are rather elaborate, especially those designed for ac—derive from this. The four resistors, R1, R2, R3 and R4 shown in A, are known as the bridge arms. For the voltmeter reading to be zero, the voltages across R3 and R4 in series must add algebraically to zero; that is, E1 must equal E2. R1 with R3 and R2 with R4 form voltage dividers across the dc source, so if

$$\frac{R1}{R3} = \frac{R2}{R4} \quad \text{then}$$

$$E1 = E2$$

The circuit is customarily drawn as shown at Fig 8B when used for resistance measurement. The equation above can be rewritten

$$R_x = R_s \frac{R2}{R1}$$

to find R_x, the unknown resistance. R1 and R2 are frequently made equal; then the calibrated adjustable resistance (the standard), R_s, will have the same value as R_x when R_s is set to show a null on the voltmeter.

Note that the resistance *ratios*, rather than the actual resistance values, determine the voltage balance. However, the values do have important practical effects on the sensitivity and power consumption. The bridge sensitivity is the readiness with which the meter responds to small amounts of unbalance about the null point; the sharper the null the more accurate the setting of R_s at balance.

The Wheatstone bridge is rarely used by amateurs for resistance measurement, the ohmmeter being the favorite instrument for that purpose. However, it is worthwhile to understand its operation because it is the basis of more complex bridges.

Electronic Voltmeters

It has been pointed out that for many purposes the resistance of a voltmeter must be extremely high in order to avoid "loading" errors caused by the current that necessarily flows through the meter. The use of high-resistance meters tends to cause difficulty in measuring relatively low voltages (under perhaps 1000 volts) because progressively smaller multiplier resistance is needed as the voltage range is lowered.

The voltmeter resistance can be made independent of the voltage range by using vacuum tubes or field-effect transistors as electronic dc amplifiers between the circuit being measured and the indicator, which may be a conventional meter movement or a digital display. As the input resistance of the electronic devices is extremely high—hundreds of megohms—they have essen-

tially no loading effect on the circuit to which they are connected. They do, however, require a closed dc path in their input circuits (although this path can have very high resistance). They are also limited in the amplitude of voltage that their input circuits can handle. Because of this, the device actually measures a small voltage across a portion of a high-resistance voltage divider connected to the circuit being measured. Various voltage ranges are obtained by appropriate taps on the voltage divider.

In the design of electronic voltmeters it has become standard practice to use a voltage divider having a resistance of 10 megohms, tapped as required, in series with a 1-megohm resistor incorporated in the meter. The total voltmeter resistance, including probe, is therefore 11 megohms. The 1-megohm resistor serves to isolate the voltmeter circuit from the active circuit.

Hybrid Digital Meters

Over the years since digital meters (DMMs) came in to the price range of experimenters, the one argument against DMMs has been the lack of a facility for peaking and nulling circuits. The conversion and display times of most digital meters are slow enough that slight peaks and dips are almost impossible to observe without going to ridiculous extremes. Because of this, most experimenters still have a good old VOM or VTVM on the bench. In the past few years, manufacturers have taken various steps to make the digital meter more attractive to the user who needs peaking and nulling facilities. Three approaches have come into use:

1) Putting a small analog meter on the front panel of the DMM to give trends indication. This meter tends to be coarsely calibrated. This method allows the user to see the peaks and nulls and still have the accuracy of the digital display available in one package. Examples of this type of meter are the Heath IM-2264 and the Simpson 460-3.

2) Addition of a bar graph on the display. This bar graph usually takes the form of 20 to 40 dots in a row which are lit in proportion to the percentage of the range that is inputted. For example, if the meter had a 2-V range and a 20-dot bar graph, a dot would light for every 0.1-V increase. So if this meter has an input of 0.6 V on its leads, the six left-hand dots would light. This may seem a bit coarse, but it is useful in practice. Examples of this type of meter are the Fluke 70 series (sold by Heath as the SM77) and the Simpson 467.

3) The third method is using peak and null LEDs. In this case, there are two LEDs: one is for peaking and one is for nulling. As a signal is peaked, the peaking LED will glow brighter. When the peak occurs, both LEDs will glow equally. As the peak is passed, the null LED will brighten. This is probably the most difficult to interpret of the three methods. An

example of this type of meter is the Sencore DVM56A.

The individual will have to decide which method best suits his or her personal taste and budget, when selecting this type of meter. But, with the steady improvement in measurement technology, the day may not be far off when mechanical meter movements may completely disappear from electrical and electronic applications.

AN IC VOLTMETER

One of the fundamental troubleshooting techniques is to check the voltage present at various points in the circuit. The measured voltage is compared with the voltage we expect to find at that point. Knowing what to expect is important. If we don't have some idea of what the voltage should be, measuring it won't tell us very much. Often the voltages we measure in a circuit will not agree exactly with our expected values. Small variations are normal and do not mean the circuit is not operating as it should. Component tolerance and meter errors are the primary causes for these variations.

Shown in Figs 9 through 12 is an easy-to-build, high-impedance dc voltmeter, first described by George Collins, KC1V, in January 1982 *QST*. All of the parts for the voltmeter are readily available. Calibration is simple and the cost is low. Construction of this meter can be considered as an easy weekend project. The input-impedance is 11 MΩ, and accuracy is better than 15%. With the RF probe shown later in this chapter, this meter can be used to make reasonably accurate RF voltage measurements at frequencies up to 30 MHz.

Circuit Details

The input impedance of the meter is determined by the total resistance of the range-selector voltage divider (R1 through R8). The values of the individual resistors have been selected to provide the desired full-scale voltage ranges and a total resistance of 11 MΩ. Some of the resistance values needed for the divider are not found in the standard series of 5%-tolerance resistor values. To avoid having to buy expensive (and hard to find) 1% resistors, two 5% units are used in series for each of the nonstandard values.

To keep the meter movement from loading the 11-MΩ divider, an operational amplifier (op amp) with JFET inputs is used to drive the meter. The LF353N IC (U1) contains two of these op amps in the same package. U1B drives the meter movement, while U1A serves as an adjustable voltage reference point. Both of the op amps are connected as voltage followers. This means that the input and output voltages are the same (a gain of 1). What makes the voltage follower useful is that the output can supply several milliamperes of current while the input draws a very small current (the input impedance is high).

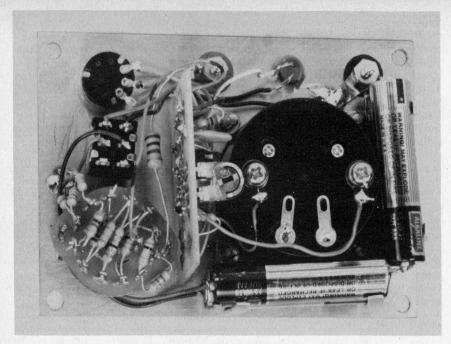

Fig 11—Inside view of the IC voltmeter. This version was built from an available parts kit. Other components and construction styles can be used as well.

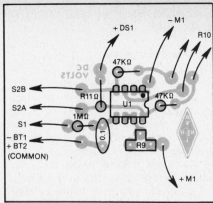

Fig 12—Parts-placement guide for the FET voltmeter. Parts are placed on the nonfoil side of the board; the shaded area represents an X-ray view of the copper pattern. A full-size etching pattern can be found at the back of this book.

By varying the voltage at pin 3 of U1 with R10, the zero setting of the meter can be adjusted to compensate for changes in battery voltage and room temperature. The fact that both op amps are in the same package helps reduce drift caused by temperature changes. R10 is mounted on the front panel so the operator can adjust it easily. R9 is the calibration control; it adjusts the meter sensitivity. R9 does not require frequent adjustment, so it is mounted inside the case.

Two batteries are used to power the meter circuit. Any battery potential between 3 and 9 V can be used without changes in the circuit. In the unit shown, four AAA penlight cells are used. These give the needed 3 V and have long life. Two 9-V transistor radio batteries will also be suitable.

Construction

Almost any type of case can be used to house the voltmeter. The exact size needed will depend on the dimensions of the batteries, meter movement and switches used. A plastic case, only 2-7/8 × 4 × 1-5/8 inches, houses the meter shown in the photographs. If a larger meter movement is used, an enclosure measuring 2-5/8 × 5-1/6 × 1-5/8 inches will be more satis-

factory. If you use a case with a metal panel, it is best if the negative jack (J2) is not connected to the panel. This permits voltage measurements below ground potential without having a voltage on the meter case.

The voltage-divider resistors are mounted on the range selector switch S1, as shown in Fig 11. If the switch has any spare lugs, they can be used as tie points for the series-connected resistors. If no lugs are available, simply solder the leads together; the remaining leads will support the resistors. The other components can be mounted on a small printed-circuit board,* although any method of wiring can be used. A quick and simple way of wiring the IC is to use a general-purpose IC-prototyping board, such as the Radio Shack 276-159.

The builder should be aware that very minute leakage currents between pins of the op-amp IC can cause improper operation

*A printed-circuit board and parts for the voltmeter, including a small circuit board for an RF probe, are available from Circuit Board Specialists, P.O. Box 951, Pueblo, CO 81002-0951.

of the instrument. If a circuit board is used, be sure to clean the rosin flux from around these pins after soldering is completed. Use a suitable solvent, such as denatured alcohol. Further, the circuit board should not be allowed to rest on any type of supporting material; even nonconducting foam may yield unexpected results.

With the resistor values shown in Fig 10, the highest full-scale range is 500 V. If this range is included, be sure the input connectors (J1 and J2) and the range switch (S1) are rated for 500 V or more. J1 and J2 should also be able to handle 500 volts safely. Only thin, fiber washers are used to insulate some type of jacks from the panel. These are fine for up to 100 V, but are not recommended for higher voltages. If the 500-V range is not needed, R7 and R8 can be connected in series or replaced by a 50-kΩ resistor (the same as R6).

Calibration

Only the sensitivity control, R9, needs to be adjusted before the meter can be used. A good method of calibration is to use two fresh carbon-zinc batteries in series to form a source of known potential. Each cell, when new, should produce 1.54 V. To adjust R9, turn the meter on, and set it to the 5-V range. With the meter leads shorted together, adjust the zero control (R10) so that the meter shows zero. Connect the two cells to the meter, and adjust R9 so the meter reads 3.1 V. This completes the voltmeter, and it is ready to use.

AC Instruments and Circuits

Although purely electromagnetic instruments that operate directly from alternating current are available, they are seen infrequently in present-day amateur equipment. For one thing, their use is not feasible above power-line frequencies.

Common instruments for audio and radio frequencies generally use a dc meter movement in conjunction with a rectifier. Voltage measurements suffice for nearly all test purposes. Current, as such, is seldom measured in the AF range. When RF current is measured the instrument used is a thermocouple milliammeter or ammeter.

The Thermocouple Meter

In a thermocouple meter the alternating current flows through a low-resistance heating element. The power lost in the resistance generates heat that warms a thermocouple, a junction of certain dissimilar metals that has the property of developing a small dc voltage when heated. This voltage is applied to a dc milliammeter calibrated in suitable ac units. The heater-thermocouple-dc meter combination is usually housed in a regular meter case.

Thermocouple meters can be obtained in ranges from about 100 mA to many amperes. Their useful upper frequency limit is in the neighborhood of 100 MHz. Their principal value in amateur work is in measuring current into a known load resistance for calculating the RF power delivered to the load.

Rectifier Instruments

The response of a rectifier-type meter is proportional (depending on the design) to either the peak amplitude or average amplitude of the rectified ac wave, and never directly responsive to the RMS value. The meter therefore cannot be calibrated in RMS without preknowledge of the relationship that happens to exist between the real reading and the RMS value. This relationship, in general, is not known, except in the case of single-frequency ac (a sine wave). Many practical measurements involve nonsinusoidal wave forms, so it is necessary to know what kind of instrument you have, and what it is actually reading, in order to make measurements intelligently.

Peak and Average with Sine-Wave Rectification

Fig 13 shows the relative peak and average values in the output of half- and full-wave rectifiers (see Chapter 6 for further details). As the positive and negative half cycles of the sine wave have the same shape, A, half-wave rectification of either the positive half, B or the negative half, C, gives exactly the same result. With full-wave rectification, D, the peak is still the same, but the average is doubled, since

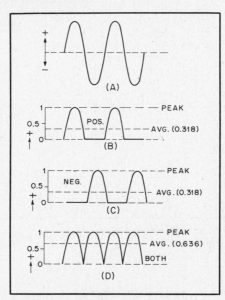

Fig 13—Sine-wave alternating current or voltage, A, with half-wave rectification of the positive half cycle, B, and negative half cycle, C. At D—full-wave rectification. Average values are shown with relation to a peak value of 1.

there are twice as many half cycles per unit of time.

Unsymmetrical Wave Forms

A nonsinusoidal waveform is shown in Fig 14A. When the positive half cycles of this wave are rectified, the peak and average values are shown at B. If the polarity is reversed and the negative half cycles are rectified, the peak value is different but the average value is unchanged. The fact that the average of the positive side is equal to the average of the negative side is true of all ac waveforms, but different waveforms have different averages. Full-wave rectification of such a lopsided wave doubles the average value, but the peak reading is always the same as it is with the half cycle that produces the highest peak in half-wave rectification.

Effective-Value Calibration

The actual scale calibration of commercially made rectifier-type voltmeters is very often (almost always, in fact) in terms of RMS values. For sine waves this is satisfactory, and useful since RMS is the standard measure at power-line frequency. It is also useful for many RF applications where the waveform is often close to sinusoidal. But in other cases, particularly in the AF range, the error may be considerable when the waveform is not pure.

Turn-Over

From Fig 14 it is apparent that the calibration of an average-reading meter will

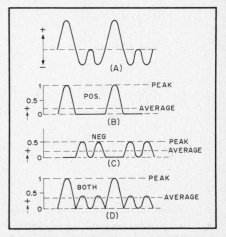

Fig 14—Same as Fig 13 for an unsymmetrical waveform. The peak values are different with positive and negative half-cycle rectification.

be the same whether the positive or negative sides are rectified. A half-wave peak-reading instrument, however, will indicate different values when its connections to the circuit are reversed (turn-over effect). Very often readings are taken both ways, in which case the sum of the two is the peak-to-peak (P-P) value, a useful figure in much audio and video work.

Average- and Peak-Reading Circuits

The basic difference between average- and peak-reading rectifier circuits is that the output is not filtered for averaged readings, while a filter capacitor is charged up to the peak value of the output voltage for measuring peaks. Fig 15A shows typical average-reading circuits, one half-wave and the other full-wave. In the absence of dc filtering the meter responds to wave forms such as are shown at B, C and D in Figs. 14 and 15, and since the inertia of the pointer system makes it unable to follow the rapid variations in current, it averages them out mechanically.

In Fig 15A, D1 actuates the meter; D2 provides a low-resistance dc return in the meter circuit on the negative half cycles. R1 is the voltmeter multiplier resistance. R2 forms a voltage divider with R1 (through D1), which prevents more than a few ac volts from appearing across the rectifier-meter combination. A corresponding resistor can be used across the full-wave bridge circuit.

In these two circuits no provision is made for isolating the meter from any dc voltage that may be on the circuit under measurement. The error caused by this can be avoided by connecting a large capacitor (not electrolytic!) in series with the hot lead. The reactance must be low compared with the meter impedance (see next section) in order for the full ac voltage to be applied

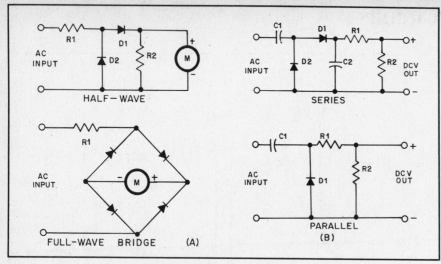

Fig 15—At A, half-wave and full-wave rectification for an instrument intended to operate on average values. At B, half-wave circuits for a peak-reading meter.

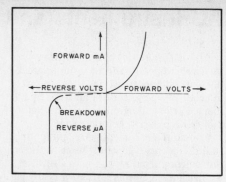

Fig 16—Typical semiconductor diode characteristic. Actual current and voltage values vary with the type of diode, but the forward-current curve would be in its steep part with only a volt or so applied. Note the change in the current scale for reverse current. Breakdown voltage, again depending on diode type, may range from 15 or 20 volts to several hundred.

to the meter circuit. As much as 1 μF may be required at line frequencies with some meters. The capacitor is not usually included in a VOM.

Series and shunt peak-reading circuits are shown in Fig 15B. C1 isolates the rectifier from dc voltage on the circuit under measurement. In the series circuit (which is seldom used) the time constant of the C2-R1-R2 combination must be very large compared with the period of the lowest ac frequency to be measured; similarly with C1-R1-R2 in the shunt circuit. The reason is that the capacitor is charged to the peak value of voltage when the ac wave reaches its maximum, and then must hold the charge (so it can register on a dc meter) until the next maximum of the same polarity. If the time constant is 20 times the ac period, the charge will have decreased by about five percent by the time the next charge occurs. The average drop will be smaller, so the error is appreciably less. The error will decrease rapidly with increasing frequency, assuming no change in the circuit values, but will increase at lower frequencies.

In Fig 15B, R1 and R2 form a voltage divider that reduces the peak dc voltage to 71 percent of its actual value. This converts the peak reading to RMS on sine-wave ac. Since the peak-reading circuits are incapable of delivering appreciable current without considerable error, R2 is usually the 11-megohm input resistance of an electronic voltmeter. R1 is therefore approximately 4.7 megohms, making the total resistance approach 16 megohms. A capacitance of 0.05 μF is sufficient for low audio frequencies under these conditions. Much smaller values of capacitance suffice for radio frequencies, obviously.

Voltmeter Impedance

The impedance of the voltmeter at the frequency being measured may have an effect on the accuracy similar to the error

caused by the resistance of a dc voltmeter, as discussed earlier. The ac meter acts like a resistance in parallel with a capacitance, and since the capacitive reactance decreases with increasing frequency, the impedance also decreases with frequency. The resistance is subject to some variation with voltage level, particularly at very low voltages (on the order of 10 volts or less) depending on the sensitivity of the meter movement and the kind of rectifier used.

The ac load resistance represented by a diode rectifier is approximately equal to one-half its dc load resistance. In Fig 15A the dc load is essentially the meter resistance, which is generally quite low compared with the multiplier resistance R1, so the total resistance will be about the same as the multiplier resistance. The capacitance depends on the components and construction, test lead length and disposition, and other such factors. In general, it has little or no effect at lower line and low audio frequencies, but the ordinary VOM loses accuracy at the high audio frequencies and is of little use at RF. For radio frequencies it is necessary to use a rectifier having very low inherent capacitance.

Similar limitations apply to the peak-reading circuits. In the parallel circuit the resistive component of the impedance is smaller than in the series circuit because the dc load resistance, R1/R2, is directly across the circuit being measured, and is therefore in parallel with the diode ac load resistance. In both peak-reading circuits the effective capacitance may range from 1 or 2 to a few hundred picofarads. Values of the order of 100 pF are to be expected in electronic voltmeters of customary design and construction.

Linearity

Fig 16, a typical current/voltage characteristic of a small semiconductor rectifier, indicates that the forward dynamic

resistance of the diode is not constant, but rapidly decreases as the forward voltage is increased from zero. The transition from high to low resistance occurs at considerably less than 1 volt, but is in the range of voltage required by the associated dc meter. With an average-reading circuit the current tends to be proportional to the square of the applied voltage. This crowds the calibration points at the low end of the meter scale. For most measurement purposes, however, it is far more desirable for the output to be linear; that is, for the reading to be directly proportional to the applied voltage.

To obtain linearity it is necessary to use a relatively large load resistance for the diode—large enough so this resistance, rather than the diode's own resistance, will govern the current flow. A linear or equally spaced scale is thus gained at the expense of sensitivity. The amount of resistance needed depends on the type of diode; 5000 to 50,000 ohms usually suffices for a germanium rectifier, depending on the dc meter sensitivity, but several times as much may be needed for silicon. The higher the resistance, the greater the meter sensitivity required; i.e., the basic meter must be a microammeter rather than a low-range milliammeter.

Reverse Current

When voltage is applied in the reverse direction there is a small leakage current in semiconductor diodes. This is equivalent to a resistance connected across the rectifier, allowing current to flow during the half cycle that should be completely nonconducting and causing an error in the dc meter reading. This back resistance is so high as to be practically unimportant with silicon diodes, but may be less than 100 kΩ with germanium.

The practical effect of semiconductor back resistance is to limit the amount of resistance than can be used in the dc load. This in turn affects the linearity of the

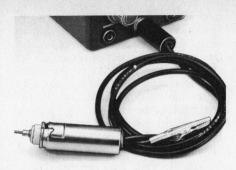

Fig 17—RF probe for use with an electronic voltmeter. The case of the probe is constructed from a seven-pin ceramic tube socket and a 2¼-inch tube shield. A half-inch grommet at the top of the tube shield prevents the output lead from chafing. A flexible copper-braid grounding lead and alligator clip provide a low-inductance return path from the test circuit.

meter scale. The back resistance of vacuum-tube diodes is infinite, for practical purposes.

RF Voltage

Special precautions must be taken to minimize the capacitive component of the voltmeter impedance at radio frequencies. If possible, the rectifier circuit should be installed permanently at the point where the RF voltage is to be measured, using the shortest possible RF connections. The dc meter can be remotely located, however.

For general RF measurements an RF probe is used in conjunction with an 11-MΩ electronic voltmeter, substituted for the dc probe mentioned earlier. The circuit of Fig 18, essentially the peak-reading shunt circuit of Fig 15B, is generally used. The series resistor, installed in the probe close to the rectifier, prevents RF from being fed through the probe cable to the electronic voltmeter. RF is further reduced from reaching the voltmeter by the cable capacitance. This resistor, in conjunction with the 10-MΩ divider resistance of the electronic voltmeter, also reduces the peak rectified voltage to a dc value equivalent to the RMS of the RF signal, to make the RF readings consistent with the regular dc calibration.

Of the diodes readily available to amateurs, the germanium point-contact or Schottky type is preferred for RF applications. It has low capacitance (on the

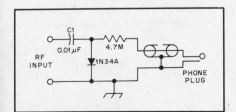

Fig 18—The RF probe circuit.

order of 1 pF) and in the high-back-resistance types the reverse current is not serious. The principal limitation is that its safe reverse voltage is only about 50-75 volts, which limits the RMS applied voltage to 15 or 20 volts. Diodes can be connected in series to raise the overall rating.

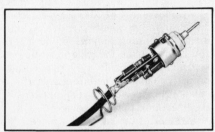

Fig 19—Inside the probe. The 1N34A diode, calibration resistor and input capacitor are mounted tight to the terminal strip with the shortest possible leads. Spaghetti tubing is placed on the diode leads to prevent accidental short circuits. The tube-shield spring and flexible copper grounding lead are soldered to the cable braid (the cable is RG-58 coax). The tip can be either a phone tip or a short pointed piece of stiff wire.

AN RF PROBE FOR ELECTRONIC VOLTMETERS

The isolation capacitor, C1, crystal diode and filter/divider resistor are mounted on a bakelite five-lug terminal strip, as shown in Fig 20. One end lug should be rotated 90 degrees so it extends off the end of the strip. All other lugs should be cut off flush with the edge of the strip. Where the inner conductor connects to the terminal lug, unravel the shield three-quarters of an inch, slip a piece of spaghetti tubing over it and then solder the braid to the ground lug on the terminal strip. Remove the spring from the tube shield, slide it over the cable, and crimp it to the remaining quarter inch of shield braid. Solder both the spring and a 12-inch length of flexible copper braid to the shield.

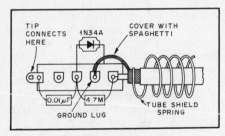

Fig 20—Component mounting details.

Next, cut off the pins on a seven-pin miniature shield-base tube socket. Use a socket with a cylindrical center post. Crimp the terminal lug previously bent out at the end of the strip and insert it into the center post of the tube socket from the top. Insert the end of a phone tip or a pointed piece of heavy wire into the bottom of the tube

socket center post, and solder the lug and tip to the center post. Insert a half-inch grommet at the top of the tube shield, sliding it over the cable and flexible braid down into the tube socket. The spring should make good contact with the tube shield to assure that the tube shield (probe case) is grounded. Solder an alligator clip to the other end of the flexible braid and mount a phone plug on the free end of the shielded wire.

Mount components close to the terminal strip, to keep lead lengths as short as possible and minimize stray capacitance. Use spaghetti tubing over all wires to prevent accidental shorts.

The phone plug on the probe cable plugs into the dc input jack of the electronic voltmeter and RMS voltages are read on the voltmeter negative dc scale.

The accuracy of the probe is within ± 10 percent from 50 kHz to 250 MHz. The approximate input impedance is 6000 ohms shunted by 1.75 pF (at 200 MHz).

RF Power

Power at radio frequencies can be measured by means of an accurately calibrated RF voltmeter connected across the load in which the power is being dissipated. If the load is a known pure resistance, the power, by Ohm's Law, is equal to E^2/R, where E is the RMS value of the voltage.

If the load is not a pure resistance, the above method indicates only apparent power. The load can be a terminated transmission line, tuned with the aid of bridge circuits such as are described in the next section to act as a known resistance. An alternative load is a dummy antenna, a known pure resistance capable of dissipating the RF power safely.

AC Bridges

In its simplest form, the ac bridge is exactly the same as the Wheatstone bridge discussed earlier. However, complex impedances can be substituted for resistances, as suggested by Fig 21A. The same bridge equation holds if Z is substituted for R in each arm. For the equation to be true, however, the phase angles as well as the numerical values of the impedances must balance; otherwise, a true null voltage is impossible to obtain. This means that a bridge with all "pure" arms (pure resistance or reactance) cannot measure complex impedances; a combination of R and X must be present in at least one arm aside from the unknown.

The actual circuits of ac bridges take many forms, depending on the type of measurement intended and on the frequency range to be covered. As the frequency is raised, stray effects (unwanted capacitances and inductances) become more pronounced. At radio frequencies, special attention must be paid to minimizing them.

Most amateur-built bridges are used for

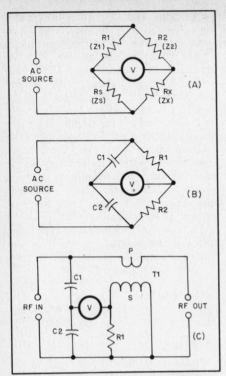

Fig 21—(A) Generalized form of bridge circuit for either ac or dc. (B) One form of ac bridge frequently used for RF measurements. (C) SWR bridge for use in transmission lines. This circuit is often calibrated in watts rather than volts.

RF measurements, especially SWR measurements on transmission lines. The circuits at Fig 21B and C are favorites for this purpose.

Fig 21B is useful for measuring both transmission lines and lumped constant components. Combinations of resistance and capacitance are often used in one or more arms; this may be required for eliminating the effects of stray capacitance.

The bridge shown in Fig 21C is used only on transmission lines, and only on those lines having the characteristic impedance for which the bridge is designed.

SWR Measurement—the Reflectometer

In measuring the standing-wave ratio, advantage is taken of the fact that the voltage on a transmission line consists of two components traveling in opposite directions. The power going from the transmitter to the load is represented by one voltage (designated incident or forward) and the power reflected from the load by the other. Because the relative amplitudes and phase relationships are definitely established by the line's characteristic

impedance, its length and the load impedance in which it is terminated, a bridge circuit can separate the incident and reflected voltages for measurement. This is sufficient for determining the SWR. Bridges designed for this purpose are frequently called reflectometers.

Refer to Fig 21A. If R1 and R2 are made equal, the bridge will be balanced when $R_x = R_s$. This is true whether R_x is an actual resistor or the input resistance of a perfectly matched transmission line, provided R_s is chosen to equal the characteristic impedance of the line. Even if the line is not properly matched, the bridge will still be balanced for power traveling outward on the line, since outward-going power sees only the Z_0 of the line until it reaches the load. However, power reflected back from the load does not see a bridge circuit, and the reflected voltage registers on the voltmeter. From the known relationship between the incident and reflected voltages the SWR is easily calculated:

$$SWR = \frac{V_o + V_r}{V_o - V_r}$$

where
V_o = incident voltage
V_r = reflected voltage

The Reflected Power Meter

Fig 21C makes use of mutual inductance between the primary and secondary of T1 to establish a balancing circuit. C1 and C2 form a voltage divider in which the voltage across C2 is in the same phase as the voltage at that point on the transmission line. The relative phase of the voltage across R1 is determined by the phase of the current in the line. If a pure resistance equal to the design impedance of the bridge is connected to the RF OUT terminals, the voltages across R1 and C2 will be out of phase and the voltmeter reading will be minimum. If the amplitudes of the two voltages are also equal (they are made so by bridge adjustment) the voltmeter will read zero. Any other value of resistance or impedance connected to the RF OUT terminals will result in a finite voltmeter reading. When used in a transmission line this reading is proportional to the reflected voltage. To measure the incident voltage, the secondary terminals of T1 should be reversed. To function as described, the secondary-winding reactance of T1 must be very large compared to the resistance of R1.

Instruments of this type are usually designed for convenient switching between forward and reflected power, and are often calibrated to read watts in the specified characteristic impedance. As the power

calibration is made at an SWR of 1:1, the actual power transmitted to the load must be calculated from the difference between the forward and reflected readings when the SWR is other than 1:1. A reading of 400 watts forward and 300 watts reflected means that 400 W − 300 W = 100 W has been delivered to the load. The SWR can be calculated from the formula

$$SWR = \frac{1 + \sqrt{\dfrac{\text{reflected power}}{\text{forward power}}}}{1 - \sqrt{\dfrac{\text{reflected power}}{\text{forward power}}}}$$

$$SWR = \frac{1 + \sqrt{\dfrac{300}{400}}}{1 - \sqrt{\dfrac{300}{400}}} = 13.9$$

Both of these calculations assume that the loss in the transmission line is very small; information on correcting for line loss is given in Chapter 16.

Sensitivity and Frequency

In all of the circuits in Fig 21 the sensitivity is independent of the applied frequency, within practical limits. Stray capacitances and couplings generally limit the performance of all three at the high-frequency end of the useful range. Fig 21A will work right down to dc, but the low-frequency performance of Fig 21B is degraded when the capacitive reactances become so large that voltmeter impedance becomes low in comparison. (In all these bridge circuits, it is assumed that the voltmeter impedance is high compared with the impedance of the bridge arms.) In Fig 21C the performance is limited at low frequencies by the fact that the transformer reactance decreases with frequency, so that eventually the reactance is not very high in comparison with the resistance of R1.

The Monimatch

A type of bridge that is simple to make, but in which the sensitivity rises directly with frequency, is the Monimatch and its various offspring. The circuit cannot be described in terms of lumped constants, as it makes use of the distributed mutual inductance and capacitance between the center conductor of a transmission line and a wire placed parallel to it. The wire is terminated in a resistance approximating the characteristic impedance of the transmission line at one end and feeds a diode rectifier at the other.

Frequency Measurement

The regulations governing amateur operation require that the transmitted signal be maintained inside the limits of certain bands of frequencies.[1] The exact frequency need not be known, as long as it is not outside the limits. On this last point there are no tolerances: It is up to the individual amateur to see that he stays safely inside.

This is not difficult to do, but requires some simple apparatus and the exercise of some care. The apparatus commonly used is the frequency-marker generator, and the method involves use of the station receiver, as shown in Fig 22.

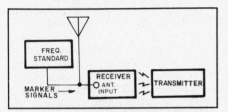

Fig 22—Setup for using a frequency standard. It is necessary that the transmitter signal be weak in the receiver—of the same order of strength as the marker signal from the standard. This requirement can usually be met by turning on just the transmitter oscillator, leaving all power off any succeeding stages. In some cases it may also be necessary to disconnect the antenna from the receiver.

The Frequency Marker

The marker generator in its simplest form is a high-stability oscillator generating a series of signals which, when detected in the receiver, mark the exact edges of the amateur assignments. It does this by oscillating at a low frequency that has harmonics falling on the desired frequencies.

Most US amateur band limits are exact multiples of 25 kHz, whether at the extremes of a band or at points marking the subdivisions between types of emission, license privileges, and so on. A 25-kHz fundamental frequency therefore will produce the desired marker signals if its harmonics at the higher frequencies are strong enough. But since harmonics appear at 25-kHz intervals throughout the spectrum, along with the desired markers, the problem of identifying a particular marker arises. This is easily solved if the receiver has a reasonably good calibration. If not, most marker circuits provide for a choice of fundamental outputs of 100 and 50 kHz as well as 25 kHz, so the question can be narrowed down to initial identification of 100-kHz intervals. From these, the desired

25-kHz (or 50-kHz) points can easily be spotted. Coarser frequency intervals are rarely required; there are usually signals available from stations of known frequency, and the 100-kHz points can be counted off from them.

Transmitter Checking

In checking one's own transmitter frequency, the signal from the transmitter is first tuned in on the receiver and the dial setting at which it is heard is noted. Then the nearest marker frequencies above and below the transmitter signal are tuned in and identified. The transmitter frequency is obviously between these two known frequencies.

If the marker frequencies are accurate, this is all that needs to be known—except that the transmitter frequency must not be so close to a band (subband) edge that sideband frequencies, especially in phone transmission, will extend over the edge.

If the transmitter signal is inside a marker at the edge of an assignment, to the extent that there is an audible beat note with the receiver BFO turned off, normal CW sidebands are safely inside the edge. (This statement does not take into account abnormal sidebands such as those caused by clicks and chirps.) For phone the safety allowance is usually taken to be about 3 kHz, the nominal width of one sideband. A frequency difference of this order can be estimated by noting the receiver dial settings for the two 25-kHz markers that bracket the signal and dividing 25 by the number of dial divisions between them. This will give the number of kilohertz per dial division.

Transceivers

The method described above is applicable when the receiver and transmitter are separate pieces of equipment. When a transceiver is used and the transmitting frequency is automatically the same as that to which the receiver is tuned, setting the tuning dial to a spot between two known marker frequencies is all that is required. The RIT control must be turned off.

The proper dial settings for the markers are those at which, with the BFO on, the signal is tuned to zero beat—the spot where the beat disappears as the tuning makes the beat tone progressively lower. Exact zero beat can be determined by a very slow rise and fall of background noise, caused by a beat of a cycle or less per second. In receivers with high selectivity it may not be possible to detect an exact zero beat, because low audio frequencies from beat notes may be prevented from reaching the speaker or headphones.

Frequency-Marker Circuits

The basic frequency-determining element in most amateur frequency markers is a

100-kHz or 1-MHz crystal. Although the marker generator should produce harmonics at 25-kHz and 50-kHz intervals, crystals (or other high-stability devices) for frequencies lower than 100 kHz are expensive and difficult to obtain. However, there is really no need for them, since it is easy to divide the basic frequency down to whatever one desires; 50 and 25 kHz require only two successive divisions, each by two. In the division process, the harmonic output of the generator is greatly enhanced, making the generator useful at frequencies well into the VHF range.

Simple Crystal Oscillators

Fig 23 illustrates two of the simpler circuits. C1 in both circuits is used for exact adjustment of the oscillating frequency to 100 kHz, which is done by using the receiver for comparing one of the oscillator's harmonics with a standard frequency transmitted by WWV, WWVH or a similar station.

Fig 23A shows a field-effect transistor analog of a vacuum-tube circuit. However, it requires a 10-mH coil to operate well, and since the harmonic output is not strong at the higher frequencies the circuit is given principally as an example of a simple transistor arrangement. A much better oscillator is shown at B. This is a cross-connected pair of transistors forming a multivibrator of the free-running or asynchronous type, locked at 100 kHz by using the crystal as one of the coupling elements. While it can use two separate bipolar transistors as shown, it is much simpler to use an integrated-circuit dual gate, which contains all necessary parts except the crystal and capacitors. Further, it is considerably less expensive and more compact than the separate components.

Frequency Dividers

Electronic division is accomplished by a bistable flip-flop or cross-coupled circuit that produces one output change for every two impulses applied to its input circuit, thus dividing the applied frequency by two. All division therefore must be in terms of some power of two. In practice this is no handicap since with modern integrated-circuit flip-flops, circuit arrangements can be worked out for division by any desired number.

As flip-flops and gates in integrated circuits come in compatible series—meaning that they work at the same supply voltage and can be directly connected together—a combination of a dual-gate version of Fig 23B and a dual flip-flop makes an attractive simple combination for the marker generator. A circuit using a similar arrangement appears in Chapter 34.

There are several different basic types of flip-flops, the variations having to do with methods of driving (dc or pulse operation)

---

[1]These limits depend on the type of emission and class of license held, as well as on international agreements. See the latest edition of *The FCC Rule Book* for current status.

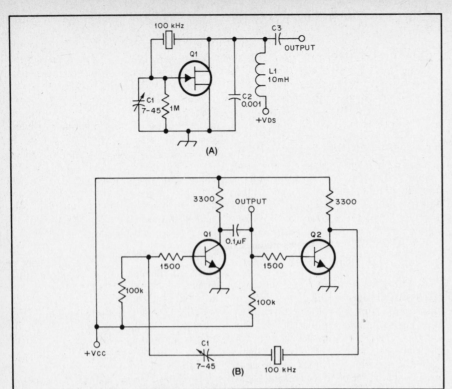

Fig 23—Two simple 100-kHz oscillator circuits. B is the most suitable of available transistor circuits (for marker generators) and is recommended where solid-state construction is to be used. In both circuits C1 is for fine frequency adjustment. The output coupling capacitor, C3, is generally small—20 to 50 pF—a compromise to avoid loading the oscillator by the receiver antenna input while maintaining adequate coupling for good harmonic strength.

and control of the counting function. Information on the operating principles and ratings of a specific type usually can be obtained from the manufacturer. The counting-control functions are not needed in using the flip-flop in a simple marker generator, although they come into play when dividing by some number other than a power of two.

Other Methods of Frequency Checking

The simplest possible frequency-measuring device is an absorption frequency-meter—a parallel LC circuit, tunable over a desired frequency range, with its tuning dial calibrated in frequency. It can be used only for checking circuits in which at least a small amount of RF power is present, because the energy required to give a detectable indication is not available in the LC circuit itself; it has to be extracted from the circuit being measured. Hence the name absorption frequency meter. What is actually measured is the frequency of the RF energy, not the frequency to which the circuit under test may be tuned.

The measurement accuracy of such an instrument is low, compared with the accuracy of a marker generator, because the Q of a practicable LC circuit is not high enough to make precise dial readings possible. Also, any two circuits coupled together react on each other's tuning. (This can be minimized by using the loosest

coupling that will give an adequate indication.)

The absorption frequency meter has one useful advantage over the marker generator—it will respond only to the frequency to which it is tuned, or to a band of frequencies very close to it. Thus there is no harmonic ambiguity, as there sometimes is with a marker generator.

Absorption Circuit

A typical absorption frequency-meter circuit is shown in Fig 24. In addition to the adjustable tuned circuit, L1-C1, it includes a pickup coil, L2, wound over L1, a high-frequency semiconductor diode, D1, and a microammeter or low-range (usually not more than 0-1 mA) milliammeter. A phone jack is included so the device can be used for listening to the signal.

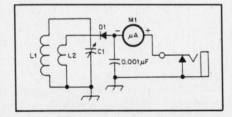

Fig 24—Absorption frequency-meter circuit. The closed-circuit phone jack may be omitted if listening is not wanted, in which case the positive terminal of M1 goes to common ground.

The sensitivity of the frequency meter depends on the sensitivity of the dc meter movement and the size of L2 in relation to L1. There is an optimum size for this coil that must be found by experimentation. An alternative is to make the rectifier connection to an adjustable tap on L1, in which case there is an optimum tap point. In general, the rectifier coupling should be a little below (that is, less tight than) the point that gives maximum response, since this will make the indication sharper.

Calibration

The absorption frequency meter must be calibrated by taking a series of readings from circuits carrying RF power at known frequencies. The frequency of the RF energy may be determined by means such as a marker generator and receiver. The setting of the dial that gives the highest meter indication is the calibration point for that frequency. This point should be determined by tuning through it with loose coupling to the circuit being measured.

Frequency Standards

The difference between a marker generator and a frequency standard is that special pains are taken to make the frequency-standard oscillator frequency as stable as possible. Variations in temperature, humidity, line voltage and other factors can cause a small change in frequency.

While there are no definite criteria that distinguish the two in this respect, a circuit designated as a standard for amateur purposes should be capable of maintaining frequency within at least a few parts per million in ambient conditions, without adjustment. Commercially made receivers that employ synthesizers use frequency standards with a tolerance of 1 part in 100,000,000. A simple marker generator using a 100-kHz crystal can be expected to have frequency variations 10 times (or more) greater under similar conditions. It can of course be adjusted to exact frequency at any time the WWV (or equivalent) signal is available.

The design considerations of high-precision frequency standards are outside the scope of this chapter, but information is available from time to time in periodicals.

Frequency Counters

One of the most accurate means of measuring frequency is the frequency counter. This instrument is capable of numerically displaying the frequency of the signal supplied to its input. For example, if an oscillator operating at 8.244 MHz is connected to the counter input, 8.244 would be displayed. At present, there are counters that are usable well up into the gigahertz range. Most counters that are used at high frequencies make use of a prescaler ahead of a basic low-frequency counter. Essentially, the prescaler divides the high-frequency signal by 10, 100, 1000 or some other amount so the low-frequency counter can display the operating frequency.

The accuracy of the counter depends on an internal crystal reference. The more accurate the crystal reference, the more accurate the readings will be. Crystals for frequency counters are manufactured to close tolerances. Most counters have a trimmer capacitor so the crystal can be set exactly on frequency. Crystal frequencies of 1 MHz, 5 MHz or 10 MHz have become more or less standard. Harmonics of the crystal can be compared to WWV or WWVH signals at 5, 10, 15 MHz, etc., and adjusted for zero beat.

Many frequency counters offer options to increase the accuracy of the counter's timebase; this directly increases the counter's accuracy. These options usually employ temperature-compensated crystal oscillators (TCXOs) or crystals mounted in constant temperature ovens that keep the crystal from being affected by changes in ambient (room) temperature. Counters with these options are capable of obtaining accuracies of 0.1 ppm (part per million) or better. For example, a counter with a timebase with an accuracy of 5.0 ppm and a second counter with a TCXO with an accuracy of 0.1 ppm are available to check a 436-MHz CW transmitter for satellite use. The counter with the 5-ppm timebase could have a frequency error of as much as 2.18 kHz, while the possible error of the counter with the 0.1-ppm timebase is only 0.0436 kHz.

A 500-MHz FREQUENCY COUNTER

The advent of LSI (large-scale-integration) integrated-circuit technology has reduced the construction costs of many electronic products. The frequency counter shown in Figs 25 through 34 was designed and built in the ARRL lab by Ed Hare, KA1CV and Tom Miller, NK1P. It uses an Intersil ICM7216C 10-MHz frequency counter to obtain a lot of performance for relatively little cost. With some careful shopping it can be built for about $125.00, including the cabinet. It is usable from audio through UHF, and its sensitivity of 15 mV makes it a useful station accessory and troubleshooting tool.

Two input connectors are provided. The 10-MHz input has a high impedance, with a usable range of 20 Hz to over 10 MHz. The sensitivity is reduced to about 50 mV at the low-frequency end. The 500-MHz input has a 50-ohm impedance, and it is used for frequencies from 400 kHz to over 500 MHz. A SENSITIVITY control allows this input to accommodate signals of up to 2 V RMS.

The project incorporates a prescaler that works with the range switch to automatically select a divide-by-10 or divide-by-100 function. This prescaler can also be built as a separate project to extend the range of an existing frequency counter. The ½-inch digital readout is easy to read in almost any lighting conditions; a GATE

Fig 25—Exterior view of the 500-MHz frequency counter. Clear red plastic has been put in front of the LED display to make it easier to read in bright light.

LED is included for those who like to watch "das-blinkenlights."

Technical Details

This project was designed in modules. Fig 26 shows the schematic of the prescaler. The 10 MHz input uses Q1 and its associated circuitry as a high-input impedance preamplifier, with input protection provided by D1 and D2. U1 is a high-speed comparator used to obtain fast rising square waves from the output of Q1. The output of this stage is coupled into Q2 to obtain the TTL voltage levels that are required by U5.

U2 is an MMIC preamplifier stage for the 500-MHz input. Other MMICs were tried successfully at U2, requiring only a change in the value of R10. (For more information on MMIC devices, see February and March 1987 QST). An 11C90 is used at U3 as a high-speed divide-by-10 stage. R12, R13 and R14 are optional components; they allow a fine adjustment of the VREF voltage output from U3, increasing the counter sensitivity. If the resistors are left out of the circuit the sensitivity will decrease to about 50 mV.

The output of U3 feeds U4 and U5, another divide-by-10 stage. A 74S160 binary counter is used in this stage. A 74LS160 can be substituted here, but this limits the counter response to 350 MHz.

U5 selects either the 10-MHz direct input or the 500-MHz input divided by 10 or 100. This selection is accomplished through the appropriate A, B or C inputs of the 74LS151. If the prescaler is built as a separate module these signals must be provided. Table 1 shows a truth table for this stage. The output of this stage is a TTL voltage level.

The prescaler module requires +4.75 to +5.25 V at 450 mA and +12 to +18 V at 120 mA. The +12 to +18 V is regulated to +10 V by D5. Extensive power-supply bypassing has been used throughout.

The power-supply schematic is shown in Fig 27. A full-wave bridge rectifier is used to generate +12 to +18 V. This is regulated to +5 V by U1. The builder has quite a bit of latitude when selecting components for the power-supply circuit. Any

12.6-V transformer with at least a 450 mA current rating may be used at T1. U2 can be rated at 50 V or higher, 1 A or better. The values for C1 and C3 are approximate and could easily vary by 50%. The capacitor voltage ratings shown are the minimum that should be used, however.

The 10-MHz counter section has been designed to make full use of the features available in the Intersil ICM7216C frequency counter IC. The schematic for this module is shown on Fig 28. The internal 10-MHz oscillator can be adjusted to zero-beat with WWV. It is important to use a high-quality capacitor at C103; a mica or polystyrene capacitor is best. If a ceramic capacitor is used, the oscillator frequency is apt to be fairly sensitive to temperature changes, and this degrades counter accuracy. C102 can be any small trimmer capacitor that has a capacitance of at least 50 pF (a 6 to 70 pF unit was used in the ARRL counter). The RESET switch allows the user to restart the counter, a useful feature on longer gate times. The HOLD feature stops the counting and maintains the last reading on the LED display. The display uses GI MAN6710 dual displays, and the display PC board is designed to accommodate this (or an equivalent) display. Experienced builders could substitute a different LED display, but the ICM7216C is specified to sink a maximum of 20 mA, so the substitution must be made accordingly. The gate LED (DS101) lights when the counter is sampling the input signal.

The counter features six ranges. Gate times are 10 s, 1 s, 100 ms and 10 ms, with 100 ms used on the two prescaled ranges. Ultimate resolution is 0.1 Hz. The left decimal point indicates that the input signal is over-range. The external decimal point mode of the ICM7216C is used, automatically switching the decimal point to accommodate Hz, kHz and the two prescaled MHz ranges.

The ICM7216C accomplishes range and decimal point changes by connecting the RANGE and DP IN inputs to the various digit-multiplex outputs and internally evaluating the multiplex relationship. The TTL circuitry of U102, U103 and U104 is

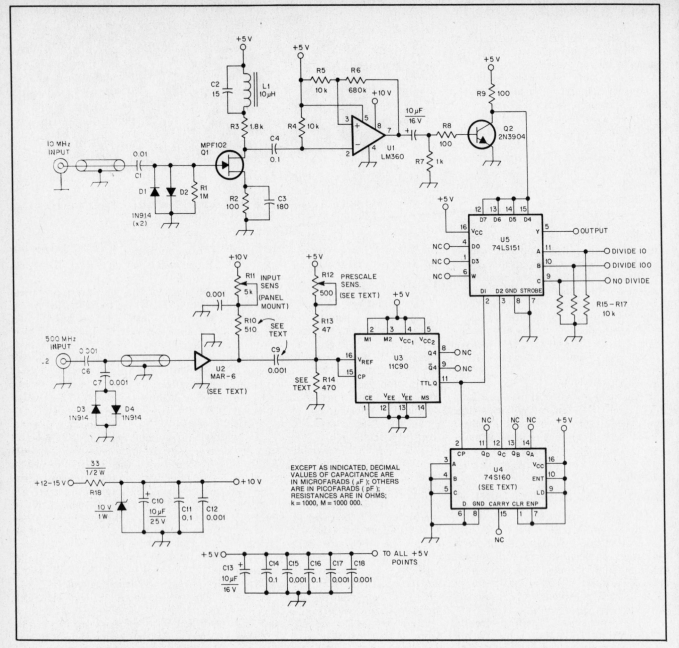

Fig 26—Schematic diagram of the prescaler module. All resistors are ¼ W, 5% tolerance, unless noted. Polarized capacitors are electrolytic.

R11—5-kΩ linear panel-mount potentiometer.
R12—500-Ω single-turn trimmer potentiometer.
Q1—MPF102 JFET.
Q2—2N3904 NPN transistor.
U1—LM360 operational amplifier.

U2—MAR-6 MMIC (see text).
U3—11C90 Prescaler (available from Circuit Specialists, see Chapter 35).
U4—74S160 Schottky decade counter (see text).

U5—74LS151 8-input multiplexer.
D1-D4—Small-signal diode: 1N4148, 1N914 etc.
D5—10-V, 1-W Zener diode.

designed to automatically generate the signals required to switch the prescaler into its various modes. U102 is configured as a flip-flop, changing state as the range switch is set to the 100-MHz or 500-MHz range by detecting a simultaneous occurrence of DP IN and digit 6 for divide-by-10 or digit 5 for divide-by-100. U103 and U104 set the 10 MHz output high on all other ranges. U102 and U103 must be CMOS, 74C00 or 74HC00. The use of a low-power Schottky device here will result in excessive loading of the multiplexing lines. C104 and

Table 1
Prescaler Control Logic

Inputs			Result
A	B	C	
H	L	L	divide by 10
L	H	L	divide by 100
X	X	H	direct (no prescale)

H = > 2.7 V
L = < 0.8 V
X = don't care

C105 are used to smooth out multiplexing glitches that would otherwise result in false switching of the flip-flop outputs. U104 can be CMOS or low-power Schottky. The ARRL version uses a 74C74 here.

Construction Details

The counter and prescaler should be constructed from the PC artwork that is found at the back of this book. Circuit layout is important to the proper operation of the 10 MHz squaring circuit and the 11C90 divide-by-10 IC. Fig 29 shows the

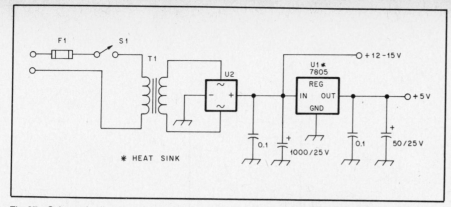

Fig 27—Schematic diagram of the power supply. Polarized capacitors are electrolytic.

S1—SPST switch.
T1—Power transformer, 120-V primary, 12.6-V 450-mA secondary.

U1—7805 voltage regulator IC.
U2—50-V, 1-A bridge rectifier.

component-side parts-placement diagram for the counter/prescaler board. Use insulated wire jumpers at the places marked W on the component-placement diagram.

Some additional components must be mounted on the bottom side of the board, and it is easier to do this after all of the component-side parts have been installed. U2 is a surface-mount MMIC, and the bypass capacitors are installed on pads on the PC board. Refer to Fig 30 for the solder-side parts-placement diagram. Fig 31 shows how the bypass capacitors have been mounted using heat-shrink tubing to keep the component leads from shorting to any of the PC board traces.

A 16-pin DIP socket may be installed in the socket for J101 if the builder wishes to use the dip header and ribbon cable seen in Fig 33. Sockets are recommended for all of the DIP ICs because it is a good idea to verify proper voltages at the power pins of the ICs before they are installed. Sockets also make IC replacement easier.

If the PC board is used for the power supply, the parts should be installed in the board as shown in the power-supply component-placement diagram, Fig 32. Pay close attention to the orientation of U1 (mount it with the tab flush against the PC board—part number facing up) and to the polarity markings on the electrolytic capacitors and the bridge rectifier. U1 requires a small heat sink similar to that shown in Fig 33.

The component-placement diagram for the display board is shown in Fig 34. A socket may be installed at J201 if desired, or the wires from J101 may be directly soldered through the appropriate holes. Pin 1 on J101 should be connected to pin 1 on J201. Install wire jumpers at the places marked W on the parts-placement diagram.

The display board mounts directly to the front panel. If ½-inch spacers are used to mount the display board, the LEDs will be flush with the front panel. For those builders who wish to exactly duplicate the ARRL version of the counter, a dimensional diagram and actual-size layout of the front panel lettering are available from the ARRL Technical Department secretary for an SASE and $1.00.

The ARRL version of the counter was built in a Bud AC-406 chassis (2 × 9 × 7 inches HWD) with a BPA-1593 bottom plate. The HOLD and RESET switches and the gate LED are mounted between the display board and the front panel, so these must be mounted before the display board is installed. It is probably easier to install the wires on the PC board and solder them to the front-panel components after they are mounted. C6, C7, D3 and D4 are mounted off-board, and should be installed directly at J2 using short leads. Short pieces of RG-174 coaxial cable are used to connect J1 and J2 to the appropriate points on the PC board. The sensitivity control should be wired so that clockwise rotation

decreases the resistance. The remainder of the layout is not critical, so the builder can easily determine the layout best suited to the components used. The interior photograph, Fig 33 shows the arrangement used in the ARRL version.

Alignment and Troubleshooting

There is little alignment that needs to be performed on this circuit. Lab tests indicate that the counter has completely stabilized after a warmup time of 60 minutes, showing 170 Hz drift at 10 MHz during the warmup period. To align the counter, allow an hour for warmup and tune in WWV (10 MHz or 20 MHz) on a general-coverage receiver. Next, tune in the signal from the counter's internal oscillator on the receiver. It may be necessary to use an external antenna placed near the counter circuitry to pick up enough signal from the ICM7216C oscillator. Adjust C102 so that the oscillator is zero-beat with the WWV carrier.

A second calibrated frequency counter could also be used to set the 10-MHz oscillator if desired. Couple a signal from the ICM7216 10-MHz oscillator by placing the second counter's probe near the oscillator components. Adjust C102 for a 10-MHz reading on the second counter.

If a variable-output VHF/UHF generator is available, the prescale sensitivity control (R12), may be adjusted for maximum sensitivity. Set the front-panel sensitivity control fully clockwise and start with about 100 mV at 500 MHz into the 500 MHz input. Gradually decrease the output from the signal generator while adjusting R12 to maintain an accurate reading. With somewhere around 20 mV input the prescale sensitivity control will not give further improvement. R12 may be left at this setting; it should not require any further adjustment. It is best to use an external attenuator to reduce strong signals to about 100 mV or so, but the front-panel sensitivity control can be used for input signals up to 2 V RMS. The 10-MHz input can withstand input voltages of about 20 V RMS. The front-panel sensitivity control has no effect on the 10 MHz input.

If the circuit fails to function refer to Chapter 26 for general troubleshooting information. The signals are divided by 10 at U3 and U4; keep this in mind if signal tracing is being performed.

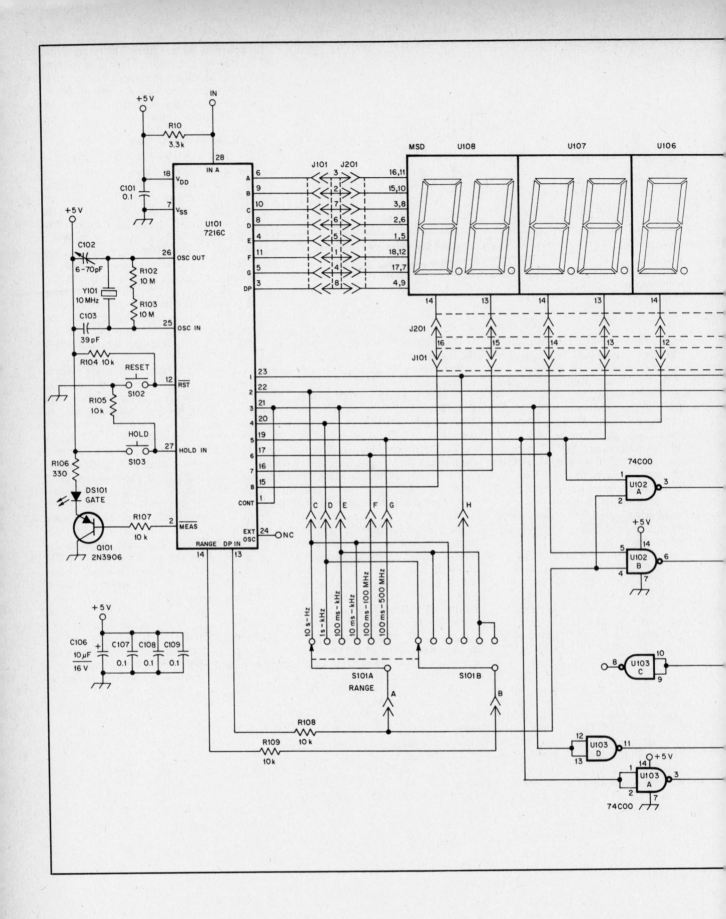

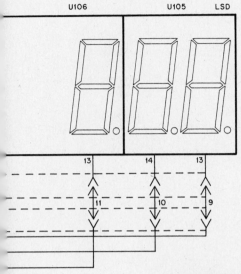

U106　　　U105　LSD

Fig 27—Schematic diagram of the 10-MHz counter module. All resistors are ¼ W, 5% tolerance. Polarized capacitors are electrolytic.

C102—6-70 pF ceramic trimmer (see text).
C103—39-pF 50-V silver mica or polystyrene.
S101—Double-pole six-position non-shorting rotary switch.
S102,S103—SPST momentary contact panel-mount push-button switch.
Q101—2N3906 PNP transistor.
U101—Intersil ICM7216C common-cathode frequency counter. (available from Digikey; see Chapter 35).
U102,U103—74C00 CMOS quad-NAND gate (see text).
U104—74C74 CMOS dual D flip-flop (see text).
U104-U108—Dual 7-segment LED display; General Instrument MAN6710, Panasonic LN524RA or ECG3074 (available from Digikey or Mouser; see Chapter 35).
Y101—10-MHz crystal, 30 pF load.

EXCEPT AS INDICATED, DECIMAL VALUES OF CAPACITANCE ARE IN MICROFARADS (μF); OTHERS ARE IN PICOFARADS (pF); RESISTANCES ARE IN OHMS; k = 1000, M = 1000 000.

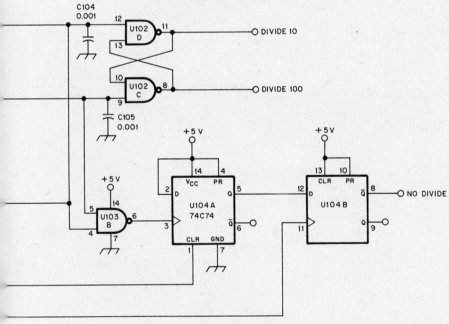

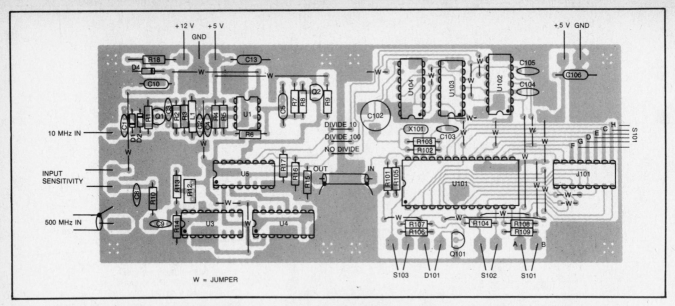

Fig 29—Parts-placement diagram for the prescaler and counter board as seen from the component side of the board. Additional parts are mounted on the solder side of the board. A full-size etching pattern can be found at the back of this book.

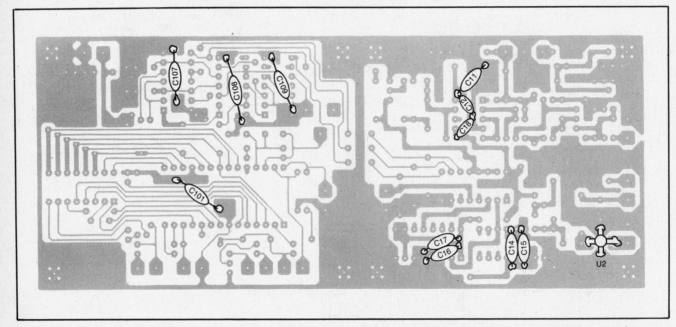

Fig 30—Parts-placement diagram for the prescaler and counter board as seen from the foil side of the board. Ensure that U2 is mounted as shown.

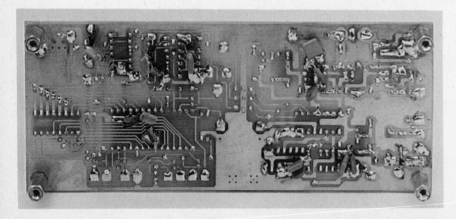

Fig 31—View of the solder side of the prescaler and counter board. Heat-shrink tubing has been used to ensure that the capacitor leads will not short circuit to the PC board traces.

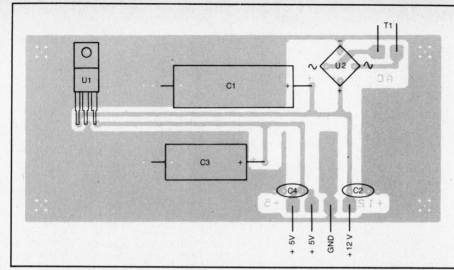

Fig 32—Parts-placement diagram for the power supply board as seen from the component side of the board. A full-size etching pattern can be found at the back of this book.

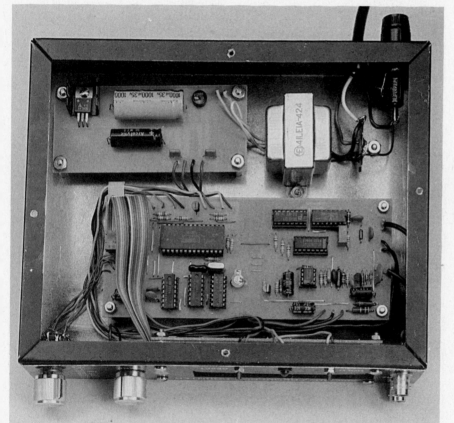

Fig 33—Interior view of the completed 500-MHz frequency counter. Ribbon cable and IDC headers have been used to connect the display board to the prescaler and counter board.

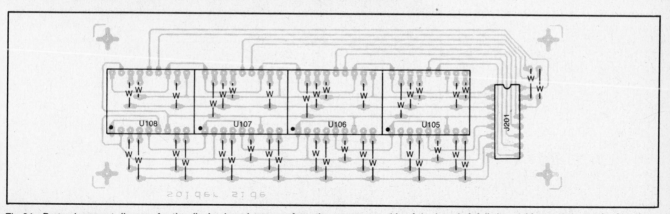

Fig 34—Parts-placement diagram for the display board as seen from the component side of the board. A full-size etching pattern can be found at the back of this book.

Other Instruments and Measurements

Many measurements require a source of ac power of adjustable frequency (and sometimes adjustable amplitude) in addition to what is already available from the transmitter and receiver. RF and AF test oscillators, for example, provide signals for purposes such as receiver alignment, testing of phone transmitters, and so on. Another valuable adjunct to the station is the oscilloscope, essential for checking waveforms.

RF Oscillators for Circuit Alignment

Receiver testing and alignment, covered in an earlier chapter, uses equipment common to ordinary radio service work. Inexpensive RF signal generators are available, both complete and in kit form. However, any source of signal that is weak enough to avoid overloading the receiver usually will serve for alignment work. The frequency marker generator is a satisfactory signal source. In addition, its frequencies, although not continuously adjustable, are known far more precisely, since the usual signal-generator calibration is not highly accurate. For rough work a gate-dip or grid-dip meter is sufficient.

AUDIO-FREQUENCY OSCILLATORS

Tests requiring an audio-frequency signal generally call for one that is a reasonably good sine wave. The best oscillator circuits for this use are RC-coupled, operating as close to a class-A amplifier as possible. Variable frequencies covering the entire audio range are needed for determining frequency response of audio amplifiers.

For most phone-transmitter testing, and for simple trouble shooting in AF amplifiers, an oscillator generating one or two frequencies with good waveform is adequate. A two-tone (dual) oscillator is particularly useful for testing sideband transmitters and adjusting them for on-the-air use.

The circuit of a simple RC oscillator that is useful for general test purposes is given in Fig 35. This Twin-T arrangement gives a waveform that is satisfactory for most purposes, and by choice of circuit constants the oscillator can be operated at any frequency in the audio range. R1, R2 and C1 form a low-pass type network, while C2, C3 and R3 form a high-pass network. As the phase shifts are opposite, there is only one frequency at which the total phase shift from collector to base is 180 degrees; oscillation will occur at this frequency. Optimum operation results when C1 is approximately twice the capacitance of C2 or C3, and R3 has a resistance about 0.1 that of R1 or R2 (C2 = C3 and R1 = R2). Output is taken across C1, where the harmonic distortion is least. A relatively high impedance load should be used—100 kΩ or more.

A small-signal AF transistor is suitable for Q1. Either NPN or PNP types can be

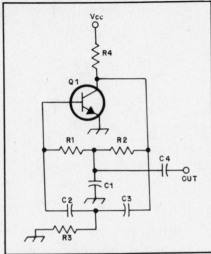

Fig 35—Twin-T audio oscillator circuit. Representative values for R1-R2 and C1 range from 18 kΩ and 0.05 μF for 750 Hz to 15 kΩ and 0.02 μF for 1800 Hz. For the same frequency range, R3 and C2-C3 vary from 1800 ohms and 0.02 μF to 1500 ohms and 0.01 μF. R4 should be approximately 3300 ohms. C4, the output coupling capacitor, can be 0.05 μF for high-impedance loads.

used, with due regard for supply polarity. R4, the collector load resistor, must be large enough for normal amplification, and may be varied somewhat to adjust the operating conditions for best waveform.

A WIDE-RANGE AUDIO OSCILLATOR

A wide-range audio oscillator that will provide a moderate output level can be built from a single 741 operational amplifier (Fig 36). Power is supplied by two 9-volt batteries, from which the circuit draws 4 mA. The frequency range is selectable from 8 Hz to 150 kHz. Distortion is approximately one percent. The output level under a light load (10 kΩ) is 4 to 5 volts. This can be increased by using higher battery voltages, up to a maximum of plus and minus 18 volts, with a corresponding adjustment of R_f.

Pin connections shown are for the TO-5 case. If another package configuration is used, the pin connections may be different. R_f (220 ohms) is trimmed for an output level about five percent below clipping. This should be done for the temperature

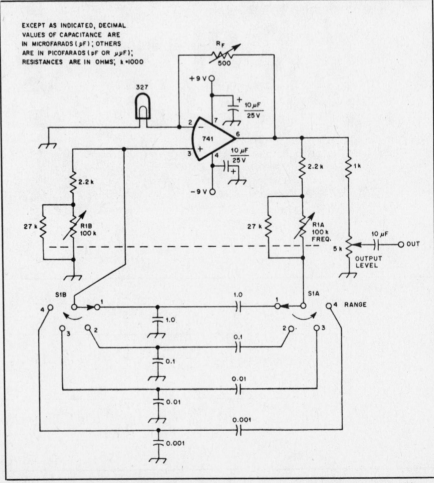

Fig 36—A simple audio oscillator that provides a selectable frequency range.

at which the oscillator will normally operate, as the lamp is sensitive to ambient temperature. This unit was originally described by Schultz in November 1974 *QST*; it was later modified by Neben as reported in June 1983 *QST*.

A TWO-TONE AUDIO GENERATOR

The audio frequency generator shown in Figs. 37 and 38 makes a very convenient signal source for testing the linearity of a single-sideband transmitter. To be suitable for transmitter evaluation, a generator of this type must produce two nonharmonically related tones of equal amplitude. The level of harmonic and intermodulation distortion must be sufficiently low so as not to confuse the measurement. The frequencies used in this generator are 700 and 1900 Hz, both well inside the normal audio passband of an SSB transmitter. Spectral analysis and practical application with many different transmitters has shown this generator to meet all of the requirements mentioned above. While designed specifically for transmitter testing, it is also useful any time a fixed-frequency, low-level audio tone is needed. Details on distortion measurement and the two-tone test can be found in Chapter 18.

Circuit Details

Each of the two tones is generated by a separate Wein bridge oscillator, U1B and U2B. The oscillators are followed by RC active low-pass filters, U1A and U2A. Because the filters require nonstandard capacitor values, provisions have been made on the circuit board for placing two capacitors in parallel in those cases where standard values cannot be used. The oscillator and filter capacitors should be polystyrene or Mylar film types if available. Two tones are combined at op amp U3A. This amplifier has a variable resistor, R4, in its feedback loop which serves as the ouput LEVEL control. While R4 varies both tones together, R3, the BALANCE control, allows the level of tone A to be changed without affecting the level of tone B. This is necessary because some transmitters do not have equal audio response at both frequencies. Following the summing amplifier is a step attenuator; S3 controls the output level in 10-dB steps. The use of two output level controls, R4 and S3, allows the output to cover a wide range and still be easy to set to a specific level. The remaining op amp, U3B, is connected as a voltage follower and serves to buffer the output while providing a high-impedance load for the step attenuator. Either a high or a low output impedance can be selected by S4.

Fig 37—Exterior view of the two-tone audio generator.

The values shown are suitable for most transmitters using either high- or low-impedance microphones.

Construction and Adjustment

Component layout and wiring are not critical, and any type of construction can be used with good results. For those who wish to use a printed-circuit board, a parts-placement guide is shown in Fig 39. Because the generator will normally be used near a transmitter, it should be enclosed in some type of metal case for shielding. Battery power was chosen to reduce the possibility of RF entering the unit through the ac line. With careful shielding and filtering, the builder should be able to use an ac power supply in place of the batteries.

The only adjustment required before use is the setting of the oscillator feedback trimmers, R1 and R2. These should be set so that the output of each oscillator, measured at pin 7 of U1 and U2, is about 0.5 volt RMS. A VTVM or oscilloscope can be used for this measurement. If neither of these is available, the feedback should be adjusted to the minimum level that allows the oscillators to start reliably and stabilize quickly. When the oscillators are first turned on, they take a few seconds before they will have a stable output amplitude. This is caused by the lamps, DS1 and DS2, used in the oscillator feedback circuit. This is normal and should cause no difficulty. The connection to the transmitter should be through a shielded cable.

Resistors at Radio Frequencies

Measuring equipment often requires essentially pure resistance in some part of its circuit—that is, resistance exhibiting only negligible reactive effects on the fre-

quencies at which measurement is intended. Of the resistors available to amateurs, this requirement is met only by small composition (carbon) resistors. The inductance of wire-wound resistors makes them useless for amateur frequencies.

The reactances to be considered arise from the inherent inductance of the resistor itself and its leads, and from small stray capacitances from one part of the resistor to another and to surrounding conductors. Although both the inductance and capacitance are small, their reactances become increasingly important as the frequency is raised. Small composition resistors, properly mounted, show negligible capacitive reactance up to 100 MHz or so in resistance values up to a few hundred ohms; similarly, the inductive reactance is negligible in values higher than a few hundred ohms. The optimum resistance region in this respect is 50 to 200 ohms.

Proper mounting includes reducing lead length as much as possible, and keeping the resistor separated from other resistors and conductors. Care must also be taken in some applications to ensure that the resistor, with its associated components, does not form a closed loop into which a voltage could be induced magnetically.

So installed, the resistance is essentially pure. In composition resistors the skin effect is very small, and the RF resistance up to VHF is nearly the same as the dc resistance.

Dummy Antennas

A dummy antenna is simply a resistor that can be substituted for an antenna or transmission line for test purposes. The dummy must exhibit the same resistance as the antenna or line it is to replace. It permits leisurely transmitter testing without radiating a signal. (Amateur regulations strictly limit the amount of on-the-air testing that may be done.) A dummy antenna is useful in testing receivers, in that electrically it resembles an antenna, but does not pick up external noise and signals, a desirable feature in some tests.

For transmitter tests the dummy antenna must be capable of dissipating safely the entire power output of the transmitter. For most testing it is desirable that the dummy simulate a perfectly matched transmission line; it should be a pure resistance of approximately 50 or 75 ohms. This is a severe limitation in home construction, because nonreactive resistors rated at more than a few watts' safe dissipation are very difficult to obtain. Tubular carborundum nonreactive resistors can often be found in military surplus transmitters and antenna tuners. These resistors usually are rated at

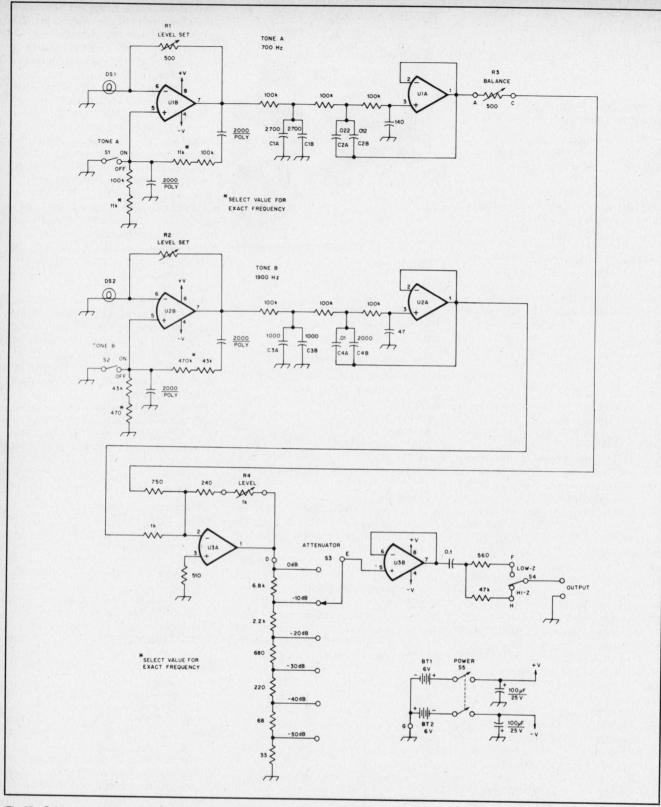

Fig 38—Schematic diagram of the two-tone audio generator. All resistors are ¼-W composition type.

BT1, BT2—Four AA cells.
C1A, B—Total capacitance of 0.054 µF, ±5%.
C2A, B—Total capacitance of 0.034 µF, ±5%.
C3A, B—Total capacitance of 0.002 µF, ±5%.
C4A, B—Total capacitance of 0.012 µF, ±5%.

DS1, DS2—12-V, 25-mA lamp.
R1, R2—500-Ω, 10-turn trim potentiometer.
R3—500-Ω, panel-mount potentiometer.
R4—1-kΩ, panel-mount potentiometer.
S1, S2—SPST toggle switch.

S3—Single-pole, 6-position rotary switch.
S4—SPDT toggle switch.
S5—DPDT toggle switch.
U1, U2, U3—Dual JFET op amp, type LF353N or TL082.

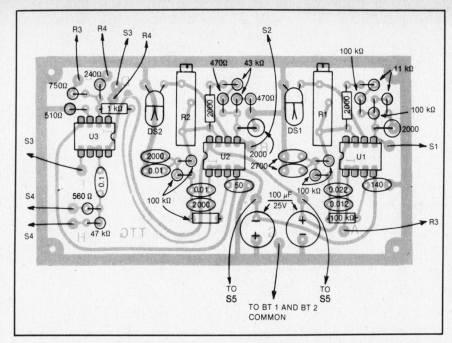

Fig 39—Parts-placement diagram for the two-tone audio generator, shown from the component side of the board. A full-size etching pattern can be found at the back of this book.

50 to 100 W. Check surplus catalogs. (There are, however, dummy antenna kits available that can handle up to a kilowatt.)

For receiver and minipower transmitter testing an excellent dummy antenna can be made by installing a 51- or 75-ohm composition resistor in a PL-259 fitting, as shown in Fig 40. Sizes from one-half to two watts are satisfactory. The disc at the end helps reduce lead inductance and completes the shielding. Dummy antennas made in this way have good characteristics through the VHF bands as well as at all lower frequencies.

Increasing Power Ratings

More power can be handled by using a number of 2-watt resistors in parallel, or series-parallel, but at the expense of introducing some reactance. Nevertheless, if some departure from the ideal impedance characteristics can be tolerated, this is a practical method for getting increased dissipation in a dummy antenna. The principal problem is stray inductance, which can be minimized by mounting the resistors on flat copper strips or sheets, as indicated in Fig 41.

The power rating on resistors is a continuous rating in free air. In practice, the maximum power dissipated can be increased in proportion to the reduction in duty cycle. Thus with keying, which has a duty cycle of about one half, the rating can be doubled. With sideband the duty cycle is usually not over about one-third. The best way of judging is to feel the resistors occasionally (with power off); if too hot to touch, they may be dissipating more power than they are rated for.

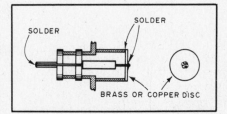

Fig 40—Dummy antenna made by mounting a composition resistor in a PL-259 coaxial plug. Only the inner portion of the plug is shown; the cap screws on after assembly is completed.

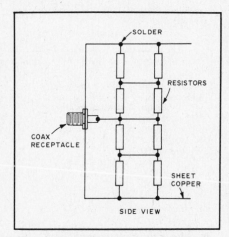

Fig 41—Use of resistors in series-parallel to increase the power rating of a small dummy antenna. Mounted in this way on pieces of flat copper, inductance is reduced to a minimum. Eight 100-ohm, 2-watt composition resistors in two groups, each four resistors in parallel, can be connected in series to form a 50-ohm dummy. The open construction shown permits free air circulation.

The dummy load shown in Fig 42 was constructed to test the feasibility of using a large number of 2-watt resistors to form a high-power load. This example uses 66 carbon resistors of 3.3 kΩ. The load will handle power levels up to 100 watts for reasonable lengths of time, and the SWR is less than 1.3 to 1 at frequencies up to 30 MHz. Because variations in construction may affect the stray reactances, it is recommended that the builder follow the general layout shown in the photograph. Complete details can be found in January 1981 QST.

A SIMPLE INDUCTANCE METER

The inductance meter shown in Figs 43 through 47 measures the relative Q of a coil as well as its inductance. The Q/L meter is based on a design by Doug DeMaw, W1FB; this version was built by Zachary Lau, KH6CP, in the ARRL Lab. Simple LC Hartley oscillators replace the crystal oscillators used in the original design, eliminating the need for low-pass filters and reducing circuit complexity and cost. The accuracy of the meter is approximately ± 10 percent using the dial layout shown. Greater accuracy can be obtained by calibrating the meter against known high-accuracy standards.

Circuit Details

The schematic of the Q/L meter is shown in Fig 43. A low-level RF signal is generated using an oscillator and an amplifier (Q1 through Q5). This signal is then applied to a tuned circuit consisting of the inductor under test and a calibrated capacitor. The voltage across the tuned circuit is measured using Q6 and displayed on M1, a 200-μA meter. When the capacitor is adjusted for resonance at the frequency of the RF signal, the voltage across the tuned circuit is maximized, resulting in a peak indication on the meter.

The crystal oscillators used in the original versions were replaced with Hartley LC oscillators as a cost-cutting measure. While the Hartley oscillators do not have the great stability of crystal oscillators, they are adequate for this application. The output of the LC oscillators is relatively clean, and no additional filtering is needed. The potentiometers from source to ground

Fig 42—100-watt dummy antenna made up of 66, 2-watt carbon resistors.

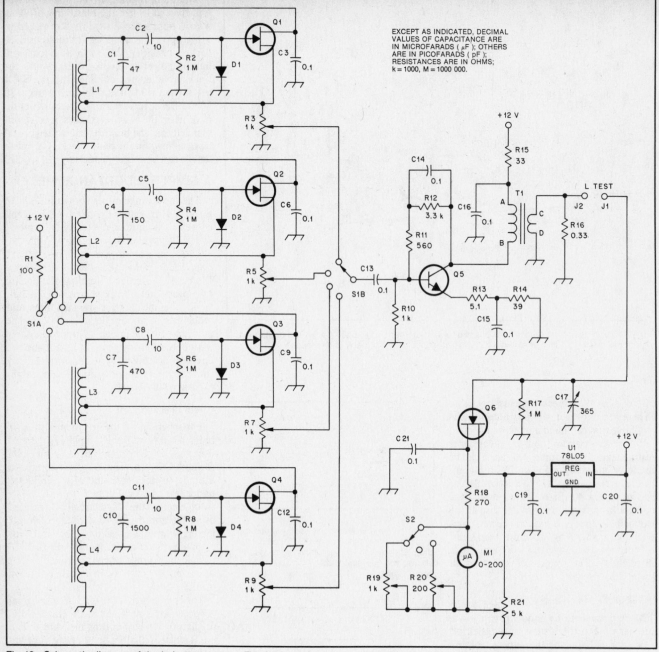

Fig 43—Schematic diagram of the inductance meter. Fixed-value capacitors are disk ceramic unless otherwise noted. Fixed-value resistors are ¼ or ½ W units.

C1, C2, C4, C5, C7, C8, C10 and C11—NP0 disc, silver mica or polystyrene.
C17—365 pF air variable, Radiokit BC-01 with Jackson Brothers 4511/DAF vernier or equiv.
D1-D4—High-speed silicon diode; 1N914 or 1N4148.
L1—17t no. 26 enam wire on an Amidon T-30-6 core. Tap 5 turns from ground.

L2—28t no. 28 enam wire on an Amidon T-30-2 core. Tap 7 turns from ground.
L3—47t no. 28 enam wire on an Amidon T-50-2 core. Tap 12 turns from ground.
L4—58t no. 28 enam wire on an Amidon T-50-1 core. Tap 14 turns from ground.
M1—200-μA meter.
Q1-Q4, Q6—2N5486 or 2N5484 JFET.

Q5—2N3866 or 2N5109 NPN transistor.
R16—Three 1-Ω ¼ or ½ W resistors in parallel.
S1—2-pole, 4-position rotary switch.
S2—SPDT center-off toggle switch.
T1—23t no. 26 enam wire on an Amidon FT-50-43 core. 1 turn no. 22 hookup wire secondary.

allow control of the power to the buffer amplifier.

The buffer amplifier is a broadband class-A amplifier with shunt and emitter feedback. T1 is used to match the 0.33-ohm load to the amplifier, as the load tends to swamp out the frequency dependence of the circuit.

A voltage-to-current amplifier (Q6) is used to determine the amount of RF present at J1. For best sensitivity, a small amount of current should always flow through the meter. Two meter shunts are used to select different Q ranges, and the 78L05 voltage regulator supplies Q6 with a constant voltage.

Construction

Construction of the Q/L meter is fairly straightforward. The parts-placement diagram is shown in Fig 45. The toroids must be wound carefully; the total number of turns on the coils is determined by counting the number of times the wire passes through the hole in the toroid.

Fig 44—This inductance meter features a direct readout over an inductance range from 0.12 μH to 1.8 mH.

Fig 46—Interior view of the inductance meter. An L-shaped bracket supports the variable capacitor.

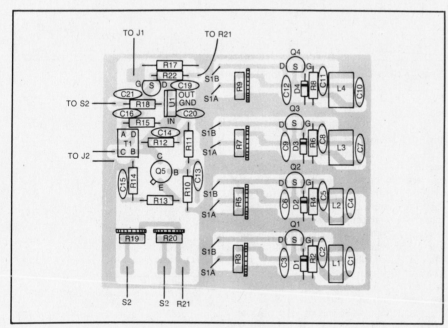

Fig 45—Parts-placement diagram for the inductance meter.

The most critical construction detail is the mounting of the capacitor and the terminals for the test inductor. See Fig 46. The ungrounded capacitor lead is connected to J1 with a piece of no. 14 wire.

The distance from the capacitor to J1 in the version shown here is 2¼ inches, and the dial layout shown in Fig 47 may be used if this terminal spacing is duplicated. It may be possible to position the capacitor and

test inductor closer together and use the given dial layout if a series compensation inductor is used. The accuracy of the dial shown here may differ because of minor differences in construction and components.

The capacitor is mounted three inches from the terminals on an L-shaped mounting bracket below a vernier drive. The 6-to-1 vernier drive is used to prevent the tuning rate from being too fast for high Q inductors. A piece of plastic scored with a sharp knife is used as a dial pointer. The cabinet used is a Ten-Tec SE-7.

Testing

An RF probe or oscilloscope should be used to verify that the oscillators are working. The peak-to-peak voltage at the input of the buffer amplifier should not exceed 0.40 V to allow the buffer amplifier to operate linearly. The potentiometers in the oscillator circuit, (R3, R5, R7 and R9), are used to adjust the oscillator output levels. They can be adjusted while measuring the oscillator output, or simply adjusted for an adequate meter deflection while measuring an inductor. If multiple peak indications are noted, the potentiometer is set too high and should be adjusted for less meter deflection. The oscillator output voltage depends somewhat on the FETs used; it may be as low as 0.20 V. R19, the high-Q shunt, should be set for a much lower meter reading than the low-Q shunt, R20.

Operation

After connecting the unknown inductor to the test jacks, the sensitivity control is adjusted for a small quiescent meter current between 5 and 25 microamperes. If this is not done, it may be impossible to find a peak, as the detector may be essentially turned off! The low-Q setting is most useful in finding the value of an unknown inductor. The capacitor is then adjusted for a peak meter reading. If no peak can be found, a different range should be tried.

Limitations

The meter will not be very accurate for inductances below 0.3 μH; at inductances this low, the measurements are greatly affected by lead length and parts layout. It may be possible to calibrate the unit in terms of Q; the meter is most useful as a measure of *relative* Q, however. The unit shown here was not calibrated for Q, as the meter reading is affected by the setting of the quiescent point. In addition, Q should always be measured at the frequency of interest, as it varies markedly with frequency. The inductance measurements are not absolute; it is not unusual for the inductance of a ferrite-core inductor to vary with frequency. Despite these limitations, this Q/L meter is quite useful in obtaining approximate inductance values.

OSCILLOSCOPES

Most engineers and technicians will tell you that the most useful single piece of test and design equipment is the modern triggered-sweep oscilloscope (commonly called just a "scope"). This is because not only can the oscilloscope measure voltage, but it can display voltage relative to time, showing the waveform pictures we are all familiar with from electronics textbooks. Over the past 10 years, the advances in semiconductor technology have brought the price of a good high-frequency scope within the affordable price range for most hams.

Fig 48 shows a simplified diagram of a triggered-sweep oscilloscope. The heart of the scope is the cathode ray tube (CRT) display. The CRT allows the visual display of an electronic signal by taking two electric signals and using them to deflect a beam of electrons. This beam is then deflected to a particular place on the phosphorescent screen of the CRT where it causes a small spot to glow. The particular location of the spot is determined by the value of the voltage applied to the vertical and horizontal inputs.

All the other circuits in the scope are used to take the real world signal and convert it to a form usable by the CRT. Let's take a look at how a signal travels through the oscilloscope circuitry. We will assume the trigger select switch is in the INTERNAL position (see Fig 48).

The first thing the input signal comes to is the input coupling switch. The switch allows the coupling of low-level ac signals that are superimposed on a large dc signal (this is often the case in audio amplifiers). In the AC position the dc component is blocked from reaching the vertical amplifier chain and saturating it (this would

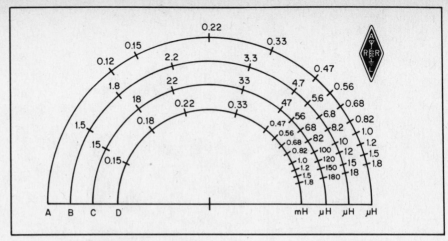

Fig 47—Full-size dial template.

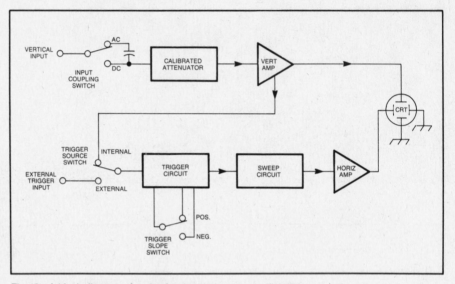

Fig 48—A block diagram of a simple triggered-sweep oscilloscope.

cause distortion that would make the displayed ac signals meaningless). It is important to note that at frequencies below 30 Hz it is not advisable to use ac coupling because the value of the blocking capacitor represents a considerable series impedance to very low-frequency signals. After the coupling switch, the signal next comes to a calibrated attenuator which is used to reduce the signal to a level that can be tolerated by the scope's vertical amplifier. The vertical amplifier amplifies the signal to a level that can drive the CRT and also adds a bias component to properly position the waveform on the screen.

A small sample of the signal from the vertical amplifier is sent to the trigger circuitry. The trigger circuit feeds a start pulse to the sweep generator when the input signal reaches a certain level. The sweep generator gives a precisely timed signal that looks like a triangle (see Fig 49). This triangular signal causes the scope trace to sweep from left to right, with the zero voltage point representing the left side of the screen and the maximum voltage representing the right side of the screen.

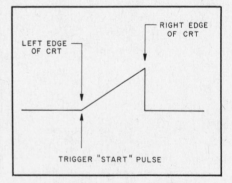

Fig 49—The sweep circuit produces a ramp pulse that is used to sweep the CRT electron beam from side to side.

The sweep circuit feeds the horizontal amplifier, which in turn drives the CRT. It is also possible to trigger the sweep system from an external source (such as the system clock in a digital system). This is done by using an external input jack with the trigger select switch in the EXTERNAL position.

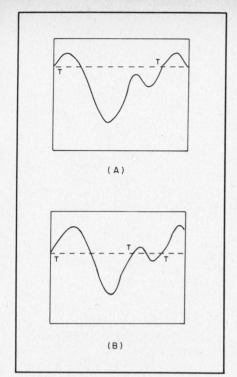

(A)

(B)

Fig 50—Selection of the trigger point is very important. Selecting the trigger point in A produces a stable display, but the trigger shown at B will produce a display that "jitters" from side to side.

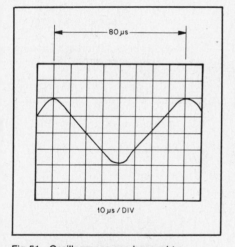

10 μs / DIV

Fig 51—Oscilloscopes can be used to measure frequency. Here the displayed wave has a period of 80 microseconds (8 divisions × 10 microseconds per division) and a frequency of 12.5 kHz (1/80 μs).

An explanation of the operation of the trigger system may be useful at this point. Basically, the trigger system looks at the reference signal (internal or external) to determine if it is positive- or negative-going and to see if the signal has passed a particular level. Fig 50A shows a typical signal, the dotted line on the figure representing the trigger level. If we assume that the trigger slope switch is set to the POSITIVE position it is obvious that the only points where a trigger will occur are the points

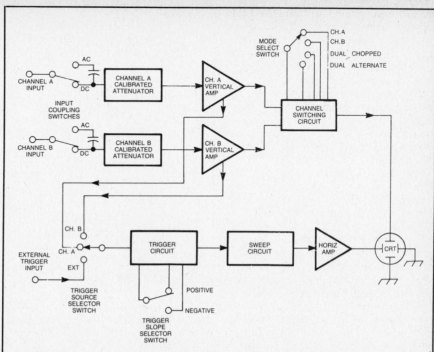

Fig 52—Block diagram of a modern dual-trace oscilloscope.

marked "T." It is important to note that once a trigger circuit is "fired" it cannot fire again until the sweep has completed one complete traversal of the screen from left to right. It is also important that the trigger level be chosen properly so the display "locks" in place.

Fig 50B shows a situation where the level has not been properly selected. Because there are two points during a single cycle of the waveform that meet the triggering requirements, the trigger circuit will have a tendency to jump from one trigger point to another. This will cause the waveform to appear to jitter from left to right. Simple adjustment of the trigger level control will solve this problem.

Up to now we have not discussed what the travel across the horizontal axis represents. This axis is calibrated in units of time. This is especially helpful because knowing the time of one cycle can give us the frequency of the waveform. In Fig 51, for example, if we know that the sweep speed selector is set at 10 μs/division and we count the number of divisions (vertical bars) between peaks of the waveform (or any similar well-defined points that occur once per cycle) we can determine the period of one cycle. In this case it is 80 μs. This means that the frequency of the waveform is 12,500 Hz (1/80 μs). The accuracy of the measured frequency depends on the accuracy of the scope's sweep oscillator (usually approximately 5%) and the linearity of the ramp generator. Obviously, this accuracy cannot compete with even the least expensive frequency counter, but the scope can still be used to determine that a circuit is functioning properly.

Unfortunately, if we wanted to compare two signals (like the input and the output of an amplifier) we would have to alternately connect the scope input to each of the two points to be measured. This inconvenience can be solved by using a *dual-trace* oscilloscope. This scope has two vertical input channels which can be displayed either individually or simultaneously.

Fig 52 shows a simplified block diagram of a dual-trace triggered-sweep oscilloscope. The only differences between this scope and our previous example are the additional vertical amplifier and the "channel switching circuit." This block selects whether we display channel A, channel B or both (simultaneously). The dual display is not a true dual display (there is only one electron gun in the CRT) but the dual traces are synthesized in the scope. True dual-trace dual-beam scopes are rare and very expensive.

There are two methods of synthesizing a dual-trace display from a single-beam scope. These two methods are referred to as "chopped mode" and "alternate mode." In the chopped mode a small portion of the channel A waveform is written to the CRT, then a corresponding portion of the channel B waveform is written to the CRT. This procedure is continued until both waveforms are completely written on the CRT. The chopped mode is especially useful where an actual measure of the time difference between the two waveforms is required. The chopped mode is usually most useful on slow sweep speeds (times greater than a few microseconds per division).

In the alternate mode, the complete

channel A waveform is written to the CRT followed immediately by the complete channel B waveform. This occurs so quickly that it appears that the waveforms are displayed simultaneously. This mode of operation is used to measure the absolute phase difference between two signals by triggering both sweeps from the same vertical input signal. Unfortunately, this mode of operation is not useful at very slow sweep speeds, but is good at most other sweep speeds.

Most dual-trace oscilloscopes also have a feature called "X-Y" operation. This feature allows one channel to drive the horizontal amplifier of the scope (called the X channel) while the other channel (called Y in this mode of operation) drives the vertical amplifier. Some oscilloscopes also have an external Y input. X-Y operation allows the scope to display Lissajous patterns for frequency and phase comparison, and to use specialized test adapters such as curve tracers, spectrum analyzer front ends, etc. Because of frequency limitations of most scope horizontal amplifiers the X channel is usually limited to a 5- or 10-MHz bandwidth.

The oscilloscope has limits, like every other device. For most purposes the voltage range of the scope can be expanded by the use of appropriate probes (which we will cover later). The most important limiting factor for most oscilloscopes is the *frequency response* (also called the *bandwidth*) of the oscilloscope. At the specified maximum response frequency, the response will be down 3 dB (0.707 voltage). For example, if we put a 100-MHz 1-V sine wave into a 100-MHz bandwidth scope we will read approximately 0.707 V on the scope display. The same scope at frequencies below 30 MHz (down to dc) should be accurate to about 5%.

Directly related to bandwidth is a parameter called *rise time*. This term refers to the fact that if a very sharp and square waveform is put into a scope input the display will appear to take a finite amount of time to rise to a specified fraction of the input voltage level. The rise time is usually defined as the time required for the display to show a transition from the 10% to 90% points of the input waveform, as shown in Fig 53. The mathematical definition of rise time is given by:

$$t_r = \frac{0.35}{BW}$$

where

t_r = rise time (μs)
BW = bandwidth (MHz)

It is also important to note that all but the most modern (and expensive) scopes are not designed for precise measurement of either time or frequency. At best they will not have better than 5% accuracy in these applications. This does not diminish the usefulness of even a moderately priced oscilloscope, however.

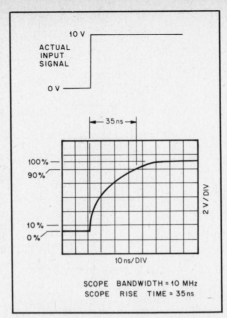

Fig 53—Oscilloscope bandwidth has a limiting effect on the rise time of the signals displayed on the scope.

Oscilloscope Probes

Connecting the signal to be measured to the input of the oscilloscope requires some form of electrical connection, preferably a shielded connection. Your first thought would probably lead you to a piece of small-diameter coax cable. Unfortunately, as we go higher in frequency the capacitance of the cable would produce a capacitive reactance considerably lower than the one-megohm input impedance of the oscilloscope. In addition, the scope has a certain built-in capacitance at its input terminals (usually between 5 and 35 pF). These two capacitances cause problems if we are probing a circuit with a relatively high impedance.

The simplest method of connecting the signal to the scope is to use a specially designed probe. The most common scope probe is a "×10" probe (called a "times ten probe"). This probe forms a 10 to 1 voltage divider using the built-in resistance of the probe and the input resistance of the scope. When you use a times ten probe you must multiply the value of the calibrated attenuator by 10. For example, if your scope is on the 1-V/div range and you were using an ×10 probe the actual readings on the scope face would be 10 V/div.

Unfortunately, if we just put a resistor in series with the scope input it seriously degrades the scope's rise-time performance and therefore its bandwidth. This is because the scope's input looks like a parallel RC circuit and the series resistor feeding it causes a significant reduction in available charging current from the source. We can correct this by using a compensating capacitor in parallel with the series resistor. This in fact forms two dividers: one resistive voltage divider and one capacitive voltage divider. With these two dividers connected in parallel with the RC relationships shown in Fig 54, the probe and scope should have a flat response curve through the whole bandwidth of the scope.

To account for manufacturing tolerances in the scope and the probe the com-

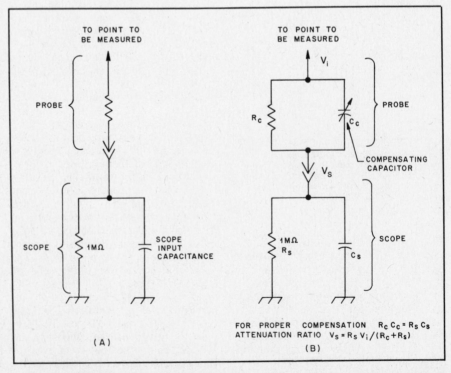

Fig 54—The probe at A is uncompensated; the compensating capacitor in the probe shown at B can be used to "tweak" the probe to compensate for circuit capacitances (see text for details).

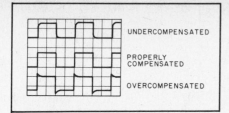

Fig 55—Oscilloscope displays showing a squarewave input and undercompensated, overcompensated and properly compensated probes. Refer to the text for a discussion of the figure.

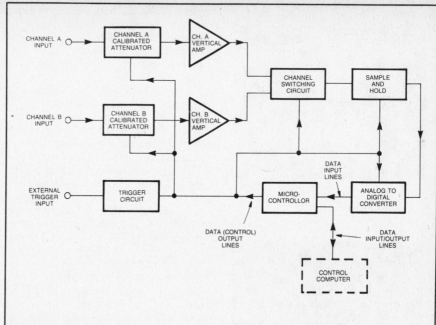

Fig 56—Block diagram of a dual-trace digital oscilloscope.

pensating capacitor is made variable and is used to "tweak" the probe and scope for proper response when they are connected together. Most scopes provide a "calibrator" output that produces a known-frequency square wave for the purpose of adjusting the compensating capacitor in a probe. Fig 55 shows possible responses when the probe is connected to the oscilloscope's calibrator jack. We can see that the leading edge shown in the top and bottom traces of Fig 55 is incorrect. If you get these pictures from the calibrator jack, adjust the compensating capacitor until you get a perfectly square display as is shown in the middle trace of the figure.

If the probe cable is too short, do not attempt to extend the length of the cable by adding a piece of common coaxial cable. There are two reasons for this. First, the compensating capacitor is chosen to compensate for the provided length of cable. It usually will not have enough range to compensate for extra lengths. Additionally, the cable used to make a scope probe is of specialized construction and electrical characteristics. Its requirements are significantly different than normal coax cable.

Another important point in scope probe use is to use the shortest ground lead possible. Long ground leads are inductors at high frequencies. In these circuits they cause ringing and other undesirable effects.

Digital Oscilloscopes

Up to now we have discussed versions of the classic oscilloscope that has existed for over 30 years. As we have noted, the cost of the oscilloscope has dropped with the widespread use of semiconductor technology. This is not the only change that semiconductors have brought to the oscilloscope market, however. In the past five years the *digital oscilloscope* has appeared in laboratories around the world. This type of oscilloscope has quickly gone from a laboratory oddity to an affordable tool in reach of the active experimenter. The design of a digital oscilloscope relies heavily on microcomputer technology, and the prominence and affordability of digital scopes has roughly paralleled the microcomputer market. We will not spend significant space on this type of scope only because almost any details we give will be

quickly overtaken by technological advances. A basic primer may be of interest, however.

A block diagram of the basic digital oscilloscope is given in Fig 56. The only thing we have previously not seen are the blocks labeled *sample and hold* (S/H), *analog to digital converter* (ADC) and *microcontroller*. These blocks take the analog signal and convert it to a sequence of digital bits which can be used to represent the waveform. Processing the signals in this way means that the information can then be passed to a computer where it can be stored (on floppy disks) or analyzed using advanced mathematical techniques (such as Fast Fourier Transforms). In fact, most digital oscilloscopes that are in the price range of the

experimenter do not even have a CRT; they use the computer terminal to display their result. Measurement parameters are adjusted via the computer keyboard, rather than with front-panel knobs and switches. Fig 57 shows a picture of a digital oscilloscope of this type. In the parlance of the engineer they are sometimes referred to as *waveform digitizers*, or *scope front ends*.

Referring to the block diagram, we can discuss the operation of the digital oscilloscope. Like most digital devices, the digital scope has a microprocessor "brain." The microprocessor commands the sample and hold circuit to sample the amplitude of the input signal at regular intervals. The sample and hold circuit then holds this value while the analog to digital converter converts the sampled value to its digital

Fig 57—A digital "oscilloscope front end" for a personal computer. Other digital oscilloscopes take the form of a circuit board that plugs into an expansion slot in the computer (photo courtesy of Heath Co).

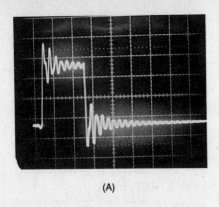

(A)

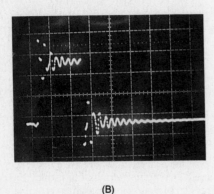

(B)

Fig 58—Two displays of the same waveform. The display at A was produced with an analog storage scope. The display at B was produced with a digital oscilloscope. Notice that the digital display is not continuous, leaving the actual shape of the rising and falling edges open to interpretation.

equivalent. This digital equivalent and information representing the time at which the sample was taken can either be sent to internal RAM for later use or can be sent out to the host microcomputer. The whole process is repeated a number of times until the waveform is duplicated.

Unfortunately, if the sample rate is not high enough, very fast signal perturbations between sampling points do not appear on the display. For example Fig 58A and 58B show the same signal as measured on an analog and digital scope. As you can see, there are large spikes in the analog scope that do not seem to show up on the digital scope. Some of the more expensive digital scopes use special methods to overcome this problem, but these methods are unlikely to come to lower priced units for a few years.

Some digital oscilloscopes are now appearing on the market in a portable form; these scopes employ a liquid-crystal display (LCD) very similar to that used on many lap-top computers. This type of digital scope requires no outboard computer and is usually battery operated. It is important to note that these scopes display the waveform as a series of unconnected dots. In certain very complex waveforms this

may cause some difficulty in interpretation. Also, as a consequence of other circuitry restrictions these hand-held scopes usually have very limited bandwidth (in the 100 to 500 kHz range) which means they are only useful in audio and slow digital work. Undoubtedly as technology advances, the performance of these scopes will improve considerably.

As with all new technology items, specifications are sometimes misleading because of lack of standard definitions for terms. If you are considering buying a digital scope we recommend you either attempt to see the unit you are interested in in operation or consult with a local ham in the electronics business who can offer some experienced advice.

Storage Oscilloscopes

Storage scopes are now showing up on the surplus markets at prices available to many amateurs. Storage scopes are analog instruments with special long-persistence phosphor CRTs that allow them to capture single-shot events. For example, if you wanted to determine the switching time of an antenna sequencer you would trigger the scope from the control signal input to the sequencer and monitor the contacts with the vertical input to determine when the relay had switched. With a regular oscilloscope the trace would be gone from the CRT before you could determine the switching time, but with the storage scope's long-persistence CRT for a reasonable time (15 to 120 seconds) you can observe the switching action before the trace fades from the screen.

A newer version of the storage scope is the *digital storage oscilloscope* (DSO). The DSO uses a high-speed semiconductor memory which is continually read back to the display so that the waveform can be viewed for as long as necessary. The DSO is a relatively straightforward extension of the digital oscilloscopes we have previously discussed.

Lissajous Figures

When sinusoidal ac voltages are applied to both sets of deflecting plates in the oscilloscope, the resultant pattern depends on the relative amplitudes, frequencies and phases of the two voltages. If the ratio between the two frequencies is constant and can be expressed in integers, a stationary pattern will be produced.

The stationary patterns obtained in this way are called Lissajous figures. Examples of some of the simpler Lissajous figures are given in Fig 59. The frequency ratio is found by counting the number of loops along two adjacent edges. Thus in C there are three loops along a horizontal edge and only one along the vertical, so the ratio of the vertical frequency to the horizontal frequency is 3:1. Similarly, in E there are four loops along the horizontal edge and three along the vertical edge, giving a ratio of 4:3.

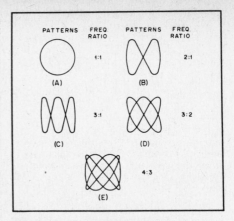

Fig 59—Lissajous figures and corresponding frequency ratios for a 90-degree phase relationship between the voltages applied to the two sets of deflecting plates.

Assuming that the known frequency is applied to the horizontal plates, the unknown frequency is

$$f2 = \frac{n2}{n1} f1$$

where

f1 = known frequency applied to horizontal plates

f2 = unknown frequency applied to vertical plates

n1 = number of loops along a vertical edge

n2 = number of loops along a horizontal edge.

An important application of Lissajous figures is in the calibration of audio-frequency signal generators. For very low frequencies the 60-Hz power-line frequency is held accurately enough to be used as a standard in most localities. The medium audio-frequency range can be covered by comparison with the 440- and 600-Hz modulation on WWV transmissions. It is possible to calibrate over a 10:1 range, both upward and downward from these frequencies and thus cover the audio range useful for voice communication.

An oscilloscope having both horizontal and vertical amplifiers is desirable, since it is convenient to have a means for adjusting the voltages applied to the deflection plates to secure a suitable pattern size.

Oscilloscope Bibliography

vanErk, R., *Oscilloscopes, Functional Operation and Measuring Examples*, McGraw-Hill Book Co, New York 1978.

Bunze, V., "Probing in Prospective-Application Note 152," Hewlett-Packard Co, Colorado Springs, CO 1972 (Pub No. 5952-1892)

The XYZs of Using a Scope, Tektronix, Inc, Portland, OR 1981 (Pub No. 41AX-4758)

"Basic Techniques of Waveform Measurement (Parts 1 and 2)," Hewlett-Packard Co, Colorado Springs, CO 1980 (Pub No. 5953-3873)

Millman, J. and Taub, H., *Pulse Digital and Switching Waveforms*, McGraw-Hill Book Co, New York 1965, pp 50-54.

Martin, V., "A Detailed Look at Probes," *Ham Radio*, September 1985, pp 75-86.

Operation and Service Manual for Model IC-4802 Computer Oscilloscope, Heath Co, Benton Harbor, MI 1986 (Pub No. 595-3458-1)

HF ADAPTER FOR NARROW-BANDWIDTH OSCILLOSCOPES

Fig 60 shows the circuit of a simple piece of test equipment that will allow you to display signals that are beyond the normal bandwidth of an inexpensive oscilloscope. This circuit was built to monitor modulation of a 10-meter signal on a scope that has a 5-MHz upper-frequency limit. This design features a Mini-Circuits Laboratory SRA-1 mixer. Any stable oscillator or VFO with an output of 10 dBm can be used for the local oscillator (LO), which mixes with the HF signal to produce an IF in the range of the oscilloscope.

The mixer can handle RF signal levels up to −3 dBm without clipping, so this was set as an upper limit for the RF input. A toroidal-transformer coupler is constructed by winding a 31-turn secondary of no. 28-AWG wire on a 3E2A core, which has a 0.038-inch diameter. The Amidon FT-37-75 is an equivalent core. The primary is a piece of coaxial cable through the core center. The coupler gives 30-dB attenuation and has a flat response from 0.5 to 100 MHz. An additional 20-dB attenuation was added for a total of 50 dB before the mixer. One-watt resistors will do fine for the attenuator. The completed adapter should be built into a shielded box.

This circuit, with the 25-MHz LO frequency, is useful on frequencies in the 20- to 30-MHz range with transmitters of up to 50-W power output. By changing the frequency of the LO, any frequency in the range of the coupler can be displayed on a 5-MHz-bandwidth oscilloscope. More attenuation will be required for higher-power transmitters. This circuit was described by Kenneth Stringham Jr., AE1X in the Hints and Kinks columm in February 1982 *QST*.

A GATED NOISE SOURCE

The circuit of Fig 61 may be used to construct a simple low-cost device to optimize a converter for best noise figure. The simplicity of this system makes effective alignment possible without a lot of test equipment.

Numerous articles have described units where noise-figure tests may be made. With the exception of certain thermal-limited diodes (5722, for example), an absolute value of noise figure is not obtainable with these units; this device is no exception.

Anyone using a classic noise-figure meter soon learns that the tune-up of a system is a cut-and-try procedure where an adjustment is made and its influence is observed by calibrating the system. Then the excess-noise source is applied and the effect evaluated. This is basically an after-the-fact method of testing after an adjustment is made, and is consequently time consuming.

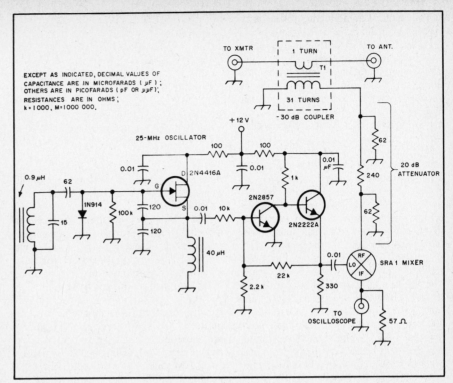

Fig 60—Schematic diagram of an adapter to display RF signals on a narrow-bandwidth oscilloscope, including a 10-dBm 25-MHz LO, −30-dB coupler, 20-dB attenuator and diode-ring mixer.
T1—See text.

The gated noise source doesn't require a special detector, or any detector at all other than your ear. By turning the noise source on and off at an audio rate, the ratio of noise contributed by the system to noise of the system plus excess noise appears as an audio note. The louder the note, the greater the differential in levels. Hence, the greater the influence of the excess noise, the better the noise figure.

If greater precision is desired than that obtained by subjectively listening to the signal, an oscilloscope may be used. Connect the scope vertical input to any point in the audio system of the receiver, such as the speaker terminals. Adjust the scope for a display of several multiples of the train of square pulses. Proceed by adjusting the device(s) being tested for greatest vertical deflection.

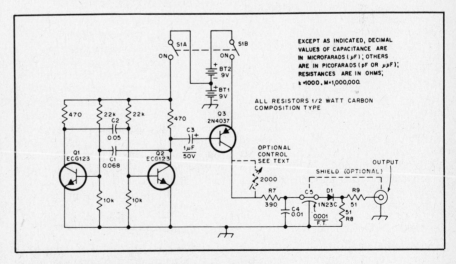

Fig 61—Schematic diagram of the gated noise source.
BT1, BT2—9-volt battery, Eveready 216 or equiv.
C5—0.001-μF feedthrough capacitor, Sprague BH-340.
S1—Double-pole, single-throw miniature toggle.

The result of an adjustment is instantly visible as an increase or decrease in the recovered audio. This method of noise evaluation is by no means new. Most modern automatic noise-figure meters turn the excess-noise source on and off and then, through rather sophisticated methods, evaluate the results. This technique is sometimes called Y-parameter testing.

While the method and circuit described here are not exceptional, they represent a fresh approach to noise evaluation. This approach does not require long-term integrating detectors and tedious "twice-power" measurements that, without absolute calibration, can result in no more than simply optimizing the system.

In some cases the available noise generated by this unit may be too great. The output may be reduced by inserting attenuators between the generator output and the device under test or by adding a 2000-ohm potentiometer at the point marked in Fig 61. The use of an attenuator is preferred because it reduces the apparent output VSWR of the generator by increasing the return loss. If a control is used it must be returned to its minimum insertion-loss position when starting a test or no signal may be heard.

This circuit uses readily available junk-box parts and may be easily duplicated. The lead placement in and around the diode itself should follow good VHF practices with short leads and direct placement.

Theory of Operation

Q1 and Q2 are used in a cross-coupled multivibrator circuit, operating at approximately 700 Hz. The value of C1 is greater than C2 to cause the duty cycle to favor the conduction of Q2 slightly. When Q2 conducts, the pulse is coupled to Q3 via C3, turning on Q3 and causing current flow through R7, D1 and R8.

The diode generates broadband noise that is passed through R9 to the output. R7, C4 and C5 form a low-pass filter to prevent high-order harmonics of the switching pulses from appearing in the output.

The influence of stray RF signals entering the device under test through the generator may be minimized by shielding the components shown. A simple box may be built by using PC-board scraps. For best match, this source should be connected directly to the input of the device under test; therefore, the unit is equipped with a male connector. This matching becomes a greater consideration as the frequency of interest increases.

Operation on Other Than AM

Some contemporary receivers and transceivers cannot be operated in the AM mode, and consequently the noise source seems unusable. The detection of noise is the process by which the noise source operates; therefore, it will not work through an FM detector, nor will it work

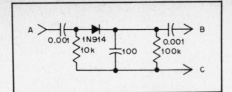

Fig 62—A simple detector that can be used when aligning SSB and FM receivers (see text for details).

through a product detector since one of the terms of the detection (the noise) is not coherent.

The SCOPE jack on most receivers is loosely coupled to the IF amplifier, preceding the detector. A wide-band scope connected to this point will show the train of pulses and eliminate the need for aural detection. The alignment of the later IF stages of a system should have the least impact on the noise performance, and maximum signal response will always occur at the same setting. Therefore, the simple detector of Fig 62 will generally work for aural AM detection. Connect point A to the last IF amplifier plate or collector. Connect point B to the audio amplifier, at or near the volume control, and ground point C. With this arrangement the normal detector output is turned down with the volume control, and the temporary detector provides AM detection.

The gated noise source has been used for literally hundreds of applications and has proved to be a powerful yet simple addition to the test bench. While no guarantee of duplication may be made, these units develop approximately 18 dB of excess noise in the region of 50-300 MHz. This unit was originally described by Hartsen in January 1977 *QST*.

NOISE BRIDGE FOR 160 THROUGH 10 METERS

The noise bridge, sometimes referred to as an antenna (RX) noise bridge, is an instrument that will allow the user to measure the impedance of antennas or other electrical circuits. The unit described here, designed for use in the 160- through 10-meter range, provides adequate accuracy for most measurements. Battery operation and small physical size make this unit ideal for remote-location use. Tone modulation is applied to the wide-band noise generator as an aid for obtaining a null indication. A detector, such as the station receiver, is required for operation of the unit.

The Circuit

The noise bridge consists of two parts—the noise generator and the bridge circuitry. See Fig 64. A 6.8-volt Zener diode serves as the noise source. U1 generates an approximate 50-percent duty cycle, 1000-Hz, square-wave signal, which is applied to the cathode of the Zener diode. The 1000-Hz

Fig 63—Interior and exterior views of the noise bridge. The unit is finished in red enamel. Press-on lettering is used for the calibration marks. Note that the potentiometer must be isolated from ground.

modulation appears on the noise signal and provides a useful null-detection enhancement effect when used with an AM receiver. The broadband-noise signal is amplified by Q1, Q2 and associated components to a level which produces an approximate S-9 signal in the receiver. Slightly more noise is available at the lower end of the frequency range, as no frequency compensation is applied to the amplifier. Roughly 20 mA of current is drawn from the 9-volt battery, thus ensuring long-battery life—as long as the power is switched off after use!

The bridge portion of the circuit consists of T1, C1, C2 and R1. T1 is a trifilar-wound transformer with one of the windings used to couple noise energy into the bridge circuit. The remaining two windings are arranged so that each one is in an arm of the bridge. C1 and R1 complete one arm and the UNKNOWN circuit along with C2 comprise the remainder of the bridge. The terminal labeled RCVR is for connection to the detector.

Construction

The noise bridge is contained in a homemade aluminum enclosure that measures 5 × 2-3/8 × 3-3/4 inches. Many of the circuit components are mounted on a circuit board that is fastened to the rear wall of the cabinet. The circuit-board layout is such that the lead lengths to the board from the bridge and coaxial connectors are at a minimum. A parts-placement guide is shown in Fig 65.

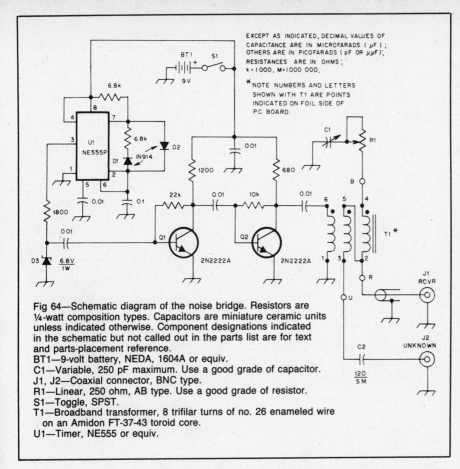

EXCEPT AS INDICATED, DECIMAL VALUES OF CAPACITANCE ARE IN MICROFARADS (μF); OTHERS ARE IN PICOFARADS (pF or μμF); RESISTANCES ARE IN OHMS; k=1000, M=1000 000.

*NOTE: NUMBERS AND LETTERS SHOWN WITH T1 ARE POINTS INDICATED ON FOIL SIDE OF PC BOARD.

Fig 64—Schematic diagram of the noise bridge. Resistors are ¼-watt composition types. Capacitors are miniature ceramic units unless indicated otherwise. Component designations indicated in the schematic but not called out in the parts list are for text and parts-placement reference.
BT1—9-volt battery, NEDA, 1604A or equiv.
C1—Variable, 250 pF maximum. Use a good grade of capacitor.
J1, J2—Coaxial connector, BNC type.
R1—Linear, 250 ohm, AB type. Use a good grade of resistor.
S1—Toggle, SPST.
T1—Broadband transformer, 8 trifilar turns of no. 26 enameled wire on an Amidon FT-37-43 toroid core.
U1—Timer, NE555 or equiv.

Care must be taken when mounting the potentiometer. For accurate readings, the potentiometer must be well insulated from ground. In the unit shown this was accomplished by mounting the control to a piece of Plexiglas, which in turn was fastened to the chassis with a piece of aluminum angle stock. Additionally, a ¼-inch control-shaft coupling and a length of phenolic rod were used to further isolate the control from ground where the shaft passes through the front panel. A high-quality potentiometer is a must if good measurement results are to be obtained.

Mounting the variable capacitor is not a problem since the rotor is grounded. As with the potentiometer, a good grade of capacitor is important. If you must cut corners to save money, look elsewhere in the circuit. Two BNC-type female coaxial fittings are provided on the rear panel for connection to a detector (receiver) and to the UNKNOWN circuit. There is no reason other types of connectors can't be used. One should avoid the use of plastic insulated phono connectors, however, as these might influence accuracy at the higher frequencies. As can be seen from the photograph, a length of miniature coaxial cable (RG-174) is used between the RCVR connector and the appropriate circuit board foils. Also, C2 has one lead attached to the circuit board and the other connected directly to the UNKNOWN circuit connector.

Calibration and Use

Calibration of the bridge is straightforward and requires no special instruments. A receiver tuned to any portion of the 15-meter band is connected to the RCVR terminal of the bridge. The power is switched on and a broadband noise with a 1000-Hz note should be heard in the receiver. Calibration of the resistance dial should be performed first. This is accomplished by inserting small composition resistors of appropriate values across the UNKNOWN connector of the bridge. The resistors should have the shortest lead lengths possible in order to mate with the connector. Start with 25 ohms of resistance (this may be made up of series- or parallel-connected units). Adjust the capacitance and resistance dials for a null of the signal as heard in the receiver. Place a calibration mark on the front panel at that location of the resistance dial. Remove the 25-ohm resistor and insert a 50-ohm resistor, a 100-ohm unit and so on until the dial is completely calibrated.

The capacitance dial is calibrated in a similar manner. Initially, this dial is set so that the plates of C1 are exactly half meshed. If a capacitor having no stops is used, orient the knob so as to unmesh the plates when the knob is rotated into the positive-capacitance region of the dial. A 50-ohm resistor is connected to the UNKNOWN terminal and the resistance control is adjusted for a null. Next, the reactance dial is adjusted for a null and its position is noted. If this setting is significantly different than the half-meshed position, the value of C2 will need to be changed. Unit-to-unit value variations of 120-pF capacitors may be sufficient to provide a suitable unit. Alternatively, other values can be connected in series or parallel and tried in place of the 120-pF capacitor. The idea is to have the capacitance dial null as close as possible to the half-meshed position of C1.

Once the final value of C2 has been determined and the appropriate component installed in the circuit, the bridge should be adjusted for a null. The zero-reactance point can be marked on the face of the unit.

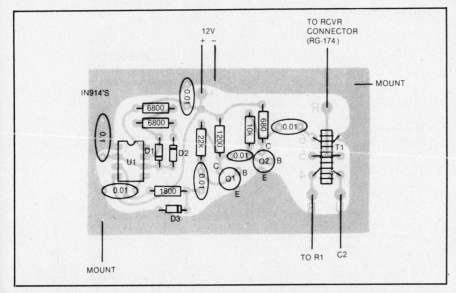

Fig 65—Parts-placement guide for the noise bridge as viewed from the component side of the board. A full-size etching pattern can be found at the back of this book.

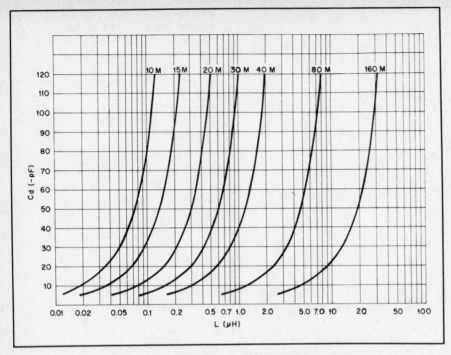

Fig 66—Graph for determining actual inductance from the calibration marks on the negative portion of the dial. These curves are accurate only for bridges having 120 pF at C2.

The next step is to place a 20-pF capacitor in series with the 50-ohm load resistor. Use a good grade of capacitor such as a silver-mica type, and keep the leads as short as possible. Null with the capacitance dial and make a calibration mark at that point. Remove the 20-pF capacitor and insert a 40-pF unit in series with the 50-ohm resistor. Again null the bridge and make a calibration mark for 40 pF. Continue in a similar manner until that half of the dial is calibrated.

To calibrate the negative half of the scale, the same capacitors may be used. This time they must be placed temporarily in parallel with C2. Connect the 50-ohm resistor to the UNKNOWN terminal and the 20-pF capacitor in parallel with C2. Null the bridge and place a calibration mark on the panel. Remove the 20-pF unit and temporarily install the 40-pF capacitor. Again null the bridge and make a calibration mark at that point. It should be pointed out that the exact resistance and capacitance values used for calibration can be determined by the builder. If resistance values of 20, 40, 60, 80, 100 ohms and so on are more in line with the builder's needs, the scale may be calibrated in these terms. The same is true for the capacitance dial. The accuracy of the bridge is determined by the components that are used in the calibration process.

Many amateurs use a noise bridge simply to find the resonant (nonreactive) impedance of an antenna system. For this service it is necessary only to calibrate the zero-reactance point of the capacitance dial. This simplification relaxes the stringent quality requirement for the bridge capacitors, C1 and C2.

Operation

The resistance dial is calibrated directly in ohms, but the capacitance dial is calibrated in picofarads of capacitance. The +C half of the dial indicates that the load is capacitive, and the −C portion is for inductive loads. To find the reactance of the load when the capacitance reading is positive, the dial setting must be applied to the standard capacitive reactance formula

$$X = \frac{1}{2\pi fC}$$

The result will be a capacitive reactance.

Inductance values corresponding to negative capacitance dial readings may be taken from the graph of Fig 66. The reactance is then found from the formula

$$X = 2\pi fL$$

When using the bridge, remember that the instrument measures the impedance of loads as connected at the UNKNOWN terminal. This means that the actual load to be measured must be directly at the connector rather than being attached to the bridge by a length of coaxial cable. Even a short length of cable will transform the load impedance to some other value. Unless the electrical length of line is known and taken into account, it is necessary to place the bridge at the load. An exception to this would be if the antenna were to be matched to the characteristic impedance of the cable. In this case the bridge controls may be preset for 50-ohms resistance and 0-pF capacitance. With the bridge placed at any point along the coaxial line, the load

(antenna) may be adjusted until a null is obtained. If the length of line is known to be an even multiple of a half-wavelength at the frequency of interest, the readings obtained from the bridge will be accurate.

Interpreting the Readings

A few words on how to interpret the measurements are in order. For example, assume the impedance of a 40-meter inverted-V antenna fed with a half-wavelength of cable was measured. The antenna had been cut for roughly the center of the band (7.150 MHz), and the bridge was nulled with the aid of a receiver tuned to that frequency. The results were 45 ohms of resistance and 70 picofarads of capacitance. The 45-ohm resistance reading is as close to 50 ohms as would be expected for this type of antenna. The capacitive reactance calculates to be 318 ohms from the equation

$$X = \frac{1}{2\pi(7.15 \times 10^6)(70 \times 10^{-12})}$$

$$= 318 \; \Omega$$

When an antenna is adjusted for resonance, the capacitive or inductance reactance will be zero. The antenna in question is a long way from that mark. Since the antenna looks capacitive it is too short; wire should be added to each side of the antenna. An approximation of how much wire to add can be made by tuning the receiver higher in frequency until a point is reached where the bridge nulls with the capacitance dial at zero. The percentage difference between this new frequency and the desired frequency indicates the approximate percentage that the antenna should be lengthened. The same system will work if the antenna has been cut too long. In this case the capacitance dial would have nulled in the −C region, indicating an inductive reactance. This procedure will work for most antennas.

SIGNAL GENERATORS FOR RECEIVER TESTING

The oscillator shown in Figs 67 and 68 was designed for testing high-performance receivers. Parts cost for the oscillator has been kept to a minimum by careful design. While the stability is slightly less than that of a well-designed crystal oscillator, the stability of the unit should be adequate for measurements of most amateur receivers. In addition, the ability to shift frequency is important when dealing with receivers that have spurious responses. More importantly, LC oscillators with high-Q components often have superior phase noise performance compared to crystal oscillators, because of power limitations in the crystal oscillators (crystals are easily damaged by excessive power).

The circuit is a Hartley oscillator

followed by a class-A buffer amplifier. To keep the frequency of the oscillator somewhat independent of the supply voltage, a 5-V regulator is used. The amplifier is cleaned up by a seven-element Chebyshev low-pass filter which is terminated by a 6-dB attenuator. The attenuator insures that the filter works properly, even with a receiver that has an input impedance other than 50 ohms. A receiver designed to work with a 50-ohm system may not have a 50-ohm input impedance. The +4 dBm of output is adequate for most receiver measurements. It may even be too much for some receivers; the potential exists for damage to narrow crystal filters following amplifiers. It is assumed that the user will use a step attenuator to lower the output level to an appropriate value.

Construction

This unit was built in a box made of double-sided circuit board. The box used in the prototype has inside dimensions of 1 × 2.2 × 5 inches (HWD). The copper foil of the circuit board makes an excellent shield, while the fiberglass helps temperature stability. Capacitor C2 should be soldered directly across L1 to insure high Q. As with any RF circuit, the component leads should be kept short. While silver mica capacitors have slightly better Q, NPØ capacitors may offer better stability. Mounting the three inductors orthoganally (axis of the inductors 90 degrees from each other) reduced the 2nd order harmonic by 2 dB compared to an initial prototype.

Alignment and Testing

The output of the regulator should be +5 V. The output of the oscillator should

Fig 67—This LC oscillator for receiver measurements offers good performance at relatively low cost.

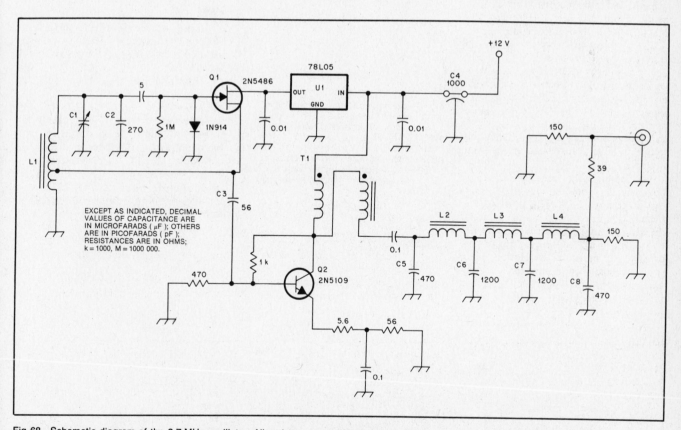

Fig 68—Schematic diagram of the 3.7-MHz oscillator. All resistors are ¼ W, 5% units. Except as indicated, decimal values of capacitance are in microfarads (μF); others are in picofarads (pF); resistances are in ohms; K = 1000, M = 1,000,000.

C1—1.4 to 9.2 pF air trimmer (value and type not critical)
C2—270 pF silver-mica or NPØ capacitor. Value may be changed slightly to compensate for variations in L1.
C3—56 pF silver mica or NPØ capacitor. Value may be changed to adjust output power.

C4—1000 pF solder-in feedthrough capacitor. Available from Microwave Components of Michigan (see Chapter 35). C5-C8—Silvermica, NPØ disc or polystyrene capacitor.
D1—1N914, 1N4148.
L1—31t no. 18 enam wire on an Amidon or Palomar Engineers T-94-6 core. Tap 8 turns

from ground end.
L2, L4—21t no. 22 enam wire on a T-50-2 core. Calculated value of these coils is 3.0 μH.
Q1—2N5486 JFET. MPF102 may give reduced output.
Q2—2N5109.
U1—78L05 low-current 5-V regulator.

be +4 dBm (2.5 milliwatts) into a 50-ohm load. Increasing the value of C3 will increase the power output to a maximum of about 10 milliwatts. The frequency should be around 3.7 MHz. Adding additional capacitance across inductor L1 (in parallel with C2) will lower the frequency if desired, while the trimmer capacitor (C1) specified will allow adjustment to a specific frequency. The drift of one of the prototypes was 5 Hz over 25 minutes after a few minutes of warm up. If the warm-up drift is large, changing C2 may improve the situation somewhat. For most receivers, a drift of 100 Hz during the measurement period has a negligible effect on measurements.

HYBRID COMBINERS FOR SIGNAL GENERATORS

Many receiver performance measurements require two signal generators to be attached simultaneously to a receiver. A combiner that isolates the two signal generators is necessary to keep one generator from being frequency or phase modulated by the other. Commercially made hybrid combiners are available from Mini-Circuits, PO Box 350166, Brooklyn NY 11235-0003. Hybrid combiners are not difficult to construct, however, as the following project shows.

The combiners described here provide 40 to 50 dB of isolation between ports while attenuating the desired signal paths (each input to output) by 6 dB. A second feature of these combiners is that of maintaining the 50-ohm impedance of the system—very important if accurate measurements are to be made.

The combiners are constructed in small boxes made from double-sided circuit-board material. Each piece is soldered to the adjacent one along the entire length of the seam. This makes an essentially RF-tight enclosure. BNC coaxial fittings are used on the units shown. However, any type of coaxial connector can be used. Leads must be kept as short as possible and precision resistors (or matched units from

the junk box) should be used. The circuit diagram for the combiners is shown in Fig 70.

Fig 69—Exterior view of the two hybrid combiners. The one on the left is designed to cover the 1 to 50 MHz range; the one on the right 50 to 500 MHz.

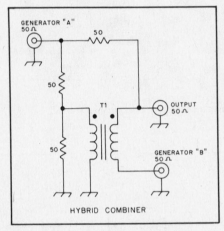

Fig 70—Schematic diagram of the hybrid combiners. For the 1- to 50-MHz model, T1 is 10 turns of no. 30 enameled wire bifilar wound on an FT-23-72 ferrite core. For the 50 to 500 MHz model, T1 consists of 10 turns of no. 30 enameled wire bifilar wound on an FT-23-63 ferrite core. Keep all leads as short as possible when constructing these units.

LOW-POWER STEP ATTENUATORS

A step attenuator is a useful and important tool for receiver testing. Step attenuators may often be found at flea

markets for reasonable prices; in addition, the construction of an attenuator is well within the capabilities of most builders. Described here is a simple low-power step attenuator suitable for receiver front-end protection, and as a calibrated attenuator for receiver performance evaluation. This attenuator uses double-pole, double-throw slide switches to select different amounts of attenuation. Coaxial fittings are used at each end of the attenuator.

Description

Fig 71 is the schematic diagram of the attenuator. Eight pi-network resistive sections are employed; the attenuation is variable in 1-dB steps. A total attenuation value of 81 dB is available with all the sections switched in. The maximum attenuation of any single section is limited to 20 dB because leak-through would probably degrade the effect of higher attenuation sections and result in inaccuracy.

This is a low-power attenuator; it is not designed for use at power levels exceeding ¼ watt. If for some reason the attenuator will be connected to a transceiver, a means of bypassing the unit during transmit periods must be devised.

Parts

All the switches are DPDT standard size slide types. Stackpole S-5022CD03-0 units are used here. Other switch types may work as well, but have not been tested. The use of subminiature switches should be avoided.

Carbon-composition or film, ¼-watt, 5%-tolerance resistors are used. Ideally, the resistors should be selected using a reliable ohmmeter; this will ensure accuracy.

Double-sided PC board is used for the enclosure. Dimensions for the model described here are given in Fig 72. The attenuator has identification lettering etched into the top surface (or front panel) of the unit. This adds a nice touch, and is a permanent means of labeling. Of course, rub-on transfers or Dymo tape labels could be used as well.

Female BNC single-hole, chassis-mount connectors are used at each end of the

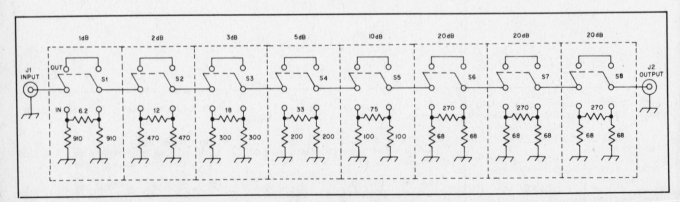

Fig 71—Schematic diagram of the attenuator. Resistors are ¼-W, 5%-tolerance, carbon-composition or film types. Resistances are given in ohms.

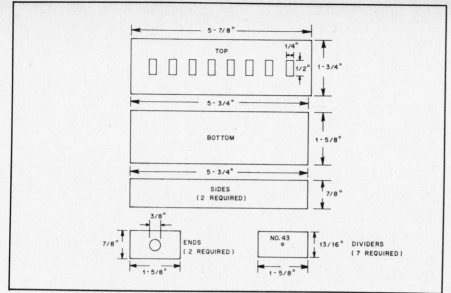

Fig 73—Close-up of the switch detail.

Fig 72—Mechanical dimensions of the attenuator enclosure.

enclosure. These connectors are small and easy to mount, have excellent RF qualities, and provide a means of easily connecting and disconnecting the attenuator by a simple twist of the wrist.

Construction

After all the box parts are cut to size and the necessary holes are made, scribe light lines to locate the inner partitions. Carefully tack-solder all partitions in position. Dress any PC-board parts that do not fit squarely. Once everything is in proper position, run a solder bead all the way around the joints. Caution: Do not use an excessive amount of solder, as the switches must later be fit flat inside the sections. The top, sides, ends and partitions can be completed. Dress the outside of the box to suit your taste. For instance, you might wish to bevel the box edges. Buff the copper with steel wool, add lettering, and finish off the work

with a coat of clear lacquer or polyurethane varnish.

Using a little lacquer thinner or acetone (and a lot of caution), soak the switches to remove the grease that was added during the manufacture. When dry, spray the inside of the switches lightly with a TV-tuner cleaner/lubricant. Using a sharp drill bit (about 3/16 inch will do), countersink the mounting holes on the actuator side of the switch mounting plate. This ensures that the switches will fit flush against the top plate. At one end of each switch, bend the two lugs over and solder them together. Cut off the upper halves of the remaining switch lugs. (A look at Fig 73 will help clarify these steps.)

Solder the horizontal members of the pi sections between the appropriate switch lugs. Try to keep the lead lengths as short as possible, and do not overheat the resistors. Now solder the switches in place

to the top section of the enclosure by flowing solder through the mounting holes and onto the circuit-board material. Be certain that you place the switches in their proper positions; correlate the resistor values with the degree of attenuation. Otherwise, you may wind up with the 1-dB step at the wrong end of the box.

Once the switches are installed, thread a piece of no. 18 bare-copper wire through the center lugs of all the switches, passing it through the holes in the partitions. Solder the wire at each switch terminal. Cut the wire between the poles of each individual switch, leaving the wire connecting one switch pole to that of the neighboring one on the other side of the partition, as shown in Fig 74.

At each of the two end switch terminals, leave a wire length of approximately 1/8 inch. Install the BNC connectors, and solder the wire pieces to the connector center conductors.

Now install the resistors that comprise the grounded legs of each pi section. Use short lead lengths. Remember that physical

Fig 74—An inside view of the completed attenuator. Use of short, direct leads enhances the performance of the unit. Brass nuts soldered at each of the four corners allow machine screws to be used to secure the bottom cover. File one corner of each nut to permit a flat, two-sided fit within the enclosure.

symmetry is conducive to good performance. Do not use excessive amounts of heat when soldering.

Solder a no. 4-40 brass nut at each inside corner of the enclosure. Recess the nuts approximately 1/16 inch from the bottom edge of the box to allow sufficient room for the bottom panel to fit flush. Secure the bottom panel with four no. 4-40, 1/4-inch machine screws and you're done!

This circuit was described by Shriner and Pagel in September 1982 *QST*. Resistance values for pi- and T-network resistive attenuators are given in Tables 2 and 3. Values are for use in 50-ohm unbalanced systems. These values can be scaled for use in other systems by dividing the resistance values by 50 and then multiplying by the system resistance. The resistances shown in Fig 71 are the nearest standard values to those resistances appearing in Tables 2 and 3, taking 5% tolerances into account. Although some of the values are a few ohms off, they are quite acceptable for amateur work.

MEASURING RECEIVER PERFORMANCE

Comparing the performance of one receiver to another is difficult at best. The features of one receiver may outweigh a second, even though its performance under some conditions is not as good as it should be. Although the final decision on which receiver to own will more than likely be based on personal preference, there are ways to compare receiver performance characteristics. The most important parameters are noise floor, intermodulation distortion, blocking (gain compression) and cross modulation.

Table 2
Pi-Network Resistive Attenuator (50 Ω)

dB Atten.	R1 (Ohms)	R2 (Ohms)
1	870.0	5.8
2	436.0	11.6
3	292.0	17.6
4	221.0	23.8
5	178.6	30.4
6	150.5	37.3
7	130.7	44.8
8	116.0	52.8
9	105.0	61.6
10	96.2	71.2
11	89.2	81.6
12	83.5	93.2
13	78.8	106.0
14	74.9	120.3
15	71.6	136.1
16	68.8	153.8
17	66.4	173.4
18	64.4	195.4
19	62.6	220.0
20	61.0	247.5
21	59.7	278.2
22	58.6	312.7
23	57.6	351.9
24	56.7	394.6
25	56.0	443.1
30	53.2	789.7
35	51.8	1405.4
40	51.0	2500.0
45	50.5	4446.0
50	50.3	7905.6
55	50.2	14,058.0
60	50.1	25,000.0

Table 3
T-Network Resistive Attenuator (50 Ω)

dB Atten.	R1 (Ohms)	R2 (Ohms)
1	2.9	433.3
2	5.7	215.2
3	8.5	141.9
4	11.3	104.8
5	14.0	82.2
6	16.6	66.9
7	19.0	55.8
8	21.5	47.3
9	23.8	40.6
10	26.0	35.0
11	28.0	30.6
12	30.0	26.8
13	31.7	23.5
14	33.3	20.8
15	35.0	18.4
16	36.3	16.2
17	37.6	14.4
18	38.8	12.8
19	40.0	11.4
20	41.0	10.0
21	41.8	9.0
22	42.6	8.0
23	43.4	7.1
24	44.0	6.3
25	44.7	5.6
30	47.0	3.2
35	48.2	1.8
40	49.0	1.0
45	49.4	0.56
50	49.7	0.32
55	49.8	0.18
60	49.9	0.10

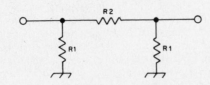

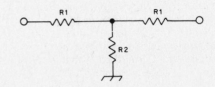

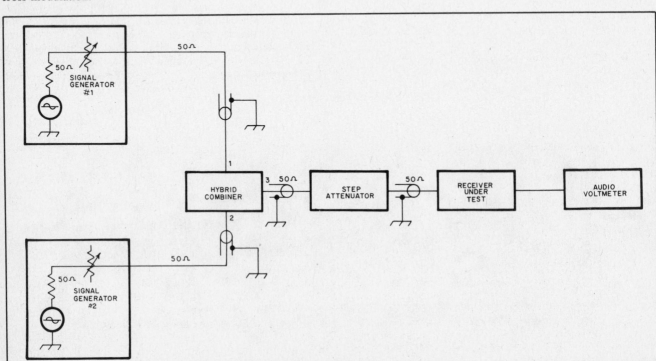

Fig 75—General test setup used for evaluating receiver performance. Two signal generators (calibrated), a hybrid combiner, a step attenuator and an audio voltmeter are required for the measurements.

The general test setup is shown in Fig 75. Two calibrated signal generators are required, along with a hybrid combiner, a step attenuator and an ac voltmeter. A hybrid combiner is essentially a unit with three ports. This device is used to combine the signals from a pair of generators. It has the characteristic that signals applied at ports 1 or 2 appear at port 3, and are attenuated by 6 dB. However, a signal from port 1 is attenuated 30 or 40 dB when sampled at port 2. Similarly, signals applied at port 2 are isolated from port 1 some 30 to 40 dB. The isolating properties of the box prevent one signal generator from being frequency or phase modulated by the other. A second feature of a hybrid combiner is that a 50-ohm impedance level is maintained throughout the system. A commercial example of a directional bridge of this kind is an HP-8721A.

The signal generators used in the test setup must be calibrated accurately in dBm or microvolts. The generators should have extremely low leakage. That is, when the output of the generator is disconnected, no signal should be detected at the operating frequency with a sensitive receiver. Ideally, at least one of the signal generators should be capable of amplitude modulation. A suitable lab-quality piece would be the HP-8640B.

While most signal generators are calibrated in terms of microvolts, the real concern is not with the voltage from the generator but with the power available. The fundamental unit of power is the watt. However, the unit which is used for most low-level RF work is the milliwatt, and power is often specified in decibels with respect to 1 milliwatt (dBm). Hence, 0 dBm would be 1 milliwatt. The dBm level, in a 50-ohm load, can be calculated with the aid of the following equation

$$dBm = 10 \log_{10} [20 (V_{RMS})^2]$$

where dBm is the power with respect to 1 milliwatt and V is the RMS voltage available at the output of the signal generator.

The convenience of a logarithmic power unit like the dBm becomes apparent when signals are amplified or attenuated. For example, a −107-dBm signal that is applied to an amplifier with a gain of 20 dB will result in an output of −107 dBm + 20 dB, or −87 dBm. Similarly, a −107-dBm signal applied to an attenuator with a loss of 10 dB will result in an output of −107 dBm − 10 dB, or −117 dBm.

Noise-Floor Measurement

A generator that is tuned to the same frequency as the receiver is used for noise-floor measurements. Output from the generator is increased until the ac voltmeter at the audio-output jack of the receiver shows a 3-dB increase. This measurement indicates the minimum discernible signal (MDS) that can be detected with the receiver. This level is defined as that which

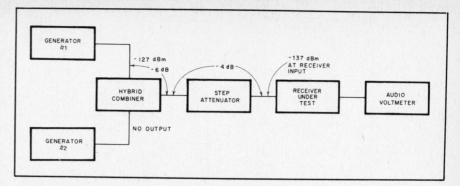

Fig 76—General test setup for measuring receiver noise floor. Signal levels for a hypothetical measurement are indicated. See text for a detailed discussion.

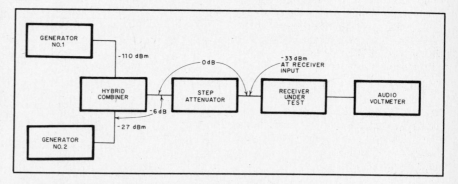

Fig 77—Test setup for measuring receiver blocking performance. Hypothetical measurements are included on the drawing.

will produce the same audio-output power as the internally generated receiver noise. Hence, the term "noise floor." As a hypothetical example, say the output of the signal generator is −127 dBm, the loss through the combiner is fixed at 6 dB and the step attenuator is set to 4 dB. The equivalent noise floor can then be calculated as follows:

$$\text{Noise floor} = −127 \text{ dBm} − 6 \text{ dB} − 4 \text{ dB}$$
$$= −137 \text{ dBm}$$

where the noise floor is the power available at the receiver antenna terminal, −6 dB is the loss through the coupler and −4 dB is the loss through the attenuator. Refer to Fig 76.

Blocking

Blocking concerns gain compression. Both signal generators are used. One is set for a weak signal of roughly −110 dBm and the receiver is tuned to this frequency. The other generator is set to a frequency of 20 kHz away and is increased in amplitude until the receiver output drops by 1 dB, as measured with the ac voltmeter. A blocking measurement is indicative of the signal level that can be tolerated at the receiver antenna terminal before desensitization will occur.

As an example, say that the output of the generator is −27 dBm, the loss through the combiner is fixed at 6 dB and there is 0 dB attenuation through the attenuator (effec-

tively switched out of the line). See Fig 77. The signal level at the receiver terminal that will cause gain compression is calculated as follows:

$$\text{Blocking level} = −27 \text{ dBm} − 6 \text{ dB}$$
$$= −33 \text{ dBm}$$

This can be expressed as dynamic range when this level is referenced to the receiver noise floor that was calculated earlier. This term can be called "receiver blocking dynamic range." Calculate it as follows:

Blocking dynamic range
$$= \text{noise floor} − \text{blocking level}$$
$$= −137 \text{ dBm} − (−33 \text{ dBm})$$
$$= −104 \text{ dB}$$

This value is usually taken in terms of absolute value and would be referred to as 104 dB.

Two-Tone IMD Test

One of the most significant parameters that can be specified for a receiver is a measure of the range of signals that can be tolerated while producing essentially no undesired spurious responses. The evaluation is made with a two-tone intermodulation-distortion (IMD) test. It is generally a conservative evaluation for other effects, such as blocking, which will occur only for signals well outside the IMD dynamic range of the receiver.

Two signals of equal level spaced 20 kHz

apart are injected into the input of the receiver. Call these frequencies f1 and f2. The so-called third-order intermodulation-distortion products will appear at frequencies of (2f1-f2) and (2f2-f1). Assume that the two input frequencies are 14.040 and 14.060 MHz. The third-order products will be at 14.020 and 14.080 MHz.

The step attenuator will be useful in this experiment. Adjust the two generators for an output of −10 dBm each at frequencies spaced 20 kHz. Tune the receiver to either of the third-order IMD products. Adjust the step attenuator until the IMD product produces an output 3 dB above the noise level as read on the ac voltmeter.

For an example, say the output of the generator is −10 dBm, the loss through the combiner is 6 dB and the amount of attenuation used is 40 dB. See Fig 78. The signal level at the receiver antenna terminal that just begins to cause IMD problems is calculated as:

$$\text{IMD level} = -10 \text{ dBm} - 6 \text{ dB} - 40 \text{ dB}$$
$$= -56 \text{ dBm}$$

This can be expressed as a dynamic range when this level is referenced to the noise floor. This term is referred to as "IMD dynamic range" and can be calculated

IMD dynamic range
$$= \text{noise floor} - \text{IM level}$$
$$= -137 \text{ dBm} - (-56 \text{ dBm})$$
$$= -81 \text{ dB}$$

Therefore, the IMD dynamic range of this receiver would be 81 dB.

Evaluating the Data

Thus far, a fair amount of data has been gathered with no mention of what the numbers really mean. It is somewhat easier to understand exactly what is happening by arranging the data in a form something like that in Fig 79. The base line is just a power line with a very small level of power at the left and a higher level (0 dBm) at the right.

The noise floor of the hypothetical receiver is drawn in at −137 dBm, the IMD level (the level at which signals will begin to create spurious responses) at −56 dBm and the blocking level (the level at which signals will begin to desense the receiver) at −33 dBm. As can be seen, the IMD dynamic range is some 23 dB smaller than the blocking dynamic range. This means IMD products will be heard across the band long before the receiver will begin to desense − some 23 dB sooner.

The figures for the hypothetical receiver represent those which would be expected from a typical communications receiver on the market today. It is interesting to note that it is possible for the home constructor to build a receiver that will outperform commercially available units (even the high-priced ones).

THE SPECTRUM ANALYZER

The spectrum analyzer is like an

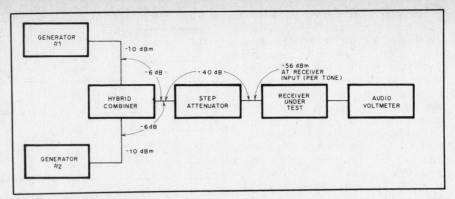

Fig 78—Receiver IMD performance test setup. Signal levels for a hypothetical measurement are given. A detailed discussion of this measurement is given in the text.

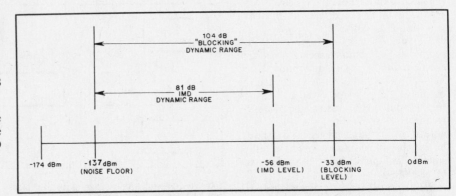

Fig 79—This graph displays the performance of a hypothetical (though typical) receiver under test. The noise floor is −137 dBm, blocking level is −33 dBm and the IMD level is −56 dBm. This corresponds to a receiver blocking dynamic range of 104 dB and an IMD dynamic range of 81 dB.

oscilloscope in that both present an electrical signal through graphical representation. The oscilloscope is used to observe electrical signals in the time domain (amplitude as a function of time). However, not all signals can be represented properly in only the time domain. Amplifiers, mixers, oscillators, detectors, modulators and filters are best characterized in terms of their frequency response. This information is obtained by viewing electrical signals in the frequency domain (amplitude as a function of frequency). One instrument that can display the frequency domain is the spectrum analyzer.

Time and Frequency Domain

To better understand the concepts of time and frequency domain, see Fig 80. The three-dimensional coordinates show time (as the line sloping toward the bottom right), frequency (as the line rising toward the top right) and amplitude (as the vertical axis). The two discrete frequencies shown are harmonically related, so we'll refer to them as f_1 and $2f_1$.

In the representation of time domain at Fig 80B, all frequency components of a signal are summed together. In fact, if the two discrete frequencies shown were applied to the input of an oscilloscope, we would see the solid line (which corresponds to $f_1 + 2f_1$) on the display.

In the frequency domain, complex signals (signals composed of more than one frequency) are separated into their individual frequency components. Additionally, a measurement is made as to the power level at each discrete frequency. The display depicted at Fig 80 C is typical of that obtained with a spectrum analyzer.

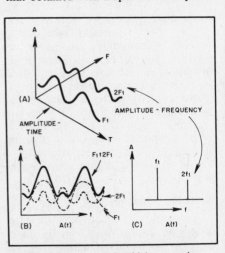

Fig 80—Different ways in which a complex signal may be characterized. At A is a three-dimensional display of amplitude, time and frequency. At B, this information is shown only in the time domain as would be seen on an oscilloscope. At C the same information is shown in the frequency domain as it would be viewed on a spectrum analyzer.

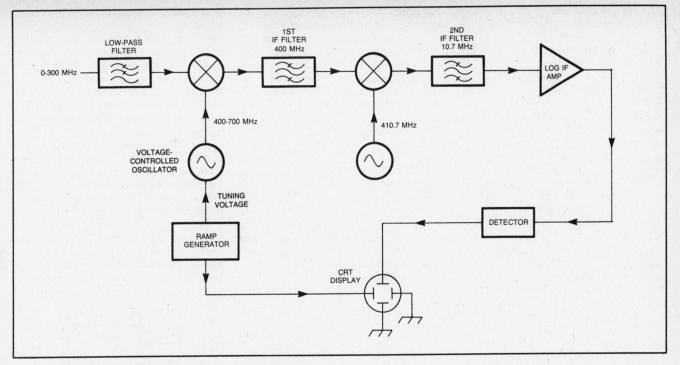

Fig 81—A block diagram of a superheterodyne spectrum analyzer.

The frequency domain can contain information not found in the time domain and therefore the spectrum analyzer offers advantages over the oscilloscope for certain measurements. As might be expected, there are some measurements that require data gathering to be done in the time domain. In these cases, the oscilloscope is an invaluable instrument.

There are several different types of spectrum analyzers, but by far the most common is nothing more than an electronically-tuned superheterodyne receiver. The receiver is tuned using a ramp voltage; this ramp voltage performs two functions: First, the ramp voltage sweeps the frequency of the analyzer local oscillator. Second, the ramp is used to deflect the horizontal axis of a CRT display, as shown in Fig 81. The vertical axis of the CRT display is provided deflection from the signal strength of the received signal. In this way, the CRT display has frequency on the horizontal axis and signal strength on the vertical axis.

Most spectrum analyzers use the up-converting technique so that a fixed-tuned input filter is capable of removing the image. Only the first local oscillator need be tuned to tune the receiver. In the up-conversion design, a wideband input is converted to an IF higher than the highest input frequency. As with most up-converting communications receivers, it is not easy to achieve the desired ultimate selectivity at the first IF, because of the rather high frequency. For this reason, dual or higher orders of conversion are used to generate an IF low enough that the desired selectivity can be achieved practically. In the example of Fig 81 dual conversion is

used; the first IF is at 400 MHz with a second IF at 10.7 MHz.

In the example spectrum analyzer, the first local oscillator is swept from 400 MHz to 700 MHz; this converts the input (from nearly 0 MHz to 300 MHz) to the first IF of 400 MHz. The usual rule of thumb for varactor-tuned oscillators is that the maximum practical tuning ratio (the ratio of the highest frequency to the lowest frequency) is an octave or a two-to-one ratio. As it can be seen from the example spectrum analyzer, the tuning ratio of the first local oscillator is 1.75 to one, which meets this specification.

The image frequency spans 800 MHz to 1100 MHz and is easily eliminated using a low-pass filter with a cut-off frequency around 300 MHz. The 400-MHz first IF is converted to 10.7 MHz where the ultimate selectivity of the analyzer is obtained. The image of the second conversion, (421.4 MHz), is eliminated by the first IF filter. The attenuation of the image should be rather great, on the order of 60 to 80 dB. This requires a first IF filter with a high Q; this is achieved by using helical resonators, SAW resonators or cavity-type filters. Another method of eliminating the image problem is to use triple conversion; converting first to an intermediate IF such as 50 MHz and then to 10.7 MHz. As with any receiver, an additional frequency conversion requires added circuitry and produces further potential spurious responses.

Most of the signal amplification takes place at the lowest IF; in the case of the example analyzer this is 10.7 MHz. This is where the communications receiver and the spectrum analyzer differ. A communi-

cations receiver demodulates the incoming signal so that the modulation can be heard or further demodulated for RTTY or packet or other mode of operation. In the spectrum analyzer, only the signal strength is needed.

In order for the spectrum analyzer to be most useful, it should be convenient to display signals of widely different levels. As an example, signals differing by 60 dB, which is a thousand to one difference in voltage or a million to one in power, would be difficult to display. This would mean that if power were displayed, one signal would be one million times larger than the other (in the case of voltage one signal would be a thousand times larger). In either case it would be difficult to display both signals on a CRT. The solution to this problem is to use a logarithmic display and to display the relative signal levels in decibels. Using this technique, a one-thousand-to-one ratio of voltage reduces to a sixty-decibel difference.

The conversion of the signal to a logarithm is usually performed in the IF amplifier or detector, resulting in an output voltage proportional to the logarithm of the input RF level. This output voltage is then used to drive the CRT display.

Spectrum Analyzer Performance Specifications

The performance parameters of a spectrum analyzer are specified in terms similar to those used for radio receivers, in spite of the fact that there are many differences between a receiver and a spectrum analyzer.

The *sensitivity* of a receiver is often specified as the minimum detectable signal, which means the smallest signal that can be heard. In the case of the spectrum analyzer, it is not the smallest signal that can be heard, but the smallest signal that can be seen. The *dynamic range* of the spectrum analyzer determines the largest and smallest signals that can be viewed on the analyzer. As with a receiver, there are several factors that can affect dynamic range, such as intermodulation distortion, 2nd and 3rd order distortion and blocking. *IMD dynamic range* is the maximum difference in signal level between the minimum detectable signal and the level of two signals of equal strength that generate an intermodulation product equal to the minimum detectable signal.

This is a good point to explain another of the significant differences between a communications receiver and a spectrum analyzer. Although the communications receiver is an excellent example to introduce the spectrum analyzer, there are several differences such as the previously explained lack of a demodulator. Unlike the communications receiver, the spectrum analyzer is not a sensitive radio receiver. To preserve a wide dynamic range, the spectrum analyzer often uses passive mixers for the first and second mixers. Therefore, referring to Fig 81, the noise figure of the analyzer is no better than the losses of the input low pass filter plus the first mixer, the first IF filter, the second mixer and the loss of the second IF filter. This often results in a combined noise figure of more than 20 dB. With that kind of noise figure the spectrum analyzer is obviously not a communications receiver for extracting very weak signals from the noise but a measuring instrument for the analysis of frequency spectrum.

The *selectivity* of the analyzer is called the *resolution bandwidth*. This term refers to the minimum frequency separation of two signals of equal level that can be resolved so that there is a 3-dB dip between the two. The IF filters used in a spectrum analyzer differ from a communications receiver in that the filters in a spectrum analyzer have very gentle skirts and rounded passbands, rather than the flat passband and very steep skirts used on an IF filter in a high-quality communications receiver. This rounded passband is necessary because the signals pass into the filter passband as the spectrum analyzer scans the desired frequency range. If the signals suddenly pop into the passband (as they would if the filter had steep skirts), the filter tends to ring; a filter with gentle skirts is less likely to ring. This ringing, called *scan loss*, distorts the display and requires that the analyzer not be swept in frequency at an excessive rate. All this means that the scan rate must be checked periodically to be certain that the signal amplitude is not affected by excessive rate.

Spectrum Analyzer Applications

When and where is a spectrum analyzer used? Spectrum analyzers are used in situations where the signals to be analyzed are

very complex and an oscilloscope display would be an undecipherable jumble. The spectrum analyzer is also used when the frequency of the signals to be analyzed is very high. Although high-performance oscilloscopes are capable of operation into the UHF region, moderately-priced spectrum analyzers can be used well into the GHz region.

The spectrum analyzer can also be used to view very low-level signals. For an oscilloscope to display a VHF waveform, the bandwidth of the oscilloscope must extend from zero to the frequency of the waveform. If harmonic distortion and other higher-frequency distortions are to be seen the bandwidth of the oscilloscope must exceed the fundamental frequency of the waveform. This broad bandwidth can also admit a lot of noise power. The spectrum analyzer, on the other hand, analyzes the waveform using a narrow bandwidth; thus

it is capable of reducing the noise power admitted.

Probably the most common application of the spectrum analyzer is the measurement of the harmonic content and other spurious signals in the output of a radio transmitter. Fig 82 shows two ways to connect the transmitter and spectrum analyzer. The method shown at A should not be used for wide-band measurements since most line-sampling devices do not exhibit a constant-amplitude output over a broad frequency range. Using a line sampler is fine for narrow-band measurements, however. The method shown at B is used in the ARRL Lab.

The attenuator must be capable of dissipating the transmitter power, and it must have sufficient attenuation to protect the spectrum analyzer input. Many spectrum analyzer mixers can be damaged by only a few milliwatts, so most analyzers

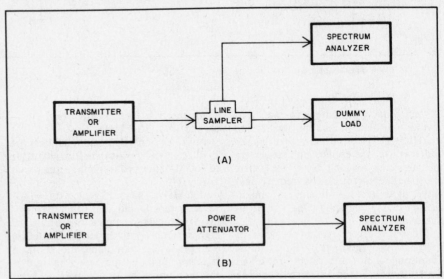

Fig 82—Two commonly used setups for observing the output of a transmitter or amplifier on a spectrum analyzer. The system at A uses a line sampler to pick off a small amount of the transmitter or amplifier power. At B, the majority of the transmitter power is dissipated in the power attenuator.

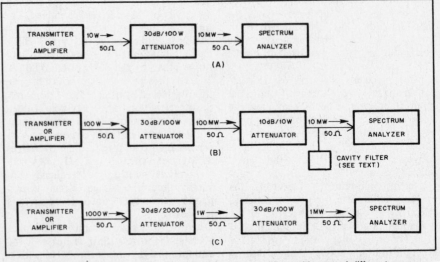

Fig 83—ARRL laboratory test setup for transmitters or amplifiers with several different power levels.

have an adjustable input attenuator that will provide a reasonable amount of attenuation to protect the sensitive input mixer from damage. The power limitation of the attenuator itself is usually on the order of a watt or so, however. This means that 20 dB of additional attenuation is required for a 100 watt transmitter, 30 dB for a 1000 watt transmitter, and so on, to limit the input to the spectrum analyzer to one watt. There are specialized attenuators that are made for transmitter testing; these attenuators provide the necessary power dissipation and attenuation in the 20 to 30 dB range.

The test setup used in the ARRL laboratory for measuring an HF transmitter or amplifier is shown in Fig 83. As can be seen, different power levels dictate different amounts of attenuation between the transmitter or amplifier and the spectrum analyzer.

When using a spectrum analyzer it is very important that the maximum amount of attenuation be applied before a measurement is made. In addition, it is a good practice to view the entire spectrum of a signal before the attenuator is adjusted. The signal being viewed could appear to be at a safe level, but another spectral component, which is not visible, could be above the damage limit. It is also very important to limit the input power to the analyzer when pulse power is being measured. The *average* power may be small enough so that the input attenuator is not damaged, but the *peak* pulse power, which may not be readily visible on the analyzer display, can destroy a mixer, literally in microseconds.

Fig 84 shows the broadband spectrum of a transmitter, showing the harmonics in the output. The horizontal (frequency) scale is 5 MHz per division; the main output of the transmitter at 7 MHz can be seen about 1.5 major divisions from the left of the trace. A very large apparent signal is seen at the extreme left of the trace. This occurs at what would be zero frequency and it is cause by the first local oscillator frequency being exactly the first IF. All up-converting superheterodyne spectrum analyzers have this IF feedthrough; in addition, this signal is occasionally accompanied by smaller spurious signals generated within the analyzer. To determine what part of the displayed signal is a spurious response caused by IF feedthrough and what is an actual input signal, simply remove the input signal and observe the trace.

It is not necessary or desirable that the transmitter be modulated for this broadband test. Investigating the sidebands from a modulated transmitter requires a narrow band spectrum analysis and produces displays similar to that shown in Fig 85. The test setup used to produce this display is shown in Fig 86. In this example a two-tone test signal is used to modulate the transmitter. Fig 85 shows the two test tones plus some of the intermodulation produced by the single-sideband transmitter.

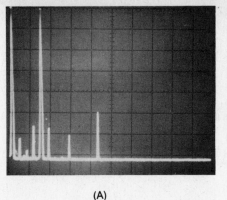 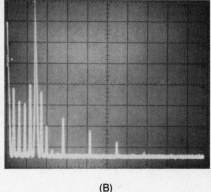

(A)   (B)

Fig 84—Spectrum analyzer display of two different transmitters on the 40-meter band. Each horizontal division represents 5 MHz and each vertical division is 10 dB. The photograph at A shows a clean transmitted signal; the transmitter at B shows more spurious signal content. Both transmitters are legal, however, according to current FCC spectral purity requirements.

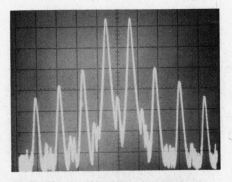

Fig 85—Spectrum analyzer photograph showing the result of an SSB transmitter two-tone test. Each horizontal division represents 1 kHz and each vertical division is 10 dB. The third-order products are 30 dB below the PEP (top line), the fifth-order products are down 37 dB and seventh-order products are down 44 dB. This represents acceptable (but not ideal) performance.

When using a spectrum analyzer it is necessary to ensure that the analyzer does not generate additional spurious signals that are then attributed to the system under test. Some of the spurious signals that can be generated by a spectrum analyzer are harmonics and intermodulation distortion. If it is desired to measure the harmonic levels of a transmitter at a level below the spurious level of the analyzer itself, a notch filter can be inserted between the attenuator and the spectrum analyzer as shown in Fig 87. This reduces the level of the fundamental signal and prevents that signal from generating harmonics within the analyzer, while still allowing the harmonics from the transmitter to pass through to the analyzer without attenuation. Caution should be used when using this technique; detuning the notch filter or inadvertently changing the transmitter frequency will allow potentially high levels of power to enter the analyzer. In addition, care must be exercised with the choice of filters; some filters (such as cavity filters) will respond not only to the fundamental but will notch out odd harmonics as well.

It is good practice to check for the generation of spurious signals within the spectrum analyzer. When a spurious signal is generated by a spectrum analyzer, adding attenuation at the analyzer input will cause the internally generated spurious signals to decrease by an amount greater than the added attenuation. If attenuation added ahead of the analyzer causes all of the visible signals to decrease by the same amount, this indicates a spurious-free display.

The input impedance for most RF spectrum analyzers is 50 ohms; not all circuits

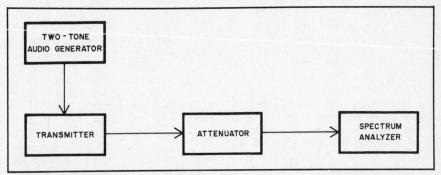

Fig 86—The test setup used in the ARRL laboratory for measuring the IMD performance of transmitters and amplifiers.

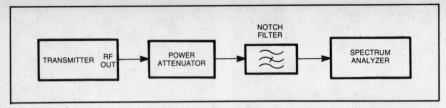

Fig 87—A notch filter can be used to reduce the level of a transmitter's fundamental signal so that the fundamental does not generate harmonics within the analyzer.

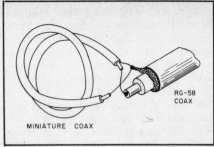

Fig 89—An inductive pickup loop, called a "sniffer" probe. This probe does not load the circuit under test. See text for details.

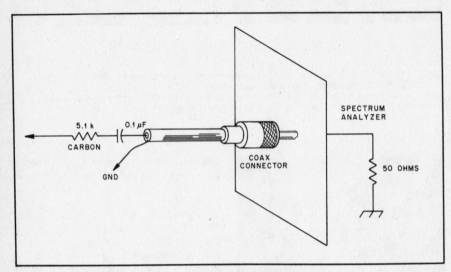

Fig 88—A voltage probe designed for use with a spectrum analyzer.

have convenient 50-ohm connections that can be accessed for testing purposes, however. Using a probe such as the one shown in Fig 88 allows the analyzer to be used as a troubleshooting tool. The probe can be used to track down signals within a transmitter or receiver, much like an oscilloscope is used. The probe shown offers a 100-to-one voltage reduction and provides a loading to the circuit of 5000 ohms.

A different type of probe is shown in Fig 89. This inductive pickup coil (sometimes called a "sniffer") is very handy for troubleshooting. The coil is used to couple signals from the radiated magnetic field of a circuit into the analyzer. A short length of miniature coax is wound into a pick-up loop and soldered to a larger piece of coax. The use of the coax shields the loop from coupling energy from the electric field component. The dimensions of the loop are not critical, but the smaller the dimensions of the loop, the more accurate the loop will be in locating the source of radiation of RF energy. The shield of the coax provides a complete electrostatic shield without introducing a shorted turn.

The sniffer allows the spectrum analyzer to sense RF energy without contacting the circuit being analyzed. If the loop is brought near an oscillator coil, the oscillator can be tuned without directly contacting (and thus disturbing) the circuit. The oscillator can then be checked for reliable starting and the generation of spurious sidebands. With the coil brought near the

tuned circuits of amplifiers or frequency multipliers, those stages can be tuned using a similar technique.

Even though the sniffer does not contact the circuit being evaluated, it does extract some energy from the circuit. for this reason, the loop should be placed as far from the tuned circuit as is practical. If the loop is placed too far from the circuit, the signal will be too weak or the pick-up loop will pick up energy from other parts of the circuit and not give an accurate indication of the circuit under test.

One very handy application of the sniffer is to locate sources of RF leakage. By probing the shields and cabinets of RF generating equipment (such as transmitters) egress and ingress points of RF energy can be identified by indications on the analyzer display.

One very powerful characteristic of the spectrum analyzer is the instrument's capability to measure very low-level signals. This characteristic is very advantageous when very high levels of attenuation are measured. Fig 90 shows the set-up for tuning the notch and bandpass of a VHF duplexer. The spectrum analyzer, being capable of viewing signals well into the low microvolt region, is capable of measuring the insertion loss of the notch cavity more than 100 dB below the signal generator output. Making a measurement of this sort requires care in the interconnection of the equipment and a well-designed spectrum analyzer and signal generator. RF energy

leaking from the signal generator from the cabinet, the line cord or even the coax itself, can leak into the spectrum analyzer through similar paths and corrupt the measurement. This leakage can make the measurement look either better or worse than the actual attenuation depending on the phase relationship of the leaked signal.

Extensions of Spectral Analysis

What if a signal generator is connected to a spectrum analyzer so that the signal generator output frequency is exactly the same as the receiving frequency of the spectrum analyzer? It would certainly appear to be a real convenience not to have to continually reset the signal generator to the desired frequency. It is, however more than a convenience. A signal generator connected in this way is called a *tracking generator* because the output frequency tracks the spectrum analyzer input frequency. The tracking generator makes it possible to make sweep-frequency measurements of the attenuation characteristics of circuits, even when the attenuation involved is large as the bandpass of narrow filters such as the cavity described in the previous example.

Fig 91 shows the connection of a tracking generator to a circuit under test. In order for the tracking generator to create an output frequency exactly equal to the input frequency of the spectrum analyzer, the internal local oscillator frequencies of the spectrum analyzer must be known; this is the reason for the interconnections between the tracking generator and the spectrum analyzer. The test set-up of Fig 91 will measure the gain or loss of the circuit under test. Only the magnitude of the gain or loss is available; in some cases, the phase angle between the input and output would also be an important and necessary parameter.

The spectrum analyzer is not sensitive to the phase angle of the tracking generator output. In the process of generating the tracking generator output, there were no guarantees that the phase of the tracking generator would be either known or constant. This is especially true of VHF spectrum analyzers/tracking generators where a few inches of coaxial cable represents a

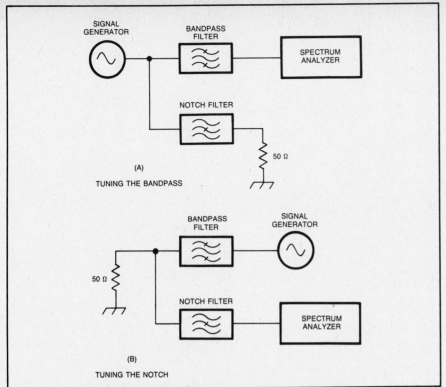

(A)

TUNING THE BANDPASS

(B)

TUNING THE NOTCH

Fig 90—A spectrum analyzer and signal generator can be used to tune the bandpass and notch filters of a duplexer. Leakage around the duplexer must be kept to a minimum.

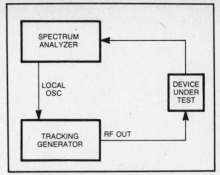

Fig 91—Filter response over a range of frequencies can be plotted by tracking a signal generator with the local oscillator of the spectrum analyzer.

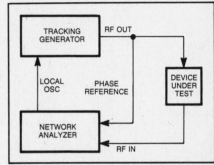

Fig 92—A network analyzer can measure both the phase and magnitude of the filter input and output signals. See text for details.

number of degrees of phase shift.

One effective way of measuring the phase angle between the input and output of a device under test is to sample the phase of the signal applied to the device under test and apply the sample to a phase detector. The phase of the output of the device under test is also sampled and applied to the phase detector. Fig 92 shows a block diagram of this technique. An instrument that can measure both the magnitude and phase

of a signal is called a *vector network analyzer*, or simply a network analyzer.

The magnitude and phase can be displayed either separately or together. When the magnitude and phase are displayed together the two can be presented as two separate traces similar to the two traces on a dual-trace oscilloscope. A much more useful method of display is to present the magnitude and phase as a polar plot where the locus of the points of a vector having

the length of the magnitude and the angle of the phase are displayed. Very sophisticated network analyzers can display all of the S parameters of a circuit in either a polar format or a Smith chart format.

Spectrum Analyzer Bibliography

Helfrick, A D, "An Inexpensive Spectrum Analyzer for The Radio Amateur," *QST*, Nov 1985.
Helfrick, A D, "A Pocket-Sized Spectrum Analyzer," *RF Design*, Jan 1988.

Chapter 26

Troubleshooting and Repair

Traditionally, the radio amateur has maintained a working knowledge of station equipment. This knowledge, and the resourcefulness to make repairs with whatever supplies are available, keeps amateur installations operating when all other communications fail. The radio amateur's ability to troubleshoot and repair equipment is not only a tradition; it is fundamental to the existence of the service.

Every amateur must make a decision when faced with equipment failure. Not everyone is an electronics wizard; your set may end up at the repair shop. The theory you learned for the FCC examinations and the information in this handbook can help you make an intelligent choice. If the problem is something simple (and most are), why not avoid the effort of shipping the radio to the manufacturer? Although the saving of time and money is gratifying, the experience and confidence gained may prove far more valuable. Especially when a problem appears during an emergency.

Troubleshooting and repair of electronic gear need not be a mystical subject. It is more like detective work. A knowledge of complex math is not required. However, one must have, or develop, the ability to read a schematic diagram and to visualize signal flow through the circuit. A working knowledge of resistors, capacitors, inductors and semiconductor devices, and how they affect the dc resistance of a network, is also necessary. The fundamental concepts are covered elsewhere in this book.

We are all a bit nervous when we enter new territory. Do not be intimidated by the intricate circuits of a modern radio. Even the most complex transceiver is made up of relatively simple circuits. As you gain experience, you will develop a first-hand understanding of circuit functions. This insight, and the confidence gained from experience, distinguish the skillful troubleshooter.

SAFETY FIRST

Always! Death is permanent. Before we

Table 1
Safety Rules

1) Keep one hand in your pocket when working on "live" circuits or checking to see that capacitors are discharged.
2) Include a conveniently located ground-fault-interrupter/circuit breaker in the workbench wiring.
3) Use grounded plugs and receptacles on all equipment that is not double insulated.
4) Use double-insulated equipment when working:
   a) outdoors.
   b) on a concrete or dirt floor.
   c) near standing water.
5) Use an isolation transformer when working on ac/dc devices.
6) Switch off the power, ground the positive lead from the supply and discharge capacitors when making circuit changes.
7) Do not subject electrolytic capacitors to:
   a) excessive voltage.
   b) ac.
   c) reverse voltage.
8) Test leads should be well insulated.
9) Do not work alone!
10) Wear safety glasses for protection against sparks and metal fragments.
11) Always use a safety belt when working above ground level.
12) Wear shoes with nonslip soles that will support your feet when climbing.
13) Wear a hard hat when someone is working above you.
14) Be careful with tools that may cause short circuits.
15) Replace fuses only with those having proper ratings.

begin considering the equipment and techniques of troubleshooting, we must discuss the preservation of the worker and his tools.

Electricity can be lethal. Only 50 mA flowing through the body is painful; 100 to 500 mA is usually fatal. Under certain conditions, as little as 24 V can kill. Keep one hand in your pocket while working on "live" circuits to prevent the possibility of resting your free hand on a ground. This reduces the risk of electricity passing

through your chest (the most fatal current path). A list of safety rules can be found in Table 1.

TEST EQUIPMENT

Test equipment is required for efficient troubleshooting. We cannot see electrons flow. Electrons do affect various devices in our equipment, however, with results that we can see.

Many fear that repairing their own ham gear means purchasing a bench full of expensive test instruments. This is not so! In fact, you probably already own the most important instruments. Some others may be purchased inexpensively, rented, borrowed or built at home. A shortage of test gear may limit the kind of repairs you attempt, but do not let it stop you from making repairs that *are* within your capabilities.

Most hams elect to buy test gear. Before buying, consider whether your use of the item merits the expense. Instruments that do not give sufficient return for an individual, may for a club. Expensive test instruments look more reasonable when the cost is split among the members of a club (and a pool of test gear may also serve to pull a sleepy club out of the doldrums).

I mentioned that you already own some of the most important instruments. We each have some of these natural test instruments.

- Eyes — Use them to search for evidence of heat or arcing. Often these conditions are indicated by discoloration or charring of components and circuit boards.
- Ears — Gross distortion can be detected by ear. The "snaps" and "pops" of arcing or the sizzling of a burning component may help you track down circuit faults.
- Nose — Arcing in a circuit produces ozone (what you sometimes smell in the air during a thunderstorm). Smoke from hot components can usually be smelled before it is seen. Some components emit a "hot" smell when overheated. This

Table 2

Sources of Equipment Manuals

HI Manuals, Inc.†
P.O. Box Q-802
Council Bluffs, IA 51502

Howard Sams Publications
4300 W. 62nd St.
Indianapolis, IN 46205

Collins Telecommunications Products
Division,
Rockwell International
Cedar Rapids, IA 52406

R. L. Drake Co.
P.O. Box 112
911 Springboro Pike
Miamisburg, OH 45342
513-866-2421

Hallicrafters Manuals
Ardco Electronics††
P.O. Box 95, Dept. Q
Berwyn, IL 60402

Hammarlund Manuals
Irving J. Abend
P.O. Box 426
Bergenfield, NJ 07621
201-384-7589

Hammarlund Manuals
Pax Manufacturing Co.
Attn: Peter Kjeldsen
100 East Montauk Hwy.
Lindenhurst, NY 11757

ICOM America, Inc.
2380-116th Ave. NE
Bellevue, WA 98004
206-454-8155

Swan Division of Cubic Communications
305 Airport Rd.
Oceanside, CA 92054
619-757-7525

Ten-Tec, Inc.
Hwy. 411 East
Sevierville, TN 37862
615-453-7172

Kenwood USA Corp.
2201 East Dominguez St.
Long Beach, CA 90810
213-639-4200

U.S. Government Surplus
General Services Administration
National Archives and Record Services
Washington, DC 20408

U.S. Government Surplus
Slep Electronics Co.††
P.O. Box 100
Otto, NC 28763

Yaesu Electronics Corp.
17210 Edwards Rd.
Cerritos, CA 90701
213-404-2700

†HI Manuals will not answer requests for information unless a $5 research fee is included. Order their catalogue ($1) to see if they have the manual you need.

††Write for manual prices (specify model).

---

smell remains long after power is removed. Nerves in the nose are extremely sensitive; heat that generates no smell can be felt when brought within 1 or 2 inches of the nose.

• Fingers — Use your fingers to measure low heat levels in components. Solid-state devices are temperature sensitive; if you can hold your finger on them while they are operating (be careful!), they are cool enough. If an operating device is too hot to hold, check for excessive current flow; additional heat sinks may be required.

• Brain — More troubleshooting problems have been solved with a VOM and a brain than with the most expensive spectrum analyzer. Use your brain to assemble information and analyze data collected by other instruments. (The magnifying glass did not make Sherlock Holmes famous; it was his mind!)

Information

There is a sea of information available (a bibliography is included at the end of this chapter). Sometimes it is difficult to find information on a specific item, but remember that Amateur Radio offers a great resource to each of us. Check into a few roundtable QSOs and ask for help. There are over 400,000 hams in the U.S.; chances are someone else has had the same trouble you are experiencing. Look in

Amateur Radio publications; ownership clubs for particular radios may advertise.

Before beginning work, obtain a manual for the radio. An owner's manual or schematic is mandatory; a service manual is better. Some sources of manuals are listed in Table 2.

ARRL members have other resources. ARRL Hq. may be able to provide copies of manuals that are in the ARRL files, and equipment-modification references for back issues of *QST*.[1] Place a "HamAds" listing in *QST* when searching for a rare manual.

The ARRL Technical Coordinator (TC) program may also be able to help. TCs and Assistant Technical Coordinators (ATCs) are volunteers who are willing to help hams with technical questions. For the name of a local ATC, contact your Section Manager (listed on page 8 of any issue of *QST*).

ELECTRONIC TEST EQUIPMENT

Simple test gear is frequently included in the equipment under repair. Nearly all receivers include a speaker. An S meter is usually connected ahead of the audio chain. If the S meter shows signals, the RF chain is probably functioning. The antenna jack of a receiver can serve as a crude signal in-

[1]Write to ARRL Headquarters for information on costs and availability of specific manuals, copies and back issues.

jector. (Noise should be heard from the speaker when a test lead is rubbed against the center conductor of the antenna jack.) Some receivers include a calibrator. The calibrator signal, which is rich in harmonics, is injected in the RF chain close to the antenna jack and may be used for signal tracing and alignment. Every transmitter has an oscillator that may be used as a signal generator.

Spare components enable you to test by substitution. A lot of troubleshooting may be done without what is normally called test equipment.

Here is a summary of test instruments and their applications. Some items serve several purposes and may substitute for others on the list. The list does not cover all equipment available, only the most common and useful instruments.

• Multimeters — This group includes vacuum-tube voltmeters (VTVMs), volt-ohm-milliammeters (VOMs), field-effect-transistor VOMs (FETVOMs), digital multimeters (DMMs). Meter operation and use is explained in Chapter 25.

Look for a meter with an input impedance of 20 kΩ/V or better. Reasonably priced models are available with 30 kΩ/V ($35) and 50 kΩ/V ($40).

The high-Z input characteristic of FETVOMs, VTVMs and other electronic voltmeters makes them the preferred instruments for voltage measurements. DMMs are usually high-Z meters, but they may be sensitive to RF. Most technicians keep an analog-display VOM handy for use near RF equipment.

Multimeters are used to read bias voltage, circuit resistance, presence and level of signal (with an appropriate probe). They can test resistors, capacitors (within certain limitations), diodes and transistors.

• Test Leads — Keep an assortment of leads that terminate in medium and small alligator clips. It is best to make your own leads because ready-made leads use small wire that is not soldered to the clips.

Open wire leads (Fig. 1A) can pick up RFI that disturbs measurements. RFI is reduced somewhat if the leads are twisted together (Fig. 1B). A coaxial cable lead (Fig. 1C) *increases* RFI problems *if* the shield is *not* grounded. The cable shield must be grounded to reduce RFI.

Low capacitance, or ×10, probes (Fig. 1D) are used to reduce the effect of instrument-input and shield capacitance on the circuit under test. A network in the probe serves as a 10:1 divider and compensates for frequency distortion in the cable and test instrument. The trimmer capacitor on ×10 and ×100 oscilloscope probes should be adjusted for the best-looking square wave from the internal calibrator. See Fig. 1H.

Demodulator probes (see Fig. 1E and Chapter 25) convert ac voltages to dc for measurement. They are also used to allow scope triggering from a modulation envelope.

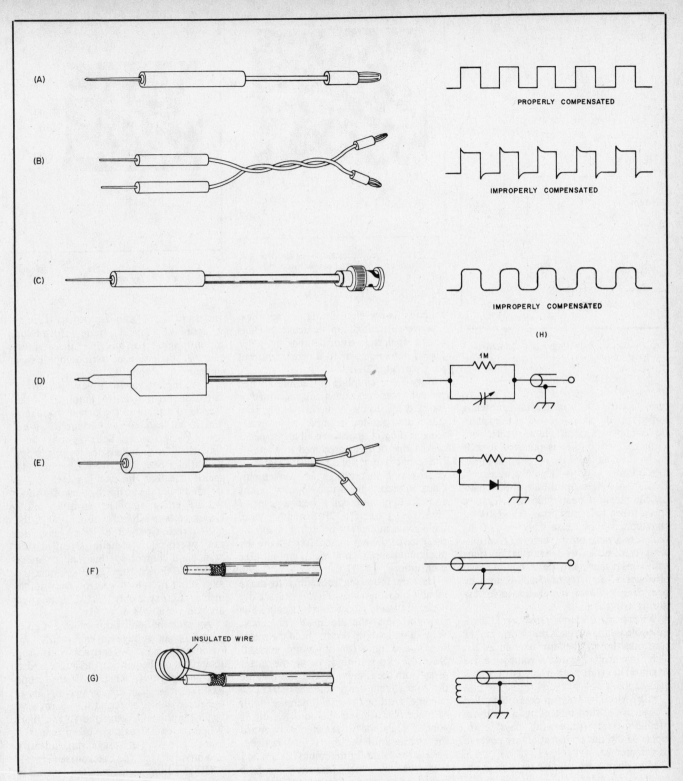

Fig. 1 — An array of test probes for use with various test instruments.

A capacitively coupled probe (Fig. 1F) can be made from a connector and a length of coax cable. Remove about ¾ inch of the cable jacket at one end. Trim the shield to ½ inch from the end and tin it. Use some RTV (room-temperature-vulcanizing) adhesive to seal the end against moisture. Install the connector at the other end of the cable.

Capacitive coupling increases with proximity and potential at the test point. Be very careful with this probe around high voltage. (The dielectric strength of air is about 80 V per 0.001 inch.)

A probe for inductive coupling (Fig. 1G) can be made in similar fashion. Connect a two- or three-turn loop across the center conductor and shield before sealing the

end. The inductive pick up is useful for coupling to high-current points.

• SWR Meter — Every shack should have one! It can be the first indicator of antenna trouble. The SWR meter can also be used between an exciter and power amplifier to spot an impedance mismatch.

Simple meters indicate relative SWR and are fine for Transmatch adjustment and

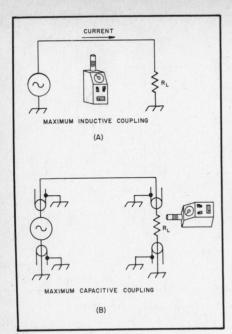

Fig. 2 — Optimum inductive and capacitive dip-meter coupling.

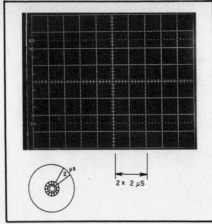

Fig. 3 — An oscilloscope display showing the relationship of the time-base setting and graticule lines.

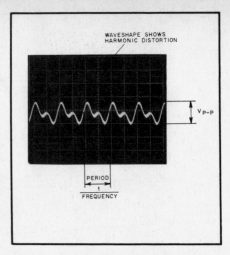

Fig. 4 — Information available from a typical oscilloscope display of a waveform.

line monitoring. However, if you want to make accurate measurements of antenna impedance or calculate SWR at the antenna feed point, a calibrated wattmeter with a directional coupler is required (one is shown in Chapter 25).

• Dummy Load — This is a necessity! Do not put a signal on the air while repairing equipment. Courtesy demands that we spare other amateurs from the QRM of tests until we are reasonably sure of the repair. More important, before a transmitter is fixed, we have no idea what spurious emissions it may generate. A dummy load also works to the technician's benefit by providing a known, matched load for use during adjustments.

When buying a dummy load, avoid used, oil-cooled dummy loads unless you can be sure that the oil does not contain PCBs. This biologically hazardous compound was common in transformer oil until a few years ago.

• Dip Meter — This device is often called a transistor dip meter or a grid-dip oscillator (from vacuum-tube days). Principles of dip-meter operation are covered in Chapter 25.

Most dip meters can also serve as an absorption frequency meter (in this mode, measurements are read at the current peak, rather than the dip). Further, some meters have a connection for headphones. The operator can usually hear signals that do not register on the meter. Because the dip meter is an oscillator, it can be used as a signal generator in certain cases where high accuracy or stability are not required.

A dip meter may be coupled to a circuit either inductively or capacitively. Inductive coupling results from the magnetic field generated by current flow. Therefore, inductive coupling should be used when a

conductor with relatively high current flow is convenient. Maximum inductive coupling results when the axis of the pick-up coil is placed perpendicular to a nearby current path (see Fig. 2A).

Capacitive coupling is required when current paths are magnetically confined or shielded. (Toroidal inductors and coaxial cables are common examples of magnetic self shielding.) Capacitive coupling depends on the electric field produced by voltage. Use capacitive coupling when a point of relatively high voltage is convenient. Capacitive coupling is maximum when the end of the pick-up coil is near a point of high voltage (see Fig. 2B). In either case, the circuit under test is affected by the presence of the dip meter. Always use the minimum coupling that yields a noticeable indication.

Use the following procedure to make reliable measurements. First, bring the dip meter gradually closer to the circuit while slowly varying the dip-meter frequency. When a current dip occurs, hold the meter steady and tune for minimum current. Once the dip is found, move the meter away from the circuit and confirm that the dip comes from the circuit under test (the current reading should increase with distance from the circuit until the dip is gone). Finally, move the meter back toward the circuit until the dip is just noticed. Retune the meter for minimum current and read the dip-meter frequency with a calibrated receiver or frequency meter.

The current dip of a good measurement is smooth and symmetrical. A non-symmetrical dip indicates that the dip-meter oscillator frequency is being significantly influenced by the test circuit. Such conditions do not yield usable readings.

When purchasing a dip meter, look for one that is mechanically and electrically stable. The coils should be in good condition. A headphone connection is helpful. Battery-operated models are easier to use for antenna measurements.

• Oscilloscope — A picture is worth a

thousand words, and the moving trace of a 'scope can give us more information about a signal than any other piece of test gear. A discussion of oscilloscope basics appears in Chapter 25.

The simplest way to display a waveform is to connect the vertical amplifier of the 'scope to a point in the circuit through a simple test lead. (When viewing RF, use a low-capacitance probe that has been "matched" to (compensated for) the 'scope amplifier.) See Fig. 1F. First, check to make sure that the circuit under test is isolated from the ac line by a transformer. Set the vertical-amplifier sensitivity so the circuit voltage will cover about half of the CRT screen. Connect the 'scope leads. Set the sweep-rate (sometimes called the TIME-BASE control so that one or more cycles are shown (see Fig. 3). Adjust the SYNC and SYNC-LEVEL controls for a stable display. Line sync only "locks" signals that are some multiple of 60 Hz.

An examination of the waveform gives peak-to-peak voltage (if calibrated), approximate period (frequency is the reciprocal of the period) and a rough idea of signal purity (see Fig. 4). If the 'scope has dual trace (can display two signals at once) capability, a second waveform may be displayed and compared to the first. When the two signals are taken from the input and output of a stage, stage linearity and phase shift can be checked (see Fig. 5).

Other display techniques are covered in previous chapters and later in this chapter. For more information, see the bibliography at the end of this chapter.

A primary performance characteristic of a 'scope is amplifier bandwidth. This specification tells us the frequency at which amplifier response has dropped 3 dB. The instrument will display higher frequencies, but cannot be used for accurate measurement of them. Expect an accuracy of no better than 3% when a 'scope is used well below its rated bandwidth. Most 'scope applications in Amateur Radio, however, do not require accurate measurements. Thus,

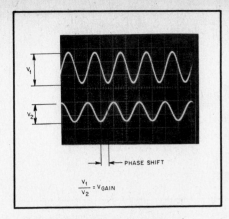

Fig. 5 — A dual-trace scope display of amplifier input and output waveforms.

'scopes are useful to hams well beyond the specifications. Do not believe that transmitter emissions are clean based only on a 'scope display. A bandwidth of 250 to 500 MHz is required to display all significant harmonics from an HF transmitter.

When buying a 'scope, get the greatest bandwidth you can afford. Old and wide is better than new and narrow, with respect to bandwidth. Old Hewlett Packard or Tektronix scopes of great bandwidth are occasionally sold at hamfests. Keep to well-known manufacturers when buying a used scope, or any test equipment for that matter.

• Signal Generators — Although signal generators are used for sophisticated circuit tests, their role in troubleshooting is limited to use in signal injection and alignment techniques. An AF/RF signal-injector schematic is shown in Fig. 6. For those instances where more accuracy is needed, the crystal-controlled signal source of Fig. 7 can be used. The AF/RF circuit provides usable harmonics up to 30 MHz, while the crystal-controlled oscillator will function with crystals from 1 to 15 MHz. These two projects are not meant to compete with standard signal generators, but they are adequate for signal injection.[2] A better generator is required for receiver alignment or for receiver quality testing.

When shopping for a signal generator, look for one with a pure sine-wave signal. A good signal generator is double or triple shielded against leakage. Fixed-frequency audio should be available for modulation of the RF and for injection into audio stages. The frequency control of a good generator does not slip or exhibit backlash. Signal level should be controlled by a switch, labeled in dBm, not a potentiometer. The output jack should be a coaxial connector (usually a BNC), not the kind used for microphone connections. Some older, high-quality units are common. Look for World War II surplus units of the

[2]More information about the signal injector and signal source appears in, Some Basics of Equipment Servicing," Feb. 1982 *QST* (Feedback, May 1982).

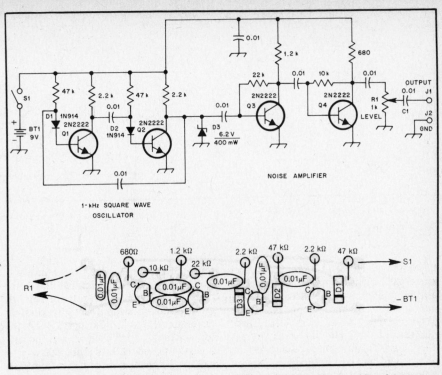

Fig. 6 — Schematic and parts-placement diagram of the AF/RF signal injector. All resistors are ¼ W, 5% carbon types, and all capacitors are disc ceramic. A full-size etching pattern can be found at the back of this book.

BT1 — 9-V transistor radio battery.
D1, D2 — Silicon switching diode, 1N914 or equiv.
D3 — 6.2-V, 400 mW Zener diode.
J1, J2 — Banana jack.

Q1-Q4 — General purpose silicon NPN transistors, 2N2222 or equiv.
R1 — 1-kΩ panel-mount control.
S1 — SPST toggle switch.

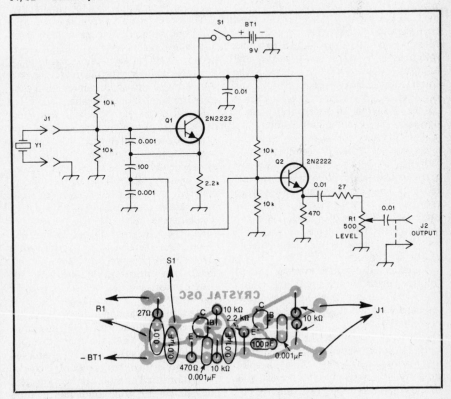

Fig. 7 — Schematic and parts-placement diagram of the crystal-controlled signal source. All resistors are ¼ W, 5% carbon types, and all capacitors are disc ceramic. A full-size etching pattern can be found at the back of this book.

BT1 — 9-V transistor radio battery.
J1 — Crystal socket to match the crystal type used.
J2 — RCA phono jack or equiv.
Q1, Q2 — General purpose silicon NPN

transistor, 2N2222 or equiv.
R1 — 500-Ω panel-mount control.
S1 — SPST toggle switch.
Y1 — 1- to 15-MHz crystal.

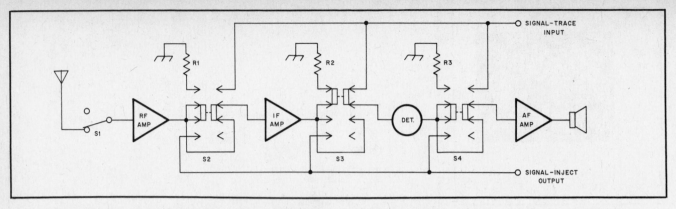

Fig. 8 — A typical radio converted for signal-tracer use. R1, R2 and R3 must be chosen to provide appropriate loads for the preceding stages.
S1 — SPST switch.   S2, S3, S4 — 2P3T slide switches.

URM series, Hewlett Packard/Boonton model 202, Measurements Inc. model 80 and 65, military units TS-497 and BC-221 and Hewlett Packard model 606. (The Technical Department at ARRL Headquarters has a bibliography of articles about the BC-221 that includes a solid-state conversion and calibration technique.) Home-built signal generators may be quite good, but make sure to check construction techniques, level control and shielding quality. Some 1000-Hz square-wave generators produce harmonics that are useful up to 10 MHz.

• Signal Tracers — Signals can be traced with a voltmeter and an RF probe, a dip meter or an oscilloscope, but there are some devices made especially for signal tracing. Most convert the traced signal to audio (often using a voltmeter RF probe) and apply the amplified audio to a speaker. The tracer must function as a receiver and detector for each frequency range in the test circuit. A high-impedance tracer input is necessary to prevent circuit loading.

An old radio can be converted into a signal injector/tracer for use with common IFs (Fig. 8). Modify the radio so a test probe can be connected to each functional block. If the stages preceding the test probe are disabled, the set is a signal tracer; if the preceding stages are operating, the set is a signal injector. (Use a step attenuator to adjust injected-signal level.) Avoid ac/dc radios for this use.

A general-coverage receiver can be used to trace signals in radios with uncommon IFs. Most receivers, however, have a low-impedance input that severely loads the test circuit. To minimize loading, use a capacitive probe. When the probe is held near the circuit, signals will be picked up and carried to the receiver. It may also pick up stray RF; make sure you are listening to the correct signal by switching the circuit under test on and off while listening.

Dedicated signal tracers are limited test instruments. All things considered, a dip meter, with an absorptive mode and a headphone connection, is a much better value.

• Tube Tester — Two types of tube testers are common. The first is the emission tester. As the name implies, this tester checks the tube by measuring the amount of electron emission from the cathode (cathode current, I_c). Each grid is shorted to the plate through a switch and the current is observed while the tube operates as a diode. By opening the switches from each grid to the plate (one at a time) we can check for opens and shorts. If the plate current does not drop slightly as a switch is opened, the element connected to that switch is either open or shorted to another element. We cannot tell an open from a short with this test. The emission tester does not indicate the ability of a tube to amplify.

The other common tube tester checks transconductance. Some transconductance testers read plate current with a fixed bias network. Others use an ac signal to drive the tube while measuring plate current. Both of these tests are meant to indicate the transconductance (amplification) of a tube.

Leakage is another tube fault that testers often check. Contamination inside the tube envelope may result in current leakage between elements. The paths can have high resistance (250 kΩ), and may result from gas or deposits of electron-emitting material that have been transferred from the cathode. Leakage can be checked with an ohmmeter using the ×1M range, depending on the actual spacing of tube elements. Tube testers use a moderate voltage to check for leakage. Beware: This voltage may *cause* shorts in tubes with very close element spacings (6AK5, GL-2C51 and GL-5670).

Use the same caution in the purchase of a tube tester that you would with any other used equipment. Know the testing method, make sure the manual is included and see that it checks the tubes you want to check.

• Transistor Tester — Transistor testers are similar to transconductance tube testers. Collector current is measured while the device is biased into conduction or while an ac signal is applied at the control terminal.

Most transistor failures appear as either an open or shorted junction. Opens and shorts can be found easily with an ohm-meter; a special tester is not required. Transistor gain characteristics vary widely, however, even between units with the same device number.

Testers are useful to determine the characteristics of a particular transistor. A tester that uses no ac signal can read the transistor dc alpha and beta, while testers that apply an ac signal show the ac alpha or beta. The tester can help you decide if a particular transistor has sufficient gain for use as a replacement. It may also help when matched transistors are required.

Transistor testers are inexpensive, so purchase of a used instrument is unnecessary. The functions of a dc tester can be performed with a power supply and a meter (details are shown later in this chapter under "Components"). Construction of a suitable ac tester for bipolar transistors, JFETs and MOSFETs is covered in Chapter 25.

• Frequency Meter — Various instruments for frequency measurement, their construction and operation are covered in Chapter 25.

Almost any digital frequency counter is good enough for troubleshooting and repair work. No special temperature compensation is necessary unless you require accuracy greater than 1 part per million. Used equipment should be in good working order with no signs of damage or abuse.

• Power supplies — A well-equipped test bench should include a means of varying the ac-line voltage, a variable-voltage regulated dc supply and an isolation transformer. If you want to work on vacuum-tube gear, the maximum voltage available from the dc supply should be high enough to serve as a bias supply for common tubes.

Ac-line voltage varies slightly with load. An autotransformer with a movable tap lets you boost or reduce the line voltage slightly. This is helpful when testing equipment supply-voltage tolerance.

A variable-voltage dc supply may be used to power various small items under repair and as a variable bias supply for testing active devices. Construction details for a laboratory power supply appear in Chapter 27.

As mentioned earlier, ac/dc radios must be isolated from the ac line during testing and repair. Keep an isolation transformer handy if you want to work on table-model broadcast radios or television sets. Some transceivers of the 1960s also used the ac/dc power supply.

• Accessories — There are a few small items that may be used in troubleshooting. You may want to keep them handy.

A cold source is useful when searching for intermittent connections. Most parts suppliers carry some kind of cooling agent in a spray can. Poor connections frequently become open circuits when cooled.

A heat source helps locate components that fail only when hot. A small incandescent lamp can be mounted in a large piece of sleeve insulation to produce localized

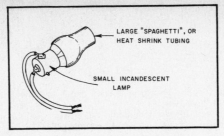

Fig. 9 — A source of local heat made from insulating sleeve and a small dial lamp.

heat for test purposes (see Fig. 9).

A stethoscope (with the pickup removed — Fig. 10) or a long piece of sleeve insulation can be used to listen for arcing or sizzling in a circuit.

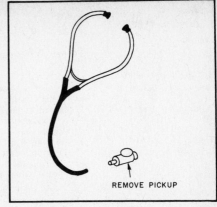

Fig. 10 — A stethoscope, with the pickup removed, used to listen for arcing in crowded circuits.

Troubleshooting Procedures

Connector faults are more common than component troubles. Consider poor connections as prime suspects in our troubleshooting detective work. Begin every job with an inspection of the connections. Is the antenna connected? How about the speaker, fuses and TR switch? Are transistors and ICs firmly seated in their sockets? Are all interconnection cables sound and securely connected? After checking the obvious, begin an in-depth investigation.

To begin that investigation, define the problem accurately. Ask yourself these questions:

1) What functions of the equipment do not work as they should; what does not work at all?

2) What kind of performance can you realistically expect?

3) Is there anything in the operational history of the equipment that could have caused the trouble?

4) Has the trouble occurred in the past? (Keep a record of troubles, maintenance and repair in the owner's manual or log book.)

Write the answers to the questions. The information will help with your work, and help service personnel if their advice or professional service is required.

SEARCHING AT THE BLOCK LEVEL

The block diagram is a road map. It shows the signal paths for each circuit function. These paths may run together, they may cross occasionally, or not at all. Those blocks that are not in the paths of faulty functions can be eliminated as suspects. Sometimes the symptoms point to a single block, and no further search is necessary.

In cases where more than one block is suspect, several approaches may be used. Each requires that a block, or stage, be tested. Signal injection, signal tracing or a

combination of both techniques may be used for this purpose.

THE INSTINCTIVE APPROACH

First, consider an instinctive approach. The check for connector problems mentioned at the beginning of this section is an

example of this procedure. Experience has shown connector faults to be so common that they should be checked even before a systematic approach begins.

When instinct springs from a base of experience, search by instinct may be the fastest procedure. If instinct is correct,

Table 3
Symptoms and Their Causes for All Electronic Equipment

Symptom	Cause
Power Supply Problems	
No output voltage	Open circuit (usually a fuse or transformer winding).
Hum or ripple	Faulty regulator, capacitor or rectifier, low-frequency oscillation.
Amplifiers	
Low gain	Transistor, coupling capacitors, emitter-bypass capacitor, AGC component, alignment.
Noise Transistors, Coupling capacitors, resistors	
Oscillations	Dirt on variable capacitor or chassis, shorted op-amp input.
Untuned (oscillations do not change with frequency)	Audio stages.
Tuned	RF, IF and mixer stages.
Squeal	Open AGC-bypass capacitor.
Static-like crashes	Arcing trimmer capacitors, poor connections.
Static in FM receiver	Faulty limiter stage, open capacitor in ratio detector, weak RF stage, weak incoming signal.
Intermittent noise	All components and connections, band-switch contacts, potentiometers (especially in dc circuits), trimmer capacitors, poor antenna connections.
Distortion (constant)	Oscillation, overload, faulty AGC, leaky transistor, open lead in tab-mount transistor, dirty potentiometer, leaky coupling capacitor, open bypass capacitors, imbalance in tuned FM detector, IF oscillations, RF feedback (cables).
Distortion (strong signals only)	Open AGC line, open AGC diode.
Frequency change	Physical or electrical variations, dirty or faulty variable capacitor, broken switch, loose compartment parts, poor voltage regulation, oscillator tuning (trouble when switching bands).
No Signals	
All bands	Dead VFO or heterodyne oscillator.
One band only	Defective crystal, oscillator out of tune, band switch.
No function control	Faulty switch, poor connection, defective switching diode or circuit.
Improper Dial Tracking	
Constant error across dial	Dial drive.
Error grows worse along dial	Circuit adjustment.

Table 4
Transmitter Problems

Symptom	Cause
Key clicks	Keying filter, distortion in stages after keying.
Modulation problems	
Loss of modulation	Broken cable (microphone, PTT, power), open circuit in audio chain, defective modulator.
Distortion on transmit	Defective microphone, RF feedback from: lead dress, modulator imbalance, bypass capacitor, improper bias, excessive drive.
Arcing	Dampness, dirt, improper lead dress.
Low output	Incorrect control settings, improper carrier shift (CW signal outside of passband), audio oscillator failure, transistor or tube failure, SWR protection circuit.
Antenna Problems	
Poor SWR	Damaged antenna element, matching network, feed line, balun failure (see below), resonant conductor near antenna, poor connection at antenna.
Balun failure	Excessive SWR, weather or cold-flow damage in coil choke, broken wire.
RFI	Arcing or poor connections anywhere in antenna system or nearby conductors.

Table 5
Transceiver Problems

Symptom	Cause
Inoperative S meter	Faulty relay
PA noise in receiver	
Excessive current on receive	
Arcing in the PA tank	
Reduced signal strength on transmit and receive	IF failure
Poor VOX operation	VOX amplifiers and diodes
Poor VOX timing	Adjustment, component failure in VOX timing circuits or amplifiers
VOX consistently tripped by receiver audio	AntiVOX circuits or adjustment

repair time and effort may be reduced substantially. As experience and confidence grow, the merits of the instinctive approach grow with them. However, the inexperienced technician who chooses this approach is at the mercy of chance.

As an example, suppose an AM receiver suddenly begins producing weak and distorted audio. An inexperienced person frequently suspects poor alignment as a common problem. Even though the manufacturer's instructions and the proper equipment are not available, our friend begins twiddling the transformer cores. Before long, the set is hopelessly misaligned. Now our misguided ham must send the radio to a shop for an alignment that was not needed before repairs were attempted.

Alignment does not shift suddenly. A normal signal-tracing procedure would have shown that the signal was good up to the detector, but badly distorted after that stage. The defective detector diode that caused the problem would have been easily found and quickly replaced.

Tables 3 through 6 list common symptoms, and their probable causes. Background information about the items in the tables and symptoms that cannot be itemized appear in the text. These servicing hints have been collected from many sources and may save some time if similar troubles are encountered. The tables list only problems that occur often in troubleshooting and repair. If a problem matches one listed, proceed to testing at the component level. If a particular symptom is not found in the tables, continue at the block level with a systematic approach.

The Power Supply

Many equipment failures are caused by trouble in the power supply. Fortunately, most power-supply problems are easy to find and repair (see Fig. 11). First, use a voltmeter to measure output. Loss of output voltage is usually caused by an open circuit. (A short circuit draws excessive current that opens the fuse, thus becoming an open circuit.) If the fuse has opened, turn off the power, replace the fuse and measure the load-circuit dc resistance. The measured resistance should be consistent with the power-supply ratings:

$$R \geq \frac{E}{I}$$

where

R = load-circuit resistance
E = supply voltage
I = maximum supply current

If R is too low, check the load circuit with an ohmmeter to locate the trouble. (Nominal circuit resistances are included in most equipment manuals.) If the load-circuit resistance is normal, look for a shorted filter capacitor or defective regulator IC. Electrolytic capacitors fail with long (two years) disuse; the electrolytic layer may be re-formed as explained later in this chapter.

IC regulators are often victims of oscillations in their high-gain amplifiers. Such oscillations are normally prevented by capacitors that connect the IC input and output pins to the adjustment or ground pin. Check or replace these capacitors whenever a regulator has failed.

Ac ripple, or hum, is caused by feedback, poor regulation or defective filter components. Look for a defective

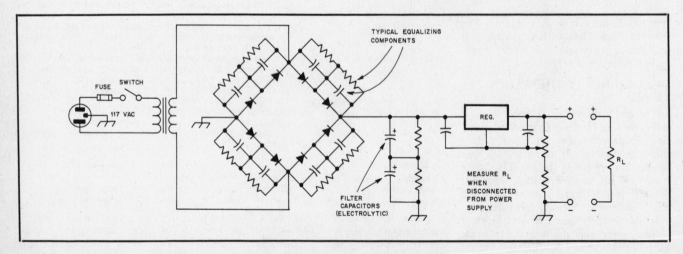

Fig. 11 — A schematic of a typical power supply showing the components mentioned in the text.

Table 6
Receiver Problems

Symptom	Cause
Low sensitivity	Semiconductor contamination weak tube, alignment
Signals and calibrator heard weakly (low S-meter readings) (strong S-meter readings)	RF chain
	AF chain, detector
No signals or calibrator heard, only hissing	RF oscillators
Distortion	
On strong signals only	AGC fault
AGC fault	Active device cut off or saturated
Difficult tuning	AGC fault
Inability to receive	Detector fault
AM weak and distorted	Poor detector, power or ground connection
CW/SSB unintelligible	BFO off frequency or dead
FM distorted	Open detector diode

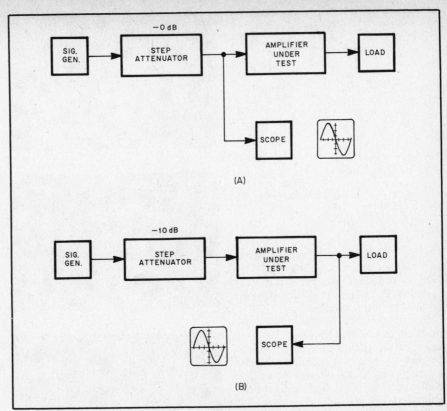

Fig. 12 — Equipment set up for amplifier gain testing. Connect the equipment as shown in A and adjust the 'scope display so that the waveform peaks are at graticule lines. Move the 'scope probe to the amplifier output and increase attenuation until the envelope size matches that of the input signal. Stage gain, independent of 'scope calibration, is shown on the step attenuator.

regulator, a shorted choke or open capacitor in the filter. In new construction projects make sure RF is not getting into the power supply.

Once the faulty component is found, inspect the surrounding circuit and consider what may have caused the problem. Referred symptoms (where a faulty component is the result of another failure, rather than the cause) are often encountered. For example, a shorted filter capacitor will increase current flow and burn out a rectifier diode. The defective diode is easy to find but the capacitor may show no visible damage. When a component has been subjected to excess current or voltage, as a result of another failure, check it.

Amplifiers

Amplifiers are the most common circuits in electronics. The output of an ideal amplifier would match the input signal in every respect except magnitude: No distortion or noise would be added. Real amplifiers *always* add noise and distortion.

Gain

Gain is the measure of amplification and it is usually expressed in decibels (dB) over a specified frequency range, known as the bandwidth or passband of the amplifier. When an amplifier is used to provide a stable load for the preceding stage, or as an impedance transformer, there may be little or no power gain. Amplifier failure usually results in a loss of gain or excessive distortion at the amplifier output. In either case, check external connections first. Is there power to the stage? Has the fuse opened? Check the speaker and leads in audio output stages, the microphone and push-to-talk (PTT)

line in transmitter audio sections.

Excess voltage, excess current or thermal runaway lead to sudden failure of semiconductors. The failure may appear as either a short, or open, circuit of one or more PN junctions.

Thermal runaway is characteristic of bipolar transistors. If degenerative feedback (the emitter resistor reduces base-emitter voltage as conduction increases) is insufficient, thermal runaway will allow excessive current flow and device failure. Check transistors by substitution, if possible.

Faulty coupling components can reduce amplifier output. Look for component failures that would increase series, or decrease shunt, impedance in the coupling network. Coupling faults can be located by signal tracing or parts substitution. Other passive-component defects reduce amplifier output by shifting bias or causing active-device failure. These failures are evident when the dc operating voltages are measured.

A fault in the AGC loop may force a transistor into cutoff or saturation. Open the AGC line to the device and substitute a variable voltage for the AGC signal. If amplifier action varies with voltage, suspect the AGC components; otherwise, suspect the amplifier.

Gain bandwidth measurements can be

made with a stable signal source, step attenuator and oscilloscope. See Fig. 12.

Noise

A slight hiss is normal in all electronic circuits. This noise is produced whenever current flows through a conductor that is warmer than absolute zero. Noise is compounded and amplified by succeeding stages. Repair is necessary only when noise threatens to obscure normally clear signals.

Semiconductors can produce hiss in two ways. First, faulty devices frequently produce excessive hiss under normal conditions. Second, any semiconductor junction produces white noise (hiss) when reverse biased. Both conditions worsen with increased temperature, so localized heat may help you find the source. Hiss can also be traced with an oscilloscope. Hissing from a resistor may increase when the component is tapped with a pencil or other insulator.

Typical sources of impulse noise are resistors, capacitors, semiconductors, vacuum tubes and connections. Poor connections and arcing are the most common causes of impulse noise.

Use signal-tracing techniques and substitution to locate noise sources. Physical shock and temperature increase noise from faulty components. A stethoscope lets you hear arcing in crowded

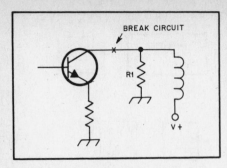

Fig. 13 — An in-circuit test for noise in active devices/coupling components. R1 should duplicate the collector-ground impedance of the active device.

circuits. Check connections at cables, sockets and switches. Look for dirty variable-capacitor wipers and potentiometers. Procedures for cleaning parts are given later in this chapter.

Rotary switches may be tested by bridging the contacts with a clip lead. Loose contacts may sometimes be repaired by gluing them to the switch deck. Operate variable components through their range while observing the noise level at the circuit output.

Potentiometers are particularly noise prone when used in dc circuits. Clean them with spray cleaner and rotate the shaft several times.

Mica trimmer capacitors often sound like lightning when arcing occurs. Test them by installing a series 0.01-μF capacitor. If the noise disappears, replace the trimmer.

Noise in transformers and loads can be spotted by substituting a suitable current-limiting resistor for the active device (Fig. 13). If the noise stops, the active component is at fault.

Oscillations

An amplifier oscillates whenever there is sufficient positive feedback. Oscillation may occur at any frequency from a slow audio buzz (often called "motorboating") up to the frequency where active-device gain is zero. Unwanted oscillations are usually the result of changes in the active device (increased junction or interelectrode capacitance), failure of an oscillation-suppressing component (open decoupling or bypass capacitors, neutralizing components) or new feedback paths (improper lead dress, dirt on the chassis or components). A shift in bias or drive levels may aggravate oscillation problems.

Oscillations that occur in audio stages do not change as the radio is tuned because the operating frequency, and therefore the component impedances, do not change. However, RF and IF oscillations vary in amplitude as operating frequency is changed.

Oscillation stops when the output circuit of the offending stage is changed enough to upset the feedback amplitude or phase. An additional bypass capacitor may pro-

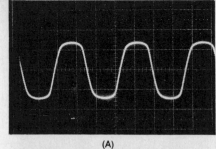

(A)

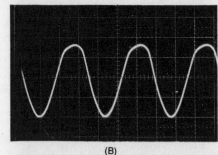

(B)

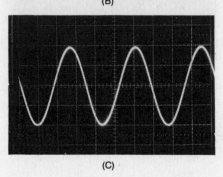

(C)

Fig. 14 — Examples of distorted wave forms. The result of clipping is shown in A and nonlinear amplification at B. A pure sine wave is shown in C for comparison. An example of harmonic distortion is shown in Fig. 4, earlier in this chapter.

duce such a change. The defective stage can be found more reliably with a signal tracer or oscilloscope.

Distortion

Distortion is the product of nonlinear operation. The resultant waveform contains not only the input signal, but new signals, at other frequencies as well. All of the frequencies combine to produce the distorted waveform. Distortion in a transmitter gives rise to splatter, harmonics and RFI.

Fig. 14 shows some typical cases of distortion. Clipping (also called flat-topping — Fig. 14A) is the consequence of excessive drive. The corners on the waveform show that harmonics are present. (A square wave contains the fundamental and all odd harmonics.) These odd harmonics would be evident as splatter above and below the operating frequency, possibly outside the amateur band. Key clicks are similar to clipping.

Harmonic distortion produces radiation at frequencies far removed from the fundamental. Harmonic distortion is a major cause of RFI; it is the product of improper neutralization or ineffective harmonic suppression.

Incorrect bias brings about unequal amplification of the positive and negative wave sections. Consequences similar to those of clipping or harmonic distortion may be present.

In most cases, the amateur's ability to detect and measure distortion is limited by available test equipment. (Instruments used for distortion measurement are the distortion meter, noise meter, spectrum analyzer and phase meter.) However, some rough measurements of distortion can be made with a function generator and an oscilloscope.

Distortion Measurement

First, set the generator for a square wave and view it on the 'scope. Use a low-capacitance probe. The wave should show square corners and a flat top. Next, inject a square wave at the amplifier input and again view the input wave on the 'scope. Any new distortion is a result of the test circuit loading the generator output. (If the wave shape is severely distorted, the test is not valid.) Now, move the test probe to the test circuit output and read the waveform. Refer to Fig. 15 to estimate circuit distortion and its cause.

The above applies only to audio amplifiers without frequency tailoring. In RF gear, the transmitter may have a very narrow audio passband, so inserting a square wave into the mic input may result in an output that is difficult to interpret. The frequency of the square wave will have a significant effect.

All components listed as sources of impulse noise may also produce distortion. Check variable capacitors and resistors, switches, connectors and sockets for dirt or corrosion. Distortion that only affects strong signals usually indicates an AGC defect.

Leaky transistors produce distorted signals. A vacuum tube with an interelectrode short also causes distortion. These semiconductor and vacuum-tube conditions may mimic AGC trouble.

IF oscillations may produce distortion. They cause constant, full AGC action. (This is evident on a 'scope display or in bias measurements.) The S meter shows strong signals even with the antenna disconnected. AGC and conduction levels in RF sections are nearly constant as the operating frequency is varied.

Improper bias often results from an overheated or open resistor. Heat causes resistor values to permanently increase. Leaky, or shorted capacitors and RF feedback can also produce distortion by disturbing bias levels. Distortion is also caused by circuit imbalance in Class AB or B amplifiers.

Oscillators

While the loss of oscillator output

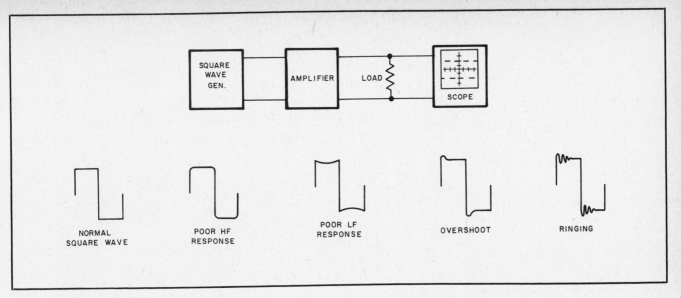

Fig. 15 — Square wave distortion and probable causes.

usually has drastic effects on equipment operation, the effects do not point directly to oscillator failure. Whenever there is weakening or complete loss of signal, check oscillator operation and frequency. There are several methods:

• Use a receiver with a coaxial probe to listen for the oscillator signal.

• A dip meter can be used to check oscillators. In the absorptive mode, tune the dip meter to within ±15 kHz of the oscillator, couple it to the circuit, and listen for a beat note in the dip-meter headphones.

A dip meter in the active mode can be set to the oscillator frequency and coupled to the oscillator output. If the oscillator is dead, the dip-meter signal will restore circuit operation. Tune the dip meter slowly, or you may pass stations so quickly that they sound like ''birdies.''

• Oscillator waveform shows on a 'scope. However, operating frequency cannot be determined accurately. Use a low-capacitance (10X) probe for oscillator observations.

• Tube oscillators usually have no grid bias at startup, but negative bias when oscillating. Use a high-impedance voltmeter to measure grid bias. The bias also changes with frequency.

• Emitter current varies with frequency in transistor oscillators. Use a sensitive, high-impedance voltmeter across the emitter resistor to observe the current level.

• Control voltage changes with oscillation (for both tubes and transistors) in a crystal oscillator. Measure the control (either grid, or base) voltage while shunting the crystal with a 0.001-μF capacitor; if the stage is oscillating, the level will vary when the capacitor is added. Do not short the crystal!

Stability

Judge stability by the standards in practice when the circuit was made. Drift of several kilohertz per hour was quite normal years ago. Your needs may demand more stability, and modifications may be in order, but drift that is consistent with the equipment design is not a defect. This applies to new equipment as well as old. It is normal for some digital displays to flash back and forth between two values for the least-significant digit.

Drift is caused by variations in the oscillator. Poor voltage regulation and heat are the most common culprits. Check regulation with a voltmeter (use one that is not affected by RF). Regulators are usually part of the oscillator circuit. Check them by substitution. Check bypass capacitors by bridging them with good capacitors.

Solid-state replacements for tubes may improve oscillator stability by reducing heat and, therefore, temperature change in the circuit.[3] Long warm-up times can be avoided by keeping the oscillator compartment warm constantly. A small lamp, such as that used in a night light, mounted in the oscillator compartment keeps components near normal operating temperature. Wire the lamp so that it is on when the receiver is off, and off when the receiver is on.

Chirp is a form of rapid drift that is usually caused by excessive oscillator loading or poor power-supply regulation. When troubleshooting chirp, look for defects in buffer stages that affect the oscillator load. (For example, a shorted coupling capacitor increases loading drastically.) Also check lead dress, tubes and switches for new feedback paths (feedback defeats buffer action).

Frequency instability may also result from defects in feedback components. Too much feedback may produce spurious

signals, while too little makes oscillator startup unreliable.

Sudden frequency changes are frequently the result of physical variations. Loose components or connections are probable causes. Check for arcing or dirt on printed-circuit boards, trimmers and variable capacitors, loose switch contacts, bad solder joints or loose connectors.

Many older rigs used rivets to hold terminal strips, tube mounting rings and transistor sockets against ground. Rivets made a fine connection to chassis ground when the rigs were new. But, as time goes on, corrosion and mechanical weakening cause intermittent or high-impedance ground connections. To locate this type of problem, use a short piece of flexible, insulated wire (such as the braid from a piece of RG-174) with a piece of insulated tubing over it. Attach an alligator clip to each end. Connect one alligator clip to the suspect terminal and the other clip to the nearest chassis ground point. If the problem disappears, drill out the offending rivet and replace it with a machine screw, lockwasher and nut of the appropriate size.

A similar problem occurs when the screws around the edge of a PC board are used to connect the board to the chassis ground. If you run into this problem, remove the mounting screws and lightly sand the edge of the circuit board with fine sandpaper. Also clean the portions of the chassis that make contact with the board and its mounting screws. Remount the board to the chassis with machine screws, lockwashers and nuts.

Dial Tracking

Dial tracking errors may be associated with oscillator operation. Misadjustments in the frequency-determining components make dial accuracy worse at the ends of the dial. Tracking errors that are constant everywhere in the passband arise from slippage in the dial drive. This is usually cured

[3]Background design tips are in the reference for Sartori at the end of this chapter.

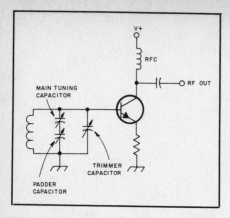

Fig. 16 — A schematic of a simple oscillator showing the locations of trimmer and padder capacitors.

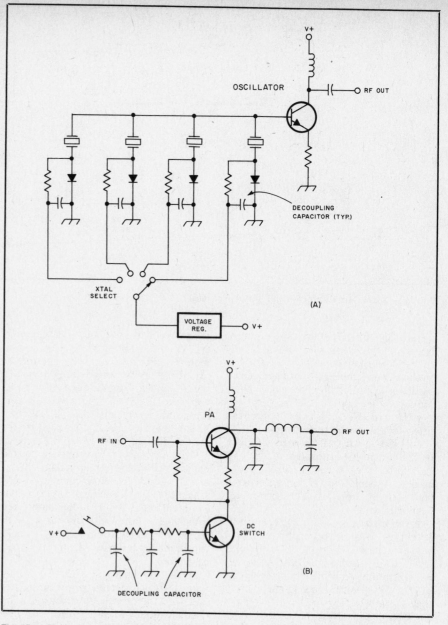

Fig. 17 — Diode switching selects oscillator crystals at A. A transistor switch is used to key a power amplifier at B.

by a simple mechanical or calibration adjustment.

Tracking at the high-frequency end of the dial is controlled by trimmer capacitors. A trimmer is a variable capacitor connected in parallel with the main tuning capacitor (see Fig. 16). The trimmer represents a higher percentage of the total capacitance at the high end of the tuning range. It has little effect on tuning characteristics at the low-frequency end of the dial.

Low-end tracking is adjusted by a padder capacitor. A padder is a variable capacitor that is connected in series with the main tuning capacitor. Padder capacitance has a greater effect at the low-frequency end of the dial. The padder capacitor is often eliminated to save money. In that case, the low-frequency tracking is adjusted by the main tuning coil.

Controls

Semiconductors have made it practical to place a transistor or diode at each point where switching is required and run only dc leads to that point. This eliminates many problems caused by long leads.

Mechanical switches are relatively rugged. They can withstand substantial voltage and current surges. The environment does not drastically affect them, and there is usually visible damage when they fail.

Semiconductor switching offers inexpensive, high-speed operation. When subjected to excess voltage or current, most transistors and diodes silently expire. Occasionally, one sends up a smoke signal to mark its passing.

Temperature alters semiconductor characteristics. A normally adequate control signal may not be effective when transistor beta is lowered by a cold environment. Heat may cause a control voltage regulator to produce an improper control signal.

The control signal is actually a bias for the semiconductor switch. Forward-biased diodes and transistors act as closed switches while reverse biased components simulate

open switches. If the control (bias) signal is not strong enough to completely saturate the semiconductor, conduction may not continue through a full ac cycle. Severe distortion results.

When dc control leads provide unwanted feedback paths, switching transistors may become modulators or mixers. Additionally, any reverse-biased semiconductor junction is a potential source of white noise.

Semiconductor switching usually reduces the cost and complexity of switching components. Switching speed is increased, contact corrosion and breakage are eliminated. In exchange, troubleshooting is complicated by additional support components such as voltage regulators and decoupling capacitors (see Fig. 17). The technician must consider many more components and symptoms when working with diode- and

transistor-switched circuits.

Another kind of control circuit is becoming common in amateur equipment. Nearly every new transceiver is controlled by a miniature computer. Microprocessor control is beyond the scope of this chapter. For successful repair of microprocessor-controlled circuits, the technician should have the knowledge and test equipment necessary for computer repair. Familiarity with machine-language programming may also be desirable.

Transmitters

Many transmitter faults have already been covered. However, there are few techniques used to ensure stable operation of RF amplifiers in transmitters that are not covered elsewhere.

A parasitic choke frequently consists of

a 51- to 100-Ω non-inductive resistor with a coil wound around the body and connected to the leads. It is used to prevent VHF and UHF oscillations in a vacuum-tube amplifier. The coil has a low impedance at the operating frequency, but effectively blocks VHF and UHF oscillations. The choke is placed in the plate lead, close to the plate connection. Parasitic chokes often fail from excessive current flow. In these cases, the resistor is charred. Occasionally, physical shock or corrosion will produce an open circuit in the coil; test for continuity with an ohmmeter. Low-frequency oscillations are usually not a problem with vacuum-tube amplifiers.

Transistor amplifiers are protected against parasitic oscillations by low-value resistors or ferrite beads in the base or collector leads. Resistors are used only at low power levels (about 0.5 W), and both methods work best when applied to the base lead. Negative feedback is used to prevent oscillations at lower frequencies. An open component in the feedback loop may cause low-frequency oscillation, especially in broadband amplifiers.

Keying

The simplest form of modulation is on/off keying. Although it may seem that there cannot be much trouble with such an elementary form of modulation, two very important transmitter faults are the result of keying problems.

Key clicks are produced by fast rise and decay times of the keying waveform. Most transmitters include an RF filter across the key connections to shape the waveform and prevent clicks. In order to prevent clicks, all amplifiers following the last keyed stage must maintain linear operation. When clicks are experienced, check the keying filter components first, then the succeeding stages. An improperly biased power amplifier, or a Class C amplifier that is not keyed, may produce key clicks even though the keying waveform is correct. Clicks caused by a linear amplifier are a sign of low-frequency parasitic oscillations.

The other modulation problem associated with on/off keying is called backwave. Backwave is a condition in which the signal is heard (usually at a reduced level) even when the key is up. This occurs when the oscillator signal feeds through a keyed amplifier; neutralization of the keyed amplifier may be necessary.

Arcing

Although the sound and smell of arcs may conjure up nostalgic memories of the days when radios were made of wood, and operators of steel, it is a sign of trouble in modern gear. Arc sites are usually easy to find because an arc that generates visible light or noticeable sound also pits and discolors conductors.

Arcing is often caused by dampness, dirt or lead dress. If the dampness is temporary, dry the area and resume operation. Dirt

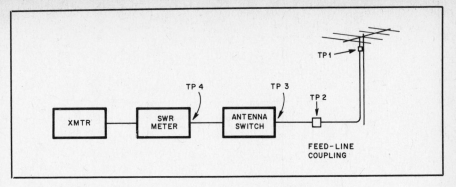

Fig. 18 — Test points used when tracing an antenna system circuit.

may be cleaned from the chassis with a residue-free cleaner. Arrange leads so high-voltage conductors are isolated. Keep them away from sharp corners and screw points.

Operating practice occasionally causes some arcing in control components. Arcing occurs in capacitors when the working voltage is exceeded. Air-dielectric variable capacitors can sustain occasional arcs without damage, but arcing indicates operation beyond circuit limits. Transmatches working beyond their ability may suffer from arcing.

Low Output Power

Some transmitters automatically reduce power in the TUNE mode. Check the owner's manual to see if the condition is normal. Check the control settings. Transmitters that use broad-bandwidth amplifiers require so little effort from the operator that control settings are seldom noticed. The CARRIER (or DRIVE) control may have been bumped. Remember to adjust tuned amplifiers after a significant change in operating frequency (usually 50 to 100 kHz).

Low power output in a transmitter may also spring from a misadjusted carrier oscillator or defective SWR protection circuit. If the carrier oscillator is set to a frequency well outside the transmitter passband, there may be no measurable output. Output power will increase steadily as the frequency is moved into the passband.

Complete loss of signal occurs when the carrier oscillator stops, although loss of the tone oscillator signal (in transmitters that use an audio tone to generate CW) may have the same effect. The tone oscillator may also be used to produce a CW sidetone. Loss of sidetone together with a loss of RF output indicates a defective tone oscillator.

Transistors may fail if the SWR protection circuit malfunctions. An open circuit in the "reflected" side of the sensing circuit leaves the transistors unprotected, a short "shuts them down."

Antennas

Begin antenna troubleshooting with a visual inspection of the antenna and surrounding area. Look for resonant conduc-

tors within a few wavelengths of the antenna. (Conductors that are perpendicular to the antenna are not significant.) If there are no obvious problems, test the antenna system.

To test the antenna system, transfer the antenna end of the feed line from the antenna to a dummy load (see Fig. 18). If the SWR appears normal, the problem must lie with the antenna itself. Make a closer inspection; look for corroded connections or an element that has slipped or broken. Check the components in any matching network at the feedpoint. Water in traps, and insects in variable capacitors or coils are common culprits.

If the SWR is still poor, move the dummy load to the next feed-line coupling that is closer to the transmitter and check the SWR again. Repeat this procedure until the SWR returns to normal. The faulty system component is the last one disconnected. Don't forget that SWR meters may have defects also. When in doubt, confirm readings with a second meter.

Balun Failures

Symptoms of balun failure include burned insulation and arcing between turns. Replace the damaged balun and avoid high-SWR operation.

A coaxial choke (usually 10 turns, 6-inch diameter for RG-58/U, larger cables require larger diameters) serves the same purpose as a balun. Falling branches, weather and so on may damage coaxial-choke baluns. Another threat to the coaxial choke is a phenomenon known as cold flow. The action is very similar to a piece of wire slicing through cheese. The center conductor slowly moves, cutting through the core dielectric as a result of the stress in the bent cable. This problem is most common in foam-dielectric cables, and is aggravated by warm weather.

RFI

RFI can be produced in any part of the antenna system. The cause is usually poor connections. Almost any metal-to-metal joint can form a diode junction. When RF is induced across the junction, rectification and strong harmonics result. Luckily, only conductors that are of a significant elec-

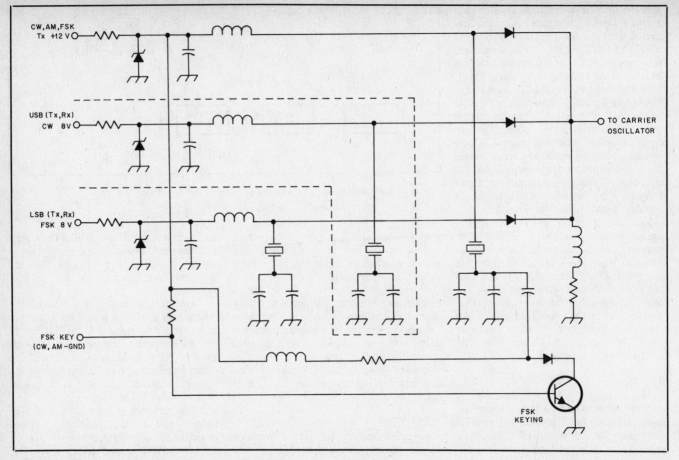

Fig. 19 — Partial schematic of a transceiver oscillator. The symptoms described in the text are caused by one or more components inside the dashed lines or a faulty USB/CW control signal.

trical length develop enough power to radiate the harmonics produced. The most likely troublemakers are tower sections, TV line amplifiers, antenna and feed-line connections, video recorders, guy wires and metal gutters. Less obvious sources can be located with direction-finding techniques. As you get closer to the RFI source, it is possible to hear the noise on higher harmonics. Switch to a higher band to prevent overload as you draw near the source. Once the source is located, correct the problem by electrically bonding the parts together. More information about RFI problems and cures appears in Chapter 40.

Transceivers

Elaborate switching schemes are used in transceivers for signal control. Many transceiver malfunctions can be attributed to relay or switching problems. Suspect the switching controls when:

- The S meter is inoperative.
- There is arcing in the tank circuit.
- Plate current is high during reception.
- There is PA noise in the receiver.
- Received signals are erratic or weak.

Since transceiver circuits are shared, stage defects frequently affect both the transmit and receive modes, although the symptoms may change with mode.

Oscillator problems usually affect both transmit and receive modes, but different oscillators, or frequencies, may be used for different emissions. Check the block diagram.

For example, one particular transceiver uses a single carrier oscillator with three different crystals (see Fig. 19). One crystal sets the carrier frequency for CW, AM and FSK transmit. Another sets USB transmit and USB/CW receive, and a third sets LSB transmit and LSB/FSK receive. This radio showed a strange symptom. After several hours of CW operation, the receiver produced only a light hiss on USB and CW. Reception was good in other modes, and the power meter showed full output during CW transmission. An examination of the block diagram and schematic showed that only one of the crystals (and seven support components) was capable of causing the problem.

Another potential trouble area is VOX control. If there is difficulty in switching to transmit, check the VOX-SENSITIVITY and ANTI-VOX control settings. Next, see if the PTT and manual (MOX) transmitter controls work. If the PTT and MOX controls function, examine the VOX-control diodes and amplifiers. Test the switches, control lines and control voltage if the transmitter does not respond to other TR controls. VOX SENSITIVITY and ANTI-VOX settings

should also be checked when the transmitter switches on in response to received audio. Suspect the anti-VOX circuitry next. Unacceptable VOX timing results from a poor VOX-delay adjustment, or a bad resistor or capacitor in the timing circuit or VOX amplifiers.

Alignment

The mixing scheme of the modern SSB transceiver is complicated. The signal passes through many mixers, oscillators and filters. Satisfactory SSB communication requires accurate adjustment of each stage. Do not attempt any alignment without a copy of the manufacturer's instructions and the necessary test equipment.

Receivers

The important characteristics of a receiver are selectivity, sensitivity, stability and fidelity. Receiver malfunctions ordinarily affect one or more of these areas.

Selectivity

Tuned transformers may develop a shorted turn, capacitors can fail and alignment is required occasionally. Such defects are accompanied by a loss of sensitivity. Filters may fall victim to physical damage from external forces. Cavities and resonators are subject to similar hazards. Except in cases of catastrophic failure (where either the filter passes all signals, or none), it is difficult to spot a loss of selectivity. Bandwidth and insertion-loss measurements are necessary to judge filter performance.

Sensitivity

A gradual loss of sensitivity results from gradual failure of an active device or faulty alignment. Sudden, partial sensitivity reductions are the effect of low RF-amplifier output, or a shift in the frequency of an oscillator or a filter passband. Complete and sudden loss of sensitivity is caused by an open circuit anywhere in the signal path or by a "dead" oscillator.

Stability

The stability of a receiver depends on its oscillators. See the section on oscillators earlier in this chapter.

Distortion

Causes of distortion are covered earlier in this chapter. Distortion may be the effect of poor connections or faulty components in the signal path. AGC circuits produce many receiver defects that appear as distortion or insensitivity.

AGC

Distortion that affects only strong signals is the normal symptom of AGC failure. All stages operate at maximum gain when the AGC influence is removed. An S meter can help diagnose AGC failure because it is operated by the AGC loop.

An open AGC bypass capacitor forces feedback through the loop and usually a squeal (oscillation). Changes in the loop time constant affect tuning. If stations consistently blast, or are too weak, the time constant is too fast. An excessively slow

time constant makes tuning difficult, and stations fade after tuning. When loop problems are extreme, AGC-controlled stages may be forced into cut off or saturation.

Some AGC circuit designs provide a shunt Zener diode to clamp loop voltage and prevent cut off or saturation. The protection diode is called an AGC gate.

Detector Problems

Detector trouble usually appears as distortion of the received signal. AM, SSB and CW signals may be weak and unintelligible, FM signals distorted. Suspect an open circuit in the detector near the detector diodes. If tests of the detector parts indicate no trouble, look for a poor connection in the power-supply or ground leads. A BFO that is "dead" or off frequency prevents SSB and CW reception.

Alignment

Unfortunately, IF transformers are as enticing to the neophyte technician as a carburetor is to a shade-tree mechanic. In truth, radio alignment (and for that matter, carburetor repair) is seldom required. Circuit alignment may be justified under the following conditions:
• The set is very old and has not been adjusted in many years.
• The circuit has been subject to abusive treatment or environment.
• There is obvious misalignment from a previous repair.
• Tuned-circuit components or crystals have been replaced.
• There is a malfunction, but *all* other circuit conditions are normal. (Faulty transformers can be located because they will not tune.)

Even if one of the above conditions is

met, do not attempt alignment unless you have the proper equipment.

Receiver alignment should progress from the detector to the antenna terminals. When working on an FM receiver, align the detector first, then the IF and limiter stages, and finally the RF and local oscillator stages. For an AM receiver, align the IF stages first, then the RF and oscillator stages.

Both AM and FM receivers can be aligned in much the same manner:

1) Set the receiver RF GAIN to maximum, BFO control to zero (or center) and tune to the high end of the receiver passband.

2) Disable the AGC.

3) Set the signal source to the center of the IF passband, with no modulation and minimum signal level.

4) Connect the signal source to the input of the IF section.

5) Connect a voltmeter to the IF output as shown in Fig. 20.

6) Adjust the signal-source level for a slight indication on the voltmeter.

7) Peak each transformer in order, from the meter to the signal source.

The adjustments interact; repeat steps 6 and 7 until adjustment brings no noticeable improvement.

8) Remove the signal source from the IF-section input, reduce the level to minimum, set the frequency to that shown on the receiver dial and connect the source to the antenna terminals.

9) Adjust the signal level to give a slight reading on the voltmeter.

10) Adjust the trimmer capacitor of the RF amplifier for a peak reading of the test signal. (Verify that you are reading the correct signal by switching the source on and off.)

11) Reset the signal source and the

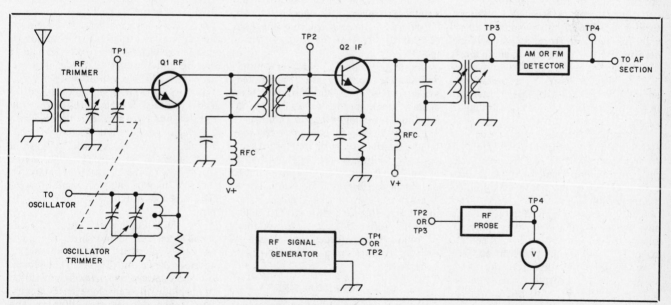

Fig. 20 — Typical receiver alignment test points. To align the entire radio, connect a dc voltmeter at TP4. Inject an IF signal at TP2, and adjust the IF transformers. Move the signal generator to TP1 and inject RF for alignment of the RF and oscillator stages. To align a single stage, place the generator at the input and an RF voltmeter at the output: TP1/TP2 for RF, TP2/TP3 for IF.

receiver tuning for the low end of the passband.

12) Adjust the local-oscillator padder for peak reading.

13) Repeat steps 8 through 11 until the settings do not change.

Digital Equipment

Few ham shacks across the country show no signs of the digital revolution. The inscrutable number crunchers have brought automated precision to everything from desk clocks to computer-controlled EME antenna arrays. Although every aspect of their operation may be resolved to a simple 1 or 0, the symptoms of their failure are far more complicated. As with other equipment:

• Observe the operating characteristics.
• Study the block diagram and the schematic.
• Test.
• Replace defective parts.

Problems in digital circuits have two elementary causes. First, the circuit may give false counts because of electrical noise at the input. Second, the gates may lock in one state.

False counts from noise are especially likely in a ham shack. (A 15- to 20-μsec voltage spike can trigger a TTL flip-flop.) Amateur Radio equipment often switches heavy loads; the attendant transients can follow the ac line or radiate directly to nearby digital equipment. Ringing in filters or leads or oscillation in the digital circuit can also produce false counts.

The effects of extra counts are determined by the design of the digital equipment: A station clock may run fast, but a microprocessor-controlled transceiver may "decide" that it is only a receiver. It might even be difficult to determine that there is a problem without a logic analyzer or multitrace 'scope and a thorough understanding of circuit operation.

Begin by removing the faulty equipment from RF fields. If the symptoms cease with the removal from RF, it will be necessary to shield the equipment against such fields. When it is possible to move the device, distance may be a permanent answer. In other cases, bypass or filter the power, input and output lines of the digital circuit.

Microprocessors use clock speeds of 1 to 10 MHz; it may be impossible to filter RF from the lines when the RF is near the clock frequency. In these cases, the best approach is to shield the digital circuit and all lines running to it.

RFI to nearby equipment may also be a sign that the station is radiating spurious signals. Inspect the station ground and all connections when such problems appear. Also check for radiation from chassis or ac lines with an absorptive wavemeter or a field-strength meter with a capacitive probe.

Strong RF fields may also cause lock up. Deal with this problem in the same

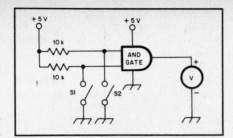

Fig. 21 — Simple digital circuits can be tested with a few components. In this case, an AND gate is tested. Open and close S1 and S2 while comparing the VOLTS reading with a truth table for the device.

manner as with RF-induced noise. Lockup may also be the result of a short, or open, circuit.

Check for the correct voltages at the pins of each chip. (Make the measurements at the pins of the IC, not the socket pins.) Next, smell and feel each IC for heat. Excessive heat is often evidence of a shorted IC.

Repair any power faults, replace overheated components and check for restored operation. If there is still a malfunction, test the circuit board and sockets for open or shorted connections.

Record the placement and orientation of any active devices installed in sockets, then remove them from the circuit board to prevent damage during testing. (Follow the handling precautions for MOS devices.) Consider the maximum voltage specifications of any devices that cannot be removed before testing. Some devices can be damaged by 0.6 V on certain pins. After these precautions, proceed to test the board with an ohmmeter.

Last, lockup can be the effect of a failed logic gate or IC. In clocked circuits, listen for the clock signal with a coax probe and a suitable receiver. If the signal is found at the clock chip, trace it to each of the other ICs, to be sure that the clock system is intact.

If a suitable scope (15-MHz bandwidth, with storage capability and a x10 probe) is available, check pulse timing and duration against circuit specifications. Pay special attention at analog-to-digital converters.

Logic gates, flip-flops and counters may be tested (see Fig. 21) by triggering them, manually, with a power supply (4 to 5 V is a safe level). Diodes may be checked with an ohmmeter. Testing of more complicated ICs requires the use of a logic analyzer, multitrace scope or a dedicated IC tester.

Check repaired circuits for restored operation. Test frequency counters against another counter or a calibrated signal generator.

SYSTEMATIC APPROACHES

The instinctive approach works well for those with years of troubleshooting experience. Those of us who are new to this game need some guidance. A systematic approach is a disciplined procedure that allows us to tackle problems in unfamiliar equipment with a reasonable hope of success.

There are two common systematic approaches to troubleshooting at the block level. The first is signal tracing; the second is signal injection. The two techniques are very similar. Differences in test equipment and the circuit under test determine which method is best in a given situation.

Signal Tracing

This approach is best for testing transmitters, solid-state and high-impedance circuits. Signal tracing is the best way to check transmitter blocks because the necessary signals are present by design. Most signal generators cannot supply the wide range of signal levels required to test a transmitter.

Signal tracing requires that a steady signal be present in the circuit passband. (An off-the-air signal may be used when testing a receiver. When the receiver is completely disabled, the signal tracer may be connected at the output of the first RF stage as an indicator of incoming signals.) A signal generator or other *stable* signal source may be used. Use a signal of −80 dBm (22 μV) for most amateur receivers. Higher signal levels (7 dBm, 0.5 V) may damage sensitive components.

Signal Tracing Equipment

A voltmeter, with an RF probe, is the most common instrument used for signal tracing. Low-level signals cannot be measured accurately with this instrument. Signals that do not exceed the junction drop of the diode, in the probe, will not register at all, but the presence, or absence, of larger signals can be observed.

A dedicated signal tracer can also be used. It is a simple instrument that may not have adjustable sensitivity. Signal information is usually presented as audio. An experienced technician may be able to judge the level and distortion of the signal by ear; only gross changes will be noticeable.

Absorptive wavemeters may be used for signal tracing if they are sufficiently sensitive and have a suitable probe. Use a wavemeter with headphones to trace weak signals. The frequency selectivity of a wavemeter is an advantage in signal tracing. Tracing a crowded circuit, however, may be impossible unless the instrument has provision for a capacitive probe.

An oscilloscope is the most versatile signal tracer. It offers high input impedance, variable sensitivity, and a constant display of the traced waveform. A demodulator probe permits observation of the modulation envelope in RF circuits. Dual-trace scopes can display the waveforms present at the input and output of a circuit simultaneously. Gross distor-

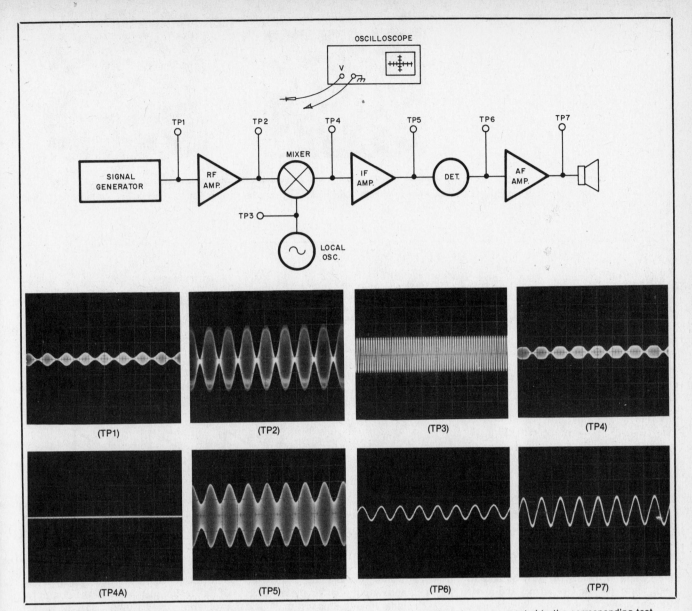

Fig. 22 — Signal tracing in a simple receiver. Photos TP1 through TP6 show typical patterns for a 'scope connected to the corresponding test points. TP1 through TP4 show 0.5 V per vertical division; TP5 through TP7 show 2.0 V per division. TP4A is TP4 with signal source removed. All patterns were made with a 10X probe and a 2-mS-per division sweep rate.

tion and phase relationship are visible on a dual-trace display.

Signal-Tracing Procedure

First, make sure that the circuit under test and test instruments are isolated from the ac line by transformers. Set the signal source to an appropriate level and frequency. Switch on power to the test circuit and connect the signal-source output to the test-circuit input. Place the tracer probe at the circuit input and check for presence of the test signal; observe the characteristics of the signal if you are using a 'scope (see Fig. 22). Compare the detected signal to the source signal during tracing. Move the tracer probe to the output of the next stage and observe the signal. Signal level should increase in amplifier stages and may decrease slightly in other stages. The signal will not be present at the output of a "dead" stage.

Low-impedance test points may not provide sufficient signal to drive a high-impedance signal tracer; tracer sensitivity is important. Also, the circuit/probe impedance relationship may make the output level appear low in stages where there is a radical impedance change from input to output (see Fig. 23). It is important to note that the circuit in Fig. 23 is a current amplifier, and that within reason the voltages at TP1 and TP2 are equal and in phase. Remember that there are two signals (the test signal and the local oscillator signal) present in a mixer stage. Switch the signal source on and off repeatedly to make sure that the tracer reading varies (it need not disappear) with source switching.

Signal Injection

Like signal tracing, signal injection is particularly suited to some situations. Signal injection is a good choice for receiver troubleshooting because a detector is already provided as part of the design. Also, the required signal-injection levels vary less widely in a receiver than in a transmitter. Signal injection is also better for vacuum-tube circuits. Low-impedance circuits may not produce enough signal to drive available signal tracers; signal injection may then be the best technique.

Signal-Injection Equipment

If you are testing equipment that does not include a suitable detector as part of the circuit, some form of signal detector is required. Any of the instruments used for signal tracing are adequate. Remember to consider the signal level at the test point when choosing an instrument.

The signal source used for injection must

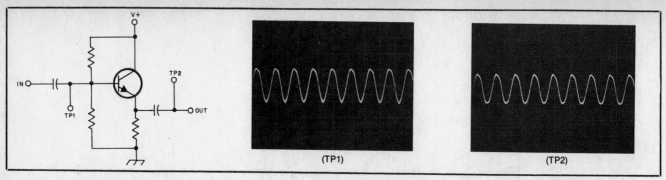

Fig. 23 — The effect of circuit impedance on a 'scope display. Although the circuit functions as an amplifier, the change in impedance from TP1 to TP2 results in the traces shown.

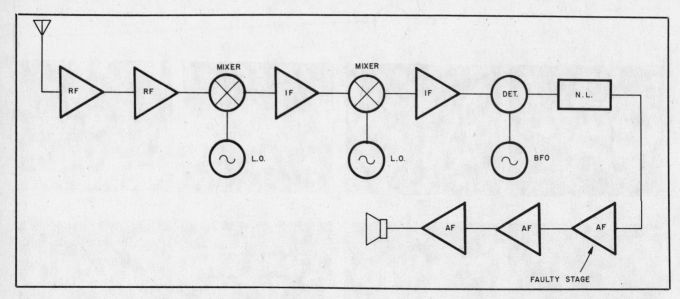

Fig. 24 — The 18-tube receiver diagnosed by the "divide and conquer" technique.

meet many demands. It must be able to supply appropriate frequencies and levels for each stage to be tested. For example, a typical superhet receiver requires AF, IF and RF signals that vary from 6 volts at AF, to 0.2 μV at RF. Each conversion stage used in a receiver requires another IF from the signal source.

Signal-Injection Procedure

If an external detector is required, set it to the proper level and connect it to the test circuit. Set the signal source for AF and inject a signal directly into the signal detector to test operation of the injector and detector.

Move the signal source to the input of the preceding stage and observe the signal. Continue moving the signal source to the inputs of successive stages. (Use a 50-Ω resistor, in series with the signal-source lead to inject a signal at the input of an inverting amplifier.) The detected signal disappears when the source is at the input of a defective stage. Prevent stage overload by reducing the level of the injected signal as testing progresses through the circuit. Remember to use suitable frequencies for each tested stage.

Make a rough check of stage gain by using the same injected-signal strength at the input and output of an amplifier stage. As with signal tracing, this test may yield misleading results if there is a radical difference in impedance from stage input to output. Understand circuit operation before testing!

Mixer stages present a special problem because they have two inputs, rather than one. A lack of signal at the test detector may result from either a faulty mixer or a faulty local oscillator (LO). Check oscillator operation with a 'scope, absorption wavemeter or by listening on another receiver. If none of these instruments are handy, inject the frequency of the LO at the LO output. If a dead oscillator is the only problem, this should restore operation.

If the oscillator is operating, but off frequency, a multitude of spurious responses will appear. A simple signal injector that produces many frequencies simultaneously is not suitable for this test. Use a well-shielded signal generator set to an appropriate level at the LO frequency.

Divide and Conquer

Both signal tracing and signal-injection procedures may be speeded by application of a technique known in computer programming as a binary search. Rather than check each stage sequentially, check a point halfway through the nonfunctioning section. As an example:

A surplus 18-tube receiver is not working. All of the filaments light, but there is absolutely no response from the speaker.

First, substitute a suitable speaker — still no sound. Next, check the high-voltage supply — no problem there. No clues indicate any particular stage. Signal tracing or injection must provide the answer.

Get out the signal generator and switch it on. Once the signal is stable, set the generator for a low-level RF signal, switch the RF OFF and connect the output to the receiver. Switch the RF ON again and place a high-Z signal-tracer probe at the lug of the antenna connection. Instantly, the tracer emits a strong audio note. Good; the test equipment is functioning.

Now, move the probe to the input of the receiver detector. As the tracer probe touches the circuit the familiar note sounds. Next, set the tracer for audio and place the probe halfway through the audio chain. It is silent! Move the probe halfway back to the detector, and the note appears once again. Yet, no signal is present at the output of the stage. The third audio stage is faulty.

Saving Time

Under certain conditions, the block search may be speeded by testing at the middle of successively smaller circuit sections. Each test limits the fault to one half of the remaining circuit (see Fig. 24). The receiver has 18 stages and the fault is in stage 12. This approach requires only four tests to locate the faulty stage, a substantial saving of time.

The divide-and-conquer tactic cannot be used in equipment in which the signal path splits during processing. Test readings taken inside feedback loops are misleading unless you understand the circuit and the waveform to be expected at each point in the test circuit. It is best to consider all stages within a feedback loop as a single block during the block search.

TESTING WITHIN A STAGE

The search is now nearing an end. A few dc measurements will usually pinpoint one or more specific components that need adjustment or replacement. First, check the active device. If the device can be tested easily in the circuit, do so; otherwise, go on to check dc voltages.

Look in the service manual to determine if the circuit is controlled by AGC. The service manual should either give test procedures for the stage with the AGC active, or have instructions for disabling the AGC.

Check the parts in the circuit against the schematic diagram to be sure that they are reasonably close to the design values (someone may have changed them during repairs). Next, make some calculations to see what the voltage should be. It is important to consider the effect of the meter in these calculations. Remember, however, that the calculated voltages are nominal; measured voltages may vary from the calculations. When making measurements, remember the following points:
- Make measurements at device leads, not at circuit-board traces or socket lugs.
- Use small test probes to prevent accidental shorts.
- Never connect or disconnect power to solid-state circuits with the switch on.

Symptoms and Faults in a Typical Amplifier Circuit

Fig. 25 is a schematic of a common-emitter transistor amplifier. The emitter, base and collector leads are labeled e, b and c, respectively. Important dc voltages are measured at these points and designated V_e, V_b and V_c. Similarly, the important currents are I_e, I_b and I_c. V+ indicates the supply voltage. Emitter-base internal resistance, R_b, is approximated by the formula:

$$R_b = \frac{26 \beta}{I_e} \qquad \text{(Eq. 2)}$$

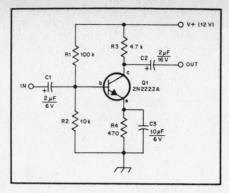

Fig. 25 — A typical common-emitter amplifier (audio).

where I_e is expressed in milliamperes. The "junction drop," V_j, is the potential measured across a semiconductor junction that is conducting. V_j is typically 0.6 V for silicon and 0.2 V for germanium transistors.

R1 and R2 form a voltage divider that supplies dc bias (V_b) for the transistor. Normally, V_e, will be equal to V_b less the e-b junction drop. R4 provides degenerative dc bias, while C3 provides a low-impedance path for the signal. From this information, normal operating voltages can be calculated:

$$V_b = \frac{(V+)R2}{(R1+R2)} = 1.09 \text{ V} \qquad \text{(Eq. 3)}$$

$$V_e = V_b - V_j = 0.49 \text{ V} \qquad \text{(Eq. 4)}$$

$$I_e = \frac{V_e}{R4} = 0.001 \text{ A} \qquad \text{(Eq. 5)}$$

$$I_c = \frac{I_e\beta}{(\beta+1)} = 0.001 \text{ A} \qquad \text{(Eq. 6)}$$

$$V_c = (V+) - (I_c \times R3) = 7.14 \text{ V} \qquad \text{(Eq. 7)}$$

Refer to Fig. 25 while reading the symptoms associated with failure of various components in the circuit.

- R1 open:
$V_b = 0$, $V_e = 0$, $V_c = 12$

V_b drops to ground potential because current through the voltage divider stops. When $V_b < V_e + V_j$, the transistor stops conducting. V_c rises to V+.

- R1 short:
$V_b = 12$, $V_e = 11.4$, $V_c < V_b$

Since the resistance of R1 is zero, $V_b = V+$. This biases the transistor into saturation (full on), usually resulting in destruction of the transistor, R3 or R4.

- R2 open:
$V_b = 1.7$, $V_e = 1.1$, $V_c < V_b$

The voltage divider becomes R1 in series with R4 and R_b. The combination of R1, R4 and R_b sets V_b at saturation, so that V_c is approximately equal to V_e.

- R2 short:
$V_b = 0$, $V_e = 0$, $V_c = 12$

A short in R2 has the same effect as an open R1.

- R3 open:
$V_b = 0.62$, $V_e = 0.02$, $V_c < V_b$

No output. Since R3 is open there is no I_c. This reduces I_e and V_e. The difference between V_e and V_b forward biases the e-b junction and the series combination of R4 + R_b appears in parallel with R2 divider. The circuit stabilizes with V_e close to zero and V_b equal to V_e plus V_j. V_c is approximately equal to V_e.

- R3 short:
$V_b = 1.09$, $V_e = 0.49$, $V_c = 12$

No output. If R3 = 0, then the output voltage $V_c = (V+) - (IC \times 0) = V+$.

- R4 open:
$V_b = 1.09$, $V_e = 0$, $V_c = 12$

No I_e can flow, therefore $I_b = I_c = 0$. V_c rises to V+. V_b changes imperceptibly. Attempts to measure V_e will indicate V_e slightly less than V_b because the meter provides a high-Z path for a small I_e.

- C1 or C2 open:
$V_b = 1.09$, $V_e = 0.49$, $V_c = 7.14$

Bias is normal; but the signal is gone. Loss of signal with normal bias can only be caused by a coupling capacitor. A 'scope or signal tracer can be used to show presence of signal and locate the faulty capacitor.

- C3 open:
$V_b = 1.09$, $V_e = 0.49$, $V_c = 7.14$

When C3 opens, impedance to the signal is increased and gain is reduced. The voltage gain becomes R3/R4.

- C3 short:
$V_b = 0.7$, $V_e = 0$, $V_c < V_b$

This shorts R4, so V_e becomes 0 V. I_b increases until V_b is equal to V_j. The transistor is saturated, so V_c is approximately equal to V_e.

- c-b junction open:
$V_b = 0.62$, $V_e = 0.02$, $V_c = 12$

No I_c flows, so V_c rises to V+. Other symptoms simulate an open R3.

- c-b short:
$V_b = 1.64$, $V_e = 1.04$, $V_c = 1.64$
$V_c = V_b$ because of the short.
$I_e = V+/(R3+R4)$; $V_e = R4 \times I_e$.
$V_b = V_e + V_j$.

- e-b open:
$V_b = 1.09$, $V_e = 0$, $V_c = 12$

No I_b and, therefore, no I_e can flow. $V_c = V+$; $V_e = 0$. V_b is normal.

- e-b short:
$V_b = 0.05$, $V_e = 0.05$, $V_c = 12$

$V_b = V_e$; V_e is at a low value because R4 is in parallel with R2, thus lowering V_b. Since the e-b junction is shorted, R_b is 0; all emitter current flows through the base.

There is no I_c; $V_c = V+$.

• e-c short:
$V_b = 1.09$, $V_e = 1.09$, $V_c = 1.09$, $V_c = V_e$. Since the emitter-collector path has no resistance, all I_e flows to the collector and there is no I_b. V_b is normal, but $V_e = V_c$ as determined by the R3/R4 voltage divider.

Voltage Levels

There are two voltage levels in most circuits. Most component failures (opens and shorts) will force one of the dc measurements very close to one of these levels. Typical failures that show up as incorrect dc voltages include: open coupling transformers; shorted capacitors; open, shorted or overheated resistors and open or shorted semiconductors.

For example, consider the circuit of Fig. 25. If R2 failed open, the voltage-divider action at the base would stop. V_b would rise to 12 V (R1 becomes a pull-up resistor) and bias the e-b junction on. With the e-b junction conducting, a new divider network is formed of R1 in series with the internal base resistance of the transistor and R4. V_b will stabilize based on this new divider — if the transistor is not destroyed while saturated. Now proceed to test the suspected parts and make the repair.

Components

Each electronic component has a function. This section acquaints you with the functions, failure modes and test procedures of resistors, capacitors, inductors and other components. Test the components implicated by symptoms and stage-level testing. In most cases, a particular faulty component will be located by these tests. If a faulty component is not indicated, check the circuit adjustments. As a last resort, use a shotgun approach. Replace all parts in the problem area with components that are known to be good.

Wires and Connectors

Short circuits are caused by physical damage to insulation or by conductive contamination. Damaged insulation is usually apparent during a close visual inspection of the conductor or connector. Look carefully where conductors come close to corners or sharp objects. Consider areas where wires and connectors are subject to stress.

Conductive contaminants range from water to metal filings. Most can be removed by a thorough cleaning. Any of the residue-free cleaners can be used, but remember that the cleaner may also be conductive. Do not apply power to the circuit until the area is completely dry. Keep the cleaners away from the variable-capacitor plates, transformers and parts that may be harmed by the chemical. The most common conductive contaminant is solder.

Increased resistance and open connections are most often the outgrowth of repeated flexing. Nearly everyone has broken a wire by bending it back and forth, and broken wires are usually easy to detect. Connector failure is more insidious. Most connectors maintain contact as a result of spring tension that forces two conductors together. As the parts age, they become brittle and lose tension.

Any connection may deteriorate because of nonconductive corrosion at the contacts. Solder helps prevent this problem but even soldered joints suffer from corrosion when exposed to weather.

Since poor connections increase joint resistance, the dissipated power (I^2R) in the joint increases. Signs of excess heat often surround poor connections in circuits that carry moderate current.

Check for short and open circuits with an ohmmeter or continuity tester. Repair worn insulation by replacing the wire or securing an insulating sleeve (spaghetti) or heat-shrink tubing over the worn area. Clean those connections that short as a result of contamination. Replace broken wires. Occasionally, corroded connectors may be repaired by cleaning, but replacement of the conductor/connector is usually required. Solder all connections that may be subject to harsh environments and protect them with acrylic enamel, RTV compound, or a similar coating.

When replacing conductors, use the same material and size, if possible. Substitute only wire of greater cross-sectional area (smaller gauge number) or material of greater conductivity. Insulated wire should be rated at the same, or higher, temperature and voltage as the wire it replaces. Replacement connectors should be chosen with consideration given to voltage and current ratings. Use connectors with symmetrical pin arrangements only where correct insertion is not critical.

Resistors

Resistors are used to limit or divide current and to divide voltage. They fail by increasing resistance (often becoming an open circuit) or developing leakage. Excessive heat causes most resistors to permanently change value. Such heat may come from external sources or from power dissipated within the resistor. Sufficient heat burns the resistor until it becomes an open circuit. Resistors also fracture (become an open circuit) as a result of physical shock.

Leakage is only a problem with resistors of very high value (100 kΩ or more). Leakage current actually flows through dirt paths on the outside of the resistor body, mounts or circuit board, not through the resistor itself.

Resistors that have permanently changed value may be replaced, or repaired by connecting an appropriate resistor in parallel with the faulty part. Leakage is cured by cleaning the resistor body and surrounding area.

Potentiometers suffer from noise (especially in dc circuits) in addition to the problems of fixed-value resistors. Dirt often causes intermittent contact between the wiper and resistive element. To cure the problem, spray cleaner into the potentiometer (through holes in the case) and rotate the shaft a few times. The resistive element in wire-wound potentiometers eventually wears and breaks from the sliding action of the wiper.

Replacement resistors should be of the same value, tolerance, type and power rating as the original. The value should stay within tolerance: Suppose a 10-kΩ resistor with a 20% tolerance, fails. Resistors of closer tolerance need not have the same nominal value as the original, as long as their value, plus or minus their tolerance, falls within the tolerance range (10,000 × 0.8 = 8000 Ω to 10,000 × 1.2 = 12,000 Ω) of the original. A 9100-Ω, 5%-tolerance resistor should have a value between 9100 × 0.95 = 8645 Ω and 9100 × 1.05 = 9555 Ω. Thus, the 5% resistor is a suitable replacement.

Replacement resistors may be of a different type than the original, if the characteristics of the replacement are consistent with circuit requirements. Some resistor families and their characteristics are:

• Carbon composition — Carbon-composition resistor values are moderately stable from 0° to 60° C (their resistance increases above *and below* this temperature range). They are not inductive, but they are relatively noisy, and have wide tolerances.

• Wire wound — These resistors are very stable in value, and are made to close tolerances. They are, however, small coils of wire and exhibit inductance when used in ac circuits.

• Metal film — Metal film resistors are exceptionally stable in value. They are made to close tolerances, are noninductive and create little noise. High-power (above 3 W) versions are not available.

• Carbon film — Carbon film resistors

are not as stable, nor are their tolerances as close, as those of metal-film resistors. They are noninductive. Carbon-film resistors are especially useful because they have a *negative* temperature characteristic. That is, their resistance decreases with temperature. They are often used in temperature-compensation circuits.

Substitute resistors may have a greater power rating than the original, except in high-power emitter circuits where the resistor also acts as a fuse.

Potentiometers should be replaced with the same type (carbon or wire wound) and taper (linear, log, reverse log, and so on) as the original. Keep the same, or better tolerance, but be mindful of the power rating. Potentiometer power ratings are specified for the entire resistive element. A 10-W potentiometer can dissipate 10 W only when set for maximum resistance. The allowable dissipation for a segment of the resistor is proportional to the resistance of that segment, depending on taper. If the wiper of a 10-W unit were set to 30% of the total resistance, it could only dissipate 3 W in that portion of the element.

In all cases, mount high-temperature resistors away from heat-sensitive components. Also, keep carbon resistors away from heat sources. This will extend their life and assure minimum resistance variations.

Fuses

Always use exact replacement fuses. It is important for maximum safety that the current, voltage and timing (fast-, normal- or slow-blow) be identical to the original. Also, the replacement fuse should be the same physical size. Never attempt to force a fuse that is almost the right physical size to fit into the fuse-holder. The substitution of a bar, wire or penny for a fuse invites a "smoke party."

Capacitors

Capacitors are used to store energy (in the electrostatic field between the plates) or block the flow of dc. They fail as a result of excess current, voltage or temperature. Dielectric failure or a permanent shift of capacitance are symptoms of current overload. Voltage overload may produce internal corona or puncture the dielectric. Excess heat may come from nearby heat-generating components, excessive current, an operating frequency well beyond the capacitor design frequency or a lossy dielectric. Leakage occurs directly, through the dielectric, or through leakage paths (dirt) across the component body or chassis surface.

Testing

The easiest way to test capacitors (out of circuit) is with an ohmmeter. In this test, the resistance of the meter forms a timing circuit with the capacitor to be checked. Capacitors from 0.01 μF to 1.0 μF can be tested with common ohmmeters. Set the meter to its highest range and connect the test leads across the discharged capacitor. When the leads are connected, current

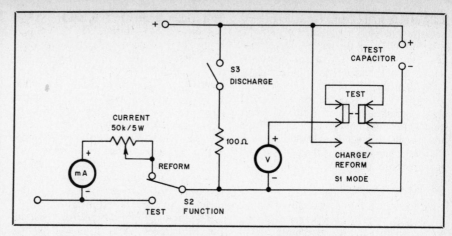

Fig. 26 — Fixture for testing capacitors and reforming the dielectric or electrolytic capacitors. Use 12 V for testing and the capacitor working voltage for reformation.

begins to flow. The capacitor passes current easily when discharged, but less easily as the charge builds. We see this on the meter as a low resistance that builds, over time, to infinity. The speed of the resistance build up corresponds to capacitance, small capacitance values seem to achieve infinite resistance instantly. A 0.01-μF capacitor checked with a 11-MΩ FETVOM would increase from zero to a two-thirds scale reading in 0.11 second, while a 1-μF unit would require 11 seconds to reach the same reading. If the tested capacitor does not reach infinity within five times the period taken to reach the two-thirds point, it has excess leakage. If the meter reads infinite resistance immediately, the capacitor is open. (Aluminum electrolytic capacitors cannot be measured by this method because they often give erroneous high-leakage readings.)

Fig. 26 shows a circuit that may be used to test capacitors. Make sure that the power supply is off, set S1 to CHARGE and S2 to TEST, then connect the capacitor to the circuit. Switch on the power supply and allow the capacitor to charge until the voltmeter reading stabilizes. Next, switch S1 to TEST and watch the meter for a few seconds. If the capacitor is good, the meter will show no potential. (In fact, there will be some voltage drop, but the meter will not be able to measure it.) Any appreciable voltage indicates excess leakage. After testing, set S1 to CHARGE, switch off the power supply, and press the DISCHARGE button until the meter shows 0 V, then remove the capacitor from the test circuit.

Capacitance can be measured with a capacitance meter, an RX bridge or a dip meter. An RX bridge (see "A Noise Bridge for 160 through 10 Meters," Chapter 25) can be used to measure capacitance by placing a 50-Ω resistor in series with the capacitor. The procedure is similar to that explained for calibrating the bridge.

To determine capacitance with a dip meter, a parallel-resonant circuit should be constructed using the capacitor of unknown value and an inductor of known value. Measure the resonant frequency of the circuit, and calculate the value of the

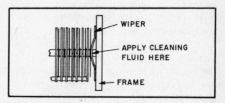

Fig. 27 — Partial view of an air-dielectric variable capacitor. Apply cleaning fluid where the wiper contacts the rotor plates.

unknown component, using the formula:

$$unknown = \frac{\left(\dfrac{1000}{2\pi f_o}\right)^2}{known} \quad \text{(Eq. 9)}$$

where

f_o = resonant frequency in MHz
known = the value (L in μH, C in pF) of the known component

This formula works for unknowns of either C or L.

It is best to keep a collection of known components that have been measured on accurate L or C meters. Alternatively, a "standard" value might be taken as the average value of several (10 or more) components with the same marked value (disregard parts with values that differ greatly from the rest), or that of a section of coaxial cable. (Capacitance per unit length is usually specified.) Prewound, air-dielectric coils, such as Miniductor®, Polycoil®, and Air-dux® are high-Q inductors with reliable values. The accuracy of tests made with any of these alternatives depends on the accuracy of the "standard" value component. Further information on this technique appears in Bartlett's article, "Calculating Component Values," in the Nov. 1978 *QST*.

Cleaning

The only variety of common capacitor that can be repaired is the air-dielectric variable capacitor. Electrical connection to the moving plates is made through a spring-wiper arrangement (see Fig. 27). Dirt normally builds on the contact area, and cleaning is occasionally required.

Before cleaning the wiper/contact, use gentle air pressure and a soft brush to remove all dust and dirt from the capacitor plates. Then use a screwdriver blade to lift the wiper gently from the rotor plate and apply some cleaning fluid. Rotate the shaft quickly several times to work in the fluid and establish contact. Use the cleaning fluid sparingly, and keep it off the plates except at the contact point.

Replacement Capacitors

Replacement capacitors should match the original in value, tolerance, dielectric, working voltage and temperature coefficient. If exact replacements are not available, substitutes may vary from the original part in the following respects.

Bypass capacitors may vary from one to three times the capacitance of the original. Coupling capacitors may vary from one-half to twice the value of the original. Capacitance values in tuned circuits (especially filters) must be held constant, and any replacement will probably require circuit alignment.

If the same type of capacitor is not available, use one with better dielectric characteristics; however, do not use mica capacitors to block high dc voltages. Also, do not substitute polarized capacitors for nonpolarized parts. Capacitors with a higher working voltage may be used, unless the replacement is an electrolytic. When choosing substitute capacitors, consider the characteristics of each type:

• Air dielectric — Air-dielectric capacitors are large when compared to other types of the same value. They are very stable over a wide temperature range. Losses are low, and a high Q can be obtained. Humidity and atmospheric pressure can cause arcing. Consider plate shape when choosing replacement air variable capacitors; correct plate shape assures proper dial tracking.

• Ceramic — There are two types of ceramic capacitors. Those with a low dielectric constant are relatively large, but very stable and nearly as good as mica capacitors at HF.

High dielectric constant ceramic capacitors are physically small for their capacitance, but their value is not as stable. Their dielectric properties vary with temperature, applied voltage and operating frequency. Use them only in coupling and bypass roles. Tolerances are usually +100% and −20%.

Ceramic capacitors are produced with values from 100 pF to 0.1 μF. They are not polarized.

• Electrolytic — Electrolytic capacitors are popular because they provide high capacitance values in small packages at a reasonable cost. Leakage is high, as is inductance, and they are polarized. Internal inductance restricts foil electrolytics to low-frequency applications. They are available with values from 1 to 500,000 μF.

A dielectric is "formed" by the voltage applied to the capacitor. As a consequence, electrolytics should not be used where the dc potential is well below the capacitor working voltage. It is best to slowly walk the voltage up, increasing about 10% every 10-15 minutes.

The dielectric must be re-formed when the capacitor has been stored, or operated at low voltage for long periods (two years or more). Re-formation is accomplished by applying full working voltage to the capacitor, with current limited, until the dielectric strength is restored.

The circuit shown in Fig. 26 can be used for this purpose. Set S2 to REFORM, S1 to REFORM and R2 to maximum resistance. Place the capacitor in the circuit, turn on the power supply and set it to the capacitor working voltage. Adjust CURRENT for a 10-mA reading and leave the power on until the voltmeter shows the full supply voltage across the capacitor. It may take several days to complete the re-formation; if full supply voltage is not achieved, the capacitor must be replaced.

Tantalum electrolytic capacitors perform better than aluminum units but their cost is higher. They are smaller, lighter and more stable, with less leakage and inductance than their aluminum counterparts. Re-formation problems are less frequent, but working voltages are not as high as with aluminum units.

• Paper dielectric — Paper capacitors are inexpensive; capacitances from 500 pF to 50 μF are available. High working voltages are possible, but paper-dielectric capacitors have high leakage rates and tolerances are no better than 10 to 20%.

Paper-dielectric capacitors are not polarized; however, the body of the capacitor is usually marked with a color band at one end. The band indicates the terminal that is connected to the outermost plate of the capacitor. This terminal is represented by a curved line in the schematic symbol for a capacitor; it should be connected to ground, or the side of the circuit with least potential, as a safety precaution.

• Plastic film — Capacitors with plastic-film (polystyrene, polyethelyne or Mylar) dielectrics have the same characteristics as paper capacitors in most respects. They are more expensive than paper capacitors, but have much lower leakage rates (even at high temperatures). Capacitance is more stable than that of paper capacitors; low temperature coefficients are possible. Plastic-film capacitors are not polarized.

• Mica — The capacitance of mica capacitors is very stable with respect to time, temperature and electrical stress. Leakage and losses are very low. Values range from 1 pF to 0.1 μF, with tolerances from ±1% to ±20%. High working voltages are possible, but they must be derated severely as operating frequency increases. Mica capacitors are not polarized.

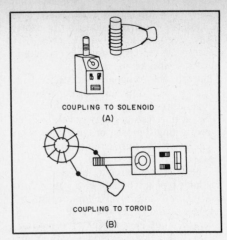

COUPLING TO SOLENOID
(A)

COUPLING TO TOROID
(B)

Fig. 28 — Dip-meter coupling to inductors.

Silver mica capacitors are made by depositing a thin layer of silver on the mica dielectric. This makes the value even more stable, but it presents the possibility of silver migration through the dielectric. The migration problem worsens with increased dc voltage, temperature and humidity. Avoid using silver-mica capacitors under such conditions.

Inductors and Transformers

Inductors store energy (in a magnetic field) or restrict the flow of ac. They can fail in several ways. The most common failure is a broken conductor. A short circuit across one or more turns of a coil will impair operation. Inductors that are used to block ac signals may "leak." (The ac signal may couple magnetically or capacitively through a faulty shield can, poor lead dress or distributed capacitance in the inductor.) Inductance may change as a result of shorted turns or physical damage to the coil or core material. Excessive tank-circuit current generates enough heat to melt plastics used as coil forms.

Inductors may be checked for open-circuit failure with an ohmmeter. Dc resistance will rarely exceed a few ohms. Shorted turns and other changes in inductance show only during alignment or inductance measurement. The procedure for measurement of inductance with a dip meter is the same as that given for capacitance measurement, except that a capacitor of known value is used in the resonant circuit. Fig. 28 shows the proper dip-meter orientation with respect to various coil configurations.

Replacement inductors must have the same inductance as the original, but that is only the first requirement. They must also carry the same current, withstand the same voltage and present nearly the same Q as the original part. Given the original as a pattern, the amateur can duplicate these qualities for many inductors. Note that inductors with cores are frequency-sensitive, so the replacement must have the same core material.

If the coil is of simple construction, with the form and core undamaged, carefully observe (and record) the number of turns and their placement on the form. Also note how the coil leads are arranged and connected to the circuit. Then determine the wire size and insulation used. (There is little hope of matching coil characteristics unless the wire is duplicated exactly in the new part.) Next, remove the old winding (be careful not to damage the form) and apply a new winding in its place. Be sure to dress all coil leads and connections in exactly the same manner as the original. Apply Q dope to hold the finished winding in place. Follow the same procedure in cases where the form or core is damaged, except that a suitable replacement form or core (same dimensions and permeability) must be found.

Ready-made inductors may be used as replacements if the characteristics of the original and the replacement are known and compatible. Unfortunately, many inductors are poorly marked. Thus, some comparisons and measurements are necessary.

Wire diameter, insulation and turn spacing are critical to the current and voltage ratings of an inductor. Next, eliminate those parts that bear no physical resemblance to the original part. This may seem an odd procedure, but the Q of an inductor depends on its physical dimensions and the permeability of the core material. Inductors of the same value, but of vastly different size or shape, will likely have a great difference in Q.

The Q of the new inductor can be checked by installing it in the circuit, aligning the stage and performing the manufacturer's passband tests. Although this practice is all right in a pinch, it does not yield an accurate Q measurement.

A better measurement of effective, unloaded Q can be made with a dip meter and an RF voltmeter (or a dc voltmeter with an RF probe). Make a parallel-resonant circuit using the replacement inductor and a capacitance equal to that of the tuned circuit under repair. Connect the voltmeter across this parallel combination and measure the resonant frequency. Adjust the dip-meter/circuit coupling for a convenient reading on the voltmeter, then maintain this dip-meter/circuit relationship for the remainder of the test. Vary the dip-meter frequency until the voltmeter reading drops to 0.707 times that at resonance. Note the frequency of the dip meter and repeat the process, this time varying the frequency on the opposite side of resonance. The difference between the two dip meter readings is the test-circuit bandwidth. This can be used to calculate the circuit Q:

$$Q = \frac{f_o}{Bw} \qquad \text{(Eq. 10)}$$

where

f_o = operating frequency
Bw = measured bandwidth in the same units as the operating frequency.

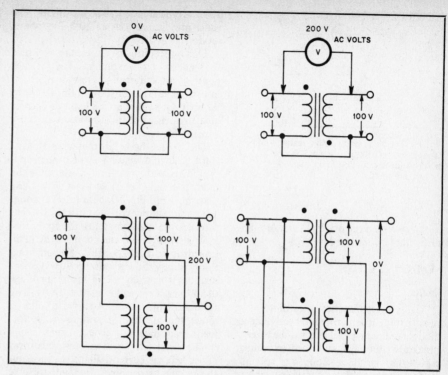

Fig. 29 — Test circuits to determine the winding "'sense" of transformers. Observe the sense dots.

Once the replacement inductor is found, install it in the circuit. Duplicate the placement, orientation and wiring of the original. Ground-lead length and arrangement should not be changed.

Isolation and magnetic shielding can be improved by replacing solenoid inductors with toroids; however, it is likely that many circuit adjustments will be needed to compensate for reduced coupling and mutual inductance. Alignment is required whenever a tuned-circuit component is replaced.

A transformer consists of two inductors that are magnetically coupled. Transformers are used to change voltage and current levels (this changes impedance also). Failure usually occurs as an open circuit or short circuit of one or more windings.

Amateur testing of power transformers is limited to ohmmeter tests for open circuits and voltmeter checks of secondary voltage. Make sure that the power-line voltage is correct, then check the secondary voltage against that specified. There should be less than 10% difference between open-circuit and full-load secondary voltage.

Replacement transformers must match the original in voltage, volt-ampere (VA), duty cycle and operating frequency ratings. (All transformer windings should be insulated for the full power-supply voltage.) Transformer windings are color coded; see Chapter 35 for more information. If it is necessary to determine the phase relationship of transformer leads, the tests illustrated in Fig. 29 can be used.

Relays

Although relays are being replaced by semiconductor switching in low-power circuits, they are still used extensively for control of high-power RF in transceivers and amplifiers. Relay action may become sluggish. Ac relays can buzz (with adjustment becoming impossible). A binding armature or weak springs can cause intermittent switching. Excessive duty cycle ruins contacts and shortens relay life.

Relays can be tested with a voltmeter or by jumping across contacts with a test lead (power on, in circuit) or with an ohmmeter (out of circuit). Look for erratic readings across the contacts, open or short circuits at contacts or an open circuit at the coil.

Most failures of simple relays can be repaired by a thorough cleaning. Clean contacts and mechanical parts with a residue-free cleaner. Keep it away from the coil and plastic parts that may be damaged. Dry the contacts with lint-free paper, such as a business card; then burnish them with a smooth steel blade. Do not use a file to clean contacts.

Replacement relays should match or exceed the original specifications for voltage, current, switching time and impedance (impedance is significant in RF circuits only). Many relays used in transceivers are specially made for the manufacturer. Substitutes may not be available from any other source.

Before replacing a multicontact relay, make a drawing of the relay, its position, the leads and their routings through the surrounding parts. This drawing allows you to complete the installation properly, even if you are distracted during the delicate operation. RF leads must be dressed exactly

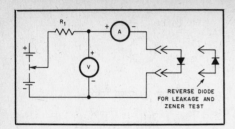

Fig. 30 — A diode conduction, leakage and Zener-point test fixture. The ammeter should read mA for conduction and Zener-point, μA for leakage tests.

SEMICONDUCTORS

Diodes

The primary function of a diode is to pass current in one direction only. Zener diodes are made with a predictable reverse-breakdown voltage and used as voltage regulators. Varactor diodes are specially made for use as voltage-controlled variable capacitors. (Any semiconductor diode may be used as a voltage-variable capacitance, but the value will not be as predictable as of a varactor.) A Diac is a special-purpose diode that passes only pulses of current in each direction.

Rectifier diodes fail by becoming an open circuit, a short circuit, by developing excess reverse leakage or by losing the ability to switch on and off quickly. Zener diodes may also fail by shifting their avalanche (reverse-breakdown) point. Varactor diodes fail in the same fashion as capacitors.

There are three basic tests for most diodes. First, is it a diode? That is, does it conduct in one direction and block current flow in the other? An ohmmeter is suitable for this test in most cases.

Before tests begin, make sure the meter uses a voltage of more than 0.7 V and less than 1.5 V to measure resistance. Next, use a good diode to determine the meter polarity: Set the meter to read ohms ×100 and connect the test probes across the diode. When the negative terminal of the ohmmeter battery is connected to the cathode, the meter will show 200 to 300 Ω (forward resistance) for a good silicon diode, 200 to 400 Ω for a good germanium diode. Reverse the lead polarity and set the meter to ohms ×1M (times one million, or the highest scale available on the meter) to measure diode reverse resistance. Good diodes should show 100- to 1000-MΩ and 100-k to 1-MΩ for silicon and germanium, respectively. When done, mark the meter lead polarity for future reference.

To check a suspected diode, disconnect one end of the diode from the circuit, then measure the forward and reverse resistance. Diode-quality is shown by the ratio of reverse to forward resistance. A ratio of 100:1 is typical for small-signal diodes. The ratio may go as low as 10:1 for power diodes.

This procedure measures the junction resistances at low voltage. If the Zener point of a rectifier diode has shifted to a low value, it may not show in this test. The defect might show in the leakage-current measurement, but substitution is the only dependable test.

The second diode test measures the forward voltage drop across the diode junction. This measurement is made while the diode is conducting. Use a test circuit like that shown in Fig. 30, connect the diode, adjust the supply voltage until the current through the diode matches the manufacturer's specification and compare the junction drop to that specified. Silicon junctions usually show about 0.6 V, while germanium is typically 0.2 V. Junction voltage drop increases with current flow. The results of this test can be used to choose diodes that are matched with respect to forward resistance at a given current level.

The final simple diode test measures diode leakage current. Place the diode in the circuit used above, but with reverse polarity. Set the specified reverse voltage and read the leakage current on the ammeter. (The currents and voltages measured in the junction voltage-drop and leakage tests vary by several orders of magnitude.)

The major criterion of Zener diodes is the Zener voltage. The Zener-voltage test also uses the circuit of Fig. 30. Place the diode in the circuit (so that it is reverse biased), increase the voltage until there is a large increase in current flow, then read the Zener voltage from the voltmeter.

Replacement rectifier diodes should have the same current and peak inverse voltage (PIV) as the original. Series combinations of diodes are often used in high-voltage rectifiers. Good engineering practice demands that the voltage be distributed equally across each diode. This is accomplished by placing a resistor and capacitor across each diode. When a diode fails, check the associated components also.

Switching diodes may be replaced with diodes that have equal or greater current ratings and a PIV greater than twice the peak-to-peak voltage of the signal they switch. Switching time requirements are not critical except in logic and some keying circuits. RF switching diodes used near resonant circuits must have exact replacements, as the diode resistance and capacitance will affect the tuned circuit. Logic circuits require exact replacements to assure compatible switching and load characteristics.

Voltage, current and capacitance characteristics must be considered when replacing varactor diodes. Once again, exact replacements are best.

Zener diodes should be replaced with parts having the same Zener voltage and equal or better current, power, impedance and tolerance specifications. Check the associated current-limiting resistor when replacing a Zener diode.

Bipolar Transistors

The fundamental purposes of transistors are to switch and to amplify. Transistor failures occur as an open junction, a shorted junction, excess leakage or a change in amplification performance.

Transistor tests are not performed at the planned operating frequency. Tests are made at dc or a low frequency (usually 1000 Hz). This practice is based on the assumption that the semiconductor meets all of its specifications equally well. That is, if it functions at low frequencies, it functions equally well at high frequencies. When special RF circuits (such as oscillators) are planned, build a circuit to test prospective devices at the operating frequency. The circuit under repair is the best test of a potential replacement part.

The simplest and probably the most telling of bipolar-transistor tests is performed with a test lead, a 10-kΩ resistor and a voltmeter, while the transistor is in a circuit with the power on. Connect the voltmeter across the emitter/collector leads and read the voltage. Then use the test lead to connect the base and emitter leads (Fig. 31A). Under these conditions, conduction of a good transistor will be cut off and the meter should show nearly the entire supply voltage across the emitter/collector leads. Next, remove the clip lead and connect the 10-kΩ resistor from the base to the collector. This should bias the transistor into conduction and the emitter/collector voltage should drop. (Fig. 31B) (This test indicates transistor response to changes in bias voltage.)

Transistors can be tested (out of circuit) with an ohmmeter in the same manner as diodes. Look up the device characteristics before testing and consider the consequences of the ohmmeter-transistor circuit. Limit junction current to 1 to 5 mA for small-signal transistors. Transistor destruction or inaccurate measurements may result from careless testing.

Use the ohms ×100 and ohms ×1000 ranges for small-signal transistors. For high-power transistors use the ohms ×1 or ohms ×10 ranges. The reverse-to-forward resistance ratio for good transistors may vary from 30:1 to 100:1. Germanium transistors sometimes show low reverse resistance when tested with an ohmmeter.

Bipolar transistor leakage may be specified from the collector to the base, emitter to base or emitter to collector (with the junction reverse biased in all cases). The specification may be identified as I_{cbo}, I_{bo}, collector cutoff current or collector leakage for the base-collector junction, I_{ebo}, and so on for other junctions.[4]

---

[4]The term, "I_{cbo}" means, "Current from collector to base with emitter open." The subscript notation indicates the status of the three device terminals. The terminals concerned in the measurement are listed first, with the remaining terminal specified as "s," shorted, or "o" open.

---

as they were on the original part, or alignment may be required.

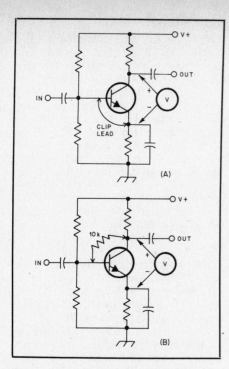

Fig. 31 — An in-circuit semiconductor test with a clip lead, resistor and voltmeter. The meter should read V+ at (A), and anything from a slight variation to 0 V at (B).

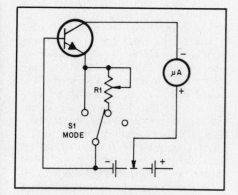

Fig. 32 — A test circuit for measuring collector-base leakage with the emitter shorted, open or shorted through a (variable) resistance, depending on the setting of S1. See the transistor manufacturer's instructions for test conditions and the value of R1 (if used). Reverse battery polarity for PNP transistors.

Manufacturers may require that tests be made with the emitter lead unconnected, shorted to the base or connected to the base through a resistor. Temperature is critical to this test. Allowable leakage current increases with junction temperature.

A suitable test fixture for base-collector leakage measurements is shown in Fig. 32. Make the required connections and set the voltage as stated in the transistor specifications and compare the measured leakage current with that specified. Small-signal germanium transistors exhibit I_{cbo} and I_{ebo} leakage currents of about 15 μA. Leakage increases to 90 μA or more in high-power

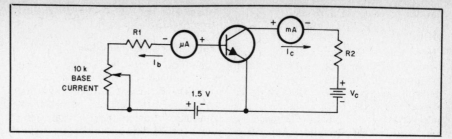

Fig. 33 — A test circuit for measuring transistor beta. Values for R1 and R2 will vary widely with the transistor tested. Reverse the battery polarities for PNP transistors.

components (germanium power transistors can develop excess leakage while sitting on the shelf.) Leakage currents for silicon NPN transistors are seldom more than 1 μA. Leakage current tends to double for every 10° C increase above 25° C.

Breakdown-voltage tests actually measure leakage at a specified voltage, rather true breakdown voltage. Breakdown voltage is known as BV_{cbo}, BV_{ces} (emitter shorted to base) or BV_{ceo}. Use the same test fixture shown for leakage tests, adjust the power supply until the specified leakage current flows, and compare the junction voltage against that specified.

A circuit to measure dc current gain is shown in Fig. 33. Current gain is also known as:

- (alpha) $= \dfrac{\Delta I_c}{\Delta I_e}$

in a common-base amplifier

- (beta) $= \dfrac{\Delta I_c}{\Delta I_b}$

in a common-emitter amplifier; "collector-to-base current gain," "forward current-transfer ratio,"
- $h_{FE} =$ (ac current gain).
- $h_{fe} =$ (dc current gain).

Conditions for gain testing are dictated by the manufacturer's data sheet. When testing, do not exceed the voltage, current (especially in the base circuit) or dissipated-power rating of the transistor. Make sure that the load resistor is capable of dissipating the power generated in the test. When I_c is high, measure the emitter-collector leakage current (with emitter shorted to base) first, and subtract the leakage from I_c before calculating gain. Do not neglect the effect of the meter resistance on the circuit.

RF devices should be tested at RF. Most component manufacturers include a test-circuit schematic on the data sheet. The test circuit is usually an RF amplifier that operates near the high end of the device frequency range.

Semiconductor failure is nearly always the result of environmental conditions. Open junctions, excess leakage (except with germanium transistors) and changes in amplification performance result from overload or excessive current. Shorted junctions are caused by voltage spikes.

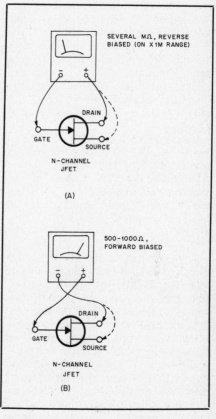

Fig. 34 — Ohmmeter tests of a JFET. The junction is reverse biased at A and forward biased at B.

Check surrounding parts for the cause of the transistor's demise, and correct the problem before installing a replacement.

JFETs

Junction FETs can be tested, out of circuit, with an ohmmeter in much the same way as bipolar transistors (see Fig. 34). Connect the negative ohmmeter lead to the gate of an N-channel JFET and check the resistance to both source and drain. The JFET is reverse biased by the ohmmeter battery, and resistance across each junction should be several megohms or more (using the ohms X1M scale). Then reverse the lead polarity to measure the forward resistance of each junction, which should be 500 to 1000 Ω. P-channel JFETs can be checked in the same manner with the test-lead polarity reversed for both tests.

MOSFETs

MOS (metal-oxide semiconductor) layers are extremely fragile. Normal body static is enough to puncture this delicate insulator. Even gate-protected (a diode is placed across the MOS layer to clamp voltage) MOSFETs may be destroyed by 6 V. Make sure the power is off, capacitors discharged and the leads of a MOSFET are shorted together before installing or removing it from a circuit. Use a voltmeter to be sure the chassis is near ground potential, then touch the chassis before and during MOSFET installation and removal. This assures that there is no difference of potential between your body and the MOSFET leads. Follow this equalization procedure whenever a MOSFET may contact another object. Ground the soldering-iron tip with a clip lead when soldering MOS devices. The source should be the first lead connected and the last disconnected to a circuit.

The insulating layers in MOSFETs prevent testing with an ohmmeter. They can be checked for operation with the transistor tester shown in Chapter 25. The test, however, is only a general check for operation, substitution is the only practical means for amateur testing of MOSFETs.

FET Considerations

Replacement FETs should be of the same kind as the original part. That is, JFET or MOSFET, P-channel or N-channel, enhancement or depletion characteristics should be chosen to match the original.

Next consider the breakdown voltage required by the circuit. Always allow for twice to four times the signal voltage in amplifiers. Provision should be made for transients of 10 times the line voltage in power supplies. Breakdown voltages are usually specified as $V_{(BR)GSS}$ or $V_{(BR)GDO}$.

Gate-voltage specifications indicate the bias or signals required for operation. The specification usually gives the gate voltage required to cut off or initiate channel current (depending on the mode of operation). Gate voltages are usually listed as $V_{GS (OFF)}$, V_p (pinch off), V_{TH} (threshold) or I_D (ON) or I_{TH}.

Gate leakage is a measure of FET input impedance. MOSFET gate leakage is extremely small; JFET leakage is greater. Leakage does not increase appreciably with temperature in either device. The characteristic may be called I_{GSS} (leakage from gate to source with drain shorted to source), I_{GSO} or I_{GDO}. (The first term is the preferred measurement, as the GSS configuration results in greater leakage.)

Dual-gate MOSFET characteristics are complicated by interaction of the two gates. Cutoff voltage, breakdown voltage and gate leakage are the important traits of each gate. $V_{G1S (OFF)}$ is the cutoff voltage designation for gate 1, while $V_{(BR)G1SS}$ and I_{G1SS} are the gate-1 breakdown voltage and leakage current designators, respectively.

Semiconductor Substitution

In all cases try to obtain exact replacement semiconductors. Specifications vary slightly from one manufacturer to the next. Also, device specifications are minimums; some units far exceed them. Obtain replacements for PA devices (and others that are not common) from the equipment manufacturer. Some parts are specially selected for a particular characteristic by the maker.

Crossover equivalents are useful, but not infallible. Before using an equivalent, check the specifications against those for the original part. When choosing a replacement, consider:

• Is it silicon or germanium?
• Is it a PNP or an NPN?
• Operating frequency and input/output capacitance.
• Dissipated power (often less than $V_{max} \times I_{max}$).
• Will it fit the original mount?
• Unusual circuit demands (low noise and so on).
• The gain-bandwidth product.

Remember that crossover equivalents are listed without testing, and manufacturer's numbers are not necessarily unique. Derate power specifications, as recommended by the manufacturer, for high-temperature operation.

INTEGRATED CIRCUITS

The basics of integrated circuits are covered in Chapter 4. Amateurs seldom have the sophisticated equipment required to test ICs. Even a multitrace 'scope can view only their simplest functions. We must be content to check every other possible cause, and only then assume that the problem lies with an IC.

There are two major classes of ICs: Linear and digital. Linear ICs are best replaced with identical units. Do not overlook equipment manufacturers as a source of replacements; they are the only source with a reason to stockpile obsolete or custom-made items. If substitution of an IC is unavoidable, look in one of the many books written about ICs, their use and traits. Understanding and assessing the necessary characteristics can be as difficult as redesigning the entire circuit. In fact, extensive modifications of the coupling circuits and wiring may be required to conform to the new device connections. The concept of the electronic building block, which implies some degree of interchangeability, is fundamental to IC production, but the intricacies of comparison make substitution difficult.

Substitution of digital and array ICs is somewhat easier than that of linear ICs because there are fewer significant characteristics to consider. Digital signals, because of their on/off nature, demand only speed and slight power, not fidelity, from a device.

First, review the explanation of the various logic families and their electrical characteristics in Chapter 8. Interface circuits and their limitations are covered there also. Consider interface and logic requirements, switching time and fan-out when choosing substitute parts. Check supply voltages, at the lead to the IC pin, once again before installing any new IC. It is a good idea to check the potential on each trace to the bad component. The old IC may have "died" as a result of a lethal voltage. Measure twice — repair once! Of course, in-circuit performance is the final test of any substitution.

TUBES

The most common tube failures in amateur service are cathode depletion and gas contamination. Whenever a tube is operated, the coating on the cathode loses some of its ability to produce electrons. It is time to replace the tube when electron production (cathode current, I_c) falls to 50 to 60% of that exhibited by a new tube. This kind of tube failure is accelerated by exceeding the recommended duty cycle, as is often the case when sweep tubes are pressed into amateur service. High-gain power-amplifier tubes often fail from cathode depletion when filament voltage is not adequately regulated.

A gas-contaminated tube can be spotted because there is often a glow *between* the elements during operation. The gas reduces tube resistance and leads to runaway plate current evidenced by a red glow from the anode, interelectrode arcing or a blown power-supply fuse. Runaway plate current is most likely to occur during long transmissions at higher operating frequencies.

Less common tube failures include an open filament, broken envelope and interelectrode shorts. Use an ohmmeter to check for an open filament (remove the tube from the circuit first). A broken envelope is visually obvious, although a cracked envelope may appear as a gassy tube. Interelectrode shorts are evident during voltage checks on the operating stage. Any two elements that show the same voltage are shorted. (Remember that some interelectrode shorts, such as the cathode-suppressor grid, are normal.) Confirm that the tube is shorted, and not the circuit, by removing the tube.

Substitution is the best test of a tube. First, check the tube for opens and shorts on a tester. If the tube has an open element, check the surrounding circuits for shorts, then try a new tube. When you suspect that tubes are weak, check them by substitution. Place a new tube in the circuit and leave it in place only if there is a noticeable improvement. In power-supply circuits, measure the B+ with the old tube in place. If a new tube increases the B+ by 10% or more, leave the replacement in the circuit. Remember that no tube tester can adequately test a tube for use as an oscillator.

The "life test" is meant to determine if a tube will fail in the near future. Connect the tube to an emission tester and reduce the filament voltage by 20% from normal. If the plate current drops below that of a good tube, the service life of the tube is nearly over. Also, the cathodic emission of a strong tube falls slowly when filament voltage is removed; emission in a weak tube stops immediately.

Tube testers are covered earlier in this chapter. Look there for more information about testers and tips on their use.

Generally, a tube may be replaced with another that has the same basic model number, but look at data sheets for the respective tubes to verify their compatibility. Consider the base configuration and pinout, interelectrode capacitances (a small variation is okay except for tubes in oscillator service), dissipated power ratings of the plate and screen grid and current limitations (both peak and average).

For example, the 6146A may be replaced with a 6146B (heavy duty), but the B-suffix tube should not be replaced with an A-suffix version. (See the tetrode and pentode transmitting tube specifications in Chapter 35: A 6146B can dissipate 35 W safely, while a 6146A is rated at only 25 W.) In some cases, the difference denotes differences in filament voltages, or even base styles, so check *all* specifications before making a replacement.

As usual, the best course is an exact replacement from the equipment manufacturer; especially for tubes used in RF-power and oscillator circuits. (These are sometimes hand-picked or close-tolerance versions of standard models. Even tubes of the same model number, prefix and suffix vary slightly, in some respects, from one supplier to the next.)

CHECKOUT

Inspect each circuit visually when repairs are complete. Look for cold solder joints and signs of damage incurred during the repair. Use a bright light close behind printed circuit boards to check for breaks in the copper foil. Double check the position, leads and polarity of diodes, transistors and electrolytic capacitors that were removed or replaced. Make sure that all ICs have pin 1 in the proper spot. Test fuse continuity with an ohmmeter and verify that its current rating matches the circuit specification.

Observe the lead dress. Place wires clear of screw points and sharp edges. Separate the leads that carry dc, RF, input and output as much as possible. Plug-in circuit boards should be firmly seated with screws tightened and star washers installed if so specified. Shields and ground straps should be installed just as they were on the original.

For Transmitters Only

Since the signal produced by an HF transmitter can be heard the world over,

a thorough check is demanded after any service. Follow the procedure below to make sure the transmitter operates properly. Do not exceed the transmitter duty cycle while testing. Limit transmissions to 10 to 20 seconds unless otherwise specified by the owner's manual.

1) Set all controls as specified in the operation manual, or at mid scale.

2) Connect a dummy load and a power meter to the transmitter output.

3) Set the drive or carrier control for low output.

4) Switch the power on.

5) Transmit and quickly set the final-amplifier bias to specifications.

6) Slowly tune the output network through resonance. The current dip should be smooth and repeatable. It should occur simultaneously with the maximum power output. If these characteristics are not present, the amplifier is unstable. Adjust the neutralization circuit (see details in Chapter 15) if one is present or check for oscillation. An amplifier requires neutralization whenever active devices, components or lead dress (that affect the output/input capacitance) are changed.

7) Check to see that the output power is consistent with the amplifier class used in the PA (efficiency should be about 25% for Class A, 50 to 60% for Class AB or B, and 70 to 75% for Class C.)

8) Repeat steps 4 through 6 for each band of operation from lowest to highest frequency.

9) Check the carrier balance (in SSB transmitters only) and adjust for minimum power output with maximum RF drive and no microphone gain.

10) Adjust the VOX controls.

11) Measure the passband and distortion levels if equipment (wide-bandwidth 'scope or spectrum analyzer) is available.

All Other Repaired Circuits

After completing all of the preliminary checks, remove any test equipment, set the circuit controls per the manufacturer's specifications (or to midrange if specifications are not available), and switch the power on. Watch and smell for smoke, and listen for odd sounds such as arcing or hum. Operate and observe the circuit for a few minutes, consistent with allowable duty cycle. Vary all operating controls through their full range (consistent with operating instructions); see that they have the desired effects. Measure the gain and distortion of repaired amplifiers.

If the circuit appears to be operating normally, vary the power-supply voltage ± 10% while checking for intermittent connections. Check for intermittents by subjecting the circuit to heat, cold and slight flexure. Also, tap or jiggle the chassis lightly with an alignment tool or other insulator. If the equipment is meant for mobile or portable service, operate it through an appropriate temperature range. Many mobile radios do not work on cold

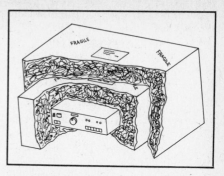

Fig. 35 — Ship equipment packed securely in a box within a box.

mornings, or on hot afternoons, because a temperature-dependent intermittent was not found during repairs.

FACTORY REPAIRS

This chapter does not tell how to perform all repairs. Those repairs that deal with very complex and temperamental circuits, or that require sophisticated test equipment, should be passed on to a professional.

You should have opened a dialog with the manufacturer early in the repair work: If you did not, do so now. Notify the repair center before shipping any equipment. Get authorization for shipping and an identification name or number for the package.

Begin the packing procedure by blocking and tying down all heavy components. Large vacuum tubes should be wrapped in packing material or shipped separately. Make sure that all circuit boards and parts are firmly attached. Enclose a statement of the trouble, a short history of operation and any test results that may help the service technician.

Use a box within a box for shipping. (See Fig. 35) Place the equipment and some packing material inside a box. Seal the box with tape, address it, and mark it "Fragile — Electronic Equipment." Place that box inside another that is at least six inches larger in each dimension. Fill the gap with packing material, seal, address and mark the outer box. Choose a good freight carrier and insure the package.

Used Equipment

There is a certain thrill we all feel as we enter the flea market at our favorite hamfest. Vikings, "Gooney birds," Warriors and the venerable S-Line appeal to almost everyone. Equally tempting is the allure of a bargain; few hams can resist one.

Keep an eye on the classified section of radio magazines. Know the going price for radios that interest you and never pay that price without a demonstration. Try to buy equipment that was of high quality when it was new. It will serve you well when it is old.

Any gear that has been in storage may need repair. When you are looking for a bargain, gear that needs repair is usually the best buy unless:

- It is heavily damaged.
- Expensive test gear is needed for repair.
- No manual or parts are available.

Recently Built Equipment

First, do not buy unless you know the manufacturer. Know the equipment and whether parts and manuals are available. See that the manual is included in the sale — manuals are difficult to locate and expensive.

When a good prospect is found, open the cabinet. Look for missing parts, evidence of fire or damage. Purchase modified equipment if you trust the seller, or if the modifications can be easily removed. Modifications are as good as the skill of the person who installed them. Check the meter functions; the readings can tell you much.

Older Gear

Gear made before 1970 is usually suitable only for CW operation. Once again, do not overpay, watch for modifications and get the manual. Check for missing parts, and do not waste time on equipment that was inferior when new.

Beware of power supplies that do not use an isolation transformer. Such gear is difficult to modify. Connection of accessories, such as a keyer, keyboard or Q multiplier, is dangerous.

Restoration

First, open the cabinet and replace any parts that show signs of physical damage. Electrolytic capacitors often fail with age. Visual failure signs include bulges and leakage of a thick fluid or a white, gray or brownish paste or powder. Even if none of these signs are present, an electrolytic capacitor may be destroyed when power is first applied after long disuse. This may be prevented by re-forming the dielectric layer before applying power (see earlier section of this chapter).

Paper capacitors (wax or plastic coated) fail by leakage. They exude fluid or develop cracks in their coating. Replace faulty paper capacitors with dipped-Mylar units of the same, or higher, working voltage.

Mica and ceramic capacitors survive storage better than their tubular counterparts, but they require more exact replacements. Replacements must match not only the capacitance and working voltage of the original, but also the tolerance and temperature coefficient in all but bypass applications.

Remove dust and dirt with a small brush and a vacuum cleaner while inspecting the circuitry. Use extreme care around variable capacitors, inductors and rotary switches. Clean potentiometers and rotary switches with residue-free spray cleaner. Service air-dielectric variable capacitors as described earlier in this chapter. An eraser will clean stubborn dirt from rotary switch contacts,

but be very gentle and do not disturb lead dress around the switch. If leads to RF circuits are disturbed during cleaning, alignment may be affected.

After all visible problems have been corrected, disconnect one of the transformer secondary leads and measure the dc resistance from the power-supply output to ground. If the resistance is not consistent with the power ratings of the supply, locate the reason before applying power. If all is well, reconnect the transformer secondary and use a variable-voltage transformer to apply power gradually to the circuit. Listen and watch for arcs in the circuit as power is applied. Equipment that has been in storage should be "burned in" before proceeding. Operate the circuit (within duty-cycle limits) several hours each day for a week. Stay near the gear during this period to make sure that the "burn in" does not become a "burn down."

Once the basic clean up is completed, note any troubles as instructed earlier in this chapter and begin the normal repair routine. Once the equipment is repaired, clean the cabinet, instrument panel and knobs with household or automotive cleaner and a small brush. Smooth surfaces may be brightened by an application of automotive wax.

A SPECIAL CASE — NEW CONSTRUCTION

In most repair work, the technician is aided by the knowledge that the circuit once worked. It is only necessary to find the faulty part(s) and replace it. This is not so with newly constructed equipment. Repair of equipment with no working history is a special, and difficult, case.

Try to tackle the project one stage at a time. Consider each stage as two halves: Input and output. Look for any active device that is either saturated or cut off. Problems with the bias circuit are usually evident even with no input signal.

Use this procedure to set the bias of a high-gain amplifier. Set up the amplifier with a suitable dummy load and an oscilloscope on the output. Feed the input with a sine wave through an appropriate matching network. Increase the drive until the output waveform is clipped to a nearly square wave (See Fig. 36). Adjust the bias so that the positive and negative sections of the wave have equal duty cycles. Decrease the drive level slowly while making fine bias adjustment to maintain the duty cycle.

If equal clipping is not possible, suspect the component values or design. Nonlinear operation can be cured by adjusting values in the bias network, unless the active device is unsuitable for the circuit. When the circuit will only operate near saturation or cut off, try reducing positive feedback.

Once the bias is adjusted, check carefully for oscillations or noise. Oscillations are most likely to start with maximum gain and the amplifier input shorted. Any noise that

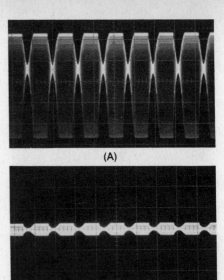

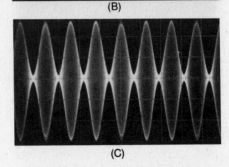

(A)

(B)

(C)

Fig. 36 — Waveforms seen while adjusting amplifier bias. The amplifier is overdriven at A, adjusted for proper duty cycle at B and operating normally at C.

is induced by 60-Hz sources can be seen with a 'scope synchronized to the ac line. Test with the supply voltage varied by ±10% (high-gain circuits are sensitive to supply-voltage changes).

RF oscillations should be cured with changes of lead dress or circuit components. Separate input leads from output leads; use coaxial cable to carry RF between stages; neutralize inter-element or junction capacitance. Ferrite beads on the control element of the active device often stop unwanted oscillations.

Low-frequency oscillations ("motor-boating") indicate poor stage isolation or inadequate power-supply filtering. Try a better lead-dress arrangement and/or increase decoupling capacitance. Use larger capacitors at the power-supply lead, increase the number of capacitors or use separate decoupling capacitors at each stage.

Coupling capacitors that are too low in value cause poor low-frequency response. Poor response to high frequencies is frequently caused by transistor input capacitance. The input capacitance of a transistor decreases with frequency; this lowers stage gain. Reducing the stage gain at lower frequencies will help, as will using a better transistor.

References

Bartlett, J., "Calculating Component Values," *QST*, Nov. 1978.

Carr, J., *How to Troubleshoot and Repair Amateur Radio Equipment*, Blue Ridge Summit, PA: Tab Books Inc., 1980.

DeMaw, D., "Understanding Coils and Measuring their Inductance," *QST*, Oct. 1983.

Gibson, H., *Test Equipment for the Radio Amateur*, London, England: Radio Society of Great Britain, 1974.

Gilmore, C., *Understanding and Using Modern Electronic Servicing Test Equipment*, TAB Books, Inc., 1976.

Glass, F., *Owner Repair of Amateur Radio Equipment*, Los Gatos, CA: RQ Service Center, 1978.

Goodman, R., *Practical Troubleshooting with the Modern Oscilloscope*, TAB Books, Inc., 1979.

Haas, A., *Oscilloscope Techniques*, New York, NY: Gernsback Library, Inc., 1958.

Hallmark, C., *Understanding and Using the Oscilloscope*, TAB Books, Inc., 1973.

Helfrick, A., *Amateur Radio Equipment Fundamentals*, Englewood Cliffs, NJ: Prentice-Hall, Inc., 1982.

Henney, K., and Walsh, C., *Electronic Components Handbook*, New York, NY: McGraw-Hill Book Company, 1957.

Klein, L., and Gilmore, K., *It's Easy to Use Electronic Test Equipment*, New York, NY: John R. Rider, Publisher, Inc. (A division of Hayden Publishing Company), 1962.

Lenk, J., *Handbook of Electronic Test Procedures*, Prentice-Hall, Inc., 1982.

Loveday, G., and Seidman, A., *Troubleshooting Solid-State Circuits*, New York, NY: John Wiley and Sons, 1981.

Margolis, A., *Modern Radio Repair Techniques*, Tab Books Inc., 1971.

Neben, H., "An Ohmmeter with a Linear Scale," *QST*, Nov. 1982.

Neben, H., "A Simple Capacitance Meter You Can Build," *QST*, Jan. 1983.

Noble, F., "A Simple L-C Meter," *QST*, Feb. 1983.

Priedigkeit, J., "Measuring Inductance and Capacitance with a Reflection-Coefficient Bridge," *QST*, May 1982.

Sartori, H., "Solid Tubes — A New Life for Old Designs," *QST*, Apr. 1977; "Questions on Solid Tubes Answered," Technical Correspondence, *QST*, Sept. 1977.

Wedlock, B., and Roberge, J., *Electronic Components and Measurements*, Prentice-Hall, Inc., 1969.

"Some Basics of Equipment Servicing," series, *QST*, Dec. 1981, Jan.-Mar. 1982, feedback May 1982.

Chapter 27

Power Supply Projects

Construction of a power supply can be one of the most rewarding projects undertaken by a radio amateur. Whether it's a charger for the NiCds in a VHF hand-held transceiver, a low-voltage, high-current monster for a new 100-watt solid-state transceiver, or a high-voltage supply for a new linear amplifier, a power supply is basic to all of the radio equipment we enjoy and operate. Power-supply projects are especially inviting to amateurs new to rolling their own. Unlike transmitters and receivers, layout and wiring of power supplies is not particularly important to proper operation. Parts for many power-supply projects are readily available in many areas, either through local retail outlets or at flea markets and hamfests. The power-supply project requiring a special, hard-to-find part is the exception, rather than the rule. In addition, final testing and adjustment of most power-supply projects requires only a voltmeter, and perhaps an oscilloscope — tools commonly available to most amateurs.

Additional information on power-supply design and theory may be found in Chapter 6. General construction techniques that may be helpful in building the projects in this chapter are outlined in Chapter 24. Chapters 2, 3, 4 and 5 contain basic information on the components that make up power supplies.

Safety must always be carefully considered during design and construction of any power supply. Power supplies contain potentially lethal voltages, and care must be taken to guard against accidental exposure. For example, electrical tape, in-

sulated tubing (spaghetti) or heat-shrink tubing is recommended for covering exposed wires, component leads, component solder terminals and tie-down points. Whenever possible, connectors used to mate the power supply to the outside world should be of an insulated type designed to prevent accidental contact.

Connectors and wire should be checked for voltage and current ratings. Always use wire with an insulation rating higher than the working voltages in the power supply. Special high-voltage wire is available for use in B+ supplies. Chapter 35 contains a table showing the current-carrying capability of various wire sizes. Scrimping on wire and connectors to save money could result in a flashover, meltdown or fire.

All fuses and switches should be placed in the hot leg(s) only. The neutral leg should not be interrupted. Use of a three-wire (grounded) power connection will greatly reduce the chance of accidental shock. The proper wiring color code for 117-V circuits is: black — hot; white — neutral; and green — ground.

Power Supply Primary Circuit Connector Standard

The International Commission on Rules for the Approval of Electrical Equipment (CEE) standard for power supply primary circuit connectors for use with detachable cable assemblies is the CEE-22 (see Fig. 1). The CEE-22 has been recognized by the ARRL and standards agencies of many countries. Rated for up to 250 V, 6 A at 65° C, the CEE-22 is the most commonly used three-wire (grounded), chassis-mount

Fig. 1 — CEE-22 connectors are available with built-in line filters and fuse holders.

primary circuit connector for electronic equipment in North America and Europe. It is often used in Japan and Australia as well.

When building a power supply requiring 6 A or less from the primary supply, a builder would do well to consider using a CEE-22 connector and an appropriate cable assembly, rather than a permanently installed line cord. Use of a detachable line cord makes replacement easy in case of damage. CEE-22 compatible cable assemblies are available with a wide variety of power plugs including most types used overseas.

Some manufacturers even supply the CEE-22 connector with a built-in line filter. These connector/filter combinations are especially useful in supplies that are operated in RF fields. They are also useful in digital equipment to minimize conducted interference to the power lines.

CEE-22 connectors are available in many styles for chassis or PC-board mounting. Some have screw terminals; others have solder terminals. Some styles even contain built-in fuse holders.

A 5-A Switching Supply

This section describes a practical 13.8-volt, 5-ampere switching power supply suitable for use with transceivers in the 25-watt-output class. Circuit simplicity and easy parts acquisition were the major design goals. The supply was built and tested in the ARRL laboratory by Greg Bonaguide, WA1VUG. A 10-watt-output 220-MHz FM transceiver was used as a load with no apparent degradation of the RF output spectral purity. The line and load regulation are acceptable, and the unit

is slightly more efficient than an equivalent linear supply. With the heat sink specified, the unit can withstand 3-minute FM transmissions interspersed with 1-minute listening periods. Because of the simple circuit configuration, conventional preregulation components are used. Therefore, this

Fig. 2 — Front panel of the switching power supply. The cabinet is by Apollo and has a built-in power switch.

Fig. 3 — Interior view of the switching power supply. The main switch transistor heat sink and fuse are mounted on the rear panel.

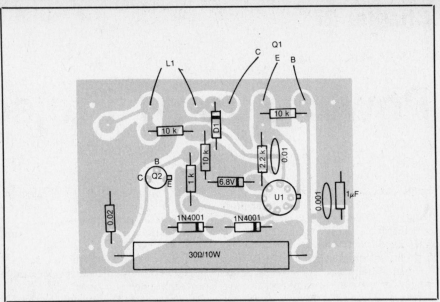

Fig. 5 — Parts-placement guide for the PC board of the switching-regulator supply control board. The component side is shown with an X-ray view of the foil. A full-size etching pattern can be found at the back of this book.

power supply offers no particular economic advantage over one using linear regulation. This project was intended to demonstrate the switching principle as applied to voltage regulation, and it serves that purpose well. However, it is not in any way represented to be a state-of-the-art design.

A more nearly technically correct name for this project might be "A 5-A Power Supply Using a Switching Regulator." However, it's not necessary to belabor the distinction between supplies and regulators in casual work.

Fig. 2 shows the cosmetic aspects of the front panel, while Fig. 3 reveals one satisfactory arrangement of internal com-

ponents. The schematic diagram is given in Fig. 4. Stud-mount rectifiers are used in a full-wave bridge. A PC board for the diodes is visible in Fig. 3, but the etching pattern is not printed here because of its simplicity. The component-placement guide for the electronics board is given in Fig. 5. A full-size etching pattern can be found at the back of this book.

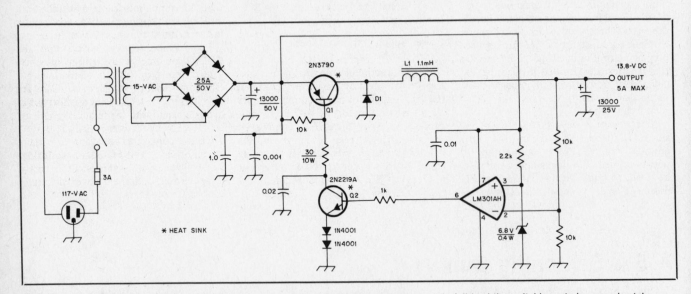

Fig. 4 — A switching power supply suitable for use with medium-power amateur transceivers. At full load the switching rate is approximately 50 kHz. Resistance values are in ohms, and capacitance values are in microfarads.

D1 — 6 A, 600 PIV, anode common to case, ECG 5863 or equiv.

L1 — 33 turns no. 14 enam. on Amidon FT 240-43 toroidal core.

Q1 Heat sink — 4 × 3 inches with ½-inch fins extruded from flat surfaces.
Q2 Heat sink — TO-5 clip-on unit.

An RF-Proof 30-Amp Supply

Solid-state technology has made possible circuit miniaturization that was only dreamed about 20 years ago. This new technology has afforded the amateur community medium-power equipment compact enough to use in even the most cramped mobile environment. This equipment is normally designed to operate from a source of 13.8-V dc and is well-suited for direct connection to the automobile electrical system. In the interest of size and weight conservation, however, an ac-operated power supply (necessary for fixed-station operation), is not often built in. If home operation from the ac mains is contemplated, an external source of 13.8-V dc is required. Matching commercial units are often costly and are not normally designed for high-duty-cycle modes, such as RTTY or SSTV.

The power supply described here (Figs. 6 through 13), designed by George Woodward, W1RN, and built in the ARRL lab by Mark Wilson, AA2Z, boasts a 30-A continuous current rating. Cost of construction will depend on the builder's junk box, but should be significantly less than commercially available units. This design features complete output metering, overvoltage shutdown and foldback current limiting. Additionally, this supply exhibits excellent immunity to RF fields, even with the cover removed. During testing, a 5-W, 2-meter transceiver was keyed with the antenna just inches from the regulator circuitry, with no loss of regulation. Also, a 100-W HF transceiver was loaded into a piece of wire that was attached to a fluorescent bulb. There was enough RF present to illuminate the bulb, but the power supply performed perfectly when the bulb was brought near the output terminals.

Design Information

The schematic diagram of the RF-proof supply is given in Fig. 7. A custom transformer, available from Avatar Magnetics, is used in this design. The hefty 25.4-V center-tapped secondary, wound with a pair of no. 11 wires, is continuous-duty rated at 34 A. At full load, the secondary voltage is down only 100 mV from nominal. Transformer output voltage is an important consideration in high-power applications: If the transformer voltage is too high, the pass transistors dissipate excessive power because of the large voltage drop across each device. Conversely, low transformer output voltage can cause loss of regulation because of insufficient voltage differential across the regulator circuitry. A separate 15-V, 1.5-A secondary winding (connected to U3) powers the regulator and associated circuitry so that the high-current secondary voltage can be kept lower.

In high-current applications, a low-voltage condition can also be caused by marginal transformer current ratings.

Fig. 6 — The completed RF-proof power supply features voltage and current metering and remote sensing.

Under no-load conditions, the current through the transformer is low, hence the $I \times R$ drop through the secondary (from the dc winding resistance) is small. When the transformer is called upon to deliver more current, however, the voltage drop through the secondary increases, causing insufficient voltage at the rectifier and regulator circuits. This is not a problem with the AV-399 transformer used here!

C1, a 190,000-μF electrolytic capacitor, provides a clean dc waveform to the regulator and pass transistors. This high value is necessary to keep ripple to a minimum. The filtered secondary voltage is close to the desired output voltage to keep pass-transistor dissipation to a minimum. This, however, means that ripple must be minimized so that under load the input voltage remains above the desired regulated output voltage. The energy stored in this capacitor is tremendous, so a bleeder resistor must be used even though the voltage is low. The 1-kΩ resistor here draws 14 mA — not enough to heat up the inside of the already-warm cabinet, but enough to safely discharge C1.

The large value chosen for C1 places a high peak-current demand on the rectifiers. To charge this capacitor, several hundred amperes of peak current will flow through the rectifiers over short periods of each cycle. Peak-current demands must be considered carefully when selecting rectifiers and wiring for this portion of the circuit. D1 and D2 have surge-current ratings of 500 A and can handle 40 A continuously. These are stud-mounted rectifiers, and they require a substantial heat sink.

Voltage regulation is handled by a 723-type IC regulator, U1. This IC generates the reference voltage for the regulated output. R3, a 25-turn potentiometer, is used to set the output voltage to the desired value; the supply shown here was set to precisely 13.80 V.

U1 is capable of supplying about 150 mA — not enough to drive the high-current pass transistors. In this circuit, the output of U1 drives a control loop consisting of U2B and Q1. Almost any NPN Darlington that can handle 5 A or more will work for Q1. The 15-A 2N6576 used here was chosen because it was available locally.

Q1 drives four 2N3055 NPN power transistors (Q3-Q6). For best current balance and efficiency, these pass transistors should all be from the same manufacturing lot. A 0.1-Ω, 5-W spreading resistor is placed in series with the emitter lead of each 2N3055 to help balance current distribution. The emitter resistor values should be within 1% of each other. Matching can be accomplished by placing low-value (1.0- to 2.2-Ω), ¼-W resistors in parallel with each 0.1-Ω unit until the desired tolerance is obtained.

If you use a different heat sink or a higher input voltage, you will probably have to use additional pass transistors to keep the junction temperature to an acceptable level. See Chapter 6 for complete information on how to calculate the proper heat sink and number of pass transistors needed for your application.

Current limiting, provided by U2A, U2D and associated components, protects the supply from overcurrent damage. An operational amplifier, U2A, monitors the voltage across the current-sense resistor (R2) and reduces the regulator drive when output current exceeds 32 A. If a short circuit should appear across the supply output terminals, the overcurrent circuit activates, reducing output voltage to next to nothing at a current level selected by the fold-back resistor, R1.

R1 is actually made from two separate resistors in series. This makes it possible to change the value slightly (by experimenting with the smaller resistor) until the desired foldback current is obtained. With the value shown, fold-back current is limited to 3 A. This value was chosen so that the M1 deflects slightly so that the operator can tell that the problem is overcurrent of some kind rather than a loss of input voltage, yet it is low enough that wires won't melt. When the supply folds back, the OVER-CURRENT lamp turns on, warning the operator of an overload.

A change in the value R1 as small as a few hundred ohms will make a significant difference, so resistor tolerances can affect the foldback current of your version of this supply. You may have to experiment with different resistors to get an acceptable foldback current. Of course, you can set the foldback current to any value you wish. Also, the value of V_{IN} has a profound effect, so if you use a different transformer and V_{IN} is much different from 20.5 V (the measured value of V_{IN} on the ARRL lab version), you should recalculate the

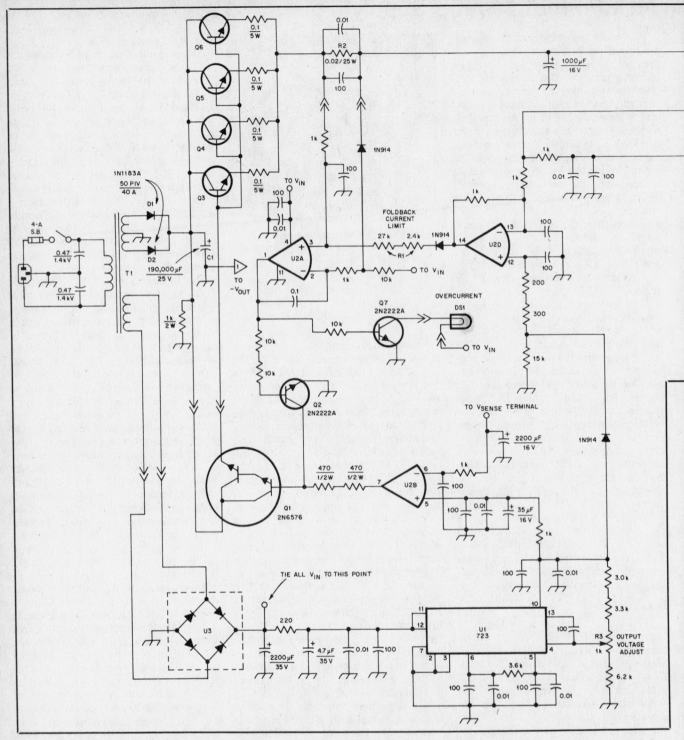

Fig. 7 — Schematic diagram of the high-current, RF-proof power supply. All resistors are ¼-W, 5% types unless noted. Capacitors are monolithic ceramic types unless noted. Capacitors marked with polarity are electrolytic. Numbered components not appearing in the parts list are for text reference.

C1 — Electrolytic capacitor, 190,000 µF, 25 V (Sprague 194G025DJ2A or equiv.).

D1, D2 — Rectifier diode, 50 PIV, 40 A (1N1183A or equiv.).

DS1 — 12-V pilot light or LED with 680-Ω dropping resistor.

M1 — Dc ammeter, 0-30 A.

M2 — Dc voltmeter, 0-15 V.

Q1 — 5 A or greater Darlington transistor (see text).

Q3-Q6 — NPN power transistor (2N3055, 2N5303 or equiv.).

Q8 — 25-A SCR, TO-220 package (ECG-5550 or equiv.).

R2 — Five 0.1-Ω, 5-W resistors in parallel (see text).

R3 — Multiturn PC-mount potentiometer, 1 kΩ, 0.5 W.

R4 — Multiturn PC-mount potentiometer, 10 kΩ, 0.5 W.

T1 — Power transformer. Primary, 117-V ac; secondary, 25.4-V CT, 34 A and 15 V, 1.5 A (Avatar AV-399 or equiv. Available from Avatar Magnetics, 1147 North Emerson, Indianapolis, IN 46219).

U1 — Adjustable voltage regulator IC, 14-pin DIP package (LM723, MC1723 or equiv.).

U2 — Quad 2-input operational amplifier (LM324).

U3 — Bridge rectifier, 50 PIV, 1.5 A.

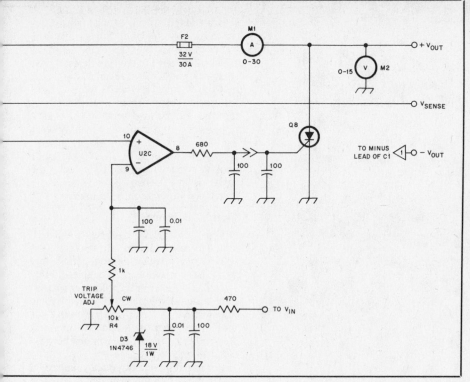

value of R1 from the equation

$$R_{FB} = \frac{V_{IN}}{0.7 - (I_{FB} \times 0.02)} - 1$$

where

  R_{FB} = the value of the foldback resistor in kΩ
  V_{IN} = the regulator board supply voltage in volts
  I_{FB} = the desired foldback current in amperes

This supply also includes a crowbar overvoltage protection circuit, consisting of U2C, Q8 and associated circuitry. Should the output voltage exceed the threshold — somewhere slightly above the desired output voltage — set by the parallel combination of D3/R4, Q8 will conduct, short-circuiting the output terminals of the supply. R4, a multi-turn potentiometer, is used to set the trip voltage. This lights DS1 and triggers the previously described overcurrent protection circuitry.

F2 is included in case of a drastic supply failure. Normally, the supply would fold back if a short circuit appeared at the output. Should something happen to the regulator board and the foldback current limiting circuitry fail to function, full supply current — in excess of 30 A — would be available at the output terminals. This is enough current to do serious damage to the load and/or interconnecting wires. F2 should blow if this happens.

Remote sensing, another important feature in high-current supplies, is included. When heavy loads are connected through long cables, the voltage drop degrades load regulation. Remote sensing allows the drop associated with long cable runs to be included in the regulator feedback loop. For normal operation, a jumper wire must be connected between the +V_{OUT} and V_{SENSE} terminals to complete the regulator feedback loop. The supply will not regulate if this jumper is omitted! To use the remote-sensing feature, remove the jumper wire between the +V_{OUT} and V_{SENSE} terminals, and connect a wire between the V_{SENSE} terminal and the positive side of the load.

Care has been taken to make this design immune to the RF that floats around most amateur stations. Each RF-sensitive component is heavily bypassed with both 100-pF and 0.01-μF capacitors to ensure radio-frequency energy a low-impedance path to ground. In some places, several resistors are used in series, rather than a single unit, to break up long leads and potential "antennas" for RF.

Construction

One possible assembly method can be seen in Fig. 12, although other methods should perform just as well. The supply is built in a Hammond enclosure. T1, C1 and the regulator circuit board mount inside the cabinet. The heat sinks for Q3-Q6 and D1 and D2 mount on the exterior of the rear panel, with the fins vertical, to take advantage of natural convection cooling. Do not mount the heat sinks with the fins horizontal! See Chapter 6 for complete information on thermal design considerations.

Components that don't generate appreciable heat are contained on a double-sided PC board (Figs. 9 and 10). One side of this board is etched. The foil on the remaining side is unaltered, to act as a ground plane. All component leads that go to ground are soldered to the ground plane. Where a component lead is to pass through the board, the hole is drilled from the etched side, and foil around the hole on the component side removed to prevent shorting to ground. The foil can be cleared away with a 1/8-inch drill bit turned by hand. Do not clear away foil from holes for component leads that go to ground. A parts-placement guide for this circuit board is shown in Fig. 10.

To avoid ground loops, this supply uses a single-point ground. PC board common does not contact the cabinet; instead, all power supply grounds (except those on the primary) are tied to the minus terminal of C1.

The pass transistors, Q3-Q6, inclusive, are mounted on a Wakefield model 441K heat sink. In this version, the transistors are mounted directly to the heat sink with thermal compound, but without insulating hardware. The transistor case is tied to the collector, so the dc potential of the heat sink is 13.8 V. Wakefield insulating standoffs are used between the sink and the cabinet. The emitter current-spreading resistors are placed on terminal strips that are mounted on the heat sinks adjacent to each pass transistor. Connections are made by soldering directly to the transistor pins. Another builder may wish to use insulating hardware and/or TO-3 sockets. A full-size drilling template for mounting four TO-3 transistors on the Wakefield 441 is given in Fig. 11, and the wiring for the pass transistors is shown in Fig. 8.

D1 and D2 are mounted on a Wakefield 421K heat sink that bolts to the rear panel. The rectifiers used in the prototype have the cathode connected to the stud. The rectifiers are mounted "upside down" on the heat sink; that is, the anode pins point toward the chassis. See Fig. 13. To avoid exposed wiring, a copper strip is mounted between the cathodes of D1 and D2 and the heat sink. Connection to the cathodes is made to this strip. Use heat sink compound between the diodes and the copper strip and between the copper strip and the heat sink. As before, these devices are not insulated from the heat sink, so the heat sink is mounted on insulating standoffs manufactured by Wakefield for this purpose.

Q8 is mounted to the chassis on a small heat sink. Because the foldback current in this supply is limited to 3 A, only a small heat sink is required. If, however, you change the foldback current to a higher value, you must use a larger heat sink.

R2, the 0.02-Ω, 25-W current-sense resistor consists of five 0.1-Ω, 5-W resistors in parallel. The resistors used here have built-in mounting brackets, so long no. 4 machine screws are used to mount them to the chassis. The sand-type power resistors

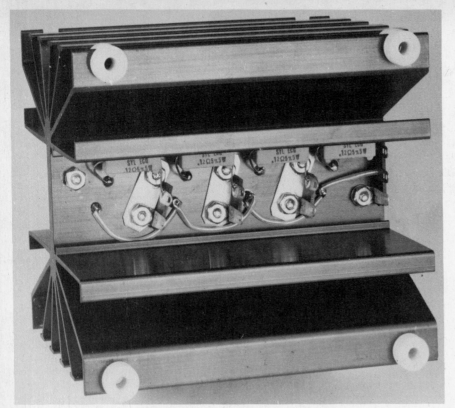

Fig. 8 — Connections to Q3-Q6 are made on the underside of the heat sink.

Fig. 9 — The regulator circuitry is mounted on a double-sided PC board. Component leads that connect to ground are soldered to the ground plane on top of the board. Copper foil is removed from the component side of the board where leads pass through to traces on the other side. See text and Fig. 10.

(like those used at the emitters of Q3-Q6) are much more common; the best way to mount sand-type resistors is between two terminal strips.

Liberal use of terminal strips simplifies the wiring and troubleshooting, should it be necessary. In a supply of this current capability, it is imperative that you use wire heavy enough for the job. If you choose a different cabinet and/or layout from the one shown here, try to keep the high-current runs between T1, D1, D2, C1 and Q3-Q6 as short as possible. Even no. 12 wire is rated for only 23 A continuous in applications such as this, so no. 10 is in order here. No. 10 wire can be difficult to obtain and work with, so you can do what we did in the prototype supply: Use several pieces of smaller wire for each run. The high-current secondary wires from T1 connect directly to the anodes of D1 and D2; the center tap connects directly to the C1 minus terminal. Three pieces of no. 12 wire are used between the rectifiers and C1. A separate length of no. 14 wire is used between the collector of each pass transistor and C1. Likewise, a separate piece of no. 14 is used between the emitter of each pass transistor and R2. Three no. 12s are used between R2 and M1, and between M1 and the positive output terminal. Likewise, three no. 12s are used between C1 and the negative output terminal. This may seem like overkill, but at 30 or more amperes, I × R losses in even short lengths of hookup wire are significant. The rest of the wiring is low current, so any no. 22 or larger hookup wire may be used.

Testing and Setup

After construction has been completed, it will be necessary to adjust the output voltage and crowbar threshold point before using the supply as a power source. All that is needed is an accurate voltmeter (preferably a digital voltmeter, if available) and a screwdriver to turn the OUTPUT VOLTAGE and TRIP VOLTAGE potentiometers. Remember to install the jumper between $+V_{OUT}$ and V_{SENSE}.

For initial setup, temporarily replace R2 with a resistor of approximately 0.5 Ω. This resistor can be made from several low-value, 1-W resistors in parallel. The actual value is not critical; the purpose is to prevent unnecessary fireworks by limiting supply current to around 1 A for initial testing and setup. Also, disconnect the gate of Q8 from U2C.

Before applying power to the regulator board and pass transistors, verify the operation of T1, D1, D2 and C1. No-load voltage on C1 should be approximately 18 V. Remove power and connect the pass transistors and the regulator board. Begin with R4 fully counterclockwise (for maximum trip voltage) and R3 fully clockwise (for minimum output voltage).

Turn the supply on. M2 should deflect upscale 10 to 12 V. If the voltmeter does not indicate any output voltage, or if the OVERCURRENT lamp is illuminated, quickly turn the supply off, unplug the line cord and check for wiring errors. If all is well, connect a voltmeter across the supply output terminals and slowly rotate R3 clockwise until the meter reads 13.8 V. Do not be surprised if you have to turn R3 many turns; the adjustable voltage range of this supply has been preset by the fixed resistors, and R3 is intended for trimming the supply voltage to the desired output, rather than for large voltage excursions. Connect a resistive load of approximately 15-Ω across the supply output terminals. This should draw approximately 1 A, and M1 should deflect slightly. Short-circuit the supply output terminals through a meter capable of reading approximately 0-1 A. Supply voltage should drop drastically, the foldback current through the ammeter should be less than 500 mA and DS1 should light to indicate OVERCURRENT. The

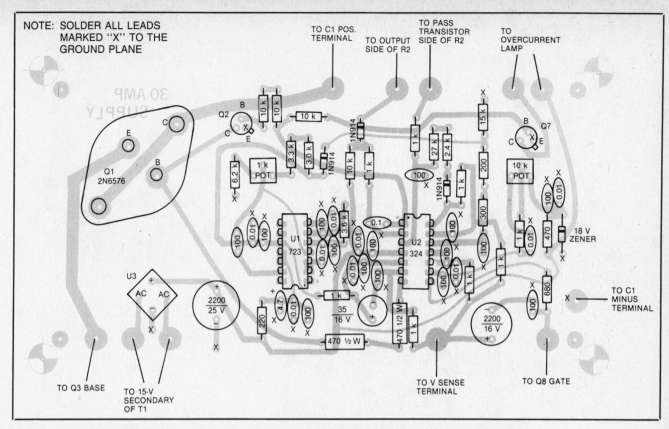

Fig. 10 — Parts-placement guide for the regulator board. A full-size etching pattern appears at the back of this book.

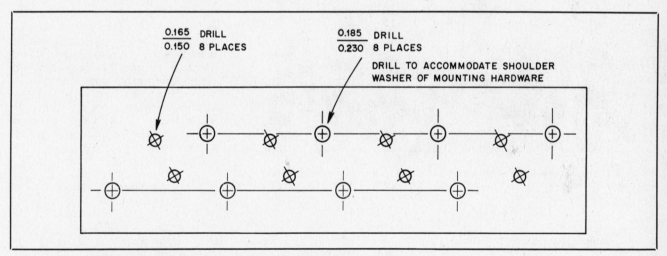

Fig. 11 — Full-size drilling template for the Wakefield 441K heat sink. See text for details.

foldback current should be less than half of the maximum supply current (set to approximately 1 A by the temporary R2).

Next, set the crowbar to fire at the desired voltage. Connect a voltmeter to the lead that will be used to connect the 680-Ω resistor on the regulator board to the gate of Q8. Set the voltmeter to a convenient range above 20 V (the value of V_{IN}). There should be no voltage present at this point until the crowbar trips. Rotate R3 until the supply output voltage reaches the desired trip voltage. A trip voltage of

14.5 V was selected for the prototype supply. Next, adjust R4 until the crowbar trips and the voltmeter indicates the value of V_{IN}. The crowbar is now set to fire at 14.5 V. Turn R3 until the supply voltage drops below the trip voltage (the voltmeter will return to zero) and connect the gate lead to Q8. Turn R3 to increase supply voltage until the crowbar trips. This time, Q8 will fire, shorting the supply output. The supply should go into foldback current limiting and behave as it did when you shorted the output terminals as described

above. Remove power and turn R3 to reduce the supply output voltage below the trip point. Apply power again and adjust R3 until supply output voltage returns to 13.8 V, or whatever you want the output voltage to be for normal operation. This completes initial testing.

Now it's time for the real smoke test. Turn off the supply and remove the temporary resistor used at R2; install the 0.02-Ω unit. Locate some low-value resistors to test the supply at various current levels. We started testing the prototype with a 5-Ω,

Fig. 12 — An internal view of the high-current power supply. The rectifiers and pass transistors are mounted to heat sinks on the rear panel. Parts are arranged to keep the high-current runs as short as possible.

Fig. 13 — D1 and D2 are mounted to the heat sink with the anode posts pointing toward the chassis. The cathodes are connected by a copper strip that is sandwiched between the devices and the heat sink. Connection to the cathodes is made to a lug that is soldered to the copper strip. See text.

200-W resistor (approximately a 3-A load) and added additional similar resistors in parallel until we reached the desired current level. To make a full-current test of the supply, you'll need a combination of resistors that totals approximately 0.5 Ω at 450 W. If you don't have access to suitable power resistors, automobile headlights are an excellent, commonly-available high-current 13.8-V load. At some point in your testing, you'll notice that although the supply voltage at the output terminals of the supply is still holding steady at 13.8 V, the voltage at the load will be somewhat less than that because the drop in the connecting wires is not included in the regulator feedback loop. This is where remote sensing comes into play. Disconnect the jumper between the $+V_{OUT}$ and V_{SENSE} terminals and connect the V_{SENSE} terminal directly to the load. The voltage at the load should now be within a few hundred millivolts of the no-load supply output voltage.

After you have verified supply performance under load, you should verify that the foldback current level is indeed the desired value. Reconnect the jumper wire between the $+V_{OUT}$ and V_{SENSE} terminals. Screw up your courage and briefly short the $+V_{OUT}$ terminal to the $-V_{OUT}$ terminal. M1 should indicate the selected value of foldback current (in this case, 3 A) and M2 should indicate almost no voltage. If this is not the case, you will have to experiment with different values for the smaller of the two resistors that make up R1. Try changing the value a few hundred ohms at a time until you reach the desired foldback current.

Congratulations! The supply is now ready to use.

A 28-V, High-Current Power Supply

Many modern high-power transistors used in RF power amplifiers require 28-V dc collector supplies, rather than the traditional 12-V supply. By going to 28 V (or even 50 V), designers significantly reduce the amount of current required for an amplifier in the 100-W or higher output class. The power supply shown in Figs. 14 through 18 is conservatively rated for 28 V at 10 A (enough for a 150-W output amplifier) — continuous duty! It was designed with simplicity and readily-available components in mind. Mark Wilson, AA2Z, built this project in the ARRL lab.

Circuit Details

The schematic diagram of the 28-V supply is shown in Fig. 15. T1 was designed by Avatar Magnetics specifically for this project. The primary requires 117-V ac, but a dual-primary (117/234 V) version is available. The secondary is rated for 32 V at 15 A, continuous duty. The primary is bypassed by two 0.01-μF capacitors and protected from line transients by an MOV.

U1 is a 25-A bridge module available

Fig. 14 — The front panel of the 28-V power supply sports only a power switch, pilot lamp and binding posts for the voltage output. There is room for a voltmeter, should another builder desire one.

from Radio Shack or a number of other suppliers. It requires a heat sink in this application. Filter capacitor C1 is a computer-grade 22,000-μF electrolytic. Bleeder resistor R1 is included for safety because

of the high value of C1; bleeder current is about 12 mA.

There is a trade-off between the secondary voltage of the transformer and the value of the filter capacitor. To maintain regulation, the minimum supply voltage to the regulator circuitry must remain above approximately 31 V. Ripple voltage must be taken into account. If the voltage on the bus drops below 31 V in ripple valleys, regulation may be lost.

In this supply, the transformer secondary voltage was chosen to allow use of a commonly-available filter value. The builder found that 50-V electrolytic capacitors of up to about 25,000 μF were common and the prices reasonable; few dealers stocked capacitors above that value, and the prices increased dramatically. If you have a larger filter capacitor, you can use a transformer with a lower secondary voltage; similarly, if you have a transformer in the 28- to 35-V range, you can calculate the size of the filter capacitor required. Chapter 6 contains the formulas needed to calculate ripple for different filter

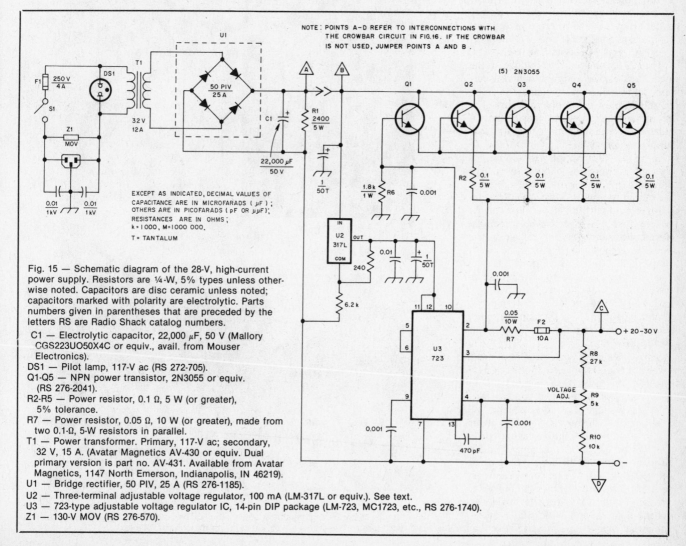

Fig. 15 — Schematic diagram of the 28-V, high-current power supply. Resistors are ¼-W, 5% types unless otherwise noted. Capacitors are disc ceramic unless noted; capacitors marked with polarity are electrolytic. Parts numbers given in parentheses that are preceded by the letters RS are Radio Shack catalog numbers.

C1 — Electrolytic capacitor, 22,000 μF, 50 V (Mallory CGS223UO50X4C or equiv., avail. from Mouser Electronics).
DS1 — Pilot lamp, 117-V ac (RS 272-705).
Q1-Q5 — NPN power transistor, 2N3055 or equiv. (RS 276-2041).
R2-R5 — Power resistor, 0.1 Ω, 5 W (or greater), 5% tolerance.
R7 — Power resistor, 0.05 Ω, 10 W (or greater), made from two 0.1-Ω, 5-W resistors in parallel.
T1 — Power transformer. Primary, 117-V ac; secondary, 32 V, 15 A. (Avatar Magnetics AV-430 or equiv. Dual primary version is part no. AV-431. Available from Avatar Magnetics, 1147 North Emerson, Indianapolis, IN 46219).
U1 — Bridge rectifier, 50 PIV, 25 A (RS 276-1185).
U2 — Three-terminal adjustable voltage regulator, 100 mA (LM-317L or equiv.). See text.
U3 — 723-type adjustable voltage regulator IC, 14-pin DIP package (LM-723, MC1723, etc., RS 276-1740).
Z1 — 130-V MOV (RS 276-570).

and transformer values.

The regulator circuitry takes advantage of commonly-available parts. The heart of the circuit is U3, a 723 voltage regulator IC. The values of R8, R9 and R10 were chosen to allow the output voltage to be varied from 20 to 30 V. The 723 has a maximum input voltage rating of 40 V, somewhat lower than the filtered bus voltage. U2 is an adjustable 3-terminal regulator; it is set to provide approximately 35 V to power U3. U3 drives the base of Q1, which in turn drives pass transistors Q2-Q5. This arrangement was selected to take advantage of common components. At first glance, the number of pass transistors seems high for a 10-A supply. Input voltage is high enough that the pass transistors must dissipate about 120 W (worst case), so thermal considerations dictate the use of four transistors. See Chapter 6 for a complete discussion of thermal design. If you use a transformer with a significantly different secondary potential, refer to the Chapter 6 thermal tutorial to verify the size heat sink required for safe operation.

R9 is used to adjust supply output voltage. Since this supply was designed primarily for 28-V applications, R9 is a "set and forget" control mounted internally. A 25-turn potentiometer is used here to allow precise voltage adjustment. Another builder may wish to mount this control, and perhaps a voltmeter, on the front panel to easily vary the output voltage.

The 723 features current foldback if the load draws excessive current. Foldback current, set by R7, is approximately 14 A, so F2 should blow if a problem occurs. The output terminals, however, may be shorted indefinitely without damage to any power-supply components.

If the regulator circuitry should fail, or if a pass transistor should short, the unregulated supply voltage will appear at the output terminals. Most 28-V RF transistors would fail with 40-plus volts on the collector, so a prospective builder might wish to incorporate the overvoltage protection circuit shown in Fig. 16 in the power supply. This circuit is optional. If you choose to use the "crowbar," make the interconnections as shown. Note that R11 and F3 of Fig. 16 are added between points A and B of Fig. 15. If the crowbar is not used, place a wire jumper between points A and B of Fig. 15.

This crowbar circuit is taken from Motorola application notes for the MC3423 overvoltage sensor IC. The chip contains a 2.5-V reference and two comparators. When the voltage at pin 2 (sense terminal) reaches 2.5 V, the output goes high (from 0.0 V to the positive bus voltage) to drive the gate of Q6, a high-current SCR. Q6 turns on, shorting the supply output terminals. This circuit is inexpensive and easy to implement, yet it allows the builder to precisely set the trip voltage. It provides excellent gate drive to the SCR, and is somewhat quicker and more reliable than

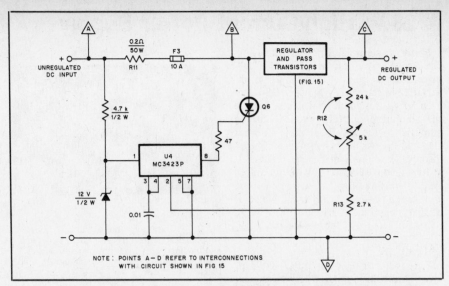

Fig. 16 — Schematic diagram of the over-voltage protection circuit. Resistors are ¼-W, 5% carbon types unless noted.

Q6 — 20-A, 100-V stud-mount SCR (RCA SK6502 or equiv.).
R12 — 5-kΩ, 10-turn potentiometer in series with a 24-kΩ, ¼-W resistor.
U4 — Overvoltage sensor IC, Motorola MC3423 (RS 276-1717).

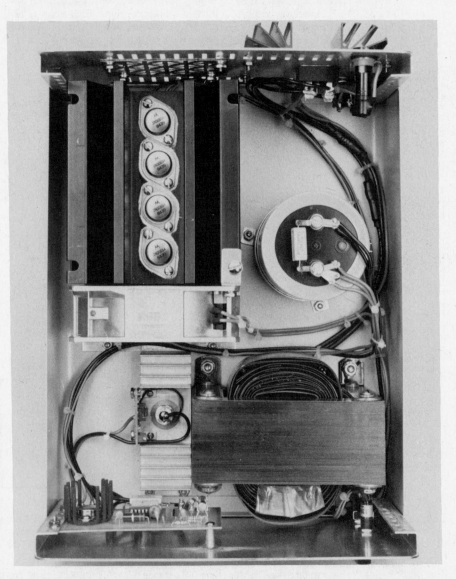

Fig. 17 — Interior of the 28-V, high-current power supply. The cooling fan is necessary only if the pass transistors and heat sink are mounted inside the cabinet. See text.

crowbar circuits that use Zener diodes. R11 and F3 protect the pass transistors from damage in case of high-current transients.

The trip voltage is determined from the equation

$$V_{TRIP} \approx 2.5 \left(1 + \frac{R11}{R12}\right)$$

The application notes recommend that R13 be less than 10 kΩ for minimum drift and suggest a value of 2.7 kΩ. In this version, R12 is preset with a 24-kΩ resistor, and a 5-kΩ, 10-turn potentiometer dials in the precise trip point.

Construction

Fig. 17 shows the interior of the 28-V supply. It is built in a Hammond 1401K enclosure. All parts mount inside the box. The regulator components are mounted on a small PC board that is attached to the rear of the front panel. See Fig. 18. Most of the parts were purchased at local electronics stores or from suppliers listed in Chapter 35. Many parts, such as the heat sink, pass transistors, 0.1-Ω power resistors and filter capacitor can be obtained from scrap computer power supplies found at flea markets.

Q2-Q5 are mounted on a Wakefield model 441K heat sink. A drilling template can be found in Fig. 11. The transistors are mounted to the heat sink with insulating washers and thermal heat-sink compound to aid heat transfer. Radio Shack TO-3 sockets make electrical connections easier. The heat sink surface under the transistors must be absolutely smooth. Carefully deburr all holes after drilling and lightly sand the edges with fine emery cloth.

A five-inch fan circulates air past the heat sink inside the cabinet. Forced-air cooling is necessary only because the heat sink is mounted inside the cabinet. If the heat sink was mounted on the rear panel with the fins vertical, natural convection would provide adequate cooling and no fan would be required.

U1 is mounted to the inside of the rear panel with heat-sink compound. Its heat sink is bolted to the outside of the rear panel to take advantage of convection cooling.

U2 may prove difficult to find. The 317L is a 100-mA version of the popular 317-series 1.5-A adjustable regulator. The 317L is packaged in a TO-92 case, while the normal 317 is usually packaged in a larger TO-220 case. Many of the suppliers listed in Chapter 35 sell them, and RCA SK7644 or Sylvania ECG1900 direct replacements are available from many local electronics shops. If you can't find a 317L, you can use a regular 317 (available from Radio Shack, among others).

R7 is made from two 0.1-Ω, 5-W resistors connected in parallel. These resistors get warm under sustained operation, so they

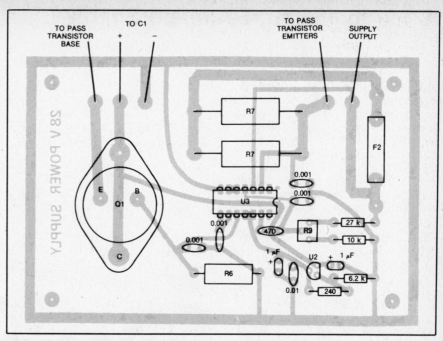

Fig. 18 — Parts-placement diagram for the 28-V power supply. A full-size etching pattern appears at the back of this book.

are mounted approximately 1/16 inch above the circuit board to allow air to circulate and to prevent the PC board from becoming discolored. Similarly, R6 gets warm to the touch, so it is mounted away from the board to allow air to circulate. Q1 becomes slightly warm during sustained operation, so it is mounted to a small TO-3 PC board heat sink.

Not obvious from the photograph is the use of a single-point ground to avoid ground-loop problems. The PC-board ground connection and the minus lead of the supply are tied directly to the minus terminal of C1, rather than to a chassis ground.

The crowbar circuit is mounted on a small heat sink near the output terminals. Q6 is a stud-mount SCR and is insulated from the heat sink. An RCA replacement SCR was used here because it was readily available, but any 20-A or greater SCR that can handle at least 50 V will work. The other components are mounted on a small circuit board that is attached to the heat sink with angle brackets. U4 is available from Radio Shack.

Although the output current is not extremely high, no. 14 or no. 12 wire should be used for all high-current runs, including the wiring between C1 and the collectors of Q2-Q5; between R2-R5 and R7; between F2 and the positive output terminal; and between C1 and the negative output terminal. Similar wire should be used between the output terminals and the load.

Testing

First, connect T1, U1 and C1 and verify that the no-load voltage is approximately 44-V dc. Then, connect unregulated voltage to the PC board and pass transistors. Leave the gate lead of Q6 disconnected from pin 8 of U4 at this time. You should be able to adjust the output voltage between approximately 20 and 30 V. Set the output to 28 V.

Next, set the crowbar to fire at 29 V or whatever trip voltage you desire. Set the potentiometer of R12 for maximum resistance. Connect a voltmeter to U4, pin 8. The voltmeter should indicate 0.0 V. Increase supply voltage to 29 V and adjust R12 until pin 8 goes high, to approximately 28 V. Back off the supply voltage; pin 8 should go low again. Connect the gate lead of Q6 and again increase supply voltage to 29 V. The crowbar should fire, shorting the supply output. Slight adjustment of R12 may be necessary. Remove power and turn R9 to reduce supply voltage below the trip point. Apply power and reset the output voltage to 28 V.

Next, short the output terminals to verify that the current foldback is working. Voltage should return to 28 when the shorting wire is disconnected. This completes testing and setup.

The supply shown in the photographs dropped approximately 0.1 V between no load and a 12-A resistive load. During testing in the ARRL lab, this supply was run for four hours continuously with a 12-A resistive load on several occasions without any difficulty.

A 1.2- to 15-Volt, 5-Ampere Supply

The power supply shown in Figs. 19 to 23 is intended for general-purpose, test-bench applications. The output is adjustable from 1.2 to 17 volts at currents up to 6 amperes. Metering is provided for voltage levels up to 15 volts and current levels up to 5 amperes. Most of the components used in this supply are of the junk-box variety with the possible exception of U2, the three-terminal voltage regulator. The circuit will tolerate fairly wide component substitutions and still offer good performance. The majority of the circuit components are mounted on a 2¾ × 4½-inch circuit board. All controls, including the mains fuse are located on the front panel for easy access.

The Circuit

As shown in Fig. 20, two power transformers are used in parallel to feed U1, the full-wave bridge rectifier assembly. The transformers specified are rated at 2 amperes each. The prospective builder might question the wisdom of using only

Fig. 19 — A simple 1.2- to 15-volt, 5-ampere power supply. All controls are mounted on the front panel for easy access. The milliammeter reading is multiplied by 100 to obtain the true output current.

Fig. 21 — Inside view of the power supply. Component placement is not at all critical; however, the layout shown here provides a neat appearance.

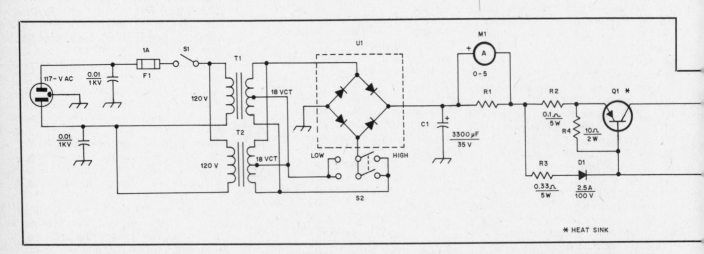

Fig. 20 — Schematic diagram of the 5-ampere power supply. Component designations on the schematic diagram but not shown in the parts list are for text or placement guide reference only.

C1 — 3300-μF, 35-volt, axial leads.
J1, J2 — Binding posts.
M1 — 0-50 mA, Calectro DI-914 or equiv.
M2 — 0-15 V, Calectro DI-920 or equiv.
Q1 — Silicon PNP power, Radio Shack 276-2043 or equiv.

R1 — Meter shunt, 13 inches no. 22 enameled wire wound on a high-value, 1-watt resistor. (Resistor used only as a form for the wire.)
R6 — 2500 ohms, 2 watts, panel mount.
S1 — SPST, toggle.

S2 — DPDT, toggle. Both sections connected in parallel.
T1, T2 — 117-V primary, 18-V CT secondary. Radio Shack 273-1515 or equiv.
U1 — Bridge-rectifier assembly, 50 V, 25 A.
U2 — Regulator, LM-317K.

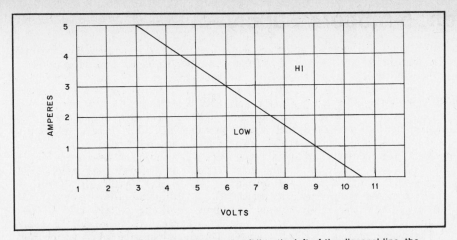

Fig. 22 — For voltage and current requirements that fall to the left of the diagonal line, the power supply may be operated in the LO mode. Pass-transistor dissipation will be reduced when the supply is operated in this manner.

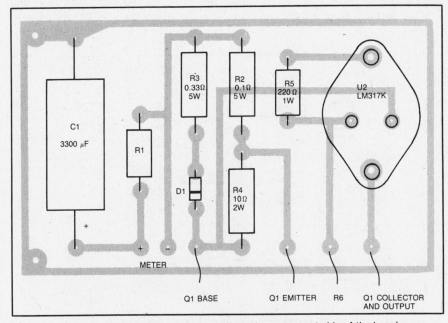

Fig. 23 — Circuit-board layout pattern as viewed from the component side of the board. A full-size etching pattern can be found at the back of this book.

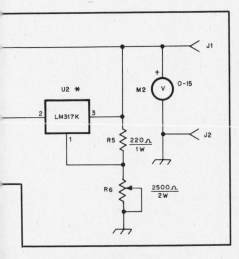

in selecting the HI or LO mode of operation.

The regulator consists of a pass transistor "wrapped around" an adjustable voltage regulator. Circuit operation can be understood by noting the values of R3 and R2. The majority of the three-terminal regulator current will flow through R3 and D1. The offset voltage in D1 is approximately equal to the emitter-base potential of Q1. Because of this, the voltage drop across R3 will be the same as that across R2. Since the ohmic value of R2 is 0.33 of R3, three times as much current will flow in Q1 as in U2. The net result is that the current capability of the overall circuit is increased by a factor of four. In addition, the current-limiting characteristics of the three-terminal regulator are transferred directly to the composite circuit.

M1 and its associated shunt resistor are placed at the input to the regulator circuit so that the voltage drop across the resistor will not adversely alter the supply voltage regulation. The relatively small current drawn by the regulator circuitry does not seriously affect the meter accuracy. M2 measures supply output voltage.

Construction

The power supply is housed in a homemade enclosure that was fabricated from sheet aluminum. Fig. 21 shows the parts layout. Dimensions of the enclosure are 5½ × 6 × 8 inches, although any cabinet that will house the components may be used. Circuit board layout information is given in Fig. 23. The completed circuit board is mounted vertically to the chassis using space lugs and no. 6 hardware. A small heat sink for the LM317K regulator was made from a scrap piece of aluminum. The pass transistor is mounted to a larger heat sink which is bolted to the rear panel of the power supply. Here, a Motorola MS-10 was used. Bear in mind that the transistor must be insulated from the sink. Use a small amount of heat-sink compound between the transistor and the sink for a good thermal bond.

Since the power supply can deliver up to 6 amperes, fairly heavy wire should be used for those runs carrying the bulk of the current. No. 18 wire was used in this unit and it appeared to be adequate.

The completed power supply may be "crowbarred" without worry of regulator or pass transistor destruction. Perhaps the only precaution that should be mentioned is that of the exposed collector of the pass transistor. Although no damage will occur if the case is shorted to ground, it will cause the loss of output voltage. This could occur if the power supply is mounted on a test bench with a number of leads dangling behind the unit. A simple fix for this would be to mount a plastic TO-3 transistor cover over the case.

4 amperes worth of transformer in a 5-ampere supply. This is a valid question. With a 5-ampere load connected to the output of the supply, the transformers deliver more than their rated secondary voltage and do not become unreasonably warm to the touch even after continuous-duty operation. If transformers of different manufacture are used, it might be wise to select units having a higher current rating — just to be sure.

S2 is included in the design so that either half or all of the secondary voltage may be applied to U1. This feature was included so that the dissipation of the pass transistor may be reduced when using the supply with low-voltage, high-current loads. The graph displayed in Fig. 22 can be used as a guide

A Deluxe 5- to 25-Volt, 5-Ampere Supply

The power supply illustrated in Figs. 24 and 26 and schematically at Fig. 25 might be termed a rich person's power supply. The unit shown can supply voltages from 5 to 25 at currents up to 5 amperes. With thermal and short-circuit protection, it is virtually destruction proof. A digital panel meter is used to monitor voltage and current, selectable by a front-panel switch. Although we termed this a "rich person's supply," it will cost far less to build than a ready-made supply with the same features. The most expensive single item in the supply is the digital panel meter, which sells in single lot quantities for around $50 as this is written.

The digital readout, however, is not much more expensive than two high-quality meters. The prospective builder should consider this when choosing between the digital panel meter and two analog panel meters. Voltage measurements are read directly from the panel meter in volts. Current is measured in amperes; a reading of 0.05 is equal to 50 mA.

Circuit Details

The circuit diagram of the power supply is shown in Fig. 25. T1 is a 36-volt, center-tapped transformer rated at 6 amperes. D1 and D2 are used in a full-wave rectifier providing dc output to the filter capacitor, C3, a 34,000-μF, 50-volt electrolytic of the computer-grade variety. The unregulated voltage is fed to U1, a National LM338 regulator, the heart of the supply. This chip is rated for 5-A continuous duty when used with an adequate heat sink. R1 and R2 form a voltage divider which sets the output voltage of the supply. R2 is a 10-turn potentiometer. U1 is bypassed with 2.2-μF tantalum capacitors directly at

Fig. 24 — Front view of the deluxe 5- to 25-volt, 5-ampere power supply.

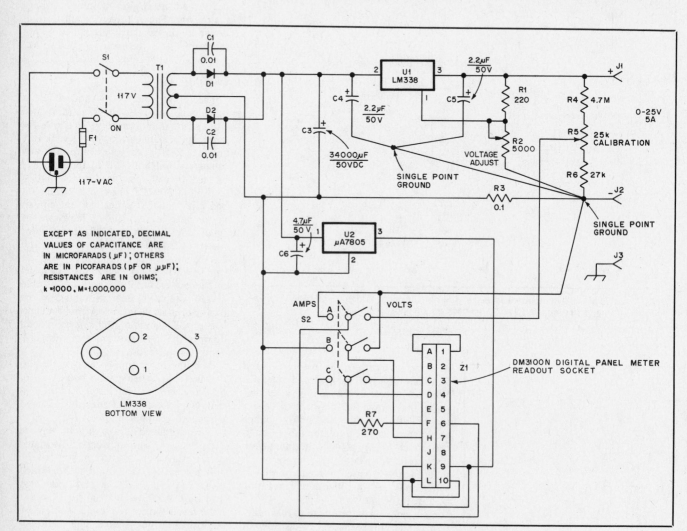

Fig. 25 — Schematic diagram of the deluxe power supply. All resistors are half-watt carbon types unless noted otherwise. Capacitors are disc ceramic unless noted otherwise. Numbered components not appearing in the parts list are for text reference only.

C3 — Electrolytic capacitor, 34,000 μF, 50 V. Sprague 36D343G050DF2A or equiv.
C4, C5 — Tantalum capacitor, 2.2 μF, 50 V.
C6 — Tantalum capacitor, 4.7 μF, 50 V.
D1, D2 — Silicon rectifier, 100 V, 12 A.
F1 — Fuse, 2 A.
J1-J3 — Binding post.

R2 — Potentiometer, 5 kΩ, linear, 10 turn.
R3 — Resistor, 0.1 Ω, circuit board mount.
S1 — Toggle switch, DPST.
S2 — Toggle switch, 3PDT.
T1 — Power transformer; primary 117 V,

secondary 36 V CT, 6 A. Stancor P-8674 or equiv.
U1 — Regulator, National LM338 or equiv. (5 A, adj.)
U2 — Regulator, μA7805 or equiv. (5 V, 1 A)
Z1 — Digital panel meter. Datel DM3100N or equiv.

Fig. 26 — Interior view of the deluxe power supply.

the input and output pins.

Z1, as outlined earlier, is a digital panel meter. Connections to the meter are made through a special edge connector supplied with the readout. U2 is used to supply a regulated 5 volts for powering the digital panel meter. The input and ground leads of this regulator are attached to the input (non regulated) side of U1.

R4, R5 and R6 form a divider circuit to supply the digital meter with an output voltage reading. R5 is made adjustable so that the meter can be calibrated. R3 is a current-sensing resistor which is placed in the negative lead of the supply. This resistor is used on the input side of the regulator (U1) so as not to affect the voltage regulation of the power supply at high load currents. Any voltage dropped across the resistor will be made up by the regulator, so the output voltage will remain unchanged. Notice that U2 is placed to the left or at the input side of the regulator. This is so the current drawn by the readout will not affect current readings taken at the load. Sections A and B of S2 are used to switch the meter between the voltage and current sensors. S2C is used to switch the decimal point in the digital panel meter to read correctly for both voltage and current.

As shown in the schematic, a single-point ground is used for the supply. Used in many commercial supplies, this technique provides better voltage regulation and stabilization than the "ground it anywhere" attitude. In this supply, the single-ground point is at the front panel binding post labeled MINUS. All leads that are to be connected to ground should go only to that point.

Construction

The deluxe power supply is housed in a homemade enclosure that measures $9 \times 11 \times 5\frac{1}{4}$ inches (Fig. 26). U1 is mounted to a large heat sink ($3 \times 5 \times 2$ inches) attached to the rear apron of the supply. The front panel sports the digital-panel meter, power switch, binding posts, fuse holder, voltage-adjust potentiometer and meter-selector switch. Although a circuit board is shown in the photograph as supporting R4, R5, R6, R7, D1, D2, C1 and C2, these items could just as well be mounted on terminal strips. For this reason a board pattern is not supplied.

The front and rear panels are painted white, and the cover is blue. Dymo labels are used on the front panel to identify each of the controls. Cable lacing of the various leads adds to the clean appearance of the supply.

A Multiple-Voltage Bench Supply

Meeting the voltage needs of today's electronic devices may pose more than a minor inconvenience to the experimenter. Having individual supplies available to power TTL, CMOS, ECL and MOSFET devices, as well as relay windings, can be annoying — particularly if the breadboarded project requires them all simultaneously. The power supply shown in Figs. 27 through 32 provides eight popular fixed voltages (± 5, 12, 15 and 24 V), as well as two continuously variable voltages (± 1.3 to 28 V) — enough to handle almost any electronic situation.

Simplicity, flexibility, low cost and ease of parts procurement were major design goals. By analyzing the transformer, rectifier/filter capacitor and regulator separately, a prospective builder should find enough information to confidently duplicate, modify or completely redesign this supply to fit his or her particular needs. The completed unit, shown in Fig. 27, was designed and built in the ARRL lab by Greg Bonaguide, WA1VUG.

Design Overview

Fig. 28 is a block diagram of a power supply. The rectifier/filter block receives a stepped-down ac voltage from the transformer secondary and delivers a filtered positive or negative voltage to the bus lines. These lines in turn supply the regulators with positive or negative input voltages. Internal regulator circuitry then drops and stabilizes this input voltage at a particular value.

Transformer Considerations

The transformer chosen for this supply, shown schematically in Fig. 29, features a 117-V primary and a 48-V, 4-A center tapped secondary. Both positive and negative voltages may be obtained by using a bridge rectifier and grounding the transformer center tap. The maximum dc voltage expected on either bus can be computed from the equation

$$V_{dc, max} = \left(\frac{1.414 \times V_{RMS}}{2}\right) - V_f$$

where

V_{RMS} = the transformer secondary RMS voltage

V_f = the forward drop across a single rectifier diode (usually 0.8 V)

The transformer used for this project yields $V_{dc, max} = 33.14$ V, the peak voltage that the dc bus filter capacitors will charge to under no-load conditions. Any applied load will tend to discharge the capacitors, producing a periodic ripple voltage. The peak-to-peak ripple value will be determined by load current, capacitance and ripple frequency. A properly designed power supply should have little or no detectable output

Fig. 27 — The multiple-voltage bench supply features a number of popular voltages used in solid-state projects.

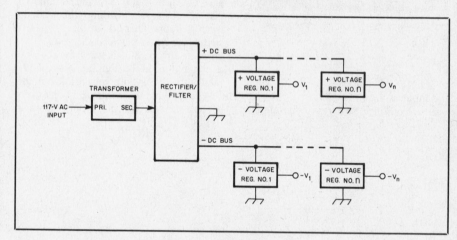

Fig. 28 — Block diagram of the multiple-voltage bench supply.

ripple, regardless of the regulator P-P input ripple voltage.

Filter Capacitor/Rectifier

In general, a 3-terminal regulator requires an input voltage at least 3-V higher than its output voltage. This value, called dropout voltage, is the minimum input voltage that the device needs to regulate. In this design, the highest voltage any regulator is expected to deliver is 28 V, so the filter capacitor should never permit the bus to drop below 31 V (28 + 3). From the earlier calculation, the no-load maximum bus voltage will never exceed 33.14 V. The difference, 2.14 V, represents the maximum peak-to-peak ripple voltage that can be allowed on either bus (Fig. 30). Additionally, each bus is expected to draw 2 A, half the available transformer current.

Equipped with the above information, the required filter capacitance may be calculated using the equation

$$C = \frac{I\sqrt{2}}{F_r \, V_{p-p}\sqrt{3}}$$

where

C = filter capacitance in farads

I = maximum current drawn from the bus in amperes

F_r = ripple frequency in hertz (120 for full-wave rectification)

V_{p-p} = peak-to-peak ripple voltage

In this case, C works out to be 6356 μF. The next closest available combination of commercial values, 6500 μF, is used.

The bridge rectifier is made from four 200-PIV, 3-A diodes. These devices provide necessary protection against the large surge current that initially charges the filter capacitors the instant the supply is switched on.

Regulators

The 3-terminal regulators selected for this project are simple to work with and have several particularly attractive features. Built-in thermal shutdown, overvoltage protection and current foldback virtually guarantee destruction-proof operation — even in the event of accidental short-circuiting of power supply output terminals

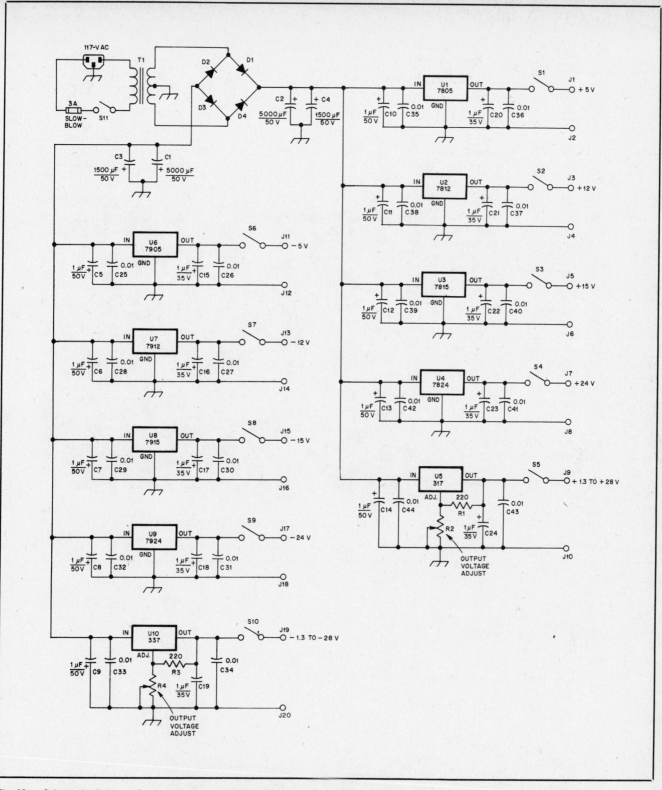

Fig. 29 — Schematic diagram of the multiple-voltage bench supply. All capacitor values are in microfarads and resistor values are in ohms. Resistors are ¼ W.

C1, C2 — Electrolytic capacitor, 5000 µF, 50 V (Mallory CG502U050U3C or equiv.).
C3, C4 — Electrolytic capacitor, 1500 µF, 50 V (Aerovox AFH 1-14-50 or equiv.).
C5-C14 — Tantalum capacitor, 1 µF, 50 V.
C15-C24 — Tantalum capacitor, 1 µF, 35 V.
D1-D4 — Silicon diode, 200 PIV, 3 A (Radio Shack 276-1143 or equiv.).
J1-J20 — Binding post.
R2-R4 — Potentiometer, 5000 Ω, 2 W.
S1-S10 — Toggle switch, SPST.
S11 — Rocker switch, SPST.

T1 — Power transformer; primary 117 V, secondary 48-V CT, 4 A (Stancor P-8619 or equiv.).
U1 — Voltage regulator, +5 V, 1 A (LM7805 or equiv.).
U2 — Voltage regulator, +12 V, 1 A (LM7812 or equiv.).
U3 — Voltage regulator, +15 V, 1 A (LM7815 or equiv.).
U4 — Voltage regulator, +24 V, 1 A (LM7824 or equiv.).
U5 — Voltage regulator, +1.2 to 37 V, 1.5 A

(LM317 or equiv.).
U6 — Voltage regulator, -5 V, 1 A (LM7905 or equiv.).
U7 — Voltage regulator, -12 V, 1 A (LM7912 or equiv.).
U8 — Voltage regulator, -15 V, 1 A (LM7915 or equiv.).
U9 — Voltage regulator, -24 V, 1 A (LM7924 or equiv.).
U10 — Voltage regulator, -1.2 to -37 V, 1.5 A (LM337 or equiv.).

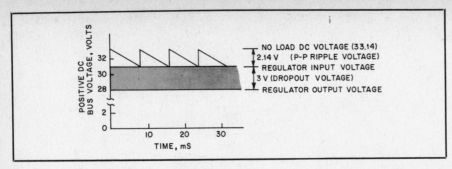

Fig. 30 — Voltage relationships on the positive dc bus.

conductive silicone heat-sink compound and nonconductive fasteners must be used.

Bypassing

Three-terminal regulators are active devices containing high-gain error-sensing amplifiers, so great care is necessary to keep them stable over a wide range of load impedances. Input and output leads should be bypassed with both 0.01-μF disc-ceramic and 1.0-μF tantalum capacitors. All bypassing should be performed as close to the regulator as possible to keep RF from

or excessive load demands. Three-terminal regulators are capable of supplying currents up to 8 A, simplifying construction and reducing overall cost by eliminating series-pass transistors and associated circuitry. The supply described here uses readily available 1.0-A regulators in TO-220-style packages.

Several design precautions must be taken when working with these regulators, however. Maximum input voltage, as well as maximum and minimum differential voltages, must conform to manufacturers' ratings for safe, trouble-free operation. Adequate heat-sinking must also be employed. In this supply, individual fin-type heat sinks cool each regulator. In some three-terminal devices, the case is also the input or output terminal, so care must be taken to ensure nothing contacts the heat sinks. The TO-220-style cases also permit installation against the side of the chassis, using it as a heat sink. However, if the regulator case isn't also the ground connection, appropriate mica washers, thermally

Fig. 32 — Interior view of the multiple-voltage bench supply.

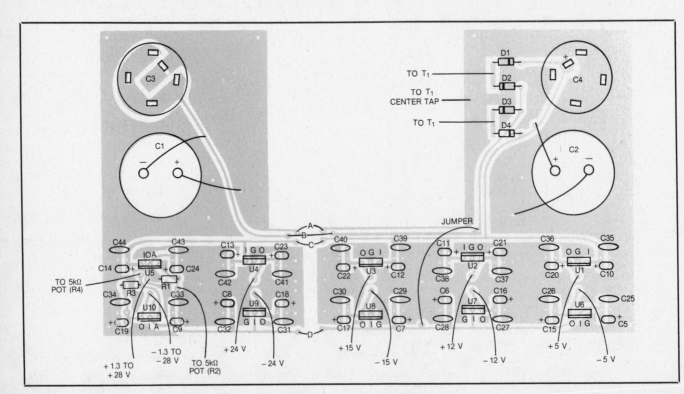

Fig. 31 — Parts placement for the multiple-voltage bench supply. A full-size etching pattern appears at the back of this book.

entering the device. Generally, use of double-sided circuit board material is recommended. The unetched component side acts as a ground plane. All component leads should be kept as short as possible. Ferrite beads on all lines leading to and from the regulators may be necessary in some cases.

Construction

Most of the components in this supply are mounted on two single-sided PC boards. Parts placement is given in Fig. 31. The boards are mounted just above the bottom of the 4 × 6¾ × 12-inch cabinet (Fig. 32) and are cut to clear the transformer. Individual heat sinks are used for each regulator to ensure adequate cooling.

Output voltages from the regulators are fed to binding posts via toggle switches. The potentiometers for the two adjustable voltages are also mounted on the front panel adjacent to the binding posts. Ac line voltage enters the supply through a standard CEE-22, 6-A connector.

The high initial surge current needed to charge the capacitors at turn-on will blow an ordinary fuse, so use of a 3-A slow blow fuse is necessary. This fuse will conduct 4 A for about five seconds, allowing the capacitors to charge while guarding against drastic circuit malfunction. No. 18 wire is used for the ac connections, while no. 22 is used for the dc interconnections. The cover of the enclosure has a black wrinkle finish that attractively complements the natural-aluminum front panel.

A Smorgasbord Supply

A smorgasbord offers foods and dishes to suit a variety of tastes. Similarly, the circuits and techniques presented here cover a wide range of low-voltage amateur power-supply applications. This smorgasbord, shown in Figs. 33 to 45 was prepared by Greg Bonaguide, WA1VUG, in the ARRL lab.

Anatomy of a Power Supply

There are four major building blocks to any linear dc supply: The transformer, rectifier, filter and output regulator. However, before components can be specified and parts ordered, certain information must be provided. The designer needs to know the line voltage he has to work with (typically either 117 V, as here, or 234-V ac) and the desired output voltage and current the supply will deliver.

Defining Output Voltages and Currents

It's handy to have ± 12 and ± 5 V at the workbench. While negative voltages are called upon to provide small currents, the + 5-V tap must be capable of supplying 5 A for power-hungry logic boards. Similarly, many medium-power VHF FM rigs, low-power amplifiers, and QRP HF transceivers require at least 5 A from a steady 12-V tap. There are also times when a continuously variable, *tracking* voltage source (single-knob adjustment of positive and negative voltages) is indispensable. The variable supply should have a range of approximately 2 to 28 V at half an amp. The smorgasbord supply was designed with these goals in mind. Enough information is presented here to enable the amateur to

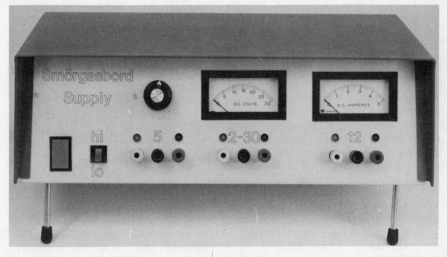

Fig. 33 — Front-panel view of the Smorgasbord Supply.

adapt these circuits and techniques to his own needs.

Examining Regulator Characteristics

A three-terminal adjustable voltage regulator, such as the LM317, is a natural choice for this supply. Besides providing excellent regulation of output voltage over the range of 1.5 to 28 V, it offers other attractive features such as thermal-overload protection and current-foldback limiting. To make any three-terminal regulator work well, however, the designer must operate the device above a minimum input voltage, called the dropout voltage (Fig. 34A). If the input voltage dips below this manufacturer-specified minimum (usually 3.0 V above the regulator's specified output voltage), regulation is lost.

Given this information, one would be tempted to apply an input voltage well above dropout. However, as the input voltage increases, so does the amount of power that the regulator must dissipate for any given output voltage and current. Three-terminal regulators have dissipation ratings that must be observed. Power dissipation can be calculated by multiplying the input-to-output voltage differential by the load current (Fig. 34B). Fig. 34C shows how input-output differential affects maximum available current. Under extreme

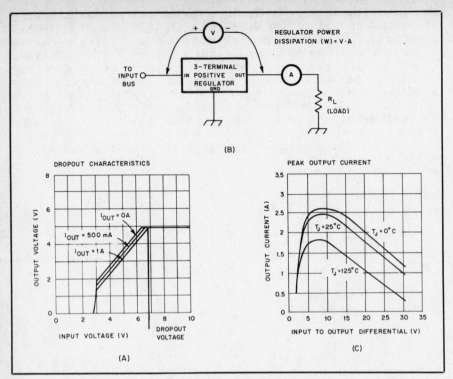

REGULATOR POWER
DISSIPATION (W) = V·A

TO INPUT BUS

3–TERMINAL POSITIVE REGULATOR

R_L (LOAD)

(B)

DROPOUT CHARACTERISTICS

$I_{OUT} = 0A$
$I_{OUT} = 500 mA$
$I_{OUT} = 1A$

OUTPUT VOLTAGE (V)

INPUT VOLTAGE (V)

DROPOUT VOLTAGE

(A)

PEAK OUTPUT CURRENT

OUTPUT CURRENT (A)

$T_J = 25°C$
$T_J = 0°C$
$T_J = 125°C$

INPUT TO OUTPUT DIFFERENTIAL (V)

(C)

Fig. 34 — Important three-terminal regulator operating characteristics. See text for details.

operating conditions, regulator power dissipation may become so great that the device shuts down to keep from burning up, a condition known as current foldback.

This leads to an important design consideration. To deliver 28 V at 1.0 A, the regulator needs 31 V at its input terminal to satisfy dropout requirements. But, when the regulator is adjusted to deliver only 2.0 V, (minimum output voltage), the input voltage will remain at 31 V, and the input-output differential will be 31 − 2, or 29 V! From the chart in Fig. 34C, we see that this condition results in severe current limiting. Fortunately, we don't need much current from the supply at this low voltage level. We will, however, need to be able to draw appreciable current at 28 V (for powering relays, and so forth), where, fortunately, a 3-V input-output differential presents no dissipation problems.

Choosing Filter Capacitors

Now that we have calculated the minimum permissible bus voltage (31 V), based on the dropout requirements of the supply's highest output voltage (28 V), we can investigate various transformer, rectifier and filter capacitor combinations.

There are several reasons for choosing designs that minimize filter capacitance. Electrolytic capacitors are bulky. They also tend to be expensive. Supplies that use large electrolytic capacitors sometimes require special turn-on sequencing to safely handle inrush current. Also, these large capacitors present a shock hazard unless they are properly discharged when the

supply is switched off. Fortunately, we can use an equation to find ways of minimizing filter capacitance requirements. It interrelates power-supply current, ripple amplitude, ripple frequency and filter capacitance:

$$C = \frac{I \sqrt{2}}{F_r V_{p-p} \sqrt{3}} \qquad \text{Eq. 1}$$

where

I = supply current in amperes
F_r = ripple frequency in hertz
V_{p-p} = peak-to-peak ripple in volts

This equation immediately tells us that we can reduce the filter capacitance by increasing the ripple frequency or by allowing a greater peak-to-peak ripple voltage to exist on the regulator input line (the bus). If we increase the ripple frequency, we will permit rectified dc pulses to replenish the energy stored in the filter capacitor at a faster rate. This is important, because the load is constantly draining this stored charge from the capacitor.

The alternative is to allow more ripple. This may seem confusing, since we have an intuitive notion that *all* power supply ripple is undesirable. Keep in mind, however, that ripple is only undesirable when it shows up *after* the regulator, where the voltage should be steady, pure dc. *Pre*-regulator ripple is an entirely different consideration. Fig. 35 helps to illustrate this point. Imagine we have a dc source of 40 V, with a 4-V p-p sawtooth waveform superimposed. Using an oscilloscope, we'd see the waveform of Fig. 35A. The voltage

begins at 40 V and decreases to 36-V valleys. If we hooked this up to a 28-V regulator, it would regulate perfectly because the bus voltage valleys never drop below the 31-V dropout value. In fact, we could let this sawtooth ripple waveform increase in amplitude until it was 9 V p-p, and we'd still have a perfectly regulated output! (Fig. 35B). As soon as the input voltage dips below 31 volts on valleys, however, output regulation is lost (Fig. 35C).

In any linear power supply, pre-regulator ripple will increase if we remove some filter capacitance or increase the load. Here, we need only enough filter capacitance to ensure that the input voltage remains above the regulator dropout voltage under full load.

Returning to the power-supply design, we can cut filter capacitance requirements in half if we use full-wave (120-Hz ripple) rather than half-wave (60-Hz ripple) rectification. Also, the amount of capacitance required can be reduced through the use of a transformer with a higher secondary voltage. With a higher bus voltage, more ripple (deeper valleys) can exist on the dc bus without dropout problems. However, as mentioned before, power dissipation becomes a problem at higher bus voltages, so all trade-offs must be considered.

The transformer chosen for the supply has a 48-V RMS, center-tapped secondary winding. The maximum dc voltage that can be expected from this supply, using a full-wave bridge rectifier and capacitive-input filter, is found from

$$V_{dc, max} = \sqrt{2} \, V_{RMS} - 2V_d \qquad \text{(Eq. 2)}$$

where

V_{RMS} = transformer RMS secondary voltage in volts
V_d = the voltage drop across a single rectifier diode, usually 0.7 V.

This equation gives 66.3 V as the maximum rectified voltage that can be expected across the full secondary. Since we want positive *and* negative voltages, divide this result by two, for 33.15 V. The peak-to-peak ripple on this line must be limited to 33.15 − 31.0 = 2.15 V, since the bus voltage must remain above dropout (31.0 V).

Since the total transformer is rated at 4 A, assume that the positive bus and negative bus will each require half that current. Using Eq. 1, each bus will require 6329 µF of filter capacitance. The next greater standard value, 6800 µF at 50-V dc, is used for each bus. Observe proper capacitor polarity!

Controlling the Adjustable-Output Regulators

A builder could install a stand-alone negative adjustable voltage regulator, such as an LM337, in the negative bus. This is a simple, reliable method for obtaining an adjustable, regulated negative voltage.

The smorgasbord supply, however, uses

a different technique to give single-knob control of both positive and negative voltages. See Fig. 36. U4, U5, Q2 and Q3 work together in a feedback loop to keep the negative output locked to the positive output. The 3.9-kΩ resistors form a voltage divider which, under normal conditions, sets pin 2 of U5 at zero volts. If the negative supply should begin to rise towards zero (as might happen if the load suddenly increases), pin 2 becomes positive with respect to pin 3 (ground), forcing pin 6 of U5 low. This forces Q2 into harder conduction, supplying more base current to Q3. Then, more load current flows through Q3 and, consequently, the load. This condition continues until a suitably negative output voltage returns zero volts to U5, pin 2. If too much current flows through Q3, the output voltage becomes too negative, and the voltage at the op-amp input drops below zero volts. This condition forces the output (pin 6) high, turning Q2 off, and removing Q3 base drive until the correct output voltage is established. Proper regulation of the positive output ensures regulation of the negative output. Whatever happens on the positive side also happens on the negative side; for example, if U4 goes into current limiting and the positive voltage drops, the negative voltage changes by the same amount. U5 is located on the rectifier board (Fig. 37).

Obtaining a High-Current 5-V Output

There are several ways to obtain regulation for the +5-V output. The simplest method requires only a three-terminal, 5-V regulator, such as the LM7805 1-A device or the LM323 5-A device. See Fig. 38. In this supply, the regulator could be fed from the 31-V bus or from the 25-V bus used for the high-current 12-V supply (described later). This arrangement is less than satisfactory because the input-output voltage differential is high enough to cause power-dissipation problems, as described earlier, which severely restrict maximum current. A larger heat sink or a heftier device might help, but output current would still be limited. Another option is to install yet another transformer/rectifier/filter combination for the 5-V circuit.

Instead, the smorgasbord supply uses a step-down switching regulator (Fig. 39). By chopping an input dc voltage the switcher converts the output to a desired lower voltage at high efficiency. In fact, the switching regulator used here operates at 80% efficiency. It draws only 1.2 A from the 31-V bus, yet provides 5 V at 5 A. The circuit is designed around a Lambda LAS6350 IC.

Because of the high switching rate (approximately 60 kHz), relatively small values of inductance and capacitance are used. D1 is a special Schottky diode specifically designed for high-speed switching applications. It is available from Varo Semiconductors, as noted in the parts list. Do

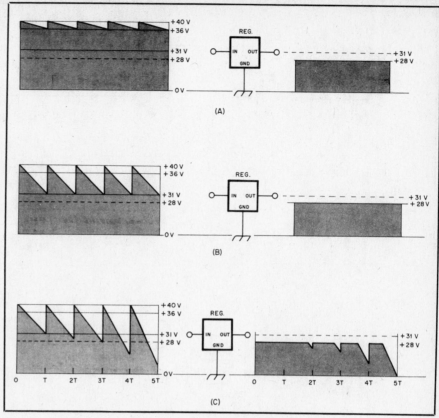

Fig. 35 — Effects of ripple on three-terminal regulator performance. See text.

not substitute an ordinary silicon diode!

In this switching regulator, care must be taken to prevent undesirable modes of oscillation. Bypassing is very important. Double-sided circuit board should be used with the layout pattern shown in Fig. 40.

Negative 5- and 12-V Considerations

Since little current is required from the negative 5-V supply, a standard 7905 three-terminal −5-V regulator will suffice. Voltage input for the 7905 is taken from the output of the negative 12-V regulator, however, to reduce the 7905 input-output differential. The negative 12-V regulator is similar; a 7912 provides the −12-V regulated output.

Designing a High Current, 12-V Regulator Subsystem

While a separate switching regulator could provide the higher current required for the positive 12-V output, a different method is used here. Switching regulators are fine for constant loads, such as heating logic chips, but they sometimes have trouble powering RF equipment. There are several reasons. First, most transceivers present widely varying loads — especially when switching between transmit and receive. Second, any RF that sneaks back into the switcher can raise havoc.

There is one other design goal that cannot be easily met by a switcher — high and

low current limiting. The smorgasbord supply includes provisions for high (5 A) and low (1 A) current settings for the 12-V supply. Shown in Figs. 36 and 41, this circuit consists of a standard three-terminal regulator with a current-boosting wraparound pass transistor operating as a dc current amplifier for the high-current setting. U8 provides sensing and voltage-error correction and transfers these characteristics to Q1. As regulator current demands increase, more base current is drawn by the transistor. This causes the transistor to conduct harder, supplying greater collector current to the load.

Details of the design are presented here to allow the builder to calculate different current levels or to substitute a pass transistor with different characteristics. Refer to Fig. 41. I_c (collector current) is equal to $I_2 - I_b$ (base current). The current through U8, I_a, is equal to $I_1 + I_b$. The total load current, I_L, is equal to $I_c + I_a$, or $(I_2 - I_b) + (I_1 + I_b) = I_1 + I_2$.

The voltage drop between V_{in} and V_b includes one diode drop, no matter which path (I_1 or I_2) is chosen: I_2 flows through the internal base-emitter drop of Q1, while I_1 flows through D1. This forces the same voltage drop across R1 and R2; so, $I_1R_1 = I_2R_2$, and

$$\frac{I_2}{I_1} = \frac{R_1}{R_2}$$

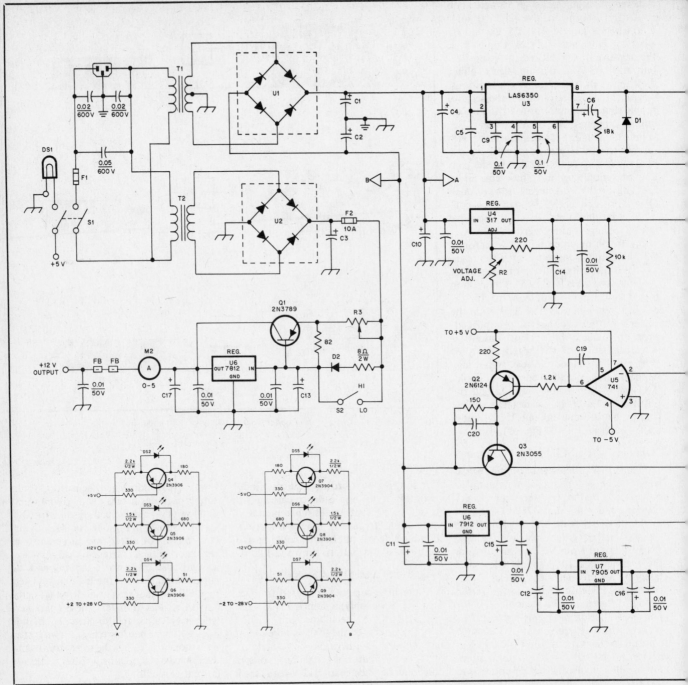

Fig. 36 — Schematic diagram of the Smorgasbord Supply. Resistors are ¼-W, 5% carbon types unless noted. Capacitors are 50-V disc ceramic types unless noted; capacitors marked with polarity are electrolytic.

C1-C3 — 6800-µF, 50-V electrolytic.
C4 — 1000-µF, 50-V electrolytic.
C5,C6 — 0.47-µF, 50-V tantalum.
C7 — 2200-µF, 50-V electrolytic.
C8 — 10-µF, 50-V electrolytic.
C9 — 0.0047-µF, 50-V mylar.
C10-C13 — 1.0-µF, 50-V tantalum.
C14-C17 — 1.0-µF, 35-V tantalum.
C18 — 1.0-µF, 50-V electrolytic.
C19, C20 — 1.0-µF, 50-V disc ceramic.
D1 — Schottky diode, VSK64. (Avail. from Varo Semiconductors, P.O. Box 40676, 1000 N. Shiloh, Garland, TX 75040. Call 214-271-8511 for nearest distributor).
D2 — Silicon diode, 1N5401 or equiv.
DS1 — 5 V-pilot lamp (see text).
DS2-DS4 — Red subminiature LED (Radio Shack 276-068 or equiv.).
DS5-DS7 — Yellow subminiature LED (Radio Shack 276-073 or equiv.).
F1 — 3-A slow-blow fuse.

F2 — 32-V, 10-A fuse.
L1 — 150-µH toroidal inductor (38t no. 16 enam. on Amidon T-157-26 core).
L2 — 1-µH toroidal inductor (10t no. 18 enam. on Amidon T-68-1 core).
M1 — Dc voltmeter, 0 to 30 V (Triplett type 220-G or equiv.).
M2 — Dc ammeter, 0 to 5 A (Triplett type 220-G or equiv.).
Q1 — PNP power transistor, 2N3789 or equiv.
Q2 — PNP power transistor, 2N6124 or equiv.
Q3 — NPN power transistor, 2N3055 or equiv.
Q4-Q6 — PNP transistor, 2N3906 or equiv.
Q7-Q9 — NPN transistor, 2N3904 or equiv.
R1 — 5-kΩ trimmer potentiometer.
R2 — 5-kΩ potentiometer, "S" taper (Clarostat 53C1 or equiv.).
R3 — 1-Ω, 25-W wire wound resistor (Clarostat VP25KA or equiv.).
S1 — DPST switch, 125-V ac, 5 A.

S2 — SPST switch, 3 A.
T1 — Power transformer; 117-V primary, 48-V, 4-A center-tapped secondary (Stancor P-8619 or equiv.).
T2 — Power transformer; 117-V primary, 18-V, 8 A secondary (Stancor P-8686 or equiv.).
U1,U2 — Bridge rectifier, 50 PIV, 6 A.
U3 — Lambda LAS6350 switching regulator IC (Available from Lambda Semiconductors, 121 International Dr., Corpus Christi, TX 78410, tel.800-255-9606).
U4 — Adjustable voltage regulator, 1.5 to 30 V, 1.5 A (LM317 or equiv.).
U5 — Type 741 op amp.
U6 — Voltage regulator, −12 V, 1 A (LM7912 or equiv.).
U7 — Voltage regulator, −5 V, 1 A (LM7905 or equiv.).
U8 — Voltage regulator, 12 V, 1 A (LM7812 or equiv.).

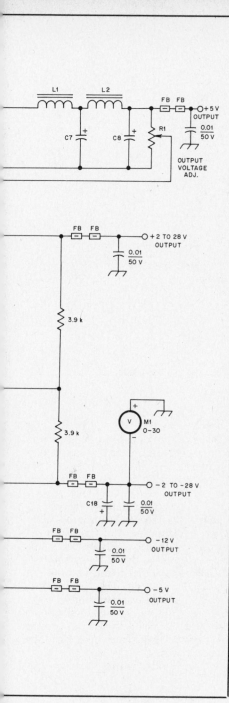

If we calculate the values for I_1 and I_2, we can find the values needed for R1 and R2. Assume that this circuit will be called upon to supply a worst-case maximum current of 8 A, to allow for momentary overloads. There will be two components of current through the regulator: I_b and I_1. Good design requires that U8 be run at somewhat less than its maximum rated current, so first assume $I_a = 0.85$ A. Then, $I_c = 8 - 0.85$, or 7.15 A through the collector of Q1. From the collector vs. base current curves for Q1 (Fig. 42), we find that

$I_b = 600$ mA for a collector current of 7.15 A. Thus, total current through R1 will be $I_1 = I_a - I_b = 0.85 - 0.6 = 0.25$ A.

Because of the high current demand, a separate transformer is used for this output. It is rated at 18 V RMS at 8 A. The maximum dc bus voltage will be approximately 25 V. A 6800-μF filter capacitor is also used on this bus, yielding a maximum ripple of 8 V (from Eq. 2). The minimum voltage on the bus will occur during ripple valleys ($25 - 8 = 17$ V). The minimum regulator input voltage to avoid dropout is 15 V. A diode drops 0.7 V, so we must add this value to the minimum regulator requirement to make up for the drop across D1. The voltage across R1 and R2 must be $17 - 15.7 = 1.3$ V.

Thus,

$$R1 = \frac{V_{R1}}{I_{R1}} = \frac{1.3 \text{ V}}{0.25 \text{ A}} = 5.2 \ \Omega$$

$$R2 = \frac{V_{R2}}{I_{R2}} = \frac{1.3 \text{ V}}{7.75 \text{ A}} = 0.168 \ \Omega$$

The power dissipated by R1 will be

$$I^2 R1 = 0.25^2 \times 5.2 = 0.325 \text{ W}$$

The power dissipated by R2 will be

$$I^2 R2 = 7.75^2 \times 0.167 = 10.03 \text{ W}$$

The final design uses two 16-ohm, 1-watt resistors in parallel for R1, and a 1-ohm, 25-watt adjustable wirewound resistor for R2.

To implement high/low current limiting (Fig. 36), S2 shorts out the combination of R1 and D1. This effectively sets the emitter and base of Q1 at the same potential, making it incapable of turning on and passing current. Thus, in the low current setting, U8 supplies all the current to the load.

Other Design Features

The LED indicators (Fig. 43) are designed to turn on when the proper voltage appears at the power supply output terminals and to extinguish if the voltage falls by 1 V or more. For the two adjustable-output voltage regulators, the LEDs will stay lit unless a short-circuit condition occurs. Any or all of these indicators may be omitted, if desired.

Precautions

Three-terminal regulators contain high-gain circuitry and may show signs of instability unless design safeguards are taken. Manufacturers urge that 0.1-μF bypass capacitors be installed on all input and output leads, preferably right at the regulator pins. In practice, the amount of bypassing will vary from installation to installation. Additionally, all ground leads should be returned to a single point. This helps avoid difficult-to-cure ground loops.

Sometimes, one can actually hear regulator instability as a high-pitched whine. More often than not, however, the oscilla-

tions can only be detected with an oscilloscope. Different values and types of bypass capacitors should be tried until a combination is found that makes the oscillations disappear. A regulator may be considered stable when it can be operated at 10, 50 and 110% of maximum rated load without signs of oscillation, even when rapidly switched off and on.

Where wrap-around current-boosting regulators are used, oscillations can often be squelched by bypassing the base of the pass transistor to ground with several hundred ohms. In extreme cases, collector or emitter bypassing may be required.

Once the power supply passes these tests operating into resistive loads, it's time to power up the HF or VHF rig. Operating in the presence of strong RF fields can open up entirely new oscillation problems. One very important, yet basic cure is to enclose the supply in a metal cabinent. Wires leading into or out of the cabinent should be RF bypassed. Do not overlook the line cord when searching for potential sources of RF ingress. In most supplies, 0.02 to 0.05 μF, 600-V capacitors across the line provide sufficient RF rejection. Ferrite beads on all outputs will also discourage RF from feeding back into the regulators.

F2 is located at the output of U2. It serves as a "safety valve" in case the regulator or pass transistor malfunctions. A 10-A automotive fuse may be used.

An 82-ohm, ¼ W resistor placed across the Q1 emitter and base guards against transient operating conditions that could destroy the device. This is a very inexpensive form of semiconductor insurance!

Construction

The interior of the smorgasbord supply can be seen in Fig. 44. A 4½ × 15 × 10-inch (HWD) steel cabinet (Bud model TV2155) houses the supply. The front panel slopes inward towards the top, making meter viewing easy. Half-height meter bezels for the Triplett meters further enhance the clean appearance of the front panel. The main power switch was primarily designed for computer applications and features a built-in pilot lamp. Power for the lamp comes from the +5 V switcher output.

For convenience, the rectifiers and filter capacitors, as well as U5, are mounted on a single-sided PC board. A parts-placement diagram is shown in Fig. 37.

The 5-V switching regulator is mounted on its own PC board (Fig. 40). The component side of the board remains unetched. Those component leads that are connected to ground are soldered to the top of the board as well as the bottom. The copper foil on the top of the board must be cleared away from component leads that are not grounded. This can be accomplished with a countersink or a 1/8-inch drill bit.

The LED drivers and associated circuitry

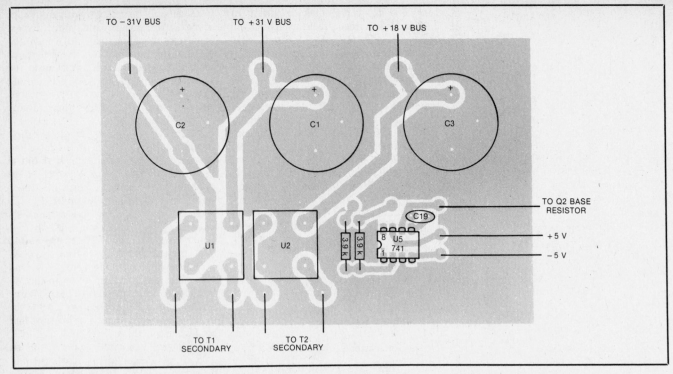

Fig. 37 — Parts-placement guide for the rectifier board. A full-size etching pattern appears at the back of this book.

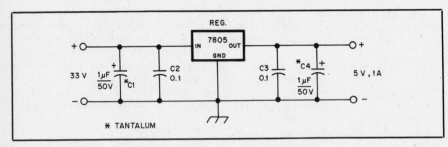

Fig. 38 — Schematic diagram of a simple 5-V, three-terminal regulator.

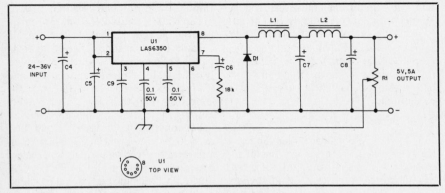

Fig. 39 — Schematic diagram of a 5-V, 5-A switching regulator.

(Q4-Q9) mount on a small PC board that attaches to the rear of the front panel. Parts placement is shown in Fig. 43.

All regulators and power transistors mount on the back panel (Fig. 45). Individual heat sinks improve thermal dissipation. TO-3 sockets are used to facilitate device replacement. The transistors and regulators are electrically insulated from the heat sinks and chassis by mica washers; use silicone heat-sink grease for better thermal conductivity. Fig. 45A shows the location of regulators and transistors as viewed from the inside of the cabinet. Fig. 45B is a top view of the power supply showing layout of the major assemblies. Use point-to-point wiring to interconnect circuit board assemblies and regulators.

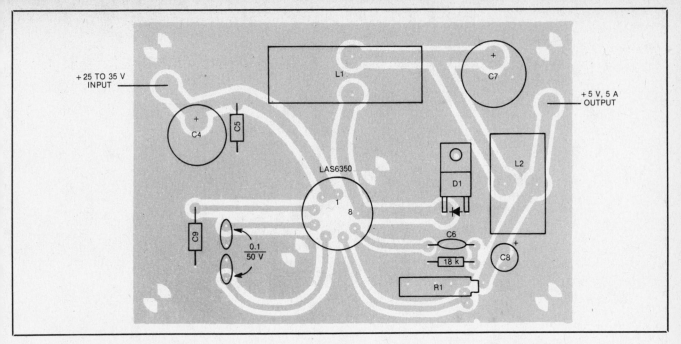

Fig. 40 — Parts placement guide for the 5-V, 5-A switching regulator. A full-size etching pattern appears at the back of this book.

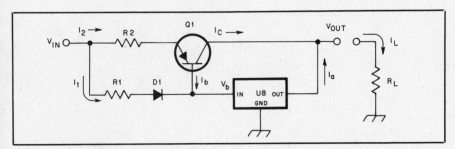

Fig. 41 — For higher output currents, a pass transistor can be "wrapped around" a three-terminal regulator. See text for discussion.

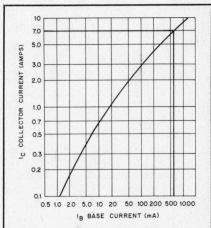

Fig. 42 — Collector vs. base current for the 2N3789 pass transistor.

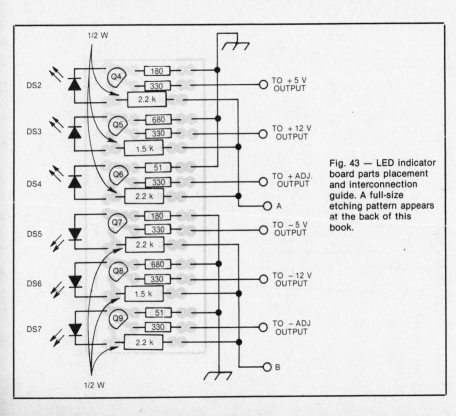

Fig. 43 — LED indicator board parts placement and interconnection guide. A full-size etching pattern appears at the back of this book.

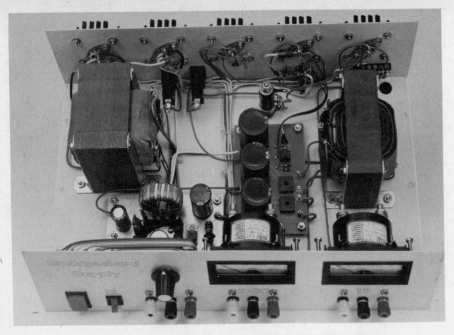

Fig. 44 — Interior view of the Smorgasbord Supply.

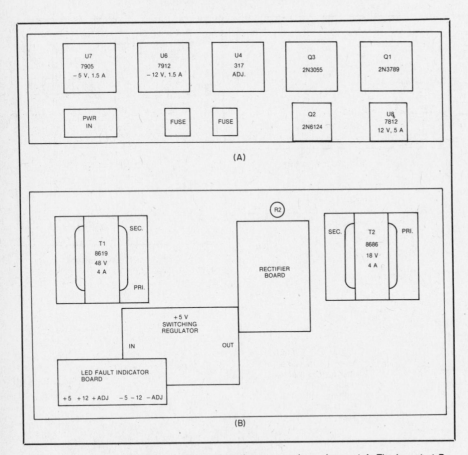

Fig. 45 — Transistor and regulator placement on the rear panel are shown at A. The layout at B provides comfortable access to all wiring and assemblies.

The Perfect 10: A Battery Eliminator for FM Portables

In an effort to reduce the size of FM portable equipment, some manufacturers have opted to reduce the number of batteries in the battery pack. Now, a typical portable rig operates from a pack containing eight cells (nominal 9.6-V potential). Operating these rigs from an external supply, typically 12-V from an automobile electrical system or commonly-used home power supply, can be risky. Pete O'Dell, KB1N, presents the following solution for those amateurs wishing to power their portables from a source other than internal batteries. The "Perfect 10," shown in Figs. 46 through 50, was originally presented in March 1984 *QST*.

The Regulator Circuit

Fig. 47 shows a circuit for use with the 12-V electrical system of an automobile (or any other 12-V dc source capable of supplying about 500 mA). Both the input and the output are fused for added safety. The mounting tab of U1 (internally connected to pin 2) is bolted to the metal cabinet that houses the circuit. Heat-sink compound is applied between the tab and the cabinet to aid heat transfer. The metal of the cabinet should be more than adequate as a heat sink because the input voltage is only 2 to 4 volts above the output voltage.

Two alternatives are available for U1. The 10-V fixed regulator is not very common, but is available from the source in the parts list. You could also use an LM317 adjustable regulator (available from Radio Shack and other sources). Adjust the 2-kΩ potentiometer for a supply output of 10 V.

D1 provides protection for the regulator chip in the event that the output terminal potential of U1 rises above that at the input terminal. Without D1, the regulator can be destroyed. That condition can occur easily. The capacitor on the input of U1 charges (assuming S1 is open). DS1 lights, then slowly extinguishes as the capacitor discharges. U1 may survive the small reverse current, but why take chances?

The two capacitors should be mounted as close to the regulator IC as possible.

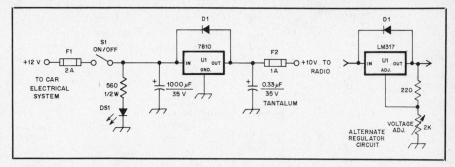

Fig. 47 — Schematic diagram of the regulator circuit. The resistor is a carbon-composition type. Capacitance is in microfarads. The input capacitor is electrolytic. Parts numbers in parentheses are from Radio Shack.

D1 — Silicon power diode, 1N4001 or equiv.
DS1 — Green LED (276-022).
F1,F2 — Fast-acting fuse, 1 A (270-1273).
S1 — SPST toggle (275-602).
U1 — 10-V, 1-A, three-terminal regulator, Texas

Instruments 7810 or equiv. (available from Active Electronic Sales Corp., P.O. Box 8000, Westboro, MA 01581, tel. 617-366-0500). Or 1-A adjustable three-terminal regulator (LM317 or equiv.).

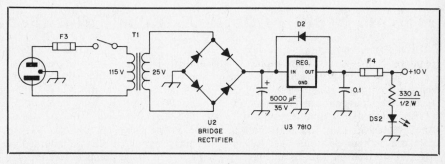

Fig. 48 — Schematic diagram of the base-station version of the battery eliminator. Components used in the regulator circuit are identical with those in Fig. 47. Consult the Fig. 47 parts list for component values.

T1 — Filament transformer; primary: 117-V ac, 60 Hz; secondary: 12.6-V ac, 1.2 A (173-1505).

U2 — Bridge rectifier, 4 A, 100 PIV (276-1171).

Fig. 46 — Two versions of the Perfect 10. The mobile unit bracket with ON/OFF switch attaches to a convenient spot on the dashboard, while the box housing the regulator mounts out of sight.

Fig. 49 — Internal view of the regulated supplies. Construction is straightforward; layout is not critical.

Some manufacturers recommend lead lengths no greater than ¼ inch. U1, D1 and the capacitors mount in a small aluminum box. A mounting tab on the box permits it to be secured under the dash, out of sight. F1 and F2 are housed in in-line fuse holders. S1 and DS1 are

mounted on a small home-made bracket that is attached to the dash.

A Base-Station Supply

Fig. 48 shows a version of the circuit that operates directly from 117-V ac. The regulator circuit is identical to that of Fig. 47. DS2 is located on the output and provides a quick visual check of the output status. The fuse holders are the panel-mounted variety instead of the in-line type. All the components are mounted inside a small aluminum box (Fig. 49). Again, the cabinet serves as the heat sink for the regulator chip.

Initially, a 1000-μF capacitor was tried for the input of the regulator. No problems were encountered when using the supply with a 2.5-W portable. However, at the 5-W level, some hum was noticeable on the signal. A 5000-μF unit cured the problem.

Making Connections

Fig. 50 shows one method of connecting the battery eliminator to a portable rig. J1 is a miniature plastic-enclosed, closed-circuit phone jack (Radio Shack 274-296 or similar). P1 is a matching plug. As a phone

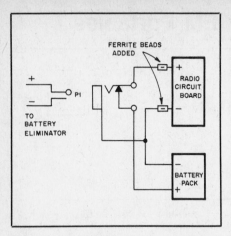

Fig. 50 — Block diagram of the connections between the battery pack, radio circuit board and battery eliminator for the Tempo S5 and similar portable units.

plug is inserted into a jack, the tip will cause a short, initially, between the elements of the jack. This results in a momentary short from the positive terminal of the battery to ground, which will not damage the battery pack (as long as the short is momentary). The event may cause a fuse in series with the battery to blow. If your radio incorporates a fuse in series with the battery pack, you should consider a different plug-and-jack system.

Finding a place to mount the jack in a small radio can be a chore. Carefully measure the openings and distances before drilling the mounting hole. Use liberal amounts of heat-shrink tubing to cover bare wires or terminals.

This method should work for most portables on the market — those that do not have slide-on battery packs. For rigs with the slide-on packs, the easiest approach is to obtain an empty pack and construct the circuit inside it.

Two ferrite beads are added to the power leads near the circuit board in Fig. 50. These minimize the possibility of RF getting back into the portable-rig circuit through the battery eliminator. If the eliminator is constructed in a slide-on pack, the beads should be placed on the output wires that are soldered to the terminal inside the pack.

Before applying power, double check all wiring. With the radio ON/OFF switch in the OFF position, apply power to the eliminator circuit and check the voltage at the power input of the radio. If everything is all right, proceed with final assembly. There is nothing to adjust. Just use and enjoy.

A Solid-State Regulator for Alternator-Battery Systems

Until very recently, automotive voltage regulators have been electromechanical devices. In some systems they can be reliable for many years, while in other systems they cause noise and fail frequently. Serious mobile operators should consider replacing their electromechanical regulator with an electronic one. Advantages of an electronic unit include greatly reduced hash and whine, tight load regulation independent of shaft speed, temperature stability and precise adjustment. Most modern radio equipment is powered from 13.8-V dc, so a battery-alternator system makes more sense for portable operation than does a 117-V ac generator and an accessory ac power supply.

The circuit shown in Fig. 51 is designed for alternators that are controlled in the positive (ungrounded) field lead. Alternator field current (up to 3 amperes) is regulated by Q1, which is driven by Q2 and U1. The "armature" terminal is also connected to the battery, so that when K1 is energized, the field becomes part of a feedback loop. U2 is a reference generator. When the alternator has sufficient speed, the loop maintains the armature/battery voltage at a level that causes no potential between the differential inputs of U1.

R1 sets the output voltage at which equilibrium is established. D1 absorbs negative-going transients generated by the field inductance that might otherwise damage Q1. In an automobile having a discharge indicator light the field is excited through the light when the ignition switch is turned on. As soon as the engine starts and the alternator stator (common connection of the 3-phase windings) develops about 5 volts, K1 closes and full alternator/battery voltage is available to drive the field. A system using an ammeter rather than an indicator light would have the "stator" ter-

minal connected to the ignition switch (through a current-limiting resistor if necessary) and the "indicator" terminal would not be connected.

The circuit can be built into a stock voltage regulator enclosure, which can serve as the heat sink for Q1. The remaining components can be mounted on a PC board or tie points. Vibration and shock can be severe in automotive service,

so clear silicone rubber compound is recommended as a cushion for the components and leads. R1 can be a PC-mount trimmer or a chassis-mount potentiometer with the bushing and shaft protruding through the regulator cover. Adjustment is greatly facilitated by the latter arrangement, but the knob should be anchored to the cover to prevent voltage changes caused by vibration or accidental contact.

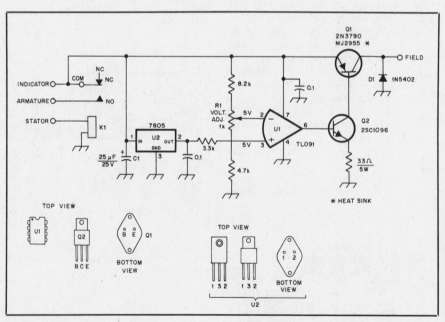

Fig. 51 — An electronic automotive voltage regulator for mobile and portable use, designed by W1RN.

C1 — 25-μF, 25-V aluminum electrolytic or 1-μF, 25-V tantalum.
D1 — Silicon diode, 3 A, 25 PIV.
K1 — 3-A, 117 V-ac SPDT contacts, 5-V dc, 70-Ω coil (RS 275-246 or similar).
Q1 — PNP silicon power transistor, 10 A, 40 V, 100 W (RS 276-2043 or similar).

Q2 — NPN silicon power transistor, 1 A, 40 V, 10 W (RS 276-2052 or similar).
R1 — 1-kΩ linear taper potentiometer (see text).
U1 — NFET operational amplifier (RS 276-1745 or similar).
U2 — 5-V, 1-A voltage regulator.

The AA6PZ Power Charger

The battery charger shown in Figs. 52 through 58 combines the best features of constant-current and constant-voltage NiCd chargers. This charger was originally described by Paul Zander, AA6PZ, in December 1982 *QST*.

The Power Charger

As shown in Fig. 53, when the battery voltage is low, the current available from the charger is high enough to keep the transmitter on the air. As the battery voltage increases, the charging current decreases in order to avoid overcharging the battery. Finally, as the battery approaches full charge the current tapers off more slowly with increasing voltage. This action provides a topping-off charge with some latitude for variations in battery voltage.

Fig. 54 shows what happens when a discharged battery is connected to the Power Charger. The initial current is high, charging the battery to 25 to 30% of its capacity in the first hour. The current drops off as the battery is charged. After approximately 6 hours the battery is fully charged. Then, as continued charging pushes the battery voltage higher, the current decreases to a few milliamperes. If you forget to disconnect the Power Charger after the battery is charged, the final charge rate is actually less stressful to the battery than if you were using a constant-current trickle charger.

What happens if you operate with the Power Charger connected? When the transmitter draws current from the battery the battery voltage drops, and more current is available from the Power Charger.

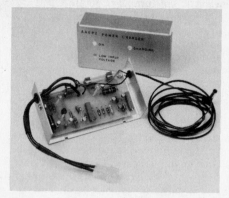

Fig. 52 — Interior view of the Power Charger. The three LEDs are mounted on long leads so they protrude through the top cover.

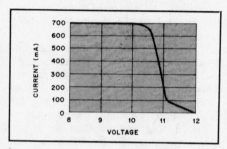

Fig. 53 — Output characteristics of the Power Charger circuit.

When the battery is less than three quarters charged the charger supplies most, if not all, of the transmitter current.

Rapid-charging capability, by itself, is of limited utility if the charger must be plugged into a wall outlet. The Power Charger can be operated with a wide variety of sources that supply from 12 to 30 volts. For mobile operation the car battery can supply the power from the cigarette lighter. For extended portable or emergency operation, lead-acid storage batteries may be useful. For fixed-station use, a simple ac-operated supply can be used.

Circuit Description

A simplified schematic diagram of the Power Charger is shown in Fig. 55. A full schematic is shown in Fig. 56. A PNP series-pass transistor, Q1, is used to control the charging current, and a 1-Ω resistor (R1) is used to monitor that current. When Q1 has enough base drive it will have 0.2 V or less between collector and emitter. This, combined with the voltage drop across R1, allows the circuit to have a minimal difference between the input and output voltages. Although it might be possible to use an IC voltage regulator in place of Q1 and some of the other parts, most IC regulators have a minimum of 2 or 3 V between the input and output. This makes them unsuited for mobile and portable operation where there is not much voltage to spare.

Q2 and Q3 control the base drive to Q1. Q3 receives input from Q4 if the charging current is too high — and from Q5 if the output voltage is too high. In response to either of these inputs, Q3 reduces the Q2 base current. This, in turn, reduces the base drive to Q1 so that the proper output voltage and current are maintained.

Q4 and the associated components monitor the output current. The base-

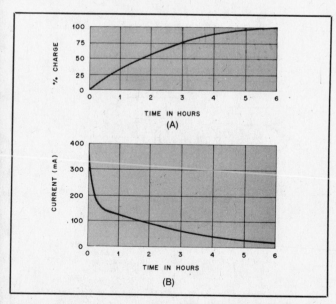

Fig. 54 — Typical charging performance of the Power Charger. Shown at A is the percentage of full charge obtained as a function of charging time. The charging current, as a function of time, is shown at B.

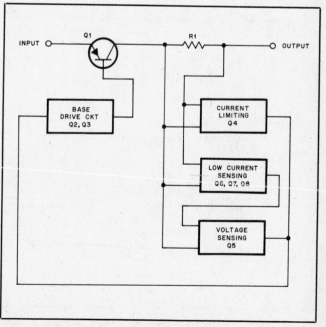

Fig. 55 — Block diagram of the Power Charger.

emitter junction of Q4 is connected across R1. When the current is more than 700 mA, Q4 is turned on. This sends a signal through the other transistors to reduce the base drive to Q1 so the current cannot go higher. In practice, this much current will only be drawn by the transceiver if you are transmitting while the battery is discharged. If the 700-mA current limit is not high enough for your transmitter it can be increased by connecting R5 between the emitter and base of Q4. The formula

$$R5 = \frac{70}{R1 \times I_{MAX} - 0.7}$$

can be used to determine the correct value. For example, a 680-Ω resistor would increase the current limit to 800 mA. Any variation in the tolerance of R1 and other resistors will affect the current limit markedly, so measure to value of your particular resistor and compute the value of R5 accordingly.

During most of the charging cycle Q5 and the 4.7-V Zener diode act as a voltage-sensing circuit. The combination of Q1, Q2, Q3 and Q5 regulate the collector voltage of Q1 at 11.2 V. There is, however, a resistor (R1) between that point and the battery. When the output current is 500 mA there is 0.5 V across R1, so the output is only 10.7 volts. At 200 mA, the drop across R1 is 0.2 volt, and the output is 11.0 volts. Viewed from the battery terminals, as the voltage increases the charging current decreases.

As we have just seen, R1 determines the slope of the voltage-current relation shown in Fig. 53. A value of 1 Ω works well with NiCd batteries having a capacity of 450 mAh. For batteries having a smaller capacity, the value of R1 should be increased. For example, when one is charging 250-mAh batteries, R1 should be 1.8 Ω.

Using a 330-Ω resistor for R5 would give a maximum current of 500 mA.

The remainder of the circuit is used to provide the topping-off part of the charging cycle. Q6 and Q7 form an amplifier that senses the voltage drop caused by charging current flowing through R1. When the current approaches 100 mA, Q7 starts to turn off and Q6 begins to turn on. The collector output of Q6 is connected to the base circuit of Q5. This causes the charger voltage to rise slightly. The output voltage is allowed to increase more as the current drops further. Eventually, a point of equilibrium is reached at which the current is only a few milliamperes. This current is unable to push the battery voltage higher. With this method, the battery can be fully charged safely and quickly.

As the output voltage increases at the end of the charging cycle, the base drive to Q8 is reduced. This causes the CHARGING LED to dim and eventually go out when charging is complete.

If the input voltage is too low the LOW-VOLTAGE LED turns on to signal that you will be unable to fully charge the battery. This is most likely to occur when using an external storage battery to supply the Power Charger. A similar situation can occur when using the Power Charger with rigs that do not have an internal diode in series with the battery. If the power-charger input is accidentally disconnected, Q1 might conduct some current in the reverse direction. The LOW VOLTAGE LED will light as a warning. If, under these circumstances, the input terminals should be accidentally short-circuited, Q2 and Q1 turn off as a safety measure to avoid damaging the radio. The residual current drain on the NiCd battery will be only a few milliamperes.

The last items in the schematic diagram

are the input and output fuses. As we have just seen, the power-charger circuit has current limiting, so perhaps the fuses are not necessary. However, consider that a fully charged NiCd battery can deliver many amperes to a short circuit, and that an automobile battery is designed to deliver peak currents of 200 A or more. In this environment a couple of 20-cent fuses seem like a good idea to prevent unwanted fireworks!

Construction

In designing the Power Charger circuit, consideration was given to using parts that are readily available. The only critical part is Q1, which should be an 2N4918. The fuse clips are Littlefuse no. 102071 or a similar part by another manufacturer. Alternatively, fuses with wire leads can be soldered directly to the board.

All of the components can be mounted on a 1.9- × 3.9-inch etched circuit board. An etching pattern for this board appears at the back of this book. This size fits comfortably inside a 2 × 4 × 1-inch aluminum box. A parts-placement diagram is shown in Fig. 58. Q1 is mounted with an insulating washer between it and the box. The mounting provides heat sinking for Q1 and mechanical support for one end of the circuit board. The other end of the board is supported by a no. 4-40 machine screw and three nuts. The first nut holds the screw securely to the box. The remaining nuts go above and below the board to hold it level in the box. The head of the same screw can be used to mount a rubber foot on the outside of the box.

Probably the only construction difficulty you will encounter is determining the proper charging connector for your transceiver. There seem to be as many different connectors as there are transceivers. Then there

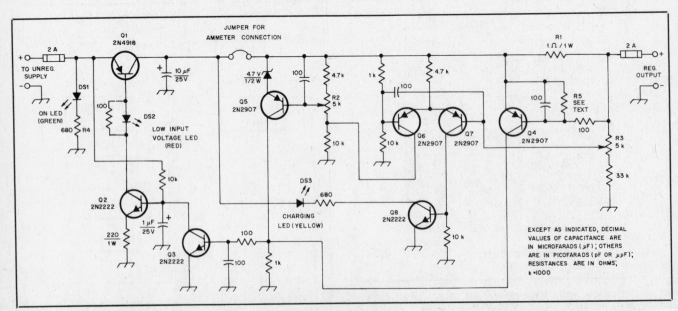

Fig. 56 — Schematic diagram of the Power Charger. Except as indicated all resistors are ¼-W, 5% carbon types. Capacitors are disc ceramic units rated at 50 V or greater. Polarized capacitors are 25-V electrolytic types.

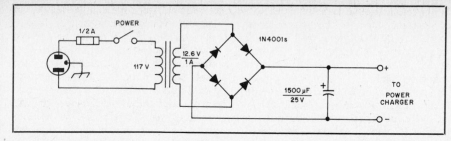

Fig. 57 — A simple ac adapter, such as the one shown here, can be used to power the charger for fixed-station use.

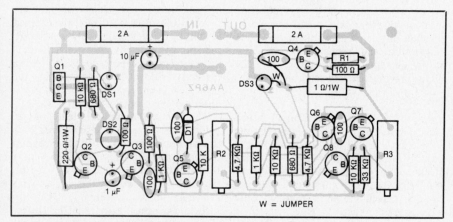

Fig. 58 — Parts-placement guide for the Power Charger. This view is from the component side of the board. Gray areas represent an X-ray view of the unetched foil. The etching pattern appears at the back of this book.

W = JUMPER

are several transceivers that are intended for use with a drop-in charger. A mating connector can be made from pieces of plastic and screws for the contacts. The challenge with the drop-in connector is to maintain contact when you transmit. Otherwise, the Power Charger cannot supply power directly to the transmitter, but can only recharge the battery when you set the transceiver back in the charger.

Another concern arises when you try to use one Power Charger with two transceivers. If the transceivers have a different number of cells in their battery packs, then the charging voltage must be different for each transceiver. It might be possible to build the Power Charger for the transceiver with the greater number of cells, and use an adapter containing two series silicon diodes for the other transceiver. The pair of diodes has a drop of approximately 1.4 V, which is similar to that of one charged NiCd cell. A compromise adjustment of the output voltage would then allow use of the Power Charger with either transceiver.

Adjustment

There are two variable resistors to adjust for proper operation of the Power Charger. Make sure the input voltage is high enough that the LOW INPUT VOLTAGE LED remains off while these adjustments are being performed. The initial adjustment procedure is based on voltage measurements. The final adjustment procedure measures the specific charging current.

For the initial adjustment, turn R2 fully counterclockwise and R3 fully clockwise. Connect the Power Charger input to a convenient dc supply and the output to a dc voltmeter. Both the ON LED and the CHARGING LED should be lit. Adjust R2 for 11.2 volts (1.4 volts per cell). If your transceiver has an internal diode, increase the voltage by 0.7 volt to compensate for the voltage drop in the diode. If you are unsure about the accuracy of your voltmeter it is advisable to adjust R2 for a voltage reading that is a little on the low side. Next, slowly turn R3 counterclockwise. The CHARGING LED should become dim and go out. Adjust R3 to the point where the CHARGING LED is so dim you can barely see it. Then turn R3 a quarter turn counterclockwise. The output should increase by about 1 volt as R3 is adjusted.

The final adjustment takes care of small errors in setting the voltage, and adjusts the Power Charger to compensate for a diode or other components that may be between the transceiver charging connector and the battery.

Next, connect a milliammeter in place of the circuit board jumper between the CHARGING LED and R1. This is the best place to measure the charging current accurately. If, instead, the meter were connected directly to the output of the Power Charger, it is likely that the voltage drop caused by the internal resistance of the meter would adversely affect the adjustment. A meter connected in place of the jumper will measure the current drawn by Q5, Q6, Q7 and the associated parts. This current, which is about 10 mA, can be measured when the Power Charger output is disconnected. When the battery is connected the true charging current can be found by subtracting the transistor current from the meter reading.

It may be tempting to measure the Power Charger input current instead, but an ammeter connected at the input would read the current drawn by the ON LED, the CHARGING LED and the base of Q1. This current could be from 20 to 100 mA, thus making an accurate determination of the charging current impossible.

Now, with the ammeter connected and the Power Charger turned on, it is time for the big moment. Connect the transceiver containing the charged NiCd battery to the Power Charger. The initial current surge may be 100 mA or more, but the current should fall quickly to less than 45 mA, which is the trickle-charge rate. Adjust R2 for a charging current between 20 and 30 mA. Adjust R3 until the CHARGING LED is so dim you can barely see it. Some interaction between these adjustments is normal, so you will probably have to repeat them several times until the current and the brightness of the CHARGING LED are correct.

Since continued charging will cause the current to decrease slowly with time, the adjustments should be completed within a minute or two of connecting the battery. If this can't be done disconnect the transceiver for a few minutes to allow the battery to recover, then try again. If you plan to use your Power Charger with more than one battery, a compromise adjustment of R2 may be necessary to compensate for construction differences in the batteries.

If you encounter difficulty with this adjustment you may have a bad component or a wiring error. Power resistors in values from 10 to 200 Ω may be used as loads to verify that your Power Charger has an output similar to that shown in Fig. 53.

Sam Bases, K2IUV, offered these additional thoughts on calibration of the Power Charger in Technical Correspondence in September 1983 QST.

Calibration of the output voltage is extremely critical, and most inexpensive VOMs are definitely not accurate enough to do the job. Furthermore, typical milliammeters have a relatively high series resistance and will affect the charge current when removed from the circuit. Here's why: To charge the battery to 100% of capacity, the individual cell voltage must reach a minimum of 1.42 V at a charge rate of 0.1C, or 11.36 V for an 8-cell battery without an internal diode. A set point of 11.20 V, or 1.40 V per cell (as suggested), will result in only 15-37% of full charge. I experienced an apparent loss of battery capacity, which was actually due to insufficient charge current at 11.20 V. Similarly, voltmeter errors as little as 0.16 V or milliammeter-induced voltage drops of the same amount will result in failure to charge the battery properly.

The charge rate must taper to the normal charge rate of 0.05C to 0.1C at 1.42 to 1.44 V per cell, or 22.5 to 45 mA for a 450-mAh battery. This charge rate may be left on indefinitely. A charge rate of 0.02C is a trickle charge, and it is used only to maintain a battery that previously has been fully charged. The power charger should not taper below 0.05C, or the battery will never reach 100% capacity; 0.1C is a better minimum value.

You may wish to calibrate the charger with a digital voltmeter as follows:

1) Using the slow-rate "wall charger," give your battery a 16-hour charge.

2) Connect the battery to the power charger for at least one-half hour. Monitor the current with a meter having a resistance of less than 0.2 Ω. You may use a current-sampling resistor of 0.1 Ω in series with the charger, and measure the voltage drop across it with a *good* DVM. For example, a reading of 0.0045 V indicates a charging current of 45 mA. (This measurement is usually beyond the means of anything but laboratory-grade instruments.)

3) If a laboratory-grade DVM is not available, connect the charger directly to the battery, without any series resistor or ammeter, using the leads intended to be permanent. Measure the battery voltage with a DVM or laboratory-grade analog meter. Adjust R2 for a battery voltage of 1.42-1.44 V per cell, adding 0.7 V if your battery pack has an internal diode. Adjust R3 so the yellow LED barely glows, then readjust R2 (because of interaction). Alternatively, measure the voltage across R1 and adjust R2 and R3 to give a reading corresponding to 0.05C-0.1C. V = R1 × 0.1C, so for a 450-mAh, 8-cell battery, V = 1.0 × 0.045 A = 0.045 V.

As a final thought, I strongly suggest the use of a multifingered or other effective heat sink and thermally conductive grease together on Q1. Without them, excessive temperatures and thermal cycling will eventually damage the transistor, possibly damaging your battery from overcharge.

Operation

Operation of the Power Charger is simple. Connect it to a convenient power source between 12 and 30 V dc. If the polarity is correct, the ON LED will light. If the voltage is high enough, the LOW INPUT VOLTAGE LED will be off. Connect the transceiver and the CHARGING LED will come on. When charging is complete the CHARGING LED will go out. Disconnect the Power Charger first from the transceiver and then from the power source. Fig. 57 shows a suggested supply if you don't have a suitable ac power supply. It has some ac ripple, but that is no problem since the charger acts as an electronic voltage regulator.

If a supply of 20 V or more is used to power the charger, an external 10- or 20-Ω power resistor should be used to reduce the power dissipation in Q1. The value of R4 should also be increased to keep the current through the ON LED below 20 mA.

The Power Charger can be used to charge the battery or to maintain the charge in the battery while the transceiver is being used. It is recommended that the latter mode of operation not be used on an exclusive basis, however. If the NiCd batteries are not allowed to cycle occasionally between a fully discharged and fully charged condition, they will suffer a temporary reduction in capacity. Although it is not necessary, a good procedure is to let the battery run down before connecting the Power Charger. This will maintain maximum battery capacity.

The last consideration is temperature. NiCd batteries are not designed for wide temperature extremes. With the exception of some special cells, most NiCds vent if they are allowed to exceed 45 °C (113 °F). Especially avoid leaving them in a closed car on a sunny day. Also, at low temperatures the chemical activity in the battery slows down. NiCds can be used below freezing, but the capacity is reduced. For these reasons, charging of any type should be minimized when the battery is very warm or very cold.

A Controlled Battery Charger for Emergency Power Supplies

It's essential to use proper charging methods to maintain emergency battery power supplies in a ready-to-be-used condition. It's possible to keep a lead-acid battery continuously connected to a charger — provided the charger incorporates features to prevent the battery from overcharge! This project, presented by Dick Jansson, WD4FAB, describes the design and construction of a battery charger that may be left connected to a lead-acid battery at all times, whether in storage or at an emergency site. The battery is kept fully charged and ready for use if commercial power fails. Figs. 59 through 63 accompany this project. A number of these chargers have been built by the Orange County, FL, ARES group.

The Circuit

Today's commonly available voltage-regulator ICs make the construction of a

Fig. 59 — The controlled battery charger can be built in several configurations. Size depends on the intended application.

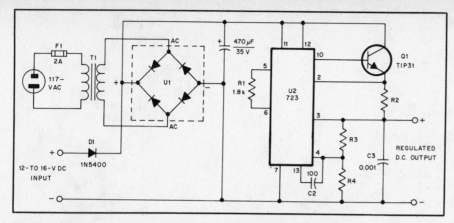

Fig. 60 — Schematic diagram of the controlled battery charger. Radio Shack part numbers are given in parentheses.

C1 — Electrolytic capacitor, 470 µF, 35 V (272-1018).
C2 — Ceramic disc capacitor, 100 pF (272-123).
C3 — Ceramic disc capacitor, 0.001 µF (272-126).
D1 — Silicon diode, 50 PIV, 3 A (1N5400 or equiv. 276-1141).
F1 — Fuse, 125 V, 2 A (270-1275) in a chassis-mount holder (270-307).
Q1 — Power transistor, NPN, 40 W, TO-220 case (TIP31 or equiv. 276-2017).

R1 — Resistor, 1.8 kΩ, ¼ W (271-1324).
R2 — Resistor, ½ or 1 W. See text and Table 1.
R3 — Resistor, ¼ W, 1%. See text and Table 1.
R4 — Resistor, 7.15 kΩ, ¼ W, 1%. See text.
T1 — Filament transformer. Primary, 117 V; secondary 12.6 V at 300 mA or 1.2 A. See text. (273-1505 or 273-1385).
U1 — Bridge rectifier, 50 PIV, 3 A (276-1151).
U2 — Regulator IC, type 723 (276-1740).

Table 1
Limiting Resistors for Controlled Battery Charger

Current Limit (mA)	R2 Value (Ohms)
1000	0.62
280	2.2
100	6.2
26	24.0

Voltage Limit (V)	R3 Value (Ohms)
10.5	3.40 k
11.9	4.75 k
13.8	6.80 k

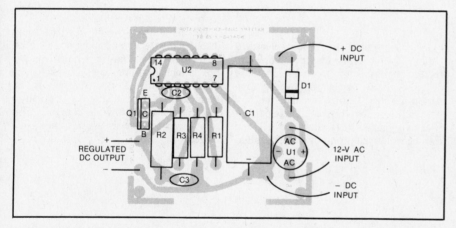

Fig. 61 — Parts-placement guide for the controlled battery charger. A full-size etching pattern appears at the back of this book.

sophisticated charger relatively easy. Fig. 60 is a schematic diagram of the charger. The unit is constructed from parts available from Radio Shack, except for the precision resistors needed to set the values for voltage limiting (R3 and R4). Several values for voltage- and current-limiting resistors are given in Table 1. The builder may customize the charger for a specific need based on the information given. For example, a builder may wish to build one charger for lead-acid batteries and another to provide controlled charging for a NiCd battery pack.

The charger has two possible inputs. For operation from 117-V ac, T1 provides 12.6-V ac to U1, a full-wave bridge rectifier, and filter C1. Filtered output is approximately 18-V dc. The charger may also be powered from a 12- to 16-V dc source connected to the DC INPUT terminals. This feature is especially handy for charging a NiCd battery pack from a car battery. U2, a 723-type (LM723, µA723, MC1723CP and so on) voltage-regulator IC, regulates the dc and controls voltage and current limits. If the battery voltage falls below the voltage limit set by the ratio of R3 and R4, the charger provides a constant-current output whose value is set by R2. When the battery charges to the voltage limit, the current "folds over" to hold that voltage limit. U2 can safely provide currents up to 100 mA. For applications requiring more current, up to 1 A, pass transistor Q1 is needed. If you do not use Q1, connect pins 2 and 10 of U2. Higher-current applications can be adapted by using a properly sized

transformer and an appropriate pass transistor with an adequate heat sink.

Construction

Although this simple circuit does not require a PC board, one was designed to simplify construction of a number of these chargers for the local ARES group. Fig. 61 is a parts-placement guide; a full-size etching pattern appears at the back of this book. Two versions of the finished unit are shown in Fig. 62. Both are built into small aluminum boxes purchased at Radio Shack. One unit was built with a 1.2-A transformer and is used for charging lead-acid batteries. Q1 is used here, and the tab is thermally connected to the box with insulating hardware. Q1 gets only comfortably warm, so no other heat sink is necessary.

The other charger uses two of the regulator boards to charge two different sizes of NiCd battery packs for hand-held transceivers. A 300-mA transformer is used for this application, and only one rectifier/filter (located on the lower PC board) is needed for both regulated outputs. Q1 is not used here, so a wire jumper is used to connect the PC board pads normally connected to the Q1 base and emitter. Insert the jumper between U2, pin 2 and U2, pin 10.

Adjustment

The most critical value to be adjusted is the output limiting voltage. With no battery load connected, turn on the charger and measure the open-circuit voltage with an accurate voltmeter. If possible, use a digital voltmeter with 0.1% accuracy. If the voltage is not exactly what you need, you will have to experiment with the value of R3. Major voltage adjustments should be made by changing the value of this resistor. Minor output voltage trimming may be accomplished by shunting R3 with a high-ohmage carbon resistor to lower the voltage, or by shunting R4 to raise the voltage.

For a six-cell lead-acid battery, set the voltage to approximately 13.75 V. This value was chosen to hold electrolyte vaporization to an acceptable level. Voltage limits for NiCd battery packs are nominally set to 1.43 V per cell. Some commercially

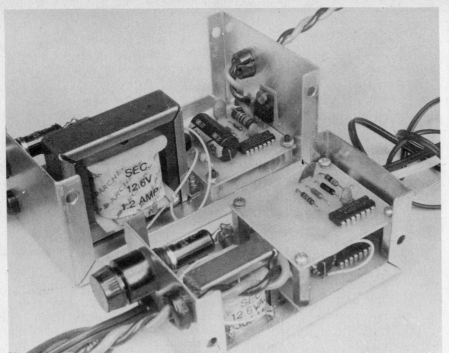

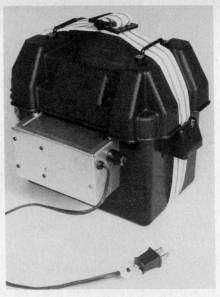

Fig. 63 — The controlled battery charger, attached to a deep-discharge lead-acid battery in a waterproof enclosure, is ready to be transported to an emergency site.

Fig. 62 — Two versions of the controlled battery charger. The unit at the left is designed to charge a lead-acid storage battery. The one at the right is designed to charge two different low-current NiCd packs. See text for a complete explanation.

made packs have built-in protective series diodes, so 0.7 V must be added to the total battery voltage to compensate for the drop across the diode.

About Batteries

The best type of battery for long-term emergency use is the deep-cycle lead-acid battery. Designed to power electrical fishing boat motors and accessories in recreational vehicles, these batteries are designed to be discharged over a long period at a steady rate. Normal automotive batteries are designed to discharge partially at a very high rate (to start the car), and then be recharged promptly while the alter-

nator is also carrying the load.

Fig. 63 shows a Sears no. 9601 deep-discharge battery in its companion waterproof case. The charger is housed in the box attached to the case. In an emergency, the entire 25-pound assembly can be picked up by the handle and taken to the field complete with charger. This setup gives reliable service at a moment's notice.

Emergency Gelled-Electrolyte Battery Pack

Since loss of commercial power is often encountered in an emergency, it's a good idea to have a fully charged prime power source on hand. The battery pack and charger described here is designed to operate a 10-W (output) transceiver for 8 hours, 80% of which is standby, 10% receiving and 10% transmitting. This project, shown in Figs. 64 through 68, was designed and built by Warren Dion, N1BBH.

Calculations indicate that a 2 ampere-hour (Ah) battery should be adequate. A battery with twice that capacity was selected, however for the following reasons:

1) Published ampere-hour ratings are based on a steady 20-hour drain. Occasional high-current bursts required for transmitting lower capacity.

2) A battery left on FLOAT (see below)

will not be charged to full capacity.

3) Some 12-V gear won't work well at the battery manufacturer's 10.5 end-voltage.

4) Aging reduces battery capacity.

Accordingly, the following project uses

Fig. 64 — Both the battery pack and companion charger are built in identical, compact enclosures.

a 4-Ah battery. While larger batteries may be substituted, smaller ones are not recommended.

Lead-acid batteries with a gelled electrolyte were selected for this project. These batteries may be left on a trickle charge for extended periods without damage, and they may be deep-discharged if necessary. The gelled electrolyte helps the user avoid leaking liquids and escaping gas normally associated with liquid-electrolyte lead-acid batteries. In addition, these cells may be operated in any position without the worry of leakage through vent holes.

The Battery Pack

The battery pack, shown in Fig. 65, is little more than two batteries in a case. A few notes apply, however. The batteries are held together with two cable ties connected end to end to provide enough length. Blocks of wood (or any smooth material) locate the batteries and keep them away from protruding screws, cables, fuse holder, etc. The blocks are braced against the sides of the case and glued to keep them from shifting. Do not expect the glue alone to keep the blocks in place. These batteries fit the selected case closely, so that its cover clamps them securely.

Lead-acid batteries easily produce currents of insulation-melting proportions. The 5-A fuse will limit damage caused by a dead short. The connector, an Anderson Power Products 1300 series, is one chosen

Fig. 65 — Interior of the battery pack. Care is taken to secure the batteries to keep them from moving.

for its performance and versatility. Being hermaphroditic and polarized, the same connector is used on both the battery and charger cables.

The Charger

The charger is a "bulletproof" design offering the following features:

1) Voltage limiting to prevent overcharging.
2) Current limiting to protect batteries that are heavily discharged.
3) A worst-case current limit of 1.5 A.
4) A thermal limit to protect the regulator.
5) FULL and FLOAT charging options.
6) Isolation from the power line.

A battery in any state of charge may be connected and left on charge without risk of damage to the charger or battery. With the FULL-FLOAT switch on float, the battery may be left on charge indefinitely, trading off a little capacity for this convenience. The LED turns on to indicate that the battery is fully charged.

The Circuit

The charger schematic diagram is shown in Fig. 66. Most components are available from Radio Shack. Since the final charging voltage must be held to within 0.1 volt, and precision components are impractical, 5% tolerances are prescribed throughout. The two trimmer potentiometers, R4 and R8, provide sufficient range to set the charging voltage and indicator light precisely.

Regulator U1 is the key component, maintaining the final charging voltage, and

in cooperation with Q1, limiting the charging current. The output voltage is approximately $1.25 \times (R7 + R8)/R6$. R9 is in series with the charging current. When this current reaches about 0.6 A, the voltage across R9 is sufficient to forward bias the base-emitter junction of Q1, which conducts, shunting the R7 + R8 combination. This pulls down the output voltage to a value that will just maintain a 0.6-A charging current.

The *absolute maximum* charging current is ¼ the rated ampere-hours, in this case 1.0 A. The recommended charging current is 0.15 to 0.20 times the Ah rating, 0.6 to 0.8 A in this case. If a faster charging rate is desired, or a larger battery is used, the charging rate can best be adjusted by shunting R9. Refer to the following table:

R9A	Current Limit
None	0.6 A
6.2-6.8	0.7
3.0-3.3	0.8
2.0-2.2	0.9
1.0	1.2

Full vs. Float

When a "12-V" battery is fully charged, its terminal voltage reaches 14.4, while accepting a low maintenance current from the charger. Long periods in this state will damage the battery. Therefore, if the battery is to be left on charge for extended periods, the voltage must be reduced by about 0.1 V per cell. In a "12-V" battery this figures out to 13.8 V, which is, conveniently, 0.6 V lower than full charge. A

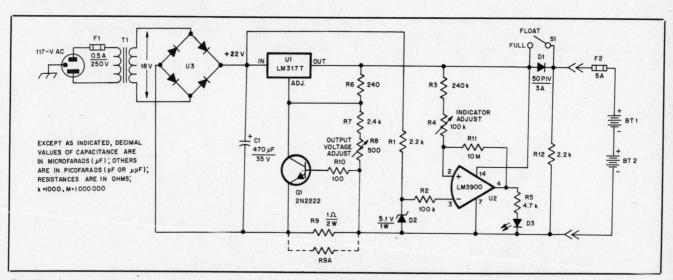

Fig. 66 — Schematic diagram of the battery pack and charger. All resistors are ¼ W, 5% tolerance, unless noted. Parts numbers given in parentheses are Radio Shack catalog numbers.

BT1, BT2 — Gelled-electrolyte, lead-acid battery, 6 V, 4 Ah (Yuasa NP4-6 or equiv. Available from Glynn Electronics, Box 800, Middleboro, MA 02346).
C1 — Electrolytic capacitor, 470 μF, 35 V (272-1030).
D1 — Silicon rectifier, 50 PIV, 3 A (276-1141).
D2 — Zener diode, 5.1 V, 1 W, 5% tolerance (276-565).
D3 — Red LED (276-062).

F1 — Fuse, 0.5 A, 250 V (270-1271).
F2 — Fuse, 5 A, 250 V (270-1278).
Q1 — Transistor, NPN, 2N2222 (276-1617).
R4 — Trimmer potentiometer, 100 kΩ. PC-board mount (271-220).
R8 — Trimmer potentiometer, 500 Ω, PC-board mount (271-226).
R9 — Resistor, 1 ohm, 2 to 5 W.
R9A — Optional resistor, see text.
S1 — Slide switch, SPST (275-401).

T1 — Transformer; primary 117 V, secondary 18 V, 1.2 A (modified Radio Shack 273-1480, see text; or Radio Shack 273-1515, with 18-V secondary).
U1 — Adjustable three-terminal regulator, LM317T (276-1778).
U2 — Quad op amp, LM3900 (276-1713).
U3 — Full-wave bridge rectifier, 100 PIV, 4 A (276-1171).

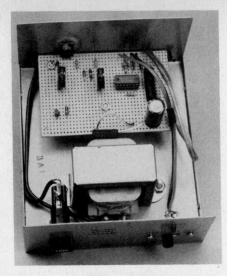

Fig. 67 — Interior view of the battery charger. Although perforated board is used in this version, a PC-board layout is shown in Fig. 68.

silicon-junction rectifier in series with the output supplies the needed drop, and requires only one adjustment for both the voltage limit and indicator light settings.

Charger Indicator

Full charge is signaled by the state of two variables, namely, terminal voltage and end-of-charge current. The current is affected by battery condition and temperature; therefore, terminal voltage is the chosen indicator. The LM3900 op amp (U2) used in this portion of the circuit is a current-sensitive device. The voltage from any forward-biased pin to common is an almost constant 0.6 V. For this application Zener D2 provides a reference voltage. The current through R2 [$(5.1 - 0.6)/100,000$

= 45 μA] is fed to pin 3. When this exceeds the current into pin 2, U2 output is pulled down and D3 is not lit. When the charger output voltage reaches 14.4, trimmer R4 is adjusted so that the current into pin 2 is slightly greater than 45 μA. At this point U2 toggles, and the light turns on. R11 provides positive feedback to sharpen the switching point.

Construction Notes

Construction of the charger, shown in Fig. 67, is conventional, and parts location not critical. Perforated board construction obviates the need for a printed circuit, but the layout is directly transferable to a one-sided etched board. For builders wishing to use a PC board, a full-size etching pattern may be found at the back of this book. Parts placement is shown in Fig. 68. It is best to socket the LM3900. The LM317T requires a heat sink. In this design it is screwed to the cabinet using a mica washer insulator. Radio Shack 3 × 5 × 5-7/8-in cabinets (No. 270-253) are used for both battery pack and charger.

Reworking T1

Transformers with 18-V secondaries are not as easy to find as the popular 12.6 and 25.2 styles. The 12.6-V types will not supply enough voltage to bias the regulator properly, and will cause the regulator to overheat. A secondary value of 16-18 V RMS is ideal. This can be obtained by pruning the secondary of the recommended 25.2-V transformer, if a suitable 18-V transformer is not available.

Place the transformer on its back and pry up the four tabs that lock the "U" frame mounting strap. Pull the legs of the frame apart and lift out the core with its windings. The core is built up of laminations shaped as letters E and I. Run a thin-bladed

knife between one of the outside I laminations and the stack. Remove the lamination and save it. Next run the knife between the abutting E lamination and the stack, especially where it passes through the coil window. Next, place the stack on the edge of the bench with the legs of the E facing upward and the loosened lamination just clear of the edge. Place a screwdriver blade against the center leg of the E, and tap gently on the handle until the lamination slides free. Carefully pull the lamination out of the stack.

In the same manner, remove the rest of the laminations and save them. Bent laminations can be straightened by placing them on a smooth, flat surface and *tapping* gently with a hammer.

Next, unwind the plastic insulating tapes and save them. Find the outside turn of the secondary and remove it from its solder lug. Unwind *55 turns*. Cut off the excess wire, clean the end and resolder it to its lug. Replace the tapes.

Reassemble the core alternating E laminations end-for-end, and interleaving the I laminations. It will be necessary to squeeze the stack a little to get the last lamination into the coil window. Place the stack on a flat surface and tap the edges of the stack with a hammer, turning so that all four sides are tapped. This will make the stack square and even.

Wrap the mounting strap back around the core and clamp the legs tightly against the sides. Bend in the four lugs, and hammer them down well. Connect the primary to 117 V, 60 Hz and measure the secondary voltage. It should be between 16 and 20 V RMS.

Initial Tests and Adjustments

Before connecting the battery, do the following:

1) Check the circuit for correctness.
2) Connect a 1000-ohm resistor across the output.
3) Turn trimmer R8 until the output voltage is exactly 14.4.
4) Turn trimmer R4 until the light turns off and then back R4 off until the light just comes back on.
5) Repeat 3 and 4, above.

Most moving-pointer (D'Arsonval) meters are not accurate enough for these adjustments. Use a digital meter in good condition with at least 2½ (preferably 3½) digits.

To Charge a Battery

Plug the battery pack into the charger and the charger into the power line. The charging light should be on. Select FULL or FLOAT, as required. If fully discharged, a recharge should take 6 to 8 hours. If partially discharged, it will take about the same number of hours that the battery has been in service since its last charge. When the light turns on, disconnect the battery. If the battery is to be left on charge, be sure it is switched to FLOAT.

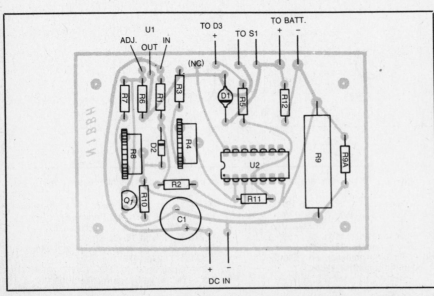

Fig. 68 — Parts-placement guide for the emergency battery pack. A full-size etching pattern is given at the back of this book.

A 2000-Volt Power Supply

The power supply shown in Figs. 69 through 72 is capable of supplying 2000 volts at 1 ampere in conservative operation. Many ceramic tubes, such as the 8874, 3CX800A7 and 4CX250 family, require 2-kV plate supplies. This project is intended to illustrate various design techniques used in high-voltage (HV) supplies. Although all of the parts used here are readily available from commercial suppliers, sufficient information is presented for a builder to adapt the project to suit individual needs and to incorporate components on hand. This power supply was designed and built in the ARRL lab by Mark Wilson, AA2Z.

Circuit Details

The heart of this supply, shown schematically in Fig. 70, is an Avatar Magnetics AV-415 transformer. Although the AV-415 has a dual primary, use of 234 V is essential to good regulation. The secondary has taps for 1375, 1515 and 1650 V to allow the builder to optimize the output voltage for the filter capacitor used. A full-value secondary was chosen, rather than a lower voltage secondary and voltage-doubler circuit, to aid voltage regulation. The transformer is rated for 1.2 A, continuous duty.

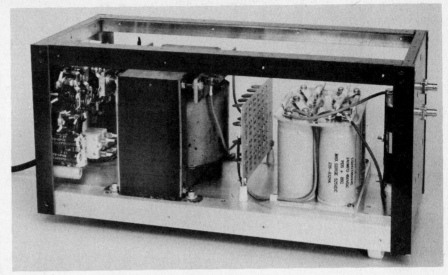

Fig. 69 — The completed 2000-V, 1-A power supply.

This supply is intended to be placed away from the operating position in an inconspicuous spot, perhaps under the operating table, so it may be activated remotely by J1. If J1 is jumpered, the supply may be activated by S1. When S1 and J1 are both closed, the coil of K1 is energized, closing the contacts and applying primary power. R1 and R2 in the primary feed to T1 provide inrush current protection. At turn-on, the filter capacitors are discharged and look like a dead short. The

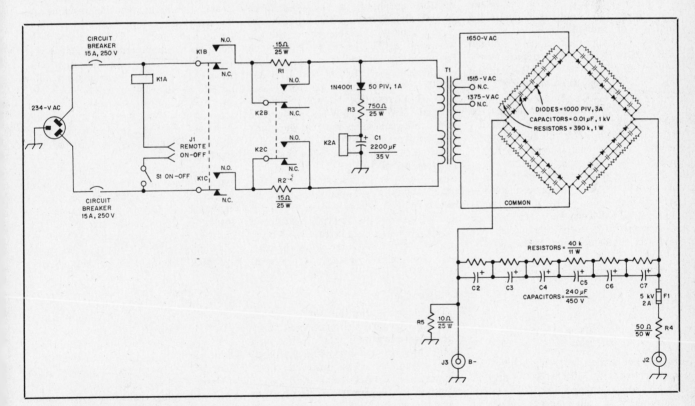

Fig. 70 — Schematic diagram of the 2-kV, 1-A supply.

C2-C7 — Electrolytic capacitor, 240 µF, 450 V (Mallory CG241T450V4C or equiv.).
F1 — High-voltage fuse, 5 kV, 2 A (Buss HVJ2).
J1 — Two-conductor socket (Cinch S-402-AB or equiv.).
J2,J3 — Chassis-mount female MHV connector (Amphenol 27000).
K1 — Relay, DPDT, 240-V ac coil, 240-V, 25-A contacts (Potter and Brumfield PRD11AYO-240V or equiv.).
K2 — Relay, DPDT, 24-V dc coil, 240-V, 25-A contacts (Potter and Brumfield PRD11DYO-24V or equiv.).
S1 — Switch, SPST, 240-V.
T1 — Power transformer; primary 117/234 V, secondary 1650-V, 1.2-A Continuous Commercial Service (CCS) (Avatar Magnetics AV-415 or equiv. Available from Avatar Magnetics, 1147 North Emerson, Indianapolis, IN 46219).

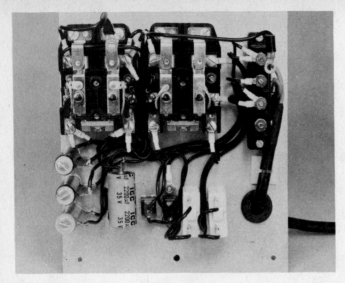

Fig. 71 — The primary components of the 2-kV supply are mounted on the rear panel.

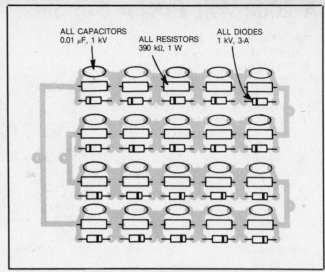

Fig. 72 — Parts placement for the rectifier board. The full-size etching pattern appears at the back of this book.

charging current surge is high enough that the rectifier diodes may rupture, so R1 and R2 are necessary to limit the inrush current to a safe value. After a 0.6-second time delay (determined by C1 and R3), K2 closes and shorts out R1 and R2. Full primary power is now available. K1 and K2 both have high-current contacts.

A full-wave bridge rectifier follows the transformer. The bridge is made from a number of individual diodes with a 1000-PIV rating. Each diode is shunted with a 0.01-μF capacitor to suppress transient voltage spikes and a 390-kΩ equalizing resistor. Although three of the 1000-PIV diodes would allow an ample safety margin for a 2-kV supply, the limiting factor is the 500-V breakdown rating of 1-W metal film resistors. Using five diode/capacitor/resistor sets in each leg of the bridge presents 400 V to each resistor.

Filtering is accomplished with a string of six 240-μF, 450-V capacitors connected in series. The nominal rating of this string is 40 μF at 2.7 kV. This value of capacitance was chosen to keep the ripple under 3% and was calculated using the formula

$$C = \frac{1}{2\sqrt{3} \, f_r \, R_L \, (\text{Ripple Factor})}$$

where
  C = capacitance in farads
  f_r = ripple frequency in hertz
  R_L = load resistance in ohms

In this application, F_r = 120 (full-wave rectification), R_L = 2000 (2000 V divided by 1 A) and the ripple factor = 0.03 (3%). Plugging these values into the equation yields 40.1 μF for the target value. The 240-μF capacitors (6 in series = 40 μF total) were the closest readily available value. The total 2.7-kV rating includes a safety margin.

A 40-kΩ, 11-W resistor shunts each filter capacitor. The actual value of electrolytic capacitors may vary as much as ±20% of the rated value, so these resistors are necessary to equalize the voltage across each capacitor in the string. The shunt resistance should be about 100 ohms per volt of supply output. In this case, 2000 V × 100 Ω/V = 200,000 ohms total. Six resistors are used, so each should be 200,000 divided by 6, or 33.3 kΩ each. The closest readily available value, 40 kΩ, was selected. The power rating of each resistor is determined from Ohm's Law (E^2 divided by R). In this case, E = 333 (2000 V divided by 6 resistors), and R = 40 kΩ, resulting in a power requirement of 2.8 W. The 11-W resistors chosen give an adequate safety margin. The equalizing resistors also form the bleeder.

Two protective devices follow the filter. F1 is a special 5-kV, 2-A fuse made by Buss. This fuse is 5 inches long and has a ceramic body. It will open if sustained excess current is drawn. R4 is a current-limiting resistor used to protect the power supply and amplifier in case of a high-voltage arc. R5 keeps the negative lead a few ohms above ground for current-metering purposes in the amplifier.

Construction

The final weight of this supply is about 60 pounds, so construction on an ordinary lightweight aluminum chassis is less than optimum. To avoid chassis-flexing problems, a framework was constructed from aluminum angle stock. Top and side panels then bolt to the framework. The bottom panel is made from 1/8-in-thick aluminum, adding to the rigidity of the structure. The power supply measures 9 × 8½ × 20 inches (HWD). The framework is easily fabricated with hand tools, and the angle stock is available from most hardware stores. The side and top panels may be cut and filed by hand, but many sheet metal suppliers will shear panels to the correct size for little or no additional cost.

All of the primary circuitry, including K1, K2, J1, S1, R1-R3 and C1 are all mounted on the rear panel (Fig. 71). Liberal use of solder lugs adds to a neat appearance. All primary wiring was done with no. 12 wire.

The bridge rectifier diodes and associated components are mounted on a glass-epoxy circuit board. The diodes and resistors are mounted slightly above the board to allow airflow to all sides for cooling. The PC board is mounted vertically on ceramic standoff insulators. Parts placement for the rectifier board is shown in Fig. 72, and a full-size etching pattern may be found at the back of this book.

Filter capacitors and equalizing resistors are mounted on a sandwich made from acrylic sheet. In addition to securing the capacitors, the acrylic provides additional insulation between the capacitors and chassis. Copper straps interconnect the capacitors, and the 40-kΩ resistors are mounted far enough above the capacitors to allow adequate airflow for cooling.

F1 and R4 are mounted on ceramic standoffs bolted to the front panel. Amphenol MHV high-voltage receptacles, similar to BNC connectors but designed for use up to 5 kV at 5 A, are used for the output. No. 18 test lead wire with insulation rated to 5 kV is used for all secondary wiring.

Care in mounting and wiring the components in a high-voltage supply is essential for safety. Any components and connections carrying high voltage should be kept well away from the chassis to avoid accidental arcs. High-voltage supplies should be fully enclosed to prevent accidental contact. Safety and common sense are the watchwords — *HV is lethal*!

A 3500-Volt Power Supply

The high-voltage power supply shown in Figs. 73 through 75 will loaf along during legal-power-limit operation. Capable of powering amplifiers employing such tubes as the 3-1000Z, 3CX1200A7, 4CX1000, and 8877, this supply is rated for 3500 V at 1 A output. Mark Wilson, AA2Z, built this project in the ARRL lab.

Circuit Details

Fig. 73 is the schematic diagram of this supply. The transformer used here is manufactured by the Peter W. Dahl Co. It features a 220/240-V primary and a 2650-V secondary rated at 1.5 A Intermittent Commercial and Amateur Service (ICAS). The core material is hipersil, which represents a significant weight saving over conventional material. One of the design goals was good regulation under heavy loads, so a 117-V primary wasn't even considered. Similarly, use of a full-value secondary and bridge rectifier, rather than a voltage-doubler scheme, aids in good regulation. Of course, a center-tapped secondary (2650-0-2650 V) would work just as well.

This supply was built to power several amplifiers requiring different voltages, so a variable autotransformer is used on the primary. Supply output can be varied between 0 and approximately 4 kV.

High-voltage block-type rectifiers are used here, rather than strings of individual 1-kV devices. These rectifiers are rated at 14 kV at 1 A and have a 250-A surge-

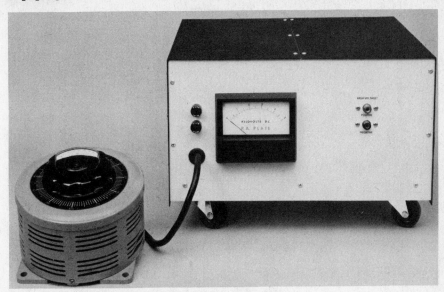

Fig. 73 — Front view of the completed 3500-V power supply.

current rating. If a builder were to purchase new the individual diodes, resistors and capacitors necessary to make a bridge rectifier suitable for this power supply, the cost could easily exceed that of the block rectifiers. Additionally, these HV rectifiers are compact and much easier to replace should they fail.

The filter capacitor used here is a 25-μF, 4-kV oil-filled unit made by Condensor

Products Corp. This capacitor does not contain any of the carcinogenic chemical PCB, but many oil-filled capacitors available at flea markets do, so be wary if another capacitor is substituted. Oil-filled capacitors have some advantages over HV filter capacitors made from strings of individual electrolytics. Oil-filled capacitors are sealed and will not dry out after years of service. They may be used all the way

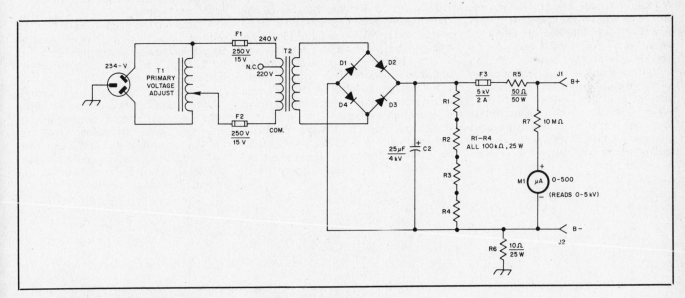

Fig. 74 — Schematic diagram of the 3500-V power supply.

C2 — Filter capacitor, 25 μF, 4 kV (Condenser Products KMOC 4M25ES-1 or equiv. Available from Condenser Products Corp., P.O. Box 997, Brooksville, FL 33512).

D1-D4 — Rectifier diode, 14 kV, 1A (K2AW HV14-1 or equiv. Available from K2AW's Silicon Alley, 175 Friends La., Westbury, NY 11590).

F3 — High-voltage fuse, 5 kV, 2 A (Buss HVJ2 or equiv.).

J1,J2 — High-voltage connector (Millen 37001 or equiv.).

R7 — Meter shunt made from five 2-MΩ, 2-W resistors in series.

T1 — Variable autotransformer, 0-280 V, 10 A (Superior Electric 236B Powerstat® or equiv.).

T2 — Power transformer; primary, 220/240 V, secondary 2650 V, 1.5-A ICAS (Dahl ARRL-001 or equiv. Available from Peter W. Dahl Co., 4007 Fort Blvd., El Paso, TX 79930).

Fig. 75 — Interior view of the 3500-V power supply.

up to their maximum voltage ratings, and they often take up less space than a stack of electrolytics. Oil-filled capacitors are somewhat more expensive initially — in this case about 30% more than a comparable string of electrolytics; however, they do last a long time and are definitely worth considering.

R1-R4 act as a bleeder. Four 100-kΩ, 25-W units are used to minimize the voltage across any individual resistor. If one large resistor was used, arcing could occur between the turns of wire that make up the resistor.

Like the 2000-V supply, this circuit employs a high-voltage fuse and a series resistor in the HV output lead to protect against overcurrent and arcs. A homemade fuse can be made from a piece of no. 30 tinned wire suspended on ceramic stand-offs. The minus lead is kept a few ohms above ground for metering purposes in the companion amplifier.

A 0-5-kV meter is included. The basic movement is 0-500 μA, so a 10-MΩ shunt resistor is necessary. Be careful to mount the resistors away from the chassis. Here,

they are mounted on a circuit board behind the meter.

Construction

This supply is built on a homemade chassis and cabinet measuring 9 × 17 × 14 inches (HWD). The bottom plate, which supports most of the components, is made from 0.125-inch-thick aluminum sheet. Aluminum angle brackets bolted to the bottom plate provide additional strength, as well as allow for attachment of the panels. The front and rear panels are made from 0.062-inch-thick aluminum sheet. The top and side panels are made from two pieces of aluminum sheet, each bent 90 degrees. The two top/side halves meet in the center where they are screwed to a bracket running between the front and rear panels. The entire chassis and cabinet could be made with hand tools, but a trip to the local aluminum supplier with a list of needed panel dimensions will probably yield the nicely sheared, square panels necessary for a professional-looking project.

As shown in Fig. 75, all components mount on the front and bottom panels. Even the line cord passes through the front panel, leaving the rear panel empty so the supply may be placed on the floor against a wall in the shack. The line cord and primary wiring are no. 12 wire. The transformer is securely bolted to the bottom plate by ¼-in bolts. The block rectifiers are mounted to the bottom panel with silicone heat-sink compound to aid in cooling.

F3 and R1-R6 all mount on a circuit board that is mounted on ceramic stand-offs above C2. Be sure that this board is mounted well-away from the chassis or other components.

All secondary wiring is done with no. 18 test lead wire with insulation rated at 5000 V. The output connectors, mounted on the front panel, are of the Millen 37001 type and are rated for 7000 V.

As with any project incorporating high voltage, lead dressing is important. High-voltage can bridge small air gaps, so care must be taken to keep component terminals and wires away from the chassis. Only wire with suitable insulation should be used, because HV will arc through insulation designed for lower voltages.

Chapter 28

Audio and Video Equipment

The story of radio begins with amateurs transmitting and receiving dots and dashes. Soon hams were communicating by voice. Amateurs were transmitting pictures 20 years before commercial TV came into being. Hams still communicate by Morse code, voice and TV. The techniques they use have changed and other modes have been added. All that communication activity has created a demand for a wide variety of electronic equipment. This chapter describes a number of projects for audio and video.

A Simple, High-Performance CW Filter

This inductor-capacitor CW filter uses one stack of the familiar 88-mH inductors and two 44-mH inductors in a five-resonator circuit that gives high performance at low cost. The center frequency is fixed at 750 Hz because most transceivers use this sidetone frequency, but sidetones between 700 and 800 Hz can be received with less than 1 dB attenuation relative to the center frequency. Ed Wetherhold, W3NQN, designed and built the filter presented here. The author can provide parts for this project at nominal cost. Write E. E. Wetherhold, W3NQN, 1426 Caplyn Place, Annapolis, MD 21401 for more information. If you need a design for a different center frequency, the author can provide that as well. Be sure to include a self-addressed, stamped 9½- × 4-inch envelope with your request.

One feature of this filter is a 3-dB bandwidth of 236 Hz. This bandwidth is narrow enough to give good selectivity, and yet broad enough for easy tuning with no ringing. Five high-Q resonator circuits provide good skirt selectivity that is equal to or better than most commercial active filters costing more than $80. In comparison, this CW filter can be built for less than $15. Simple construction, low cost and good performance make this filter an ideal first project for anyone interested in putting together a useful station accessory.

Design

Fig 1 shows the filter schematic diagram and component values. These values were selected for a center frequency of 750 Hz and for a filter impedance level of 230 ohms. The filter sees a 230-ohm source impedance consisting of the 200-ohm source (transformed from 8 ohms), a 22-ohm transformer winding resistance and an 8-ohm inductor resistance. In a similar way, the filter sees a load impedance of 230 ohms. This design was selected so that only one turn needs to be removed from both windings of a standard 44-mH inductor to give the required L2 and L4 values.

Construction

Fig 2 is a pictorial diagram showing the filter wiring. Note the 44-mH lead connection, as well as the connections between the capacitor leads, the 88-mH stack terminals and the 44-mH inductor leads. Fig 3 shows the finished filter installed in an aluminum box. Before beginning construction, obtain one 88-mH five-inductor stack with a mounting clip and two 44-mH inductors, and then follow steps 1 to 5.

1) Remove one turn from each of the two windings of one 44-mH inductor to get 43.5 mH (total turns removed is two). Carefully scrape off the film insulation and connect the start lead (with sleeve) of one

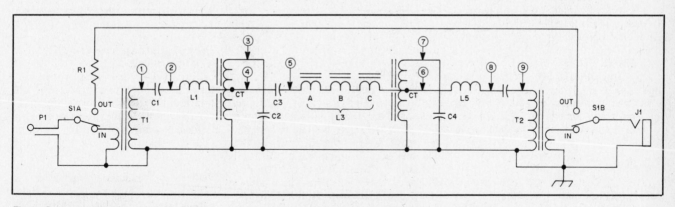

Fig 1—Schematic diagram of 750-Hz CW filter. Use 1% tolerance capacitors for best results.

C1, C5—0.512 μF capacitor.
C2, C4—1.036 μF capacitor.
C3—170.7 nF capacitor.
J1—Phone jack, or jack to match your headphones.
L1, L5—88-mH toroid (part of toroid stack, see text).

L2, L4—43.5 mH toroid (modified 44-mH toroid, see text).
L3—264-mH toroid (part of toroid stack, see text).
P1—Phone plug, or plug to match your receiver.

R1—Zero to 220-ohm, ½-W, 10% resistor (see text).
S1—DPDT switch.
T1, T2—8-ohm to 200-ohm impedance-matching transformer, 0.4-W.

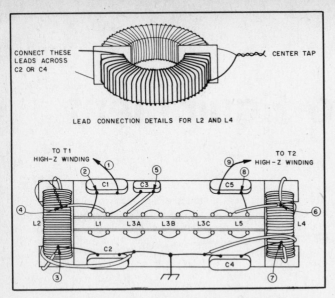

Fig 2—Pictorial diagram showing L2 and L4 lead connection and wiring of inductor stack (L1, L3, L5).

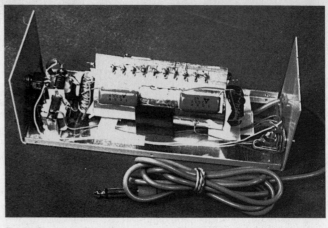

Fig 3—The assembled filter is shown installed in a CR-800 aluminum box. The thru/bypass switch (S1) and input/output transformers (T1, T2) are at the left end of the box.

winding to the finish lead (no sleeve) of the adjacent winding to make the center tap as shown in Fig 2. Do the same for the second 44-mH inductor.

2) Fasten both of the 43.5-mH inductors to opposite ends of the 88-mH stack using clear silicone-rubber sealant, available from most hardware stores.

3) Position the 43.5-mH inductors so their leads can be easily connected to the rest of the circuit. Solder the capacitor leads to the stack terminals as shown in Fig 2.

4) Obtain a suitable box and make holes for the inductor mounting clip, the DPDT switch, and the phone jack and phone cord. First, install matching transformers T1 and T2 and the inductor stack with capacitors. Fasten the transformers (with leads pointing up) to the bottom of the box with silicone rubber sealant. Secure the stack to the bottom of the box with a 1-3/8-inch component mounting clip and two no. 6-32 × 5/16-inch screws. Instead of the 8 × 3 × 2¾-inch aluminum box shown in Fig 3 (Mouser Stock No. 537-CR-800), a small cardboard box may be used to minimize cost.

5) Complete the wiring of the transformers, the DPDT switch with resistor R1, and the phone jack and phone plug. Then check the correctness of your wiring by measuring and comparing the filter node-to-node resistances with the values listed in Table 1.

Installation

T1 and T2 match the filter to the receiver low-impedance audio output and to an 8-ohm headset or speaker. If your headset is high impedance, T2 may be omitted. In this case, connect a 10%, ½ W resistor from node 9 (C5 output lead) to ground. Choose the resistor value so the parallel combination of the headset and resistor gives the correct filter termination impedance (within 10 percent of 230 ohms).

Performance

The measured 30-dB and 3-dB bandwidths are about 511 and 235 Hz, respectively, and the 30-dB/3-dB shape factor is 2.17. This factor can be used to compare the performance of this filter with others. The measured insertion loss at 750 Hz is less than 3 dB and is typical of passive filters of this type. This small loss is compensated by slightly increasing the receiver audio gain. R1 helps to maintain a constant audio level when the filter is switched out of the circuit. The correct value of R1 for your audio system should be determined by experiment. Start with a short circuit for R1 and then gradually increase the resistance until the audio level appears to be the same with the filter in or out of the circuit.

More than 700 hams have constructed this five-resonator filter (using either the 2-stack or the newer 1-stack arrangement) and many have commented on its excellent performance and lack of hiss and ringing.

References

Wetherhold, "Modern Design of a CW Filter using 88- and 44-mH Surplus Inductors," *QST*, Dec. 1980 and Feedback, *QST*, Jan. 1981, p. 43.

Wetherhold, "High Performance CW Filter," *Ham Radio*, Apr 1981.

Radio Handbook, 23rd edition, W. Orr, ed., Howard W. Sams & Co, 1987 (1-Stack CW Filter), p. 13-4.

Table 1

Node-to-Node Resistances for the CW Audio Filter

Nodes From	To	Components	Resistance (ohms)
1	GND	T1 hi-Z winding	12
2	GND	L1 and ½ L2	10
3	GND	L2	4
4	GND	½ L2	2
5	GND	L3 and ½ L4	26
6	GND	½ L4	2
7	GND	L4	4
8	GND	L5 and ½ L4	10
9	GND	T2 hi-Z winding	12
2	4	L1	8
5	6	L3	24
6	8	L5	8
2	3	L1 and ½ L2	10
8	7	L5 and ½ L4	10

Notes

1) See Figs 1 and 2 for the filter node locations.

2) Check your wiring using the resistance values in Table 1. If there is a significant difference between your measured values and the table values, you have a wiring error that must be corrected.

3) For accurate measurements, use a digital VOM or an analog VOM (such as a Triplett Model 630) that has a scale center of about 5 ohms on the ×1 ohmmeter range.

A Passive Audio Filter for SSB

While audio filters are most often used during CW reception, the SSB operator can also benefit from their use. Shown in Figs 4 and 5 is a passive band-pass filter designed by Ed Wetherhold, W3NQN, for phone operation. This filter was described in Dec. 1979 *QST*.

All of the inductors are the surplus 88-mH toroidal type with their windings wired either in series or parallel to get the required 88 or 22 mH of inductance. The series connection is shown in Fig 2. The 0.319-μF capacitors were selected from several 0.33-μF capacitors that were about 3 percent on the low side. The 0.638-μF value was obtained with a single 0.68-μF capacitor that was about 6 percent on the low side. The 1.276-μF values were obtained by paralleling selected 1-μF and 0.33-μF capacitors.

Fig 5 shows the measured and calculated attenuation responses of the filter. The difference between the measured and calculated responses at the low frequency side of the passband is probably caused by the much lower Q of the inductors at these frequencies.

The necessary termination resistance of this filter is 206 ohms. While this is not a standard value, it should not be too difficult for most amateurs to accommodate. If low-impedance headphones are used, a matching transformer can be used to provide the correct termination. A suitable transformer is available from Mouser Electronics (see Chapter 35 parts-suppliers list). The part number is 42TU200, and it is a 200-ohm CT to 8-ohm CT unit.

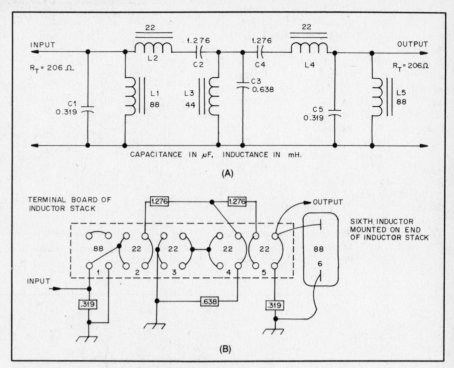

Fig 4—Schematic diagram of the SSB band-pass filter (A). Shown in B is a pictorial wiring diagram of the terminal board on the inductor stack.

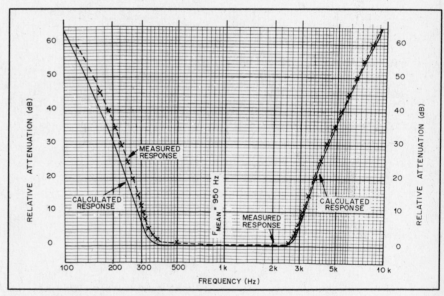

Fig 5—Response curves of the SSB band-pass filter.

RC Active Audio Filters

The active audio filter is more popular than the passive type shown in Figs 1 and 4. The primary advantages of active filters are (1) unity gain or greater (passive filters have some insertion loss), and (2) that they are more compact than LC filters. Another advantage of RC active filters is that they can be made with variable Q and variable center or cutoff frequencies. These two features can be controlled at the front panel of the receiver by means of potentiometers.

Most RC active filters are designed for a gain of 1 to 5. A recommended gain is 2 for most amateur applications. The more filter sections placed in cascade, the better the skirt selectivity. The maximum number of usable RC filter sections is typically four. The minimum acceptable number is two for CW work, but a single-section RC active filter is often suitable for SSB reception in

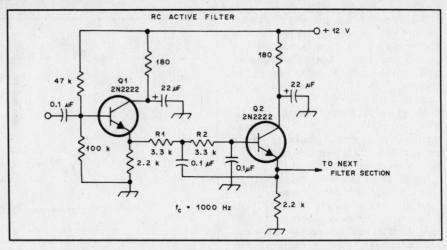

Fig 6—Circuit example of one pole section of an RC active audio filter that uses discrete active devices, Q1 and Q2.

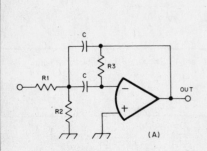

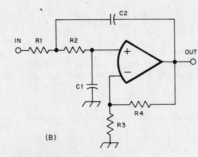

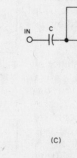

(A) (B) (C)

BAND-PASS FILTER

Pick $H_oQ,\omega_o = 2\pi f_c$
where f_c = center freq.
Choose C

Then $R1 = \dfrac{Q}{H_o\omega_oC}$

$R2 = \dfrac{Q}{(2Q^2 - H_o)\omega_oC}$

$R3 = \dfrac{2Q}{\omega_oC}$

If $H_o = 2$, $f_o = 800$ Hz, $Q = 5$ and
C = 0.022 μF
R1 = 22.6 kΩ (use 22k)
R2 = 942 Ω (use 1000)
R3 = 90.4 kΩ (use 91k or 100k)

Fig 7—Equations for designing a band-pass RC active audio filter are given at A. B and C show design information for low-pass and high-pass Butterworth filters.

LOW-PASS FILTER

$C_1 \leq \dfrac{\left[a^2 + 4b(K - 1)\right]C_2}{4b}$

$R_1 = \dfrac{2}{\left[aC_2 + \sqrt{[a^2 + 4b(K - 1)]C_2{}^2 - 4bC_1C_2}\right]\omega_c}$

$R_2 = \dfrac{5}{bC_1C_2R_1\omega_c{}^2}$

$R_3 = \dfrac{K(R_1 + R_2)}{K - 1}$ (K > 1)

$R_4 = K(R_1 + R_2)$

where

K = gain
f_c = −3 dB cutoff point
$\omega_c = 2\pi f_c$
a = 0.765367 (see text)
b = 1
C2 = a standard value near 10 / f_c μF
Note: For unity gain, short R4 and omit R3.

Example:
K = 2
f = 4000 Hz
ω_c = 25,132.74
C2 ≈ 0.0025 μF
C1 ≤ 0.003 μF
R1 = 19,254 Ω (use 20 kΩ)
R2 = 11,278 Ω (use 11 kΩ)
R3 = 61,066 Ω (use 62 kΩ)
R4 = 61,066 Ω (use 62 kΩ)

HIGH-PASS FILTER

$R_1 = \dfrac{4b}{\left[a + \sqrt{a^2 + 8b(K - 1)}\right]\omega_cC}$

$R_2 = \dfrac{b}{\omega_c{}^2C^2R_1}$

$R_3 = \dfrac{KR_1}{K - 1}$ (K > 1)

$R_4 = KR_1$

where

K = gain
f_c = −3 dB cutoff point
$\omega_c = 2\pi f_c$
a = 0.765367 (see text)
b = 1
C = a standard value near 10 / f_c μF
Note: For unity gain, short R4 and omit R3.

Example:
K = 4
f = 250 Hz
ω_c = 1570.8
C ≈ 0.04 μF (use 0.039 μF)
R1 = 11,407 Ω (use 11 kΩ)
R2 = 23,358 Ω (use 24 kΩ)
R3 = 15,210 Ω (use 15 kΩ)
R4 = 45,630 Ω (use 47 kΩ)

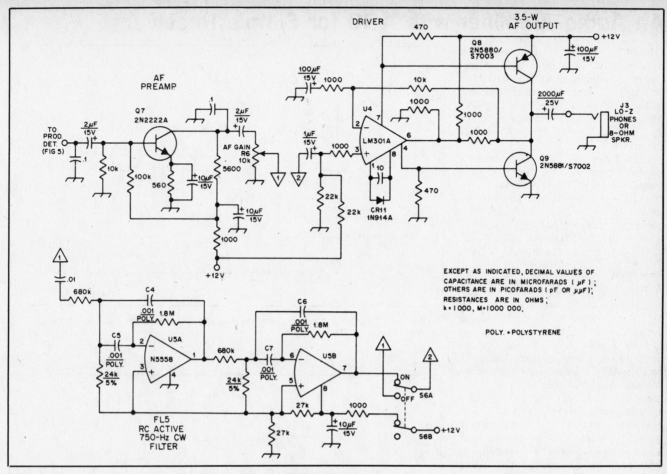

Fig 8—Practical circuit for a two-pole CW RC active filter, showing how it can be switched into and out of the audio channel of a receiver.

simple receivers. As the Q and number of filter sections increases there is a strong tendency toward "ringing." This appears in the speaker or earphones as a howling sound, which can be most unpleasant to hear. The same is true of passive audio filters that have extremely high loaded-Q values.

Op amp ICs are used as the active devices in most RC active filters. The 741, LM301 and 747 ICs are suggested for that application. However, discrete devices can be used with equal success if the builder so desires. Fig 6 shows one section of an active filter that uses transistors. Q1 serves as a source-follower at the input and Q2 is one section of the filter. Additional cascaded filter sections would consist of the circuit that is common to Q2. The values of R1 and R2 would be changed to modify the center frequency (f_c) of the filter. The lower the resistance value the higher the f_c. A dual potentiometer could be used in place of R1 and R2 to provide variable frequency response.

Design data for a band-pass active filter using a single op amp is shown in Fig 7A. Only one pole is shown for clarity. The term K represents the desired gain of the filter. Gains of 1 (unity) and 2 are most common. The desired filter Q can be calculated by dividing the center frequency by the bandwidth (delta f) at the −3 dB points.

Design data for second-order Butterworth VCVS low-pass and high-pass filters is shown in Figs 7B and 7C respectively. The desired filter gain is designated K, and the frequency at the −3 dB points is designated f. To design one of these filters, start by selecting the gain (K) and cutoff frequency (f_c). Then start plugging the variables into the formulas in the order given. Second-order calculations are shown, and these give approximately 40 dB per decade of attenuation outside the passband. If greater attenuation is desired, cascade another second-order section, this time substituting 1.847759 for a in the second section. This second section attenuates unwanted frequencies an additional 40 dB per decade.

High-Q, stable capacitors are imperative to proper filter performance. Polystyrene capacitors are recommended for use at C of Figs 6 and 7. The frequency-determining resistors and capacitors should be as close to the design values as possible. Variations greater than 5 percent in resistance and capacitance in a multipole filter will widen the 3-dB bandwidth and cause dips in the nose of the response curve. In other words, f_c should be exactly the same for *all* filter sections in an ideal example.

A practical example of a two-pole RC active filter that uses a dual op-amp IC is given in Fig 8. It is switched in and out of the audio amplifier by means of S1. As shown, the filter represents the minimum acceptable design for most CW work. A three- or four-section filter of this type would be more desirable for CW work under adverse band conditions (QRM or weak signals).

An Audio Amplifier with AGC for Simple Receivers

Many excellent articles describing simple receivers have appeared in the amateur literature. In many respects these receivers perform well. However, most of the designs lack an automatic gain control (AGC) circuit. This amplifier module was first described in April 1983 *QST* by Rick Littlefield, K1BQT.

There are several ICs on the market that satisfy amateur receiver audio requirements. The LM-386 is well suited for amateur use. This chip features low distortion, plenty of gain (up to 200), low current drain and high stability. With a 9-V supply it delivers 200 mW of audio power, more than sufficient to drive a speaker.

Automatic Gain Control

IF-derived AGC is probably the best. However, deriving a control signal from the IF stage may not be practical in simple designs in which the IF signal levels are low and BFO leakage is common. Audio-derived AGC is usually a safer choice. But deriving AGC from a receiver that employs a single high-gain audio chip can also present a problem. The only voltage-sampling point before the volume control is at the output of the product detector, where the voltage level is very low. Thus, a functional AGC circuit would require the addition of high-gain circuitry to develop a usable control voltage.

The simplest alternative is to derive the control voltage at the audio-amplifier output, where the signal level is high. The AGC circuit then keeps the amplifier from being driven past a predetermined output level. Limiting is set somewhere below the maximum undistorted output level of the audio amplifier at a comfortable "maximum" volume level for the radio. A drawback to this approach is that AGC action does not occur at very low gain settings. This should not be a problem in quiet listening environments — the human ear can adapt easily to changes of 20 dB or more in normal conversation. In noisy environments the radio would be operated at a higher volume level, and heavy AGC action would keep the output more constant. This is a compromise compared to a 90-dB AGC system operating independently of the audio gain control; but it works and does not add much to the complexity of the receiver circuit.

Circuit Information

The circuit shown in Fig 9 has been used successfully in several small receivers. The LM386 audio section is compact and stable. Miniature 220-μF, 16-V capacitors (available from Radio Shack) contribute to the small size of the layout. The circuit is designed to drive an 8-ohm speaker. Operation with headphones is possible by adding a simple attenuator similar to those used in transistor radios. A 10-ohm resistor

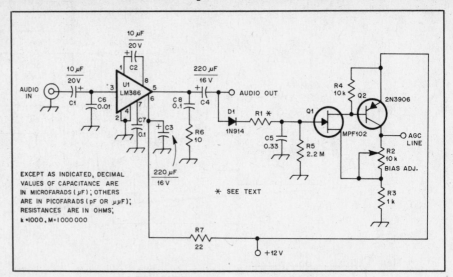

Fig 9—Circuit diagram of a simple audio amplifier and AGC module for use with homemade receivers.

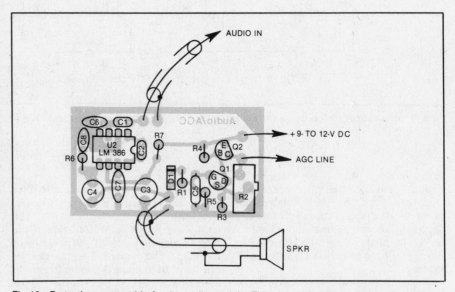

Fig 10—Parts-placement guide for the audio module. This view is from the component side of the board. Gray areas represent an X-ray view of the unetched foil. A full-size etching pattern can be found in the back of this book.

replaces the speaker, and a 470-ohm series resistor drops the output to a comfortable listening level.

The AGC design was borrowed from a more conventional audio-derived configuration and adapted for this application. This circuit has some interesting features that make it universal. First, it samples a wide range of audio levels while remaining virtually transparent to the circuit because of the high input impedance. Second, it can be used with either a bipolar or an IC-based IF section that requires a positive bias-voltage swing, or with MOSFET stages, which employ a negative bias-voltage swing.

Since Q1 will respond to either positive or negative voltage, the polarity of D1 determines the direction of the bias-voltage swing. R2 sets the resting-bias level. Adjust this level to the point where AGC action begins to lower the receiver background noise on a quiet frequency. R1 is used to adjust the drive. It should be set for smooth response. Overdrive will cause a pronounced overshoot or a cracking sound when strong signals come on. Lack of drive will cause insufficient limiting, and the audio amplifier will be driven into distortion. A substitution box can be used to determine a value for R1 that will work best with your radio. Or, start with a 50-kΩ potentiometer and measure the required resistance.

An Expandable Headphone Mixer

From time to time, active amateurs find themselves wanting to listen to two or more rigs simultaneously with one set of headphones. For example, a DXer might want to comb the bands looking for new ones while keeping an ear on the local 2-meter DX repeater. Or, a contester might want to work 20 meters in the morning while keeping another receiver tuned to 15 meters waiting for that band to open. There are a number of possible uses for a headphone mixer in the ham shack.

The mixer shown in Figs 11 through 13 will allow simultaneous monitoring of up to three rigs. Level controls for each channel allow the audio in one channel to be prominent, while the others are kept in the background. Although this project was built for operation with three different rigs, the builder may vary the number of input sections to suit particular station requirements. This mixer was built in the ARRL lab by Mark Wilson, AA2Z.

Circuit Details

The heart of the mixer is an LM386 low-power audio amplifier IC. This 8-pin device is capable of up to 400 mW output at 8 ohms — more than enough for headphone listening. The LM386 will operate from 4- to 12-V dc, so almost any station power supply, or even a battery, will power it.

As shown in Fig 12, the input circuitry for each channel consists of an 8.2-ohm resistor (R1-R3) to provide proper termination for the audio stage of each transceiver, a 5000-ohm level control (R4-R6) and a 5600-ohm resistor (R7-R9) for isolation between channels. C1 sets the gain of the LM386 to 46 dB. With pins 1 and 8 open, the gain would be 26 dB. Feedback resistor R10 was chosen experimentally for minimum amplifier total harmonic distortion

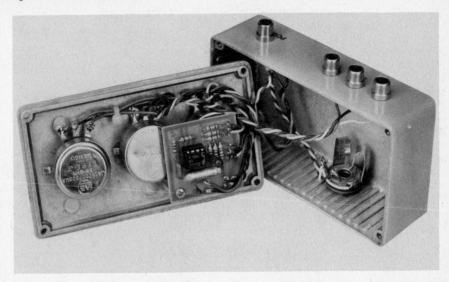

Fig 11—The 3-channel headphone mixer is built on a small PC board. Note that the lead length was kept to a minimum to aid stability.

(THD). C2 and R11 form a "snubber" to prevent high-frequency oscillation, adding to amplifier stability. None of the parts values are particularly critical, except R1-R3, which should be as close to 8 ohms as possible.

Construction

Most of the components are arranged on a small PC board. Parts placement is shown in Fig 13, and a full-size etching pattern is shown in the back of this book. Perfboard will work fine also, but some attention to detail is necessary because of the high gain of the LM386. Liberal use of ground connections, short lead lengths and a bypass capacitor on the power-supply line

all add to amplifier stability.

The mixer was built in a small diecast box. Tantalum capacitors and ¼-W resistors were used to keep size to a minimum. The '386 IC is available from Radio Shack (cat. no. 276-1731). A 0.01-μF capacitor and a ferrite bead on the power lead help keep RF out of the circuit. In addition, shielded cable is highly recommended for all connections to the mixer. The output jack is wired to accept stereo headphones.

Output power is about 250 mW at 5% THD into an 8-ohm load. The output waveform faithfully reproduces the input waveform, and no signs of oscillation or instability are apparent.

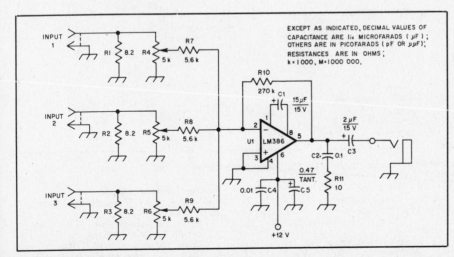

Fig 12—Schematic diagram of the LM386 headphone mixer. All resistors are ¼ W. Capacitors are disc ceramic unless noted.

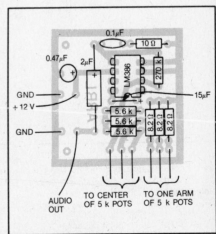

Fig 13—Parts-placement diagram for the LM386 headphone mixer. A full-size etching pattern appears in the back of this book.

An ATV Monitoring Instrument

To maintain high-quality video images, amateur television enthusiasts must monitor their transmissions. RF sampling is often used to siphon a portion of the transmitter's RF output from the transmission line for monitoring purposes. This relatively easy-to-build video sampler, shown in Figs 14 through 21, is based on this technique, but goes several steps further. By employing a directional coupler, it not only provides low-level output to a TV monitor, but also permits percentage modulation and relative SWR measurements. Biagio Presti, of Aptron Laboratories, Inc., designed and built this instrument. John Gereke of Southwest Sound also contributed to this project.

Circuit Description

Fig 14 is a schematic diagram of the unit. The stripline directional coupler provides detection of both forward and reverse power. When S1 is in the VIDEO position, modulation is detected and routed to chopper IC U1. The IC switches inside U1 are connected as shown in Fig 15.

If S3 is also placed in the VID position, then IC switch 1-2 is always closed, and IC switches 3-4 and 9-8 are always open. Detected video is then routed to the input of the video amplifier consisting of Q2, Q3 and Q4. The output from Q4 is capable of driving a TV monitor input and must be terminated in 75 ohms. For percentage modulation measurements, this output is connected to an oscilloscope.

If S3 is placed in the CHOP position, then IC switches 3-4 and 9-8 are turned off and on at a 60-Hz rate by Q1. In this configuration, the connections of switch 9-8 form a 60-Hz inverter that drives IC switch 1-2. This produces a ground reference on the face of an oscilloscope connected to the video output terminal and set to view the signal at the horizontal TV rate.

Placing S1 in the SWR position allows making SWR measurements.

PC Board Assembly

Fig 16 shows the PC-board parts placement. The board is made from 1/16-inch, double-sided glass-epoxy stock, so the

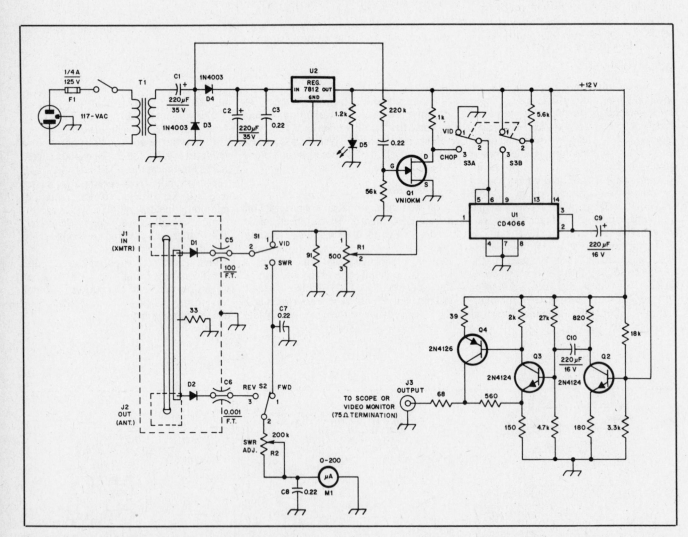

Fig 14—Schematic diagram of the ATV monitoring instrument. Resistors are ¼ W, 5% types unless otherwise noted. Capacitors are disc ceramic unless noted. Capacitors marked with polarity are electrolytic.

C1, C2 — 220-μF, 35-V miniature PC-mount electrolytic capacitor.
C3, C4, C7, C8 — 0.22-μF epoxy-coated monolithic ceramic capacitor.
C5 — 100-pF feedthrough capacitor.
C6 — 0.001-μF feedthrough capacitor.
C9, C10 — 220-μF, 16-V miniature electrolytic capacitor, axial leads.

D1, D2 — 1N34 diode.
D3, D4 — 1N4003 diode.
D5 — LED in a panel-mount holder.
J1, J2 — Chassis-mount female Type-N connector.
J3 — Chassis-mount female BNC connector.
M1 — Microammeter, 200 μA full scale.
Q1 — VN10KM FET.
Q2, Q3 — General-purpose NPN silicon

transistor, 2N4124 or equiv.
Q4 — Small-signal PNP amplifier transistor, 2N4126 or equiv.
T1 — Power transformer. Primary, 117-V ac. Secondary, 12.6-V ac, 100 mA.
U1 — Quad bilateral switch IC, CD4066.
U2 — 12-V, 1.5-A three-terminal regulator (LM7812 or equiv.).

component side acts as a groundplane. If you can make double-sided boards, use the etching pattern provided for the component side to clear away unwanted copper from around component holes. If you cannot make double-sided boards, leave the component side unetched. After drilling all lead holes, clear away the component-side copper from holes for component leads that are not grounded. A 1/8-inch drill bit turned by hand works well for clearing holes.

The PC board mounts vertically to the rear plate of the case used. The Type-N connector center pins connect directly to the PC board, so the board must be mounted very close to the panel. This means some components must be mounted on the circuit side of the PC board. All of the resistors, small bypass capacitors, rectifier diodes and U1 mount on the component side. The remainder of parts and wiring mount on the circuit side. Any component leads that are grounded should be soldered to the groundplane of the board as well as the circuit trace.

When mounting the parts on the circuit side, be sure to orient U2 and the transistors properly. Leave the body of the electrolytics elevated above the board to allow soldering of the leads to the pads as shown in Fig 17A. Remember to clip leads on the opposite side to avoid shorts.

The case used should be at least three inches deep to allow room for the switches and meter to mount to the front panel. The power transformer can be mounted to the case bottom plate. Use sensible wiring

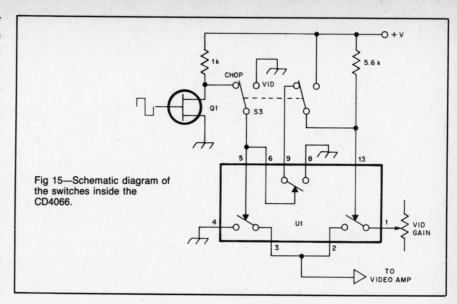

Fig 15—Schematic diagram of the switches inside the CD4066.

practice in wiring the ac leads to the switch and fuse.

A short piece of 75-ohm coaxial cable should be used to connect the video from the board to J3. A grounding lug should be used on the video connector to bring the coax shield to ground without relying on the case metal for ground.

When mounting the stripline diodes (D1 and D2) and the 33-ohm resistor, be careful not to allow solder blobs to accumulate. The leads for the diode and resistor to the stripline should be very short to the body of the component, and the leads should not extend beyond the stripline area as shown in Fig 17B.

The 1N34 diodes should be a matched pair for best results. You can match such diodes adequately by using an ohmmeter and measuring the forward drop of a number of diodes until two are found that read the same or nearly so.

Metal Parts

The area of construction requiring some care is the mechanical placement of the Type-N connectors to the rear panel and the metal shield parts. The reason is that

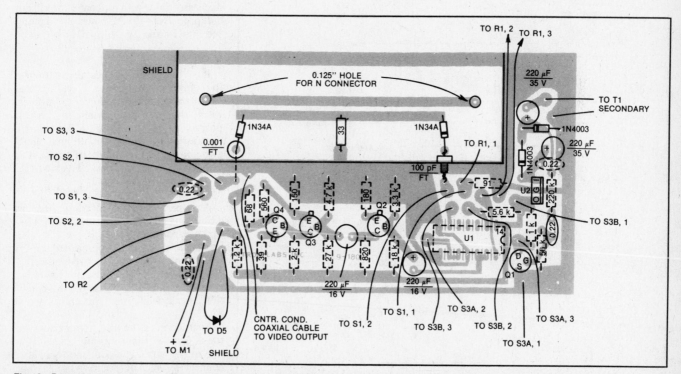

Fig 16—Parts-placement diagram of the ATV monitoring instrument. Components that mount on the circuit side of the board are drawn with solid lines. Components represented by dashed lines mount on the component side of the board. See text. A full-size etching pattern appears at the back of this book.

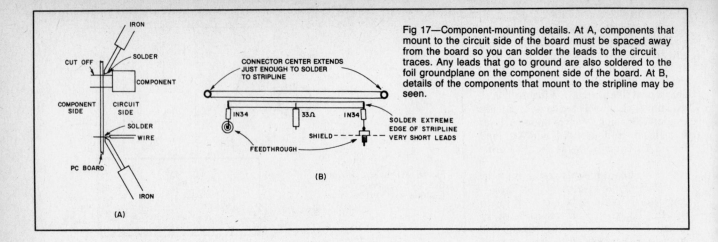

Fig 17—Component-mounting details. At A, components that mount to the circuit side of the board must be spaced away from the board so you can solder the leads to the circuit traces. Any leads that go to ground are also soldered to the foil groundplane on the component side of the board. At B, details of the components that mount to the stripline may be seen.

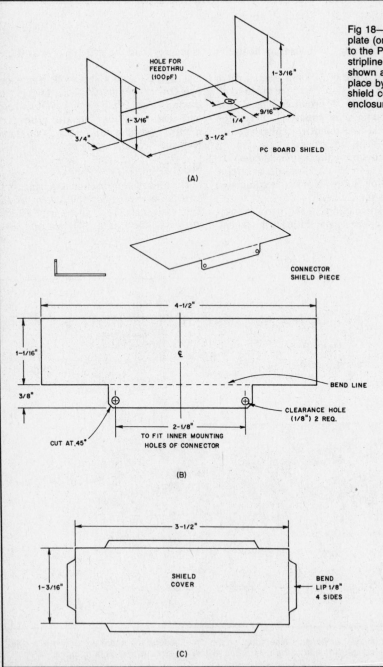

Fig 18—The pieces of the PC board shield are made from tin plate (or thin sheet brass or copper). The piece at A solders to the PC board circuit side to form three sides of the stripline compartment shield. The connector shield piece, shown at B, forms the fourth side of the shield. It is held in place by the connector-mounting hardware. See Fig 19C. The shield cover piece shown at C completes the stripline enclosure.

once the connectors are mounted, along with the shield piece associated with the connector mounting, the nuts holding these pieces are no longer reachable when the PC board is slipped into position. The PC mounting holes for the center conductor of the connectors must then slip perfectly into place or the PC board will not fit, and the shield associated with the connector will not butt properly to the PC board shield piece. See Figs 18 and 19.

How to Use the Instrument

It is a simple matter to use the instrument to monitor video. First, connect the video output to a monitor terminated in 75 ohms. Place the VID/CHOP switch in the VID position. Adjust video gain for suitable contrast. If you have a multiburst signal available, then the quality of transmitter performance can be checked as well.

To Check for Percentage Modulation

A properly designed TV transmitter has a clamping circuit that establishes the sync tip to a constant value. Because the RF amplitude of the signal approaches zero for "white-going" signals, any overmodulating white peaks will bring the carrier to zero, causing "sync buzz." To avoid this condition, peak white modulation is set to 87.5%.

First, connect the transmitter output to the IN Type-N connector and the antenna line to the OUT Type-N connector. Next, connect the video output to an oscilloscope with sufficient bandwidth to view video (10 MHz) and terminate the output in

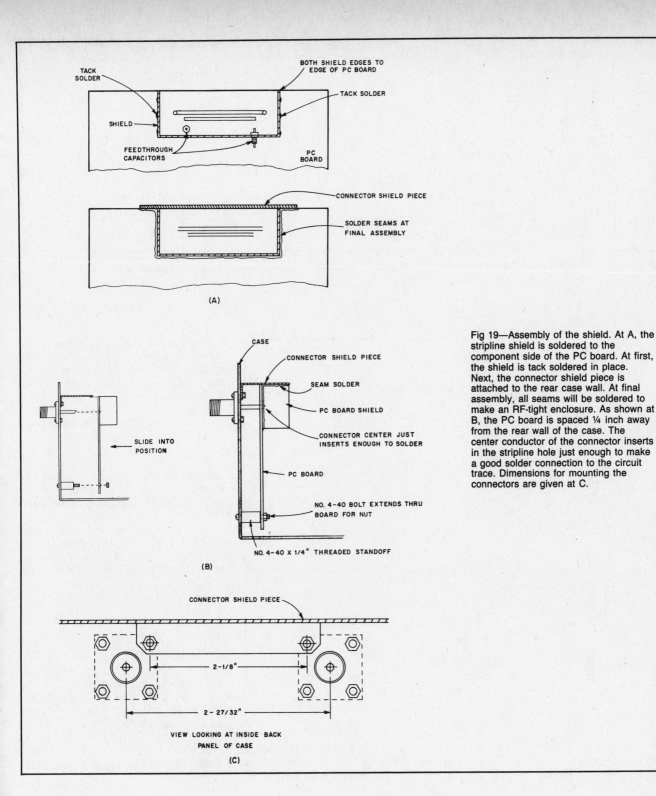

Fig 19—Assembly of the shield. At A, the stripline shield is soldered to the component side of the PC board. At first, the shield is tack soldered in place. Next, the connector shield piece is attached to the rear case wall. At final assembly, all seams will be soldered to make an RF-tight enclosure. As shown at B, the PC board is spaced ¼ inch away from the rear wall of the case. The center conductor of the connector inserts in the stripline hole just enough to make a good solder connection to the circuit trace. Dimensions for mounting the connectors are given at C.

75 ohms. You can make an adequate termination by using a ¼-watt, 5%, 75-ohm resistor with one lead soldered to the center pin of a BNC male connector and the opposite end of the resistor soldered to the shell. By using a BNC T connector, you can then make the connection to the scope. TV monitors generally provide 75-ohm terminations, but if not, then the same scheme can be used for the monitor as well. Then, set the scope horizontal sweep rate to allow two or three TV lines to be viewed.

The ideal signal to use for modulating the transmitter is a test-generator stairstep signal since the "white" step is clearly defined. If, however, you do not have a test signal, a black card with a white square in the center can be used in conjunction with your camera. The idea is to have a clear white video reference with which to set the percentage modulation.

Set the VID/CHOP switch to CHOP. Set the VID/SWR switch to VID. With every-

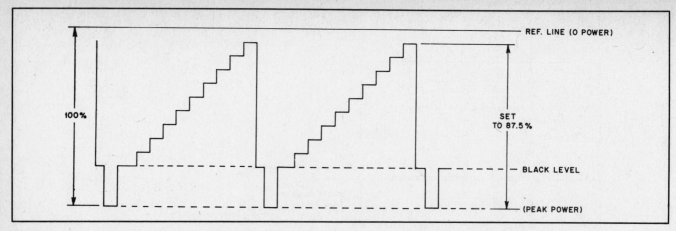

Fig 20—When you have set up the monitoring instrument with a test generator to check percentage modulation, the scope display should look like this.

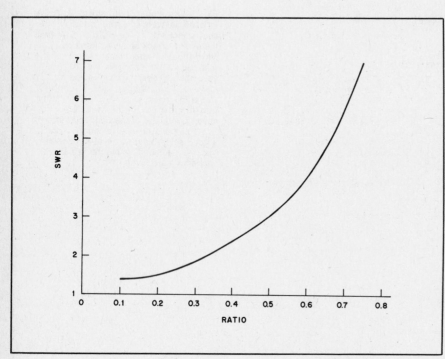

Fig 21—Use this graph to determine SWR from the meter readings. See text.

flection from the chopper reference bar to the tip of sync, then each cm of deflection represents 25%. Setting the tip of sync to video white at 3.5 cm is equivalent to 87.5% modulation.

Adjust the *transmitter* video level and sync stretch for proper sync to video ratio and percentage modulation.

To Monitor SWR

Set the VID/SWR switch to SWR and the FWD/REV switch to FWD. Adjust the SET control for full-scale reading on the meter (200 µA).

Now switch the FWD/REV switch to REV and read the meter. The ratio of the reverse to forward reading is used by referring to the SWR chart in Fig 21 or calculate using

$$\frac{F + R}{F - R}$$

where

F = forward reading
R = reverse reading

For example, if the reverse reading is 40 µA, then the ratio is 40/200 or 0.2. Referring to Fig 21, read 0.2 on the RATIO scale to the curve and then to the SWR scale and read 1.5 SWR.

The instrument will handle more than 50 watts and can be left in place without degradation to the quality of transmission or power output.

thing turned on, the scope display should appear as in Fig 20.

Adjust the scope vertical gain and the instrument video gain to a calibrated vertical deflection. For example, if the scope display is set to a 4-cm vertical de-

A Weather Satellite Display System Using A Dedicated CRT

A system for display of automatic picture transmission (APT) weather satellite imagery using a dedicated CRT is described here. This method has been used by amateur and professional satellite users for many years, although commercial users have now largely abandoned it in favor of more complex (and much more costly) imaging systems. This system, shown in Figs 22-38, was designed and built by Grant Zehr, WA9TFB.

Using a dedicated CRT system, the electron beam is intensity modulated as it is swept across the CRT screen. With proper timing the satellite image is traced out on the CRT screen. Because of the long times required to draw a complete image, the screen must be photographed (in a darkened room) for the image to be seen in detail.

While there are some limitations to a dedicated CRT system, it is still popular among amateurs because it offers high-resolution, photographic-quality imagery at low cost. The only other amateur systems with similar resolution and photographic image quality are those built around photographic drum machines.[1] These machines produce excellent imagery but offer the complexity of drums, motors, glow lamps and traverse mechanisms.

The principles of CRT operation are discussed in Chapter 7 and are well worth reviewing before beginning this project. The CRT for this project is used in an entirely conventional manner except that the sweep rates are much slower than those typical for standard monitor use.

System Overview

Operation of the system is relatively straightforward. During the time a satellite is within range of the receiving station, the signal from the satellite is recorded on the right channel of a stereo tape recorder (see Fig 22). At the same time, a highly accurate 2400 Hz signal of constant strength is recorded on the left channel of the tape recorder. During playback, the satellite signal (right channel) is fed into the analog processing circuit and the resulting signal used to intensity modulate the CRT using a high-voltage transistor amplifier. The synchronizing tone (left channel) is fed into the phase-locked loop/digital countdown circuit. These countdown circuits generate properly timed digital pulses to reset the analog sweep circuits and blank the screen during retrace of the CRT beam.

Circuit Description

The 2400 Hz local oscillator shown in Fig 23 is a TTL circuit similar to that used by many amateurs in crystal calibrators. The 2.4 MHz TTL signal is divided by ten in a 7490 decade counter then divided by

[1]Emilani, G. and M. Righini, "Printing Pictures for "Your Weather Geostationary Satellite," *QST*, Apr 1981, pp 20-25.

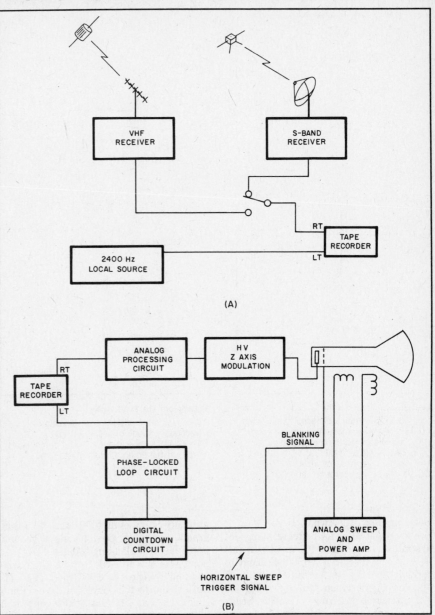

Fig 22—Block diagram of a satellite imaging system. Part A shows the recording mode, and part B shows the playback and display mode.

ten again in a CMOS 4017 decade counter. The last divide by ten is done with a CMOS 4018 in a digital sine-wave generator configuration. The secondary winding of a small audio transformer serves as a choke to further smooth the output of the local oscillator to provide a signal that can be recorded without excessive distortion. Because the harmonics have been greatly reduced by the sine-wave generation step, the playback signal is easier to use later for synchronization of the horizontal sweep.

After the satellite pass is complete, the signal can be played back and an image produced. During playback the satellite signal from the right channel is routed to the analog processing circuit and applied to the input of a 600 Hz high pass active filter (U5 of Fig 24) to reduce 60 Hz interference. The output from this filter is detected by a precision rectifier (U6A, B). The 10 kΩ potentiometer allows adjustment for symmetrical detection. The output from the detector is then filtered to remove the 2400 Hz subcarrier (U7, 8). The resulting dc level follows the video subcarrier envelope. An offset amplifier (U9) allows for adjustment of the amplitude and dc offset before the signal is applied to the high voltage driver amplifier.

In some situations, a stage of logarithmic amplification will produce a more satisfactory image. With logarithmic amplification the black level remains black, the grey level is somewhat brighter than with linear amplification and the whites will

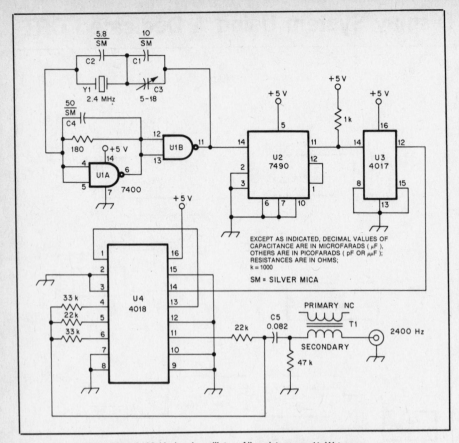

Fig 23—Schematic of the 2400 Hz local oscillator. All resistors are ¼ W types.

C1—10 pF silver-mica capacitor (see text).
C2—5.8 pF silver-mica capacitor (see text).
C3—Ceramic or foil trimmer capacitor, approximately 5-18 pF.
C4—50 pF silver-mica capacitor.
C5—0.082 µF disc-ceramic capacitor.

T1—Any small audio transformer with a secondary winding of about 2 kΩ
Y1—2.4-MHz general-purpose, fundamental-type crystal (International Crystal type no. 432115 or equiv.).

be compressed somewhat. In some situations this will allow for better resolution of ground detail in the APT image. In this display unit, an Intech A-730 hybrid circuit module is used to achieve logarithmic shaping of the video signal. See Fig 25. Other integrated circuits or discrete circuits could be used for this purpose. Even greater flexibility is possible using digital image enhancement at this point.[2]

The video signal, either linear or logarithmic, is applied to the contrast potentiometer shown in Fig 26. Adjusting the brightness control allows for a fairly wide range of voltages on the cathode of the CRT while still maintaining forward bias on the emitter-base junction of Q1 and class-A operation of the amplifier stage.

The CRT is used in a conventional cathode drive configuration. Electro-magnetic deflection is used with a matching yoke, and electrostatic focus is used. All voltages supplying the CRT are regulated (except the 9 kV anode supply). If unregulated voltages are used, the circuits will operate satisfactorily but the images will show objectionable ac "ripple" that degrades the image. Grid 2 of the CRT

[2]Grant Zehr, "The VIP: a VIC Image Processor," QST, Aug 1985, pp 25-31.

normally operates at 100 V dc; grid 3 is at ground potential. Grid 1 rests at ground potential until blanking is desired. At that point −15 V is applied to grid 1 to produce good blanking of the CRT trace. The TTL pulse signaling the timing for the blanking is amplified at U11 and further amplified with care taken to preserve the correct polarity of the final blanking pulse. This circuitry could probably be simplified, again taking care to preserve the option to blank three out of four sweeps of the CRT beam (see discussion below).

During playback of the satellite signal the 2400 Hz sine wave recorded on the left channel is fed into the phase-locked loop circuitry shown in Fig 27. First the 2400 Hz signal is amplified in two stages (U12 A, B). The signal is next filtered in a 2400 Hz bandpass filter to reduce harmonic content and noise (U13). The signal is then fed into a comparator to achieve a square wave (U14A) and reduced in amplitude with an amplifier having less than unity gain (U14B). After rectification, a 5 V square wave is available for the phase-locked loop to lock onto. This allows the monitor to track the satellite on the right channel even though the level of the satellite signal drops to zero. Systems locking on the satellite subcarrier will "unlock" if the signal level

drops below threshold resulting in loss of horizontal synchronization of the CRT trace. A 4049 CMOS buffer is used to obtain a satisfactory level to drive the TTL ICs used in the "countdown" circuitry.

The countdown circuitry is shown in Fig 28. This circuit divides the 2400 Hz reference down to either 8, 4 or 2 Hz. (U18, 19, 20). One half of a 74123 (U21A) is used to generate a short pulse at these slower frequencies. This pulse is used to reset the horizontal sweep on the CRT sweep drive (see sweep circuit discussion below). This pulse is also applied to the line blanking circuit to blank the CRT while the retrace occurs. The other half of the 74123 (U21B) is used to trigger on the output from the first half, pause until the sweep is complete, and finally blank the CRT so that damage to the phosphor does not occur. This feature is needed only when the sweep rate is increased (for example, to expand a small feature by expanding the portion of the image of interest). Two 7476 flip flops (U22, 23) are used to provide the option of blanking either alternate lines or three lines out of four. The alternate line blanking is used for "normal" processing of the Tiros image so that either the visible or IR image portion of the image can be shown alone. Using this option, the vertical and horizontal sweep rates must be increased so the image fills the entire screen and the proper aspect ratio is maintained. By blanking three lines out of four the typical Tiros image can be "enlarged" on the CRT so that small details on the image can be seen. The sweep rates must again be increased when this option is used.

The 74LS368 (U17) tri-state buffer allows the option of phasing the picture either to the right or left during operation of the monitor. This is a very convenient feature during operation of the monitor. Pressing the horizontal phase (right) button causes the 2400 Hz input to be interrupted for a short pulse (determined by U21) each sweep cycle. This results in fewer than 2400 cycles being counted and allows the picture to slew to the right on the screen. Pressing the left phase button allows 2400 Hz to be applied briefly to U18 at pin one instead of 1200 Hz (from pin 12). This results in more than 2400 Hz being counted and allows the picture to slew to the left. In an effort to reduce the amount of digital noise around the monitor, TTL gates (U24-27) were used so that panel switching simply involves grounding one pin of the appropriate gate. Direct switching of the TTL lines can be employed but will result in more "noise" in the linear video circuits described above because of the longer lead lengths required. The various blanking inputs are combined and switched at Q4. The horizontal reset pulse from U21A is routed to the sweep circuits described below.

The video sweep circuits for the CRT are shown in Fig 29. An op amp connected as an integrator is used to generate the voltage ramp to drive the sweep circuit (U28A, 29A). The vertical sweep is reset manually

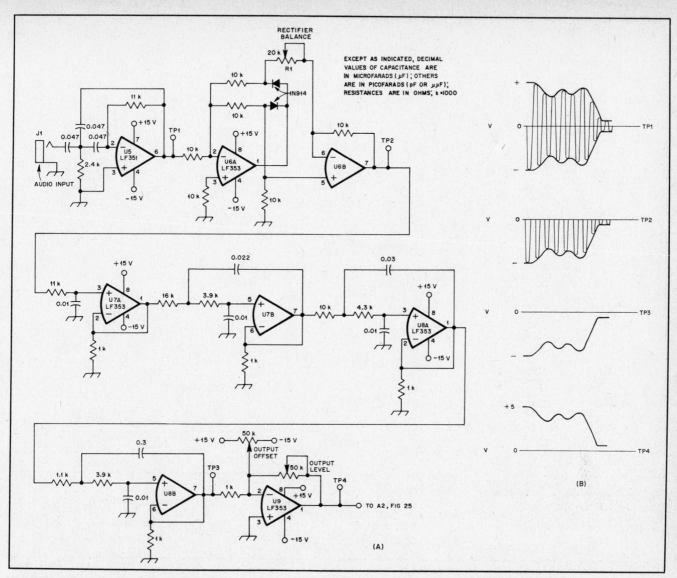

Fig 24—Schematic diagram of the analog processing circuit (A), and waveforms that should be expected at test points (B). Resistors are ¼ W types unless noted. Capacitors are disc-ceramic types.

at the end of each frame. The horizontal sweep is reset from the countdown circuit described above (U21A). A second op amp (U28B, 29B) allows adjustment of centering and size of the trace to fill the CRT screen. Power transistors are required to drive the yoke, which was originally designed for use with a computer terminal CRT.

Construction and Alignment

The circuits were built in four separate modules on separate chassis. These in-cluded the power supply, the CRT chassis (including the CRT high-voltage supply and high-voltage amplifier), the linear/digital chassis, and the local oscillator chassis. The linear and digital circuits were built on four separate plug-in experimenter boards. Four individual 4.5- × 4.5-inch boards were used for the PLL circuits, the analog processing circuits, the digital/countdown circuits, and the sweep circuits. These four boards were mounted in a "bus" con-figuration on the "analog/digital" chassis with appropriate controls mounted on the front panel. To avoid congestion, especially for the digital circuits, 4.5- × 6-inch boards could be used instead of the 4.5-inch-square size. See Figs 30, 31 and 32.

The local oscillator should be built in a separate chassis with RF shielding and, preferably, its own enclosed power supply. You can leave the oscillator running on a continuous basis. At room temperature, the frequency stability is satisfactory. Accurate calibration of the 2400 Hz local oscillator

Fig 25—Schematic diagram of the logarithmic video amplifier circuit. All resistors are ¼ W types. Z1 is an Intech A-730 hybrid circuit module.

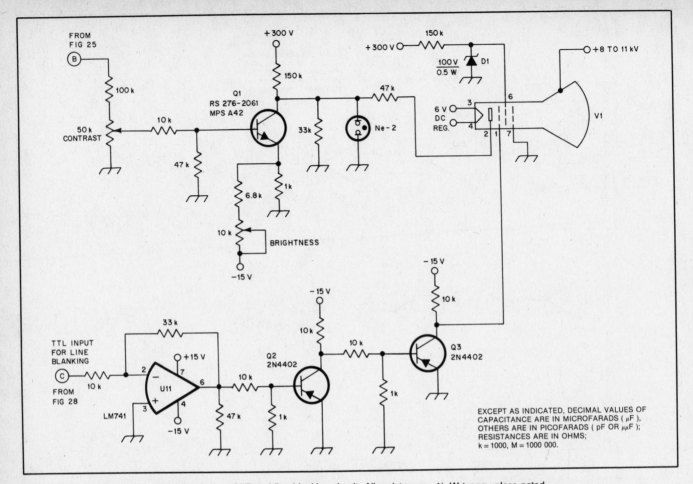

Fig 26—Schematic diagram of the video driver, CRT and line blanking circuit. All resistors are ¼ W types unless noted.

D1—100 V, 500 mW Zener diode, 1N5271 or equiv.
Q1—NPN high-voltage, small-signal tran-

sistor, RS 276-2061 or equiv. See text.
Q2, Q3—Small-signal PNP transistor, 2N4402, or equiv.

U11—741-type or LF351 op amp.
V1—9- to 12-inch CRT, P31 or P39 phosphor. See text.

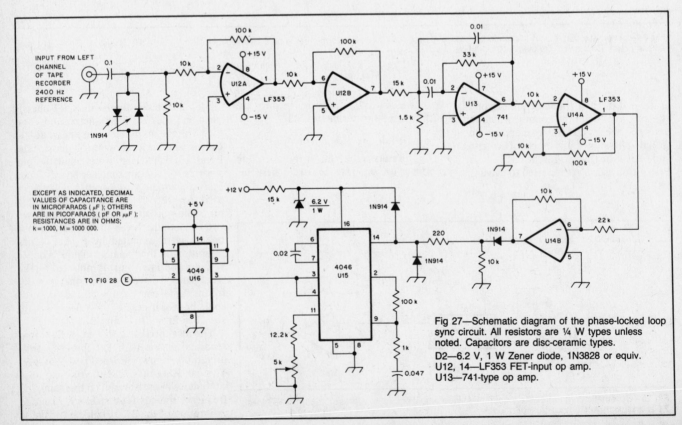

Fig 27—Schematic diagram of the phase-locked loop sync circuit. All resistors are ¼ W types unless noted. Capacitors are disc-ceramic types.

D2—6.2 V, 1 W Zener diode, 1N3828 or equiv.
U12, 14—LF353 FET-input op amp.
U13—741-type op amp.

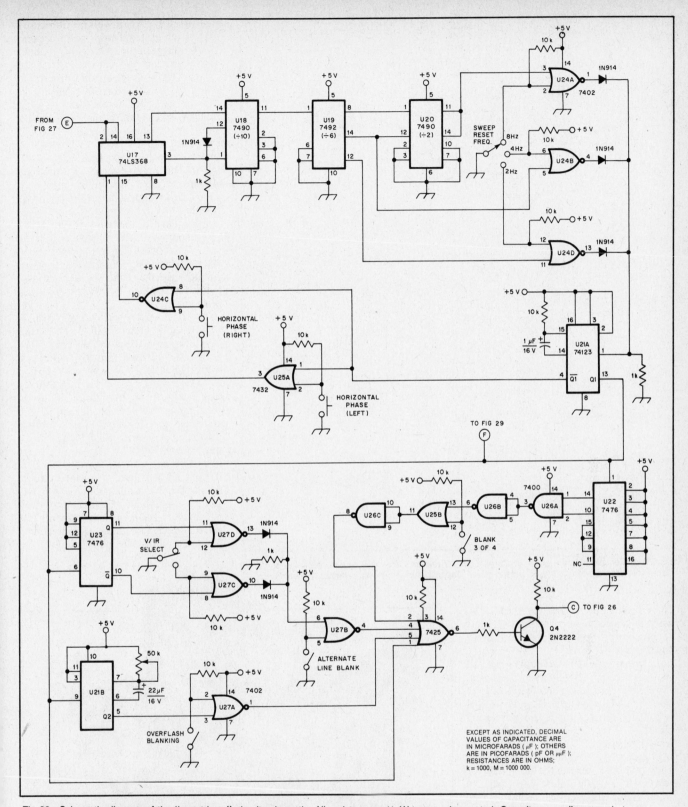

Fig 28—Schematic diagram of the "countdown" circuit schematic. All resistors are ¼ W types unless noted. Capacitors are disc-ceramic types unless noted. Capacitors marked with polarity are electrolytic.

is important so the images will have straight edges. The initial adjustment of the oscillator can be checked with a frequency counter. If a highly accurate frequency counter is available, this may be adequate. Depending on the exact crystal used, the values of C1 and C2 may have to be changed slightly to achieve centering at exactly 2.4 MHz. An inexpensive way to achieve accurate adjustment is to calibrate the local oscillator against the satellite generated 2400 Hz subcarrier. For this an oscilloscope with external trigger capability is needed. First the vertical input is connected to the receiver audio connection and the sweep rate of the oscilloscope is adjusted so the 2400 Hz sine wave can be seen while the satellite passes overhead. Next the external trigger input of the oscilloscope is connected to the output of the local oscillator. By adjusting C3 while one of the

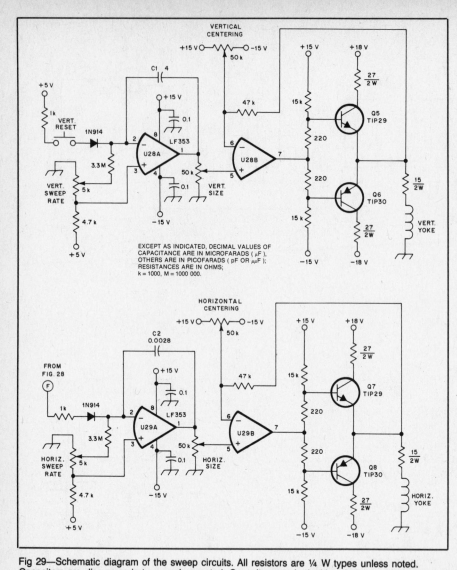

Fig 29—Schematic diagram of the sweep circuits. All resistors are ¼ W types unless noted. Capacitors are disc-ceramic types unless noted. Capacitors marked with polarity are electrolytic.

EXCEPT AS INDICATED, DECIMAL VALUES OF CAPACITANCE ARE IN MICROFARADS (μF), OTHERS ARE IN PICOFARADS (pF OR μμF); RESISTANCES ARE IN OHMS; k = 1000, M = 1000 000.

small-signal NPN high-voltage transistors might be substituted. During normal operation the voltage at the emitter of Q1 should be between about − 1 and − 2 V. With the Sylvania CRT used in the prototype, + 54 V at the collector of Q1 results in black on the CRT screen, and + 40 V results in white. Initial adjustments should be made on the high side so the trace will be too dark rather than too white (which can burn the phosphor coating on the CRT screen). The brightness and contrast controls should be panel mounted since they will require frequent adjustment during operation of the display unit.

In the phase-locked loop circuit (Fig 27) the 5 kΩ potentiometer should be adjusted so the phase-locked loop runs free at about 2400 Hz. There is no need to offset the free-running frequency, and the only adjustment is to approach 2400 Hz as closely as possible.

The CRT used here was a Sylvania 9ST4716AP39. Other small (9 or 12 inch) green phosphor (P31 or P39 phosphor) CRTs could be used for this project. A 9-inch tube is easier to work with from a mechanical standpoint and requires somewhat less power-supply current. Select a tube using electromagnetic deflection and electrostatic focusing, and be sure to obtain a matching yoke with the CRT. A high-impedance yoke (50-75 ohms) is easier to use in this application if available. If a different CRT is used a manufacturer's data sheet should be obtained to be sure the "pin-outs" are the same and the operating voltages described here are appropriate for the CRT selected. With the widespread use of computer terminals, new surplus CRTs

NOAA satellites passes overhead, the satellite-generated sine wave can be made to march off the oscilloscope screen either toward the right or left. By carefully adjusting the capacitor, the sine wave will remain very nearly stationary on the screen, indicating that the local oscillator and the satellite signal are at the same frequency.

Construction of the audio processing circuitry is straightforward. An audio signal generator is useful to check the operation of this section (although playing back a tape recording of the local oscillator works well as a substitute). With a 2400 Hz signal input, the rectifier balance can be adjusted for symmetrical detection at the output of U6B. Output offset should be adjusted so that low-amplitude input (black) gives an output of about zero volts. Maximum signal level (white) should be around + 5 V initially, but may need to be adjusted depending on the other circuit adjustments.

The video amplifier (Fig 26) uses a Radio Shack 276-2061 high-voltage transistor. If this transistor cannot be located the commercial equivalent is the SE7056. Other

Fig 30—The completed CRT and high-voltage supply are mounted on a small homemade chassis.

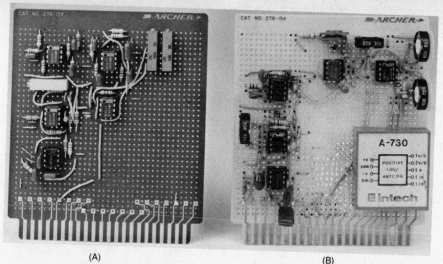

(A)   (B)

Fig 31—The linear video processing circuit of Fig 24 is shown at A, while the board at B also contains the logarithmic video amplifier of Fig 25.

are often available at low cost at hamfests and computer gatherings. The experimenter is encouraged to purchase two tubes initially if an inexpensive source is located. During the testing stages some phosphor burns are almost inevitable. Later when the unit is working properly, the "fresh" CRT can be used to provide images free of phosphor defects.

In the sweep circuits the timing capacitors used to generate the ramps (C1 and C2 of Fig 29) should be the highest quality Mylar or polyester type that can be obtained. Values slightly different from those listed could be used since the sweep rate can be adjusted by changing the setting of the sweep rate controls. These controls should be panel mounted. The drive transistors should be mounted using heat sinks. In a high-current application such as this, it may be wise to substitute a Darlington-type power transistor. One

possibility is to substitute a TIP 122 NPN Darlington transistor for the TIP 29 shown and to substitute a TIP 127 PNP Darlington transistor for the TIP 30 shown. These devices have approximately twice the current rating of the sweep transistors used here.

The yoke was modified by separating the windings at the yoke feed and connecting all the horizontal coils in series rather than parallel. The vertical windings were also reconnected in series. This raised the resistance of the yoke from about 1 Ω to 6-8 Ω. A 2 W resistor was connected in series with the yoke to provide some current limiting. These modifications are not needed if a yoke in the 50-75 Ω impedance range is available. Raising the resistance of the yoke reduces the power requirement in the drive circuit.

The power supply for this project is conventional, with regulated voltages

supplied throughout. See Figs 33 to 39. Simple 1 A regulators (with heat sinks) are suitable for the 5 V, 6 V (filament supply), +15 V and −15 V supplies. The 6 V regulator operates near its maximum rating and should have a generous heat sink or be chassis mounted. Because of the large power requirements in the sweep circuits, separate regulated 18 V, 2.25 A power supplies are used. Although multiple-secondary transformers could be used, separate transformers for each supply are an effective way to keep noise and "ripple" generated in the digital circuits from appearing in the sensitive linear video processing and sweep circuits.

The CRT high-voltage supply was built using an NE 555 (U33) oscillator to drive a power transistor (see Fig 36) which drives a television-type flyback transformer. The NE 555 oscillator is adjusted to run at about 15 kHz. Be sure to use a high grade polyester capacitor for C1. The value of C2 may need to be adjusted depending on the flyback transformer used. Values from 0.1 to 0.47 μf will usually be appropriate. It is important that this supply be built within a separate shielded chassis to avoid electrical shock. The 2N3055 power transistor should be mounted using a heat sink. Depending on the construction used, the wall of the shielding chassis may serve as a mounting point and heat sink. The 500 MΩ, 20 kV resistor and 500 pF, 20 kV capacitor provide some degree of regulation of the high voltage. By inserting a 470 kΩ, 1/2 W carbon resistor at the ground end of the 500 MΩ resistor, a test point is created. About 20 V here corresponds to an output voltage of about 10 kV. If a separate HV module can be obtained, it may be preferable to use a commercially available 8-11 kV supply. A Venus brand LU 15 high-voltage module, located at a hamfest, has been substituted for the unit shown here.[3]

[3]Venus Scientific Inc., 399 Smith St, Farmingdale, New York 11735.

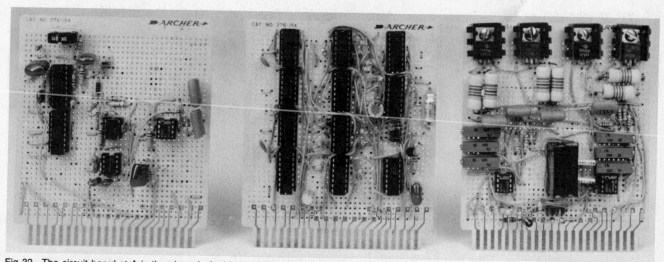

Fig 32—The circuit board at A is the phase-locked loop of Fig 27. At B is the countdown circuitry of Fig 28, and the sweep circuits of Fig 29 are on the circuit board shown at C.

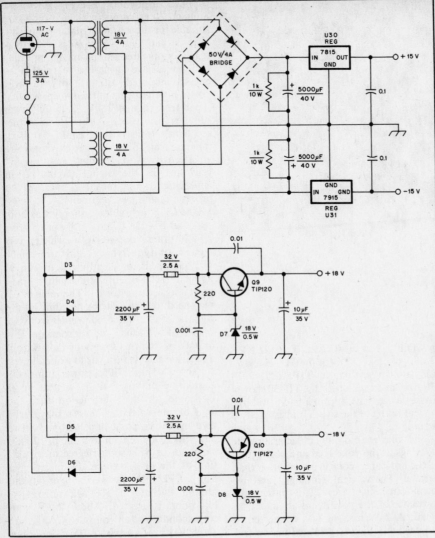

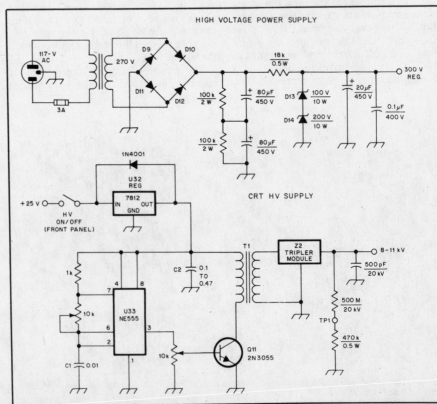

▲Fig 33—Schematic diagram of the low-voltage power supply circuitry. All resistors are ¼ W types unless noted. Capacitors are disc-ceramic types unless noted. Capacitors marked with polarity are electrolytic.

D3-D6—100 PIV, 3 A silicon rectifier, 1N5401 or equiv.

D7, D8—18 V, 500 mW Zener diodes, 1N4746 or equiv.

Q9—NPN power transistor, TIP 120 or equiv.

Q10—PNP power transistor, TIP 127 or equiv.

▼Fig 34—Schematic diagram of the high-voltage power supplies. All resistors are ¼ W types unless noted. Capacitors are disc-ceramic types unless noted. Capacitors marked with polarity are electrolytic.

D9-D12—1000 PIV, 1 A silicon rectifier, 1N4007 or equiv.

D13—100 V, 10 W Zener diode, 1N3005 or equiv.

D14—200 V, 10 W Zener diode, 1N3015 or equiv.

T1—Flyback transformer, see text.

Z2—Tripler module, Heath 57-83, GE 525, RCA SK3301 or equiv. See text.

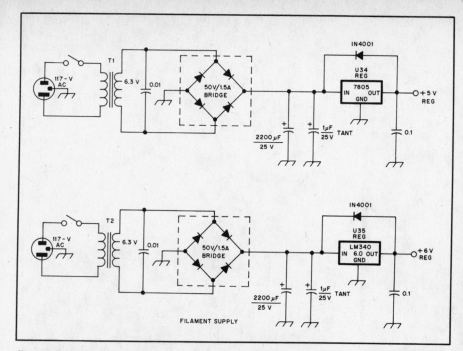

Fig 35—Schematic diagram of the +5- and +6 V power supplies. All resistors are ¼ W types unless noted. Capacitors are disc-ceramic types unless noted. Capacitors marked with polarity are electrolytic.

T1, T2—117 V primary, 6.3 V, 2 A secondary.

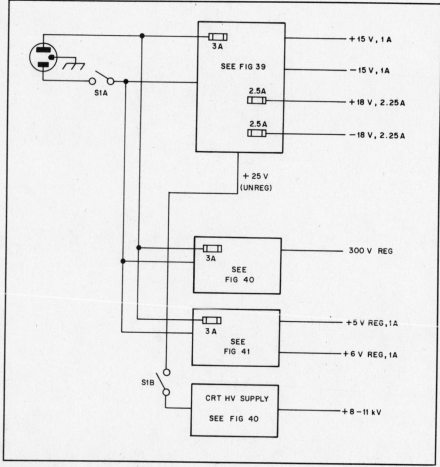

Fig 36—Power supply block diagram.

Another alternative is to obtain a CRT complete with HV supply and use the original HV supply circuitry. A separate front panel switch was included to shut off the high voltage supply independently.

The 300 V supply was built using high-voltage capacitors from surplus flash equipment. Zener diodes can be combined back to back to obtain the required values.

The importance of regulating these power supplies cannot be overemphasized, if annoying ripple is to be avoided. It is also important to build the power supply and sweep circuitry on a chassis which can be moved an inch or two away from the CRT if needed to control residual ripple.

Operation

The image is obtained by darkening the room and photographing the CRT screen using long exposure times and a closeup lens. A "macro" lens is ideal, but less expensive screw-on lenses have been used at this location with satisfactory results. A 35mm SLR camera is mounted on a tripod in front of the monitor and the shutter is opened with a cable release, holding the shutter open until the entire image has been drawn on the screen. The shutter is then released and the film processed and printed. The results may be seen in Figs 37 and 38.

During photography sessions the contrast can be reduced somewhat from what appears ideal to the eye. The camera film will respond better, in most cases, to a CRT image that appears only moderately bright and almost too low in contrast to the eye. Care must be taken to focus on the CRT trace itself rather than the glass screen of the CRT or the film will show an image which is slightly out of focus. Experimentation is required to find the best camera settings. Good results have been obtained with Kodak Plus-X film (ASA 125) developed in Kodak D-76 and exposed at about f4. Remember that exposure times may change if a different developer is used.

While construction of this monitor is a major project, some experimenters will find it easier to manage than the mechanical drum type construction. The major disadvantage of the CRT method is the inconvenience of processing film before viewing the image. In addition, because of the geometry of the CRT, there will almost always be some slight "bowing" of the image margins. Advantages of the CRT method include simple adjustment of the aspect ratio, adaptability to new sweep rates, and the option of rapid sweeping of the CRT to "blow up" a small part of the image for an enlarged view. If immediate results are needed, Polaroid films can be used. If 35mm film is used, the final image can be printed many times from a single favorite negative. With careful voltage regulation to eliminate ac "ripple" in the images this unit can serve as a versatile, reliable, and inexpensive imaging instrument for the amateur APT station.

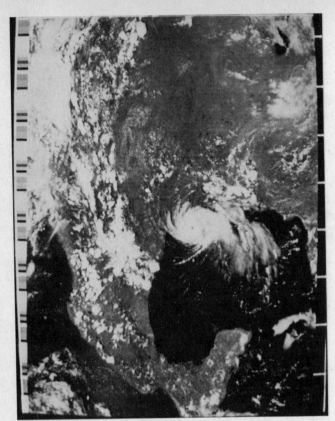

Fig 37—This photograph, made from the dedicated CRT screen, shows Hurricane Alicia on Aug. 17, 1983. This a full frame from NOAA 7.

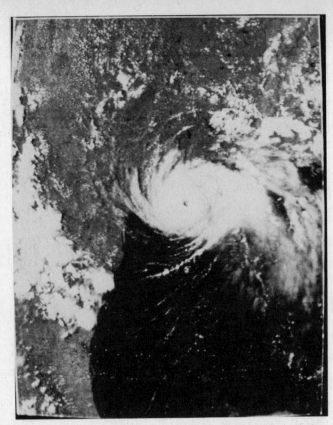

Fig 38—Using the dedicated CRT in the expanded mode (by sweeping the beam faster), it is possible to "blow up" an image. This is part of the image shown in Fig 43, but much more detail is available.

Chapter 29

Digital Equipment

Digital techniques are finding more and more application in Amateur Radio with each passing year. In addition to enhancing projects for the "standard" modes, such as CW and Baudot RTTY, digital technology is leading amateurs into new areas such as AMTOR and packet radio. As time passes, amateurs are thinking of new ways to interface their radio equipment with computers for even more enjoyment, creating the need for modems specifically designed for radio applications. Chapters 8 and 19 contain related information that may be helpful in understanding the projects in this chapter.

A Simple CMOS Iambic Keyer

The keyer shown in Figs 1 through 3 features iambic operation, dot and dash memories and internal sidetone generation. The keyer was designed and built by Paul Newland, AD7I, and originally appeared in June 1988 *QST*.

Circuit Details

Fig 2 shows the schematic of the CMOS keyer. The clock circuit was designed by Roy Lewallen, W7EL, and was used in his keyer project that first appeared in an issue of *SPRAT*—a QRP journal—several years ago. This clock has the advantage of good asynchronous starting without elongating the first clock period. Exclusive-OR gates U1B and U1C are used for the dot and dash memories. A Schmitt-trigger gate (U2D) configured as an AF oscillator serves as a handy, low-cost, minimum parts-count sidetone generator.

The keyer circuit is composed of five major sub-circuits. They are:

• Dot timer
• Dash timer
• Dot memory
• Dash memory
• Asynchronous clock

First, let's focus on the dot and dash timers. The heart of each timer is a CMOS 4017 decade counter. This chip's count advances on the rising edge of the clock input when both the reset and inhibit inputs are low. If the inhibit line is high, clock signals are ignored. When the reset line is high, the counter immediately goes to the zero state. The outputs of this counter are decoded; when the counter output is in the zero state, the zero output is high and all other outputs are low. When the counter is in the two state, the two output is high and all other outputs are low, and so on.

The dot and dash timers operate in a similar way. For clarity, we'll cover just the dash timer. When no dashes are desired,

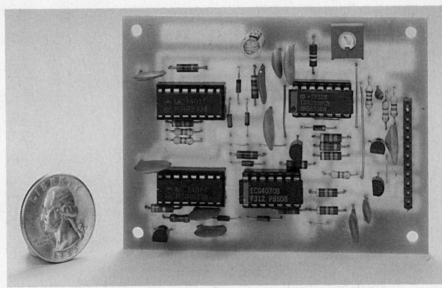

Fig 1—This simple CMOS keyer uses four common ICs, seven diodes, four transistors, 21 resistors, 13 capacitors and two potentiometers.

the timer is in the four state and is held there because the inhibit input is tied to the four output. If both the dot and dash timers are inhibited, both inputs of U2A are high and the clock is disabled. When in the idle state, the outputs of both the dot memory (U1B) and dash memory (U1C) are high. Note that when the dot memory is idle, D5 is reverse biased, and when the dash timer is idle, D6 is reverse biased.

The only thing keeping the dash timer from being reset is that D4 is forward biased, holding the dash timer's reset input low (inactive). When the dash paddle is closed, the output of the dash memory formed by U1C, R6 and R7 goes low (C4 bypasses stray RF). This signal is inverted by U1D, and D4 is then reverse biased. Because all diodes at the reset input of the

dash timer are reverse biased, the reset signal goes high, forcing the counter to go from four to zero.

When the dash timer is in the zero state, the keying transistor, Q1, is biased on by the current flowing through R9. Because of the low base-emitter voltage drop of Q1, most of R9's current goes into the base-emitter junction of Q1. When the state-four output of the dash timer goes low (when the timer leaves the four state), the clock signal from U2B goes from high to low and quickly back to high. (The speed of this process is determined by how fast C7 discharges through D7.) The time constant provided by R8 and C6 ensures that the dash timer remains reset long enough to ignore the first rising edge of the clock. (If this time constant is too short, the first dash

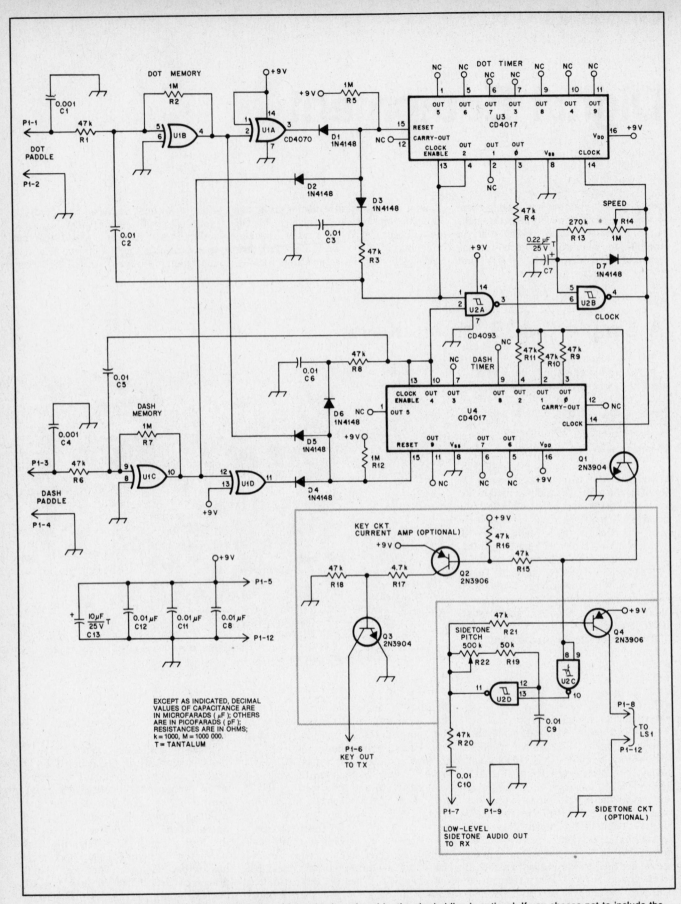

Fig 2—Schematic of the simple CMOS keyer. The portion of the circuit enclosed by the shaded line is optional. If you choose not to include the sidetone circuit, be sure to ground the unused inputs of U2 (pins 8, 9, 12 and 13).

C7—0.22 μF tantalum or ceramic.
J1—¼-in. three-conductor jack.
LS1—8-Ω or 100-Ω speaker.

Q1, Q3—2N3904, 2N2222 or equiv.
Q2, Q4—2N3906, 2N2907 or equiv.
R14—1 MΩ potentiometer.

U1—CD4070.
U2—CD4093.
U3, U4—CD4017.

would only be two Morse elements in length instead of the requisite three.) On the second rising edge of the clock, the dash timer moves from state zero to state one. Now the bias current for Q1 is provided by R10 instead of R9. Similarly, R11 provides bias current for state two. State three provides the space for the inter-element timing period.

The next rising edge of the clock moves the dash timer into the idle state (state four) and disables the clock. A reset pulse from the dash timer's state-four output also tries to reset the dash memory (U1C) via C5. If the paddle is open, this pulse clears the memory to a high state. If the paddle is still closed, the pulse momentarily clears the dash memory, but because of the remaining key closure, the memory is set again as soon as the charge on C5 is dissipated through R6.

If the dot paddle is closed while a dash is being sent, the dot memory is set before the dash timer reaches state four. During the period when C5 is trying to reset the dash memory, the dot timer is reset (D4, D5 and D6 all reverse biased). This makes the keyer send a dot followed by a space immediately after the dash/space combination is finished. Voila—iambic operation! During the time when the dot timer is active, the dash timer is held idle by forward-biased D5, until the dot-timer cycle finishes.

Construction

The prototype keyer was built on a piece of perf board. A PC board has since been developed; Fig 3 shows the parts-placement diagram. The PC board and a kit of parts are available from A&A Engineering[1]. Nothing about the layout is particularly critical, so the layout shown in Fig 3 can be modified to suit your needs. Just about any construction method is fine; dead bug, wire wrap, point-to-point and etched PC-board wiring are all fine. One caution you need to observe is the sensitivity of CMOS ICs to static electricity. Always use a grounding wrist strap or its equivalent to protect the ICs during handling. Also, if you elect not to use the PC-board layout shown in Fig 3, be sure to place the bypass capacitors as close as possible to their respective IC pins. The signal levels in the keyer are very small, and poor RF bypassing can make the keyer do some mighty strange things.

A word of caution about C7, the clock capacitor: the keyer uses CMOS ICs to provide keying at minimal current drain.

[1]PC boards and kits of parts are available from A & A Engineering, 2521 W La Palma Ave, Unit K, Anaheim, CA 92801, tel 714-952-2114.

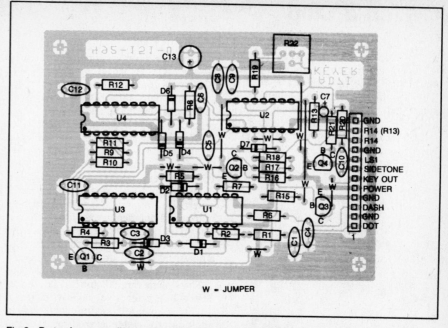

Fig 3—Parts-placement diagram for the keyer. Parts are mounted on the non-foil side of the board; the shaded area represents an X-ray view of the copper pattern. A full-size etching pattern appears at the back of the book.

Because of this, no on/off switch is included. Be sure to use a low-leakage capacitor for C7 (either ceramic or tantalum). Don't use a garden-variety electrolytic here if the keyer will be battery powered; these are often very leaky. The keyer will work fine if you use a leaky cap for C7, but battery life will be markedly reduced.

Optional Circuitry

The optional circuitry enclosed by the shaded line in Fig 2, including the keying amplifier and the sidetone circuit, is not essential. CMOS gates were used because of their flexibility, as well as the low current drain already mentioned. These devices can be powered from anything between 3 and 15 V (74HC-series parts could also be used, but over a norrower supply-voltage range). The problem is that these devices can only supply a few hundred microamps of current from their outputs for a guaranteed logic level. If these gates are used to drive a transistor with a gain of 50 or so, only a small amount of keying current will be available to drive a transmitter. Often this is too little current for anything but QRP rigs. Q2 and Q3 (with associated resistors) form the keying amplifier, and increase available current to levels suitable for keying almost any transmitter. The value of drive current can be controlled by changing the values of R17 and R18. In general, decreasing the value of R17 will increase the available keying current.

The second optional item is the sidetone generator formed by a Schmitt-trigger gate (U2D) with an RC circuit for feedback. The sidetone frequency can be changed by varying R22. As the resistance of R22 is increased the frequency of the sidetone decreases. Low-level audio output is available to drive an audio amplifier (such as the one in a transceiver), and the high-level output is suitable for driving a small speaker. Try to avoid using the high-level output to drive a speaker if you are running the keyer from a 9-V battery. The sidetone circuit itself doesn't draw much current, but a speaker sure does! Whenever possible, let the rig provide the sidetone, or you'll be feeding the keyer an awful lot of batteries.

Operation

The keyer operates just like any other electronic keyer. The circuit has one anomaly that you should be aware of, however. When power is first applied (when you connect a battery, or turn on your rig if the keyer is built into it), a reset pulse is not always sent to the dot or dash counter chips. The lack of a reset signal can cause more than one of the counters to be in active states at the same time, and the keyer will stick in the key-down mode. All it takes to "un-stick" the keyer is to squeeze the paddle after power is applied. This anomaly generally isn't a problem because battery life is long, and the power is only cycled when changing cells. If you build the keyer into a rig, however, this key-down bug can be more objectionable. In this case, try a different set of 4017 counters at U3 and U4.

CW on a Chip

Fig 4 shows a single-chip electronic keyer that can easily be built in a weekend, thanks to the Curtis keyer-on-a-chip. This IC has made it possible to construct a quality, flexible keying system with a minimum of additional discrete components. This project first appeared in an article by Paul Pagel, N1FB, and Bob Shriner, WA0UZO, in December 1983 *QST*.

Fig 4—View of the single-IC iambic keyer. The unit shown here was built around the Curtis 8044M integrated circuit, and features an analog speed readout.

The Curtis Chip

There are four versions of the Curtis 8044 CMOS IC. Two of these (8044/8044B) are contained in a 16-pin package and the others (8044M/8044BM) in an 18-pin package.[1] The additional pins are connected to internal circuitry that provides a keyer sending speed monitoring function by means of a meter and a few other external components. The whole family of ICs features contact debouncing, RF immunity and self-completing character generation. A weight control, sidetone output and dot memory are also included. The memory function helps to prevent dot loss if the operator "leads" the keyer. With a quiescent current drain of about 50 μA, an on/off switch is not really required.

The "plain vanilla" (no suffix) and B-suffix ICs offer two slightly different iambic (squeeze) keying methods in addition to single-lever (non-squeeze) keying.[2,3] With the no-suffix IC, a dot or dash being sent when the paddles are released is completed and nothing else is sent. The B-suffix IC completes the dot or

dash being sent upon paddle release, and then sends an opposite element; that is, a dot after a dash or a dash after a dot. Many squeeze-key operators prefer the latter method of iambic operation. If only single-lever paddle operation is desired, either IC should suffice.

Board Design

To make the keyer as universal in application as possible, the board is patterned so that any of the ICs mentioned earlier can be used with or without some of their inherent capabilities.[4,5] The board is single sided, and is small enough to fit inside almost any transmitter or transceiver.

Any or all of the variable controls may be mounted on the board or brought out for external adjustment. Two on-board output keying options are provided: reed relay output (with or without arc suppression components across the contacts) or a transistor-keyed output that can be con-

[1]Curtis Electro Devices, Inc. generously supplied the ICs used in this project.
[2]L. Fay, "The Iambic Gambit," July 1981 *QST*.
[3]Curtis Electro Devices Lil' Bugger, Product Review, March 1981, *QST*.
[4]An etched circuit board and a complete kit of parts is available from Circuit Board Specialists, P.O. Box 969, Pueblo, CO 81002.
[5]The keyer ICs are available from Curtis Electro Devices, Inc., Box 4090, Mountain View, CA 94040. Be sure to specify IC type when ordering.

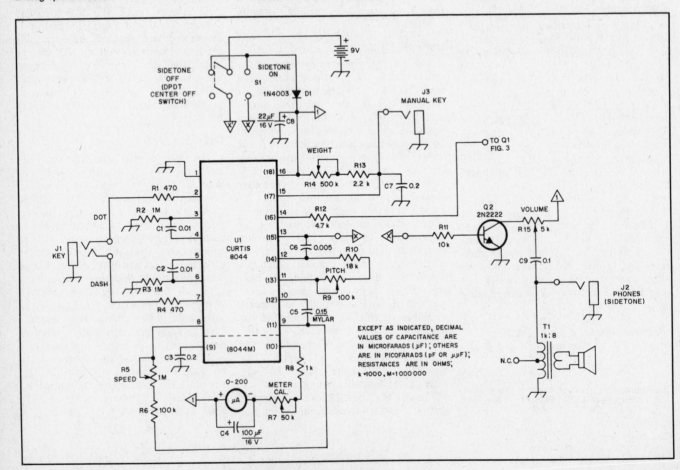

Fig 5—Schematic diagram of the keyer. All resistors shown are ¼-W, 5% types.

figured to fit station keying requirements.

Assembly

Construction details are given in Figs 5 and 6. A parts-placement guide appears in Fig 7.[6] All components except the IC should be soldered to the PC board. The integrated circuit should be installed last. A socket is recommended to avoid over-heating the chip. Use an 18-pin socket, so either the 16- or 18-pin version may be used. If the 16-pin chip is used, the two extra positions on the socket are left empty; just be sure to install the IC correctly.

The keyer should be built to suit the user's requirements. Any components associated with unwanted features may be eliminated. These may include: sidetone output, level and pitch control, and the speed-meter function. If transistor-output keying is desired, the relay arc-suppression network may be deleted. A 5.6-kΩ resistor between pins 15 and 16 of U1 will eliminate the WEIGHT control, if a variable unit is not needed. If the weighting appears to be too heavy, reduce the value of C6 or eliminate it altogether. The MANUAL KEY input can be used as a TUNE function; an SPST switch that brings the line to ground will create a key-down condition.

If an IC with the speed-meter function is chosen, any meter with a full-scale deflection range of 50 to 500 μA (with a linear scale) can be used. In the unit shown, a modified VU meter was used. A new meter scale was made with two-words-per minute increments between 0 and 100. With the 100-kΩ resistor shown in series with the SPEED potentiometer in Fig 2, the maximum speed of the keyer is about 50 WPM. Alter the value of the fixed resistor to modify the speed range. The top-end speed for this unit is about 80 WPM.

A switch is included to turn the keyer and/or the transistor audio amplifier (Q2) on and off. To increase battery life, the audio amplifier can be left off. The side-tone oscillator will probably not be required, since most modern transmitters and transceivers have built-in sidetone monitoring circuits. The keyer monitor does serve as a good indicator of battery condition: as the battery becomes depleted, the note will become quite chirpy.

Relay-Contact Arc Suppression

Some transmitter keying lines may require the inclusion of an arc-suppression network across the keying relay contacts. Most modern transmitters and transceivers should not need this network (C10, R19), as they are usually operated at low voltage levels. But keying some transmitters and transceivers using tubes in the final amplifier may require the relay contact

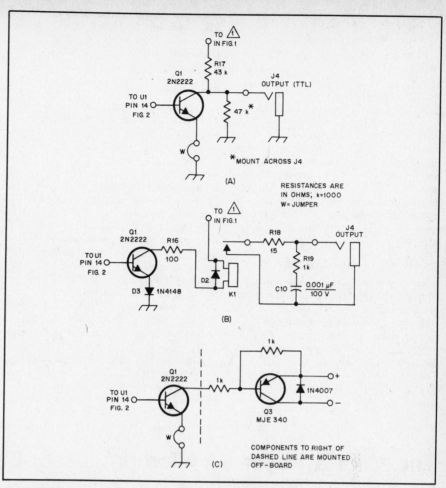

Fig 6—Some possible output circuit arrangements. The jumper (W) shown at (A) and (C) is inserted in place of D3. D3 is included to ensure rapid energization of K1; it may not be needed. Q3 and the components to the right of the dashed line in (C) are not mounted on the keyer circuit board.

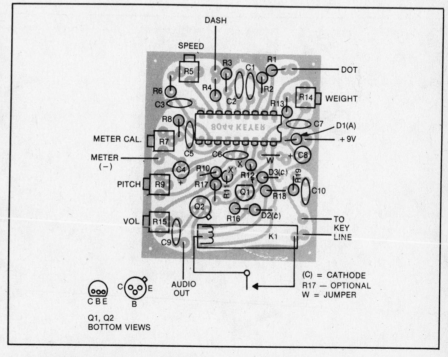

Fig 7—Parts-placement guide for the Curtis-IC keyer. Parts are placed on the nonfoil side of the board; the shaded area represents an X-ray view of the copper pattern. A full-size etching pattern appears at the back of this book.

[6]Templates for the sloping-panel keyer cabinet are availabe from ARRL HQ for a business-size envelope and $2 to cover template and postage costs.

Table 1

Table 1
Series Resistor Values

Voltage	Value (ohms)
70	100
100	200
150	450
200	800

protection. The prototype keyer has been successfully used with a Kenwood TS-820S transceiver without the arc-suppression network; no problems resulted. The key line voltage in that rig is −65-V dc. The information presented in Table 1 should assist prospective users to determine the required network values.

Calibrating the Speed Control

Sending speed can be determined by counting the number of dashes (dahs) sent in a five-second period. This number is the code speed in words per minute. A more refined measurement method uses an oscilloscope or a frequency counter connected to pin 12 of U1.
The formula is

$$speed = \frac{f}{1.2}$$

where speed is in WPM and f is the dot frequency measured at pin 12 of U1. Thus, for speeds of 10, 25 and 50 WPM, the frequency counter should display dot frequencies of 8.33, 20.8 and 41.6 Hz, respectively.

Keyer Case Construction

Two enclosure styles have been tried. The first—the rectangular configuration—is easy to assemble. The control labels may be difficult to read at some angles, however. The sloping-panel version is more attractive, but a bit more difficult to put together. Either case style is available from Circuit Board Specialists; case style desired should be mentioned when ordering.

Double-sided PC board material is used for the enclosure. Once the box parts have been cut to size, they should be burnished with fine steel wool. The edges must be beveled for a good fit. Use caution to avoid removing too much material from the parts.

The speaker should be laid on a flat surface and the grill pressed into position. Quick-drying epoxy cement secures the grill.

After the box parts are prepared, they should be tack soldered together and checked for alignment and correct fit. If all is well, lay a bead of solder around each seam. A 25- to 45-W soldering iron should be sufficient.

A light coat of clear polyurethane varnish over most of the box will help maintain a pleasant sheen. The prototype cabinet was accented with light-blue epoxy spray paint.

Install the panel mounted controls, jacks, meter and speaker. The keyer board should be mounted with the meter-calibration potentiometer on the bottom.

Appendix

Most modern Amateur Radio transmitters, particularly solid-state designs, do not require additional relay-contact protection. When using this keyer with such a low-voltage transmitter, R18 can be removed or jumpered.

Some grid-block-keyed transmitters will require additional relay-contact protection in the form of an added series-connected resistor if the key-line voltage exceeds 70 and a large-value bypass capacitor is tied between the key line and ground. The appropriate resistor value can be determined from Table 1. Resistor power rating can be calculated by multiplying the resistor value by the square of the key-down circuit current in amperes. This resistor can often be placed within the body of the key-line plug. A silicon diode should be connected across the contacts of K1 (in place of C10, R19) to absorb the "kick" when keying inductive loads (such as another relay).

The Mike & Keyer—A Microprocessor-Based Memory Keyer

Microprocessors are everywhere in electronics today. Most modern commercial radio equipment contains at least one microprocessor, so it is only natural to find home-built projects following this electronic evolution. Figs 8 through 11 show the Mike and Keyer, an electronic memory keyer designed around the Zilog Z80® microprocessor. The keyer was designed and built by Michael Dinkelman, WA7UVJ. In addition to eight 2-kbit memories it includes such standard features as iambic operation, dot/dash memories and paddle-controlled message abort. In addition, optional enhancements include a

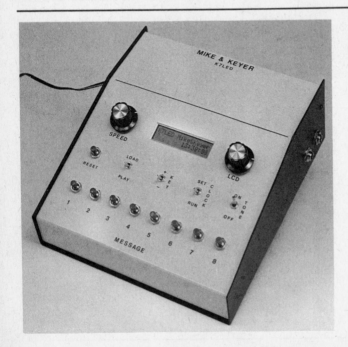

(A)

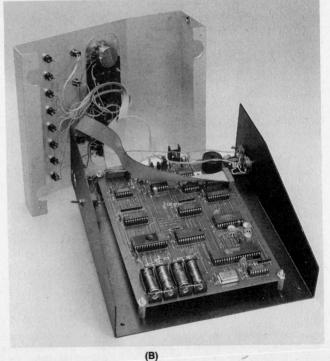

(B)

real-time clock, liquid-crystal display and loop options allowing several message positions to perform repetitive tasks automatically. Those dismayed by the finicky storage operations of some memory keyers will find the Mike and Keyer a breeze to load. All these features are available in a package requiring only 10 integrated circuits. The key to all this versatility is that the Mike and Keyer is, in reality, a small computer. Source code for the operating software is available, and modification of the software should be easy for anyone with a knowledge of Z80 assembly language and access to an EPROM programmer.

The Mike and Keyer can be built in two versions. The simplest is a basic memory keyer. This version contains the standard features listed above, with load/send message indicators displayed by discrete LEDs. The addition of one IC (an Intersil 7170 real-time clock), an LCD module and several discrete parts allows you to bring all the bells and whistles on board.

Why call it the "Mike and Keyer?" Just a small way of saying thanks to all the members of the Mike and Key ARC in Seattle who provided immeasurable support. Special thanks to Ray Leslie, WA7EKH, who designed the PC board.

Circuit Description

Fig 9 shows the schematic diagram of the Mike and Keyer. Most apparent are the VLSI (Very Large Scale Integration) chips that are the heart of the Mike and Keyer. The keyer is controlled by U2, the Z80 CPU. The CPU receives its instructions from the 2764 ROM, U4. The ROM contains approximately 3 kbytes of instructions and the LCD message formats.

The 6116L, U5, is the system RAM. A small portion of this chip is reserved for the Z80's internal needs, but the majority is message storage for the keyer. Message memory is organized into eight sections, each one bit wide. The 6116L can store 2 kbytes, so there are approximately

(C)

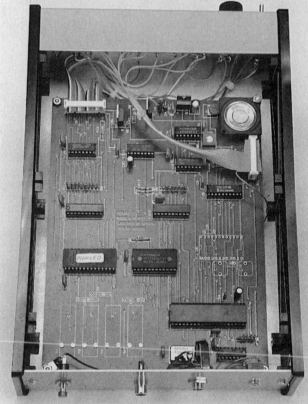

(D)

Fig 8—The Mike and Keyer, a Z80-based memory keyer, can be built in two versions. The "full dress" version shown at A and B, has an LCD and a real-time clock, along with options in the software that allow repetitive looping of memories. The LCD version is built in a Ten Tec SW-8 enclosure. The "no frills" version (shown at C and D) uses LEDs as indicators and has no real-time clock or looping options. The LED version is built in a Ten Tec FG-7 enclosure. Notice in the interior view at D that the backup batteries and real-time clock have been omitted. Batteries N3 and N4 are not required in the LED version, and N1 and N2 (backup power for the memory) are optional in either version.

(Next page) Fig 9—Schematic of the Mike and Keyer. The circuitry inside the dotted lines is used for the LED version, and the ICM7170 is not required. Unless otherwise noted, all resistors are ¼ W 5% units and all capacitors are disc ceramic. Printed-circuit boards are available from Leslie PCB[1].

D1-D3—Small-signal silicon diode, 1N914 or similar.
D4—1N4001 or 1N4002.
C1-C15, C21—0.01 µF.
C16—0.047 µF.
C17, C18—7-25 pF trimmer.
C19—4.7 µF, 16 V electrolytic.
C20—10 µF, 25 V electrolytic.
C22—100 µF, 25 V electrolytic.
N1-N4—1.5-V type N battery.
H1—16-pin header.
H2—20-pin header.
H3—3-pin header.
LCD1—Seiko M1632 16-character × 2-line LCD (see text).
Q1—TIP31.
Q2—TIP30 (choose Q1 and Q2 for your transmitter—see text).
Q3—MJE803 Darlington transistor.
R1—10 kΩ SIP (5 resistors, 6 pins).
R2—2.2 kΩ SIP (7 resistors, 8 pins).
R3—2.2 kΩ SIP (9 resistors, 10 pins).
R4—100 kΩ.
R5—22 kΩ.
R6, R7, R11, R12—10 kΩ.
R8, R13—2.2 kΩ.
R9—100 kΩ trimmer potentiometer.
R10—100 Ω trimmer potentiometer.
R14—1 kΩ.
R15—25 kΩ panel-mount speed control potentiometer.
S1-S9—SPST momentary switch.
S10—SPDT toggle switch.
S11—DPDT toggle switch.
S12, S13—SPST toggle switch.
U1—TTL-compatible oscillator module (see text).
U2—Z80 microprocessor.
U3, U7—74LS138.
U4—27C64 8k × 8-bit EPROM[1].
U5—6116L 2k × 8-bit RAM.
U6—Intersil ICM7170 real-time clock.
U8, U13—74LS74.
U9—74LS273.
U10, U12—74LS244.
U14—74HC132.
Y1—32.7680 kHz clock crystal.

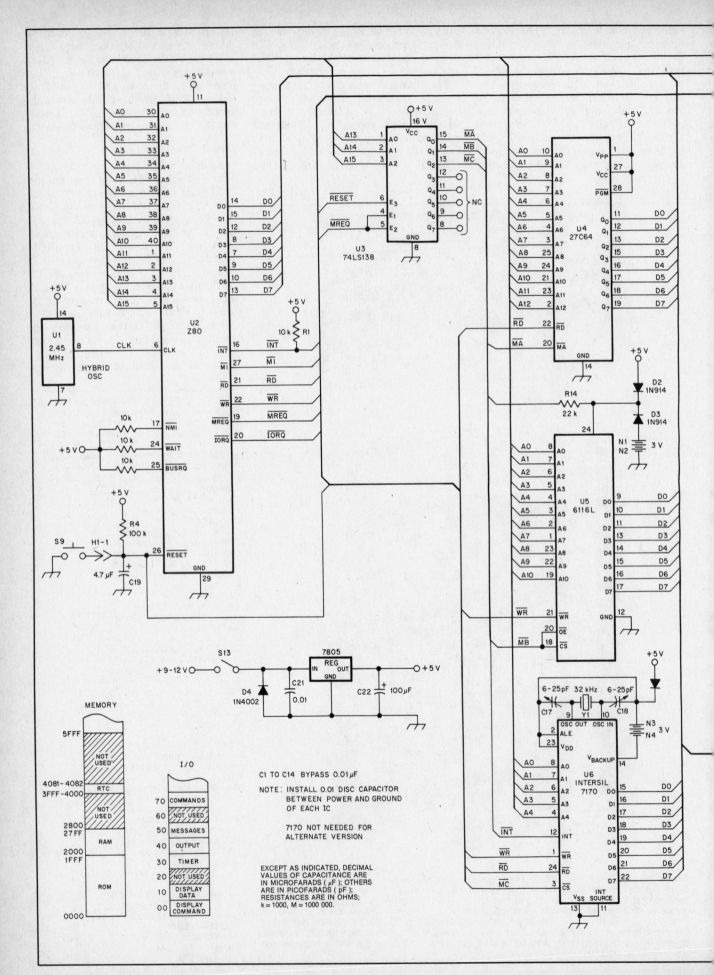

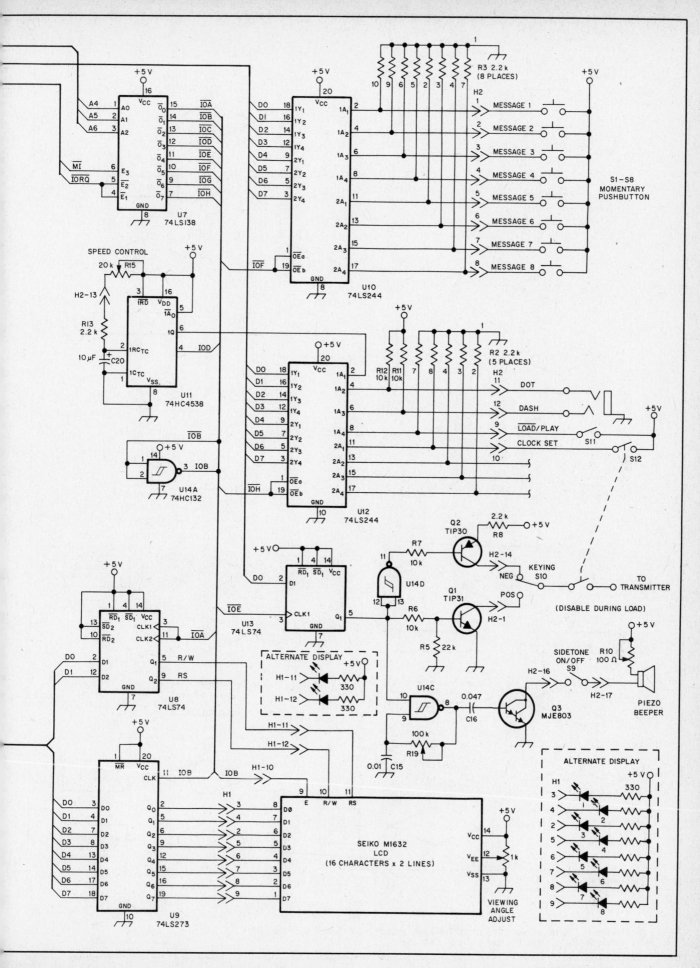

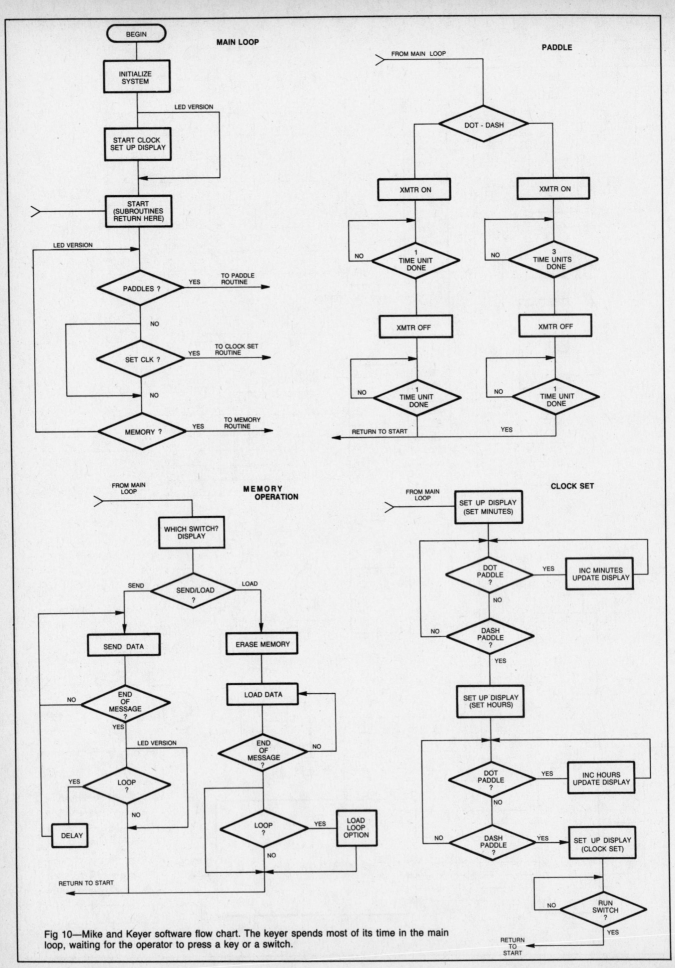

Fig 10—Mike and Keyer software flow chart. The keyer spends most of its time in the main loop, waiting for the operator to press a key or a switch.

2000 bits for each of the eight memory storage areas. Since the typical character takes 10 bits, up to 200 characters can be stored in each memory position. Batteries N1 and N2 enable the memory to retain information while the keyer is off.

U6, an Intersil 7170, is a special device. It is a real-time clock (RTC) designed for easy interfacing with microprocessors. Most important, it has the ability to generate interrupts for the Z80. These periodic "taps on the shoulder" allow the keyer to display a continuous clock on the LCD and to create special repetitive loops with several of the messages. The RTC has its own battery back-up circuit, so it will continue to operate even while the power is off.

The remaining ICs make up the "support circuitry." U1 is the master clock for the CPU. U3 and U7 help select sections of memory or I/O. For inputs, U10 buffers the message select switches, while U12 provides access to the dot/dash paddles and function switches. For output, the LCD or LEDs are controlled by U8 and U9, while U13 and U14 provide interfacing to the radio. Finally, U11 provides the basic timing for the dots and dashes.

Functional/Operational Description

It would be redundant to describe how microprocessor systems function. Several good books cover the subject as does Chapter 8 of this *Handbook*. The flow chart in Fig 10 shows the Mike and Keyer program logic. After power is applied to the circuit, the unit scans for input from the message or function switches and the paddles. This is how the keyer spends most of its time, waiting for you to press the key or a switch.

Touching a paddle starts a simple but basic sequence of events. To understand these events it is necessary to review the structure of Morse code. A character in Morse code can be broken up into simple time units. A dot is one unit long, a dash is three units long. The space between dots and dashes within a character equals one unit and the spaces between words are seven units long. U11 creates "units" for the Mike and Keyer. Every time U11 is triggered, it turns on for a specified time period determined by its RC network (R13, R15 and C20). This time is monitored by the CPU.

The number of times U11 is triggered is determined by the Morse element to be generated. For a dot, the CPU turns on the transmitter, then triggers U11. When U11 is finished, the CPU recognizes the event, turns off the transmitter, and triggers U11 a second time. This second trigger creates the necessary space required between marks, exactly the same unit length as the dot. A dash works similarly, but once the transmitter is turned on U11 is triggered three times before the radio is unkeyed. After the transmitter is off, U11 is triggered once more to create the mandatory space

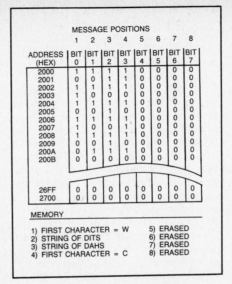

ADDRESS (HEX)	BIT 0	BIT 1	BIT 2	BIT 3	BIT 4	BIT 5	BIT 6	BIT 7
2000	1	1	1	1	0	0	0	0
2001	0	0	1	1	0	0	0	0
2002	1	1	1	1	0	0	0	0
2003	1	0	0	0	0	0	0	0
2004	1	1	1	1	0	0	0	0
2005	0	1	1	0	0	0	0	0
2006	1	1	1	1	0	0	0	0
2007	1	0	0	1	0	0	0	0
2008	1	1	1	1	0	0	0	0
2009	0	0	1	1	0	0	0	0
200A	0	1	1	1	0	0	0	0
200B	0	0	0	0	0	0	0	0
26FF	0	0	0	0	0	0	0	0
2700	0	0	0	0	0	0	0	0

MESSAGE POSITIONS 1 2 3 4 5 6 7 8

MEMORY

1) FIRST CHARACTER = W   5) ERASED
2) STRING OF DITS   6) ERASED
3) STRING OF DAHS   7) ERASED
4) FIRST CHARACTER = C   8) ERASED

Fig 11—Mike and Keyer memory organization chart. A one in a memory location means "transmitter on," a zero means "transmitter off." See text for details.

unit. Adjusting the length of U11's timing period with the external speed control (R15) changes the speed of the Morse code generated by the keyer.

Loading code sequences into memory is a little more complicated. Fig 11 shows how the data is stored in the RAM. A byte consists of eight bits, and each byte has a unique address. The starting address for RAM in the Mike and Keyer is 2000 (hex) and the last address reserved for message memory is 2700 (hex). Each individual bit can be defined by a data line (D0 - D7) and its address. Each memory position uses one of the bits in each memory byte. If the transmitter should be keyed when the memory is played back, a "1" is placed in the appropriate bit of the current memory location. If a space between words or characters is required, the keyer stores a "0." Dots and dashes are predefined as discussed above (a dot is a one followed by a zero, a dash is three ones followed by a zero).

Spaces, between dots and dashes or between letters and words, are a bit more difficult; the CPU cannot differentiate between character and word spaces. A free-running clock would load the proper spaces into memory, but this technique is wasteful if the operator pauses while loading a memory. The Mike and Keyer uses a limited run count for spaces in memory. When a memory load is started, spaces are placed in memory until a maximum of ten spaces are loaded. The memory load will then stop until a paddle is touched or the LOAD/RUN switch is set to RUN. Touching a dot or dash paddle will load the appropriate data into memory, reset the spaces counter, and start the memory loading again. Having the memory counter run a maximum of ten spaces between characters allows for a little "swing" in the operator's sending style. Switching the

LOAD/RUN switch to RUN during a pause in loading will complete the loading process. In the unlikely event that the memory is overrun, a warning message will occur and the memory must be reloaded.

"Playing" the memory is a much simpler operation. Starting at address 2000 (hex) the CPU simply examines the bit corresponding to the message requested and turns the transmitter on if it finds a one, or off if it finds a zero. The keyer then triggers U11 and waits for it to time out. Again, the speed of the generated Morse code is controlled by varying the period of U11 with the speed control. Each address is examined in turn until the end of message is reached.

The end of message memory is determined by counting the number of adjacent spaces in the memory. When the message was first loaded, the CPU quickly erased each bit corresponding to the selected memory in all the RAM locations by placing a zero in each bit location. In addition, the keyer will only load 10 spaces (zeroes) before pausing for input and checking the LOAD/RUN switch. All this means that there should never be more than 10 spaces in a row before the end of a message. If the keyer finds more than 10 spaces in a row, we have reached the end of stored memory.

If the RTC and LCD are installed, interrupts every second will update the clock and display. These interrupts occur in the background to the main program and are not noticed by the user.

Options

Including the real-time clock and liquid crystal display enables certain features of the Mike and Keyer. Some of the memories may be repeated at predetermined intervals. The original Mike and Keyer design provided for four looping options in memory locations six through eight, as well as normal single runs. These loops are set for 15, 30, 120 and 150 seconds. UHF/VHF aficionados will notice that these times correspond to the standard time intervals used in meteor-scatter and moonbounce activities.

A second version of the software provides an option of looping messages six through eight continuously with short pauses, with the pauses adjustable from two to eight seconds in two-second increments. This version may be more useful for HF operations such as calling CQ for DX or during contests. In both these versions any 16-character message can be placed on the first line of the LCD when the function/instructional messages are not needed. This message is permanently installed in ROM, locations 1000-100F (hex). If you like, you can personalize your keyer by having your call sign directly on the display.

The third software version is for the basic keyer without the RTC/LCD installed. It provides for limited function display with

LEDs; no looping or clocks are supported.

Construction

The Mike and Keyer lends itself to easy home-brew construction. A double-sided PC board is available[1] and is probably the easiest method to avoid wiring errors. If you prefer, wire-wrapping is an option and easily accomplished with a couple evenings work. A wire-wrapped prototype was built by the *Handbook* editor, and the unit worked perfectly from first power-up. As always with IC projects, sockets are a good idea to facilitate troubleshooting and repairs.

U1 is a TTL-compatible clock oscillator. The clock frequency is not critical; clock frequencies from 1 MHz to 2.5 MHz were used successfully during testing. The keyer spends most of its time waiting for the operator, so a fast clock is not required.

The keyer can key a voltage to ground (through Q1) or a 5-V keyed output can be supplied if required (through Q2). The keying transistors should be selected for your transmitter's keying voltage, current and polarity requirements. The keying transistors shown will work with most rigs.

The layout of parts is not critical, but do not skimp on the bypass capacitors that help eliminate transients on the power bus. Digital devices are sensitive to power fluctuations, so a well-regulated and spike-suppressed power supply is mandatory. A metal enclosure should be used to prevent RFI both to and from the keyer—microprocessor systems make good transmitters.

Several CMOS devices are used in the Mike and Keyer. These devices are very sensitive to static charges. Observe good handling procedures when installing these and the other integrated circuits.

A printed listing of the software (in Z80 assembly language) is available from the ARRL Technical Department Secretary for an SASE and $3. Programmed PROMs or floppy disk listings are also available from WA7UVJ[2].

Suitable liquid-crystal display modules are manufactured by a number of companies. Versions of the Mike and Keyer have been built with LCDs from Seiko, Stanley and Phillips; all worked well. Other companies that make suitable displays include Sharp and AND. These modules all use the same basic chip and commands for controlling the LCD matrix, but be aware that the pinouts usually vary from company to company. The schematic covers the Seiko M1632 LCD module—if you purchase another device, get a data sheet with it! Probably the easiest source for the module is from Digi-Key (model AMX216-ND; see Chapter 35 for the address), but almost any large regional electronics supply house stocks a version of this display. The LCD module should consist of two lines of 16 characters in a 5 × 7 format, operate from +5 V dc, and contain its own character set. It must be serially controlled and have eight-bit capability. Normally, the control chips in the module are made by Hitachi.

Testing

Nothing is more frustrating than a circuit that doesn't work the first time. Most of the problems with prototype Mike and Keyer units were caused by careless assembly processes. Check for cold solder joints, solder bridges or incorrect wiring connections. Be careful when you insert the ICs into their sockets; it takes only a fraction of a second to kill a chip installed backwards. It is most helpful to have a friend check your work for you; many times a different pair of eyes will see mistakes you have missed.

Adjustment

The LED version of the keyer requires no adjustments. With the LCD/RTC

[1]PC boards and parts kits are available from Leslie PCB, 13517 117 Ave NE, Kirkland, WA 98034. Send SASE for information.
[2]Programmed EPROMS and disk listings only (IBM® PC or Commodore 128™ CP/M®) are available from Michael Dinkelman, 637 2nd Ave South, Kent, WA 98032. Send SASE for information.

version, the real-time clock oscillator frequency should be adjusted for most accurate timing. Do not try to measure the crystal directly as loading will give erroneous results. Place a frequency counter or oscilloscope on the Z80 interrupt line (INT) and adjust the trimmer capacitors (C17 and C18) for a 1-Hz reading. Setting the clock time is easy and straightforward. After you place the CLOCK SET/RUN switch to SET, the dot paddle is used to insert the correct values for minutes and hours the dash paddle places the values into memory. The software will prompt you on the LCD. After you set the hours and minutes, setting the clock switch to RUN will start the clock running at zero seconds.

Operation

The Mike and Keyer is simple to operate. After power is applied, the keyer functions as a straight iambic keyer. To load a memory, switch the LOAD/PLAY switch to LOAD and press the appropriate memory pushbutton. Start sending the code; remember that the keyer automatically stops loading the memory after ten spaces, so this is the longest pause that can be stored in memory. When you are done loading the memory, switch the LOAD/PLAY switch back to PLAY. Playing a memory is accomplished by pressing the appropriate memory pushbutton with the LOAD/PLAY switch in the PLAY position. To abort memory playback, simply press either side of the paddle. Pressing the memory pushbutton again after an abort restarts the memory playback *at the beginning of the memory*.

Conclusion

As mentioned above, the beauty of the microprocessor approach is its flexibility. The designer of this project hopes that others who like to dabble in software will program new enhancements for the Mike and Keyer. New programs could be written for more effective contesting, DXing, or other CW activities. Please write to WA7UVJ about any projects you undertake (enclose an SASE if you'd like a reply).

A Deluxe Memory Keyer

Here it is—a memory keyer with "all the features" that doesn't cost a fortune. The memory keyer described here and shown in Figs 12 through 15 is designed around a Curtis 8044-series keyer chip and features the following:

- Optional remote-mounting of the memory pushbuttons
- Ultra-low power consumption—less than 5 mA average
- MOSFET keying outputs to accommodate virtually any rig
- Single, low-profile circuit board
- Uses commonly available, inexpensive ICs
- Can use a nonvolatile memory
- Powered by four AA-size NiCd batteries

One of the major design goals of this keyer was to include as many features as possible while keeping power consumption low and price moderate. This goal was accomplished through the use of 4000 series CMOS logic, a Curtis 8044 series keyer IC and a low-power memory IC. The Curtis keyer IC helps reduce circuit complexity; it is used for dot and dash generation (iambic operation is optional) and weighting (positive or negative). The optional M-series ICs contain a speed meter interface for indication of keyer speed. This keyer was designed and built in the ARRL Lab by Thomas Miller, NK1P.

Circuit Details

To better understand the circuit, let's break it down into some basic sections. Fig 13 shows the keyer schematic diagram. The first section is the clock and its control circuitry. The master clock, generated by U19, operates at 16 times the actual speed of the dots. It runs at about 130 to 1100 Hz to achieve a speed range of 7 to 80 WPM (calculated by the Paris method). This master clock is divided to the dot speed by U10 and is synchronized to the rest of the circuit by U5A, U13 and U8A. Upon closing either the dot or dash paddle, U10 is reset to zero and starts counting, supplying U3 with a pulse train. The resulting dot or dash from U3 is fed back via U4A and, upon completion, resets U10. The circuit returns to idle until another paddle closure. The master clock is also used synchronously for read and write cycles of the memory circuitry.

The second section is that of the memory and control, which includes most of the remaining circuitry. One of the most significant features of the read/write control is that of the "mouse" (the push buttons and circuitry that select the memory), which is basically a 1-of-8 encoder. This encoder is used to select and start one of the eight memory strings. It can be used remotely (next to the paddle), or it can be built inside the enclosure.

The mouse schematic is shown in Fig 14.

Fig 12—The deluxe memory keyer can be built with the memory pushbuttons inside the box or mounted in an external "mouse" that can be placed next to the paddle.

Note that only six wires are required to connect the mouse to the main circuit board. This is made possible by U1 and U2, which are used to generate and latch a three-bit word according to the pushbutton pressed. The three-bit word ultimately selects the memory to be used. U3 and its associated components are used as visual feedback. This circuit can be disabled by shorting pins 1 and 2 of P6 when low-power operation is required. Of course, this can also be done using a SPDT switch.

The remainder of the read/write control circuitry is made up of U11, U12 and associated parts. The control circuit allows for the inclusion of switches for starting a read or write function and for automatic repeat of stored messages after a user-defined delay.

An inexpensive, commonly available 2-k × 8 memory IC (U6) is used for storage. Since the memory is 8 bits wide, it is necessary to write to one of the bits while retaining the others for other messages. This is accomplished by using a feedback loop on the data I/O pins consisting of U1, U2 and feedback resistors R1-R8. Data to be written to the memory is routed through U2 (pins 9 and 10) to the appropriate data line on U1 by the mouse. All other outputs of U2 remain in the high-impedance state, allowing data at the selected address in memory to be fed back (by the feedback resistors) and rewritten to the same location.

The remaining block of the keyer is the interface to the outside world. The sidetone oscillator/amplifier was designed in its simplest form and adds less than 15 mA to the total current consumption at full output. The other part of the interface circuitry is that of the keying outputs. Both grid-block and positive outputs are provided, so this keyer will key almost any transmitter. The keying interface is extremely efficient in operation because of the use of MOSFETs. The manufacturer specifies 8 ohms of on-state resistance, resulting in reliable performance over the long haul.

Construction

The keyer can be constructed in a variety of ways, such as wire wrap and point to point. Fig 15 shows two printed-circuit board versions that were built by the author. One was built with the mouse contained inside the cabinet. This option can be incorporated in a variety of ways. It is not necessary to include all eight of the memory selection buttons and LEDs, just as many as you need. Connection points for unused pushbuttons can be left open.

PC boards and parts kits for this project are available from A&A Engineering.[1] Use of one of these PC boards is highly recommended. The board measures 5 × 7 inches and is double sided with plated-through holes. All connections to the board are made through connectors to facilitate easy servicing and modification. Table 1 lists the connections to each connector strip.

A template package for this keyer, including PC board parts overlay and drilling information for the front and rear panels, is available from the ARRL Technical Department secretary for an SASE and $2.00.

A number of features were added to the PC board to aid in operation/construction flexibility of the keyer. For example, the board is designed to accommodate any of the 8044 series ICs. The 8044ABM is recommended if you don't already have a keyer chip. Strapping options for the 8044ABM allow you to disable or enable iambic paddle operation (Fig 13, connection X1) and to select positive or negative weight control (Fig 13, connection X4). Also available as an option is a meter for relative indication of speed ("M" series ICs only). See Fig 13. The usable range of the meter needs to be 0-50 microamps, so just about any 0-50 or 0-100 microamp meter will suffice.

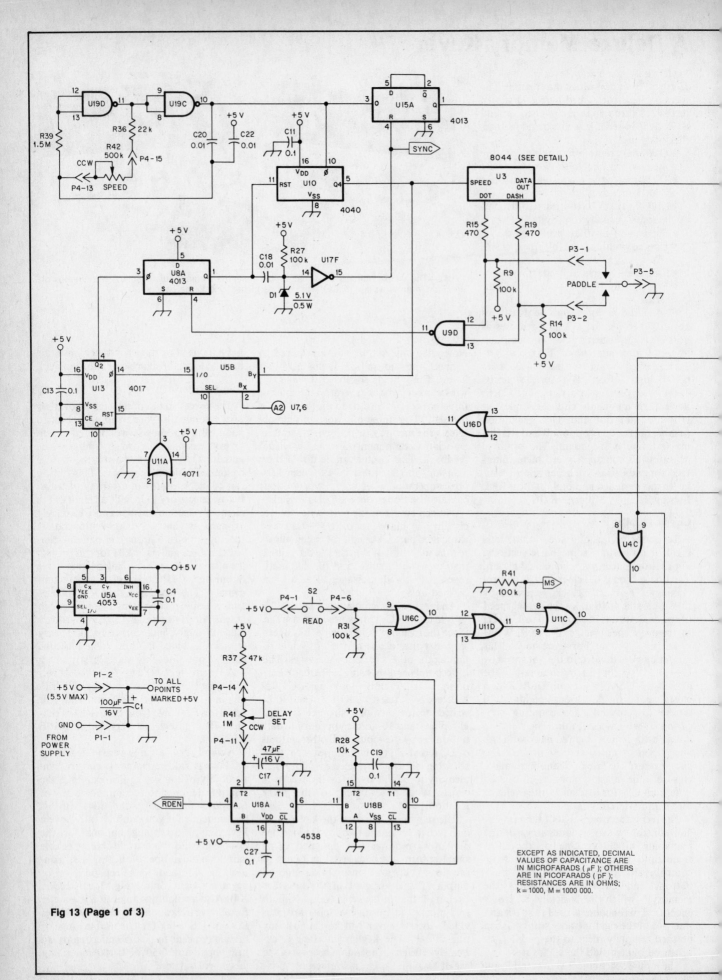

EXCEPT AS INDICATED, DECIMAL
VALUES OF CAPACITANCE ARE
IN MICROFARADS (μF); OTHERS
ARE IN PICOFARADS (pF);
RESISTANCES ARE IN OHMS;
k = 1000, M = 1000 000.

Fig 13 (Page 1 of 3)

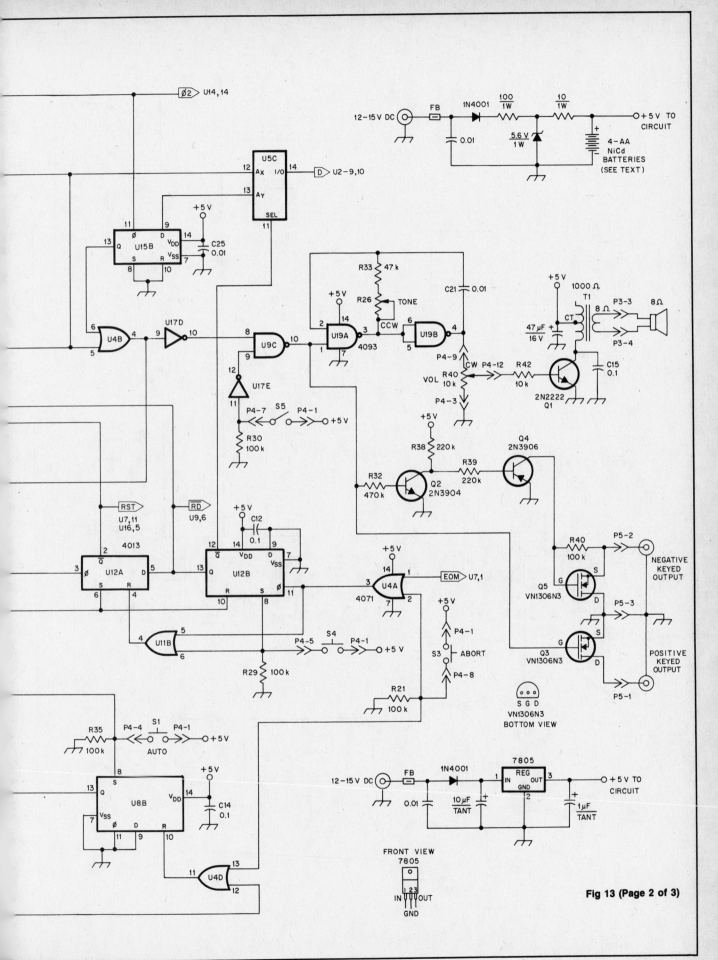

Fig 13 (Page 2 of 3)

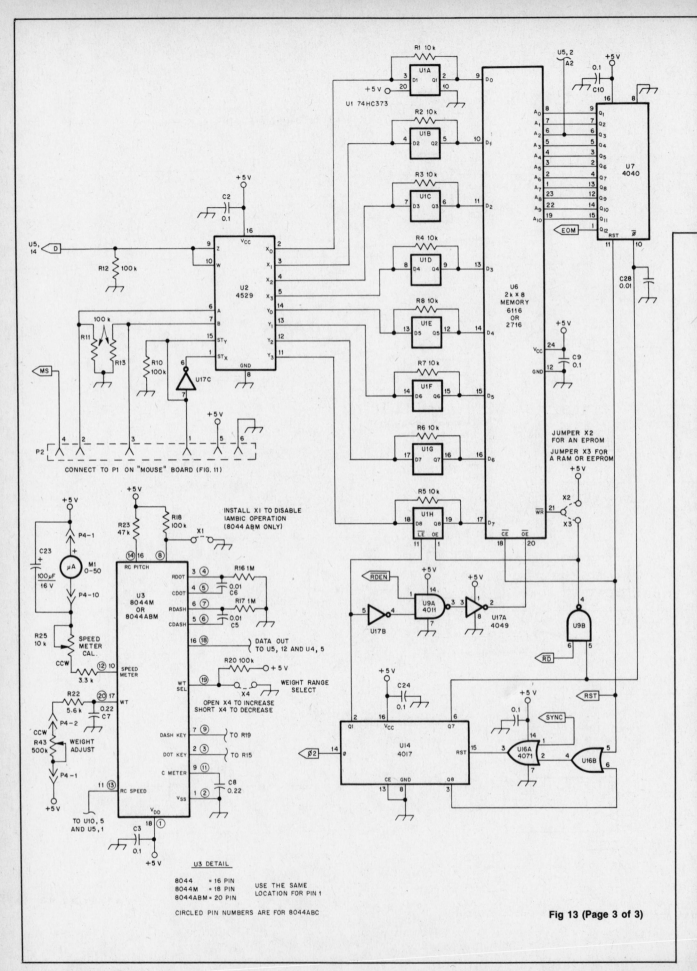

Fig 13 (Page 3 of 3)

Fig 13—Schematic diagram of the deluxe memory keyer. All resistors are ¼-W, 10% units unless noted. Capacitors are disc ceramic units unless noted. Capacitors marked with polarity are electrolytic. PC boards and all parts for this project are available from A&A Engineering (see text).

C1, C23—100-µF, 16-V radial lead electrolytic capacitor.
C5, C6, C18—0.01-µF mylar capacitor.
C7, C8—0.22-µF monolithic ceramic capacitor.
C16—16-µF, 16-V radial lead electrolytic capacitor.
C17—47-µF, 16-V tantalum capacitor.
C20, C21, C22—0.01-µF polystyrene capacitor.
D1—5.1-V, 0.5-W Zener diode (1N5230B or equiv.)
P1—2-pin, single-row header, pins spaced on 0.156-inch centers.
P2—6-pin, single-row header, pins spaced on

0.1-inch centers.
P3—5-pin, single-row header, pins spaced on 0.1-inch centers.
P4—15-pin, single-row header, pins spaced on 0.1-inch centers.
P5—3-pin, single-row header, pins spaced on 0.1-inch centers.
Q3, Q5—60-V, 0.5-A, N-channel enhancement-mode power FET (Supertex VN1306N3 or equiv.).
R25—10-kΩ PC-board mount potentiometer.
R26—10-kΩ PC-board mount potentiometer.
R40—10-kΩ panel-mount potentiometer.

R41—1-MΩ panel-mount potentiometer.
R42, R43—500-kΩ panel-mount potentiometer.
S1-S4—SPST, momentary contact pushbutton switch.
S5—SPST toggle switch.
T1—Audio transformer, 500-Ω to 8-Ω (or half of a 1000-Ω to 8-Ω unit).
U3—Curtis 8044 series keyer chip (see text).
U6—CMOS RAM, 6116-LP-4 or equiv. (see text).
X1, X4—2-pin, single-row header, pins spaced on 0.1-inch centers.
X2, X3—3-pin, single-row header, pins spaced on 0.1-inch centers.

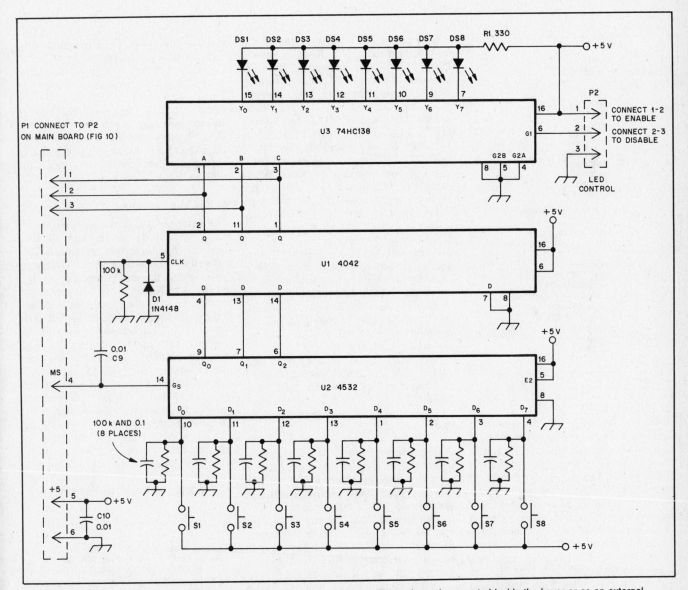

Fig 14—Schematic diagram of the memory pushbutton and indicator LED unit. This unit can be mounted inside the keyer or as an external "mouse." All resistors are ¼-W, 10% units unless noted. Capacitors are disc ceramic units; all capacitance values are in µF.

DS1-DS8—Red LED.
P1—6 pin, single-row header, pins spaced on 0.1-inch centers.

P2—3 pin, single-row header, pins spaced on 0.1-inch centers.

S1-S8—SPST, momentary contact pushbutton switch.

Fig 15—Interior of the deluxe memory keyer. The version at A has the memory pushbutton board mounted behind the front panel and uses the internal NiCd battery pack and charger option. The version at B has the memory pushbutton board mounted in an external "mouse" and uses a 7805 regulator for power from a 12-15-V source.

Table 1
Keyer Connections

Connector	Function
P1-1	Ground
-2	+5 V
P2	Connect to P1 on mouse board
P3-1	Dot paddle
-2	Dash paddle
-3	Speaker
-4	Speaker
-5	Ground
P4-1	+5 V
-2	WEIGHT adjust control
-3	Ground
-4	AUTO switch
-5	WRITE switch
-6	READ switch
-7	TUNE switch
-8	ABORT switch
-9	VOLUME control
-10	Speed meter (optional)
-11	DELAY control
-12	VOLUME control
-13	SPEED control
-14	DELAY control
-15	SPEED control

Another feature of the PC board is a provision to enable the keyer to read from an EPROM (Fig 13, connections X2/X3). With J2 installed, a 2716 (or equivalent) can be installed to play back a "canned" message. EPROMs can be read, but not programmed, by the keyer, so those applications will require at least an EPROM programmer. Data is stored in memory as a string of variable length, one bit wide. This enables the keyer to store as many as eight messages, each 2 k long. If you wanted to store a series of messages in EPROM, it would be necessary to first make a listing of each message in ones and zeros (3:1 dash-to-dot ratio). Then you could combine all of the messages into eight-bit words to be stored in the EPROM.

Connection X3 is used to enable the keyer to read and write to RAM. A Hitachi HM6116P-3 is recommended, but just about any 2-k × 8 RAM will work as long as the pin-out is the same. Some memory ICs that have been tried include TMM2016P-1 and TC5517APL (Toshiba) and the XLS2816AC-350 EEPROM (Exel). It is desirable to use RAM ICs with "P" or "LP" suffixes, as they require substantially lower standby currents. When using these ICs, total current draw of the keyer in standby is less than one milliamp!

The circuit will support EEPROM (Electronically Erasable Programmable Read Only Memory) which is the only true nonvolatile RAM. Since EEPROM does not require power to retain memory, it is perfectly suitable for use where battery operation is required, reducing cost by eliminating the need for batteries to back up the RAM. See Chapters 6 and 27 for more information on building an ac-operated 5-V supply.

Keep in mind that all of the integrated circuits used are CMOS and are sensitive to static discharge. Use standard CMOS handling precautions. A metal enclosure will help prevent RF from getting into the unit.

All of the parts are quite common (with the exception of the 8044) and can be obtained from most surplus houses. Many suitable suppliers are listed in Chapter 35.[2] Capacitors can be inexpensive ceramics or electrolytics with the exceptions of C17 (tantalum) and C20, C21 and C22 (polystyrene) for stability. The audio output transformer is a general-purpose device, with a 1-kΩ center-tapped (or 500-ohm) primary, and an 8-Ω secondary.

The circuit runs from 5-V dc and can be powered for considerable lengths of time by a NiCd battery pack. Please note that you should *never apply more than 5.5 V* to the circuit; greater voltage will damage the RAM and other ICs. Fig 13 shows two

possible power-supply configurations. One uses a 7805 three-terminal regulator to provide the necessary voltage. The other incorporates a constant-current charger and battery backup. If you choose to use the NiCd battery pack, check the voltage before connecting the batteries to the keyer. If the battery pack has been charged to more than 5.6 V, let the the pack discharge through the Zener for 24 hours or until the voltage drops below 5.6 V.

Testing and Operation

Once the unit has been built, the fun can begin! Start by double checking all connections. This may seem a little tedious, but it is well worth it in the long run. Once you are sure that the unit is wired correctly, apply power. Rotate the volume control to maximum, and upon closing the paddles, you should hear a repetition of dots and dashes. If not, try rotating the volume control in the opposite direction (it might have been wired backwards). Once you hear dots and dashes, rotate the tone and speed controls to the desired positions. Send a few words to get used to the "feel." If you have included the speed meter option, R25 on the PC board is used to set the relative position.

When all of the basic keyer functions are working, test the memory. If you are using an external mouse, connect it at this time. Press the pushbutton for the memory that you want to program (1 to 8). Press WRITE

[1]Contact A&A Engineering, 2521 W. LaPalma Ave., Unit K, Anaheim, CA 92801, tel 714-952-2114, for the latest information.

[2]In addition to the suppliers listed in Chapter 35, the following may be of help:
Curtis Electro Devices, Box 4090, Mountain View, CA 94040, tel. 415-964-3846.
Jameco Electronics, 1355 Shoreway Rd., Belmont, CA 94002, tel. 415-592-8097.
JDR Microdevices, 1224 S. Bascom Ave., San Jose, CA 95128, tel. 408-995-5430.

once and then key in a short CQ or your call. Remember that data is not written to the memory until the first paddle closure. Do not leave too much space between words; the memory will interpret this as the end of the message. If you make a mistake, simply press WRITE again and start over. It is not necessary to press a mouse pushbutton again unless you intend to write to a different memory. Once you have completed the message to be stored, wait the equivalent of five dashes or more and press ABORT.

To play back the stored message, simply press the mouse pushbutton that corresponds to the memory you just programmed, and the message should be played back exactly as sent. If the mouse is not connected, you can play back the memory by pressing READ. If your

message stops in the middle, chances are that the keyer interpreted a long space as the end of the message; this is how it achieves variable message length.

To use the auto-repeat feature, rotate the DELAY pot counter clockwise and press AUTO. The message currently selected by the mouse will be played, then there will be a pause (set by the DELAY pot), and the message will be played again. This will continue until ABORT is pressed or one of the paddles is closed.

Next, test the keying outputs by connecting the appropriate output to a radio. Be sure to check your owner's manual for the specification of the voltage and current present on the key terminals, as the maximum drain-to-source rating of the keying MOSFETs (Q3 and Q5) is 60 V at 400 mA.

If you are not sure which type of keying output is required, again check the owner's manual. This completes the testing. CQ DX CQ DX CQ DX DE...

If you have used the unit for a few QSOs and feel that the paddles just don't feel quite right in either iambic or non-iambic modes, a slight capacitor value change might be in order. There are two one-shots on the 8044 (one for the dot, one for the dash) that are used to briefly inhibit their respective paddles. This feature, sometimes called "anti echo," prevents insertion of extra dots or dashes. To adjust the feel, alter the value of C5 (for the dash paddle) or C6 (for the dot paddle) by an order of magnitude. This is strictly trial and error, and be sure to take your time if you decide to try this.

The W2IHY Audio Memory Keyer

A digital voice recorder (called a DVR, or audio keyer) is the audio equivalent of a CW memory keyer. A CW memory keyer stores Morse code messages for retransmitting. A DVR stores digitized audio messages for retransmission. Digital voice recorders are gaining popularity with SSB/FM contesters. With a DVR, memory segments can be loaded with short, frequently used messages (such as CQs and contest exchanges). These messages can then be easily retransmitted. Other uses for the audio keyer include record and play-back of received audio during a QSO and voice identification for a repeater. In a few years audio keyers will be as common a station accessory as CW memory keyers are today.

The audio keyer shown in Figs 16 through 18 was designed and built by Julius Jones, W2IHY.

W2IHY thanks Herb and Barbara Sweet, (K2GBH and WA2KCL) for their assistance with the project writeup. The audio keyer has the following features:

1) Four message memories each capable of storing up to 10 seconds of digitized audio
2) A message write/read (record/send) indicator
3) A microphone gain indicator
4) A built-in audio monitor
5) The ability to repeatedly read (send) a selected memory
6) Adjustable delays between repeats
7) Electronic control of transmitter push to talk (PTT)
8) Adjustable PTT sensitivity and delay controls
9) Voice-activated reset (VAR)
10) Adjustable VAR sensitivity control
11) Battery back-up capability

The audio keyer is designed using commonly available components, including 74HC series high-speed CMOS, 256k × 1-bit dynamic random-access memory

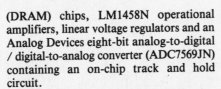

Fig 16—The W2IHY Audio Memory Keyer can store four messages, each containing up to 10 seconds of audio.

(DRAM) chips, LM1458N operational amplifiers, linear voltage regulators and an Analog Devices eight-bit analog-to-digital / digital-to-analog converter (ADC7569JN) containing an on-chip track and hold circuit.

The audio keyer can be constructed, using all new parts, for about $185. A single-message version can be built for about $130 by replacing the 256k DRAM ICs with 64k ICs (150 ns or 120 ns). The single-message keyer may be later upgraded to full capacity simply by installing 256k ICs. All the parts are available from Jameco Electronics[1] and Mouser Electronics[2] except the A-to-D converter which is available from Analog Devices[3].

The keyer has been used in high RF

[1] Jameco Electronics, 1355 Shoreway Road, Belmont, CA 94002, tel 415-592-8121.
[2] Mouser Electronics, 2401 Hwy 287 North, Mansfield, TX 76063, tel 1-800-346-6873.
[3] Analog Devices Inc, Rt 1 Industrial Park, Norwood, MA 02062-9106, tel 617-329-4700. (AD7569JN may be phone ordered).

environments without any problems. It has also been used in a multi-band SSB contest. On-the-air tests indicate that other stations find it difficult to distinguish between live signals and keyer-generated signals.

Circuit Description

The schematic of the DVR is shown in Fig 17, and a block diagram of the keyer is shown in Fig 18. Audio from the microphone is amplified through two cascaded operational amplifiers. The audio microphone level control (R49) sets the gain of the second stage of amplification.

The output of the microphone amplifier is sent to a six-pole unity-gain low-pass filter and to an analog audio switch. (The audio switch will be discussed later.) The low-pass filter removes high frequencies present in the amplified microphone signal. The 3-dB cut-off frequency of the filter is 1.8 kHz.

The output of the microphone low-pass filter is sent to the analog-to-digital converter (ADC), microphone-level indica-

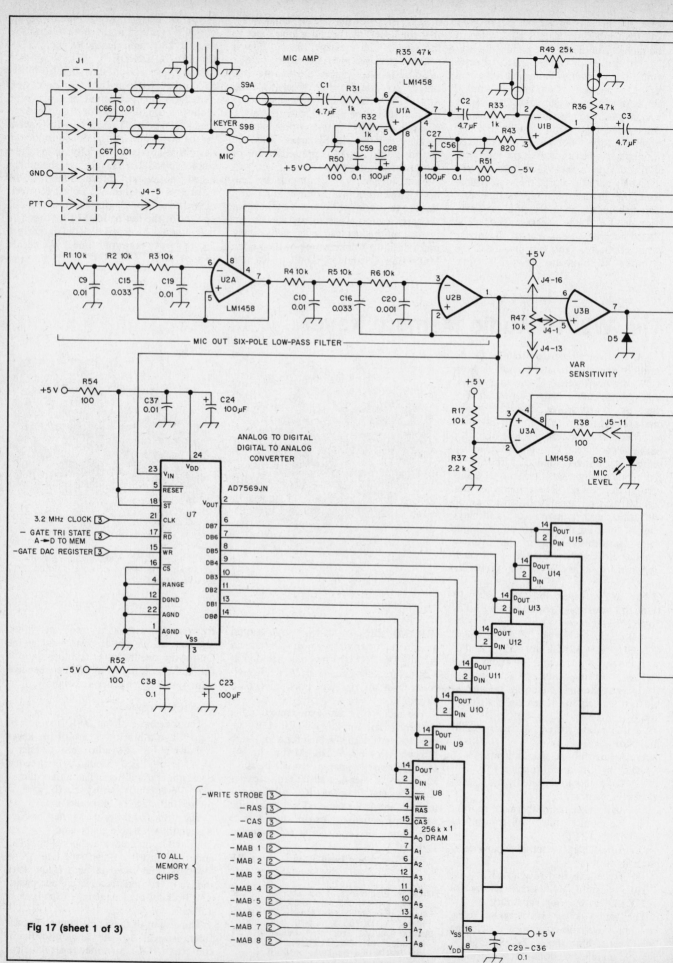

Fig 17 (sheet 1 of 3)

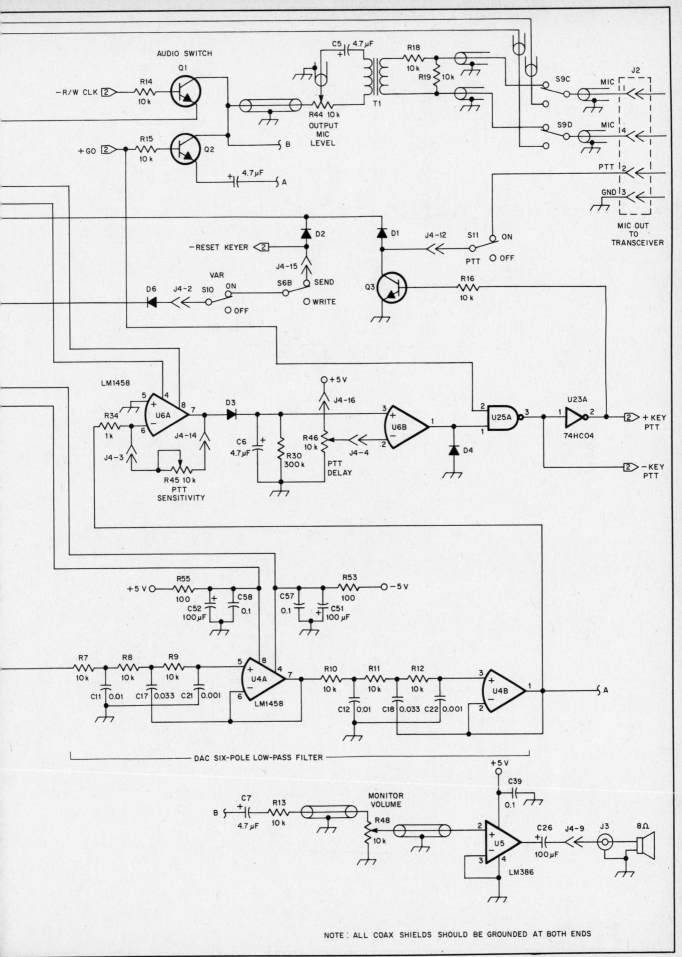

NOTE: ALL COAX SHIELDS SHOULD BE GROUNDED AT BOTH ENDS

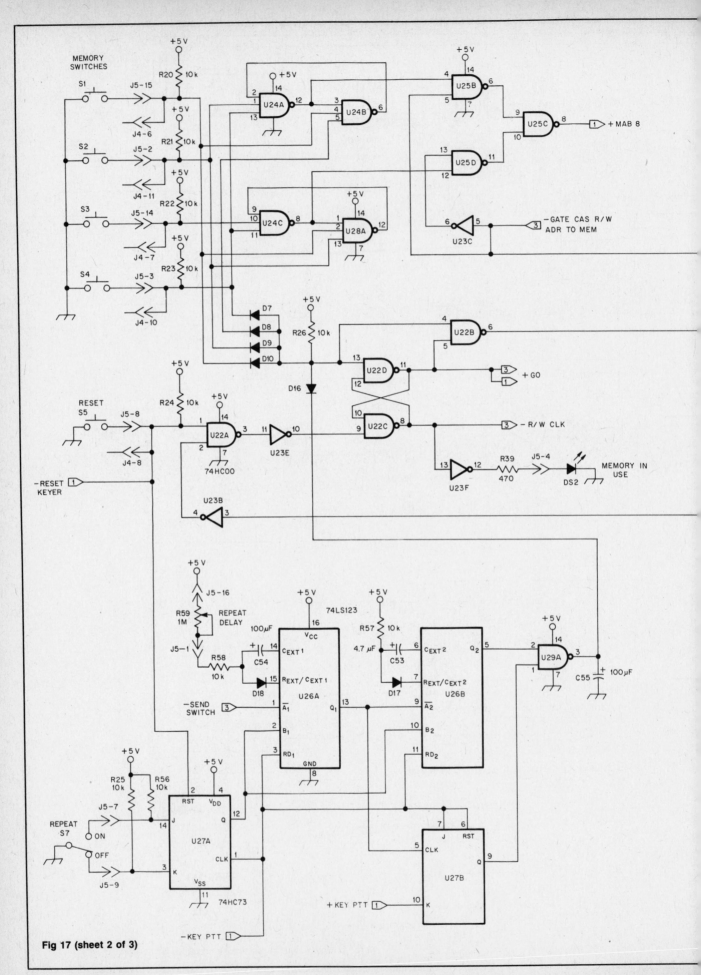

Fig 17 (sheet 2 of 3)

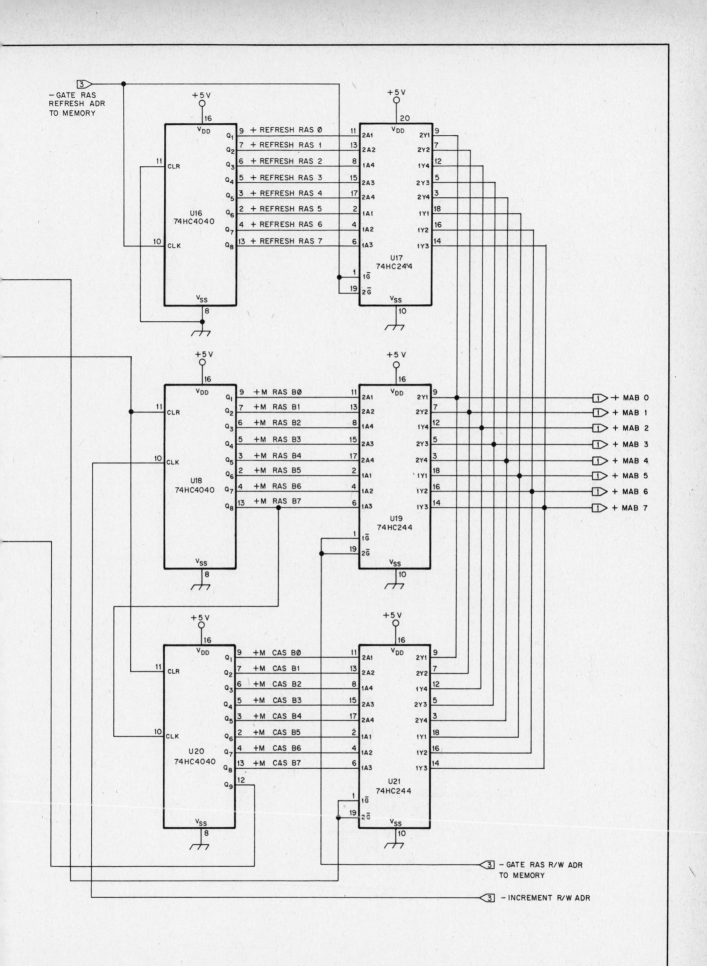

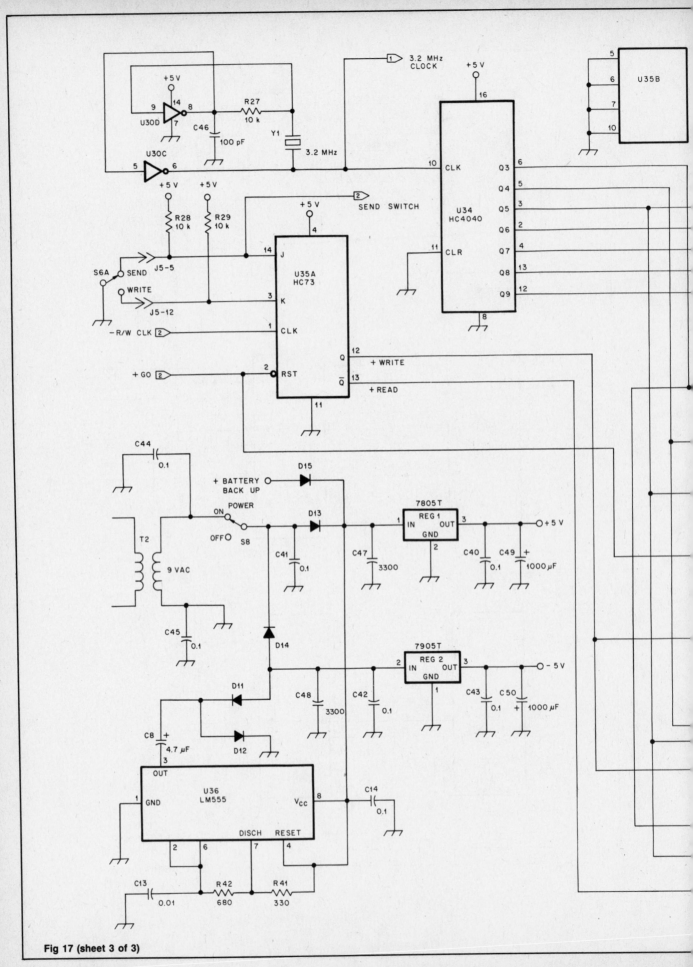

Fig 17 (sheet 3 of 3)

29-24   Chapter 29

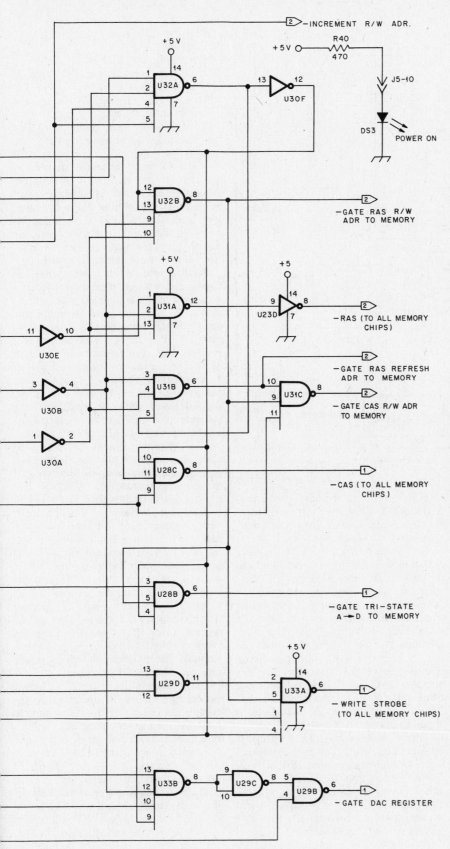

Fig 17—Schematic diagram of the W2IHY audio keyer. All resistors are ¼ W, 5% units unless noted. Capacitors are disc ceramic units unless noted. Capacitors marked with polarity are radial electrolytic. PC boards for this project are available from W2IHY[4].

C1-C8, C53—4.7 µF, 16 V radial electrolytic.
C9-C12, C66, C67—0.01 µF, 50 V Mylar 10%.
C13, C14, C29-C45, C56-C65—0.1 µF, 25 V Mylar 10%.
C46—100 pF, 50 V disc ceramic.
C23,C24,C26-C28,C51,C52,C54,C55—100 µF, 16 V radial electrolytic.
C15-C18—0.033 µF, 50 V Mylar 10%.
C19-C22—0.001 µF, 50 V Mylar 10%.
C47, 48—3300 µF, 25 V radial electrolytic.
C49,C50—1000 µF, 25 V radial electrolytic.
D1-D10, D16-D18—Silicon diode (1N914 or 1N4148).
D11-D15—1N4001.
J1,J2—Female microphone connector (choose to match your equipment).
J3—RCA phono plug.
J4,J5—16-pin IC socket.
R1-R29, R56-R58—10 kΩ.
R31- R34—1 kΩ.
R43—820 Ω.
R38, R50-R55—100 Ω.
R41—330 Ω.
R39,R40—470 Ω.
R37—2.2 kΩ.
R30—300 kΩ.
R35—47 kΩ.
R36—4.7 kΩ.
R42—680 Ω.
R49—25 kΩ potentiometer.
R44-R48—10 kΩ potentiometer, ½ W.
R59—1 MΩ potentiometer.
S1-S5—Single-pole momentary pushbutton switch.
S6,S8—DPDT toggle switch.
S7—SPDT toggle switch.
S9—4PDT toggle switch.
S10, S11—SPST toggle switch.
T1—Audio transformer, 600 Ω 1:1 (Mouser 42TM016).
T2—117 V ac primary, 9 V ac 500 mA secondary.
Q1-Q3—2N2222A.
U1-U4, U6—LM1458N.
U5—LM386N.
U7—AD7569JN A/D and D/A converter IC; Analog Devices Inc[3].
U8-U15—41256 256k × 1-bit dynamic RAM.
U16, U18, U20, U34—74HC4040.
U17, U19, U21—74HC244.
U22, U25, U29—74HC00.
U23, U30—74HC04.
U24, U28, U31—74HC10.
U26—74LS123.
U27, U35—74HC73.
U32, U33—74HC20.
U36—LM555.
Y1—3.2768 MHz crystal.

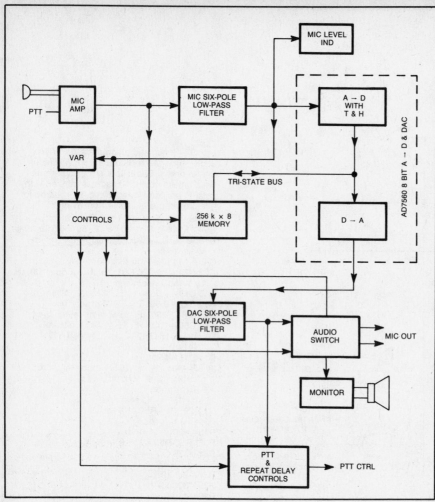

Fig 18—Block diagram of the audio keyer. See text for a discussion of this diagram.

tion circuitry and VAR circuitry. The ADC samples the low-pass filter output and converts it into an 8-bit data word.

Signals sampled by the ADC exceeding ±1.25 volts will be clipped to ±1.25 volts. The microphone level LED (DS1) provides visual indication when the input level to the ADC is near +1.25 V. The microphone-level indicator is turned on when the ADC is sent a signal from the microphone low-pass filter exceeding 0.9 V. The microphone low-pass filter output is also sent to the voice-activated reset (VAR) circuitry. When activated, VAR allows the audio keyer to be reset when you speak into the microphone. This is analogous to resetting a CW memory keyer by tapping the paddle. VAR sensitivity potentiometer R47 establishes the microphone low-pass filter level that will reset the keyer.

The output of the ADC is sent to eight 256k × 1 DRAM chips. The control circuits allow the digitized microphone input to be stored in the memory.

To send (play back) data stored in memory, an eight-bit word is read from memory and sent to the DAC, where the data is restored to analog form.

The output of the DAC is sent to a six-pole low-pass filter. (The DAC low-pass filter is identical to the microphone low-pass filter.) This filter removes high-frequency components from the DAC-generated analog signal.

The output of the DAC low-pass filter is sent to the analog audio switch and to the push-to-talk controls. During playback, audio in analog form is gated from the DAC low-pass filter through the analog switch (transistor Q2). When the keyer is not sending or writing, the analog switch selects audio from the microphone amplifier (transistor Q1). During a write operation neither the microphone amplifier nor the DAC low-pass filter are gated through the analog audio switch.

Audio gated through the analog switch is sent to output level control R44 and to the speaker monitor amplifier. T1 provides isolation between the keyer audio output and the microphone input of the transceiver. R44 is adjusted to drive the attached transceiver to the appropriate audio level. The output of the analog switch is also sent to the speaker monitor. Control R48 adjusts the monitor volume.

The PTT controls detect audio sent from the DAC low-pass filter. When audio is detected, transistor Q3 grounds the transceiver's PTT. Adjustment of the PTT sen-

sitivity control (R45) determines the DAC low-pass filter level that will ground the transceiver's PTT. The delay (PTT) control (R46) determines the length of time Q3 is grounded after the absence of detected audio from the DAC low-pass filter.

The control circuits provide the following functions:

1) crystal control timebase
2) refresh control for the dynamic RAMS
3) interfaces to various front-panel controls including memory select pushbuttons (S1-S4), reset pushbutton (S5), repeat switch (S7), repeat delay (R59) and send/write switch (S6).

When a memory button is pressed, the position of the send/write switch (S6) determines whether the selected memory is loaded with microphone audio or played to the transceiver. During a write operation, the VAR and repeat functions are disabled and the audio being recorded is not gated through the audio switch.

Sending recorded data with repeat on causes the selected memory to be continuously read. Sending (with or without repeat on) may be stopped by pushing the reset button, pushing (enabling) the microphone PTT or using the voice activated reset (VAR). The repeat delay function allows the user to select the amount of time between delays. The repeat delay time-out begins when the microphone PTT is disabled (transceiver PTT unkeyed). Enabling the microphone PTT prior to completion of the repeat time-out resets the timer.

The following example is meant to clarify the operation of repeat delay. Each memory may contain up to 10 seconds of audio. Suppose that memory number one contains six seconds of audio followed by four seconds of no audio. Memory number two contains four seconds of audio followed by two seconds of no audio, followed by three seconds of audio and one second of no audio. Repeat is on and repeat delay is set to six seconds. If memory number one is read, the message will be repeated every 12 seconds. If memory number two is read, the message will be repeated every 15 seconds. The minimum delay in a repeat time-out is a few milliseconds. The maximum delay in a repeat time-out is about 20 seconds.

The Memory In Use LED (DS2) is turned on during a send/write operation. At the completion of reading from or writing to memory, the LED is turned off. During a repeat operation that takes more than 10 seconds, the Memory In Use LED stays on for only the first 10 seconds of the operation (only when the memory is being read). Referring to the previous example, if memory number one is read with repeat on, the Memory In Use LED will remain on for 10 seconds and will be turned off for two seconds. If memory number two was read, the Memory In Use LED would remain on for 10 seconds and would be turned off for five seconds.

Construction

Point-to-point or wire-wrap methods can be used to construct the keyer. A glass-epoxy double-sided board with plated-through holes, solder mask, and a silk-screened parts-placement pattern is available from W2IHY[4]. With this board it is recommended that sockets be used for all of the ICs. Also, all components should be soldered to the board with a low-wattage soldering iron (25 to 40 watts). The board measures 5.5 by 9.375 inches.

Wiring most switches and potentiometers to the PC board is accomplished through 16-pin IC sockets J4 and J5. Two single-ended 24-inch 16-pin DIP jumpers are connected between the various panel LEDs, potentiometers and switches and the sockets on the PC board. The schematic diagram and Table 1 provide details of the connections.

The memory select and reset pushbuttons S1 through S5 can be connected to the board through either IC socket (J4 or J5). If one socket is used to connect front-panel switches to the board, the other socket may be connected to an external source (through a connector) to provide remote control of the keyer. It is important to use shielded cables as shown on the schematic. Decoupling capacitors must be used at the power connectors (0.1 μF) and at the microphone input connections (0.01 μF) to the keyer. Spurious signals will be heard if the capacitors are omitted. Power (ac or dc) entering the keyer must first be grounded on the PC board to prevent ground loops. In the prototype, this was accomplished by using a three-conductor miniature stereo plug and jack for the power connectors.

Power Supply

The audio keyer may be powered by either a 9 to 12 V ac supply or a 10 to 20 V dc supply. Battery backup may also be used for short periods. If only 12 V dc or battery back up operation is desired, diode D14 may be removed. If only ac operation (without battery backup) is desired, components D11, D12, D15, U36, R41, R42, C8, C13 and C14 are not necessary. The battery backup voltage should never exceed the voltage on capacitor C47.

Crystal Frequency Selection

The crystal frequency should be selected to optimize audio bandwidth and message size while considering the minimum refresh rate. All memory cells must be refreshed within 4 milliseconds. The minimum crystal frequency that will do this is 2.048 MHz. The following equation defines (for a given crystal frequency) the message length in seconds.

[4]PC boards and information available from: Julius D. Jones (W2IHY), 15 Vanessa Lane, Staatsburg, NY 12580, tel 914-889-4933.

Table 1
J4 and J5 Pin-outs

J4 Pin	Signal	J5 Pin	Signal
1	VAR sensitivity R47	1	Repeat delay potentiometer R59
2	VAR reset to switch S10	2	Memory pushbutton S2
3	PTT sensitivity potentiometer R45	3	Memory pushbutton S4
4	PTT delay potentiometer R46	4	Memory In Use LED DS2
5	PTT from microphone	5	Send/write switch S6
6	Memory pushbutton S1	6	Ground
7	Memory pushbutton S3	7	Repeat switch S7
8	Keyer pushbutton S5	8	Keyer Reset pushbutton S5
9	Audio out to speaker	9	Repeat switch S7
10	Memory pushbutton S4	10	Power on LED DS3
11	Memory pushbutton S2	11	Microphone Level LED DS1
12	PTT out to transceiver (to S11)	12	SEND/WRITE switch S6
13	Ground	13	Not used
14	PTT sensitivity potentiometer R45	14	Memory pushbutton S3
15	VAR reset (from switch S16)	15	Memory pushbutton S1
16	+5 V VAR sensitivity potentiometer R47	16	+5 V to repeat delay potentiometer R59

Message Time (in seconds) =

$$\frac{\text{Memory Size} \times 1024}{2 \times \text{Crystal Frequency (in hertz)}}$$

Using a 3.2768 MHz crystal, it will take 10.24 seconds to load one 65,536 8-bit word memory segment. It is recommended that a crystal between the frequencies of 3.0 MHz and 3.6 MHz be used. Using a crystal with a frequency below 3 MHz will require modifying the response of the low-pass filters.

Wiring Options

Microphone Impedance—The microphone amplifier shown in the schematic is designed to work with a low-impedance microphone. A high-impedance microphone may be used if the following component values are used: R31 = 47k, R35 = 470k, R32 = 47k.

Microphone Level—Control of the microphone amplifier gain is provided by potentiometer R49. Increasing the value of R49 increases the microphone amplifier gain.

PTT Sensitivity—Potentiometer R45 controls the sensitivity of the PTT circuitry. Increasing the resistance of R45 increases the PTT sensitivity.

PTT Delay—Potentiometer R46 controls the hold time of the push to talk circuit. Decreasing the voltage on the wiper of R46 increases the PTT hold time.

Audio Output Level—Potentiometer R44 controls the audio keyer output level. Increasing the resistance of R44 increases the keyer analog output level.

Monitor Volume—Potentiometer R48 controls the monitor audio output volume. Increasing the resistance of R48 will increase the monitor volume.

Voice Activated Reset Sensitivity—VAR sensitivity controls the microphone audio level that resets the audio keyer. As the voltage on the wiper of R47 is increased, higher microphone audio levels are required to reset the keyer.

Repeat Delay—Potentiometer R59 controls the delay between reading and reread-

ing the memory. Increasing the resistance of R59 increases the delay between repeats.

Operating the Audio Keyer

When power is first applied to the keyer it may be sending or writing. Press the reset button or microphone PTT to cancel these operations. This will return the keyer to a known state.

Message Writing

While you talk into the microphone, adjust the microphone level control (R49) until the microphone level indicator (DS1) begins to light. Next set the send/write switch (S6) to the write position. Push one of 4 memory pushbuttons to load (write) a memory. *Do not press your PTT button at this point*; this will reset the keyer. The Memory In Use indicator (DS2) will stay on until the selected memory is full. It takes about 10 seconds to load a particular memory. During write operations the repeat, monitor output and VAR functions are automatically disabled.

Message Sending

Set the send/write switch (S6) to the send position. Turn off the repeat and VAR switches. Press one of the four memory pushbuttons to begin the send operation. The Memory In Use indicator (DS2) will stay on until the memory read operation is completed. Again, do not press the PTT switch at this point; this will reset the keyer.

PTT Sensitivity and PTT Delay

Follow the message sending procedure. Adjust the PTT delay to mid scale. Adjust the PTT sensitivity (R45) until the transceiver's push to talk is enabled. Adjust the PTT delay potentiometer (R46) for the desired hold time, and readjust the PTT sensitivity control if necessary.

Repeat and Repeat Delay

PTT sensitivity and PTT delay *must* be adjusted for desired operation. Turn on the repeat switch (S7). Follow the message sending procedure. Adjust R59 to obtain the desired spacing between messages.

VAR and VAR Sensitivity

Turn on the Voice Activated Reset (VAR) switch (S10). Follow the message sending procedure. Talk into the microphone. Adjust the VAR sensitivity control (R47) until the keyer is reset.

Transceiver PTT Operation

Transceiver PTT operations may be activated by turning on switch S11 and either sending a message from the keyer or by pushing the microphone PTT. When the microphone PTT is depressed, the keyer is reset. Audio from the microphone is amplified and sent to the transceiver and the transceiver PTT is grounded (activated). Switch S11 should be turned off if VOX operation is desired.

Keyer Bypass

A four-pole double-throw switch (S9) is used to switch either the audio keyer or the microphone to the audio output jack. If S9 has switched the microphone directly to the transceiver, the keyer need not be powered on for the microphone and its PTT to be used with the transceiver.

Audio Output

The audio output control (R44) should be adjusted to drive the transceiver to the same audio level that is obtained from a directly connected microphone.

BOB—An RS-232-C Break-Out Box

The EIA RS-232-C interface standard greatly simplifies the interconnection of digital devices. When something goes wrong at an RS-232-C interface, however, the 25 signal lines and two equipment configurations—data terminal equipment (DTE) and data-circuit terminating equipment (DCE)—can easily confuse the troubleshooter. The RS-232-C break-out box (BOB) is a piece of test equipment used to simplify RS-232-C troubleshooting. Jeff Ward, K8KA, designed and built the break-out box described here in the ARRL lab.

Break-out boxes can be simple or very complex. In its simplest form, a break-out box displays the polarity of the voltage on an RS-232-C line. Deluxe break-out boxes monitor and change the voltages on several RS-232-C lines and alter the wiring at the interface between two RS-232-C devices. The deluxe break-out box is powerful, but expensive. Fortunately, many of the problems that amateurs will encounter can be diagnosed with a relatively simple break-out box.

BOB

This project (Figs 19-23) describes the construction and use of BOB, an average break-out box. BOB is inexpensive, yet useful in many situations. It has the following features: a male and a female DB-25 connector at each end; a patch bay that allows any of the 7 most important lines at one end to be patched to any of the lines at the other end; the ability to set any line to +12 V or −12 V; and bi-color LEDs to indicate the polarity of the voltage on each line.

The circuit (Fig 20) is simple. Each of the signal lines (2,3,4,5,6,8 and 20) is brought from the DB-25 connectors to the DIP socket in the center of the BOB. The DIP socket acts as a patch bay. Any line from one pair of connectors can be patched to any line on the other pair. Each line is attached to a DPDT center-off toggle switch. One position of the switch sets the line to +12 V; the other position sets the line to −12 V. The OFF position lets the line "float" at the voltage applied to it by devices attached to BOB. Bi-color, polarity-

Fig 19—The final version of the RS-232-C break-out box uses PC-board construction. The prototype, built on a perforated board using wire-wrap techniques, performed just as well.

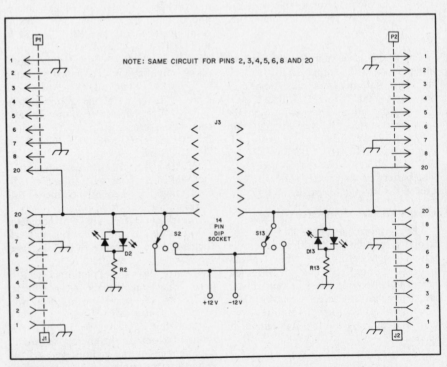

Fig 20—Schematic diagram for the RS-232-C break-out box. The same circuit is used for pins 2, 3, 4, 5, 6, 8 and 20.

D1-D14—Bi-color LEDs (Jameco #XC5491 or equiv.).
J1, J2—DB-25-F (AMP 206584-2 or equiv.).
J3—14-pin DIP socket (3M 214-3339 or equiv.).
P1, P2—DB-25-M (AMP 206604-1 or equiv.).
R1-R14—1.2 kΩ, ¼ W.

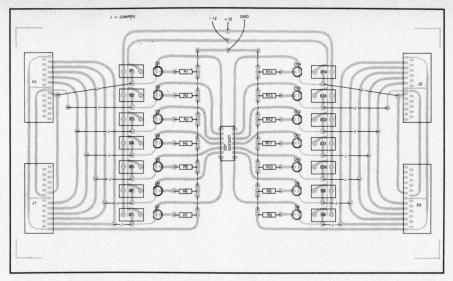

Fig 21—Parts-placement guide for the RS-232-C break-out box, shown from the component side of the board. A full-size etching pattern appears at the back of this book.

indicating LEDs with current limiting resistors display the polarity of each line.

If BOB is to be used to set voltage levels on RS-232-C lines, an external ±12-V power supply is needed. In this case, the ground from BOB should be connected to the power supply ground. Otherwise, BOB should be grounded to the unit(s) under test.

Construction

The prototype for BOB was built on a perforated-board chassis using wire-wrap techniques. The LEDs were mounted on the top of the board, while the switches were mounted through the board using standard mounting hardware. Although not as refined as later PC-board models, the prototype performed well in the ARRL lab.

To avoid the complicated, time-consuming wire-wrap process, a printed circuit board was designed for BOB. All components, including the switches, LEDs and DB-25 connectors, are mounted directly on top of the board. Because the board is single sided, some jumpers are necessary. Parts placement is shown in Fig 21. Be careful to insert all of the LEDs in the same direction, with their longer leads toward the center of the board. A full-size etching pattern for this board can be found at the back of this book.

Partial Construction

If you do not want to invest in a complete set of parts for BOB, you can construct part of the circuit. If you populate one-half of the board, you will be able to use your BOB for everything but in-line monitoring. Installing only the LEDs results in a fine RS-232-C monitoring device. Of course, you can leave one of the DB-25 connectors off of each end and

simply have a less versatile BOB. If you initially construct a partial BOB, you can add the remaining components at any time.

The Patch Bay

The prototype had a female DB-25 connector in the center of the board for use as a patch bay. The PC-board version uses a 14-pin zero-insertion-force (ZIF) DIP socket (Fig 22). The DIP socket, although smaller and more delicate than the DB-25, is the better choice. A 7-circuit DIP switch can be inserted into the socket and used to switch through or interrupt any of the 7 RS-232-C lines. DIP component headers can also be used to wire various patch bay configurations. Although any 14-pin DIP socket will work, one with zero- or low-insertion force makes changing patches or switches much easier.

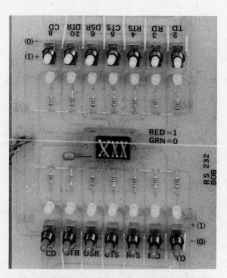

Fig 22—The zero-insertion-force DIP socket in the center of the break-out box can hold DIP headers with various jumper configurations.

Potential Problems

Before you use BOB, you should understand two circuit shortcomings. First, BOB draws current from the unit under test. Second, unless you are careful, you can use BOB to create short circuits.

BOB draws current from the circuit under test to light the polarity-indicating LEDs. Using current from the circuit being monitored eliminates the need for active buffers, transistor switches, and a power supply. However, it means that BOB loads the RS-232-C device hooked to it. In extreme cases, this loading may interfere with circuit testing. If you suspect this is the case, use an oscilloscope or VOM to check circuit voltages. If positive voltages are much below 12 V or negative voltages are much above – 12 V, the circuit is overloaded.

It is possible to switch a line to + 12 V on one side of the patch bay, patch it across to the other side, and then switch it to – 12 V. This will short-circuit your power supply. You will also cause a short circuit if you attempt to reverse the polarity of an RS-232-C line which is being driven by the equipment under test. Be sure that you do not attempt to set one line to two levels, and be sure that the equipment under test is not setting a line that you are trying to set with BOB. If you are careful, you will not be bothered by the economies in BOB's design.

Using The Break-out Box

You should become familiar with the material in Chapter 19 concerning the RS-232-C interface. While only the numbers for the various lines will be referred to here, the full names of the lines can often give insight into their functions. It is also valuable to know which type of device, DCE or DTE, generates each signal.

When you are familiar with the RS-232-C interface, and familiar enough with BOB to understand its limitations, you will be able to begin using it. It is impossible to anticipate every configuration of BOB that will be useful. However, descriptions of a few common problems and how BOB is used to solve them can help you learn how to use it.

Activating DCE

It is often necessary to test a modem or other piece of data-circuit terminating equipment (DCE) without having it attached to a computer. To activate a DCE, lines 20 and 4 must be set to + 12 V. To do this, use BOB with no jumpers in the patch bay and a power supply attached. Put all of BOB's switches in the OFF position. Connect the DCE to one of BOB's DB-25 connectors, and set the switches for pins 4 and 20 to the + 12 position (Fig 23A). Be sure to use the switches on the same side of the patch bay as the DB-25 that you are using. Now, assuming that the DCE is a modem, you should see a carrier at its analog output.

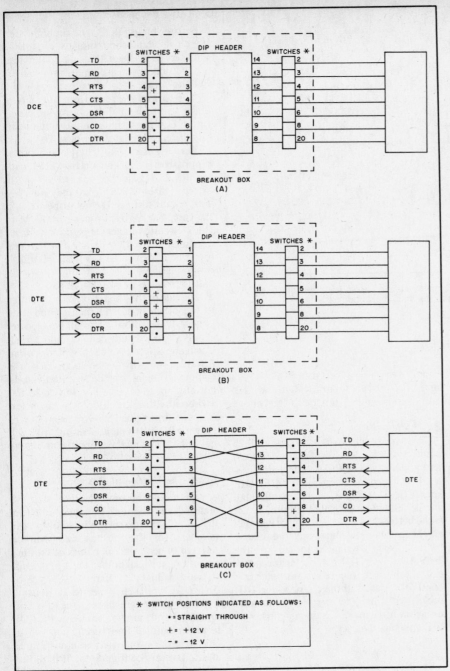

Fig 23—A shows how to use the break-out box to activate a DCE device. To activate a DTE device, use the switch settings shown at B. The jumpers and switch settings shown at C configure the break-out box for connecting two DTE devices.

modem might never send any data.

If the modem does not respond correctly, then you have learned that it is not operating as a standard DCE device. Further troubleshooting is in order. If the modem does respond as predicted, you know that any problems you have are caused by some other piece of equipment.

Activating DTE

BOB can also be used to activate a data terminal equipment (DTE) device, such as a terminal, to see if it is operating correctly. When BOB is configured as shown in Fig 23B, data sent by the DTE should appear as flickering of the LED on line 2.

If the data does appear, there are still some things you should use BOB to determine: Does the DTE set line 20 to + 12 V? Many DCE devices must have this line set to + 12 V. Does the DTE under test set line 4 to + 12 V before sending data? How does the DTE react to changes on lines 5, 6 and 8? You can answer these questions quickly by changing switches and observing LEDs. The answers will come in handy when you wish to interface the DTE to other equipment.

Connecting Like Devices

It is sometimes necessary to connect two RS-232-C devices with the same configuration (DCE or DTE). If two terminals (DTE devices) are simply wired together, they would both attempt to transmit on line 2 and to receive on line 3. The transmitted signals would collide and no received signals would be heard. Also, both units would be setting line 4 to + 12 V, but lines 5, 6 and 8 would be set by neither. This would not work.

With the aid of BOB, and a DIP header wired as shown in Fig 23C, two DTE devices can be connected with no problems. In this configuration BOB is called a "null modem," since it replaces the modems that would usually connect two terminals. Two DCE devices can be connected in a similar manner.

Conclusion

These are only a few of the things you can do with BOB. BOB is a powerful device for monitoring and changing signals at an RS-232-C interface. If you are careful to avoid short circuits, BOB can teach you about RS-232-C interfaces and help solve confusing problems.

By setting the switch for line 2 (transmitted data) to + 12 V you should be able to change the tone coming from the modem.

LEDs should indicate that the modem is setting pins 5 and 6 to + 12 V. These lines are used to activate equipment attached to the modem. If they are inactive, or remain at − 12 V, then a device attached to the

An LSI Modem for Amateur Radio

Whether you are trying to get on the air with a complex packet-radio terminal node controller or simply hooking up your home computer to copy RTTY, you will need an interface to change the analog output of your radio to digital input for your computer, and to change the digital output from your computer to analog input for your radio. This interface MODulates input to the radio and DEModulates output from the radio, so it is called a *modem*. This project, shown in Figs 24 through 27, presents a modem designed around an Advanced Micro Devices Am7910 IC.[1] Use of this large scale integrated (LSI) circuit device greatly simplifies construction of an inexpensive and simple, yet versatile and noise-immune, modem. Jeff Ward, K8KA designed and built this project in the ARRL lab.

[1]Devices, application notes and information are available from Advanced Micro Devices, 901 Thompson Pl, PO Box 3453, Sunnyvale, CA 94088.

The Am7910

The Am7910 is a complete modem in a 28-pin DIP package. The chip requires no external filters or control circuits. Fig 24 shows that all of the control and data lines for a full-duplex FSK modem are available right on the chip. Five "mode-control" pins allow the selection of various FSK tone pairs and filter parameters. External circuitry is needed only to buffer the input and output signals and to provide operating voltage to the Am7910.

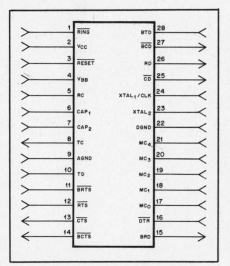

Fig 24—The Am7910 is a complete, full-duplex modem in a 28-pin DIP package.

The modulator and demodulator in the chip employ both analog and digital signal-processing circuits. The incoming signal first passes through an analog prefilter, then is converted by an analog-to-digital (A/D) converter, then passes through digital band-pass filters, and finally is digitally demodulated. The modulator is a digital sine synthesizer, followed by digital band-pass filters, a digital-to-analog converter (DAC), and an analog post filter. As a result of this mixture of analog and digital circuitry, the Am7910 has a clean, stable modulator and a selective demodulator.

Modes and Mode Control

The mode-control pins (pins 17-21) on the Am7910 deserve some explanation. These pins should be thought of as five binary bits. Changing the voltages on the pins changes the binary word which they represent. Different "control words" cause the modem to enter different modes.

Four of the available modes are used in radio communication. Table 1 shows the switch settings and binary codes for these modes. Bell 202 is a 1200-baud FSK protocol found on most packet-radio networks. The only difference between the two Bell 202 implementations is in demodulator filter response.

Bell 103 is a 200-Hz shift FSK protocol. It is used for packet radio on the HF bands, and is also very close to the 170-Hz shift

used by standard amateur RTTY stations. Bell 103 is a full-duplex protocol—the modem receives on one pair of tones while simultaneously transmitting on another pair. Unfortunately, for Amateur Radio use, we need to have the modem receive and transmit on the same tone pair. On most modems, this means changing a switch setting every time the station is switched from transmit to receive or vice versa. However, the "loopback" modes on the Am7910 solve this problem. These modes are usually used for testing the modem. For that purpose, the modem transmitter and receiver are put on the same tone pair. Since this is what is desired in the half-duplex radio environment, the loop-back modes are used for normal Amateur Radio operation.

Circuit Description

Using discrete components to build a modem as versatile as the Am7910 would be a costly and delicate task. In contrast, only 10 components and a power supply are needed to put the Am7910 in operation. The circuit presented here (Fig 25) is not the bare-bones implementation of the Am7910. To make the modem easy to use, this circuit adds two ICs and a few other components to the minimum configuration.

While the Am7910 uses 0 V for a logic low and +5 V for a logic high, computer equipment which adheres to the RS-232-C interface standard uses +12 V for logic high and −12 V for logic low. U2 and U3, a 1489 RS-232-C line receiver and a 1488 RS-232-C line driver, make the modem RS-232-C compatible.

A single DIP package contains S1 through S5, the mode-control switches. These switches make it easy to use any of the modes available on the AM7910. The mode-control pins are TTL compatible and draw little current. When one of the switches (S1-S5) is open, the corresponding mode-control pin is set to +5 V (logic high) via one of resistors R1-R5. When one of the switches is closed, the corresponding pin is grounded (logic low). In the grounded position, the high value chosen for R1-R5 reduces current drain from the +5 V supply. Of course, the mode of operation could be hard-wired into the circuit, but why sacrifice versatility?

The rest of the chip is wired as specified in Advanced Micro Devices application notes for the Am7910, ignoring all but one

of the "back-channel" connections. The one back-channel control line which must be connected is Back Request To Send (BRTS), pin 11. This pin must be set to +5 V, unasserting BRTS; otherwise the back channel would take control of the modem. A copy of the applications notes for the Am7910 will help you to better understand the circuit.

Power Supply

Since the Am7910 runs on ±5 V and the RS-232-C interface demands ±12 V, the circuit includes a four-voltage regulated power supply. Regulators rated at 100 mA are used on all but the +5 V supply. The +5-V supply must provide more than 100 mA, so a 1-A regulator and a small heat-sink are employed. To keep the supply small, a toroidal power transformer was used. If you do not intend to mount your modem in a small box, any transformer with a 25-V, center-tapped secondary could be used. The combination of C25 and R10 causes a 5-ms rise time on the +5-V supply, as required by the Am7910. The rest of the power supply design is standard, with special attention paid to the selection and placement of bypass capacitors. For a complete discussion of regulated power supplies, see Chapter 6.

Construction Technique

Although the circuit can be built on a perforated board using point-to-point wiring, it is recommended that you use the circuit-board layout provided (Fig 26). Use of the circuit board not only speeds construction and trouble-shooting, but results in a stable, reproducible circuit. Sockets are used for all ICs, and for the DIP switch

---

Table 1
Am7910 Modes Used in Amateur Radio

Switch Setting†

S5	S4	S3	S2	S1	
MC4	MC3	MC2	MC1	MC0	Mode
1	0	0	0	0	Bell 103 originate loop-back.
1	0	0	0	1	Bell 103 answer loop-back.
1	0	0	1	0	Bell 202 main loop-back.
1	0	0	1	1	Bell 202 with equalizer loop-back.

†1 indicates an open switch; 0 a closed switch.

Table 2
Pin-outs for P1 and P2

P1		P2	
Pin	Signal	Pin	Signal
1	GND	1	Transmit carrier (TC)
2	TD	2	GND
3	RD	3	GND
4	RTS	4	Receive carrier (RC)
5	CD		
6	CTS		
7	NC		
8	DTR		

P1 has pin 1 toward the power supply, while P2 has pin 4 toward the power supply.

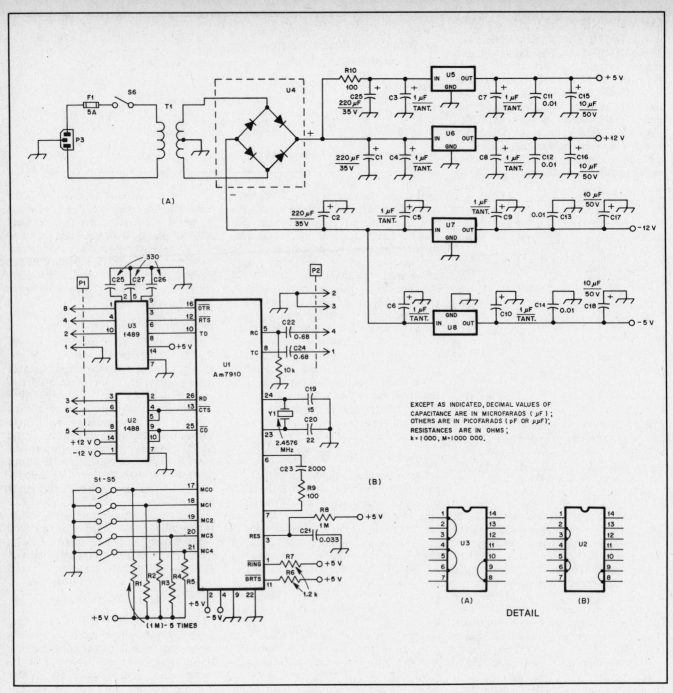

Fig 25—Schematic diagram for the Am7910-based modem and power supply. All resistors are ¼ W.

C1,C2,C25—220 µF, 35-V electrolytic, radial leads.
C3-C10—1 µF tantalum.
C11-C14—0.01µF disc ceramic.
C15-C18—10 µF, 50-V electrolytic, axial leads.
C19—22 pF silver mica.
C20—15 pF silver mica.
C21—0.033 µF disc ceramic.
C22,C24—0.68 µF disc ceramic.
C23—2000 pF disc ceramic.
P1—8-pin locking header (GC-41-088 or equiv.).
P2—4-pin locking header (GC-41-084 or equiv.).

P3—CEE-22 male, chassis mount.
R10—100 Ω, 2 W.
S1-S5—SPST, contained on a single DIP switch.
S6—SPST.
T1—Power transformer; 117-V primary, 30-V CT secondary (Cat. no. 102.152; available from Toroid Corp. of Maryland, 4720-Q Boston Way, Lanham, MD 20706).
U1—Advanced Micro Devices Am7910 (see text).
U2—1488 RS-232-C line driver. For TTL interfacing, use jumpers shown at A in detail. See text.
U3—1489 RS-232-C line receiver. For TTL

interfacing, use jumpers shown at B in detail. See text.
U4—Bridge rectifier (Radio Shack 276-1146 or equiv.).
U5—LM7805 +5-V, 1-A voltage regulator.
U6—LM78L12 +12-V, 100-mA voltage regulator.
U7—LM79L12 -12-V, 100-mA voltage regulator.
U8—LM79L05 -5-V, 100-mA voltage regulator.
Y1—2.4576-MHz crystal (M-Tron MP-2 or equiv. Available from M-Tron Industries Inc., P.O. Box 630, Yankton, SD 57078.).

Fig 26—Using the toroidal power transformer, the modem and the power supply fit neatly on a small PC board.

(S1-S5). Locking, polarized connectors are used for all digital and analog signals to and from the modem. See Table 2. Parts placement for the PC board is shown in Fig 27, and a full-size etching pattern appears at the back of this book.

The circuit board is mounted on the bottom plate of an $8 \times 10 \times 2$-in chassis. Appropriate digital, audio and power connectors, and an ON/OFF switch are mounted on the chassis itself, and the chassis is then attached to the bottom plate, over the circuit board. Of particular interest is the use of a CEE-22 power connector with integral fuse holder, which saves some precious rear-panel space.

Circuit Check-Out

Once you have mounted all of the components on your PC board, make some tests before putting any of the ICs in their sockets. Check for properly regulated voltages at the ungrounded ends of C15, C16, C17 and C18. Check to see that there is $+5$ V on pin 2 of U1 and pin 14 of U2. Check that -5 V is on pin 4 of U1. Pin 1 of U3 should be at -12 V, and pin 14 of U3 should be at $+12$ V. Once you have made these measurements, turn off the supply and insert U2 and U3 in their sockets. Turn on the supply, and measure the voltage at every pin on the socket for U1. None of these pins should have a voltage greater than $+5$ V or less than -5 V. If any of these pins are outside of the ± 5 V range, check to see that U2 and U3 are in the correct sockets, and that all PC-board jumpers are installed correctly. Only when you have ensured that no voltages too high or too low are being applied to the socket for U1 should you install U1 and attempt to use the modem. It may be hard to resist the temptation to simply plug everything in at once and "go for it"—but remember, the Am7910 is the single most expensive component on the board. Its protection is worth a few minutes delay.

Modem Operation

Once you have completed the preliminary check-out of your modem, connect it to a computer or (preferably) an RS-232-C break-out box. Set the mode-control switches to Bell 202 main loopback (switches 1, 3 and 4 in the ON position; switches 2 and 5 OFF). Now set RTS and DTR to $+12$ V. The modem should respond by setting CTS to $+12$ V. CD should remain at -12 V. If you now connect pins 1 and 4 on the audio connector (connecting the transmit output to the receive input) CD should go to $+12$ V, and whatever voltage is on TD should also appear at RD. If all goes as predicted, hook the modem to your rig, and you're on the air!

Trouble Shooting

Although the Am7910 is a complex chip, and not all possible problems can be covered here, a few common problems should be mentioned.

If the modem operates well on the test bench, but erratically when connected to a radio, you have RFI problems. If the modem is not in a box, put it in one. Ground the box, and put ferrite beads on all lines into and out of the modem. Shield all of the I/O cables. Unless you have unusual amounts of RF in your shack, these precautions should eliminate the RFI.

If CTS does not go to $+12$ V when you set RTS and DTR, momentarily ground the case of the crystal (Y1). If this solves the problem, your crystal is not starting to oscillate when you turn on the modem. Although there is no sure fix for this problem, try varying the values of C19 and C20. Experiment with asymmetrical values between 10 and 20 pF. A high value (1-MΩ) resistor across the leads of Y1 may also result in consistent oscillator start-up. If oscillator start-up does not seem to be the problem, check to see that you have indeed sent the correct voltages to U1: pin 12 and pin 16 should see 0.0 V (resulting from the $+12$-V signals on RTS and DTR being inverted and buffered by U2). Pin 11 and pin 1 (BRTS and RING) should be at $+5$ V, disabling the back channel and enabling the modem. If all of these conditions are met, and your modem still does not work, get the application notes from the manufacturer and start troubleshooting.

Non-RS-232-C Operation

Some home computers do not have true RS-232-C interface signals. These machines commonly use $+5$ V and 0 V for I/O. Since the Am7910 uses these same signal levels, such computers can be interfaced to the modem without U2 and U3, the RS-232-C transmitter and receiver. U2 and U3 should be replaced with jumpers as shown in Fig. 19. Remember that control signals will now be considered ON when they are at 0 V and OFF when they are at $+5$ V. Check to see that this is how your computer's control signals are defined. If they send $+5$ V when they are ON you will have to invert them before sending them to the Am7910.

Experiments

If you're interested in experimentation, there are some interesting things that can be tried with the Am7910. Both of these experiments should allow the modem to be used on 170-Hz RTTY. The simplest way to use the Am7910 on 170-Hz RTTY is to use one of the 200-Hz shift Bell 103 loopback modes. The 30-Hz difference between

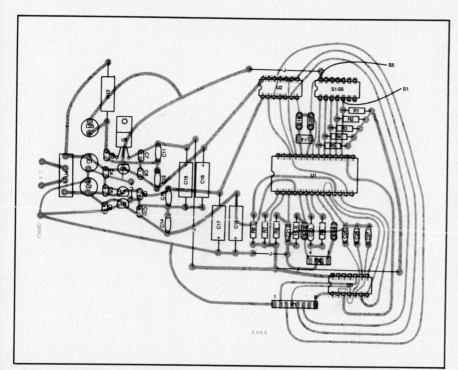

Fig 27—Parts-placement guide for the Am7910-based modem and power supply, shown from the component side of the board. A full-size etching pattern appears at the back of this book.

the amateur tones and the Bell 103 tones does not degrade performance very much. Try it and see!

Another way to put the Am7910 on 170-Hz shift would be to change the frequency of Y1. Since Y1 is the frequency reference for the entire modem, lowering the frequency of Y1 by 15% should change the 200-Hz shift to 170-Hz. This modification has not been tried yet, but that is what

home-brewing is all about! If you are going to use your Am7910 modem on HF, be sure you understand what RF tones will result if you use an SSB transmitter with the AF tones from the modem. You will want to be sure that you are listening and transmitting on the same frequencies.

Conclusion

The Am7910 modem, using the circuit presented here, has been in use for some time on a packet-radio repeater at ARRL Headquarters. If you are interested in many facets of digital communication, this modem, with its many modes, could be the solution to your interface problems. For a state-of-the-art project that's easy to build and useful in the shack, the Am7910 "World Chip" fills the bill.

Chapter 30

HF Radio Equipment

In this chapter you will find descriptions of radio equipment that you can build for the amateur bands below 30 MHz. The projects fall into three broad categories: receivers, transmitters and power amplifiers.

There is much personal pride and satis-faction in building and using a piece of radio equipment. The knowledge you gain of the workings on both sides of the front panel is another source of pleasure.

The theory necessary to understand the operation of the equipment described in this chapter is found elsewhere in this book. Some of the chapters you may wish to review are: Chapter 10, Radio Frequency Oscillators and Synthesizers; Chapter 11, Radio Transmitting Principles; Chapter 12, Radio Receiving Principles and Chapter 15, Radio Frequency Power Amplifiers.

A Band-Imaging CW Receiver for 10 and 18 MHz

Band imaging has long been used in Amateur Radio as a means of making a stable local oscillator (LO) do double duty. Instead of building equipment using only one LO-to-RF relationship for frequency conversion—with, for instance, only the *difference* between the LO and higher-frequency incoming signals giving output at the IF—*two* of several LO-to-RF relationships can be exploited for two-band coverage. A band-imaging receiver appeared in every edition of this *Handbook* from 1953 through 1966, from "A Two-Band Four-Tube Superheterodyne" in 1953 to "The HB-65 Five-Band Receiver" in 1966. Each of these receivers converted the 80- and 40-meter amateur bands to a 1.7-MHz IF by means of a 5.2-5.7-MHz LO. On 80 meters, the conversion relation-ship in such a receiver is LO − RF = IF; on 40, the relationship is RF − LO = IF. Both bands "tune in the same direction" with this system: The received frequencies of 3.5 and 6.9 MHz correspond to the lower limit of the LO tuning range.

Band imaging can also be used to cover the 80- and 20-meter amateur bands: A 5.0-5.5-MHz LO is used to convert each band to a 9-MHz IF. In such a system, the LO-to-RF relationship on 80 meters is RF + LO = IF; on 20, RF − LO = IF. The drawback to this band-imaging system is that the lower band "tunes backwards": The lower limit of the LO tuning range corresponds to 4.0 MHz on 80 meters and 14.0 MHz on 20. Nonetheless, the 80/20 band-imaging system has also been popular with radio amateurs because of the inherent sideband inversion between the image bands: The BFO-to-IF relationship that affords LSB reception on 80 meters demodulates USB on 20.

Fig. 1 — The band-imaging receiver covers the CW segments of the 10- and 18-MHz amateur bands with good stability, sensitivity and single-signal selectivity. Larger controls are (l-r) TUNING, IF GAIN and AF GAIN. The two smaller controls, separated by the PHONES jack, are BAND and SIDETONE LEVEL. The TUNING scale is drawn on contact paper applied to an aluminum disc.

With this overview of band-imaging techniques in place, we present a band-imaging CW receiver for 10 and 18 MHz (see Fig. 1). Using a 14-MHz LO, it converts the entire 10-MHz amateur band, and the CW portion of the 18-MHz amateur band, to a 4-MHz IF. Both bands tune in the same direction. At 4 MHz, a four-crystal ladder filter provides single-signal selectivity. The design emphasizes good basic receiver performance with an eye toward compactness; hence, features such as a digital frequency display, AGC and active audio filtering have been omit-ted. Alignment and checkout of the band-imaging receiver requires only (1) a 51-ohm resistor; (2) a receiver capable of CW recep-tion at 14.0-14.2 MHz and 4 MHz ±1 kHz with an S-meter and frequency display resolution of 1 kHz or greater; and (3) a

crystal-controlled marker generator capable of providing 10-kHz markers. The perfor-mance measurements given later in this article were obtained from a receiver aligned *by ear* with such test equipment. You need not have access to a radio lab to enjoy similar results. David Newkirk, AK7M, designed and built this project in the ARRL lab.

Circuit Description: RF Amplifiers

A separate 40673 RF amplifier is used for each band. (See Fig. 2). The circuit is electrically identical to that used for the RF amplifiers in "A High-Performance Com-munications Receiver," presented later in this chapter. Several other circuits in the band-imaging receiver are based on the K5IRK/W7ZOI high-performance design. To simplify alignment of the band-imaging

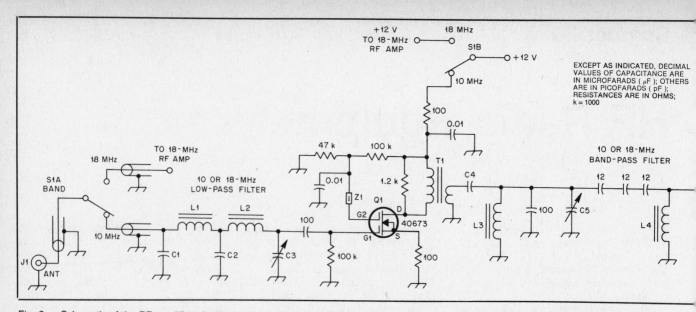

Fig. 2 — Schematic of the RF amplifiers for the band-imaging receiver. A separate amplifier is used for each band. The low- and band-pass filters may be aligned with the aid of a crystal-controlled marker generator; see text. Capacitors are disc ceramic unless otherwise noted. Capacitors marked with polarity are electrolytic. All resistors are ¼-W, 10% units unless otherwise noted.

C1, C2, C4, C7 — Silver mica, polystyrene or ceramic capacitor; see Table 1 for values.
C3, C5, C6 — Ceramic or mica compression trimmer. Mouser Electronics ceramic trimmer 24AA067 (12-100 pF) used for 100 pF; Mouser 24AA064 (5-45 pF) used for 45 pF.
J1 — Coaxial RF connector.

L1-L4 — Wound on Amidon T-50-6 powdered-iron toroid core or equiv. All inductors use no. 22 enameled wire with one exception: For L2 at 10 MHz, use no. 24 enameled wire. See Table 1 for number of turns.
Q1 — 40673 dual-gate MOSFET.
S1 — 3PDT toggle (Radio Shack 275-661 or equiv).

T1 — Transformer wound with no. 28 enameled wire on Amidon FT-37-43 ferrite toroid core or equiv. Primary (168 µH), 20 turns; secondary (6.7 µH), 4 turns.
Z1 — Ferrite bead on Gate 2 lead of Q1, Amidon FB-43-101 or equiv.

Table 1

Component Values for the Band-Imaging Receiver RF Amplifiers

MHz	C1	L1	C2	L2	C3	C4, C7	L3, L4	C5, C6
10	300	13	680	29	100	33	17	100
18	180	10	390	22	45	22	10	100

Values listed for capacitors are capacitance in pF. Values listed for inductors are number of turns of wire required.

receiver, the variable coupling capacitor (C15 in Ch. 30, Fig. 12) between the two sections of the output filter is replaced by three 12-pF capacitors in series. Gain of this circuit is 12 to 15 dB, depending on alignment and the characteristics of Q1. Band changing is accomplished by switching RF input, RF output and dc connections between the 10- and 18-MHz amplifier boards via S1, a 3PDT toggle. Input and output (I/O) impedances of each RF amplifier board are 50 ohms.

Mixer, IF Filter and IF Amplifiers

See Fig. 3. The band-imaging receiver uses a Mini-Circuits SBL-1 doubly balanced diode-ring mixer (U1) followed by a strong bipolar-transistor IF amplifier (Q2).[1] This is the circuit used in the K5IRK/W7ZOI receiver, with several modifications. In the band-imaging receiver, the bifilar 4:1 collector transformer in the original design has been replaced with a toroidal monofilar choke,

[1]Mini-Circuits, PO Box 166, Brooklyn, NY 11235, tel. 212-934-4500.

RFC1. The supply end of the 1-kΩ Q2 base bias resistor is now connected directly to the 12-V dc line at the cold end of RFC1. This removes the RF feedback present in the original circuit. Surprisingly, this feedbackless configuration results in better sensitivity *and* two-tone 3rd-order IMD dynamic range than the unmodified circuit affords *at a 4-MHz IF*. The original circuit, intended for use at an IF of 9 MHz, did not provide a comparable performance even when the inductance of its 4:1 collector transformer was scaled for 4 MHz.

The post-mixer amplifier feeds a four-crystal ladder filter via a 6-dB pad. The I/O impedances of the crystal filter are 200 ohms. Because this is a good match for the collector impedance of Q2, the step-down transformer in the original post-mixer amplifier circuit is not required. The 50-ohm 6-dB attenuator of the original circuit has been scaled to 200 ohms. This pad should *not* be replaced with one of lower attenuation: It assures a nonreactive wideband termination for Q2 *and* the crystal filter. Less attenuation here results in reduced IMD dynamic range, as confirmed by lab tests.

The crystal filter was designed using Hayward's technique (see "Simple Cohn Crystal Filters," *QST*, July 1987, pp 24-29). Measured selectivity of the prototype filter was 405 Hz at −6 dB and 1850 Hz at −60 dB, resulting in a −60 dB/−6 dB shape factor of 4.57. Insertion loss was 2 dB, and passband ripple was less than 0.4 dB. As is characteristic of simple ladder crystal filters, the upper passband slope is steeper. Because of this, the BFO must be set on the *upper* side of the filter for best single-signal reception. With the BFO set to provide a 550-Hz beat note for signals at IF center, rejection of the audio image in the prototype receiver was 73 dB. Ultimate attenuation was 90 dB.

No filter adjustment is necessary, but it *is* important that you use the specified crystals if you intend to duplicate the post-mixer-amplifier/pad/filter arrangement shown in Fig. 3. Substitutions at Y1-Y4 will require filter capacitors of other than 300 pF, resulting in I/O impedances of other than 200 ohms. Hayward states that the series-resonant frequencies of the four filter crystals must fall within a spread of no more than 10% of the desired *3-dB* filter bandwidth. We chose to evaluate the performance of the filter in the more popular terms of −60 and −6 dB bandwidths; it follows that 10% spread is too generous where a given *−6 dB* filter bandwidth is the target. Experiments with various new and surplus 4-MHz microprocessor-clock crystals in the ARRL lab showed that the new International Crystal Mfg. (ICM) crystals provided the best

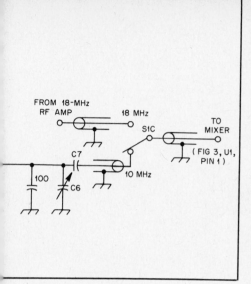

performance overall. Shape factors (−60 dB/−6 dB) for the clock-crystal filters were rarely less than 5, and sometimes more than 6. I/O impedances were between 300 and 400 ohms. Several times, four crystals within a suitably narrow frequency spread could be found only by grading *10 or more* clock crystals. Custom-ground crystals offer the added advantage of resonating —within tolerance, of course—on the frequency you specify. Their unit price is higher, but they come closest to guaranteeing that your filter will perform as predicted.

Post-filter IF amplification is provided by U2, an MC1350P video amplifier IC. The 200-ohm resistor between pins 4 and 6 of U2, in conjunction with the 0.1 μF bypass capacitor at pin 6, terminates the crystal filter output. Manual gain control is achieved by applying a variable positive voltage to pin 5 of U2 through a 27-kΩ

resistor and IF GAIN control R1. Receiver muting is accomplished by means of Q3: Grounding the MUTE terminal (center conductor at J2) applies maximum gain-reduction voltage to U2. The supply voltage (nominally 12) appears across J2 with the receiver unmuted; current through the grounded MUTE line is 5 mA. IF output (Z ≈ 50 ohms) is available at the secondary of T2.

Local Oscillator

The schematic of the band-imaging receiver LO is shown in Fig. 4. An MPF102 JFET, Q4, operates as a Colpitts oscillator. The oscillator signal is amplified by Q5, a 40673 dual-gate MOSFET. Bandspread is achieved by tapping the tuning capacitor, C9, down on LO tank inductor L5. Tuning range of the circuit is approximately 14.060 to 14.153 MHz. Air-dielectric trimmer C10 shifts this range for dial calibration.

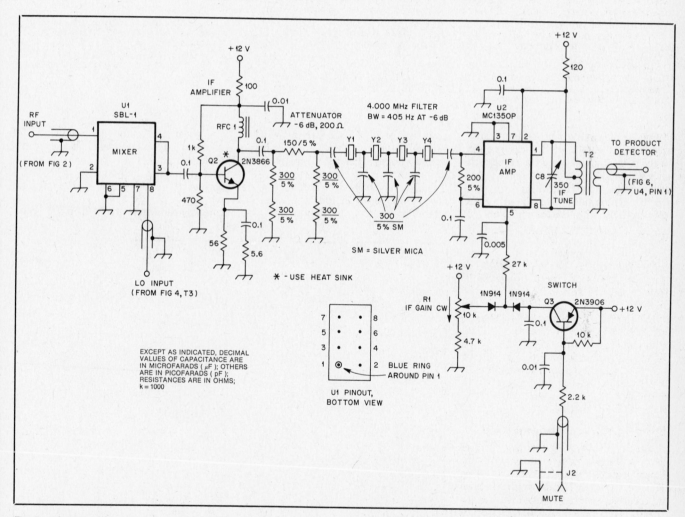

Fig. 3 — Schematic of the mixer, crystal filter and IF amplifier stages of the band-imaging receiver. Capacitors are disc ceramic unless otherwise noted. Capacitors marked with polarity are electrolytic. All resistors are ¼-W, 10% units unless otherwise noted.

C8 — 350-pF compression trimmer (Arco 428 or equiv).
J2 — Phono jack.
Q2 — 2N3866 or 2N5109. Use a small heat sink on this transistor.
Q3 — 2N3906.
R1 — 10-kΩ linear potentiometer.
RFC1 — 95 μH; 15 turns no. 24 enameled

wire on Amidon FT-37-43 ferrite toroid core.
T2 — Transformer wound on Amidon T-68-1 powered-iron toroid core, or equiv. Primary (12.9 μH): 36 turns no. 26 enameled wire, center-tapped; secondary (0.9 μH): 3 turns no. 26 enameled wire over center of primary.

U1 — Mini-Circuits SBL-1 doubly balanced diode-ring mixer.
U2 — MC1350P video amplifier IC.
Y1-Y4 — 4.000000-MHz custom-etched crystal, 26°C calibration temperature, grade CS-1 (0.001% tolerance), F-700 holder, series resonant. International Crystal Mfg Co type 433340. See text.

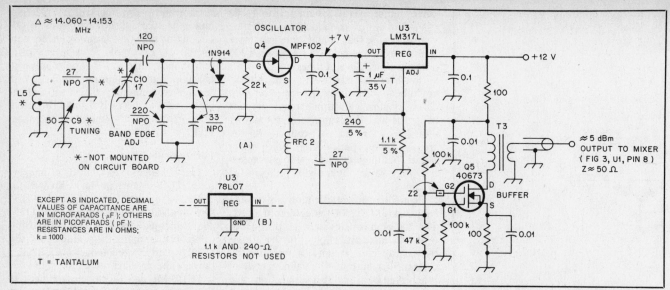

Fig. 4 — Schematic of the band-imaging receiver LO and buffer circuit. Capacitors are disc ceramic unless otherwise noted. Capacitors marked with polarity are electrolytic. All resistors are ¼-W, 10% units unless otherwise noted. At A, an LM317L adjustable regulator is used at U3. The inset at B shows connections for an 78L07 regulator at U3. For best stability, use only NP0 (C0G) capacitors in the circuitry associated with the gate and source leads of Q4. Space L5 by at least its diameter from other components and the LO shield box. See text and Fig. 8.

C9 — 50-pF air variable (Jackson Bros 4667-50 or equiv).
C10 — 17-pF air trimmer (Johnson 189-506-5 or equiv).
L5 — 1.4 µH: 11 turns no. 22 tinned wire, 24 turns per inch (Barker & Williamson 3038 Miniductor). Tap at 2 or 3 turns from ground end. See text and Figs. 8 and 9B.

Q4 — MPF102 JFET.
Q5 — 40673 dual-gate MOSFET.
RFC2 — 39 µH: Miller 70F395AI, or 24 turns no. 26 enameled wire on Amidon FT-50-61 ferrite toroid core.
T3 — Transformer wound on Amidon FT-37-43 ferrite toroid core or equiv.

Primary (50 µH): 11 turns no. 26 enameled wire. Secondary (3.8 µH): 3 turns no. 26 enameled wire.
U3 — Voltage regulator, LM317LH, LM317LZ or (with circuit changes shown at inset B) 78L07.
Z2 — Ferrite bead on Gate 2 lead of Q2, Amidon FB-43-101 or equiv.

Despite the relatively high LO operating frequency, stability is good. Measured drift of the point-to-point-wired prototype oscillator was − 530 Hz in the 45-minute period after turn-on, 460 Hz of which occurred in the first ten minutes. Over the next three hours, this oscillator drifted approximately − 20 Hz. Stability was even better with the circuit rebuilt on an etched circuit board: Drift for the ten minutes after turn-on was only − 256 Hz. *The key to this stability is the use of NP0 (C0G) ceramic units for all fixed capacitors associated with the gate and source of Q4.* Although silver-mica or polystyrene capacitors may be hand-picked for low drift, only NP0 capacitors offer minimum drift "off the shelf." Oscillator stability is further improved by the use of a three-terminal regulator to stabilize Q4's drain supply, and by enclosing LO and buffer in a shield box to slow the effect of changes in ambient air temperature.

BFO, Detector and Audio Stages

The K5IRK/W7ZOI crystal-controlled BFO is used in this receiver with one modification: The secondary of T4 in Fig. 5 carries only RF and no dc. Y5 is an inexpensive 4-MHz microprocessor clock crystal. Every such crystal we tried worked well in this circuit; a custom-ground crystal is unnecessary here.

The detector and AF stages of the band-imaging receiver are shown in Fig. 6. The product detector (U4) is a Mini-Circuits SBL-1 doubly balanced diode-ring mixer. RFC3 and the 0.001-µF capacitor provide RF filtering ahead of the AF preamp, U5, an NE5534 low-noise audio op amp. The

parts list for Fig. 6 specifies a "wind-it-yourself" toroidal choke for RFC3; pi-wound chokes tried here were prone to pickup of 60-Hz harmonics.

U6, an LM380N-8, serves as the AF power amplifier. Its output is connected to a front-panel stereo headphone jack, J4,

and a rear-panel phono connector, J5. J4 is wired to accept stereo headphones; monaural phones may be used if inserted no farther than the first detent. The 1-kΩ resistor from the output lead to ground serves to charge U6's 470-µF output coupling capacitor at power-up if a head-

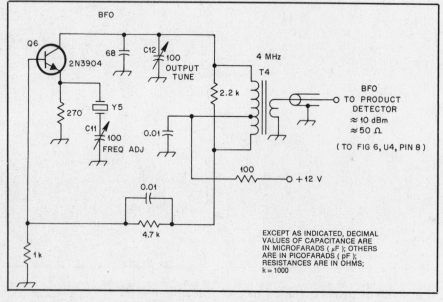

Fig. 5 — Schematic of the band-imaging receiver BFO. Capacitors are disc ceramic unless otherwise noted. Capacitors marked with polarity are electrolytic. All resistors are ¼-W, 10% units unless otherwise noted.

C11, C12 — 100-pF ceramic or mica compression trimmer. Mouser Electronics 24AA067 (12-100 pF) suitable.
Q6 — 2N3904.
T4 — Transformer wound on Amidon T-68-2

powdered-iron toroid core or equiv. Primary (19.8 µH): 59 turns no. 28 enameled wire, tapped at 12 turns. Secondary (0.98 µH): 9 turns no. 28 enameled wire over tap end of primary.

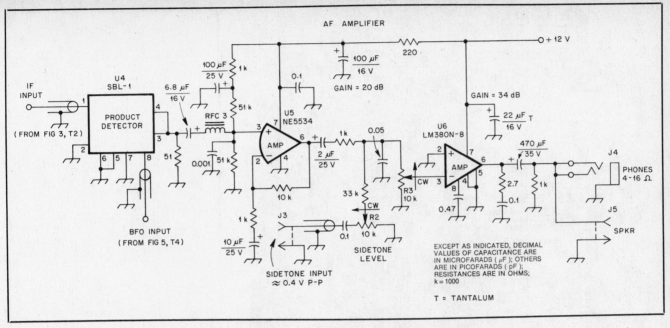

Fig. 6 — Schematic of the product detector and audio amplifiers for the band-imaging receiver. Capacitors are disc ceramic unless otherwise noted. Capacitors marked with polarity are electrolytic. All resistors are ¼-W, 10% units unless otherwise noted.

J3, J5 — Phono jack.
J4 — Stereo headphone jack.
R2, R3 — 10-kΩ audio-taper potentiometer.

RFC3 — 1-mH RF choke: 34 turns no. 30 enameled wire on Amidon FT-37-72 ferrite toroid core or equiv.
U4 — Mini-Circuits SBL-1 doubly balanced

diode-ring mixer.
U5 — NE5534 low-noise audio op amp.
U6 — LM380N-8 audio power amp.

phone or speaker load has not already been installed at J4 or J5. Without this resistor, the capacitor would charge on connection of the audio transducer, resulting in a loud thump.

As mentioned earlier, no active audio filtering is included in the band-imaging receiver. The higher audio components in detected IF amplifier hiss are reduced by the 0.05-μF capacitor connected between the hot end of the AF GAIN control, R3, and ground.

Sidetone can be injected into the audio chain at J3. Sidetone level is adjusted from the front panel by R2. Setting R2 to minimum shunts the AF GAIN control with a 33-kΩ resistor; this reduces overall audio gain by less than 1 dB. A 400-mV signal at J3 provides more than enough sidetone audio at normal AF GAIN settings.

This receiver requires, but does not include, a regulated dc power supply capable of providing a maximum of 220 mA at 12 V. See Chapter 27, Power Supply Projects, for suitable circuits.

Construction

The receiver was prototyped using point-to-point and "dead bug" modular construction (see Fig. 7). Later, circuit boards were designed and debugged.[2] You may use either method for building your receiver, with good results. The following construction hints are based on the circuit-board version of the receiver, but much of the information here will be of use to builders using either style of construction.

Parts for this receiver are available from

[2] Circuit-board templates are available from the ARRL Technical Department secretary for $3.00 and a no. 10 s.a.s.e.

Fig. 7 — The band-imaging receiver prototype. The LO is in the left foreground; just behind it, the 10- and 18-MHz RF amplifiers. At center, the mixer/filter/IF amplifier module, with the detector/AF amplifier module at far right. The smallest module, upper right, is the BFO. Performance was good even though etched circuit boards were not used.

a number of sources. Virtually everything can be obtained from RadioKit, Mouser, Radio Shack, DigiKey and Circuit Specialists. See the parts suppliers list at the end of Chapter 35 for addresses and telephone numbers of these suppliers.

See Fig. 8. The receiver is housed in a Hammond 1590F diecast aluminum box (approximately 7¼ × 7¼ × 2½ inches). Threaded standoffs are used to mount all circuit boards except the detector/audio board; spade lugs are used to mount this module vertically. Miniature 50-ohm coaxial cable (RG-174) is used for all RF connections between modules except the LO-mixer line. Here, miniature Teflon®-dielectric cable is used because of

Teflon's high melting point (see Fig. 9A). RG-174 is also used to connect J3, SIDETONE INPUT, to the detector/audio board. Connections from this board to the AF GAIN and SIDETONE LEVEL controls are made with stranded hookup wire in three colors. This makes for more compact wiring than miniature coax allows and causes no problems with hum or crosstalk. The IF GAIN control and audio output connections are also made in this way. Dc wiring is stranded hookup wire; binding posts are used to bring dc into the receiver.

We recommend that you build, test and install the band-imaging receiver modules in this order: (1) LO; (2) detector/audio and BFO; (3) mixer/filter/IF amplifier; (4)

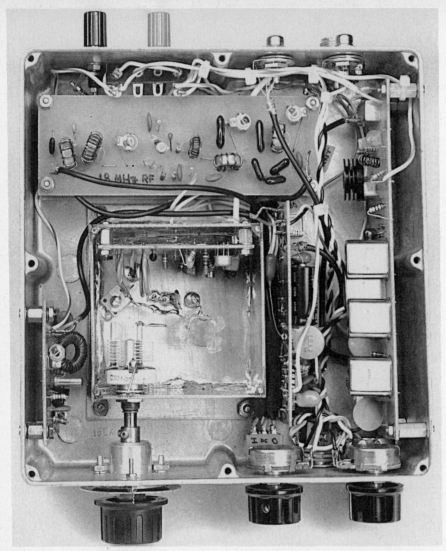

Fig. 8 — The band-imaging receiver just fits into a Hammond 1590F diecast aluminum box. In this top view, the LO (in its shield box, shown here with cover removed) dominates the layout. The RF amplifier boards are at top, one above the other. The BFO is to the left of the LO, with the detector/audio amplifier board (edge-on) to the right. At far right, the mixer/filter/IF amplifier board. The two capacitors and resistor beneath the mixer/AF board have since been incorporated into the board design. For a similar view of the band-imaging receiver in full color, see the cover of July 1987 *QST*.

Fig. 9 — The detail drawing at A shows installation of the LO output lead. Miniature Teflon coax is recommended here because its dielectric will not melt during soldering. The shield must be flared and soldered to the *inner* wall of the LO shield box. At B, details of the LO inductor, L5, are shown. Taps are made by pushing in adjacent turns of the coil stock. The untrimmed plastic bars preserve the Q of L5 by holding the Miniductor away from the shield box bottom. Before counting the 11 turns necessary for the coil, unwind enough turns from the uncut Miniductor stock to allow L5 to stand above the shield-box bottom by its diameter. The plastic bars are fragile; they may be impregnated with Duco or similar cement for greater strength. The completed L5 is cemented to the shield-box bottom after the LO tuning range has been set.

RF amplifiers. The LO comes first because its installation entails the majority of the metalwork necessary to build the receiver. The sequence allows you to use completed modules as part of your test equipment for the modules later in the sequence.

The LO shield box is made of double-sided copper-clad circuit board. The 10:1 epicyclic reduction drive is a Jackson Bros 5857. Because the sides of the Hammond diecast aluminum box are not perpendicular to the bottom, special construction techniques are needed to ensure that the TUNING capacitor shaft is perpendicular to the front panel of the box. The following construction sequence resulted in a smooth-tuning, no-backlash LO installation in the ARRL lab version of the band-imaging receiver:

1) Mount the 10:1 reduction drive on the front panel.

2) Build the LO shield box (four sides and bottom), soldering only the side and rear pieces into place on the bottom. The front and rear pieces of the shield box must butt the shield box sides as shown in Fig. 8. *Tape* the front side into place. Drill four mounting holes in the shield box bottom plate.

3) Chisel the molded-in printing from the center of the diecast-box bottom to smooth the box floor. Sanding may also be necessary to achieve this.

4) Locate the C9 (TUNING capacitor) mounting hole in the LO shield box front by pushing the box up against the reduction-drive coupler. Size this hole slightly larger than the capacitor mounting bushing. This allows later adjustment of C9's position. Temporarily mount the capacitor in the front side of the shield box and keep this assembly taped to the rest of the box.

5) Place the LO box in the diecast box so that C9 is inserted into the reduction-drive coupler. By feel, be sure that the

capacitor is about 1/16-inch short of full insertion into the coupler sleeve. This allows leeway for later adjustment. Mark the diecast box to pass the LO mounting screws through the holes in the shield-box bottom plate. Drill these holes now.

6a) Remove the taped-on front of the shield box. Build and install the LO/buffer circuit board, including the LO output cable and 12-V dc line, in the partially completed shield box. The output cable consists of a 6-inch piece of miniature Teflon coaxial cable (see Fig. 9A) Fig. 9B shows how to prepare LO inductor L5 from a length of B & W Miniductor.

6b) Temporarily install C9, C10, L5 and the 27-pF NP0 LO-tuned-circuit capacitor to the LO/buffer board by short leads. Terminate the LO output cable with a 51-ohm resistor. Verify operation of the LO/buffer board by applying dc power and tuning in the LO signal on the 14-MHz test receiver. Adjust C9 and C10 as necessary to bring the signal into your receiver's tuning range. You may need to add or remove fixed capacitors in the LO tuned circuit. Don't spend time now on setting the LO tuning range; that comes later.

6c) Once LO performance has been verified, disconnect C9, C10, L5 and the fixed tuned-circuit capacitor from the LO/buffer board. Install the board into the LO shield box.

7) Install C10 flat to the shield box floor by soldering down its rotor tab. Be sure to allow clearance for C9. Bend the C10 stator tab up to clear the box bottom. Connect C10 to the LO/buffer circuit board with tinned no. 18 solid wire. Solder L5 into the circuit; it will be cemented to the box floor later, but do not do this yet.

8) Bolt the LO into the diecast box. Loosely mount C9 in the front side of the LO box. Slide the front LO box side into place, and at the same time, slide the C9 shaft into the reduction-drive coupler to about 1/16 inch short of full insertion. *Do not* tape the LO box front into place as before.

9) Adjust the reduction drive to bring its coupler worm screws to approximately 10 and 2 o'clock. Set C9 to maximum capacitance without disturbing the reduction drive. Now, with C9 loose in its mounting hole, tighten the worm screws in the reduction-drive coupler.

10) Tighten C9 to the front of the LO shield box.

11) Using the reduction drive, turn C9 back and forth through its range several times to settle the LO box front into position. Depending on how tightly the front is held in place by the LO box sides, you may need to push the sides apart slightly to free the front piece. By eye, the front of the LO box should appear parallel to the front of the diecast box. If all looks well,

12) Solder the front side of the LO box into position.

Final tuning-range and anti-backlash adjustments will be made during alignment and testing of the receiver.

The circuit board placement shown in Fig. 8 works well. Although the position of the LO shield box left little choice as to the placement of the rest of the circuit boards, maximum spacing between the BFO and mixer/filter/IF amplifier boards was decided on *beforehand* to keep the BFO signal out of the IF amplifier circuitry.

Alignment

Test equipment necessary for aligning the band-imaging receiver is a 51-ohm resistor, a receiver capable of CW reception at 14.0-14.2 MHz and 4 MHz ±1 kHz with an S-meter and frequency display resolution of 1 kHz or greater, and a crystal-controlled marker generator capable of

providing 10-kHz markers. Equip the coaxial input of the test receiver with a short test cable terminated with alligator clips.

Detector/audio amplifier and BFO. The audio amplifiers require no adjustment. Align the BFO as follows: Without connecting the BFO to the detector, connect a 51-ohm resistor across the secondary of T4. Set C11 (FREQ ADJ) and C12 (OUTPUT TUNE) to midrange. Apply 12 V dc to the BFO. Set the test receiver for CW reception at 4000 kHz and attach the shield clip of its test cable to the BFO ground foil. Leave the center-conductor clip unconnected. Next, tune in the BFO on the test receiver. Adjust C12 for maximum received signal as indicated by the test receiver S-meter. Adjust C11 to put the BFO at approximately 4000.5 kHz. This completes alignment of the BFO for now. Remove the 51-ohm resistor from the T4 secondary and connect the BFO to the detector with RG-174 cable.

Mixer, filter and IF amplifier. The IF amplifier requires only one adjustment: With 12 V applied to the mixer/filter/IF amplifier board and later stages, adjust C8, IF TUNE, for maximum noise in the speaker or headphones.

Local oscillator tuning range. Connect the LO output cable to the mixer, and apply 12 V to the LO. Tune C9 to the low end of its range, and set the test receiver to 14.060 MHz. Connect the test cable shield clip to the LO box, but leave the center-conductor clip unconnected. Adjust C10 until you hear the LO in the test receiver. Be sure that the unconnected test-cable lead is far enough from the LO tuned circuit to have no effect on the LO frequency. Set the test receiver to 14.155 MHz. Tune the LO upward in frequency until you hear it in the test receiver. With luck, the TUNING capacitor will be nearly at minimum capacitance. Depending on the exact values of the capacitors in the Q4 gate circuitry, however, your LO may not have enough tuning range, requiring that you search downward for it with the test receiver even with the TUNING capacitor at minimum capacitance. If this is so, move the tap on L5 from 2 to 3 turns above ground and readjust the 14.060-MHz band edge with C10. This will increase the tuning range. (You may need to add capacitance in parallel with the 27-pF LO tuned-circuit capacitor to allow C9 to hit the band edge.) C10's tuning range is much larger than that of the TUNING capacitor, so adjust it carefully. With experimentation, you should be able to achieve a TUNING range of between 90 and 150 kHz. Remember that you'll need to make your final band-edge adjustment *after* installation of the LO box cover; be sure to provide a hole in the cover for this purpose, but leave the cover off for now. After you have set the LO tuning range, cement the base of each L5 pillar to

the shield-box bottom with Duco® (or similar) cement.

RF amplifiers. Install the 10-MHz RF amplifier in the receiver, and solder a 51-ohm resistor from the center conductor of J1 to ground. Connect the crystal calibrator, set for 10-kHz markers, to J1. Set the BAND switch to 10 MHz. Tune in a marker near the center of the tuning range and adjust C3 for maximum signal. Tune in the lowest marker in the range; adjust C5 for maximum signal. Tune in the highest marker in the range; adjust C6 for maximum signal. Because the C5 and C6 adjustments interlock somewhat, repeat them several times for good measure. Now, install the 18-MHz RF amplifier board and repeat this procedure at 18 MHz with C3 (at band center), C5 (at the lowest marker) and C6 (at the highest marker). This completes alignment of the RF amplifier boards.

Anti-backlash adjustment. With luck, the TUNING control will turn freely and require the same input torque across the tuning range. Backlash should be imperceptible throughout the range. If backlash is present, try loosening the reduction-drive coupler screws and tightening them again. Backlash in the ARRL lab version of this receiver was done away with by loosening and retightening the tuning capacitor in its mounting hole, *and* by slipping the TUNING capacitor several degrees to one side in the drive coupling sleeve before retightening the coupler worm screws.

Dial calibration. Calibrate the tuning dial *after* the tuning range has been set *and* any backlash has been ironed out. In the model shown in Fig. 1, calibration of the 10- and 18-MHz TUNING scales differs by the width of a dial marking. The left edge of each mark is used during 18-MHz tuning (18 MHz L); the right edge is used during 10-MHz reception (10 MHz R). Calibration of the full TUNING capacitor rotation (360°; 180° for each band) would make this unnecessary, but one band would tune "backward" relative to the other.

Performance

Measured performance of the band-imaging receiver at 10 MHz: Minimum discernible signal (MDS), −140.5 dBm; two-tone 3rd-order IMD dynamic range (20-kHz spacing), 89.5 dB; blocking dynamic range, 134 dB; image rejection, 74 dB. At 18 MHz: MDS, −140.0 dBm; two-tone 3rd-order IMD dynamic range, (20-kHz spacing), 90.0 dB; blocking dynamic range, 131 dB; image rejection, 82 dB. With a signal tuned in on the 10-MHz band, dropping the receiver three inches to the operating table produced *no discernible shift* in the pitch of the received signal. Maximum audio output was 0.66 W into an 8.2-Ω test load. Current drain at 12 V dc was 95.1 mA with no input signal, 220 mA at maximum audio output.

A High-Performance Communications Receiver

The receiver described here is the work of W7ZOI and K5IRK, and was presented in November 1981 *QST*. A progressive system was used in the *QST* version, starting with a simple (but very useable) direct conversion receiver and concluding with a multiband superheterodyne. This approach is highly recommended for those who may lack construction experience.

(A)

(B)

Fig. 10 — A 5-band, SSB version of the high-performance communications receiver constructed by K5IRK. In the top view, the VFO is located in the center with the input filter, mixer and IF boards to the left. The board at the far right contains the product detector and audio stages. The two boards to the right of the VFO are active audio filters (A). The bottom view shows the converter and oscillator boards. The BFO is contained in the shielded box at the left and the mixer board is at the lower center (B).

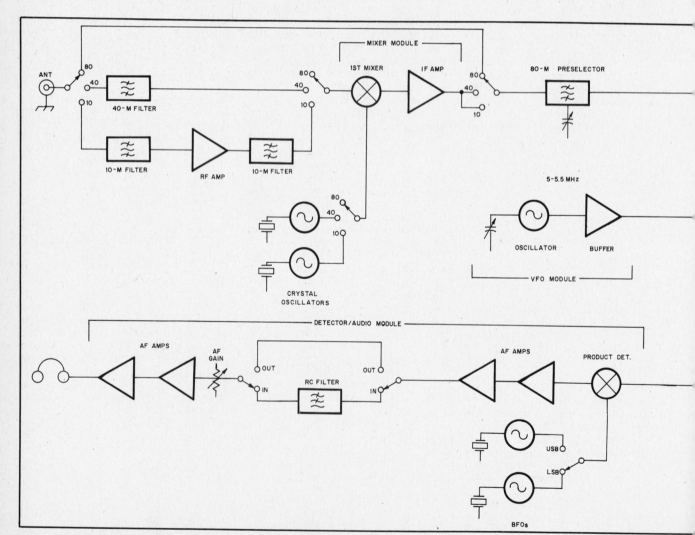

Fig. 11 — Block diagram of the W7ZOI/K5IRK communications receiver.

Shown in Figs. 10 through 19 is the receiver in its final form. The overall layout is shown in the block diagram, Fig. 11. On 80 meters the receiver functions as a single-conversion superheterodyne, with reception of the higher bands provided by higher-performance crystal-controlled converters. All of the converters use the same mixer module, switching only the converter filter and crystal oscillator modules. As shown, the 40-meter converter does not use an RF amplifier — it is unnecessary and it degrades dynamic range. The lower noise figure obtained with the RF stage is required on the bands above 7 MHz, however.

Many criteria were used in the design of this receiving system, but first and foremost were simplicity and ease of duplication. To this end, readily available components were used throughout. Alternative components are suggested where appropriate, and the circuits are insensitive to transistor type, allowing freedom in substitution. This does not in any way imply that the performance has been compromised; indeed, this receiver can equal the strong-signal performance of many of the high-priced receivers

on the market today.

Circuit Description: The Converter, Filter and RF Amplifier

Preselection for the individual converter sections is provided by the circuit shown in Fig. 12. The optional RF amplifier is shown in A, and the version without the amplifier is shown in B. The same circuit board layout can be used for both versions.

Each filter module uses two types of filters. The first is a 5-pole low-pass, necessary to prevent spurious reponses from VHF TV and FM broadcast signals. The second filter provides the majority of the front-end selectivity. It is a double-tuned circuit comprised of L7, L8 and their related capacitors. A variable capacitor is used at C15 because the small, non-standard values required here are difficult to obtain; a 1- to 5-pF variable is readily available and can be preset to the value given in Table 1 or adjusted during alignment.

The RF amplifier uses a dual gate MOSFET and modified-input low-pass filter. The first section is a simple low-pass filter, while the second section is a pi net-

work that transforms from 50 to 2000 ohms with a Q of 10. This provides an optimum driving impedance for the amplifier. The output uses a broadband transformer to provide a 50-ohm output impedance, ensuring proper termination for the following double-tuned circuit.

The filters may be aligned with a signal generator or crystal calibrator. If a calibrator is used, the input of the receiver should be terminated with a 50-ohm resistor. Initially, C15 is set near minimum capacitance, and the receiver is tuned to the center of the band. C14 and C17 are then adjusted for maximum response. C15 is then partially meshed, and C14 and C17 are again peaked for maximum response. The filter bandwidth is estimated by observing the reponse as the receiver is tuned toward band edges. This procedure is repeated until the desired bandwidth is realized. The input pi network used with the RF amplifier is adjusted by setting C22 for maximum response at the center of the band.

Mixer Module

The two mixer modules used in the composite receiver are identical, each being comprised of a doubly balanced diode-ring mixer, U2 of Fig. 13, followed by a 9-MHz IF amplifier. The IF amplifier, Q9, is one of the more critical stages in the receiver. It must have a reasonable noise figure, low IMD, and a 50-ohm input and output impedance. A bipolar transistor with negative feedback is used to establish the gain and impedances. The 6-dB pad at the output preserves the input and output match of the stage. The moderately high bias current used ensures low distortion.

The transistor type used for Q9 *is critical*. It should have an F_t of at least 500 MHz. The 2N5109, 2N3866 and 2SC1252 are all suitable.

Amplifier gain, including the loss of the pad, is about 16 dB. The mixer has a loss of about 6 dB, leaving a net module gain of 10 dB. The amplifier output intercept is about +30 dBm. Careful measurements have shown that a diplexer is not required between the mixer and the amplifier.

Crystal Oscillator Module

Shown in Fig. 14 is the circuit used for all of the crystal oscillators in the receiver. One or two of the modules are used for the BFO, and one module is used with each converter. The circuit is a Hartley oscillator with the crystal in series with the feedback tap from the coil. A trimmer capacitor in series with the crystal adjusts the operating frequency of the oscillator. If a single BFO is used, the 12-volt operating bias is applied through the output link as shown in Fig. 14B. When more than one module is used, as with the converters, operating voltage is applied through the bandswitch, Fig. 14C. Only the oscillator in use has power applied to it.

This circuit will deliver an output power

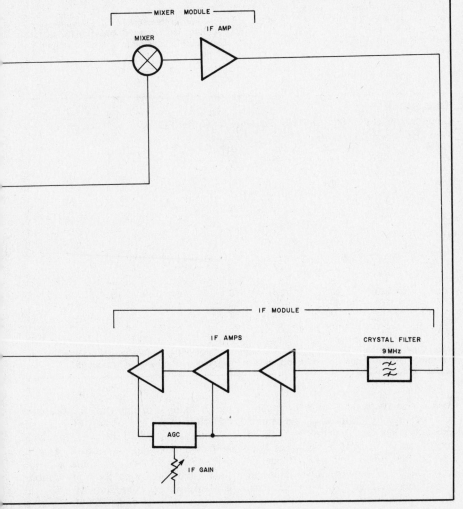

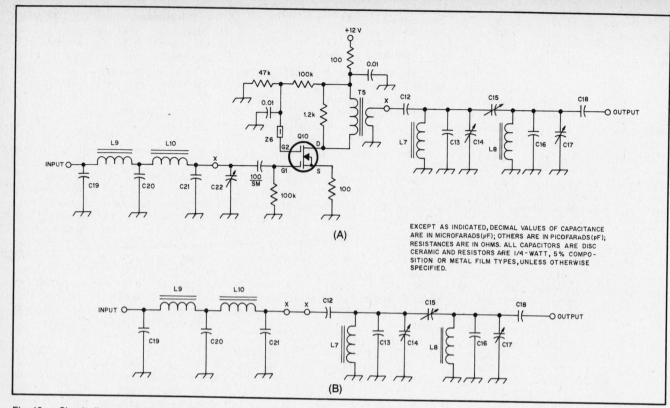

Fig. 12 — Circuit diagram of the converter filter and optional RF amplifier used in the W7ZOI/K5IRK receiver. The circuit using the RF amplifier is shown in A, while the configuration without the amplifier is shown in B.

C12, C13, C16, C18-21 — Silver mica or ceramic, see Table 1 for values.
C14, C17, C22 — Mica compression trimmer or similar variable, see Table 1 for values.
C15 — Air variable, 1 to 5 pF.

L7-10 — No. 22 enamel wire wound on Amidon T50-6 core, see Table 1 for number of turns.
Q10 — Dual gate MOSFET, 40673, 3N211,

3SK40 or similar.
T5 — Ferrite transformer, 20 turns primary, 4 turns secondary, on Amidon FT37-43 core.
Z6 — Ferrite bead on lead of Q10. Amidon type FB43-101.

of about +10 dBm, which is more than enough to drive the diode mixers. Adjustment of the oscillators should be done with the mixer attached. C11 is tuned for maximum output and proper starting of the oscillator. The series capacitor is then adjusted for the correct operating frequency. This capacitor may be eliminated in those modules used with the converters. In these cases, the crystal is connected directly to ground.

80-Meter Preselector Filter

The 80-meter preselector filter is shown in Fig. 15. It consists of two cascaded filters: The first is a 7-pole, high-pass (3-MHz cutoff), composed of the components between the two 650-pF capacitors. This filter suppresses spurious responses from AM broadcast signals. The second part of the filter, while basically a low-pass type, was designed for a very pronounced peak, resulting in a sharp, bandpass-like response. C6 is a 365-pF broadcast capacitor mounted on the front panel.

VFO Module

The variable frequency oscillator, Fig. 16, uses a JFET in a Hartley circuit, followed by a dual-gate MOSFET buffer. For best temperature stability, type SF material (Amidon -6 code) is used for the inductor

Table 1

Component Data for W7ZOI/K5IRK Receiver Converter Filters

Parts for filter without amplifier:

F,MHz	C19 C21	C20	L9 L10	C12 C18	C13 C16	C14 C17	C15	L7 L8
7.1	430	860	17	42	50	180	4.6	25
10.6	300	600	13	32	50	180	4.1	17
14.2	220	430	12	20	—	180	2.3	17
18.2	180	360	10	22	50	180	3.9	10
21.2	150	300	10	18	—	180	3.0	10
24.2	130	270	9	14	—	180	2.1	10
28.5	110	220	8	12	—	180	1.6	10

Parts for filter with RF amplifier:

F,MHz	C19	C20	C21	C22	L9	L10
10.6	300	680	33	50	13	29
14.2	220	500	22	50	12	25
18.2	180	390	—	50	10	22
21.2	150	330	—	50	10	20
24.2	130	300	—	50	9	19
28.5	110	250	—	50	8	17

Note: Other filter parts are identical to that without the amplifier. Values listed for capacitors are capacitance in pF. Values listed for inductors are the number of turns of wire required.

core, as this material has a lower temperature coefficient than the usual slug-tuned inductor. All of the capacitors in the tuned circuit should be NP0 ceramics, as they have the lowest temperature coefficient of any readily available type. Silver mica and polystyrene types should not be used in this circuit.

The resonator (tuned circuit) should be

lightly loaded; to this end, the coupling capacitor to the gate of the JFET is kept as small as possible. If the specified 2.7-pF NP0 ceramic cannot be obtained, a small air variable of similar value can be substituted. Following these precautions will ensure excellent stability. Typical warm-up drift is under 200 Hz over a period of 10 minutes. After warm-up, drift

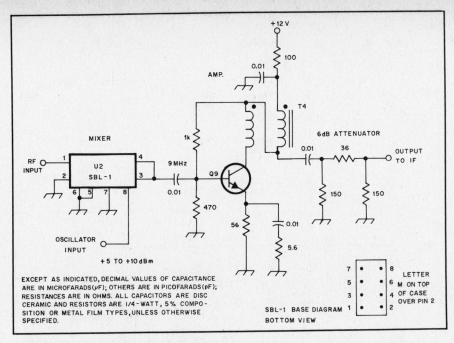

Fig. 13 — Diagram of the mixer module. Two of these modules are used in the completed receiver. One is for 80-meter input, with its output at 9 MHz. The second is used with the converters; its output is at 80 meters.

Q9 — TO-39 CATV type bipolar transistor, F_T = 1 GHz or greater. 2N5109, 2SC1252 2SC1365 or 2N3866 suitable. A small heat sink is used on this transistor.

T4 — Broadband ferrite transformer, 10 bifilar turns no. 28 enamel on Amidon FT37-43

core.

U2 — Mini Circuits Lab doubly balanced mixer, type SBL-1. Type SRA-1 is also suitable, as are similar units from other manufacturers.

is no more than 10 or 20 Hz in a 5-minute period.

The buffer stage, Q7, is conventional with the exception of the broadband output transformer, T2. The buffer provides good isolation for the oscillator and an output power of between +5 and +8 dBm.

Intermediate Frequency Amplifier

The heart of this receiver is its IF section. This design uses an IF of 9 MHz, with selectivity provided by a crystal filter of the builder's choice. The circuitry shown in Fig. 17 is designed for a filter requiring 500-ohm input and output terminations.

The input is a pi network which transforms the 50-ohm source impedance of the mixer module to the 500 ohms required by the filter. The filter output is terminated in a 560-ohm resistor.

The majority of the IF gain is provided by two dual-gate MOSFETs, Q11 and Q12. The bias on these stages is shifted upward by a pair of silicon diodes. This extends the gain control range as the gate 2 bias is altered. The last IF stage is a differential pair of PNP transistors, Q13 and Q14. Outputs are available from each of the collectors. The one from Q14 is routed through coaxial cable to the product detector.

The other IF output drives the detector,

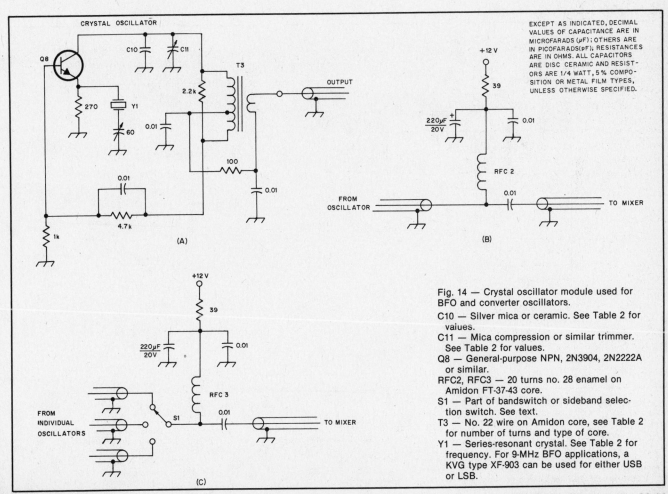

Fig. 14 — Crystal oscillator module used for BFO and converter oscillators.

C10 — Silver mica or ceramic. See Table 2 for values.

C11 — Mica compression or similar trimmer. See Table 2 for values.

Q8 — General-purpose NPN, 2N3904, 2N2222A or similar.

RFC2, RFC3 — 20 turns no. 28 enamel on Amidon FT-37-43 core.

S1 — Part of bandswitch or sideband selection switch. See text.

T3 — No. 22 wire on Amidon core, see Table 2 for number of turns and type of core.

Y1 — Series-resonant crystal. See Table 2 for frequency. For 9-MHz BFO applications, a KVG type XF-903 can be used for either USB or LSB.

D1. When large signals are present, the detected voltage from D1 appears at the base of Q16, discharging the timing capacitor, C24. The voltage change at C24 is coupled to the line through a diode, and reduces the gain of Q11 and Q12. R10, the "AGC set," is adjusted for a dc potential of 0.4- to 0.5-volt at the base of Q16. This adjustment is made with the AGC on, with no signals present. When measured with a high-impedance voltmeter, the AGC line should show about 6 volts at maximum gain.

Two transistor switches are contained in the amplifier. Q17 is used to defeat the

Table 2

Component Data for W7ZOI/K5IRK Receiver Crystal Oscillator Module

Y1	Band	C10	C11	T3			
				Core type	primary turns	tap turns	secondary turns
3.3 MHz	40	100 pF	90 pF	T68-2	65	13	10
9	BFO	56	60	T50-6	35	7	6
10.5	20	56	60	T50-6	30	7	6
11	20/40	22	60	T50-6	30	7	6
17.5	15	33	60	T50-6	23	5	4
24.5	10/15	—	60	T50-6	20	4	4
32	10	—	60	T50-6	15	3	3
6.5	30	100	60	T50-6	35	7	6
14.5	17	33	60	T50-6	23	5	4
20.5	12	—	60	T50-6	20	4	4

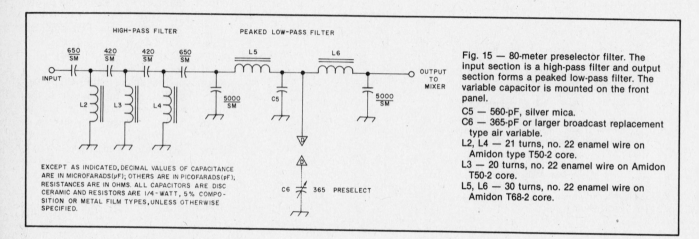

Fig. 15 — 80-meter preselector filter. The input section is a high-pass filter and output section forms a peaked low-pass filter. The variable capacitor is mounted on the front panel.

C5 — 560-pF, silver mica.
C6 — 365-pF or larger broadcast replacement type air variable.
L2, L4 — 21 turns, no. 22 enamel wire on Amidon type T50-2 core.
L3 — 20 turns, no. 22 enamel wire on Amidon T50-2 core.
L5, L6 — 30 turns, no. 22 enamel wire on Amidon T68-2 core.

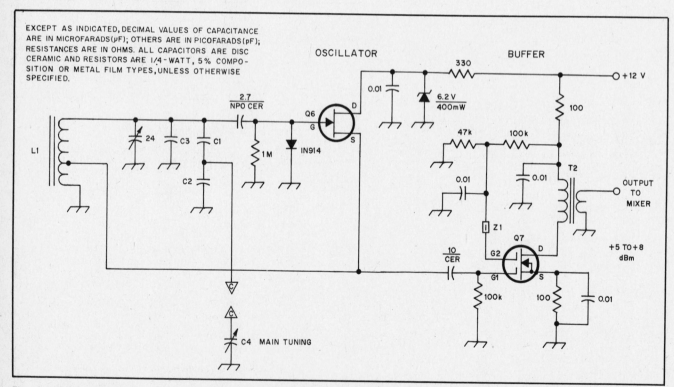

Fig. 16 — Schematic diagram of the VFO used in the W7ZOI/K5IRK receiver. This circuit will function well from 2.5 to 10 MHz. The tuning range using the components listed is 5.0 to 5.5 MHz.

C1 — 100-pF NP0 ceramic.
C2 — 82-pF NP0 ceramic.
C3 — 126-pF NP0 ceramic.
C4 — Air variable, 365-pF broadcast replacement type.
L1 — 35 turns no. 28 enamel wire on Amidon

T50-6 core. Tap 8 turns from ground. Approximately 4.9-μH total inductance.
Q6 — General-purpose JFET, MPF-102, 2N4416 TIS-88, 2SK19GR or similar.
Q7 — Dual gate MOSFET, 40673, 3N140, 3N211, 3SK40 or similar.

T2 — Ferrite transformer, 18-turn primary, 5-turn secondary, no. 28 enamel wire on Amidon FT37-43 core.
Z1 — Ferrite bead on lead of Q7. Amidon FB43-101 or similar.

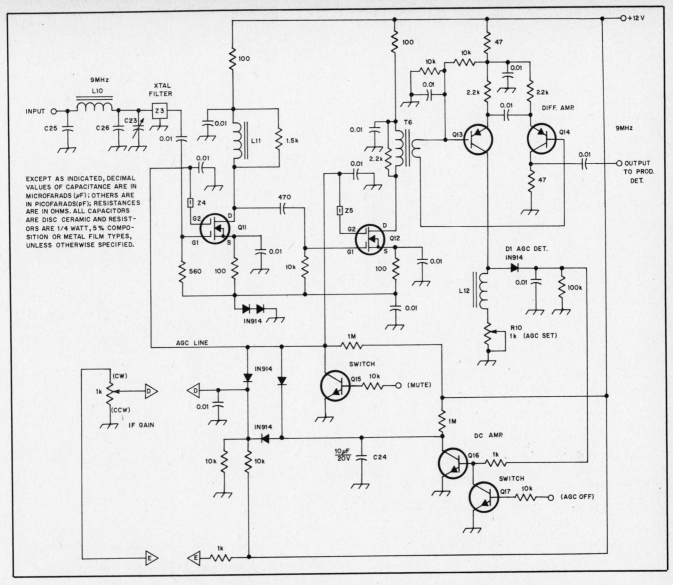

Fig. 17 — Schematic diagram of the 9-MHz IF amplifier.

C23 — 90-pF compression trimmer or similar variable.
C25 — 470-pF silver mica or ceramic.
C26 — 130-pF silver mica or ceramic.
L10 — 23 turns, no. 22 enamel wire on Amidon T50-6 core.
L11, L12 — 15 turns, no. 22 enamel wire on Amidon FT37-43 core.
Q11, Q12 — Dual-gate MOSFET, 40673,

3N140, 3N211, 3SK40 or similar.
Q13, Q14 — PNP silicon transistor, 2N2907A, 2N3906, 2N4403 or similar.
Q15-17 — NPN silicon transistor, 2N2222A, 2N3904, 2N4401 or similar.
R10 — 1-kΩ PC-board mount variable resistor linear taper.
T6 — Ferrite transformer, 15-turn primary,

3-turn secondary, no. 28 enamel wire on Amidon FT37-43 core.
Z3 — 9-MHz crystal filter, 500 ohm impedance level: For SSB, KVG type XF-9B, Yaesu XF-92A or Fox-Tango Corp., eq. For CW, KVG type XF-9M or XF-9NB, Yaesu XF-90C or Fox-Tango Corp., eq.
Z4, Z5 — Ferrite bead on lead of MOSFET, Amidon FB43-101 or similar.

AGC. It is activated by a positive voltage applied to the "AGC off" line. The other switch, Q15, is attached directly to the AGC line. A positive voltage applied to its input shorts the AGC line to ground, muting the receiver. The extra diodes allow muting to occur quickly while not discharging C24. The IF returns rapidly to full gain after muting periods.

The AGC response is more than adequate and overshoot is minimal. Recovery time is relatively independent of signal level. The recovery time may be shortened by decreasing the value of C24 or the associated 1-megohm resistor.

Detector Audio Module

The IF section of the receiver is followed

by the detector and audio amplifiers, shown in Fig. 18. The detector used here is a doubly balanced diode-ring mixer, a Mini-Circuits Labs SBL-1. Mixers from other manufacturers or homemade equivalents will work as well. The excellent balance provided by this type of mixer helps eliminate problems with the AGC system caused by BFO leakage.

The detector output is applied to a diplexer network formed by RFC1 and related components. This network ensures that the detector is properly terminated at all frequencies from audio to VHF.

The first audio amplifier is somewhat unusual in that it uses the common-base configuration. When biased for an emitter current of about 0.5 mA, it provides the

50-ohm input impedance necessary to properly terminate the detector. The second audio stage, Q2, is a direct-coupled PNP amplifier. The receiver may be muted by shorting the collector of Q2 to ground. This is done by applying a positive voltage to the muting input, saturating Q5. The output of Q2 drives the audio gain control, which is mounted on the front panel. If the optional RC active filter is used, it is connected between the output of Q2 and the gain control.

Q3 functions as a common-emitter amplifier, while Q4 is an emitter follower. Q4, biased for an emitter current of about 30 mA, provides sufficient audio output to drive low-impedance (4 to 16 ohms) headphones. If high-impedance headphones are

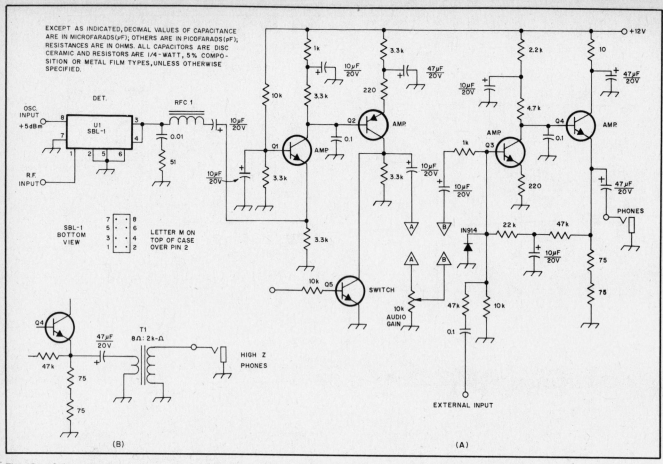

Fig. 18 — Schematic diagram of the product detector and audio amplifer. If high-impedance headphones are to be used, the output circuit shown in B is recommended.

Q1, Q3 — Low-noise NPN, 2N3565 or similar.
Q2 — General-purpose PNP, 2N3906 or similar.

Q4 — TO-5 or TO-39 NPN, 2N3053 or similar with small heat sink.
R3 — 10-kΩ audio taper.

RFC1 — 20 turns no. 28 enamel wire on Amidon FT37-43 ferrite toroidal core.

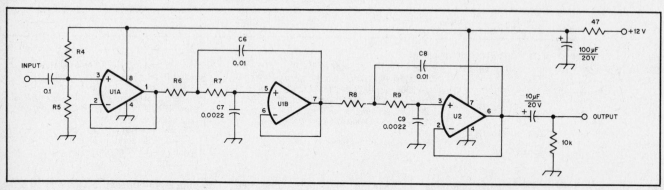

Fig. 19 — Schematic diagram of the optional RC active audio filter. A 4-pole low-pass filter is shown here. Additional sections may be added for improved performance.

C6, C8 — 0.01-μF, 10% or better tolerance, ceramic or polystyrene.

C7, C9 — 0.0022-μF, 10% or better tolerance, ceramic or polystyrene.
R4-9 — 33-kΩ for CW or 15-kΩ for SSB filter.

U1 — 1458 or similar, dual op-amp.
U2 — 741, or similar, op-amp.

to be used, a step-up transformer should be used to increase the system voltage gain. This is shown in Fig. 18B. An auxiliary input to the audio amplifier is provided for injection of a sidetone signal for CW monitoring.

The optional RC active filter shown in Fig. 19 can be used to improve selectivity during CW reception, especially when the

IF filter is of bandwidth suitable for SSB operation. As shown in Fig. 19, the filter has a single pole of high-pass filtering and four poles of low-pass response. The cutoff frequency is about 1-kHz for CW or 2-kHz for SSB. The filter bandwidth is determined through proper choice of resistor values. Values for both bandwidths are given in Fig. 19. The filter may be expanded to

many more sections for improved skirt selectivity.

Construction

Mechanical details, such as the type of cabinet and dial drive, are left to the discretion of the builder. Band switching is not critical, as all of the switched points occur at low impedance levels. A multi-wafer

Table 3

Measured and Calculated Receiver Performance Characteristics

Circuit	RF Amp.	Bandwidth, Hz	NF, dB	IP$_{in}$, dBm	MDS, dBm	DR, dB
Single Conv.*	no	500	16	+ 18	− 131	99
Single Conv.	no	2500	16	+ 18	− 124	94
Single Conv.	yes	500	5	+ 2	− 142	96
Single Conv.	yes	2500	5	+ 2	− 135	92
Dual Conv.	no	500	18	+ 12	− 129	94
Dual Conv.*	no	2500	18	+ 12	− 122	89
Dual Conv.	yes	500	6	− 2	− 141	92
Dual Conv.*	yes	2500	6	− 2	− 141	88

RF amplifier assumed to have a 3 dB noise figure, a 15 dB gain and a + 22 dBm output intercept. Circuits marked with (*) are measured cases. All measurements done at 14 MHz.

rotary switch will work well. Small coaxial cable, such as RG-174/U, should be used for all signal lines.

The version shown here was built by K5IRK. The bulk of the 80-meter part of the receiver is mounted above the chassis. The VFO is contained in an aluminum box, providing both shielding and mechanical strength. The BFOs are below the chassis in a box constructed of scrap circuit-board material. The converters are mounted below the chassis.

Homemade, etched circuit boards were used in the K5IRK model, while "ugly-but-quick" breadboards were utilized by W7ZOI in the construction of his receiver.[1,2] "Ugly boards" are easily built, with scraps of unetched circuit-board material serving as ground foil. The circuitry is supported by those components that are normally grounded. Additional support is provided by suitable tie points. Large-value resistors serve well for this purpose, especially in RF circuits where the impedance level is low. All of the circuitry for the receiver was initially breadboarded using this method. While not as "professional" in appearance, the performance was identical to that of later circuit board versions. In few cases where a performance difference could be detected, the ugly breadboards were superior, usually a result of improved grounding.

Performance

It should be emphasized that, although relatively simple, this receiver is not a toy. It features excellent stability, selectivity consistent with the filter used by the builder, adequate sensitivity, and a dynamic range that rivals or exceeds that of many commercially available equivalents. The only major compromise is the utilization of dual conversion on the higher bands. The penalty is small, because the gain distribution has been carefully planned.

System measurements were made on the receiver at various stages of development.

This data is summarized in Table 3. Both measured data and calculated results are presented to give the prospective builder some feel for the performance to be expected. The table shows the system noise figure, input intercept, minimum detectable signal and the two-tone dynamic range. Both CW and SSB bandwidths are considered in both single- and dual-conversion designs, with and without an RF amplifier. Measurements and calculations generally agree within 1 dB.

Table 3 reveals no surprises. The nature of the trade-off between single and dual conversion is well illustrated, as is the effect of adding an RF amplifier. The system showing the largest dynamic range is the single-converson design without an RF amplifier. It should be noted that this data pertains only to the modules described. Changes in gain, noise figure or intercept of any stage will change the results. The table should not be used for general comparisons.

The dual-conversion systems are about 5 dB "weaker" than the single conversion ones; this is typical. It should be understood that this observation applies only to dual-conversion systems with a wide bandwidth first IF. Modern systems using a crystal filter at the first IF will display performance much the same as a single-conversion receiver, even if they utilize several conversions. No gain compression was measureable in the single-conversion model with no RF amplifier, even with an input signal of − 10 dBm. VFO phase noise was measured to be − 152 dBc/Hz at a spacing of 10 kHz.

Operationally, the receiver is pleasing to use, offering a clean, crisp sound that is not always found in commercial equivalents. Of greatest significance is that the receiver should be easily duplicated at a cost well under that of commercial units.

[1] Circuit board templates are available from ARRL for $1 and a large s.a.s.e.
[2] Circuit boards, negatives and many parts for this receiver are available from Circuit Board Specialists, P.O. Box 969, Pueblo, CO 81002.

High-Performance Receiver Design Concepts

Amateur Radio design technology is changing so rapidly today that it is impossible to publish a high-performance receiver circuit that remains timely when the work is committed to print. As new components and active devices are introduced to the market, better designs become possible. These advances make obsolete many of the circuits found in current amateur literature. Therefore, this section is devoted to design objectives, circuit techniques and some practical examples. This will serve as the basis for individual designs which can be carried out by the more experienced amateur.

The interest in building homemade receivers of the more complex variety has waned tragically during the past decade. This has been brought on by an increased interest in operating, and through the availability of sophisticated receivers and transceivers found on the commercial market. For this reason it seems prudent to devote this portion of the *Handbook* to design approaches. The information given here is based on circuit and performance investigations in the ARRL laboratory. It is slanted toward the practical side of design and application in order to be of use to amateurs who have no formal background in electronics.

Performance Objectives

What should an amateur look for in terms of high performance when building a receiver? A subjective outlook would call for "bells and whistles" with which to play, but a discerning operator is interested in performance under all of the adverse conditions one might encounter in the course of operating an amateur station. The following are representative of the major considerations in receiver performance:

1) High dynamic range. This is the ability of the receiver to perform well in the presence of strong signals within and outside the amateur band of interest. Poor dynamic range results in cross-modulation effects, receiver desensitization and spurious responses from the mixer, which appear in the tuning range as additional signals (mixer IMD).

2) Good selectivity. This feature includes the receiver front end (RF amplifier and mixer) along with the IF and audio selectivity. The object is to have the receiver pass only those frequencies to which it is tuned, while rejecting all others. This Utopian goal cannot be realized, but it can be approached closely enough to ensure good performance.

3) Low noise figure. The noise figure should be such that it is somewhat below the level of the receiver antenna noise under typical "quiet" band conditions. This means that the noise generated within the receiver — notably the early stages — should be kept to an absolute minimum so that it does not mask weak incoming signals.

4) High order of stability. All of the receiver oscillators, whether crystal-controlled or LC types, should, ideally, be drift free. Since this is practically impossible, maximum drift should not exceed 50 to 100 Hz in a good design. The greater the IF selectivity, the more important the oscillator stability. Self-oscillations should not be allowed to take place in any part of a receiver.

5) Wide-range AGC. The AGC circuit should engage at low signal levels and hold the receiver output at a plateau over a wide range of input-signal levels. For example, the audio output should remain constant in amplitude over a range of input signal from less than a microvolt to better than 10,000 μV, depending on the external noise level which reaches the receiver front end. The AGC attack time should be set so that "pumping" and "clicking" is not noted when strong signals are received.

6) Local oscillator. Not only must the local oscillator be stable, it needs to have low noise and good spectral purity at the output. Ideally, the LO noise floor should be 80 to 100 dB below the peak output voltage. Spurs and harmonics in the output should fall at least 50 dB below peak output. LO output energy must be confined to the mixer by means of appropriate shielding and filtering.

7) IF amplifiers. An IF amplifier strip needs to have sufficient gain to drive the detector and provide ample excitation to the audio channel. The design should include active devices which can operate with a collective 80 dB or greater AGC swing over the input-signal range mentioned in item 5.

Wideband noise is generated within most IF amplifier chains. An improvement in receiver "noise bandwidth" can be realized by adopting the W7ZOI filter "tailending" scheme which calls for use of a second IF filter immediately after the last IF amplifier. The second filter can have slightly greater bandwidth than the filter used ahead of the first IF amplifier. The tailend filter will reduce wideband noise components. This technique is discussed in greater detail in *Solid State Design for the Radio Amateur*, by the ARRL.

8) IF filters. In order for a filter to function, there must be some insertion loss (IL) if a passive network is being used. The IL is typically highest when a mechanical filter is used. This factor must be taken into account when planning the receiver gain distribution. Most mechanical filters have an IL of 8 dB or greater, whereas a well designed crystal-lattice filter has a characteristic IL of less than 5 dB.

In some designs the IF filter is the limiting factor in obtaining high performance in mixer IMD. This is because of the movement of the mechano-electrical contacts within the filter, which generate IM products which are independent of those in the mixer. Laboratory investigations indicate that mechanical filters were somewhat worse than crystal filters in this respect, limiting the receiver IMD profile to roughly 95 dB. The crystal holders in lattice filters must be able to provide positive electrical contact with the quartz element and the circuit to minimize the generation of IM products.

Careful attention must be given to correct filter termination and input/output resonance to ensure minimum passband ripple. Most filters require a specified external terminal capacitance to resonate the input and output transformers within the filter module. Similarly, each filter has a characteristic input and output impedance which must be matched to the source and load.

9) Detector and audio channel. An otherwise excellent receiver can be spoiled by an inferior product detector or audio-amplifier strip. The detector must be able to handle the highest output signal from the IF strip without saturating. Although active product detectors are sometimes used, they are the most prone to this malady. The preference of most designers is a passive diode detector of the singly or doubly balanced variety. This type of detector can handle high signal levels with large amounts of BFO injection. Since the detector is the lowest-level part of the audio channel, hum and noise should be minimal at this point in the circuit. Passive detectors do not need operating voltages; hence one primary source of hum is avoided.

A low-noise audio preamplifier should follow the detector in a quality design. It should be able to withstand the maximum output from the detector at peak receiver input signal without operating in a nonlinear manner. The audio gain control can be used to the best advantage when it is located after the AF preamplifier. If is is used ahead of the preamplifier, the noise figure of the audio channel may be degraded at low settings of the gain control.

Audio shaping is normally applied to the AF channel to provide a low-frequency rolloff at some frequency well above 60 or 120 Hz. This greatly reduces the chance of power-supply ripple appearing in the audio output. Also, there is little need for low-frequency response below, say, 300 Hz in a communications receiver. Similarly, the high-frequency response should be restricted so that rolloff starts around 2000 Hz. A satisfactory tailoring of the audio passband can often be done by proper selection of the R and C components in that part of the circuit.

All of the audio stages must operate linearly up to peak signal levels. This will minimize distortion and aid weak-signal reception greatly. It is prudent, therefore, to use an output stage which is capable of delivering greater undistorted power output than will ever be needed. Cross-over distortion is also to be avoided. The effects of this are most apparent under weak-signal conditions. The signal has a "fuzzy" sound when this type of distortion is present. Some of the audio-power ICs have significant cross-over distortion which can not be corrected. This is because the biasing is done within the IC, and it can't be changed. For this reason it is helpful to use discrete devices in the audio channel. This enables the designer to bias the amplifiers for minimum distortion.

Tuned audio amplifiers can be used to provide steep skirts outside the desired passband. An example of a simple application of this, using a single pot-core inductor with variable Q, was described by K1TX in April 1979 QST. Various types of passive LC filters can be used to obtain CW or SSB selectivity at audio.

RC active audio filters with variable Q and adjustable peak frequency offer an excellent means for limiting the audio bandwidth, minimizing wide-band noise and reducing QRM. Ideally, these filters should be contained in the low-level part of the audio channel rather than at the receiver output. This will prevent overloading of the filter, which can impair the performance and introduce intolerable amounts of distortion.

10) Structural considerations. There can be considerable latitude in the mechanical approach one takes when laying out a high-performance receiver. Aesthetics have no place in this discussion. We will address the matter of structure versus performance and leave the beauty of the front panel to the builder.

The major points of concern are rigidity of the overall assembly and shielding against incidental pickup and radiation. The chassis and panels should be strong enough to prevent undue stress on the PC boards during flexing or vibration. In a like manner, the local oscillator should be relatively immune to any mechanical stress which is imposed on the receiver.

An excellent assembly technique is one that uses a modular approach for the various key circuit assemblies in the receiver. Each module is contained in its own shielded box. All signal leads entering and leaving the various modules are made from RG-174/U or similar coaxial cable. The shield braid is grounded at each end of each cable. Leads which carry dc are decoupled where they leave the module

shield. LC or RC decoupling networks are suitable in most instances. Feed-through capacitors can be mounted on the box walls of each module to serve as terminal connections for the dc voltages, while also functioning as parts of the decoupling networks. Since 50-Ω miniature cable is suggested for interconnecting leads, the points to which they connect in the circuit should be designed for a similar impedance.

This form of modular construction and shielding greatly reduces the chances for "birdies" by keeping RF energy where it belongs. It also prevents unwanted external signals from being picked up by low-level parts of the circuit. Miniboxes or die-cast aluminum boxes are excellent for use in modular work. Homemade enclosures can be fashioned inexpensively from pieces of double-sided PC board. Modular construction permits the amateur to try new circuits within the receiver without disrupting the remainder of the circuitry.

11) "Bells and whistles." This discussion does not include such themes as synthesizers, IF passband tuning, noise blankers, computer-programmed functions and digital frequency readout. These are primarily subjective matters, however useful they might be.

For reasons of practicality, the builder must decide whether he will use analog or digital readout of the receiver frequency. There are two disadvantages with analog systems: (1) Quality dial mechanisms are scarce and highly expensive. (2) Readout resolution is usually poor if more than 200 kHz of any band is covered. The major advantages of analog frequency readout are reduced circuit complexity, lower cost (sometimes) and less current drain from the receiver power supply. Heating is diminished also — a definite benefit to stability.

A frequency counter and a digital display, on the other hand, permit 500-kHz frequency spreads with good resolution. A shaft encoder is needed for synthesized LO systems to avoid thumb-wheel frequency selection. But, it is easy to use parts of the synthesizer circuit for the frequency counter, thereby making the two circuits compatible. In this type of system, or in one which has a conventional LO and a counter, a 10:1 vernier drive without detectable backlash is almost mandatory to keep the tuning rate within practical limits.

It should be stressed here that counters can create noise and spurious responses if they aren't designed and used correctly. Careful shielding and filtering must be applied to prevent the counter from affecting the other parts of the receiver circuit.

The same general considerations apply to synthesizers. The design must be carried out with care to minimize phase noise, which can degrade the mixer noise figure and the ultimate IF selectivity.

Three QST articles are offered as

references on high-performance receivers. They contain information which will be of value to the amateur designer.[3]

RF Amplifiers

When it is deemed necessary to use an RF amplifier ahead of the receiver mixer, thought must be given to gain, linearity, signal-handling ability and noise figure. The choice between bipolar transistors and FETs is another consideration. An RF amplifier should not be necessary in a properly designed receiver, even if a passive mixer is used, provided the input networks are not highly lossy and are matched to the source and load. This rationale applies to frequencies up to approximately 14 MHz. At 20 meters and higher, an RF amplifier may be needed to ensure an acceptable receiver noise figure.

As a rule, the designer should use no more gain in the RF stage than is necessary to obtain an acceptable noise figure. The higher the stage gain, the greater the sensitivity. But, more gain than is needed will degrade the receiver dynamic range markedly, by virtue of the mixer being fed larger amounts of input signal than if no RF amplifier was used. So, even at the very early part of a receiver it is vital to pay attention to *gain distribution*. This fundamental rule applies from stage to stage through the receiver.

There should be sufficient selectivity ahead of the RF amplifier (and in most instances between it and the mixer) to restrict passage of signals outside the amateur band of interest. This will greatly reduce the probability of unwanted images in the tuning range. Furthermore, it will help prevent very strong out-of-band commercial signals from entering the receiver front end and impairing performance. This form of selectivity is called "preselection." It can take the form of LC circuits which are very narrow in bandwidth, and tracked manually from the front panel. Alternatively, fixed-tuned LC filters can be used to provide selective circuits. A band-pass filter or tuned circuit is the choice of most designers because rejection is offered above and below the frequency band of interest.

The choice between small-signal FETs and bipolar transistors in an RF preamplifier is more than arbitrary. FETs exhibit low noise figures at HF and they consume less dc power than bipolars for an equivalent output intercept. Generally speaking, FETs are less subject to blocking in the presence of strong input signals.

[3]Wes Hayward, W7ZOI, "A Competition-Grade Receiver," QST, March and April 1974. Doug DeMaw, W1FB, "His Eminence the Receiver," QST, June and July 1976. Jay Rusgrove, W1VD, "Human Engineering the Station Receiver," QST, January 1979.

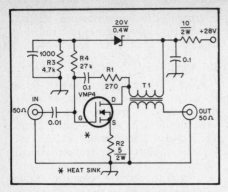

Fig. 20 — Diagram of the Class A large-signal RF amplifier which uses a VMOS power FET. T1 has 9½ turns of no. 30 enamel wire on a Stackpole no. 57-9130 ferrite balun core.

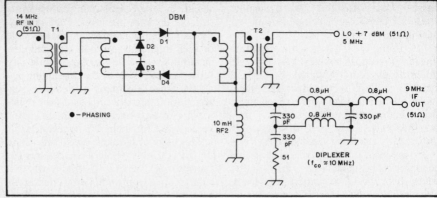

Fig. 21 — Practical circuit for a doubly balanced diode-ring mixer. The components are discussed in the text.

Bipolar transistors, on the other hand, have rather well defined input and output impedances and can be used more easily with negative feedback than can FETs. These features make them ideal for ensuring a proper and constant filter termination. A common-source FET which operates in the HF spectrum can not meet the foregoing condition. The use of negative or degenerative feedback in a bipolar-transistor preamplifier makes possible a low noise figure and a good input and output match.

Bipolar transistors which are designed for CATV and UHF oscillator work, such as the 2N5179 (biased for about 20 mA) and the 2N5109 (biased for roughly 50 mA) are excellent for use as RF amplifiers ahead of a mixer.

Fig. 20 shows a high-performance RF amplifier which uses a VMOS VHF power FET. (Oxner, May 1979 *QST*, p. 23). The circuit is designed with feedback and is structured for a source and load impedance of 50 Ω. The stage gain is determined by the designer's needs. Once this parameter is chosen, the values for R1 and R2 can be obtained from

$$Q = \frac{\sqrt{R_S R_L}}{2} \qquad \text{(Eq. 1A)}$$

$$X = \frac{\sqrt{G}(R_S + R_L)}{\sqrt{2} \ R_S R_L} \qquad \text{(Eq. 1B)}$$

$$R1 = Q \times [\sqrt{G} + \sqrt{G + 4(1 + X)}] \qquad \text{(Eq. 1C)}$$

$$R2 = \left(\frac{R_S R_L}{R1}\right) - \frac{1}{G_M} \qquad \text{(Eq. 2)}$$

where

G is the desired stage gain for the amplifier, and

G_M is the forward transconductance value of the transistor expressed in siemens (Y_{21} real).

The equations don't yield standard resistance values in most instances. In an amateur application the nearest standard value will often suffice. R2 of Fig. 20 consists of six 30-Ω resistors in parallel to obtain 5 Ω. Three are soldered from one source tab to ground, and the other three go from the remaining source tab to ground. This helps reduce stray inductance in that part of the circuit. A power gain of approximately 13 dB results with the component values shown. Noise figure is 4 dB at 30 MHz. A 1-dB saturation power output of 3.7 watts was observed, indicating the suitability of this type of circuit for high signal-handling applications.

The VMP4 is a fairly expensive M/A COM transistor. It is likely that one of the lower-priced devices, such as the VN66AK, would provide good service at HF in the circuit of Fig. 20.

High Performance Mixers

Doubly balanced diode-ring mixers (DBMs) of the type discussed earlier are often used to obtain high dynamic range. Among the advantages are low noise (diode mixers generate very little noise) and broadband characteristics. The mixer noise figure is approximately the conversion loss of the diode ring — typically 7 to 8 dB. The balanced mixer circuit provides port-to-port isolation which is not possible with single-ended or singly balanced mixers. This feature can aid the mixer IMD and help to minimize spurious responses resulting from the LO energy entering other parts of the receiver circuit.

The main shortcomings of diode mixers are the high level of LO injection needed (approximately +7 dBm for most) and the necessity of proper mixer termination, especially at the IF output port. This type of mixer is subject to harmonic mixing — another trait the designer must deal with.

Some high-level diode-ring mixers are available commercially. They require a high amount of injection power (+17 dBm for acceptable performance). Laboratory analysis suggests that high-level mixers misbehave as a result of diode imbalance at specific current levels. The effect is one of the IMD not dropping 3 dB when the input tones are lowered 1 dB in level. This phenomenon could be caused in part by saturation of the broadband input and output transformers at specific power levels. A thorough discussion concerning diode mixers and their behavior is presented in the League's book, *Solid State Design for the Radio Amateur*, Chapter 6. Fig. 21 shows a practical circuit for a DBM. It includes a diplexer at the IF port to establish a 51-Ω termination for the mixer. This offers an improvement to the IMD level by a few decibels over a similar mixer with no diplexer.

The diodes can be HP2800 hot-carrier types. Carefully matched 1N914s are sometimes used as substitutes. T1 and T2 are broadband toroidal transformers (baluns). For wideband use in the HF spectrum the cores should have a high permeability. A 0.37-inch-diameter ferrite core (Amidon FT37-43) with a mu of 950 will work nicely. Ten trifilar turns of no. 30 enamel wire can be used for the windings. Output intercept for this circuit is typically +13 dBm with the LO injection at +7 dBm. This provides an input intercept of 20 dBm (output intercept plus the 7 dB conversion loss = 20 dBm). Calculations for a high-level diode mixer, assuming a +17 dBm LO level (recommended), indicate the output intercept will be +23 dBm. Again, assuming a 7-dB conversion loss, the input intercept becomes quite desirable: +30 dBm. This is based on the respective performances of the commercially available SRA-1 and SRA-1H DBMs. It can be seen from the foregoing that better mixer performance can be realized at the higher LO-injection levels. The actual LO power applied will depend upon the current-handling ability of the diodes.

160-Meter Linear QSK Transverter

The 160-m QSK transverter shown in Figs. 22 through 28 was designed and built in the ARRL lab by Zachary Lau, KH6CP/1. A prototype was used in the ARRL 160-m contest to make 56 contacts from a second floor apartment. It is designed for use with a 7-MHz QSK CW/SSB rig such as the Ten-Tec Argonaut, although any rig that operates on 7 MHz can be adapted for use. It features very low distortion; the unit pictured has IMD products at least 48 dB down at its rated output of 2-W PEP. The circuit is easy to duplicate: Several versions were constructed with good results. With just an RF probe and multimeter for measurements, an IMD performance of −42 dB can be obtained.

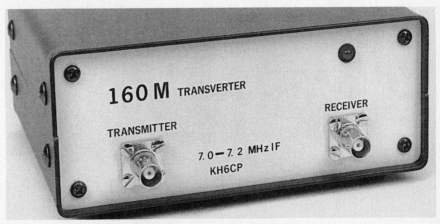

Fig. 22 — This QRP 160-m transverter features QSK and low distortion.

Circuit Details

The schematic diagram of the transverter is shown in Figs. 23 and 24. A pair of Mini-Circuits SBL-1 mixers is used to convert between 7 and 1.8 MHz. Q1 generates the 5.2-MHz local-oscillator signal that feeds the splitter circuit composed of T1 and R4.

After the 1.8-MHz signal is generated using U1, the 7-MHz IF signal, and the local oscillator, it is amplified by a pair of broadband amplifiers with roughly 44 dB of total gain. Q3 runs fairly warm, since it dissipates about 1.3 W. The large standing current is necessary to obtain the excellent IMD performance. A 2-pole band-pass filter is used to remove the unwanted mixer products. Better IMD performance may be obtainable by using filters at both the input and output of the amplifier strip, but more parts are required and the adjustment procedure may be more involved. Care must be taken to properly terminate the mixer if this approach is taken.

A matched pair of MRF476 transistors is used to take the low-level output of roughly 20 mW up to 2 W PEP. The biasing is slightly unusual in that an LM317T current source is used. It is essential that diode D3 be thermally coupled to the transistors, since it is supposed to draw current away from the transistors as the circuit heats up. Surprisingly, it is the driver stage and not the output transistors that limits the IMD performance. An IMD performance of −54 dB has been measured at a power level of 4-W PEP output when the final stage was driven by a pair of signal generators. See Fig. 25.

A 7-element low-pass filter removes the harmonics from the output signal and eliminates the receiver image signals. A series-resonant circuit is used to pick off the receive signal. This technique has been described by Roy Lewallen, W7EL.[1] Despite the simplicity of this circuit, the unwanted RF at the receiver during transmit is only −15 dBm. As one might expect, the third order input intercept of the mixer is degraded to +2.5 dBm, but this is more than adequate for casual QRP work. The output of the mixer feeds the receiver directly, since an IF receiver with a tuned input will reject most unwanted mixer products.

Construction

Construction of the transverter is straightforward. It requires stuffing two PC boards, attaching the connectors, and mounting everything in a shielded box. Figs. 26 and 27 show the parts placement for these boards. When you are drilling the amplifier board, don't forget to drill the no. 33 or 1/8-inch holes to mount the heat sink.

The box used is a Ten-Tec BU-625. Its design allows the final amplifier to be well shielded from the driver stages. This shielding is necessary because of the very high gain on a single frequency. It is necessary to use heat sinks for all of the transistors except Q1. Q3 runs fairly warm and requires a hefty heat sink such as a Thermalloy 2215B. The heat sink for the MRF476 transistors is a strip of aluminum that also acts as a mounting bracket. Nylon screws and insulating washers are used to electrically isolate the MRF476 transistors from the grounded aluminum strip. Greaseless TO-220 insulators seem to work well in this application, although standard mica insulators would work if heat-sink compound was used. D3 should be mounted so as to ensure good thermal contact with the MRF476 transistors. Figs. 28 and

[1]Roy W. Lewallen, "An Optimized QRP Transceiver," QST, August 1980, pp. 14-19.

29 show the mounting arrangement.

Testing

A good preliminary check is a measurement of the current drawn by the four stages. Q1 should draw 9 mA; Q2 draws 40 mA; Q3 draws a hefty 110 mA. These values were measured on the prototype and may vary slightly because of component tolerances. If the current for Q3 is significantly higher and operation at 13.8 V is intended, R14 should be increased from 15 to 18 ohms. R5 should be adjusted so that the MRF476 transistors have a bias current of 40 mA; The current through jumper W2 should be 40 mA with no signal applied at the amplifier input.

An RF probe or oscilloscope should be used to adjust the oscillator trimmer (C3) for maximum output measured at point A of T1, consistent with quick oscillator starting.

A wattmeter or dummy load/RF probe can be used to align the band-pass filter trimmers (C15 and C17). With a signal of roughly 2 mW applied to the transverter input, alternately peak the trimmers for maximum output at 1833 and 1965 kHz (7033 and 7165 kHz input, respectively). It usually helps to use one trimmer for peaking at the low end and the other for peaking at the high end of the band. Approximately 2 W output should be noted. Lower passband ripple can be obtained by peaking at 1826 and 1956 kHz, but the output will fall off rapidly past 1983 kHz.

Operation

The transverter requires an input of roughly 2 mW PEP. If desired, R5, R6, and R7 can be changed to make the unit as much as 30 dB more sensitive. This lowers the requirement to 2 µW PEP. Like any linear amplifier, the transverter must not be overdriven when used on SSB or the

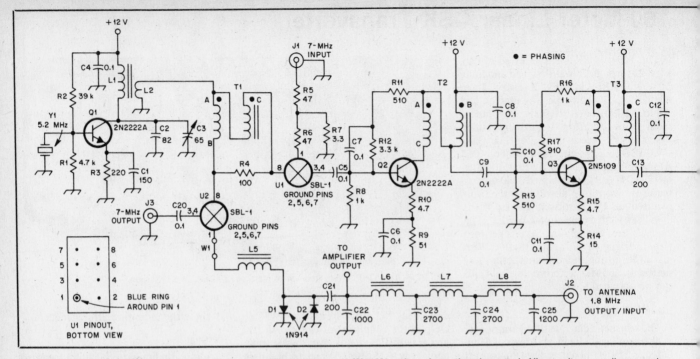

Fig. 23 — Schematic diagram of the 160-m transverter. All resistors are ¼-W, 10% units unless otherwise noted. All capacitors are disc-ceramic units unless otherwise marked. Capacitors marked with polarity are electrolytic.

C3, C15, C17 — 9- to 62-pF trimmer capacitor.

C13, C14, C16, C18, C19, C21-C25 — Silver-mica, NP0 disc or polystyrene capacitor.

D1, D2 — High-speed silicon diode (1N914 or 1N4148). Do not use Schottky diodes.

J1-J3 — Female chassis-mount BNC connector.

L1 — 40t no. 28 enam wire on an Amidon T-37-2 core.

L2 — 5t no. 28 enam wire over the +12 V side of L1.

L3, L4 — 21.6 µH, Qu = 224 @ 1.9 MHz; 59t no. 26 enam on an Amidon T-68-2 core.

L5 — 61t no. 28 enam wire on an Amidon T-50-1 core.

L6, L8 — 5.4 µH; 34t no. 28 enam wire on an Amidon T-37-2 core.

L7 — 6.4 µH; 37t no. 28 enam wire on an Amidon T-37-2 core.

R14 — 15-Ω, ½-W resistor. See text.

Q1, Q2 — 2N2222A.

Q3 — 2N5109.

T1 — 6t no. 26 enam wire bifilar wound on an Amidon FT-37-43 or FT-23-43 core. Palomar F-37-43 or F-23-43 cores should also be suitable.

T2, T3 — 8t no. 26 enam wire bifilar wound on an Amidon FT-50-43 core.

U1, U2 — Mini-Circuits SBL-1 doubly balanced diode mixer.

Y1 — 5.2-MHz fundamental crystal, 32-pF load capacitance (International Crystal no. 433115 or equiv.).

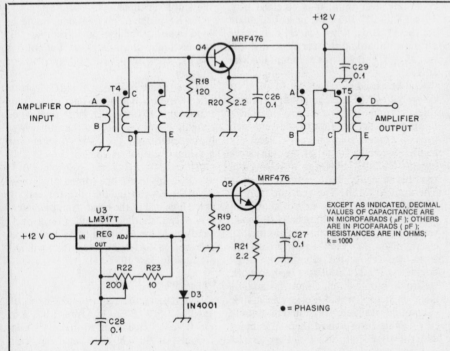

Fig. 24 — Schematic diagram of the 160-m transverter power amplifier. All resistors are ¼-W, 10% units unless otherwise noted. All capacitors are disc-ceramic units unless otherwise marked. Capacitors marked with polarity are electrolytic.

D3 — 1-A rectifier diode. Voltage rating not important (1N4001 or equiv.). See text for mounting instructions.

Q4, Q5 — Matched pair of MRF476 transistors.

R18-R21, R23 — ½-W resistors

R22 — 200-Ω PC-mount potentiometer. May also use a 100-Ω trimmer and choose R23 appropriately.

T4 — Primary: 2t no. 22 enam wire on an Amidon BLN-73-302 core. Secondary: 2t no. 22 enam wire bifilar wound.

T5 — Primary: 2t no. 22 enam wire bifilar wound on an Amidon BLN-73-302 core. Secondary: 4t no. 22 enam. wire.

U3 — LM317T adjustable regulator.

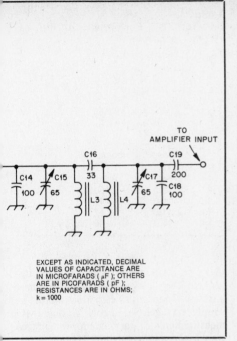

TO AMPLIFIER INPUT

EXCEPT AS INDICATED, DECIMAL VALUES OF CAPACITANCE ARE IN MICROFARADS (µF); OTHERS ARE IN PICOFARADS (pF); RESISTANCES ARE IN OHMS; k = 1000

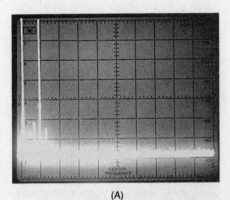

(A)

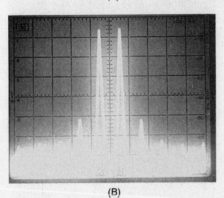

(B)

Fig. 25 — Spectral analysis of the 160-m transmitting converter. Display at A shows the fundamental, the second harmonic, and several close-in mixing products. Each vertical division is 10 dB; each horizontal division is 2 MHz. All spurious signals are at least 52 dB down. The two-tone IMD spectrum is shown in B. Each vertical division is 10 dB; each horizontal division is 2 kHz. The third-order products are 52 dB below PEP.

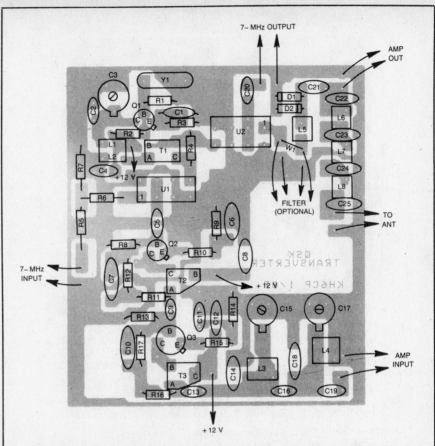

Fig. 26 — Parts-placement guide for the 160-m transverter. A full-sized etching pattern is shown at the back of this book.

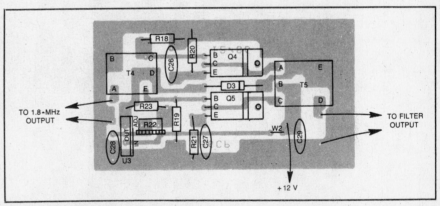

Fig. 27 — Parts-placement guide for the 160-m transverter amplifier board. A full-sized etching pattern is shown at the back of this book.

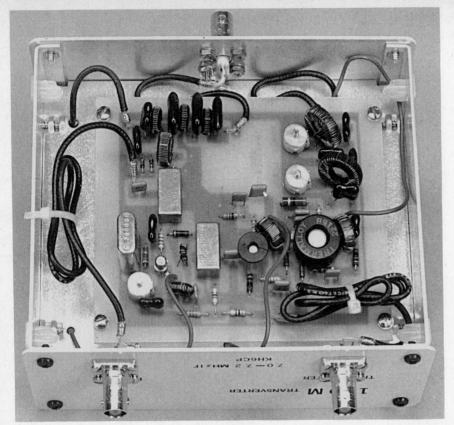

Fig. 28 — Bottom interior view of the 160-m transverter.

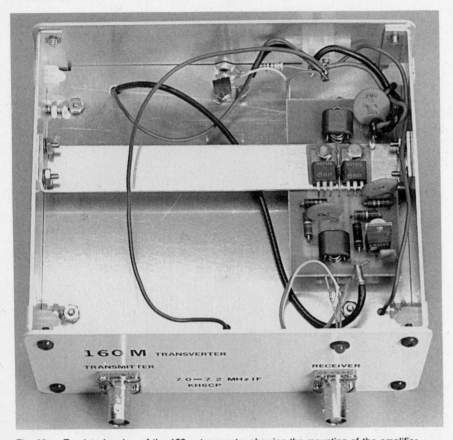

Fig. 29 — Top interior view of the 160-m transverter showing the mounting of the amplifier board with a long aluminum heat sink.

Table 1
Two-Tone IMD Performance of the 160-m Transverter

P_{OUT} (W)	3rd-order IMD (dB)
2.0	−52
2.5	−47
3.0	−45
3.5	−40
3.8	−32
4.0	−28

P_{OUT} (W)	Voltage (V)	3rd-order IMD (dB)
2.0	14.0	−52
2.0	13.0	−50
2.0	12.0	−48
2.0	11.0	−47
1.9	10.0	−46

IMD performance will be severely degraded. The transverter IMD performance is relatively insensitive to fluctuations in supply voltage. See Table 1. The IMD measurements at varying supply voltages were made with a constant RF input level. The IMD characteristics also vary with frequency, as the driver IMD performance worsens as the band-pass filter becomes more lossy. As a result, the IMD varies between −48 and −52 dB.

QSK is obtained by connecting the transmitter port of the transverter to the output of a QSK rig such as a Ten-Tec Argonaut and connecting the receiver port directly to an internal receiver connection after the T/R switch. It would also be possible to use a separate receiver and transmitter, but receiver muting during transmit would probably be required.

Although the receive performance was adequate despite a broadcast station less than a mile away, jumper W1 may be replaced with a filter to improve AM broadcast-band rejection. Unfortunately, little can be done to eliminate interference from the second harmonic of a broadcast station if it falls within the 160-m band.

A High-Performance AGC Circuit

Fig. 30 contains the circuit of an IF strip and AGC chain that offers excellent performance. This circuit was designed by W7ZOI for use in his Competition-Grade CW Receiver. The complete receiver circuit is found in *Solid State Design for the Radio Amateur*, Chapter 9.

This AGC circuit employs a full "hang" action. The AGC is defeated by means of S1. The time constant is selectable at S2. R1 at U3 should be set for +5 volts at pin 6 of U3 with the AGC off. With Q5 and Q6 as part of the circuit, the receiver is practically silent after a strong signal disappears from the passband. But, after a timing period associated with network C1-R2, the receiver will return to full gain in roughly 50 milliseconds. This is very advantageous when loud pulses of noise enter the receiver. The effect is similar to that of a noise blanker. A detailed description of this type of AGC circuit can be found in Chapter 5 of *Solid State Design for the Radio Amateur*, by the ARRL.

A less complex AGC circuit for use with RCA CA3028A IF amplifiers is provided in Chapter 12. It does not incorporate the hang feature used in Fig. 30.

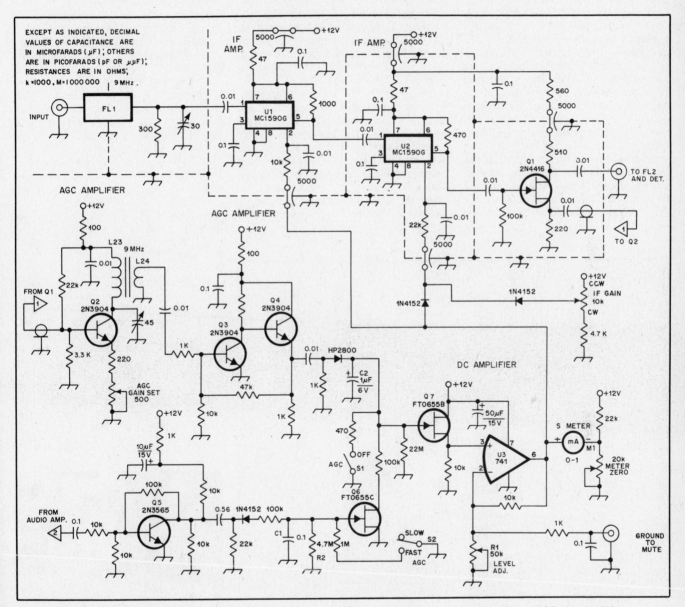

Fig. 30—Circuit details of the W7ZOI IF amplifier and high-performance AGC system. This circuit provides full-hang AGC characteristics.

A BC-Band Energy-Rejection Filter

Inadequate front-end selectivity, or poorly performing bipolar-transistor RF amplifier and mixer stages, can result in unwanted cross-talk and overloading from adjacent commercial or amateur stations. A simple cure for this problem is to install a filter between the antenna and receiver to attenuate the out-of-band signals but pass those signals of interest with little or no attenuation.

If the receiver is designed for reception of frequencies below and above the broadcast band, a 550- to 1600-kHz band-stop filter will be required. However, if reception is desired only below or above the broadcast band, then a less complex low- or high-pass filter will suffice. Because a majority of ham receivers are used only for reception above 1600 kHz, a high-pass filter will generally be preferable to the band-reject filter. For the same number of components, the high-pass filter performance is superior to that of the band-reject type.

Since the power level of broadcast stations can be quite high, the stop-band attenuation of the high-pass filter should also be high, preferably in excess of 60 dB. The cutoff frequency should be selected so less than 1 dB of attenuation occurs above 1800 kHz, the start of the 160-meter band. Receivers are generally designed to present a 50-ohm load to the antenna. The filter should also be designed for the same impedance level. The rate of attenuation rise, SWR, passband ripple, and number of filter components are all interrelated and many design choices are possible. In the high-pass design to be discussed, the maximum SWR of the filter was selected to be 1.353. To obtain adequate stop-band attenuation and a reasonable rate of attenuation rise, a filter of 10 elements was necessary. Finally, to simplify construction, only those designs permitting the use of standard-value capacitors were considered.

Building the Filter

The filter layout, schematic diagram, response curve, component values used and toroidal-inductor winding specifications are shown in Fig. 31. The standard-value capacitors used are listed under the filter schematic diagram. Note that all standard values are within 2.8 percent of the design values. Since the maximum deviation between the actual capacitance used and the design value will be only 5.3 percent, there should be little or no difficulty in obtaining the desired reponse. If the attenuation peaks (f2, f4 and f6) at 0.677, 1.293 and 1.111 MHz are not obtained, a slight squeezing or separating of the toroidal-inductor windings should be all that is required to tune the series-resonant circuits.

Note that series circuit C6-L6 should resonate at f6 = 1.111 MHz, but from the response curve it actually resonated at about 1.130 MHz. This frequency error of about 2 percent is small enough to ignore. The A_S value was selected to be 58.3 dB, and examination of the response curve shows the measured filter response to be in good agreement. The measured values of cutoff frequency (at the attenuation level of 0.0988 dB) and the measured value of fA_S (the frequency where A_S is first reached) are also in good agreement with the calculated values. The measured passband loss was less than 0.8 dB from 1.8 to 10 MHz. Between 10 and 100 MHz, the insertion loss of the filter gradually increased to 2 dB. The measured input impedance versus frequency was in good agreement with the calculated input impedance between 1.7 and 4.2 MHz. (The frequency range above 4.2 MHz was not tested.) Over the range tested, the input impedance of the filter remained within the 37 to 67.7 ohms input-impedance window (equivalent to a maximum SWR of 1.353).

Construction of the filter is relatively simple, as shown in Fig. 32, and no difficulty should be experienced if the Mallory SXM polystyrene capacitors are used. These capacitors have a standard tolerance of 2.5 percent and are available through all Mallory distributors. The Micrometals powdered iron T50-2 toroidal cores are available through either Amidon or Palomar Engineers. This material originally appeared in a February 1978 *QST* article by Ed Wetherhold, W3NQN.

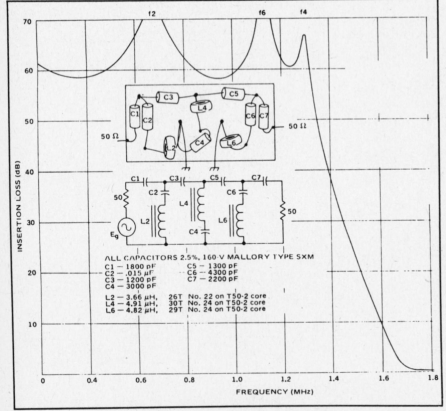

Fig. 31—Filter-response curve, insertion loss, layout and schematic diagram. Terminal impedance is 50 ohms for this 1.7-MHz, high-pass filter.

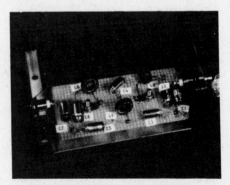

Fig. 32—The filter is built on perfboard in a 2 × 2 × 5-inch Minibox. The filter can be made smaller if desired, and phono connectors can be used in place of the BNC fittings shown here.

A Computer Controlled Shortwave Converter

Imagine a computer-controlled, general-coverage receiver that can be tuned in 10-Hz increments, hold an unlimited number of frequencies in memory and be programmed to suit individual needs. The project described here is a general-coverage receiving converter that can turn any 10-meter amateur receiver or transceiver into a 100-kHz to 20-MHz general-coverage receiver. It uses a synthesizer that is controlled by an Apple II personal computer. As shown in Fig. 33, this compact receiving converter makes extensive use of surface-mounted device (SMD) technology. It was designed and built by Lee Snook, W1DN.

The converter design goals included:

1) Tuning that is controlled from a flexible BASIC language computer program

2) High resolution: 50 Hz for CW and SSB

3) Simplicity: no counters, BFO, IF or AGC

4) Good strong-signal handling and sensitivity.

The final design is a "black box" converter and synthesizer that tunes the spectrum from 100 kHz to 20 MHz in 10-Hz steps. (It can tune the entire range from 0 to 20 MHz, but the local oscillator (LO) is the only signal heard below 100 kHz.) The converter receives signals and converts them to 28.620 MHz, where they can be heard on virtually any 10-meter amateur receiver.

The only connections to the converter are for antenna input, 28.620 MHz output, power (12-V dc at 250 mA) and computer control. There are no front-panel switches. The computer does all the work. Since the converter/synthesizer is connected to the computer through a single 16-wire ribbon cable (only 6 lines are used), it can be put in any out-of-the-way location in the shack.

Operation is simple. Tune the transceiver to 28.62 MHz and boot the computer with the converter program disk. The monitor displays a menu of options. A box in the center of the screen shows the current operating frequency in Hz. The user can scan through any range of frequencies at several rates, or disk options can be selected to store or recall a maximum of 800 frequencies in memory under 20 different file names. The screen format used here limits the maximum number of files to 20, with 40 stations per file. (More could be stored using another format.) More on this later.

CIRCUIT DESCRIPTION

The circuit can be broken down into two distinct parts: the three-loop synthesizer and the converter. Each will be discussed in the following sections.

Synthesizer

For a description of basic synthesizer

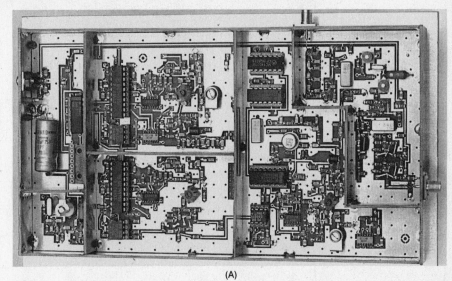

(A)

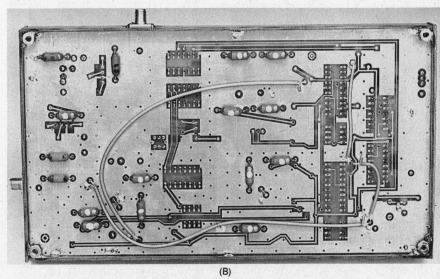

(B)

Fig. 33 — The photo at A shows the top of the synthesized receiving converter; the photo at B shows the reverse side. Most of the components are surface mounted on the top side. PC board material is used to shield the various compartments. Starting at the left of the board: The 16.38-MHz reference oscillator is in the lower left corner. Above it is the power supply and Apple II interface circuitry. Next are the coarse and fine synthesizer loops, then the output loop. Along the upper right edge is the converter circuitry. The only components on the bottom of the board are RF chokes and a few interconnecting cables. Partitions similar to those on the top of the board will help reduce birdies and other spurious signals.

operation see references 1, 2 and 3. Fig. 34 shows the synthesizer block diagram. The heart of the circuit is the Motorola MC145159 sample-and-hold synthesizer IC. Two of these are used (for the coarse and fine tuning loops), along with a third loop employing the phase detector in an RCA CD4046. A 16.38-MHz crystal oscillator is used as the reference for the synthesizers, as well as for the second injection frequency for the converter. This oscillator is not temperature compensated, so its accuracy is approximately ±20 ppm with variations in temperature. The overall frequency accuracy is about ±400 Hz at 20 MHz. This is well within the accuracy of the analog dial on the FT-101E the

author used as an IF receiver. A temperature-compensated crystal oscillator (TCXO) or another standard could be used for greater frequency accuracy.

The MC145159 is the latest in a series of 20-pin DIP synthesizer chips from Motorola. It features serial data loading and a sample-and-hold phase detector. The MC145159 has a programmable 14-bit reference (R) counter, as well as programmable 10-bit divide-by-N and 7-bit divide-by-A counters. See Fig. 35. The counters are programmed serially through a common data input and latched into the appropriate counter latch. If the last bit (the control bit) is a logic high, then all three counters are latched. If the last bit is

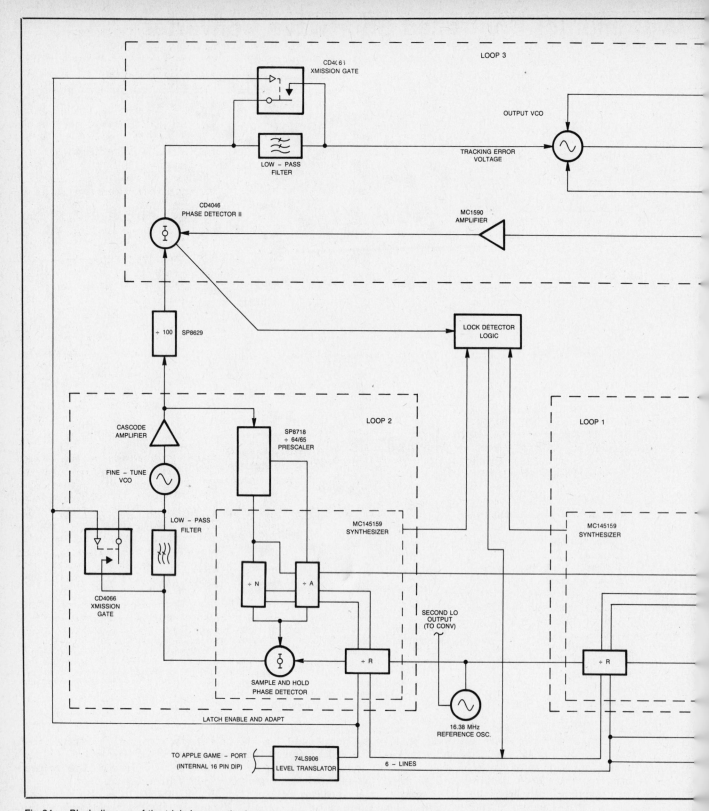

Fig 34 — Block diagram of the triple-loop synthesizer.

a zero, then only the N and A counters are latched. This is done to speed up data entry if the R counter never changes. In this case, the R counter is loaded every time.

When this synthesizer chip is used with an adaptive loop filter, it has fast frequency/phase acquisition and good reference suppression. The adaptive loop filter uses a CD4066 transmission gate to bypass the low-pass filter when changing frequencies or when the loop loses lock. The CD4066 is nothing more than four single-pole switches in an IC package. The switches are activated by a logic high on their control lines. In the adapt mode, with the low-pass filter bypassed, the filter circuit's cutoff frequency is greatly increased for a period of 10 ms. This allows the output of the phase detector to sweep the VCO (voltage-controlled oscillator) to the desired frequency.

When combined with a loop filter and VCO, the MC145159 can provide all the remaining functions for a PLL frequency synthesizer operating up to the device's frequency limit (approximately 15 MHz). For higher-frequency operation, a down

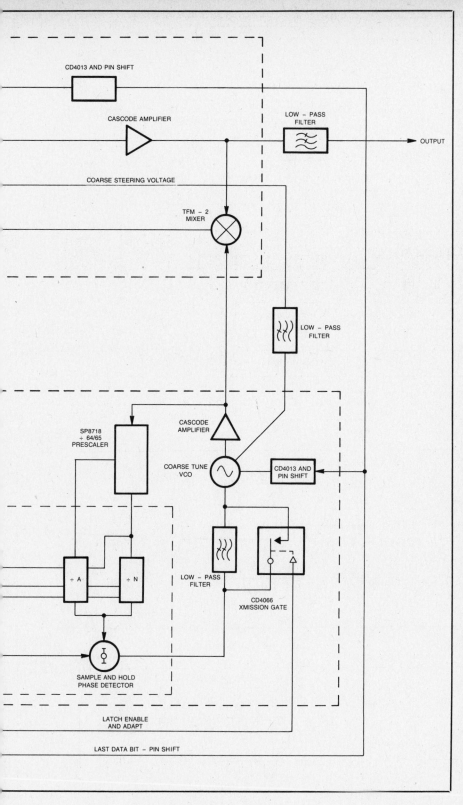

with the data lines. The LATCH ENABLE line is also connected to the control pins of the transmission gate. This allows the LATCH ENABLE to function as an adapt pulse, switching the bandwidth of the loop filters as required. The DATA line is also connected to a CD4013 edge-triggered flip-flop. One extra bit at the end of the data stream sets the flip-flop either high or low. This extra bit does not enter the MC145159 because it is not accompanied by a clock pulse. The flip-flops are used to turn a PIN diode switch on and off. This switch shifts the VCO frequency to extend its range.

The gain of the sample-and-hold phase detector is set with two external components, R_r (pin 20) and C_r (pin 15). See reference 4. A graph and formula given in the data sheet were used to calculate the values used for this project.

The MC145159's internal sample-and-hold phase detector operates in two states. One is the linear state where the loop is very near its final settling frequency and is trying to lock. In this case, the voltage from the phase detector is essentially a transient dc voltage that pulls the VCO phase closer to that of the reference.

In the second state, where the loop is not locked (the two inputs to the phase detector are not at equal frequencies), the output of the detector is a sawtooth waveform with a frequency equal to the difference between the two inputs. This ramp serves to sweep the VCO closer to its lock frequency. The problem is that the low-pass filter between the detector and the VCO attenuates this ramp because the filter has a very low cutoff frequency. This is the reason for including the transmission gate to form an adaptive loop filter to increase the cutoff frequency of the low-pass filter while the PLL is trying to lock.

Pin 9 provides a lock detector output. When locked, it is in a high state. When it is not locked, it produces a pulsed wave with a frequency equal to the difference between the two phase detector inputs.

Dual Modulus Prescaler

As mentioned previously, the input frequency to the MC145159 synthesizer is limited to about 15 MHz. The VCO operates from 45 to 65 MHz, so it must be converted to a lower frequency by mixing or by means of a prescaler. This circuit uses a dual modulus prescaler. A dual modulus prescaler is nothing more than two high-frequency dividers with different divide ratios that are connected in parallel in a single package. (Note: The terms divider and counter are used interchangeably in this article). A logic level on the modulus control line determines which divider is used.

The prescaler used in this project is a Plessey SP8718 divide-by-64/65 prescaler IC. (See reference 5.) To illustrate how and why it was used, an example follows.

Suppose we wish to generate frequencies

mixer or a dual modulus prescaler can be used between the VCO and the MC145159.

The data is loaded serially at pin 12. Each bit is advanced into the register on a low-to-high transition of the clock, pin 11. The last bit allows the reference counter information to be loaded into its 14-bit latch. If the clock continues pulsing after the 32 bits are loaded, the data is shifted

out of pin 14. This arrangement allows for a serial connection of two or more synthesizers that can be loaded from a single data line. The data in each register is then latched on the positive edge of the pulse on pin 13 (LATCH ENABLE). Fig. 36 shows the data format used to load the two synthesizers for this project.

Two additional functions are performed

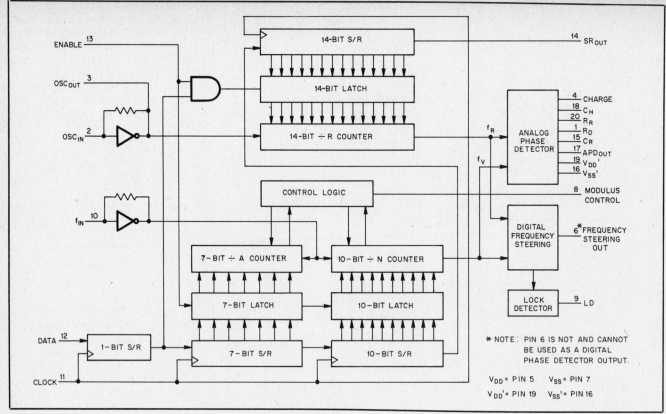

Fig. 35 — Internal block diagram of the Motorola MC145159 synthesizer IC.

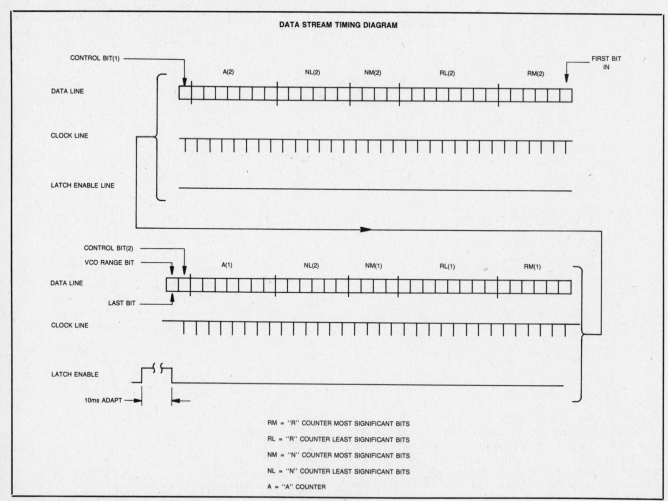

Fig. 36 — Data stream timing for the synthesizer.

from 46.1 to 47.1 MHz in 10-kHz increments. Since the synthesizer output frequency (F) is equal to the reference frequency (F_r) multiplied by an integer number N_t,

$$F = F_r \times N_t$$

or,

$$N_t = \frac{F}{F_r}$$

If the MC145159's internal dividers could divide signals from a VCO operating from 46.1 to 47.1 MHz, all we would have to do is program the dividers to divide by numbers between 4610 and 4710. The MC145159 can only operate to 15 MHz, however. High-frequency dividers could be placed between the VCO and the MC145159, but they would have to be programmed, adding to the complexity of the circuit. Why not use a fixed divider?

Let's say we use a divide-by-10. Now the programmable dividers in the MC145159 will see 4.61 to 4.71 MHz, well within range. What will the dividers in the MC145159 have to be programmed to?

$$N_t = \frac{(F/10)}{F_r}$$

$$N_t(min) = \frac{(46.1 \text{ MHz}/10)}{10 \text{ kHz}} = 461$$

$$N_t(max) = \frac{(47.1 \text{ MHz}/10)}{10 \text{ kHz}} = 471$$

Can we still program 10-kHz steps? If we increase the N count by one, what will the output be?

$$F = F_r \times (10 \ N_t)$$
$$F = 10 \text{ kHz} \times [10 \times (461 + 1)]$$
$$= 46.2 \text{ MHz}$$

The frequency has changed by 100 kHz instead of 10 kHz. With a fixed prescaler, the resolution is decreased by the value of the fixed divisor.

The alternative is to use two fixed prescalers with different divide ratios and then switch between them at the appropriate time to give a fractional divide ratio. Inside the MC145159 is a control circuit that uses the N and A counters to decide when to switch. Its output is the modulus control line (pin 8).

How can we divide by 4610 using this method? Let us say we are using a divide-by-64/65 prescaler. If we divide by 65 for A times and divide by 64 for (N − A) times, we will get

$$4610 = (65 \times A) + [64 \times (N - A)]$$
(Eq. 1)

Now let us find the values of N and A. Let A = 0, since we have one equation and two unknowns.

$$4610 = 64 \text{ N}$$

or

$$N = 72.03125$$

A divider can only divide by an integer number, so let N = 72. Substituting N = 72 into Eq. 1 and solving for A, A = 2.

This means that if we divide by 65 for 2 times and then divide by 64 for (72 − 2) times, we will have effectively divided by 4610. Have we still maintained the 10-kHz channel spacing? If we increment A by one count, we have

$$N_t = (65 \times 3) + [64 \times (72 - 3)] = 4611$$

and

$$F = N_t \times F_r = 4611 \times 10 \text{ kHz}$$
$$= 46.11 \text{ MHz}$$

The increment is 10 kHz.

The equation for N_t can be simplified to

$$N_t = [(P + 1) \times A] + [P \times (N - A)]$$

or

$$N_t = (N \times P) + A$$

In these equations, P is the lower prescaler number (in this case, 64). We have accomplished a great deal with the dual modulus prescaler and the MC145159 internal control logic. The MC145159 N divider has the capability to divide between 16 and 1023, and the A divider between 0 and 127 (provided that A is less than N). So,

$$From \ N_t = (N \times P) + A$$
$$N_t(min) = (16 \times 64) + 0 = 1024$$
$$N_t(max) = (1023 \times 64) + 127 = 65599.$$

Therefore,

$$From \ F = F_r \times N_t$$
$$F(min) = 10 \text{ kHz} \times 1024 = 10.24 \text{ MHz}$$
$$F(max) = 10 \text{ kHz} \times 65599$$
$$= 655.99 \text{ MHz}!$$

This range is with 10-kHz channel spacing! Of course, the VCO and prescaler must be able to cover such a great range.

The price? We need a machine to calculate all the N, A, and R values, given an output frequency. It must also convert these numbers to serial data, pulse the adapt line, know when to switch the VCO to a different range and know when to flag an unlocked synthesizer. That's what the Apple II computer does.

Triple Loop Approach

Why can't we use just one loop to generate 10-Hz resolution? One reason is that the N divider would have to be very large to multiply the reference (10 Hz) up to the 45- to 65-MHz range. Remember: $F = F_r P N$. If $F_r = 10$ and P = 64, then N = 101,562. This is clearly larger than the allowed maximum of 65,599.

The second reason is that the loop filter would have to have a cutoff frequency of much less than 10 Hz to attenuate the reference leakage through the phase detector before it reaches the VCO. This causes many problems. The loop natural frequency would have to be so low that the loop would be susceptible to microphonics.

To ensure stability, the loop gain would have to be low, thereby causing slow lockup time.

(Fig. 37 appears on next page)

Fig. 37 — Schematic diagram of the coarse and fine synthesizer loops. Components marked with an asterisk (*) are surface-mounted devices.

C100, 304, 310, 404, 410 — *0.1 μF.
C101, 102, 303, 314, 403, 414 — *4.7 μF, 15 V (Sprague 195D475X0015V5K77 or equiv.).
C103 — 470 μF, 25 V axial.
C300, 307, 308, 312, 313, 315, 322-324, 329, 330-332, 400, 408, 409, 412, 413, 415, 422-426 — *0.001 μF.
C301, 401, 419 — *33 pF.
C302 — *330 pF.
C305, 306, 309, 311, 405-407, 411 — *0.01 μF.
C316 — *3.3 pF.
C317, 421 — *10 pF.
C318, 319, 325 — *22 pF.
C320, 326 — *56 pF.
C321 — *5.1 pF.
C327, 328 — *100 pF.
C333 — 3.5-30 pF ceramic piston trimmer (Johanson 8093).
C402 — *2200 pF.
C416-418 — *30 pF.
C420 — *5.6 pF.
D300, 400 — *MMBZ5226 3.3-V Zener diode (Motorola).
D301-304, 401 — *MMBV109 varactor diode (Motorola).
D305 — *MMBV3401 PIN diode (Motorola).
J100 — 16-pin DIP socket.
L300-303, 400-402 — 3.9 μH axial molded RF choke.
L304 — 0.175 μH; 9.5t no. 26, close wound, 0.15 in. diam.
L305, 403 — 130-320 nH slug-tuned coil.
L306 — *0.13 μH.
Q300, 302, 401, 402 — *MMBT3904 NPN transistor (Motorola).
Q301, 400 — *MMBT3906 PNP transistor (Motorola).
Q303, 403 — *U310 JFET transistor (Motorola).
Q304, 305, 404, 405 — *MMBR930 NPN transistor (Motorola).
R100, 101, 405 — *5.1 kΩ.
R300, 303, 400, 407 — *100 kΩ.
R301 — *220 kΩ.
R302, 406 — *470 Ω.
R304, 305, 408 — *33 kΩ.
R306, 409 — *51 kΩ.
R307, 309, 402, 411 — *4.7 kΩ.
R308 — *560 Ω.
R310, 403 — *10 kΩ.
R311, 404 — *11 kΩ.
R312 — *1 kΩ.
R313-315, 413-415 — *2.2 kΩ.
R316-319, 321 — *180 Ω.
R320 — *30 Ω.
R322, 324 — *150 Ω.
R323 — *36 Ω.
R325 — *100 Ω.
R401 — *120 kΩ.
R410 — *56 kΩ.
R412 — *330 Ω.
R416 — 220 Ω.
R417, 419 — *82 Ω.
R418 — *91 Ω.
U100 — 74C906 open-drain buffer (National Semiconductor).
U101 — 7808 three-terminal 8-V regulator, TO-220 case.
U300, 400 — MC145159 synthesizer (Motorola).
U301, 401 — SP8718 divide-by-64/65 prescaler (Plessey).
U302, 402 — *14066B quad bilateral switch (Motorola).
U303 — *14013 dual-D flip-flop (Motorola).

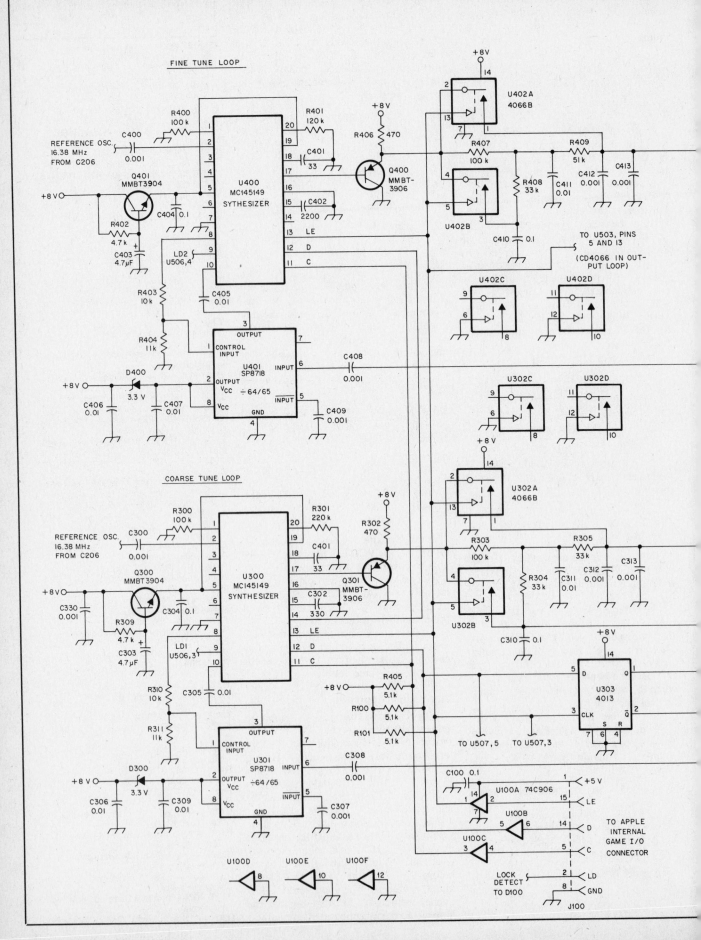

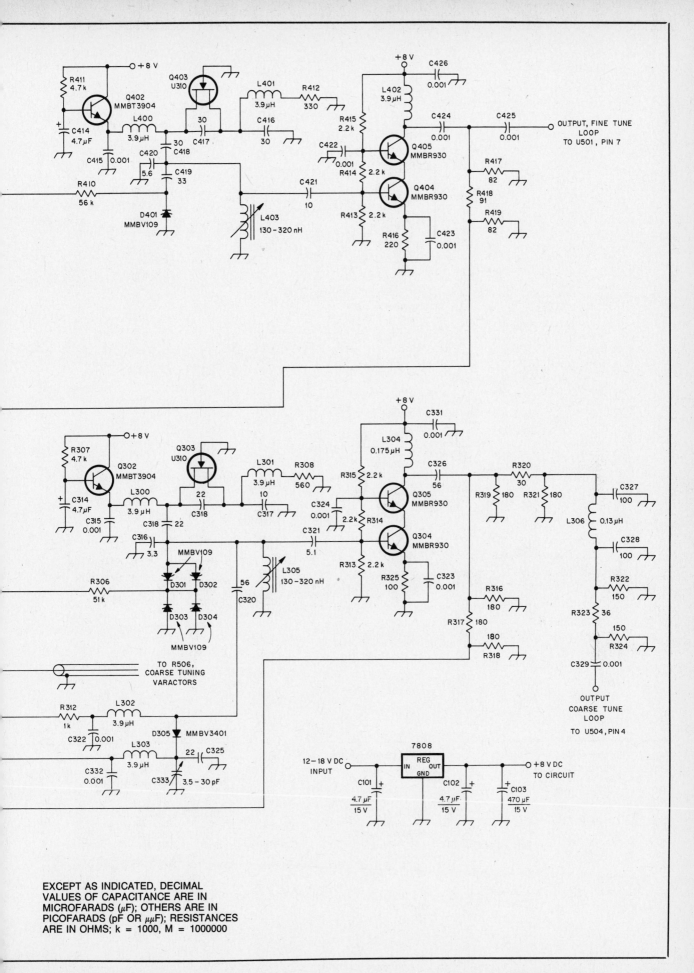

EXCEPT AS INDICATED, DECIMAL
VALUES OF CAPACITANCE ARE IN
MICROFARADS (μF); OTHERS ARE IN
PICOFARADS (pF OR μμF); RESISTANCES
ARE IN OHMS; k = 1000, M = 1000000

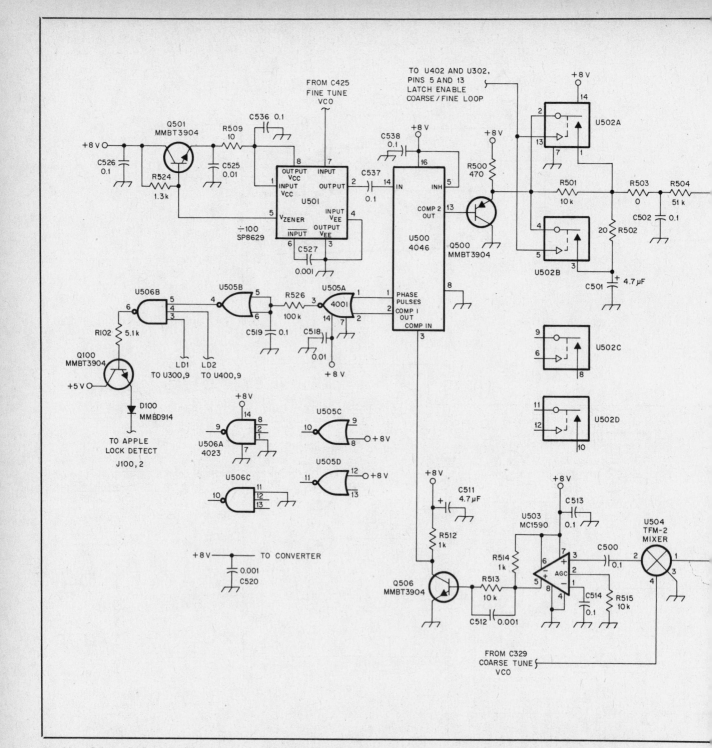

Fig. 38 — Schematic diagram of the synthesizer output loop. Components marked with an asterisk (*) are surface-mounted devices.

C500, 502, 513, 514, 519, 526, 533, 536-538 — *0.1 μF.
C501, 511, 524 — *4.7 μF, 15 V (Sprague 195D475X0015V5K77 or equiv.).
C503, 505 — *22 pF.
C504, 506 — *10 pF.
C507 — *5.1 pF.
C508-510, 512, 517, 518, 520, 527, 528, 531, 532, 535 — *0.001 μF.
C515, 516, 521, 522 — *100 pF.

C523, 529 — *51 pF.
C525 — *0.01 μF.
C530 — *33 pF.
C534 — 3.5-30 pF ceramic piston trimmer (Johanson 8093).
D100 — *MMBD914 general-purpose silicon diode (Motorola).
D500-503, 505 — *MMBV109 varactor diode (Motorola).
D504 — *MMBV3401 PIN diode (Motorola).

L500-503 — 3.9 μH axial molded RF choke.
L504 — 130-320 nH slug-tuned coil.
L505 — 0.175 μH; 9.5t no. 26, close wound, 0.15 in. diam.
L506, 507 — *0.13 μH.
Q100, 501, 502, 506 — *MMBT3904 NPN transistor (Motorola).
Q500 — *MMBT3906 PNP transistor (Motorola).
Q503 — *U310 JFET transistor (Motorola).

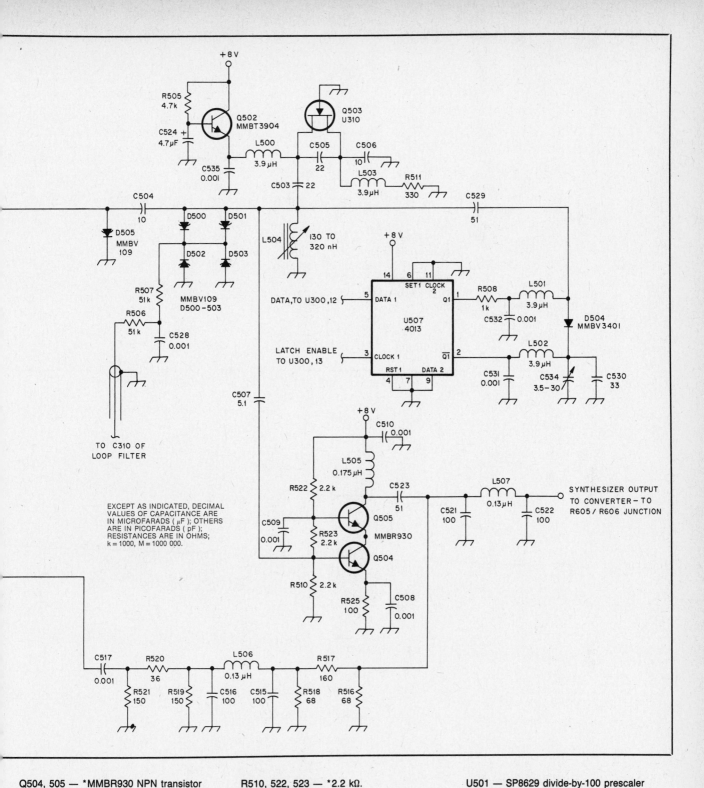

Q504, 505 — *MMBR930 NPN transistor
(Motorola).
R102 — *5.1 kΩ.
R500 — *470 Ω.
R501, 513, 515 — *10 kΩ.
R502 — *20 Ω.
R503 — *0 Ω.
R504, 506, 507 — *51 kΩ.
R505 — *4.7 kΩ.
R508, 512, 514 — *1 kΩ.
R509 — *10 Ω.

R510, 522, 523 — *2.2 kΩ.
R511 — *330 Ω.
R516, 518 — *68 Ω.
R517 — *160 Ω.
R519, 521 — *150 Ω.
R520 — *36 Ω.
R524 — *1.3 kΩ.
R525 — *100 Ω.
R526 — *100 kΩ.
U500 — CD4046BCN phase-locked loop
(RCA).

U501 — SP8629 divide-by-100 prescaler
(Plessey).
U502 — *14066B quad bilateral switch
(Motorola).
U503 — MC1590G op amp (Motorola).
U504 — Mini-Circuits TFM-2 doubly balanced
mixer.
U505 — CD4001BCN quad NOR gate (RCA).
U506 — MC14023B quad NAND gate
(Motorola).
U507 — *14013 dual-D flip-flop (Motorola).

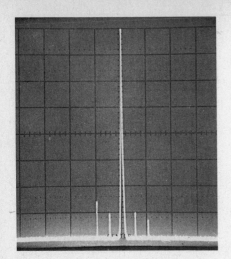

Fig. 39 — Spectrum of synthesizer output. The vertical scale is 10 dB per division; the horizontal scale is 1 MHz per division. Leakage of the 500-kHz reference signal (and its 1 MHz second harmonic) through the balanced mixer is better than 65 dB down from the carrier.

A multiple-loop synthesizer solves these problems. See Figs. 34, 37 and 38. Two MC145159 ICs are used to form independent serially loaded synthesizer loops, one coarse tuned (U300) and one fine tuned (U400). The third loop, the output loop, tracks the coarse loop with an offset from the fine loop. A CD4046 phase detector (U500) is used to generate the control voltage for the output loop.

The coarse loop tunes from 44.5 to 64.5 MHz in 10-kHz steps. Its control voltage is used to steer the output VCO about 500 kHz above the coarse VCO. The construction of both the coarse and output VCOs should be as close as possible to allow good tracking. The output of the tracking VCO is amplified and mixed with the coarse loop VCO output in U504, a Mini-Circuits TFM-2 doubly balanced diode-ring mixer. The difference signal at the mixer output is amplified by the MC1590 (U503), clipped and used as one input to pin 3 of U500, the CD4046. The

fine tune loop, divided by 100 in U501, an SP8629, feeds the other phase detector input (pin 14) with a signal of 0.50 to 0.50999 MHz in 10-Hz steps. An extra 40 dB of 1-kHz reference suppression is obtained by dividing by 100.

When it is locked, the output VCO generates a frequency that, when mixed with the coarse VCO, produces a mixer output that is equal to the fine loop divided by 100. What we end up with is the output frequency equal to the coarse loop frequency plus the fine loop frequency divided by 100.

$$F = F1 + \frac{F2}{100}$$

where

F = output frequency
F1 = coarse loop frequency
F2 = fine loop frequency.

The output VCO frequency should track the coarse-tune VCO plus the fine-tune

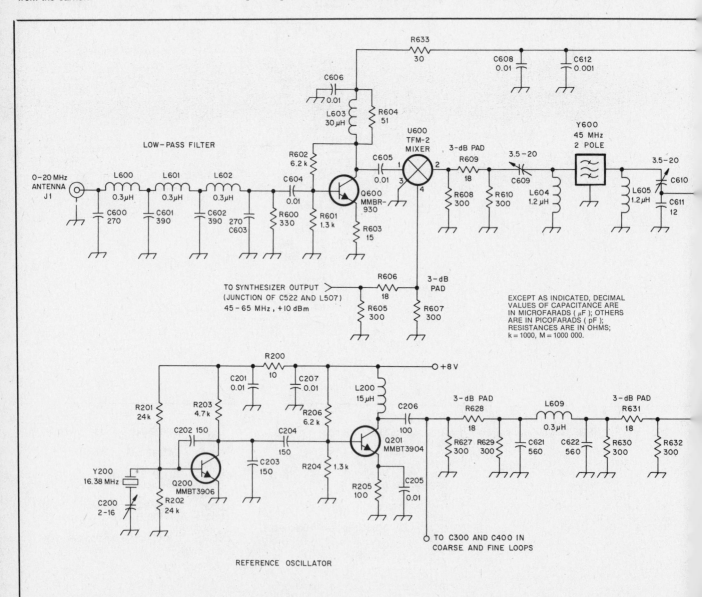

frequency. Any error in tracking is compensated for by the CD4046 phase detector. The detector produces a waveform whose average value is a dc level when filtered by the low-pass filter. This voltage is applied to a varactor that pulls the VCO closer to the correct frequency/phase. To extend the range of the VCOs in the output and coarse loop, the tank circuits are shunted by a PIN switching diode in series with a variable capacitor. The capacitor is switched into the tank when the output frequency is to go below 53 MHz.

A lock detector is formed by a CD4001 NOR gate (U505) and a CD4023 three-input NAND gate (U506). If any of the three loops lose lock, the detector causes the PB(0) switch input on the Apple game port (see Fig. 41 later in this writeup) to go low through a bipolar switch (Q100) pulled up by a 220-Ω resistor to +5 V. The PB(0) input is in parallel with the "open apple" key internally pulled to ground through a 560-Ω resistor. D100, at the output of the bipolar switch, keeps the lock detector from loading the "open apple" line when the receiver is not in use. A low voltage at PB(0) (out of lock) causes a value of less than 127 to appear in location −16287. The BASIC program periodically scans this location and reloads the synthesizers if an "out of lock" is detected. At the same time it displays "unlock" on the screen.

Colpitts grounded-gate oscillators (Q303, Q403 and Q 503) are used in all three VCOs to give excellent phase noise characteristics, approximately −80 dBc at 1 kHz from the carrier in a 1-Hz bandwidth. This configuration adds fixed capacitance in parallel with the varactor tuning, making the tuning bandwidth narrower than with other oscillator configurations. Cascode buffer amplifiers (Q304/305, Q404/405 and Q504/505) are used after the VCOs for approximately 40 dB of reverse isolation.

All three loops use adaptive filtering to bypass the low-pass filters when changing frequency. Although the CD4046 phase detector can acquire lock without the adaptive filter, its lock time is sped up with the filter. The output frequency is within 1 kHz of the settling frequency in approximately 10 ms after a frequency jump of 20 MHz.

Keeping the spurious signal levels down in the synthesizer was a problem. The coarse VCO leaked through the balanced mixer and appeared at the output as 500 kHz spurs. These spurs were attenuated by increasing the level from the output VCO and then padding this output to the balanced mixer. The pad not only provided some isolation between the mixer and the output, but provided a better match to the mixer. The better match results in improved port-to-port isolation. Spurious products are at least 60 dB below the carrier. See Fig. 39. The other buffer stages provided more output power than was needed, so 50-Ω pads were used to terminate the inputs and outputs of each low-pass filter.

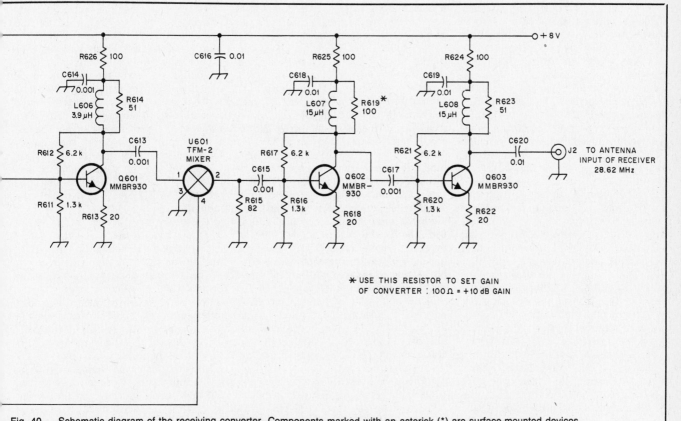

Fig. 40 — Schematic diagram of the receiving converter. Components marked with an asterisk (*) are surface-mounted devices.

C200 — 1.9-15.7 pF air variable (Johnson C187-0109-005 or equiv.).
C201, 205, 207, 604-606, 608, 616, 618-620 — *0.01 μF.
C202-204 — *150 pF.
C206 — *100 pF.
C600, 603 — *270 pF.
C601, 602 — *390 pF.
C609, 610 — 3.5-20 pF air variable (Johnson C274-20-105 or equiv.).
C611 — *12 pF.
C612-615, 617 — *0.001 μF.
C621, 622 — *560 pF.
L200, 607, 608 — 15-μH axial molded inductor.
L600-602, 609 — *0.3 μH.

L603 — 30-μH axial molded inductor.
L604, 605 — 1.2-μH axial molded inductor.
L606 — 3.9-μH axial molded inductor.
Q200 — *MMBT3906 PNP transistor (Motorola).
Q201 — *MMBT3904 NPN transistor (Motorola).
Q600-603 — *MMBR930 NPN transistor (Motorola).
R200 — *10 Ω.
R201, 202 — *24 kΩ.
R203 — *4.7 kΩ.
R204, 601, 611, 616, 620 — *1.3 kΩ.
R205, 619, 624-626 — *100 Ω.
R206, 602, 612, 617, 621 — *6.2 kΩ.

R600 — *330 Ω.
R603 — *15 Ω.
R604, 614, 623 — *51 Ω.
R605, 607, 608, 610, 627, 629, 630, 632 — *300 Ω.
R606, 609, 628, 631 — *18 Ω.
R613, 618 — *20 Ω.
R615 — *82 Ω.
R622 — *20 kΩ.
R633 — *30 Ω.
U600, 601 — Mini-Circuits TFM-2 doubly balanced mixer.
Y200 — 16.38-MHz crystal, 32-pF load capacitance, anti resonant, HC-18 holder.
Y600 — 45-MHz, 2-pole crystal filter, Piezo Technology model 2843A or 2844A.

Converter

Not much time was spent designing the converter, as the emphasis was placed on the synthesizer and the program. Although a converter with better performance could be designed, this one has good enough dynamic range to work within the wide front-end passband of the converter and not be overloaded by the crowded RF environment below 20 MHz.

Fig. 40 is a schematic diagram of the converter. Two stages of conversion are used in an upconverting design. The first stage converts the input from the antenna (100 kHz to 20 MHz) to 45 MHz by mixing with the 45- to 65-MHz output of the synthesizer. This occurs in U600, a TFM-2 doubly balanced diode-ring mixer. The second stage, U601 (another TFM-2), mixes the 45-MHz first IF with the 16.38-MHz reference-oscillator signal (from Y200/Q200/Q201) to produce the 28.62-MHz output to the IF receiver. See reference 6 for further discussion of the upconverting technique.

A 7-pole, 3-dB ripple low-pass filter (L600-602, C600-603) is used prior to the first RF amplifier (Q600) for attenuation of any 45-MHz signals and for image rejection above 90 MHz. The filter response is 40 dB down at 45 MHz and greater than 80 dB down at 90 MHz.

The gain of the first RF amplifier, including losses in the low-pass filter, is 8 dB. The RF-amplifier input impedance is shunted with a 330-Ω resistor to set the filter passband ripple to 3 dB. The RF amplifier has a noise figure of about 6 dB and a third-order intercept of +20 dB relative to its output. The RF amplifier presents a broadband 50 Ω output for the mixer input, and the mixer output is terminated with a 50-Ω 3-dB pad (R608-610). A 50-Ω 3-dB pad (R605-607) is used on the mixer LO port. The output from the synthesizer is +10 dBm, so the pad reduces this to the correct LO level of +7 dBm. There is significant loss in this part of the circuit. After going through the mixer, pad and matching network, the conversion gain from antenna to the input of the 45-MHz crystal filter (Y600) is approximately 0 dB.

The crystal filter is terminated in 2600 ohms at its input by the matching network and at its output by the bias resistor divider of the next stage. An 8-dB amplifier (Q601) follows the crystal filter to compensate for losses in the second mixer and filter. The conversion gain at the output of the Q601 amplifier circuit is 6 dB. The second mixer is terminated in the same fashion as the first.

Two more stages of gain (Q602, Q603) were added, but they are probably not required. Their total gain is adjusted by changing the value of R619, which is across the collector choke L607 on Q602. In the prototype, a value of 100 Ω for R619 yielded a total converter gain of 12 dB. This brings the atmospheric noise level just above the input noise floor of my trans-

ceiver. Higher resistances increase the gain and vice-versa.

All the amplifiers used in the converter have fairly high overload characteristics and do not require broad-band matching transformers. The receiver sensitivity averages 1 μV for a 6-dB S/S + N ratio using a 5-kHz IF filter for a 100% modulated AM signal.

PROGRAMMING

The receiver program consists of a basic program called DUAL1 and assembly language program SYN1. DUAL1 is approximately 200 lines and SYN1 is 55 lines long. SYN1 was written using the Apple's mini-assembler, which is described in reference 7. Listings for these programs are available from ARRL Hq. as part of the template package for this project.[1] Alternatively, a disk with all of the software needed to control this converter is available from the author.[2]

The primary function of DUAL1 is to calculate the values for the N and A counters once a frequency has been selected. The rest of the program formats the display, waits for command characters, and handles files. Lines 360 through 520 scan the keyboard, waiting for an entry that directs the program to change, save, recall a frequency or call a disk subroutine.

Once a frequency has been selected, lines 520 through 550 calculate the coarse-loop frequency. If the coarse-loop frequency is greater than 52 MHz, it sets the VCO range bit high. Lines 580 and 590 find the fine-tune loop frequency. Then, lines 610 through 640 calculate the proper N and A values for both loops.

Since the N values comprise 10 bits, they are broken into two bytes by lines 650 to 680. The A values are only 7 bits long so they can fit into one byte. The R value does not change. Its bit representation has been pre-calculated and is POKED directly into the binary program. RM(1) and RL(1) are POKED into locations 9 and 26 (decimal) respectively. RM(2) and RL(2) are POKED into locations 235 and 236 (decimal) respectively. Lines 690 to 780 multiply the appropriate bytes by powers of two to position the most significant bit next to the carry and then POKE all of the values into 10 vacant memory locations.

After all 10 bytes are POKED into memory, the basic program calls the binary program, SYN1, located at 38000 to 38135 (decimal) or 9470 to 94F7 (hex). This program has the job of turning on and off the data, clock, and latch enable output ports in the correct sequence. See Fig. 41 for a description of the game port pin-outs

and the address to toggle each port. The binary program has five bytes to look at for each synthesizer. Each byte is loaded into the accumulator and then rotated one bit to the left into the carry bit. The X register is loaded with the number of bits to be rotated and is decremented until all the bits are sent for each byte. At the end of each sequence, one more high bit is sent representing the control bit.

If the carry is high, a subroutine is called which sets the data line to a 1 and vice-versa. The line is set high by loading the accumulator (LDA) to $C059 and low by loading to $C058. (See Fig. 41.) A clock pulse is sent with each data bit by loading accumulator (LDA) to $C040. The clock (strobe) goes low for one instruction cycle. The value in the accumulator is irrelevant when loading these addresses. After all 64 bits have been sent, 32 per synthesizer, the VCO range shift bit, stored at $946F, is loaded into the accumulator. The data line is either set high or low if this byte is a 1 or 0. No clock pulse is sent with this bit, so it does not enter the synthesizer, but remains on the data line which is also connected to the VCO range flip-flop. The latch enable is now set high, by loading the accumulator to $C059. It is high for a period determined by the wait subroutine at address $FCA8. The length of the wait is determined by the value loaded into the accumulator, in our case $3D. This gives us a wait of approximately 10 ms. The line is then turned off by loading the accumulator to $C058. Data gets latched into the synthesizers on the positive edge of this pulse and its duration determines how long the adaptive filter stays in its wide-band mode.

The binary program then returns control to DUAL1 which then starts the process over again by scanning the keyboard, waiting for another command.

The other part of the program requiring comment is the method of creating files of frequencies on disk. File names are stored as text files. The Apple catalog command won't work to display only the text-file names in the right hand corner of the screen. In this program, a separate file (FMEN$) contains the filenames. By using this technique, the filenames can be pulled out and placed anywhere on the screen. After initializing a new disk, you must run FCREATE to open the FMEN file.

CONSTRUCTION

Fig. 33 shows the completed receiving converter. Although the author's unit relies heavily on surface-mounted devices, regular axial components can be used as well. All of the components mount on a single 4.25- × 7.5-inch PC board. The board material is double-sided, 0.062-inch glass-epoxy stock. Circuit traces are etched on both sides, and the holes are plated through. Unetched circuit board material is soldered directly to the PC board to form shielded compartments. The sides are 1.75

[1] A template package, including program listings, etching patterns and parts placement guide is available from the ARRL Technical Department Secretary for $5.00.
[2] Program disks and etched PC boards are available from the author. Contact Lee Snook, W1DN, RR 1, Box 126, Janesville, MN 56048 for details.

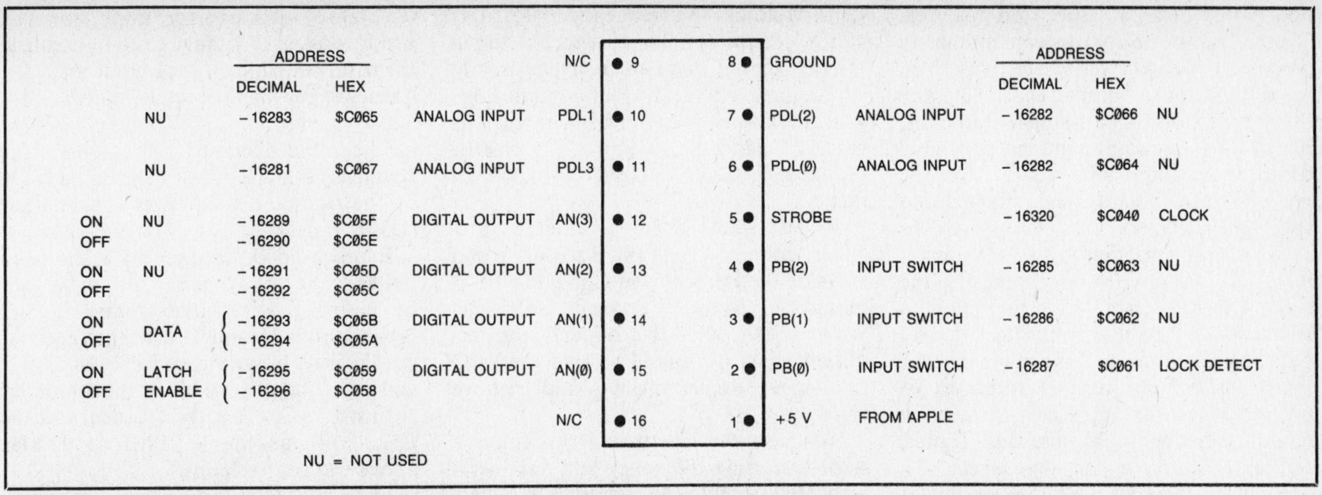

	ADDRESS					pin	pin				ADDRESS		
		DECIMAL	HEX									DECIMAL	HEX
					N/C	9	8	GROUND					
NU		−16283	$C065	ANALOG INPUT	PDL1	10	7	PDL(2)	ANALOG INPUT		NU	−16282	$C066
NU		−16281	$C067	ANALOG INPUT	PDL3	11	6	PDL(0)	ANALOG INPUT		NU	−16282	$C064
ON	NU	−16289	$C05F	DIGITAL OUTPUT	AN(3)	12	5	STROBE		CLOCK		−16320	$C040
OFF		−16290	$C05E										
ON	NU	−16291	$C05D	DIGITAL OUTPUT	AN(2)	13	4	PB(2)	INPUT SWITCH	NU		−16285	$C063
OFF		−16292	$C05C										
ON	DATA	−16293	$C05B	DIGITAL OUTPUT	AN(1)	14	3	PB(1)	INPUT SWITCH	NU		−16286	$C062
OFF		−16294	$C05A										
ON	LATCH ENABLE	−16295	$C059	DIGITAL OUTPUT	AN(0)	15	2	PB(0)	INPUT SWITCH	LOCK DETECT		−16287	$C061
OFF		−16296	$C058										
					N/C	16	1	+5 V	FROM APPLE				

NU = NOT USED

Fig. 41 — Pin outs of the Apple II internal game I/O port. Be careful of common-mode voltages when measuring levels on the game port — you could damage the I/O chip. Be sure that all test equipment is properly grounded to the computer.

inches high. Etched PC boards are available from the author. See note 2. A template package with top and bottom etching patterns, drilling information and a parts-placement guide is available from ARRL Hq. See note 1.

Surface-mounted components are available from a growing number of suppliers. If you have trouble finding specific components, the author may be able to assist. The surface-mounted ICs, transistors and diodes are available from Motorola. For information on the nearest distributor, write Motorola Semiconductors, Box 20912, Phoenix, AZ 85036. The chip resistors used here were made by Dale Electronics, 2064 12th Ave., Columbus, NE 68601. The Kemet chip capacitors are from Union Carbide, Electronics Div., P.O. Box 5928, Greenville, SC 29606. The inductors are made by CoilCraft, 1102 Silver Lake, Cary, IL 60013. Other good sources of chip resistors and capacitors are Mouser Electronics and Newark Electronics.

A minimum of construction details will be given since the PC board reduces assembly to soldering parts in the right location. Once you overcome the fear of handling and soldering SMDs, they offer several advantages over conventional construction. Most of the circuitry and printed-circuit traces are on one side of the board, leaving the other for a good ground plane. The board can be laid flat in the work area, and you won't be continually turning it over to solder and clip component leads. Another advantage is time saved by not having to form the leads on conventional components before stuffing the board.

The key to SMD construction is soldering technique. It's mandatory to have the right tools for the job. Use small-gauge silver solder (available from standard suppliers such as Kester). Regular tin/lead solder will eventually leach or bridge across component terminals, resulting in a short across the component over time. Invest in a low-temperature, fine-tipped soldering iron. You'll also need a higher wattage iron for soldering to ground planes and shields.

The surface-mounted resistors and capacitors used for this project measure 0.06 × 0.12 inches and can be held in place with fine-tipped tweezers. Place a dab of solder on the iron and temporarily tack one end of the component to its trace. This will hold the component in place. Now you can have both hands free to heat the other end of the component and feed it solder for a good connection. Then, return to the first end and resolder it to make a good connection. Don't overdo the amount of solder. A small magnifying glass is helpful for checking the connection.

Although no standard coding scheme has been adopted for chip capacitors and resistors, the two most common schemes are given in Table 1. Avoid buying surface-mounted components that don't have identifying markings. Such components do exist, but use of them can make troubleshooting difficult.

The small surface-mounted ICs are probably the hardest parts to solder. A fine-tipped iron is mandatory, and the leads should be carefully checked for solder bridges. Pin number one is in the lower left corner when the IC is positioned so you can read the part number.

The axial RF chokes are mounted on the bottom side of the PC board. The exception to this is L605, which should be mounted on the top side for better isolation between the input and output of the crystal filter.

Good shielding around each compartment and around the whole board is required in the noisy environment of the computer and the four oscillators. Double-sided unetched PC board was soldered directly to the main PC board making compartments around the reference oscillator, synthesizers, and the converter. Nickel-plated brass nuts were soldered in each corner for attaching a top and bottom lid.

Table 1
Surface-Mounted Capacitor Markings

Letter/ Number	Capacitance (pF)	
	American	Japanese
A	10	1.0
B	11	1.1
C	12	1.2
D	13	1.3
E	15	1.5
F	—	1.6
G	—	1.8
H	16	2.0
I	18	—
J	20	2.2
K	22	2.4
L	24	2.7
M	—	3.0
N	27	3.3
O	30	—
P	—	3.6
Q	—	3.9
R	33	4.3
S	36	4.7
T	39	5.1
U	—	5.6
V	43	6.2
W	47	6.8
X	51	7.5
Y	56	8.2
Z	62	9.1
3	68	—
4	75	—
7	82	—
9	91	—

Multiplier Color (American)		Multiplier Number (Japanese)	
Orange	× 0.1	0	X 1
Black	× 1.0	1	X 10
Green	× 10	2	X 100
Blue	× 100	3	X 1000
Violet	× 1000	4	X 10,000
Red	× 10,000	5	X 100,000

Tune-Up Procedure

Probably the most difficult part of building a synthesizer is getting the voltage-controlled oscillator to operate properly. For testing high-quality VCOs, a spectrum analyzer and a means of measuring phase noise are essential. A good VCO has phase-noise components a few kilohertz away from the carrier which are far below the noise floor of an RF spectrum analyzer. These phase-noise components ultimately determine the receiver's selectivity with regard to strong adjacent signals. One way this phase noise can be measured is to phaselock to the source and mix the signal down to an audio level where it can be viewed with a high-dynamic-range audio spectrum analyzer.

Since most hams don't have these instruments, we will have to be content with the signal purity we get. As a minimum, a frequency counter and an oscilloscope will be required to tune up this project. If your frequency counter reading seems to be unstable while making a measurement, it could indicate high spurious signal, reference signal, or harmonic levels.

Refer to the OPERATION section of this article. Start the Apple II computer control program and enter any starting frequency. Set the increment step size to 10 Hz. Remember, the starting frequency is in MHz and the increments are in Hz. Press the up-arrow or down-arrow key and look at the waveforms on pins 11, 12, 13 of U300 and U400 with your scope. These are the CLOCK, DATA, and LATCH ENABLE signals sent from the computer through buffer U100. They should resemble the waveforms in Fig 36. The CLOCK and DATA pulses are very short, so a sweep rate of 0.2 ms/div is required.

Measure the voltage on pin 1 of U303 and U507. It should be 0 V when frequencies greater than 8 MHz are entered into the computer and about 8 V when frequencies less than 8 MHz are entered. The 10-ms LATCH ENABLE pulse should appear on pins 5 and 13 of U302, U402 and U502 during frequency changes. This pulse is used to turn on the transmission gates that bypass the loop filters during frequency acquisition.

Be sure to remove power from the circuit when removing and replacing ICs and components in the following steps. First, tune up the coarse-tune loop. Remove U506 (to disable the lock-detect feature) and U300. Connect a variable voltage source to pin 17 of the socket for U300. This can be accomplished by using a small multi-turn trimmer potentiometer with one leg in pin 19, the other in pin 16 and the slider leg in pin 17 as the variable voltage source. Adjust this voltage source until 1 V appears on pin 17. Connect your frequency counter to the output of the coarse-tune loop at C329. Enter a frequency of greater than 8 MHz (increment step size can be any value) and adjust L305 until the counter reads approximately 52 MHz with 1 V on pin 17. Now adjust the voltage source for 7 V on pin 17. You should measure about 67 MHz on the counter. If you don't get this frequency span, the VCO does not have the required gain, probably caused by a misplaced part in the oscillator circuit.

Enter a frequency of less than 8 MHz and repeat the procedure. This time, use

C333 to adjust the frequency to 45 MHz with 1 V on pin 17. Increase to 7 V on pin 17; this should increase the frequency to approximately 57 MHz. Don't retune L305 while making this second adjustment. The value of padder capacitor C325 can be changed if C333 doesn't have enough range.

To test U301, the dual modulus prescaler, connect U300, pin 8 to +8 V and measure the frequency at U300, pin 10. It should be 1/64 the VCO output frequency. When U300, pin 8 is grounded, the frequency at pin 10 should be 1/65 the VCO frequency. Replace U300 and remove U400.

Now test the fine-tune loop. Enter a 5-MHz starting frequency and any increment step size into the computer. Connect a variable voltage source to the C410-R408 junction and adjust for 1 V. Attach the counter to C425 and adjust L403 for a reading of 48 MHz. Set the voltage source to 7 V. The frequency should now be about 53 MHz. Check U401, the dual modulus prescaler, in the same manner as U301. Replace U400. This completes the initial VCO tune-up.

Tune the reference oscillator. The frequency should be measured at pin 3 of U300 and U400. This pin is a buffered output that will not load the reference oscillator during the measurement. Adjust C200 until the counter reads 16.3800 MHz.

Check the coarse-tune loop. With U300 and U400 back in their sockets, set the starting frequency to 8 MHz with a step size of 1 MHz. With the up-arrow key, increment the frequency to 20 MHz while using your scope to observe the voltage at the C310-R304 junction. The voltage should start at about 1 V and jump in 0.3 V steps to about 6 V. If this is not the case, retune L305 as described previously.

Set the starting frequency to 1 MHz and repeat the procedure while incrementing from 1 to 7 MHz. This time, the voltage should start at about 1 V and jump in 0.5 V steps to 5 V. If this is not the case, retune C333 as described previously. Remember to tune only L305 when the frequency of operation is above 8 MHz and to tune only C333 when the frequency is below 8 MHz.

The fine-tune loop can be checked in a similar fashion. Set the starting frequency to 5 MHz and the step size to 1000 Hz. The voltage at the C410-R408 junction should start at approximately 3.6 V and increase to 5 V. Then it should drop back to 3.6 V and again increase to 5 V. This pattern should repeat every 10 kHz.

Adjustment of the output loop is probably the most critical and important step in the procedure. This is because the output VCO must track the coarse-tune VCO fairly closely, and its frequency should be greater than the coarse VCO − 500 kHz. The output of the phase/frequency detector in U500 will compensate for about

0.5 MHz of tracking error. It will, however, produce an error voltage to compensate in the wrong direction if the output VCO frequency is less than the coarse-tune VCO − 500 kHz.

Check the divide-by-100 counter. The frequency at pin 2 of U501 should be 1/100 of the fine-tune synthesizer loop output (U501, pin 7).

Remove U500. Connect a 4-V source to pin 13 of the U500 socket. This can be done by inserting a resistive voltage divider between pins 16 and 7, with the center at pin 13. Enter a starting frequency of 8 MHz and a step size of 1 MHz into the computer. Adjust L504 so that the frequency at the C522-L507 junction is 53 MHz (500 kHz above the coarse loop).

Enter a 1-MHz starting frequency and a 1-MHz step size. Adjust C534 so the frequency at C522-L507 is 46 MHz. Now, step to 20 MHz and record the frequencies. The frequencies should never exceed 500 kHz less than or 1 MHz greater than the coarse-tune VCO frequency. Variations in varactor characteristics and differences between the parallel capacitances across the varactors can cause differences in tracking. If the output VCO changes frequency faster the coarse-tune VCO (that is, if the output VCO frequency is higher than the coarse-tune VCO frequency by more than 1 MHz), decrease the value of C316 by 1 or 2 pF and repeat the alignment and test procedure. Increase the value of C316 if the output VCO frequency is lower than the coarse-tune VCO frequency by more than 500 kHz. Note that the coarse loop will not cover 20 MHz if the value of C316 is increased too much.

Replace U500. Set the starting frequency to 1 MHz and step from 1 to 20 MHz in 1 MHz steps while observing the voltage at the C501-R502 junction. The voltage should remain between 3 and 7 V. If it does not, the VCOs are not tracking or the correction VCO does not have enough gain. If the tracking cannot keep the oscillators within acceptable limits, as a last resort, the compensation loop gain can be increased by increasing C504 by a small amount. Increasing C504 too much will cause degradation of the tracking characteristics.

Monitor the synthesizer output frequency at the R605-R606 junction. Step through various frequencies. The output should follow the computer display, plus 45 MHz.

Tune the 45-MHz crystal filter, Y600. This can be done by sweeping a signal generator across the passband or by tuning the receiver across a strong constant signal; check the converter output for a flat passband. If more conversion gain is desired, decrease the value of R619. For less conversion gain, increase the value of R619. In the prototype, 100 Ω for R619 set the conversion gain to about 12 dB.

The lock detector output from the converter is in parallel with the computer's open-apple key. If the converter does not lock (the display will indicate LOCKED or UNLOCKED) when applying power, depress the open-apple key to force load the synthesizers.

Note that the sidebands will be reversed because of the relationship of the RF and LO signals during the conversions. Tune the receiver to 28.62 MHz and have fun!

OPERATION

You'll need to make up an interface ribbon cable to go between the receiving converter's I/O DIP socket and the Apple II computer's internal game port. Note that only six lines are used. You'll also need to connect the converter to an antenna and make a cable to go between the converter output and the receiver antenna input. The last step is to connect a 12-V dc, 250-mA supply to the converter.

Turn on the supply voltage to the receiving converter and tune the station receiver to 28.62 MHz. Make up a disk for your Apple II that has the operating system and the receiving converter control program files. These files are HELLO, DUAL1, SYN1, FCREATE, FMEN$ and B.TIME BY INTERRUPT. Turn on the computer with the program disk inserted. The Apple's disk operating system will run the start-up program called HELLO. This program does the following: loads a date/clock program, sets the maximum high memory address (HIMEM) to make room for the binary loader program (SYN1), loads the binary loader program, turns on the 80-column display and runs the BASIC receiver program (DUAL1). If you do not have a date/time card in your computer, the top two lines will be blank.

Remember that the core of the program DUAL1 is the synthesizer loader routine located at lines 520 to 840. It does most of the work to calculate and load the serial data stream to the synthesizers. The rest of the program just performs the frills for incrementing the frequencies, selecting options, and saving and recalling files from disk. What you would like your receiver display to look like and do is probably different than mine, so don't be afraid to start with the core program and change it to suit your needs. (Don't forget to make a copy of the master disk first!)

After booting from the disk, the screen will go to the 80-column display mode. You'll be prompted "Create a File Y/N." At this point, you can either create a file or go right into the converter control screen. You can create up to 20 files; each file can hold up to 40 station frequencies. Each station frequency can have an eight-letter comment. For instance, you can create a file called "HAMSSB" which contains your favorite frequencies in the 160 through 20-meter voice segments. Or you could create files of all the international broadcast segments with each segment containing BBC, Radio Moscow, HCJB and so on. Note that files and frequencies can also be entered while the

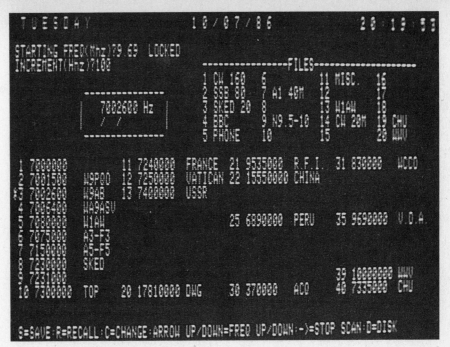

Fig. 42 — The computer receiver display.

program is in the listening mode.

If you want to create a file, enter Y for yes. You will now be asked how many stations you want to enter in the file, with a maximum of 40. For example, let's create a file with the 5, 10 and 15-MHz WWV frequencies. Enter 3 for the number of stations. You will now be prompted for the frequency (in MHz) of the first station. For this example, enter 5.0. Next, you will be prompted for comments. Enter WWV. Now you will be prompted for the frequency of the second station, then for comments. When you have entered the information on the last station, you will be prompted for a file number for this information to be stored under. For example, enter 1. Next, you will be prompted for a file name. In this case, WWV is appropriate. Now the file information will be stored on disk.

After you are done entering file information (or if you answer N(o) to the "Create a file" prompt), the receiver frequency display will be loaded to the screen. See Fig. 42. This screen contains important information. A box with the current operating frequency appears at the left center of the screen. On the upper right-hand side of the screen is a list of FILES. There are file numbers (1 to 20) and names. If you created the example file described previously, the file name WWV would be displayed for file 1. The words LOCKED or UNLOCKED are displayed in the upper center of the screen to indicate the condition of all three synthesizer loops. UNLOCKED will be displayed if one or more of the loops loses lock.

A prompt will ask you to enter the starting frequency in MHz. Then you will be prompted to enter the frequency increment step size in Hz. Increments as small as

10 Hz can be entered. When you have entered the frequency and step size, the selected frequency (displayed down to the 1 Hz) will be shown in the frequency display box. The converter will automatically be set to that frequency.

At the bottom of the screen is a menu of commands to help you tune the converter. Three commands make use of the cursor-control arrow keys. Pressing up-arrow causes the converter to scan higher in frequency; down-arrow makes it scan lower in frequency; and right-arrow stops the scanning. Other commands are called by entering the first letter of each command. These are Save, Recall, Change and Disk.

To begin scanning from the starting frequency, press the up-arrow or down-arrow keys. The default scanning rate is approximately two increment steps per second. To increase the scan rate, press the greater than (>) key; to decrease the scan rate, press the less than (<) key. To stop scanning, press the right-arrow key. The scan rate could be sped up by compiling the BASIC program or writing it in assembly language.

Putting a frequency into memory and then into a disk file can easily be accomplished directly from the receiver display screen. Simply enter the desired frequency into the display box or scan to it, and then press S for save. The computer will ask for a memory location (1-40) and for a comment (call sign or whatever). After entering the memory number, frequency and comment, this information will be displayed about halfway down the screen. Up to 40 of these station frequencies can be stored in a disk file under one file name.

To access files on disk, press D. The program will ask you if you want to Save,

Recall or Delete a file. The file name must begin with a letter and be no longer than 8 characters. To retrieve a file, simply press D, R (for recall) and the file number (from the file menu). For example, you could recall file 1 (WWV) from the previous example by pressing D, R and 1. All of the frequencies in the file, along with numbers and names, will be displayed in four columns across the lower part of the screen. In this case, the frequency list would read

1 5000000 WWV
2 10000000 WWV
3 15000000 WWV

Any of these frequencies can be loaded by pressing R and the corresponding frequency number. For example, press R 2 to load the 10.0 MHz WWV frequency. An asterisk appears next to the selected memory number and the frequency is placed in the dial box. You can scan up or down from this frequency at the current step increment by pressing the appropriate arrow key.

If at any time you want to change the frequency or increment, press C (for change) and the cursor will move to the starting frequency location. Then enter the new starting frequency and step size information.

As written, the BASIC program (DUAL1) does not have error handling routines or elaborate schemes for coping with keyboard entry errors. Outside of the synthesizer control part of the program, there is nothing special here. The BASIC code is relatively straightforward and can be refined if you wish. If the program bombs, it is best to re-boot the disk and start over. Hopefully, you will have saved your files on disk before this happens.

FUTURE PROJECTS

Fig. 41 shows the Apple II I/O game port. Two digital outputs AN(2)-AN(3), three analog inputs PDL(0)-PDL(1)-PDL(2), and two digital inputs PB(1)-PB(2) are not used on the game port. These lines could be used for extra features not implemented in the existing configuration.

The output line could switch in an attenuator, turn the power on and off, provide a blanking pulse to quiet the receiver during the time the synthesizer changes frequency, or control a TR switch. It should be noted that precautions should be taken not to transmit into the converter if it is connected to a transceiver.

An AGC voltage could be used to feed the analog input line for carrier detection in the scan mode. The scan could be programmed to stop or pause if a carrier above a fixed level is detected. Multiple listening frequencies for a foreign broadcast could be programmed, and when the carrier drops below a level, the computer could search for a stronger carrier at another frequency. This same method might be used to plot the spectrum while scanning.

A joystick can be used to scan up or down at various rates as a function of the stick displacement. Also, the two unused inputs might be connected to a digital shaft encoder for tuning the receiver.

If you have a time of day clock in your computer, a short program could tune the bands at a particular time of day for recording. Another future project is to design an RS-232-D interface that could accept the inputs from any computer.

With the flexible computer controller, the tasks you can make this receiver do are only limited by your imagination.

References

1. Rohde, Ulrich L., *Digital PLL Frequency Synthesizers, Theory and Design*, Prentice Hall, Inc., Englewood Cliffs, NJ 07638.

2. Gardner, Floyd M., *Phaselock Techniques*, 2nd ed, John Wiley and Sons, Inc.

3. Best, Roland E., *Phase-Lock Loops, Theory, Design, and Applications*, McGraw-Hill Book Co.

4. *CMOS/NMOS Special Functions Data Book*, MOS Logic Marketing, Motorola Inc., PO Box 6000, Austin, TX 78762.

5. *Integrated Circuit Databook*, Plessey Solid State, 3 Whitney, Irvine, CA 92714.

6. Helfrick, Albert D., "A Modern Up-converting General Coverage Receiver," *QST*, Dec 1981, pp 15-22.

7. *Apple II Reference Manual*, Apple Computer, Inc. 20525 Mariani Ave. Cupertino, CA 95014.

A Modified Cubic-Incher

This transmitter is a modified version of the 40-m Cubic Incher by Dennis Monticelli, AE6C, which appeared in July 1982 *QST*. This version, shown in Figs. 43 through 46, can be built for the 80- or 40-meter band and features improved harmonic rejection. It was developed in the ARRL lab by Greg Bonaguide, WA1VUG.

Circuit Description

T1 keeps the design simple by taking the place of several separate inductors and transformers. C1 and C2 form a capacitive voltage divider to limit the amount of feedback in the oscillator loop. C3, the transmitter tuning control, is adjusted to obtain the cleanest CW note. D1 serves to reduce harmonics in the output waveform. C4 bypasses RF on the dc line, and C7 helps to eliminate any residual power supply ripple which might otherwise become superimposed on the output signal. Start-up resistor R1 delivers about 1 mA to the base of Q1, a 4-watt NPN transistor that can handle infinite SWR when operated at 12 volts or less. Half-wave filter components C5, L1 and C6 are used to clean up the RF before it reaches antenna jack J1. Design values for C1, C2, C3, C5 and C6, plus winding information for L1, can be found in Table 1. Crystal Y1 is a fundamental cut type chosen for the desired operating frequency.

Construction

As can be seen in Fig. 44, the transmitter uses point-to-point wiring on double-sided epoxy circuit board material. Layout is not critical, except that component lead lengths should be kept short (1 inch or less). Since

Table 1

Component Values for the Modified Cubic-Incher

	80 meters	40 meters
C1	0.001-μF silver mica capacitor	470-pF silver mica capacitor
C2	100-pF silver mica capacitor	56-pF silver mica capacitor
C3	80-600-pF mica compression trimmer	18-100-pF mica trimmer capacitor
C5, C6	820 pF disc ceramic capacitor	470-pF disc ceramic capacitor
L1	24 t. no. 26 wire on a T-37-2 core	17 t. no. 26 wire on a T-37-2 core

Fig. 43 — The Modified Cubic-Incher.

Fig. 44 — Bottom view showing construction details. T1 is centrally located. Q1 is electrically isolated from the side of the metallic box by a nylon screw and mica spacer.

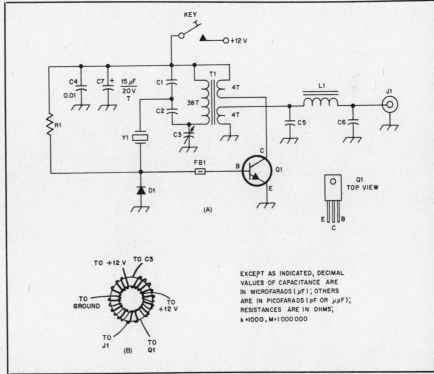

Fig. 45 — Schematic diagram of the Modified Cubic-Incher (A). Winding details for T1 appear at (B).

C1, C2, C3, C5, C6 — See Table 1.
D1 — High-speed silicon switching diode, 1N914 or equiv.
J1 — RCA phono jack.
Q1 — NPN medium-power RF transistor, MRF472 or equiv.
R1 — 10-kΩ, ¼-watt carbon.

L1 — See Table 1.
RFC1 — Ferrite bead, FB43-101 or equiv.
T1 — Toroidal transformer wound with no. 26 enamel wire on T50-2 core. Primary, 38 turns. Secondaries, 4 turns each.
Y1 — Fundamental-cut crystal in an FT-243 or HC-6 holder.

toroids are self-shielding devices, they can be located close together.

It is a good idea to first construct the transmitter on a piece of unetched circuit board to check operation before installing all the parts into a tight space. In this way, component changes and circuit voltage checks can easily be made.

Alignment

Alignment requires a dummy load (two 2-watt, 100-ohm resistors in parallel will work fine), a 0-1 A meter, and a monitoring receiver. Connect the ammeter in the + 12-volt line, being sure to observe polarity markings. Next, attach the dummy load, insert a crystal, turn the transmitter power supply on and tune the monitoring receiver around the crystal frequency. If nothing is heard, and 250- to 450-mA of current is being drawn, adjust C3 until the crystal begins to oscillate. Next, attach a straight key in series with the 12-volt line, send a string of dits and dahs, and adjust C3 for the best sounding CW note. It should be stable and chirp free. Finally, attach an antenna and readjust C3 slightly. The frequency of the transmitter will shift whenever anything connected to the oscillator varies. This includes not only C3, but environmental changes affecting the antenna, such as wind and rain, as well!

This circuit is an easy way to get a homebrew QRP signal on the air. Provided the operator takes care in adjusting the oscillator/transmitter for a pure-sounding CW note, he or she should get miles of enjoyment from this tiny 2-watter.

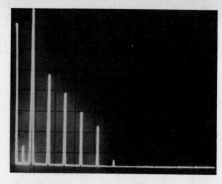

Fig. 46 — Spectral photograph of the Modified Cubic-Incher, tuned for 1.6-W output. Each horizontal division represents 5 MHz. Vertical divisions represent 10 dB. The second harmonic is 32 dB below the carrier. This transmitter meets FCC spectral purity requirements.

A VXO-Controlled CW Transmitter for 3.5 to 21 MHz

The 6-W CW transmitter shown in Figs. 47 to 50 can be built in a few evenings and will provide hours of on-the-air enjoyment. It features a variable-crystal oscillator (VXO) to generate a highly stable, adjustable-frequency signal. With the circuit shown here, frequency spans of 5 kHz or more can be realized. See Table 1. Only a few crystals are necessary for coverage of the popular CW frequencies. This single-band transmitter may be built for any one band from 80 through 15 meters. Since most crystals for frequencies above 25 MHz are overtone types, and this transmitter requires fundamental-type crystals, there is no provision for 10-meter operation.

Circuit Description

The schematic diagram of the transmitter is shown in Fig. 48. Q1 and associated components form a Colpitts variable-frequency crystal oscillator. C1 is used to adjust the frequency of the oscillator, and C2 is used to limit the span of the oscillator. If no limit is provided, the oscillator can operate "on its own" and no longer be under the control of the crystal. This is undesirable. On the 30-, 40- and 80-meter bands, C2 is not necessary and is omitted from the circuit. Supply voltage is fed to the oscillator only during transmit and spot periods. This prevents the oscillator from interfering with received stations operating on the same frequency.

Output energy from the oscillator is routed to Q2, a grounded-base amplifier. This stage provides some gain, but more important, it offers a high degree of isolation between the oscillator and the driver stage. Oscillator pulling and chirp are virtually nonexistent.

The driver stage uses a broadband amplifier that operates class A. This stage is keyed by grounding the base and emitter resistors. C10 is used to shape the keying waveform. Although the keying is rather hard, there is no evidence of clicks.

Two MRF476 transistors are used in parallel for the power amplifier. These transistors were designed for the Citizens Band service and work nicely at HF frequencies. Each transistor is rated for 3 W output. The original transmitter design used MRF472 output transistors, but Motorola no longer manufactures these devices. They are still available from many surplus outlets, however. L2 is used as a dc ground for the bases, making the transistors operate class C.

The low output impedance at the collectors of the output transistors is stepped up to 50 ohms by broadband transformer T3. A five-element Chebyshev low-pass filter is used to assure a clean output signal. This transmitter exceeds current FCC spectral purity specifications (see Fig. 49). D2 is used to clamp the collector voltage waveform to protect the output transistors if the transmitter is operated into an open circuit or high-SWR antenna system. The transmitter is designed to operate into a load that is close to 50 ohms resistive. S1 is used as the transmit/receive switch. One section transfers the antenna to an accompanying receiver or to the output of the transmitter. Another section is used to activate the VXO during transmit and the third section is provided for receiver muting purposes.

D3 and the associated components form an RF output driver for M1. This circuitry

Table 1
Component Values for the VXO-Controlled, 6-Watt Transmitter

Band	C1 (pF)	C2 (pF)	C3, C4 (pF)	C6 (pF)	C17, C18 (pF)	L1	L3, L5	L4	VXO Range
80 M	365	*	220	100	820	47 Turns T50-2	25 Turns T50-2	32 Turns T50-2	3- 5 kHz
40 M	365	*	100	100	470	36 Turns T50-2	17 Turns T50-2	21 Turns T50-2	5- 8 kHz
30 M	150	*	68	50	330	27 Turns T50-2	14 Turns T50-2	16 Turns T50-2	8-10 kHz
20 M	50	10	50	50	240	30 Turns T50-6	14 Turns T50-6	17 Turns T50-6	10-12 kHz
15 M	50	10	33	33	150	23 Turns T50-6	11 Turns T50-6	14 Turns T50-6	12-14 kHz

*Not used

Fig. 47 — The completed 6-W VXO controlled transmitter is housed in a small aluminum enclosure. RG-174 is used for all interconnections carrying RF. This version uses the older MRF472 output transistors, which are mounted flat on the board.

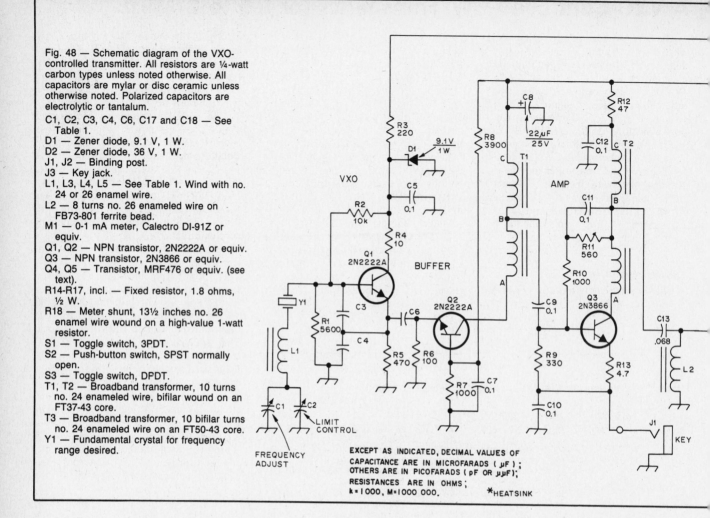

Fig. 48 — Schematic diagram of the VXO-controlled transmitter. All resistors are ¼-watt carbon types unless noted otherwise. All capacitors are mylar or disc ceramic unless otherwise noted. Polarized capacitors are electrolytic or tantalum.

C1, C2, C3, C4, C6, C17 and C18 — See Table 1.
D1 — Zener diode, 9.1 V, 1 W.
D2 — Zener diode, 36 V, 1 W.
J1, J2 — Binding post.
J3 — Key jack.
L1, L3, L4, L5 — See Table 1. Wind with no. 24 or 26 enamel wire.
L2 — 8 turns no. 26 enameled wire on FB73-801 ferrite bead.
M1 — 0-1 mA meter, Calectro DI-91Z or equiv.
Q1, Q2 — NPN transistor, 2N2222A or equiv.
Q3 — NPN transistor, 2N3866 or equiv.
Q4, Q5 — Transistor, MRF476 or equiv. (see text).
R14-R17, incl. — Fixed resistor, 1.8 ohms, ½ W.
R18 — Meter shunt, 13½ inches no. 26 enamel wire wound on a high-value 1-watt resistor.
S1 — Toggle switch, 3PDT.
S2 — Push-button switch, SPST normally open.
S3 — Toggle switch, DPDT.
T1, T2 — Broadband transformer, 10 turns no. 24 enameled wire, bifilar wound on an FT37-43 core.
T3 — Broadband transformer, 10 bifilar turns no. 24 enameled wire on an FT50-43 core.
Y1 — Fundamental crystal for frequency range desired.

is optional as there are no power-output tuning adjustments. M1 is also used to monitor transmitter current consumption.

Construction

The majority of the circuit components are mounted on a double-sided PC board. One side of the board is etched with the circuit pattern, and the other side is left unetched as a ground plane. A small amount of copper is removed from around each hole on the ground-plane side of the board to prevent leads from shorting to it. A test transmitter was built in the ARRL lab using single-sided board and the transmitter seemed to function normally with no instability. No long-term testing was performed, however. A parts-layout guide and photo of the finished board appear in Fig. 50.

Affixed to the front panel are the transmit/receive switch, spot switch and the tuning capacitor. The rear apron supports the antenna and mute jacks, key jack and binding posts.

A homemade cabinet measuring 3 × 6 × 8¼ inches was used in the construction of this transmitter. The builder may elect to build a cabinet from sheet aluminum or circuit-board material. The layout is not critical except that the lead from the circuit board to C1 should be kept as short as possible — an inch or two is fine.

A bent aluminum heat sink was attached to the output transistors. Commercial TO-220 heat sinks could also be used. If MRF472 transistors are available, they can be mounted flat on the circuit board and screws passed through the center of the transistors to hold them down. The ground plane will act as a heat sink sufficient for short key-down periods.

The only adjustment needed is that of setting the VXO limit capacitor (C2), and even this adjustment is not needed for the 80-, 40- and 30-meter transmitters. This adjustment can be done with the aid of a receiver. With a fundamental crystal in the circuit, adjust C2 for a maximum frequency spread that approximates the value shown in Table 1. If too much frequency spread is available, increase the amount of capacitance. Make a final check with the receiver by listening to the keyed signal from the transmitter. It should be steady and chirp free.

To provide wider frequency coverage, several crystals may be used. A crystal socket may be mounted on the front panel, or several sockets can be mounted on a separate circuit board and a simple rotary switch used to connect the desired crystal into the circuit. This option is shown in Fig. 48. Any number of crystals may be used, depending on the number of positions on the rotary switch. With crystals spaced 10-kHz apart, the circuit can provide continuous coverage of 50-60 kHz of the 20-meter band.

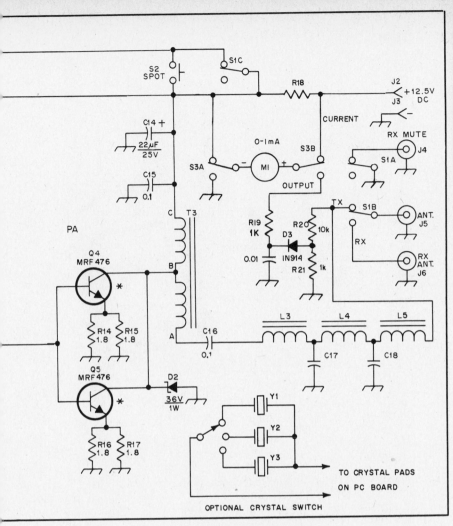

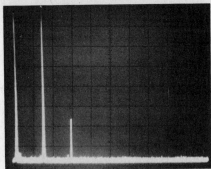

Fig. 49 — Spectral display of the VXO-controlled transmitter. Here the transmitter is operated in the 20-meter band. The second harmonic is down 56 dB relative to the fundamental output. Similar presentations were obtained on each of the other bands. This transmitter complies with the current FCC specifications regarding spectral purity.

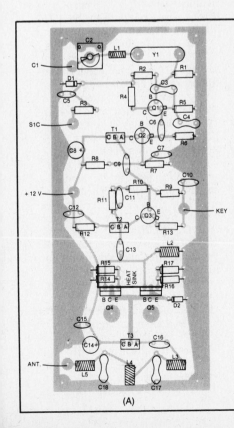

Fig. 50 — The component-placement diagram for the 6-W transmitter PC board is shown at A. The component side of the board is shown, with an X-ray view of the circuit foil. A full-size etching pattern appears at the back of this book. At B is a photo of the transmitter circuit board built using MRF476 transistors. The devices are mounted upright on the board with a heat sink attached to the metal tabs on the transistors.

A 1500-W-Output Amplifier for 160 Meters

Until a few years ago, most commercially available power amplifiers included no provision for top-band operation. With low sunspot numbers, 160 meters has become a great DX band. Sometimes a barefoot transceiver just isn't enough, and an amplifier is needed to make the QSO. The project described here and shown in Figs. 51 through 57 can enhance DX possibilities for those amateurs with an older five-band HF amplifier. It uses an EIMAC 3CX1500A7/8877 in grounded-grid configuration and will develop 1500 watts PEP output with 2500 volts on the plate. The amplifier is easy to drive, 50 to 60 W for full output, so any of the 100-W transceivers will do the job. Dick Stevens, W1QWJ, designed and built this project.

Circuit Details

The schematic diagram is shown in Fig. 52. The RF deck includes everything except the high-voltage power supply. Several suitable supplies are described elsewhere in this book. See Chapters 27 and 31, and the six-band 8877 HF amplifier described later in this chapter for ideas on building a high-voltage supply.

The input circuit (C7, L1, C8) is a low-Q pi network. No filament choke is needed with the 8877, which further simplifies the circuit.

The plate circuit is a pi network designed for a Q of 12. The tuning capacitor consists of a 300-pF fixed value (C17) in parallel with a 241-pF variable (C18). L2 is a section of standard Barker and Williamson coil stock. The loading capacitor consists of a 1600-pF fixed value (C20) in parallel with a 1000-pF variable (C19).

C13 bypasses the B+ line. Although an 8-kV unit was used here, any value above 4 kV should work fine. RFC2 is a Millen 1-mH, 1-A transmitter plate choke. RFC3 is included for safety to ground the high voltage that would appear at the antenna connector if dc blocking capacitor C16 fails.

Filament inrush current protection is provided by R1 and K1 in the T1 primary. After the initial current surge, K1A shorts out R1, applying full filament voltage. Power is applied to DS2, indicating that the filaments are on. T1 is a 5.0-V, 10.5-A transformer made by Avatar Magnetics. The primary is tapped to allow for precise adjustment of the filament voltage at the tube pins.

A grid overcurrent trip circuit is included to protect the 8877 from damage. The cost of an 8877 makes protection circuitry a must! Grid current is sensed across R5. If the grid current exceeds a preset value, Q1 conducts and actuates K2. Contact K2A, in the K3 coil supply line, opens and removes power from K3, placing the ampli-

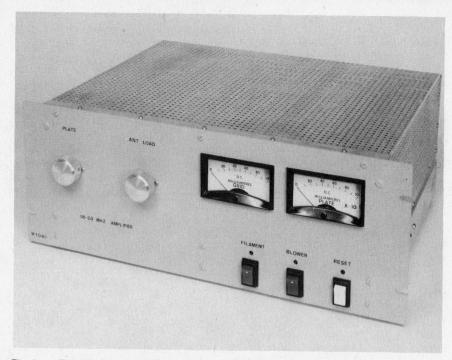

Fig. 51 — The 160-meter amplifier is built behind a 19- × 7-inch rack panel. Meters with bezels add to the clean appearance of the unit.

fier in STANDBY. To restore normal circuit operation, turn S3 to the off position; this will reset K2. S3 is "on" for normal operation. This switch provides an additional function: It controls the VOX circuit and antenna bypass relay circuit, so the amplifier can be switched in and out of the line between antenna and transceiver.

Zener diode D5 provides operating bias for the 8877. During receive periods, R6 is in the circuit, and the tube is cut off. For transmit, K3A shorts out R6.

All metering is done in the negative return lead for safety. C5, C6, D3 and D4 protect the meters. The antenna bypass relay (K4) is located externally but is powered from the amplifier.

Construction

The amplifier uses parts that are not too hard to find and are often available at flea markets at a very reasonable price. All components are, however, available commercially if the builder chooses that route. Most of the parts used in this project were purchased from Radio Shack, Electronic Emporium and RadioKit. See Chapter 35 for addresses.

Although an EIMAC SK-2200 series socket could be used, the socket used for this amplifier was homemade. It is built around a Johnson type 247 and an aluminum plate. Complete details of this plate, along with a full-size template, are shown in Fig. 53. This socket assembly can be made with hand tools — there is no fancy

or complex metal work required.

Figs. 54 and 55 show the top and bottom of the chassis of the completed amplifier. Construction of a single-band amplifier is significantly easier than that of a multiband unit. Layout is not critical, and there is plenty of room to spare.

The amplifier is built on a 2 × 17 × 13-inch (HWD) aluminum chassis (Bud AC-419). This chassis is bolted to a standard 19 × 7-inch rack panel (Hammond PBPA 19 007). The top and side panels are formed from a single piece of perforated sheet metal often found at hardware stores.

Fig. 54 shows the top of the chassis. The 8877 is located along the back edge, near the center. C18 is mounted along the left-hand edge of the chassis, and the fixed capacitors that make up C17 mount directly behind C18. Details of the mounting plate for C17 are given in Fig. 56. C19 is mounted to the right of C18, and the capacitors that make up C20 mount to the chassis behind C19. L2 is mounted on an acrylic strip that is supported above the chassis by two ceramic standoff insulators. J3 and J5 mount to an aluminum bracket that bolts to the rear of the chassis.

The blower and filament transformer may be seen along the right-hand edge of the chassis. M1 and M2 mount to the front panel near the filament transformer. K1 and K3 mount in sockets between the filament transformer and the front panel.

Fig. 55 shows the underside of the chassis.

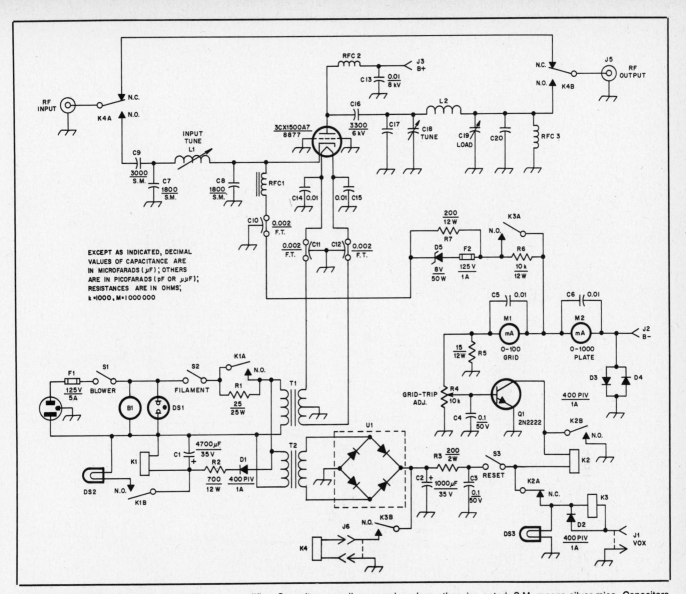

Fig. 52 — Schematic diagram of the 160-meter amplifier. Capacitors are disc ceramic unless otherwise noted. S.M. means silver-mica. Capacitors marked with polarity are electrolytic.

B1 — Dayton 4C012A 54-CFM blower.
C7, C8 — 1800-pF silver-mica capacitor; made from a 1500-pF unit in parallel with a 300-pF unit.
C9 — 3000-pF, 400-V silver-mica capacitor.
C10, C11, C12 — 0.002-μF, 8-kV feedthrough capacitor (Erie 1202-005 or equiv.).
C13 — 0.01-μF, 8-kV ceramic capacitor.
C16 — 3300-pF, 6-kV ceramic doorknob capacitor.
C17 — Three 100-pF, 5-kV ceramic doorknob capacitors in parallel.
C18 — 241-pF, 4.5-kV air variable capacitor (Cardwell 154-16 or equiv.).
C19 — 1000-pF, 1.5-kV air variable capacitor (Cardwell 154-30 or equiv.).

C20 — Two 800-pF, 5-kV ceramic doorknob capacitors in parallel.
DS1 — 117-V ac pilot light.
DS2, DS3 — 12-V dc pilot light.
J1 — Female chassis-mount phono jack.
J2, J3 — Millen 37001 high-voltage connector.
J4, J5 — Chassis-mount SO-239 connector.
J6 — Two-pin chassis-mount Cinch-Jones connector.
K1 - DPDT relay, 12-V dc coil, 10-A contacts, socket-mount.
K2 — Subminiature DPDT relay, 12-V dc coil.
K3 — DPDT relay, 12-V dc coil, 3-A contacts.
L1 — 4.4 μH; 25t no. 24 enam. on a ½-inch-OD, slug-tuned ceramic coil form.
L2 — 20 μH; 17t no. 12 tinned, 3-inch ID, 6 turns per inch (B&W 3033 miniductor stock

or equiv.).
M1 — 0-100 mA meter.
M2 — 0-1000 mA meter.
RFC1 — 100-μH RF choke.
RFC2 — 1-mH, 1000-mA transmitting type RF choke (Miller 4534 or equiv.).
RFC — 2.5 mH, 200-mA RF choke (Miller 4537 or equiv.).
T1 — Filament transformer. 117-V ac primary; 5-V, 10.5-A center-tapped secondary (Avatar AV-434 or equiv.). Available from Avatar Magnetics, 1147 N. Emerson St., Indianapolis, IN 46219.
T2 — Power transformer. 117-V ac primary; 12.6-V, 1.2-A secondary.
U1 — 50-PIV, 4-A bridge rectifier.

An enclosure is fashioned around the input circuit for shielding and to direct the air flow up through the socket. The blower draws air from behind the amplifier and blows it into the input compartment. A homemade chimney, fashioned from

Teflon sheet, directs air through the tube cooler. A suitable chimney, part number SK-2216, is available from EIMAC.

L1 mounts to the rear chassis wall so the input match can be adjusted with the covers on. Filament and bias voltage enter the in-

put compartment through feedthrough capacitors. D5 mounts to a chunk of scrap aluminum bar for heat sinking. The grid overcurrent components (Q1, K2) are arranged on a small piece of perforated board. R4 is mounted to the chassis near

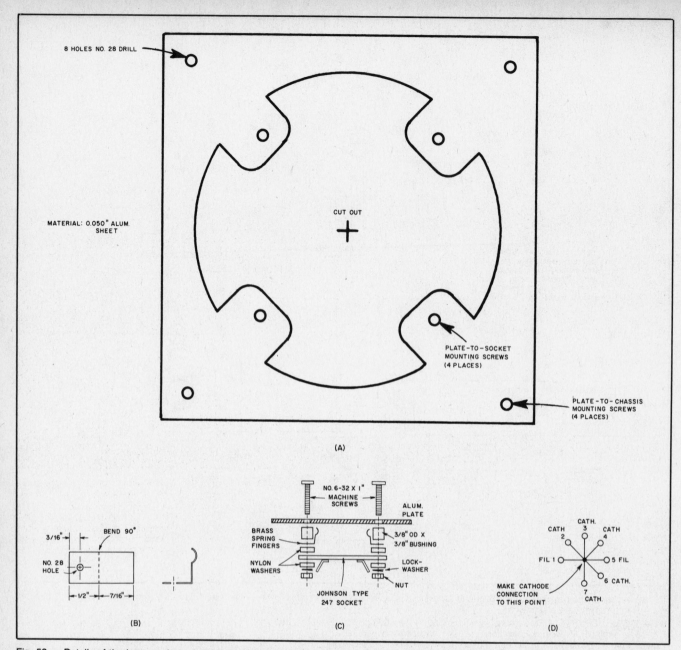

Fig. 53 — Details of the homemade socket for the 8877 tube. All hardware is no. 6-32. A full-size template for the aluminum plate is shown at A. Connection to the grid is made through four spring fingers made from spring brass or copper. Details of these clips are shown at B. Assembly of the socket is shown at C. Assemble as follows. Insert four 1-inch no. 6-32 machine screws through the plate. Slip a 3/8-inch by 3/8-inch-OD aluminum bushing over each screw. Place a spring finger on each screw, followed by a nylon washer. Slide the Johnson 247 socket onto the screws, and secure each screw with a nylon washer, metal lock washer and nut. Align the spring fingers for good contact with the grid ring and tighten. Connections to the bottom of the socket are shown at D.

the meters and is adjustable from the top. Liberal use of terminal strips simplifies wiring. There is plenty of room, and wiring is not critical at all, except that the input network should be mounted near the tube socket.

Testing and Tune Up

Connect the amplifier output to a dummy load through a wattmeter capable of reading 1500 W. Connect an HF transmitter capable of supplying at least 60 W to the amplifier input through an SWR/watt-meter. Connect a high-voltage supply, preferably with a variable autotransformer on the ac line to make the B+ continuously variable, to J2 and J3. The power supply should be able to deliver approximately 2500 V at 1 A. Be careful: The high voltage in this amplifier can be lethal!

Apply 117-V ac to the RF deck and operate the blower switch. The blower should come on, and the blower indicator lamp will light. You should feel air exhausting through the tube cooler.

Operate the filament switch. After a short delay, the filament indicator will light showing that the inrush protection circuit has cycled. Operate the reset switch, and the indicator lamp will light showing the amplifier is in the "go" condition. EIMAC recommends that the 8877 filaments be on for three minutes before high voltage is applied. No delay circuit for plate voltage control has been provided here, so wait three minutes before turning on the plate supply.

Apply plate voltage gradually. Do not apply drive yet. Watch and listen for any

Fig. 54 — Top-side view of the 160-meter amplifier chassis.

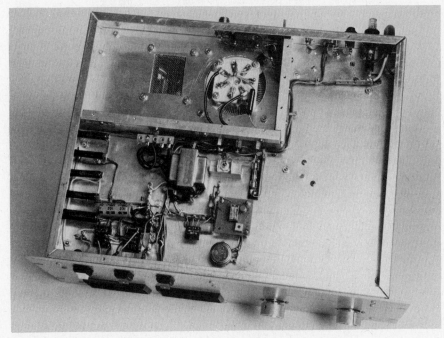

Fig. 55 — View of the underside of the 160-meter amplifier chassis.

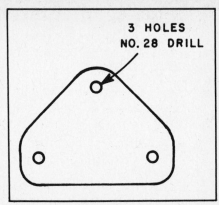

Fig. 56 — Full-size template of the plate used to connect the three doorknob capacitors that make up C17 to C18.

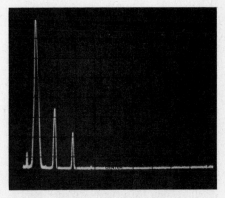

Fig. 57 — Spectral output of the 160-meter amplifier. Power output is 1500 W on 1.8 MHz. Each vertical division is 10 dB and each horizontal division is 2 MHz. All harmonics and spurious emissions are greater than 44 dB down. This amplifier complies with current FCC spectral-purity requirements.

signs of arcing as you bring the plate voltage up to full operating potential. There should be no indication on M1 or M2. If all appears to be okay, lower the plate voltage to about 1000 V and apply about 10-W drive. Adjust L1 for lowest input SWR.

Increase drive and adjust the pi network for maximum output. Most likely the grid overcurrent trip circuit will operate, placing the amplifier in STANDBY. Decrease the sensitivity by adjusting R4. This control is near high-voltage points, so take care and use an insulated screwdriver. Adjust this potentiometer so the grid protection circuit trips at 60 to 70 mA.

Apply full plate voltage and gradually increase drive. Adjust C18 and C19 for maximum output. These controls are somewhat interactive. Do not exceed 1 ampere of plate current. Recheck the input SWR when you reach full output power. Some additional adjustment of L1 may be necessary. Output power should be 1500 W with a plate voltage of approximately 2600 at 1 ampere and 60-W drive. Fig. 57 shows the amplifier spectral output.

A Legal-Limit Amplifier for 160 through 10 Meters

The amplifier described here is an update of the 1500-W output legal-limit amplifier described in the 1985 and 1986 *ARRL Handbooks*. It uses the EIMAC 8877 (3CX1500A7) high-mu triode in grounded-grid service. The 8877 will easily deliver 1500-W RF output with the drive levels afforded by most 100-W-class transceivers. Operating at 1500-W output, tube dissipation is well within the manufacturer's specification.

Shown in Figs. 58 through 68, this amplifier covers 160 through 10 meters, not including the WARC bands. Operation on the 24-MHz WARC band is possible with the band switch in the 28-MHz position, but the input SWR will be a little higher than on the other bands. Many of the components used were purchased at flea markets and from mail-order dealers. The names and addresses of the parts suppliers are detailed at the end of Chapter 35. Amplifier parts are expensive. The builder is encouraged to substitute components from surplus dealers or flea markets where possible if cost is a problem. Be sure, however, to use only components with adequate ratings. The dc and RF power levels found in this project are capable of destroying parts with insufficient ratings.

Dick Stevens, W1QWJ, designed and built this project. The power supply and RF deck are built on separate chassis because of the size and weight of the power supply. Standard 19-inch rack panels are used for the front panels, so both pieces may be rack mounted if desired.

Power Supply

The high-voltage power supply delivers 1 A continuously at user-selectable voltages of 2.5, 3 or 3.5 kV. During normal operation, only the 3-kV position is used, but a lower level is available for tuneup, and a higher level is available if line sag is a problem. Circuits are provided for protection, automatic start and PTT line control. Fig. 59 is the power-supply schematic diagram, and the unit is pictured in Figs. 60 and 61.

Circuit Details

The secondary side of the power supply is conventional. T1 is a Hipersil transformer with a 234-V primary and secondary taps at 1770, 2125 and 2475 V. S3 is a high-voltage switch with large contacts and ceramic insulation used to select among the taps. The 234-V primary and full-value secondary contribute to good voltage regulation. D2 through D5 are 14-kV, 1-A rectifiers connected in a full-wave bridge. C2 is a surplus 53-μF, 3.5-kV oil-filled capacitor. If the builder cannot locate one of these units, a suitable replacement should be at least 25 μF at a similar voltage rating. R4, the 200-kΩ bleeder resistor, is made up of eight 25-kΩ, 25-W units connected in series.

Fig. 58 — The 8877 HF amplifier uses a standard rack-type front panel. The tuning and loading capacitors are controlled by venier drives with turns counters. Once the settings for each band are determined, they can be recorded for quick band changes.

A high-voltage meter (M1) is included in the supply. The value of R_M, the multiplier resistor, will depend on the value of the basic meter movement. See Chapter 25 for information on meter multipliers and shunts. The B– lead floats above chassis ground. R3 is included to keep the B– near ground potential. The floating B– allows plate current to be metered in the negative lead for safety. C4 and C5 are bypass capacitors.

Power enters the supply from the 234-V line through a 20-A, 2-pole circuit breaker. DS1 will light when the circuit breaker is closed. With S1 in AUTOMATIC and S2 in START, a 24-V signal at J4 from the amplifier control circuitry will energize K1. K1A closes, energizing K2. DS2 lights. K2A and K2B apply reduced primary voltage to T1 through R2. After a second or so, C1 charges and energizes K3. K3A closes, shorting across R2, to apply full primary voltage to T1. This circuit protects D2-D5 from damage by effectively limiting inrush current through the rectifiers while C2 charges. R1 adjusts the charging time for C1. DS3 lights, signaling that the power supply is ready for service. An additional protective feature is that K1B is in the exciter PTT line to prevent keying the amplifier without high voltage present. Note that the supply can be activated manually (not by the amplifier control circuit) by placing S1 in the OFF position and using S2 to turn the supply on and off. PTT line protection is lost in the manual

mode, however.

J1 is a grounded, polarized 117-V ac receptacle used to provide power to the RF deck. Correct 117-V ac wiring must be used here. The ground (green) wire is connected to the chassis ground. The neutral (white) wire is connected to the 234-V line neutral. The hot side (black) is connected to one side of the 234-V line. Likewise, standard wiring must be used in the RF deck.

J1 and J2 are Millen connectors rated at 7 kV. Test prod wire (5-kV insulation) is used for the B+ and B– runs to the RF deck. A braid ground strap bonds the power supply and RF deck chassis together for added protection.

Construction

The power supply is built behind a standard 12½-inch-high, 19-inch-wide rack panel. It is designed to fit inside a Bud Prestige series cabinet. T1 needs a good base to support the weight, so the bottom plate is made from a piece of ¼-inch-thick aluminum plate that is approximately 14 inches square. The size and shape of the power supply enclosure will depend on the actual components used. Be sure that it is sturdy, and allow sufficient room for air to circulate.

T1 and C2 are bolted to the bottom plate. The rectifiers, control relays and other parts mount on a shelf above C2. The eight 25-W resistors that make up the bleeder are mounted to two vertical strips of phenolic stock.

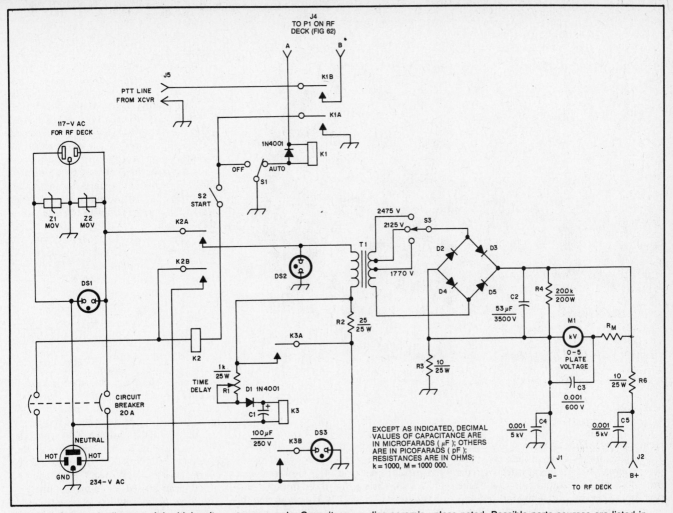

Fig. 59 — Schematic diagram of the high-voltage power supply. Capacitors are disc ceramic unless noted. Possible parts sources are listed in parentheses. See Chapter 35 for the addresses of the parts suppliers.

C2 — Oil-filled capacitor, 53 µF, 3.5 kV (Electronic Emporium cat. no. 4W308T or equiv.).

D2-D5 — Rectifier, 14 kV, 1 A (K2AW's Silicon Alley HV-14 or equiv.).

DS1-DS5 — Neon pilot lamp, 120 V, with built-in dropping resistor (Radio Shack).

J1, J2 — High-voltage connector (Millen 37001 or equiv.).

J3 — Chassis-mount three-wire 117-V ac socket.

J4 — Two-conductor Jones socket.

J5 — Chassis-mount phono jack.

K1 — 4PDT relay, 24-V dc coil, 2-A contacts. (Potter & Brumfield KH4703-1 or equiv.).

K2, K3 — DPST relay, 120-V ac coil, 20-A contacts. (Potter & Brumfield PRD7AYO or equiv.).

M1 — 0 to 50 µA meter with suitable multiplier resistor (R_M). See text.

R_M — Meter multiplier resistor. Exact value will depend on meter movement used. See text and Chapter 35 for information on meter multipliers.

R1 — Adjustable resistor, slider type, 1-kΩ, 25 W.

R4 — 200-kΩ, 200-W bleeder made from eight 25-kΩ, 25-W resistors in series (see text).

S1 — SPDT switch, 117-V, 3-A contacts.

S2 — SPST switch, 117-V, 3-A contacts.

S3 — High-voltage switch (see text).

T1 — Power transformer; 234-V primary, secondary taps at 1770, 2125 and 2575 V (Peter W. Dahl Co. no. PT-2475 or equiv.).

Z1, Z2 — 130-V metal-oxide varistor (Radio Shack).

Fig. 60 — Front panel view of the 8877 amplifier high-voltage supply. It is built on a 12½-inch high rack panel and mounted in a Bud Prestige cabinet. Output voltages of 2.5, 3 and 3.5 kV are switch-selectable from the front panel.

RF Deck

The RF deck houses the 8877 and associated RF circuitry as well as the amplifier-control circuit board. This amplifier uses a tuned input circuit and a pi-L output circuit. The filament transformer (T1) and antenna relay (K6) are built in. Cooling is provided by blower B1. Fig. 62 is a schematic diagram of the control circuitry, and Fig. 63 shows the RF circuitry.

Control Circuits and Metering

The control circuits are somewhat involved and are designed to protect the expensive 8877 tube from excessive fila-

Fig. 61 — Interior of the high-voltage power supply cabinet.

lays (K6, K7). K6 has a slight time delay, provided by C5. This feature assures that K7 will close before K6 to prevent hot switching.

Protection from excessive grid current is provided by Q1 and associated components. When the grid current exceeds a certain amount (set to approximately 70 mA by R5), Q1 fires and operates K4. K4C lights the TRIP LED. K4B opens, shutting off the HV supply and opening the PTT circuitry to protect the tube. To reset the amplifier, switch S2 off and then on again and everything will restart automatically. S2 can also be used to place the amplifier in standby if operation with the transceiver only is desired.

M1 and M2 of Fig. 63 monitor grid current and plate current. Plate current is monitored in the B− lead, rather than in the B+ lead, for safety. A pair of back-to-back diodes and 0.01-μF capacitors protect the meters from transients and RF voltages.

Input Circuitry

The RF input appears at J1 of Fig. 63. If the amplifier is turned off or bypassed by S2 of the control circuit, the input signal is routed directly to the output connector, J2. When the amplifier is in operation, the input signal goes to the tuned input circuitry and then to the 8877 cathode. Although the 8877 input impedance is close to 50 ohms and RF drive could be connected directly to the 8877 cathode, the tuned input is used for improved linearity and lower drive requirements.

There is a separate pi network (C1, L1, C2) for each band. S1 selects the appropriate network for the band in use. The input networks were designed with a Q of approximately 2 to be broadly resonant on each band. For clarity, only one pi network is shown in Fig. 63. Table 1 presents the complete input network data.

The coil (L1) for each band is wound on a 3/8-inch slug-tuned form. The slugs were purchased from a surplus dealer, but any 3/8-inch slug-tuned form with a red core should work. Table 1 gives both calculated capacitor values and the real-world values used in this amplifier.

Output Circuitry

B+ enters the RF deck through J3 and is routed through RFC2 and RFC3. The B+ line is bypassed by C8 and C9. C7 is a dc blocking capacitor that keeps dc voltages out of the tank circuit and off the feed

ment inrush current, excessive grid current, and application of drive in the absence of plate voltage. A three-minute delay circuit prevents keying the amplifier until the tube gets up to operating temperature, and another time-delay circuit keeps the blower running for a short while after power is removed.

When S1 of Fig. 62 (POWER) is closed, several things happen. First, 117-V ac is applied through R2 to the primary of filament transformer T1. R2 limits inrush current and is in the line until C2 charges, energizing K2. K2A shorts across R2, applying full primary voltage to T1. K2B lights the FILS LED. Note that T1 has several primary taps to allow adjustment of the filament voltage (measured at the socket) to 5.0 V, ±0.25 V, as specified by the tube manufacturer. At the same time filament power is applied, blower B1 starts and time-delay relay K1 starts to heat. K1A does nothing at this time. When power is turned off, however, K1A will stay closed for several minutes to run the blower and cool the tube after filament voltage is removed. K1 does not control the blower

directly, but rather does so through K8, a relay with contacts heavy enough to handle the current needed by the blower motor.

S1 also starts K3, a three-minute time-delay relay. When K3 energizes, the OPERATE LED lights. Finally, S1 applies primary voltage to T2 of the 24-V dc control-relay supply. When S2, RESET, is closed, 24 V is applied to various points in the control circuit. A 24-V signal is sent through relay contacts K4B and K3A to energize K1 in the high-voltage power supply. K3A also allows 24 V to reach K5, the PTT relay, so that the amplifier may be keyed from the PTT line. Note that since K3A is in the line to the HV-supply control relay and the PTT relay, the HV supply cannot be turned on and the PTT line cannot be keyed until after the three-minute warmup period.

The transceiver PTT control connects to the HV power supply, and K1B in the power supply activates K5 in the RF deck to prevent keying the amplifier in the absence of B+. K5A applies operating bias to the 8877, and K5B operates the antenna TR re-

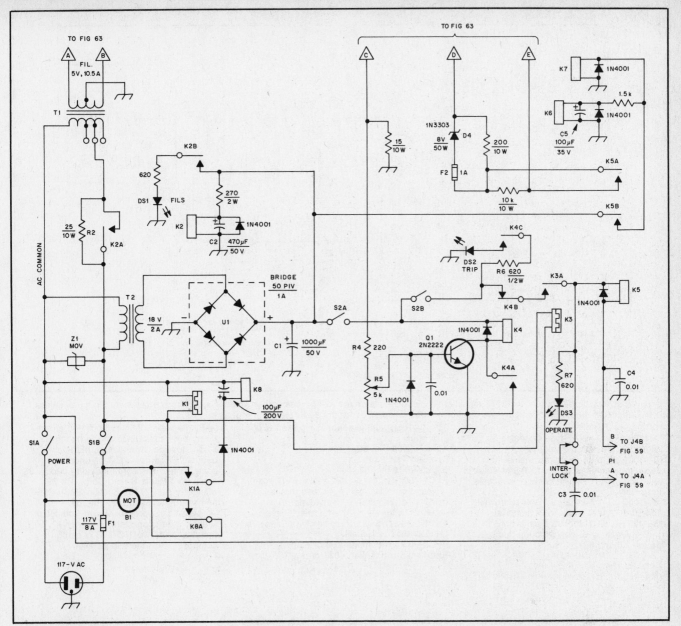

Fig. 62 — Schematic diagram of the 8877 amplifier control circuitry. Resistors are ¼-W types unless noted. Capacitors are disc-ceramic types unless noted. Capacitors marked with polarity are electrolytic. See Chapter 35 for the addresses of the parts suppliers.

B1—Blower, 54 CFM (Dayton 4C012A or equiv.).
DS1-DS3—LED.
D4—Stud-mount Zener diode, 8 V, 50 W (1N3303 or equiv.).
K1—Time-delay relay with a 2-3 minute delay on release, 115-V ac coil (Amperite 115NO69 or equiv.).
K2—DPDT relay, 12-V dc coil with 120-V, 10-A contacts (Radio Shack).

K3—Time-delay relay with a 3-minute delay on operate, 115-V ac coil (Amperite 115NO180 or equiv.).
K4, K5—4PDT relay, 24-V dc coil with 2-A contacts (Potter & Brumfield KH4703-1 or equiv.).
K6, K7—Dow Key Model 60 coaxial antenna relays with 24-V dc coils.
K8—DPDT relay, 120-V ac coil with 125-V ac, 10-A contacts (Radio Shack).

T1—Filament transformer. Primary: 120-V with taps for ±4% and ±8%; secondary: 5 V, 10.5 A (Avatar Magnetics AV-454 or equiv.).
P1—Two-conductor Jones plug to match J4 in the HV power supply (Fig. 59).
T2—Power transformer. Primary: 117-V; secondary: 18 V, 2 A.
U1—Bridge rectifier module, 50 PIV, 1 A (Radio Shack).
Z1—130-V metal-oxide varistor (Radio Shack).

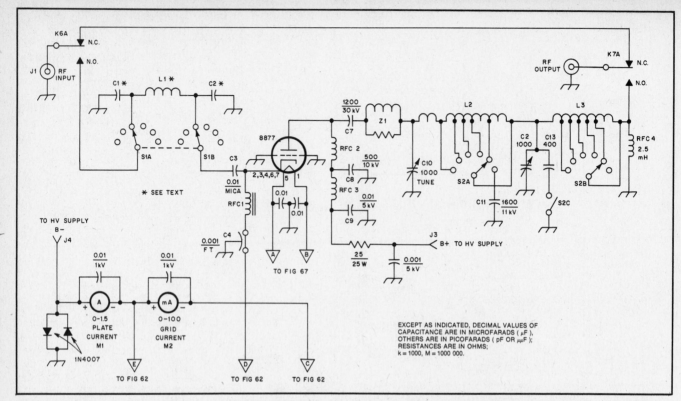

Fig. 63 — Schematic diagram of the 8877 amplifier RF deck. Resistors are ¼-W types unless noted. Capacitors are disc-ceramic types unless noted. Capacitors marked with polarity are electrolytic. See Chapter 35 for the addresses of the parts suppliers.

C1, C2—See Table 1.
C7—1200-pF, 30-kV RF-type doorknob capacitor (Electronic Emporium).
C10—1000-pF, 5-kV vacuum variable capacitor (Jennings UCSL 1000-5 or equiv.).
C11—Two 800-pF, 5-kV doorknob capacitors in parallel.
C12—1000-pF, 1.5-kV air variable (Cardwell 154-30 or equiv., avail. from RadioKit).
C13—400-pF, 5-kV doorknob capacitor.
J1, J2—Female UHF connector (in this case, part of K6 and K7).

J3, J4—High-voltage connector (Millen 37001 or equiv.).
K6, K7—See Fig. 62.
L1—See Table 1.
L2, L3—See Fig. 64.
RFC1—50 µH; 39T no. 24 enam. wire close wound on a ¼- inch powdered iron core.
RFC2—No. 24 enam. close wound on a 1-inch Teflon form. The form is 5 inches long; the winding is 4 inches long. Measured inductance is 195 µH.
RFC3—8.2 µH (Ohmite Z50 or equiv.).
RFC4—2.5 mH (Millen 4537 or equiv.).

S1—Single wafer, 2-pole, 6-position ceramic rotary switch with indexing to match S2.
S2—Six position, two section, high-power ceramic RF switch with non-shorting contacts and 60-degree indexing (Model 86, avail. from Radio Switch Corp., Rt. 79, Marlboro, NJ 07746). Section S2C is a homemade addition. See Fig. 68.
Z1—Parasitic suppressor. Made from 2T of ½-inch wide copper strap, ½-inch ID. Shunt the strap with two 100-Ω, 2-W carbon-composition resistors in parallel.

Table 1
Input Coil Data

Band	L1 (µH) (Calculated)	L1 Winding Information†	C14, C15 (pF) (Calculated)	C14, C15 (pF) (Values Used)
160	4.2	27T no. 24	1640	1120 + 560 = 1680
80	2.07	16T no. 22	820	820
40	1.18	9T no. 20	430	210 + 210 = 420
20	0.59	7T no. 20	220	210
15	0.39	5T no. 20	150	150
10	0.30	4T no. 20	100	100

†All coils wound on 3/8-in-diameter form; slug-tuned with red core. Available from Electronic Emporium, part number PLS5-B-2.

line and antenna. Should C22 short, placing dangerous voltage on the output connector, RFC4 will short the B+ line to ground and blow the power-supply circuit breaker.

This amplifier uses a pi-L tank circuit for better suppression of harmonic energy. To get a better Q and improved efficiency on 10 and 160 meters, a 1000-pF vacuum-variable capacitor was chosen for C10 (TUNE).

C12 (LOAD), a 1000-pF air variable, does not have enough capacitance for operation on 160 and 80 meters. A spare contact on S2A switches C11 into the circuit for 160-meter operation, while S2C switches C13 into the circuit for 80-meter operation.

L2 is made from a section of ¼-inch copper tubing for 10 meters and a section of B&W Pi Dux stock for the other bands. The L coil, L3, is also made from a small section of Pi Dux stock. S2 switches the taps for use on the various bands. Fig. 64 is a detailed diagram of the tank coil connections.

Construction

The completed RF deck is shown in Figs. 65 and 66. The main support for the components is a standard 4- × 17- × 10-inch (HWD) chassis (Bud AC-427). Like the power supply, the chassis is mounted

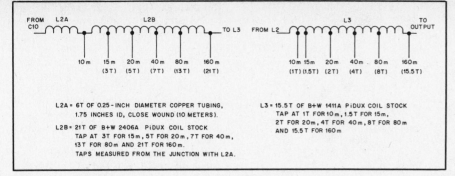

Fig. 64 — Details of the pi-L coils, L2 and L3 of Fig. 63.

Fig. 65 — Underside of the 8877 amplifier chassis.

(RCX-25) are available from Small Parts, Inc. See the parts suppliers list in Chapter 35. Align the chain drive with both switches in the 10-meter position. Leave some slack in the chain to allow the detents in both switches to function effectively.

The underside of the chassis is shown in Fig. 65. A single circuit board in the lower half of the photo contains the bias and control circuitry. The antenna relays are mounted to the rear chassis wall, to the right of the control circuitry. The filament and control circuit transformers are mounted at the top left of the photo, near the front panel, and the tube socket is to the right of the transformers. A Dayton blower pressurizes the enclosed cathode compartment, and air exits through the tube anode cooler. A Teflon chimney directs the exhaust air.

The tube socket is homemade from a Johnson 247 ceramic socket and miscellaneous hardware. Complete details for fabrication of this socket are shown earlier in this chapter, associated with a 160-meter 8877 amplifier. The recommended EIMAC socket is part number SK-2200.

The relative positions of the output-network components is shown in Fig. 66. The tuning capacitor is mounted close to the tube, and the loading capacitor is on the other side of the chassis. Both of these capacitors use Groth turns-counter dials with vernier drives for smooth tuning. These drives were purchased from Robert H. Bauman Sales, P.O. Box 122, Itasca, IL 60143.

The band switch is in the center of the chassis. S2C, to switch in additional loading capacitance for 80 meters, is a homemade addition to the band switch. It consists of two Barker and Williamson high-power switch contacts (available from B&W or RadioKit — see Chapter 35) and a wafer made from a piece of double-sided PC-board material. See Fig. 68. S2C is attached to the main band switch shaft.

The pi coil, L2, may be seen to the left of the 8877. The two coil pieces are supported by a rectangular piece of glass-epoxy circuit board material with all copper removed. The coil assembly is mounted above the chassis with ceramic standoffs. The L coil is mounted with a similar arrangement to the left of the band switch and perpendicular to the pi coil. The two coils must be mounted at right angles to prevent undesired coupling. All plate circuit connections are made with copper straps.

In this amplifier, it was necessary to provide a ½-inch wide ground strap between the tuning and loading capacitors to prevent ground-loop currents. Connect the frame of the loading capacitor to the lead-screw mechanism of the vacuum variable. Do not ground this strap to the chassis.

to a 19-inch-wide, 12½-inch-high rack panel. For RF shielding, sidewalls 4¾ inches high are mounted to the chassis with aluminum angle stock. A top cover bolts to aluminum angle stock that is connected to the sidewalls. Similar aluminum-sheet-and-angle-stock construction is used to enclose the tube socket underneath the chassis.

A separate enclosure that is bolted to the amplifier top cover houses the input circuitry. This enclosure is positioned so that its band switch is directly above the main band switch in the output network. A simple chain drive from the main band switch makes both input and output switches turn in unison from a single front-panel control. The sprockets (part no. RCS-12) and chain

Fig. 66 — Interior of the 8877 amplifier RF deck. The input compartment, which was attached to the top cover, has been removed. The chain-drive sprocket on the main band-switch shaft is visible between the chassis and the front panel.

Tuneup and Operation

Check the high-voltage power supply for proper operation before connecting it to the RF deck. Check all circuit paths in the B+ supply and RF deck for possible short circuits. Verify the operation of all control circuitry with the high-voltage off and with the tube removed from the socket. If everything works, apply power to the filament transformer, blower and control circuitry, and measure the filament voltage at the tube socket. If all is well, remove power from the RF deck and install the covers. Connect the high-voltage supply. Connect an exciter capable of delivering about 100 W to the input through a wattmeter and connect the amplifier output to a suitable dummy load through a second wattmeter.

Place the power-supply switches in AUTOMATIC and START. Close the RESET switch on the RF deck and turn the POWER switch on. The blower will start and filament power will be applied. The FILS lamp will light. After three minutes, K3 will close and the high-voltage supply will turn on. Press the transceiver PTT switch. M2 should indicate about 150 mA resting plate current.

Apply about 10 to 20 W of drive and adjust the TUNE and LOAD controls until the grid-current meter reads about 70 mA. Adjust R5 until the grid-trip circuit functions. Reset the circuit and verify that the grid-trip circuit functions at about 70 to 80 mA.

Apply about 10-W drive and adjust the slugs of the coils in the input pi networks for lowest SWR at the center of each band. Adjust the TUNE and LOAD controls for maximum output. Tuning of a pi-L circuit is similar to that for a conventional pi circuit, except that the setting of the LOAD capacitor is much more critical. Gradually increase drive until the power output reaches 1500 W. Recheck the input SWR on each band with the amplifier running at full output.

Table 2 shows the operating parameters for this amplifier as measured in the ARRL lab. Fig. 67 shows the exceptionally clean spectral output from the pi-L network.

Table 2
8877 Amplifier Operating Characteristics

Band	Power Output (W)	Plate Voltage (V)	Plate Current (mA)	Grid Current (mA)	Drive Power (W)	Input SWR
160	1500	2700	1000	30	65	1.1:1
80	1500	2800	900	35	65	1.2:1
40	1500	2800	900	40	65	1.1:1
20	1500	2800	1000	40	65	1.2:1
15	1500	2800	1000	45	65	1.2:1
12	1500	2800	1000	50	65	1.5:1
10	1500	2800	980	50	65	1.4:1

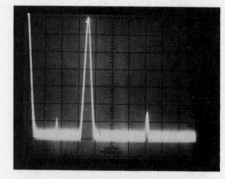

Fig. 67 — Worst-case spectral photograph of the 8877 amplifier. Power output is 1500 W on 3.5 MHz. Each vertical division is 10 dB, and each horizontal division is 1 MHz. All harmonics and spurious emissions are at least 54 dB down. This amplifier complies with current FCC spectral-purity requirements.

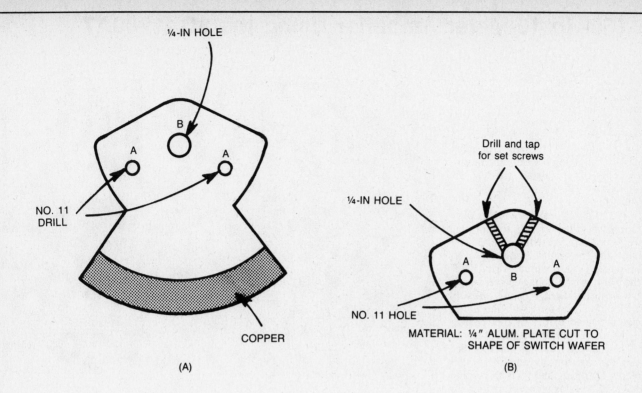

¼-IN HOLE

B

A   A

NO. 11
DRILL

COPPER

(A)

Drill and tap
for set screws

¼-IN HOLE

A   A
B

NO. 11 HOLE

MATERIAL: ¼" ALUM. PLATE CUT TO
SHAPE OF SWITCH WAFER

(B)

4-1/4" SQ. X 1/4" THICK PLEXIGLAS SHEET

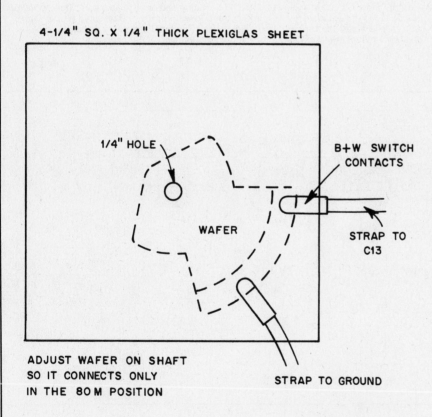

1/4" HOLE

WAFER

B+W SWITCH
CONTACTS

STRAP TO
C13

ADJUST WAFER ON SHAFT
SO IT CONNECTS ONLY
IN THE 80M POSITION

STRAP TO GROUND

(C)

Fig. 68 — Details of the homemade switch
used for S2C. At A is a full-size template of
the wafer. Material is double-sided circuit
board. Remove all copper but that indicated
by the shaded area; the pattern should be the
same on each side of the board. The
aluminum piece that mates the switch wafer
to the band-switch shaft is shown at B. At C,
the entire switch is assembled on a Plexiglas
sheet.

A 160- to 10-Meter Amplifier Using the 3CX1200A7

Billed as a replacement for the glass 3-1000Z, the EIMAC 3CX1200A7 metal/-ceramic triode is the basis of the 160- to 10-meter amplifier described here and shown in Figs 69 through 78. The amplifier will produce 1500 W output for about 100 W drive, making it a natural companion for today's solid-state transceivers. Dick Stevens, W1QWJ, designed and built this project.

The 3CX1200A7 requires a higher plate voltage than other ceramic transmitting tubes. In this case, the design is for 3500 V at full load. The amplifier will, however, still produce 1000 W output with 2500 V on the plate. A high-voltage power supply design is not included with this writeup. This amplifier was used with the power supply described earlier in this chapter along with a 160- to 10-meter 8877 amplifier. The high-voltage supply was modified slightly for use with this amplifier: C1 is increased to 200 μF, and R1 is changed to 1.5 kΩ to allow a longer charging time for C2. Another suitable high-voltage power supply design appears in Chapter 27.

A large volume of air is required to adequately cool the 3CX1200A7. Use of

Fig 69—The 3CX1200A7 160- to 10-meter amplifier is built behind a standard 19 × 10½-inch rack panel. Vernier drives are used to control the loading and tuning capacitors. The top meter can be switched to read plate voltage, grid current, forward power and reflected power. (The meter switch is above the LOAD and PLATE dial drives.) The bottom meter is dedicated to plate current. An external low-pass filter is mounted on the left side of the chassis.

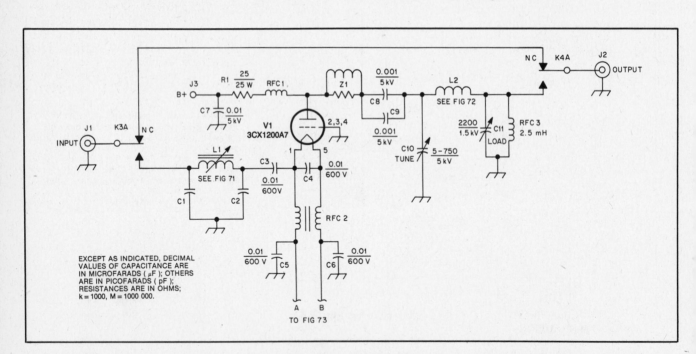

Fig 70—Schematic diagram of the 3CX1200A7 amplifier RF deck. Details of the input and plate-tank circuits are shown in Figs 71 and 72.

C1, C2—See Fig 71.
C3, C4—0.01-μF, 600-V mica capacitor.
C5, C6—0.01-μF, 600-V disc-ceramic capacitor.
C7—0.01-μF, 6-kV disc-ceramic capacitor.
C8, C9—0.001-μF, 10-kV RF-type doorknob capacitor.
C10—5 to 750-pF, 5-kV vacuum-variable capacitor.
C11—2200-pF, 1.5-kV air-variable capacitor.

J1, J2—UHF female chassis-mount connector.
J3—High-voltage connector (Millen 37001 or equiv).
K3, K4—See Fig 73.
L1—Input pi-network tunable coil. See Fig 71 for details.
L2—Plate-tank coil. See Fig 72 for details.
RFC1—90-μH, 1-A RF choke (Hammond 1521 or equiv).

RFC2—Bifilar wound 30-A filament choke, 28 bifilar turns no. 10 enam wire on a ½-in. diam × 7½-in. long ferrite core (Peter W. Dahl Co).
RFC3—2.5 mH RF choke (Millen 4537 or equiv).
Z1—Parasitic suppresser, four 220-Ω, 2-W metal-film resistors in parallel, mounted on a 5-inch long, ½-inch wide copper strap bent in a U shape.

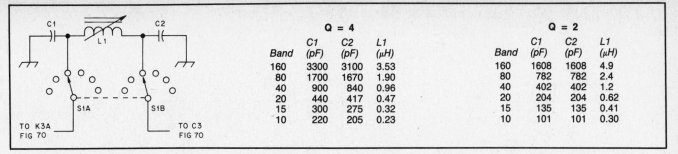

	Q = 4				Q = 2		
Band	C1 (pF)	C2 (pF)	L1 (µH)	Band	C1 (pF)	C2 (pF)	L1 (µH)
160	3300	3100	3.53	160	1608	1608	4.9
80	1700	1670	1.90	80	782	782	2.4
40	900	840	0.96	40	402	402	1.2
20	440	417	0.47	20	204	204	0.62
15	300	275	0.32	15	135	135	0.41
10	220	205	0.23	10	101	101	0.30

Fig 71—Details of the 3CX1200A7 amplifier input circuit. Component values are shown for Q = 4 and Q = 2. The amplifier was built using the values for Q = 4 (see text).

C1, C2—500-V silver-mica capacitors. See table for values.

L1—The coils for each band are wound on JW Miller 42A000CBI tunable ceramic coil forms (3/8-inch diam with red core). The number of turns required for each band:

10 meters—4 turns no. 22 wire spaced to occupy 3/8 in.

15 meters—6 turns no. 22 wire spaced to occupy ½ in.

20 meters—7 turns no. 22 wire spaced to occupy ½ in.

40 meters—9 turns no. 22 wire spaced to occupy ½ in.

80 meters—11 turns no. 22 wire close wound.

160 meters—20 turns no. 24 wire close wound.

S1—Double wafer, 6-position ceramic rotary switch with 60° indexing to match plate-tank bandswitch S2 (Centralab 2551 or equiv).

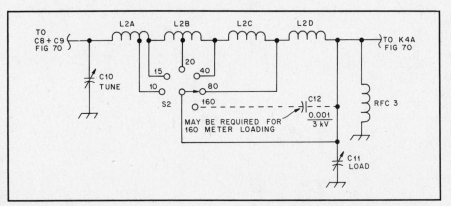

Fig 72—Details of the plate-tank circuit for the 3CX1200A7 amplifier. See Fig 70 for the rest of the circuit.

C10, C11—See Fig 70.

L2A—5 turns of ¼-inch diam copper tubing, 2¼ inches ID, 2½ inches long. See text. Tap at 4 turns for 10 meters. Tap at junction of L2A and L2B for 15 meters.

L2B—6¼ turns of B&W 2404T coil stock. Tap at 3 turns for 20 meters. Tap at junction of L2B and L2C for 40 meters.

L2C—6½ turns of B&W 2406T coil stock. Tap at junction of L2C and L2D for 80 meters.

L2D—10 turns of B&W 2408T coil stock. For 160 meters, use L2A, L2B, L2C and L2D.

RFC3—See Fig 70.

S2—Six position, one section, high-power ceramic RF rotary switch with nonshorting contacts and 60° indexing (Radio Switch Corp Model 86 or equiv).

the proper EIMAC air system socket (part no. SK410) and chimney (part no. SK436) is essential. The blower is a Dayton 4C440, rated at 60 CFM.

This project makes extensive use of parts purchased at flea markets and from surplus dealers. Although most components could be purchased new (at considerable expense), there are many good parts available surplus. The builder is encouraged to modify the design shown here when possible to make use of whatever surplus parts are available.

RF Circuits

Fig 70 is a schematic of the amplifier RF circuitry. For clarity, some details of the input and output circuits are omitted—these details are shown in Figs 71 and 72.

Details of the tuned input circuit are shown in Fig 71. Individual pi networks are used for each band. A tunable coil and two fixed capacitors are used for each band.

These circuits are designed for a Q of 4, based on EIMAC's recommendation. Tubes requiring high values of grid drive require a higher Q than the usual 2 to prevent flattopping. Tuning of the input circuits is sharp, but if they are adjusted for an SWR of 1:1 at the center of each band, the SWR at the band edges is less than 1.5:1.

Unfortunately, you won't be able to "fudge" the 10-meter input circuit to reach 12 meters—the Q is too high. Component values are given for input networks with a Q of 2 for those builders wishing to reach 12 meters.

The amplifier plate tank circuit is a standard pi network with a calculated Q of 10. Details are shown in Fig 72. Use of a vacuum variable (C10) for the tuning capacitor makes it easy to get a reasonable Q at 160 and 10 meters. The 3CX1200A7 has a high internal capacitance: tube capacitance and stray circuit capacitance

are almost enough to resonate the plate circuit at 10 meters. The minimum capacitance of an air variable suitable for tuning on 160 meters is too great to allow a reasonable Q on 10 meters. So, the vacuum variable, with its low minimum capacitance, is ideal for this application. An air-variable capacitor (C11) is used for loading. As shown in Fig 72, an additional 1000-pF fixed capacitor (C12) may be switched in if more loading range is needed for 160 meters.

The location of RFC1 differs from many other designs. Normally, the plate RF choke is connected to the junction of the parasitic suppressor (Z1) and plate blocking capacitor (C8 and C9). In this case, a low-frequency parasitic was noticed during testing, and moving RFC1 to the position shown corrected this problem.

Switches for the input (S1) and output (S2) circuits are connected by a chain drive so only one front-panel bandswitch control is needed. S1 and S2 *must* have the same indexing. All of the parts necessary for making the chain drive arrangement are available from Small Parts, Inc (see Chapter 35).

Metering and Control Circuits

Control and metering circuits are shown in Figs 73 through 75. Circuitry is included to prevent amplifier operation without plate voltage, and filament inrush protection is also provided. No ALC arrangement is used because modern transceivers will not overdrive the amplifier. Metering includes grid current, plate current, plate voltage and forward and reverse power.

Refer to Fig 73. Closing S1 starts the blower and lights DS1. Closing S2 applies 120 V to T2 through R3, the inrush-current-limiting resistor. Simultaneously, C1 charges through R2, and after a second or so K1 operates and contacts K1A short out R3, applying full primary voltage to T2. K1B contacts cause DS2 to light, showing that the filament inrush protection has functioned. Closing S2 also applies 120 V

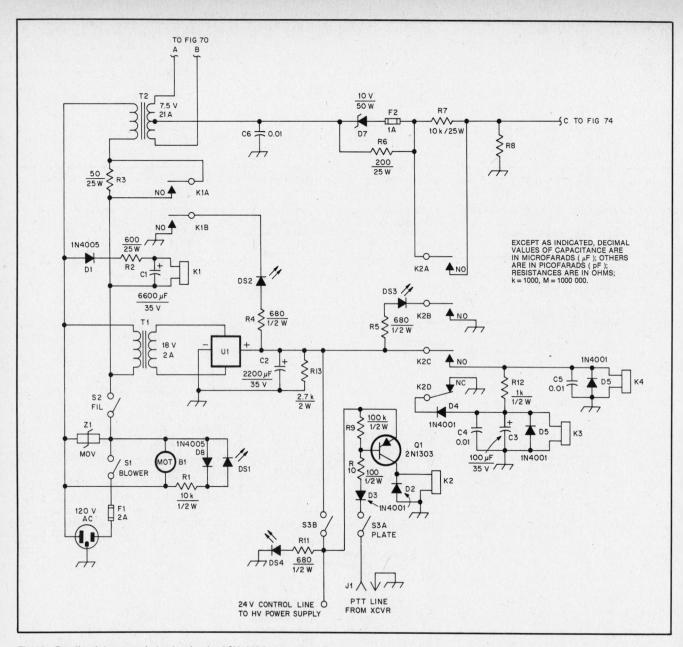

Fig 73—Details of the control circuitry for the 3CX1200A7 amplifier. Capacitors are disc-ceramic, 50-V types unless noted. Capacitors marked with polarity are electrolytic.

B1—60 CFM blower (Dayton 4C440 or equiv).
C1—6600-μF, 35-V electrolytic made from two 3300-μF, 35-V capacitors in parallel.
D1, D8—600-PIV, 1-A diode (1N4005).
D2-D6—50-PIV, 1-A diode (1N4001).
D7—10-V, 50-W Zener diode (1N2808 or equiv).
DS1-DS4—LED.
J1—Phono jack.
K1—DPDT relay with 10-A contacts and 12-V

dc coil.
K2—4PDT relay with 3-A contacts and 24-V dc coil.
K3—RF input relay (see Fig 70). SPST relay with 3-A contacts and 24-V dc coil.
K4—Antenna relay (see Fig 70). SPST relay with 10-A contacts and 24-V dc coil. Dow-Key antenna relay preferred; must be able to handle 1.5 kW of RF.
R8—Grid-current meter shunt. Value depends

on meter movement used. See text.
S1, S2—SPST switch.
S3—DPDT switch.
T1—Control circuit transformer. Primary, 120 V; secondary, 18 V at 2 A.
T2—Filament transformer. Primary, 120 V; secondary, 7.5 V at 21 A (Peter W. Dahl Co).
U1—100-PIV, 1.5-A bridge rectifier.
Z1—130-V metal-oxide varistor.

to T1. T1 and related components supply 24 V dc for other control circuits.

Closing S3 sends 24 V dc to the high-voltage power supply. The high-voltage supply is activated by a 24-V dc relay in the primary. DS3 lights when S3 is closed. You may choose a different method of activating the high-voltage supply, depending on your individual needs.

Q1 is the amplifier keying transistor. Many modern transceivers have amplifier keying circuits with insufficient current

and/or voltage ratings to close a relay coil. They can, however, key Q1 which in turn keys the relays in the amplifier. When Q1 conducts, K2 operates and completes several functions. K2A shorts out R7, which changes the tube bias from cutoff to operating bias. DS3 lights to show that the tube is ready to operate. K2C activates K3 and K4, the antenna relays. K4, the output relay, closes immediately. There is a slight delay in the closure of K3, caused by R12 and C3, to prevent hot switching K4.

When the amplifier is unkeyed to return to receive, K2D discharges C3 so that there is no additional delay as K3 returns to the receive position.

Operating bias is provided by D7. F2 protects the grid from excessive current. R8 is a grid milliammeter shunt. Its exact value depends on the meter used.

This amplifier includes metering for plate current, grid current, plate voltage, forward power and reverse power. See Fig 74. M2 is a dedicated plate-current meter. M1

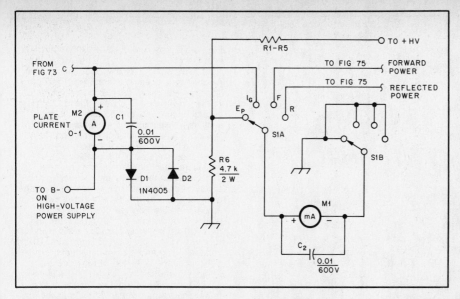

Fig 74—Schematic of the metering circuitry for the 3CX1200A7 amplifier. Note that S1B may be left out and the negative lead from M1 may be returned directly to ground.

M1—0 to 1-mA meter movement with scales to read 0-2 and 0-5. See text.
M2—0 to 1 A meter.
R1-R5—HV meter multiplier, 5 MΩ at 10 W.

Made from five 1-MΩ, 2-W (2% tolerance) resistors in series.
S1—2-pole, 4-position rotary switch.

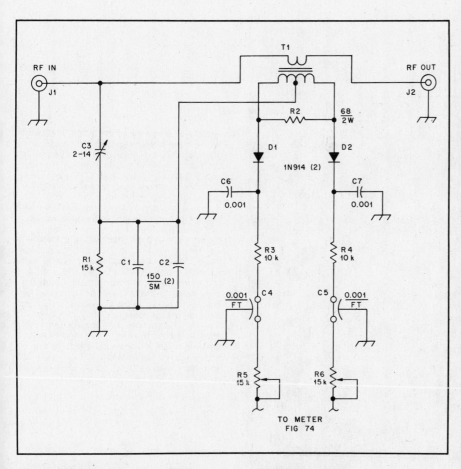

Fig 75—Schematic of the wattmeter circuitry for the 3CX1200A7 amplifier.

C1, C2—150-pF silver-mica capacitor.
C3—Miniature ceramic or air trimmer, approx 2-14 pF.
J1, J2—SO-239 chassis-mount connector.
T1—Secondary: 20 turns no. 22 enam on

Amidon T-50-2 core, center tapped.
Primary: Section of RG-8 through center of secondary core; connect the RG-8 center conductor between the center pins of J1 and J2. Ground the shield.

is a 0-1 mA movement with scales that read 0-2 and 0-5. S1 is used to switch in the appropriate shunt or meter multiplier resistor. Depending on switch position, M1 reads 0-500 mA grid current, 0-5 kV plate voltage, 0-2000 W forward power and 0-200 W reflected power. R6 prevents the voltage at the end of multiplier string R1-R5 from going to supply voltage when S1 is set to positions other than plate voltage. Note that S1 can be a single-wafer switch. S1B is not needed—the negative end of M1 may be grounded and S1B eliminated. Fig 75 shows the wattmeter. This circuit is similar to many others that have appeared over the years.

Construction

Figs 76 and 77 show the amplifier with the covers removed. Detailed dimensions are not given because many of the parts used are surplus. It would be difficult to duplicate this amplifier exactly, so you should consider the information presented here as guidelines, rather than as "cookbook" directions.

The input circuit is built on a 3 × 17 × 4-inch (HWD) chassis (Bud AC-432) that runs along the back of the amplifier chassis assembly. The 3CX1200A7 socket is mounted to this chassis, and underneath are the filament choke, input bandswitch, individual pi-network input circuits for each band, coupling and bypass capacitors, and the input relay and associated circuitry. Originally, the filament transformer was mounted on this chassis as well, but it was moved to cure an instability problem. (The metal case of the filament transformer did not provide a good shield). B1 blows air into the sealed input-circuit compartment, and air is exhausted into the output-circuit compartment.

The filament transformer, control circuits and bias components are built in a 3 × 9.25 × 7-inch (HWD) chassis that is bolted to the input chassis. (This is a standard Bud AC-408 chassis shortened to 9.25 inches.)

Output-circuit components, including the tuning and loading capacitors, tank coil and bandswitch, are mounted on a piece of 1/8-inch aluminum sheet. A heavy copper strap is used for grounding the tuning and loading capacitors. Use of this strap is necessary to prevent ground-loop currents.

Four separate pieces make up the plate-tank coil. See Figs 72 and 76 for details. A length of ¼-inch diameter soft-drawn copper tubing is the basis for the 10-meter coil. Clean the tubing well and wind 6 turns on a 1¾-inch diameter form. (This is one more turn than needed, but gives some extra material to work with.) Flatten one end of the tubing and drill it to clear a no. 6-32 screw. Remove the excess turn, flatten the other end and drill it to pass a no. 6-32 screw. Spread or compress the coil so that there are 2½ inches between the centers of the holes drilled in the ends of

Fig 76—Top view of the 3CX1200A7 amplifier with the cover removed. The blower, at the right rear, is bolted to the input chassis that runs along the rear of the amplifier assembly. The filament transformer is mounted in front of the blower, on the chassis that holds the control circuits. K1 and K2 are mounted in sockets to the right of the filament transformer. The wattmeter adjustment potentiometers (R5 and R6, Fig 75) are mounted on a bracket on the side of the plate-circuit enclosure near the filament transformer. The plate-current meter and multimeter are mounted on the front panel (lower right corner of photo), and the multimeter rotary switch is mounted to the front panel in front of the plate-circuit enclosure. The 3CX1200A7 is mounted on the input-circuit chassis at the right rear of the plate circuit enclosure. High-voltage wiring and bypass capacitors, and the RF choke, are to the left of the tube. The Minibox in the left rear corner of the plate-circuit compartment houses K4, the output relay. The vacuum-variable tuning capacitor is mounted at the right front of the plate compartment, and the loading capacitor is at the left front. The plate-coil assembly and plate-circuit bandswitch (under the plate coil) are mounted between the two capacitors.

Fig 77—Bottom view of the 3CX1200A7 amplifier with the covers removed. The amplifier chassis assembly consists of three parts: (1) the input chassis (left side of photo) runs the entire width of the amplifier; (2) the control circuit chassis (lower right); and (3) the plate compartment (upper right) built on an aluminum sheet. The tuned input circuits are visible in the input chassis, above the tube socket. The pi networks are built on an unetched piece of circuit board material, with the coil adjustment shafts protruding through the rear panel. The input bandswitch is mounted to the opposite chassis wall, with its shaft protruding into the plate-circuit compartment. Below the tube socket is the filament choke. A small circuit board bolted to the rear wall to the left of the tube socket holds the input relay and associated components. In the control-circuit chassis, most components are mounted on terminal strips. K1 and K2 are mounted in sockets in the lower center of the chassis. Q1 and associated parts are mounted on a small circuit board in the lower right corner.

the coil. The finished coil is 5 turns, spaced 2½ inches.

A 2-7/8 × 8¼-inch piece of ¼-inch-thick Plexiglas® serves as the main plate-tank coil support. Drill a mounting hole in each corner of the Plexiglas sheet, ¼ inch in from the ends. Drill two holes for the 10-meter coil 1-5/8 inch in from one end. Mount the 10-meter coil above the Plexiglas sheet on 3/8-inch spacers.

Next, cut three pieces of B&W coil stock. You'll need 6¼ turns of 2404T, 6½ turns of 2406T and 10 turns of 2408T. Slide the 2404T onto the Plexiglas support, followed by the 2406T and then the 2408T. Using a thin piece of brass shim stock, connect one end of the 2404T to one of the mounting screws for the 10-meter coil. Using more shim stock to make a good mechanical bond, splice the other end of the 2404T to the 2406T, and the 2406T to the 2408T. Solder all connections. The free end of the 2408T goes to the loading capacitor.

The output-circuit shield is made of sheet aluminum and angle stock. The top has a screened hole for hot air exhaust.

K4, the output antenna relay, is in the Minibox mounted to the inside rear wall of the output compartment, directly behind the loading capacitor. The wattmeter RF-sensing circuitry is inside the Minibox mounted on the outside of the back panel, near the tube.

Problem Areas

During original tests, the amplifier operated very poorly above 20 meters. It had severe LF and VHF parasitics, and none of the common cures worked. By accident, it was found that the case of the filament transformer was hot with RF energy, and in fact an arc could be drawn from the transformer mounting bolts. An NE2 neon lamp, mounted on an insulated rod and placed near the tube or filament transformer, glowed bright purple. (Normally,

Table 1
Typical Operating Conditions for the 3CX1200A7 Amplifier

Plate Voltage	3500 V
Plate Current	800 mA
Grid Current	180 mA
Power Output	1500 W
Drive Power	100 W
Input SWR	<1.5:1

the neon lamp glows orange in the presence of HF energy. If it glows purple, VHF energy is present.) Removing the filament transformer from the shielded plate compartment cured the VHF parasitic problem, but there was still a trace of a low-frequency parasitic. Moving the plate RF choke (RFC1) to the position shown in Fig 70 cured this problem.

Tune-Up

Check the high-voltage supply for proper operation before connecting it to the RF deck. Check all circuit paths in the power supply and RF deck for possible short circuits. Verify the operation of all control circuits with the high-voltage off and with the tube removed from the socket. When everything works, place the tube in the circuit and apply power to the blower, filament transformer and control circuitry. Measure the filament voltage at the tube socket; it should be 7.5 V. If all is well, remove power from the RF deck and install the covers. Connect the high-voltage supply. Connect an exciter capable of delivering about 100 W to the input through a wattmeter and connect the amplifier output to a suitable dummy load through a second wattmeter.

Close the the BLOWER and FILS switches. The blower will start and filament power will be applied. The BLOWER and FILS lamps will light. Close the PLATE switch,

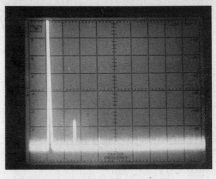

Fig 78—Spectral photograph of the 3CX1200A7 amplifier operating through the low-pass filter. Power output is 1500 W at 14 MHz. Each horizontal division is 10 MHz; each vertical division is 10 dB. All harmonics and spurious emissions are at least 56 dB below peak output.

and the high-voltage will come on. Plate voltage should register with M1 in the E_P position. Press the transceiver PTT switch. M2 should indicate about 150 mA resting plate current.

Apply about 10-W drive and adjust the slugs of the coils in the input pi networks for lowest SWR at the center of each band. Adjust the TUNE and LOAD controls for maximum output. Gradually increase drive until the power output reaches 1500 W. With 3500 V on the plate, grid current should be about 200 mA, and plate current should be about 800 mA. Recheck the input SWR on each band with the amplifier running at full output.

Table 1 shows the operating parameters for this amplifier. Although it has no problem meeting FCC specifications for spectral purity, a B&W low-pass filter is bolted to the side of the plate-tank compartment for added harmonic suppression. Fig 78 shows the spectral output of the amplifier with the low-pass filter.

Chapter 31

VHF Radio Equipment

The bands at 50, 144 and 220 MHz each contain more spectrum space than the total of *all* bands below 50 MHz! Little wonder that the VHF bands are popular favorites with many hams.

This chapter contains projects for the VHFer. In it you will find preamplifiers and converters for receiving. You will also find transmitting converters and transverters (used for transmitting and receiving). Most hams will be interested in the RF power amplifiers. Perhaps you will find what you are looking for.

For a discussion of RF amplifier theory see Chapter 15. Transmitter and receiver theory is covered in Chapters 11 and 12.

Dual-Gate MOSFET Preamplifiers for 28, 50, 144 and 220 MHz

The performance of many older receivers can be improved with a simple preamplifier in the front end. The simple design shown in Figs. 1 through 5 uses a dual-gate MOSFET to provide about 20-dB gain with associated noise figures of 1 to 1.5 dB on 28, 50, 144 and 220 MHz. They can be used for weak-signal or satellite work, and they can help put some life into an older 144- or 220-MHz FM transceiver. This is a simple, stable design. When built with a 3SK48, 3SK51 or 3N204, it will hold its own with all but the best GaAsFET designs. Kent Britain, WA5VJB, designed and built these preamplifiers.

Circuit Details

The basic circuit for the 28- and 50-MHz designs is shown in Fig. 1. Fig. 2 shows the 144- and 220-MHz versions. Q1 is a VHF dual-gate MOSFET. Many devices will work in this circuit. The author has successfully used the following devices (listed in order of preference): 3SK48, 3SK51, 3N204, 3SK40, 3N211. L1 and C1 tune the input. C6 and L2 tune the output of the 144- and 220-MHz versions. The 28- and 50-MHz preamplifiers use an untuned resistive output. R1 and R2 provide gate-2 bias.

Construction

The preamps shown here were built

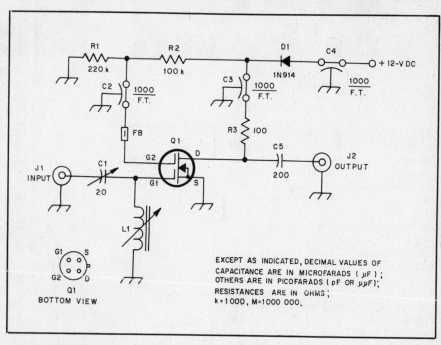

Fig. 1 — Schematic diagram of the 28- and 50-MHz dual-gate MOSFET preamplifiers. Resistors are ¼-W carbon types.

C1 — Ceramic or piston trimmer capacitor, 20- to 25-pF max.
C2, C3 — 500- to 1000-pF feedthrough capacitor, solder-in type preferred.
C4 — 500- to 1000-pF feedthrough capacitor.
C5 — 100- to 200-pF silver-mica or ceramic capacitor.

J1, J2 — Female chassis-mount RF connectors, BNC preferred.
L1 — 28 MHz: 23t no. 24 enam. on ¼-inch slug-tuned coil form. 50 MHz: Same, but 10t.
Q1 — Dual-gate MOSFET such as 3SK51 or 3N204. See text.

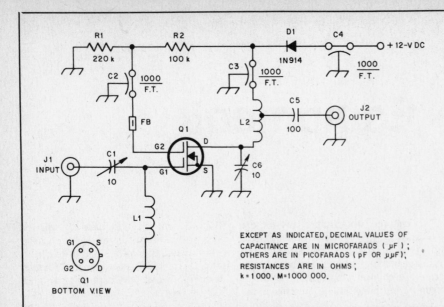

Fig. 2 — Schematic diagram of the 144- and 220-MHz dual-gate MOSFET preamplifiers. Resistors are ¼-W carbon types.

C1, C6 — Ceramic or piston trimmer capacitor, 10- to 15-pF max.
C2, C3 — 500- to 1000-pF feedthrough capacitor, solder-in type preferred.
C4 — 500- to 1000-pF feedthrough capacitor.
C5 — 50- to 200-pF silver-mica or ceramic capacitor.
J1, J2 — Female chassis-mount RF connectors, BNC preferred.
L1 — 144 MHz: 8t no. 18 wire, 3/16-inch ID, 1/2-inch long. 220 MHz: 5t no. 18 wire, 3/16-inch ID, 1/2-inch long.
L2 — 144 MHz: 8t no. 18 wire, 3/16-inch ID, 1/2-inch long. Tap at 1t. 220 MHz: 5t no. 18 wire, 3/16-inch ID, 1/2-inch long. Tap at 1/2t.
Q1 — Dual-gate MOSFET such as 3SK51 or 3N204. See text.

Fig. 3 — RF component side of the board of a 2-meter version of the dual-gate MOSFET preamplifier.

Fig. 4 — Another view of the 2-meter MOSFET preamp showing the dc components mounted on the other side of the board.

"dead bug" style on a piece of unetched double-sided circuit-board material. The foil acts as a ground plane, and all ground connections are made directly to it. The finished preamp fits inside a small Minibox. See Figs. 3 and 4. Fig. 5 shows the general layout. Dc components, including C4, D1, R1 and R2 mount on one side of the board. Voltage is fed to the transistor through C2 and C3. Parts placement is not at all critical. Solder the case of Q1 to the ground plane.

Parts for this project are available from many of the suppliers listed in Chapter 35. One source for 3SK51 transistors is Fuji-Svea, P.O. Box 3375, Torrance, CA 90510.

Adjustment is easy. Simply tune in a weak signal and adjust C1 and L1 (28- and 50-MHz models) or C1 and C6 (144- and 220-MHz models) for best signal-to-noise ratio.

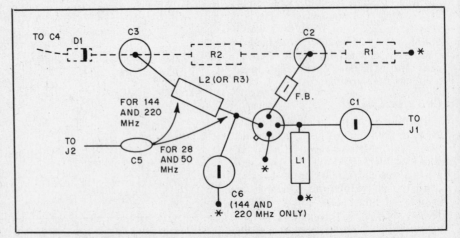

Fig. 5 — Suggested board layout for the MOSFET preamplifier. There are slight differences between the 28- and 50-MHz versions and the 144- and 220-MHz units. See text. Parts outlined with dashed lines are mounted on the back of the board. Component numbers refer to Figs. 1 and 2.

Dual-Gate GaAsFET Preamplifiers for 28, 50, 144 and 220 MHz

Dual-gate GaAsFET devices offer builders the opportunity to experiment with the latest technology — low-noise GaAsFETs — for a modest investment. The preamplifiers described here use GaAsFETs that cost less than $5 each, yet the preamplifiers offer noise figures well under 1 dB with gains ranging from 20 to 24 dB. Kent Britain, WA5VJB, designed and built the preamplifiers shown in Figs. 6 through 10.

Circuit Details

The basic circuit for the 28- and 50-MHz designs is shown in Fig. 6. Fig. 7 shows the 144- and 220-MHz versions. Q1 is a dual-gate GaAsFET. Many devices will work in this circuit, including the NEC NE41137, Motorola MRF966/967 and Mitsubishi MGF1100. Parts values are given for all of these devices so the builder can use whatever transistor is easiest to find. L1, C1 and C2 tune the input. C5 and L2 tune the output of the 144- and 220-MHz versions. The 28- and 50-MHz preamplifiers use an untuned resistive output. R1 and R2 provide gate-2 bias.

Variable capacitors can be miniature ceramic trimmer capacitors or piston trimmers. The exact value is not critical; any trimmer with a maximum value around that specified will do. C7, the output coupling capacitor, can be any small silver-mica around the specified value. For

stability, it is important to use good-quality capacitors for the gate 2 and source bypasses. Ceramic chip or leadless disc-ceramic types are necessary to provide a low-impedance ground.

Construction

The dual-gate GaAsFET preamplifiers are built "dead bug" style on a piece of double-sided PC board material. See Figs. 8 and 9. The circuit board acts as a low-impedance ground plane, and all ground connections are made directly to it. J1, J2 and C9 also secure the circuit board to the lid of the enclosure. The enclosure can be any aluminum diecast box (Bud CU-124; Hammond 1490B) or Mini-box (Bud CU-3002A) approximately 4 by 2 inches, or you can make one by soldering pieces of circuit board material together.

Q1 is supported above the ground plane by its gate 2 and source leads, which are soldered directly to C3 and C4. The other ends of these capacitors are soldered to the ground plane. R4 should be soldered in place with essentially no lead length. Placement for the rest of the parts is not critical, but keep the leads short. A layout that has proven successful is shown in Fig. 10.

GaAsFETs are static-sensitive, so handle them accordingly. See Chapter 24 for information on working with static-sensitive devices. It's best to assemble the rest of the circuit and install Q1 last. Use a grounded iron if possible, and solder the leads quickly.

Parts for this project are available from Microwave Components of Michigan and Applied Invention, both listed in Chapter 35. NEC transistors are available from California Eastern Laboratories, 3260 Jay St., Santa Clara, CA 95050.

Table 1

L1 and L2 Values for 144- and 220-MHz Dual-Gate GaAsFET Preamplifiers

Device	L1 144 MHz	L1 220 MHz	L2 144 MHz	L2 220 MHz
NE41137	7t	5t	7t, tap at 2t	5t, tap at 1½t
3SK97 and MGF1100	5t	3½t	7t, tap at 2t	5t, tap at 1½t
MRF966/967	12t	8t	8t, tap at 2t	5t, tap at 1½t

Note: All coils are no. 20 to 24 wire, 3/16-inch ID, spaced one wire dia.

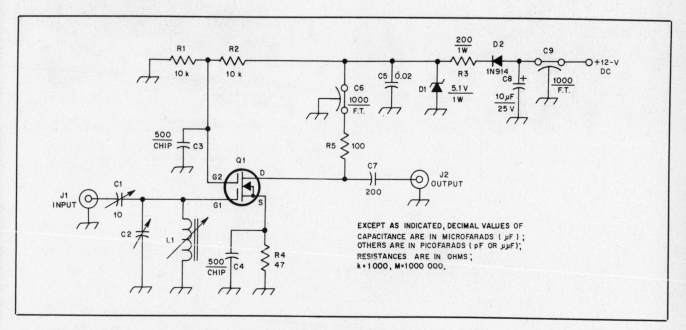

Fig. 6 — Schematic diagram of the 28- and 50-MHz dual-gate GaAsFET preamplifiers. Resistors are ¼-W carbon types unless noted.

C1 — Ceramic or piston trimmer capacitor, 10-pF max.
C2 — Ceramic or piston trimmer capacitor, 20- to 25-pF max.
C3, C4 — 200- to 1000-pF chip capacitor.
C6, C10 — 500- to 1000-pF feedthrough capacitor, solder-in type preferred.

C7 — 200- to 500-pF silver-mica or disc-ceramic capacitor.
C8 — 5- to 25-μF electrolytic capacitor.
C9 — 500- to 1000-pF feedthrough capacitor.
D1 — 5.1-V, 1-W Zener diode (1N4733 or equiv.).
D2 — 1N914, 1N4148 or any diode with ratings

of 25 PIV and 50 mA or greater.
J1, J2 — Female-chassis mount RF connectors, BNC preferred.
L1 — 28 MHz: 13t no. 24 enam. on ¼-inch slug-tuned coil form. 50 MHz: Same, but 9t.
Q1 — Dual-gate GaAsFET, NEC NE41137, Mitsubishi MGF1100 or 3SK97.

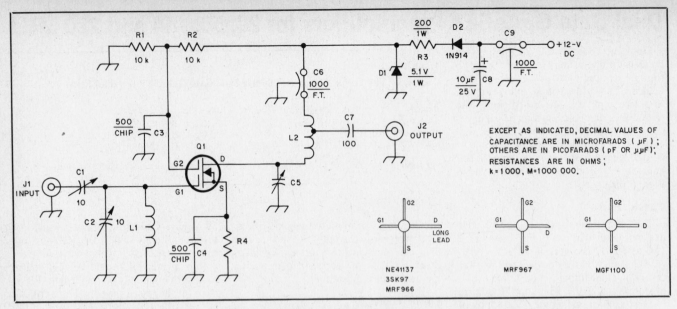

Fig. 7 — Schematic diagram of the 144- and 220-MHz dual-gate GaAsFET preamplifiers. The same circuit is used for all versions, but some parts values are different. See Table 1. Resistors are ¼-W carbon-composition types unless otherwise noted.

C1, C2 — 10-pF (max.) ceramic or piston trimmer capacitor.
C3, C4 — 200- to 1000-pF ceramic chip capacitor or leadless disc-ceramic capacitor.
C5 — For 144 MHz: 20 pF trimmer capacitor. For 220 MHz: 10-pF trimmer capacitor.
C6, C9 — 400- to 1000-pF feedthrough capacitor.
C7 — 50- to 100-pF silver-mica capacitor.

C8 — 1- to 25-µF, 25-V electrolytic capacitor.
D1 — 5.1-V, 1-W Zener diode (1N4733 or equiv.)
D2 — 1N914, 1N4148 or any diode with ratings of 25 PIV and 50 mA or greater.
J1, J2 — Female chassis-mount BNC or Type-N connector.
L1 — See Table 1.
L2 — See Table 1.

Q1 — NEC NE41137, Motorola MRF966/967, Mitsubishi MGF 1100 or 3SK97 dual-gate GaAsFET. See text.
R1 — For NE41137, MGF1100, 3SK97: 10 kΩ; for MRF966/967: 4.7 kΩ.
R2 — 10 kΩ for all devices.
R3 — 150- to 250-ohm, 1- or 2-W resistor.
R4 — For NE41137, MGF1100, 3SK97: 47 Ω; for MRF966/967: 100 Ω.

Fig. 8 — This version of the 50-MHz dual-gate GaAsFET preamplifier is built in a small diecast box. Construction of the 28-MHz preamp is identical. A suggested parts placement is shown in Fig. 10.

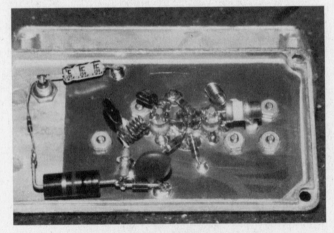

Fig. 9 — This 144-MHz dual-gate GaAsFET preamp uses an NE41137 and is built on the lid of an aluminum diecast box. The 220-MHz version is similar. See Fig. 10 for a suggested parts placement.

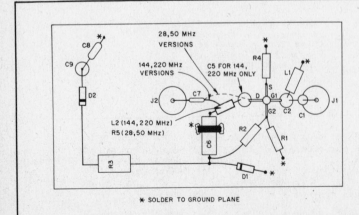

∗ SOLDER TO GROUND PLANE

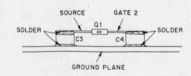

Fig. 10 — Suggested board layout for the dual-gate GaAsFET preamplifier. There are slight differences between the 28- and 50-MHz versions and the 144- and 220-MHz units. See text. Component numbers refer to Figs. 6 and 7.

GaAsFET Preamplifiers for 144 and 220 MHz

GaAsFET devices enable amateurs to build preamplifiers with very low noise figures for terrestrial, satellite and EME work. In addition, GaAsFETs offer superior strong-signal IMD performance — a characteristic that is important on our crowded bands. Described here and shown in Figs. 11 through 13 are preamplifiers for 144 and 220 MHz that use readily available Mitsubishi MGF1402 or MGF1202 devices. Other single-gate GaAsFETs such as the Dexcel 1500 and 2500 series, or the NEC 700 series, also work well. Gain is typically 20 to 24 dB with noise figures of about 0.5 dB. Kent Britain, WA5VJB, designed and built these preamplifiers.

Circuit Details

Fig. 11 is a schematic diagram of the preamplifier. The circuit is the same for both bands, although some component values are different. L1, C1 and C2 tune the input. C5 and L2 tune the output. Variable capacitors can be miniature ceramic trimmer capacitors or piston trimmers; piston types are recommended. The exact value is not critical; any trimmer with a maximum value around that specified will do. C6, the output coupling capacitor, can be any small silver-mica around the specified value. For stability, it is important to use good-quality capacitors for the source bypasses. Note that there are two

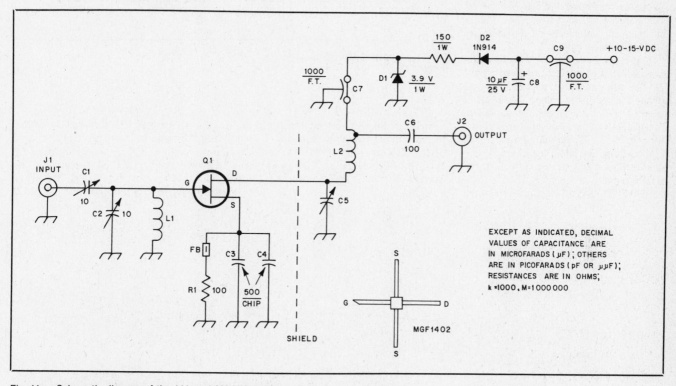

Fig. 11 — Schematic diagram of the 144- and 220-MHz MGF1402 GaAsFET preamplifiers.

C1, C2 — 10-pF (max.) ceramic or piston trimmer capacitor (piston trimmer preferred).
C3, C4 — 200- to 1000-pF ceramic chip capacitor or leadless disc-ceramic capacitor.
C5 — For 144 MHz: 20 pF trimmer capacitor. For 220 MHz: 10-pF trimmer capacitor (piston trimmer preferred).
C6 — 100- to 200-pF silver-mica capacitor.
C7, C9 — 400- to 1000-pF feedthrough capacitor.

C8 — 1- to 25-μF, 25-V electrolytic capacitor.
D1 — 3.9-V, 1-W Zener diode (1N4730 or equiv.).
D2 — 1N914, 1N4148 or any diode with ratings of 25 PIV and 50 mA or greater.
J1, J2 — Female chassis-mount BNC or Type-N connector.
L1 — For 144 MHz: 9t no. 20 wire, 3/16-inch ID, spaced 1 wire dia. For 220 MHz: same as

144 MHz, but 6t.
L2 — For 144 MHz: 6t no. 20 wire, 3/16-inch ID, spaced 1 wire dia. Tap at 1½t. For 220 MHz: 4t no. 20 wire, 3/16-inch ID, spaced 1 wire dia. Tap at 1t.
Q1 — Mitsubishi MGF1402 GaAsFET. Others usable (see text).
R1 — 100-Ω, 1/4- or 1/8-W resistor.
R2 — 150- to 250-ohm, 1- or 2-W resistor.

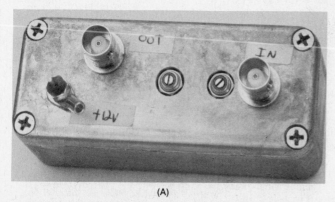

(A)

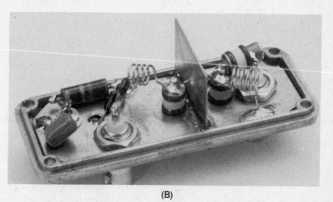

(B)

Fig. 12 — This 220-MHz GaAsFET preamplifier is built in a Hammond 1590A diecast box. All ground connections are made to the circuit-board ground plane. A suggested parts placement is shown in Fig. 13.

source leads and that each has a bypass capacitor. Ceramic chip or leadless disc-ceramic types are necessary to provide a low-impedance ground.

Construction

The GaAsFET preamplifiers are built "dead bug" style on a piece of double-sided PC board material. See Fig. 12. The circuit board acts as a low-impedance ground plane, and all ground connections are made directly to it. J1, J2 and C9 also secure the circuit board to the lid of the enclosure. The enclosure can be any aluminum diecast box (Bud CU-124; Hammond 1590B) or Minibox (Bud CU-3002A) approximately 4 by 2 inches, or you can make one by soldering pieces of circuit board material together.

Q1 is mounted to the shield by its source leads, which are soldered directly to C3 and C4. The other ends of these capacitors are soldered to the ground plane. R4 should be soldered in place with essentially no lead length. A layout that has proven successful is shown in Fig. 13. To help prevent L1 and L2 from coupling, wind one coil in the left-hand direction and the other in the right-hand direction.

GaAsFETs are static-sensitive, so handle

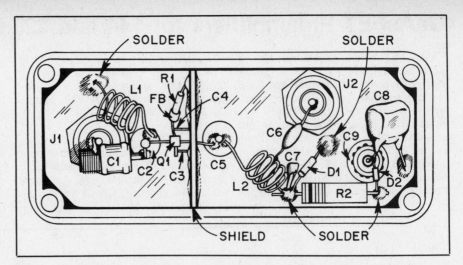

Fig. 13 — Suggested board layout for the MGF1402 144- and 220-MHz GaAsFET preamplifier. Component numbers refer to Fig. 11.

them accordingly. See Chapter 24 for information on working with static-sensitive devices. It's best to assemble the rest of the circuit and install Q1 last. Use a grounded iron if possible, and solder the leads quickly.

Parts for this project are available from Microwave Components of Michigan and Applied Invention, both listed in Chapter 35. MGF1402 transistors are available from Advanced Receiver Research, Box 1242, Burlington, CT 06013.

Remote Preamplifier-Switching System

The preamplifier-switching system described here and shown in Figs. 14 through 16 is intended primarily for EME applications. Serious VHF and UHF operators may wish to consider similar systems for terrestrial work, as a tower-mounted preamplifier usually means a noticeable reduction in system noise figure.

The Circuit

The relay-switching system is separated into two parts. One section is mounted at the tower, and the other, the control circuitry and power supply, is mounted at the station. A length of four-conductor, TV-type rotator cable can be used to inter-connect the two units. The package mounted at the tower consists of two RF coaxial relays and a preamplifier. As shown in Fig. 15, K1 is used to switch the antenna between the transmit line and a line that connects K1 with K2. K2 switches the preamplifier to either the antenna or a 50-ohm termination. With the system shown, it is impossible to transmit accidentally into the preamplifier.

K1 must be able to handle the full transmitter power. Relays rated for high power at VHF and UHF include the Dow-Key DK-60 series and the Transco Y and D series. These relays are expensive when purchased new, but they often show up at

Fig. 14 — A preamplifier relay switching system suitable for EME work. The relay box is mounted at the antenna and the control box is located in the station. A length of four-conductor wire connects the two units.

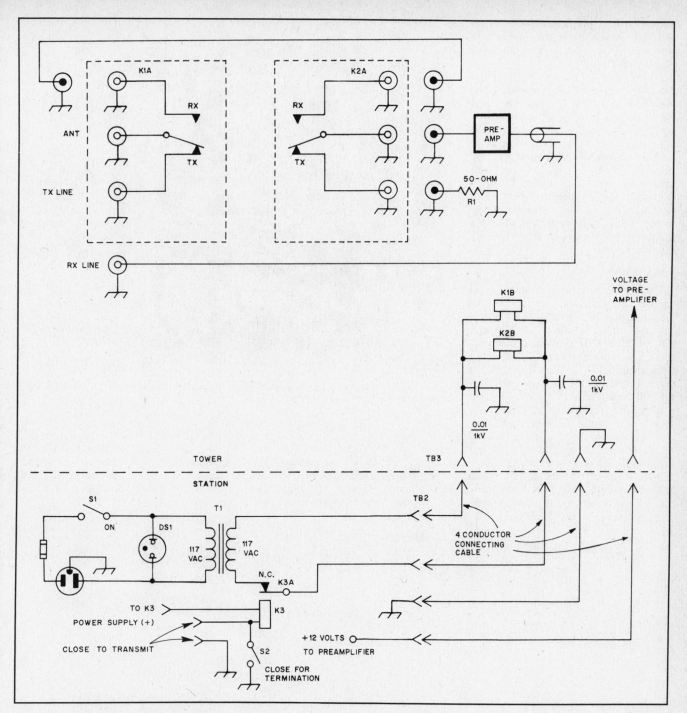

Fig. 15 — Schematic diagram of the preamplifier switching system. The diagram is divided into two parts; the top portion for the circuitry at the antenna and the bottom for use in the station.

DS1 — Neon indicator light with built-in dropping resistor.
K1, K2 — RF-coaxial relays suitable for the frequency range to be used.
K3 — SPDT relay with contacts to carry

voltage/current needed for coils of K1, K2.
R1 — Termination, 50 ohms, noninductive. Resistor built into a PL-259 connector.
S1, S2 — Toggle SPST.

T1 — Isolation transformer, 117-V ac primary, 117-V ac secondary, 15 VA, Stancor P-6411 or equiv.
TB1-TB3, incl. — Terminal block, screw connection, four terminals.

flea markets. K2 is used only to switch the receive line, so it may be a low-power relay. Typical surplus relays have BNC connectors. The project shown here uses two Dow-Key relays.

Relay isolation is an important consideration. High power levels can leak through the unused relay contacts and damage the receive preamplifier, so the use of two relays helps keep the level of RF that appears at the preamp to a minimum.

Generally, no more than 10 mW or RF should be allowed to reach the preamp input. Use of a 0.1- to 0.25-wavelength section of cable between K1 and K2, instead of the double-male connector shown here, would substantially increase isolation.

The control circuit is set up so that the relays must be energized to receive. This offers two advantages. When the station is not in use, the preamp is terminated in a 50-ohm load rather than the antenna. This

protects the preamp from static charges from nearby lightning strikes. Also, if the tower-mounted preamp or one of the relays should fail, the transmit line would be connected directly to the antenna. Another TR relay can be installed at the station end of the transmit line and operation could continue (although without the benefit of a tower-mounted preamp).

S2 allows the operator to switch the relays without putting the station into transmit. The preamp input can be switched between the 50-ohm load and the antenna. This allows for sun-noise measurements.

A method of sequencing the relays and the other components is recommended so that the relays are not "hot switched" with RF power applied. A TR sequencer is described elsewhere in this chapter.

The portion of the system mounted at the station is essentially a power supply and control circuitry. A line-isolation transformer is used to power the 117-V relay coils. A pilot light, fuse and ON/OFF switch are provided in this design. Many surplus relays have 26-V dc coils, so a suitable dc power supply will be necessary in some cases.

Construction

The items to be mounted at the tower are enclosed in an ordinary chassis and bottom plate assembly. Fig. 16 indicates the general layout. Short lead lengths are used throughout. Bulkhead UHF feed-through connectors are used to ensure an RF- and watertight enclosure. The item shown at the bottom center of the enclosure is a commercial preamplifier. Power for the preamplifier (12-V dc) is fed through the fourth wire of the four-wire cable that connects the two modules (tower and station). When all components are mounted properly, the chassis is sealed with silicone rubber (RTV). A terminal block provides for connection to the four-conductor cable. Two 0.01-μF capacitors are mounted across

Fig. 16 — Interior view of the package mounted at the antenna. The object at the bottom center of the chassis is a commercial preamplifier.

the relay coils at terminal block TB3.

The station circuitry is mounted in a small aluminum cabinet. A neon indicator, on/off power and termination switch are

mounted to the front panel. The fuse and interconnection terminal blocks TB1 and TB2 are mounted on the rear apron. Component layout is not at all critical.

TR Time-Delay Generator

If you've ever blown up your new GaAsFET preamp or hard-to-find coaxial relay, or are just plain worried about it, this transmit/receive (TR) time-delay generator is for you. This little circuit makes it simple to put some reliability into your present station or to get that new VHF or UHF transverter on the air fast, safe and simple. Its primary application is for VHF/UHF transverter, amplifier and antenna switching, but it can be used in any amplifier-antenna scheme. An enable signal to the TR generator will produce sequential output commands to a receive relay, a TR relay, an amplifier and a transverter — automatically. All you do is sit back and work DX! This project, shown in Figs. 17 through 23, was designed and built by Chip Angle, N6CA.

Why Sequence?

In stations using transverters, extra power amplifiers and external antenna-mounted TR relays, several problems may arise. The block diagram of a typical station is shown in Fig. 17. When the HF exciter is switched into transmit by the PTT or VOX line, it immediately puts out a ground (or in some cases a positive voltage) command for relay control and an RF signal.

If voltage is applied to the transverter, amplifier and antenna relays simultaneously, RF can be applied as the relay contacts bounce. In most cases, RF will be applied before a relay can make full closure. This can easily arc contacts on dc and RF relays and cause permanent damage. In addition, if the TR relay is not fully closed before RF from the power amplifier is applied, excessive RF may leak into the receive side of the relay. The likely result — preamplifier failure!

The TR time-delay generator supplies commands, one after another, going into transmit and going back to receive from transmit, to turn on all station relays in the right order, eliminating the problems just described.

Circuit Details

Here's how it works. See the schematic diagram in Fig. 19. Assume we're in receive and are going to transmit. A ground command to Q2 (or a positive voltage command to Q1) turns Q2 off. This allows C1 to charge through R1 plus 1.5 kΩ. This rising voltage is applied to all positive (+) inputs of U1, a quad comparator. The ladder network on all negative (−) inputs of U1 sets the threshold point of each comparator at a successively higher level. As C1 charges up, each comparator, starting with U1A, will sequentially change output states.

The comparator outputs are fed into U2, a quad exclusive-OR gate. This was in-

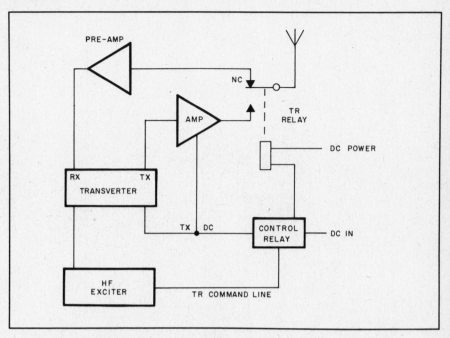

Fig. 17 — A typical VHF or UHF station arrangement with transverter, preamp and power amp. As shown, most TR relays change at the same time.

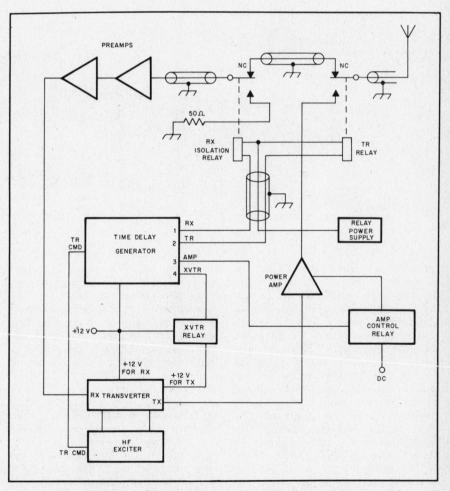

Fig. 18 — Block diagram of a VHF/UHF station with a remote-mounted preamp and antenna relays. The TR time-delay generator makes sure that everything turns on in the right order.

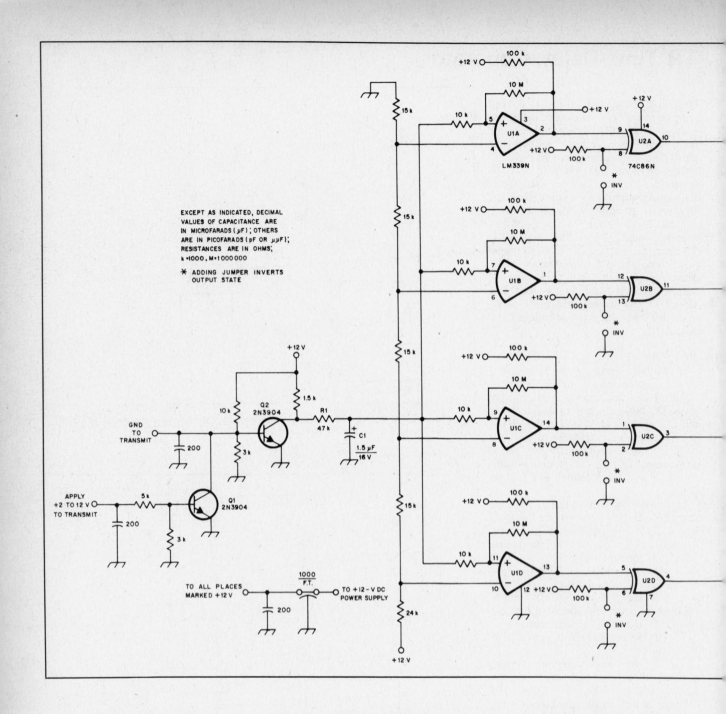

EXCEPT AS INDICATED, DECIMAL
VALUES OF CAPACITANCE ARE
IN MICROFARADS (μF); OTHERS
ARE IN PICOFARADS (pF OR μμF);
RESISTANCES ARE IN OHMS;
k =1000, M=1000000

✱ ADDING JUMPER INVERTS
OUTPUT STATE

cluded in the design to allow "state programming" of the various relays throughout the system. Because of the wide variety of available relays, primarily coaxial, you may be stuck with a relay that's exactly what you need — except its contacts are open when it's energized! To use this relay, you merely invert the output state of the delay generator by inserting a jumper between the appropriate OR-gate input and ground. Now, the relay will be "on" during receive and "off" during transmit. This might seem kind of strange; however, high-quality coaxial relays are hard to come by and if "backwards" relays are all you have, you'd better use them.

The outputs of U2 drive transistors Q3-Q6, which are "on" in the receive mode. Drive from the OR gates turns these transistors "off." This causes the collectors of Q3-Q6 to go high, allowing the base-to-emitter junctions of Q7-Q10 to be forward-biased through the LEDs to turn on the relays in sequential order. The LEDs serve as built-in indicators to check performance and sequencing of the generator. This is convenient if any state changes are made.

When the output transistors (Q7-Q10) are turned on, they pull the return side of the relay coils to ground. These output transistors were selected because of their high beta, very low saturation voltage (V_{CE}) and low cost. They can switch (and have been tested at) 35 V at 600 mA for many days of continuous operation. If substitutions are planned, test one of the new transistors with the relays you plan to

use to be sure that the transistor will be able to power the relay for long periods.

To go from transmit to receive, the sequencing order is reversed. This gives additional protection to the various system components. C1 discharges through R1 and Q2 to ground.

Fig. 20 shows the relative states and duration of the four output commands when enabled. With the values specified for R1 and C1, there will be intervals of 30 to 50 milliseconds between the four output commands. Exact timing will vary because of component tolerances. Most likely everything will be okay with the values shown, but it's a good idea to check the timing with an oscilloscope just to be sure. Minor changes to the value of R1 may be necessary.

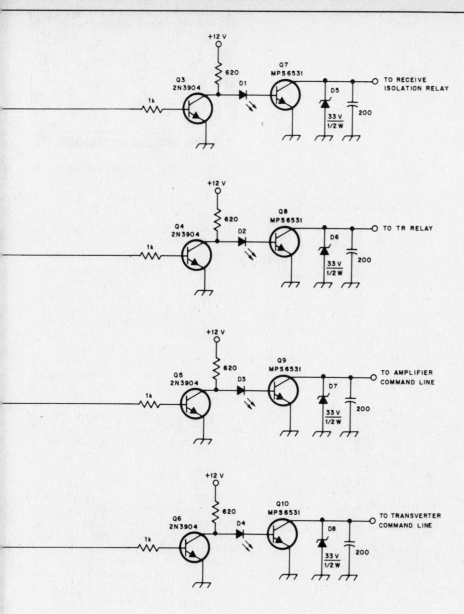

Fig. 19 — Schematic diagram of the TR time-delay generator. Resistors are ¼ W. Capacitors are disc ceramic. Capacitors marked with polarity are electrolytic.

D1-D4 — Red LED (MV55, HP 5082-4482 or equiv.).
D5-D8 — 33-V, 500-mW Zener diode (1N973A or equiv.).
C1 — 1.5-μF, 16-V or greater, axial-lead electrolytic capacitor. See text.
Q1-Q6 — General purpose NPN transistor (2N3904 or equiv.).
Q7-Q10 — Low-power NPN amplifier transistor, MPS6531 or equiv. Must be able to switch up to 35 V at 600 mA continuously. See text.
R1 — 47-kΩ, ¼-W resistor. This resistor sets the TR delay time constant and may have to be varied slightly to achieve the desired delay. See text.
U1 — Quad comparator, LM339 or equiv.
U2 — Quad, 2-input exclusive OR gate (74C86N, CD4030A or equiv.).

Most relays, especially coaxial, will require about 10 ms to change states and stop bouncing. The 30-ms delay will give adequate time for all closures to occur.

Construction and Hookup

One of the more popular antenna changeover schemes uses two coaxial relays — one for actual TR switching and one for receiver/preamplifier protection. See

Fig. 20 — The relative states and durations of the four output commands when enabled. This diagram shows the sequence of events when going from receive to transmit and back to receive. The TR delay generator allows about 30 to 50 ms for each relay to close before activating the next one in line.

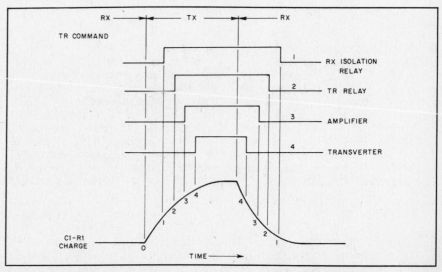

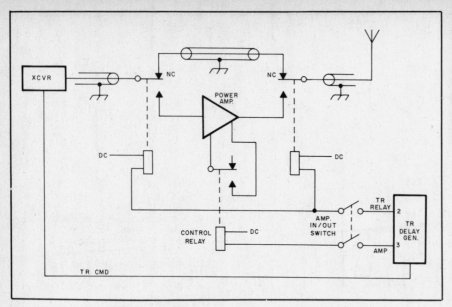

Fig. 21 — The TR time-delay generator can also be used to sequence the relays in an HF power amplifier.

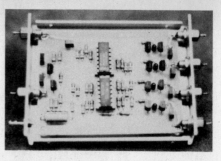

Fig. 23 — The completed time-delay generator fits in a small aluminum box.

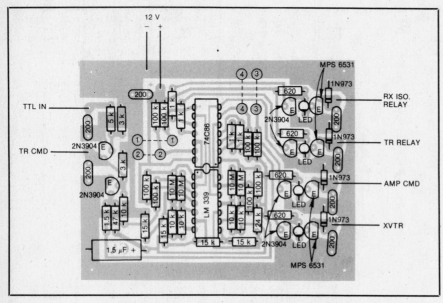

Fig. 22 — Parts placement diagram of the TR time-delay generator. A full-size etching pattern is given at the back of this book.

Fig. 18. A project using this scheme is presented elsewhere in this chapter.

Many RF relays have very poor isolation, especially at VHF and UHF frequencies. Some of the more popular surplus relays have only 40 dB isolation at 144 MHz or higher. If you are running high power, say 1000 W ($+60$ dBm) at the relay, the receive side of the relay will see $+20$ dBm (100 mW) when the station is transmitting. This power level is enough to inflict fatal damage on your favorite preamplifier.

Adding a second relay, called the RX isolation relay here, terminates the preamp in a 50-ohm load during transmission and increases the isolation significantly. Also, in the event of TR relay failure, this extra relay will protect the receive preamplifier.

As shown in Fig. 18, both relays can be controlled with three wires. This scheme provides maximum protection for the receiver. If high-quality relays are used and

verified to be in working order, relay losses can be kept well below 0.1 dB, even at 1296 MHz. The three-conductor cable to the remote relays should be shielded to eliminate transients or other interference.

By reversing the RX-TX state of the TR relays (that is, connecting the transmitter hardline and 50-ohm preamp termination to the normally open relay ports instead of the normally closed side), receiver protection can be provided. When the station is not in use and the system is turned off, the receiver preamplifier will be terminated in 50 Ω instead of being connected to the antenna. The relays must be energized to receive. This might seem a little backward; however, if you are having static-charge-induced preamplifier failures, this may solve your problem.

Most coaxial relays aren't designed to be energized continuously. Therefore, adequate heat sinking of coaxial relays must be considered. A pair of Transco Y relays can be energized for several hours when mounted to an aluminum plate 12 inches square and ¼ inch thick. Thermal paste will give better heat transfer to the plate. For long-winded operators, it is a good idea to heat sink the relays even when they are energized only in transmit.

Fig. 21 shows typical HF power amplifier interconnections. In this application, amplifier in/out and sequencing are all provided. The amplifier will always have an antenna connected to its output before drive is applied.

Many TR change-over schemes are possible depending on system requirements. Most are easily satisfied with this TR delay generator.

The TR delay generator is built on a 2½ × 3¼ inch PC board. See Fig. 22. The board can be mounted in a small aluminum case, as shown in Fig. 23. Connections to the rest of the system are made through feedthrough capacitors. Do not use feedthrough capacitors larger than 2000 pF because peak current through the output switching transistors may be excessive.

Three-Watt Transmitting Converter for 6 Meters

The linear transmitting converter shown in Figs. 36 through 39 is a simple, low-cost way to extend the coverage of any 28-MHz transmitter to the 6-meter band. Output power is 3 watts PEP and the 28-MHz drive requirement is 1 mW. This drive level is compatible with the transverter outputs found on many current HF transceivers. By selecting the appropriate resistor values, the input attenuator can easily be adjusted for drive levels as high as 100 mW.

Particular care was taken during the design stage to ensure that spurious emissions are at a minimum. The IMD performance is such that this unit can be used to drive a high-power linear amplifier without generating excessive adjacent-channel interference.

Circuit Description

LO injection is supplied to the mixer by a crystal oscillator operating at 22 MHz. The output of the oscillator is filtered to reduce harmonics to approximately -40 dBc and is then applied to a power splitter, T2. One port of the splitter feeds the mixer, and the other output is connected to J1. This output can be used to supply LO to a receiving converter for transceiver operation. A 3-dB attenuator between the splitter and the mixer provides a wideband termination for the mixer LO port. The power delivered to the mixer is $+7$ dBm.

A commercial diode-ring module is used for the mixer. The excellent balance of this type of mixer reduces the band-pass filter requirements following the mixer. The IMD performance is also very good; the third-order products are 40 dB below each tone of the two-tone output signal when the input signal is at the recommended level of -10 dBm (each tone). The attenuator at the IF port prevents overdriving the mixer. The 20-dB pad shown in Fig. 37 allows a driving signal of 30 to 40 mW (PEP) to be used. This attenuator should be adjusted if other drive levels are desired.

The mixer is followed by a broadband amplifier, 6-dB pad and a 3-pole band-pass filter. The amplifier has a gain of nearly 20 dB at 50 MHz and provides a good termination for the mixer. Signal level at the filter output is about -10 dBm. Two more stages of class-A amplification follow the band-pass filter, bringing the signal level up to the $+16$ dBm necessary to drive the power-amplifier stage.

The power amplifier uses a Motorola MRF476 transistor operating class AB. The MRF476, a low-cost device in a plastic TO-220 package, will deliver a minimum of 3 watts PEP with good linearity at 50 MHz. A CW output of up to 4 watts can be obtained from this converter, but the key-down time should be limited to 60 seconds. This is because of the relatively small heat sink used in this unit. The power-amplifier input and output networks

Fig. 36 — Six-meter transmitting converter and 3-W PEP amplifier.

are both designed to match the transistor to 50 ohms. Their design was based on data supplied in the *Motorola RF Data Manual* (2nd edition). The amplifier output is filtered by a 7-pole low-pass filter designed for a cutoff frequency of 56 MHz and a ripple factor of 0.17 dB.

Construction

The converter is constructed in two sections: the low-level section consisting of the LO, mixer, band-pass filter and class-A amplifiers and the power-amplifier/low-pass filter section. No printed circuit board is needed for the low-level section; rather, the components are mounted directly on a piece of etched copper-clad board. The component leads going to ground support the components above the board. Among the many advantages to this style of construction is the ease with which the circuits can be modified. Also, if a printed-circuit board is used, the components used to lay out the board, such as variable capacitors,

are likely to be the only types that can be used. This makes it difficult for the builder to use junk box components.

Note that the LO and its filter are completely enclosed by a shield made of double-sided circuit-board material. This was found to be necessary to ensure that 22-MHz LO energy does not find its way into the amplifier stages. It was expected that a 1-inch-high shield around the LO would provide sufficient isolation — this is not the case; a complete enclosure is required. It is recommended that a shield enclosure be constructed around the remaining portion of the low-level section. This will prevent feedback from the power-amplifier board to the class-A stages. It was not necessary to do this in the prototype shown in the photographs, but is good practice when the two boards are mounted in the same enclosure (such as an aluminum chassis).

The power-amplifier board is constructed on a printed board (Fig. 38). It is

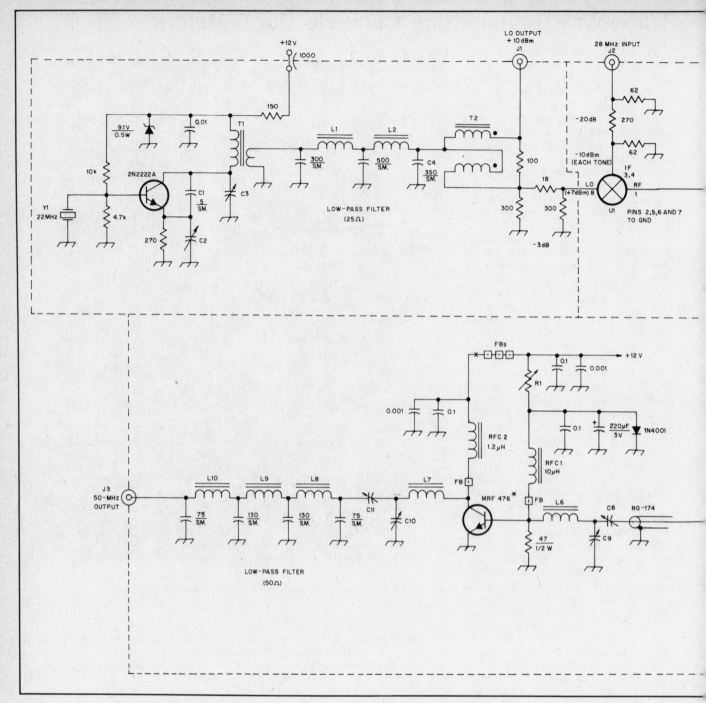

Fig. 37 — Schematic diagram of the 6-meter transmitting converter.

C2, C3, C5-C7, incl. — 2- to 25-pF air or foil-
   dielectric trimmer.
C8, C9, C11 — 15-115 pF mica trimmer (Arco
   406 or equiv.)
C10 — 1.5- to 20-pF mica trimmer (Arco 402 or

   equiv.)
FB — Ferrite bead, Amidon FB101-43.
L1, L2 — 7 turns, no. 28 enamel on T25-6 core.
L3-L5, incl. — 12 turns, no. 28 enamel on
   T37-6 core.

L6 — 6½ turns, no. 26 enamel on T37-12 core.
L7 — 9½ turns, no. 22 enamel on T50-12 core.
L8, L10 — 8 turns, no. 22 enamel on T50-12
   core.
L9 — 8 turns, no. 22 enamel on T50-12 core.

not necessary that the board be etched. The layout shown can be used as a guide and the copper clad cut away with a sharp X-acto® knife. After cutting through the copper with the knife, applying heat from a soldering iron allows the foil to be lifted off easily. The heat sink for the MRF476 is made from a 3½ × 1-inch strip of 1/16-inch aluminum folded into a U shape. Use a small amount of heat-sink compound between the transistor body and the heat

sink. The MRF476, unlike many RF transistors, has the collector connected to the mounting tab; thus the heat sink is hot for both RF and dc. The center pin of the transistor is also connected to the collector, but it is not used. Connection to the collector is made through the mounting tab. The unused pin should be cut off at the point where it becomes wider (about 1/8 inch from the transistor body). The other pins are bent down to make connection to the

circuit board at this same point. Connection between the two boards can be made with a short length of RG-174 coaxial cable.

Tune-Up and Operation

The low-level stages are aligned before connecting the power-amplifier stage. A 50-ohm resistor is connected through the 0.01-μF blocking capacitor to the output of the last class-A amplifier and 12 volts is ap-

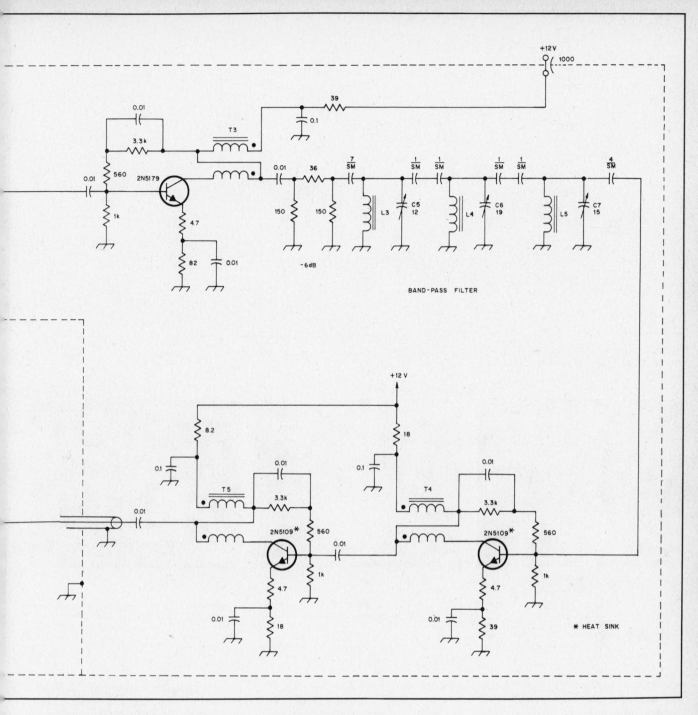

BAND-PASS FILTER

* HEAT SINK

Compress winding to cover half of core.
R1 — 220-ohm, 1-W resistor in series with a
  500-ohm, ½-W trimmer.
RFC1 — 10-μH solenoid type RF choke (Miller
  74F105AP or equiv.).
RFC2 — 1.2-μH solenoid type RF choke (Miller

74F126AP or equiv.).
T1 — 27 turns primary, 5 turns secondary,
  no. 28 enamel on T37-6 core.
T2 — 5 bifilar turns, no. 26 enamel on FT37-61
  core.

T3 — 7 bifilar turns, no. 28 enamel on FT23-43
  core.
T4, T5 — 5 bifilar turns, no. 26 enamel on
  FT37-61 core.
U1 — Mini-Circuits SRA-1 doubly balanced
  mixer.

plied to the oscillator and class-A amplifiers. The oscillator is adjusted by setting C2 at mid-range. Next, C3 is tuned for maximum output as measured with a VTVM and RF probe connected to pin 8 of the mixer. C2 can then be adjusted to bring the frequency to exactly 22 MHz. If a frequency counter with a high-impedance input is used, it should be connected to pin 8 of the mixer. If the counter has a 50-ohm input it can be connected to J1.

The adjustment of C2 and C3 will interact so the procedure should be repeated. J1 should always be connected to a load close to 50 ohms. If a receive converter is not being used, a 50-ohm resistor, mounted in a BNC connector, should be connected to J1.

The band-pass filter is aligned by applying a 28.7-MHz CW signal to the input of the converter and adjusting C5, C6 and C7 for maximum output at the output of the

last amplifier. Check the frequency response of the filter by varying the frequency of the input signal from 28 to 29 MHz. The output power should not vary more than 1 dB within the passband of the filter. The 50-ohm resistor at the output of the last class-A amplifier should now be removed and the power amplifier connected through a short length of coaxial cable.

To align the input and output networks

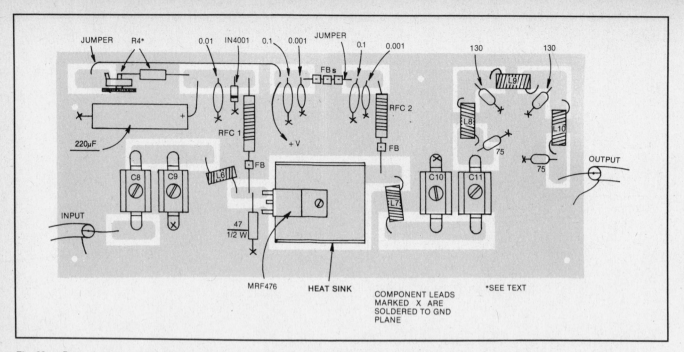

Fig. 38 — Parts-placement guide for the 6-meter transmitting converter. All components mount on the foil side of the board. A full-size etching pattern can be found at the back of this book.

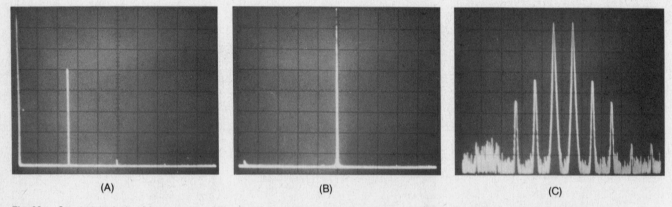

Fig. 39 — Spectral analysis of 6-meter transmitting converter. Display at A shows the fundamental and second harmonic. The harmonic is 73 dB below the 3-watt carrier. The fundamental has been notched 30 dB to prevent overload of the spectrum analyzer. Each horizontal division is 20 MHz. A close-in display is shown in B: Each horizontal division is 2 MHz and the center frequency is 50 MHz. The 2×28-MHz spurious product (which falls in channel 2) is attenuated more than 75 dB relative to the carrier power. The two-tone IMD spectrum is shown in C; the third-order products are 34 dB below PEP. Each horizontal division is 2 kHz. In all three displays each vertical division is 10 dB.

of the final, connect the amplifier output to a wattmeter and 50-ohm load. Attach the amplifier to a 12-volt, 1-A power supply (maximum current drain is approximately 750 mA for the complete converter) and place a milliammeter in the collector supply line at the point marked $\times$ in Fig. 37. With no drive applied, turn on the dc power and adjust R1 for a collector current of 35 to 40 mA. Apply just enough 29-MHz drive to the input of the converter to get an indication on the wattmeter and peak C8 through C11 for maximum output. Now increase the drive slowly, keeping the networks peaked, until the output power is 4 watts. Do not maintain the output at 4 watts any longer than is necessary. It is important that the output network be adjusted while the output is at the 3.5- to 4-watt level. If it is adjusted at a lower level the IMD performance will be degraded. When the output is adjusted as described above, the third-order IMD products should be at least 32 dB below the 3-watt-PEP level (Fig. 39C).

Linear Transverters for 144 and 220 MHz

The CW and SSB portions of the 144- and 220-MHz bands have swelled with activity during the past few years. Although there is plenty of commercially made CW and SSB gear for 144 MHz, there is little available for 220. Paul Drexler, WB3JYO, designed and built the linear transverters described here and shown in Figs. 40 through 62. These projects enable the use of a standard 28-MHz transceiver as a tunable IF for 144- or 220-MHz operation. Construction is less complicated than building a complete transceiver, and all of the good features of the HF rig (such as a clean SSB source, stable VFO and good crystal filters) are incorporated. Chapters 11 and 12 contain additional information on transverter theory. Although these transverters may be tuned for any segment of operation, they were designed mainly to cover the lower portion of each band (144 to 145 MHz and 220 to 221 MHz). A 12-V power supply and an antenna are the only other equipment necessary to complete the VHF station.

The complete transverter design includes a minimum of hard-to-find components and should be easily reproducible. Although the text and illustrations center around the 220-MHz transverter, component values are given for the 144-MHz unit as well. Except for the local oscillator (LO), all circuits are common to both designs.

The receive converter has a 0.6-dB noise figure and an overall conversion gain of 28 dB. These figures were verified on an HP8970A noise-figure meter with an HP346A noise source. Transmit-converter power output is a conservative 1 W under linear operation. The companion amplifier produces 8 to 10 W of linear output power. Much care was taken to make the transmit chain as clean as possible, and the receive converter incorporates techniques to maximize sensitivity and dynamic range.

CIRCUIT DESCRIPTION

Fig. 41 shows the transverter block diagram. The main difference between the 144- and 220-MHz versions is the LO. Although transceive operation is depicted, the experimenter may choose to limit construction to either transmit only or receive only. LO energy is injected into a high-level (+17 dBm) doubly balanced mixer during receive. Received signals are amplified by a GaAsFET preamplifier and then filtered before entering the mixer. The 28-MHz

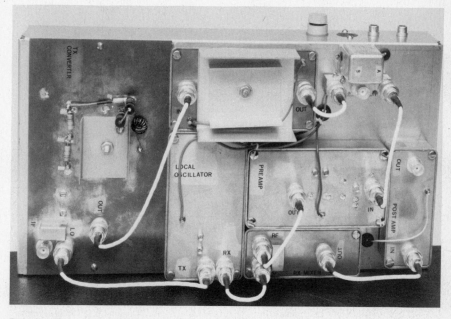

Fig. 40 — The 144- or 220-MHz transverter is built using a modular circuit approach. Each circuit is boxed and mounted on a chassis with interconnections of short lengths of 50-ohm coaxial cable.

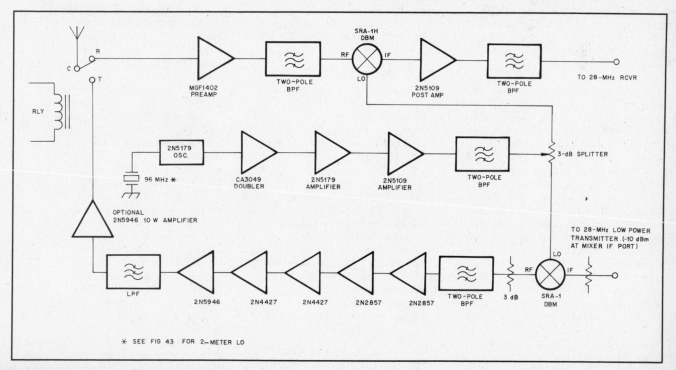

Fig. 41 — Block diagram of the 144- and 220-MHz transverters. All blocks but the local oscillator are common to both bands.

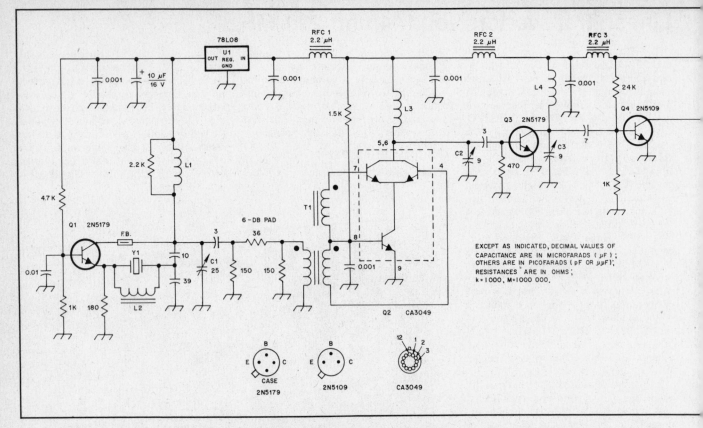

Fig. 42 — Schematic diagram of the 192-MHz local oscillator. All resistors are ¼-W carbon-composition types unless otherwise noted. Capacitors are silver-mica or miniature monolithic ceramic types unless otherwise noted. Capacitors marked with polarity are electrolytic.

C1 — 25-pF (max.) miniature ceramic trimmer.
C2, C3, C4 — 9-pF (max.) miniature ceramic trimmer.
C5, C7 — 10-pF (max.) ceramic piston trimmer or miniature ceramic trimmer.
C6 — 0.5-pF ceramic chip or gimmick capacitor. See text.

J1, J2 — Chassis-mount female BNC connector.
L1 — 10t no. 24 enam., 0.100-inch ID, close wound.
L2 — 15t no. 28 enam. on T25-6 toroid core.
L3, L4 — 5t no. 24 enam., 0.125-inch ID, close wound.

L5 — 4t no. 24 enam., 0.125-inch ID, close wound.
L6, L7 — 5t no. 18 tinned, 0.250-inch ID, spaced one wire dia. Tap at 1t from ground.
Q2 — RCA CA3049 transistor array. CA3054 or Motorola MC3346 are acceptable substitutes.
RFC1-RFC4 — 2.2-µH molded miniature RF

signal at the output port of the mixer is sent to a post amplifier and band-pass filter, and then to the 28-MHz IF receiver.

On transmit, LO energy and a low-level RF signal from the 28-MHz IF transceiver are fed to a standard-level (+7 dBm) doubly balanced mixer. The mixer output is filtered to eliminate the image and other unwanted responses, and then the desired signal is amplified by five stages to reach the 1-W level. The output of the 1-W transmit converter is further filtered to meet FCC rules and regulations. An optional 8- to 10-W amplifier is described for those desiring greater power output.

This transverter is built in a modular fashion. All of the major circuit blocks are built into separate enclosures and interconnected by 50-ohm coaxial cable. Modular construction lends itself to on-the-air experimentation and development, as well as simple troubleshooting. Any stage may be removed from the circuit for modification, or a new stage may be substituted. Although commercially available diecast boxes (BUD CU-123 and CU-124 or Hammond 1590A and 1590B) are used here, double-sided PC-board enclosures or Miniboxes may be used. Mounting all cir-

cuits in individual boxes is highly recommended for versatility and shielding. All component connections are made using direct, point-to-point construction techniques.

Local Oscillator

A schematic diagram of the local oscillator for the 220-MHz transverter is shown in Fig. 42. A frequency of 96 MHz has been chosen so that only one stage of multiplication is necessary to obtain the needed 192-MHz LO frequency. The crystal, Y1, is a series-resonant, fifth-overtone type in an HC-18/U package. The oscillator used is a common-base circuit derived from an article by Joe Reisert, W1JR.[1] Most crystal oscillators tend to oscillate at a frequency slightly higher than the crystal's fundamental mode of operation. In this circuit, L2 cancels the C_o crystal capacitance, thus bringing the oscillator down to the desired frequency.

The author chose to use a 95.95-MHz crystal (for a 191.9-MHz LO); an IF fre-

[1]Reisert, Joe, "VHF/UHF Receivers," *Ham Radio*, March 1984, pp. 42-46.

quency of 28.1 MHz corresponds to an operating frequency of 220.0 MHz. Strong signals from stations operating on 10 meters may be picked up by the transverter interconnecting cables, interfering with signals being received from 220 MHz. Also, when more than one transverter with a 28-MHz IF is operated in the same room (at a multioperator VHF contest station, for example), it is not uncommon to hear signals from the other IF transceivers.

Oscillator output is fed into an active doubler constructed with a CA3049 transistor array. This circuit is based on a suggestion by Bill Strunk, K3ZMA. The doubler features as much as 7-dB gain and excellent harmonic suppression because of its balanced input circuit. A CA3054 or MC3346 transistor array may be used in place of the CA3049.

Output energy from the doubler is further amplified by Q3 and Q4 to bring the LO to the required mixer input level. Filtering is accomplished at the LO output by a lightly coupled double-tuned filter. All spurious and harmonic energy is better than 55 dB below the desired signal. A clean LO is necessary to avoid interference problems caused by undesired mixing products

RFC 4
2.2 µH

+ 12.5 —
V DC

0.001
F.T.

TO RX
MIXER

+ 19 dBm

2.2 µF
16 V

0.001

L5

BAND PASS FILTER
C6   0.5 pF

J1

16

C4   9

5

L6   C5   C7   L7

10   10

16

16

20   33

J2

+ 7 dBm
TO TX
MIXER

choke.

T1 — 7t no. 30 enam., trifilar wound on T25-12
core or Mini-Circuits T4-1 transformer. See
text.

U1 — Three-terminal voltage regulator, 8 V,
100 mA (LM78L08 or equiv.).

Y1 — Fifth-overtone, 96-MHz series-resonant
crystal, HC-18/U holder.

during receive and transmit.

After the filter, a resistive power divider is used to drive the transmit and receive mixers. Resistive pads attenuate the LO signal to the proper level while providing a 50-Ω impedance for the mixer input.

Although a similar lineup could be employed for 144 MHz, a three-stage 116-MHz oscillator is shown schematically in Fig. 43. The 144-MHz transverter uses a 116-MHz series-resonant crystal (Y1) to avoid frequency multipliers. Oscillator output is amplified by Q2 and Q3 until it reaches the level necessary to drive both the transmit and receive mixers. LO output is filtered by a double-tuned band-pass filter. After the filter, a resistive power divider is used to drive the transmit and receive mixers; again, pads attenuate the signal to the proper level.

With the filters properly adjusted, all spurious outputs from the LO are about 58 dB below the fundamental. Fig. 44 shows the spectral output from the LO receive port.

Receive Converter

The 144- and 220-MHz receive converters are identical, except for the tuned

circuits in the front end. In each case, the converter consists of a GaAsFET preamplifier, a mixer/diplexer circuit and an optional 28-MHz post amplifier. The mixer/diplexer and post amplifier circuits are the same, regardless of band. Each of the three receive converter blocks is built into a separate module. This was done to facilitate experimentation and development of each stage. Of course, it is possible to build all three circuits in one box. This subject will be addressed in the construction portion of this article.

Mixer

The heart of the receive converter is a Mini-Circuits SRA-1H high-level, doubly balanced mixer (DBM). See Fig. 45. This mixer requires an LO injection level of +17 dBm, compared with the +7 dBm injection level required for standard mixers. The high-level mixer offers superior strong-signal handling characteristics while maintaining the port-to-port isolation, image suppression and simplicity inherent in a DBM. The SRA-1H is modestly priced and available in small quantities directly from the manufacturer.

Reactive terminations can ruin the excellent IMD characteristics of a DBM.[2,3] The IF port, in particular, is most sensitive to a nonresistive 50-ohm termination. Anything short of a 20-dB resistive pad at the IF port will result in increased IMD products and a lower third-order-intercept point. Feeding the output of a DBM directly into a narrowband amplifier will decrease the mixer's third-order intercept point as compared to a purely resistive termination. The diplexer circuit shown in Fig. 45 represents one solution to the problem of proper mixer termination. The diplexer's low-pass response presents a 20-dB return loss at 28 MHz and terminates higher frequencies into 50 ohms.

RF Preamplifier

A low-noise, high-dynamic-range GaAsFET preamplifier is used in front of the mixer to overcome mixer conversion loss. The GaAsFET device offers exceptional performance, compared with most bipolars and MOSFETs, and designs abound.[4,5] The circuit in Fig. 46 has proven reliable during many hours of on-the-air operation. This simple design offers a noise figure of 0.4 dB, as measured on an HP8970A noise-figure meter with the HP346A noise source. This noise figure is much lower than the feed line loss preceding the preamplifier; performance is exceptional for all applications short of in-

tensive EME receiving. A 30-mA bias current achieves optimum signal-handling capability. The third-order-intercept point is +25 dBm. Gain is 24 dB.

The double-tuned filter between the preamp and mixer provides a reasonable degree of filtering. A trap (L2) is used to attenuate the 164-MHz (188 MHz for the 2-meter version) image. Fig. 47A shows the swept frequency response of the 220-MHz version. A comb line or helical filter might be used if greater selectivity is required.

28-MHz Post Amplifier

For most amateur applications, a 28-MHz post amplifier is not necessary. It serves to amplify the 28-MHz IF signal to increase S-meter readings. The author lives among several of the "big gun" VHF stations in southeastern Pennsylvania, so high dynamic range is essential to avoid overload problems. The receive converter operates nicely without any post amplification, thereby preserving the IF receiver's dynamic range.

The 28-MHz post amplifier shown in Fig. 48 has been included here for those operators fortunate enough to live away from strong in-band signals. The 2N5109 is readily available and provides good performance at low cost. In this circuit, the device is biased to provide 13-dB gain with a third-order-intercept point of +26 dBm. The design features a tuned input circuit and a broadband output transformer. A double-tuned band-pass filter at the output assures a clean signal for the IF receiver. Fig. 47B shows the swept frequency response of the post amplifier.

If you live in an area with loud local signals, yet want to use a post amplifier, a pad may be used between the post amplifier and the IF receiver to reduce the converter gain to a level that the IF receiver can handle. The value of attenuation will depend on the IF receiver's ability to handle large signals. When you first connect the receive converter to the IF receiver, you will probably notice that the S-meter on the receiver moves up to S3 or higher (a lot depends on the nature of your specific receiver), even with no signals present. To determine the right pad value for your application, place a variable step attenuator in the line between post amp and IF receiver and increase the attenuation until the IF receiver S-meter is just above zero. If you want to leave the step attenuator in the line, fine. If not, you can build a pad with the correct value from the attenuator tables given in Chapter 25 of this *Handbook*.

Transmit Converter

A schematic diagram for the 1-W transmit converter is shown in Fig. 49. The 192-MHz LO (116-MHz LO for the 144-MHz version) and 28-MHz signals are mixed in a Mini-Circuits SRA-1 standard-level DBM. A pad is necessary to limit the 28-MHz input to a maximum level of −10 dBm, ensuring good linearity and

[2]Will, Peter, "Reactive Loads — The Big Mixer Menace," *Microwaves*, April 1971, pp. 38-42.
[3]Cheadle, Dan, "Selecting Mixers for Best Intermod Performance," *Microwaves*, Nov. and Dec. 1973.
[4]Kraus, Geoff, "VHF and UHF Low Noise Preamplifiers," *QEX*, Dec. 1981, p. 3.
[5]Reisert, Joe, "Low Noise GaAsFET Technology," *Ham Radio*, Dec. 1984, pp. 99-112.

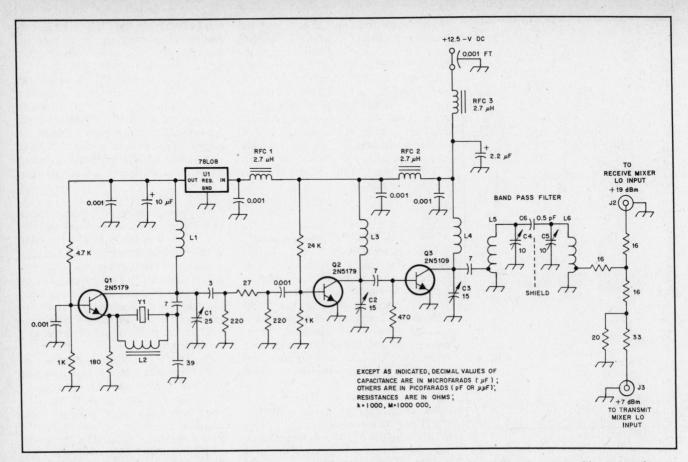

Fig. 43 — Schematic diagram of the 116-MHz local oscillator for the 2-meter transverter. All resistors are ¼-W, carbon-composition types unless otherwise noted. Capacitors are silver-mica or miniature monolithic ceramic types unless otherwise noted. Capacitors marked with polarity are electrolytic.

C1 — 25-pF (max.) miniature ceramic trimmer.
C2, C3 — 15-pF (max.) miniature ceramic trimmer.
C4, C5 — 10-pF (max.) ceramic piston trimmer or miniature ceramic trimmer.
C6 — 0.5-pF ceramic chip or gimmick capacitor. See text.

J1, J2 — Chassis-mount female BNC connector.
L1 — 8t no. 24 enam., 0.100-inch ID, close wound.
L2 — 13t no. 28 enam. on T25-6 toroid core.
L3, L4 — 7t no. 24 enam., 0.125-inch ID, close wound.
L5, L6 — 8t no. 18 tinned, 0.250-inch ID,

spaced one wire dia. Tap at 1t from ground.
RFC1-RFC3 — 2.7-μH molded miniature RF choke.
U1 — Three-terminal voltage regulator, 8 V, 100 mA (LM78L08 or equiv.).
Y1 — Fifth-overtone, 116-MHz series-resonant crystal, HC-18/U holder.

spectral purity. No parts values are shown for the IF pad; the exact resistor values will depend on the amount of 28-MHz drive available from the transverter output of your IF transceiver. For example, if your IF rig delivers 20 mW (+13 dBm) at the transverter output, you would need to build a 23-dB pad. See Chapter 25 for tables listing resistor values for different levels of attenuation.

Mixer output is fed through a resistive pad for proper termination, and then filtered by a double-tuned band-pass filter to reduce the image and other undesired mixing products. Two 2N2857 amplifier stages follow the filter, followed by two 2N4427 stages. The final amplifier stage is a 2N5946. All stages are biased for linear operation. The 2N5946 may be substituted with a lower-power 2N5945 or 2N5944 device; if you substitute, you may have to alter the input and output matching, as well as the bias circuit. A 7-element Chebyshev low-pass filter (Fig. 50) follows the 2N5946. Swept filter response is shown in Fig. 51B. The output is exceptionally clean; a spec-

tral plot is shown in Fig. 51A.

Although some designers may question the use of five stages to achieve 1-watt output, there are several good reasons for doing so. This transmit converter is rated

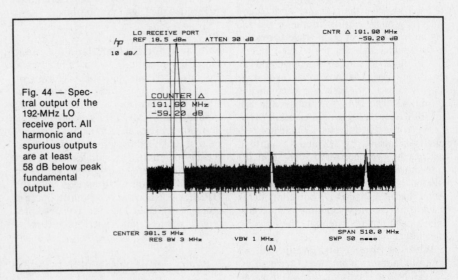

Fig. 44 — Spectral output of the 192-MHz LO receive port. All harmonic and spurious outputs are at least 58 dB below peak fundamental output.

for conservative operation at a healthy, filtered 1-W output. It is not "almost" a watt, like some four-stage designs. All stages are run below their maximum possible output level, ensuring clean, linear

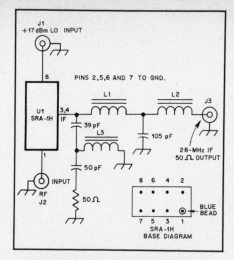

Fig. 45 — Schematic diagram of the receive mixer and diplexer filter. Capacitors are silver-mica. Resistors are ¼-W carbon-composition types.

J1-J3 — Chassis-mount female BNC connector.
L1 — 0.35 µH; 6t no. 24 enam. on T50-6 toroid core.
L2 — 0.205 µH; 5t no. 24 enam. on T50-6 toroid core.
L3 — 0.18 µH; 5t no. 24 enam. on T37-6 toroid core.
U1 — Mini-Circuits SRA-1H high-level, doubly balanced mixer. Available directly from the manufacturer. See text.

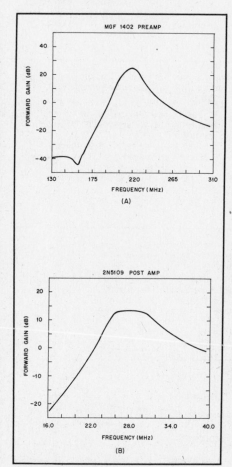

Fig. 47 — Swept response of the 220-MHz preamplifier (A) and the 28-MHz post amplifier (B). The image filter at the output of the 220-MHz preamplifier produces a deep null at 164 MHz.

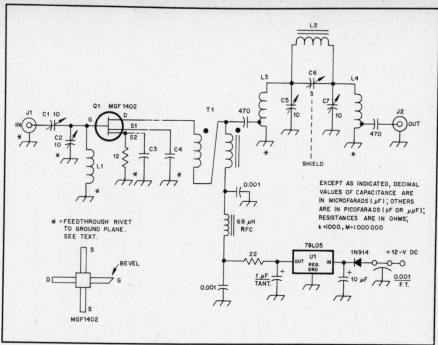

Fig. 46 — Schematic diagram of the 144- or 220-MHz GaAsFET preamplifier. All resistors are ¼-W carbon composition types unless otherwise noted. Capacitors are silver-mica or miniature monolithic ceramic types unless otherwise noted. Capacitors marked with polarity are electrolytic.

C1, C2, C5, C7 — 10-pF (max.) ceramic piston trimmer or miniature ceramic trimmer.
C3, C4 — 500-pF ceramic chip capacitor.
C6 — 0.3 to 3 pF ceramic piston trimmer.
J1, J2 — Chassis-mount female BNC connector.
L1 — 220 MHz: 3t no. 18 enam., 0.250-inch ID, close wound. 144 MHz: 5t.
L2 — 220 MHz: 13t no. 24 enam. on T44-10 toroid core; 144 MHz: 27t no. 26 enam. on T37-10 toroid core.

L3, L4 — 220 MHz: 5t no. 18 tinned wire, 0.250-inch ID, spaced one wire dia. Tap at 1t from ground. 144 MHz: same as 220 MHz, except coil is 7t.
Q1 — Single-gate GaAsFET, MGF1402.
RFC1 — 68-µH molded miniature RF choke.
T1 — 4t no. 28 enam. bifilar wound on T25-12 toroid core.
U1 — Three-terminal voltage regulator, 5 V, 100 mA (LM78L05 or equiv.).

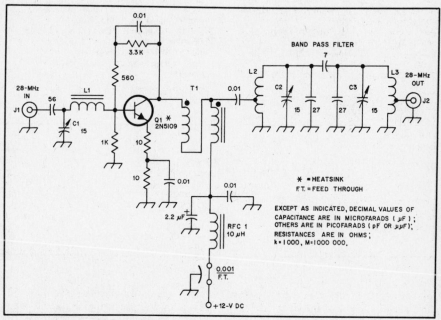

Fig. 48 — Schematic diagram of the 28-MHz post amplifier. All resistors are ¼-W carbon-composition types unless otherwise noted. Capacitors are silver-mica or miniature monolithic ceramic types unless otherwise noted. Capacitors marked with polarity are electrolytic.

C1, C2, C3 — 15-pF (max.) miniature ceramic trimmer.
J1, J2 — Chassis-mount female BNC connector.
L1 — 12t no. 26 enam. on T37-6 toroid core.

L2, L3 — 15t no. 26 enam. on T37-6 toroid core. Tap at 4t from ground.
RFC1 — 10-µH miniature molded RF choke.
T1 — 20t no. 30 enam. bifilar wound on T37-10 toroid core.

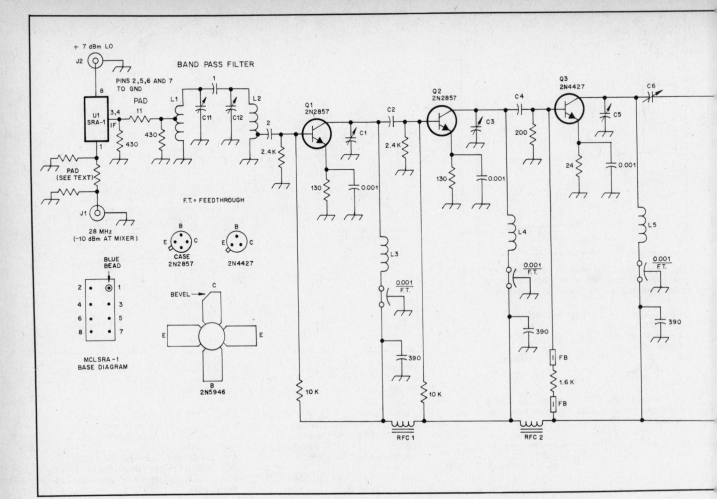

Fig. 49 — Schematic diagram of the 144- and 220-MHz transmit converter. All resistors are ¼-W carbon-composition types unless otherwise noted. Capacitors are silver-mica or miniature monolithic ceramic types unless otherwise noted. Capacitors marked with polarity are electrolytic.

C1, C3, C5, C6, C7 — Miniature ceramic trimmer, 220 MHz: 8 pF (max.); 144 MHz: 25 pF (max.).
C2 — Silver-mica capacitor, 220 MHz: 2 pF; 144 MHz: 3 pF.
C4 — Silver-mica capacitor, 220 MHz: 3 pF; 144 MHz: 5 pF.
C8 — Miniature ceramic trimmer, 220 MHz:

15 pF (max.); 144 MHz: 25 pF (max.).
C9, C10 — 25-pF (max.) miniature ceramic trimmer (same for both bands).
C11, C12 — Ceramic piston trimmer or miniature ceramic trimmer, 220 MHz: 10 pF (max.); 144 MHz: 25 pF (max.).
J1-J3 — Chassis-mount female BNC connector.

L1, L2 — 220 MHz: 4t no. 20 tinned, 0.250-inch ID, spaced one wire dia. Tap at 1t from ground. 144 MHz: Same as 220 MHz except coil is 6t.
L3, L4, L5 — 220 MHz: 3t no. 20 tinned, 0.250-inch ID, spaced one wire dia. 144 MHz: Same as 220 MHz except coil is 5t.
L6 — 220 MHz: 1t no. 20 tinned, 0.250-inch ID.

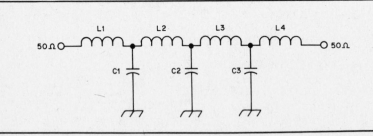

Fig. 50 — Schematic diagram of 7-element Chebyshev low-pass filter. Capacitors are silver-mica types.

C1, C3 — 220 MHz: 15 pF; 144 MHz: 27 pF.
C2 — 220 MHz: 18 pF; 144 MHz: 30 pF.
L1, L4 — 220 MHz: 3t no. 22 enam., 3/16-inch ID, close wound; 144 MHz also 3t.
L2, L3 — 220 MHz: 4t no. 22 enam., 3/16-inch ID, close wound; 144 MHz: 5t.

operation. The gain from the extra stage also allows the use of the filters after the mixer and at the output. The author lives in an area where the VHF bands are congested, so clean, linear operation is a must for sound relations with other amateurs sharing the band.

Power Amplifier

An optional 8- to 10-W linear power

amplifier is shown schematically in Fig. 52. This amplifier uses another 2N5946 transistor. The design is relatively simple. Input matching is accomplished by C1, C2 and L1. L2, C3 and C4 match the output. The only differences between the 220-MHz and 144-MHz versions are the values of L1 and L2.

The bias circuit, suggested by Dave Mascaro, WA3JUF, uses an LM317 ad-

justable regulator to provide a stiff bias supply. The LM317 circuit is capable of providing 30- to 40-mA of stable bias current.

Switching

Fig. 53 is a schematic diagram of the transverter switching circuitry. K1 is used to switch power to the transverter modules in transmit and receive. When the trans-

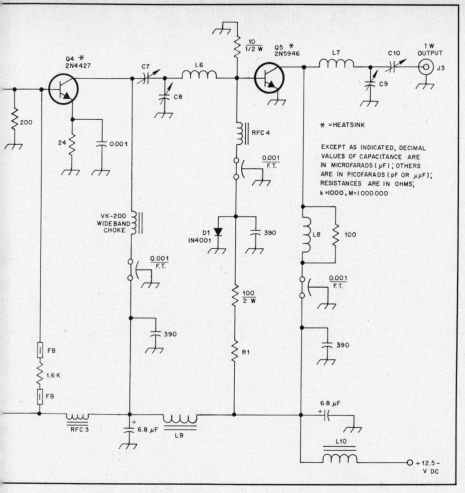

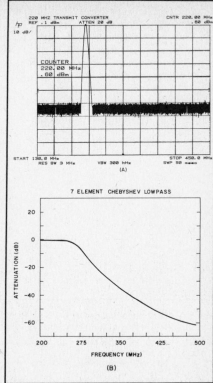

Fig. 51 — The plot at A shows the spectral output of the 1-W, 220-MHz transmit converter after filtering. All harmonics and spurious emissions are at least 60 dB below the fundamental output. This transmit converter meets current FCC spectral-purity specifications. The plot at B is the swept frequency response of the low-pass filter shown in Fig. 50.

144 MHz: 2t no. 20 tinned, 0.250-inch ID, spaced one wire dia.
L7 — 220 MHz: 1t no. 20 tinned, 0.250-inch ID. 144 MHz: 3t no. 20 tinned, 0.250-inch ID, spaced one wire dia.
L8 — 220 MHz: 6t no. 20 enam., 0.250-inch ID, close wound. 144 MHz: Same as 220 MHz except coil is 8t.
L9, L10 — 15 t no. 22 enam. on T44-10 toroid core (same for both bands).
R1 — Start with 68-ohm, 2-W resistor; vary for proper bias. See text.
RFC1-RFC4 — 220 MHz: 1.8-μH miniature molded RF choke; 144 MHz: 2.2-μH miniature molded RFC.
U1 — Mini-Circuits SRA-1 standard level, doubly balanced mixer. Available directly from the manufacturer. See text.

verter is powered on, 12.5-V dc is applied directly to the LO and through K1 to the VHF preamplifier and 28-MHz post amplifier. When J2 is closed, K1 removes power from the preamplifier and post amplifier and applies 12.5-V dc to the transmit converter and power amplifier. The LO, used for transmit and receive, is always on.

When J2 is closed, K2 is also energized. K2, an RF relay, switches the antenna between the VHF preamplifier input and the power amplifier output. Relays of this type are generally available at flea markets at modest prices. Since most HF transceivers have separate transmitter input and output connections, no relay switching for the IF rig is included. The transmit converter IF input is connected directly to the transverter output on the HF transceiver, while the post amplifier output is connected directly to the transverter input on the HF transceiver.

Relays with 26-V dc coils were used here

since most surplus coaxial relays require this voltage. While relays with any coil voltage could be used, it is a good idea to run relays and electronics from separate power supplies to avoid possible problems caused by voltage transients that occur when the relay coils are switched. The diode and capacitor connected across the relay coil power line help to alleviate transients.

CONSTRUCTION TECHNIQUES

Although this project is not intended for a first-time effort, anyone having a reasonable amount of VHF construction experience should encounter no difficulty. As with all VHF circuits, a certain amount of construction care is required.

Proper grounding techniques, RF bypassing and shielding will ensure stable operation. Feedthrough grounding is used to provide a low-inductance ground return to both sides of the PC board. Basically, this means drilling a hole through the PC

board at key points where components must have a good RF ground and installing a rivet or piece of no. 20 tinned wire soldered to both sides of the board. See Fig. 54. Check the schematic diagrams for each circuit and install ground feedthroughs accordingly.

High-quality, low-inductance capacitors ensure a good RF bypass. Ceramic chip capacitors work best. These can be expensive, however, so they are used only where absolutely necessary. Epoxy encapsulated, miniature, monolithic ceramic capacitors work quite well as bypass capacitors.

Tight shielding between the input and output of each stage of the transmit converter eliminates the likelihood of feedback. Shielding was only found necessary on the transmit converter.

The variable capacitors used in each stage are miniature ceramic or piston trimmers. The value of these capacitors is not critical as long as you use capacitors with maximum values close to those specified in the schematic diagram. For example, there are many capacitors available with a range from 2 to 8 pF or 2 to 12 pF. Any of these will work fine in circuits that call for 9- or 10-pF-maximum variables. Ceramic piston trimmers are most convenient for building the double-tuned band-pass filters. Johan-

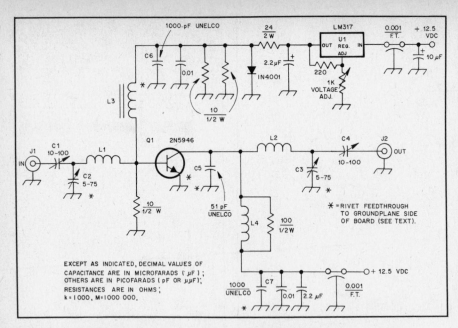

Fig. 52 — Schematic diagram of the 8- to 10-W power amplifier. Capacitors are monolithic ceramic types unless otherwise noted. Capacitors marked with polarity are electrolytic.

C1, C4 — 10-100 pF mica trimmer (Arco 406 or equiv.).
C2, C3 — 5-75 pF mica trimmer (Arco 405 or equiv.).
C5 — 51 pF metal-clad book-mica capacitor. See text.
C6, C7 — 1000 pF metal-clad book-mica capacitor. See text.
J1, J2 — Chassis-mount female BNC connector.
L1 — 220 MHz: ½t no. 20 tinned, 0.250-inch ID ("hairpin loop"); 144 MHz: 1t no. 20 tinned,

0.250-inch ID.
L2 — 220 MHz: 1t no. 20 tinned, 0.250-inch ID; 144 MHz: 2t no. 20 tinned, 0.250-inch ID, spaced one wire dia.
L3 — VK-200 wideband RF choke.
L4 — 8t no. 20 enam., 0.250-inch ID, close wound.
Q1 — 2N5946 or Thompson-CSF SD1160 RF power transistor.
U1 — Three-terminal adjustable voltage regulator, 1 A (LM317 or equiv.).

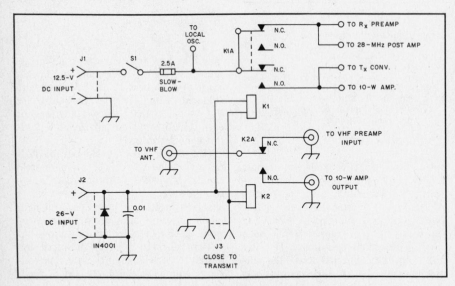

Fig. 53 — Schematic diagram of the transverter switching arrangement.

J1, J2 — Chassis-mount female Cinch-Jones power plug.
J3 — Chassis-mount phono jack.
K1 — DPDT power relay, 3-A contacts, 26-V dc

coil.
K2 — SPDT coaxial antenna changeover relay, BNC connectors, 26-V dc coil.
S1 — SPDT switch, 5-A contacts.

son, Trimtronics or Voltronics all manufacture 8- to 10-pF maximum piston trimmers suitable for these filters.

"Dead bug" layout is best suited for VHF/UHF construction. As shown in the accompanying photographs, components are supported by their leads above the

ground plane. In most cases, component leads are soldered directly to the leads of other components, keeping the length of each interconnection to a minimum. A low-inductance RF ground is achieved since the component leads are soldered directly to the ground plane. This construction

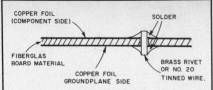

Fig. 54 — Detail of the feedthrough grounding used to achieve a good RF ground. See text for details.

method eliminates the need for designing an etched PC board, a task not easily accomplished at VHF and above.

Each circuit, or "module," is built on a piece of double-sided circuit board stock and mounted in a separate metal enclosure. In the transverter shown in the photographs, each module is built into a small Hammond 1590 series diecast box. In each case, the box cover has been discarded. The circuit board holding the components has been shaped to fit the box tightly in place of the cover. Each module has BNC connectors for RF input and output. Voltage is supplied through a feedthrough capacitor. The result is an RF-tight enclosure for each part of the circuit. Since the 1-W transmit converter board is large, and a suitable diecast box is expensive, the transmit converter is mounted under the chassis that supports the rest of the modules.

Finding Parts

These days, it is becoming increasingly difficult to find parts. This transverter makes use of parts available from a number of sources. The parts suppliers table in Chapter 35 lists many possible sources, and still others advertise in *QST* and other amateur magazines. Some parts sources are listed below. You can find addresses and other information in Chapter 35 for companies that are listed by name only here. This list is by no means complete; these are sources that the author knows of that will sell to individuals in small quantities.

- Advanced Receiver Research, Box 1242, Burlington, CT 06013 (MGF1402 GaAsFETs)
- Amidon Associates (toroid cores, ferrite beads, VK-200 chokes)
- Applied Invention (piston trimmers, chip caps)
- MHz Electronics, 3802 N. 27th Ave., Phoenix, AZ 85017 (transistors, book-mica capacitors)
- Microwave Components of Michigan (trimmers, piston trimmers, GaAsFETs, chip caps, feedthrough caps, diecast boxes)
- Mini-Circuits, P.O. Box 166, Brooklyn, NY 11235 (mixers, T1 in 192-MHz LO)
- Mouser Electronics (resistors, capacitors, ceramic trimmers, molded RF chokes, diecast boxes)
- Radiokit (toroid cores, ferrite beads, Arco variables, diecast boxes)

•R.F. Gain, Ltd. (transistors)

Local Oscillator

Build the local oscillator first; you'll need LO energy to test most of the other stages. Before soldering anything, gather all of the necessary components and lay them out on the board in order to plan adequate room before constructing. Follow the schematic diagram and refer to Fig. 55 for an idea of how to lay out the board. Although the 192-MHz LO for the 220-MHz transverter is shown, layout is similar for the 116-MHz version.

Once the general layout scheme has been achieved, the components may be soldered together, beginning with Q1. Keep component leads short, but leave enough room so that you can change components if necessary. Make sure that the leads of Y1 are as short as possible and that the crystal will not touch the box that the LO will be installed in. A dab of silicone sealant will hold the crystal in place.

Q1, Q2 and Q3 are mounted "belly up" with their leads sticking into the air. Note that Q1 and Q3 have a lead that is connected to the case. Solder this lead to ground. The collector of Q4 is tied to the case, so it is mounted with the leads facing the ground plane. Make sure that the case of Q4 does not touch the ground plane or other components.

Dc power for the LO enters the case through a feedthrough capacitor. Miniature molded RF chokes and bypass capacitors decouple the power line at each stage. In Fig. 55, these RFCs and capacitors are arranged in a horizontal line that traverses the top of the board. Q1 is powered through an 8-V, three-terminal regulator IC. Mount the associated bypass capacitors as close to the IC body as possible.

Q2 has several unused sections. Cut off all unused leads before soldering it into the circuit. T1 is a trifilar-wound transformer on a T25-12 toroid core. See Fig. 56 for details. Although the transformer used here was wound by hand, a Mini-Circuits model T4-1 may also be used.

The output filter uses two ceramic piston trimmer capacitors mounted through the circuit board. These piston trimmers make it convenient to mount L6 and L7; common ceramic trimmers like those used in the rest of the circuit will also work here. A shield made from a piece of double-sided circuit-board material separates L6 and L7, eliminating overcoupling and ensuring a "clean" filter response. Although a 0.5-pF ceramic chip capacitor was used to couple the two filter sections, a ¼-inch-long gimmick capacitor made from two pieces of no. 28 enameled wire twisted tightly together will work here.

One of the advantages of dead bug construction — one that you should take advantage of — is that each stage can be tested individually. After you build the basic oscillator (Q1 and associated components), you can test for 96 (or 116) MHz

Fig. 55 — The 192-MHz local oscillator is laid out in the order it is drawn on the schematic diagram. Q1 is at the far left; Q4 is near the center of the board. The band-pass filter is at the right, next to J1 and J2. The resistors used for the power divider and pad are soldered into the circuit with virtually no lead length.

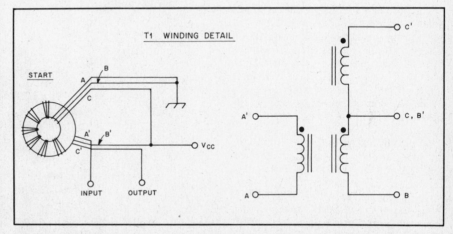

Fig. 56 — Winding details for T1 in the 192-MHz LO. Label one end of three 5-inch pieces of enameled wire as shown. Holding all three wires flat, in parallel, begin threading the toroid starting with the unlabeled end of the three wires. Carefully wind the toroid by feeding the entire group of wires in parallel. Once seven turns have been wound, use an ohmmeter to trace the unlabeled ends. Label the end of wire A by A', and so on. After the windings have been properly labeled, connect T1 as shown.

Fig. 57 — Closeup of the VHF RF preamplifier. The input components at the left of the board are mounted with as little lead length as possible, to minimize losses.

output by capacitively coupling into a receiver or spectrum analyzer from the 50-ohm pad at the output of the stage. When you are sure Q1 is working, build the next stage and test it. Proceed alternately building and testing each stage until you reach the final output stage.

When all of the LO stages are complete, check the operating frequency with a counter or spectrum analyzer. Tune each stage for maximum output. Tuning is somewhat interactive, so recheck each stage after you have done the initial tuneup. Each variable capacitor should have a definite peak. If you have a method of checking low power levels, check the output at the receive and transmit ports. Power output should be as indicated on the drawings.

All other stages are constructed in a like manner. Lay the board out in advance; start at the input and work toward the output. Test each stage after you build it, and fix any problems before continuing to the next stage.

Preamplifier

The GaAsFET transistor requires special handling care because it is especially sensitive to static electricity. Solder the transistor into the circuit last; use a grounded-tip, low-temperature soldering iron. If a static-free work station is unavailable, ground yourself before removing the MGF1402 from its protective package to prevent static buildup from destroying the device.

Although an MGF1402 is specified, you can use other devices if you change the biasing resistors accordingly. Consult the references listed at the end of this project writeup before attempting a substitution. The MGF1402 is a fairly common transistor and is available from several of the suppliers listed in Chapter 35.

The general layout is shown in Fig. 57. Although BNC connectors were used here, Type-N or SMA connectors may be employed. A number of ground feedthroughs are used at points indicated on the schematic diagram. These feedthroughs are necessary for stable operation and optimum performance; they must be used. See Fig. 54.

Ceramic chip capacitors are mandatory for the source bypass on the MGF1402. Do not attempt to substitute low-grade capacitors here! Chip capacitors provide a low-impedance source ground; this is of particular importance for stable operation with high-gain devices such as the microwave GaAsFET used here. The MGF1402 is mounted directly to the source bypass capacitors by its source leads. First, solder one end of each chip capacitor to the ground plane. Then solder one source lead to each chip capacitor. See the preamplifier projects that appear earlier in this chapter for complete details of this mounting scheme.

The output filter is similar to the one described in the local-oscillator section. In

Fig. 58 — Closeup of the 28-MHz post amplifier. Q1 uses a push-on finned heat sink. The transistor case is tied to the collector, so the heat sink must be positioned away from the case and nearby components.

Fig. 59 — Closeup of the receive mixer and diplexer filter. RG-174 cable is used between the mixer and input connectors.

this case, however, the coupling capacitor is a 0.3- to 3-pF trimmer. A toroidal coil is added for the 164-MHz image trap.

28-MHz Post Amplifier

The 2N5109 post amplifier shown in Fig. 58 requires little special care. Note, however, the use of a push-on, finned heat sink. Keep the heat sink away from the circuit board and other components since the 2N5109 case is tied to the transistor collector.

Receive Mixer

The receive mixer and diplexer filter may be housed in one enclosure, as shown in Fig. 59. The SRA-1H is best mounted on top of the circuit board with the pins protruding through to the component side. Carefully mark and drill eight holes using a no. 59 drill bit. All holes on top of the board are deburred with a 1/8-inch drill bit; turn the bit by hand to remove the copper from around the hole. This allows for ample clearance where leads are close to touching the board. Deburr the holes for pins 1, 3, 4 and 8 on the component side of the board. The other pins are grounded and should be soldered directly to the ground plane.

Transmit Converter

Construction of the transmit converter requires a little extra planning for component layout because of the number of stages involved. Position the SRA-1 and the five transistors to allow sufficient room for the remaining components. It is better to allow extra room than to be cramped for space. Refer to Fig. 60 for ideas on board layout.

Fig. 60 — The transmit converter incorporates extensive bypassing and shielding. J1 and J2 are at the bottom left. Q1, Q2 and Q3 are arranged along the lower edge of the board, below the horizontal shield. Q4 and Q5 are above that shield; Q4 is to the right. The two emitter leads of Q5 are soldered directly to the ground plane near the center of the board. The low-pass filter is between Q5 and the output connector.

Drill any mounting holes as you go along. Start by mounting the SRA-1 using the same technique described above. Then mount J1 and J2. Build the mixer bandpass filter first, and then work stage-by-stage toward the output.

Q1, Q2 and Q3 should be shielded by placing a small piece of sheet metal or double-sided circuit-board material over each transistor. Cut a U-shaped notch in the shield to clear the transistor case. Each shield may be soldered directly to the ground plane. Q1 and Q2 are mounted "belly up," while Q3 and Q4 are mounted "right side up." Cut the Q3 and Q4 emitter lead to ¼ inch before soldering the transistor in place. Note that Q1 and Q2 have four leads; one is tied directly to the transistor case and must be soldered to the ground plane. The case of Q3, however, is connected to the collector, so be sure to leave adequate clearance between the case and the shield. Q4 must have a push-on, finned heat sink. The heat sink should not touch any other components. Again, keep all leads as short as possible.

Q5 is a stud-mounted power transistor. Drill a hole just large enough to pass the threaded stud and transistor base through the circuit board. The emitter leads should lay almost flat on the component side of the board. Cut the collector and base leads to half of the original length, while leaving the emitter leads full length. The heat sink for Q5 is made from a U-shaped piece of brass sheet (the same material used for the circuit-board shields). Mount D1 on top of the heat sink for good thermal contact;

Fig. 61 — Space is a bit tight, but the 10-W amplifier will fit inside a small diecast box. Q1 is mounted at the center of the board. Input circuitry is to the left, output to the right. U1 and associated bias circuitry are behind the shield adjacent to J1.

solder the ground end of D1 directly to the heat sink.

The value of R1 must be determined experimentally. Start with a 68-ohm, 2-W resistor. Measure the quiescent current of Q5 by inserting a milliammeter in the circuit at the cold end of L8. The quiescent current should be between 20 and 40 mA; adjust the value of R1 until proper bias is achieved.

The transmit converter is tuned by applying 12.5-V dc, LO energy and a 28-MHz signal attenuated to provide −10 dBm at the mixer input. Peak the double-tuned filter for maximum 220-MHz (or 144-MHz) output. If a spectrum analyzer is available, tune for a balance between maximum 220-MHz (or 144-MHz) energy and minimum spurious output. Next, peak C1 through C3. Alternately peak C5 and C6 for maximum output. C7, C8, C9 and C10 are adjusted in the same manner. Although a wattmeter may be used for tuning purposes, a spectrum analyzer tells the full story. Tune each stage for a compromise between output power and spectral purity.

Harmonics are 60-dB down after filtering. After you're through adjusting the transmit converter, power output should be 1 watt.

Power Amplifier

The 10-W amplifier shown in Fig. 61 is mounted in its own diecast box. Ground feedthroughs are used beneath the 2N5946 emitter leads and at all variable capacitor grounds. The LM317 regulator IC must be attached to a heat sink. Bias should be adjusted for a quiescent current of 40 to 60 mA.

C5, C6 and C7 are Unelco metal-clad book-mica capacitors. These capacitors provide an excellent low-impedance RF ground and are designed to work at high-current points. For stable operation, it is important that you use book-mica capacitors at these points.

The heat sink is fashioned from two U-shaped pieces of aluminum sheet. Be careful when mounting the heat sink; lateral pressure on the 2N5946 stud may break the transistor.

Summary

The transverter modules are arranged on a chassis as shown in Figs. 40 and 62. Short runs of 50-ohm coaxial cable interconnect the units. Most of the dc power wiring is done underneath the chassis.

The 144/220-MHz transverter represents a low-cost, modern approach to getting on the VHF bands. Circuit construction is straightforward, and the design makes use

Fig. 62 — The transmit converter is mounted in a cutout portion of the chassis, component side down. Dc for the transmit and receive chains is brought in to K1 and from there to a barrier strip. Wires are routed from the barrier strip to each module.

of commercially available parts. The modular construction approach offers flexibility for easy troubleshooting and experimentation.

The author wishes to thank Ron Whitsel, WA3AXV, and other members of the Mount Airy VHF Radio Club "Packrats"

whose encouragement made this project possible.

References
Hardy, *High Frequency Circuit Design*. Reston, VA: Reston Publishing Co., Inc., 1979.
Hayward, *Introduction to Radio Frequency Design*. Englewood Cliffs, NJ: Prentice Hall, Inc., 1982.

144- or 220-MHz Band-Pass Filter

Spectral purity is necessary during transmitting. Tight filtering in a receiving system ensures the rejection of out-of-band signals. Those unwanted signals may otherwise lead to receiver overload and increased intermodulation-distortion (IMD) products that manifest themselves as annoying in-band "birdies."

The double-tuned band-pass filter is not new. Filter designs involving this circuit abound in amateur designs because they are simple, are easy to construct, offer a relatively narrow passband and are quite forgiving of resonator variations. This filter was designed by Paul Drexler, WB3JYO.

The filter shown in Fig. 77 includes a resonant trap coupled between the resonators to provide increased rejection of undesired frequencies. Many popular VHF conversion schemes use a 28-MHz intermediate frequency (IF).

Proper filtering of the image frequency is often overlooked in amateur designs. The low-side injection frequency used in 144-MHz mixing schemes is 116 MHz and the image frequency, 88 MHz, falls in the

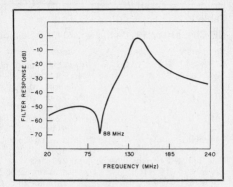

Fig. 78 — Frequency-response of the 144-MHz band-pass filter. Image-reject notch is for a 28-MHz IF.

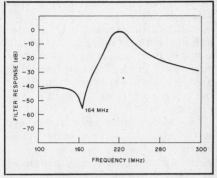

Fig. 79 — Frequency-response of the 220-MHz band-pass filter. Image-reject notch is for a 28-MHz IF.

Channel 6 television band. Inadequate rejection of a broadcast carrier at this frequency results in a strong, wideband signal present in the low end of the two-meter band. A similar problem exists on the transmit side that can cause TVI. The band-pass filters described have been effectively used to suppress undesired mixing products. See Figs. 78 and 79.

Circuit construction is easily accomplished using a double-sided copper-clad circuit board. Observing good construction techniques, component lead lengths must be kept to a minimum to eliminate resistive losses and unwanted stray coupling. The piston trimmers are mounted through the board with the coils soldered to the opposite end, parallel to the board. Placing a shield between L1 and L3 decreases mutual coupling and improves the frequency response. Peak C1 and C3 for optimum response. Component values are given in Table 2.

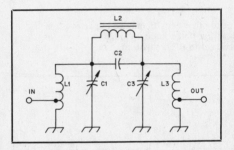

Fig. 77 — Schematic diagram of the double-tuned band-pass filter. Component values are given in Table 2.

Table 2

Component Values

	144 MHz	220 MHz
C2	1 pF	1 pF
C1, C3	1-7 pF piston	1-7 pF piston
L2	27t no. 26 enam. on T37-10	15t no. 24 enam. on T44-10
L1, L3	7t no. 18, ¼-in ID, tap 1½t	4t no. 18, ¼-in ID, tap 1½t

L1, C1, L3 and C3 form the tank circuits that resonate at the desired frequency. C2 and L2 reject the undesired energy while allowing the desired signal to pass. The tap points on L1 and L3 are for 50-ohm matching and may be adjusted for optimal energy transfer. Several filters have been constructed using a miniature variable capacitor in place of C2 so that the notch frequency could be varied.

RF Chokes for the VHF Bands

Many of the construction projects in this chapter call for fabricating one or more RF chokes. Specific instructions are given in each case, but additional guidance is often sought by those embarking on projects of their own design. The information presented here is a brief tutorial in theory and practice for those interested in constructing VHF equipment.

Distributed capacitance limits the range over which an RF choke will work. A coil wound with some spacing between the adjacent turns exhibits less distributed capacitance than one in which turns are close. This makes the space-wound choke superior to the close-wound one. A minimum of cement on the windings is also desirable. The space-wound 50-MHz choke in Table 3 and shown in the upper right of Fig. 80 is as good as you can make for that band, and better than most chokes you could buy. It is good at 144 MHz as well, and even serviceable at 220 MHz. A close-wound choke of fine wire, heavily doped with lacquer, might be usable on only one VHF band, and very likely it would not be too good even there.

To construct RF chokes you will need some wire: no. 22 enamel (Nycald or Formvar preferred); no. 28 enamel, silk or cotton covered; and no. 30 or 32, of any similar insulation. Silk- or cotton-covered wires take cement nicely, but enamel is okay otherwise, and it is usually most readily available.

High-value ½- or 1-watt carbon resistors make good winding forms for use at 144 MHz and higher. A 2-watt resistor is big enough for a 50-MHz choke, but Teflon or Nylon rod stock is better. Do not use polystyrene or Lucite, if any heat is to be involved. These materials will melt in the heat of an average transmitter enclosure.

Table 3

RF Chokes for 50-, 144- and 220-MHz Service

Frequency	Inductance	Description
50 MHz	7.8 to 9.5 µH	B&W miniductor no. 3004, 1-3/8 to 1-9/16 inch long.*
50 MHz	8.3 µH	No. 28 double silk covered (dsc), space-wound on ½-inch Teflon rod. Winding 1¾ inch long. See text.
50 MHz	7.2 µH	No. 28 dsc, close-wound on ¼-inch Teflon rod. Winding 1-7/16 inch long.
144 MHz	2.15 µH	No. 22 Nyclad, close-wound 1-3/16 inch on ¼-inch Teflon rod.
144 MHz	1.42 µH	31 turns no. 28 dsc, space-wound on ¼-inch dia self-supporting.
(The 144-MHz chokes work well on 220 MHz.)		
220 MHz	0.6 µH	13 turns no. 22 Nyclad on ¼-inch Teflon rod.
220 MHz	0.75 µH	17 turns no. 28 dsc space-wound on ¼-inch Teflon rod. Winding 5/8 inch long.
220 MHz	0.52 µH	22 turns no. 22 Nyclad close-wound on no. 24 drill, self-supporting.

*Excellent for use except where high temperatures are involved.

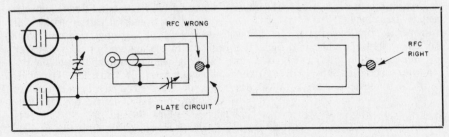

Fig. 81 — How a choke is positioned with respect to other circuits may be important. The choke at the top is coupled to the plate line of the transmitter tuned circuit. Outside the loop, as shown at the bottom, makes the choke far less subject to RF breakdown.

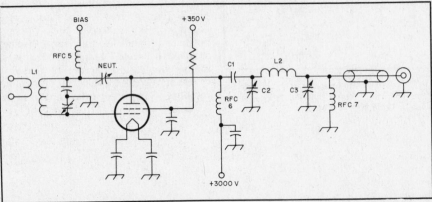

Fig. 82 — Transmitter applications for RF chokes vary markedly in regard to the quality of choke needed. In the grid circuit, RFC5 has no difficult job to do, and any choke suitable for low-power use is suitable. The shunt-feed choke, RFC6, must meet severe requirements, especially in high-powered amplifiers. It is effectively connected across the transmitter tank circuit, and is subjected to high temperature, current and voltage. The output choke, RFC7, is mainly a safety device, and it operates under much less stringent circumstances.

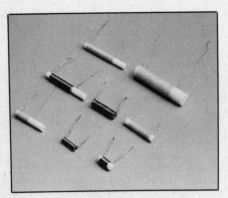

Fig. 80 — Typical handmade VHF chokes. At the rear are close-wound and space-wound chokes for 50 MHz wound on ¼-inch and ½-inch Teflon rod, drilled and tapped for end-mounting. Three 144-MHz chokes are shown in the center; the two at the left being excellent for high-current applications. In front are similar types to these, but for the 220-MHz band.

Teflon rod can be found in plastics supply houses, in ¼- and ½-inch diameter. It drills and taps nicely, it won't melt, and its insulating quality is excellent. Bakelite rod or even wood dowelling is good enough for the less-critical choke applications. The smallest-diameter prepared coil stock is usable for 50-MHz chokes, but it won't stand much heat.

Space-winding RF chokes is easy. First drill through the rod at spacings indicated under winding length in the table. Now measure off slightly more than a half wavelength of wire for the design frequency band. Double it back on itself and feed the end through one of the holes in the rod. Now wind the coil as if it were to be bifilar. If you clamp the other end of the double wire in a vise, or tie it down firmly otherwise, this can be done easily. Keep the wires under tension, and be sure they are not twisted at any point. Wind tightly and then feed the end through the other rod hole.

Now remove one of the wires by unwinding carefully, keeping it under tension throughout. The remaining wire will be space-wound as neatly as if done by machine. Apply a thin coating of polystyrene cement, using a bit more around the lead holes, and your choke is done.

Self-supporting chokes of excellent quality can be made by winding no. 22 or 24 wire tightly on various drill sizes, and then slipping the drill or other winding form out. If wound under tension the coil will hold its shape when slipped off the form. Turns can be spaced by running a thin knife blade between them.

You can tell a good choke from an inferior one easily enough. Connect it across your driver-stage tuned circuit, and see what it does to your final-stage grid current. Also note how much you have to retune the driver circuit to restore resonance. A perfect choke would have no harmful effects, and it would not heat up. You won't find one that good, but a well-designed choke will come close. If the choke is not a good one, don't run the test too long at any appreciable power level, or you won't have to look for indications — you'll smell them!

6-Meter Amplifier Using the 3CX800A7

The amplifier described here uses the new EIMAC 3CX800A7 in a grounded-grid circuit. With 2200 V on the plate and 25-W drive, it will provide about 700-W output — enough for weak-signal work such as meteor scatter. The 3CX800A7 is well suited for this application. It combines low drive requirements normally associated with tetrodes, yet it offers the stability of grounded-grid triodes. Figs. 83 through 88 depict the amplifier. Dick Stevens, W1QWJ, designed and built this project.

Circuit Details

Fig. 84 is a schematic diagram of the amplifier. The RF deck includes everything except the high-voltage power supply. Several suitable supplies are described elsewhere in this chapter and in Chapter 27. This amplifier requires 2000- to 2200-V at about half an ampere. The plate tank circuit is a pi-L network consisting of C11, C12, L3 and L4. This configuration gives somewhat better harmonic suppression than a standard pi network. The cathode input circuit is a T network consisting of L1, L2 and C8. A filament choke is recommended above 50 MHz to keep the filament above ground for RF. RFC2 is bifilar wound on a ferrite rod. According to sources at EIMAC, use of a center-tapped transformer at T1 is not necessary. If you have a transformer of the correct voltage, but without a center tap, go ahead and use it.

C13 and C14 bypass the B+ line. RFC3 is a hand-wound choke on a Teflon form, while RFC4 is an Ohmite Z-144. RFC5, an

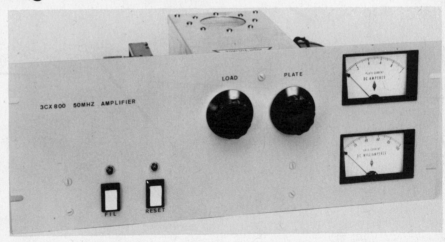

Fig. 83 — The 3CX800A7 6-meter amplifier is mounted on a rack panel. Meters are mounted behind the panel with bezels.

Ohmite Z-50, is included for safety to prevent high voltage from appearing at the antenna connector if dc blocking capacitor C15 fails.

R1, in the B+ line, will protect the tube and power supply in case of a high-voltage arc. In this case, R1 was mounted in the power supply. It doesn't matter if you mount it in the power supply or the RF deck, but the use of this resistor is highly recommended.

Cathode bias is provided by a 5.1-V Zener diode, D5. R3 prevents the cathode voltage from soaring if D5 fails. F2 will blow if excessive cathode current is drawn. R4 nearly cuts the tube off on receive.

All metering is done in the negative return lead for safety. C9, C10, D3 and D4 protect the meters. The antenna relay is located externally but is powered from the amplifier.

A grid overcurrent trip circuit is included to protect the 3CX800A7 from damage. The cost of the tube makes protection circuitry a must! Grid current is sensed across R5. If the grid current exceeds a preset value, Q1 conducts and actuates K1. Contact K1A, in the K2 coil supply line, opens and removes power from K2, placing the amplifier in STANDBY. To restore normal circuit operation, turn S2 to the off position; this will reset K1. S2 is "on" for normal operation. This switch provides an additional function: It controls the VOX

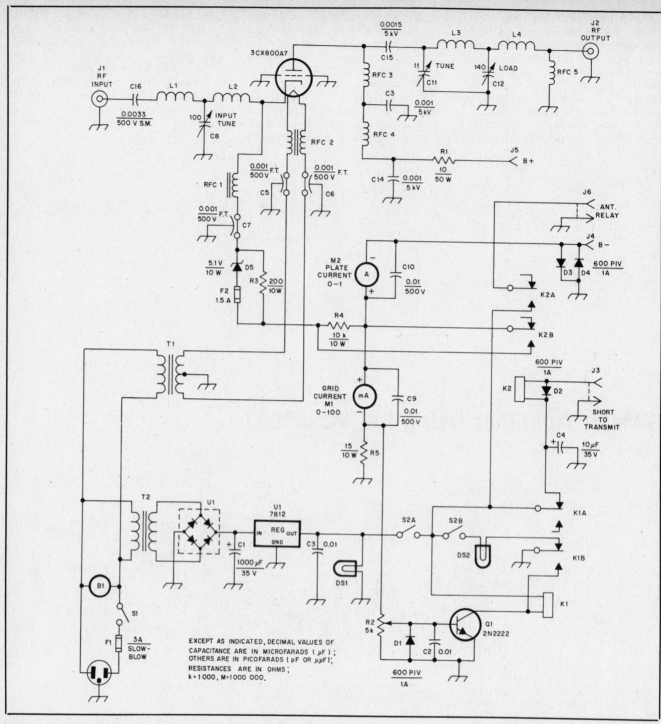

Fig. 84 — Schematic diagram of the 6-meter amplifier using the 3CX800A7. Resistors are ¼-W carbon types unless noted. Capacitors are disc ceramic types unless noted. SM means silver-mica. Capacitors marked with polarity are electrolytic.

B1 — Dayton 4C012A 54-CFM blower.
C8 — 100-pF air variable (Millen 22100 or equiv.).
C11 — Modified air variable, 4 to 11 pF. See text.
C12 — 8- to 140-pF air variable, receive spacing (Cardwell 149-6 or equiv.).
C13, C14 — 1000-pF, 5-kV disc-ceramic capacitor.
C15 — 1500-pF, 5-kV RF-type doorknob capacitor.
C16 — 3300-pF, 500-V silver-mica capacitor.
D1-D4 — 600-PIV, 1-A silicon diode.
D5 — 5.1-V, 10-W Zener diode (1N3996 or equiv.).
DS1, DS2 — 12-V dc pilot light.
J1 — Chassis-mount female BNC connector.
J2 — Chassis-mount female Type-N connector.

J3, J6 — Chassis-mount phono jack.
J4, J5 — Millen 37001 high-voltage connector.
K1 — DPDT miniature relay, PC-board mount, 12-V dc coil (RS 275-213).
K2 — DPDT relay, 12-V dc coil, 10-A contacts.
L1, L2 — 10t no. 18 wire, 5/8-inch ID, 1 inch long.
L3 — 5t 1/8-inch copper tubing, 1-1/8 inches ID, 2½ inches long.
L4 — 5t no. 12 wire, 1/2-inch ID, 1-1/8 inches long.
M1 — 0-100 mA dc meter (Simpson Model 1327, avail. from Larson Instrument Co., Greenbush Rd., Orangeburg, NY 10962).
M2 — 0-1 A dc meter (Simpson Model 1327, avail. from Larson Instrument Co.).
Q1 — 2N2222 or equiv.
RFC1 — 10-μH RF choke. 40t no. 22 enam. on

a ¼-inch ferrite form or a Radio Shack no. 273-101.
RFC2 — 14t no. 16 enam. on a ½-inch-OD ferrite rod. See text.
RFC3 — 40t no. 22 enam. closewound on a 2-inch-long by ¾-inch OD Teflon rod. See text.
RFC4 — Ohmite Z-144 RF choke.
RFC5 — Ohmite Z-50 RF choke.
T1 — Filament transformer. 117-V ac primary; 13.5-V, 1.5-A center-tapped secondary (Avatar AV-440 or equiv. Available from Avatar Magnetics, 1147 N. Emerson St., Indianapolis, IN 46219).
T2 — Power transformer. 117-V ac primary; 12.6-V, 1.2-A secondary.
U1 — 50-PIV, 4-A bridge rectifier.
U2 — 12-V, 1.5-A three-terminal regulator (LM7812 or equiv.).

circuit and antenna bypass relay circuit, so the amplifier can be switched in and out of the line between antenna and transceiver.

Construction

The amplifier is built on a 10 × 12 × 3-inch chassis (LMB 10123C). For the best possible RF shielding, the plate circuitry is built inside its own 5 × 10 × 3-inch chassis (LMB 5103C) that is mounted on top of the main chassis. The chassis is attached to a standard 19 × 7 inch aluminum rack panel (Hammond PBPA 19 007). The meters are mounted to the front panel with bezels.

Most of the parts used in this project were purchased from Radio Shack, Semiconductor Surplus, 2822 N. 32nd St., Phoenix, AZ 85008 and RadioKit, P.O. Box 411, Greenville, NH 03048.

Considerable care must be given to the tank circuit construction to keep stray capacitance to a minimum. To this end, the plate tuning capacitor (C11) is rebuilt from a receiving type. All available commercial capacitors had too high a minimum capacitance.

The first step in construction is to fabricate the plate tuning capacitor. The capacitor is made from a stock Cardwell ER25AS split-stator capacitor having a maximum capacitance of 25 pF per section, with four stator and five rotor plates. The capacitor is 1-5/8 inches long between the end bearing plates and is made up of individual stator and rotor plates held together with spacers, washers and nuts.

Carefully disassemble the capacitor. Be careful not to lose any of the small parts. Two new no. 4-40 by 2-1/8-inch-long tie rod bolts are needed. Reassemble the capacitor as a single section with four stator and three rotor plates with three spacers between plates. The completed capacitor will have a spacing adequate for 3.5-kV breakdown. The minimum capacitance is 4 pF and the maximum is 11 pF. Any capacitor of this type construction in the same general capacitance range should work. Alternatively, a small vacuum variable will work.

The tuning and loading capacitors, as well as L3 and C15, are mounted on an aluminum subassembly, as shown in Fig. 85. C12, the plate loading capacitor, is a standard value and required no modification. L3 is made of 1/8-inch copper tubing and is suspended between C11 and C12. The bottom of L3 should be about ½ inch above the bottom of the subassembly. Additional details of the plate compartment may be seen in Fig. 86.

The subassembly is made from a piece of aluminum sheet that measures 4¾ × 5¾ inches. Fold up 2 inches on the longer side at a right angle to form an L bracket, as shown in Fig. 85. The capacitor shafts are mounted 1 inch in from each end, in the center of the 2-inch lip.

RFC3, the plate choke, is wound on a 2-inch length of ¾-inch-OD Teflon rod. Drill each end to a depth of ½ inch and

Fig. 85 — The output network mounts on a separate subassembly. C15 connects to the rest of the circuit with wide copper straps.

Fig. 86 — Top view of the 3CX800A7 6-meter amplifier with the plate-compartment cover removed. The blower pressurizes the plate compartment and air exhausts through a chimney made from Teflon sheet (not shown). A Dow-Key 260B is used here as a TR relay. T1 and K2 are mounted on top of the chassis.

tap for no. 10-32 hardware. Screw a 1-inch length of no. 10-32 threaded rod (or a bolt with the head cut off) into each end. Starting at ½ inch from one end of the rod, wind 40 turns of no. 22 enameled wire. Leave extra wire at each end for making connections. The winding should be about 1 inch long and is centered on the form.

Secure both ends of the winding with epoxy cement. Slip a no. 10 solder lug over one of the pieces of threaded rod and solder one end of the winding to it. This will be the end that attaches to C15.

The plate compartment is pressurized, and a chimney made from Teflon sheet that fits around the tube anode forces the air

Fig. 87 — The underside of the chassis is divided in half by a partition to shield the input circuit.

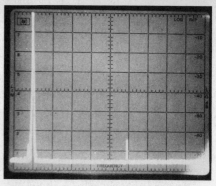

Fig. 88 — Spectral output of the 3CX800A7 6-meter amplifier. Power output is 700 W on 50 MHz. Each vertical division is 10 dB and each horizontal division is 10 MHz. All harmonics and spurious emissions are greater than 64 dB down. This amplifier complies with current FCC spectral purity requirements.

Table 4

Typical Operating Conditions for the 3CX800A7 6-Meter Amplifier

Plate supply voltage	2200 V
Zero-signal plate current	65 mA
Single-tone plate current	500 mA
Single-tone grid current	40 mA
Drive power	25 W
Output power	700 W
Efficiency	62%

up through the tube to an exhaust hole cut in the top of the plate compartment. The hole is shielded with aluminum screen. Details of this type of chimney and cooling system may be found in the 2-meter 8930 amplifier project presented later in this chapter.

Fig. 87 shows the underside of the amplifier chassis. A partition down the center of the chassis shields the input circuitry. The tube socket, input network and filament choke mount on one side of the partition. The dc components and control circuitry mount on the other. Operating voltages reach the tube through feedthrough capacitors in the partition.

To insure stability, the grid grounding straps should go from pin 4 to pin 7, and from pin 4 to pin 11. Ground pin 4 to the chassis at a solder lug at one of the tube-socket mounting screws. These pin numbers are for the EIMAC SK-1900 socket. Alternatively, EIMAC can supply a collet (part no. 720359) to ground the grid. The cathode straps should go between pins 1-8, 2-9 and 3-10. Solder these straps together where they cross at the center of the socket and make the cathode RF connection to this point.

Filament choke RFC2 is wound on a ferrite core. Cut two 36-inch lengths of no. 16 enameled wire and bifilar wind them on a ½-inch wooden dowel. Dress the ends to make the winding 2 inches long. Cut a 3-inch length of ferrite rod and file the ends so there are no sharp edges. Carefully slide the bifilar winding off the dowel and onto the ferrite rod. After the winding is centered on the rod, give it a coat of epoxy cement to hold it in place. See Chapter 24 for more details on this type of RF choke.

The grid protection circuitry is built on a small piece of Radio Shack part no. 276-159 prototyping board. One half the board is used. Most of the other components mount to the chassis or to terminal strips.

Initial Tune-Up and Adjustment

Connect an SWR meter between a 6-meter transmitter capable of supplying 20 W and the amplifier input. Connect a wattmeter and a good 50-ohm dummy load capable of handling at least 700 W to the output connector. Close S1 and allow the 3CX800A7 to warm up for at least 3 minutes before applying plate voltage. Caution: never apply filament voltage without the blower running. Never apply drive without plate voltage and a load connected to the tube, or the tube grid-dissipation rating may be exceeded.

Apply low plate voltage, about 1 kV, and apply a watt or two of drive. Adjust C8 for minimum input SWR. It might be necessary to spread or squeeze turns on L2 to get a 1.1:1 SWR.

Adjust C11 and C12 for maximum output. There is not much leeway with L3. It has to be the correct value to resonate with the output capacitance of the 3CX800A7 and the low capacitance of C11. Chances are that you may have to shut the amplifier off, disconnect the high voltage and spread the turns on L3 slightly to get the output circuit to resonate properly. The circuit should resonate with C11 set at ½ mesh.

When you get the output circuit to resonate at low power, bolt down all the covers and apply full plate voltage. Zero-signal plate current should be 65 to 70 ma. Apply a couple of watts drive and tune the amplifier for maximum output. Slowly increase the drive, keeping the output tuned for maximum. With 10-W drive from a typical multi-mode transceiver, output should be 400 to 500 W. With 25-W drive, output should be approximately 700 W. Table 4 lists typical operating conditions. Check the input SWR at full power and readjust as necessary.

Most likely the grid overcurrent trip circuit will operate at some point during the initial tune-up, placing the amplifier in STANDBY. Decrease the sensitivity by adjusting R2, mounted on the grid-trip circuit board. A hole is cut in the chassis bottom plate for access to this control. With the values shown, the circuit should trip at 20 mA with R2 set to a minimum. Adjust R2 slightly until the circuit holds at 40 to 50 mA of grid current. The tube's maximum grid-current rating is 60 mA.

A spectral photograph of the amplifier output is shown in Fig. 88. It meets current FCC spectral-purity requirements.

A 2-kW PEP Amplifier for 50 to 54 MHz

A number of manufacturers sell low-power-output (10-watt) transceivers and transverters for 6-meter operation. This particular power-output level is not in line with that required to drive the popular grounded-grid amplifiers. Even the high-mu triodes, such as those of the 8874 family, require at least 25 watts of drive for 1 kilowatt of input. Over 50 watts of drive may be required for 2-kilowatt service.

This leaves two options for owners of 10-watt-output rigs. One solution is to use a solid-state "brick" amplifier to drive the grounded-grid amplifier. This results in complex relay-switching systems not to mention the cost of the amplifier and a high-current supply to power it. The other approach is to use a tetrode such as the 4CX1000A in a grounded-cathode amplifier (see Figs. 89 through 97). In this circuit, 10 watts of drive will provide 2 kilowatts of input power, thereby eliminating an intermediate amplifier stage.

Circuit Description

The schematic diagram of the amplifier is shown in Fig. 90. As the 4CX1000A is a high-mu tube, care must be taken in the design and construction of the amplifier to prevent instability. In the circuit shown, this was accomplished without the need for neutralization — one of the drawbacks commonly associated with the use of these tubes.

The input network is series tuned. A link couples power from the exciter to the tube. Series tuning was chosen primarily because of the large input capacitance of the tube (roughly 100 pF). The tube socket and strays add an additional 20 pF. This tuning method places the tube input capacitance in series with the tuning capacitor, effectively reducing the tuning capacitance. A number of other networks were tried, including one somewhat exotic current-feed method, but none proved as simple and effective as the one shown. Bias voltage is shunt fed to the grid through R1. This resistor also heavily swamps the grid circuit and helps to provide amplifier stability without the need for neutralization. Should it be desired to increase the gain of the amplifier for use with lower power rigs, the value of R1 can be increased and the network components changed as appropriate. The cathode is grounded through short, heavy leads at the base of the SK-800B tube socket. A screen bypass capacitor is an integral part of this socket. The capacitor is 1500 pF, is rated for 400 volts and is of the mylar-film variety. Capacitors are connected between ground and each of the three filament-connection points.

The plate circuit used in this amplifier is of the common pi-network variety, with two exceptions. First, the internal tube plate capacitance, which is on the order of

10 pF, comprises what would be considered the "tuning" capacitor of the network. Since the tube capacitance is fixed, the inductor is used as an adjustable element. A compressible/expandable coil is used here. This arrangement allows the amplifier to cover the entire 6-meter band, whereas other designs using shorted-turn and slug-tuned techniques will cover little more than 1 MHz of tuning range. Additionally, the compressing and expanding of the coil does not have an adverse effect on the inductor Q, as typically occurs with the other systems. The tube capacitance dictates the circuit Q, which is roughly 14.

A single, multisecondary transformer provides power for the filament, bias and screen supplies. Although the filament winding is rated for 6.3 V at 8.8 A, the transformer delivers 6.0 V at 9.0 A — precisely that required by the 4CX1000A. The transformer was run for a period of 10 hours under these conditions and no appreciable heating of the wires or core was noted. Transformers of other manufacture or different voltage/current ratings should be carefully checked.

The high-voltage secondary of the transformer provides energy for the bias and screen supplies. Zener diodes limit the dissipation of the bias potentiometer. R6 cuts the amplifier off during receive periods. J5 is shorted for transmit.

The transformer secondary also feeds the screen supply. Here, a full-wave rectifier provides the dc energy. Q1, D3, R1 and R2 form a regulator that limits the current that can be drawn through the transistor to 40 mA. This extra circuitry is desirable in the event of high-voltage supply failure. No more than 40 mA of current can be drawn through the regulator, ensuring that the screen is protected. Without such a safety device, a failure of the high-voltage supply would mean that the screen dissipation might be pushed well above the specified 12-watt maximum rating. D4-D9, inclusive, make up the screen-voltage regulator.

M1, located in the −HV lead, continuously monitors cathode current. Meter shunts are provided in the bias and screen supplies. M2 can be switched between grid and screen by means of S2. This meter reads 0 to 10 mA for grid and 0 to 100 mA for screen current.

The amplifier requires a high-voltage supply. The design and construction of such supplies is covered in Chapter 27.

Construction

Many of the construction details can be seen in Fig. 91. The general layout is not especially critical, although good isolation between the grid and plate circuitry is a must. To this end, the Eimac SK-800B socket is used. The built-in screen-bypass capacitor is an important factor in amplifier stability. Other sockets without this

Fig. 89 — Front-panel view of the 2-kW, 6-meter amplifier.

capacitor should be avoided.

The amplifier is built on a 10 × 12 × 3-inch aluminum chassis that is attached to an 8¾ inch-high, 19-inch-wide rack panel. Sheet-aluminum panels are made so that the overall height of the amplifier enclosure is 7½ inches.

The grid compartment is constructed from a second aluminum chassis that measures 7 × 7 × 2 inches. It is easiest to cut the hole for the tube socket with the smaller chassis bolted in place. In this manner exact alignment can be guaranteed. A heavy-duty hand nibbler tool was used to make the large hole. All leads entering the grid compartment pass through 1000-pF feedthrough capacitors to ensure an RF-tight enclosure. The amplifier input connection is through a UG-625B (female) chassis-mount BNC connector, directly into the grid compartment. The general placement of the input-network components can be seen in the photograph. A perforated aluminum cover is placed over the smaller grid-compartment chassis. This piece must allow adequate air flow for proper tube cooling. It is a good idea to check air flow with and without this piece in place in order to determine whether the material is suitable.

The remainder of the bottom chassis houses the screen and bias power-supply components and the control circuitry. Many of the parts are contained on a single-sided circuit board. The parts-layout diagram is shown in Fig. 96. Zener diodes are mounted on an aluminum plate and the circuit board. While the aluminum provides some heat-sinking, cooling for the diodes is supplied primarily by the system blower, as the diode stack is mounted in the air flow path. Cooling for the tube is by means of a blower that is mounted external to the chassis. A plastic flange (Newtone 366, used with central vacuum cleaning systems) is suitable for use with 2-inch automobile heater/defroster hose. A length of this hose connects the pressurized chassis to the blower, which can be mounted at some convenient location. For ultra-quiet operation the blower can be mounted in a closet or in an adjoining room. Copper or brass screening is used to cover the flange opening, thus maintaining an RF-tight chassis. Information on how to select a

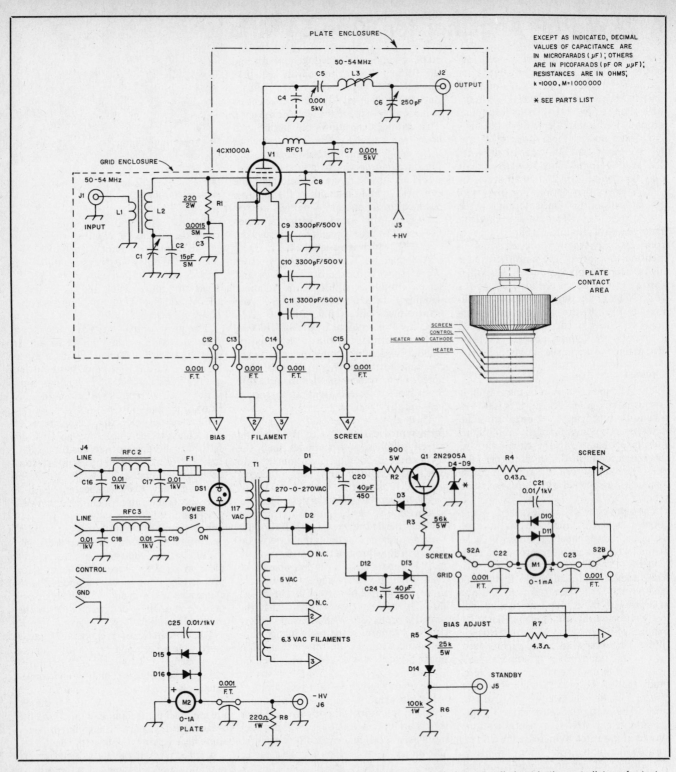

Fig. 90 — Schematic diagram of the 6-meter amplifier. Parts designations shown on the schematic but not called out in the parts list are for text reference only.

C1 — Miniature variable, panel mount, 32 pF maximum.
C2 — Silver mica, 15 pF, mounted directly across C1.
C4 — Tube internal plate capacitance.
C5, C7 — Transmitting capacitor, 0.001 µF, 5000 volt. Centralab 858S-1000 or equiv.
C6 — Transmitting variable, 250 pF maximum.
C8 — Screen bypass built into SK-800B socket.
C20, C24 — Electrolytic, 40 µF, 450 volt.
D1, D2, D10-D12, incl., D15, D16 — Silicon, 2.5 ampere, 1000 volt.
D3 — Zener diode, 45 volt, 1 watt.
D4-D9, incl. — Zener diode, stud mount, 56 volt, 10 watt. Connect in series.

D13 — Zener diode, 200 volt, 5 watt.
D14 — Zener diode, 30 volt, 1 watt.
DS1 — Neon indicator built into S1.
F1 — Fuse, 1 ampere.
J1 — Coaxial connector, BNC chassis mount, UG-625B or equiv.
J2 — Coaxial connector, type N chassis mount, UG-58 or equiv.
J3 — High-voltage connector, Millen 37001 or equiv.
J4 — 4 conductor.
J5, J6 — Phono connector.
L1 — 2 turns no. 24 enam. wire wound over L2.
L2 — 10 turns no. 24 enam. wire on a T-50-12 core.

L3 — See text and drawing.
M1 — 0.1 mA, Simpson 15070 or equiv.
M2 — 0.1 A, Simpson 15101 or equiv.
Q1 — 2N2905A or equiv.
R4 — Meter shunt, 19½ in of no. 34 enam. wire wound on a high-value, ½-watt resistor.
R5 — Potentiometer, 25k-ohm, 5 watt.
R7 — Meter shunt, 4.3 ohm, ¼ watt.
RFC1 — 36 turns no. 24 enam. wire on a 1-in-diameter Teflon rod.
RFC2, RFC3 — 6 turns no. 22 enam. wire on an FT-50-43 core.
S1 — SPST, rocker type with built-in neon indicator.
S2 — Rotary, 2 pole, 2 position.
T1 — Stancor P8356 or equiv. See text.

Fig. 91 — Underside of the amplifier. The grid-compartment chassis is normally enclosed by a perforated aluminum cover which was removed for this photograph.

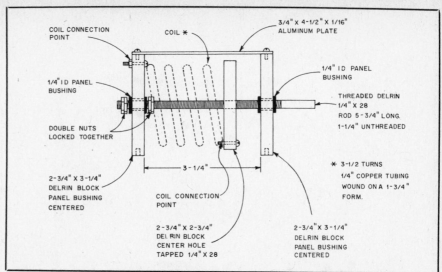

Fig. 93 — Dimensional drawing of the inductor assembly. Delrin can be obtained from most plastic supply houses. Check the Yellow Pages of your local telephone directory for dealers.

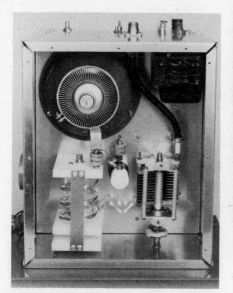

Fig. 92 — Plate compartment of the amplifier. Details of the inductor are given in the text and in additional photographs and drawings.

Fig. 94 — Photograph of the assembled adjustable inductor. Make certain to use only plated brass hardware for pieces that come into contact with the coil.

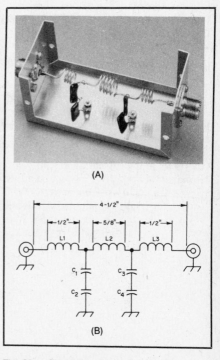

Fig. 95 — Photo (A) and schematic diagram (B) of the low-pass filter that is used after the amplifier. The filter is built to fit inside a 2¼ × 2¼ × 5-in box. Good quality capacitors such as silver-mica types should be used.

C1-C4 — 110 pF silver mica, 1000 V.
L1, L3 — 4 turns no. 14 wire, 5/16-inch ID.
L2 — 5 turns no. 14 wire, 7/16-inch ID.

suitable blower is given in Chapter 15 of this *Handbook*. Connection to the meters is through 1000-pF feedthrough capacitors.

The plate compartment is somewhat less busy than the underside of the amplifier. See Fig. 92. The only component that needs explanation is the inductor. Details of the construction are shown in Fig. 93, and a photograph is shown in Fig. 94. Two pieces of 3/8-inch Delrin plate are used as the stationary end pieces. A ¼-inch control bushing is used at the center of each piece. A third piece of Delrin stock is used for the movable plate. This piece is tapped to accommodate the ¼-28 thread of the shaft that is used to move the plate. This shaft

is double nutted on each side of the rear support to prevent the shaft from moving in relation to either the front or rear supports. As the front-panel knob is turned, only the movable plate changes position to compress or expand the coil. The exact dimension of the coil is fairly critical, so it should be made as close to specification as possible. Once the coil has been wound it can be silver plated to prevent oxidation.

As a point of interest, the amplifier front panel was painted orange. Black press-on lettering was used to label the front panel controls. A light coat of clear lacquer was applied to protect the lettering.

Operation

It is a good idea initially to check out the various tube-element voltages without the tube in the socket. The screen voltage may

turn out to be different than the nominal 336 because of poor Zener-diode voltage tolerances. As long as the voltage is under 375 no problems should be encountered. Operation of the protective circuitry can be verified by loading the output of the screen supply with appropriate high-wattage resistors. Ensure that the bias control will allow adjustment between the values shown on the schematic. Set the voltage to roughly −150 so that the tube will be cut off when

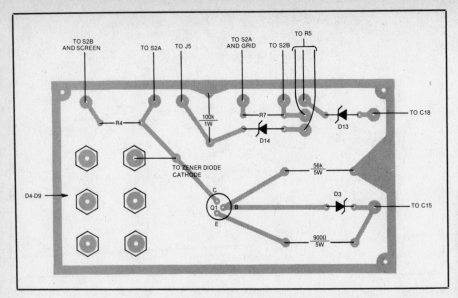

Fig. 96 — Parts-placement guide for the printed-circuit board as shown from the component side. A full-size etching pattern can be found at the back of this book.

Table 5

Operating Parameters

	1 kW	2 kW
Plate voltage	2100 V dc	3000 V dc
Plate current (single tone)	480 mA	667 mA
Plate current (idling)	50 mA	50 mA
Power input	1000 W	2000 W
Power output	620 W	1250 W
Efficiency	62 %	62.5 %
Drive power	4 W	9 W

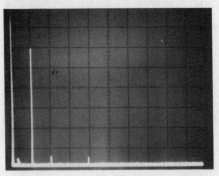

Fig. 97 — Spectral photograph of the amplifier and filter adjusted for 1-kilowatt input. Each horizontal division represents 50 MHz and each vertical division is 10 dB. This amplifier complies with current FCC requirements for spectral purity.

the amplifier is first turned on. Eimac recommends that the 4CX1000A heater voltage be applied for a period of not less than three minutes before plate voltage and current are applied.

Connect a transmitter capable of supplying 10 watts of power to J1 through an SWR indicator. Also, connect a dummy load and power meter to J2. After the amplifier has warmed up for several minutes, short J5 and adjust the bias control for an idling current of 50 mA. Apply a small amount of drive power and adjust L3 and C6 for maximum power output. Next, adjust C1 for minimum reflected power as indicated by the input SWR indicator. Apply additional drive power and

continue adjusting L3 and C6 for maximum power output. Use the operating parameters given in Table 5 as a guide. For the amplifier described here an adjustable high-voltage supply was used. The voltage and current levels for 1-kW and 2-kW input level were chosen to equalize the plate impedance, thus requiring only a minimal change in the settings of L3 and C6 for changing power level. If the fixed-voltage supply is used to power this amplifier, similar efficiencies should be attained.

As with many amplifier designs for VHF operation, additional low-pass filtering is required to meet the FCC spectral-purity requirements (all spurious and harmonic emissions − 60 dB or greater below peak

power). A suitable filter circuit is shown in Fig. 95. The unit was constructed in a small aluminum box and is mounted directly at the output connector by means of a double-male adapter. With this filter the amplifier easily exceeds the FCC requirements. A spectral photograph is shown in Fig. 97.

A Medium-Power 144-MHz Amplifier

The typical 2-meter rig these days puts out 10 watts or so. While 10 W is adequate for local communications, it is not enough for reliable long-haul work. Although occasional band openings and enhanced conditions allow long-distance contacts with low power, something more is desirable for weak-signal work. The amplifier pictured in Figs. 98 through 110 will supply more than 300-W output for 10-W of drive. It is clean, easy to build and reliable. If the builder shops wisely, the entire amplifier can be built for less than the cost of a commercially manufactured 150-W "brick" amplifier and associated 20-A, 13.8-V regulated supply. This project was designed by Clarke Greene, K1JX, and built by Mark Wilson, AA2Z, in the ARRL lab.

Circuit Details

An 8930 tetrode (350-W plate dissipation) is used in this design. The 8930 is identical electrically to the 4CX250R, but it features a larger anode cooler for higher dissipation with lower air flow and pressure drop requirements. If the builder must purchase a new tube, the 8930 is a good choice because it is more rugged than the 4CX250. However, a commonly available 4CX250B will work fine in this circuit. Only minor mechanical changes are necessary to accommodate the 4CX250's smaller anode cooler. The amplifier and power supply are shown schematically in Figs. 99 and 100.

The input circuit consists of L2, which is series tuned by C2 and the tube input capacitance. C3 is used in parallel with C2 to preset the tuning range for a smooth vernier action. A single 15- or 20-pF variable may be used at C2, but the tuning will be more critical. Power is coupled to L2 through link L1.

Bias voltage is fed to the tube through R1. This resistor, which consists of three 820-ohm, 2-W carbon composition resistors in parallel, swamps the grid heavily. Heavy swamping is important for amplifier stability. It also raises the drive requirement to about 8 W for full output. The amplifier can be driven to full output with about 2 W if the grid swamping resistor is changed to 2000 ohms or so and the input matching is adjusted accordingly. Excessive drive, however, will make the tube run in the nonlinear region, causing splatter up and down the band. It is impossible to adequately control the drive power by adjusting the exciter mic gain control. An input circuit designed for the drive level available from the exciter makes it difficult to overdrive the amplifier, ensuring a clean signal and peaceful co-existence with nearby amateurs sharing the band.

The output circuitry consists of a quarter-wavelength strip line, L3, which is tuned by C9. A tuned link, L4 and C10, couples power to the output jack. High

voltage is fed to the tube at the low-impedance (ground) end of the line through RFC1. C8 is the plate-blocking capacitor.

The filament, bias and screen voltages are all supplied by T1. EIMAC specifies filament voltage at 6.0 ±0.3 V, so R11 is included to drop the transformer voltage to the proper level.

T1 has a secondary winding of 270-0-270 V which is used for the screen and bias supplies. Screen voltage regulation is accomplished by five 60-V Zener diodes, D7 through D11. The voltage from the supply is dropped to the Zener stack voltage by current-limiting resistor R5. R7, a screen bleeder resistor, is connected in parallel with the Zener stack to allow for negative screen current developed under certain tube operating conditions.

A tetrode should not be operated without plate voltage and a load. Otherwise, the screen would act like the anode and draw excessive current. Should this happen accidentally, F2 will blow and protect the screen. F2 is placed between the Zener stack and the bleeder, rather than between the bleeder and the tube. If the tube were operated without a load on the screen, the screen potential would rise to the anode voltage, destroying the screen bypass capacitor. Should the screen voltage rise above 390 V, Z1, a metal-oxide varistor, will conduct and clamp the screen voltage to ground. The power-supply fuse should blow before any damage occurs.

Bias voltage is taken from the same secondary winding as the screen voltage. When the amplifier is in standby, −200 V is applied to the grid to cut the tube off. In transmit, D12 is switched in, supplying operating bias. R9 is used to adjust operating bias from −40 to −60 V to allow for differences among tubes. D15 is included to minimize Zener voltage drift as D14 changes temperature.

The 5-V winding on T1 supplies coil voltage for K1. Contacts K1B switch the bias from standby to operate. K1C provides "dry" contacts for antenna relay switching.

S1 applies voltage to the blower motor, while S2 energizes T1. These two switches are connected so that the blower must be

switched on before filament voltage is applied. The blower may be left on after the amplifier has been turned off.

The high-voltage supply is designed to deliver 2000 V under load. The heart of the supply is a heavy-duty Hammond transformer that will run cool even under continuous use. The 234-V primary and full-wave bridge rectifier in the secondary aid voltage regulation.

The B+ supply is turned on from the amplifier front panel, so the supply may be located away from the operating position. K1 is a heavy-duty power relay with contacts large enough to handle the supply current. R1 and R2 provide inrush current protection for the rectifiers. These resistors are shorted out by the contacts of K2, applying full primary voltage after a short delay. This delay is about 0.6 second, and is determined by R3 and C1.

D1 through D4 are high-voltage rectifiers provided by K2AW. They eliminate the need for a PC board full of individual 1000-PIV units with associated capacitors and equalizing resistors. C2 is a surplus oil-filled capacitor. Individual electrolytic capacitors would work fine, if enough units are connected in series to give about 40 μF at a voltage rating of at least 2500 V.

R4 through R6 make up the bleeder resistor. Several individual resistors are used, rather than one large unit, to keep the voltage across each resistor within ratings. F2 is a high-voltage fuse that will protect the power supply and amplifier in case of a short circuit. The negative lead of the power supply floats above chassis ground so that plate current may be monitored in the negative lead in the RF deck. R8 keeps the B− lead close to chassis potential. R9, in series with the B+ lead, protects the tube and power supply from high-voltage flashovers.

Plate current is monitored by a 0- to 500-mA meter in the negative lead. A 0- to 5-mA meter monitors grid current and screen current. Shunt R6 increases the movement to a 50-mA range for the screen. The high-voltage supply has a separate 0- to 4-kV meter.

S3, used to turn on the high-voltage

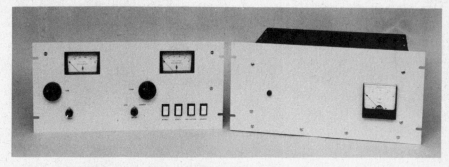

Fig. 98 — The 8930 amplifier and accompanying high-voltage power supply are built behind standard 19-in rack panels.

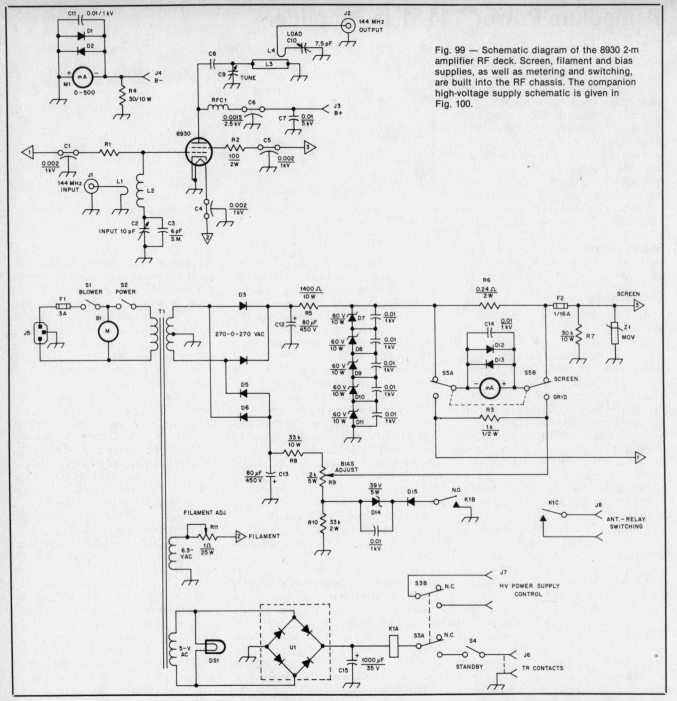

Fig. 99 — Schematic diagram of the 8930 2-m amplifier RF deck. Screen, filament and bias supplies, as well as metering and switching, are built into the RF chassis. The companion high-voltage supply schematic is given in Fig. 100.

B1 — 54 CFM blower (Dayton 4C012A or equiv.).
C1, C4, C5 — Feedthrough capacitor, 2000 pF, 1 kV (Erie 1202-005 or equiv.).
C2 — Miniature air variable, 1.8 to 8.7 pF (Cardwell 160-104 or equiv.).
C6 — Feedthrough capacitor, 1500 pF, 2.5 kV (Erie 1280-060 or equiv.).
C8 — Plate dc blocking capacitor; part of plate line assembly. See text and Fig. 104.
C9 — Flapper capacitor. See text and Fig. 104.
C10 — Air variable, 2.8 to 7.5 pF (Millen 22006 or equiv.).
D1-D6, D12, D13, D15 — Silicon diode, 1 kV, 1 A (1N4007).
D7-D11 — Zener diode, 60 V, 10 W (RCA SK192).
D14 — Zener diode, 39 V, 5 W (1N5366).
DS1 — Meter illumination lamps (see text).
J1 — BNC chassis mount.

J2 — Type N chassis mount.
J3, J4 — High-voltage connector (Millen 37001 or equiv.).
J5 — CEE-22 chassis mount power connector.
J6 — Chassis-mount phono jack.
J7 — Two-conductor socket (Cinch S-402AB or equiv.).
J8 — Terminal strip, two conductor.
K1 — DPDT relay, 6-V dc coil.
L1 — 1 turn no. 16 enam. wound over center of L2.
L2 — 4 turns no. 16 tinned wire, 3/8-in ID, 5/8-in long.
L3 — Plate line. See text and Fig. 104.
L4 — Output coupling link. See text and Fig. 104.
M1 — Plate current meter, 0-500 mA (Simpson model 1327 with custom face; Cat. no. 10552, avail. from Larson Instrument Co., Greenbush Road, Orangeburg, NY 10962).

M2 — Grid and screen current meter, 0-5 mA movement (Simpson model 1327 with custom face; Larson Instruments Co., cat. no. 10551.
R1 — Three 820-Ω, 2 W carbon composition resistors in parallel. See text.
RFC1 — 1.72 μH, 600 mA RF choke (J. W. Miller RFC-144 or equiv.).
S1, S2, S4 — SPST rocker switch.
S3 — DPDT rocker switch.
S5 — Rotary switch, 2 circuit, 2 position, non-shorting (Mallory 3222J or equiv.).
T1 — Power transformer, 117-V ac primary; secondary: 270-0- 270-V at 120 mA, 6.3-V CT at 3.5 A, 5 V at 3 A (Stancor PC-8405 or equiv.).
U1 — Bridge rectifier, 50 V, 1A.
Z1 — Metal-oxide varistor, 370-V dc continuous, 390-470 V peak (Zenamic Z275LA40A or equiv.).

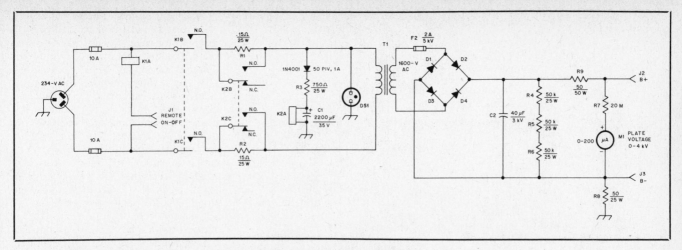

Fig. 100 — Schematic diagram of the high-voltage power supply for the 8930 2-m amplifier.

D1-D4 — High-voltage rectifier, 14 kV, 1 A (K2AW HV14-1 or equiv.).

F2 — High-voltage fuse, 5 kV, 2 A (Buss HVJ-2 or equiv.).

J1 — Two-conductor socket (Cinch S-402AB or equiv.).

J2, J3 — High-voltage connector (Millen 37001 or equiv.).

K1 — Power relay, DPST, 234-V ac coil, 277-V ac, 25-A contacts (Potter & Brumfield PRD7AYO-240V or equiv.).

K2 — Power relay, DPDT, 24-V dc coil, 277-V ac, 25-A contacts (Potter & Brumfield PRD11DYO-24V or equiv.).

T1 — Power transformer; primary 117/234-V; secondary 1600 V, 1.2 A (Hammond Mfg. 126270 or equiv.).

supply from the RF deck, is wired so it is impossible to key the amplifier until the B+ is applied. When S2 is thrown, the filaments, bias and screen supplies all come on together. If operating bias was placed on the tube (by keying the amplifier) and no B+ was present, the screen would draw excessive current, as described before. As long as the tube is cut off, having screen voltage present without plate voltage is not a problem.

Construction

The 8930 amplifier is built using three separate chassis (see Figs. 101 and 102). The main chassis measures 10 × 17 × 3-in and is bolted to a standard 19-in-wide, 8¾-in-high rack panel. The plate compartment is a 13 × 5 × 3-in chassis, while the input compartment is a 4 × 5 × 2-in chassis. Use of separate enclosures ensures good shielding.

Amplifier cooling is accomplished using a split airflow, rather than the conventional base-to-anode flow. The blower pressurizes the plate compartment. A special chimney, fabricated from Teflon sheet and phenolic tubing (see Fig. 105), seals the area between the tube cooler and the plate compartment top cover. Air enters the plate compartment and exits through the anode cooler. Some of the air exits through the tube socket to cool the tube seals and input compartment. Measurements by ARRL TA Dick Jansson, WD4FAB, indicate that this cooling system reduces blower back-pressure requirements by about 50 percent with this tube. The blower chosen is capable of supplying 25 cubic feet per minute at 0.4-in of static column pressure — more than enough to cool the amplifier, even at full dissipation.

An EIMAC SK-620A socket is used in this project. The SK-620A and SK-630A sockets feature built-in low-inductance screen bypass capacitors and screen

Fig. 101 — Top view of the 8930 2-m amplifier. The tube and plate circuitry are mounted in a separate chassis that bolts to the main chassis. The plate compartment top cover has been removed for clarity, although it must be installed during amplifier operation.

shielding rings. These features are essential to amplifier stability. If possible, an SK-630A should be used because it features grounded cathode pins. The SK-620A was used here because one was on hand. The four cathode pins are bent in toward the socket center ring and soldered to it, assuring a solid, low-inductance cathode ground.

The input compartment is centered on the tube socket. Filament, bias and screen voltages enter the input box through feedthrough capacitors. All leads are kept as short as possible. The swamping resistor is connected between the feedthrough capacitor and the grid terminal. Likewise, the 100-ohm screen isolation resistor connects the screen pin and the feedthrough capacitor. L2 is supported by C2 and the

socket grid pin. L1 is mounted on a small ceramic standoff insulator near L2. J1 is a panel-mount BNC connector attached to the rear panel and connected to L1 by a short length of 50-ohm coaxial cable.

The plate compartment is mounted along the rear of the main chassis. Fig. 103 shows the relationships of the pieces inside the compartment. Details of L3, L4, C8 and C9 are given in Fig. 104.

L3 is a ¼-wavelength strip line made from brass and aluminum sheet. Complete details are shown in Fig. 104. One end bolts to the chassis, while the other end is silver-soldered to a copper strap that clamps around the anode cooler. C8, the plate blocking capacitor, is a "sandwich" capacitor formed by the two pieces of the plate line. The dielectric is a piece of 0.01-in

Fig. 102 — Underside of the 8930 2-m amplifier. The input circuitry is enclosed in a separate chassis. The screen and bias supplies are built on small printed-circuit boards. D7-D11, the screen regulator Zener diodes, are mounted on a piece of ¼-in-thick aluminum plate for adequate heat-sinking.

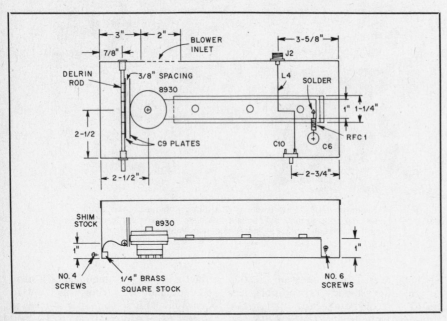

Fig. 103 — Component mounting information for the plate compartment.

thick Teflon sheet. Three 6-32 brass screws hold the sandwich together. The top half of the line has B+ on it, and the bottom half is grounded, so Teflon shoulder washers are used to insulate the screws from the top half of the line. The shoulder washers used here were cut on a lathe from ¾-in-diameter Teflon rod. A light coating of Dow Corning DC-4 silicone grease is spread on each side of the Teflon sheet to fill in any irregularities.

In this amplifier, the bottom half of the plate line is formed from a single piece of aluminum sheet bent in a sheet metal brake. With patience and a good bench vise and hammer, it should be possible to get a clean bend. If the builder does not have sheet metal bending capability, the bottom half of the line may be made from two separate pieces bolted together. The plate line may be made from copper or brass if this method is used. The top half of the sandwich must be made from copper or brass because it is soldered to the tube clamp.

C9 is made from two brass plates. Fig. 104 gives complete mechanical details of the capacitor assembly. One plate is silver soldered to the anode clamp. The movable plate is attached to a ¼-in Delrin rod supported by bushings in the chassis walls. A wide, low-inductance strap made from 0.003-in thick brass shim stock connects the movable capacitor plate to the chassis. A length of ¼-in-square brass bar is drilled and tapped to hold the shim stock securely against the chassis. A Jackson 6:1 ball drive (available from RadioKit — see Chapter 35 parts suppliers list) is used on the capacitor support shaft to ensure smooth tuning. A 6-32 machine screw tapped into the Delrin shaft ensures a solid mechanical stop. The two capacitor plates must *not* be allowed to touch — one plate is at ground potential, while the other is at 2 kV.

The copper, brass, Teflon, Delrin and aluminum stock necessary for this project are available from Small Parts, Inc. (see Chapter 35 parts suppliers list). Brass or plated brass hardware should be used throughout the plate compartment. Steel is unacceptable for use in strong RF fields. After assembly, the anode clamp, C9 plates and brass half of the plate line were treated with rub-on silver plating.

L4 is made from a piece of no. 14 tinned wire suspended over the cold end of the plate line. Details are shown in Fig. 104. It is supported on the ends by J2 and C10. Greatest coupling occurs when the section of the link that runs parallel to the strip line is placed over the center of the line. In this amplifier, best coupling occurred with the link spaced about ¼-in above the line, but the builder should experiment with different spacings to find the optimum distance for best efficiency.

High-voltage enters the plate compartment through feedthrough C6. RFC1 is soldered between C6 and the cold end of the line. Carefully clean away all flux from the plate line after soldering. Any flux left near the end of the line may offer a path for a high-voltage arc across the sandwich capacitor.

Brass screen, soldered to a brass plate, covers the blower inlet to ensure good shielding. Similarly, a piece of screen soldered to a plate shields the hot air exhaust. A liberal number of machine screws are used to secure the plate compartment top cover to the chassis to keep the RF in. The machine screws thread into aluminum captive fasteners pressed into the chassis. These fasteners are available from Penn Engineering and Mfg. Co., Danboro, PA 18916.

All parts for the screen and bias supplies except D7 through D11 and their 0.01-μF bypass capacitors are mounted on printed-circuit boards. Parts layouts for these boards are given in Fig. 106, and full-size etching patterns are given at the back of this book. The resistors tend to run warm, so they are spaced about ¼-in off the board to allow airflow on all sides. D7

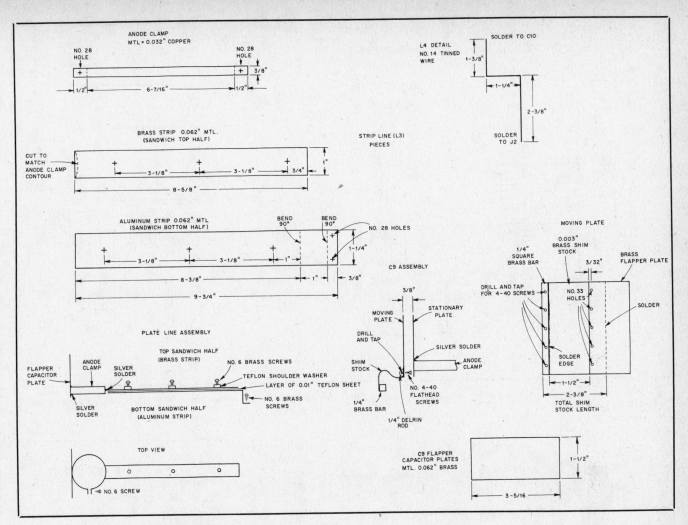

Fig. 104 — Construction and assembly of L3, L4, C8 and C9.

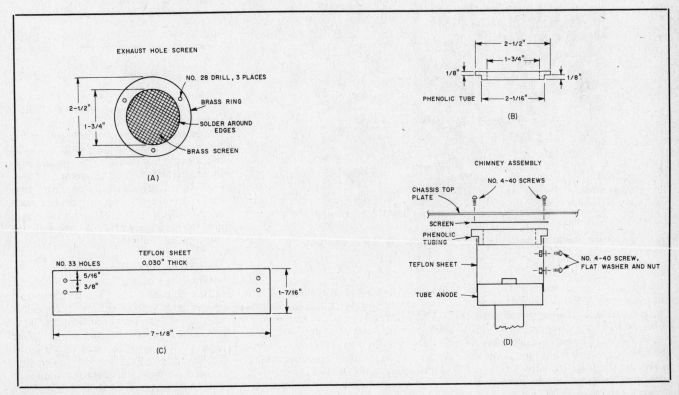

Fig. 105 — Chimney construction details.

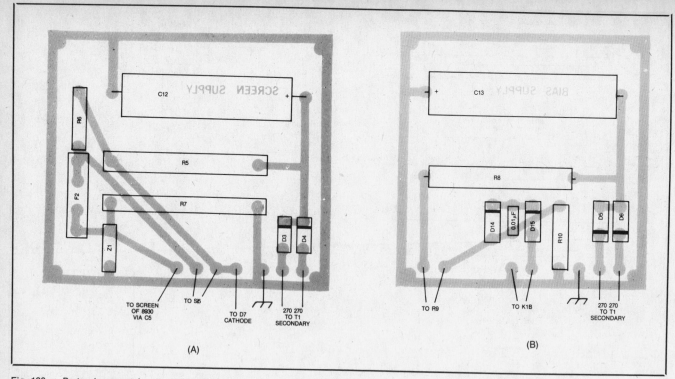

Fig. 106 — Parts placement for the screen (A) and bias (B) power supply boards as seen from the component side. Full-size etching patterns are given at the back of this book.

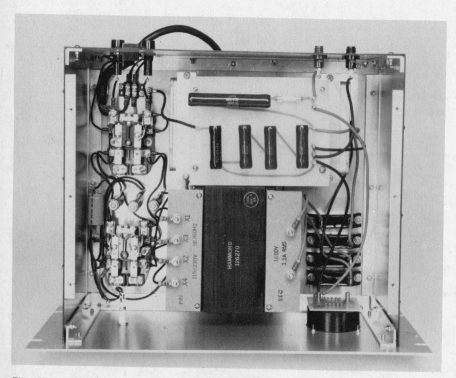

Fig. 107 — Interior view of the high-voltage power supply for the 8930 2-meter amplifier.

Table 6

Operating Parameters, 8930 144-MHz Amplifier

Plate voltage	2000-V dc
Plate current (single tone)	290 mA
Plate current (idling)	90 mA
Screen current	15 mA
Power input	580 W
Power output	340 W
Efficiency	59%
Drive power	6 W

8½-in-high rack panel (see Fig. 107). The major power supply components are bolted to a 1/8-in-thick plate. The front, rear and side panels are attached to the bottom plate by pieces of aluminum angle stock. All connections are made via the rear panel. Only the high-voltage meter (to show that the secondary is on) and a pilot lamp (to show that primary energy is applied) are mounted on the front panel. The high-voltage projects in Chapter 27 show full details on this type of construction.

Tune Up and Operation

When the wiring is complete, check and recheck all connections. The voltages present in this amplifier are potentially dangerous. Disconnect the filament, bias and screen supplies at the feedthrough capacitors on the input compartment. Turn on T1, if possible through a variable autotransformer, and check the voltages. The filament should be about 6-V ac. The screen supply should be around 300 V, plus

through D11 are mounted with insulating hardware on a ¼-in-thick aluminum plate that is bolted to the chassis for heat sinking.

M1 and M2 are Simpson Model 1327 3½-in meters. They are mounted behind the front panel using Simpson mounting bezels and illumination kits. Larson Instruments designed custom meter scales drawn for this project. M2 has two scales;

the GRID CURRENT scale reads 0 to 5 mA, while the SCREEN CURRENT scale reads −10 0 10 20 30 40 mA. The screen supply bleeder resistor draws 10 mA, so the zero for the SCREEN scale is offset 10 mA. Ordering information for these meters is included in the parts list in Fig. 99.

The high-voltage supply is built on a separate chassis, also mounted behind an

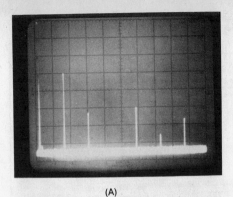

(A)

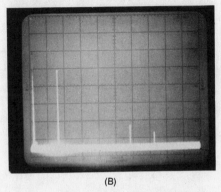

(B)

Fig. 108 — Spectral photograph of the 8930 2-meter amplifier is given at A. Power output is 340 W. Horizontal divisions are each 100 MHz; vertical divisions are each 10 dB. A shorted ¼-wavelength stub (Fig. 109) is necessary to make the amplifier meet current FCC spectral-purity requirements. The spectral photograph at B, taken with the stub installed, shows that all spurious and harmonic emissions are at least 60 dB down. In both photos, the fundamental has been reduced in amplitude by 30 dB by means of notch filters to prevent analyzer overload.

or minus a few volts. This may change some as the Zeners heat up to their operating temperature. The bias voltage should be about − 200 V. Close the TR relay, K1. Bias voltage should change to somewhere around − 50 V. If everything checks out, unkey K1 and turn the power off.

Connect the filament voltage to the feed-through capacitor on the input box. Turn on the blower and check for airflow. Turn on T1 (remember, no voltages other than filament are applied to the tube). Use an accurate voltmeter to check the voltage at the socket filament pins. Adjust R11 until the filament voltage is exactly 6.0 volts. Turn off the filament voltage and blower, and connect the screen and bias supplies to the appropriate feedthrough capacitors. Put the cover on the input compartment.

Connect the high-voltage supply to the RF deck. Connect a dummy load to J2 through an accurate VHF wattmeter, such as a Bird Model 43. Connect a 2-meter rig capable of supplying about 10 W to J1 through another VHF wattmeter. If only one wattmeter is available, connect it to the input first for initial adjustments and con-

nect the output directly to a dummy load.

Start the blower and check for airflow. Turn on T1, again preferably through a variable autotransformer in case of trouble. The exhaust air should be slightly warm from the filament. The screen current meter should read 0 (10 mA upscale from the current drawn by R7). The plate and grid current meters should not move.

After a minute or so, gradually apply plate voltage. Bring the voltage up to 1000 V or so. Listen and watch for any signs of high-voltage arcing. If everything seems okay, key the amplifier at J6. Adjust the idling plate current to approximately 90 mA using R9. Screen and grid current should stay at zero.

Apply a few watts of drive and adjust C2 for minimum reflected power on the SWR meter connected to the input. If the SWR is not better than 1.2 or 1.3, shut off the amplifier. Disconnect the high voltage, open up the input compartment and adjust the position of the link. It may take several tries, but it is important that the input match be as good as possible for best IMD characteristics. If moving the link does not bring the input SWR close to 1.2, try adjusting the spacing of L2 slightly.

When the input is matched, turn on the amplifier and bring the plate voltage up to 2000 V. If no arcing is noticed, apply a small amount of drive and tune C9 and C10 for maximum output. Increase drive slowly, and readjust C9 and C10 for maximum output. Watch the screen and plate current, and do not exceed the manufacturer's ratings. Table 6 lists the normal operating parameters for this amplifier. Six watts of drive should produce about 340-W output. If the efficiency is very much different from that shown in the table, try adjusting the position of the link with respect to the plate line.

This amplifier is extremely stable. With operating voltages applied (and no drive), the controls were tuned through their ranges with no hint of instability. This amplifier is not neutralized, so maximum output may not occur with the plate dip. The grid, tuning and loading controls work smoothly. Their adjustment is not critical — in fact, tuning is smoother than on some commercially manufactured HF amplifiers.

Spectral photographs are shown in Figs. 108 and 110. In Fig. 108A the second and fourth harmonics are about 50 dB below the amplitude of the fundamental. The sixth is 58 dB down. By increasing the value of C10 and reducing the size of L4, it is possible to raise the link Q to the point where the amplifier meets FCC spectral-purity requirements (all spurious and harmonic emissions at least 60 dB down); however, the high-Q link makes retuning necessary for small excursions in the 2-meter band.

Addition of a simple shorted stub cut to an electrical quarter wavelength at 144 MHz (Fig. 109) results in the spectral

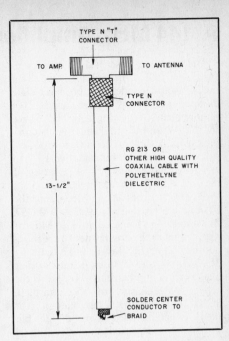

Fig. 109 — Details of the ¼-wavelength stub.

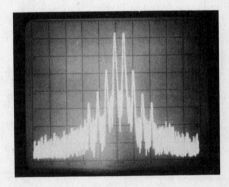

Fig. 110 — Two-tone spectral photograph of the 8930 2-meter amplifier. Power output is 340 W PEP.

photograph shown in Fig. 108B. At the transmitter end, the stub looks like an open circuit for RF at 144 MHz, but it looks like an RF short circuit at the even harmonics. The only visible harmonic energy is at the fourth, and this is just out of the analyzer noise floor at − 62 dB.

The shorted stub serves a second purpose as well. In the unlikely event that the link detaches from its mounting and falls on the top of the plate line, placing high voltage on the output connector, the stub will short the dc to ground and blow the power supply fuse.

A two-tone test of this amplifier is shown in Fig. 110. The test setup consisted of two RF tones, one at 144.200 MHz and the other at 144.201 MHz fed into a combiner and then into the amplifier input. Except for the third-order products, which are down 28 dB from PEP, all IMD products generated within the amplifier are lower in amplitude than those found in most 10-W, solid-state, 2-m transceivers.

A 144-MHz Amplifier Using the 3CX800A7

The amplifier described here and pictured in Figs. 111 through 118 is based on the 3CX800A7 triode. This tube has a plate dissipation of 800 W with modest cooling requirements. Requiring only 19-W drive for 700-W output, this amplifier is compatible with some of today's low-power, solid-state multi-mode transceivers.

This amplifier was originally presented in April 1984 *QST* by David D. Meacham, W6EMD. It is based on a design for the 8874 tube presented in January 1972 *QST* by Raymond F. Rinaudo, W6ZO. Most of the changes to the original design are to accommodate the larger size of the 3CX800A7 and its attendant higher capacitances and currents. Fig. 112 is the schematic diagram of the complete 3CX800A7 amplifier. A plate-current meter is not included in this design, but

Fig. 111 — Front-panel view of the 3CX800A7 2-meter amplifier.

there is room on the front panel if the builder wishes to add one.

Construction

The amplifier chassis is mounted in a standard 19-inch-wide, 5¼-inch-high aluminum rack panel (Bud No. SFA-1833). A 5 × 13 × 3-inch aluminum chassis (Bud No. AC-422) is spaced 1¾ inches behind the panel by two aluminum end brackets.

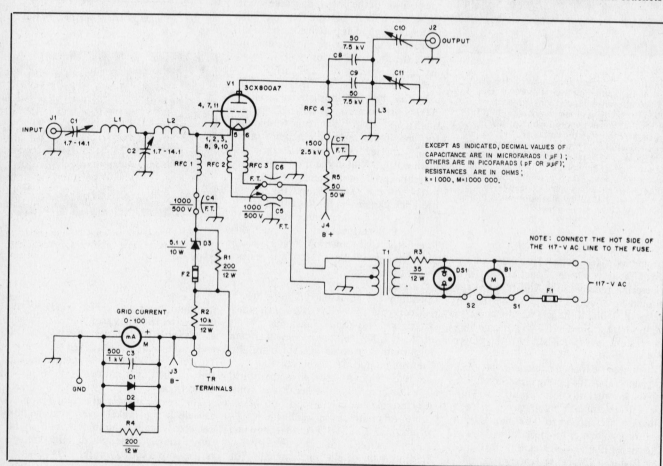

Fig. 112 — Schematic diagram of the 3XC800A7 amplifier.

B1 — Blower (Dayton 4C012 or equiv.); see text.
C1, C2 — 1.7-14.1 pF air variable (E. F. Johnson 189-505-4 or equiv.).
C4, C5, C6 — Ceramic feedthrough, 1000 pF, 500 V (Erie 357- 001).
C7 — EMI feedthrough filter, 1500 pF, 2.5 kV (Erie 1280-060).
C8, C9 — 50 pF, 7.5 kV, NP0 (HEC HT-50 852).
C10, C11 — See text.
D1, D2 — 1N4001.
D3 — Zener diode, 5.1 V, 10 W (1N3996A).
DS1 — NE51 in holder with built-in dropping resistor.
F1 — Slow-blow fuse, 2 A (Buss MDL-2 or equiv.).
F2 — 1.5 A fuse (Buss AGC 1½ or equiv.).
J1 — BNC bulkhead feedthrough connector (UG-492 A/U).
J2 — Type N female connector (part of C10 assembly — see text).
J3, J4 — High-voltage connector (Millen 37001 or equiv.).
L1 — 8 turns no. 16 tinned wire, 3/8-in ID, 11/16 in long.
L2 — 6 turns no. 16 tinned wire, 3/8-in ID, 11/16 in long.
L3 — Plate line, 1-5/16 in wide, 8-3/16 in long. See text and Fig. 115.
R1, R4 — 200 ohms, 12 W.
R2 — 10 kΩ, 12 W.
R3 — 35 ohms, 12 W.
R5 — 50 ohms, 50 W.
RFC1 — 1.0 µH choke, 300 mA (Miller 4602).
RFC2, RFC3 — 11 turns no. 20 HF wire on 3/8-in Noryl rod.
RFC4 — 10 turns no. 16 tinned wire, ½-in ID, 1-3/16 in long.
T1 — Filament transformer, 14.0-V, 2-A secondary (Triad F-251X or equiv.).

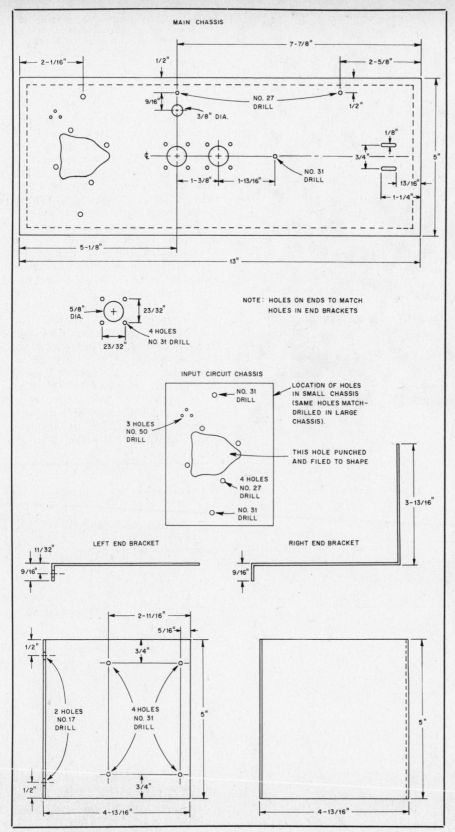

Fig. 113 — Mechanical details of the chassis and end brackets. The end brackets are made from 0.032-in aluminum, 5052 alloy or softer.

cabinet chosen is a lightweight aluminum unit made by Ten-Tec (No. 19-0525).

Input Circuit

In the cathode-driven configuration, the input impedance of the 3CX800A7 appears as a nominal capacitance of 26.5 pF in parallel with a resistive component that varies with operating conditions but is typically about 49 ohms. The computer-designed input circuit of this amplifier operates at a loaded Q of about 2.1. It can be set anywhere in the 2-meter band for an SWR of less than 1.3:1. C1 and C2 are predominantly matching and tuning controls, respectively; there is some interaction between them, however. When the input is tuned at 144.5 MHz, the input SWR will be less than 1.4:1 from 144 to 145 MHz.

The tube socket is an EIMAC SK-1900 or a Johnson part No. 124-311-100. Its center is mounted 2 inches from the end of the input box, adjacent to the mounting bracket. The tube pins and input circuit are cooled by a small amount of air admitted to the input box from the pressurized output box. Three holes, made with a no. 50 drill, provide adequate flow. These holes are spaced in a close triangle and are located diagonally across from the variable capacitors. Air exhausts through the tuning holes. Fig. 114 shows details of the input circuitry.

The heater circuit includes two chokes, feedthrough capacitors, the filament transformer, a voltage-dropping resistor and a switch. The chokes are wound with a turn-to-turn spacing of about one-half the wire diameter. They are self-resonant (parallel resonant) just above the 2-meter band. Nominal heater voltage for the 3CX800A7 is 13.5 V. The closest available commercial transformer has a 14.0-V secondary, so R3 is used in the primary. Switching is set up so the blower must be on before the heater can be activated. Conversely, this arrangement allows the blower to be left on after switching off the heater — a highly recommended practice.

Cathode bias is provided by a 5.1-V Zener diode, D3. R1 prevents the cathode voltage from soaring if the Zener fails. F2 will blow if excessive cathode current is drawn. R2 nearly cuts off plate current on receive. Because the grid is at dc ground, the negative supply lead must be kept above ground for grid-current metering by M1. R4 keeps the negative side of the plate supply from rising if M1, D1 and D2 all open up. D1, D2 and C3 protect the meter from transients and RF voltages.

Output Circuit

The output tank circuit is a silver-plated quarter-wave strip line (Fig. 115) fore-shortened by the tube, loading and tuning capacitances at, and near, its open end. It operates at a loaded Q of approximately 20. The silver-plated anode collet (Fig. 115) is made of 0.062-inch-thick brass sheet with Tech-Etch 134B finger stock soldered on

A 3½ × 1-inch aluminum chassis (Bud No. AC-1402) houses the input circuitry and is mounted on the larger chassis between it and the front panel. The right-hand end bracket has a large lip on the rear for mounting the heater transformer and connectors. Fig. 113 shows mechanical details of the chassis and end brackets. The

Fig. 114 — Close-up of the input circuitry and tube-socket wiring. Keep all leads as short as possible. Although L2 is shown here with 5 turns, the final version uses 6 turns, as described in the parts list.

the inside. It is supported by two Teflon® standoffs 1½ inch in diameter and 1 inch long. At the far end of the line, a silver-plated shorting block contacts the chassis and the strip line. This block (Fig. 115) is made from 7/16-inch-thick brass 1-5/16 inches wide and 1 inch high. EIMAC CF-800 finger stock is screwed and then soldered on both top and bottom. The chassis and strip line are both slotted to allow 5/16 inch of shorting-block travel to set the tuning range of the capacitive-tuning paddle. A 1-inch-long Teflon standoff supports the center of the line. Construction details may be seen in Fig. 117.

The tuning (C11) and loading (C10) paddles are 1-3/16-inch-diameter discs made from silver-plated, 1/16-inch-thick brass. They are spaced 1-3/8 inches center-to-center. The output loading paddle is nearest the tube, and its center is 1-13/16 inches from the tube cooler surface. Spacing between the paddle and line during operation is about 0.135 inch. The output tuning boss, EIMAC part no. 720362, is tapped

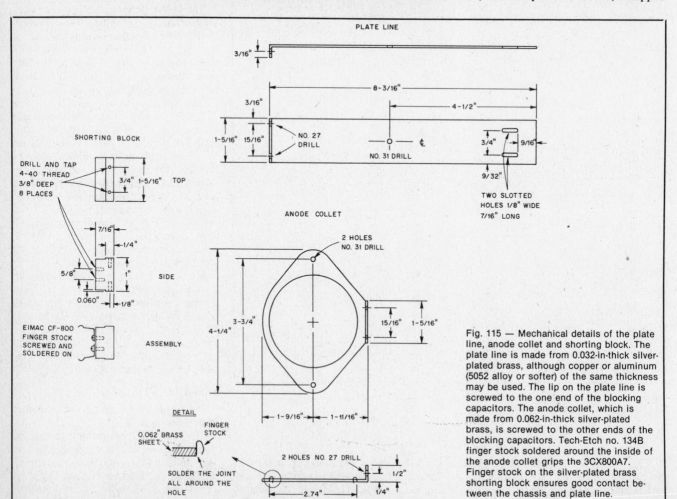

Fig. 115 — Mechanical details of the plate line, anode collet and shorting block. The plate line is made from 0.032-in-thick silver-plated brass, although copper or aluminum (5052 alloy or softer) of the same thickness may be used. The lip on the plate line is screwed to the one end of the blocking capacitors. The anode collet, which is made from 0.062-in-thick silver-plated brass, is screwed to the other ends of the blocking capacitors. Tech-Etch no. 134B finger stock soldered around the inside of the anode collet grips the 3CX800A7. Finger stock on the silver-plated brass shorting block ensures good contact between the chassis and plate line.

for a ¼-28 threaded rod. The output loading control and output connector are two separate EIMAC assemblies combined into one unit. They are available from EIMAC as Support Assembly No. 720361 and Sliding Probe Assembly No. 720407. For this amplifier, the outer conductor was lengthened to reach the front panel. The grid collet is also available from EIMAC as part No 720359.

In addition to RFC4 and feedthrough capacitor C7, an energy-absorbing resistor (R5) is wired in series with the dc plate supply. This resistor will protect the tube and power supply in the event of a high-voltage arc. It provides necessary protection while dissipating only 12.5 W at 500-mA dc plate current.

Cooling

The blower specified provides a measured 25 CFM airflow through the output box and tube cooler at 0.42 inch of water-column static pressure. This amount of cooling is sufficient for 800 W of plate dissipation at sea level with inlet air temperatures up to 35° C. It is adequate for the same dissipation at 5000 feet of altitude with inlet-air temperatures up to 25° C.

Not evident from the photographs is the care taken to separate outgoing hot air from incoming cool air. This is accomplished by the addition of a simple dividing wall inside the cabinet running from the rear chassis cover of the output box to the rear of the cabinet. The material used is rigid fiberglass insulation — the kind you

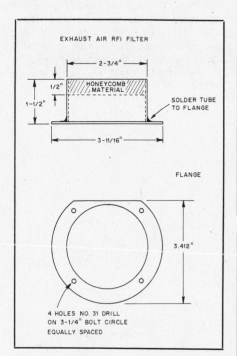

Fig. 116 — Mechanical details of the exhaust air RFI filter. The honeycomb material is soldered inside a 2¾-in OD, 0.065- in wall brass tube. The flange, which bolts to the chassis, is made from 0.125-in-thick brass sheet.

Fig. 117 — Various views of the completed 3CX800A7 amplifier.

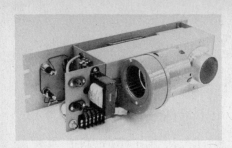

Fig. 118 — Rear view of the amplifier with the plate-compartment cover in place. Note the honeycomb RFI filter at the hot-air exhaust.

cut with a saw. It provides a measure of blower-noise suppression in addition to assuring cool inlet air.

Hot air leaving the anode cooler is directed to a homemade "honeycomb" RFI filter through a chimney made from rolled-up Teflon sheet. The filter is made of brass tubing and sheet with the honeycomb material soldered in place. See Figs. 116 and 118 for construction details. This filter acts as a waveguide-beyond-cutoff having high attenuation at 144 MHz.

The equation for calculating attenuation by this type of filter is

$$A_a = 32 \frac{D}{d}$$

where

A_a = aperture attenuation (dB)
D = length of pipe
d = inside diameter of pipe.

In this case, the "pipe" is each cell of the honeycomb. The basic material for the honeycomb is cadmium-plated brass heat-radiator core.[1] Each hexagonal-shaped cell has a width between flat sides (diameter) of 0.100 inch. The material is ½ inch thick.

[1] One source for the honeycomb radiator core material is G-O-Metric, c/o Marty Manion, 909 Norwich Ave., Delran, NJ 08075.

Plugging these values into the equation yields an A_a of 160 dB — more than enough! From a practical stand-point, it is usually sufficient to make cell length at least three times the nominal cell diameter.

Tune-up

Initial work may be done with a dip meter, particularly on the output circuit. With the tube in place, but with no voltages applied, the shorting block on the end of the plate line should be set to give a paddle-tuning range that straddles the desired operating frequency. Bear in mind that when the tube is "hot," the resonant frequency will be somewhat lower than it is without electron flow.

The input-tuning capacitor can be dip-meter resonated with the matching capacitor set at one-half mesh for a start. Further work here must be done "hot" with an SWR measuring device on the input.

Set the initial spacing between the output paddle and the plate line to 1/8 inch. Connect the output to a 50-ohm dummy load capable of handling at least 700 W at 144 MHz through an accurate VHF watt-meter, such as a Bird model 43. Connect a driver capable of delivering about 20 W to the input through an SWR-measuring device.

The heater of the 3CX800A7 should be run for at least three minutes before applying plate voltage. After the warm-up, short the TR terminals and apply about 1000 V to the plate. This should result in a small amount of idling plate current. Next, apply enough drive to produce a rise in plate current, and adjust the plate tuning for a peak in output power. Now tune and match the input circuit for minimum input SWR. Next, adjust the output loading paddle for maximum power output while keeping the plate current dipped with the plate tuning paddle. Apply full plate voltage and

Table 7

Operating Conditions for the 3CX800A7 Amplifier

Plate supply voltage	2200 V
Zero-signal plate current †	65 mA
Single-tone plate current	500 mA
Grid bias (Zener bias)	−5.1 V
Single-tone grid current †	40 mA
Driving power	18.5 W
Output power	707 W
Gain	15.8 dB
Efficiency ††	64%

†Values may vary considerably from tube to tube.
††Actual tube efficiency is about one percent higher because of power loss in the 50-ohm series resistor in the plate lead (R5).

higher drive power, and then repeak the output tuning and loading controls. Touch up the input tuning and matching, and the amplifier is ready for service.

A few words of caution are in order. Remember that the heater voltage must never be applied without the blower running, and that the heater must warm up at least three minutes before applying plate voltage. Never exceed 60-mA dc grid current, even during tune-up. Also, because of the relatively low grid dissipation of the 3CX800A7, RF drive must *never* be applied unless plate voltage is applied to the tube and a suitable load is connected to the output. Following these simple rules will substantially increase tube life and amplifier reliability.

Typical Operating Conditions

With Zener-diode bias, the 3CX800A7 is best operated in Class AB_2 for linear service. The data in Table 7 represent measured performance in linear service at 144 MHz. Complete data sheets are available from EIMAC. This amplifier is easily capable of conservative operation at 700-W output. Two of these tubes will run at the 1500-W output legal limit.

A 2-kW PEP Amplifier for 144 MHz

Large external-anode triodes in a cathode-driven configuration offer outstanding reliability, stability and ease in obtaining high power at 144 MHz.

The techniques employed in the design and construction of the cathode-driven 3CX1500A7/8877 amplifier described here (Figs. 119 to 122) have removed many of the mechanical impositions of other designs. Those interested in obtaining complete construction details should refer to the two-part article by Edward L. Meade Jr., K1AGB, appearing in December 1973 and January 1974 *QST*.

Output Circuit

The plate tank operates with a loaded Q on the order of 40 at 2-kW PEP input and 80 at 1 kW. Typical loaded Q values of 10 to 15 are used in HF amplifiers. In comparison, we are dealing with a relatively high loaded Q, so losses in the strip-line tank-circuit components must be kept very low. To this end, small-diameter Teflon rods are used as mechanical drives for the tuning capacitor and for physical support as well as mechanical drive for the output-coupling capacitor. The tuning vane or flapper capacitor is solidly grounded through a wide flexible strap of negligible inductance, directly to the chassis in close proximity to the grid-return point. A flexible-strap arrangement, similar to that of the tuning capacitor, is used to connect the output-coupling capacitor to the center pin of a type-N coaxial connector mounted in the chassis base.

Ceramic (or Teflon) pillars, used to support the air strip line, are located under the middle set of plate-line dc isolation bushings. This places these pillars well out of the intense RF field associated with the tube, or high-impedance end of the line. In operation, plate tuning and loading is quite smooth and stable, so a high-loaded Q is apparently not bothersome in this respect.

In this amplifier, output coupling is accomplished by the capacitive probe method. Major advantages of capacitive-

Fig. 120 — The placement of input-circuit components and supporting bracket may be seen in this bottom view. When the bottom cover is in place, the screened air inlet allows the blower to pull air in, pressurizing the entire under-chassis area. The Minibox on the rear apron is a housing for the input reflectometer circuit.

Fig. 121 — The tube and plate line is in place with the top and side of the compartment removed for clarity. The plate-tuning vane is at bottom center. A bracket is attached to the side panel to support the rear of the Teflon rod supporting the tuning vane. The coil at the opposite end of the plate line is RFC1, connected between the high-voltage-bypass plate and the top section of the plate-line sandwich. Items outside the tube enclosure include the filament transformer, blower motor, relays and a power supply to operate a VOX-controlled relay system.

Fig. 119 — Front-panel layout of the 2-meter kilowatt amplifier.

Table 8
Performance Data

	1 kW	1.6 kW
Power input, watts	1000	1600
Plate voltage	2600	2450
Plate current (single tone)	385 mA	660 mA
Plate current (idling)	50 mA	50 mA
Grid bias	− 10 V	− 10 V
Grid current (single tone)	35 mA	54 mA
Drive power, watts	18	41
Efficiency (apparent)	59.5%	61.8%
Power gain (apparent)	15.2 dB	13.9 dB
Power output, watts	595	1000

probe coupling are loading linearity and elimination of moving-contact surfaces.

Capacitive-probe coupling is a form of reactive transformation matching whereby the feed-line (load) impedance is transformed to the tube resonant-load impedance (Z_o) of 1800 ohms (at the 2-kW level) by means of a series reactance (a capacitor in this case).

Support Electronics

An RF-output monitor is a virtual necessity in VHF amplifiers. In this amplifier, one of these built-in circuits is achieved quite handily. The circuit consists of a 10:1 resistive voltage divider, diode rectifier, filter and adjustable indicating instrument. Two 7500-ohm, 2-watt carbon resistors are located in the plate compartment connected between the type-N RF-output connector and a BNC connector, inside a Minibox, with the 1500-ohm, 1-watt composition resistor and the rectifier diode joined at this point. Relative output voltage is fed, via feedthrough capacitors, to the level-setting potentiometer and multimeter switch.

Testing and Operation

The amplifier is unconditionally stable, with no parasitics. To verify this, a zero-bias check for stability was made. This involved shorting out the Zener diode in the cathode return lead, reducing bias to essentially zero volts. Plate voltage was applied, allowing the tube to dissipate about 885 watts. The input and output circuits were then tuned through their ranges with no loads attached. There was no sign of out-

put on the relative output meter and no change in the plate and grid currents. As with most cathode-driven amplifiers, there is a slight interaction between grid and plate currents during normal tune-up under RF-applied conditions. This should not be misconstrued as amplifier instability.

Tolerances of the Zener diode used in the cathode return line will result in values of bias voltage and idling plate currents other than those listed in Table 8. The 1N3311, a 20-percent-tolerance unit, is rated at 12 volts nominal but actually operates at 10 volts in this amplifier (within the 20-percent tolerance).

To commence routine operation, the variable capacitor in the input circuit should be set at the point where lowest input SWR was obtained during the cold-tube initial tune-up. The ability of the plate tank to resonate at 144-145 MHz with the top cover in place should be verified with

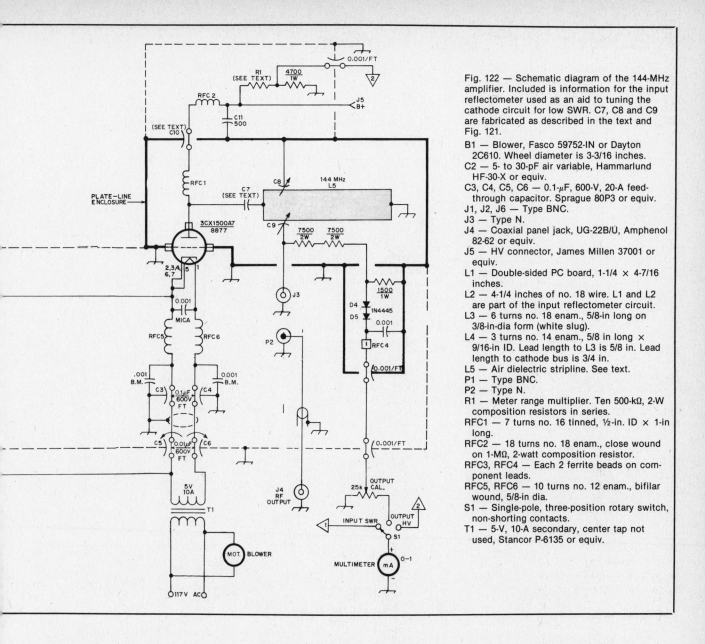

Fig. 122 — Schematic diagram of the 144-MHz amplifier. Included is information for the input reflectometer used as an aid to tuning the cathode circuit for low SWR. C7, C8 and C9 are fabricated as described in the text and Fig. 121.

B1 — Blower, Fasco 59752-IN or Dayton 2C610. Wheel diameter is 3-3/16 inches.

C2 — 5- to 30-pF air variable, Hammarlund HF-30-X or equiv.

C3, C4, C5, C6 — 0.1-μF, 600-V, 20-A feed-through capacitor. Sprague 80P3 or equiv.

J1, J2, J6 — Type BNC.

J3 — Type N.

J4 — Coaxial panel jack, UG-22B/U, Amphenol 82-62 or equiv.

J5 — HV connector, James Millen 37001 or equiv.

L1 — Double-sided PC board, 1-1/4 × 4-7/16 inches.

L2 — 4-1/4 inches of no. 18 wire. L1 and L2 are part of the input reflectometer circuit.

L3 — 6 turns no. 18 enam., 5/8-in long on 3/8-in-dia form (white slug).

L4 — 3 turns no. 14 enam., 5/8 in long × 9/16-in ID. Lead length to L3 is 5/8 in. Lead length to cathode bus is 3/4 in.

L5 — Air dielectric stripline. See text.

P1 — Type BNC.

P2 — Type N.

R1 — Meter range multiplier. Ten 500-kΩ, 2-W composition resistors in series.

RFC1 — 7 turns no. 16 tinned, 1/2-in. ID × 1-in long.

RFC2 — 18 turns no. 18 enam., close wound on 1-MΩ, 2-watt composition resistor.

RFC3, RFC4 — Each 2 ferrite beads on component leads.

RFC5, RFC6 — 10 turns no. 12 enam., bifilar wound, 5/8-in dia.

S1 — Single-pole, three-position rotary switch, non-shorting contacts.

T1 — 5-V, 10-A secondary, center tap not used, Stancor P-6135 or equiv.

a grid-dip meter. Use a one-turn link attached to the RF output connector. Top and bottom covers are then secured. As with all cathode driven amplifiers, *excitation should never be applied when the tube heater is activated and plate voltage is removed.* Next, turn on the tube heater and blower simultaneously, allowing 90 seconds for warm-up. A plate potential between 2400 and 3000 volts then may be applied and its presence verified on the multimeter. The power supply should be able to deliver

800 mA or so. With the VOX relay actuated, resting current should be indicated on the cathode meter.

A small amount of drive is applied and the plate-tank circuit tuned for an indication of maximum relative power output. The cathode circuit can now be resonated, tuning for minimum reflected power on the reflectometer, and not for maximum drive power transfer. Tuning and loading of the plate-tank circuit follows the standard sequence for any cathode driven amplifier.

Resonance is accompanied by a moderate dip in plate/cathode current, a rise in grid current and a considerable increase in relative power output. Plate-current dip is not absolutely coincident with maximum power output, but it is very close. Tuning and output-loading adjustments should be for maximum efficiency and output as indicated on the output-meter. Final adjustment for lowest SWR at amplifier input should be done when the desired plate input-power level has been reached.

A 220-MHz High-Power Amplifier

This amplifier employs the 3CX1500A7/8877 in a cathode-driven circuit. The grid is grounded directly to the chassis, adding to the stability. The amplifier (Figs. 123 to 127) is unconditionally stable — more so than some amplifiers built for the HF region.

Circuit Details

Fig. 124 is a schematic diagram of the amplifier. The input circuit consists of a T network. Medium values of Q were chosen to provide high efficiency. Both the cathode and the heater are operated at the same RF potential; the heater is held above

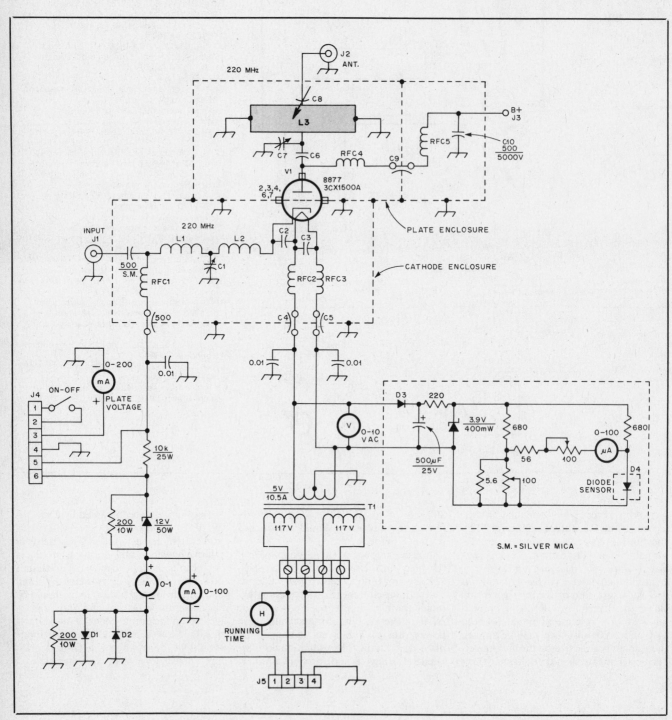

Fig. 124 — Schematic diagram of the 220-MHz amplifier. Unless otherwise specified, all capacitors are disc ceramic and resistors are 1/2-watt carbon composition.

C1 — Air variable, 15 pF.
C2, C3 — Button mica, 500-pF, 500-V rating.
C4-C9, inclusive — Teflon capacitor (use 10-mil Teflon sheet).
C10 — Doorknob capacitor, 500 pF, 5-kV rating.
D1-D4 — 1000 PIV, 3A.

J1, J2 — Coaxial receptacle, type N.
J3 — High-voltage connector (Millen).
L1 — 3 turns no. 14, 1/4-inch ID, 3/4-inch long.
L2 — 1/4-inch-wide, 2-3/8-inch-long copper flashing strap.
L3 — Plate inductor (see Fig. 125).
RFC1 — 8 turns no. 18 enam. 1/2-inch dia,

3/4-inch long.
RFC2, RFC3 — 10 turns no. 18 enam. bifilar wound on 3/4-inch Teflon rod close wound.
RFC4 — 5 turns no. 16 enam. wound on 1-MΩ, 2-watt composition resistor.
T1 — Filament transformer, 5.0 V at 10.5 A.

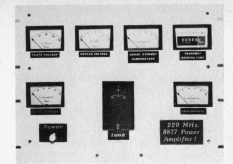

Fig. 123 — Front-panel layout of the 220-MHz kilowatt.

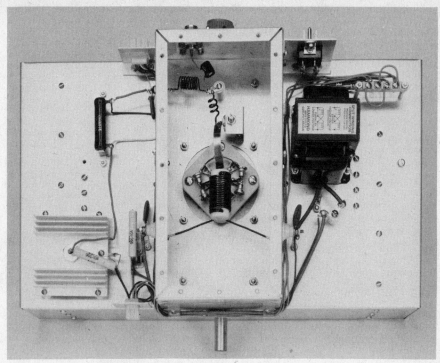

Fig. 125 — Bottom view of the amplifier. RFC2 and RFC3 can be seen above tube socket (bifilar winding). Copper strap is L2 shown connected to C1. Small coil is L1 and larger coil is RFC1. The grid of the tube should be grounded to the chassis with finger stock similar to that used in the plate line. Component mounted on the heat sink at left is the Zener diode used for biasing purposes.

RF ground by the impedance of the filament choke. The plate tank is a pair of quarter-wavelength striplines placed symmetrically about the tube. This arrangement permits a more uniform flow of current through the anode, preventing "hot spots" on the anode conducting surface. Additionally, tube output capacitance is effectively halved, as one-half the tube capacitance (13 pF) is used to load each stripline. Striplines act as low-pass circuit elements even with the high unloaded-Q conditions found at 220 MHz. Linear inductors also offer control of odd-mode harmonics. No spurious responses could be found in this amplifier up through the 900-MHz region.

A strip-line impedance can be varied by changing its width and relation to its ground planes. Physical dimensions of the tube limit the position of the stripline above one ground plane. In order to use a commercially available chassis, the stripline was placed 1¼ inch above one side of an inverted 4-inch-high chassis. This means that approximately 75 percent of the RF current flows through the chassis, but only 25 percent flows through the top shield cover. The small percentage flowing through the top reduces the effect of any mechanical anomalies associated with a removable cover.

For quarter-wavelength lines, the ratio of line impedance to reactance should be between 1.5 and 2.0 for the best bandwidth. Taking stray capacitance into account, expected tuning capacitance and tube output capacitance gives a value of 55 ohms for X_C. Values of line impedance versus line length for resonance at 222 MHz were computed on a programmable calculator for impedances between 30 and 100 ohms. These were plotted on a graph. Final dimensions were determined using this system, choosing dimensions that fell into the middle of the graph, thus allowing for any unpredicted effects.

Construction details are given in Figs. 125, 126 and 127. The plate blocking capacitor consists of a sandwich of brass plate and the stripline, with Teflon sheet as the dielectric. This forms a very-low-loss, high-voltage capacitor. The plate bypass capacitor is built along the same principles. A piece of circuit board was sandwiched with Teflon sheet to the side wall of the chassis. This technique is used effectively throughout as an inexpensive bypass or feedthrough capacitor at VHF.

Amplifier output is coupled through a capacitive probe. Transformation of the load impedance to the tube resonant-load impedance is achieved by means of a series reactance (the loading capacitor). The tuning capacitor is solidly grounded by means of a flexible strap of negligible inductance.

A rather elaborate metering system is employed. Although all of the meters provide useful data, only the plate and grid meters are necessary for proper amplifier use.

The anode exhaust-temperature metering circuit takes advantage of a thermal property of semiconductors. As the temperature changes, the forward resistance of a diode changes in a nearly linear manner. The diode sensor is made a part of a bridge circuit, allowing calibrated operation. Calibration may be determined by packing the diode in ice for the low point (0° C) and immersing it in boiling water for the high point (100° C). The amount of heat dissipated by the tube is inversely proportional to the efficiency for a given power input. Low heat dissipation yields longer tube life.

High-power amplifiers require considerable attention to cooling. The plate compartment is pressurized by air from an external blower. Holes in the chassis allow a portion of this air to pass through the grid and cathode structure. Most of the air

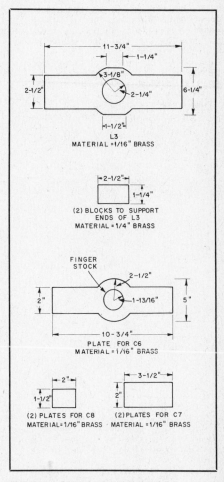

Fig. 126 — Construction details of the 220-MHz plate line and associated components.

flows through the anode, a handmade Teflon chimney, then out the top cover. Aluminum screening is tightly bonded around these two openings. No radiated RF could be detected around the chassis except within one inch of the anode exhaust hole.

To commence operation, the input should be adjusted for minimum SWR with no voltages applied. The covers should be in place whenever voltage is present. Drive should *never* be applied without plate voltage and a load connected if the filament is energized. Cooling air must always be supplied whenever the filament is turned on.

After a 60-second warmup, small amounts of drive may be applied. The plate circuit is then tuned for maximum output indication. The drive level is then increased. Tuning and loading follow the normal procedure for any cathode-driven amplifier: Adjustments are made for maximum output and efficiency. When the desired plate output power has been reached, the input circuit should be adjusted for minimum input SWR.

Fig. 127 — Interior view of the 220-MHz plate compartment. Contact of the tube with "hot side" of C6 is accomplished with suitable finger stock (available from the tube manufacturer). This conductor, in conjunction with a similar one separated by the Teflon insulator, forms the L3/C6 combination. The entire assembly is sandwiched together by means of four insulated bushings (approximately ¾-inch diameter). Placement of bushings is not critical. RFC4 can be seen at the right connected to C9. C8 is at the right of the photo and has a nominal spacing of 1 inch to a similar plate soldered to L3. Tuning capacitor C7 can be seen at the left connected to C6. Drive mechanism can be of builder's choice.

A Legal-Limit 2-Meter Tetrode Amplifier

With today's typical 2-meter multimode transceiver or transverter, increasing transmit output power beyond a few hundred watts is a difficult proposition. High-level transistor "bricks" are available to raise the 10-W output from the modern transceiver to 200 W or so; an amplifier based on a member of the 4CX250 family is capable of 300 W. Going beyond this power plateau to the kilowatt level or the legal limit often requires cascades of amplifiers. The most popular types of tubes used at and beyond the kilowatt level at 144 MHz are the 8877, pairs of 8874s, and recently the 3CX800A7. While each of these tubes is of the high-mu design, with power gains approaching 16 dB, none of them is capable of producing significantly more output power than the 4CX250 when driven with only 10 W. A very popular approach is to drive a "brick" from the multimode transceiver; the brick in turn drives the triode amplifier to the desired power output.

A cleaner approach (both spectrally and in complexity) is to use a tetrode amplifier that has enough gain to develop the desired output from the 10 W available from the transceiver. The EIMAC 4CX1000A and 4CX1500B tetrodes are both capable of delivering over 1500 W output when operated class AB1. Typical realizable gain for either tube at 144 MHz is in excess of 25 dB. The amplifier shown in Figs. 128 through 150 and described here uses a 4CX1000A or 4CX1500B (the tubes are interchangeable in this design) to develop 1500-W output in linear service. The drive required to attain this output level is less than 10 W. Clarke Greene, K1JX, and Jay Rusgrove, W1VD, designed and built this project.

Design Information

One of the main reasons for the popularity of grounded-grid triode amplifiers is the assumed inherent stability claimed for the grounded-grid circuit. There is an element of truth to that legend; since the gain of a grounded grid amplifier is almost always lower than a comparable grid-driven tetrode design, there is a larger "margin of error" available for sloppy design and construction techniques. That doesn't mean that grounded-grid triode amplifiers will not oscillate or that it is humanly impossible to make a tetrode work at high frequencies. The key consideration in maintaining stability in any kind of amplifier circuit is control of undesired feedback paths. To prevent oscillation, the gain of the amplifier in a forward direction (the desired path) must be substantially greater than the "gain" in the reverse direction. If there is sufficient leakage from the output back to the input, the stage will probably oscillate. Generally, the isolation between output and input

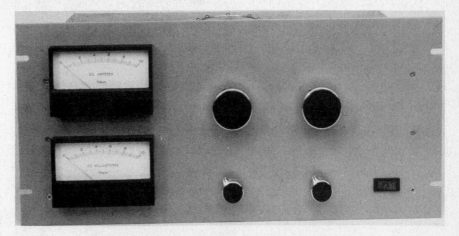

Fig. 128 — Front-panel view of the 2-meter legal-limit tetrode amplifier. Front-panel controls are for power, plate tuning and loading, meter switching and input tuning.

Fig. 129 — View of the plate compartment with the covers removed. Note the liberal use of machine screws to secure the top cover to ensure tight shielding.

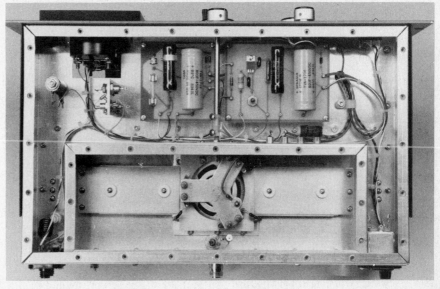

Fig. 130 — View of the amplifier with the bottom covers removed.

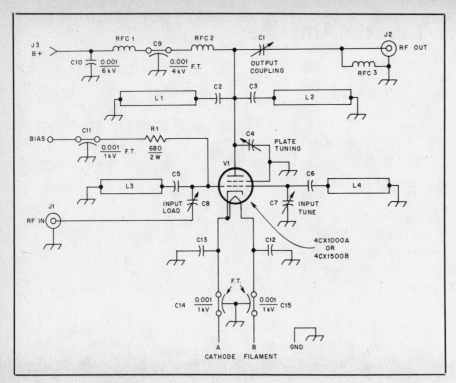

Fig. 131 — Schematic diagram of the 2-meter amplifier RF deck.

C1 — Output coupling capacitor, homemade from brass sheet. See text.
C2, C3 — Plate blocking capacitor, homemade from brass sheet with 10-mil Teflon dielectric. See text.
C4 — Plate tuning capacitor, homemade from brass sheet. See text.
C5, C6 — Grid blocking capacitor, homemade from brass sheet with 10-mil Teflon dielectric. See text.
C7 — 8.7-pF miniature air variable (Cardwell 160-104) with 10-pF silver-mica in parallel.
C8 — 14.2-pF miniature air variable (Cardwell 160-107) with 10-pF silver-mica in parallel.
C9 — 0.001-μF, 4-kV feedthrough capacitor (Erie 2498 or equiv.).
C10 — 0.001 μF, 6-kV disc ceramic (Centralab DD60 series or equiv.).
C11, C14, C15 — 0.001-μF, 1-kV feedthrough

capacitor.
C12 — 500-pF, 500-V silver-mica, DM-15 type.
C13 — Three 500-pF, 500-V silver-mica capacitors, DM-15 type, between the cathode pins and the brass socket plate. See text.
J1, J2 — UG-58 female chassis-mount Type-N connector.
J3, J4 — Millen 37001 high-voltage connectors.
L1, L2 — Stripline inductor made from brass sheet. See text.
L3, L4 — Stripline inductor made from brass sheet. See text.
RFC1, RFC2 — 7t no. 16 wire, ½-inch ID, 1 inch long.
RFC3 — 18t no. 18 enam. wire close-wound on a 1-MΩ, 2-W carbon-composition resistor.
V1 — EIMAC 4CX1000A or 4CX1500B. See text.

impedance path to ground is needed for best shielding effect. Any inductance in series between the screen connection and ground allows the screen potential to rise above ground and permits undesired coupling between anode and control grid. A (very expensive) version of the 4CX1000, the 4CX1000K, has a special screen ring built in to provide direct, low-impedance grounding of the screen grid through a matching special socket (also very expensive). Instead of using this high-priced combination, a homemade socket was fabricated to accommodate a 4CX1000A. Parts from a used standard socket were used extensively. The resulting socket, shown in the photographs and described in detail later, yields a measured isolation between amplifier output and input of greater than 40 dB at 144 MHz. Since the expected stage gain is about 25 dB, adequate isolation is available.

Fig. 131 is a schematic diagram of the amplifier RF deck. Fig. 132 shows the filament, grid and screen supplies, while Fig. 133 is a diagram of the high-voltage supply. To directly ground the screen, the tube cathode must be left floating for dc. The cathode is actually operated below dc ground by the desired tube screen potential. The cathode is grounded for RF by standard silver-mica capacitors. Since circuit circulating current is low at that point, less than perfect RF grounding causes no ill effects; any RF potential between the cathode and ground acts as degenerative feedback, only raising the amplifier's drive requirement for full output.

Both the output tank circuit and the input tank circuit are formed from dual quarter-wavelength striplines (L1, L2 and L3, L4). The circuit configuration, as described in Chapter 15, reduces the effect of tube input and output capacitances. Particularly in the input circuit where the circuit capacitance is very high, splitting the capacitance between the two lines allows for a physically manageable line, not possible with only a single quarter-wavelength line.

Historically, many amplifier builders have chosen to use a half-wavelength line at this point, letting the tube input capacitance "absorb" a large portion of the line. Tuning is then accomplished by varying a tuning capacitor at the open end of the line. Unfortunately, at frequencies well below the amplifier design frequency, the impedance to ground presented by the tuning capacitor becomes so large as to vanish. The transmission line portion of the circuit also effectively vanishes, since its inductive reactance is almost zero at low frequencies. As a result, the control grid tends to float toward infinite impedance at low frequencies. The amplifier's gain at low frequencies then becomes significantly higher than the isolation between output and input at low frequencies, and the amplifier tends to oscillate.

Several published designs have exhibited

from feedback paths must be at least 10 dB greater than the stage gain for stable operation. For a 25-dB-gain amplifier, at least 35 dB of isolation between output and input is necessary to prevent oscillation. The grounded-grid triode amplifier, with its 16-dB stage gain, requires only about 26 dB of isolation between input and output for stability; a lot of sloppiness can be successfully tolerated.

The paths for feedback primarily take two forms: coupling between input and output caused by leakage paths, and coupling within the amplification device itself. Good construction practices can eliminate the first path through proper shielding and line bypassing. However, achieving minimal coupling through the tube itself is a challenging task. In any practical vacuum tube, parasitic reactances exist between every tube element and their connection to the outside world. The small-diameter interconnecting wires are significant fractions of a wavelength long at 144 MHz and above, and they may act as either induc-

tive elements or capacitive elements. Additionally, the interelement capacitances (for example, screen grid to control grid), while only a few picofarads, represent very low impedances at these high frequencies. A common approach to cancelling these effects is neutralization; unfortunately, most builders of high-power VHF and UHF amplifiers find neutralization a form of "black art" and are unsuccessful in attaining complete and unconditional stability as a result of their considerable efforts.

A different approach is taken here. Every effort is made to minimize feedback paths through sound design and good construction techniques. As a result, this amplifier is completely stable even without neutralization.

The first notable aspect of the amplifier is the direct grounding of the screen-grid connection to the chassis. Since the screen acts as a shield between the tube's anode and control grid (the output and input), the proper use of the screen can add significantly to output-input isolation. A very low

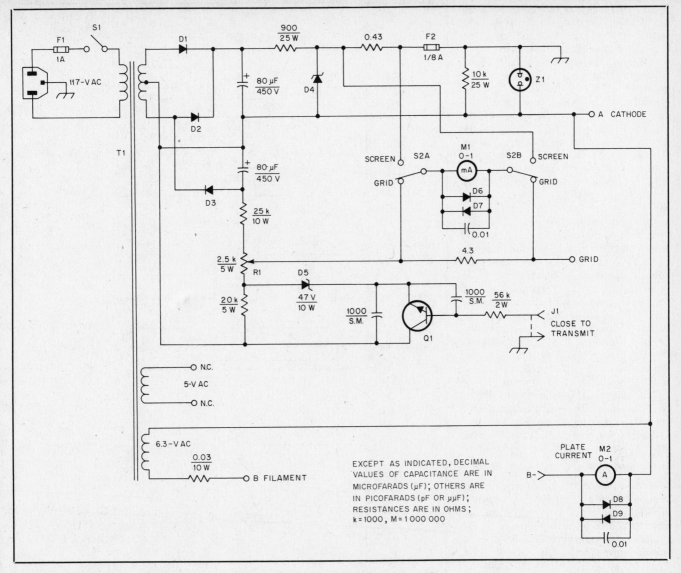

Fig. 132 — Schematic diagram of the filament and bias supplies for the tetrode 2-meter amplifier.

D1-D3, D6-D9 — Silicon rectifier, 1000 PIV, 2.5 A.
D4 — Five 56-V, 10-W Zener diodes in series.
D5 — 47-V, 10-W Zener diode.
J1 — Chassis-mount phono jack.

M1 — 0-1 mA dc meter, Simpson 15070 or equiv.
M2 — 0-1 A dc meter, Simpson 15101 or equiv.
Q1 — High-voltage NPN power transistor, 500-V, 1 A, TO-220 case (RCA SK3220 or equiv.).

T1 — Power transformer. Primary, 117-V ac; secondary, 260-0-260 V at 260 mA, 6.3 V at 8.8 A (Stancor P8356 or equiv.).
Z1 — 470-V MOV or surge-voltage protector (Siemens CG470L or equiv.).

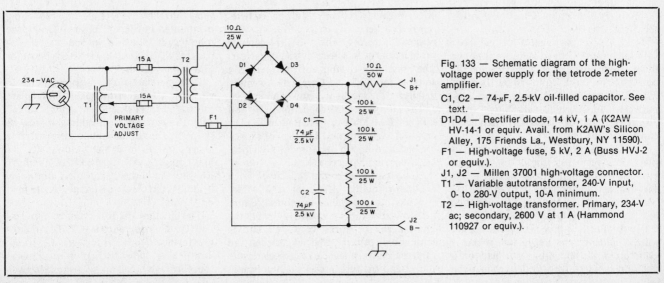

Fig. 133 — Schematic diagram of the high-voltage power supply for the tetrode 2-meter amplifier.

C1, C2 — 74-μF, 2.5-kV oil-filled capacitor. See text.
D1-D4 — Rectifier diode, 14 kV, 1 A (K2AW HV-14-1 or equiv. Avail. from K2AW's Silicon Alley, 175 Friends La., Westbury, NY 11590).
F1 — High-voltage fuse, 5 kV, 2 A (Buss HVJ-2 or equiv.).
J1, J2 — Millen 37001 high-voltage connector.
T1 — Variable autotransformer, 240-V input, 0- to 280-V output, 10-A minimum.
T2 — High-voltage transformer. Primary, 234-V ac; secondary, 2600 V at 1 A (Hammond 110927 or equiv.).

this tendency. The only effective cure is the use of a swamping resistor across the tube control grid to reduce the amplifier's input impedance at low frequencies. The unfortunate side effect is an increase in drive level required for full output. With the grounded quarter-wavelength lines as used here, the amplifier input impedance naturally decreases below the operating frequency. At HF, the control grid is essentially grounded. No tendencies toward oscillation at any frequency have been detected in the amplifier under any conditions.

Much of the control and dc power supply circuitry is similar to that used in the Medium-Power 144-MHz Amplifier described elsewhere in this chapter. The main difference is the reference point for the various bias supplies. Since the screen is grounded for dc, both the cathode and the control grid must be operated below dc ground. Additionally, the negative side of the plate supply must be connected to the tube cathode, rather than ground. In reality, this is only a slight variation on the usual dc supply schemes. The main difference is that the reference point for zero dc potential has been changed from ground to the tube cathode (which is normally at dc ground potential). Q1 allows switching of the tube cut-off bias to be switched with reference to ground. Grounding the center contact of J1 removes cutoff bias from the tube.

Construction Details

At first glance, this amplifier seems to require a lot of specialty components, especially the brass, copper, Teflon and G-10 stock used in the striplines and for the tuning and output coupling capacitor drive mechanisms. Fortunately, all of these materials are available in small quantities from Small Parts, Inc. The address is listed in Chapter 35.

Drawings for the required metal work are shown in Figs. 134 through 145. The main chassis that virtually everything is mounted on is 3 × 10 × 17 inches (Bud AC-416 or equiv.). The grid compartment is a similar chassis, 2.5 × 5 × 13.5 inches (Bud AC-1410 or equiv.), mounted inside the main chassis.

The plate compartment is formed from sheet aluminum bent to form walls above the main chassis. See Fig. 135. Standard construction techniques described in Chapter 24 can be used. None of the compartment dimensions is very critical from an electrical standpoint, so only mechanical considerations apply.

The tube socket is fabricated from a plate of sheet brass (Fig. 136), the grid stripline (also made from brass — see Fig. 137) and components from an EIMAC SK-800 series socket. Only the ceramic spacers, contact fingers and bayonet from the socket are used here. The predominant failure mode for a power-tetrode socket is puncture of the screen-bypass capacitor, a part not required for this design since the

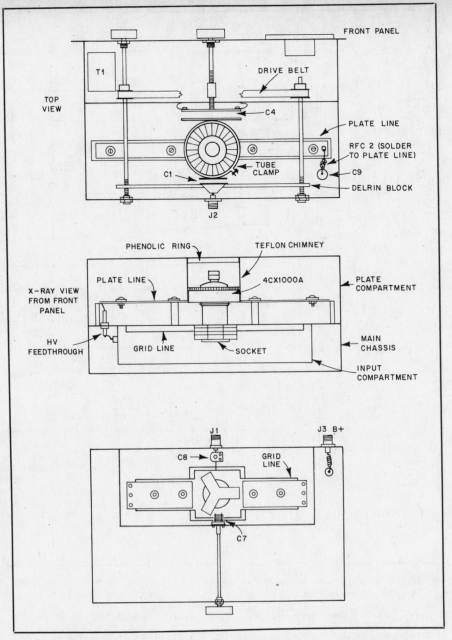

Fig. 134 — General layout of the 2-meter amplifier, shown from various angles.

screen is going to be directly grounded here. Surplus sockets can often be found at very reasonable prices, with the only damage being a destroyed screen-bypass capacitor. The remaining components are perfectly suitable for use here. Complete assembly details for the socket are shown in Fig. 138.

The cathode bypass capacitors, C12 and C13, are DM-15 silver-mica units with their leads cut to 0.25 inches. The type of capacitor and lead length are somewhat critical as the lead inductances and the parasitic inductances internal to the capacitors form a series-resonant circuit at 144 MHz, yielding low impedance to ground. Other types of capacitors may be suitable, but the type suggested works very well and is reasonably priced. The capacitors are soldered between the cathode tabs on the socket and the screen contact brass piece

(Fig. 139). C8, the grid loading capacitor, is mounted in the air between J1 and the tube socket. This is a set-and-forget component. C7 is adjustable from the panel for grid tuning.

Both the plate-circuit striplines (L1, L2) and grid-circuit striplines (L3, L4) are fabricated from sheet brass (Figs. 140 and 141). Brass or nickel-plated brass hardware should be used throughout the stripline assemblies wherever the hardware comes into contact with a surface that conducts RF. Steel and other hardware will instantly heat up from flowing RF current and melt or "spark."

The blocking capacitors for both the input (C5, C6) and output (C2, C3) circuits are formed from a sandwich of 10-mil Teflon sheet between the two brass pieces that form the tuned lines. Details are given in Figs. 140 and 141. Care should be taken when assembling these components to ensure that no metal chips embed themselves into the Teflon. Dow Corning DC-4 silicon grease should be used to coat the mating surfaces between Teflon and brass to fill in any imperfections in the surface that might allow eventual dielectric breakdown.

The plate tuning (C4) and output coupling (C1) capacitors are homemade double-plate units, also fabricated from brass sheet (Figs. 134 and 142). The flexible sections used for connecting the tuning capacitor to ground and for connecting the coupling capacitor to the output connector are made from 3-mil beryllium-copper sheet. All soldering to the plate tank circuit was done using silver solder to minimize RF and thermal resistance.

The drive mechanism for the plate-tuning capacitor (shown in Figs. 134 and 143) is made from a short section of threaded Delrin rod. The rod threads into a mating piece held fixed on the movable plate of the tuning capacitor. As the tuning control is turned, the capacitor plate moves in relation to the plate line.

The output coupling capacitor is similar in principle, except that for mechanical reasons, the movable capacitor plate is attached to a piece of 0.25-inch-thick Teflon sheet. Assembly details are shown in Figs. 134 and 144. As the two belt-coupled drive shafts rotate in parallel, the movable coupling-capacitor plate moves in relation to the fixed plate. Threaded G-10 fiberglass rods are used for the coupling-capacitor control shafts because Delrin lacked sufficient mechanical rigidity. Since the rods cross the plate tank circuit near the low-impedance ends of the plate line, no discernible heating has been observed. Mechanical stops in the form of locked pairs of nuts on the drive shafts limit the travel for the capacitor plates, preventing possible shorting of the plate voltage to ground.

The amplifier is mounted to a standard 19-inch-wide rack panel that is 8¾ inches high. Complete fabrication information for the front and rear amplifier panels is given in Fig. 145.

Most of the dc supply and control components are mounted on two etched circuit boards. Parts-placement information for these boards is given in Fig. 146. The Zener diodes for regulating the screen supply are mounted on a small aluminum plate supported by an aluminum bracket. No additional heat sinking is required. The protective components for the two panel meters (D6-D9 and 0.01-μF disc-ceramic capacitors) are mounted directly on their respective meter terminals.

Cooling for the tube is provided by an airstream that pressurizes the anode compartment. Air enters the chassis through a plastic flange (Nutone 366 or equiv.), screened with brass or aluminum screening. The majority of air passes through the tube anode cooler, through a chimney formed from 1/16-inch Teflon tape and out

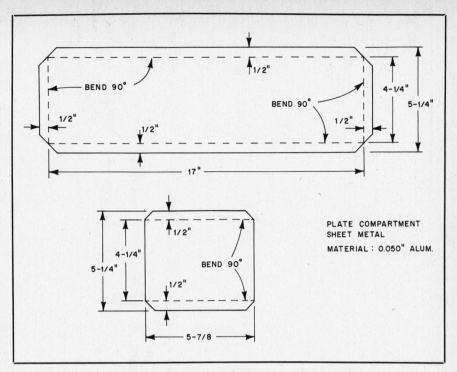

Fig. 135 — Details of the aluminum sheet that makes up the plate compartment.

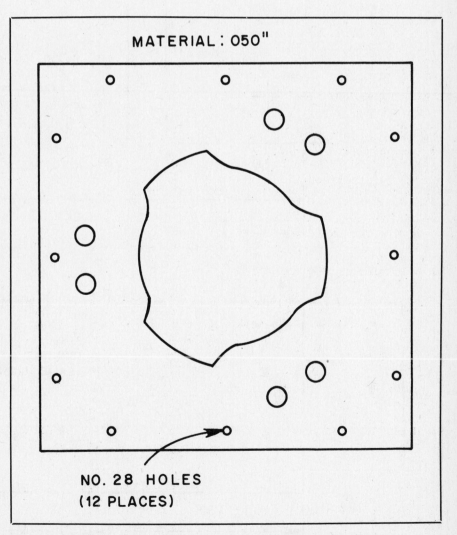

Fig. 136 — Template for the homemade brass plate for the socket assembly. This template is reproduced full size.

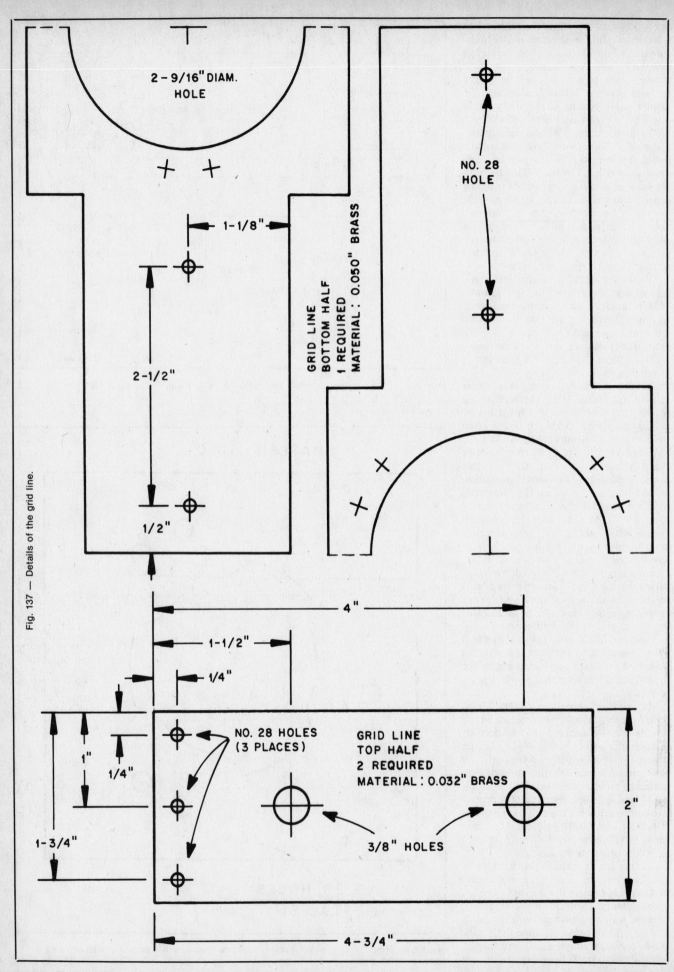

Fig. 137 — Details of the grid line.

2-9/16" DIAM. HOLE

NO. 28 HOLE

GRID LINE
BOTTOM HALF
1 REQUIRED
MATERIAL: 0.050" BRASS

1-1/8"

2-1/2"

1/2"

4"

1-1/2"

1/4"

1"

1/4"

1-3/4"

NO. 28 HOLES
(3 PLACES)

GRID LINE
TOP HALF
2 REQUIRED
MATERIAL: 0.032" BRASS

3/8" HOLES

2"

4-3/4"

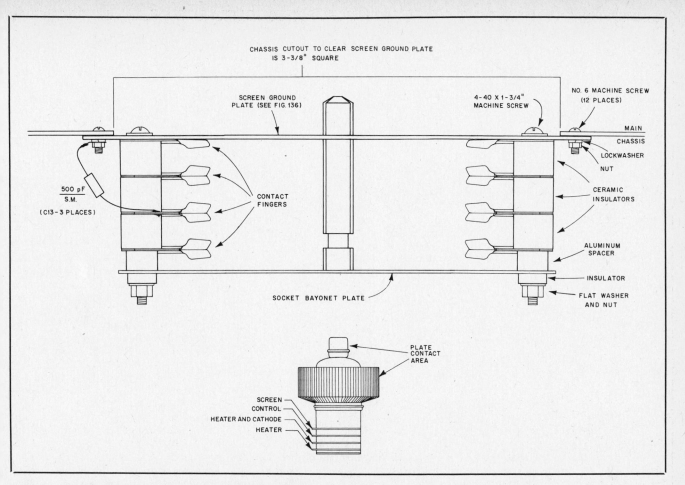

Fig. 138 — Assembly details of the homemade 4CX1000A socket.

Fig. 139 — Component placement around the socket. See text.

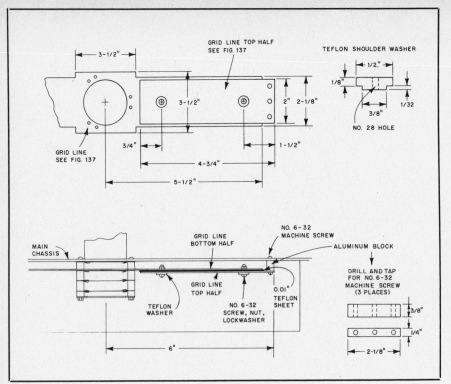

Fig. 140 — Assembly details for the grid compartment stripline.

Table 9

Operating Conditions for the 2-Meter 4CX1000A Amplifier

Plate voltage	3000 V
Plate current (single tone)	800 mA
Plate current (idling)	250 mA
Screen current	40 mA
Power output	1410 W
Efficiency	59%
Drive power	10 W

through a screened hole in the cabinet top cover. Some of the air flows past the tube base through slots in the tube socket into the grid compartment. Eight 0.25-inch holes in the grid compartment cover plate allow sufficient air to reach the tube base. Chimney details are shown in Fig. 147. A blower could be attached directly to the back of the cabinet if the builder so desires. See Chapter 15 for information on selecting an appropriate blower.

The high-voltage supply is shown in Fig. 148. The massive Hammond transformer is bolted to the ¼-inch-thick plate that forms the cabinet bottom. D1-D4 are mounted with thermal compound so that the chassis acts as a heat sink. C1 and C2 are mounted with aluminum straps. These capacitors were purchased surplus. Similar oil-filled capacitors are often available at flea markets at reasonable prices. Many surplus oil-filled capacitors use oil containing the carcinogenic chemical PCB, so avoid any units that are leaking. Note that the B– lead must float above dc ground in this design. Make connections between the power supply and amplifier with appropriate high-voltage wire, and ground the two chassis together for safety.

Tune-Up and Operation

After checking all the wiring and inspecting various RF components for solid connections, a resistance check should be made of the dc blocking capacitors in the input and output circuits. The resistance (with the power supplies off and disconnected) should be infinite. Any deviation

from this high value indicates a fault somewhere in the dielectric. Any fault like this will definitely cause a failure when voltage is applied. Clear the fault before applying any voltage.

Once it has been established that there are absolutely no dc leakage paths, attach the cover plate. Connect the supply of cooling air supply (or turn on the blower) and apply filament voltage. Do not connect any other voltages at this time! After five minutes of warm-up time, the air stream exiting through the top cover should be slightly warmer than room temperature.

At this point, the input circuit can be adjusted. Attach the bottom cover and connect a 2-meter exciter to the amplifier input through an SWR indicator. About 5 W of 144-MHz drive should be applied. Adjust C7 and C8 for minimum reflected power. C7 should tune rather broadly. A near perfect match should be possible. Once the operation is complete, drive should be removed. Never apply drive to the amplifier when screen voltage is present without plate voltage or when plate voltage is present and no antenna is connected.

The two bias supplies can now be connected. These are hazardous voltages, so considerable caution should be exercised. When S2 is set so that M1 is indicating screen current, the meter should indicate approximately 30 mA. This is a normal condition. The 4CX1000A often operates under conditions of negative screen current, so a 10-kΩ load resistor is connected from the screen supply to ground to maintain screen regulation during negative screen

current conditions. This load resistor draws approximately 30 mA. As a side benefit, having the meter's zero reference lie 30 mA upscale allows for meaningful measurements of negative screen current.

Measure the filament voltage and bias voltages at the tube base. Filament voltage should be 6.0-V ac. If not, experiment with the value of the series resistor until the voltage is correct. Cathode voltage relative to ground should be approximately –280 V dc. Control grid voltage should be in the area of –500 V dc. Shorting J4 should change the control-grid voltage to about –330 V dc, variable with R1. If all the voltages are within reasonable range, final tests can be made. Turn off all power to the amplifier.

Connect the high-voltage supply to the amplifier. Connect a dummy load to J2 through a VHF wattmeter such as a Bird Model 43. Connect the exciter to J1 through another VHF wattmeter. Make sure the covers are fastened in place, and connect the air supply. Turn the filament and bias supplies on and allow the filament to warm up for 5 minutes. Assuming all is still well, gradually raise the plate voltage. Plate current as indicated by M2 should remain at zero, and neither wattmeter should show any indication. For initial testing, a plate potential of 2500 V should be used. Short J4 and adjust R1 for a resting plate current of 250 mA.

Drive may now be gently applied. Increase drive to about 1 W. Plate current should increase noticeably. Pay close attention to screen current during initial tune-up. Excessive screen current will damage a tube much more quickly than excessive plate current. Adjust C7 for maximum plate current, and then alternately adjust C1 and C4 for maximum amplifier output. Gradually increase drive while adjusting C1 and C4. Do not key the amplifier for more than about 30 seconds at a time, particularly before C1 and C4 are properly adjusted. Once full output is achieved, C8 can be readjusted for zero reflected power as seen by the exciter at full drive. Once this control is properly set, it will not require further adjustment. Proper operating conditions for the amplifier are shown in Table 9.

Spectral output of the amplifier is shown

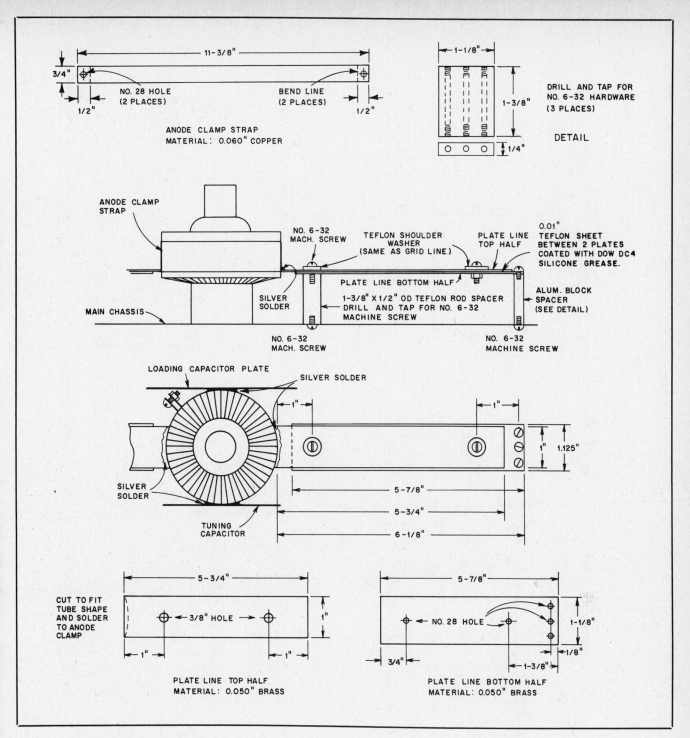

Fig. 141 — Plate line assembly details.

in Fig. 149. The filter shown in Fig. 150 is necessary to make the amplifier meet current FCC spectral-purity requirements for commercial equipment. This amplifier has provided stable, trouble-free operation at the legal limit for extended periods during several VHF contests and meteor showers. Although construction is a bit involved, the results are well worth the effort. This amplifier provides all the power necessary for weak-signal 2-meter work, including EME. It is easily driven by a 10-W transverter, and the resulting signal is exceptionally clean as verified by extensive on-the-air testing with local amateurs.

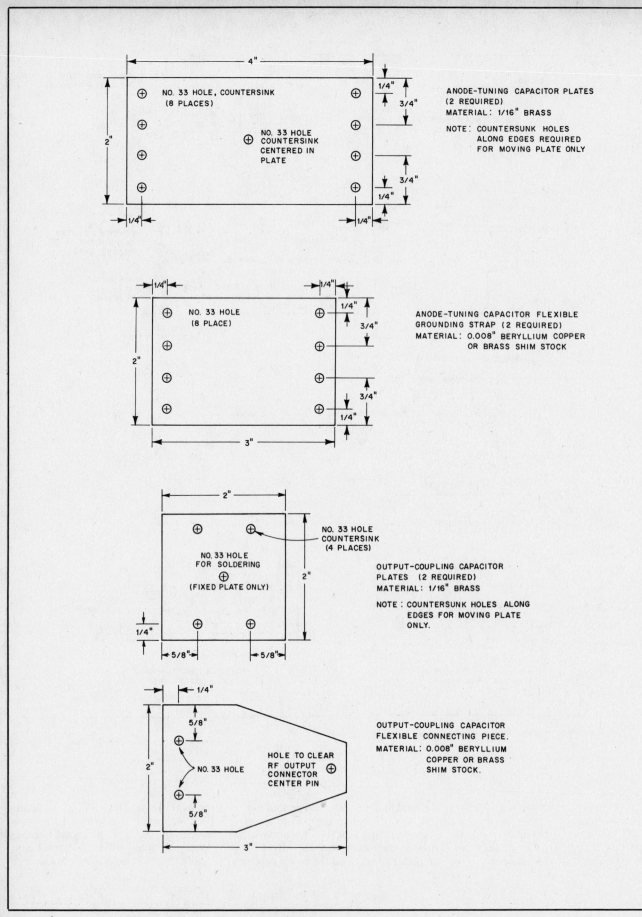

Fig. 142 — Construction details for the tuning and loading capacitors.

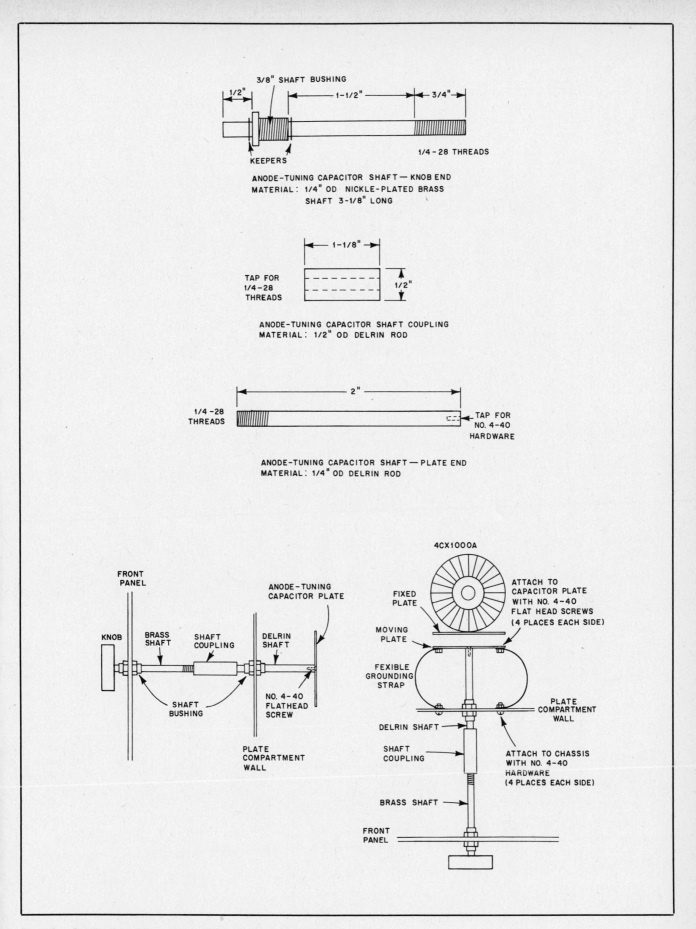

Fig. 143 — Details of the drive mechanism for the plate tuning capacitor.

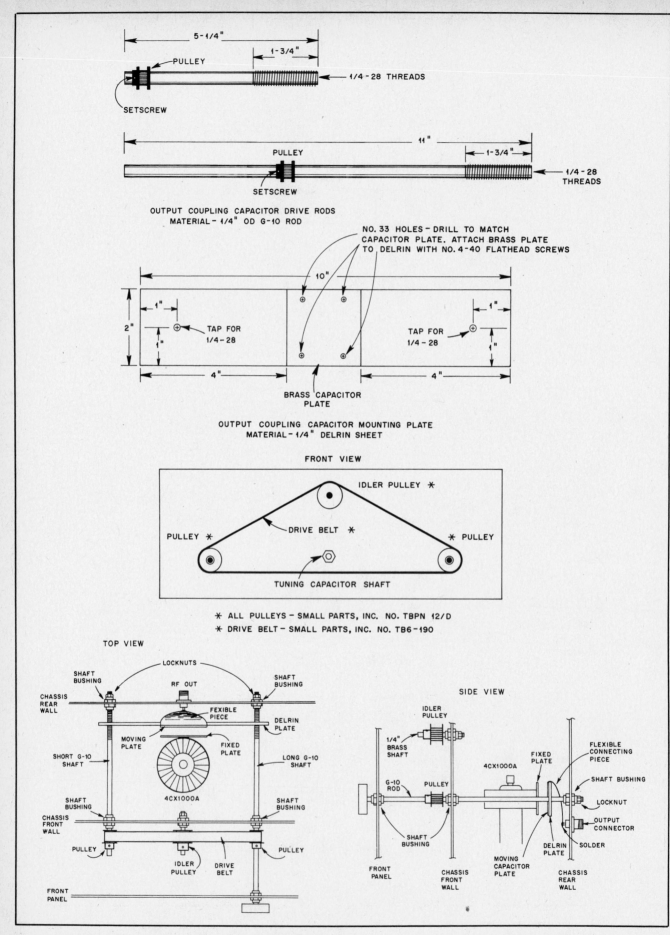

Fig. 144 — Details of the drive mechanism for the output coupling capacitor.

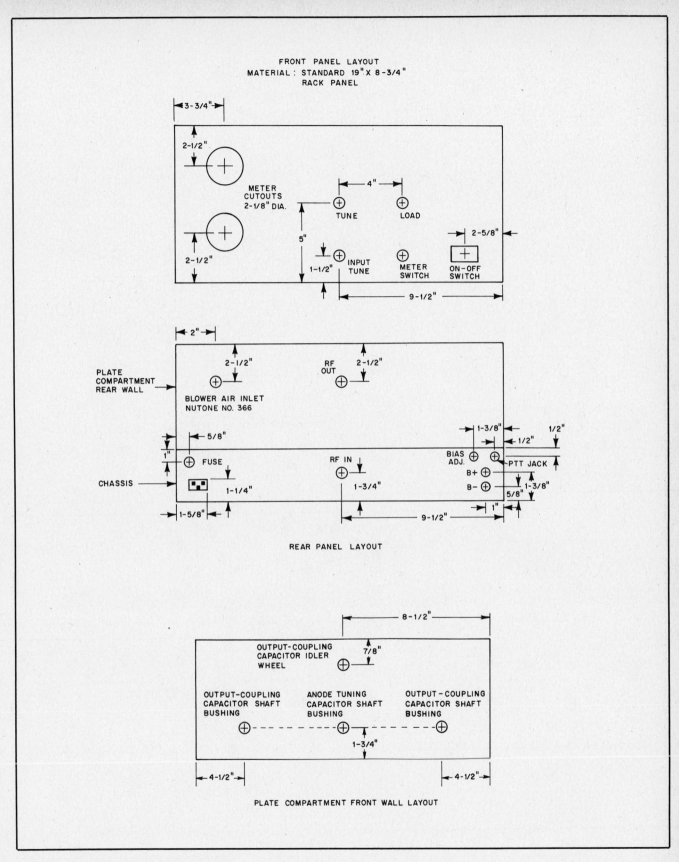

Fig. 145 — Layout for the front and back panels.

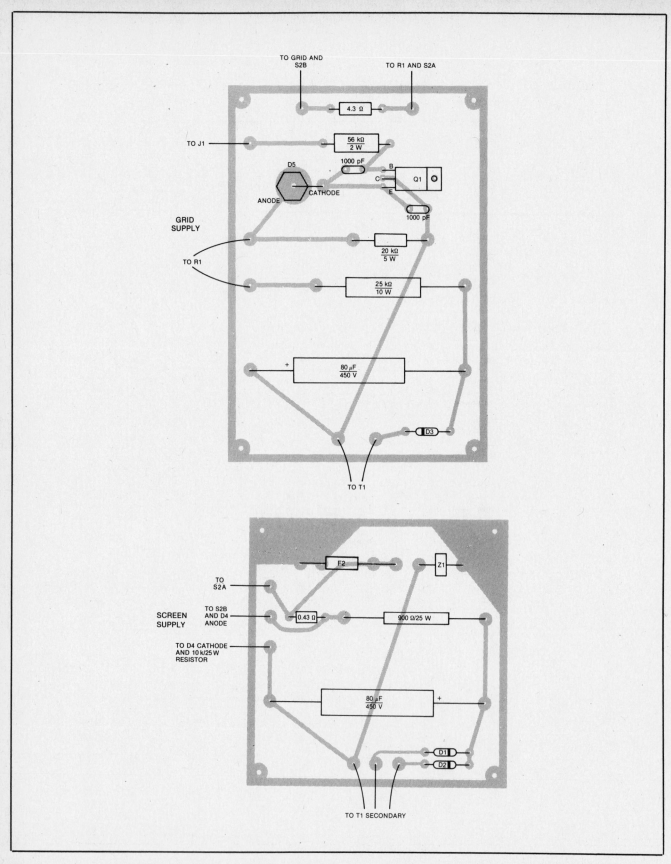

Fig. 146 — Parts placement diagrams for the grid and screen power supply boards. Full-size etching patterns are given at the back of this book.

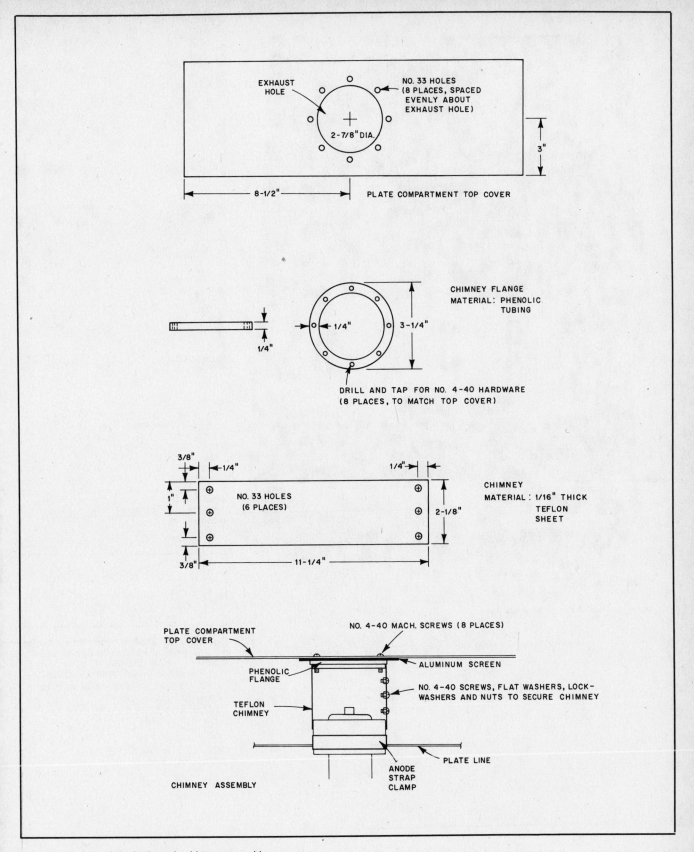

Fig. 147 — Details of the homemade chimney assembly.

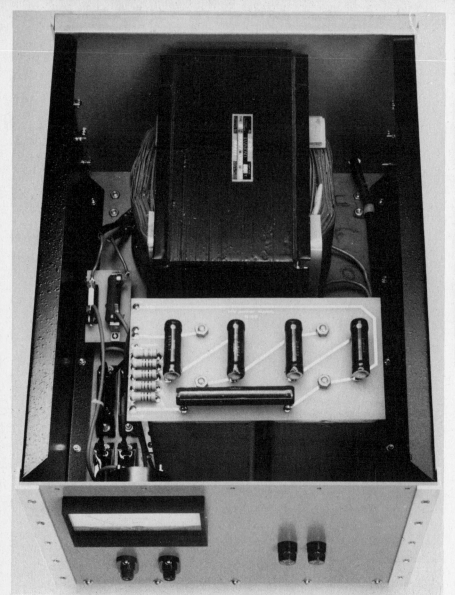

Fig. 148 — Interior of the high-voltage power supply for the tetrode 2-meter amplifier.

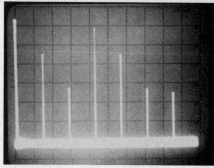

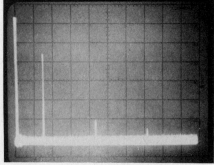

(B)

Fig. 149 — Spectral output of the tetrode 2-meter amplifier without filter (A) and after filtering (B). All harmonic and spurious emissions are at least 60 dB down after filtering. Vertical divisions are each 10 dB; horizontal divisions are each 100 MHz. The amplifier was being run at 1410-W output at 144 MHz. The fundamental is reduced in amplitude approximately 26 dB by means of notch filters to avoid spectrum-analyzer overload. This amplifier meets current FCC spectral-purity requirements.

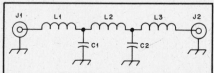

Fig. 150 — Schematic diagram of the filter for the legal-limit 2-meter tetrode amplifier. The filter is built inside a 3-1/4 × 2-1/8 × 1-1/8-inch Minibox (Bud CU-3017A or equiv.).

C1, C2 — 27-pF Centralab 850 series ceramic transmitting capacitor.
J1, J2 — Female chassis-mount Type-N connector (UG-58 or equiv.).
L1, L3 — 2t no. 14 wire, 0.3125 inch ID, 0.375 inch long.
L2 — 3t no. 14 wire, 0.3125 inch ID, 0.4375 inch long.

Chapter 32

UHF and Microwave Equipment

U HF and up represents a challenge to the experimenter and builder. The techniques used at HF and VHF do not always work here. Tuned circuits for this range rarely contain coils and capacitors as they do at HF and VHF. Sections of transmission line are used instead. Sometimes those line sections are in the form of cavity resonators. At other times, a section of line is a trace that has been etched on a PC board.

In this chapter you will find new techniques, and old, illustrated in the various construction projects. Those projects range from simple to sophisticated.

Dual-Gate GaAsFET Preamplifiers for 432 MHz

Dual-gate GaAsFET devices in plastic packages offer amateurs the opportunity to build inexpensive preamplifiers with good noise performance. Described here and shown in Figs. 1 through 3 are two preamplifiers that use GaAsFET devices costing only a few dollars each. These preamplifiers were designed and built by Kent Britain, WA5VJB.

Circuit Details

The basic circuit is shown in Fig. 2. In-formation is given so that the NEC NE41137, Motorola MRF966/967, Mitsubishi MGF1100 or 3SK97 devices may be used. C1, C2 and C5 can be miniature ceramic trimmer capacitors or piston trimmers. The exact value is not critical; any trimmer with a maximum value of 8 to 10 pF will do. C7, the output coupling capacitor, can be any small silver-mica unit between 50 and 100 pF. For stability, it is important to use good-quality capacitors for the gate 2 and source bypasses. Ceramic chip or leadless disc-ceramic types are necessary to provide a low-impedance ground.

Construction

The dual-gate GaAsFET preamplifiers are built "dead bug" style on a piece of double-sided PC board material. See Fig. 1. The circuit board acts as a low-impedance ground plane, and all ground connections are made directly to it. J1, J2 and C9 also secure the circuit board to the lid of the

Fig. 1 — The dual-gate GaAsFET 432-MHz preamplifier is built "dead bug" style on a piece of copper-clad circuit board material.

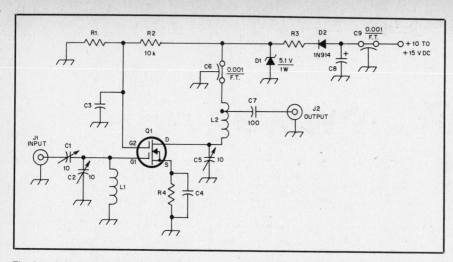

Fig. 2 — Schematic diagram of the 432-MHz dual-gate GaAsFET preamplifiers. The same circuit is used for either version, but some parts values are different. Resistors are ¼-W carbon composition types unless otherwise noted.

C1, C2, C5 — 10-pF (max.) trimmer capacitor. Piston trimmer preferred, but miniature ceramic trimmer will work fine.
C3, C4 — 200- to 1000-pF ceramic chip capacitor or leadless disc-ceramic capacitor.
C6, C9 — 400- to 1000-pF feedthrough capacitor.
C7 — 50- to 100-pF silver-mica capacitor.
C8 — 1- to 25-µF, 25-V electrolytic capacitor.
D1 — 5.1-V, 1-W Zener diode (1N4733 or equiv.).
D2 — 1N914, 1N4148 or any diode with ratings of 25 PIV and 50 mA or greater.
J1, J2 — Female chassis-mount BNC or Type-N connector.
L1 — For NE41137: 2½t no. 20 copper wire, 3/16-inch ID, spaced one wire dia. For

MRF966/967: 4t no. 20 copper wire, 3/16-inch ID, spaced one wire dia. For 3SK97 or MGF1100: 2t no. 20 copper wire, 3/16-inch ID, spaced 1 wire dia.
L2 — For NE41137, MGF1100 or 3SK97: 2t no. 20 copper wire, 3/16-inch ID, spaced one wire dia. Tap at 1t. For MRF966/967: 2t no. 20 copper wire, 3/16-inch ID, spaced one wire dia. Tap at ½t.
Q1 — NEC NE41137, Motorola MRF966/967, Mitsubishi MGF1100 or 3SK97. See text.
R1 — For NE41137, MGF1100 or 3SK97: 10 kΩ; for MRF966/967: 4.7 kΩ.
R2 — 10 kΩ for all versions.
R3 — 150- to 250-ohm, 1- or 2-W resistor.
R4 — For NE41137, MGF1100 or 3SK97: 47 Ω; for MRF966/967: 100 Ω.

enclosure. The enclosure can be any aluminum diecast box (Bud CU-124; Hammond 1490B) or Minibox (Bud CU-3002A) approximately 4 by 2 inches, or you can make one by soldering pieces of circuit-board material together.

Q1 is supported above the ground plane by its gate 2 and source leads, which are soldered directly to C3 and C4. The other ends of these capacitors are soldered to the ground plane. R4 should be soldered in place with essentially no lead length. Placement for the rest of the parts is not critical, but keep the leads short. A layout that has proven successful is shown in Fig. 3.

GaAsFETs are static-sensitive, so handle them accordingly. See Chapter 24 for information on working with static-sensitive devices. It's best to assemble the rest of the circuit and install Q1 last. Use a grounded iron if possible, and solder the leads quickly.

Parts for this project are available from Microwave Components of Michigan and Applied Invention, both listed in Chapter 35. NEC transistors are available from California Eastern Laboratories, 3260 Jay St., Santa Clara, CA 95050.

Peak the tuning capacitors on a weak signal and adjust for best signal-to-noise ratio. For best results, you should have access to a noise-figure meter. This preamplifier typically measures 0.7- to 0.9-dB noise figure and 22- to 24-dB gain with any of the listed devices.

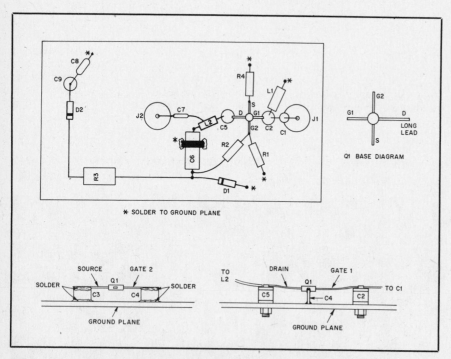

Fig. 3 — Suggested layout for the dual-gate GaAsFET 432-MHz amplifier. The circuit board is approximately 1¾ × 3¾ inches. J1 and J2 are approximately 1¾ inches apart.

GaAsFET Preamplifier for 70 cm

The preamplifier described here and shown in Figs. 4 through 6 offers good noise performance and gain, and is suitable for terrestrial, EME and satellite applications at 432 and 435 MHz. Gain is approximately 15 dB and the noise figure is 0.55 dB, measured on an HP8970A noise-figure meter with the HP346A noise source. This preamplifier is easy to build and offers stable operation with little adjustment. It was designed by Chip Angle, N6CA. The version shown here was built and tested in the ARRL lab by Mark Wilson, AA2Z.

Circuit Details

The schematic diagram of the 70-cm GaAsFET preamplifier is shown in Fig. 5. The circuit was originally designed for an NEC NE21889 GaAsFET, but a Mitsubishi MGF1402 or MGF1202 device will work fine. The version shown here uses an MGF1402, but most single-gate GaAsFETs will work.

This preamplifier uses source feedback to bring the input impedance close to 50 ohms. Input and output return loss is typically better than 15 dB. This means that a band-pass filter can be used ahead of the preamplifier without introducing mismatch loss or instability.

The design uses plenty of decoupling capacitors and RF chokes. Good-quality bypass capacitors are a must for stable operation, so use chip capacitors where specified. D1 and D2 are included for reverse and overvoltage protection. Regulator U1 allows virtually any voltage greater than 9-V dc to be used.

Construction

The preamplifier is built on a G-10 glass-epoxy printed-circuit board using surface-mounting techniques. Layout is shown in Fig. 6. All components mount to the circuit traces. The other side of the board is unetched copper except around the connector center pins, and acts as a ground plane. All grounded pads on the etched side are connected to the ground plane with eyelets or pieces of tinned wire that pass through holes in the board and are soldered to both sides.

The board is mounted to the lid of a die-cast box (Hammond 1590B or Bud CU-124) and is designed to accommodate Type-N connectors. Aluminum spacers for the connectors must be fabricated, as shown in Fig. 6. These spacers mount between the box lid and the ground plane of the PC board.

Q1 has two source leads. Only one is used on this board, and the extra lead may be clipped off. The ferrite bead on the drain lead mounts in a small hole cut in the board. The drain lead passes through this bead and is soldered to a pad.

GaAsFETs are static-sensitive, so handle them accordingly. See Chapter 24 for in-

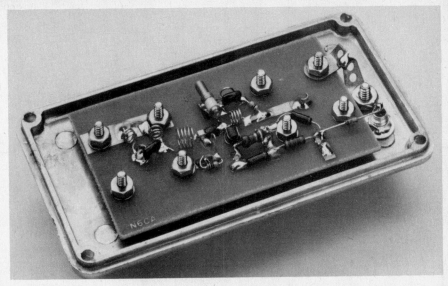

Fig. 4 — The 70-cm GaAsFET preamplifier is built on a PC board that is mounted to the lid of a diecast box. This version uses a Zener diode instead of the 3-terminal regulator shown in Figs. 5 and 6. Connector mounting hardware also holds the board in place. Component layout is shown in Fig. 6.

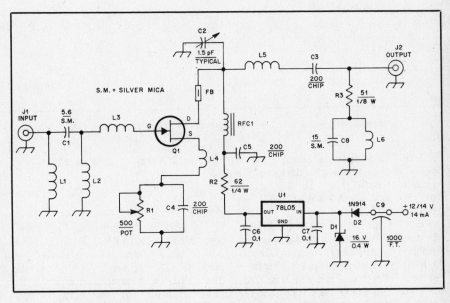

Fig. 5 — Schematic diagram of the 70-cm GaAsFET preamplifier. Resistors are carbon-composition types. Resistor values are given in ohms; capacitor values are given in pF.

C1 — 5.6-pF silver-mica capacitor or same as C2.
C2 — 0.6- to 6-pF ceramic piston trimmer capacitor (Johanson 5700 series or equiv.).
C3, C4, C5 — 200-pF ceramic chip capacitor.
C6, C7 — 0.1-µF disc ceramic capacitor, 50 V or greater.
C8 — 15-pF silver-mica capacitor.
C9 — 500- to 1000-pF feedthrough capacitor.
D1 — 16- to 30-V, 500-mW Zener diode (1N966B or equiv.).
D2 — 1N914, 1N4148 or any diode with ratings of at least 25 PIV at 50 mA or greater.
J1, J2 — Female chassis-mount Type-N connectors, PTFE dielectric (UG-58 or equiv.).
L1, L2 — 3t no. 24 tinned wire, 0.110-inch ID, spaced 1 wire dia.
L3 — 5t no. 24 tinned wire, 3/16-inch ID,

spaced 1 wire dia. or closer. Slightly larger diameter (0.010 inch) may be required with some FETs.
L4 — 1t no. 24 tinned wire, 1/8-inch ID.
L5 — 4t no. 24 tinned wire, 1/8-inch ID, spaced 1 wire dia.
L6 — 1t no. 24 tinned wire, 1/8-inch ID.
Q1 — Mitsubishi MGF1402 (see text).
R1 — 200- or 500-Ω cermet potentiometer set to midrange initially.
R2 — 62-Ω, 1/4-W resistor.
R3 — 51-Ω, 1/8-W carbon composition resistor, 5% tolerance.
RFC1 — 5t no. 26 enam. wire on a ferrite bead.
U1 — 5-V, 100-mA 3-terminal regulator (LM78L05 or equiv. TO-92 package).

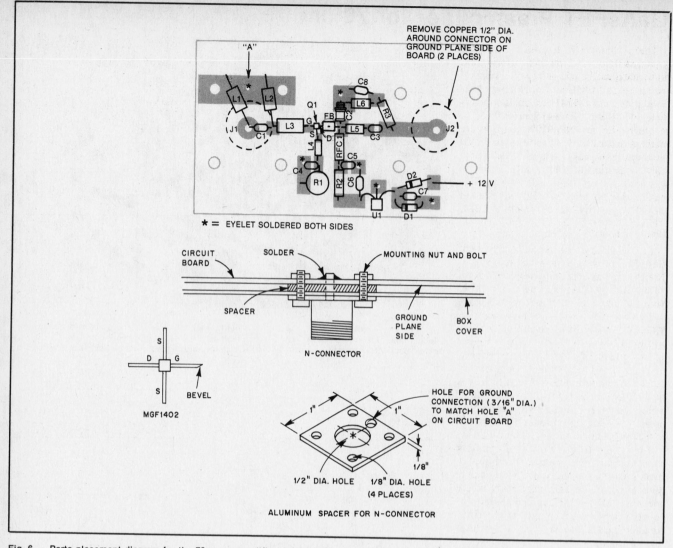

Fig. 6 — Parts placement diagram for the 70-cm preamplifier. A full-size etching pattern may be found at the back of this book.

formation on working with static-sensitive devices. It's best to assemble the rest of the circuit and install Q1 last. Use a grounded iron if possible, and solder the leads quickly.

Parts for this project are available from Microwave Components of Michigan and Applied Invention, both listed in Chapter 35. MGF1402 transistors are available from Advanced Receiver Research, Box 1242, Burlington, CT 06013. NEC transistors are available from California Eastern Labora-tories, 3260 Jay St., Santa Clara, CA 95050.

Adjustment

The preamplifier shown in Fig. 4 was assembled in the ARRL lab and tested before any adjustments were made. For starters, C2 was set at midrange. As assembled, the preamp measured 13 dB gain with a 0.57-dB noise figure — per-fectly acceptable performance for a device such as this. After adjusting C2 and bend-ing L3 and L5, the gain increased to 15.25 dB with a 0.54-dB noise figure. This is acceptable for all but perhaps the most demanding EME applications. No amount of spreading or compressing the coils could make the gain fall below 12 dB or the noise figure rise above 0.65 dB. At no time did the preamp display signs of instability. In short, this is a reproducible project. You can assemble it without test equipment and be reasonably sure of the performance.

Dual-Gate GaAsFET Preamplifiers for 1296 MHz

Dual-gate GaAsFET devices in plastic packages offer amateurs the opportunity to build inexpensive UHF preamplifiers with noise figures that were unattainable a few years ago. Described here and shown in Figs. 7 through 9 are two 1296-MHz preamplifiers that use GaAsFET devices costing only a few dollars each. These preamplifiers were designed and built by Kent Britain, WA5VJB.

Circuit Details

The basic circuit is shown in Fig. 8. Information is given so that either the NEC NE41137 or the Motorola MRF966/967 devices may be used. C1 is a piston trimmer. C2, C3 and C6 are ceramic chip capacitors. L1 and L2 are striplines made from brass or copper strips, or copper wire. For stability, it is important to use good-quality capacitors for the gate 2 and source bypasses. Ceramic chip capacitors are necessary to provide a low-impedance ground and help to avoid lead inductance problems. The value of these chip capacitors is not critical; any value between 200 and 1000 pF should work.

Construction

The dual-gate GaAsFET preamplifiers are built "dead bug" style on a piece of double-sided PC board material. See Fig. 7. The circuit board acts as a low-impedance ground plane, and all ground connections are made directly to it. J1, J2 and C8 also secure the circuit board to the lid of the enclosure. The enclosure can be any aluminum diecast box (Bud CU-124; Hammond 1490B) or Minibox (Bud CU-3002A) approximately 4 by 2 inches, or you can make one by soldering pieces of circuit board material together.

Dimensions for the striplines should be followed carefully, especially the height above the ground plane. Note that the lengths given for L1 do not include the 3/16-inch lip that is soldered to the ground plane. Also, keep in mind that the length of L2 includes transistor lead length and should be measured from the case of Q1 to C6. The brass or copper stock for L1 and L2 should be available at hobby stores. Alternatively, it is available from Small Parts, Inc. (see Chapter 35). Although 0.020-inch-thick stock was used here, 0.010- to 0.030-inch stock will work.

Q1 is supported above the ground plane by its gate 2 and source leads, which are soldered directly to C2 and C3. The other ends of these capacitors are soldered to the ground plane. R4 should be soldered in place with essentially no lead length. For best results, follow the layout shown in Fig. 9. The most important trick is to mount J1 and J2 with their center pins 2-3/8 inches apart.

GaAsFETs are static-sensitive, so handle them accordingly. See Chapter 24 for in-

Fig. 7 — The dual-gate GaAsFET 1296-MHz preamplifier is built "dead bug" style on a piece of copper-clad circuit board material in the lid of a diecast box. See Fig. 9 for parts-placement information.

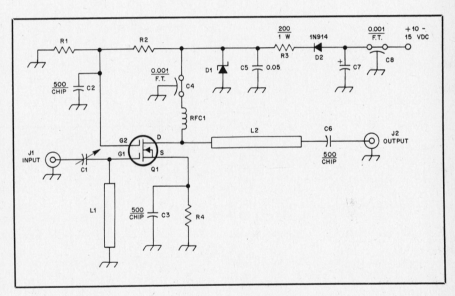

Fig. 8 — Schematic diagram of the 1296-MHz dual-gate GaAsFET preamplifiers. The same circuit is used for either version, but some parts values are different. Resistors are ¼-W carbon-composition types unless otherwise noted.

C1 — 10-pF (max.) piston trimmer capacitor (Johanson 5200 series, Voltronics EQT-9 or equiv.).
C2, C3, C6 — 200- to 1000-pF ceramic chip capacitor.
C4, C8 — 200- to 1000-pF feedthrough capacitor.
C5 — 0.05-μF disc-ceramic capacitor.
C7 — 1- to 20-μF, 25-V electrolytic capacitor.
D1 — For NE41137: 4.3-V, 1-W Zener diode (1N4731 or equiv.). For MRF966/967: 5.1-V, 1-W Zener diode (1N4733 or equiv.).
D2 — 1N914, 1N4148 or any diode with ratings of 25 PIV and 50 mA or greater.
J1, J2 — Female chassis-mount BNC, SMA or Type-N connector.
L1 — For NE41137: Brass or copper strip, 0.020-inch thick, ¼-inch wide, 1-inch long (plus 3/16-inch lip), mounted 3/16-inch above the ground plane. For MRF966/967: Brass or copper strip, 0.020-inch thick, ¼-inch wide,

7/8-inch long (plus 3/16-inch lip), mounted 3/16-inch above the ground plane.
L2 — For NE41137: Piece of no. 22 copper wire, 1-1/8 inches long (measured from the body of Q1 to C6), mounted ¼ inch above the ground plane. For MRF966/967: Brass or copper strip, 0.020-inch thick, 1/8-inch wide, 1¼ inches long, mounted ¼ inch above the ground plane.
Q1 — NEC NE41137 or Motorola MRF966/967. See text.
R1 — For NE41137: 10 kΩ; for MRF966/967: 2.2 kΩ.
R2 — For NE41137: 10 kΩ; for MRF966/967: 4.3 kΩ.
R3 — 150- to 250-ohm, 1- or 2-W resistor.
R4 — For NE41137: 47 Ω; for MRF966/967: 100 Ω.
RFC1 — 6t no. 22 enam., 1/8-inch ID, close wound.

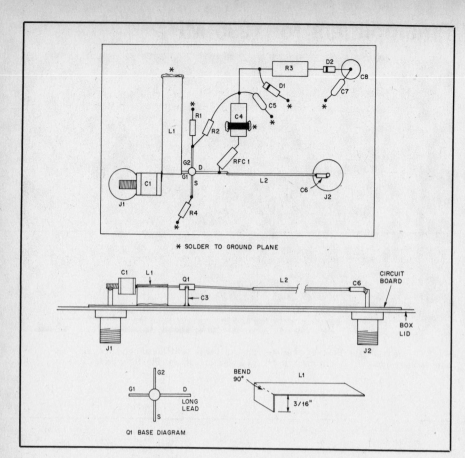

✱ SOLDER TO GROUND PLANE

Q1 BASE DIAGRAM

formation on working with static-sensitive devices. It's best to assemble the rest of the circuit and install Q1 last. Use a grounded iron if possible, and solder the leads quickly.

Parts for this project are available from Microwave Components of Michigan and Applied Invention, both listed in Chapter 35. NEC transistors are available from California Eastern Laboratories, 3260 Jay St., Santa Clara, CA 95050. MRF966 transistors are available from MHz Electronics, 3802 North 27th Ave., Phoenix, AZ 85017.

Peak C1 on a weak signal and adjust for best signal-to-noise ratio. For best results, you should have access to a noise-figure meter. This preamplifier typically provides 13-dB gain with any of the listed devices. Noise figure for the NE41137 is typically 2.5- to 3.0-dB; for the MRF966/967, noise figure typically runs 1.5- to 2.0-dB.

Fig. 9 — Suggested layout for the dual-gate GaAsFET 1296-MHz amplifier. The circuit board is approximately 1¾ × 3 inches. J1 and J2 are approximately 2-3/8 inches apart.

A GaAsFET Preamplifier for 1296 MHz

Single-gate GaAsFET devices are readily available to amateurs. They are capable of providing the low noise figures needed for weak-signal operation, including EME and terrestrial work, at 1296 MHz. Described here and shown in Figs. 10 through 12 is a 1296-MHz preamplifier that uses the popular Mitsubishi MGF1402 transistor. Dexcel 1500 or 2500 series and NEC 700 series GaAsFETs should work as well. This project was designed and built by Kent Britain, WA5VJB.

Circuit Details

The basic circuit is shown in Fig. 11. C1 is a piston trimmer. It must be capable of tuning to 0.5 pF or less. Typically C1 tunes between 0.5- and 1-pF at resonance. L1 and L2 are striplines made from brass or copper strips, or copper wire. C2, C3 and C4 are ceramic chip capacitors. For stability, it is important to use good-quality capacitors for the source bypasses. The MGF1402 has two source leads, so a ceramic chip capacitor is used on each lead to provide a low-impedance ground. The value of these chip capacitors is not critical; any value between 200 and 1000 pF should work.

Construction

The GaAsFET preamplifier is built "dead bug" style on a piece of double-sided PC board material. See Fig. 10. The circuit board acts as a low-impedance ground plane, and all ground connections are made directly to it. J1, J2 and C8 also secure the circuit board to the lid of the enclosure. For best performance, Type-N or SMA connectors are recommended for J1. J2 may be a BNC connector. The enclosure can be any aluminum diecast box (Bud CU-124; Hammond 1490B) or Minibox (Bud CU-3002A) approximately 4 by 2 inches, or you can make one by soldering pieces of circuit board material together.

Dimensions for the striplines should be followed carefully, especially the height above the ground plane. The brass or copper stock for L1 should be available at hobby stores. Alternatively, it is available from Small Parts, Inc. (see Chapter 35). Although 0.020-inch-thick stock was used here, 0.010- to 0.030-inch stock will work. If you have the equipment and inclination to experiment with GaAsFETs other than the MGF1402, you may have to cut L1 a bit long and use a piece of brass foil as a sliding short to determine the correct length. Likewise, L2 can be changed by using finer wire for more inductance or thicker wire (or a brass strip) for less inductance.

Fig. 10 — The 1296-MHz preamplifier with MGF1402 is built "dead bug" style on a piece of copper-clad circuit board material in the lid of a diecast box. See Fig. 12 for parts-placement information.

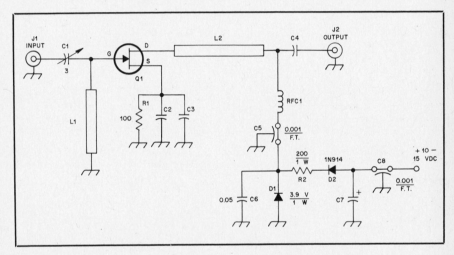

Fig. 11 — Schematic diagram of the MGF1402 1296-MHz GaAsFET preamplifier.

C1 — 0.3- to 3-pF piston trimmer capacitor (Johanson 5800 or Giga-trim series; Voltronics EQ or M series; Erie MVM-003 series; or Johnson Teflon trimmer).
C2, C3, C4 — 200- to 1000-pF ceramic chip capacitor.
C5, C8 — 200- to 1000-pF feedthrough capacitor.
C6 — 0.05-μF disc-ceramic capacitor.
C7 — 1- to 20-μF, 25-V electrolytic capacitor.
D1 — 3.9-V, 1-W Zener diode (1N4730 or equiv.).
D2 — 1N914, 1N4148 or any diode with ratings of 25 PIV and 50 mA or greater.
J1, J2 — Female chassis-mount BNC, SMA or

Type-N connector.
L1 — Brass or copper strip, 0.020-inch thick, 1/8-inch wide, 1¼ inches long (plus 1/8-inch lip), mounted 1/8 inch above the ground plane.
L2 — Piece of no. 24 copper wire, ¾-inch long, mounted 1/8 inch above the ground plane.
Q1 — Mitsubishi MGF1402. See text.
R1 — 100-Ω, 1/4- or 1/8-W, carbon-composition resistor.
R2 — 150- to 250-ohm, 1- or 2-W resistor.
RFC1 — 6t no. 24 enam., 1/8-inch ID, close wound.

Q1 is supported above the ground plane by its source leads, which are soldered directly to C2 and C3. The other ends of these capacitors are soldered to the ground plane. Because L1 must be mounted so close to the ground plane, C2 and C3 will probably have to be mounted at an angle.

R1 should be soldered in place with essentially no lead length. For best results, follow the layout shown in Fig. 12. The most important trick is to mount J1 and J2 with their center pins 1¾ inches apart.

GaAsFETs are static-sensitive, so handle them accordingly. See Chapter 24 for in-

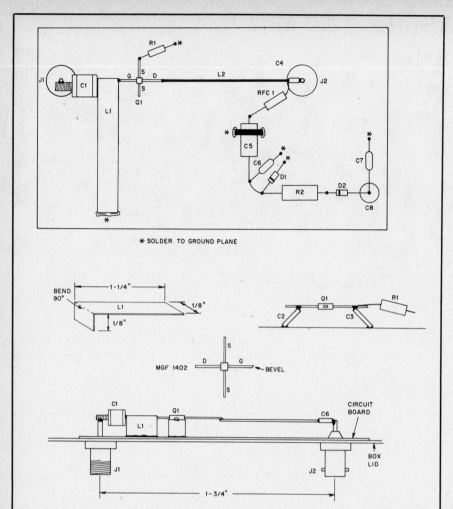

✻ SOLDER TO GROUND PLANE

MGF 1402

Fig. 12 — Suggested layout for the dual-gate GaAsFET 1296-MHz amplifier. The circuit board is approximately 1¾ × 3 inches. J1 and J2 are approximately 1¾ inches apart.

formation on working with static-sensitive devices. It's best to assemble the rest of the circuit and install Q1 last. Use a grounded iron if possible, and solder the leads quickly.

Parts for this project are available from Microwave Components of Michigan and Applied Invention, both listed in Chapter 35. MGF 1402 transistors are available from Advanced Receiver Research, Box 1242, Burlington, CT 06013. NEC transistors are available from California Eastern Laboratories, 3260 Jay St., Santa Clara, CA 95050.

Peak C1 on a weak signal and adjust for best signal-to-noise ratio. For best results, you should have access to a noise-figure meter. This preamplifier typically provides 13- to 15-dB gain. Noise figure for this preamplifier is typically 0.8 to 0.9 dB, measured on an HP8970 noise-figure meter.

Cost-Effective Preamp for 1296 MHz

Obtaining a low noise figure in amateur microwave receiving systems is no longer the expensive proposition it once was. The preamplifier shown in Figs. 13, 14 and 15 is taken from a catalog of designs published by Geoffrey Krauss, WA2GFP, in June 1982 *QST*. Featuring a simple circuit using a commonly available transistor, the unit boasts a noise figure of less than 2.5 dB with a 10-dB power gain.

Construction information is given in Figs. 14 and 15, with additional instructions in the captions. All controls should be adjusted for minimum noise figure (maximum signal-to-noise ratio with a weak signal). Start with all capacitors at minimum capacitance, adjusting the output network first. Repeat the adjustments several times because they interact. R2 sets the collector current. Minimum noise figure will occur in the 3- to 7-mA range.

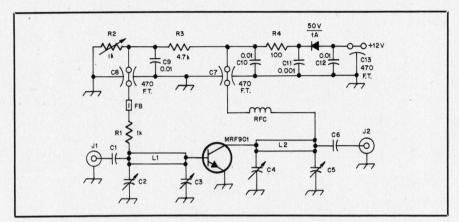

Fig. 13 — Circuit diagram of the 1296-MHz preamplifier.

C1, C6 — Microwave capacitor, ceramic chip or leadless disc, 100 pF or greater.
C2-C5 — Piston trimmer capacitor, 0.8 to 10 pF.
J1,J2 — Coaxial connector suitable for frequency range (type BNC acceptable).
L1,L2 — Copper strips; see Fig. 15.
RFC — 8 turns no. 28 enam. wire, closewound. 0.1-in-dia. air core.

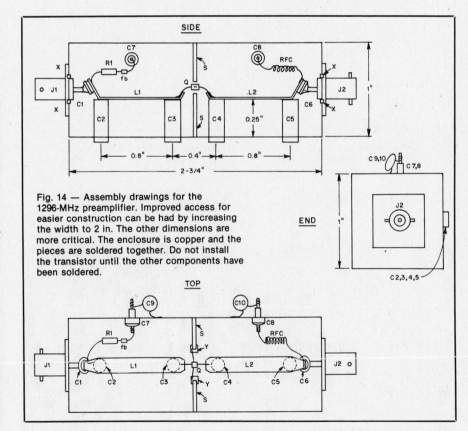

Fig. 14 — Assembly drawings for the 1296-MHz preamplifier. Improved access for easier construction can be had by increasing the width to 2 in. The other dimensions are more critical. The enclosure is copper and the pieces are soldered together. Do not install the transistor until the other components have been soldered.

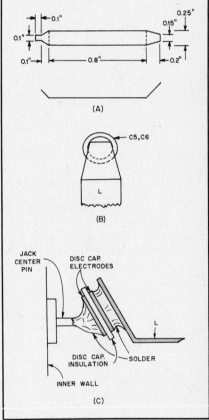

Fig. 15 — (A) Dimensional drawing of the inductors. (B) Detail of tapered end of inductor. (C) Method for installing a leadless disc capacitor at C1 or C6.

Interdigital Converter for 1296 or 2304 MHz

In a world where RF spectrum pollution is becoming more serious, even into the microwave region, it is almost as important to keep unwanted signals out of a receiver as it is to prevent radiation of spurious energy. An interdigital filter was described some years ago, featuring low insertion loss, simplicity of construction, and reasonable rejection to out-of-band signals. It could be used in either transmitters or receivers.

This twice-useful principle has now been put to work again — as a mixer. Again, the ease of construction and adaptation leads many to wonder that it had not been thought of before. It was first described by W2CQH in January 1974 *QST*.

A Filter and Mixer

A layout of the microwave portions of both converters is shown in Fig. 16. The structure consists of five interdigitated round rods, made of 3/8-inch-OD brass or copper tubing. They are soldered to two sidewalls and centrally located between two ground planes made of 1/16-inch sheet brass or copper-clad epoxy fiberglass. One ground plane is made larger than the microwave assembly and thus provides a convenient mounting plate for the remainder of the converter components.

The sidewalls are bent from 0.032-inch-thick sheet brass or they can be made from ¼ × ¾-inch brass rod. One edge of each sidewall is soldered to the larger ground plane. The other edge is fastened to the smaller ground plane by 4-40 machine or self-tapping screws, each located over the center line of a rod. The sidewall edges should be sanded flat, before the ground plane is attached, to assure continuous electrical contact. Note that no end walls are required since there are no electric fields in these regions.

Electrically, rods A, B and C comprise a one-stage, high-loaded-Q (Q_L = 100), interdigital filter[1] that is tuned to the incoming signal frequency near 1296 or 2304 MHz. The ungrounded end of rod A

---

[1]Fisher, "Interdigital Bandpass Filters for Amateur VHF/UHF Applications," *QST*, March 1968.

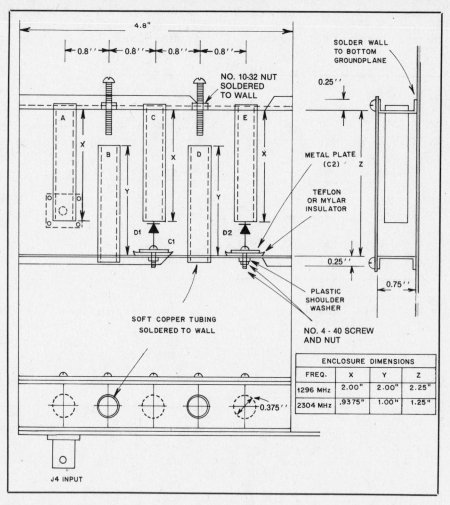

ENCLOSURE DIMENSIONS			
FREQ.	X	Y	Z
1296 MHz	2.00"	2.00"	2.25"
2304 MHz	.9375"	1.00"	1.25"

Fig. 16 — Dimensions and layout for the filter and mixer portions of the interdigital converters. The signal input is to the left rod, labeled A. Local-oscillator injection is through the diode to rod E. D1 is the mixer diode, connected to the center rod in the assembly.

Fig. 17 — The converter for 1296 MHz. This unit was built by R. E. Fisher, W2CQH. While the mixer assembly (top center) in this model has solid brass walls, it can be made from lighter material, as explained in the text and shown in Fig. 16. The IF amplifier is near the center, just above the mixer-current-monitoring jack, J1. A BNC connector at the lower left is for 28-MHz output. The local oscillator and multiplier circuits are to the lower right. Note that L6 is very close to the chassis, just above the crystal. The variable capacitor near the crystal is an optional trimmer to adjust the oscillator to the correct frequency.

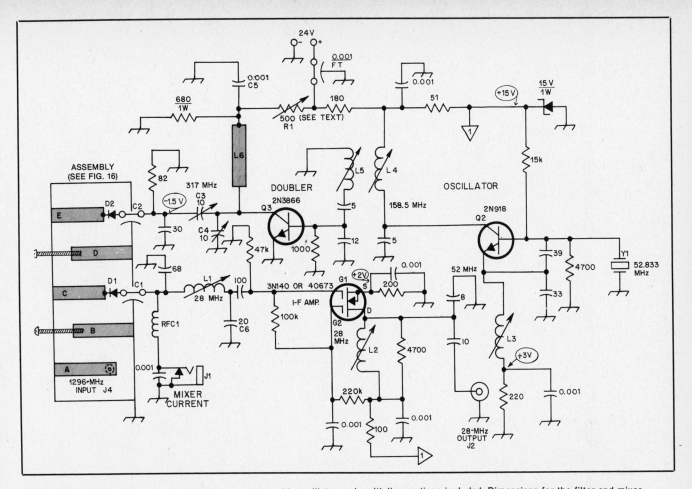

Fig. 18 — Schematic diagram of the 1296-MHz converter with oscillator and multiplier sections included. Dimensions for the filter and mixer assembly are given in Fig. 16.

C1, C2 — 30-pF homemade capacitor. See
   text and Fig. 16.
C3, C4 — 0.8- to 10-pF piston trimmer,
   Johanson 5200 series or equiv.
C5 — 0.001-µF button mica.
C6 — 2- to 20-pF air variable, E. F. Johnson
   189-507-004 or equiv.

D1 — Hewlett-Packard 5082-2817 or 5082-2835.
D2 — Hewlett-Packard 5082-2811 or 5082-2835.
J1 — Closed-circuit jack.
J2 — Coaxial connector, type BNC
   acceptable.
L1, L2 — 18 turns no. 24 enam. on ¼-inch-

OD slug-tuned form (1.5 µH nominal).
L3 — 10 turns like L1 (0.5 µH).
L4, L5 — 6 turns like L1 (0.2 µH).
L6 — Copper strip, ½ inch wide × 2½ inches
   long. See text and photographs.
RFC1 — 33 µH, J. W. Miller 74F33SAI or equiv.

is connected to a BNC coaxial connector filter input. Rod B is the high-Q resonator and is tuned by a 10-32 machine screw. Rod C provides the filter output-coupling section to the mixer diode, D1.

The original mixer diode was a Hewlett-Packard 5082-2577 Schottky-barrier type that is no longer available. The 5082-2817 and MA-4853 (Microwave Associates) are recommended substitutes. The cheaper 5082-2853 can be used instead, but this substitution will increase the 2304-MHz mixer noise figure by approximately 3 dB.

One pigtail lead of the mixer diode is tack-soldered to a copper disc on the ungrounded end of rod C. Care should be taken to keep the pigtail lead as short as possible. If rod C is machined from solid brass stock, then it is feasible to clamp one of the mixer-diode leads to the rod end with a small setscrew. This alternative method facilitates diode substitution and was used in the mixer model shown in the photograph.

Fig. 16 also shows that the other end of D1 is connected to a homemade 30-pF

Table 1

Converter Specifications

	1296 MHz	2304 MHz
Noise figure	5.5 dB	6.5 dB
Conversion gain	20 dB	14 dB
3-dB bandwidth	2 MHz	7 MHz
Image rejection	18 dB	30 dB
IF output	28 MHz	144 MHz

bypass capacitor, C1, which consists of a ½-inch-square copper or brass plate clamped to the sidewall with a 4-40 machine screw. The dielectric material is a small sheet of 0.004-in-thick Teflon® or Mylar®. A 4-40 screw passes through an oversize hole and is insulated from the other side of the wall by a small plastic shoulder washer.

In the first converter models constructed by the author and shown in the photographs, C1 was a 30-pF button mica unit soldered to the flange of a 3/8-in-diameter threaded panel bearing (H. H. Smith No. 119). The bearing was then screwed into a threaded hole in the sidewall. This provi-

sion made it convenient to measure the insertion loss and bandwidth of the interdigital filters since the capacitor assembly could be removed and replaced with a BNC connector.

Rods C, D and E comprise another high loaded-Q ($Q_L = 100$) interdigital filter tuned to the local oscillator (LO) frequency. This filter passes only the fourth harmonic (1268 or 2160 MHz) from the multiplier diode, D2. The two filters have a common output-coupling section (rod C) and their loaded Qs are high enough to prevent much unwanted coupling of signal power from the antenna to the multiplier diode and LO power back out to the antenna.

The multiplier diode is connected to the driver circuitry through C2, a 30-pF bypass capacitor identical to C1. D2 is a Hewlett-Packard 5082-2811, although the 5082-2835 works nearly as well. Fifty milliwatts drive at one-quarter of the LO frequency is sufficient to produce 2 mA of mixer diode current, which represents about 1 mW of local-oscillator injection. A Schottky-barrier

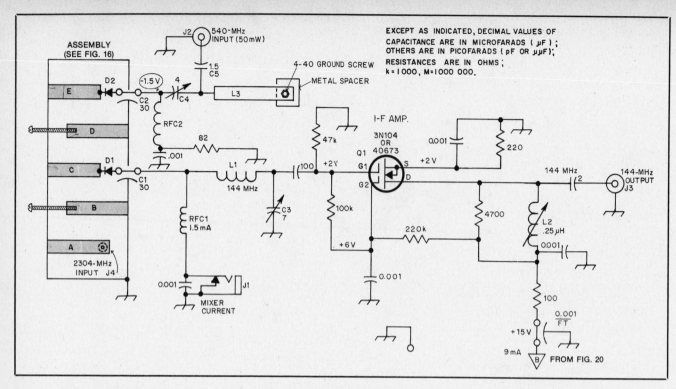

Fig. 19 — Schematic diagram of the 2304-MHz version of the converter, with the IF amplifier. The oscillator and multiplier circuits are constructed separately.

C1, C23 — 30-pF homemade capacitor (see text).
C3, C4, C5 — 0.8- to 10-pF piston trimmer, Johanson 5200 series or equiv.
D1 — Hewlett-Packard 5082-2817 or 5082-2835.
D2 — Hewlett-Packard 5082-2811 or 5082-2835.

J1 — Closed-circuit jack.
J2, J3, J4 — Coaxial connector, type BNC.
L1 — 5 turns no. 20 enam. ¼-inch ID × ½-inch long.
L2 — 6 turns no. 24 enam. on ¼-inch OD slug-tuned form (0.25 µH).
L3 — Copper strip ½ × 2-11/16 inches. See text and photographs.
RFC1 — Ohmite Z-144 or equiv.
RFC2 — Ohmite Z-460 or equiv.

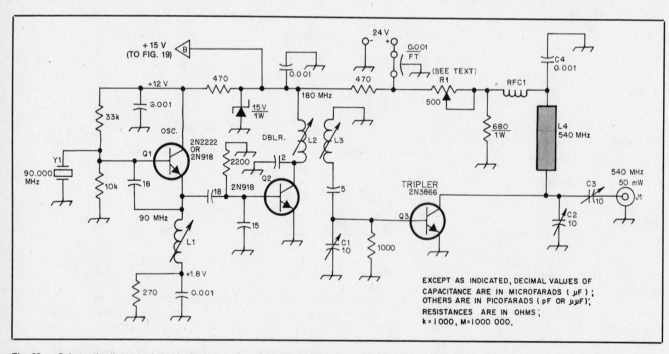

Fig. 20 — Schematic diagram of the oscillator and multiplier for the 2304-MHz converter. As explained in the text, a fixed- value resistor may be substituted for R1 after the value that provides proper performance has been found.

C1, C2, C3 — 0.8- to 10-pF piston trimmer, Johanson 5200 series or equiv.
C4 — 0.001-µF button mica.
J1 — Coaxial connector, type BNC or equiv.

L1 — 10 turns no. 24 enam. on ¼-in-OD slug-tuned form.
L2, L3 — 3 turns no. 24 enam. on ¼-in-OD slug-tuned form.

L4 — Copper strip ½ × 1½ in. Space 1/8 inch from chassis.
RFC1 — 10 turns no. 24 enam. 1/8-in-ID, close wound.

diode was chosen over the more familiar varactor diode for the multiplier because it is cheaper and more stable, and requires no idler circuit.

Fig. 18 shows the schematic diagram of the 1296 to 28 MHz converter. All components are mounted on a 7 × 9-inch sheet of brass or copper-clad epoxy-fiberglass board. As mentioned earlier, this mounting plate also serves as one ground plane for the microwave mixer. When completed, the mounting plate is fastened to an inverted aluminum chassis which provides a shielded housing.

Oscillator and Multipliers

The nonmicrowave portion of the converter is rather conventional. Q1, a dual-gate MOSFET, was chosen as the 28-MHz IF amplifier since it can provide 25 dB of gain with a 1.5 dB noise figure. The mixer diode is coupled to the first gate of Q1 by a pi-network matching section. It is most important that the proper impedance match be achieved between the mixer and IF amplifier if a low noise figure is to be obtained. In this case, the approximately 30-ohm output impedance of the mixer must be stepped up to about 1500 ohms if Q1 is to yield its rated noise figure of 1.5 dB. It is for this reason that a remote IF amplifier was not employed, as is the case with many contemporary UHF converters.

Q2 functions in an oscillator-tripler circuit that delivers about 10 mW of 158.5-MHz drive to the base of Q3. The emitter coil, L3, serves mainly as a choke to prevent the crystal from oscillating at its fundamental frequency. Coils L4 and L5, which are identical, should be placed close together so their windings almost touch.

Q3 doubles the frequency to 317 MHz, providing about 50 mW drive to the multiplier diode. It is important that the emitter lead of Q3 be kept extremely short; ¼ inch is probably too long. L6, the stripline inductor in the collector circuit of Q3, consists of a ½ × 2½-inch piece of flashing copper spaced 1/8-inch above the ground plane. The cold end of L6 is bypassed to ground by C5, a 0.001-μF button mica capacitor.

The multiplier circuits are tuned to resonance in the usual manner by holding a wavemeter near each inductor being tuned. Resonance in the Q3 collector circuit is found by touching a VTVM probe (a resistor must be in the probe) to C2 and adjusting the Johanson capacitors until about −1.5 volts of bias is obtained. The 317- to 1268-MHz multiplier cavity is then resonated by adjusting the 10-32 machine screw until maximum mixer current is measured at J1. When resonance is found, R1 should be adjusted so that about 2 mA of mixer current is obtained. As an alter-

native to mounting a potentiometer in the converter, once a value of resistance has been found that provides correct performance it can be measured and the nearest standard fixed-value resistor substituted. Some means of adjusting the collector voltage of the multiplier stage must be provided initially to allow for the non-uniformity of transistors.

A 2304-MHz VERSION

Figs. 19 and 20 show the schematic diagrams of the 2304-MHz converter and multiplier. The mixer and IF preamplifier were built on a separate chassis since, at the time of their construction, a multiplier chain from another project was available. An IF of 144 MHz was chosen, although 50 MHz would work as well. An IF output of 28 MHz, or lower, should not be used since this would result in undesirable interaction between the mixer and multiplier interdigital filters.

The 2304-MHz mixer and IF amplifier section, shown in Fig. 20, is very similar to its 1296-MHz counterpart. Q1, the dual-gate MOSFET, operates at 144 MHz and thus has a noise figure about 1 dB higher than that obtainable at 28 MHz.

The multiplier chain, Fig. 20, has a separate oscillator for improved drive to the 2N3866 output stage. Otherwise the circuitry is similar to the 1296-MHz version.

1296-MHz Transverter

Presented here is a basic 1296-MHz transverter system which, with some additions, could serve in a 1296-MHz moonbounce system. The basic unit coupled with a good antenna is capable of reasonable terrestrial performance as well. Construction requires a minimum of special skills and test equipment.

The established design specifications are listed below:

1) The unit forms the basis of a 1296-MHz weak-signal/moonbounce system.

2) The unit must be easy to build and duplicate.

3) It must be as inexpensive as possible.

4) It must use a minimum of special microwave components.

5) It must have a 28-MHz tunable IF.

6) The receiver portion should exhibit a noise figure of 3 dB or less and sufficient gain to override noise in the tunable IF. Reasonable dynamic range is also a requirement.

7) The transmitter output should be 250 mW with all spurs and harmonics down at least 40 dB.

8) The LO should produce at least +15-dBm output with all spurs and harmonics down 30 dB or more. It must be easy to align without special instruments.

9) At low microwave frequencies lumped constants are not very effective, so microstrip techniques are used wherever possible. The microstrip portions of the circuit can be constructed easily with nothing more than a sheet of copper, some single-sided 0.062-inch G-10 (glass epoxy) board, a sharp pair of scissors, and a few tubes of Superglue! This eliminates the mess of etching PC boards.

The 1296 transverter presented here, designed and constructed by Dave Eckhardt, W6LEV, fulfills all of these requirements. The total price tag from start

Fig. 23 — The completed 1296-MHz transverter consists of three circuit boards. The boards may be stacked as shown and mounted in a convenient enclosure.

to finish should be less than half the cost of comparable commercial units.

Theory of Operation

Fig. 22 presents the block diagram of the transverter. It is broken into three major functional units. From top to bottom these are the receive chain, local oscillator (LO) and transmit chain. A photograph of the finished unit is shown in Fig. 23.

A good LO is the key to a clean transverter design. The LO used here is based on a design originally presented in *Ham Radio*.[1] This reference is all but required reading before tuning up the LO without special instruments. An International Crystal OE-5 modular oscillator forms the heart of the LO chain. It oscillates at 105.658333 MHz, which places 1296.000 MHz at 28.100 MHz on the tunable IF. This frequency was chosen because the band-pass filters in some receivers exhibit loss below 28 MHz. It also allows for some error in the exact LO frequency.

The price of the module is reasonable in single quantities. The output link of the OE-5 module resonates at 105 MHz, and

[1]Shuch, H. P., "Compact and Clean L-Band Local Oscillators," December 1979 *Ham Radio*, p. 40.

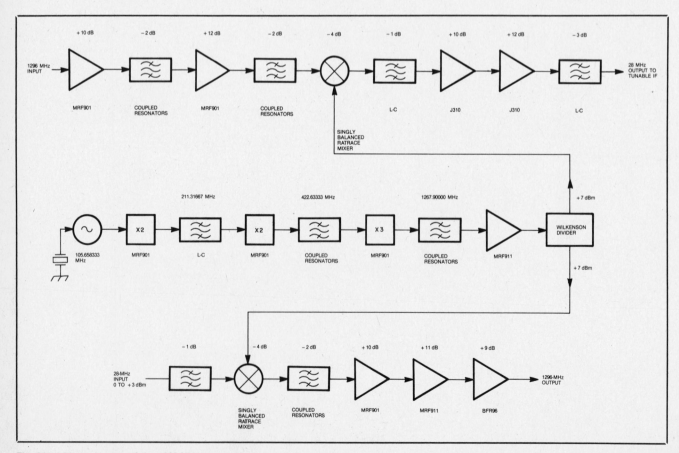

Fig. 22 — Block diagram of the 1296-MHz transverter.

feeds the first doubler. This stage tunes to 211 MHz with a single parallel-tuned resonant circuit with the inductor tapped down to feed the second doubler. The second doubler feeds a band-pass filter consisting of two inductively coupled, capacitively shortened quarter-wave stripline resonators tuned for 422.6 MHz. Output from the 422.6 MHz filter is fed to a parametric tripler which has both its input and output tuned to 1268 MHz. This technique yields about 4 dB more gain and a cleaner output than a conventional tripler. The 1268-MHz energy then passes through a 3-stage filter constructed of 0.025-inch-thick brass sheet and UT-141 semi-rigid coax. This filter was designed by Chip Angle, N6CA. The output of the filter feeds an MRF911 amplifier where the level is brought up to approximately +15 dBm. From there, the 1268-MHz signal is applied to an in-phase (Wilkinson) power divider. Each of the two output ports of the divider supply roughly +7 dBm of 1268-MHz energy to the mixers in the receive and transmit chains.

All spurs and harmonics of the LO are down 40 dB from the 1268-MHz output when tuned with the aid of a spectrum analyzer. The builder can achieve −30 dB on spurs and harmonics by following the procedure (outlined in the referenced *Ham Radio* article) for alignment without special instruments. The LO is easy to tune and achieves a clean output. It is, however, the most critical of any of the circuits in the transverter. Meticulous adherence to the outlined tune-up procedure is a must!

The receiver nominally exhibits a 3-dB noise figure with 32 dB of conversion gain. The antenna feeds the first MRF901 amplifier, which is optimized for best noise figure. Builders may want to precede the first stage with a low-loss band-pass filter, T/R switching and a GaAsFET preamplifier. Output from the noise-optimized first stage is fed through a band-pass filter identical to the 1268-MHz filter in the LO. The signal then goes to the second MRF901 amplifier, which is optimized for gain. Another band-pass filter couples the second stage to the mixer. The singly balanced ratrace mixer also accepts +7 dBm of 1268-MHz energy from one of the ports of the power divider located at the output of the LO. This type of mixer functions well with as little as +1 to +2 dBm of LO energy, but linearity and dynamic range are enhanced with +7 to +10 dBm of LO injection.

A terminating network follows the IF port of the mixer and also serves as a low-Q band-pass filter for the required difference frequency of 28 MHz. This network presents the IF port of the mixer with a constant 50-ohm load over a wide frequency range. Postconversion amplification is by means of two J310 JFETs in a common-gate configuration, with the first stage optimized for noise performance. If

the builder has a sensitive 28-MHz receiver, both stages of postconversion gain may not be required. The final block in the receiver chain is a tight band-pass filter centered on 28.5 MHz, which precedes the tunable IF.

The gain-optimized MRF901 amplifier and the two stages of postamplification gain are unconditionally stable. The noise-optimized MRF901 stage may be open- or short-circuited at its input with no damage to the device.

The transmit mixer accepts inputs from the second port of the LO power divider and 0 to +3 dBm of 28-MHz energy from the tunable IF. A terminating network/BPF identical to that in the receive chain is used at the IF port of the mixer. This mixer is identical to the receive mixer, except that the sum frequency is used. The filter that follows the RF port is, therefore, tuned to 1296 MHz, and is identical to the filters in the LO and receiver chain. The filter response is typically 15 dB down at 1268 MHz (the LO frequency) when tuned to 1296 MHz. If more attenuation is needed, two filters can be cascaded, or a high-Q open quarter-wave line cut for 1268 MHz can be installed in shunt on the filter port away from the mixer. Three stages of amplification, each supplying roughly 10 dB of unconditionally stable gain, are fed by the filter. Final output from the transmitter is between 200 and 300 mW.

SPECIAL COMPONENTS AND TECHNIQUES

Several special techniques are used in building this transverter, but don't be alarmed! They are easy to master, and may be adapted to other UHF and microwave projects. Also, application of several special components greatly simplifies the

process of building any piece of microwave gear.

Components

All microwave circuits require some special components, and this project is no exception. However, since most amateurs have a limited bank account (that's one reason we build), the number of special components has been kept to a minimum.

Variable capacitors that work well at microwave frequencies are quite expensive, so another method of tuning was devised. Tuning involves use of transmission-line sections and chip capacitors. The builder should have about 30 to 50 good microwave chip capacitors on hand. Not all of these will be used, but allow 30% more than needed to account for breakage or metallization separation during soldering. Don't skimp! In many cases these capacitors establish the low- and high-frequency stability of the circuits. They should *not* be the "series resonant" type, but should present a very low capacitive reactance from dc to at least 5 GHz. American Technical Ceramics makes a good line of microwave chips. Although a value of 0.01 μF is shown in the schematics, values ranging from 0.01 μF to 1000 pF should work well. Several of the ads in *QST* routinely carry good microwave chip capacitors ranging from $.50 to $1.50 each.

Special care must be exercised in soldering the chip capacitors. Fig. 24 illustrates a good technique. All surfaces should be clean and lightly tinned before you attempt to solder the chip. Position the chip in place and hold it with a toothpick. Do not use a screwdriver or tweezers, since the metal can easily damage the ceramic base of the capacitor. Lay the chisel side

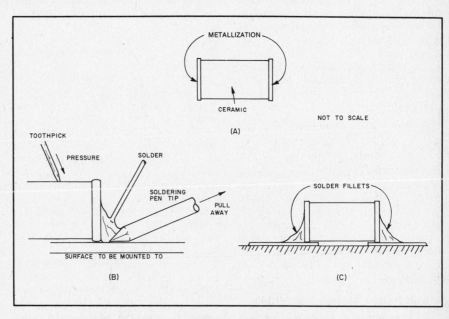

Fig. 24 — A typical chip capacitor is shown at A. The proper technique for holding and soldering the chip is shown at B. At C the final appearance of the capacitor in the circuit. Check for good solder flow, no metallization separation and no cracks or fractures in the ceramic.

of the tip of a 15- to 20-watt soldering pen on the surface to which the capacitor is to be soldered, with the tip of the chisel just touching the chip capacitor.

Touch and flow a minimum amount of solder between the chip and the soldering pen tip and pull the tip away at a low angle to the surface. Repeat the procedure on the other side of the chip. If you don't dump too much heat into the chip while soldering the second end, you may not have to hold it in place. Fig. 24C illustrates how the connection should look when done. Inspect the solder and the chip capacitor with a magnifier for good solder flow and lack of fractures and cracks. Don't overheat the chip or the metallization may separate from the ceramic and you'll have to discard the capacitor. Several attempts may be necessary, but the technique can be mastered after a few tries.

In addition to chip capacitors, several high-quality microwave variable capacitors are needed. These devices are also available from some of the advertisers in *QST*. Johanson or Triko units may be used. The Johansons are more expensive, but mechanically superior. If you can tune the circuit to "perfection" with only 5 to 8 tries use the Trikos. However, they will not withstand repeated tweaking. (Who can resist?) Don't make too many substitutions of critical components unless you have some good microwave test gear.

Several feet of UT-141 semi-rigid Teflon dielectric coax is also required for resonators in the filters and interconnection between boards. Roughly 2 feet is required for the filters. In 2-foot lengths, UT-141 costs between $1.35 to $3 per foot. A few advertisers in *QST* offer it for sale. It is also available from most microwave parts/cabling and connector supply houses.

Order the OE-5 oscillator module directly from International Crystal. The builder may start with a good precision crystal and roll his own oscillator, but the cost difference between a high-accuracy crystal and the module is small. Drift and temperature characteristics of the OE-5 are good. The formula for the module frequency is:

Frequency (OE-5) = (1296.000 − IF)/12

All frequencies are in megahertz. This should be considered if the builder desires 1296.000 MHz to fall at some frequency other than 28.100 MHz. Although I have not built this unit for the OSCAR transponder, a change in LO frequency could bring 1269 MHz in range of a 28-MHz tunable IF. All circuits should have a wide enough tuning range to accommodate the slightly lower LO and operating frequencies.

Circuit Boards, Microstrip and Tuning

As mentioned in a previous section, much of the RF circuitry uses microstrip transmission-line sections. This was done because lumped-constant matching networks are not very effective at microwave frequencies. Their size is a significant portion of a wavelength, so they no longer "look like" their lumped dc equivalents. Lumped elements are available to the microwave industry, but they are generally quite expensive. The use of transmission-line sections offers a versatile, simple and inexpensive way of impedance matching and routing RF around a board. Microstrip transmission line is especially well suited because it may be etched from a double-sided copper-clad board or cut out with a sharp knife or a pair of scissors from sheet copper and mounted above the ground plane with an adhesive.

Fig. 25 shows microstrip parameters; characteristic impedance is given by Schneider.[2] This project assumes a dielectric constant of 4.5 for G-10 board and 1.0 for air. The width of the strip, spacing above the ground plane, and dielectric material are all variables that determine the impedance of the line and its electrical length. If a circuit is designed with this in mind, it can be tuned by tweaking the spacing or width of a line. In addition, a shunt line may be shortened or lengthened by sliding a chip capacitor up and down the line! This is the technique used to tune the transverter. It may sound odd to the uninitiated, but it works well and doesn't involve expensive gold-plated components! Most of the circuits that need tuning are designed with a shunt line and a series line with the 50-ohm port located at the end of the series line. The matching circuit variables are chosen so that the normal scatter of S-parameters from device variation is compensated for by sliding the chip capacitor up and down the shunt line until the stage is optimized. Remember to use a toothpick to hold the capacitor in place.

In the design stages of this project, this method was modeled on a Hewlett-Packard HP-41CV calculator with the aid of the Smith Chart. The method looked promising. When the original air-microstrip version was constructed, no stage took more than 5 minutes of sliding chip capacitors up and down the shunt line and "fine tuning" the spacing of the microstrip elements to optimize. When the G-10 version was constructed, the "sliding capacitor" technique proved quite adequate. This method may be applied to just about any large- or small-signal RF circuitry that uses transmission-line sections for matching.

Several characteristic line impedances are required. Table 2 presents these impedances with the required line widths and spacings. Absolute precision is not needed in cutting the line widths, but reasonable care is required. "Eyeballing" them with a ruler graduated in 0.01-inch increments is adequate. In the case of G-10 board, the spacing is always that of the board, 0.062

[2]Hewlett Packard, *HP-41CV Circuit Analysis Pac.*

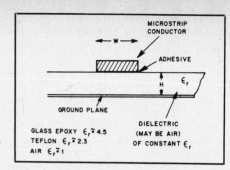

Fig. 25 — Microstripline physical parameters. See Table 2.

Table 2
Microstrip Parameters

Z_o	Air (ϵ_r = 1)		G-10 (ϵ_r = 4.5)	
	w	h	w	h = 0.062
10	1.0000	0.0313		
16	1.2813	0.0625	0.5000	
18	1.1250	0.0625	0.4375	
50	0.3125	0.0625	0.1000	
70	0.3750	0.1250	0.0600	
86				0.0370
90	0.1250	0.0625	0.030	
115	0.1563	0.125	0.016	
125	0.1250	0.1250	0.013	

inches. This project may be built with air-microstrip lines, or the lines may be mounted directly on the dielectric side of the single-sided G-10 board. In the latter case, the line length for a given electrical length of air line must be shortened by the square root of the dielectric constant. For a given electrical line length, the G-10 line is 0.471 times ($1/\sqrt{4.5}$) the physical length of the corresponding air line. Therefore, for a given electrical length of line, the G-10 version occupies less than half the board space of an air-microstrip version.

In either case, the lines are simply cut to the appropriate width and length from a sheet of 1- or 2-ounce copper. If the G-10 version is built, the line is simply glued in place with Superglue. The air line version requires small spacers of 0.032, 0.062 or 0.125 inches. These are made from small (1/8 × 1/8 inch) squares of clean (no copper) 0.032, 0.062 or 0.125-inch-thick glass-epoxy board. The spacers and air line are then glued in place. A minimum number of spacers should be used. What could be easier! The ground plane on the "business side" of the G-10 board is cut from the same copper sheet to fit around and frame the microstrip elements. It is also glued in place with Superglue. It is necessary to curl the top ground over the edge of the board and solder it to the bottom ground plane. Small holes to pass no. 18 wire are drilled through both ground planes at intervals of 1/4 to 1/2 inch. This is especially important on ground-plane edges that frame microstrip elements. Short pieces of no. 18 wire (or metal grommets) are then passed through each hole, crimped with pliers, and soldered to both sides of the board.

Active Devices

The transverter circuitry uses only four active device types. These are the MRF901, MRF911, BFR96 and J-310. The first three are bipolar transistors and are used in the microwave circuitry. The J-310 is a JFET and is used only in the post-conversion amplifier of the receiver chain. The bipolars require a few comments.

The proper way to mount microwave devices is a bit different than for HF devices. All leads must be as short as possible and ground returns must be as direct as possible with minimum series impedance. Figs. 26 and 27 illustrate the proper way to mount the microwave devices. A hole is first drilled in the board where the transistor is to be mounted. The hole should be slightly larger than the plastic "pill." Bend the emitter lead(s) of the active device toward the bottom of the device at the pill edge and at right angles to the pill. Insert the active device into the hole. Orient the base and collector leads properly, and bend the portion of the emitter lead(s) that protrude through the bottom of the board flat with the bottom ground plane. Solder the emitter leads to the bottom ground plane.

The dc parameters of a bipolar RF device generally do not depend on the RF parameters, but the RF parameters are

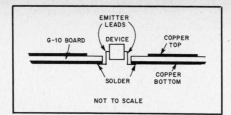

Fig. 26 — Mounting technique for microwave transistors. Collector and base leads are straight; emitter leads are bent and soldered to the ground plane. See text.

Fig. 27 — Photograph showing mounting technique for microwave transistors.

Table 3
Transistor DC Bias Parameters

Device	Noise Match		Power Match	
	V_{cc} (Volts)	I_c (mA)	V_{cc} (Volts)	I_c (mA)
MRF901	6	6	10	15
MRF911			10	20
BFR 96			10	50

dependent on the dc bias parameters. Each device is characterized by the manufacturer under specific bias conditions. Although the RF parameters at this bias level (specific V_{cc} and I_c) are closely controlled by the manufacturer, the dc parameters (h_{fe} or β) of the RF devices generally are not as closely controlled. Since dc gain may vary greatly from device to device, the bias circuit must be flexible enough to set the same dc quiescent bias point in all cases. In some cases noise and gain optimization of a device require slightly different bias conditions to accommodate different high-frequency parameters. This is another way of saying "best noise performance generally does not coincide with maximum gain."

Table 3 summarizes the dc bias conditions for each bipolar transistor used in this project. These should be adhered to closely. The bias resistors shown in the schematics are those used at W6LEV. Since dc gain is not a controlled parameter, however, some minor adjustments to the bias resistors may have to be made. The dc characteristics of the devices may be determined with dc instruments and no RF drive. It is a good idea, however, to terminate each stage as it is built and during set-up with a 50-ohm resistor at input and output. Remember to block dc with a capacitor. The important parameters are collector current and collector voltage (see Table 3). The dc current gain of each device may even be measured before installing it into the circuit and the bias network "tweaked" for this gain. Remember that bipolar transistors are current-controlled, not voltage-controlled, devices. The only stage that may present a biasing problem is the BFR96. Establish the bias parameters for temperature stability.

Allow the device to warm for a few seconds after application of power to reach the stable bias point given in Table 3. This technique is preferable to letting it thermally drift through that point and run away.

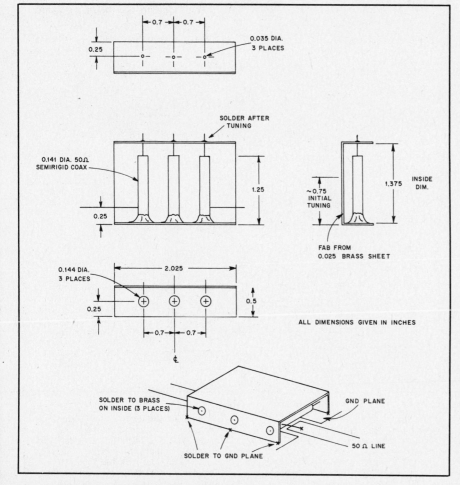

Fig. 28 — 1268- or 1296-MHz band-pass filter. See text.

Fig. 29 — Details of the band-pass filter shown in Fig. 28.

CONSTRUCTION

3-Element 12xx-MHz Filter

Figs. 28 and 29 illustrate the mechanical details of the 1268/1296-MHz filter that is used in four places in the transverter circuitry. The filter consists of three coupled, capacitively loaded quarter-wave resonators. The center conductor of the coax forms a tunable cylindrical capacitor at the Hi-Z end of each resonator. The only tools required are a large pair of tin snips capable of cutting 0.025-inch-thick brass sheet, a ruler, vise and 0.035- and 0.144-in drill bits. A small hammer will aid in bending the brass, and a sharp X-acto knife can be used to cut the semi-rigid UT-141 coax. All dimensions are in inches and are shown in Fig. 33. The dimensions should be followed closely — to a few hundredths of an inch is desirable.

The brass sheet should be marked and cut to size. The holes are drilled next. Do *not* bend the brass before drilling. When the brass work is finished, cut the appropriate lengths of UT-141 semi-rigid coax. It is critical to cut the correct length. Straighten a length of the coax and carefully mark off the required 1.250 inches. With the X-acto or other sharp knife, score the coax around its perimeter at the mark by rolling it on a flat surface while applying moderate pressure to the cutting edge. Once the cut has been defined around the periphery, the pressure is increased while rolling the coax until the copper tubing is cut. Be careful not to cut too deeply into the Teflon, since the center conductor should not be cut at this point. Carefully cut through the Teflon.

To remove the coax from the bulk of the straightened portion and the center conductor, grab the length of coax to be removed lightly with a pair of pliers and pull carefully while slowly twisting it. Don't break the center conductor, since it is used to tune the resonators. Once the length of tubing and Teflon is removed, cut off the center conductor and lay both pieces aside. Repeat this procedure until the required number of resonators has been cut; three are required for each filter. Set the brass filter box on edge on a flat surface with the large holes downward and insert the lengths of semi-rigid coax into the large holes. Lightly coat the center conductors with silicone grease and insert them through the top (small) holes of the structure and into the Teflon of each resonator. The silicone grease aids in sliding the conductor in and out of the Teflon when tuning each filter. Do not insert it more than a half inch. This will hold the resonator in place while the resonators are soldered to the brass structure. Confirm that the copper of each resonator element is flush with the outer bottom surface of the brass "box." Use a 150-watt soldering gun, and solder each resonator in place from the inside of the box, making sure that the elements are parallel with the box side (the large surface) and each other. Do *not* solder the center conductors at this time.

To install the filter, set it in place on the ground plane. Make sure the filter ground plane forms a complete sheet of copper under the filter assembly. When installed, the ground plane will form the fourth side of the box. Carefully solder the box to the ground plane continuously along each edge. Do not add so much heat that the resonators slump in position within the filter. Do not solder the center conductors.

Input and output connections to the filters are formed by soldering a short piece of large-gauge wire or a copper strip from the 50-ohm microstripline to the resonator at a point 0.250 inch from the cold end.

Filter tuning is straightforward. It is assumed that the builder has some method of sensing the amplitude of the required signal at either 1268 or 1296 MHz. This may be a selective absorption wavemeter (resonant cavity or quarter-wave line and RF probe) or a receiver (converter and tunable IF). Start with each center conductor pushed about 0.5 inch into each resonator. As the filter is tuned it is imperative that the center conductors keep good ground contact with the brass structure where they pass through it! If a good connection is not maintained, the filter will either not tune or will change tuning once the center conductors are soldered.

The builder may want to introduce a slight large radius bend in each center conductor to keep friction between the brass and each tuning element. Be sure that the center conductors are coated with silicone grease before inserting them into the resonators. Starting with the input reso-

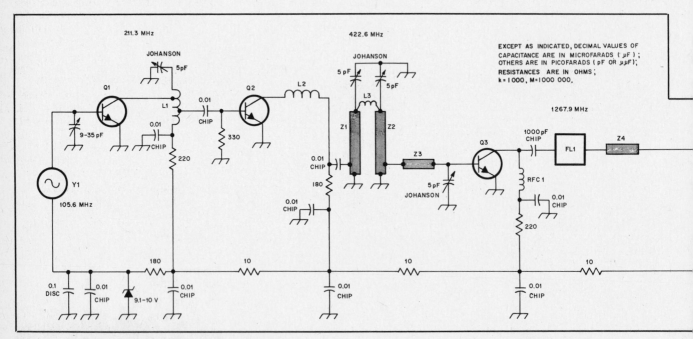

Fig. 30 — Schematic diagram of the local oscillator.

FL1 — 1268-MHz BPF. See Fig. 28 and text.
L1 — 90 nH, 4 turns no. 18 enameled wire, 7/32 in ID turns spaced one wire diameter. Collector tap 2½ turns, output tap 2 turns from cold end.
L2 — 23 nH, 2 turns no. 18 enameled wire,

3/16 in ID, closewound.
L3 — 0.33 μH, 12 turns no. 24 enameled wire, 3/16 in ID, closewound.
Q1, Q2, Q3 — MRF901.
Q4 — MRF911.
RFC1, RFC2 — 8 turns no. 24 enameled wire, 1/8 in ID, closewound.

Y1 — International Crystal Manufacturing Company, Inc. Local Oscillator module. Model no. OE-5. Note: International Crystal no longer manufactures these modules. A suitable oscillator circuit is shown in Chapter 31, Fig. 43.

nator, move the tuning element (center conductor) *slowly* in and out to peak the filter. Repeat for the center and output resonator. The center resonator is relatively critical, so don't use an overdamped meter movement to indicate output. Repeat the process several times, since the adjustments are interactive. *Carefully* solder each tuning element in place and cut off the excess.

LO and Power Splitter

Fig. 30 shows the schematic of the LO and power splitter, while Fig. 31 presents the layout used in the prototype. If the layout is modified, try to stick relatively close to a left-to-right progression. This is important for a clean output. You can construct the LO a stage at a time, tune it and then proceed to the next stage until the unit is completed, or complete the unit and then tune it up. The latter is preferred since many of the adjustments interact. It is recommended that all the microstrip elements be laid out first. Then cut and glue the ground plane in place. Before mounting any components, solder the top and bottom ground planes together as described earlier with copper tape and no. 18 wire (or metal grommets). The lengths of microstripline given in Fig. 30 are for G-10 board, 0.062 inch thick.

Remove the aluminum cover from the OE-5 module and discard it. Holes must be pre-drilled in the circuit board for this component. Then mount the OE-5 module on the back of the board. A copper box (with a hole punched in it for the single adjustment) should be fabricated to enclose the oscillator. This box should be soldered to the ground plane.

The LO is the only place where high quality Johanson trimmer capacitors are used. Triko units can be used, but they are not as rugged.

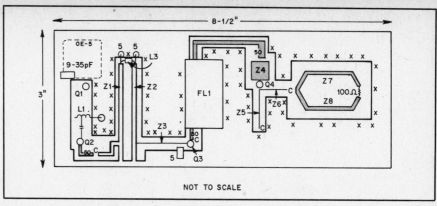

Fig. 31 — Local oscillator board layout. Microstripline sections marked "50" are 50 ohms, any length. C indicates location of a chip capacitor; X indicates a jumper from top to bottom ground planes. Wrap board edges with copper and solder top and bottom.

When the unit is finished the tune-up procedure of footnote 1 should be followed. This LO will *not* perform properly when simply tuned for maximum output on a broadband power meter! Careful alignment is a must! Resist the temptation to go back and tweak. Connection to the receiver and transmitter boards should be made with short lengths of semi-rigid Teflon cable, since RG-58 foam coax is leaky at 1268 MHz.

The power divider is a standard Wilkinson in-phase divider. No alignment is necessary. The mismatch resistor between the ports should be either a 1/8-watt carbon composition unit with extremely short leads or a chip resistor.

Mixer

The same rat-race singly balanced mixer and IF port terminator are used in both the receiver and transmitter. See Figs. 32 and 33. The LO (1268 MHz) is injected ¼ and ¾ wavelength from each diode (the diodes are 180 degrees out of phase with each other). The non-microstrip side of the diodes are placed at 1296-MHz ground, but at a high impedance for the 28-MHz IF. The IF is fed through a network which theoretically maintains 50 ± j0 ohms everywhere, but is transparent within its passband. It keeps the IF port at a constant impedance, which is important for proper mixer operation. The design equations for this network are given in Fig. 34. The RF is fed midway (¼ wavelength) from each diode.

The hot-carrier Schottky diodes should be matched. This may be accomplished in a rough but satisfactory manner by choosing two diodes which exhibit the same forward dc voltage drop when biased with 5 to 10 mA of forward current.

Nominal values are given for the tuned components of the IF-port terminator. Since dip mica and chip capacitor values may vary by 5%, each tuned circuit should be adjusted to resonance with a dip meter or other means before installation. The inductor in each case may be adjusted by expanding or compressing turns. Be sure to tune each circuit with the same lead lengths that will be used when finally installed. The circuit for 28 MHz was modeled on an HP-41CV using the GNAP option of the Circuits Pac. Table 4 presents the results. Actual results on the high-frequency end will depend on component quality.

Postconversion Amplifier and BPF

Fig. 35 shows the two stages of postconversion amplification and output BPF. The first amplifier stage may be noise-matched to 50 ohms, if desired. As previously stated, the builder may opt to incorporate only one stage if sufficient gain and a low noise figure are present in the 28-MHz tunable IF.

The output BPF tuned components are critical. They should be tuned individually to resonance with a dip meter and frequency counter or some other method. Be sure to tune them with the same lead lengths that will be used when finally installed in the circuit. The shunt elements are especially critical. This filter may be used on other converter projects as well. For reference, Fig. 36 gives the modeled response of the

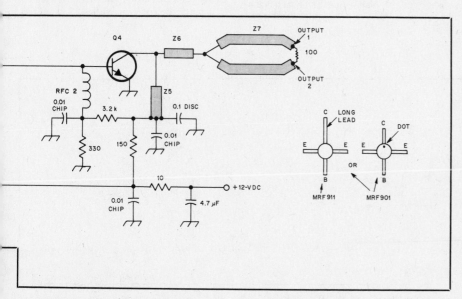

Z1, Z2 — 50-Ω microstripline, 1.5 × 0.10 in (length × width). Tapped at 0.3 in from cold end. Z1 to Z2 spacing is 0.25 in.
Z3 — 90-Ω microstripline, 1.0 × 0.03 in.
Z4 — 16-Ω microstripline, 0.397 × 0.50 in.

Z5 — 90-Ω microstripline, 1.044 × 0.03 in. Tune for maximum gain with sliding chip capacitor.
Z6 — 115-Ω microstripline, 0.78 × 0.016 in.
Z7, Z8 — 70-Ω microstripline, 1.228 × 0.06 in.

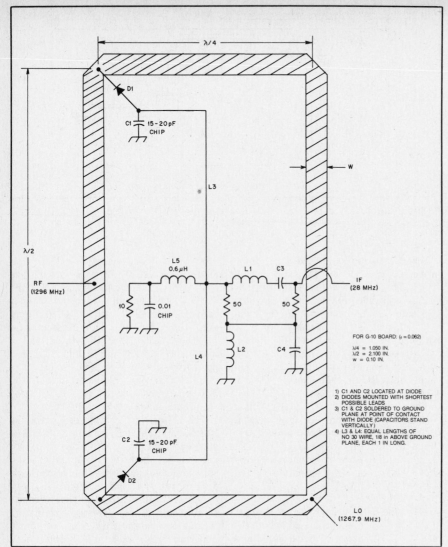

Fig. 32 — Rat-race mixer and IF-port terminator.
C1, C2 — 15- to 20-pF chip capacitor.
C3 — 22.7-pF silver mica capacitor.
C4 — 568-pF silver mica capacitor.
D1, D2 — Schottky diodes — see text.
L1 — 1.42 μH.

L2 — 56.8 nH.
L3, L4 — 1.0 in length of no. 30 wire, 1/8-in above ground plane. Must be equal lengths.
L5 — 0.6 μH, 10 turns no. 24 enameled wire on a T-37-6 core.

Fig. 33 — Photograph of the singly balanced rat-race mixer, see Fig. 32.

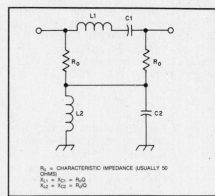

Fig. 34 — IF-port terminator diagram and design equations. A good choice for Q is approximately 5.

filter (HP-41CV with the LNAP option of the Circuits Pac). Actual response is very close to ideal.

Transmitter

The transmitter board is prepared in the same manner as the LO board. See Figs. 37 and 38. It is desirable to construct the transmitter stage by stage, since problems are easier to isolate this way. The mixer, IF port terminator and first amplifier (MRF901) are constructed first. Always ter-

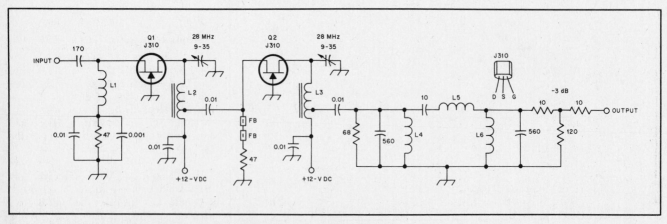

Fig. 35 — Postconversion amplifier and band-pass filter. Total gain is 23 dB, NF is 1 dB.
FB — Ferrite bead.
L1 — 0.613 μH, 14 turns of no. 22 enameled wire, ¼ in ID, closewound.
L2, L3 — 1.35 μH, 20 turns of no. 24

enameled wire on a T-37-6 core. L2 tapped 3.5 turns from cold end; L3 tapped 1.5 turns from cold end.

L4, L6 — 56 nH, 4 turns of no. 18 enameled wire, 5/32-in ID, closewound.
L5 — 3.1 μH, 27 turns of no. 24 enameled wire on a T-50-6 core.

minate the output of the last constructed stage with a "leadless" 51-ohm ¼-watt carbon composition resistor. First apply dc power and set the bias resistors for the collector current previously noted. Be sure it is stable with temperature. Then apply RF and adjust the sliding capacitor shunt line for maximum gain. An absorption wavemeter (cavity or quarter-wave line plus RF probe) or spectrum analyzer may be used. *Always* solder a chip capacitor at the far end of the tuned shunt line before applying power. Leave this chip in place and use another chip to fine-tune the line. Be sure to include the multiple bypass capacitors, since a 0.01-µF chip capacitor does not supply sufficient decoupling for LF and HF.

Receiver

If the builder has a signal source and a means of quantitative measurement, he may opt to construct the receiver stage by stage. See Figs. 39 and 40. If not, it's best to construct the entire circuit and then do the tune-up. The tuned-shunt lines are peaked in the same manner as the transmitter board. The 1296-MHz filters are tuned as previously described. You may want to vary the LO injection level. Generally, somewhere between + 3 and + 10 dBm injection yields quietest mixer operation. If the receiver is preceded by a GaAsFET preamplifier, the difference will be slight but noticeable over the MRF901 alone.

Conclusion

This transverter is relatively easy and in-

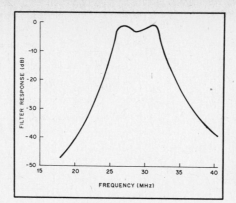

Fig. 36 — Calculated response of the 28-MHz output filter.

expensive to construct. It is also relatively easy to get operating, provided good RF and microwave construction practices are followed. Various views of the completed unit are shown in Figs. 41, 42 and 44.

Useful test equipment to go along with this project include a low-pass filter and a signal source. The filter shown in Fig. 43 is taken from the RSGB *Radio Communication Handbook*. This unit has a measured insertion loss of 0.35 dB. Response is down 3 dB at 12 MHz either side of 1296 MHz and 23 dB at 1152 MHz. The filter could be built into the transverter — a lower loss cavity filter would be better if you use a GaAsFET preamplifier.

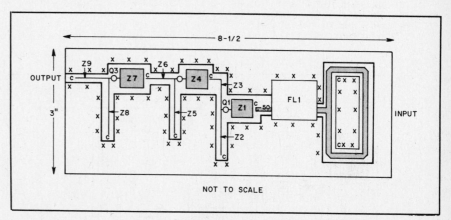

Fig. 37 — Transmitter board layout. Microstripline section marked "50" is 50 ohms, any length. C indicates location of a chip capacitor; X indicates a jumper from top to bottom ground planes. Wrap board edges with copper and solder top and bottom.

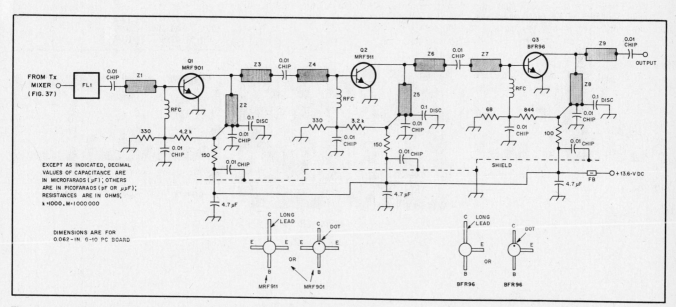

Fig. 38 — Schematic diagram of the transmitter section. Z2, Z8 and Z9 are tuned for maximum gain with "sliding" chip capacitors.

FL1 — 1296-MHz filter. See Fig. 28 and text.
FB — Ferrite bead.
RFC — 8 turns of no. 24 enameled wire, 1/8-in ID, closewound.
Z1 — 16-Ω microstripline, 0.62 × 0.50 in

(length × width).
Z2 — 90-Ω microstripline, 1.113 × 0.03 in.
Z3 — 125-Ω microstripline, 0.779 × 0.013 in.
Z4 — 16-Ω microstripline, 0.397 × 0.50 in.
Z5 — 90-Ω microstripline, 1.044 × 0.03 in.

Z6 — 115-Ω microstripline, 0.78 × 0.016 in.
Z7 — 16-Ω microstripline, 0.384 × 0.50 in.
Z8 — 90-Ω microstripline, 0.974 × 0.03 in.
Z9 — 86-Ω microstripline, 0.992 × 0.037 in.

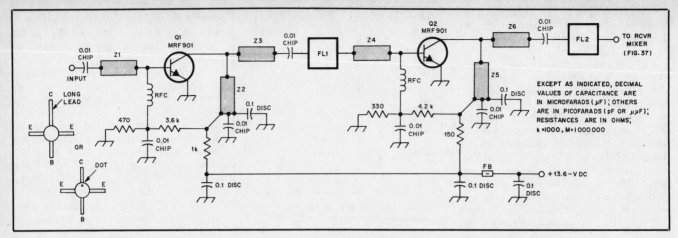

Fig. 39 — Schematic diagram of the receiver section. Z2 and Z5 are tuned for maximum gain with "sliding" chip capacitors.

FL1, FL2 — 1296-MHz band-pass filter — see Fig. 28 and text.
RFC — 8 turns of no. 24 enameled wire, 1/8-in ID, closewound.

Z1 — 18-Ω microstripline, 0.868 × 0.438 in (length × width).
Z2 — 90-Ω microstripline, 0.904 × 0.03 in.
Z3 — 125-Ω microstripline, 0.708 × 0.013 in.

Z4 — 16-Ω microstripline, 0.620 × 0.50 in.
Z5 — 90-Ω microstripline, 1.113 × 0.03 in.
Z6 — 125-Ω microstripline, 0.779 × 0.013 in.

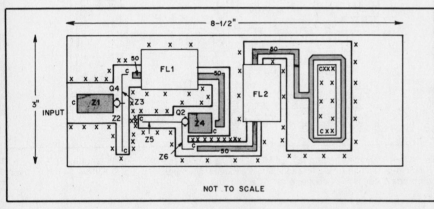

Fig. 40 — Receiver board layout. Microstripline sections marked "50" are 50 ohms, any length. C indicates location of a chip capacitor; × indicates a jumper from top to bottom ground planes. Wrap board edges with copper and solder top and bottom.

Table 4
IF-Port Terminator Performance

Freq. (MHz)	Z_{in}	Phase Angle (Degrees)
5	49.87	0.09
10	49.39	0.29
15	48.11	1.04
20	44.44	5.08
25	44.76	43.19
26	60.90	61.13
27	115.69	76.65
28	831.0	88.26
29	173.3	− 81.63
30	78.6	− 70.18
31	52.25	− 56.38
35	39.62	− 15.78
40	44.00	− 4.38
45	46.35	− 1.89
50	47.53	− 1.03
100	49.63	− 0.10
200	49.92	− 0.03

Fig. 41 — Photograph of the transmitter board.

Fig. 42 — Photograph of the receiver board.

Fig. 44 — Bottom view of the receiver board. This view shows the jumpers that are soldered from top to bottom ground planes.

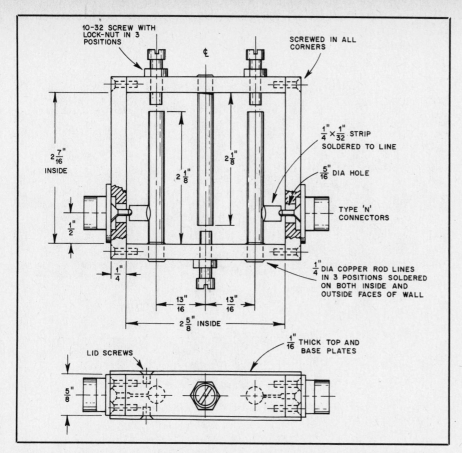

Fig 43 — Interdigital band-pass filter for 1296 MHz. From *Radio Communication Handbook* by RSGB. Walls and plates in the W6LEV version were made of 0.062-in double-sided, copper-clad PC board.

1296- to 144-MHz Transverter

This section describes the design and construction of a complete 1296-MHz transverter for use with any multimode 144-MHz transceiver. Integral dc and RF switching is included, so complete transverter TR control is accomplished by merely activating the transceiver PTT. The transverter is modularized to allow the individual to build just what suits his needs. This allows the builder to get on with the "barebones" transverter and build up. This project, shown in Figs. 51 through 76, was designed and built by Al Ward, WB5LUA.

BLOCK DIAGRAM

The basic transverter is shown as the dashed area in the block diagram in Fig. 52. The unit centers around a single doubly balanced mixer (DBM) that is used as an upconverter and downconverter. The conventional approach generally has been to use separate mixers for transmit and receive. Local oscillator (LO) output is derived from a power divider. The approach taken here is to use a single mixer with both IF port and RF port switching for the up and down conversion processes. The LO starts with a 96-MHz oscillator and multiplies up to supply about + 6 dBm of power at 1152 MHz to drive the mixer.

The antenna jack of the 144-MHz transceiver is connected to an IF switch that uses low-cost PIN diodes. On transmit, the switch directs the nominal 10-W 2-meter signal to an adjustable 40-dB attenuator. The attenuator output applies a nominal 1-mW (0-dBm) signal to the DBM. The 144-MHz signal is then mixed with the 1152-MHz LO signal to produce RF signals at 1296 MHz (RF = LO + IF) and 1008 MHz (RF = LO − IF). A cavity filter provides 28 dB of image rejection at 1008 MHz while contributing only 0.4-dB loss at 1296 MHz. After the filter, a PIN-diode RF switch directs the signal to the low-level amplifiers. The power level available to drive the low-level amplifiers is about 0.1 mW (− 10 dBm).

At this point there are several options open to the builder, depending on how much power output is desired. All options shown in Fig. 52 use state-of-the-art, low-cost broadband monolithic microwave integrated circuit amplifiers (MMICs) recently released by Avantek. The MSA-0204 driving the MSA-0304 can deliver + 10 dBm, which is enough to drive four MSA-0404s in a parallel configuration to 40 mW (option 1). If desired, an NEC NE41620

Fig. 51 — The various modules that make up the 1296-MHz transverter are mounted in a chassis and interconnected with short pieces of coaxial cable.

bipolar transistor amplifier can be used instead to reach the 100-mW level (option 2). The MSA-0204 and MSA-0304 are in inexpensive plastic packages and cost about $3 each in small quantities. The MSA-0235 and MSA-0335, encased in the micro-x ceramic package, offer slightly more gain and power output, but they cost about three times as much. The 0235/0335 combination could drive four 0404s slightly harder, producing 50 to 60 mW.

The best driver for both the parallel 0404 amplifier and the NE41620 would be a single MSA-0404 running at + 13 dBm, as shown in options 3 and 4. Since all of the gain available from MSA-0404 is not required, the IF drive level to the mixer should be reduced several decibels to avoid overdriving the transmitter strip.

The MSA-0404 costs about $3.25 in single quantities, making the component price of the parallel amplifier $13. The NEC NE41620 costs about $16 each in small quantities. The advantage of the parallel MSA-0404 amplifier is that no tuning whatsoever is required to attain the power level. The NE41620 offers higher power that can be used to drive additional transistor amplifiers to higher output. Ac-

tual MMIC performance will be discussed at length later.

On the receive side, the HXTR-3101 RF preamplifier helps establish front-end noise figure. The noise figure of the preamplifier by itself is 2 dB. Preamplifier output is routed to the PIN diode RF switch. On receive, the signal from the preamplifier is sent through the RF filter, where 28 dB of image rejection is obtained. The RF is then down converted in the mixer to the IF frequency of 144 MHz. The PIN diode IF switch directs the signal to the low-noise 144-MHz IF amplifier. Output from this amplifier is then switched via another PIN diode switch to the antenna jack of the 2-meter tranceiver used as the IF.

If the HXTR-3101 front-end RF amplifier were not used, the SSB noise figure would be approximately 11 to 12 dB depending on cable losses and type of doubly balanced mixer used. Adding the preamplifier brings the overall receiver noise figure down to a reasonable 5 to 6 dB, which is typical of the many commercial transverters available on the market today. Adding a good low noise GaAsFET preamplifier will bring the noise figure down to under 1 dB.

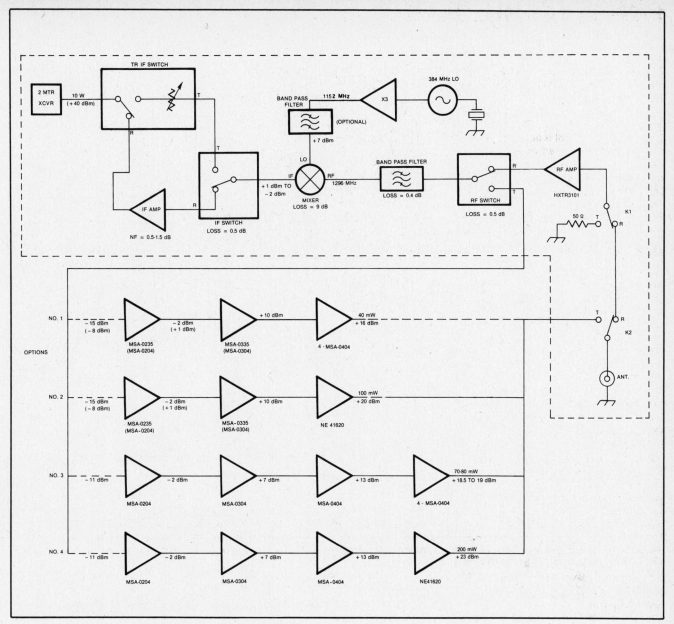

Fig. 52 — Block diagram of the 1296-MHz transverter.

Each section of the transverter is covered separately and completely with respect to description of operation, design, construction and tune-up. This should enhance the ability of the builder to construct and tune up each module separately.

1152-MHz LOCAL OSCILLATOR

The LO shown in Fig. 53 produces +6 to +8 dBm at 1152 MHz with inexpensive devices. The power level is adequate to drive any of the popular commercially available low-power doubly balanced mixers.

For ease of tune-up, the tripler should be built as a separate module. The 384-MHz unit produces an output power level of +7 dBm into a 50-ohm termination. When tuned up with the aid of a spectrum analyzer, all harmonics are 40 dB below the desired output. When a spectrum analyzer is not available and all tuned circuits are peaked for maximum 384-MHz output, the undesired harmonics will be approximately 30 dB below the desired signal. This is still quite acceptable because the harmonics will be suppressed even further

by the addition of the tripler stage.

The 384-MHz-to-1152-MHz tripler was designed by John Stankus, KN5N. The design goal was simplicity and conversion efficiency. This was accomplished by using a Motorola MRF901 in the circuit shown in Fig. 53. The multiplier has a power output of +6 dBm to +8 dBm into a 50-ohm termination when driven with +4 dBm at 384 MHz. The multiplier uses a parallel tuned output network and a low-pass filter input network. Without any additional filtering, the simple network will deliver the harmonic rejection shown below:

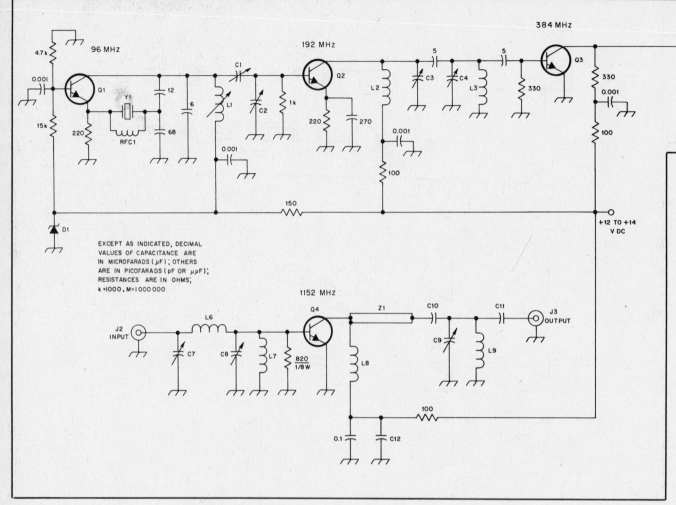

Fig. 53 — Schematic diagram of the local oscillator chain for the 1296-MHz transverter. Capacitors are silver-mica types unless noted. Resistors are ¼-W carbon-composition types unless noted.

C1-C9 — 0.8- to 16-pF piston trimmer capacitor.
C10, C12 — 30-pF chip or UNELCO bookmica capacitor.
C11 — 1- to 2-pF chip or NP0 ceramic capacitor (see text).
D1 — 9-V, 500-mW Zener diode (1N757 or equiv.).
J1-J3 — Chassis-mount female SMA or BNC connectors.
L1 — 4t no. 24 enam. wire, closewound on a

0.25-inch-OD adjustable coil form with white core (suitable for use in the 50-200 MHz range).
L2, L3 — 3t no. 14 wire, 0.25 inch ID, spaced 1 wire dia.
L4 — 1t no. 14 wire, 0.25 inch ID.
L5 — 1t no. 14 wire, 0.25 inch ID, tap at ½t.
L6 — 5t no. 24 wire, 0.2 inch ID, 0.36 inch long.
L7 — 7t no. 24 wire, 0.2 inch ID, 0.36 inch long.
L8 — 3t no. 24 wire, 0.2 inch ID, 0.6 inch

long.
L9 — 0.3-inch length of no. 24 wire between top of C9 and ground.
Q1, Q2 — 2N918, 2N3570, MPS3563 or equiv.
Q3, Q4 — MRF901.
RFC1 — 0.38-µH miniature molded RF choke.
Y1 — 96-MHz series-resonant overtone crystal.
Z1 — ¼-wavelength, 100-ohm transmission line. See text.

Frequency	Rejection
384 MHz	− 16 dBc
768 MHz	− 30 dBc
1536 MHz	− 23 dBc

The results show what can be achieved with very simple tuned circuits. If additional rejection is deemed necessary, a cavity filter similar to that used on the RF port of the mixer can be inserted between the multiplier and the LO port of the mixer. The author has used the LO chain described here without additional filtering for

two years with no problems from spurious signals.

Circuit Description

For 1296-MHz operation, an LO injection frequency of 1152 MHz is required. The basic oscillator uses a 96.0-MHz overtone crystal. For satellite Mode L operation at 1269 MHz, a 93.6667 MHz overtone crystal will be required, assuming an IF of 145 MHz is used. If you want to operate on both band segments, you can build the LO with a crystal for each and use a miniature DPDT toggle switch to switch between

them. If desired, PIN diodes can be used to switch crystals. The mechanical switch was chosen for simplicity, and it created no stability problems. Voltage to the oscillator is regulated at 9 V by D1.

The crystal frequencies are close enough that one setting of L1 produces maximum power output and easy crystal starting. Q1 can be a 2N918, MPS3563 or equivalent device. Any lower subharmonic of the crystal frequency is not suggested as this will only make adequate harmonic rejection harder to obtain.

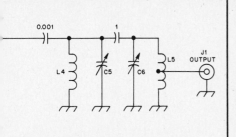

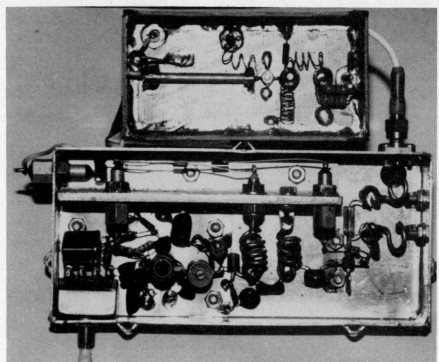

Fig. 54 — The 384-MHz LO and tripler are built in separate enclosures.

Capacitive divider C1 and C2 matches the input of Q2, which again can be a 2N918 or MPS3563. The interstage matching network between Q2 and Q3 is a two-pole filter composed of L2-C3 and L3-C4 tuned to 192 MHz. The last doubler uses an MRF901 with another two-pole filter composed of C5-L4 and C6-L5 tuned to 384 MHz. The output is tapped halfway up on L5 to maximize output into a 50-ohm termination.

The tripler (Q4) also uses an MRF901. The input network is a standard low-pass filter to match to 50 ohms. L7 acts as an RF choke, and the 820-ohm resistor insures stability. Network C9-L9 forms a parallel-tuned filter at 1152 MHz. The quarter wave 100-ohm line (Z1) and series capacitor C11 transform the required 50-ohm output impedance up to a level required to enhance harmonic output of the MRF901. Depending on the lead inductance of C11, the output may be overcoupled and the power output at 1152 MHz low. If this is the case, change C11 to a 1-pF chip capacitor and retune the output.

Construction

The basic 384-MHz local-oscillator assembly is built in a 2-3/8 × 5-3/8 × 1-inch box that was found surplus. The tripler was built in a 1¾ × 3½ × ¾-inch box made from circuit board material. The builder may use any aluminum diecast box or Minibox of appropriate dimensions. SMA connectors were used for size considerations. BNC connectors should work as well.

As with all UHF circuitry, use the shortest lead lengths possible. There are two options available to the builder for producing Z1. The first option, which was used by KN5N in his original design, is a microstripline 0.050 inches wide by 1.9 inches long etched on 0.062-inch thick Duroid 5880 dielectric material (Er = 2.2). The second option, used by the author, is a 1.7-inch length of 0.141-inch, 50-ohm semi-rigid miniature hardline (UT-141). To make this line the right impedance, remove the center conductor and replace it with a 0.010-inch-diameter (no. 30) copper wire of similar length. The small-diameter wire raised the coaxial impedance to approxi-

mately 100 ohms. L9 is simply a 0.3-inch length of no. 24 wire soldered between the top of piston trimer C7 and the ground plane. The finished LO is shown in Fig. 54.

Tune-Up

Initially, the 96-MHz crystal oscillator is best adjusted by peaking L1 with a dip meter operating as a detector. Start with C1 and C2 near minimum capacitance. After L1 is peaked, C1 and C2 are adjusted to maximize the voltage across the 220-ohm resistor in the emitter lead of Q2. In a similar fashion, maximize Q3 collector current by adjusting C3 and C4. Finally, C5 and C6 are adjusted for maximum output at 384 MHz. Harmonics should be 30 dB down from the 384-MHz signal. With the aid of a spectrum analyzer, harmonic rejection of 40 dB can be achieved. When the basic local oscillator is running properly, typical Q1 collector current is 6-10 mA for +7 dBm of output power. Total current drawn by Zener diode D1 and Q1 will be approximately 32 mA. Q2 collector current will be 9-10 mA typically. Q3 collector current will be 17-20 mA, typically, for +7 dBm output power. If you experience low 384-MHz output power, try the following. Change the 1-pF capacitor between L4, C5 and C6, L5 to a 2-pF silver-mica or ceramic unit and retune the output network.

When the 384-MHz assembly is operating properly, insert a 6-dB attenuator between the 384-MHz output and the tripler

input. This attenuator value can be optimized later, as the tripler typically requires +1 to +4 dBm to drive it to full output. The MRF901 is capable of continuous collector current of only 30 mA. Typically, 20 to 24 mA is the maximum desired. With this in mind, tune C7 and C8 to increase Q4 collector current, then tune C9 for maximum output at 1152 MHz. About +6 to +8 dBm of 1152-MHz power will be obtained with 20 to 24 mA of collector current.

Adjust the input attenuator as required to insure that the MRF901 collector current will not exceed 20 to 24 mA. With minimal adjustment, the LO can be tuned for equivalent power levels at 1152 MHz (1296-MHz operation) and 1124 MHz (1269-MHz operation).

DOUBLY BALANCED MIXER

One of the main reasons for using a commercially available doubly balanced mixer was size. The typical DBM is quite a bit smaller than the commonly used singly balanced rat-race mixer. A DBM can easily be mounted with SMA connectors in a small 1.4- × 2.25-inch diecast aluminum box.

Most DBMs have a broad frequency response because of the inherently broadband nature of the toroidal transformers used. A schematic diagram of a typical DBM is shown in Fig. 55. No tuning is required. Fig. 56 shows the mixer used here.

The balanced nature of the DBM is the characteristic that determines the magnitude of the isolation between each of the three ports. Typically, 15 to 25 dB of isolation is the minimum desired to reduce interaction at each of the mixer ports. Typical conversion losses of 7 to 10 dB are possible with SWRs of 2:1 to 3:1.

The main drawback to a commercial DBM is cost. Most DBMs that are rated to 1300 MHz cost at least $20 in single quantities. Careful research with several mixer manufacturers has revealed that several DBMs that are advertised to operate up to 1 GHz can still produce adequate performance at 1300 MHz.

Table 6 summarizes mixers available from several manufacturers. All mixers except the Tele-Tech MT55L require +7 dBm of LO power. Even the mixers that are running at a higher frequency than advertised still produce acceptable conversion loss performance at 1300 MHz. Most likely, the isolation and SWR characteristics are slightly worse. Two mixers stand out as high performers for low cost — the Mini-Circuits SBL-1X and the Tele-Tech MT57. The SBL-1X has been tried by W5ETG with good results, so 9-dB conversion loss should be attainable. The MT57 has not been tried but should work as well. The SBL-1X is available in the standard 8-pin relay package, while the MT-57 is available in a TO-5 package.

Mini-Circuits and Tele-Tech have no minimum-order requirement; both will deal factory-direct. Just be sure to include enough extra money for postage and handling. Synergy sells direct but presently has a $50 minimum-order requirement.

TRANSMIT/RECEIVE IF SWITCH

The IF switching arrangement described here performs the important function of mating the 2-meter transceiver to the doubly balanced mixer. See Fig. 57. In the transmit mode, the 10-W 2-meter signal is switched to an attenuator through D1, a high-power solid-state TR switch. The attenuated signal is routed through a low-power PIN diode switch (D2 and D3) to feed the IF port of the mixer. In the receive mode, the IF signal from the mixer is fed through the low-power PIN diode switch to a low noise, dual-gate FET IF amplifier. The output of the IF amplifier is fed to the 2-meter transceiver via the high-power PIN diode switch.

The high-power MA-8334 series multi-throw switches use high-breakdown-voltage PIN diode chips. Two package styles are available, along with several switch configurations. Two types are applicable for this project. A 10-W version (MA-8334-100) is available in a 0.350-inch-diameter can. A 100-W version (MA-8334-001) is available in a package that lends itself to microstripline applications. The -100 style is adequate for the power levels encountered here.

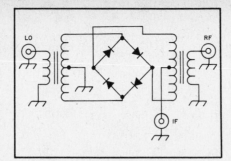

Fig. 55 — Schematic diagram of the circuit inside a typical doubly balanced mixer module.

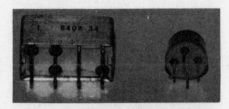

Fig. 56 — Typical DBMs and method of mounting with SMA connectors.

The low-power glass PIN diodes are either Microwave Associates part number MA47047, 47110, 47123 or Hewlett Packard HP5082-3379. These diodes cost about $2 in small quantities and are available from local distributors.

For a typical doubly balanced mixer running at +7 dBm LO power, a good rule of thumb is to not exceed +1 dBm IF input power when the mixer is run as an up-converter. With the 2-meter transceiver running at full power (typically 10 W or so), adjust C1 for 0 dBm (1 mW) at the mixer.

The receive side of the IF switch includes a low-noise IF amplifier using an inexpensive 3N204 or 3N211 dual-gate FET device. The circuit is as shown in Fig. 57. An alternate circuit using an inexpensive NEC NE411 is shown in Fig. 58. The 3N204-type devices typically deliver 1.5- to 2.0-dB noise figures, while the NE411 typically delivers a 0.5- to 0.7-dB noise figure.

A low-pass type filter network was chosen for the input circuit to enhance rejection of the unwanted sideband (1152 MHz + 1296 MHz) generated by the doubly balanced mixer and to reflect any LO power coming out of the IF port. The input circuit should be adjusted for minimum noise figure, and the output network can be optimized for maximum gain. If a gain adjustment is desired, the voltage of gate 2 can be further decreased by making the 51-kΩ resistor variable.

For both the input and output PIN diode switches, bias current is limited to 50-60 mA by the 220-ohm resistor in the common-port leg of each switch. In addition, this produces 11 V at the cathode to reverse bias the "off" section diode. The 100-kΩ resistors allow leakage current to flow through the "off" diode. The reverse bias condition tends to increase the isolation in the "off" state by decreasing diode capacitance.

The TR IF switch is a good basic building block that can be applied to any VHF or UHF transverter design. It is best built in three separate enclosures — high-power IF switch and attenuator, low-power IF switch and IF preamplifier. Since the high-power

Table 6
Doubly Balanced Mixers

Manufacturer	Model	LO/RF Freq. (MHz)	IF Freq. (MHz)	Conversion Loss (dB)
Mini-Circuits	SRA-4	5-1250	5-500	9
P.O. Box 166	SRA-5	5-1500	10-600	8.5
Brooklyn, NY 11235	SBL-1X	10-1000	5-500	9
tel. 718-934-4500	TFM-2	1-1000	dc-1000	8
Synergy	S-4	10-1000	5-1000	10
483 McLean Blvd. &	CLP-211	50-1500	dc-1000	7
18th Ave.				
Paterson, NJ 07504				
tel. 201-881-8800				
Tele-Tech Corp.	MT-55L†	1-1500	dc-1000	9
2050 Fairway Dr.	MT-55	1-1500	dc-1000	9
Box 1827	MT85	5-1500	dc-1000	9
Bozeman, MT 59715	MT57	10-2000	dc-1000	9
tel. 406-586-0291				

†requires +3 dBm LO

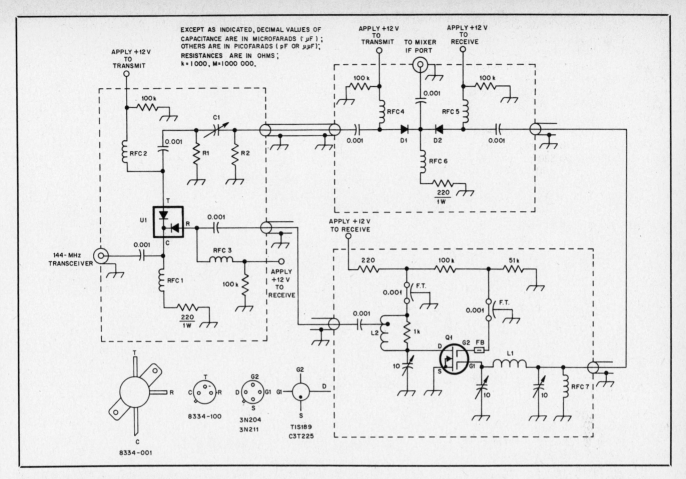

Fig. 57 — Schematic diagram of the high-power IF switch, low-power IF switch and IF preamplifier. Resistors are ¼-W carbon-composition types unless noted. Capacitors are silver-mica types unless noted.

C1 — Typically 1 to 2 pF. Adjust for desired drive level at mixer. See text.
D1, D2 — PIN switching diode (MA47047, 47110, 47123, HP 5082-3379 or equiv.).
L1 — 6t no. 18 wire, 0.25 inch ID, spaced 1 wire dia.

L2 — 5t no. 18 wire, 0.25 inch ID, spaced 1 wire dia., tapped 1t from cold end.
Q1 — Dual-gate MOSFET (TIS189, 3N204, 3N211 or equiv.).
R1 — 50-Ω, 10-W noninductive resistor. May be made from a combination of 2-W carbon-

composition resistors in parallel.
R2 — 51-Ω, ½-W carbon-composition resistor.
RFC1-RFC7 — 1-µH miniature molded RF choke.
U1 — Microwave Associates MA8334 series TR switch. See text.

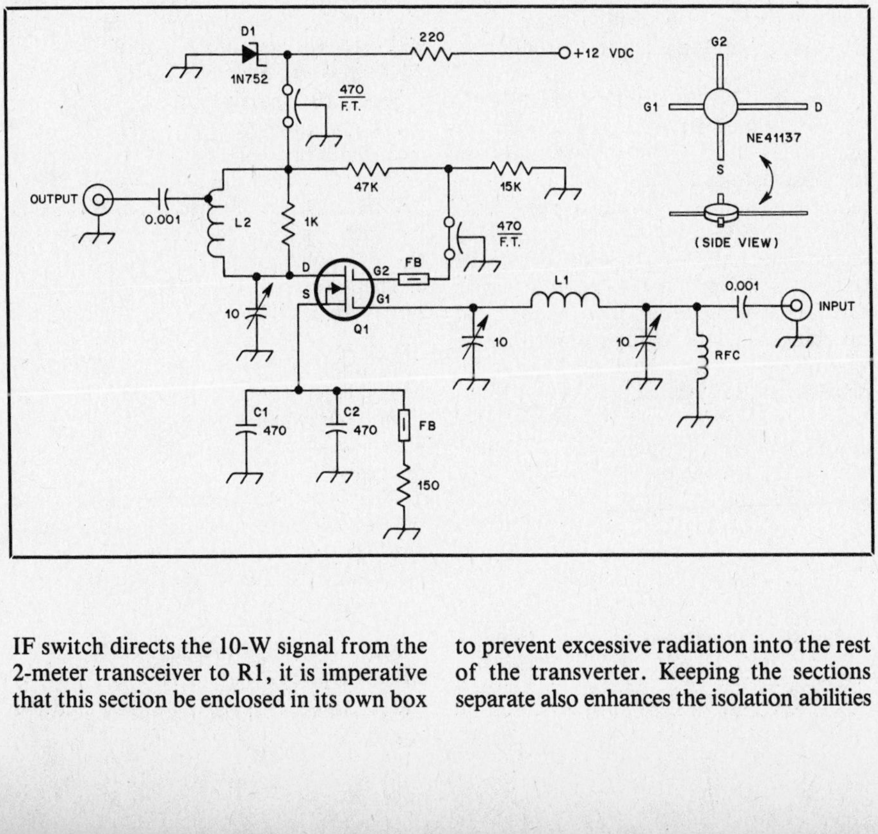

Fig. 58 — Schematic diagram of the dual-gate GaAsFET IF preamplifier. Resistors are ¼-W carbon-composition types. Capacitors are disc-ceramic or silver-mica types unless noted.

C1, C2 — 470- to 1000-pF feedthrough, button or ceramic chip capacitors.
C3, C4 — 470 to 1000-pF feedthrough capacitor.
D1 — 5.6-V, 500-mW Zener diode (1N752 or equiv.).
FB — Ferrite bead.
L1 — 6t no. 24 wire, 0.25 inch ID, spaced 1 wire dia.
L2 — 5t no. 24 wire, 0.25 inch ID, spaced 1 wire dia., tapped 1t from cold end.
Q1 — NEC NE41137 dual-gate GaAsFET.
RFC1 — 1-µH miniature molded RF choke.

of the PIN diode switches. See Fig. 59.

Use good VHF wiring practice when building the IF preamplifier. A good shield between the input and output is suggested, as well as a ferrite bead on gate 2 of the device to reduce tendency to oscillate. If stability is a problem under certain load conditions, decrease the value of the resistor in parallel with L2.

IF switch directs the 10-W signal from the 2-meter transceiver to R1, it is imperative that this section be enclosed in its own box to prevent excessive radiation into the rest of the transverter. Keeping the sections separate also enhances the isolation abilities

Fig. 59 — The high-power IF switch is fully shielded to prevent RF from entering other stages.

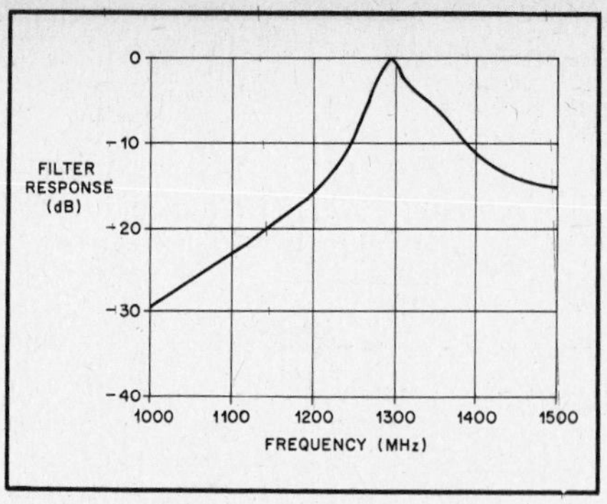

Fig. 60 — Frequency response of the capacitively coupled cavity filter.

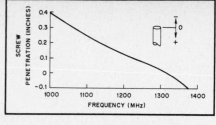

Fig. 62 — Approximate resonant frequency of the cavity filter for various degrees of screw penetration.

1296-MHz RF BAND-PASS FILTER

The primary function of the RF band-pass filter is to achieve adequate image rejection during receive and adequate unwanted sideband rejection during transmit. In both cases, the undesired product is 1152 MHz − 144 MHz = 1008 MHz. The simple quarter-wave filter described here achieves 28 dB of rejection at 1008 MHz while introducing only 0.4-dB loss at 1296 MHz. In addition, 20 dB of rejection is achieved at the 1152 MHz LO frequency. The complete frequency response is shown in Fig. 60. The 1-dB bandwidth is 18 MHz, while the 3-dB bandwidth is 40 MHz. SWR at resonance is 1.15:1.

Construction details are given in Fig. 61. Three of the four sides and the bottom of the filter are made from 1/16-inch-thick double-sided circuit board material. The inner dimensions of the the cavity should be 0.75 inch square by 1.9 inches long. A piece of sheet brass covers the top of the cavity and also serves as a mounting point for the no. 6-32 brass screw.

A 1.8-inch length of 0.25-inch-diameter hobby brass tubing is centered on the cavity floor and aligned with the tuning screw. Make sure the tuning screw never makes contact with the brass tubing over the entire tuning range.

Connectors should be SMA type because of their small size. The author used SMA connectors with four small mounting tabs that soldered to the cavity inner wall. Whatever type SMA connectors you decide to use, make sure that the connector ground connection is made directly to the inner cavity wall; otherwise, RF performance will suffer. The capacitive probes that are soldered to the connector center conductors should be made out of thin sheet brass and formed as shown in Fig. 61. With the cavity assembled, adjust the spacing between the tab and center conductor to 0.050 inch.

After all components are assembled inside the cavity, install a piece of sheet brass over the front of the cavity to seal the unit. Press the sheet brass slightly into the cavity (about 0.020 inch) so it can be soldered directly to the inner cavity walls at the edges.

If additional out-of-band rejection is desired, two filters can be cascaded. Other

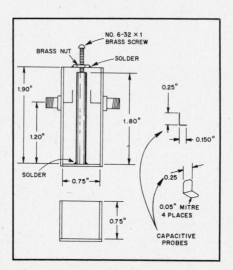

Fig. 61 — Construction details of the capacitively coupled cavity filter. See text for additional information.

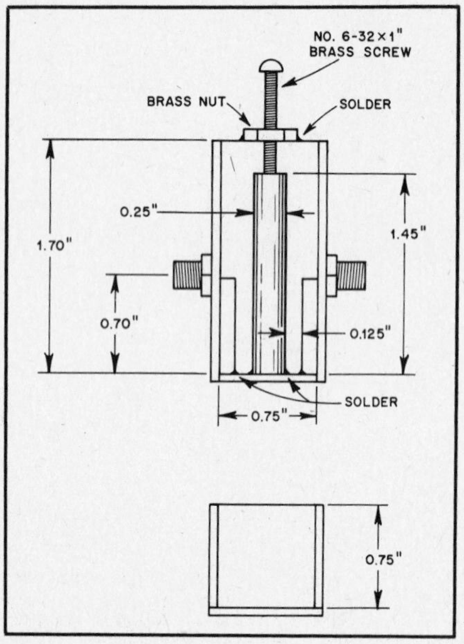

options are to decrease the size of the coupling capacitors and/or increase the impedance of the cavity from 70 ohms to about 100 ohms. Loss will increase slightly. Another option open to the builder is to build a multi-element interdigital filter as described by Fisher and Hinshaw[1,2].

As an added feature, the filter can be used as a wavemeter to aid in tuning up local oscillators. Fig. 62 is a plot of resonant frequency versus depth of penetration of the no. 6-32 screw into the 0.250-inch diameter tubing. Positive numbers indicate penetration; 0 means flush; and negative numbers indicate distance from open end of tubing. With the tuning screw completely removed, the cavity will resonate at approximately 1375 MHz.

The first band-pass filter tried by the author was a scaled W1JR 432 MHz cavity filter using inductive coupling[3]. It does not offer the image rejection of the

[1]R. Fisher, "Interdigital Bandpass Filter for Amateur VHF/UHF Applications," *QST*, March 1968, p. 32.
[2]J. Hinshaw, "Computer-aided Interdigital Bandpass Filter Design," *Ham Radio*, January 1985, p. 12.
[3]J. Reisert, "Ultra Low Noise UHF Preamplifier," *Ham Radio*, March 1975, p. 18.

capacitively coupled filter just described, but it does have a characteristic that may be advantageous to someone who operates at both 1296 and 1269 MHz. When adjusted for minimum loss (0.5 dB) at 1296 MHz, the insertion loss at 1269 MHz is only 3 dB. The capacitively coupled filter rolls off about 6 dB at 1269 MHz. The 1-dB bandwidth of the inductively coupled filter is 30 MHz, and the 3-dB bandwidth is 56 MHz. The image rejection at 1008 MHz is 18 dB, compared to 28 dB for the capacitively coupled design.

The inductively coupled filter construction is shown in Fig. 63. Construction and materials are very similar for both filters. The inductive loops are made from no. 20 enameled wire. The filter tunes to 1296 MHz when the tuning screw penetrates 0.28 inch into the brass-tubing resonator.

The filters just discussed represent the simplest designs that can be used effectively at 1296 MHz. They both offer adequate rejection of unwanted signals when used in conjunction with a doubly balanced mixer.

SINGLE-POLE, DOUBLE-THROW RF SWITCH

At first glance, the requirement for a

Fig. 63 — Construction details of the inductively coupled cavity filter.

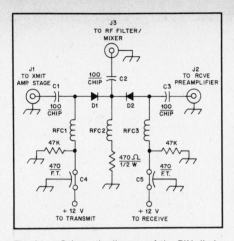

Fig. 64 — Schematic diagram of the PIN diode RF switch.

C1-C3 — 100-pF ceramic chip capacitor.
C4, C5 — 100- to 470-pF feedthrough capacitor.
D1, D2 — HP 5082-3379 PIN diode.
J1-J3 — Female SMA connector.
RFC1-RFC3 — 6t no. 24 enam. wire, 0.125 inch ID, closewound.

low-loss, low-cost RF switch would seem to dictate that the switch be electro mechanical. But not so! Hewlett Packard manufactures a PIN diode, part number HP5082-3379, which has extremely low loss and is inexpensive (a couple of dollars each in small quantities). The circuit shown in Fig. 64 is a simple SPDT switch using a series diode in each leg. "Off" isolation in enhanced by using the positive voltage that is created across the 470-ohm resistor when either series diode is forward biased. The positive voltage reverse biases the opposite diode because reverse leakage current is allowed to flow through the 47-kΩ resistors. The RF switch has only 0.5 dB loss at 1296 MHz. The "off" port isolation at 1296 MHz has been measured at 21 dB.

Having additional isolation is not extremely important in this transverter design since only the receive or transmit section is powered up at any one time. The isolation as measured is adequate to eliminate loading effects on the normal "through" port performance.

The circuit is laid out on a 1.2- × 2.0-inch, 0.062-inch-thick, double-sided, glass-epoxy circuit board. One side is unetched and acts as a ground plane. Parts placement is shown in Fig. 65 and a photo of the finished unit is shown in Fig. 66. Transmission lines of 50 ohms (0.1 inch wide) enhance the high-frequency response. Holes 0.25 inch in diameter are drilled through the board to allow the PIN diode to be mounted with the leads flat on the circuit trace. C1, C2 and C3 can be either 0.05- or 0.1-inch square and are soldered directly to the circuit trace in the center of the line. SMA type connectors (2 or 4 hole flange mount) are recommended and should be soldered to the ground plane. In addition, run small screws, metal tabs or wire through the SMA connector holes and solder both sides to tie the top and bottom ground planes together.

1296-MHz PREAMPLIFIER

The simple preamplifier shown in Fig. 67 uses an inexpensive Hewlett Packard HXTR-3101 bipolar device. Noise figure is just under 2 dB with an associated gain of 12 dB. The preamp as described here will decrease the SSB mixer noise figure from 12 dB to 4 dB — a significant improvement in sensitivity!

Circuit Design

The input match is accomplished with a conventional pi network that is tuned by C1 and C2 for best noise figure. The output match is accomplished by a high-pass L network that requires no tuning. The bias circuit is set up for V_{ce} of 10 V with an I_c of 10 mA to obtain the lowest possible noise figure. If necessary, the 7.5-kΩ base-bias resistor can be varied to obtain 10-mA idling current.

Construction

The preamplifier is built in a 1 × 2 × 3-inch enclosure made from double-sided circuit-board stock. With the layout shown in Fig. 67, no shield between the input and output circuits should be required. Both Q1 emitter leads are soldered to the ground plane with leads as short as possible. Connectors can be any convenient type — N, BNC or SMA.

The input microstripline is 0.25 inch wide. It is mounted 0.2 inch above the ground plane and suspended between C1 and C2. These piston trimmers are mounted on 0.75 inch centers. The actual length of L1 will be approximately 0.5 inch.

LOW-POWER RF AMPLIFIER

After the IF signal is mixed up to 1296 MHz, filtered and passed through the RF switch, the power level, if all goes well, is in the region of −10 dBm. The requirement now exists for a stage that can amplify the −10 dBm (0.1 mW) signal up to +10 to +13 dBm (10 to 20 mW). Typically, many builders turn to the common MRF901 preamplifier circuits and operate them as small-signal class-A amplifiers. This is fine, but the circuits must be tuned by expensive piston trimmers and these are best reserved for higher power or preamplifier stages.

The two-stage amplifier shown in Fig. 68 requires no tuning and is unconditionally stable. It is built in a 1.2- × 2.0-inch enclosure. Compare this to the MRF901 stages which are typically 2 × 4 inches each, require tuning and are sometimes difficult to get running. The amplifier used here is designed around two silicon mono-lithic microwave integrated circuits (MMICs) manufactured by Avantek. The first stage uses an MSA-0235-21, while the second stage is an MSA-0335-21. Inside each device is a Darlington-connected transistor pair that uses both series and shunt feedback to obtain broad-band gain usable to above 3 GHz. The devices used here are part of a series of ICs that are rated at different gain and power levels.

The MMICs have been optimized for use in a 50-ohm system. The input and output of each device are dc connected and require an external bias decoupling choke. The only other parts needed for a complete gain stage are a series current-limiting resistor and dc blocking capacitors. Compare this to the many associated parts needed for an MRF901 stage.

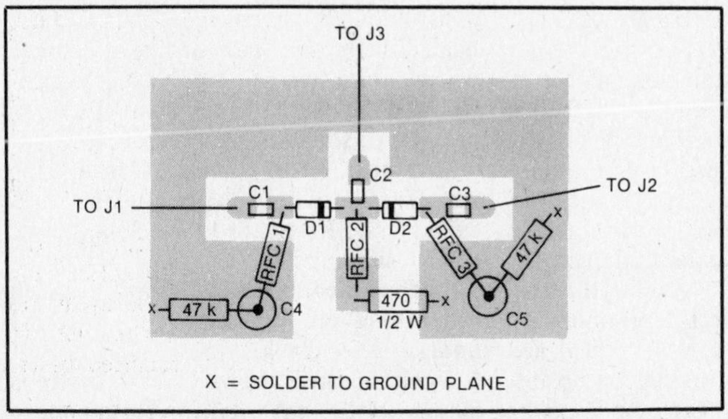

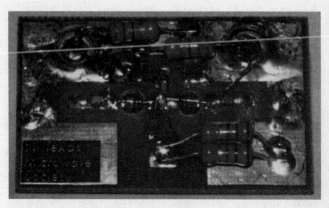

Fig. 65 — Parts placement for the PIN diode RF switch board. A full-size etching pattern appears at the back of this book.

X = SOLDER TO GROUND PLANE

Fig. 66 — Photograph of the SPDT PIN diode RF switch.

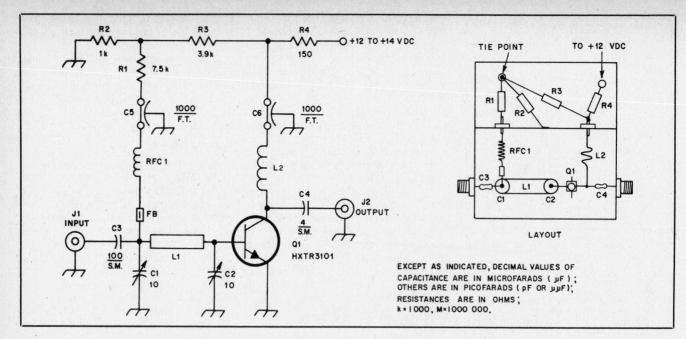

Fig. 67 — Schematic diagram of the HXTR-3101 1296-MHz preamplifier. Resistors are ¼-W carbon-composition types. Capacitors are silver-mica types unless noted.

C1, C2 — 10-pF piston trimmer capacitor.
C5, C6 — 470- to 1000-pF feedthrough capacitor.
FB — Ferrite bead.

J1, J2 — Female chassis-mount RF connector, type SMA preferred.
L1 — 0.25-inch-wide microstripline suspended between C1 and C2. See text.
L2 — 2t no. 24 wire, 0.125 inch ID, spaced 1 wire dia.

Q1 — Hewlett Packard HXTR-3101 bipolar microwave transistor.
RFC1 — 6t no. 24 enam. wire, 0.125-inch ID, closewound.

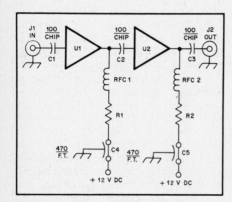

Fig. 68 — Schematic diagram of the two-stage, low-power MMIC amplifier.

C1-C3 — 100-pF chip capacitor.
C4, C5 — 100- to 470-pF feedthrough capacitor.
J1, J2 — Female SMA connector.
R1, R2 — Use values given in Table 7 for initial setup.
RFC1, RFC2 — 6t no. 24 enam. wire, 0.125-inch ID, closewound.
U1, U2 — Avantek MMIC. See text and Table 8.

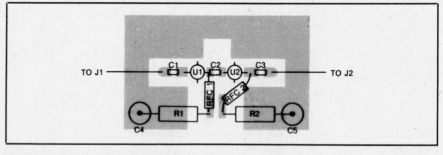

Fig. 69 — Parts placement diagram for the low-power MMIC board. A full-size etching pattern may be found at the back of this book.

The bias circuitry was optimized for operation at 1296 MHz, although the two-stage amplifier is still quite usable at 2304 MHz:

Frequency   Gain   1-dB Compression Pt.
1296 MHz   24 dB   +13 dBm
2304 MHz   14 dB   + 7 dBm

The amplifier circuit was etched on 0.062-inch-thick, double-sided, glass-epoxy circuit-board material. The etching pattern is a modified version of the SPDT switch design, except the center piece of 50-ohm microstripline was cut in the middle to accommodate C2. One side is unetched and acts as a ground plane. See Fig. 68. The etched board fits nicely into a Bud CU-123 diecast box. See Fig. 69.

The input and output connectors are 2- or 4-hole flange-type SMAs and are soldered to the unetched ground plane. On the ground-plane side, use an end mill or drill bit approximately 0.136 inch in diameter to remove the copper around the SMA center pin. This minimizes the discontinuity between the SMA connector and board by insuring that the 50-ohm impedance is maintained when they are mated. It is desirable to connect the top and bottom ground planes together at the connectors with screws or wire feed throughs that are soldered to each side.

Drill a 0.25-inch-diameter hole through the board to clear the body of each MMIC. Carefully bend the ground leads of each device downward through the hole and solder them to the unetched ground plane. Use a 1/32-inch-diameter lead as a tool to assist in bending the fragile leads. Do not bend directly at the case, or they may break off. Use care in soldering the chip capacitors and the devices to the board. A 15-W pencil-type soldering iron is sufficient. Excessive heat will only burn off what little metallization is available on the chip capacitors. Keep the RF chokes and current-limiting resistors away from the input side of the MMICs and away from each other to avoid problems with external undesired feedback.

Typical current drawn by each device is 50 mA, or 100 mA total for the two-stage amplifier. If you are so inclined to run this amplifier as a preamplifier, the noise

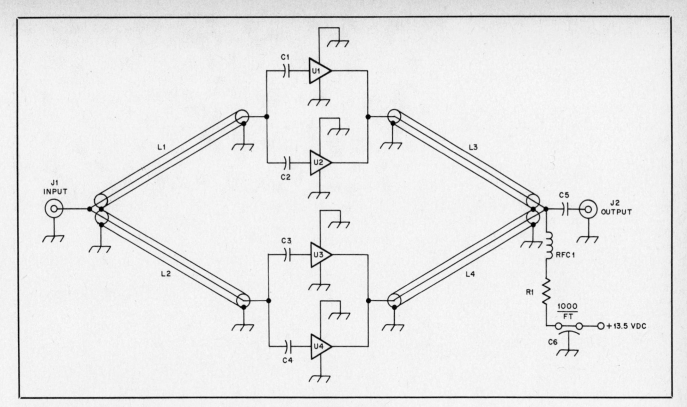

Fig. 70 — Schematic diagram of the 80-mW amplifier using four MMICs in parallel.

C1-C5 — 100- to 820-pF chip capacitor (value not critical).
C6 — 470- to 1000-pF feedthrough capacitor.
J1, J2 — Female SMA connector.
L1-L4 — ¼-wavelength section of 50-ohm

0.141-inch semi-rigid miniature cable (UT-141 or equiv.). The outer shield of each piece is 1.6 inches long from end to end.
R1 — 40-Ω, 2-W resistor (made from several 1-W units in parallel). Alter value as

necessary to limit current to 200 mA. See text.
RFC1 — 6t no. 24 wire, 0.125-inch ID, spaced 1 wire dia.
U1-U4 — Avantek MSA-0404 MMIC. See text.

Table 7
MMIC Performance at 1300 MHz

Device	I_c (mA)	Bias Resistor for V_{cc} = 13 V Ohms	P_d (W)	P_{out} @ 1-dB C.P. (dBm)	Gain @ Low Level (dB)
MSA-0104	30	267	0.24	+ 1	13
MSA-0135-21	30	267	0.24	+ 1	16
MSA-0204	30	267	0.24	+ 5	9
MSA-0235-21	50	160	0.40	+ 10	13
MSA-0304	40	200	0.32	+ 10	9
MSA-0335-21	50	160	0.40	+ 12	12
MSA-0404	50	150	0.38	+ 13	7
MSA-0435	60	125	0.45	+ 13	8
MSA-0420	90	72	0.59	+ 15	8

Table 8
Cascaded MMIC Performance

Lineup (MSA-####)	P_{out} @ 1 dB C.P. (dBm)	Gain @ Low Level (dB)	P_{in} @ 1 dB C.P. (dBm)
0235-0335	+ 12	25	− 13
0204-0304	+ 10	20	− 10
0304-0404	+ 13	16	− 3
0204-0304-0404	+ 13	25	− 12

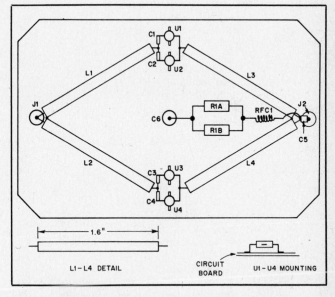

Fig. 71 — Parts placement diagram for the 80-mW amplifier using four MSA-0404 MMICs.

figures at 1296 MHz and 2304 MHz are typically 4.7 dB and 5.3 dB, respectively.

The current-limiting resistors suggested in Table 7 should be used as a guide when initially setting up the MMICs. The values should be adjusted accordingly to draw the desired current. No tune-up is required!

Avantek has recently released an entire line of plastic package equivalents to the "micro-x" package devices shown here. Typical 1300-MHz performance for both package styles is shown in Table 7. Several suggested device line ups are characterized in Table 8. Probably the best approach for the money is the 0204/0304/0404 line up. The 1-dB compression point is good, and gain is high enough that additional mixer and or filter loss could be tolerated if required.

80-mW AMPLIFIER

The amplifier stage described here is capable of producing greater than 100 mW of output power. The 1-dB compression point is 80 mW (+ 19 dBm). The schematic diagram is given in Fig. 70. Gain has been measured at 5.5 to 6.0 dB in its linear operating range. This amplifier requires no tuning whatsoever. Apply 12- to 13.5-V dc and RF drive, and that's it.

Circuit Details

The heart of the amplifier is the Avantek MSA-0404 MMIC with usable gain from dc through 3 GHz. This design uses four of these devices. Two 0404s are connected in parallel, and then the two parallel-connected pairs are combined with Wilkinson power dividers on the input and output. Because they are connected in parallel, the input and output impedance of each pair of 0404s is 25 ohms. The 25-ohm impedances are transformed up to 100 ohms through quarter-wavelength sections of 50-ohm coaxial cable. At the input and output connectors, the 100-ohm impedances from each pair are paralleled to obtain the desired 50 ohms for the input and output. Blocking capacitors are used on the input to each MMIC for dc isolation. The outputs of each circuit are paralleled for dc, and bias is applied through one common series bias resistor and RFC.

Construction

The amplifier is built on a 2.05- × 4.05-inch piece of 0.062-inch-thick, glass-epoxy, double-sided circuit board. This board installs nicely into an aluminum diecast box (Bud CU-124, Hammond 1490B). Recommended component layout is shown in Fig. 71. SMA connectors are suggested. The transmission lines are made from miniature 50-ohm semi-rigid coaxial cable (UT-141 or equiv.). Cut each piece so that the outer jacket is 1.6 inches end-to-end. All four cables should be carefully cut to the same length within ±0.20 inch. This is a little over ±1 electrical degree at 1296 MHz. Tack solder the outer jacket of each piece of semi-rigid cable to the ground plane at the end of each line.

Be very careful when handling the MMICs; the leads may break off if they are bent directly at the case. Use the lead from a 1-W resistor as a tool to form the leads of each device as shown in Fig. 71. Use a 15-W soldering iron to tack solder the ground leads to the ground plane and solder to the input and output leads. Mount the devices so that the center-to-center distance is 5/16 inch maximum. This way, the 0.1- × 0.1-inch chip capacitors can be mounted between the semi-rigid center conductor and the device input. A very short length of no. 24 wire can be used to tie the MMIC outputs and the semi-rigid center conductor together. When laying out the semi-rigid coax be sure to leave enough room near the output connector for C5. See Fig. 72.

NE41620 AMPLIFIER

The NE41620 is a rugged stud-mount device that will deliver 250 mW at the 1-dB

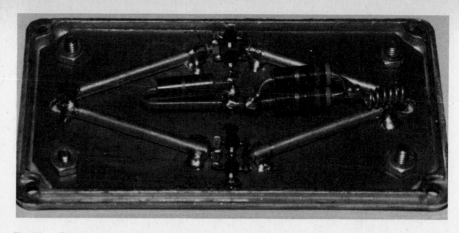

Fig. 72 — The 80-mW amplifier fits inside a standard Bud or Hammond diecast box.

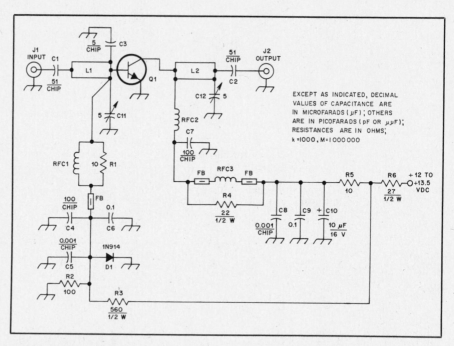

Fig. 73 — Schematic diagram of the NE41620 power amplifier. Resistors are ¼-W carbon-composition types unless noted.

C1, C2 — 51-pF chip capacitor.
C3 — 5-pF chip capacitor.
C4, C7 — 100-pF chip capacitor.
C5, C8 — 1000-pF chip capacitor.
C6, C9 — 0.1-μF disc-ceramic capacitor.
C10 — 10-μF electrolytic capacitor.
C11, C12 — 5-pF variable capacitor.
D1 — 1N4001 or 1N914.
FB — Ferrite bead.
L1 — 27-Ω microstripline, ¼ wavelength long. See text.

L2 — 51-Ω microstripline, ¼ wavelength long. See text.
Q1 — NEC NE41620 power transistor.
RFC1 — 4t no. 24 enam. wire, 0.125 inch ID, 0.375 inch long.
RFC2 — 3t no. 24 enam. wire, 0.125 inch ID, 0.25 inch long.
RFC3 — 1-μH RF choke; 18t no. 24 enam. wire close spaced on Amidon T50-10 toroid core.

compression point when used in the circuit shown in Fig. 73. Gain is 8 to 10 dB at 200-mW output, so the + 13-dBm power level available from the low-level MSA-204/0304/0404 driver is just right.

Design and Construction

This circuit uses L networks for input and output matching. The series elements are microstriplines. The characteristic impedance of the input line is 27 ohms, and the output is 51 ohms. The shunt matching elements are variable capacitors C11 and C12; they can be either miniature piston

trimmers or PC-board mount parallel-plate trimmers. Solder C3 directly across the Q1 base and emitter leads.

Q1 idling current is set by a bias network consisting of R2, R3 and D1. RFC1 and RFC2 were chosen for optimum RF decoupling performance at 1296 MHz, while RFC3 and R4 were chosen to reduce the transistor's ability to generate low-frequency spurious signals. C4 and C7 provide a good bypass at the frequency of operation, and the remaining bypass capacitors provide a low-impedance path to ground for lower frequencies.

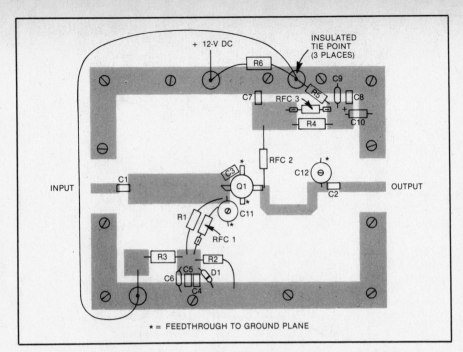

Fig. 74 — Parts placement diagram for the NE41620 amplifier. A full-size etching pattern appears at the back of this book.

Fig. 75 — The NE41620 amplifier is mounted to an aluminum block and the RF connectors are mounted to the ends of the block.

The NE41620 amplifier is built on a double-sided, 0.062-inch-thick, glass-epoxy PC board. Parts placement is shown in Fig. 74. One side of the board is left unetched to act as a ground plane. The PC board is mounted on a 0.25-inch-thick aluminum block to facilitate the mounting of four-hole flange-type SMA connectors for the RF input and output. If desired, the amplifier can be mounted in a suitable enclosure and coaxial cable can be run between the circuit board and RF connectors. Type-N or BNC connectors could also be

used. Be sure to use screws to tie the top ground plane to the bottom ground plane and aluminum plate, as shown in Fig. 74. This is especially important near the critical bypass areas. Additional details may be seen in Fig. 75.

Tune-up

For initial tune-up, start with C11 and C12 at minimum capacitance and connect 50-ohm loads to both the input and output. Slowly apply 12 to 14 V and monitor the collector current through R5. Idling

current should be approximately 20 mA. If adjustment is necessary, change the value of R3 accordingly (that is, increase the resistance value to decrease collector idling current). It has been observed with various devices that R3 may need to be as high as 1000 ohms to keep idling collector current at 20 mA. This is normal.

Apply 10-mW drive and tune C11 and C12 for maximum power output. The output power should be 70 to 100 mW. Increase drive and readjust C11 and C12 so the output power achieved is 200 mW. Collector current should be approximately 70 mA. Saturated power output has been measured at 300 mW. The measured 1-dB compression point of the amplifier is 250 mW, but it is suggested that the amplifier be run at the 200-mW level.

Do not omit R5 and R6 in an attempt to attain greater power output; collector current will be excessive. Do not exceed the absolute maximum collector current rating of 100 mA as suggested by the manufacturer. R5 also provides a convenient point at which to monitor collector current.

TR SEQUENCER

The need for a sequencing circuit to control TR switching became rather obvious after an IF amplifier was blown out by excessive RF from the 2-meter rig. When the transverter was used for SSB operation, RF was available at the IF TR switch before it had completely switched to the transmit circuits. This RF could potentially destroy RF preamplifiers and coaxial TR relays. The situation is even further compounded during FM operation since a carrier is sent to the transverter at the same time the coaxial switch and the IF and RF PIN-diode switches are in the process of switching.

The desired sequence of events when going from receive to transmit is as follows:

1) Transverter receive section turns off.
2) Transverter TR relay transfers to the transmit state.
3) Transverter transmitter stages turn on.
4) Power amplifier turns on, or feedback to IF transceiver to turn on power stage.

Conversely, when going from transmit to receive, the opposite sequence is desired:

1) Power amplifier turns off, or IF transceiver power stages turn off.
2) Transverter transmitter stages turn off.
3) Transverter TR relay transfers to receive state.
4) Transverter receive section turns on.

The circuit shown in Fig. 76 fulfills all of these requirements.

The sequencing circuit used here was inspired by work done by Chip Angle, N6CA. The N6CA version is presented as a stand-alone project in Chapter 31. The circuit shown here uses an inexpensive LM324 quad op-amp. Each section is used as a differential amplifier. A string of resistors produces a progressively higher

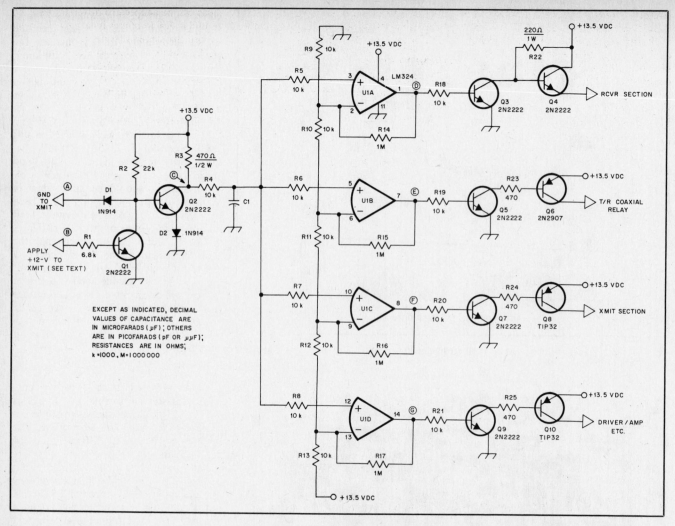

Fig. 76 — Schematic diagram of the TR sequencer. Resistors are ¼-W carbon types unless noted.

C1 — 20-μF electrolytic (see text).
D1, D2 — 1N914 or equiv.

Q1-Q5, Q7, Q9 — 2N2222A or equiv. (RS 276-2009).
Q6 — 2N2907 or equiv. (RS 276-2023).

Q8, Q10 — TIP32 or equiv. (RS 276-2027).
U1 — LM324 quad op-amp.

reference level on each section. The RC circuit (R4, C1) controls the timing and the voltage applied to the + input of each section.

The sequencer is set up for two input options. The A input is set up for the ground-to-transmit mode and can be paralleled with the microphone PTT on most 2 meter multi-mode radios. Input B is activated by applying 9- to 12-V dc. Since an isolation diode is used in series with port A input, an extra control diode is used in the emitter lead to ensure that Q2 can be turned off so point C can go high to charge the R4/C1 combination.

As the R4/C1 charges, each op-amp turns on in sequence, causing points D through G to sequence on. When point D goes high (12.5 V), Q3 turns on, causing Q4 to turn off; this turns off the receive section. When point E goes high (12.5 V), Q5 turns on, causing Q6 to turn on; this triggers the coaxial TR relay. In a similar fashion, the bias circuits on the transmit power stages are turned on via Q7 and Q8. Q9 and Q10 can be used to turn on an ex-

ternal driver/final amplifier, or this switch can be used to turn on the IF transceiver transmit stages if desired.

When the PTT switch is opened to return to the receive mode, point C returns low. This causes C1 to discharge through R4, Q2 and D2. Each of the op-amps turns off in sequence. This turns off the load connected to Q10, then the transmit section, then the TR relay goes back to receive, and finally the receiver circuits are turned on. The value of C1 can be set up for any charging/discharging rate. With 20 μF, the transverter sequences in less than several hundred milliseconds. The gain of the op-amps is reduced to an acceptable level by means of feedback resistors R14 through R17. This reduces the susceptibility to noise and RF. The 10-kΩ series resistors (R18 through R21) also help minimize RF susceptibility of the unit. Q3, Q5, Q7 and Q9 are buffer stages which minimize the current drawn from the LM324.

When a 2N2222 is used at Q4, 200 to 300 mA can be drawn in the receive mode. A heat sink is recommended. A 2N2907 at

Q6 allows up to 300 mA to be pulled safely. This should be adequate for most coaxial relays. If TIP32s are used at Q8 and Q10, 2 A can be pulled safely from each device. Q8 can be used to switch on all of the transmit circuits previously discussed. If higher powered amplifier stages are installed, it is suggested that V_{cc} be applied to each power stage at all times. Use Q8 and Q10 to apply bias for transmit. Otherwise the current-handling requirement of Q8 and Q10 would have to increase significantly.

FINDING PARTS

Many small parts, such as resistors and capacitors, are available from the suppliers listed in Chapter 35. Applied Invention and Microwave Components of Michigan are sources of piston trimmer capacitors, chip capacitors, feedthrough capacitors and other related components. NEC transistors are available from California Eastern Laboratories, 3260 Jay St., Santa Clara, CA 95050. Avantek MMICs are available from stocking distributors around the country. Contact Avantek, 3175 Bowers Ave., Santa Clara, CA 95051 for the name of the nearest distributor. Hewlett Packard devices are available from three major distributors that have offices around the country: Hall-Mark, Hamilton Avnet and Schweber. If you have trouble locating an HP distributor, call the sales office nearest you and ask an engineer in the electronic components group to steer you in the right direction. Microwave Associates parts are available through Zeus in Port Chester, New York (tel. 914-937-7400) or Anaheim, California (tel. 714-632-6880).

CONCLUSION

It is hoped that the ideas and concepts presented here will inspire you to build your own 1296-MHz transverter. With the availability of state-of-the-art MMIC amplifiers, a lot of work has been minimized. These ideas can be adapted to different designs. Modular design makes it easy to pick and choose the stages you need.

The author wishes to thank his wife Emily, K5DOI, KN5N, WB5AFY, WA5TKU, W5ETG, WA5VJB and K5GW for their assistance in bringing this project to fruition. Microwave Associates, Hewlett Packard, California Eastern Laboratories, Avantek and Tele-Tech provided components.

1296-MHz Solid-State Power Amplifiers

With the ever-increasing number of commercial 1296 MHz transverters available on the market today, there exists an even greater demand for a simple and economical way to generate higher power. Most of the transverters presently available produce 0.5 to 1 watt of output. If this power level is used to drive a typical 2C39/7289 stripline or cavity amplifiers, the low power output is often disappointing. If the tube-type amplifier is run with at least 1 kV on the plate, typical gain is around 10 dB. This means that the 0.5 to 1 W is now at best 5 to 10 W output. One alternative is to cascade two of these amplifiers. We now could expect 50 to 100 W of output power.

This section describes two solid-state amplifiers that can replace the 2C39/7289 as the driver and provide 10-20 W — enough to drive a two-tube amplifier to full output. These amplifiers, shown in Figs. 77 through 81 were designed and built by Al Ward, WB5LUA.

The NEC NEL1306 and the NEL1320 were recently introduced as an economical solid state approach to generating moderate power (10-20 W) at 1269 and 1296 MHz. These amplifiers can be used for terrestrial or satellite work. When OSCAR 10 was designed, it was thought that 10 W into a modest gain antenna (20 dBi) would pro- duce usable signals from the satellite. Unfortunately, there were some problems and the sensitivity of AO-10 was not as originally expected for the Mode L uplink. These amplifiers can, however, be used as a driver for a higher-powered tube amplifier for Mode L service. If all goes according to plan, with the launch of AMSAT-OSCAR Phase IIIC in mid 1986, 10 to 20 W with a 20-dBi gain antenna will produce acceptable downlink signals.

The NEL1306 is rated for 6-W output at 1296 MHz at the 1-dB compression point; the NEL1320 is rated at 20 W. One of the advantages of these devices is that they were designed for a V_{cc} of 12- to 13.6-V dc, making them ideal for portable and mobile operation. The NEL1306 is in the $26 price range, while the NEL1320 costs about $42. These are single-quantity prices, making the devices attractive to experimenters.

The performance of these amplifiers as built and tested by the author is shown in Table 9. The NEL1306 is a good buy. With 1.5 W input, 8 W output can be achieved. When tuned up at lower power levels, power gain can be as high as 10 dB. With 0.2 W drive from the transverter described earlier in this chapter, an output power level of 2.0 W is obtainable.

Table 9

Typical Operating Conditions for the 1296-MHz Solid-State Power Amplifiers

Device	NEL130681-12	NEL132081-12
P_{out} (1-dB C.P.)	7 W	18 W
Gain (1-dB C.P.)	6 dB typ.	5 dB typ.
Collector efficiency	40-50%	40-50%
Idling current	50 mA	150 mA
I_c @ 1-dB C.P.	1.1 A	3.0 A
V_{cc}	13.5 V	13.5 V
Power input	14.9 W	40.5 W

Power gains as high as 17 dB are possible with a two-stage amplifier (an NEL1306 driving an NEL1320). With a mere 200 mW of drive, 10-W output can be obtained. When driven with 1 W, the 1-dB compression point of 18 W will be achieved.

Circuit Details

The basic design, shown schematically in Fig. 77, is an adaption of a circuit described in the NEL1300 series data sheet. The design incorporates 30-ohm quarter-wavelength microstriplines on both the input and output. Shunt capacitors C3, C4, C7 and C8 form a pi network designed to match the low input impedance of the

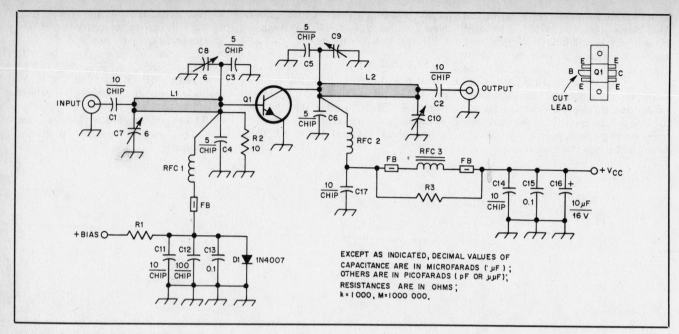

Fig. 77 — Schematic diagram of the NEL1306 and NEL1320 1296-MHz solid-state power amplifiers. The schematic is identical for both versions. Component values are the same except as noted.

C1, C2, C11, C17 — 10-pF chip capacitor.
C3, C4, C5, C6 — 3.6- to 5.0-pF chip capacitor.
C7, C8 — 1.8- to 6.0-pF miniature trimmer capacitor (Mouser 24AA070 or equiv.).
C9, C10 — Same as C7 and C8 for the NEL1306 amplifier. For the NEL1320 version, 0.8- to 10-pF piston trimmers are used (Johanson 5221 or 8053 or equiv.).
C12, C14 — 100-pF chip capacitor.

C13, C15 — 0.1-μF disc ceramic capacitor.
C16 — 10-μF electrolytic capacitor.
D1 — 1N4007 diode.
L1, L2 — 30-ohm microstripline, ¼ wavelength long (see text).
Q1 — NEC NEL130681-12 (6 W) or NEL132081-12 (18 W) transistor.
R1 — 82- to 100-Ω resistor, 2-W minimum. Vary for specified idling current.

R2 — 10-Ω, ¼-W carbon-composition resistor with "zero" lead length. See text.
R3 — 15-Ω, 1-W carbon-composition resistor.
RFC1 — 3t no. 24 wire, 0.125 inch ID, spaced 1 wire dia.
RFC2 — 1t no. 24 wire, 0.125 inch ID, spaced 1 wire dia.
RFC3 — 1-μH RF choke; 18t no. 24 enam. close-spaced on a T50-10 toroid core.

device to 50 ohms. C5, C6, C9 and C10 and the 30-ohm transmission line form an output pi network that matches the device to 50 ohms. C10 is not always necessary, depending on variations among devices and PC-board material. Glass-epoxy circuit-board material 0.031 inch thick was chosen for the dielectric material. A 30-ohm line in this dielectric equates to a line width of 0.121 inches, which is equivalent to the width of both the collector and base leads of the NEL1300 series devices. This minimizes the discontinuity among L1, L2 and Q1.

Bias is provided by R1, R2 and D1. R1 can be optimized, if desired, to adjust the collector idling current.

RFC1 and RFC2 were selected by choosing the lowest possible reactance that will not affect power gain or output power. The RF chokes and the 10-pF bypass capacitors afford adequate decoupling at the frequency of operation. The values of RFC1 and RFC2 were purposely made different to avoid oscillations caused by bias choke coupling.

Keeping the high-frequency RF chokes in the collector as small as possible allows the parallel R3/RFC3 combination, in conjunction with bypass capacitors C14, C15, C16, to reduce the tendency of the transistor to generate low-frequency spurious signals. The 1-μH RF choke must be capable of handling 1 A for the NEL1306 and 3 A for the NEL1320. Special RF

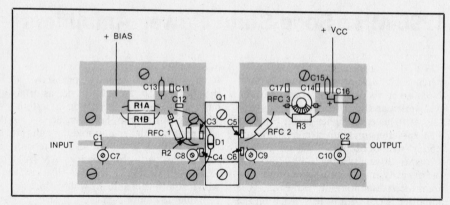

Fig. 78 — Parts-placement diagram for the solid-state 1296-MHz power amplifiers. The same PC boards are used for each version. A full-size etching pattern may be found at the back of this book.

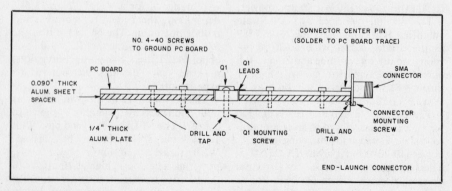

Fig. 79 — Construction details for the solid-state 1296-MHz power amplifiers. See text for additional information.

Fig. 80 — This NEL1306 amplifier was built inside a diecast box. Miniature coaxial cable runs to the connectors.

Fig. 81 — The NEL1320 amplifier is virtually identical to the NEL1306 version, except for the capacitors on the output stripline. C10 was not necessary on this version.

chokes were built to withstand the current.

Construction

The circuit is etched on 0.031-inch thick, double-sided, glass-epoxy circuit board. The layout is given in Fig. 78. Two separate boards are actually used — one for the input side and one for the output — and they are mirror images of each other. The copper is retained on the bottom side and serves as the ground plane. In addition, the grounded areas on the top side must be held at RF ground. This insures low-inductance grounding for the transistor emitter leads, matching capacitors and bias circuitry.

Several effective methods of grounding of the top ground plane to the bottom ground plane are summarized here.

1) Plated-through holes at the critical grounding areas mentioned above.

2) Use of pins or screws that penetrate through the circuit board and into the aluminum base plate at the critical areas.

3) Use of "wrap-around" foils on all edges of the ground plane.

Plated-through holes are often used in the commercial and military electronic marketplace but are not so easily reproduced in the average builder's circuit-board shop. The best technique is to use the "wrap-around" foils on all edges and use no. 4-40 screws to tie the circuit board to the base plate at the critical areas shown in Fig. 78. Solder the brass or copper foil to both ground planes.

It will be necessary to mill out a 0.240-inch-wide slot in the aluminum base plate to allow the transistor flange to be mounted 0.090 inch below the main surface, or to build up the input and output printed circuit boards with a 0.090-inch-thick aluminum plate. This is required since the transistor leads protrude from the device 0.165 inch above the bottom of the flange, and the PC boards are only 0.031 inch high. These dimensions allow a wraparound foil and solder buildup of 0.040 inch maximum.

One-inch screws are used to hold the transistor to the aluminum base plate. The base plate was drilled and tapped to accept the screws. The extra screw length that protrudes from the bottom of the base plate allows the use of an external heat sink, which is highly suggested if continous operation is desired.

The clearance holes in the NEL1300 devices are for no. 4-40 hardware. The holes were drilled out to accept no. 6-32 screws to make the assembly more rugged, although this may not be necessary. Be careful if you decide to drill out the holes to accept no. 6-32 hardware: the base plate is soft copper, and you could damage the device. Use a small vise to hold the transistor flange during the drilling operation.

The transistor leads should be soldered in place only after the circuit boards and transistor have been firmly bolted down to the base plate. This is necessary to minimize any buildup of stress in the transistor leads. Chip capacitors C3, C4, C5 and C6 should be soldered directly onto the leads of Q1 to ensure the shortest possible lead length.

In the original design of these amplifiers, Johanson piston trimmer capacitors (part no. 5221 or 8053) were used for C7, C8, C9 and C10. These capacitors are fairly large, and in some instances the coupling between the bodies of C8 and C9 was enough to cause an in-band oscillation. Smaller variable capacitors should be used if at all possible. Mouser Electronics (see Chapter 35) part number 24A070, a 1.8- to 6.0-pF miniature trimmer capacitor, has been used successfully in the input and output networks of the NEL1306 with no performance degradation. The Johanson piston trimmers were used in the output network of the NEL1320 amplifier because of the high RF currents involved.

The original printed circuit board was designed to allow end-launch SMA-type connectors to be installed. Fig. 79 illustrates this technique. Four-hole, flange-mount SMA connectors can be mounted to the edge of the 0.250-inch-thick base plate using two of the four mounting holes. Drill

and tap the base plate for no. 2-56 hardware. Extreme care must be exercised — it's easy to accidentally cross-thread or over-torque the no. 2-56 hardware.

An alternative approach is to mount the amplifier in an aluminum diecast box (Bud CU-124 or Hammond 1590B) and run miniature 50-ohm coaxial cable such as RG-174 from the amplifier board to the connector. The amplifiers shown in Figs. 80 and 81 use the standard SMA connectors mounted to the walls of the metal box. BNC or Type-N connectors should work equally as well. When preparing each end of the coaxial cable, try to keep the pigtail leads as short as possible (1/8 inch or less); otherwise the mismatch will be difficult to tune out.

A performance comparison was made on one amplifier with end-launch connectors and another using the approach just described. No difference in gain or 1-dB compression point could be measured. Etched PC boards and parts kits are available from A + A Engineering.[3]

Tune-up and Operation

Initial setup of each amplifier is begun by terminating the input and output in good 50-ohm loads. Start with all capacitors at their minimum value. Apply 12- to 13.5-V dc to the V_{cc} and bias terminals. The collector idling current should be as shown in Table 9. R1 can be varied as required to obtain the correct idling current. For initial tuneup, it is suggested that a fuse be used in the collector lead to protect the device.

Measuring RF output power at this frequency can be difficult. The author used a calibrated 20-dB directional coupler along with enough attenuator pads to allow power to be read with a Hewlett Packard HP430C power meter and an HP477B thermistor mount. The HP431 power meter with its associated HP478A thermistor mount is a newer version of the HP430C and does not suffer from temperature drift problems. Bird offers several low-power slugs for the popular model 43 in this frequency range. The 400-1000 MHz slugs can be used with decreased accuracy.

For the NEL1306 amplifier, start out with 50 to 100 mW of drive. Adjust the output network for maximum power output and then peak the input network for output power. Increase drive and repeak both matching networks for rated performance as shown in Table 9. Similarly, start out with approximately 1 W of drive for the NEL1320 and follow the same procedure. After a minute or two of operation at maximum power output, remove RF drive power and check to see that the collector idling current has not increased more than 25% over the initial setting. Keep D1 close to Q1. Thermal compound will enhance heat transfer to D1 to insure minimal drift in idling current with temperature changes.

Considerable effort was put forth to make sure that the amplifiers are stable. The devices have fairly high gain at the frequency of operation, so layout and good construction practices are very important. Here are some construction hints that can help ensure amplifier performance and stability.

1) Use the smallest (physical) size variable capacitors that will still handle the RF current.

2) Use wrap-around ground foils as noted. Grounding screws are required at the critical RF ground areas near the shunt variable capacitors, shunt bypass capacitors and Q1 emitters.

3) Connect braids from the coaxial cable jumpers to same ground as the shunt variable capacitors.

4) Use as little lead length as possible on R2 — less than 0.125 inch.

5) In some instances where the large piston trimmers are used, a shield approximately 0.75 inch high mounted on top of Q1 and grounded via the mounting screws can improve isolation between C8 and C9.

If transistor switching is being considered as a method of applying dc power to the power amplifier stages during transmit, the following technique is recommended. Keep 13.5-V dc applied to the V_{cc} terminal during receive and transmit. Use a series transistor switch to apply 12- to 13.5-V dc to the bias terminal during transmit. A power transistor capable of carrying only a few hundred milliamperes of bias current, as opposed to several amperes of collector current, will be required. More important,

the voltage drop across the transistor switch in the V_{cc} line will be eliminated. This will ensure maximum power output of the NEL1300 devices by keeping V_{cc} at 13.5-V dc.

Several amplifiers of this design are currently in use. The NEL1300 series amplifiers offer a simple and efficient means of generating medium power on 1296 MHz. The 13.5-V dc requirement makes them ideal for mobile and portable operation.

Parts for this project are available from Applied Invention, Microwave Components of Michigan and Mouser Electronics (see Chapter 35). NEC transistors are available from California Eastern Laboratories, 3260 Jay St., Santa Clara, CA 95050.

The author wishes to thank everyone who offered technical advice, especially WA5TKU for helping with the construction and evaluation of the prototype amplifiers.

[3]A + A Engineering, 7970 Orchid Dr., Buena Park, CA 90620, tel. 714-521-4160.

A Quarter Kilowatt 23-cm Amplifier

The amplifier project described here and shown in Figs. 99 through 123 offers the following features.

1) Covers 1240 to 1300 MHz.

2) Linear operation.

3) Grounded grid 7289/2C39 cavity amplifier, single tube.

4) Power gain ranges from 12 dB to 20 dB depending on output power, input power, loading, anode voltage and grid voltage.

5) 50-ohm input and output — no stub tuner required.

6) Power output greater than 200 W with about 12-W drive.

This project was designed and built by Chip Angle, N6CA. It was originally presented in March and April 1985 *QST*. All of the information given in the original article is reproduced here, including details of the design and construction of the RF deck and power supply, a practical water cooling system, testing and operation. The amplifier described here is a tried and proven design. It works well, is reliable and can be duplicated.

General Design Approach

A cavity amplifier is similar to a conventional amplifier designed for lower frequencies. The tube anode excites a resonant circuit, and power is in turn coupled into a load, usually 50 ohms. Instead of using coils and capacitors, as at lower frequencies, the cavity provides the resonant circuit necessary to tune the amplifier output.

The anode cavity of this amplifier is a squat cylinder. Cylinder height is set by mechanical tube requirements. The inside diameter of the cylinder sets the highest resonant frequency. Any capacitance added from the top to the bottom of the cavity will lower its resonant frequency, as will increasing the cavity diameter.

This amplifier uses 1/8-inch-thick copper plates for the cavity top and bottom and a thick-wall aluminum ring, cut from tubing, for the walls. This heavy construction virtually eliminates all resonant frequency variations caused by thermal and mechanical changes.

Fig. 100 is the schematic diagram of the cavity amplifier. The circuit is simple. Filament voltage enters the RF deck through C4, C5, RFC1 and RFC2. High voltage is fed to the anode through RFC3. C8, the anode bypass capacitor, is homemade from Teflon® dielectric sandwiched between a copper plate and the chassis.

The input pi network easily tunes the entire band at any power level. It is made from two Johanson piston trimmer capacitors and a "coil" made from copper wire. An input cavity is not necessary at 23 cm.

Output coupling is through a rotatable loop that serves as a variable loading control. This allows amplifier-tuning flexibility; it may be tuned for maximum gain

or for maximum power. Light loading can produce stable power gains of up to 20 dB.

Amplifier tuning is accomplished with a homemade cylindrical coaxial capacitor with Teflon dielectric (C6). There are no moving metal parts to cause erratic performance. The Teflon rod/tube screws in and out of the coaxial capacitor, increasing or decreasing the capacitance by changing the amount of Teflon dielectric inside the cylinder. With the rod all the way in, the dielectric is all Teflon; with the rod all the way out, the dielectric is all air.

Teflon has a relative dielectric constant (relative to air = 1) of 2.05, which means that the value of the capacitor with the Teflon rod all the way in is twice the value of the capacitor with the rod all the way out. Full capacitance will pull the resonant frequency of the amplifier down to 1240 MHz. Use of only one anode tuning adjustment means the amplifier will have more gain because shunt capacitance has been minimized.

Thermal Considerations

The cavity walls are formed by a thick-wall aluminum ring, which is sandwiched between two thick copper plates. RF and thermal properties of these two metals are reasonably close, whereas brass is rather poor in both respects. The 7289/2C39 tube used in this amplifier is being run at 2 to 2½ times its normal dissipation rating; therefore it's important to have a cavity that remains thermally stable.

Most previously described amplifiers

have used sheet brass in their construction. This has usually meant constant retuning of resonance to maintain output power at or near maximum.

The copper and aluminum construction in this amplifier has solved all thermal stability problems. The amplifier can easily be run key down for over an hour at 200-W output without retuning. This, of course, is obtained only with a good tube and water cooling.

Water cooling keeps the internal structure of the tube thermally stable. When air cooling is used for output levels of 100 to 150 W, output power fluctuations are a direct result of internal tube changes. These changes vary from tube to tube and must be tested for. In some cases, perfectly good RF tubes have had poor thermal stability. Such tubes can make good drivers at lower power levels.

Construction

Hand tools are great if you are skilled and patient. Most people want to hurry up and finish their new project. If that's you, then have a machine shop make all of the parts, leaving only the final assembly up to you. It should cost you about $200. The parts are not difficult to fabricate, but the process is time consuming. If you have the time and patience to do it yourself, this amplifier can be very inexpensive.

Gathering the Materials

All of the materials used in this amplifier are fairly common and should be available

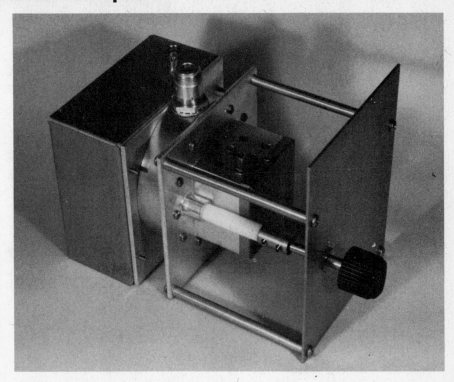

Fig. 99 — The completed 23-cm amplifier.

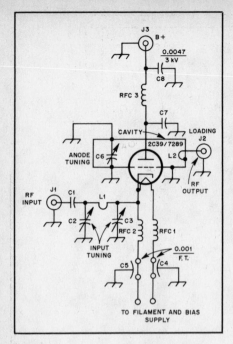

Fig. 100 — Schematic diagram of the 23-cm amplifier.

C1 — 3-pF dipped mica capacitor.
C2, C3 — 1- to 10-pF piston trimmer capacitor (Johanson no. 3957, 5201 or equiv.).
C6 — Anode-tuning capacitor. See text and Fig. 110.
C7 — Anode-bypass capacitor, 90 pF. Homemade from copper plate and Teflon sheet.
C8 — Disc ceramic, 0.0047-μF, 3-kV capacitor.
J1 — 5-mm SMA connector, chassis mount, female.
J2 — Modified Type-N connector. See text and Fig. 106.
J3 — Female chassis-mount BNC connector.
L1 — Loop of no. 18 bus wire soldered between C2 and C3. See Fig. 114.
L2 — Output-coupling loop. Part of output-connector assembly. See text and Fig. 106.
RFC1, RFC2 — 5t no. 20 tinned, 3/16-inch ID.
RFC3 — 3t no. 20 tinned wound on a 20-ohm, 1-W carbon-composition resistor.

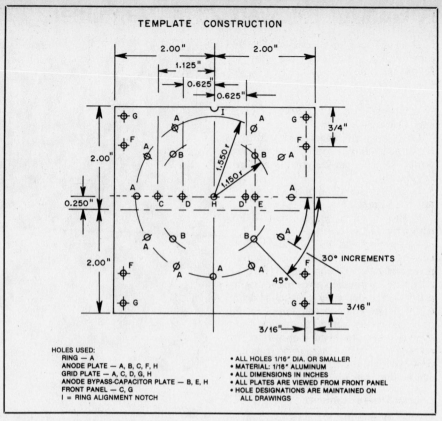

Fig. 101 — Complete dimensions for the aluminum template.

from suppliers in most metropolitan areas. Some suppliers have "short sale" racks where they sell odd pieces cut off standard lengths or sheets at reduced prices. The parts for this project are small enough that they may be fashioned from cutoff stock. Surplus-metal houses have some great buys, so start there if one is nearby.

The key to successfully completing this project is careful layout work before cutting or drilling any parts. Invest in a can of marking dye, a sharp scribe, an accurate rule, vernier calipers and several center punches. These tools are available at any machinists' supply shop. The marking dye will make cutting and filing lines much easier to see. Measure all dimensions as carefully as you can and then recheck them before cutting. Mark with a sharp scribe because the sharper the scribe, the finer the marked line, and the finer the marked line, the closer your cut will be to where it should be. Remember — the accuracy of your

drilled holes is only as good as your center-punching ability, so use a fine punch for the first mark and then a bigger one to enlarge the mark enough for drilling.

Access to a drill press is a must. It's extremely difficult to drill holes accurately with a hand drill. Although not absolutely necessary, you should have access to a lathe or milling machine to do the best possible job.

Other tools that will aid you with this project are a nibbling tool, a set of files and some sharp drill bits. If you don't already have one, purchase a file card to clean aluminum and copper shavings out of your files as you work. Clean, sharp files are faster and more accurate to work with. You'll also need an assortment of sandpaper for the final finishing work.

The Template Approach

It's best to fabricate a single template for marking and drilling the anode plate, anode bypass capacitor, cavity ring, grid plate and front panel. The template shown in Fig. 101 has all of the holes for these parts. If you use the template, you'll only have to make the careful measurements once — after that, it's simple to mark and drill the rest of the parts.

The template approach offers several other advantages. A template makes it much easier to maintain accuracy between the anode plate, cavity ring, grid plate and front panel; these parts will fit perfectly because they were all drilled from the same

master. The template approach also makes it possible to set up a small production line if you decide to build more than one of these amplifiers and combine them for higher power, or if a friend wants to build an amplifier along with you.

See Fig. 101 for complete template dimensions. Start with a piece of 1/16-inch-thick aluminum stock that is larger than you need and degrease it with soap and water. Dry it off and spray it with marking dye. Scribe a 4-inch square on the stock and cut the template to size. A shear will make this job much easier, but it can be cut with hand tools and filed to size.

Carefully measure and scribe all holes. Note that holes A and B are on the circumference of circles. Use a compass to scribe the circles, and then locate the holes. After you have marked and checked all holes, centerpunch and drill them. The holes should be drilled with a 1/16-inch or smaller bit. Recheck all measurements. If you goof, start again. The time you spend making the template as perfect as you can will save you much time and aggravation when you make and assemble the other parts.

When you finish the template, mark the front side for future reference. All plates that will be made from the template are marked and drilled from the front side (as viewed from the front panel).

Making the Copper Plates

Once you have completed the template,

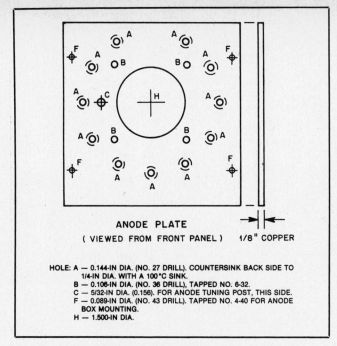

ANODE PLATE
(VIEWED FROM FRONT PANEL)   1/8" COPPER

HOLE: A — 0.144-IN DIA. (NO. 27 DRILL). COUNTERSINK BACK SIDE TO
1/4-IN DIA. WITH A 100°C SINK.
B — 0.106-IN DIA. (NO. 36 DRILL), TAPPED NO. 6-32.
C — 5/32-IN DIA. (0.156). FOR ANODE TUNING POST, THIS SIDE.
F — 0.089-IN DIA. (NO. 43 DRILL). TAPPED NO. 4-40 FOR ANODE
BOX MOUNTING.
H — 1.500-IN DIA.

Fig. 102 — Drilling details for the anode plate. See Fig. 101 for
additional information on hole location.

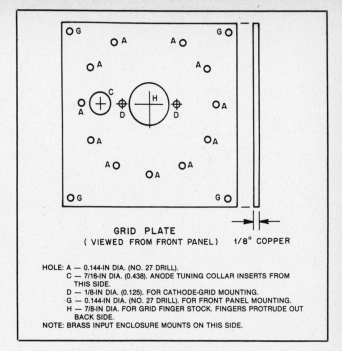

GRID PLATE
(VIEWED FROM FRONT PANEL)   1/8" COPPER

HOLE: A — 0.144-IN DIA. (NO. 27 DRILL).
C — 7/16-IN DIA. (0.438). ANODE TUNING COLLAR INSERTS FROM
THIS SIDE.
D — 1/8-IN DIA. (0.125). FOR CATHODE-GRID MOUNTING.
G — 0.144-IN DIA. (NO. 27 DRILL). FOR FRONT PANEL MOUNTING.
H — 7/8-IN DIA. FOR GRID FINGER STOCK. FINGERS PROTRUDE OUT
BACK SIDE.
NOTE: BRASS INPUT ENCLOSURE MOUNTS ON THIS SIDE.

Fig. 103 — Drilling details for the grid plate. See Fig. 101 for
additional information on hole location.

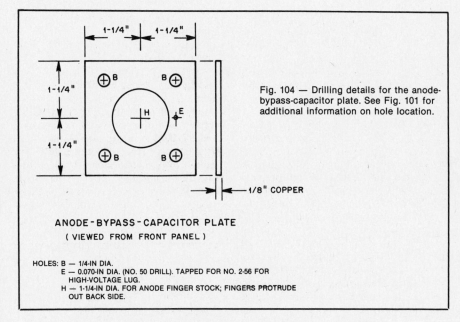

ANODE-BYPASS-CAPACITOR PLATE
(VIEWED FROM FRONT PANEL)

HOLES: B — 1/4-IN DIA.
E — 0.070-IN DIA. (NO. 50 DRILL). TAPPED FOR NO. 2-56 FOR
HIGH-VOLTAGE LUG.
H — 1-1/4-IN DIA. FOR ANODE FINGER STOCK; FINGERS PROTRUDE
OUT BACK SIDE.

Fig. 104 — Drilling details for the anode-
bypass-capacitor plate. See Fig. 101 for
additional information on hole location.

bit and be sure to grind it symmetrically. Modified drill bits can still be used on aluminum and other metals.

Always start with a smaller drill and work up to the final hole size. It's safer and more accurate. The larger holes can be cut with a flycutter, or you can drill a series of smaller holes around the inside of a larger hole and file to finish. Either way is fine. Use lots of cutting fluid to lubricate the drill bit, and wear safety glasses and an old shirt. Remember, some cutting fluids are not to be used on aluminum.

Start with a no. 50 (0.070-inch) or smaller bit and drill pilot holes at each of your punched marks. The details for finishing each hole are listed in the drawings. Some holes are countersunk or tapped. Pay attention to the details and take your time.

When you are through drilling, you must deburr each hole. Copper is soft, so it tends to rise up around the hole during drilling and deburring. Use a flat file for the initial cut, and then remove any remaining material with a countersink. File the copper plates flat again because a flush fit on both sides of the aluminum ring is very important.

When all copper work is done, you should be able to stack the plates and see all pertinent holes align correctly. Enough tolerance is included in the dimensions to accommodate minor errors. After the holes are drilled, it can be difficult to tell which side of each plate is which, so mark the front side of each plate with a permanent marker.

Machining the Ring

The aluminum ring that forms the cavity

it will be easy to make the copper plates. The anode plate, grid plate and anode-bypass-capacitor plate are all made from 1/8-inch-thick copper. See Figs. 102, 103 and 104 for the dimensions of these pieces.

Measure and cut the three plates to the proper dimensions. Carefully break (deburr) all sharp edges to avoid small cuts to your fingers and hands.

Clean the plates with alcohol and spray them with marking dye. Clamp the aluminum template to each plate and carefully scribe the correct holes. Remember that all plates do not have the same holes. The anode plate uses holes A, B, C, F and H; the grid plate uses holes A,

C, D, G and H. The anode-bypass-capacitor plate uses holes B, E and H.

Use a small center punch to punch all holes accurately and lightly. If they then look accurate, enlarge them enough for drilling.

Copper isn't the easiest metal to work with. It's very stringy, and drilling it can be frustrating. You'll need the proper drill bits for best results. Special drills can be purchased, but that's not really necessary. You can use a grinder to carefully remove the sharp points on the outer edge of the cutting surface of each side of a standard drill bit. This will eliminate any tendency for the copper to grab. Practice on an old

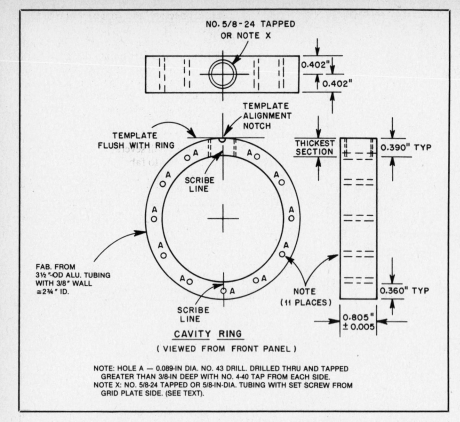

NO. 5/8-24 TAPPED
OR NOTE X

0.402"
0.402"

TEMPLATE
ALIGNMENT
NOTCH

TEMPLATE
FLUSH WITH RING

THICKEST
SECTION

0.390" TYP

SCRIBE
LINE

A

A

A

A

A

A

A

A

A

A

A

FAB. FROM
3½"-OD ALU. TUBING
WITH 3/8" WALL
≅ 2¾" ID.

0.360" TYP

NOTE
(11 PLACES)

SCRIBE
LINE

0.805"
± 0.005

CAVITY RING
(VIEWED FROM FRONT PANEL)

NOTE: HOLE A — 0.089-IN DIA. NO. 43 DRILL. DRILLED THRU AND TAPPED
GREATER THAN 3/8-IN DEEP WITH NO. 4-40 TAP FROM EACH SIDE.
NOTE X: NO. 5/8-24 TAPPED OR 5/8-IN-DIA. TUBING WITH SET SCREW FROM
GRID PLATE SIDE. (SEE TEXT).

Fig. 105 — Details of the cavity ring. See Fig. 101 for additional information on hole location.

wall is cut (sliced) from a length of 3½-inch OD tubing with a 3/8-inch wall thickness. See Fig. 105. The tubing ID is approximately 2¾ inches. The dimensions of the ring are the most critical in this amplifier. Tolerance of the ring thickness is ±0.005 inch to maintain full band coverage.

The ring can be hacksawed or bandsawed out of the tubing, but take extreme care to be accurate. Cutting tubing straight isn't easy. Clamp the tubing to prevent rotating on the bandsaw. The final finish cut is best done on a lathe or milling machine, but careful filing will work.

Once the ring is the correct thickness, deburr the sharp edges and spray it with marking dye. Notice that the outside and inside diameters are not concentric. This is normal for large tubing and is not a problem. Lay the ring flat and find the thickest wall section. Scribe a line across the wall at this point, across the center of the ring, and across the wall on the other side. The scribed lines on each side of the ring will be used to align the template. The output connector will be placed at the thick wall section.

Carefully align notch I on the template with the line scribed on the thickest wall section on the ring. Clamp the template onto the ring. Mark each of the 11 holes labeled A on the template. After you mark the holes and remove the template, check alignment with the copper plates just in case. If everything lines up, center punch all eleven holes on one side of the ring only, and drill each hole completely through the

ring. Use lots of cutting fluid. File the ring flat before and after deburring, taking care not to change the wall thickness. Tap each hole to accept no. 4-40 machine screws. Each hole will have to be tapped to a depth of at least 3/8-inch from both sides because long taps don't exist. The inside of the ring doesn't need to be polished.

The hole for mounting the output connector can now be drilled. There are two ways to mount this connector, and either scheme works fine. Read ahead to the section on making the output connector for more information. The first method of mounting the output connector involves tapping the ring with a no. 5/8-24 tap and using a lathe to cut matching threads on the output connector coupling sleeve. Large taps are expensive, but both a tap and die for Type-N connectors come in handy if you do a lot of building.

If you don't have access to a lathe or a large tap, the second method is easier. Make the output connector coupling sleeve from 5/8-inch-OD brass or copper tubing and drill the ring to just clear it. Then drill and tap the grid-plate side of the ring above the output connector to accept a setscrew. Use the setscrew to secure the output connector.

Output Connector

A standard Type-N chassis-mount female connector (silver plated) is used for the output probe/connector. See Fig. 106. First, remove the flange with a hacksaw and file flush with the connector body.

Next, make the output coupling sleeve that is right for your application (threaded or unthreaded, depending on how you fabricated the ring). The sleeve will be the same length in either case. The output coupling loop is fashioned from a piece of 0.032-inch-thick copper sheet that is 5/32 inch wide. Bend it to the dimensions shown in Fig. 106. We will solder the output connector together later.

Grid Compartment

The grid compartment measures 2 inches square by 1½ inches high. See Fig. 107. It is made from brass and can be sawed out of square tubing or bent from sheet stock. The cover can be made from any sheet metal.

Two small PC boards (Fig. 108) hold the finger stock that makes contact with the filament pin and cathode ring on the 2C39 tube. These boards are cut from 1/16-inch-thick, double-sided G-10 glass-epoxy stock. The copper pattern is identical for both sides of each piece. Mark and drill or file the holes first and then cut the boards to size. Small boards are difficult to hold while drilling them. Mark each side of each board and score the copper foil with a sharp knife.

The unwanted copper can be removed easily by heating the foil with a soldering iron and lifting it off. Use a flat file to deburr the boards. Do not use a countersink because the copper foil must be as close to the holes as possible to facilitate soldering the finger stock in place.

The input connector is a 5-mm SMA type. This is an excellent RF connector, especially for low-power UHF applications. Although an SMA is recommended, any small, screw-on connector will do. If you really feel you have to use a BNC then do so, but it's a lousy connector at frequencies above 200 MHz. Remember to move

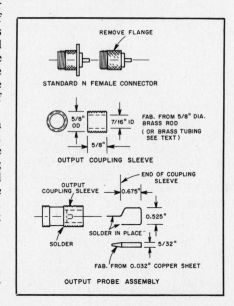

REMOVE FLANGE

STANDARD N FEMALE CONNECTOR

5/8"
OD

7/16" ID

5/8"

FAB. FROM 5/8" DIA.
BRASS ROD
(OR BRASS TUBING
SEE TEXT)

OUTPUT COUPLING SLEEVE

OUTPUT
COUPLING SLEEVE

END OF COUPLING
SLEEVE

0.675"

0.525"

SOLDER IN PLACE

5/32"

SOLDER

FAB. FROM 0.032" COPPER SHEET

OUTPUT PROBE ASSEMBLY

Fig. 106 — Output-probe/connector assembly details.

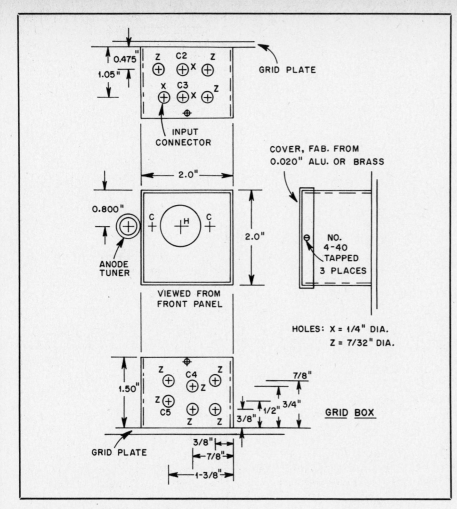

HOLES: X = 1/4" DIA.
Z = 7/32" DIA.

Fig. 107 — Input compartment details.

the connector hole to accommodate its larger size.

The input connector must be as close as possible to the first input capacitor. Lead length of the input dc blocking capacitor must be as short as possible. The 3-pF capacitor is series resonant at 1200 MHz only with short (1/16-inch or less) leads.

Miscellaneous Bits and Pieces

There are still several small, but very important parts to fabricate. The front panel is shown in Fig. 109. It is made from a piece of 1/8-inch-thick aluminum sheet. Some builders may wish to mount the amplifier on a rack panel. Wash and dry your front-panel material and spray it with marking dye. Clamp it to the template and mark the holes. Check the hole alignment with the

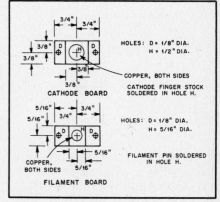

Fig. 108 — Cathode and filament PC-board details.

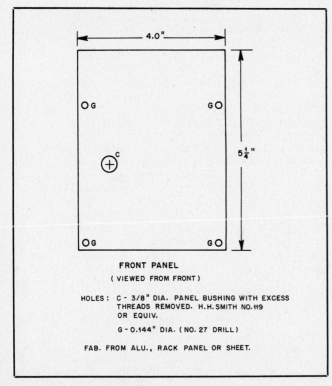

FRONT PANEL
(VIEWED FROM FRONT)

HOLES : C - 3/8" DIA. PANEL BUSHING WITH EXCESS THREADS REMOVED. H.H. SMITH NO. 119 OR EQUIV.

G - 0.144" DIA. (NO. 27 DRILL)

FAB. FROM ALU., RACK PANEL OR SHEET.

Fig. 109 — Front-panel details.

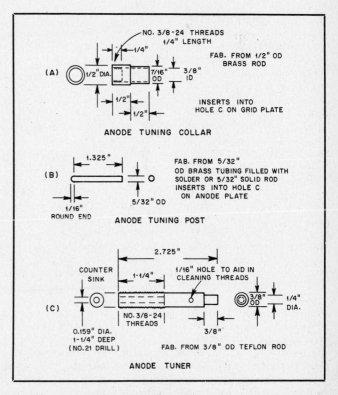

Fig. 110 — Anode-tuning capacitor details.

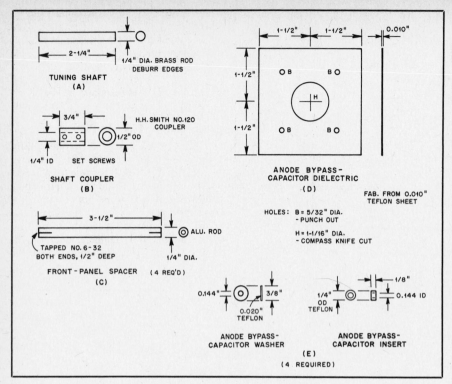

Fig. 111 — Miscellaneous parts necessary to complete the amplifier.

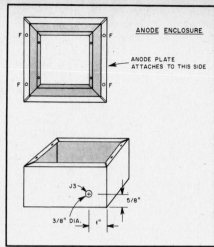

Fig. 112 — Anode enclosure details.

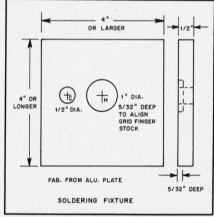

Fig. 113 — Dimensions of the soldering fixture. See Fig. 101 for more information on hole location.

copper grid and anode plates. If all lines up correctly, center punch and drill the holes. The only front-panel control is for the anode tuning capacitor, which is adjusted by a ¼-inch shaft protruding through a 3/8-inch panel bushing in hole C.

The anode tuning collar, shown in Fig. 110A, is made from a piece of ½-inch-OD brass rod. This rod has a 3/8-inch hole drilled through its center, and it is turned down to 7/16-inch OD for half its length. The inside of the ½-inch-OD end is tapped to a depth of ¼ inch to accept 3/8-24 threads. This collar will be inserted into hole C on the grid plate.

Fig. 110B also shows the anode tuning post. It is simply a length of 5/32-inch-OD brass rod that inserts into hole C on the copper anode plate. This rod will form one plate of the anode tuning capacitor; the cavity wall is the other plate.

The anode tuner (Fig. 110C) is machined from a piece of 3/8-inch-OD Teflon rod. One end of the rod is drilled out with a no. 21 drill. The outer wall of this end is threaded with a no. 3/8-24 tap. This is the end that will thread into the anode tuning collar and slip over the anode tuning post. The other end is turned down to fit inside a ¼-inch shaft coupler.

Fig. 111 shows the remaining parts. The tuning shaft (A) is made from a piece of ¼-inch brass rod. A coupler (B) to connect the tuning shaft to the anode tuner may be purchased or made. This also applies to the front-panel spacers (C). The Teflon dielectric for the anode bypass capacitor (D) is made from 0.010-inch-thick Teflon sheet.

Use the template to locate holes B and H. Teflon washers and inserts (E) are used to insulate the mounting hardware for the anode bypass capacitor from the chassis. The inserts are made from ¼-inch-OD Teflon rod. The washers are made from Teflon sheet. Sharpen a piece of 3/8-inch OD tubing and chuck it up in a drill press. This tool will cut neat, round washers from the sheet.

The box that encloses the anode compartment (Fig. 112) is fabricated from a Bud AU-1083 utility cabinet. Clean the chassis and spray it with marking dye. Secure the template to the side of the enclosure that contacts the anode plate and scribe the holes labeled F. Make sure that these holes line up with the holes on the copper anode plate. If they do, center punch and drill them to size. If air cooling is used, the blower will mount to this box.

Soldering the Subassemblies

Once all copper and brass parts are drilled and deburred, they should be cleaned with alcohol and Scotch-Brite®, a nonmetallic pot cleaner, and washed in alcohol again. Set the pieces aside and avoid touching them. Fingerprints will inhibit soldering.

The best way to solder the heavy brass and copper parts is to first build the soldering fixture shown in Fig. 113. This soldering fixture, made from ½-inch-thick aluminum plate, will evenly heat the entire assembly to be soldered. Even heating will allow you to do a much better soldering job than you could otherwise.

The aluminum soldering fixture should be preheated on a stove or hot plate until bits of solder placed on its surface just melt. At this point, reduce the heat slightly. Avoid excessive heat. If the copper parts placed on the fixture suddenly turn dark, the heat is too high.

Solder the grid plate assembly first. You will need the copper grid plate, grid finger stock, anode tuning collar and brass input compartment.[1] Look at the drawings again to be sure that you know which parts go where. Insert the grid finger stock into hole H on the grid plate. As viewed from the front-panel side, the curved fingers will protrude out the back side, away from you. Apply liquid or paste flux and set the grid plate in the soldering fixture. The finger

[1]The finger stock for this project is available from Instrument Specialties, P.O. Box A, Delaware Water Gap, PA 18327. Contact them for the name of the closest distributor. The part numbers for this amplifier are: anode bypass capacitor plate, no. 97-70A; grid plate, no. 97-74A; cathode board, no. 97-420A; filament board, no. 97-280A.

stock will fit in hole H in the fixture, allowing the grid plate to rest flush with the surface of the fixture. Next, apply flux to the anode tuning collar and insert it in hole C of the grid plate. Part of the tuning collar will slip into hole C in the soldering fixture. Make sure the collar seats flush with the grid plate. The flux should start to bubble at this point. Carefully apply solder directly to the joints of the installed parts. The solder should melt almost immediately and flow bright and smooth. Next, place the square brass input compartment in place and apply flux. In a few seconds, it can be soldered by running solder around the joints, inside and outside. If you have trouble getting it to flow on both sides, merely tap the brass box aside (1/16 inch) and return it to its original position.

Now comes the hard part — getting the soldered assembly away from the heat without disturbing the alignment. A pair of forceps is recommended, but long pliers will do. Carefully lift the assembly off the soldering fixture and set on a cooling rack. Do this without moving any part. The cooling rack can be any two pieces of metal that will allow clearance for the protruding parts. You can expedite cooling by using an ordinary hair dryer in the "cool" position to gently blow air across the assembly.

While the grid assembly is cooling, assemble the output connector. See Fig. 106. Place the modified Type-N female connector, threaded end down, on the soldering fixture. Apply flux to the top and install the output coupling sleeve. Allow both parts to heat before applying solder. Carefully remove the soldered output connector from the fixture. When it has cooled, solder one end of the loop to the center pin of the N connector and the other to the output coupling sleeve.

Now place the anode plate on the soldering fixture and allow to heat. Apply flux to hole C. Insert the anode tuning post (5/32-inch-OD brass tube) and allow to heat; apply solder. Remove the parts and cool as before.

This completes the work with the soldering fixture. Be sure to let it cool off before handling! Save the fixture for future construction; you never know when you might want it again.

The anode plate and the anode-bypass-capacitor plate must be filed and then sanded flat on their butt surfaces to assure that there are no solder bumps or sharp points to puncture the Teflon dielectric. This must be done after soldering. The Teflon sheet is adequate insulation for many times the anode potential of this amplifier, but only if the surfaces it separates are smooth!

Next, clean the cathode and filament PC boards. Install the finger stock in hole H of the cathode board. Apply flux to both sides of the board. Heat with a hot iron and apply solder around the circumference of hole H, soldering the finger stock on both

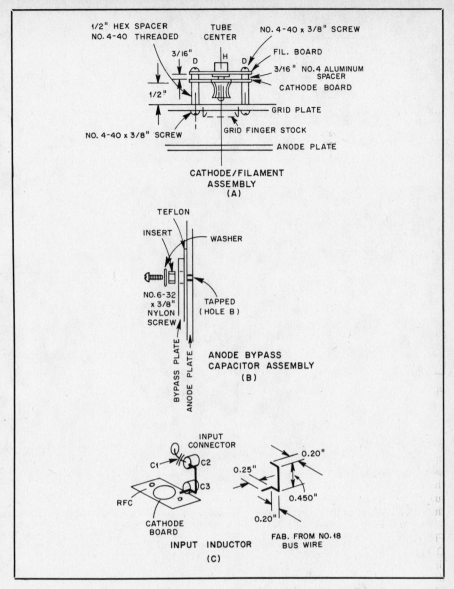

Fig. 114 — Assembly details for the filament and cathode boards (A), the anode-bypass capacitor (B) and the input pi network (C).

sides of the board. Use the same technique to install the filament pin.

After all parts have cooled, use a spray can of flux remover to clean them. Slight scrubbing with "Scotch-Brite" pot cleaner will finish them nicely. Congratulations: You have finished the pieces and are now ready to bolt the amplifier together.

Silver Plating

Over the years, many people have pushed silver plating as the only way to go. You may wish to silver plate the amplifier components before soldering them together, but it is not necessary. The RF skin conductivity of aluminum and copper is pretty good at 23 cm; these materials are much better than brass. In actual testing with four amplifiers, there was no difference in performance among tin-plated, silver-plated and unplated versions. A nickel-plated amplifier exhibited 3-dB less gain than the others.

Assembly

After fabrication of all parts, assembly is simple. Figs. 114 through 116 show assembly details. Loosely fasten the grid and anode plates to the ring. Mount the input connector and capacitors on the input compartment. Loosely install the cathode and filament boards and their respective spacers. See Fig. 114A.

Now insert a 7289/2C39 tube. This will center up all finger stock. Place the Teflon anode tuner in its collar on the grid plate and screw it most of the way in. Now tighten all of the screws. The 7289/2C39 tube should slide in and out snugly, and the anode tuner should screw in and out smoothly.

The Teflon sheet and anode bypass

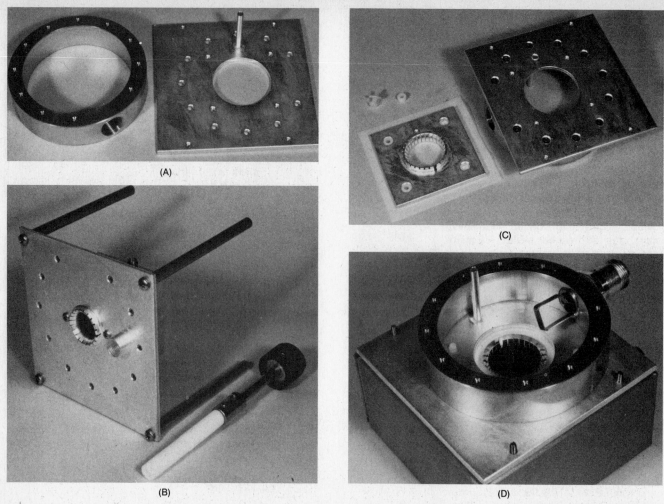

Fig. 115 — The completed cavity ring and anode plate with anode tuning post soldered in place are shown at A. The photo at B shows the grid plate with finger stock, input compartment and anode tuning collar soldered in place. The completed anode tuner is at the right. C shows the cavity ring attached to the anode plate. The anode-bypass capacitor is ready for installation. At D, the interior of the cavity as seen from the grid plate side is visible. The output probe/connector assembly is installed. The anode bypass capacitor and anode enclosure have been installed on the anode plate.

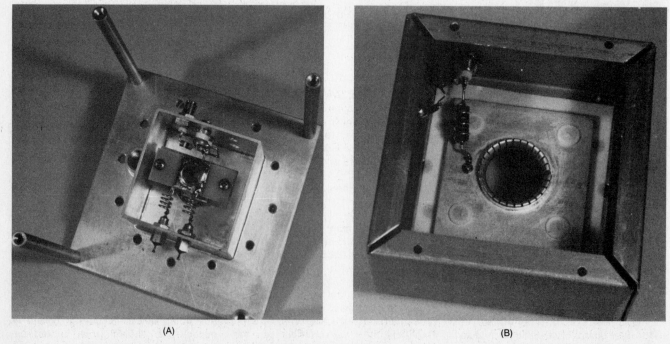

Fig. 116 — At A, the interior of the completed input compartment is visible. The photo at B shows the interior of the anode compartment with the anode bypass, RFC3, C8 and J3 installed.

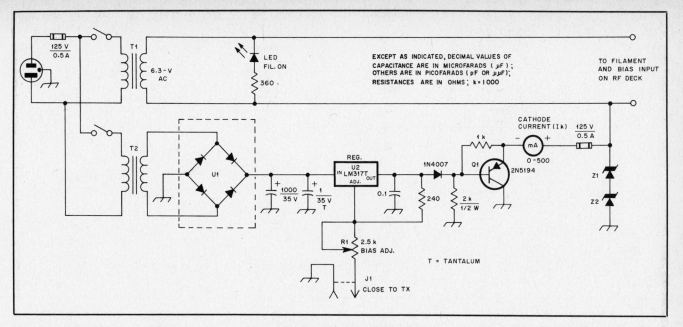

Fig. 117 — Schematic diagram of the cavity amplifier filament and bias supplies. All resistors are ¼-W carbon types unless otherwise noted.

J1 — Female chassis-mount phono connector.
T1 — Filament transformer. Primary, 117-V; secondary, 6.3 V at 1 A.
T2 — Power transformer. Primary, 117-V; secondary, 24 to 28 V at 50 mA or greater.
U1 — Bridge rectifier, 50 PIV, 1 A.
U2 — Adjustable 3-terminal regulator (LM317T or equiv.).
Z1, Z2 — 20-V unipolar metal-oxide varistor (General Semiconductor SA20 or equiv.) or two 20-V, 1-W Zener diodes.

capacitor plate can be installed now (Fig. 114B). Assemble the remaining input components, the filament feed-through capacitors and RFCs (Fig. 114C). Screw the output probe into the cavity ring (or push in the probe and tighten the set screw, depending on which method you chose). Install the high-voltage connector and other parts in the anode box. Mount the amplifier on the front panel and install the anode tuner shaft. This completes the assembly.

Power Supplies

The filament and bias supplies for the cavity amplifier are shown schematically in Fig. 117. The manufacturer's specification for the 7289/2C39 filament is 6.0-V ac at 1 A. The use of a standard 6.3-V ac, 1-A transformer only slightly increases the tube emission without much loss of tube life. The filament should be allowed to warm up before operating the amplifier, so the filament, bias and high-voltage supplies incorporate separate primary switches.

Biasing

Many biasing schemes have been published for grounded-grid amplifiers. Fig. 117 shows a bias network that satisfies all of the following operating requirements:

1) External bias supply referenced to ground.
2) Low-power components.
3) Variable bias to accommodate tube-to-tube variations.
4) TR switchable with relay contact or transistor to ground.
5) Bias-supply protection in case of a defective or shorted tube.

U2 provides a variable bias-voltage source, adjustable by R1. The output of U2 drives the base of Q1, which is used to increase the current-handling capability of the bias supply. Q1 must be mounted on a heat sink. J1 is connected to the station TR switching system so that R1 is grounded on transmit and disconnected on receive. The approximate range of the bias supply is 6 to 20 V. Z1 and Z2 provide protection for R1 in case of a shorted tube. The amplifier can be run without Z1 and Z2 if you keep the anode voltage below 1100 V.

High-Voltage Power Supply

A safe, reliable high-voltage power supply is described here. Of course, you can use any readily available HV supply; keep in mind, however, that the 7289/2C39 anode potential should never exceed 1400-V dc at full load and that the amplifier will withstand 1900-V dc at low cathode current and cut-off-bias conditions. For maximum power output, assuming adequate drive power is available, anode voltage under full load should be about 1200- to 1400-V dc.

Fig. 118 is a schematic diagram of the high-voltage supply. A power transformer (T2) that delivers 900- to 1050-V ac is ideal. The type of rectifier circuit used will depend on the type of transformer chosen. Each leg of the rectifier is made from two 1000-PIV, 3-A silicon diodes connected in series. Each diode is shunted with a 0.01-μF capacitor to suppress transient voltage spikes, and a 470-kΩ equalizing resistor.

Filtering is accomplished with a string of four 360-μF, 450-V electrolytic capacitors connected in series. R3-R6 equalize the voltage across each capacitor in the string and serve as bleeder resistors. Of course, a single oil-filled capacitor may be used here if available. Whatever type of filter you use, the total capacitance should be about 80 μF at a voltage rating of at least 1500-V dc. This value allows adequate "droop" of the anode voltage under high-current loads to protect the amplifier in case of RF overdrive or a defective tube.

Protective Circuitry

Some type of start-up protection should be incorporated in the primary. Fully discharged filter capacitors look like a dead short at supply turn-on. Initial surge current (until the capacitors charge) may be high enough to destroy the rectifiers. R1 and R2 provide some surge-current limiting, but either of the two primary configurations shown in Fig. 118 should be used. T1, a variable autotransformer (Variac and Powerstat are two common trade names), is ideal. In addition to allowing you to bring the primary up slowly, (and charging the capacitors gradually), it also allows full control of amplifier output power by varying anode voltage.

The second method, a "step-start" system, uses a resistor in the T2 primary to limit the turn-on surge current. When the capacitors have charged, K1 is energized, shorting out R11 and applying full voltage to the T2 primary.

F1 and R7 protect against high-voltage arc-overs or short circuits. If sustained overcurrent is drawn, F1 will open and remove B+ from the RF deck. Use a high-voltage fuse here; standard fuses may arc when blown and not interrupt the B+. R7

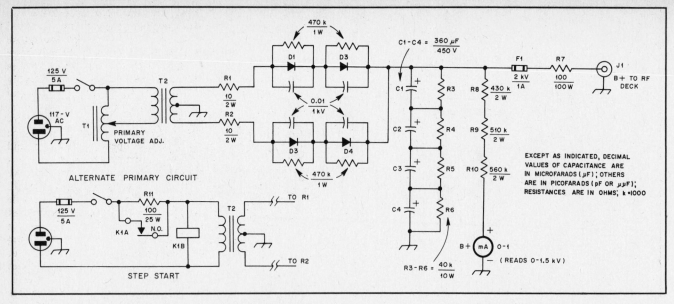

Fig. 118 — Schematic diagram of the amplifier high-voltage supply.

C1-C4 — Electrolytic capacitor, 360 μF, 450 V.
D1-D4 — Silicon rectifier, 1000 PIV, 3 A.
F1 — High-voltage fuse, 2 kV, 1 A.

J1 — Chassis mount female BNC or
  MHV connector.
R3-R6 — Wirewound resistor, 40 kΩ, 11 W.

T1 — Variable autotransformer, 500 VA.
T2 — High-voltage transformer. Primary,
  117 V; secondary, 900 to 1050 V at 500 mA.

provides current limiting to protect the amplifier and power supply in case of a high-voltage arc.

Safety

An HV meter should always be used to monitor the status of the power supply. The values for R8-R10 shown in Fig. 118 will give a 1500-V dc full-scale reading on a 0-1 mA meter. RG-58 or -59 coaxial cable should be used for the high-voltage interconnection between the power supply and the RF deck. Ground the shield at both ends for safety and a good dc return.

Safety must be observed when working with all power supplies. These voltages are lethal! Always disconnect ac power and then discharge the filter capacitors before working on the power supply. Never guess or make assumptions about the status of a power supply. Assume it is hot.

Metering

Cathode-current monitoring is all that's really necessary for observing amplifier dc performance. Cathode current (I_K) is the sum of the plate (I_P) and grid (I_G) currents. Normally, when this amplifier is driven to 300- or 400-mA I_K, the grid current will be around 40 to 50 mA. The inclusion of a grid-current meter is not really necessary and only makes biasing and TR switching complicated.

Cooling

Desired output power and the level of drive power available will dictate what type of cooling to use. For intermittent duty (SSB, CW) at output levels less than 50 W, air cooling is satisfactory. Any small blower may be easily mounted to the aluminum box surrounding the tube anode. For high-duty-cycle modes and/or output levels

greater than 50 W, water cooling is highly recommended. Greater than twice the normal air-cooled output power can be obtained from a water-cooled tube, and water cooling is quiet.

Tube Modification and Water Jacket

The first step is to remove the air radiator from the tube. The air radiator screws on, so it may simply be unscrewed without damage to the tube.

First, place a hose clamp around the tube anode. Secure the radiator fins in a vise and grip the hose clamp with a pair of large pliers. Gently unscrew the tube from the radiator. If the hose clamp slips slightly, tighten it.

Some 7289/2C39 tubes use an air radiator that is attached with set screws. To remove the radiator, simply remove the set screws and pull the radiator off.

The air radiator will be replaced with a water jacket that allows water to be circulated past the tube anode and through a heat exchanger, where it is cooled and circulated past the tube anode again. Two different types of water jackets are described here.

The water jacket shown in Fig. 119 will work with any type of 7289/2C39. It is fabricated from a 1-inch-OD copper tubing cap and two short pieces of 9/32-inch OD brass tubing. The copper tubing cap should be available from a local hardware store or plumbing supply house. Brass tubing is available from many hobby stores and metal supply houses.

Mark and drill the copper cap so that the brass tubing is a snug fit. Thoroughly clean the parts until they shine. Push the tubing into the holes in the end cap and degrease the assembly with alcohol. Use plenty of flux and solder the seam around each sec-

tion of tubing. Allow the jacket assembly to cool.

Meanwhile, thoroughly clean the 7289/2C39 anode to a bright finish. Check the water jacket for fit. In some cases, you'll have to use a 0.005- to 0.010-inch-thick copper shim to fill the gap between the copper cap and the tube anode. This shim helps eliminate pin holes in the solder.

Using plenty of flux, solder the water jacket to the tube anode. Solder it quickly with a hot, high-wattage iron. Allow the tube to cool in the air after soldering to avoid thermal shock and possible breakage. After the tube has cooled, use plenty of alcohol to remove all traces of flux from the tube and water jacket.

The second type of water jacket is shown in Fig. 120. This jacket will work only with 7289/2C39 tubes that have a screw-on air radiator. It is designed to thread onto the tube anode just like the air radiator did. This jacket is machined from a piece of 1¼-inch aluminum rod. The water inlet

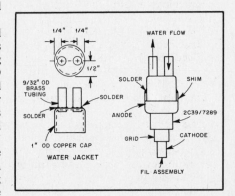

Fig. 119 — Details of the solder-on water jacket.

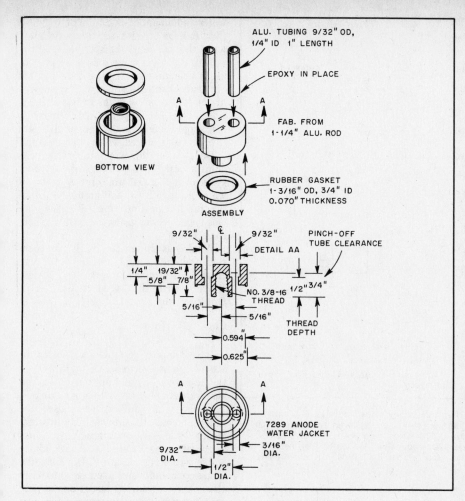

ALU. TUBING 9/32" OD,
1/4" ID 1" LENGTH

EPOXY IN PLACE

FAB. FROM
1-1/4" ALU. ROD

BOTTOM VIEW

RUBBER GASKET
1-3/16" OD, 3/4" ID
0.070" THICKNESS

ASSEMBLY

DETAIL AA

PINCH-OFF
TUBE CLEARANCE

NO. 3/8-16
THREAD

THREAD
DEPTH

7289 ANODE
WATER JACKET

Fig. 120 — Details of the screw-on water jacket.

and outlet tubes are made from 9/32-inch-OD, 1/4-inch-ID aluminum tubing that is epoxied in place. A rubber gasket seals the jacket against leaks.

If you have access to a lathe, you should have no trouble duplicating the jacket. You could have one made up at a local machine shop. Complete screw-on water jackets are also available from Angle Linear.[2]

After you unscrew the air radiator from the 7289/2C39, check for and remove any burrs from the tube anode. The anode surface must be flat if the rubber gasket is to be effective. Screw the water jacket onto the tube. Tighten by hand only. Do not use any tools, or you could damage the tube or jacket! Do not use the water inlet and outlet tubes for leverage — they have thin walls and break easily.

Water System

Fig. 121 depicts the complete water-cooling system. Recommended pumps and accessories that have proven reliable and effective are listed in the caption.

Any small pump, such as a fountain pump, that can deliver 160 to 200 gallons

[2]25309 Andreo Ave., Lomita, CA 90717.

per hour can be used here. Most inexpensive pumps are not self-priming, which means that they won't pump water if they have air in the rotor. Although water can be forced through the pump for the initial prime, my system uses gravity priming. The water reservoir is a 2-foot length of 3-inch-OD plastic pipe that is available from hardware or plumbing stores. It is usually sold for use in residential sewer systems. The outlet is at the bottom and the inlet about halfway up the column. The inlet is located here to eliminate aeration that ionizes the water and reduces its effectiveness. The outlet feeds the pump directly. The pump and the reservoir outlet port should be mounted in the same plane. The pump should be oriented so that air bubbles will rise into the impeller output port and can be blown out once the pump starts running.

Flow Indicator and Heat Exchanger

Water cooling is best described as "super quiet." There is no noisy fan to hear to reassure you that the tube is receiving adequate cooling. If water flow is reduced or cut off during amplifier operation, tube damage is virtually assured.

Flow interlocks and switches to shut down the amplifier if water flow is reduced are hard to find and expensive. Flow in-dicators, however, are inexpensive and reliable. A flow indicator has a spoked rotor that turns as water passes through the unit. If the wheel is turning, there is water flow; if not, you have a problem. Changes in flow rate can be observed by watching for speed changes in the rotor. A small lamp illuminates the flow indicator, making it easy to see rotation. The flow in-dicator should be mounted where it can be seen from the operating position and monitored during operation.

Heat exchangers, or radiators, remove the heat from water as it passes through. For this application, a small automobile transmission-oil cooler works great. Most auto-parts stores and speed shops have a good selection. Some come with mounting brackets. Look for a cooler with the input and output ports on the top so air bubbles will rise to the top and move on without becoming trapped. Trapped air degrades cooler performance.

If you use the amplifier for high-duty-cycle modes such as ATV or FM, or for long, slow-speed CW transmissions (EME, for example), you should use a small axial whisper fan to increase the effectiveness of the heat exchanger. A fan isn't necessary during normal operation, or even for sustained operation at moderate power levels, but is highly recommended if you plan prolonged operation at maximum power. Locate the fan so the warm exhaust air won't heat up other equipment.

Hoses and Fittings

Most hardware stores carry a complete line of brass fittings and adapters that can be used for this project. Brass, however, will eventually corrode and pollute the water supply. Plastic fittings are cheaper and don't corrode, but they are harder to find. Recreational vehicle suppliers are my main source for these parts. They are used extensively in drinking water systems for mobile homes and travel trailers. Procure the fittings when you have the rest of the parts in hand, as there are many variables to consider.

You can use any relatively soft, thin-wall vinyl tubing for all water lines. The main runs are made from 3/8-inch-ID hose, while 1/4-inch-ID stock is used to connect to the 7289/2C39 water jacket. The 1/4-inch-ID tubing fits snugly over the 9/32-inch-OD inlet and outlet tubes on the water jacket, so no clamps are required. All other hose connections should be secured with stainless-steel clamps to prevent leaks. Any leaks mean air in the system and deterioration of cooling performance.

Safety

The anode of the tube, and hence the water jacket and water, are in direct contact with the high-voltage supply, so some safety precautions must be observed. Approximately 12 to 18 inches of tubing should run between the 7289/2C39 jacket and any other component in the cooling

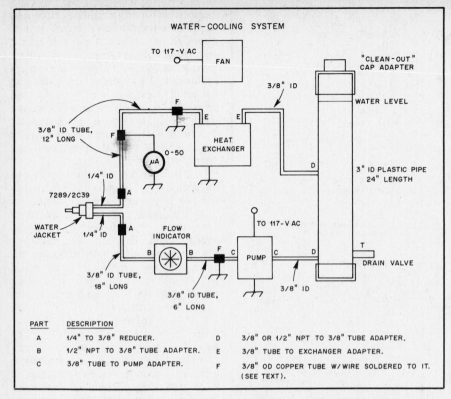

WATER-COOLING SYSTEM

PART	DESCRIPTION		
A	1/4" TO 3/8" REDUCER.	D	3/8" OR 1/2" NPT TO 3/8" TUBE ADAPTER.
B	1/2" NPT TO 3/8" TUBE ADAPTER.	E	3/8" TUBE TO EXCHANGER ADAPTER.
C	3/8" TUBE TO PUMP ADAPTER.	F	3/8" OD COPPER TUBE W/WIRE SOLDERED TO IT. (SEE TEXT).

Fig. 121 — Details of the water-cooling system. Recommended pumps are: (1) Little Giant Pump Co. Model 1-42A or larger, available from most hardware stores; or (2) Calvert Engineering, Cal Pump Model 875S (160 gal/h), available from Calvert, 7051 Hayvenhurst Ave., Van Nuys, CA 91406, tel. 818-781-6029. The flow indicator (Model 15C; requires two ½-inch NPT connectors) is available from Proteus Industries, 240 Polaris Ave., Mountain View, CA 94043, tel. 415-964-4163.

system. This will allow enough resistance in the water to provide adequate current limiting, should the water contact any components that are grounded.

It is best to ground the water supply at the pump. This can be accomplished by replacing a short section of the tubing that runs to the flow indicator with a piece of brass or copper tubing. Solder a wire to this metal tubing and connect the other end of the wire to your station ground. Use at least 24 inches of vinyl tubing between the anode

cooling jacket and the ground point.

On the warm-water side of the 7289/2C39, run 12 inches of vinyl tubing to a small metal fitting or short section of metal tubing, and then another 12 inches of vinyl tubing to a grounded point (this can be at the heat exchanger). You can measure the water leakage current to ground by placing a microammeter between the metal fitting that connects the two vinyl hoses and ground. Leakage current should be less than 10 μA with clean

water and an anode potential of 1 kV. As the water ages, the leakage current will rise; when this happens, replace the water.

Grocery stores carry distilled water for use in steam irons. It may be deionized and not truly distilled, but it works fine for about four to six months in this application. Filters can be purchased from scientific supply houses, but it's not really worth it because deionized water is so cheap.

Do not use tap water under any circumstances! When you turn on the water system for the first time, run a gallon of water through it for half an hour to wash out fabrication impurities. Replace with clean water before using the system to cool the amplifier.

Water was chosen because it's inexpensive, nontoxic, nonflammable, and easy to clean up if you have a leak. Better liquid coolants are available, but they are toxic. Don't use them!

Cooling Performance

Fig. 122 is a graph of several transmit/receive cycles on a water-cooled, 500-W output, 23-cm power amplifier. For this test, two of the amplifiers described here were coupled with a hybrid combiner. This particular cooling system used one gallon of water. Experiments indicate that, during extended operation, the water temperature rises only 30 to 35 °F above ambient room temperature. Typically, the tube anode and water average 10 to 15 °F above ambient during casual operating.

Flow rates in this system are typically ⅓ gallon per minute per tube, which is more than adequate. At this rate, more than 300 W of dissipation from a single inefficient 7289/2C39 were required to boil the water in the water jacket. The water should not be allowed to boil because this will heat the rubber gasket.

Tubes

It is not really necessary to buy a new 7289/2C39. Used tubes can be found surplus for a few dollars and, in many cases, will perform as well as a new tube. Most used tubes have been sitting around for several years, so it's a good idea to run them through the dishwasher to clean them up and then run the filaments for about 24 hours. This will restore operation in many cases.

If you buy a new tube, you should be aware that the 7289/2C39 is being run far in excess of its ratings in this amplifier. The manufacturer's warranty will not cover tubes run in this application.

Contrary to popular opinion, glass tubes will work. Physically, they are not as rugged as the ceramic version, but the glass-to-metal seal seems to provide better shelf life than the ceramic seal. The glass tubes make great driver tubes and will work fine for power levels up to 100-W output. Pulse tubes (7815, 7211) are not recommended because of their poor thermal stability at high power levels. Also, they generally are

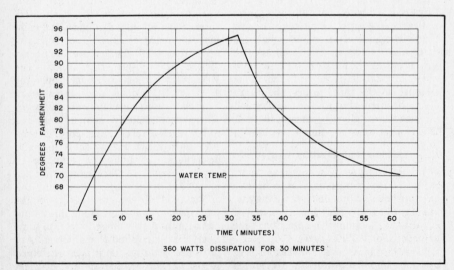

Fig. 122 — Performance graph of the water-cooling system.

Microwave Radiation Safety

Intense RF radiation concentrated on body tissues can produce heat damage; the extent and penetration will depend on the radio frequency in use and on exposure duration. You should be aware of the approximate intensity of RF radiation of transmitting equipment and antennas that you come in contact with.

RF intensity is commonly expressed in milliwatts per square centimeter (mW/cm^2), which is the power flowing away from a source through a unit sampling or interception area at some specified distance. Although the United States as yet has no federal RF protection standard, a useful interim guide is the 1982 standard of the American National Standards Institute (ANSI '82). The most stringent level in this standard is 1 mW/cm^2 for frequencies between 30 and 300 MHz. Above 300 MHz, the protection level rises until it reaches 5 mW/cm^2 at 1500 MHz. Beyond 1500 MHz, the recommended level remains at 5 mW/cm^2. These levels represent the average power density allowed over any six-minute period and are for the sum of all polarizations from a given source.

At 1296 MHz, where one wavelength (λ) equals 23 cm, a thick resonant dipole feeding a calibrated power meter with matched coaxial cable (itself free of pickup) may be used to obtain an indication of power density. A reasonably lossless resonant dipole has an effective aperture of $\lambda/8$; at 23 cm this is 66 cm^2. The power meter reading in milliwatts, divided by 66, is the indicated power density. For this to be a reliable indication, the dipole must be positioned far enough from the RF source to be in its far field. For a small source, the distance should be at least $\lambda/2$, and here that would be about 12 cm (4.5 inches). The dipole should be oriented for alignment with the dominant polarization. Note that the power meter must be capable of readings well below 1 mW.

This arrangement would be useful for checking leaks along the coaxial route that the high power (here 250 W) takes to a load, be it dummy load or antenna. Cable connectors may not be tightly secured, or they may be faulty. For equipment operating in the SHF region, waveguide flanges may not be clamped properly.

Direct measurement of electric field strength near an antenna (with a calibrated instrument, preferably one with the indicating meter shielded and possibly positioned at the center of the sampling dipole) is another way to check for adequate protection. A field strength of 60 V per meter (V/m) corresponds to 1 mW/cm^2; 134 V/m corresponds to 5 mW/cm^2. At a distance 60 cm (2 feet) from an isolated dipole fed with 26 watts, the field strength would be about 60 V/m. This is a far-field field strength for all frequencies where the half wavelength is less than 60 cm, or for frequencies above 250 MHz. For full 250 watts applied to the dipole, the 60 V/m level occurs at a distance of 1.8 meters (6 feet) and at this distance this holds for all frequencies above 80 MHz.

With SSB or CW keying, the fields during Amateur Radio operation are highly intermittent, and usually include considerable pauses or intervals for listening. These factors reduce the average power density over the six-minute averaging period.

Further information on RF safety and protection estimates can be found in Chapter 7 of *The Satellite Experimenter's Handbook,* published by ARRL. The following rules of good practice for RF protection are recommended:

• Never operate an RF amplifier with equipment shielding removed.
• Never handle antennas with RF power applied.
• Never guess that RF levels are safe. Take the time to consult a reliable reference for an estimate, or measure levels carefully. Allow a cushion of about 6 dB (factor of 4 in power density). If possible, borrow an RF radiation monitor (after learning how to use it), or consult with a ham who is well informed on RF protection.
• Never look into an open end of a power waveguide; never point a powered directive antenna (a beam or a paraboloid, for example) toward people. Keep all VHF and UHF transmitting antennas up as high as possible, distant from human activity.
• Use good-quality, well-constructed, coaxial cable and connectors to avoid RF leaks.
• Think RF and electrical safety first; test later!
• Watch *QST* for news on RF measurement techniques and progression, protection standards and proposed federal and state RF regulations. — *David Davidson, W1GKM*

---

30 to 40 MHz lower in resonant frequency in this amplifier compared to the 7289/2C39. Some 7289 tubes can be as much as 30 MHz lower in frequency. Minor length adjustment of the anode-tuning post may be required to accommodate amplifier and tube differences.

Tube Insertion

Extreme care must be exercised when inserting the 7289/2C39 tube. Never force the tube in place as damage (bending) of the cathode finger stock may result. Observe the layout of the finger stock to get an idea of how the tube inserts. Carefully position the tube so it is straight as you gently push. It should slide in snugly without any solid resistance.

Testing

After you have completed all of the parts for the amplifier, it's time to test everything before hooking it all together. Test the water-cooling system by turning it on and watching for steady water flow as indicated on the flow meter. The tube and water jacket can be removed from the cavity amplifier for this test.

Check all of the power-supply voltages first without connecting them to the RF deck. Then, without the tube in place, hook the bias and filament supplies to the cavity and check the voltages again at the tube finger-stock connections. Connect the high-voltage supply to the RF deck and bring the voltage up slowly with a variable autotransformer. Monitor the high-voltage on the anode bypass capacitor plate, and look and listen for any possible arcing between the anode-bypass-capacitor plate and ground. Use extreme care when measuring and testing the high-voltage supply. If everything looks okay with the power supplies, shut them off and disconnect them.

You can make a safe, low-power test of the cavity resonance without applying any voltage. With the tube in place, insert a 2-inch-long coupling loop on the end of a piece of coaxial cable between the spring fingers of the anode down into the cavity. Connect the amplifier output probe/connector to a device capable of detecting low-level RF at 23 cm (for example, a spectrum analyzer or microwattmeter). Feed a signal from an L-band signal generator into cable attached to the wire coupling loop that you inserted into the cavity. Set the signal generator for various frequencies in the 23-cm band and tune the amplifier anode tuner. There will be a sharp peak in output at cavity resonance.

This testing method can be used to determine cavity tuning range, anode-bypass-capacitor effectiveness and resonance of various tube types for use in this amplifier. Any cavity amplifier can be tested completely without ever applying high voltage. The better your test equipment, the easier the amplifier is to test. If all dimensions were followed strictly, the amplifier will tune as designed.

Amplifier Hookup

Installation and operation of this amplifier is relatively straightforward, but as with any amplifier, several precautions must be followed. If these are adhered to, it will provide years of reliable service.

The amplifier is designed to be operated in a 50-ohm system and should never be turned on without a good 50-ohm load connected to the output connector. Never operate it into an antenna that has not been tuned to 50 ohms!

Drive power to the amplifier should never exceed 15 W. Never apply drive power in excess of 1 W unless all operating voltages are present and the tube is biased on. Otherwise, the tube grid-dissipation rating will be exceeded and you will probably ruin it.

As in all TR-switched systems, some type of interlock or sequencing of transmit and receive functions should be incorporated. In most systems, the sequence for going into transmit is something like this: First, switch the antenna changeover relay from the receiver to the power amplifier. Next, bias the power amplifier on. Last, key the exciter and apply drive to the amplifier. To go to receive, unkey the exciter, remove operating bias from the amplifier and switch the antenna relay back to the receiver. A TR sequencer is described in Chapter 31.

If the antenna relays are switched while the power amplifier is operating and putting out power, damage to the relay contacts and/or the amplifier is likely. If there is a momentary removal of the antenna while the power amplifier is biased on, it may oscillate. This can damage the TR relay, the tube, or even the receive preamplifier.

Tune Up and Operation

This is it — the big moment when you will see your project come to life! Connect

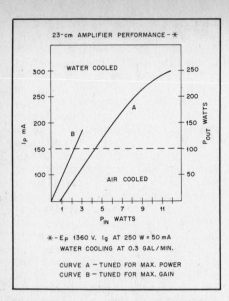

23-cm AMPLIFIER PERFORMANCE - ✳

＊ - Ep 1360 V. Ig AT 250 W = 50 mA

WATER COOLING AT 0.3 GAL/MIN.

CURVE A — TUNED FOR MAX. POWER
CURVE B — TUNED FOR MAX. GAIN

Fig. 123 — Performance of the cavity amplifier under different drive and plate-current conditions.

an accurate UHF power meter and a 50-ohm antenna or load to the amplifier output connector. A Bird Model 43 wattmeter with a 100- or 250-W, 400- to 1000-MHz slug will give reasonable accuracy, depending on the purity of the drive signal. Apply filament power and tube cooling, and allow 3 to 5 minutes for the filaments to warm up. Turn on bias supply (the amplifier will draw maximum current if the anode voltage is applied without bias). Apply 300 to 400 V to the anode. There should be no current flowing in the tube as indicated on the cathode-current

meter. Ground J1 on the bias supply to apply transmit bias and observe cathode current. As R1, the bias control, is turned clockwise, quiescent idling current should increase. Set for about 25 mA.

Apply 1 W of RF drive power. Turn the anode tuner while observing the RF output power meter and tune for maximum output. The output should go through a pronounced peak at cavity resonance. Adjust C2 and C3 on the input tuning network for maximum amplifier output. If possible, use a directional wattmeter between the driver and the amplifier input to check that best input SWR and maximum amplifier output occur at roughly the same setting.

Depending on the amount of drive power available, you may want to tune the amplifier for maximum power output or maximum gain. Fig. 123 shows what you can expect from different drive levels.

Once the amplifier is tuned for best input SWR and maximum output with 1 W of drive, anode voltage and drive power can be increased. Increase both in steps; be sure to keep the anode tuner peaked for maximum output power. When you get to the 100-W output level, very carefully readjust the input circuit for maximum output. The input capacitor closest to the cathode is critical and should need to be rotated less than 90 degrees maximum. Maximum output power will be roughly coincident with best input SWR.

Increase the drive power and keep the anode tuner peaked for maximum output. Increase the drive until you reach the desired output level, but do not exceed 400-mA I_K! At 1300-V dc and 350-mA I_K,

output power with a good tube should be about 230 to 250 watts. At lower anode voltages, I_K will be higher for the same output power. Higher anode voltages result in higher gain, lower drive levels, lower grid current and lower plate current for a given output power.

The anode tuner's tuning rate is approximately 5 MHz per turn. Clockwise rotation of the tuner lowers the resonant frequency of the cavity. This control will require readjustment as you make large frequency excursions within the 23-cm band (for example, if you go from 1296 weak-signal work to the 1269-MHz satellite segment). You should also check the input SWR if you move more than 15 MHz. Generally, amplifier tuning does not change much after initial setup. You should be able to turn it on and use it without retuning as it heats up. Slight adjustments may be necessary, however, depending on cooling, inherent thermal differences from tube to tube and duty cycle of the operating mode. Always keep the anode tuner peaked for maximum output, and check it from time to time, especially while you are first learning how the amplifier operates.

The output loading control is the output connector and probe assembly. Loading is changed by minor rotational adjustment of the Type-N connector. First loosen the jam-nut (or set screw) slightly. While observing output power and keeping the anode tuner peaked, rotate the loading control ±30 degrees maximum for greatest output power. This should be done only once and should not need repeating unless another tube is installed. Even then it may not be required.

Loop Yagis for 23 CM

Described here and shown in Figs. 124 through 128 are loop Yagis for the 23-cm band. Several versions are described, so the builder can choose the boom length and frequency coverage desired for the task at hand. Mike Walters, G3JVL, brought the original loop Yagi design to the amateur community in the 1970s. Since then, many versions have been developed with different loop and boom dimensions. Chip Angle, N6CA, developed the antennas shown here.

Three sets of dimensions are given. Good

performance can be expected if you follow the dimensions carefully. Recheck all dimensions before you cut or drill anything! The 1296-MHz version is intended for weak-signal operation at 1296 MHz, while the 1270-MHz version is optimized for FM and Mode L satellite work. The 1283-MHz antenna provides acceptable performance from 1280 to 1300 MHz.

These antennas have been built on 6- and 12-foot booms. Results of gain tests at VHF conferences and by individuals around the country peg the gain of the

6-footer at about 18 dBi, while the 12-foot version provides about 20.5 dBi. Swept measurements indicate that gain is about 2 dB down from maximum gain at ±30 MHz from the design frequency. SWR, however, deteriorates within a few megahertz on the low side of the design center frequency.

The Boom

The dimensions given here apply only to a ¾-inch-OD boom. If you change the boom size, the dimensions must be scaled

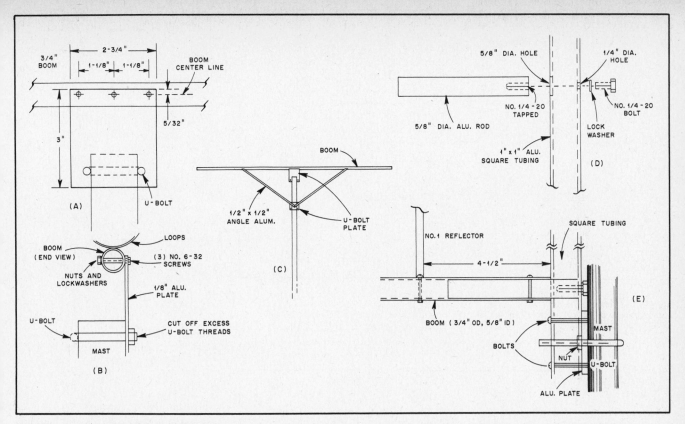

Fig. 124 — Boom-to-mast plate details are given at A. B shows how the Yagi is mounted to the mast. A boom support for long antennas is shown in C. The arrangement shown in D and E may be used to rear mount antennas up to 6 or 7 feet long.

accordingly. Many hardware stores carry aluminum tubing in 6- and 8-foot lengths, and that tubing is suitable for a short Yagi. If you plan a 12-foot antenna, you should find a piece of more rugged material, such as 6061-T6 grade aluminum. Do not use anodized tubing. The 12-foot antenna must be supported to minimize boom sag. The 6-foot version can be rear mounted. For rear mounting, allow 4.5 inches of boom behind the last reflector to eliminate SWR effects from the support.

The antenna is mounted to the mast with a gussett plate. This plate mounts at the boom center. See Fig. 124. Drill the plate-mounting holes perpendicular to the element-mounting holes, assuming the antenna polarization is to be horizontal. Elements will be mounted to the boom with no. 4-40 machine screws, so a series of no. 33 (0.113-inch) holes must be drilled along the center of the boom to accommodate this hardware. Fig. 125 shows the element spacings for different parts of the band. Tolerances should be followed as closely as possible.

Parasitic Elements

The reflectors and directors are cut from 0.0325-inch-thick aluminum sheet and are ¼ inch wide. Fig. 126 indicates the lengths for the various elements. These lengths apply only to elements cut from the specified material. For best results, the element strips should be cut with a shear. If you

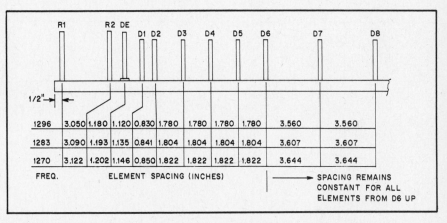

FREQ.	R1	R2	DE	D1	D2	D3	D4	D5	D6	D7	D8
1296	3.050	1.180	1.120	0.830	1.780	1.780	1.780	1.780	3.560		3.560
1283	3.090	1.193	1.135	0.841	1.804	1.804	1.804	1.804	3.607		3.607
1270	3.122	1.202	1.146	0.850	1.822	1.822	1.822	1.822	3.644		3.644

ELEMENT SPACING (INCHES) — SPACING REMAINS CONSTANT FOR ALL ELEMENTS FROM D6 UP

Fig. 125 — Boom drilling dimensions. Pick the version you want and follow these dimensions carefully. Spacing is the same for all directors after D6. Use as many as needed to fill up your boom.

leave the edges sharp, birds won't sit on the elements!

Drill the mounting holes as shown in Fig. 126. Measure carefully! After the holes are drilled, you must form each strap into a circle. This is easily done by wrapping the element around a round form (a small juice can works great).

Mount the loops to the boom with no. 4-40 × 1-inch machine screws, lock washers and nuts. See Fig. 127. It's best to use stainless-steel or plated-brass hardware for everything. Although the initial cost is

higher than for ordinary plated-steel hardware, stainless or brass hardware won't rust and need replacement after a few years. Unless the antenna is painted, it will definitely deteriorate.

Driven Element

The driven element is cut from 0.0325-inch copper sheet and is ¼ inch wide. Drill three holes in the strap, as detailed in Fig. 126. Trim the ends as shown and form the strap into a loop similar to the other elements. This antenna is like a

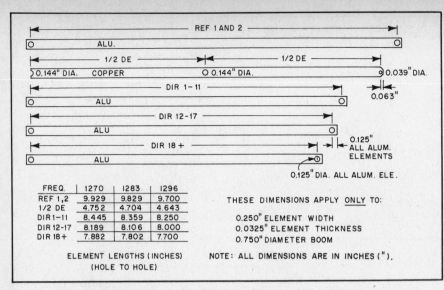

Fig. 126 — Parasitic elements are made from aluminum sheet. The driven element is made from copper sheet. The dimensions given are for ¼ inch wide by 0.0325-inch-thick elements only. Lengths specified are hole-to-hole distances; the holes are located 0.125 inch in from each element end.

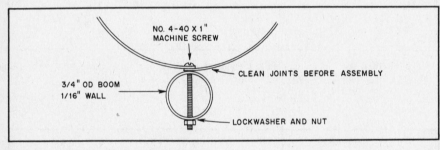

Fig. 127 — Element-to-boom mounting details.

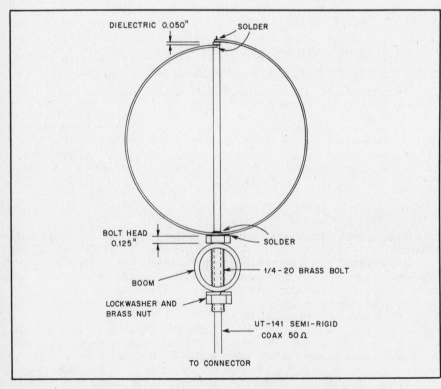

Fig. 128 — Driven element details. See Fig. 126 and the text for more information.

quad; if the loop is fed at the top or bottom, it will be horizontally polarized.

Driven-element mounting details are shown in Fig. 128. A mounting fixture is made from a ¼ 20 × 1¼ inch brass bolt. File the bolt head to a thickness of 0.125 inch. Bore a 0.144-inch (no. 27 drill) hole lengthwise through the center of the bolt. A piece of 0.141-inch semi-rigid Hardline (UT-141 or equiv.) will mount through this hole and connect to the driven loop. The point at which the UT-141 passes through copper loop and brass mounting fixture should be left unsoldered at this time to allow for matching adjustments when the antenna is completed, although the range of adjustment is not that great.

The UT-141 can be any convenient length. Attach the connector of your choice (preferably Type N). Use a short piece of low-loss RG-8 size cable (or better yet, ½-inch Hardline) for the run down the boom and mast to the main feed line. For best results, your main feed line should be the lowest-loss 50-ohm cable obtainable. Good 7/8-inch Hardline measures at 1.5 dB per 100 feet and virtually eliminates the need for remote mounting of the transmit amplifier.

Tuning the Driven Element

If you built the antenna carefully to the dimensions given, the SWR should be fine. Just to be sure, check the SWR if you have access to test equipment. You must be sure that your signal source is clean, however; wattmeters are often fooled by dirty signals and can give erroneous readings. If you have a problem, recheck all dimensions. If they look okay, a minor improvement may be realized by changing the shape of the driven element. Slight bending of reflector 2 may also help optimize SWR. When you have obtained the desired match, solder the point where the UT-141 jacket passes through the loop and brass bolt.

Quite a few people believe that practices common on lower frequencies can be used on 1296 MHz. That's the biggest reason that they don't work all the DX that there is to be worked!

First, when you have a design that's proven, copy it exactly — don't change it! This is especially true for antennas. Use the best feed line you can get. Why build an antenna if you are going to use lossy cable?

Here are some realistic measurements of common coaxial cables at 1296 MHz (loss per 100 feet): RG-8, 213, 214 — 11 dB; ½ inch foam/copper Hardline — 4 dB; 7/8 inch foam/copper Hardline — 1.5 dB.

Mount the antenna(s) to keep feed line loss to an absolute minimum. Antenna height is less important than keeping the line losses low! Do not allow the mast to pass through the elements, like on antennas for lower frequencies! Cut all U-bolts to the minimum length needed — a quarter wavelength at 1296 MHz is a little over 2 inches! Avoid any unnecessary metal around the antenna.

10-GHz Gunnplexer Communications

Communications on the amateur 10-GHz band have been simplified greatly with the advent of the Microwave Associates Gunnplexer™ transceivers. It is interesting to note that a similar communications system as little as 10 years ago would have required, literally, a rack full of equipment. The Gunnplexer transceiver shown in Fig. 129 will fit conveniently in your hand and is operated from a single 12-V power supply, either ac line operated or batteries! This makes the Gunnplexer ideal for fixed or portable operation.

The Gunnplexer is most often used in wide-band FM systems and is ideal for audio, full-color video and data transfer. With suitable peripheral equipment, systems for full-duplex audio, color video and up to several megabit data transfer are possible. Such high data rates should allow direct computer-to-computer memory transfer that would not be practical (or legal) on other lower-frequency amateur bands.

The Gunnplexer

The heart of the Gunnplexer is a Gunn diode oscillator, named after its inventor, John Gunn of IBM. A more detailed discussion of the Gunn diode is presented in Chapter 4. Refer to the cutaway drawing of the Gunnplexer, Fig. 130, for this discussion.

The Gunn diode is mounted with a varactor diode in a resonant cavity. When a regulated voltage is applied to the Gunn diode it oscillates; the frequency of oscillation is determined by the capacitance of the varactor diode and two mechanical tuning screws. The mechanical tuning screws can be likened to coarse tuning controls and are factory set for the appropriate tuning range. The voltage applied to the varactor diode (1 to 20 V dc) tunes the frequency electronically a minimum of 60 MHz. Power is coupled out of the cavity through a small iris that has been designed as somewhat of a compromise between maximum power output and isolation from changes in diode impedance and load.

The Gunn oscillator is also used to provide the local-oscillator signal for the detector diode. A ferrite circulator couples an appropriate amount of energy into the low-noise Schottky mixer diode and isolates the transmitter and receiver. Because the Gunn oscillator functions as both the transmitter and the receiver local oscillator, the IF at each end must be at the same frequency. Furthermore, the frequencies of the Gunn-

Fig. 129 — View of the Microwave Associates Gunnplexer. A protective resistor and diode are connected from the detector out-put to ground. A 50-μF electrolytic capacitor is connected from the Gunn diode terminal to ground. The Gunnplexer is designed to mate a UG-39/U waveguide flange when the horn antenna is removed.

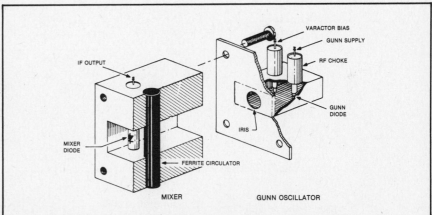

Fig. 130 — Drawing of the Gunnplexer assembly. The Gunn oscillator assembly consists of a resonant cavity in which is mounted the Gunn diode and varactor diode. The cylinders contain the quarter-wave choke sections, which connect to the diodes. The Gunn oscillator assembly bolts to the mixer assembly, which contains the detector diode and ferrite circulator. A horn antenna bolts to the front of the mixer assembly. (Reproduced with permission from *Ham Radio* Magazine, January 1979.)

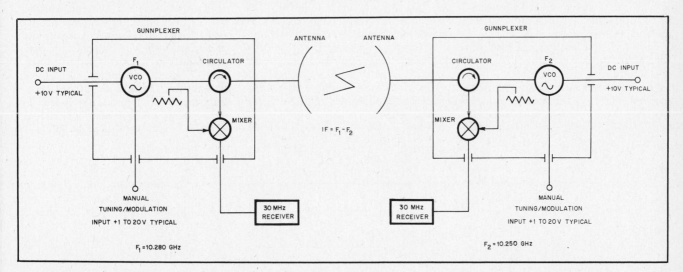

Fig. 131 — This drawing depicts a Gunnplexer communications system running full duplex. The VCOs are offset by the desired IF, which in this case is 30 MHz.

Table 12

Gunnplexer Specifications @ T_A = 25° C

Electrical Characteristics

RF Center Frequency	10.250 GHz[1]
Tuning	
Mechanical	± 50 MHz
Electronic	60 MHz min.
Linearity	1 to 40%
Frequency Stability	− 350 kHz/°C max.
RF Power vs.	
Temperature and	
Tuning Voltage	6 dB max.
Frequency Pushing	15 MHz/V max.
Input Requirements	
Dc Gunn Voltage	
Range[2]	+ 8.0 to + 10.0-V dc[2]
Maximum Operating	
Current	500 mA
Tuning Voltage	+ 1 to + 20 volts
Noise Figure[3]	<12 dB

RF Output Power[1]

Model	P out (mW)
MA87141-1	10 min. 15 typ.
MA87141-2	20 min. 25 typ.
MA87141-3	35 min. 40 typ.

Notes

[1] Tuning voltage set at 4.0 volts.
[2] Operating voltage specified on each unit.
[3] 1.5 dB IF NF at 30 MHz.

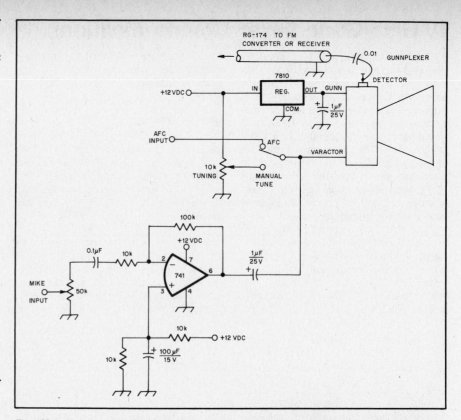

Fig. 132 — Here is a complete bare-minimum communications system using the Gunnplexer transceiver. An inexpensive automobile FM converter or receiver is used for the receiver IF. A 7810 voltage regulator is used for the Gunn-diode supply, and a simple 741 op amp serves as the microphone stage. The frequency of operation is set by the potentiometer that biases the varactor diode. A 10-turn potentiometer provides a comfortable tuning rate.

plexers must be separated by the IF. This is illustrated in Fig. 131. Intermediate frequencies of 30 MHz are more or less standard for audio work in the U.S. Both 45 and 70 MHz are used for video and high-speed data work.

As can be seen from Fig. 131, the Gunnplexer communications system is full duplex. In other words, both parties can talk and listen at the same time, without throwing switches. This is something that may take a while to get used to as most amateurs are programmed for VOX or PTT operation. In short, it is the ultimate break-in system!

One detail of the Gunnplexer that does require some specific attention is frequency control. The Gunnplexer has a frequency-stability specification of − 350 kHz frequency change per degree Celsius increase. This does not pose much of a problem with wide-bandwidth applications such as video or data transfer. However, for relatively narrowband audio work (200 kHz and less) some form of AFC, a phase-lock or other frequency-control scheme is required. In most cases simple AFC circuitry is suffi-

cient and quite easy to implement. The electrical characteristics of the Gunnplexer are given in Table 12.

Bare Minimum Audio Communications System

The simplest of communications systems using the Gunnplexer transceivers can be formed with two 88- to 108-MHz FM receivers, two Gunnplexers, two microphones and associated amplifiers, and two sources of 12 V dc. A diagram of such a system is shown in Fig. 132. Some of the low-cost FM converters and receivers for automotive use make good IF strips for this communications system. The AFC signal developed in the converter or receiver can be routed to the Gunnplexer varactor diode to lock the two units together. The microphone amplifier can be a single 741 operational amplifier as shown in the diagram.

There are two shortcomings with this

system. The first involves the use of the FM broadcast band as the IF. If mountaintop DXing is planned, it is likely that strong FM broadcast stations will be received no matter how short the lead between the detector and the FM converter or receiver is made. It will be necessary to select that part of the band where there are no strong signals present.

The second item involves AFC. Since it is possible to tune the Gunnplexers over a 60-MHz range with the varactor diode tuning, it is possible to tune them on either side of each other. This means that the single-polarity AFC system in the receiver or converter will work for only one combination. If the units are operated so that the AFC polarity is incorrect, the AFC will push the received signal out of the receiver passband. If this happens, simply tune the two Gunnplexers to produce the other IF signal. AFC lock should then be obtained.

A High-Performance Audio Communications System

The high-performance Gunnplexer audio system described here was developed by Advanced Receiver Research, and is sold as an amateur/commercial product.[1] The information is included here for those who wish to construct their own system.

This Gunnplexer support system has been specifically designed for use with the Microwave Associates Gunnplexers. The board contains a complete 30-MHz FM receiver, diode-switched IF filters, dual-polarity AFC system, Gunn-diode regulator and modulators for phone and CW. The system is suitable for fixed, portable or mobile operation. Power-supply requirements are a nominal 13 volts at 250 mA (this includes the current drawn by the Gunnplexer assembly). The circuit is specified for operation over the temperature range −25 to +65° C.

Theory of Operation

The circuit diagram is shown in Fig. 135, and an interconnection wiring guide in Fig. 134. Signal energy arriving at the IF input (pins 24 and 25) is routed to Q1, a low-noise RF amplifier. A band-pass filter, with a 3-dB bandwidth of 2 MHz, is located between the RF amplifier and mixer, Q2. LO injection for the mixer is provided by a crystal-controlled 40.7-MHz oscillator stage. Output from the mixer, at 10.7 MHz, is applied to either of two diode-switched ceramic filters. Supply voltage applied to pin 23 selects FL1 (supplied) and supply voltage applied to pin 22 selects FL2 (optional). Output from the filter is fed to IF amplifier stage Q3. Amplified 10.7-MHz energy is routed to the FM subsystem chip, U1, a CA3189E. Detected audio passes through an AF gain control (pins 16, 17 and 18) to U2, the audio amplifier. This stage has sufficient power to drive headphones and/or a speaker.

Squelch control is available by U5A. U5B inverts the amplified AFC signal so that either polarity AFC may be selected by S3. Overall AFC gain is controlled by R_c. U5C sums the AFC information along with the manual tuning voltage. This composite signal is applied to the varactor terminal of the Gunnplexer. U5D functions as a meter driver. Its input can be switched for center-tuning (discriminator zero) and manual-tuning voltage. Input to the driver is selected by S1.

[1]Advanced Receiver Research, Box 1242, Burlington, CT 06013, tel. 203-584-0776.

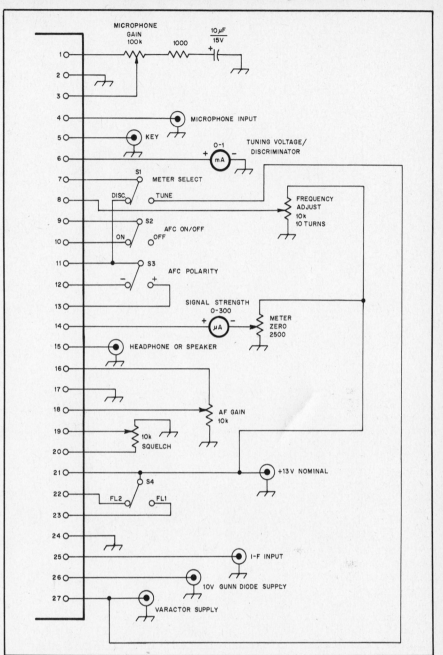

Fig. 134 — Interconnection diagram for the high-performance audio Gunnplexer communications system.

U3 is a microphone amplifier boosting the output from a microphone to a level suitable for modulating the Gunnplexer. U4 generates an approximate 500-Hz tone for MCW operation. Voice and MCW information is applied to the varactor of the Gunnplexer along with the AFC/tuning voltage. U6 is a three-terminal positive voltage regulator that provides an accurate 10-volt supply for the Gunn diode. A 1N4001 diode is included to protect against inadvertent application of reverse-polarity voltage.

It is advisable to mount the board in a well-shielded enclosure since the receiver is quite sensitive. The Gunnplexer unit can be

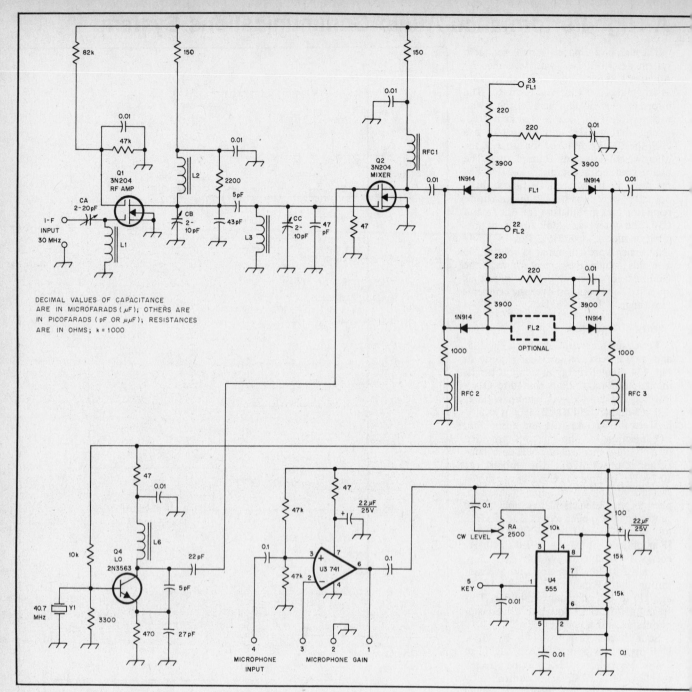

Fig. 135 — Schematic diagram of the high-performance audio communications system. All capacitors specified in picofarads are silver mica, DM5 variety. All resistors are ¼-watt film types.

FL1 — Murata SFJ10. 7 mA, 10.7 MHz.
L1 — 25 turns no. 30 enam. wire on a T25-10 core.
L2, L3 — 13 turns no. 28 enam. wire on a T25-6 core.

L4, L5 — 25 turns no. 30 enam. wire on a T25-20 core.
L6 — 16 turns no. 28 enam. wire on a T25-10 core.
RFC1 — 10 turns no. 28 enam. wire on an FT23-43 core.

RFC2, RFC3 — 10 turns no. 28 enam. wire on an FT23-72 core.
RFC4 — 11 turns no. 28 enam. on an FT23-43 core.
Y1 — 40.7 MHz, third overtone.

mounted atop a tower and connected to the board through three lengths of coaxial cable. Although the Gunn diode and varactor line carry only dc and low-level audio signals, shielded cable is a must as even small-voltage pickups can produce relatively large modulating voltages, which are undesirable. In cases of severe pickup, small RF chokes may be required directly at the Gunn diode and varactor connections

of the Gunnplexer. Lengths of cable of up to several hundred feet between the Gunnplexer and the board have been used with no problems. For long IF connection runs it is desirable to install a 30-MHz preamplifier at the Gunnplexer to prevent the long cable run from adversely affecting the system noise figure.

The Gunnplexer system is capable of complete duplex communications, so good

separation between the microphone and speaker is mandatory. The use of headphones for this type of communications is generally advisable.

Alignment

1) Connect a signal generator capable of delivering a 30-MHz signal to the IF input connection.

2) Adjust the output level of the

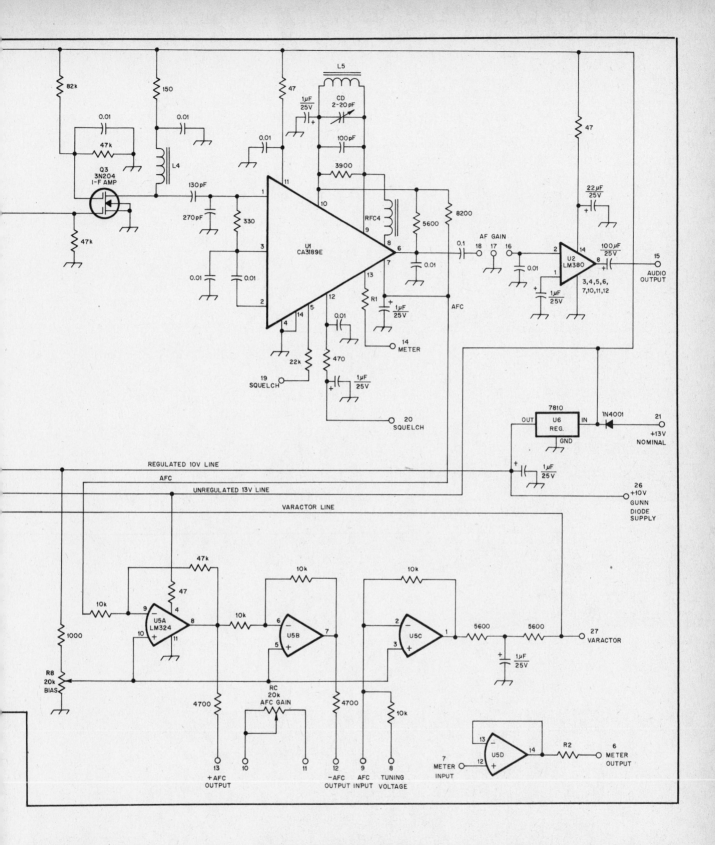

generator to a point where the signal-strength meter just begins to move up scale.

3) Adjust C_a, C_b and C_c for maximum indication on the meter.

4) Temporarily disconnect the signal generator from the input.

5) Place S1 in the discriminator position.

6) While switching S3 between the + and − AFC positions, adjust R_b so that the same meter reading is obtained in both

switch positions. This reading should be mid scale on the discriminator meter.

7) Reconnect the signal generator to the input and center the signal in the receiver passband.

8) While switching S1 between the AFC ON and OFF positions, adjust C_d so that the same meter reading is obtained for both positions. This reading should be mid scale on the discriminator meter.

9) Adjust R_c for the desired AFC gain/locking range by shifting the signal-generator frequency away from the passband center.

10) R_c is adjusted while in communication with another station. Advance the level of this control to the point where distortion occurs. Back off the setting of the control to the point where distortion is no longer present.

Chapter 33

Antenna Projects

Since the dawn of Amateur Radio, hams have been fascinated by antennas. Almost every ham has built at least one antenna, be it a dipole for one of the HF bands or a directional antenna for a VHF frequency. Through the years amateurs have tried countless possible configurations — everything from rain-gutter long wires, to balloon-supported wires, to full-size 80-meter beams. A quick scan of any of the ham bands will yield several discussions that attest to the popularity of antenna experimentation among amateurs.

This chapter presents projects for amateurs wishing to construct their own antennas. Related information may be found in Chapters 16 and 17. Chapter 37 tells how to get antennas into the air and keep them there safely.

Construction of Wire Antennas

There are many different types of wire antennas. The virtues of each, and the formulas for computing the right length for the desired operating frequency, are well documented in Chapter 17. The purpose of this section is to offer information on the actual physical construction of wire antennas. Because the dipole, in one of its configurations (Fig. 1), is probably the most common amateur wire antenna, it is used in the following examples. However, the techniques described here to enhance the reliability and safety of your antenna installations apply to all wire antennas.

Wire

Choosing the right type of wire for the project at hand is the key to a successful antenna — the kind that gets out like gangbusters and stays up in the face of a winter ice storm or a gusty spring wind storm. What gauge of wire to use is the first question to settle, and the answer depends on strength, ease of handling, cost, availability and visibility. Generally, antennas that are expected to support their own weight, plus the weight of the feed line, should be made from no. 12 wire. Horizontal dipoles, Zepps, some long wires and the like fall into this category. Antennas supported in the center, such as inverted-V dipoles and delta loops, may be made from lighter material, such as no. 14 wire — the minimum size called for in the National Electrical Code.

The type of wire to be used is the next important decision. Table 1 details several

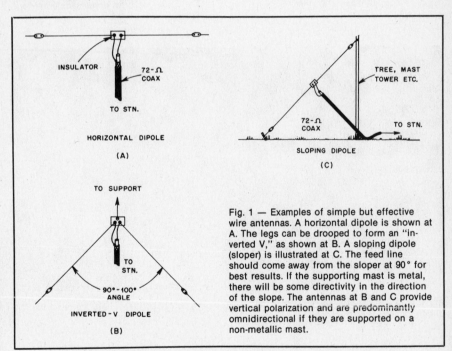

Fig. 1 — Examples of simple but effective wire antennas. A horizontal dipole is shown at A. The legs can be drooped to form an "inverted V," as shown at B. A sloping dipole (sloper) is illustrated at C. The feed line should come away from the sloper at 90° for best results. If the supporting mast is metal, there will be some directivity in the direction of the slope. The antennas at B and C provide vertical polarization and are predominantly omnidirectional if they are supported on a non-metallic mast.

popular wire styles and sizes. The strongest wire suitable for antenna service is copper-clad steel, also known as Copperweld® The copper coating is necessary for RF service because steel is a relatively poor conductor. Practically all of the RF current is confined to the copper because of skin effect. Copper-clad steel is outstanding for permanent installations, but it can be difficult to work with. Kinking, which severely weakens the wire, is a constant threat when handling any solid conductor.

Solid-copper wire, either hard drawn or soft drawn, is another popular material. Easier to handle than copper-clad steel, solid copper is available in a wide range of sizes. It is generally more expensive, however, because it is all copper. Soft-

Table 1

Stressed Antenna Wire

American Wire Gauge	Recommended Tension[1] (pounds)		Weight (pounds per 1000 feet)	
	Copper-clad steel[2]	Hard-drawn copper	Copper-clad steel[2]	Hard-drawn Copper
4	495	214	115.8	126
6	310	130	72.9	79.5
8	195	84	45.5	50
10	120	52	28.8	31.4
12	75	32	18.1	19.8
14	50	20	11.4	12.4
16	31	13	7.1	7.8
18	19	8	4.5	4.9
20	12	5	2.8	3.1

[1]Approximately one-tenth the breaking load. Might be increased 50 percent if end supports are firm and there is no danger of ice loading.
[2]"Copperweld," 40 percent copper.

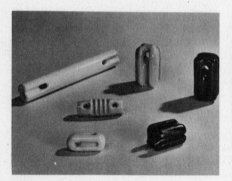

Fig. 2 — Various types of commercially made insulators.

Fig. 3 — Some ideas for homemade antenna insulators.

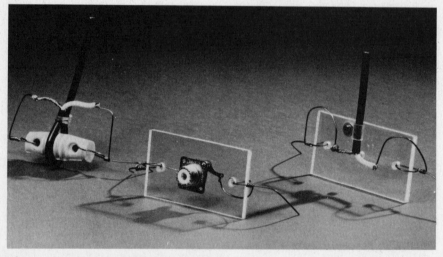

Fig. 4 — Some dipole center insulators have built-in SO-239 connectors. Others are designed for direct connection to the feed line.

drawn tends to stretch under tension, so periodic pruning of the antenna may be necessary in some cases.

Enamel-coated "magnet-wire" is a good choice for experimental antennas because it is easy to manage, and the coating protects the wire from the weather. Although it stretches under tension, the wire may be prestretched before final installation and adjustment. A local electric motor rebuilder might be a good source for magnet wire.

Hook-up wire, speaker wire or even ac lamp cord are suitable for temporary installations. Almost any copper wire may be used, as long as it is strong enough for the demands of the installation. Steel wire is a poor conductor and should be avoided. Aluminum may be suitable in some cases, although it is not usually very strong and is difficult to adequately connect a feed line to.

It matters not (in the HF region at least)

whether the wire chosen is insulated or bare. However, the wire should be either enameled or insulated to prevent corrosion. If exposed, the bare copper surface will corrode, causing a resistive loss in current, which flows only on the surface due to skin effect. If insulated wire is used, a 3 to 5% shortening beyond the standard 468/f length will be required to obtain resonance at the desired frequency, because of the increased distributed capacitance resulting from the dielectric constant of the plastic insulating material. The actual length for resonance must be determined experimentally by pruning and measuring because the dielectric constant of the insulating material varies from wire to wire. Portions that might come into contact with humans or animals should be insulated to reduce the chance of shock or burns. Several suppliers of wire suitable for the projects in this chapter are listed in Chapter 35.

Insulators

Wire antennas must be insulated at the ends. Generally, commercially available insulators are made from ceramic, glass or plastic. Some of the more common types are pictured in Fig. 2. Insulators like these are available from many Amateur Radio dealers. Radio Shack and local hardware stores are other possible sources.

In a pinch, acceptable homemade insulators may be fashioned from a variety of material including (but not limited to) acrylic sheet or rod, PVC tubing, wood, fiberglass rod or even stiff plastic from a discarded container. Fig. 3 shows some homemade insulators. Ceramic or glass insulators will usually outlast the wire, so they are highly recommended for a safe, reliable, permanent installation. Other materials may tear under stress or break down in the presence of sunlight. Many types of plastic do not weather well.

Many types of wire antennas require an insulator at the feed point. Although there are many ways to connect the feed line, there are few things to keep in mind. If you feed your antenna with coaxial cable, you have two choices. You can install an SO-239 connector on the center insulator and use a PL-259 on the end of your coax, or you can separate the center conductor from the braid and connect the feed line directly the antenna wire. Although the latter method is initially less costly, the former method offers several advantages. Coaxial cable braid soaks up water like a sponge. If you do not adequately seal the antenna end of the feed line, water will find its way into the braid. Water in the feed line will lead to contamination, rendering the coax useless long before its normal lifetime is up. It is not uncommon for water to drip from the end of the coax inside the shack after a year or so of service if the antenna connection is not properly waterproofed. Use of a PL-259/SO-239 combination (or connector of your choice) makes the task of waterproofing connections much easier.

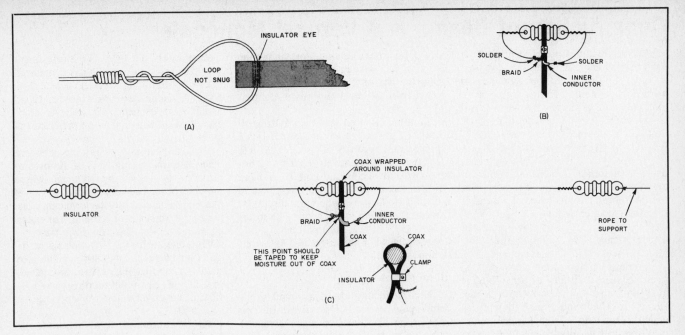

Fig. 5 — Details of the dipole antenna. The end insulator connection is shown at A, while B illustrates the center insulator. Part C illustrates the completed antenna.

Another advantage to using the PL-259/SO-239 combination is that feed line replacement is much easier, should that become necessary.

Whether you use coaxial cable, open-wire "ladder line" or plastic-encased transmitting-type twin lead to feed your antenna, an often-overlooked consideration is the mechanical strength of the connection. Wire antennas and feed lines tend to move a lot in the breeze, and unless the feed line is attached securely, the connection will weaken with time. The resulting failure can range from a frustrating intermittent electrical connection to a complete separation of feed line and antenna. Fig. 4 illustrates several different ways of attaching the feed line to the antenna.

Putting It Together

Fig. 5 shows some of the finer points of antenna construction. Although a dipole is used for the examples, the techniques illustrated here apply to any type of wire antenna.

How well you put the pieces together is second only to the ultimate strength of the materials used in determining how well your antenna will work over the long term. Even the smallest details, such as how you connect the wire to the insulators (Fig. 5A), contribute significantly to antenna longevity. By using plenty of wire at the insulator and wrapping it tightly, you will decrease the possibility of the wire pulling loose in the wind. There is no need to solder the wire once is it wrapped. There is no electrical connection here, only mechanical. The high heat needed for soldering can anneal the wire, significantly weakening it at the solder point.

Similarly, the feed-line connection at the center insulator should be made to the antenna wires after they have been secured to the insulator (Fig. 5B). This way, you will be assured of a good electrical connection between the antenna and feed line without compromising the mechanical strength. Do a good job of soldering the antenna and feed-line connections. Use a heavy iron or a torch, and be sure to clean the materials thoroughly before starting the job. Proper planning should allow you to solder indoors at a workbench, where the best possible joints may be made. Poorly soldered or unsoldered connections will become a headache as the wire oxidizes and the electrical integrity degrades with time. Besides degrading your antenna performance, poorly made joints can be a cause of TVI because of rectification.

If made from the right materials, the dipole illustrated in Fig. 5C should give the builder years of maintenance-free service — unless of course a tree falls on it! As you build your antenna, keep in mind that if you get it right the first time, you won't have to do it again for a long time.

Construction of Beams and Vertical Monopoles

Most beams and verticals are made from sections of aluminum tubing that have been extruded or drawn. Compromise beams have been fashioned from less-expensive materials such as electrical conduit (steel) or bamboo poles wrapped with conductive tape or aluminum foil. The steel conduit is heavy, is a poor conductor and is subject to rust. Similarly, bamboo with conducting material attached to it will deteriorate rapidly in the weather. Aluminum tubing is by far the best material for building antennas in the HF and VHF regions. Copper or brass may be used in some VHF and UHF applications, and this idea is discussed later in this chapter.

Aluminum tubing comes in a variety of sizes, detailed in Table 2. Such tubing is available in many metropolitan areas. Dealers may be found in the Yellow Pages under "Aluminum." Tubing usually comes in 12-foot lengths, although 20-foot lengths are available in some sizes. Your aluminum dealer will probably also sell aluminum plate in various thicknesses needed for boom-to-mast and boom-to-element connections.

Aluminum is rated according to its hardness. The most common material used in antenna construction is grade 6061-T6. This material is relatively strong and has good workability. In addition, it will bend without taking a "set," an advantage in antenna applications where the pieces are constantly flexing in the wind. The softer grades (5051, 3003, etc.) will bend much more easily, while harder grades (7075, etc.) are more brittle.

Beam Elements

Beam elements are generally made from telescoping sections of aluminum tubing. The prime consideration when choosing tubing sizes is mechanical strength. Ob-viously, a 66-foot-long 40-meter element would be made from much larger material than an 18-foot-long 10-meter element. Fig. 6 shows generic element designs for 10, 15 and 20-meter elements. The exact length of each element will depend on the electrical design.

Wall thickness is of primary concern when selecting tubing. It is of utmost importance that the tubing fits snugly where the element sections join. Sloppy joints will make a mechanically unstable antenna. The magic wall thickness is 0.058 in. For example (from Table 2), 1-in outside diameter (OD) tubing with a 0.058-in wall has an inside diameter (ID) of 0.884 in. The next smaller size of tubing, 7/8 in, has an OD of 0.875 in. The 0.009-in difference provides just the right amount of clearance for a snug fit.

Fig. 7 shows several methods of fastening antenna element sections to-

Table 2
Standard Sizes of Aluminum Tubing

6061-T6 (61S-T6) round aluminum tube in 12-foot lengths

OD (in.)	Wall Thickness in.	Wall Thickness stubs ga.	ID (in.)	Approx. Weight (lb) per ft	per length	OD (in.)	Wall Thickness in.	Wall Thickness stubs ga.	ID (in.)	Approx. Weight per ft	per length
3/16	.035	no. 20	.117	.019	.228	1	.083	no. 14	.834	.281	3.372
	.049	no. 18	.089	.025	.330	1 1/8	.035	no. 20	1.055	.139	1.668
1/4	.035	no. 20	.180	.027	.324		.058	no. 17	1.009	.228	2.736
	.049	no. 18	.152	.036	.432	1 1/4	.035	no. 20	1.180	.155	1.860
	.058	no. 17	.134	.041	.492		.049	no. 18	1.152	.210	2.520
5/16	.035	no. 20	.242	.036	.432		.058	no. 17	1.134	.256	3.072
	.049	no. 18	.214	.047	.564		.065	no. 16	1.120	.284	3.408
	.058	no. 17	.196	.055	.660		.083	no. 14	1.084	.357	4.284
3/8	.035	no. 20	.305	.043	.516	1 3/8	.035	no. 20	1.305	.173	2.076
	.049	no. 18	.277	.060	.720		.058	no. 17	1.259	.282	3.384
	.058	no. 17	.259	.068	.816	1 1/2	.035	no. 20	1.430	.180	2.160
	.065	no. 16	.245	.074	.888		.049	no. 18	1.402	.260	3.120
7/16	.035	no. 20	.367	.051	.612		.058	no. 17	1.384	.309	3.708
	.049	no. 18	.339	.070	.840		.065	no. 16	1.370	.344	4.128
	.065	no. 16	.307	.089	1.068		.083	no. 14	1.334	.434	5.208
1/2	.028	no. 22	.444	.049	.588		*.125	1/8"	1.250	.630	7.416
	.035	no. 20	.430	.059	.708		*.250	1/4"	1.000	1.150	14.823
	.049	no. 18	.402	.082	.984	1 5/8	.035	no. 20	1.555	.206	2.472
	.058	no. 17	.384	.095	1.040		.058	no. 17	1.509	.336	4.032
	.065	no. 16	.370	.107	1.284	1 3/4	.058	no. 17	1.634	.363	4.356
5/8	.028	no. 22	.569	.061	.732		.083	no. 14	1.584	.510	6.120
	.035	no. 20	.555	.075	.900	1 7/8	.058	no. 17	1.759	.389	4.668
	.049	no. 18	.527	.106	1.272	2	.049	no. 18	1.902	.350	4.200
	.058	no. 17	.509	.121	1.452		.065	no. 16	1.870	.450	5.400
	.065	no. 16	.495	.137	1.644		.083	no. 14	1.834	.590	7.080
3/4	.035	no. 20	.680	.091	1.092		*.125	1/8"	1.750	.870	9.960
	.049	no. 18	.652	.125	1.500		*.250	1/4"	1.500	1.620	19.920
	.058	no. 17	.634	.148	1.776	2 1/4	.049	no. 18	2.152	.398	4.776
	.065	no. 16	.620	.160	1.920		.065	no. 16	2.120	.520	6.240
	.083	no. 14	.584	.204	2.448		.083	no. 14	2.084	.660	7.920
7/8	.035	no. 20	.805	.108	1.308	2 1/2	.065	no. 16	2.370	.587	7.044
	.049	no. 18	.777	.151	1.810		.083	no. 14	2.334	.740	8.880
	.058	no. 17	.759	.175	2.100		*.125	1/8"	2.250	1.100	12.720
	.065	no. 16	.745	.199	2.399		*.250	1/4"	2.000	2.080	25.440
1	.035	no. 20	.930	.123	1.476	3	.065	no. 16	2.870	.710	8.520
	.049	no. 18	.902	.170	2.040		*.125	1/8	2.700	1.330	15.600
	.058	no. 17	.884	.202	2.424		*.250	1/4	2.500	2.540	31.200
	.065	no. 16	.870	.220	2.640						

*These sizes are extruded; all other sizes are drawn tubes. Shown here are standard sizes of aluminum tubing that are stocked by most aluminum suppliers or distributors in the United States and Canada.

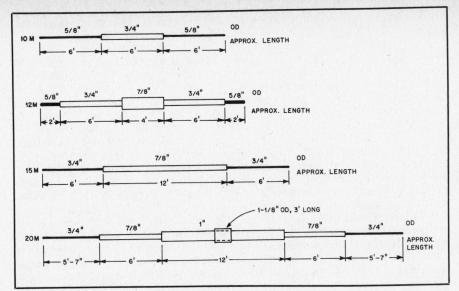

Fig. 6 — Element designs for Yagi antennas. (Use 0.058-in-wall aluminum tubing except for end pieces where thinner-wall tubing may be used.)

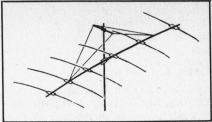

Fig. 8 — A long boom needs both vertical and horizontal support. The crossbar mounted above the boom can support a double truss, which will help keep the antenna in position.

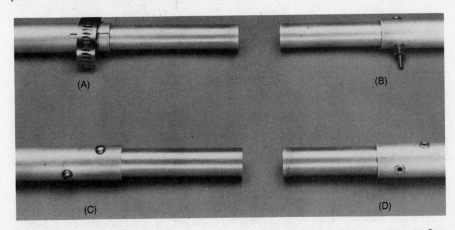

Fig. 7 — Some methods of connecting telescoping tubing sections to build beam elements. See text for a discussion of each method.

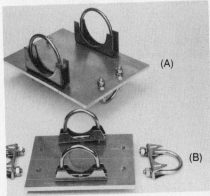

Fig. 9 — The boom-to-element plate at A uses muffler-clamp-type U-bolts and saddles to secure the round tubing to the flat plate. The boom-to-mast plate at B is similar to the boom-to-element plate. The main difference is the size of materials used.

gether. The slot and hose clamp method shown in Fig. 7A is probably the best for joints where adjustments are needed. Generally, one adjustable joint per element half is sufficient to tune the antenna. Stainless steel hose clamps (beware — some "stainless steel" models do not have a stainless screw and will rust) are recommended for longest antenna life.

Figs. 7B, 7C and 7D show possible fastening methods for joints that are not adjustable. At B, machine screws and nuts hold the elements in place. At C, sheet metal screws are used. At D, rivets secure the tubing. If the antenna is to be assembled permanently, rivets are the best choice. Once in place, they are permanent. They will never work free, regardless of vibration or wind. If aluminum rivets with aluminum mandrels are employed, they will never rust. Also, being aluminum, there is no danger of corrosion from interaction between dissimilar metals. If the antenna is to be disassembled and moved periodically, either B or C will work. However, if machine screws are used, take all possi-

ble precautions to keep the nuts from vibrating free. Use of lock washers, lock nuts and flexible adhesive such as silicon bathtub sealant will keep the hardware in place.

Use of a conductive grease at the element joints is essential for long life. Left untreated, the aluminum surfaces will oxidize in the weather, resulting in a poor connection. Some trade names for this conductive grease are Penetrox, Noalox and Dow Corning Molykote 41. Many electrical supply houses carry these products.

Boom Material

The boom size for a rotatable Yagi or quad should be selected to provide stability to the entire system. The best diameter for the boom depends on several factors, but mostly the element weight, number of elements and overall length. Tubing diameters of 1¼ inches can easily support 3-element 10-meter arrays and perhaps a 2-element 15-meter system. For larger 10-meter antennas or for harsh weather conditions, and for antennas up to 3 ele-

ments on 20 meters or 4 elements on 15 meters, a 2-inch-diameter boom will be adequate. Two-inch-diameter booms should not be made any longer than 24 feet unless additional support is given to reduce both vertical and horizontal bending forces. Suitable reinforcement for a long 2-inch boom can consist of a truss or a truss and lateral support, as shown in Fig. 8.

A boom length of 24 feet is about the point where a 3-inch diameter begins to be very worthwhile. This dimension provides a considerable amount of improvement in overall mechanical stability as well as increased clamping surface area for element hardware. The latter is extremely important if heavy icing is commonplace and rotation of elements around the boom is to be avoided. Pinning an element to the boom with a large bolt helps in this regard. On the smaller diameter booms, however, the elements sometimes work loose and tend to elongate the pinning holes in both the element and the boom. After some time the elements shift their positions slightly (sometimes from day to day!) and give a ragged appearance to the system, even though this may not harm the electrical performance.

A 3-inch-diameter boom with a wall thickness of 0.065 inch is very satisfactory for antennas up in size to about a 5-element, 20-meter array that is spaced on a 40-foot-long boom. A truss is recommended for any boom longer than 24 feet. One possible source for large boom material is irrigation tubing sold at farm supply houses.

Putting It Together

Once you assemble your boom and elements, the next step is to fasten the elements to the boom securely and then fasten the boom to the mast or supporting structure. Fig. 9A shows the most practical method of attaching the elements to the boom. Of course, if the home constructor has access to a machine shop, there are any number of different ways to accomplish this. The size of the mounting plate will depend on the size of the boom and element materials used. Be sure to leave plenty of material on either side of the U-bolt holes. The U-bolts selected should be a snug fit for the tubing. If possible, purchase the muffler-clamp type U-bolts that come with saddles.

The boom-to-mast plate shown in Fig. 9B is similar to the boom-to-element plate. The size of the plate and number of U-bolts used will depend on the size of the antenna. Generally, antennas for the bands up through 20 meters require only two U-bolts each for the mast and boom. Longer antennas for 15 and 20 (35-ft booms and up) and most 40-meter beams should have four U-bolts each for the boom and mast because of the torque that the long booms and elements exert as the antennas move in the wind. When tightening the U-bolts, be careful not to crush the tubing. Once the wall begins to collapse, the connection begins to weaken. Many aluminum suppliers sell ¼-in plate just right for this application. Often they will shear pieces to the correct size on request. As with tubing, the relatively hard 6061-T6 grade is a good choice for mounting plates.

The antenna should be put together with good-quality hardware. Stainless steel is best for long life. Rust will attack plated steel hardware after a short while, making nuts difficult, if not impossible, to remove. If stainless muffler clamps are not available, the next best thing is to have them plated. If you can't get them plated, then at least paint them with a good zinc-chromate primer and a finish coat or two. Good-quality hardware is more expensive initially, but if you do it right the first time, you won't have to take the antenna down after a few years and replace the hardware. Also, when repairing or modifying an installation, nothing is more frustrating than fighting rusty hardware at the top of a tower.

Portable Trap Dipoles

Vacationers, campers, sales people and QRPers take note! You need not carry a large multiband trap dipole afield if your transmitter is in the 150-W-output class, or lower. You can construct your own traps inexpensively with ordinary materials, and they can be made quite small without becoming poor performers. This project describes some easy techniques for fabricating homemade antenna traps. Additional hints are offered for keeping the bulk and weight of portable antennas within reason. Doug DeMaw, W1FB, originally presented this material in June 1983 QST.

A Review of the Trap Concept

A "trap" is exactly what the term implies. It traps an RF signal to prevent it from passing beyond a specific point along an electrical conductor. At some other frequencies, however, it no longer acts as a trap, and permits the passage of RF energy.

An antenna trap is designed for a particular operating frequency, and there may be several traps in the overall system — each designed for a specific frequency. Therefore, a 40- through 10-meter trap dipole might contain traps for 10, 15, 20 and 30 meters. On 40 meters, all of the traps are "absorbed" into the system to become part of the overall 40-meter dipole. Owing to the loading effect of the traps, the 40-meter portion of the antenna will be somewhat shorter than a full-size 40-meter dipole with no traps. The antenna bandwidth will be narrower when traps are used. Fig. 10A illustrates the general format for a multiband dipole.

A trap style of antenna is not as efficient as a full-size dipole. This is because there will always be some losses in the traps. But the losses in a well-designed system are usually so low that they are hard to measure by simple means. The losses represent a small trade-off for the convenience of being able to accommodate many ham bands with one radiator and a single feed line. Yagi antennas contain traps in the parasitic elements (directors and reflectors) as well as in the driven element. Therefore, a multielement antenna of that type may have as many as 12 traps.

Electrical Characteristics

An antenna trap is a parallel-resonant L-C circuit. Therefore, it is similar to the tuned circuit in a transmitter or receiver. A resonator of this kind, if designed correctly, has a moderate Q and a fairly narrow bandwidth. This means the trap capacitor should have a high Q and the trap coil should contain wire that is reasonably large in cross section. These traits will help to reduce losses.

Fig. 10B shows the equivalent circuit for an antenna trap. Once this network is adjusted to resonance in the desired part of an amateur band, it will not be affected significantly by the attachment of the wires that comprise the antenna. A well-designed and -constructed trap should not change frequency by any great amount when the temperature or humidity around it varies. Therefore, it is important to use a stable capacitor, a rigid coil and some type of sealant.

Mini Trap Using a Toroid Core

In an effort to scale down the size of antenna traps during a design exercise for a portable antenna, small toroidal cores on which to wind the coils were tried. Ferrite cores were ruled out because they aren't as stable as powdered-iron ones. Furthermore, the powdered-iron material has a much greater flux density than an equivalent-size ferrite core, which means that the core will not saturate as easily at moderate RF power levels.

Development work started with Micrometals Corporation T50-6 toroids, which are sold by Amidon Associates, Palomar Engineers and RadioKit (see Chapter 35). The first effort resulted in a pair of very small 20-meter traps. A silver-mica capacitor was chosen for the parallel-tuned circuit. Ceramic capacitors were not used because of previous experiences with changes in value under temperature ex-

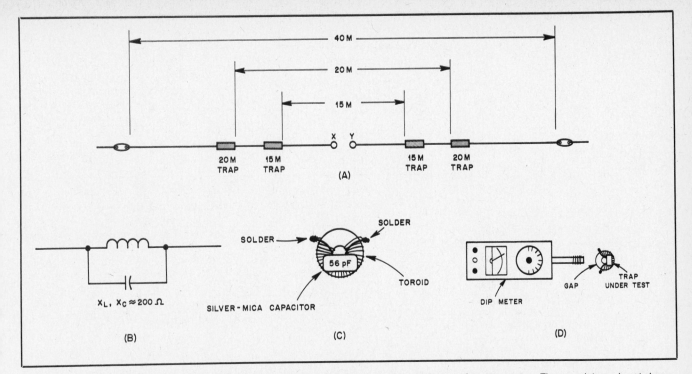

Fig. 10 — At A, a representation of a three-band trap dipole antenna. B shows the electrical equivalent of an antenna trap. The ac resistance is not shown. A suitable reactance value for the coil and capacitor is 200 ohms. C illustrates the physical arrangement for one of the toroidal L-C traps. Put spaghetti tubing over the capacitor leads to prevent them from shorting to the turns on the toroid. D shows one test method for finding the resonant frequency of a trap. Different points around the toroid will yield better dip indications. Experiment with the position of the dipper coil.

tremes; better results were obtained with dipped silver-mica units.

The rule of thumb for choosing the coil and capacitor values for traps is based on a reactance of approximately 200 ohms, although values up to 300 have also yielded good results. Using 200 ohms as the basis for the design, the calculated capacitor value was very close to a standard one — 56 pF for trap resonance at 14.100 MHz. This was obtained from

$$C\ (\mu F) = \frac{1}{2\pi f(MHz)X_c}$$

Hence

$$C = \frac{1}{6.28 \times 14.1 \times 200}$$

$$= 0.0000564\ \mu F\ (56\ pF)$$

Since X_c and X_L are equal at resonance, the coil was calculated from

$$L\ (\mu H) = \frac{X_L}{2\pi f(MHz)}$$

Hence

$$L = \frac{200}{6.28 \times 14.1} = 2.25\ \mu H\ (approx.)$$

The value of the coil will have to be adjusted slightly after the trap is assembled to allow for capacitor tolerance and stray capacitance, which accounts for the term "approximate."

The Amidon toroid tables were consulted to learn the A_L factor of a T50-6 core (½-inch-diameter toroid). The value is 40.

From this, the number of turns was calculated from

$$Turns = 100\ \sqrt{L_{\mu H}/A_L}$$

Hence

$$Turns = 100\ \sqrt{2.25/40} = 23.7$$

For practical reasons a 24-turn winding was used: A partial turn is not convenient on a toroid form.

The same procedure was used for the remaining traps in the antenna. The equations can be useful to those who have not previously designed resonant circuits or used toroidal cores.

Toroidal-Trap Adjustment

It's best to use the largest size wire that will fit easily on the toroid core. The stiffness of the heavier magnet wire will help to keep the coil turns in place, thereby minimizing detuning. This design used no. 24 enameled wire.

The capacitor leads and coil "pigtails" should be kept as short as possible. Fig. 10C illustrates the layout. The leads at each end of the mica capacitor are soldered to the related coil leads before final adjustment is made.

A dip meter can be used to determine the resonant frequency of the trap, as shown in Fig. 10D. Although a prominent feature of a toroidal coil is the self-shielding characteristic, which makes it difficult for us to get ample coupling with a dip meter, it is possible to read a dip. By inserting the dip-meter coil into the area of the winding

gap on the tuned circuit (Fig. 10D) a dip can be obtained. By approaching the trap from different angles, it should be easy to find a spot where a dip can be read on the meter. Once the dip is found, back off the instrument until the dip is barely discernible (the minimum-coupling point). Monitor the dip-meter signal on a calibrated receiver to learn the resonant frequency of the trap.

Select a part of the related amateur band for trap resonance, perhaps the center of the frequency spread most often used. For example, set 20-meter traps for resonance at 14.025 MHz for 14.000 to 14.050 MHz. For phone-band coverage, 14.275 MHz is a good choice for the trap frequency. A compromise frequency for phone and CW operation would be 14.100 MHz. Owing to the trap Q, coverage of an entire band is not possible without having an SWR of 2:1 or greater at the band-edge extremes. The absolute bandwidth will depend on the trap Q and the Q of the antenna itself.

If the trap is not on the desired frequency, move the turns of the toroidal coil farther apart to raise the frequency. Push them closer together to lower the frequency. An alternative method for finding the trap resonance is shown in Fig. 11. The trap being tested is connected to terminals x and y. The coupling is very light in order to prevent the test-circuit capacitance from appearing in parallel with the trap. For this reason the coupling capacitors are only 2 pF. The station transmitter is adjusted for the lowest power output that will provide a reading on M1. The VFO is then swept manually across the band. When the resonant frequency of the trap is located, the meter (M1) will deflect upward sharply, indicating resonance. Adjust the trap for a frequency that is approximately 5% lower than the desired one. This will compensate for the shunt capacitance presented by the 2-pF coupling capacitors.

When the coil turns are set in the correct manner, spread a bead of fast-drying epoxy cement across the turns on the two flat sides of the toroid. This will prevent unwanted position changes that could cause a shift in resonance later on from handling.

Housing the Mini Trap

A 7/8-inch-OD PVC plumbing coupling, 1¼ inches long, will serve nicely as a housing for the toroidal traps. A ridge inside the couplings at the center can be filed out easily to provide clearance for the trap. A rat-tail file does the job quickly. Fig. 12A shows a breakaway view of how the trap is assembled. Slices of dowel rod are used for end plugs. A knot is tied in the antenna wire that enters the trap housing; this prevents strain on the trap coil.

After the antenna wire has been soldered to the trap at each end, add a layer of epoxy glue to the outer perimeter of one of the dowel plugs, then insert it into the PVC coupling until it is flush. Fill the coupling with noncorrosive sealant such as aquarium cement. Finally, place epoxy glue on the remaining end plug and insert it in the PVC coupling. Allow the trap to set for 48 hours, until the sealant has hardened. Fig. 12B is a photograph of a mini trap, along with a dipole center insulator made from a PVC T-coupling. The coupling is filled with sealant after the wires are soldered to the coaxial feed line. Long plugs are used to close the three open ends of the T connector. A closed nylon loop, made from strong spaghetti tubing, was fed through two small holes at the top of the T-coupling to permit erecting the dipole as an inverted V. A small eye bolt and nut could have been used instead.

There was a minor downward shift in trap resonance after the sealant hardened. Both 20-meter traps shifted roughly 30 kHz lower. No doubt this was caused by increased distributed capacitance across the coil turns with the sealant in place. This seemed to have no effect on the trap quality; it had a measured parallel resistance of 25 kΩ before and after encapsulation (using the laboratory RX meter for tests). Generally, anything greater than 10 kΩ is suitable for an antenna trap.

Mini Coaxial-Cable Traps

Two very interesting articles concerning antenna traps appeared in the amateur literature during 1981.[1,2] Some advantages

[1]G. O'Neil, "Trapping the Mysteries of Trapped Antennas," *Ham Radio*, Oct. 1981, p. 10.
[2]R. Johns, "Coaxial Cable Antenna Traps," *QST*, May 1981, p. 15.

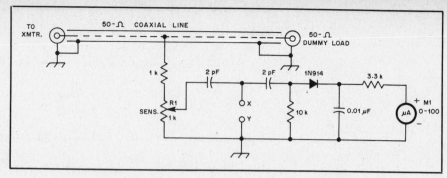

Fig. 11 — Test fixture suitable for checking trap resonance with the station transmitter. Use the least amount of power necessary for meter deflection.

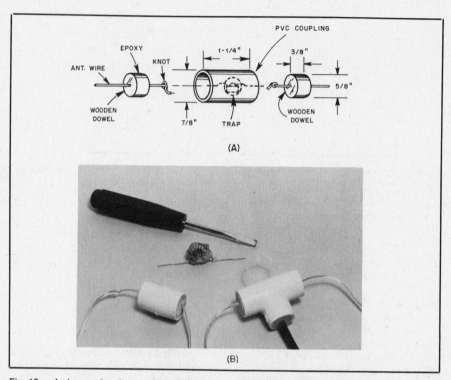

Fig. 12 — A shows a breakaway view of a toroidal mini trap. The knots in the wire prevent stress on the tuned circuit. B illustrates a toroidal mini trap, an encapsulated toroid and a PVC T-coupling for use as a center insulator. RG-58 cable is shown in this example (see text).

over the usual coil/capacitor style of trap were described by the authors: (1) The traps were not especially frequency sensitive to changes in temperature and climate; (2) the coaxial trap offers greater effective bandwidth; and (3) parallel resistance is quite high — on the order of 50 kΩ.

The articles under discussion contained practical information about the use of RG-58 and RG-8 cable for the trap coils. A small, lightweight trap was desired, so miniature cable — RG-174 — was tried. A completed mini coaxial trap for 20 meters is shown in Fig. 13A.

The principle of operation is covered well by O'Neil (note 1). Since this article deals with the practical aspects of traps, we won't delve too deeply into the electrical characteristics of the coaxial trap. However, a diagram showing how it is hooked up is offered in Fig. 13C. A length of coaxial line is wound on a coil form, and the inner conductor at one end is attached to the outer conductor at the opposite end.

The distributed capacitance of the two conductors and the inductance of the coil combine to provide a resonant circuit. An acceptable Q results, and the trap can accommodate considerable RF voltage and current without being damaged. A parallel resistance of 50 kΩ was measured for the 20-meter trap of Fig. 13A. The bandwidth at the 10-kΩ points was somewhat greater than with the toroidal trap.

Coaxial-Trap Assembly

PVC plumbing pipe, 5/8-in OD, is an acceptable and low-cost material for the coaxial traps. Fig. 13B shows a typical trap assembly. End plugs made from 1/2-inch wooden dowel fit snugly inside the PVC pipe. The completed trap contains a length of bus wire inside it for connecting the braid and center conductor of the cable together, as discussed earlier. The ends of the bus wire and the related cable ends are routed outside the PVC tubing through small holes, then soldered. Aquarium

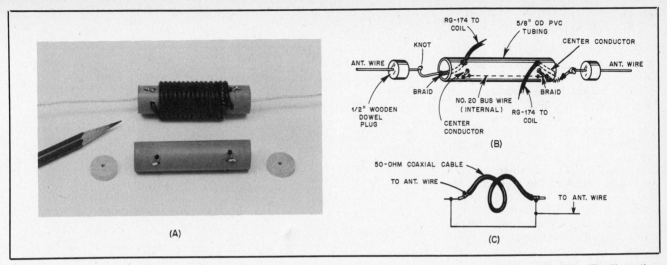

Fig. 13 — A shows completed 20-meter coaxial trap with miniature RG-174 coaxial cable. At B is a breakaway view of a coaxial trap. The illustration at C shows the electrical connections for a coaxial trap.

cement was applied to the sides of the wooden plugs before inserting them into the tubing. A layer of vinyl electrical tape can be wound over the coaxial coil if desired, although this should not be necessary. If weather protection is desired, a coating of exterior polyurethane varnish can be applied to the completed close-wound coil. This will keep the turns affixed in the desired position after final adjustment. Tune-up is carried out in the same manner as prescribed for the toroidal traps, using a dip meter or the test fixture described in Fig. 12.

The length of the coaxial cable used will have to be determined experimentally. The 20-meter coaxial trap contains 15 close-wound turns of RG-174 cable (36 inches, 89 pF) to provide resonance at 14.100 MHz. Final adjustment was done by moving the three outer turns at one end until the desired frequency was noted. The coil form for the 20-meter trap is 2½ inches long. The wooden end plugs are 3/8 inch thick. The inside of this trap is not filled with sealant, but it could be if desired. Avoiding the use of filler will make the traps lighter, thereby permitting the use of lighter-gauge wire for the antenna sections.

Trap Performance

Both trap styles were subjected to RF power tests to determine whether they could handle the output of a typical 150-W class transceiver. A Bird wattmeter was connected between the trap and the transmitter. A 50-ohm dummy load was attached to the opposite end of the trap. Next, 40- and 80-meter RF energy was applied (in separate tests) gradually while the reflected power was observed, which, of course, was not conducive to providing an SWR of 1:1 with the trap in the line. Neither trap showed signs of heating or breakdown at power levels up to 150 W. A key-down period of five minutes was tried during the tests, using a linear amplifier adjusted for 150-W

output. Still no sign of power limitations. The SWR did not change under these conditions. Although the traps were not tested beyond 150 W, it's safe to conclude they could sustain substantially more power without damage. This may not be true of the toroidal trap.

Toward a Lightweight Dipole

Having solved the problem of light-weight, small traps the next step was reducing the bulk of the remainder of the multiband dipole. The cost of materials was an important factor in the selection of wire and end insulators. Strong and light in weight, suitable wire is available from Radio Shack and similar outlets for use as speaker cable. It has a clear plastic outer covering and contains a no. 22 conductor. This roll yields 200 feet of wire because the parallel conductors can be pulled apart easily without harming the outer insulation. In addition to the insulation aiding the strength of wire portions of the antenna, it protects the copper wire from corrosion. This can be especially beneficial in areas where salt water and industrial pollutants affect the atmosphere.

Although RG-58 coaxial cable is less offensive in terms of loss per 100 feet than RG-174, the smaller cable is much more portable. Normally, a 50-foot length of feeder cable is adequate for portable work. Measured loss for 50 feet of RG-174 in decibels is as follows: 3.5 MHz — 1.19; 7.0 MHz — 1.42; 14.0 MHz — 1.67; 21 MHz — 1.93; 29 MHz — 2.0. Therefore, in a worst-case situation (10 meters), a 100-W power input to the cable would result in an antenna feed-point power of 63 W. RG-58, on the other hand, would have a 1-dB loss at 29 MHz, which would mean an antenna feed-point power of 79.4 W. This is not too significant when operating in the 50- to 150-W range, but it can be important when using a QRP rig with only a few watts or milliwatts of output power. The end insulators for the trap dipole discussed in this

article were fashioned from inch-long pieces of 5/8-inch-diameter PVC tubing. Holes were drilled to accommodate the dipole wires and nylon guy wires.

Summary Comments

The overall length of any dipole section in a trap type of antenna will be less than if the dipole were cut for a single band without traps. The exception is the first dipole section after the feed point (out to the first set of traps). The following percentages (approximate) were typical in a coaxial-trap dipole built for use from 40 through 10 meters, compared to the length of a full dipole (100%) for each band: 10 meters — 100%; 15 meters — 92.4%; 20 meters — 88.8%; 40 meters — 83.6%. The shortening becomes more pronounced as the frequency is lowered, owing to the cumulative loading effects of the traps.

These percentages can be applied during initial structuring of the antenna. Starting with the highest band, the dipole sections for each frequency of interest are trimmed or lengthened for the lowest attainable SWR. After the exact dimensions are known, continuous lengths of wire can be used between the traps. This will add strength to the antenna by avoiding breaks in the speaker-wire insulation, if that type of conductor is used. The percentage reductions listed above are not necessarily applicable to antennas that use toroidal or other coil/capacitor traps. The wire diameter and insulation may also affect the final dimensions of the dipole.

For long-term installations, use some type of sealer (spar varnish or polyure-thane) over the wooden end plugs of the traps. All trap holes need to be sealed securely to prevent moisture from building up inside them.

Miniature antenna traps and lightweight trap dipole antennas are practical and inexpensive to build. Try one during your next vacation or business trip.

A Practical Two-Band Vertical

Fig. 14 shows a two-band trap vertical (20/15 meters) which can be collapsed to 39 inches for easy transportation on holidays, DXpeditions or camping trips. All of the tubing sections except B telescope together to make a compact package. The trap and base plate will be separate from the remainder of the antenna during storage or transport.

If portability is not a requirement, a single section of aluminum tubing can be used below the trap, although two sections (telescoping) are recommended to facilitate adjustment of the 15-meter portion of the system. Similarly, two telescoping tubes can be used above the trap (as shown) to permit adjustment for 20-meter operation.

Table 3 contains data on the starting lengths of the tubing sections, plus approximate dimensions for resonance on a variety of band pairings. Final adjustment is done for the lowest SWR attainable in the chosen part of each band (resonance). The adjustment must be done while the antenna is mounted for use with the ground system in place.

Fig. 15 shows the details of a simple trap with the tubing sections keyed to the nomenclature of Fig. 14. The ID of the PVC tubing is too small to accommodate the ½-inch-OD tubing. Therefore, a hacksaw is used to cut four slots at the ends of sections B and C so they will compress and fit into the PVC tubing. The wooden dowel plugs permit a tight bond when the hose clamps are compressed over the ends of the PVC tubing. Innovative builders can find other methods for mounting the trap in the antenna.

Copper straps G are slid into the PVC tubing to provide electrical contact with tubing sections B and C. The straps are bent (as shown) so they will fit under the hose clamps that affix the trap to the tubing. The trap capacitor and coil leads are soldered to the copper straps.

A thin layer of conductive silicon grease between the copper straps and the tubing sections will prevent undue oxidation. Similarly, the grease is applied to the mating surfaces of all of the tubing sections.

If coax cable is used for the trap capacitor, it can be taped (after soldering) to the upper tubing section (B) of Fig. 14. If a fixed-value transmitting capacitor is used, it should be located at the trap coil with stiff wire leads, which are soldered to copper tabs G. If a fixed-value capacitor is used, the trap can be brought to resonance by adjusting the coil turns. The exposed ends of the coax capacitor should be weather sealed.

Ground System

If the antenna is mounted at ground

Table 3

Tubing-Section Lengths for 2-Band Vertical

	Band (MHz)	A	B	C	D	E	F	C1 (pF)[1]	L1 (approx.) (µH)
Tubing length (inches)	21/28	25	16	25	25	25	33	18	1.70
	14/21	38	33	37	37	37	33	25	2.25
	10/14	42	42	54	54	54	49	39	3.25
Tubing length at resonance (approx. inches)[2]	21/28	20	16	21.5	21.5	21.5	33	—	—
	14/21	33	33	33	33	33	33	—	—
	10/14	37	37	49.6	49.6	49.6	49	—	—

[1]See text
[2]Midband dimensions $X_c 1$, $X_L 1$ = 300 Ω

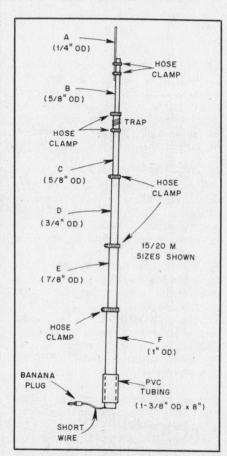

Fig. 14 — Details of the two-band trap vertical, which telescopes to 39 inches when dismantled. Stainless-steel hose clamps are used to hold the tubing sections together and to affix the trap to the tubing. A short length of flexible wire and a banana plug are connected to the base of the antenna for joining the antenna to the coax connector of Fig. 16.

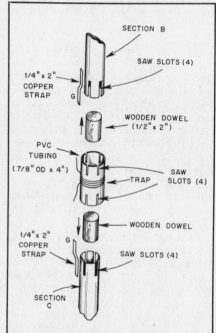

Fig. 15 — Close-up details of the trap construction and how it connects to tubing sections B and C of Fig. 14.

level, section F of Fig. 14 should be as close to ground as possible. At least 20 radials are recommended. They need not be longer than 20 feet, and should be buried from 2 to 4 inches below the surface of the soil or lawn.

If an above-ground installation is planned, use at least four radials per band. The wires are cut to one-quarter wavelength for each of the bands, although some amateurs prefer to make them 5 percent longer than one-quarter wavelength to increase the antenna bandwidth slightly.

The slope of the above-ground radials can be changed to help provide a match to 50-ohm feed line. The greater the angle between the vertical element and the radial wires, the higher the feed impedance. The feed impedance will be approximately 30 ohms when the radials are at right angles to the vertical element. If such an installation is contemplated, a 1.6:1 broadband toroidal transformer can be used at the feed point to effect a matched condition.

Universal Mounting Plate

Fig. 16 illustrates a mounting plate that will satisfy a host of conditions one might encounter when operating from an unfamiliar place. The hole size and spacing will depend on the U-bolts or muffler clamps used with the antenna. The lower set of holes (except the bottom-most two) permit using a supporting mast that is either vertical or horizontal. The holes in the top half of the plate permit the antenna to be mounted vertically or at 45° angle. Hole B is for a female-to-female bulkhead connector. The feed line attaches to one side of the fitting, and the banana plug of Fig.14 plugs into the center hole of the opposite end of the connector. This permits easy disconnection when disassembling the antenna. The radials are bolted to the two holes marked C, at the left and right center of the plate. The two holes (C) at the bottom of the plate are for bolting an iron or aluminum angle stock to the plate. A second angle-stock piece is cut to the same size as the first and is used with the mounting hardware when it is convenient to clamp the mounting plate to a porch railing, window sill and so on. A pair of large C clamps can be used for this mounting technique.

The plate is made from ¼-inch aluminum. Brass or copper material could be used equally well if available.

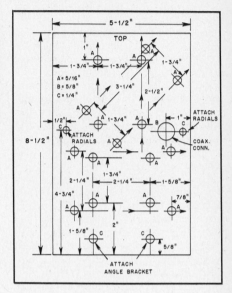

Fig. 16 — Layout details for a universal mounting plate. The hole sizes and spacing will depend on the type of U-bolts used (see text).

Adapting Commercial Trap Verticals to the WARC Bands

The frequency coverage of a multiband vertical antenna can be modified simply by altering the lengths of the tubing sections and/or adding a trap. Several companies manufacture trap verticals covering 40, 20, 15 and 10 meters. Many amateurs roof-mount these antennas because an effective ground radial system isn't practical, to keep the children away from the antenna or to clear metal-frame buildings. On the three highest bands, the tubing and radial lengths are convenient for rooftop installations, but 40 meters sometimes presents problems. Prudence dictates erecting an antenna with the assumption that it will fall down. When the antenna falls, it and the radial system must clear any nearby power lines. Where this consideration rules out 40-meter operation, careful measurement may show that 30-meter dimensions will allow adequate safety. The antenna is resonated by pruning the tubing above the 20-meter trap and installing tuned radials.

Several new frequency combinations are possible. The simpler ones, 12/10, 17/15/10 and 40/30/20/15/10 meters, are shown in Fig. 17 applied to the popular ATV series of trap verticals manufactured by Cushcraft. Operation in the 30-meter band requires an additional trap — use Fig. 15 as a guide for constructing this component.

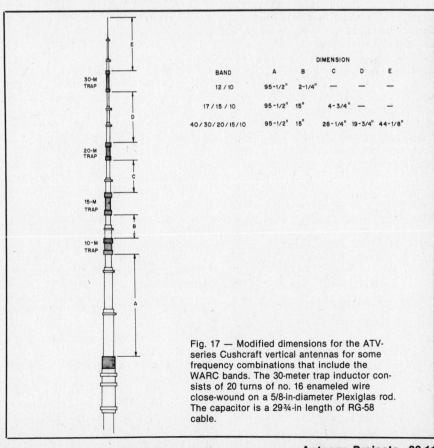

BAND	A	B	C	D	E
12 / 10	95-1/2"	2-1/4"	—	—	—
17 / 15 / 10	95-1/2"	15"	4-3/4"	—	—
40 / 30 / 20 / 15 / 10	95-1/2"	15"	28-1/4"	19-3/4"	44-1/8"

Fig. 17 — Modified dimensions for the ATV-series Cushcraft vertical antennas for some frequency combinations that include the WARC bands. The 30-meter trap inductor consists of 20 turns of no. 16 enameled wire close-wound on a 5/8-in-diameter Plexiglas rod. The capacitor is a 29¾-in length of RG-58 cable.

A Three-Band Quad Antenna System

Quads have been popular with amateurs during the past few decades because of their light weight, their relatively small turning radius, and their unique ability to provide good DX performance when mounted close to the earth. A two-element three-band quad, for instance, with the elements mounted only 35 feet above the ground, will give good performance in situations where a triband Yagi will not. Fig. 18 shows a large quad antenna that can be used as a basis for design of either smaller or larger arrays.

Five sets of element spreaders are used to support the 3-element 20-meter, 4-element 15-meter, and 5-element 10-meter wire-loop system. The spacing between elements has been chosen to provide optimum performance consistent with boom length and mechanical construction. See Fig. 19. Each of the parasitic loops is closed (ends soldered together) and requires no tuning. All of the loop sizes are listed in Table 4 and are designed for a center frequency of 14.1, 21.1 and 28.3 MHz. Since quad antennas are broad-tuning devices, excellent performance is achieved in both CW and SSB segments of each band (with the possible exception of the very high end of 10 meters). Changing the dimensions to favor a frequency 200 kHz higher in each band to create a "phone" antenna is not necessary.

One question that comes up often is whether to mount the loops in a diamond or a square configuration. In other words, should one spreader be horizontal to the earth, or should the wire be horizontal to the ground (spreaders mounted in the fashion of an X)? From the electrical point of view, it is probably a trade-off. While the square configuration has its lowest point higher above ground than a diamond version (which may lower the angle of radiation slightly), the top is also lower than that of a diamond-shaped array. Some authorities indicate that separation of the current points in the diamond system gives

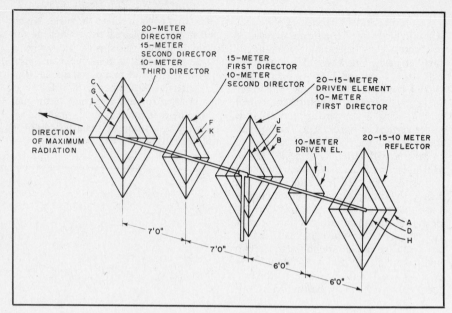

Fig. 19 — Dimensions of the three-band cubical quad. See Table 4 for the dimensions of the lettered wires.

Table 4

Three-Band Quad Loop Dimensions

Band	Reflector	Driven Element	First Director	Second Director	Third Director
20 meters	(A) 72'8"	(B) 71'3"	(C) 69'6"	—	—
15 meters	(D) 48'6½"	(E) 47'7½"	(F) 46'5"	(G) 46'5"	—
10 meters	(H) 36'2½"	(I) 35'6"	(J) 34'7"	(K) 34'7"	(L) 34'7"

Letters indicate loops identified in Fig. 19.

slightly more gain than is possible with a square layout. It should be pointed out, however, that there never has been any substantial proof in favor of one or the other, electrically.

Spreader supports (sometimes called spiders) are available from many different manufacturers. The builder who is keeping the cost at a minimum should consider building his or her own. The expense is about half that of a commercially manufactured equivalent and, according to some authorities, the homemade arm supports described below are less likely to rotate on the boom as a result of wind pressure.

A 3-foot-long section of 1-inch-per-side steel angle stock is used to interconnect the pairs of spreader arms. The steel is drilled at the center to accept a muffler clamp of sufficient size to clamp the assembly to the boom. The fiberglass is attached to the steel angle stock with automotive hose clamps, two per pole. Each quad-loop spreader frame consists of two assemblies of the type shown in Fig. 20.

Fig. 18 — The assembled and installed three-band cubical-quad beam antenna.

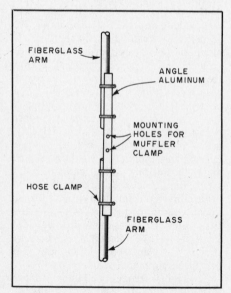

Fig. 20 — Details of one of two assemblies for a spreader frame. The two assemblies are jointed to form an X with a muffler clamp mounted at the position shown.

An Optimum-Gain Two-Band Yagi Array

If optimum performance is desired from a Yagi, the dual four-element array shown in Fig. 21 will be of interest. This antenna consists of four elements on 15 meters interlaced with the same number for 10. Wide spacing is used, providing excellent gain and good bandwidth on both bands. Each driven element is fed separately with 50-ohm coax; gamma-matching systems are employed. If desired, a single feed line can be run to the array and then switched by a remotely-controlled relay.

The element lengths shown in Fig. 21 are for the phone portions of the band,

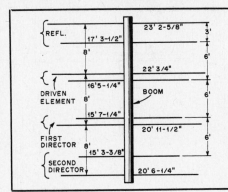

Fig. 21 — The element lengths shown are for the phone sections of the bands. Table 5 provides the dimensions for CW frequencies.

Table 5
Element Lengths for 20, 15 and 10 Meters, Phone and CW

Freq. (kHz)	Driven Element A	B	Reflector A	B	First Director A	B	Second Director A	B
14,050	33' 5-3/8"	33' 8"	35' 2-1/2"	35' 5-1/4'	31' 9-3/8"	31' 11-5/8"	31' 1-1/4"	31' 3-5/8"
14,250	32' 11-3/4"	33' 2-1/4"	34' 8-1/2"	34' 11-1/4"	31' 4"	31' 6-3/8"	30' 8"	30' 10-1/2"
21,050	22' 4"	22' 5-5/8"	23' 6"	23' 7-3/4"	21' 2-1/2"	21' 4"	20' 9-1/8"	20' 10-7/8"
21,300	22' 3/4"	22' 2-3/8"	23' 2-5/8"	23' 4-1/2"	20' 11-1/2"	21' 1"	20' 6-1/4"	20' 7-3/4"
28,050	16' 9"	16' 10-1/4"	17' 7-5/8"	17' 8-7/8"	15' 11"	16'	15' 7"	15' 9-1/2"
28,600	16' 5-1/4"	16' 6-3/8"	17' 3-1/2"	17' 4-3/4"	15' 7-1/4"	15' 8-1/2"	15' 3-3/8"	15' 4-1/2"

```
     A
| 0.2 | 0.2 | 0.2 |
```

```
         B
| 0.15 | 0.15 | 0.15 |
```

These lengths are for 0.2- or 0.15-wavelength element spacing.

centered at 21,300 and 28,600 kHz. If desired, the element lengths can be changed for CW operation, using the dimensions given in Table 5. The spacing of the elements will remain the same for both phone and CW.

The elements are supported by commercially made U-bolt assemblies. Muffler clamps also make excellent element supports. The boom-to-mast support (Fig. 22) is also a manufactured item that is designed to hold a 2-inch-diameter boom and that can be used with mast sizes up to 2½ inches in diameter (Table 6). Another feature of this device is that it permits the beam to be tilted after it is mounted in place on the tower, providing access to the elements if they need to be adjusted once the beam has been mounted on the tower.

Fig. 22 — The boom-to-mast fixture that holds the two 12-foot boom sections together. The unit is made by Hy-Gain Electronics.

Table 6
Materials for Two-Band Yagi

Quantity	Length (ft)	Diameter (in)	Reynolds No.
2	8	1	9A
4	8	3/4	8A
1	8	1¼	10A
1	6	7/8	4231

Two U-bolts, TV antenna to mast type, 1 variable capacitor, 150 pF maximum, any type, 1 plastic freezer container, approx. 5 × 5 × 5 inches, to house gamma capacitor.

Gamma rod, 3/8- to ½-inch diameter aluminum tubing, 36 inches long. (Aluminum curtain rod or similar.)

Simple Antennas for HF Portable Operation

The typical portable HF antenna is a random-length wire flung over a tree and end-fed through a Transmatch. QRP Transmatches can be quite compact, but each additional piece of equipment necessary makes portable operation less attractive. The station can be simplified by using resonant impedance-matched antennas for the bands of interest. Perhaps the simplest antenna of this type is the half-wave dipole, center-fed with 50- or 75-ohm coax. Unfortunately, RG-58, RG-59 or RG-8 cable is heavy and bulky for backpacking, and the miniature cables such as RG-174 are too lossy. A practical solution to the coax problem is to use folded dipoles made from lightweight TV twin lead. The characteristic impedance of this type of dipole is near 300 ohms, but it can be transformed to a 50-ohm source or load by means of a simple matching stub.

Fig. 23 illustrates the construction method and important dimensions for the twin-lead dipole. A silver-mica capacitor is shown for the reactive element, but an open-end stub of twin lead can serve as well, provided it is dressed at right angles to the transmission line for some distance. The stub method has the advantage of easy adjustment of the system resonant frequency.

To preserve the balance of the feeder, a 1:1 balun must be used at the end of the feed line. In most applications the balance is not important, and the twin lead can be connected directly to a coaxial output jack, one lead to the center contact, and one lead to the shell. To preserve its low loss, the twin lead must be kept dry and free from bunching or coiling. Because of its higher

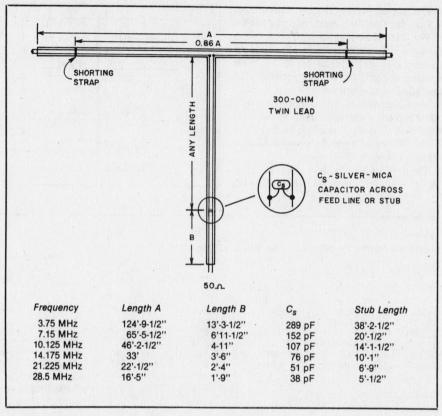

Frequency	Length A	Length B	C_s	Stub Length
3.75 MHz	124'-9-1/2"	13'-3-1/2"	289 pF	38'-2-1/2"
7.15 MHz	65'-5-1/2"	6'-11-1/2"	152 pF	20'-1/2"
10.125 MHz	46'-2-1/2"	4'-11"	107 pF	14'-1-1/2"
14.175 MHz	33'	3'-6"	76 pF	10'-1"
21.225 MHz	22'-1/2"	2'-4"	51 pF	6'-9"
28.5 MHz	16'-5"	1'-9"	38 pF	5'-1/2"

Fig. 23 — A twin-lead folded dipole makes an excellent portable antenna that is easily matched to 50-ohm stations.

impedance, a folded dipole exhibits a wider bandwidth than a single-conductor type. The antennas described here are not as broad as a standard folded dipole because the impedance transformation mechanism is frequency selective. However, the bandwidth should be adequate. An antenna cut for 14.175 MHz, for example, will present an SWR of less than 2:1 over the entire 20-meter band.

HF Mobile Antennas

The antenna is perhaps the most important item in the successful operation of a mobile installation. Mobile antennas, whether designed for single or multiband use, should be securely mounted to the automobile, as far from the engine compartment as possible (for reducing noise pickup), and should be carefully matched to the coaxial feed line connecting them to the transmitter and receiver. All antenna connections should be tight and weatherproof. Mobile loading coils should be protected from dirt, rain and snow if they are to maintain their Q and resonant frequency. The greater the Q of the loading coil, the better the efficiency, but the narrower the bandwidth of the antenna system.

Though bumper-mounted mobile antennas are favored by some, it is better to place the antenna mount on the rear deck of the vehicle, near the rear window. This locates the antenna high and in the clear, assuring less detuning of the system when the antenna moves to and from the car body. *Never use a base-loaded antenna on a bumper mount.*

The choice of base or center loading a mobile antenna has been a matter of controversy for many years. In theory, the center-loaded whip presents a slightly higher base impedance than does the base-loaded antenna. However, with proper impedance-matching techniques employed there is no discernible difference in performance between the two methods. A base-loading coil requires fewer turns of wire than one for center loading, and this is an electrical advantage because of reduced coil losses. A base-loaded antenna is more stable during wind loading and sway. If a homemade antenna system is contemplated, either system will provide good results, but the base-loaded antenna may be preferred for its mechanical advantages (Table 7).

Loading Coils

There are many commercially built antenna systems available for mobile operation, and some manufacturers sell the coils as separate units. Air-wound coils of large wire diameter are excellent for use as loading inductors. Large Miniductor coils can be installed on a solid phenolic rod and used as loading coils. Miniductors, because of their turns spacing, are easy to adjust when resonating the mobile antenna and provide excellent Q. Phenolic-impregnated paper or fabric tubing of large diameter is suitable for making homemade loading coils. It should be coated with liquid fiberglass, inside and out, to make it weatherproof. Brass insert plugs can be installed in each end, their centers drilled and tapped for a standard 3/8 × 24 thread to accommodate the mobile antenna sections. After the coil winding is pruned to

Table 7
Approximate Values for 8-foot Mobile Whip
Base Loading

f(kHz)	Loading L(µH)	RC(Q50) Ohms	RC(Q300) Ohms	RR Ohms	Feed R* Ohms	Matching L(µH)
1800	345	77	13	0.1	23	3
3800	77	37	6.1	0.35	16	1.2
7200	20	18	3	1.35	15	0.6
14,200	4.5	7.7	1.3	5.7	12	0.28
21,250	1.25	3.4	0.5	14.8	16	0.28
29,000	—	—	—	—	36	0.23

Center Loading

1800	700	158	23	0.2	34	3.7
3800	150	72	12	0.8	22	1.4
7200	40	36	6	3.0	19	0.7
14,200	8.6	15	2.5	11.0	19	0.35
21,250	2.5	6.6	1.1	27.0	29	0.29

RC = Loading-coil resistance; RR = radiation resistance.
*Assuming loading coil Q = 300, and including estimated ground-loss resistance.
Suggested coil dimensions for the required loading inductance are shown in a following table.

Table 8
Suggested Loading-Coil Dimensions

Required L (µH)	No. Turns	Wire Size	Dia In	Length In
700	190	22	3	10
345	135	18	3	10
150	100	16	2½	10
77	75	14	2½	10
77	29	12	5	4¼
40	28	16	2½	2
40	34	12	2½	4¼
20	17	16	2½	1¼
20	22	12	2½	2¾
8.6	16	14	2	2
8.6	15	12	2½	3
4.5	10	14	2	1¼
4.5	12	12	2½	4
2.5	8	12	2	2
2.5	8	6	2-3/8	4½
1.25	6	12	1¾	2
1.25	6	6	2-3/8	4½

resonance (Table 8) it should be coated with a high-quality, low-loss compound to hold the turns securely in place and to protect the coil from the weather. Liquid polystyrene is excellent for this. Hobby stores commonly stock this material for use as a protective film for wall plaques and other artwork. Details for making a home-built loading coil are given in Fig. 24.

Impedance Matching

Fig. 25 illustrates the shunt-fed method of obtaining a match between the antenna and the coaxial feed line. For operation on 75 meters with a center-loaded whip, L2 will have approximately 18 turns of no. 14 wire, spaced one wire thickness between turns, and wound on a 1-inch-diameter form. Initially, the tap will be approximately five turns above the ground end of L2. Coil L2 can be inside the car body, at the base of the antenna, or it can be located at the base of the whip, outside the car

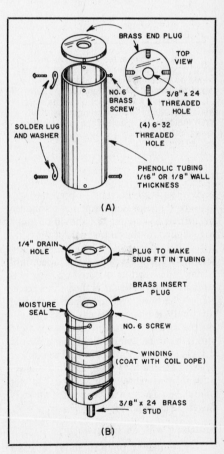

Fig. 24 — Details for making a home-built mobile loading coil. A breakdown view of the assembly is given at A. Brass end plugs are snug-fit into the ends of the phenolic tubing, and each is held in place by four 6-32 brass screws. Center holes in the plugs are drilled and tapped for 3/8-24 thread. The tubing can be any diameter from 1 to 4 inches. The larger diameters are recommended. Illustration B shows the completed coil. Resonance can be obtained by installing the coil, applying transmitter power, then pruning the turns until the lowest SWR is obtained. Pruning the coil for maximum field-strength-meter indication will also serve as a resonance indication.

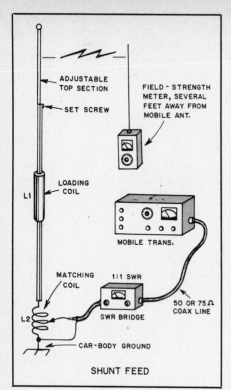

Fig. 25 — A mobile antenna using shunt-feed matching. Overall antenna resonance is determined by the combination of L1 and L2. Antenna resonance is set by pruning the turns of L1, or adjusting the top section of the whip, while observing the field-strength meter or SWR indictor. Then, adjust the tap on L2 for lowest SWR.

body. The latter method is preferred.

Since L2 helps determine the resonance of the overall antenna, L1 should be tuned to resonance in the desired part of the band with L2 in the circuit. The adjustable top section of the whip can be telescoped until a maximum reading is noted on the field-strength meter. The tap is then adjusted on L2 for the lowest reflected-power reading on the SWR bridge. Repeat these two adjustments until no further increase in field strength can be obtained; this point should coincide with the lowest SWR. The number of turns needed for L2 will have to be determined experimentally for 40- and 20-meter operation. There will be proportionately fewer turns required.

Matching with an L Network

Any resonant mobile antenna that has a feed-point impedance less than the characteristic impedance of the transmission line can be matched to the line by means of a simple L network, as shown in Fig. 26. The network is composed of C_M and L_M. The required values of C_M and L_M may be determined from

$$C_M = \frac{\sqrt{R_A(Z_0 - R_A)}}{2\pi f \text{ (kHz)} \, R_A Z_0} \times 10^9 \text{ pF}$$

and

$$L_M = \frac{\sqrt{R_A(Z_0 - R_A)}}{2\pi f \text{ (kHz)}} \times 10^3 \, \mu H$$

where

R_A = the antenna feed-point resistance
Z_0 = the characteristic impedance of the transmission line.

As an example, if the feed-point resistance is 20 ohms and the line is 50-ohm coaxial cable, then at 4000 kHz,

$$C_M = \frac{\sqrt{20 (50 - 20)}}{(6.28)(4000)(20)(50)} \times 10^9$$

$$= \frac{\sqrt{600}}{(6.28)(4)(2)(5)} \times 10^4$$

$$= \frac{24.5}{251.2} \times 10^4 = 975 \text{ pF}$$

$$L_M = \frac{\sqrt{20 (50 - 20)}}{(6.28)(4000)} \times 10^3$$

$$= \frac{\sqrt{600}}{25.12} = \frac{24.5}{25.12} = 0.97 \, \mu H$$

The chart of Fig. 27 shows the capacitive reactance of C_M and the inductive reactance of L_M necessary to match various antenna impedances to 50-ohm coaxial cable. The chart assumes the antenna element has been resonated.

In practice, L_M need not be a separate inductor. Its effect can be duplicated by adding an equivalent amount of inductance to the loading coil, regardless of whether the loading coil is at the base or at the center of the antenna.

Adjustment

In adjusting this system, at least part of C_M should be variable, the balance being made up of combinations of fixed mica capacitors in parallel as needed. A small, one-turn loop should be connected between C_M and the chassis of the car, and the loading coil should then be adjusted for resonance at the desired frequency as indicated by a dip meter coupled to the loop at the base. Then the transmission line should be connected, and a check made with an SWR indicator connected at the transmitter end of the line.

With the line disconnected from the antenna again, C_M should be readjusted and the antenna returned to resonance by readjustment of the loading coil. The line should be connected again, and another check made with the SWR bridge. If the SWR is less than it was on the first trial, C_M should be readjusted in the same direction until the point of minimum SWR is found. Then the coupling between the line and the transmitter can be adjusted for proper loading.

It will be noticed from Fig. 27 that the inductive reactance varies only slightly over the range of antenna resistances likely to be encountered in mobile work. Therefore, most of the necessary adjustment is in the capacitor. The one-turn loop at the base

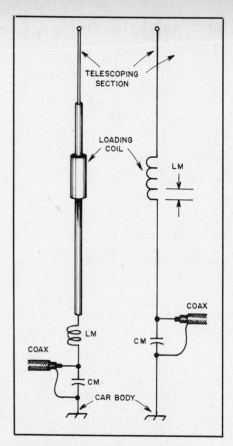

Fig. 26 — A whip antenna may also be matched to coax line by means of an L network. The inductive reactance of the L network can be combined in the loading coil, as indicated at the right.

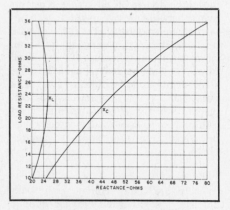

Fig. 27 — Curves showing inductive and capacitive reactances required to match a 50-ohm coax line to a variety of antenna resistances.

should be removed at the conclusion of the adjustment and slight compensation made at the loading coil to maintain resonance.

For manufactured loading coils that have no means for either adjusting or tapping into the inductance, here is a simple L-network matching method that requires no series inductor, and needs only a measurement of SWR to determine the value of the shunt matching capacitor C_M. By extending the whip slightly beyond its natural resonant length, the series induc-

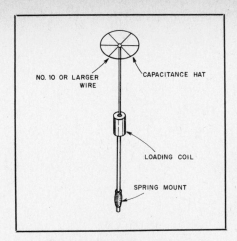

Fig. 28 — A capacitive hat can be used to improve the performance of base- or center-loaded whips. A solid metal disc can be used in place of the skeletal disc shown here.

tance required for the L network appears in the feed-point impedance. At the correct length, the resulting parallel reactance component, X_L, causes the parallel resistance component, R_A, to equal the feed line characteristic impedance, Z_0 ohms. The match is accomplished by cancelling the parallel inductive reactance component, X_L, with shunt capacitor C_M, of equal but opposite reactance.

To perform the matching operation, first resonate the antenna at the desired frequency by adjusting the whip length for minimum SWR without the capacitor. The approximate value of X_L with the whip lengthened to make R_A equal to Z_0, may now be found from

$$X_L = Z_0 \times \sqrt{SWR}/(SWR - 1)$$

where SWR is that obtained at resonance. The reactance, X_C, of the shunt matching capacitor C_M is the negative of X_L. The capacitance of C_M may be determined for the desired frequency, f, in MHz from

$$C_M = 1 \times 10^6/2\pi f X_C \text{ pF}.$$

Form a capacitor, C_M, from a combina-

tion of fixed-mica capacitors in parallel, as needed, and connect the combination across the antenna input terminals as shown in Fig. 26. Finally, increase the whip length in small increments until the minimum SWR is reached — it should be very low.

If a lower SWR is desired, a trimmer capacitor may be added to C_M, and by alternate adjustment of trimmer and whip length, a perfect 1:1 match can be obtained. Once C_M has been established for a given band, the antenna can then be matched at other frequencies in the band by simply adjusting the whip length for minimum SWR. If an inductive rather than a capacitive shunt element is preferred, replace the capacitor with an inductor having the same absolute value of reactance, and shorten the whip instead of lengthening it from the natural resonant length.

Top-Loading Capacitance

Because the coil resistance varies with the inductance of the loading coil, the resistance can be reduced, beneficially, by reducing the number of turns on the coil. This can be done by adding capacitance to that portion of the mobile antenna that is *above* the loading coil (Fig. 28). To achieve resonance, the inductance of the coil is reduced proportionally. Capacitive hats can consist of a single stiff wire, two wires or more, or a disc made up from several wires like the spokes of a wheel. A solid metal disc can also be used. The larger the capacitive hat, in terms of surface area, the greater the capacitance. The greater the capacitance, the smaller the amount of inductance needed in the loading coil for a given resonant frequency.

There are two schools of thought concerning the attributes of center-loading and base-loading. It has not been established that one system is superior to the other, especially in the lower part of the HF spectrum. For this reason both base and center-loading schemes are popular. Capacitive-hat loading is applicable to either system. Since more inductance is required for

center-loaded whips to make them resonant at a given frequency, capacitive hats should be particularly useful in improving their efficiency.

The capacitance that is increased by the hat is the total antenna-to-ground and vehicle capacitance, which provides the path for the antenna currents to return to the generator or transmitter. To obtain optimum effectiveness, the hat should be placed at the top of the whip, which may require special mechanical considerations to keep the whip upright with wind loading. If the hat is large enough to be effective, only a negligible increase in capacitance will result from extending the whip above the hat. It is important to know that with center-loaded whips, the hat actually becomes more detrimental than helpful if placed directly above the loading coil. This incorrect placement causes an undesirable increase in hat-to-coil capacitance, which adds to the distributed capacitance of the coil, lowering the Q and increasing the coil resistance — just the opposite of the hat's intended purpose. Although often seen in this position, it is usually because of the mistaken notion that increased radiation results from the lower resonant SWR obtained by this incorrect placement. Such a placement does reduce feed-line load mismatch and SWR at resonance, but for the wrong reason. The total antenna-circuit load resistance is increased, bringing it closer to the 50-ohm impedance of the feed line. However, since this increase in load resistance is from the increased coil resistance, and not from an increase in radiation resistance, reducing the SWR in this manner increases the loss instead of the radiation. It is also for this reason that loading coils of the highest Q should be favored above the coils of lower Q used by some manufacturers to enable them to advertise a lower SWR. The rule to follow then, is that, since the coil with the highest Q has the lowest loss resistance, the coil that yields the highest SWR at resonance (without any matching circuitry) will produce the greatest radiation.

The VHF, UHF Yagi

The small size of VHF and, especially, UHF arrays opens up a wide range of construction possibilities. Finding components is somewhat difficult for home constructors of ham gear, but it should not hold back antenna work. Radio and TV distributors have many useful antenna parts and materials. Hardware stores, metal suppliers, lumber yards, welding-supply and plumbing-supply houses and even junkyards should not be overlooked. With a little imagination, the possibilities are endless.

Boom Materials

Wood is very useful in antenna work. It is available in a great variety of shapes and sizes. Rug poles of wood or bamboo make fine booms. Round wood stock (doweling) is found in many hardware stores in sizes suitable for small arrays. Square or rectangular boom and frame materials can be cut to order in most lumber yards if they are not available from the racks in suitable sizes.

There is no RF voltage at the center of a half-wave dipole or parasitic element, so no insulation is required in mounting elements that are centered in the support, whether the latter is wood or metal. Wood is good for the framework of multibay arrays for the higher bands, as it keeps down the amount of metal in the active area of the array.

Wood used for antenna construction should be well-seasoned and free of knots or damage. Available materials vary, depending on local sources. Your lumber dealer can help you better than anyone else in choosing suitable materials. Joining wood members at right angles is often done advantageously with gusset plates. These can be made of thin outdoor-grade plywood or Masonite. Round materials can be handled in ways similar to those used with metal components, with U clamps and with other hardware.

Metal booms have a small "shorting effect" on elements that run through them. With materials sizes commonly employed, this is not more than one percent of the element length, and may not be noticeable in many applications. It is just perceptible with ½-inch tubing booms used on 432 MHz, for example. Formula lengths can be used as given, if the matching is adjusted in the frequency range one expects to use. The center frequency of an all-metal array will tend to be 0.5 to 1 percent higher than a similar system built of wooden supporting members.

Element Materials

Antennas for 50 MHz need not have elements larger than ½-inch diameter, though up to 1 inch is used occasionally. At 144 and 220 MHz the elements are usually 1/8 to 1/4 inch in diameter. For 420

MHz, elements as small as 1/16 inch in diameter work well, if made of stiff rod. Aluminum welding rod, 3/32 to 1/8 inch in diameter is fine for 420-MHz arrays, and 1/8 inch or larger is good for the 220-MHz band. Aluminum rod or hard-drawn wire works well at 144 MHz. Very strong elements can be made with stiff-rod inserts in hollow tubing. If the tubing is slotted and tightened down with a small clamp, the element lengths can be adjusted experimentally with ease.

Sizes recommended above are usable with formula dimensions given in Table 9. Larger diameters broaden frequency response; smaller ones sharpen it. Much smaller diameters than those recommended will require longer elements, especially in 50-MHz arrays.

Element and Boom Dimensions

Tables 9 through 12 list element and boom dimensions for several Yagi configurations for operation on 50, 144, 220 and 432 MHz. These figures are based on information contained in the *National Bureau of Standards Technical Note 688,* which offers element dimensions for maximum-gain Yagi arrays as well as other types of antennas. The original information provides various element and boom diameters. The information shown in the tables represents a highly condensed set of antenna designs, however, making use of standard and readily available material. Element and boom diameters have been chosen so as to produce lightweight, yet very rugged, antennas. Fig. 29 shows various methods of attaching the elements to the boom.

Since these antennas are designed for maximum forward gain, the front-to-back pattern ratios may be a bit lower than those for some other designs. Ratios on the order of 15 to 25 dB are common for these antennas and should be more than adequate for most installations. Additionally, the patterns are quite clean, with the side lobes well suppressed. The driven-element lengths for the antennas represent good starting-point dimensions. The type of feed system used on the array may require longer or shorter lengths, as appropriate. Full details of the various methods for feeding a Yagi array are given in Chapter 17 and in the *ARRL Antenna Book.* Generally speaking, a balanced feed system is preferred in order to prevent pattern skewing and the possibility of unwanted side lobes, which can occur with an unbalanced feed system.

Element spacing for the various arrays is presented in Fig. 30 in terms of the wavelength of the boom, as noted in the first column of Tables 9 through 12. The 0.4, 0.8, 2.2 and 3.2-wavelength boom antennas have equally spaced elements for both reflector and directors; 1.2- and 4.2-wavelength boom antennas have dif-

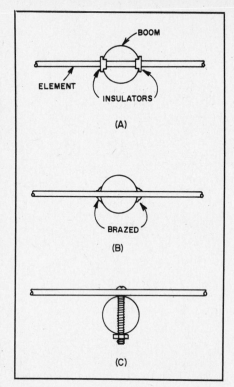

Fig. 29 — Various techniques for mounting Yagi elements.

ferent reflector and director spacings. As all of the antenna parameters are interrelated, changes in element diameter, boom diameter and element spacing will require that a new design be worked out. As the information presented in the *NBS Technical Note 688* is straightforward, the serious antenna experimenter should have no difficulty designing antennas with different dimensions from those presented in the tables. For antennas with the same element and boom diameters, but for different frequencies within the band, standard scaling techniques may be applied.

Stacking Yagis

Where suitable provision can be made for supporting them, two Yagis mounted one above the other and fed in phase may be preferable to one long Yagi having the same theoretical or measured gain. The pair will require a much smaller turning space for the same gain, and their lower radiation angle can provide interesting results. On long ionospheric paths a stacked pair occasionally may show an *apparent* gain much greater than the 2 to 3 dB that can be measured locally as the gain due to stacking.

Optimum spacing for Yagis of five elements or more is one wavelength, but this may be too much for many builders of 50-MHz antennas to handle. Worthwhile results can be obtained with as little as one half-wavelength (10 feet) and 5/8 wavelength (12 feet). The difference be-

Table 9
NBS 50.1-MHz Yagi Dimensions

Boom Length	Boom Diameter	Element Diameter	Insulated Elements	Ref.	Driven	Dir. 1	Dir. 2	Dir. 3	Dir. 4	Dir. 5	Dir. 6	Dir. 7	Dir. 8	Dir. 9	Dir. 10
7'10"(0.4 λ)	1-1/4"	1/2"	YES	9'7"	9'1-3/4"	9'5/8"									
15'8-1/2"(0.8 λ)	2"	3/4"	YES	9'7-3/4"	9'1-3/4"	9'1-3/8"	8'8-3/8"	8'9-1/8"							
			NO	9'6-1/2"	9'1-3/4"	8'9-1/8"	8'9-5/8"	8'10-1/4"	8'9-1/8"						
23'6-7/8"(1.2 λ)	2"	3/4"	YES	9'7-3/4"	9'1-3/4"	8'10-1/4"	8'8-7/8"	8'7-3/4"	8'7-3/4"						
			NO	9'6-1/2"	9'1-3/4"	8'9-1/8"	8'7"	8'8-7/8"	8'10-1/4"						
39'3-3/8"(2.2 λ)	2"	3/4"	YES	9'6-1/2"	9'1-3/4"	8'9-7/8"	8'8-7/8"	8'5-3/8"	8'3-1/2"	8'1-3/4"	8'1-3/4"	8'1-3/4"	8'1-3/4"	8'3-1/2"	8'5-3/8"
			NO	9'7-3/4"	9'1-3/4"	8'11"	8'8-1/8"	8'6-1/2"	8'4-5/8"	8'3"	8'3"	8'3"	8'3"	8'4-5/8"	8'6-1/2"

Table 10
NBS 144.1-MHz Yagi Dimensions

Boom Length	Boom Diameter	Element Diameter	Insulated Elements	Ref.	Driven	Dir. 1	Dir. 2	Dir. 3	Dir. 4	Dir. 5	Dir. 6	Dir. 7	Dir. 8	Dir. 9	Dir. 10	Dir. 11	Dir. 12	Dir. 13	Dir. 14	Dir. 15
5'5-9/16"(0.8 λ)	1"	3/16"	YES	3'4"	3'2-3/16"	3'7/8"	3'11/16"	3'7/8"												
			NO	3'4-5/8"		3'1-1/2"	3'1-3/8"	3'1-1/2"												
8'2-5/16"(1.2 λ)	1"	3/16"	YES	3'4"		3'7/8"	3'7/16"	3'7/16"	3'7/8"											
			NO	3'4-5/8"		3'1-1/2"	3'1-1/8"	3'1-1/8"	3'1-1/2"											
15'1/4"(2.2 λ)	1-1/4"	3/16"	YES	3'4"		3'1-1/2"	3'1-1/8"	2'11-13/16"	2'11-1/4"	2'10-9/16"	2'10-9/16"	2'10-9/16"	2'10-9/16"	2'11-1/4"	2'11-13/16"					
			NO	3'4-13/16"		3'1-15/16"	3'1-1/8"	3'5/8"	3'	2'11-3/8"	2'11-3/8"	2'11-3/8"	2'11-3/8"	3'	3'5/8"					
21'10-1/16"(3.2 λ)	1-1/2"	3/16"	YES	3'4"		3'7/8"	3'9/16"	3'5/8"	3'	2'10-7/8"	2'10-5/16"	2'10-5/16"	2'10-5/16"	2'10-5/16"	2'10-5/16"	2'10-5/16"	2'10-5/16"	2'10-5/16"	2'10-5/16"	2'10-5/16"
			NO	3'5-1/16"		3'7/8"	3'1-3/8"	3'13/16"	3'3/16"	2'11-1/4"	2'11-1/8"	2'11-3/8"	2'11-3/8"	2'11-3/8"	2'11-3/8"	2'11-3/8"	2'11-3/8"	2'11-3/8"	2'11-3/8"
28'8-1/8"(4.2 λ)	1-1/2"	3/16"	YES	3'3-3/8"		3'9/16"	3'9/16"	3'3/8"	2'11-5/8"	2'11-1/2"	2'10-13/16"	2'10-13/16"	2'10-9/16"	2'10-9/16"	2'10-9/16"	2'10-9/16"	2'10-9/16"	2'10-5/16"	2'10-5/16"
			NO	3'4-1/2"		3'1-5/8"	3'1-5/8"	3'1-7/16"	3'11/16"	3'9/16"	3'3/16"	2'11-7/8"	2'11-5/8"	2'11-5/8"	2'11-5/8"	2'11-5/8"	2'11-5/8"	2'11-3/8"	2'11-3/8"

Table 11
NBS 220.1-MHz Yagi Dimensions

Boom Length	Boom Diameter	Element Diameter	Insulated Elements	Ref.	Driven	Dir. 1	Dir. 2	Dir. 3	Dir. 4	Dir. 5	Dir. 6	Dir. 7	Dir. 8	Dir. 9	Dir. 10	Dir. 11	Dir. 12	Dir. 13	Dir. 14	Dir. 15
3'6-15/16"(0.8 λ)	1"	3/16"	YES	2'2-1/16"	2'1"	1'11-13/16"	1'11-11/16"	1'11-13/16"												
			NO	2'2-3/4"		2'1/2"	2'3/8"	2'1/2"												
5'4-3/8"(1.2 λ)	1"	3/16"	YES	2'2-1/16"		1'11-13/16"	1'11-9/16"	1'11-9/16"	1'11-3/16"											
			NO	2'2-3/4"		2'1/2"	2'1/2"	2'1/4"	2'1/2"											
9'10"(2.2 λ)	1"	3/16"	YES	2'2-1/16"		2'1/2"	2'1/4"	2'1/4"		1'10-1/2"	1'10-1/8"	1'10-1/8"	1'10-1/8"	1'10-1/2"	1'10-15/16"					
			NO	2'2-3/4"		2'3/4"	2'1/16"	1'10-15/16"	1'11-3/16"	1'10-1/8"	1'10-1/8"	1'10-7/8"	1'10-7/8"	1'11-1/4"	1'11-5/8"					
14'3-11/16"(3.2 λ)	1-1/4"	3/16"	YES	2'2-1/16"		1'11-13/16"	1'11-9/16"	1'11-7/8"	1'11-1/4"	1'10-7/8"	1'10-7/8"	1'9-7/8"	1'9-7/8"	1'10-1/2"	1'10-7/8"	1'9-7/8"	1'9-7/8"	1'9-7/8"	1'9-7/8"	1'9-7/8"
			NO	2'3"		2'3/4"	2'7/16"	1'11-7/8"	1'11-7/16"	1'11-1/4"	1'10-1/2"	1'10-13/16"	1'10-13/16"	1'10-13/16"	1'10-13/16"	1'10-13/16"	1'10-13/16"	1'10-13/16"	1'10-13/16"
18'9-5/16"(4.2 λ)	1-1/2"	3/16"	YES	2'1-11/16"		1'11-5/8"	1'11-5/8"	1'11-7/16"	1'10-7/8"	1'10-3/4"	1'10-1/2"	1'10-5/16"	1'10-1/8"	1'9-7/8"	1'10-1/8"	1'10-1/8"	1'10-1/8"	1'10-1/8"	1'10-1/8"	1'9-7/8"
			NO	2'2-3/4"		2'11/16"	2'11/16"	2'1/2"	2'	1'11-13/16"	1'11-9/16"	1'11-3/8"	1'11-3/16"	1'11-3/16"	1'11-3/16"	1'11-3/16"	1'11-3/16"	1'11-3/16"	1'10-13/16"

Table 12
NBS 432.1-MHz Yagi Dimensions

Boom Length	Boom Diameter	Element Diameter	Insulated Elements	Ref.	Driven	Dir. 1	Dir. 2	Dir. 3	Dir. 4	Dir. 5	Dir. 6	Dir. 7	Dir. 8	Dir. 9	Dir. 10	Dir. 11	Dir. 12	Dir. 13	Dir. 14	Dir. 15
2'8-13/16"(1.2 λ)	1"	3/16"	YES	1'1-3/16"	1'23/32"	11-13/16"	11-5/8"	11-5/8"	11-13/16"											
			NO	1'1-15/16"		1'17/32"	1'11/32"	1'11/32"	1'17/32"											
5'1/8"(2.2 λ)	1"	3/16"	YES	1'1-3/16"		11-29/32"	11-7/16"	11-1/4"	11"	10-13/16"	10-13/16"	10-11/16"	10-13/16"	11"	11-1/4"					
			NO	1'1-15/16"		1'21/32"	13/16"	1"	11-3/4"	11-17/32"	11-17/32"	11-17/32"	11-17/32"	11-3/4"	1'					
7'3-15/32"(3.2 λ)	1"	3/16"	YES	1'1-3/16"		11-27/32"	11-5/8"	11-1/4"	11"	10-29/32"	10-13/16"	10-11/16"	10-11/16"	10-11/16"	10-11/16"	10-11/16"	10-11/16"	10-11/16"	10-11/16"	10-11/16"
			NO	1'1-15/16"		1'9/16"	1'11/32"	1"	11-3/4"	11-5/8"	11-17/32"	11-13/32"	11-13/32"	11-13/32"	11-13/32"	11-13/32"	11-13/32"	11-13/32"	11-13/32"
9'6-25/32"(4.2 λ)	1"	3/16"	YES	1'1"		11-22/32"	11-22/32"	11-19/32"	11"	11-5/32"	11"	10-29/32"	10-13/16"	10-13/16"	10-13/16"	10-13/16"	10-13/16"	10-13/16"	10-13/16"	10-13/16"
			NO	1'1-3/4"		1'7/16"	1'7/16"	1'11/32"	1'	11-7/8"	11-3/4"	11-5/8"	11-17/32"	11-17/32"	11-17/32"	11-17/32"	11-17/32"	11-17/32"	11-13/32"

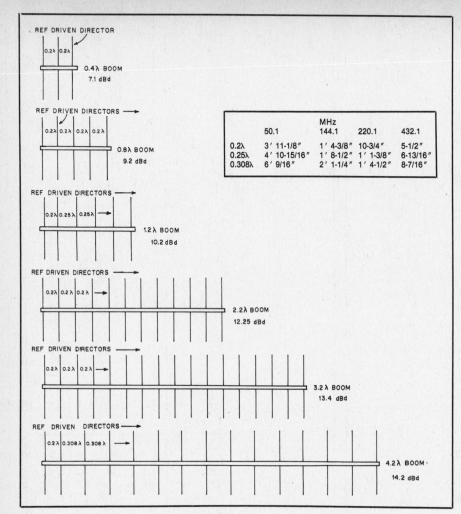

		MHz		
	50.1	144.1	220.1	432.1
0.2λ	3′ 11-1/8″	1′ 4-3/8″	10-3/4″	5-1/2″
0.25λ	4′ 10-15/16″	1′ 8-1/2″	1′ 1-3/8″	6-13/16″
0.308λ	6′ 9/16″	2′ 1-1/4″	1′ 4-1/2″	8-7/16″

Fig. 30 — Element spacing for the various arrays, in terms of boom wavelength.

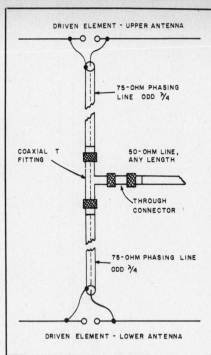

Fig. 31 — A method for feeding a stacked Yagi array.

tween 12 and 20 feet may not be worth the added structural problems involved in the wider spacing, at 50 MHz, at least. The closer spacings give lower measured gain, but the antenna patterns are cleaner than will be obtained with one-wavelength spacing. The extra gain realized with wider spacings is usually the objective on 144 MHz and higher bands, where the structural problems are not severe.

One method for feeding two 50-ohm antennas, as might be used in a stacked Yagi array, is shown in Fig. 31. The transmission lines from each antenna to the common feed point must be equal in length and an odd multiple of a quarter wavelength. This line acts as an impedance transformer and raises the feed impedance of each antenna to 100 ohms. When the two antennas are connected in parallel at the coaxial "T" fitting, the resulting impedance is close to 50 ohms.

A 5-Element Yagi for 50 MHz

The antenna described here (Figs. 32 through 36) was designed from information contained in Table 9 and Fig. 30. This antenna has a theoretical gain of 11.3 dB over an isotropic antenna (dBi) and should exhibit a front-to-back ratio of roughly 18 dB. The pattern is quite clean, with side lobes well suppressed. A hairpin matching system is used, and if the dimensions are followed closely no adjustment should be necessary. The completed antenna is rugged, yet lightweight, and should be easy to install on any tower or mast.

Mechanical Details

Construction details of the antenna are given in Figs. 33 and 34. The boom of the antenna is 17 feet long and is made from a single piece of 2-inch aluminum irrigation tubing that has a wall thickness of 0.047 inch. Irrigation tubing is normally supplied in 20 foot sections, so several feet may be removed from the length.

Elements are constructed from ¾-inch-OD aluminum tubing of the 6061-T6 variety, with a wall thickness of 0.058 inch. Each element, with the exception of the driven element, is made from a single length of tubing. The driven element is split in the center and insulated from the boom to provide a balanced feed system. The reflector and directors have short lengths of 7/8-inch aluminum tubing telescoped over the center of the elements for reinforcement purposes. Boom-to-element clamps (Fig. 36) were fashioned from 3/16-inch-thick aluminum-plate stock, as shown in the photographs.

Two muffler clamps hold each plate to the boom, and two U-bolts affix each element to the plate. Exact dimensions of the plate are not critical, but should be great enough to accommodate the two muffler clamps and two U-bolts. The element and clamp structure may seem to be a bit over-engineered. However, the antenna is

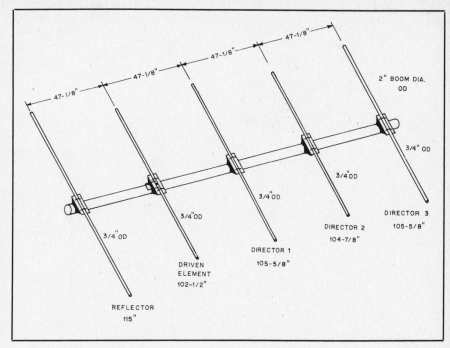

Fig. 33 — Dimensional drawing of the 5-element, 50-MHz Yagi antenna described in the text.

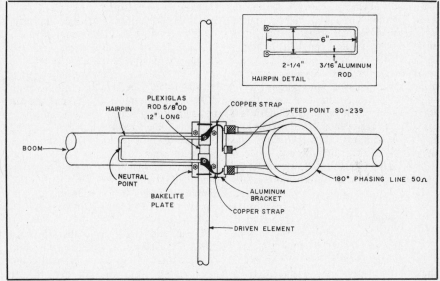

Fig. 34 — Detailed drawing of the feed system used with the 50-MHz Yagi. Phasing-line lengths are: for cable with 0.80 velocity factor — 7 ft 10-3/8 in; for cable with 0.66 velocity factor — 6 ft 5-3/4 in.

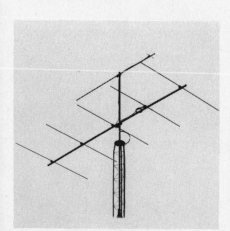

Fig. 32 — Photograph of the 5-element, 50-MHz Yagi. A 432-MHz beam is mounted above.

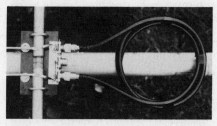

Fig. 35 — Closeup of the driven element and feed system. The phasing line is coiled and taped to the boom.

designed to withstand the severe weather conditions common to New England. This antenna has withstood many wind storms and several ice storms, and no maintenance has been required.

The driven element is mounted to the boom on a Bakelite plate of similar dimension to the reflector and director element-to-boom plates. A piece of 5/8-inch Plexiglas rod, 12 inches in length, is inserted into each half of the driven element. The Plexiglas piece allows the use of a single clamp

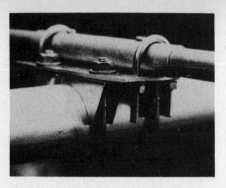

Fig. 36 — Photograph of the element-to-boom clamp. U-bolts are used to hold the element to the plate, and 2-inch plated muffler clamps hold the plates to the boom.

on each side of the element and also seals the center of the elements against moisture. Self-tapping screws are used for connection to the driven element. A length of ¼-inch polypropylene rope is inserted into each element, and end caps are placed on the elements. The rope damps element vibrations that could lead to element or hardware fatigue.

Feed System

Details of the feed systems are shown in Figs. 34 and 35. A bracket fashioned from a piece of scrap aluminum is used to mount the three SO-239 connectors to the driven

element plate. A half-wavelength phasing line connects the two element halves, providing the necessary 180-degree phase difference between them. The "hairpin" is connected directly across the element halves. It should be noted that the exact center of the hairpin is electrically neutral and may be fastened to the boom or allowed to hang. Phasing-line lengths are: for cable with 0.80 velocity factor — 7′ 10-3/8″; for cable with 0.66 velocity factor — 6′ 5¾″. It will be noted that the driven element is the shortest element in this array. While this may seem a bit unusual, it is necessary with the hairpin matching system.

A Portable 144-MHz Antenna for Emergency Use

Much local emergency work is done on VHF FM these days. VHF FM equipment lends itself to portable operation, and many amateurs have 2-meter capability at home, in their cars or portable with hand-held transceivers. But how does an emergency-powered station located in a remote shelter filled with evacuees communicate with the emergency operations center (EOC) when the repeater has been knocked off the air? Effective simplex communications are needed when that high-altitude repeater antenna is not available. One way to get the needed ERP for simplex communications is the addition of a high-power amplifier. A power amplifier, however, requires a healthy chunk of battery power, and in an emergency situation battery power may be at a premium.

In an emergency situation with many shelters in operation, some as many as 25 miles away, communications are usually controlled from a central EOC. The shelters need not hear each other, but they all must be able to communicate reliably with that EOC. Often, these communications are conducted with the help of a repeater; however, in an emergency, the repeater may not always be operational. At times, the stations in remote shelters may have to maintain simplex contact with the EOC for several days using very low transmitter power to conserve whatever battery power is available. Some type of gain antenna that can be set up in a hurry solves the problem of getting enough ERP to maintain communications while not drawing excessive battery power. Dick Jansson, WD4FAB, describes here the portable Yagi

that the Orange County, Florida, ARES group developed for 2-meter communications with its remote emergency shelters.

The antenna illustrated in Figs. 37 through 39 fits in a package no larger than 4 × 8 × 48 inches (the original shipping cartons taped together) and can be set up in less than five minutes. The antenna can be as high as 10 feet off the ground. Fig. 37 shows the erected antenna (not extended to its full 10-foot height). Fig. 38 shows the pieces as they were shaken out of the boxes. This antenna system was made from commercially manufactured items with a few simple additions to speed assembly. It is stored partially assembled and requires no tools for setup.

The antenna itself is a Cushcraft model A147-4 four-element Yagi. The screws and brackets that hold the elements to the boom are premounted to the elements with the original hardware supplied with the antenna. A visit to the local hardware store yielded wing nuts that allow speedy assembly of the prepared brackets and elements to the boom. To reduce setup time to a minimum, each element is color coded with spray paint, and the boom is coded accordingly. No instructions are necessary. The operator just shakes the parts out of the box, matches up the colors, and

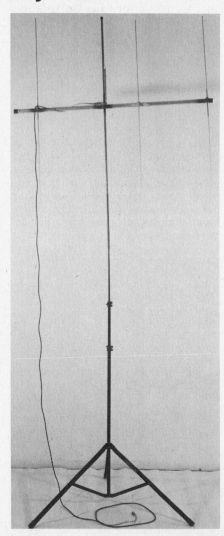

Fig. 37 — The complete emergency antenna takes less than five minutes to set up, yet provides gain equal to a power amplifier without draining battery power. The driven element doubles as a boom support, and the tripod mount is a standard photographic-studio light stand.

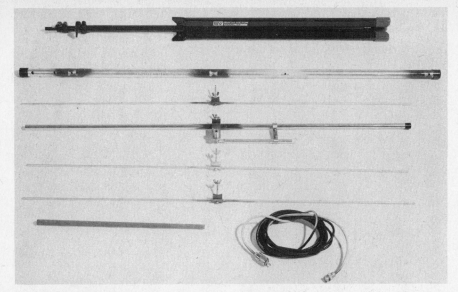

Fig. 38 — The antenna breaks down into several compact pieces for transportation. Element hardware is preassembled and color coded for speedy assembly.

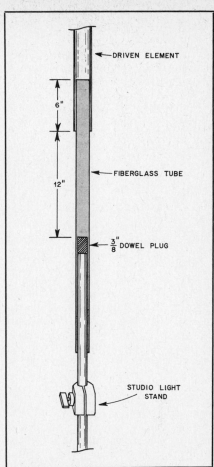

Fig. 39 — Details of the driven-element-to-support mount.

tightens down the wing nuts by hand. Assembly is as simple as possible.

The antenna driven element is made from tubing, rather than rod. This is a distinct advantage, because the driven element can double as an extension of the light stand, providing support for the boom. This arrangement further reduces weight and parts count. A section of tapered tubular-fiberglass fishing-rod stock that was purchased from a local sporting goods shop insulates the driven element tip from the support mast. The rod was selected so that one end slipped snugly into the driven element, and the other slipped over the support mast. See Fig. 39. A small piece of 3/8-inch-OD wooden dowel was jammed about 6 inches inside the large end of the fiberglass tubing to keep the tubing from sliding too far over the support stand tip.

Use a fiberglass section that is at least 12 inches long to keep the metal mast out of the field of the beam. This practice will help you to avoid unwanted coupling between the metal parts (which would detune the antenna).

The antenna support is a photographic-studio light stand that folds into a compact package for transportation. The support shown here is a Smith-Victor model S3. This model and other similar stands are available at many camera stores. It is used without modification.

Be sure your antenna kit has a 15-foot length of RG-58 coaxial cable with the necessary connectors for your equipment. The setup illustrated here uses a PL-259 connector for the antenna end, while the transceiver end is equipped with a UG-88 BNC connector.

This antenna adds significant range to a hand-held transceiver that normally uses a "rubber-duck" antenna, and you'll be surprised at the increase in your reliable communications range.

A 15-Element Yagi for 432 MHz

This 432-MHz Yagi antenna was designed using the information shown in Table 12 and Fig. 30. The theoretical gain for this antenna is 16.3 dBi, with a front-to-back pattern ratio of approximately 22 dB. The pattern is very sharp and quite clean, as would be expected from a well-tuned array of this size. Four of these antennas in a "box" or "H" array would serve well for terrestrial work, while eight would make a respectable EME system. The completed antenna is shown in Fig. 40.

Mechanical Details

Dimensions for the antenna are given in Figs. 41 and 42. The boom of the antenna is made from a length of 1-inch aluminum tubing. Each of the elements is mounted through the boom, and only the driven element is insulated. Auveco 8715 external retaining rings secure each of the parasitic elements in place. These rings are available at most hardware supply houses. Consult your local telephone company Yellow Pages for a hardware dealer near you.

The driven element is insulated from the boom by a length of ½-inch Teflon rod. The length of rod is drilled to accept the ¼-inch-thick aluminum tubing driven element. A press fit was used to secure the Teflon piece in the boom of the antenna. An exact fit can be achieved by drilling the hole slightly undersized and enlarging the hole with a hand reamer, a small amount at a time. Should the hole turn out to be oversized, a small amount of RTV (silicone seal) can be used to secure the Teflon in place. Although it wasn't tried with this antenna, it should be possible to use a driven element that is not insulated from the boom. Small changes in the position of the matching rods and/or clamps might be necessary.

Fig. 40 — The completed 15-element Yagi for 432 MHz, ready for installation atop the tower.

Details of the feed system are shown in Fig. 42. This is a form of the "T" match, where the driven element is shortened from its resonant length to provide the necessary capacitance to tune out the reactance of the matching rods. With this system no variable capacitors are required, as in the more conventional T-match systems used at HF. This is a definite plus in terms of

antenna endurance in harsh-weather environments.

The center pin of the UG-58A N-type connector attaches to one of the matching rods. A half wavelength of 50-ohm, foam-dielectric cable is used to provide the 180-degree phase shift from one half of the element to the other. An alternative to the large and cumbersome cable used here

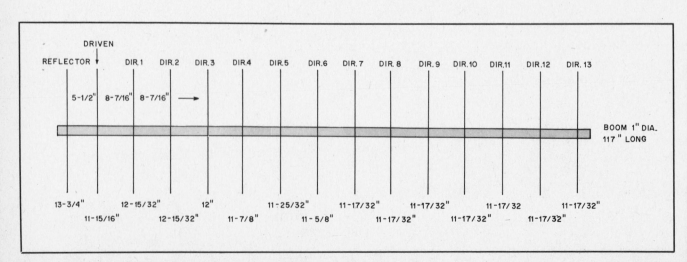

Fig. 41 — Dimensional drawing of the 15-element Yagi for 432 MHz. The balance point of the antenna is between the fifth and sixth director. The antenna was designed from the information presented in Table 12 and Fig. 30 of this chapter.

would be the miniature copper Hardline with Teflon dielectric material, such as RG-401.

Each of the matching rods is secured to two threaded steatite standoffs at the center of the antenna. These standoffs provide tie points for the ends of the phasing lines, the center pin of the coaxial connector as well as the ends of the matching rods. Solder lugs are used for each of the connections for easy assembly or disassembly. The clamps that connect the matching rods to the driven element are constructed from pieces of aluminum measuring ¼ × ½ × 1-5/16 inches. These pieces are drilled and slotted so when the screws are tightened the pieces will compress slightly to provide a snug fit. Alternatively, simple clamps could be fashioned from strips of aluminum.

Adjustment

If the dimensions given in the drawings are followed closely, little adjustment should be necessary. If adjustment is necessary, as indicated by an SWR greater than 1.5 to 1, move the clamps a short distance along the driven element. Keep in mind that the clamps should be located equidistant from the center of the boom.

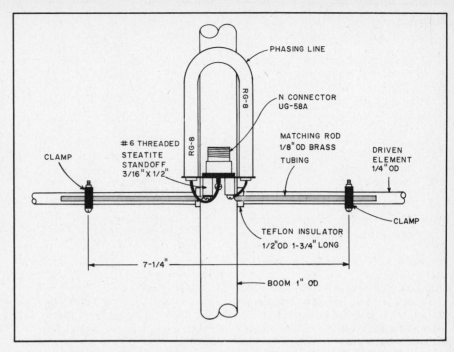

Fig. 42 — Detailed drawing of the feed system used with the Yagi. A small copper plate is attached to the coaxial-connector plate assembly for connection of the phasing line braid. As indicated, the braid is soldered to the plate. A coat of clear lacquer or enamel is recommended for waterproofing the feed system.

The VHF Quagi

First described by K6YNB in April 1977 *QST*, the quagi has become a very popular antenna for use on 144 MHz and above. The long-boom quagi was presented by K6YNB/N6NB in February 1978 *QST*.

How to Build a Quagi

There are a few tricks to quagi building The designer has mass produced as many as 16 in one day. Table 13 gives the dimensions for an 8-element quagi, while Table 14 gives the dimensions for a long-boom, 15 element, 432-MHz version.

The boom is *wood* or any other nonconductor (e.g., fiberglass). If a metal boom is used, a new design and new element lengths will be required. Many VHF antenna builders go wrong by failing to follow this rule: If the original uses a metal boom, use the same size and shape metal boom when you duplicate it. If it calls for a wooden boom, use a nonconductor. Many amateurs dislike wood booms, but in a salt-air environment they outlast aluminum

Table 13
Dimensions, Eight-Element Quagi

Element Lengths	144.5 MHz	147 MHz	222 MHz	432 MHz	446 MHz
Reflector (all no. 12 TW wire, closed)	86-5/8" (loop)	85"	56-3/8"	28"	27-1/8"
Driven element (no. 12 TW, fed at bottom)	82" (loop)	80"	53.5"	26-5/8"	25-7/8"
Directors	35-15/16" to 35" in 3/16" steps	35-5/16" to 34-3/8" in 3/16" steps	23-3/8" to 22-3/4" in 1/8" steps	11-3/4" to 11-7/16" in 1/16" steps	11-3/8"to 11-1/16 in 1/16" steps
Spacing					
R-DE	21"	20-1/2"	13-5/8"	7"	6.8"
DE-D1	15-3/4"	15-3/8"	10-1/4"	5-1/4"	5.1"
D1-D2	33"	32-1/2"	21-1/2"	11"	10.7"
D2-D3	17-1/2"	17-1/8"	11-3/8"	5.85"	5.68"
D3-D4	26.1"	25-5/8"	17"	8.73"	8.46"
D4-D5	26.1"	25-5/8"	17"	8.73"	8.46"
D5-D6	26.1"	25-5/8"	17"	8.73"	8.46"
Stacking Distance Between Bays					
	11'	10'10"	7'1-1/2"	3'7"	3'5-5/8"

Table 14

432-MHz, 15-Element, Long-Boom Quagi Construction Data

Element Lengths — Inches		Interelement Spacing — Inches	
R — 28″ loop	D7 — 11-3/8	R-DE — 7	D6-D7 — 12
DE — 26-5/8″ loop	D8 — 11-5/16	DE-D1 — 5-1/4	D7-D8 — 12
D1 — 11-3/4	D9 — 11-5/16	D1-D2 — 11	D8-D9 — 11-1/4
D2 — 11-11/16	D10 — 11-1/4	D2-D3 — 5-7/8	D9-D10 — 11-1/2
D3 — 11-5/8	D11 — 11-3/16	D3-D4 — 8-3/4	D10-D11 — 9-3/16
D4 — 11-9/16	D12 — 11-1/8	D4-D5 — 8-3/4	D11-D12 — 12-3/8
D5 — 11-1/2	D13 — 11-1/16	D5-D6 — 8-3/4	D12-D13 — 13-3/4
D6 — 11-7/16			

Boom — 1 × 2-inch × 12-ft Douglas fir, tapered to 5/8 inch at both ends.
Driven element — No. 12 TW copper-wire loop in square configuration, fed at center bottom with type N connector and 52-ohm coax.
Reflector — No. 12 TW copper-wire loop, closed at bottom.
Directors — 1/8-inch rod passing through boom.

Fig. 43 — A close-up view of the feed method used on a 432-MHz quagi. This arrangement produces an excellent SWR and an actual measured gain in excess of 13 dB over an isotropic antenna with a 4-foot 10-inch boom! The same basic arrangement is used on lower frequencies, but wood may be substituted for the Plexiglas spreaders. The boom is ½-inch exterior plywood.

(and surely cost less). Varnish the boom for added protection.

The 2-meter version is usually built on a 14-foot 1 × 3 inch boom, with the boom cut down to taper it to 1 inch at both ends. Clear pine is best because of its light weight, but construction-grade Douglas fir works well. At 220 MHz, the boom is under 10 feet long and most builders use 1 × 2 or (preferably) ¾ by 1¼-inch pine molding stock. On 432 MHz the boom must be ½-inch thick or less. Most builders use strips of ½-inch exterior plywood for 432.

The quad elements are supported at the current maxima (the top and bottom, the latter beside the feed point) with Plexiglas or small strips of wood. The quad elements are made of no. 12 copper wire, commonly used in house wiring. Some builders may elect to use no. 10 wire on 144 MHz and no. 14 wire on 432 MHz, although this will change the resonant frequency slightly. Solder a type-N connector (an SO-239 is

often used at 2 meters) at the midpoint of the driven element bottom side, and close the reflector loop.

The directors are mounted through the boom. They can be made of almost any metal rod or wire of about 1/8-inch diameter. Welding rod or aluminum clothesline wire will work well if straight. (The designer used 1/8-inch stainless-steel rod secured from an aircraft surplus store.)

A TV-type U-bolt mounts the antenna on a mast. The author uses a single machine screw, washers and nut to secure the spreaders to the boom so the antenna can be quickly "flattened" for travel. In permanent installations two screws are recommended.

Construction Reminders

Here are a couple of hints based on the experiences of some who have built the quagi. First, remember that at 432 MHz even a 1/8-inch measuring error will

degrade performance. Cut the loops and elements as carefully as possible. No precision tools are needed, but be careful about accuracy. Also, make sure to get the elements in the right order. The longest director goes closest to the driven element.

Finally, remember that a balanced antenna is being fed with an unbalanced line. Fig. 43 shows details of the feed system on the 432-MHz quagi. Every balun the designer tried introduced more losses than the feed imbalance problem. Some builders have tightly coiled several turns of the feed line near the feed point to limit radiation further down the line. In any case, the feed line should be kept at right angles to the antenna. Run it from the driven element directly to the supporting mast and then up or down perpendicularly for best results.

Helical Antenna for 432 MHz

Polarization is described as "horizontal" or "vertical," but these terms have no meaning once the reference of the earth's surface is lost. Many propagation factors can cause polarization change: reflection or refraction, passage through magnetic fields (Faraday rotation) and satellite rolling, for example. Polarization of VHF waves is often random, so an antenna capable of accepting any polarization is useful. Circular polarization, generated with helical antennas or with crossed elements fed 90 degrees out of phase, has this quality.

The circularly polarized wave, in effect, threads its way through space, and it can be left- or right-hand polarized. These polarization "senses" are mutually exclusive, but either will respond to any plane polarization. A wave generated with right-hand polarization comes back with left-hand, when reflected from the moon, a fact to be borne in mind in setting up EME circuits. The same applies with satellite circuits. Stations communicating on direct paths should have the same polarization sense.

Both senses can be generated with crossed dipoles, with the aid of a switchable phasing harness. With helical arrays, both senses are provided with two antennas, wound in opposite directions.

The eight-turn helix of Fig. 44 is designed for 432 MHz, with left-hand polarization.

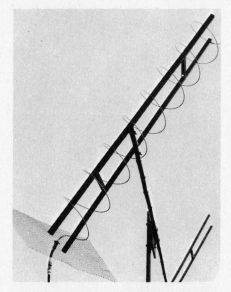

Fig. 44 — An eight-turn 432-MHz helical array, wound from aluminum clothesline wire. Left-hand polarization is shown. Each turn is one wavelength, with a pitch of 0.25 wavelength. Feed is with 50-ohm coax, through an 84-ohm Q section.

It is made from 213 inches of aluminum clothesline wire, including 6 inches that is used for cutting back to adjust the feed impedance.

Each turn is one wavelength long, and the pitch is about 0.25 wavelength. Turns are stapled to the wooden supports, which should be waterproofed with liquid fiberglass or exterior varnish. The reflecting screen is one wavelength square, with a type-N coaxial fitting soldered at its center, for connection of the required coaxial Q section.

The nominal impedance of a helical antenna is 140 ohms, calling for an 84-ohm matching section to match to a 50-ohm line. This can be approximated with copper tubing of 0.4-inch inside diameter, with no. 10 inner conductor, both 6½ inches long. With the antenna and transformer connected, apply power and trim the outer end of the helix until the reflected power approaches zero.

The support arms are made from sections of 1 × 1-inch wood and are each 60 inches long. The spacing between them is 8¼ inches outer dimension. The screen of the antenna in Fig. 44 is tacked to the support arms for temporary use. A wooden framework for the screen would provide a more rugged antenna structure. The theoretical gain of an eight-turn helical is approximately 14 decibels. Where both right- and left-hand circularity are desired, two antennas can be mounted on a common framework, a few wavelengths apart, and wound for opposite senses.

VHF Quarter-Wavelength Vertical Monopole

Ideally, the VHF vertical monopole should be installed over a perfectly flat plane reflector to assure uniform omni-directional radiation. This suggests that the center of the automobile roof is the best place to mount it. Alternatively, the flat portion of the auto rear-trunk deck can be used, but will result in a directional pattern because of car-body obstruction.

Fig. 45 illustrates at A and B how a Millen high-voltage connector can be used as a roof mount for a 144-MHz whip. The hole in the roof can be made over the dome light, thus providing accessibility through the upholstery. RG-59 and matching section L, Fig. 45C, can be routed between the car roof and the ceiling upholstery and brought into the trunk compartment, or down to the dashboard of the car. Some operators install an SO-239 coax connector on the roof for mounting the whip. The method is similar to that of Fig. 45.

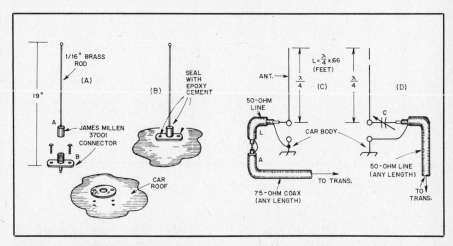

Fig. 45 — At A and B, an illustration of how a quarter-wavelength vertical antenna can be mounted on a car roof. The whip section should be soldered into the cap portion of the Millen connector, then screwed to the base socket. This handy arrangement permits removing the antenna when desired. Epoxy cement should be used at the two mounting screws to prevent moisture from entering the car. Diagrams C and D are discussed in the text.

2-Meter, 5/8-Wavelength Vertical

Perhaps the most popular vertical antenna for FM mobile and fixed-station use is the 5/8-wavelength vertical. Compared to a ¼-wavelength vertical, it has some gain over a dipole. Additionally, the so-called "picket fencing" type of flutter that results when the vehicle is in motion is greatly reduced when a 5/8-wavelength radiator is employed.

This style of antenna is suitable for mobile or fixed-station use because it is small, omnidirectional and can be used with radials or solid-plane ground (such as is afforded by the car body). If radials are used, they need be only ¼ wavelength or slightly longer. A 5 percent increase in length over ¼ wavelength is suggested for the radial wires or rods.

Construction

The antenna shown here is made from low-cost materials. Fig. 46A and B show the base coil and aluminum mounting plate. The coil form is a piece of low-loss solid rod, such as Plexiglas or phenolic. The dimensions for this and other parts of the antenna are given in Fig. 47. A length of brazing rod is used as the whip section.

The whip should be 47 inches long. However, brazing rod comes in standard 36-inch lengths, so it is necessary to solder an 11-inch extension to the top of the whip. A piece of no. 10 copper wire will suffice. Alternatively, a stainless-steel rod can be purchased to make a 47-inch whip. Shops that sell CB antennas should have such rods for replacement purposes on base-loaded antennas. The limitation one can expect with brazing rod is the relative fragility of the material, especially when the threads are cut for screwing the rod into the base-coil form. Excessive stress can cause the rod to break where it enters the coil form. The problem is complicated somewhat in this design by the fact that a spring is not used at the antenna mounting point. Innovators can find all manner of solutions to these problems by changing the physical design and using different materials when constructing the overall antenna. The main purpose of this description is to provide dimensions and tune-up data.

The aluminum mounting bracket must be shaped to fit the car with which it will be used. The bracket can be used to effect a "no-holes" mount. The inner lip of the vehicle trunk (or hood for front mounting) can be the point where the bracket attaches by means of no. 6 or no. 8 sheet-metal screws. The remainder of the bracket is bent so that when the trunk lid or car hood is raised and lowered, there is no contact between the bracket and the moving part. Details of the mounting unit are seen in Fig. 47 at B. A 14-gauge metal thickness (or greater) is recommended for best rigidity.

There are 10½ turns of no. 10 or no. 12

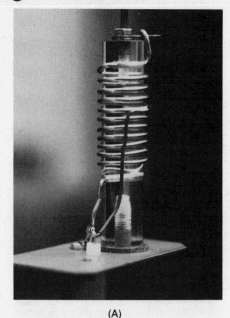

(A)

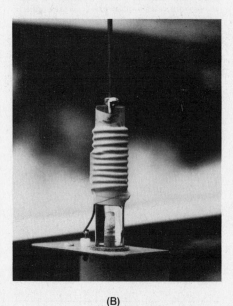

(B)

Fig. 46 — (A) A photograph of the 5/8-wavelength vertical base section. The matching coil is affixed to an aluminum bracket that screws onto the inner lip of the car trunk. (B) The completed assembly. The coil has been wrapped with vinyl electrical tape to prevent dirt and moisture from degrading performance.

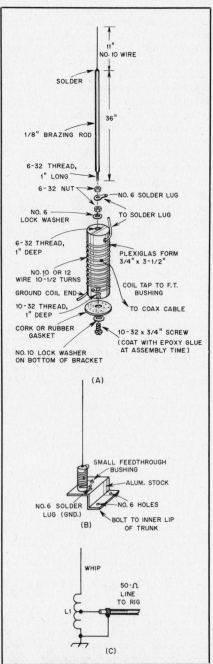

Fig. 47 — Structural details for the 2-meter antenna are provided at A. The mounting bracket is shown at B and the equivalent circuit is given at C.

copper wire wound on the ¾-inch-diameter coil form. The tap on L1 is placed approximately four turns below the whip end. A secure solder joint is imperative.

Tune-Up

After the antenna has been affixed to the vehicle, insert an SWR indicator in the 50-Ω transmission line. Turn on the 2-meter transmitter and experiment with the coil-tap placement. If the whip section is 47 inches long, an SWR of 1:1 can be obtained when the tap is at the right location. As an alternative to this method of adjustment, place the tap at four turns from the top of L1, make the whip 50 inches long and trim the whip length until an SWR of 1:1 is secured. Keep the antenna free of conductive objects or human bodies during tune-up, as conductive objects will detune the antenna and spoil the match. This antenna was described more completely in June 1979 *QST*, page 15.

A 5/8-Wavelength 220-MHz Mobile Antenna

This antenna, shown in Figs. 48 and 49, was developed to fill the gap between a homemade ¼-wavelength mobile antenna and a commercially made 5/8-wavelength model. There have been other antennas made using modified CB models. This still presents the problem of cost in acquiring the original antenna. The major cost in this setup is the whip portion, which can be any tempered rod that will spring easily.

Construction

The base insulator portion is constructed of ½-inch Plexiglas rod. A few minutes work on a lathe was sufficient to shape and drill the rod. The bottom ½-inch of the rod is turned down to a diameter of 3/8 inch. This portion will now fit into a PL-259 UHF connector. A hole, 1/8-inch diameter, is drilled through the center of the rod. This hole will contain the wires that make the connections between the center conductor of the connector and the coil tap. The connection between the whip and the top of the coil is also run through this opening. A stud is force-fitted into the top of the Plexiglas rod. This allows the whip to be detached from the insulator portion.

The coil should be wound initially on a

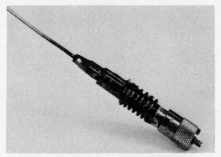

Fig. 48 — Photograph of the 220-MHz, 5/8-wavelength mobile antenna. The bottom end of the coil is soldered to the coaxial connector.

form slightly smaller than the base insulator. When the coil is transferred to the Plexiglas rod it will keep its shape and will not readily move. After the tap point has been determined, a longitudinal hole is drilled into the center of the rod. A no. 22 wire can then be inserted through the center of the insulator into the connector. This method is also used to attach the whip to the top of the coil. After the whip has been fully assembled, a coating of epoxy cement

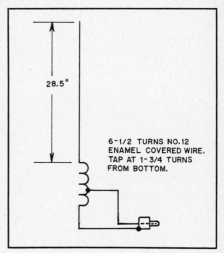

Fig. 49 — Diagram of the 220-MHz mobile antenna.

is applied. It seals the entire assembly and provides some additional strength. During a full winter's use there were no signs of cracking or mechanical failure. The adjustment procedure is the same as for the 2-meter antenna just described.

Ground-Plane Antennas for 144, 220 and 420 MHz

For the FM operator living in the primary coverage area of a repeater the ease of construction and low cost of a quarter-wavelength ground-plane antenna make it an ideal choice. Three different types of construction are detailed in Figs. 50 through 53; the choice of construction method will depend on the materials at hand and the desired style of antenna mounting.

The 2-meter model shown in Fig. 50 uses a flat piece of sheet aluminum, to which the radials are connected with machine screws. A 45-degree bend is made in each of the radials. This bend can be made with the aid of an ordinary bench vise. An SO-239 chassis connector is mounted at the center of the aluminum plate with the threaded part of the connector facing down. The vertical portion of the antenna is made of no. 12 copper wire soldered directly to the center pin of the SO-239 connector.

The 220-MHz version, Fig. 51, uses a slightly different technique for mounting

and sloping the radials. In this case the corners of the aluminum plate are bent down at a 45-degree angle with respect to the remainder of the plate. The four radials are held to the plate with matching machine screws, lockwashers and nuts. A mounting tab has been included in the design of this antenna as part of the aluminum base. A compression-type hose clamp could be used to secure the antenna to a mast. As with the 2-meter version the vertical portion of the antenna is soldered directly to the SO-239 connector.

A very simple method of construction, shown in Figs. 52 and 53, requires nothing more than an SO-239 connector and some 4-40 hardware. A small loop, formed at

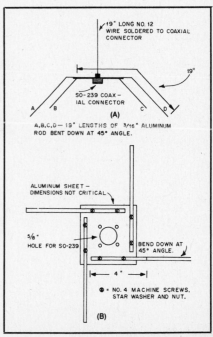

Fig. 50 — These drawings illustrate the dimensions for the 146-MHz ground-plane antenna. The radials are bent down at a 45-degree angle.

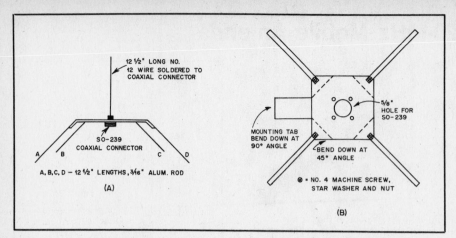

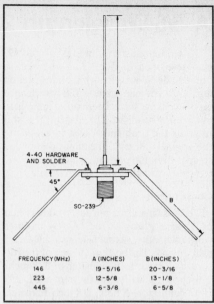

Fig. 51 — Here is the dimensional information for the 220-MHz ground-plane antenna. The corners of the aluminum plate are bent down at a 45-degree angle rather than bending the aluminum rod as in the 146-MHz model. Either method is suitable for these antennas.

FREQUENCY (MHz)	A (INCHES)	B (INCHES)
146	19-5/16	20-3/16
223	12-5/8	13-1/8
445	6-3/8	6-5/8

Fig. 52 — Simple ground-plane antenna for the 146, 220 and 440-MHz bands. The vertical element and radials are 3/32- or 1/16-inch brass rod. Although 3/32-inch rod is preferred for the 2-meter antenna, 10- or 12-gauge copper wire can also be used.

one end of the radial, is used to attach the radial directly to the mounting holes of the coaxial connector. After the radial is fastened to the SO-239 with 4-40 hardware, a large soldering iron or propane torch is used to solder the radial and the mounting hardware to the coaxial connector. The radials are bent to a 45-degree angle and the vertical portion is soldered to the center pin to complete the antenna. The antenna can be mounted by passing the feed line through a mast of ¾-inch ID plastic or aluminum tubing. A compression hose clamp can be used to secure the PL-259 connector, attached to the feed line, in the end of the mast. Dimensions for the 146, 220 and 440-MHz bands are given in Fig. 52.

If these antennas are to be mounted outside it is wise to apply a small amount of RTV sealant or similar material around the area of the center pin of the connector to prevent water from entering the connector and coax line.

Fig. 53 — A 440-MHz ground-plane antenna constructed using only an SO-239 connector, 4-40 hardware and 1/16-inch brass rod.

Chapter 34

Station Accessories

An Amateur Radio station is much more than the major pieces of equipment found in the ham shack. The minor (only in size and complexity) pieces of equipment often determine the efficiency of the station and the pleasure of the operator.

In this chapter you will find descriptions of a number of accessory items. These represent the kind of projects amateurs like to build. For the most part, they are projects that do not require a lot of time and money to build. Projects of this type help to personalize your station.

A Homemade DTMF Encoder

To be compatible with all repeaters and telephone systems, a DTMF signal must be accurate and stable in frequency, and have a nearly sinusoidal waveform. Simpler encoders than the one described here can be built, but they will not provide its high performance.

This encoder features internal voltage regulation, allowing power to be taken from the rig it is used with; there is no need to depend on separate batteries for power. When a tone pair is selected by pressing the keyboard switch, the transmitter is keyed automatically. When the key switch is released, a delay timer keeps the transmitter on long enough for the next tone pair to be selected. It's no longer necessary to hang onto the push-to-talk switch while fumbling with the DTMF pad, and there are no squelch tails between digits. It has a low-impedance audio output which is electronically disconnected from the transmitter audio system when no keyboard switches are pressed. The encoder may be connected to the mic input of transceivers having either high- or low-impedance mic inputs, with negligible loading of the transmitter audio circuitry. The audio frequencies are crystal controlled, meaning there is minimal drift. This circuit was originally described by Roy Hejhall, K7QWR, in the February 1979 issue of *QST*.

Theory of Operation

Fig. 1 is a schematic diagram of the encoder. Tone generation is performed by U1, a CMOS IC. High-frequency tones from pin 15 are mixed with their low-frequency counterparts from pin 2 and passed through the level control, R1, before reaching emitter follower Q1. Q1 performs an impedance transformation, providing the low-impedance output mentioned previously. Q2, Q4 and Q5 are used as

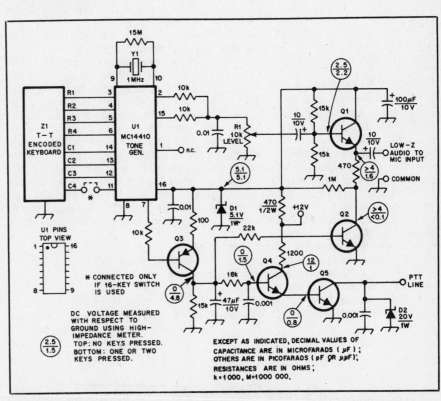

Fig. 1 — Schematic diagram of the K7QWR DTMF encoder. Any properly encoded keyboard may be used with this circuit, but the units specified will plug directly into a row of Molex pins soldered to the circuit board. If the encoder is constructed on a printed-circuit board there should be no difficulties. Should you experience problems, voltage levels at various points in the circuit are included on the schematic diagram.

D1 — 5.1-volt, 400-mW Zener diode, 1N4733, HEP Z0406 or equiv.
D2 — 20-volt, 1-watt Zener diode, 1N4747 or equiv.
Q1, Q2, Q4 — Silicon NPN transistor, 2N4123 or equiv.
Q3 — Silicon PNP transistor, 2N4125 or equiv.
Q5 — Silicon NPN transistor, 2N4401 or equiv.
R1 — Circuit-board-mounted trimmer potentiometer, 10-kΩ, linear taper.
U1 — Integrated-circuit DTMF encoder, Motorola MC14410.
Y1 — 1-MHz crystal in HC-18/U holder. Frequency tolerance is 0.1 percent; series resistance and load capacitance are typically 540 Ω and 7 pF, respectively. Available from Data Signal, Inc., 2403 Commerce La., Albany, GA 31707.
Z1 — DTMF encoding keyboard. The circuit-board layout will accommodate Digitran keyboards KL0054 (12-key) or KL0049 (16-key). They are available from distributors in single-lot quantities. For the name of the nearest distributor, contact Bob Privell at Digitran, 855 South Arroyo Pkwy., Pasadena, CA 91105, or call him at 213-449-3110.

switches. Q2 forces the audio-output impedance high when no keyboard switches are depressed, preventing the encoder from loading the transmitter mic input. Q4 and Q5 are operated as a Darlington pair, keying the transmitter push-to-talk (PTT) line when a keyboard switch is pressed. A single-package Darlington pair was originally used in this application, but its saturated collector voltage was high enough to prevent transmitter keying in some transceivers. Substituting discrete transistors solved the problem. Q2 and Q4 are driven by Q3, which is turned on by pulses from pin 7 of U1 when a keyboard switch is depressed.

U1 requires a 5-volt supply for proper operation. This is provided by the 470-ohm resistor and 1N4733 Zener diode, D1. With the exception of Q4, the remainder of the encoder circuit was also designed to operate from a 5-volt supply.

The length of time the transmitter remains keyed after a keyboard switch is released is determined by the 47-μF capacitor connected to the collector of Q3. On the prototype unit, a value of 47 μF provided a delay of just over one second. If you prefer a longer drop-out time, increase the value of this capacitor. Lowering its value will decrease drop-out time.

Construction and Testing

The prototype was built on a piece of perforated board, but a PC board is preferable. A parts-placement guide for a commercially available circuit board is shown in Fig. 2. U1 should be installed in a socket. The 0.001-μF disc capacitors connected to the base of Q4 and collector of Q5 should be installed as near the transistors as possible. Their function is to bypass RF from the transmitter, which can cause Q4 and Q5 to latch up in the keyed position.

A few simple checks will tell whether the circuit is functioning properly. The following tests may be performed before connecting the encoder to the radio, using only a 12-volt power supply, a high-impedance dc voltmeter and a scope, if one is available.

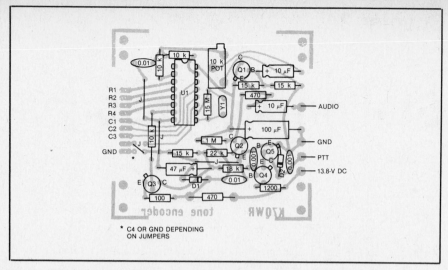

Fig. 2 — If the circuit board is used, this parts overlay will guide you when installing components. Circuit boards are available from Lea Engineering, 1230 E. Layola Dr., Tempe, AZ 85282. A full-size etching pattern can be found at the back of this book.

First ensure that D1 is regulating the encoder supply voltage at +5.1 V dc ±10 percent. U1 may be damaged if more than 6 volts is applied to pin 16.

The two operating states for the encoder are (1) no keyboard buttons depressed and (2) one or more buttons depressed. Connect the 12-V dc supply and measure the voltage at the test points shown on the schematic diagram. Voltages measured should be in accordance with those shown.

If any voltages are incorrect, look for wiring errors. If the collector voltage of Q2 is not at least 4 volts with no buttons pressed, the problem may be a leaky transistor at Q1 or Q2. If a scope is available it may be used to inspect the audio output. Pressing any one button should produce a two-tone signal, while depressing any two buttons simultaneously should produce a single tone.

Installation

The electrical portion of the installation simply involves running four wires from the encoder to the transceiver: +12 V, ground, push-to-talk (PTT) and audio output. Shielded audio cable is recommended for the audio output, which is connected to the transceiver mic input. The PTT lead is connected to the hot side of the mic PTT switch. The encoder PTT circuit is designed for rigs with an antenna relay coil connected to the +12 V bus and the PTT switch. The switch grounds the cold side of the relay coil during transmit. Ensure that your rig has this type of PTT circuit and that the relay coil draws less than 300 mA. Most of the popular VHF and UHF FM rigs have this type of PTT circuit. The mechanical details of the installation are left to the discretion of the builder.

Since the encoder will not load the audio system it should not be necessary to change the setting of any transmitter mic-level controls. Adjust only R1 in the encoder for proper tone deviation. The prototype unit has provided excellent performance on both a Tempo VHF/One 2-meter rig and a Kenwood TR-8300 UHF rig.

A Simple DTMF Decoder

A Touch Tone® (dual-tone, multifrequency, or DTMF) decoder can be a valuable link to a remotely controlled device, such as the phone patch in a repeater. Tone decoders of the past were rather large and unstable because they were constructed from discrete tone comparators such as the NE567. The '567 scheme also required rather expensive capacitors to keep the frequency even remotely accurate over a wide temperature range. Recently, Silicon Systems, Inc. introduced a new line of tone decoder ICs. They are hybrid devices, containing an audio filter, time base, tone decoders and TTL-compatible output drivers all on the same chip. Cost wise, it is less expensive to use the SSI IC as

Fig 3—The simple DTMF decoder uses a minimum of parts and fits on a compact PC board.

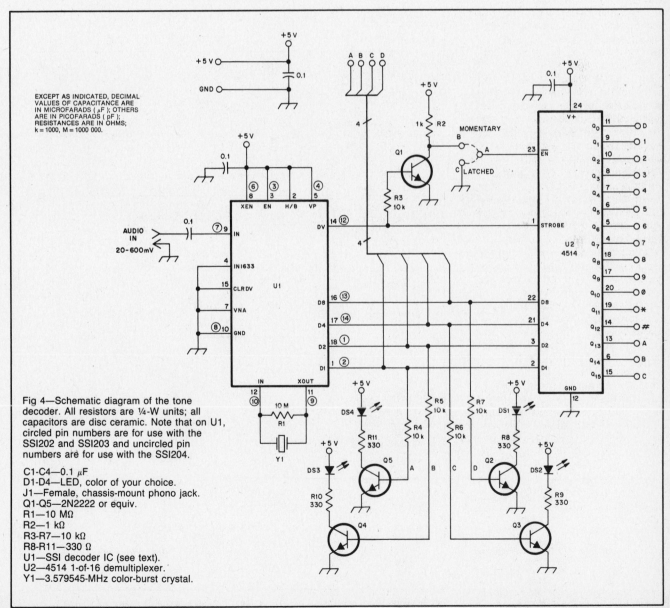

Fig 4—Schematic diagram of the tone decoder. All resistors are ¼-W units; all capacitors are disc ceramic. Note that on U1, circled pin numbers are for use with the SSI202 and SSI203 and uncircled pin numbers are for use with the SSI204.

C1-C4—0.1 µF
D1-D4—LED, color of your choice.
J1—Female, chassis-mount phono jack.
Q1-Q5—2N2222 or equiv.
R1—10 MΩ
R2—1 kΩ
R3-R7—10 kΩ
R8-R11—330 Ω
U1—SSI decoder IC (see text).
U2—4514 1-of-16 demultiplexer.
Y1—3.579545-MHz color-burst crystal.

opposed to other methods, and the finished unit is also considerably smaller.[1] The project shown in Figs. 3 through 8, designed and built by Tom Miller, KA1JQW, in the ARRL Lab, uses the SSI202, SSI203 or SSI204 as a one-of-sixteen tone decoder.

Circuit Details

The decoder described here is designed so you can tailor the circuit to suit your needs. As mentioned before, three different SSI chips can be used. All produce the same results and are available from different sources at various prices. Fig. 4 shows the decoder schematic. Audio containing the DTMF tones is fed into the circuit at the AUD IN jack. The input amplitude should be between 20 and 600 mV RMS. The signal is ac-coupled and fed directly into the decoder IC. Notice that there are two sets of pin numbers for U1. The circled numbers are for use with the SSI204, and the uncircled numbers are for use with the SSI202 and '203.

Inside the decoder chip, the audio signal is filtered to rid it of noise and 60-Hz hum, and then split into two bands, one for high tones, and the other for low tones. From there, the two bands are further filtered and fed into the tone-detection circuits. The tone detection circuits compare the frequency of the incoming signal to that of an internal clock signal. This clock signal is generated by dividing down a crystal-controlled 3.579545 MHz oscillator (from a common color-burst crystal). Once detected, the outputs send the corresponding binary word and a strobe (called DV) for use by an external circuit.

The circuit described here uses a 4514 (1-of-16 decoder) to decode the binary output into 16 separate logic-level outputs. These outputs can be momentary (by installing a jumper between points A and B) or latched (by installing a jumper between points A and C). Some sort of driver circuitry must be used to prevent damage to the 4514. Examples of driver circuits are shown in Fig. 5.

The four data outputs from U1 are also available for use, and LEDs indicate activity. The strobe from the tone decoder is not specifically designated as an output on the circuit board. It can, however, be taken off the board from pin 1 of the 4514 or the collector of Q1 if the compliment is necessary. This will allow you to strobe the binary word onto a computer bus or other control device.

The circuit requires a single 5-V supply. It can readily run from 12-V dc or 117-V ac using the suggested power supply circuits shown in Fig. 6. Maximum current consumption at 5 V is 60 mA with all four LEDs lit, and 20 mA with no LEDs lit.

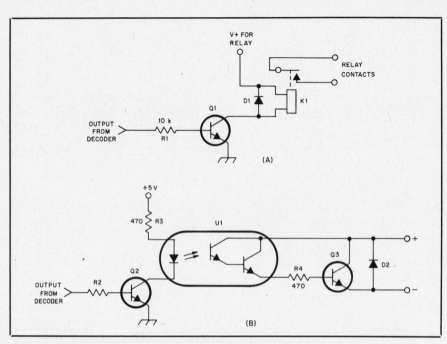

Fig. 5—Suggested driver circuits for use with logic level outputs. A typical relay or solenoid driver is shown at A, and B shows a typical high current/voltage driver with an opto-isolated output stage. Resistors are ¼-W units.

D1—1N914 or 1N4148.
D2—1N4001 or equiv.
K1—Relay (application specific).
Q1, Q2—2N2222 or equiv.
Q3—2N3055 or equiv.

R1—10 kΩ.
R2—15 kΩ.
R3, R4—470 Ω.
U1—Opto-isolator, Sylvania ECG3045 or equiv.

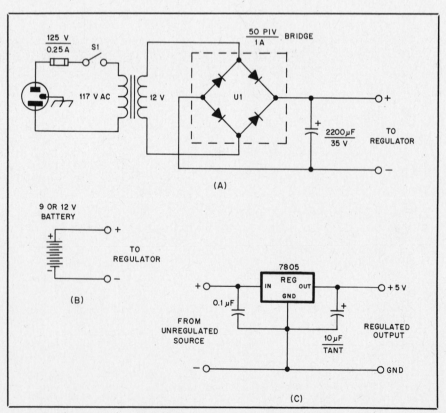

Fig. 6—Suggested power supplies and regulator for use with the decoder.

[1]The SSI decoder ICs referred to in this article cost between $10 and $25 at the time of publishing. Possible sources are:
Hall-Mark Electronics, Inc., 11333 Pagemill, Dallas, TX 75243, tel 214-343-5000
Circuit Specialists, P.O. Box 3047, Scottsdale, AZ 85257, tel 602-966-0764
Silicon Systems Inc., 14351 Myford Rd, Tustin, CA 92680, tel 714-731-7110
Your local Radio Shack store, catalog no. 276-1303

Table 1

Binary Output Table as Indicated by the LEDs

Tone Pair	Lit LEDs† D	C	B	A	Binary Word D	C	B	A
D	O	O	O	O	0	0	0	0
1	O	O	O	●	0	0	0	1
2	O	O	●	O	0	0	1	0
3	O	O	●	●	0	0	1	1
4	O	●	O	O	0	1	0	0
5	O	●	O	●	0	1	0	1
6	O	●	●	O	0	1	1	0
7	O	●	●	●	0	1	1	1
8	●	O	O	O	1	0	0	0
9	●	O	O	●	1	0	0	1
0	●	O	●	O	1	0	1	0
*	●	O	●	●	1	0	1	1
#	●	●	O	O	1	1	0	0
A	●	●	O	●	1	1	0	1
B	●	●	●	O	1	1	1	0
C	●	●	●	●	1	1	1	1

†O = not lit; ● = lit.

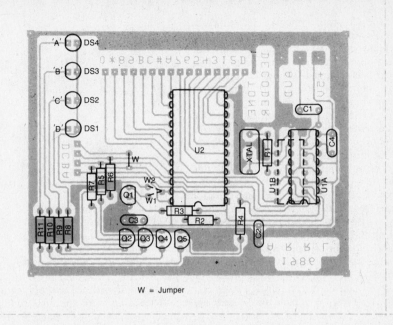

W = Jumper

Fig. 7—Parts placement guide for the DTMF decoder. This view is from the component side of the board. A full-size etching pattern appears at the back of this book.

Construction

The PC-board etching pattern is located at the back of this book. Fig. 7 shows the component layout from the component side of the board. Use sockets for the ICs, as this makes testing and repair easier. The board was designed for use with any of the SSI chips, so care should be taken when stuffing the circuit board—it is very easy to place the U1 in the wrong set of holes. If you are using the 14-pin variety decoder (SSI204), it should be located in the set of holes nearest to the crystal with pin 1 in the square pad. Conversely, if you are using one of the 18-pin decoders (SSI202 and '203) for U1, use the set of holes farthest from the crystal. Once again, be sure that pin 1 is located correctly. Fig. 8 shows two of the finished units, one using the SSI202, and the other using the SSI204. Notice the slight difference in location of the ICs; the two sets of holes are offset just enough to allow you to use one set and not the other.

If the decoder is to be used in a high-RF environment, such as that of a repeater controller, some precautions should be taken to ensure that there is no interference to the decoder. In this case, the unit should be housed in a closed aluminum chassis. Use a shielded wire and connector for the audio line, and install plenty of bypass capacitors in the power supply.

Testing/Set Up

Once the unit has been built, remove the ICs and apply 5-V dc to the power supply connections on the PC board. Then check the power-supply pins on the IC sockets with a voltmeter to be sure that the power supply is working properly and is correctly polarized. Then turn off the power and install the ICs. Although these steps might seem trivial, they can save a lot of grief from burned-out ICs.

Be sure that the right jumper (not both) is installed between point A and point B or

Fig. 8—Two finished units ready to go. Note the offset of the two decoder ICs.

C. Apply power once again. Next, you'll need a source of DTMF tones to test the unit. An easy source of the required tones is a pair of hand-held VHF transceivers. One must have a tone pad; this will be connected to a dummy load and used for transmitting the required tones. Connect the speaker output of the other hand-held transceiver to the decoder and set the audio output to low volume. Upon sending an audio tone pair to the decoder, the LEDs

should indicate the corresponding binary word for as long as the tone is present. Table 1 shows the correlation between tone pairs and which LEDs are lit. Note that the output of the decoder IC is standard binary. The LED indication is not latched when a jumper is installed between points A and C; just the 1-of-16 outputs are latched. The completed unit is now ready for installation in the device to be controlled.

A PIN-Diode TR Switch

The TR switch system described here is usable with almost any tube-final 100-W (output) power level transmitter/receiver or transceiver/receiver combination. This system is designed for flexible operation and interconnection to various pieces of commercial or homemade equipment. For the simplest setup, no modifications to the transmitter or receiver are necessary. All that's required is to plug the station equipment into the system.

The only limitation associated with this simple setup is the recovery speed of the receiver. If the receiver AGC time constant is fairly fast, it should be possible to hear signals between characters at keying speeds of up to 25 WPM. If the receiver AGC is turned off, or set for very fast recovery, signals can be heard between characters at speeds of up to 50 WPM. If you prefer, the receiver can be muted during characters. Two outputs, the + mute and − mute are provided for this purpose. Several transmitters were tried with this system, and it was possible to use them without modification, so long as the final amplifier was biased off under key-up conditions. No background hash was noticeable.

The PIN-diode TR system is not plagued by problems commonly associated with some other systems. First, "suckout" (receiver desensing) has been eliminated, as has the problem of critical interconnecting line lengths. Also, since the saturated-diode technique has been abandoned there should be no chance for TVI. No high-priced vacuum relays are used. No amplifiers are placed ahead of the receiver that could affect receiver performance. In short, the system described here provides excellent performance and suffers none of the ills of earlier designs.

The Circuit

A schematic diagram of the circuit is shown in Fig. 15. The diagram is divided into two sections, as is the actual circuitry. That portion of the circuit to the left of the dotted line is intended to be mounted at the operating position for easy access. Circuitry to the right of the line can be mounted remotely, perhaps behind the station equipment.

The transmitter is connected to the antenna through a quarter-wavelength, lumped-constant circuit. S1 selects the appropriate circuit for the frequency in use. Quarter-wave circuits are required to prevent suckout of the received signal. Suckout occurs with tube-type transmitters when the high-impedance end of the transmitter pi network becomes unloaded; during receive periods, for example. As the pi network is one type of impedance-inverting network, the high resistance presented by the non-operational tube causes the low-impedance end of the network to approach 0 ohms. The quarter-

Fig. 14 — Exterior view of the PIN Diode TR Switch. The box at the right is mounted at the operating position. The box at the left can be mounted remotely.

wavelength lumped-constant sections provided in the TR system serve to step the nearly zero impedance level of the transmitter output up to an almost infinitely high impedance that will not reduce the received signal level. As shown in the schematic diagram, the antenna is connected directly to J3 which feeds the PIN-diode switch section of the TR circuit.

The components located between J4 and J5 comprise the switch that protects the receiver from the transmitted signal. A T configuration is used, with D1 and D2 connected in series and D3 in shunt. Combination switches provide better attenuation performance than either the series or shunt elements alone. Approximately 50 dB of isolation from the antenna to the receiver connection is provided throughout the RF range. These results should be reproducible if the same type of PIN diodes are used and the board layout shown is followed closely.

The station keyer (or straight key, bug or keyboard) is connected to either J6 or J7, depending on the output keying potential. Q1 and Q2 provide a suitable signal for driving Q3 through Q9. Q3 controls D3 and turns the diode on during transmission and off during receive. Q4 and Q5 control D1 and D2, biasing the diodes on for recieve and off for transmit. At first glance it might appear that some simplification of the switching diode and control circuitry might be possible. Because it was desired to power the system from a 12-V dc source (for portable operation), and high negative voltages could not be used to reverse bias the diodes during transmit, an unusual arrangement was devised. Hence, the more complicated circuit. J8 and J9 are provided for keying the transmitter; one of the two outputs should be suitable for almost any transmitter. J10 and J11 are for muting the station receiver during transmission. Again, dual-polarity outputs are provided. Choose the one applicable to the equipment in use.

The power supply is depicted at the bottom center of the schematic diagram. Power to the system is routed through an

ON/OFF switch that is mounted at the control head. An LED indicator is included as a reminder that power is switched on. Connection from the control head to the remote unit is made via feedthrough capacitors at each box. These capacitors ensure that each cabinet remains RF-tight. A step-down transformer, full-wave rectifier, filter and three-terminal regulator provide the necessary voltage. A fuse is included in one leg of the ac line.

Construction

As mentioned earlier, the TR switch system is constructed in two enclosures. The circuitry intended for mounting at the operating position is built into a Bud Minibox that measures 3 × 4 × 5 inches. The part number of this box is CU-3005A. This enclosure houses the circuit-board mounted quarter-wavelength sections, rotary switch, ON/OFF power switch, LED indicator, coaxial connectors and feedthrough capacitors. An interior view of this unit is shown in Fig. 16. The rotary-switch wafers are positioned to line up closely with the appropriate circuit board connection points. An extra ceramic spacer is inserted in each of the switch section support rods to provide the needed separation. Short lengths of no. 18 tinned wire are used for the connections from the board to the switch contacts. RG-58A/U cable is used to make the connection from the antenna coaxial connector to the front wafer. The cable braid is soldered to ground lugs at each end.

The second enclosure used for the TR switch system is constructed from sheet aluminum and measures 2-1/8 × 4-1/8 × 7 inches. An ideal commercial enclosure would be the Bud CU-247 die-cast aluminum box. These boxes are extremely rugged and RF tight. The power supply, PIN-diode switch and control circuitry are mounted on a circuit board. Circuit-board parts-layout information is shown in Fig. 18. Double-sided circuit-board material was used, with the top side of the board left substantially unetched to act as

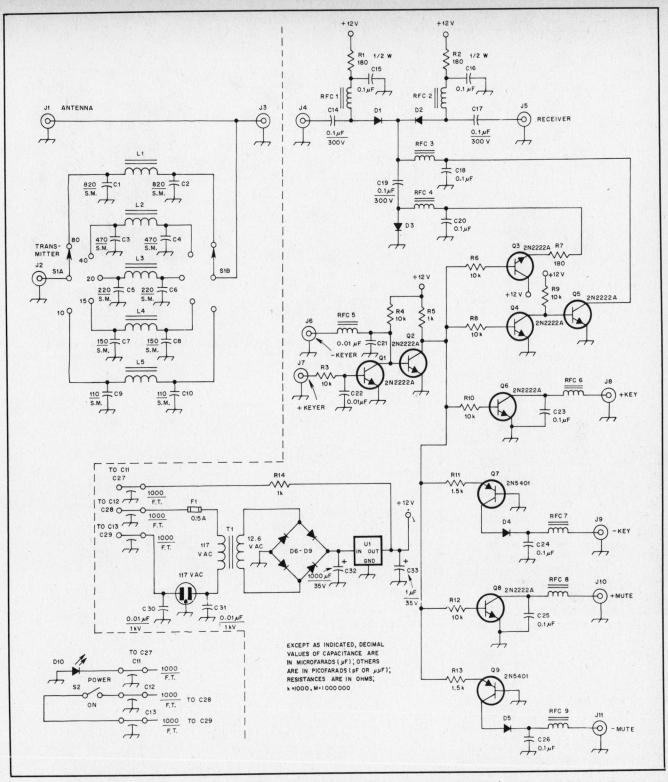

Fig. 15 — Schematic diagram of the TR system. All resistors are ¼-watt composition types. All capacitors are miniature ceramic, 50-volt types unless polarity is indicated. Polarized types are aluminum electrolytic or tantalum. Component designations listed in the schematic, but not called out in the parts list, are for text or layout reference only.

C1, C2 — Mica, 820 pF, 500 V.
C3, C4 — Mica, 470 pF, 500 V.
C5, C6 — Mica, 220 pF, 500 V.
C7, C8 — Mica, 150 pF, 500 V.
C9, C10 — Mica, 110 pF, 500 V.
C32 — Electrolytic, 1000 µF, 35 V.
C33 — Tantalum, 1 µF, 35 V.
D1-D3, incl. — PIN diode, Unitrode 1N5767 or equiv.
D4-D9, incl. — Power, 100 PIV, 1A.
D10 — Light-emitting diode.
F1 — Fuse, ½ A.
J1-J5, incl. — RF connector, female (builder's

choice).
J6, J7 — Phone, ¼ inch or builder's choice.
J8-J11, incl. — Phono or builder's choice.
L1 — Toroid, 20 turns no. 18 enam. wire on a T-80-2 core.
L2 — Toroid, 15 turns no. 18 enam. wire on a T-80-2 core.
L3 — Toroid, 11 turns no. 18 enam. wire on a T-80-6 core.
L4 — Toroid, 9 turns no. 18 enam. wire on a T-80-6 core.
L5 — Toroid, 8 turns no. 18 enam. wire on a

T-80-6 core.
RFC1, RFC2 — Toroid choke, 20 turns no. 26 enam. wire on an FT-37-75 core.
RFC3-RFC9, incl. — Toroid choke, 26 turns no. 30 enam. wire on an FT-23-75 core.
S1 — Rotary wafer, 2 sections, 5 positions, ceramic.
S2 — Toggle, SPST.
T1 — Miniature power, primary 117 V, secondary 12 V at 300 mA. Radio Shack 273-1385 or equiv.
U1 — Three-terminal regulator, 12-V output. Radio Shack RS-7812 or equiv.

Fig. 16 — Interior view of the control head. The quarter-wave sections are mounted to the single-sided PC board.

Fig. 17 — Inside view of the remotely mounted portion of the system. Short lengths of wire are used to attach the connectors to the appropriate circuit-board foils. All power supply components are mounted on the circuit board.

a ground plane. Copper must be removed from around circuit-hole locations for components that are not connected to ground. This can be accomplished in the etching process with the aid of the top-side etching pattern. Alternatively, the copper can be removed from around holes with a large drill. Do not remove copper from around holes where component leads are grounded. Many of the component ground connections are not made on the pattern (bottom) side of the board. These components must be soldered on the top side to complete the ground connection.

Garden-variety components are used, with the exception of the PIN diodes. The diodes are Unitrode 1N5767 types, which can be obtained from many supply houses. Microwave Associates makes several suitable replacements. All of the RF chokes are hand wound on small ferrite cores. Since encapsulated chokes are relatively expensive and cores are not, the time spent winding the chokes can result in reduced cost.

Circuit Checkout

Interconnection of the two modules requires four lengths of hookup wire, each long enough to reach between the two units when installed in their operating positions. The wires are twisted, cable-tied or laced together. Wires of different colors will help distinguish the connections and prevent possible surprises the first time power is applied! Connections are as follows: C11 to C27, C12 to C28 and C13 to C29. The fourth wire, ground, connects the two boxes. A coaxial cable is used to connect J3 and J4.

Connection to the station equipment is a simple matter. The keyer is plugged into either J6 or J7. If the keyer provides a positive voltage when keyed, use J7. If the keyer provides a ground, use J6. Connect

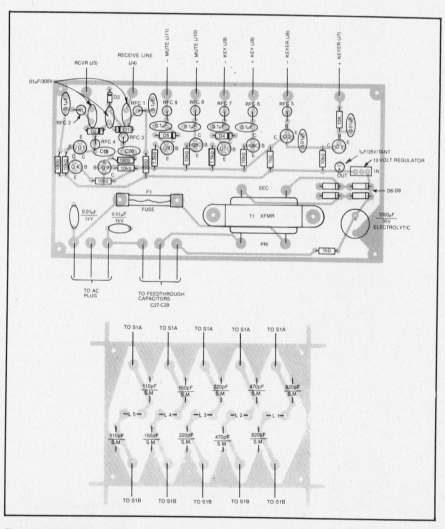

Fig. 18 — Parts-layout patterns of the two printed-circuit boards. Each board is shown from the component side with an X-ray view of the foil. Full-size etching patterns can be found at the back of this book.

the antenna to J1 and the station receiver to J5. Do not connect the transmitter at this time. A check of the system operation can now be made. If all is in order at this point, signals should be heard in the receiver. Actuating the keyer should cause the signals to become inaudible.

The exact amount of attenuation can be measured using a calibrated signal generator or a step attenuator and received signals. Attenuation should be on the order of 50 dB. If no measurement equipment is available, a received signal and the receiver S meter may be used. A strong signal should become almost completely buried in the receiver noise when the keyer is activated. Connect the transmitter output to J2 and install a cable between the transmitter key jack and J8 or J9. If muting of the receiver is desired, make the appropriate connection at J10 or J11.

QRP SWR Indicator

This compact SWR indicator is suitable for use with transmitters in the 1- to 150-watt output range. It will function properly from 1.8 to 30 MHz. The project, shown in Fig. 19, was first described by Doug DeMaw, W1FB in the Hints and Kinks column of August 1982 *QST*.

The principal difference in this circuit from some others is that a two-turn link is wound on T1 to increase the low-power sensitivity of the instrument. Normally, a single wire is passed through the center hole of the torodial transformer for sampling the 50-ohm transmission line. See Fig. 20.

Most of the components were garnered

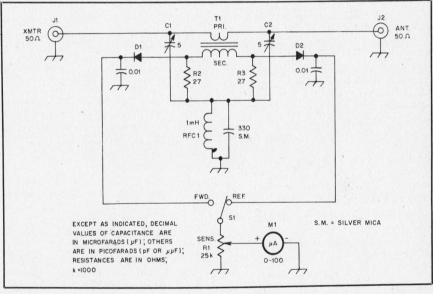

Fig. 20 — Schematic diagram of the SWR indicator. Fixed-value capacitors are disc ceramic except those marked with SM, which are silver mica. R2 and R3 are ¼-watt carbon-composition units.

C1, C2 — Miniature PC-mount air trimmer (see text).
D1, D2 — Silicon switching diode. 1N914 type, matched for equivalent forward resistance (use an ohmmeter).
J1, J2 — Single-hole-mount phono jack.
M1 — Miniature 50- to 100-µA dc meter (see text).

R1 — Linear-taper miniature control, 25 KΩ.
RFC1 — Miniature 1-mH RF choke.
S1 — Miniature SPDT slide or toggle switch.
T1 — Toroidal transformer. Secondary: 60 turns no. 30 enam. wire on an Amidon, Radiokit or Palomar T68-2 powdered-iron core. Primary is two turns over secondary winding.

Fig. 19 — Photograph of the assembled SWR indicator in the homemade PC-board material box. A commercial cabinet or a Minibox can be used to obtain a more professional effect.

at hamfest flea markets. A miniature FM tuning meter is used at M1. It has a 100-µA movement, but microampere meters of other full-scale characteristics will work nicely in this circuit. A miniature slide switch is used for S1, while nulling trimmers C1 and C2 are surplus PC-mount

trimmers. Piston trimmers can be used in place of the units shown for C1 and C2. The type chosen should be mechanically stable and capable of withstanding at least 87 volts RMS (typical maximum voltage for 150 watts at 50 ohms). Greater voltages may be present in a mismatched system.

Use care in choosing the capacitors, with special attention to the minimum capacitance available. Only 2 or 3 pF of capacitance should be needed when the bridge is nulled for 50-ohm use.

Double-sided PC-board material is used for the case. It is soldered together along the inner seam of the walls and the base plate. The top plate is tacked to the case, using one solder blob on each side. This should be done after the circuit has been adjusted and is considered ready to use.

The PC board is supported in the box by means of a single standoff post, directly under T1. The rear edge of the PC board butts firmly against the back wall, as shown in the photograph. Be sure to connect the ground foil of the PC board to the box walls. Two short lengths of bus wire can

be used for the purpose. A glob of noncorrosive RTV sealant is placed in the center hole of T1 to keep the transformer in position. Similarly, M1 is glued to the front panel by means of quick-drying contact cement. Four adhesive-backed plastic feet are attached to the bottom plate of the instrument. You can use label tape to identify the controls and the input/output jacks on the rear of the box.

Adjustment is done by connecting a 50-ohm resistive termination to the antenna jack, applying RF energy to the transmitter jack and adjusting R1 for a full-scale reading (S1 in the FWD position). Next, switch S1 to REF and adjust the trimmer that causes the meter reading to change (one of the trimmers will be unresponsive

in this setting). Set the trimmer for minimum meter deflection. It should read zero. Next, reverse the cables at J1 and J2. Put S1 in the REF position and apply RF energy. The meter should read full scale. Switch S1 to FWD and adjust the remaining trimmer for minimum meter reading (again, it should fall to zero). The bridge has now been balanced for 50 ohms. This set of adjustments should be done on 20 or 15 meters to ensure proper high-range performance.

When using the instrument, always adjust R1 for a full-scale reading with S1 in FWD mode. Adjust the antenna or antenna-matching network for the lowest reading attainable with S1 in the REF position. A zero reading in REF will be equivalent to an SWR of 1:1.

Digital PEP Wattmeter and SWR Calculator

Every radio amateur wants to measure transmitter power or transmission-line SWR at some time. The wattmeter and SWR calculator shown in Figs. 21 through 31 will allow the operator to perform these simple operations for peak-envelope power (PEP) in forward and reverse directions and automatically calculate SWR with no adjustments once the unit is calibrated. The unit reads from 0 to 2000 watts and SWR from 1.00 to 19.99. Because it measures PEP, the unit satisfies FCC requirements for monitoring your power output. The RF sampling unit is designed to operate from 160 through 6 meters. This project was designed and built by Jon Towle, WB1DNL, in the ARRL lab.

Circuit Description

The wattmeter and SWR calculator consists of four parts: An RF current sampling unit, a peak measuring and calculator board, an LCD digital display with 3½

digit A/D converter and a four-voltage power supply.

The RF sampling unit circuit is shown in Fig. 22. It consists of a wideband current transformer (T1) and a capacitive voltage divider. The circuit produces a voltage that is proportional to the power in the transmission line at each of the feed-through capacitors. These are high impedance outputs, and are connected to the op-amp inputs on the calculator board.

The primary of T1 is simply a short section of RG-8 coaxial cable soldered between the input and output connectors. The secondary is a ferrite toroid core with the appropriate number of turns. Thirty turns on the core provide 10-V output from the current transformer when operated at the 2000-W level. A 10-turn secondary would provide 10 V at the 200-W level if more accuracy is desired at low power levels. The toroid slips over the center of the outer jacket of the section of RG-8. The 1N34A diodes, 22-ohm resistors and 3-pF capacitors should be matched for best performance.

The sampling unit in the wattmeter shown here is attached to the back of the cabinet. The forward and reflected voltages from the sampling unit can also be fed to the main unit through shielded cable (such as RG-174) if you want a remote-sensing wattmeter. In this way, the sensing unit can be connected directly at the transmitter output (or directly to the antenna terminals).

The calculator board (Fig. 23) performs several functions. First, two peak detector circuits (U1) look at the maximum voltage from the RF sampling unit. A one-second time constant holds the peak voltage sufficiently long to view the power output on the display. The peak detector circuit uses a TL084 quad op-amp, which has a very

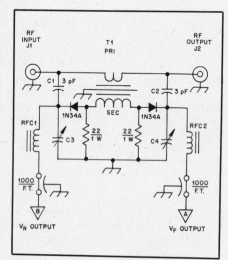

Fig. 22 — Schematic diagram of the RF sampling unit.

C1, C2 — 3-pF silver-mica capacitor.
C3, C4 — 37- to 250-pF mica compression trimmer capacitor (Arco 426 or equiv., avail. from RadioKit).
J1, J2 — SO-239 coaxial connector.
RFC1, RFC2 — 4t no. 24 enam. on 3/8-inch-OD, 850-µi ferrite bead (Amidon FB-43-2401 or equiv.).
T1 — Wideband current transformer. Primary, section of RG-8 through center of core; secondary, 30t no. 22 enam. on Amidon FT-82-61 toroid core (125 µi). See text and Fig. 26.

high input impedance, to avoid loading the output from the RF sampling unit.

A TL082 dual op-amp (U2) is used for two functions. U2A is wired as a summing amplifier, while U2B is connected as a difference amplifier. The output of the peak detector circuits is applied to both U2A and U2B to give sum and difference voltages. That is: VF + VR and VF − VR. These values are used to calculate SWR.

Fig. 21 — Front-panel view of the digital PEP wattmeter. The LCD display is mounted directly to the front panel.

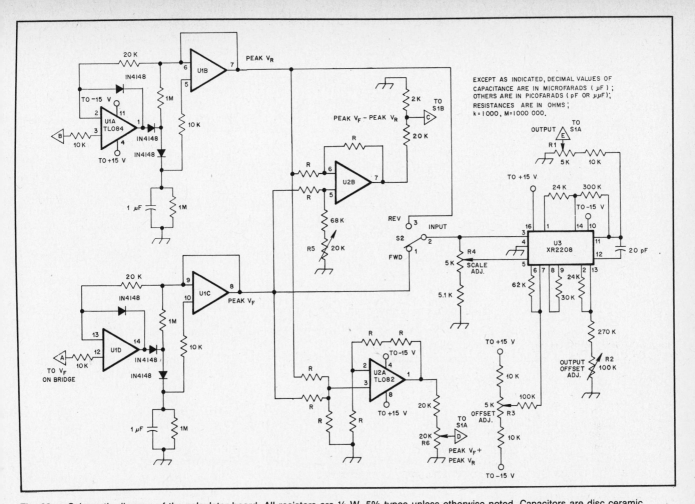

Fig. 23 — Schematic diagram of the calculator board. All resistors are ¼-W, 5% types unless otherwise noted. Capacitors are disc ceramic.

R — 82.5-kΩ, ¼ W, 1%, 50-ppm metal-film resistors. See text.
R1-R6 — Multiturn Cermet trimmers (Bourns BP3299W series or equiv.).
U1 — TL084 BIFET quad operational amplifier.
U2 — TL082 BIFET dual operational amplifier.
U3 — Exar XR2208 operational multiplier (Avail. from Jameco Electronics, 1355 Shoreway Rd., Belmont, CA 94002).

An Exar XR-2208 operational multiplier IC, U3, is wired as a squaring circuit to make the conversion to watts for the digital display. The output of either peak detector circuit can be applied to the input of the multiplier through S2. This output is scaled by R1 and is proportional to the output power in watts.

The LCD display is controlled by an Intersil ICL7106 3½ digit single-chip A/D converter. Fig. 24 is a schematic diagram of this circuit, which is wired as a 0- to 2-volt voltmeter for this application. No decimal point is used in the wattmeter mode, so the range is 0 to 2000 watts. Also, no provision is made to change scales between high and low power. This is because the full-scale range is determined by the RF sampling unit. A multi-turn potentiometer, R7, is used to set the 1.000-V reference for the ICL7106.

The ICL7106 A/D converter operates by calculating the ratio between IN HI and REF HI (pins 31 and 36). This makes it easy to compute SWR, which is the ratio of the maximum voltage to the minimum voltage in the transmission line. In the SWR mode, the difference voltage from the calculator board is applied to

REF HI, and the summed voltage is applied to IN HI. To get a useful range of SWR, R6 on the calculator board is used to scale IN HI for a range of 1.00 to 19.99. The decimal point indicates only in the SWR mode.

The power supply uses common three-terminal regulators to generate the ±15-volt supplies for the op-amps and the multipliers and the ±5-volt supplies for the A/D converter. T1 has a dual 17-V ac secondary; it is connected to give separate full-wave, center-tapped outputs for the negative and positive supplies. A complete schematic diagram is given in Fig. 25.

Construction Details

The RF sampling unit is built in a Hammond 1590B die-cast box for maximum shielding. Fig. 26 shows how the parts are arranged inside the box. The components are mounted on a PC board (Fig. 27). Surface-mounting techniques are used to make the RF sampling unit small and simple. Don't forget to solder a wire from the shield of the T1 primary to ground!

Two SO-239 connectors are attached to the outside of the box cover. The mounting screws for the connectors are also used

to secure the circuit board to the inside of the cover. Mounting the connectors on the outside allows plenty of room inside for T1. Holes drilled in the bottom of the die-cast box allow access to C3 and C4 for nulling the bridge. Only one end of RFC1 and RFC2 is soldered to the circuit board. The other end of each RFC solders to one side of a feedthrough capacitor. The feed-through capacitors mount in holes drilled in the bottom of the box.

Assembly of the calculator board shown in Fig. 28 is very simple. The resistors marked R on the schematic must be 1% resistors. The actual value used may be anywhere in the range of 75 kΩ to 100 kΩ, but all resistors used at these points must be within 1% of each other. You will see the best possible performance if you measure the values of the resistors before using them and discard those that are not within 1% of the desired value. It pays to order a few extra so they can be matched during construction. The 82.5-kΩ units used in the unit shown here are available from Mouser Electronics, Digi-Key and other sources listed in Chapter 35. The 1N34A diodes may be available from Radio Shack.

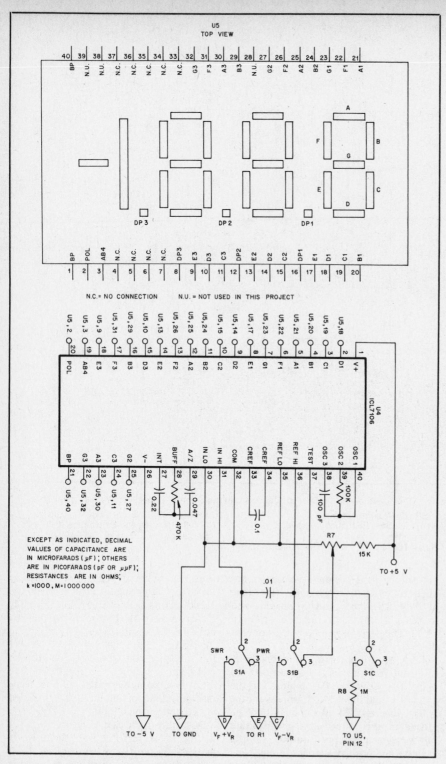

Fig. 24 — Schematic diagram of the digital display and A/D converter. Resistors are ¼-W, 5% types unless otherwise noted. Capacitors are disc ceramic.

R7 — Multi-turn 10-kΩ potentiometer.
U4 — Intersil ICL7106 3½-digit, single-chip A/D converter (Avail. from Jameco

Electronics).
U5 — FEO203D 3½-digit LCD display for ICL7106 (Avail. from Jameco Electronics).

The circuit board layout seems to be critical around U3, so changing the artwork provided may result in drifting or unstable performance. Other than this one area, layout is not critical.

The ICL7106 A/D converter and the LCD digital display are mounted on one circuit board. Layout is shown in Fig. 29.

The ICL7106 is a CMOS device and should be handled with care. The LCD display is sandwiched between thin glass plates and should also be handled carefully. Static charge may cause segments to illuminate when handled. This is not a problem. Do not apply dc to the display because this may damage the segments. The board is laid out

to accommodate low-profile sockets for both ICs. The display socket is made from a normal 40-pin socket that is cut in half lengthwise.

Power supply components, including T1, mount on a PC board. Layout for this board is given in Fig. 30. Layout is not critical; the power-supply board was made exactly the same size as the calculator board for ease of mounting, as detailed below. Heat sinks are necessary on the 15-V regulators. Remember that the tab of U2 is connected to the input, while the tab of U1 is connected to ground. Do not allow the U2 heat sink to contact the U1 heat sink, the chassis or any other components! If possible, mount U2 to its heat sink with insulating hardware.

This project has been designed to fit inside a Hammond 1426K-B equipment box. Fig. 31 shows the inside of the finished unit. It is best to complete adjustments (outlined below) before mounting the wattmeter in the enclosure. The display board mounts behind the front panel. The calculator board and power supply are stackable and mount to the cabinet bottom plate. Mount the calculator board on the bottom, upside down to allow access to the board adjustments through holes drilled in the bottom panel. The RF sampling unit attaches to the rear panel.

Setup Procedures

RF Sampling Unit — Connect a 100-W transmitter to the input and a 50-Ω dummy load to the output of the sampling unit. Connect a high-impedance voltmeter to the VR output feedthrough capacitor (the reverse side of the bridge). Apply 100 W to the sampling unit and adjust C3 until you observe a voltage null on the meter attached to the VR feedthrough capacitor. Next, reverse the RF connections, connect the voltmeter to the VF feedthrough capacitor and adjust C4 until you see a voltage null on the forward side of the bridge also.

Calculator Board — Connect the inputs of the two peak detectors together (points A and B on Fig. 23) and apply 5.0-V dc. Adjust R5 so that the output of the difference amplifier (point C on Fig. 23) is 0.0 V. The output of the summing amplifier at pin 1 of U2 should be 10.0 V.

The squaring circuit requires several iterations for proper adjustment, but it is easy just the same. Connect a voltmeter to pin 11 of U3 (the output). Apply 0 V to the input (at S2) and set the output offset adjustment (R2) to indicate 0 V at pin 11. Next, alternately apply ± 10 V to the input and set the offset adjustment (R3) so the output value at pin 11 is the same for the positive and negative voltages. This may require several tries. Finally, with 10 V applied to the input, set the scale adjustment (R4) so the output voltage at pin 11 of U3 is 10 V.

Display Board — Apply voltage to the display board. Connect a voltmeter to U4, pin 36 and adjust R7 until the voltage at

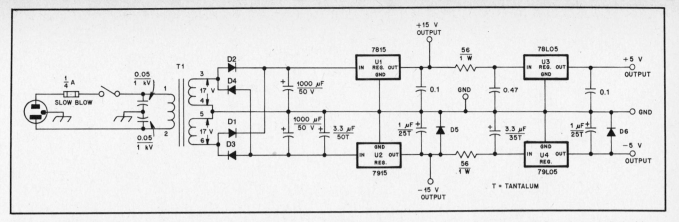

Fig. 25 — Schematic diagram of the PEP wattmeter power supply. Capacitors are disc ceramic unless otherwise noted. Capacitors marked with polarity are electrolytic.

D1-D6 — 100-PIV, 1-A diode (1N4002 or equiv.).
T1 — Dual-secondary, PC-board mount
    transformer. Primary, 117-V ac; secondary,
    17-V ac, 600 mA (Digi-Key T116-ND or equiv.)

U1 — 15-V, 1.5-A three-terminal regulator
    (LM7815 or equiv.)
U2 — Negative 15-V, 1.5-A three-terminal
    regulator (LM7915 or equiv.).

U3 — 5-V, 100-mA three-terminal regulator
    (LM78L05 or equiv.).
U4 — Negative 5-V, 100-mA three-terminal
    regulator (LM79L05 or equiv.).

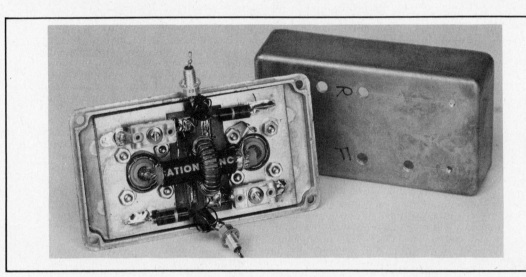

Fig. 26 — The RF sampling unit is built inside a Hammond 1590B diecast box.

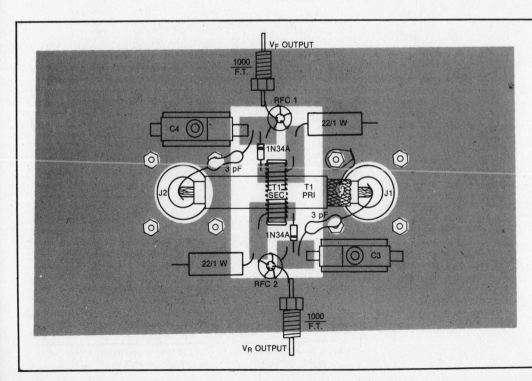

Fig. 27 — Parts placement guide for the sampling unit. All components mount on the foil side of the board. See text and Fig. 26. A full-size etching pattern appears at the back of this book.

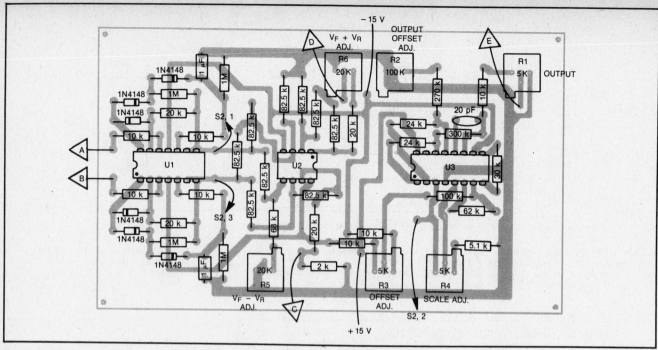

Fig. 28 — Parts placement guide for the calculator board. A full-size etching pattern appears at the back of this book.

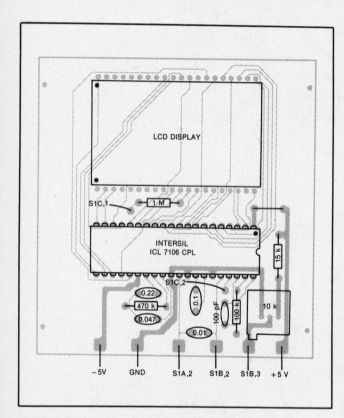

Fig. 29 — Parts placement guide for the display board. A full-size etching pattern appears at the back of this book.

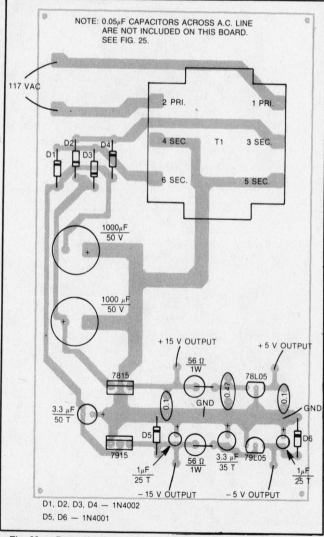

Fig. 30 — Parts placement guide for the power-supply board. A full-size etching pattern appears at the back of this book.

Fig. 31 — The power-supply and calculator boards are stacked on the bottom panel of the enclosure. The calculator board is not visible in this photograph. The RF sampling unit attaches to the rear panel, while the display board is attached to the front panel with standoff hardware. S1 and S2 also mount to the front panel.

pin 36 measures exactly 1.000 V. This sets the reference for U4.

Final Calibration — Connect the digital PEP wattmeter and a wattmeter of known accuracy in series between a transmitter and a good 50-Ω load. Apply power and adjust R1 until the PEP wattmeter reads the same power as the known wattmeter.

Disconnect VR from the input of the peak detector at point B and short the input to ground. Next, switch to the SWR mode and with RF applied, set R6 to read an SWR of 1.00. Reconnect VR to the peak detector and fully assemble the wattmeter/SWR calculator. Alternatively, the SWR reading can be adjusted by applying +5 V (from the power supply) to the V_F input (point A of Figs. 23 and 28) with the V_R input (point B) grounded. Set R6 to read an SWR of 1.00.

Operation

The wattmeter operation is controlled by S1 and S2, which are mounted on the front panel. S2 allows the operator to view either forward or reflected power. S1 selects the power or SWR mode. In the SWR position, the display will flash random values because the A/D converter lacks a reference when there is no RF in the transmission line. This is normal and serves as a visual reminder that the SWR mode is in use.

A Transmatch for Balanced or Unbalanced Lines

Most modern transmitters are designed to operate into loads of approximately 50 ohms. Solid-state transmitters produce progressively lower output power as the SWR on the transmission line increases, owing to the built-in SWR protection circuits. Therefore, it is useful to employ a matching network between the transmitter and the antenna feeder when antennas with complex impedances are used.

One example of this need can be seen in the case of an 80-meter, coax-fed dipole antenna which has been cut for resonance at, say, 3.6 MHz. If this antenna were used in the 75-meter phone band, the SWR would be fairly high. A Transmatch could be used to give the transmitter a 50-ohm load, even though a significant mismatch was present at the antenna feed point.

It is important to remember that the Transmatch will not correct the actual SWR condition; it only conceals it as far as the transmitter is concerned. A Transmatch is useful also when using a single-wire antenna for multiband use. By means of a balun at the Transmatch output it is possible to operate the transmitter into a

Fig. 32 — Exterior view of the SPC Transmatch. Radio Shack vernier drives are used for adjusting the tuning capacitors. A James Millen turns-counter drive is coupled to the rotary inductor. Green paint and green Dymo tape labels are used for panel decor. The cover is plain aluminum with a lightly grooved finish (sandpapered) that has been coated with clear lacquer. An aluminum foot holds the Transmatch at an easy access angle.

balanced transmission line, such as a 300- or 600-ohm feed system of the type that would be used with a multiband tuned

dipole, V beam or rhombic antenna.

A secondary benefit can be realized from Transmatches of certain varieties: The matching network can, if it has a band-pass response, attenuate harmonics from the transmitter. The amount of attenuation is dependent upon the loaded Q (Q_L) of the network after the impedance has been matched. The higher the Q_L, the greater the attenuation. Some Transmatches, such as the Ultimate Transmatch of Fig. 33, can exhibit a high-pass response (undesirable), depending on the transformation ratio they are adjusted to accommodate. In a worst-case condition the attenuation of harmonic currents may be as low as 3 to 6 dB. Under different (better) conditions of impedance transformation, the attenuation can be as great as 20 to 25 dB.

The SPC Transmatch described here was developed to correct for the sometimes poor harmonic attenuation of the network in the Ultimate Transmatch. The SPC (series-parallel capacitance) circuit maintains a bandpass response under load conditions of less than 25 ohms to more than 1000 ohms (from a 50-ohm transmitter).

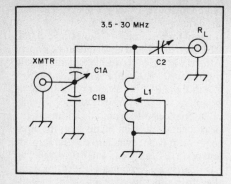

Fig. 33 — Circuit for the Ultimate Transmatch showing the network which can degenerate to a high-pass network under some conditions of transformation (see text). A T-network configuration will provide identical matching range, making a split-stator capacitor unnecessary at C1.

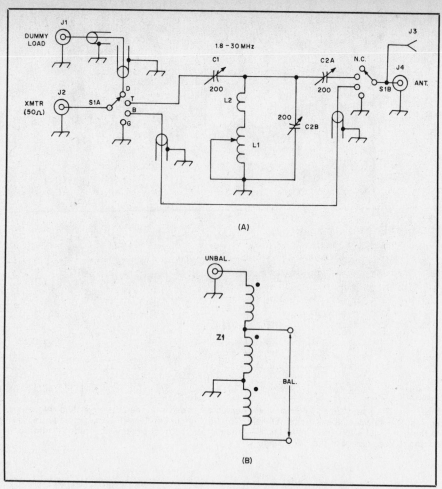

(A)

(B)

Fig. 34 — Schematic diagram of the SPC circuit. Capacitance is in picofarads.

C1 — 200-pF transmitting variable with plate spacing of 0.075 inch or greater. J. W. Miller Co. no. 2150 used here.
C2 — Dual-section variable, 200 pF per section. Same plate spacing as C1. J. W. Miller Co. no. 2151 used here. (Catalog no. 79, J. W. Miller Co., 19070 Reyes Ave., Compton, CA 90224.)
J1, J2, J4 — SO-239 style coaxial connector. J4 should have high-dielectric insulation if high-Z single-wire antennas are used at J3. Teflon insulation is recommended.

J3 — Ceramic feedthrough bushing.
L1 — Rotary inductor, 25 µH min. inductance. E. F. Johnson 229-203 or equiv.
L2 — Three turns no. 8 copper wire, 1 inch ID × 1½ inches long.
S1 — Large ceramic rotary wafer switch with heavy contacts. Two-pole, 4-position type. Surplus Centralab JV-9033 or equiv., two positions unused.
Z1 — Balun transformer. 12 turns no. 12 Formvar wire, trifilar, close-wound on 1-inch-OD phenolic or PVC-tubing form.

This is because a substantial amount of capacitance is always in parallel with the rotary inductor (C2B and L1 of Fig. 34). In comparison with the Ultimate circuit of Fig. 33, it can be seen that at high load impedances, the Ultimate Transmatch will have minimal effective output capacitance in shunt with the inductor, giving rise to a high-pass response.

Another advantage of the SPC Transmatch is its greater frequency range with the same component values used in the Ultimate Transmatch. The circuit of Fig. 34 operates from 1.8 to 30 MHz with the values shown. Only three-fourths of the available inductance of L1 is needed on 160 meters.

The notable difference in outward performance over the circuit in Fig. 33 is somewhat sharper tuning. This is because of the increased network Q. This is especially prominent at 40, 80 and 160 meters. For this reason there are vernier-drive dials on C1 and C2. They are also useful in logging the dial settings for changing bands or antennas.

Construction

Figs. 32 and 35 show the structural details of the Transmatch. The cabinet is homemade from 16-gauge aluminum sheeting. L brackets are affixed to the right and left sides of the lower part of the cabinet to permit attachment of the U-shaped cover.

The conductors which join the components should be of heavy-gauge material to minimize stray inductance and heating. Wide strips of flashing copper are suitable for the conductor straps. The center conductor and insulation from RG-59 polyfoam coaxial cable is used in this model for the wiring between the switch and the related components. The insulation is sufficient to prevent breakdown and arcing at 2 kW PEP input to the transmitter.

All leads should be kept as short as possible to help prevent degradation of the cir-

cuit Q. The stators of C1 and C2 should face toward the cabinet cover to minimize the stray capacitance between the capacitor plates and the bottom of the cabinet (important at the upper end of the Transmatch frequency range). Insulated ceramic shaft couplings are used between the vernier drives and C1 and C2, since the rotors of both capacitors are floating in this circuit. C1 and C2 are supported above the bottom plate on steatite cone insulators. S1 is at-

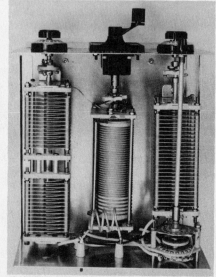

Fig. 35 — Interior view of the W1FB SPC Transmatch. L2 is mounted on the rear wall by means of two ceramic standoff insulators. C1 is on the right and C2 is at the left. The coaxial connectors, ground post and J3 are on the lower part of the rear panel.

tached to the rear apron of the cabinet by means of two metal standoff posts.

Operation

The SPC Transmatch is designed to handle the output from transmitters which operate up to 2 kW PEP input. L2 has been added to improve the circuit Q at 10 and 15 meters. However, it may be omitted from the circuit if the rotary inductor (L1) has a tapered pitch at the minimum-inductance end. It may be necessary to omit L2 if the stray wiring inductance of the builder's version is high. Otherwise, it may be impossible to obtain a matched condition at 28 MHz with certain loads.

An SWR indicator is used between the transmitter and the Transmatch to show when a matched condition is achieved. The builder may want to integrate an SWR meter in the Transmatch circuit between J2 and the arm of S1A (Fig. 34A). If this is done there should be room for an edgewise panel meter above the vernier drive for C2.

Initial transmitter tuning should be done with a dummy load connected to J1, and with S1 in the D position. This will prevent interference which could otherwise occur if tuning is done on the air. After the transmitter is properly tuned into the dummy load, unkey the transmitter and switch S1 to T (Transmatch). Never hot switch a Transmatch, as this can damage both transmitter and Transmatch. Set C1 and C2 at midrange. With a few watts of RF, adjust L1 for a decrease in reflected power. Then adjust C1 and C2 alternately for the lowest possible SWR. If the SWR cannot be reduced to 1:1, adjust L1 slightly and repeat the procedure. Finally, increase the transmitter power to maximum and touch up the Transmatch controls if necessary. When tuning, keep your transmissions brief and identify your station.

The air-wound balun of Fig. 34B can be used outboard from the Transmatch if a low-impedance balanced feeder is contemplated. Ferrite or powdered-iron core material is not used in the interest of avoiding TVI and harmonics that can result from core saturation.

The B position of S1 permits switched-through operation when the Transmatch is not needed. The G position is used for grounding the antenna system, as necessary; a quality earth ground should be attached at all times to the Transmatch chassis.

Final Comments

Surplus coils and capacitors are okay in this circuit. L1 should have at least 25 μH of inductance, and the tuning capacitors need to have 150 pF or more of capacitance per section. Insertion loss through this Transmatch was measured at less than 0.5 dB at 600 watts of RF power on 7 MHz.

A Link-Coupled Matching Network

Link coupling offers many advantages over the types of systems where a direct connection between the transmitter and antenna is required. This is particularly true on 80 meters, where commercial broadcast stations often induce sufficient voltage to cause either rectification or front-end overload. Transceivers and receivers that show this tendency can usually be cured by using only magnetic coupling between the transceiver and antenna system. There is no direct connection and better isolation results along with the inherent band-pass characteristics of magnetically coupled tuned circuits.

Although link coupling can be used with either single-ended or balanced antenna systems, its most common application is with balanced feed. The model shown here is designed for 80- through 10-meter operation.

The Circuit

The circuit shown in Fig. 37 and the accompanying photographs are those of a band-switched link coupler. L2 is the link and C1 is used to adjust the coupling. S1B

Fig. 36 — Exterior view of the band-switched link coupler. Alligator clips are used to select the proper tap positions of the coil.

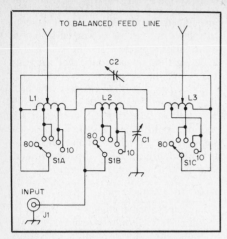

Fig. 37 — Schematic diagram of the link coupler. The connections marked as "to balanced feed line" are steatite feedthrough insulators. The arrows on the other ends of these connections are alligator clips.

C1 — 350 pF maximum, 0.0435 inch plate spacing or greater.
C2 — 100 pF maximum, 0.0435 inch plate spacing or greater.
J1 — Coaxial connector.
L1, L2, L3 — B&W 3026 Miniductor stock, 2-inch diameter, 8 turns per inch, no. 14 wire. Coil assembly consists of 48 turns. L1 and L3 are each 17 turns tapped at 8 and 11 turns from outside ends. L2 is 14 turns tapped at 8 and 12 turns from C1 end. See text for additional details.
S1 — 3-pole, 5-position ceramic rotary switch.

Fig. 38 — Interior view of the coupler showing the basic positions of the major components. Component placement is not critical, but the unit should be laid out for minimum lead lengths.

selects the proper amount of link inductance for each band. L1 and L3 are located on each side of the link and are the coils to which the antenna is connected. Alligator clips are used to connect the antenna to the coil because antennas of different impedances must be connected at different points (taps) along the coil. With many antennas it will be necessary to change taps for different bands of operation. C2 tunes L1 and L3 to resonance at the operating frequency.

Switch sections S1A and S1C select the amount of inductance necessary for each of the HF bands. The inductance of each of the coils has been optimized for antennas in the impedance range of roughly 20-600 ohms. Antennas that exhibit impedances well outside this range may require that some of the fixed connections to L1 and L3 be changed. Should this be necessary remember that the L1 and L3 sections must be kept symmetrical — the same number of turns on each coil.

Construction

The unit is housed in a homemade aluminum enclosure that measures 9 × 8 × 3½ inches. Any cabinet with similar dimensions that will accommodate the components may be used. L1, L2 and L3 are a one-piece assembly of B&W 3026 Miniductor stock. The individual coils are separated from each other by cutting two of the turns at the appropriate spots along the length of the coil. Then, the inner ends of the outer sections are joined by a short wire that is run through the center of L2. Position the wire so that it will not come into contact with L2. Each of the fixed tap points on L1, L2 and L3 is located and lengths of hookup wire are attached. The coil is mounted in the enclosure and the connections between the coil and the bandswitch are made. Every other turn of L1 and L3 is pressed in toward the center of the coil to facilitate connection of the alligator clips.

As can be seen from the schematic, C2 must be isolated from ground. This can be accomplished by mounting the capacitor on steatite cones or other suitable insulating material. Make sure the hole through the front panel for the shaft of C2 is large enough so the shaft does not come into contact with the chassis.

Tuneup

The transmitter should be connected to the input of the Transmatch through some sort of instrument that will indicate SWR. S1 is set to the band of operation and the balanced line is connected to the insulators on the rear panel of the coupler. The alligator clips are attached to the mid points of coils L1 and L3 and power is applied. Adjust C1 and C2 for minimum reflected power. If a good match is not obtained, move the antenna tap points either closer to the ends or center of the coils. Again apply power, and tune C1 and C2 until the best possible match is obtained. Continue moving the antenna taps until a 1-to-1 match is obtained.

The circuit described here is intended for power levels up to roughly 200 watts. Balance was checked by means of two RF ammeters, one in each leg of the feed line. Results showed the balance to be well within 1 dB.

A Balanced QRP Transmatch

The balanced QRP Transmatch shown in Figs 39 through 41 was designed and built by Zachary Lau, KH6CP, in the ARRL Lab. It is designed for use with balanced feed lines, although random-wire antennas can be fed if one of the antenna terminals is grounded. Unlike most Transmatches designed for use with balanced feed lines, this design features a balun at the input, rather than at the output. As a result, the balun sees impedances close to the design impedances once the Transmatch has been properly adjusted. This results in lower loss and freedom from core saturation at low power levels.

Since it is balanced currents that prevent feed-line radiation, this circuit was designed to balance currents rather than voltages. Some antenna systems use circuits that provide balanced voltages, making it necessary to make the system symmetrical in an effort to balance the currents. By going straight to balanced currents, instead of balanced voltages, it is possible to use a much simpler matching network. In addition, the actual current balance in typical amateur open-wire feed lines should be improved.

Construction

The inductor L1 and the capacitor C1 should be of the highest quality obtainable for best performance. Low-impedance loads will require a good inductor for efficient matching, while high-impedance loads will require a good capacitor.

L1 was wound with tinned copper wire to make it easier to adjust taps. It is necessary to wind the wire with spaces between turns to prevent shorts which may make the inductor lossy; no. 16 wire is heavy enough to stay in position on the toroid. The inductor used had a full-inductance Q of 420 at 7.9 MHz; the Q was 410 after the taps and switch were added. The use of clip lead taps is not recommended as they increase losses, although they may be useful in initially setting up tap positions.

Capacitor C1 should have a value of at least 250 pF, and larger capacitors will work even better, increasing the range of the Transmatch at low frequencies. Suitable capacitors are usually available at hamfests. The value of C2 and C3 should equal C1, and C4 should be twice the value of C1. If the calculated values of C2, C3 and C4 are not available, smaller values may be used.

Capacitor C1 must be insulated from the chassis, so it was mounted using ½-inch Plexiglas® with tapped screw holes. An insulated shaft coupler was used to prevent high voltage from appearing on the knob set screw. The cabinet is a Ten-Tec MW-8 with a model 91-206 matching tilt-up bail. Although it is large for a QRP Transmatch, the cabinet matches the author's QRP rig

Fig 39 — This QRP Transmatch for balanced feed lines features a balun at the input of the matching network.

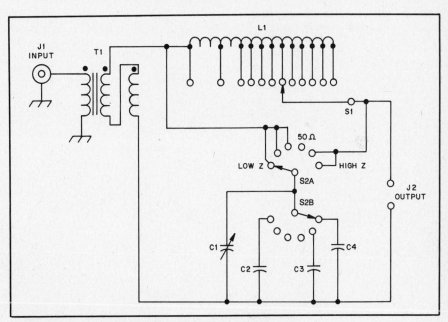

Fig 40 — Schematic diagram of the QRP Balanced Transmatch.

C1—330 pF variable capacitor, 500 V rating.
C2,C3—330 pF silver-mica capacitor, 500 V rating.
C4—600 pF silver-mica capcitor, 500 V rating.
J1—Female chassis-mount BNC connector.
J2—Two ceramic feedthrough insulators.
L1—36t no. 16 tinned wire on an Amidon T-200-6 core.

S1—Ceramic rotary switch, single wafer, 1 pole, 12 position.
S2—Ceramic rotary switch, single wafer, 2 pole, 6 position.
T1—12 trifilar turns on an Amidon FT-114-61 core; primary, no. 16 enam wire; secondary, no. 18 enam wire.

and allows the controls to be spaced apart for easy use. The logging scale is typewritten paper attached to the cabinet with a Plexiglas sheet.

T1 is a trifilar-wound transformer.

Winding details are shown in Fig 40. It is possible to wind this coil with only two windings, eliminating the solder joint. The coil should be duplicated exactly with regard to the number of turns and core material unless the transformer can be tested at the operating frequency. Testing can be done by hooking up two baluns in series and measuring the insertion loss. The matching network will compensate for a poor balun, but efficiency will probably suffer. A toroidal choke balun would be recommended for a higher power version.

Switches S1 and S2 should be ceramic. Phenolic switches are not recommended, although they should work at low power levels on the order of a few watts. The switch positions should never be changed while more than a few watts of RF is applied.

Adjustment

Adjustment of the Transmatch is much easier if the approximate impedance of the load is known. In his article in *The ARRL Antenna Compendium, Volume 1*, "Mr. Smith's 'Other' Chart and Broadband Rigs" Roger Ghormley, W0KK, details how parts values for L networks can be calculated. Alternately, received signals can be peaked up by first adjusting the inductor and then the capacitor. As with any Transmatch, low power should be used in the initial adjustment. The actual power handling capability will depend on the load. The capacitor breakdown voltage is the limiting factor on high-impedance loads; a 2000-ohm load will cause the 500 V capacitors to reach their maximum rating at 62.5 W, while the maximum rating will be reached with 625 watts into a 200-ohm load. The current-handling capability of the wire is the limit on low-impedance loads;

Fig. 41 — Interior layout of the QRP Transmatch. The variable capacitor is mounted on a Plexiglas block.

a 40-ohm load will cause a 90-W signal to generate 1.5 A through the wire, while 450 watts will generate 1.5 A if the load is 200 ohms. These values are for resistive loads; a reactive load would require higher current and voltage ratings. The unit shown here has worked well in low-power operation (up to 4.5 watts).

A 50- to 75-Ohm Broadband Transformer

Shown in Figs. 42 through 44 is a simple 50- to 75-ohm or 75- to 50-ohm transformer that is suitable for operation in the 2- to 30-MHz frequency range. A pair of these transformers is ideal for using 75-ohm CATV hardline in a 50-ohm system. In this application one transformer is used at each end of the cable run. At the antenna one transformer raises the 50-ohm impedance of the antenna to 75 ohms, thereby presenting a match to the 75-ohm cable. At the station end a transformer is used to step the 75-ohm line impedance down to 50 ohms.

The schematic diagram of the transformer is shown in Fig. 42, and the winding details are given in Fig. 43. C1 and C2 are compensating capacitors; the values shown were determined through swept return-loss measurements using a spectrum analyzer and tracking generator. The transformer consists of a trifilar winding of no. 14 enameled copper wire wound over an FT-240-61 (Q1 material) or equivalent core. As shown in Fig. 43, one winding has only half the number of turns of the other two. Care must be taken when connecting the loose ends so the proper phasing of the turns is maintained. Improper phasing will become apparent when power is applied to the transformer.

If the core has sharp edges it is a good idea either to sand the edges until they are relatively smooth or wrap the core with tape. The core shown in the photograph was wrapped with ordinary vinyl electrical tape, although glass-cloth insulating tape would be better. The idea is to prevent chafing of the wire insulation.

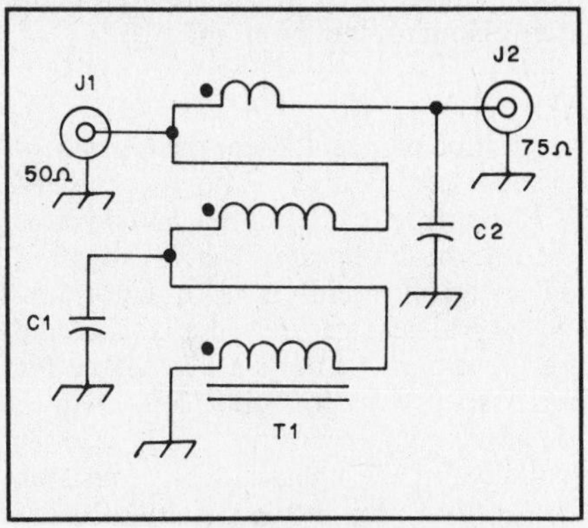

Construction

The easiest way to construct the transformer is wind the three lengths of wire on the core at the same time. Different color wires will aid in identifying the ends of the windings. After all three windings are securely in place, the appropriate wire may be unwound three turns as shown in the diagram. This wire is the 75-ohm connection point. Connections at the 50-ohm end are a bit tricky, but if the information in Fig. 43 is followed carefully no problems should be encountered. Use the shortest connections possible, as long leads will degrade the high-frequency performance.

The transformer is housed in a home-made aluminum enclosure measuring $3\frac{1}{2} \times 3\frac{3}{4} \times 1\frac{1}{4}$ inches. Any commercial cabinet of similar dimensions will work fine. In the unit shown in the photograph, several "blobs" of silicone seal (RTV) were used to hold the core in position. Alternatively, a piece of phenolic insulating material may be used between the core and the aluminum enclosure. Silicone seal is used to protect the inside of the unit from moisture. All joints and screw heads should receive a generous coating of RTV.

Checkout

Checkout of the completed transformer or transformers is quite simple. If a 75-ohm dummy load is available connect it to the 75-ohm terminal of the transformer. Connect a transmitter and SWR indicator (50 ohm) to the 50-ohm terminal of the transformer. Apply power (on each of the HF bands) and measure the SWR looking into the transformer. Readings should be well under 1.3 to 1 on each of the bands. If a 75-ohm load is not available and two transformers have been constructed they may be checked out simultaneously as follows. Connect the 75-ohm terminals of both transformers together, either directly through a coaxial adaptor or through a length of 75-ohm cable. Attach a 50-ohm load to one of the 50-ohm terminals and connect a transmitter and SWR indicator (50 ohm) to the remaining 50-ohm terminal. Apply power as outlined above and record the measurements. Readings should be under 1.3 to 1.

The transformers were checked in the ARRL laboratory under various mismatched conditions at the 1500-watt power level. No spurious signals (indicative of core saturation) could be found while viewing the LF, HF and VHF frequency range with a spectrum analyzer. A keydown, 1500-watt signal produced no increase in the temperature of the windings.

Using the Transformers

For indoor applications, the transformers can be assembled open style, without benefit of a protective enclosure. For outdoor installations, such as at the antenna feed point, the unit should be encapsulated in epoxy resin or mounted in a suitable weatherproof enclosure. A Minibox, sealed against moisture, works nicely for the latter.

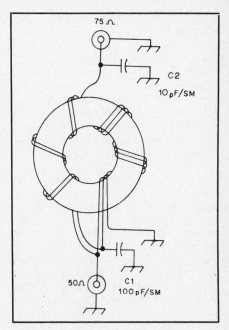

Fig. 43 — Pictorial drawing of the 50- to 75-ohm transformer showing details of the windings.

Fig. 44 — Packaging of the 50- to 75-ohm transformers.

Fig. 42 — Schematic diagram of the 50- to 75-ohm transformer described in the text. C1 and C2 are compensating capacitors.

C1 — 100 pF, silver mica.
C2 — 10 pF, silver mica.
J1, J2 — Coaxial connectors, builder's choice.
T1 — Transformer, 6 trifilar turns no. 14 enameled copper wire on an FT-240-61 (Q1 material, $\mu_i = 125$) core. One winding has one-half the number of turns of the other two.

Super Dummy

Super Dummy is an RF load for use with full-power RF amplifiers. See Figs. 45 through 49. Designed by ARRL TA Dick Jansson, WD4FAB, this dummy load will "loaf" along at 1500 watts for 10 minutes without complaint. Super Dummy is not just a copy of the similar appearing Heathkit® HN-31A Cantenna. A lot of development went into proving this design. These efforts concentrated on RF matching, heat transfer design and coolant selection.

It is not easy to design a dummy load for high-power operation. Heat must be safely dissipated and a good match maintained. Forced air cooling of 300 two-watt carbon resistors and running a Heath Cantenna with automobile antifreeze coolant were tried. Neither of these ideas proved satisfactory. Efforts were then turned to other directions. An improved dummy load evolved from this work. Super Dummy is suitable for HF and low VHF regions at 1500 watts.

The Bird Model 612 dummy load/watt-meter uses the type of construction shown in Fig. 46. This design ensures proper characteristic impedance (Z_0) along the structure. Starting with the coaxial connector, a tapered line is used to provide a constant impedance (50 ohms) to the resistor. The coaxial shield then decreases in size to

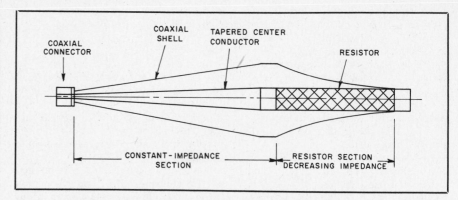

Fig. 46 — An RF load should have proper characteristic impedance along the entire structure for operation over a wide frequency range. See text.

match the diminishing impedance to ground along the resistor. Coaxial line impedance can be determined from the formula:

$$Z_0 = \frac{138}{\sqrt{\epsilon}} \log \frac{D}{d}$$

where
Z_0 = coaxial line characteristic impedance
D = ID of coaxial shell
d = OD of coaxial conductor
ϵ = dielectric constant of coolant (air = 1.0)

This equation shows that as the impedance along the resistor diminishes following a straight-line curve, the shape of the coaxial shield or shell around the resistor follows a smooth logarithmic curve. That curve can be seen in Fig. 46.

A logarithmically tapered coaxial shell is difficult to form. At HF, we don't have to worry about the shape of the coaxial shell. Unless you are a purist, it is far easier to just use the straight tubing shown in Figs. 47 and 48. A straight shell proves quite usable into the VHF region. For a 50-ohm resistor, the coaxial shell is sized for a Z_0 of about 25 ohms.

RF Performance

The SWR of Super Dummy was measured in the ARRL lab. Fig. 49 shows the results, an excellent match through the HF region (SWR less than 1.05). SWR is only 1.08 at 50 MHz, rising to a tolerable 1.4 at 144 MHz.

Construction

Construction of Super Dummy is straightforward. Two one-gallon paint cans were obtained from a painting supply house. A 5.5-inch-diameter disc is removed from the closed end of the top can and four 0.120-inch holes are drilled equally spaced through the remaining bottom rim with matching holes in the top rim of the bottom can. The two cans can now be locked together with four no. 4 screws and nuts,

forming a "two high" can assembly. Carefully solder the rims of the cans to fully seal the assembly against oil leakage. Be sure to secure the cans with the screws as solder is dependable only for sealing and not for mechanical strength.

Mount the shield tube to the cover with four aluminum angle brackets. Space the tube 1.5 to 2 inches from the cover, allowing room for the coolant to flow out of the top. Clamp the copper straps to each end of the resistor. Under the top strap be sure to include the U-shaped strap that is used to connect the resistor and the SO-239 coaxial connector. Install the resistor in the tube, making sure that the U strap slides over the SO-239 connector center pin. Center and clamp the lower end firmly with four no. 6 brass screws and lock nuts. Solder the U strap to the connector pin after clamping the lower end.

Coolant

Selection of a coolant for the resistor involved a lengthy investigation. The use of PCB transformer oils is illegal and involves a serious health hazard. The investigation did identify several modern-day transformer and heat transfer oils that could be used, but none of these are available for purchase in any quantity less than a 55-gallon drum.[1] Transformer oil has lower viscosity. That means lower resistor temperatures, but at the penalty of using an oil with a 140 to 150°C flash point (vaporization temperature). Transformer oils are also treated to remove water, but maintaining this dry condition is difficult and expensive. Calculations of the fluid heat transfer show that turbulent nature convection flow exists for rated power levels. The use of higher viscosity oil produces higher resistor temperatures, but the gain in flash point temperature is greater than the rise in resistor temperature. Oils designed for heat transfer have flash points in the 230 to 250°C region.

[1]Shell Oil Co., "Thermia"; Chevron Oil Co., "Heat Transfer Oil No. 1"; Gulf Oil Co., Security No. 53; Texas Oil Co., "Thermia".

Fig. 45 — Super Dummy is an RF load designed to handle 1.5 kW.

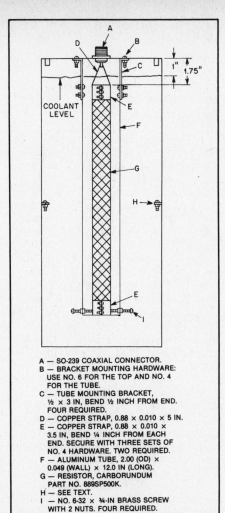

A — SO-239 COAXIAL CONNECTOR.
B — BRACKET MOUNTING HARDWARE:
USE NO. 6 FOR THE TOP AND NO. 4
FOR THE TUBE.
C — TUBE MOUNTING BRACKET,
½ × 3 IN, BEND ½ INCH FROM END.
FOUR REQUIRED.
D — COPPER STRAP, 0.88 × 0.010 × 5 IN.
E — COPPER STRAP, 0.88 × 0.010 ×
3.5 IN, BEND ¼ INCH FROM EACH
END. SECURE WITH THREE SETS OF
NO. 4 HARDWARE. TWO REQUIRED.
F — ALUMINUM TUBE, 2.00 (OD) ×
0.049 (WALL) × 12.0 IN (LONG).
G — RESISTOR, CARBORUNDUM
PART NO. 889SP500K.
H — SEE TEXT.
I — NO. 6-32 × ¾-IN BRASS SCREW
WITH 2 NUTS. FOUR REQUIRED.

Fig. 47 — X-ray view of the interior of Super
Dummy. To ensure proper cooling, coolant
level should be higher than the top of the
tube.

Fig. 48 — View of the insides of Super Dummy. Construction details are given in Fig. 47.

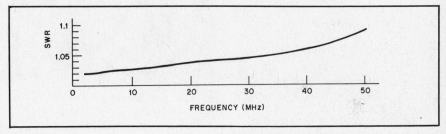

Fig. 49 — Super Dummy SWR curve. Measurements were made in the ARRL lab using swept
return-loss techniques.

Heat transfer oils are composed predominantly of mineral oil. Highly refined or medicinal mineral oil has a usably high flash point but an undesirably high cost. Heat transfer oils contain small amounts of rust and oxidation inhibitors, making them preferable to mineral oils. Heat transfer oils are not available in small quantities, but in at least one case the very same product is relabeled and sold in 5 gallon cans for service as a "turbine oil." That product is sold by TEXACO distributors as Regal Oil R&O No. 46.

Use of automotive motor and transmission oils is undesirable. They are composed of a wide mixture of oils with differing flash points. Motor oil additives will damage the metal platings on an RF resistor. There is no danger in the short term use of automotive oils, but the long-term

effects are a problem. Further, there are no economic advantages with automotive oils, as they cost as much as turbine oil.

Thermal Tests

Thermal performance tests were done with Super Dummy in mineral oil at a power level of 1.5 kW. The large resistor used in Super Dummy has a reasonable power density (3.5 watts/cm²).[2] That

[2]Carborundum 889SP500K. Available in single quantities from Radiokit, P.O. Box 973, Pelham, NH 03076 (tel 603-635-2235). Also available from The Carborundum Co., Electric Products Div., P.O. Box 339, Niagra Falls, NY 14302.

results in moderate operating temperatures and negligible resistance shift. Some tests were conducted without the coaxial shield tube. Considerable thermal stagnation of the oil was measured in these tests with temperatures in the range of 160 °C at the top and only 40 °C at the bottom of the container. The coaxial shield tube around the resistor performs another valuable function here. The tube enhances oil flow from the bottom of the can to the top, forcing better oil mixing in the can and effective use of the total thermal mass of the 2 gallons of coolant.

UHF Gallon Dummy

UHF Gallon Dummy is an RF load for use with VHF and UHF RF amplifiers up to 1.3 GHz and power outputs to 1 kW. See Figs. 50 through 58. Designed by ARRL TA Dick Jansson, WD4FAB, this dummy load is an extension of the work done with "Super Dummy," the HF legal-limit load presented earlier in this chapter. It will handle 1000 watts for five minutes without complaint. A lot of development went into proving this design, with efforts concentrated on RF matching, heat transfer design and coolant selection.

The Bird Model 612 dummy load/watt-meter uses the type of construction shown in Fig. 46. This design ensures proper characteristic impedance (Z_0) along the structure. Starting with the coaxial connector, a tapered line is used to provide a constant impedance (50 ohms) to the resistor, while expanding the transmission line size up to meet that of the resistor. The coaxial shield then decreases in size along the resistor to match the diminishing impedance-to-ground of the resistor. Coaxial line impedance can be determined from the formula

$$Z_0 = \frac{138}{\sqrt{\epsilon}} \log \frac{D}{d}$$

where

Z_0 = coaxial line characteristics impedance
D = ID of coaxial shell
d = OD of coaxial conductor
ϵ = dielectric constant of coolant (air = 1.0).

This equation shows that as the impedance along the resistor diminishes following a straight-line curve; the shape of the coaxial shield or shell around the resistor follows a smooth logarithmic curve. That curve can be seen in Fig. 46.

A logarithmically tapered coaxial shell is difficult to form, so most designers just use straight tubing at HF. This approach works just fine. For the VHF and UHF region, however, proper RF impedance matching is very important because the dimensions of the load become significantly sizable compared to the wavelengths of interest. UHF Gallon Dummy provides a close approach to the ideal shapes.

RF Performance

The SWR of the UHF Gallon Dummy was measured in the ARRL lab. Fig. 51 shows the results: An excellent match through 220 MHz (SWR less than 1.05). SWR is a tolerable 1.25 for the 70-cm band and less than 1.25 for the 23-cm band. To obtain these results, you *must* have UHF Gallon Dummy immersed in its coolant oil. The dielectric constant of the oil is approximately 2.3. UHF Gallon Dummy tested dry showed an SWR of approximately 2.0 across most of the spectrum. The change

is caused by the difference between the dielectric constant of the oil and that of air.

Deviations in performance from the target design are caused by some of the practicalities of constructing such a load. The builder can carefully watch, and avoid, such elements as the dimensional steps in the center conductor as it attaches to the connector, and at the junction with the resistor. The lack of a logarithmically shaped resistor shell will, of course, provide an element of performance error. In short, paying attention to small details is important in obtaining the best performance from this load.

Construction

Although you can build the UHF dummy load shown here from parts purchased separately from suppliers, a good way to begin is to purchase a Heathkit® HN-31A "Cantenna" dummy load and modify it. This way, you won't have to track down the power resistor, and you'll get one of the paint cans complete with a lid with a vent hole.

The housing for the UHF Gallon Dummy is made from two one-gallon paint cans obtained from a painting supply house. Remove a 5.5-inch-diameter disc from the closed (bottom) end of the top can. Drill four 0.120-inch holes, equally spaced, around the remaining bottom rim; drill matching holes in the top rim of the bottom can. The two cans can now be locked together with four no. 4 screws and nuts, forming a "two high" can assembly. Carefully solder the rims of the cans to fully seal the assembly against oil leakage. You must secure the cans with the screws, as solder is only dependable for sealing and not for mechanical strength. Prime and paint the exterior of this can assembly to resist the rigors of rust corrosion.

The interior parts of the UHF Gallon Dummy were built using hobby store brass shim stock, a 1-kW (rating with coolant) 50-ohm resistor (available from Carborundum[1] or from a Heathkit HN-31A) and

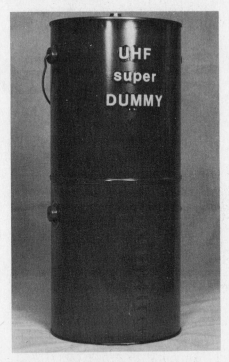

Fig. 50 — UHF Gallon Dummy is built into two one-gallon paint cans bolted and soldered together.

[1]Carborundum 886SP500K. Available from The Carborundum Co., Electric Products Div., P.O. Box 339, Niagara Falls, NY 14302.

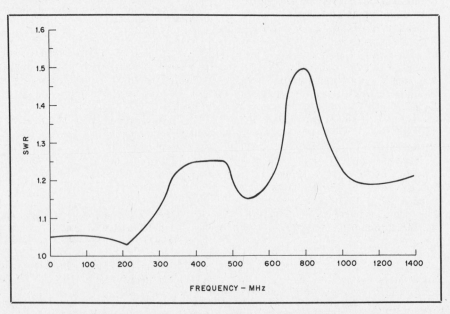

Fig. 51 — UHF Gallon Dummy SWR curve made in the ARRL lab using swept return-loss techniques. Measurements were made in the two-gallon can with turbine oil coolant.

some wooden dowels. Apart from an electric drill and a soldering iron, this double conical shape was formed entirely with hand tools. Fig. 52 is an X-ray view of the interior of the UHF Gallon Dummy showing the arrangement of the parts.

Make all of the parts before you begin assembly of any part of the load. Figs. 53 and 54 detail the flat layout of the sheet brass that will be used to make the transition and resistor cones. Fig. 55 shows the center-conductor cone, four braces and two clamp bands. These drawings are full-size templates. They may be photocopied and used directly to do the sheet-metal layout. Carefully lay out and cut these parts to size. Drill the no. 60 holes and cut the joining tabs.

Start assembly by forming the brass stock for the transition cone into a conical shape. The edges of the brass stock are lap-joined and soldered along the longitudinal seam. You will get better results if you drill some small holes in the tab where the two edges overlap and install 1/16-inch eyelets to act as rivets to hold the assembly together as you solder. Assemble the resistor cone the same way. Assembly of the conical copper center conductor is similar, except that it is just soldered along its seam (no rivets). Figs. 56 and 57 show how the pieces should look when finished and how they will attach to the paint can lid.

Drill a hole for the UG-680 hermetic Type-N connector in the center of the paint can lid. Attach the connector nut and securely tighten it in place. Solder the connector to top of the lid, as shown in Fig. 58, to seal the connection against possible coolant leakage.

Solder the small end of the brass transition cone to a shortened nut from a PL-259 connector. This nut will screw onto the back-side threads of the UG-680 connector. Be sure to drill a small vent relief hole through the PL-259 nut to allow coolant oil into the upper section of cone; see Fig. 56. Check this assembly on the UG-680 to be sure that it is perpendicular to the lid, but do not attach yet.

Clamp the large end of the conical center conductor to the resistor and check that it is straight and true. Check the small end of the center-conductor tube for fit on the UG-680 center conductor. Slide this cone-and-resistor subassembly into the brass transition shell; leave it loose. Carefully line up the tabs on the large end of the transition cone with the tabs on the large end of the resistor cone. When you are sure of the alignment, solder the tabs together to form one large double-conical shell. The cone-and-resistor subassembly will still be a loose part inside the shell assembly.

Slide the small end of the center-conductor cone through the small end of the transition cone and solder it to the center conductor of the UG-680 connector. Screw the double-cone shell assembly to the outside of the UG-680 connector. When the connector end is in place, the free end of

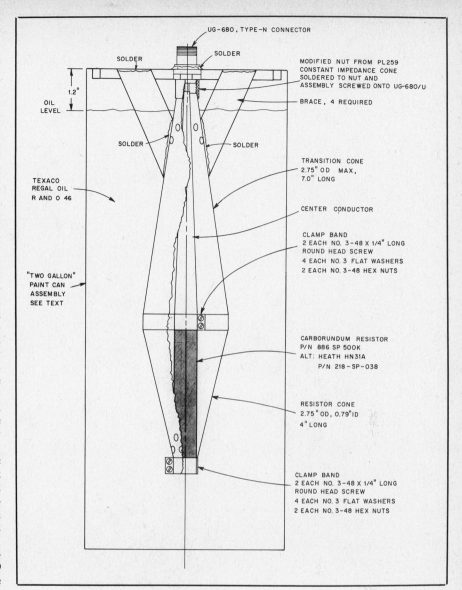

Fig. 52 — Cutaway view of UHF Gallon Dummy.

the resistor should fit into the small end of the resistor cone, where it will be secured with a brass clamp band.

Add the four braces by soldering them between the lid and the transition cone to make a more rigid structure. Check your work at this point and make any corrections.

Coolant

Selection of a coolant for the resistor involved a lengthy investigation. The use of PCB transformer oils is illegal and involves a serious health hazard. The investigation did identify several modern-day transformer and heat transfer oils that could be used, but none of these are available for purchase in any quantity less than a 55-gallon drum.[2] Transformer oil has lower

viscosity. That means lower resistor temperatures, but at the penalty of using an oil with a 140 to 150°C flash point (vaporization temperature). Calculations of the fluid heat transfer show that a turbulent natural convection flow exists for rated power levels. The use of higher viscosity oil produces higher resistor temperatures, but the gain in flash point temperature is greater than the rise in resistor temperature. Oils designed for heat transfer have flash points in the 230 to 250°C region. Transformer oils are treated to remove water, but maintaining this dry condition is difficult and expensive.

Heat transfer oils are composed mainly of mineral oil. Highly refined or medicinal mineral oil has a usably high flash point but an undesirably high cost. Heat transfer oils contain small desirable amounts of rust and oxidation inhibitors, making them preferable to mineral oils. Heat transfer oils are not available in small quantities, but in at

[2]Shell Oil Co., "Thermia"; Chevron Oil Co., "Heat Transfer Oil #1"; Gulf Oil Co., "Security #53"; Texas Oil Co., "Thermia".

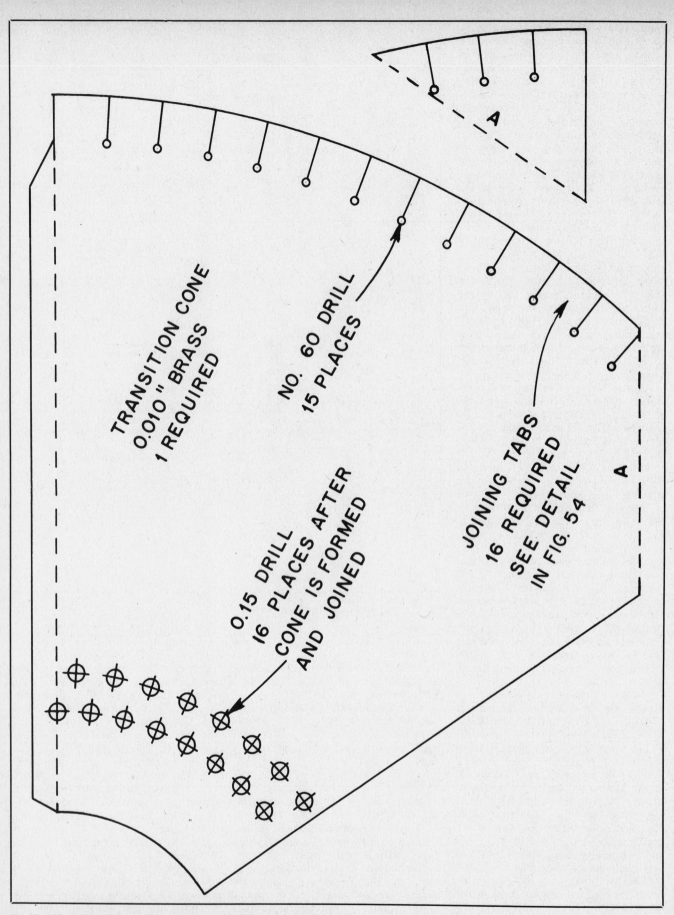

TRANSITION CONE
0.010" BRASS
1 REQUIRED

NO. 60 DRILL
15 PLACES

0.15 DRILL
16 PLACES AFTER
CONE IS FORMED
AND JOINED

JOINING TABS
16 REQUIRED
SEE DETAIL
IN FIG. 54

A

A

Fig. 53 — Full-size template of the transition cone. Make a copy of this page, cut out the two pieces and tape them together along edge A. You can use this template directly to lay out the transition cone on sheet brass.

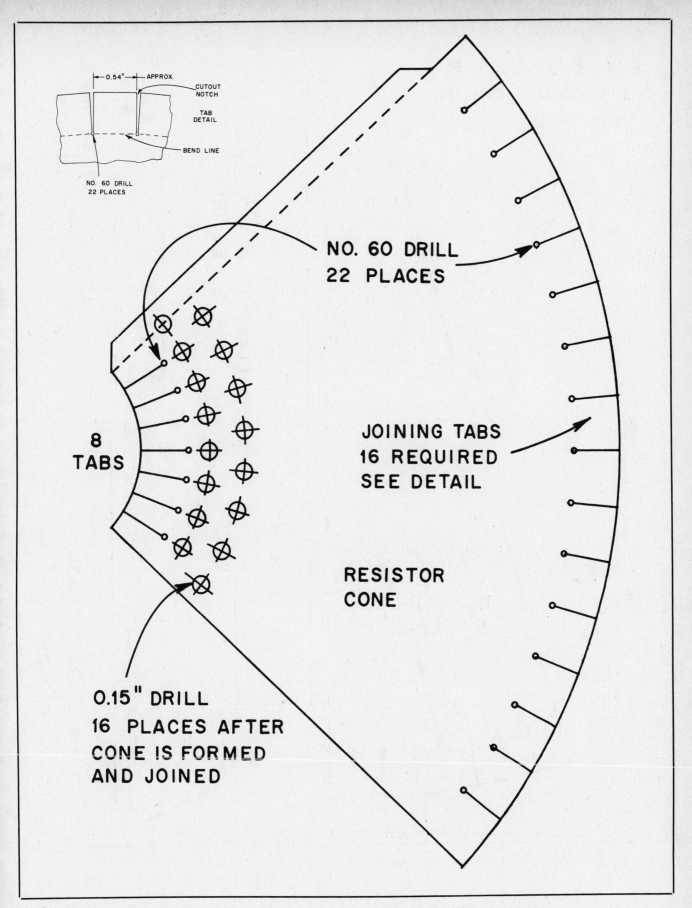

0.54" — APPROX.

CUTOUT
NOTCH

TAB
DETAIL

BEND LINE

NO. 60 DRILL
22 PLACES

NO. 60 DRILL
22 PLACES

8
TABS

JOINING TABS
16 REQUIRED
SEE DETAIL

RESISTOR
CONE

0.15" DRILL
16 PLACES AFTER
CONE IS FORMED
AND JOINED

Fig. 54 — Full-size template of the resistor cone. The tab detail applies to this cone and the transition cone of Fig. 53.

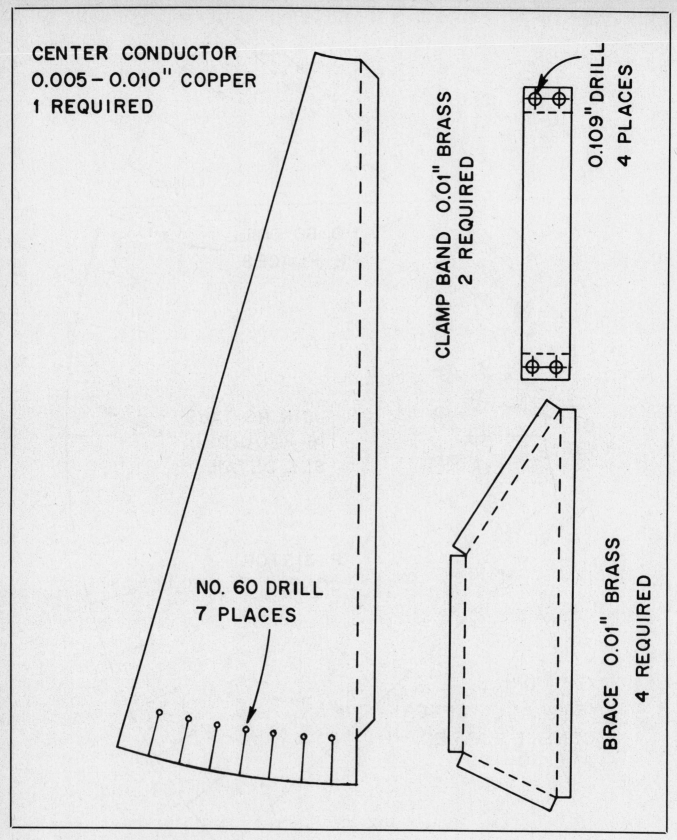

CENTER CONDUCTOR
0.005 – 0.010" COPPER
1 REQUIRED

CLAMP BAND 0.01" BRASS
2 REQUIRED

0.109" DRILL
4 PLACES

NO. 60 DRILL
7 PLACES

BRACE 0.01" BRASS
4 REQUIRED

Fig. 55 — Full-size templates for the center conductor cone, braces and clamp bands.

Fig. 56 — Four braces secure the transition cone to the lid of the paint can. Drill a hole in the PL-259 nut that solders to the transition cone and screws to the Type-N connector.

Fig. 57 — The resistor clamps to the end of the resistor cone (top of photo). At the bottom of the photo, the tabs on the resistor and transition cones are soldered together.

Fig. 58 — The UG-680 Type-N connector is soldered to the lid of the paint can to seal the hole against leakage. To the right of the connector is a vent made from a no. 6-32 machine screw, a spring and a nut.

least one case the very same product is relabled and sold in 5 gallon cans for service as a "turbine oil." That product is sold by the Texas Oil Company (TEXACO) distributors as Regal Oil R&O No. 46.

Use of automotive motor and transmission oils is undesirable. They are composed of a wide mixture of oils with differing flash points. Motor oil additives will damage the metal platings on an RF resistor. There is no danger in the short term use of automotive oils, but the long term effects are a problem. There are no economic advantages with automotive oils either; they cost just as much as turbine oil.

Thermal Tests

Thermal performance tests were done with the UHF Gallon Dummy with a 1-kW resistor (rating with coolant) in mineral oil at a power level of 1.5 kW. The tests were conducted with considerable apprehension, but no spectacular events occurred except for very vigorous convection flow of oil over the resistor. Don't use this size resistor at higher power levels than its 1 kW (cooled) rating. The resistor gets *very hot* at 1.5 kW, and substantial permanent resistance shifts have been measured.

Some tests were conducted without a coaxial shield tube. Considerable thermal stagnation of the oil was measured in these tests with temperatures in the range of 160 °C at the top and only 40 °C at the bottom of the container. The coaxial shield assembly at the top and only 40 °C at the bottom of the container. The coaxial shield assembly around the resistor performs another valuable function here. The shield enhances oil flow from the bottom of the can to the top, forcing better oil mixing in the can and effective use of the total thermal mass of the two gallons.

A Marker Generator for 100, 50 and 25 kHz

A marker generator that provides reference signals at 100-, 50-, and 25-kHz intervals is a useful addition to any ham shack. The generator can be used as a low-level signal source when testing receivers, or it can be installed inside a receiver and used to mark the edges of the amateur bands and subbands. When built into a receiver, the marker generator is usually called a crystal calibrator.

The standard design for a marker generator involves using a crystal oscillator operating at 100 kHz and then dividing this signal with flip-flops to get the 50- and 25-kHz signals. The 100-kHz crystals are rather expensive, however, and are difficult to locate.

Another way to produce marker signals is to use a crystal oscillator operating at a multiple of 1 MHz. Crystals for 1, 2 and 4 MHz are readily available at reasonable prices on the computer surplus market.

Fig. 63—Front view of the 100-, 50- and 25-kHz marker generator. A two-pole rotary switch was used in this unit; one section of the switch was used to apply power to the generator, and the other section was used to select the desired signal. Another builder may wish to use a separate power switch and/or LED power indicator.

Low-cost TTL ICs can then be used to divide the signal down to 100, 50 and 25 kHz. This is the approach taken in the marker generator shown in Figs. 63 through 66. The marker generator is simple to build, and it provides a good introduction to the principles of digital circuits. The output of the generator is rich in harmonics and provides marker signals up into the 2-meter band. This project was designed and built by Bruce Hale, KB1MW, in the ARRL Lab.

Circuit Description

The schematic of the marker generator is shown in Fig. 64. Two of the NAND gates in a 7400 quad-NAND package, U1, are used as an oscillator. The crystal, Y1, sets the operating frequency at 1, 2 or 4 MHz, and C1 is used to trim the frequency during the calibration procedure. U1C acts as a buffer between the oscillator and the rest

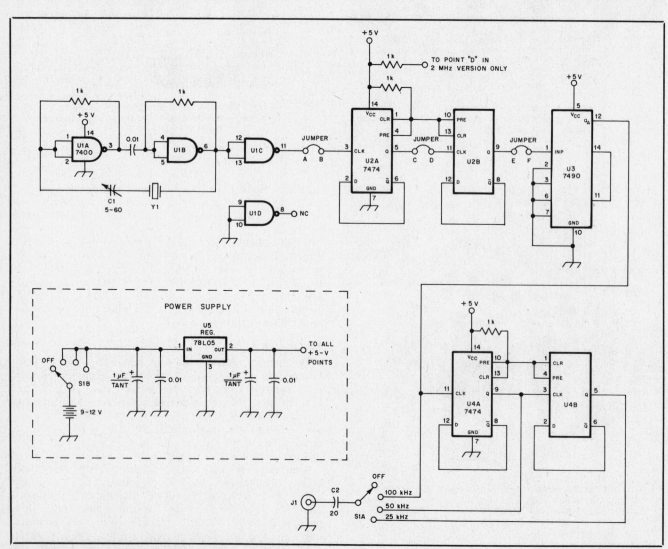

Fig. 64—Schematic diagram of the marker generator. All resistors are ¼-W, 5% types. All capacitors are disk ceramic unless noted. Capacitors marked with polarity are tantalum electrolytic.

C1—5-60 pF miniature trimmer capacitor.
C2—20 pF disk ceramic (see text).
J1 — Female, chassis-mount BNC connector, or other suitable connector of the builder's choice.
U1—7400 or 74LS00 quad NAND gate.
U2, U4—7474 or 74LS74 dual D flip-flop.
U3—7490 or 74LS90 decade counter.
U5—78L05 or 7805 5-V regulator (see text).
Y1—1, 2 or 4-MHz crystal in HC-25, HC-33 or HC-6 holder.

Fig. 65—The marker generator may be built using a 1 MHz, 2 MHz or 4 MHz crystal. With a 1 MHz crystal, U2 is not needed. The 4-MHz version was built using a 7805 regulator. All the units built in the ARRL lab could be trimmed to within 50 Hz of 100 kHz, including a prototype built on perforated board.

of the circuit; U1D is not used, so its inputs are grounded. Unused TTL gate inputs should always be connected either to ground or to V_{cc} through a pull-up resistor.

U2 is a 7474 dual D flip-flop. The two flip-flops in this package are used to divide the output of the oscillator down to 1 MHz. This circuit is designed so that crystals operating at 1, 2 or 4 MHz can be used, depending on what is available. If a 1-MHz crystal is used, U2 is not needed and can be left completely out of the circuit. If a 2-MHz crystal is used, one flip-flop of the two available is used, and the input of the second flip-flop is pulled high with a 1-kΩ resistor. Using a 4-MHz crystal requires both flip-flops for a 1-MHz output to U3. Notice the PRESET and CLEAR inputs (pins 1, 4, 10 and 13); these inputs can be pulled low to set and clear the flip-flops. During normal operation these inputs must be held high—a 1-kΩ pull-up resistor is connected from the PRESET and CLEAR inputs to V_{cc}.

U3 is a 7490 divide-by-ten counter (sometimes called a decade counter). A 1 MHz input to this IC produces a 100 kHz output signal at pin 12. We could stop here if we were building a 100-kHz signal generator.

Another 7474 dual flip-flop is used at U4 to divide the 100-kHz output of U3 down to 50 and 25 kHz. Again, the PRESET and CLEAR inputs must be pulled high. The outputs of U3, U4A and U4B are connected to switch S1A to select the desired output. C2 decouples the output jack from the circuit to prevent excessive loading.

U5 is a 78L05 low-current, 5-V regulator IC in a TO-92 package. The regulator input and the output are bypassed with 1-μF tantalum and 0.01-μF disc ceramic capacitors. A 9-V transistor battery provides the power source, and S1B is used as an on-off switch.

Construction

Construction of this project is straight-forward. A PC board was used in the final versions shown in Fig. 65, but a prototype built on a Radio Shack perforated board worked well. Wire-wrap construction could also be used.

A PC component-placement diagram is shown in Fig. 66. The PC pattern shown can be used to build the marker generator using any of the three crystal frequencies by varying the placement of three jumper wires. Locations for the jumpers are clearly labeled on the PC pattern and they should be set as shown in Table 1, depending on the crystal frequency. Provision is made on the PC board for either crystals in small

HC-25 holders or the larger HC-33 or HC-6 holders.

If low-power Schottky ICs are available, they should be used at U1, U2, U3 and U4. These ICs draw considerably less current than their standard TTL counterparts, and using them will reduce the drain on the battery. These devices are labeled with "LS" as part of their numerical designator; a low-power quad NAND gate is designated a 74LS00.

The value of the coupling capacitor, C2, is not critical. Any value between 10 and 100 pF will work, with the larger values providing a bit less signal at the output jack.

A standard 7805 regulator IC in a TO-220 package can be used at U5 if a 78L05 cannot be located. This is "overkill" in a way, since the 7805 is rated to pass at least an amp of current, and the marker generator draws only about 70 mA, even with standard TTL ICs. The 78L05 regulator should be available from the same companies that sell the TTL ICs used in this project.

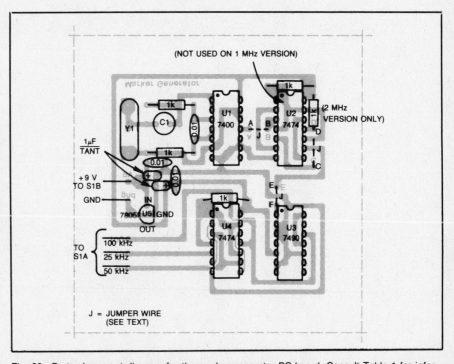

Fig. 66—Parts-placement diagram for the marker generator PC board. Consult Table 1 for information on placing the jumpers at points A, B, C, D, E and F. Extra pads in the oscillator section of the board allow for the use of a crystal in a small HC-25 holder or a larger HC-33 or HC-6 holder. Pads are also provided for different sizes of trimmer capacitor. A full-size etching pattern appears at the back of this book.

The input voltage to the regulator can be as high as 12-V dc; this may be desired if the marker generator is built inside a receiver using a 12-V power supply or some other source of 12-V dc is convenient. The low-power regulator may require a small press-on heat sink if 12-V dc is used, however.

A Radio Shack 2-pole 6-position rotary switch was used in this version of the marker generator, but any available rotary switch with at least four positions can be used. If a 2-pole switch is not available, a separate power switch can be used, but be careful to switch the generator off when it is not in use. An LED can be used as a power indicator.

Computer surplus crystals were used in all the units shown here, and all the generators could be trimmed to within 50 Hz of 100 kHz. These crystals are available from many of the electronics dealers listed in Chapter 35, such as BCD Electro or All Electronics Corporation. A crystal can also be obtained from a manufacturer such as International Crystal or JAN Crystals, but the cost will be higher.

The value of C1 (the oscillator frequency trimmer) is not critical. Trimmers are available at Radio Shack in a 5 to 60 pF range—this is fine. Too small a value may not trim the oscillator frequency properly, and too large a value may prevent the oscillator from operating.

Adjustment and Calibration

To check the power supply section of the marker generator, measure the input and output voltages at U5. Input to the IC should be at least 8 V, and the output should be between 4.8 and 5.2 V.

Use an oscilloscope to observe the output waveform as S1 is rotated to select the 100, 50 and 25-kHz marker signals. The output should approximate a 3 to 5 V peak-to-peak square wave. The larger the value of the coupling capacitor C2, the less "square" the output will be as the higher-frequency harmonics are attenuated.

The easiest way to calibrate the marker generator is with a frequency counter. With the counter connected to the output of the calibrator, adjust C1 until the counter measures the frequency to the desired precision. Greater capacitance at C1 will lower the oscillator frequency; if the oscillator will not trim with C1 at maximum capacitance, a low-value capacitor can be added in parallel with C1 to further reduce the oscillator frequency.

If a frequency counter is not available, the generator can be calibrated using WWV. Tune in WWV and disconnect the receiver antenna. Switch the receiver to the CW position (or turn on the BFO) without changing the setting of the tuning dial. Connect a short piece of wire to the output of the generator and place it near the receiver. Adjust C1 until the marker generator can be clearly heard in the receiver.

Chapter 35

Component Data

Few of us have the time or space to collect all the literature available on the many different commercially manufactured components. Even if we did, the task of keeping track of new and obsolete devices would surely be formidable. Fortunately, amateurs tend to use a limited number of component types. This chapter provides a source of information on the kinds of components most often encountered by the amateur experimenter.

Component Values

Values of composition resistors and small capacitors (mica and ceramic) are specified throughout this *Handbook* in terms of "preferred values." In the preferred-number system, all values represent (approximately) a constant-percentage increase over the next lower value. The base of the system is the number 10. Only two significant figures are used.

"Tolerance" means that a variation of plus or minus the percentage given is considered satisfactory. For example, the actual resistance of a "4700-ohm" 20-percent resistor can lie anywhere between 3700 and 5600 ohms, approximately. The permissible variation in the same resistance value with 5-percent tolerance would be in the range from 4500 to 4900 ohms, approximately.

In the component specifications in this *Handbook*, it is to be understood that when no tolerance is specified the *largest* tolerance available in that value will be satisfactory.

Table 1
Resistor-Capacitor Color Code

Color	Significant Figure	Decimal Multiplier	Tolerance (%)	Voltage Rating*
Black	0	1	–	–
Brown	1	10	1*	100
Red	2	100	2*	200
Orange	3	1,000	3*	300
Yellow	4	10,000	4*	400
Green	5	100,000	5*	500
Blue	6	1,000,000	6*	600
Violet	7	10,000,000	7*	700
Gray	8	100,000,000	8*	800
White	9	1,000,000,000	9*	900
Gold	–	0.1	5	1000
Silver	–	0.01	10	2000
No color	–	–	20	500

*Applies to capacitors only.

Values that do not easily fit into the preferred-number system (such as 500, 25,000) can be substituted. It is obvious, for example, that a 5000-ohm resistor falls well within the tolerance range of the 4700-ohm 20-percent resistor used in the example above. It would not, however, be usable if the tolerance were specified as 5 percent.

Component Markings

The values of small components such as capacitors and resistors are usually marked with standard color codes or cryptic abbreviations. If you have an ohmmeter and capacitor meter, you can quickly determine the value of an unknown component. If not, the information in the following section can help you identify the majority of components you're likely to encounter.

Table 1 shows the standard resistor-capacitor color code. From the information in Table 1, you can determine the value and tolerance of most resistors. You can also use this information to determine the value, tolerance and voltage rating of some capacitors, such as mica and molded paper types.

Table 2 shows standard values for resistors and capacitors. Note that these are the significant figures in the value and that these standard values may be multiplied by powers of 10. For example, 2.2 μF, 22 μF, 220 μF and 2200 μF are all standard capacitance values. Standard resistor values

Table 2
Standard Values for Capacitors and Resistors

*1.0	*1.5	*2.2	*3.3	*4.7	*6.8
1.1	1.6	2.4	3.6	5.1	7.5
*1.2	*1.8	*2.7	*3.9	*5.6	*8.2
1.3	2.0	3.0	4.3	6.2	9.1

Note: All values available in ±5%
 *Available in ±10%

include 1.5 Ω, 15 Ω, 150 Ω, 1.5 kΩ, 15 kΩ, 150 kΩ 1.5 MΩ and 15 MΩ.

Capacitor Markings

The value, tolerance, temperature characteristics and voltage rating of many small capacitors such as disc-ceramic, mica and tantalum types can be determined from a color code or from a combination of letters and numbers. Fig. 1 shows several popular marking systems.

In addition to the value, disc-ceramic capacitors may be marked with a letter/number combination signifying temperature characteristics. See Table 3. For example, a capacitor marked Z5U is suitable for use between +10 and +85°C, with a maximum capacitance change of −56% or +22%.

Capacitors with highly predictable temperature coefficients of capacitance are sometimes used in oscillators that must be frequency stable with temperature. If an application called for a temperature coefficient of −750 parts per million per degree Celsius (N750), a capacitor marked U2J would be suitable. These codes are listed in Table 4.

Some capacitors, such as dipped silver mica units, have a letter designating the capacitance tolerance. These letters are deciphered in Table 5.

Fixed-Value Resistors

Composition resistors (and small wire-wound units molded in cases identical to the composition type) are color-coded as shown in Fig. 2. Color bands are used on resistors having axial leads; on radial-lead resistors the colors are placed as shown in the drawing. When bands are used for color coding the body color has no significance.

Examples: A resistor of the type shown in the lower drawing of Fig. 2 has the fol-

Fig. 1 — Capacitors can be identified by color codes and markings. Shown here are identifying markings found on many common capacitor types.

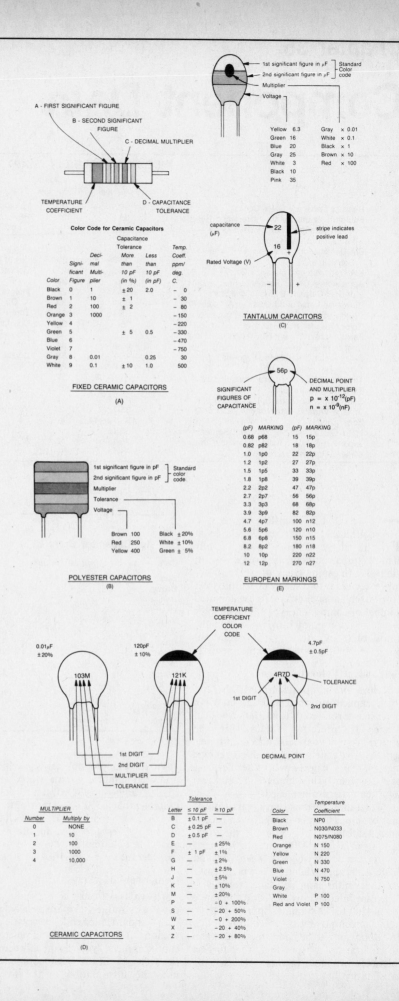

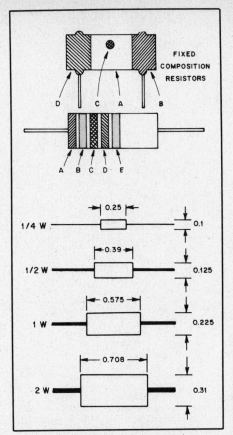

FIXED COMPOSITION RESISTORS

D C A B

A B C D E

1/4 W |←0.25→| 0.1

1/2 W |←0.39→| 0.125

1 W |←0.575→| 0.225

2 W |←0.708→| 0.31

Fig. 2 — Color coding and body size for fixed resistors. The color code is given in Table 1. The colored areas have the following significance.

A — First significant figure of resistance in ohms.
B — Second significant figure.
C — Decimal multiplier.
D — Resistance tolerance in percent. If no color is shown the tolerance is ±20%.
E — Relative percent change in value per 1000 hours of operation; Brown, 1%; Red, 0.1%; Orange, 0.01%; Yellow, 0.001%.

Table 3
EIA Temperature Characteristic Code for Disc Ceramic Capacitors

Minimum Temperature		Maximum Temperature		Max. Cap. Change Over Temp. Range	
X	−55°C	2	+45°C	A	±1.0%
Y	−30°C	4	+65°C	B	±1.5%
Z	+10°C	5	+85°C	C	±2.2%
		6	+105°C	D	±3.3%
		7	+125°C	E	±4.7%
				F	±7.5%
				P	±10%
				R	±15%
				S	±22%
				T	−33%, +22%
				U	−56%, +22%
				V	−82%, +22%

Table 4
EIA Designations for Capacitor Temperature Coefficient

Industry	EIA
NP0	C0G
N033	S1G
N075	U1G
N150	P2G
N220	R2G
N330	S2H
N470	T2H
N750	U2J
N1500	P3K
N2200	R3L

Table 5
EIA Capacitance Tolerance Codes

C	±¼ pF
D	±½ pF
F	±1 pF or ±1%
G	±2 pF or ±2%
J	±5%
K	±10%
L	±15%
M	±20%
N	±30%
P	−0%, +100% or GMV
W	−20%, +40%
Y	−20%, +50%
Z	−20%, +80%

lowing color bands: A, red; B, red; C, orange; D, no color. The significant figures are 2,2 (22) and the decimal multiplier is 1000. The value of resistance is therefore 22,000 ohms and the tolerance is ±20 percent.

A resistor of the type shown in the upper drawing of Fig. 2 has the following colors: Body (A), blue; end (B), gray; dot, red; end (D), gold. The significant figures are 6, 8 (68) and the decimal multiplier is 100, so the resistance is 6800 ohms. The tolerance is ±5 percent.

The preferred values over one decade are given in Table 2. All resistance values (from less than 1 ohm to about 22 megohms) are available in ±2% metal film and ±5% carbon-composition units.

Carbon composition and metal film resistors have standard power ratings of 1/10, 1/8, 1/4, 1/2, 1 and 2 watts. The 1/10- and 1/8-watt sizes are expensive and difficult to purchase in small quantity.

They are used only where miniaturization is essential. The 1/4-, 1/2-, 1- and 2-watt composition resistors are drawn to scale in Fig. 2. Metal film resistors are usually slightly smaller than carbon units of the same power rating. Film resistors can often be distinguished by a glossy vitreous enamel coating and an hourglass profile as opposed to a right circular cylinder. They are the most commonly available type today, but the resistive film is often deposited in the form of a solenoid. Thus, they can be highly inductive and therefore less desirable than carbon-composition resistors in RF circuits.

Power Transformers

1) Primary Leads: black
   If tapped:
   Common: black
   Tap: black and yellow striped
   Finish: black and red striped
2) High-Voltage Plate Winding; red
   Center-Tap: red and yellow striped
3) Rectifier Filament Winding: yellow
   Center-Tap: yellow and blue striped
4) Filament Winding no. 1: green
   Center-Tap: green and yellow striped
5) Filament Winding no. 2: brown
   Center-Tap: brown and yellow striped
6) Filament Winding no. 3: slate
   Center-Tap: slate and yellow striped

IF Transformers

Plate lead — *blue*
B-plus lead — *red*
Grid (or diode) lead — *green*
Grid (or diode) return — *black*
   Note: If the secondary of the IF transformer is center-tapped, the second diode plate lead is green-and-black striped, and black is used for the center-tap lead.

Audio Transformers

Plate (finish) lead of primary — *blue*
B-plus lead (this applies whether the primary is plain or center-tapped) — *red*
Plate (start) lead on center-tapped primaries. (Blue may be used for this lead if polarity is not important) — *brown*
Grid (finish) lead to secondary — *green*
Grid return (this applies whether the secondary is plain or center-tapped) — *black*
Grid (start) lead on center-tapped secondaries (Green may be used for this lead if polarity is not important) — *yellow*
   Note: These markings apply also to line-to-grid and tube-to-line transformers.

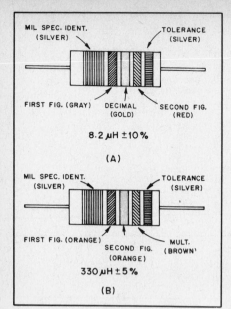

Fig. 3 — Color coding for tubular encapsulated RF chokes. At A, an example of the coding for an 8.2-μH choke is given. At B, the color bands for a 330-μH inductor are illustrated. The color code is given in Table 1.

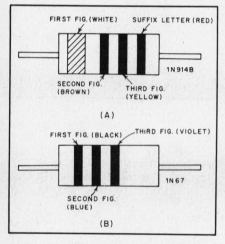

Fig. 4 — Color coding for semiconductor diodes. At A, the cathode is identified by the double-width first band. At B, the bands are grouped toward the cathode. Two-figure designations are signified by a black first band. The color code is given in Table 1. The suffix-letter code is: A — brown, B — red, C — orange, D — yellow, E — green, F — blue. The 1N prefix is understood.

Table 6

Approximate Series-Resonant Frequencies of Disc Ceramic Bypass Capacitors

Capacitance	Freq.[1]	Freq.[2]
0.01 μF	13 MHz	15 MHz
0.0047	18	22
0.002	31	38
0.001	46	55
0.0005	65	80
0.0001	135	165

[1]Total lead length of 1 inch
[2]Total lead length of 1/2 inch

Table 7

Powdered-Iron Toroid Cores — Magnetic Properties

Inductance and Turns Formula

The number of turns (N) necessary to obtain a specific inductance (L) can be calculated by

$$N = 100 \sqrt{\frac{L}{L_{100}}}$$

L = desired inductance (μH)

L_{100} = core inductance (μH per 100 turns)

Inductance Per 100 Turns ±5%

No.	MIX-41	MIX-3	MIX-15	MIX-1	MIX-2	MIX-6	MIX-10	MIX-12
T-200	755 μH	360 μH			120 μH	105 μH		
T-184	1640 μH	720 μH			240 μH			
T-157	970 μH	420 μH			140 μH	115 μH		
T-130	785 μH	330 μH	250 μH	200 μH	110 μH	96 μH		
T-106	900 μH	405 μH	330 μH	280 μH	135 μH	116 μH		
T-94	590 μH	248 μH		160 μH	84 μH	70 μH	58 μH	32 μH
T-80	450 μH	180 μH	170 μH	115 μH	55 μH	45 μH	34 μH	22 μH
T-68	420 μH	195 μH	180 μH	115 μH	57 μH	47 μH	32 μH	21 μH
T-50	320 μH	175 μH	135 μH	100 μH	50 μH	40 μH	31 μH	18 μH
T-37	240 μH	110 μH	90 μH	80 μH	42 μH	30 μH	25 μH	15 μH
T-25	200 μH	100 μH	85 μH	70 μH	34 μH	27 μH	19 μH	13 μH
T-12	90 μH	60 μH		48 μH	24 μH	19 μH	12 μH	8.5 μH

Magnetic Properties
Iron Powder Cores

Material	Color Code	Permeability	Temperature Stability	Typical Frequency Range	Optimum Frequency Range
41' HA'	Green	$\mu = 75$	975 ppm/c	10 kHz- 100 kHz	20 kHz- 50 kHz
3' HP'	Gray	$\mu = 35$	370 ppm/c	20 kHz- 2 MHz	50 kHz- 500 kHz
15' GS6'	Rd. & Wh.	$\mu = 25$	190 ppm/c	20 kHz- 5 MHz	500 kHz- 1 MHz
1' C'	Blue	$\mu = 20$	280 ppm/c	40 kHz- 5 MHz	1 MHz- 2 MHz
2' E'	Red	$\mu = 10$	95 ppm/c	200 kHz- 30 MHz	2 MHz- 10 MHz
6' SF'	Yellow	$\mu = 8$	35 ppm/c	2 MHz- 50 MHz	10 MHz- 20 MHz
10' W'	Black	$\mu = 6$	150 ppm/c	4 MHz-100 MHz	20 MHz- 40 MHz
12' Irn-8'	Gr. & Wh.	$\mu = 3$	170 ppm/c	10 MHz-200 MHz	40 MHz- 90 MHz
0' Ph'	Tan	$\mu = 1$	—	50 MHz-300 MHz	90 MHz-150 MHz

Courtesy of Amidon Assoc., N. Hollywood, CA 91607

Table 8

Powdered-Iron Toroid Cores — Dimensions

Red E Cores — 500 kHz to 30 MHz ($\mu = 10$)

No.	OD (In)	ID (In)	H (In)
T-200-2	2.00	1.25	0.55
T-94-2	0.94	0.56	0.31
T-80-2	0.80	0.50	0.25
T-68-2	0.68	0.37	0.19
T-50-2	0.50	0.30	0.19
T-37-2	0.37	0.21	0.12
T-25-2	0.25	0.12	0.09
T-12-2	0.125	0.06	0.05

Black W Cores — 30 MHz to 200 MHz ($\mu = 7$)

No.	OD (In)	ID (In)	H (In)
T-50-10	0.50	0.30	0.19
T-37-10	0.37	0.21	0.12
T-25-10	0.25	0.12	0.09
T-12-10	0.125	0.06	0.05

Yellow SF Cores — 10 MHz to 90 MHz ($\mu = 8$)

No.	OD (In)	ID (In)	H (In)
T-94-6	0.94	0.56	0.31
T-80-6	0.80	0.50	0.25
T-68-6	0.68	0.37	0.19
T-50-6	0.50	0.30	0.19
T-26-6	0.25	0.12	0.09
T-12-6	0.125	0.06	0.05

Number of Turns vs. Wire Size and Core Size

Approximate maximum number of turns — single layer wound — enameled wire.

Wire Size	T-200	T-130	T-106	T-94	T-80	T-68	T-50	T-37	T-25	T-12
10	33	20	12	12	10	6	4	1		
12	43	25	16	16	14	9	6	3		
14	54	32	21	21	18	13	8	5	1	
16	69	41	28	28	24	17	13	7	2	
18	88	53	37	37	32	23	18	10	4	1
20	111	67	47	47	41	29	23	14	6	1
22	140	86	60	60	53	38	30	19	9	2
24	177	109	77	77	67	49	39	25	13	4
26	223	137	97	97	85	63	50	33	17	7
28	281	173	123	123	108	80	64	42	23	9
30	355	217	154	154	136	101	81	54	29	13
32	439	272	194	194	171	127	103	68	38	17
34	557	346	247	247	218	162	132	88	49	23
36	683	424	304	304	268	199	162	108	62	30
38	875	544	389	389	344	256	209	140	80	39
40	1103	687	492	492	434	324	264	178	102	51

Actual number of turns may differ from above figures according to winding techniques, especially when using the larger size wires. Chart prepared by Michael J. Gordon, Jr., WB9FHC

Courtesy of Amidon Assoc., N. Hollywood, CA 91607

Table 9
Color Code for Hookup Wire

Wire Color	Type of Circuit
Black	Grounds, grounded elements and returns
Brown	Heaters or filaments, off ground
Red	Power supply B plus
Orange	Screen grids and base 2 of transistors
Yellow	Cathodes and transistor emitters
Green	Control grids, diode plates, and base 1 of transistors
Blue	Plates and transistor collectors
Violet	Power supply, minus leads
Gray	Ac power line leads
White	Bias supply, B or C minus, AGC

Note: Wires with tracers are coded in the same manner as solid-color wires, allowing additional circuit identification over solid-color wiring. The body of the wire is white and the color band spirals around the wire lead. When more than one color band is used, the widest band represents the first color.

Table 10
Ferrite Toroids—A_L Chart (mH per 1000 turns) Enameled Wire

Core Size	63-Mix $\mu = 40$	61-Mix $\mu = 125$	43-Mix $\mu = 950$	72-Mix $\mu = 2000$	75-Mix $\mu = 5000$
FT-23	7.9	24.8	189.0	396.0	990.0
FT-37	17.7	55.3	420.0	884.0	2210.0
FT-50	22.0	68.0	523.0	1100.0	2750.0
FT-82	23.4	73.3	557.0	1172.0	2930.0
FT-114	25.4	79.3	603.0	1268.0	3170.0

Number turns = $1000 \sqrt{\text{desired L (mH)} \div A_L \text{ value (above)}}$

Ferrite Magnetic Properties

Property	Unit	63-Mix	61-Mix	43-Mix	72-Mix	75-Mix
Initial perm. (μ_i)		40	125	950	2000	5000
Maximum perm.		125	450	3000	3500	8000
Saturation flux density @ 13 oer	Gauss	1850	2350	2750	3500	3900
Residual flux density	Gauss	750	1200	1200	1500	1250
Curie temp.	°C	500	300	130	150	160
Vol. resistivity	ohm/cm	1×10^8	1×10^8	1×10^5	1×10^2	5×10^2
Opt. freq. range	MHz	15-25	0.2-10	0.01-1	0.001-1	0.001-1
Specific gravity		4.7	4.7	4.5	4.8	4.8
Loss factor	$\dfrac{1}{\mu_o}$	9.0×10^{-5} @ 25 MHz	2.2×10^{-5} @ 2.5 MHz	2.5×10^{-5} @ 0.2 MHz	9.0×10^{-6} @ 0.1 MHz	5.0×10^{-6} @ 0.1 MHz
Coercive force	Oer.	2.40	1.60	0.30	0.18	0.18
Temp. Coef. of initial perm.	%/°C 20-70°C	0.10	0.10	0.20	0.60	—

Ferrite Toroids — Physical Properties

Core Size	OD	ID	Height	A_e	l_e	V_e	A_s	A_w
FT-23	0.230	0.120	0.060	0.00330	0.529	0.00174	0.1264	0.01121
FT-37	0.375	0.187	0.125	0.01175	0.846	0.00994	0.3860	0.02750
FT-50	0.500	0.281	0.188	0.02060	1.190	0.02450	0.7300	0.06200
FT-82	0.825	0.520	0.250	0.03810	2.070	0.07890	1.7000	0.21200
FT-114	1.142	0.748	0.295	0.05810	2.920	0.16950	2.9200	0.43900

OD — Outer diameter (inches)
ID — Inner diameter (inches)
Hgt — Height (inches)
A_w — Total window area (in)^2

A_e — Effective magnetic cross-sectional area (in)^2
l_e — Effective magnetic path length (inches)
V_e — Effective magnetic volume (in)^3
A_s — Surface area exposed for cooling (in)^2

Courtesy of Amidon Assoc., N. Hollywood, CA 91607

Digital ICs

Digital-logic ICs are divided into two basic families, bipolar and MOS, depending on the transistor types used. These families are further divided into subfamilies by the internal circuitry of the IC ("totem pole" output versus open collector output) or by differences in fabrication techniques (silicon gate-recessed oxide versus metal gate). The digital ICs most commonly encountered by amateurs are the TTL (bipolar) and CMOS (MOS) types.

The TTL family is divided as shown in tables presented later in this chapter. It is important to recognize the differences between the types of TTL ICs. In many digital circuits, ICs of several types are connected together. If care is not exercised when doing this, one or more of the TTL outputs may be overloaded. Be aware that the different series of TTL operate at different speeds and with different propagation delays (see Chapter 8).

To determine if a substitute TTL IC can be used in a given circuit, several questions must be answered. First, is the logic element operating at a speed below the maximum specified for the TTL type? Will any output of the IC be required to sink or source a current beyond its capabilities? Even if the answers to these questions seem to indicate that the IC can be used, a deeper analysis of the circuit may reveal that a change in propagation delay will affect the operation of the circuit. If in doubt the IC can always be tried in the circuit. TTL ICs of the "wrong" type will generally not harm anything.

The major consideration when using CMOS is operating speed. CMOS logic is inherently slow. If clock rates greater than a few megahertz are required, CMOS probably cannot be used (although higher speed CMOS families have recently been introduced by several manufacturers). The CMOS speed question is complicated by the fact that the speed of any CMOS logic element is affected by its operating voltage. A manufacturer's data book should be consulted to determine the speed of a specific IC.

Finding Parts

No chapter on components and component data would be complete without information on where to buy. Amateurs, on a dwarfed scale, must function as purchasing agents in these perplexing times. A properly equipped buyer maintains as complete a catalog file as possible. Many of the companies listed in the last table in this chapter will provide free catalogs upon written request. Others may charge a small fee for catalogs. Mail ordering, especially for those distant from metropolitan areas, is today's means to the desired end when collecting component parts for an amateur project. Prices are, to some extent, competitive. A wise buyer will study the catalogs and select his merchandise accordingly.

Delays in shipment can be lessened by avoiding the use of personal checks when ordering, especially for those distant from metropolitan areas. Personal checks often take a week to clear, thereby causing frustrating delays in the order reaching you. The parts suppliers table is updated with each new edition of this *Handbook*. Suppliers wishing to be listed in the table are urged to contact the editors.

Table 11
Copper-Wire Table

Wire Size A.W.G. (B&S)	Diam. in Mils[1]	Circular Mil Area	Turns per Linear inch (25.4 mm)[2]			Cont.-duty current[3] single wire in open air	Cont.-duty current[3] wires or cables in conduits or bundles	Feet per Pound (0.45 kg) Bare	Ohms per 1000 ft. 25° C	Current Carrying Capacity at 700 C.M. per Amp.	Diam. in mm.	Nearest British S.W.G. No.
			Enamel	S.C.E.	D.C.C.							
1	289.3	83690	—	—	—	—	—	3.947	.1264	119.6	7.348	1
2	257.6	66370	—	—	—	—	—	4.977	.1593	94.8	6.544	3
3	229.4	52640	—	—	—	—	—	6.276	.2009	75.2	5.827	4
4	204.3	41740	—	—	—	—	—	7.914	.2533	59.6	5.189	5
5	181.9	33100	—	—	—	—	—	9.980	.3195	47.3	4.621	7
6	162.0	26250	—	—	—	—	—	12.58	.4028	37.5	4.115	8
7	144.3	20820	—	—	—	—	—	15.87	.5080	29.7	3.665	9
8	128.5	16510	7.6	—	7.1	73	46	20.01	.6405	23.6	3.264	10
9	114.4	13090	8.6	—	7.8	—	—	25.23	.8077	18.7	2.906	11
10	101.9	10380	9.6	9.1	8.9	55	33	31.82	1.018	14.8	2.588	12
11	90.7	8234	10.7	—	9.8	—	—	40.12	1.284	11.8	2.305	13
12	80.8	6530	12.0	11.3	10.9	41	23	50.59	1.619	9.33	2.053	14
13	72.0	5178	13.5	—	12.8	—	—	63.80	2.042	7.40	1.828	15
14	64.1	4107	15.0	14.0	13.8	32	17	80.44	2.575	5.87	1.628	16
15	57.1	3257	16.8	—	14.7	—	—	101.4	3.247	4.65	1.450	17
16	50.8	2583	18.9	17.3	16.4	22	13	127.9	4.094	3.69	1.291	18
17	45.3	2048	21.2	—	18.1	—	—	161.3	5.163	2.93	1.150	18
18	40.3	1624	23.6	21.2	19.8	16	10	203.4	6.510	2.32	1.024	19
19	35.9	1288	26.4	—	21.8	—	—	256.5	8.210	1.84	.912	20
20	32.0	1022	29.4	25.8	23.8	11	7.5	323.4	10.35	1.46	.812	21
21	28.5	810	33.1	—	26.0	—	—	407.8	13.05	1.16	.723	22
22	25.3	642	37.0	31.3	30.0	—	5	514.2	16.46	.918	.644	23
23	22.6	510	41.3	—	37.6	—	—	648.4	20.76	.728	.573	24
24	20.1	404	46.3	37.6	35.6	—	—	817.7	26.17	.577	.511	25
25	17.9	320	51.7	—	38.6	—	—	1031	33.00	.458	.455	26
26	15.9	254	58.0	46.1	41.8	—	—	1300	41.62	.363	.405	27
27	14.2	202	64.9	—	45.0	—	—	1639	52.48	.288	.361	29
28	12.6	160	72.7	54.6	48.5	—	—	2067	66.17	.228	.321	30
29	11.3	127	81.6	—	51.8	—	—	2607	83.44	.181	.286	31
30	10.0	101	90.5	64.1	55.5	—	—	3287	105.2	.144	.255	33
31	8.9	80	101	—	59.2	—	—	4145	132.7	.114	.227	34
32	8.0	63	113	74.1	61.6	—	—	5227	167.3	.090	.202	36
33	7.1	50	127	—	66.3	—	—	6591	211.0	.072	.180	37
34	6.3	40	143	86.2	70.0	—	—	8310	266.0	.057	.160	38
35	5.6	32	158	—	73.5	—	—	10480	335	.045	.143	38-39
36	5.0	25	175	103.1	77.0	—	—	13210	423	.036	.127	39-40
37	4.5	20	198	—	80.3	—	—	16660	533	.028	.113	41
38	4.0	16	224	116.3	83.6	—	—	21010	673	.022	.101	42
39	3.5	12	248	—	86.6	—	—	26500	848	.018	.090	43
40	3.1	10	282	131.6	89.7	—	—	33410	1070	.014	.080	44

[1] A mil is 0.001 inch. A circular mil is a square mil $\times$ $\pi/4$. The circular mil (c.m.) area of a wire is the square of the mil diameter.
[2] Figures given are approximate only; insulation thickness varies with manufacturer.
[3] Max. wire temp. of 212° F (100° C) and max. ambient temp. of 135° F (57° C).
[4] 700 circular mils per ampere is a satisfactory design figure for small transformers, but values from 500 to 1000 c.m. are commonly used.

Table 12
EIA Vacuum-Tube Base Diagrams

Socket connections correspond to the base designations given in the column headed "Base" in the classified tube-data tables. Bottom views are shown throughout. Terminal designations are as follows:

A — Anode	B — Deflecting Plate	IS — Internal Shield	RC — Ray-Control Electrode
B — Beam	F — Filament	K — Cathode	Ref — Reflector
BP — Bayonet Pin	FE — Focus Elect.	NC — No Connection	S — Shell
BS — Base Sleeve	G — Grid	P — Plate (Anode)	TA — Target
C — Ext. Coating	H — Heater	P_1 — Starter-Anode	U — Unit
CL — Collector	IC — Internal Con.	P_{BF} — Beam Plates	● — Gas-Type Tube

Alphabetical subscripts D, P, T and HX indicate, respectively, diode unit, pentode unit, triode unit or hexode unit in multi-unit types. Subscript CT indicates filament or heater tap.

Generally when the No. 1 pin of a metal-type tube, with the exception of all triodes, is shown connected to the shell, the No. 1 pin in the glass (G or GT) equivalent is connected to an internal shield.

*On 12AQ, 12AS and 12CT: index = large lug; ● = pin cut off

Type	Maximum Ratings						Cathode		Capacitances			Base	Typical Operation							
	Plate Dissipation Watts	Plate Voltage	Plate Current mA	DC Grid Current mA	Freq. MHz Full Ratings	Amplification Factor	Volts	Amperes	C_{in} pF	C_{gp} pF	C_{out} pF		Class of Service	Plate Voltage	Grid Voltage	Plate Current mA	DC Grid Current mA	Approx. Driving Power Watts	P-to-P Load Ohms	Approx. Output Power Watts
5675	5	165	30	8	3000	20	6.3	0.135	2.3	1.3	0.09	Fig. 21	GG·O	120	−8	25	4	—	—	0.05
2C40	6.5	500	25	—	500	36	6.3	0.75	2.1	1.3	0.05	Fig. 11	C·T·O	250	−5	20	0.3	—	—	0.075
5893	8.0	400	40	13	1000	27	6.0	0.33	2.5	1.75	0.07	Fig. 21	C·T	350	−33	35	13	2.4	—	6.5
													C·P	300	−45	30	12	2.0	—	6.5
2C43	12	500	40	—	1250	48	6.3	0.9	2.9	1.7	0.05	Fig. 11	C·T·O	470	—	38[7]	—	—	—	9[2]
811-A	65	1500	175	50	60	160	6.3	4.0	5.9	5.6	0.7	3G	C·T	1500	−70	173	40	7.1	—	200
													C·P	1250	−120	140	45	10.0	—	135
													B·GG	1250	0	27/175	28	12	—	165
													AB$_1$	1250	0	27/175	13	3.0	—	155
812-A	65	1500	175	35	60	29	6.3	4.0	5.4	5.5	0.77	3G	C·T	1500	−120	173	30	6.5	—	190
													C·P	1250	−115	140	35	7.6	—	130
													B[2]	1500	−48	28/310	270[9]	5.0	13.2K	340
3CX100A5[6]	100	1000	125[5]	50	2500	100	6.0	1.05	7.0	2.15	0.035	—	A·GG	800	−20	80	30	6	—	27
	70	600	100[5]										C·P	600	−15	75	40	6	—	18
2C39	100	1000	60	40	500	100	6.3	1.1	6.5	1.95	0.03	—	G1C	600	−35	60	40	5.0	—	20
													C·T·O	900	−40	90	30	—	—	40
													C·P	600	−150	100[5]	50	—	—	—
AX9900/5866[6]	135	2500	200	40	150	25	6.3	5.4	5.8	5.5	0.1	Fig. 3	C·T	2500	−200	200	40	16	—	390
													C·P	2000	−225	127	40	16	—	204
													B[2]	2500	−90	80/330	350[4]	14[3]	15.68K	560
572B/T160L	160	2750	275	—	—	170	6.3	4.0	—	—	—	3G	C·T	1650	−70	165	32	6	—	205
													B·GG[2]	2400	−2.0	90/500	—	100	—	600
810	175	2500	300	75	30	36	10	4.5	8.7	4.8	12	2N	C·T	2500	−180	300	60	19	—	575
													C·P	2000	−350	250	70	35	—	380
													GMA	2250	−140	100	2.0	4	—	75
													B[2]	2250	−60	70/450	380[4]	13[3]	11.6K	725
8873	200	2200	250	—	500	160	6.3	3.2	19.5	7.0	0.03	Fig. 87	AB$_2$	2000	—	22/500	98[3]	27[3]	—	505
8875	300	2200	250	—	500	160	6.3	3.2	19.5	7.0	0.03	Fig. 87	AB$_2$	2000	—	22/500	98[3]	27[3]	—	505
833A	350	3300	500	100	30	35	10	10	12.3	6.3	8.5	Fig. 41	C·T·O	2250	−125	445	85	23	—	780
														3000	−160	335	70	20	—	800
	450[6]	4000[6]			20[6]								C·P	2500	−300	335	75	30	—	635
														3000	−240	335	70	26	—	800
													B[2]	3000	−70	100/750	400[4]	20[4]	9.5K	1650
8874	400	2200	350	—	500	160	6.3	3.2	19.5	7.0	0.03	—	AB$_2$	2000	—	22/500	98[3]	27[3]	—	505
3-400Z	400	3000	400	—	110	200	5	14.5	7.4	4.1	0.07	Fig. 3	B·GG	3000	0	100/333	120	32	—	655
3-500Z	500	4000	400	—	110	160	5	14.5	7.4	4.1	0.07	Fig. 3	B·GG	3000	—	370	115	30	5K	750
													C·T	3500	75	300	115	22		850
3CX800A7	800	2250	600	60	350	200	13.5	1.5	26	—	6.1	Fig. 87	AB$_2$·GG[7]	2200	−8.2	500	36	16	—	750
3-1000Z	1000	3000	800	—	110	200	7.5	21.3	17	6.9	0.12	Fig. 3	B·GG	3000	0	180/670	300	65	—	1360
3CX1200A7	1200	5000	800	—	110	200	7.5	21.0	20	12	0.2	Fig. 3	AB$_2$·GG	3600	−10	700	230	85	—	1500
8877	1500	4000	1000	—	250	200	5.0	10	42	10	0.1	—	AB$_2$	2500	−8.2	1000	—	57	—	1520

* Cathode resistor in ohms.

[1] KEY TO CLASS-OF-SERVICE ABBREVIATIONS

A$_1$ = Class-A$_1$ AF modulator.
AB$_1$ = Class-AB$_1$ push-pull AF modulator.
AB$_2$ = Class-AB$_2$ push-pull AF modulator.
B = Class-B push-pull AF modulator.
C·M = Frequency multiplier.

C·P = Class-C plate-modulated telephone.
C·T = Class-C telegraph.
C·T·O = Class-C amplifier-osc.
AB$_2$·GG = Grounded-grid class AB$_2$ amp.
B·GG = Grounded-grid class B amp. (single tone)
GG·O = Grounded-grid osc.
GIC = Grid-isolation circuit.

GMA = Grid-modulated amp.
[2] Values are for two tubes in push-pull.
[3] Max. signal value.
[4] Peak AF grid-to-grid volts.
[5] Max. cathode current in mA.
[6] Forced-air cooling required.
[7] Key-down CW.

Type	Plate Dissipation Watts	Plate Voltage	Screen Dissipation Watts	Screen Voltage	Freq. MHz Full Ratings	Volts	Amperes	Cin pF	Cgp pF	Cout pF	Base	Class of Service[14]	Plate Voltage	Screen Voltage	Suppressor Voltage	Grid Voltage	Plate Current mA	Screen Current mA	Grid Current mA	Approx. Driving Power Watts	P-P Load Ohms	Approx. Output Power Watts
6939[3]	7.5	275	3	200	500	6.3 / 12.6	0.75 / 0.375	6.6	0.15	1.55	Fig. 13	C-T	200	200	—	-20	60	13	2	1.0	—	7.5
												C-P	180	180	—	-20	55	11.5	1.7	1.0	—	6
												C-M	200	190	—	68K[1]	46	10	2.2	0.9	—	—
7551 / 7558	12	300	2	250	175	12.6 / 6.3	0.38 / 0.8	10	0.15	5.5	9LK	C-T	300	250	—	-55	80	5.1	1.6	1.5	—	10
												C-P	250	250	—	-75	70	3.0	2.3	1.0	—	7.5
5763 / 6417	13.5	350	2	250	50	6.3 / 12.6	0.75 / 0.375	9.5	0.3	4.5	9K	C-T	350	250	—	-28.5	48.5	6.2	1.6	0.1	—	12
												C-P	300	250	—	-42.5	50	6	2.4	0.15	—	10
												C-M[2]	300	250	—	-75	40	4	1	0.6	—	2.1
												C-M[4]	300	235	—	-100	35	5	1	0.6	—	1.3
2E26 / 6893	13.5	600	2.5	200	125	6.3 / 12.6	0.8 / 0.4	12.5	0.2	7	7CK	C-T	600	185	—	-45	66	10	3	0.17	—	27
												C-P	500	180	—	-50	54	9	2.5	0.15	—	18
												AB1	500	200	—	-25	9/45	10[7]	0	0	—	15
6360[3]	14	300	2	200	200	6.3 / 12.6	0.82 / 0.41	6.2	0.1	2.6	Fig. 13	C-T	300	200	—	-45	100	3	3	0.2	—	18.5
												C-P	200	100	—	15K[1]	86	3.1	3.3	0.2	—	9.8
												C-M[11]	300	150	—	-100	65	3.5	3.8	0.45	—	4.8
												AB2	300	200	—	-21.5	30/100	1/11.4	64[8]	0.04	6.5K	17.5
6252 / AX9910[3]	20	750	4	300	300	6.3 / 12.6	1.3 / 0.65	6.5	—	2.5	Fig. 7	C-T	600	250	—	-60	140	14	4	2.0	—	—
												C-P	500	250	—	-80	100	12	3	4.0	—	—
												B	500	250	—	-26	25/73	0.7/16	52[8]	—	20K	23.5
1614	25	450	3.5	300	80	6.3	0.9	10	0.4	12.5	7AC	C-T	450	250	—	-45	100	8	2	0.15	—	31
												C-P	375	250	—	-50	93	7	2	0.15	—	24.5
												AB1[6]	530	340	—	-36	60/160	20[7]	—	—	7.2K	50
815[3]	25	500	4	200	125	6.3 / 12.6	1.6 / 0.8	13.3	0.2	8.5	8BY	C-T-O	500	200	—	-45	150	17	2.5	0.13	—	56
												AB2	500	125	—	-15	22/150	32[7]	—	0.36[7]	8K	54
6146 / 6146A	25	750	3	250	60	6.3	1.25	13	0.24	8.5	7CK	C-T	500	170	—	-66	135	9	2.5	0.2	—	48
												C-T	750	160	—	-62	120	11	3.1	0.2	—	70
8032 / 6883						12.6	0.585					C-T[12]	400	190	—	-54	150	10.4	2.2	3.0	—	35
												C-P	400	150	—	-87	112	7.8	3.4	0.4	—	32
												C-P	600	150	—	-87	112	7.8	3.4	0.4	—	52
6159B						26.5	0.3					AB2[6]	600	190	—	-48	28/270	1.2/20	2[7]	0.3	5K	113
												AB2[6]	750	165	—	-46	22/240	0.3/20	2.6[7]	0.3	7.4K	131
												AB1[6]	750	195	—	-50	23/220	1/26	100[8]	0	8K	120
6524[3] / 6850	25	600	—	300	100	6.3 / 12.6	1.25 / 0.625	7	0.11	3.4	Fig. 76	C-T	600	200	—	-44	120	8	3.7	0.2	—	56
												C-P	500	200	—	-61	100	7	2.5	0.2	—	40
												AB2	500	200	—	-26	20/116	0.1/10	2.6	0.1	11.1K	40
807 / 807W / 5933	30	750	3.5	300	60	6.3	0.9	12	0.2	7	5AW	C-T	750	250	—	-45	100	6	3.5	0.22	—	50
												C-P	600	275	—	-90	100	6.5	4	0.4	—	42.5
												AB1	750	300	—	-35	15/70	3/8	75[8]	0	—	72
1625						12.6	0.45				5AZ	B[10]	750	—	—	0	15/240	—	555[8]	5.3[7]	6.65K	120
2E22	30	750	10	250	—	6.3	1.5	13	0.2	8	5J	C-T-O	750	250	22.5	-60	100	16	6	0.55	—	53
6146B / 8298A	35	750	3	250	60	6.3	1.125	13	0.22	8.5	7CK	C-T	750	200	—	-77	160	10	2.7	0.3	—	85
												C-P	600	175	—	-92	140	9.5	3.4	0.5	—	62
												AB1	750	200	—	-48	25/125	6.3	—	—	3.6K	61
AX-9903[3] / 5894A	40	600	7	250	250	6.3 / 12.6	1.8 / 0.9	6.7	0.08	2.1	Fig. 7	C-T	600	250	—	-80	200	16	2	0.2	—	80
829B[3] / 3E29[3]	40	750	7	240	200	6.3 / 12.6	2.25 / 1.125	14.5	0.12	7	7BP	C-T	500	200	—	-45	240	32	12	0.7	—	83
												C-P	425	200	—	-60	212	35	11	0.8	—	63
												B	500	200	—	-18	27/230	—	56[8]	0.39	4.8K	76
3D24	45	2000	10	400	125	6.3	3	6.5	0.2	2.4	Fig. 75	C-T-O	2000	375	—	-300	90	20	10	4.0	—	140
												C-T-O	1500	375	—	-300	90	22	10	4.0	—	105
8117[3]	60	750	7	300	175	6.3 / 12.6	1.8 / 0.9	11.8	3.7	0.09	Fig. 7	AB1	600	250	—	-32.5	60/212	1.9/25	—	—	1410	76
814	65	1500	10	300	30	10	3.25	13.5	0.1	13.5	Fig. 64	C-T	1500	300	—	-90	150	24	10	1.5	—	160
												C-P	1250	300	—	-150	145	20	10	3.2	—	130
4-65A	65	3000	10	600	150	6	3.5	8	0.08	2.1	Fig. 25	C-T-O	1500	250	—	-85	150	40	18	3.2	—	165
												C-T-O	3000	250	—	-100	115	22	10	1.7	—	280
												C-P	1500	250	—	-125	120	40	16	3.5	—	140
												C-P	2500	250	—	-135	110	25	12	2.6	—	230
												AB1	2500	400	—	-85	15/66	3[7]	—	—	—	100
7854[3]	68	1000	8	300	175	6.3 / 12.6	1.8 / 0.9	6.7	2.1	0.09	Fig. 7	C-T	750	260	—	-75	240	12.7	5.5	3.5	—	123
												C-P	600	225	—	-75	200	7.8	5.5	3.5	—	85
4E27 / 8001	75	4000	30	750	75	5	7.5	12	0.06	6.5	7BM	C-T	2000	500	60	-200	150	11	6	1.4	—	230
												C-P	1800	500	60	-130	135	11	8	1.7	—	178
7270 / 7271	80	1350	—	425	175	6.3 / 13.5	3.1 / 1.25	8	0.4	0.14	Fig. 84	C-T	850	400	—	-100	275	15	8	10	—	135
												AB1	665	400	—	-119	220	15	6	10	—	85
8072	100	2200	8	400	500	13.5	1.3	16	0.13	0.011	Fig. 85	C-T-O	700	200	—	-30	300	10	20	5	—	85

Table 14 continued on next page.

Table 14 (continued)

Type	Plate Dissipation Watts	Plate Voltage	Screen Dissipation Watts	Screen Voltage	Freq. MHz Full Ratings	Volts	Amperes	Cin pF	Cgp pF	Cout pF	Base	Class of Service[14]	Plate Voltage	Screen Voltage	Suppressor Voltage	Grid Voltage	Plate Current mA	Screen Current mA	Grid Current mA	Approx. Driving Power Watts	P-P Load Ohms	Approx. Output Power Watts
6816[9] / 6884	115	1000	4.5	300	400	6.3 / 26.5	2.1 / 0.52	14	0.085	0.015	Fig. 77	C·T·O	900	300	—	-30	170	1	10	3	—	80
												C·P	700	250	—	-50	130	10	10	3	—	45
												AB1[6]	850	300	—	-15	80/200	0/20	30[8]	0	7K	80
												AB2[6]	850	300	—	-15	80/335	0/25	46[8]	0.3	3.96K	140
813[13]	125	2500	20	800	30	10	5	16.3	0.25	14	5BA	C·T·O	1250	300	0	-75	180	35	12	1.7	—	170
												C·T·O	2250	400	0	-155	220	40	15	4	—	375
												AB1	2500	750	0	-95	25/145	27[7]	0	0	—	245
												AB2[6]	2000	750	0	-90	40/315	1.5/58	230[8]	0.1[7]	16K	455
													2500	750	0	-95	35/360	1.2/55	235[8]	0.35[7]	17K	650
4-125A / 4D21 / 6155	125	3000	20	600	120	5	6.5	10.8	0.07	3.1	5BK	C·T·O	2000	350	—	-100	200	50	12	2.8	—	275
												C·T·O	3000	350	—	-150	167	30	9	2.5	—	375
												AB2[6]	2500	350	—	-43	93/260	0/6	178[8]	1.0[7]	22K	400
												AB1[6]	2500	600	—	-96	50/232	0.3/8.5	192[8]	0	20.3K	330
												GG	2000	0	—	0	10/105[17]	30[17]	55[17]	16[17]	10.5K	145
4E27A / 5-125B	125	4000	20	750	75	5	7.5	10.5	0.08	4.7	7BM	C·	3000	500	60	-200	167	5	6	1.6	—	375
													1000	750	0	-170	160	21	3	0.6	—	115
803	125	2000	30	600	20	10	5	17.5	0.15	29	5J	C·T	2000	500	40	-90	160	45	12	2	—	210
												C·P	1600	400	100	-80	150	45	25	5	—	155
7094	125	2000	20	400	60	6.3	3.2	9.0	0.5	1.8	Fig. 82	C·T	1500	400	—	-100	330	20	5	4	—	340
												C·P	1200	400	—	-130	275	20	5	5	—	240
												AB1	2000	400	—	-65	30/200	35[7]	60[8]	0	12K	250
4X150A / 4X150G[15]	150[9]	2000	12	400	500	6 / 2.5	2.6 / 6.25	15.5 / 27	0.03 / 0.035	4.5 / 4.5	Fig. 75 / —	C·T·O	1250	250	—	-90	200	20	10	0.8	—	195
												C·P	1000	250	—	-105	200	20	15	2	—	140
												AB2[6]	1250	300	—	-44	475[7]	0/65	100[8]	0.15[7]	5.6K	425
8121	150	2200	8	400	500	13.5	1.3	16	0.13	0.011	Fig. 5	C·T·O	2000	200	—	-30	300	10	30	5	—	165
4-250A / 5D22 / 6156	250[8]	4000	35	600	110	5	14.5	12.7	0.12	4.5	5BK	C·T·O	2500	500	—	-150	300	60	9	1.7	—	575
												C·T·O	3000	500	—	-180	345	60	10	2.6	—	800
												C·P	2500	400	—	-200	200	30	9	2.2	—	375
												C·P	3000	400	—	-310	225	30	9	3.2	—	510
												AB2[6]	2000	300	—	-48	510[7]	0/26	198[8]	5.5[7]	8K	650
												AB1[6]	2500	300	—	-110	430[7]	0.3/13	180[8]	0	11.4K	625
4X250B	250[9]	2000	12	400	175	6	2.1	18.5	0.04	4.7	Fig. 75	C·T·O	2000	250	—	-90	250	25	27	2.8	—	410
												C·P	1500	250	—	-100	200	25	17	2.1	—	250
												AB1[6]	2000	350	—	-50	500[7]	30[7]	100[8]	0	8.26K	650
7034/[9] 4X150A	250	2000	12	300	150	6	2.6	16	0.03	4.4	Fig. 75	C·T·O	2000	250	—	-88	250	24	8	2.5	—	370
												C·P	1600	250	—	-118	200	23	5	3	—	230
7035/ 4X150D	250	2000	12	400		26.5	0.58				Fig. 75	AB2[6]	2000	300	—	-50	100/500	0/36	106[8]	0.2	8.1K	630
												AB1[6]	2000	300	—	-50	100/470	0/36	106[8]	0	8.76K	580
4CX-300A	300[9]	2000	12	400	500	6	2.75	29.5	0.04	4.8	—	C·T	2000	250	—	-90	250	25	27	2.8	—	410
												C·P	1500	250	—	-100	200	25	17	2.1	—	250
												AB1[6]	2000	350	—	-50	500[7]	30[7]	100[8]	0	8.26K	650
175A	400	4000	25	600	—	5	14.5	15.1	0.06	9.8	Fig. 86	C·T·C·P	4000	600	0	-200	350	29	6	1.4	—	960
												C·T·C·P	2500	600	0	-180	350	40	7	1.6	—	600
												AB1	2500	750	—	-143	100/350	1/35	0	0	—	570
4-400A	400[9]	4000	35	600	110	5	14.5	12.5	0.12	4.7	5BK	C·T·C·P	4000	300	—	-170	270	22.5	10	10	—	720
												GG	2500	0	—	0	80/270[17]	55[17]	100[17]	38[17]	4.0K	325
												AB1	2500	750	—	-130	95/317	0/14	0	0	—	425
8122	400	2200	8	400	500	13.5	1.3	16	0.13	0.011	Fig. 86	C·T·O	2000	200	—	-30	300	5	30	5	—	300
5-500A	500	4000	35	600	30	10	10.2	19	0.10	12	—	C·T	3000	500	0	-220	432	65	35	12	—	805
												C·T	3100	470	0	-310	260	50	15	6	—	580
												AB1	3000	750	0	-112	320	26	—	—	—	612
8166/ 4-1000A	1000	6000	75	1000	—	7.5	21	27.2	.24	7.6	—	C·T	3000	500	—	-150	700	146	38	11	—	1430
												C·P	3000	500	—	-200	600	145	36	12	—	1390
												AB2	4000	500	—	-60	300/1200	0/95	—	11	7K	3000
												GG	3000	0	—	0	100/700[17]	105[17]	170[17]	130[17]	2.5K	1475
4CX1000A	1000	3000	12	400	400	6	12.5	35	.005	12	—	AB1[6]	2000	325	—	-55	500/2000	-4/60	—	—	2.8K	2160
												AB1[6]	2500	325	—	-55	500/2000	-4/60	—	—	3.1K	2920
												AB1[6]	3000	325	—	-55	500/1800	-4/60	—	—	3.85K	3360
8295/ 172	1000	3000	30	600	—	6	8.2	38	.09	18	—	C·T	2000	500	35	-175	850	42	10	1.9	—	1155
												C·T	2500	500	35	-200	840	40	10	2.1	—	1440
												C·T	3000	500	35	-200	820	42	10	2.1	—	1770
												AB1	2000	500	35	-110	200/800	12/43	110[8]	—	2.65K	1040
												AB1	2500	500	35	-110	200/800	11/40	115[8]	—	3.5K	1260
												AB1	3000	500	35	-115	220/800	11/39	115[8]	—	4.6K	1590

[1]Grid-resistor.
[2]Doubler to 175 MHz.
[3]Dual tube. Values for both sections, in push-pull. Interelectrode capacitances, however, are for each section.
[4]Tripler to 175 MHz.
[5]Filament limited to intermittent operation.
[6]Values are for two tubes.
[7]Max.-signal value.
[8]Peak grid-to-grid volts.
[9]Forced-air cooling required.
[10]Two tubes triode connected, G_2 to G_1 through 20K Ω. Input to G_2.
[11]Triper to 200 MHz.
[12]Typical Operation at 175 MHz.
[13]± 1.5 volts.

[14]KEY TO CLASS-OF-SERVICE ABBREVIATIONS
AB1 = Class-AB1
AB2 = Class-AB2.
B = Class-B push-pull at modulator.
C·M = Frequency multiplier.
C·P = Class-C plate-modulated telephone.
C·T = Class-C telegraph.
C·T·O = Class-C amplifier-osc.
GG = Grounded-grid (grid and screen connected together).
[15]No Class-B data available.
[16]HK257B 120 MHz, full rating.
[17]Single tone.

Table 15
TV Deflection Tubes

Type	Plate Dissipation Watts	Screen Dissipation Watts	Transconductance Micromhos	Heater (6.3V) Amperes	Cin pF	Cgp pF	Cout pF	Base	Class of Service	Plate Voltage	Screen Voltage	Grid Voltage	Plate Current mA	Screen Current mA	Grid Current mA	Approx. Driving Power Watts	Approx. Output Power Watts
6DQ5	24	3.2	10.5k	2.5	23	0.5	11	8JC	C	400	200	−40	100	12	1.5	0.1	25
6DQ6B	18	3.6	7.3k	1.2	15	0.5	7	6AM									
6FH6	17	3.6	6k	1.2	33	0.4	8	6AM									
6GC6	17.5	4.5	6.6k	1.2	15	0.55	7	8JX									
6GJ5	17.5	3.5	7.1k	1.2	15	0.26	6.5	9NM	C	500	200	−75	180	15	5	0.43	63
									AB₁	500	200	−43	85	4			35
6HF5	28	5.5	11.3k	2.25	24	0.56	10	12FB	C	500	140	−85	232	12.5	8	0.76	77
									AB₁	500	140	−46	133	4.5			58
6JB6	17.5	3.5	7.1k	1.2	15	0.2	6	9QL	C	500	200	−75	180	13.3	5	0.43	63
									AB₁	500	200	−42	85	4.2			35
6JE6C	30	5	10.5k	2.5	24.3		14.5	9QL	C	500	125	−85	222	17	8	0.82	76
									AB₁	500	125	−44	110	3.9			47
6JG6A	17	3.5	10k	1.6	22	0.7	9	9QU	C	450	150	−80	202	20	8	0.75	63
									AB₁	450	150	−35	98	4.5			38
6JM6	17.5	3.5	7.3k	1.2	16	0.6	7	12FJ	C	500	200	−75	190	13.7	4	0.32	61
									AB₁	500	200	−42	85	-4.4			37
6JN6	17.5	3.5	7.3k	1.2	16	0.34	7	12FK									
6JS6C	30	5.5		2.25	24	0.7	10	12FY									
6KD6	33	5	14k	2.85	40	0.8	16	12GW	GG	800	0	−11	150			12.5	82
6LB6	30	5	13.4k	2.25	33	0.4	18	12GJ									
6LG6	28	5	11.5k	2	25	0.8	13	12HL									
6LQ6	30	5	9.6k	2.5	22	0.46	11	9QL									
6MH6	38.5	7	14k	2.65	40	1.0	20	12GW									

Note: For AB₁ operation, input data is average 2-tone value. Output power is PEP.

Table 16
Semiconductor Diode Specifications

Listed numerically by device

Device	Type	Material	Peak Inverse Voltage, PIV (Volts)	Average Rectified Current Forward (Reverse) I_o (A) (I_R(A))	Peak Surge Current, I_{FSM} 1 sec. @ 25°C (A)	Average Forward Voltage, V_F (Volts)
1N34	Signal	Germanium	60	8.5 m (15.0 μ)		1.0
1N34A	Signal	Germanium	60	5.0 m (30.0 μ)		1.0
1N67A	Signal	Germanium	100	4.0 m (5.0 μ)		1.0
1N191	Signal	Germanium	90	5.0 m		1.0
1N270	Signal	Germanium	80	0.2 (100 μ)		1.0
1N914	Fast Switch	Silicon (Si)	75	75.0 m (25.0 n)	0.5	1.0
1N1184	Rectifier, Fast Recovery (RFR)	Si	100	35 (10 m)		1.7
1N2071	RFR	Si	600	0.75 (10.0 μ)		0.6
1N3666	Signal	Germanium	80	0.2 (25.0 μ)		1.0
1N4001	RFR	Si	50	1.0 (0.03 m)		1.1
1N4002	RFR	Si	100	1.0 (0.03 m)		1.1
1N4003	RFR	Si	200	1.0 (0.03 m)		1.1
1N4004	RFR	Si	400	1.0 (0.03 m)		1.1
1N4005	RFR	Si	600	1.0 (0.03 m)		1.1
1N4006	RFR	Si	800	1.0 (0.03 m)		1.1
1N4007	RFR	Si	1000	1.0 (0.03 m)		1.1
1N4148	Signal	Si	75	10.0 m (25.0 n)		1.0
1N4149	Signal	Si	75	10.0 m (25.0 n)		1.0
1N4152	Fast Switch	Si	40	20.0 m (0.05 μ)		0.8
1N4445	Signal	Si	100	0.1 (50.0 n)		1.0
1N5400	RFR	Si	50	3.0	200	
1N5401	RFR	Si	100	3.0	200	
1N5402	RFR	Si	200	3.0	200	
1N5403	RFR	Si	300	3.0	200	
1N5404	RFR	Si	400	3.0	200	
1N5405	RFR	Si	500	3.0	200	
1N5406	RFR	Si	600	3.0	200	
1N5767	Signal	Si		0.1 (1.0 μ)		1.0
ECG5863	RFR	Si	600	6	150	0.9

Table 17
Voltage—Variable Capacitance Diodes

Listed numerically by device

Device	CT Nominal Capacitance pF ±10% @ $V_R = 4.0$ V $f = 1.0$ MHz	Capacitance Ratio 4-60 V Min.	Q @ 4.0 V 50 MHz Min.	Case Style
1N5441A	6.8	2.5	450	
1N5442A	8.2	2.5	450	
1N5443A	10	2.6	400	DO-7
1N5444A	12	2.6	400	
1N5445A	15	2.6	450	
1N5446A	18	2.6	350	
1N5447A	20	2.6	350	
1N5448A	22	2.6	350	DO-7
1N5449A	27	2.6	350	
1N5450A	33	2.6	350	
1N5451A	39	2.6	300	
1N5452A	47	2.6	250	
1N5453A	56	2.6	200	DO-7
1N5454A	68	2.7	175	
1N5455A	82	2.7	175	
1N5456A	100	2.7	175	
1N5461A	6.8	2.7	600	
1N5462A	8.2	2.8	600	
1N5463A	10	2.8	550	DO-7
1N5464A	12	2.8	550	
1N5465A	15	2.8	550	
1N5466A	18	2.8	500	
1N5467A	20	2.9	500	
1N5468A	22	2.9	500	DO-7
1N5469A	27	2.9	500	
1N5470A	33	2.9	500	
1N5471A	39	2.9	450	
1N5472A	47	2.9	400	
1N5473A	56	2.9	300	DO-7
1N5474A	68	2.9	250	
1N5475A	82	2.9	225	
1N5476A	100	2.9	200	
MV2101	6.8	2.5	450	
MV2102	8.2	2.5	450	
MV2103	10	2.0	400	TO-92
MV2104	12	2.5	400	
MV2105	15	2.5	400	
MV2106	18	2.5	350	
MV2107	22	2.5	350	
MV2108	27	2.5	300	TO-92
MV2109	33	2.5	200	
MV2110	39	2.5	150	
MV2111	47	2.5	150	
MV2112	56	2.6	150	
MV2113	68	2.6	150	TO-92
MV2114	82	2.6	100	
MV2115	100	2.6	100	

Table 18
Zener Diodes

Volts	Power (Watts)							
	0.25	0.4	0.5	1.0	1.5	5.0	10.0	50.0
1.8	1N4614							
2.0	1N4615							
2.2	1N4616							
2.4	1N4617	1N4370,A	1N4370,A 1N5221,B 1N5985,B					
2.5			1N5222B					
2.6	1N702,A							
2.7	1N4618	1N4371,A	1N4371,A 1N5223,B 1N5839, 1N5986					
2.8			1N5224B					
3.0	1N4619	1N4372,A	1N4372 1N5225,B 1N5987					
3.3	1N4620	1N746,A 1N764,A 1N5518	1N746,A 1N5226,B 1N5988	1N3821 1N4728,A	1N5913	1N5333,B		
3.6	1N4621	1N747,A 1N5519	1N747A 1N5227,B 1N5989	1N3822 1N4729,A	1N5914	1N5334,B		
3.9	1N4622	1N748,A 1N5520	1N748A 1N5228,B 1N5844, 1N5990	1N3823 1N4730,A	1N5915	1N5335,B	1N3993A	1N4549,B 1N4557,B
4.1	1N704,A							
4.3	1N4623	1N749,A 1N5521	1N749,A 1N5229,B 1N5845 1N5991	1N3824 1N4731,A	1N5916	1N5336,B	1N3994,A	1N4550,B 1N4558,B
4.7	1N4624	1N750,A 1N5522	1N750A 1N5230,B 1N5846, 1N5992	1N3825 1N4732,A	1N5917	1N5337,B	1N3995,A	1N4551,B 1N4559,B
5.1	1N4625 1N4689	1N751,A 1N5523	1N751,A, 1N5231,B 1N5847 1N5993	1N3826 1N4733	1N5918	1N5338,B	1N3996,A	1N4552,B 1N4560,B
5.6	1N708A 1N4626	1N752,A 1N5524	1N752,A 1N5232,B 1N5848, 1N5994	1N3827 1N4734,A	1N5919	1N5339,B	1N3997,A	1N4553,B 1N4561,B

Table 18 — Zener Diodes — Continued

Volts	0.25	0.4	0.5	Power (Watts) 1.0	1.5	5.0	10.0	50.0
5.8	1N706A	1N762						
6.0			1N5233B 1N5849			1N5340,B		
6.2	1N709,1N4627 MZ605, MZ610 MZ620, MZ640	1N753,A 1N821,3,5,7,9;A	1N753,A 1N5234,B, 1N5850 1N5995	1N3828,A 1N4735,A	1N5920	1N5341,B	1N3998,A	1N4554,B 1N4562,B
6.4	1N4565-84,A							
6.8	1N4099	1N754,A 1N957,B 1N5526	1N754,A 1N757,B 1N5235,B 1N5851 1N5996	1N3016,B 1N3829 1N4736,A	1N3785 1N5921	1N5342,B	1N2970,B 1N3999,A	1N2804B 1N3305B 1N4555, 1N4563
7.5	1N4100	1N755,A 1N958,B 1N5527	1N755A, 1N958,B 1N5236,B 1N5852 1N5997	1N3017,A,B 1N3830 1N4737,A	1N3786 1N5922	1N5343,B	1N2971,B 1N4000,A	1N2805,B 1N3306,B 1N4556, 1N4564
8.0	1N707A							
8.2	1N712A 1N4101	1N756,A 1N959,B 1N5528	1N756,A 1N959,B 1N5237,B 1N5853 1N5998	1N3018,B 1N4738,A	1N3787 1N5923	1N5344,B	1N2972,B	1N2806,B 1N3307,B
8.4		1N3154-57,A	1N3154,A 1N3155-57					
8.5	1N4775-84,A		1N5238,B 1N5854					
8.7	1N4102					1N5345,B		
8.8		1N764						
9.0		1N764A	1N935-9;A,B					
9.1	1N4103	1N757,A 1N960,B 1N5529	1N757,A, 1N960,B 1N5239,B, 1N5855 1N5999	1N3019,B 1N4739,A	1N3788 1N5924	1N5346,B	1N2973,B	1N2807,B 1N3308,B
10.0	1N4104	1N758,A 1N961,B 1N5530,B	1N758,A, 1N961,B 1N5240,B, 1N5856 1N6000	1N3020,B 1N4740,A	1N3789 1N5925	1N5347,B	1N2974,B	1N2808,B 1N3309,A,B
11.0	1N715,A 1N4105	1N962,B 1N5531	1N962,B 1N5241,B 1N5857, 1N6001	1N3021,B 1N4741,A	1N3790 1N5926	1N5348,B	1N2975,B	1N2809,B 1N3310,B
11.7	1N716,A 1N4106		1N941-4;A,B					
12.0		1N759,A 1N963,B 1N5532	1N759,A, 1N963,B, 1N5242,B, 1N5858 1N6002	1N3022,B 1N4742,A	1N3791 1N5927	1N5349,B	1N2976,B	1N2810,B 1N3311,B
13.0	1N4107	1N964,B 1N5533	1N964,B 1N5243,B, 1N5859 1N6003	1N3023,B 1N4743,A	1N3792 1N5928	1N5350,B	1N2977,B	1N2811,B 1N3312,B
14.0	1N4108	1N5534	1N5244B 1N5860			1N5351,B	1N2978,B	1N2812,B 1N3313,B
15.0	1N4109	1N965,B 1N5535	1N965,B 1N5245,B, 1N5861, 1N6004	1N3024,B 1N4744A	1N3793 1N5929	1N5352,B	1N2979,A,B	1N2813,A,B 1N3314,B
16.0	1N4110	1N966,B 1N5536	1N966,B, 1N5246,B, 1N5862, 1N6005	1N3025,B 1N4745,A	1N3794 1N5930	1N5353,B	1N2980,B	1N2814,B 1N3315,B
17.0	1N4111	1N5537	1N5247,B 1N5863			1N5354,B	1N2981B	1N2815,B 1N3316,B
18.0	1N4112	1N967,B 1N5538	1N967,B 1N5248,B 1N5864, 1N6006	1N3026,B 1N4746,A	1N3795 1N5931	1N5355,B	1N2982,B	1N2816,B 1N3317,B
19.0	1N4113	1N5539	1N5249,B 1N5865			1N5356,B	1N2983,B	1N2817,B 1N3318,B
20.0	1N4114	1N968,B 1N5540	1N968,B 1N5250,B 1N5866, 1N6007	1N3027,B 1N4747,A	1N3796 1N5932,A,B	1N5357,B	1N2984,B	1N2818,B 1N3319,B
22.0	1N4115	1N959,B 1N5541	1N969,B 1N5241,B 1N5867, 1N6008	1N3028,B 1N4748,A	1N3797 1N5933	1N5358,B	1N2985,B	1N2819,B 1N3320,A,B
24.0	1N4116	1N5542 1N9701B	1N970,B 1N5252,B, 1N586 1N6009	1N3029,B 1N4749,A	1N3798 1N5934	1N5359,B	1N2986,B	1N2820,B 1N3321,B
25.0	1N4117	1N5543	1N5253,B 1N5869			1N5360,B	1N2987B	1N2821,B 1N3322,B
27.0	1N4118	1N971,B	1N971 1N5254,B, 1N5870, 1N6010	1N3030,B 1N4750,A	1N3799 1N5935	1N5361,B	1N2988,B	1N2822B 1N3323,B
28.0	1N4119	1N5544	1N5255,B 1N5871			1N5362,B		
30.0	1N4120	1N972,B 1N5545	1N972,B 1N5256,B, 1N5872, 1N6011	1N3031,B 1N4751,A	1N3800 1N5936	1N5363,B	1N2989,B	1N2823,B 1N3324,B
33.0	1N4121	1N973,B 1N5546	1N973,B 1N5257,B 1N5873 1N6012	1N3032,B 1N4752,A	1N3801 1N5937	1N5364,B	1N2990,A,B	1N2824,B 1N3325,B

Table 18 — Zener Diodes — Continued

Volts	0.25	0.4	0.5	Power (Watts) 1.0	1.5	5.0	10.0	50.0
36.0	1N4122	1N974,B	1N974,B 1N5258,B 1N5874, 1N6013	1N3033,B 1N4753,A	1N3802 1N5938	1N5365,B	1N2991,B	1N2825,B 1N3326,B
39.0	1N4123	1N975,B	1N975,B, 1N5259,B 1N5875, 1N6014	1N3034,B 1N4754,A	1N3803 1N5939	1N5366,B	1N2992,B	1N2826,B 1N3327,B
43.0	1N4124	1N976,B	1N976,B 1N5260,B, 1N5876, 1N6015	1N3035,B 1N4755,A	1N3804 1N5940	1N5367,B	1N2993,A,B	1N2827,B 1N3328,B
45.0							1N2994B	1N2828B 1N3329B
47.0	1N4125	1N977,B	1N977,B, 1N5261,B 1N5877, 1N6016	1N3036,B 1N4756,A	1N3805 1N5941	1N5368,B	1N2996,B	1N2829,B 1N3330,B
50.0								1N2830B 1N3331B
51.0	1N4126	1N978,B	1N978,B, 1N5262,A,B 1N5878, 1N6017	1N3037,B 1N4757,A	1N3806 1N5942	1N5369,B	1N2997,B	1N2831,B 1N3332,B
52.0							1N2998B	1N3333
56.0	1N4127	1N979,B	1N979 1N5263,B 1N6018	1N3038,B 1N4758,A	1N3807 1N5943	1N5370,B	1N2999,B	1N2822,B 1N3334,B
60.0	1N4128		1N5264,A,B			1N5371,B		
62.0	1N4129	1N980,B	1N980 1N5265,A,B 1N6019	1N3039,B 1N4759,A	1N3808 1N5944	1N5372,B	1N3000,B	1N2833,B 1N3335,B
68.0	1N4130	1N981,B	1N981,B 1N5266,A,B 1N6020	1N3040,A,B 1N4760,A	1N3809 1N5945	1N5373,B	1N3001,B	1N2834,B 1N3336,B
75.0	1N4131	1N982,B	1N982 1N5267,A,B 1N6021	1N3041,B 1N4761,A	1N3810 1N5946	1N5374,B	1N3002,B	1N2835,B 1N3337,B
82.0	1N4132	1N983,B	1N983 1N5268,A,B 1N6022	1N3042,B 1N4762,A	1N3811 1N5947	1N5375,B	1N3003,B	1N2836,B 1N3338,B
87.0	1N4133		1N5269,B			1N5376,B		
91.0	1N4134	1N984,B	1N984 1N5270,B 1N6023	1N3043,B 1N4763,A	1N3812 1N5948	1N5377,B	1N3004,B	1N2837,B 1N3339,B
100.0	1N4135	1N985	1N985,B 1N5271,B 1N6024	1N3044,A,B 1N4764,A	1N3813 1N5949	1N5378,B	1N3005,B	1N2838,B 1N3340,B
105.0							1N3006B	1N2839,B 1N3341,B
110.0		1N986	1N986 1N5272,B 1N6025	1N3045,B 1M110ZS10	1N3814 1N5950	1N5379,B	1N3007A,B	1N2840,B 1N3342,B
120.0		1N987	1N987,B 1N5273,B 1N6026	1N3046,B 1M120ZS10	1N3815 1N5951	1N5380,B	1N3008A,B	1N2841,B 1N3343,B
130.0		1N988	1N988,B 1N5274,B 1N6027	1N3047,B 1M130ZS10	1N3816 1N5952	1N5381,B	1N3009,B	1N2842,B 1N3344,B
140.0		1N989	1N5275,B			1N5382B	1N3010B	1N3345B
150.0		1N990	1N989 1N5276,B 1N6028	1N3048,B 1M150ZS10	1N3817 1N5953	1N5383,B	1N3011,B	1N2843,B 1N3346,B
160.0		1N991	1N990 1N5277,B 1N6029	1N3049,B 1M160ZS10	1N3818 1N5954	1N5384,B	1N3012A,B	1N2844,B 1N3347,B
170.0		1N992	1N5278,B	1M170ZS10		1N5385,B		
175.0							1N3013B	1N3348B
180.0			1N991,B 1N5279,B 1N6030	1N3050,A,B 1M180ZS10	1N3819 1N5955	1N5386,B	1N3014,B	1N2845,B 1N3349,B
190.0			1N5280,B			1N5387,B		
200.0			1N992 1N5281,B 1N6031	1N3051,B 1M200ZS10	1N3820 1N5956	1N5388B	1N3015,B	1N2846,B 1N3350,B

Table 19
Metal-Oxide Varistors Transient Suppressors

Listed by voltage

Type No.	V ac$_{RMS}$	Maximum Applied Voltage V ac$_{Peak}$	Maximum Energy (Joules)	Max. Peak Current (Amps)	Max. Power (Watts)	Max. Varistor Voltage (Volts)
V180ZA1	115	163	1.5	500	0.2	285
V180ZA10	115	163	10.0	2000	0.45	290
V130PA10A	130	184	10.0	4000	8.0	350
V130PA20A	130	184	20.0	4000	15.0	350
V130LA1	130	184	1.0	400	0.24	360
V130LA2	130	184	2.0	400	0.24	360
V130LA10A	130	184	10.0	2000	0.5	340
V130LA20A	130	184	20.0	4000	0.85	340
V150PA10A	150	212	10.0	4000	8.0	410
V150PA20A	150	212	20.0	4000	15.0	410
V150LA1	150	212	1.0	400	0.24	420
V150LA2	150	212	2.0	400	0.24	420
V150LA10A	150	212	10.0	2000	0.5	390
V150LA20A	150	212	20.0	4000	0.85	390
V250PA10A	250	354	10.0	4000	4.0	670
V250PA20A	250	354	20.0	4000	7.0	670
V250PA40A	250	354	40.0	4000	13.0	670
V250LA2	250	354	2.0	400	0.28	690
V250LA4	250	354	4.0	400	0.28	690
V250LA15A	250	354	15.0	2000	0.6	640
V250LA20A	250	354	20.0	2000	0.6	640
V250LA40A	250	354	40.0	4000	0.9	640

Courtesy of General Electric Company

Table 20
Japanese Semiconductor Nomenclature

All transistors manufactured in Japan are registered with the Electronic Industries Association of Japan (EIAJ). In addition, the Japan Industrial Standard JIS-C-7012 provides type numbers for transistors and thyristors.

Each transistor type number consists of five elements.

Example:

i	ii	iii	iv	v
2	S	C	82D	A
Figure	Letter	Letter	Figure	Letter

i) Kind of device, indicating the number of effective electrical connections minus one.

ii) For a semiconductor registered with the EIAJ, this letter is always an S.

iii) This letter designates polarity and application, as follows:

Letter	Polarity and Application
A	PNP transistor, high frequency
B	PNP transistor, low frequency
C	NPN transistor, high frequency
D	NPN transistor, low frequency
E	P-gate thyristor
G	N-gate thyristor
H	N-base unijunction transistor
J	P-channel FET
K	N-channel FET
M	Bi-directional triode thyristor

iv) These figures designate the order of application for EIAJ registration, starting with 11.

v) This letter indicates the level of improvement. An improved device may be used in place of a previous-generation device, but not necessarily the other way around.

Table 21
General Purpose Transistors

Listed numerically by device

Device	Type	V_{CEO} Max. Collector-Emitter Voltage (Volts)	V_{CBO} Max. Collector-Base Voltage (Volts)	V_{EBO} Max. Emitter-Base Voltage (Volts)	I_c Max. Collector Current (mA)	P_D Max. Device Dissipation Watts	Min. DC Current Gain $I_c = 0.1$ mA h_{FE}	Min. DC Current Gain $I_c = 150$ mA h_{FE}	Current-Gain Bandwidth Product f_T* (MHz)	Noise Figure NF Max. (dB)
2N918	NPN	15	30	3.0	50	0.200	20(3 mA)	—	600	6.0
2N2102	NPN	65	120	7.0	1000	1.0	20	40	60	6.0
2N2218	NPN	30	60	5.0	800	0.8	20	40	250	
2N2218A	NPN	40	75	6.0	800	0.8	20	40	250	
2N2219	NPN	30	60	5.0	800	3.0	35	100	250	
2N2219A	NPN	40	75	6.0	800	3.0	35	100	300	4.0
2N2222	NPN	30	60	5.0	800	1.2	35	100	250	
2N2222A	NPN	40	75	6.0	800	1.2	35	100	300	4.0
2N2905	PNP	40	60	5.0	600	0.6	35	—	200	
2N2905A	PNP	60	60	5.0	600	0.6	75	100	200	
2N2907	PNP	40	60	5.0	600	0.400	35	—	200	
2N2907A	PNP	60	60	5.0	600	0.400	75	100	200	
2N3053	NPN	40	60	5.0	700	5.0	—	50	100	
2N3053A	NPN	60	80	5.0	700	5.0	—	50	100	
2N3904	NPN	40	60	6.0	200	0.625	40	—	300	5.0
2N3906	PNP	40	40	5.0	200	1.5	60	—	250	4.0
2N4037	PNP	40	60	7.0	1000	5.0	—	50		
2N4123	NPN	30	40	5.0	200	0.35	—	25(50 mA)	250	6.0
2N4124	NPN	25	30	5.0	200	0.350	120(2 mA)	60(50 mA)	300	5.0
2N4125	PNP	30	30	4.0	200	0.625	50(2 mA)	25(50 mA)	200	5.0
2N4126	PNP	25	25	4.0	200	0.625	120(2 mA)	60(50 mA)	250	4.0
2N4401	NPN	40	60	6.0	600	0.625	20	100	250	
2N4403	PNP	40	40	5.0	600	0.625	30	100	200	
2N5320	NPN	75	100	7.0	2000	10.0	—	30(1 A)		
2N5415	PNP	200	200	4.0	1000	10.0	—	30(50 mA)	15	
MM4003	PNP	250	250	4.0	500	1.0	20(10 mA)	—		
MPSA55	PNP	60	60	4.0	500	0.625	—	50(100 mA)	50	
MPS6547	NPN	25	35	3.0	50	0.625	20(2 mA)	—	600	

*Test conditions: $I_c = 20$ mAdc; $V_{CE} = 20$V; f = 100 MHz

Table 22
Low-Noise Transistors

The low-noise devices listed are produced with carefully controlled r_b and f_T to optimize device noise performance. Devices listed in the matrix are classified according to noise figure performance versus frequency.

NF dB	Frequency MHz 60	100	200	450	1000	2000	Polarity
1.5	2N5829	2N5829					PNP
	2N5031	2N5031	MRF904				NPN
2.0	2N4957	2N4957	2N5829				PNP
	2N5032	2N5032	2N5031	MRF904	MRF901		NPN
2.5	2N4958	2N4958	2N4957	2N5829			PNP
	2N5032	2N5032	2N5032	2N5031	MRF901		NPN
					2N6603		NPN
3.0	2N4959	2N4959	2N4958	2N4957	2N5829		PNP
	2N2857	2N2857	2N5032	3N5032	MRF901	MRF902	NPN
					2N6604		NPN
3.5	2N4959	2N4959	2N4959	2N4958	2N4957		PNP
	2N5179	2N5179	2N2857	3N5032	2N5031	MRF901	NPN
4.0	2N4959	2N4959	2N4959	2N4959	2N4958		PNP
	2N5179	2N5179	2N5179	2N2857	2N5031		NPN
4.5	2N4959	2N4959	2N4959	2N4959	2N4959		PNP
	2N5179	2N5179	2N5179	2N2857	2N5032		NPN

Table 23
RF Power Transistors
High Frequency, Low Voltage Amplifier Transistors
Power FETs

The transistors listed in this table are specified for operation in RF power amplifiers and are listed by ascending input power. Modulation type is given in each application heading.
Courtesy Motorola Semiconductor Prod. Inc.

Device Type	P_{in} Input Power Watts	P_{out} Output Power Watts	G_{PE} Power Gain dB Min	V_{CC} Supply Voltage Volts	Package
1.5-30 MHz SSB Transistors					
MRF476	0.1	3.0 PEP/CW	15	12.5	TO-220
MRF432*†	0.125	12.5 PEP/CW	20	12.5	211-07
MRF433*	0.125	12.5 PEP/CW	20	12.5	211-07
MRF426	0.16	25.0 PEP/CW	22	28.0	211-07
MRF426A	0.16	25.0 PEP/CW	22	28.0	145A-09
2N6367	0.36	9.0 PEP/CW	14	12.5	211-07
MRF427	0.40	25.0 PEP/CW	18	50.0	211-11
MRF427A	0.40	25.0 PEP/CW	18	50.0	145A-10
2N6370	0.62	10.0 PEP/CW	12	28.0	211-07
MRF475	1.2	12.0 PEP/CW	10	13.6	TO-220
2N5070	1.25	25.0 PEP/CW	13	28.0	TO-60
MRF401	1.25	25.0 PEP/CW	13	28.0	145A-09
MRF406	1.25	20.0 PEP/CW	12	12.5	211-07
MRF466	1.25	15.0 PEP/CW	15	28.0	211-09
MRF477	1.25	40.0 PEP/CW	15	12.5	TO-220
MRF486	1.25	40.0 PEP/CW	15	28.0	TO-220
MRF479	1.5	15.0 PEP/CW	10	12.5	TO-220
MRF485	1.5	15.0 PEP/CW	10	28.0	TO-220
2N5941†	2.0	40.0 PEP/CW	13	28.0	211-07
MRF460	2.5	40.0 PEP/CW	12	12.5	211-11
MRF463†	2.53	80.0 PEP/CW	15	28.0	211-08
MRF464	2.53	80.0 PEP/CW	15	28.0	211-09
MRF464A	2.53	80.0 PEP/CW	15	28.0	145A-10
MRF412	3.5	70.0 PEP/CW	13	13.6	211-11
MRF428	7.5	150.0 PEP/CW	13	50.0	211-11
MRF428A†	7.5	150.0 PEP/CW	13	50.0	307-01
MRF429	7.5	150.0 PEP/CW	13	50.0	211-11
MRF421	10.0	100.0 PEP/CW	10	12.5	211-11
MRF422	15.0	150.0 PEP/CW	10	28.0	211-11
MRF435	15.0	150.0 PEP/CW	10	28.0	211-11
MRF448	15.7	250.0 PEP/CW	12	50.0	211-11

Device Type	P_{in} Input Power Watts	P_{out} Output Power Watts	G_{PE} Power Gain dB Min	V_{CC} Supply Voltage Volts	Package
14-30 MHz Amateur Transistors					
MRF8003	0.05	0.5	10	12.5	TO-39
MRF476	0.10	3.0	15	12.5	TO-220
MRF449†	0.3	30.0	10	13.6	211-07
MRF8004	0.35	3.5	10	12.5	TO-39
MRF475	0.4	4.0	10	13.6	TO-220
MRF449A	1.9	30.0	12	13.6	145A-09
MRF453	3.0	60.0	13	12.5	211-11
MRF455	3.0	60.0	13	12.5	211-07
MRF455A	3.0	60.0	13	12.5	145A-09
MRF450	4.0	50.0	11	13.6	211-09
MRF450A	4.0	50.0	11	13.6	145A-09
MRF497	4.0	40.0	10	12.5	TO-220
MRF453A†	4.8	60.0	11	12.5	145A-10
MRF454	5.0	80.0	12	12.5	211-11
MRF454A	5.0	80.0	12	12.5	145A-10
MRF458	5.0	80.0	12	12.5	211-11
MRF458A	5.0	80.0	12	12.5	145A-10
MRF492	5.6	70.0	11	12.5	211-11
MRF492A	5.6	70.0	11	12.5	145A-10
106-175 MHz VHF AM Transistors					
2N3866	0.1	1.0	10	28	TO-39
2N3553	0.25	2.5	10	28	TO-39
2N5641	1.0	7.0	8.4	28	144-05
2N5642	3.0	20	8.2	28	145A-09
MRF314	3.0	30	10	28	211-07
MRF314A	3.0	30	10	28	145A-09
2N5643	6.9	40	7.6	28	145A-09
MRF315	5.7	45	9.0	28	211-07
MRF315A	5.7	45	9.0	28	145A-09
MRF316*	8.0	80	10	28	316-01
MRF317*	12.5	100	9.0	28	316-01

*PNP/NPN complements for Complementary Symmetry Driver. For matched pairs, order MK433.
†Discontinued

*Controlled "Q" transistor

Table 23 (continued)

2-175 MHz SSB TMOS Power FETs

Device Type	P_{in} Input Power Watts	P_{out} Output Power Watts	G_{PE} Power Gain dB Min	V_{CC} Supply Voltage Volts	Package
MRF138	0.6/1.2	30.0	17/14	28	211-07
MRF140	4.7/3.7	150.0	15/16	28	211-11
MRF148	0.48/0.95	30.0	18/15	50	211-07
MRF150	3.0/23	150.0	17/8	50	211-11

130-175 MHz VHF FM Transistors

Device Type	P_{in} Input Power Watts	P_{out} Output Power Watts	G_{PE} Power Gain dB Min	V_{CC} Supply Voltage Volts	Package
2N4427	0.1	1.0	10	12.0	TO-39
MRF553	0.1	1.5	13.0	12.5	317C-01
MRF604	0.1	1.0	10	12.5	TO-46
MRF607	0.12	1.75	11.5	12.5	TO-39
2N6080	0.25	4.0	12	12.5	145A-09
MRF237*	0.25	4.0	12	12.5	TO-39
MRF215**	0.33	20.0	8.2	12.5	316-01
2N5589	0.44	3.0	8.2	13.6	144B-05
2N6255	0.5	3.0	7.8	12.5	TO-39
MRF260	0.5	5.0	10	12.5	TO-220
MRF212	1.25	10.0	9.0	12.5	145A-09
2N5590	3.0	10.0	5.2	13.6	145A-09
MRF239	3.0	30.0	10.0	13.6	145A-09
MRF261	3.0	10.0	5.2	12.5	TO-220
2N6081	3.5	15.0	6.3	12.5	145A-09
MRF221	3.5	15.0	6.3	12.5	211-07
MRF262	3.5	15.0	6.3	12.5	TO-220
MRF238	3.7	30.0	9.0	13.6	145A-09
MRF240	5.0	40.0	9.0	13.6	145A-09
2N6082	6.0	25.0	6.2	12.5	145A-09
MRF222	6.0	25.0	6.2	12.5	211-01
2N6083	8.1	30.0	5.7	12.5	145A-09
MRF223	8.1	30.0	5.7	12.5	211-07
MRF216**	8.5	40.0	6.7	12.5	316-01
2N5591	9.0	25.0	4.4	13.6	145A-09
MRF264	9.1	30.0	5.2	12.5	TO-220
MRF243**	12.0	60.0	7.0	12.5	316-01
2N6084	14.3	40.0	4.5	12.5	145A-09
MRF224	14.3	40.0	4.5	12.5	211-07
MRF247**	15.0	75.0	7.0	12.5	316-01
MRF245**	18.2	80.0	6.4	12.5	316-01
MRF4070	20.0	70.0	5.0	12.5	316-01

*Grounded emitter TO-39 package   **Controlled "Q" transistor

220 MHz FM Transistors

Device Type	P_{in} Input Power Watts	P_{out} Output Power Watts	G_{PE} Power Gain dB Min	V_{CC} Supply Voltage Volts	Package
MRF525*	0.001	0.02	13.0	26.0	TO-39
MRF313	0.03	1.0	15.0	28.0	305A-01
MRF313A	0.03	1.0	15.0	28.0	305-01
2N4428	0.075	25.0	10.0	28.0	TO-39
2N3866	0.1	1.0	10.0	28.0	TO-39
MRF208	0.1	10.0	10.0	12.5	145A-09
MRF5174	0.125	2.0	12.0	28.0	244-04
MRF227*	0.13	3.0	13.5	12.5	TO-39
MRF209	0.15	1.0	8.2	12.5	TO-39
2N5160	0.16	1.0	8.0	28.0	TO-39
MRF225	0.18	1.5	9.0	12.5	TO-39
MRF321**	0.62	10.0	12.0	28.0	244-04
MRF226	1.6	13.0	9.0	12.5	145A-09
MRF331	1.6	10.0	8.0	28.0	244-04
MRF323**	2.0	20.0	10.0	28.0	244-04
MRF325**	4.3	30.0	8.5	28.0	316-01
MRF390***	6.8	60.0	9.5#	28.0	744-02
MRF5177	7.5	30.0	6.0	28.0	215-02
MRF5177A	7.5	30.0	6.0	28.0	145A-09
MRF326**	8.0	40.0	9.0	28.0	316-01
MRF209	9.1	25.0	4.4	12.5	145A-09
MRF309	10.0	50.0	7.0	28.0	316-01
2N6439	10.0	60.0	7.8	28.0	316-01
MRF327**	14.9	80.0	7.3	28.0	316-01
MRF329**	20.0	100.0	7.0	28.0	333-03

*Grounded emitter TO-39 package   ***Internal impedance matched push pull transistors
**Internal impedance matched   #Typical
†PNP

407-512 MHz UHF FM Transistors

Device Type	P_{in} Input Power Watts	P_{out} Output Power Watts	G_{PE} Power Gain dB Min	V_{CC} Supply Voltage Volts	Package
MRF581	0.023	0.6	14.0	12.5	317-01
MRF559	0.025	0.5	13.0	12.5	317-01
2N6256	0.05	0.5	10.0	12.5	249-05
MRF626	0.05	0.5	10.0	12.5	305-01
MRF627	0.05	0.5	10.0	12.5	305A-01
MRF628	0.05	0.5	10.0	12.5	249-05
MRF750	0.05	0.5	10.0	7.5	305A-01
MRF515	0.12	0.75	8.0	12.5	TO-39
2N5644	0.20	1.0	7.0	12.5	145A-09
2N3948	0.25	1.0	6.0	13.6	TO-39
2N5944	0.25	2.0	9.0	12.5	244-04
MRF629*	0.32	2.0	8.0	12.5	TO-39
MRF630*	0.33	3.0	9.5	12.5	TO-39
MRF752	0.4	2.5	8.0	7.5	249-05
MRF652	0.5	5.0	10.0	12.5	244-04
2N5945	0.64	4.0	8.0	12.5	244-04
MRF660	2.0	7.0	5.4	12.5	TO-220
MRF754	2.0	8.0	6.0	7.5	249-05
2N5946	2.5	10.0	6.0	12.5	244-04
MRF641**	3.75	15.0	7.8	12.5	316-01
MRF644**	5.9	25.0	6.2	12.5	316-01
MRF646**	13.3	40.0	4.8	12.5	316-01
MRF338**	15.0	80.0	4.3	28.0	333-03
MRF648**	22.0	60.0	4.4	12.5	316-01

*Grounded emitter TO-39 package   **Internal impedance matched

806-947 MHz UHF FM Transistors

Device Type	P_{in} Input Power Watts	P_{out} Output Power Watts	G_{PE} Power Gain dB Min	V_{CC} Supply Voltage Volts	Package
MRF581	0.06	0.6	10#	12.5	317-01
MRF816	0.075	0.75	10.0	12.5	249.05
MRF559	0.080	0.5	8.0	12.5	317-01
MRF838	0.22	1.0	6.5	12.5	305A-01
MRF838A	0.22	1.0	6.5	12.5	305-01
MRF890	0.25	2.0	9.0	24.0	305-01
MRF817	0.59	2.5	6.2	13.6	244-04
MRF841@	0.7	5.0	8.5	12.5	305A-01
MRF870	1.0	3.0	5.0	12.5	305A-01
MRF870A	1.0	3.0	5.0	12.5	305-01
MRF892 @	2.0	14.0	8.5	24.0	319-04
MRF840@	2.5	10.0	6.0	12.5	319-04
MRF842@	5.0	20.0	6.0	12.5	319-04
MRF894@	6.0	30.0	7.0	24.0	319-04
MRF844@	9.0	30.0	5.2	12.5	319-04
MRF846@	15.0	40.0	4.3	12.5	319-04

@ Common base
# Typical

Table 24
VHF and UHF Class-A Transistors

The devices listed below are excellent for class-A linear applications.
The devices are listed according to increasing current-gain bandwidth (f_T).

Device Type	Nominal Test Conditions V_{CE}/I_C Volts/mA	f_T MHz Min.	Noise Figure Max/Freq. MHz	Distortion Specifications 2nd Order IMD	3rd Order IMD	Output Level dBmV	Package
MRF501	6/5	600	4.5*/200				TO-72
MRF502	6/5	800	4.0*/200				TO-72
2N5179	6/5	900	4.5/200				TO-72
BFY90	5/2	1000	5.0/500				TO-72
BFW92	5/2	1000	4.0/500				302A-01
2N6305	5/10	1200	5.5/450				TO-72
BFX89	5/25	1200	6.5/500				TO-72
2N5109	15/50	1200	3.0*/200				TO-39
2N5943	15/50	1200	3.4/200	−50		+50	TO-39
2N6304	5/10	1400	4.5/450				TO-72
MRF511	20/80	1500	7.3*/200	−50	−65	+50	144D-04
MRF517	15/60	2200	7.5/300	−60	−72	+45	TO-39
BFR90	10/14	5000*	2.4*/500				302A-01
BFR91	5/35	5000*	1.9*/500				302A-01
BFR96	10/50	5000*	3.3*/500				302A-01

*Typ.

Table 25
Suggested Small-Signal FETs

Device No.	Type	Max. Diss. (mW)	Max. V_{DS} (volts)	$V_{GS(off)}$ volts	Min. gfs (µS)	Input C (pF)	Max. I_D (mA)*	Upper Freq. (MHz)	Noise Figure (typ.)	Case Type	Base Conn.	Mfgr. (see code)	Applications (general)
2N4416	N-JFET	300	30	−6	4500	4	−15	450	400 MHz 4 dB	TO-72	1	S, M	VHF/UHF/RF amp., mix., osc.
2N5484	N-JFET	310	25	−3	2500	5	30	200	200 MHz 4 dB	TO-92	2	M	VHF/UHF amp., mix., osc.
2N5485	N-JFET	310	25	−4	3500	5	30	400	400 MHz 4 dB	TO-92	2	S	VHF/UHF/RF amp., mix., osc.
3N200	N-Dual-Gate MOSFET	330	20	−6	10,000	4-8.5	50	500	400 MHz 4.5 dB	TO-72	3	R	VHF/UHF/RF amp., mix., osc.
3N202	N-Dual-Gate MOSFET	360	25	−5	8000	6	50	200	200 MHz 4.5 dB	TO-72	3	S	VHF amp., mixer
MPF102	N-JFET	310	25	−8	2000	4.5	20	200		TO-92	2	N, M	HF/VHF amp., mix., osc.
MPF106/ 2N5484	N-JFET	310	25	−6	2500	5	30	400	200 MHz 4 dB	TO-92	2	N, M	HF/VHF/UHF amp., mix., osc.
40673	N-Dual-Gate MOSFET	330	20	−4	12,000	6	50	400	200 MHz 6 dB	TO-72	3	R	HF/VHF/UHF amp., mix., osc.
U300	P-JFET	300	−40	+10	8000	20	−50	—	40n V/√Hz	TO-18	4	S	General-purpose amp.
U304	P-JFET	350	−30	+10		27	−50	—		TO-18	4	S	analog switch, chopper
U310	N-JFET	500	30	−6	10,000	2.5	60	450	450 MHz 3.2 dB	TO-52	5	S	common-gate VHF/UHF amp., osc., mix.
U350	N-JFET Quad	300 / 1W	30 / 25	−6	9000	5	60	100	100 MHz 7 dB	TO-99	6	S	matched JFET doubly bal. mix.
U431	N-JFET Dual	300	25	−6	10,000	5	30	100	—	TO-99	7	S	matched JFET cascode amp. and bal. mix.

*25°C   M = Motorola.   N = National Semiconductor.   R = RCA.   S = Siliconix Inc.   D = Drain.   S = Source.   G = Gate.

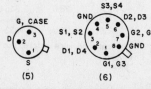

(1) (2) (3) (4) (5) (6) (7)

Table 26
Power FETs

Device No.	Type	Max. Diss. (W)	Max. V_{DS} (volts)	Max. I_D (A)	G_{fs} μmhos (typ.)	Input C C_{iss} (pF)	Output C C_{oss} (pF)	Approx. Upper Freq. (MHz)	Case Type	Base Conn. Mfgr.	Applications (general)
DV1202S	N-Chan.	10	50	0.5	100k	14	20	500	.380 SOE	1/S	RF pwr. amp., osc.
DV1202W	N-Chan.	10	50	0.5	100k	14	20	500	C-220	5/S	RF pwr. amp., osc.
DV1205S	N-Chan.	20	50	1	200k	26	38	500	.380 SOE	1/S	RF pwr. amp., osc.
DV1205W	N-Chan.	20	50	1	200k	26	38	500	C-220	5/S	RF pwr. amp., osc.
2SK133	N-Chan.	100	120	7	1M	600	350	1	TO-3	6/H	AF pwr. amp., switch (complement to 2SJ48)
2SK134	N-Chan.	100	140	7	1M	600	350	1	TO-3	6/H	AF pwr. amp., switch (complement to 2SJ49)
2SK135	N-Chan.	100	160	7	1M	600	350	1	TO-3	6/H	AF pwr. amp., switch (complement to 2SJ50)
2SJ48	P-Chan.	100	120	7	1M	900	400	1	TO-3	6/H	AF pwr. amp., switch (complement to 2SK133)
2SJ49	P-Chan.	100	140	7	1M	900	400	1	TO-3	6/H	AF pwr. amp., switch (complement 2SK134)
2SJ50	P-Chan.	100	160	7	1M	900	400	1	TO-3	6/H	AF pwr. amp., switch (complement 2SK135)
VMP4	N-Chan.	25	60	2	170,000	32	4.8	200	.380-SOE	1/S	VHF pwr. amp., rcvr front end (rf amp., mixer).
VN10KM	N-Chan.	1	60	0.5	100,000	48	16	—	TO-92	2/S	High-speed line driver, relay driver, LED stroke driver
VN64GA	N-Chan.	80	60	12.5	150,000	700	325	30	TO-3	3/S	Linear amp., power-supply switch, motor control
VN66AF	N-Chan.	15	60	2	150,000	50	50	—	TO-202	4/S	High-speed switch, hf linear amp., audio amp. line driver.
VN66AK	N-Chan.	8.3	60	2	250,000	33	6	100	TO-39	5/S	RF pwr. amp., high-current analog switching.
VN67AJ	N-Chan.	25	60	2	250,000	33	7	100	TO-3	3/S	RF pwr. amp., high-current switching
VN89AA	N-Chan.	25	80	2	250,000	50	10	100	TO-3	3/S	High-speed switching, hf linear amps., line drivers.
IRF100	N-Chan.	125	80	16	300,000	900	25	—	TO-3	3/IR	High-speed switching, audio amps., motor control, inverters.
IRF101	N-Chan.	125	60	16	300,000	900	25	—	TO-3	3/IR	Same as IRF100

*25°C S = M/A-COM IR = International Rectifier.
(case) H = Hitachi

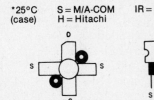
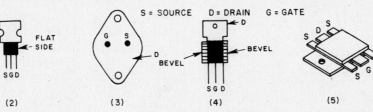

S = SOURCE   D = DRAIN   G = GATE

(1) (2) (3) (4) (5) (6)

Table 27
General Purpose Silicon Power Transistors

TO-220 CASE POWER TRANSISTORS

NPN	PNP	I_C Max. Collector Current (Amps)	V_{CEO} Max. Collector-Emitter Voltage (Volts)	hFE Min. DC Current Gain	F_T Current-Gain Bandwidth Product (MHz)	Pd Max Device Dissipation (Watts)
D44C1*		4	30	25	50	30
	D45C1*	−4	−30	25	50	30
D44C2*		4	30	100/220	50	30
	D45C2*	−4	−30	40/120	50	30
D44C3*		4	30	40/120	50	30
	D45C3*	−4	−30	40/120	50	30
D44C4*		4	45	25	50	30
	D45C4*	−4	−45	25	50	30
D44C5*		4	45	100/220	50	30
	D45C5*	−4	−45	40/120	50	30
D44C6		4	45	40/120	50	30
	D45C6*	−4	−45	40/120	50	30
D44C7*		4	60	25	50	30
	D45C7*	−4	−60	25	50	30
D44C8*		4	60	100/220	50	30
	D45C8*	−4	−60	40/120	50	30
D44C9*		4	60	40/120	50	30
	D45C9*	−4	−60	40/120	50	30
D44C10*		4	80	25	50	30
	D45C10*	−4	−80	25	50	30
D44C11*		4	80	40/120	50	30
	D45C11*	−4	−80	40/120	50	30
D44C12*		4	80	40/120	50	30
	D45C12*	−4	−80	40/120	50	30
D44H1*		10	30	20	50	50
	D45H1*	−10	−30	20	50	50
D44H2*		10	30	40	50	50
	D45H2*	−10	−30	40	50	50
TIP61*		0.5	40	15	3	15
	TIP62*	0.5	40	15	3	15
TIP61C		0.5	100	15	3	15
MJE2360T		0.5	350	40	10	30
41501		1	225	25	10	40
TIP29*		1	40	15/75	3	30
	TIP30A*	1	40	15/75	3	30
TIP29A*		1	60	15/75	3	30
	TIP30A*	1	60	15/75	3	30
TIP29B		1	80	15/75	3	30
TIP29C*		1	100	15/75	3	30
	TIP30C*	1	100	15/75	3	30
TIP47		1	250	30/150	10	40
TIP48		1	300	30/150	10	40
TIP49		1	350	30/150	10	40
TIP50		1	400	30/150	10	40
	TIP116	2	80	500	25	50
TIP31*		3	40	25	3	40
	TIP32*	3	40	25	3	40
TIP31A*		3	60	25	3	40
	TIP32A*	3	60	25	3	40
TIP31B*		3	80	25	3	40
	TIP32B*	3	80	25	3	40
TIP31C*		3	100	25	3	40
	TIP32C*	3	100	25	3	40
2N6121*		4	45	25/100	2.5	40
	2N6124*	4	45	25/100	2.5	40
2N6122		4	60	25/100	2.5	40
MJE13004		4	300	6/30	4	60
	TIP42	6	40	15/75	3	65
TIP41A		6	60	15/75	3	65
TIP41B		6	80	15/75	3	65
	2N61111	7	30	30/150	4	40
2N6290*		7	50	30/150	4	40
	2N6109*	7	50	30/150	4	40
2N6292*		7	70	30/150	4	40
	2N6107	7	70	30/150	4	40
MJE15030		8	150	20	30	50
MJE3055T*		10	60	20/70	—	75
	MJE2955T*	10	60	20/70	—	75
2N6486		15	40	20/150	5	75
2N6488		15	80	20/150	5	75

*Complementary Pairs.

Table 27 (continued)

TO-204 (TO-3) CASE

NPN	PNP	I_C Max. Collector Current (Amps)	V_{CEO} Max. Collector-Emitter Voltage (Volts)	hFE Min. DC Current Gain	F_T Current-Gain Bandwidth Product (MHz)	Pd Max Device Dissipation (Watts)
MJ12002		2.5	1500	1.11	4	75
2N3902		3.5	325	30/90	—	100
MJ410		5	200	30/90	2.5	100
MJ3029		5	250	30	—	125
MJ411		5	300	30/90	2.5	100
MJ3030		5	325	3.75	—	125
BU208A		5	1500	2.25	7.5	125
2N6307		8	300	15/75	5	125
2N6308		8	350	12/60	5	125
2N6545		8	400	7/35	6	125
MJ12005		8	1500	5	—	100
	2N3789	10	60	15	4	150
2N3715*		10	60	30	4	150
	2N3791*	10	60	30	4	150
2N5877*		10	60	20/100	4	150
	2N5875*	10	60	20/100	4	150
2N3790		10	80	15	4	150
2N3716*		10	80	30	4	150
	2N3792*	10	80	30	4	150
	2N6230	10	120	20/80	1	150
MJ15011*		10	250	20/100	—	200
	MJ15012*	10	250	20/100	—	200
MJ413		10	325	20/80	2.5	125
MJ423		10	325	30/90	2.5	125
MJ10012		10	400	100/2K	—	175
2N3055*		15	60	20/70	2.5	115
	MJ2955*	15	60	20/70	2.5	115
2N3055A		15	60	20/70	0.8	115
2N5881		15	60	20/100	4	160
	2N5880	15	80	20/100	4	160
	MJ15015	15	120	20/70	1	150
MJ150001*		15	140	25/150	2	200
	MJ15002*	15	140	25/150	2	200
2N6249		15	200	10/50	2.5	175
2N6250		15	275	8/50	2.5	175
2N6546		15	300	6/30	6-24	175
2N6251		15	350	6/50	2.5	175
MJ4033*		16	60	1K	—	150
	MJ4030*	16	60	1K	—	150
MJ4034*		16	80	1K	—	150
	MJ4031*	16	60	1K	—	150
2N5269		16	100	25/100	1	150
MJ4035*		16	100	1K	—	150
	MJ4032*	16	100	1K	—	150
2N5630*		16	120	20/80	1	200
	2N6030*	16	120	20/80	1	200
2N3773		16	140	15/60	4	200
2N5631*		16	140	15/60	1	200
	2N6031*	16	140	15/60	1	200
2N5039		20	75	20/100	60	140
2N5303		20	80	15/60	2	200
2N6284*		20	100	750/18K	—	160
	2N6287*	20	100	750/18K	—	160
MJ15003*		20	140	25/150	2	250
	MJ15004*	20	140	25/100	2	250
2N5885*		25	60	—	4	200
	2N5883*	25	60	—	4	200
2N5886*		25	80	20/100	4	200
	2N5884*	25	80	20/100	4	200
MJ15024*		25	250	15/60	5	250
	MJ15025*	25	250	15/60	5	250
2N3771		30	40	—	2	150
2N5301*		30	40	15/60	2	200
	2N4398*	30	40	15/60	2	200
2N5302*		30	60	15/60	2	200
	2N4399*	30	60	15/60	2	200
MJ802*		30	100	25/100	2	200
	MJ4502*	30	100	25/100	2	200
	2N5683‡	50	60	15/60	2	300
2N6274		50	100	30/120	30	250
MJ14002		70	80	15/100	—	300

‡Modified TO-3—60 Mil Pins.   *Complementary Pair

152 Case Power Transistors

NPN	PNP	I_C Max. Collector Current (Amps)	V_{CEO} Max. Collector-Emitter Voltage (Volts)	hFE Min. DC Current Gain	F_T Current-Gain Bandwidth Product (MHz)	Pd Max Device Dissipation (Watts)
MPSU31		0.5	65	10	—	10
MPSU10*		0.5	300	30	60	10
	MPSU60*	0.5	300	30	60	10
MPSU02*		0.8	40	30	150	10
	MPSU52*	0.8	40	30	150	10
MPSU03		1	120	40	100	10
MPSU04		1	180	40	100	10
MPSU01*		2	30	50	50	10
	MPSU51*	2	30	50	50	10
MPSU01A*		2	40	50	50	10
	MPSU51A*	2	40	50	50	10
MPSU05*		2	60	60	50	10
	MPSU55*	2	60	60	50	10
MPSU06*		2	80	60	50	10
	MPSU56*	2	80	60	50	10
MPSU07*		2	100	30	50	10
	MPSU57*	2	100	30	50	10

*Complementary Pairs.

Table 28
Package Information

TO-204 (TO-3)

CASE 22-03
TO-18

CASE 79-02, 31-03
TO-205 (TO-39), TO-5

CASE 36-03
TO-60

CASE 20-03
TO-72

CASE 029-02
TO-92

CASE 221A-01
TO-220AB

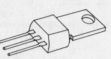

TO-202

TO-237

CASE 144B-05

CASE 144D-05

CASE 145A-09

CASE 145A-10

CASE 211-07
CASE 211-08
CASE 211-09
CASE 211-11

CASE 211-10

CASE 215-02

CASE 244-04

CASE 249-05

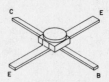

CASE 303-01
CASE 332A-01

CASE 305-01
CASE 332-04

CASE 305A-01

CASE 307-01

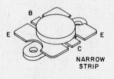

NARROW
STRIP

CASE 316-01

Macro-X
Case 317-01

Macro-T
Case 317A-01

SOT-23
CASE 318-02

CASE 319-01

CASE 328-02
(μ0.230'' PILL)

CASE 328A-01
(μ0.230'' FLANGE)

CASE 336-03

CASE 337-02
(μ0 290'' FLANGE)

Micro-X
Case 358-01

CASE 361A-01

Table 29
Integrated Circuit Operational Amplifiers

Listed numerically by device.

Device	Fabrication Technology	Notes
101A	Bipolar	General purpose
108	Bipolar	
108A	Bipolar	
124	Bipolar	Quad, low power
148	Bipolar	Quad, 741
158	Bipolar	Dual, low power
301	Bipolar	Bandwidth extendable with specific components
324	Bipolar	Quad, single supply applications
347	BIFET	Quad, high-speed
351	BIFET	
353	BIFET	
355	BIFET	
355B	BIFET	
356A	BIFET	
356B	BIFET	
357	BIFET	
357B	BIFET	
358	Bipolar	Dual-single supply applications
411	BIFET	Low offset, low drift
709	Bipolar	
741	Bipolar	
741S	Bipolar	Improved 741 for audio application
1436	Bipolar	High-voltage
1437	Bipolar	Matched, Dual 1709
1439	Bipolar	
1456	Bipolar	Dual 1741
1458	Bipolar	
1458S	Bipolar	Improved 1458 for audio applications
1709	Bipolar	
1741	Bipolar	
1747	Bipolar	Dual 1741
1748	Bipolar	Non-compensated 1741
1776	Bipolar	Micropower, programmable
3140	Bipolar/MOSFET	Strobable output
3403	Bipolar	Quad, low power
3405	Bipolar	Dual op-amp and dual comparator
3458	Bipolar	Dual, low power
3476	Bipolar	
3900	Bipolar	Quad, Norton, single supply
4558	Bipolar	Dual, wideband
4741	Bipolar	Quad 1741
5534	Bipolar	Low noise-can swing 20V p-p across 600Ω
5556	Bipolar	Equivalent to 1456
5558	Bipolar	Dual; equivalent to 1458
34001	BIFET	JFET Input
OP-27A	Bipolar	Ultra-low noise, high speed
OP-37A	Bipolar	
TL-071	BIFET	Low noise
TL-081	BIFET	
TL-084	Bipolar/JFET	Quad; high-performance audio applications

Table 30

IC Op Amp Base Diagrams (Top View — not to scale)

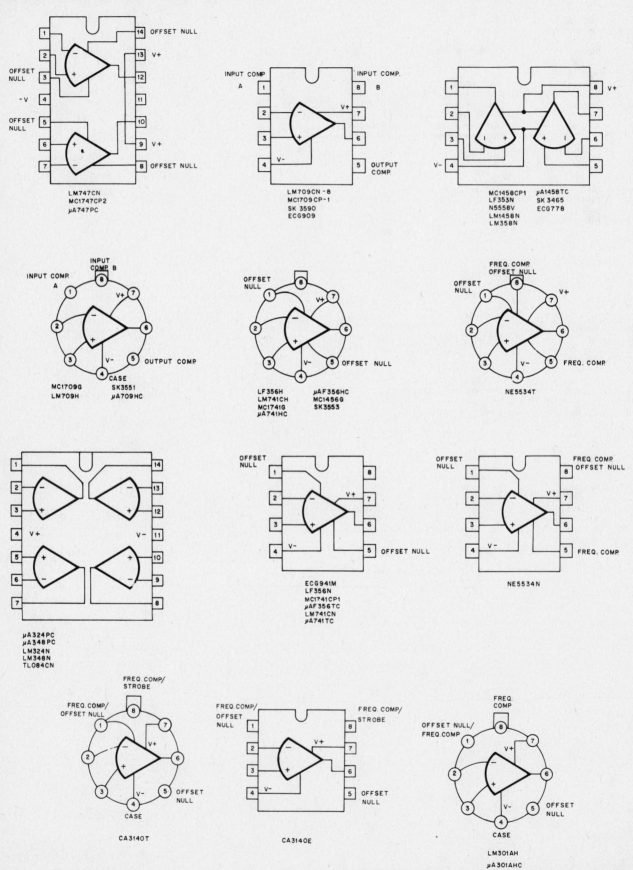

LM747CN
MC1747CP2
µA747PC

LM709CN-8
MC1709CP-1
SK 3590
ECG909

MC1458CP1   µA1458TC
LF353N   SK 3465
N5558V   ECG778
LM1458N
LM358N

MC1709G   SK3551
LM709H   µA709HC

LF356H   µAF356HC
LM741CH   MC1456G
MC1741G   SK3553
µA741HC

NE5534T

µA324PC
µA348PC
LM324N
LM348N
TL084CN

ECG941M
LF356N
MC1741CP1
µAF356TC
LM741CN
µA741TC

NE5534N

CA3140T

CA3140E

LM301AH
µA301AHC

Table 30 (continued)

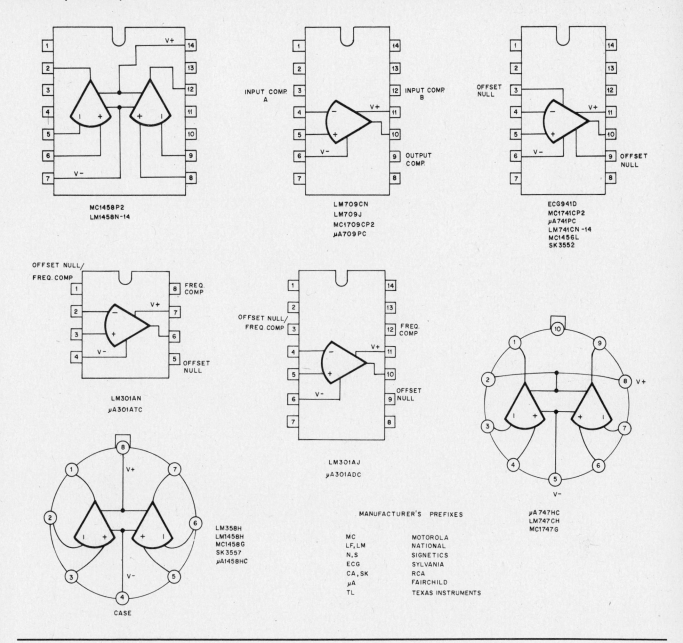

MC1458P2
LM1458N-14

INPUT COMP. A   INPUT COMP. B

OUTPUT COMP.

LM709CN
LM709J
MC1709CP2
μA709PC

OFFSET NULL

OFFSET NULL

ECG941D
MC1741CP2
μA741PC
LM741CN-14
MC1456L
SK3552

OFFSET NULL/ FREQ. COMP

FREQ. COMP

OFFSET NULL

LM301AN
μA301ATC

OFFSET NULL/ FREQ COMP

FREQ. COMP

OFFSET NULL

LM301AJ
μA301ADC

V+

V−

μA747HC
LM747CH
MC1747G

V+

V−

CASE

LM358H
LM1458H
MC1458G
SK3557
μA1458HC

MANUFACTURER'S PREFIXES

MC	MOTOROLA
LF, LM	NATIONAL
N, S	SIGNETICS
ECG	SYLVANIA
CA, SK	RCA
μA	FAIRCHILD
TL	TEXAS INSTRUMENTS

Table 31
Three-Terminal Voltage Regulators
Listed numerically by device

Device	Description	Package	Voltage	Current (Amps)
317	Adj Pos	TO-205	+1.2 to +37	0.5
317	Adj Pos	TO-204,TO-220	+1.2 to +37	1.5
317L	Low Current Adj Pos	TO-205,TO-92	+1.2 to +37	0.1
317M	Med Current Adj Pos	TO-220	+1.2 to +37	0.5
350	High Current Adj Pos	TO-204,TO-220	+1.2 to +33	3.0
337	Adj Neg	TO-205	−1.2 to −37	0.5
337	Adj Neg	TO-204,TO-220	−1.2 to −37	1.5
337M	Med Current Adj Neg	TO-220	−1.2 to −37	0.5
309		TO-205	+5	0.2
309		TO-204	+5	1.0
323		TO-204,TO-220	+5	3.0
140-XX	Fixed Pos	TO-204,TO-220	Note 1	1.0
340-XX		TO-204,TO-220		1.0
78XX		TO-204,TO-220		1.0
78LXX		TO-205,TO-92		0.1
78MXX		TO-220		0.5
78TXX		TO-204		3.0
79XX	Fixed Neg	TO-204,TO-220	Note 1	1.0
79LXX		TO-205,TO-92		0.1
79MXX		TO-220		0.5

Note 1—XX indicates the regulated voltage; this value may be anywhere from 1.2 volts to 35 volts. A 7815 is a positive 15-volt regulator, and a 7924 is a negative 24-volt regulator.

The regulator package may be denoted by an additional suffix, according to the following:

Package	Suffix
TO-204 (TO-3)	K
TO-220	T
TO-205 (TO-39)	H,G
TO-92	P,Z

For example, a 7812K is a positive 12-volt regulator in a TO-204 package. An LM340T-5 is a positive 5-volt regulator in a TO-220 package. In addition, different manufacturers use different prefixes. An LM7805 is equivalent to a μA 7805 or MC7805.

K SUFFIX
METAL TO-204 PACKAGE

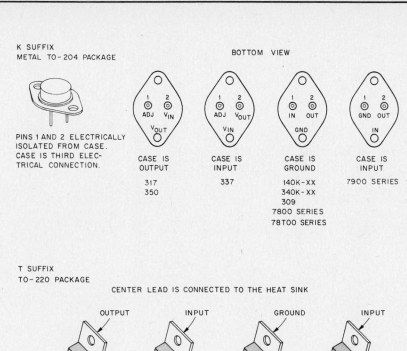

PINS 1 AND 2 ELECTRICALLY
ISOLATED FROM CASE.
CASE IS THIRD ELEC-
TRICAL CONNECTION.

BOTTOM VIEW

CASE IS
OUTPUT

317
350

CASE IS
INPUT

337

CASE IS
GROUND

140K-XX
340K-XX
309
7800 SERIES
78T00 SERIES

CASE IS
INPUT

7900 SERIES

T SUFFIX
TO-220 PACKAGE

CENTER LEAD IS CONNECTED TO THE HEAT SINK

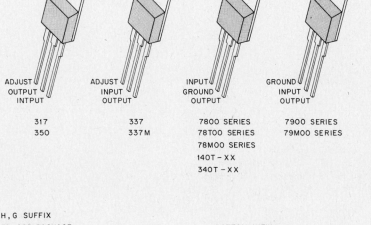

317
350

337
337M

7800 SERIES
78T00 SERIES
78M00 SERIES
140T-XX
340T-XX

7900 SERIES
79M00 SERIES

H,G SUFFIX
TO-205 PACKAGE

BOTTOM VIEW

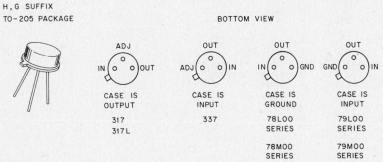

CASE IS
OUTPUT

317
317L

CASE IS
INPUT

337

CASE IS
GROUND

78L00
SERIES

78M00
SERIES

CASE IS
INPUT

79L00
SERIES

79M00
SERIES

P,Z SUFFIX
TO-92 PACKAGE

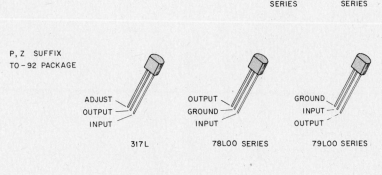

317L

78L00 SERIES

79L00 SERIES

Table 32

Coaxial Cable End Connectors

UHF Plugs — Male†

Construction	Cable RG-	Military No.
UHF-Straight plug	8, 9, 11, 13, 63, 87, 149, 213, 214, 216, 225	PL-259
UHF-Straight plug	59, 62, 71, 140, 210	UG-111

UHF Panel Receptacles — Female†

Construction	Description	Military No.
Panel receptacle (Female contacts)	Standard, mica-filled Phenolic insulation	SO-239
Bulkhead receptacle (female)	Rear mount/pressurized/ copolymer of styrene ins.	UG-266

† Except as noted, all series UHF are non-weatherproof, and have a non-constant impedance.
Freq. range = 0-500 MHz.
Voltage rating: 500 V peak

UHF Adapters

Construction	Description	Military No.
Adapter Polystyrene ins.	Straight/jack-jack	PL-258
Adapter Polystyrene ins.	Bulkhead/jack-jack	UG-363
Adapter Polystyrene ins.	Bulkhead/jack-jack	UG-224
Adapter Polystyrene ins.	Angle/jack-plug	UG-646
Adapter Polystyrene ins.	Angle/plug-jack	M-359A
Adapter Polystyrene ins.	T/jack-plug-jack	M-358

UHF Reducers

Construction	Cable RG-	Military No.
Reducing adapter	55, 58, 141, 142 (except 55A)	UG-175
Reducing adapter	59, 62, 71, 140, 210	UG-176

N Plugs and Jacks

Construction	Cable RG-	Construction Notes	Military No.
N Straight plug	8, 9, 213, 214	Impedance = 50 Ω weatherproof	UG-21
N Straight plug	11, 13, 149, 216	Impedance = 70 Ω weatherproof	UG-94A
N Straight plug	58, 141, 142	Impedance = 50 Ω weatherproof	UG-536
N Straight plug	59, 62, 71, 140, 210	Impedance = 50 Ω weatherproof	UG-603
N Straight jack	8, 9, 87, 213, 214, 225	Impedance = 50 Ω weatherproof	UG-23, B-E
N Straight jack	59, 62, 71, 140, 210	—	UG-602
N Panel jack	8, 9, 87, 213, 214, 225	—	UG-22B, D, E
N Panel jack	58, 141, 142	Impedance = 50 Ω weatherproof	UG-1052
N Panel jack	59, 62, 71, 140, 210	Impedance = 50 Ω weatherproof	UG-593
N Bulkhead jack	8, 9, 87, 213, 214, 225	Impedance = 50 Ω weatherproof	UG-160A, B, D
N Bulkhead jack	58, 141, 142	Impedance = 50 Ω weatherproof	UG-556
N Panel jack		Impedance = 50 Ω	UG-58, A
N Angle female panel jack		Impedance = 50 Ω 1¹/₁₆″ clearance above panel	UG-997A
Bulkhead female		Front mounted, hermetically sealed	M39012/04-0001
Bulkhead female		Front mounted, pressurized	UG-680

N Adapters

Construction	Adapter Ends	Construction Notes	Military No.
Straight	Jack-jack	50 Ω, TFE ins.	UG-29A
Straight	Plug-plug	50 Ω, TFE ins.	UG-57B
Straight	Jack-jack	50 Ω, TFE ins.	UG-29B
Straight	Jack-jack	50 Ω, TFE ins.	UG-29
Straight	Plug-plug	50 Ω, TFE ins.	UG-57A
Angle	Jack-plug	Mitre body	UG-27A
Angle	Jack-plug	Mitre body	UG-27B
Angle	Jack-plug	Mitre body	UG-212A
T	Jack-plug-jack	—	UG-107A
T	Jack-jack-jack	—	UG-28A
T	Jack-plug-jack	—	UG-107B

*NOTE: N connectors with gaskets are weatherproof. RF leakage: −90 dB min @ 3 GHz
Temp. limits: TFE: −67° to 390°F (−55° to 199°C). Insertion loss 0.15 dB max @ 10 GHz
Copolymer of styrene: −67° to 185°F (−55° to 85°C)
Z = 50 or 70 Ω, as noted. Freq. range: 0-11 GHz
Voltage rating: 1500 V p-p. Dielectric withstanding voltage 2500 V_{RMS}.
VSWR (MIL-C-39012 cable connectors) 1.3 max 0-11 GHz

Table 32 (continued)

HN Plugs and Jacks†

Construction	Cable	Construction Notes	Military No.
Straight plug	8, 9, 213, 214	—	UG-59A
Straight jack	8, 9, 87, 213, 214, 225	Captivated contact	UG-1214
Straight jack	8, 9, 213, 214	Copolymer of styrene ins.	UG-60A
Panel jack	8, 9, 87, 213, 214, 225	Captivated contact	UG-1215
Panel receptacle	—	Female contact	UG-560
Panel receptacle	—	Female contact	UG-496
Angle adapter	—	Jack-plug/berylium outer contact	UG-212C

†Note: Z = 50 Ω, Freq. range = 0-4 GHz, voltage rating = 1500 V p-p.
Dielectric withstanding voltage = 5000 V_{RMS} VSWR = 1.3
All HN series are weatherproof. Temp. limits: TFE: −67° to 390°F (−55° to 199°C)
Copolymer of styrene: −67° to 185°F (−55° to 85°C)

BNC Plugs and Jacks

Construction	Cable RG-	Construction Notes	Military No.
BNC straight plug	8, 9	—	UG-959
BNC straight plug	59, 62, 71, 140, 210	Rexolite insulation	UG-260,A
BNC panel jack	59, 62, 71, 140, 210	Rexolite insulation	UG-262
BNC panel jack	59, 62, 71, 140, 210	Non-weatherproof Rexolite ins.	UG-262A
BNC panel jack	55, 58, 141, 142, 223, 400	—	UG-291
BNC panel jack	55, 58, 141, 142, 223, 400	Non-weatherproof	UG-291A
BNC bulkhead jack front mount	59, 62, 71, 140, 210	Front mtg/Rexolite ins.	UG-624
BNC bulkhead receptacle	—	Standard	UG-1094A UG-625B UG-625
BNC straight adapter	Plug-plug	—	UG-491A
BNC straight adapter	Plug-plug	Beryl. outer contact	UG-491B
BNC straight adapter	Jack-jack	—	UG-914
BNC straight adapter	Plug-plug	—	UG-491
BNC angle adapter	Jack-plug	—	UG-306
BNC angle adapter	Jack-plug	Berl. outer contact	UG-306A
BNC angle adapter	Jack-plug	Berl. outer contact	UG-306B
BNC panel adapter	Jack-jack	3-56 tapped flange holes	UG-414
BNC panel adapter	Jack-jack	3-56 tapped flange holes	UG-414A
Straight	Plug-plug	—	UG-491A
Straight	Plug-plug	Beryl. outer contact	UG-491B
Straight	Jack-jack	—	UG-914
Straight	Plug-plug	—	UG-491
Angle	Jack-plug	—	UG-306
Angle	Jack-plug	Beryl. outer contact	UG-306A
Angle	Jack-plug	Beryl. outer contact	UG-306B
Panel	Jack-jack	3-56 tapped flange holes	UG-414
Panel	Jack-jack	3-56 tapped flange holes	UG-414A
T	Jack-plug-jack	—	UG-274
T	Jack-plug-jack	Beryl. outer contact	UG-274A
T	Jack-plug-jack	Beryl. outer contact	UG-274B

NOTES: Z = 50 Ω Freq. range: 0-4 GHz w/low reflection; usable to 11 GHz.
Voltage rating: 500 V p-p. Dielectric withstanding voltage 500 V_{RMS}.
SWR: 1.3 max 0-4 GHz. RF leakage −55 dB min @ 3 GHz.
Insertion loss: 0.2 dB max @ 3 GHz.
Temp. limits: TFE: −67° to 390°F (−55° to 199°C)
Rexolite insulators: −67° to 185°F (−55° to 85°C)

Table 33

Cross-Family Adapters

(adapters described by their own end construction)

Construction	Adapter Ends		Description	Impedance	Military No.
			Rear mounted		
HN to BNC	HN-plug	BNC-jack	Straight	50 Ω	UG-309
N to BNC	N-plug	BNC-jack	Straight	50 Ω	UG-201A
	N-jack	BNC-plug	Straight	50 Ω	UG-349A
	N-plug	BNC-jack	Straight	50 Ω	UG201
	N-jack	BNC-plug	Straight/copolymer of styrene-TFE ins.	50 Ω	UG-349
	N-plug	BNC-plug	Straight	50 Ω	UG-1034
N to UHF	N-plug	UHF-jack	Straight/copolymer/ non-weatherproof/ non-constant impedance	—	UG146
	N-jack	UHF-plug	Straight/copolymer/ non-weatherproof/ non-constant impedance	—	UG-83
	N-plug	UHF-plug	Straight/copolymer of styrene insulation/ non-constant impedance	—	UG-318
	N-jack	UHF-plug	Straight/non-constant impedance	—	UG-83B
UHF to BNC	UHF-plug	BNC-jack	Straight/copolymer of styrene insulation/ non-constant impedance	—	UG-273
	UHF-jack	BNC-plug	Straight/copolymer of styrene insulation	50 Ω	UG-255

Table 34
Miniature Lamp Guide

1 Submidget Flanged	2 Midget Grooved	3 Miniature Screw	4 Midget Flanged	5 Index Bayonet	6 Miniature Flanged	7 Candelabra Screw	8 Miniature Bayonet	9 S.C. Bayonet	10 D.C. Bayonet

Type	V	Amp (Design)	Life Hours	Base	Bulb
PR-2	2.38	0.50	15	7	B-3½
PR-3	3.57	0.50	15	7	B-3½
PR-4	2.33	0.27	10	7	B-3½
PR-6	2.47	0.30	30	7	B-3½
PR-12	5.95	0.50	15	7	B-3½
PR-13	4.75	0.50	15	7	B-3½
10	2.5	0.5	3000	−	G-3½
12	6.3	0.15	L	−	G-3½
13	3.7	0.30	15	8	G-3½
14	2.47	0.30	15	8	G-3½
19	14.4	0.10	1000	−	G-3½
27	4.9	0.30	30	8	G-4¼
37	14.0	0.09	1500	+	T-1¾
40	6.3	0.15	3000	8	T-3¼
43	2.5	0.50	3000	9	T-3¼
44	6.3	0.25	3000	9	T-3¼
45	3.2	0.35	3000	9	T-3¼
47	6.3	0.15	3000	9	T-3¼
48	2.0	0.06	1000	8	T-3¼
49	2.0	0.06	1000	9	T-3¼
50	7.5	0.22	1000	8	G-3½
51	7.5	0.22	1000	9	G-3½
53	14.4	0.12	1000	9	G-3½
55	7.0	0.41	500	9	G-4½
57	14.0	0.24	500	9	G-4½
63	7.0	0.63	1000	10	G-6
73	14.0	0.08	15000	+	T-1¾
74	14.0	0.10	500	+	T-1¾
82	6.5	1.05	500	11	G-6
85	28.0	0.04	7000	+	T-1¾
86	6.3	0.20	20000	+	T-1¾
88	6.8	1.9	300	11	S-8
93	12.8	1.04	500	10	S-8
112	1.2	0.22	5	8	GTL-3
130	6.3	0.15	3000	9	G-3½
131	1.3	0.10	50	8	G-3½
158	14.0	0.24	500	+	T-3¾
159	6.3	0.15	10000	+	T-3¼
161	14.0	0.19	4000	+	T-3¼
168	14.0	0.35	1500	+	T-3¼
219	6.3	0.25	L	10	G-3½
222	2.25	0.25	5	8	GTL-3
239	6.3	0.36	5000	9	T-3¼
240	6.3	0.36	5000	9	T-3¼
252	2.5	0.35	10000	4	GTL-1¾
253 X	2.5	0.35	10000	2	GTL-1¾
259	6.3	0.25	10000	+	T-3¼
261	2.5	0.35	10000	2	GTL-1¾
268	2.5	0.35	10000	4	T-1¾
334LSV	28.0	0.06	25000	2	T-1¾
305	28.0	0.56	480	10	S-8
307	28.0	0.66	300	10	S-8
308	28.0	0.67	300	11	S-8
313	28.0	0.17	500	9	T-3¼
323	3.0	0.19	350	−	T-1¼
324	3.0	0.19	350	W	T-1¼
327	28.0	0.04	7000	4	T-1¾
327AS15	28.0	0.04	7000	4	T-1¾
327AS25	28.0	0.04	7000	4	T-1¾
328	6.0	0.20	1000	4	T-1¾
330	14.0	0.08	750	4	T-1¾
331	1.35	0.06	500	4	T-1¾
334	28.0	0.04	7000	2	T-1¾
335	28.0	0.04	7000	3	T-1¾
336	14.0	0.08	750	2	T-1¾
337	6.0	0.20	1000	2	T-1¾
338	2.7	0.06	6000	4	T-1¾
342	6.0	0.04	10000	3	T-1¾
344	10.0	0.014	10000	4	T-1¾
345	6.0	0.04	10000	4	T1¾
346	18.0	0.04	10000	2	T-1¾
349	6.3	0.20	3000	4	T-1¾
370	18.0	0.04	10000	4	T-1¾
373	14.0	0.08	750	3	T-1¾
375	3.0	0.015	10000	4	T-1¾
376	28.0	0.06	25000	4	T-1¾
380	6.3	0.04	50000	4	T-1¾
381	6.3	0.20	50000	4	T-1¾
382	14.0	0.08	50000	4	T-1¾
385	28.0	0.04	50000	4	T-1¾
386	14.0	0.08	50000	2	T-1¾
387	28.0	0.04	25000	4	T-1¾
388	28.0	0.04	25000	2	T-1¾
397	10.0	0.04	5000	2	T-1¾
398	6.3	0.20	3000	2	T-1¾
399	28.0	0.04	25000	3	T-1¾
502	5.1	0.15	100	8	G-4½
515	5.0	0.115	40000	W	T-1¼
555	6.3	0.25	3000	+	T-3¼
583D	5.0	0.06	100000	W	T-1¼
656	28.0	0.06	5000	+	T-3¼
680	5.0	0.06	100000	W	T-1
682	5.0	0.06	100000	1	T-1
682AS15	5.0	0.06	100000	1	T-1
683	5.0	0.06	100000	W	T-1
683AS15	5.0	0.06	100000	W	T-1
685	5.0	0.06	100000	1	T-1
685AS15	5.0	0.06	100000	1	T-1
715	5.0	0.115	40000	W	T-1
715AS15	5.0	0.115	40000	W	T-1
718	5.0	0.115	40000	1	T-1
755	6.3	0.15	50000	9	T-3¼
756	14.0	0.08	50000	9	T-3¼
757	28.0	0.08	7000	9	T-3¼
1034	12.8	1.80	200	6	S-8
1073	12.8	1.80	200	10	S-8
1130	6.4	2.6	200	11	S-8
1133	6.2	3.91	200	10	RP-11
1141	12.8	1.44	500	10	S-8
1143	12.5	2.00	40	10	RP-11
1184	5.5	6.25	100	11	RP-11
1251	28.0	0.23	L	10	G-6
1445	18.0	0.15	250	10	G-3½
1487	14.0	0.20	3000	−	T-3¼
1488	14.0	0.15	200	9	T-3¼
1490	3.2	0.16	3000	9	T-3¼
1493	6.5	2.80	100	11	S-8
1619	6.7	1.90	500	10	S-8
1630	6.5	2.75	100	−	S-8
1648	8.0	2.0	1500	11	S-8
1691	28.0	0.61	1000	10	S-8
1750D	14.0	0.08	750	W	T-1¾
1728D	1.35	0.06	500	W	T-1¾
1730D	6.0	0.04	10000	W	T-1¾
1738	2.7	0.06	6000	W	T-1¾
1762D	28.0	0.04	7000	W	T-1¾
1764D	28.0	0.04	7000	W	T-1¾
1767	2.5	0.20	500	3	T-1¾
1768	6	0.20	1000	3	T-1¾
1775	6.3	0.075	500	3	T-1¾
1813	14.4	0.10	1000	9	T-3¼
1815	14	0.20	3000	9	T-3¼
1816	13	0.33	1000	9	T-3¼
1819	28	0.04	1000	9	T-3¼
1820	28	0.10	1000	9	T-3¼
1821	28	0.17	500	5	T-3¼
1822	36	0.10	1000	9	T-3¼
1828	37.5	0.05	3000	9	T-3¼
1829	28	0.07	1000	9	T-3¼
1835	55.0	0.05	5000	9	T-3¼
1847	6.3	0.15	20000	9	T-3¼
1850	5	0.09	1500	9	T-3¼
1864	28.0	0.17	1500	9	T-3¼
1866	6.3	0.25	20000	9	T-3¼
1869D	10.0	0.014	10000	W	T-3¼
1891	14.0	0.24	500	9	T-3¼
1892	14.4	0.12	1000	9	T-3¼
1893	14.0	0.33	3000	9	T-3¼
1895	14.0	0.27	1500	9	G-4½
2102D	18.0	0.04	10000	W	T-1¾
2107D	10.0	0.04	5000	W	T-1¾
2114D	6.0	0.06	3000	W	T-1¾
2124	2.5	0.35	10000	W	GTL-1¾
2128	3.0	0.0125	16000	W	T-1
2158	3.0	0.015	10000	W	T-1¾
2162D	14.0	0.100	10000	W	T-1¾
2169	2.5	0.350	10000	W	T-¾
2180	6.3	0.04	50000	W	T-¾

Table 34 (continued)

Type	Volts	Amps	Hours	Base	Bulb		Type	Volts	Amps	Hours	Base	Bulb
2181D	6.3	0.20	50000	W	T-1¾		8369	28.0	0.065	5000	—	T-1¾
2182D	14.0	0.08	50000	W	T-1¾		8610	6.3	0.20	5000	W	T-1¾
2187D	28.0	0.04	25000	W	T-1¾		8623	28.0	0.04	1000	—	T-1¼
6803	5.0	0.060	100000	W	T-¾		8627	28.0	0.04	1000	W	T-1¼
6833	5.0	0.060	100000	W	T-¾		8632	28.0	0.04	1000	—	T-1¼
6838	28.0	0.024	16000	W	T-1		8635	28.0	0.04	1000	—	T-1¼
6839	28.0	0.024	16000	1	T-1		8640	14	0.08	1000	W	T-1¼
7153	5.0	0.115	40000	W	T-¾		8666	5.0	0.08	15000	W	T-¾
7216	5	0.125	5000	W	T-1							
8362	14.0	0.08	50000	—	T-1¾							

W-Wire terminals.   (−)-Not illustrated.   (+)-Wedge base.   L-Long life.

BULB STYLES

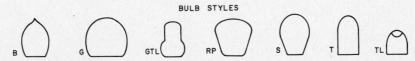

B   G   GTL   RP   S   T   TL

**Bulbs are described by a letter indicating shape and a number that is an approximation of diameter expressed in eights of an inch. For example S-8 is "S" shape, 8 eighths or 1 inch in dia.

(†) W-Wire terminals.   (*) Knurled screw.

NEON GLOW LAMPS

Type	Design Volts	Life Watts	Hours	**Base	Bulb
A1B	105-125	1/25	25000	W	T-2
A1C	105-125	1/10	25000	W	T-2
A1A(NE2)	105-125	1/17	25000	W	T-2
C7A(NE2D)	105-125	1/12	25000	4	T-2
A9A(NE2E)	105-125	1/12	25000	W	T-2
C2A(NE2H)	105-125	1/4	25000	W	T-2
C9A(NE2J)	105-125	1/4	25000	4	T-2
B6A(NE21)	110-125	1/4	7500	10	T-4½
B7A(NE45)	110-125	1/4	7500	8	T-4½
B9A(NE48)	110-125	1/4	7500	11	T-4½
B1A(NE51)	105-125	1/25	25000	9	T-3¼
B2A(NE51H)	105-125	1/7	25000	9	T-3¼
AIG	105-125	1/25	25000	4	T-2
AIH	105-125	1/10	25000	4	T-2
J5A(NE-30)	105-125	1	10000	0	S-11

INDICATOR LAMPS

Type	volts	Amps	End. Foot Candles
6PSB	6	.140	800
12PSB	12	.170	2400
24PSB	24	.073	2200
28PSB	28	.040	1900
48PSB	48	.053	1500
60PSB	60	.050	2400
120PSB	120	.025	1600

STANDARD VOLTAGE LAMPS

Type	Design Volts	Life Watts	Base Hours	**Style	Bulb
6S6	115	6	1500	8	S-6
6S6	145	6	1500	8	S-6
6S6/DC	120V	6	1500	8	S-6
6T4½/1	120V	6	1500	8	T-4½
7C7	115-125	7	1500	8	T-4½
10C7	115-125	10	—	8	C-7
10C7DC	115-125	10	3000	11	C-7
10S6/10	230	10	1500	8	S-6

CM8-SERIES LAMPS

Type	Design V.	Amps	Life Hours	Base	**Bulb
CM8-373	2.5	.35	10000	2	TL-1¾
CM8-374	2.5	.35	10000	W	TL-1¾
CM8-967	120	.02	5000	9	TL-2

W: wire terminals

CM7-SERIES BI-PIN LAMPS

Type	Design V.	Amps	Life Hours	**Bulb
CM7112	5.0	.021	10000	T-¾
CM7118	5.0	.115	40000	T-¾
CM7327	28.0	.04	7000	T-1¾
CM7328	6.0	.20	1000	T-1¾
CM7330	14.0	.08	750	T-1¾
CM7344	10.0	.014	10000	T-1¾
CM7367	10.0	.04	5000	T-1¾
CM7371	12.0	.04	10000	T-1¾
CM7373	14.0	.10	10000	T-1¾
CM7376	28.0	.065	5000	T-1¾
CM7377	6.3	.075	500	T-1¾
CM7380	6.3	.04	50000	T-1¾
CM7382	14.0	.08	50000	T-1¾
CM7515	5.0	.115	40000	T-1¼
CM7583	5.0	.06	100000	T-1¼
CM7628	6.0	.20	1000	T-1¼
CM7632	28.0	.04	1000	T-1¼
CM7646	14.0	.08	1000	T-1¼
CM7838	2.7	.06	6000	T-1¾
CM7945	6.0	.04	10000	T-1¾

Table 35
Crystal Holders

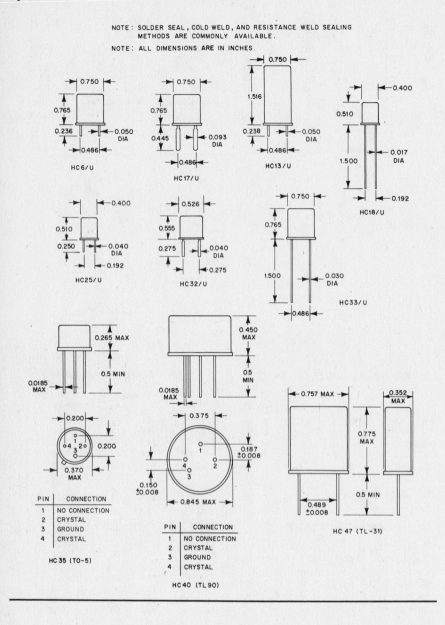

NOTE: SOLDER SEAL, COLD WELD, AND RESISTANCE WELD SEALING METHODS ARE COMMONLY AVAILABLE.

NOTE: ALL DIMENSIONS ARE IN INCHES.

Table 36
Properties of Common Thermoplastics

ASTM or UL test	Property	NYLONS (DRY, AS MOLDED) — Type					PHENOLICS — Type of compound						POLYETHYLENE			
		6/6	6	6/12	11	Castable	General purpose	impact	Non-bleeding	Electrical	Heat resistant	Special purpose*	Low density	Medium density	High density	Ultra-high molec weig
	PHYSICAL															
D792	Specific gravity	1.14	1.13	1.06	1.04	1.15-1.17	1.35-1.46	1.36-1.41	1.37-1.38	1.36-1.75	1.41-1.84	1.37-1.75	0.910-0.925	0.926-0.940	0.941-0.965	0.928-0.
D792	Specific volume (in³/lb)	24.2	24.5	25.9	26.6	23.8							30.4-29.9	29.9-29.4	29.4-28.7	29.
D570	Water absorption, 24 h. ⅛-in. thk (%)	1.2	1.6	0.25	0.4	0.9	0.6-0.7	0.6-0.9	0.8-0.9	0.05-0.20	0.30-0.35	0.20-0.40	<0.01	<0.01	<0.01	<0.
	MECHANICAL															
D638	Tensile strength (psi)	12,000	11,800	8,800	8,500	11,000-14,000	6,500-7,000	6,000-7,000	6,000-7,000	5,000-7,000	5,000-6,000	7,000-9,000	600-2,300	1,200-3,500	3,100-5,500	4,000-(
D638	Elongation (%)	60	200	150	120	10-50	11-13	12	10	17-25	14	10	90-800	50-600	20-1,000	200-(
D638	Tensile modulus (10⁵ psi)	4.2	3.8	2.9	1.8	3.5-4.5							0.14-0.38	0.25-0.55	0.6-1.8	0.20-
D785	Hardness, Rockwell ( )	121 (R)	119 (R)	114 (R)	—	112-120 (R)	70-95 (E)	82 (E)	82 (E)	75-88 (E)	94 (E)	76 (E)	10 (R)	15 (R)	65 (R)	55
D790	Flexural modulus (10⁵ psi)	4.1	3.9	2.9	1.5	—	11-14	12-25	10-12	12-25	11-23	10-19	0.08-0.60	0.60-1.15	1.0-2.0	1.0-
D256	Impact strength, Izod (ft-lb/in of notch)	1.0	0.8	1.0	3.3	0.9	0.30-0.35	0.6-1.05	0.28	0.28-0.45	0.26	0.50	No break	0.5-16	0.5-20	No br
	THERMAL															
C177	Thermal conductivity (Btu-in/hr-ft²-°F)	1.7	1.7	1.5	—	1.7	7.1†	7.9†	—	16.0†	—	8.8†	8.0†	8.0-10.0†	11.0-12.4†	11.l
D696	Coef of thermal expansion (10⁻⁵ in./in.-°F)	4.0	4.5	5.0	5.1	5.0	3.95	3.56	4.40	2.60	2.80	3.60	5.6-12.2	7.8-8.9	6.1-7.2	7.l
D648	Deflection temperature (°F) At 264 psi	194	152	194	118	300-425	275-360	270-500	370	310-400	330-380	360-430	90-105	105-120	110-130	11
	At 66 psi	455	365	356	154	400-425							100-121	120-165	140-190	17
UL 94	Flammability rating	V-2	V-2	V-2	—	—	V-1	HB	—	V-0	V-0	HB				
	ELECTRICAL															
D149	Dielectric strength (V/mil) Short time, ⅛ in. thk	600	400	400	425	500-600*	350	350-400	200	400	170	175	460-700	460-650	450-500	90(
D150	Dielectric constant At 1 kHz	3.9	3.7	4.0	3.3	3.7	5.2-5.3	5.2-5.4		4.9-6.5	11.7	7.8	2.25-2.35	2.30-2.35	2.30-2.35	
D150	Dissipation factor At 1 kHz	0.02	0.02	0.02	0.03	0.02	0.04-0.05	0.04-0.06		0.025-0.10	0.15	0.12	0.0002	0.0002	0.0003	0.00
D257	Volume resistivity (ohm-cm) At 73°F, 50% RH	10^{15}	10^{15}	10^{15}	2×10^{13}	—	10^{11}-10^{12}	10^{11}-10^{12}	10^{12}	10^{11}-10^{13}	10^{12}	10^{11}	10^{15}	10^{15}	10^{15}	10
D495	Arc resistance (s)	116	—	121	—	—	100	50	—	184	181	—	135-160	200-235	—	—
	OPTICAL															
D542	Refractive Index												1.51	1.52	1.54	—
D1003	Transmittance (%)												4-50	4-50	10-50	—

* kV/cm.
* Chemical-resistant compound. †⅛-in. thick specimens.
* 0.040 in. thick specimen
† (10⁻⁴ cal-cm/sec-cm²-°C)

ASTM or UL test	POLYPROPYLENE			POLYPHENYLENE SULFIDE[a]						POLYSTYRENE					POLYVINYL CHLORIDE	
	Unmodified resin	Glass reinforced	Impact grade	Glass reinforced		Glass and mineral filled				Polymers		Crystal clear	Copolymers		Rigid	Flexible
				R-3	R-4	R-8	R-9	R-10[b]	R-11	General purpose	Impact modified		Impact modified	10-20% (wt.) Glass reinf*		
D792	0.905	1.05-1.24	0.89-0.91	1.57	1.67	1.8	1.9	1.96-1.98	1.98	1.04-10.9	1.03-1.10	1.08-1.10	1.05-10.8	1.13-1.22	1.30-1.58	1.20-1.70
D792	30.8-30.4	24.5	30.8-30.5	–	–	–	–	–	–	26.0-25.6	28.1-25.2	–	–	–	20.5-19.1	–
D570	0.01-0.03	0.01-0.05	0.01-0.03	–	<0.05	0.03	–	–	–	0.03-0.10	0.05-0.6	0.1	0.1	0.08	0.04-0.4	0.15-0.75
D638	5,000	6,000-14,500	2,800-4,400	15,500	17,500	10,750	11,000	10,000-11,500	11,000	5,000-12,000	1,500-7,000	7,000-7,600	4,800-7,200	10,500-12,500	6,000-7,500	1,500-3,500
D638	10-20	2.0-3.6	350-500	1.1	1.25	0.47	0.5	0.5-0.6	0.6	0.5-2.0	2-60	1.4-1.7	2.0-20.0	1.3-2.0	40-80	200-450
D638	1.6	4.5-9.0	1.0-1.7	–	–	–	–	–	–	4.0-6.0	1.4-5.0	4.4-4.7	2.8-4.2	6.3-10.0	3.5-6.0	3.9-13.9
D785	80-110 (R)	110 (R)	50-85 (R)	–	123 (R)	121 (R)	–	120 (R)	–	65-80	10-90	108	80	101	65-85D (Shore)	50-100A (Shore)
D790	1.7-2.5	3.8-8.5	1.2-1.8	14	17	22	21	18	20	4.0-4.7	1.5-4.6	4.6-4.9	3.2-4.5	5.5-9.8	3.5	–
D256	0.5-2.2	1.0-5.0	1.0-15	1.0	1.1	0.59	0.7	0.6-1.0	0.8	0.2-0.45	0.5-4.0	0.3-0.5	0.5-4.4	1.8-2.6	0.4-20.0	–
C117	2.8†	–	3.0-4.0†	–	2.0	–	–	–	–	2.4-3.3	1.0-3.0	2.4-3.3	1.0-3.0	–	3.5-5.0†	3.0-4.0†
D696	3.2-5.7	1.6-2.9	3.3-4.7	–	2.2	1.6	1.1	–	–	3.3-4.4	1.9	3.5-3.7	3.5-3.7	2.0-2.2	2.8-5.6	3.9-13.9
D648	125-140 / 200-250	230-300 / 310	120-135 / 160-210	500	500	500	500	500	500	190-220 / 180-230	160-200 / 180-220	235-249	235-249	235-260	140-170 / 135-180	–
UL 94	HB[b]	HB[b]	HB[b]	V-0	V-0/5V	V-0/5V	V-0	V-0/5V	V-0	HB[b]	HB[b]	HB[b]	HB[b]	HB[b]	–	–
D149	500-660	475	500-650	–	–	–	–	–	–	500-700	300-600	500-700	300-600	–	350-500	300-400
D150	2.2-2.6	2.36	2.3	–	4.0*	4.3*	4.5*	4.8-6.1*	–	2.40-2.65	2.4-4.5	–	–	–	3.0-3.8	4.0-8.0
D150	0.0005-0.0018	0.0017	0.0003	–	0.0014*	0.016*	0.0072*	0.01-0.02*	–	0.0001-0.0003	0.0004-0.0020	–	–	–	0.009-0.017	0.07-0.16
D257	10^{17}	2×10^{18}	10^{18}	–	–	–	–	–	–	10^{17}-10^{19}	10^{16}	–	–	–	$>10^{16}$	10^{11}-10^{15}
D495	160	100	–	–	34	182	180	116-182	–	60-135	20-100	95	95	–	60-80	–
D542										1.60	–	1.59	–	–		
D1003										87-92	35.57	92	–	–		

[a]Test specimen molding conditions, 275°F mold temperature.
[b]Representative of a series of various pigmented compounds.
*At 1.0 MHz.
[b]V-2, V-1, and V-0 grades are also available.

Table 36 (continued)

Polyvinyl Chloride (PVC)

Advantages:
- can be compounded with plasticizers, fillers, stabilizers, lubricants and impact modifiers to product wide range of physical properties.
- can be pigmented to almost any color
- Rigid PVC has good corrosion and stain resistance, thermal & electrical insulation, and weatherability

Disadvantages:
- base resin can be attached by aromatic solvents, ketones, aldehydes, naphthalenes, and some chloride, acetate, and acrylate esters
- should not be used above 140° F

Applications:
- conduit
- conduit boxes
- electrical fittings
- housings
- pipe
- wire and cable insulation

Polystyrene

Advantages:
- low cost
- moderate strength
- electrical properties only slightly affected by temperature and humidity
- sparkling clarity
- impact strength is increased by blending with rubbers, such as polybutadiene

Disadvantages:
- brittle
- low heat resistance

Applications:
- capacitors
- light shields
- knobs

Polyphenylene Sulfide (PPS)

Advantages:
- excellent dimensional stability
- strong
- high-temperature stability
- chemical resistant
- Inherently completely flame retardant
- completely transparent to microwave radiation.

Applications:
R3-R5 have various glass-fiber levels that are suitable for applications demanding high mechanical and impact strength as well as good dielectric properties.
R8 and R10 are suitable for high arc-resistance applications
R9-901 is suitable for encapsulation of electronic devices

Polypropylene

Advantages:
- low density
- good balance of thermal, chemical, and electrical properties
- moderate strength (increases significantly with glass-fiber reinforcement)

Disadvantages:
- Electrical properties affected to varying degrees by temperature (as temperature goes up, dielectric strength increases and volume resistivity decreases.)
- Inherently unstable in presence of oxidative and UV radiation

Applications:
- Automotive battery cases
- blower housings
- fan blades
- fuse housings
- insulators
- lamp housings
- supports for current-carrying electrical components.
- TV yokes

Polyethylene (PE)

Advantages: Low Density PE
- Good toughness
- excellent chemical resistance
- excellent electrical properties
- low coefficient of friction
- near-zero moisture absorbtion
- easy to process
- relatively low heat resistance

Disadvantage
- susceptible to environmental and some chemical stress cracking
- wetting agents (such as detergents) accelerate stress cracking

Advantages: High Density PE
- Same as above, plus increased rigidity and tensile strength

Advantages: Ultra-High Molecular Weight PE
- outstanding abrasion resistance
- low coefficient of friction
- high impact strength
- excellent chemical resistance
- material does not break in impact strength tests using standard notched specimens

Applications:
- bearings
- components requiring maximum abrasion resistance, impact strength, and low coefficent of friction

Phenolic

Advantages:
- low cost
- superior heat resistance
- high heat-deflection temperatures
- good electrical properties
- good flame resistance
- excellent moldability
- excellent dimensional stability
- good water and chemical resistance

Applications:
- commutators and housings for small motors
- heavy duty electrical components
- rotary-switch wafers
- insulating spacers

Nylon

Advantages
- excellent fatigue resistance
- low coefficient of friction
- toughness a function of degree of crystallinity
- resists many fuels and chemicals
- good creep- and cold-flow resistance as compared to less rigid thermoplastics
- resists repeated impacts

Disadvantages:
- all nylons absorb moisture
- nylons that have not been compounded with a UV stabilizer are sensitive to UV light, and thus not suitable for extended outdoor use

Applications
- bearings
- housing and tubing
- rope
- wire coatings
- wire connectors
- wear plates

Table 37
TTL Families
Listed by ascending maximum clock frequency.

TTL Series	Input Pull-up Resistor (kΩ)	Maximum High-level Input Current (µA)	Maximum Low-level Input Current (mA)	Typical Gate Propagation Delay Time (ns)	Maximum Flip-flop Clock Frequency (MHz)
54L/74L[1]	40/8	10/20	−0.18-−0.8	33	3
54/74	4	40	−1.6	10	35
54LS/74LS	18	20	−0.4	9.5	45
54H/74H	2.8	50	−2	6	50
54ALS/74ALS	37	20	−0.1	4	70
54S/74S	2.8	50	−2	3	125
54AS/74AS	10	20	−0.5	1.7	200

[1]Series 54L/74L has two types of standard input.
Courtesy of Texas Instruments Inc.

Table 38

Fractions of an inch with Metric Equivalents

Fractions of an inch	Decimals of an inch	mm
1/64	0.0156	0.397
1/32	0.0313	0.794
3/64	0.0469	1.191
1/16	0.0625	1.588
5/64	0.0781	1.984
3/32	0.0938	2.381
7/64	0.1094	2.778
1/8	0.1250	3.175
9/64	0.1406	3.572
5/32	0.1563	3.969
11/64	0.1719	4.366
3/16	0.1875	4.763
13/64	0.2031	5.159
7/32	0.2188	5.556
15/64	0.2344	5.953
1/4	0.2500	6.350
17/64	0.2656	6.747
9/32	0.2813	7.144
19/64	0.2969	7.541
5/16	0.3125	7.938
21/64	0.3281	8.334
11/32	0.3438	8.731
23/64	0.3594	9.128
3/8	0.3750	9.525
25/64	0.3906	9.922
13/32	0.4063	10.319
27/64	0.4219	10.716
7/16	0.4375	11.113
29/64	0.4531	11.509
15/32	0.4688	11.906
31/64	0.4844	12.303
1/2	0.5000	12.700
33/64	0.5156	13.097
17/32	0.5313	13.494
35/64	0.5469	13.891
9/16	0.5625	14.288
37/64	0.5781	14.684
19/32	0.5938	15.081
39/64	0.6094	15.478
5/8	0.6250	15.875
41/64	0.6406	16.272
21/32	0.6563	16.669
43/64	0.6719	17.066
11/16	0.6875	17.463
45/64	0.7031	17.859
23/32	0.7188	18.256
47/64	0.7344	18.653
3/4	0.7500	19.050
49/64	0.7656	19.447
25/32	0.7813	19.844
51/64	0.7969	20.241
13/16	0.8125	20.638
53/64	0.8281	21.034
27/32	0.8438	21.431
55/64	0.8594	21.828
7/8	0.8750	22.225
57/64	0.8906	22.622
29/32	0.9063	23.019
59/64	0.9219	23.416
15/16	0.9375	23.813
61/64	0.9531	24.209
31/32	0.9688	24.606
63/64	0.9844	25.003
—	1.0000	25.400

Table 39

Aluminum Numbers for Amateur Use

Common Alloy Numbers

Type	Characteristic
2024	Good formability, high strength
5052	Excellent surface finish, excellent corrosion resistance, normally not heat treatable for high strength
6061	Good machinability, good weldability, can be brittle at high tempers
7075	Good formability, high strength

Common Tempers

Type	Characteristics
T0	Special soft condition
T3	Hard
T6	Very hard, possibly brittle
TXXX	Three digit tempers — usually specialized high-strength heat treatments, similar to T6

General Uses

Type	Uses
2024-T3 7075-T3	Chassis boxes, antennas, anything that will be bent or flexed repeatedly
6061-T6	Mounting plates, welded assemblies or machined parts

Table 40
Voltage-Power Conversion Table

Based on a 50-ohm system

RMS	Voltage — Peak-to-Peak	dBmV	Power — Watts	dBm
0.01 μV	0.0283 μV	− 100	2×10^{-18}	− 147.0
0.02 μV	0.0566 μV	− 93.98	8×10^{-18}	− 141.0
0.04 μV	0.113 μV	− 87.96	32×10^{-18}	− 134.9
0.08 μV	0.226 μV	− 81.94	128×10^{-18}	− 128.9
0.1 μV	0.283 μV	− 80.0	200×10^{-18}	− 127.0
0.2 μV	0.566 μV	− 73.98	800×10^{-18}	− 121.0
0.4 μV	1.131 μV	− 67.96	3.2×10^{-15}	− 114.9
0.8 μV	2.236 μV	− 61.94	12.8×10^{-15}	− 108.9
1.0 μV	2.828 μV	− 60.0	20.0×10^{-15}	− 107.0
2.0 μV	5.657 μV	− 53.98	80.0×10^{-15}	− 101.0
4.0 μV	11.31 μV	− 47.96	320.0×10^{-15}	− 94.95
8.0 μV	22.63 μV	− 41.94	1.28×10^{-12}	− 88.93
10.0 μV	28.28 μV	− 40.00	2.0×10^{-12}	− 86.99
20.0 μV	56.57 μV	− 33.98	8.0×10^{-12}	− 80.97
40.0 μV	113.1 μV	− 27.96	32.0×10^{-12}	− 74.95
80.0 μV	226.3 μV	− 21.94	128.0×10^{-12}	− 68.93
100.0 μV	282.8 μV	− 20.0	200.0×10^{-12}	− 66.99
200.0 μV	565.7 μV	− 13.98	800.0×10^{-12}	− 60.97
400.0 μV	1.131 mV	− 7.959	3.2×10^{-9}	− 54.95
800.0 μV	2.263 mV	− 1.938	12.8×10^{-9}	− 48.93
1.0 mV	2.828 mV	0.0	20.0×10^{-9}	− 46.99
2.0 mV	5.657 mV	6.02	80.0×10^{-9}	− 40.97
4.0 mV	11.31 mV	12.04	320×10^{-9}	− 34.95
8.0 mV	22.63 mV	18.06	1.28 μW	− 28.93
10.0 mV	28.28 mV	20.00	2.0 μW	− 26.99
20.0 mV	56.57 mV	26.02	8.0 μW	− 20.97
40.0 mV	113.1 mV	32.04	32.0 μW	− 14.95
80.0 mV	226.3 mV	38.06	128.0 μW	− 8.93
100.0 mV	282.8 mV	40.0	200.0 μW	− 6.99
200.0 mV	565.7 mV	46.02	800.0 μW	− 0.97
223.6 mV	632.4 mV	46.99	1.0 mW	0
400.0 mV	1.131 V	52.04	3.2 mW	5.05
800.0 mV	2.263 V	58.06	12.80 mW	11.07
1.0 V	2.828 V	60.0	20.0 mW	13.01
2.0 V	5.657 V	66.02	80.0 mW	19.03
4.0 V	11.31 V	72.04	320.0 mW	25.05
8.0 V	22.63 V	78.06	1.28 W	31.07
10.0 V	28.28 V	80.0	2.0 W	33.01
20.0 V	56.57 V	86.02	8.0 W	39.03
40.0 V	113.1 V	92.04	32.0 W	45.05
80.0 V	226.3 V	98.06	128.0 W	51.07
100.0 V	282.8 V	100.0	200.0 W	53.01
200.0 V	565.7 V	106.0	800.0 W	59.03
223.6 V	632.4 V	107.0	1000.0 W	60.0
400.0 V	1,131.0 V	112.0	3,200.0 W	65.05
800.0 V	2,263.0 V	118.1	12,800.0 W	71.07
1000.0 V	2,828.0 V	120.0	20,000 W	73.01
2000.0 V	5,657.0 V	126.0	80,000 W	79.03
4000.0 V	11,310.0 V	132.0	320,000 W	85.05
8000.0 V	22,630.0 V	138.1	1.28 MW	91.07
10,000.0 V	28,280.0 V	140.0	2.0 MW	93.01

Voltage, $V_{p-p} = V_{RMS} \times 2 \times \sqrt{2}$

Voltage, $dBmV = 20 \times Log_{10}\left[\dfrac{V_{RMS}}{0.001 \text{ V}}\right]$

Power, $Watts = \dfrac{(V_{RMS})^2}{50\Omega}$

Power, $dBm = 10 \times Log_{10}\left[\dfrac{Power \text{ (watts)}}{0.001 \text{ W}}\right]$

Table 41

Equivalent Values of Reflection Coefficient, Attenuation, SWR and Return Loss

Reflection Coefficient (%)	Attenuation (dB)	Max. SWR	Return Loss (dB)
1.000	0.000434	1.020	40.00
1.517	0.001000	1.031	36.38
2.000	0.001738	1.041	33.98
3.000	0.003910	1.062	30.46
4.000	0.006954	1.083	27.96
4.796	0.01000	1.101	26.38
5.000	0.01087	1.105	26.02
6.000	0.01566	1.128	24.44
7.000	0.02133	1.151	23.10
7.576	0.02500	1.164	22.41
8.000	0.02788	1.174	21.94
9.000	0.03532	1.198	20.92
10.000	0.04365	1.222	20.00
10.699	0.05000	1.240	19.41
11.000	0.05287	1.247	19.17
12.000	0.06299	1.273	18.42
13.085	0.07500	1.301	17.66
14.000	0.08597	1.326	17.08
15.000	0.09883	1.353	16.48
15.087	0.10000	1.355	16.43
16.000	0.1126	1.381	15.92
17.783	0.1396	1.433	15.00
18.000	0.1430	1.439	14.89
19.000	0.1597	1.469	14.42
20.000	0.1773	1.500	13.98
22.000	0.2155	1.564	13.15
23.652	0.2500	1.620	12.52
24.000	0.2577	1.632	12.40
25.000	0.2803	1.667	12.04
26.000	0.3040	1.703	11.70
27.000	0.3287	1.740	11.37
28.000	0.3546	1.778	11.06
30.000	0.4096	1.857	10.46
31.623	0.4576	1.925	10.00
32.977	0.5000	1.984	9.64
33.333	0.5115	2.000	9.54
34.000	0.5335	2.030	9.37
35.000	0.5675	2.077	9.12
36.000	0.6028	2.125	8.87
37.000	0.6394	2.175	8.64
38.000	0.6773	2.226	8.40
39.825	0.7500	2.324	8.00
40.000	0.7572	2.333	7.96
42.000	0.8428	2.448	7.54
42.857	0.8814	2.500	7.36
44.000	0.9345	2.571	7.13
45.351	1.0000	2.660	6.87
48.000	1.1374	2.846	6.38
50.000	1.2494	3.000	6.02
52.000	1.3692	3.167	5.68
54.042	1.5000	3.352	5.35
56.234	1.6509	3.570	5.00
58.000	1.7809	3.762	4.73
60.000	1.9382	4.000	4.44
60.749	2.0000	4.095	4.33
63.000	2.1961	4.405	4.01
66.156	2.5000	4.909	3.59
66.667	2.5528	5.000	3.52
70.627	3.0000	5.809	3.02
70.711	3.0103	5.829	3.01

$$\rho = \frac{SWR - 1}{SWR + 1}$$

where $\rho = 0.01 \times$ (reflection coefficient in %)

$$\rho = 10^{\frac{-RL}{20}}$$

where RL = return loss (dB)

$$\rho = \sqrt{1 - (0.1^x)}$$

where $X = A/10$ and A = attenuation (dB)

$$SWR = \frac{1 + \rho}{1 - \rho}$$

Return loss (dB) = $-8.68589 \ln(\rho)$

where ln is the natural log (log to the base e)

Attenuation (dB) = $-4.34295 \ln(1 - \rho^2)$

where ln is the natural log (log to the base e)

Table 42

ARRL Parts Suppliers List

W
*sase
A&A Engineering
2521 W LaPalma Ave, Unit K
Anaheim, CA 92801
714-952-2114

A,E,G,H,I,K,L,
M,N,X,Y
*free
**$10
All Electronics Corp
PO Box 20406
Los Angeles, CA 90006
800-826-5432

A,B,C,D,E,G,H,
I,J,K,L,M,U,X,Y
*free
Alpha Electronic Laboratories
705 Vandiver, Suite A
Columbia, MO 65202-2099
314-874-1514

B
Amidon Associates, Inc
12033 Otsego St
N Hollywood, CA 91607

A,D,F,G,H
*sase
Antennas Etc/Unadilla
(Millen components)
PO Box 215BV
Andover, MA 01810-0814
617-475-7831

N,O
Atlantic Surplus Sales
(facsimile equipment)
3730 Nautilus Ave
Brooklyn, NY 11224
718-372-0349

T,U
ATV Research, Inc
13th & Broadway
Dakota City, NE 68731
402-987-3771

Avatar Magnetics
see Ronald C Williams

A,D,H
Barker & Williamson
10 Canal St
Bristol, PA 19007
215-788-5581

A,B,E,G,H,I,J,
K,L,M,N,U,W,X,
Y
*sase
BCD Electro
PO Box 830119
Richardson, TX 75083-0119
214-343-1770

S
**$25
Bird Electronic Corporation
30303 Aurora Rd
Cleveland, OH 44139
216-248-1200

H,S
Caywood Electronics, Inc
(Millen capacitors)
PO Box U
Malden, MA 02148-0921
617-322-4455

A,B,C,D,E,G,H,
I,W
*free
Circuit Board Specialists
PO Box 951
Pueblo, CO 81002-0951
719-542-4525

A,B,C,E,I,K
*$1
**$15
(mail orders)
Circuit Specialists, Inc
Box 3047
Scottsdale, AZ 85257
602-966-0764

A,B,D,E,H,I,T,
W
*free
Communications Concepts, Inc
121 Brown St
Dayton, OH 45402
513-220-9677

L,S,tools
*free
**none
Contact East
PO Box 786
No Andover, MA 01845
617-682-2000

D,E,I,M,H,X
Peter W Dahl Co, Inc
5869 Waycross
El Paso, TX 79924
915-715-2300

A,E,I,K,L,X,Y
*free
Digi-Key Corporation
701 Brooks Ave So
PO Box 677
Thief River Falls, MN 56701
800-344-4539

A,C,D,E,F,G,H,
I,K,L,X,Y
*$25
Electro Sonic, Inc
1100 Gordon Baker Rd
Willowdale, Ontario Canada
M2H 3B3

A,B,C,D,E,G,H,
I,J,K,L,M,S,X,Y
**$10
Electronic Emporium, Inc
(Formerly Semiconductors
Surplus)
3621-21 E Wier Ave
Phoenix, AZ 85040
602-437-8633

F
*free
Elwick Supply Co
Dept 633-SS
230 Woods Lane
Somerdale, NJ 08083

C,L,W
E-Z CIRCUIT by Bishop
Graphics, Inc
5388 Sterling Center Dr
PO Box 5007
Westlake Village, CA 91359
818-991-2600

D,G,H,I,M,N,Q
*free
**$10
Fair Radio Sales Co, Inc
Box 1105
1016 E Eureka St
Lima, OH 45802
419-227-6573
24 hour FAX 419-227-1313

J
*sase
Fox-Tango Corp
(8-pole crystal filters)
PO Box 15944
W Palm Beach, FL 33416
305-683-9587

D
*free
**$25
Gregory Electronics
249 Route 46
Saddle Brook, NJ 07662
201-489-9000

A,B,I,K,M,N,T
*free
**$15
H & R Corporation
A Herbach & Rademan, Inc Co
401 E Erie Ave
Philadelphia, PA 19134
215-426-1708

I,K
Hammond Mfg Co, Inc
1690 Walden Ave
Buffalo, NY 14225

I,K   Hammond Mfg, Ltd
394 Edinburg Rd, N
Guelph, Ontario Canada
N1H 1E5

Q
*$1 Catalog
required   Hi-Manuals
PO Box 802
Council Bluffs, IA 51502

D,J,L,Q,W,Y
*sase   International Radio and
Computers, Inc
747 South Macedo Blvd
Port St Lucie, FL 33452
305-879-6868

Y
*free   JAN Crystals
2341 Crystal Dr
PO Box 06017
Ft Myers, FL 33906-6017
800-237-3063

E
*free
**$10   K2AW's Silicon Alley
175 Friends Lane
Westbury, NY 11590
516-334-7024

C
*free
**$15   Kepro Circuit Systems, Inc
630 Axminister Dr
Fenton, MO 63026-2992
800-325-3878 (out of state)
314-343-1630 (MO)

F
**$10   Kirk Electronics
Ivoryton Industrial Park
Main Street
Ivoryton, CT 06442
203-767-2104
800-243-9310

C,E,F,G,I,K,L,X   Lashen Electronics, Inc
21 Broadway
Denville, NJ 07834
201-627-3783

A,B,C,D,E,F,G,
H,I,J,K,L,M,N,
W
*free
**$10   Marlin P Jones Associates
PO Box 12685
Lake Park, FL 33403
305-848-8236

F (aluminum
and stainless
tubing and
pipe)
**$50   Metal and Cable Corp, Inc
PO Box 117
Twinsburg, OH 44087
216-425-8455

K,L   MFJ Enterprises
PO Box 494
Mississippi State, MS 39762
601-323-5869

A,B,D,E,H,M
*sase   Microwave Components of
Michigan
11216 Cape Cod
Taylor, MI 48180
313-941-8469 (evenings)

A,B,C,D,E,F,G,
H,I,J,K,L,S,U   Milo Associates, Inc
4169 Millersville Rd
Indianapolis, IN 46205
317-546-6456

A,B,C,E,F,G,H,
I,J,K,L,U,X,Y
*free
**$20   Mouser Electronics
11433 Woodside Ave
Santee, CA 92071
619-449-2222

M,N
*$1   OCTE Electronics
PO Box 276
Albury, VT 05440

T
*free   PC Electronics
2522 S Paxson Lane
Arcadia, CA 91006
818-447-4565

B
*free   Palomar Engineers
Box 455
Escondido, CA 92025
619-747-3343

A,D,X
*free   Pasternack Enterprises
PO Box 16759
Irvine, CA 92713
714-261-1920

A,D,E   RF Gain, Ltd
(RF transistors, amplifiers and
electron tubes)
116 South Long Beach Rd
Rockville Centre, NY 11570
800-645-2322

A,B,C,D,E,F,G,
H,I,J,K,L,W,X,Y
*$1   Radiokit
PO Box 973
Pelham, NH 03076
603-635-2235

D,Y   Sentry Mfg Co
Crystal Park
Chickasha, OK 73018
405-224-6780

**$10   Small Parts, Inc
(Mechanical components and
metal stock)
PO Box 381736
Miami, FL 33238
305-751-0856

A,B,C,D,E,F,G,
H,I,J,K,L,S,U,
W,X,Y
*$2
**$20   Dick Smith Electronics, Inc
PO Box 468
Greenwood, IN 46142
317-888-7265

A,D,F,J,Y   Spectrum International, Inc
PO Box 1084
Concord, MA 01742
617-263-2145

A,M,N
**$4   Star-Tronics
PO Box 683
McMinnville, OR 97128

A,B,C,D,E,F,G,
H,I,J,K,L,M,N,
P,Q,R,T,W,X,Y   Surplus Sales of Nebraska
1315 Jones St
Omaha, NE 68102
402-346-4750

K,H   Ten-Tec, Inc
Highway 411, E
Sevierville, TN 37862

I
*free   Toroid Corporation of Maryland
6000 Laurel-Bowie Rd
Bowie, MD 20715-4037
301-464-2100

B,O,Q
*sase   Typetronics
PO Box 8873
Ft Lauderdale, FL 33310

V   US Tower Corp
8975 West Goshen Ave
Visalia, CA 93291

I,Z   Ronald C Williams W9JVF
(formerly Avatar Magnetics)
1408 W Edgewood Ave
Indianapolis, IN 46217-3618
317-783-1211

A,E,F,T,U,W   Wyman Research, Inc
Box 95, RR 1
Waldron, IN 46182
317-525-6452

Chart Coding
A—New Components
B—Toroids and Ferrites
C—Etched Circuit Board Materials
D—Transmitting and Receiving Materials
E—Solid-State Devices
F—Antenna Hardware
G—Dials and Knobs
H—Variable Capacitors
I—Transformers
J—IF Filters
K—Cabinets and Boxes
L—General Supplier
M—Surplus Parts
N—Surplus Assemblies
O—RTTY Equipment and Parts
P—Surplus FM Gear and Parts
Q—Equipment Manuals
R—Service of Collins Equipment
S—Test Equipment
T—Amateur TV Cameras and Components
U—Microcomputer Peripheral Equipment
V—Towers
W—Ready-made Printed Circuit Boards
X—Wire
Y—Crystals
Z—Climbing and Safety Equipment

*Catalog Price
**Minimum Order

To the best of our knowledge the suppliers shown are willing to sell components to amateurs in small quantities by mail. This listing does not necessarily indicate that these firms have the approval of ARRL.

Chapter 36

How to Become a Radio Amateur

Why become a ham? Perhaps because you want to participate in hands-on technical experiments. Perhaps you heard about ham radio from a television or magazine advertisement. Or maybe you witnessed Amateur Radio operators providing emergency communications during a recent disaster. Or it could be the desire to talk with people easily and freely, people who may live down the street or thousands of miles away.

Whatever your reason, it's not hard to become a radio amateur. This chapter will guide you along the path. Millions of people around the world have done it; youngsters too young to ride a bicycle have gotten their amateur licenses (the record is a five year old), as have people over 95.

In the United States, Amateur Radio operation is governed by the Federal Communications Commission (FCC). The FCC rules, known as Part 97 for the section dealing specifically with Amateur Radio, describe five levels of amateur license. These follow a ladder-like progression. Each license class has a more difficult examination, but also grants greater operating privileges. The *Novice* is the introductory license level. Other amateur licenses, in order of increasing privileges, are the Technician class, General class, Advanced class and Amateur Extra class.

The examination for the Novice license covers three main areas: First is some regulatory material, which is simply common sense (and therefore easy to learn). The second deals with on-the-air operating skills. And the third covers radio theory — purposely made easy and at an introductory level. In addition, you'll be tested on the Morse code. Aha!, you say, I knew there was a catch! It's not a catch. It's not even hard.

Learning the code is often regarded as the major obstacle to getting an amateur license. Not so! Realistically, it takes less than two weeks to learn the alphabet, 10

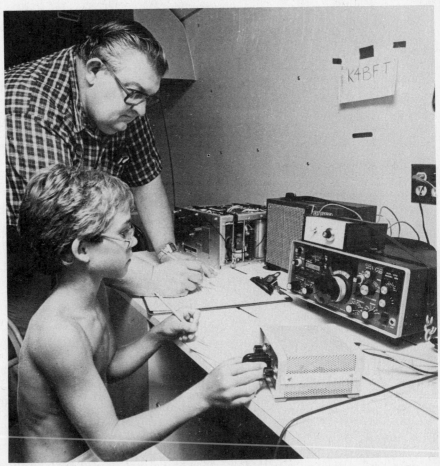

Amateur Radio — a hobby where *everyone* has fun. Here, Bill Christian, K4IKR (left), and Tim Dionne, KB4BDG, operate in the annual Field Day. During this two-day event, thousands of radio amateurs around the U.S. set up and operate portable emergency-powered stations.

numerals and punctuation; perhaps it will take you even less time, depending on your motivation. After you get the letters down, it's just a matter of regular practice until you're comfortably communicating at 5 words per minute (WPM). Higher speeds come easily, too! Again, the key is *regular*

practice. You could learn (and remember) seven Morse code letters in half an hour.

Novices can communicate worldwide using International Morse code, voice and digital modes such as RTTY and packet radio on a wide range of radio frequencies. The Novice license examination is free.

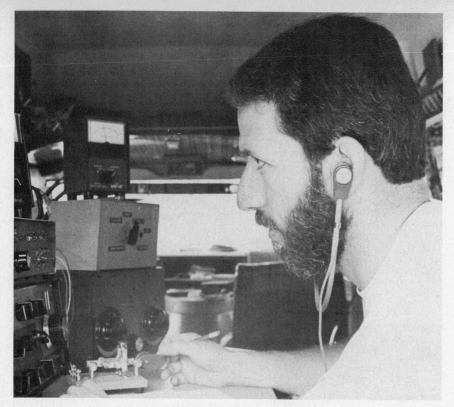

John Comella, N8AA, concentrates as he makes Morse code contacts.

Radio Wave Propagation

Q. What type of propagation involves radio signals that travel along the ground? A: Ground-wave propagation.

Amateur Radio Practice

Q. For safety purposes, how high should all portions of a horizontal wire antenna be? A: High enough so that a person cannot touch them from the ground.

Q. Why is harmonic radiation undesirable? A: It will cause interference to other stations and/or result in out-of-band signal radiation.

Q. A voltage standing wave ratio is measured with a(n) A: SWR bridge.

Electrical Principles

Q. What are the two polarities of a voltage? A: Positive and negative.

Q. Are audio frequencies higher or lower than radio frequencies? A: Lower.

Q. What is the unit of electrical current? A: The ampere.

Circuit Components

Q. What does a voltmeter measure? A: Volts.

Practical Circuits

Q. What is the unlabeled box in this block diagram?

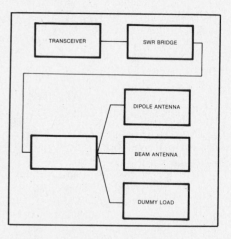

A. Antenna switch.

Signals and Emissions

Q. What does the term "chirp" mean? A: A slight shift in oscillator frequency each time a CW transmitter is keyed.

Antennas and Feedlines

Q. When a vertical antenna is lengthened, what happens to its resonant frequency? A: It decreases.

Q. What commonly available feedline can be buried directly in the ground for some distance without adverse affects? A: Coaxial cable.

Technician class licensees have Novice class privileges; they also are assigned additional frequencies in the very-high- and ultra-high-frequency (VHF and UHF) spectrums. The General class license permits even more frequency choices.

A Close Look at the Novice Examination

As was briefly explained earlier, the Novice exam covers three main topics: Rules and Regulations, Operating Procedures and Amateur Radio Theory. Volunteer-examiner coordinators have put together a list of about 300 questions that cover these different areas. Every Novice exam is a randomly selected collection of 30 of these 300 questions. Each test is compiled as described in Table 1.

To show how a typical Novice exam might be worded, we'll go over a sample test. The answers are given as short phrases.

Rules and Regulations

Q. What is a control operator? A: A control operator is a licensed operator designated to be responsible for the emissions of a particular station.

Q. What are the Novice class operator transmitting frequency privileges in the 80-meter band? A: 3700-3750 kHz.

Q. What does the term A1A emission mean? A: CW Morse code without any audio modulation of the carrier.

Q. Under what circumstances, if any, may the control operator cause false or deceptive signals or communications to be transmitted? A: Under no circumstances.

Q. With which amateur stations may an FCC-licensed Amateur Radio station communicate? A: All amateur stations, unless prohibited by the amateur's government.

Q. An Amateur Radio station must be identified at least once every A: 10 minutes.

Q. Should an Amateur Radio operator receive an Official Notice of Violation from the FCC, how promptly should he/she respond? A: Within 10 days.

The factual knowledge in this first section can be picked up easily by reading through the ARRL publication *Tune in the World With Ham Radio*.

Operating Procedures

Q. The "T" in the RST signal report stands for A: the tone of the signal.

Table 1

The Novice Class Written Examination

Rules and Regulations:	9 questions
Operating Procedures:	2 questions
Radio Wave Propagation:	2 questions
Amateur Radio Practice:	4 questions
Electrical Principles:	4 questions
Circuit Components:	2 questions
Practical Circuits:	2 questions
Signals and Emissions:	2 questions
Antennas and Feedlines:	3 questions

There is *nothing* hard about the Novice theory exam! If you want to see the rest of the 300 questions, a copy of the question

Listening to over-the-air practice transmissions is the best way to increase your speed and proficiency with the code. Members of the Mobile (Alabama) Amateur Radio Club put this philosophy to good use.

Study is more fun when you are part of a group. In an organized class you will find encouragement from your fellow students and help from a knowledgeable instructor.

pool is available free from ARRL Hq. Please include an s.a.s.e. with your request.

Finding Help

You can choose several paths when going for your amateur license. There is the personalized-instruction route, the self-study route or the class method. No matter which one you choose, the ARRL can help you.

The ARRL has a program specifically set up to help people get their ham licenses, a referral service that ensures you get the help and assistance you need, and that you get it when you need it.

Personalized Instruction

To get started toward your Amateur Radio license, write to ARRL Hq. in Newington, Connecticut. We will match you up with a Registered Amateur Radio Instructor in your area. There are thousands of Registered Instructors around the United States. Their level of instruction, names and addresses are kept in a computer file. The large number of Instructors ensures a wide geographical coverage. When you request assistance, there's a good chance you'll be linked with a Registered Instructor who lives close by.

The great benefit of the Registered Instructor system is that the Instructor offers patient instruction, tailored to the pace at which you feel comfortable. If you miss a point, your Instructor will spend as much time as is necessary to make sure you understand what's going on. This is the sort of education you can get only in the finest private schools.

Registered Instructors have altruistic reasons for helping you, the future ham. This buddy system works extremely well, and has been the path into the hobby for tens of thousands of radio amateurs. The personalized instruction and assistance provided by the Elmers, as Instructors are often called, can't be beat.

The Elmer is also the one who teaches you the Morse code, and often administers the license examination, if you are preparing for the Novice license.

Self Study

If you choose the self-study method, there's an excellent package that will guide you along the way. *Tune in the World With Ham Radio,* published by ARRL, will teach you all you need to know to pass your Novice examination. The book covers the concepts of amateur theory that are found in the Novice examination. It covers the following areas: Exploring ham radio; managing the radio spectrum (covering the regulatory and legal aspects of the hobby); learning the Morse code; understanding basic theory; setting up a Novice station; and a section on operating and operating activities. The package contains two Morse code tapes that teach you every letter, numeral and punctuation mark you need to know. The tapes build competence in the code with practice conversations.

License Study Classes

The third alternative open to you as a prospective radio amateur is to attend license classes. These classes are usually sponsored by a local Amateur Radio club, and there are over 2000 such clubs all around the United States, usually centered in and around metropolitan areas.

Here again ARRL Hq. can help you. We have a listing of over 1700 *ARRL-Affiliated* clubs around North America. Many of these clubs run licensing classes twice a year, in the fall and spring, and register these classes with ARRL Hq. If you want to get in touch with a class in your area, give us a call. The club records are kept in a constantly accessible computer file, so we can rapidly give you the information you want.

A license class usually follows a 10-week schedule. Morse code is introduced simultaneously with the elementary theory needed for the Novice license. Class size commonly ranges from 5 to 50 people. You can take your license examination any time,

You Can Do It Too!

Think Morse code is a difficult hurdle that takes weeks and weeks of hard work to master? Think again! Guy Mitchell, WDØDVX, of Buckingham, Iowa, picked up Morse code from tapes his parents were using for their Novice class. Soon, he began studying on his own, and mastered the alphabet and numbers with little difficulty. What makes this story unique is that Guy Mitchell was four years old when he passed the five-words-per-minute code test! He had had another birthday when he passed the written test, however. At five years of age, Guy was the youngest licensed ham radio operator in the U.S., and perhaps the world!

Guy Mitchell started kindergarten five months after he passed the Novice code test.

if you progress faster than the rest of the class.

If you want to get in touch with a radio amateur immediately, chances are that you'll find one right in your neighborhood. There is one ham for every 550 Americans, and a simple walk around your neighborhood might give good results: Look for houses or apartments with *large* antennas on the roof.

The Code

Morse code should not be considered a

Susan Smith, K3YL, takes a break from handling a few Amateur Radio messages.

These Chinese amateurs use directional antennas to search for a hidden transmitter during a "fox hunt."

series of dots and dashes. Rather, words descriptive of the sound of the two different Morse code elements used: A dot is called a *dit,* and a dash is called a *dah.* The dah, the long tone, is three times as long as the dit.

When two or more dits are put together, or when a dit is followed immediately by a dah, as they are for 12 of the 26 letters and most of the numbers, they are written as didit, or dididit or didah (dropping the "t" from all but the final dit).

An easy method of learning the code, and one in which you'll learn seven letters in less than an hour, is the "EISH-TMO" system. This method works by teaching you letters by association and relation, a powerful learning tool, as teachers know.

Here is the EISH-TMO method:

E	*dit*	T	*dah*
I	*didit*	M	*dahdah*
S	*dididit*	O	*dahdahdah*
H	*didididit*		

You see how these first seven letters relate to one another. The rest of the Morse code letters are just about as easy to learn. Here's the next group:

A	*didah*	N	*dahdit*
W	*didahdah*	D	*dahdidit*
J	*didahdahdah*	B	*dahdididit*

From this point on, the relations become less apparent, but you've already learned half of the alphabet in a short time. Regular practice is the key to solidifying the code characters in your mind. After you get these few letters down, think about them as you go about your daily business. A good way to practice the code is to translate the letters you've learned whenever and wherever you see them — in newspapers, on signs, even as you dial the telephone.

Here are a few more letters.

U   *dididah*
V   *didididah* (think of Beethoven's fifth symphony)
G   *dahdahdit*
Z   *dahdahdidit*
K   *dahdidah*
R   *didahdit*
P   *didahdahdit*
X   *dahdididah*

Now come the letters with no relatives, the "oddballs."

F   *dididahdit*
C   *dahdidahdit*
L   *didahdidit*
Q   *dahdahdidah*
Y   *dahdidahdah*

And that's the entire alphabet. Not too hard, and a lot easier than learning how to touch type or practicing the piano. You can take it with you wherever you go, too!

When listening to code-practice transmissions, as you hear the tones, translate them in your head and write down the letters on paper. Eventually you'll be able to skip the paper, but that shouldn't be your immediate goal. At this stage, it's important to practice the letters you've learned as often as possible. After learning how to receive and translate code in your head, you should practice sending it. The key is to first train the ear, then train the hand.

You also have to know the numerals from zero to nine for the Novice code examination. Each numeral consists of five code elements. You'll see how straightforward this is:

One	*didahdahdahdah*
Two	*dididahdahdah*
Three	*didididahdah*
Four	*dididididah*
Five	*dididididit*

You probably noticed the pattern followed by these first five digits. Six through zero follow a similar pattern.

Six	*dahdidididit*
Seven	*dahdahdididit*
Eight	*dahdahdahdidit*
Nine	*dahdahdahdahdit*
Zero	*dahdahdahdahdah*

You must learn three punctuation marks

for the Novice code test. These are: The *period (didahdidahdidah),* the *comma (dahdahdididahdah)* and the *question mark (dididahdahdidit).* The rhythm of punctuation marks makes them easy to learn.

Computer-aided Morse code instruction may be of interest to you. Programs that teach the Morse code are available for most popular home computers.

The Handi-Ham System

People with disabilities will find the world of Amateur Radio easy to enter, and for individuals restricted to the confines of a bed and four walls, Amateur Radio can be a vehicle for communication, fraternity and fellowship with other human beings in the outside world. A valuable therapeutic tool, Amateur Radio continues to hold a profound appeal for individuals with physical handicaps. It is a means of people-to-people contact on a basis of absolute equality — two minds communicating and relating on a level of mutual and equal empathy.

With today's miniaturized, transistorized transceivers (a transmitter and receiver in one package), an Amateur Radio station can be set up just about anywhere. WD8KCP can talk around the world from the kitchen.

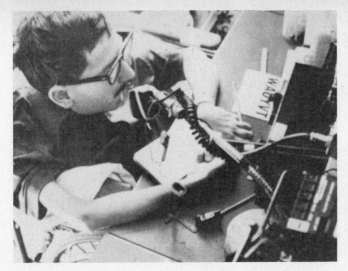

Amateur Radio is person-to-person communication, spanning miles and continents.

Amateur Radio is a hobby for young and old alike. From left to right, Brad Stave (KA0BIJ), Steve Stave (KA0AEK), and Art Stevens (K0LXI), ages 12, 11 and 88, respectively, operate K0LXI.

A special-interest organization is set up for radio amateurs with disabilities. Called the Courage HANDI-HAM system, this group is an international service organization of over 2000 handicapped and able-bodied radio amateurs working together to bring ham radio to individuals with physical handicaps. There are HANDI-HAM members in every state and in 20 other countries ready to help. They can provide textbooks (or cassette recordings thereof), code practice tapes, a key and a code practice oscillator. There are local HANDI-HAMs to assist you with studies at home. Once you receive your license, the Courage HANDI-HAM system may lend you basic ham radio equipment to get you started on the air.

For information, contact the Courage Center at 3915 Golden Valley Rd., Golden Valley, MN 55422, tel. 612-588-0811.

Putting It All Together

Assembling a station doesn't have to be expensive. Many companies that sell used equipment advertise in *QST*. First you'll need a receiver. That receiver will do three things for you. By listening to over-the-air code-practice transmissions, you can increase your skill with the code. By monitoring the ham bands you will get an idea of what happens; you'll increase your familiarity with ham radio operation. And after you get your license, you can use the receiver for making contacts. For these reasons, buy the best receiver you can afford. Think of the long-term use of the equipment. The extra dollar spent shrinks as you use the gear year after year.

Of course, you don't have to buy a receiver. You can build your own. This *Handbook* has receiver designs in Chapter 30.

W1AW Code Practice

Code practice is sent by the ARRL Hq.

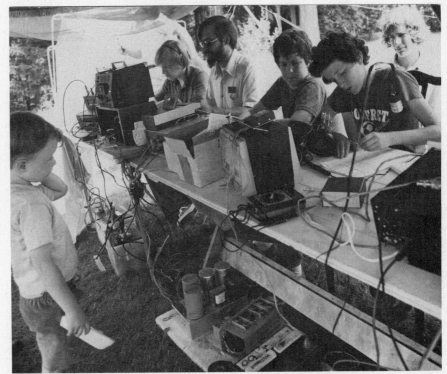

The W1FYM group uses a microcomputer to ensure they work stations only once per band. In keeping with the emergency and portable emphasis of Field Day, everything in this two-transmitter station is powered from the batteries at the bottom of the picture.

Table 2
W1AW Slow Code Practice Schedule
Speeds: 5, 7½, 10, 13, 15 WPM

EST/EDST	CST/CDST	MST/MDST	PST/PDST	Days
9 A.M.	8 A.M.	7 A.M.	6 A.M.	M W F
7 P.M.	6 P.M.	5 P.M.	4 P.M.	M W F
4 P.M.	3 P.M.	2 P.M.	1 P.M.	T Th S Sn
10 P.M.	9 P.M.	8 P.M.	7 P.M.	T Th S Sn

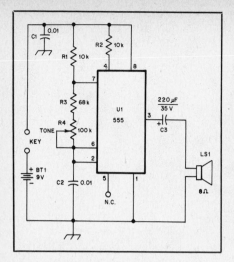

Fig. 1 — Schematic diagram of the code-practice oscillator. Radio Shack part numbers are given in parentheses.

BT1 — 9-V battery (23-553).
C1, C2 — 0.01-μF capacitor (272-131).
C3 — 220-μF, 35-V electrolytic capacitor (272-1029).
LS1 — 2-in loudspeaker, 8 Ω (40-245).
R1, R2 — 10-kΩ resistor, ¼ W (271-1335).
R3 — 68-kΩ resistor, ¼ W.
R4 — 100-kΩ potentiometer (271-220).
U1 — 555 IC timer (276-1723).

Fig. 2 — Photograph of the code-practice oscillator. This version is available as a kit (less wooden base) from: Circuit Board Specialists, P.O. Box 969, Pueblo, CO 81002. See Chapter 24 for a discussion of construction techniques that can be used to build this circuit.

station, W1AW. Transmission frequencies are approximately 1.818, 3.58, 7.08, 14.07, 21.08, 28.08, 50.08 and 147.555 MHz. Texts are from recent issues of *QST* and checking references are sent several times in each session. For practice purposes, the order of words in each line of text may be reversed during the 5-15 WPM transmissions. The transmission time schedule is shown in Table 2. See Chapter 38 for more information.

A CODE-PRACTICE OSCILLATOR

The simple oscillator (see Figs. 1 and 2) is a good introduction to electronic construction as well as a useful tool for learning to send the Morse code. The costs can be cut dramatically through intelligent substitution and liberal scrounging from a ham's junk box.

The schematic diagram and photographs show the unit in detail, but exact duplication is not necessary. Construction techniques are presented in Chapter 24, and the schematic symbols are printed near the front of this volume. The project can be assembled in an evening. This code practice oscillator is intended for beginners using hand keys. Because the circuit is keyed in the positive battery line, the oscillator may not be compatible with some electronic keyers.

Chapter 37

Assembling a Station

Although many hams never try to build a major project, such as a transmitter, receiver or amplifier, they do have to assemble the various components (say, for example, transceiver, power supply and antenna) into a working station. There are many benefits to be derived from assembling a safe, comfortable, easy-to-operate collection of radio gear, whether the shack is at home, in the car or portable in a field.

Rather than discuss electronic theory or individual construction projects, this chapter will detail some of the "how tos" of setting up a station for fixed, mobile and portable operation. Such topics as station location, safety, finding adequate power sources, station layout and cable routing are covered, along with some of the practical aspects of antenna erection and maintenance.

Two major themes recur throughout this chapter: Plan ahead, and most shortcuts aren't really, in the long run. The key to a successful installation, whether it's the installation of a mobile rig under the dash of your car or the erection of a 150-foot tower with a large monobander at the top, is proper planning. If you take the time to think the project through to its conclusion, to consider alternatives, to make measurements and rough sketches, your chances of success are much greater than if you start drilling and cutting blindly. You'll avoid 99 percent of such problems as "Oops, that guy wire has to go through the center of this tree," or "Oops, I can't drive my car with the mobile rig installed because I put the rig where my left knee has to go."

Most shortcuts will eventually come back to haunt you, so it's really worth doing a job right the first time. A "bargain" antenna mast for your tower is no bargain if it folds up in a windstorm, sending your antenna crashing to the ground. Not taking the time to weatherproof your outdoor connections is no shortcut if you have to replace a cable after six months because it filled up with water. Running your shack from an overloaded electrical service is no "deal" if the circuit breaker opens and you're off the air every time the refrigerator compressor kicks in. Plan ahead, and do

Fig. 1 — W9OBF is quite active from this compact station built into a hideaway desk. A station arranged like this can easily be tucked into the corner of a living room without disrupting the rest of the household.

the best job you can when putting together your station.

SELECTING A LOCATION

Selecting the right location for your station is the first and perhaps most important step in assembling a safe, comfortable, convenient station. The exact location will depend on the type of home or apartment you have and how much of that space can be devoted to your station. Fortunate amateurs will have a spare room to devote to housing the station; some may even have a separate building for their exclusive use. Most aren't that lucky, however, so usually a spot in the cellar or attic, or a corner of the living room is pressed into service.

Examine the possibilities from several angles. A station should be comfortable; odds are good that you'll be spending a lot of time there over the years. Some unfinished basements are damp and drafty — not an ideal environment for several hours of leisurely hamming. Attics have their

drawbacks, too; they can be stifling during warmer months. If possible, locate your station away from the heavy traffic areas of your home. Operation of your station should not interfere with family life. A night of chasing DX on 80 meters may be exciting to you, but the other members of your household may not share your enthusiasm.

Keep in mind that you must connect your station to the outside world. The location you choose should be convenient to a good power source and an adequate ground. There should be a fairly direct route to the outside for running antenna feed lines, rotator control cables and the like.

Although most homes will not have an "ideal" space meeting all requirements, the right location for you will be obvious after you scout around. The amateurs whose stations are depicted in Figs. 1 through 3 all found the right spot for them. Weigh the tradeoffs and decide which features you can do without and which are necessary for

Fig. 2 — The basement makes a good station location if it is dry. This setup belongs to WB2NPE.

Fig. 3 — Some amateurs, like PP2ZDD, are fortunate enough to have a separate room available for their stations.

your style of operation. If possible, pick an area large enough for future expansion.

SAFETY

Although the RF, ac and dc voltages in most amateur stations pose a potentially grave threat to life and limb, common sense and knowledge of safety practices will help you avoid accidents. Building and operating an Amateur Radio station can be, and is for almost all amateurs, a perfectly safe pastime. However, carelessness can lead to severe injury, or even death. The ideas presented here are only guidelines; it would be impossible to cover all safety precautions. Remember: There is no substitute for common sense.

A fire extinguisher is a must for the well-equipped amateur station. The fire extinguisher should be of the carbon-dioxide type to be effective in electrical fires. Store it in an easy-to-reach spot and check it at recommended intervals.

Family members should know how to turn the power off in your station. They should also know how to apply artificial respiration. Many community groups offer courses on cardiopulmonary resuscitation (CPR).

AC and DC Safety

The primary wiring for your station should be controlled by one master switch, and other members of your household should know how to kill the power in an emergency. All equipment should be connected to a good ground. All wires carrying power around the station should be of the proper size for the current carried and be insulated for the voltage level involved. Bare wire, open-chassis construction and exposed connections are an invitation to accidents. Remember that high-current, low-voltage power sources are just as dangerous as high-voltage, low-current sources. Possibly the most-dangerous voltage source in your station is the 117-V primary supply; it is a hazard often overlooked because it is a part of everyday life. Respect even the lowliest power supply in your station.

Whenever possible, kill the power and

unplug equipment before working on it. Discharge capacitors with an insulated screwdriver; don't assume the bleeder resistors are 100 percent reliable. In a transmitter, always short the tube plate cap to ground just to be sure. If you must work on live equipment, keep one hand in your pocket. Avoid bodily contact with any grounded object to prevent your body from becoming the return path from a voltage source to ground. Use insulated tools for adjusting or moving any circuitry. Never work alone. Have someone else present; it could save your life in an emergency.

RF Safety

An often-overlooked safety precaution is the avoidance of unnecessary exposure to RF energy. Body tissues that are subjected to large amounts of RF energy may suffer serious heat damage. At frequencies where the body's length is around 0.4 wavelength, RF energy is most efficiently absorbed; this occurs in the VHF range.

Most Amateur Radio operations use relatively low RF power, and the operations are highly intermittent. Hams spend more time listening than transmitting, and actual transmissions — like keyed CW and SSB — are inherently intermittent. The exception is RTTY, because the RF carrier is present continuously at its maximum level.

Potential exposure situations should be taken seriously. Observing these "RF awareness" guidelines will help make Amateur Radio operation safe.

• Confine RF radiation to antenna radiating elements themselves. Provide a single, good station ground (earth), and eliminate radiation from transmission lines.

• In high-power operation (several hundred watts and up) in the HF and VHF region, and particularly with RTTY, try to avoid human presence near antenna ends. With vertical monopole antennas, humans should go no closer than 10-15 feet during high-power, non-intermittent operation.

• Don't operate RF power amplifiers, especially at VHF/UHF, with the covers removed.

• In the UHF/SHF region, never look

into the open end of an activated length of waveguide or point it toward anyone. Never point a high-gain, narrow-beamwidth antenna (a paraboloid, for instance) toward people.

• With mobile rigs of 10-W RF power or more, do not power your rig if anyone is standing near the antenna.

• With hand-held transceivers with RF power output above several watts, maintain an inch or so between forehead and antenna. Keep the antenna tip away from your head.

• Don't work on antennas that have RF power applied.

• Amateur antennas up on towers, well away from people, pose no exposure problem. Make sure, however, that transmission lines are not radiating.

Further details on RF safety can be found in *The Satellite Experimenter's Handbook,* published by the ARRL. *QST* carries information regarding RF environmental regulations at the local and federal levels, and tells how these regulations may affect amateur operations.

National Electrical Code®

The *National Electrical Code®* is a comprehensive document that details safety requirements for all types of electrical installations. In addition to setting safety standards for house wiring and grounding, the Code also contains a section on Radio and Television Equipment — Article 810. Sections C and D specifically cover Amateur Transmitting and Receiving Stations. Highlights of the section concerning Amateur Radio stations follow. If you are interested in learning more about electrical safety, you may purchase a copy of *The National Electrical Code* or *The National Electrical Code® Handbook,* edited by Peter Schram, from the National Fire Protection Association, Batterymarch Park, Quincy, MA 02269.

Antenna installations are covered in some detail in the Code. It specifies minimum conductor sizes for different length wire antennas. For hard-drawn copper wire, the Code specifies no. 14 for open (unsupported) spans less than 150 feet, and no. 10 for longer spans. Copper-clad steel, bronze or other high-strength conductors may be no. 14 for spans less than 150 feet and no. 12 for longer runs. Lead-in conductors (for open-wire transmission line) should be at least as large as those specified for antennas.

The Code also says that antenna and lead-in conductors attached to buildings must be firmly mounted at least 3 inches clear of the surface of the building on nonabsorbent insulators. The only exception to this minimum distance is when the lead-in conductors are enclosed in a "permanently and effectively grounded" metallic shield. The exception covers coaxial cable.

According to the Code, lead-in conductors (except those covered by the exception) must enter a building through a rigid, noncombustible, nonabsorbent insulating tube or bushing, through an opening provided for the purpose that provides a clearance of at least 2 inches or through a drilled window pane. All lead-in conductors to transmitting equipment must be arranged so that accidental contact is difficult.

Transmitting stations are required to have a means of draining static charges from the antenna system. An antenna discharge unit (lightning arrester) must be installed on each lead-in conductor (except where the lead-in is protected by a continuous metallic shield that is permanently and effectively grounded, or the antenna is permanently and effectively grounded). An acceptable alternative to lightning arrester installation is a switch that connects the lead-in to ground when the transmitter is not in use.

Grounding conductors are described in detail in the Code. Grounding conductors may be made from copper, aluminum, copper-clad steel, bronze or similar erosion-resistant material. Insulation is not required. The "protective grounding conductor" (main conductor running to the ground rod) must be as large as the antenna lead-in, but not smaller than no. 10. The "operating grounding conductor" (to bond equipment chassis together) must be at least no. 14. Grounding conductors must be adequately supported and arranged so that they are not easily damaged. They must run in as straight a line as practical between the mast or discharge unit and the ground rod. Adequate ground systems are covered later in this chapter.

Some information on safety inside the station is covered in the Code. All conductors inside the building must be at least 4 inches away from conductors of any lighting or signalling circuit except when they are separated from other conductors by conduit or a nonconducting material. Transmitters must be enclosed in metal cabinets, and the cabinets must be grounded. All metal handles and controls accessible by the operator must be grounded. Access doors must be fitted with

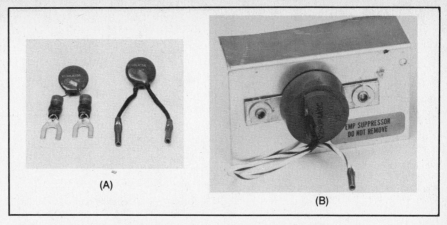

Fig. 4 — Typical LA series MOV protectors are shown at A with various connectors on the leads. A PA series MOV is shown at B as it might be mounted in a service-entrance box.

interlocks that will disconnect all voltages above 350 when the door is opened.

Lightning Protection

The National Fire Protection Association (NFPA) publishes a booklet called *Lightning Protection Code* (NFPA no. 78-1983) that should be of interest to radio amateurs. Write to the NFPA at the address given in the previous section for information on obtaining a copy. Two paragraphs of particular interest to amateurs are presented below:

"3-26 Antennas. Radio and television masts of metal, located on a protected building, shall be bonded to the lightning protection system with a main size conductor and fittings.

"3-27 Lightning arresters, protectors or antenna discharge units shall be installed on electric and telephone service entrances and on radio and television antenna lead-ins."

The best protection from lightning is to disconnect all antennas from equipment and disconnect the equipment from the power lines. Ground antenna feed lines to safely bleed off static buildup. Eliminate the possible paths for lightning strokes. Rotator cables and other control cables from the antenna location should also be disconnected during severe electrical storms.

In some areas, the probability of lightning surges entering homes via the 117/234-V line may be high. Lightning produces both electrical and magnetic fields that vary with distance. These fields can couple into power lines and destroy electronic components. Radio equipment can be protected from these surges through the use of transient-protective devices. The following information originally appeared in a February 1982 *QST* article by Gene Collick, W8LEQ and Ken Stuart, W3VVN.

Protective Devices

The General Electric Home Lightning Protector® is designed to prevent lightning surges (entering through the wiring) from damaging electrical wiring and/or appliances. The protector is a two-pole, three-wire device designed primarily for single-phase 117/234-V grounded-neutral service. It mounts via a ½-inch pipe-thread connection through a knockout in the service-entrance box preferably, or, at the weatherhead or within the meter housing. The purpose of this device is to reduce the amplitude of large transients.

TransZorbs® are silicon devices manufactured by General Semiconductor Industries for transient suppression. They contain a large-area PN junction having integral heat sinks and are capable of handling short-duration, high-power pulses (typically 1500 watts for 1 millisecond and 100,000 watts for 100 nanoseconds). The TransZorb® protects by clamping transient voltages to a safe level, with sub-nanosecond reaction time.

A metal-oxide varistor (MOV) is a bulk semiconductor device whose resistance varies with the magnitude (but not polarity) of the applied voltage. At extremely low currents, a varistor acts like a linear resistor with a resistance that can exceed hundreds of megohms. At higher currents, the voltage-current relation is nonlinear, and at extremely high currents the device acts like a constant resistor with a very small value (typically about 1 ohm). Varistors can respond in low-nanosecond times, with clamping-threshold voltages ranging from 10 to 1500 V, with continuous power from 0.5 to 5 watts and with peak energy of more than 600 joules. MOVs have a capacitance that is a device parameter, and this should be considered for high-frequency applications. GE manufactures devices with wire leads (LA series) and the heat-sink type (PA series). The PA type may be used for service-entrance box and fuse-mounting; use the LA units for coils, solenoids and such within the equipment. See Fig. 4.

Installation

The first protective device you can install is a secondary low-voltage lightning arrester rated at 650 V (a typical device is a GE

Fig. 5 — Construction of a plug-in protector. Install a V130LA10A MOV across the brass and silver terminals and another across the silver and green terminals.

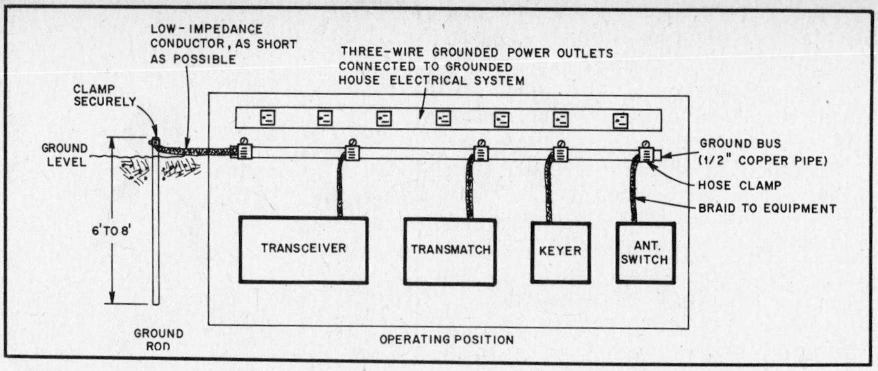

Fig. 6 — An effective station ground bonds the chassis of all equipment together with low-impedance conductors and ties into a good earth ground.

Table 1

Transient-Protective Device Selection Data

General Electric Co., Semiconductor Div., W. Genesee St., Auburn, NY 13021

"Home Lightning Protector"
GE Thyrite 9L15BCB002
"Metal Oxide Varistors"
V130LA10A
V130PA20A
V39ZA6
V56ZA2

The numbers after the V indicate the normal operating voltage (RMS) of the circuit. Voltages at a preset value greater than that will cause the device to conduct heavily, clamping the voltage.

General Semiconductor Industries, P.O. Box 3078, Tempe, AZ 85281

TransZorbs®
1.5KE6.8
1.5KE7.5 to 47
1.5KE51 to 110
1.5KE120 to 150
1.5KE160 to 200
1N6072

Bidirectional, for connection directly across 117-V ac line.
The numbers after KE are the approximate breakdown voltage of the device.

Radio Shack

MOVs for 117-V Lines
V130LA1 — Very Light Duty (276-571)
SNR14A130K — Light Duty (276-570)
SNR20A130K — Medium Duty (276-568)

Thyrite type 9L15BC002). Power should be removed from the feed to the house, and the device should be installed ahead of the service-entrance box by a qualified electrician. The black leads (2) are connected to each hot leg, with the white lead terminated to the meter box (be sure the meter box is directly connected to the service-entrance box). The reason the white lead is terminated to the box rather than the electrical circuit neutral is that the device is being installed to protect against the common-mode (phase-to-case ground) transient.

The second device to be installed is a low-voltage surge arrester such as a GE V130PA20A MOV. These should be connected from each hot leg to case ground. The arrester should be located in a way that provides minimum lead length. These can

be installed with an active power feed to the house, but with the main breaker turned off. Connect the MOVs on the load side of the main breaker.

The ac input to the transmitter can be protected by one of two methods. The transmitter may be modified by installing a V130LA10A MOV across the ac input and another one from neutral to case ground. If you do not want to modify the transmitter, a plug-in protector can be purchased or constructed (see Fig. 5). This device is then plugged into the duplex receptacle feeding the transmitter. This method should be ideal for renters, for portable operation or to provide some protection for those who can't afford an electrician to install the other devices. Receivers, transceivers, TV sets and other electronic equipment should all be protected in a similar fashion. An alternative would be to wire an outlet box or strip with protective devices.

Rotator protection consists of installing a plug-in MOV device in one side of the duplex receptacle that the rotator remote-control unit is plugged into. The rotator control cables should terminate in a grounded metal box. Each control cable conductor should have a GE V56ZA2 MOV connected across it to case ground. If the control cable is not shielded, the same type of protective devices can be installed within the rotator. If shielded cable is used, the devices inside the rotator would not be needed. Check the voltage on the control cable and select a value of MOV rated for a safe operating voltage above the control voltage.

MOVs are available from some of the suppliers listed in Chapter 35. Some may be available from local electrical supply houses. Table 1 lists some transient-protective devices of interest to radio amateurs.

GROUND

An effective ground system is a must for every amateur station. The mission of the ground system is twofold. First, it reduces the possibility of electrical shock if something in a piece of equipment should fail and the chassis or cabinet becomes "hot." If connected properly, three-wire

electrical systems ground the chassis, but the majority of amateur equipment still uses the ungrounded two-wire system. A ground system to prevent shock hazards is generally referred to as "dc ground."

The second job the ground system must perform is to provide a low-impedance path to ground for any stray RF current inside the station. Stray RF can cause equipment to malfunction and contributes to RFI problems. This low-impedance path is usually called "RF ground." In most stations, dc ground and RF ground are provided by the same system.

The first step in building a ground system is to bond together the chassis of all equipment in your station. Ordinary hookup wire will do for a dc ground, but for a good RF ground you need a low-impedance conductor. Copper strap, sold as "flashing copper," is excellent for this application, but it may be hard to find. Braid from coaxial cable is a popular choice; it is readily available, a low-impedance conductor and flexible.

Grounding straps can be run from equipment chassis to equipment chassis, but a more convenient approach is illustrated in Fig. 6. In this installation, a ½-inch-diameter copper water pipe runs the entire length of the operating bench. A thick braid (from discarded RG-8 cable) runs from each piece of equipment to a clamp on the pipe. Copper water pipe is inexpensive and available at most hardware stores and home centers. Alternatively, a strip of flashing copper may be run along the rear of the operating bench.

After the equipment is bonded to a common ground bus, the ground bus must be wired to a good earth ground. This run should be made with a heavy conductor (braid is a good choice again) and should be as short and direct as possible. The earth ground usually takes one of two forms.

In most cases, the best approach is to drive one or more ground rods into the earth at the point where the conductor from the station ground bus leaves the house. The best ground rods to use are those available from an electrical supply house. These rods are 6 to 8 feet long and are made from steel with a heavy copper plating. Do

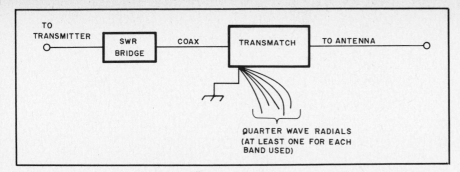

Fig. 7 — Here is an alternative to earth ground if the station is located far from the ground point and RF in the station is a problem. Install at least one ¼-wavelength radial for each band used.

not depend on shorter, thinly plated rods sold by some home electronics suppliers. These rods begin to rust almost immediately after they are driven into the soil, and they become worthless within a short time. Good ground rods, while more expensive initially, offer long-term protection. If your soil is soft and contains few rocks, an acceptable alternative to "genuine" ground rods is ½-inch-diameter copper water pipe. A 6- to 8-foot length of this material offers a good ground, but it may bend while being driven into the earth.

Once the ground rod is installed, clamp the conductor from the station ground bus to it with a clamp that can be tightened securely and will not rust. Copper-plated clamps made especially for this purpose are available from electrical supply houses, but a stainless-steel hose clamp will work too. Alternatively, drill several holes through the pipe and bolt the conductor in place. If a torch is available, solder the connection.

Another popular station ground is the cold water system in the building. To take advantage of this ready-made ground system, run a low-impedance conductor from the station ground bus to a convenient cold water pipe, preferably somewhere near the point where the main water supply enters the house. Avoid hot water pipes; they do not run directly into the earth. The advent of PVC (plastic) plumbing makes it mandatory to inspect the cold water system from your intended ground connection to the main inlet. PVC is an excellent insulator, so any PVC pipe or fittings rule out your cold water system for use as a station ground.

For some installations, especially those located above the first floor, a conventional ground system such as that just described will make a fine dc ground but will not provide the necessary low-impedance path to ground for RF. The length of the conductor between the ground bus and the ultimate ground point becomes a problem. For example, the ground wire may be about ¼ wavelength (or an odd multiple of ¼ wavelength) long on some amateur band. A ¼-wavelength wire acts as an impedance inverter from one end to the other. Since the grounded end is at a very low im-

pedance, the equipment end will be at a high impedance. The likely result is RF hot spots around the station while the transmitter is in operation. In this case, this ground system may be worse than having no ground at all.

An alternative RF ground system is shown in Fig. 7. No attempt is made to ground the station to an earth ground. Instead, the station ground bus is connected to a system of ¼-wavelength radials. Install at least one radial for each band used. Try this system if you have RF in the shack even though your equipment is connected to a conventional ground. Keep in mind, however, that while the radial ground system offers an RF ground, it will not act as a dc ground to protect you from shock hazards.

Ground Noise

Noise in ground systems can affect our sensitive radio equipment. It is usually related to one of three problems:
1) Insufficient ground conductor size
2) Loose ground connections
3) Ground loops

These matters are treated in precise scientific research equipment and certain industrial instruments by attention to certain rules. The ground conductor should be at least as large as the largest conductor in the primary power circuit. Ground conductors should provide a solid connection to both ground and to the equipment being grounded. Liberal use of lockwashers and star washers is highly recommended. A loose ground connection is a tremendous source of noise, particularly in a sensitive receiving system.

Ground loops should be avoided at all costs. A short discussion of what a ground loop is and how to avoid them may lead you down the proper path. A ground loop is formed when more than one ground current is flowing in a single conductor. This commonly occurs when grounds are "daisy-chained" (series linked). The correct way to ground equipment is to bring all ground conductors out radially from a common point to either a good driven earth ground or a cold-water system.

Ground noise can affect transmitted and received signals. With the low audio levels

required to drive amateur transmitters, and the ever-increasing sensitivity of our receivers, correct grounding is critical.

STATION POWER

Amateur Radio stations generally require a 117-V ac power source. The 117-V ac is then converted to the proper ac or dc levels required for the station equipment. Power supply theory is covered in Chapter 6. If your station is located in a room with electrical outlets, you're in luck. If your station is located in the basement, an attic, or another area without a convenient 117-V source, you will have to run a line to your operating position.

Stations with high-power amplifiers should have a 234-V ac power source in addition to the 117-V supply. Some amplifiers may be powered from 117 V, but they require current levels at that voltage that are near, or in excess of, the limits of standard house wiring. For safety, and for the best possible voltage regulation in the equipment, it is advisable to install a separate 234-V line with an appropriate current rating if you use an amplifier.

The usual line running to baseboard outlets is rated at 15 amperes. This may or may not be enough current to power your station. To determine how much current your station requires, check the ratings for each piece of gear. Usually, the manufacturer will specify the required current at 117-V; if the power consumption is rated in watts, divide that rating by 117 V to get amperes. If the total current required for your station is near 15 amperes, you need to think about running another line or upgrading the present one. Keep in mind that other rooms may be powered from the same branch of the electrical system, so the power consumption of any equipment connected to other outlets on the branch must be taken into account. Whenever possible, power your station from a separate, heavy-duty line run directly to the distribution panel.

In modern residential systems, three wires are brought in from the outside to a distribution panel (fuse panel, or circuit-breaker panel). In the three-wire system, the third wire is neutral, which is grounded. The voltage between the other two wires is usually 234, while half of this voltage appears between each wire and neutral. In a three-wire system, the three wires should be brought into the station so that the load can be distributed to keep the line balanced. The voltage across a fixed load on one side of the line will increase as the load current on the other side is increased. The rate of the increase will depend on the resistance of the neutral wire. If the resistance of the neutral wire is low, the increase will be correspondingly small. When the currents in the two circuits are balanced, no current flows in the neutral wire, and the system is operating at maximum efficiency.

If you decide to install a separate heavy-duty 117-V line or a 234-V line, consult the

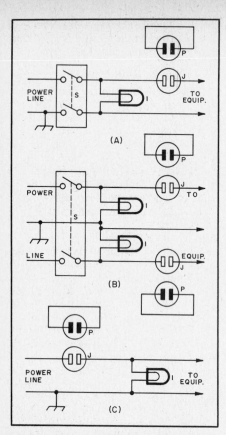

Fig. 8 — Reliable arrangements for cutting off all power to the station. S is an enclosed double-pole, double-throw power switch of appropriate ratings. J is a standard ac outlet. P is a shorted plug to fit outlet J. I is a red warning lamp. The system at A is for two-wire, 117-V lines; B is for three-wire 234-V systems; C is a simplified arrangement for low-power stations.

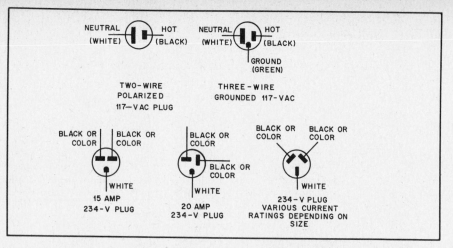

Fig. 9 — Proper wiring for power plugs found in amateur stations. Follow correct wiring procedures to minimize shock hazards. On 117-V plugs, the screw terminal or connection for one prong is brass and the other is plated ("white"). Connect the black wire to the brass prong and the white wire to the white prong. Always use 234-V connectors for 234-V systems; otherwise, you may destroy a piece of equipment by applying too much voltage.

local power company for local requirements. In some areas, this work must be performed by a licensed electrician. You may need a special building permit for the work, and even if you are allowed to do the work yourself, you might have to get a licensed electrician to inspect the job. Your best bet is to go through the system and get the necessary permits and inspections. If something should happen, such as a fire, you might run into insurance or other problems if you didn't follow the required steps, even if the accident had nothing to do with your wiring job.

If you decide to do the job yourself, study the applicable sections of the National Electrical Code and any local building codes. This information should be available from your local building inspector. Books are available at libraries or from outlets selling electrical supplies that will tell you everything you need to know to install separate ac lines for your station. Use only good-quality wire and other materials with ratings as specified in the building codes. If you have any doubts about doing the work yourself, get a licensed electrician to do the installation.

Safety Precautions

All power supplies in an installation

should be fed through a single main power-line switch so that all power may be cut off quickly, either before working on the equipment, or in case of an accident. Spring-operated switches or relays are not sufficiently reliable for this important service. Foolproof devices for cutting off all power to the transmitter and other equipment are shown in Fig. 8. The arrangements shown in Figs. 8A and B are similar circuits for two-wire (117-volt) and three-wire (234-volt) systems. S is an enclosed double-throw switch of the sort usually used as the entrance switch in house installations. J is a standard ac outlet and P a shorted plug to fit the outlet. The switch should be located prominently in plain sight, and members of the household should be instructed in its location and use. I is a red lamp located alongside the switch. Its purpose is not so much to serve as a warning that the power is on as it is to help in identifying and quickly locating the switch should it become necessary for someone else to cut the power off in an emergency.

The outlet J should be placed in some corner out of sight where it will not be a temptation for children or others to play with. The shorting plug can be removed to open the power circuit if there are others around who might inadvertently throw the switch while the operator is working on the rig. If the operator takes the plug when the station is not in use, it will prevent unauthorized persons from turning on the power. Of utmost importance is the fact that the outlet J must be placed in the ungrounded side of the line.

Those who are operating low power and feel that the expense or complication of the switch isn't warranted can use the shorted-plug idea as the main power switch. In this case, the outlet should be located prominently and identified by a signal light, as shown in Fig. 8C.

The test bench should be fed through the

main power switch, or a similar arrangement at the bench, if the bench is located remotely from the transmitter.

Three-Wire 117-V Power Cords

To meet the requirements of state and national codes, electrical tools, appliances and many items of electronic equipment now being manufactured to operate from the 117-volt line must be equipped with a three-conductor power cord. Two of the conductors carry power to the device in the usual fashion, while the third conductor is connected to the case or frame. See Fig. 9.

When plugged into a properly wired mating receptacle, the three-contact polarized plug connects this third conductor to an earth ground, thereby grounding the chassis or frame of the appliance and preventing the possibility of electrical shock to the user. All commercially manufactured items of electronic test equipment and most ac-operated amateur equipment are being supplied with these three-wire cords. Adapters are available for use where older electrical installations do not have mating receptacles. For proper grounding, the lug of the green wire protruding from the adapter must be attached underneath the screw securing the cover plate of the outlet box where connection is made, and the outlet box itself must be grounded.

Multiple Outlets

Once you have the right line or lines running from the distribution panel to your station, you will need to bring the power to each piece of equipment. All of the accessories, such as keyer, rotator control box and clock, as well as the station transceiver, need a separate outlet. Commercial outlet strips, such as that pictured in Fig. 10, will provide a grounded outlet for each piece of equipment. If the commercially available outlet strips do not suit your setup, you can make your own.

Fig. 11 shows one way to make your own

Fig. 10 — Commercially made power strips are available with different numbers of outlets. Some feature built-in fuses or circuit breakers. Three-wire strips are recommended for use in amateur stations.

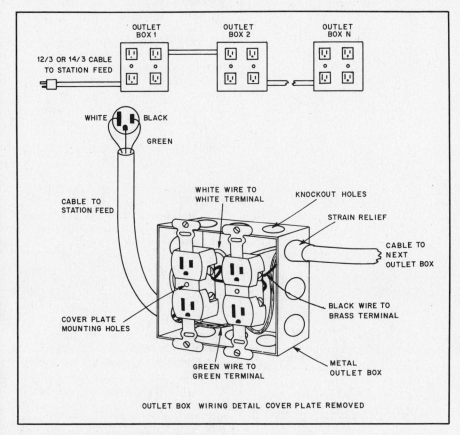

Fig. 11 — It is possible to make a custom outlet strip for your station. Space the outlets at convenient intervals for your operating position. See text for details.

outlet strip. The materials to make this type of strip are available at most hardware stores or home centers. Use wire with sufficient current rating. See Chapter 35 for a table listing the current-carrying ability of different wire sizes. No. 14 wire is rated to handle 17 amperes continuously when bundled in cables, so it is adequate for most installations. The best kind of cable to buy for this job is stranded 14/3 (three conductors, each no. 14) with a rubberized outer jacket. This type of cable is easy to work with and is flexible.

First, measure the length of cable needed to run from your operating position to the station power feed. Include enough extra so that you can run the cable neatly along the floor if necessary. Refer to Fig. 9 to see which color wire goes to which plug terminal. Locate the first outlet box near the first group of equipment. Wire the outlets in parallel, paying close attention to where the various color conductors should go. Wire any additional outlet boxes in parallel with the first. Depending on the amount of equipment you have, you may choose either two- or four-outlet boxes. The individual outlet boxes may be left loose, attached to a strip of wood of the appropriate dimensions, or bolted directly to the operating bench.

STATION LAYOUT

Station layout is largely a matter of personal taste. It will depend to a large extent on the amount of space available, the amount of gear involved and the types of operating activities engaged in. Although no two stations are exactly alike, there are some underlying considerations you should be aware of. These design considerations will help you build a convenient, easy-to-use station around your needs and means.

The Operating Table

The operating table may be an office desk, a kitchen table or a custom-made bench; what you use will depend on the available space and materials on hand. The two most important considerations are the size of the top and the height of the top above the floor. The top of most commercially manufactured desks is about 29 inches above the floor. This is a comfortable height for most adults. If the top is much lower, your knees may not clear the underside, and you will be sitting in an uncomfortable pitched-forward position. If the table top is much higher, you may not be able to reach the equipment, and writing will be awkward.

The dimensions of the top of the operating table are a prime consideration. Although the top need be only deep enough to hold the equipment (a foot or so), a much wider top greatly enhances the comfort and convenience of your operation. Leaving room for a power outlet strip at the back of the table, a ground bus, the equipment and room for writing, the depth of a comfortable operating table grows to at least 36 inches, and preferably more. For an ideal operating table, 42 to 48 inches is about the right depth. See Fig. 12. The length of the top will depend on how much equipment is used.

An office desk makes a good operating table. The top is usually 30 to 36 inches deep by 60 inches wide. The drawers can be put to good use for log books, headphones and the like.

A Homemade Operating Table

An unfinished flush-type door, measuring approximately 30 × 84 inches, makes an acceptable operating bench if the power outlet strip and ground bus are not mounted on the table top. A depth of 30 inches is just enough for the equipment and writing area. The door may be mounted on two 2-drawer file cabinets (see Fig. 13), making an easy bench with drawer space.

A flush door may also be made into an attractive homemade table. Fig. 14 shows construction details for building a simple but sturdy homemade operating table. Side members for supporting the top are made from ordinary 2 × 3 lumber which comes in standard lengths of 8 and 12 feet. Making legs that are good and solid isn't the easiest job in the world, and if the table isn't solid it will be a poor operating desk. You can get around this by using the construction technique shown in Fig. 14A and D. The 2 × 3s are formed into a square or rectangular format, as shown at A. Each

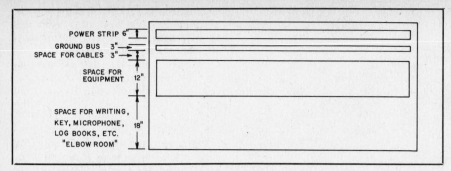

Fig. 12 — The ideal operating table is about 42 inches deep and as wide as you need to make it to fit your available space. This arrangement leaves plenty of room for equipment and cables, with lots of elbow room for comfort.

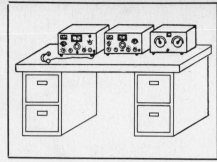

Fig. 13 — Two file cabinets and a flush-type door make an effective operating bench.

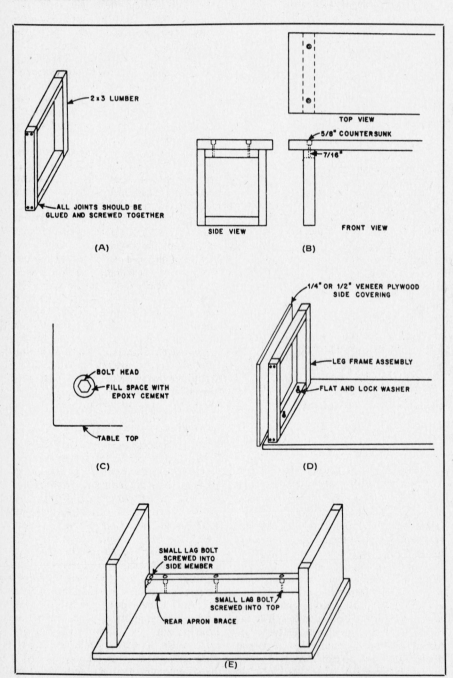

Fig. 14 — A flush door may also be used with a simple frame to make a furniture-quality operating table. (A) 2 × 3 lumber is used to form the side supports. Each joint should be glued and then screwed together. (B) Front, top and side views detailing the method of fastening the side supports to the top. (C) Completely fill the space around the bolt head with epoxy. (D) Table assembly and side covering information. (E) The rear-apron brace should be attached to both side members at the top of the desk.

joint should be glued and screwed together. Nails will not provide enough sturdiness.

Once the glue holding the 2 × 3s together has completely dried (several hours is sufficient with most types of glue), these assemblies can be drilled and made ready for mounting. If you are planning to cover the top and edges of the door with a plastic laminate, such as Formica, holes should be drilled through the door and countersunk as shown in Fig. 14B. After the holes in the door have been drilled, place the pieces in the final position. Drill through the 2 × 3s at the four bolt locations. After these holes have been drilled, the four bolts should be inserted in the holes, through the 2 × 3s and secured by flat washers and nuts. With the bolts securely in place, fill the gaps between the bolt heads and wood with epoxy cement. See Fig. 14C.

Once the epoxy has hardened (it's a good idea to let it set overnight), the nuts can be removed and the three sections taken apart. If you are planning on having the top laminated, now is the time. Most kitchen cabinet shops can do this for you, as they have all of the necessary tools. Pick a color that compliments the rest of the shack!

The table is now ready for assembly. Place the table top on its good surface with the bolts sticking up. Now the side members can be slipped over the bolts and held in place with flat washers, lock washers and nuts. Make sure to tighten the nuts securely.

At this point, the side coverings can be attached to the leg assemblies. Ordinary ¼-inch plywood or Masonite can be used if you plan on painting this portion of the table. An alternative would be to use oak-, walnut-, birch- or teak-veneered plywood. Although this material is a bit more expensive than ordinary plywood, it yields a quality furniture-like appearance. The wood can be stained and coated with clear lacquer to preserve the finish. Another alternative would be to use pieces of wood paneling for the side coverings. In any case, the coverings on the outside of the table should be glued and tacked to the frame. The inner side coverings should be tacked only to the frame so they can be removed and the table taken apart should the need arise. The front face of the side members

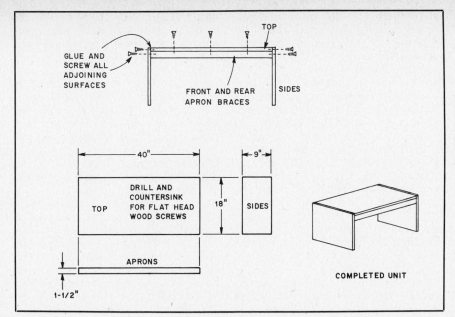

Fig. 15 — A simple but very strong equipment shelf can be built from ¾-inch plywood. See text for details.

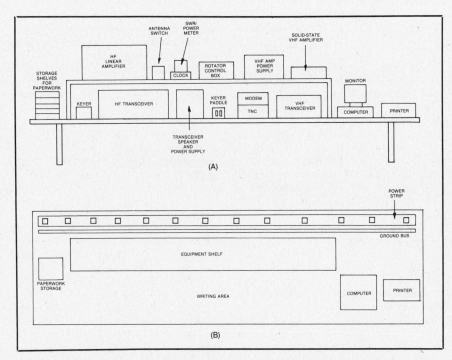

Fig. 16 — Example station layout as seen from the front (A) and the top (B). The equipment is spaced far enough apart that air circulates on all sides of each cabinet. Equipment that is adjusted frequently sits on the table top, while equipment requiring infrequent adjustment is perched on a shelf. All equipment is positioned so that the operator does not have to move the chair to reach anything at the operating position.

should be covered with a wide strip of veneer or plastic laminate of the same color used for the top of the table.

A rear-apron brace can be made from a 2 × 4. By painting this brace flat black it will not detract from the appearance of the table even if a small amount of its surface shows. The power outlet strip and ground bus can be attached to the rear-apron brace.

If you have some basic woodworking tools and knowledge, you can construct a larger top for your operating table and mount it on the type of framework just described. The homemade top is fashioned from a single 4 × 8-foot sheet of plywood or particle board. Particle board is less expensive than plywood. Its perfectly smooth, even surface makes a good base for plastic laminates, but it is substantially heavier than plywood and not as strong. Depending on your needs, you may use the entire sheet for your table top, or you may trim it to size. If you make the top at least 42 inches deep, you will have plenty of room for setting up the equipment.

The type of finish you select for your top will depend on the material used. Plastic laminates, while expensive, are attractive and durable; they also make an excellent writing surface and are easy to keep clean. Plywood may be stained and finished with a tough varnish, or it may be painted. If you paint or varnish particle board or plywood, you should use a piece of glass or other hard material as a writing surface.

Stacking Equipment

No matter how large your operating table is, some vertical stacking of equipment is necessary if you are to reach everything from your chair. Piling the equipment cabinets one on top of another is not a particularly good idea because most amateur equipment needs air flowing around it for cooling. Construction of a shelf like that shown in Fig. 15 will improve equipment layout in most installations.

The exact shelf dimensions will depend on the size of your operating table and the size of your equipment. This shelf is completely open both front and back and has no center support; access to equipment and interconnecting cables is excellent. The shelf should be made from strong material, such as ¾-inch plywood. A shelf constructed like that shown in Fig. 15 will support several hundred pounds without bowing in the center. The top, sides and apron braces are all made from ¾-inch plywood. The top and side dimensions can be varied to match your station requirements. Glue and screw all adjoining surfaces for maximum strength. Finish the shelf to match the operating table. If you build it right, you should be able to stand on the shelf without it bending.

Arranging the Equipment

When you have acquired the operating table and shelving for your station, the next task is arranging the equipment in a convenient, orderly manner. The first step is to provide power outlets and a good ground as described in a previous section. Be conservative in estimating the number of power outlets for your installation; radio equipment has a habit of multiplying with time, so plan for the future at the outset.

Fig. 16 illustrates a sample station layout. The rear of the operating table is spaced about 1½ feet from the wall to allow easy access to the rear of the equipment. This installation incorporates two separate operating positions, one for HF and one for VHF. When the operator is seated at the HF operating position, the keyer and transceiver controls are within easy reach. The keyer, keyer paddle and transceiver are the most-often adjusted pieces of equipment in the station. The speaker is positioned right in front of the operator for the best possible reception.

Fig. 17 — This station, owned by K5RC, was designed with contest operation in mind. Note the high-efficiency layout; separate rigs for each band makes for quick band changes. Control switches, keyer and rig are all within easy reach.

Fig. 18 — WA3AXV is interested primarily in VHF work. Each band requires a separate transverter and amplifier. This equipment, all of which is rarely touched once adjusted, is located above the operating table. The transceiver used as a tuneable IF and the accessories are within easy reach.

Fig. 19 — K1JX has an HF operating position and a VHF operating position arranged on a corner table.

Accessory equipment not often adjusted, including the amplifier, antenna switch and rotator control box, is located on the shelf above the transceiver. The SWR/power meter and clock, often consulted but rarely touched, are located where the operator can view them without head movement. All HF-related equipment can be reached without moving the chair.

This layout assumes that the operator is right-handed. The keyer paddle is operated with the right hand, and the keyer speed

and transceiver controls are operated with the left hand. This setup allows the operator to write or send with the right hand without having to cross hands to adjust the controls. If the operator is left-handed, some repositioning of equipment is necessary, but the idea is the same. For best results during CW operation, the paddle should be weighted to keep it from "walking" across the table. It should be oriented such that the operator's entire arm from wrist to elbow rests on the table top to prevent fatigue.

Some operators prefer to place the station transceiver on the shelf to leave the table top clear for writing. This arrangement leads to fatigue from having an unsupported arm in the air most of the time. If you rest your elbows on the table top, they will quickly become sore. If you rarely operate for prolonged periods, however, you may not be inconvenienced by having the transceiver on the shelf. The real secret to having a clear table top for logging, etc., is to make the operating table deep enough that your entire arm from elbow to wrist rests on the table with the front panels of the equipment at your fingertips. This leaves plenty of room for paperwork, even with a microphone and keyer paddle on the table.

The VHF operating position in this station is similar to the HF position. The amplifier and power supply are located on the shelf. The station triband beam and VHF beam are on the same tower, so the rotator control box is located where it can be seen and reached from both operating positions. This operator is active on packet radio on a local VHF repeater, so the computer, printer, terminal node controller and modem are all clustered within easy reach of the VHF transceiver.

This sample layout is intended to give you ideas for designing your own station. Study the photos of station layouts presented here, in other chapters of this Handbook and in QST. Visit the shacks of amateur friends to view their ideas. Station layout is always changing as you acquire new gear, dispose of old gear, change operating habits and interests, or become active on different bands. Configure the station to suit your interests, and keep thinking of ways to refine the layout. Figs. 17 through 19 show station arrangements tailored for specific purposes.

INTERCONNECTING YOUR EQUIPMENT

After you've chosen the equipment for your station and arrived at a layout, you'll have to interconnect everything. The complexity of the wiring job depends on the setup; but large or small, HF or VHF, high or low power, every shack needs some cabling between equipment. Commercially manufactured cables with connectors are available for some applications. In most instances, however, there will not be a readily available cable, so you'll have to make your

own. Making custom cables is not difficult. Most equipment manuals tell you the pinouts for microphone, key, accessory and control connectors. Many manufacturers include connectors, but they leave the cable construction up to you so you can make up the right lengths for your installation.

You should use good wire, cables and connectors in your shack. Take the time to make good mechanical and electrical connections on your cable ends. Poorly made cables are a source of trouble. Connections may break, requiring maintenance. Improperly assembled transmission-line connectors can cause RFI. More important, cables made from inferior materials or assembled sloppily are a safety hazard. Wire too small for the current load can heat up, presenting a fire hazard. Connections lacking mechanical integrity can work free, possibly shorting a power supply to ground or applying voltage where not expected. Wire with poor or underrated insulation is a shock hazard.

Wherever possible, neatly bundle your wires and cables. There are many ways to do this. Some amateurs use masking tape or cloth adhesive tape. Others use paper labels attached to the cables with string. See Fig. 20. Whatever method you use, proper labeling makes disconnecting, reconnecting or rearranging equipment much easier. Good labels will also help you avoid inadvertently putting voltage where it doesn't belong.

Types of Wire and Cable

The type of wire or cable to use depends on the job at hand. If the wire is to carry any significant current, check the wire table in Chapter 35 for the ratings of different sizes. For safety, always use wire heavy enough for the task. Also, check the insulation rating. Almost any insulation will do for low voltages, but use the right wire for higher voltages; anything you can do to minimize shock hazards around your shack is important for your safety.

Use the right size coaxial cable for connecting transmitters, transceivers, antenna switches, SWR bridges, Transmatches and the like. For example, RG-58/U cable is fine between your transceiver and SWR meter, but it is too small to use between your legal-limit amplifier and Transmatch. Chapter 16 contains information on the characteristics of popular coaxial cables.

Hookup wire may be stranded or solid. Generally, stranded wire is the better choice for most applications because solid wire will break under repeated flexing.

Many interconnections should be made from shielded wire to reduce the chances of RF getting into equipment. RG-174/U subminiature coaxial cable is a good choice for control, audio and some power leads. It is shielded, small and very flexible. Fig. 21 shows how to prepare coaxial cable for attachment to a connector.

Connectors

Connectors found in amateur stations

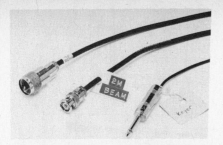

Fig. 20 — Labels on the cables make it much easier to rearrange things in the station. Labeling ideas include masking tape, cardboard labels attached with string and labels attached to fasteners found on plastic bags (such as bread bags).

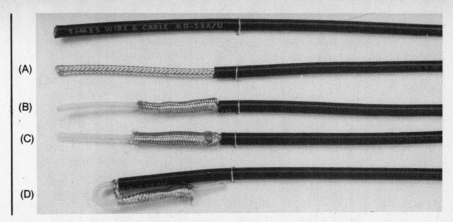

Fig. 21 — Preparing coaxial cable for use with a connector involves a few easy steps. A — Remove the outer insulation with a sharp knife or wire stripper. If you nick the braid, start over. B — Push the braid accordian-fashion against the outer jacket. C — Spread the shield strands at the point where the outer insulation ends. D — Fish the center conductor through the opening in the braid. Smooth the braid. Now strip the center conductor insulation back far enough for the type of connector you're using and tin both center conductor and shield. Be careful not to use too much solder, making the conductors inflexible. A pair of pliers used as a heat sink will help. The outer jacket removed in step A makes a perfect insulator for the braid if necessary.

run the gamut from single-conductor ring terminals for use on terminal blocks, to four-conductor microphone connectors, to 50-conductor ribbon cable connectors. Although the number of connectors in existence is mind boggling, most manufacturers of amateur gear use a few standard types. If you are involved in any group activities, such as public service and emergency preparedness work, check to see what kinds of connectors others in the group use. If group members standardize whenever possible, the equipment will be more interchangeable — a definite plus in an emergency.

Audio, Power and Control Connectors

The simplest form of connector is found on terminal blocks. Although it is possible to strip the insulation from wire and wrap it around the screw, this method leaves something to be desired. Sometimes it is difficult to hold the wire in place when tightening the screw; this is especially troublesome when the wire is large compared to the screw size, which is frequently the case. It's not always easy to get a good connection. Changing the wire on a terminal block can be a headache with the strip-and-wrap method. Also, the conductors in stranded wire tend to separate under compression, making an unsightly mess.

Fortunately, the terminal lugs illustrated in Fig. 22 eliminate these problems. These lugs may be soldered or crimped (with the right tool), or both. A good crimping tool compresses the top and bottom while keeping the sides from expanding, making a secure connection. Terminal lugs are available in different sizes; you'll get the best mechanical and electrical connection if you use the right size for your wire. Lugs are available with or without insulation. The uninsulated kind usually cost less, and you can use heat-shrink tubing after installation on the wire if insulation is necessary. The fully closed ring terminals will hold even if the screw works loose, but they require you to remove the screw completely from the block during installation. The U-shaped spade lugs are easier to install, requiring you only to remove the screw partially, but they may fall off if the

Fig. 22 — The wires on one side of this terminal block have connectors; the others do not. The connectors make it possible to secure different wire sizes to the strip and also make it much easier to change things around.

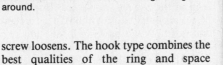

screw loosens. The hook type combines the best qualities of the ring and space terminals.

Some common multi-pin connectors are illustrated in Fig. 23. The connector shown in Fig. 23A, often called a "Cinch-Jones plug," is frequently used for cabling between equipment and outboard power supplies. Supplying from two to eight or more conductors, these connectors are keyed to go together only one way. They offer a good electrical and mechanical connection, and the pins are large enough to handle high current. If your cable is too small for the strain relief, build up the outer jacket with a few layers of tape until the strain relief clamps down securely. Properly used, the strain relief will keep your wires from breaking away from the connector because of excessive flexing of the soldered joint or from a sudden tug on the cable.

The plug shown in Fig. 23B is usually called a "molex" connector. This device consists of an insulated outer shell, resembling a honeycomb, which houses many individual male or female "fingers." Each finger is individually soldered or crimped onto a conductor of the cable and

(C)

Fig. 23 — The plug shown at A is often used to connect equipment to remote power supplies. The multipin connector at B is used for control, signal and power lines. The DIN plug at C offers shielding and is often used for connecting accessories to transceivers.

then inserted into the shell, locking into place. These connectors are available for different wire sizes and are used on many brands of amateur gear for power and accessory interconnections.

Fig. 23C shows a DIN connector. Commonly available with five, seven or eight pins, these connectors are a European standard that have found favor with amateur equipment manufacturers the world over. They are generally used for accessory connectors.

The connector shown in Fig. 24 is a ¼-inch phone plug, often just called a phone plug. Usually used on amateur equipment for headphones and Morse key lines, phone plugs have two or three circuits. They are available with plastic and metal bodies. Besides offering a shielded connection, the metal-jacketed version is more durable — it won't break when stepped on!

Fig. 24 also shows a 1/8-inch phone plug, often called a miniature phone plug. Used for earphone, external speaker, key and control lines, mini-phone plugs are available in shielded and unshielded forms. An even smaller phone plug, the 1/16-inch, or subminiature, version is not common on amateur gear.

The phono, or RCA, plug illustrated in Fig. 25 is a favorite among amateurs. Phono plugs are found on amateur equipment for everything from amplifier relay control lines, to low-voltage power lines, to low-level RF lines, to antenna lines. Several styles of phono plugs are available, but the best choice for general use is the shielded type with a screw-on metal body. These plugs are much easier to grip when plugging and unplugging; they are shielded, and more durable.

Microphone connectors seem to change with the wind, and the two most popular types used today are shown in Fig. 26. At one time, microphone connectors used three conductors — one for audio, one for the PTT switch, and one for ground. Today's microphone connectors offer four or eight conductors. In many rigs, the PTT switch does not short to a common ground, but rather closes two contacts independent of ground. Some manufacturers offer a limited amount of frequency control — UP and DOWN — from switches on the mic, while others use a preamplifier in the microphone that derives its power from the rig; hence the move to eight-conductor mic connectors. When hooking up a mic to your rig, consult the manual. You should be aware of the manufacturer's recommended wiring for best results.

Some amateurs own more than one rig but have only one mic. Rather than lopping off the connector at each rig change, you can make adapters with short pieces of cable that route the signals from one mic to the right pins for several rigs. Check your local dealer; you may be surprised at the variety of male and

female mic connectors available.

RF Connectors

There are many different types of RF connectors for coaxial cable on the market, but the three most common for amateur use are the UHF, Type N and BNC families. The type of connector used for a specific job depends on the size of the cable, the frequency of operation and the power levels involved.

The UHF connector is found on most HF and some VHF equipment. It is the only connector many hams will ever see on coaxial cable. PL-259 is another name for the UHF male, and the female is also known as the SO-239. These connectors are rated for full legal amateur power at HF. They are poor for UHF work because they do not present a constant impedance, so the UHF label is a misnomer. PL-259 connectors are designed to fit RG-8 and RG-11 size cable (0.405-in OD). Adapters are available for use with RG-58 and RG-59 size cable. UHF connectors are not weatherproof.

Fig. 27 shows how to install the solder type of PL-259 on RG-8 type cable. Proper preparation of the cable end is the key to success. Follow these simple steps. Measure back about ¾-inch from the cable end and slightly score the outer jacket around its circumference. With a sharp knife, cut through the outer jacket, through the braid, and through the dielectric, right down to the center conductor. Be careful not to score the center conductor. Cutting through all outer layers at once keeps the braid from separating. Pull the severed outer jacket, braid and dielectric off the end of the cable as one piece. Inspect the area around the cut, looking for any strands of braid hanging loose, and snip them off. There won't be any if your knife was sharp enough. Next, score the outer jacket about 5/16-in back from the first cut. Cut through the jacket lightly; do not score the braid. This step takes practice. If you score the braid, start again. Remove the outer jacket.

Tin the exposed braid and center conductor, but apply the solder sparingly and avoid melting the dielectric. Slide the coupling ring onto the cable. Screw the connector body onto the cable. If you prepared the cable to the right dimensions, the center conductor will protrude through the center pin, the braid will show through the solder holes, and the body will actually thread onto the outer cable jacket.

Solder the braid through the solder holes. Solder through all four holes; poor connection to the braid is the most common form of PL-259 failure. A good connection between connector and braid is just as important as that between center conductor and connector. Use a large soldering iron for this job. With practice, you'll learn how much heat to use. If you use too little heat, the solder will bead up, not really flowing onto the connector body. If you use too

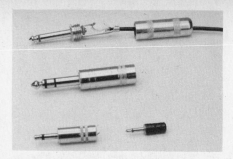

Fig. 24 — The phone-plug family. The ¼-inch type is often used for headphone and key connections on amateur equipment. The three-circuit version is used with stereo headphones. The mini phone plug is commonly used for connecting external speakers to receivers and transceivers. A submini-phone plug is shown in the foreground for comparison. The shielded style with metal barrel is more durable than the plastic style.

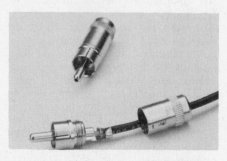

Fig. 25 — Phono plugs have countless uses around the shack. They are small and shielded; the type with the metal body is easy to grip. Be careful not to use too much heat when soldering the ground (outer) conductor — you may melt the insulation.

Fig. 26 — The four-pin mic connector is common on modern transmitters and receivers. More elaborate rigs use the eight-pin type. The extra conductors may be used for switches to remotely control the frequency or to power a preamplifier built into the mic case.

much heat, the dielectric will melt, letting the braid and center conductor touch. Most PL-259s are nickel plated, but there are available slightly more expensive connectors with silver-plated bodies. These are much easier to solder.

Solder the center conductor to the center pin. The solder should flow on the inside, not the outside of the center pin. If you wait until the connector body cools off from soldering the braid, you'll have less trouble with the dielectric melting. Trim the center conductor to be even with the end

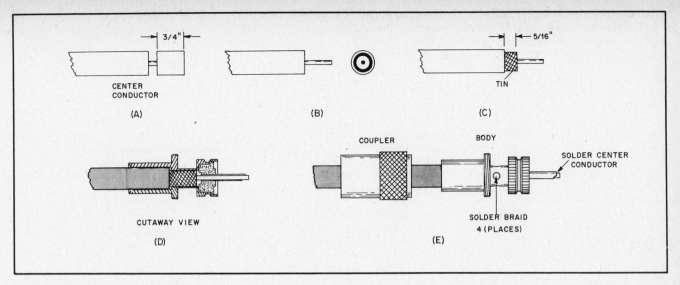

Fig. 27 — The PL-259, or UHF, connector is almost universal for amateur HF work and is-popular for equipment operating in the VHF range. Steps A through E are described in detail in the text.

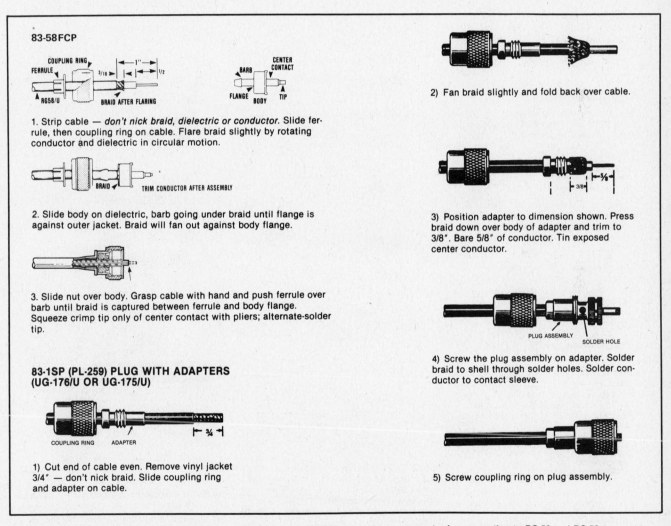

83-58FCP

1. Strip cable — *don't nick braid, dielectric or conductor*. Slide ferrule, then coupling ring on cable. Flare braid slightly by rotating conductor and dielectric in circular motion.

2. Slide body on dielectric, barb going under braid until flange is against outer jacket. Braid will fan out against body flange.

3. Slide nut over body. Grasp cable with hand and push ferrule over barb until braid is captured between ferrule and body flange. Squeeze crimp tip only of center contact with pliers; alternate-solder tip.

83-1SP (PL-259) PLUG WITH ADAPTERS (UG-176/U OR UG-175/U)

1) Cut end of cable even. Remove vinyl jacket 3/4″ — don't nick braid. Slide coupling ring and adapter on cable.

2) Fan braid slightly and fold back over cable.

3) Position adapter to dimension shown. Press braid down over body of adapter and trim to 3/8″. Bare 5/8″ of conductor. Tin exposed center conductor.

4) Screw the plug assembly on adapter. Solder braid to shell through solder holes. Solder conductor to contact sleeve.

5) Screw coupling ring on plug assembly.

Fig. 28 — Crimp-on connectors and adapters for use with standard PL-259 connectors are popular for connecting to RG-58 and RG-59 type cable. This material courtesy of Amphenol Electronic Components, RF Division, Bunker Ramo Corp.

of the center pin. Use a small file to round the end, removing any solder that built up on the outer surface of the center pin. Use a sharp knife, very fine sandpaper or steel wool to remove any solder flux from the outer surface of the center pin. Screw the coupling onto the body, and you're finished.

Fig. 28 shows two options available if you want to use RG-58 or RG-59 size cable with PL-259 connectors. The crimp-on connectors manufactured specially for the smaller cable work very well if installed correctly. The alternative method involves using adapters for the smaller cable with standard RG-8 size PL-259s. Prepare the cable as shown. Once the braid is prepared, screw the adapter into the PL-259 shell and finish the job as you would a PL-259 on RG-8 cable.

The BNC connectors illustrated in Fig. 29 are popular for low power levels at VHF and UHF. They accept RG-58 and RG-59 cable, and are available for cable mounting in both male and female versions. There are several different styles available, so be sure to use the dimensions for the type you have. Follow the installation instructions carefully. If you prepare the cable to the wrong dimensions, the center pin will not seat properly with connectors of the opposite sex. Sharp scissors are a big help for trimming the braid evenly. Properly assembled, BNC connectors are weatherproof.

The Type N connector, illustrated in Fig. 30, is a must for high-power VHF and UHF operation. N connectors are available in male and female versions for cable mounting and are designed for RG-8 size cable. Unlike UHF connectors, they are designed to maintain a constant impedance at cable joints. Like BNC connectors, it is important to prepare the cable to the right dimensions. The center pin must be positioned correctly to mate with the center pin of connectors of the opposite sex. Use the right dimensions for the connector style you have. These connectors are weatherproof.

Computer Connectors

Adding computer equipment to your shack introduces you to many potentially confusing connectors. Most of the connections between computers and their peripheral equipment are made with multi-conductor ribbon cable (Fig. 31), with 10 or more individual conductors. Although the computer industry has tried to standardize connectors, there are dozens of types of connectors in use. Some of the common ribbon cable connectors are shown in Fig. 32. Fortunately, these connectors can be divided into three categories,

---

BNC CONNECTORS

Standard Clamp

1. Cut cable and even. Strip jacket. Fray braid and strip dielectric. *Don't nick braid or center conductor.* Tin center conductor.

2. Taper braid. Slide nut, washer, gasket and clamp over braid. Clamp inner shoulder should fit squarely against end of jacket.

3. With clamp in place, comb out braid, fold back smooth as shown. Trim center conductor.

4. Solder contact on conductor through solder hole. Contact should butt against dielectric. Remover excess solder from outside of contact. Avoid excess heat to prevent swollen dielectric which would interfere with connector body.

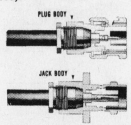

5. Push assembly into body. Screw nut into body with wrench until tight. *Don't rotate body on cable to tighten.*

Improved Clamp

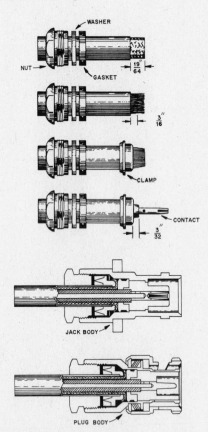

Follow 1, 2, 3 and 4 in BNC connectors (standard clamp) except as noted. Strip cable as shown. Slide gasket on cable *with groove facing clamp.* Slide clamp on cable *with sharp edge facing gasket.* Clamp *should* cut gasket to seal properly.

C. C. Clamp

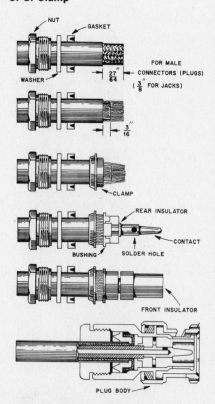

1) Follow steps 1, 2 and 3 as outlined for the standard-clamp BNC connector.

2) Slide on the bushing, rear insulator and contact. The parts must butt securely against each other, as shown.

3) Solder the center conductor to the contact. Remove flux and excess solder.

4) Slide the front insulator over the contact, making sure it butts against the contact shoulder.

5) Insert the prepared cable end into the connector body and tighten the nut. Make sure that the sharp edge of the clamp seats properly in the gasket.

Fig. 29 — BNC connectors are common on VHF and UHF equipment at low power levels. (courtesy Amphenol Electronic Components, RF Division, Bunker Ramo Corp.)

based on how they are attached to ribbon cable: insulation displacement, solder-type and crimp-type.

Insulation Displacement Connectors

An insulation displacement connector (IDC) has three parts: the body, the cover and the strain-relief clamp (Fig. 33). On the body of an IDC there are one or more rows of metal clips — one clip for each conductor in the connector. When ribbon cable is pressed on to the body of the connector, each clip cuts through the insulation around one of the ribbon cable conductors, making contact to that conductor (Fig. 34). The IDC cover applies even pressure during installation of the connector, and clamps the connector in place during use. Some IDCs are also provided with strain-relief clamps so that strain on the cable will not be transferred to the insulation-displacing clips.

To install an IDC, select the appropriate ribbon cable; place it on the back of the

---

TYPE N CONNECTORS

Standard Clamp

NUT   WASHER SPREAD   GASKET   CLAMP

FEMALE CONTACT   JACK BODY

MALE CONTACT   PLUG BODY

1) Cut cable and even. Remove 9/16″ of vinyl jacket. When using double-shielded cable remove 5/8″.

2) Comb out copper braid as shown. Cut off dielectric 7/32″ from end. Tin center conductor.

3) Taper braid as shown. Slide nut, washer and gasket over vinyl jacket. Slide clamp over braid with internal shoulder of clamp flush against end of vinyl jacket. When assembling connectors with gland, be sure knife-edge is toward end of cable and groove in gasket is toward the gland.

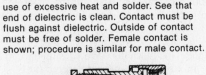

4) Smooth braid back over clamp and trim. Soft-solder contact to center conductor. Avoid use of excessive heat and solder. See that end of dielectric is clean. Contact must be flush against dielectric. Outside of contact must be free of solder. Female contact is shown; procedure is similar for male contact.

5) Slide body into place carefully so that contact enters hole in insulator (male contact shown). Face of dielectric must be flush against insulator. Slide completed assembly into body by pushing nut. When nut is in place, tighten with wrenches. In connectors with gland, knife edge should cut gasket in half by tightening sufficiently.

Improved Clamp

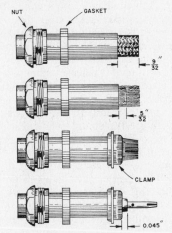

1) Follow instructions 1 through 4 as detailed in the standard clamp (be sure to use the correct dimensions).

2) Slide the body over the prepared cable end. Make sure the sharp edges of the clamp seat properly in the gasket. Tighten the nut.

C. C. Clamp

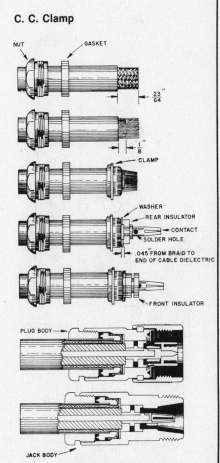

1) Follow instructions 1 through 3 as outlined for the standard-clamp Type N connector.

2) Slide on the washer, rear insulator and contact. The parts must butt securely against each other.

3) Solder the center conductor to the contact. Remove flux and excess solder.

4) Slide the front insulator over the contact, making sure it butts against the contact shoulder.

5) Insert the prepared cable end into the connector body and tighten the nut. Make sure the sharp edge of the clamp seats properly in the gasket.

Fig. 30 — Type N connectors are a must for high-power VHF and UHF operation. (courtesy Amphenol Electronic Components, RF Division, Bunker Ramo.)

Fig. 31 — Ribbon cable with various numbers of conductors is used to interconnect computer equipment.

Fig. 32 — Various computer connectors.

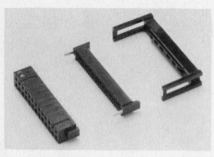

Fig. 33 — Insulation displacement type connector (IDC). See text for details.

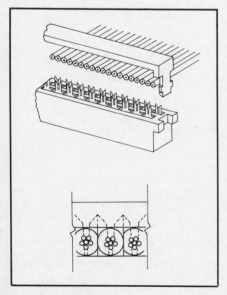

Fig. 34 — Installation of ribbon cable on an IDC. See text for details.

connector, making sure each conductor is in the jaws of one of the clips; place the cover on the connector and apply sufficient pressure to seat the cover. Use a vice to apply even pressure across the cover. If the connector has a strain-relief clamp, route the cable through the clamp, as shown in Fig. 35.

IDCs are the easiest ribbon cable connectors to install. If there is an IDC available for your application, use it.

Solder-Type Connectors

Each conductor from the ribbon cable must be soldered to a tap or post in solder-type connectors (Fig. 36). Use a razor or sidecutters to separate the conductors of the ribbon cable by cutting the insulation parallel to and between the conductors. The length of cable that should be separated into individual conductors will depend on the width of the cable and the width of the connector. Assure yourself that each conductor will reach the correct pin on the connector. Once you have separated the conductors, strip and tin each one. Then solder the conductors to the appropriate pins. Use good soldering techniques.

Crimp-Type Connectors

Each conductor in the cable is crimped to a contact; the contacts are inserted into a plastic housing (Fig. 37). The contacts can be either male (plugs) or female (sockets). Preparing ribbon cable for a crimp-type connector is the same as preparing cable for solder-type connectors, except that the stripped conductors should not be tinned. Insert each conductor into a contact and crimp the contact closed around the conductor. Properly crimped contacts should provide secure mechanical and electrical connections. Unfortunately, proper crimping is hard to achieve without the appropriate, expensive crimping tool. Therefore, many amateurs solder wires to crimp type contacts. If you choose to do this, use good soldering technique and be sure not to fill the locking tab(s) on the contact with solder.

Once the contacts are attached to the cable, use a pair of pliers to insert them in the housing. With some connectors you must be sure that the locking tab on the contact will engage the locking slot on the housing. If you insert the contacts backward, the locking tab will prevent complete insertion of the contact.

Computer cabling is often complicated. Time spent on proper selection and installation of connectors will be paid back in time saved on troubleshooting and recabling.

ROUTING ANTENNA CABLES

Once you have interconnected the equipment in your station, the next step is getting transmission lines, rotator cables and control lines outside to the devices at the other end. The exact method you use will depend on the location of your station and the construction of your house. Use the information presented here for ideas for your station.

The run from the shack to the antennas should be as short as possible to minimize feed-line loss and control-cable voltage drop. The most direct route may be to open a window and run the cables outside. Fig. 38 shows a way to do this and still seal your house against the weather (and bugs). Apply a piece of thick foam weather-stripping to the lower sash bottom rail and another to the sill. Adhesive-backed weatherstripping is available in various sizes at hardware stores. For this job, the thicker the better. The foam not only seals the area around the cables, but also protects the cables from being crushed or cut by the window.

On many double-hung windows, meeting rails on the top and bottom sashes no longer meet with the window partially open, so more weatherstripping is needed to seal the gap. You'll also have to devise another way to lock the window. Blocks of wood between the sash and sill will do the trick. Running cables through open windows leaves something to be desired in terms of weatherproofing and security, but it is easy and leaves no permanent scars on the house.

Another popular approach is to replace a glass window pane with material that can be drilled, such as acrylic or aluminum sheet. Fig. 39 shows how to run cables through a window with one glass pane replaced by acrylic sheet. Each piece of coaxial cable from the antennas connects to a double-female bulkhead connector. Bulkhead connectors are similar to double-female barrel connectors except they are longer and the outer body is threaded to accept nuts. UHF and Type N bulkhead connectors are commonly available at electronics part stores. Rotator and other control cables are run through holes in the window just large enough to clear the outer jacket. If additional weatherproofing is needed, run a bead of silicone seal around the cable jacket or use a grommet of the right dimensions. The multiconductor cable runs directly to a terminal strip; from there, another wire runs to the control box. Open-wire line or single conductors may be run safely through the window using ceramic feedthrough insulators. This method offers a neat, weatherproof way to run cables between your station and antennas. The original window pane may be reinstalled at any time, so there are no permanent scars. The only drawback is that you may lose the use of the window.

Fig. 40 shows how to use a dryer vent pipe to bring cables into a house. The vent pipe may be installed in a hole bored through the wall of the house, or it may be mounted in an acrylic or aluminum replacement window pane. The vent pipe hood keeps rain out of the tube, and a bead of caulk around the outside of the tube where it passes through the wall completes weather sealing. A standard three-inch vent

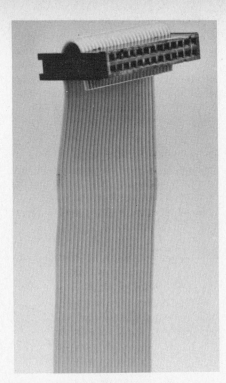

Fig. 35 — Completed IDC with ribbon cable, showing proper strain-relief.

pipe will pass more than enough cables for the average amateur station. The vent-pipe approach does not require cables to be broken, and the number of cables may be changed at any time without drilling more holes. For the VHF/UHF operator, it allows a continuous run of hardline from the antenna to the station. It's a good idea to stuff the pipe with insulation after installing the cables to keep small animals and insects out of the house and heat in.

DOCUMENTING YOUR STATION

An often neglected but very important part of putting together your station is properly documenting your work. Ideally, you should diagram your entire station from the ac power lines to the antenna on paper and keep the information in a special notebook with sections for the various facets of your installation. Having the station well documented is an invaluable aid when tracking down a problem or planning a modification. Rather than having to search your memory for information on what you did a long time ago, you'll have the facts on hand. The best time to prepare your notes is as you're building, so take a few minutes at each step to record your progress.

When documenting your shack, include information such as the type, size and length of the primary wiring, as well as the value of circuit breakers or fuses. Keep track of your cables. Write down the pinouts of the various plugs and connectors joining your equipment, including microphone cables, keyer wiring and control circuitry. If a piece of equipment has more than one possible hookup configuration (say, for example, high or low impedance for a microphone or 117 or 234 V for a transformer), note which connections you used as well as the ones you didn't. For multiconductor cables, make a list of the colors of the wires and their functions.

Documenting your antenna system is especially important. For towers, make and keep drawings of the entire installation. Include sketches with measurements of such things as the tower base and guy anchors. Write down your guy-wire lengths. If you break your guys into non-resonant lengths with insulators, note the lengths of the individual sections for each set of wires. Knowing what you have in the present installation will make repairing or upgrading the tower much easier in the future.

Make a list of the bolt sizes used for the tower, rotator, mast and antenna(s). List the wrench size required for all hardware. This way, you'll know which wrenches to take along for a specific job. For example, to remove your rotator for repair, you may need a 7/16-in wrench to remove the U-bolts securing the mast and a ½-in wrench for the bolts holding the rotator housing to the mounting plate. If you guess and take the wrong size, you'll have to make additional unnecessary trips up and down the tower. If you bring along all of your tools for a job requiring only one or two wrenches, you'll weigh yourself down and the climb will be harder than necessary.

If your diagram of the mast and antenna(s) includes the size of the mast and the spacing between antennas (if you have more than one on the same mast), you'll be better prepared to add or change antennas in the future. You'll be able to fabricate boom-to-mast clamps on the ground and hoist them up the tower with the confidence that you'll have to do the job only once — the hardware will fit perfectly. By knowing how much mast is sticking out of the top of the tower, you'll know, for example, whether or not you can replace an antenna that has a self-supporting boom with another requiring a specific number of feet of mast for a boom brace. If you know the distance between antennas, you'll be able to determine if an additional antenna on the mast is feasible. Having the measurements on hand takes the guesswork out of many jobs and saves unnecessary trips up and down the tower to make new measurements every time you want to make a change.

Cables and connectors are worth documenting as well. As you build your antenna system, note the length and type of each piece of feed line, rotator and control cable used. Document the type of connectors used. Be sure to write down the color code used for the rotator cable; if some day you disconnect the control box in the shack and don't have the codes in your notebook, you'll have to climb the tower to find out how they go. If you know the lengths of your cables and the types of

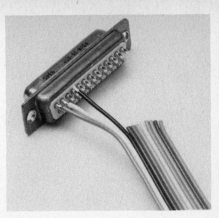

Fig. 36 — Solder-type DB-25 connector and prepared ribbon cable.

Fig. 37 — Crimp-type connector with prepared ribbon cable.

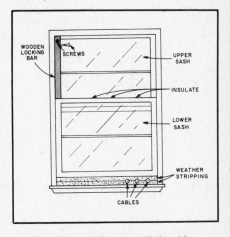

Fig. 38 — Many amateurs run their cables through an open double-hung window. The window should be sealed against the weather.

connectors used, it will be a great help in troubleshooting. You will be able to prepare test and/or replacement cables before leaving the ground.

Besides documenting the interconnections and hardware around your station, you should also keep track of the performance of your equipment. There are many different performance tests you can make, and what you do will depend on your

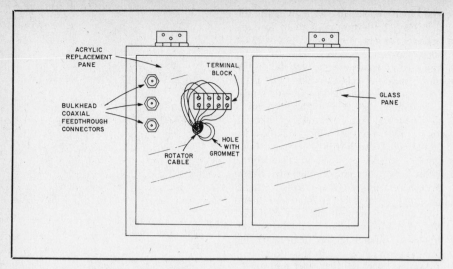

Fig. 39 — A window pane may be replaced with an acrylic sheet and then drilled to pass cables.

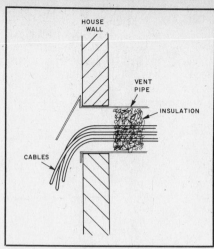

Fig. 40 — An ordinary dryer vent pipe will pass a large number of cables.

technical ability and the test equipment on hand. There are, however, some basic tests that almost every amateur can make. Each time you install a new antenna, measure the SWR at different points in the band and make a table or plot a curve. Later, if you suspect a problem, you'll be able to look in your records and compare your SWR with the original performance. You might also want to measure the loss of each piece of feed line before installation. Most feed line degrades with time, so again you'll have a benchmark to compare with if you have a problem later. This measurement is easy; just terminate the feed line with a dummy load, and compare the power measured at the transmitter end with that measured at the dummy load.

In your shack, you can measure the power output from your transmitter(s) and amplifier(s) on each band. These measurements will be helpful if you later

suspect you have a problem, such as your rig going out of alignment. They will also help you see when your tubes start to go soft. For receivers, it may be helpful to turn on the crystal calibrator and note the S-meter reading on various bands. The calibrator output is fairly constant, so future S-meter readings different from the original ones may indicate alignment or other problems. Of course, if you have access to a signal generator you can measure receiver performance for future reference.

As part of your station records, keep a maintenance log showing details of any repairs or modifications to your gear. For modifications, make a copy of the schematic and add in changes on the copy. You can scrape traces from a photocopy with a sharp knife and draw in the new lines and/or components. Also, write down the details and keep a copy of any literature

pertaining to it (a magazine article or manufacturer's update note, for example). If you do any before/after testing, note your results.

Keep a list of purchases of station equipment. Include the serial number, price paid, date of purchase, and name and address of the seller. If you purchased the gear new from a dealer, this information should appear on the invoice. These records will come in handy at resale time. It's also advisable to have this information for insurance purposes, should anything happen to your equipment.

Station documentation can be as formal or informal as you like. Some amateurs will keep a file drawer with folders for the various aspects of the station. Others will keep a three-ring notebook separated into several sections. If you keep the documentation simple, you will be more likely to do it, and good records are very helpful.

Assembling Your Antenna System

The antenna system is one area where amateurs have considerable room for experimentation. No two antenna systems are alike, and the shape yours takes will depend on the amount of land available, your resources and your interests. Your antenna system might consist of a multiband dipole, or you might want to chase DX and work contests with monoband beams installed on individual towers. The antenna system is your chance to be creative.

The information presented here explains some of the methods used to put antennas into the air and keep them there. Information on antenna design is covered in Chapter 17, while antenna construction is presented in Chapter 33. The *ARRL*

Antenna Book is another good source of information on antenna theory and construction. This section assumes that you have an antenna in mind or assembled and are looking for ideas on how to support it.

SUPPORTING WIRE ANTENNAS

The antenna can be supported at the ends by buildings, trees or anything you may be able to use. Later in this chapter, there is information on putting up poles and masts. A mast may be the real answer to holding up at least one end of the antenna. Many operators fasten one end of the antenna to the house, or to a short pole on the roof, and build a mast far enough away to permit using the desired antenna length. The

spot for the mast can be chosen so the antenna will be in the clear.

You can slope the antenna wire if your two supports aren't the same height. A sloping antenna usually will radiate best in the direction of the downslope. You can take advantage of this effect by deliberately sloping the wire to get a desired "best" direction.

Trees as Antenna Supports

From the beginning of Amateur Radio, trees have been used widely for supporting wire antennas. Trees cost nothing, of course, and will often provide a means of supporting a wire antenna at considerable height. As an antenna support, however, a tree is highly unstable in the presence of wind, unless the tree is a very large one and the antenna is suspended from a point well down on the tree trunk. As a result, the antenna must be constructed much more sturdily than would be necessary with stable supports. Even with rugged construction, it is unlikely that an antenna suspended from a tree, or between trees, will stand up indefinitely.

There are two general methods of securing a pulley to a tree. If the tree can be climbed safely to the desired level, a pulley can be wired to the trunk of the tree, as shown in Fig. 41. If, after passing the halyard through the pulley, both ends of the halyard are simply brought back down to ground along the trunk of the tree, there may be difficulty in bringing the antenna end of the halyard out where it will be clear of branches. To avoid this, one end of the halyard can be tied temporarily to the tree at the pulley level, while the remainder of the halyard is coiled up, and the coil thrown out horizontally from this level, in the direction in which the antenna will run. It may help to have the antenna end of the halyard weighted. Then, after attaching the antenna to the halyard, the other end is untied from the tree, passed through the pulley, and brought to ground along the tree trunk in as straight a line as possible. The halyard need be only long enough to reach the ground after the antenna has been hauled up, since additional rope can be tied

to the halyard when it becomes necessary to lower the antenna.

The other method consists of passing a line over the tree from ground level, and using this line to haul a pulley up into the tree and hold it there. Several ingenious methods have been used to accomplish this. The simplest method employs a weighted pilot line, such as fishing line or mason's chalk line. Grasping the line about 2 feet from the weight, the weight is swung back and forth, pendulum style, and then heaved with an underhand motion in the direction of the tree top. Several trials may be necessary to determine the optimum size of the weight for the line selected, the distance between the weight and the hand before throwing, and the point in the arc of the swing where the line is released. The weight, however, must be sufficiently large to ensure that it will carry the pilot line back to ground after passing over the tree. Flipping the end of the line up and down so as to put a traveling wave on the line often helps to induce the weight to drop down if the weight is marginal. The higher the tree, the lighter the weight and the pilot line must be. A glove should be worn on the throwing hand, because a line running swiftly through the bare hand can cause a severe burn.

If there is a clear line of sight between ground and a particularly desirable crotch in the tree, it may be possible to hit the crotch eventually after a sufficient number of tries. Otherwise, it is best to try to heave the pilot line completely over the tree, as close to the center line of the tree as possible. If it is necessary to retrieve the line and start over again, let the weighted end drop to the ground and pull the pilot line through the tree. If you pull the weight back through the tree, it may wrap the line around a small limb, making retrieval impossible.

Stretching the line out in a straight line on the ground before throwing may help to keep the line from snarling, but it places extra drag on the line, and the line may snag on obstructions overhanging the line when it is thrown. Another method is to make a stationary reel by driving eight nails, arranged in a circle, through a 1-inch board. After winding the line around the circle formed by the nails, the line should reel off readily when the weighted end of the line is thrown. The board should be tilted at approximately right angles to the path of the throw.

Other devices that have been used successfully to pass a pilot line over a tree are the bow and arrow with heavy thread tied to the arrow, and the short casting rod and spinning reel used by fishermen. Still another method that has been used where sufficient space is available is to fly a kite. After the kite has reached sufficient altitude, simply walk around the tree until the kite string lines up with the center of the tree. Then pay out string until the kite falls to the earth. This method has been used

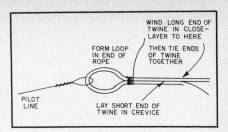

Fig. 42 — In connecting the halyard to the pilot line, a large knot that might snag in the crotch of a tree should be avoided, as shown.

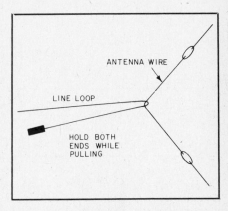

Fig. 43 — A weighted line thrown over the antenna can be used to pull the antenna to one side to avoid overhanging obstructions, such as branches of trees in the path of the antenna, as the antenna is pulled up. When the obstruction has been cleared, the line can be removed by releasing one end.

successfully to pass a line over a patch of woods between two higher supports, which would have been impossible using any other method.

The pilot line can be used to pull successively heavier lines over the tree until one of adequate size to take the strain of the antenna has been reached. This line is then used to haul a pulley up into the tree after the antenna halyard has been threaded through the pulley. The line that holds the pulley must be capable of withstanding considerable chafing where it passes through the crotch, and at points where lower branches may rub against the standing part. For this reason, it may be advisable to use galvanized sash cord or stranded guy wire for raising the pulley.

Especially with larger sizes of line or cable, care must be taken when splicing the pilot line to the heavier line to use a splice that will minimize the chances that the splice cannot be coaxed through the tree crotch. One type of splice is shown in Fig. 42.

The crotch which the line first comes to rest in may not be sufficiently strong to stand up under the tension of the antenna. However, if the line has been passed over, or close to, the center line of the tree, it will usually break through the lighter crotches and finally come to rest in one sufficiently strong lower down on the tree.

Needless to say, any of the suggested

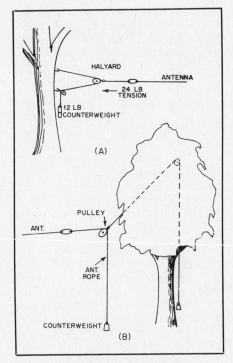

Fig. 41 — Methods of counterweighting to minimize antenna movement. The method at A limits the fall of the counterweight should the antenna break. It also has a 2 to 1 mechanical advantage, as indicated. The method at B has the disadvantage that the point of support in the tree must be higher than the end of the antenna.

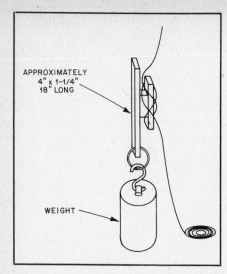

APPROXIMATELY
4" × 1-1/4"
18" LONG

WEIGHT

Fig. 44 — The cleat helps you avoid having to untie a knot that may have been weather-hardened.

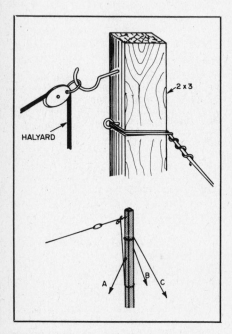

2 × 3

HALYARD

A B C

Fig. 45 — Using a length of 2 × 3 as a support for a horizontal antenna. Lengths up to 20 feet can be used with a single set of guys arranged as shown.

methods should be used with due respect to persons or property in the immediate vicinity. A child's sponge-rubber ball (baseball size) makes a safe weight for heavy thread line or fishing line.

If the antenna wire becomes snagged in lower branches of the tree when the wire is pulled up, or if branches of other trees in the vicinity interfere with raising the antenna, a weighted line thrown over the antenna and slid along to the appropriate point is often helpful in pulling the antenna wire to one side to clear the interference as the antenna is being raised, as shown in Fig. 43.

Wind Compensation

The movement of an antenna suspended

between supports that are not stable in wind can be reduced materially by the use of heavy springs, such as screen-door springs under tension, or by a counter-weight at the end of one halyard, as shown in Fig. 41. The weight, which may be made up of junk-yard metal, window sash weights, or a galvanized pail filled with sand or stone, should be adjusted experimentally for best results under existing conditions. Fig. 44 shows a convenient way of fastening the counterweight to the halyard. It avoids the necessity for untying a knot in the halyard which may have hardened under tension and exposure to the weather.

Man-Made Supports

In practice, a height of 30 to 60 feet is in the optimum region for any band above 7 MHz where the ionosphere plays a part.

You can get by with even less than 30 feet if you have to. However, nearly everyone can put an antenna at least 25 feet or so in the air. This height puts it in the class where, usually, at least one end can be supported by your house. The other end can be held up by another building (if you have permission to use it) far enough away, or by a tree, or if necessary by a pole you put up yourself. The pole doesn't have to be an elaborate structure, especially if it can be erected on top of something. A detached garage makes a good base. It should give you a start of 12 to 15 feet, and generally will have enough roof area to let you put guys on a simple wood pole that can be up to 20 feet tall. It is important to realize that antennas with horizontal polarity should be one-half wavelength or more above ground for best results in long-distance communications.

Anchoring

Wire antennas are not heavy, and even when pulled up tight the strain can easily be supported by a husky screw eye or hook, if it can be sunk 1½ or 2 inches into good solid wood. Screw eyes can be used similarly for anchoring guy wires from light poles up to 20 feet long. Often other anchorages for guy wires offer themselves; use anything that seems solid enough and which gives you the opportunity for wrapping the wire around a couple of times so it won't slip.

A length of 2 × 3 makes a good pole for heights up to 20 feet or so. It isn't heavy, but with three solidly anchored guys it will hold any wire antenna you may use with it. Fig. 45 shows a simple type of construction, using fittings that you can buy at a hardware store. The pulley and halyard make it possible to raise and lower the antenna easily. With the antenna up, only two guys are really necessary, placed 90 to 120 degrees apart back of the antenna, as shown at B and C. The third guy, A, can be part way down the pole so it will be out of the way of the pulley and halyard, and should pull off in about the same direction as the antenna. Its purpose is to hold the

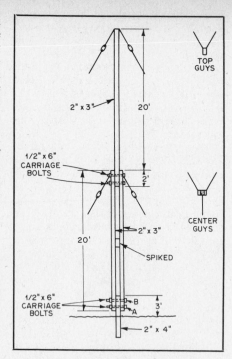

2" × 3"

20'

1/2" × 6" CARRIAGE BOLTS

2'

20'

2" × 3"

SPIKED

1/2" × 6" CARRIAGE BOLTS

B
A

3'

2" × 4"

TOP GUYS

CENTER GUYS

Fig. 46 — This mast construction is suitable for heights up to 40 feet, for supporting a horizontal antenna. Pivoting the mast on the lower bolt, A, simplifies the job of raising and lowering.

pole up when the antenna is down.

The guys can simply be wrapped around the pole near the top. One is shown in the drawing. To keep them from slipping down they can be run through a screw eye as shown. This avoids putting any strain on the screw eye. Stranded steel wire available in hardware stores (a typical type has four strands of about the same diameter as no. 18 copper) has ample strength for guying a pole of this type. Guy wire also is available from radio supply houses.

Plastic clothesline can be used for the halyard. The plastic kind stands the weather better than ordinary cotton clothesline. If you do use cotton, get the type having a steel core; it has greater strength and lasting qualities. But whatever type you use, be sure to use enough. Remember that you need to be able to reach the free end of the halyard when the antenna is down!

Simple Wooden Mast

If you have to set a mast on the ground, the structure shown in Fig. 46 is easy to build and erect. Three 20-foot lengths of 2 × 3 are used to make a mast 40 feet tall. It makes a sturdy support for a horizontal-wire antenna when properly guyed. The mast is pivoted at the base for ease in raising (and lowering, if necessary), the pivot being the lower carriage bolt (A) shown in the drawing.

The base is a length of 2 × 4 set solidly in the ground. The 2 × 4 should be buried to a depth of about 3 feet and should extend at least 3 feet above the ground level. Use a plumb line to make sure it is vertical

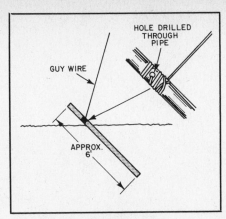

Fig. 47 — Pipe guy anchor. One-inch or 1½-inch iron pipe is satisfactory for a mast such as is shown in Fig. 46.

when setting it. Pack rocks around it before refilling the hole with dirt; this will make a solid foundation, which is helpful when the mast is being raised. After it is up and the guys are in place, the stress on the base is practically all straight down, so there is really no need for an elaborate footing.

The mast should be assembled on the ground (with the antenna halyard affixed; the method shown in Fig. 45 can be used for this) and fastened to the ground post with the lower carriage bolt. The guy wires, which should be made long enough to reach their anchorages while the mast is down, also should be attached. Push the mast straight up at the middle, letting it pivot on the bolt. When it gets up high enough, have an assistant slide a ladder under it to hold it. Keep working it up, moving the ladder down the mast as you go. Stop now and then to pull up on the guys. Keep the front guy short enough so the mast won't swing all the way over and come down on the other side, but leave it loose enough so it can't interfere as the mast goes up. When the mast approaches the vertical, pulling on the top guys will bring it up the rest of the way. Slip in the upper carriage bolt (B) and the mast will stand by itself (if the base is solid) while you adjust the lower set of guys. Don't give the top guys their final tightening until the antenna is up, as they must work against the pull of the antenna.

The guy wires should be anchored about 20 feet from the base of the mast, for a height of 40 feet. If shorter 2 × 3s are used so that the completed mast is less tall, this distance can be correspondingly reduced.

Insulating Guy Wires

Guy wires are like antenna wires — if their length happens to be near resonance at your operating frequency, they can pick up energy from the antenna field and re-radiate it. This may do no harm, but then again it might distort the radiation pattern you hope the antenna will give.

To circumvent this, guy wires are often "broken up" by inserting insulators at strategic points. The idea is to avoid letting any section of guy wire have a length that is near resonance at any operating frequency or harmonic of it. Egg-type insulators are generally used as a safety measure. With these insulators the wires are looped around each other so that if the insulator breaks the guy wire will still hold.

General practice is to put an egg in each guy a foot or so from where it is attached to the pole or mast, and another a foot or so from the point where it is anchored. Whether any more are needed depends on the remaining length. In most cases it is sufficient to insert insulators in such a way that no section of the wire will have a length in feet that can be divided evenly by 16 or 22.

Guy Anchors

Trees or buildings at the right distance from the base of a mast can be used for anchoring guys. However, they don't always "just happen" to be present. In such a case a guy anchor can be sunk into the ground.

A 6-foot length of 1-inch iron pipe driven into the ground at about a 45-degree angle makes a good anchor. It should be driven toward the mast, with 12 to 18 inches protruding from the ground for fastening the wires. Fig. 47 shows the general idea, with a suggested way of fastening the guy so it can't slip off the anchor.

Weather Protection

A wooden mast or pole needs the same protection against weather as the exterior woodwork on a dwelling. New wood should be given three coats of a good paint made for outdoor use, with the first coat thinned so it sinks in well. Follow the directions on the paint container.

Masts made from several pieces of lumber, such as the one shown in Fig. 46, should have their individual parts painted separately before assembly. The heads of nails or screws should be sunk into the surface far enough so they can be puttied over. These, too, should be painted — after assembly. Hooks, screw eyes and similar hardware should likewise have a coat of paint to prevent rusting.

Other Types of Masts

Amateur ingenuity has conjured up all sorts of mast and tower designs over the years. It is possible, for example, to make a satisfactory pole, 40 feet or so high, out of lengths of metal down-spouting. Wooden lattice construction has long been a favorite for masts and towers. A number of such designs will be found in *The ARRL Antenna Book*, and new ones keep coming along regularly in *QST*. Then there are TV masts, built up in sections of metal tubing and provided with guy rings and other hardware — readily available and not at all costly. These can be run up to heights as great as 50 feet, with proper guying.

In fact, anything you can put up, and that will stay up, will be satisfactory. Never forget that you can't afford to take any chances with safety. A falling mast can result in accidental injury, or even death, to anyone that happens to be in its path. The taller and heavier the mast, the more necessary it becomes to build in plenty of safety factor — in size of guy wire, type of anchorage, and in the strength and cross section of the materials used in the mast or tower itself.

TOWERS

A mast is fine for supporting one end of a wire antenna, or a small vertical, or even a small VHF/UHF beam, but the right way to support an HF beam or a large VHF/UHF array is with a communications tower designed for this purpose. The information in this section is intended to help you choose the right type of tower for your needs, provide information on erecting the tower, and present some ideas on getting the antenna to the top.

Working on towers and antennas is dangerous, and possibly fatal, if you do not know what you are doing. Your tower and antenna can cause serious property damage and personal injury if any part of the installation should fail. Always use the highest quality materials in your system. Follow the manufacturer's specifications, paying close attention to base pier and guying details. Do not overload the tower. If you have any doubts about your ability to work on your tower and antennas safely, contact another amateur with experience in this area or seek professional assistance.

Information on antenna theory appears in Chapter 17. Practical antenna construction projects are presented in Chapter 33. Information on making the cables to connect your antennas, rotator cables and control lines appears earlier in this chapter. *The ARRL Antenna Book* is an excellent source of antenna information, and *QST* regularly features articles on antennas.

TOOLS

Any mechanical job is easier if you have the right tools. Tower work is no exception. In addition to a good assortment of wrenches, screwdrivers and pliers, you will need some specialized tools to work safely and efficiently on a tower. You may already own some of these tools. Others will be purchased or borrowed. Do not start a job until you have assembled all of the necessary tools. Shortcuts or improvised tools can be fatal if you gamble and lose at 50 feet in the air.

Clothing

The clothing you wear when working on towers and antennas should be selected for maximum comfort and safety. Wear clothing that will keep you warm, yet allows complete freedom of movement. Long denim pants and long-sleeve shirts will protect you from scrapes and cuts. Wear work shoes with heavy soles, or better yet steel inserts in the soles, to give your feet the support they need to stand on a narrow tower rung.

Gloves are necessary for both the tower climber and all ground-crew members. Good quality leather gloves will protect hands from damage and keep them warm. They also offer protection and a better grip when you are handling rope.

Ground-crew members should have hard hats to protect them in case something falls from the tower. It is not uncommon for the tower climber to drop tools and hardware. A wrench dropped from 100 feet will bury itself several inches in soft ground; imagine what it might do to an unprotected skull.

Safety Belt and Climbing Accessories

Any amateur with a tower *must* own a high-quality safety belt, such as the one shown in Fig. 48. Do not attempt to climb a tower, even a short distance, without a belt. The climbing belt is more than just a safety device for the experienced climber. It is a tool to free up both hands for work. The belt allows the climber to lean back away from the tower to reach bolts or connections. It also provides a solid surface to lean against to exert greater force when hoisting antennas into place.

A climber must trust his life to his safety belt. For this reason, nothing less than a professional-quality, commercially made, tested and approved safety belt is acceptable. Check the equipment suppliers list in Chapter 35 and the ads in *QST* for suppliers of climbing belts and accessories. Check your belt *before each use* for defects. If the belt or lanyard (tower strap) are cracked, frayed or worn in any way, destroy the damaged piece and replace it with a new one. You should never have to wonder if your belt will hold.

Fig. 48 — A good-quality safety belt is a requirement for working on a tower. The belt should contain large steel loop for the strap snap. Leather loops at the rear of the belt are handy for holding tools. (*photo by K1WA*)

Along with your climbing belt, you should seriously consider the purchase of some climbing accessories. A canvas bucket is a great help for carrying tools and hardware up the tower. Two buckets, a large one for carrying tools and a smaller one for hardware, make it easier to find things when needed. A few extra snap hooks like those on the ends of your belt lanyard are useful for attaching tool bags and equipment to the tower at convenient spots. These hooks are better than using rope with knots in many cases because they can be hooked and unhooked with one hand.

Gorilla hooks, shown in Fig. 49, are especially useful for ascending and descending the tower. They attach to the belt and are hooked to the tower on alternating rungs as the climber progresses. With these hooks, the climber is secured to the tower at all times. Gorilla hooks were specially designed for amateur climbers by Avatar Magnetics (see Chapter 35 suppliers list).

Rope and Pulley

Every amateur who owns a tower should also own a good quality rope at least twice as long as the tower is tall. The rope is essential for safely erecting towers and installing antennas and cables. For most installations, a good quality ½-inch-diameter manila hemp rope will do the job, although a thicker rope is stronger and easier to handle. Some types of polypropylene rope are acceptable also; check the manufacturer's strength ratings. Nylon rope is not recommended because it tends to stretch and is difficult to fasten knots with.

Check your rope *before each use* for tearing or chafing. Do not attempt to use damaged rope; if it breaks with a tower section or antenna in mid-air, property damage and personal injury are likely results. If your rope should get wet, let it thoroughly air dry before putting it away.

A most-worthwhile purchase is a pulley like the one shown in Fig. 50. Use the right size pulley for your rope. Be sure that the pulley you purchase will not jam or bind as the rope passes through it.

The Gin Pole

The gin pole, shown in Fig. 51, is a handy device for working with tower sections and masts. This gin pole is designed to clamp onto one leg of Rohn No. 25 or 45 tower. The tubing, which is about 12 feet long, has a pulley on one end. A rope is routed through the tubing and over the pulley. When the gin pole is attached to the tower and the tubing extended into place, the rope may be used to haul tower sections or the mast into place. Fig. 52 shows the basic process.

A gin pole can be expensive for an individual to buy, especially for a one-time tower installation. Often, local radio clubs own a gin pole for use by their members. Stores that sell tower sections to amateurs and commercial customers frequently will

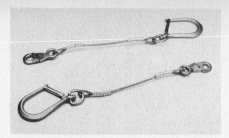

Fig. 49 — Gorilla hooks are designed to keep the climber attached to the tower at all times when ascending and descending.

Fig. 50 — A good quality rope and pulley are a must for anyone working on towers and antennas. This pulley is encased in wood so the rope cannot jump out of the pulley wheel and jam.

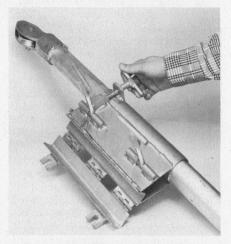

Fig. 51 — A gin pole is a mechanical device that can be clamped to a tower leg to aid in the assembly of sections as well as the installation of the mast. The aluminum tubing extends through the clamp and may be slipped into position before the tubing clamp is tightened. A rope should be routed through the tubing and over the pulley mounted at the top.

rent erection fixtures. If you attempt to make your own gin pole, use materials heavy enough for the job. Provide a means for securely clamping the pole to the tower. There are many cases on record where homemade gin poles have failed, sending tower sections crashing down amidst the ground crew.

When you use a gin pole, make every effort to keep the load as vertical as possible. Although gin poles are strong, you are asking for trouble if you apply too much lateral force.

SELECTING A TOWER

Probably the most important part of any Amateur Radio installation is the antenna

Fig. 52 — The assembly of tower sections is made simple when a gin pole is used to lift each one into position. Note that the safety belts of both climbers are fastened below the pole, thereby preventing the strap from slipping over the top section. (photo by K1WA)

Table 2

Tower Manufacturers

Aluma Tower
Box 2806
Vero Beach, FL 32960

Hy-Gain Division
Telex Communications, Inc.
9600 Adrich Ave., S.
Minneapolis, MN 55420

Unarco-Rohn
P.O. Box 2000
Peoria, IL 61601

Tri-Ex Tower Corp.
7182 Rasmussen Ave.
Visalia, CA 93277

Universal Manufacturing Company
12357 E. 8 Mile Rd.
Warren, MI 48089

Glenn Martin Engineering
P.O. Box 253
Boonville, MO 65233

E-Z Way Products, Inc.
P.O. Box 11535
Tampa, FL 33680

US Tower Corp.
8975 West Goshen Ave.
Visalia, CA 93291

system. It determines the effectiveness of the signal transmitted at a particular power level. In terms of dollar investment, the antenna provides double duty; while it can provide gain during transmitting periods (which can also be accomplished by increasing transmitter power), it has the same beneficial effect on received signals. Therefore, a tall support for a gain antenna that can be rotated is very desirable. This is especially true if DX contacts are of prime interest on the 20, 15 and 10-meter bands.

Of the two important features of an antenna system, height and antenna gain, height is usually considered the most important if the antenna is horizontally polarized. The typical amateur installation consists of a three-element triband beam (tribander) for 20, 15 and 10 meters mounted on a tower that may be as low as 25 or 30 feet, or as high as 65 or 70 feet. Some systems use large antennas on much taller towers.

The selection of a tower, its height and the type of antenna and rotator to be used all may seem like a complicated matter, particularly for the newcomer. These four aspects of an antenna system are interrelated, and one should consider the overall system before making any decisions as to a specific component. Perhaps the most important consideration for many amateurs is the effect of the antenna system on the surrounding environment. If plenty of space is available for a tower installation and there is little chance of the antenna irritating neighbors, the amateur is indeed fortunate. This amateur's limitations will be mostly financial. But for most, the size of the backyard, the effect of the system on the family members and neighbors, local ordinances, and the proximity of power lines and poles influence the overall selection of antenna components considerably.

The amateur must consider several practical limitations for installation:

1) A tower should never be installed in such a way that it could fall onto a neighbor's property.

2) The antenna must be located in such a position that it cannot possibly tangle with power lines during normal operation, or if a disastrous windstorm comes along.

3) Sufficient yard space must be available to position a guyed tower properly. The guy anchors should be between 60 and 80 percent of the tower height in distance from the base of the tower.

4) Provisions must be made to keep neighborhood children from climbing the support.

5) Local ordinances should be checked to determine if any legal restrictions are on record.

Other important considerations are:

6) The total dollar value to be invested.

7) The size and weight of the antenna desired.

8) The overall yearly climate.

9) Ability of the owner to climb a fixed tower.

The selection of a tower support is usually dictated more by circumstances than by desire.

Once a decision has been tentatively made, the next step is to write to the manufacturer (several are listed in Table 2) and request a specification sheet. Meanwhile, lay out any guy anchor points needed to ensure that they will fit on the assigned property. The specification sheet for the tower should give a wind-load capability; an antenna can then be selected that will

not overload it. If a tentative decision on the antenna type is made, a note to the antenna manufacturer giving the complete set of details for installation is not a bad idea. Be sure to give complete details of your plans, including all specifications of the antenna system planned. Remember, the manufacturer will not custom design a system directly for your needs, but may offer comments.

It is often very helpful to the novice tower installer to visit other local amateurs who have installed towers. Look over their hardware and ask questions. If possible, have a few local experienced amateurs look over your plans — before you commit yourself. They may be able to offer a great deal of help. If someone in your area is planning to install a tower and antenna system, be sure to offer your assistance. There is no substitute for experience when it comes to tower work, and your experience there may prove invaluable to you later.

The Tower

The most common variety of tower is the guyed tower made of stacked identical sections. The information in Fig. 53 is based on data taken from the Unarco-Rohn catalog. A list of tower manufacturers can be found in Table 2. Rohn calls for a maximum vertical separation between sets of guy wires of 35 feet for this series of tower. At A, the tower is 70 feet high, and there are two sets of evenly spaced guy wires. At B, the tower is 80 feet high, and there are three sets of evenly spaced guy wires. Exceeding the vertical spacing requirements could result in the tower buckling.

This may not seem like a reasonable possibility unless you understand the functioning of the guy wires. The guy wires restrain the tower against the force of the wind and translate the lateral force of the wind into a downward compression that forces the tower down onto the base. Normally, the manufacturers specify the initial tension in the guy wires. This is another force that is translated into the downward compression on the tower. If there are not enough guys and if they are not properly spaced, a heavy gust of wind may turn out to be the "straw that breaks the camel's back." Fig. 53C is an overhead view of a guyed tower. Manufacturers usually call for equal angular spacing between guys. If it is necessary to deviate from this spacing, you would be well advised to contact the engineering staff of the manufacturer or a civil engineer.

Some types of towers are not normally guyed — these are usually referred to as free-standing or self-supporting towers. The principles involved are the same regardless of the manufacturer's choice of names. The wind blowing against the side of the tower creates an overturning moment that would topple the tower if it were not for the anchoring at the base. Fig. 54 details the action and reaction involved. The tower

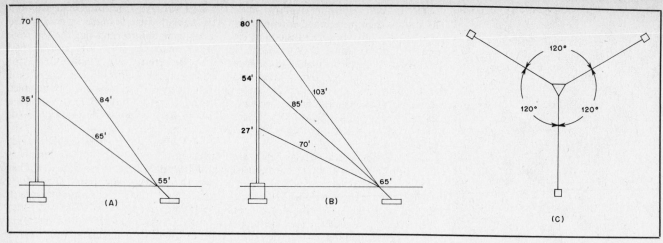

Fig. 53 — Diagram depicting proper method of installation of a typical guyed tower.

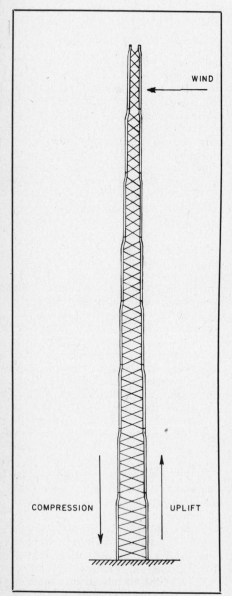

Fig. 54 — Diagram of typical free-standing (unguyed) tower. Arrows indicate the directions of the forces acting upon the structure. See text for discussion.

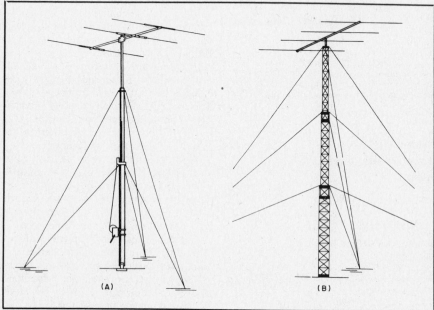

Fig. 55 — Two examples of "crank-up" towers.

is restrained by the base. As the wind blows against one side of the tower, the opposite side is compressed downward much as in the guyed-type setup. Because there are no guys to restrain the top, the side that the wind is blowing against is simultaneously being pulled up (uplift). The force of the wind is creating a moment that tends to pivot about a point in the base of the tower. The base of the guyed tower simply must hold the tower up, but the base of the free-standing tower must simultaneously hold one side of the tower up and the other side down! It should not be surprising that manufacturers often call for a great deal more concrete in the base of free-standing towers than they do in the base of guyed towers.

Fig. 55 shows two variations of another popular type of tower, the crank-up. In regular guyed or free-standing towers, each section is bolted atop the next lower section. The height of the tower is the sum of the heights of the sections (minus any overlap). Not so with the crank-up towers. The outer diameter of each section is smaller than the inner diameter of the next lower section. Instead of bolting together, the sections are attached with a complex set of cables and pulleys. The overall height of the tower is adjusted by using the pulleys and cables to "telescope" the sections together or apart.

Depending on the design, the manufacturer may or may not require guy wires. The primary advantage of the crank-up tower is that the owner must do the antenna work near the ground. A second advantage is that the tower can be kept retracted except during use, which reduces the guying

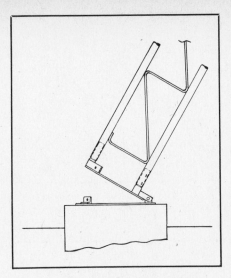

Fig. 56 — Fold-over or tilting base. There are several different variations of hinged sections permitting widely different types of installation. Great care should be exercised when raising or lowering a tilting tower.

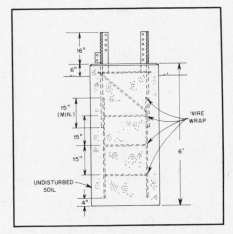

Fig. 57 — Another example of a concrete base (Tri-Ex LM-470).

needs (presumably, you would not try to extend the tower and use it during periods of high winds). The disadvantages include mechanical complexity and (usually) cost. It is extremely dangerous to climb on a crank-up tower, even if it is extended only a small amount.

Some towers have another convenience feature — a hinged section that permits the owner to fold over all or a portion of the tower. The primary benefit is in allowing antenna work to be done closer to ground level without the necessity of removing the antenna and lowering it. Fig. 56 shows a hinged base; of course, the hinged section can be designed for portions of the tower other than the base. Also, a hinged feature can be added to a crank-up tower.

Misuse of hinged sections during tower erections is a common problem among radio amateurs. Unfortunately, these episodes often end in accidents. If you do not have a good grasp of the fundamentals of physics, it might be wise to avoid hinged towers (or to consult an expert). It is often far easier (and safer) to erect a regular guyed tower or self-supporting tower with gin pole and climbing belt than it is to try to "walk up" an unwieldy hinged tower.

The Base

Each manufacturer will provide customers with detailed plans for properly constructing the base. Fig. 57 is an example of one such plan. This plan calls for a hole that is $3.5 \times 3.5 \times 6.0$ feet. Steel reinforcement bars are lashed together and placed in the hole. The bars are positioned such that they will be completely embedded in the concrete, yet will not contact any metallic object in the base itself. This is done to minimize the possibility of a direct discharge path for lightning through the base. Should such a discharge occur, the concrete base would likely explode and bring about the collapse of the tower.

A strong wooden form is constructed around the top of the hole. The hole and the wooden form are filled with concrete so that the resultant block will be 4 inches above grade. The anchor bolts are embedded in the concrete before it hardens. Usually it's easier to ensure that the base is level and properly aligned by attaching the mounting base and the first section of the tower to the concrete anchor bolts. Each manufacturer will provide specific detailed instructions for the proper mounting procedure. Fig. 58 provides a slightly different design for a tower base.

The one assumption so far is that you have normal soil. "Normal soil" is a mixture of clay, loam, sand and small rocks. A technical discussion is beyond the scope of this book, but you may want to adopt more conservative design parameters for your base (usually, more concrete) if your soil is sandy, swampy or extremely rocky. If you have any doubts about your soil, contact your local agricultural extension office and ask for a more technical description of your soil. Once you have that information, contact the engineering department of your tower manufacturer or a civil engineer.

Tower Installation

The installation of a tower is not difficult when the proper techniques are known. A guyed tower, in particular, is not hard to erect since each of the individual sections are relatively light in weight and can be handled with only a few helpers and some good-quality rope. A gin pole is a handy device for working with tower sections.

One of the most important aspects of any tower-installation project is the safety of all persons involved. Helpers should always stand clear of the tower base to prevent being hit by a dropped tool or hardware. When two people are climbing the tower, one should climb into position before the other begins climbing. The same procedure is required for climbing down a tower after the job is completed. The purpose is to have the nonclimbing person stand relatively still so as not to drop any tools or objects on the climbing person, or unintentionally obstruct his movements. When two persons are working on top of a tower, only one should change position (unbelt and move) at a time.

Attaching Guy Wires

In typical Amateur Radio installations a guy wire may experience "pulls" in excess of 1000 pounds. Under such circumstances, you do not merely twist the wires together and expect them to hold. Fig. 59 depicts the traditional method for fixing the end of a

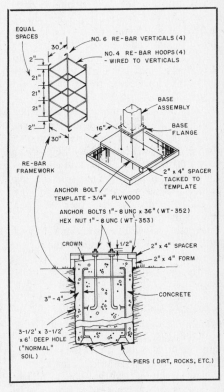

Fig. 58 — Plans for installing concrete base for Wilson ST-77B. Although the instructions and dimensions will vary from one tower to the next, this is representative of the type of concrete base specified by most manufacturers.

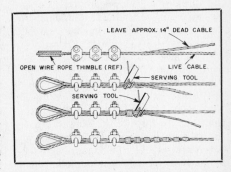

Fig. 59 — Traditional method of securing the end of a guy wire.

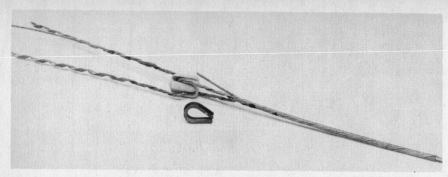

Fig. 60 — Alternative method for attaching guy wires using dead ends. The dead end on the right is completely assembled (the end of the guy wire was left extending from the grip for illustrative purposes). On the left, one side of the dead end has been partially attached to the guy wire. In front, a thimble for use where a sharp bend might cause the guy wire or dead end to break.

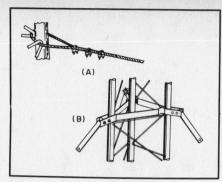

Fig. 61 — Two methods of attaching guy wires to tower. See text for discussion.

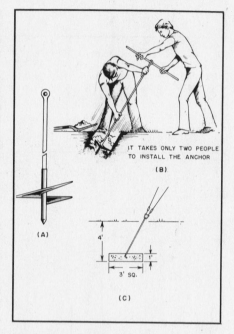

Fig. 62 — Two standard types of guy anchors. The earth screw shown at A is easy to install and widely available, but it may not be suitable for use with abnormal soil. The concrete anchor is more difficult to install properly, but it is suitable for use with a wide variety of soil conditions and will satisfy most building code requirements.

piece of guy wire. A thimble is used to prevent the wire from breaking because of a sharp bend at the point of intersection. Three cable clamps follow to hold the wire securely. As a final backup measure, the individual strands of the free end are unraveled and wrapped around the guy wire. It is a lot of work, but it is necessary to ensure a firm connection.

Fig. 60 shows the use of a device that replaces the clamps and twisted strands of wire. These devices are known as dead ends. They are far more convenient to use than are clamps. You must cut the guy wire to the proper length. The dead end is installed into whatever the guy wire is being attached to (use a thimble, if needed). One side of the dead end is then wrapped around the guy wire. The other side of the dead end follows. The savings in time and trouble more than make up for the slightly higher cost.

Guy wire comes in different sizes, strengths and types. Typically, 3/16-inch EHS (extra high strength) guy wire with a breaking strength of 3990 pounds will be adequate for the moderate tower installation found at most Amateur Radio stations. Some amateurs prefer to use 5/32-in "aircraft cable." Although this cable is somewhat more flexible than 3/16-inch EHS, it is only about 70% as strong. It is recommended that you stay with standard guy wire and that you use nothing smaller than 3/16-inch EHS.

Fig. 61 shows two different methods for attaching guy wires to towers. At A, the guy wire is simply looped around the tower leg and terminated in the usual manner. At B, a "torque bracket" has been added. There probably isn't much difference in performance for wind forces that are tending to push the tower over. If you happen to have more projected area (antennas, feed lines, etc.) on one side of the tower than the other, or a long-boom Yagi at the top, then the force of the wind will cause the tower to tend to twist into the ground. The torque bracket will be far more effective in resisting this twisting motion than will the simpler installation. If you have any doubts about how much twisting a tower

can do, go outside and carefully watch one in a windstorm. The trade-off, of course, is in terms of initial cost.

There are two main types of anchors used for guy wires. Fig. 62A depicts an earth screw. It usually measures 4 to 6 feet long. The screw blade at the bottom typically measures 6 to 8 inches in diameter. Fig. 62B illustrates two people installing the anchor. The shaft is tilted such that it will be in line with the guy wires. Earth screws are suitable for use in "normal" soil where permitted by local building codes.

The alternative to earth screws is the concrete block anchor. Fig. 62C shows the installation of this type of anchor; it is suitable for any soil condition, with the possible exception of a bed of lava rock or coral. Consult the instructions from the manufacturer for the precise method of installation.

Guy wires can receive and reradiate RF energy. Reradiation can, in turn, distort the pattern of a directional array and can sometimes contribute to RFI. These conditions are aggravated when one or more guy wires is resonant at the operating frequency or one of its harmonics. To avoid resonance, the guy wires should be broken up into nonresonant lengths with egg strain insulators. Fig. 63 shows wire lengths that fall within 10 percent of half-wave resonance in the popular HF bands. These lengths should be avoided.

Turnbuckles and associated hardware are used to attach guy wires to anchors and to provide a convenient method of adjusting tension on the guy wires. Fig. 64A shows a turnbuckle of a single guy wire attached to the eye of the anchor. Turnbuckles are usually fitted with either two eyes or one eye, and one jaw. The eyes are the oval ends, while the jaws are U-shaped with a bolt through the tips.

Fig. 64B depicts two turnbuckles attached to the eye of an anchor. The procedure for installation is to remove the bolt from the jaw, pass the jaw over the eye of the anchor and reinstall the bolt through the jaw, through the eye of the anchor, and through the other side of the jaw. For two or more guys attached to one anchor, it is

recommended that you install an equalizer plate (Fig. 64C). In addition to providing a convenient point to attach the turnbuckles, the plate will pivot slightly and tend to equalize the tension on the guy wires. Once the installation is complete, a safety wire should be passed through the turnbuckles in a "figure-eight" fashion to prevent the turnbuckle from working loose.

The Tower Shield

A tower can be legally classified as an "attractive nuisance" that could cause injuries unless some precautions are taken. The tower shield should eliminate the worry.

Generally, the "attractive nuisance" doctrine is based on the theory that one who maintains upon his premises an agency or condition that is dangerous to children of tender years by reason of their inability

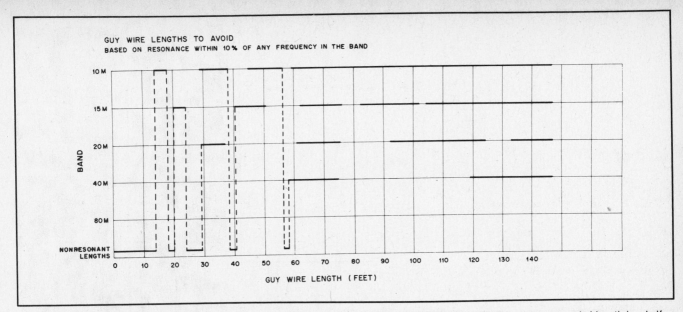

Fig. 63 — Guy wire resonant lengths. The solid lines opposite the band designations indicate that the corresponding ungrounded length is a half wavelength or an integral multiple thereof. Grounded wires will exhibit resonance at odd multiples of a quarter wavelength. Ideal guy wire lengths are 0 to 14 ft, 18 to 20 ft, 24 to 29 ft, 39 to 40 ft and 57 to 58 ft. This chart was prepared by Jerry Hall, K1TD.

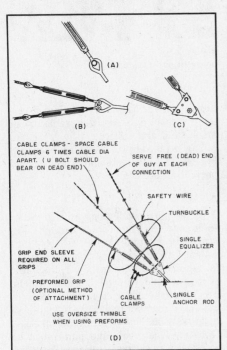

Fig. 64 — Variety of means available for attaching guy wires and turnbuckles to anchors.

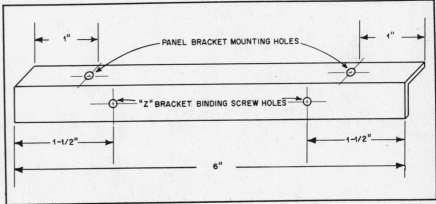

Fig. 65 — Z-bracket component pieces.

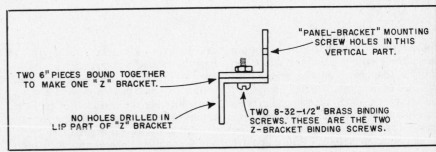

Fig. 66 — Assembly of the Z bracket.

to appreciate danger and which may reasonably be expected to attract children to the premises, is under a duty to exercise reasonable care to protect them against dangers of the attraction.

The tower shield is simply composed of panels that enclose the tower and make climbing practically impossible. These panels are 5 feet in height and are wide enough to fit snugly between the tower legs and flat against the rungs. A height of 5 feet is sufficient in most every case. The panels are constructed from 18-gauge galvanized sheet metal obtained and cut to

proper dimensions from a local sheet-metal shop. A lighter gauge could probably be used, but the extra physical weight of the heavier gauge is an advantage if no additional means of securing the panels to the tower rungs are used. The three types of metals used for the components of the shield are supposedly rust proof and nonreactive. The panels are galvanized sheet steel, the brackets aluminum, and the screws and nuts are brass. The tower shield

consists of three panels, one for each of the three sides, supported by two brackets. These brackets are constructed from 6-inch pieces of thin aluminum angle stock. Two of these pieces are bolted together to form a Z bracket (see Figs. 65, 66 and 67). The Z brackets are bolted together with binding-head brass machine screws.

The panels are laid flat for measuring, marking and drilling. The first measurement is from the top of the upper mounting

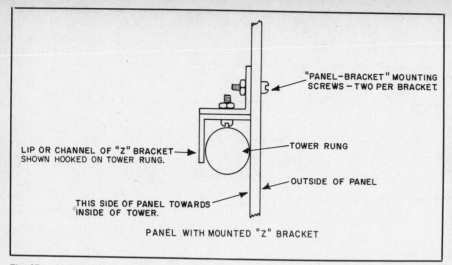

Fig. 67 — Installation of the shield on a tower rung.

Fig. 69 — Installed tower shield. Note the holes for using the handles.

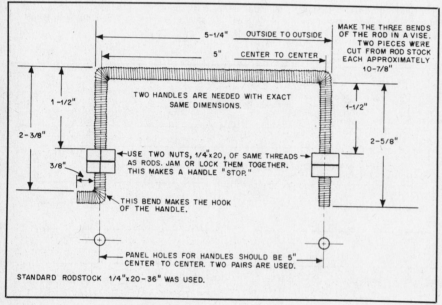

Fig. 68 — Removable handle construction.

rung on the tower to the top of the bottom rung. These mounting rungs are selected to position the panel on the tower. This distance from rung to rung is then marked on the panel. Using the same size brass screws and nuts, the top vertical portion of each Z bracket is bolted to the panel. The mounting-screw holes are drilled about 1 inch from the end of the Z brackets so that an offset clearance occurs between the Z-bracket binding-screw holes and the panel-bracket mounting-screw holes. The panel noles are drilled to match the Z-bracket holes.

The panels are held on the tower by their own weight. They are not easy to grasp because they fit snugly between the tower legs. If the need exists for added safety against deliberate removal of the panels, this can be accomplished by means of tie wires. A small hole can be drilled in the panel just above, just below, and in the

center of each Z bracket. Run a piece of heavy galvanized wire through the top hole, around the Z bracket, and then back through the hole just below the Z bracket. Twist together the two ends of the wire. One tie wire should be sufficient for each panel, but use two if desired.

The completed panels are rather bulky and difficult to handle. A feature that is useful if the panels have to be removed often for tower climbing or accessibility is a pair of removable handles. The removable handles can be constructed from one threaded rod and eight nuts (see Fig. 68). The two pairs of handle holes were drilled in the panels a few inches below the top Z bracket and several inches above the bottom Z bracket. For panel placement or removal, the handles are hooked in these holes in the panels. The hook, on the top of the handle, fits into the top hole of each pair of the handle holes. The handle is op-

tional, but for the effort required it certainly makes removal and replacement much safer and easier.

The installed tower shield can be seen in Fig. 69. This relatively simple device could prevent an accident.

ANTENNA INSTALLATION

All antenna installations are different in some respects. Therefore, thorough planning is the most important first step in installing any antenna. At the beginning, before anyone climbs the tower, the whole process should be thought through. The procedure should be discussed to be sure each crew member understands what is to be done. Plan how to work out all bugs. Consider what tools and parts must be assembled and what items must be taken up the tower. Extra trips up and down the tower can be avoided by using forethought.

Getting ready to raise a beam requires planning. Done properly, the actual work of getting the antenna into position can be done quite easily with only one person at the top of the tower. The trick is to let the ground crew do all the work and leave the person on the tower free to guide the antenna into position.

Before the antenna can be hoisted into position, the tower and the area around it must be prepared. The ground crew should clear the area around the base while someone climbs the tower to remove any wire antennas or other objects that might get in the way. The first person to climb the tower should also rig the rope and pulley that will

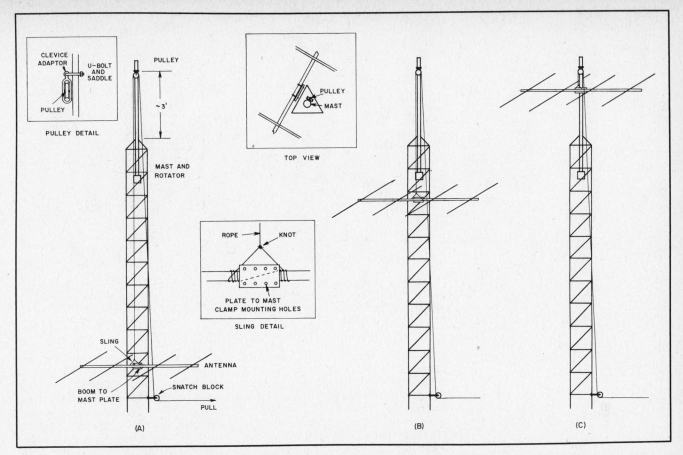

Fig. 70 — Installing antennas on free-standing towers is relatively simple. At A, the pulley is rigged to the mast about 3 feet above the antenna mounting point. A snatch block at the tower base is used so the ground crew can pull horizontally, while keeping the strain on the tower in the vertical plane. The Yagi is suspended in a rope sling at the tower base. The ground crew pulls the antenna up (B), and the antenna will remain parallel to the ground if rigged properly. When the antenna reaches the tower top, the tower climber simply bolts it into place (C).

be used to raise the antenna. The time to prepare the tower is before the antenna leaves the ground, not after it becomes hopelessly entwined with your 80-meter dipole.

Raising the Antenna

If you have a free-standing or crank-up tower, the task of actually getting the antenna from the ground to the tower top is simple. Rig a pulley to the mast as shown in Fig. 70. The pulley should be attached about 2 to 3 feet above the point where the antenna will bolt to the mast. If the antenna is heavy, it is advisable to rig a second pulley, or "snatch block," at the tower base. This arrangement allows the ground crew to use their entire body weight when pulling, rather than just the strength in their arms, making for a smoother and easier installation. The snatch block also keeps the load on the tower vertical, rather than lateral. Although many towers can take some lateral strain, it is better to be safe than sorry.

A properly designed and constructed antenna attaches to the mast at the balance point on the boom. If you rig a sling around the boom as shown in Fig. 70, you should be able to lift the antenna horizontally from the ground to the tower top. Be sure to rig the rope so that it does not obstruct the boom-to-mast bolt holes. Do not rig the rope so that it will come between the boom-to-mast plate and the mast. Test your rigging by lifting the beam a few feet above the ground. If it tilts in any direction, let it back down and rig it again. It should be possible to balance the antenna in the horizontal plane.

When the beam is balanced and safely rigged, the ground crew starts pulling the antenna up the tower. In some installations, it will be possible to bring the antenna to the top of the tower without anyone leaving the ground. In most, however, the tower climber will have to follow the antenna up the tower to keep it from getting caught. When the antenna reaches the tower top, the climber simply has to bolt it into place.

Avoiding Guy Wires

Although the same basic methods of installing a Yagi apply to any tower, guyed towers pose a special problem. Steps must be taken to avoid snagging the antenna on the guy wires. Let's consider, therefore, some ways in which this difficulty can be circumvented. With proper precautions, even large antennas can be pulled to the top of a tower, even if the mast is guyed at several levels.

Sometimes one of the top guys can pro- vide a track to support the antenna as it is pulled upward. Insulators in the guys, however, may offer obstructions that could cause the beam to get hung up. A better track made with rope is an alternative, provided there is sufficient rope available. One end of the rope is secured outside the guy anchors. The other end is passed over the top of the tower and back down to an anchor near the first anchor. So arranged, the rope forms a narrow V track strung outside the guy wires. Once the V track is secured, the antenna may simply be pulled up the track.

Another method is to tie a rope to the back of the antenna (but within reach of the center). The ground crews then pull the antenna out away from the guys as the antenna is raised. With this method, some crew members are pulling up the antenna to raise it while others are pulling down and out to have the beam clear the guys. Obviously, the opposing crews must be coordinated or they can literally tear the antenna apart. The beam is especially vulnerable when it begins to tip into the horizontal position. If the crew pulling out and down continues to pull against the antenna, the boom can be broken. Another problem with this approach is that the antenna may rotate on the axis of the boom as it is raised. To prevent such rotation, long

Fig. 71 — The PVRC mount, boom-to-mast plate, mast and rotator ready to go. The mast and rotator are installed first.

Fig. 72 — Close-up of the PVRC mount. Two of the four locking pins (bolts) may be seen at the midline of the left hand vertical plate. The other two pins are located along the axis of the short pipe section; the head of the right-hand bolt blends in with the U-bolt lock nut to the rear.

lengths of twine may be tied to an element, one piece on each side of the boom. Ground personnel may then use these to stabilize the antenna. Where this is done, provision should be made for untying the twine once the antenna is in place.

A third method is to tie the halyard to the center of the antenna. A crew member, wearing a safety belt, walks the antenna up the tower as the crew on the ground raises it. Because the halyard is tied at the center balance point, the man on the tower can rotate the elements around the guys. A tag line can be tied to the bottom end of the boom so that a man on the ground can help move the antenna around the guys. The tag line must be removed while the antenna is still vertical and you can reach it.

THE PVRC MOUNT

The above methods may not be satisfactory for large arrays on guyed towers. The best way to handle large Yagis is to assemble them on top of the tower. The way to do that is by using the "PVRC mount." Many members of the Potomac Valley Radio Club have used this method successfully to install their large antennas. Simple and ingenious, the idea involves offsetting the boom from the mast to permit the boom to tilt 360° and rotate axially 360°. This permits the entire length of the boom to be brought alongside the tower, allowing the elements to be attached one by one.

See Figs. 71 through 75. The mount itself consists of a short length of pipe of the same (or greater) diameter as the rotating mast, a steel plate and the hardware to hold it all together. The plate is drilled for eight U-bolts: four to attach the plate to the pipe, and four to attach the pipe to the antenna boom-to-mast plate. Additionally, four pinning bolts are used to ensure that the antenna will end up level and parallel to the ground. Two pinning bolts pass through the mast, two through the horizontal pipe. When the horizontal pipe pinning bolts are removed and the U-bolts are loosened, the boom-to-mast plate can be tilted 360°, allowing either half of the boom to come alongside the tower.

After all critical dimensions have been marked, the antenna elements are removed from the boom. Once the rotator and mast have been installed on the tower, a gin pole is used to bring the adapter plate and pipe to the top of the tower. There, the "top crew" unpins the horizontal pipe and tilts the antenna boom-to-mast plate to place it in the vertical plane. The boom is attached to the boom-to-mast plate at the balance point of the assembled antenna. It is important that the boom be rotated axially so the bottom side of the boom is closest to the tower. This will ensure that the antenna elements will be parallel to the side of the tower during installation, allowing the boom to be tilted without the elements striking the tower.

During installation it may be necessary to remove one guy wire temporarily to allow for tilting of the boom. As a safety precaution, a temporary guy should be mounted to the same leg of the tower just low enough so that the antenna will not hit it.

The elements are assembled on the boom starting with those closest to the center of the boom, working out alternately to the farthest director and reflector. This procedure must be followed. If all the elements are put first on one half of the boom, it will be dangerous (if not impossible) to put on the remaining elements. By starting at the middle and working outward the antenna weight will never be so far removed from the balance point that tilting of the boom becomes impossible.

When the last element is attached, the boom is brought parallel to the ground, the horizontal pipe is pinned and the U-bolts tightened. Now, all the antenna elements are positioned vertically. Next, loosen the U-bolts that hold the boom and rotate the boom axially 90°, bringing the elements parallel to the ground. Then tighten the boom U-bolts and double-check all the hardware.

Many long-boom Yagis employ a truss to prevent boom sag. With the type of mount just described the truss must be attached to a pipe that is independent of the rotating mast. A short length of pipe is attached to the boom as close as possible to the balance point. The truss will now move along with the boom whenever the boom is tilted or twisted.

ROTATOR SYSTEMS

There are not many choices when it comes to antenna rotators for the amateur antenna system. Choosing the right rotator is very important if trouble-free operation is desired. There are basically four grades of rotators available to the amateur. The lightest-duty rotator is the TV type, typically used to turn TV antennas. These rotators will handle, without much difficulty, a small three-element tribander (20-, 15- and 10-meter) array or a single 15- or 10-meter monoband three-element antenna.

The important consideration with a TV rotator is that it lacks braking or holding capability. High winds will turn the rotator

Fig. 73 — Working at the 70-foot level. A gin pole makes pulling up and mounting the boom to the boom-to-mast plate a safe and easy procedure.

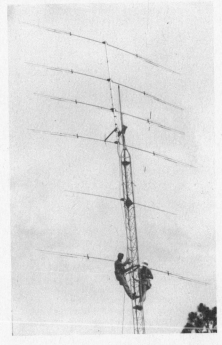

Fig. 74 — Mounting the last element prior to positioning the boom in a horizontal plane.

motor via the gear train in a reverse fashion, sometimes resulting in broken gears. The next grade up from the TV class of rotator usually includes a braking arrangement whereby the antenna is held in place when power is not applied to the

rotator. Generally speaking, the brake will prevent gear damage on windy days. If adequate precautions are taken, this group of rotators is capable of holding and turning stacked monoband arrays, or up to a five-element 20-meter system. The next step up in strength is more expensive. This class of rotator will turn just about anything the most demanding amateur might want to install.

A description of antenna rotators would not be complete without the mention of the prop-pitch class. This rotator system consists of a surplus aircraft propeller-blade pitch motor coupled to an indicator system as well as a power supply. There are installation problems, however. It has been said that a prop-pitch rotator system, properly installed, is capable of turning a house. This is no doubt true! Perhaps in the same class as the prop-pitch motor but with somewhat less capability is the electric motor of the type used for opening garage doors. These have been used successfully in turning large arrays.

Proper installation of the antenna rotator can provide many years of trouble-free service; sloppy installation can cause problems such as a burned-out motor, slippage, binding and casting breakage. Most rotators are capable of accepting mast sizes of different diameters, and suitable precautions must be taken to shim an undersized mast to ensure dead-center rotation. It is very desirable to mount the rotator inside and as far below the top of the tower as possible. The mast will absorb the torque developed by the antenna during high winds, as well as during starting and stopping. A mast length of 10 feet or more between the rotator and the antenna will add greatly to the longevity of the entire system. Some amateurs have used a long mast from the top to the base of the tower. Installation and service can be accomplished at ground level. Another benefit of mounting the rotator 10 or more feet below the antenna is that any misalignment among the rotator, mast and the top of the tower is less significant. A tube at the top of the tower through which the mast protrudes will almost completely eliminate lateral forces on the rotator casting. All the rotator need do is support the downward weight of the antenna system and turn the array. While the normal weight of the antenna and the mast is usually not more than a couple of hundred pounds, even with a large system, one can ease this strain on the rotator by installing a thrust bearing at the top of the tower. The bearing is then the component that holds all the weight, and the rotator need perform only the rotating task.

A problem often encountered in the

Fig. 75 — The mast-to-pipe U-bolts are loosened and the boom is turned to a horizontal position. This puts the elements in a vertical plane. Then the pipe U-bolts are tightened and pinning bolts secured. The boom U-bolts are then loosened and the boom turned axially 90°.

amateur installation is that of misalignment between the direction indication of the rotator control box and the heading of the antenna. This is caused by mechanical slippage in the system caused by loose bolts or antenna boom-to-mast hardware. Many texts suggest that the boom be pinned to the mast with a heavy-duty bolt and, likewise, the rotator be pinned to the mast. There is a trade-off here. The amateur might not like to climb the tower and straighten out the assembly after each heavy windstorm. However, if there is sufficient wind to cause slippage in the couplings, the wind could break a casting. The slippage will act as a clutch release, which may prevent serious damage.

Completing the Installation

Once you've completed the tower, mast, antenna and rotator installation, verify that everything works. After you've done this, it is important to put on the finishing touches. Waterproof any coaxial cable connections and control lead connections. Use plenty of electrical tape, some heat-shrink tubing, or putty. Tape the cable along the antenna boom and down the mast. At the tower top, leave a loop large enough that the antenna can turn 360 degrees without binding the coaxial cable. Then secure any cables to the tower leg all the way to the ground. Cables that flap in the wind may be damaged under stress.

Maintenance

It's a good idea to inspect your tower and antenna system several times during the year. Make a checklist and run through it from top to bottom. Look for loose or missing hardware, rust, structural defects and worn cables. Periodically check the tension of your guy wires and adjust them if necessary. Catch problems before they become serious.

Mobile and Portable Stations

A major justification for the existence of Amateur Radio in the USA is to provide a pool of experienced operators in time of national or community need. When the call for emergency communications is voiced by cities, towns, counties, states or the federal government, mobile and portable radio equipment is pressed into service where needed. Aside from the occasional disaster and emergency type of communications, a great deal of pleasure and challenge can reward the amateur when operating portable or mobile under normal conditions.

Most mobile operation is carried out today by means of narrow-band FM and repeaters. The major repeater frequencies are in the 146, 220 and 440-MHz bands. It is expected that this reliable service mode will soon include widespread occupancy of the 1240-MHz band and higher.

Mobile HF-band operation still appeals to numerous amateurs because it eliminates the constrictions imposed by VHF and UHF repeaters, their operators and their normal coverage contours. When operating mobile on SSB or CW with HF transceivers, worldwide contacts are possible for those who enjoy that style of communication.

MOBILE STATIONS

Layout of a mobile station is considerably easier than that of a fixed station. In most vehicles, space is limited so there are only one or two possible spots for locating the equipment. Whenever possible, it should be positioned so that the operator can reach the front-panel controls without compromising the ability to operate the automobile safely. Driving safely is *always* the first priority; operating radio equipment is secondary.

While most mobile operation is done at low power levels, such as 10 watts from a VHF FM transceiver, high-power mobile operation is entirely practical from the 13.6-V dc battery system. There are many VHF power amplifiers on the market today at power levels up to 150 watts that require 13.6-V dc. Also, many HF rigs in the 100-watt-output class will operate directly from the automobile power system. Dynamotors, vibrator packs and dc-to-dc converters are things of the past, making mobile operation much more attractive.

If the amateur transceiver or amplifier requires more than a few amperes at 13.6 volts, it is best to run a heavy cable directly to the automobile battery. Few automobile electrical systems are set up to allow you to safely draw the 20 amperes required for a 150-watt transmitter from the accessory connection on the fuse panel. Consult the wire table in Chapter 35 for the safe size wire for your installation. Keep in mind that the larger the wire, the lower the voltage drop under load; a larger conductor will help you to get the most from your

equipment. Adequately fuse your power cable. Automobile fires are costly and dangerous.

A variety of mobile antennas are available. Chapter 33 of this Handbook contains information on building and using mobile antennas for HF and VHF.

Electrical-Noise Elimination

One of the most significant deterrents to effective signal reception during mobile or portable operation is electrical impulse noise from the automotive ignition system. The problem also arises during the use of gasoline-powered portable ac generators. This form of interference can completely mask a weak signal, thus rendering the station ineffective. Most electrical noise can be eliminated by taking logical steps toward suppressing it. The first step is to clean up the noise source itself, then utilize the receiver's built-in noise-reducing circuit as a last measure to knock down any noise pulses from passing cars, or from other man-made sources.

Spark-Plug Noise

Most vehicles manufactured prior to 1975 were equipped with simple Kettering inductive-discharge ignition systems. A variety of noise-suppression methods were devised for these systems, including resistor spark plugs, clip-on suppressors, resistive high-voltage cable and even complete shielding. Resistive high-voltage cable and resistor plugs provide the greatest noise reduction for the least expense and effort. While almost all vehicles produced after 1960 had resistance cable as standard equipment, such cable develops microscopic cracks in the insulation and segmentation of the conductor after a few years of service. These defects can defeat the suppression ability of the cable before the engine performance degrades noticeably. Two years is a reasonable replacement interval for spark-plug cable. Earlier editions of this *Handbook* described complete shielding methods for inductive-discharge ignitions.

Some late-model automobiles employ sophisticated high-energy electronic ignition systems in an attempt to reduce exhaust pollution and increase fuel mileage. With increased sophistication comes greater sensitivity to modification — solutions to RFI caused by older Kettering systems cannot be uniformly applied to the modern electronic ignitions.

Such fixes may be ineffective at best, and at worst may impair the engine performance. One should thoroughly understand an ignition system before attempting to modify it. One of the significant features of capacitive discharge systems, for example, is extremely rapid voltage rise, which combats misfire caused by fouled spark plugs. Rapid voltage rise depends on a low

Fig. 76 — This may represent the ultimate in portable VHF operation. The aggregate antenna gain respresented by this setup could be very effective in time of emergency. This installation was built and operated by N6NB. The antenna shown is effective also for EME work.

Fig. 77 — Field Day is many things to many people, but everyone who participates does so away from commercial power. These gents, members of the Murgas ARC of Wilkes-Barre, Pennsylvania, hoisted a 15-meter beam atop a truck for their 1981 effort. (*photo by Mike Benish, K3SAE*)

RC time constant presented to the output transformer. For this reason, high-voltage suppression cable designed for capacitive-discharge systems is wound with monel wire. It exhibits a distributed resistance of only about 600 ohms per foot, as contrasted with 10 kΩ per foot for carbon-center cable used with inductive-discharge systems. Increasing the RC product by shielding or installing improper spark-plug cable could seriously compromise the capacitive-discharge circuit operation.

Ferrite beads represent a possible means

for RFI reduction in newer vehicles. Both primary and secondary ignition leads are candidates for beads. Install them liberally, then load test the engine for adequate spark energy.

The plane sheet metal surfaces and cylindrical members such as exhaust pipes often exhibit resonance in one of the amateur bands. Such resonances encourage reradiation of spark impulse energy. Bonding these structural members with heavy braid can reduce the level of spark noise inside the vehicle and out. Other types of noise to be described later can also be helped by bonding. Here are the main areas to bond. (1) Engine to frame; (2) air cleaner to engine block; (3) exhaust lines to frame; (4) battery ground terminal to frame; (5) hood to fire wall; (6) steering column to frame; (7) bumpers to frame; and (8) trunk lid to frame.

On the whole, modern automobiles are cleaner from an RFI standpoint than those of 20 years ago. The interference problem, however, at least at VHF and UHF, persists because present-day receivers are about 10 dB more sensitive than their predecessors.

Charging-System Noise

Noise from the vehicular battery-charging circuit can interfere with both transmission and reception of radio signals. The charging system of a modern automobile consists of a belt-driven, three-phase alternator and a solid-state voltage regulator. Solid-state regulators are a great improvement over the electromechanical vibrating contact units of earlier times. Interference from the charging system can affect a receiver in two ways: RF radiation can be picked up by the antenna, and noise can be conducted directly into the circuits via the power cable. "Alternator whine" is a common form of conducted interference. It has the greatest effect on VHF FM communications, because synthesized carrier generators and local oscillators are easily frequency modulated by power supply voltage fluctuations. The alternator ripple is most noticeable when transmitting, because the machine is more heavily loaded in that condition. If the ripple amplitude is great enough, alternator whine will be imparted to all incoming signals by the LO.

Conducted noise can be minimized by connecting the radio power leads directly to the battery, as this is the point in the electrical system having the lowest impedance. If the regulator is adjustable, set the voltage no higher than is necessary to ensure complete battery charging. Radio equipment manufacturers combat voltage variations by internally regulating critical circuits wherever possible.

Both conducted and radiated noise can be suppressed by filtering the alternator leads. Coaxial capacitors (about 0.5 μF) are suitable, but don't connect a capacitor to the field. The field lead can be shielded

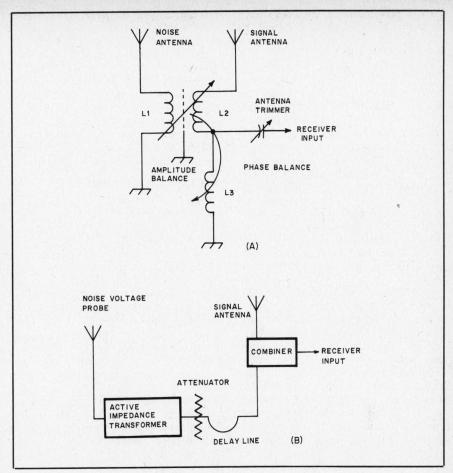

Fig. 78 — Automotive noise-cancelling systems. At A, the circuit used in the BC-342 HF receiver. At B, a suggested broad-band noise-cancelling scheme.

or loaded with ferrite beads if necessary. A parallel-tuned LC trap in this lead may be effective against radiated noise. Such a trap in the output lead must be made of no. 10 wire or larger, as some alternators conduct up to 100 amperes. The alternator slip rings should be kept clean to prevent excess arcing. An increase in "hash" noise may indicate that the brushes need to be replaced.

Instrument Noise

Some automotive instruments are capable of creating noise. Among these gauges and senders are the heat- and fuel-level indicators. Ordinarily, the addition of a 0.5-μF coaxial capacitor at the sender element will cure the problem.

Other noise-generating accessories are turn signals, window-opener motors, heating-fan motors and electric windshield-wiper motors. The installation of a 0.25-μF capacitor will usually eliminate their interference.

Corona-Discharge Noise

Some mobile antennas are prone to corona build-up and discharge. Whip antennas that come to a sharp point will sometimes create this kind of noise. This is why most mobile whips have steel or plastic balls at their tips. But, regardless of the structure of the mobile antenna, corona buildup will frequently occur during or just before a severe electrical storm. The symptoms are a high-pitched "screaming" noise in the mobile receiver, which comes in cycles of one or two minutes' duration, then changes pitch and dies down as it discharges through the front end of the receiver. The condition will repeat itself as soon as the antenna system charges up again. There is no cure for this condition, but it is described here to show that it does not originate within the electrical system of the automobile.

Electronic Noise Reduction

When all electrical noise generated within a vehicle has been eliminated, the mobile operator can be annoyed by RFI from passing vehicles. Some measures can be taken in the receiver to reduce or reject impulse noise. (Noise limiters and noise blankers are discussed in another chapter.) The placement of a noise blanker in the receiver is important. The blanking circuit must be placed ahead of the sharp selectivity; otherwise, the IF filter will stretch the noise pulses, and they cannot be blanked without destroying a major portion of the received intelligence.

The Soviet "woodpecker" over-the-horizon radar has inspired some serious development work on noise blankers that don't degrade receiver dynamic range. Receivers for VHF FM service are generally designed for optimum noise figure at the expense of resistance to overload. Recent advances in RF amplifier design have proven that low noise figure and high dynamic range are not mutually exclusive. A high-performance noise blanker is useless if the front end of the receiver overloads on the noise pulses. A helical resonator at the receiver input affords some protection against noise overload because it restricts the total noise energy delivered to the front end.

Some FM receivers suffer from impulse noise because of inadequate AM rejection. The cure for this ailment is to ensure hard limiting in the IF stages and to use a detector that is inherently insensitive to amplitude variations.

Particularly troublesome vehicular impulse noise can sometimes be cancelled at the receiver input. The technique involves sampling the noise voltage from a separate "noise antenna" and adjusting its phase and amplitude to cancel the noise delivered by the "signal antenna." For this system to be effective, the signal antenna must be positioned to provide the best possible signal-to-noise ratio, and the noise antenna located close to the noise source and effectively shielded from the desired signal.

Fig. 78A shows the noise cancellation circuit used in some models of the BC-342, a WW II receiver. The Faraday shield between L1 and L2 ensures that the coupling is purely magnetic. The coupling between L2 and L3 is purely electrostatic. Adjusting the coil coupling causes the noise to null. The block diagram of Fig. 78B illustrates a more modern broadband approach to noise cancellation. A short wire near the ignition coil couples impulse energy into the active impedance transformer, which is simply an FET source follower stage. The amplitude and phase of the noise are controlled by the attenuator and delay line, respectively. The signal combiner can be a hybrid ferrite transformer at HF or a transmission line multi-coupler at VHF.

PORTABLE STATIONS

Many amateurs experience the joys of portable operation once each year in the annual emergency exercise known as Field Day. Setting up an effective portable station requires organization, planning and some experience. For example, some knowledge of propagation is essential to picking the right band or bands for the intended communications link(s). Portable operation is difficult enough without dragging along excess equipment and antennas that will never be used.

Some problems encountered in portable operation that are not normally experienced in fixed-station operation include finding an appropriate power source and erecting an effective antenna. The equipment used should be as compact and lightweight as possible. A good portable setup is simple. Although you may bring gobs of gear to Field Day and set it up the day before, during a real emergency speed is of the essence. The less equipment to set up, the faster it will be operational.

PORTABLE AC POWER SOURCES

There are two popular sources of ac power for use afield. The first is what is referred to as a dc-to-ac converter, or more commonly, an inverter.

The ac output voltage is a square wave. Therefore, some types of equipment cannot be operated satisfactorily from the inverter. Certain types of motors are among those items that require a sine-wave output. Fig. 79 shows a picture of one style of commercial inverter. Heat sinks are used to cool the switching transistors. It delivers 117 volts of ac at 175 watts continuous power rating. The primary voltage is 6 or 12 dc.

When sine-wave output is required from a portable ac power supply, gasoline-engine alternators are used. They are available with ratings of several kilowatts, or as little as 500 watts. One of the larger units is shown in Fig. 80 where WB9QPI has just completed a maintenance run for the WØOHU/Ø Field Day group.

Alternators powered by internal-combustion engines have been used for years to supply 117/234-V ac independent of the commercial mains. Such combinations range from tiny units powered by two-cycle or four-cycle gasoline engines in the low-wattage class to giant multicylinder diesels capable of supplying megawatts of power. Perhaps the most practical power range for most purposes is in the neighborhood of 2 kW. Larger units tend to become too heavy for one person to lift and handle easily while smaller generators lack sufficient power output for many applications. A 2-kW alternator is quite heavy but is capable of supplying power for just about any large power tool. It is roughly the equivalent of having a single 15-A outlet in an ordinary electric service. Of course, it will handle moderate-power amateur equipment with ease.

Maintenance Checklist

Although more complicated maintenance chores should be performed by qualified service personnel, many simple measures which will prolong the life of the alternator can be done at home or afield. Perhaps the best plan is to log the dates the unit was used and the operating time in hours. Also include dates of maintenance and type of service performed. Oil changes, when gasoline was purchased for emergency purposes, and

Fig. 79 — Photograph of a commercial dc-to-ac inverter that operates from 6 to 12 volts dc and delivers 117 volts ac (square wave) at 175 watts.

Fig. 80 — Large gasoline generators of the kilowatt and higher class are excellent for powering several amateur stations from a complex field site. Maintenance, as discussed in the text, is a vital matter to ensure reliable operation. Here, WB9QPI has just finished a maintenance check of the group's Field Day power plant.

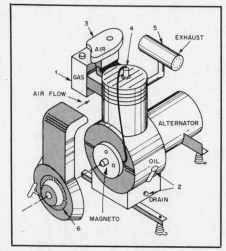

Fig. 81 — The numbers indicate the primary maintenance points of a large power generator (see text for details).

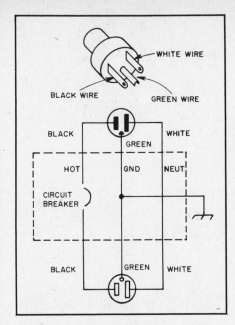

Fig. 82 — A simple accessory that provides overload protection for generators that do not have such provisions built in.

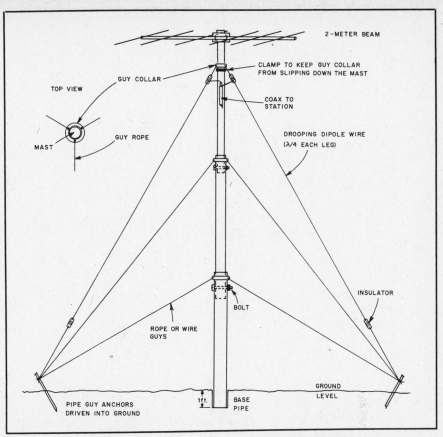

Fig. 83 — A simple portable antenna mast may be built from telescoping tubing sections. The mast is assembled and bolted together on the ground, and the antennas attached. The mast is then pushed into the air and guyed in place. If slip-ring guy collars are used, the entire mast may be rotated by hand from ground level. A dipole may be used as part of the guy system.

similar data would fall under this category.

Important points common to all types of generators are indicated for a typical one in Fig. 81. (Consult the manufacturer's manual for additional instructions that might apply to a particular model.) The following checklist relates to the numbers on the drawing.

1) Use the proper grade of fuel. Check the owner's manual to determine whether oil must be mixed with the gasoline. While two-cycle models require an oil-gas mixture, most generators have a four-cycle engine that burns ordinary gasoline with no extra additives. Gasoline for emergency purposes should be stored only in small amounts and rotated on a regular basis. Older stock can be burned in a car (that uses the same grade of gas as the generator) since storing gasoline for any length of time is inadvisable. The more volatile components evaporate, leaving excess amounts of a varnish-like substance that will clog carburetor passages. Also, be sure gasoline containers are of an approved type with a clean interior, free of rust or other foreign matter. Similar considerations apply to the gas tank on the engine itself.

The majority of difficulties with small engines are related to fuel problems in some way. Dirty fuel or water in the gasoline is one source, with carburetor trouble because of the use of old gas being another common cause. Except for minor adjustments recommended in the instruction manual, it is seldom necessary to touch the carburetor controls. Avoid the temptation to make such adjustments in the case of faulty operation. Follow the recommendations in this guide so that more complicated maintenance procedures (such as carburetor overhaul) are not required.

2) Another important factor often neglected in maintenance of alternator engines is oil. While lubrication is one job oil has to perform, there are other considerations as well. The engine oil in the crankcase also collects a large amount of solid combustion products, bits of metal worn away by the moving parts, and any dust or other foreign matter that enters the carburetor intake. For instance, it is especially important to observe the manufacturer's recommendations concerning the length of time the engine may be operated before an oil change is required during the break-in period. If you ever have the opportunity to examine the oil from a new engine, you will note a metallic sheen to it. This is from the excessive amount of metal that is worn away. After the break-in period, much less metal is abraded and the oil doesn't have to be replaced as often.

The oil level should be checked frequently during engine operation. Each time fuel has to be added the oil should be checked also. When storing an alternator, it is also wise to drain the oil and replace it with fresh stock. This is because one of the combustion products is sulfur, which forms sulfuric acid with water dispersed in the oil. The acid then attacks the special metal in the bearing surfaces causing pitting and the need for premature replacement.

Also note the grade and weight of oil recommended by the manufacturer. Unlike their larger counterparts in automobiles, most small engines do not have oil filters, which is another reason required changes are more frequent. Some manufacturers recommend a high-detergent oil that comes in various service grades such as MS and SD. Examine the top or side of the cans in which the oil is sold and see if the letters correspond to those recommended by the engine manufacturer.

3) The carburetor mixes gasoline with air, and the mixture is then burned in the engine. Before entering the carburetor, the air must be filtered so it is free of dust and other foreign matter that might otherwise be drawn into cylinder(s). Particles that do get by the air filter are picked up by the oil. (Oil should be changed more often if the alternator is operated in a dusty location.) Also, it is important to clean the air filter frequently. It contains a foam-like substance that can be cleaned in kerosene and then soaked in fresh motor oil. Squeeze excess oil from the filter before replacing. Also, consult the instruction manual for further recommendations.

4) Once the gas/air mixture enters the cylinder, it is compressed by the piston into a very small volume and ignited by the spark plug. During the rapid burning that then occurs, the expansion caused by the resulting heat forces the piston down and

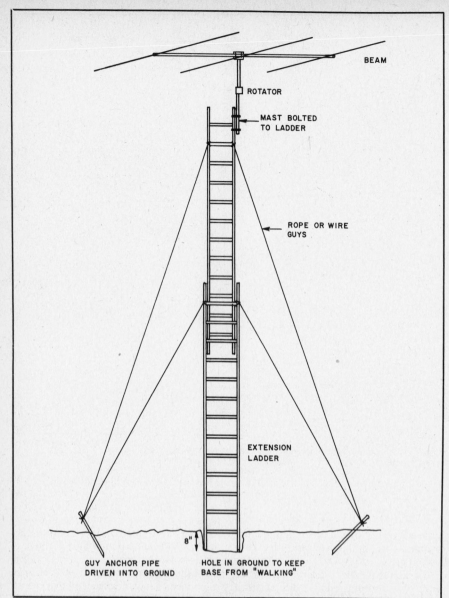

Fig. 84 — An aluminum extension ladder makes a simple but sturdy portable antenna support. Attach the antenna and feed lines to the top ladder section while it is nested and laying on the ground. Push the ladder vertical, attach the bottom guys and extend the ladder. Attach the top guys. Do not attempt to climb this type of antenna support.

(B)

Fig. 85 — The portable tower mounting system described by WA7LYI in the text. At A, a truck is "parked" on the homemade base plate to weigh it down. At B, the antennas, mast and rotator are mounted before the tower is pushed up. Do not attempt to climb a temporary tower installation.

delivers the mechanical power to the alternator.

As might be expected, proper operation of the ignition system is an important factor in engine performance. Power for the spark is supplied by a device called a magneto that is normally installed on the front of the engine. The magneto seldom requires servicing and such work should only be done by those qualified to do so. (This is one reason the magneto is often located under a flywheel that is difficult to remove by the inexperienced.)

Faulty spark plugs are the usual cause of ignition problems. Special equipment is required to test a spark plug properly, but an easier solution is to have a new one handy. In fact, keep two spare plugs on hand. Spark plug life can be notoriously short on occasion. However, repeated plug failure is also abnormal; other causes such as a poor gas/air mixture might be the culprit.

Replace the spark plug with a type similar to the one that came with the alternator or a substitute recommended by the manufacturer. Some models have resistor-type plugs, which are desirable for ignition-noise suppression. Resistor plugs are usually indicated by an R prefix. For instance the resistor version of a Champion CJ-8 would be an RC-J8.

5) Little exhaust system maintenance is required. In some forested areas, a spark-arrester type of muffler is required, so be sure your unit is so equipped before operating in such a location. "Quiet hours" may also be imposed in some places during the nighttime hours if generator exhaust noise is too loud.

Two very important safety precautions should be observed with regard to the exhaust system. Never operate an alternator in closed surroundings such as a building. Dangerous gases are emitted from the exhaust. Second, never refuel an engine while it is running or if the exhaust system is still very hot. Unfortunately, this last precaution is disregarded by many, which is extremely foolish. (Experienced service station operators will refuse to refuel an automobile with the motor running, a practice that is often prohibited by law.) Don't become an unnecessary statistic.

6) Most alternators are air-cooled as opposed to the water-cooled radiator system of the automobile. A fan on the front of the engine forces air over the cylinder. Avoid operating the alternator in areas where obstruction to this flow might result (such as in tall grass). Alternators should be operated such that a sufficient amount of air circulation is present for cooling, carburetion and exhaust.

Storage

Proper maintenance of an alternator when it is not being used is just as important as during the time it is in operation. The usual procedure is to run the engine dry of gasoline, drain the crankcase and fill it with fresh oil, and remove the spark plug. Then pour a few tablespoons of oil into the cylinder and turn the engine over a few times with the starter and replace the plug. But never crank the engine with the plug removed and the ignition or start switch in the on or run position. The resulting no-

(A)

(B)

(C)

Fig. 86 — The portable mast and tripod described by WA7LYI in the text. At A, the tripod is clamped to stakes driven into the ground. The rotator is attached to a homemade pipe mount. At B, rocks piled on the rotator mount keep the rotator from twisting and add weight to stabilize the mast. At C, a 10-foot mast is inserted into the tripod/rotator base assembly. Four 432-MHz quagis are mounted at the top.

load high voltage might damage the magneto. It is also a good idea to ground the spark-plug wire to the engine frame with a clip lead in case the switch is accidentally activated.

Moisture is the greatest enemy of an iron product — such as a generator — in storage. The coating of oil helps retard rust formation that might actually weld the two surfaces when the engine is restarted, resulting in premature wear. Consequently, it is important to store the alternator in an area of low humidity.

Although the maintenance procedures outlined may seem like a chore, the long-term benefits include low repair costs and like-new performance. Engines for alternator combinations must be able to handle a variety of loads while maintaining a constant speed in order to keep the output frequency constant. A mechanical governor performs this latter function by metering the fuel supplied to the engine under different load conditions. However, the system cannot function properly with an engine in poor mechanical condition because of lack of proper maintenance.

Grounds

Newer generators are supplied with a three-wire outlet, and the ground connection should go to the plug, as shown in Fig. 82. On older types, the ground would have to be connected separately to the generator frame and then to the common terminal in the junction box. A pipe or rod can then be driven into the ground and a wire connection made to either a clamp supplied with the rod or by means of a C-clamp for larger sizes of pipe. From an ignition-noise-

suppression standpoint, the ground is desirable along with safety considerations when power tools are being used.

The ground connection goes to the green wire in commercially made three-wire conduit. Conduit purchased from an electrical store comes with a color-coded insulation and the colored wires should be connected as shown in Fig. 82. Consult the owner's manual for the generator for further details on power hookup that might apply to your particular model.

PORTABLE ANTENNA CONSIDERATIONS

An effective antenna system is essential to all types of operation. Effective portable antennas, however, are more difficult to devise than their fixed-station counterparts. A portable antenna must be light, compact and easy to assemble. It is also important to remember that the portable antenna may be erected at a variety of sites, not all of which will offer ready-made supports. Strive for the best antenna system possible because operations in the field are often restricted to low power by power supply and equipment considerations. Some antennas suitable for portable operation are described in Chapter 33.

Antenna Supports

While some amateurs have access to a truck or trailer with a portable tower, most are limited to what nature supplies, along with simple push-up masts. If possible, select a portable site as high and clear as possible. Elevation is especially important if your operation involves VHF. Trees, buildings, flagpoles, telephone poles and

the like can be pressed into service to support wire antennas. Usually, drooping dipoles are chosen over horizontal dipoles because they require only one support.

A portable mast up to 30 feet high can be constructed from telescoping tubing sections, as shown in Fig. 83. A mast of this type can be used as a wire-antenna support, or it could support a small triband beam or VHF array. If the guy collars allow the mast to slip, the entire mast may be rotated by hand. Some of the guy wires may be used for wire antennas. The limiting factor in terms of height with a mast of this type is the number of people required to push it into place.

An aluminum extension ladder makes an effective antenna support, as shown in Fig. 84. In this installation, a mast, rotator and beam are attached to the top of the second ladder section with the ladder near the ground. The ladder is then pushed vertical and the lower set of guy wires attached to the guy anchors. When the first set of guy wires is secured, the ladder may be extended and the top guy wires attached to the anchors. Do not attempt to climb a temporarily guyed ladder.

Figs. 85 and 86 illustrate two methods for mounting portable antennas described by Terry Wilkinson, WA7LYI. Although the antennas shown are used for VHF work, the same principles can be applied to small HF beams as well.

In Fig. 85A, a 3-foot section of Rohn 25 tower is welded to a pair of large hinges, which in turn are welded to a steel plate measuring approximately 18 × 30 inches. One of the rear wheels of a pickup truck is "parked" on the plate, ensuring that it will not move. In Fig. 85B, a quad array of antennas for 144 and 220 MHz are mounted on a Rohn 25 top section, complete with rotator and feed lines. The tower is then pushed up into place using the hinges, and guy ropes anchored to stakes driven into the ground complete the installation. This method of portable tower installation offers an exceptionally easy-to-erect, yet sturdy, antenna support. Towers installed in this manner may be 30 or 40 feet high; the limiting factor is the number of "pushers" and "rope pullers" needed to get it into the air. A portable station located in the bed of the pickup truck completes the installation.

The second method of mounting portable beams described by WA7LYI is shown in Fig. 86. This support is intended for use with small or medium sized VHF and UHF arrays. The tripod is available from any dealer selling television antennas; tripods of this type are usually mounted on the roof of a house. Open the tripod to its full size and drive a pipe into the ground at each leg. Use a hose clamp or small U-bolt to anchor each leg to its pipe.

The rotator mount is made from a 6-inch-long section of 1½-inch-diameter pipe welded to the center of an "X" made from two 2-foot-long pieces of concrete reinforcing rod (rebar). The rotator clamps onto the pipe, and the whole assembly is placed in the center of the tripod. Large rocks placed on the rebar hold the rotator in place, and the antennas are mounted on a 10 or 15-foot mast section. This system is easy to make and set up.

Tips for Portable Antennas

Any of the antennas described in Chapter 33 or available from commercial manufacturers may be used for portable operation. Generally, though, big or heavy antennas should be passed over in favor of smaller arrays. The couple of decibels of gain a 5-element, 20-meter beam may have over a 3-element version is insignificant compared to the mechanical considerations. Stick with arrays of reasonable size that are easily assembled.

Wire antennas should be cut to size and tuned prior to their use in the field. Be careful when coiling these antennas for transport, or you may end up with a tangled mess when you need an antenna in a hurry. The coaxial cable should be attached to the center insulator with a connector for speed in assembly. Use RG-58 for the low bands and RG-8X for higher-band antennas. Although these cables exhibit higher loss than standard RG-8, they are far more compact and weigh much less for a given length.

Beam antennas should be assembled and tested before taking them afield. Break the beam into as few pieces as necessary for transportation and mark each joint for speed in reassembly. Hex nuts can be replaced with wing nuts to reduce the number of tools necessary.

Chapter 38

Operating a Station

This Handbook serves primarily the technical phases of Amateur Radio — and indeed radio/electronics in general. As far as hamming is concerned, only about half the game deals with how you handle the soldering iron. The other half is concerned with what you do with your radio equipment once you get it operating. This is the subject of this chapter, and the real fun of Amateur Radio!

The *ARRL Operating Manual* is the definitive source of information on all phases of on-the-air activity for radio amateurs and goes into more detail than we can here. You can purchase a copy from ARRL Hq. or from your local dealer. Newcomer and veteran alike will profit from in-depth treatment of everything from DXing to digital communications. This publication is available at your local dealer or direct from ARRL Hq.

Amateur Radio is enriched by the influx of enthusiastic young operators such as Mike, KA3HID.

Operating Standards

Through the years, Amateur Radio has developed a number of operating standards and procedures. Some of these are borrowed from other services, such as commercial or military. Some have been coined by amateurs for our particular use, as a part of "ham jargon." Still others have been innovated by the League to fulfill a need. All of them together make up a "standard operating procedure" for amateurs that differs, at least in part, from the procedures used in any other radio service. ARRL recommends standard procedures that are based on our particular needs. If we all use different procedures, we will have difficulty, at times, in communicating with each other. Let's all use the same standard and be a single organized, operating service.

Establishing Contact

If you are looking for a contact with anyone, then you may want to call "CQ." Before your signal uses valuable spectrum space, however, listen to see if the frequency appears to be in use. Follow that

up with, "Is the frequency in use?" on voice, or QRL? on CW. You don't need a long CQ — just a 3 × 3. If still no answer, it means one of two things: Either no one heard you or no one listening wants to contact you. Try it again, another 3 × 3. If still no answer, that's enough. Move to another (clear) frequency before you try it again. A CQ call on CW might go like this: CQ CQ CQ DE AB1U AB1U AB1U K. On voice, "CQ CQ, this is AB1U, Alfa Bravo One Uniform, go ahead."

If you call CQ, it means you are willing to talk to anyone. If you want to be fussy, don't call CQ: Find someone you would like to talk to and call that person. In this case, you can observe the same principle; zero beat his frequency and give him a short call, such as: K1MM K1MM K1MM DE VE3GT VE3GT AR, or even shorter. On voice, "K1MM K1MM from VE3GT Victor Echo Three Golf Tango, over."

Notice the "ending signals." These aren't just happenstance; each one of them means something. On CW for example, K at the end of a transmission means "anyone go ahead," while AR means "I

have just called another station and want only him to reply. On voice, "go" or "go ahead" refers to anybody listening, while "over" refers to a specific station. And so on. There is a complete list of ending prosigns and prowords. Use them properly, even if others you hear do not.

Now about the QSO itself. You are introducing yourself to a brand-new acquaintance. Don't bore him. The usual procedure is to give your name and location. After that, tell him the things about yourself that you would like to know about him, but make your transmissions short. Most voice operators use VOX or push to talk, which makes it possible to talk back and forth rapidly, much like a telephone conversation. Identification of your station during each transmission is unnecessary, but be sure to sign your call at least once every 10 minutes during the course of the QSO and at the conclusion of the QSO.

The ARRL QSL Bureaus

Only one thing is left to do before you can consider the contact complete — confirm it with a QSL card. The QSL is con-

Table 1

Some Facts About Time Conversion

This chart has been arranged to show UTC and the time zones used by most amateurs in North America. Universal Coordinated Time is the standard reference time used throughout the world. ARRL recommends that all amateur logging be done in UTC.

All times shown are in 24-hour time for convenience. To convert to 12-hour time: For times between 0000 and 0059, change the first two ciphers to 12, insert a colon and add A.M.; for times between 1200 and 1259, insert a colon and add P.M.; for times between 1300 and 2400, subtract 12, insert a colon and add P.M.

Time zone letters may be used to identify the kind of time being used. For example, UTC is designated by the letter Z, EDT/AST by the letter Q, CDT/EST by R, MDT/CST by S, PDT/MST by T, PST by U; thus, 1200R would indicate noon in the CDT/EST zone, which would convert to 1700 UTC or 1700Z.

In converting from one time to another, be sure the day or date corresponds to the new time. That is, 2100R (EST) on January 1 would be 0200Z (UTC) on January 2; similarly, 0400Z on January 2 would be 2000U (PST) on January 1.

A good method is to use UTC (Z) for all amateur logging, schedule-making, QSLing and other amateur work. Otherwise, confusion with all the different time zones is inevitable. Leave your clock on UTC. Most of Alaska and Hawaii use W time (UTC + 10 hrs.)

Time changes one hour with each change of 15° in longitude. The five time zones in the U.S. proper and Canada roughly follow these lines.

0000* and 2400 are interchangeable; 2400 is associated with the date of the day ending, 0000 with the day just starting.

UTC	AST/EDT	EST/CDT	CST/MDT	MST/PDT	PST
0000*	2000	1900	1800	1700	1600
0100	2100	2000	1900	1800	1700
0200	2200	2100	2000	1900	1800
0300	2300	2200	2100	2000	1900
0400	0000*	2300	2200	2100	2000
0500	0100	0000*	2300	2200	2100
0600	0200	0100	0000*	2300	2200
0700	0300	0200	0100	0000*	2300
0800	0400	0300	0200	0100	0000*
0900	0500	0400	0300	0200	0100
1000	0600	0500	0400	0300	0200
1100	0700	0600	0500	0400	0300
1200	0800	0700	0600	0500	0400
1300	0900	0800	0700	0600	0500
1400	1000	0900	0800	0700	0600
1500	1100	1000	0900	0800	0700
1600	1200	1100	1000	0900	0800
1700	1300	1200	1100	1000	0900
1800	1400	1300	1200	1100	1000
1900	1500	1400	1300	1200	1100
2000	1600	1500	1400	1300	1200
2100	1700	1600	1500	1400	1300
2200	1800	1700	1600	1500	1400
2300	1900	1800	1700	1600	1500
2400	2000	1900	1800	1700	1600

---

sidered the final courtesy of a contact and is an Amateur Radio tradition. Most amateurs have printed cards, some personalized (at greater cost), some using standard setups provided by printers for the purpose. Whatever form you adopt, be sure your QSL card shows very clearly the correct call of the station contacted, the date (including year) and UTC time, the band on which contact was made, and the mode. Most awards based on QSL cards require at least these essentials, along with, of course, your street address, city, state or province, and country. Some awards require your county as well. Other interesting data might include some details of your equipment, antennas, former calls held, class of license, signal report and any friendly comments.

If everybody waited to receive a card before sending one, there would be no QSLing. Admittedly, printing and postage are expensive. If you are very active on the air, QSLing every contact may seem like a needless expense. If you want a QSL from your contact, be sure to send your card; and as a matter of common courtesy, send a card to everyone who sends you one.

Receiving DX QSLs

Within the U.S. and Canada, the ARRL DX QSL Bureau System is made up of 22 call area bureaus. Most of the cards from DX bureaus go directly to the individual bureaus.

At the individual bureaus, the incoming cards are sorted by the first letter of the suffix. This sorting divides the work load into portions that can be handled by a single individual.

To claim your cards, send a 5 × 7½ inch, self-addressed stamped envelope to the bureau serving your district. (Addresses for the U.S. and Canadian bureaus are shown twice a year in QST in the "QSL Corner" column, or send an s.a.s.e. to ARRL Hq. for QSL bureau information.) These envelopes should have your call sign printed neatly in the upper-left corner of the envelope to assist the sorter of your cards. Some bureaus will sell envelopes or postage credits as well as handling s.a.s.e.'s. The bureau will provide the proper size envelope and affix appropriate postage upon prepayment of a certain fee. The exact arrangement of your area bureau can be obtained by sending the bureau an s.a.s.e. with your inquiry.

Since many of the overseas stations use the bureau system, this area bureau can be very important to someone who works DX. It is a complex volunteer arrangement requiring good cooperation on the part of the DXers to function properly.

Sending Your DX QSLs

Each month, every member of the ARRL (except family and sightless members) is mailed a copy of QST. The address label on the wrapper of QST is the member's "ticket" for use of the Overseas QSL Service. Twelve times per year, an ARRL member may send any number of QSL cards for amateurs overseas. With each mailing the member must include the address label from the current copy of QST and $1 (check or money order) for each pound or part of a pound of cards; approximately 155 cards weigh one pound. QSLs must be presorted by prefix. Include only the cards, address label and money in the package. Wrap the package securely and address it to ARRL-Membership Overseas QSL Service, 225 Main St., Newington, CT 06111.

Family members of the ARRL may send cards in the same package but must include $1 per pound for each member sending cards and indicate that the QST address label includes a "family membership."

Sightless members, who do not receive QST, need only include $1 per pound with a note indicating that the cards are from a sightless member. Associate (unlicensed) members may use the Overseas QSL Service to send SWL reports to overseas amateur stations. No cards will be sent to individual QSL managers.

Additional information is available from ARRL Hq. Send an s.a.s.e. and request the QSL Bureau reprint.

Table 2

ARRL Communications Procedures

Voice	Code	Situation
Go ahead	K	Used after calling CQ, or at the end of a transmission, to indicate any station is invited to transmit.
Over	AR	Used after a call to a specific station before the contact has been established.
	KN	Used at the end of any transmission when only the specific station contacted is invited to answer.
Stand by or wait	AS	A temporary interruption of the contact.
Roger	R	Indicates a transmission has been received correctly and in full.
Clear	SK	End of contact. SK is sent before the final identification.
Leaving the air or closing station	CL	Indicates that a station is going off the air, and will not listen for or answer any further calls. CL is sent after the final identification.

Annotations: See inside front cover. — Output in Watts. — UTC recommended — RST. See back inside cover. — This column may also be used for contest exchange info. received.

	FIXED				VARIABLE				COMMENTS			QSL	
DATE	FREQ.	MODE	POWER	TIME	STATION WORKED	SENT	REC'D	TIME OFF	QTH	NAME	QSL VIA	S	R
28 JUL	146.52	FM	10	0430	WA1CCR				Wallingford	Eric	New converter works!		
3 OCT	7.0	CW	150	2319	WA6VEF	001	322	CONTRA COS	CALIFORNIA QSO PARTY				
				22	N6OJ	002	157	SONO					
				24	K6NA	003	331	SD					
				31	N6OP/M	004	117	CALAV					
9 OCT	28.6	SSB	1 KW	0301	JA1OCA	59	57		Tokyo	Isao	BURO	✓	
	21	CW		1545	EA9GD	559	579		Melilla	Jose	Box 348	✓	✓
				56	6ØØDX	599	599		Somalia		I2YAE		✓
5 NOV	3.8	SSB	150.	0030	W9NA	59+	59+	0117	Wausau, WI	Reno			
9 NOV	21	CW	10	1642	G4BUE	339	449	1657		1 watt!			

The ARRL Log is adaptable for all types of operating — ragchewing, contesting, DXing. References are to pages in the ARRL Log.

Your Station Record — Logging

Although the FCC does not require you to keep a log, an accurate, complete and neat log book can be a matter of personal pride. It can also be a strong form of protection for you against possible claims by others of intentional interference, or against troubles caused by unauthorized use of your call by others.

A log should be convenient to use — a bound one is best, so pages will not be lost. Your log becomes a written diary of your amateur operation and should include everything that will be of interest to you in years to come. Most amateurs retain their logs indefinitely as a historical record.

The Voice Modes

The use of proper procedure to get best results is very important. Voice operators say what they want to have understood, while CW operators have to spell it out or abbreviate. Since on phone the speed of transmission is generally between 150 and 200 words per minute, the matter of readability and understandability is critical to good communications. The good voice operator uses operating habits that are beyond reproach.

Phone Operating Practices

Listen with care. It is very natural to answer the loudest station that calls, but with a little digging, if need be, answer the best signal instead. Not all amateurs can run a kilowatt, but there is no reason every amateur cannot have a signal of the highest quality. Do not reward the operator who cranks up the transmitter gain and splatters by answering him if another station is calling.

Use VOX or push-to-talk. If you use VOX, don't defeat its purpose by saying "aah" to keep the relay closed. If you use push-to-talk, let go of the mic button every so often to make sure you are not "doubling" with the other station. Don't be a monologist.

Listen before transmitting. Make sure the frequency isn't being used before you come barging onto it. If you don't hear any station on the frequency, make this announcement: "Is the frequency in use? WB5LUA." If it is still clear, you are ready to make your call.

Interpose your call frequently, in distinct, measured tones. Use approved phonetics if your call sign is hard to understand or if conditions are poor. Remember you can be cited for improper identification if it cannot be understood.

Keep microphone (mic) gain constant. Don't "ride" the mic gain. Try to speak in an even amplitude the same distance from the microphone, keeping the gain down to eliminate room noise. Follow the manufacturer's instructions for use of the microphone; some require close-talking, while some need to be turned at an angle to the speaker's mouth.

The speed of radiotelephone transmission (with perfect accuracy) depends almost entirely on the skill of the two operators concerned. Use a rate of speech that allows perfect understanding as well as permitting the receiving operator to record the information.

Voice Operating Hints

1) Listen before calling.
2) Make short calls with breaks to listen. Avoid long CQs; do not answer overlong CQs.
3) Use push-to-talk or voice control (VOX). Give data concisely in first transmissions.
4) Make reports honest. Use definitions of strength and readability for reference. Make your reports informative and useful.
5) Limit transmission length. Two minutes or less will convey much information. When three or more stations converse in roundtables, brevity is essential.
6) Display sportsmanship and courtesy. Bands are congested — make transmissions meaningful — give others a break.
7) Check transmitter adjustment — avoid splatter. Check speech-processor adjustment. Do not transmit when moving VFO frequency. Complete testing before busy hours — use a dummy load!

Repeater Operating

A repeater is a device that receives a signal on one frequency and simultaneously transmits (repeats) it on another frequency. Often located atop a tall building or high mountain, VHF and UHF repeaters greatly extend the operating coverage of amateurs using mobile and hand-held transceivers.

To use a repeater you must have a transceiver with the capability of transmitting on the repeater's input frequency (the frequency that the repeater listens on) and receiving on the repeater's output frequency (the frequency the repeater transmits on). This capability can be acquired by dialing the correct frequency and selecting the proper offset (frequency difference between input and output).

When you have the frequency capability, all you need do is key the microphone button and you will turn on (access) the repeater. Some repeaters have limited access, requiring the transmission of a subaudible tone, a series of tones or bursts in order to gain access. Most repeaters briefly transmit a carrier after a user has stopped transmitting to inform the user that he is actually accessing a repeater.

After acquiring the ability to access a repeater, you should become acquainted with the operating practices that are inherent to this unique mode of Amateur Radio:

Table 3

International Telecommunication Union Phonetics

A — Alfa	J — Juliett	S — Sierra
B — Bravo	K — Kilo	T — Tango
C — Charlie	L — Lima	U — Uniform
D — Delta	M — Mike	V — Victor
E — Echo	N — November	W — Whiskey
F — Foxtrot	O — Oscar	X — X-Ray
G — Golf	P — Papa	Y — Yankee
H — Hotel	Q — Quebec	Z — Zulu
I — India	R — Romeo	

1) Monitor the repeater to become familiar with any notable features in its operation.

2) To initiate a contact simply indicate that you are on frequency. Various geographical areas have different practices on making yourself heard, but, generally, "This is W1XX monitoring" will suffice. One practice that is usually looked upon with disfavor throughout the U.S. and Canada is calling CQ on a repeater.

3) Identify legally. You must transmit your call sign at the end of each contact and every 10 minutes during each contact. It is illegal to "kerchunk" a repeater (transmit an unmodulated carrier) without identifying, as this constitutes an unidentified transmission.

4) Pause between transmissions. This allows other hams to use the repeater (someone may have an emergency). On most repeaters a pause is necessary to reset the timer.

5) Keep transmissions short and thoughtful. Your monologue may prevent someone with an emergency from using the repeater. If your monologue is long enough, you may time-out the repeater. Your transmissions are being heard by many listeners, including nonhams with public service band monitors and scanners; don't give a bad impression of Amateur Radio.

6) Use simplex whenever possible. If you can complete your QSO or contact on a direct frequency, there is no need to tie up the repeater and prevent others from using it.

7) Use the minimum amount of power necessary to maintain communications. This FCC regulation minimizes the possibility of accessing distant repeaters on the same frequency.

8) Don't break into a contact unless you have something to add. Interrupting is no more polite on the air than it is in person.

9) Many repeaters are equipped with autopatch facilities which, when properly accessed, connect the repeater to the telephone system to provide a public service. The FCC forbids using an autopatch for anything that could be construed as business communications. An autopatch should not be used to avoid a toll call. Do not use an autopatch where regular telephone service is available. Abuses of autopatch privileges may lead to their loss.

10) All repeaters are assembled and

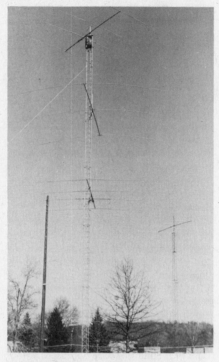

This W1AW 120-foot tower holds phased 4-element, 20-meter Yagis at 60 and 120 feet and a 2-element, 40-meter Yagi at 90 feet.

maintained at considerable expense and inconvenience. Usually an individual or a group is responsible, and those who are regular users of a repeater should support the efforts of keeping the repeater on the air.

The ARRL *Repeater Directory*, listing approximately 10,000 amateur repeaters, is published annually by ARRL.

CW Operating

If you spend an entire Amateur Radio career on phone once you have mastered enough CW to pass the necessary tests, you are missing out on at least 50 percent of the fun of hamming. Mastering the art of CW communication is 10 times easier than learning to talk, and you did that when you were two years old. All it takes is some basic learning principles, then practice. This is not drudgery as you might think, because you can combine learning with listening to actual signals on the band, and even with operating, since the Novice class license requires only 5 WPM.

But listening is the best way to go, in the

beginning. You don't need a license for that. Once you have learned the basic sounds of code, you will soon start recognizing common words — the, and, CQ, DE (from), etc. Copying calls is an excellent means of getting prelicense practice. You can even get good sending practice by pretending to call the CQing station on your code practice oscillator. Sending practice is important. Start with a simple "straight" key; later you can switch to an electronic keyer.

There are many pitfalls to developing into a finished CW operator, not the least of which is the acquisition of bad habits. Many of these come from mimicking your peers or elders, some of whom are themselves the victims of bad CW habits. Don't let undesirable operating habits rub off on you.

Probably the worst of these is carelessness about spacing. Your early CW training should have taught you that spacing length is just as important as dit and dah length. One way to improve your spacing is to practice sending "in step" with W1AW using a code practice oscillator (but not on the air!). If you can send in step with W1AW, your sending is perfect. The source of each W1AW code practice test is sent several times during the transmission to enable you to check your copy.

On phone it is unnecessary and therefore improper to use jargon and abbreviations, but on CW abbreviations are a necessity. Without them, it takes a long time to say what you want to say, especially at beginner speeds. Most of the abbreviations hams use have developed within the fraternity; some of them are borrowed from or are carryovers from old-time telegraphy abbreviations. Which is which doesn't matter; it is all ham radio to us. Learn to use ham CW abbreviations liberally and you will get much more said in much less time. When you reach a high-proficiency level, you will find that CW is almost as fast as talking, thus all but eliminating one of its principal disadvantages while still retaining its many advantages.

Despite the fact that learning CW is easier than learning to talk, nearly everybody can talk but few people can communicate by CW. Thus, there is considerable pride of accomplishment in CW operating. Regardless of your level of proficiency, chances are that you still have something to learn about CW operating, additional proficiency goals you can achieve. ARRL offers a series of awards in this field, starting at 10 WPM and progressing all the way through 40 WPM, in 5-WPM increments. You will find more details elsewhere in this chapter.

Copying CW and comprehending it are not the same thing. The word "copy" implies something written, so "if you don't put it down, that ain't copying." On the other hand, for conversational CW purposes, copying really isn't necessary and can be quite cumbersome. Most beginners on CW learn by copying everything down, and some find that this habit is hard to break as they achieve higher levels of proficiency. However, it does involve a "translation" process that can and should be eliminated for conversational purposes. Sooner or later, in order to realize the full value of CW communication, you must learn just to listen to it, as you do the spoken word, rather than to "copy" it. It should not be necessary to translate the CW into written copy and then translate the written copy into intelligence by reading it. Eventually, the sound of code should directly trigger your consciousness just as the spoken word does, and then the "copy" and understanding functions are reversed; that is, you understand it first, then you copy it.

CW is not just something that has been imposed on us to make passing the amateur test a little more difficult. It is an entirely different method of communication, and a long way from obsolete. Learn it well and your enjoyment of Amateur Radio will be enriched.

Amateur Television Operating

Amateur television (ATV) has much in common with commercial image-transmission systems. Amateur fast-scan television (FSTV) has borrowed equipment and techniques from broadcast television. Commercial freeze-frame systems are to some extent a spin-off of amateur slow-scan television (SSTV) developments. There are some differences that make ATV distinctive. ATV is a two-way operation, which gives it the potential of being highly interactive. Owing to transmitter power limitations and the insatiable desire of amateurs to work DX, ATV operators are usually willing to sacrifice quality and signal-to-noise ratio for range.

FSTV and SSTV are primarily *image-*transmission media, often with an audio channel. Many amateurs who start out in ATV point a camera at themselves, their shack, their household pet (wearing earphones, of course) and then run out of ideas. After everyone on the repeater has had a few nights of this, there is inevitably a discussion of how to get some "new blood" in the group. ATV is a visual medium, and the artistic aspects of operating deserve some attention.

What can you do to make ATV useful and *interesting*? Well, the FCC rules say you can't transmit music or broadcast to the general public. That leaves a tremendous number of possibilities for ATV. It takes some planning, a reasonable amount of equipment, and some follow-through to make full use of the medium. Much can be learned from still photography and home video, which is largely supplanting home movies. These are tools that can be used to prepare material for ATV transmission. Character generators and special effects can spice up an ATV transmission. FSTV lends itself to transmission of prerecorded video tapes on subjects of interest to radio amateurs, such as technical talks and training courses. Portable and mobile operating adds to the fascination of FSTV. Emergency communications, two-way TV third-party conversations possibly involving sign language for the deaf, and assisting in coordination of special events, are but a few FSTV applications that have been tried. Some spectacular color still photography is being transmitted via SSTV these days. The creative side of ATV is only beginning.

For either FSTV or SSTV, it is useful to have a good selection of test patterns for measuring the quality of both equipment and the communications channel. Several test charts are available from the Electronic Industries Association (EIA), the International Radio Consultative Committee (CCIR), and the International Telegraph and Telephone Consultative Committee (CCITT) for FSTV and facsimile.[1] In addition, you may want to make your own, either mechanically or electronically. Mechanical test charts can be made on poster board by combining line drawings, hand or press-on lettering, press-on shading and colored paper with photographs. Test charts can be generated electrically with a personal computer that can produce letters, and black-and-white graphics. Computers with a color capability can be used to make color bars, a gray scale and other useful patterns.

Fast-Scan Television Operating

Amateur fast-scan television (FSTV) is

[1]Contact EIA at 2001 Eye St., N.W., Washington, DC 20006. CCIR and CCITT materials are available from the U.N. Bookstore, United Nations Building, New York, NY 10017.

Table 4
Q Signals

These Q signals are the ones whose meanings most often need to be expressed with brevity and clearness in CW amateur work. (Q abbreviations take the form of questions only when followed by a question mark.)

QRG Will you tell me my exact frequency (or that of ___)? Your exact frequency (or that of ___) is ___ kHz.

QRH Does my frequency vary? Your frequency varies.

QRI How is the tone of my transmission? The tone of your transmission is ___ (1. Good; 2. Variable; 3. Bad).

QRK What is the intelligibility of my signals (or those of ___)? The intelligibility of your signals (or those of ___) is ___ (1. Bad; 2. Poor; 3. Fair; 4. Good; 5. Excellent.)

QRL Are you busy? I am busy (or I am busy with ___). Please do not interfere.

QRM Is my transmission being interfered with? Your transmission is being interfered with ___ (1. Nil; 2. Slightly; 3. Moderately; 4. Severely; 5. Extremely.)

QRN Are you troubled by static? I am troubled by static ___

QRO Shall I increase power? Increase power.

QRP Shall I decrease power? Decrease power.

QRQ Shall I send faster? Send faster (___ WPM).

QRS Shall I send more slowly? Send more slowly (___ WPM).

QRT Shall I stop sending? Stop sending.

QRU Have you anything for me? I have nothing for you.

QRV Are you ready? I am ready.

QRW Shall I inform ___ that you are calling him on ___ kHz. Please inform ___ that I am calling him on ___ kHz.

QRX When will you call me again? I will call you again at ___ hours (on ___ kHz).

QSA What is the strength of my signals (or those of ___)? The strength of your signals is ___ (1. Scarcely perceptible; 2. Weak; 3. Fairly good; 4. Good; 5. Very good).

QSB Are my signals fading? Your signals are fading.

QSD Are my signals defective? Your signals are defective.

QSG Shall I send ___ messages at a time? Send ___ messages at a time.

QSK Can you hear me between your signals and if so can I break in on your transmission? I can hear you between my signals; break in on my transmission.

QSL Can you acknowledge receipt? I am acknowledging receipt.

QSM Shall I repeat the last message which I sent you, or some previous message? Repeat the last message you sent me [or message(s) number (s) ___].

QSN Did you hear me (or ___) on ___ kHz? I did hear you (or ___) on ___ kHz.

QSO Can you communicate with ___ direct or by relay? I can communicate with ___ direct (or by relay through ___).

QSP Will you relay to ___? I will relay to ___

QSU Shall I send or reply on this frequency (or on ___ kHz)? Send or reply on this frequency (or ___ kHz).

QSV Shall I send a series of Vs on this frequency (or ___ kHz)? Send a series of Vs on this frequency (or on ___ kHz).

QSW Will you send on this frequency (or on ___ kHz)? I am going to send on this frequency (or on ___ kHz).

QSX Will you listen to ___ on ___ kHz? I am listening to ___ on ___ kHz.

QSY Shall I change to transmission on another frequency? Change to transmission on another frequency (or on ___ kHz).

QSZ Shall I send each word or group more than once? Send each word or group two (or ___ times).

QTA Shall I cancel message number ___? Cancel message number ___.

QTB Do you agree with my counting of words? I do not agree with your counting of words; I will repeat the first letter or digit of each word or group.

QTC How many messages have you to send? I have ___ messages for you (or for ___).

QTH What is your location? My location is ___.

QTR What is the correct time? The time is ___.

Special Q signal adopted by ARRL:

QST General call preceding a message addressed to all amateurs and ARRL members.

steadily gaining popularity. This growth has been largely brought about by UHF ATV repeaters, which greatly extend the communications range of an individual ATV station. U.S. and Canadian ATV repeaters are listed in the ARRL *Repeater Directory*.

Because an FSTV signal can occupy as much as 6 MHz, only the 70 cm and higher-frequency bands are wide enough to ac-commodate this type of operation. Specific ATV channels are designated in the ARRL band plans, which are included in the *Repeater Directory*.

U.S. and Canadian FSTV stations generally follow Electronic Industries Association (EIA) RS-170 standard for black-and-white transmissions and National Television Standards Committee (NTSC) recommendations for color televi-sion. Amateurs are not limited to using in-dustry standards, but their use assures the compatibility of a wide variety of commer-cially available TV cameras, sync generators and receiving equipment. Most home TV receivers require only a frequency converter to translate the 70- or 23-cm ATV frequencies to a normal broadcast TV channel. Some UHF-TV tuners can receive 70-cm ATV by retuning broadcast channel

---

Table 5

U.S. Amateur Radio Frequency Allocations

Frequency band (kHz)	Emissions	Limitations
1800-2000	A1A,F1B,A3E, F3E,G3E,A3C, F3C,A3F,F3F, H3E,J3E,R3E	3,5,21
3500-3750	A1A,F1B	1,3,32,33
3750-4000	A1A,A3E,F3E, G3E,A3C,A3F, F3C,F3F,H3E, J3E,R3E	
5167.5	J3E,R3E	2
7000-7075	A1A,F1B	3,32
7075-7100	A1A,F1B,H3E, J3E,R3E	3,32,34
7100-7150	A1A,F1B	1,3,32,33
7150-7300	A1A,A3E,F3E, G3E,A3C,F3C, A3F,F3F,H3E, J3E,R3E	3,32
10100-10150	A1A,F1B	28,32
14000-14150	A1A,F1B	32
14150-14350	A1A,A3E,F3E, A3C,F3C,A3F, F3F,H3E,J3E, R3E	32
21000-21200	A1A,F1B	1,32,33
21200-21450	A1A,A3E,F3E, A3C,F3C,A3F, F3F,H3E,J3E, R3E	32
24890-24930	A1A,F1B	29,32
24930-24990	A1A,A3E,F3E, G3E,A3C,F3C, A3F,F3F,H3E, J3E,R3E	29,32
28000-29700	A1A	
28000-28300	A1A,F1B	
28300-29500	A1A,A3E,F3E, G3E,A3C,F3C, A3F,F3F,H3E, J3E,R3E	35
29500-29700	A1A,A3E,F2A, F3E,G3E,A3C, F3C,A3F,F3F, H3E,J3E,R3E	
(MHz)		
50.0-50.1	A1A	3
50.1-51.0	A1A,A2A,A2B, A3E,A3C,A3F, F1B,F2B,F3E, G3E,F3C,F3F, H3E,J3E,R3E	3
51.0-54.0	NØN,A1A,A2A, A2B,A3E,A3C, A3F,F1B,F2B, F3E,G3E,F3C, F3F,H3E,J3E, R3E	3
144.0-144.1	A1A	3,32
144.1-148.0	NØN,A1A,A2A, A2B,A3E,A3F, F1B,F2B,F3E, G3E,F3C,F3F, H3E,J3E,R3E	3,32
220-225*		3,4,5,36
420-450*		3,5,6,7,10,30
902-928**		3,5,8,9
1240-1300*		5,11,22,37
2300-2310*†		3,5,12,13
2390-2450*†		3,5,13,14
(GHz)		
3.3-3.5*†		3,5,15,17
5.650-5.925*†		3,5,18,19,20
10.0-10.5*		5,21,22,31
24.00-24.25*†		3,5,22,24,26
47.0-47.2*†		
75.5-81*†		5,21,22
119.98-120.02*†		15,25
142-149*†		5,15,21,22
241-250*†		5,21,22,27
above 300*†		15

*Amateur stations are authorized the following emissions on this band: NØN, A1A, A2A, A2B, A3E, A3C, A3F, F1B, F2B, F3E, G3E, F3C, F3F, H3E, J3E, R3E.
**F8E emissions may also be used in the 902-928 MHz band.
†PØN may also be used on this band.

Limitations:
The following list contains those limitations from Section 97.61 considered most important to amateurs in the US. For a complete list of the limitations, see *The FCC Rule Book*.

(1) Novice and Technician class radio operators are limited to the use of international Morse code when communicating in this band.

(2) This band may only be used by Amateur stations in the State of Alaska or within fifty nautical miles of the State of Alaska for emergency com-munications with other stations authorized to use this band in the State of Alaska. This frequency band is shared with licensees in the Alaska-private fixed service who may use it for certain non-emergency purposes.

(3) Where, in adjacent regions or subregions, a band of frequencies is allocated to different services of the same category, the basic principle is the equality of right to operate. Accordingly, the stations of each service in one region or subregion must operate so as to not cause harmful interference to services in the other regions or subregions. (See International Telecommunication Union Radio Regulations, RR 346 [Geneva, 1979].)

(4) This band is allocated to the amateur, fixed and mobile services in the United States on a co-primary basis. The basic principle which applies is the equality of right to operate. Amateur, fixed and mobile stations must operate so as not to cause harmful interference to each other.

(5) Amateur stations in the 1900-2000 kHz, 220-225 MHz, 420-450 MHz, 902-928 MHz, 1240-1300 MHz, 2300-2310 MHz, 2390-2450 MHz, 3.3-3.5 GHz, 5.650-5.925 GHz, 10.0-10.5 GHz, 24.05-24.25 GHz, 76-81 GHz, 144-149 GHz and 241-248 GHz bands must not cause harmful inter-ference to stations in the Government radiolocation service and are not protected from interference due to the operation of stations in the Government radio-location service.

(6) No amateur station shall operate north of Line A (see § 97.3[i]) in the 420-430 MHz band.

(7) The 420-430 MHz band is allocated to the Amateur service in the United States on a secondary basis, but is allocated to the fixed and mobile (except aeronautical mobile) services in the International Table of Allocations on a primary basis. Therefore, amateur stations in this band must not cause harmful interference to stations authorized by other nations in the fixed and mobile (except aeronautical mobile) services and are not protected from interference due to the operation of stations authorized by other nations in the fixed and mobile (except aeronautical mobile) services.

(8) In the 902-928 MHz band, amateur stations shall not operate within the States of Colorado and Wyoming, bounded by the area of: latitude 39° N to 42° N, and longitude 103° W to 108° W. This band is allocated on a secondary basis to the Amateur Service subject to not causing harmful interference to the operations of Government stations authorized in this band or to Automatic Vehicle Monitoring (AVM) systems. Stations in the Amateur service are not pro-tected from any interference due to the operation of industrial, scientific and medical (ISM) devices, AVM systems or Government stations authorized in this band.

(9) In the 902-928 MHz band, amateur stations shall not operate in those portions of the States of Texas and New Mexico bounded on the south by latitude 31° 41' N, on the east by longitude 104° 11' W, on the north by latitude 34° 30' N, and on the west by longitude 107° 30' W.

(10) The 430-440 MHz band is allocated to the Amateur service on a secondary basis in ITU Regions 2 and 3. Amateur stations in this band in ITU Regions 2 and 3 must not cause harmful inter-ference to stations authorized by other nations in the radiolocation service and are not protected from interference due to the operation of stations authorized by other nations in the radiolocation service. In ITU Region 1 the 430-440 MHz band is allocated to the Amateur service on a co-primary basis with the radiolocation service. As between these two services in this band in Region 1 the basic principle which applies is the equality of right to operate. Amateur stations authorized by the United States and radiolocation stations authorized by other nations in Region 1 must operate so as not to cause harmful interference to each other.

(11) In the 1240-1260 MHz band amateur stations must not cause harmful interference to stations authorized by other nations in the radionavigation-satellite service and are not protected from inter-ference due to the operation of stations authorized by other nations in the radionavigation-satellite service.

(12) In the United States, the 2300-2310 MHz band is allocated to the Amateur service on a co-secondary basis with the Government fixed and mobile services. In this band, the fixed and mobile services must not cause harmful interference to the Amateur service.

(13) In the 2300-2310 MHz and 2390-2450 MHz band, the Amateur service is allocated on a secon-dary basis in all ITU Regions. In ITU Regions 2 and 3, stations in the Amateur service must not cause harmful interference to stations authorized by other nations in the fixed, mobile and radiolocation services, and are not protected from interference due to the operation of stations authorized by other nations in the fixed, mobile and radiolocation services.

(14) Amateur stations in the 2400-2450 MHz band are not protected from interference due to the opera-tion of industrial, scientific and medical devices on 2450 MHz.

(15) Amateur stations in the 3.332-3.339 GHz, 3.3458-3.3525 GHz, 119.98-120.02 GHz, 144.68-144.98 GHz, 145.45-145.75 GHz, 146.82-147.12 GHz and 343-348 GHz bands must not cause harmful interference to stations in the radio astronomy service. Amateur stations in the 300-302 GHz, 324-326 GHz, 345-347 GHz, 363-365 GHz and 379-381 GHz bands must not cause harmful inter-ference to stations in the space research service (passive) or Earth exploration-satellite service (passive).

(17) In the United States the 3.3-3.5 GHz band is allocated to the Amateur service on a co-secondary basis with the non-government radiolocation service.

(18) In the 5.650-5.725 GHz band, the Amateur service is allocated in all ITU regions on a co-secondary basis with the space research (deep space) service. In the 5.725-5.850 GHz band the Amateur service is allocated in all ITU regions on a secondary basis. In the 5.650-5.850 GHz band amateur stations must not cause harmful interference to stations authorized by other nations in the radio-location service, and are not protected from inter-ference due to the operation of stations authorized by other nations in the radiolocation service. In the 5.850-5.925 GHz band the Amateur service is al-located in ITU Region 2 on a co-secondary basis with the radiolocation service. In the 5.850-5.925 GHz band amateur stations must not cause harmful inter-ference to stations authorized by other nations in the fixed, fixed-satellite and mobile services, and are not protected from interference due to the operation of stations authorized by other nations in the fixed, fixed-satellite and mobile services.

(19) In the United States, the 5.850-5.925 GHz band is allocated to the Amateur service on a secondary basis to the non-government fixed-satellite service. In the 5.850-5.925 GHz band amateur sta-tions must not cause harmful interference to stations in the non-government fixed-satellite service and are not protected from interference due to the operation of stations in the non-government fixed-satellite service.

(20) Amateur stations in the 5.725-5.875 GHz band are not protected from interference due to the opera-tion of industrial, scientific and medical devices on 5.8 GHz.

(21) Amateur stations in the 1900-2000 kHz, 10.45-10.50 GHz, 76-81 GHz, 144-149 GHz and 241-248 GHz bands must not cause harmful inter-ference to stations in the non-government radio-location service and are not protected from interference due to the operation of stations in the non-government radiolocation service.

(22) Amateur stations in the 1240-1300 MHz, 10.0-10.5 GHz, 24.05-24.25 GHz, 76-81 GHz, 144-149 GHz and 241-248 GHz bands must not cause harmful interference to stations authorized by other nations in the radiolocation service and are not protected from interference due to the operation of stations authorized by other nations in the radiolocation service.

(24) In the United States, the 24.05-24.25 GHz band is allocated to the Amateur Service on a co-secondary basis with the non-government radio-location and Government and non-government Earth

exploration-satellite (active) services.

(25) The 119.98-120.02 GHz band is allocated to the Amateur service on a secondary basis. Amateur stations in this band must not cause harmful interference to stations operating in the fixed, inter-satellite service and mobile services, and are not protected from interference caused by the operation of stations in the fixed, inter-satellite and mobile services.

(26) Amateur stations in the 24.00-24.25 GHz band are not protected from interference due to the operation of industrial, scientific and medical devices on 24.125 GHz.

(27) Amateur stations in the 244-246 GHz band are not protected from interference due to the operation of industrial, scientific and medical devices on 245 GHz.

(28) Amateur stations in the 10100-10150 kHz band must not cause harmful interference to stations authorized by other nations in the fixed service. Amateur stations shall make all necessary adjustments (including termination of transmission) if harmful interference is caused.

(29) Until July 1, 1989, amateur stations in this band must not cause harmful interference to stations authorized by other nations in the fixed and mobile services. Amateur stations must make all necessary adjustments (including termination of transmission) if harmful interference is caused.

(30) Amateur stations in the 449.5-450 MHz band must not cause interference to and are not protected from interference due to the operation of stations in the space operation service, the space research service, or for space telecommand.

(31) In the United States, the 10.0-10.5 GHz band is allocated to the Amateur service on a co-secondary basis with the non-government radiolocation service.

(32) Amateur stations in these bands may be used for communications related to relief operations in connection with natural disasters. See Appendix 6 to this Part.

(33) Novice and Technician class radio operators may not use F1B emissions in this band.

(34) Amateur stations located in Regions 1 and 3, and amateur radio stations located within Region 2 which are west of 130° West longitude may also use A3E, F3E and G3E emissions.

(35) Novice and Technician class radio operators are only permitted to use A1A and J3E emissions from 28300-28500 at a maximum power output of 200 watts.

(36) Novices are permitted all authorized modes from 222.1 to 223.9 MHz at a maximum power output of 25 watts.

(37) Novices are permitted all authorized modes from 1270 to 1295 MHz at a maximum power output of 5 watts.

Note

The types of emission referred to in the amateur rules are as follows:

Type N0N—Steady, unmodulated pure carrier.
Type A1A—Telegraphy without use of modulating audio frequency.
Type A2A, A2B—Amplitude tone-modulated telegraphy.
Type A3E—Double-sideband AM telephony.
Type J3E—Single-sideband, suppressed-carrier AM telephony.
Type R3E—Reduced-carrier SSB telephony.
Type H3E—Full-carrier SSB telephony.
Type A3C—Facsimile.
Type A3F—Television.
Type F1B—Frequency-shift telegraphy.
Type G1B—Phase-shift telegraphy.
Type F2B—Audio frequency-shift telegraphy.
Type G2B—Audio phase-shift telegraphy.
Type F3E—Frequency-modulated telephony.
Type G3E—Phase-modulated telephony.
Type F3C—FM facsimile.
Type G3C—PM facsimile.
Type F3F—FM television.
Type G3F—PM television.
Type P0N—Pulse.

Table 6

Canadian Amateur Bands*

Band (limitations)	Frequency (MHz)	Emissions
80 meters	3.500-3.725	A1, F1
(1, 3, 4, 5)	3.725-4.000	A1, A3, F3
40 meters	7.000-7.050	A1, F1
(1, 3, 4, 5)	7.050-7.100	A1, A3, F3
	7.100-7.150	A1, F1
	7.150-7.300	A1, A3, F3
30 meters	10.100-10.150	A1, F1
20 meters	14.000-14.100	A1, F1
(1, 3, 4, 5)	14.100-14.350	A1, A3, F3
17 meters (16)	18.068-18.168	A1, F1, A3, F3, A4, F4, A5, F5
15 meters	21.000-21.100	A1, F1
(1, 3, 4, 5)	21.000-21.450	A1, A3, F3
12 meters (16)	24.890-24.990	A1, F1, A3 F3, A4, F4, A5, F5
10 meters	28.000-28.100	A1, F1
(2, 3, 4, 5)	28.100-29.700	A1, A3, F3
6 meters	50.000-50.050	A1
(3, 4)	50.050-51.000	A1, A2, A3, F1, F2, F3
	51.000-54.000	A0, A1, A2, A3, A4, F1, F2, F3, F4
2 meters	144.000-144.100	A1
(3, 4)	144.100-145.500	A0, A1, A2, A3, A4, F1, F2, F3, F4
(3, 4, 7)	144.500-145.800	P0, P1, A0, A1, A2, A3, A4, F1, F2, F3, F4
(3, 4)	145.800-148.000	A0, A1, A2, A3, A4, F1, F2, F3, F4
(3, 4)	220.000-220.100	A0, A1, A2, A3, A4, F1, F2, F3, F4
(9, 10, 13, 15)	220.100-220.500	
(9, 12, 13, 15)	220.500-221.000	
(11, 13, 14, 15)	221.000-223.000	
(9, 12, 13, 15)	223.000-223.500	
(3, 4)	223.500-225.000	A0, A1, A2, A3, A4, F1, F2, F3, F4
(4, 6)	430.000-433.000	A0, A1, A2, A3, A4, A5, F1, F2, F3 F4, F5
(12, 13, 14, 15)	433.000-434.000	
(3, 4, 8)	434.000-434.500	P0, P1, P2, P3, A0, A1, A2, A3, A5, F1, F2, F3, F4, F5
(4, 6)	434.500-450.000	A0, A1, A2, A3, A4, A5 F1, F2, F3, F4, F5
	902.000-928.000	A3, F3
	1215.000-1300.000	A0, A1, A2, A3, A4, A5, F1, F2, F3, F4, F5
	2300.000-2450.000	†
	3300.000-3500.000	†
	5650.000-5925.000	†
	10,000.000-10,500.000	†
(9, 14, 15)	24,000.000-24,010.000	†
	24,010.000-24,250.000	†

†Amateur stations are authorized the following emissions on this band: A0, A1, A2, A3, A4, A5, F1, F2, F3, F4, F5, P0, P1, P2, P3, P4, P5, P9.

Limitations

1) Phone privileges are restricted to holders of advanced Amateur Radio Operators Certificates, and of Commercial Certificates.

2) Phone privileges are restricted as in footnote 1, and to holders of Amateur Radio Operators Certificates, whose certificates have been endorsed for operation on phone in these bands.

3) Amplitude modulation (A2, A3, A4) shall not exceed ±3 kHz (6A3).

4) Frequency modulation (F2, F3, F4) shall not produce a carrier deviation exceeding ±3 kHz, (6F3) except that in the 52-54 MHz and 144.148 MHz bands and higher the carrier deviation shall not exceed ±15 kHz (30F3).

5) Slow-scan television (A5), permitted by special authorization shall not exceed a bandwidth greater than that occupied by a normal single-sideband voice transmission.

6) Television (A5), permitted by special authorization, shall employ a system of standard interface and scanning with a bandwidth of not more than 4 MHz.

7) Pulse modulation with any mode of transmission shall not produce signals of a bandwidth exceeding 15 kHz.

8) Pulse modulation with any mode of transmission shall not produce signals of a bandwidth exceeding 30 kHz.

9) Any mode may be used.

10) Packet transmissions shall not produce signals exceeding 10 kHz.

11) Packet transmissions shall not produce signals exceeding 25 kHz.

12) Packet transmissions shall not produce signals exceeding 100 kHz.

13) Licensees performing an Amateur Experimental Service may use such modulation techniques or types of emission for packet transmission as they may select by experimentation on conditions that they do not exceed the bandwidths established in 10, 11 and 12.

14) Only packet transmissions shall be used.

15) Final RF output power used for packet transmissions shall not exceed 100 watts peak power and 10 watts average power.

16) On 17 meters and 12 meters, all emissions may have a maximum bandwidth of 6 kHz. Holders of Amateur Radio Operators Certificates may use A1 or F1 emissions only. Holders of advanced Amateur Radio Operators Certificates may use all authorized modes.

Operation in frequency band 1.800-2.000 MHz shall be limited to the area as indicated in the following table and shall be limited to the indicated maximum dc power input to the anode of the final radio frequency stage of the transmitter during day and night hours respectively; for the purpose of this table "day" means the hours between sunrise and sunset, and "night" means the hours between sunset and sunrise. A1, A3, and F3 emissions are permitted.

	A	B	C	D	E	F	G	H
British Columbia	3¹	3	3	1	0	0	0	0
Alberta	3¹	3	3	3	1	0	0	1
Saskatchewan	3	3	3	3	3	1	1	3
Manitoba	3¹	2	2	2	2	2	2	3
Ontario North of 50° N.	3	1	1	1	0	0	0	2
Ontario South of 50° N.	3¹	2	1	0	0	0	0	1
Province of Quebec North of 52° N.	1	0	0	1	1	0	0	2
Province of Quebec South of 52°N.	3	2	1	0	0	0	0	0
New Brunswick	3	2	1	0	0	0	0	0
Novia Scotia	3	2	1	0	0	0	0	0
Prince Edward Island	3	2	1	0	0	0	0	0
Newfoundland (Island)	3	1	1	0	0	0	0	0
Newfoundland (Labrador)	2	0	0	0	0	0	0	0
Yukon Territory	3	3	3	1	0	0	0	0
District of MacKenzie	3	3	3	3	1	0	0	1
District of Keewatin	3	1	1	3	2	0	0	2
District of Franklin	0	0	0	0	1	0	0	1

¹The power levels 500 day/100 night may be increased to 1000 day/200 night when authorized by a radio inspector of the Department of Communications.

Frequency Band

A	1.800-1.825 MHz	E	1.900-1.925 MHz
B	1.825-1.850 MHz	F	1.925-1.950 MHz
C	1.850-1.875 MHz	G	1.950-1.975 MHz
D	1.875-1.900 MHz	H	1.975-2.000 MHz

Power Level—Watts

0—Operation not permitted
1— 25 night   125 day
2— 50 night   250 day
3—100 night   500 day

*At presstime there was no official word on the implementation of the new WARC emissions designators, but we understand that the Department of Communications has the matter under consideration.

P5 — Excellent

P4 — Good

P3 — Fair

P2 — Poor

P1 — Barely perceptible

ATV picture quality reporting system.

14, possibly with slight modification. Some of the newer "cable-ready" TV receivers and video cassette recorders (VCRs) have wide-range tuners that can tune to channels in the 70-cm amateur band.

As of June 15, 1983, the FCC discontinued the requirement for FSTV stations to identify by CW or voice, thus permitting FSTV stations to identify by video. Please note that this applies only to transmissions employing U.S. 525-line standards, legally "those which conform, at a minimum, to the monochrome transmission standards of Section 73.682(a)(6) through Section 73.682(a)(13), inclusive (with the exception of Section 73.682(a)(9)(iii) and Section 73.682(a)(9)(iv)." The FCC also requires that the characters be "readily legible." It follows that ATV operators should use an ordinary type style for identification and make their call signs cover enough of the screen to be readable even under weak-signal conditions.

The RST signal reporting system used for CW and phone was not designed for images. Instead, ATVers report picture quality by the letter P followed by a single digit, as follows: 1-barely perceptible, 2-poor, 3-fair, 4-good, and 5-excellent. Some ATV operators make a permanent still-photograph or video-tape record of received pictures for their station log and to mail to the transmitting station.

An off-air monitor is a handy station accessory to have. It will help you judge picture modulation as well as let you know when there is a malfunction.

Slow-Scan Television Operating

The popularity of SSTV is rising as a result of increased availability of equipment, some of which can transmit and receive color pictures. The slow-scan technique permits the video signal to be transmitted in a normal voice bandwidth. This makes it possible to use the HF amateur bands for SSTV. Popular frequencies for SSTV operation are:

U.S. SSTV Freq. (kHz)	Minimum License	IARU Region 1 (Europe/Africa) Freq. (kHz)
3845	General	3730-3740
7171	Advanced	7035-7045
14,230	General	14,225-14,235
21,340	General	21,335-21,345
28,680	General	28,675-28,685

Before transmitting, first listen around the calling frequencies, then either respond on voice to a station calling CQ SSTV or call CQ SSTV yourself after ascertaining that the frequency is not in use. Call in the following manner: "CQ SSTV CQ SSTV CQ SSTV this is W9NTP W9NTP W9NTP over." Once voice contact is established and it is clear to both stations which SSTV standard is being used, the video transmission can start. You should identify your video as you start your transmission and as you end transmission of video as follows: "W9NTP this is WØLMD, video follows. (SSTV pictures sent). Okay, how

do you like that one? W9NTP this is WØLMD over."

When you are operating SSTV, be sure to carefully monitor your transmissions to make sure your signal does not exceed normal voice bandwidth limitations.

SSTV signals can be stored on audio cassette tape. It's nice to be able to play back a rare SSTV DX QSO to visitors in your shack when there is no on-the-air SSTV activity to demonstrate.

Digital Communications Procedures

Baudot radioteletype (RTTY) has been a regular part of Amateur Radio since the late 1940s. Federal Communications Commission (FCC) rules now permit three specified codes on any frequency where digital communication is permitted. These three specified codes are: ITA2 (Baudot/Murray), the American National Standard Code for Information Interchange (ASCII) and CCIR 476-2 and 476-3, also known as Amateur Teleprinting Over Radio (AMTOR). Above 50 MHz, any digital code is permitted. For a detailed discussion of these codes see Chapter 19.

Frequencies

On the HF bands, the digital communications (RTTY) frequencies are usually found at the top of the CW portion of each band. Traditionally, the 20-meter RTTY subband has received the heaviest use during daylight hours; the vicinity of 3.6 MHz is active at night. The increased popularity of RTTY has brought about increased use of the 40-meter RTTY subband. See the Radio-Frequency Allocations Tables in this chapter for a list of bands where RTTY is authorized. Suggested operating frequencies for RTTY are:

U.S. RTTY Frequencies (kHz)	IARU Region 1 (Europe/Africa) Band Plan (kHz)
3590 RTTY DX 3610-3630	3580-3620
7040 RTTY DX 7080-7100	7035-7045
10,140-10,150	10,140-10,150
14,075-14,100	14,080-14,100
18,100-18,110*	18,100-18,110
21,090-21,100	21,080-21,120
24,920-24,930	24,920-24,930
28,090-28,100	28,050-28,150

*Pending FCC approval for U.S. amateur use.

Station Identification

FCC rules permit identification of a digital communication station in the digital code used for communication when ITA2, AMTOR or ASCII are used, and also by Morse code or voice (where radiotelephone is permitted). Above 50 MHz when using other digital codes, identification must be by one of the methods just mentioned.

ITA2 Operation

ITA2 (Baudot/Murray) has been the predominant RTTY code used by amateurs. It became popular because of the availability of inexpensive, surplus teleprinters that use 5-unit code and operate at speeds of 60, 67, 75 or 100 WPM, corresponding to 45, 50, 56 and 75 bauds, respectively. Gradually, starting in the late 1970s, these mechanical teleprinters were replaced with computer-based systems. Some of these systems are based on personal computers programmed for ASCII-ITA2 code conversion. Others are commercially designed electronic RTTY terminals.

When transmitting via ITA2, you are usually in control of the receiving station's copy format. Because of the variety of mechanical and electronic devices in use, your message may be displayed on a video-display terminal (VDT) or a printer.

At the beginning of each transmission, it is desirable to make sure that the receiving station's printer or video display is set to the first printing position of a new line and in letters case. This can be done by sending two carriage returns (2CR), a line feed (LF), and a letters (LTRS) function. This 2CR LF LTRS combination is also used as an end-of-line (EOL) sequence whenever you wish to begin a new line. While only one CR is needed for newer printers, two CRs are necessary for older teleprinters to give their massive carriages enough time to return to the first printing position of the line. Only one LF is sent to conserve the receiving station's supply of printer paper. The LTRS function ensures that the printer is in letters case.

For years, RTTY operators were accustomed to teleprinters which could print up to 72 characters per line. It is no longer that simple. There are video display terminals (VDTs) and personal computers which may display from 20 to 80 characters per line, sometimes as many as 132. Printers are available with anywhere from 40 to 132 characters per line. So it is difficult to know how many characters to type per line and still have it fit the other station's screen or printer format. If you know that the other station is using a printer or VDT with 40 characters per line, you can keep that in mind when you type. However, as a general rule for normal teleprinter-to-teleprinter communication, it is desirable not to exceed 69 characters per line, the CCITT standard for both printers and video displays.

Some VDTs and printers automatically go to the first printing position of the new line when the previous line is full. This feature may be called *automatic new line* or *wrap around*. Not all mechanical printers have this feature, however. When they don't, characters can pile up at the end of the line and not be readable. To accommodate the older printers, an end-of-line sequence of 2CR LF LTRS must be sent at the end of each line.

When you are typing, precede all figures-case characters with a FIGS function. After finishing the figures-case characters, be sure to return to letters case by sending LTRS before typing anything else. Some, but not all, teleprinters have a feature known as "unshift on space" which automatically inserts a LTRS function upon receipt of a space function. This feature can be a blessing at times but also can be a nuisance when printing a page full of numbers. To be positive that the different display equipment will print your figures correctly, it is good practice to precede each number group with a FIGS function and include a LTRS function when shifting to the letters case.

At the end of each transmission, it is considered courteous to transmit an end-of-line sequence (2CR LF LTRS) so that both stations' printers will start the next transmission on a new line and in letters case.

In ITA2, it is safe to assume that all the letters-case characters will print correctly on the other station's printer or diplay. This is not necessarily true of the figures-case characters, particularly for FIGS D, F, G, H and J, which are subject to national variation. The teleprinters which many U.S. amateurs have used also differs from ITA2 for FIGS V and Z. Avoid these figures-case characters when talking to DX RTTY stations. A few stations are using teleprinters with weather symbols in figures case. If you want to play it safe, avoid figures-case characters by spelling out most punctuation, using figures case only for numbers, period and slant.

ITA2 has a limited character set that provides only capital letters. Unlike ASCII, which has a variety of control characters, ITA2 has only a few machine functions: space, carriage return, line feed, figures, letters and blank. Because of the lack of control characters, commercial and government services have evolved a number of sequences of combinations used for special purposes. Some of these sequences have found their way into Amateur Radio RTTY. One that you may have seen is NNNN, which is an end-of-message signal. Another that has appeared on the ham bands is ZCZC, the start-of-message signal. Both of these sequences are from CCITT Recommendation S.4, which lists some others that you could encounter sometime:

CCCC	for remotely switching on a reperforator or equivalent device
SSSS	for remotely switching on a terminal
FFFF	for remotely turning off a reperforator
KKKK	ready-for-test signal
KLKL	for remotely switching on a reader
XXXXX	error signal

When RTTY equipment is tested, lines of RY sequences are sometimes used. The RY sequence is excellent for exercising mechanical teleprinters because R and Y have complementary bit patterns, causing most mechanical parts to move each character. However, it does not test to see whether every different character can be printed. For that purpose, many RTTY operators use a test message text such as one of the following:

THE QUICK BROWN FOX JUMPS OVER THE LAZY DOG 1234567890

VOYEZ LE BRICK GEANT QUE J'EXAMINE PRES DU WHARF 1234567890

In most other respects ITA2 RTTY procedure is similar to that used on CW. All the CW abbreviations and Q signals are used.

AMTOR Operation

For a technical overview of AMTOR, see Chapter 19. The basic reason for using AMTOR in preference to ITA2 is that AMTOR has error-reduction features.

At the moment, most AMTOR activity is around 14,075 and 3637.5 kHz, which are used as calling frequencies. Operators usually move off the calling frequency to free it for others. However, there are some automatically controlled stations that just stay on these frequencies. At this writing, automatic control of an AMTOR station in the U.S. requires an FCC Special Temporary Authority.

In most respects, AMTOR operation is the same as for ITA2. However, AMTOR has several modes of operation.

In Mode A, the transmitting station sends groups of three characters. If a group is received correctly, the receiving station acknowledges, and the transmitting station sends the next three. If the receiving station does not acknowledge, the transmitting station keeps sending the three characters until acknowledged. This system is known as Automatic Repeat Request (ARQ) and is limited to contacts between two stations.

Mode B employs Forward Error Correction (FEC). Collective Mode B is used to transmit to more than one station, as in a CQ, bulletin transmission, or when talking to a net. Selective Mode B uses FEC but is intended for a single station or perhaps a group of stations.

ASCII Operation

ASCII operation differs from ITA2 operation in that it has a larger character set. Besides a number of control characters, ASCII also provides both upper- and lower-case letters and a more complete set of punctuation marks.

ASCII does away with letters and figures cases that sometimes cause ITA2 characters to print in the wrong case. ASCII is a character set that was designed for both computer and data communications uses.

Because of newer printers and elimination of the LTRS/FIGS problem, only CR LF need be sent as an end-of-line sequence. There are other ASCII *format effectors* (control signals which effect where print occurs on the page) if you want to make use of its entire capability. There are also control characters for start of message, start of text, device control, etc. which may be part of a communications protocol.

To take advantage of the expanded character set, ASCII test messages could be as follows:

ThE QuicK BrowN FoX JumpS OveR ThE LazY Dog 1234567890
VoyeZ Le BricK GeanT QuE J'ExaminE PreS Du WharF 1234567890

When ASCII is used, it is possible that the way a transmitting station formats a message will not be the way that the message is presented to the receiving operator. Computer-based systems may be used to provide split-screen displays, simply file messages away for later use, or process incoming messages according to an agreed protocol.

Packet-Radio Protocols

From a user's viewpoint, the above information about ASCII operation also applies to packet radio. However, in packet radio, there are seven different protocol layers, as outlined in Chapter 19. The Application and Presentation Layers affect what the operator sees on a VDT or printer, depending on the design of the software. However, the Physical, Link and Network layers govern what is sent on the air. Although there is some standardization at the lower protocol layers, revolutionary changes are expected in packet-radio protocol development over the years ahead.

The use of the RTTY subbands is encouraged for routine packet operations on all HF bands. The frequencies listed below are exceptions to this general rule to provide usable channels for automatic message forwarding:

3594.3 kHz	Intercontinental message forwarding
3607.3 kHz	North American message forwarding
7038.3 kHz	Intercontinental message forwarding
7091.3 kHz	North American message forwarding
10145.3 kHz	Intercontinental message forwarding
10147.3 kHz	North American message forwarding

Note: These frequencies are subject to noninterference with fixed stations outside the United States.

| 14102.3 kHz | Intercontinental message forwarding (experimental) |
| 14108.3 kHz | Intracontinental message forwarding (experimental) |

Note: The lowest frequency provides sufficent protection to receivers using CW bandwidths to receive 14100-kHz beacons.

The frequencies listed below conform to the RTTY subbands and are suggested for automatic message forwarding when propagation is favorable.

1802.3 kHz
18106.3 and 18108.3 kHz (pending FCC approval)
21096.3 and 21098.3 kHz
24926.3 and 24928.3 kHz
28102.3 and 28104.3 kHz
Note: These 28-MHz frequencies may be considered for both automatic message forwarding and network entry points for Novices and Technicians.

For more information, see Moved and Seconded, Sept. 1987 *QST*, p. 54.

Computer-Based Message Systems

Since the mid 1970s, a number of Computer-Based Message Systems (CBMSs) have appeared on the ham bands. Possibly you have heard them called MSO (Message Storage Operation), bulletin board or mailbox. They will automatically respond to calls on their operating frequency if the calling station uses the correct character sequence. A remote station can write messages to be stored in the system. Stations can subsequently read remotely entered messages plus any bulletin messages.

Message handling in this manner is third-party traffic, and the system operator

US AMATEUR SUBBAND ALLOCATIONS, 1.8 to 1300 MHz

At all times, transmitter power should be kept down to that necessary to carry out the desired communications. Power is rated in watts PEP output. Unless otherwise stated, the maximum power output is 1500 W. Power for all license classes is limited to 200 W in the 10,100-10,150 kHz band and in all Novice subbands below 28,100 kHz. Novices and Technicians are restricted to 200 W in the 28,100-28,500 kHz subband. In addition, Novices are restricted to 25 W in the 222.1-223.91 MHz subband and 5 W in the 1270-1295 MHz subband.

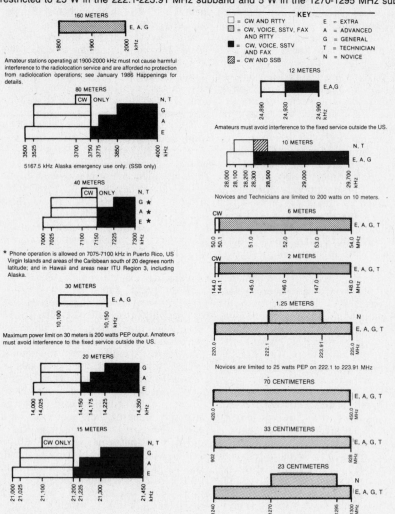

(SYSOP) is required to observe appropriate rules concerning message content. Also the SYSOP is reponsible for maintaining control of the transmitter and removing it from the air in the event of malfunction.

Working DX

Most amateurs at one time or another make working DX a major aim. As in every other phase of amateur operating, there are right and wrong ways to go about getting best results in working foreign stations. This section will outline a few of them.

The ham who has trouble contacting DX stations readily may find that poor transmitter efficiency is not the reason. He may find that his sending is poor, his call ill-timed, or his judgment in error. Working DX requires the know-how that comes with experience. If you just call CQ DX you may get a call from a foreign station, but it isn't likely to be a "rare one." On the other hand, unless you are

experienced enough to know that conditions are right, your receiver is sensitive and selective enough, and your transmitter and antenna properly tuned and oriented, you may get no calls at all and succeed only in causing some unnecessary QRM.

The call CQ DX means slightly different things to amateurs on different bands:

a) On VHF, CQ DX is a general call ordinarily used only when the band is open, under favorable propagation conditions. For VHF work, such a call is used for looking for new states, countries and grid squares, as well as for distances beyond the customary "line-of-sight" range on most VHF bands.

b) CQ DX on our HF bands may be taken to mean "general call to any foreign station." The term "foreign station" usually refers to any station on a different continent. If you do call CQ DX, remember that it implies you will answer any DX station that calls. If you don't mean "general call to any DX station," then listen and call the station you do want.

Codes and Ethics

One of the most effective ways to work DX is to know the operating habits of the DX stations sought, and to abide by the procedures they use. Know when and where to call, and for how long, and when to remain silent while awaiting your chance. DXing has certain understood codes of ethics and procedures that will make this popular amateur pursuit more fun for everybody if everybody follows them. One of the sad things about DXing is to listen to some of the abuse that goes on, mostly by stations on "this" side, as they trample on each other trying to raise their quarry. DX stations have been known to go off the air in disgust at some of the tactics. If W and VE stations will use the procedure in the "DX Operating Code" detailed elsewhere on these pages, we can all make a good impression on the air.

Snagging the Rare Ones

Once in a while a CQ DX will result in snagging a rare DX contact, if you're lucky. This seldom happens, however; usually, what you have to do is listen — and listen — and then listen some more. If everybody transmits, nobody is going to hear anything. Be a snooper. Usually, unless you are lucky enough to be among the first to hear him, a rare DX station will be found under a pileup, with stations swarming all over him like worker bees over a queen. The bedlam will subside when the DX station is transmitting (although some stations keep right on calling him), and you can hear him. Don't immediately join the pack; be a little cagey. Listen a while, get an idea of his habits, find out where he is listening (if not on his exact frequency), bide your time, and await your chance.

Make your calls short, snappy and distinct. No need to repeat his call (he knows it very well; all he needs to know is that you are calling him), but send your own call a couple of times. Try to find a time when few stations are calling him and he is not transmitting; then get in there! With experience, you'll learn all kinds of tricks, some of them clever, some just plain dirty. You'll have no trouble discerning which is which. Learn to use the clever ones, and shun the dirty ones.

Choosing Your Band

If it does nothing else in furthering your education, striving to work DX will certainly teach you a few things about propagation. You will find that four principal factors determine propagation characteristics: (1) the frequency band on which you do your operating, (2) the time of day or night, (3) the season of the year and (4) the sunspot cycle. For example, the 3.5- to 4.0-MHz band at high noon in the summertime at the peak of the sunspot cycle is the poorest possible choice, while the same band at midnight during the wintertime at

Table 7
The R-S-T System

Readability
1 — Unreadable
2 — Barely readable, occasional words distinguishable.
3 — Readable with considerable difficulty.
4 — Readable with practically no difficulty.
5 — Perfectly readable.

Signal Strength
1 — Faint signals, barely perceptible.
2 — Very weak signals.
3 — Weak signals,
4 — Fair signals.
5 — Fairly good signals.
6 — Good signals.
7 — Moderately strong signals.
8 — Strong signals.
9 — Extremely strong signals.

Tone
1 — Sixty-cycle ac or less, very rough and broad.
2 — Very rough ac, very harsh and broad.
3 — Rough ac tone, rectified but not filtered.
4 — Rough note, some trace of filtering.
5 — Filtered rectified ac but strongly ripple-modulated.
6 — FIltered tone, definite trace of ripple modulation.
7 — Near pure tone, trace of ripple modulation.
8 — Near perfect tone, slight trace of modulation.
9 — Perfect tone, no trace of ripple or modulation of any kind.

If the signal has the characteristic steadiness of crystal control, add the letter x to the RST report. If there is a chirp, the letter c may be added to so indicate. Similarly for a click, add κ. See FCC Regulations 97.73, purity of emissions. This reporting system is used on both CW and voice, except that the "tone" report is left out on voice.

the low part of the cycle might produce some very exciting DX. Similarly, you will learn by experience when to operate on which band for the best DX by juggling these factors using both long-range and other indication of band conditions. WWV transmissions can also be helpful in indicating both current and immediate-forecast band conditions.

On some bands, such as 10 and 6 meters, beacons have been established to give an indication of band openings. Listen between 28.2 and 28.3 MHz on 10 meters and around 50.110 MHz on 6 meters. Commercial stations near ham-band edges are also a fair indication of openings. But remember that many of these run many times the maximum amateur power, and consequently may be heard well before skip improves to the point necessary to sustain amateur communications.

Conditions in the transmission medium often make it possible for the signals from low-powered transmitters to be received at great distances. In general, the higher the frequency band, the less important power considerations become, for occasional DX work. This accounts in part for the relative popularity of the 14-, 21- and 28-MHz bands among amateurs who like to work DX.

DX Operating Code (for W/VE Amateurs)

If observed by all W/VE amateurs, these suggestions will go a long way toward making DX more enjoyable for everybody.

1) Call DX only after he calls CQ or QRZ?, signs S̄K, or the phone equivalent.

2) Do not call a DX station:

a) on the frequency of the station he is working until you are sure the QSO is over. This is indicated by the ending signal S̄K on CW and any indication that the operator is listening, on phone.

b) because you hear someone else calling him.

c) when he signs K̄N, ĀR, CL or phone equivalents.

d) after he calls a directional CQ, unless of course you are in the right direction or area.

3) Keep within your frequency-band limits.

4) Observe calling instructions given by DX stations. 10U means call 10 kHz up from his frequency, 15D means 15 kHz down, etc.

5) Give honest reports. Many stations depend on signal reports for adjustment of station equipment.

6) Keep your signal clean. Key clicks, chirp, hum or splatter give you a bad reputation and may get you a citation from the FCC.

7) Listen for and call the station you want. Calling CQ DX is not the best assurance that the rare DX will reply.

8) When there are several W or VE stations waiting to work a DX station, avoid asking him to "listen for a friend." Let your friend take his chances with the rest. Also avoid engaging DX stations in rag chews against their wishes.

Awards

League-sponsored operating activities have useful objectives and provide much enjoyment for members of the fraternity. Achievement in Amateur Radio is also recognized by various awards offered by ARRL. Basic rules require that sufficient funds be included with all submissions of cards to ensure their safe return. A basic fee schedule for return postage is included with each award application. Applicants in the U.S., its possessions and Canada must be ARRL members to participate in the WAS and DXCC programs. DX stations are exempted from this requirement.

DX Century Club Award

The DXCC is one of the most popular and sought-after awards in all of Amateur Radio, and among the more difficult to acquire. Its issuance is carefully supervised at ARRL Hq.

To obtain DXCC, an amateur must make two-way contact with 100 "countries" on the ARRL DXCC list. Written confirmations are required for proof of contact. These must show clearly your call sign, and the date, time, frequency and mode. Such confirmations must be sent to ARRL Hq., where each one is carefully scrutinized to make sure it actually con-

firms a contact with the applying amateur, that it is not altered, and that the "country" claimed is actually on the ARRL list. Further safeguards are applied to maintain the high standards of this award. A handsome, king-size certificate and DXCC lapel pin are sent to each amateur qualifying.

The term "country" for DXCC purposes does not necessarily agree with the dictionary definition. Many bodies of land not having independent status politically are classified as countries. For example, the states of Alaska and Hawaii are considered "countries" because of their distance from the U.S. mainland. There are over 300 such designations on the ARRL DXCC list. Once a basic DXCC is issued, the certificate can be endorsed, by sticker, for additional countries by sending the additional cards to ARRL Hq. for checking. Separate DXCC awards are available for mixed modes, all phone, all CW, RTTY, 160 meters and satellites.

Before applying, familiarize yourself with the rules. Application forms (CD-253) may be obtained from ARRL Hq. Also, the complete ARRL DXCC Countries List, containing the current DXCC countries, along with those countries served by the ARRL Membership Overseas QSL Service, ITU and CQ Zones for each country, third-party agreements, and band/mode check-offs for record keeping, is available from the League for $1. [Note: By action of the ARRL Board of Directors, 10-MHz confirmations are not creditable for ARRL awards.]

Five-Band DXCC

Entirely separate from DXCC, ARRL

also offers a Five-Band DXCC (5BDXCC) Award for those amateurs who submit written proof of having made two-way contact with 100 or more countries on each of five bands since January 1, 1969. For a copy of the complete rules, send an s.a.s.e. to ARRL Hq., 225 Main St., Newington, CT 06111.

WAC Award

The ever-popular Worked All Continents Award (WAC), sponsored by the International Amateur Radio Union (IARU), can be yours by simply submitting proof of contact with each of the six continents. Two-way confirmed contacts must be made with amateurs in each of six continental areas of the world: Africa, Asia, Europe, North America, Oceania and South America. Certificates are available for the following: Basic award, CW, phone, SSTV, RTTY, FAX, satellite and 5-band. Endorsement stickers are available for these achievements: 6-band (for 5-band awards), QRP (5-W output or less), 1.8 MHz, 3.5 MHz, 50 MHz, 144 MHz and 430 MHz. Contacts on 10, 18 and 24 MHz are void for the 5-band certificate and 6-band sticker. Confirmations must be submitted for any awards claimed.

Amateurs residing in the USA or its possessions may obtain full details about the WAC Awards from ARRL Hq. QSLs (not copies) must be sent to ARRL Hq. for checking. ARRL membership is required of all US applicants. Applicants in other countries must send QSLs to their IARU amateur society. That society will certify their eligibility to the IARU Secretariat (ARRL) for issuance of the award on behalf of the Union. Applicants in countries not belonging to the IARU may send their QSLs direct to ARRL Hq. for checking.

WAS Award

"WAS" means Worked All States. This award is offered to any amateur who contacts and receives a QSL from each U.S. state. Contacts may be made over any period of time on any or all amateur bands.

Endorsements to the basic award are available for WAS on CW, SSB, Novice, QRP and any single band. Specialty awards (numbered separately) are available for

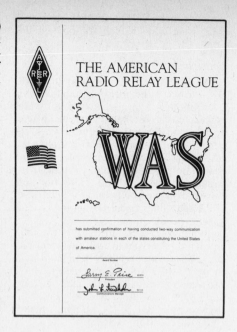

OSCAR, satellite, SSTV, RTTY, 432 MHz, 220 MHz, 144 MHz, 50 MHz and 160 meters. Confirmations must clearly state that contact took place under the circumstances of the desired endorsement or specialty award.

Complete rules and application forms (MCS-217) are available from ARRL Hq. for an s.a.s.e. Confirmations and application forms may be submitted to an ARRL Special Service Club HF Awards Manager for verification. You can find out the name of the nearest awards manager from ARRL Hq. If you cannot find an awards manager in your area, you may send your cards and application to ARRL Hq. for checking. Be sure to include sufficient postage for their safe return.

5BWAS Awards

A handsome certificate will be issued to all amateurs who submit original proof of contact with all 50 states on each of five amateur bands, made after January 1, 1970 (only contacts made after that day will be eligible). Rules require applicants in the U.S., its possessions and Canada to be ARRL full members. Standard WAS rules apply. Write to ARRL Hq. for the application, rules and postage fees for the return of your cards.

VUCC

The VHF/UHF Century Club, VUCC, is an ARRL-sponsored achievement award for working grid squares measuring 2° longitude by 1° latitude on frequencies above 50 MHz. Individual awards are issued per band, with initial qualifying levels as follows: 50 MHz — 100; 144 MHz — 100; 220 MHz — 50; 432 MHz — 50; 902 MHz — 25; 1296 MHz — 25; 2.3 GHz — 10; 3.4 GHz — 5; 5.7 GHz — 5; 10 GHz — 5. Each award is endorsable in increments of 25 for 50 and 144 MHz, 10 for 220 and 432 MHz, and 5 for 902 MHz and above. Certificates offered for 220 and 432

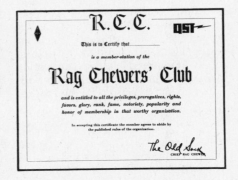

MHz indicate membership in the Half Century Club; for higher frequencies, the Quarter Century Club is appropriate. Only contacts made on January 1, 1983 and after count for VUCC. QSL cards are certified by an approved ARRL VHF Awards Manager. Please send ARRL Hq. an s.a.s.e. for complete application materials.

A-1 Operator Club

The A-1 Operator Club should include in its ranks every good operator. To become a member one must be nominated by two persons who already belong. General keying (not speed) or voice technique, procedure, copying ability, judgment and courtesy all count in the rating of candidates under the club rules. Aim to make yourself a fine operator and one of these days you will be pleasantly surprised when your letter carrier arrives at your QTH with your certificate of membership in the A-1 Operator Club.

Old Timers Club

If you held an Amateur Radio license 20 or more years ago and are licensed at the present time, you are eligible to become a member of the Old Timers Club. Lapses in activity during intervening years are permitted. An s.a.s.e. (business size, at least 10 × 4 inches) will expedite your certificate.

Rag Chewers Club

Your first contact as a licensed amateur may very well earn your first award. The Rag Chewers Club is designed to encourage friendly contacts and discourage the "contest" type of QSO with nothing more than an exchange of calls, signal reports and so on. It furthers fraternalism through Amateur Radio.

Membership certificates are awarded to amateurs who report a fraternal-type contact with another amateur lasting a half hour or longer. This does not mean a half hour spent trying to work a rare DX station, but a solid half hour of pleasant "visiting" with another amateur, discussing subjects of mutual interest. If you nominate someone for RCC, please send the information to the nominee who will (in turn) apply to ARRL Hq. for membership. Or if you know you qualify for the RCC, just report the conversation to ARRL Hq. (c/o RCC) and back will come your membership certificate. An s.a.s.e. is appreciated when requesting this award.

Code Proficiency Award

The Code Proficiency Award permits each amateur to prove himself as a proficient CW operator, and sets up a system of achievement for code skill. It enables every amateur to check individual code proficiency, to better that proficiency, and to receive a certification of receiving speed.

This program is a lot of fun. The League will award a certificate to any interested individual who demonstrates perfect copy for at least one minute, plain-language. Continental code at 10, 15, 20, 25, 30 or 35 words per minute, as transmitted twice monthly from W1AW and once a month from W6OWP. Special 40-WPM qualifying runs are sent by W1AW in February, June and October and by W6OWP in April, August and December. Neither an amateur license nor ARRL membership is required to participate.

As part of the ARRL Code Proficiency program, W1AW transmits plain-language practice material several times daily at speeds from 5 to 35 WPM, occasionally in reverse order. All amateurs are invited to use these transmissions to increase their code-copying ability. Nonamateurs are invited to utilize the lower speeds, 5, 7½ and 10 WPM, which are transmitted for the benefit of persons studying the code in preparation for the amateur license examination. Check the W1AW material in this chapter or refer to *QST* for details.

Contesting

Contesting is to Amateur Radio what the Olympic Games are to worldwide amateur athletic competition: a showcase to display talent and learned skills, as well as a stimulus for further achievement through competition. Increased operating skills and greater station efficiency are the predominant end results of Amateur Radio contesting, whether the operator is a serious contender or a casual participant.

Don't believe it? Tune across the band, any band, and listen for the most efficient operators. Chances are better than even that they are avid contesters or at least have contesting as one of their favorite Amateur Radio activities. How can one tell who is a contester just by listening to a particular operator's style? It is easier to tell who is not interested in contesting by listening. The contester is not likely to be the one, who (while thousands on the frequency are gnashing their teeth in anger) asks the operator of the rarest DXpedition in two decades what the weather is like in "Lower Slobbovia." The contester is not likely to

Kurt, NI6W, contests from Santa Ana, California.

Section leaders earn this handsome certificate in the annual Sweepstakes contest. There are similar awards for other contests.

NK7U is none other than former Major League baseball player Joe Rudi. Joe is active on all bands, and particularly enjoys contests, operating from his home in Oregon.

be the operator, who when working a much-sought-after station on one of the many award nets, punctuates his repeating of the needed exchange 37 times with a long series of "uhhs" and assorted other noises for the lack of anything better to say. The contest operator knows from experience that conciseness aids in efficient and courteous operating.

The contest operator is also likely to have one of the better signals on the band — not necessarily the most elaborate station equipment, but a signal enhanced by the most efficient use of station components available. Contest operation encourages optimization of station and operator efficiency.

The ARRL contest program is so diverse that it holds appeal for almost every operator — the beginning contester and the old hand, the newest Novice and oldest Extra-Classer, "Top Band" buff and microwave enthusiast.

A thumbnail sketch of many of the contests sponsored by the ARRL follows. Complete entry rules and details appear in *QST,* usually the month before the contest occurs.

January

VHF Sweepstakes. Premiere VHF operating event. All bands, 50 MHz and up. ARRL affiliated-club competition, based on members' aggregate total scores.

February

Novice Roundup. Competition geared for the beginning (Novice and Technician) amateur. Increase code speed through operating, work stations needed for WAS, and other achievements. Awards for ARRL Section winners. Fun for all.

International DX Contest, CW. W/VE amateurs work the rest of the world for individual section, country and ARRL affiliated club honors; single-band, QRP and multioperator categories also, with many plaques awarded.

March

International DX Contest, Phone.

June

VHF QSO Party. One of two VHF QSO parties. This one (and the September party) lends itself to multioperator mountaintop expedition operation. Use all bands above 50 MHz.

Field Day. The number one operating event of the year. More than 25,000 participants take to the fields to operate some 1500 emergency stations for informal competition, a score listing in *QST* and an all-around good time. Don't miss this one.

July

IARU HF World Championship. Worldwide competition. Everybody works everybody else for ITU zone, DXCC country and U.S. state honors. Varying point scale; ITU zones and IARU member-society headquarters stations are scoring multipliers. Some of those hard-to-work DXCC countries turn out for this contest.

August

UHF Contest. Similar to the VHF contests, but utilizes the bands at 220 MHz and above. Scoring mulipliers are grid squares. The UHF bands come alive for this contest weekend.

September

VHF QSO Party. Second of two VHF QSO parties (also see June).

November

Sweepstakes. The most prestigious domestic contest. Two weekends (actually separate contests and listings); one weekend for phone and one for CW. Twenty-four-hour time limit on each mode. W and VE operators work each other. ARRL sections are the scoring multipliers. Awards for both high- and low-power ARRL section winners. An ARRL-affiliated club competition highlights the Sweepstakes activity.

December

160-Meter Contest. A gathering of "top band" enthusiasts. W/VE types work each other and DX stations for contest credit.

10-Meter Contest. A 10-meter operator's

dream come true as 28 MHz springs to life and everyone, worldwide, tries to work everyone else for top score (in country, continent and ARRL section) honors.

That's the ARRL contest program in a nutshell. Of course, more detailed rules and descriptions of the awards structure (certificates and plaques awarded to designated top scorers) are announced in *QST* for each of these events. The monthly "Contest Corral" column of *QST* also details the entry rules for many contests other than those sponsored by ARRL, including the very popular state QSO parties and most other major contests.

Public Service

Tens of thousands of U.S. amateurs are involved with public service. Where do you fit in? The emergency preparedness and third-party-traffic handling facets of Amateur Radio beckon. There's a place for every ham in the League's Amateur Radio Emergency Service (ARES), an emergency-preparedness group of approximately 25,000 amateurs who have signed up voluntarily to keep Amateur Radio in the forefront of public service operating. The ARRL National Traffic System (NTS) functions as a message-handling network operating 365 days a year for the systematic handling of third-party traffic. Together, ARES and NTS form a major part of the League's extensive volunteer Field Organization.

Also recognized by ARRL as a part of the organized public service effort is the Radio Amateur Civil Emergency Service (RACES), a part of the Amateur Service serving civil defense under a separate subpart of the amateur regulations; the Military Affiliate Radio System (MARS)

Operating Aids for Public Service

ARRL Hq. makes available the following free operating aids for public service communications:

Public Service Communications Manual	FSD-235
ARRL numbered radiograms	FSD-3
Sample emergency plan	FSD-27
ARES registration form	FSD-98
Amateur message form	FSD-218
Emergency Reference information	FSD-255
Field Organization Brochure	FSD-300

This entire Public Service Package can be obtained by sending a large (9 × 12-inch) envelope with First Class postage for 6 ounces.

---

sponsored by the armed forces to provide military communications training for amateurs; and the numerous amateur groups organized into nets or monitoring services by individuals, clubs or other amateur entities for public service. The detailed workings of the League's emergency and traffic programs are covered briefly here and in more depth in the *Public Service Communications Manual,* available free from ARRL Hq. (please provide an s.a.s.e.). Additional information can be found in the *ARRL Operating Manual,* a for-sale publication available at your local radio dealer or direct from Hq.

ARES and NTS — How They Apply to You

As a member of the local ARES group, you'll be training to provide communications at the city or county level. Each group is headed by an ARRL Emergency Coordinator (EC). Most ARES activities are centered on 2-meter FM, so it's advantageous to have your own emergency-powered VHF gear. However, you really don't need any equipment to join; the training and practice are most important. All you really need is an interest in serving your community through Amateur Radio and participation in periodic tests as time permits. These tests run the gamut from serious simulated emergencies to providing communications for parades and walkathons, or conducting a message-handling service at a shopping center during the Christmas season. Many hams have trained with the National Weather Service to become tornado and storm spotters. All of these activities exist, so that when a flood or an ice storm disrupts the community, experienced hams will know exactly what to do.

Becoming involved is as simple as requesting a registration card (FSD-98) from ARRL Hq. and filling it out. These cards are turned over to the local EC, who registers you in the local ARES organization. Should your community not have an EC, why not volunteer yourself? You qualify if you are a licensed amateur of Technician class or higher, an ARRL member, have a sincere interest in public service and a willingness to put in the time

and effort to fulfill the appointment. If this sounds like you, contact your Section Manager — whose name, address and telephone number appear on page 8 of each issue of *QST.*

The bulk of localized emergency communications is handled on VHF. Much involves repeater operation. The reason is simple. Repeaters can be accessed with low-cost and lightweight equipment, hand-held or mobile. Best of all, they provide clear, reliable communications up to 100 miles or so. Many repeaters have emergency-power capabilities as well, making them the mainstay in any widespread emergency.

What's the National Traffic System all about, then? NTS serves a dual purpose: The rapid movement of long-haul traffic from origin to destination and the training of amateur operators in the handling of

formal radiogram traffic in efficient directed nets. A rundown of the NTS schedule of nets and functions can be found in the *Public Service Communications Manual.* In the overview, however, NTS can be visualized as a pony express of the airwaves, with assigned amateurs carrying traffic to and from the next higher (or lower) level in the system. NTS operations are concentrated mainly on the high frequencies (HF), but local nets on 2 meters have become more and more popular as the ideal place to distribute traffic for delivery.

Most ARRL sections have section nets on 80 meters, both phone and CW. You can find the traffic and emergency nets that service your area in the annual ARRL *Net Directory.* Directed net procedures, especially on CW, take a little getting used to, but if you consult the League's

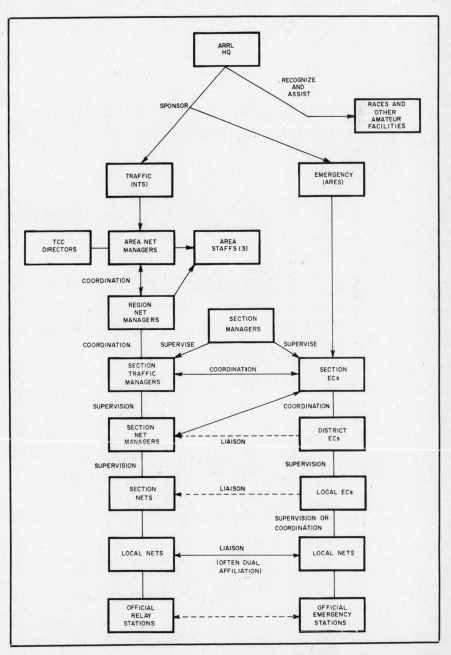

AMATEUR MESSAGE FORM

Every formal radiogram message originated and handled should contain the following component parts in the order given.

I   PREAMBLE
  a. Number (begin with 1 each month or year)
  b. Precedence (R, W, P or EMERGENCY)
  c. Handling Instructions (optional, see text) .
  d. Station of Origin (first amateur handler)
  e. Check (number of words/groups in text only)
  f. Place of Origin (not necessarily location of station of origin)
  g. Time Filed (optional with originating station)
  h. Date (must agree with date of time filed)

II   ADDRESS (as complete as possible, include zip code and telephone number)

III TEXT (limit to 25 words or less, if possible)

IV   SIGNATURE

CW MESSAGE EXAMPLE

I   NR 1 R HXG W1AW 8 NEWINGTON CONN 1830Z July 1
    a b c   d   e   f   g   h

II DONALD SMITH $\overline{AA}$
    164 EAST SIXTH AVE $\overline{AA}$
    NORTH RIVER CITY MO 00789 $\overline{AA}$
    733 4968 $\overline{BT}$

III HAPPY BIRTHDAY X SEE YOU SOON X LOVE $\overline{BT}$

IV   DIANA $\overline{AR}$

Note that X, when used in the text as punctuation, counts as a word.

CW: The prosign $\overline{AA}$ separates the parts of the address. $\overline{BT}$ separates the address from the text and the text from the signature. $\overline{AR}$ marks end of message; this is followed by B if there is another message to follow, by N if this is the only or last message. It is customary to copy the preamble, parts of the address, text and signature on separate lines.

RTTY: Same as cw procedure above, except (1) use extra space between parts of address, instead of $\overline{AA}$; (2) omit cw procedure sign $\overline{BT}$ to separate text from address and signature, using line spaces instead; (3) add a CFM line under the signature, consisting of all names, numerals and unusual works in the message in the order transmitted.

PHONE: Use *prowords* instead of prosigns, but it is not necessary to name each part of the message as you send it. For example, the above message would be sent on phone as follows: "Number one routine HX Golf W1AW eight Newington Connecticut one eight three zero zulu July one Donald Smith *Figures* one six four East Sixth Avenue North River City Missouri zero zero seven eight nine *Telephone* seven three three four nine six eight *Break* Happy birthday X-ray see you soon X-ray love *Break* Diana *End of Message* Over.   "End of Message" is followed by "More" if there is another message to follow, "No More" if it is the only or last message. Speak clearly using VOX (or pause frequently on push-to-talk) so that the receiving station can get fills. Spell phonetically all difficult or unusual words — *do not* spell out common words. Do not use cw abbreviations or Q-signals in phone traffic handling.

PRECEDENCES

The precedence will follow the message number. For example, on cw 207R or 207 EMERGENCY. On phone, "Two Zero Seven, Routine (or Emergency)."

EMERGENCY — Any message having life and death urgency to any person or group of persons, which is transmitted by Amateur Radio in the absence of regular commercial facilities. This includes official messages of welfare agencies during emergencies requesting supplies, materials or instructions vital to relief of stricken populace in emergency areas. During normal times, it will be *very rare*. On cw, this designation will always be spelled out. When in doubt, *do not* use it.

PRIORITY — Important messages having a specific time limit. Official messages not covered in the Emergency category. Press dispatches and other emergency-related traffic not of the utmost urgency. Notification of death or injury in a disaster area, personal or official. Use the abbreviation P on cw.

WELFARE — A message that is either a) an inquiry as to the health and welfare of an individual in the disaster area or b) an advisory or reply from the disaster area that indicates all is well should carry this precedence, which is abbreviated W on cw. These messages are handled *after* Emergency and Priority traffic but before Routine.

ROUTINE — Most traffic normal times will bear this designation. In disaster situations, traffic labeled Routine (R on cw) should be handled *last*, or not at all when circuits are busy with Emergency, Priority or Welfare traffic.

Handling Instructions (Optional)

HXA — (Followed by number.) Collect landline delivery authorized by addressee within .... miles. (If no number, authorization is unlimited.)
HXB — (Followed by number.) Cancel message if not delivered within .... hours of filing time; service originating station.
HXC — Report date and time of delivery (TOD) to originating station.
HXD — Report to originating station the identity of station from which received, plus date and time. Report identity of station to which relayed, plus date and time, or if delivered report date, time and method of delivery.
HXE — Delivering station get reply from addressee, originate message back.
HXF — (Followed by number.) Hold delivery until .... (date).
HXG — Delivery by mail or landline toll call not required. If toll or other expense involved, cancel message and service originating station.

For further information on traffic handling, consult the *Public Service Communications Manual* or the *ARRL Operating Manual*, both published by ARRL.

FSD-218 (585)

ARRL HQ., 225 Main St., Newington, CT 06111

---

reference material and do a little monitoring beforehand, you shouldn't have much of a problem. In fact, the many slow-speed nets that meet on the Novice frequencies are a boon to learning the simple procedures.

Each day, hams enjoy the challenge of these activities. Why not check into a net and check it out!

What You Should Do

Before an emergency occurs, prepare for it by keeping your station and emergency power supply (if you have one) in good working order. Participate in the annual nationwide Simulated Emergency Test, contests and Field Day.

Register your station with your local EC. During an emergency, report to the EC at once and follow the EC's directions.

Monitor your local emergency net frequency, but don't transmit unless you are specifically requested to or are certain you can be of assistance.

Copy special W1AW bulletins for latest developments.

Use your receiver more, your transmitter less. Interference can be intense during an emergency.

After an emergency, tell your EC or net manager of your activities, so a timely report is submitted to ARRL Hq. Each month *QST* chronicles Amateur Radio's emergency communications efforts.

The Amateur Radio Service has been a vital part of emergency communications for more than 50 years, whether it be relaying medical traffic into an earthquake-ravaged village in South America, answering "Mayday" from a ship in the Pacific Ocean, or finding out if a neighbor's relative survived a blizzard in the Midwest.

Why not become a part of it!

Volunteer Monitoring

Another significant role for the ARRL Field Organization is in volunteer monitoring. The Communications Amendments Act of 1982 (commonly referred to as Public Law 97-259) made it possible for the FCC to formally use volunteer Amateur Radio operators to monitor the ham bands for rules violations. This is a crucial factor in maintaining order and the traditional high standards of the Amateur Radio Service, as governmental belt-tightening no longer provides the FCC's Field Operations Bureau with the necessary resources to monitor the amateur bands to any great extent. Therefore, the FCC created the Amateur Auxiliary to its Field Operations Bureau, implemented by the ARRL Field Organization.

The Amateur Auxiliary addresses both day-to-day maintenance monitoring (conducted by the League's well-established Official Observer program) and amateur-to-amateur interference. Maintenance monitoring is carried out through an enhanced ARRL Official Observer program, while amateur-to-amateur interference is handled by specifically authorized Local Interference Committees. ARRL sections are the focal point for liaison and the primary communications channel between organized amateurs and the FCC's Field Operations Bureau. In effect, the Amateur Auxiliary will be administered through the League's Section Managers. See page 8 of *QST* for a listing of Section Managers.

The FCC has publicly acknowledged that the Amateur Radio Service has traditionally been the most self-regulated and disciplined service over which it has jurisdiction. The Amateur Auxiliary is a direct response to self-regulation and the confidence that is placed in amateurs by the FCC. Such self-regulation increases in importance as the trend of federal deregulation accelerates. The basic message is that amateur problems are for amateurs to solve, by the experts: The Amateur Auxiliary. For more information on volunteer monitoring see Chapter 39.

W1AW: ARRL Hq. Station

The Maxim Memorial Station, W1AW, is dedicated to serve the amateur fraterni-

All W1AW CW code practice, Baudot, CW, AMTOR and ASCII bulletin transmissions are now generated by computer.

The W1AW visitors' operating position is on the left, and the satellite ground station is on the right.

ty. It is adjacent to the ARRL Hq. offices and is operated by the Headquarters Operators Club. Operating hours are 8:30 A.M. to 1 A.M. Monday through Friday and 3:30 P.M. to 1 A.M. Saturday and Sunday. The station is open to visitors at all times it is in operation. If you wish to operate W1AW while visiting, the period between 1 and 4 P.M., Monday through Friday, is available. Be sure to bring a copy of your FCC license with you if you plan to operate.

W1AW Code Practice

Code practice is sent on approximately 1.818, 3.58, 7.08, 14.07, 21.08, 28.08, 50.08 and 147.555 MHz. Texts are from recent issues of QST and checking references are sent several times during each session. For practice purposes, the order of words in each line of text may be reversed during the 5-15 WPM transmissions. Two code proficiency qualifying runs are sent each month; dates are in the "Contest Corral" section of QST. Here is the code practice schedule:

Speeds
5, 7½, 10, 13, 15 WPM

EST/EDST	PST/PDST
9 A.M. MWF	6 A.M. MWF
7 P.M. MWF	4 P.M. MWF
4 & 10 P.M. TTHSSU	1 & 7 P.M. TTHSSU

Speeds
35, 30, 25, 20, 15, 13, 10 WPM

EST/EDST	PST/PDST
9 A.M. TTH	6 A.M. TTH
7 P.M. TTHSSU	4 P.M. TTHSSU
4 & 10 P.M. MWF	1 & 7 P.M. MWF

Bulletins

CW bulletins at 18 WPM are sent daily at 5, 8 and 11 P.M. and Monday through Friday at 10 A.M EST/EDST. Frequencies are the same as those used for code practice.

Baudot radioteletype bulletins, at 60 WPM with 170-Hz shift, are sent daily at 6, 9 and 12 P.M. and Monday through Friday at 11 A.M. EST/EDST. Each Baudot bulletin is followed by a repeat on 110-baud

W1AW Special Emergency Bulletin Schedule

Phone on the hour: 1890, 3990, 7290, 14,290, 21,390, 28,590, 50,190, 147,555 kHz.
RTTY 15 minutes past the hour: 3625, 7095, 14,095, 21,095, 28,095, 147,555 kHz.
CW on the half hour: 1818, 3580, 7080, 14,070, 21,080, 28,080, 50,080, 147,555 kHz.

ASCII. Frequencies are 3.625, 7.095, 14.095, 21.095, 28.095 and 147.555 MHz. Transmissions on AMTOR Mode B will follow the Baudot/ASCII bulletins when time permits.

Voice bulletins are daily at 9:30 P.M. and 12:30 A.M. EST/EDST on 1.890, 3.990, 7.290, 14.290, 21.390, 28.590, 50.19 and 147.555 MHz.

A complete W1AW schedule is available from ARRL Hq. for an s.a.s.e.

General Operation

W1AW is equipped for operation on all bands from 1.8 to 144 MHz, and for teleprinter, and satellite communications as well as for CW, SSB and FM.

Allocation of International Call Signs

Call Sign Series	Allocated to
AAA-ALZ	United States of America
AMA-AOZ	Spain
APA-ASZ	Pakistan (Islamic Republic of)
ATA-AWZ	India (Republic of)
AXA-AXZ	Australia
AYA-AZZ	Argentine Republic
A2A-A2Z	Botswana (Republic of)
A3A-A3Z	Tonga (Kingdom of)
A4A-A4Z	Oman (Sultanate of)
A5A-A5Z	Bhutan (Kingdom of)
A6A-A6Z	United Arab Emirates
A7A-A7Z	Qatar (State of)
A8A-A8Z	Liberia (Republic of)
A9A-A9Z	Bahrain (State of)
BAA-BZZ	China (People's Republic of)
CAA-CEZ	Chile
CFA-CKZ	Canada
CLA-CMZ	Cuba
CNA-CNZ	Morocco (Kingdom of)

COA-COZ	Cuba
CPA-CPZ	Bolivia (Republic of)
CQA-CUZ	Portugal
CVA-CXZ	Uruguay (Oriental Republic of)
CYA-CZZ	Canada
C2A-C2Z	Nauru (Republic of)
C3A-C3Z	Andorra (Principality of)
C4A-C4Z	Cyprus (Republic of)
C5A-C5Z	Gambia (Republic of the)
C6A-C6Z	Bahamas (Commonwealth of the)
C7A-C7Z*	World Meteorological Organization
C8A-C9Z	Mozambique (People's Republic of)
DAA-DRZ	Germany (Federal Republic of)
DSA-DTZ	Republic of Korea
DUA-DZZ	Philippines (Republic of the)
D2A-D3Z	Angola (People's Republic of)
D4A-D4Z	Cape Verde (Republic of)
D5A-D5Z	Liberia (Republic of)
D6A-D6Z	Comoros (Federal and Islamic Republic of the)
D7A-D9Z	Republic of Korea

EAA-EHZ	Spain
EIA-EJZ	Ireland
EKA-EKZ	Union of Soviet Socialist Republics
ELA-ELZ	Liberia (Republic of)
EMA-EOZ	Union of Soviet Socialist Republics
EPA-EQZ	Iran (Islamic Republic of)
ERA-ESZ	Union of Soviet Socialist Republics
ETA-ETZ	Ethiopia
EUA-EWZ	Byelorussian Soviet Socialist Republic
EXA-EZZ	Union of Soviet Socialist Republics
FAA-FZZ	France
GAA-GZZ	United Kindgom of Great Britain and Northern Ireland
HAA-HAZ	Hungarian People's Republic
HBA-HBZ	Switzerland (Confederation of)
HCA-HDZ	Ecuador
HEA-HEZ	Switzerland (Confederation of)
HFA-HFZ	Poland (People's Republic of)
HGA-HGZ	Hungarian People's Republic
HHA-HHZ	Haiti (Republic of)

HIA-HIZ	Dominican Republic	
HJA-HKZ	Colombia (Republic of)	
HLA-HLZ	Republic of Korea	
HMA-HMZ	Democratic People's Republic of Korea	
HNA-HNZ	Iraq (Republic of)	
HOA-HPZ	Panama (Republic of)	
HQA-HRZ	Honduras (Republic of)	
HSA-HSZ	Thailand	
HTA-HTZ	Nicaragua	
HUA-HUZ	El Salvador (Republic of)	
HVA-HVZ	Vatican City State	
HWA-HYZ	France	
HZA-HZZ	Saudi Arabia (Kingdom of)	
H2A-H2Z	Cyprus (Republic of)	
H3A-H3Z	Panama (Republic of)	
H4A-H4Z	Solomon Islands	
H6A-H7Z	Nicaragua	
H8A-H9Z	Panama (Republic of)	
IAA-IZZ	Italy	
JAA-JSZ	Japan	
JTA-JVZ	Mongolian People's Republic	
JWA-JXZ	Norway	
JYA-JYZ	Jordan (Hashemite Kingdom of)	
JZA-JZZ	Indonesia (Republic of)	
J2A-J2Z	Djibouti (Republic of)	
J3A-J3Z	Grenada	
J4A-J4Z	Greece	
J5A-J5Z	Guinea-Bissau (Republic of)	
J6A-J6Z	Saint Lucia	
J7A-J7Z	Dominica	
J8A-J8Z	St. Vincent and the Grenadines	
KAA-KZZ	United States of America	
LAA-LNZ	Norway	
LOA-LWZ	Argentina (Republic of)	
LXA-LXZ	Luxembourg	
LYA-LYZ	Union of Soviet Socialist Republics	
LZA-LZZ	Bulgaria (People's Republic of)	
L2A-L9Z	Argentina (Republic of)	
MAA-MZZ	United Kingdom of Great Britain and Northern Ireland	
NAA-NZZ	United States of America	
OAA-OCZ	Peru	
ODA-ODZ	Lebanon	
OEA-OEZ	Austria	
OFA-OJZ	Finland	
OKA-OMZ	Czechoslovak Socialist Republic	
ONA-OTZ	Belgium	
OUA-OZZ	Denmark	
PAA-PIZ	Netherlands (Kingdom of the)	
PJA-PJZ	Netherlands Antilles	
PKA-POZ	Indonesia (Republic of)	
PPA-PYZ	Brazil (Federative Republic of)	
PZA-PZZ	Suriname (Republic of)	
P2A-P2Z	Papua New Guinea	
P3A-P3Z	Cyprus (Republic of)	
P4A-P4Z	Netherlands Antilles	
P5A-P9Z	Democratic People'e Republic of Korea	
QAA-QZZ	(Service abbreviations)	
RAA-RZZ	Union of Soviet Socialist Republics	
SAA-SMZ	Sweden	
SNA-SRZ	Poland (People's Republic of)	
SSA-SSM	Egypt (Arab Republic of)	
SSN-STZ	Sudan (Democratic Republic of the)	
SUA-SUZ	Egypt (Arab Republic of)	
SVA-SZZ	Greece	
S2A-S3Z	Bangladesh (People's Republic of)	
S6A-S6Z	Singapore (Republic of)	
S7A-S7Z	Seychelles (Republic of)	
S9A-S9Z	Sao Tome and Principe (Democratic Republic of)	
TAA-TCZ	Turkey	
TDA-TDZ	Guatemala (Republic of)	
TEA-TEZ	Costa Rica	
TFA-TFZ	Iceland	
TGA-TGZ	Guatemala (Republic of)	
THA-THZ	France	
TIA-TIZ	Costa Rica	
TJA-TJZ	Cameroon (United Republic of)	
TKA-TKZ	France	
TLA-TLZ	Central African Republic	
TMA-TMZ	France	
TNA-TNZ	Congo (People's Republic of the)	
TOA-TQZ	France	
TRA-TRZ	Gabon Republic	

TSA-TSZ	Tunisia	
TTA-TTZ	Chad (Republic of)	
TUA-TUZ	Ivory Coast (Republic of the)	
TVA-TXZ	France	
TYA-TYZ	Benin (People's Republic of)	
TZA-TZZ	Mali (Republic of)	
T2A-T2Z	Tuvalu	
T3A-T3Z	Kiribati Republic	
T4A-T4Z	Cuba	
T5A-T5Z	Somali Democratic Republic	
T6A-T6Z	Afghanistan (Democratic Republic of)	
T7A-T7Z	San Marino (Republic of)	
UAA-UQZ	Union of Soviet Socialist Republics	
URA-UTZ	Ukrainian Soviet Socialist Republic	
UUA-UZZ	Union of Soviet Socialist Republics	
VAA-VGZ	Canada	
VHA-VNZ	Australia	
VOA-VOZ	Canada	
VPA-VSZ	United Kingdom of Great Britain and Northern Ireland	
VTA-VWZ	India (Republic of)	
VXA-VYZ	Canada	
VZA-VZZ	Australia	
V2A-V2Z	Antigua and Barbuda	
V3A-V3Z	Belize	
V4A-V4Z	St. Christopher and Nevis	
WAA-WZZ	United States of America	
XAA-XIZ	Mexico	
XJA-XOZ	Canada	
XPA-XPZ	Denmark	
XQA-XRZ	Chile	
XSA-XSZ	China (People's Republic of)	
XTA-XTZ	Upper Volta (Republic of)	
XUA-XUZ	Democratic Kampuchea	
XVA-XVZ	Viet Nam (Socialist Republic of)	
XWA-XWZ	Lao People's Democratic Republic	
XXA-XXZ	Portugal	
XYA-XZZ	Burma (Socialist Republic of the Union of)	
YAA-YAZ	Afghanistan (Democratic Republic of)	
YBA-YHZ	Indonesia (Republic of)	
YIA-YIZ	Iraq (Republic of)	
YJA-YJZ	New Hebrides	
YKA-YKZ	Syrian Arab Republic	
YLA-YLZ	Union of Soviet Socialist Republics	
YMA-YMZ	Turkey	
YNA-YNZ	Nicaragua	
YOA-YRZ	Romania (Socialist Republic of)	
YSA-YSZ	El Salvador (Republic of)	
YTA YUZ	Yugoslavia (Socialist Federal Republic of)	
YVA-YYZ	Venezuela (Republic of)	
YZA-YZZ	Yugoslavia (Socialist Federal Republic of)	
Y2A-Y9Z	German Democratic Republic	
ZAA-ZAZ	Albania (Socialist People's Republic of)	
ZBA-ZJZ	United Kingdom of Great Britain and Northern Ireland	
ZKA-ZMZ	New Zealand	
ZNA-ZOZ	United Kingdom of Great Britain and Northern Ireland	
ZPA-ZPZ	Paraguay (Republic of)	
ZQA-ZQZ	United Kingdom of Great Britain and Northern Ireland	
ZRA-ZUZ	South Africa (Republic of)	
ZVA-ZZZ	Brazil (Federative Republic of)	
Z2A-Z2Z	Zimbabwe (Republic of)	
2AA-2ZZ	United Kingdom of Great Britain and Northern Ireland	
3AA-3AZ	Monaco	
3BA-3AZ	Mauritius	
3CA-3CZ	Equatorial Guinea (Republic of)	
3DA-3DM	Swaziland (Kingdom of)	
3DN-3DZ	Fiji	
3EA-3FZ	Panama (Republic of)	
3GA-3GZ	Chile	
3HA-3UZ	China (People's Republic of)	
3VA-3VZ	Tunisia	
3WA-3WZ	Viet Nam (Socialist Republic of)	
3XA-3XZ	Guinea (People's Revolutionary Republic of)	
3YA-3YZ	Norway	
3ZA-3ZZ	Poland (People's Republic of)	
4AA-4CA	Mexico	

4DA-4IZ	Philippines (Republic of the)	
4JA-4LZ	Union of Soviet Socialist Republics	
4MA-4MZ	Venezuela (Republic of)	
4NA-4OZ	Yugoslavia (Socialist Federal Republic of)	
4PA-4SZ	Sri Lanka (Democratic Socialist Republic of)	
4TA-4TZ	Peru	
4UA-4UZ*	United Nations Organization	
4VA-4VZ	Haiti (Republic of)	
4WA-4WZ	Yemen Arab Republic	
4XA-4XZ	Israel (State of)	
4YA-4YZ*	International Civil Aviation Organization	
4ZA-4ZZ	Israel (State of)	
5AA-5AZ	Libya (Socialist People's Libyan Arab Jamahiriya)	
5BA-5BZ	Cyprus (Republic of)	
5CA-5GZ	Morocco (Kingdom of)	
5HA-5IZ	Tanzania (United Republic of)	
5JA-5KZ	Colombia (Republic of)	
5LA-5MZ	Liberia (Republic of)	
5NA-5OZ	Nigeria (Federal Republic of)	
5PA-5QZ	Denmark	
5RA-5SZ	Madagascar (Democratic Republic of)	
5TA-5TZ	Mauritania (Islamic Republic of)	
5UA-5UZ	Niger (Republic of the)	
5VA-5VZ	Togolese Republic	
5WA-5WZ	Western Samoa	
5XA-5XZ	Uganda (Republic of)	
5YA-5ZZ	Kenya (Republic of)	
6AA-6BZ	Egypt (Arab Republic of)	
6CA-6CZ	Syrian Arab Republic	
6DA-6JZ	Mexico	
6KA-6NZ	Republic of Korea	
6OA-6OZ	Somali Democratic Republic	
6PA-6SZ	Pakistan (Islamic Republic of)	
6TA-6UZ	Sudan (Democratic Republic of the)	
6VA-6WZ	Senegal (Republic of the)	
6XA-6XZ	Madagascar (Democratic Republic of)	
6YA-6YZ	Jamaica	
6ZA-6ZZ	Liberia (Republic of)	
7AA-7IZ	Indonesia (Republic of)	
7JA-7NZ	Japan	
7OA-7OZ	Yemen (People's Democratic Republic of)	
7PA-7PZ	Lesotho (Kingdom of)	
7QA-7QZ	Malawi (Republic of)	
7RA-7RZ	Algeria (Algerian Democratic and Popular Republic)	
7SA-7SZ	Sweden	
7TA-7YZ	Algeria (Algerian Democratic and Popular Republic)	
7ZA-7ZZ	Saudi Arabia (Kingdom of)	
8AA-8IZ	Indonesia (Republic of)	
8JA-8NZ	Japan	
8OA-8OZ	Botswana (Republic of)	
8PA-8PZ	Barbados	
8QA-8QZ	Maldives (Republic of)	
8RA-8RZ	Guyana	
8SA-8SZ	Sweden	
8TA-8YZ	India (Republic of)	
8ZA-8ZZ	Saudi Arabia (Kingdom of)	
9AA-9AZ	San Marino (Republic of)	
9BA-9DZ	Iran (Islamic Republic of)	
9EA-9FZ	Ethiopia	
9GA-9GZ	Ghana	
9HA-9HZ	Malta (Republic of)	
9IA-9JZ	Zambia (Republic of)	
9KA-9KZ	Kuwait (State of)	
9LA-9LZ	Sierra Leone	
9MA-9MZ	Malaysia	
9NA-9NZ	Nepal	
9OA-9TZ	Zaire (Republic of)	
9UA-9UZ	Burundi (Republic of)	
9VA-9VZ	Singapore (Republic of)	
9WA-9WZ	Malaysia	
9XA-9XZ	Rwanda (Republic of)	
9YA-9ZZ	Trinidad and Tobago	

Note

The series of call signs with an asterisk indicate the international organization to which they are allocated.

Chapter 39

Monitoring and Direction Finding

Radio direction finding is almost as old as radio itself. Frequently called RDF, the location of radio transmitters with direction-finding techniques is considered by many to be either an art or a science — or perhaps a bit of both. Although sophisticated and complex equipment pushing the state of the art has been developed for use by governments and commercial enterprises, relatively simple equipment can be built at home to offer the radio amateur an opportunity to RDF.

In many countries of the world, the hunting of hidden transmitters takes on the atmosphere of a sport, as participants wearing jogging togs or track suits dash toward the area where they believe the transmitter is located. The sport is variously known as fox hunting, bunny hunting, ARDF (Amateur Radio direction finding) or simply transmitter hunting. In North America, most hunting of hidden transmitters is conducted from automobiles, although hunts on foot are gaining popularity.

There is a more serious side to RDF, as well. Although not directly related to Amateur Radio, radio navigation is one application of RDF. The locating of downed aircraft is another, and one in which amateurs often lend their skills. Or sometimes a stolen amateur rig will be placed into operation by a person who is not familiar with Amateur Radio, and by being lured into making frequent transmissions, the operator unsuspectingly permits himself to be located by hams using RDF equipment. Jammers of repeaters, traffic nets and other amateur operations can also be located with RDF equipment. Indeed, it has many useful applications.

Volunteer Monitoring

An important aspect of the Communications Amendments Act of 1982, commonly known as Public Law 97-259, is one that authorized the FCC to formally enlist the use of amateur volunteers for monitoring the airwaves and for rules violations. To implement volunteer monitoring, the FCC's Field Operations Bureau has created an Amateur Auxiliary, administered by the League's Section Managers.

The Auxiliary is concerned with both maintenance monitoring and amateur-to-amateur interference. Maintenance monitoring is conducted through an enhanced ARRL Official Observer program, while amateur-to-amateur interference is handled by specifically authorized Local Interference Committees. To locate offenders, such committees must rely in a large part on RDF techniques. Thus, RDF becomes a federally sanctioned activity in Amateur Radio.

DIRECTION-FINDING SYSTEMS

Required for any RDF system are a directive antenna and a device for detecting the radio signal. In Amateur Radio applications the signal detector is usually a receiver; for convenience it will have a meter to indicate signal strength. At very close ranges a simple diode detector and dc microammeter may suffice for the signal detector.

The receiver may be a small portable, a mobile or the regular fixed-station instrument. In government or commercial installations the "receiver" is usually a system of electronics coupled to a cathode-ray tube that indicates azimuth. Antennas used in these systems use electronic beam forming, discussed in a later section. But for amateur use, a standard, unmodified, commercially available receiver works well. Some enterprising amateurs have designed and built their own receivers for RDF work, but their primary goal has usually been portability that is not offered in commercially made equipment. The point is this: If you want to try your hand at RDF, going out to buy a special RDF receiver is *not* one of the considerations.

Antennas for RDF work, on the other hand, are not generally the types used for normal two-way communications. Directivity is a prime requirement, and here the word directivity takes on a somewhat different meaning than is commonly applied to antennas. Normally we associate directivity with gain, and we think of the ideal antenna pattern as one having a long, thin main lobe. Such a pattern may be of value for coarse measurements in RDF work, but precise bearing measurements are not possible. There is always a spread of a few (or perhaps many) degrees on the "nose" of the lobe where a shift of antenna bearing produces no detectable change in signal strength. In RDF measurements, it is desirable to correlate an exact bearing or compass direction with the position of the antenna. To do this accurately, an antenna exhibiting a null in its pattern is used. A null can be very sharp in directivity, to within a half degree or less.

Loop Antennas

A simple antenna for RDF work is a small loop tuned to resonance with a capacitor. Several factors must be considered in the design of an RDF loop. The loop must be small compared with the wavelength. In a single-turn loop, the conductor should be less than 0.08 wavelength long. For 28 MHz, this represents a length of less than 34 inches (diameter of approximately 10 inches). Maximum response from the loop antenna is in the plane of the loop, with nulls exhibited at right angles to that plane.

To obtain the most accurate bearings, the loop must be balanced electrostatically with respect to ground. Otherwise, the loop will exhibit two modes of operation. One is the mode of a true loop, while the other is that of an essentially non-directional vertical antenna of small dimensions ("antenna effect"). The voltages introduced by the two modes are not in phase and may add or subtract, depending on the direction from which the wave is coming.

The theoretical true loop pattern is illustrated in Fig. 1A. When properly balanced, the loop exhibits two nulls that are 180° apart. Thus, a single null reading with a small loop antenna will not indicate the exact direction toward the transmitter — only the line along which the transmitter lies. Ways to overcome this ambiguity are discussed in later sections.

When the antenna effect is appreciable and the loop is tuned to resonance, the loop may exhibit little directivity, as shown in Fig. 1B. However, by detuning the loop to shift the phasing, a pattern similar to C may be obtained. Although this pattern is not symmetrical, it does exhibit a null. The null, however, may not be as sharp as that obtained with a loop that is well balanced, and may not be at exact right angles to the plane of the loop.

By suitable detuning, the unidirectional pattern of Fig. 1D may be approached. This adjustment is sometimes used in RDF work to obtain a unidirectional bearing, although there is no complete null in the pattern. In most cases, however, the loop is adjusted for the best null.

An electrostatic balance can be obtained by shielding the loop, as shown in Fig. 2. The shield is represented by the broken lines in the drawing, and eliminates the antenna effect. The response of a well-constructed shielded loop is quite close to the ideal pattern of Fig. 1A.

For the low-frequency amateur bands, single-turn loops of convenient physical size for portability are generally unsatisfactory for RDF work. Therefore, multi-turn loops are generally used instead. Such a loop is shown in Fig. 3. This loop may also be shielded, and if the total conductor length remains below 0.08 wavelength, the directional pattern is that of Fig. 1A.

Ferrite Rod Antennas

The development of magnetic core materials has led to the use of loop antennas with such cores. These antennas consist essentially of many turns of wire around a ferrite rod. They are known as loopstick antennas and also as ferrite-rod antennas. Probably the best-known example of this type of antenna is that used in small portable AM broadcast receivers. The advantage of ferrite-rod antennas is their reduced size. For portable work, loopsticks are used almost exclusively for frequencies below 150 MHz.

As implied in the earlier discussion of shielded loops in this chapter, the true loop antenna responds to the magnetic field of the radio wave, and not to the electrical field. The voltage delivered by the loop is proportional to the amount of magnetic flux passing through the coil, and to the number of turns in the coil. The action is much the same as in the secondary winding of a transformer. For a given size of loop, the output voltage can be increased by increasing the flux density, and this is done with a ferrite core of high permeability. A ½-inch-diameter, 7-inch rod of Q2 ferrite ($\mu_i = 125$) is suitable for a loop core from

the broadcast band through 10 MHz. For increased output, the turns may be wound on two rods that are taped together, as shown in Fig. 4. Loopstick antennas for construction are described later in this chapter.

Maximum response of the loopstick antenna is broadside to the axis of the rod as shown in Fig. 5, whereas maximum response of the small loop is in the plane of the loop. Otherwise the performances of the ferrite-rod antenna and of the ordinary loop are similar. The loopstick may also be shielded to eliminate the antenna effect, such as with a U-shaped or C-shaped channel of aluminum or other form of "trough." The length of the shield should equal or slightly exceed the length of the rod.

Sensing Antennas

Because there are two nulls that are 180° apart in the directional pattern of a loop or a loopstick, an ambiguity exists as to which one indicates the true direction of the station being tracked. For example, assume you take a bearing measurement and the result indicates the transmitter is somewhere on a line running approximately east and west from your position. With this single reading, you have no way of know-

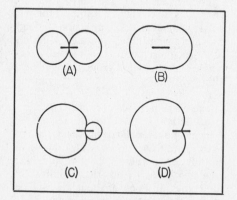

Fig. 1 — Small-loop field patterns with varying amounts of antenna effect — the undesired response of the loop acting merely as a mass of metal connected to the receiver antenna terminals. The heavy lines show the plane of the loop.

ing for sure if the transmitter is east of you or west of you.

If there is more than one receiving station taking bearings on a single transmitter, or if a single receiving station takes bearings from more than one position on the transmitter, the ambiguity may be worked out by a method called triangulation. (Further detail on triangulation is

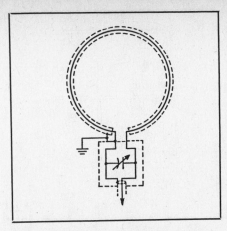

Fig. 2 — Shielded loop for direction finding. The ends of the shielding turn are not connected, to prevent shielding the loop from magnetic fields. The shield is effective against electric fields.

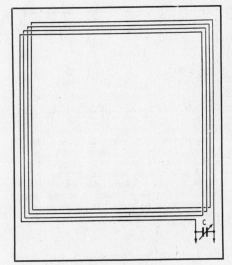

Fig. 3 — Small loop consisting of several turns of wire. The total conductor length is very much less than a wavelength. Maximum response is in the plane of the loop.

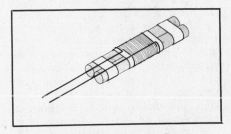

Fig. 4 — A ferrite-rod or loopstick antenna. Turns of wire may be wound on a single rod or, to increase the output from the loop, the core may be two rods taped together, as shown here. The type of core material must be selected for the intended frequency range of the loop. To avoid bulky windings, fine wire such as no. 28 or no. 30 is often used, with larger wire for the leads.

given in a later section.) However, it is sometimes desirable to have a pattern with only one null, so there is no question about

whether the transmitter in the above example would be east or west from your position.

A loop or loopstick antenna may be made to have a single null if a second antenna element is added. The element is called a sensing antenna, because it gives an added sense of direction to the loop pattern. The second element must be omnidirectional, such as a short vertical. When the signals from the loop and the vertical element are combined with a 90° phase shift between the two, a cardioid pattern results. The development of the pattern is shown in Fig. 6A.

Fig. 6B shows a circuit for adding a sensing antenna to a loop or loopstick. R1 is an internal adjustment and is used to set the level of the signal from the sensing antenna. For the best null in the composite pattern, the signals from the loop and the sensing antenna must be of equal amplitude, so R1 is adjusted experimentally during setup. In practice, the null of the cardioid is not as sharp as that of the loop, so the usual measurement procedure is to first use the loop alone to obtain a precise bearing reading, and then to add the sensing antenna and take another reading to resolve the ambiguity. (The null of the cardioid is 90° away from the nulls of the loop.) For this reason, provisions are usually made for switching the sensing element in and out of operation.

Phased Arrays

Phased arrays are also used in amateur RDF work. Two general classifications of phased arrays are end-fire and broadside configurations. Depending on the spacing and phasing of the elements, end-fire patterns may exhibit a null off one end of the axis of the two or more elements. At the same time, the response is maximum off the other end of the axis, in the opposite direction from the null. A familiar arrangement is two elements spaced ¼ wavelength apart and fed 90° out of phase. The resultant pattern is a cardioid, with the null in the direction of the leading element. Other arrangements of spacing and phasing for an end-fire array are also suitable for RDF work. One of the best known is the Adcock array, discussed in the next section.

Broadside arrays are inherently bidirectional, which means there are always at least two nulls in the pattern. Ambiguity therefore exists in the true direction of the transmitter, but depending on the application, this may be no handicap. Broadside arrays are seldom used for amateur RDF applications.

Adcock Arrays

One of the most popular types of end-fire phased arrays is the Adcock. This system was invented by F. Adcock and patented in 1919. The array consists of two vertical elements fed 180° apart, and

mounted so the system may be rotated. Element spacing is not critical, and may be in the range from 1/10 to 3/4 wavelength. The two elements must be of identical lengths, but need not be self-resonant. Elements that are shorter than resonant are commonly used. Because neither the element spacing nor the length is critical in terms of wavelengths, an Adcock array may be operated over more than one amateur band.

The radiation pattern of the Adcock is shown in Fig. 7A. The nulls are in directions broadside to the axis of array, and become sharper with greater element spacings. However, with an element spacing greater than ¾ wavelength, the pattern begins to take on additional nulls in the directions off the ends of the array axis. At a spacing of 1 wavelength the pattern is that of Fig. 7B, and the array is unsuitable for RDF applications.

Short vertical monopoles are often used in what is sometimes called the U-Adcock, so named because the elements with their feeders take on the shape of the letter U. In this arrangement the elements are worked against the earth as a ground or counterpoise. If the array is used only for reception, earth losses are of no great consequence. Short, elevated vertical dipoles are also used in what is sometimes called the H-Adcock. The Adcock array is often used at HF for skywave RDF work. In this application, it is not considered a portable system.

The Adcock array, with two nulls in its pattern, has the same ambiguity as the loop and the loopstick. Adding a sensing element to the Adcock array has not met with great success. Difficulties arise from mutual coupling between the array elements and the sensing element, among other things. Because Adcock arrays are

Fig. 5 — Field pattern for a ferrite rod antenna. The dark bar represents the rod on which the loop turns are wound.

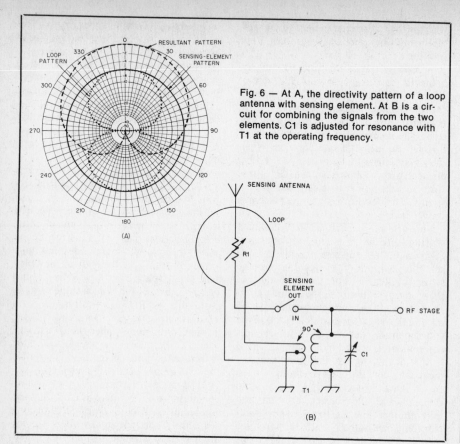

Fig. 6 — At A, the directivity pattern of a loop antenna with sensing element. At B is a circuit for combining the signals from the two elements. C1 is adjusted for resonance with T1 at the operating frequency.

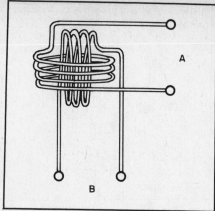

Fig. 8 — An early type of goniometer that is still used today in some RDF applications. This device is a special type of RF transformer that permits a movable coil in the center (not shown here) to be rotated and determine directions even though the elements connected at A and B are stationary.

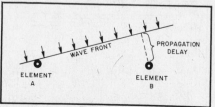

Fig. 9 — This diagram illustrates one technique used in electronic beam forming. By delaying the signal from element A by an amount equal to the propagation delay, the two signals may be summed precisely in phase, even though the signal is not in the broadside direction. Because this time delay is identical for all frequencies, the system is not frequency sensitive.

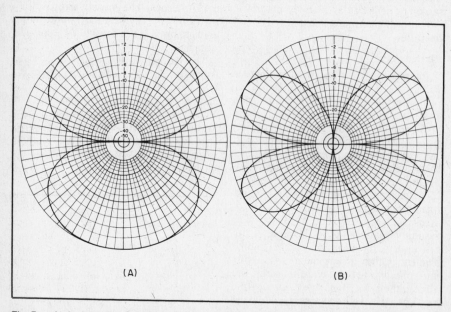

Fig. 7 — At A, the pattern of the Adcock array with an element spacing of ½ wavelength. In these plots the elements are aligned with the vertical axis. As the element spacing is increased beyond ¾ wavelength, additional nulls develop off the ends of the array, and at a spacing of 1 wavelength the pattern at B exists. This pattern is unsuitable for RDF work.

used primarily for fixed-station applications, the ambiguity presents no serious problem. The fixed station with an Adcock is usually one of a group of stations in an RDF network.

Loops vs. Phased Arrays

Although loops can be made smaller than suitable phased arrays for the same frequency of operation, the phased arrays are preferred by some for a variety of reasons. In general, sharper nulls can be obtained with phased arrays, but this is also a function of the care used in constructing and feeding the individual antennas, as well as of the size of the phased array in terms of wavelengths. The primary constructional consideration is the shielding and balancing of the feed line against unwanted signal pickup, and the balancing of the antenna for a symmetrical pattern.

Loops are not as useful for skywave RDF work because of random polarization of the received signal. Phased arrays are somewhat less sensitive to propagation effects, probably because they are larger for the same frequency of operation and therefore offer some space diversity. In general, loops and loopsticks are used for mobile and portable operation, while phased arrays are used for fixed-station operation. However, phased arrays are used successfully above 144 MHz for portable and mobile RDF work. Practical examples of both types of antennas are presented later in this chapter.

The Goniometer

Most fixed RDF stations for government and commercial work use antenna arrays of stationary elements, rather than mechanically rotatable arrays. This has been true since the earliest days of radio. The device used in those early days to permit finding directions without moving the elements is called a radiogoniometer, or simply a goniometer. Various types of goniometers are still used today in many installations, and offer the amateur many possibilities.

The early style of goniometer is a special form of RF transformer, as shown in Fig. 8. It consists of two fixed coils mounted at right angles to one another. Inside the fixed coils is a movable coil, not shown in Fig. 8 to avoid cluttering the diagram. The pairs of connections marked A and B are connected respectively to two elements in an array, and the output to the detector or receiver is taken from the movable coil. As the inner coil is rotated, its coupling to one fixed coil increases while that to the other decreases. Both the amplitude and the phase of the signal coupled into the pickup winding are altered with rotation in a way that corresponds to actually rotating the array itself. Therefore, the rotation of the inner coil can be calibrated in degrees to correspond to bearing angles from the station location.

In the early days of radio, the type of goniometer just described saw frequent use with fixed Adcock arrays. A refinement of that system employed four Adcock elements, two arrays at right angles to each other. With a goniometer arrangement, RDF measurements could be taken in all compass directions, as opposed to none off the ends of a two-element fixed array. However, resolution of the four-element system was not as good as with a single pair of rotatable elements. To overcome this difficulty, a few systems of eight elements were installed.

Various other types of goniometers have been developed over the years, such as commutator switching to various elements in the array. A later development is the diode switching of capacitors to provide a commutator effect. As mechanical action has gradually been replaced with electronics to "rotate" stationary elements, the word goniometer is used less frequently these days. However, it still appears in many engineering reference texts. The more complex electronic systems of today are often called beam-forming networks.

Electronic Antenna Rotation

With an array of many fixed elements, beam rotation can be performed electronically by sampling and combining signals from various individual elements in the array. Contingent on the total number of elements in the system and their physical arrangement, almost any desired antenna pattern can be formed by summing the sampled signals in appropriate amplitude and phase relationships. Delay networks are used for some of the elements before the summation is performed. In addition, attenuators may be used for some elements to develop patterns such as from an array with binomial current distribution.

One system using these techniques is the Wullenweber antenna, used primarily in government and military installations. The Wullenweber consists of a very large number of elements arranged in a circle, usually outside of (or in front of) a circular reflecting screen. Depending on the installation, the circle may be anywhere from a few hundred feet to more than a quarter of a mile in diameter. Although the Wullenweber is not one that would be constructed by an amateur, some of the techniques it uses may certainly be applied to Amateur Radio.

For the moment, consider just two elements of a Wullenweber antenna, shown as A and B in Fig. 9. Also shown is the wavefront of a radio signal arriving from a distant transmitter. As drawn, the wavefront strikes element A first, and must travel somewhat farther before it strikes element B. There is a finite delay before the wavefront reaches element B.

The propagation delay may be measured by delaying the signal received at element A before summing it with that from element B. If the two signals are combined directly, the amplitude of the resultant signal will be maximum when the delay for element A exactly equals the propagation delay. This results in an in-phase condition at the summation point. Or if one of the signals is inverted and the two are summed, a null will exist when the element-A delay equals the propagation delay; the signals will combine in a 180° out-of-phase relationship. Either way, once the time delay is known, it may be converted to distance. Then the direction from which the wave is arriving may be determined by trigonometry.

By altering the delay in small increments, the peak of the antenna lobe (or the null) can be steered in azimuth. This is true without regard to the frequency of the incoming wave. Thus, the system is not frequency sensitive, other than for the frequency range that may be covered satisfactorily by the array elements themselves. Delay lines used in such systems can be lumped-constant networks if the system is used only for receiving. Rolls of coaxial cable of various lengths are used in installations for transmitting. In this case, the lines are considered for the time delay they provide, rather than as simple phasing lines. The difference is that a phasing line is ordinarily designed for a single frequency (or for an amateur band), while a delay line offers essentially the same delay at all frequencies.

By combining signals from other Wullenweber elements appropriately, the broad beamwidth of the pattern from the two elements can be narrowed, and unwanted sidelobes can be suppressed. Then, by electronically switching the delays and attenuations to the various elements, the beam so formed can be rotated around the compass. The package of electronics designed to do this, including delay lines and electronically switched attenuators, is the beam-forming network. However, the Wullenweber system is not restricted to forming a single beam. With an isolation amplifier provided for each element of the array, several beam-forming networks can be operated independently. Imagine having an antenna system that offers a dipole pattern, a rhombic pattern and a Yagi beam pattern, all simultaneously and without frequency sensitivity. One or more may be rotating while another is held in a particular direction. The Wullenweber was designed to fulfill this type of requirement.

One feature of the Wullenweber antenna is that it can operate at 360° around the compass. In many government installations, there is no need for such coverage, as the areas of interest lie in a single azimuth sector. In such cases an in-line array of elements with a backscreen or curtain reflector may be installed broadside to the center of the sector. By using the same techniques as the Wullenweber, the beams formed from this array may be slewed left and right across the sector. The maximum sector width available will depend on the installation, but beyond 70 to 80° the patterns begin to deteriorate to the point that they are unsatisfactory for precise RDF work.

Direction Finding Techniques

Locating a transmitter with RDF techniques is a skill that is acquired only with practice. Familiarity with the antenna being used and especially with its limitations are probably the most important criteria. Of course, knowing the measuring equipment is also a requirement. But in addition to this, one must know how radio signals behave at different frequencies and in different kinds of terrain. Here is where experience becomes the best teacher, although reading about the experiences of others and talking with others who are active in RDF also helps.

GROUND-WAVE RDF

Most amateur RDF activity is conducted with ground-wave signals, where the transmitter is seldom more than a dozen or so miles away. Even if the position of a transmitter has been localized with sky-wave measurements, it is usually not possible to determine the precise transmitter location without taking some ground-wave readings. The considerations for these measurements are discussed in this section. Some of these considerations also apply to sky-wave measurements, but a later section discusses the peculiarities of sky-wave RDF work.

Before setting out to locate a signal source, it is a good idea to note some general information about the signal itself. Is its frequency constant, or does it drift? Drifting means you may have to retune the receiver often during the hunt. Do transmissions occur at rather regular intervals, or are they sporadic? Are transmissions very short in duration, modest in length, or is there a continuous carrier? The most difficult sources to locate are those which transmit at irregular intervals and are on the air for very brief periods, such as a few seconds. Finding these transmitters becomes a real challenge and requires extreme patience. It also requires quick action when the transmitter is on the air, to obtain a null reading in those few seconds. Once you've determined the general signal characteristics, you can begin the search.

Best accuracy in determining a bearing to a signal source is obtained when the propagation path is over homogeneous terrain, and when only the vertically polarized component of the ground wave is present. If a boundary exists such as between land and water, the different conductivities of the two mediums under the ground wave cause bending (refraction) of the wave front. In addition, reflection of RF energy from vertical objects such as mountains or buildings can add to the direct wave and cause RDF errors.

The effects of refraction and reflection are shown in Fig. 10. At A, the signal is actually arriving from a different bearing than the true direction to the transmitter.

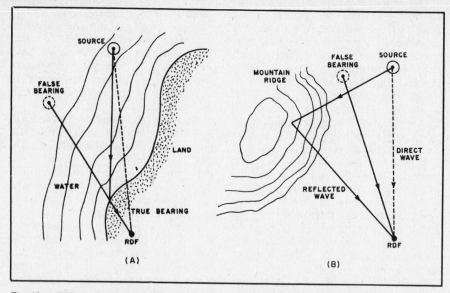

Fig. 10 — RDF errors caused by refraction (A) and reflection (B). At A a false reading is obtained because the signal actually arrives from a direction that is different from that to the source. At B a direct signal from the source combines with a reflected signal from the mountain ridge. The two are averaged at the antenna, giving a false bearing to a point somewhere between the two apparent sources.

This happens because the wave is refracted at the shoreline. Even the most sophisticated of measuring equipment will not indicate the true bearing in this situation, as the equipment can only show the direction from which the signal is arriving. Knowing how radio signals behave at different frequencies and in different kinds of terrain is important in this situation. If the measuring equipment is portable, the experienced RDF enthusiast will probably move to the shoreline to take a measurement with less chance of error. Even so, depending on the direction to the source compared to the direction of the shoreline, a precise reading may still be impossible to obtain.

In Fig. 10B there are two apparent sources for the incoming signal — a direct wave from the source itself, and another wave that is reflected from the mountain ridge. In this case the two waves add at the antenna of the RDF equipment. Without care being used in performing the measurement, the uninitiated observer would obtain a false bearing in a direction somewhere between the directions to the two sources. The experienced RDF enthusiast might notice that the null reading in this situation is not as sharp as it usually is, or perhaps not as deep. But these indications would be subtle, and easy to overlook. Here is where knowing the antenna and the equipment becomes very helpful.

Water towers, tall radio towers and similar objects can also lead to false bearings. The effects of these objects become significant when they are large in terms of a wavelength. In fact, if the direct path to the transmitter is masked by intervening terrain or other radio "clutter," it is possible that a far stronger signal may be received by reflection than from the true direction to the source. Such towers atop hills are prime sources of strong "pinpoint" reflections. Triangulation may even "confirm" that the transmitter is at that high elevation, unless a reading is taken from in the clear, where the true radio path is unmasked by the clutter.

Local objects also tend to distort the field, such as buildings of concrete and steel construction, power lines and the like. It is important that the RDF antenna be in the clear, well away from such surrounding objects. Trees with foliage may also create an adverse effect, especially at VHF and above. If it is not possible to avoid such objects, the best procedure is to take readings at several different positions, and to average the readings for the result.

Triangulation

If two RDF bearing measurements are taken from locations that are separated by a significant distance, the bearing lines may be drawn from those positions as represented on a map. The point where the two lines cross (assuming the two bearings are not the same nor are 180° apart) will indicate a "fix," or the approximate position of the transmitter.

The word "approximate" is used here because there is always some degree of uncertainty in the readings obtained from the measurement. This uncertainty arises from equipment limitations (antenna null width, for example). Propagation effects

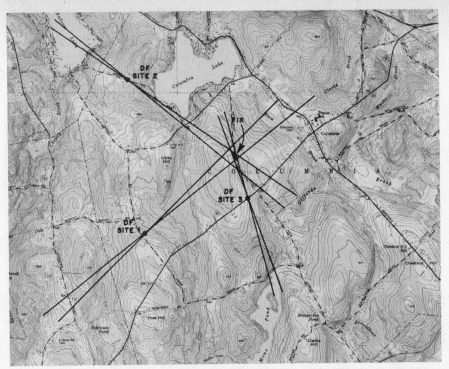

Fig. 11 — Bearing sectors from three RDF positions drawn on a map. This method is known as triangulation. Note here that sensing antennas are not required at any of the DF sites; antennas with two null indications 180° apart are quite acceptable.

may increase the degree of uncertainty. In order to obtain a more precise fix, a bearing measurement from yet a third position is helpful. When the positions and bearing lines for all three measurements are placed on a map, the three lines seldom intersect at a single point. Instead, they form a small triangle on the map. The unknown source is presumed to be located inside the triangle. Thus, the name triangulation.

In order to best indicate the probable area of the transmitter location on the map, the bearings from each position should be drawn as a narrow sector instead of as a single line. The sector width should represent the amount of uncertainty. A portion of a map marked in this manner is shown in Fig. 11. Here it becomes evident graphically how the measurement from DF Site 3 has narrowed down the probable area of the transmitter position. The small black section indicates where the transmitter is likely to be found, a significantly smaller quadrangle than that of the sectors from DF Sites 1 and 2 alone.

Several types of maps are suitable for triangulation. Ordinary road maps, however, usually lack sufficient detail, and are not generally satisfactory. Contour maps of the area like that shown in Fig. 11 are preferred for open country. Aeronautical charts are also suitable. City maps and street maps are usually acceptable. Your local city or county engineer can help you find a source for suitable maps, perhaps even through his office.

Closing in on the Fix

In the final phases of locating an

unknown source, portable RDF equipment is used to pinpoint the exact location. As the equipment is brought near the transmitter, signals become very strong. Overloading of the front-end circuits of the receiver may lead to confusing and inconsistent readings. One way to combat this problem is to insert an attenuator in the transmission line from the antenna to the receiver. Attenuators are discussed in another chapter of this book.

Even with an attenuator in the line, in the presence of a strong electromagnetic field there may be some energy coupled directly into the receiver circuitry. This effect is noted as an apparent nondirectional characteristic of the antenna. In other words, the received signal strength changes only slightly or perhaps not at all as the RDF antenna is rotated in azimuth. The only cure for this effect is to shield the receiving equipment. Something as simple as placing the receiver in a bread pan or cake pan covered with a piece of copper or aluminum screening, securely fastened at several points, may provide sufficient shielding to eliminate this problem. Of course more elaborate precautions may also be taken.

An alternative to protecting the receiver from overload is to switch to a simple diode detector as you are closing in on the fix. The detector should have a sensitive meter indicator with some kind of sensitivity control, and may be made tuneable in frequency, if desired. It might even include a meter-amplifier circuit to increase its sensitivity. It should, however, be constructed inside a metal enclosure

that affords shielding, to avoid unwanted signal pickup in the environment of a strong RF field.

Depending on the nature of the area surrounding the fix, the transmitter may be easy or difficult to locate physically. If the emission is a spurious output from a communications transmitter, a transmitting antenna will probably stand out. On the other hand, a hidden transmitter that is jamming a repeater may be very difficult to find. Sometimes such transmitters are buried in the ground, and the antenna may be a thin wire concealed in the bushes. It may not be spotted by an observer standing right next to it. Sharp eyes and some astute observations are needed here.

A Test or Target Transmitter

Some situations may arise where repeated inconsistent RDF readings simply won't yield a reliable fix. This can happen in mountainous or hilly terrain, or in a canyon, where many signal reflections occur. It also happens in the "concrete jungle" of a metropolitan area where canyons are formed by tall buildings. (In this case it may be advantageous to take a reading from atop one of those buildings.)

A simple aid to locating the source of such a signal is to use a test or target transmitter. This is a second transmitter that is temporarily placed in the general area where the unknown source is thought to be located, and operated on a slightly different frequency. RDF readings from the unknown source may then be compared with readings from the target transmitter to gain a better idea of the unknown location. The target transmitter may even be moved about until the readings coincide, to obtain a reliable fix. Successfully tracking down an unknown transmitter involves detective work — examining all the clues, weighing the evidence and using one's best judgment. For many amateurs, RDF presents an interesting challenge.

SKY-WAVE RDF

Many factors make it difficult to obtain accuracy in sky-wave RDF work. In the previous section, mention was made that the vertical component of the signal is used in taking ground-wave measurements. Because of Faraday rotation during propagation, sky-wave signals are received with random polarization — sometimes the vertical component is strongest, and at other times the horizontal. This necessitates somewhat different techniques for sky-wave work, because the vertical component of the signal should still be the one used for bearing measurements. During periods when the vertical component is weak, the signal will appear to fade. At these times, determining an accurate signal null becomes difficult or impossible.

One important thing to keep in mind with sky-wave measurements is the degree of uncertainty. Accuracy to within 1 or 2°

is the exception rather than the rule. Errors of as much as 3 to 5° are common. An error of 3° at a thousand miles represents a distance of 52 miles. Even with every precaution taken in measurements, don't expect to pinpoint a location to more than a county, a corner of a state or a large metropolitan area. What you *can* expect is to determine where an RDF group should begin making a local search.

Great-Circle Radio Paths

It is generally assumed that a sky-wave signal follows a great-circle path in its travel from a transmitter to a receiver. This is not always the case, however. For example, if the signal is refracted in a tilted layer of the ionosphere, it could arrive from a direction that is several degrees away from the true great-circle bearing. Weather fronts and other meteorological conditions may also result in false readings. Some scientists believe that the earth's magnetic field causes a departure of wave travel from the great-circle path, and propagation conditions during geomagnetic disturbances are known to have this effect.

Another cause of signals arriving off the great-circle path is termed sidescatter. It is possible that, at a given time, the ionosphere does not support great-circle propagation of the signal from the transmitter to the receiver because the frequency is above the MUF for that path. However, at the same time, propagation may be supported from both ends of the path to some mutually accessible point off the great-circle path.

The signal from the source may propagate to that point on the earth's surface and hop in a sideways direction to continue to the receiver. For example, signals from Central Europe have been known to propagate to New England by hopping from an area in the Atlantic Ocean off the northwest coast of Africa, whereas the great-circle path puts the reflection point in the North Atlantic, off the southern coast of Greenland. Readings in error by as much as 50° or more may result from this type of propagation.

Indeed, many factors do combine to make it difficult to obtain accurate sky-wave RDF measurements. The effect may be that the source of the signal seems to wander somewhat over a few minutes of time, or it may be weak and fluttery. At other times, however, there may be no tell-tale signs to indicate that the readings are suspect.

Antennas for Sky-Wave RDF

Loop antennas and loopsticks are generally unsatisfactory for sky-wave RDF because of not only the random polarization shift, but also because the vertical wave angle from which the signal arrives may not be constant. The Adcock or other type of phased vertical array is used most often. Such arrays offer some degree of space diversity over small loops, and

therefore are less subject to fading effects. Greatest accuracy for sky-wave signals is obtained when the vertical angle of arrival is small.

At high arrival angles, even the Adcock array becomes less reliable because its response to high-angle signals falls off. (It will not respond at all to a signal arriving from directly overhead.) Some experimenters have attempted to overcome this difficulty by tilting the array, but little improvement in reliability seems to result. A tilted loop sometimes yields results equivalent to those of a vertical Adcock array for high-angle signals. Signals from high angles often seem to arrive not from a single point, but from a small area in space. This phenomenon is probably associated with scattering of the signal in the ionosphere.

Sky-Wave Triangulation

Triangulation mapping with sky-wave signals is a bit more complex than with ground waves because the paths are great-circle routes. Commonly available world maps are difficult to work with, as the triangulation lines must be curved lines, rather than straight. Depending on the method used to create the map, one showing a country or similar large area may be suitable.

Many U.S. highway maps are usable for triangulation work. Most are some form of conical projection, such as the Lambert conformal conic system. The method of developing such a map is shown in Fig. 12. A cone is made to coincide with the earth's surface along two parallels of latitude, and the axis of the cone coincides with that of the earth. The two latitude lines are called standard parallels, and are chosen for the area to be covered for minimum distortion over the entire projection. The standard parallels are located inside the limits of the projection, as shown. This system of coordinates maintains the accuracy of angular representation, but the distance scale is not constant over the entire projection.

A type of map that can be pressed into service for worldwide areas is an azimuthal-equidistant projection, as shown in Fig. 13. This particular map is prepared for Wichita, Kansas, but they may be obtained for any location. Such maps are available for a nominal fee from Bill Johnston, N5KR, 1808 Pomona Dr., Las Cruces, NM 88001. His maps are custom prepared by computer for a specific location.

With an azimuthal-equidistant map, true bearings for great-circle paths are shown as straight lines from the center to all points on the earth. Maps for three or more different RDF sites may be compared to gain an idea of the general geographic area for an unknown source.

A special type of map has been developed that is well suited for sky-wave triangulation. This map is known as a gnomonic projection, one on which *all* great-circle paths are represented by

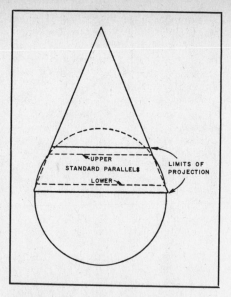

Fig. 12 — The method of obtaining a Lambert conformal conic projection. To make the average distance errors as small as possible, the two standard parallels are chosen for the area to be mapped. Accuracy of angular representations is maintained.

straight lines. To understand how this is possible, see Fig. 14. The drawing shows a transparent sphere with latitude and longitude lines. A small lamp is positioned at the exact center of the sphere and illuminated.

If a large flat card or sheet of paper is touched against the globe at some point as shown in Fig. 14, the surface features will be projected as shadows on that flat surface. This is the gnomonic projection. All meridians and any lines representing great-circle paths will appear as straight lines on the projection. This happens because the lamp is in the plane of each of the great circles, and cannot therefore throw a curved shadow.

Unfortunately, gnomonic charts are not readily available. They are prepared especially for government agencies and military use. Such a projection can be prepared at home if one wants to take the trouble. The work is tedious. Instructions for making a gnomonic projection are given in pages 749 through 758 of R. Keene (see the bibliography at the end of this chapter).

RDF SYSTEM CALIBRATION AND USE

Once an RDF system is initially assembled, it should be "calibrated" or checked out before actually being put into use. Of primary concern is the balance or symmetry of the antenna pattern. A lopsided figure-8 pattern with a loop, for example, is undesirable; the nulls are not 180° apart nor are they at exact right angles to the plane of the loop. If this fact was not known in actual RDF work, measurement accuracy would suffer.

Initial checkout can be performed with a low-powered transmitter at a distance of a few hundred feet. It should be within visual range and must be operating into a vertical antenna. (A quarter-wave vertical or a loaded whip is quite suitable.) The site must be reasonably clear of obstructions, especially steel and concrete or brick buildings, large metal objects, nearby power lines, and so on. If the system operates above 30 MHz, trees and large bushes should also be avoided. An open field makes an excellent site.

The procedure is to "find" the transmitter with the RDF equipment as if its position were not known, and compare the RDF null indication with the visual path to the transmitter. For antennas having more than one null, each null should be checked.

If imbalance is found in the antenna system, there are two options available. One is to correct the imbalance. Toward this end, pay particular attention to the feed line. Using a coaxial feeder for a balanced antenna invites a nonsymmetrical pattern, unless an effective balun is used. A balun is not necessary if the loop is shielded, but a nonsymmetrical pattern can result with misplacement of the break in the shield itself. The builder may also find that the presence of a sensing antenna upsets the balance slightly. Experimenting with its position with respect to the main antenna may lead to correcting the error. You will also note that the position of the null shifts by 90° as the sensing element is switched in and out, and the null is not as deep. This is of little concern, however, as the purpose of the sensing antenna is only to resolve ambiguities. The sensing element should be switched out when accuracy is desired.

The second option is to accept the imbalance of the antenna and use some kind of indicator to show the true directions of the nulls. Small pointers, painted marks on the mast or an optical sighting system might be used. Sometimes the end result of the calibration procedure will be a compromise between these two options, as a perfect electrical balance may be difficult or impossible to attain.

The discussion above is oriented toward calibrating portable RDF systems. The same general suggestions apply if the RDF array is fixed, such as an Adcock. However, it won't be possible to move it to an open field. Instead, the array is calibrated in its intended operating position through the use of a portable or mobile transmitter. Because of nearby obstructions or reflecting objects, the null in the pattern may not appear to indicate the precise direction of the transmitter. Do not confuse this with imbalance in the RDF array. Check for imbalance by rotating the array 180° and comparing readings.

Once the balance is satisfactory, you should make a table of bearing errors noted in different compass directions. These error values should be applied as corrections when actual measurements are made. The

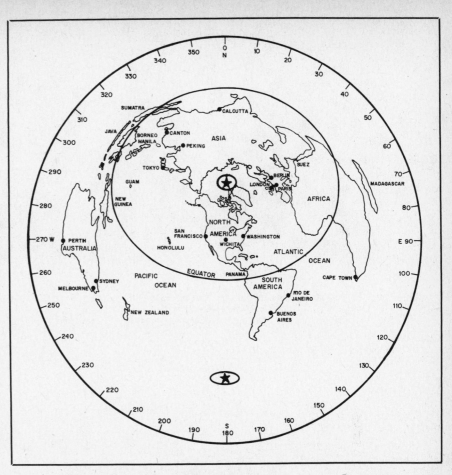

Fig. 13 — Azimuthal map centered on Wichita, Kansas. With such a map for each site in an RDF network, the general geographic area for an unknown source can be obtained by comparison of the maps. *Copyright by Rand McNally & Co., Chicago. Reproduction License no. 394.*

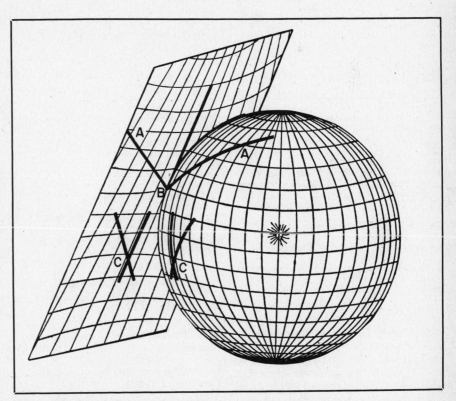

Fig. 14 — Optical means of showing the development of a gnomonic projection. On such a map, all great-circle routes appear as straight lines, and angular measurements with respect to meridians are true.

Fig. 15 — The linear field-strength meter. The control at the upper left is for C1 and the one to the right for C3. At the lower left is the band switch. The zero-set control for M1 is located directly below the meter.

Fig. 16 — Inside view of the field-strength meter. At the upper right is C1 and to the left, C3. The dark leads from the circuit board to the front panel are the shielded leads described in the text.

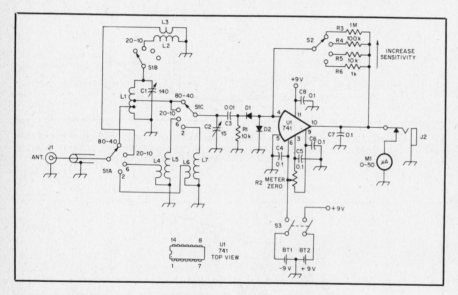

Fig. 17 — Circuit diagram of the linear field-strength meter. All resistors are ¼- or ½-watt composition types.

C1 — 140-pF variable.
C2 — 15-pF variable
D1, D2 — 1N914 or equiv.
L1 — 34 turns no. 24 enam. wire wound on an Amidon* T-68-2 core, tapped 4 turns from ground end.
L2 — 12 turns no. 24 enam. wire wound on T-68-2 core.
L3 — 2 turns no. 24 enam. wire wound at ground end of L2.
L4 — 1 turn no. 26 enam. wire wound at ground end of L5.
L5 — 12 turns no. 26 enam. wire wound on T-25-12 core.
L6 — 1 turn no. 26 enam. wire wound at

ground end of L7.
L7 — 1 turn no. 18 enam. wire wound on T-25-12 core.
M1 — 50 or 100 μA dc.
R2 — 10-kΩ control, linear taper.
S1 — Rotary switch, 3 poles, 5 positions, 3 sections.
S2 — Rotary switch, 1 pole, 4 positions.
S3 — DPST toggle.
U1 — Type 741 op amp. Pin nos. shown are for a 14-pin package.

*Amidon Associates, 12033 Otsego St., North Hollywood, CA 91607

mobile or portable transmitter should be at a distance of 2 or 3 miles for these measurements, and should be in as clear an area as possible during transmissions. The idea is to avoid conduction of the signal along power lines and other overhead wiring from the transmitter to the RDF site. Of course the position of the transmitter

must be known accurately for each transmission.

Using RDF Antennas for Communications

Because of their directional characteristics, RDF antennas would seem to be useful for two-way communications. It has not

been mentioned earlier that the efficiency of receiving loops is poor. The radiation resistance is very low, and the resistance of wire conductors by comparision is significant. For this reason it is common to use some type of preamplifier with receiving loops. Small receiving loops can often be used to advantage in a fixed station, to null out either a noise source or unwanted signals.

A loop that is small in terms of a wavelength may also be used for transmitting, but a different construction technique is necessary. A thick conductor is needed at HF, several inches in diameter. The reason for this is to decrease the ohmic losses in the loop. Special methods are also required to couple power into a small loop, such as links. A small loop is highly inductive, and the inductance may be canceled by inserting a capacitor in series with the loop itself. The capacitor must be able to withstand the high RF currents that flow during transmissions.

The Adcock antenna and other phased arrays have been used extensively for transmitting. In this application maximum response is off the ends of the Adcock, which is 90° away from the null direction used for RDF work.

A PORTABLE FIELD-STRENGTH METER

Few amateur stations, fixed or mobile, are without need of a field-strength meter. An instrument of this type serves many useful purposes during antenna experiments and adjustments, and may be used for closing in on an RDF fix. When work is to be done from many wavelengths away, a simple wavemeter lacks the necessary sensitivity. Further, its linearity leaves much to be desired.

The field-strength meter described here takes care of these problems. Additionally, it is small, measuring only 4 × 5 × 8 inches. The power supply consists of two 9-volt batteries. Sensitivity can be set for practically any amount desired. However, from a usefulness standpoint, the circuit should not be too sensitive or it will respond to unwanted signals. This unit also has excellent linearity with regard to field strength, which is very helpful in RDF work. (The field strength of a received signal varies directly with the distance from the source, all other things being equal.) The frequency range includes all amateur bands from 3.5 through 148 MHz, with band-switched circuits, thus avoiding the use of plug-in inductors. All in all, it is a quite useful instrument

The unit is pictured in Figs. 15 and 16, and the schematic diagram is shown in Fig. 17. A type 741 op-amp IC is the heart of the unit. The antenna is connected to J1, and a tuned circuit is used ahead of a diode detector. The rectified signal is coupled as dc and amplified in the op amp. Sensitivity of the op amp is controlled by inserting R3 through R6 in the circuit by means of S2.

With the circuit shown, and in its most sensitive setting, M1 will detect a signal from the antenna on the order of 100 μV. Linearity is poor for approximately the first 1/5 range of M1, but then is almost straight-line from there to full-scale deflection. The reason for the poor linearity at the start of the readings is because of nonlinearity of the diodes at the point of first conduction. If gain measurements are being made this is of no real importance, however, as accurate gain measurements can be made in the linear portion of the readings. For RDF measurements, just remember that the indicated strength is not proportional to the distance at the low end of the scale.

The 741 op amp requires both a positive and a negative voltage source. This is obtained by connecting two 9-volt batteries in series and grounding the center. One other feature of the instrument is that it can be used remotely by connecting an external meter at J2. This is handy if one wants to adjust an antenna and observe the results without having to leave the antenna site.

L1, the 80/40 meter coil, is tuned by C1. The coil is wound on a toroid form. For 20, 15 or 10 meters, L2 is switched in parallel with L1 to cover the three bands. L5 and C2 cover approximately 40 to 60 MHz, and L7 and C2 from 130 MHz to approximately 180 MHz. The two VHF coils are also wound on toroid forms.

Construction Notes

The majority of the components may be mounted on an etched circuit board. A shielded lead should be used between pin 4 of the IC and S2. The same is true for the leads from R3 through R6 to the switch. Otherwise, parasitic oscillations may occur in the IC because of its very high gain.

For the unit to cover the 144-MHz band, L6 and L7 should be mounted directly across the appropriate terminals of S1, rather than on a circuit board. The extra lead length adds too much inductance to the circuit. It isn't necessary to use toroid forms for the 50- and 144-MHz coils. They were used in the version described here simply because they were available. Air-wound coils of the appropriate inductance can be substituted.

Calibration

The field-strength meter can be used "as is" for a relative-reading device for RDF work. A linear indicator scale will serve admirably. However, it will be a much more useful instrument for antenna work if it is calibrated in decibels, enabling the user to check relative gain and front-to-back ratios. If one has access to a calibrated signal generator, it can be connected to the field-strength meter and different signal levels can be fed to the device for making a calibration chart. Signal-generator voltage ratios can be converted to decibels by using the equation

$$dB = 20 \log (V1/V2) \qquad (Eq. 1)$$

where V1/V2 is the ratio of the two voltages, and log is the common logarithm.

Let's assume that M1 is calibrated evenly from 0 to 10. Next, assume we set the signal generator to provide a reading of 1 on M1, and that the generator is feeding a 100-μV signal into the instrument. Now we increase the generator output to 200 μV, giving us a voltage ratio of 2 to 1. Also let's assume M1 reads 5 with the 200-μV input. From the equation above, we find that the voltage ratio of 2 equals 6.02 dB between 1 and 5 on the meter scale. M1 can be calibrated more accurately between 1 and 5 on its scale by adjusting the generator and figuring the ratio. For example, a ratio of 126 μV to 100 μV is 1.26, corresponding to 2.0 dB. By using this method, all of the settings of S2 can be calibrated. In the instrument shown here, the most sensitive setting of S2 with R3, 1 megohm, provides a range of approximately 6 dB for M1. Keep in mind that the meter scale for each setting of S1 must be calibrated similarly for each band. The degree of coupling of the tuned circuits for the different bands will vary, so each band must be calibrated separately.

Another method for calibrating the instrument is using a transmitter and measuring its output power with an RF wattmeter. In this case we are dealing with power rather than voltage ratios, so this equation applies:

$$dB = 10 \log (P1/P2) \qquad (Eq. 2)$$

where P1/P2 is the power ratio.

With most transmitters the power output can be varied, so calibration of the test instrument is rather easy. Attach a pickup

antenna to the field-strength meter (a short wire a foot or so long will do) and position the device in the transmitter antenna field. Let's assume we set the transmitter output for 10 watts and get a reading on M1. We note the reading and then increase the output to 20 watts, a power ratio of 2. Note the reading on M1 and then use Eq. 2. A power ratio of 2 is 3.01 dB. By using this method the instrument can be calibrated on all bands and ranges.

With the tuned circuits and coupling links specified in Fig. 17, this instrument has an average range on the various bands of 6 dB for the two most sensitive positions of S2, and 15 dB and 30 dB for the next two successive ranges. The 30-dB scale is handy for making front-to-back antenna measurements without having to switch S2.

A SHIELDED LOOP WITH SENSING ANTENNA FOR 28 MHz

Fig. 18 shows the construction and mounting of a simple, shielded, 10-meter loop. The loop is made from an 18-inch length of RG-11/U (solid or foamed dielectric) secured to an aluminum box of any convenient size, with two coaxial cable hoods (Amphenol 83-1HP/U). The outer shield must be broken at the exact center. C1 is a 25-pF variable capacitor, and is connected in parallel with a 33-pF mica padder capacitor, C3. C1 must be tuned to the desired frequency while the loop is connected to the receiver in the same way as it will be used for RDF. C2 is a small differential capacitor used to provide electrical symmetry. The lead-in to the receiver is 67 inches of RG-59/U cable (82 inches if the cable has foamed dielectric).

The loop can be mounted on the roof of the car with a rubber suction cup. The builder might also fabricate some kind of bracket assembly to mount the loop temporarily in the window opening of the automobile, allowing for loop rotation. Reasonably true bearings may be obtained through the windshield when the car is pointed in the direction of the hidden transmitter. More accurate bearings may be obtained with the loop held out the window and the signal coming toward that side of the car.

Sometimes the car broadcast antenna may interfere with accurate bearings. Disconnecting the antenna from the broadcast receiver may eliminate this trouble.

Sensing Antenna

A sensing antenna can be added to the loop to check on which of the two directions indicated by the loop is the correct one. Add a phono jack to the top of the aluminum case shown in Fig. 18. The insulated center terminal of the jack should be connected to the side of the tuning capacitors connected to the center conductor of the RG-59/U coax feed line. The jack then takes a short vertical antenna rod of a diameter to fit the jack, or a piece of

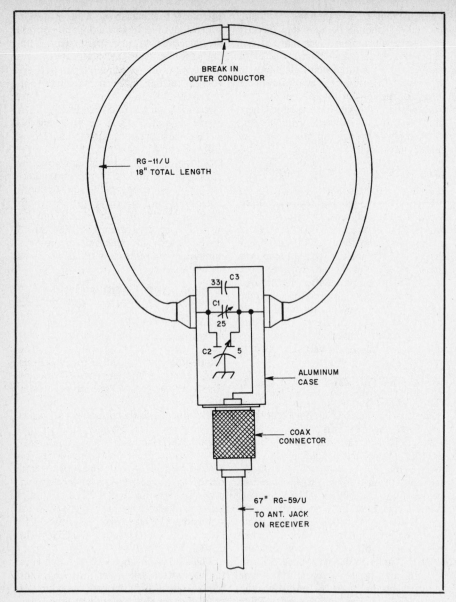

BREAK IN
OUTER CONDUCTOR

RG-11/U
18" TOTAL LENGTH

33 C3

C1

25

C2   5

ALUMINUM
CASE

COAX
CONNECTOR

67" RG-59/U
TO ANT. JACK
ON RECEIVER

Fig 18 — Sketch showing constructional details of the 28-MHz RDF loop. The outer braid of the coax loop is broken at the center of the loop. The gap is covered with waterproof tape, and the entire assembly is given a coat of acrylic spray.

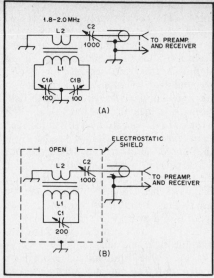

1.8-2.0 MHz
L2   C2
1000
TO PREAMP.
AND RECEIVER

L1

C1A   C1B
100   100

(A)

OPEN   ELECTROSTATIC
SHIELD
L2   C2
1000
TO PREAMP.
AND RECEIVER

L1
C1
200

(B)

Fig. 19 — Diagram of a ferrite loop (A). C1 is a dual-section air variable. The circuit at B shows a rod loop contained in an electrostatic shield channel (see text). A low-noise preamplifier is shown in Fig. 22.

A photograph of a shielded rod loop is offered in Fig. 20. It was developed experimentally for 160 meters and uses two 7-inch ferrite rods that were glued end-to-end with epoxy cement. The longer core resulted in improved sensitivity during weak-signal reception.

Obtaining a Cardioid Pattern

Although the bidirectional pattern of loop antennas can be used effectively in tracking transmitters down by means of triangulation, an essentially unidirectional loop response will help to reduce the time spent when doing so. Adding a sensing antenna to the loop is simple to do, and it will provide the desired cardioid response.

Fig. 21 shows how this can be accomplished. The link from the rod loop is connected via coaxial cable to the primary of T1, which is a tuned toroidal transformer with a split secondary winding. C3 is adjusted for peak signal response at the frequency of interest (as is C4), then R1 is adjusted for minimum back response of the loop. It will be necessary to readjust C3 and R1 several times to compensate for the interaction of these controls. The adjustments are repeated until no further null depth can be obtained. Tests show that null depths as great as 40 dB can be obtained with the circuit of Fig. 21 on 75 meters. A near-field weak-signal source was used during the tests.

The greater the null depth the lower the signal output from the system, so plan to include a preamplifier with 25 to 40 dB of gain. Q1 of Fig. 21 will deliver approximately 15 dB of gain. The circuit of Fig. 22 can be used following T2 to obtain an additional 24 dB of gain. In the interest of maintaining a good noise figure, even at 1.8 MHz, Q1 should be a low-noise

no. 12 or 14 solid wire may be soldered to the center pin of a phono plug for insertion in the jack. The sensing antenna can be plugged in as needed. Starting with a length of about four times the loop diameter, the length of the sensing antenna should be pruned until the pattern is similar to Fig. 1D.

LOOPSTICK ANTENNA FOR 1.8 MHz

Fig. 19 contains a diagram for a rod loop. The winding (L1) has the appropriate number of turns to permit resonance with C1 at the operating frequency. L1 should be spread over approximately 1/3 of the core center. Litz wire will yield the best Q, but Formvar magnet wire can be used if desired. A layer of 3M Company glass tape (or Mylar tape) is recommended as a covering for the core before adding the

wire. Masking tape can be used if nothing else is available.

L2 functions as a coupling link over the exact center of L1. C1 is a dual-section variable, although a differential capacitor might be better toward obtaining optimum balance (not tried). The loop Q can be controlled by means of C2, which is a mica compression trimmer.

Electrostatic shielding of rod loops can be effected by centering the rod in a U-shaped aluminum, brass or copper channel that extends slightly beyond the ends of the rod loop (1 inch is suitable). This idea is shown in Fig. 19B. The open side (top) of the channel can't be closed, as that would constitute a shorted-turn condition and render the antenna useless. This can be proved by shorting across the center of the channel with a screwdriver blade when the loop is tuned to an incoming signal.

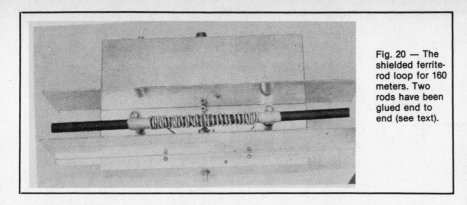

Fig. 20 — The shielded ferrite-rod loop for 160 meters. Two rods have been glued end to end (see text).

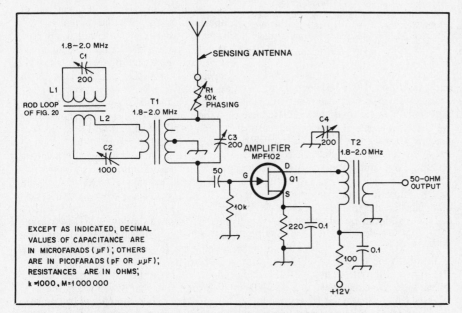

Fig. 21 — Schematic diagram of a rod-loop antenna with a cardioid response. The sensing antenna, phasing network and a preamplifier are shown also. The secondary of T1 and the primary of T2 are tuned to resonance at the operating frequency of the loop. T68-2 or T68-6 Amidon toroid cores are suitable for both transformers. Amidon also sells the ferrite rods for this type of antenna.

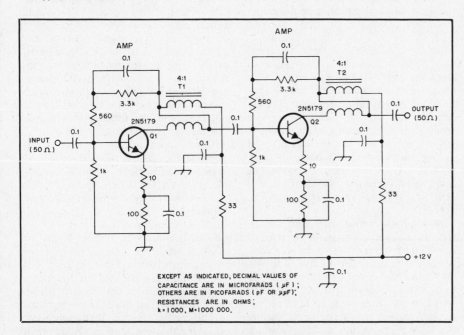

Fig. 22 — Schematic diagram of a two-stage broadband amplifier. T1 and T2 have a 4:1 impedance ratio and are wound on FT-50-61 toroid cores (Amidon) that have a μ_i of 125. They contain 12 turns of no. 24 enam. bifilar wound wire. The capacitors are disc ceramic. This amplifier should be built on double-sided circuit board for best stability.

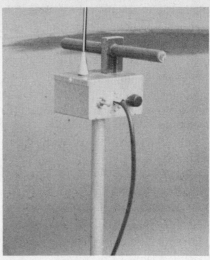

Fig. 23 — Unidirectional 75-meter RDF using ferrite-core loop with sensing antenna. Adjustable components of the circuit are mounted in the aluminum chassis supported by a short length of tubing.

device. A Siliconix U310 JFET would be ideal in this circuit, but a 2N4416, an MPF102 or a 40673 MOSFET would also be satisfactory.

The sensing antenna can be mounted 6 to 15 inches from the loop. The vertical whip need not be more than 12 to 20 inches long. Some experimenting may be necessary in order to obtain the best results. It will depend also on the operating frequency of the antenna.

A LOOPSTICK FOR 3.5 MHz

Figs. 23 through 25 show an RDF loop suitable for the 3.5-MHz band. It uses a construction technique that has had considerable application in low-frequency marine direction finders. The loop is a coil wound on a ferrite rod from a broadcast-antenna loopstick. Because it is possible to make a coil of high Q with the ferrite core, the sensitivity of such a loop is comparable to a conventional loop that is a foot or so in diameter. The output of the vertical-rod sensing antenna, when properly combined with that of the loop, gives the system the cardioid pattern shown in Fig. 1D.

To make the loop, remove the original winding on the ferrite core and wind a new coil as shown in Fig. 24. Other types of cores than the one specified may be substituted; use the largest coil available and adjust the winding so the circuit resonates in the 75-meter band within the range of C1. The tuning range of the loop may be checked with a dip meter.

The sensing system consists of a 15-inch whip and an adjustable inductance that will resonate the whip as a quarter-wave antenna. It also contains a potentiometer to control the output of the antenna. S1 is used to switch the sensing antenna in and out of the circuit.

The whip, the loopstick, the inductance L1, the capacitor C1, the potentiometer R1

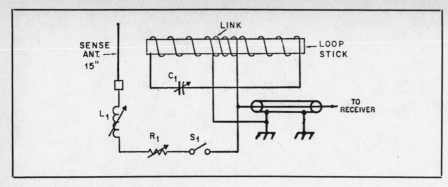

Fig. 24 — Circuit of the 75-meter direction finder.

C1 — 140 pF variable (125-pF ceramic trimmer in parallel with 15-pF ceramic fixed).
L1 — Approx. 140 μH adjustable (Miller No. 4512 or equiv.)
R1 — 1-kΩ carbon potentiometer.
S1 — SPST toggle.

Loopstick — Approx. 15 μH (Miller 705-A, with original winding removed and wound with 20 turns of no. 22 enam.). Link is two turns at center. Winding ends are secured with Scotch electrical tape.

Fig. 25 — Components of the 75-meter RDF are mounted on the top and sides of a Channel-lock type box. In this view R1 is on the left wall at the upper left and C1 is at the lower left. L1, S1 and the output connector are on the right wall. The loopstick and whip mount on the outside.

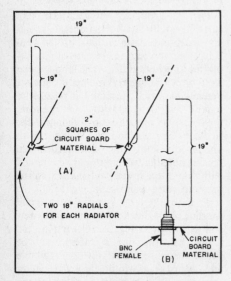

Fig. 26 — At A is a simple configuration that produces a cardioid pattern. At B is a convenient way of fabricating a sturdy mount for the radiator using BNC connectors.

and the switch S1 are all mounted on a 4 × 5 × 3-inch box chassis as shown in Fig. 25. The loopstick may be mounted and protected inside a piece of ½-inch PVC pipe. A section of ½-inch electrical conduit is attached to the bottom of the chassis box and this supports the instrument.

To produce an output having only one null there must be a 90° phase difference between the outputs of the loop and sensing antennas, and the signal strength from each must be the same. The phase shift is obtained by tuning the sensing antenna slightly off frequency by means of the slug in L1. Since the sensitivity of the whip antenna is greater than that of the loop, its output is reduced by adjusting R1.

Adjustment

To adjust the system, enlist the aid of a friend with a mobile transmitter and find a clear spot where the transmitter and RDF receiver can be separated by several hundred feet. Use as little power as possible at the transmitter. (Remove your own transmitter antenna before trying to make any loop adjustments and remember to leave it off during transmitter hunts.) With the test transmitter operating on the proper frequency, disconnect the sensing antenna with S1, and peak the loopstick using C1, while watching the S meter on the receiver. Once the loopstick is peaked, no further adjustment of C1 will be necessary. Next, connect the sensing antenna and turn R1 to minimum resistance. Then vary the adjustable slug of L1 until a maximum reading of the S meter is again obtained. It may be necessary to turn the unit a bit during this adjustment to obtain a larger reading than with the loopstick alone. The last turn of the slug is quite critical, and some hand-capacitance effect may be noted.

Now turn the instrument so that one side (not an end) of the loopstick is pointed toward the test transmitter. Turn R1 a complete revolution; if the proper side was chosen a definite null should be observed on the S meter for one particular position of R1. If not, turn the RDF 180° and try again. This time leave R1 at the setting that produces the minimum reading. Now adjust L1 very slowly until the S-meter reading is reduced still further. Repeat this several times, first R1, and then L1 until the best minimum is obtained.

Finally, as a check, have the test transmitter move around the RDF and follow it by turning the RDF. If the tuning has been done properly the null will always be broadside to the loopstick. Make a note of the proper side of the RDF for the null, and the job is finished.

A PHASED ARRAY FOR 144-MHz RDF WORK

Although there may be any number of different antennas that will produce a cardioid pattern, the simplest design is depicted in Fig. 26. Two ¼-wavelength vertical elements are spaced one ¼-wavelength apart and are fed 90° out of phase. Each radiator is shown with two radials approximately 5 percent shorter than the radiators.

During the design phase of this project a personal computer was used to predict the impact on the antenna pattern of slight alterations in its size, spacing and phasing of the elements. The results suggest that this system is a little touchy and that the most significant change comes at the null. Very slight alterations in the dimensions caused the notch to become much more shallow and, hence, less usable for RDF. Early difficulties in building a working model bore this out.

This means that if you decide to build this antenna, you will find it advantageous to spend a few minutes to tune it for the deepest null. If it is built using the techniques presented here, this should prove to be a small task that is well worth the extra effort. Tuning is accomplished by adjusting the length of the vertical radiators, the spacing between them and, if necessary, the lengths of the phasing harness that connects them. Tune for the deepest null on your S meter when using a signal source such as a moderately strong repeater. This should be done outside, away from buildings and large metal objects. Initial indoor tuning on this project was tried in the kitchen, which revealed that reflections off the appliances were producing spurious readings. Beware too of distant water towers, radio towers, and large office or apartment buildings. They can reflect the signal and give false indications.

Construction is simple and straightforward. Fig. 26B shows a female BNC connector (Radio Shack 278-105) that has been mounted on a small piece of PC-board material. The BNC connector is held "upside down" and the vertical radiator is soldered to the center solder lug. A 12-in piece of brass tubing provides a snug fit over the solder lug. A second piece of tubing, slightly smaller in diameter, is telescoped inside the first. The outer tubing is crimped slightly at the top after the inner tubing is installed. This provides positive contact between the two tubes. For 146 MHz the length of the radiators is calculated to be about 19 in.

You should be able to find small brass tubing at a hobby store. If none is available

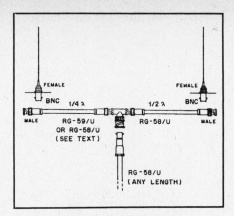

Fig. 27 — The phasing harness for the 144-MHz RDF array. The phasing sections must be measured from the center of the T connector to the point that the vertical radiator emerges from shield portion of the upside-down BNC female. Don't forget to take the length of the connectors into account when constructing the harness. If care is taken and coax with polyethylene dielectric is used, you should not have to prune the phasing line. With this phasing system, the null will be in a direction that runs along the boom, on the side of the ¼-wavelength section.

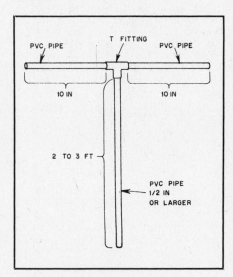

Fig. 28 — A simple mechanical support for the DF antenna made of PVC pipe and fittings.

in your area, consider brazing rods. This is often available in hardware sections of discount stores. It will probably be necessary to solder a short piece to the top since these come in 18-in sections. Also, tuning will not be quite as convenient. Two 18-inch radials are added to each element by soldering them to the board. Two 36-in pieces of heavy brazing rod were used in this project.

The Phasing Harness

As shown in Fig. 27, a T connector is used with two different lengths of coaxial line to form the phasing harness. This method of feeding the antenna is superior to other simple systems toward obtaining equal current in the two radiators. Un-equal currents tend to reduce the depth of the null in the pattern, all other factors being equal.

With no radials or with two radials perpendicular to the vertical element, it was found that a ¼-wavelength section made of RG-59/U, 75-Ω coax produced a deeper notch than a ¼-wavelength section made of RG-58/U, 50-Ω line. However, with the two radials bent downward somewhat, the RG-58/U section seemed to outperform the RG-59/U. There will probably be enough variation from one antenna to the next that it will be worth your time and effort to try both sections and determine which works best for your antenna. The ½-wavelength section can be made from either RG-58/U or RG-59/U because it should act as a 1-to-1 transformer.

The most important thing about the coax for the harness is that it be of the highest quality (well shielded and with a polyethylene dielectric). The reason for avoiding foam dielectric is that the velocity factor can vary from one roll to the next — some say it varies from one foot to the next in the same roll. Of course, it can be used if you have test equipment available that will allow you to determine its electrical length. Assuming you do not want to or cannot go to that trouble, stay with solid polyethylene-dielectric coax. Avoid coax that is designed for the CB market or the do-it-yourself cable-TV market. (A good choice is Belden 8240 for the RG-58/U or Belden 8241 for the RG-59/U.)

Both RG-58 and RG-59 with poly-ethylene dielectric have a velocity factor of 0.66. Therefore, for 146 MHz a quarter wavelength of transmission line will be 19 in × 0.66 = 12.5 in. A half-wavelength section will be twice this length or 25 in. One thing that you must take into account is that the transmission line is the total length of the cable *and the connectors*. Depending on the type of construction and the type of connectors that you choose, the actual length of the coax by itself will vary somewhat. You will have to determine that empirically.

Y connectors that mate with RCA phono plugs are widely available and the phono plugs are easy to work with. Avoid the temptation to substitute these for the T and BNC connectors. Phono plugs and a Y connector were tried. The results with that system were not satisfactory. The performance seemed to change from day to day and the notch was never as deep as it should have been. Although they are more difficult to find, BNC T connectors will provide superior performance and are well worth the extra effort. If you must make substitutions, it would be preferable to use UHF connectors (type PL-259).

Fig. 28 shows a simple support for the antenna. PVC tubing is used throughout. Additionally, you will need a T fitting, two end caps and possibly some cement. (By not cementing the PVC fittings together, you will have the option of disassembly for transportation.) Cut the PVC for the dimensions shown with a saw or a tubing cutter. A tubing cutter is preferred because it produces smooth, straight edges and is less messy. Drill a small hole through the PC board near the female BNC of each element assembly. Measure 19 inches along the boom (horizontal) and mark the two end points. Drill a small hole vertically through the boom at each mark. Use a small nut and bolt to attach the element assembly to the boom.

Tuning

The dimensions given throughout this section are those for approximately 146 MHz. If the signal you will be tracking is above that frequency, then the measurements will probably need to be a bit shorter. If you are to operate below that frequency, then they will need to be a little longer.

Once you have built the antenna to the rough size, the fun begins. You will need a signal source near the frequency that you will be using for your RDF work. Adjust the length of the radiators and the spacing between them for the deepest null on your S meter. Make changes in increments of ¼ in or less. If you must adjust the phasing line, make sure that the ¼-wavelength section is exactly one-half the length of the half-wavelength section. Keep tuning until you have a satisfactorily deep null on your S meter.

AN ADCOCK ANTENNA

While loops are adequate in applications where only the ground wave is present, the question arises of what can be done to improve the performance of an RDF system for sky-wave reception. One type of antenna that has been used successfully for this purpose is the Adcock antenna. There are many possible variations, but the basic configuration is shown in Fig. 29.

The operation of the antenna when a vertically polarized wave is present is very similar to a conventional loop. As can be seen from Fig. 29, currents I1 and I2 will be induced in the vertical members by the passing wave. The output current in the transmission line will be equal to their difference. Consequently, the directional pattern will be identical to the loop with a null broadside to the plane of the elements and with maximum gain occurring in end-fire fashion. The magnitude of the difference current will be proportional to the spacing, d, and the length of the elements. Spacing and length are not critical, but somewhat more gain will occur for larger dimensions than for smaller ones. In an experimental model, the spacing was 21 feet (approximately 0.15 wavelength on 40 meters) and the element length was 12 feet.

Response of the Adcock antenna to a horizontally polarized wave is considerably different from that of a loop. The currents induced in the horizontal members (dotted

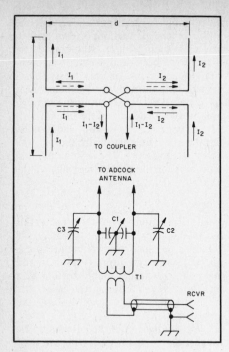

Fig. 29 — A simple Adcock antenna and suitable coupler (see text).

arrows in Fig. 29) tend to balance out regardless of the orientation of the antenna. This effect seemed to be borne out in practice, since good nulls were obtained with the experimental model under sky-wave conditions that produced only poor nulls with small loops (both conventional and ferrite-loop models). Generally speaking, the Adcock antenna has very attractive properties with regard to amateur RDF applications. Unfortunately, its portability leaves something to be desired, and it is more suitable to fixed or semi-portable applications. While a metal support for the mast and boom could be used, wood is preferable because of its nonconducting properties. Less distortion of the pattern should result.

Since a balanced feed system is used, a coupler is required to match the unbalanced input of the receiver. It consists of T1 which is an air-wound coil with a two-turn link wrapped around the middle. The combination is then resonated to the operating frequency with C1. C2 and C3 are null-clearing capacitors. A low-power signal source is placed some distance from the Adcock antenna and broadside to it. C2 and C3 are then adjusted until the deepest null is obtained.

The coupler can be placed on the ground below the wiring-harness junction on the boom and connected by means of a short length of 300-ohm twin-lead. A piece of aluminum tubing was used as a mast, but in a practical application it could be replaced with a length of PVC tubing extending to the ground. This would facilitate rotation and would provide a means of attaching a compass card for obtaining bearings.

Bibliography

Bonaguide, G., "HF DF — A Technique for Volunteer Monitoring," *QST,* March 1984.

Bond, D. S., *Radio Direction Finders,* 1st edition (1944), McGraw-Hill Book Co., New York, NY.

DeMaw, D., "Maverick Trackdown," *QST,* July 1980.

DeMaw, D., "Beat the Noise with a Scoop Loop," *QST,* July 1977.

Dorbuck, T., "Radio Direction-Finding Techniques," *QST,* August 1975.

Jasik, H., *Antenna Engineering Handbook,* 1st edition, McGraw-Hill Book Co., New York, NY.

Keene, R., *Wireless Direction Finding,* 3rd edition (1938), Wireless World, London.

Kraus, J., *Antennas* (1950): McGraw-Hill Book Co., New York, NY.

Kraus, J., *Electromagnetics,* McGraw-Hill Book Co., New York, NY.

McCoy, L., "A Linear Field-Strength Meter," *QST,* January 1973.

O'Dell, P., "Simple Antenna and S-Meter Modification for 2-Meter FM Direction Finding," *QST,* March 1981.

Ramo and Whinnery, *Fields and Waves in Modern Radio,* John Wiley & Sons, New York, NY.

Terman, F., *Radio Engineering,* McGraw-Hill Book Co., New York, NY.

Chapter 40

Interference

Radio Frequency Interference (RFI): It is a jungle out there! Not a jungle of wildlife and tangled undergrowth, however. It is a jungle of electrical and electronic equipment, each piece of which is a potential source or victim of RFI, or both! Today, a host of new devices are creating an increasingly complex RF environment. Each new gadget brings with it a new set of problems to the residential community.

The Electronic Home

Let's look around a perhaps "not-so-average" home. In the kitchen the blender whirrs and the microwave oven digital timer counts down to the blinking of the Public Service scanner. The wireless intercom emits a mild protest to the electronic activity. In the bathroom a family member is using a hair dryer and a portable radio. Both items are plugged into a ground-fault interrupter outlet.

In a quiet room supplied with mood lighting from an electronically controlled lamp, a color TV set is showing a movie that is delivered by cable. A video cassette recorder (VCR) records another program for later viewing.

One of the older children is in a bedroom doing homework on a microcomputer. Music is coming from the stereo record player in the corner. In the next bedroom a younger family member finds amusement in the video game connected to a portable TV set.

In the hall a pager sits in the battery charger just below a smoke detector. Someone had decided to place a call on the cordless telephone from this secluded spot.

Down in the basement, the water heater has turned on under control of the thermostat. Another thermostat, located upstairs, has called for hot air. The fuel oil gun is running in the furnace (as is the continuous ignition system) — the air is just now hot enough that a sensor starts the circulation blower. Mounted on the wall next to the power distribution box is the doorbell transformer.

Out in the garage the power tools and lawn mower silently wait for someone to use them. The garage door opener waits for a signal to be decoded by the receiver. Outside a loud sound will activate a fixture, turning on an outdoor light. When the family is away, the premises are guarded by an electronic security system. The system has a microwave intrusion detector; alarms are electronically reported over the telephone line.

Every device mentioned in this account has the potential of being either a source or victim of RFI — or both! Many other possibilities exist in the residential environment these days. Even the plumbing and wiring can conduct, radiate or rectify RF emanations from whatever source. And every month some new electronic "whiz bang" finds its way into our homes.

A Technical Problem

Interference is a technical problem. The causes and cures are often straightforward, but they can be mysterious and complicated. Interference side effects have economic and political impact. Experience has shown that nearly all RFI problems experienced with home-entertainment devices result from basic design deficiencies in these devices. The few small components or filters that would prevent RFI are often left out of otherwise well-designed products as manufacturers attempt to reduce costs, and hence to reduce the prices of their products.

Interference is primarily a matter of emanation (sending) and interception (receiving). Unwanted emanations occur by radiation (as from an antenna) or conduction (as along a wire). How and where to treat unwanted emanation or interception will depend on where and how it occurs. If interfering harmonics are generated in a transmitter and are radiated by the station antenna, usually a filter in the feed line will solve the problem. If a TV set lacks immunity to the radio frequency environment, various types of filters can be used to cure the problem.

The source of the energy is not always the cause of interference; the actual cause is sometimes difficult to determine. Interfering harmonics or spurious emissions of an Amateur or CB Radio signal may be generated in the transmitter, in a bad (rectifying) connection in the antenna system, in the affected receiver RF amplifier (preamplifier) or in any poor metallic connection in the vicinity of the transmitter antenna. For that reason, the cause is usually discovered through a process of elimination. Usually, more than one action will be necessary to cure the problem.

The usual procedure for dealing with harmonically related interference calls for filters to be installed at the transmitter and the receiver. That treatment usually solves the problem; if it doesn't, try using direction-finding techniques on a harmonic of the offending signal.

Radio amateurs have traditionally been involved with the analog world — a realm of sine waves and distinct frequencies. Digital equipment can be found in almost every home and ham shack. The digital domain is a realm of square waves.

The frequency spectra of sine and square waves of the same fundamental frequency are quite different, as shown in Fig. 1. The sine wave is ideally represented by a single frequency. In contrast, an ideal square wave consists of a fundamental frequency plus an infinite series of odd harmonics. It is these harmonics that require proper treatment in digital circuits to prevent RF emanations. Treatment, which must be applied to the digital circuit and enclosure, generally consists of proper shielding, filtering and bypassing. The measures that help minimize the possibilities that a digital device will be a source of interference will also protect it from being a victim of interference.

What's Being Done About RFI?

Along with the growth of new appliances in the electronic home come new regulations to deal with accompanying RFI

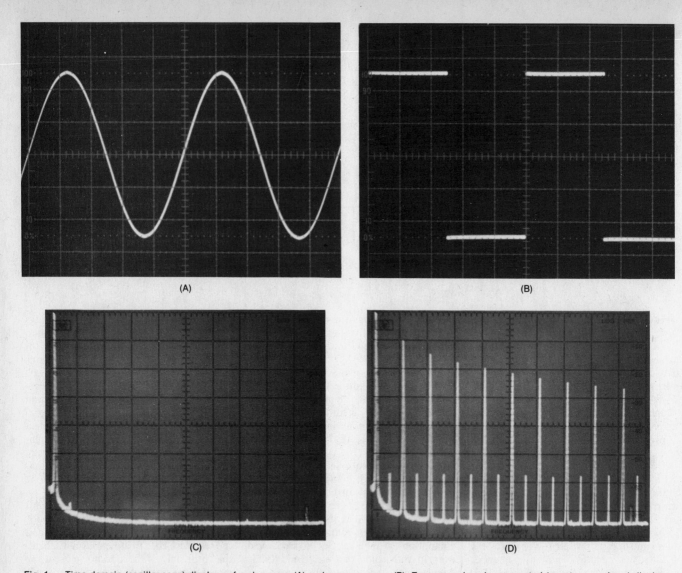

Fig. 1 — Time domain (oscilloscope) displays of a sine wave (A) and a square wave (B). Frequency-domain or spectral (spectrum-analyzer) displays of a sine wave (C) and a square wave (D). Vertical divisions in the spectral displays are 10 dB. With second-harmonic energy down 67 dB, the sine wave can be considered spectrally clean. By contrast, the square wave shows considerable energy in odd-order harmonics.

troubles. FCC rules now deal specifically with such services and devices as CATV, computers, cordless telephones and security devices. Although Public Law 97-259 gives the FCC authority to set standards for home-entertainment devices, the current approach looks toward voluntary compliance in place of legislation. Television sets and Amateur Radio equipment of recent vintage are, for the most part, superior to earlier models. Some VCRs still show a lack of RF immunity.

Cable television (CATV) is regulated by Parts 15 and 76 of the FCC rules. The Commission defines a CATV system as a "nonbroadcast facility" — a plant consisting of cables that carry television programming to subscribers. Cable television offers the amateur service several advantages over conventional reception. Signal levels on the cable are at a higher level than normally received "off air." This effectively eliminates fringe area (weak

signal) reception. In addition, these signals are supplied on a shielded cable, which further reduces the susceptibility to interference.

In the field, however, signals may leak from these ostensibly "closed" systems. When this leakage occurs, harmful interference to over-the-air services can occur. The FCC has specified the maximum allowable leakage from cable systems.

Section 76.613 regulates interference from CATV systems. Paragraph (a) defines harmful interference as "any emission, radiation or induction which endangers the functioning of a radio-navigation service or of other safety services or *seriously degrades, obstructs or repeatedly interrupts a radiocommunication service operating in accordance with this chapter*" (emphasis added). Paragraph (b) says "the operator of a cable television system that causes harmful interference shall promptly take appropriate measures

to eliminate the harmful interference."

Section 76.605(a)(12) spells out the limits for allowable radiation from a CATV system:

Frequencies	Radiation Limit ($\mu V/m$)	Distance (Feet)
Up to and including 54 MHz	15	100
Over 54 up to and including 216 MHz	20	10
Over 216 MHz	15	100

RF emanating from digital devices such as computers, and other incidental radiation devices such as power lines, industrial machines and electrical fences, are regulated in Part 15 of the FCC rules. Communication devices such as cordless telephones, FM wireless microphones and

Form A (ARRL RFI immunity complaint report):

COMPLAINANT'S NAME | CALL | ARRL HQ. CASE NO.

ADDRESS | PHONE NO.

CITY | STATE/PROV. | ZIP/PC

DATE (YEAR/MONTH/DAY) | LOCATION OF RFI (INCLUDING ZIP OR PC) | ARRL SECTION (IF NOT KNOWN, LIST COUNTY)

OWNER'S NAME | PHONE NO.

ADDRESS

CITY | STATE/PROV. | ZIP/PC

	TYPE (VCR, ETC.)	MANUFACTURER	MODEL NO.	NATURE OF INTERFERENCE
EMANATING EQUIPMENT				
SUSCEPTIBLE EQUIPMENT				

DISTANCE BETWEEN EMANATING & SUSCEPTIBLE EQUIPMENT (FT.) | SEVERITY ☐ BARELY PERCEPTIBLE ☐ WEAK ☐ MODERATE ☐ STRONG ☐ OVERWHELMING | COUPLING MECHANISM ☐ NOT DETERMINED ☐ CONDUCTED ☐ RADIATED

COMPLAINT RESOLVED? ☐ YES ☐ NO | DESCRIBE TECHNICAL SOLUTION

PROBLEM FIXED BY ☐ AMATEUR ☐ REPAIRMAN ☐ MFGR. OF EMANATING EQUIPMENT ☐ MFGR. OF SUSCEPTIBLE EQUIPMENT ☐ OTHER

WHO WAS INFORMED ☐ UTILITY ☐ CABLE OPERATOR ☐ FCC ☐ EIA ☐ MFGR. OF EMANATING EQUIP. ☐ MFGR. OF SUSCEPTIBLE EQUIPMENT ☐ OTHER

REMARKS

RFI IMMUNITY COMPLAINT REPORT TD 1/84

(A)

Form B (CATVI Complaint Report Form):

I-D: Serial_____ State_____

Date:

Name of cable company

Address

Phone Number:

Name & Title of Person Contacted:

Areas Served by System:

Nature of Interference:

Method of Observation (List of equipment used):

Dates of Contact With:

A. Cable operator

B. FCC

C. Controller of franchise (town, county, etc.)

Steps Taken by Cable Operator:

Does interference still exist?

Your name: | Callsign:

Address:

Phone:

FCC Involvement (if any):

(For ARRL use only)

NCTA contact:

(B)

Fig. 2 — At A, ARRL RFI immunity complaint report. At B, CATVI Complaint Report Form.

other low-power appliances are also covered in Part 15. These devices may not cause harmful interference to other radio service communications. In addition, specifications for limits on amounts of RF energy emanating from a given device are included. For example, the wireless microphones mentioned earlier in this paragraph may not exceed certain field strengths.

Also, units (such as most 1.7-MHz cordless telephones) that operate under a waiver of the field strength requirements are controlled in another manner: These units must not exceed standards for RF current in the power and telephone lines.

Radiation limits also apply to personal computers manufactured after January 1, 1981. In many cases, labeling, indentification and FCC certification are required for devices covered in Part 15. These requirements help ensure that the device operator understands the nature of the unit, and his or her obligations under the rules. For applicable excerpts of Part 15 or Part 76 of the Commission's rules, contact ARRL Headquarters.

Government regulation is not a panacea, however. We are in a period of shrinking government involvement in the life of the society it governs. Thus, we have seen deregulation on the part of an FCC determined to let marketplace forces resolve their differences. And, with the FCC budget greatly reduced, we are witnessing a Commission less able to enforce the rules. Government rules are a help, but not the sole solution to RFI.

Help is available from other quarters as well. Manufacturers and professional trade associations are aware of the RFI phenomenon. While these organizations are not in positions to dictate policies to members, they can promote the need for responsibility in manufacturing and operations. Associations often publish pamphlets and manuals for their members explaining how they can make products less susceptible to RFI problems. The same associations lobby the government on behalf of their members.

The ARRL, as the principal membership organization of the amateur community, is ready to assist. The League publishes *Radio Frequency Interference: How to Identify and Cure It*, a guide with practical ideas and tips to bring harmony back to a neighborhood troubled with RFI. ARRL Headquarters is also the home of the Technical Information Service (TIS). Members can call or write the TIS with their RFI questions. See Fig. 2A. CATVI complaints should be submitted on the form shown at Fig. 2B. *QST* publishes articles with timely information on the evolving RFI picture.

ARRL is also active on the legislative front. The League worked hard for passage of the Goldwater-Wirth measure (P.L. 97-259). In January 1982, ARRL filed a petition for rulemaking with the FCC to have CATV vacated from channels coinciding with amateur frequencies. This action was in response to membership concern over the problem of leakage and interference from "closed" cable TV systems.

An Official Observer (OO) program has been part of the ARRL field organization for more than half a century. First conceived in 1926, the OO Program was designed to "help brother amateurs by calling attention to violations of good practice...in the right way...in better operating...and ham enjoyment."

One of the significant aspects of Public Law 97-259 (known as the Communications Amendments Act of 1982 before being signed into law in September 1982)

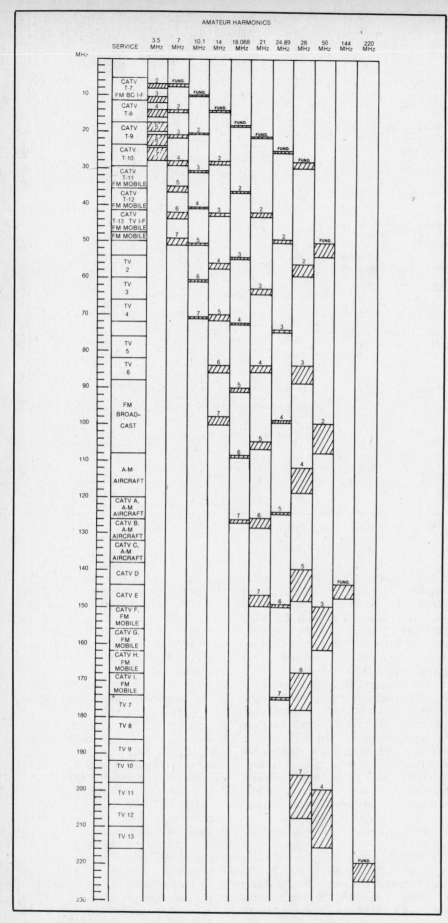

AMATEUR HARMONICS

Fig. 3 — Relationships of HF and VHF amateur bands to frequencies used in consumer electronic devices. The CATV channels are used in closed systems, but experience has shown these systems to have poor isolation from outside signals.

is one that authorizes the FCC to formally enlist the use of amateur volunteers in monitoring the airwaves for rules discrepancies or violations. (The same legislation paved the way for the Volunteer Examiner program.) Before approving the volunteer monitoring program, FCC's Field Operations Bureau (FOB) fully recognized the value of the organized and disciplined Amateur Radio community through the ARRL.

To fully utilize the ARRL Field Organization structure, the FOB has specifically indicated that the volunteer-monitoring program should be compatible with the amateur organizational structure. To this end, FOB created an Amateur Auxiliary (similar to those of the Civil Air Patrol and the Coast Guard Auxiliary). The Amateur Auxiliary/OO program is administered by the League's Section Managers and OO/RFI Coordinators, with support from ARRL Hq. For more information on the Amateur Auxiliary, see "The Amateur Auxiliary for Volunteer Monitoring," in August 1984 *QST*.

What Can You Do?

Be informed — read, report, experiment and test. As an ARRL member, you will have access to publications that contain a wealth of information about RFI. Further, a skilled Headquarters staff stands ready to assist you.

Report — January 1, 1984 marked the beginning of the ARRL interference reporting system. All RFI reports/complaints are placed into a computer data base, to help locate and cure chronic interference problems. Use the format shown in Fig. 2 to document your report. Call or write the League's Technical Information Service with all requested information. If a chronic problem exists, we will probably have a cure on file. Similarly, if you experience a problem and find the cure, let your League know about it! Additional report forms are available from Headquarters for an s.a.s.e.

Experiment and test — RFI problems and solutions are not always simple. Try the usual; then try the unusual. It is also possible that more than one action will be necessary to cure the problem. This chapter will help you with ideas.

Be cooperative — the first step is to ensure that your station is assembled and operated according to "good engineering and amateur practice." Check your transmitter; try using a low-pass filter. Then, check your own TV set for problems. Put your own house in order before checking on your neighbor's.

When approaching your neighbor, bear in mind that he or she likely knows little of radio or RFI. Proceed accordingly. Explain your interpretation of the situation in simple terms. Then, politely make your recommendations. Neighborhood RFI disputes are settled when a cooperative atmosphere exists.

Be communicative — contact the manufacturer of the offending equipment. Many responsible manufacturers have a policy of supplying filters for eliminating television interference when such cases are brought to their attention. A list of those manufacturers, and a more thorough treatment of the RFI problem, can be obtained by writing the ARRL. If a given manufacturer is not listed, it is still possible that he can be persuaded by writing either directly to the manufacturer or to the Electronic Industries Association (EIA).[1]

Similarly, if the problem lies with a CATV operator, public utility or other facility, let the responsible party know about it. You will often find assistance in your efforts to track down an RFI source.

If an RFI problem appears to be a violation of FCC rules, and there is no response to your requests for corrective action, contact your nearest FCC district office. See *The FCC Rule Book* or the *License Manual*, both published by ARRL, for the address.

Don't forget to communicate your RFI problems and solutions to League Headquarters. The ARRL RFI Task Group needs a complete and accurate picture of the ever-changing RFI scene so it can make appropriate policy recommendations to the ARRL Board of Directors.

Dealing With Interference

Many interference problems are caused by harmonics. Fig. 3 is a chart showing the frequency relationships between broadcast channels and amateur harmonics.

The visible effects of interference vary with the type and intensity of the interference. Blackout, where the picture and sound disappear completely, leaving the screen dark, occurs only when the transmitter and receiver are close together. Strong interference ordinarily causes the picture to be broken up, leaving a jumble of light and dark lines, or turns the picture "negative" — the normally white parts of the picture turn black and the normally black parts turn white.

Cross-hatching — diagonal bars or lines in the picture — often occurs as well, and also represents the most common type of less severe interference. The bars are the result of the beat between the harmonic frequency and the picture carrier frequency. They are broad and relatively few in number if the beat frequency is comparatively low — near the picture carrier — and are numerous and very fine if the beat frequency is very high — toward the upper end of the channel. Typical cross-hatching is shown in Fig. 4A.

Whether or not cross-hatching is visible, an amplitude-modulated transmitter may cause sound bars in the picture. These are

shown in Fig. 4B. They result from the variations in the intensity of the interfering signal when modulated. Under most circumstances modulation bars will not occur if the amateur transmitter is frequency- or phase-modulated. With these types of modulation the cross-hatching will wiggle from side to side with the modulation.

Except in the more severe cases, there is seldom any effect on the sound reception when interference shows in the picture, unless the frequency is close to the sound carrier. In this event the sound may be interfered with even though the picture is clean.

Reference to Fig. 3 will show whether or not harmonics of the frequency in use will fall in any television channels that can be received in the locality. It should be kept in mind that not only harmonics of the final frequency may interfere, but also harmonics of any frequencies that may be present in mixer or frequency-multiplier stages. In the case of 144-MHz transmitters, frequency-multiplying combinations that require a doubler or tripler stage to operate on a frequency in a low-band VHF channel in use in the locality should be avoided.

Harmonic Suppression

Effective harmonic suppression has three separate phases:

1) Reducing the amplitude of harmonics generated in the transmitter. This is a matter of circuit design and operating conditions.

2) Preventing stray radiation from the transmitter and associated wiring. This requires adequate shielding and filtering of all circuits and leads from which radiation can take place.

3) Preventing harmonics from being fed into the antenna.

It is impossible to build a transmitter that will not generate some harmonics. It is advantageous to reduce harmonic strength through circuit design and choice of operating conditions before attempting to prevent harmonics from being radiated. Harmonic radiation from the transmitter itself or from its associated wiring obviously will cause interference just as readily as radiation from the antenna. Measures taken to prevent harmonics from reaching the antenna will not reduce TVI if the transmitter itself is radiating harmonics. But once it has been found that the transmitter itself is free from harmonic radiation, devices for preventing harmonics from reaching the antenna can be expected to produce results.

Reducing Harmonic Generation

Since reasonably efficient operation of RF power amplifiers always is accompanied by harmonic generation, good judgment calls for operating all frequency-multiplier stages at a very low power level. When the final output frequency is reached, it is desirable to use as few stages as possible

(A)

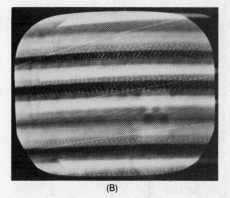

(B)

Fig. 4 — At A, cross-hatching caused by the beat between the picture carrier and an interfering signal inside the TV channel. At B, sound bars or modulation bars accompanying amplitude modulation of an interfering signal. In this case the interfering carrier is strong enough to destroy the picture, but in mild cases the picture is visible through the horizontal bars. Sound bars may accompany modulation even though the unmodulated carrier gives no visible cross-hatching.

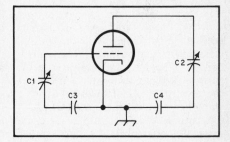

Fig. 5 — A VHF resonant circuit is formed by the tube capacitance and the lead inductances through the tank and blocking capacitors. Regular tank coils are not shown, since they have little effect on such resonances. C1 is the grid tuning capacitor and C2 is the plate tuning capacitor. C3 and C4 are the grid and plate blocking or bypass capacitors, respectively.

in building up to the final output power level and to use devices that require minimum drive power.

Circuit Design and Layout

Harmonic currents of considerable

[1]Electronic Industries Association, 2001 Eye St., N.W., Washington, DC 20006. Attention: Director of Consumer Affairs.

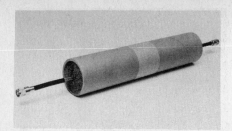

Fig. 6 — This coax shield decoupler is made with steel wool stuffed into an IBM copier tube. This size was selected for photographic purposes, but to be effective, the device should be about twice as long.

amplitude flow in both the input and output circuits of RF power amplifiers, but they will do relatively little harm if they can be efficiently bypassed to the cathode, emitter or source of the active device. Fig. 5 shows the paths followed by harmonic currents in an amplifier circuit. Because of the high reactance in the tank coil, there is little harmonic current in it, so the harmonic currents simply flow through the tank capacitor, the input or output blocking capacitor and the device interelectrode capacitances. The lengths of the leads forming these paths is of great importance, since the inductance in this circuit will resonate with the interelectrode capacitance in the VHF range. Generally, only the interelectrode capacitance is considered in this respect, because of the relatively large values of capacitance contained in the blocking and tank circuitry. If such a resonance happens to fall near the same frequency as one of the transmitter harmonics, the effect is the same as though a harmonic tank circuit had been deliberately introduced; the harmonic at that frequency would be tremendously increased in amplitude.

Such resonances are unavoidable, but by keeping the path from plate to cathode and from grid to cathode (and between corresponding terminals in solid-state devices) as short as possible, the resonant frequency can usually be raised above 100 MHz in amplifiers of medium power. This places the harmonic energy where it will cause minimal interference.

It is easier to place grid-circuit VHF resonances where they will do no harm when the amplifier is link-coupled to the driver stage, since this generally permits shorter leads and more favorable conditions for bypassing the harmonics than is the case with capacitive coupling. Link coupling also reduces the coupling between the driver and amplifier at harmonic frequencies, thus preventing driver harmonics from being amplified.

The inductance of leads from the tube to the tank capacitor can be reduced not only by shortening but by using flat strap instead of wire conductors. It is also better to use the chassis as the return from the blocking capacitor or tuned circuit to the

cathode, since the chassis path will have less inductance than almost any other form of connection.

The VHF resonance points in amplifier tank circuits can be found by coupling a dip meter covering the 50-250 MHz range to the grid and plate leads. If a resonance is found in or near a TV channel, methods such as those described above should be used to move it well out of the TV range. The dip meter also should be used to check for VHF resonances in the tank coils, because coils made for 14 MHz and below usually will show such resonances. In making the check, disconnect the coil entirely from the transmitter and move the dip meter coil along it while exploring for a dip in the 54- to 88-MHz band. If a resonance falls in a TV channel that is in use in the locality, changing the number of turns will move it to a less-troublesome frequency.

Operating Conditions

Grid bias and grid current have an important effect on the harmonic content of the RF currents in both the grid and plate circuits. In general, harmonic output increases as the grid bias and grid current are increased, but this is not necessarily true of a particular harmonic. The third and higher harmonics, especially, will go through fluctuations in amplitude as the grid current is increased, and sometimes a high value of grid current will minimize one harmonic as compared with a low value. This characteristic can be used to advantage where a particular harmonic is causing interference; however, operating conditions that minimize one harmonic may greatly increase another.

Suppression Practices

Complete elimination of TVI is often not a simple process. It seldom happens that a single measure such as installing a high-pass filter at the TV set will cure the problem. Rather, a number of methods must be applied simultaneously. The principal factor in any TVI situation is the ratio of TV signal strength to interference level. This includes interference of all types such as ignition noise, random or thermal noise (which isn't really interference but sets the minimum signal that permits snow-free reception), and unwanted signals that fall within the TV channel. A signal-to-interference ratio greater than approximately 35 to 40 dB is required for good picture quality.

In this regard, an area frequently overlooked in TVI difficulties is the TV antenna. A poor antenna with little gain in the direction of the TV station, old and corroded wire, and connections (which can cause the harmonic generation by rectification of a clean signal generated in a nearby amateur transmitter), may result in a TVI situation that is impossible to solve. Generally speaking, if the picture quality on the TV set experiencing the interference

is poor to begin with, even sophisticated suppression measures are likely to prove futile. In such cases, the only solution is replacement of the defective TV antenna system components. Corroded and cracked lead-in cables should be replaced, and the antenna terminals cleaned. If the TV antenna shows signs of corrosion and/or deterioration, it should also be replaced. Note: The amateur has no legal or financial responsibility to pay for the replacement of the antenna system — such replacement is the responsibility of the TV owner.

Grounds

Grounding of equipment has long been considered a first step in eliminating interference. While the method is very effective in the MF range and below, for all practical purposes it is useless in suppressing VHF energy. This is because even short lengths of wire have considerable reactance at VHF. For instance suppose a length of wire by itself has an inductance of 1 μH. At 550 kHz, the reactance would be about 3.46 ohms. On the other hand, the same wire would have a reactance of over 300 ohms at 56 MHz, which is in the frequency range of TV channel 2. (Actually, the impedance of a wire becomes a more complicated entity to define at VHF. The delay effects along the wire are similar to those on the surface of an antenna. Consequently, the wire might even appear as an open circuit rather than as a ground as the electrical length approaches a quarter wavelength.)

From a shock-hazard point of view, grounding is important. However, never connect a ground *for any reason* to the chassis of a TV set. This is because many TV sets derive their operating voltages directly from the ac-service line. Although a schematic diagram of a TV set may indicate a power transformer is being used, caution should be exercised to be sure it is actually being employed for this purpose. Often, the only voltage the transformer is supplying is for the TV picture tube filament.

Shielding

Effective shielding is perhaps the single most important measure in preventing or curing any RFI problem. However, unwanted RF energy must be dissipated. The task becomes harder to perform when the spacing between the source of energy and the boundaries of the shield diminish. Consequently, the use of a double shield is one way of reducing residual radiation from the primary shielding surface.

In order to obtain maximum effectiveness of a particular shielding measure, no breaks or points of entry should be permitted. Small holes for ventilation purposes usually do not degrade shielding effectiveness. But even here, a honeycomb type of duct is often employed when maximum isolation is required. (A parallel bundle of small tubing has very high attenuation since

Fig. 7 — Winding the cable on a ferrite toroid is an effective shield current suppressor in some cases. Reversing the winding as pictured allows more turns with less shunt capacitance. RG-58 will suffice for moderate power applications. The most important property of the cable is complete shielding — avoid "bargain" cable having less than 95-percent braid coverage.

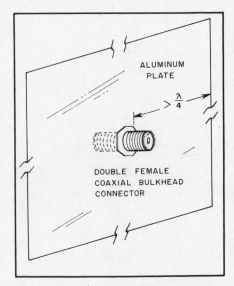

Fig. 8 — A large metallic baffle inhibits waves propagating on the outside of a coaxial cable. For VHF TV channel 2, the smallest effective baffle is a 9-foot-diameter circle, but the required size decreases linearly with frequency.

each tube by itself acts as a waveguide below cutoff.)

The isolation of a coaxial cable can be degraded considerably unless the ends of the shield are terminated properly. A braid should be soldered so it completely encloses the inner conductor(s) at the connector junction. For instance, the practice of twisting the braid and point soldering it to the base of a connector may result in a 20-dB degradation in isolation. Normally, this effect is not serious if the cable is run through an area where sensitive circuits don't exist. However, the isolation afforded by a filter can be reduced considerably in circuits where such cable breaks occur.

One instance where a shield break causes a serious problem is in the connection between the antenna terminals on a TV set and the tuner. Newer sets have a 75-ohm coaxial input along with a balun for 300-ohm line. However, because many TV sets have direct connections to the ac line, a decoupling network is used. The shielded lead to the tuner is broken and a capacitor is connected in series with the braid. This provides a low-impedance path for RF energy while presenting a high impedance at 60 Hz. Consequently, because of the cable break, high-pass filters at the antenna input terminals are not as effective as those built into the tuner itself.

Coax Shield Chokes

As mentioned previously, VHF currents flowing on the outside of coaxial cables are frequently the cause of RFI. Figs. 6, 7 and 8 show techniques for reducing or eliminating conducted chassis radiation from coaxial cables. The cardboard tube stuffed with steel wool in Fig. 6 works on the absorption principle. The steel wool is very lossy and dissipates the RF energy on the shield. The tube pictured is 18 inches long. A longer tube would be even more effective.

Fig. 7 shows a choke wound on a ferrite toroidal core. Another coax radiation-suppression device is illustrated in Fig. 8. If the plate is at least a half wavelength (at the harmonic frequency to be suppressed) on its smallest dimension, it will provide a very effective barrier. Large pieces of sheet metal are expensive, so the baffle can be made from a sheet of cardboard or Masonite® covered with aluminum foil. The ideal placement of any of these chokes will vary with the standing wave pattern on the coax shield, but in general they should be close to the transmitter. Like all RFI remedies, the effectiveness of these devices varies widely with each interference situation. Therefore, one should not expect miracles. A coax shield choke installed at a TV receiver prevents signals picked up on the coax braid from reaching the tuner.

Capacitors at RF

Capacitors are common elements found in almost any piece of electronic gear. However, some precautions are necessary when they are employed in RFI-preventive devices such as filters and in bypassing applications. In particular, lead inductance may be sufficient to resonate with the capacitor and cause the entire combination to have a high inductive reactance rather than the desired capacitive reactance.

This effect is illustrated in the accompanying photographs. The response curve shown in Fig. 9A is for a 10-MHz low-pass filter arranged in a "pi" configuration. However, this particular circuit realization required some large-value capacitors. Using ordinary capacitor types resulted in an unwanted resonance as evidenced by the sharp dip in the response curve at approximately

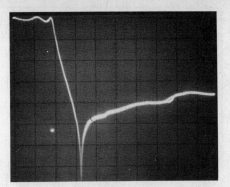

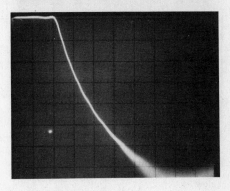

Fig. 9 — Stray lead inductance of a capacitor can degrade filter performance.

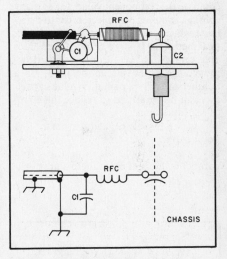

Fig. 10 — Additional lead filtering for harmonics or other spurious frequencies in the high VHF TV band (174-216 MHz).

C1 — 0.001-μF disc ceramic.
C2 — 500- or 1000-pF feedthrough bypass (Centralab FT-1000. Above 500 volts, substitute Centralab 858S-500).
RFC — 14 inches no. 26 enam. close-wound on 3/16-inch-dia form or composition resistor body.

15 MHz. However, by going to the equivalent T configuration (see the section on filters in Chapter 2, on electrical laws and circuits), a circuit realization for the desired response required much smaller capacitance values. The curve shown in Fig. 9B approximated this response closely, and no

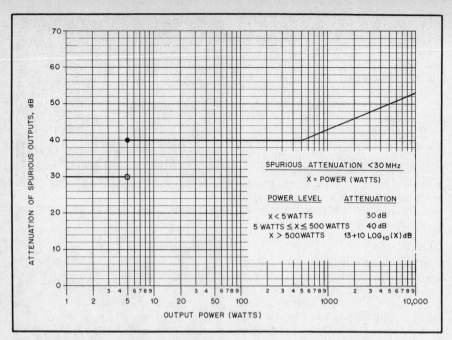

Fig. 11 — The FCC specifies that the spurious signals generated by transmitting equipment must be reduced well below the level of the fundamental. This graph illustrates exactly how far the spurious components must be reduced. This applies to amateur transmitters operating below 30 MHz.

effects of parasitic inductance were noticeable. To design a filter, it is advisable to compute the component values for as many configurations as possible in order to determine which one results in the most practical elements. If large capacitance values are unavoidable, either special low-inductance types should be used, or a number of ordinary smaller-value capacitors can be paralleled to reduce the effect of lead inductance.

A very desirable capacitor from an RFI point of view is C2 in Fig. 10. Instead of having two or more plates arranged in a parallel fashion, the conductors are coaxial and are separated by the dielectric. Such feedthrough capacitors are highly recommended for conducting leads in and out of circuits where the radiation of harmonic energy is possible. In addition, the RFC illustrated in Fig. 10 could consist of either a small coil wound over a composition resistor as shown or a ferrite bead on a straight piece of wire.

Decoupling from the AC Line

Direct feedback of RF energy into the ac power service is usually not a problem with modern transmitting equipment. However, currents induced on the antenna feed line

may flow in the transmitter chassis and back into the ac line. A rig hot with RF or even the presence of broadcast harmonics while receiving may indicate a problem of this sort. In the case where you are using an antenna that requires a ground (such as an end-fed wire), never use any part of the ac conduits, water systems or other conductors in a building. It is always advisable to have a separate ground system for the antenna itself.

It is also good practice to use an antenna-matching network with no direct connection between the transmitter and antenna feed line. Any matching network that uses mutual-magnetic coupling exclusively will fulfill this requirement. Antenna pattern is another factor to consider; try to choose a type that directs the minimum possible signal into other dwellings. For instance, ground-mounted vertical antennas have considerable low-angle radiation, while a dipole directs energy at angles below the horizontal plane. A ground plane or beam mounted on as high a tower as practical will generally be better from an RFI and TVI standpoint than antennas closer to the ground.

FCC Rules Concerning RFI

Part 97.73 of the U.S. amateur regulations specifies the amateur's responsibility for signal purity:

(a) Except for a transmitter or transceiver built before April 15, 1977 or first marketed before January 1, 1978, the mean power of any spurious emission or radiation from an amateur transmitter, transceiver, or external radio frequency power amplifier being operated with a carrier frequency below 30 MHz shall be at least 40 decibels below the mean power of the fundamental without exceeding the power of 50 milliwatts. For equipment of mean power less than five watts, the attenuation shall be at least 30 decibels.

(b) Except for a transmitter or transceiver built before April 15, 1977 or first marketed before January 1, 1978, the mean power of any

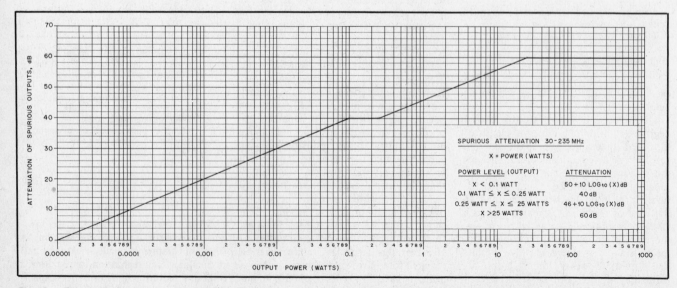

Fig. 12 — This graph illustrates to what level spurious-output energy must be reduced for equipment designed to operate in the 30- to 225-MHz range.

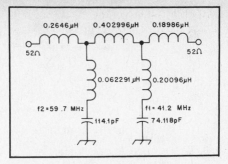

Fig. 13 — Schematic diagram showing component values of an experimental elliptic-function filter.

MHz are shown in Fig. 13. The filter is supposed to provide an attenuation of 35 dB above 40 MHz. An experimental model was built and the response is shown in Fig. 14. As can be seen, the filter came close to the design goals. Unfortunately, as with most of the designs in this section, alignment of the more complicated filters requires some sort of sweep-generator setup. This is the only practical way of tweaking a filter to the desired response. While building a sweep setup is not beyond the talents of an advanced experimenter, the lack of one is not an obstacle in the home construction of filters.

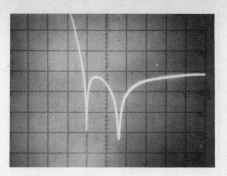

Fig. 14 — Response curve of the filter shown in Fig. 13. Vertical scale represents 10 dB/div, and horizontal scale is 10 MHz/div.

AN ABSORPTIVE FILTER

The filter shown in Fig. 15 not only provides rejection by means of a low-pass section, it also includes circuitry that absorbs harmonic energy. A high-pass section consisting of L1, L2, C1 and C2 is terminated in a 50-ohm idler load. This combination absorbs harmonics. The advantages of this technique are that degradation of filter rejection caused by antenna mismatch at the harmonic frequency is not as severe, and the transmitter is terminated in a resistive load at the harmonic.

Construction and Test Techniques

If good performance above 100 MHz is not a necessity, this filter can be built using conventional fixed capacitors. Teflon® -dielectric PC board can be purchased from Microwave Components of Michigan, listed in Chapter 35. Regular fiberglass-insulated board is satisfactory for low power. One such filter has been used with an SSB transceiver running 100 watts PEP output. Although the Q of the fiberglass capacitors will be lower than that of Teflon® -dielectric capacitors, this should not greatly affect the type of filter described here.

Test equipment needed to build this filter at home includes a reasonably accurate dip oscillator, an SWR bridge, a reactance chart or the ARRL L/C/F Calculator, a 50-ohm dummy load and a transmitter.

Once the value of a given capacitor has been calculated, the next step is to determine the capacitance per square inch of the double-clad circuit board you have. This is done by connecting one end of a coil of known inductance to one side of the circuit board, and the other coil lead to the other side of the circuit board. Use a dip meter, coupled lightly to the coil, to determine the resonant frequency of the coil and the circuit-board capacitor. When the frequency is known, the total capacitance can be determined by working the calculator or by looking the capacitance up on a reactance chart. The total capacitance divided by the number of square inches on one side of the circuit board gives the capacitance per square inch. Once this figure is determined, capacitors of almost any value can be laid out with a ruler!

High voltages can be developed across capacitors in a series-tuned circuit, so the

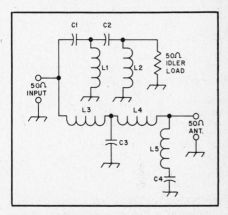

Fig. 15 — Schematic diagram of the absorptive filter. The pc-board used is MIL-P-13949D, FL-GT-062 in, C-2/2-11017, Class 1, Grade A. Polychem Bud Division. Capacitance between copper surfaces is 10-pF per square inch. Values are as follows for a design cutoff frequency of 40 MHz and rejection peak in TV channel 2:

C1 — 52 pF	L2 — 0.52 μH
C2 — 73 pF	L3 — 0.3 μH
C3 — 126 pF	L4 — 0.212 μH
C4 — 15 pF	L5 — 0.55 μH
L1 — 0.125 μH	

spurious emission or radiation from an amateur transmitter, transceiver, or external radio frequency power amplifier being operated with a carrier frequency above 30 MHz but below 225 MHz shall be at least 60 decibels below the mean power of the fundamental. For a transmitter having a mean power of 25 watts or less, the mean power of any spurious radiation supplied to the antenna transmission line shall be at least 40 decibels below the mean power of the fundamental without exceeding the power of 25 microwatts, but need not be reduced below the power of 10 microwatts.

(c) Paragraphs (a) and (b) of this section notwithstanding, all spurious emission or radiation from an amateur transmitter, transceiver, or external radio frequency power amplifier shall be reduced or eliminated in accordance with good engineering practice.

(d) If any spurious radiation, including chassis or power line radiation, causes harmful interference to the reception of another radio station, the licensee may be required to take steps to eliminate the interference in accordance with good engineering practice.

NOTE: For the purposes of this section, a spurious emission or radiation means any emission or radiation from an amateur transmitter, transceiver, or external radio frequency power amplifier which is outside of the authorized Amateur Radio Service frequency band being used.

The numerical limits cited in 97.73 are interpreted graphically in Figs. 11 and 12. Note, however, that paragraphs (c) and (d) go beyond absolute limits in defining the amateur's obligation.

Filters and Interference

The judicious use of filters, along with other suppression measures such as shielding, has provided solutions to interference problems in widely varying applications. As a consequence, considerable attention has been given to the subject over the years resulting in some very esoteric designs. Perhaps the most modern approach is to design filters with a digital computer. However, there are a number of other types with component values presented in tabular form. Of these, the most important ones are the so-called Chebyshev and elliptic-function filters. (Butterworth filters are often considered a special case of Chebyshev types with a ripple factor of zero.)

Elliptic-function filters might be considered optimum in the sense that they provide the sharpest rolloff between the passband and stopband. Computed values for a low-pass filter with a 0.1-dB ripple in the passband and a cutoff frequency of 30.6

copper material should be trimmed back at least 1/8 inch from all edges of a board, except those that will be soldered to ground, to prevent arcing. This should not be accomplished by filing, because the copper filings would become imbedded in the board material and just compound the problem. The capacitor surfaces should be kept smooth, and sharp corners should be avoided.

If the filter box is made of double-clad fiberglass board, both sides should be bonded together with copper stripped from another piece of board. Stripped copper foil may be cleaned with a razor blade before soldering. To remove copper foil from a board, use a straight edge and a sharp scribe to score the thin copper foil. When the copper foil has been cut, use a razor blade to lift a corner. This technique of bonding two pieces of board can also be used to interconnect two capacitors when construction in one plane would require too

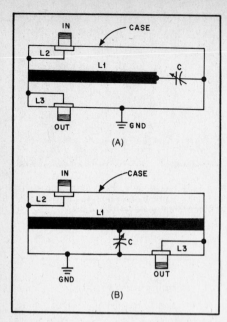

Fig. 16 — Equivalent circuits for the strip-line filters. At A, the circuit for the 6- and 2-meter filters are shown. L2 and L3 are the input and output links. These filters are bilateral, permitting interchanging of the input and output terminals. At B, the representative circuit for the 220- and 432-MHz filters. These filters are also bilateral.

much area. Stray inductance must be minimized and sufficient clearance must be maintained for arc-over protection.

Capacitors with Teflon® dielectric have been used in filters passing up to 2 kW PEP. One further word of caution: No low-pass filter will be fully effective until the transmitter with which it is used is properly shielded and all leads filtered.

The terminating loads for the high-pass section of the filter can be made from 2-watt, 10-percent-tolerance composition resistors. Almost any dissipation rating can be obtained by suitable series-parallel combinations. A 16-watt, 50-ohm load should handle the harmonic energy of a signal with peak fundamental power of 2 kilowatts. With this load, the harmonic energy will see an SWR of less than 2:1 up to 400 MHz. For low power (<300 watts PEP), a pair of 2-watt, 100-ohm resistors is adequate.

This filter was originally described by Weinrich and Carroll in November 1968 *QST*. The component values given here were calculated by Keith Wilkinson, ZL2BJR.

VHF TVI CAUSES AND CURES

The principal causes of TVI from VHF transmitters are:

1) Interference in channels 2 and 3 from 50 MHz.

2) Fourth harmonic of 50 MHz in channels 11, 12 or 13, depending on the operating frequency.

3) Radiation of unused harmonics of the oscillator or multiplier stages. Examples are 9th harmonic of 6 MHz, and 7th harmonic

of 8 MHz in channel 2; 10th harmonic of of 8 MHz in channel 6; 7th harmonic of 25-MHz stages in channel 7; 4th harmonic of 48-MHz stages in channel 9 or 10; and many other combinations. This may include IF pickup, as in the cases of 24-MHz interference in receivers having 21-MHz IF systems, and 48-MHz trouble in 45-MHz IFs.

4) Fundamental blocking effects, including modulation bars, usually found only in the lower channels, from 50-MHz equipment.

5) Image interference in channel 2 from 144 MHz, in receivers having a 45-MHz IF.

6) Sound interference (picture clear in some cases) resulting from RF pickup by the audio circuits of the TV receiver.

There are other possibilities, but nearly all can be corrected completely. The rest can be substantially reduced.

Items 1, 4 and 5 are receiver faults, and no amount of filtering at the transmitter can eliminate them. The only cure is to reduce the amount of 50-MHz energy reaching the receiver, such as by reducing power. In mild cases, increasing the separation between the transmitting and TV antenna systems may reduce or eliminate the problem. Item 6 is also a receiver fault, but it can be alleviated at the transmitter by using FM or CW instead of SSB.

Treatment of the various harmonic troubles, Items 2 and 3, follows the standard methods detailed elsewhere in this Handbook. The prospective builder of new VHF equipment should become familiar with TVI prevention techniques and incorporate them in new construction projects.

Use as high a starting frequency as possible to reduce the number of harmonics that might cause trouble. Select crystal frequencies that do not have harmonics in local TV channels. Example: The 10th harmonic of 8-MHz crystals used for operation in the low part of the 50-MHz band falls in channel 6, but 6-MHz crystals for the same band have no harmonic in that channel.

If TVI is a serious problem, use the lowest transmitter power that will do the job at hand. Keep the power in the multiplier and driver stages at the lowest practical level, and use link coupling in preference to capacitive coupling. Plan for complete shielding and filtering of the RF sections of the transmitter, should these steps become necessary.

Use coaxial line to feed the antenna system, and locate the radiating portion of the antenna as far as possible from TV receivers and their antenna systems.

A complete discussion of the problems and cures for interference is in the ARRL publication, *Radio Frequency Interference*.

FILTERS FOR VHF TRANSMITTERS

High rejection of unwanted frequencies is possible with the tuned-line filters of Fig. 16. Examples are shown for each band from 50 through 450 MHz. Construction is relatively simple, and the cost is low.

Fig. 17 — Interior of the 50-MHz strip-line filter. Inner conductor of aluminum strip is bent into U shape, to fit inside a standard 17-inch chassis.

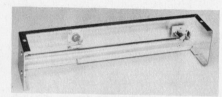

Fig. 18 — The 144-MHz filter has an inner conductor of ½-inch (5/8-inch OD) copper tubing 10 inches long, grounded to the left end of the case and supported at the right end by the tuning capacitor.

Fig. 19 — A half-wave strip line is used in the 220-MHz filter. It is grounded at both ends and tuned at the center.

Standard boxes are used for ease of duplication.

The filter of Fig. 17 is selective enough to pass 50-MHz energy and attenuate the seventh harmonic of an 8-MHz oscillator that falls in TV channel 2. With an insertion loss at 50 MHz of about 1 dB, it can provide up to 40 dB of attenuation to energy at 57 MHz in the same line.

The filter uses a folded line in order to keep it within the confines of a standard chassis. The case is a 6 × 17 × 3-inch chassis (Bud AC-433) with a cover plate that fastens in place with self-tapping screws. An aluminum partition down the middle of the assembly is 14 inches long, and the full height of the chassis is 3 inches.

The inner conductor of the line is 32 inches long and 13/16-inch wide, of 1/16-inch brass, copper or aluminum. This was made from two pieces of aluminum spliced together to provide the 32-inch length. Splicing seemed to have no ill effect on the circuit Q. The sides of the "U" are 2-7/8 inches apart, with the partition at the center. The line is supported on ceramic standoffs. These were shimmed with sections of hardwood or Bakelite rod, to give the required 1½-inch height.

The tuning capacitor is a double-spaced variable (Hammarlund HF-30-X) mounted 1½ inches from the right end of the chassis. Input and output coupling loops are of no. 10 or 12 wire, 10 inches long. Spacing from the line is adjusted to about ¼ inch.

The 144-MHz model shown in Fig. 18 is housed in a 2¼ × 2½ × 12-inch Minibox® (Bud CU-2114-A).

One end of the tubing is slotted ¼-inch deep with a hacksaw. This slot takes a brass angle bracket 1½ inches wide, ¼ inch high, with a ½-inch mounting lip. The ¼-inch lip is soldered into the tubing slot, and the bracket is then bolted to the end of the box, so as to be centered on the end plate.

The tuning capacitor (Hammarlund HF-15-X) is mounted 1¼ inches from the other end of the box, in such a position that the inner conductor can be soldered to the two stator bars.

The two coaxial fittings (SO-239) are 11/16 inch in from each side of the box, 3½ inches from the left end. The coupling loops are no. 12 wire, bent so that each is parallel to the center line of the inner conductor, and about 1/8 inch from its surface. Their cold ends are soldered to the brass mounting bracket.

The 220-MHz filter (Fig. 19) uses the same size box as the 144-MHz model. The inner conductor is 1/16-inch brass or copper, 5/8-inch wide, just long enough to fold over at each end for bolting to the box. It is positioned so that there will be a 1/8-inch clearance between it and the rotor plates of the tuning capacitor. The latter is a Hammarlund HF-15-X, mounted slightly off-center in the box, so that its stator plates connect to the exact mid-point of the line. The 5/16-inch mounting hole in the case is 5½ inches from one end. The SO-239 coaxial fittings are 1 inch in from opposite sides of the box, 2 inches from the ends. Their coupling links are no. 14 wire, 1/8 inch from the inner conductor of the line.

The 420-MHz filter is similar in design, using a 1-5/8 × 2 × 10-inch Minibox (Bud CT-2113-A). A half-wave line is used, with the disc tuning at the center. The discs are 1/16-inch brass, 1¼-inch diameter. The fixed one is centered on the inner conductor, the other is mounted on a no. 6 brass lead-screw. This passes through a threaded bushing, which can be taken from the end of a discarded slug-tuned form. An advantage of these is that usually a tension device is included. If there is none, use a lock nut.

Type-N coaxial connectors were used on the 420-MHz model. They are 5/8 inch in from each side of the box, and 1-3/8 inches in from the ends. Their coupling links of no. 14 wire are 1/16 inch from the inner conductor.

Adjustment and Use

If you want the filter to work on both transmitting and receiving, connect the filter between antenna line and SWR in-

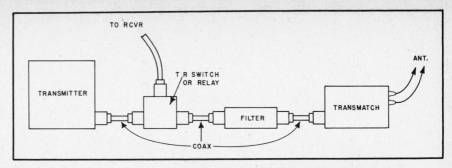

Fig. 20 — The proper method of installing a low-pass filter between the transmitter and a Transmatch. If the antenna is fed through coax, the Transmatch can be eliminated, but the transmitter and filter must be completely shielded. If a TR switch is used, it should be installed between the transmitter and low-pass filter. TR switches can generate harmonics themselves, so the low-pass filter should follow the TR switch.

dicator. With this arrangement you need merely adjust the filter for minimum reflected power reading on the SWR bridge. This should be zero, or close to it, if the antenna is well matched. The bridge should be used, as there is no way to adjust the filter properly without it.

When the filter is properly adjusted (with the SWR bridge), you may find that reception can be improved by retuning the filter. Don't do it if you want the filter to work best for the job it was intended to do: The rejection of unwanted energy, transmitting or receiving. If you want to improve reception with the filter in the circuit, work on the receiver input circuit. To get maximum power out of the transmitter and into the line, adjust the transmitter output coupling, not the filter. If the effect of the filter on reception bothers you, connect it in the line from the antenna relay to the transmitter only.

Summary

The methods of harmonic elimination outlined here have been proven beyond doubt to be effective even under highly unfavorable conditions. It must be emphasized once more, however, that the problem must be solved one step at a time, and the procedure must be in logical order. It cannot be done properly without two items of simple equipment: A dip meter and wavemeter covering the TV bands, and a dummy antenna.

To summarize:

1) Take a critical look at the transmitter on the basis of the design considerations outlined under Reducing Harmonic Generation.

2) Check all circuits, particularly those connected with the final amplifier, with the dip meter to determine whether there are any resonances in the TV bands. If so, rearrange the circuits so the resonances are moved out of the critical frequency region.

3) Connect the transmitter to the dummy antenna and check for the presence of harmonics on leads with the wavemeter and around the transmitter enclosure. Seal the weak spots in the shielding and filter the leads until the wavemeter shows no indication at any harmonic frequency.

4) At this stage, check for interference with a TV receiver. If there is interference, determine the cause by the methods described previously and apply the recommended remedies until the interference disappears. More than one remedy may be required.

5) When the transmitter is completely clean on the dummy antenna, connect it to the regular antenna and check for interference on the TV receiver. If the interference is not bad, a Transmatch or matching circuit, installed as shown in Fig. 20, should clear it up. Alternatively, a low-pass filter may be used. If neither the Transmatch nor filter makes any difference in the interference, the evidence is strong that the interference, at least in part, is being caused by receiver overloading because of the strong fundamental-frequency field about the TV antenna and receiver. A Transmatch and/or filter, installed as described above, will invariably make a difference if the interference is caused by transmitter harmonics alone.

6) If there is still interference after installing the Transmatch and/or filter, and the evidence shows that it is probably caused by a harmonic, more attenuation is needed. A more elaborate filter may be necessary. However, it is well at this stage to assume that part of the interference may be caused by receiver overloading. Take steps to alleviate such a condition before trying highly elaborate filters and traps on the transmitter.

Harmonics by Rectification

Even though the transmitter is completely free of harmonic output, it is still possible for interference to occur because of harmonics generated outside the transmitter. These result from rectification of fundamental-frequency currents induced in conductors in the vicinity of the transmitting antenna. Rectification can take place at any point where two conductors are in poor electrical contact, a condition that frequently exists in plumbing, downspouting, BX cables crossing each other, and numerous other places in an ordinary residence. It can also occur at any exposed circuitry in the station that may not be shielded from RF. Poor joints anywhere in

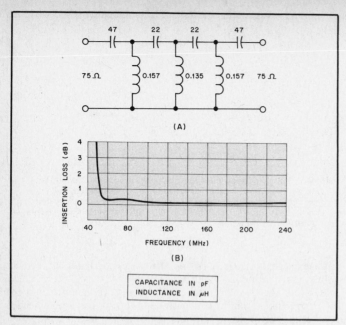

Fig. 21 — The schematic diagram of a 75-ohm Chebyshev filter assembled on PC board is shown at A. At B, the passband response of the 75-ohm filter. Design inductances: 0.157 μH: 12 turns no. 24 wire on T44-0 core. 0.135 μH: 11 turns no. 24 wire on T44-0 core. Turns should be evenly spaced, with approximately ¼ inch between the ends of the winding. If T37-0 cores are used, wind 14 and 12 turns, respectively.

Fig. 22 — Photo showing construction of the 75-ohm unbalanced filter.

the antenna system are especially bad, and rectification may take place in the contacts of antenna-changeover relays. Another rectification problem is caused by overloading the front end of the communications receiver when it is used with a separate antenna. This can result in the radiation of harmonics generated in the first stage of the receiver.

Rectification of this sort will not only cause harmonic interference, but also is frequently responsible for cross-modulation effects. It can be detected to some degree in most locations, but fortunately the harmonics thus generated are not usually of high amplitude. However, they can cause considerable interference in the immediate vicinity in fringe areas, especially when operation is in the 28-MHz band. The amplitude decreases rapidly with the order of the harmonic, the second and third being the worst. It is ordinarily found that even in cases where destructive interference results from 28-MHz operation, the interference is comparatively mild from 14 MHz, and is negligible at still lower frequencies.

Nothing can be done at either the transmitter or receiver when rectification occurs in other objects. The remedy is to find and eliminate the poor contact either by separating the conductors or bonding them together. A crystal wavemeter (tuned to the fundamental frequency) is useful for hunting the source by showing which conductors are carrying RF and, comparatively, how much.

Interference of this kind is frequently intermittent since the rectification efficiency will vary with its susceptibility to vibration, weather and so on. The possibility of cor-

roded contacts in the TV receiving antenna should not be overlooked, especially if it has been up for a year or more.

TV Receiver Deficiencies

When a television receiver is located close to the transmitter, the intense RF signal from the transmitter's fundamental may overload one or more of the receiver circuits to produce spurious responses that cause interference.

If the overload is moderate, the interference is of the same nature as harmonic interference; it is caused by harmonics generated in the early stages of the receiver and, since it occurs only on channels harmonically related to the transmitting frequency, it is difficult to distinguish from harmonics actually radiated by the transmitter. In such cases, additional harmonic suppression at the transmitter will do no good, but any means taken at the receiver to reduce the strength of the amateur signal reaching the first stage will improve performance. With very severe overloading, interference also will occur on channels not harmonically related to the transmitting frequency, so such cases are easily identified.

Intermodulation

Under some circumstances, overloading will result in cross modulation or mixing of the amateur signal with that from a local FM or television station. For example, a 14-MHz signal can mix with a 92-MHz FM station to produce a beat at 78 MHz, the *difference* between the two frequencies (92 − 14 = 78). Since 78 MHz falls in TV channel 5, interference to television recep-

tion would occur on that channel. Similarly, a 14-MHz signal could mix with a TV broadcast station operating on channel 6 to produce a beat at 99 MHz, the *sum* of the two frequencies (14 + 85 = 99). Neither of the broadcast channels interfered with is in harmonic relationship to 14 MHz, and *both* signals must be on the air for the interference to occur. Eliminating either at the receiver will eliminate the interference.

There are many combinations of this type, depending on the band in use and the local frequency assignments of FM and TV stations. As noted earlier, the interfering frequency is equal to the amateur fundamental frequency either added to, or subtracted from, the frequency of some local station. Whenever interference occurs in a frequency that is not harmonically related to the amateur transmitter frequency, the possibilities in such frequency combinations should be investigated.

IF Interference

Some TV receivers do not have sufficient selectivity to prevent strong signals in the intermediate-frequency range from forcing their way through the front end and getting into the IF amplifier. The third harmonic of 14 MHz and second harmonic of 21 MHz fall into the television IF, as do some of the local-oscillator frequencies used in a heterodyne type of transmitter or transceiver. If these frequencies are breaking through the TV tuner, a high-pass filter can improve the situation significantly. Even so, the amateur is responsible for keeping his or her radiation in the TV IF region within the limits defined by FCC rules and

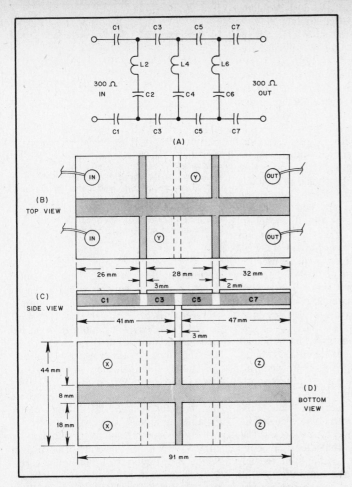

Fig. 23 — Schematic and pictorial diagrams of the 300-ohm balanced elliptical high-pass filter with PC-board capacitors. Shaded areas indicate where copper has been removed. Dimensions are given in millimeters for ease of measurement. L2-C2 connects between the points marked x, L4-C4 connects between the points marked Y and L6-C6 connects between the points marked z on the pictorial. C1 = 28.0 pF, C3 = 14.0 pF, C5 = 14.8 pF, C7 = 34.2 pF, C2 = 162 pF, C4 = 36.0 pF, and C6 = 46.5 pF. Design inductances: L2 = 0.721 μH: 14 turns no. 26 wire evenly wound on a T44-10 core. L4 = 0.766 μH: 14 turns no. 26 wire bunched as required on a T44-10 core. L6 = 0.855 μH: 15 turns evenly wound on a T44-10 core. These coils should be adjusted for resonance at 14.7, 30.3 and 25.2 MHz.

(A)

(B)

Fig. 24 — A top view of the 300-ohm elliptic filter using PC-board capacitors is at A. Twin-lead is tack soldered at the left and right ends of the board. At B, bottom view of the filter.

good engineering practice.

A form of IF interference peculiar to 50-MHz operation near the low edge of the band occurs with some receivers having the standard 41-MHz IF, which has the sound carrier at 41.25 MHz and the picture carrier at 45.75 MHz. A 50-MHz signal that forces its way into the IF system of the receiver will beat with the IF picture carrier to give a spurious signal on or near the IF sound carrier, even though the interfering signal is not actually in the normal passband of the IF amplifier.

There is a type of IF interference unique to the 144-MHz band in localities where certain UHF TV channels are in operation. It affects only those TV receivers in which double-conversion type plug-in UHF tuning strips are used. The design of these strips involves a first intermediate frequency that varies with the TV channel to be received and, depending on the particular strip design, this first IF may be in or close to the 144-MHz amateur band. Since there is comparatively little selectivity in the TV

signal-frequency circuits ahead of the first IF, a signal from a 144-MHz transmitter will ride into the IF, even when the receiver is at a considerable distance from the transmitter. The channels that can be affected by this type of IF interference are 20-25, 51-58, 82 and 83. If the receiver is not close to the transmitter, a trap of the type shown in Fig. 25 will be effective. However, if the separation is small, the 144-MHz signal will be picked up directly on the receiver circuits. The best solution is to readjust the strip oscillator so that the first IF is moved to a frequency not in the vicinity of the 144-MHz band. This should be done only by a competent technician.

IF interference is easily identified since it occurs on all channels — although sometimes the intensity varies from channel to channel — and the cross-hatch pattern it causes will rotate when the receiver's fine-tuning control is varied. When the interference is caused by a harmonic, overloading or cross modulation, the structure of the interference pattern does not change

(its intensity may change) as the fine-tuning control is varied.

High-Pass Filters

In all of these cases the interference can be eliminated if the fundamental signal strength is reduced to a level that the receiver can handle. To accomplish this with signals on bands below 30 MHz, the most satisfactory device is a high-pass filter having a cutoff frequency just below 54 MHz installed at the tuner input terminals of the receiver.

Fig. 21 shows the schematic diagram of a filter designed for use with 75-ohm coaxial cable. Double-sided 1/16-inch FR-4 epoxy-glass PC board is used as a base for the filter components. A section of copper on the top is stripped away on both sides of center to approximate a 75-ohm microstrip line about 3/32-inch wide (see Fig. 21). Both sides of the top copper foil (at the edges) are connected to the ground plane foil underneath.

Slice off the extruded insulation around

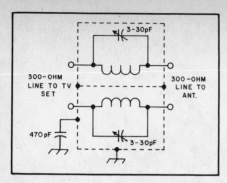

Fig. 25 — Parallel-tuned traps for installation in the 300-ohm line to the TV set. The traps should be mounted in an aluminum Minibox with a shield partition between them, as shown. For 50 MHz, the coils should have nine turns of no. 16 enamel wire, close-wound to a diameter of ½ inch. The 144-MHz traps should contain coils with a total of six turns of the same type wire close-wound to a diameter of ¼ inch. Traps of this type can be used to combat fundamental-overload TVI on the lower-frequency bands as well.

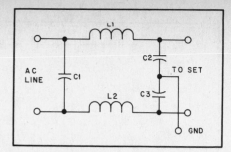

Fig. 26 — Brute-force ac line filter for receivers. The values of C1, C2 and C3 are not generally critical; capacitances from 0.001 to 0.01 μF can be used. L1 and L2 can be a 2-inch winding of no. 18 enameled wire on a ½-inch-diameter form. In making up such a unit for use external to the receiver, make sure there are no exposed conductors to offer a shock hazard.

the solder pins on two type-F coaxial connectors (Radio Shack 278-212). Butt the connectors directly against the PC board. Solder the connector shells to the bottom ground plane and the center pins to the micro-strip line. Cut the micro-strip line in four equally spaced places. The capacitors should be mounted across the spaces; inductors can be connected between the capacitor junctions and the ground plane on the top of the board. Use NP0 ceramic or silver mica capacitors. Inductors are wound on toroidal powdered-iron cores; winding details are given in Fig. 22.

Fig. 22 shows the schematic and pictorial diagrams of a 300-ohm balanced elliptical high-pass filter that uses PC-board capacitors. Use double-sided 1/32 inch FR-4 glass-epoxy PC board. Thicker board will require more area for the desired capacitances. C2, C4 and C6 should be NP0 ceramic or silver-mica capacitors. It is easier to strip away rather than etch copper to form the series of capacitive elements. Mark the edges by cutting with a sharp knife; heating with a hot soldering iron will help lift the strips more easily. Top and bottom views of the filter are shown in Fig. 24.

Neither of the high-pass filters described requires a shielded enclosure. For mounting outside the receiver, some kind of protective housing is desirable, however. These filters were presented in "Practical 75- and 300-Ohm High-Pass Filters," QST, February 1982, by Ed Wetherhold, W3NQN.

Simple high-pass filters cannot always be applied successfully in the case of 50-MHz transmissions, because they do not have sufficiently sharp cutoff characteristics to give both good attenuation at 50-54 MHz and no attenuation above 54 MHz. A more elaborate design capable of giving the required sharp cutoff has been described (Ladd, "50-MHz TVI — Its Causes and Cures," QST, June and July 1954). This article also contains other information

useful in coping with the TVI problems peculiar to 50-MHz operation.

As an alternative to such a filter, a high-Q wave trap tuned to the transmitting frequency may be used, suffering only the disadvantage that it is quite selective and therefore will protect a receiver from overloading over only a small range of transmitting frequencies in the 50-MHz band. A trap of this type is shown in Fig. 25. These suck-out traps, while absorbing energy at the frequency to which they are tuned, do not affect the receiver operation otherwise. The assembly should be mounted near the input terminals of the TV tuner and its case should be RF grounded to the TV set chassis by means of a small capacitor. The traps should be tuned for minimum TVI at the transmitter operating frequency. An insulated tuning tool should be used for adjustment of the trimmer capacitors, since they are at a hot point and will show considerable body-capacitance effect.

High-pass filters are available commercially at moderate prices. In this connection, it should be understood by all parties concerned that while an amateur is responsible for *harmonic* radiation from his transmitter, it is no part of his responsibility to pay for or install filters, wave traps or other devices that may be required at the receiver to prevent interference caused by his *fundamental* frequency. Proper installation usually requires that the filter be installed at the input terminals of the RF tuner of the TV set and not merely at the external antenna terminals, which may be at a considerable distance from the tuner. The question of cost is one to be settled between the set owner and the organization with which he deals. Don't overlook the possibility that the manufacturer of the TV receiver may supply a high-pass filter free of charge.

If the fundamental signal is getting into the receiver by way of the line cord, a line filter such as those shown in Fig. 26 may help. To be most effective it should be installed inside the receiver chassis at the point where the cord enters, making the ground connections directly to the chassis at this point. It may not be so helpful if placed between the line plug and the wall socket unless the RF is actually picked up on the house wiring rather than on the line cord itself.

Antenna Installation

Usually the transmission line between the TV receiver and the antenna will pick up a great deal more energy from a nearby HF transmitter than the television receiving antenna itself. The currents induced on the TV transmission line in this case are of the parallel type, where the phase of the current is the same in both conductors. The line simply acts like two wires connected together to operate as one. If the receiver antenna input circuit were perfectly balanced it would reject these parallel or common-mode signals and respond only to the true transmission-line (push-pull or dif-

ferential mode) currents. That is, only signals picked up on the actual antenna would cause a receiver response. However, no receiver is perfect in this respect, and many TV receivers will respond strongly to such common-mode currents. The result is that the signals from a nearby amateur transmitter are much more intense at the first stage in the TV receiver than they would be if the receiver response were confined entirely to energy picked up on the TV antenna alone.

A simple common-mode choke can be formed by winding several turns of TV twin-lead through an FT-114 ferrite core. Best results will be obtained if you use twin-lead with an oval cross-sectional profile. The situation can also be improved by using coaxial cable or shielded twin-lead. For best results, coax line should terminate in a coaxial fitting on the receiver chassis. A balun can be used between the coax and the 300-ohm balanced input terminals of a receiver having no coaxial connector. The foil of shielded twin-lead should be connected to the chassis near the antenna terminals through a small capacitor (470 pF). RF currents on the outside of the shield can be dealt with effectively by using a shield choke as described earlier in this chapter.

In most TV receiving installations the transmission line is much longer than the antenna itself, and is consequently exposed to more of the harmonic fields from the transmitter. Much of the harmonic pickup, therefore, is on the receiving transmission line when the transmitter and receiver are located close together. Shielded line, plus relocation of either the transmitting or receiving antenna to take advantage of directive effects, often will reduce overloading and harmonic pickup to a level that does not interfere with reception.

UHF Television

Harmonic TVI in the UHF-TV band is far less troublesome than in the VHF band. Harmonics from transmitters operating below 30 MHz are of such high order that they would normally be expected to be weak; in addition, the components, circuit conditions and construction of low-

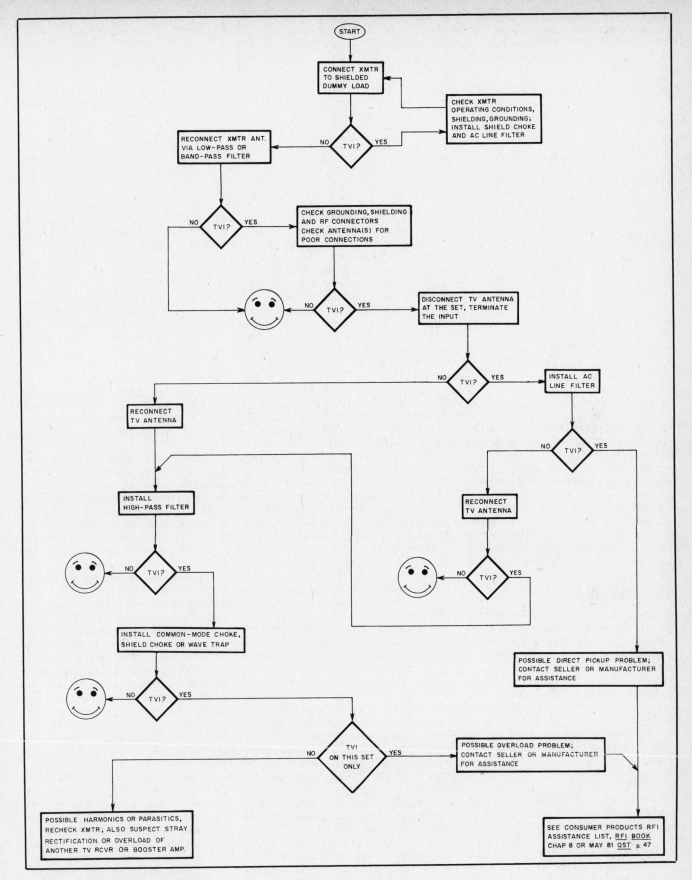

Fig. 27 — TVI troubleshooting flowchart.

frequency transmitters tend to prevent very strong harmonics from being generated in this region. However, this is not true of amateur VHF transmitters, particularly those working in the 144-MHz and higher bands. Here the problem is similar to that of the low VHF-TV band with respect to transmitters operating below 30 MHz.

There is one highly favorable factor in UHF TV that does not exist in most of the VHF-TV band: If harmonics are radiated, it is possible to move the transmitter fre-

Amateur Band	Harmonic	Fundamental Freq. Range	Channel Affected
144 MHz	4th	144.0-144.5	31
		144.5-146.0	32
		146.0-147.5	33
		147.5-148.0	34
	5th	144.0-144.4	55
		144.4-145.6	56
		145.6-146.8	57
		146.8-148.0	58
	6th	144.0-144.33	79
		144.33-145.33	80
		145.33-146.33	81
		146.33-147.33	82
		147.33-148.00	83
220 MHz	3rd	220-220.67	45
		220.67-222.67	46
		222.67-224.67	47
		224.67-225	48
	4th	220-221	82
		221-222.5	83
420 MHz	2nd	420-421	75
		421-424	76
		424-427	77
		427-430	78
		430-433	79
		433-436	80

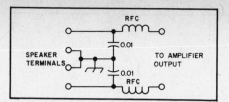

Fig. 28 — A method for removing RF current from loudspeaker leads. The chokes should be near the output terminals, preferably within the amplifier cabinet. The RF chokes can be 24 turns of no. 18 wire closewound on a pencil.

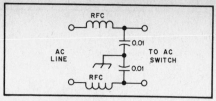

Fig. 29 — Ac line filter for audio amplifiers. The chokes are the same as those described in Fig. 28. The capacitors must be rated for ac service (or 1.4 kV dc).

quency sufficiently (within the amateur band being used) to avoid interfering with a channel that may be in use in the locality.

The harmonics from amateur bands above 50 MHz span the UHF channels as shown in Table 1. Since the assignment plan calls for a minimum separation of six channels between any two stations in one locality, there is ample opportunity to choose a fundamental frequency that will move a harmonic out of range of a local TV frequency.

TVI Troubleshooting

Determination and experimentation are essential for success in tracking down and eliminating troublesome TVI. The variables involved are usually so complicated as to make the task more of an art than a science. To ensure a logical and systematic approach, use the TVI troubleshooting flowchart found in Fig. 27. The best place to begin is in your own home. Lessons learned there will assist you when dealing with problems elsewhere in the neighborhood.

Stereo Interference

Since the introduction of stereo receivers, interference to this type of home-entertainment device has become a severe problem for amateurs. Unfortunately, most of the stereo equipment now being sold has little or no filtering to prevent RF interference. In most cases, corrective measures must be taken at the stereo installation.

Stereo Equipment

Stereo gear can consist of a simple amplifier, with record or tape inputs and speaker outputs. The more elaborate installations may also include a tape deck, turntable, AM/FM tuner, amplifier and two or more speakers. These units are usually connected by means of shielded leads, and in most cases the speakers are positioned some distance from the amplifier and connected with long leads. When such a setup is operated within a few hundred feet of an amateur station, there are two important paths through which RF energy can reach the stereo equipment and cause interference.

The first step is to try to determine how the interference is getting into the unit. If the volume control has no effect on the level of interference, or has a very slight effect, the audio rectification of the amateur signal is taking place past the volume control, or on the output end of the amplifier. This is by far the most common type. It usually means that the amateur signal is being picked up on the speaker leads, or possibly on the ac line, and is then being fed back into the amplifier.

Experience has shown that most of the RF gets into the audio system via the speaker leads, or the ac line, most often the speaker leads. If the speaker leads happen to be resonant near an amateur band in use, there is likely to be an interference problem. The speaker lead will act as a resonant antenna and pick up the RF. One easy cure is illustrated in Fig. 28. RF chokes rated to carry the load current are installed at the amplifier output terminals. Capacitors may be used on the *load* side of the chokes, but should not be placed on the amplifier side. Most solid-state audio amplifiers have high values of loop gain, which makes them prone to supersonic oscillation when working into capacitive loads. Such oscillation can destroy the output transistors in short order.

In particularly stubborn cases, use shielded wire for the speaker leads, grounding the shields at the amplifier chassis and still using the bypasses on the terminals. All chassis used in the stereo installation should be bonded together and connected to a good earth ground if at all possible. It has been found that grounding sometimes eliminates the interference. On the other hand, don't be discouraged if grounding doesn't appear to help.

Fig. 29 shows a method for filtering the ac line at the input of the amplifier chassis. Be sure that the capacitors are rated for ac because the dc types have been known to short out.

Antenna Pickup

RF may also get into the audio equipment by way of the FM antenna, so precautions should be taken here. A TV-type high-pass filter can prove effective in some cases.

Turntables and Tape Decks

In the more elaborate setups, several assemblies may be connected by means of patch cords. It is a good idea when checking for RFI to disconnect the units, one at a time, observing any changes in the interference. Not only disconnect the patch cords connecting the pieces, but also unplug the ac line cord for each item as you make the test. This will help you determine which section is the culprit.

Patch cords are usually, but not always, made of shielded cable. The lines should be shielded, which brings up another point. Many commercially available patch cords have poor shields. Some have wire spirally wrapped around the insulation, covering the main lead, rather than braid. This method provides poor shielding and could be the reason for RFI problems.

Record-player tone-arm connections to the cartridge are usually made with small clips. The existence of a loose clip, particularly if oxidation is present, offers an excellent invitation to RFI. Also, the leads from the cartridge and those to the amplifier are sometimes resonant at VHF, providing an excellent receiving antenna for RF. One cure for unwanted RF pickup is to install ferrite beads, one on each cartridge lead. Check all patch-cord connections for looseness or poor solder joints. Inferior connections can cause rectification and RFI.

Tape decks should be treated the same as turntables. Loose connections and bad solder joints can cause trouble. Ferrite beads can be slipped over the leads to the recording and play-back pickup heads. Bypassing of the tone-arm or pickup-head leads is also effective, but sometimes it is difficult to install capacitors in the small area available. Disc capacitors (0.001 µF) should be used as close to the cartridge or pickup head as possible. Keep the capacitor leads as short as possible.

Preamplifiers

One or more preamplifiers are usually used in a stereo amplifier. The inputs to these stages can be very susceptible to RFI. Fig. 30 illustrates a typical preamplifier circuit. In this case the leads to the bases of the transistors are treated for RFI with ferrite beads by the addition of RFC2 and RFC4. This is a very effective method for stopping RFI when VHF energy is the source of the trouble.

The emitter-base junction of a transistor is a common RFI offender. This junction operates as a forward-biased diode, with the bias set so that a change of base current with signal will produce a linear but amplified change in collector current. Should RF energy reach the junction, the bias could increase, causing nonlinear amplification and distortion as the result. If the RF level is high it can completely block (saturate) a transistor, causing a complete loss of gain. Therefore, it may be necessary to reduce the transmitter power output to pinpoint the particular transistor stage that is affected.

In addition to ferrite beads placed over the base lead, it may be necessary to bypass the base of the transistor to chassis ground, C1 and C2, Fig. 30. A suitable value is 100 pF, and keep the leads short! As a rule, the capacitor value should be as large as possible without degrading the high-frequency response of the amplifier. Values up to 0.001 µF can be used. In severe cases, a series inductor (RFC1 and RFC3) may be required, such as the Ohmite Z-50 or Z-144, or their equivalents (7 and 1.8 µH, respectively). Fig. 30 shows the correct placement for an inductor, bypass capacitor and ferrite bead. Also, it might help to use a ferrite bead in the B+ lead to the preamplifier stages (RFC5 in Fig. 30). Keep in mind that Fig. 30 represents only one preamplifier of a stereo set. Both channels may require treatment.

FM Tuners

Much of the interference to FM tuners is caused by fundamental overloading of the first stage (or stages). The cure is the installation of a high-pass filter, the same type used for TVI. The filter should be installed as close as possible to the antenna input of the tuner. The high-pass filter will attenuate the amateur fundamental signal, thus preventing overloading of the front end.

Shielding

Lack of shielding on the various components in a stereo installation can permit RF to get into the equipment. Many units have no bottom plates, or are installed in plastic cases. One easy method of providing shielding is to use aluminum foil. Make sure the foil doesn't short-circuit the com-

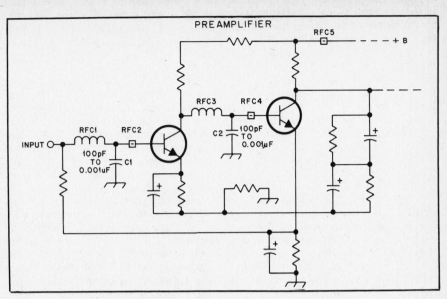

Fig. 30 — Typical circuit of a solid-state preamplifier.

ponents, and connect it to chassis ground.

Interference with Medium-Wave Broadcast Reception

Transmitter Defects

Out-of-band radiation is something that must be cured at the amateur transmitter. Parasitic oscillations are a frequently unsuspected source of such radiation, and no transmitter can be considered satisfactory until it has been thoroughly checked for both low- and high-frequency parasitics. Very often, parasitics show up only as transients, causing key clicks in CW transmitters and "splashes" or "burps" on modulation peaks in SSB transmitters. Methods for detecting and eliminating parasitics are discussed in the transmitter chapter.

In CW transmitters the sharp make and break that occurs with unfiltered keying causes transients that, in theory, contain frequency components through the entire radio spectrum. Practically, they are often strong enough in the immediate vicinity of the transmitter to cause serious interference to broadcast reception. Key clicks can be eliminated by proper wave shaping. See Chapter 19.

BCI is frequently made worse by radiation from the power wiring or the RF transmission line. This is because the signal causing the interference is radiated from wiring that is nearer the broadcast receiver than the transmitting antenna. Much depends on the method used to couple the transmitter to the antenna, a subject that is discussed in the chapters on transmission lines and antennas. If at all possible, the antenna itself should be placed some distance from house wiring, telephone and power lines, and similar conductors.

The BC Receiver

Most present-day receivers use solid-state

active components rather than tubes. A large number of the receivers in use are battery powered. This is to the amateur's advantage because much of the BC interference an amateur encounters is caused by ac line pickup. In the case where the BC receiver is powered from the ac line, whether using tube or solid-state components, the amount of RF pickup must be reduced or eliminated. A line filter such as is shown in Fig. 26 often will help accomplish this. The values used for the coils and capacitors are in general not critical. The effectiveness of the filter may depend considerably on the ground connection used, and it is advisable to use a short ground lead to a cold-water pipe if at all possible. The line cord from the set should be bunched up to minimize the possibility of pickup on the cord. To get satisfactory operation, it may be necessary to install the filter inside the receiver, so the filter is connected between the line cord and the set wiring.

Cross Modulation

With phone transmitters, there are occasionally cases where the noise is heard whenever the broadcast receiver is tuned to a BC station, but there is no interference when tuned between stations. This is cross modulation, a result of rectification in one of the early stages of the receiver. Receivers that are susceptible to this trouble usually also get a similar type of interference from regular broadcasting if there is a strong local BC station and the receiver is tuned to some other station.

The remedy for cross modulation in the receiver is the same as for images and oscillator-harmonic response — reduce the strength of the amateur signal at the receiver by means of a line filter.

The trouble is not always in the receiver. Cross modulation can occur in any nearby rectifying circuit — such as a poor contact in water or gutter pipes, and other conductors in the strong field of the transmitting antenna. Locating the cause may be difficult, and is best attempted with a battery-powered portable broadcast receiver used as a probe to find the spot where the interference is most intense. When such a spot is located, inspection of the metal structures in the vicinity should indicate the cause. The remedy is to make a good electrical bond between the two conductors having the poor contact.

Handling BCI Cases

Tune the receiver through the broadcast band to see whether the interference tunes like a regular BC station. If so, image or oscillator-harmonic response is the cause. If there is interference only when a BC station is tuned in, but not between stations, the cause is cross modulation. If the interference is heard at all settings of the tuning dial, the trouble is pickup in the audio circuits. In this case, the receiver volume control may or may not affect the strength of the interference, depending on how your signal is being rectified.

Organs

The electronic organ is an RFI problem area. All of the techniques outlined for audio gear hold true in getting rid of RFI in an organ. Two points should be checked — the speaker leads and the ac line. Many organ manufacturers have special service guides for taking care of RFI. However, to get this information you or the organ owner must contact the manufacturer, not the dealer or distributor. Don't accept the statement from a dealer or serviceman that nothing can be done about the interference.

Public Address Systems

The cure for RFI in public address systems is almost the same as that for audio gear. The one thing to watch for is RF on the leads that connect the various stations

in a public address system. These leads should be treated the same as speaker leads, and filtering should be done at both ends of the lines. Also, watch for ac-line pickup of RF.

Telephone Interference

Because of a change in FCC rules, subscribers increasingly own their telephone instruments, leasing only the lines from the telephone company. Interference-prevention measures to the instrument are the owner's responsibility. If a fault occurs in the line, the telephone company must make the necessary repairs. Responsible instrument manufacturers should provide necessary modification to minimize RFI.

Telephone interference may be cured by connecting a bypass capacitor (about 0.01 μF) across the microphone unit in the telephone handset. Telephone companies have capacitors for this purpose, but will only service their own instruments. When such a case occurs, get in touch with the repair department of the phone company, giving the particulars. If you have purchased your instrument from the phone company, you will be billed for the service call if the warranty has expired. Section 500-150-100 of the Bell System Practices *Plant Series* gives detailed instructions; for the General Telephone System, refer to General System Practices *Engineering — Plant Series* Section 471-150-200. This section discusses causes and cures of telephone interference from radio signals. It points out that interference can come from corroded connections, unterminated loops and other sources. It correctly points out that the RF can be picked up on the drop wire coming into the house, and also on the wiring within the house, but the RF detection usually occurs inside the telephone instrument. The detection usually takes place at the varistors in the compensation networks, and/or at the receiver noise suppressor and the carbon microphone.

But interference suppression should be handled two ways: Prevent the RF from getting to the phone, and prevent it from being rectified.

AT&T markets RF filter modules that plug into modular telephone jacks. Model Z100A, for desk phones, is available at local AT&T phone centers (stock number SKU-52610). A second filter, the Z101A for wall-mounted telephones, is available only by mail from the AT&T National Service Center in St. Louis, Missouri (tel 800-222-3111). The stock number for the Z101A is SKU-57293.

Additional Information

In response to the many hundreds of thousands of RFI-related complaints it has received in recent years, the FCC has produced a booklet designed to show how to solve common RFI problems before they become serious. Entitled *How to Identify and Resolve Radio-TV Interference Problems,* it is available for $5 from the Government Printing Office, Washington, DC 20402. Make your check payable to "Superintendent of Documents." The ARRL publication, *Radio Frequency Interference,* which sells for $3, covers all aspects of RFI and includes the complete FCC booklet.

Additional information can be found in the sources listed below.

Consumer Electronics Service Technician Interference Handbook — Audio Rectification (Washington, DC: CES, n.d.)*
Consumer Electronics Service Technician Interference Handbook — Television Interference (Washington, DC: CES, n.d.)*
Nelson, W. R., *Interference Handbook* (Wilton, CT: Radio Publications, 1981).

*Single copies of the interference handbooks for audio rectification and television interference may be obtained by writing to: Director of Consumer Affairs, Consumer Electronics Group, Electronic Industries Association, 2001 Eye St., N.W., Washington, DC 20006.

Printed-Circuit Board Etching Patterns

The following pages contain printed-circuit-board etching patterns for the projects in this book. All patterns are full size and are shown from the foil side. Black areas represent unetched copper. The companion parts layout for each pattern is shown with the project description.

More information on making PC boards may be found in Chapter 24. A list of suppliers of PC board and related materials may be found in Chapter 35.

In the interest of making best use of the space available, the patterns are not arranged in the same order the projects appear in the book. The contents list that follows will help you locate the pattern(s) for your project. In addition, each pattern is clearly labeled with the chapter and figure number of the parts-placement diagram.

Etching Pattern Contents

CAUTION:

Thoroughly check the printed pattern before you begin. In case of broken traces, continuity can be restored with a fine-point pen. Any ink bridges can be removed by careful scraping with a sharp, thin knife. Take time to be sure that these potential problems do not exist before you begin — it can save you much time later.

The scraping and drawing techniques described above for knife and pen can also be used to modify or customize patterns. If you feel uneasy, practice on a photocopy to sharpen your skill.

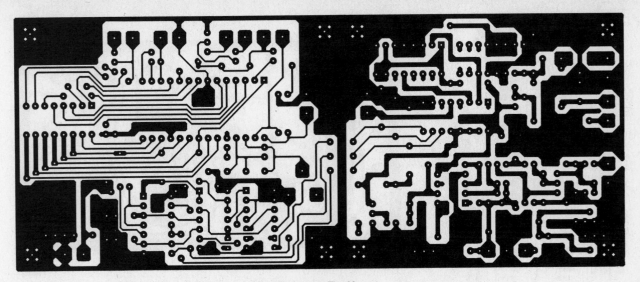

Chapter 25, Fig. 30

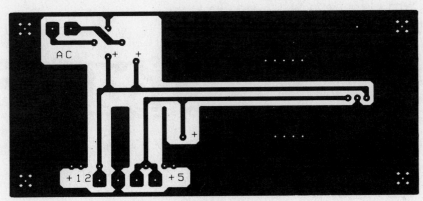

Chapter 25, Fig. 32

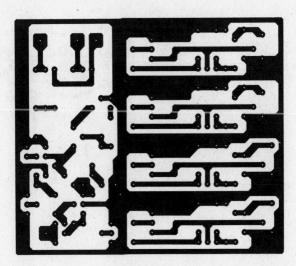

Chapter 25, Fig. 45

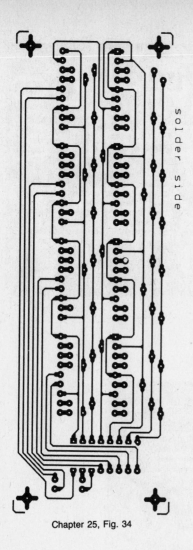

solder side

Chapter 25, Fig. 34

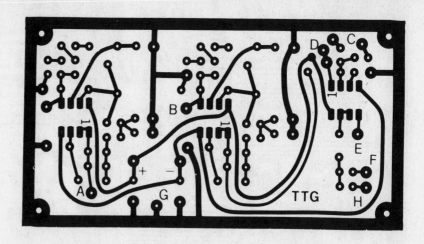

Chapter 25, Fig. 51

Chapter 26, Fig. 6

Chapter 25, Fig. 12

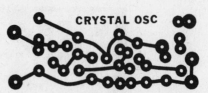

Chapter 26, Fig. 7

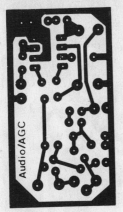

Chapter 28, Fig. 15

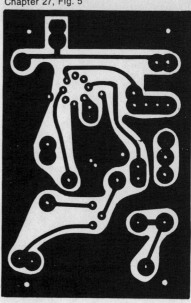

Chapter 27, Fig. 5

Chapter 27, Fig. 23

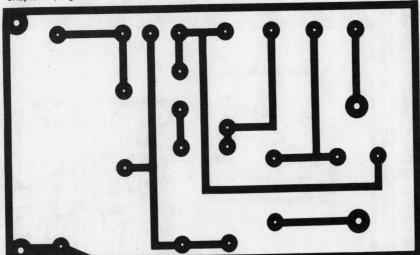

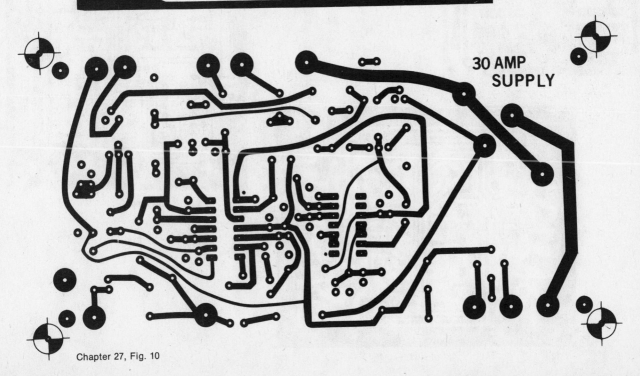

30 AMP
SUPPLY

Chapter 27, Fig. 10

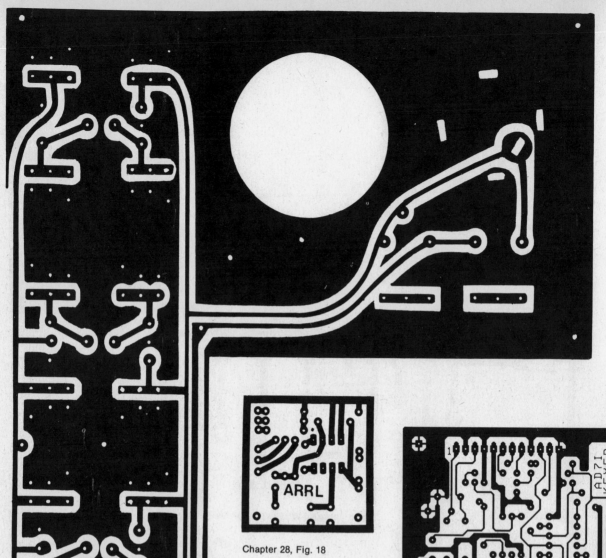

Chapter 27, Fig. 31

Chapter 28, Fig. 18

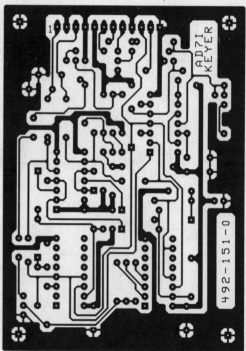

Chapter 29, Fig. 2

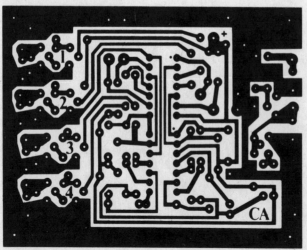

Chapter 31, Fig. 22

6   Printed-Circuit Board Etching Patterns

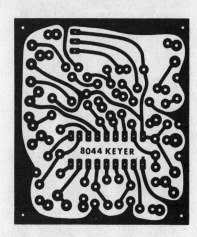

Chapter 29, Fig. 4

Chapter 32, Fig. 65

Chapter 32, Fig. 69

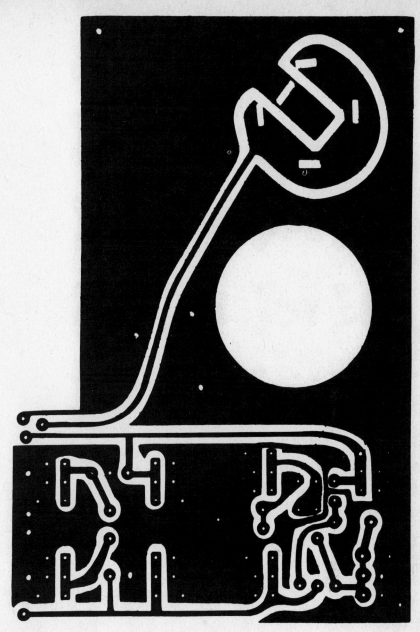

Chapter 27, Fig. 31

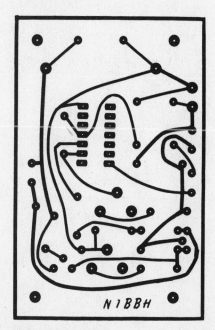

Chapter 27, Fig. 68

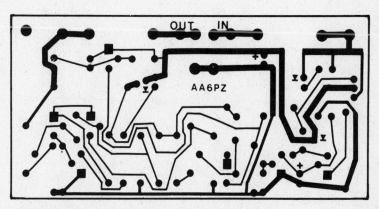

Chapter 27, Fig. 58

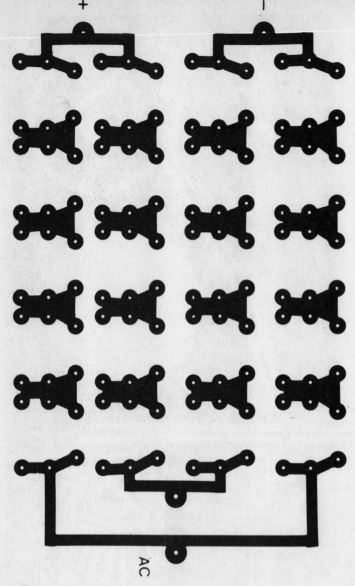

Chapter 27, Fig. 72

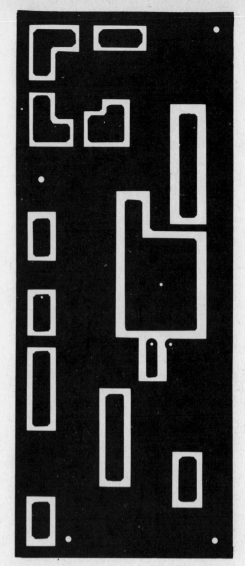

Chapter 31, Fig. 38

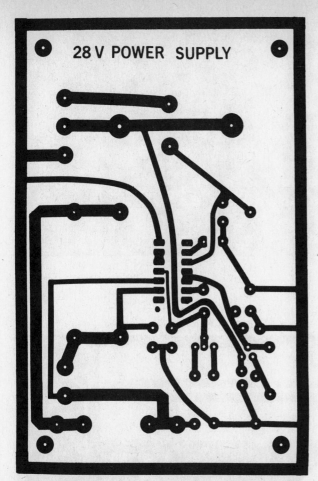

28 V POWER SUPPLY

Chapter 27, Fig. 18

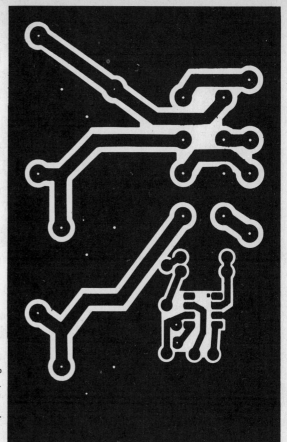

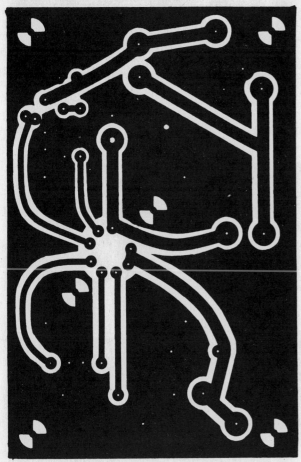

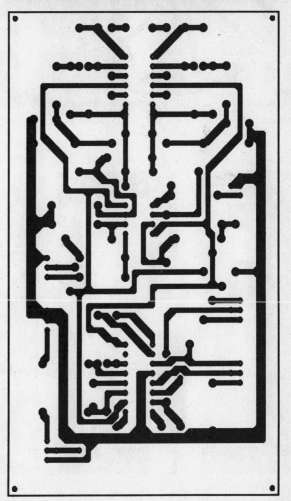

Chapter 34, Fig. 28

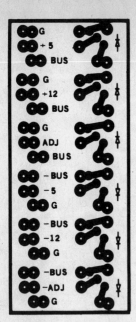

Chapter 27, Fig. 43

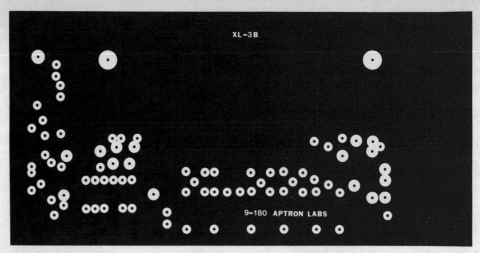

Chapter 28, Fig. 22 (Top)

Chapter 27, Fig. 61

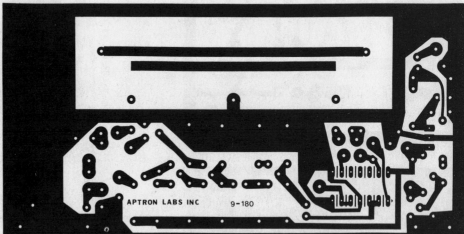

Chapter 28, Fig. 22 (Bottom)

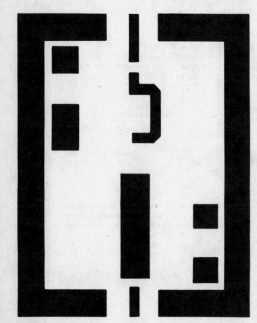

Chapter 32, Fig. 74

Chapter 34, Fig. 27

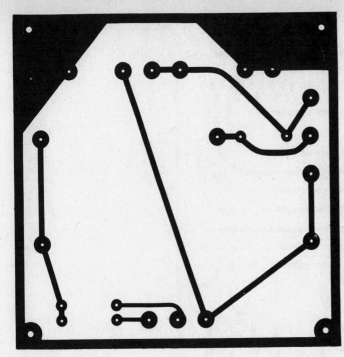

Chapter 31, Fig. 146A

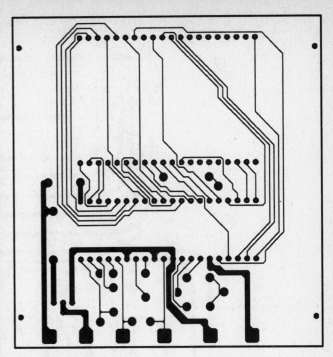

Chapter 34, Fig. 29

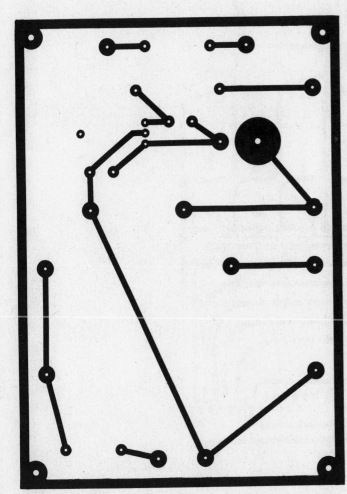

Chapter 31, Fig. 146B

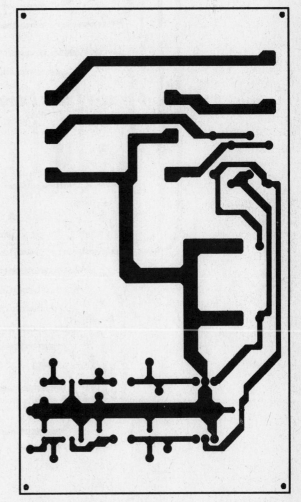

Chapter 34, Fig. 30

Printed-Circuit Board Etching Patterns 11

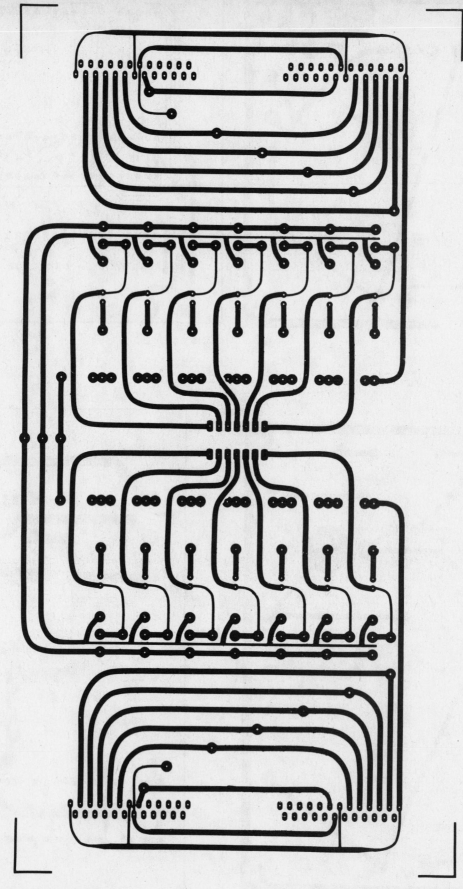

Chapter 29, Fig. 15

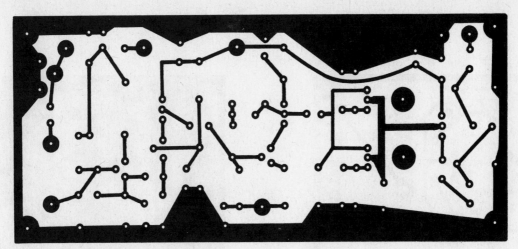

Chapter 30, Fig. 50

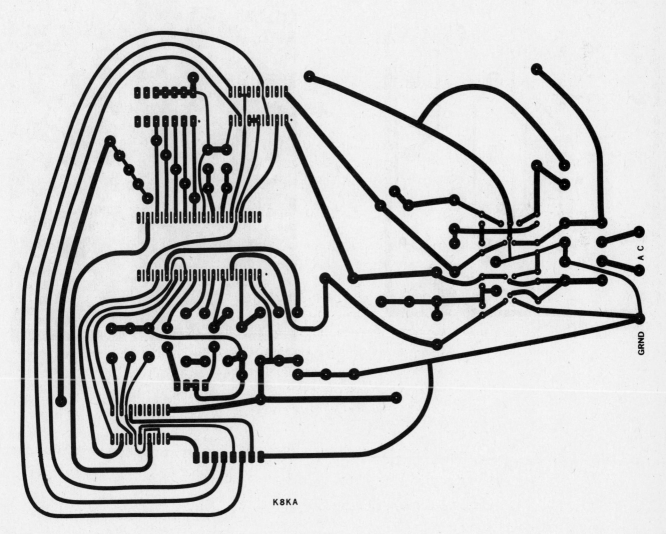

Chapter 29, Fig. 21

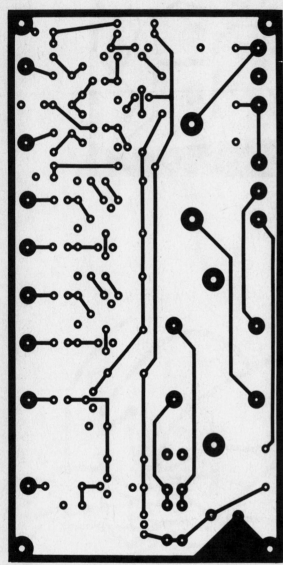

Chapter 34, Fig. 18 (Bottom)

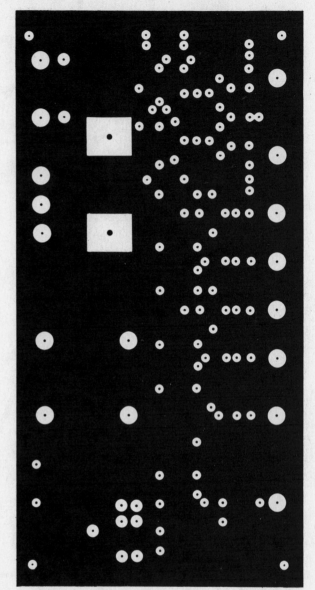

Chapter 34, Fig. 18 (Top)

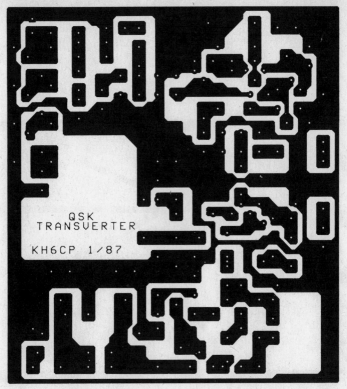

Chapter 30, Fig. 26

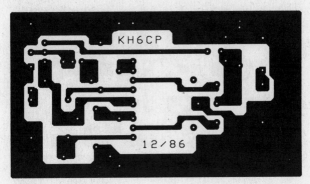

Chapter 30, Fig. 27

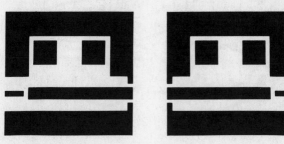

Chapter 32, Fig. 78

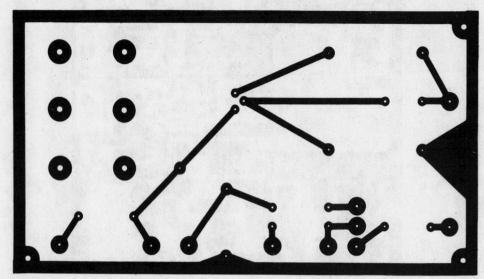

Chapter 31, Fig. 96

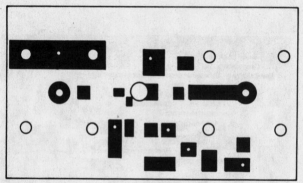

Chapter 32, Fig. 6

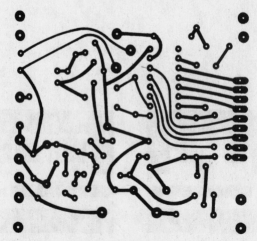

Chapter 34, Fig. 2

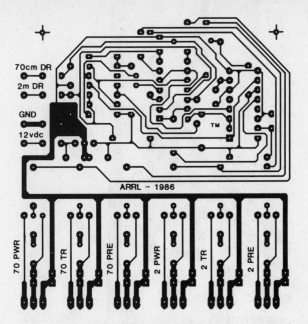

Chapter 23, Fig. 26

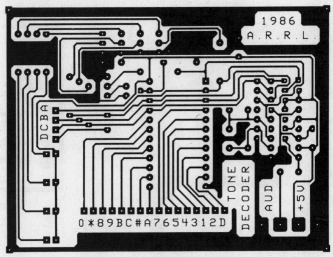

Chapter 34, Fig. 7

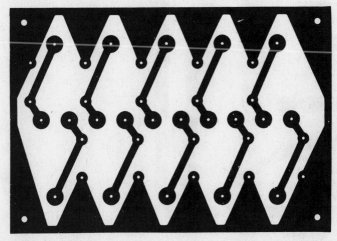

Chapter 34, Fig. 18

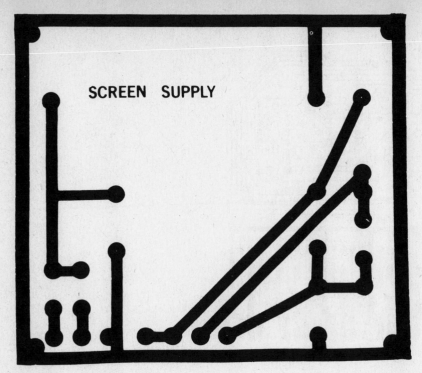

Chapter 31, Fig. 106A

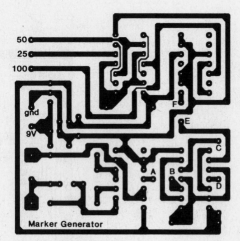

Marker Generator

Chapter 34, Fig. 66

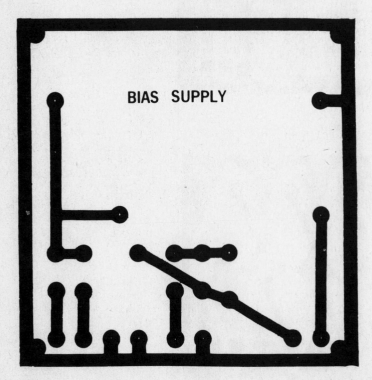

Chapter 31, Fig. 106B

Abbreviations List

These abbreviations, revised to conform with contemporary electronics and communications standards, appear in League publications.

a — atto (prefix for 10⁻¹⁸)
a — atto (prefix for 10^{-18})
A — ampere (unit of electrical current)
ac — alternating current
ACC — Affiliated Club Coordinator
ACSSB — Amplitude-compandored single sideband
A/D — analog-to-digital
ADC — analog-to-digital converter
AF — audio frequency
AFC — automatic frequency control
AFSK — audio frequency-shift keying
AGC — automatic gain control
Ah — ampere hour
AIRS — ARRL Interference Reporting System
ALC — automatic level control
AM — amplitude modulation
AMTOR — Amateur Teleprinting Over Radio
ANT — antenna
ARA — Amateur Radio Association
ARC — Amateur Radio Club
ARES — Amateur Radio Emergency Service
ARQ — Automatic repeat request
ARS — Amateur Radio Society (Station)
ASCII — American National Standard Code for Information Interchange
ASSC — Amateur Satellite Service Council
ATC — Assistant Technical Coordinator
ATV — amateur television
AVC — automatic volume control
AWG — American wire gauge
az-el — azimuth-elevation

B — bel; blower
balun — balanced to unbalanced (transformer)
BC — broadcast
BCD — binary-coded decimal
BCI — broadcast interference
Bd — baud (bit/s in single-channel binary data transmission)
BER — bit error rate
BFO — beat-frequency oscillator
bit — binary digit
bit/s — bits per second
BM — Bulletin Manager
BPF — band-pass filter
BPL — Brass Pounders League
BT — battery
BW — bandwidth

c — centi (prefix for 10^{-2})
C — coulomb (quantity of electric charge); capacitor

CAC — Contest Advisory Committee
CATVI — cable-television interference
CB — Citizens Band (radio)
CBMS — computer-based message system
CCTV — closed-circuit television
CCW — coherent CW
ccw — counterclockwise
CD — civil defense
cm — centimeter
CMOS — complementary-symmetry metal-oxide semiconductor
coax — coaxial cable
COR — carrier-operated relay
CP — code proficiency (award)
CPU — central processing unit
CRT — cathode-ray tube
CT — center tap
CTCSS — continuous tone-coded squelch system
cw — clockwise
CW — continuous wave

d — deci (prefix for 10^{-1})
D — diode
da — deca (prefix for 10)
D/A — digital-to-analog
DAC — digital-to-analog converter
dB — decibel (0.1 bel)
dBi — decibels above (or below) isotropic antenna
dBm — decibels above (or below) 1 milliwatt
DBM — doubly balanced mixer
dBV — decibels above/below 1V (in video, relative to 1 V P-P)
dBW — decibels above/below 1 watt
dc — direct current
D-C — direct conversion
DEC — District Emergency Coordinator
deg — degree
DET — detector
DF — direction finding; direction finder
DIP — dual in-line package
DPDT — double-pole double-throw (switch)
DPSK — differential phase-shift keying
DPST — double-pole single-throw (switch)
DS — direct sequence (spread spectrum); display
DSB — double sideband
DTMF — dual-tone multifrequency
DVM — digital voltmeter
DX — long distance; duplex
DXAC — DX Advisory Committee
DXCC — DX Century Club

E — voltage
e — base of natural logarithms (2.71828)
EC — Emergency Coordinator

ECAC — Emergency Communications Advisory Committee
ECL — emitter-coupled logic
EHF — extremely high frequency (30-300 GHz)
EIRP — effective isotropic radiated power
ELF — extremely low frequency
EMC — electromagnetic compatibility
EME — earth-moon-earth (moonbounce)
EMF — electromotive force
EMI — electromagnetic interference
EMP — electromagnetic pulse
EPROM — erasable programmable read-only memory

f— femto (prefix for 10^{-15}); frequency
F — farad (capacitance unit); fuse
FAX — facsimile
FD — Field Day
FET — field-effect transistor
FL — filter
FM — frequency modulation
FSK — frequency-shift keying
ft — foot (unit of length)

g — gram (unit of mass)
G — giga (prefix for 10^9)
GaAs — gallium arsenide
GDO — grid- or gate-dip oscillator
GHz — gigahertz
GND — ground

h — hecto (prefix for 10^2)
H — henry (unit of inductance)
HF — high frequency (3-30 MHz)
HFO — high-frequency oscillator
HPF — highest probable frequency; high-pass filter
Hz — hertz (unit of frequency)

I — current, indicating lamp
IC — integrated circuit
ID — identification; inside diameter
IF — intermediate frequency
IMD — intermodulation distortion
in — inch (unit of length)
in/s — inch per second (unit of velocity)
I/O — input/output
IRC — international reply coupon
ITF — Interference Task Force

j — operator for complex notation, as for reactive component of an impedance ($+j$ inductive; $-j$ capacitive)
J — joule (kg m²/s²) (energy or work unit); jack
J — joule (kg m^2/s^2) (energy or work unit); jack
JFET — junction field-effect transistor

k — kilo (prefix for 10^3); Boltzmann's constant (1.38×10^{-23} J/K)

K — kelvin (used without degree symbol) (absolute temperature scale); relay
kBd — 1000 bauds
kbit — 1024 bits
kbit/s — 1000 bits per second
kbyte — 1024 bytes
kg — kilogram
kHz — kilohertz
km — kilometer
kV — kilovolt
kW — kilowatt
kΩ — kilohm

l — liter (liquid volume)
L — lambert; inductor
lb — pound (force unit)
LC — inductance-capacitance
LCD — liquid crystal display
LED — light-emitting diode
LF — low frequency (30-300 kHz)
LHC — left-hand circular (polarization)
LO — local oscillator; League Official
LP — log periodic
LS — loudspeaker
LSB — lower sideband
LSI — large-scale integration

m — meter; milli (prefix for 10^{-3})
M — mega (prefix for 10^6); meter
mA — milliampere
mAh — millamperehour
MDS — Multipoint Distribution Service; minimum discernible (or detectable) signal
MF — medium frequency (300-3000 kHz)
mH — millihenry
mho — mho (use siemens)
MHz — megahertz
mi — mile, statute (unit of length)
mi/h — mile per hour
mi/s — mile per second
mic — microphone
min — minute (time)
MIX — mixer
mm — millimeter
MOD — modulator
modem — modulator/demodulator
MOS — metal-oxide semiconductor
MOSFET — metal-oxide semiconductor field-effect transistor
MS — meteor scatter
ms — millisecond
m/s — meters per second
MSI — medium-scale integration
MUF — maximum usable frequency
mV — millivolt
mW — milliwatt
MΩ — megohm

n — nano (prefix for 10^{-9})
NBFM — narrow-band frequency modulation
NC — no connection; normally closed
NCS — net-control station; National Communications System
nF — nanofarad
NF — noise figure
nH — nanohenry
NiCd — nickel cadmium

NM — Net Manager
NMOS — N-channel metal-oxide silicon
NO — normally open
NPN — negative-positive-negative (transistor)
NR — Novice Roundup (contest)
ns — nanosecond
NTS — National Traffic System

OBS — Official Bulletin Station
OD — outside diameter
OES — Official Emergency Station
OO — Official Observer
op amp — operational amplifier
ORS — Official Relay Station
OSC — oscillator (schematic diagram abbrev.)
OTC — Old Timer's Club
OTS — Official Traffic Station
oz — ounce (force unit, 1/16 pound)

p — pico (prefix for 10^{-12})
P — power; plug
PA — power amplifier
PAM — pulse-amplitude modulation
PC — printed circuit
PEP — peak envelope power
PEV — peak envelope voltage
pF — picofarad
pH — picohenry
PIA — Public Information Assistant
PIN — positive-intrinsic-negative (transistor)
PIO — Public Information Officer
PIV — peak inverse voltage
PLL — phase-locked loop
PM — phase modulation
PMOS — P-channel (type) metal-oxide semiconductor
PNP — positive-negative-positive (transistor)
pot — potentiometer
P-P — peak to peak
ppd — postpaid
PRAC — Public Relations Advisory Committee
PROM — programmable read-only memory
PSHR — Public Service Honor Roll
PTO — permeability-tuned oscillator
PTT — push to talk

Q — figure of merit (tuned circuit); transistor
QRP — low power (less than 5-W output)

R — resistor (schematic diagram abbrev.)
RACES — Radio Amateur Civil Emergency Service
RAM — random-access memory
RC — resistance-capacitance
R/C — radio control
RCC — Rag Chewers' Club
RF — radio frequency
RFC — radio-frequency choke
RFI — radio-frequency interference
RHC — right-hand circular (polarization)
RIT — receiver incremental tuning

RLC — resistance-inductance-capacitance
RM — rule making (number assigned to petition)
r/min — revolution per minute
RMS — root mean square
ROM — read-only memory
r/s — revolution per second
RST — readability-strength-tone
RTTY — radioteletype
RX — receiver, receiving

s — second (time)
S — siemens (unit of conductance); switch
SASE — self-addressed stamped envelope
SEC — Section Emergency Coordinator
SET — Simulated Emergency Test
SGL — State Government Liaison
SHF — super-high frequency (3-30 GHz)
SM — Section manager; silver mica (capacitor)
S/N — signal-to-noise (ratio)
SPDT — single-pole double-throw (switch)
SPST — single-pole single-throw (switch)
SS — Sweepstakes; spread spectrum
SSB — single sideband
SSC — Special Service Club
SSI — small-scale integration
SSTV — slow-scan television
STM — Section Traffic Manager
SX — simplex
sync — synchronous, synchronizing
SWL — shortwave listener
SWR — standing-wave ratio

T — tera (prefix for 10^{-2}); transformer (schematic diagram abbrev.)
TA — Technical Advisor
TC — Technical Coordinator
TCC — Transcontinental Corps
TD — Technical Department (ARRL HQ)
tfc — traffic
TR — transmit/receive
TTL — transistor-transistor logic
TTY — teletypewriter
TV — television
TVI — television interference
TX — transmitter, transmitting

U — integrated circuit
UHF — ultra-high frequency (300 MHz to 3 GHz)
USB — upper sideband
UTC — Coordinated Universal Time
UV — ultraviolet

V — volt; vacuum tube (schematic diagram abbrev.)
VCO — voltage-controlled oscillator
VCR — video cassette recorder
VDT — video-display terminal
VE — Volunteer Examiner
VEC — Volunteer Examiner Coordinator
VFO — variable-frequency oscillator

VHF — very-high frequency
(30-300 MHz)
VLF — very-low frequency (3-30 kHz)
VLSI — very-large-scale integration
VMOS — vertical metal-oxide
semiconductor
VOM — volt-ohm meter
VOX — voice operated switch
VR — voltage regulator
VRAC — VHF Repeater Advisory
Committee
VSWR — voltage standing-wave ratio
VTVM — vacuum-tube voltmeter
VUAC — VHF/UHF Advisory
Committee
VUCC — VHF/UHF Century Club
VXO — variable crystal oscillator

W — watt (kg m^{2}s^{-3}, unit of power)
WAC — Worked All Continents
WARC — World Administrative Radio
Conference
WAS — Worked All States
WBFM — wide-band frequency
modulation
Wh — watthour
WPM — words per minute
WVDC — working voltage, direct
current

X — reactance
XCVR — transceiver
XFMR — transformer
XO — crystal oscillator
XTAL — crystal
XVTR — transverter

Y — crystal (schematic diagram
abbrev.)
YIG — yttrium iron garnet

Z — impedance; also see UTC

5BDXCC — Five-Band DXCC
5BWAC — Five-Band WAC
5BWAS — Five-Band WAS
6BWAC — Six-Band WAC

° — degree (plane angle)
°C — degree Celsius (temperature)
°F — degree Fahrenheit (temperature)
α — (alpha) angles; coefficients,
attenuation constant, absorption
factor, area, common-base forward
current-transfer ratio of a bipolar
transistor
β — (beta) angles; coefficients, phase
constant current gain of common-
emitter transistor amplifiers
γ — (gamma) specific gravity, angles,
electrical conductivity, propagation
constant
Γ — (gamma) complex propagation
constant
δ — (delta) increment or decrement,
density angles
Δ — (delta) increment or decrement
determinant, permittivity
ϵ — (epsilon) dielectric constant,
permittivity, electric intensity
ζ — (zeta) coordinates, coefficients
η — (eta) intrinsic impedance,
efficiency, surface charge density,
hysteresis, coordinate

θ — (theta) angular phase displacement,
time constant, reluctance, angles
ι — (iota) unit vector
K — (kappa) susceptibility, coupling
coefficient
λ — (lambda) wavelength, attenuation
constant
Λ — (lambda) permeance
μ — (mu) permeability, amplification
factor, micro (prefix for 10^{-6})
μC — microcomputer
μF — microfarad
μH — microhenry
μP — microprocessor
ξ — (xi) coordinates
π — (pi) 3.14159
ρ — (rho) resistivity, volume charge
density, coordinates, reflection
coefficient
σ — (sigma) surface charge density,
complex propagation constant,
electrical conductivity, leakage
coefficient, deviation
Σ — (sigma) summation
τ — (tau) time constant, volume
resistivity, time-phase displacement,
transmission factor, density
ϕ — (phi) magnetic flux, angles
Φ — (phi) summation
χ — (chi) electric susceptibility, angles
ψ — (psi) dielectric flux, phase
difference, coordinates, angles
ω — (omega) angular velocity 2πf
Ω — (omega) resistance in ohms, solid
angle

INDEX

[Editor's Note: Except for commonly used phrases and abbreviations, topics are indexed by their noun name. Many topics are also cross-indexed, especially when noun modifiers appear (such as "voltage amplifier" and "amplifier, voltage"). For topics having extended coverage, only the first page of presentation is listed. Numerous terms and abbreviations pertaining to Amateur Radio but not contained in this index may be found in the glossaries indexed herein.]

HAM RADIO IS FUN!

Fun to learn
Fun to operate

Tune In The World With Ham Radio has put the fun back into learning what Amateur Radio is all about. Enhanced Novice class privileges have brought the fun back into operating. Now beginners with their Novice licenses no longer have to spend all of their time on the air using only Morse Code. Novices can now use voice communications on 10-meters and use VHF and UHF repeaters. The new privileges include the use of digital communications so that home computers can be linked through packet radio networks. The FCC requires that Novices know something about their new privileges and that's where the expanded *Tune In The World With Ham Radio* text comes in. You'll find what you need to know explained in clear, concise bite-sized chunks of information. You'll find all 300 possible questions that may appear on the 30-question Novice exam with their distractors and answer key.

Besides improving the text, we've added almost three times the code practice material to the package in the form of two C-90 tape cassettes. One tape teaches the code, the other provides practice. They are recorded in stereo so you can switch off the voice portion for even more practice. These new tapes make learning the code a snap!

Tune In The World With Ham Radio is available at your dealer or from ARRL for $15.00 plus $3.50 for UPS shipping and handling.

THE AMERICAN RADIO RELAY LEAGUE, INC.
225 MAIN STREET
NEWINGTON, CT 06111

A Great New ARRL Antenna Book

Twice the size of the previous edition
Over 700 pages
987 figures
Edited by Gerald L. Hall, K1TD

QST Associate
Technical Editor

The 15th Edition of *The ARRL Antenna Book* has been dramatically expanded in a similar fashion to recent editions of the *ARRL Handbook* and the *ARRL Operating Manual*. We've drawn on material produced by the ARRL Technical Department and from 16 well-known outside authors who have done much to contribute to the state-of-the-art in antenna and transmission line design. Available in softcover only for $18 at your dealer or directly from ARRL (shipping and handling: $2.50, $3.50 for UPS)

CHAPTER LINEUP:

The number of pages appears in parentheses after the chapter title. Page counts may vary slightly. Safety First (18), Antenna Fundamentals (42), The Effects of the Earth (14), Selecting Your Antenna System (30), Loop-Antennas (16), Multielement Arrays (42), Broadband Antennas (12), Log Periodic Arrays (24), Yagi Arrays (26), Quad Arrays (14), Long Wire and Traveling Wave Antennas (18), Direction Finding Antennas (26), Portable Antennas (10), Mobile and Maritime Antennas (30), Repeater Antenna Systems (20), VHF and UHF Antenna Systems (44), Antennas for Space Communications (32), Spacecraft Antennas (8), Antenna Materials and Accessories (20), Antenna Supports (22), Radio Wave Propagation (26), Transmission Lines (26), Coupling the Transmitter to the Line (18), Coupling the Line to the Antenna (28), Antenna and Transmission-Line Measurements (36), Smith Chart Calculations (16), Topical Bibliography on Antennas (32), Glossary and Abbreviations (4), Contents, Index, etc (16).

ARRL 225 MAIN ST., NEWINGTON, CT 06111

ARRL BOOKSHELF

Prices are subject to change without notice. Shipping and handling: add $2.50 for book rate. For $3.50, shipments to US addresses can be made by insured parcel post or UPS. Please specify. The minimum charge order is $7.50. Payment must be in US funds.

ARRL, 225 MAIN STREET, NEWINGTON, CT 06111

THE 1989 ARRL HANDBOOK

This is the most comprehensive edition since the *Handbook* was first published in 1926. It is updated yearly to present the cutting edge of rf communication techniques while presenting hundreds of projects the average Amateur Radio operator can build. The 66th edition is packed with information on digital communication modes as well as new power supplies and amplifiers. Ready-to-use etching patterns are provided for many projects. This *Handbook* belongs in every ham shack.

Hardcover only #1662 $21 US, $23 elsewhere

ANTENNA BOOKS

THE ARRL ANTENNA BOOK represents the best and most highly regarded information on antenna fundamentals, transmission lines, design and construction of wire antennas as well as yagis and quads for HF. You'll find chapters on VHF/UHF antennas, test equipment and propagation. The new 15th edition has over 700 pages of practical antenna information.
©1988, Softcover .... #2065 $18

Novice Antenna Notebook is written for the beginner or experienced amateur who wants practical information on basic antenna designs and construction.
©1988. Softcover .... #2073 $ 8

W1FB's Antenna Notebook Practical wire and vertical antenna designs .... #0488 $ 8

TRANSMISSION LINE TRANSFORMERS, cover baluns, use of ferrites, and other aspects of antenna transmission line design and operation. 128 pages .... ©1987 #0471 $10

ANTENNA COMPENDIUM Packed with new material on quads, yagis and other interesting topics.
©1985 178 pages #0194 $10 US, $11 elsewhere

HF ANTENNAS FOR ALL LOCATIONS
G6XN's look at antennas with practical construction data.
©1982 264 pages #R576 $15

YAGI ANTENNA DESIGN by Dr. James L. Lawson, W2PV. Over 210 pages of practical theory and design information.
©1986 #0410 $15

PASSING POWER! - THESE PUBLICATIONS WILL HELP YOU THROUGH THE EXAMS

Beginning with **Tune in the World with Ham Radio** for the Novice and progressing through the critically acclaimed **ARRL License Manual Series** for the Technician through Extra Class; you will find passing each exam element a snap! There are accurate text explanations of the material covered along with FCC question pools and answer keys. The latest edition of **The FCC Rule Book** is invaluable as a study guide for the regulatory material found on the exams and as a handy reference. *Every* amateur needs an up-to-date copy. **The ARRL Code Kit** has a booklet and two C-60 cassettes to take you from 5 to 13 WPM quickly. **Morse Code the Essential Language** has tips on learning the code, high speed operation and history. If you have a Commodore 64™ or C 128 computer, **Morse University*** provides hours of fun and competition in improving your code proficiency. **First Steps in Radio** from QST presents electronic principles for the beginner.
*MORSE UNIVERSITY is a trademark of AEA, Inc.

Tune in the World with Ham Radio 1987 edition
Kit with book and cassettes ..... #0380 $15
Book only .... #0399 $12
Cassettes $10
License Manual Series
Technician/General Class .... #0143 $ 5
Advanced Class................. #016X $ 5
Extra Class for exams thru
10/31/88 .... #0178 $ 5
Extra Class for exams after
11/1/88 .... #0763 $ 5
FCC Rule Book 7th Ed. ....#0453 $ 5
Code Proficiency
Code Kit .... #5501 $ 8
Morse University .... #0259 $40
GGTE Morse Tutor .... #2081 $20
C-60 Code Practice Cassettes
30 min. each at 5 and 7½ WPM*...#1030 $ 5
30 min. each at 10 and 13 WPM*....#1040 $ 5
30 min. each at 15 and 20 WPM .. #2050 $ 5
*Same tapes included in *Code Kit*
Morse Code: The Essential Language covers sending, receiving, high speed operation and history ©1986 .... #0356 $ 5
First Steps in Radio .... #2286 $ 5

OPERATING

The ARRL Operating Manual 688 pages packed with information on how to make the best use of your station, including: interfacing home computers, OSCAR, VHF-UHF, contesting, DX traffic/emergency matters and shortwave listening.
©1987 3rd ed. #1086 $15

The ARRL Repeater Directory, 1988-89
#0437 .... $ 5

The ARRL Net Directory-free shipping ... #0275 $1

HOLA CQ Learn to communicate with Spanish-speaking radio amateurs. 90 min. cassette and 15 page text. #901N $7
The RSGB Operating Manual The third edition published in 1985 is packed with practical operating tips, techniques and tables................. #R69X $14

Passport To World Band Radio 350 pages of information and listings of shortwave broadcast stations with frequency, times and languages. 1988 ed. $15

PACKET RADIO/COMPUTERS

Computer Networking Conferences 1-4 from 1981-1985 Pioneer Papers on Packet Radio .. #0224 $18
5th Computer Networking Conference Papers
©1986 .... #033X $10
6th Computer Networking Conference Papers
©1987 .... #2022 $10
7th Computer Networking Conference Papers
©1988 .... #2138 $12
AX.25 Link Layer Protocol .... #0119 $8
Get***Connected to Packet Radio ..... #Q221 $13
RSGB Amateur Radio Software Contains 85 BASIC programs, 6 in assembly language covering CW, RTTY, Amtor, Packet, Antenna Design, Satellite Predictions, Distances, Bearings and Locators $18
Gateway to Packet Radio How to get started, equipment you need and more ....#2030 $10

DX

The Complete DX'er by W9KNI .... #2083 $10
DX Power by K5RSG .... #T740 $10
DXCC Countries List — free shipping .. #0291 $ 1
Low Band Dxing ©1987 .... #047X $10

QRP

QRP Notebook by Doug DeMaw, W1FB. An exciting book for the low power enthusiast and experimenter. Copyright 1986, 70 pages .... #0348 $ 5

VHF-UHF, MICROWAVE, SPACE

RSGB VHF/UHF Manual .... #630 $30
RSGB Microwave Newsletter Col. ..... #R000 $18
21st Central Sts. VHF Conf. $10
22nd Central States VHF Conf. .... #209X $12
Microwave Update 1987 Conf. $10
Microwave Update 1988 Conf. $12
Mid-Atlantic VHF Conference $10
The Satelite Experimenter's Handbook by Martin Davidoff, K2UBC, 208 pages, copyright 1985
#0046 $10 US, $11 elsewhere
AMSAT NA 5th Space Symposium $12
Satellite Anthology .... $ 5

INTERFERENCE/DFing

Interference Handbook (Radio Pubs)......... $10
Transmitter Hunting (Tab) $18

OTHER PUBLICATIONS

Fifty Years of ARRL .... #0135 $ 4
GIL: Collection of cartoons from QST .. #0364 $ 5
Oscarlocator .... #3037 $8.50 US, $9.50 elsewhere
200 Meters and Down .... #0011 $ 4
Solid State Design for the Radio Amateur. First published in 1977; just reprinted by popular demand #0402 .... $12
RSGB Radio Communications Hndbk. .. #R584 $35
RSGB Buyer's Guide .... #R680 $15
RSGB Data Book .... #R673 $18

FOR INSTRUCTORS

Written for those teaching classes using
ARRL License Manuals or *Tune In The World*
General Class Instructor's Guide .... $ 5
Technician Instructor's Guide .... $ 5
Novice Instructor's Guide .... $ 5

ADVENTURE

Grand Canyon QSO ... (Tompkins) ....#5048 $ 5
SOS at Midnight.. (Tompkins) ....#5005 $ 5
CQ Ghost Ship... (Tompkins)..........#5013 $ 5
DX Brings Danger(Tompkins)..........#5021 $ 5
Death Valley QTH (Tompkins).........#503X $ 5
Set of 5 Tompkins books ....#1490 $20

Notes

 # ORDER FORM

USE THIS FORM OR PHOTOCOPY

Please allow 1 week for us to receive your order, 1 week for processing and 1 to 3 weeks shipping time after your order leaves ARRL.

BOOK#	QUANTITY	TITLE	AMOUNT
Shipping/Handling ☐ by Mail $2.50 ☐ by Insured Mail $3.50 ☐ by UPS $3.50			
Payment must be in US funds drawn on a US bank		TOTAL	

$7.50 minimum all credit card orders

Charge to ☐ VISA   ☐ Mastercard
☐ AMEX   ☐ Discover

Name _____

Call _____

Street _____   _____
                                     Card Number

City _____   Card good from _____

_____   Card good to _____
                              Expiration Date

Signature _____

HDBK

ARRL   225 Main Street   Newington, CT 06111 USA

Notes

Please use this form to give us your comments on this book and what you'd like to see in future editions.

Name _____ Call sign _____

Address_____ Daytime Phone ( ) _____

City _____ State/Province_____ ZIP/Postal Code _____

From _____

_____

_____

Editor, 1989 Handbook
American Radio Relay League
225 Main Street
Newington, CT 06111
USA

please fold and tape